True Text-Art Integration

visual anatomy & physiology

2nd edition

MARTINI / OBER / NATH / BARTHOLOMEW / PETTI

A Complete Learning Solution

visual
anatomy &
physiology
2nd
MARTINI / OBER / NATH / BARTHOLOMEW / PETTI

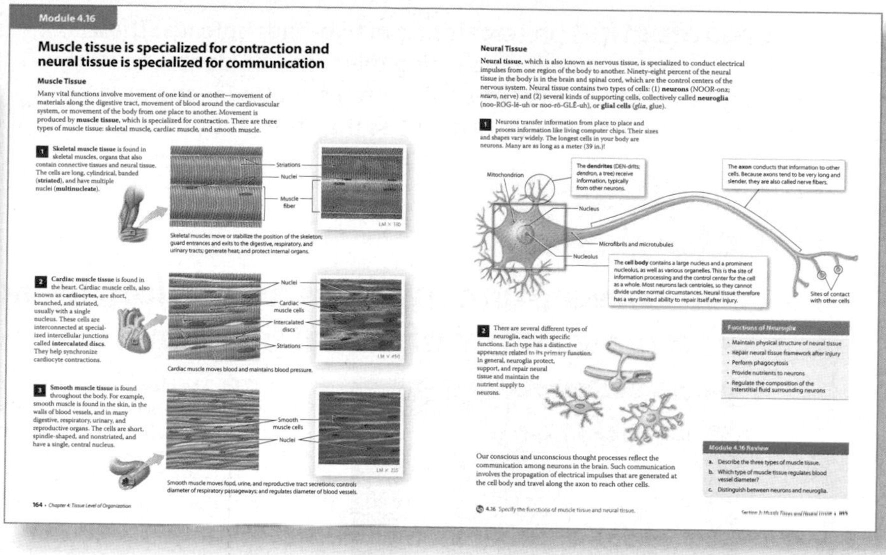

Module 4.16
Muscle tissue is specialized for contraction and neural tissue is specialized for communication
Muscle Tissue
Neural Tissue
Functions of Neuroglia
Module 4.16 Review

Visual Anatomy & Physiology, Second Edition combines a visual approach with a modular organization to deliver an easy-to-use and time-efficient book that uniquely meets the needs of today's students—without sacrificing the coverage of A&P topics required for careers in nursing and other allied health professions.

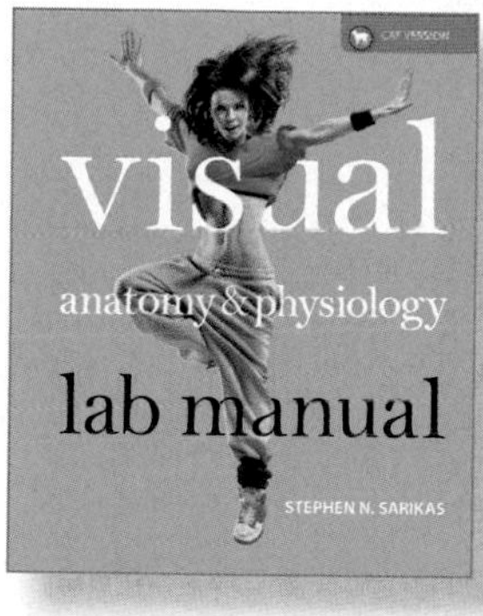

visual
anatomy & physiology
lab manual
STEPHEN N. SARIKAS

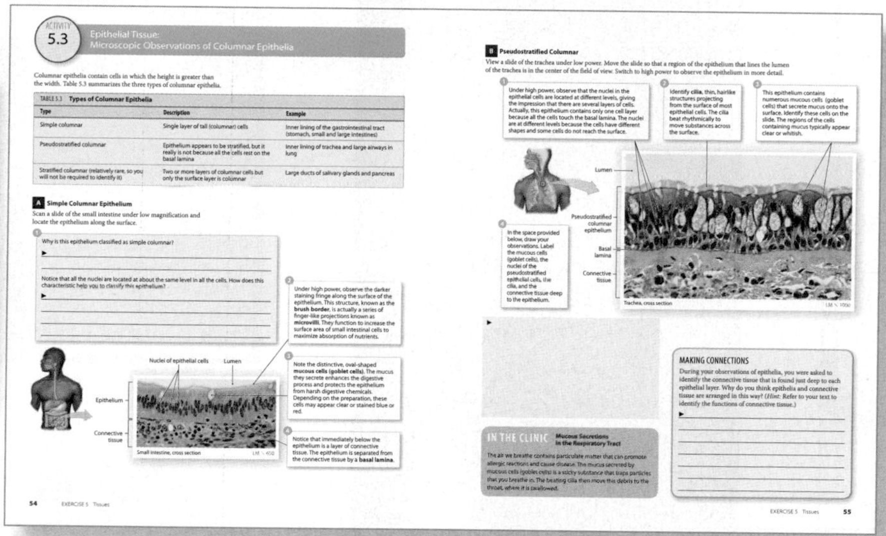

ACTIVITY
5.3
Epithelial Tissue: Microscopic Observations of Columnar Epithelia
Simple Columnar Epithelium
Pseudostratified Columnar
IN THE CLINIC
MAKING CONNECTIONS

Visual Anatomy & Physiology Lab Manual brings all of the strengths of the revolutionary *Visual Anatomy & Physiology* book to the lab. This lab manual combines a visual approach with a modular organization to maximize learning. The lab practice consists of hands-on activities in the lab manual and assignable content in MasteringA&P. Main, Cat, and Pig versions are available.

MasteringA&P®

MasteringA&P is an online learning and assessment system proven to help students learn and designed to help instructors teach more efficiently.

- Lets instructors easily assign media that is automatically graded
- Provides students with personalized coaching through answer-specific feedback and hints
- Motivates students to come to class prepared
- Easily captures data to demonstrate assessment outcomes

The Modular Organization

The time-saving modular organization presents topics in two-page spreads. These two-page spreads give students an efficient organization for managing their time. Students can study each module during the limited time they have in their busy schedules—ten minutes for one module now, ten minutes for another module later—checking off each module as they complete it.

First, the top left page begins with a full-sentence topic heading that teaches the major point of the module. (These topic headings are correlated by number to the learning outcomes on the chapter-opening page and at the bottom of each module. The learning outcomes are derived from the learning outcomes recommended by the Human Anatomy & Physiology Society.)

Next, the red-boxed numbers guide students through the presentation of the topic.

Then, instead of long columns of narrative text that refer to visuals, brief text is built right into the visuals. Students read while looking at the corresponding visual, which means:

- No long paragraphs
- No page flipping
- Everything in one place

Module 15.19

Photoreception involves activation, bleaching, and reassembly of visual pigments

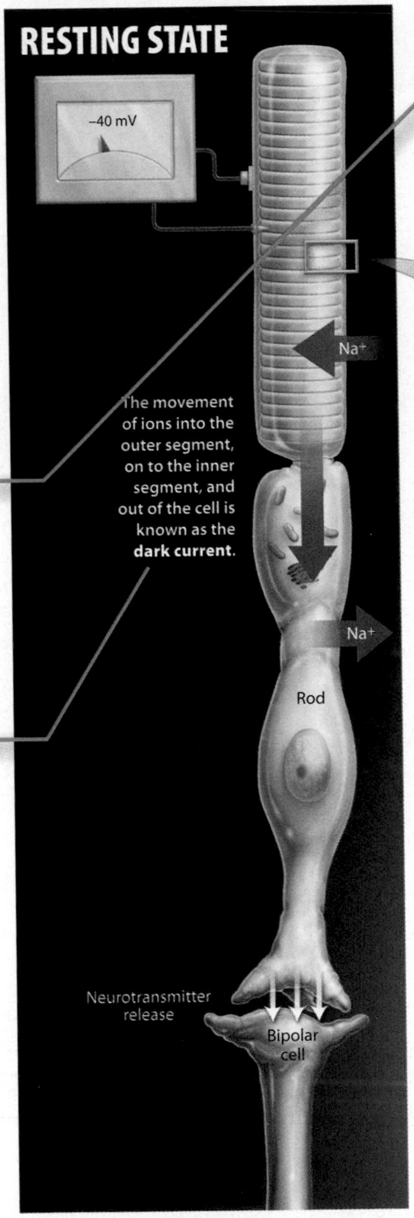
RESTING STATE
–40 mV
Na+
The movement of ions into the outer segment, on to the inner segment, and out of the cell is known as the **dark current**.
Na+
Rod
Neurotransmitter release
Bipolar cell

1 The plasma membrane in the outer segment of the photoreceptor contains chemically gated sodium ion channels. In darkness, these gated channels are kept open in the presence of cGMP (cyclic guanosine monophosphate), a derivative of the high-energy compound guanosine triphosphate (GTP). Because the channels are open, the membrane potential is approximately –40 mV, rather than the –70 mV typical of resting neurons. At the –40 mV membrane potential, the photoreceptor is continuously releasing neurotransmitters across synapses to bipolar cells. The inner segment also continuously pumps sodium ions (Na^+) out of the cytosol.

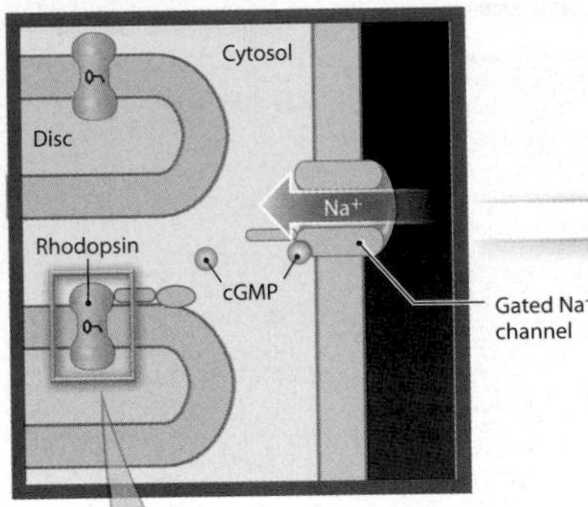
Cytosol
Disc
Na+
Rhodopsin
cGMP
Gated Na+ channel

2 The bound retinal molecule in rhodopsin has two possible configurations: the bent or curved 11-*cis* form and the more linear 11-*trans* form. Normally, in the dark, the molecule is in the 11-*cis* form. On absorbing light it changes to the 11-*trans* form. This change activates the opsin molecule.

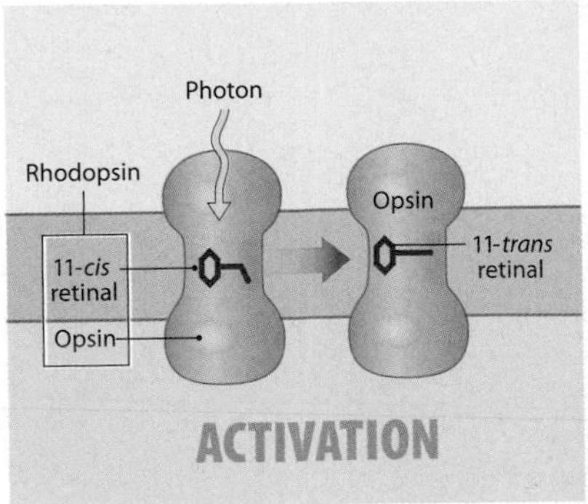
Photon
Rhodopsin
11-*cis* retinal
Opsin
Opsin
11-*trans* retinal
ACTIVATION

3 Opsin then activates **transducin**, a G protein bound to the disc membrane. The transducin in turn activates the enzyme **phosphodiesterase (PDE)**.

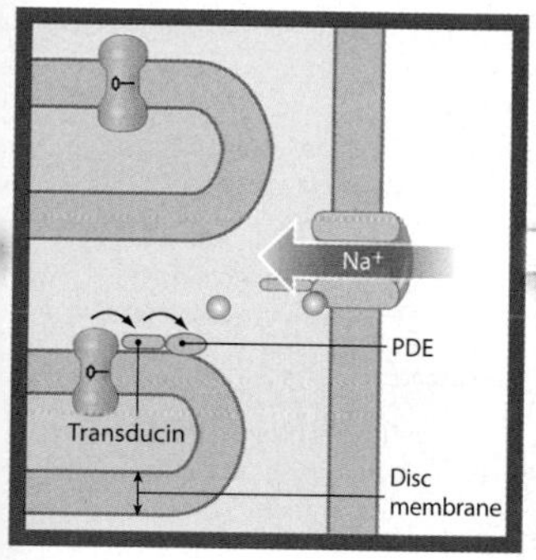

Na+
PDE
Transducin
Disc membrane

4 Phosphodiesterase is an enzyme that breaks down cGMP. The removal of cGMP from the gated sodium channels results in their inactivation. The rate of Na^+ entry into the cytosol then decreases.

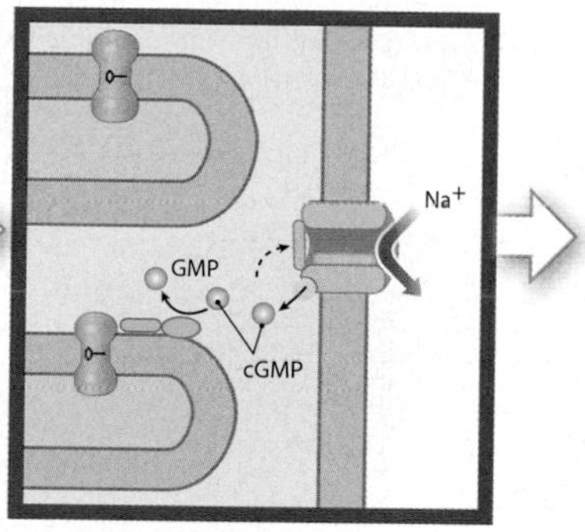

Na+
GMP
cGMP

ACTIVE STATE

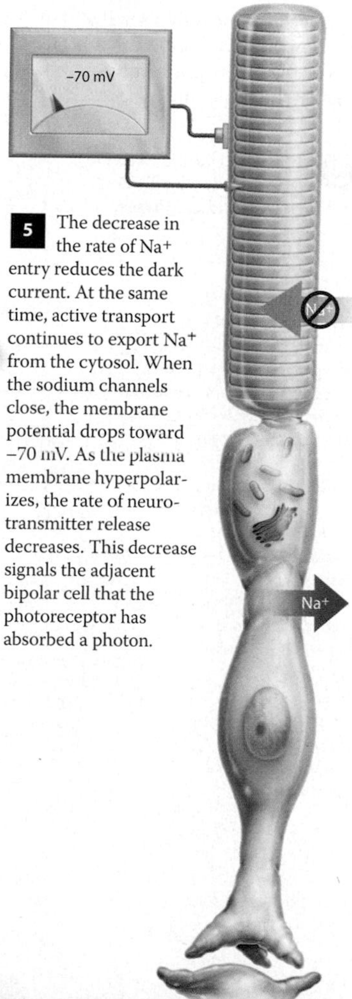

–70 mV
Na+
Na+

5 The decrease in the rate of Na^+ entry reduces the dark current. At the same time, active transport continues to export Na^+ from the cytosol. When the sodium channels close, the membrane potential drops toward −70 mV. As the plasma membrane hyperpolarizes, the rate of neurotransmitter release decreases. This decrease signals the adjacent bipolar cell that the photoreceptor has absorbed a photon.

6 Rhodopsin cannot respond to additional photons until its retinal component regains its original shape. It does not spontaneously revert to the 11-*cis* form. Instead, the entire rhodopsin molecule must be broken down into retinal and opsin in a process called **bleaching**. The retinal is then converted to its original *cis* shape. This conversion requires energy in the form of ATP. Opsin and 11-*cis* retinal are reassembled and the rhodopsin molecule is ready to repeat the cycle.

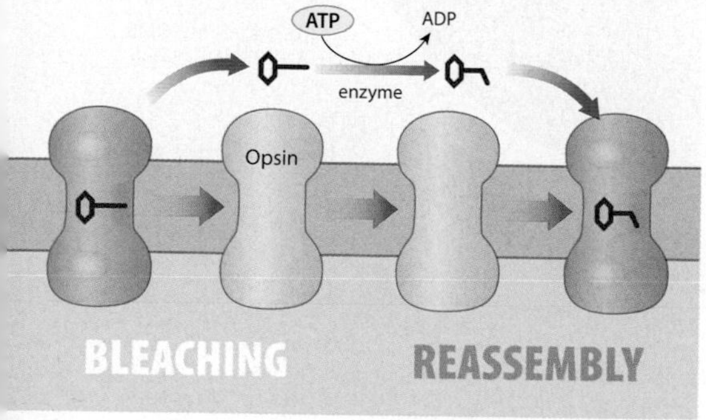

ATP
ADP
enzyme
Opsin
BLEACHING
REASSEMBLY

Module 15.19 Review

a. Visual pigments undergo which three changes during photoreception?

b. What are the two configurations of retinal?

c. When during photoreception is ATP required?

15.19 Explain photoreception and how visual pigments are activated.

Finally, each module ends with a set of Module Review questions that help students check their understanding before moving on.

The Visual Approach

The unique visual approach allows the illustrations to be the central teaching and learning element, with the text built directly around them—creating true text-art integration. This approach matches how students naturally want to use their A&P textbook. Our extensive research with A&P students—via student reviews, student focus groups, and student class tests—reveals that A&P students go first to the visuals and then to the corresponding text.

Module 3.19

Watch
MasteringA&P®
A&PFlix
DNA Replication

During interphase, the cell prepares for cell division

Most cells spend only a small part of their time actively engaged in cell division. **Somatic** (*soma*, body) **cells** spend most of their functional lives in a state known as **interphase**. During interphase, a cell performs all its normal functions and, if necessary, prepares for cell division.

1 In a cell preparing to divide, interphase can be divided into the **G_1, S,** and **G_2 phases.**

When the G1 phase is complete, the cell enters the S phase. Over the next 6–8 hours, the cell duplicates its chromosomes. This involves DNA replication and the synthesis of histones and other proteins in the nucleus.

A cell that is ready to divide first enters the G_1 phase. In this phase, the cell makes enough mitochondria, cytoskeletal elements, endoplasmic reticula, ribosomes, Golgi membranes, and cytosol for two functional cells. Centriole replication begins in G_1 and commonly continues until G_2. In a cell dividing at top speed, G_1 may last just 8–12 hours. Such a cell pours all its energy into mitosis, and all other activities cease. If G_1 lasts for days, weeks, or months, preparation for mitosis occurs as the cell performs its normal functions.

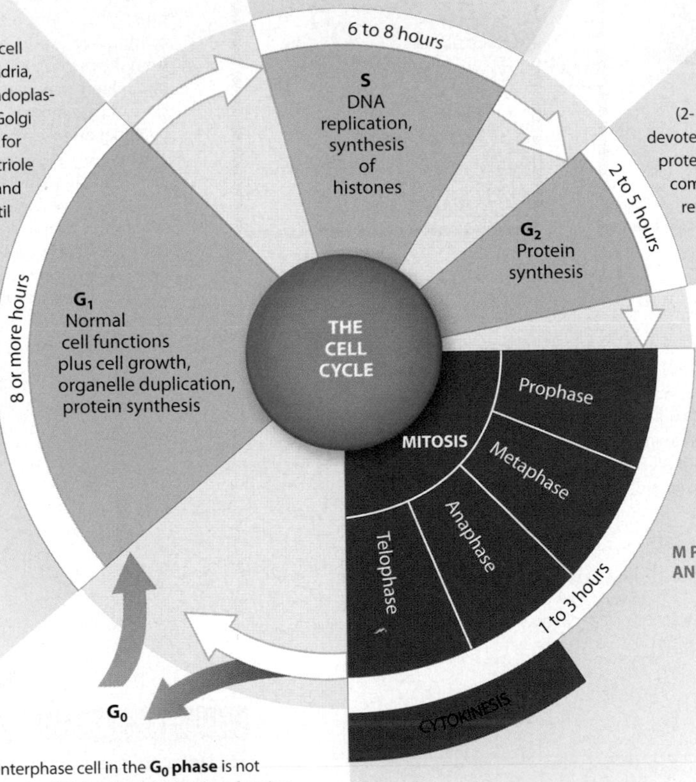

6 to 8 hours
S
DNA replication, synthesis of histones
2 to 5 hours
G₂
Protein synthesis
8 or more hours
G₁
Normal cell functions plus cell growth, organelle duplication, protein synthesis
THE CELL CYCLE
MITOSIS
Prophase
Metaphase
Anaphase
Telophase
1 to 3 hours
CYTOKINESIS
G₀

When DNA replication ends, there is a brief (2- to 5-hour) G_2 phase devoted to last-minute protein synthesis and to completing centriole replication.

M Phase – MITOSIS AND CYTOKINESIS

An interphase cell in the **G_0 phase** is not preparing for division, but is instead performing all of the other functions appropriate for that particular cell type. Some mature cells, such as skeletal muscle cells and most neurons, remain in G_0 indefinitely and never divide. In contrast, **stem cells**, which divide repeatedly with very brief interphase periods, never enter G_0.

Descriptions and key terminology are embedded in the art.

2 During the S phase of the cell cycle, DNA is replicated. The goal of DNA replication is to copy the genetic information in the nucleus. The process occurs in cells preparing to undergo either mitosis or meiosis.

1 **DNA replication** begins when **DNA helicase** enzymes unwind the strands and disrupt the hydrogen bonds between the bases. As the strands unwind, molecules of **DNA polymerase** bind to the exposed nitrogenous bases. This enzyme (1) promotes bonding between the nitrogenous bases of the DNA strand and complementary DNA nucleotides in the nucleoplasm and (2) links the nucleotides by covalent bonds.

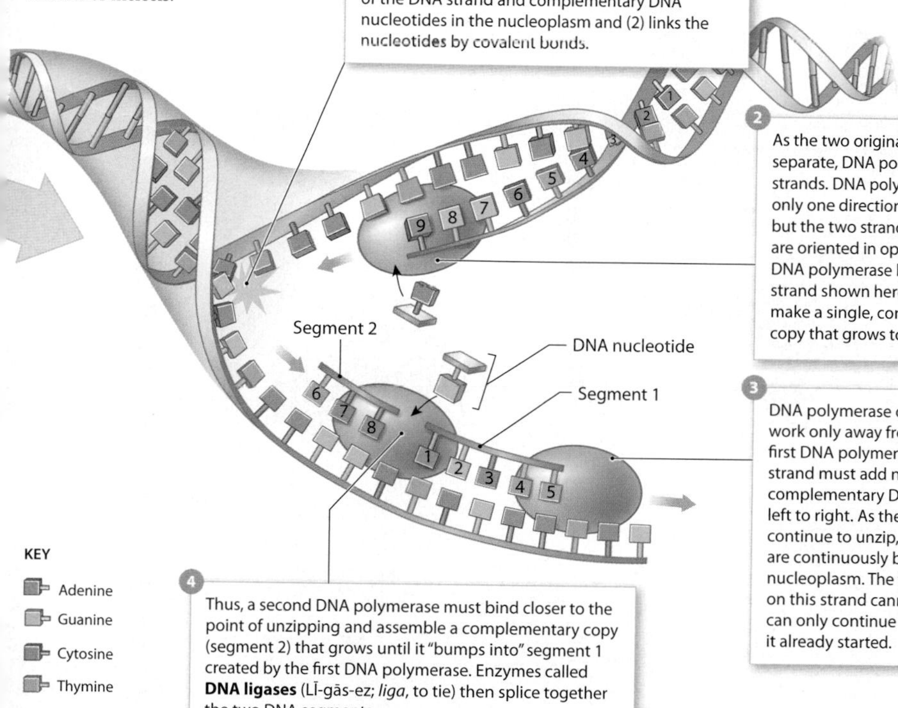

Segment 2
DNA nucleotide
Segment 1
KEY
Adenine
Guanine
Cytosine
Thymine

2 As the two original strands gradually separate, DNA polymerase binds to the strands. DNA polymerase can work in only one direction along a strand of DNA, but the two strands in a DNA molecule are oriented in opposite directions. The DNA polymerase bound to the upper strand shown here adds nucleotides to make a single, continuous complementary copy that grows toward the "zipper."

3 DNA polymerase on the lower strand can work only away from the zipper. So the first DNA polymerase to bind to this strand must add nucleotides and build a complementary DNA strand moving from left to right. As the two original strands continue to unzip, additional nucleotides are continuously being exposed to the nucleoplasm. The first DNA polymerase on this strand cannot go into reverse. It can only continue to elongate the strand it already started.

4 Thus, a second DNA polymerase must bind closer to the point of unzipping and assemble a complementary copy (segment 2) that grows until it "bumps into" segment 1 created by the first DNA polymerase. Enzymes called **DNA ligases** (LĪ-gās-ez; *liga*, to tie) then splice together the two DNA segments.

3 Eventually, the unzipping completely separates the original strands. The copying ends, the last splicing is done, and two identical DNA molecules have formed. Once the DNA is replicated, the centrioles duplicated, and the necessary enzymes and proteins synthesized, the cell leaves interphase and is ready to proceed to mitosis.

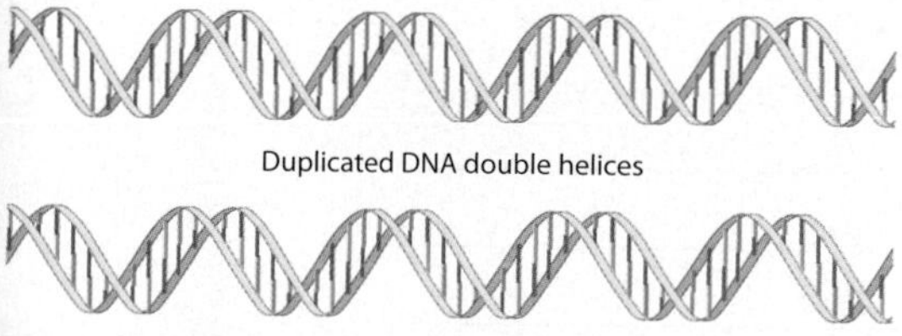

Duplicated DNA double helices

Module 3.19 Review

a. Describe interphase, and identify its stages.

b. What enzymes must be present for DNA replication to proceed normally?

c. A cell is actively manufacturing enough organelles to serve two functional cells. This cell is probably in what phase of interphase?

LO 3.19 Describe interphase, and explain its significance.

Step numbers and manageable "chunks" of information that are linked to visuals guide students through complex processes.

Frequent Practice

Three predictable places to stop and check understanding help students pace their learning throughout the chapter.

Module Reviews
appear at the end of every module for frequent and consistent self-assessment.

Module 11.4 Review

a. Identify the neuroglia of the central nervous system.

b. Which glial cell protects the CNS from chemicals and hormones circulating in the blood?

c. Which type of neuroglia would increase in the brain tissue of a person with a CNS infection?

Section Reviews
appear after groups of related modules and include "workbook-style" review activities, such as labeling and concept mapping.

SECTION 1 Review

Labeling

Label each of the structures in the following diagram of a neuron.

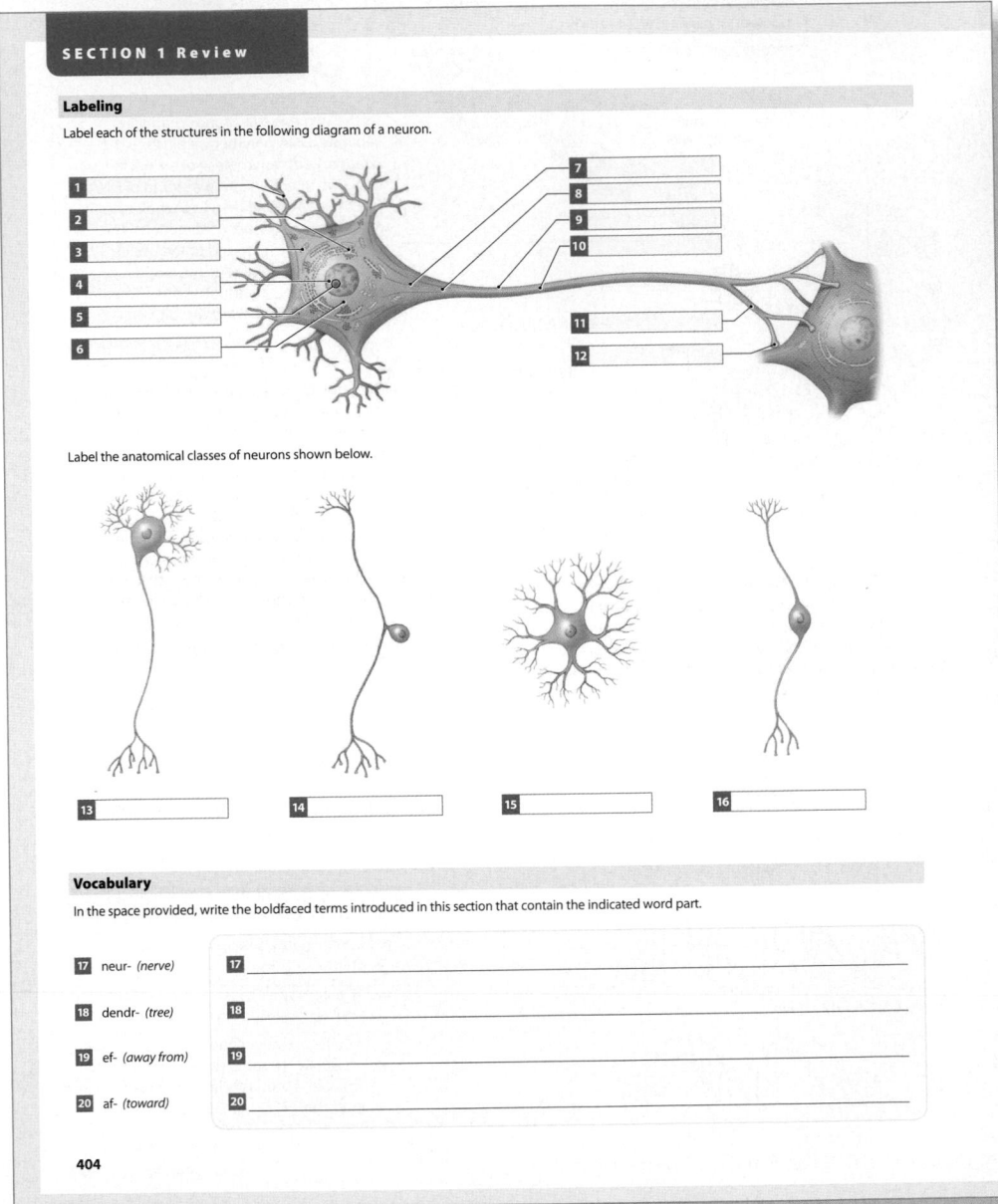

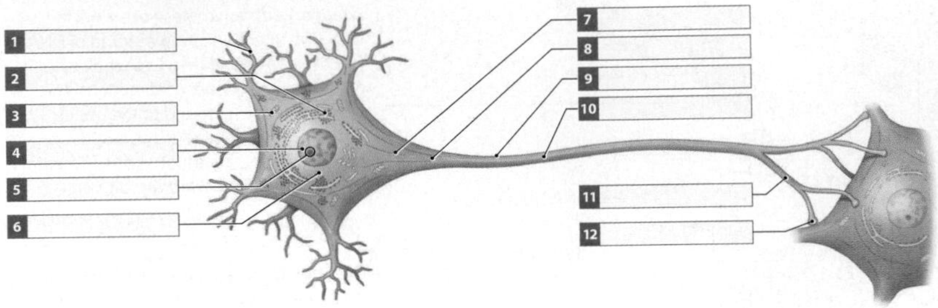

1
2
3
4
5
6
7
8
9
10
11
12

Label the anatomical classes of neurons shown below.

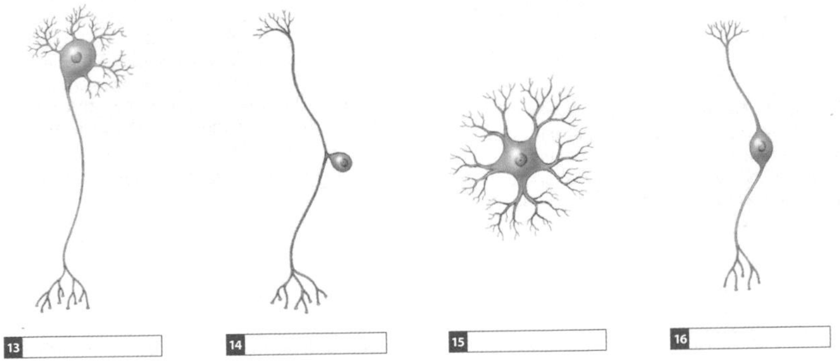

13
14
15
16

Vocabulary

In the space provided, write the boldfaced terms introduced in this section that contain the indicated word part.

17 neur- *(nerve)* 17 ______

18 dendr- *(tree)* 18 ______

19 ef- *(away from)* 19 ______

20 af- *(toward)* 20 ______

404

NEW Chapter Reviews

include brand-new narrative Study Outlines. Each Study Outline entry begins with the module number and title and then summarizes the module content.

All-new Chapter Review Questions include comprehensive questions, such as labeling, true/false, and multiple choice.

In the Chapter Integration section, one or two clinical scenarios are followed by critical thinking questions that help students tie important concepts together.

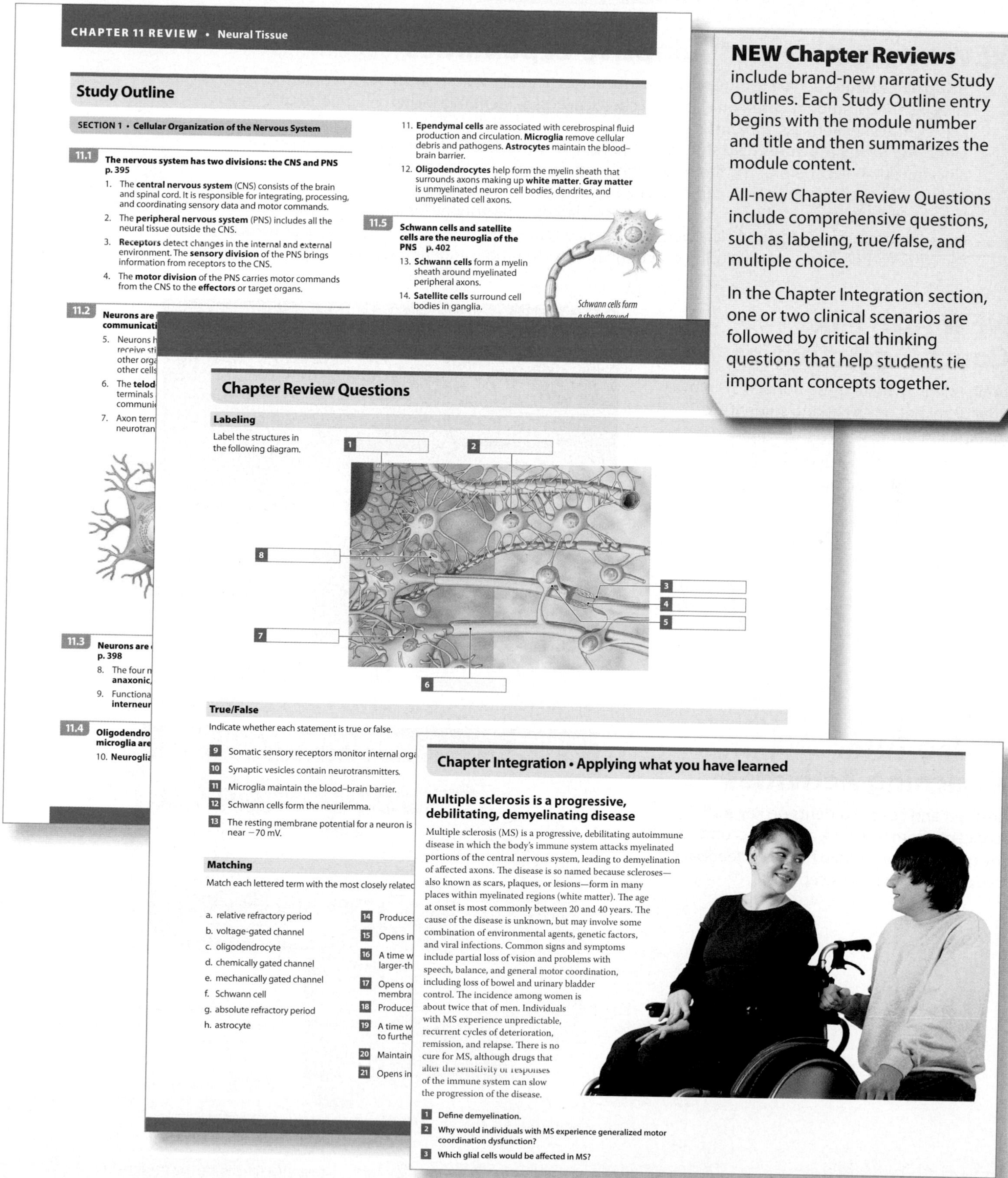

CHAPTER 11 REVIEW • Neural Tissue
Study Outline
SECTION 1 • Cellular Organization of the Nervous System
11.1 The nervous system has two divisions: the CNS and PNS p. 395
1. The central nervous system (CNS) consists of the brain and spinal cord. It is responsible for integrating, processing, and coordinating sensory data and motor commands.
2. The peripheral nervous system (PNS) includes all the neural tissue outside the CNS.
3. Receptors detect changes in the internal and external environment. The sensory division of the PNS brings information from receptors to the CNS.
4. The motor division of the PNS carries motor commands from the CNS to the effectors or target organs.
11. Ependymal cells are associated with cerebrospinal fluid production and circulation. Microglia remove cellular debris and pathogens. Astrocytes maintain the blood–brain barrier.
12. Oligodendrocytes help form the myelin sheath that surrounds axons making up white matter. Gray matter is unmyelinated neuron cell bodies, dendrites, and unmyelinated cell axons.
11.5 Schwann cells and satellite cells are the neuroglia of the PNS p. 402
13. Schwann cells form a myelin sheath around myelinated peripheral axons.
14. Satellite cells surround cell bodies in ganglia.
Schwann cells form
Chapter Review Questions
Labeling
Label the structures in the following diagram.
1
2
3
4
5
6
7
8
True/False
Indicate whether each statement is true or false.
9 Somatic sensory receptors monitor internal orga
10 Synaptic vesicles contain neurotransmitters.
11 Microglia maintain the blood–brain barrier.
12 Schwann cells form the neurilemma.
13 The resting membrane potential for a neuron is near −70 mV.
Matching
Match each lettered term with the most closely related
a. relative refractory period
b. voltage-gated channel
c. oligodendrocyte
d. chemically gated channel
e. mechanically gated channel
f. Schwann cell
g. absolute refractory period
h. astrocyte
Chapter Integration • Applying what you have learned
Multiple sclerosis is a progressive, debilitating, demyelinating disease
Multiple sclerosis (MS) is a progressive, debilitating autoimmune disease in which the body's immune system attacks myelinated portions of the central nervous system, leading to demyelination of affected axons. The disease is so named because scleroses—also known as scars, plaques, or lesions—form in many places within myelinated regions (white matter). The age at onset is most commonly between 20 and 40 years. The cause of the disease is unknown, but may involve some combination of environmental agents, genetic factors, and viral infections. Common signs and symptoms include partial loss of vision and problems with speech, balance, and general motor coordination, including loss of bowel and urinary bladder control. The incidence among women is about twice that of men. Individuals with MS experience unpredictable, recurrent cycles of deterioration, remission, and relapse. There is no cure for MS, although drugs that alter the sensitivity or responses of the immune system can slow the progression of the disease.
1 Define demyelination.
2 Why would individuals with MS experience generalized motor coordination dysfunction?
3 Which glial cells would be affected in MS?

MasteringA&P® Assignable Content

NEW Interactive and Adaptive Capabilities

- **Adaptive Follow-up Assignments** allow instructors to easily assign personalized content for each individual student based on strengths and weaknesses identified by his or her performance on MasteringA&P parent assignments.
- **Dynamic Study Modules** help students acquire, retain, and recall information faster and more efficiently than ever before. The flashcard-style modules are available as a self-study tool or can be assigned by the instructor. They can be easily accessed with smartphones.
- **Learning Catalytics** is a "bring your own device" (laptop, smartphone, or tablet) student engagement, assessment, and classroom intelligence system. With Learning Catalytics, instructors can assess students in real time using open-ended tasks to probe student understanding.

NEW A&P Flix™ Coaching Activities

bring interactivity to these popular 3D movie-quality animations by asking students to manipulate the visuals.

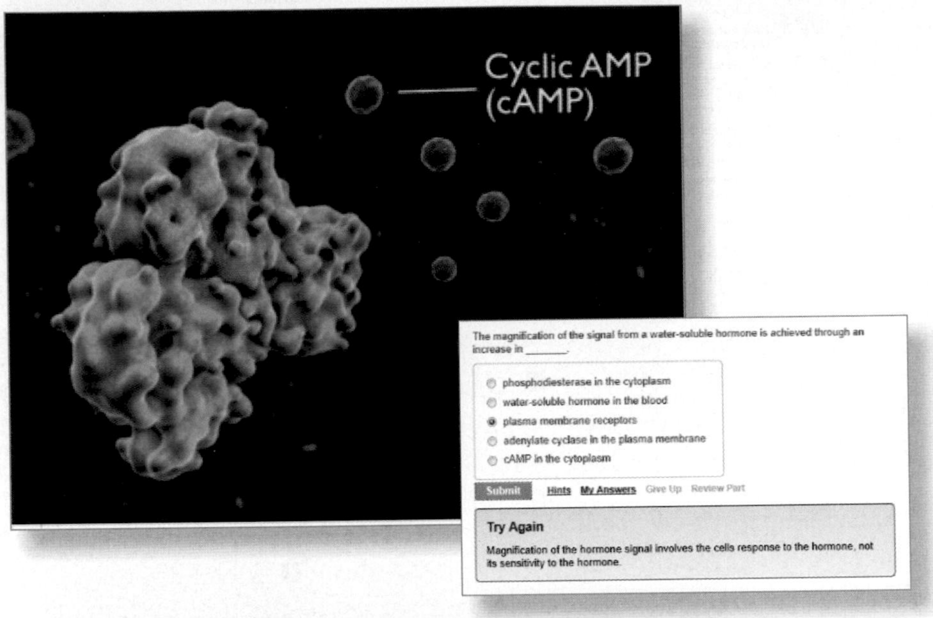

Cyclic AMP (cAMP)
The magnification of the signal from a water-soluble hormone is achieved through an increase in ______.
phosphodiesterase in the cytoplasm
water-soluble hormone in the blood
plasma membrane receptors
adenylate cyclase in the plasma membrane
cAMP in the cytoplasm
Submit
Hints
My Answers
Give Up
Review Part
Try Again
Magnification of the hormone signal involves the cells response to the hormone, not its sensitivity to the hormone.

Video Tutor Coaching Activities

instruct and coach students on key A&P concepts from the book and are accompanied by questions with video hints and feedback specific to their misconceptions.

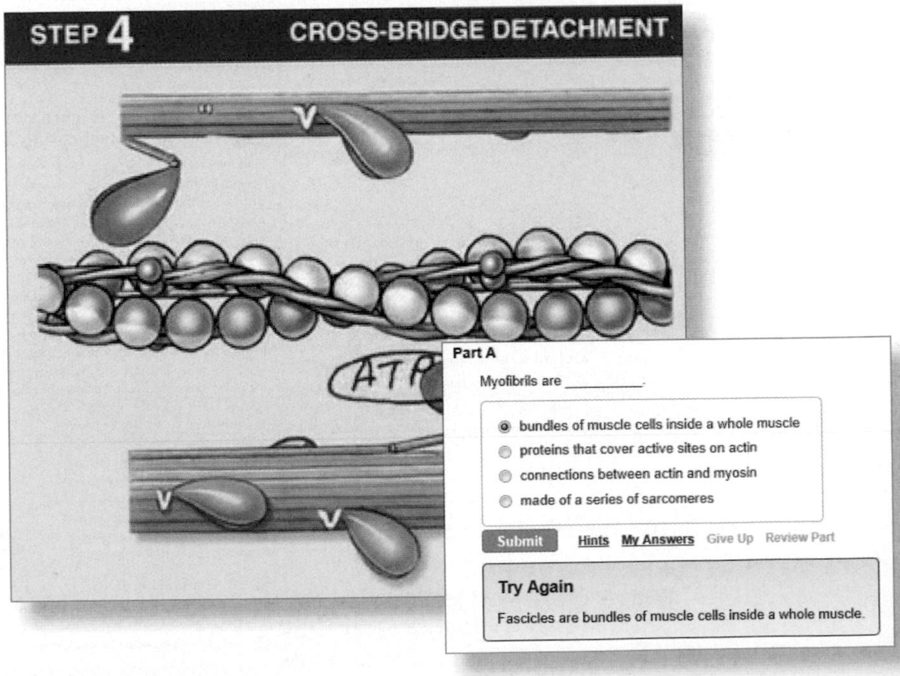

STEP 4
CROSS-BRIDGE DETACHMENT
Part A
Myofibrils are ______.
bundles of muscle cells inside a whole muscle
proteins that cover active sites on actin
connections between actin and myosin
made of a series of sarcomeres
Submit
Hints
My Answers
Give Up
Review Part
Try Again
Fascicles are bundles of muscle cells inside a whole muscle.

NEW Tough Topic Coaching Activities

are highly visual, assignable activities designed to bring interactivity to select two-page modules in the book. These multi-part activity items include the ranking and sorting types that ask students to manipulate the visuals.

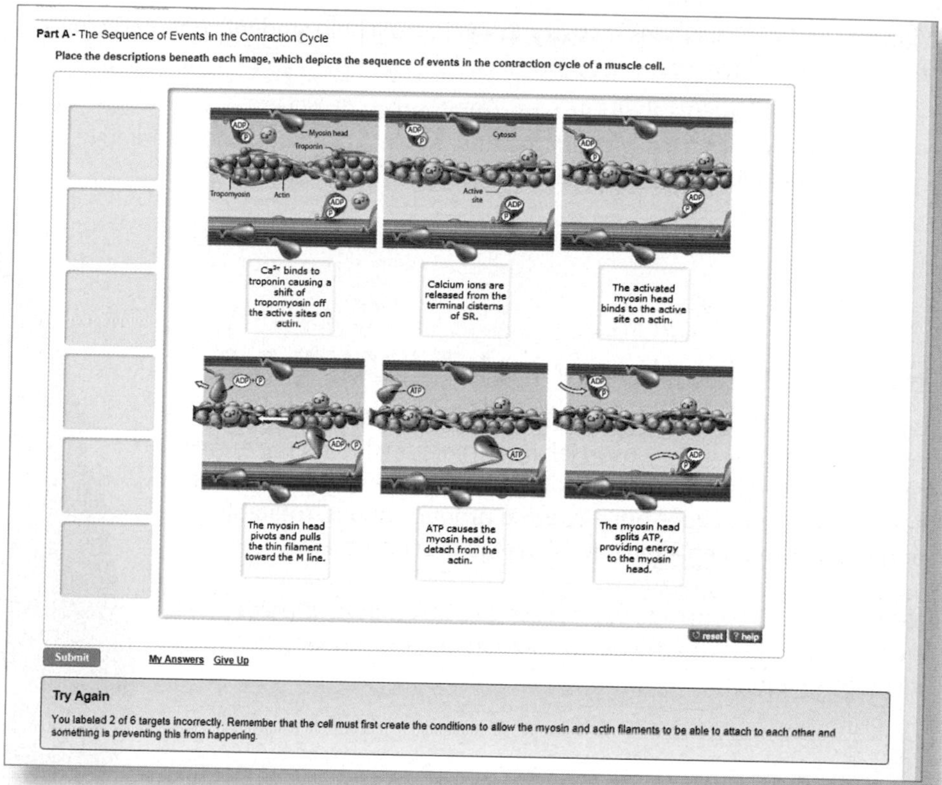

Part A - The Sequence of Events in the Contraction Cycle
Place the descriptions beneath each image, which depicts the sequence of events in the contraction cycle of a muscle cell.
Myosin head
Troponin
Tropomyosin
Actin
Cytosol
Active site
ADP
P
ATP
Ca^{2+} binds to troponin causing a shift of tropomyosin off the active sites on actin.
Calcium ions are released from the terminal cisterns of SR.
The activated myosin head binds to the active site on actin.
The myosin head pivots and pulls the thin filament toward the M line.
ATP causes the myosin head to detach from the actin.
The myosin head splits ATP, providing energy to the myosin head.
reset
help
Submit
My Answers
Give Up
Try Again
You labeled 2 of 6 targets incorrectly. Remember that the cell must first create the conditions to allow the myosin and actin filaments to be able to attach to each other and something is preventing this from happening.

Interactive Physiology® Coaching Activities

help students dive deeper into complex physiological processes using the Interactive Physiology tutorial program.

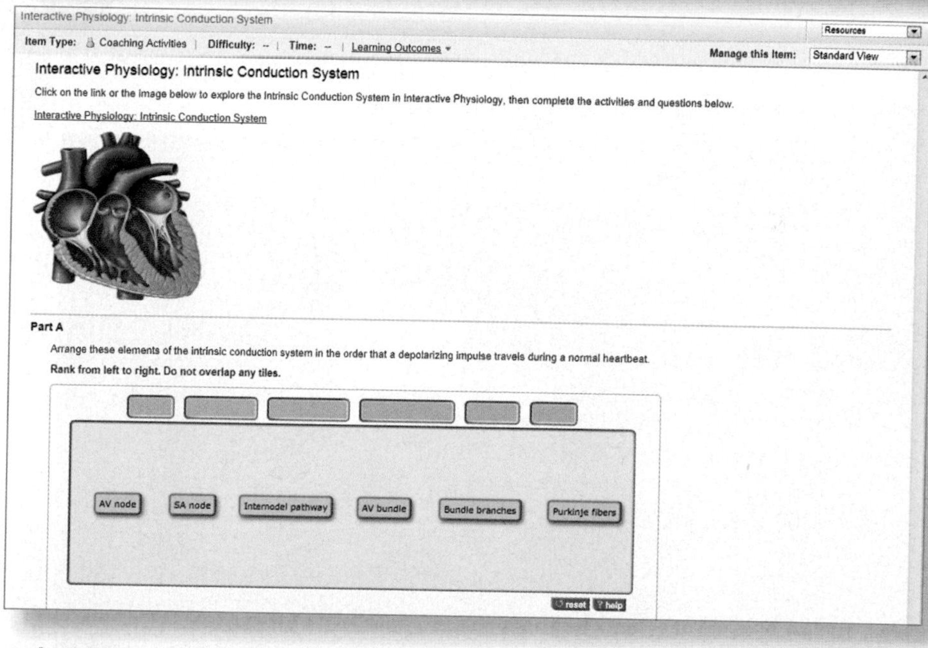

Interactive Physiology: Intrinsic Conduction System
Resources
Item Type: Coaching Activities | Difficulty: -- | Time: -- | Learning Outcomes
Manage this Item: Standard View
Interactive Physiology: Intrinsic Conduction System
Click on the link or the image below to explore the Intrinsic Conduction System in Interactive Physiology, then complete the activities and questions below.
Interactive Physiology: Intrinsic Conduction System
Part A
Arrange these elements of the intrinsic conduction system in the order that a depolarizing impulse travels during a normal heartbeat.
Rank from left to right. Do not overlap any tiles.
AV node
SA node
Internodal pathway
AV bundle
Bundle branches
Purkinje fibers
reset
help

Also Assignable in MasteringA&P®:

- **Art-labeling Activities** are drag and drop activities that allow students to assess their knowledge of terms and structures.
- **Art-based Questions** are conceptual questions related to art and instruct students with wrong-answer feedback.
- **Chemistry Review Activities** reinforce chemistry concepts necessary for an understanding of A&P.
- **PAL 3.0** and assessments
- **PhysioEx™ 9.1** and assessments
- **Reading Quiz Questions**
- **Chapter Test Questions**
- **Test Bank Questions**

MasteringA&P® Study Area

MasteringA&P® includes a **Study Area** that will help students get ready for tests with its simple three-step approach. Students can:

1. **Take a pre-test** and obtain a personalized study plan.
2. **Learn and practice** with animations, labeling activities, and interactive tutorials.
3. **Self-test** with quizzes and a chapter practice test.

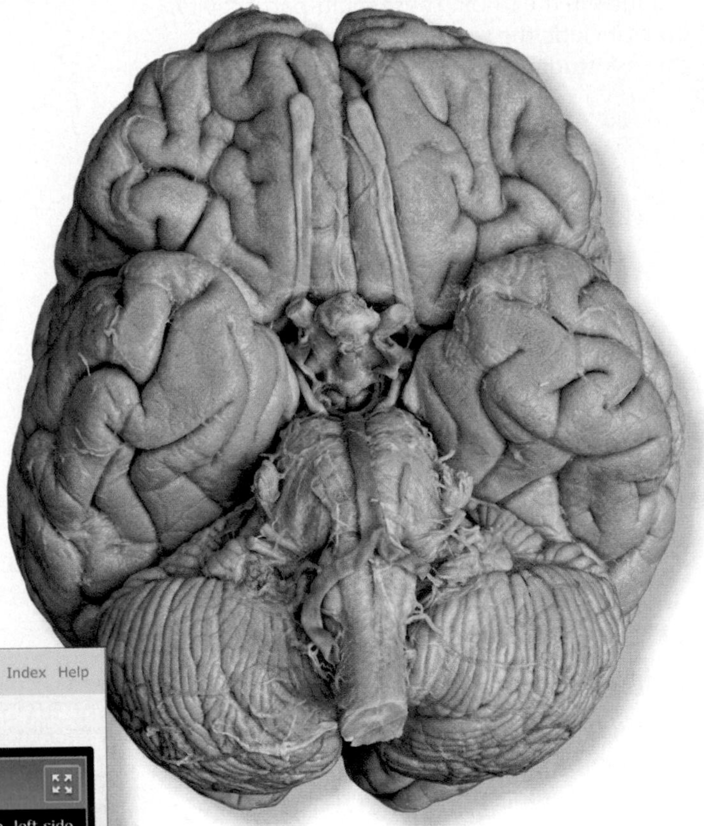

Practice Anatomy Lab™ (PAL™) 3.0

is a virtual anatomy study and practice tool that gives students 24/7 access to the most widely used lab specimens, including the human cadaver, anatomical models, histology, cat, and fetal pig. PAL 3.0 is easy to use and includes built-in audio pronunciations, rotatable bones, and simulated fill-in-the-blank lab practical exams.

NEW The PAL 3.0 App lets you access PAL 3.0 on **your iPad or Android tablet**. With the pinch-to-zoom feature, images can be instantly enlarged.

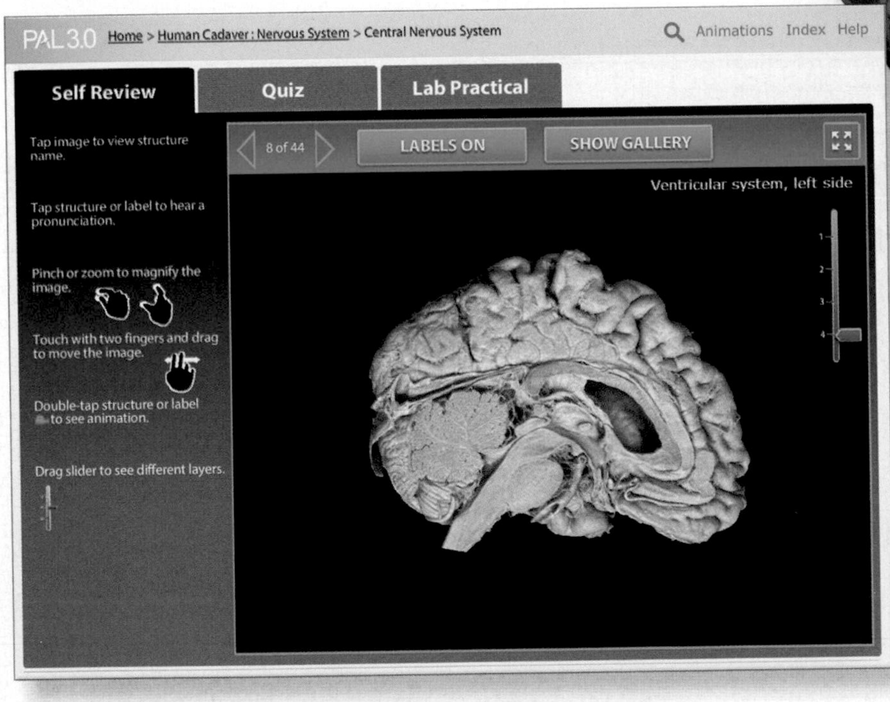
PAL 3.0
Home > Human Cadaver : Nervous System > Central Nervous System
Animations Index Help
Self Review
Quiz
Lab Practical
Tap image to view structure name.
Tap structure or label to hear a pronunciation.
Pinch or zoom to magnify the image.
Touch with two fingers and drag to move the image.
Double-tap structure or label to see animation.
Drag slider to see different layers.
8 of 44
LABELS ON
SHOW GALLERY
Ventricular system, left side

Also Available in the Study Area:

- eText
- PhysioEx™ 9.1
- Video Tutors

A&P Flix™

are 3D movie-quality animations with self-paced tutorials and gradable quizzes that help students master the toughest topics in A&P:

- **NEW** Protein Synthesis
- Membrane Transport
- DNA Replication
- Mitosis
- **NEW** Endochondral Ossification
- Events at the Neuromuscular Junction
- The Cross Bridge Cycle
- Excitation-Contraction Coupling
- Resting Membrane Potential
- Generation of an Action Potential
- Propagation of an Action Potential
- **NEW** The Stretch Reflex
- **NEW** The Mechanism of Hormone Action: Second Messenger cAMP
- Origins, Insertions, Actions, and Innervations
  Over 60 animations on this topic
- Group Muscle Actions and Joints
  Over 50 animations on this topic

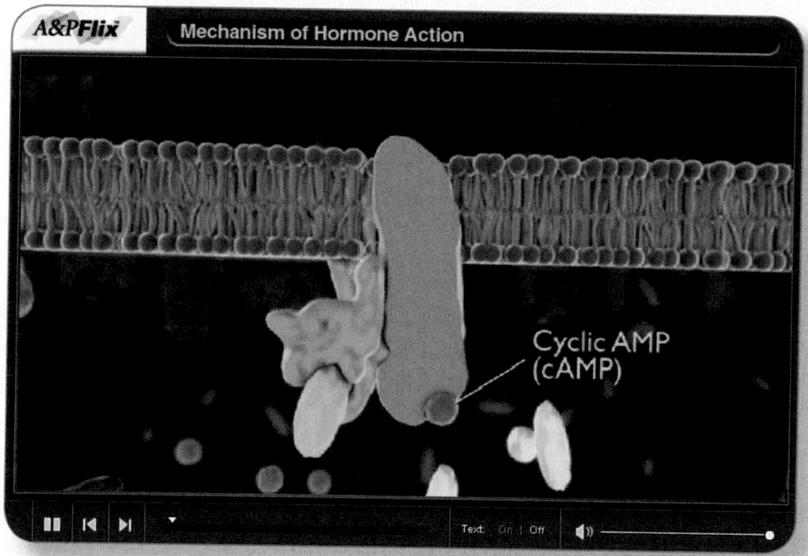
A&PFlix
Mechanism of Hormone Action
Cyclic AMP (cAMP)
Text: On | Off

Interactive Physiology® (IP)

helps students understand the hardest part of A&P: physiology. Fun, interactive tutorials, games, and quizzes give students additional explanations to help them grasp difficult concepts.

Modules:

- Muscular System
- Nervous System I
- Nervous System II
- Cardiovascular System
- Respiratory System
- Urinary System
- Fluids & Electrolytes
- Endocrine System
- Digestive System
- Immune System

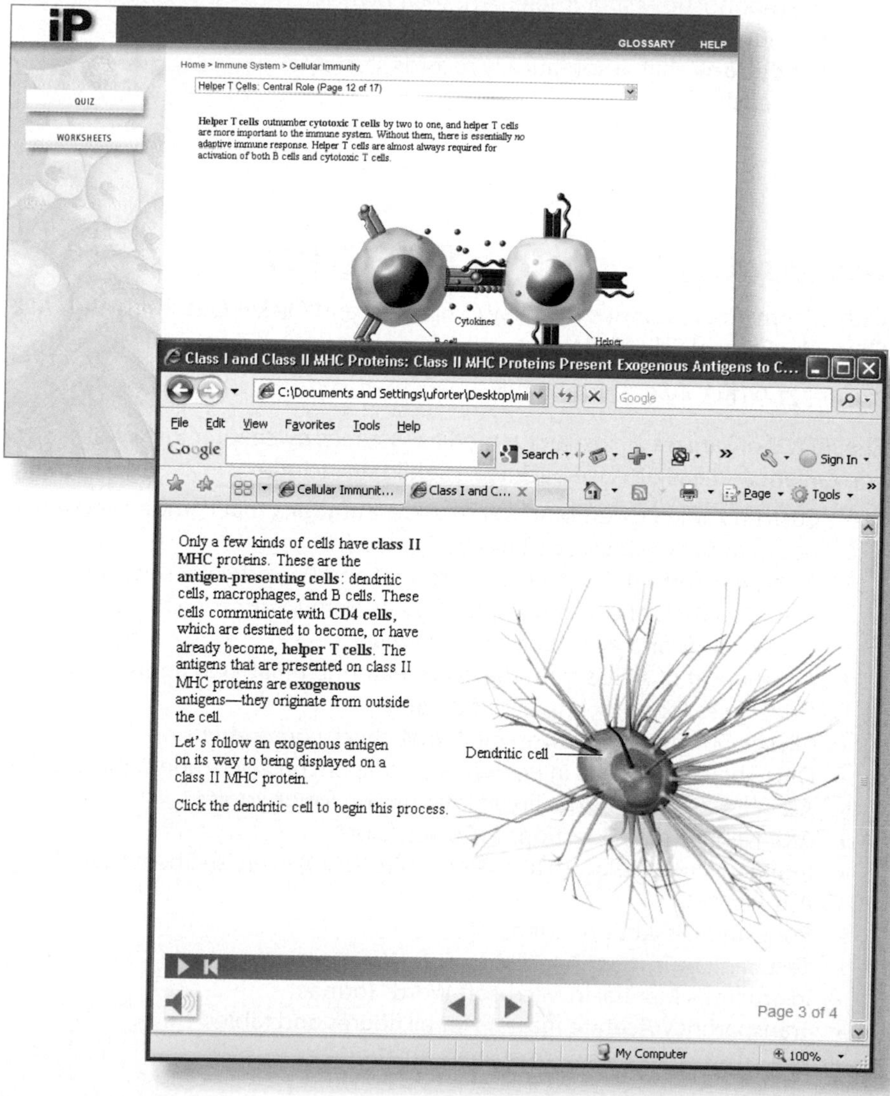
iP
GLOSSARY
HELP
QUIZ
WORKSHEETS
Home > Immune System > Cellular Immunity
Helper T Cells: Central Role (Page 12 of 17)
Helper T cells outnumber cytotoxic T cells by two to one, and helper T cells are more important to the immune system. Without them, there is essentially no adaptive immune response. Helper T cells are almost always required for activation of both B cells and cytotoxic T cells.
Cytokines
Class I and Class II MHC Proteins: Class II MHC Proteins Present Exogenous Antigens to C...
Only a few kinds of cells have class II MHC proteins. These are the antigen-presenting cells: dendritic cells, macrophages, and B cells. These cells communicate with CD4 cells, which are destined to become, or have already become, helper T cells. The antigens that are presented on class II MHC proteins are exogenous antigens—they originate from outside the cell.
Let's follow an exogenous antigen on its way to being displayed on a class II MHC protein.
Click the dendritic cell to begin this process.
Dendritic cell
Page 3 of 4
My Computer
100%

Support for Instructors

MyReadinessTest™

by Lori K. Garrett

MyReadinessTest for A&P is a powerful online system that gets students prepared before their course starts. It assesses students' proficiency in study skills and foundational science and math concepts and provides coaching in core areas where students need additional practice and review. It offers:

- **Student online access** upon registration for their course
- **Diagnostic Test and Cumulative Test** based on learning outcomes from the widely used A&P primer, *Get Ready for A&P*
- **Personalized Study Plan** based on students' test results that includes practice questions with Tutorials
- **Flexible Testing** that allows instructors to edit the Diagnostic Test or implement their own placement test or exit exam
- **Gradebook** that automatically records students' test results

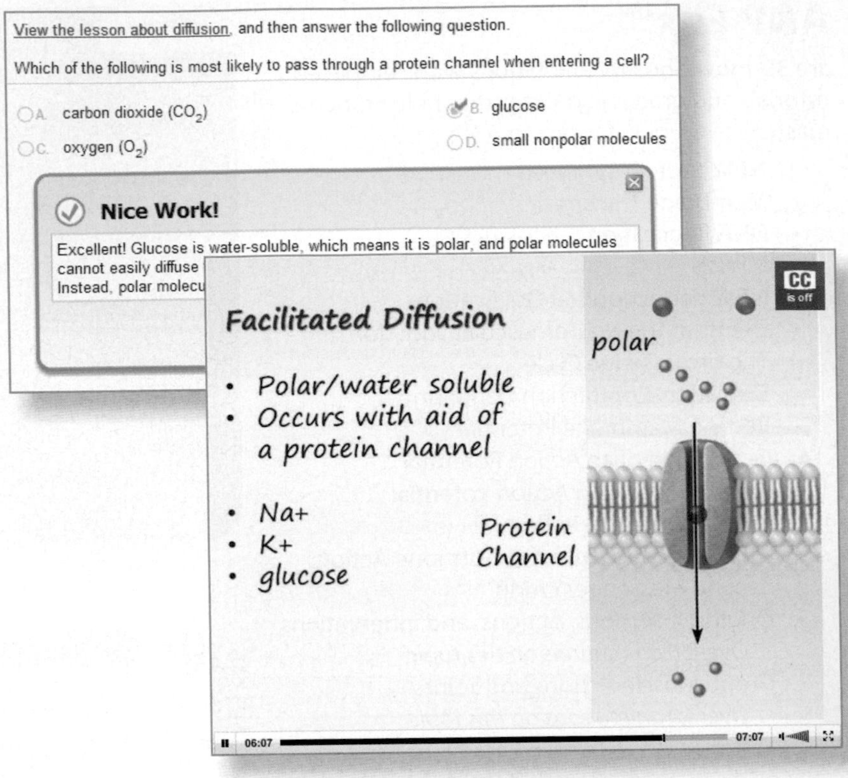

View the lesson about diffusion, and then answer the following question.
Which of the following is most likely to pass through a protein channel when entering a cell?
A. carbon dioxide (CO_2)
B. glucose
C. oxygen (O_2)
D. small nonpolar molecules
Nice Work!
Excellent! Glucose is water-soluble, which means it is polar, and polar molecules
cannot easily diffuse
Instead, polar molecu
Facilitated Diffusion
Polar/water soluble
Occurs with aid of a protein channel
Na+
K+
glucose
polar
Protein Channel
CC is off
06:07
07:07

Instructor Resource DVD (IRDVD)

with Lecture Presentations by Betsy C. Brantley and Clicker Questions and Quiz Shows by Samuel Schwarzlose

978-0-321-95143-4 / 0-321-95143-3

The IRDVD organizes all instructor media resources by chapter into one convenient and easy-to-use package. Highlights include:

- Customizable PowerPoint® Lecture Presentations that combine lecture notes, figures, tables, and links to animations
- All figures from the book in JPEG format and PowerPoint® slides (with editable labels and without) plus figures from *Martini's Atlas of the Human Body* and *A&P Applications Manual*
- Another set of JPEGs from the book featuring unlabeled figures *with leader lines* for quick and easy quizzing
- Clicker Questions in PowerPoint® that check comprehension
- Quiz Show Questions in PowerPoint® that encourage student interaction
- A&P Flix™ 3D movie-quality animations on tough topics
- A&P Flix™ Clicker Questions in PowerPoint®
- Interactive Physiology® 10-System Suite (IP-10) Exercise Sheets and Answer Key
- Bone and Dissection Videos
- Test Bank in TestGen® and Microsoft Word® formats
- Instructor's Manual in Microsoft Word® format
- Transparency Acetate masters for all figures and tables
- The IRDVD for Practice Anatomy Lab™ (PAL™) 3.0

INSTRUCTOR RESOURCE DVD
visual anatomy & physiology
2nd edition
MARTINI / OBER / NATH / BARTHOLOMEW / PETTI

eText with Whiteboard Mode

The *Visual Anatomy & Physiology* eText comes with Whiteboard Mode, allowing instructors to use the eText for dynamic classroom presentations. Instructors can show one-page or two-page views from the book, zoom in or out to focus on select topics, and use the Whiteboard Mode to point to structures, circle parts of a process, trace pathways, and customize their presentations.

Instructors can also add notes to guide students, upload documents, and share their custom-enhanced eText with the whole class.

Instructors can find the eText with Whiteboard Mode on MasteringA&P.

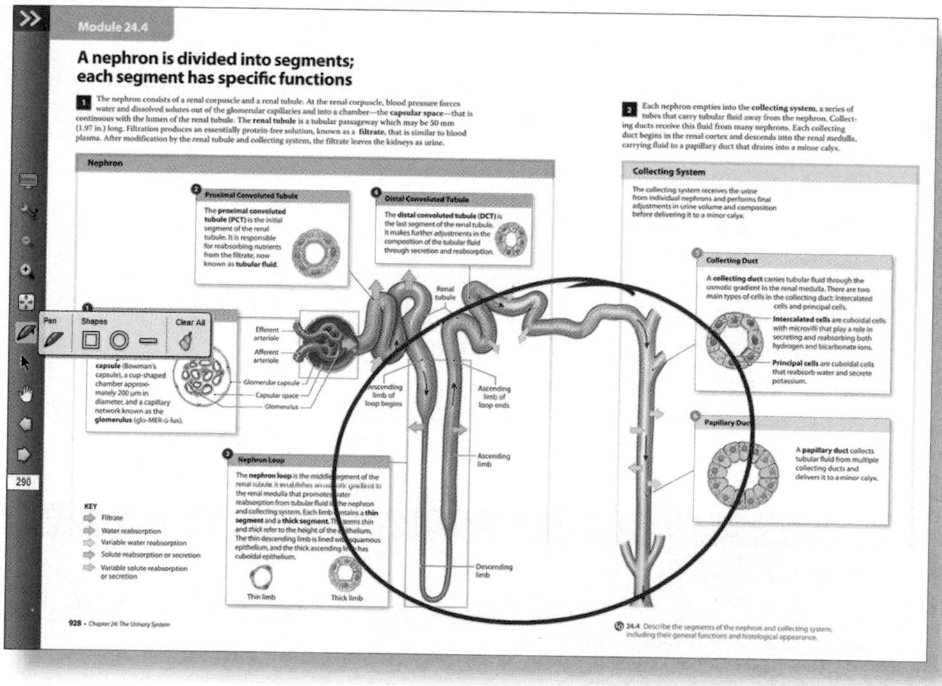
Module 24.4
A nephron is divided into segments; each segment has specific functions
Nephron
Collecting System
Pen
Shapes
Clear All
290

INSTRUCTOR'S MANUAL
visual
anatomy &
physiology
2nd
MARTINI / OBER / NATH / BARTHOLOMEW / PETTI

Instructor's Manual
by Jeff Schinske

978-0-321-96256-0 / 0-321-96256-7

This useful resource includes a wealth of materials to help instructors organize their lectures, such as lecture ideas, visual analogies, suggested classroom demonstrations, vocabulary aids, applications, and common student misconceptions/problems.

TEST BANK
visual
anatomy &
physiology
2nd
MARTINI / OBER / NATH / BARTHOLOMEW / PETTI

Printed Test Bank
by Alexander G. Cheroske and Jason LaPres

978-0-321-96268-3 / 0-321-96268-0

The test bank of more than 3,000 questions tied to the Learning Outcomes in each chapter helps instructors design a variety of tests and quizzes. The test bank includes text-based and art-based questions. This supplement is the print version of TestGen® that is in the IRDVD package.

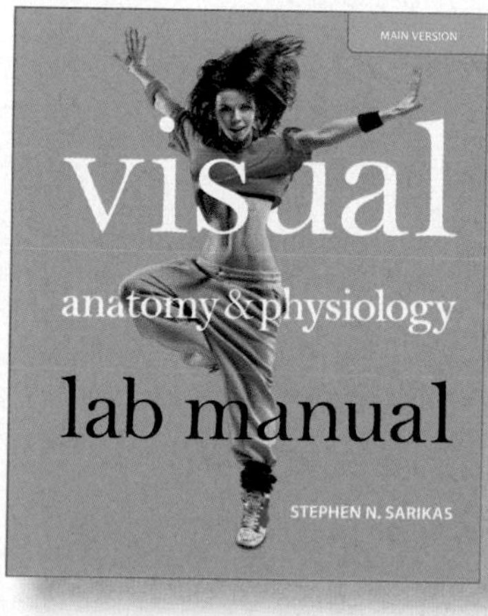
MAIN VERSION
visual
anatomy & physiology
lab manual
STEPHEN N. SARIKAS

Visual Anatomy & Physiology Lab Manual
by Stephen N. Sarikas

978-0-321-92854-2 / 0-321-92854-7

The *Visual Anatomy & Physiology Lab Manual* brings all of the strengths of the revolutionary *Visual Anatomy & Physiology* book to the lab. This lab manual combines a visual approach with a modular organization to maximize learning. The lab practice consists of hands-on activities in the lab manual and assignable content in MasteringA&P. Main, Cat, and Pig versions are available.

Support for Students

eText

MasteringA&P® includes an eText. Students can access their textbook wherever and whenever they are online. eText pages look exactly like the printed text yet offer additional functionality. Students can:

- Create notes.
- Highlight text in different colors.
- Create bookmarks.
- Zoom in and out.
- View in single-page or two-page view.
- Click hyperlinked words and phrases to view definitions.
- Link directly to relevant animations.
- Search quickly and easily for specific content.

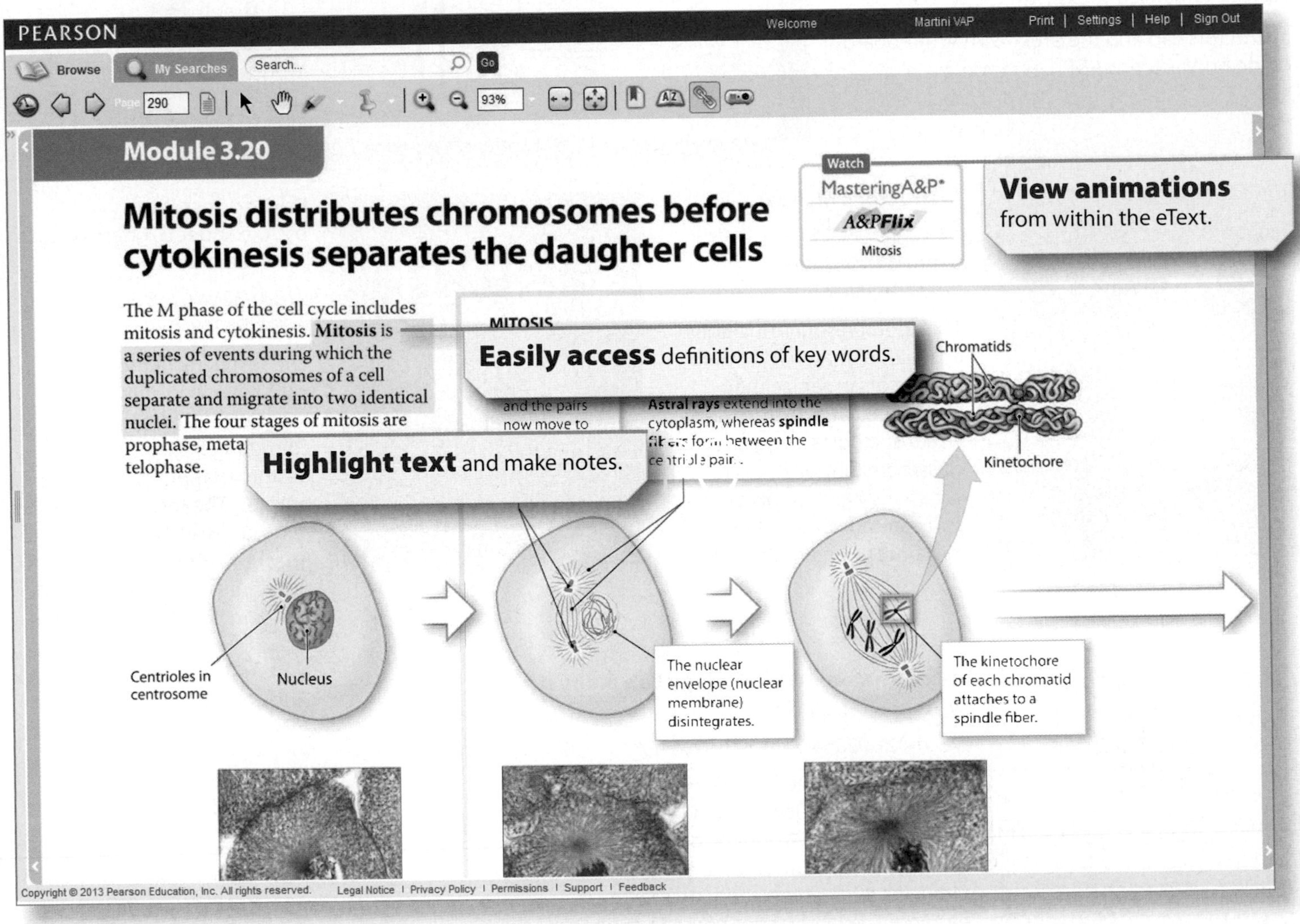

PEARSON
Welcome
Martini VAP
Print | Settings | Help | Sign Out
Browse
My Searches
Search...
Go
290
93%
Module 3.20
Watch
MasteringA&P®
A&PFlix
Mitosis
View animations from within the eText.
Mitosis distributes chromosomes before cytokinesis separates the daughter cells
The M phase of the cell cycle includes mitosis and cytokinesis. Mitosis is a series of events during which the duplicated chromosomes of a cell separate and migrate into two identical nuclei. The four stages of mitosis are prophase, meta
telophase.
Easily access definitions of key words.
Highlight text and make notes.
MITOSIS
Chromatids
Kinetochore
Centrioles in centrosome
Nucleus
The nuclear envelope (nuclear membrane) disintegrates.
The kinetochore of each chromatid attaches to a spindle fiber.
Copyright © 2013 Pearson Education, Inc. All rights reserved.
Legal Notice | Privacy Policy | Permissions | Support | Feedback

Every item can be packaged with the main student text.

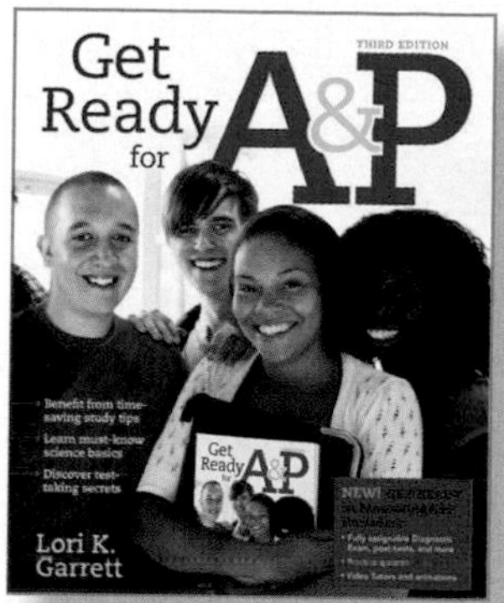
THIRD EDITION
Get Ready for A&P
Lori K. Garrett

Get Ready for A&P
by Lori K. Garrett

978-0-321-81336-7 / 0-321-81336-7

This book and online component were created to help students be better prepared for their A&P course. Features include pre-tests, guided explanations followed by interactive quizzes and exercises, and end-of-chapter cumulative tests. Also available in the Study Area of MasteringA&P.

STUDENT WORKSHEETS
visual anatomy & physiology
2nd
MARTINI / OBER / NATH / BARTHOLOMEW / PETTI

Student Worksheets for Visual Anatomy & Physiology
by Frederic H. Martini, William C. Ober, Judi L. Nath, Edwin F. Bartholomew, and Kevin Petti

978-0-321-95631-6 / 0-321-95631-1

This booklet contains all of the Section Review pages from the book for students who would prefer to mark their answers on separate pages rather than in the book itself.

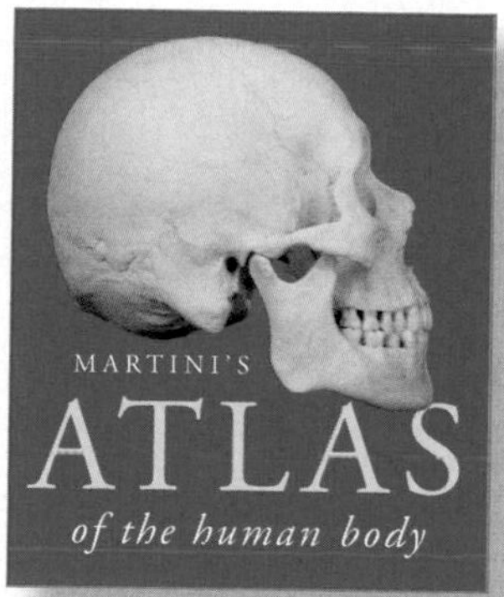
MARTINI'S
ATLAS
of the human body

Martini's Atlas of the Human Body
by Frederic H. Martini

978-0-321-94072-8 / 0-321-94072-5

The Atlas offers an abundant collection of anatomy photographs, radiology scans, and embryology summaries, helping students visualize structures and become familiar with the types of images seen in a clinical setting.

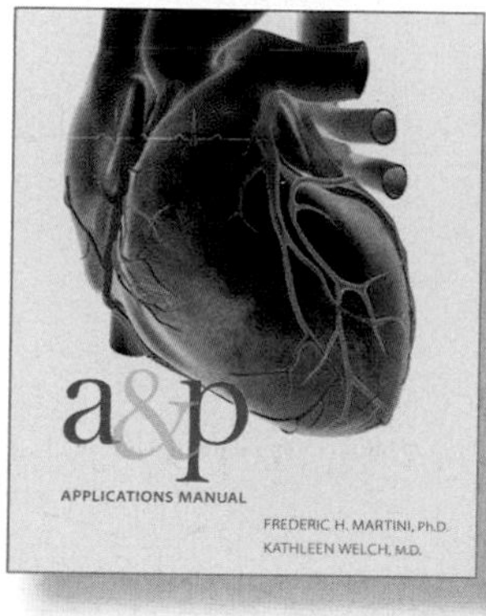
a&p
APPLICATIONS MANUAL
FREDERIC H. MARTINI, Ph.D.
KATHLEEN WELCH, M.D.

A&P Applications Manual
by Frederic H. Martini and Kathleen Welch

978-0-321-94973-8 / 0-321-94973-0

This manual contains extensive discussions on clinical topics and disorders to help students apply the concepts of anatomy and physiology to daily life and their future health professions.

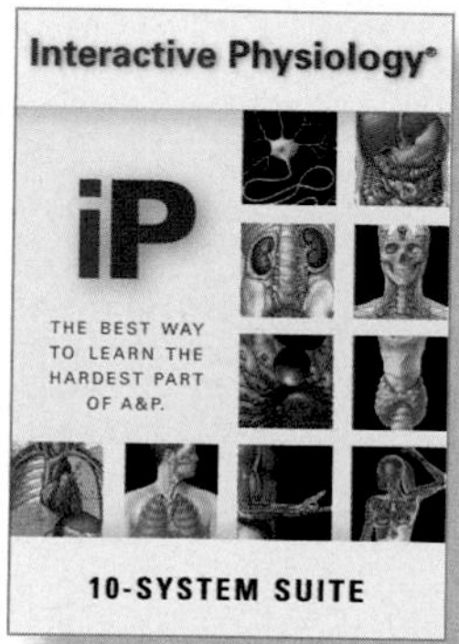
Interactive Physiology®
iP
THE BEST WAY TO LEARN THE HARDEST PART OF A&P.
10-SYSTEM SUITE

Interactive Physiology® 10-System Suite (IP-10) CD-ROM
978-0-131-36275-8 / 0-131-36275-5

IP-10 helps students understand the hardest part of A&P: physiology. Fun, interactive tutorials, games, and quizzes give students additional explanations to help them grasp difficult physiological concepts.

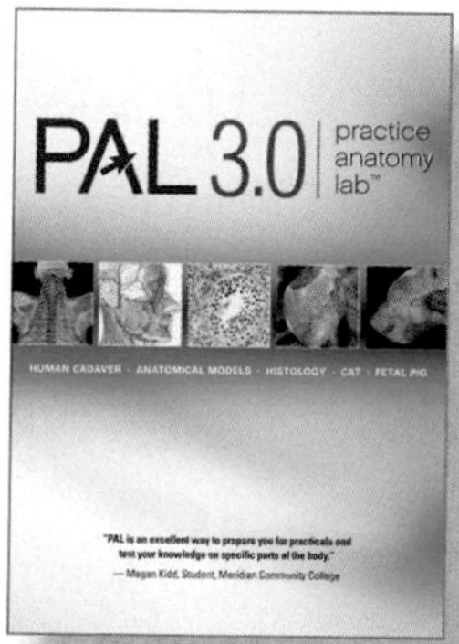
PAL 3.0
practice anatomy lab™

Practice Anatomy Lab™ (PAL™) 3.0 DVD
by Ruth Heisler, Nora Hebert, Jett Chinn, Karen Krabbenhoft, and Olga Malakhova

978-0-321-68211-6 / 0-321-68211-4

PAL 3.0 is an indispensable virtual anatomy study and practice tool that gives students 24/7 access to the most widely used lab specimens, including the human cadaver, anatomical models, histology, cat, and fetal pig.

Also available: PAL 3.0 Lab Guide
978-0-321-84025-7 / 0-321-84025-9

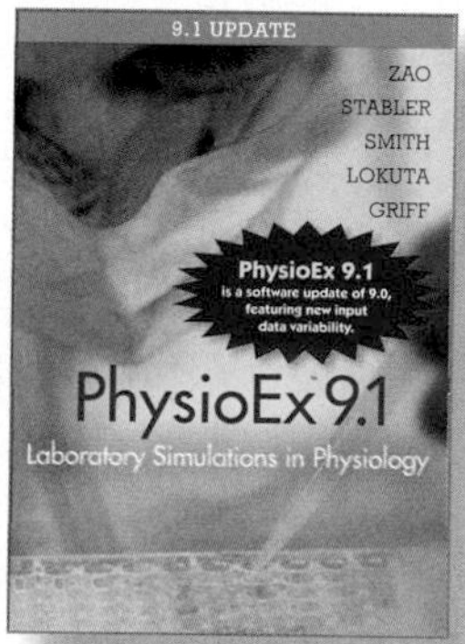
9.1 UPDATE
ZAO
STABLER
SMITH
LOKUTA
GRIFF
PhysioEx 9.1
PhysioEx 9.1
Laboratory Simulations in Physiology

PhysioEx™ 9.1 Laboratory Simulations in Physiology
by Peter Zao, Timothy Stabler, Lori A. Smith, Andrew Lokuta, and Edwin Griff

978-0-321-92964-8 / 0-321-92964-0

This easy-to-use laboratory simulation software and lab manual consists of 12 exercises containing 63 physiology lab activities that can be used to supplement or substitute wet labs safely and cost-effectively. Now with input data variability.

visual anatomy & physiology

2nd edition

Frederic H. Martini, Ph.D.
University of Hawaii at Manoa

William C. Ober, M.D.
Washington and Lee University

Judi L. Nath, Ph.D.
Lourdes University, Sylvania, Ohio

Edwin F. Bartholomew, M.S.

Kevin Petti, Ph.D.
San Diego Miramar College

Claire E. Ober, R.N.
Illustrator

Kathleen Welch, M.D.
Clinical Consultant

Ralph T. Hutchings
Biomedical Photographer

PEARSON

Boston Columbus Indianapolis New York San Francisco Upper Saddle River
Amsterdam Cape Town Dubai London Madrid Milan Munich Paris Montréal Toronto
Delhi Mexico City São Paulo Sydney Hong Kong Seoul Singapore Taipei Tokyo

Executive Editor: *Leslie Berriman*
Associate Project Editor: *Lisa Damerel*
Assistant Editor: *Cady Owens*
Editorial Assistant: *Sharon Kim*
Director of Development: *Barbara Yien*
Development Editor: *Molly Ward*
Managing Editor: *Mike Early*
Assistant Managing Editor: *Nancy Tabor*
Director of Digital Product Development: *Lauren Fogel*
Executive Content Producer: *Liz Winer*
Senior Content Producer: *Aimee Pavy*
Production Management and Composition: *S4Carlisle Publishing Services, Inc.*
Copyeditor: *Michael Rossa*
Design Manager: *Mark Ong*
Interior Designer: *Gibson Design Associates*
Cover Designer: *tani hasegawa*
Art House: *Precision Graphics*
Contributing Illustrators: *imagineeringart.com; Anita Impagliazzo*
Photo Permissions Management: *Bill Smith Group*
Photo Researchers: *Stefanie Ramsay; Luke Malone; Cordes Hoffman*
Associate Director of Image Management: *Travis Amos*
Senior Procurement Specialist: *Stacey Weinberger*
Senior Anatomy & Physiology Specialist: *Derek Perrigo*
Senior Marketing Manager: *Allison Rona*

Cover Photo Credit: Alexander Yakovlev/Shutterstock

Credits and acknowledgments for materials borrowed from other sources and reproduced, with permission, in this textbook appear on page C-1.

Notice: Our knowledge in clinical sciences is constantly changing. The authors and the publisher of this volume have taken care that the information contained herein is accurate and compatible with the standards generally accepted at the time of the publication. Nevertheless, it is difficult to ensure that all information given is entirely accurate for all circumstances. The authors and the publisher disclaim any liability, loss, or damage incurred as a consequence, directly or indirectly, of the use and application of any of the contents of this volume.

Library of Congress Cataloging-in-Publication Data
Martini, Frederic, author.
Visual anatomy & physiology / Frederic H. Martini, William C. Ober, Judi L. Nath, Edwin F. Bartholomew, Kevin Petti, Claire E. Ober; illustrator, Kathleen Welch. — Second edition.
p.; cm.
Visual anatomy and physiology
Includes index.
ISBN-13: 978-0-321-91894-9
ISBN-10: 0-321-91894-0
I. Title. II. Title: Visual anatomy and physiology.
[DNLM: 1. Anatomy. 2. Physiological Phenomena. QS 4]
QP31.2
612—dc23
2013036685

ISBN 10: 0-321-91894-0 (Student edition)
ISBN 13: 978-0-321-91894-9 (Student edition)
ISBN 10: 0-321-95132-8 (Instructor's Review Copy)
ISBN 13: 978-0-321-95132-8 (Instructor's Review Copy)

PEARSON

www.pearsonhighered.com

2 3 4 5 6 7 8 9 10—RRW—17 16 15 14

To my son, PK, for convincing me it was time to look at teaching and learning in new ways, and to the A&P students and instructors who helped shape the resulting text.

— Ric Martini

To my sons, Todd and Carl, whose warmth and humor have enriched my life in countless ways.

— Bill Ober

To my students and students everywhere, who make writing textbooks worthwhile. And, as always and in all ways, to my husband, Mike.

— Judi Nath

To my daughters Ivy and Kate, grandchildren Awley, Rhyan, Finna, and Raya, and former students, who have given me the opportunity to touch the future.

— Ed Bartholomew

To Coreen, my bride of over 25 years, and to Olivia and Dominic, the light of my life.

— Kevin Petti

Frederic (Ric) H. Martini, Ph.D.
Author

Dr. Martini received his Ph.D. from Cornell University in comparative and functional anatomy for work on the pathophysiology of stress. In addition to professional publications that include journal articles and contributed chapters, technical reports, and magazine articles, he is the lead author of ten undergraduate texts on anatomy and physiology or anatomy. Dr. Martini is currently affiliated with the University of Hawaii at Manoa and has a long-standing bond with the Shoals Marine Laboratory, a joint venture between Cornell University and the University of New Hampshire. He has been active in the Human Anatomy and Physiology Society (HAPS) for over 20 years and was a member of the committee that established the course curriculum guidelines for A&P. He is now a President Emeritus of HAPS after serving as President-Elect, President, and Past President over 2005–2007. Dr. Martini is also a member of the American Physiological Society, the American Association of Anatomists, the Society for Integrative and Comparative Biology, the Australia/New Zealand Association of Clinical Anatomists, the Hawaii Academy of Science, the American Association for the Advancement of Science, and the International Society of Vertebrate Morphologists.

Judi L. Nath, Ph.D.
Author

Dr. Judi Nath is a biology professor at Lourdes University, where she teaches anatomy and physiology, pathophysiology, and medical terminology. She received her Bachelor's and Master's degrees from Bowling Green State University and her Ph.D. from the University of Toledo. Dr. Nath is devoted to her students and strives to convey the intricacies of science in captivating ways that are meaningful, interactive, and exciting. She has won the Faculty Excellence Award—an accolade recognizing effective teaching, scholarship, and community service—multiple times. She is active in many professional organizations, notably the Human Anatomy and Physiology Society (HAPS), where she has served several terms on the board of directors. Dr. Nath is a coauthor of *Fundamentals of Anatomy & Physiology, Visual Essentials of Anatomy & Physiology,* and *Anatomy & Physiology* (all published by Pearson), and she is the sole author of *Using Medical Terminology*. Her favorite charities are those that have significantly affected her life, including the local Humane Society, the Cystic Fibrosis Foundation, and the ALS Association. On a personal note, Dr. Nath enjoys family life with her husband and their dogs.

Edwin F. Bartholomew, M.S.
Author

Edwin F. Bartholomew received his undergraduate degree from Bowling Green State University in Ohio and his M.S. from the University of Hawaii. Mr. Bartholomew has taught human anatomy and physiology at both the secondary and undergraduate levels and a wide variety of other science courses (from botany to zoology) at Maui Community College and at historic Lahainaluna High School, the oldest high school west of the Rockies. He is a coauthor of *Fundamentals of Anatomy & Physiology, Essentials of Anatomy & Physiology, Visual Essentials of Anatomy & Physiology, Structure and Function of the Human Body,* and *The Human Body in Health and Disease* (all published by Pearson). Mr. Bartholomew is a member of the Human Anatomy and Physiology Society (HAPS), the National Association of Biology Teachers, the National Science Teachers Association, the Hawaii Science Teachers Association, and the American Association for the Advancement of Science.

Kevin Petti, Ph.D.
Author

Dr. Petti is a professor at San Diego Miramar College. He teaches courses in human anatomy and physiology, human dissection, and health education. He is a President Emeritus of the Human Anatomy and Physiology Society (HAPS) and holds a Ph.D. from the University of San Diego. Dr. Petti believes that weaving art and culture into the fabric of an anatomy class is an effective educational tool. This approach is well received by his students and allows them to acquire an interdisciplinary perspective. As a dual U.S./Italian citizen, he regularly leads excursions to Italy, the fountainhead of the Renaissance as well as modern anatomical studies, to consider the relationship between art and science. Touring university museums in Rome, Florence, Bologna, and Padua, his visits explore dissection theaters and wax anatomical models that date back hundreds of years. Dr. Petti is often invited to speak at conferences and universities about the rich heritage of anatomy as a science and its influence on medicine, art, and the humanities.

William C. Ober, M.D.
Author and Illustrator

Dr. Ober received his undergraduate degree from Washington and Lee University and his M.D. from the University of Virginia. He also studied in the Department of Art as Applied to Medicine at Johns Hopkins University. After graduation, Dr. Ober completed a residency in Family Practice and later was on the faculty at the University of Virginia in the Department of Family Medicine and in the Department of Sports Medicine. He also served as Chief of Medicine of Martha Jefferson Hospital in Charlottesville, VA. He is currently a Visiting Professor of Biology at Washington and Lee University, where he has taught several courses and led student trips to the Galápagos Islands. He was on the Core Faculty at Shoals Marine Laboratory for 24 years, where he taught Biological Illustration every summer. Dr. Ober has collaborated with Dr. Martini on all of his textbooks in every edition.

Claire E. Ober, R.N.
Illustrator

Claire E. Ober, R.N., B.A., practiced family, pediatric, and obstetric nursing before turning to medical illustration as a full-time career. She returned to school at Mary Baldwin College, where she received her degree with distinction in studio art. Following a five-year apprenticeship, she has worked as Dr. Ober's partner in Medical & Scientific Illustration since 1986. She was on the Core Faculty at Shoals Marine Laboratory and co-taught the Biological Illustration course with Dr. Ober for 24 years. The textbooks illustrated by Medical & Scientific Illustration have won numerous design and illustration awards.

Kathleen Welch, M.D.
Clinical Consultant

Dr. Welch received her B.A. from the University of Wisconsin–Madison, her M.D. from the University of Washington in Seattle, and did her residency in Family Practice at the University of North Carolina in Chapel Hill. Participating in the Seattle WWAMI rural medical education program, she studied in Fairbanks, Anchorage, and Juneau, Alaska, with time in Boise, Idaho, and Anacortes, Washington, as well. For two years, she served as Director of Maternal and Child Health at the LBJ Tropical Medical Center in American Samoa and subsequently was a member of the Department of Family Practice at the Kaiser Permanente Clinic in Lahaina, Hawaii, and on the staff at Maui Memorial Hospital. She has been in private practice since 1987 and is licensed to practice in Hawaii and Washington State. Dr. Welch is a Fellow of the American Academy of Family Practice and a member of the Maui County Medical Society and the Human Anatomy and Physiology Society (HAPS). With Dr. Martini, she has coauthored both a textbook on anatomy and physiology and the *A&P Applications Manual*. She and Dr. Martini were married in 1979, and they have one son.

Ralph T. Hutchings
Biomedical Photographer

Mr. Hutchings was associated with Royal College of Surgeons for 20 years. An engineer by training, he has focused for years on photographing the structure of the human body. The result has been a series of color atlases, including the *Color Atlas of Human Anatomy*, the *Color Atlas of Surface Anatomy*, and *The Human Skeleton* (all published by Mosby-Yearbook Publishing). For his anatomical portrayal of the human body, the International Photographers Association has chosen Mr. Hutchings as the best photographer of humans in the twentieth century. He lives in North London, where he tries to balance the demands of his photographic assignments with his hobbies of early motor cars and airplanes.

Preface

Visual Anatomy & Physiology is a comprehensive textbook for the two-semester A&P course. It combines a visual approach with a modular organization to deliver subject matter in an easy-to-use and time-efficient manner that uniquely meets the needs of today's students—without sacrificing the coverage of A&P topics required for careers in nursing and other allied health professions.

For the Second Edition, prior to revising or creating a module, we asked ourselves three questions: (1) How can we best make this information meaningful, manageable, and comprehensible? (2) Does the module spark interest and encourage students to read it? (3) Will students be able to answer "Why is this important?" after the module?

In essence, we want students to be excited about learning human anatomy and physiology. During the revision process, our team of content experts, medical illustrators, award-winning teaching professionals, academic authors, and publishing specialists worked together to write and design this academic text. We scrutinized every sentence, visual, and layout, ensuring that the narrative made sense, the content was accurate, and the combinations of text and visuals flowed together seamlessly over the one- and two-page module presentations. We read countless reviews and listened to our own students in the classroom. This end product is the culmination of the very best all involved had to offer.

To help improve future editions, we encourage you to send any pertinent information and remarks about the organization or content of this textbook to us directly, using the e-mail addresses below. We warmly welcome comments and suggestions and will carefully consider them in the preparation of the Third Edition.

New to the Second Edition

These are the key changes in this new edition:

- **Increased physiology coverage** in select modules gives students a better understanding of tough physiology topics.
- **The conversion of Section Openers to modules** allows their content to be linked (by number) to Learning Outcomes and supported by new Module Reviews for additional in-the-book practice.
- **New end-of-chapter study and practice material** includes a new narrative Study Outline and new comprehensive Chapter Review Questions (labeling, true/false, matching, multiple choice, fill-in, and short answer) to help students learn and integrate the chapter content.
- **The repetition of the chapter-opening Learning Outcomes on the module spreads** underscores the connection between the HAPS-based Learning Outcomes and the associated teaching points. Author Judi Nath sat on the Human Anatomy and Physiology Society (HAPS) committee that developed the HAPS Learning Outcomes, recommended to A&P instructors, and the Learning Outcomes in this book are based on them. Additionally, the assessments in MasteringA&P are organized by these HAPS-based Learning Outcomes.
- **New assignable MasteringA&P activities** include the following:
  - **New Tough Topic Coaching Activities** are highly visual, assignable activities designed to bring interactivity to select modules in the book. Multi-part activities include the ranking and sorting types that ask students to manipulate the visuals.
  - **New Adaptive Follow-up Assignments** allow instructors to easily assign personalized content for each individual student based on strengths and weaknesses identified by his or her performance on MasteringA&P parent assignments.
  - **New Dynamic Study Modules** help students acquire, retain, and recall information quickly and efficiently. The modules are available as a self-study tool or can be assigned by the instructor. They can be easily accessed with smartphones.
- **New *Visual Anatomy & Physiology Lab Manual*** uses the same visual approach and modular organization to help students succeed in the lab.

Frederic (Ric) H. Martini
martini@pearson.com

William C. Ober

Judi L. Nath
nath@pearson.com

Edwin F. Bartholomew
bartholomew@pearson.com

Kevin Petti

Chapter-by-Chapter Changes in the Second Edition

This annotated table of contents provides select examples of revision highlights in each chapter of the Second Edition.

Chapter 1: An Introduction to Anatomy and Physiology

- All Section Openers have been converted to Modules (now 1.1, 1.5, 1.12, and 1.14) linked by number to Learning Outcomes and supported by new Module Reviews.
- New Module 1.9 introduces the major organs/structures of the integumentary, skeletal, muscular, and nervous systems.
- New Module 1.10 introduces the major organs/structures of the endocrine, cardiovascular, lymphatic, and respiratory systems.
- New Module 1.11 introduces the major organs/structures of the digestive, urinary, and reproductive systems.
- Revised Module 1.14 (formerly Section 4 Opener) includes a table with eponyms and equivalent terms, and includes information about the history of the study of human anatomy in a university setting.
- Revised Module 1.16 (formerly 1.9) incorporates content on directional terms and sectional planes that was formerly in tables into the art, and includes new art for sectional planes.
- Revised Module 1.17 (formerly 1.10) includes updated discussion of body cavities.
- New Study Outline with select illustrations replaces the Visual Outline with Key Terms to give students a narrative chapter summary.
- New Chapter Review Questions give students comprehensive practice at the end of the chapter.
- New Chapter Integration with critical thinking questions has been added.

Chapter 2: Chemical Level of Organization

- All Section Openers have been converted to Modules (now 2.1, 2.6, 2.10, and 2.13) linked by number to Learning Outcomes and supported by new Module Reviews.
- Revised Module 2.17 (formerly 2.13) includes updated art on protein structures (ribbon models).
- Revised Module 2.20 (formerly 2.16) includes updated art on phosphate-nitrogenous base structure.
- New Study Outline with select illustrations replaces the Visual Outline with Key Terms to give students a narrative chapter summary.
- New Chapter Review Questions give students comprehensive practice at the end of the chapter.
- New Chapter Integration with critical thinking questions replaces the one in the previous edition.

Chapter 3: Cellular Level of Organization

- All Section Openers have been converted to Modules (now 3.1, 3.8, 3.13, and 3.18) linked by number to Learning Outcomes and supported by new Module Reviews.
- Revised Module 3.7 (formerly 3.6) includes information on mitochondrial DNA.
- Revised Module 3.10 (formerly 3.8) includes new reference to A&P Flix: Protein Synthesis, and reference to new genetic code (mRNA codons) table in the Appendix.
- Revised Module 3.12 (formerly 3.10) includes updated art and text to describe the three phases of translation: initiation, elongation, and termination.
- Revised Module 3.13 (formerly Section 3 Opener) includes new reference to A&P Flix: Membrane Transport.
- Revised Module 3.19 (formerly 3.15) includes new reference to A&P Flix: DNA Replication.
- Revised Module 3.20 (formerly 3.16) includes new reference to A&P Flix: Mitosis.
- New Study Outline with select illustrations replaces the Visual Outline with Key Terms to give students a narrative chapter summary.
- New Chapter Review Questions give students comprehensive practice at the end of the chapter.
- New Chapter Integration with critical thinking questions has been added.

Chapter 4: Tissue Level of Organization

- All Section Openers have been converted to Modules (now 4.1, 4.9, and 4.15) linked by number to Learning Outcomes and supported by new Module Reviews.
- New Module 4.2 describes microscopy techniques used to study cells and tissues.
- New Study Outline with select illustrations replaces the Visual Outline with Key Terms to give students a narrative chapter summary.
- New Chapter Review Questions give students comprehensive practice at the end of the chapter.
- New Chapter Integration with critical thinking questions replaces the one in the previous edition.

Chapter 5: The Integumentary System

- All Section Openers have been converted to Modules (now 5.1 and 5.6) linked by number to Learning Outcomes and supported by new Module Reviews.
- Revised Module 5.3 (formerly 5.2) describes and illustrates the subpapillary plexus.
- Revised Module 5.4 (formerly 5.3) clarifies presentation on touch receptors of the skin.
- New Module 5.5 describes the classification of burns and the types of skin grafts.
- Revised Module 5.6 (formerly Section 2 Opener) describes and illustrates embryonic development of accessory structures derived from the epidermis.
- New Study Outline with select illustrations replaces the Visual Outline with Key Terms to give students a narrative chapter summary.

- New Chapter Review Questions give students comprehensive practice at the end of the chapter.
- New Chapter Integration with critical thinking questions replaces the one in the previous edition.

Chapter 6: Osseous Tissue and Bone Structure

- All Section Openers have been converted to Modules (now 6.1 and 6.10) linked by number to Learning Outcomes and supported by new Module Reviews.
- Revised Module 6.7 (formerly 6.6) includes new art and photograph of epiphyseal line, and new reference to A&P Flix: Endochondral Ossification.
- Revised Module 6.8 (formerly 6.7) contains new art and a more detailed description of intramembranous ossification.
- Revised Module 6.9 (formerly 6.8) includes information on congenital talipes equinovarus, and a new photograph for pituitary growth failure.
- New Study Outline with select illustrations replaces the Visual Outline with Key Terms to give students a narrative chapter summary.
- New Chapter Review Questions give students comprehensive practice at the end of the chapter.
- New Chapter Integration with critical thinking questions replaces the one in the previous edition.

Chapter 7: The Skeleton

- All Section Openers have been converted to Modules (now 7.1 and 7.14) linked by number to Learning Outcomes and supported by new Module Reviews.
- Revised Module 7.8 (formerly 7.7) contains a new internal ear figure, and includes labels for the auditory ossicles in the Petri dish.
- Revised Modules 7.10 and 7.11 (formerly 7.9 and 7.10) include a vertebral arch arrow to clarify location on the art.
- Revised Module 7.17 (formerly 7.15) contains a new illustration of carpal bones that includes the articular cartilages.
- Revised Module 7.19 (formerly 7.17) clarifies differences in the male pelvis and female pelvis by incorporating information from a bulleted list into the art.
- Revised Module 7.21 (formerly 7.19) includes an x-ray image of a dancer's fracture.
- New Study Outline with select illustrations replaces the Visual Outline with Key Terms to give students a narrative chapter summary.
- New Chapter Review Questions give students comprehensive practice at the end of the chapter.
- New Chapter Integration with critical thinking questions has been added.

Chapter 8: Joints

- All Section Openers have been converted to Modules (now 8.1 and 8.6) linked by number to Learning Outcomes and supported by new Module Reviews.
- Revised Module 8.4 (formerly 8.3) includes further clarification for movements at the ankle.
- Revised Module 8.7 (formerly 8.5) differentiates between a bulging disc and a herniated disc.
- Revised Module 8.8 (formerly 8.6) clarifies the differences between a shoulder dislocation and a shoulder separation, and between a hip dislocation and a hip fracture.
- Revised Module 8.10 (formerly 8.8) provides additional examples of arthritis, and new photographs of artificial joints.
- New Study Outline with select illustrations replaces the Visual Outline with Key Terms to give students a narrative chapter summary.
- New Chapter Review Questions give students comprehensive practice at the end of the chapter.
- New Chapter Integration with critical thinking questions has been added.

Chapter 9: Skeletal Muscle Tissue

- All Section Openers have been converted to Modules (now 9.1 and 9.8) linked by number to Learning Outcomes and supported by new Module Reviews.
- New Module 9.5 discusses the electrical nature of cells, what an action potential is, and how an action potential is propagated along the axon of a neuron or sarcolemma of a skeletal muscle fiber.
- Revised Module 9.6 (formerly 9.4) includes new reference to A&P Flix: Events at the Neuromuscular Junction.
- Revised Module 9.7 (formerly 9.5) includes enhanced art, and new reference to A&P Flix: The Cross-Bridge Cycle.
- Revised Module 9.8 (formerly Section 2 Opener) contains new art with integrated text broken into steps, and includes new reference to A&P Flix: Excitation-Contraction Coupling.
- Revised Module 9.11 (formerly 9.8) includes a new photograph illustrating an isotonic concentric contraction.
- Revised Module 9.14 (formerly 9.11) includes updated Properties of Skeletal Muscle Fiber Types table to parallel the order of topic sequences in the text.
- Revised Module 9.15 (formerly 9.12) includes information on muscular dystrophies.
- New Study Outline with select illustrations replaces the Visual Outline with Key Terms to give students a narrative chapter summary.
- New Chapter Review Questions give students comprehensive practice at the end of the chapter.
- New Chapter Integration with critical thinking questions replaces the one in the previous edition.

Chapter 10: The Muscular System

- All Section Openers have been converted to Modules (now 10.1, 10.5, and 10.12) linked by number to Learning Outcomes and supported by new Module Reviews.

- Revised Module 10.1 (formerly Section 1 Opener) includes new reference to A&P Flix: Origins, Insertions, Actions, and Innervations and A&P Flix: Group Muscle Actions & Joints.
- Revised Module 10.2 (formerly 10.1) includes updated art of rectus femoris muscle to clarify its bipennate structure, and updated art illustrating third-class levers.
- Revised Module 10.4 (formerly 10.3) includes an adjusted leader line position for serratus anterior muscle.
- Revised Module 10.8 (formerly 10.6) includes updated labeling of the muscles of the floor of the mouth.
- Revised Module 10.14 (formerly 10.11) includes expanded discussion of trapezius muscle.
- Revised Module 10.16 (formerly 10.13) includes additional signs and symptoms of carpal tunnel syndrome.
- Revised Module 10.21 (formerly 10.18) includes fibularis tertius muscle.
- New Study Outline with select illustrations replaces the Visual Outline with Key Terms to give students a narrative chapter summary.
- New Chapter Review Questions give students comprehensive practice at the end of the chapter.
- New Chapter Integration with critical thinking questions has been added.

Chapter 11: Neural Tissue

- All Section Openers have been converted to Modules (now 11.1 and 11.6) linked by number to Learning Outcomes and supported by new Module Reviews.
- Revised Module 11.2 (formerly 11.1) includes discussion of initial segment, and uses *axon terminal* as primary term and *synaptic terminal* as secondary term.
- Revised Module 11.4 (formerly 11.3) includes ependymocytes and tanycytes in the discussion of ependymal cells.
- Revised Module 11.5 (formerly 11.4) integrates numbered step boxes and explanatory text into the PNS axon myelination art.
- Revised Module 11.6 (formerly Section 2 Opener) clarifies the use of *resting membrane potential* and *resting potential* as synonymous terms.
- Revised Module 11.7 (formerly 11.5) clarifies equilibrium potential, and includes new reference to A&P Flix: Resting Membrane Potential.
- Revised Module 11.8 (formerly 11.6) describes chemically gated channels also as ligand-gated channels, and includes new art showing the distribution of gated channels on a neuron.
- Revised Module 11.10 (formerly 11.8) includes new reference to A&P Flix: Generation of an Action Potential.
- Revised Module 11.11 (formerly 11.9) includes new reference to A&P Flix: Propagation of an Action Potential.
- New Study Outline with select illustrations replaces the Visual Outline with Key Terms to give students a narrative chapter summary.
- New Chapter Review Questions give students comprehensive practice at the end of the chapter.

Chapter 12: The Spinal Cord, Spinal Nerves, and Spinal Reflexes

- All Section Openers have been converted to Modules (now 12.1 and 12.10) linked by number to Learning Outcomes and supported by new Module Reviews.
- Revised Module 12.2 (formerly 12.1) includes new color-coded art illustrating spinal nerves and spinal segments.
- Revised Module 12.12 (formerly 12.10) includes new reference to A&P Flix: The Stretch Reflex.
- Revised Module 12.14 (formerly 12.12) clarifies reflex reinforcement by explaining the Jendrassik maneuver.
- New Study Outline with select illustrations replaces the Visual Outline with Key Terms to give students a narrative chapter summary.
- New Chapter Review Questions give students comprehensive practice at the end of the chapter.
- New Chapter Integration with critical thinking questions has been added.

Chapter 13: The Brain, Cranial Nerves, and Sensory and Motor Pathways

- All Section Openers have been converted to Modules (now 13.1 and 13.15) linked by number to Learning Outcomes and supported by new Module Reviews.
- Term for *aqueduct of the midbrain* is now *cerebral aqueduct*.
- Revised Module 13.3 (formerly 13.2) includes updated choroid plexus art showing microvilli on ependymal cells, and enhanced art on flow across the choroid plexus.
- Revised Module 13.5 (formerly 13.4) includes updated discussion of cerebellar cortex.
- Revised Module 13.6 (formerly 13.5) includes a new cadaver photograph of the midbrain.
- Revised Module 13.7 (formerly 13.6) clarifies the status of pulvinar as a thalamic nucleus.
- Revised Module 13.8 (formerly 13.7) includes *amygdala* as secondary term for *amygdaloid body*.
- Revised Module 13.15 (formerly Section 2 Opener) clarifies sensory receptor transduction.
- Revised Module 13.17 (formerly 13.15) clarifies tactile disc structure.
- Revised Module 13.18 (formerly 13.16) includes updated art to specify the position of the primary sensory cortex.
- Revised Module 13.20 (formerly 13.18) includes a diagram summarizing levels of motor complexity.
- Revised Module 13.21 (formerly 13.19) includes new images for rabies and Alzheimer's plaques.
- New Study Outline with select illustrations replaces the Visual Outline with Key Terms to give students a narrative chapter summary.
- New Chapter Review Questions give students comprehensive practice at the end of the chapter.
- New Chapter Integration with critical thinking questions replaces the one in the previous edition.

Chapter 14: The Autonomic Nervous System

- All Section Openers have been converted to Modules (now 14.1 and 14.7) linked by number to Learning Outcomes and supported by new Module Reviews.
- Revised Module 14.2 (formerly 14.1) includes new autonomic nervous system art.
- Revised Module 14.4 (formerly 14.3) clarifies classification and description of splanchnic nerves.
- Revised Module 14.5 (formerly 14.4) contains updated text and art on alpha and beta receptor stimulation including the roles of G proteins.
- Revised Module 14.10 (formerly 14.8) includes new art on baroreceptor and chemoreceptor locations.
- New Study Outline with select illustrations replaces the Visual Outline with Key Terms to give students a narrative chapter summary.
- New Chapter Review Questions give students comprehensive practice at the end of the chapter.
- New Chapter Integration with critical thinking questions has been added.

Chapter 15: The Special Senses

- All Section Openers have been converted to Modules (now 15.1, 15.5, and 15.12) linked by number to Learning Outcomes and supported by new Module Reviews.
- Revised Module 15.2 (formerly 15.1) includes clarification of olfactory receptor cell structure and the role of G proteins in olfactory reception.
- Revised Module 15.3 (formerly 15.2) includes new text and art on foliate papillae, and uses the term *vallate papillae* instead of *circumvallate papillae*.
- Revised Module 15.4 (formerly 15.3) includes updated text and art to clarify physiology of salt and sour channels.
- Revised Module 15.5 (formerly Section 2 Opener) uses the term *internal ear* instead of *inner ear*.
- Revised Module 15.6 (formerly 15.4) includes updated labyrinth art indicating orientation of maculae.
- Revised Module 15.8 (formerly 15.6) includes updated art showing orientation of the macula in the saccule, and uses the term *otolithic membrane* instead of *statoconia*.
- Revised Module 15.11 (formerly 15.9) includes updated text and art regarding hearing pathways.
- Revised Module 15.14 (formerly 15.11) uses *fibrous*, *vascular*, and *inner layers* of the eye instead of *fibrous*, *vascular*, and *neural tunics*. This revised Module also clarifies the usage of *vitreous body* and *vitreous humor*, and uses the term *ciliary zonule* instead of *suspensory ligaments*.
- Revised Module 15.18 (formerly 15.15) includes a color blindness chart.
- New Module 15.19 discusses photoreception and details how visual pigments are activated and reassembled.
- New Study Outline with select illustrations replaces the Visual Outline with Key Terms to give students a narrative chapter summary.
- New Chapter Review Questions give students comprehensive practice at the end of the chapter.
- New Chapter Integration with critical thinking questions has been added.

Chapter 16: The Endocrine System

- All Section Openers have been converted to Modules (now 16.1 and 16.13) linked by number to Learning Outcomes and supported by new Module Reviews.
- Revised Module 16.3 (formerly 16.2) includes updated art on second messengers, and new reference to A&P Flix: Mechanism of Hormone Action: Second Messenger cAMP.
- Revised Module 16.7 (formerly 16.6) includes updated art and text on conversion of iodide ions to iodine atoms in thyroid hormone production.
- Revised Module 16.8 (formerly 16.7) includes updated blood calcium homeostasis flowchart with simpler, standardized terms.
- Revised Module 16.10 (formerly 16.9) includes updated blood glucose homeostasis flowchart with simpler, standardized terms.
- Revised Module 16.14 (formerly 16.12) includes updated blood pressure and volume homeostasis flowchart with simpler, standardized terms, and uses the term *renin-angiotensin-aldosterone system (RAAS)* instead of *renin-angiotensin system*.
- Revised Module 16.16 (formerly 16.14) uses updated terminology in the Overview of Endocrine Disorders table.
- New Study Outline with select illustrations replaces the Visual Outline with Key Terms to give students a narrative chapter summary.
- New Chapter Review Questions give students comprehensive practice at the end of the chapter.

Chapter 17: Blood

- To reduce organizational complexity, the combined Blood and Blood Vessels chapter in the previous edition has been split into two separate chapters: Chapter 17: Blood and Chapter 18: Blood Vessels and Circulation.
- All Section Openers have been converted to Modules (now 17.1 and 17.4) linked by number to Learning Outcomes and supported by new Module Reviews.
- Revised Module 17.3 (formerly 17.7) includes additional text on the production of formed elements, and uses *hematopoietic stem cells* as the primary term and *hemocytoblasts* as a secondary term.
- New Module 17.4 defines hematology, describes the elements of a complete blood count, and gives examples of red blood cell lab tests.
- Revised Module 17.6 (formerly 17.3) clarifies that reticulocytes are released into the bloodstream, not fully mature red blood cells.
- Revised Module 17.7 (formerly 17.4) includes an additional column in the blood types chart to elucidate blood type compatibility.
- Revised Module 17.9 (formerly 17.6) clarifies that lymphocytes make up 20–40% of a differential count.

- Revised Module 17.10 (formerly 17.8) defines vascular spasm as part of the vascular phase of hemostasis.
- New Study Outline with select illustrations replaces the Visual Outline with Key Terms to give students a narrative chapter summary.
- New Chapter Review Questions give students comprehensive practice at the end of the chapter.
- New Chapter Integration with critical thinking questions has been added.

Chapter 18: Blood Vessels and Circulation

- To reduce organizational complexity, the combined Blood and Blood Vessels chapter in the previous edition has been split into two separate chapters: Chapter 17: Blood and Chapter 18: Blood Vessels and Circulation.
- All Section Openers have been converted to Modules (now 18.1 and 18.5) linked by number to Learning Outcomes and supported by new Module Reviews.
- New Module 18.5 explores the embryonic development of blood vessels, defines vasculogenesis and angiogenesis, and discusses the formation of the aortic arch and venae cavae.
- Revised Module 18.12 (formerly 17.19) explicitly defines the hepatic portal system.
- New Study Outline with select illustrations replaces the Visual Outline with Key Terms to give students a narrative chapter summary.
- New Chapter Review Questions give students comprehensive practice at the end of the chapter.
- New Chapter Integration with critical thinking questions wraps up the chapter.

Chapter 19: The Heart and Cardiovascular Function

- All Section Openers have been converted to Modules (now 19.1, 19.9, and 19.17) linked by number to Learning Outcomes and supported by new Module Reviews.
- Revised Module 19.3 (formerly 18.2) clarifies description of the pericardial sac.
- Revised Module 19.4 (formerly 18.3) contains a new cadaver heart photograph.
- Revised Module 19.11 (formerly 18.9) contains new art showing the distribution of the SA node action potential by the conduction system of the heart.
- Revised Module 19.13 (formerly 18.11) contains new art clarifying autonomic nervous system distribution to the heart.
- Revised Module 19.15 (formerly 18.13) includes updated art clarifying the factors affecting heart rate and stroke volume.
- Revised Module 19.19 (formerly 18.16) includes updated y-axis values on the graph of average blood pressure (now 0–120 mm Hg instead of 0–100 mm Hg).
- Revised Module 19.23 (formerly 18.20) includes updated art with a bidirectional arrow to clarify separate increasing and decreasing stimuli.
- New Study Outline with select illustrations replaces the Visual Outline with Key Terms to give students a narrative chapter summary.
- New Chapter Review Questions give students comprehensive practice at the end of the chapter.
- New Chapter Integration with critical thinking questions replaces the one in the previous edition.

Chapter 20: The Lymphatic System and Immunity

- All Section Openers have been converted to Modules (now 20.1, 20.8, and 20.13) linked by number to Learning Outcomes and supported by new Module Reviews.
- Revised Module 20.1 (formerly Section 1 Opener) classifies lymphoid tissues and organs as primary (sites where lymphocytes are formed and matured) and secondary (sites where lymphocytes are activated and cloned).
- Revised Module 20.4 (formerly 19.3) clarifies destinations of maturing T cells produced and selected in the thymus.
- Revised Module 20.5 (formerly 19.4) clarifies preliminary role of dendritic cells in an immune response.
- Revised Module 20.6 (formerly 19.5) includes information on thymic epithelial cells (TECs).
- Revised Module 20.8 (formerly Section 2 Opener) defines immunity and clarifies its two forms—innate and adaptive.
- Revised Module 20.11 (formerly 19.9) states that the number of plasma proteins of the complement system is over 30.
- Revised Module 20.13 (formerly Section 3 Opener) includes an updated, color-coded flowchart to clarify different forms of adaptive immunity.
- Revised Module 20.15 (formerly 19.12) clarifies that CD8 T cells provide cell-mediated immunity.
- Revised Module 20.16 (formerly 19.13) includes updated, color-coded art to differentiate CD4 T cells from B cells.
- Revised Module 20.20 (formerly 19.17) uses *innate* and *adaptive immunity* instead of *specific* and *nonspecific defenses.*
- New Study Outline with select illustrations replaces the Visual Outline with Key Terms to give students a narrative chapter summary.
- New Chapter Review Questions give students comprehensive practice at the end of the chapter.

Chapter 21: The Respiratory System

- All Section Openers have been converted to Modules (now 21.1 and 21.8) linked by number to Learning Outcomes and supported by new Module Reviews.
- Revised Module 21.2 (formerly 20.1) includes expanded discussion of cystic fibrosis.
- Revised Module 21.3 (formerly 20.2) includes updated art with nasal bones added, and with a label and leader line for nasopharyngeal meatus.
- Revised Module 21.4 (formerly 20.3) contains an updated definition of glottis that includes vocal folds and rima glottidis.

- Revised Module 21.5 (formerly 20.4) features a new title that clarifies the Module content.
- Revised Module 21.6 (formerly 20.5) includes updated bronchial tree art that color codes the bronchopulmonary segments with lobe of lung, and includes art and text on the root of the lung.
- Revised Module 21.7 (formerly 20.6) clarifies the structure of the respiratory membrane.
- Revised Module 21.10 (formerly 20.8) clarifies tidal volume terminology.
- Revised Module 21.11 (formerly 20.9) contains new lung art to clarify anatomic dead space.
- Revised Module 21.12 (formerly 20.10) includes an example of the calculation of partial pressures of gases in the atmosphere to clarify Dalton's law.
- Revised Module 21.13 (formerly 20.11) includes updated hemoglobin art (ribbon model).
- Revised Module 21.16 (formerly 20.14) includes pre-Bötzinger complex.
- Revised Module 21.18 (formerly 20.16) includes a simplified graph of the incidence of lung cancer in males and females.
- New Study Outline with select illustrations replaces the Visual Outline with Key Terms to give students a narrative chapter summary.
- New Chapter Review Questions give students comprehensive practice at the end of the chapter.

Chapter 22: The Digestive System

- All Section Openers have been converted to Modules (now 22.1, 22.5, and 22.18) linked by number to Learning Outcomes and supported by new Module Reviews.
- Revised Module 22.1 (formerly Section 1 Opener) contains an updated presentation of the major organs of the digestive tract and accessory organs of the digestive system.
- Revised Module 22.7 (formerly 21.5) includes an updated discussion of molars.
- Revised Module 22.8 (formerly 21.6) includes a new LM of esophageal mucosa.
- Revised Module 22.10 (formerly 21.8) includes a new cadaver photograph of stomach and greater omentum.
- Revised Module 22.12 (formerly 21.10) uses the term *circular folds* instead of *plicae circulares*.
- Revised Module 22.13 (formerly 21.11) includes three new cadaver photographs of duodenum, jejunum, and ileum, and uses the term *circular folds* instead of *plicae circulares*. This revised Module also includes a mnemonic for remembering the order of the small intestine segments from proximal to distal.
- Revised Module 22.15 (formerly 21.13) clarifies the naming of the phases of gastric secretion.
- Revised Module 22.20 (formerly 21.17) clarifies information on the porta hepatis and lobes of the liver by grouping labels and text in boxed headings.
- Revised Module 22.21 (formerly 21.18) includes updated discussion of portal area.
- New Study Outline with select illustrations replaces the Visual Outline with Key Terms to give students a narrative chapter summary.
- New Chapter Review Questions give students comprehensive practice at the end of the chapter.

Chapter 23: Metabolism and Energetics

- All Section Openers have been converted to Modules (now 23.1, 23.5, and 23.15) linked by number to Learning Outcomes and supported by new Module Reviews.
- Revised Module 23.1 (formerly Section 1 Opener) includes updated discussion of metabolic turnover and the nutrient pool.
- Revised Module 23.2 (formerly 22.2) now appears before the discussion of the citric acid cycle.
- Revised Module 23.3 (formerly 22.1) contains new art illustrating the details of the citric acid cycle.
- New Module 23.4 describes the electron transport system and how it establishes a proton gradient used to make ATP.
- Revised Module 23.8 (formerly 22.5) includes new art and text on very low density lipoproteins (VLDLs).
- Revised Module 23.10 (formerly 22.7) clarifies reactants in the urea cycle.
- Revised Module 23.11 (formerly 22.8) includes updated labeling in art to clarify the role of insulin.
- Revised Module 23.12 (formerly 22.9) includes a cross reference to minerals in Module 25.3, and uses *hypovitaminosis* instead of *avitaminosis*.
- Revised Module 23.13 (formerly 22.10) includes USDA's MyPlate icon instead of the older MyPyramid icon.
- Revised Module 23.17 (formerly 22.13) includes a new photograph showing mechanisms of heat transfer.
- Revised Module 23.18 (formerly 22.14) includes information on the Lewis wave (hunter's response).
- New Study Outline with select illustrations replaces the Visual Outline with Key Terms to give students a narrative chapter summary.
- New Chapter Review Questions give students comprehensive practice at the end of the chapter.

Chapter 24: The Urinary System

- All Section Openers have been converted to Modules (now 24.1, 24.6, and 24. 15) linked by number to Learning Outcomes and supported by new Module Reviews.
- Revised Module 24.2 (formerly 23.1) includes a mnemonic for remembering the retroperitoneal organs.
- Revised Module 24.4 (formerly 23.3) includes new art and text on the intercalated and principal cells of the collecting duct.
- Revised Module 24.10 (formerly 23.8) includes an additional transport processes key under the diagram on the right-hand page for clarification.
- Revised Modules 24.11, 24.12, and 24.13 (formerly 23.9, 23.10, and 23.11) include updated art indicating that the thin ascending limb is impermeable to water.

- Revised Module 24.13 (formerly 23.11) includes updated art and art key to increase comprehension of renal function.
- New Study Outline with select illustrations replaces the Visual Outline with Key Terms to give students a narrative chapter summary.
- New Chapter Review Questions give students comprehensive practice at the end of the chapter.

Chapter 25: Fluid, Electrolyte, and Acid–Base Balance

- All Section Openers have been converted to Modules (now 25.1 and 25.6) linked by number to Learning Outcomes and supported by new Module Reviews.
- Revised Module 25.8 (formerly 24.6) includes updated art and text to clarify the discussion of protein buffer systems.
- New Study Outline with select illustrations replaces the Visual Outline with Key Terms to give students a narrative chapter summary.
- New Chapter Review Questions give students comprehensive practice at the end of the chapter.

Chapter 26: The Reproductive System

- All Section Openers have been converted to Modules (now 26.1 and 26.8) linked by number to Learning Outcomes and supported by new Module Reviews.
- Revised Modules 26.2 and 26.4 (formerly 25.1 and 25.3) include updated text to indicate there are several seminiferous tubules per testicular lobule.
- Revised Module 26.3 (formerly 25.2) includes diploid and haploid state in all art, along with the duration of each stage of spermatogenesis. This revised Module uses *acrosome* as the primary term, and *acrosomal cap* as the secondary term.
- Revised Module 26.4 (formerly 25.3) includes updated art that clarifies the position of blood–testis barrier.
- Revised Module 26.5 (formerly 25.4) clarifies the definition of seminal fluid versus semen.
- Revised Module 26.6 (formerly 25.5) includes *erectile dysfunction (ED)* as an alternate term for *impotence*.
- Revised Module 26.10 (formerly 25.8) contains updated art on oogenesis that clearly illustrates when formation of second polar body occurs.
- Revised Module 26.13 (formerly 25.11) contains updated art on vulva that shows position of the vestibular bulbs.
- Revised Module 26.15 (formerly 25.13) contains art with enhanced detail and color.
- Revised Module 26.17 (formerly 25.15) includes a new photograph and art on testicular cancer, and includes a new label indicating site of breast cancer with calcification.
- New Study Outline with select illustrations replaces the Visual Outline with Key Terms to give students a narrative chapter summary.
- New Chapter Review Questions give students comprehensive practice at the end of the chapter.

Chapter 27: Development and Inheritance

- All Section Openers have been converted to Modules (now 27.1 and 27.13) linked by number to Learning Outcomes and supported by new Module Reviews.
- Term for *embryological* is now *embryonic*.
- Revised Module 27.2 (formerly 26.1) includes the metabolic changes involved on oocyte activation.
- Revised Module 27.3 (formerly 26.2) includes a label for the zona pellucida on the advanced morula, and notes the loss of the zona pellucida in the blastocyst text. This revised Module also contains art that has been enhanced to show endometrial capillaries.
- Revised Module 27.8 (formerly 26.7) includes a new ultrasound image after 6 months of gestation, and includes information on multiple births.
- Revised Module 27.11 (formerly 26.10) includes enhanced art on the milk let-down reflex, and includes a new photograph to illustrate the neonatal period.
- Revised Module 27.13 (formerly Section 2 Opener) includes the definition of epigenetics following a discussion of phenotype.
- Revised Module 27.15 (formerly 26.13) clarifies sex-linked inheritance and provides examples of both X- and Y-linked genes.
- Revised Module 27.16 (formerly 26.14) includes discussion of the human epigenome.
- New Study Outline with select illustrations replaces the Visual Outline with Key Terms to give students a narrative chapter summary.
- New Chapter Review Questions give students comprehensive practice at the end of the chapter.

Supplemental Material

- New Appendix includes the following:
  - Periodic table of elements
  - Normal physiological values tables ("The Composition of Minor Body Fluids" and "The Chemistry of Blood, Cerebrospinal Fluid, and Urine")
  - Genetic code (mRNA codons) table
- New end sheets include helpful information on common abbreviations used in science, and foreign word roots, prefixes, suffixes, and combining forms.

Acknowledgments

We feel very fortunate that the success of the First Edition of *Visual Anatomy & Physiology* has provided us with the opportunity to improve on our initial efforts with a Second Edition. Like the First Edition, this Second Edition required us to generate manuscript in a nontraditional format and working protocol. We are grateful that Leslie Berriman, Executive Editor; Frank Ruggirello, Vice President and Editorial Director of Applied Sciences; and Paul Corey, President of Pearson Science, Business, and Technology, gave unwavering support to this effort—despite the many challenges it continues to pose. We thank Frank and Paul for working closely with Leslie to ensure we had the resources necessary to publish what today's students need to succeed. We cannot imagine writing this book without Frank's and Paul's support and without Leslie at the helm.

Throughout this process, Associate Project Editor, Lisa Damerel, and Assistant Editor, Cady Owens, our in-house team at Pearson, kept the wheels turning by coordinating reviews, tracking manuscripts, assisting with photo research, and attending to a million other details so that the authors could write. Lisa and Cady are by far the best at what they do, and their work made this book possible.

We would also like to thank our Development Editor, Molly Ward, who assisted us by serving as both a professional editor and "surrogate student." She helped organize material and provided constructive feedback so that the final product would be the best it could be.

Director of Development, Barbara Yien, oversaw many aspects of the project and helped keep the book on track, even when it sometimes seemed that time was not on our side.

Assistant Managing Editor, Nancy Tabor, expertly shepherded this unique book through production and into print. Mary Tindle led the skilled team at S4Carlisle in moving the book smoothly through composition.

Michael Rossa's careful attention to detail in his copyedit resulted in many important changes for improved consistency and accuracy.

Thanks to Anita Impagliazzo for her creative work on the illustrations and preliminary layouts.

The easy-to-navigate interior design was created by Jim Gibson and Gibson Design Associates, who also adjusted and finalized the layouts for optimal flow.

The striking cover design was created by tani hasegawa.

Thanks also to our photo researchers at Bill Smith Group, Stefanie Ramsay, Luke Malone, and Cordes Hoffman, for finding the perfect photographs.

We are grateful to Cady Owens for her dedicated work on the print and media supplements, and to Shannon Kong and Dorothy Cox for shepherding them smoothly through production. Thanks also to Stacey Weinberger for her expert handling of the physical manufacturing of the book.

Thanks are due to Sharon Kim, Editorial Assistant, who coordinated the administrative details of the entire textbook program and who helped with key editorial aspects of the book.

We are grateful to Aimee Pavy, Senior Content Producer, and Liz Winer, Executive Content Producer, for their expert management of the media resources for instructors and students, especially MasteringA&P.

And, a round of applause goes to Allison Rona, the dedicated Senior Marketing Manager for A&P, and our biggest cheerleader, Derek Perrigo, Senior A&P Specialist.

We would also like to thank Judi's colleagues at Lourdes University—Christine W. Boudrie, Anjali D. Gray, Clayton McKenzie, and Elizabeth T. Wise—for their constructive reviews and discussions of the text.

We are indebted to the following instructors for their comments and suggestions throughout the development of the First Edition and the Second Edition:

Reviewers of the Second Edition

Marianne Crocker, *Ozarks Technical Community College*
Miranda Dunbar, *Southern Connecticut State University*
Sharon Ellerton, *Queensborough Community College – CUNY*
Greg Erianne, *County College of Morris*
Bruce Fisher, *Roane State Community College*
Aaron Fried, *Mohawk Valley Community College*
Lori K. Garrett, *Parkland College*
Jared Gilmore, *San Jacinto College – Central*
P. Michele Glass, *Northeast State Community College*
Noah Henley, *Rowan Cabarrus Community College*
Julie Huggins, *Arkansas State University – Jonesboro*
Judy Jiang, *Triton College*
Robert S. Kellar, *Northern Arizona University*
Jay Koepke, *Edison State College*
Stephen H. Lambert, *Jefferson Davis Community College*
Dale P. Ledford, *Northeast State Community College*
Dan Lykins, *University of Colorado – Colorado Springs*
David Maldon, *Northeast State Community College*
Bruce Maring, *Daytona State College*
L. Shannon G. Meadows, *Roane State Community College*
Jennifer Menon, *Johnson County Community College*
Apryl Nenortas, *Clovis Community College*
Tammy Oliver, *Eastfield College*
Deborah Palatinus, *Roane State Community College*
Saeed Rahmanian, *Roane State Community College*
Hope Sasway, *Suffolk County Community College – Grant*
Dee Ann Sato, *Cypress College*
Samuel Schwarzlose, *Amarillo College*
Joanne Settel, *Baltimore City Community College*
Jane Slone, *Cedar Valley College*
Delbert Stallwood, *Metropolitan Community College*
Delon Washo-Krupps, *Arizona State University*

Reviewers of the First Edition

Benja Allen, *El Centro College*
Kara Battle, *Durham Technical Community College*
Nina Beaman, *Bryant & Stratton College*
Felicia Brenoe, *Glendale Community College*
Janet Brodsky, *Ivy Tech Community College – Lafayette*
Steve Byrne, *St. Petersburg College*
Maura Cavanagh-Dick, *Salem Community College*
Alexander Cheroske, *Mesa Community College – Red Mountain*
Robert Clark, *Ozarks Technical Community College*
Gerard Cronin, *Salem Community College*
Lynnette Danzl-Tauer, *Rock Valley College*
Martha Dixon, *Diablo Valley College*
Sharon Ellerton, *Queensborough Community College – CUNY*
Seema Endley, *Blinn College – Bryan*
Linda Falkow, *Mercer County Community College*
Carl Frailey, *Johnson County Community College*
Patrick Galliart, *North Iowa Area Community College*
William Gressett, *Holmes Community College*
Michael Harman, *Lone Star College – North Harris*
Lisa Hawthorne, *Brown Mackie College*
Chris Hazzi, *State College of Florida – Manatee-Sarasota*
Stuart Hill, *Blinn College – Bryan*
Dale Horeth, *Tidewater Community College*
Julie Huggins, *Arkansas State University – Jonesboro*
Jason Hunt, *Brigham Young University – Idaho*
Alexander Ibe, *Weatherford College*
Jeba Inbarasu, *Metropolitan Community College*
Jason Jennings, *Southwest Tennessee Community College – Union Avenue*
Thomas Jordan, *Pima Community College – Northwest*
Leslie King, *University of San Francisco*
William Kleinelp, *Middlesex County College*
Chad Knights, *Northern Virginia Community College – Alexandria*
Michael LaPointe, *Indiana University – Northwest*
Thomas McDonald, *Pima Community College – East & West*
Abraham Miller, *University of Tampa*
Claire Miller, *Community College of Denver*
Michele N. Moore, *Ivy Tech Community College – East Central*
David Moyer, *Piedmont Virginia Community College & University of Virginia School of Medicine*
Hong Nguyen, *Northern Virginia Community College – Alexandria*
Phillip Nicotera, *St. Petersburg College*
Margaret (Betsy) Ott, *Tyler Junior College*
Thomas Pilat, *Illinois Central College – East Peoria*
Jacqueline Quiros, *Suffolk County Community College – Grant*
Susan Rohde, *Triton College*
Hiranya Roychowdhury, *New Mexico State University*
Karla Rues, *Ozarks Technical Community College*
Hope Sasway, *Suffolk County Community College – Grant*
Jeff Schinske, *De Anza College*
Donald Shaw, *The University of Tennessee at Martin*
Marilyn Shopper, *Johnson County Community College*
Mark Slivkoff, *Collin College*
Scott Smidt, *Laramie County Community College – Albany County*
Phillip Snider, *Gadsden State Community College*
Delbert Stallwood, *Metropolitan Community College*
Asha Stephens, *College of the Mainland*
Shelia Taylor, *Ozarks Technical Community College*
Keti Venovski, *Lake-Sumter Community College – South Lake*
Patricia Visser, *Jackson Community College*
Delon Washo-Krupps, *Arizona State University*
Alan Wasmoen, *Metropolitan Community College*
Jen Wortham, *University of Tampa*
Patricia Wu, *Chabot College*
Janice Yoder Smith, *Tarrant County College – Northwest*

Class Testers of the First Edition

Melissa Bailey, *Emporia State University*
Verona Barr, *Heartland Community College*
Claudia Barreto, *University of New Mexico*
Janet Brodsky, *Ivy Tech Community College – Lafayette*
Robert Brozanski, *Community College of Allegheny County – North*
Peter Bushnell, *Indiana University – South Bend*
Nickolas Butkevich, *Schoolcraft College*
Zinnia Callueng, *Central Florida Community College*
Alexander Cheroske, *Mesa Community College – Red Mountain*
Ron Clark, *Mercer County Community College*
Debra Claypool, *Mid Michigan Community College*
Jan Clifton, *Ivy Tech Community College – Muncie*
Vicki Clouse, *Montana State University – Northern*
Judy Cunningham, *Montgomery County Community College*
Lynnette Danzl-Tauer, *Rock Valley College*
Mario De La Haye, *Kennedy-King College*
Robin Dodson, *Parkland College*
Patricia Dolan, *Pacific Lutheran University*
Sondra Dubowsky, *McLennan Community College*
Sharon Ellerton, *Queensborough Community College – CUNY*
Theresia Elrod (now Whelan), *State College of Florida – Manatee-Sarasota*
Seema Endley, *Blinn College – Bryan*
Jeff Engel, *Western Illinois University*
Greg Erianne, *County College of Morris*
Linda Falkow, *Mercer County Community College*
Alyssa Farnsworth, *Ivy Tech Community College – Muncie*
Carol Flora, *Indiana University – Purdue University Indianapolis*
Lori K. Garrett, *Parkland College*
Jennifer Gibbs, *Hinds Community College – Raymond*
John Gillen, *Hostos Community College – CUNY*
Theresa Gillian, *Virginia Polytechnic Institute and State University*
Evan Goldman, *Philadelphia University*
Suzanne Gould, *Ivy Tech Community College – Muncie*
Joanna Greene, *Ivy Tech Community College – Muncie*
David Griffith, *Ferris State University*
Rebecca Harris, *Hinds Community College – Raymond*
Amy Harwell, *Oregon State University*
Susan Holland, *Wilson Community College*
Kevin Holt, *Northeast Alabama Community College*
Amanda Huffstutler, *Chattahoochee Valley Community College*
Alexander Ibe, *Weatherford College*
Renu Jain, *Houston Community College*
Anthony Jones, *Tallahassee Community College*
Philip Jones, *Manchester Community College*
Warren Jones, *Loyola University*
Leslie King, *University of San Francisco*
David Klarberg, *Queensborough Community College – CUNY*
William Kleinelp, *Middlesex County College*
Chad Knights, *Northern Virginia Community College – Alexandria*
Jeff Laborda, *State College of Florida – Manatee-Sarasota*
Michael LaPointe, *Indiana University – Northwest*
Ray Larsen, *Bowling Green State University*
Andrey Lebed, *Lee College*
Curtis Lee, *Dallas Baptist University*
Carlos Liachovitzky, *Bronx Community College – CUNY*
Mitch Lockhart, *Valdosta State University*
Jodi Long, *Santa Fe College*
James Ludden, *College of DuPage*
John Moore, *Parkland College*
Tammy Oliver, *Eastfield College*
Margaret (Betsy) Ott, *Tyler Junior College*
Keith Overbaugh, *Northwestern Michigan College*
Paul Passalacqua, *Owens Community College*
Krya Perry, *Mesa Community College – Red Mountain*
Harry Pierre, *Keiser University*
Thomas Pilat, *Illinois Central College – East Peoria*
Brandon Poe, *Springfield Technical Community College*
Eugenie Pool, *Lee College*
Julie Porterfield, *Tulsa Community College – Southeast*
Faina Riftina, *Hostos Community College – CUNY*
Mark Robbins, *Ivy Tech Community College – Marion*
Dan Roberts, *Owens Community College*
Susan Rohde, *Triton College*
Amanda Rosenzweig, *Delgado Community College*
Nick Roster, *Northwestern Michigan College*
Thomas Ruehlmann, *College of DuPage*
Chrisanna Saums, *Hinds Community College – Raymond*
Steve Schenk, *Truckee Meadows Community College*
Ralph Schwartz, *Hostos Community College – CUNY*
Joanne Settel, *Baltimore City Community College*
Donald Shaw, *The University of Tennessee at Martin*
Marilyn Shopper, *Johnson County Community College*
Mark Slivkoff, *Collin College*
Jane Slone, *Cedar Valley College*
Scott Smidt, *Laramie County Community College – Albany County*
Joy Smoots, *Cape Fear Community College*
Phillip Snider, *Gadsden State Community College*
Julian Stark, *Queensborough Community College – CUNY*
Olga Steinberg, *Hostos Community College – CUNY*
Yung Su, *Florida State University*
James Timbilla, *Queensborough Community College – CUNY*
Corinne Ulbright, *Indiana University – Purdue University Indianapolis*
Peter Van Dyke, *Walla Walla Community College*
Patricia Visser, *Jackson Community College*
Jane Walden, *Chattahoochee Valley Community College*
Delon Washo-Krupps, *Arizona State University*
Patricia Wu, *Chabot College*
Janice Yoder Smith, *Tarrant County College – Northwest*
Robert Zdor, *Andrews University*

Contents

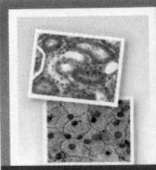

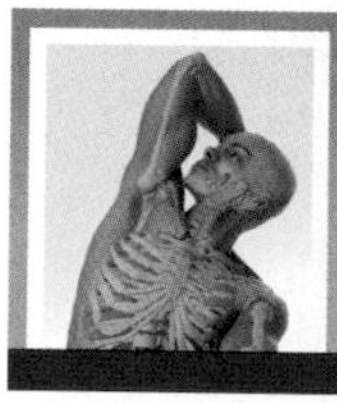

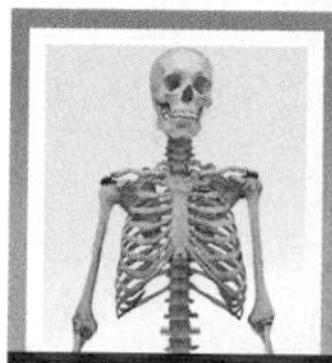

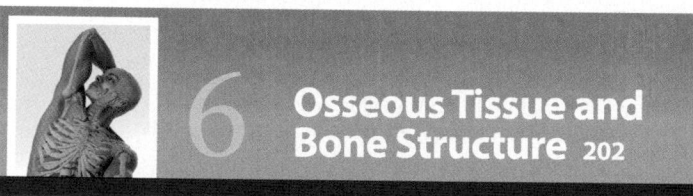

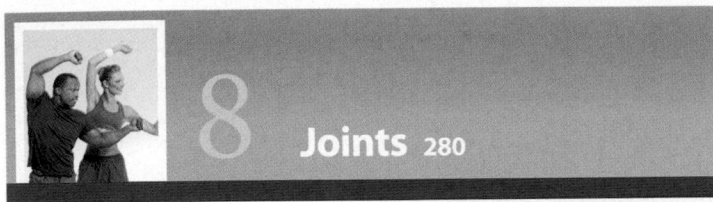

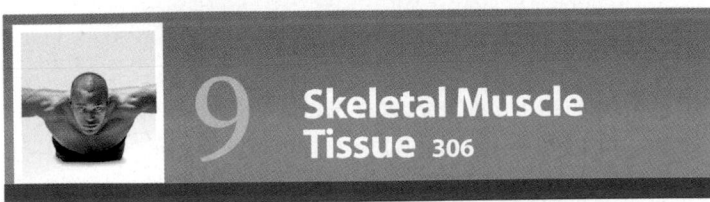

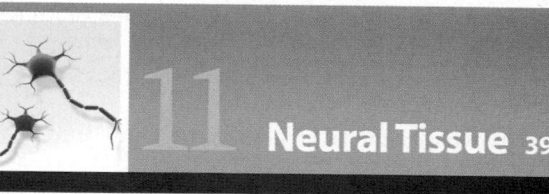

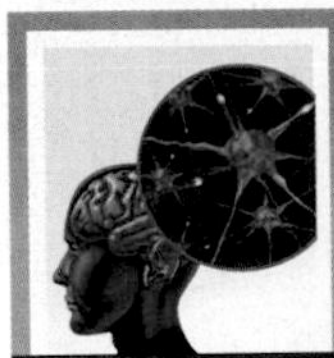

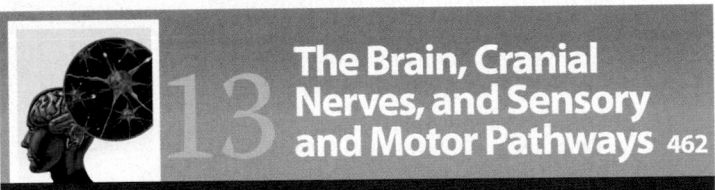

13 The Brain, Cranial Nerves, and Sensory and Motor Pathways 462

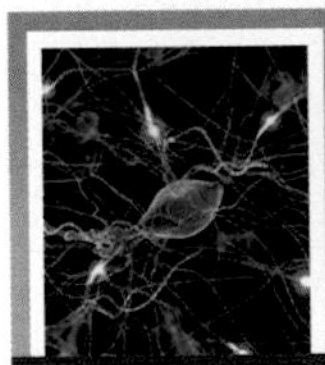

14 The Autonomic Nervous System 508

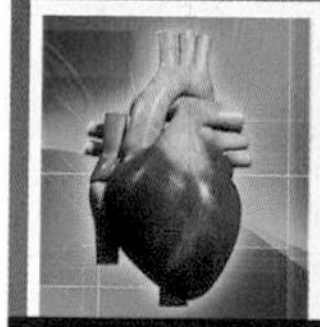

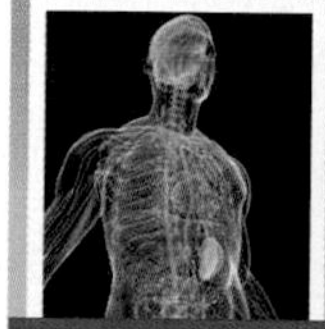

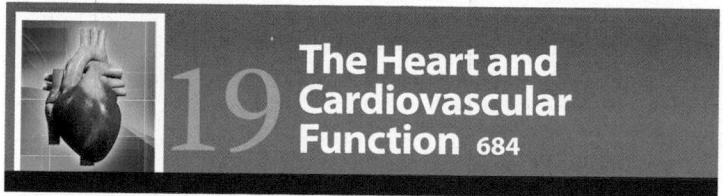

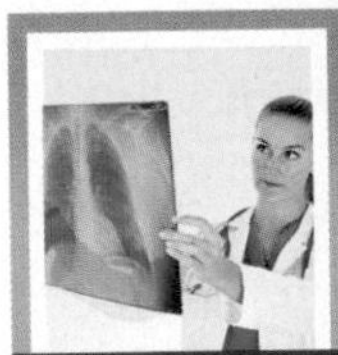

21 The Respiratory System 786

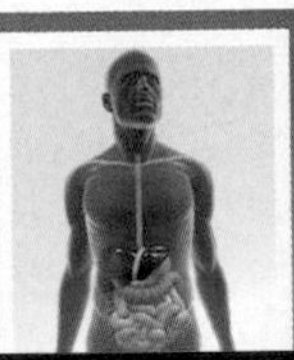

22 The Digestive System 828

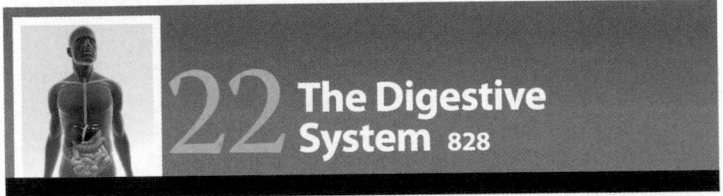

27 Development and Inheritance 1026

visual anatomy & physiology

2nd edition

1 An Introduction to Anatomy and Physiology

LEARNING OUTCOMES

These Learning Outcomes correspond by number to this chapter's modules and indicate what you should be able to do after completing the chapter.

SECTION 1 • A&P in Perspective

- **1.1** Describe homeostasis and identify basic study skill strategies to use in this course.
- **1.2** Describe the universal characteristics of living things.
- **1.3** Define anatomy and physiology, and describe macroscopic and microscopic anatomy.
- **1.4** Explain the relationship between structure and function.

SECTION 2 • Levels of Organization

- **1.5** Describe the various levels of organization in the human body.
- **1.6** Describe various types of cells in the human body and explain the basic principles of the cell theory.
- **1.7** Define histology and explain the interrelationships among the various types of tissues.
- **1.8** Identify the 11 organ systems of the human body, and describe the major functions of each.
- **1.9** Describe the major organs of the integumentary, skeletal, muscular, and nervous systems and briefly describe their functions.
- **1.10** Describe the major organs of the endocrine, cardiovascular, lymphatic, and respiratory systems and briefly describe their functions.
- **1.11** Describe the major organs of the digestive, urinary, and reproductive systems and briefly describe their functions.

SECTION 3 • Homeostasis

- **1.12** Describe the mechanisms of homeostatic regulation.
- **1.13** Discuss the roles of negative feedback and positive feedback in maintaining homeostasis.

SECTION 4 • Anatomical Terms

- **1.14** Describe the history of anatomical terminology.
- **1.15** Use correct anatomical terms to describe superficial and regional anatomy.
- **1.16** Use correct directional terms and sectional planes to describe relative positions and relationships among body parts.
- **1.17** Identify the major body cavities of the trunk and the subdivisions of each.

LO

Learning Outcomes are repeated at the bottom of each module.

Focused study is important for learning anatomy and physiology

Human anatomy and physiology considers how the human body performs the functions that keep you alive and alert. You will learn many interesting and important facts about the human body as we proceed. However, the approach you learn and the attitude you develop will be at least as important as the things you memorize. The basic approach in A&P can be summed up as "What is that structure, and how does it work?" The complexity of the answer depends on the level of detail you need. In science, if we know what something does but we don't know how, it's usually called a "Black Box." The more you learn, the smaller (and more numerous) those Black Boxes become. That is, the more you learn, the more you realize how much you don't know.

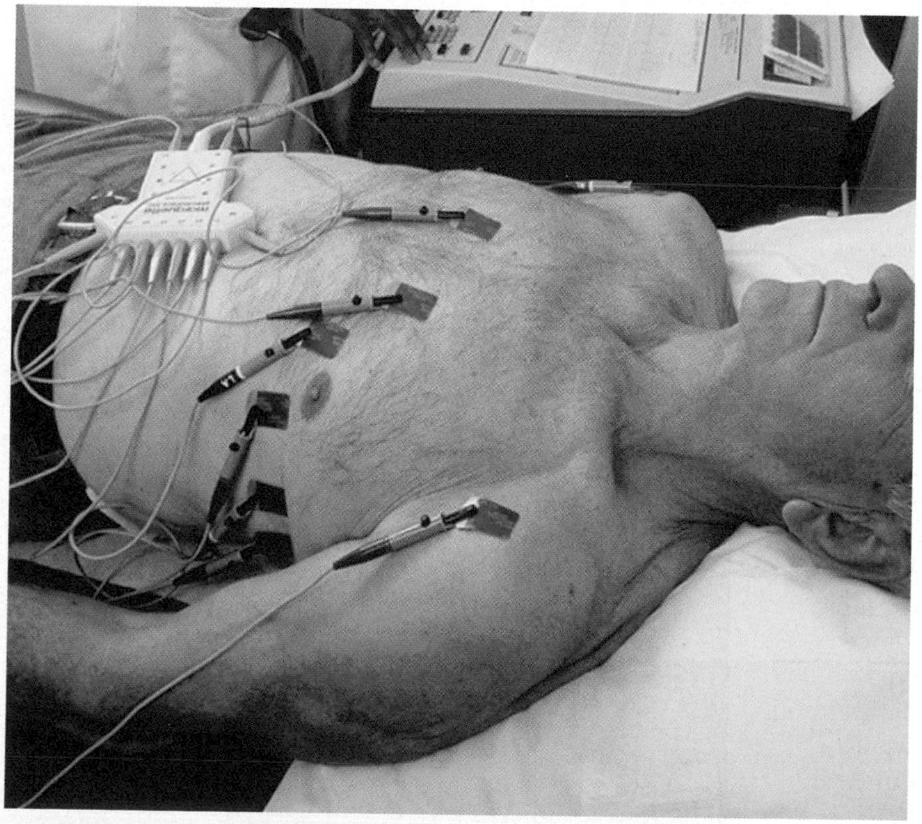

We will devote considerable time to explaining how the body responds to normal and abnormal conditions and maintains **homeostasis**, a relatively constant internal environment. As we proceed, you will see how your body's anatomy and physiology work together to cope with injury, disease, or anything else that threatens homeostasis.

Tips on How to Succeed in Your A&P Course

- **Approach the information in different ways.** For example, you might visualize the information, talk it over with or "teach" a fellow student, or spend additional time in lab asking questions of your lab instructor.
- **Set up a study schedule** and stick to it.
- **Devote a block of time each day** to your A&P course.
- **Practice memorization.** Memorization is an important skill, and an integral part of the course. You are going to have to memorize all sorts of things—among them muscle names, directional terms, and the names of bones and brain parts. Realize that this is an important study skill, and that the more you practice, the better you will be at remembering terms and definitions. We will try to give you handles and tricks along the way, to help you keep the information in mind.
- **Avoid shortcuts.** Actually there are no shortcuts. (Sorry.) You won't get the grade you want if you don't put in the time and do the work. This requires preparation throughout the term.
- **Attend all lectures, labs, and study sessions.** Ask questions and participate in discussions.
- **Read your lecture and lab assignments** before coming to class.
- **Do not procrastinate!** Do not do all your studying the night before the exam! Actually STUDY the material several times throughout the week. Marathon study sessions are often counterproductive. There is no easy button; you must push yourself.
- **Seek assistance immediately if you have a problem understanding the material.** Do not wait until the end of the term, when it is too late to salvage your grade.

Module 1.1 Review

a. Identify several strategies for success in this course.

b. Explain the purpose of the learning outcomes.

c. What do scientists mean when they use the term "Black Box"?

LO 1.1 Describe homeostasis and identify basic study skill strategies to use in this course.

Module 1.2

Biology is the study of life

The world around us contains a variety of living organisms with different appearances and lifestyles. Despite this diversity, all living things perform the same basic functions:

1. Living things **respond** to changes in their immediate environment—Plants orient to the sun, you move your hand away from a hot stove, and your dog barks at passing strangers.
2. Organisms show **adaptability**—Their internal operations and responses to stimulation can vary from moment to moment.
3. Over time, organisms **grow** and **develop**, and **reproduce**—This creates subsequent generations of similar, but not identical, organisms.
4. Many organisms are capable of some degree of **movement**. If that movement takes them from one place to another, we call the process **locomotion**.

Responsiveness, adaptability, growth and development, reproduction, and locomotion are active processes that require energy. This energy must continually be replaced as it is used. For animals, energy capture typically involves oxygen absorption from the atmosphere through respiration and the absorption of various chemicals from the surrounding environment. Each living organism also generates and discharges waste products into the environment in the process of excretion. These are the basic characteristics of living things, both plant and animal.

For very small organisms, absorption, respiration, and excretion involve simply transferring materials across exposed surfaces. But for larger creatures like dogs, cats, or human beings, this is not possible. For example, human beings cannot absorb steaks or ice cream without processing them first. That processing, called **digestion**, occurs in specialized areas where complex foods are broken down into simpler components that can be easily absorbed. Finally, because absorption, **respiration**, and **excretion** are performed in different portions of the body, most animals have an internal distribution system, or **circulation**, that transports materials from one place to another.

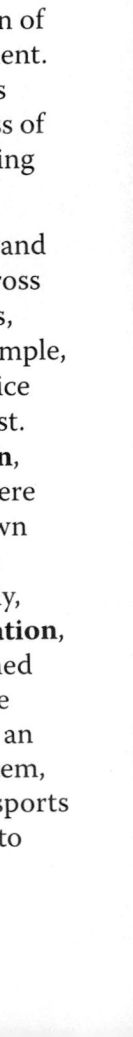

Characteristics of Living Organisms

Characteristic	Importance	Notes
Responsiveness	Indicates that the organism recognizes changes in its internal or external environment	Required for adaptability
Adaptability	Changes the organism's behavior, capabilities, or structure	Required for survival in a constantly changing world
Growth and development	Inherited patterns for growth (an increase in size) and development (changes in structure and function) produce organisms characteristic of their species	Growth and development to maturity is controlled by inherited instructions in the form of DNA
Reproduction	Produces the next generation	Sexual reproduction between two parents produces offspring with varied characteristics
Movement and locomotion	Distributes materials throughout large organisms; changes orientation or position of a plant or immobile animal; moves mobile animals around the environment	Animals show locomotion at some point in their lives
Respiration*	Usually refers to oxygen absorption and utilization, and carbon dioxide generation and release	Oxygen is required for chemical processes that release energy in a usable form; carbon dioxide is released as a waste product
Circulation*	Movement of fluid within the organism; may involve a pump and a network of special vessels	The circulation provides an internal distribution network
Digestion*	The chemical breakdown of complex materials for absorption and use by the organism	The chemicals released can be used to generate energy or to support growth
Excretion*	The elimination of chemical waste products generated by the organism	The waste products are often toxic, so their removal is essential

* The mechanics of the process depend on the size and complexity of the organism.

In the next 26 chapters we will consider the mechanics of each of these vital processes. Although we will examine the functions of the human body, the basic concepts have broad application in biology.

Module 1.2 Review

a. Define biology.

b. List the basic functions shared by all living things.

c. Explain why most animals have an internal circulation system that transports materials from place to place.

LO 1.2 Describe the universal characteristics of living things.

Module 1.3

Anatomy is the study of structure . . .

Anatomy, which means "a cutting open," is the study of internal and external structures of the body and the physical relationships among body parts. Here is an overview of the anatomy of the heart, with the walls opened so that you can see the complexity of its internal structure.

1 **Gross anatomy**, or **macroscopic anatomy**, involves the examination of relatively large structures and features usually visible with the unaided eye. This illustration of a dissected heart is an example of gross anatomy.

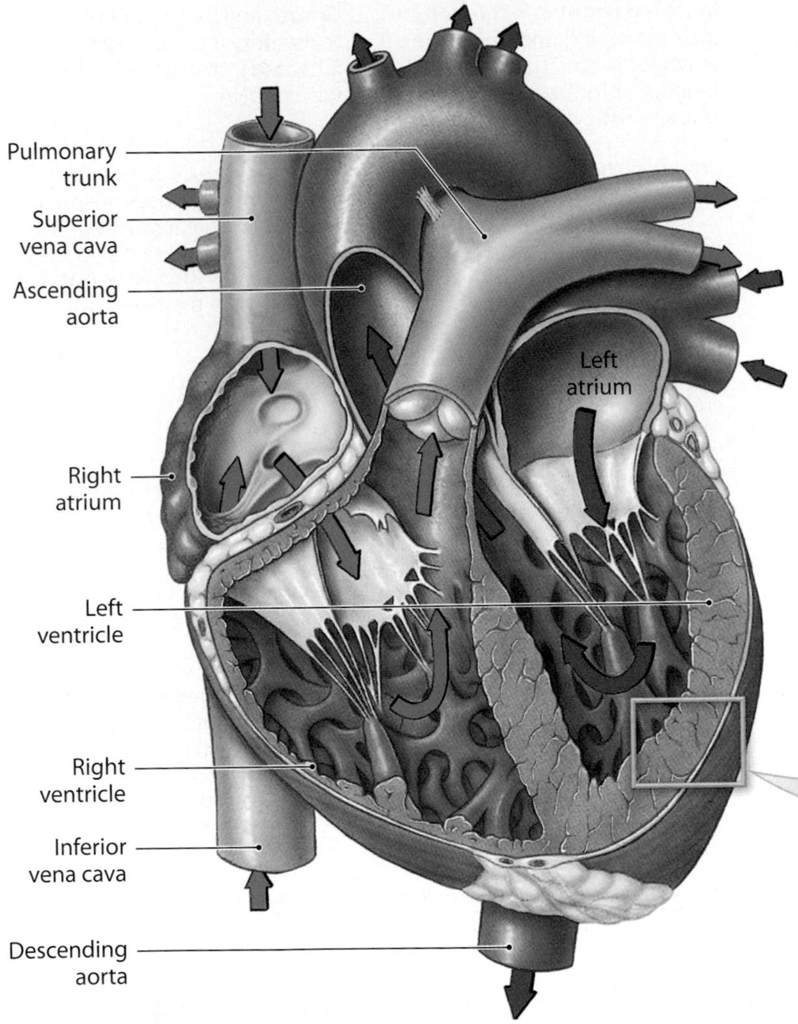

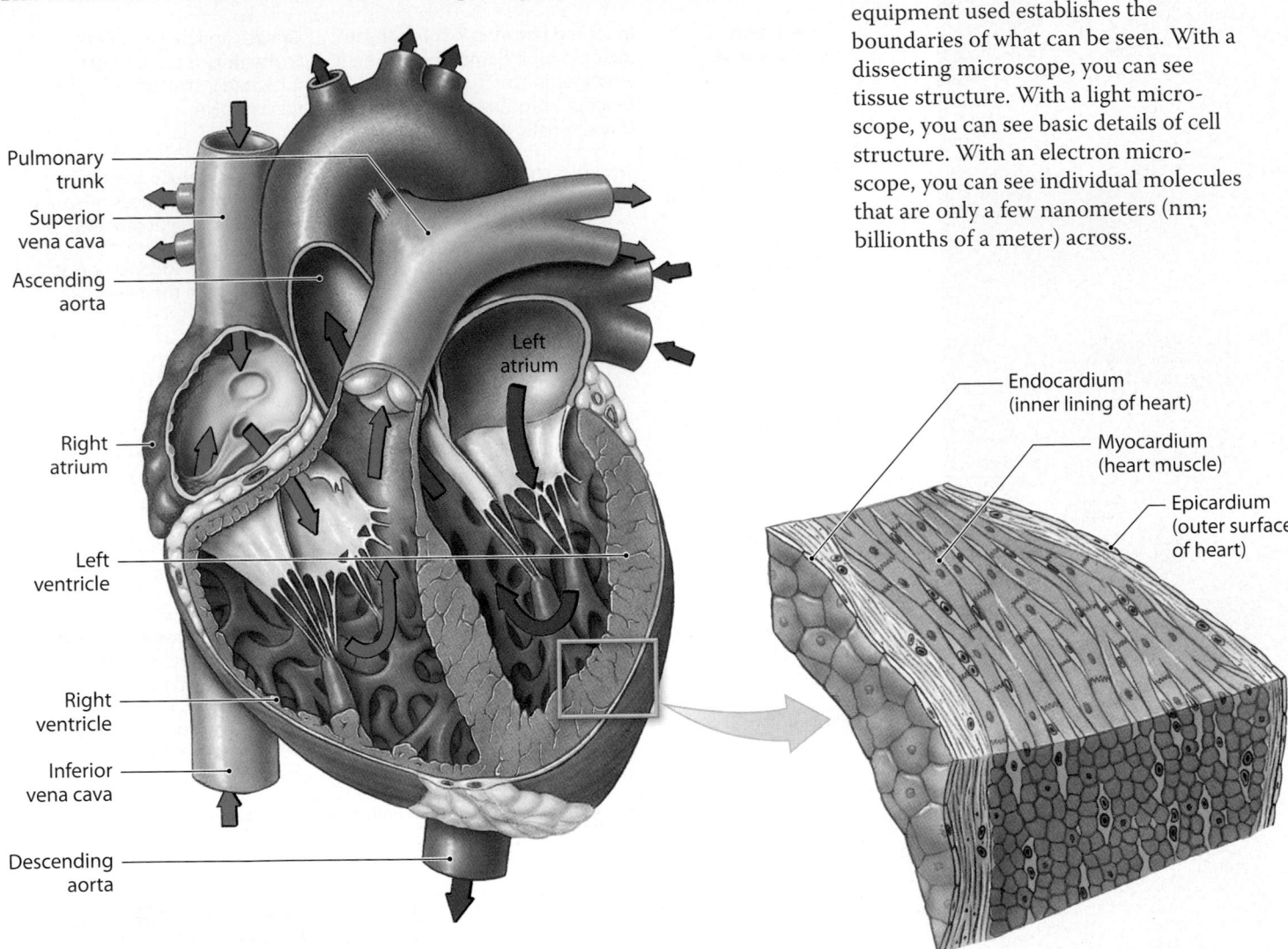

Pulmonary trunk
Superior vena cava
Ascending aorta
Right atrium
Left ventricle
Right ventricle
Inferior vena cava
Descending aorta
Left atrium

2 **Microscopic anatomy** deals with structures that cannot be seen without magnification, and thus the equipment used establishes the boundaries of what can be seen. With a dissecting microscope, you can see tissue structure. With a light microscope, you can see basic details of cell structure. With an electron microscope, you can see individual molecules that are only a few nanometers (nm; billionths of a meter) across.

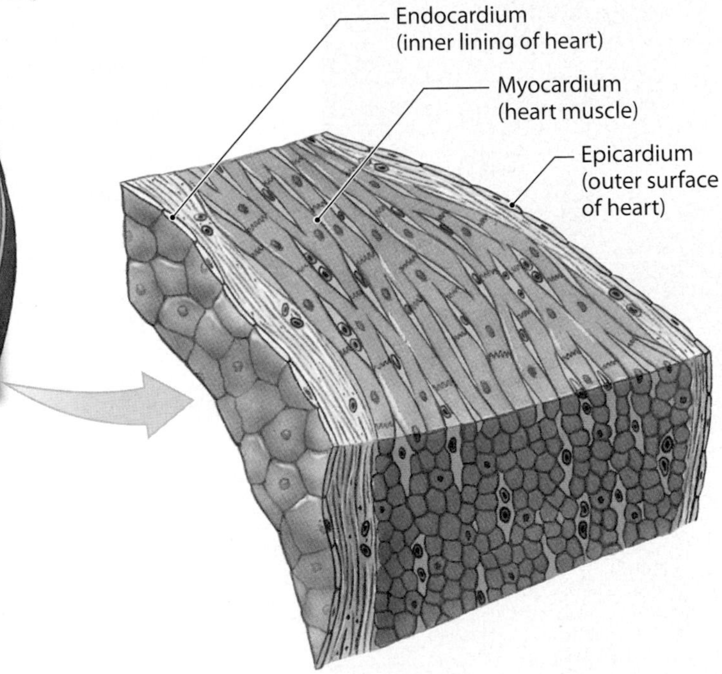

Endocardium (inner lining of heart)
Myocardium (heart muscle)
Epicardium (outer surface of heart)

All specific functions are performed by specific structures. The link between structure and function is always present, but not always understood. For example, although the anatomy of the heart was clearly described in the 15th century, almost 200 years passed before the heart's pumping action was demonstrated.

. . . and physiology is the study of function

Physiology is the study of function and how living organisms perform their vital functions. These functions are complex and much more difficult to examine than most anatomical structures. A physiologist looking at the heart focuses on its functional properties, such as the timing and sequence of the heartbeat, and its effects on blood pressure in the major arteries.

3 The heartbeat is coordinated by electrical events within the heart muscle. Those electrical events can be detected by monitoring electrodes placed on the body surface. A record of these electrical events is called an electrocardiogram, or ECG.

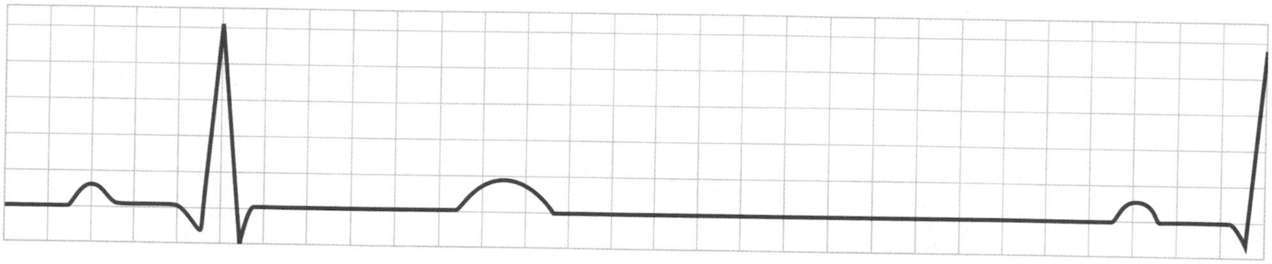

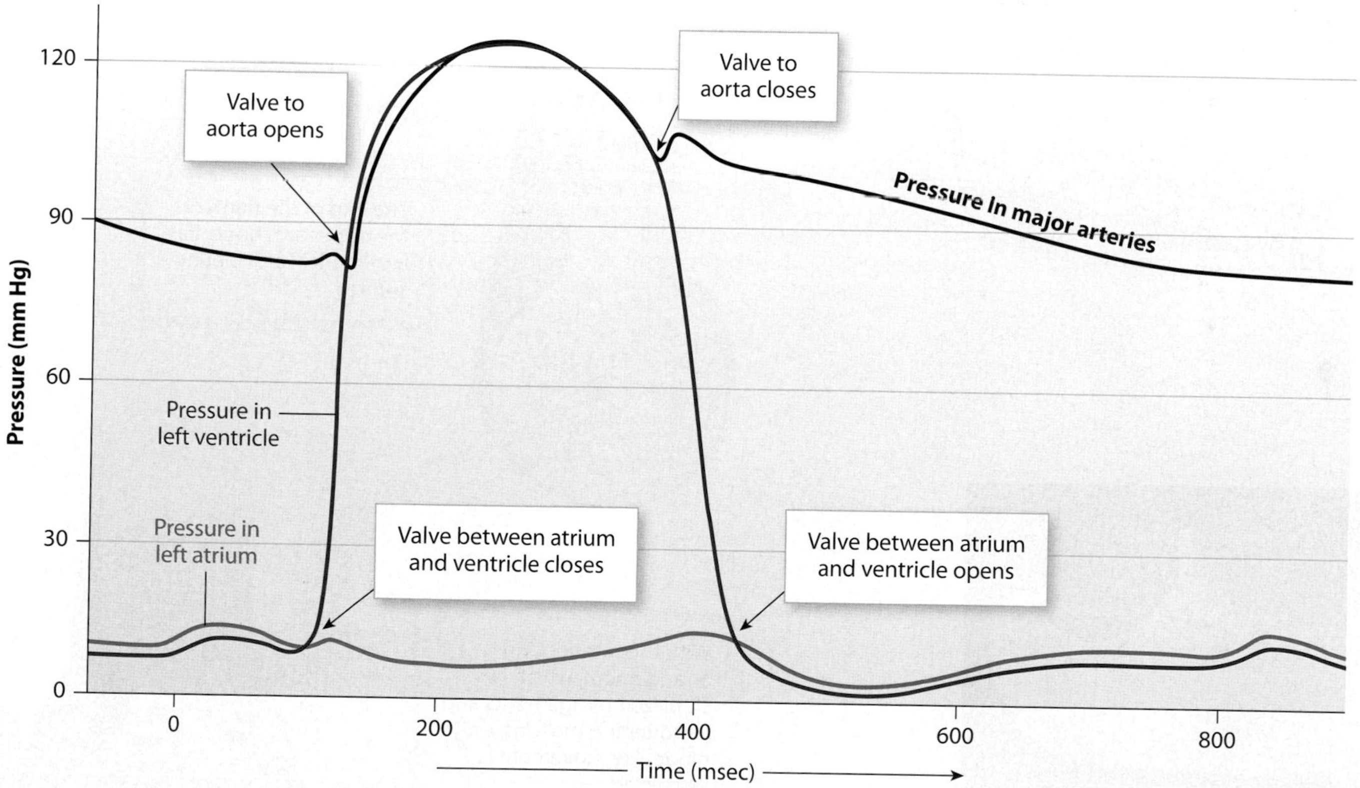
120
90
60
30
0
Pressure (mm Hg)
Valve to aorta opens
Valve to aorta closes
Pressure In major arteries
Pressure in left ventricle
Pressure in left atrium
Valve between atrium and ventricle closes
Valve between atrium and ventricle opens
0
200
400
600
800
Time (msec)

4 As the heart beats, pressure rises and falls within the major arteries and the chambers of the heart. Blood pressure in the major arteries must be maintained within normal limits to prevent vessel damage (from high pressures) or vessel collapse (from low pressures).

Module 1.3 Review

a. Define anatomy and physiology.
b. What are the differences between gross anatomy and microscopic anatomy?
c. Explain the link between anatomy and physiology.

LO **1.3** Define anatomy and physiology, and describe macroscopic and microscopic anatomy.

Module 1.4

Structure and function are interrelated

Physiology and anatomy are closely interrelated both theoretically and practically. Anatomical details are significant only because each has an effect on function, and physiological mechanisms can be fully understood only in terms of the underlying structural relationships.

1 This relationship is easily understood at the gross anatomical level. You are well aware that your elbow joint functions like a hinge. It lets your forearm move toward or away from your shoulder, but it does not allow twisting at the joint. These functional limits are imposed by the internal structure of the joint.

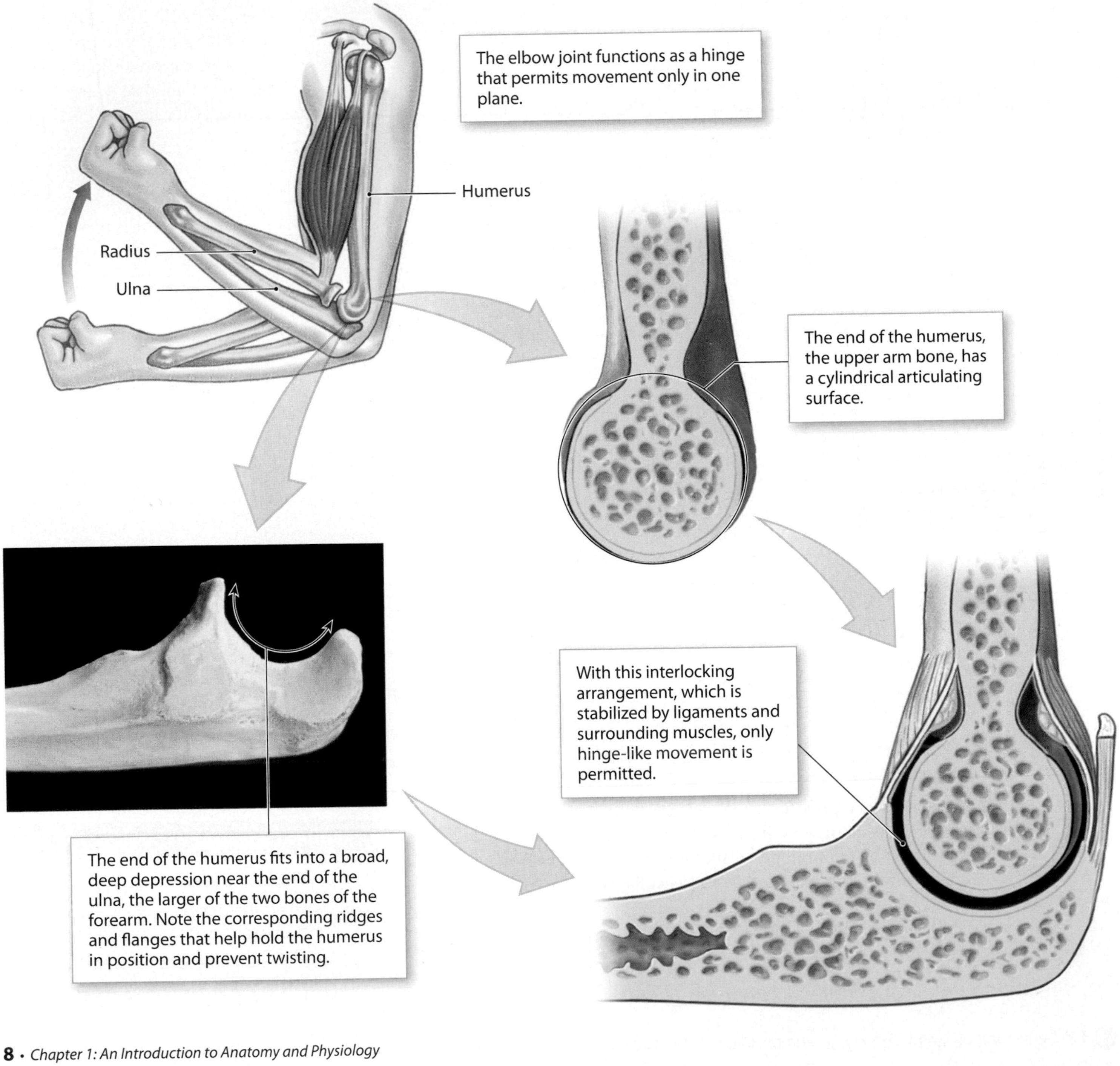

The elbow joint functions as a hinge that permits movement only in one plane.
Humerus
Radius
Ulna
The end of the humerus, the upper arm bone, has a cylindrical articulating surface.
With this interlocking arrangement, which is stabilized by ligaments and surrounding muscles, only hinge-like movement is permitted.
The end of the humerus fits into a broad, deep depression near the end of the ulna, the larger of the two bones of the forearm. Note the corresponding ridges and flanges that help hold the humerus in position and prevent twisting.

2 The relationship between structure and function also applies at the chemical level. Cells throughout your body communicate with one another through the use of chemical messengers, which you will learn more about in later chapters. The detection of and response to these messengers usually involves the attachment of the chemical messenger released by one cell to a receptor molecule at another cell. That attachment depends in large part on the three-dimensional shapes of the messenger and the receptor, and how well they fit together.

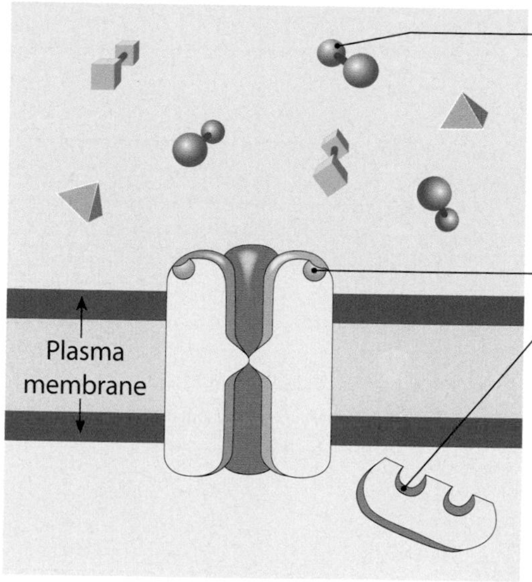
Plasma
membrane

Each chemical compound has a specific size and three-dimensional shape. Chemical messengers come in a variety of sizes and shapes. In general, they are released by one cell to affect other cells.

Receptor molecules can be on the outer surface or inside the cell. This figure shows that one receptor extends across the plasma membrane (boundary of a cell). The plasma membrane is also known as the cell membrane.

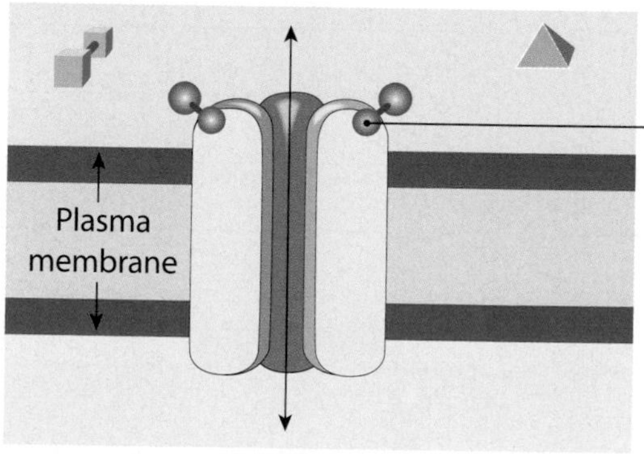
Plasma
membrane

Chemical messengers are detected when they attach, or bind, to a receptor that has the proper shape. Binding creates a new structure—messenger and receptor—and the entire complex often changes shape as a result. This can change the function of the receptor. In this case, the messenger binding opens a passageway through the receptor molecule, permitting substances to cross the plasma membrane.

It's important to realize that no mysterious forces are involved in the workings of the body. Although our knowledge is incomplete, it is quite clear that living systems are subject to the same laws of physics and chemistry as buildings, oceans, and mountain ranges. In fact, many advances in our understanding of the human body came only after advances in one of the physical or applied sciences. For example, the action and purpose of the heart valves remained a mystery until the 1600s, when pumps containing valves were developed to remove the water from flooded coal mines. An English physician, William Harvey, was then astute enough to demonstrate that those design principles explained the function of the heart and the circulation of the blood.

Module 1.4 Review

- **a.** Describe how structure and function are interrelated.
- **b.** Compare the functioning of the elbow joint with a door on a hinge.
- **c.** Predict what would happen to the function of a structure if its anatomy were altered.

LO **1.4** Explain the relationship between structure and function.

SECTION 1 Review

Vocabulary

For each of the following descriptions, write the appropriate characteristic of living things in the corresponding blank.

1 Usually refers to the absorption and utilization of oxygen and the generation and release of carbon dioxide

2 Produces organisms characteristic of its species

3 Changes in the behavior, capabilities, or structure of an organism

4 Movement of fluid within the body; may involve a pump and a network of special vessels

5 Elimination of chemical waste products generated by the body

6 Chemical breakdown of complex structures for absorption and use by the body

7 Transports materials around the body of a large organism; changes orientation or position of a plant or immobile animal; moves mobile animals around the environment (locomotion)

8 Indicates that the organism recognizes changes in the internal or external environment

1 ____________________

2 ____________________

3 ____________________

4 ____________________

5 ____________________

6 ____________________

7 ____________________

8 ____________________

Write each of the following terms under the proper heading.

- Right atrium
- Myocardium
- Valve to aorta opens
- Left ventricle
- Valve between left atrium and left ventricle closes
- Pressure in left atrium
- Electrocardiogram
- Endocardium
- Superior vena cava
- Heartbeat

9 Anatomy

____

____

____

____

____

10 Physiology

____

____

____

____

____

Short answer

Briefly describe how the relationship of form and function of a house key and its front door lock are both similar to and different from a chemical messenger and its receptor molecule.

11 ____________________

____

____

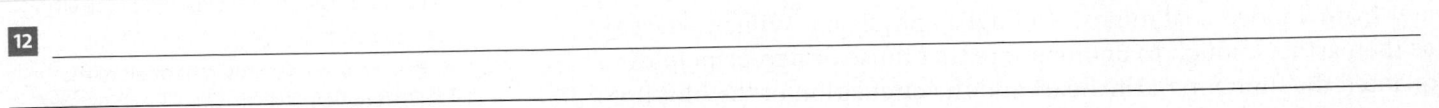

Section integration

How might a large organism's survival be affected by an inadequate internal circulation network?

12 ____________________

____

The human body has multiple interdependent levels of organization

The human body is complex, but the apparent complexity represents multiple **levels of organization**. Each level is more complex than the underlying one, but all can be broken down into similar chemical and cellular components.

Organism Level. An **organism**—in this case, a human—is the highest level of organization. All organ systems of an organism's body work together to maintain life and health.

Organ System Level. (Chapters 5–27) Organs interact in organ systems. Each time it contracts, the heart pumps and pushes blood into a network of blood vessels. Together, the heart, blood, and blood vessels form the cardiovascular system, one of 11 **organ systems** in the body.

Organ Level. An **organ** consists of two or more tissues working to perform several functions. Layers of heart muscle tissue, in combination with connective tissue (another type of tissue), form the bulk of the wall of the heart, a hollow internal organ.

Tissue Level. (Chapter 4) A **tissue** is a group of cells and cell products working together to perform one or more specific functions. Heart muscle cells, or cardiac muscle cells (*cardium*, heart), form cardiac muscle tissue.

Cellular Level. (Chapter 3) **Cells** are the smallest living units in the body. Their functions depend on organelles, intracellular structures composed of complex molecules. Each organelle has a specific function; for example, one type provides the energy that powers the contractions of muscle cells in the heart.

Chemical Level. (Chapter 2) **Atoms**, the smallest stable units of matter, can combine to form **molecules** with complex shapes. The functional properties of a particular molecule are determined by its unique shape and atomic components.

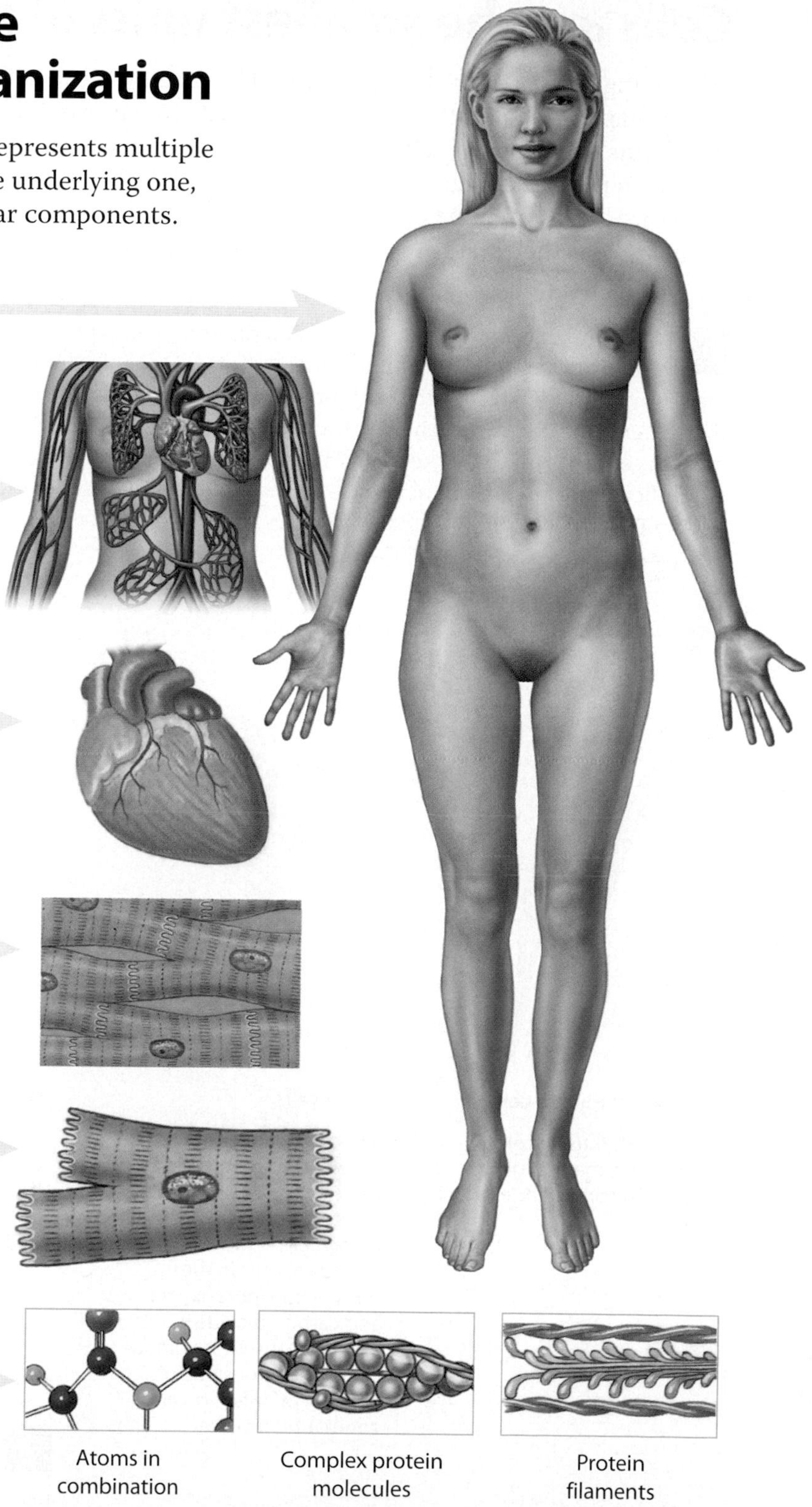
Atoms in combination
Complex protein molecules
Protein filaments

Module 1.5 Review

a. Define organ.

b. Name the lowest level of organization that includes the smallest living units in the body.

c. List the levels of organization between cells and an organism.

LO 1.5 Describe the various levels of organization in the human body.

Cells are the smallest units of life

Free-living **cells** are the smallest living structures, with all of the characteristics described earlier in the chapter. Most of the plants and animals you are familiar with are multicellular, consisting of thousands to billions of cells. These cells do not exist as independent entities. They work together, each with its own characteristics and functions. Cells are the living building blocks of our bodies. There are literally trillions of cells in your body, but there are only an estimated 200 different types of cells. Nevertheless, those 200 types show remarkable diversity in appearance and function.

1 The human body contains about 200 different cell types. The dimensions of cells are usually given in terms of micrometers (μm). One micrometer is one-millionth of a meter, or approximately 1/25,000th of an inch. All the cells illustrated here are shown with the dimensions they would have if they were magnified about 1500 times. The skeletal muscle cells that give you the ability to move around are too large to illustrate here. At this magnification a large skeletal muscle cell would have the diameter of a small dinner plate and be over 300 m long.

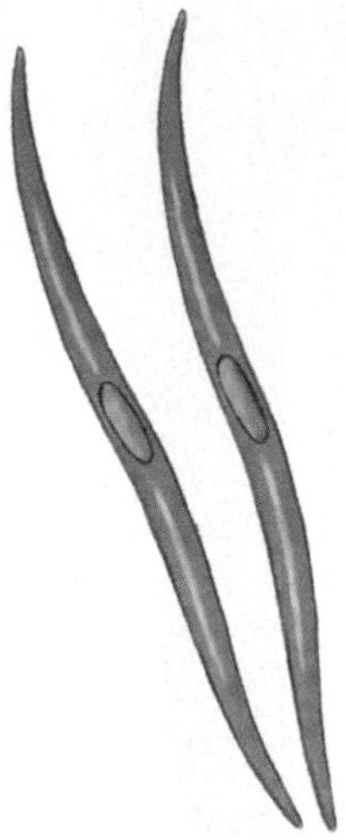

Smooth muscle cells

Smooth muscle cells, found in many organs, are long and slender.

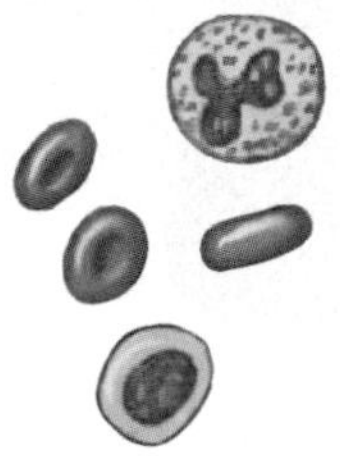

Blood cells

Blood cells are either flattened discs (red blood cells) or roughly spherical (white blood cells). Red blood cells—the most abundant cells in the body—transport oxygen and carbon dioxide in the bloodstream. White blood cells are responsible for fighting off infection and combating disease.

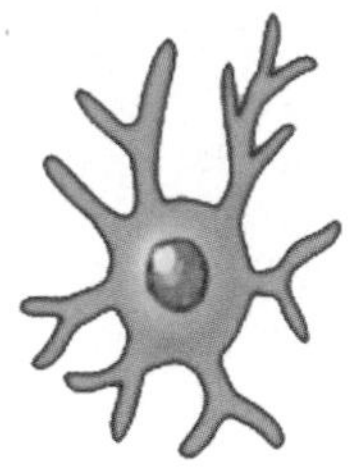

Bone cells

Bone cells reside within small cavities inside the mass of a bone. These cells maintain bone and recycle the calcium and phosphate stored there.

Fat cells

Fat cells are spherical storage containers. Whenever we take in more energy than we expend, the excess energy obtained from the food gets stored as fat, and these cells get larger and more numerous.

2 The importance of cells is apparent in the **cell theory**, one of the foundations of modern biology. The basic principles of the cell theory are as follows:

Basic Principles of the Cell Theory

- Cells are the structural building blocks of all plants and animals.
- Cells are produced by the divisions of pre-existing cells.
- Cells are the smallest structural units that perform all vital functions.

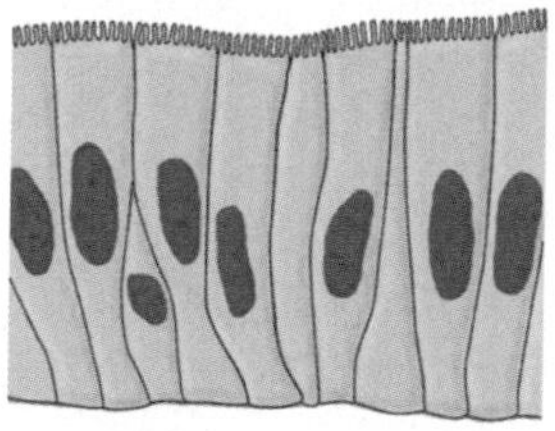

Cells lining the digestive tract

Cells lining the digestive tract are relatively delicate. These cells absorb the nutrients, vitamins, minerals, and water we need.

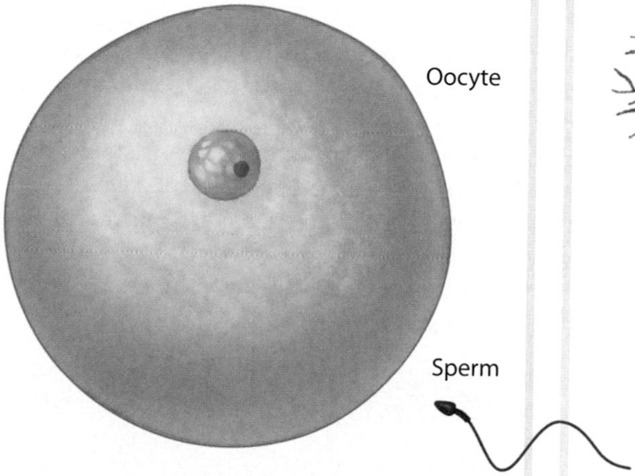
Oocyte
Sperm

Reproductive cells (sex cells)

Cells involved in sexual reproduction are called sex cells. Women produce relatively large oocytes in very small numbers, usually at monthly intervals. Males continuously produce relatively tiny sperm in enormous numbers.

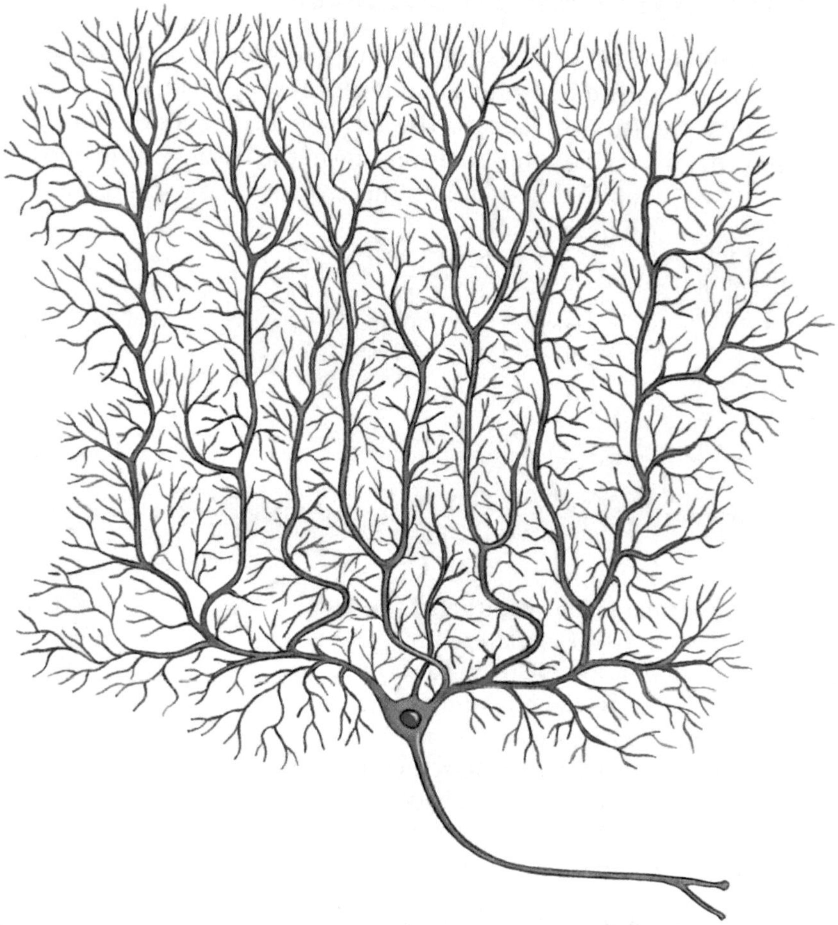

Nerve cells (neurons)

Nerve cells, or neurons, are the equivalent of computer chips—they process information. Thought, memory, consciousness, and muscle control are all based on the actions of, and interactions among, neurons. There are many different types and shapes of neurons. This is a neuron from a part of the brain involved with the control of balance and movement. The extensive branching provides a huge surface area for communicating with other neurons.

The cells of the body work together, and our lives ultimately depend on their actions. If they don't do the right thing at the right time, we're in trouble. If our cells can't survive, we're doomed. Yet each individual cell remains unaware of its role in the "big picture"—it simply responds and adapts to changes in its local environment. How the responses of cells in different parts of the body are coordinated and controlled is obviously a key question, and we will spend considerable time in later chapters considering the answers.

Module 1.6 Review

a. Name and define the unit used to measure cell size.

b. List the three basic principles of the cell theory.

c. Relate the functions of a fat cell and a neuron to their shapes.

LO 1.6 Describe various types of cells in the human body and explain the basic principles of the cell theory.

Module 1.7

Tissues are specialized groups of cells and cell products

The roughly 200 different cell types in the body combine to form **tissues**, collections of cells and cell products that perform specific functions. **Histology** (*histos*, tissue) is the study of tissues. This module introduces the four **primary tissue types** that, in various combinations, form the tissues of the body: epithelial tissue, connective tissue, muscle tissue, and neural tissue.

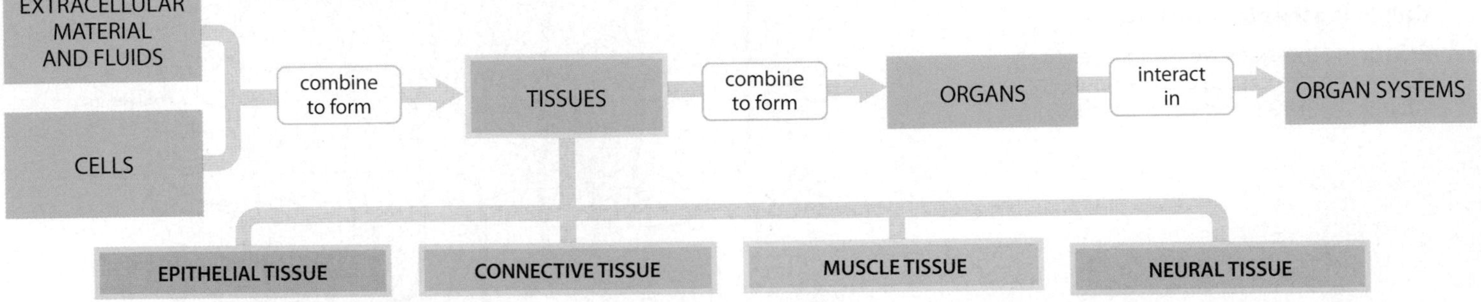
EXTRACELLULAR MATERIAL AND FLUIDS
CELLS
combine to form
TISSUES
combine to form
ORGANS
interact in
ORGAN SYSTEMS
EPITHELIAL TISSUE
CONNECTIVE TISSUE
MUSCLE TISSUE
NEURAL TISSUE

1 The most common type of **epithelial** (ep-i-THĒ-lē-ul) **tissue** is a layer of cells that forms a barrier with specific properties. Epithelia cover every exposed body surface; line the digestive, respiratory, reproductive, and urinary tracts; surround internal cavities such as the chest cavity or the fluid-filled chambers in the brain, eye, and inner ear; and line the inner surfaces of the blood vessels and heart.

EPITHELIAL TISSUE

- Covers and protects exposed surfaces
- Lines internal passageways and chambers
- Produces glandular secretions

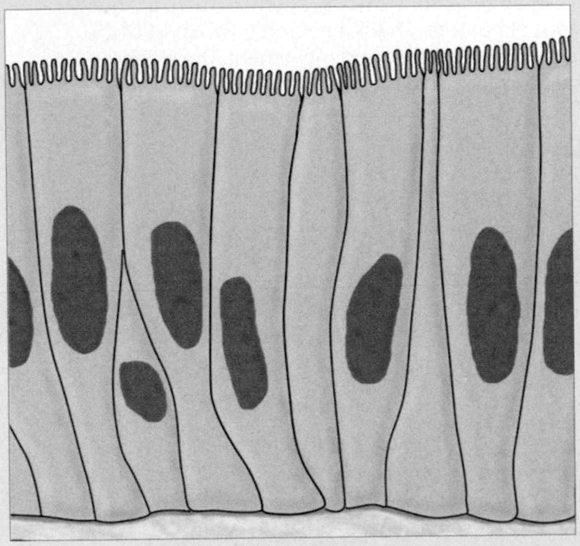

2 **Connective tissue** is quite diverse in appearance. All forms of connective tissue contain cells and an extracellular **matrix** that consists of protein fibers and a liquid known as the ground substance. The amount and consistency of the matrix depend on the particular type of connective tissue. In blood, the cells are suspended in a watery matrix called plasma. Bone has a more durable matrix, with crystals of calcium salts organized around a fibrous framework, and very little ground substance.

CONNECTIVE TISSUE

- Fills internal spaces
- Provides structural support
- Stores energy

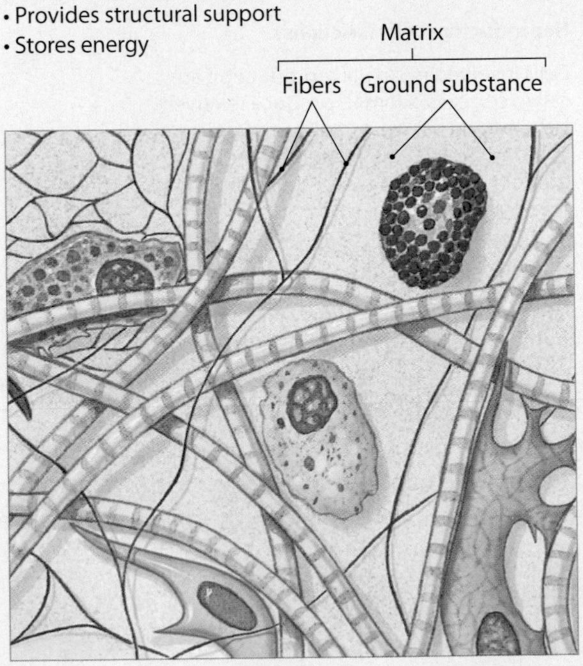
Matrix
Fibers
Ground substance

3 **Muscle tissue** is unique because individual muscle cells can contract forcefully. Major functions of muscle tissue include skeletal movement, soft tissue support, maintenance of blood flow, movement of materials along internal passageways, and the stabilization of normal body temperature. There are three different types of muscle tissue.

MUSCLE TISSUE

- Contracts to produce active movement

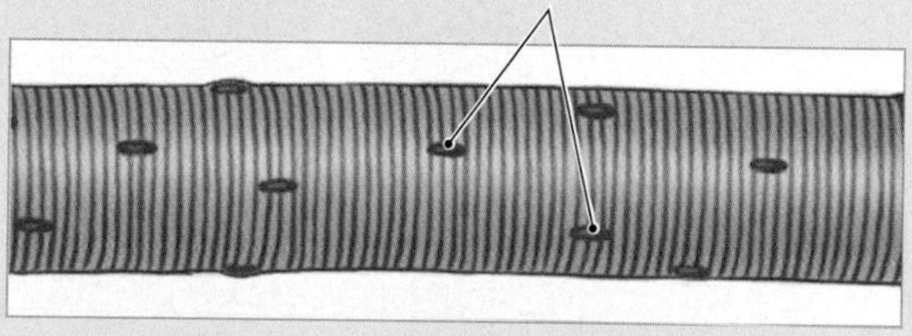

Nuclei

Skeletal muscle tissue is usually directly or indirectly attached to the skeleton. When it contracts, it moves or stabilizes the position of bones or internal organs.

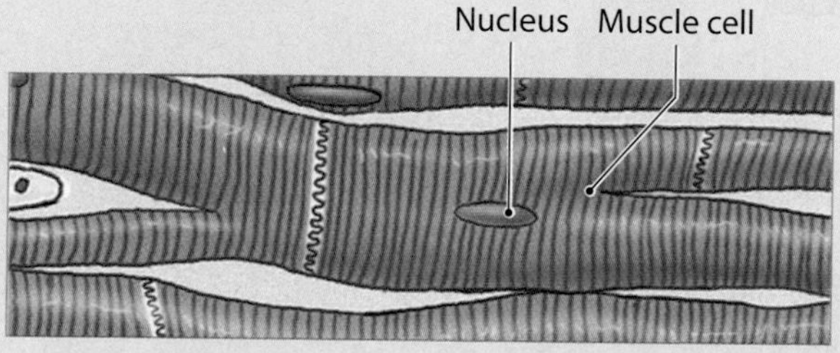

Nucleus
Muscle cell

Cardiac muscle tissue is found in the heart, where its coordinated contractions propel blood through the blood vessels.

Smooth muscle tissue is found in the walls of blood vessels, within glands, and along the respiratory, circulatory, digestive, and reproductive tracts.

4 **Neural tissue** is specialized to carry information or instructions from one place in the body to another. Two basic types of cells are present: nerve cells, or **neurons** (NOO-rons; *neuro*, nerve), and supporting cells, or **neuroglia** (noo-RŌG-lē-uh; *glia*, glue). Neurons transmit information in the form of electrical impulses. Neuroglia isolate and protect neurons while forming a supporting framework. The neural tissue in the body can be divided on anatomical grounds into the **central nervous system**, which consists of the brain and spinal cord, and the **peripheral nervous system**, which includes the nerves connecting the central nervous system with other tissues and organs.

NEURAL TISSUE

- Conducts electrical impulses
- Carries information

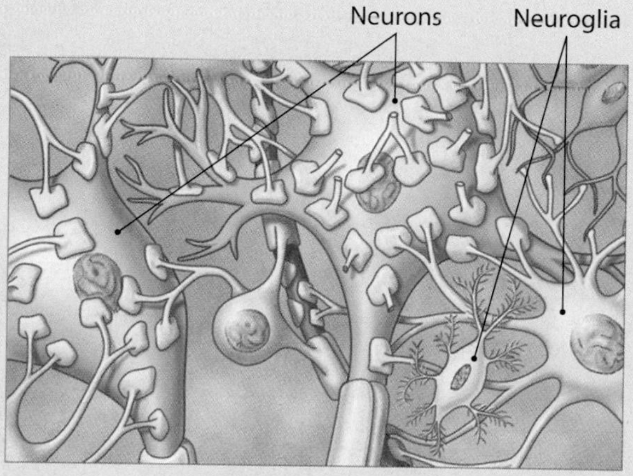

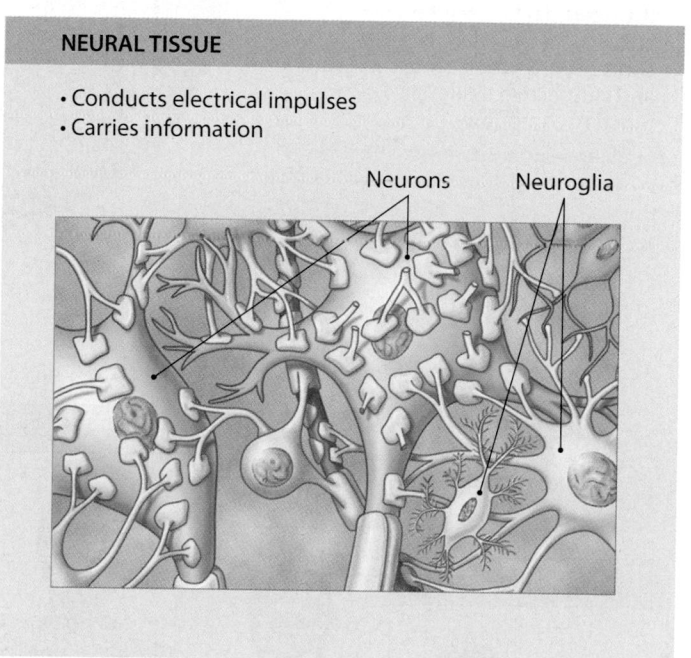

Neurons
Neuroglia

Module 1.7 Review

a. Define histology.

b. Identify the four primary tissue types.

c. Explain the functions of each of the primary tissue types.

Module 1.8

Organs and organ systems perform vital functions

1 An **organ** is a functional unit composed of more than one tissue type. The particular combination and organization of tissues within an organ both determines and limits the organ's functions. For example, an organ with a flattened shape can provide protection (like a sheet of cardboard on a table surface), and an organ with a three-dimensional shape can house additional structures (like a cardboard box and its contents). An **organ system** consists of organs that interact to perform a specific range of functions, often in a coordinated fashion.

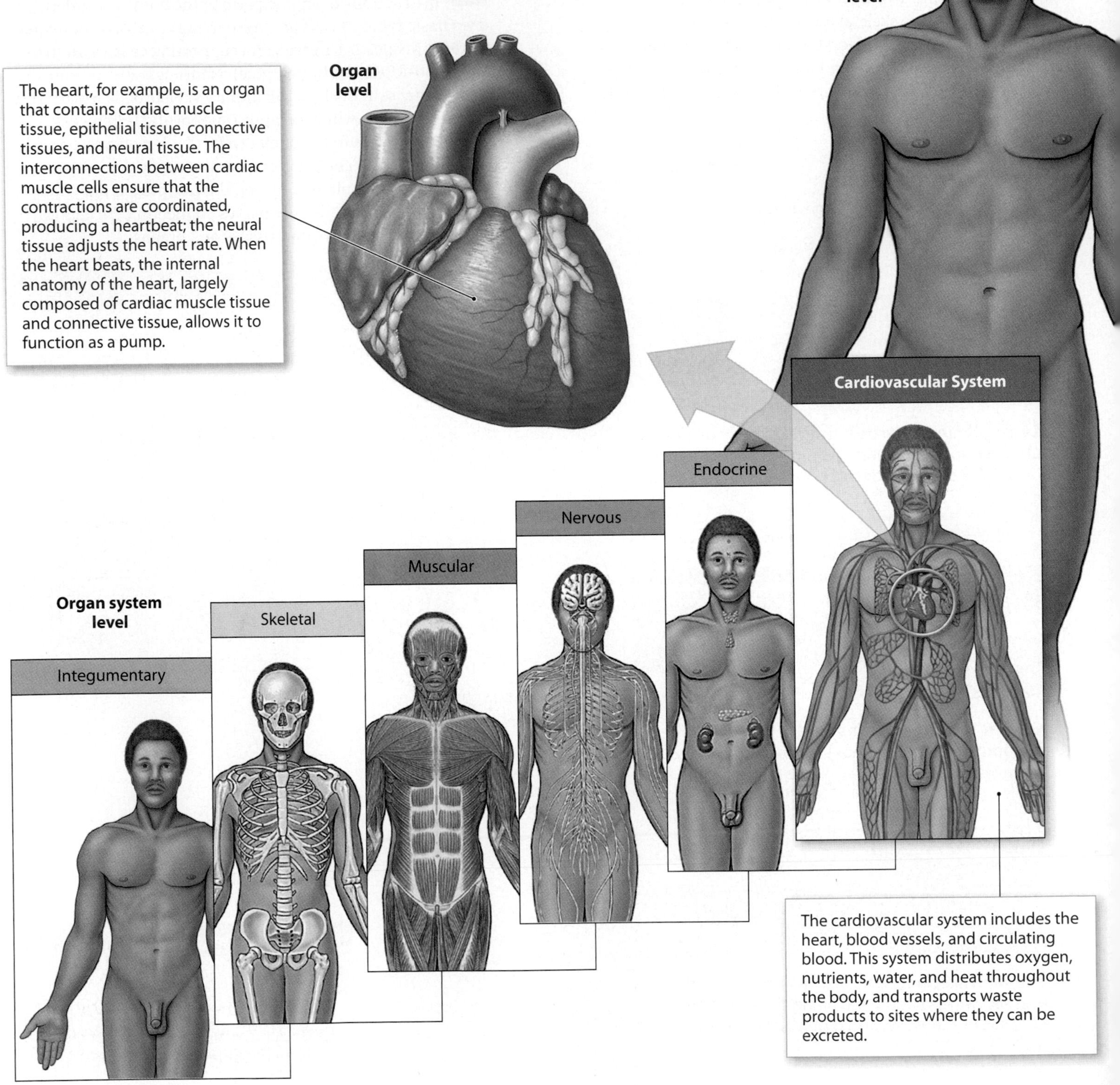

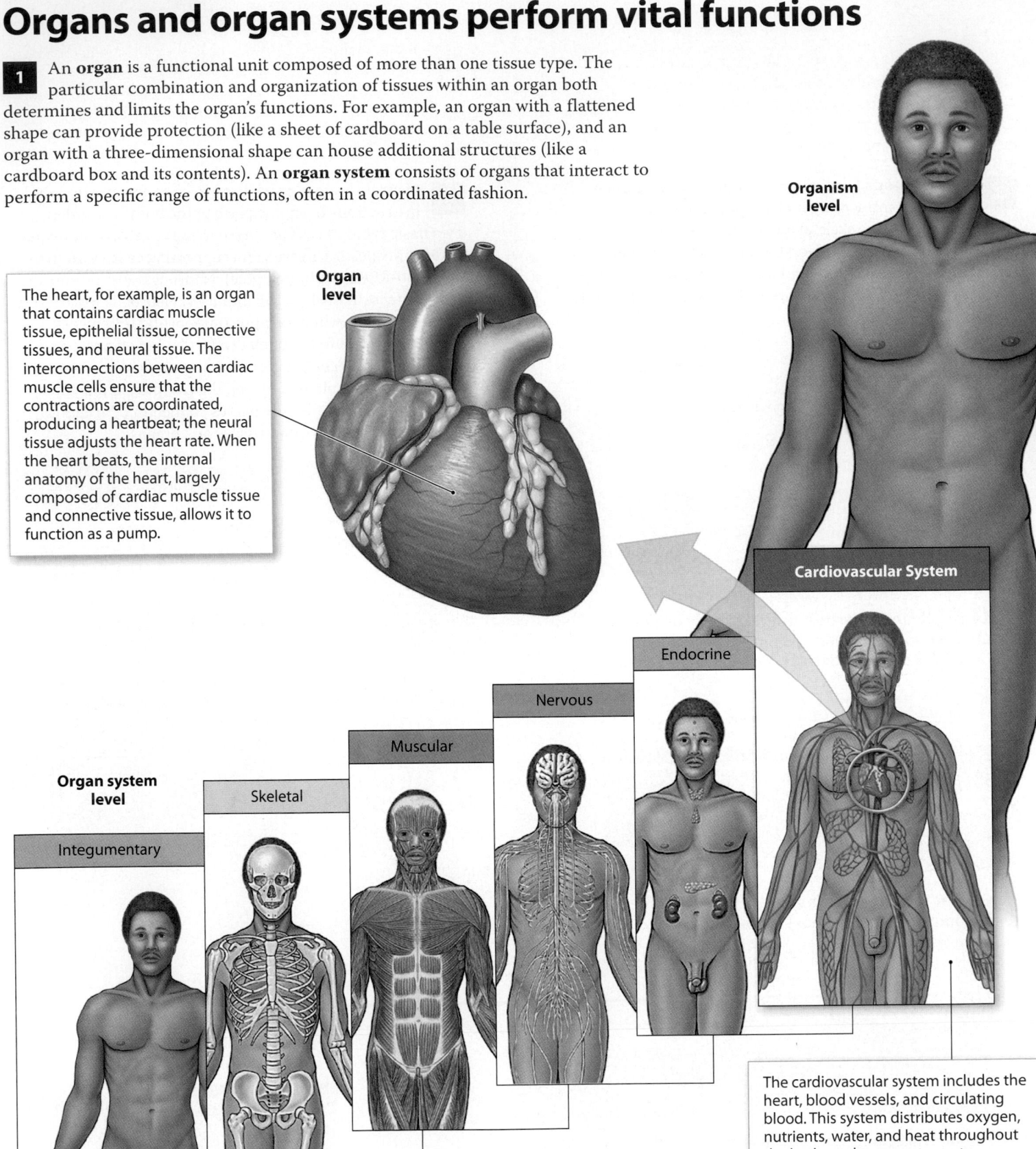

Organism level
Organ level
The heart, for example, is an organ that contains cardiac muscle tissue, epithelial tissue, connective tissues, and neural tissue. The interconnections between cardiac muscle cells ensure that the contractions are coordinated, producing a heartbeat; the neural tissue adjusts the heart rate. When the heart beats, the internal anatomy of the heart, largely composed of cardiac muscle tissue and connective tissue, allows it to function as a pump.
Organ system level
Integumentary
Skeletal
Muscular
Nervous
Endocrine
Cardiovascular System
The cardiovascular system includes the heart, blood vessels, and circulating blood. This system distributes oxygen, nutrients, water, and heat throughout the body, and transports waste products to sites where they can be excreted.

2 The table at right lists the 11 organ systems in the human body. Although this categorization is a convenient way to organize information, the concept of separate "organ systems" is artificial and somewhat misleading. Nothing in the body functions in isolation—not cells, not tissues, not organs, and certainly not organ systems. Organs and organ systems are interdependent, and something that affects one organ will affect the functioning of the body as a whole. For example, the heart cannot pump blood effectively after massive blood loss. If the heart cannot pump and blood cannot flow, oxygen and nutrients cannot be distributed. Very soon, cardiac muscle tissue begins to break down as individual muscle cells die from oxygen and nutrient starvation. These changes will not be restricted to the cardiovascular system. All cells, tissues, and organs in the body will be damaged, with potentially fatal results.

Organ Systems	Major Functions
Integumentary system	Protects against environmental hazards; helps control body temperature
Skeletal system	Provides support; protects tissues; stores minerals; forms blood cells
Muscular system	Produces movement; provides support; generates heat
Nervous system	Directs immediate responses to stimuli, usually by coordinating the activities of other organ systems
Endocrine system	Directs long-term changes in other organ systems
Cardiovascular system	Transports cells and dissolved materials, including nutrients, wastes, and gases
Lymphatic system	Defends against infection and disease; returns tissue fluid to the bloodstream
Respiratory system	Delivers air to sites where gas exchange occurs between the air and circulating blood; produces sound
Digestive system	Processes food and absorbs nutrients (carbohydrates, fats, proteins, vitamins, minerals, and water)
Urinary system	Eliminates excess water, salts, and wastes; controls pH
Reproductive system	Produces sex cells and hormones; supports embryonic development from fertilization to birth (female)

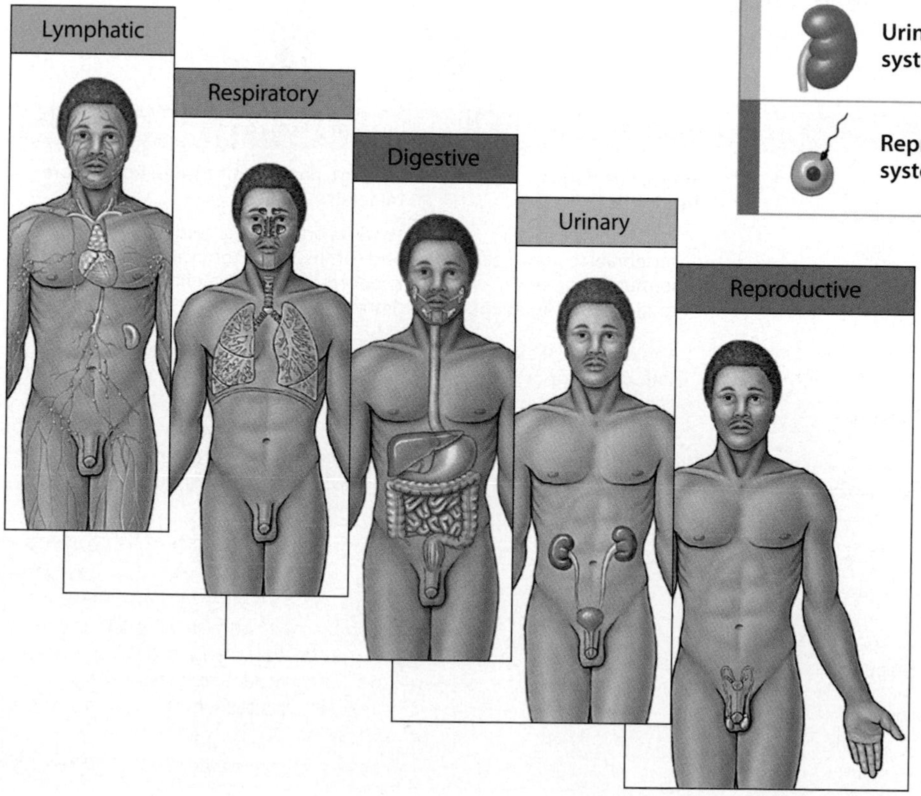
Lymphatic
Respiratory
Digestive
Urinary
Reproductive

Module 1.8 Review

a. List the 11 organ systems of the body.

b. Explain the relationship between the skeletal system and the digestive system.

c. Using the table as a reference, describe how falling down a flight of stairs could affect at least six of the organ systems.

LO 1.8 Identify the 11 organ systems of the human body, and describe the major functions of each.

Module 1.9

Organs of the integumentary, skeletal, and muscular systems support and move the body . . .

Integumentary System

Protects against environmental hazards; helps control body temperature

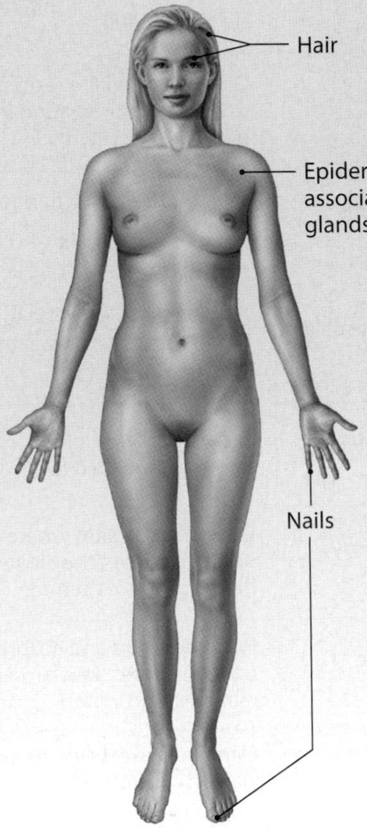

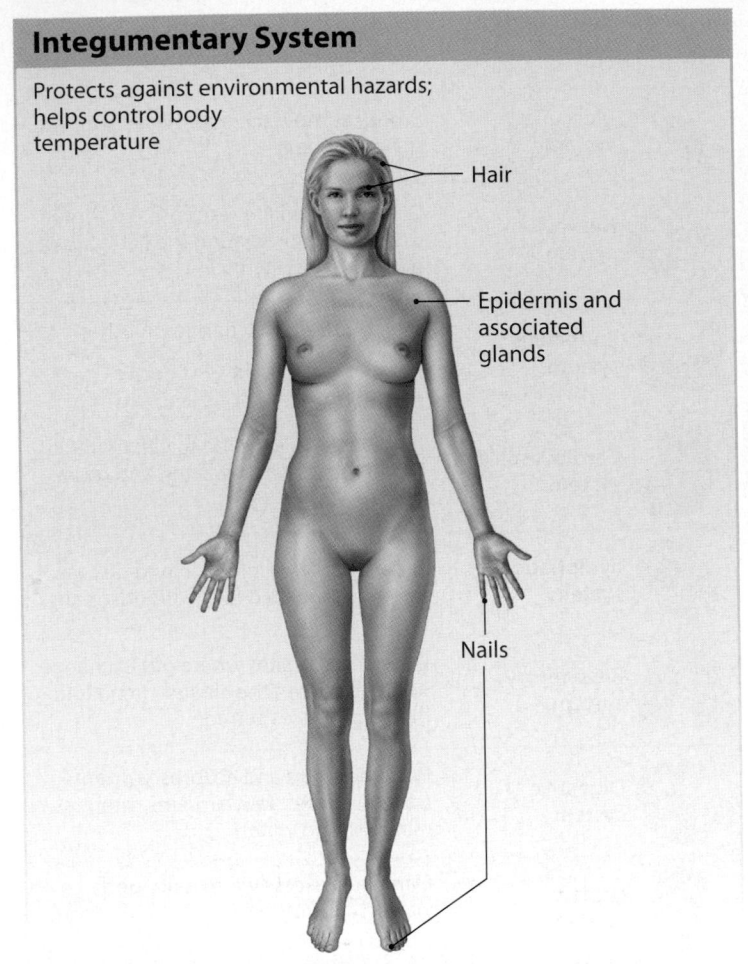

Hair
Epidermis and associated glands
Nails

Organ/Structure	Primary Function
Cutaneous Membrane	
Epidermis	Covers surface; protects deeper tissues
Dermis	Nourishes epidermis; provides strength; contains glands
Hair Follicles	Produce hair; innervation provides sensation
Hairs	Provide protection for head
Sebaceous glands	Secrete lipid coating that lubricates hair shaft and epidermis
Sweat Glands	Produce perspiration for evaporative cooling
Nails	Protect and stiffen distal tips of digits
Sensory Receptors	Provide sensations of touch, pressure, temperature, pain
Hypodermis	Stores lipids; attaches skin to deeper structures

Skeletal System

Provides support; protects tissues; stores minerals; forms blood cells

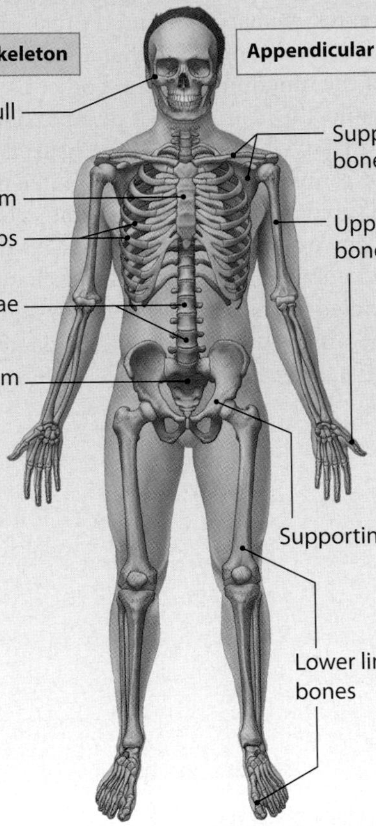

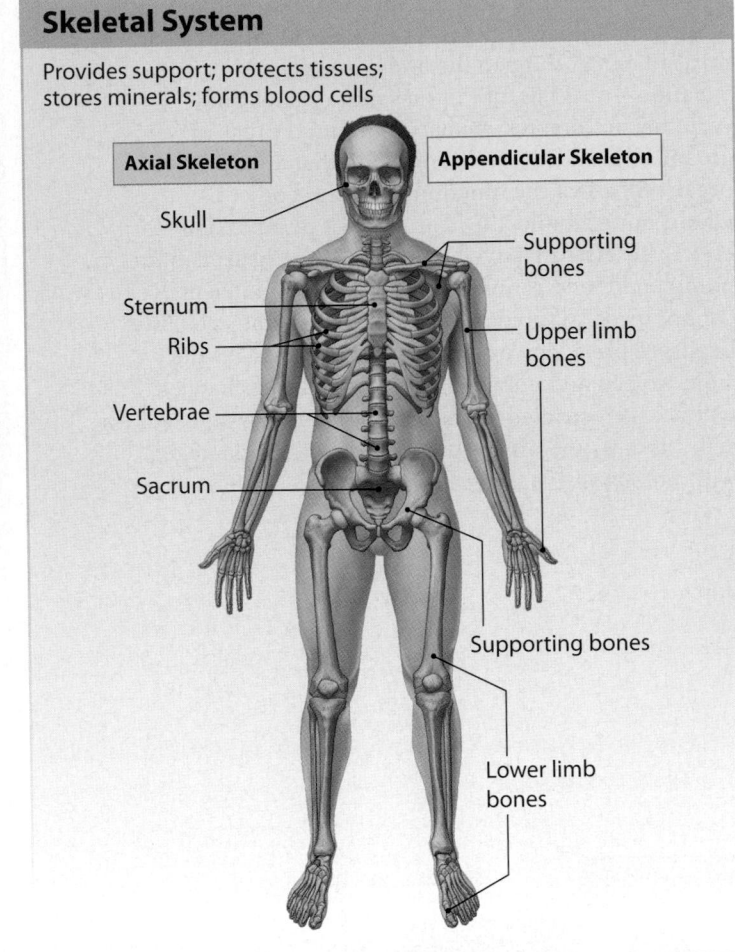

Axial Skeleton
Appendicular Skeleton
Skull
Sternum
Ribs
Vertebrae
Sacrum
Supporting bones
Upper limb bones
Supporting bones
Lower limb bones

Organ/Structure	Primary Function
Bones, Cartilages, and Joints	Support, protect soft tissues; bones store minerals
Axial skeleton (skull, vertebrae, sacrum, coccyx, sternum, supporting cartilages and ligaments)	Protects brain, spinal cord, sense organs, and soft tissues of thoracic cavity; supports the body weight over lower limbs
Appendicular skeleton: limbs and supporting bones and ligaments	Provides internal support and positioning of the limbs; supports and moves axial skeleton
Bone Marrow	Primary site of blood cell production (red marrow); stores of energy in fat cells (yellow marrow)

... and organs of the nervous system provide rapid control and regulation

Muscular System

Produces movement; provides support; generates heat

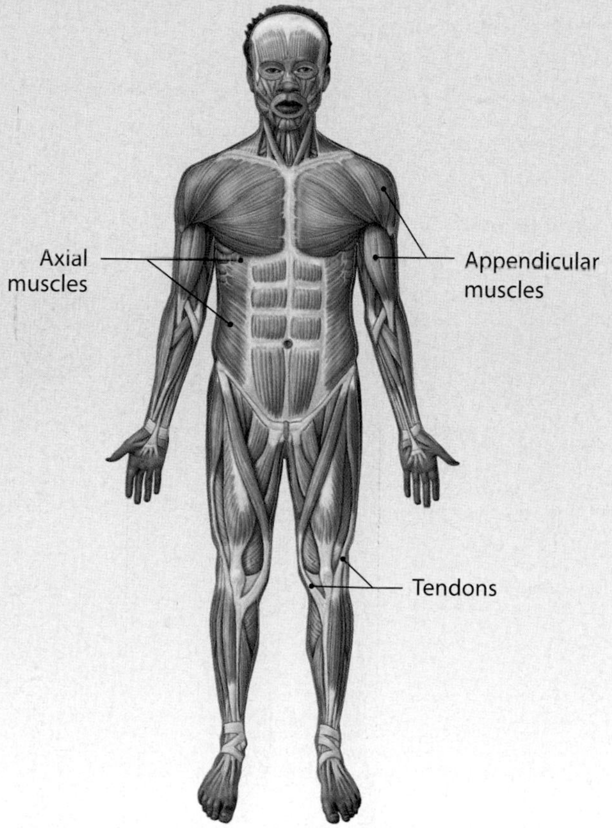

Axial muscles
Appendicular muscles
Tendons

Organ/Structure	Primary Functions
Skeletal Muscles	Provide skeletal movement; control entrances to digestive and respiratory tracts and exits from digestive and urinary tracts; produce heat; support skeleton; protect soft tissues
Axial muscles	Support and position axial skeleton
Appendicular muscles	Support, move, and brace limbs
Tendons, Aponeuroses	Use forces of contraction to perform specific tasks

Nervous System

Directs immediate responses to stimuli, usually by coordinating the activities of other organ systems

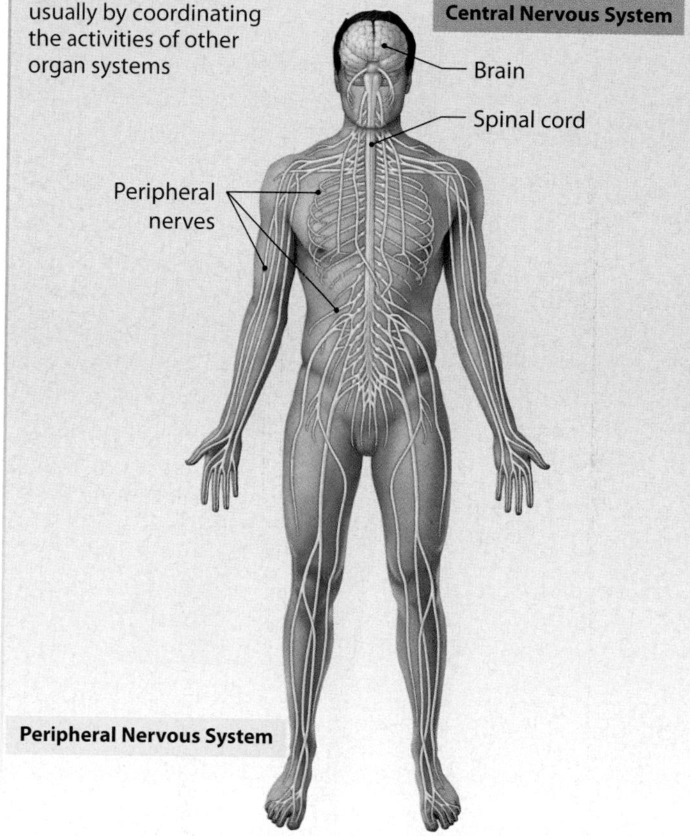

Central Nervous System
Brain
Spinal cord
Peripheral nerves
Peripheral Nervous System

Organ/Structure	Primary Function
Central Nervous System (CNS)	Acts as control center for nervous system; processes information; provides short-term control over activities of other systems
Brain	Performs complex integrative functions; controls both voluntary and involuntary activities
Spinal cord	Relays information to and from brain; performs less-complex integrative activities
Special senses	Provide sensory input to the brain relating to sight, hearing, smell, taste, and equilibrium
Peripheral Nervous System (PNS)	Links CNS with other systems and with sense organs

Module 1.9 Review

a. Identify the major organs of the integumentary, skeletal, muscular, and nervous systems.

b. Explain the functions of each of these systems.

c. How would a nervous system disorder affect the muscular system?

LO **1.9** Describe the major organs of the integumentary, skeletal, muscular, and nervous systems and briefly describe their functions.

Module 1.10

Organs of the endocrine system secrete chemicals that are carried by organs of the cardiovascular system . . .

Endocrine System

Directs long-term changes in other organ systems

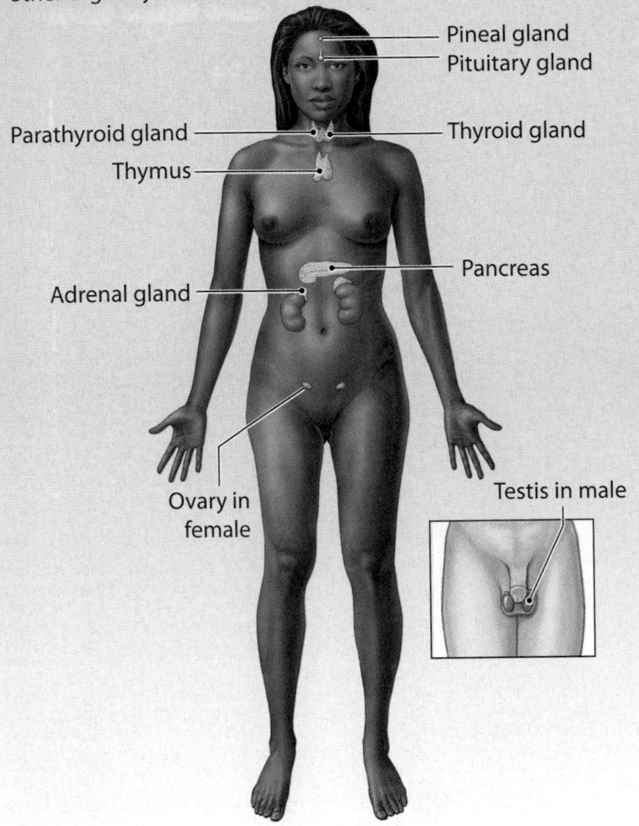
Pineal gland
Pituitary gland
Parathyroid gland
Thyroid gland
Thymus
Pancreas
Adrenal gland
Ovary in female
Testis in male

Organ/Structure	Primary Function
Pineal Gland	May control timing of reproduction and set day-night rhythms
Pituitary Gland	Controls other endocrine glands; regulates growth and fluid balance
Thyroid Gland	Controls tissue metabolic rate; regulates calcium levels
Parathyroid Glands	Regulate calcium levels (with thyroid gland)
Thymus	Controls maturation of lymphocytes
Adrenal Glands	Adjust water balance, tissue metabolism, cardiovascular and respiratory activity
Kidneys	Control red blood cell production, elevate blood pressure, and assist in calcium homeostasis
Pancreas	Regulates blood glucose levels
Gonads	
Testes	Support male sexual characteristics and reproductive functions (Module 1.11)
Ovaries	Support female sexual characteristics and reproductive functions (Module 1.11)

Cardiovascular System

Transports cells and dissolved materials, including nutrients, wastes, and gases

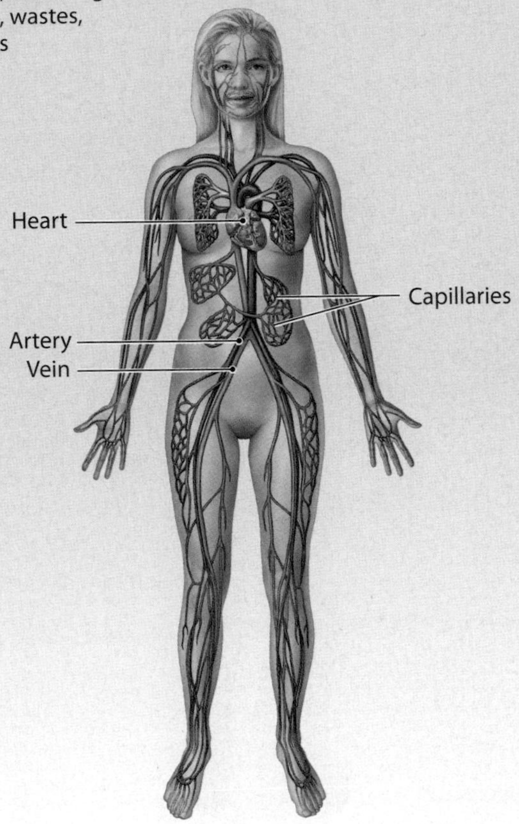
Heart
Capillaries
Artery
Vein

Organ/Structure	Primary Function
Heart	Propels blood; maintains blood pressure
Blood Vessels	Distribute blood around the body
Arteries	Carry blood from the heart to capillaries
Capillaries	Permit diffusion between blood and interstitial fluids
Veins	Return blood from capillaries to the heart
Blood	Transports oxygen, carbon dioxide, and blood cells; delivers nutrients and hormones; removes wastes; assists in temperature regulation and defense against disease

... organs of the lymphatic system defend the body, and organs of the respiratory system exchange vital gases

Lymphatic System

Defends against infection and disease; returns tissue fluid to the bloodstream

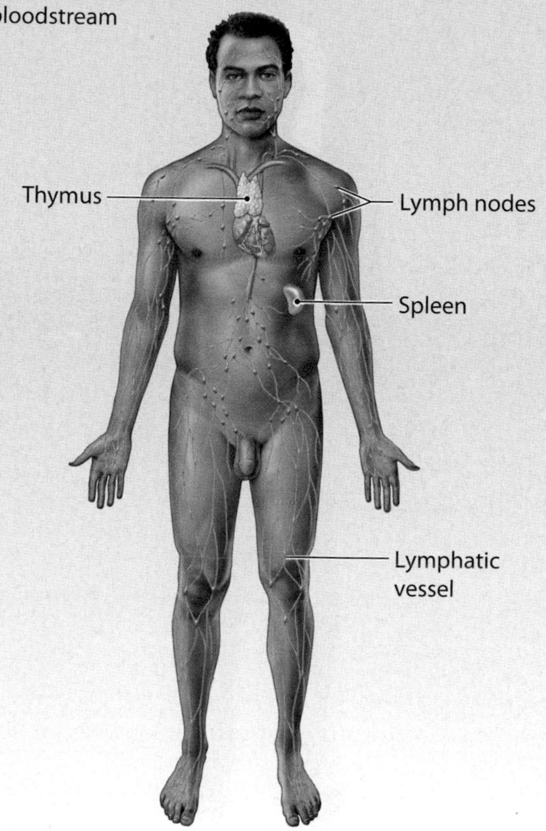

Thymus
Lymph nodes
Spleen
Lymphatic vessel

Organ/Structure	Primary Function
Lymphatic Vessels	Carry lymph (water and proteins) and lymphocytes from peripheral tissues to veins of the cardiovascular system
Lymph Nodes	Monitor the composition of lymph; engulf pathogens; stimulate immune response
Spleen	Monitors circulating blood; engulfs pathogens and recycles red blood cells; stimulates immune response
Thymus	Controls development and maintenance of one class of lymphocytes (T cells)

Respiratory System

Delivers air to sites where gas exchange occurs between the air and circulating blood; produces sound

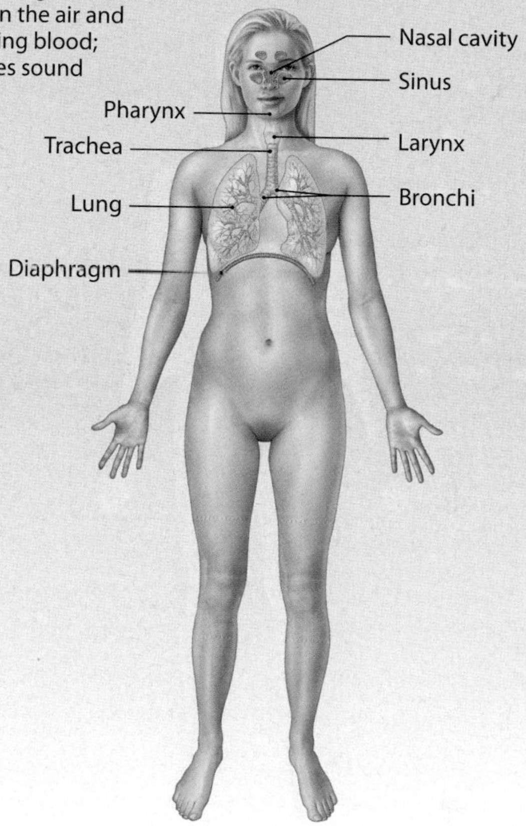

Nasal cavity
Sinus
Pharynx
Larynx
Trachea
Bronchi
Lung
Diaphragm

Organ/Structure	Primary Function
Nasal Cavities and Paranasal Sinuses	Filter, warm, humidify air; detect smells
Pharynx	Conducts air to larynx; a chamber shared with the digestive tract
Larynx	Protects opening to trachea and contains vocal cords
Trachea	Filters air; cartilages keep airway open
Bronchi	Conducts air between trachea and lungs
Lungs	Responsible for air movement; alveoli within the lungs are sites of gas exchange between air and blood

Module 1.10 Review

a. Identify the major organs of the endocrine, cardiovascular, lymphatic, and respiratory systems.

b. Explain the functions of each of these systems.

c. How would a lymphatic system disease affect the cardiovascular system?

LO **1.10** Describe the major organs of the endocrine, cardiovascular, lymphatic, and respiratory systems and briefly describe their functions.

Module 1.11

Organs of the digestive system make nutrients available and with the urinary system excrete wastes …

Digestive System

Processes food and absorbs nutrients

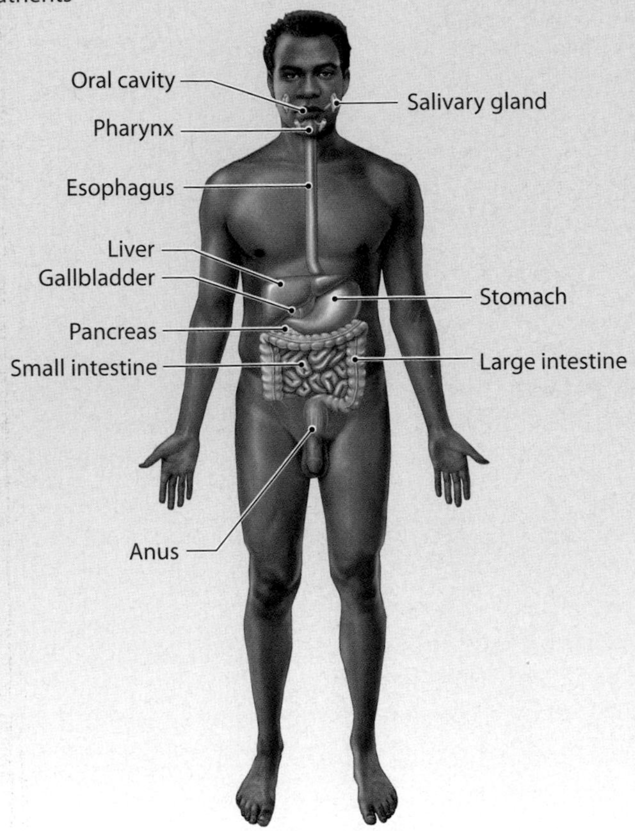

Oral cavity
Pharynx
Esophagus
Liver
Gallbladder
Pancreas
Small intestine
Anus
Salivary gland
Stomach
Large intestine

Organ/Structure	Primary Function
Oral Cavity (Mouth)	Cavity for food; works with associated structures (teeth, tongue) to break up food and pass food and liquids to pharynx
Salivary Glands	Provide buffers and lubrication; produce enzymes that begin digestion
Pharynx	Conducts solid food and liquids to esophagus; chamber shared with respiratory tract
Esophagus	Delivers food to stomach
Stomach	Secretes acids, enzymes, and hormones
Small Intestine	Secretes digestive enzymes, buffers, and hormones; absorbs nutrients
Liver	Secretes bile; regulates nutrient composition of blood
Gallbladder	Stores and concentrates bile for release into small intestine
Pancreas	Secretes digestive enzymes and buffers; contains endocrine cells
Large Intestine	Removes water from feces; stores wastes

Urinary System

Eliminates excess water, salts, and wastes

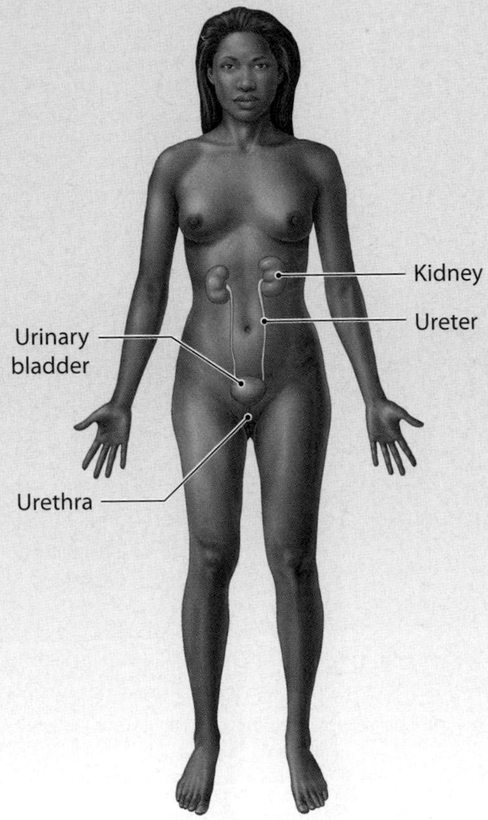

Kidney
Ureter
Urinary bladder
Urethra

Organ/Structure	Primary Function
Kidneys	Form and concentrate urine; regulate blood pH and ion concentrations; perform endocrine functions
Ureters	Conduct urine from kidneys to urinary bladder
Urinary Bladder	Stores urine for eventual elimination
Urethra	Conducts urine to exterior

... and organs of the male and female reproductive systems provide for the continuity of life

Female Reproductive System

Produces sex cells and hormones; supports embryonic development from fertilization to birth

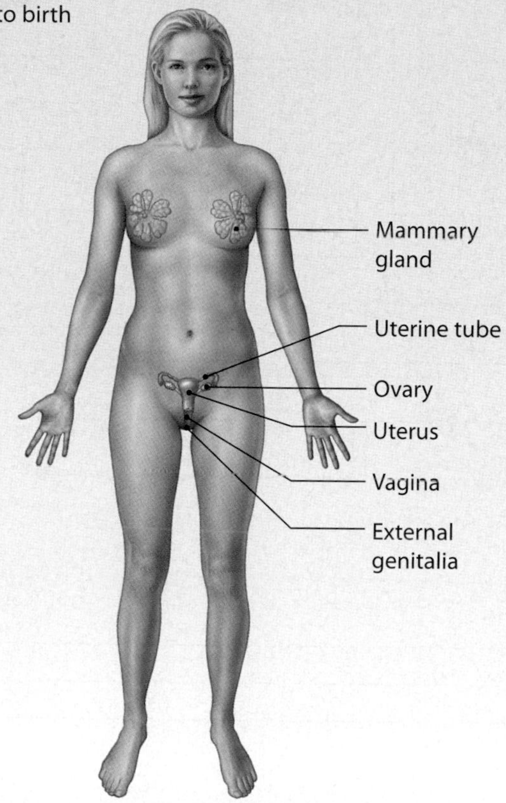

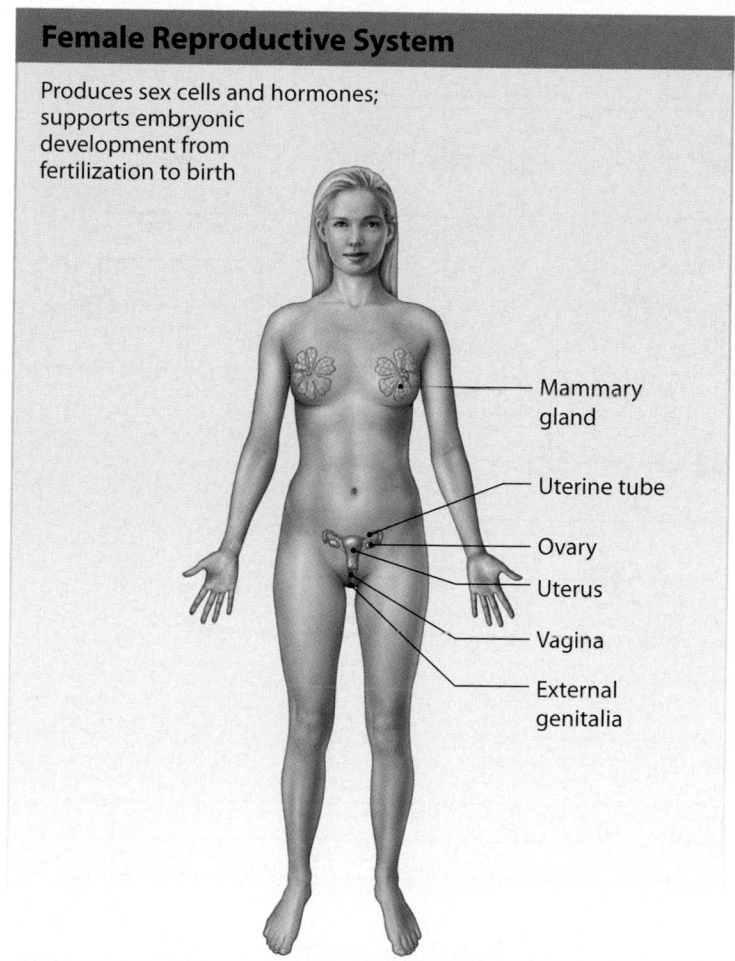

Mammary gland
Uterine tube
Ovary
Uterus
Vagina
External genitalia

Organ/Structure	Primary Function
Ovaries	Produce oocytes and hormones
Uterine Tubes	Deliver oocyte or embryo to uterus; normal site of fertilization
Uterus	Site of embryonic development and exchange between maternal and fetal bloodstreams; sheds lining during menstruation
Vagina	Site of sperm deposition; acts as a birth canal during delivery; provides passageway for fluids during menstruation
External Genitalia	
Clitoris	Contains erectile tissue; provides pleasurable sensations during sexual activities
Labia	Contain glands that lubricate entrance to vagina
Mammary Glands	Produce milk that nourishes newborn infant

Male Reproductive System

Produces sex cells and hormones

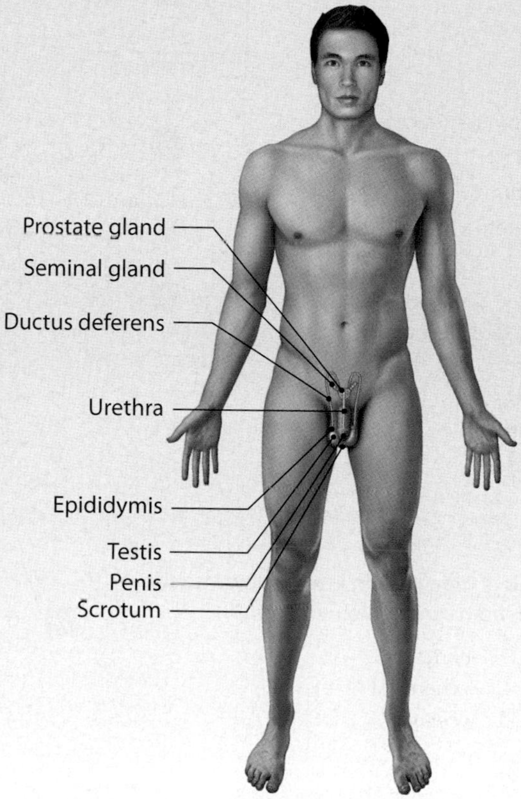

Prostate gland
Seminal gland
Ductus deferens
Urethra
Epididymis
Testis
Penis
Scrotum

Organ/Structure	Primary Function
Testes	Produce sperm and hormones
Accessory Organs	
Epididymis	Acts as site of sperm maturation in each testis
Ductus deferens	Conducts sperm from the epididymis
Seminal glands	Secrete fluid that contributes to semen
Prostate gland	Secretes fluid and enzymes
Urethra	Conducts semen to exterior
External Genitalia	
Penis	Deposits sperm in vagina of female
Scrotum	Surrounds the testes and controls their temperature

Module 1.11 Review

a. Identify the major organs of the digestive, urinary, and reproductive systems.

b. Explain the functions of each of these systems.

c. How would a reproductive system disorder affect the urinary system?

LO **1.11** Describe the major organs of the digestive, urinary, and reproductive systems and briefly describe their functions.

Concept map

Use each of the following terms once to fill in the blank boxes to correctly complete the map.

- organs
- epithelial tissue
- cells
- connective tissue
- muscle tissue
- neural tissue
- organ systems
- external and internal surfaces
- matrix
- glandular secretions
- bones of the skeleton
- neuroglia
- blood
- materials within digestive tract
- protein fibers
- ground substance
- movement

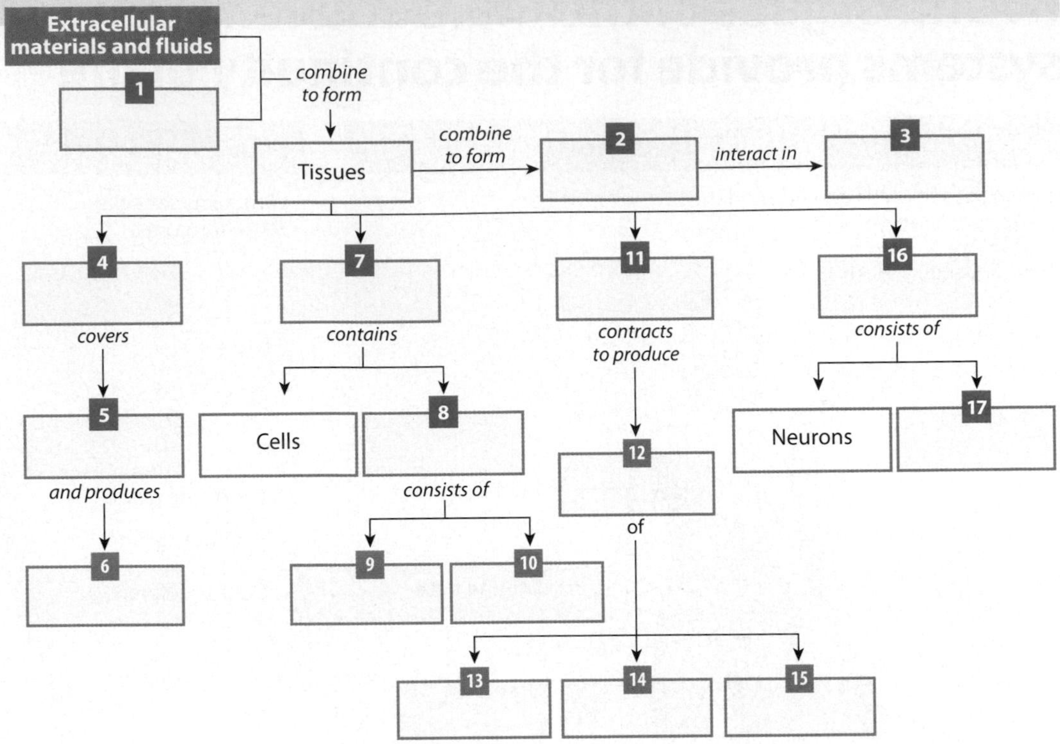

Extracellular materials and fluids
1
combine to form
Tissues
combine to form
2
interact in
3
4
covers
5
and produces
6
7
contains
Cells
8
consists of
9
10
11
contracts to produce
12
of
13
14
15
16
consists of
Neurons
17

Vocabulary

Reorder the levels of organization listed below into the correct sequence from simplest to most complex.

- organ system
- tissue
- organism
- organ
- chemical
- cellular

18 ______________________________

Short answer

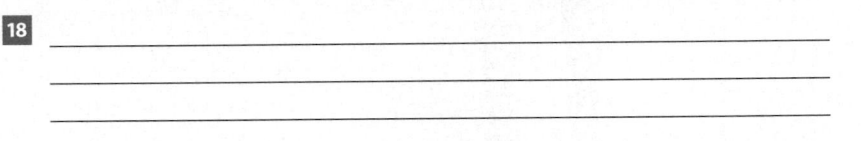

Summarize the major functions of each of the following organ systems.

19 Integumentary
20 Skeletal
21 Muscular
22 Nervous
23 Endocrine
24 Cardiovascular
25 Lymphatic
26 Respiratory
27 Digestive
28 Urinary
29 Reproductive

19 ______________________________
20 ______________________________
21 ______________________________
22 ______________________________
23 ______________________________
24 ______________________________
25 ______________________________
26 ______________________________
27 ______________________________
28 ______________________________
29 ______________________________

Section integration

For five different organ systems in the human body, identify a specialized cell type found in that system.

30 ______________________________

Homeostatic regulation relies on a receptor, a control center, and an effector

Homeostasis (hō-mē-ō-STĀ-sis; *homeo*, unchanging + *stasis*, standing) is the presence of a stable internal environment. Maintaining homeostasis is absolutely vital to an organism's survival. Failure to maintain homeostasis soon leads to illness or even death. The principle of homeostasis is the central theme of this text and the foundation of all modern physiology. **Homeostatic regulation** is the adjustment of physiological systems to preserve homeostasis in environments that are often inconsistent, unpredictable, and potentially dangerous. An understanding of homeostatic regulation is crucial to making accurate predictions about the body's responses to both normal and abnormal conditions.

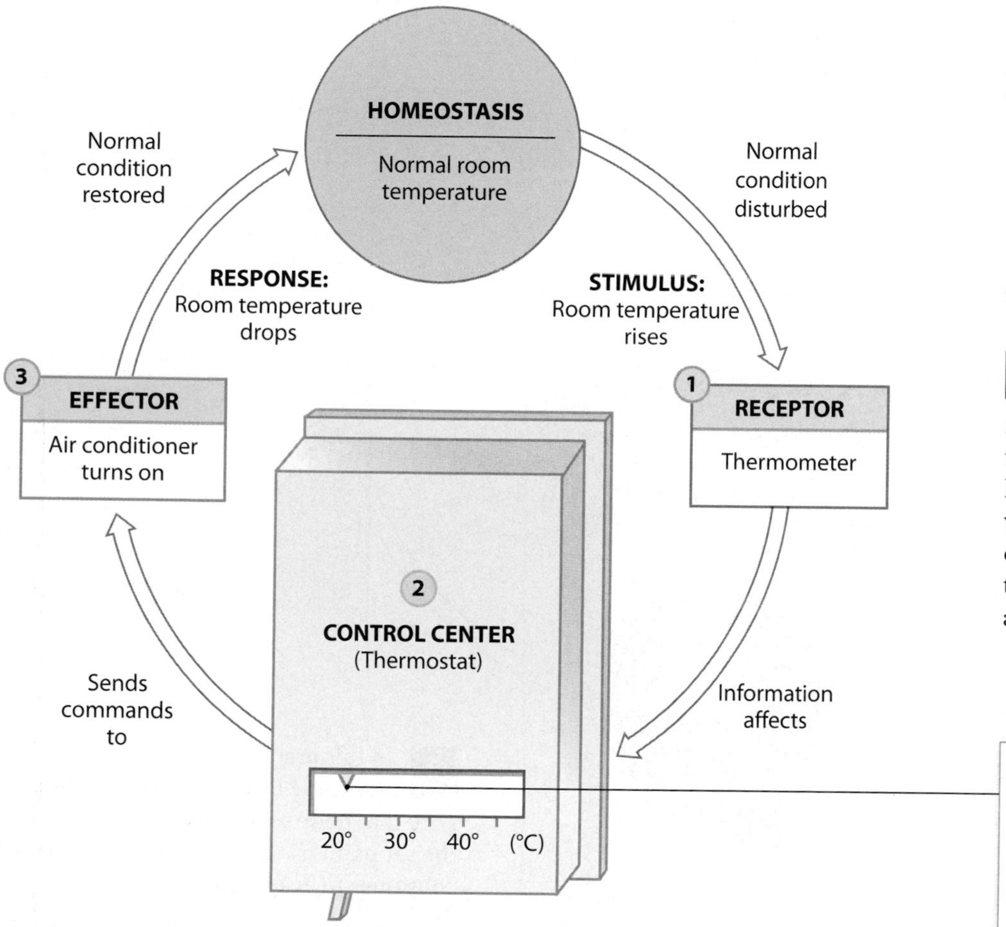
HOMEOSTASIS
Normal room temperature
Normal condition disturbed
STIMULUS:
Room temperature rises
1
RECEPTOR
Thermometer
Information affects
2
CONTROL CENTER
(Thermostat)
20° 30° 40° (°C)
Sends commands to
3
EFFECTOR
Air conditioner turns on
RESPONSE:
Room temperature drops
Normal condition restored

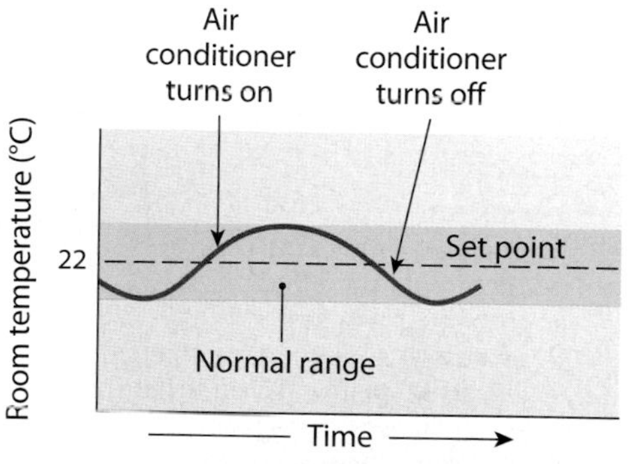
Air conditioner turns on
Air conditioner turns off
Room temperature (°C)
22
Set point
Normal range
Time

2 Homeostatic control is not precise—it maintains a normal range rather than an absolute value. The same is true for controlling room temperature—a house may have the thermostat on one wall of one room, and the air conditioning outlets at multiple locations. Over time, the temperature in the house will oscillate around the set point.

The setting on a thermostat establishes the **set point**, or desired value, which in this case is the temperature you select. (In our example, the set point is 22°C, or about 72°F.) The function of the thermostat is to keep room temperature within acceptable limits, usually within a degree or so of the set point.

1 Maintaining a relatively constant temperature in your living space is a familiar example of homeostasis. Like all homeostatic regulatory mechanisms, it consists of (1) a **receptor** or sensor—in this case, a thermometer—that is sensitive to a particular environmental change, or stimulus; (2) a **control center** or integration center—in this case, a thermostat—which receives and processes the information supplied by the receptor, and which sends out commands; and (3) an **effector**—in this case, an air conditioner—which responds to these commands by opposing the stimulus. The net effect is that any variation outside normal limits triggers a response that restores normal conditions.

Module 1.12 Review

a. Define homeostasis.

b. Why is homeostatic regulation important to an organism?

c. Describe the three parts necessary for homeostatic regulation.

LO 1.12 Describe the mechanisms of homeostatic regulation.

Module 1.13

Negative feedback provides stability . . .

Feedback occurs when receptor stimulation triggers a response that changes the environment at the receptor. In the case of temperature control by a thermostat, temperature variation outside the desired range triggers an automatic response that corrects the situation. This method of homeostatic regulation is called **negative feedback**, because an effector activated by the control center opposes, or negates, the original stimulus. Negative feedback thus tends to minimize change, keeping variation in key body systems within limits compatible with our long-term survival.

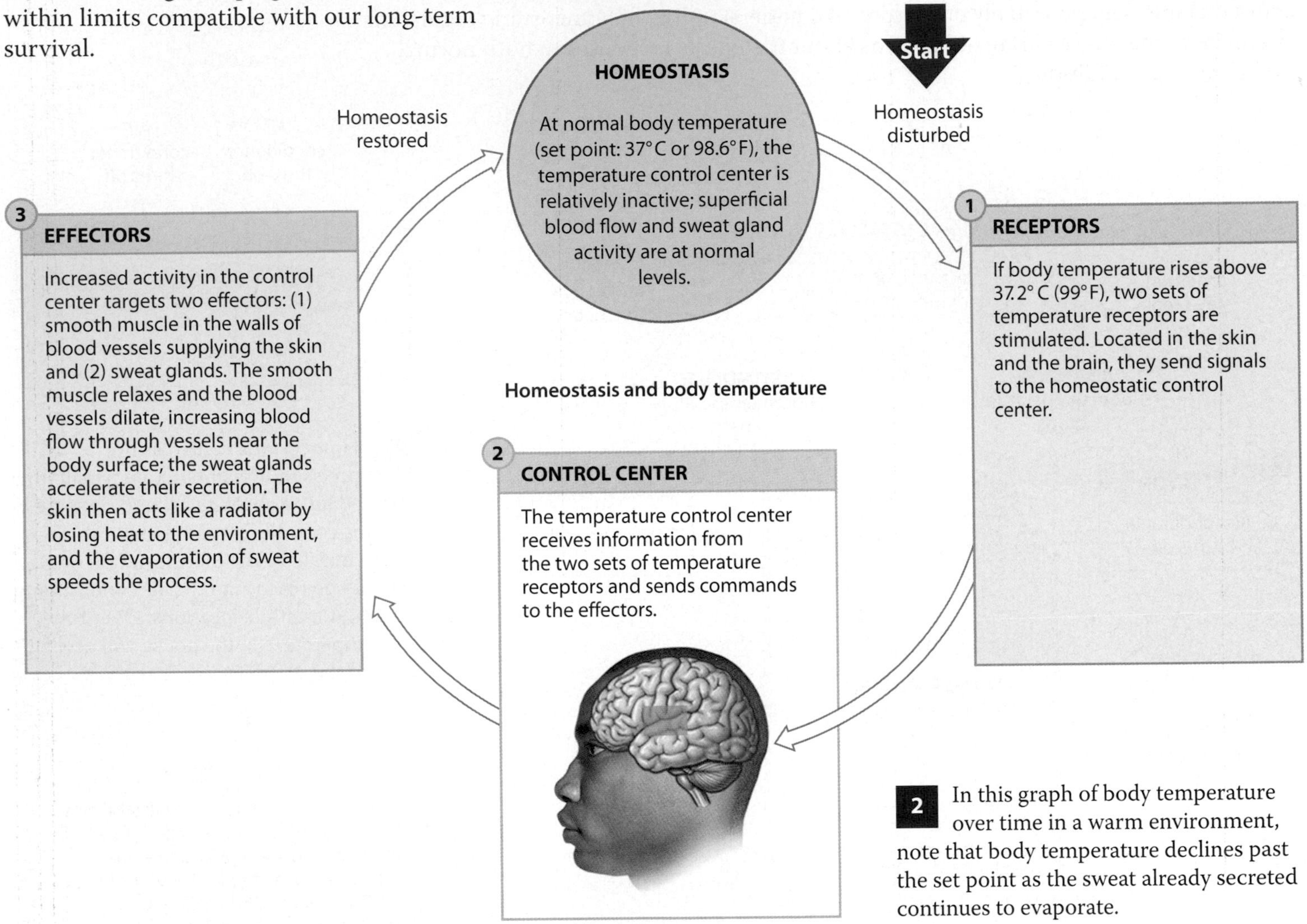

Start
Homeostasis disturbed
HOMEOSTASIS
At normal body temperature (set point: 37°C or 98.6°F), the temperature control center is relatively inactive; superficial blood flow and sweat gland activity are at normal levels.
Homeostasis restored
1
RECEPTORS
If body temperature rises above 37.2°C (99°F), two sets of temperature receptors are stimulated. Located in the skin and the brain, they send signals to the homeostatic control center.
2
CONTROL CENTER
The temperature control center receives information from the two sets of temperature receptors and sends commands to the effectors.
3
EFFECTORS
Increased activity in the control center targets two effectors: (1) smooth muscle in the walls of blood vessels supplying the skin and (2) sweat glands. The smooth muscle relaxes and the blood vessels dilate, increasing blood flow through vessels near the body surface; the sweat glands accelerate their secretion. The skin then acts like a radiator by losing heat to the environment, and the evaporation of sweat speeds the process.

Homeostasis and body temperature

1 Negative feedback is the primary mechanism of homeostatic regulation, and it provides long-term control over the body's internal conditions and systems. Homeostatic mechanisms using negative feedback normally ignore minor variations, and they maintain a normal range rather than a fixed value. The regulatory process itself is dynamic, because the set point may vary with changing environments or differing activity levels. For example, when you are asleep, your thermoregulatory set point is lower, whereas when you work outside on a hot day (or when you have a fever), it is higher. Thus, body temperature can vary from moment to moment or from day to day for any individual, due to either small oscillations around the set point or changes in the set point. Comparable variations occur in all other aspects of physiology.

2 In this graph of body temperature over time in a warm environment, note that body temperature declines past the set point as the sweat already secreted continues to evaporate.

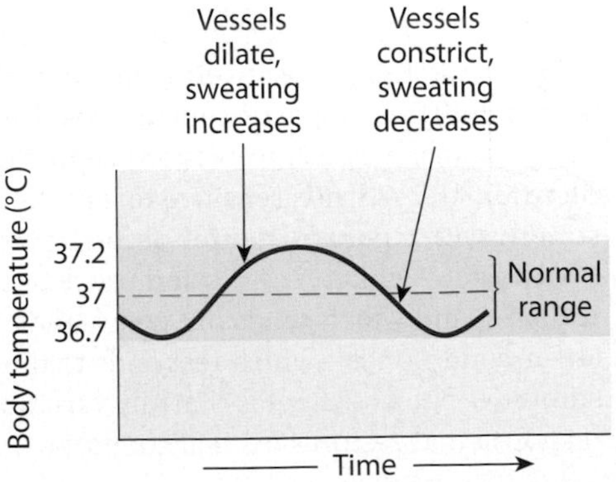

Vessels dilate, sweating increases
Vessels constrict, sweating decreases
Body temperature (°C)
37.2
37
36.7
Normal range
Time

. . . and positive feedback accelerates a process to completion

In **positive feedback**, an initial stimulus produces a response that exaggerates or enhances the change in the original conditions, rather than opposing it. You seldom encounter positive feedback in your daily life, simply because it tends to produce extreme responses. For example, suppose that the thermostat in your house was accidentally connected to a heater rather than to an air conditioner. Now, when room temperature exceeds the set point, the thermostat turns on the heater, causing a further rise in room temperature. Room temperature will continue to increase until someone switches off the thermostat, turns off the heater, or intervenes in some other way. This kind of escalating cycle is often called a **positive feedback loop**.

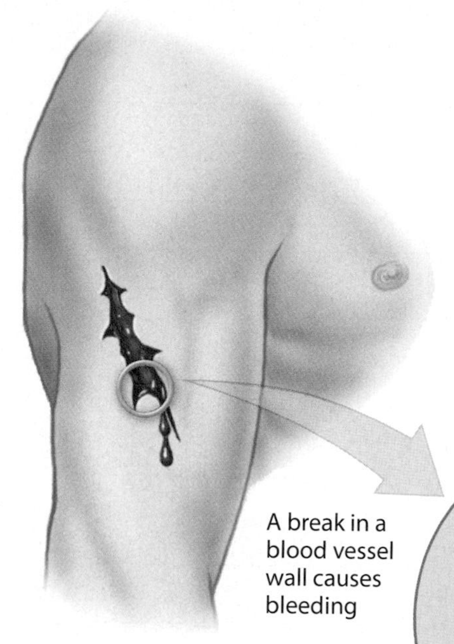

A break in a blood vessel wall causes bleeding

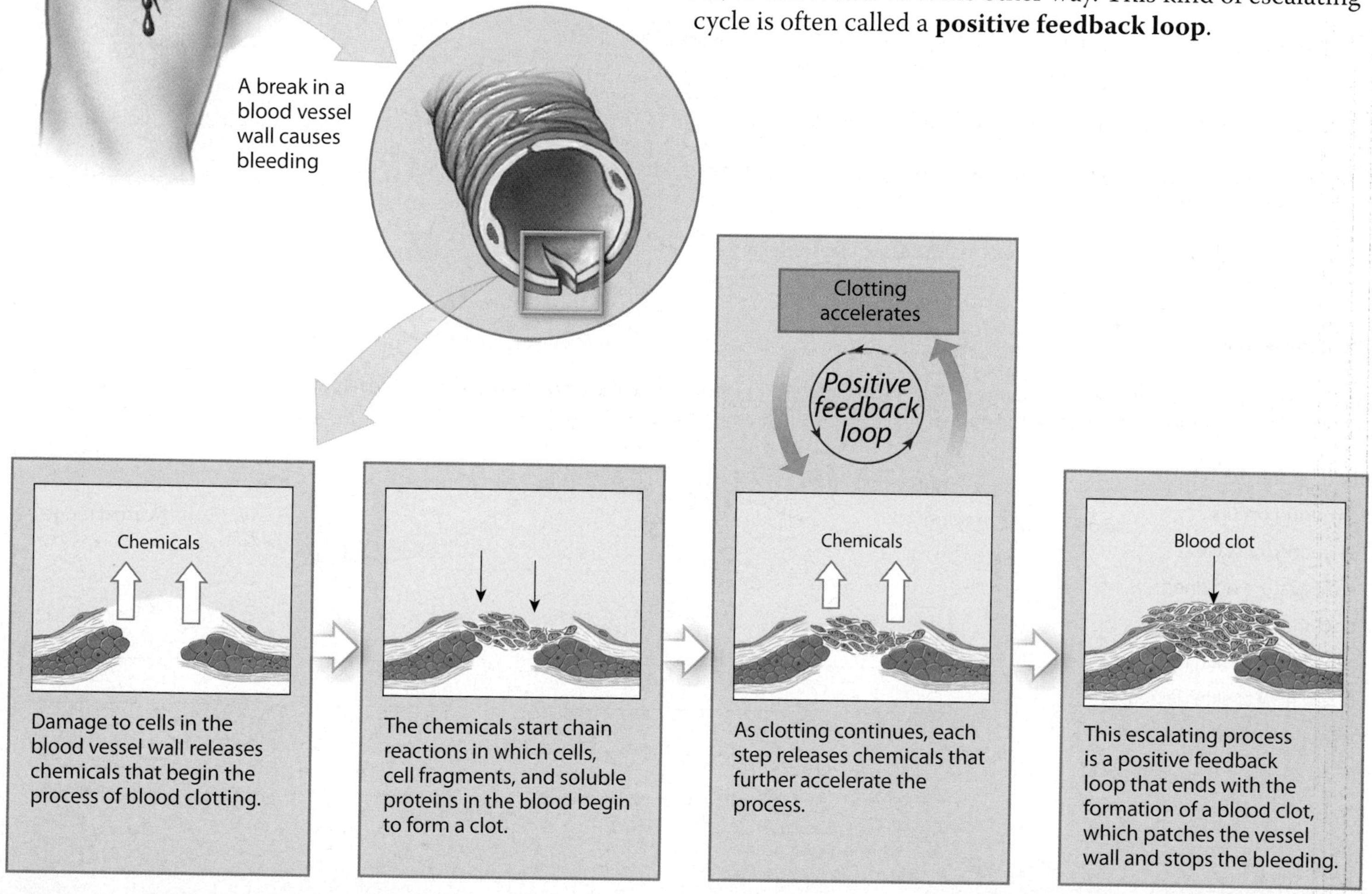

Clotting accelerates
Positive feedback loop
Chemicals
Chemicals
Blood clot
Damage to cells in the blood vessel wall releases chemicals that begin the process of blood clotting.
The chemicals start chain reactions in which cells, cell fragments, and soluble proteins in the blood begin to form a clot.
As clotting continues, each step releases chemicals that further accelerate the process.
This escalating process is a positive feedback loop that ends with the formation of a blood clot, which patches the vessel wall and stops the bleeding.

3 In the body, positive feedback loops are typically found when a potentially dangerous or stressful process must be completed quickly before homeostasis can be restored. For example, the immediate danger from a severe cut is blood loss, which can lower blood pressure and reduce the heart's efficiency.

Module 1.13 Review

a. Provide an example of negative feedback homeostatic regulation in the body.

b. Explain the function of negative feedback systems.

c. Why is positive feedback helpful in blood clotting but unsuitable for regulating body temperature?

LO 1.13 Discuss the roles of negative feedback and positive feedback in maintaining homeostasis.

Vocabulary

Write the term for each of the following descriptions in the space provided.

1 Mechanism that increases a deviation from normal limits after an initial stimulus

2 Adjustment of physiological systems to preserve homeostasis

3 The maintenance of a relatively constant internal environment

4 Homeostatic regulatory component that detects changes

5 Corrective mechanism that opposes or cancels a variation from normal limits

1 ______
2 ______
3 ______
4 ______
5 ______

Indicate whether each of the following processes represents negative feedback or positive feedback.

6 A rise in the level of calcium dissolved in the blood stimulates the release of a hormone that causes bone cells to deposit more of the calcium in bone.

7 Labor contractions become increasingly forceful during childbirth.

8 An increase in blood pressure triggers a nervous system response that results in lowering the blood pressure.

9 Blood vessel cells damaged by a break in the vessel release chemicals that accelerate the blood clotting process.

6 ______
7 ______
8 ______
9 ______

Short answer

Assuming a normal body temperature range of 36.7°–37.2°C (98°–99°F), identify from the graph below what would happen if there were an increase or decrease in body temperature beyond the normal limits. Use the following descriptive terms to explain what would happen at (10) and (11) on the graph.

- body surface cools
- shivering occurs
- sweating increases
- temperature declines
- body heat is conserved
- blood flow to skin increases
- blood flow to skin decreases
- temperature rises

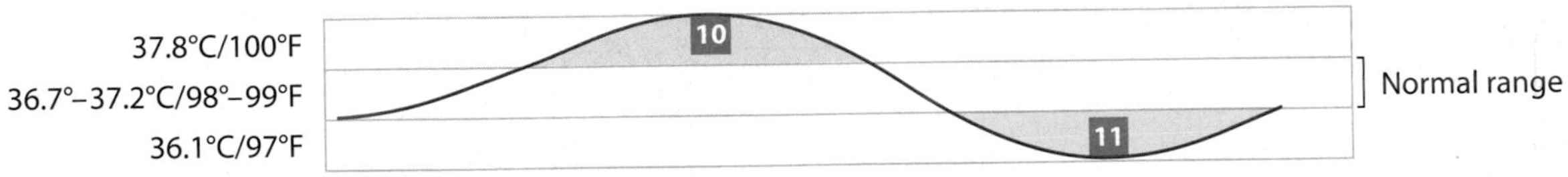
37.8°C/100°F
36.7°–37.2°C/98°–99°F
36.1°C/97°F
10
11
Normal range

10 ______

11 ______

Section integration

It is a warm day and you feel a little chilled. On checking your temperature, you find that your body temperature is 1.5 degrees below normal. Suggest some possible reasons for this situation.

12 ______

Anatomical terms have a long and varied history

Early anatomists created maps of the human body, and we still rely on maps for orientation. The landmarks are prominent anatomical structures; distances are measured in centimeters or inches; and we use specialized directional terms. In effect, anatomy uses a special language that must be learned almost at the start. Many terms are based on Latin or Greek words used by ancient anatomists. However, Latin and Greek terms are not the only ones that have been imported into the anatomical vocabulary over the centuries, and the vocabulary continues to expand. Many anatomical structures and clinical conditions were initially named after either the discoverer or, in the case of diseases, the most famous victim. Most of these commemorative names, or **eponyms**, have been replaced by more precise terms, but a few persist.

1 The table below lists some eponyms you may already know, along with their equivalent anatomical or medical terms.

Eponym	Equivalent Term
Achilles tendon	Calcaneal tendon
Broca's area	Speech center
Eustachian tube	Auditory tube
Krebs cycle	Citric acid cycle

The study of cadaver-based human anatomy by medical professionals in a university can be traced back to Medieval Europe. Founded in 1088, the University of Bologna in Italy is the first university in Europe, and one of the oldest in the world. The most famous anatomist at Bologna was Mondino dei Liuzzi. He wrote the *Anatomia,* which is perhaps the first anatomy text for university students. Mondino was so revered, that images of him presiding over dissections were included in other anatomy texts for many years.

Anatomy as a science was dramatically improved during the Renaissance at the University of Padua. Established in 1222 in Italy's Venetian Republic, Padua's most famous anatomist was Andreas Vesalius, who conducted frequent, detailed dissections. The result was a visually beautiful and amazingly accurate anatomical text titled *De Humani Corporis Fabrica.* Published in 1543, this text corrected the mistakes of previous anatomists and served as an early model for modern anatomy education.

As you begin your anatomy studies, consider yourself as the latest generation of college and university anatomy students. You are studying an area of science that is centuries old and rich with history.

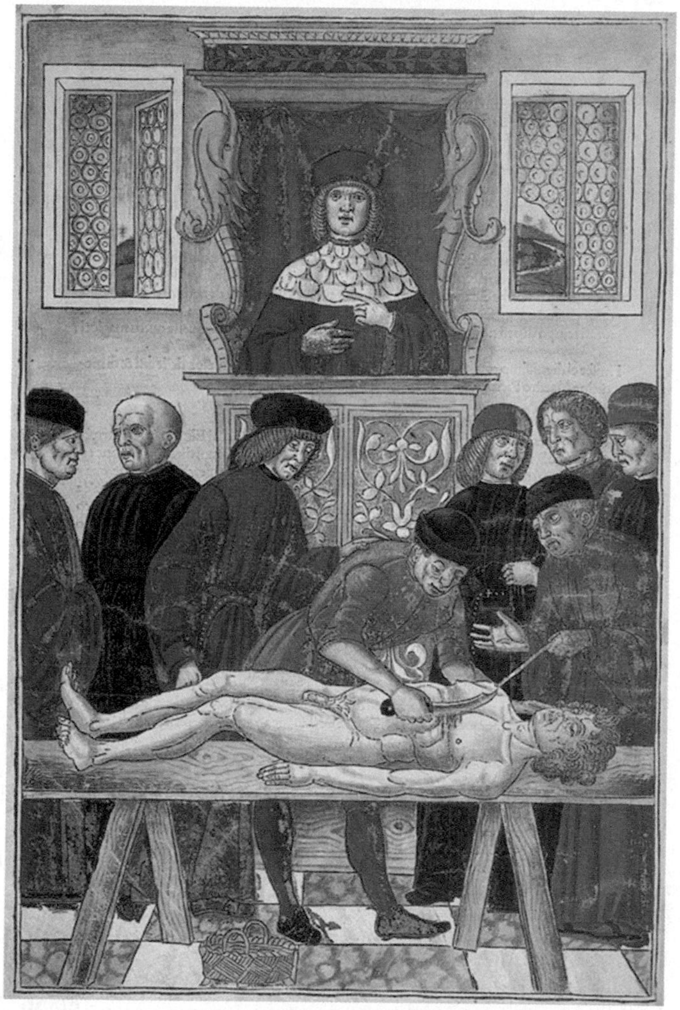

2 Frontispiece to John of Ketham's ***Fasciculus Medicinae***, 1493 showing Mondino dei Liuzzi, the lofty figure adorned with the ornate collar, presiding over a dissection.

Module 1.14 Review

a. Which languages are the source of many modern anatomical terms?

b. Define the word eponym.

c. In what country was cadaver-based anatomy established as a discipline studied by medical professionals?

LO 1.14 Describe the history of anatomical terminology.

Module 1.15

Superficial anatomy and regional anatomy indicate locations on or in the body

1 This illustration shows the body in the **anatomical position**. In this position, the hands are at the sides with the palms facing forward, and the feet are together. Unless otherwise noted, all descriptions in this text refer to the body in the anatomical position. A person lying down in the anatomical position is said to be **supine** (soo-PĪN) when face up, and **prone** when face down.

2 Why is it important to learn these terms? We need common anatomical terms to communicate effectively in a medical setting. For example, stating that a patient has a "bump on the back" does not give very precise information about its location. So, anatomists created maps of the body, naming superficial anatomical structures and identifying regional landmarks to help locate the exact point of that "bump on the back."

TUMMY ACHE?
MAYBE AN EENSY KIDNEY STONE?
...OR, WAY INSIDE, DOWN AROUND THE BELLY BUTTON, SOME KIND OF BOO-BOO...

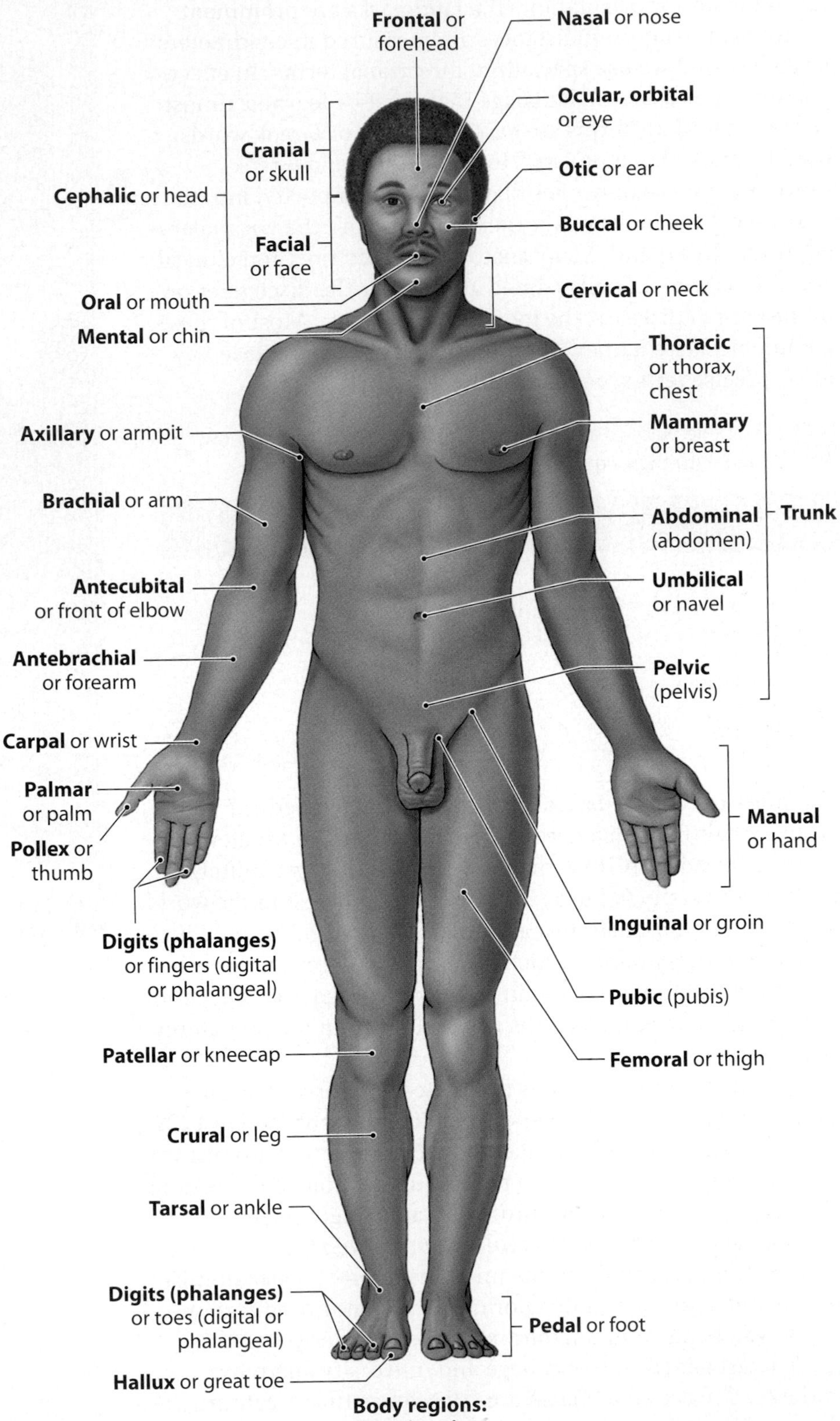

Frontal or forehead
Nasal or nose
Ocular, orbital or eye
Cranial or skull
Cephalic or head
Otic or ear
Buccal or cheek
Facial or face
Oral or mouth
Cervical or neck
Mental or chin
Thoracic or thorax, chest
Axillary or armpit
Mammary or breast
Brachial or arm
Abdominal (abdomen)
Trunk
Antecubital or front of elbow
Umbilical or navel
Antebrachial or forearm
Pelvic (pelvis)
Carpal or wrist
Palmar or palm
Pollex or thumb
Manual or hand
Inguinal or groin
Digits (phalanges) or fingers (digital or phalangeal)
Pubic (pubis)
Patellar or kneecap
Femoral or thigh
Crural or leg
Tarsal or ankle
Digits (phalanges) or toes (digital or phalangeal)
Pedal or foot
Hallux or great toe

Body regions: Anterior view

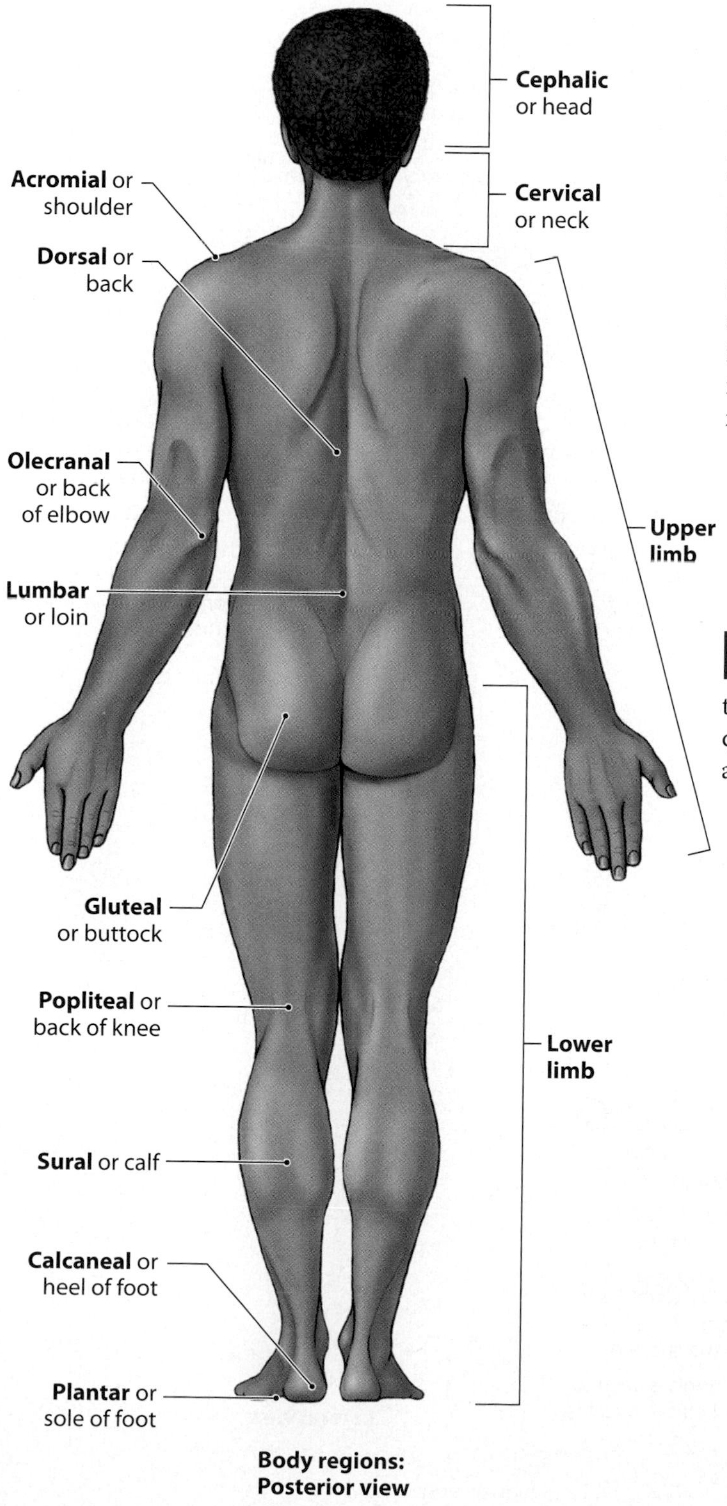

Cephalic or head
Cervical or neck
Acromial or shoulder
Dorsal or back
Olecranal or back of elbow
Lumbar or loin
Upper limb
Gluteal or buttock
Popliteal or back of knee
Lower limb
Sural or calf
Calcaneal or heel of foot
Plantar or sole of foot

Body regions: Posterior view

3 Clinicians refer to four **abdominopelvic quadrants** formed by a pair of imaginary perpendicular lines that intersect at the umbilicus (navel). This simple method provides useful references for the description of aches, pains, and injuries. The location can help physicians determine the possible cause.

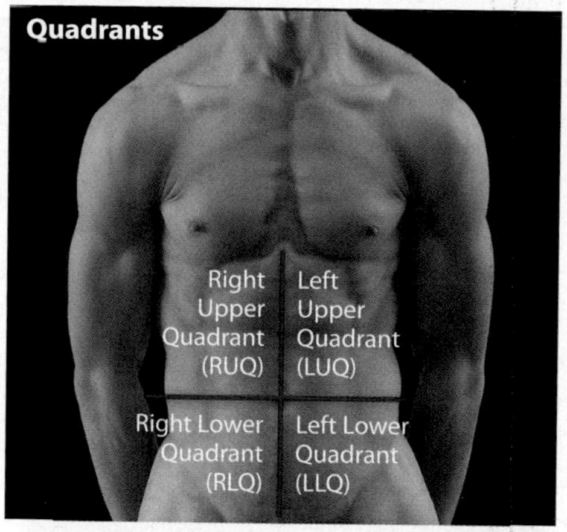

Quadrants
Right Upper Quadrant (RUQ)
Left Upper Quadrant (LUQ)
Right Lower Quadrant (RLQ)
Left Lower Quadrant (LLQ)

4 Anatomists prefer more precise terms to describe the location and orientation of internal organs. Anatomists use nine **abdominopelvic regions**.

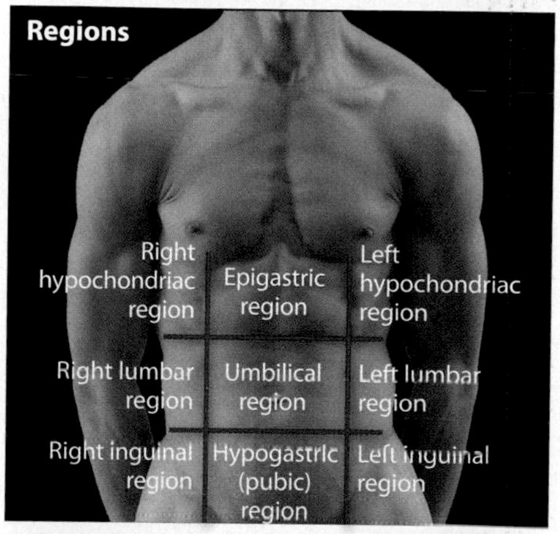

Regions
Right hypochondriac region
Epigastric region
Left hypochondriac region
Right lumbar region
Umbilical region
Left lumbar region
Right inguinal region
Hypogastric (pubic) region
Left inguinal region

5 The image at the lower right shows the relationships among quadrants, regions, and internal organs.

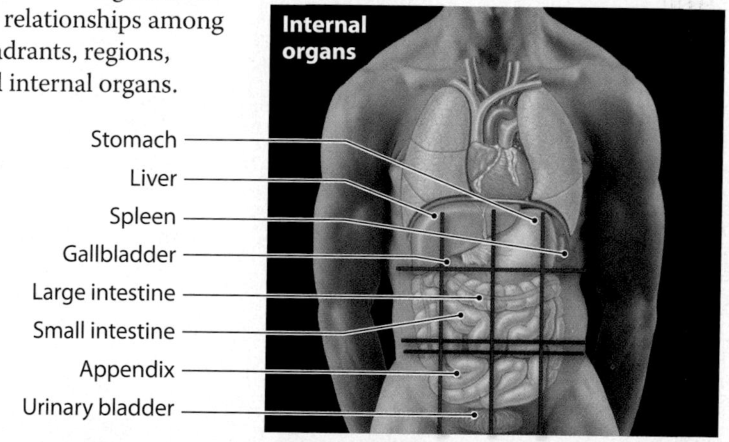

Internal organs
Stomach
Liver
Spleen
Gallbladder
Large intestine
Small intestine
Appendix
Urinary bladder

Module 1.15 Review

a. Describe a person in the anatomical position.

b. Contrast the descriptions used by clinicians and anatomists when referring to the positions of injuries or internal organs of the abdomen and pelvis.

c. A massage therapist often begins a massage by asking clients to lie face down with their arms at their sides. Which anatomical term describes that position?

LO 1.15 Use correct anatomical terms to describe superficial and regional anatomy.

Directional terms and sectional planes describe specific points of reference

1 The figures on this page introduce the principal directional terms and examples of their use. There are many different terms, and some can be used interchangeably. As you learn these directional terms, it is important to remember that all anatomical directions utilize the anatomical position as the standard point of reference.

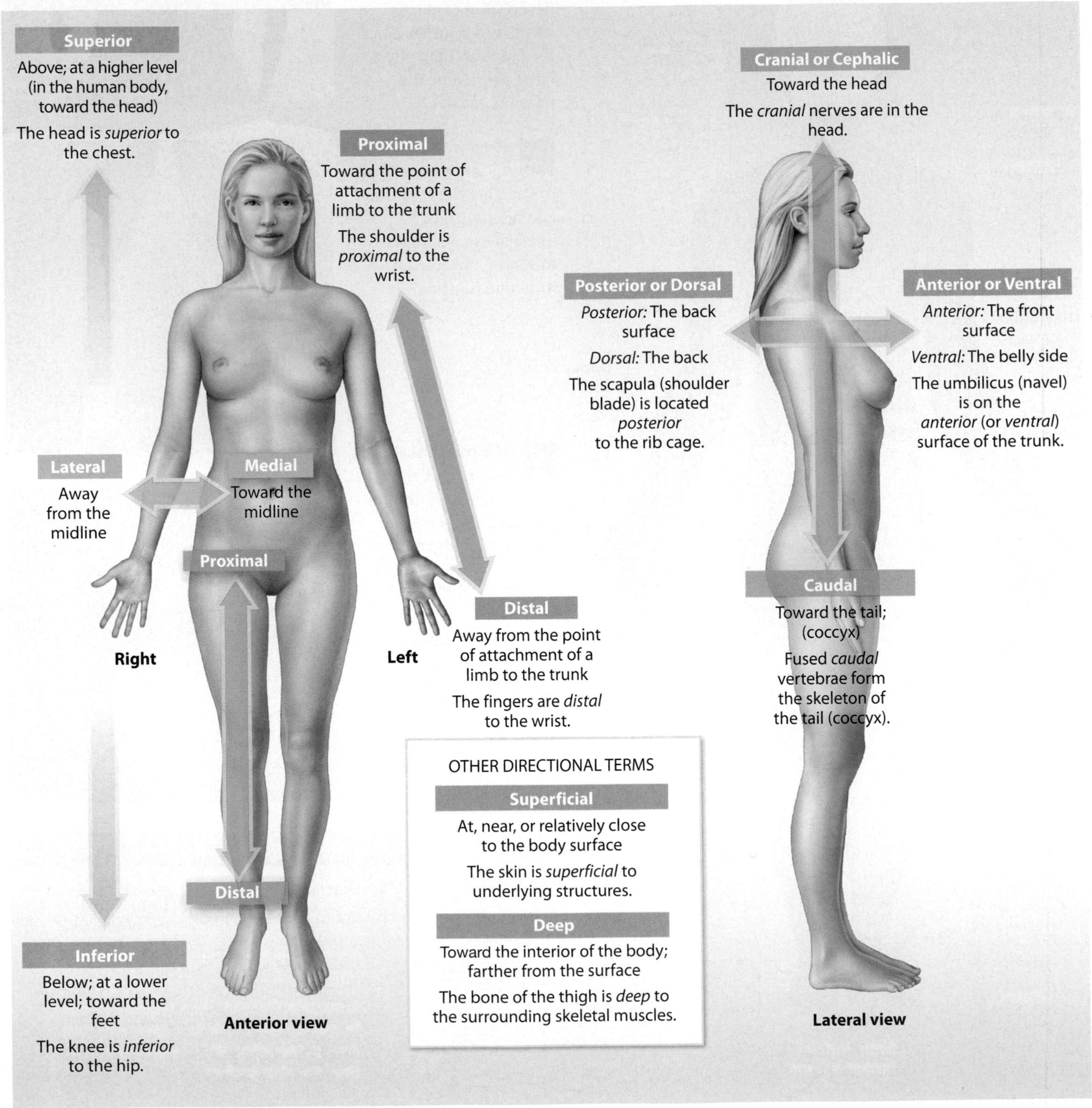
Superior
Above; at a higher level (in the human body, toward the head)
The head is *superior* to the chest.
Proximal
Toward the point of attachment of a limb to the trunk
The shoulder is *proximal* to the wrist.
Cranial or Cephalic
Toward the head
The *cranial* nerves are in the head.
Posterior or Dorsal
Posterior: The back surface
Dorsal: The back
The scapula (shoulder blade) is located *posterior* to the rib cage.
Anterior or Ventral
Anterior: The front surface
Ventral: The belly side
The umbilicus (navel) is on the *anterior* (or *ventral*) surface of the trunk.
Lateral
Away from the midline
Medial
Toward the midline
Proximal
Caudal
Toward the tail; (coccyx)
Fused *caudal* vertebrae form the skeleton of the tail (coccyx).
Right
Left
Distal
Away from the point of attachment of a limb to the trunk
The fingers are *distal* to the wrist.
OTHER DIRECTIONAL TERMS
Superficial
At, near, or relatively close to the body surface
The skin is *superficial* to underlying structures.
Deep
Toward the interior of the body; farther from the surface
The bone of the thigh is *deep* to the surrounding skeletal muscles.
Distal
Inferior
Below; at a lower level; toward the feet
The knee is *inferior* to the hip.
Anterior view
Lateral view

2 A presentation in sectional view is sometimes the only way to illustrate the relationships between the parts of a three-dimensional object. The development of medical imaging techniques has made it even more important to understand sectional planes and terms.

Frontal, or **coronal, plane**

Plane is oriented parallel to long axis

A *frontal*, or *coronal, section* separates anterior and posterior portions of the body. Coronal usually refers to sections passing through the skull.

Directional term: frontally or coronally

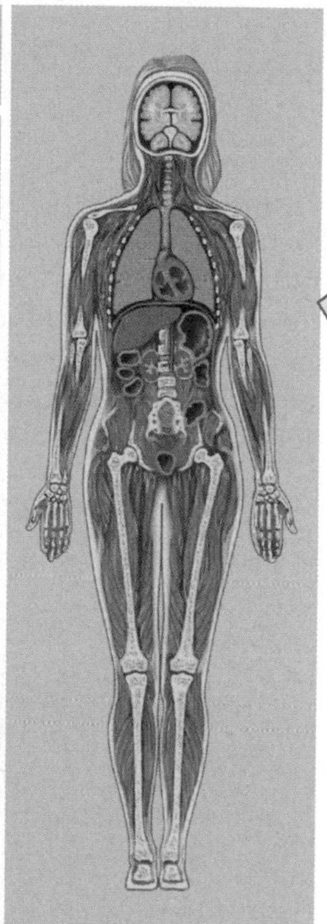

Frontal plane

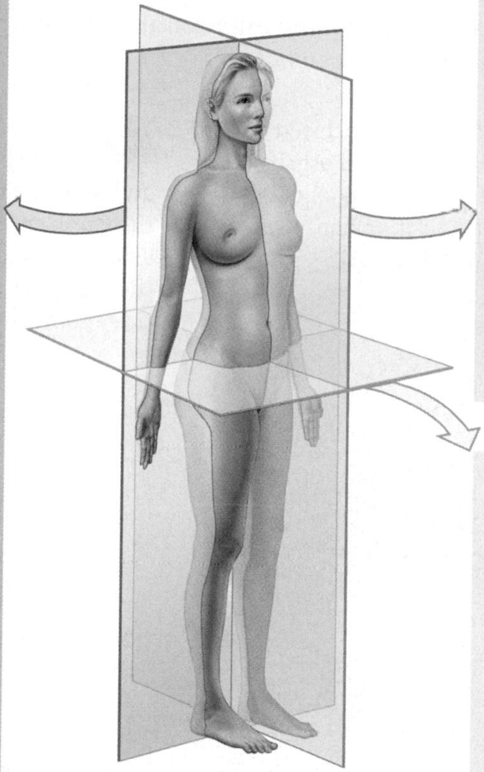

Sagittal plane

Plane is oriented parallel to long axis

A *sagittal section* separates right and left portions. You examine a sagittal section, but you section sagittally.

In a *midsagittal section*, the plane passes through the midline. It separates the body into equal right and left sides.

A *parasagittal section* misses the midline. It separates the body into unequal right and left sides.

Directional term: sagittally

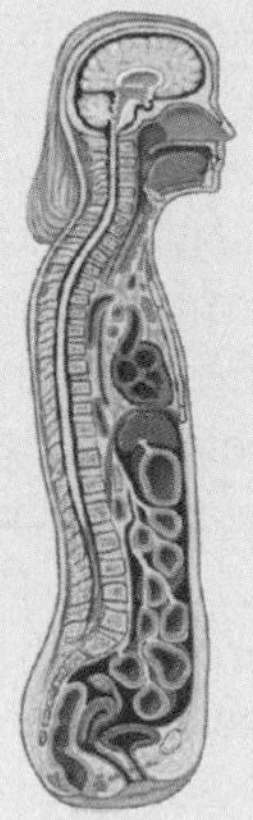

Midsagittal plane

Transverse, or **horizontal, plane**

Plane is oriented perpendicular to long axis

A *transverse*, or *horizontal*, *section* separates superior and inferior portions of the body. A cut in this plane is also called a **cross section**.

Directional term: transversely or horizontally

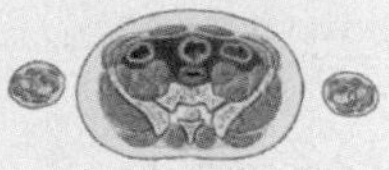

Transverse plane (inferior view)

3 Sectional planes are used for visualization purposes. Here, we are serially sectioning a bent tube, which looks like a piece of elbow macaroni. Notice how the sectional views change as you approach the curve. Keep in mind these effects of sectioning when you are looking at slides under the microscope. These sectional views also affect the appearance of internal organs when seen in a CT or MRI scan. For example, the small intestine is a simple tube, but it can look like a pair of tubes, a dumbbell, an oval, or a solid, depending on where the section was taken.

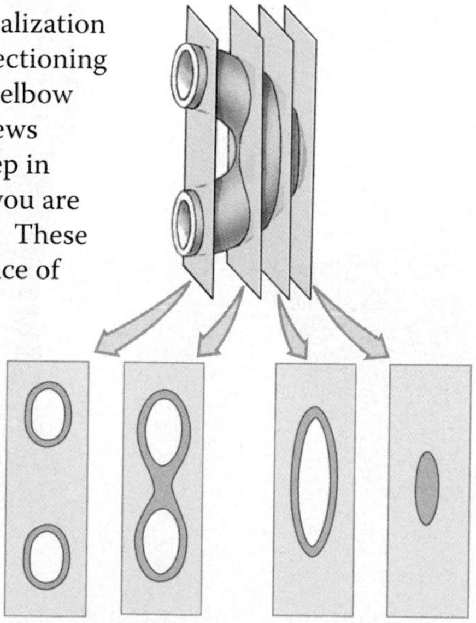

Module 1.16 Review

a. What is the purpose of directional and sectional terms?

b. In the anatomical position, describe an anterior view and a posterior view.

c. What type of section would separate the two eyes?

LO **1.16** Use correct directional terms and sectional planes to describe relative positions and relationships among body parts.

Body cavities protect internal organs and allow them to change shape

The interior of the trunk of the body is often subdivided into regions established by the body wall. For example, everything deep to the chest wall is considered to be within the **thoracic cavity**, and all of the structures deep to the abdominal and pelvic walls are said to lie within the **abdominopelvic cavity**. Many vital internal organs within these regions are suspended within closed, fluid-filled chambers that are true **body cavities**. True body cavities are lined by a serous membrane and share a common embryonic origin. They have two essential functions: (1) They protect delicate organs from shocks and impacts; and (2) they permit significant changes in the size and shape of internal organs.

1 The internal organs that are partially or completely enclosed by body cavities are called **viscera** (VIS-e-ruh) or visceral organs. Viscera do not float within the body cavities—they remain connected to the rest of the body. To understand the physical relationships, we will examine the **pericardial cavity** that surrounds the heart.

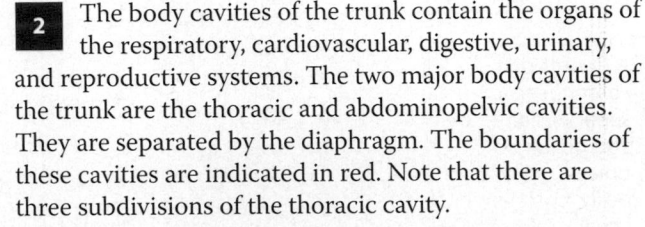

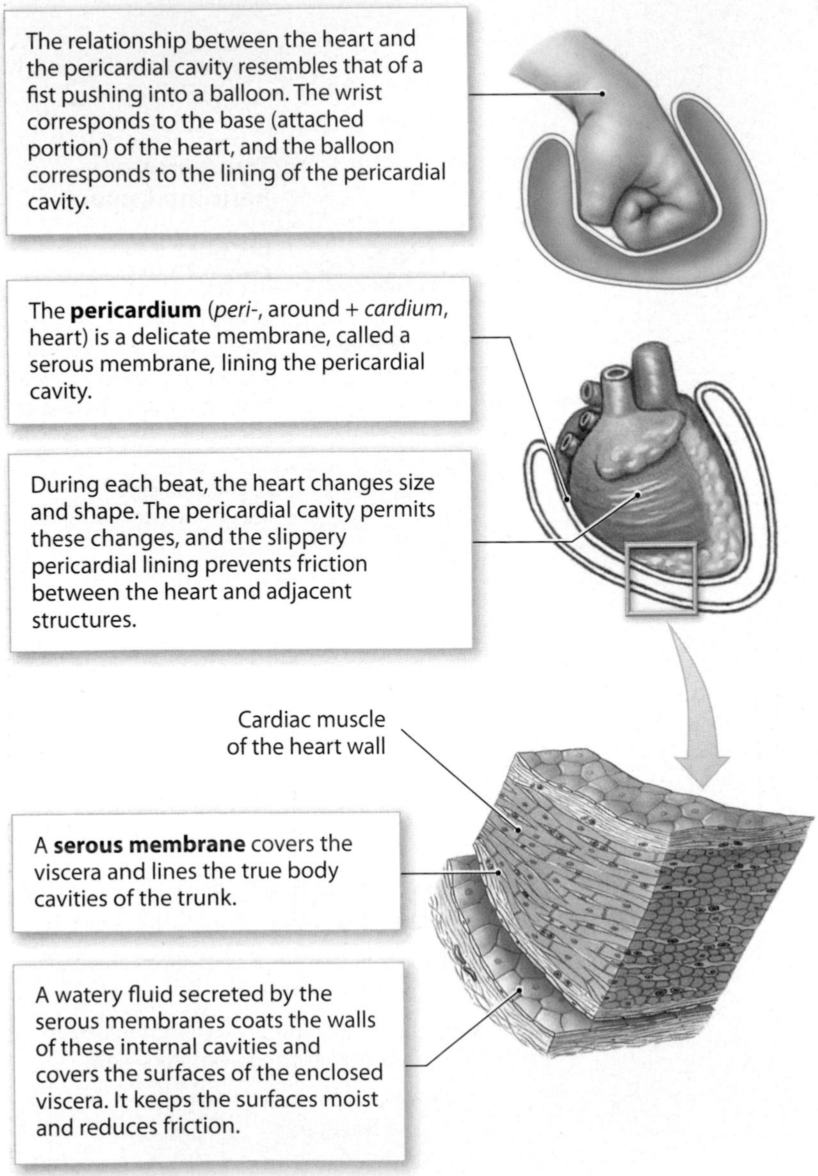
The relationship between the heart and the pericardial cavity resembles that of a fist pushing into a balloon. The wrist corresponds to the base (attached portion) of the heart, and the balloon corresponds to the lining of the pericardial cavity.
The **pericardium** (*peri-*, around + *cardium*, heart) is a delicate membrane, called a serous membrane, lining the pericardial cavity.
During each beat, the heart changes size and shape. The pericardial cavity permits these changes, and the slippery pericardial lining prevents friction between the heart and adjacent structures.
Cardiac muscle of the heart wall
A **serous membrane** covers the viscera and lines the true body cavities of the trunk.
A watery fluid secreted by the serous membranes coats the walls of these internal cavities and covers the surfaces of the enclosed viscera. It keeps the surfaces moist and reduces friction.

2 The body cavities of the trunk contain the organs of the respiratory, cardiovascular, digestive, urinary, and reproductive systems. The two major body cavities of the trunk are the thoracic and abdominopelvic cavities. They are separated by the diaphragm. The boundaries of these cavities are indicated in red. Note that there are three subdivisions of the thoracic cavity.

BODY CAVITIES OF THE TRUNK

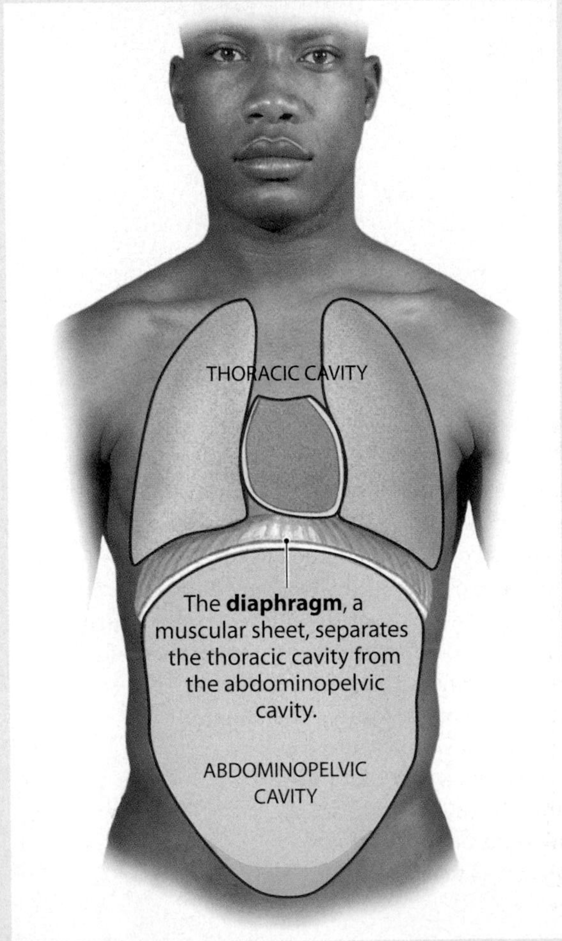
THORACIC CAVITY
The **diaphragm**, a muscular sheet, separates the thoracic cavity from the abdominopelvic cavity.
ABDOMINOPELVIC CAVITY

3 The thoracic cavity contains the lungs, heart, and other structures. Its boundaries are established by the chest wall and diaphragm.

THORACIC CAVITY

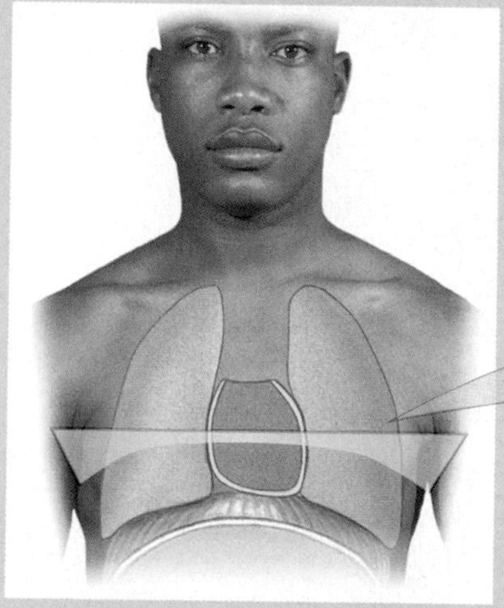

A horizontal section through the thoracic cavity shows the relationship between its three subdivisions.

Each lung is enclosed within a **pleural cavity**, lined by a shiny, slippery serous membrane called the **pleura** (PLUR-uh).

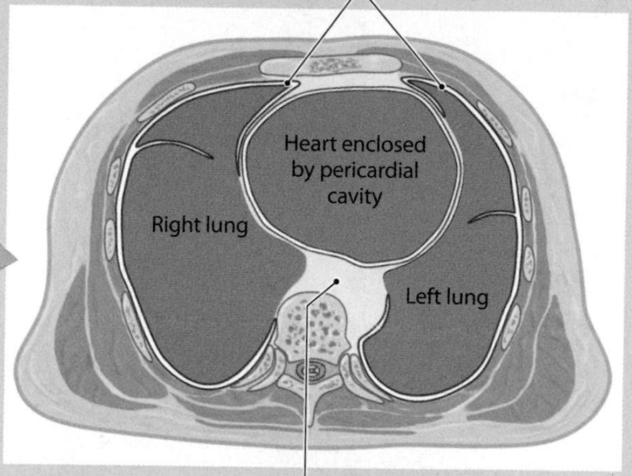

Heart enclosed by pericardial cavity
Right lung
Left lung

Note the orientation of the section. Unless otherwise noted, all cross sections are shown as if the viewer were standing at the feet of a supine person and looking toward the head.

The pericardial cavity is embedded within the **mediastinum**, a mass of connective tissue that separates the two pleural cavities and stabilizes the positions of embedded organs and blood vessels.

ABDOMINOPELVIC CAVITY

The abdominopelvic cavity encloses the **peritoneal** (per-i-tō-NĒ-al) **cavity**, a chamber lined by a serous membrane known as the **peritoneum** (per-i-tō-NĒ-um). A few organs, such as the kidneys and pancreas, lie between the peritoneal lining and the muscular wall of the abdominal cavity. Those organs are said to be **retroperitoneal** (re-trō-per-i-tō-NĒ-al; *retro*, behind). The peritoneum covers the ovaries and the uterus in females, as well as the superior portion of the urinary bladder in both sexes. Organs such as the urinary bladder and the distal portions of the ureters and large intestine, which extend inferior to the peritoneal cavity, are said to be **infraperitoneal**.

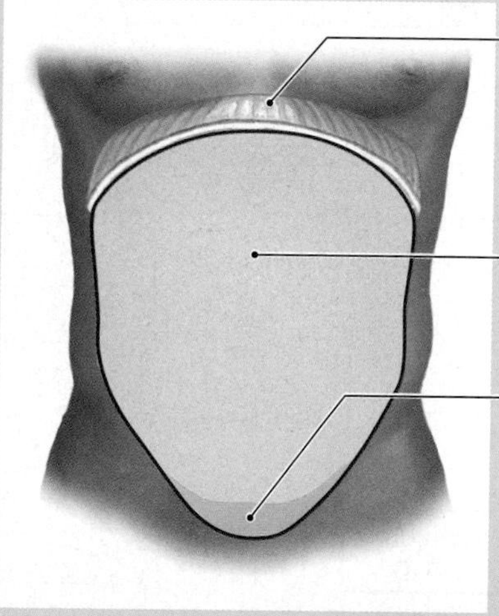

Diaphragm

Peritoneum (red) showing the boundaries of the peritoneal cavity

The abdominal cavity contains many digestive glands and organs

Retroperitoneal area

The pelvic cavity contains the urinary bladder, reproductive organs, and the last portion of the digestive tract. Many of these structures lie posterior to, or inferior to, the peritoneal cavity.

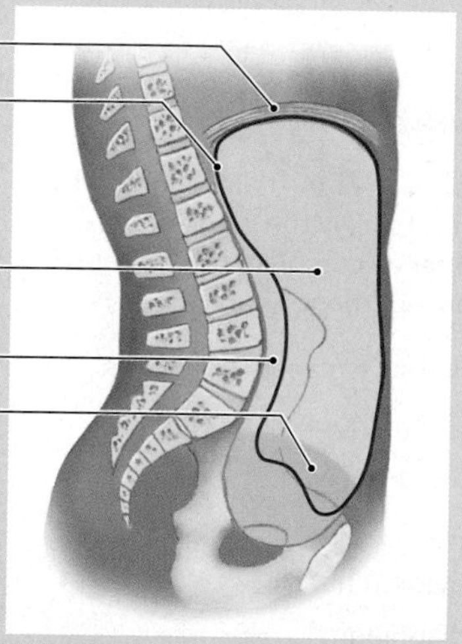

4 The boundaries of the abdominopelvic cavity are established by the diaphragm, the muscles of the abdominal wall, the trunk muscles and inferior portions of the vertebral column, and the bones and muscles of the pelvis. It may be subdivided into the **abdominal cavity** and the **pelvic cavity**.

Module 1.17 Review

a. Describe two essential functions of true body cavities.

b. Identify the body cavities of the trunk.

c. If a surgeon makes an incision just inferior to the diaphragm, what body cavity will be opened?

LO **1.17** Identify the major body cavities of the trunk and the subdivisions of each.

Labeling

Label the directional terms in the figures at right.

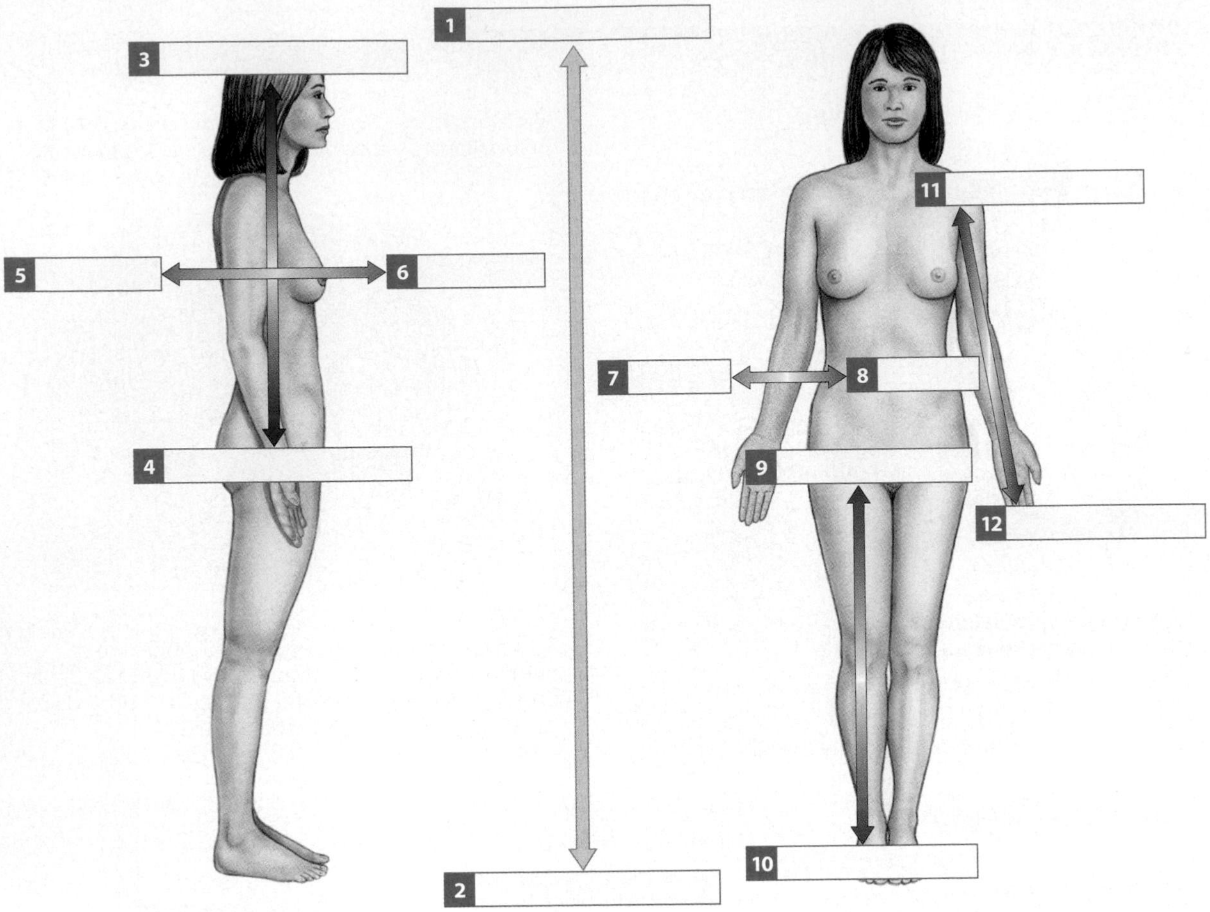
1
2
3
4
5
6
7
8
9
10
11
12

Concept map

Use each of the following terms once to fill in the blank boxes to correctly complete the body cavities concept map.

- digestive glands and organs
- abdominopelvic cavity
- thoracic cavity
- heart
- mediastinum
- diaphragm
- pelvic cavity
- trachea, esophagus
- reproductive organs
- left lung
- peritoneal cavity

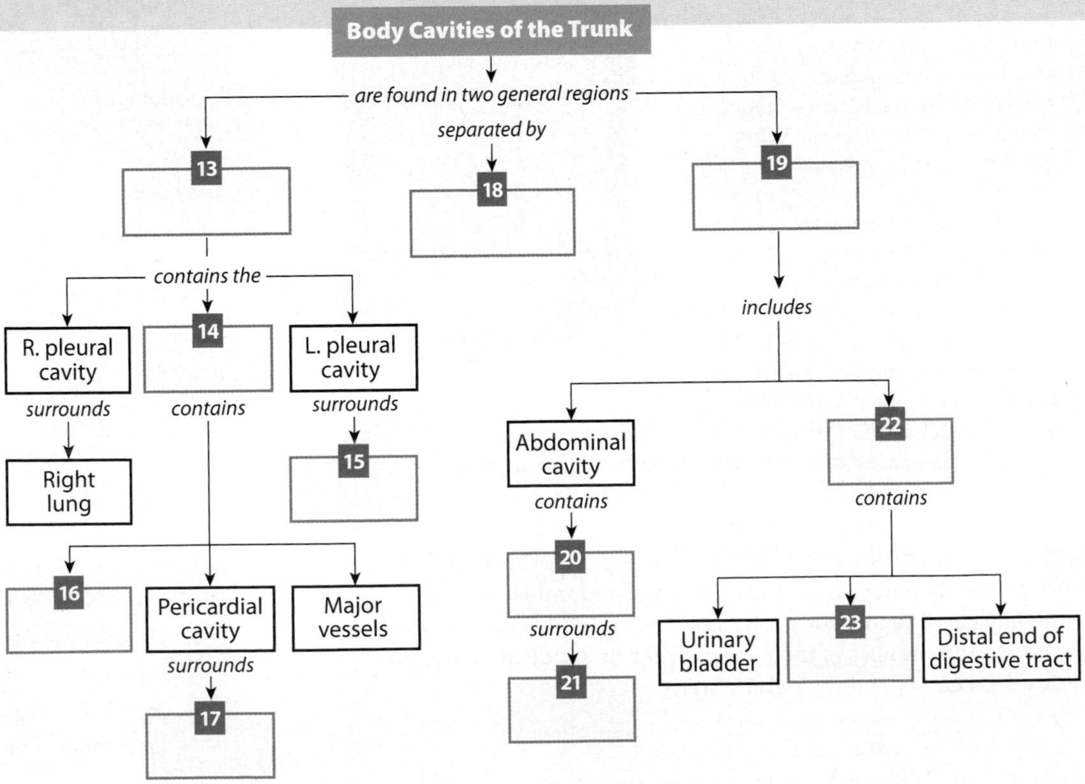
Body Cavities of the Trunk
are found in two general regions
separated by
13
18
19
contains the
R. pleural cavity
14
L. pleural cavity
surrounds
contains
surrounds
Right lung
15
16
Pericardial cavity
Major vessels
surrounds
17
includes
Abdominal cavity
22
contains
contains
20
surrounds
21
Urinary bladder
23
Distal end of digestive tract

Study Outline

SECTION 1 • A&P in Perspective

1.1 Focused study is important for learning anatomy and physiology p. 3

1. The basic approach in A&P can be summed up as "What is that structure and how does it work?"
2. **Homeostasis** is a relatively constant internal environment.

1.2 Biology is the study of life p. 4

3. All living things perform the same basic functions: **respond** to the environment, **adapt** to stimuli, **grow** and **reproduce**, **movement** or **locomotion**, **digestion**, **respiration**, **excretion**, and **circulation**.
4. The basic functions of life are active processes that require energy.

1.3 Anatomy is the study of structure, and physiology is the study of function p. 6

5. **Gross** or **macroscopic anatomy** examines relatively large structures that are visible with the unaided eye.
6. **Microscopic anatomy** deals with structures that cannot be seen without magnification.
7. **Physiology** is the study of the complex functions of the human body.

1.4 Structure and function are interrelated p. 8

8. An example of how structure and function are related is how the anatomical shape of the end of the humerus allows the elbow to function as a hinge joint.
9. An example of physiological mechanisms is how chemical messengers depend on their three-dimensional shapes to bind with cellular receptors.

SECTION 2 • Levels of Organization

1.5 The human body has multiple interdependent levels of organization p. 11

10. Progressing from smallest to largest, the levels of organization are: **chemical**, **cellular**, **tissue**, **organ**, **organ system**, and **organism.**

1.6 Cells are the smallest units of life p. 12

11. Cells are the smallest independent organisms and they are remarkably diverse.
12. **Cell theory** states that cells are the structural building blocks of plants and animals, divisions of pre-existing cells produce them, and they are the smallest structural units that perform all vital functions.

1.7 Tissues are specialized groups of cells and cell products p. 14

13. **Histology** is the study of tissues, which are groups of cells that perform specific functions.
14. There are four **primary tissue types**: **epithelial**, **connective**, **muscle**, and **neural**.

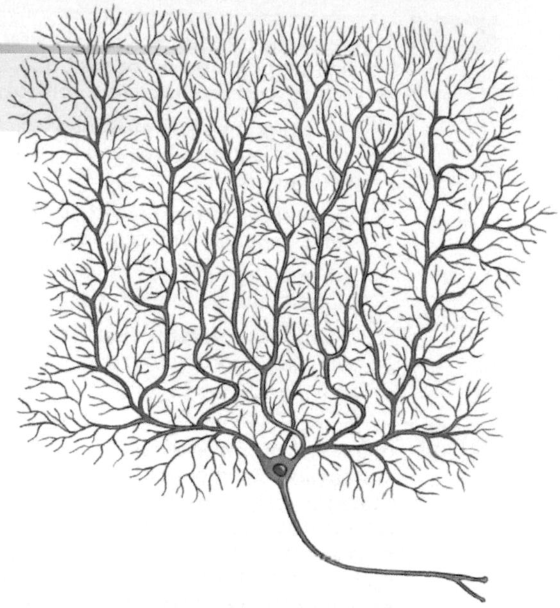

The many different kinds of cells in the human body each have their own characteristics and functions. This nerve cell (neuron) is from a part of the brain involved with the control of balance and movement. The extensive branching provides a huge surface area for communication with other neurons.

1.8 Organs and organ systems perform vital functions p. 16

15. An **organ** is a functional unit composed of more than one tissue type. An **organ system** consists of organs that interact to perform a specific range of functions.
16. The 11 organ systems of the body are: **integumentary**, **skeletal**, **muscular**, **nervous**, **endocrine**, **cardiovascular**, **lymphatic**, **respiratory**, **digestive**, **urinary**, and **reproductive**.

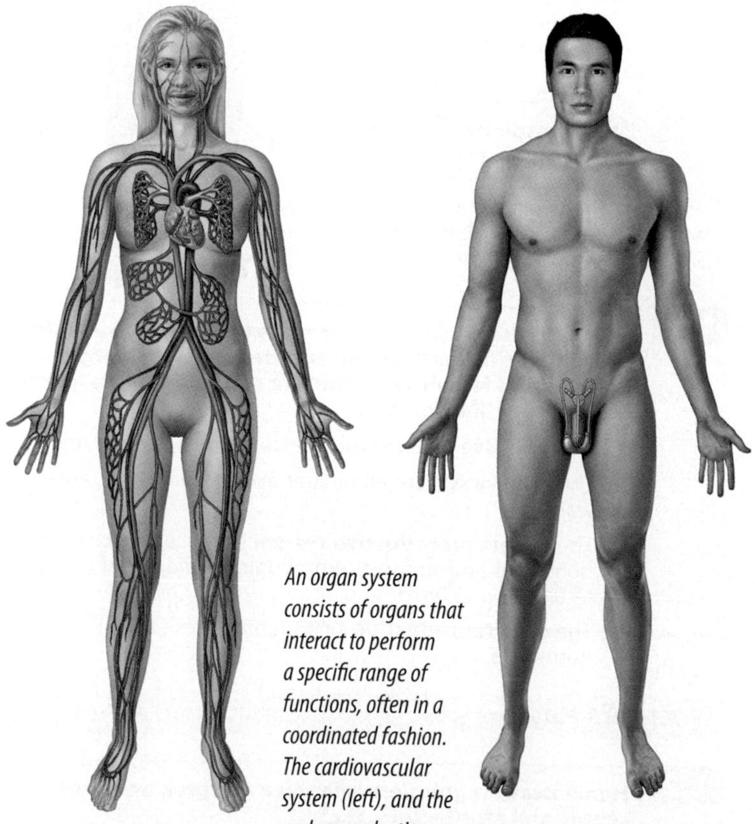

An organ system consists of organs that interact to perform a specific range of functions, often in a coordinated fashion. The cardiovascular system (left), and the male reproductive system are shown here.

1.9 **Organs of the integumentary, skeletal, and muscular systems support and move the body, and organs of the nervous system provide rapid control and regulation** p. 18

17. The **integumentary system** protects against environmental hazards and helps control body temperature.
18. The **skeletal system** provides support, protects tissues, stores minerals, and forms blood cells.
19. The **muscular system** produces movement, provides support, and generates heat.
20. The **nervous system** directs immediate responses to stimuli by coordinating the activities of the other organ systems.

1.10 **Organs of the endocrine system secrete chemicals that are carried by organs of the cardiovascular system, organs of the lymphatic system defend the body, and organs of the respiratory system exchange vital gases** p. 20

21. The **endocrine system** directs long-term changes in other organ systems.
22. The **cardiovascular system** transports cells and dissolved materials, including nutrients, wastes, and gases.
23. The **lymphatic system** defends against infection and disease and returns tissue fluid to the bloodstream.
24. The **respiratory system** delivers air to sites where gas exchange occurs between the air and circulating blood, and produces sound.

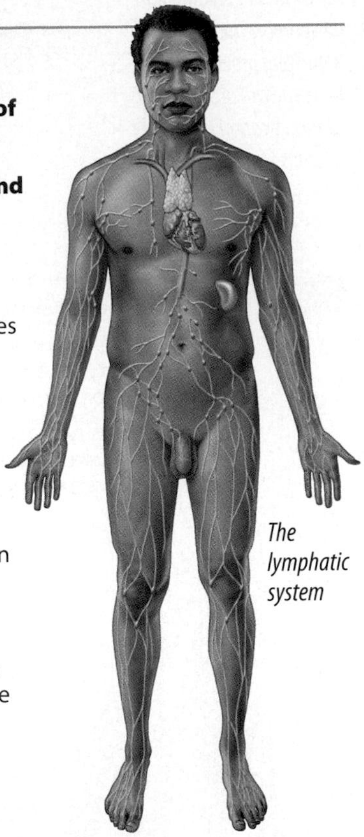

The lymphatic system

1.11 **Organs of the digestive system make nutrients available and with the urinary system excrete wastes, and organs of the male and female reproductive systems provide for the continuity of life** p. 22

25. The **digestive system** processes food and absorbs nutrients.
26. The **urinary system** eliminates excess water, salts, and wastes.
27. The **female reproductive system** produces sex cells and hormones and supports embryonic development from fertilization to birth.
28. The **male reproductive system** produces sex cells and hormones.

SECTION 3 • Homeostasis

1.12 **Homeostatic regulation relies on a receptor, a control center, and an effector** p. 25

29. **Homeostasis** is the presence of a stable internal environment.
30. **Homeostatic regulation** is the adjustment of physiological systems to preserve homeostasis and consists of three necessary parts: a **receptor** or sensor, a **control center**, and an **effector**. Homeostasis functions around a desired value or **set point**.

1.13 **Negative feedback provides stability, and positive feedback accelerates a process to completion** p. 26

31. **Negative feedback** regulates homeostasis by correcting any variation away from set point. It is the primary mechanism of homeostatic regulation.
32. In **positive feedback**, a stimulus produces a response that exaggerates change, creating a **positive feedback loop**. Positive feedback is seldom encountered in our daily lives because it produces extreme responses.

SECTION 4 • Anatomical Terms

1.14 **Anatomical terms have a long and varied history** p. 29

33. Many anatomical structures are named after the discoverer, whereas many diseases are named after the most famous victim of that disease. Such names, called **eponyms**, have been replaced with more precise terms.
34. Anatomy as a discipline studied by medical professionals began at the University of Bologna in Italy, which is considered the first modern university.

1.15 **Superficial anatomy and regional anatomy indicate locations on or in the body** p. 30

35. In the **anatomical position**, the hands are at the sides with the palms facing forward and the feet together. A person lying down in the anatomical position is said to be **supine** when face up and **prone** when face down.
36. To describe the location of internal organs, clinicians refer to four **abdominopelvic quadrants.** Anatomists use nine more precise **abdominopelvic regions**.

1.16 **Directional terms and sectional planes describe specific points of reference** p. 32

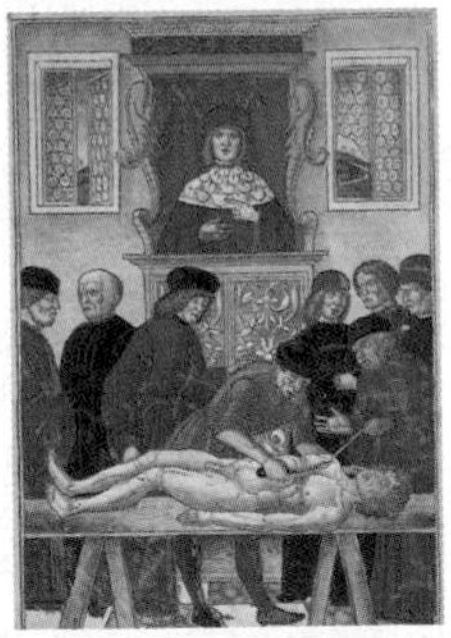

37. Directional terms utilize the anatomical position as a point of reference. The principal directional terms are superior, inferior, proximal, distal, medial, lateral, cranial (or cephalic), caudal, anterior (or ventral), posterior (or dorsal), superficial, and deep.
38. The sectional planes used to illustrate the relationship between body structures are **frontal (coronal)**, **sagittal**, and **transverse (horizontal)**.

1.17 **Body cavities protect internal organs and allow them to change shape** p. 34

39. Body structures deep to the chest wall are within the **thoracic cavity;** body structures deep to the abdominal and pelvic walls lie within the **abdominopelvic cavity**.
40. True **body cavities** are lined by a serous membrane and share a common embryonic origin. They function to protect organs from shocks and impacts, and permit significant changes in the size and shape of internal organs.

41. **Viscera** are internal organs that are partially or completely enclosed by body cavities. The heart, for example, is enclosed by the **pericardial cavity**, which is lined with a **serous membrane** called the **pericardium**.

42. The thoracic cavity contains the lungs, heart, and other structures. Its boundaries are the chest wall and a muscular sheet called the **diaphragm**.

43. The thoracic cavity consists of two **pleural cavities**, each surrounding a lung and each lined with a thin membrane known as the **pleura**. A tissue mass known as the **mediastinum** separates the pleural cavities. Within the mediastinum is the pericardial cavity that surrounds the heart.

44. The boundaries of the abdominopelvic cavity are the diaphragm, the abdominal wall muscles, the trunk muscles and inferior portions of the vertebral column, and the bones and muscles of the pelvis. It can be subdivided into the **abdominal cavity** and **pelvic cavity**.

45. The abdominal cavity and pelvic cavity contain the **peritoneal cavity**, a chamber lined by the **peritoneum**, a serous membrane.

46. Some organs lie between the peritoneal lining and the muscular wall of the abdominal cavity, and are said to be **retroperitoneal**. Other organs extend inferior to the peritoneal cavity and are said to be **infraperitoneal**.

Chapter Review Questions

Labeling

Identify the body cavities as well as the organs they enclose.

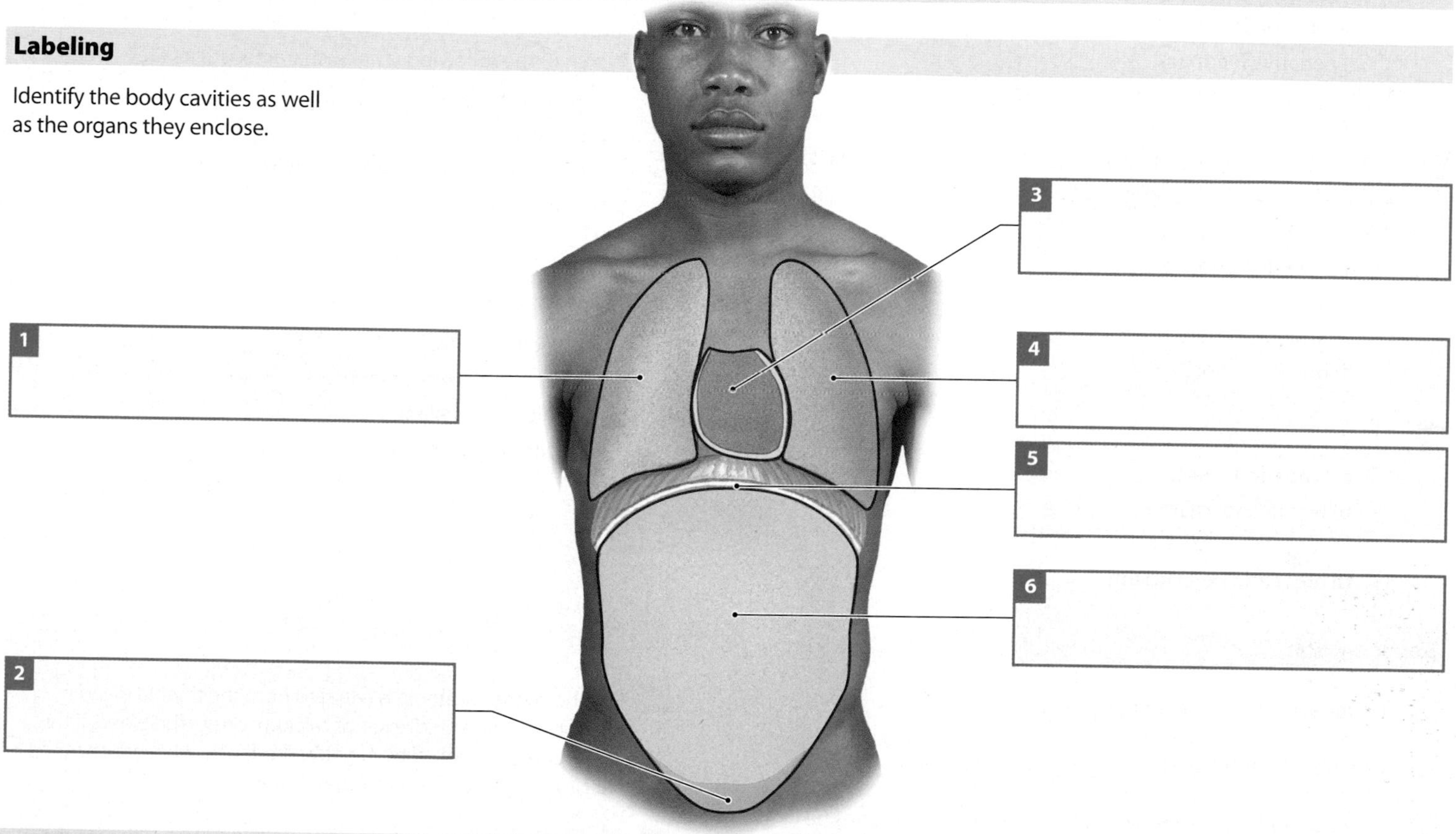

1
2
3
4
5
6

Matching

Match each lettered term with the most closely related description.

a. cytology
b. physiology
c. histology
d. anatomy
e. homeostasis
f. muscle
g. heart
h. endocrine
i. temperature regulation
j. labor and delivery
k. supine
l. prone
m. abdominopelvic cavity
n. pericardium

7 ________ Study of tissues
8 ________ Constant internal environment
9 ________ Face up position
10 ________ Study of functions
11 ________ Positive feedback
12 ________ Organ system
13 ________ Study of cells
14 ________ Negative feedback
15 ________ Serous membrane
16 ________ Study of internal and external body structures
17 ________ Diaphragm tissue
18 ________ Peritoneal cavity
19 ________ Organ
20 ________ Face down position

Multiple choice

Select the correct answer from the list provided.

21 What is the correct order, from simplest to most complex, of the six levels of organization that make up the human body?

- ☐ a) cell, chemical, tissue, organ, organ system, organism
- ☐ b) chemical, cell, tissue, organ, organ system, organism
- ☐ c) chemical, cell, tissue, organ system, organ, organism
- ☐ d) chemical, cell, organ, tissue, organ system, organism
- ☐ e) cell, tissue, chemical, organ, organism, organ system

22 The increasingly forceful labor contractions during childbirth are an example of

- ☐ a) receptor activation.
- ☐ b) effector shutdown.
- ☐ c) negative feedback.
- ☐ d) positive feedback.

23 A plane through the body that passes perpendicular to the long axis of the body and divides the body into a superior and an inferior section is a

- ☐ a) sagittal section.
- ☐ b) transverse section.
- ☐ c) coronal section.
- ☐ d) frontal section.

24 The mediastinum is the region between the

- ☐ a) lungs and heart.
- ☐ b) two pleural cavities.
- ☐ c) chest and abdomen.
- ☐ d) heart and pericardium.

25 The unit used to measure cell dimensions is the

- ☐ a) centimeter (cm).
- ☐ b) millimeter (mm).
- ☐ c) micrometer (μm).
- ☐ d) kilometer (km).

26 The two major body cavities of the trunk are the

- ☐ a) pleural cavity and pericardial cavity.
- ☐ b) pericardial cavity and peritoneal cavity.
- ☐ c) pleural cavity and peritoneal cavity.
- ☐ d) thoracic cavity and abdominopelvic cavity.

27 Which of the following is *not* a characteristic of life?

- ☐ a) responsiveness
- ☐ b) movement
- ☐ c) manipulation of external environment
- ☐ d) reproduction

28 Which sectional plane would divide the body so that the face remains intact?

- ☐ a) sagittal plane
- ☐ b) frontal (coronal) plane
- ☐ c) midsagittal plane
- ☐ d) parasagittal plane

Short answer

29 Define anatomy. Define physiology.

30 Describe the three basic principles of the cell theory.

31 In which body cavity would each of the following organs be enclosed? Heart, small intestine, large intestine, lung, kidneys.

32 Identify each of the four primary tissue types and give an example of where in the body that tissue would be found.

33 The hormone calcitonin is released from the thyroid gland in response to increased levels of calcium ions in the blood. If this hormone is controlled by negative feedback, what effect would calcitonin have on blood calcium levels?

34 A stroke occurs when there is a disruption in blood flow to the brain, causing brain cells to die. Predict the kind of symptoms a stroke patient would have. Apply your knowledge of organ system function.

Chapter Integration • Applying what you have learned

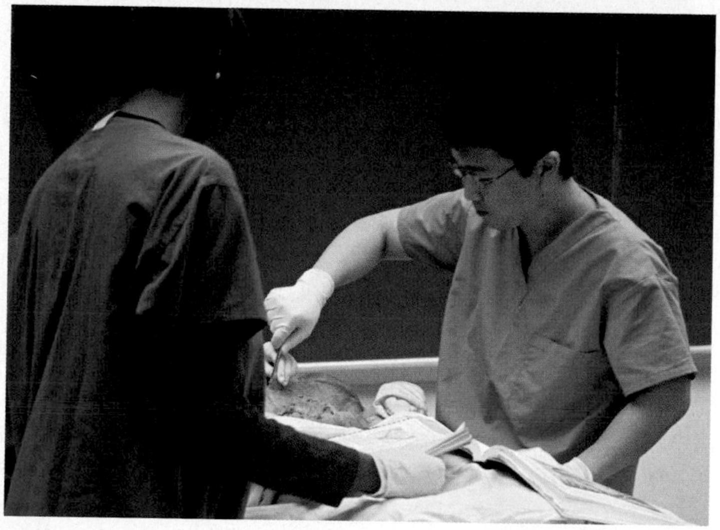

The amazing experience of studying the human body firsthand

The school year and your A&P studies are just beginning. The reading workload is massive, the terms seem like a foreign language, and the instructor seems to think that your entire life should revolve around her class. But you really do love anatomy and are lucky enough to be able to experience the study of the human body firsthand. At your last class you were introduced to the human anatomy lab, which you now approach with a little trepidation, as you have never had to study a deceased human body. Body donation for scientific purposes is a noble gesture on the part of the deceased, and you fully understand your instructor's expectations for showing complete respect to the cadaver.

Your instructor probably has you put on your gloves, apron, and eye protection before presenting the cadaver to you. Your fears prove groundless as this amazing experience begins. Your instructor explains that this cadaver was the victim of a gunshot wound to the lower right abdomen, with no exit wound. No autopsy was performed, so the anatomical/physiological cause of death has yet to be determined. The information on the body donor card provides the following information: male, 45 years old, no known medical history, and occupation unspecified. As a beginning anatomy student, use your knowledge to answer the following questions.

1 **Upon superficial examination, you discover that the bullet entered about 2 cm inferior to the umbilicus (navel). Using anatomical terminology, describe the location.**

2 **Before opening the abdominal cavity, predict what organs may have been affected.**

3 **Propose a plausible cause of death.**

Aaron has been extremely tired, thirsty, and urinating more than normal

Aaron is an active and apparently healthy 14-year-old boy. Lately, however he has been extremely tired, thirsty, and has been urinating more than what would seem normal. His parents take him to the doctor and Aaron is diagnosed as having type 1 (insulin dependent) diabetes. The physician explains to Eddie and Patty, his concerned parents, that Aaron's blood glucose (sugar) is much higher than it should be. She goes on to explain that after meals when his blood glucose rises, that it should then be absorbed into his body's tissues. Insulin, she continues, is the hormone responsible for this absorption and it is normally produced by the pancreas. Aaron is no longer producing insulin and despite high levels of blood glucose, it is not going into his tissues. Subsequently his body is being starved of glucose, explaining his lack of energy. Furthermore, this excess blood glucose is being eliminated in his urine which is why he is frequently urinating. Considering what you have read in this chapter, can you answer any of these questions?

4 **Which one of Aaron's organ systems is failing? List other systems that could be affected.**

5 **Applying what you learned about homeostasis (Module 1.12) and negative feedback (Module 1.13), predict the process that normally controls blood sugar, and then conclude which physiological mechanism is failing.**

6 **Which dietary and medical interventions would you judge most effective in treating his type 1 diabetes?**

2 Chemical Level of Organization

LEARNING OUTCOMES

These Learning Outcomes correspond by number to this chapter's modules and indicate what you should be able to do after completing the chapter.

SECTION 1 • Atoms and Molecules

- **2.1** Define an atom, and describe the properties of its subatomic particles.
- **2.2** Describe an atom and how atomic structure affects the mass number and atomic weight of the various chemical elements.
- **2.3** Explain the relationship between electrons and energy levels.
- **2.4** Compare the ways in which atoms combine to form molecules and compounds.
- **2.5** Describe the three states of matter and the importance of hydrogen bonds in liquid water.

SECTION 2 • Chemical Reactions

- **2.6** Define metabolism and distinguish between work, kinetic energy, and potential energy.
- **2.7** Use chemical notation to symbolize chemical reactions.
- **2.8** Distinguish among the major types of chemical reactions that are important for studying physiology.
- **2.9** Describe the crucial role of enzymes in metabolism.

SECTION 3 • Water in the Body

- **2.10** Describe four important properties of water and their significance in the body.
- **2.11** Explain how the chemical properties of water affect the solubility of inorganic and organic molecules.
- **2.12** Discuss the importance of pH and the role of buffers in body fluids.

SECTION 4 • Organic Compounds

- **2.13** Describe the common elements of organic compounds and how functional groups modify the properties of organic compounds.
- **2.14** Discuss the structures and functions of carbohydrates.
- **2.15** Discuss the structures and functions of lipids.
- **2.16** Discuss the structures and diverse functions of eicosanoids, steroids, phospholipids, and glycolipids.
- **2.17** Discuss protein structure and the essential functions of proteins within the body.
- **2.18** Explain how enzymes function within the body.
- **2.19** Discuss the structure and function of high-energy compounds.
- **2.20** Compare and contrast the structures and functions of DNA and RNA.

200
150
100
50

Learning Outcomes are repeated at the bottom of each module.

Atoms and molecules are the basic particles of matter

Our study of the human body begins at the chemical level of organization. Chemistry is the science that studies the structure of **matter**, which is defined as anything that takes up space and has mass. **Mass** is the quantity of matter in an object. (You can physically touch matter.) The more matter an object contains, the greater its mass. Within the Earth's gravitational field, the mass of an object determines its weight. However, the two are not always equivalent: In orbit you would be weightless, but your mass would remain unchanged.

1 **Atoms** are the smallest stable units of matter. They are composed of **subatomic particles**, only three of which are important for understanding the basic chemical properties of matter.

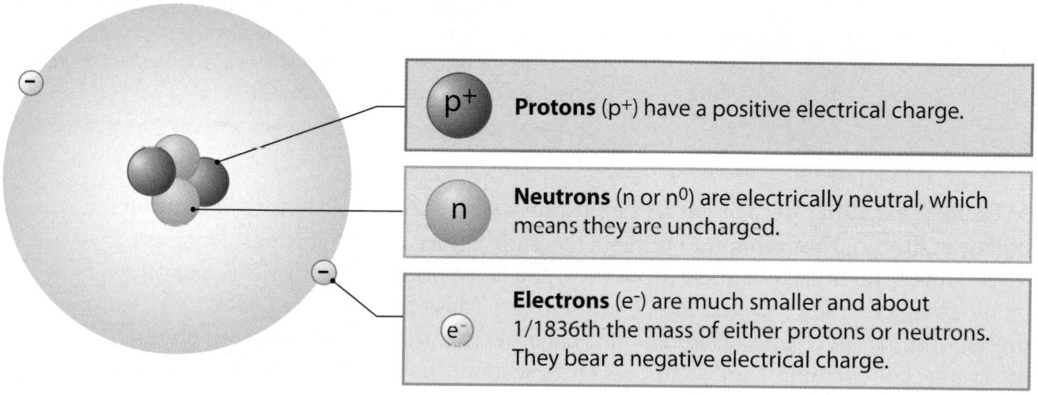
p+
Protons (p+) have a positive electrical charge.
n
Neutrons (n or n^0) are electrically neutral, which means they are uncharged.
e-
Electrons (e^-) are much smaller and about 1/1836th the mass of either protons or neutrons. They bear a negative electrical charge.

2 Atoms can be subdivided into the nucleus and the electron cloud.

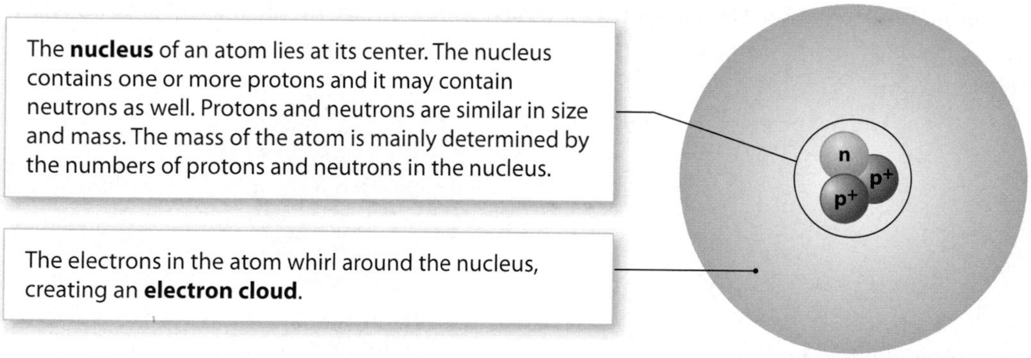
The **nucleus** of an atom lies at its center. The nucleus contains one or more protons and it may contain neutrons as well. Protons and neutrons are similar in size and mass. The mass of the atom is mainly determined by the numbers of protons and neutrons in the nucleus.
The electrons in the atom whirl around the nucleus, creating an **electron cloud**.
n
p+
p+

3 A molecule forms when atoms interact and produce larger, more complex structures.

p+
p+

Everything around us is composed of atoms in varying combinations. The unique characteristics of each object, living or nonliving, result from the types of atoms involved and the ways those atoms combine and interact.

Module 2.1 Review

a. Define atom.

b. Which subatomic particles have a positive charge? Which are uncharged?

c. Describe the subatomic particle not in the nucleus.

LO 2.1 Define an atom, and describe the properties of its subatomic particles.

Module 2.2

Typical atoms contain protons, neutrons, and electrons

Atoms normally contain equal numbers of protons and electrons. The number of protons in an atom is known as the **atomic number**; the total number of both protons and neutrons is its **mass number**. An **element** is a pure substance consisting only of atoms with the same atomic number.

1 **Hydrogen (H)** is the simplest atom, with an atomic number of 1. Thus, an atom of hydrogen contains one proton and one electron. A hydrogen atom's proton is located in the center of the atom and forms the nucleus.

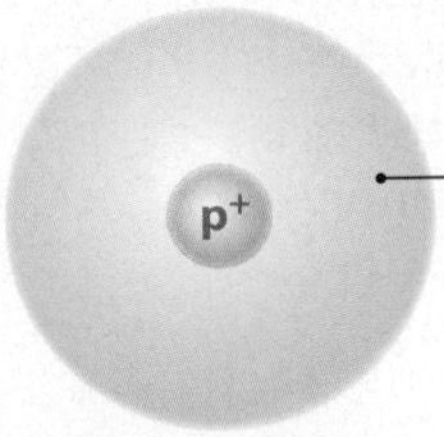
p+

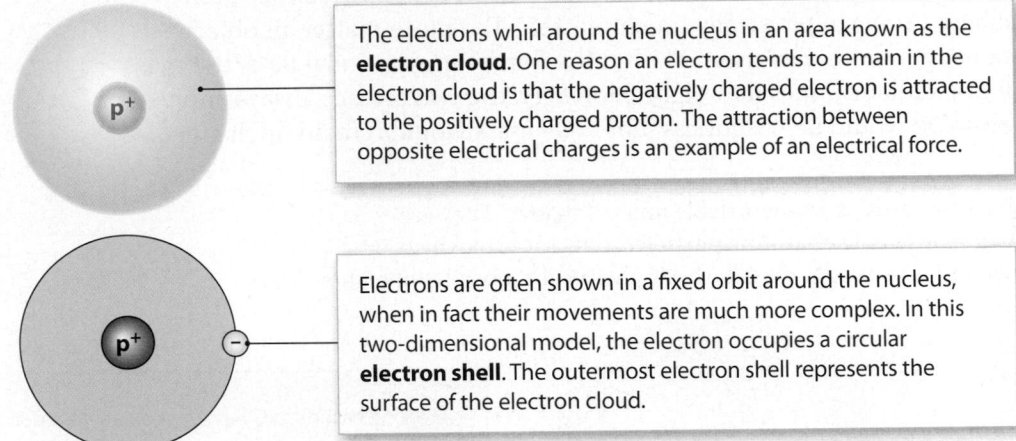

The electrons whirl around the nucleus in an area known as the **electron cloud**. One reason an electron tends to remain in the electron cloud is that the negatively charged electron is attracted to the positively charged proton. The attraction between opposite electrical charges is an example of an electrical force.

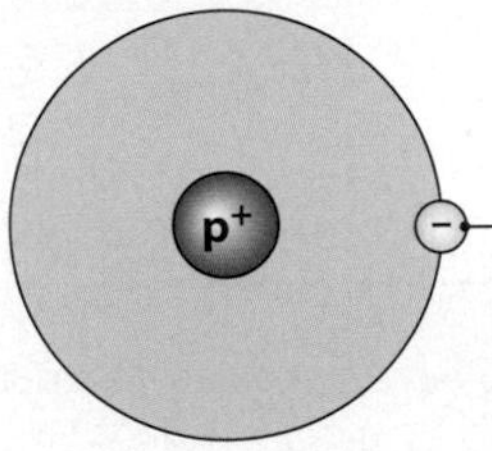
p+
−

Electrons are often shown in a fixed orbit around the nucleus, when in fact their movements are much more complex. In this two-dimensional model, the electron occupies a circular **electron shell**. The outermost electron shell represents the surface of the electron cloud.

2 The atoms of a single element can differ in the number of neutrons in the nucleus. Atoms whose nuclei contain the same number of protons, but different numbers of neutrons, are called **isotopes**. Different isotopes of an element have essentially identical chemical properties, and so are indistinguishable except on the basis of mass. The mass number is therefore used to designate isotopes. Mass numbers are useful because they tell us the number of subatomic particles in the nuclei of different atoms. However, they do not tell us the actual mass of the atoms. For example, they do not take into account the masses of the electrons or the slight difference between the mass of a proton and that of a neutron. The actual mass of an atom of a specific isotope is known as its **atomic mass**. The unit used to express atomic mass is the **atomic mass unit (amu)**, or dalton. By international agreement, 1 amu is equal to one-twelfth the mass of a carbon-12 atom.

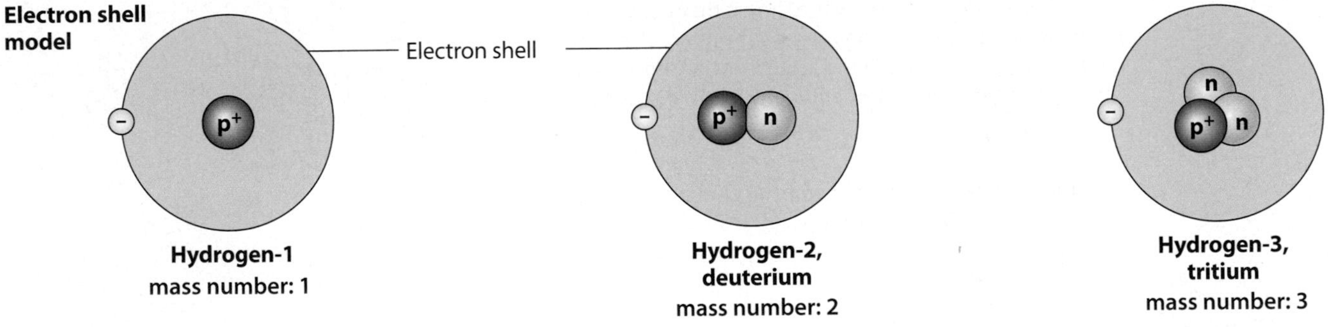
Electron shell model
Electron shell
−
p+
n
Hydrogen-1
mass number: 1
Hydrogen-2, deuterium
mass number: 2
Hydrogen-3, tritium
mass number: 3

3 The **atomic weight** of an element is an average of the different atomic masses and proportions of its different isotopes. As a result, the atomic weight of an element is very close to the mass number of the most common isotope of that element. For example, the mass number of the most common isotope of hydrogen is 1, but the atomic weight of hydrogen is 1.0079, primarily because some hydrogen atoms (0.015 percent) have a mass number of 2, and even fewer have a mass number of 3. (In most cases, the periodic table of the elements in the Appendix shows the atomic weight of each element rounded to its nearest one-hundredth decimal place. For example, the atomic weight of hydrogen is shown as 1.01, rather than 1.0079).

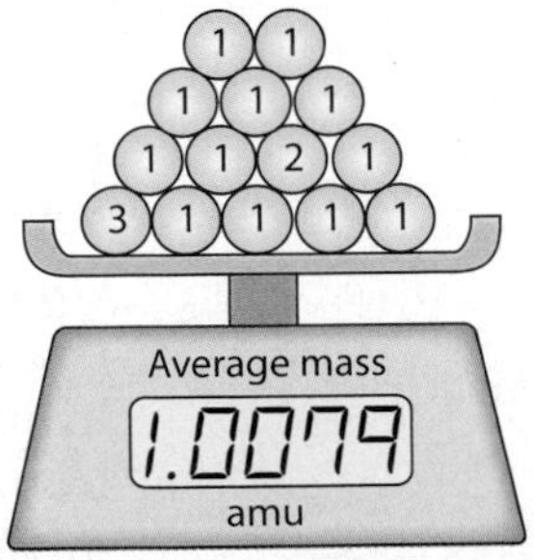
1 1
1 1 1
1 1 2 1
3 1 1 1 1
Average mass
1.0079
amu

Atomic weight of hydrogen = 1.0079

Principal Elements of the Human Body

Element (% of total body weight)	Significance
Oxygen, O (65)	A component of water and other compounds; gaseous form is essential for respiration
Carbon, C (18.6)	Found in all organic molecules
Hydrogen, H (9.7)	A component of water and most other compounds in the body
Nitrogen, N (3.2)	Found in proteins, nucleic acids, and other organic compounds
Calcium, Ca (1.8)	Found in bones and teeth; important for cell membrane function, nerve impulses, muscle contraction, and blood clotting
Phosphorus, P (1.0)	Found in bones and teeth, nucleic acids, and high-energy compounds
Potassium, K (0.4)	Important for proper cell membrane function, nerve impulses, and muscle contraction
Sodium, Na (0.2)	Important for blood volume, cell membrane function, nerve impulses, and muscle contraction
Chlorine, Cl (0.2)	Important for blood volume, cell membrane function, and water absorption
Magnesium, Mg (0.06)	A cofactor* for many enzymes
Sulfur, S (0.04)	Found in many proteins
Iron, Fe (0.007)	Essential for oxygen transport and energy capture
Iodine, I (0.0002)	A component of hormones of the thyroid gland
Trace elements: silicon (Si), fluorine (F), copper (Cu), manganese (Mn), zinc (Zn), selenium (Se), cobalt (Co), molybdenum (Mo), cadmium (Cd), chromium (Cr), tin (Sn), aluminum (Al), boron (B), and vanadium (V)	Some function as cofactors*; the functions of many trace elements are poorly understood

* A cofactor is a mineral or non-protein compound. It acts with proteins called enzymes to speed up chemical reactions in living things.

4 Our bodies consist of many elements, and the 13 most abundant elements are listed in this table. The human body also contains atoms of another 14 elements—called **trace elements**—that are present in very small amounts. Only 92 elements exist in nature, although about two dozen additional elements have been created through nuclear reactions in research laboratories. Every element has a **chemical symbol**, an abbreviation recognized by scientists everywhere. Most of the symbols are easily connected with the English names of the elements (O for oxygen, N for nitrogen, C for carbon, and so on), but a few are abbreviations of their names in other languages. For example, the symbol for sodium, Na, comes from the Latin word *natrium*.

Module 2.2 Review

a. Define atomic number and mass number.

b. How is it possible for two samples of hydrogen to contain the same number of atoms yet have different weights?

c. Describe trace elements.

LO **2.2** Describe an atom and how atomic structure affects the mass number and atomic weight of the various chemical elements.

Electrons occupy various energy levels

Atoms are electrically neutral; every positively charged proton is balanced by a negatively charged electron. Thus, with each increase in atomic number there is a comparable increase in the number of electrons traveling around the nucleus. Within the electron cloud, electrons occupy an orderly series of energy levels. Although the electrons in an energy level may travel in complex patterns around the nucleus, for our purposes the patterns are diagrammed as a series of concentric electron shells. The first electron shell (the one closest to the nucleus) corresponds to the lowest energy level. The number of electrons in the outermost electron shell or energy level determines the chemical properties of the element.

Reactive elements

1 The outermost energy level forms the "surface" of the atom. It is also called the **valence shell**. Atoms with unfilled energy levels, such as hydrogen and lithium, will react with other atoms, usually in ways that give them full outer energy levels. An atom with a filled outermost energy level is stable and does not readily react with other atoms.

Noble gases

2 Elements that do not readily participate in chemical processes are said to be **inert**. Inert elements, such as helium and neon, have filled outermost energy levels, and their atoms neither react with one another nor combine with atoms of other elements. Helium and neon are two of six such elements, called **noble gases**.

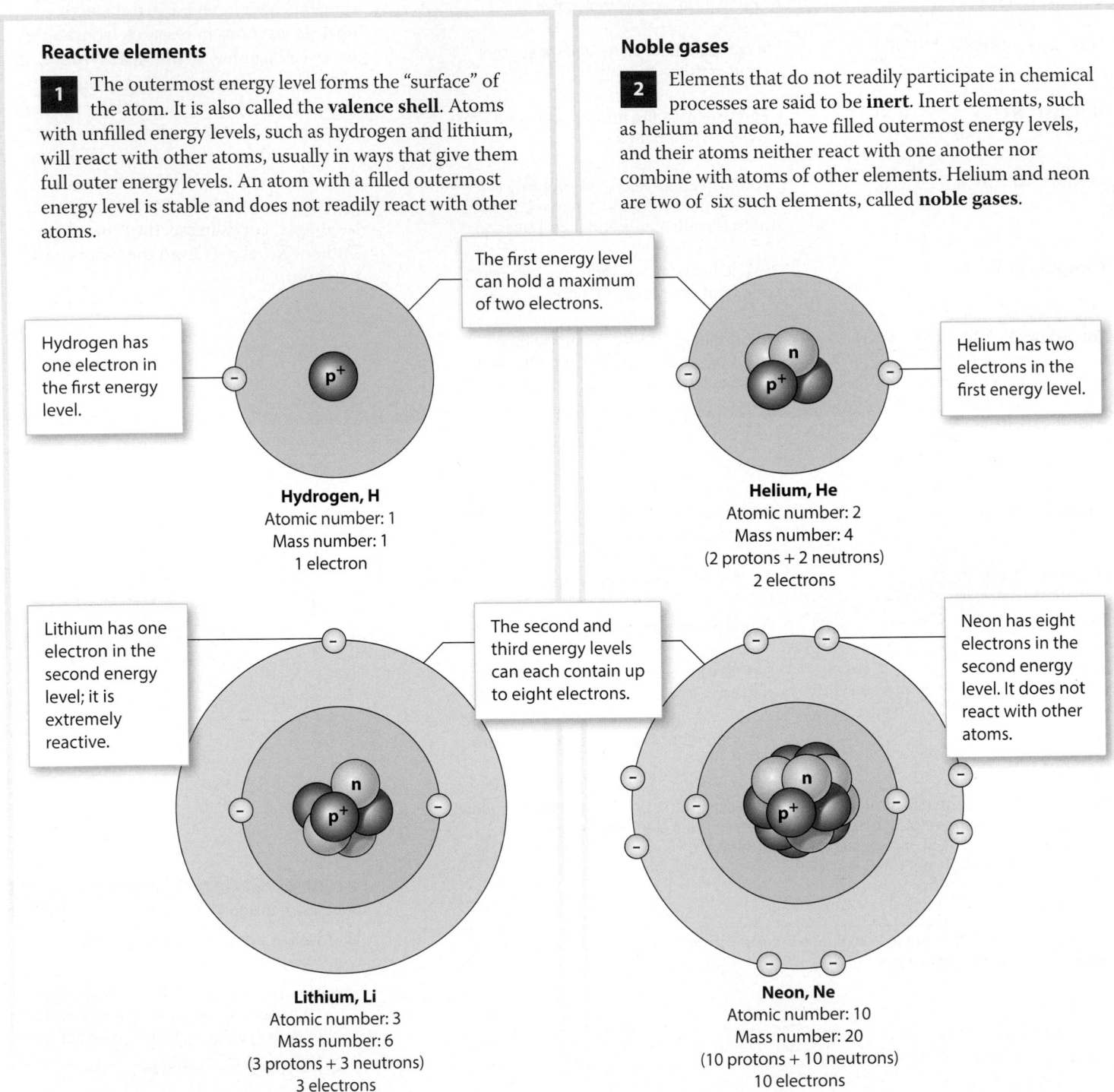

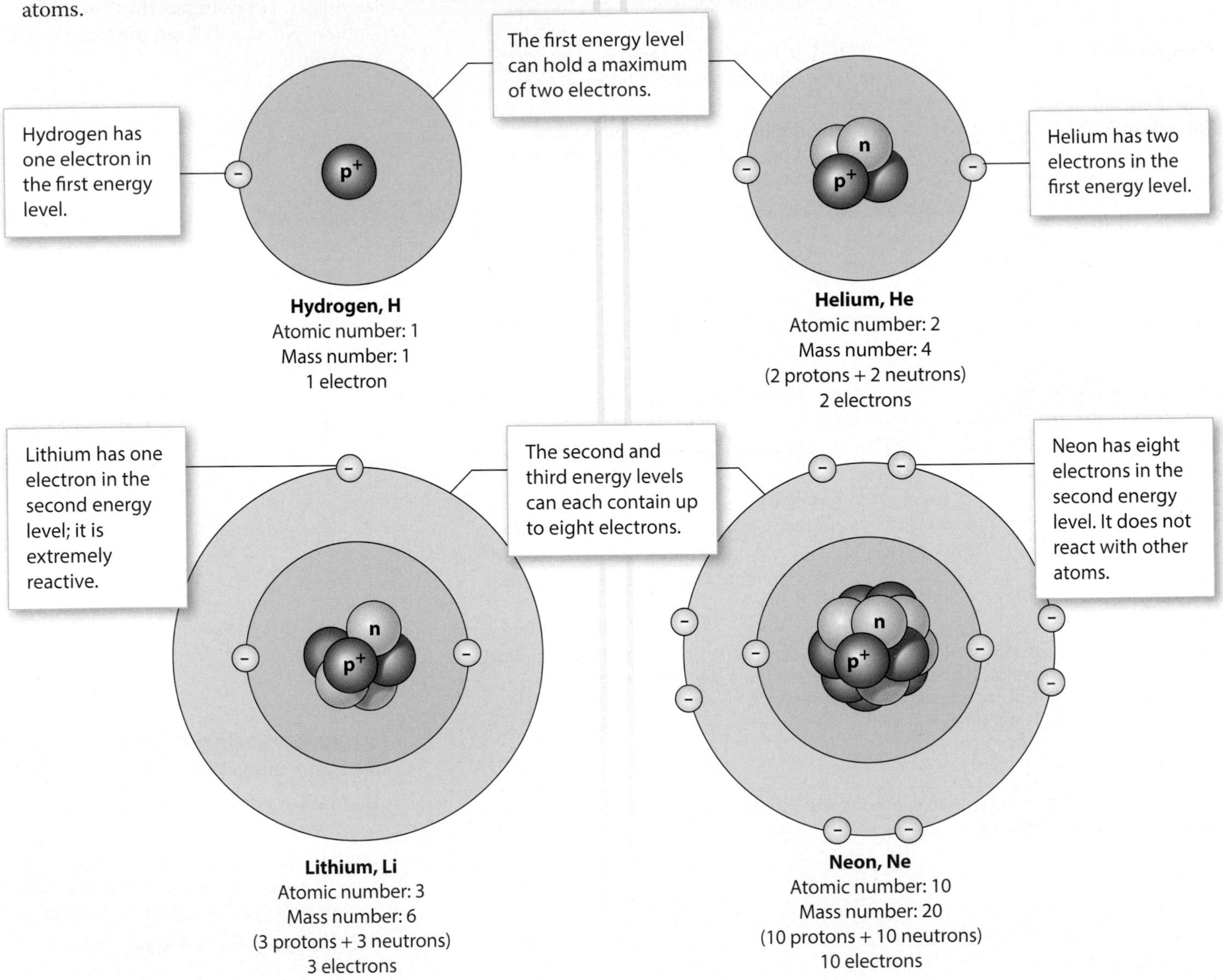
The first energy level can hold a maximum of two electrons.
Hydrogen has one electron in the first energy level.
p+
Helium has two electrons in the first energy level.
n
p+
Hydrogen, H
Atomic number: 1
Mass number: 1
1 electron
Helium, He
Atomic number: 2
Mass number: 4
(2 protons + 2 neutrons)
2 electrons
Lithium has one electron in the second energy level; it is extremely reactive.
The second and third energy levels can each contain up to eight electrons.
Neon has eight electrons in the second energy level. It does not react with other atoms.
Lithium, Li
Atomic number: 3
Mass number: 6
(3 protons + 3 neutrons)
3 electrons
Neon, Ne
Atomic number: 10
Mass number: 20
(10 protons + 10 neutrons)
10 electrons

3 Elements with unfilled outermost energy levels, such as hydrogen, lithium, or sodium, are called **reactive**, because they readily interact or combine with other atoms. Reactive atoms become stable by gaining, losing, or sharing electrons to fill their outermost energy level. When atoms gain or lose electrons, they are no longer electrically neutral and they become **ions**. Atoms which lose electrons from the outer energy level have more protons than electrons. The atom has a net positive charge and is called a positive ion or **cation**. A single missing electron gives the ion a charge of +1. Some ions carry charges of +2, +3, or +4, depending on how many electrons are lost to become stable.

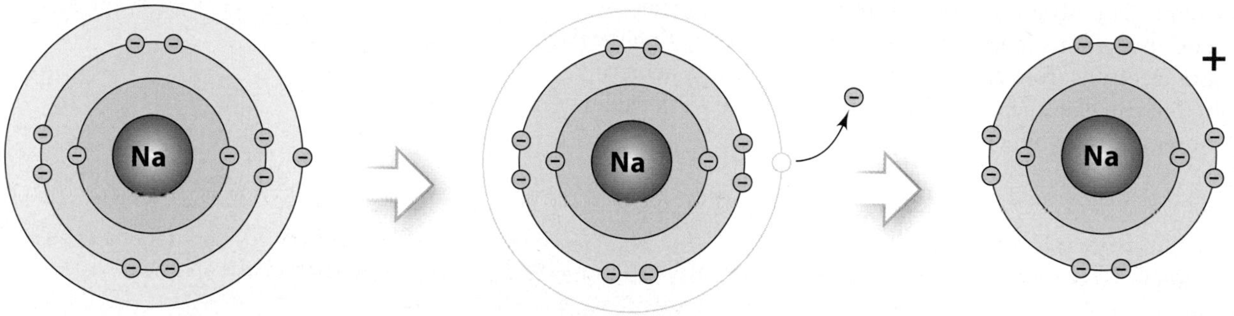

Na
Na
Na
+

Sodium atom, Na (reactive)

Sodium ion, Na^+ (stable)

4 At other times atoms become stable by filling their outer energy level with electrons obtained from other atoms. This also creates an atom that is no longer electrically neutral—it has more electrons than protons. The atom now has a net negative charge and is called a negative ion or **anion**. A single extra electron gives the ion a charge of –1. Some ions carry charges of –2, –3, or –4, depending on how many electrons are needed to become stable.

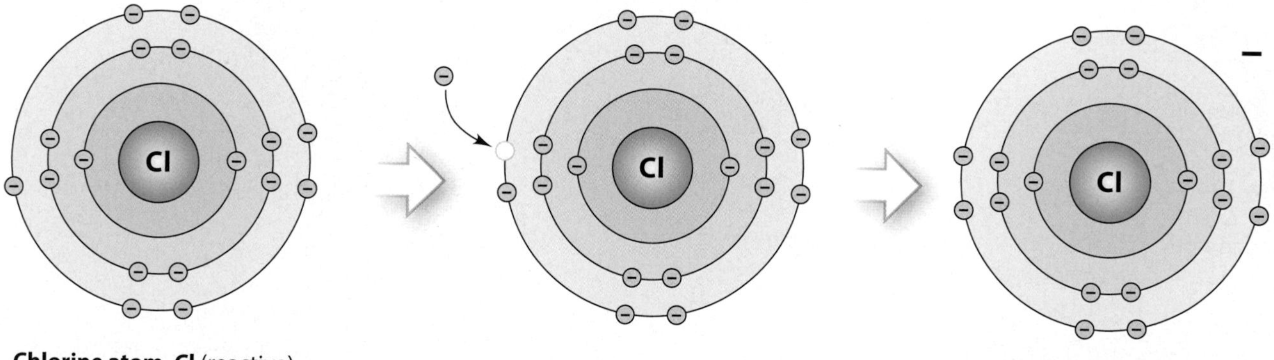

Cl
Cl
Cl
–

Chlorine atom, Cl (reactive)

Chloride ion, Cl^- (stable)

The interactions that stabilize the outer energy levels of atoms often result in **chemical bonds**. These bonds hold the participating atoms together when the reaction has ended.

Module 2.3 Review

a. Indicate the maximum number of electrons that can occupy each of the first three energy levels (electron shells) of an atom.

b. Explain why the atoms of inert elements do not react with one another or combine with atoms of other elements.

c. Explain how cations and anions form.

LO **2.3** Explain the relationship between electrons and energy levels.

Module 2.4

The most common chemical bonds are ionic bonds and covalent bonds

Chemical bonding forms new chemical entities called compounds and molecules. A **compound** is a chemical substance made up of atoms of two or more different elements in a fixed proportion, regardless of the type of bond joining them.

1 **Ionic bonds** are chemical bonds created by the electrical attraction between cations (positive ions) and anions (negative ions). Ionic bond formation involves the transfer of one or more electrons from an atom that can lose them to achieve stability, to another atom that can gain them to achieve stability. Here we consider formation of the ionic compound, sodium chloride (table salt).

Step 1
Sodium and chloride ion formation. The sodium atom loses an electron to the chlorine atom. This produces two stable ions with filled outer energy levels.

Step 2
Ionic bond formation. Because these ions form close together and have opposite charges, they are attracted to one another. This creates NaCl, an ionic compound.

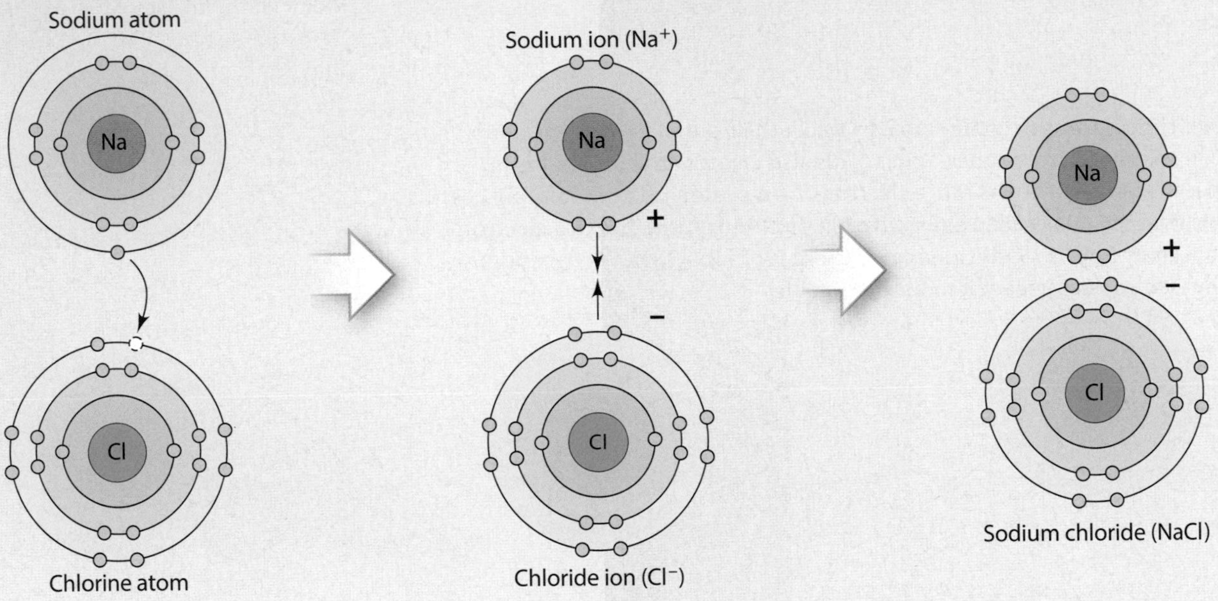

Sodium atom
Na
Cl
Chlorine atom
Sodium ion (Na+)
Na
+
−
Cl
Chloride ion (Cl−)
Na
+
−
Cl
Sodium chloride (NaCl)

2 A crystal of sodium chloride contains a large number of sodium and chloride ions packed closely together. The packed orientation of the sodium and chloride ions forms cube-shaped crystals.

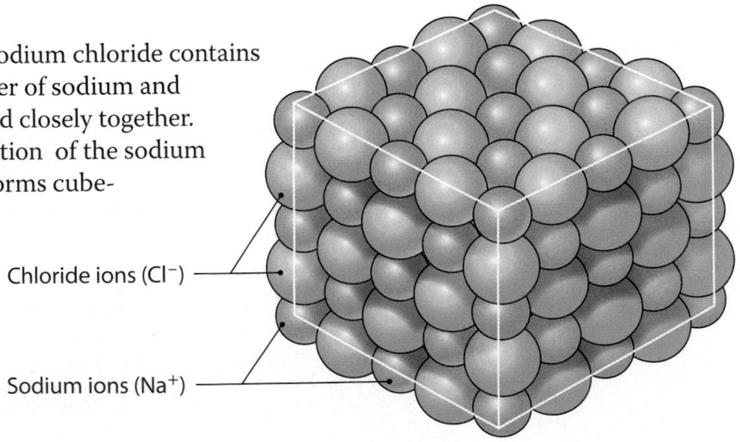

Chloride ions (Cl−)
Sodium ions (Na+)

Sodium chloride crystals ×0.5

3 Some atoms can complete their outer electron shells not by gaining or losing electrons, but by sharing electrons with other atoms. Such sharing creates **covalent** (kō-VĀ-lent) **bonds** between the atoms involved. A **molecule** is a chemical structure consisting of atoms of one or more elements held together by covalent bonds.

Molecule	Description
Hydrogen (H_2) 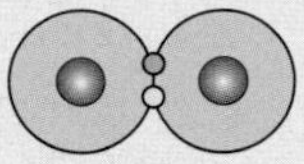	Hydrogen atoms are not found as single atoms. They exist as molecules, each containing a pair of hydrogen atoms. When the two hydrogen atoms share their electrons, each electron whirls around both nuclei. The sharing of one pair of electrons creates a **single covalent bond.**
Oxygen (O_2) 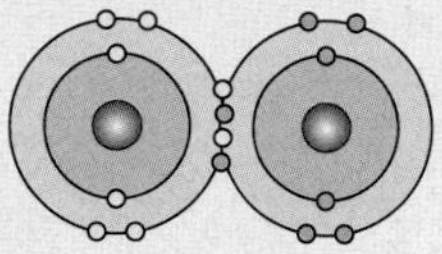	An oxygen atom has 6 electrons in its outer energy level. By sharing two pairs of electrons, it forms a **double covalent bond** with another oxygen atom, and an oxygen molecule is created with a stable outer energy level.
Carbon dioxide (CO_2) 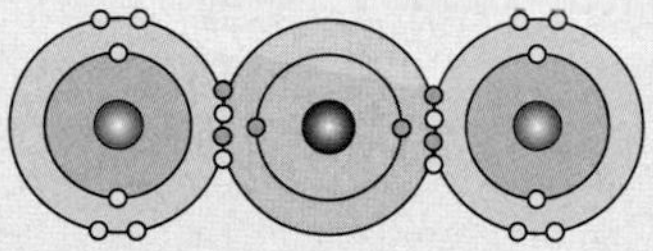	A carbon atom has 4 electrons in its outer energy level, so it needs to gain 4 from other atoms to achieve stability. In a molecule of carbon dioxide, a carbon atom shares a pair of electrons with each of two oxygen atoms and forms two double covalent bonds.

4 These are space-filling models of oxygen and carbon dioxide molecules. The spherical diameters of the atoms are shown in proportion to each other in these models. In a typical covalent bond, the participating atoms share the electrons equally, and there is no electrical charge on the molecule. Due to this lack of electrical charge, such molecules are called **nonpolar molecules**.

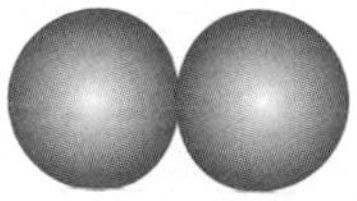

Oxygen (O_2)

Carbon dioxide (CO_2)

5 Some molecules, however, are formed by covalent bonds that involve an unequal sharing of electrons. This electron shell or energy level model shows the formation of a water molecule.

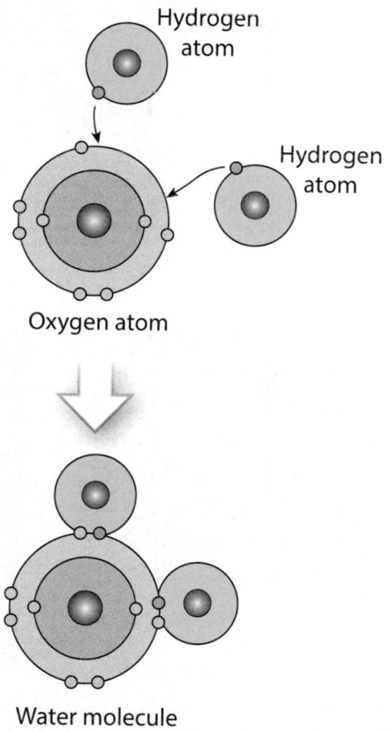
Hydrogen atom
Hydrogen atom
Oxygen atom
Water molecule

6 In a water molecule, the electron clouds of the hydrogen atoms are distorted because the 8 protons in the oxygen atom have a much stronger attraction for the electrons than do the single protons of the hydrogen atoms. As a result, each hydrogen atom carries a slightly positive charge (δ^+), and the oxygen atom carries a slightly negative charge ($2\delta^-$). (The Greek delta symbol is used to denote partial charges.) This creates an asymmetrical **polar molecule.** Covalent bonds that produce polar molecules are called **polar covalent bonds.**

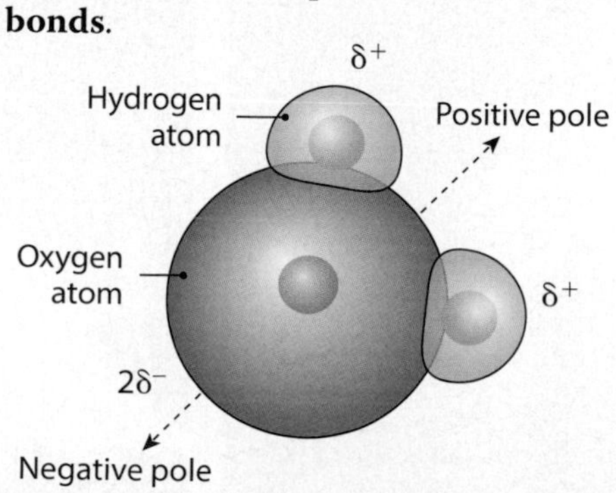
δ+
Hydrogen atom
Positive pole
Oxygen atom
δ+
2δ−
Negative pole

Module 2.4 Review

a. Name and distinguish between the two most common types of chemical bonds.

b. Describe the kind of bonds that hold the atoms in a water molecule together.

c. Explain why we can use the term *molecule* for the smallest particle of water but not for that of table salt.

LO 2.4 Compare the ways in which atoms combine to form molecules and compounds.

Matter may exist as a solid, a liquid, or a gas

Most matter in our environment exists in one of three states: solid, liquid, or gas. Whether a particular substance is a solid, a liquid, or a gas depends on the degree of motion or vibration due to its thermal energy and the type of interaction among its atoms or molecules. The particles of a solid move slowest and are held tightly together, while those of a liquid less so, and the particles of a gas are independent of each other.

1 Solids maintain their volume and their shape at ordinary temperatures and pressures. A lump of granite, a brick, and a textbook are solid objects.

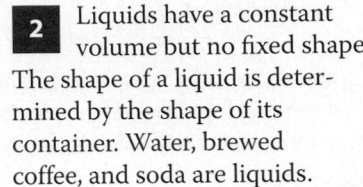

2 Liquids have a constant volume but no fixed shape. The shape of a liquid is determined by the shape of its container. Water, brewed coffee, and soda are liquids.

3 A gas has neither a constant volume nor a fixed shape. Gases can be compressed or expanded; unlike liquids they will fill a container of any size. The air of our atmosphere is the gas with which we are most familiar.

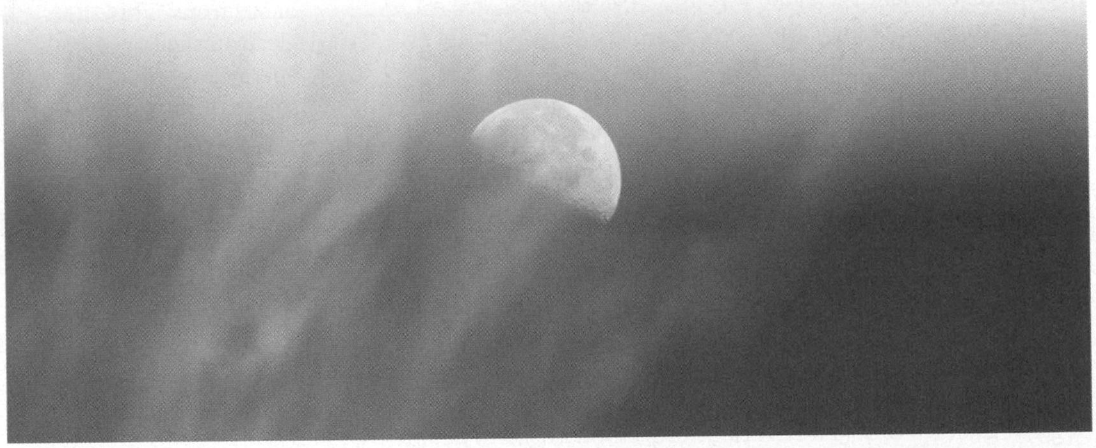

4 Water is the only substance that occurs as a solid (ice), a liquid (water), and a gas (water vapor) at temperatures compatible with life. Water exists as a liquid over a broad range of temperatures primarily because of hydrogen bonding among the water molecules.

5 The small positive charges on the hydrogen atoms of one polar molecule can be attracted to the negative charges on another polar molecule, and this can change the shapes of the molecules or pull adjacent molecules together. This weak attractive force is called a **hydrogen bond**. In the liquid state, hydrogen bonds between adjacent water molecules are continually forming and breaking. When water freezes into ice, the hydrogen bonds lock the molecules in a lattice that occupies more space than did the liquid water. This accounts for the expansion of water during freezing, and because ice is less dense than liquid water, it floats. Water becomes a vapor or gas when all of the hydrogen bonds between adjacent water molecules are broken.

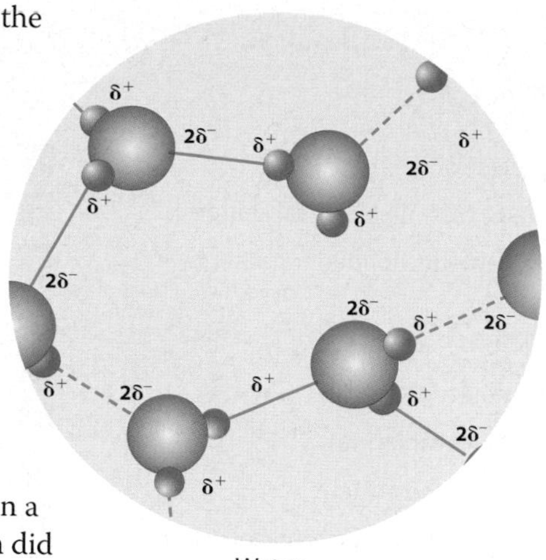

δ^+
$2\delta^-$

Water

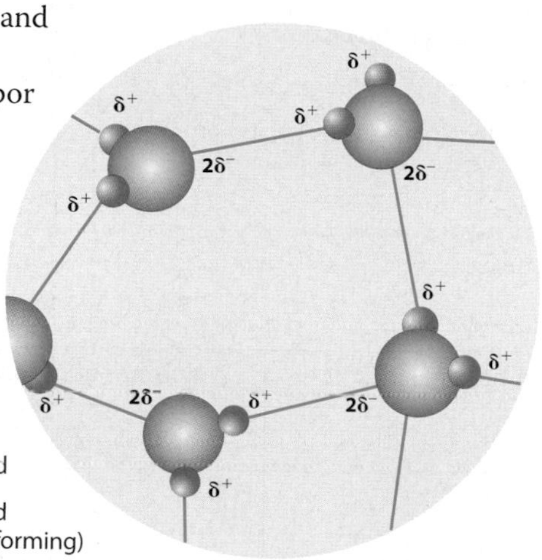

δ^+
$2\delta^-$

Ice

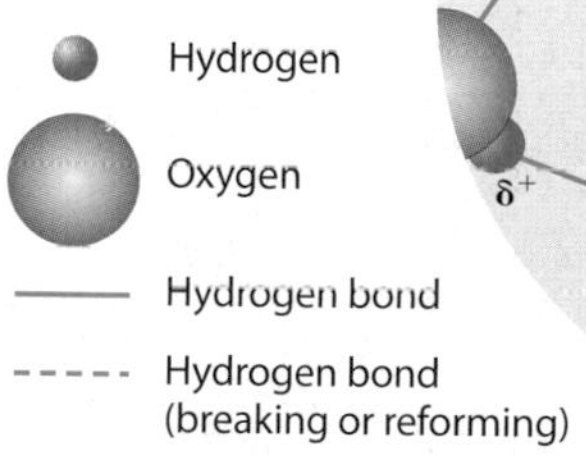

KEY
Hydrogen
Oxygen
Hydrogen bond
Hydrogen bond (breaking or reforming)

6 At the water surface, the hydrogen bonds between water molecules slow the rate of evaporation and form what is known as **surface tension**. Surface tension acts as a barrier that keeps small objects from entering the water. For example, it allows insects to walk across the surface of a pond or puddle. Similarly, the surface tension in a layer of tears on the eye prevents dust particles from touching the surface of the eye.

7 The polar charges on water molecules give water the ability to disrupt the ionic bonds of a variety of inorganic compounds and cause them to dissolve. Almost all naturally occurring elements are found in seawater, and at least 29 elements are dissolved in our body fluids.

Module 2.5 Review

a. Describe the different states of matter in terms of shape and volume.

b. By what means are water molecules attracted to each other?

c. Explain why small insects can walk on the surface of a pond, and tears protect the surface of the eye from dust particles.

LO 2.5 Describe the three states of matter and the importance of hydrogen bonds in liquid water.

Matching

Match each lettered term with the most closely related description.

a. atomic number
b. electrons
c. protons
d. neutrons
e. isotopes
f. ions
g. ionic bond
h. covalent bond
i. mass number
j. element
k. compound
l. hydrogen bond

1 Atoms that have gained or lost electrons
2 Subatomic particles in the nucleus, have no electrical charge
3 Atoms of two or more different elements bonded together in a fixed proportion
4 The number of protons in an atom
5 Attractive force between water molecules
6 Type of chemical bond within a water molecule
7 The number of subatomic particles in the nucleus
8 Substance composed only of atoms with same atomic number
9 Subatomic particles in the nucleus, have an electrical charge
10 Atoms of the same element with different masses
11 Type of chemical bond in table salt
12 Subatomic particles outside the nucleus, have an electrical charge

1 ____
2 ____
3 ____
4 ____
5 ____
6 ____
7 ____
8 ____
9 ____
10 ____
11 ____
12 ____

Fill-in

Fill in the missing information in the following table.

Element	Number of protons	Number of electrons	Number of neutrons	Mass number
Helium	13	2	2	14
Hydrogen	1	15	16	1
Carbon	6	17	6	18
Nitrogen	19	7	20	14
Calcium	21	22	20	40

Indicate which of the following molecules are also compounds.

H_2 (hydrogen)

23 ____

H_2O (water)

24 ____

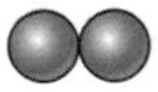
O_2 (oxygen)

25 ____

CO (carbon monoxide)

26 ____

Section integration

Describe how the following pairs of terms concerning atomic interactions are similar and how they are different.

27 Inert element/reactive element
28 Polar molecules/nonpolar molecules
29 Covalent bond/ionic bond

27 ____
____
28 ____
____
29 ____
____

Chemical reactions and energy transfer are essential to cellular functions

Cells remain alive and functional by controlling chemical reactions. In a chemical reaction, new chemical bonds form between atoms, or existing bonds between atoms are broken. These changes take place as atoms in the reacting substances, called **reactants**, are rearranged to form different substances, or **products**. All of the reactions under way in the cells and tissues of the body at any given moment make up its **metabolism** (me-TAB-ō-lizm).

1 In effect, each cell is a chemical factory. Growth, maintenance and repair, secretion, and contraction all involve complex chemical reactions. Cells also use chemical reactions to provide the energy they need to maintain homeostasis and to perform essential functions.

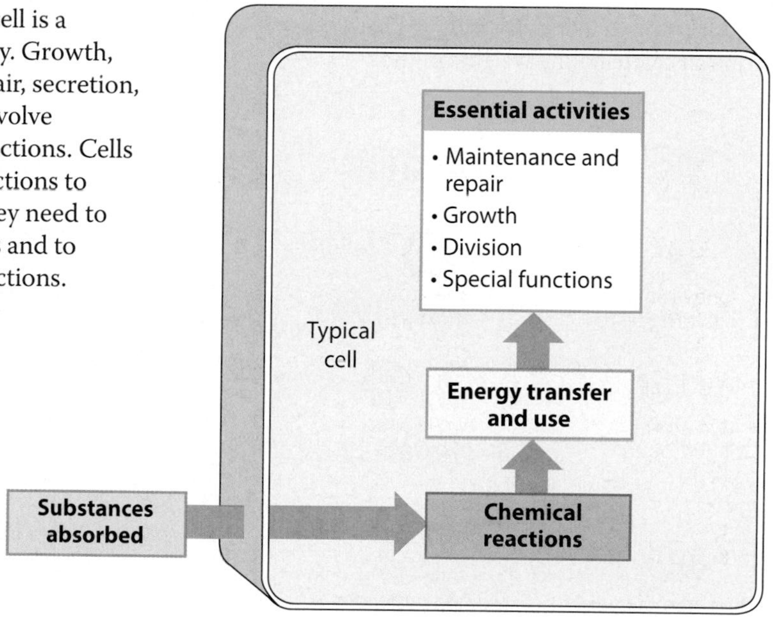
Essential activities
• Maintenance and repair
• Growth
• Division
• Special functions
Typical cell
Energy transfer and use
Substances absorbed
Chemical reactions

2 **Work** is the movement of an object or a change in the physical structure of matter. In your body, work includes movements like walking or running, and also the synthesis of molecules and the conversion of liquid water to water vapor (evaporation).

Energy is the capacity to perform work, and movement or physical change cannot take place without energy. **Kinetic energy** is the energy of motion, energy that can be transferred to another object and do work. **Potential energy** is stored energy, energy that has the potential (capability) to do work. It may derive from an object's position (you standing on a ladder) or from its physical or chemical structure (a stretched spring or a charged battery).

Cells do work as they synthesize complex molecules and move materials into, out of, and within the cell. The cells of a skeletal muscle at rest, for example, contain potential energy in the form of the positions of protein filaments and the covalent bonds between molecules within the cells. When a muscle contracts, it performs work, and potential energy is converted into kinetic energy. Such a conversion is never 100 percent efficient. Each time an energy exchange or transfer occurs, some of the energy is released as heat. The amount of heat is proportional to the amount of work done. As a result, when you exercise, your body temperature rises.

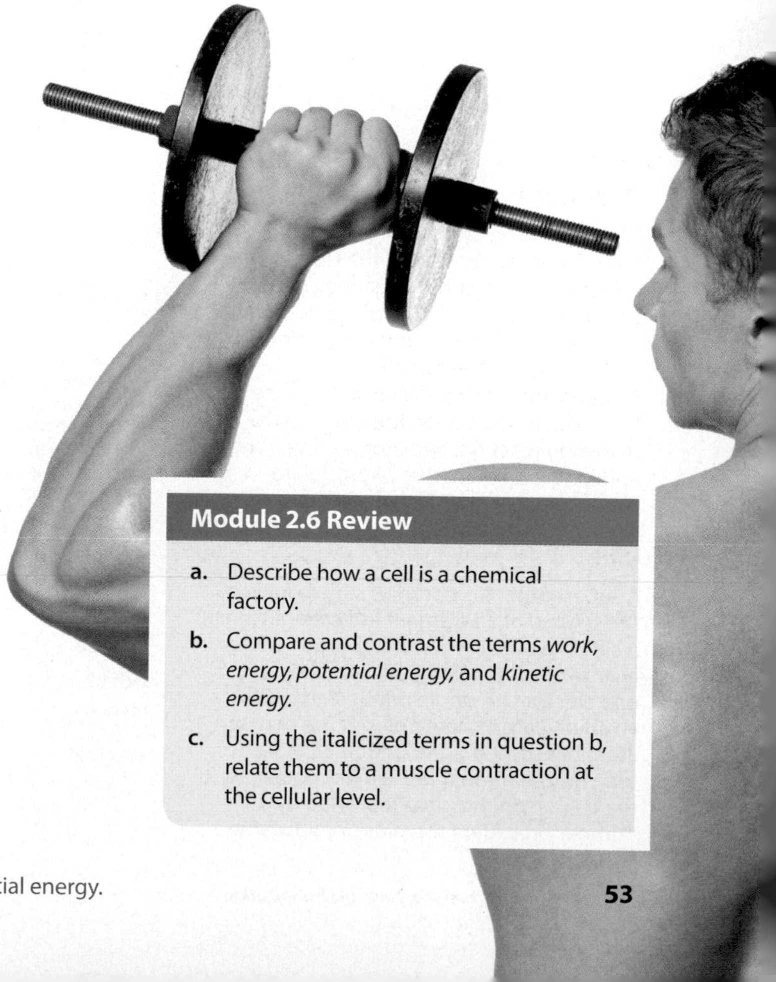

Module 2.6 Review

a. Describe how a cell is a chemical factory.

b. Compare and contrast the terms *work, energy, potential energy,* and *kinetic energy.*

c. Using the italicized terms in question b, relate them to a muscle contraction at the cellular level.

LO 2.6 Define metabolism and distinguish between work, kinetic energy, and potential energy.

Chemical notation is a concise method of describing chemical reactions

Before we can consider the specific compounds found in the human body, we must be able to describe chemical compounds and reactions effectively. Using sentences to describe chemical structures and events often leads to confusion. A simple form of "chemical shorthand" makes communication much more efficient. The chemical shorthand we use is known as **chemical notation**.

1 Chemical notation allows us to describe complex events briefly and precisely. It is relatively easy to use chemical notation to calculate the weights of the reactants involved in a particular reaction.

VISUAL REPRESENTATION = **CHEMICAL NOTATION**

Atoms

The symbol of an element indicates one atom of that element. A number preceding the symbol of an element indicates more than one atom of that element.

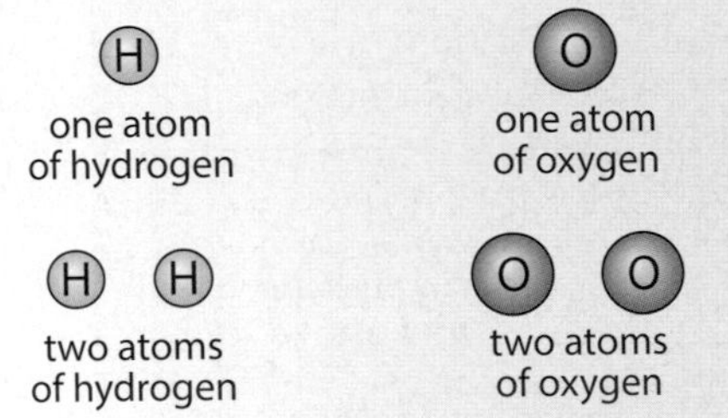
H
one atom of hydrogen
O
one atom of oxygen
H H
two atoms of hydrogen
O O
two atoms of oxygen

=

H — one atom of hydrogen; O — one atom of oxygen

$2\,H$ — two atoms of hydrogen; $2\,O$ — two atoms of oxygen

Molecules

A subscript following the symbol of an element indicates a molecule with that number of atoms of that element.

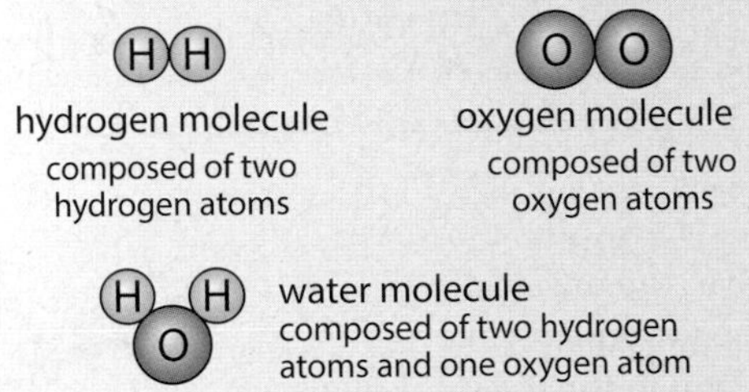
H H
hydrogen molecule
composed of two hydrogen atoms
O O
oxygen molecule
composed of two oxygen atoms
H H O
water molecule
composed of two hydrogen atoms and one oxygen atom

=

H_2 — hydrogen molecule; O_2 — oxygen molecule

H_2O — water molecule

Reactions

In a description of a chemical reaction, the participants at the start of the reaction are called reactants, and the reaction generates one or more products. Chemical reactions are represented by chemical equations. An arrow indicates the direction of the reaction, from reactants (usually on the left) to products (usually on the right). In the following reaction, two atoms of hydrogen combine with one atom of oxygen to produce a single molecule of water.

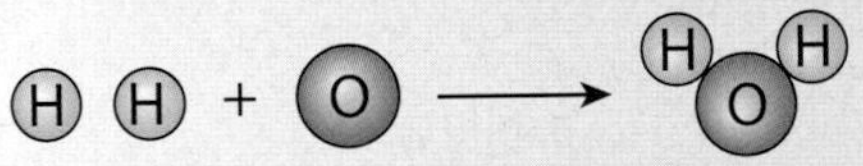
H H + O
H O H

Chemical reactions neither create nor destroy atoms; they merely rearrange atoms into new combinations. Therefore, the numbers of atoms of each element must always be the same on both sides of the equation for a chemical reaction. When this is the case, the equation is balanced.

=

$2\,H + O \longrightarrow H_2O$
Balanced equation

$2\,H + 2\,O \longrightarrow H_2O$
Unbalanced equation

Ions

A superscript plus or minus sign following the symbol of an element indicates an ion. A single plus sign indicates a cation with a charge of +1. (The original atom has lost one electron.) A single minus sign indicates an anion with a charge of −1. (The original atom has gained one electron.) If more than one electron has been lost or gained, the charge on the ion is indicated by a number preceding the plus or minus sign.

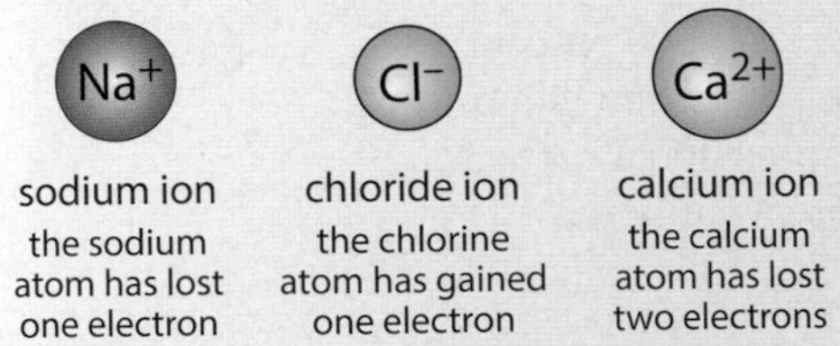
Na+
sodium ion
the sodium atom has lost one electron
Cl−
chloride ion
the chlorine atom has gained one electron
Ca2+
calcium ion
the calcium atom has lost two electrons

=

Na^+ — sodium ion; Cl^- — chloride ion; Ca^{2+} — calcium ion

2 A **mole** (abbreviated mol) is a quantity with a weight in grams equal to an element's atomic weight. One mole of a given element always contains the same number of atoms as 1 mole of any other element. The atomic weight of oxygen is 16 and the atomic weight of hydrogen is 1. So a mole of oxygen will weigh 16 grams and contain the same number of atoms as a mole of hydrogen, which weighs 1 gram.

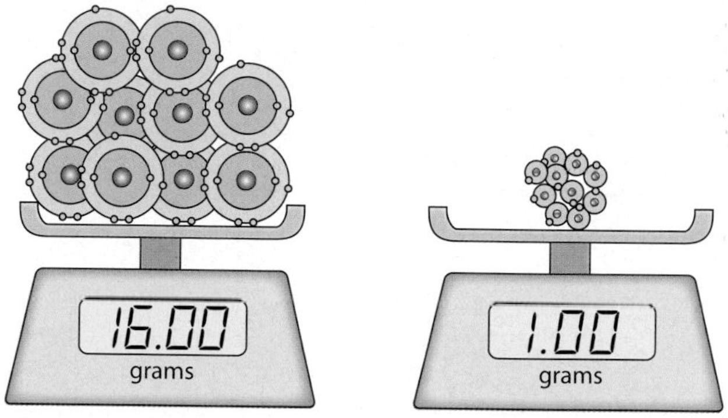
16.00
grams
1.00
grams

1 mole of oxygen   1 mole of hydrogen

3 The **molecular weight** of a molecule or compound is the sum of the atomic weights of its component atoms. For ionic compounds which do not form molecules, the term formula weight is used instead. Molecular and formula weights are important because you can neither handle individual molecules or ions nor easily count the billions of molecules or ions involved in chemical reactions in the body. To simplify our calculations of moles and molecular weights, we are rounding the atomic weights to the nearest whole number.

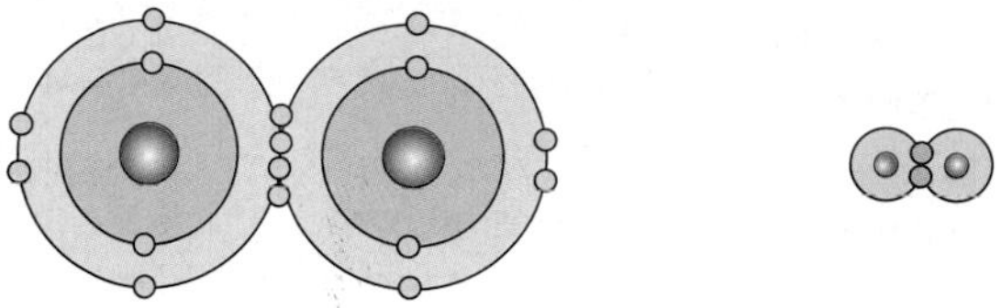

Molecular weight of O_2 = 32   Molecular weight of H_2 = 2

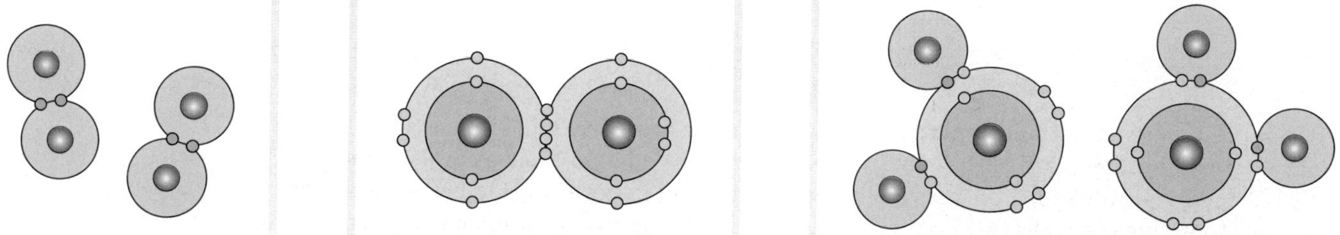

$2H_2$ + O_2 → $2H_2O$

Atomic weight of H = 1
Molecular weight of H_2 = 2
Two moles of H_2 weigh 4 g

Atomic weight of O = 16
Molecular weight of O_2 = 32
One mole of O_2 weighs 32 g

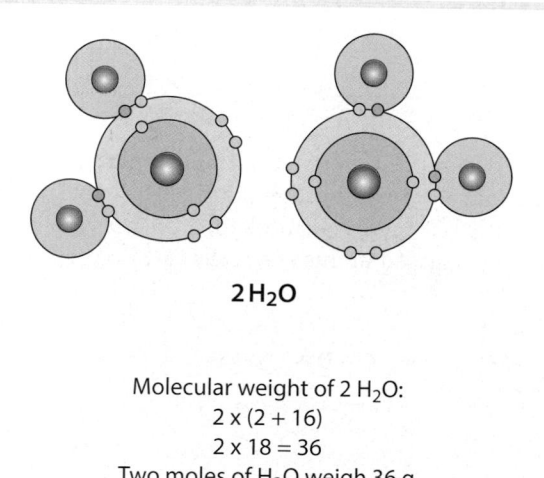

Molecular weight of 2 H_2O:
2 x (2 + 16)
2 x 18 = 36
Two moles of H_2O weigh 36 g

4 Using molecular weights, you can calculate the quantities of reactants needed for a specific reaction and determine the amount of product generated. For example, suppose you want to form water from hydrogen and oxygen according to this chemical equation: $2 H_2 + O_2 \rightarrow 2 H_2O$. The first step is to calculate the molecular weights involved. As detailed above, combining 4 g of hydrogen with 32 g of oxygen yields 36 g of water. Although by convention we use grams, you could also work with ounces, pounds, or tons, as long as the proportions remained the same. Notice that when you do the calculation correctly, the molecular weights of reactants and products are balanced.

Module 2.7 Review

a. The chemical shorthand used to describe chemical compounds and reactions effectively is known as ________________.

b. Using the rules of chemical notation, write the molecular formula for glucose, a compound composed of 6 carbon (C) atoms, 12 hydrogen (H) atoms, and 6 oxygen (O) atoms.

c. Calculate the weight of 1 mole of glucose. (The atomic weight of carbon = 12.)

LO **2.7** Use chemical notation to symbolize chemical reactions.

Three basic types of chemical reactions are important for understanding physiology

Decomposition Reactions

A **decomposition** reaction breaks a molecule into smaller fragments. A simple decomposition reaction is diagrammed below:

$$AB \longrightarrow A + B$$

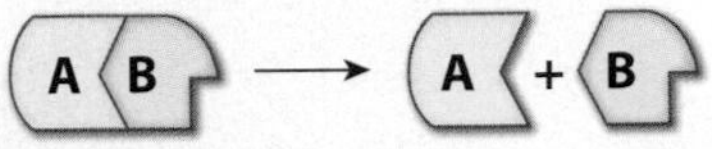
A
B
A
+
B

Decomposition reactions occur outside cells as well as inside them. For example, a typical meal contains molecules of fats, sugars, and proteins that are too large and too complex to be absorbed and used by your body. Decomposition reactions in the digestive tract break these molecules down into smaller fragments so the body can absorb them.

Decomposition reactions involving water are important in the breakdown of complex molecules in the body. In **hydrolysis** (hī-DROL-i-sis; *hydro-*, water + *lysis*, a loosening), one of the bonds in a complex molecule is broken, and the components of a water molecule (H and OH) are added to the resulting fragments:

$$A\text{-}B + H_2O \longrightarrow A\text{-}H + OH\text{-}B$$

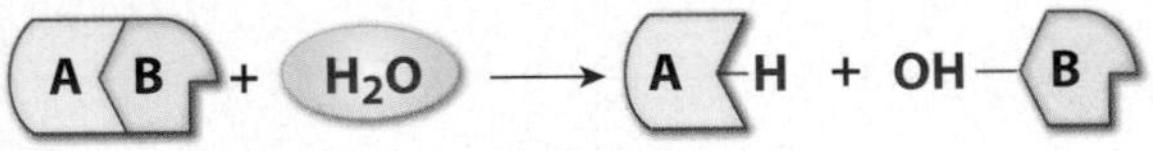
A
B
+
H2O
A
H
+ OH
B

The decomposition reactions of complex molecules within the body's cells and tissues are referred to collectively as **catabolism** (ka-TAB-ō-lizm; *katabole*, a throwing down). When a covalent bond—a form of potential energy—is broken, it releases kinetic energy that can perform work. By harnessing the energy released in this way, cells carry out vital functions such as growth, movement, and reproduction.

$$CD \longrightarrow C + D + \text{ENERGY}$$

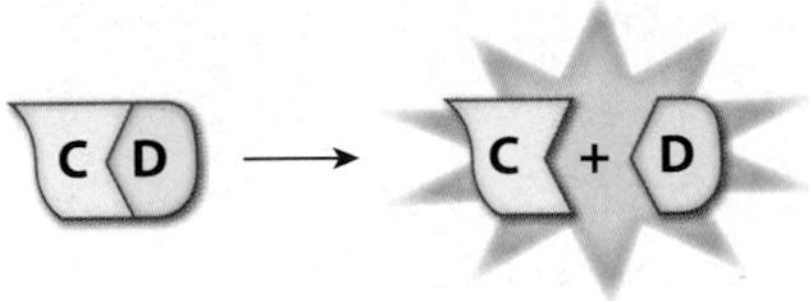
C
D
C
+
D

Synthesis Reactions

Synthesis (SIN-the-sis) is the opposite of decomposition. A synthesis reaction assembles smaller molecules into larger molecules. A simple synthetic reaction is diagrammed here:

$$A + B \longrightarrow AB$$

A
+
B
A
B

Synthesis reactions may involve combining atoms or molecules to form even larger products. Water formation from hydrogen and oxygen molecules is a synthesis reaction. Synthesis always forms new chemical bonds, whether the reactants are atoms or molecules.

Dehydration synthesis, or condensation reaction, forms a complex molecule by removing a water molecule:

$$A\text{-}H + OH\text{-}B \longrightarrow A\text{-}B + H_2O$$

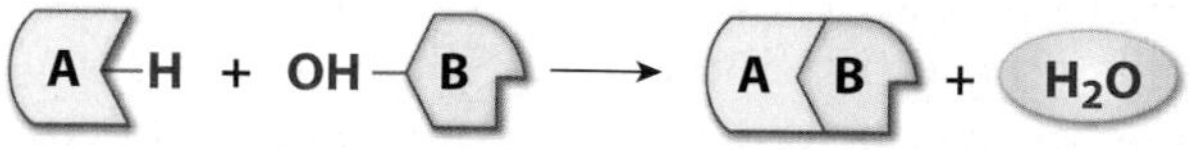
A
H
+
OH
B
A
B
+
H₂O

Dehydration synthesis is the opposite of hydrolysis. We will see examples of both reactions in later sections.

Synthesis of new molecules within the body's cells and tissues is known collectively as **anabolism** (a-NAB-ō-lizm; *anabole*, a throwing upward). Anabolism is usually considered an "uphill" process because it takes energy to create a chemical bond (just as it takes energy to push something uphill). Cells must balance their energy budgets, with catabolism providing the energy to support anabolism and other vital functions.

Chemical reactions are reversible (at least theoretically), so if

$A + B \longrightarrow AB$, then $AB \longrightarrow A + B$

Many important biological reactions are freely reversible. Such reactions can be represented as an equation:

$$A + B \rightleftharpoons AB$$

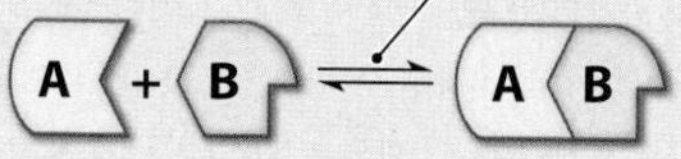
A
+
B
A
B

This equation indicates that, in a sense, two reactions are taking place at the same time. One is synthesis and the other decomposition. At **equilibrium**, the rates at which the two reactions proceed are in balance: As fast as one molecule of AB forms, another degrades into A + B.

Exchange Reactions

In an **exchange reaction**, parts of the reacting molecules are shuffled around to produce new products:

$$AB + CD \longrightarrow AD + CB$$

A
B
+
C
D
A
D
+
C
B

The reactants and products contain the same components (A, B, C, and D), but those components are present in different combinations. In an exchange reaction, the reactant molecules AB and CD must break apart (a decomposition) before they can interact with each other to form AD and CB (a synthesis).

Module 2.8 Review

a. Identify and describe three types of chemical reactions important in human physiology.

b. Distinguish the roles of water in hydrolysis and dehydration synthesis reactions.

c. In cells, glucose, a six-carbon molecule, is converted into two three-carbon molecules by a reaction that releases energy. What is the source of the energy?

LO 2.8 Distinguish among the major types of chemical reactions that are important for studying physiology.

Enzymes lower the activation energy requirements of chemical reactions

Most chemical reactions do not take place spontaneously, or if they do, they occur so slowly that they would be of little value to cells. Before a reaction can proceed, enough energy must be provided to activate the reactants. The amount of energy required to start a reaction is called the **activation energy**. Many reactions can be activated by changes in temperature or acidity (low pH), but such changes are deadly to cells. Instead, your cells use special proteins called **enzymes** to perform most of the complex synthesis and decomposition reactions in your body.

1 Enzymes promote chemical reactions by lowering the activation energy required. In doing so, they make it possible for chemical reactions, such as the breakdown of sugars, to proceed under conditions compatible with life. Enzymes belong to a class of substances called **catalysts** (KAT-uh-lists; *katalysis*, dissolution), compounds that speed up chemical reactions without themselves being permanently changed or consumed. Enzymatic reactions, which are generally reversible, proceed until an equilibrium is reached.

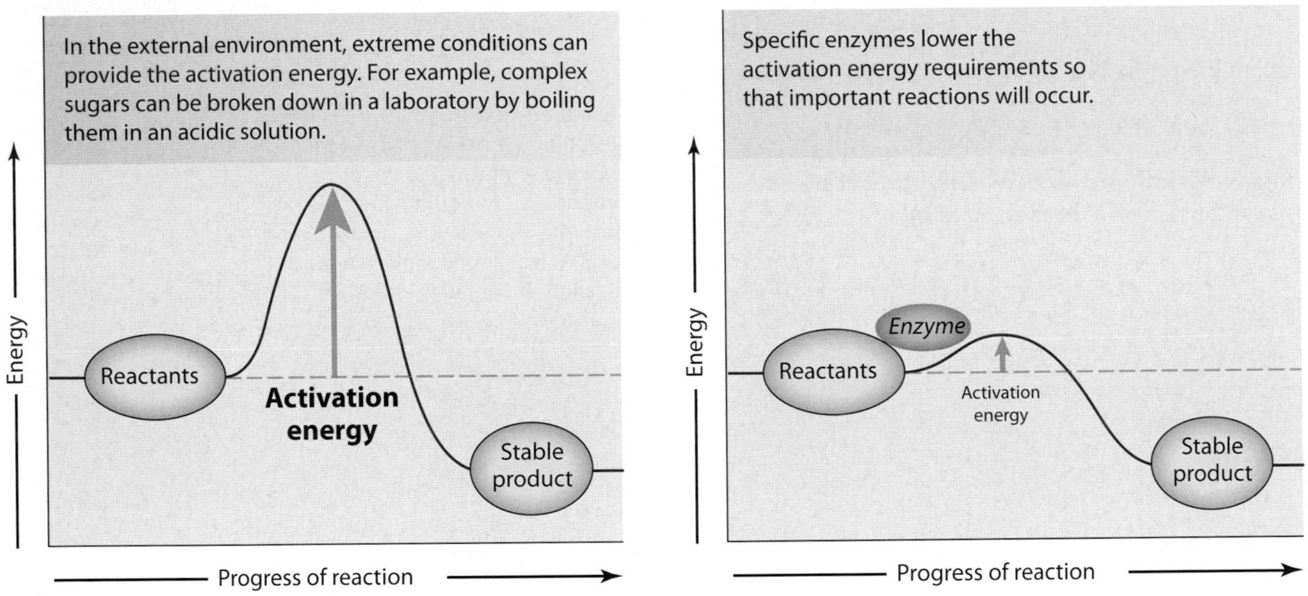
In the external environment, extreme conditions can provide the activation energy. For example, complex sugars can be broken down in a laboratory by boiling them in an acidic solution.
Energy
Reactants
Activation energy
Stable product
Progress of reaction
Specific enzymes lower the activation energy requirements so that important reactions will occur.
Energy
Enzyme
Reactants
Activation energy
Stable product
Progress of reaction

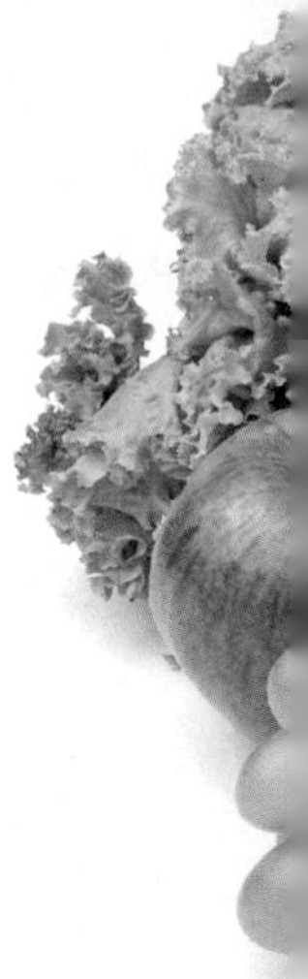

2 The complex reactions that support life proceed in a series of interlocking steps, each controlled by a specific enzyme. Such a reaction sequence is called a **metabolic pathway**. A synthetic pathway can be diagrammed as:

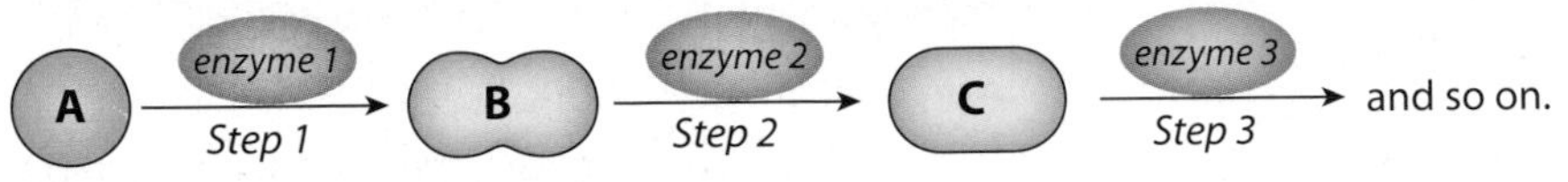
A
enzyme 1
Step 1
B
enzyme 2
Step 2
C
enzyme 3
Step 3
and so on.

It takes activation energy to start a chemical reaction, but once it has begun, the reaction as a whole may absorb or release energy as it proceeds to completion. Reactions that release energy are said to be **exergonic** (*exo-*, outside). If more energy is required to begin the reaction than is released as it proceeds, the reaction is called **endergonic** (*endo-*, inside). Exergonic reactions are relatively common in the body. They generate the heat that maintains your body temperature.

3 Enzymatic reactions are essential to the processing of **metabolites** (me-TAB-ō-līts; *metabole*, change), which are substances that can be synthesized or broken down by chemical reactions inside our bodies. **Nutrients** are essential metabolites that are normally obtained from the diet. Nutrients and metabolites can be broadly categorized as either *organic* or *inorganic*.

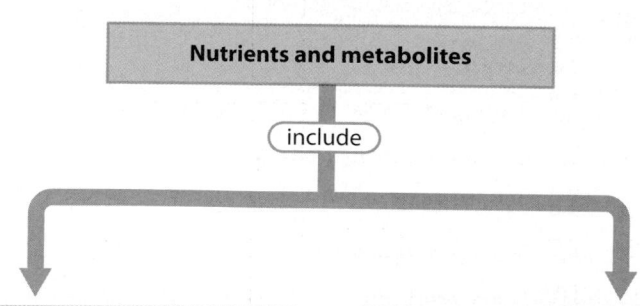

Nutrients and metabolites

include

Organic compounds

Organic compounds always contain carbon and hydrogen atoms as their primary structural ingredients.

Examples: sugars, fats, proteins, and nucleic acids (RNA, DNA), which are produced by living organisms

Inorganic compounds

Inorganic compounds generally do not contain carbon and hydrogen atoms as their primary structural ingredients.

Examples: carbon dioxide, oxygen, water, acids, bases, and salts

Module 2.9 Review

a. What is an enzyme?

b. Why do our cells need enzymes?

c. Explain the differences between metabolites and nutrients.

LO **2.9** Describe the crucial role of enzymes in metabolism.

Matching

Match each lettered term with the most closely related description.

a. exergonic
b. activation energy
c. organic compounds
d. exchange reaction
e. hydrolysis
f. endergonic
g. reactants
h. enzyme

1 Catalyst
2 Starting substances in a chemical reaction
3 Chemical reaction involving water
4 Reactions that absorb energy
5 Shuffles parts of reactants
6 Primary components are carbon and hydrogen
7 Reactions that release energy
8 Requirement for starting a chemical reaction

1 ______
2 ______
3 ______
4 ______
5 ______
6 ______
7 ______
8 ______

Short answer

Using chemical notation, write the formula of each of the following:

9 One molecule of hydrogen
10 Two atoms of hydrogen
11 Six molecules of water
12 One molecule of sucrose (in this order: 12 atoms of carbon, 22 atoms of hydrogen, and 11 atoms of oxygen)

9 ______
10 ______
11 ______
12 ______

Write the chemical equation for the following chemical reaction: one molecule of glucose combined with six molecules of oxygen produce six molecules of carbon dioxide and six molecules of water.

13 ______

Indicate which of the following is a hydrolysis reaction and which is a dehydration synthesis reaction.

14 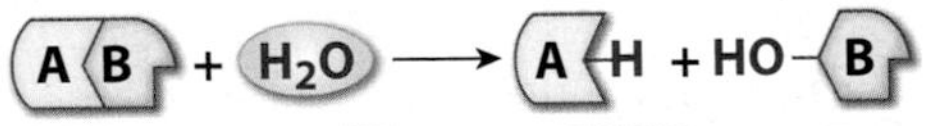
A B
+ H2O
A H
+ HO
B

15 
A H
+ HO
B
A B
+ H2O

14 ______
15 ______

Section integration

In a metabolic pathway that consists of four steps, how would decreasing the amount of enzyme that catalyzes the second step affect the amount of product at the end of the pathway?

16 ______

Water has several important properties

Water (H_2O) is the most important substance in the body. It accounts for up to two-thirds of total body weight. A change in the body's water content can be fatal because virtually all physiological systems will be affected. Although water is familiar to everyone, it has some highly unusual properties.

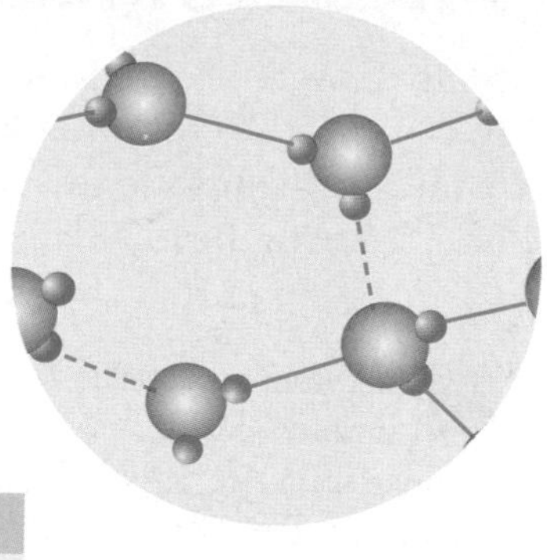

Important Properties of Water

Lubrication

Water is an effective lubricant because there is little friction between water molecules. Thus even a thin layer of water between two opposing surfaces will greatly reduce friction between them; water reduces friction within joints and in body cavities.

SER

Chemical reactant

In our bodies, chemical reactions occur in water, and water molecules are also participants in some reactions, including hydrolysis and dehydration synthesis.

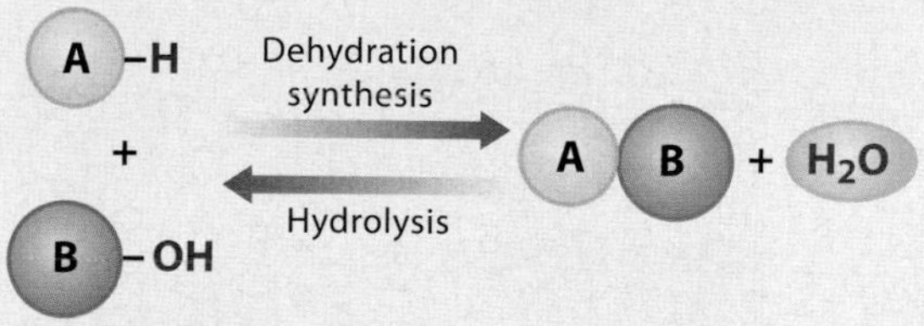
A –H
+
B –OH
Dehydration synthesis
Hydrolysis
A B + H₂O

High heat capacity

Heat capacity is the quantity of heat required to raise the temperature of a unit mass of a substance 1°C. Water has an unusually high heat capacity, because water molecules in the solid and the liquid state are attracted to one another through hydrogen bonding. Some features of a high heat capacity are listed at right:

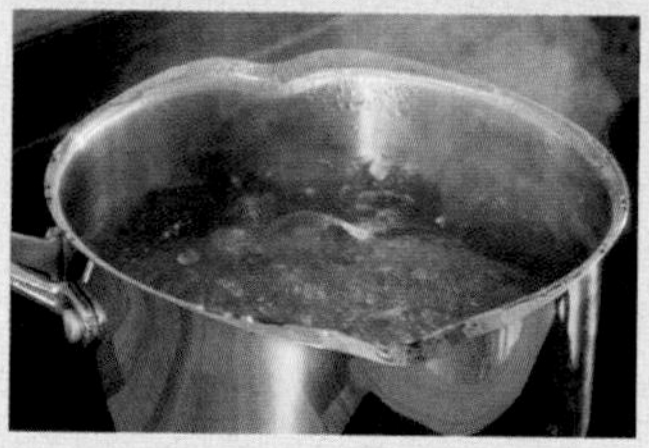

- Water temperature must be high for individual molecules to develop enough kinetic energy to break all the hydrogen bonds to become water vapor, a gas.
- Water carries a great deal of heat away with it when it changes from a liquid to a gas. This feature accounts for the cooling effect of perspiration on the skin.
- A large mass of water changes temperature very slowly. This property is called **thermal inertia**.

Solubility

A remarkable number of inorganic and organic molecules will dissolve in water. The individual particles become dispersed within the water, and the result is a **solution**—a uniform mixture of two or more substances. The liquid in which other atoms, ions, or molecules are distributed is called the **solvent**. The dissolved substances are the **solutes**. In **aqueous solutions**, water is the solvent.

Module 2.10 Review

Predict how an exercising student's body could use water for:

a. lubrication
b. reactivity
c. cooling and solubility

LO 2.10 Describe four important properties of water and their significance in the body.

Module 2.11

Physiological systems depend on water

Many inorganic compounds are held together partially or completely by ionic bonds. In water, these compounds undergo **dissociation** (di-sō-sē-Ā-shun) or **ionization** (ī-on-ī-ZĀ-shun). In this process, ionic bonds are broken as the individual ions interact with the positive or negative poles of polar water molecules.

1 A water molecule is said to be polar because it has positive and negative poles. This polarity is due to the asymmetrical positions of the hydrogen atoms that are attached by polar covalent bonds.

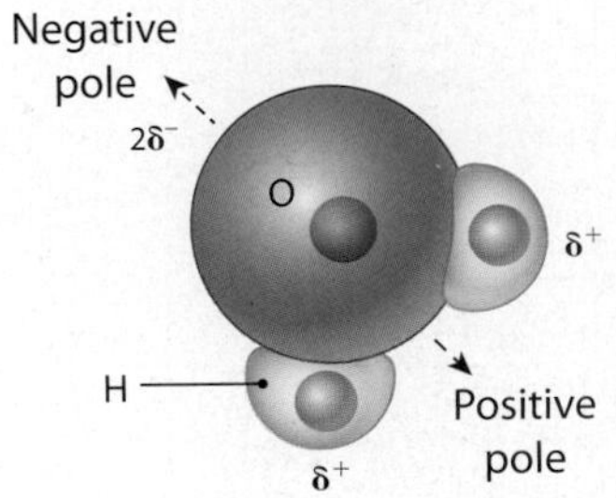

Negative pole
2δ⁻
O
δ+
H
Positive pole
δ+

2 In solution, an ionic compound dissociates as water molecules break them apart. The anions are surrounded by the positive poles of water molecules, and the cations are surrounded by the negative poles of water molecules. The sheath of water molecules around an ion in solution is called a **hydration sphere**.

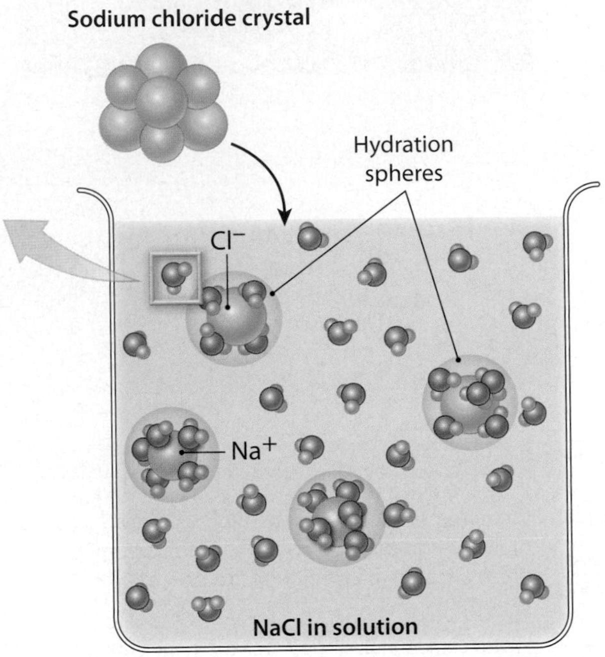

Sodium chloride crystal
Hydration spheres
Cl⁻
Na⁺
NaCl in solution

3 Hydration spheres also form around an organic molecule containing polar covalent bonds. If the molecule binds water strongly, as does glucose, it will be carried into solution—in other words, it will dissolve. Molecules that interact readily with water molecules in this way are called **hydrophilic** (hī-drō-FIL-ik; *hydro-*, water + *philos*, loving). Glucose, an important soluble sugar, is one example.

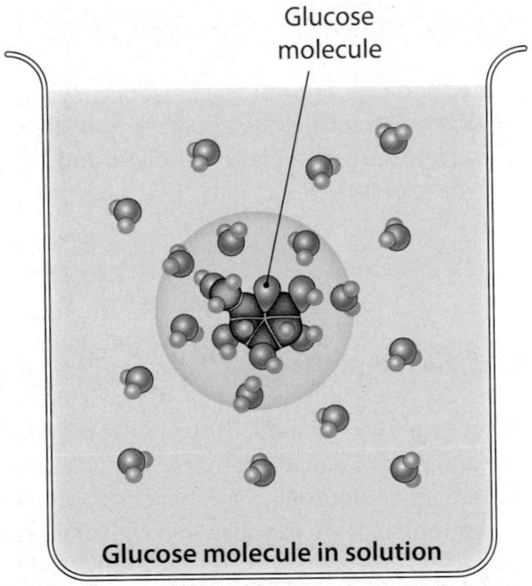

Glucose molecule
Glucose molecule in solution

4 An aqueous solution containing anions and cations will conduct an electrical current. Soluble inorganic substances whose ions will conduct an electrical current in solution are called **electrolytes** (e-LEK-trō-līts). Sodium chloride is an important electrolyte in body fluids. In an electrical field, cations in solution will move toward the negative side, or negative terminal, and anions will move toward the positive terminal. Electrical forces across plasma membranes affect the functioning of all cells. Small electrical currents carried by ions are essential to muscle contraction and nerve function, two topics that will be discussed in later chapters.

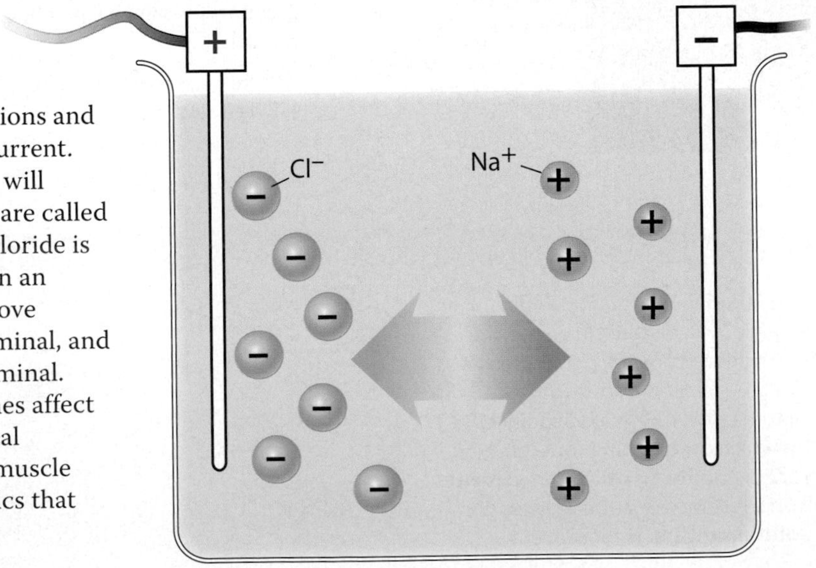

+
−
Cl⁻
Na⁺

5 The table at right lists the most important electrolytes and their dissociation products in the body. The dissociation of electrolytes in blood and other body fluids releases a variety of ions. Changes in the concentrations of electrolytes in body fluids will disturb almost every vital function. For example, declining potassium levels will lead to a general muscular paralysis, and rising concentrations will cause weak and irregular heartbeats. The concentrations of ions in body fluids are carefully regulated, primarily by the coordination of activities at the kidneys (ion excretion), the digestive tract (ion absorption), and the skeletal system (ion storage or release).

Important Electrolytes That Dissociate in Body Fluids

Electrolyte		Ions Released
NaCl (sodium chloride)	→	$Na^+ + Cl^-$
KCl (potassium chloride)	→	$K^+ + Cl^-$
$CaPO_4$ (calcium phosphate)	→	$Ca^{2+} + PO_4^{2-}$
$NaHCO_3$ (sodium bicarbonate)	→	$Na^+ + HCO_3^-$
$MgCl_2$ (magnesium chloride)	→	$Mg^{2+} + 2\ Cl^-$
Na_2HPO_4 (sodium hydrogen phosphate)	→	$2\ Na^+ + HPO_4^{2-}$
Na_2SO_4 (sodium sulfate)	→	$2\ Na^+ + SO_4^{2-}$

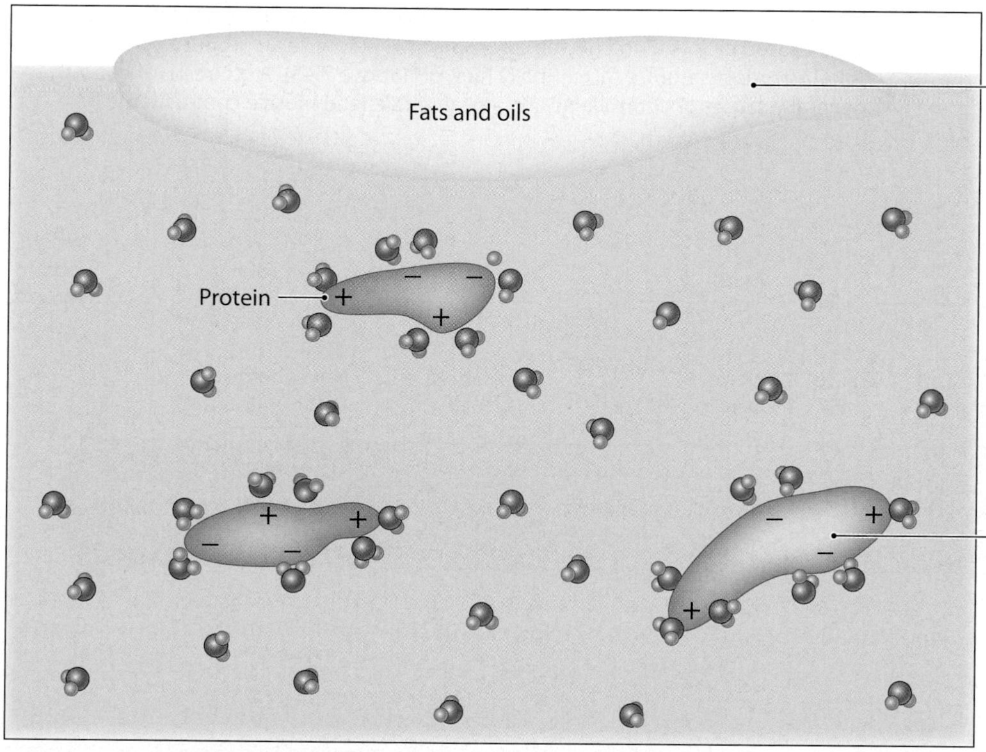

Fats and oils
Protein

Many organic molecules either lack polar covalent bonds or have very few. Such molecules do not have positive and negative poles and are thus nonpolar. When nonpolar molecules are exposed to water, hydration spheres do not form and the molecules do not dissolve. Molecules that do not readily interact with water are called **hydrophobic** (hī-drō-FŌ-bik; *hydro-*, water + *phobos*, fear). Fats and oils of all kinds are some of the most familiar hydrophobic molecules.

Body fluids typically contain large and complex organic molecules, such as proteins, that are held in solution by their association with water molecules.

6 A solution containing dispersed proteins or other large molecules is called a **colloid**. The particles or molecules in a colloid will remain in solution indefinitely. Liquid Jell-O is a familiar, viscous (thick) colloid. In contrast, a **suspension** contains large particles in solution, but if undisturbed, its particles will settle out of solution due to the force of gravity. Whole blood is a temporary suspension, because the blood cells are suspended in the blood plasma. If clotting is prevented, the cells in a blood sample will gradually settle to the bottom of the container.

Module 2.11 Review

- **a.** Explain how the ionic compound sodium chloride dissolves in water.
- **b.** Define electrolytes.
- **c.** Distinguish between hydrophilic and hydrophobic molecules.

LO 2.11 Explain how the chemical properties of water affect the solubility of inorganic and organic molecules.

Regulation of body fluid pH is vital for homeostasis

A hydrogen atom involved in a chemical bond or participating in a chemical reaction can easily lose its electron to become a **hydrogen ion, H^+**. Hydrogen ions are extremely reactive in solution. In excessive numbers, they will break chemical bonds, change the shapes of complex molecules, and generally disrupt cell and tissue functions. As a result, the concentration of hydrogen ions in body fluids must be regulated precisely.

1 A few hydrogen ions are normally present even in a sample of pure water, because some of the water molecules dissociate spontaneously, releasing a hydrogen ion, H^+, and a **hydroxide** (hī-DROK-sīd) **ion, OH^-**.

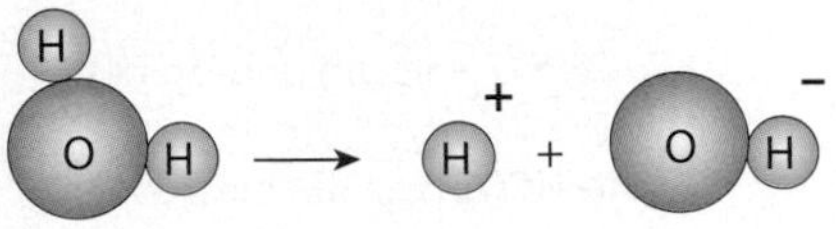
H
O H
H
+
O H
−

2 The hydrogen ion concentration in body fluids is so important to physiological processes that a special shorthand is used to express it. The **pH** of a solution is defined as the negative logarithm* of the hydrogen ion concentration in moles per liter (mol/L). For common liquids, the pH scale ranges from 0 to 14.

Blood

The pH of blood normally ranges from 7.35 to 7.45. Abnormal fluctuations in pH can damage cells and tissues by breaking chemical bonds, changing the shapes of proteins, and altering cellular functions. **Acidosis** is an abnormal physiological state caused by low blood pH (below 7.35). A pH below 7 can produce coma. **Alkalosis** results from an abnormally high pH (above 7.45). A blood pH above 7.8 generally causes uncontrollable and sustained skeletal muscle contractions.

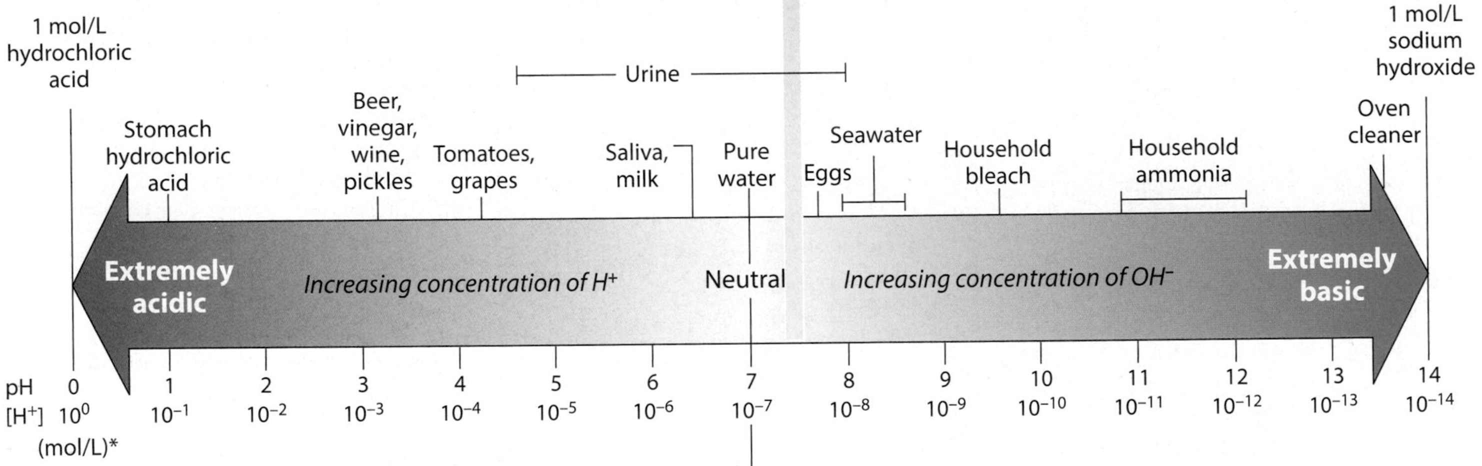
1 mol/L hydrochloric acid
Stomach hydrochloric acid
Beer, vinegar, wine, pickles
Tomatoes, grapes
Urine
Saliva, milk
Pure water
Eggs
Seawater
Household bleach
Household ammonia
Oven cleaner
1 mol/L sodium hydroxide
Extremely acidic
Increasing concentration of H+
Neutral
Increasing concentration of OH−
Extremely basic
pH 0 1 2 3 4 5 6 7 8 9 10 11 12 13 14
[H+] 10^0 10^−1 10^−2 10^−3 10^−4 10^−5 10^−6 10^−7 10^−8 10^−9 10^−10 10^−11 10^−12 10^−13 10^−14
(mol/L)*

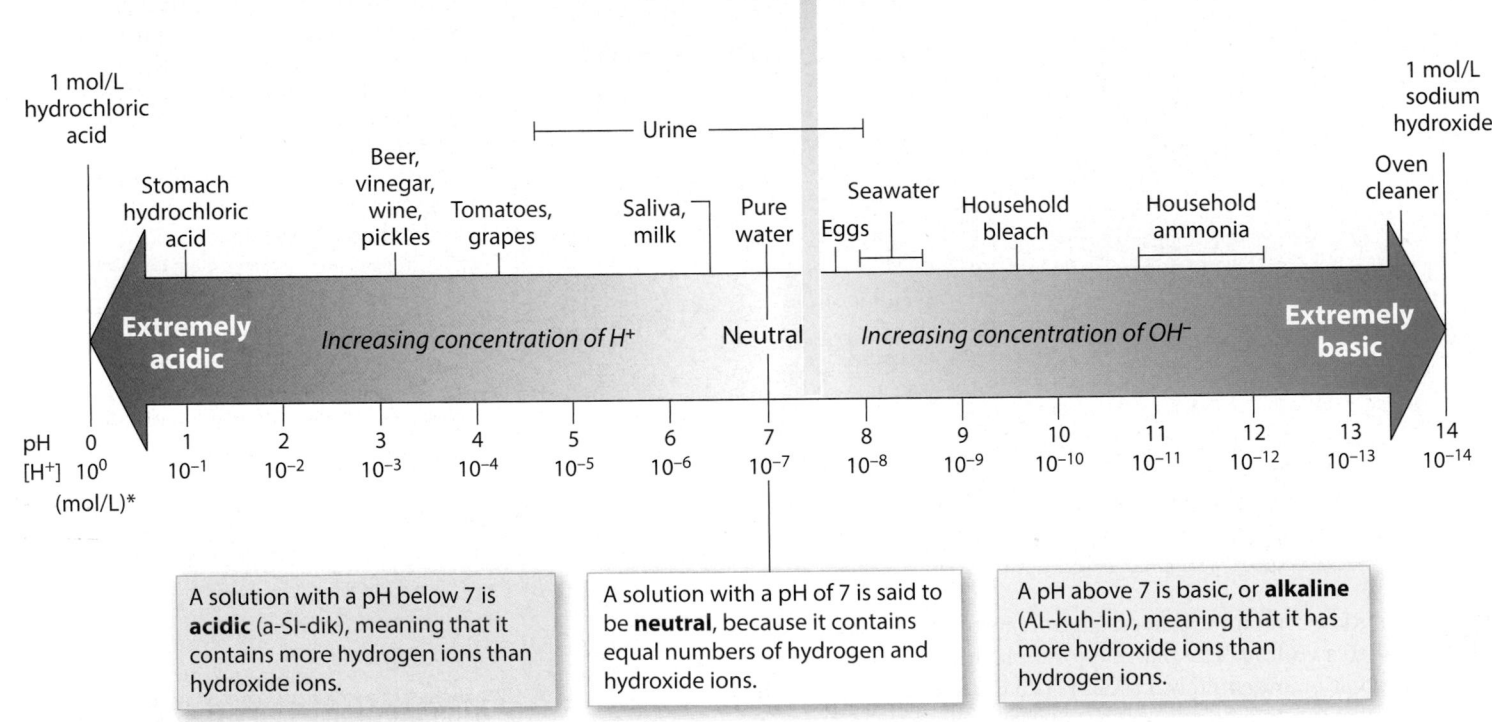

A solution with a pH below 7 is **acidic** (a-SI-dik), meaning that it contains more hydrogen ions than hydroxide ions.

A solution with a pH of 7 is said to be **neutral**, because it contains equal numbers of hydrogen and hydroxide ions.

A pH above 7 is basic, or **alkaline** (AL-kuh-lin), meaning that it has more hydroxide ions than hydrogen ions.

* One liter of pure water contains about 0.0000001 mol of hydrogen ions and an equal number of hydroxide ions. In other words, the concentration of hydrogen ions in a solution of pure water is 0.0000001 mol per liter. This can be written as $[H^+] = 10^{-7}$ mol/L. The brackets around the H^+ signify "the concentration of," another example of chemical notation. The negative logarithm of 10^{-7} is equal to $(-)(-7) = +7$.

Acids and Bases

3 An **acid** is any solute that dissociates in solution and releases hydrogen ions, thereby lowering the pH. Because a hydrogen atom that loses its electron consists solely of a proton, hydrogen ions are often referred to simply as protons, and acids as proton donors. A strong acid dissociates completely in solution, and the reaction occurs essentially in one direction only. **Hydrochloric acid** (HCl) is a representative strong acid; in water, it ionizes as follows:

$$\mathbf{HCl \longrightarrow H^+ + Cl^-}$$

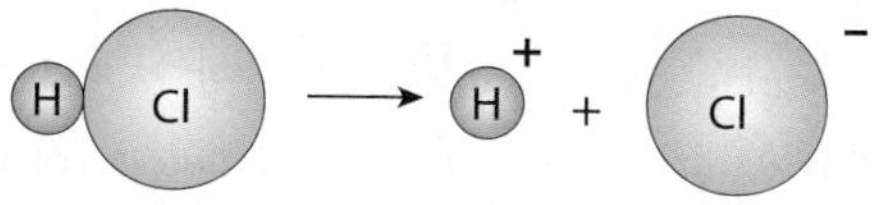
H
Cl
H
+
Cl

4 A **base** is a solute that removes hydrogen ions from a solution and thereby acts as a proton acceptor. In solution, many bases release a hydroxide ion. Hydroxide ions have a strong affinity for hydrogen ions and react quickly with them to form water molecules. A strong base dissociates completely in solution. **Sodium hydroxide** (NaOH) is a strong base. In solution, it releases sodium ions and hydroxide ions:

$$\mathbf{NaOH \longrightarrow Na^+ + OH^-}$$

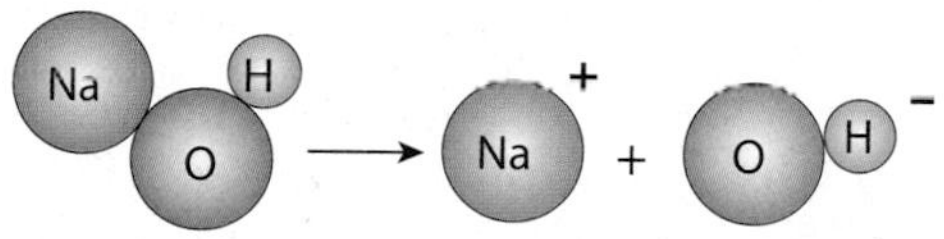
Na
H
O
Na
+
O
H

5 Weak acids and weak bases fail to dissociate completely. At equilibrium, a significant number of molecules remain intact in the solution. For a given number of molecules in solution, weak acids and weak bases therefore have less of an impact on pH than do strong acids and strong bases. **Carbonic acid** (H_2CO_3) is a weak acid found in body fluids. In solution, carbonic acid reversibly dissociates into a hydrogen ion and a **bicarbonate ion** (HCO_3^-).

$$\mathbf{H_2CO_3 \rightleftharpoons H^+ + HCO_3^-}$$

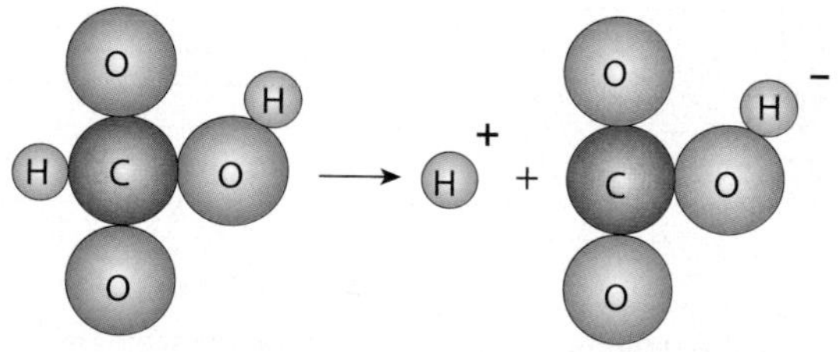
O
H
H
C
O
O
H
+
O
H
C
O
O

Salts

6 A **salt** is an ionic compound consisting of any cation except a hydrogen ion and any anion except a hydroxide ion. Because they are held together by ionic bonds, many salts dissociate completely in water, releasing cations and anions. For example, sodium chloride (table salt) dissociates immediately in water, releasing Na^+ and Cl^-, the most abundant ions in body fluids. The ionization of sodium chloride does not affect the local concentrations of hydrogen ions or hydroxide ions, so NaCl, like many salts, is a "neutral" solute. Other salts may indirectly affect the concentrations of H^+ and OH^-, making a solution slightly acidic or slightly basic.

$$\mathbf{NaCl \longrightarrow Na^+ + Cl^-}$$

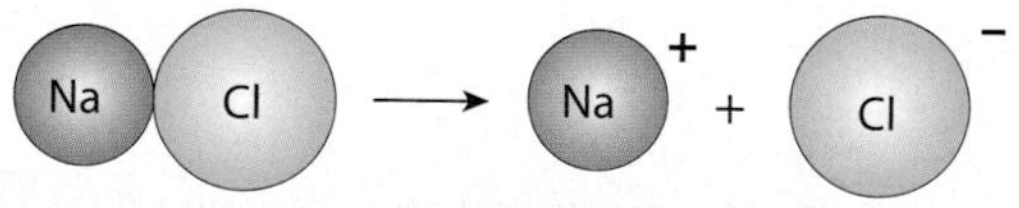
Na
Cl
Na
+
Cl

Buffers

7 **Buffers** are compounds that stabilize the pH of a solution by removing or replacing hydrogen ions. **Buffer systems** typically involve a weak acid and its related salt, which functions as a weak base. For example, the body's carbonic acid–bicarbonate buffer system consists of carbonic acid (H_2CO_3) and sodium bicarbonate, ($NaHCO_3$), otherwise known as baking soda. Buffers and buffer systems in body fluids help maintain pH within normal limits.

Module 2.12 Review

a. Define pH.

b. What is the significance of pH in physiological systems?

c. Explain the differences among an acid, a base, and a salt.

LO **2.12** Discuss the importance of pH and the role of buffers in body fluids.

Matching

Match each lettered term with the most closely related description.

a. solvent
b. water
c. buffers
d. hydrophilic
e. inorganic compounds
f. hydrophobic
g. acid
h. solute
i. alkaline
j. salt

1 HCl, NaOH, and NaCl
2 A dissolved substance
3 A solution with a pH greater than 7
4 Molecules that readily interact with water
5 Fluid medium of a solution
6 Ionic compound not containing hydrogen ions or hydroxide ions
7 Compounds that stabilize pH in body fluids
8 Solution with a pH of 6.5
9 Molecules that do not interact with water
10 Makes up two-thirds of human body weight

1 ____
2 ____
3 ____
4 ____
5 ____
6 ____
7 ____
8 ____
9 ____
10 ____

Short answer

List four properties of water important to the functioning of the human body.

11 ____
12 ____
13 ____
14 ____

Identify the regions 15, 16, and 17 on the pH scale below:

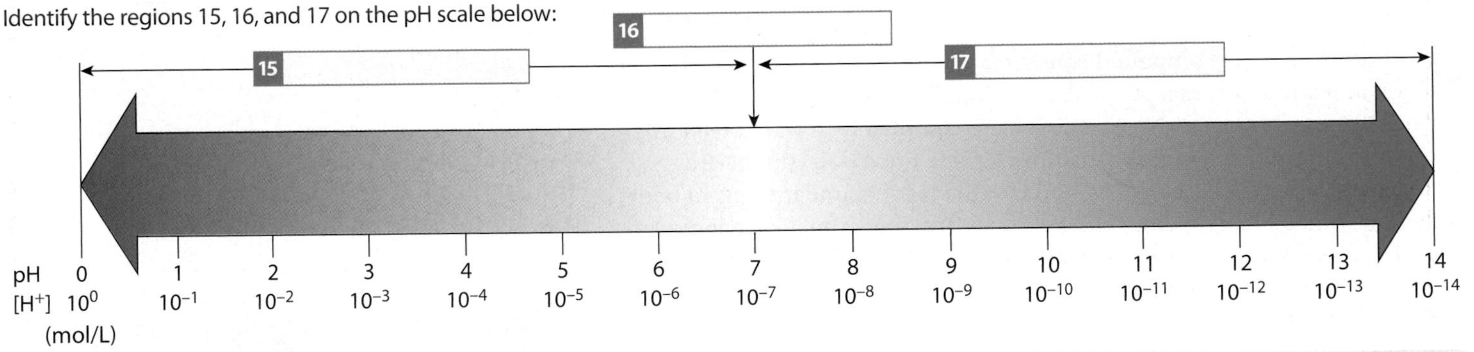
16
15
17
pH 0 1 2 3 4 5 6 7 8 9 10 11 12 13 14
[H+] (mol/L) 10^0 10^{-1} 10^{-2} 10^{-3} 10^{-4} 10^{-5} 10^{-6} 10^{-7} 10^{-8} 10^{-9} 10^{-10} 10^{-11} 10^{-12} 10^{-13} 10^{-14}

18 How much more or less acidic is a solution of pH 3 compared to one with a pH of 6?

19 Describe three negative effects of abnormal pH fluctuations in the human body.

18 ____

19 ____

Section integration

The addition of table salt to pure water does not result in a change in its pH. Why?

20 ____

All organic compounds contain carbon and hydrogen atoms

Organic compounds always contain the elements carbon and hydrogen, and generally oxygen as well. Many organic molecules are made up of long chains of carbon atoms linked by covalent bonds. The carbon atoms typically form additional covalent bonds with hydrogen or oxygen atoms and, less commonly, with nitrogen, phosphorus, sulfur, iron, or other elements. Many organic molecules are soluble in water. Although organic compounds are diverse, certain groupings of atoms occur again and again, even in very different types of molecules. These **functional groups** greatly influence the properties of any molecule in which they occur.

Important Functional Groups of Organic Compounds

Functional Group	Structural Formula *†	Importance	Examples
Amino group, $-NH_2$	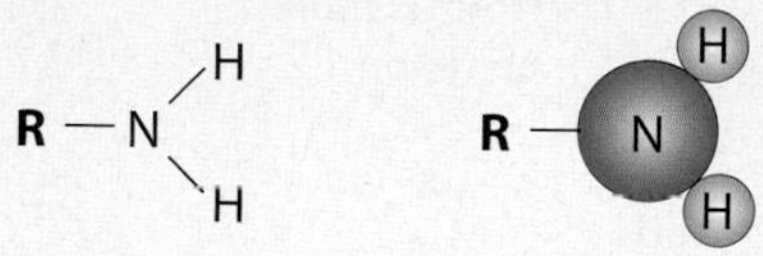	Acts as a base, accepting H^+, depending on pH; can form bonds with other molecules	• Amino acids
Carboxyl group, $-COOH$	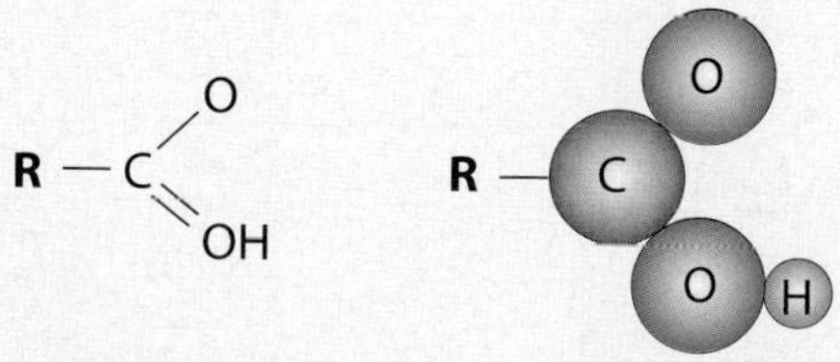	Acts as an acid, releasing H^+ to become $R\text{–}COO^-$	• Fatty acids • Amino acids
Hydroxyl group, $-OH$		May link molecules through dehydration synthesis; hydrogen bonding between hydroxyl groups and water molecules affects solubility	• Carbohydrates • Fatty acids • Amino acids • Alcohols
Phosphate group, $-PO_4$	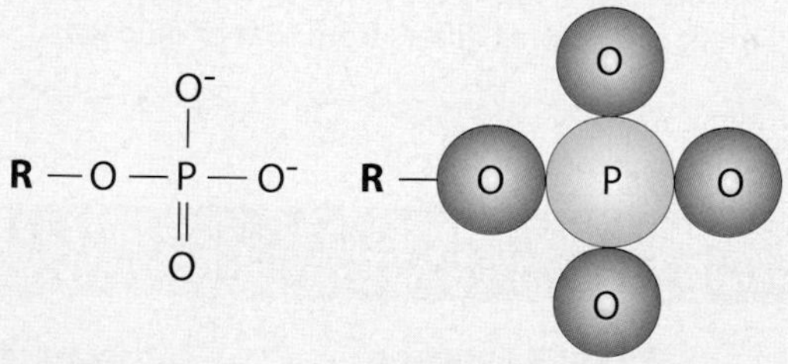	May link other molecules to form larger structures; may store energy	• Phospholipids • Nucleic acids • High-energy compounds

* A structural formula shows the covalent bonds within a molecule or functional group. For example, a single covalent bond is represented by a single line (—), a double bond by two parallel lines (=). Although not shown here, a triple bond is represented by three parallel lines (≡).

† **R** represents R group, which denotes the rest of the molecule to which a functional group is attached.

In this section we will introduce the major classes of organic compounds. We will also consider how enzymes facilitate essential reactions within living cells, and how cells capture and transfer energy with special **high-energy compounds**.

Module 2.13 Review

a. List the elements that construct organic compounds.

b. What is a functional group?

c. Identify the important functional groups of organic compounds.

2.13 Describe the common elements of organic compounds and how functional groups modify the properties of organic compounds.

Carbohydrates contain carbon, hydrogen, and oxygen, usually in a 1:2:1 ratio

A **carbohydrate** is an organic molecule that contains carbon, hydrogen, and oxygen in a ratio near 1:2:1. Familiar carbohydrates include the sugars and starches that make up about half of the typical U.S. diet. Carbohydrates typically account for less than 1.5 percent of total body weight. Although they may have other functions, carbohydrates are most important as energy sources that are catabolized rather than stored.

Monosaccharides

1 A simple sugar, or **monosaccharide** (mon-ō-SAK-uh-rīd; *mono-*, single + *sakcharon*, sugar), is a carbohydrate containing from three to seven carbon atoms. A monosaccharide can be called a triose (three-carbon), tetrose (four-carbon), pentose (five-carbon), hexose (six-carbon), or heptose (seven-carbon). The hexose **glucose** (GLOO-kōs) is the most important metabolic "fuel" in the body.

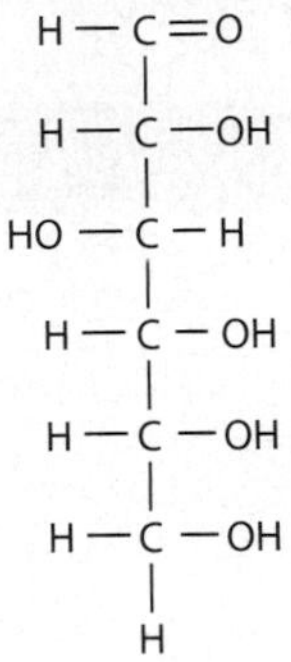

The atoms in a glucose molecule may form either a straight chain or a ring. In the body, the ring form is more common.

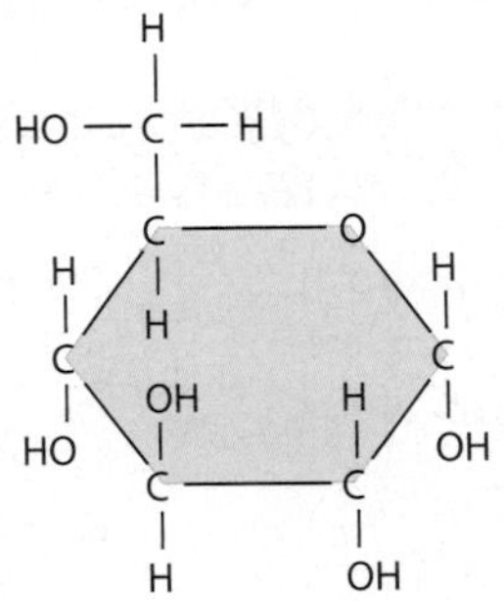

2 The three-dimensional structure of an organic molecule is an important characteristic, because it usually determines the molecule's fate or function. Some molecules have the same molecular formula—in other words, the same types and numbers of atoms—but different structures. Such molecules are called **isomers**. The body usually treats different isomers as distinct molecules.

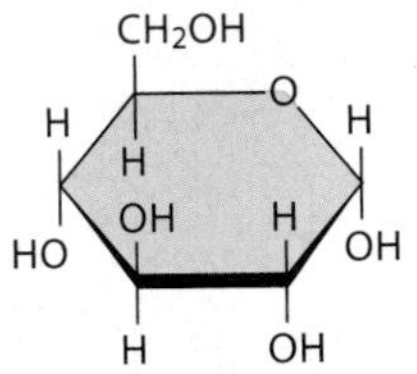

Glucose

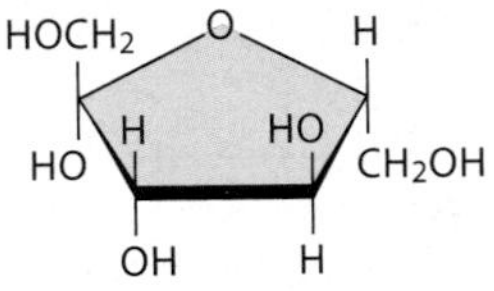

Fructose

The monosaccharides glucose and **fructose** are isomers. Fructose is a hexose found in many fruits. Although its chemical formula, $C_6H_{12}O_6$, is the same as that of glucose, the arrangement of its atoms differs from that of glucose.

Carbohydrates in the Body

Structural Class	Examples	Primary Function	Remarks
Monosaccharides (simple sugars)	Glucose, fructose	Energy source	Manufactured in the body and obtained from food; distributed in body fluids
Disaccharides	Sucrose, lactose, maltose	Energy source	Sucrose is table sugar, lactose is in milk, and maltose is malt sugar; all must be broken down to monosaccharides before absorption
Polysaccharides	Glycogen	Glucose storage	Glycogen is in animal cells; other starches and cellulose are within or around plant cells

Disaccharides

Two monosaccharides joined together form a **disaccharide** (dī-SAK-uh-rīd; *di-*, two). Disaccharides such as **sucrose** (table sugar) have a sweet taste and, like monosaccharides, are quite soluble in water. The formation of sucrose involves dehydration synthesis.

3 During dehydration synthesis, two molecules are joined by the removal of a water molecule.

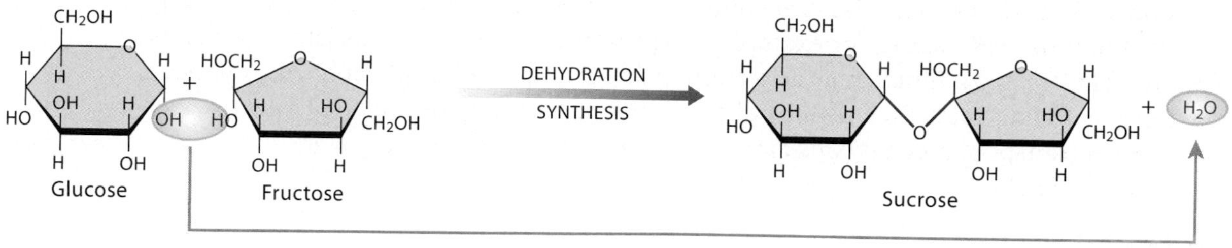

Glucose
Fructose
DEHYDRATION SYNTHESIS
Sucrose
H_2O

4 Hydrolysis reverses the steps of dehydration synthesis. A complex molecule is broken down by the addition of a water molecule.

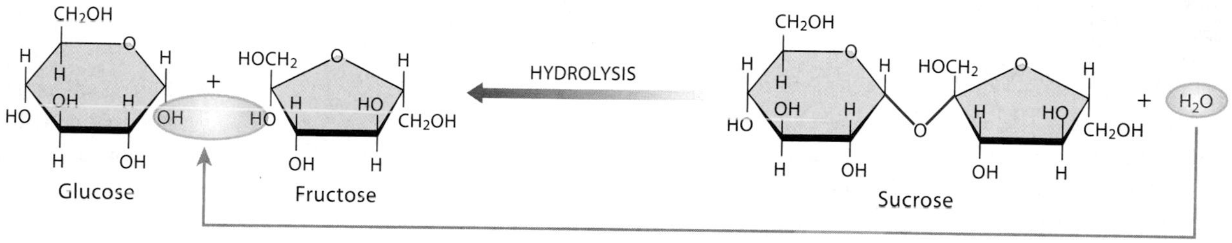

Glucose
Fructose
HYDROLYSIS
Sucrose
H_2O

Polysaccharides

5 More complex carbohydrates result when repeated dehydration synthesis reactions add additional monosaccharides or disaccharides. These large molecules are called **polysaccharides** (pol-ē-SAK-uh-rīdz; *poly-*, many). **Starches** are large polysaccharides formed from glucose molecules. Your digestive tract can break these molecules into monosaccharides. Starches such as those in potatoes and grains are a major dietary energy source.

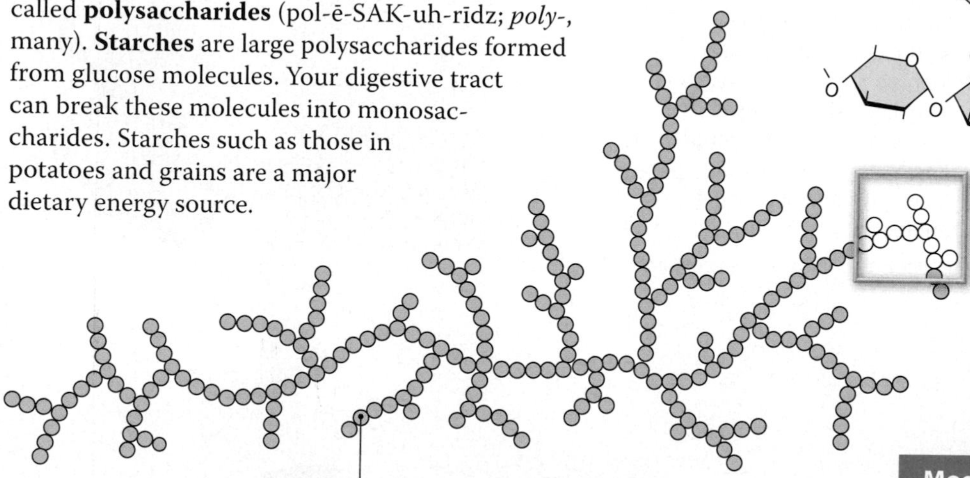

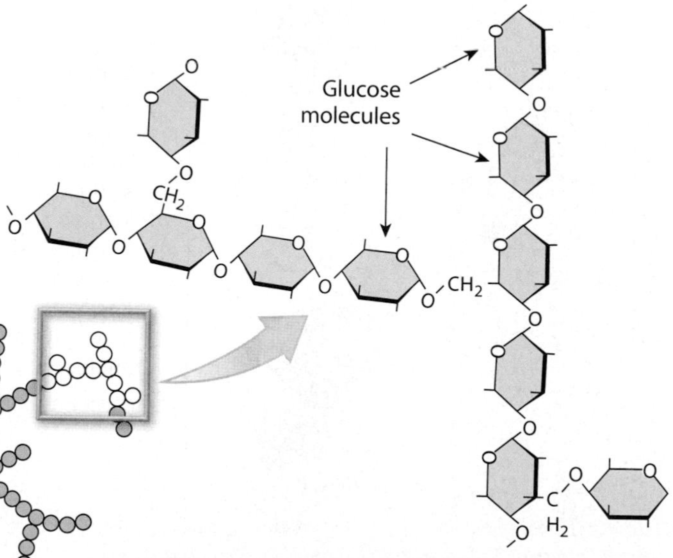

Glucose molecules

The polysaccharide **glycogen** (GLĪ-kō-jen), or animal starch, has many side branches consisting of chains of glucose molecules. Muscle cells make and store glycogen. When these cells have a high demand for glucose, glycogen molecules are broken down. When the need is low, they absorb glucose from the bloodstream and rebuild glycogen reserves.

Module 2.14 Review

a. A food contains organic molecules with the elements C, H, and O in a ratio of 1:2:1. What class of compounds do these molecules belong to, and what are their major functions in the body?

b. List the three structural classes of carbohydrates, and give an example of each.

c. Predict the reactants and the type of chemical reaction involved when muscle cells make and store glycogen.

LO 2.14 Discuss the structures and functions of carbohydrates.

Module 2.15

Lipids often have a carbon-to-hydrogen ratio of 1:2

Like carbohydrates, **lipids** (*lipos*, fat) contain carbon, hydrogen, and oxygen, and the carbon-to-hydrogen ratio is typically near 1:2. However, lipids contain much less oxygen than do carbohydrates with the same number of carbon atoms. The hydrogen-to-oxygen ratio is therefore very large. For example, a representative lipid, such as lauric acid, has a formula of $C_{12}H_{24}O_2$. Lipids may also contain small quantities of phosphorus, nitrogen, or sulfur. Familiar lipids include fats, oils, and waxes. Most lipids are insoluble in water, but special transport mechanisms carry them into the bloodstream.

Fatty Acids

1 **Fatty acids** are long carbon chains with hydrogen atoms attached. One end of the carbon chain, called the head, is attached to a **carboxyl** (kar-BOK-sil) **group**: —COOH.

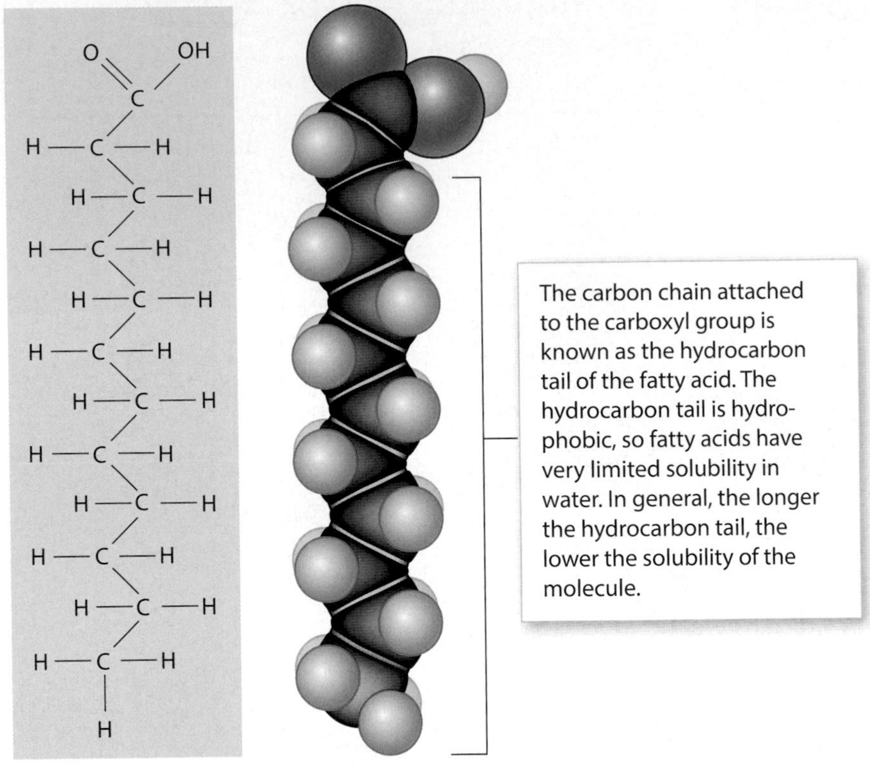
The carbon chain attached to the carboxyl group is known as the hydrocarbon tail of the fatty acid. The hydrocarbon tail is hydrophobic, so fatty acids have very limited solubility in water. In general, the longer the hydrocarbon tail, the lower the solubility of the molecule.

Lauric acid ($C_{12}H_{24}O_2$)

2 In a **saturated fatty acid**, each carbon atom in the tail has four single covalent bonds.

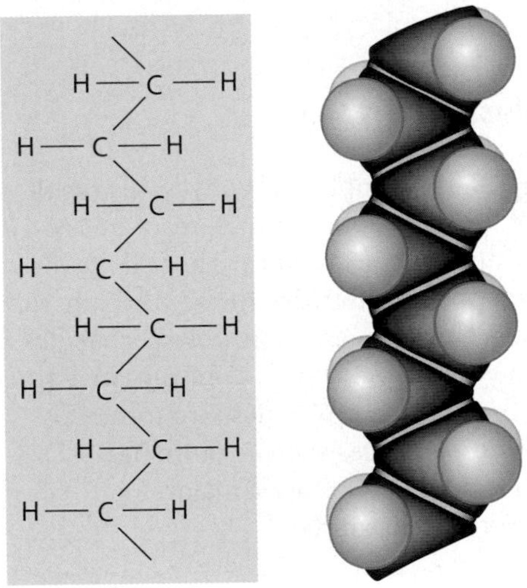

3 In an **unsaturated fatty acid**, one or more of the single covalent bonds between carbon atoms has been replaced by a double covalent bond. As a result, each carbon atom involved will bind only one hydrogen atom rather than two. This changes both the shape of the hydrocarbon tail and the way the fatty acid is metabolized. A monounsaturated fatty acid has a single double bond in the hydrocarbon tail. A polyunsaturated fatty acid contains multiple double bonds.

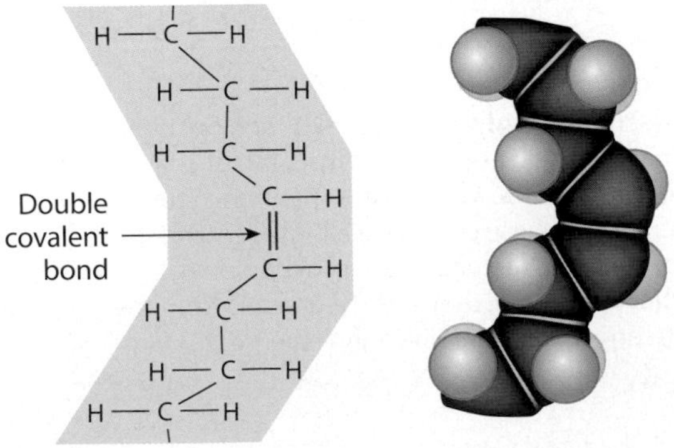
Double covalent bond

Representative Lipids in the Body

Lipid Type	Examples	Primary Functions	Remarks
Fatty acids	Lauric acid	Energy sources	Absorbed from food or synthesized in cells; transported in the blood
Glycerides	Monoglycerides, diglycerides, triglycerides	Energy sources, energy storage, insulation, and physical protection	Stored in fat deposits; must be broken down to fatty acids and glycerol before they can be used as an energy source
Eicosanoids (see Module 2.16)	Prostaglandins, leukotrienes	Chemical messengers coordinating local cellular activities	Prostaglandins are produced in most body tissues
Steroids (see Module 2.16)	Cholesterol	Structural components of cell membranes, hormones, digestive secretions in bile	All steroids have the same carbon ring framework
Phospholipids, glycolipids (see Module 2.16)	Lecithin (a phospholipid)	Structural components of cell membranes	Derived from fatty acids and nonlipid components

Glycerides

4 Individual fatty acids cannot be strung together in a chain by dehydration synthesis. But they can be attached to another compound, **glycerol** (GLIS-er-ol), through a similar reaction. The result is a lipid known as a **glyceride** (GLIS-er-īd).

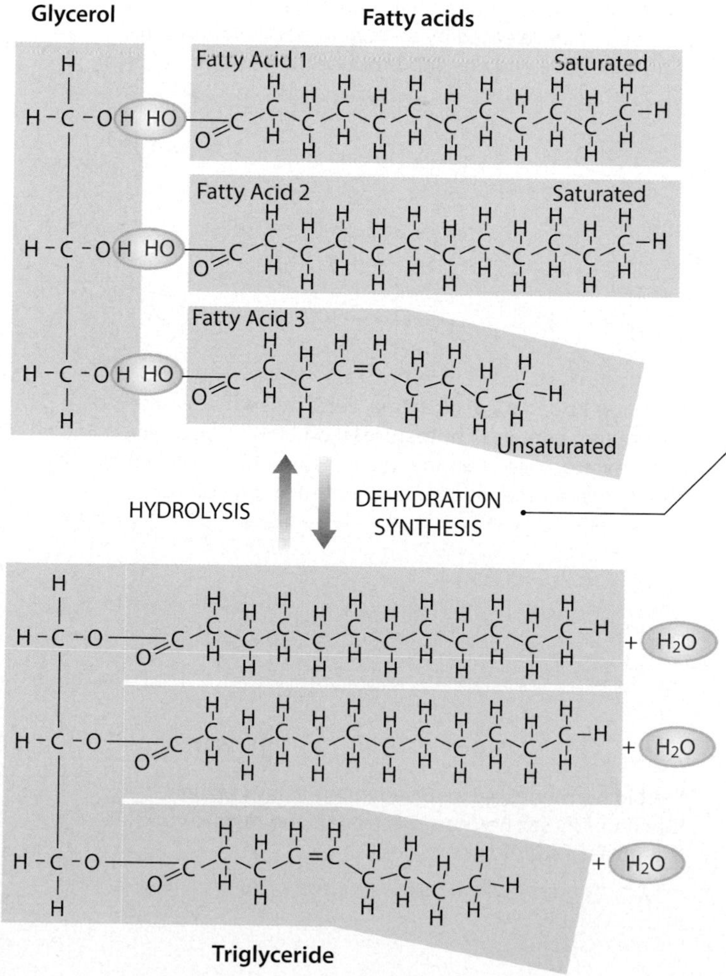

Glycerol
Fatty acids
Fatty Acid 1
Saturated
Fatty Acid 2
Saturated
Fatty Acid 3
Unsaturated
HYDROLYSIS
DEHYDRATION SYNTHESIS
Triglyceride

Dehydration synthesis can produce a **monoglyceride** (mon-ō-GLI-ser-īd), consisting of glycerol + one fatty acid. Subsequent reactions can yield a **diglyceride** (glycerol + two fatty acids) and then a **triglyceride** (glycerol + three fatty acids). Triglycerides are also known as triacylglycerols or neutral fats. Hydrolysis breaks the glycerides into fatty acids and glycerol.

Lipids form essential components of all cells. In addition, lipid deposits are important as energy reserves. On average, lipids provide twice as much energy as carbohydrates do, gram for gram, when broken down in the body. Lipids normally account for 12–18 percent of total body weight of adult men, and 18–24 percent in adult women. The human body cannot synthesize all the lipids it needs, and several fatty acids must be obtained from the diet.

Module 2.15 Review

a. Describe lipids.

b. Describe the structures of saturated and unsaturated fatty acids.

c. In the hydrolysis of a triglyceride, what are the reactants and the products?

LO 2.15 Discuss the structures and functions of lipids.

Eicosanoids, steroids, phospholipids, and glycolipids have diverse functions

Lipids are important as chemical messengers and as components of cellular structures. **Structural lipids** help form and maintain a cell's surrounding membrane and its intracellular membranes. At the cellular level, membranes are sheets or layers composed primarily of hydrophobic lipids. Functionally, a membrane is an effective barrier that can separate two aqueous solutions of differing composition.

Eicosanoids

1 **Eicosanoids** (ī-KŌ-sa-noydz) are lipids derived from arachidonic (ah-rak-i-DON-ik) acid, a fatty acid that must be absorbed from food because it cannot be synthesized by the body. **Leukotrienes** (lū-kō-TRĪ-ēnz) are produced primarily by cells involved with coordinating the responses to injury or disease, and they will be considered in later chapters.

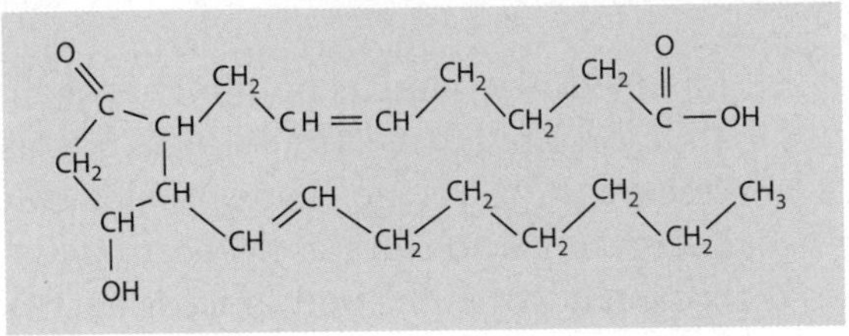

Prostaglandins (pros-tuh-GLAN-dinz) are short-chain fatty acids in which five of the carbon atoms are joined in a ring. These compounds are released by cells to coordinate or direct local cellular activities, and they are extremely powerful, even in minute quantities. Small amounts of prostaglandins released by damaged tissues, for example, stimulate nerve endings and produce the sensation of pain.

Steroids

2 **Steroids** are large lipid molecules that share a distinctive carbon-ring framework. They differ in the functional groups that are attached to the basic ring structure.

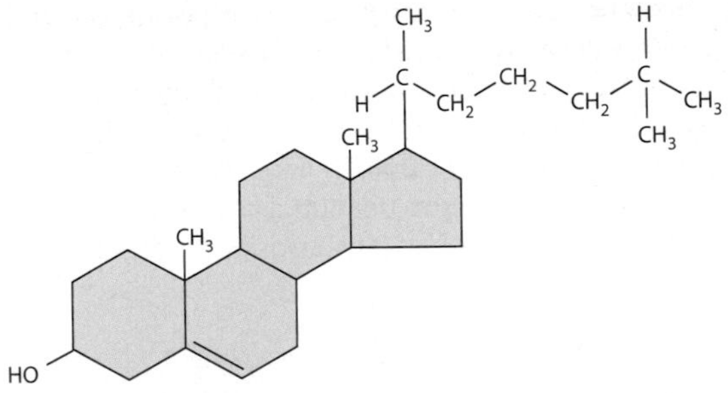

Cholesterol

The outer boundary of all animal cells, called the plasma membrane, contains **cholesterol** (kōh-LES-ter-ol; *chole-*, bile + *stereos*, solid). Cells need cholesterol to maintain their plasma membranes and for cell growth and division.

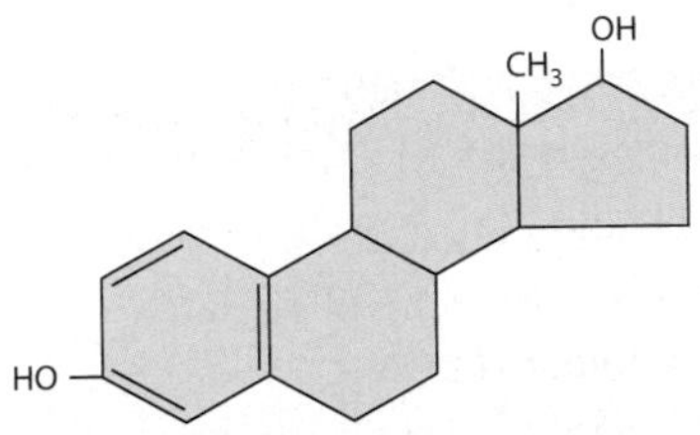

Estrogen

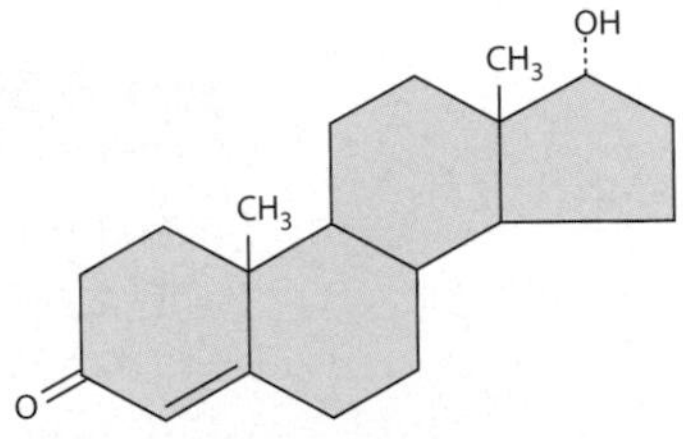

Testosterone

Steroid hormones are involved in regulating sexual function. The sex hormones estrogen and testosterone are examples of steroid hormones.

Phospholipids and Glycolipids

3 **Phospholipids** (FOS-fō-lip-idz) and **glycolipids** (GLĪ-kō-lip-idz) are structurally related, and our cells can synthesize both types of lipids, primarily from fatty acids. Cholesterol, phospholipids, and glycolipids are considered structural lipids because of their roles in membrane structure.

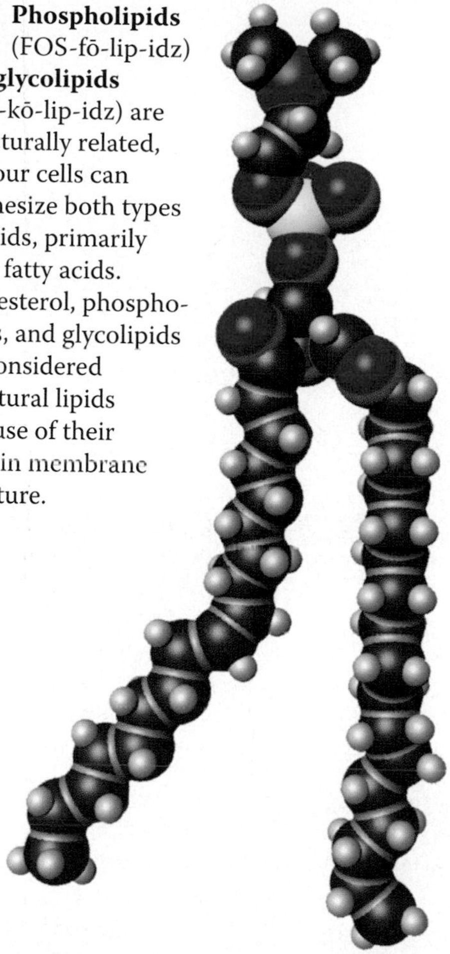

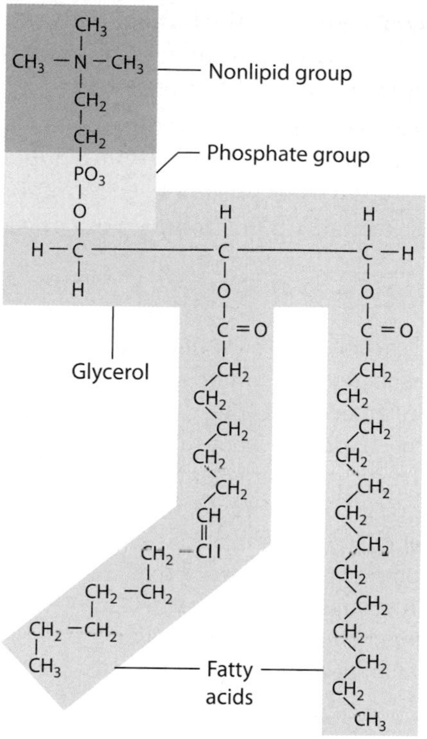

Nonlipid group
Phosphate group
Glycerol
Fatty acids

In a phospholipid, a phosphate group links a diglyceride to a nonlipid group.

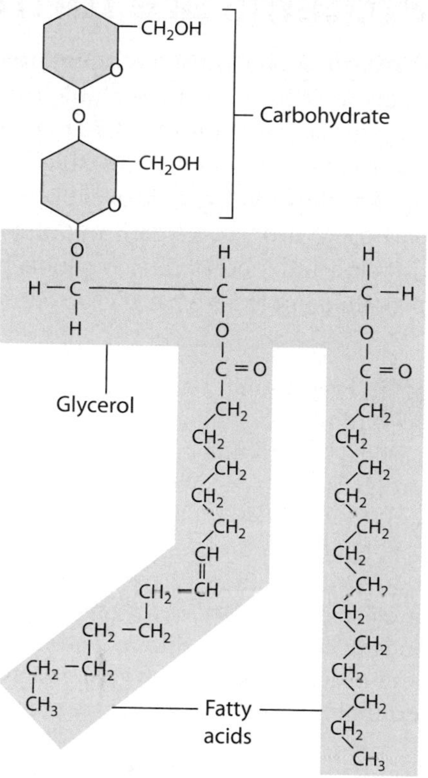

Carbohydrate
Glycerol
Fatty acids

In a glycolipid, a carbohydrate is attached to a diglyceride.

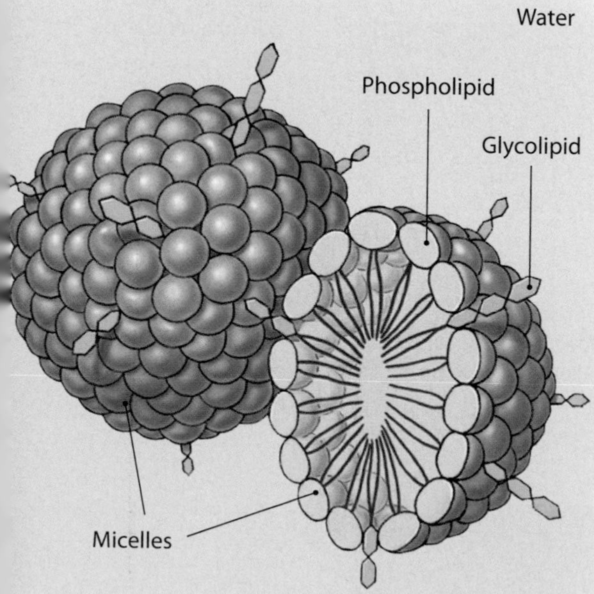

Water
Phospholipid
Glycolipid
Micelles

4 The long hydrocarbon tails of phospholipids and glycolipids are hydrophobic, but the opposite ends, the nonlipid heads, are hydrophilic. In water, large numbers of these molecules tend to form droplets, or **micelles** (mī-SELZ), with the hydrophilic portions on the outside. Most meals contain a mixture of lipids and other organic molecules, and micelles form as the food breaks down in the digestive tract.

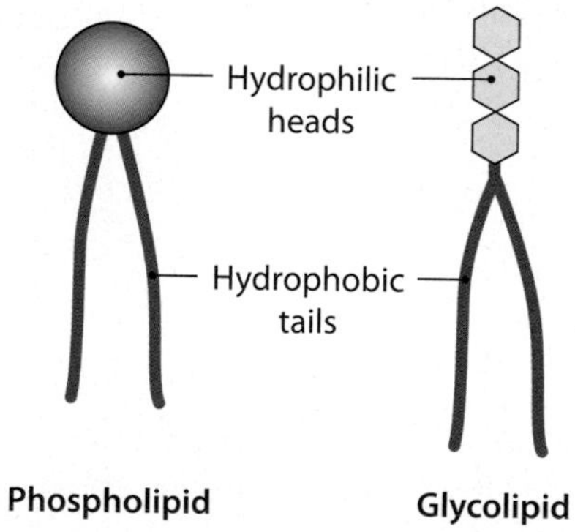

Hydrophilic heads
Hydrophobic tails
Phospholipid
Glycolipid

Module 2.16 Review

a. Describe the basic functions of eicosanoids, steroids, phospholipids, and glycolipids.

b. Why is cholesterol necessary in the body?

c. Describe the orientations of phospholipids and glycolipids when they form a micelle.

LO 2.16 Discuss the structures and diverse functions of eicosanoids, steroids, phospholipids, and glycolipids.

Module 2.17

Proteins are formed from amino acids

Proteins are the most abundant organic components of the human body, and in many ways the most important. The human body contains many different proteins, and they account for about 20 percent of total body weight. Proteins consist of long chains of organic molecules called **amino acids**. Twenty different amino acids are used as the building blocks of proteins in the human body. All amino acids contain carbon, hydrogen, oxygen, and nitrogen. Sulfur is also present in two amino acids, cysteine and methionine. A typical protein contains 1000 amino acids. The largest protein complexes have 100,000 or more.

1 Every amino acid consists of a central carbon atom to which four different groups are attached: a hydrogen atom, an amino group, a carboxyl group, and a variable side group designated as R. The carboxyl group can act as an acid by releasing a hydrogen ion to become a carboxyl ion (COO^-). The amino group can act as a base by accepting a hydrogen ion, to become an amino ion ($-NH_3^+$). The result is a molecule, with both positive and negative charges, but a net charge of zero. Such molecules are called *zwitterions*, derived from the German word that means "hybrid." The different atoms of the R groups distinguish one amino acid from another, giving each its own chemical properties. For example, different R groups are polar, nonpolar, or electrically charged.

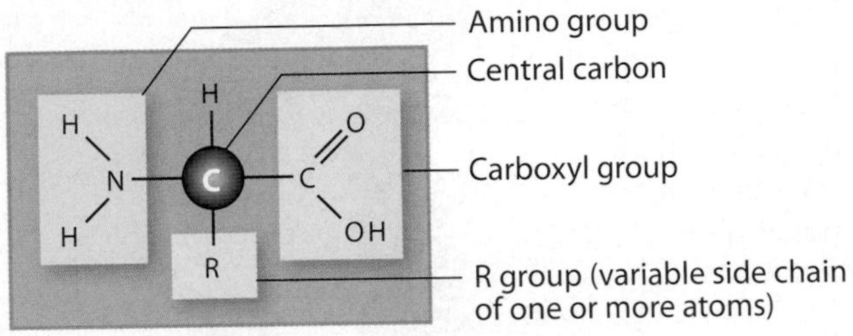

Amino group
Central carbon
Carboxyl group
R group (variable side chain of one or more atoms)

2 Two amino acids can be linked together by dehydration synthesis.

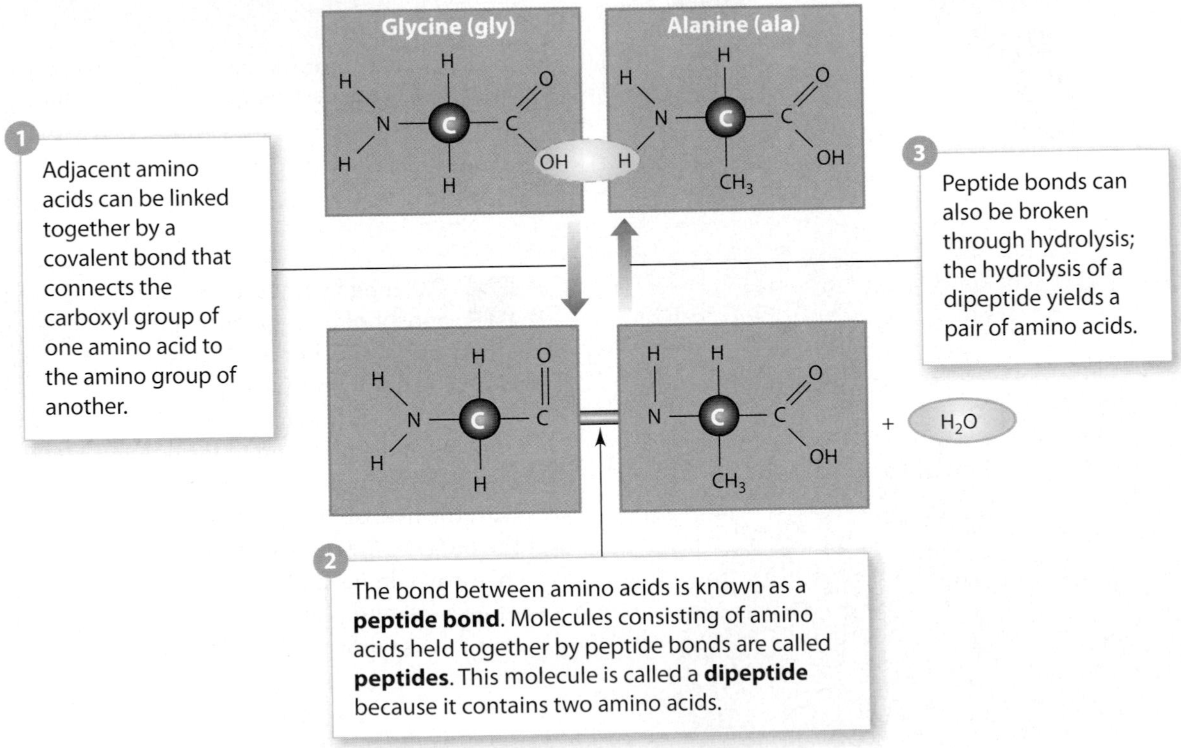

Glycine (gly)
Alanine (ala)
1 Adjacent amino acids can be linked together by a covalent bond that connects the carboxyl group of one amino acid to the amino group of another.
3 Peptide bonds can also be broken through hydrolysis; the hydrolysis of a dipeptide yields a pair of amino acids.
2 The bond between amino acids is known as a **peptide bond**. Molecules consisting of amino acids held together by peptide bonds are called **peptides**. This molecule is called a **dipeptide** because it contains two amino acids.

3 The chain can be lengthened by the addition of more amino acids with peptide bonds. Attaching a third amino acid produces a tripeptide.

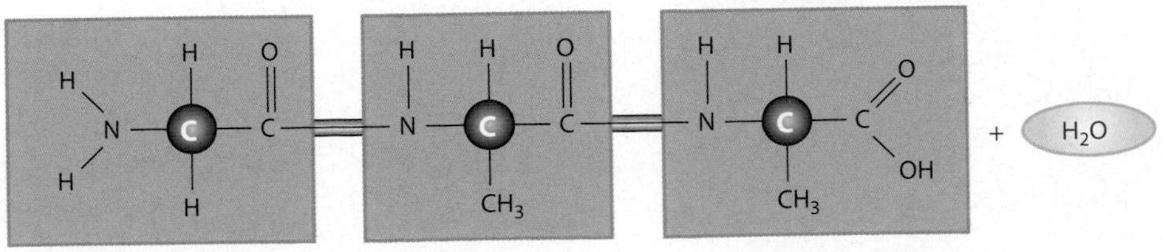

4 Tripeptides and larger peptide chains are called **polypeptides**. Polypeptides containing more than 100 amino acids are called proteins, which can have up to four levels of structural complexity. The first level of complexity is called the primary structure.

Primary Structure

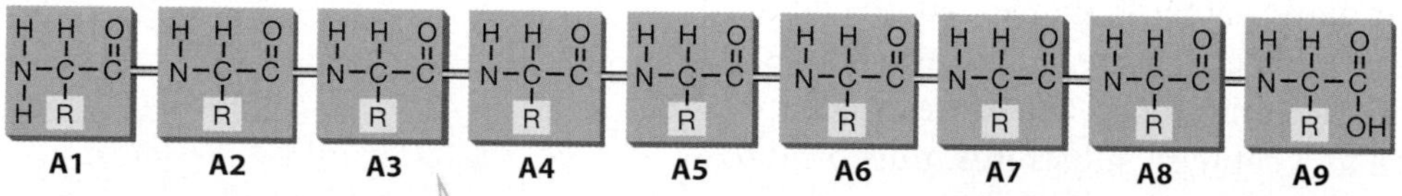

H H O
N-C-C
H R
R OH
A1
A2
A3
A4
A5
A6
A7
A8
A9

Primary structure results from the sequence of amino acids bonded together in a linear chain.

Secondary Structure

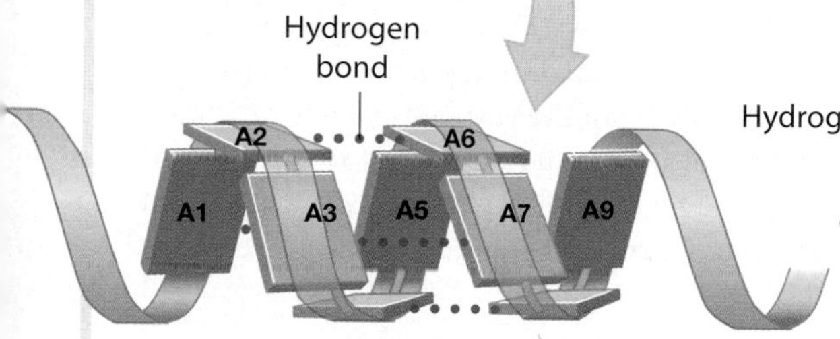

Hydrogen bond
A1
A2
A3
A5
A6
A7
A9

Alpha helix

or

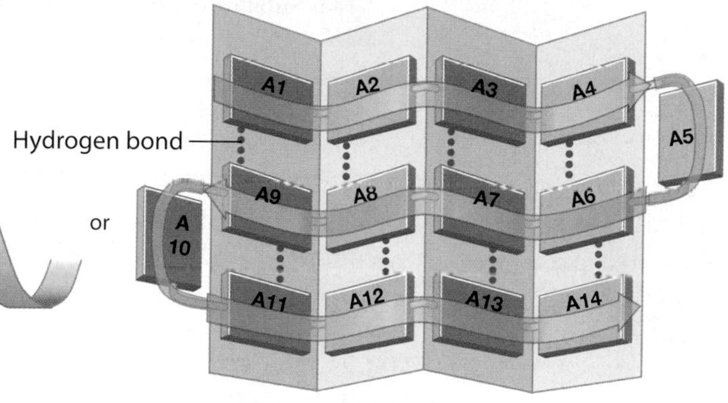

Hydrogen bond
A1
A2
A3
A4
A5
A6
A7
A8
A9
A 10
A11
A12
A13
A14

Beta sheet

Secondary structure results from bonds between atoms at different parts of the polypeptide chain. Hydrogen bonding, for example, may create either a simple spiral, called an alpha helix (α helix) or a flat pleated sheet known as a beta sheet (β sheet). The alpha helix is the most common form, but a given polypeptide chain may have both helical and pleated sections.

Tertiary Structure

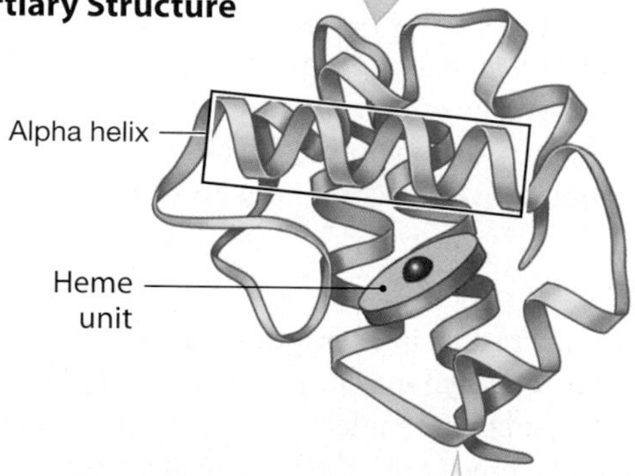

Alpha helix
Heme unit

Tertiary structure results from the complex coiling and folding that gives a protein its final three-dimensional shape. Tertiary structure results primarily from interactions between the polypeptide chain and the surrounding water molecules, and to a lesser extent from interactions between the R groups of amino acids in different parts of the molecule. This subunit of hemoglobin contains a heme unit that can bind to oxygen.

Quaternary Structure

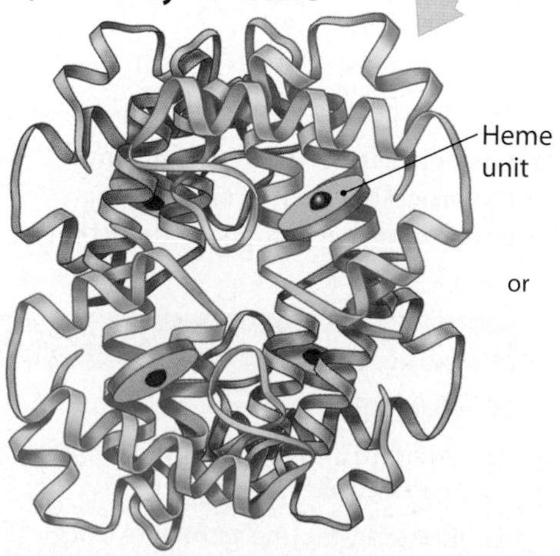

Heme unit

Hemoglobin (globular protein)

or

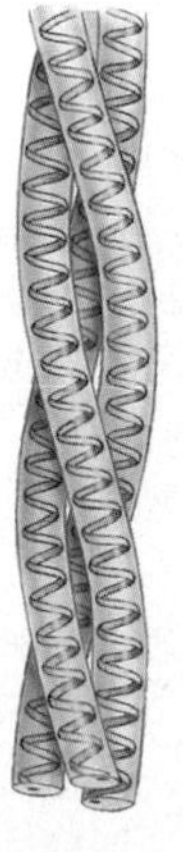

Collagen (fibrous protein)

Quaternary structure results from the interaction between individual polypeptide chains to form a protein complex. The protein **hemoglobin** contains four polypeptide subunits. Hemoglobin is found within red blood cells, where it binds and transports oxygen. It is an example of a globular protein. In **collagen**, three linear subunits intertwine, forming a fibrous protein. The three-dimensional shape of a protein plays an essential role in determining its functional properties.

As temperatures rise, protein shape changes and enzyme function deteriorates. Eventually the protein undergoes **denaturation**, a change in tertiary or quaternary structure that makes it nonfunctional. Death occurs at very high body temperatures (above 43°C, or 110°F) because the denaturation of structural proteins and enzymes causes irreparable damage to organs and organ systems.

Module 2.17 Review

a. Describe proteins.

b. What kind of bond forms during the dehydration synthesis of two amino acids?

c. Why does boiling a protein affect its structural and functional properties?

Module 2.18

Enzymes are proteins with important regulatory functions

Almost everything that happens inside the human body does so because a specific enzyme makes it possible. Enzymes are organic catalysts that speed up cellular reactions (p. 58). The reactants in enzymatic reactions are called **substrates**. As in other types of chemical reactions, the interactions among substrates yield specific products. Before an enzyme can function as a catalyst, substrates must bind to a specific region of the enzyme.

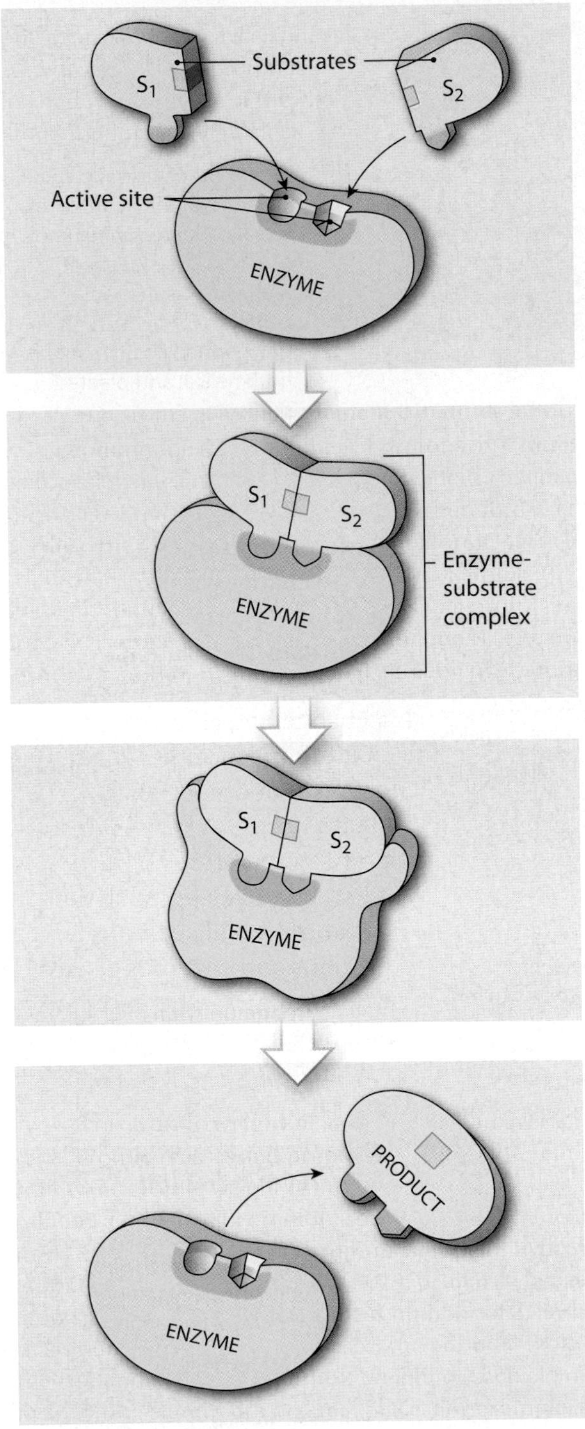
Substrates
S_1
S_2
Active site
ENZYME
S_1
S_2
ENZYME
Enzyme-substrate complex
S_1
S_2
ENZYME
PRODUCT
ENZYME

1 Substrate binding occurs at the **active site**, typically a groove or pocket into which one or more substrates nestle, like a key fitting into a lock. Weak electrical attractive forces, such as hydrogen bonding, reinforce the physical fit. The tertiary or quaternary structure of the enzyme molecule determines the shape of the active site. Each enzyme catalyzes only one type of reaction, a characteristic called **specificity**. An enzyme's specificity is determined by the ability of its active sites to bind only to substrates with particular shapes and charges.

2 Substrate binding produces an **enzyme-substrate complex**. Each cell contains an assortment of enzymes, and any particular enzyme may be active under one set of conditions and inactive under another. Virtually anything that changes the tertiary or quaternary shape of an enzyme can turn it "on" or "off" by changing the properties of the active site and preventing formation of an enzyme-substrate complex. Because the change is immediate, enzyme activation or inactivation is an important method of short-term control over reaction rates and pathways.

3 Substrate binding typically results in a temporary, reversible change in the enzyme's shape; this change may further the reaction by placing physical stresses on the substrate molecules. The enzyme then promotes product formation. In some cases, the change in enzyme shape that accompanies substrate binding is sufficient to catalyze the reaction. In other cases, an external source must provide the activation energy required.

4 The completed product then detaches from the active site, and the enzyme is free to repeat the process. When every enzyme molecule is cycling through its reaction sequence at top speed, further increases in substrate concentration will not affect the rate of reaction. The substrate concentration required for the maximum reaction rate is called the **saturation limit**.

Module 2.18 Review

a. Define active site.

b. What are the reactants in an enzymatic reaction called?

c. Relate an enzyme's structure to its reaction specificity.

LO 2.18 Explain how enzymes function within the body.

High-energy compounds may store and transfer a portion of energy released during enzymatic reactions

1 Enzymes may catalyze synthesis, decomposition, or exchange reactions. When product formation requires an energy donor, that donor is typically a **high-energy compound**. High-energy compounds contain **high-energy bonds**, covalent bonds whose breakdown releases energy under controlled conditions. The most common high-energy compound is **adenosine triphosphate**, or **ATP**.

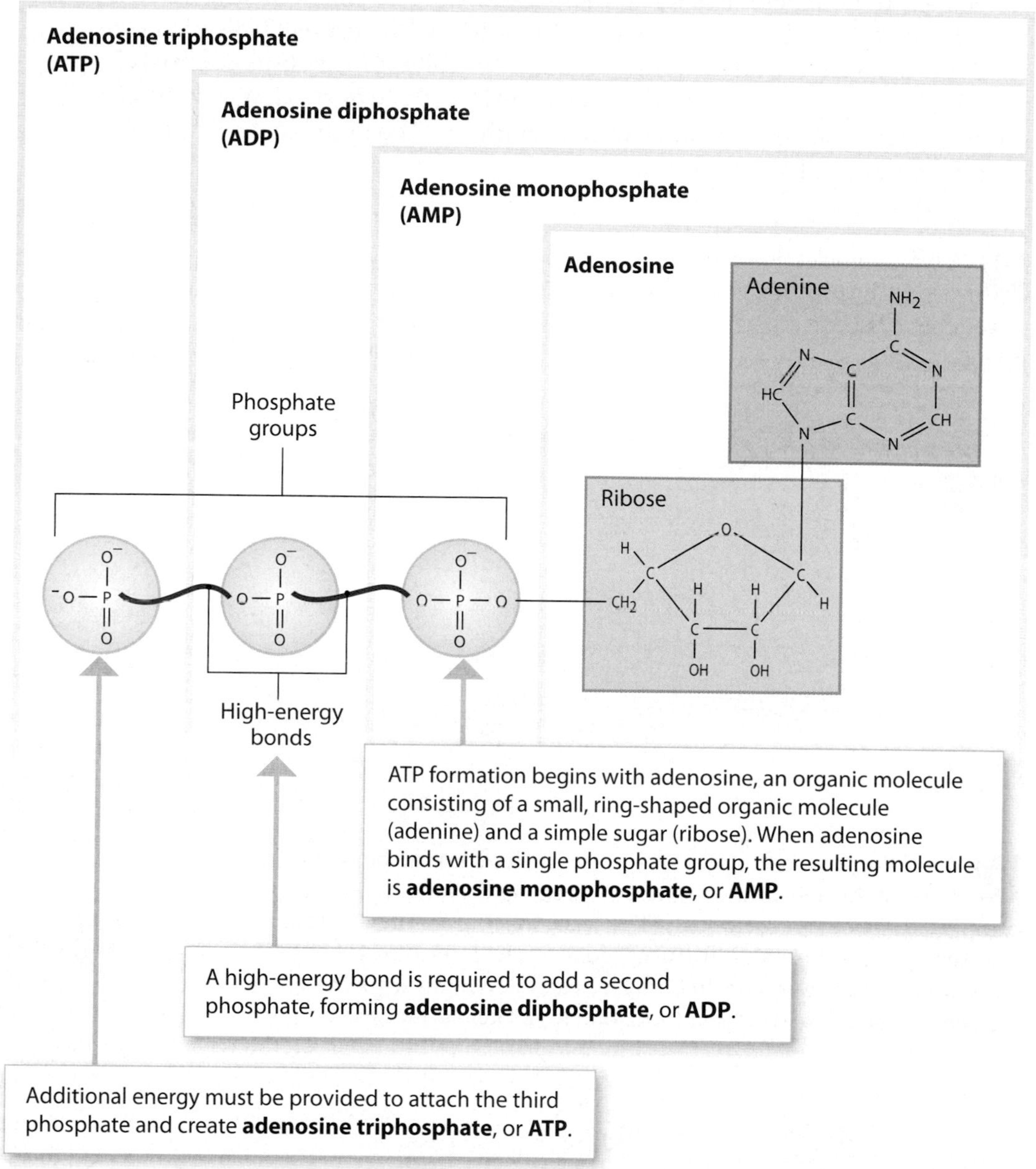

Adenosine triphosphate (ATP)
Adenosine diphosphate (ADP)
Adenosine monophosphate (AMP)
Adenosine
Adenine
Ribose
Phosphate groups
High-energy bonds
ATP formation begins with adenosine, an organic molecule consisting of a small, ring-shaped organic molecule (adenine) and a simple sugar (ribose). When adenosine binds with a single phosphate group, the resulting molecule is adenosine monophosphate, or AMP.
A high-energy bond is required to add a second phosphate, forming adenosine diphosphate, or ADP.
Additional energy must be provided to attach the third phosphate and create adenosine triphosphate, or ATP.

2 The formation of ATP from ADP is a reversible reaction. The energy stored when ATP forms is released when it breaks down to ADP. Cells can synthesize ATP in one location and then break it down in another, harnessing the energy released to power essential activities.

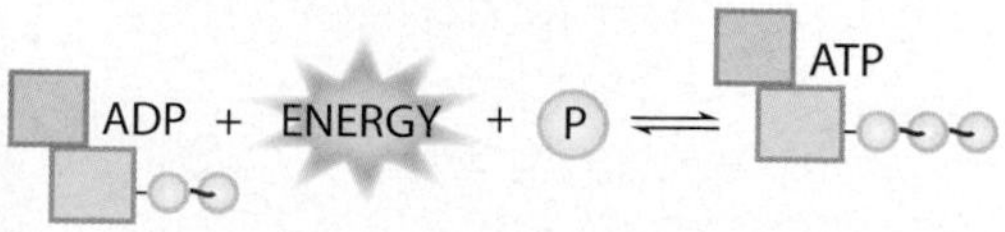

ADP + ENERGY + P ⇌ ATP

ATP and related high-energy compounds provide the energy to power many vital functions, including muscle contraction and the enzymatic reactions responsible for synthesizing proteins, carbohydrates, and lipids.

Module 2.19 Review

a. Where do cells obtain the energy needed for their vital functions?

b. Describe ATP.

c. Compare AMP with ADP.

LO 2.19 Discuss the structure and function of high-energy compounds.

DNA and RNA are nucleic acids

Nucleic (noo-KLĀ-ik) **acids** are large organic molecules composed of carbon, hydrogen, oxygen, nitrogen, and phosphorus. The two classes of nucleic acid molecules are **deoxyribonucleic** (dē-oks-ē-rī-bō-noo-KLĀ-ik) **acid**, or **DNA**, and **ribonucleic** (rī-bō-noo-KLĀ-ik) **acid**, or **RNA**. The primary role of nucleic acids is to store and transfer information, specifically, information essential to cellular protein synthesis. A nucleic acid consists of one or two long chains of subunits that are formed by dehydration synthesis. The individual subunits of a nucleic acid are called **nucleotides**.

1 A typical nucleotide consists of a phosphate group, a 5C sugar (deoxyribose or ribose), and an organic molecule known as a **nitrogenous base** that may be either a **purine** or a **pyrimidine**. Adenosine monophosphate (AMP) is an example of a nucleotide that you have already encountered.

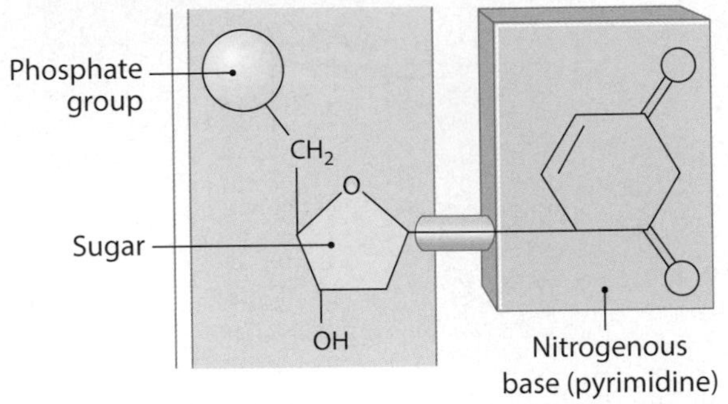
Phosphate group
CH_2
O
Sugar
OH
Nitrogenous base (pyrimidine)

2 The phosphate and sugar of adjacent nucleotides can be strung together by dehydration synthesis, creating the long chains that comprise functional nucleic acids. The "backbone" of this molecule is a linear sugar-to-phosphate-to-sugar sequence, with the nitrogenous bases projecting to one side. In both DNA and RNA, this sequence of nitrogenous bases carries the information for protein synthesis.

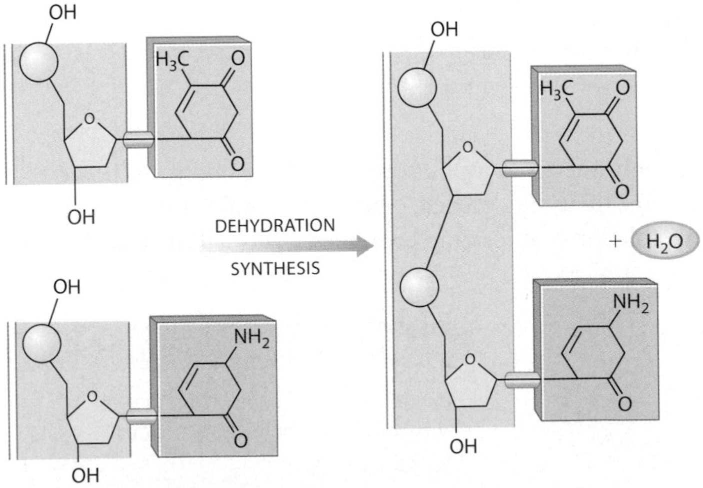
OH
H_3C
O
OH
DEHYDRATION
SYNTHESIS
NH_2
+ H_2O

Nitrogenous bases

The purines **adenine** and **guanine** are found in both DNA and RNA.

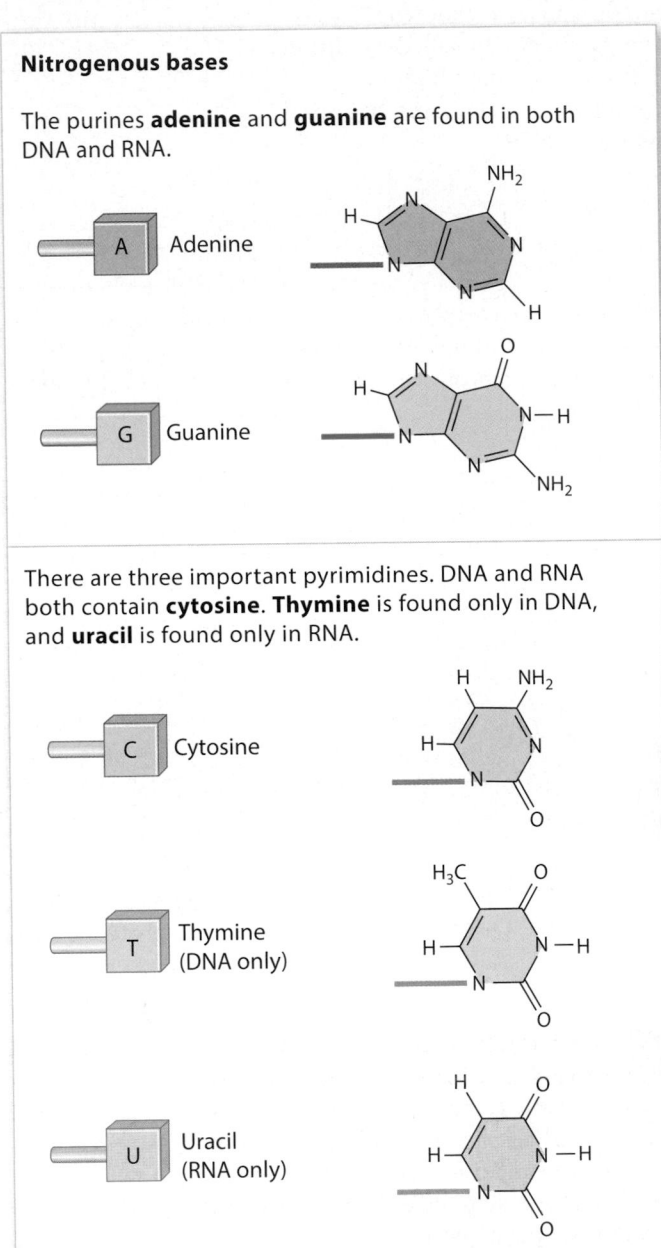

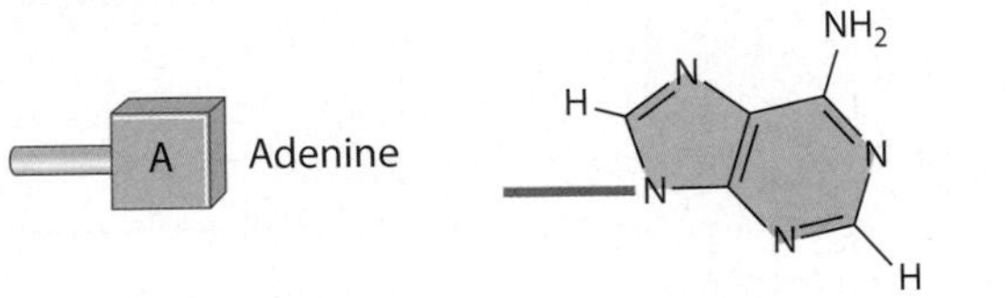
A
Adenine
NH_2
H
N

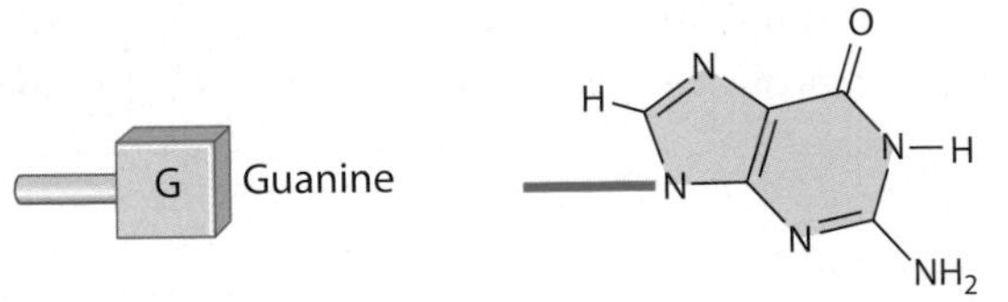
G
Guanine
O
N—H
NH_2

There are three important pyrimidines. DNA and RNA both contain **cytosine**. **Thymine** is found only in DNA, and **uracil** is found only in RNA.

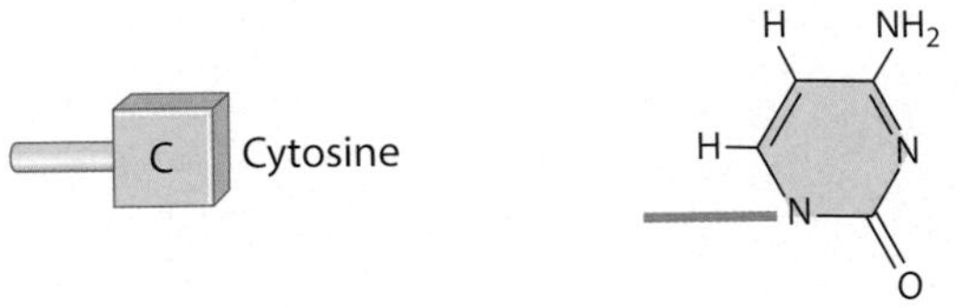
C
Cytosine
H
NH_2
N
O

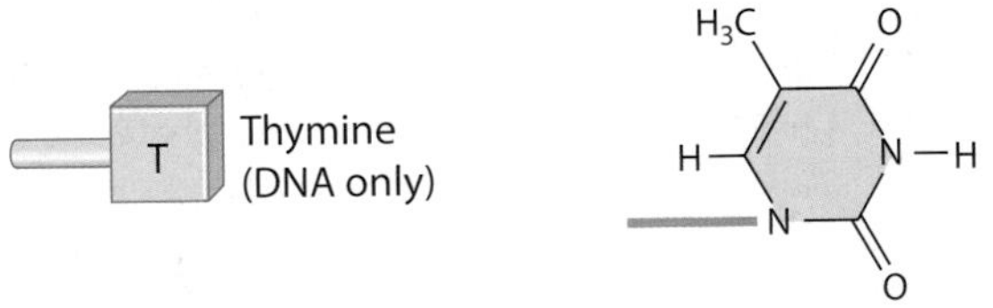
T
Thymine
(DNA only)
H_3C
O
N—H
N

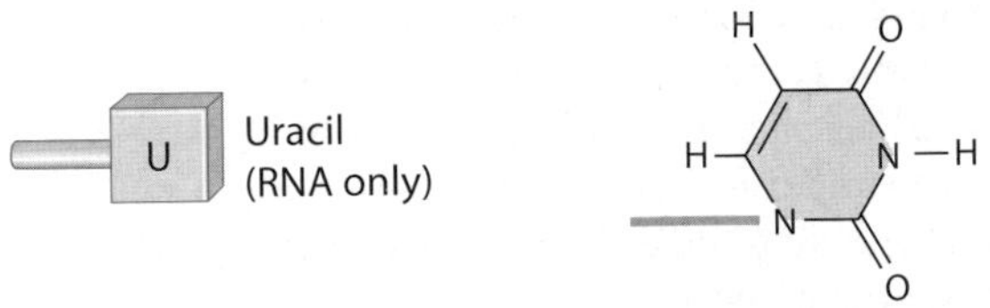
U
Uracil
(RNA only)
H
O
N—H
N

A Comparison of DNA with RNA

Characteristic	DNA	RNA
Sugar	Deoxyribose	Ribose
Nitrogenous bases	Adenine (A), guanine (G), cytosine (C), thymine (T)	Adenine, guanine, cytosine, uracil (U)
Number of nucleotides in typical molecule	Always more than 45 million	Varies from fewer than 100 to about 50,000
Molecular shape	Paired strands coiled in a double helix	Varies with hydrogen bonding along the length of the strand of each of the three main types (mRNA, tRNA, rRNA)
Function	Stores genetic information that controls protein synthesis	Performs protein synthesis as directed by DNA

3 A DNA molecule consists of a pair of nucleotide chains. Hydrogen bonding between opposing nitrogenous bases (a purine on one strand and a pyrimidine on the other) holds the two strands together. The shapes of the nitrogenous bases allow adenine to bond only to thymine, and cytosine to bond only to guanine. As a result, the combinations adenine–thymine (A-T) and cytosine–guanine (C-G) are known as **complementary base pairs**, and the two nucleotide chains of the DNA molecule are known as **complementary strands**.

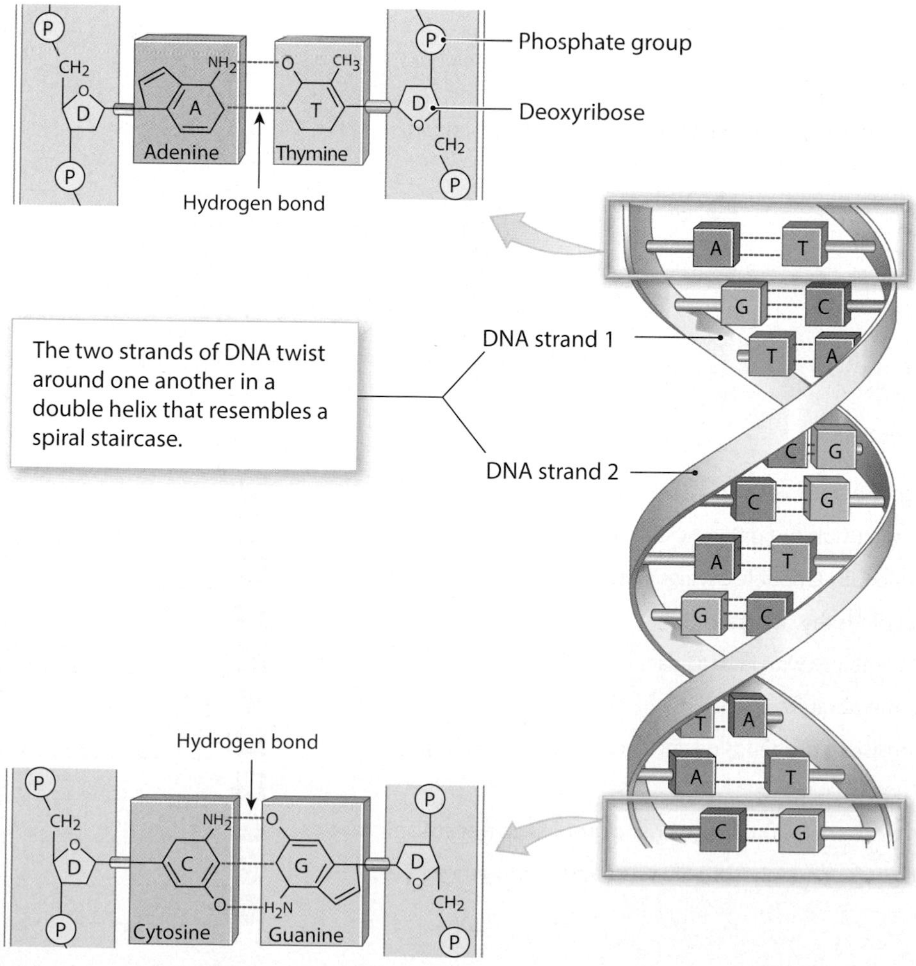
Phosphate group
Deoxyribose
Adenine
Thymine
Hydrogen bond
The two strands of DNA twist around one another in a double helix that resembles a spiral staircase.
DNA strand 1
DNA strand 2
Hydrogen bond
Cytosine
Guanine

4 A molecule of RNA consists of a single chain of nucleotides. Its shape, and thus its function, depends on the order of the nucleotides and the interactions among them. Our cells have three types of RNA: (1) **messenger RNA (mRNA)**, (2) **transfer RNA (tRNA)**, and (3) **ribosomal RNA (rRNA)**. Their specific functions will be discussed in Chapter 3.

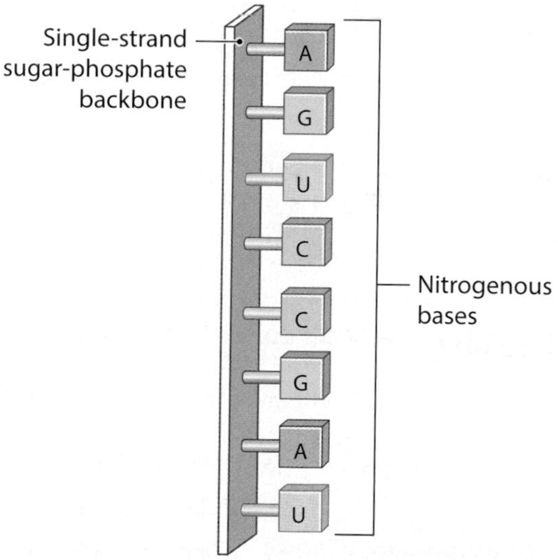
Single-strand sugar-phosphate backbone
Nitrogenous bases
A
G
U
C
C
G
A
U

Module 2.20 Review

a. Describe nucleic acids.

b. A large organic molecule composed of ribose, nitrogenous bases, and phosphate groups is which kind of nucleic acid?

c. Explain how the complementary strands of DNA are held together.

LO 2.20 Compare and contrast the structures and functions of DNA and RNA.

Concept map

Use each of the following terms once to fill in the blank boxes to correctly complete the organic compounds concept map.

- lipids
- carbohydrates
- nucleic acids
- disaccharides
- RNA
- fatty acids
- phosphate groups
- glycerol
- polysaccharides
- proteins
- monosaccharides
- ATP
- amino acids
- DNA
- nucleotides

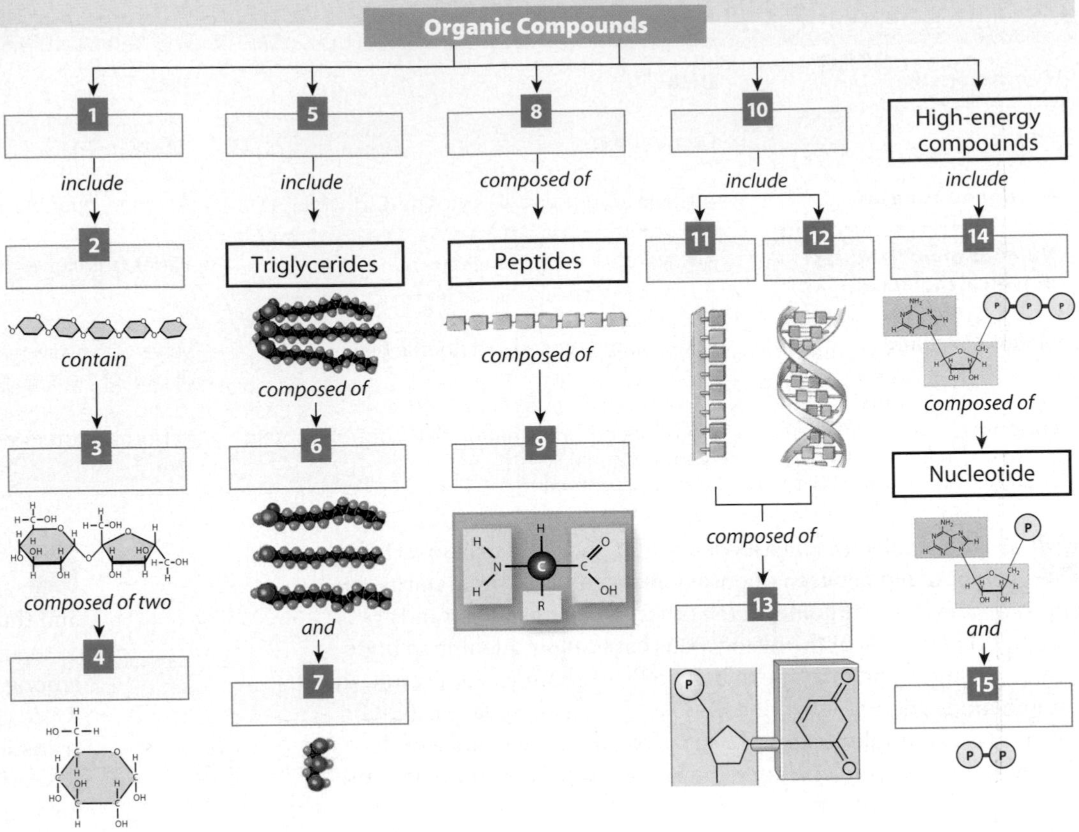
Organic Compounds
1
include
2
contain
3
composed of two
4
5
include
Triglycerides
composed of
6
and
7
8
composed of
Peptides
composed of
9
10
include
11
12
composed of
13
High-energy compounds
include
14
composed of
Nucleotide
and
15

Vocabulary

In the space provided, write the boldfaced terms introduced in this section that contain the indicated word part.

16 poly- *(many)* 16 ______________________

17 tri- *(three)* 17 ______________________

18 di- *(two)* 18 ______________________

19 glyco- *(sugar)* 19 ______________________

Matching

Match each lettered term with the most closely related description.

a. monosaccharide
b. ATP
c. polyunsaturated
d. glycerol
e. cholesterol
f. isomers
g. glycogen
h. active site
i. nucleotide
j. RNA
k. peptide

20 Polysaccharide with an energy-storage role in animal tissues 20 ______
21 Molecules with same chemical formula but different structure 21 ______
22 A fatty acid with more than one C-to-C double covalent bond 22 ______
23 The region of an enzyme that binds the substrate 23 ______
24 Three-carbon molecule that combines with fatty acids 24 ______
25 A steroid essential to plasma membranes 25 ______
26 A high-energy compound consisting of adenosine and three phosphate groups 26 ______
27 A nucleic acid that contains the sugar ribose 27 ______
28 The covalent bond between the carboxylic acid and amino groups of adjacent amino acids 28 ______
29 Organic molecule consisting of a sugar, a phosphate group, and a nitrogenous base 29 ______
30 A simple sugar 30 ______

Study Outline

SECTION 1 • Atoms and Molecules

2.1 Atoms and molecules are the basic particles of matter p. 43

1. **Matter** is anything that has mass and occupies space. **Mass** is the amount of material in matter.
2. **Atoms** are the smallest stable units of matter. They are composed of **subatomic particles**, called **protons**, **neutrons** and **electrons**. Protons and neutrons reside in the **nucleus**. Electrons whirl around the nucleus, forming an **electron cloud**. A molecule forms when atoms interact to form larger, more complex structures.

2.2 Typical atoms contain protons, neutrons, and electrons p. 44

3. The number of protons in an atom is known as the **atomic number**. The total combined number of protons and electrons in an atom is its **mass number**. **Isotopes** are atoms of the same **element** whose nuclei contain different numbers of neutrons
4. Electrons occupy an orderly series of energy levels, also called **electron shells**.

2.3 Electrons occupy various energy levels p. 46

5. The electrons in the outermost energy level, or **valence shell**, determine the chemical property of an element.
6. **Reactive** elements have atoms with an unfilled outermost energy level. **Inert** elements have filled outermost energy levels and are also called **noble gases**.
7. An **ion** is an atom with an electrical charge. A **cation** has a positive charge. An **anion** has a negative charge.

2.4 The most common chemical bonds are ionic bonds and covalent bonds p. 48

8. **Ionic bonds** are formed by the electrical attraction between cations and anions.
9. Atoms that share electrons to complete their outer electron shells create **covalent bonds**. A sharing of one pair of electrons is a **single covalent bond**. A sharing of two pairs is a **double covalent bond**. A bond with unequal sharing of electrons is a **polar covalent bond**.

2.5 Matter may exist as a solid, a liquid, or a gas p. 50

10. Solids maintain their volume, liquids have a constant volume but no fixed shape, and gases have neither constant volume nor a fixed shape.
11. Water is the only substance that occurs as a solid, liquid, or gas at temperatures consistent with life.
12. A **hydrogen bond** is a weak attractive force between polar molecules, and gives water many of its properties, such as **surface tension**.

SECTION 2 • Chemical Reactions

2.6 Chemical reactions and energy transfer are essential to cellular functions p. 53

13. A chemical reaction occurs when **reactants** are rearranged to form one or more **products**. **Metabolism** is all the chemical reactions in the body.
14. **Work** is the movement of an object or a change in the physical structure of matter. **Energy** is the capacity to perform work. **Kinetic energy** is energy in motion. **Potential energy** is stored energy.

2.7 Chemical notation is a concise method of describing chemical reactions p. 54

15. The rules of **chemical notation** are used to describe chemical reactions.
16. A **mole** (abbreviated as mol) is a quantity with a weight in grams equal to an element's atomic weight.

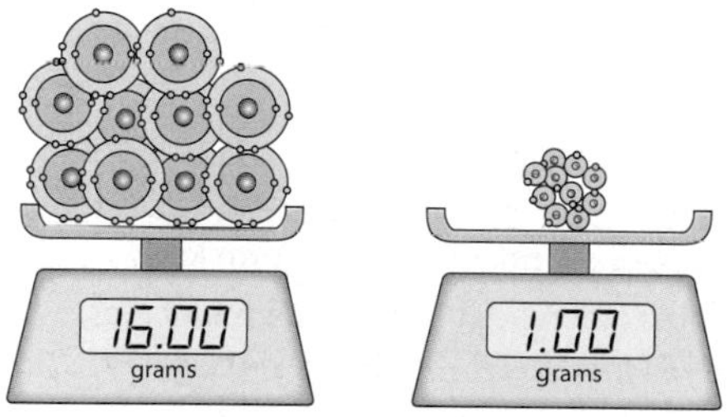

16.00
grams
1.00
grams

One mole of a given element always contains the same number of atoms as one mole of any other element.

17. The **molecular weight** of a molecule or compound is the sum of the atomic weights of its component atoms.

2.8 Three basic types of chemical reactions are important for understanding physiology p. 56

18. A chemical reaction is classified as a **decomposition**, a **synthesis**, or an **exchange reaction**.
19. In **hydrolysis**, one of the bonds in a complex molecule is broken and the components of a water molecule are added to the resulting fragments. Decomposition reactions within the body's cells and tissues are called **catabolism**.
20. **Dehydration synthesis** is the formation of a complex molecule by the removal of a water molecule. The synthesis of new molecules within the body's cells and tissues is called **anabolism**.
21. At **equilibrium**, the rates at which the two reactions proceed are in balance.

2.9 Enzymes lower the activation energy requirements of chemical reactions p. 58

22. **Activation energy** is the amount of energy required to start a chemical reaction. **Enzymes** are **catalysts**, compounds that speed up chemical reactions without themselves being permanently changed or consumed. Enzymes promote chemical reactions, or **metabolic pathways**, by lowering the activation energy requirements.

23. **Exergonic** reactions release energy; **endergonic** reactions require more energy than is released.
24. **Metabolites** are molecules that can be synthesized or broken down by chemical reactions inside our bodies. **Nutrients** are the essential metabolites obtained from the diet. Nutrients and metabolites can be broadly categorized as either **organic** or **inorganic compounds**.

SECTION 3 • Water in the Body

2.10 Water has several important properties p. 61

25. Water is the most important substance in the body. The important properties of water are **lubrication**, **reactivity**, **high heat capacity**, and **solubility**.

A remarkable number of inorganic and organic molecules will dissolve in water. The individual particles become dispersed within the water, and the result is a solution—a uniform mixture of two or more substances.

2.11 Physiological systems depend on water p. 62

26. Many inorganic compounds, called **electrolytes**, undergo **dissociation**, or **ionization**, in water to form ions.
27. Molecules that interact readily with water molecules are called **hydrophilic**; those that do not are called **hydrophobic**.
28. A solution containing dispersed proteins or other large molecules is called a **colloid**, and these particles will remain in solution indefinitely. A **suspension** contains large particles in solution but if undisturbed will settle out due to gravity.

2.12 Regulation of body fluid pH is vital for homeostasis p. 64

29. The **pH** of a solution indicates the concentration of **hydrogen ions (H^+)** it contains. Solutions are classified as **neutral**, **acidic**, or **basic (alkaline)** on the basis of pH. An **acid** releases hydrogen ions in solution; a **base** removes hydrogen ions from a solution.
30. A **salt** is an electrolyte whose cation is not a hydrogen ion (H^+) and whose anion is not a **hydroxide ion (OH^-)**. **Buffers** remove or replace hydrogen ions in solution. **Buffer systems** maintain pH within normal limits.

SECTION 4 • Organic Compounds

2.13 All organic compounds contain carbon and hydrogen atoms p. 67

31. Although organic compounds are diverse, certain groupings of atoms called **functional groups** repeatedly occur. The important functional groups are **amino groups**, **carboxyl groups**, **hydroxyl groups**, and **phosphate groups**.

2.14 Carbohydrates contain carbon, hydrogen, and oxygen, usually in a 1:2:1 ratio p. 68

32. The three major types of carbohydrates are **monosaccharides** (simple sugars), **disaccharides**, and **polysaccharides**. Disaccharides and polysaccharides form from monosaccharides by dehydration synthesis.

2.15 Lipids often have a carbon-to-hydrogen ratio of 1:2 p. 70

33. **Fatty acids** are long carbon chains with hydrogen atoms attached. Fatty acids can be **saturated** or **unsaturated fatty acids**.
34. Long chains of fatty acids can be strung together by **glycerol**, resulting in a lipid known as a **glyceride**.

2.16 Eicosanoids, steroids, phospholipids, and glycolipids have diverse functions p. 72

35. **Structural lipids** form and maintain intracellular and plasma membranes. Four classes of lipids are **eicosanoids, steroids, phospholipids,** and **glycolipids**.

2.17 Proteins are formed from amino acids p. 74

36. Proteins are formed by 20 different **amino acids** arranged in long chains and complex shapes. The four levels of protein structure are **primary structure** (amino acid sequence), **secondary structure** (amino acid interactions by hydrogen bonds), **tertiary structure** (complex coiling and folding), and **quaternary structure** (formation of protein complexes from individual subunits).

Proteins can have up to four levels of structural complexity. Hydrogen bonding in this secondary structure forms an alpha helix.

2.18 Enzymes are proteins with important regulatory functions p. 76

37. The reactants in an enzymatic reaction, called **substrates**, interact to yield a product by binding to the enzyme's **active site**. Each enzyme catalyzes only one type of reaction, a characteristic called **specificity**.

2.19 High-energy compounds may store and transfer a portion of energy released during enzymatic reactions p. 77

38. Cells store and transfer energy in the **high-energy bonds** of **high-energy compounds**. The most common high-energy compound is **adenosine triphosphate (ATP)**.
39. ATP formation begins by adding a phosphate group to **adenosine**, forming **adenosine monophosphate (AMP)**. Additional energy is required to attach a second phosphate group to form **adenosine diphosphate (ADP)**. Even more energy is required to attach a third phosphate group to form ATP.

2.20 DNA and RNA are nucleic acids p. 78

40. **Nucleic acids** are chains of **nucleotides**. Each nucleotide contains a sugar, a phosphate group, and a **nitrogenous base**. The sugar is ribose in RNA and deoxyribose in DNA.

41. DNA is a two-stranded double helix containing the nitrogenous bases **adenine**, **guanine**, **cytosine**, and **thymine**. RNA consists of a single strand and it contains **uracil** instead of thymine.
42. A DNA molecule consists of a pair of nucleotide chains. Hydrogen bonds hold together opposing nitrogenous bases. The **complementary strands** combine by means of **complementary base pairing**: adenine–thymine (A-T) and cytosine–guanine (C-G).
43. Human cells contain three types of RNA: **messenger RNA (mRNA)**, **transfer RNA (tRNA)**, and **ribosomal RNA (rRNA)**.

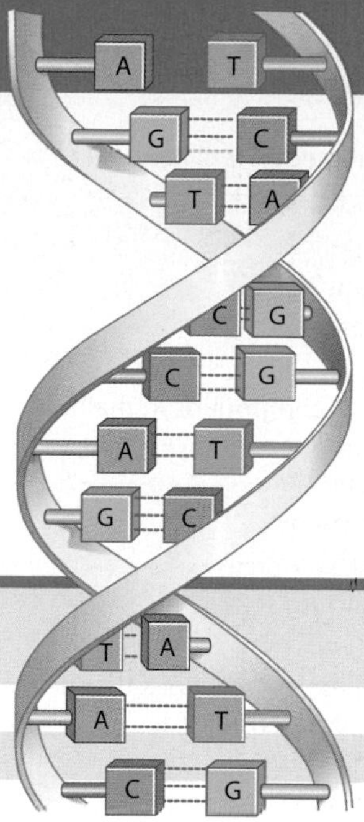

A
T
G
C
T
A
C
G
C
G
A
T
G
C
T
A
A
T
C
G

A DNA molecule consists of a pair of nucleotide chains. Hydrogen bonding between opposing nitrogenous bases holds the two strands together.

Chapter Review Questions

Multiple choice

Select the correct answer from the list provided.

1 If a polypeptide contains 10 peptide bonds, how many amino acids does it contain?
- ☐ a) 9
- ☐ b) 10
- ☐ c) 11
- ☐ d) 12

2 A dehydration synthesis reaction between glycerol and a single fatty acid would yield a(n)
- ☐ a) micelle.
- ☐ b) omega-3 fatty acid.
- ☐ c) monoglyceride.
- ☐ d) diglyceride.

3 An atom of calcium has 20 protons and 20 neutrons. What is its atomic number?
- ☐ a) 10
- ☐ b) 20
- ☐ c) 40
- ☐ d) 60

4 In an exergonic reaction,
- ☐ a) large molecules are broken down into smaller ones.
- ☐ b) small molecules are assembled into larger ones.
- ☐ c) molecules are rearranged to form new molecules.
- ☐ d) molecules move from reactants to products and back.
- ☐ e) energy is released during the reaction.

5 The hydrogen bonding that occurs in water is responsible for all of the following, except
- ☐ a) the high boiling point of water.
- ☐ b) the low freezing point of water.
- ☐ c) the ability of water to dissolve nonpolar substances.
- ☐ d) the ability of water to dissolve inorganic salts.
- ☐ e) the high surface tension of water.

6 The subatomic particle with the least mass
- ☐ a) carries a negative charge.
- ☐ b) carries a positive charge.
- ☐ c) plays no part in the atom's chemical reactions.
- ☐ d) is found only in the nucleus.

7 A(n) __________ forms when atoms interact to produce larger, more complex structures.
- ☐ a) isotope
- ☐ b) enzyme
- ☐ c) molecule
- ☐ d) nucleus

8 Isotopes of an element differ from each other in the number of
- ☐ a) protons in the nucleus.
- ☐ b) neutrons in the nucleus.
- ☐ c) electrons in the outer shells.
- ☐ d) a, b, and c are all correct.

9 The number and arrangement of electrons in an atom's outer energy level determine the atom's
- ☐ a) atomic weight.
- ☐ b) atomic number.
- ☐ c) molecular weight.
- ☐ d) chemical properties.

10 A __________ is a quantity with a weight in grams equal to an element's atomic weight.
- ☐ a) mole
- ☐ b) molecule
- ☐ c) compound
- ☐ d) synthesis

11 Energy in motion is called
- [ ] a) end product.
- [ ] b) kinetic.
- [ ] c) transfer work.
- [ ] d) potential.

12 All organic compounds in the human body contain all of the following elements except
- [ ] a) hydrogen.
- [ ] b) oxygen.
- [ ] c) carbon.
- [ ] d) calcium.

13 All the chemical reactions that occur in the human body are collectively referred to as
- [ ] a) anabolism.
- [ ] b) catabolism.
- [ ] c) metabolism.
- [ ] d) homeostasis.

14 A pH of 7.8 in the human body typifies a condition referred to as
- [ ] a) acidosis.
- [ ] b) alkalosis.
- [ ] c) dehydration.
- [ ] d) homeostasis.

15 A(n) ___________ is a solute that dissociates to release hydrogen ions, and a(n) ___________ is a solute that removes hydrogen ions from solution.
- [ ] a) base, acid
- [ ] b) salt, base
- [ ] c) acid, salt
- [ ] d) acid, base

16 Special organic catalysts that control chemical reactions in the human body are called
- [ ] a) enzymes.
- [ ] b) cytozymes.
- [ ] c) cofactors.
- [ ] d) activators.
- [ ] e) cytochromes.

17 Complementary base pairing in DNA includes the pairs
- [ ] a) adenine–uracil and cytosine–guanine.
- [ ] b) adenine–thymine and cytosine–guanine.
- [ ] c) adenine–guanine and cytosine–thymine.
- [ ] d) guanine–uracil and cytosine–thymine.

18 When the energy stored in ATP is released, it is broken down into
- [ ] a) adenosine + energy.
- [ ] b) AMP + P + energy.
- [ ] c) P + P + P + energy.
- [ ] d) ADP + P + energy.

Short answer

19 What are the three stable subatomic particles in atoms?

20 What four major classes of organic compounds are found in the body?

21 List three functions performed by lipids in the body.

22 Identify the structural characteristics of a protein.

23 a) What three components make up a nucleotide of DNA?
b) What three components make up a nucleotide of RNA?

24 Explain how enzymes function in chemical reactions.

25 What is a salt? How does a salt differ from an acid or a base?

26 Explain the differences among nonpolar covalent bonds, polar covalent bonds, and ionic bonds.

27 An organic molecule has the following constituents: carbon, hydrogen, oxygen, nitrogen, and phosphorus. Is the molecule more likely to be a carbohydrate, a lipid, a protein, or a nucleic acid?

28 Explain how an insect can walk across the top of a pond without falling through the surface.

29 A student eats a dinner of tomato salad with vinegar dressing and a glass of wine. Shortly thereafter he complains of an upset stomach. Considering what you know about pH and foods, can you predict why his stomach is upset? Could you give any suggestions as to the kinds of foods he could eat to alleviate his symptoms?

Sketching exercise

30 An oxygen atom has eight protons. a) Sketch in the arrangement of electrons around the nucleus of the oxygen atom. b) How many more electrons will it take to fill the outermost energy level?

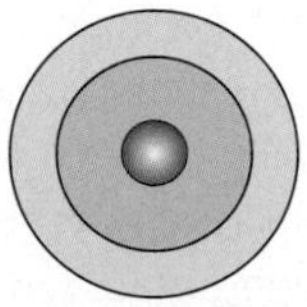

Oxygen atom

MasteringA&P®

Access more chapter study tools online in the MasteringA&P Study Area:

- **Chapter Quizzes, Chapter Practice Test, Art-labeling Activities, Animations, MP3 Tutor Sessions, and Clinical Case Studies**

- **Practice Anatomy Lab** — PAL™
- **Interactive Physiology** — iP®
- **A&P Flix** — A&PFlix™
- **PhysioEx** — PhysioEx™

Chapter Integration • Applying what you have learned

The chemistry of an after-dinner drink

Claire has just graduated college and her mother, Leslie, takes her to Italy as a graduation present. In Tuscany they enjoy a meal of vegetables, pasta, and lean meats at Massimo's family-owned osteria (a simple restaurant). Massimo provides his customers, locals and tourists alike, a culinary experience rich in cultural tradition. He takes pride in serving dinner slowly and offering dishes that are prepared simply and feature fresh ingredients.

When the women have finished their meals, Massimo offers them a *digestivo*, explaining that this after-dinner drink is part of the dining experience and necessary for proper digestion. Leslie orders an anise-flavored liqueur called sambuca, while Claire requests an amaro, a bitter, herbal drink. The two slowly sip their spirits and enjoy the flavorful end to their meal.

After they finish their drinks, they continue to sit and chat about their wonderful dining experience. Leslie comments that despite having eaten quite a bit of food, she doesn't feel overly full. Claire, while removing her jacket, agrees that her dinner, too, is indeed digesting comfortably.

Based on what you have learned about basic chemistry, answer the following questions about this scenario.

1 **If Massimo's *digestivo* was responsible for the women's assisted digestion, which type of chemical reaction is occurring?**

2 **Explain why Claire might be removing her jacket.**

3 **If the *digestivo* is assisting in the chemistry of digestion, which type of molecule would you predict to be present in the drink?**

4 **The reactions facilitated by the molecules in the *digestivo* are breaking down the complex molecules of the meals into simpler molecules. Using your knowledge of organic compounds, identify the simple molecules to which these nutrients will be reduced.**

3 Cellular Level of Organization

LEARNING OUTCOMES

These Learning Outcomes correspond by number to this chapter's modules and indicate what you should be able to do after completing the chapter.

SECTION 1 • Introduction to Cells

3.1 Describe the cell theory and the process of cellular differentiation.

3.2 Describe the cell and its organelles, including the structure and function of each.

3.3 Describe the structural and functional features of the plasma membrane.

3.4 Differentiate among the structures and functions of the cytoskeleton.

3.5 Describe the ribosome, smooth and rough endoplasmic reticula, and indicate their specific functions.

3.6 Describe the Golgi apparatus and indicate its specific functions.

3.7 Describe the structure of a mitochondrion, and explain the significance of mitochondria to cellular function.

SECTION 2 • Structure and Function of the Nucleus

3.8 Describe the role of the nucleus in maintaining homeostasis at the cellular level.

3.9 Describe the functions of the cell nucleus, and distinguish between chromatin and a chromosome.

3.10 Discuss the nature of the genetic code, and summarize the process of protein synthesis.

3.11 Summarize the process of transcription.

3.12 Summarize the process of translation.

SECTION 3 • How Substances Enter and Leave the Cell

3.13 Contrast permeable, selectively permeable, and impermeable membranes.

3.14 Explain the process of diffusion, and identify its significance in the body.

3.15 Explain the process of osmosis, and identify its significance in the body.

3.16 Describe carrier-mediated transport and its role in the absorption and removal of specific substances.

3.17 Describe vesicular transport as a mechanism for facilitating the absorption or removal of specific substances from cells.

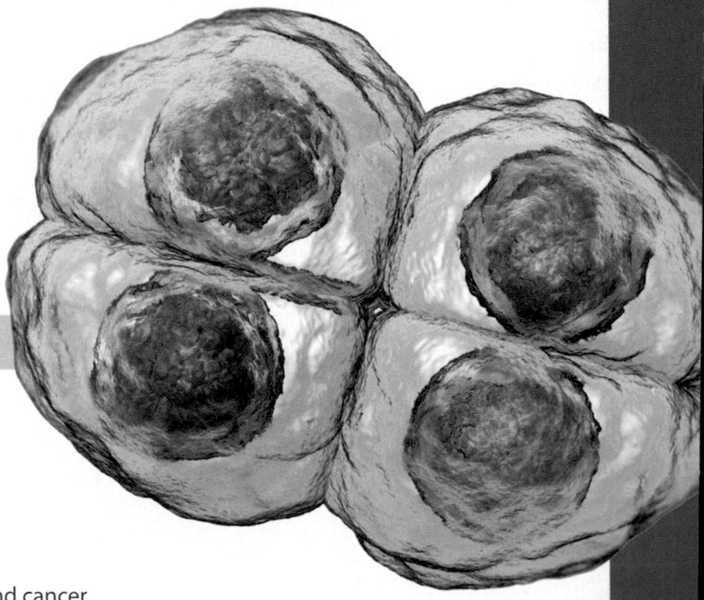

SECTION 4 • Cell Life Cycle

3.18 Distinguish between interphase and cell division in the cell cycle.

3.19 Describe interphase, and explain its significance.

3.20 Describe the process of mitosis, and its role in the cell life cycle.

3.21 Discuss the relationship between cell division and cancer.

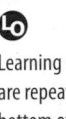

Learning Outcomes are repeated at the bottom of each module.

Cells differentiate for cellular specialization

1 A typical cell—the smallest living unit in the human body—is only about 0.1 mm in diameter. As a result, no one could examine the structure of a cell until relatively effective microscopes were invented in the 17th century. Research over time has produced the **cell theory**, which is summarized below:

Cells are the building blocks of all plants and animals.

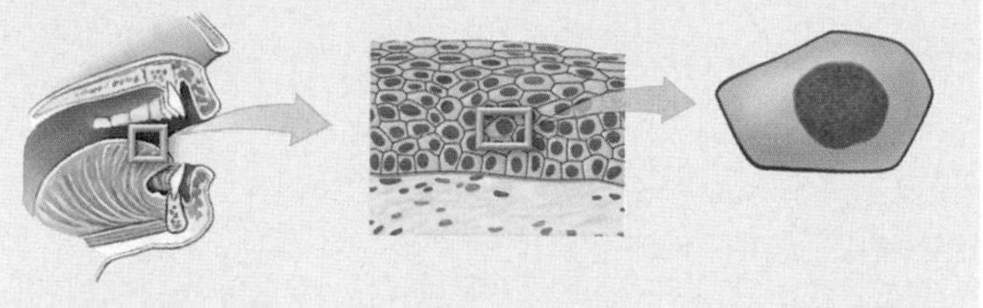

All new cells come from the division of pre-existing cells.

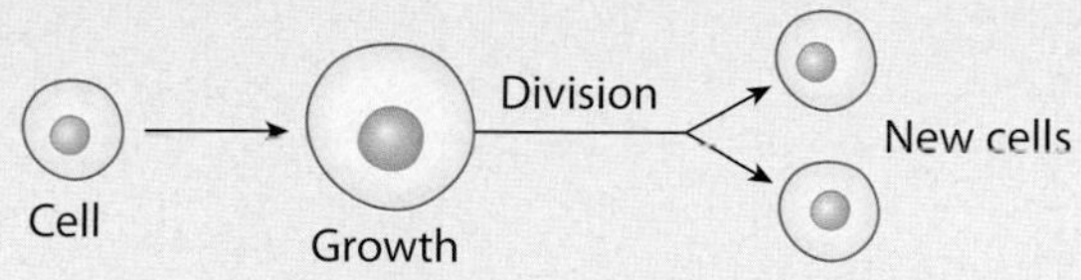
Cell
Growth
Division
New cells

Cells are the smallest structural units that carry out all vital physiological functions.

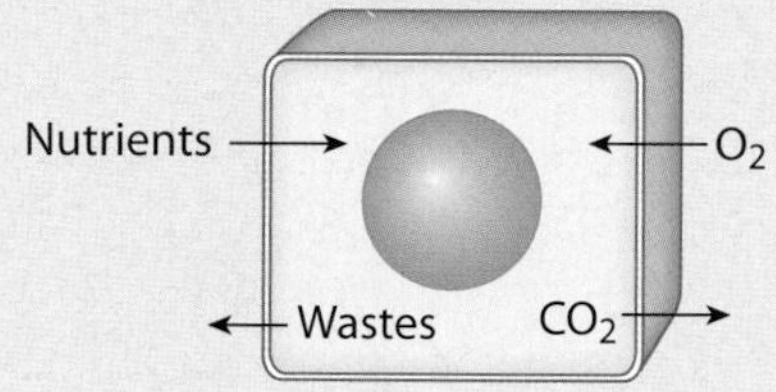
Nutrients
O_2
Wastes
CO_2

2 Each cell maintains homeostasis at the cellular level, but it requires the combined and coordinated actions of many cells to achieve homeostasis at higher levels of organization. Although cells of the human body vary widely in size, shape, and function, all are the descendants of a single cell: the fertilized ovum.

At fertilization, the fertilized ovum—which is very large—contains the genetic potential to become any cell in the body.

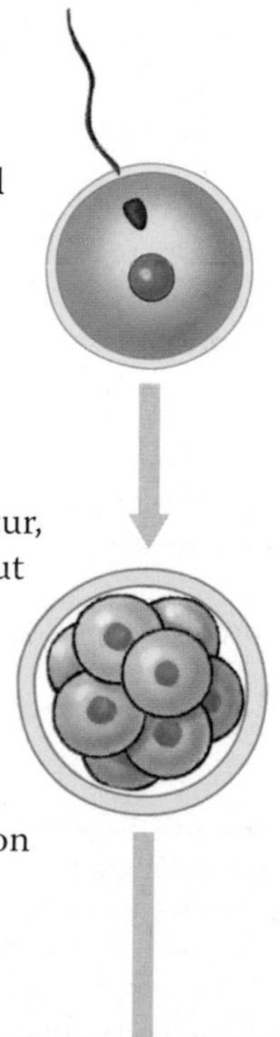

As the first cell divisions occur, the new cells do not grow, but subdivide the ovum cytoplasm into smaller parcels. These parcels differ from one another because there were regional differences in the composition of the ovum cytoplasm at fertilization.

The cytoplasmic differences affect the DNA of the new cells, turning specific genes on or off. The descendant, or daughter, cells begin to develop specialized structural and functional characteristics. This process of gradual specialization is called **differentiation**.

Differentiation produces the specialized cells that form the tissues of the body.

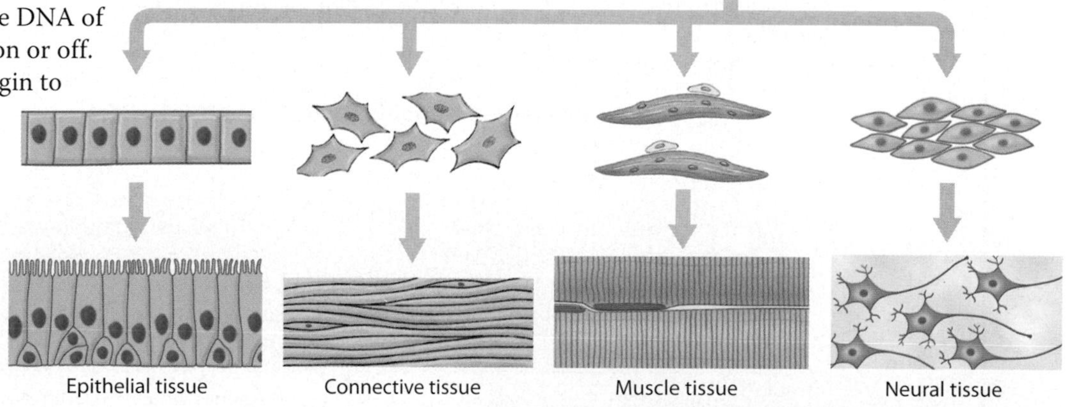
Epithelial tissue
Connective tissue
Muscle tissue
Neural tissue

Module 3.1 Review

a. Describe the cell theory.

b. Identify the cell from which all cells are descendants.

c. Define differentiation.

LO 3.1 Describe the cell theory and the process of cellular differentiation.

Module 3.2

Cells are the smallest living units of life

Our body cells are surrounded by a watery medium known as the **extracellular fluid**. The extracellular fluid in most tissues is called **interstitial** (in-ter-STISH-ul) **fluid** (*interstitium*, space between cells in a tissue). This module introduces the major components of our cells, although not all of these components are found in every cell of the body.

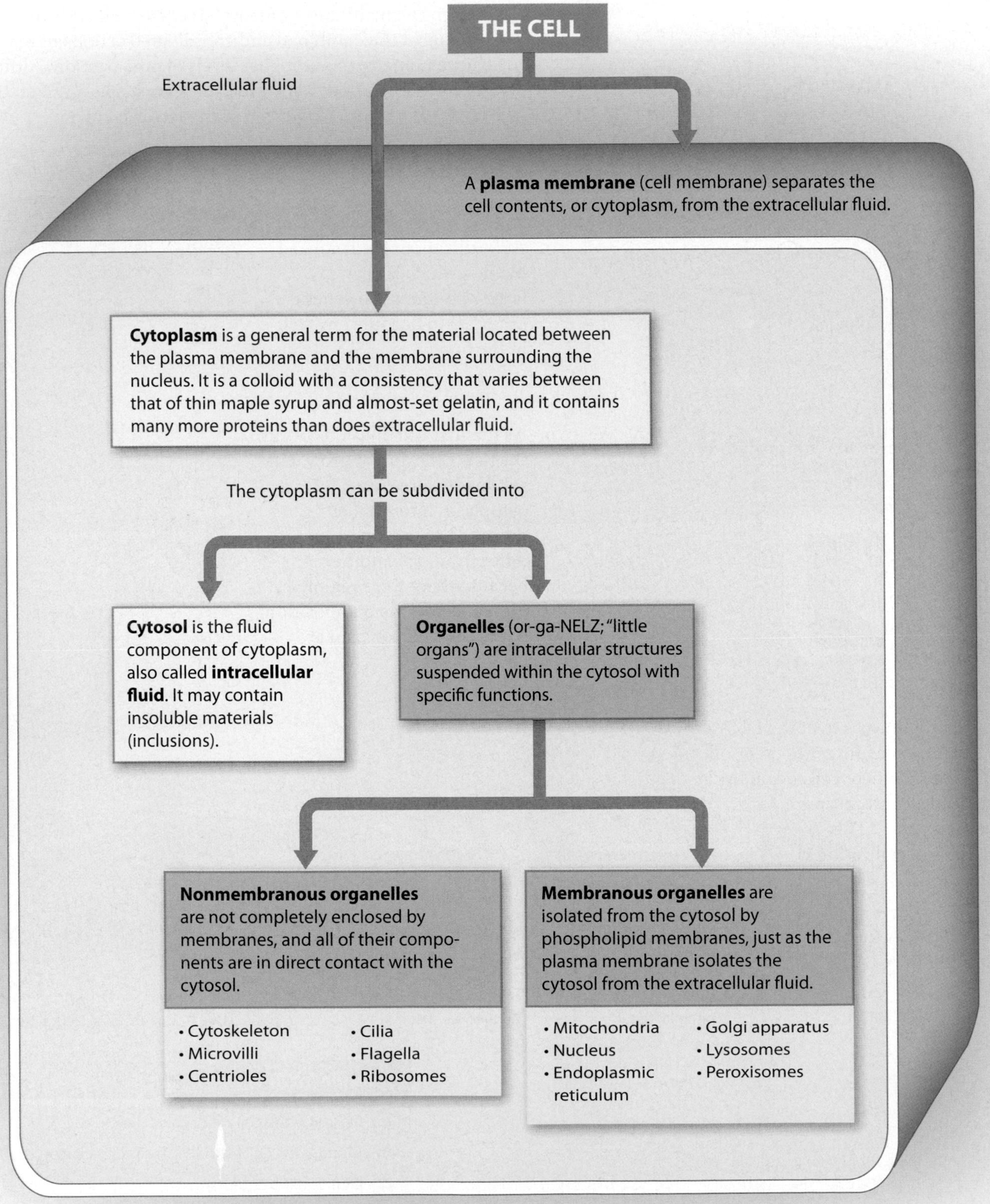

THE CELL
Extracellular fluid
A **plasma membrane** (cell membrane) separates the cell contents, or cytoplasm, from the extracellular fluid.
Cytoplasm is a general term for the material located between the plasma membrane and the membrane surrounding the nucleus. It is a colloid with a consistency that varies between that of thin maple syrup and almost-set gelatin, and it contains many more proteins than does extracellular fluid.
The cytoplasm can be subdivided into
Cytosol is the fluid component of cytoplasm, also called **intracellular fluid**. It may contain insoluble materials (inclusions).
Organelles (or-ga-NELZ; "little organs") are intracellular structures suspended within the cytosol with specific functions.
Nonmembranous organelles are not completely enclosed by membranes, and all of their components are in direct contact with the cytosol.
• Cytoskeleton
• Microvilli
• Centrioles
• Cilia
• Flagella
• Ribosomes
Membranous organelles are isolated from the cytosol by phospholipid membranes, just as the plasma membrane isolates the cytosol from the extracellular fluid.
• Mitochondria
• Nucleus
• Endoplasmic reticulum
• Golgi apparatus
• Lysosomes
• Peroxisomes

Peroxisome
STRUCTURE: Vesicles (membrane-bound sacs) containing degradative enzymes.
FUNCTION: Breakdown of organic compounds; neutralization of toxic compounds generated in the process

Lysosome
STRUCTURE: Vesicles containing digestive enzymes.
FUNCTION: Breakdown of organic compounds and damaged organelles or pathogens

Microvilli
STRUCTURE: Membrane extensions containing microfilaments. FUNCTION: Increase surface area to facilitate absorption of extracellular materials

Golgi apparatus
STRUCTURE: Stacks of flattened membranes (cisternae) containing chambers. FUNCTION: Stores, alters, and packages synthesized products

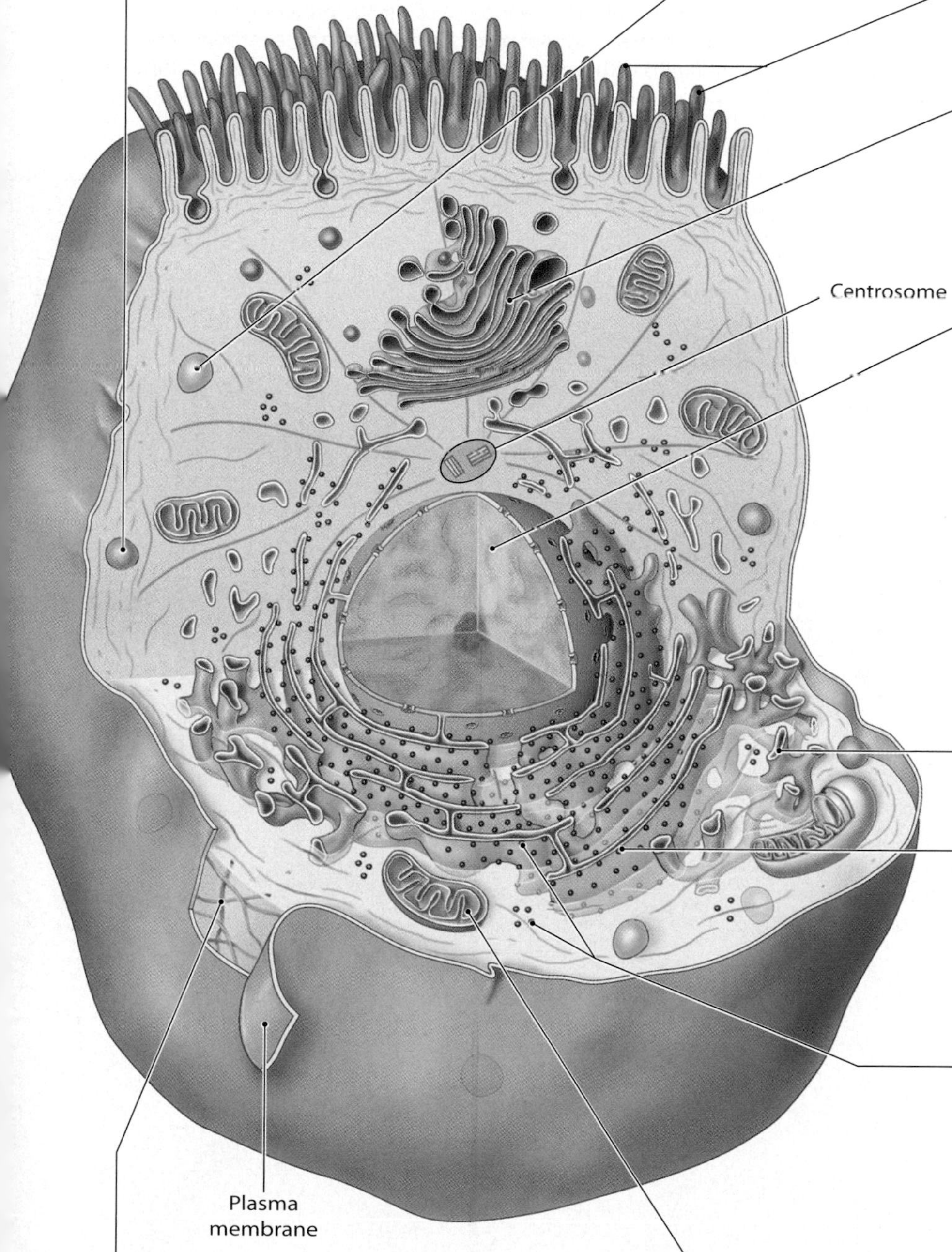
Centrosome
Plasma membrane

Nucleus
STRUCTURE: A fluid nucleoplasm containing enzymes, proteins, DNA, and nucleotides; surrounded by a double membrane, the **nuclear envelope.**
FUNCTION: Controls metabolism, stores and processes genetic information, controls protein synthesis

Endoplasmic reticulum (ER)
STRUCTURE: Network of membranous sheets and channels extending throughout the cytoplasm. FUNCTION: Synthesis of secretory products; intracellular storage and transport; detoxification of drugs or toxins

- **Smooth ER**, which has no attached ribosomes, synthesizes lipids and carbohydrates.
- **Rough ER**, which has ribosomes bound to the membranes, modifies and packages newly synthesized proteins.

Ribosomes
STRUCTURE: RNA and proteins; fixed ribosomes bound to rough ER, free ribosomes scattered in cytoplasm.
FUNCTION: Protein synthesis

Cytoskeleton
STRUCTURE: Proteins organized in fine filaments or slender tubes; organizing center located at the **centrosome**, a cytoplasmic region that contains a pair of centrioles.
FUNCTION: Strengthens and supports cell, aids in movement of cellular structures and materials

Mitochondrion
STRUCTURE: Double membrane, with inner membrane folds enclosing important metabolic enzymes.
FUNCTION: Produces 95% of the ATP required by the cell

Module 3.2 Review

a. Distinguish between the cytoplasm and cytosol.

b. Identify the membranous organelles and describe their functions.

c. Describe the functions of the cytoskeleton.

3.2 Describe the cell and its organelles, including the structure and function of each.

Module 3.3

The plasma membrane isolates the cell from its environment and performs varied functions

1 The **plasma membrane** is a physical barrier that separates the inside of the cell from the surrounding extracellular fluid. It is a selectively permeable barrier that controls the entry of ions and nutrients, such as glucose; the elimination of wastes; and the release of secretions.

Superficial membrane carbohydrates form a layer known as the **glycocalyx** (glī -kō-KĀ-liks; *calyx*, cup). Carbohydrates account for about 3 percent of the weight of a plasma membrane. They are components of complex molecules such as proteoglycans (carbohydrate with some protein attached), glycoproteins (protein with some carbohydrate attached), and glycolipids (lipids with carbohydrates attached). The glycocalyx is important in cell recognition, binding to extracellular structures, and lubrication of the cell surface.

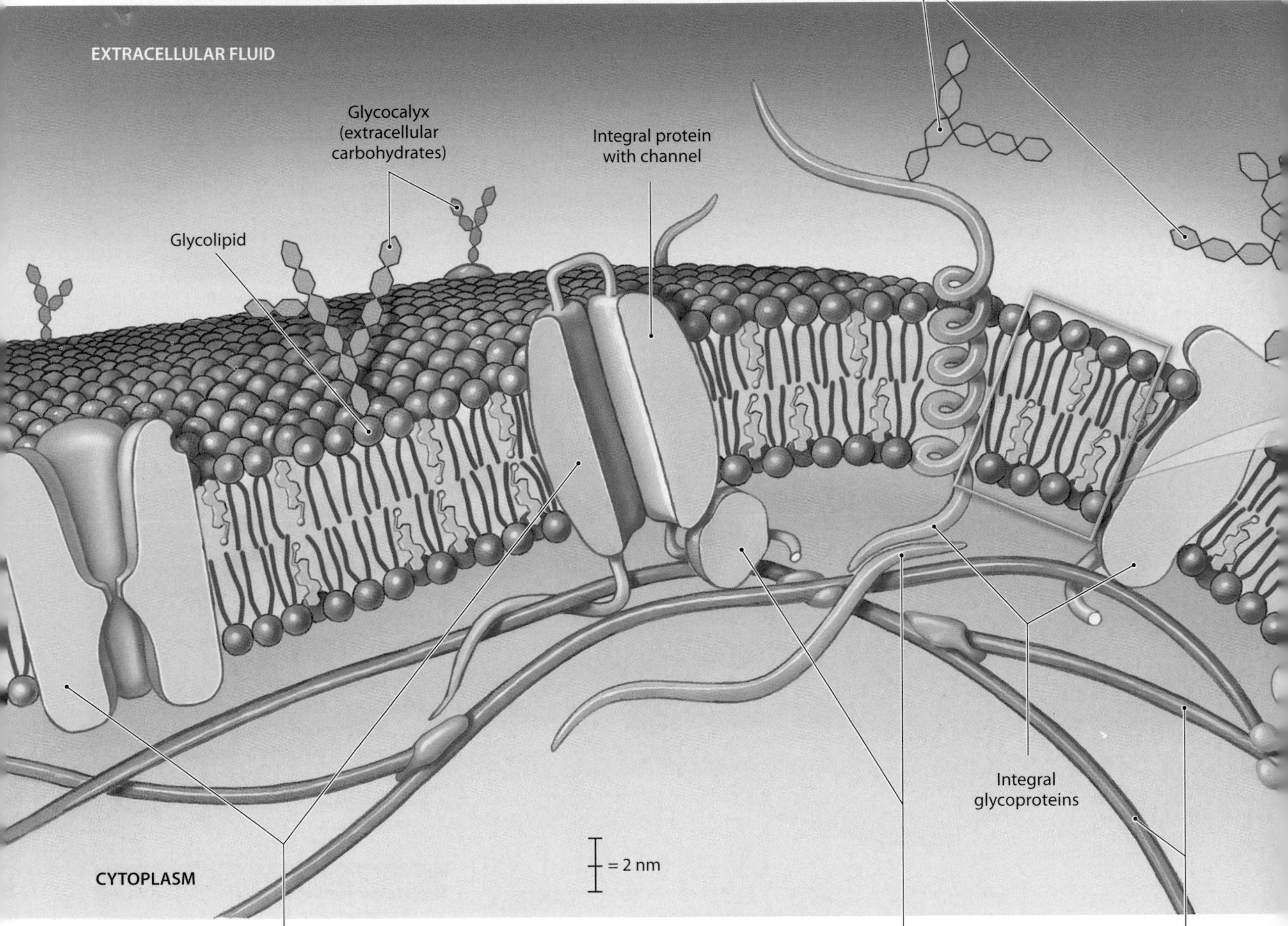

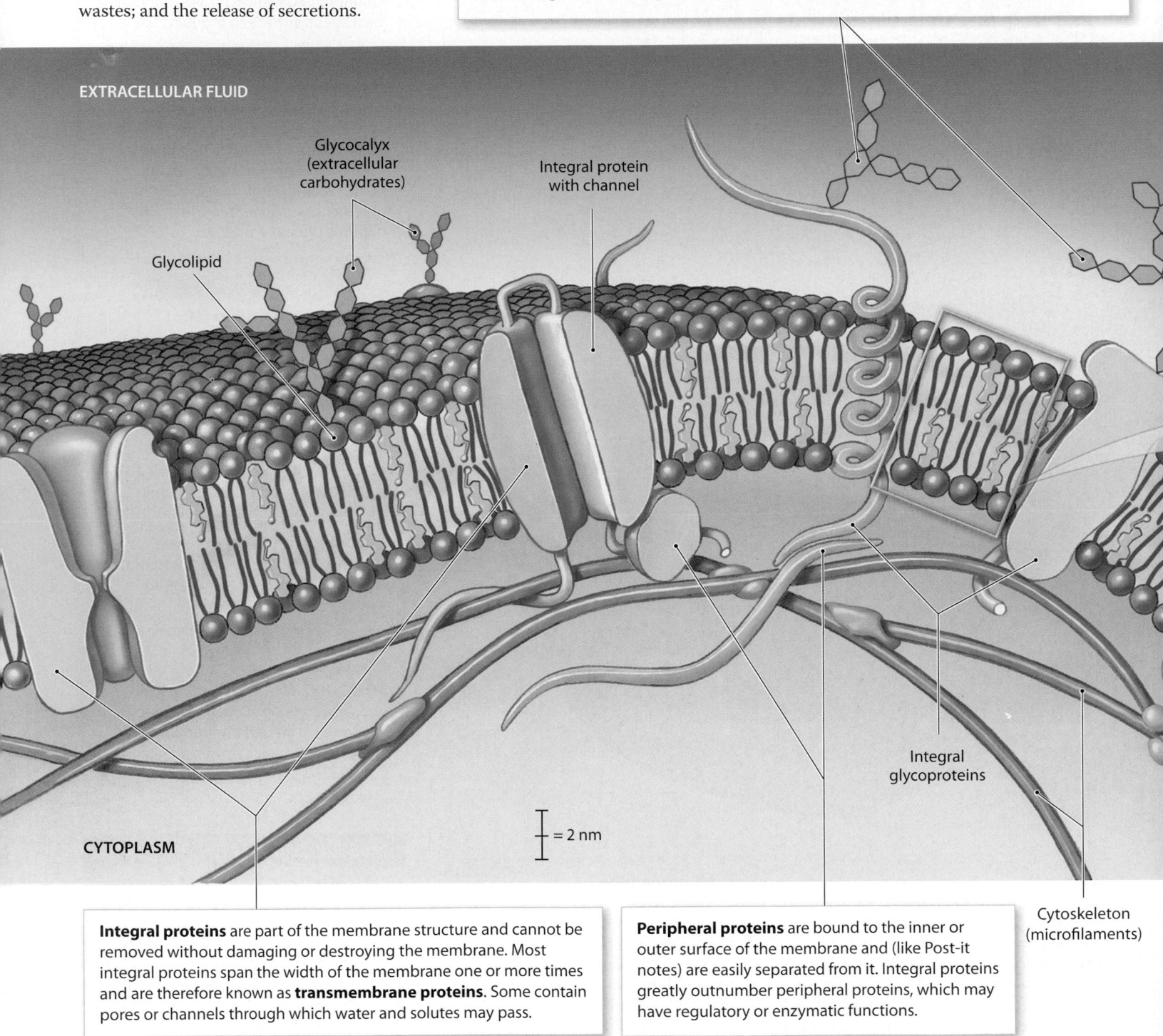

EXTRACELLULAR FLUID
Glycocalyx (extracellular carbohydrates)
Integral protein with channel
Glycolipid
Integral glycoproteins
= 2 nm
CYTOPLASM
Cytoskeleton (microfilaments)

Integral proteins are part of the membrane structure and cannot be removed without damaging or destroying the membrane. Most integral proteins span the width of the membrane one or more times and are therefore known as **transmembrane proteins**. Some contain pores or channels through which water and solutes may pass.

Peripheral proteins are bound to the inner or outer surface of the membrane and (like Post-it notes) are easily separated from it. Integral proteins greatly outnumber peripheral proteins, which may have regulatory or enzymatic functions.

2 The plasma membrane is extremely thin (6–10 nm) and very delicate. It is called a **phospholipid bilayer**, because the different phospholipid molecules in it form two layers. In each half of the bilayer, the phospholipids lie with their hydrophilic heads at the membrane surface and their hydrophobic tails on the inside, just as in a micelle (Module 2.16, p. 73). The hydrophobic layer in the center of the membrane isolates the cytoplasm from the extracellular fluid. Such isolation is important because the composition of cytoplasm is very different from that of extracellular fluid.

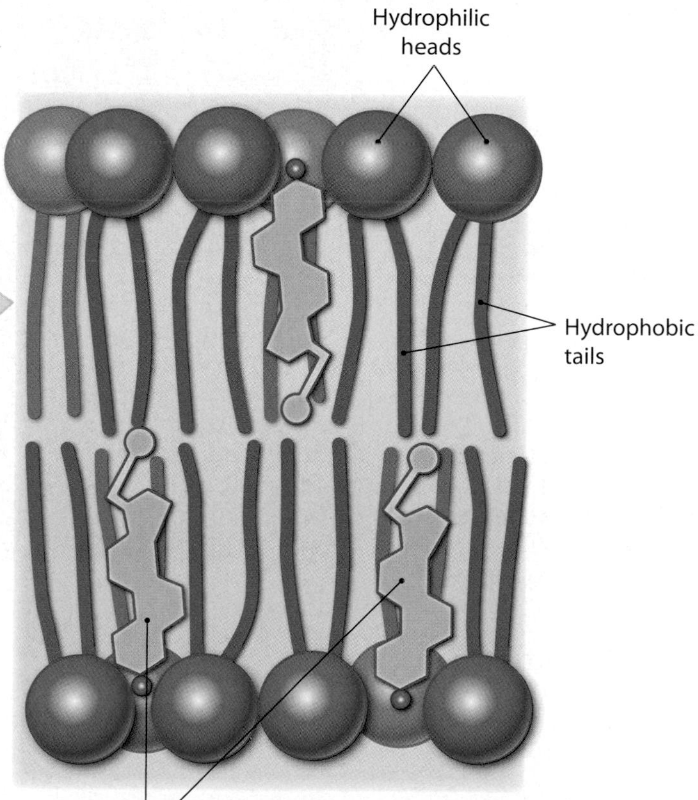
Hydrophilic heads
Hydrophobic tails

Cholesterol is an important component of plasma membranes, with almost one cholesterol molecule for each phosholipid molecule. Like phospholipid molecules, cholesterol is an **amphipathic** molecule, which means that it has both hydrophobic and hydrophilic portions. Its polar hydroxyl group aligns with the hydrophilic heads of the phospholipid molecule and its rigid steroid ring and nonpolar tail lie within the hydrophobic layer. Cholesterol "stiffens" the plasma membrane, making it less fluid and less permeable.

3 The general functions of the plasma membrane include physical isolation, regulation of exchange with the environment, sensitivity to the environment, and structural support. The lipid bilayer provides isolation, and the membrane proteins perform most of the other functions. The five major functional classes of plasma membrane proteins are described in the table below.

Functional Classes of Membrane Proteins

Anchoring proteins attach the plasma membrane to other structures and stabiliize its position. Inside the cell, membrane proteins are bound to the cytoskeleton, a network of supporting filaments in the cytoplasm.

Recognition proteins are detected by cells of the immune system.

Enzymes in plasma membranes may be integral or peripheral proteins.

Receptor proteins bind to specific extracellular molecules called **ligands** (LĪ-gandz). A ligand can be anything from a small ion like calcium, to a relatively large and complex hormone like insulin.

Carrier proteins bind solutes and transport them across the plasma membrane.

Channels are integral proteins containing a central pore (channel) that forms a passageway completely through the plasma membrane. The channel permits the passage of water and small solutes that cannot otherwise cross the lipid bilayer of the plasma membrane.

Module 3.3 Review

a. List the general functions of the plasma membrane.

b. Which structural component of the plasma membrane is mostly responsible for its ability to isolate a cell from its external environment?

c. Which type of integral protein allows water and small ions to pass through the plasma membrane?

LO 3.3 Describe the structural and functional features of the plasma membrane.

The cytoskeleton plays both a structural and a functional role

The **cytoskeleton** functions as the cell's skeleton. It provides an internal protein framework that gives the cytoplasm strength and flexibility.

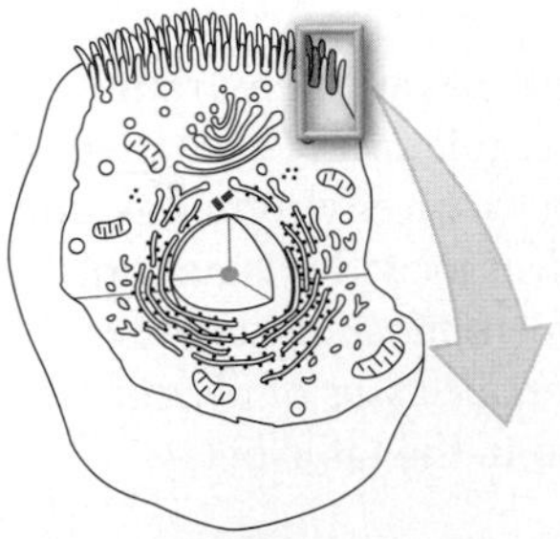

1 The cytoskeleton of all cells includes microfilaments, intermediate filaments, and microtubules.

Microvilli (singular, *microvillus*) are finger-shaped extensions of the plasma membrane of some cells. A core of microfilaments stiffens each microvillus and anchors it to the cytoskeleton at the terminal web. Microvilli greatly increase the surface area of the cell and enhance its ability to absorb materials from the extracellular fluid.

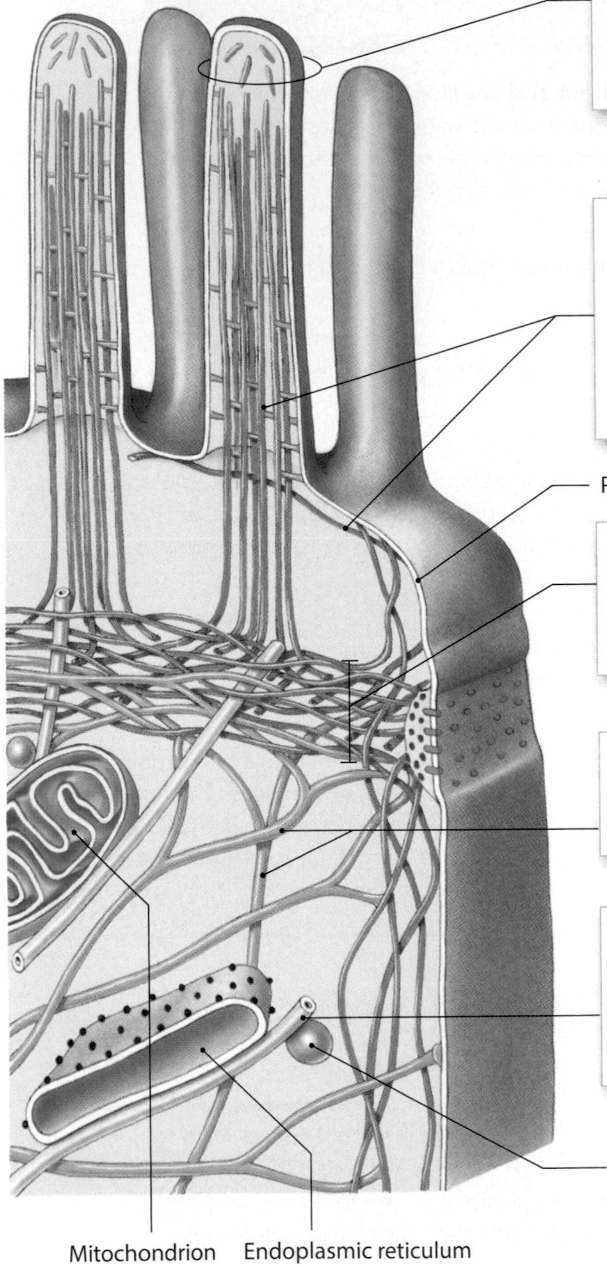
Plasma membrane
Secretory vesicle
Mitochondrion
Endoplasmic reticulum

The smallest of the cytoskeletal elements are the **microfilaments**. These protein strands are generally less than 6 nm in diameter. Typical microfilaments are composed of the protein **actin**. They are common in the periphery of the cell, but relatively rare in the region immediately surrounding the nucleus.

The **terminal web** is a layer of microfilaments just inside the plasma membrane at the exposed surface of a cell that forms a layer or lining, such as that of the intestinal tract.

Intermediate filaments, which range from 7 to 11 nm in diameter, are the strongest and most durable cytoskeletal elements.

Microtubules are the largest components of the cytoskeleton, with diameters of about 25 nm. Microtubules extend outward into the periphery of the cell from a region near the nucleus called the centrosome.

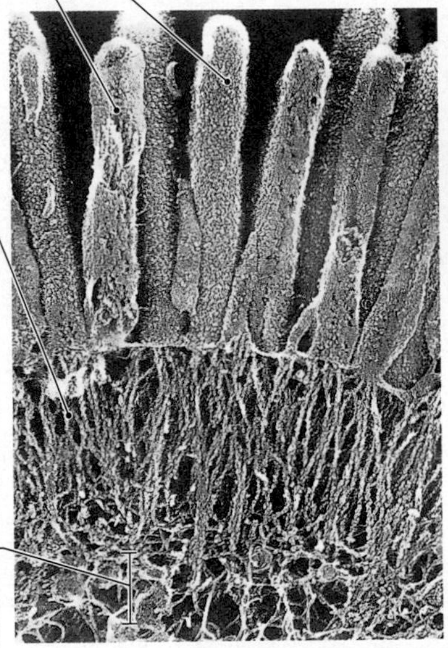

Microvilli SEM × 30,000

2 This table summarizes the general components of the cytoskeleton.

The Cytoskeleton

Structure	Remarks	Location	Functions
Microfilaments	Present in most cells; best organized in skeletal and cardiac muscle cells	In bundles beneath the plasma membrane and throughout the cytoplasm	Provide strength, alter cell shape, bind the cytoskeleton to the plasma membrane, tie cells together, involved in muscle contraction
Intermediate filaments	Present in most cells; at least five types known	In cytoplasm	Provide strength, move materials through cytoplasm
Thick filaments	Found in skeletal and cardiac muscle cells	In cytoplasm	Interact with actin microfilaments to produce muscle contraction
Microtubules	Present in most cells	In cytoplasm radiating away from centrosome	Provide strength, move organelles
Centrioles	Nine groups of microtubule triplets form a short cylinder	In centrosome near nucleus	Organize microtubules in the spindle to move chromosomes during cell division
Cilia	Nine groups of microtubule doublets form a cylinder. Motile cilia have a central pair (9 + 2)	Extensions of plasma membrane	Multiple motile cilia propel fluids or solids across cell suface. The single primary cilium of a cell detects environmental stimuli
Flagella	Nine groups of microtubule doublets form a cylinder around a central pair (9 + 2)	Extension of plasma membrane	A flagellum propels sperm

3 **Centrioles** are cylindrical structures composed of short microtubules. The microtubules form nine groups, three in each group. Each of these nine "triplets" is connected to its nearest neighbors on either side. Two centrioles are located in a region known as the **centrosome**. During cell division, the centrioles are associated with the movement of DNA strands. Cells that lack centrioles, such as red blood cells and skeletal muscle cells, cannot divide.

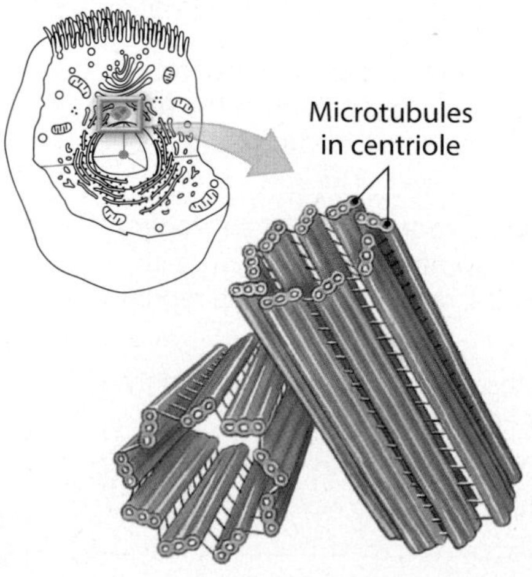
Microtubules
in centriole

4 **Cilia** (singular, *cilium*) are long, slender extensions of the plasma membrane. Multiple **motile cilia** are found on cells lining portions of the respiratory tract and the reproductive tract. Motile cilia resemble centrioles but have nine pairs of microtubules (rather than triplets) surrounding a central pair. The microtubules are anchored to a compact **basal body** just beneath the cell surface. A **primary cilium** lacks a central pair of microtubules and functions as a sensor of a cell's surroundings. **Flagella** (sing., *flagellum*) have the same microtubule structure as motile cilia, but are much longer and beat in a wavelike fashion. The only human cell with a flagellum is a sperm cell.

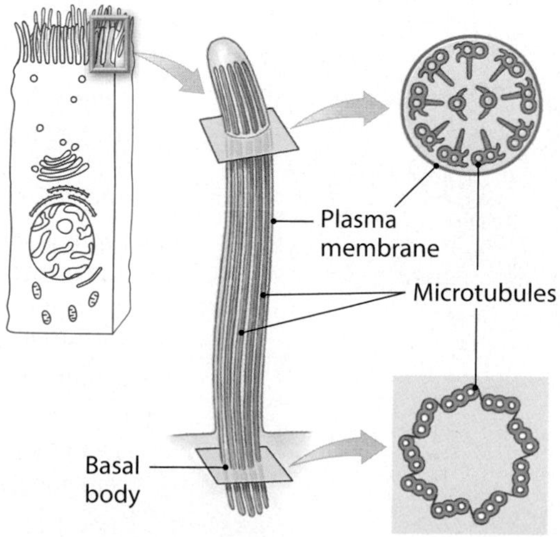
Plasma
membrane
Microtubules
Basal
body

5 Motile cilia beat rhythmically to move fluids or secretions across the cell surface. The ciliated cells lining the trachea beat their cilia in synchronized waves to move sticky mucus and trapped dust particles toward the throat and away from delicate respiratory surfaces.

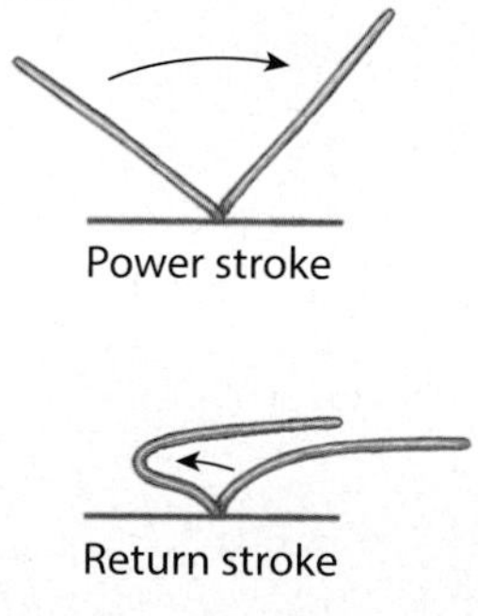
Power stroke
Return stroke

Module 3.4 Review

a. List the three basic components of the cytoskeleton.

b. Which cytoskeletal component is common to both centrioles and cilia?

c. What is the function of cilia?

LO 3.4 Differentiate among the structures and functions of the cytoskeleton.

Ribosomes are responsible for protein synthesis and are often associated with the endoplasmic reticulum

Ribosomes are the organelles responsible for protein synthesis. The more proteins a cell synthesizes, the more ribosomes it has. Liver cells have many; fat cells have relatively few. Free ribosomes are scattered throughout the cytoplasm. The proteins they manufacture enter the cytosol. Ribosomes synthesizing proteins with destinations other than the cytosol become temporarily bound, or fixed, to the endoplasmic reticulum (ER), a membranous organelle.

1 A functional ribosome consists of two subunits that are normally separate and distinct. These subunits contain special proteins and ribosomal RNA (rRNA), one of the RNA types introduced in Module 2.20 (p. 79). Before protein synthesis can begin, large and small ribosomal subunits must join together, as shown here.

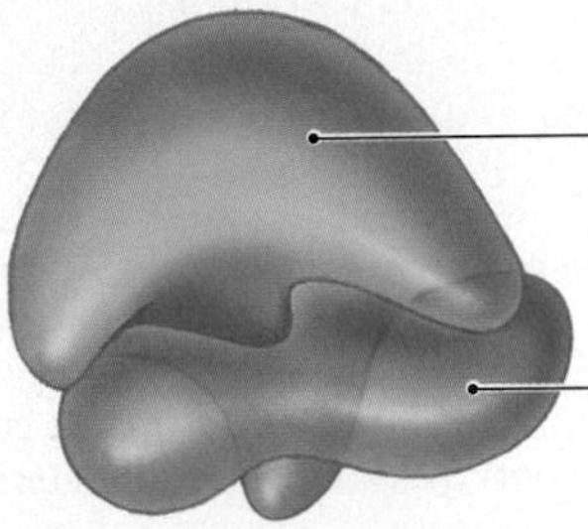
Large ribosomal subunit
Small ribosomal subunit

2 The **endoplasmic reticulum** (en-dō-PLAZ-mik re-TIK-ū-lum), or **ER**, is a network of intracellular membranes continuous with the nuclear envelope, which surrounds the nucleus.

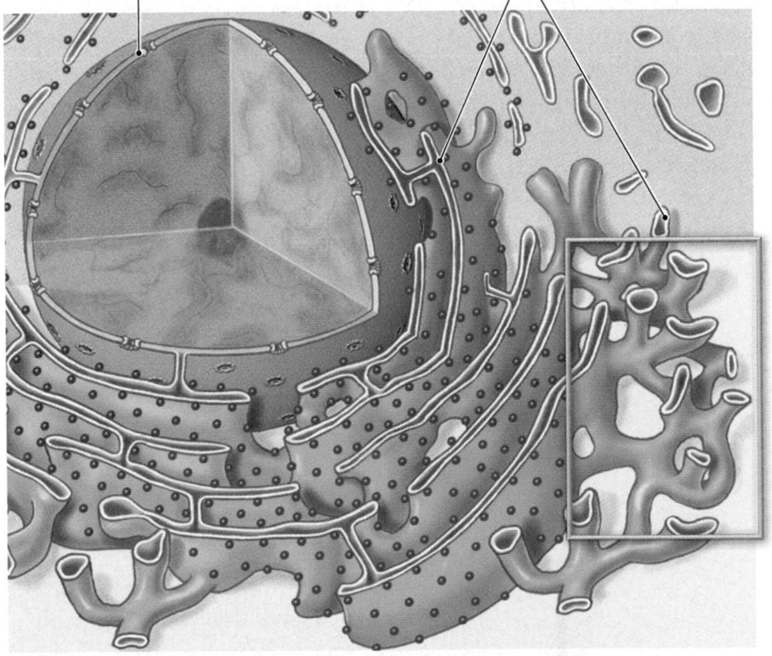
Nuclear envelope
The ER forms hollow tubes, flattened sheets, and chambers called **cisternae** (sis-TUR-nē; singular, *cisterna*, a reservoir for water).

3 The **smooth endoplasmic reticulum**, or **SER**, lacks ribosomes, and the cisternae are often tubular.

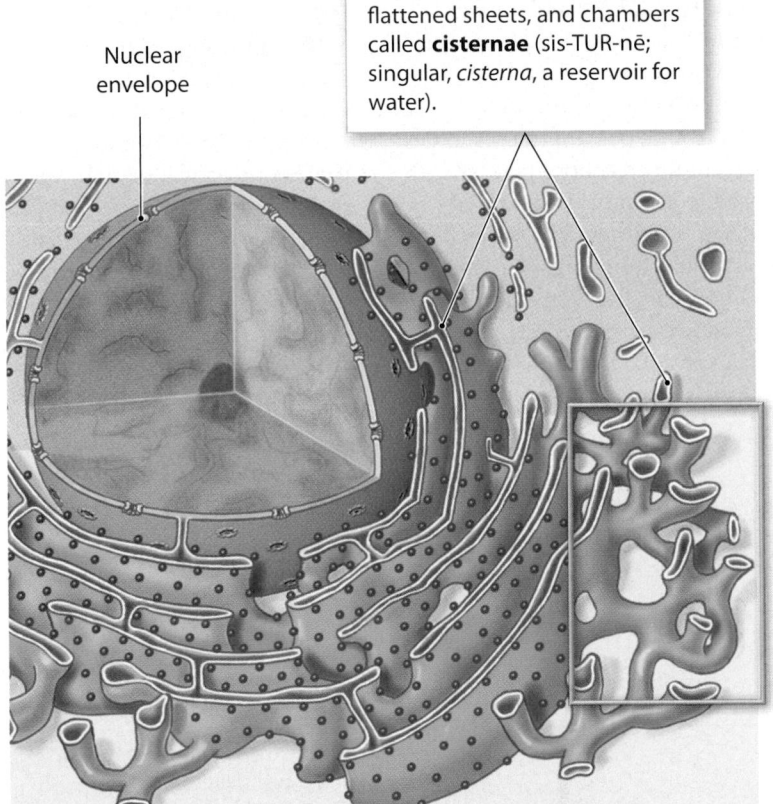

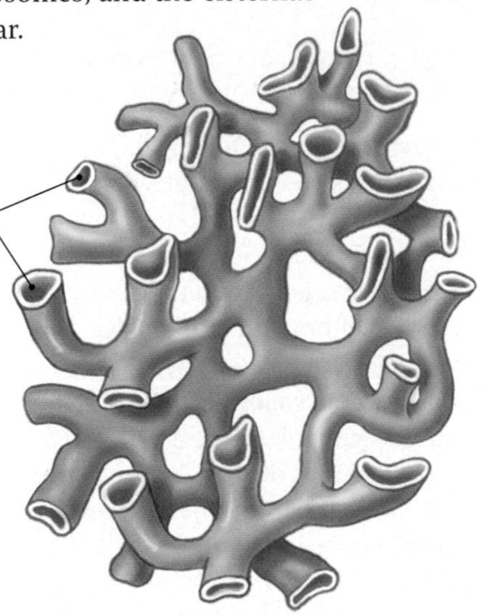
Tubular cisternae

The Functions of the SER

- Synthesize the phospholipids and cholesterol needed for maintenance and growth of the plasma membrane, ER, nuclear envelope, and Golgi apparatus
- Synthesize steroid hormones, such as androgens and estrogens (the dominant sex hormones in males and in females, respectively) in the reproductive organs
- Synthesize and store glycerides, especially triglycerides, in liver cells and fat cells
- Synthesize and store glycogen in skeletal muscle cells and liver cells

4 The **rough endoplasmic reticulum (RER)** functions as a combination workshop and shipping warehouse. It is where many newly synthesized proteins are chemically modified and packaged for export to their next destination, the Golgi apparatus, described in the next module.

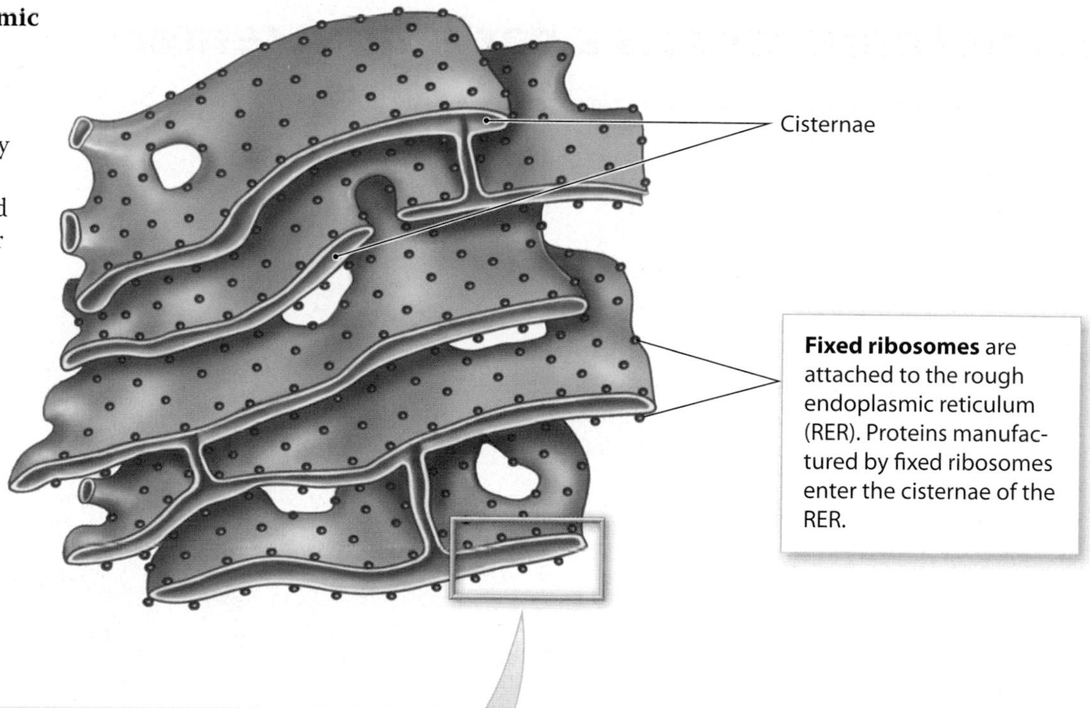

Cisternae
Fixed ribosomes are attached to the rough endoplasmic reticulum (RER). Proteins manufactured by fixed ribosomes enter the cisternae of the RER.

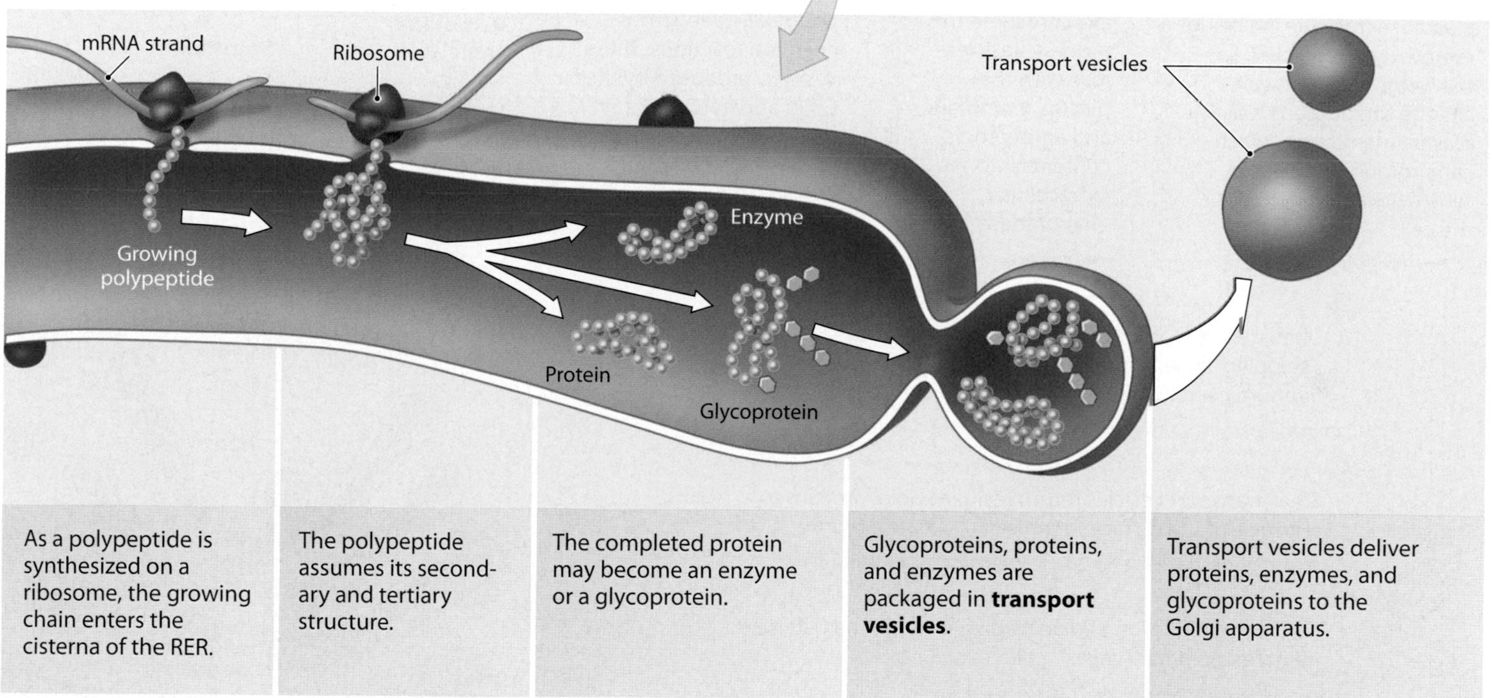

mRNA strand
Ribosome
Transport vesicles
Enzyme
Growing polypeptide
Protein
Glycoprotein

As a polypeptide is synthesized on a ribosome, the growing chain enters the cisterna of the RER.

The polypeptide assumes its secondary and tertiary structure.

The completed protein may become an enzyme or a glycoprotein.

Glycoproteins, proteins, and enzymes are packaged in **transport vesicles**.

Transport vesicles deliver proteins, enzymes, and glycoproteins to the Golgi apparatus.

The amount of endoplasmic reticulum and the proportion of RER to SER vary with the type of cell and its ongoing activities. For example, pancreatic cells that manufacture digestive enzymes contain an extensive RER, but their SER is small. The situation is just the reverse in reproductive system cells that synthesize steroid hormones.

Module 3.5 Review

a. Describe the immediate cellular destinations of newly synthesized proteins from free ribosomes and fixed ribosomes.

b. Describe the structure of smooth endoplasmic reticulum.

c. Why do certain cells in the ovaries and testes contain large amounts of smooth endoplasmic reticulum (SER)?

LO **3.5** Describe the ribosome, smooth and rough endoplasmic reticula, and indicate their specific functions.

The Golgi apparatus is a packaging center

The **Golgi apparatus**, or **Golgi complex**, (1) renews or modifies the plasma membrane; (2) modifies and packages secretions, such as hormones or enzymes, for release outside the cell through a transport process called exocytosis; and (3) packages special enzymes within vesicles for use in the cytosol.

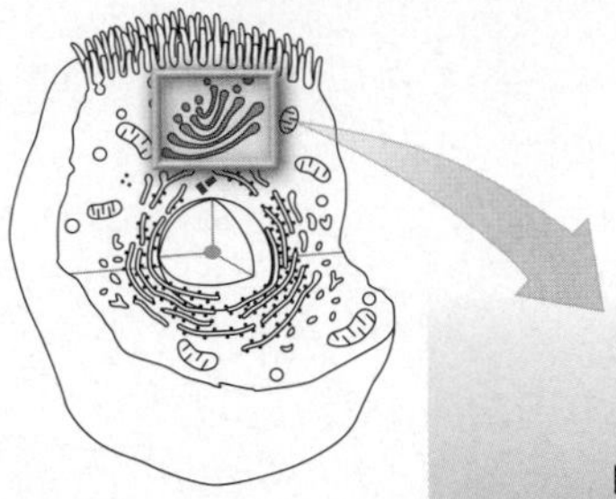

1 The Golgi apparatus typically consists of five or six flattened membranous discs called cisternae. A single cell may contain more than one Golgi apparatus, typically located near the nucleus.

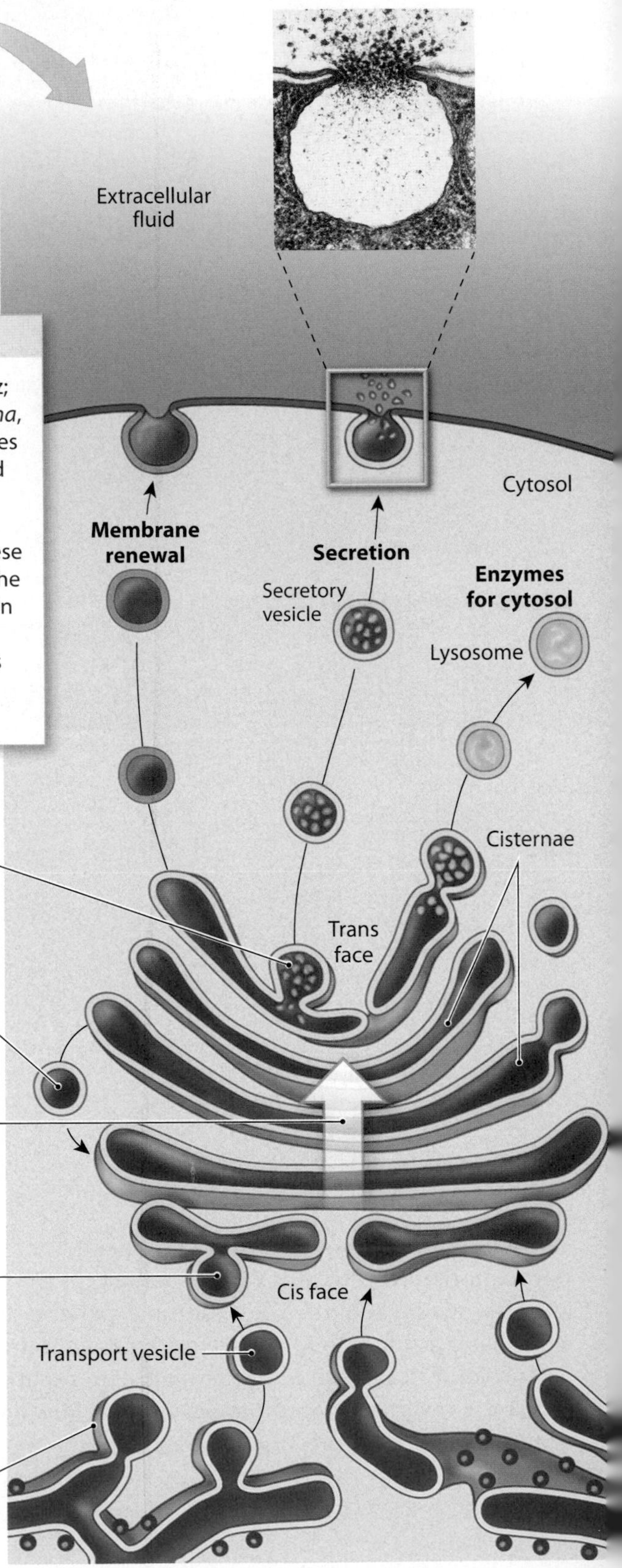
Extracellular fluid
Cytosol
Membrane renewal
Secretion
Secretory vesicle
Enzymes for cytosol
Lysosome
Cisternae
Trans face
Cis face
Transport vesicle

Membrane renewal

Membrane renewal vesicles add to the surface area of the plasma membrane. At the same time, other areas of the plasma membrane are being removed and recycled. So, the Golgi apparatus can change the properties of the plasma membrane, which can profoundly alter the sensitivity and functions of the cell.

Secretion

Secretory vesicles contain products that will be discharged from the cell by **exocytosis**. In this process, vesicles fuse with the plasma membrane and empty their contents into the extracellular environment.

Enzymes for cytosol

Lysosomes (LĪ-sō-sōmz; *lyso-*, a loosening + *soma*, body) are special vesicles that provide an isolated environment for potentially dangerous chemical reactions. These vesicles, produced by the Golgi apparatus, contain digestive enzymes whose varied functions are described on the facing page.

5 Ultimately, the product arrives at the **trans face** ("shipping" side), which is usually oriented toward the free surface of the cell.

4 Small transport vesicles return resident Golgi proteins to cisternae of the cis face for reuse.

3 Further modification and packaging occur as the cisternae migrate to the trans face.

2 Multiple transport vesicles combine to form cisternae on the cis face. Inside the Golgi apparatus, enzymes modify the arriving proteins and glycoproteins. For example, the enzymes may change the carbohydrate structure of a glycoprotein, or they may attach a phosphate group, sugar, or fatty acid to a protein.

Start 1 Some proteins and glycoproteins synthesized in the rough endoplasmic reticulum (RER) are delivered to the Golgi apparatus by **transport vesicles**. The vesicles generally arrive at a cisterna known as the **cis face** ("receiving" side).

2 Cells often need to break down and recycle large organic molecules, and even complex structures like organelles. The breakdown process requires powerful enzymes and often generates toxic chemicals that could damage or kill the cell. Lysosomes isolate those chemical reactions from the rest of the cytoplasm. The three basic functions of the lysosomes are shown below.

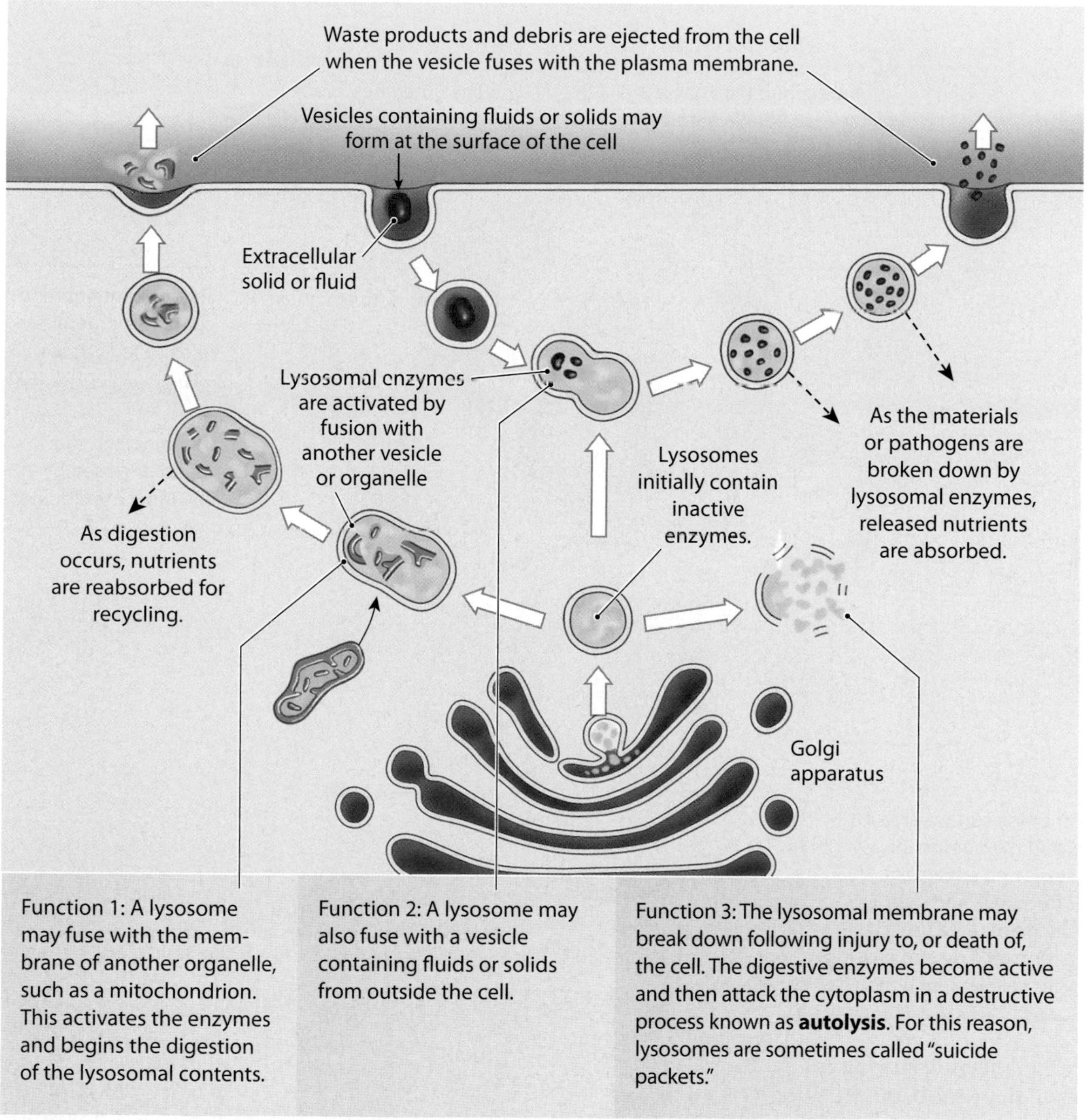

Waste products and debris are ejected from the cell when the vesicle fuses with the plasma membrane.
Vesicles containing fluids or solids may form at the surface of the cell
Extracellular solid or fluid
Lysosomal enzymes are activated by fusion with another vesicle or organelle
As digestion occurs, nutrients are reabsorbed for recycling.
Lysosomes initially contain inactive enzymes.
As the materials or pathogens are broken down by lysosomal enzymes, released nutrients are absorbed.
Golgi apparatus
Function 1: A lysosome may fuse with the membrane of another organelle, such as a mitochondrion. This activates the enzymes and begins the digestion of the lysosomal contents.
Function 2: A lysosome may also fuse with a vesicle containing fluids or solids from outside the cell.
Function 3: The lysosomal membrane may break down following injury to, or death of, the cell. The digestive enzymes become active and then attack the cytoplasm in a destructive process known as **autolysis**. For this reason, lysosomes are sometimes called "suicide packets."

With the exception of mitochondria, all membranous organelles in the cell are either interconnected or in communication through the movement of vesicles. This continuous movement and exchange is called **membrane flow**. In an actively secreting cell, an area equal to the entire membrane surface may be replaced *each hour*. Membrane flow is an example of the dynamic nature of cells. It gives cells a way to change the characteristics of their plasma membranes—the lipids, receptors, channels, anchors, and enzymes—as they grow, mature, or respond to a specific environmental stimulus.

Module 3.6 Review

a. List the three major functions of the Golgi apparatus.

b. The Golgi apparatus produces lysosomes. What do these lysosomes contain?

c. Describe three functions of lysosomes.

LO **3.6** Describe the Golgi apparatus and indicate its specific functions.

Mitochondria are the powerhouses of the cell

The cells of all living things require energy to carry out the functions of life. The organelles that produce energy are the **mitochondria** (mī-tō-KON-drē-uh; singular, mitochondrion; *mitos*, thread + *chondrion*, granule).

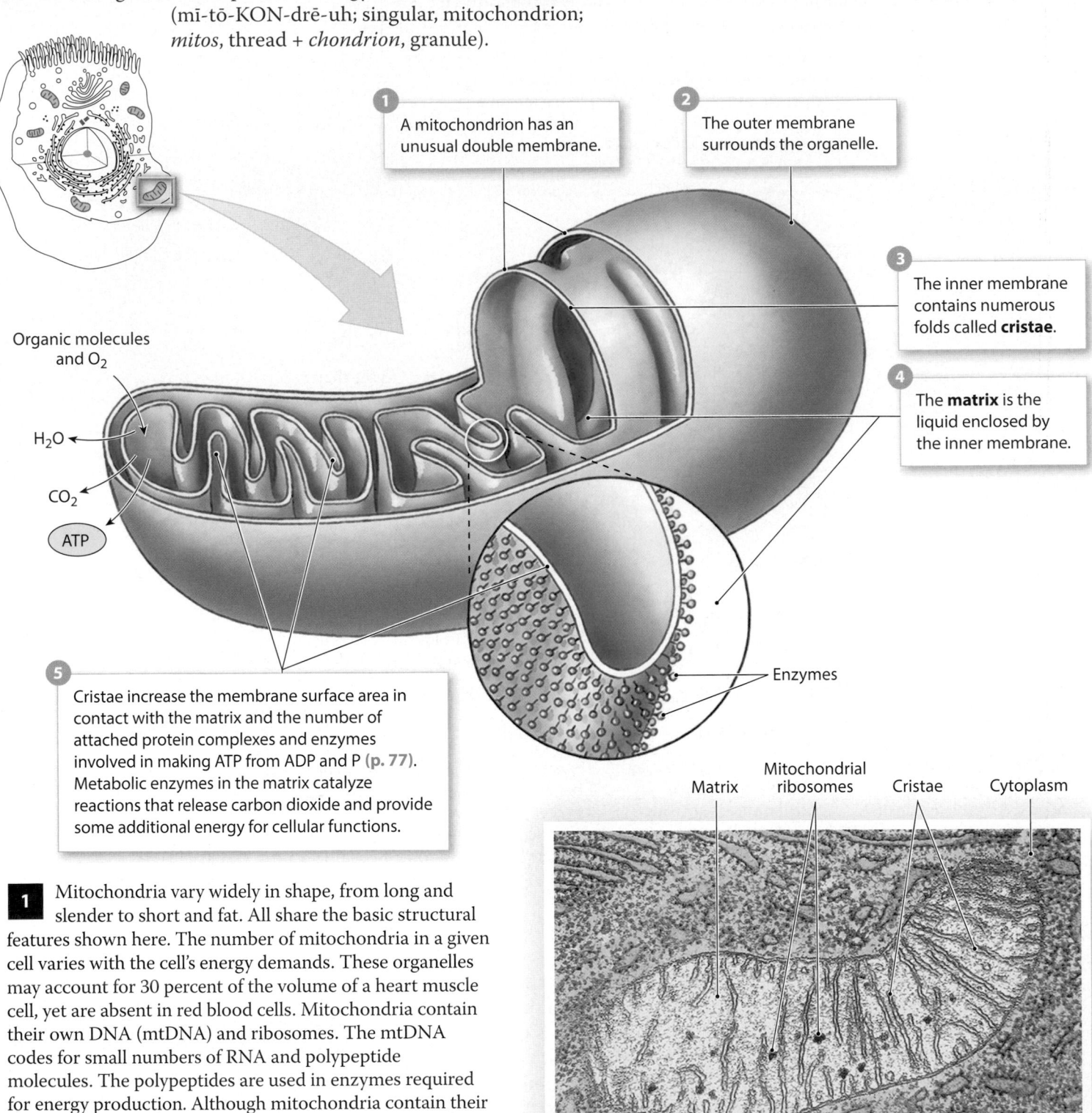

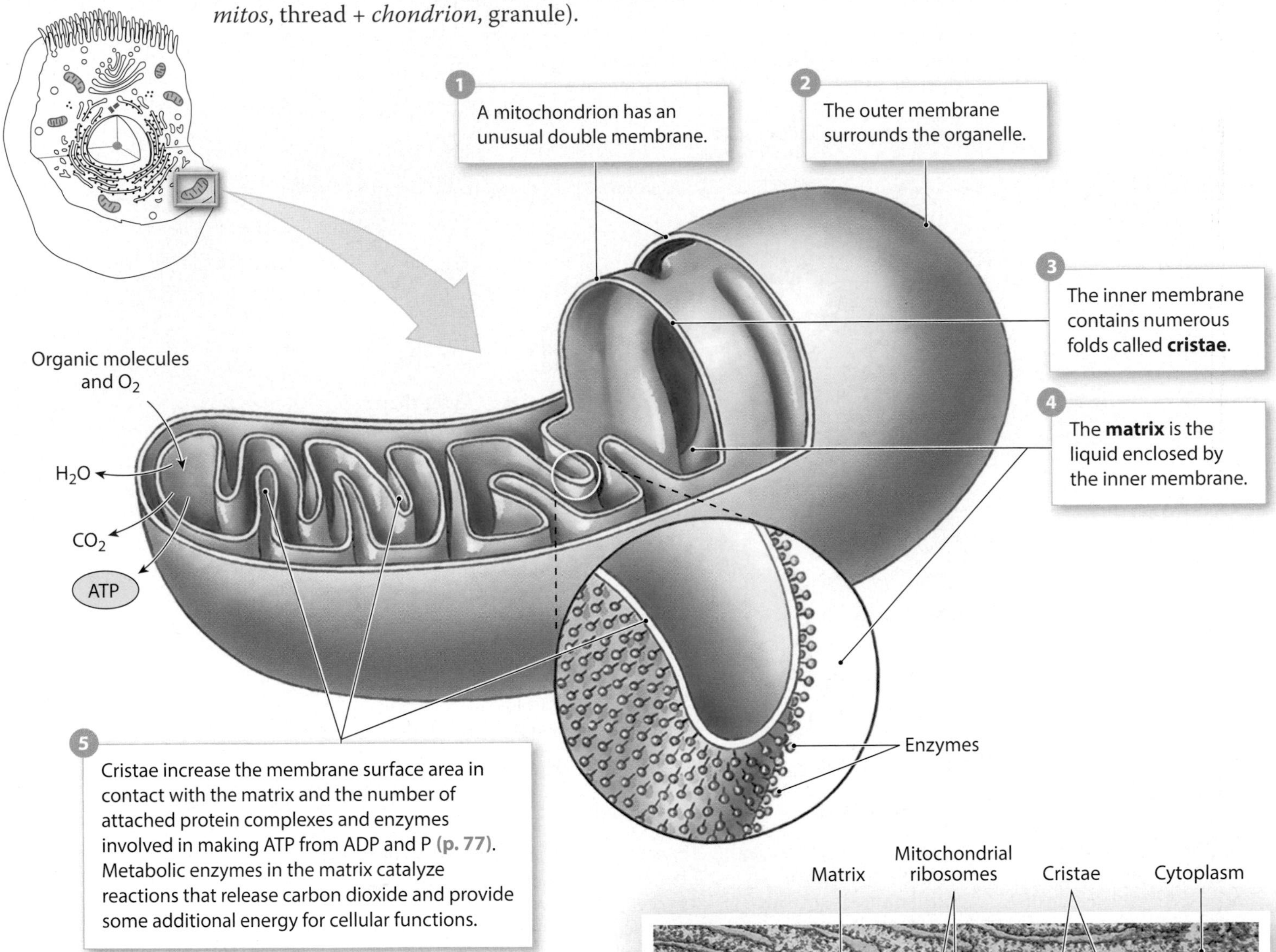
1
A mitochondrion has an unusual double membrane.
2
The outer membrane surrounds the organelle.
3
The inner membrane contains numerous folds called **cristae**.
4
The **matrix** is the liquid enclosed by the inner membrane.
Organic molecules and O_2
H_2O
CO_2
ATP
Enzymes
5
Cristae increase the membrane surface area in contact with the matrix and the number of attached protein complexes and enzymes involved in making ATP from ADP and P (p. 77). Metabolic enzymes in the matrix catalyze reactions that release carbon dioxide and provide some additional energy for cellular functions.

1 Mitochondria vary widely in shape, from long and slender to short and fat. All share the basic structural features shown here. The number of mitochondria in a given cell varies with the cell's energy demands. These organelles may account for 30 percent of the volume of a heart muscle cell, yet are absent in red blood cells. Mitochondria contain their own DNA (mtDNA) and ribosomes. The mtDNA codes for small numbers of RNA and polypeptide molecules. The polypeptides are used in enzymes required for energy production. Although mitochondria contain their own genetic system, their functions depend on imported proteins coded by nuclear DNA.

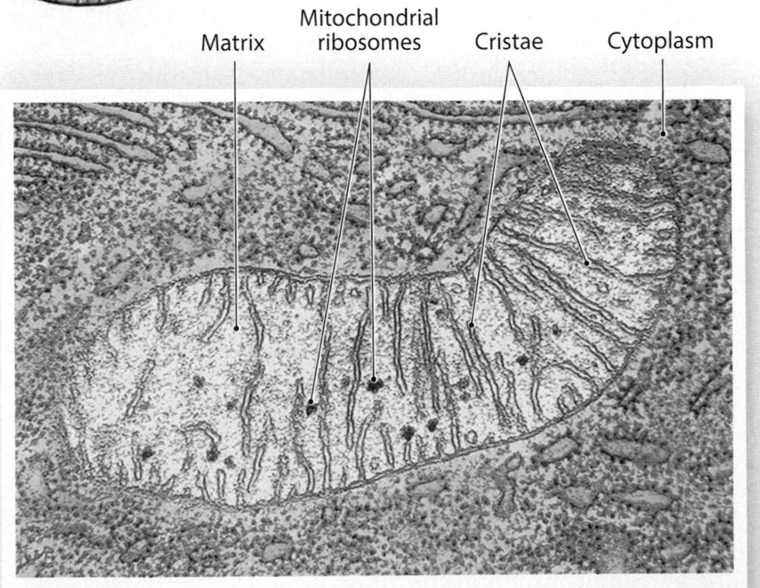
Matrix
Mitochondrial ribosomes
Cristae
Cytoplasm
Mitochondrion TEM × 50,000

2 Most cells generate ATP and other high-energy compounds as they break down carbohydrates, especially glucose. The major steps in the process are shown here.

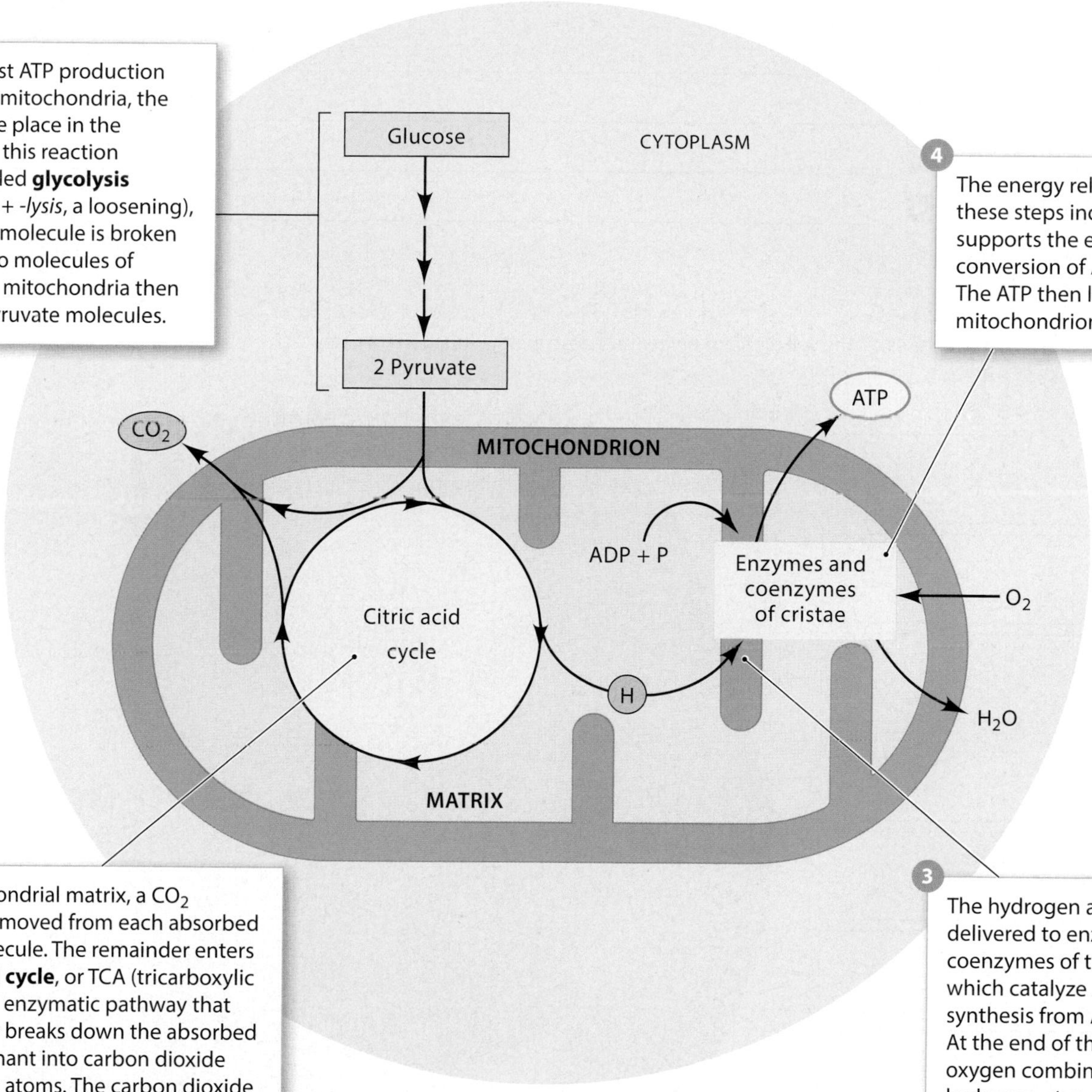

Glucose
CYTOPLASM
2 Pyruvate
CO_2
MITOCHONDRION
ATP
ADP + P
Enzymes and coenzymes of cristae
O_2
Citric acid cycle
H
H_2O
MATRIX

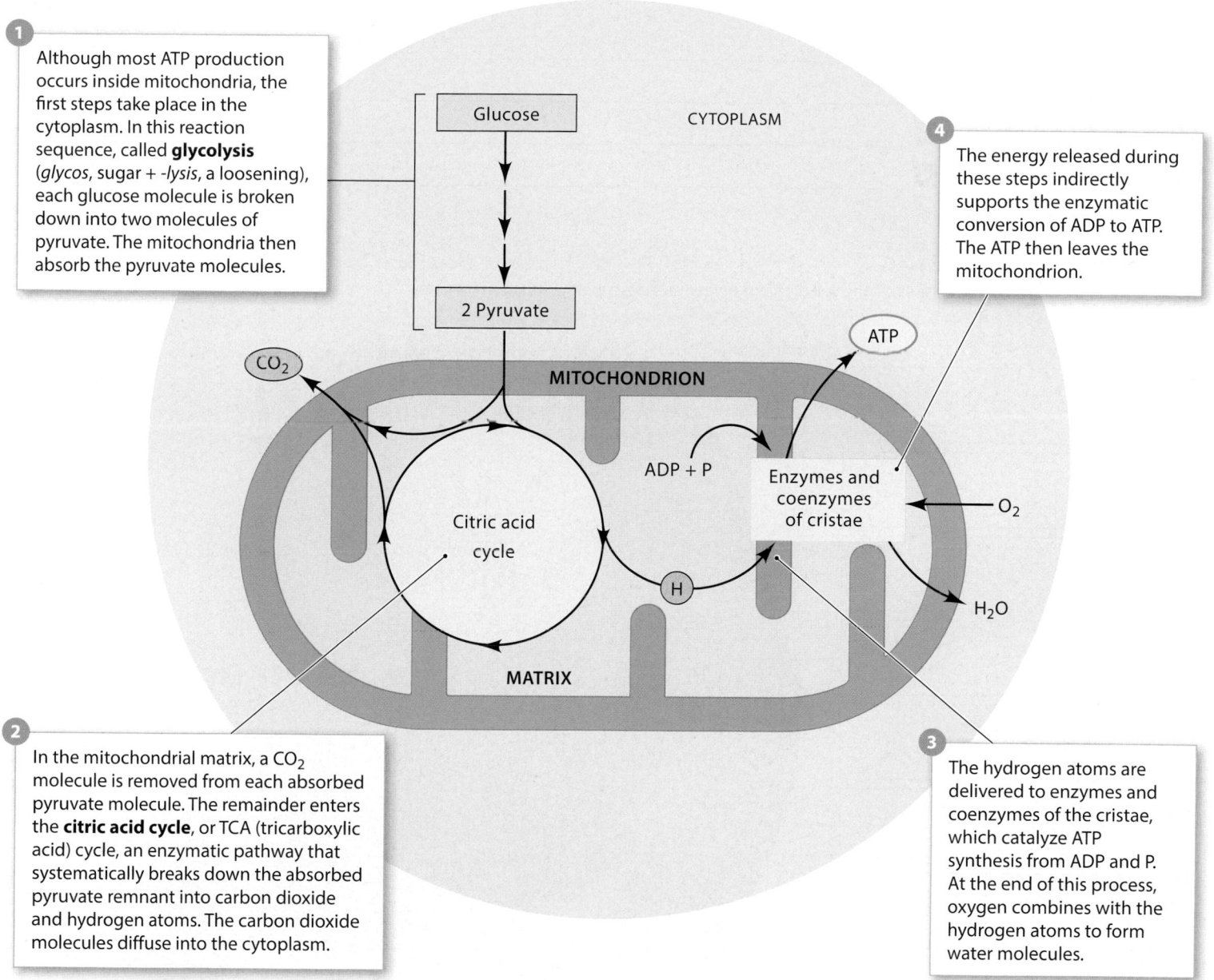

1 Although most ATP production occurs inside mitochondria, the first steps take place in the cytoplasm. In this reaction sequence, called **glycolysis** (*glycos*, sugar + *-lysis*, a loosening), each glucose molecule is broken down into two molecules of pyruvate. The mitochondria then absorb the pyruvate molecules.

2 In the mitochondrial matrix, a CO_2 molecule is removed from each absorbed pyruvate molecule. The remainder enters the **citric acid cycle**, or TCA (tricarboxylic acid) cycle, an enzymatic pathway that systematically breaks down the absorbed pyruvate remnant into carbon dioxide and hydrogen atoms. The carbon dioxide molecules diffuse into the cytoplasm.

3 The hydrogen atoms are delivered to enzymes and coenzymes of the cristae, which catalyze ATP synthesis from ADP and P. At the end of this process, oxygen combines with the hydrogen atoms to form water molecules.

4 The energy released during these steps indirectly supports the enzymatic conversion of ADP to ATP. The ATP then leaves the mitochondrion.

The amount of ATP generated during glycolysis is very small compared with the amount produced by the breakdown of pyruvate within mitochondria. Because mitochondrial activity requires oxygen, this method of ATP production is known as **aerobic metabolism** (*aer*, air + *bios*, life), or *cellular respiration*. Aerobic metabolism in mitochondria produces about 95 percent of the ATP needed to keep a cell alive. (Enzymatic reactions in the cytoplasm produce the rest.) Note that although the chemical reactions that release energy occur in the mitochondria, most of the cellular activities that require energy occur in the surrounding cytoplasm. So mitochondria are like little cellular batteries or fuel cells that provide the energy needed to power cellular functions.

LO 3.7 Describe the structure of a mitochondrion, and explain the significance of mitochondria to cellular function.

Module 3.7 Review

a. Describe the structure of a mitochondrion.

b. Most of a cell's ATP is produced within its mitochondria. What gas do mitochondria require to produce ATP?

c. What does the presence of many mitochondria imply about a cell's energy requirements?

SECTION 1 Review

Vocabulary

In the space provided, write the boldfaced terms introduced in this section that contain the indicated word part.

1. glycos- *(sugar)* 1 ______
2. aero- *(air)* 2 ______
3. micro- *(small)* 3 ______
4. lyso- *(a loosening)* 4 ______

Short answer

Correctly label the indicated structures on the cell diagram below, and describe the functions of each.

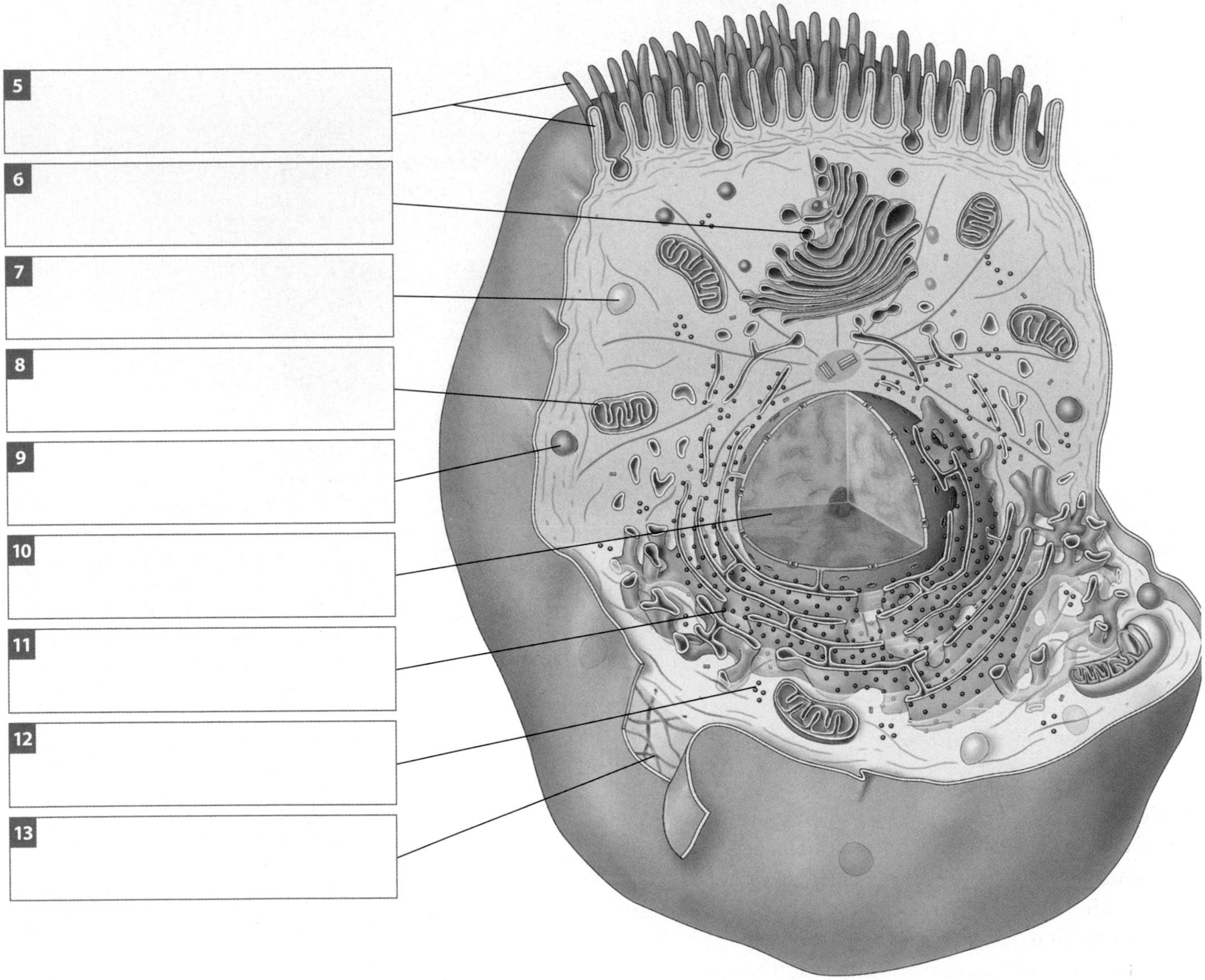

5
6
7
8
9
10
11
12
13

Section integration

What is the advantage of having some of the cellular organelles enclosed by a membrane similar to the plasma membrane?

14 ______

The nucleus is the control center for cellular homeostasis

The nucleus is usually the largest and most conspicuous structure in a cell. If you look at a human cheek cell under a light microscope, it is often the only organelle visible. The nucleus is the control center for cellular operations. A single nucleus stores all the information needed to direct the synthesis of more than 100,000 different proteins in the human body. This genetic information is coded in the sequence of nucleotides in DNA. The nucleus determines the structure of the cell and what functions it can perform by controlling which proteins are synthesized, under what circumstances, and in what amounts. Most cells contain a single nucleus, but exceptions exist. For example, skeletal muscle cells have many nuclei, whereas mature red blood cells have none. Because a cell without a nucleus cannot repair itself, a red blood cell will disintegrate within 3 or 4 months.

1 This is an overview of the role the nucleus plays in preserving homeostasis at the cellular level. When the extracellular environment changes, the cell has mechanisms that can provide rapid short-term adjustments. If the extracellular fluid (ECF) quickly returns to normal, those short-term adjustments may be sufficient to save the cell. But if the environmental stresses persist, the cell must adapt, and this involves long-term changes in its biochemical and physical characteristics. Dramatic, long-term changes in cell structure and function also occur as part of growth, development, and aging.

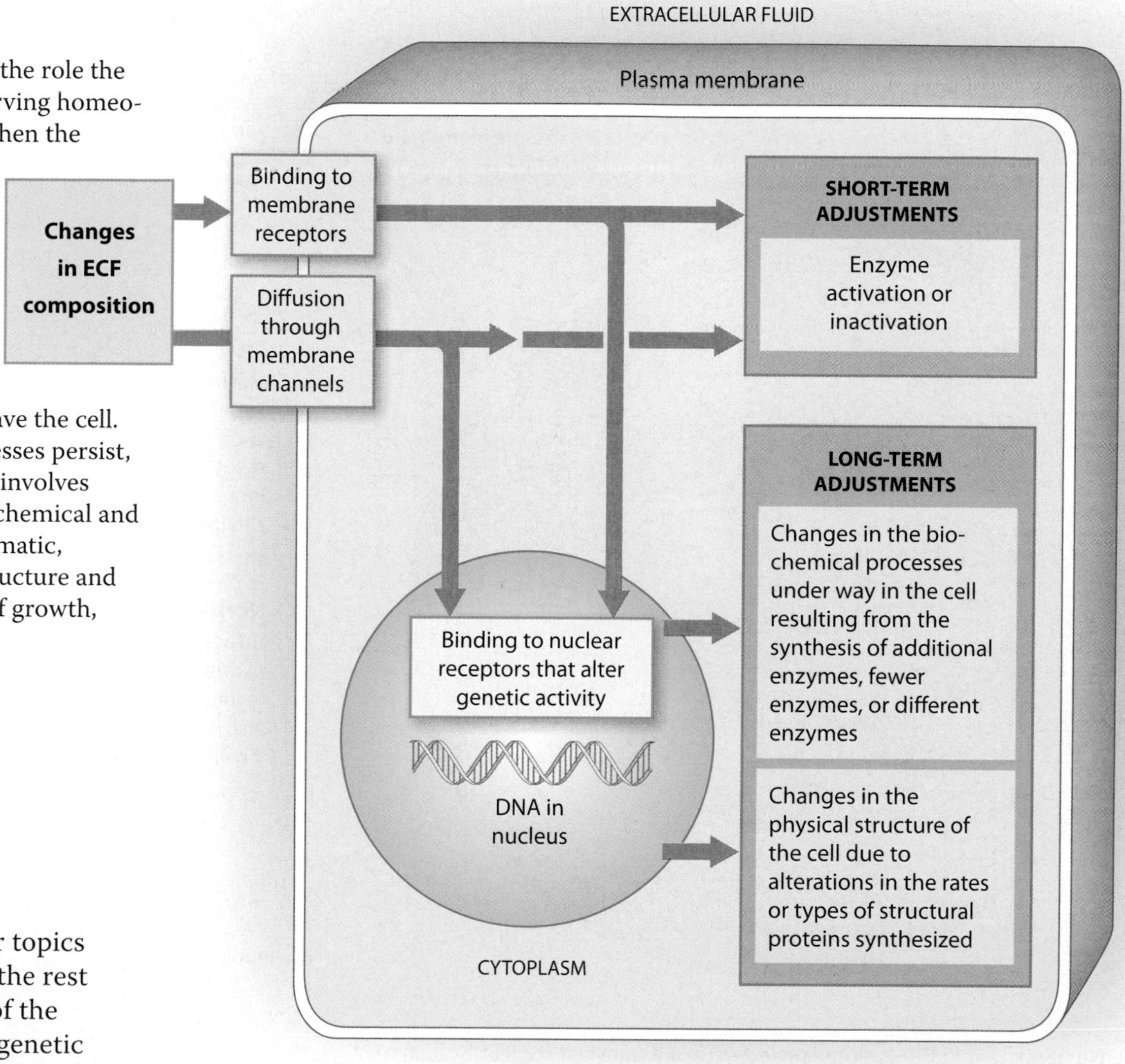
EXTRACELLULAR FLUID
Plasma membrane
Changes in ECF composition
Binding to membrane receptors
Diffusion through membrane channels
SHORT-TERM ADJUSTMENTS
Enzyme activation or inactivation
LONG-TERM ADJUSTMENTS
Changes in the biochemical processes under way in the cell resulting from the synthesis of additional enzymes, fewer enzymes, or different enzymes
Changes in the physical structure of the cell due to alterations in the rates or types of structural proteins synthesized
Binding to nuclear receptors that alter genetic activity
DNA in nucleus
CYTOPLASM

In this section we consider topics of extreme importance to the rest of this text: the structure of the nucleus, the nature of the genetic material, how protein synthesis is controlled and directed, and how genetic information is partitioned during cell division.

Module 3.8 Review

a. Identify the functions of the nucleus.

b. How many nuclei do most cells contain?

c. How is genetic information coded in the cell?

3.8 Describe the role of the nucleus in maintaining homeostasis at the cellular level.

Module 3.9

The nucleus contains DNA, RNA, organizing proteins, and enzymes

The **nucleus** is often the most prominent and visible organelle. In standard histological sections and micrographs, the nucleus stains darkly and is easily identified.

1 The nucleus is the control center for cellular processes. The surrounding nuclear envelope, with its nuclear pores, controls chemical communication between the cytoplasm and the nucleus.

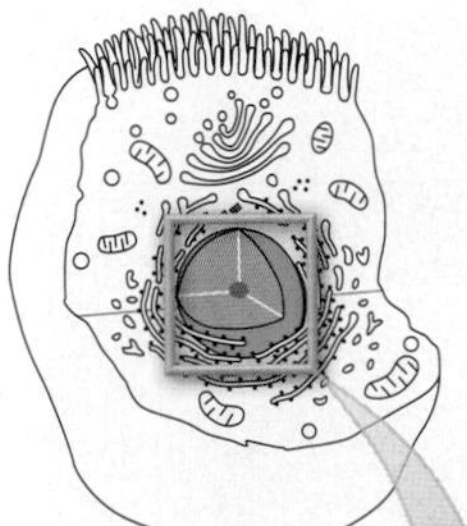

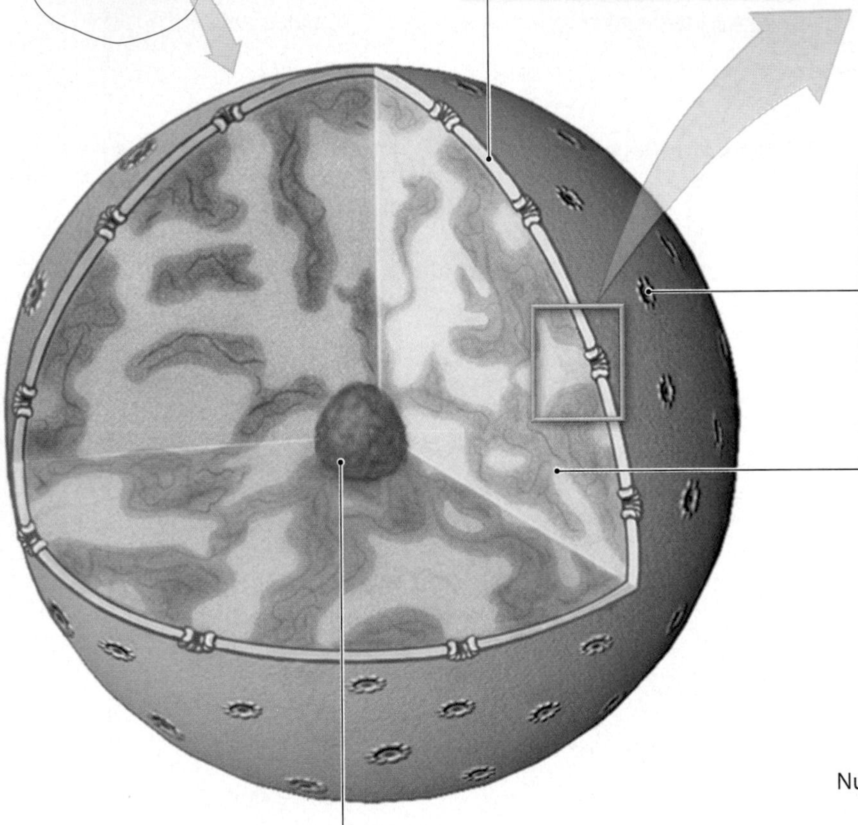

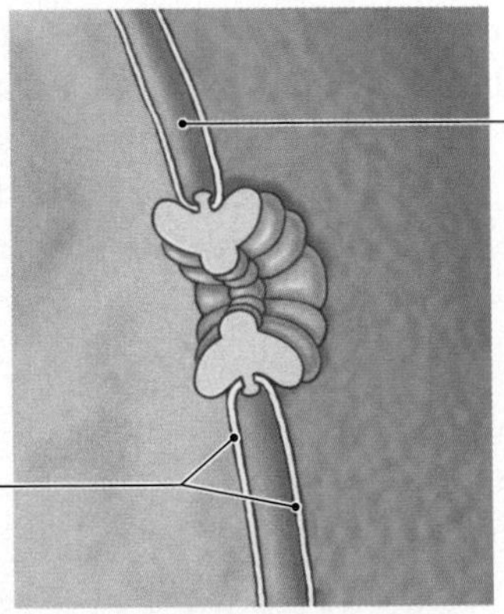

The narrow **perinuclear space** (*peri-*, around) separates the two layers of the nuclear envelope.

The **nuclear envelope**, which surrounds the nucleus and separates it from the cytoplasm, is a double membrane.

Nuclear pores, which account for about 10 percent of the surface of the nucleus, are passageways that permit chemical communication between the nucleus and the cytosol. Proteins at the pores regulate the movement of ions and small molecules, and neither proteins nor DNA can freely cross the nuclear envelope.

Nucleoplasm is the fluid, gel-like substance of the nucleus. It contains the nuclear matrix, a network of fine filaments that provides structural support and may be involved in regulating genetic activity. The nucleoplasm also contains ions, enzymes, RNA and DNA nucleotides, small amounts of RNA, and DNA.

Nucleoli (noo-KLĒ-ō-lī; singular, *nucleolus*) are transient nuclear organelles that synthesize ribosomal RNA. They also assemble the ribosomal subunits, which enter the cytoplasm by carrier-mediated transport at the nuclear pores. Nucleoli are composed of RNA, enzymes, and proteins called **histones**. When the instructions for producing ribosomal proteins and RNA are being carried out, nucleoli form around the portions of DNA containing those instructions. Nucleoli are most prominent in cells that manufacture large amounts of proteins, such as liver, nerve, and muscle cells.

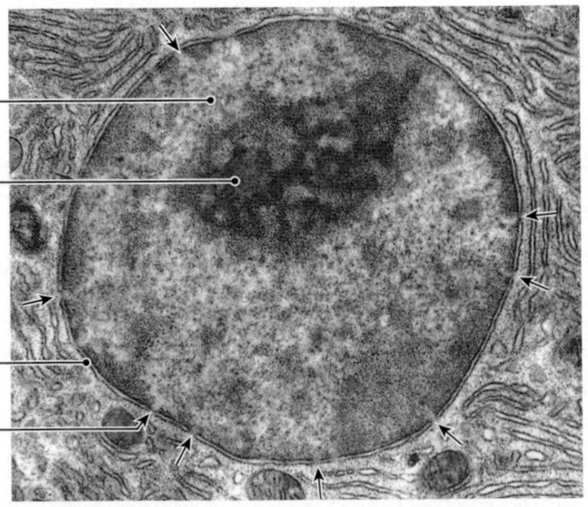
Nucleoplasm
Nucleolus
Nuclear envelope
Nuclear pore

Nucleus TEM × 34,800

2 The DNA in the nucleus stores the instructions for protein synthesis. In the nucleus, the DNA strands are coiled, rather than straight, allowing a great deal of DNA to be packaged in a relatively small space. The coils wrap around histone molecules, forming complexes known as **nucleosomes**. The coiling can be tight or loose, and the entire chain of nucleosomes may coil around other proteins.

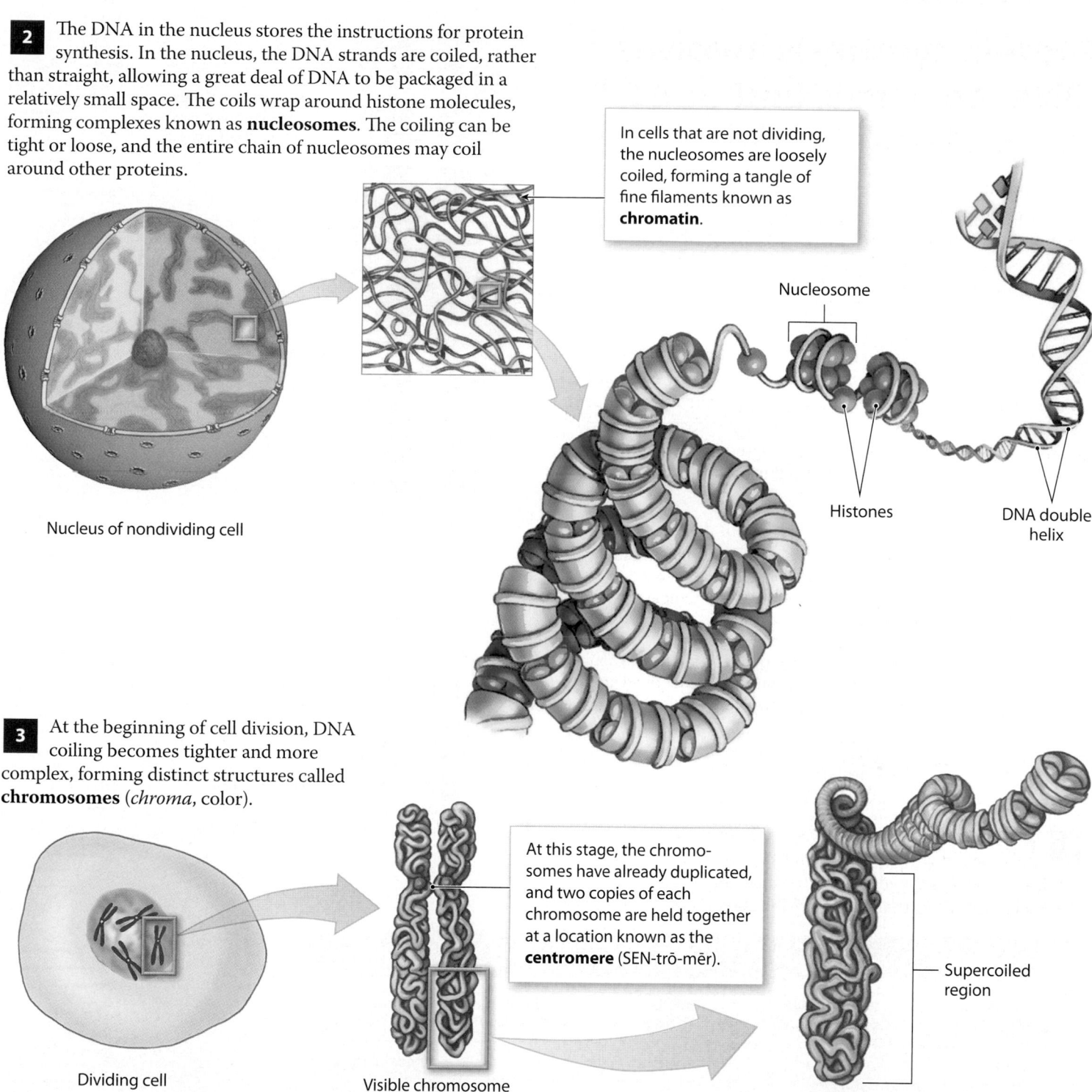

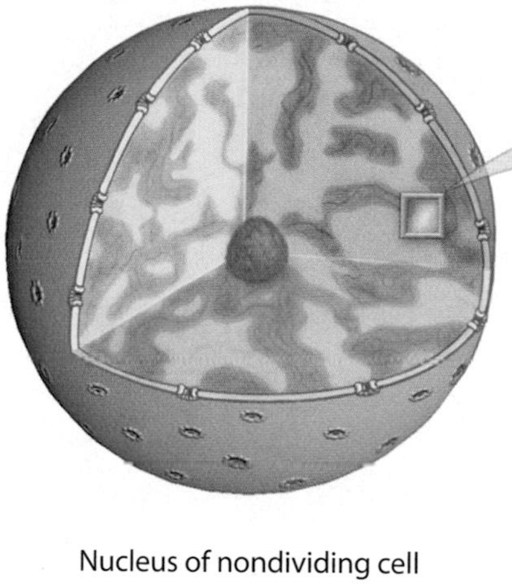
Nucleus of nondividing cell

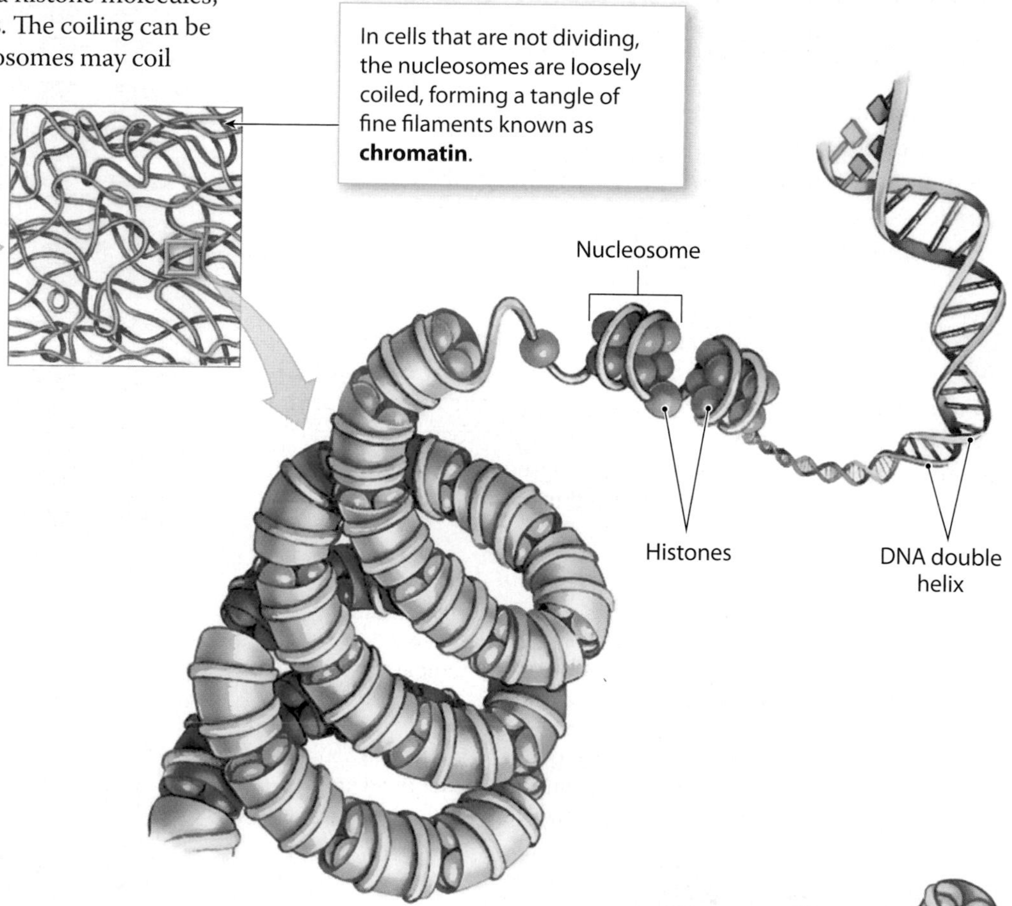
In cells that are not dividing, the nucleosomes are loosely coiled, forming a tangle of fine filaments known as **chromatin**.
Nucleosome
Histones
DNA double helix

3 At the beginning of cell division, DNA coiling becomes tighter and more complex, forming distinct structures called **chromosomes** (*chroma*, color).

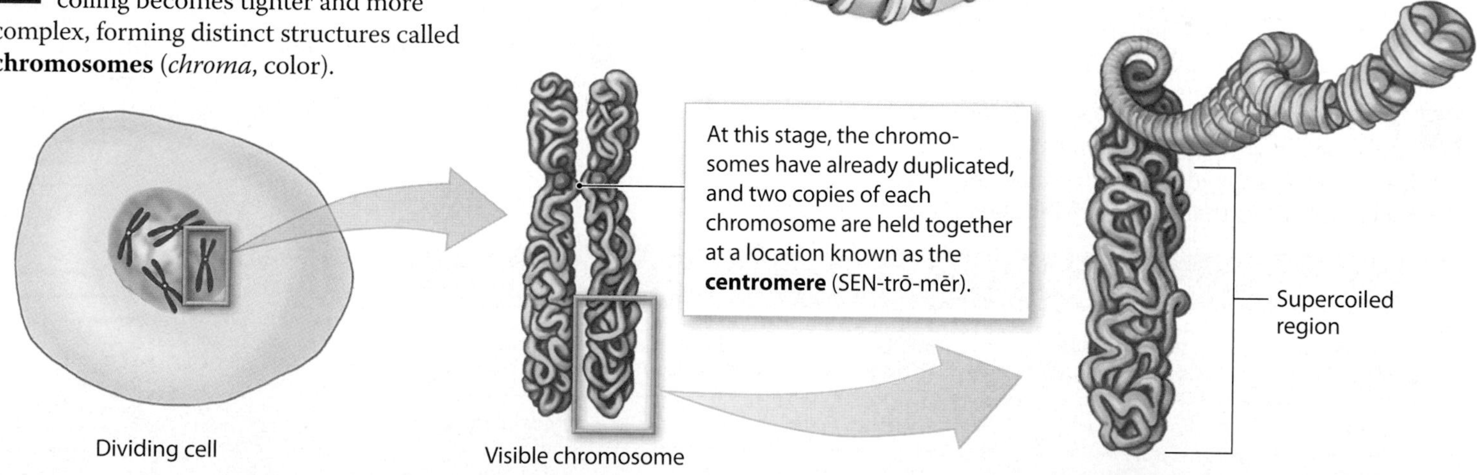
At this stage, the chromosomes have already duplicated, and two copies of each chromosome are held together at a location known as the **centromere** (SEN-trō-mēr).
Supercoiled region
Dividing cell
Visible chromosome

In humans, the nuclei of somatic cells (general body cells, as opposed to sperm and oocytes, which are called sex cells) contain 23 pairs of chromosomes. One member of each pair is derived from the mother and one from the father. The DNA in these chromosomes carries the instructions for synthesizing both proteins and RNA. In addition, some DNA segments have a regulatory function, and still others have as yet, no known function.

Module 3.9 Review

a. Describe the contents and the structure of the nucleus.

b. What molecule in the nucleus contains instructions for making proteins?

c. How many chromosomes are contained within a typical somatic cell?

LO **3.9** Describe the functions of the cell nucleus, and distinguish between chromatin and a chromosome.

Protein synthesis involves DNA, enzymes, and three types of RNA

Watch
MasteringA&P®
A&PFlix™
Protein Synthesis

In this module we introduce the key components and events of protein synthesis, a process we examine in detail in Modules 3.11 and 3.12.

1 DNA consists of long, parallel chains of nucleotides. The two adjacent DNA chains are held together by hydrogen bonding between complementary base pairs. The nitrogenous bases involved are adenine (A), thymine (T), cytosine (C), and guanine (G). Information is stored in the sequence of base pairs. The chemical "language" the cell uses is known as the **genetic code**.

A **gene** is the functional unit of heredity. It contains all the DNA nucleotides needed to produce specific proteins. The number of nucleotides in a gene depends on the size of the polypeptide represented. A short polypeptide chain might need fewer than 300 nucleotides, but the instructions for building a large protein might involve 3000 or more nucleotides.

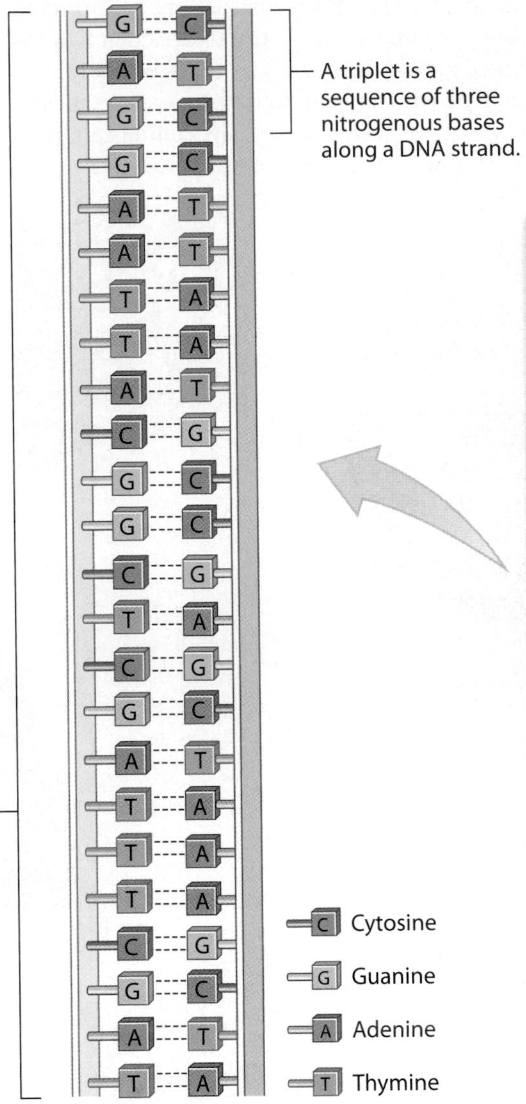

A triplet is a sequence of three nitrogenous bases along a DNA strand.
Cytosine
Guanine
Adenine
Thymine

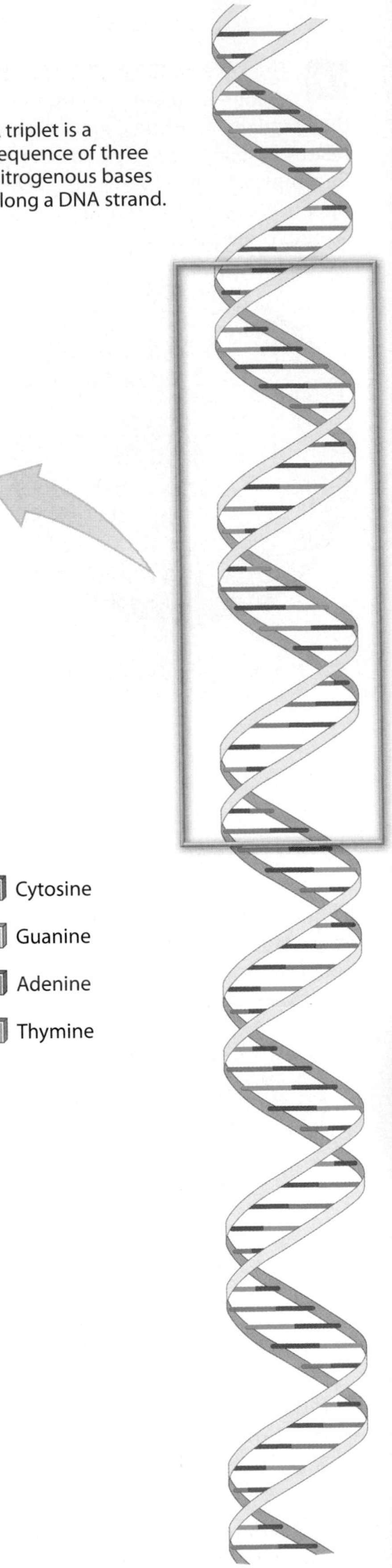

2 The genetic code is also called a **triplet code**, because a sequence of three nitrogenous bases, or **triplet**, specifies the identity of a single amino acid. We discuss this in further detail in Module 3.11.

Examples of the Genetic Code

DNA Triplet Template Strand	DNA Triplet Coding Strand	mRNA Codon	tRNA Anticodon	Amino Acid
AAA	TTT	UUU	AAA	Phenylalanine
AAT	TTA	UUA	AAU	Leucine
ACA	TGT	UGU	ACA	Cysteine
CAA	GTT	GUU	CAA	Valine
TAC	ATG	AUG	UAC	Methionine
TCG	AGC	AGC	UCG	Serine
GGC	CCG	CCG	GGC	Proline
CGG	GCC	GCC	CGG	Alanine

3 Before a gene can affect a cell, the portion of the DNA molecule containing that gene must be uncoiled, and the histones temporarily removed. The factors controlling this process, called **gene activation**, are only partially understood. We know, however, that every gene contains segments responsible for regulating its own activity.

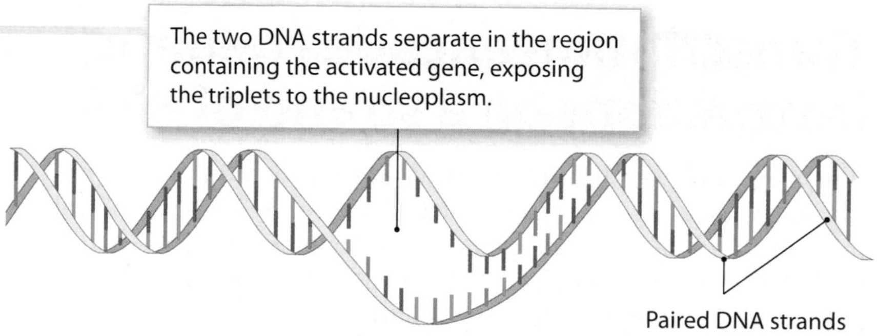

The two DNA strands separate in the region containing the activated gene, exposing the triplets to the nucleoplasm.
Paired DNA strands

4 Enzymes then assemble a strand of **messenger RNA (mRNA)** by interconnecting complementary RNA nucleotides (A, G, C, U). This produces an RNA strand that contains information in triplets of RNA nucleotides. Those triplets are called **codons**. Each codon codes for a specific amino acid. The different codons and their respective amino acids are shown in the Appendix.

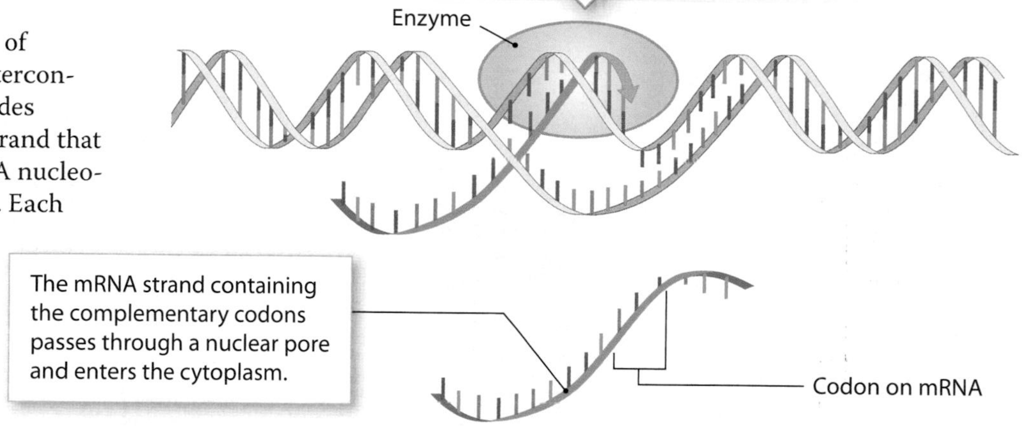

Enzyme
The mRNA strand containing the complementary codons passes through a nuclear pore and enters the cytoplasm.
Codon on mRNA

5 At a ribosome in the cytoplasm, nucleotide triplets called **anticodons** on **transfer RNA (tRNA)** bind to the mRNA codons. There are many different types of tRNA, and each type carries a specific amino acid.

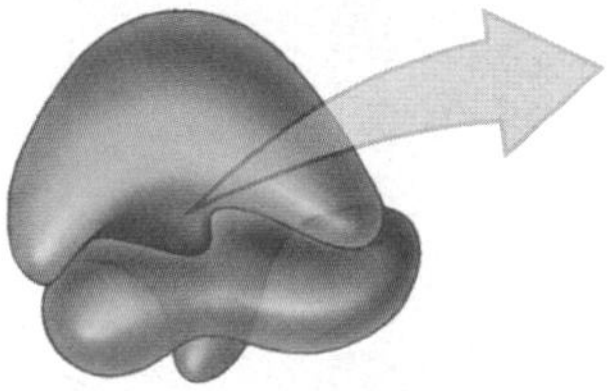

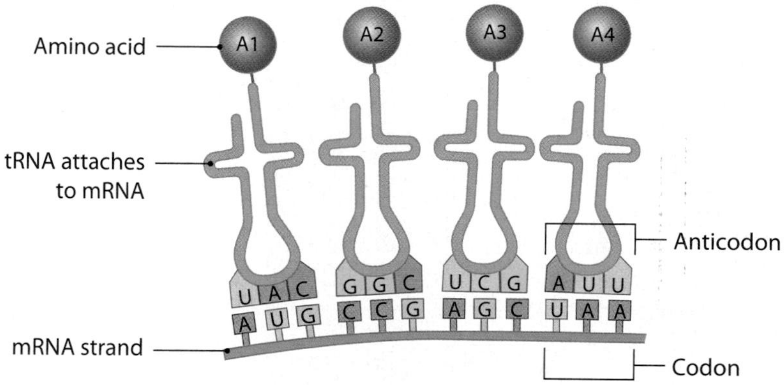

Amino acid
A1
A2
A3
A4
tRNA attaches to mRNA
Anticodon
U A C
G G C
U C G
A U U
A U G
C C G
A G C
U A A
mRNA strand
Codon

6 The **ribosomal RNA (rRNA)** of the ribosome then strings amino acids together to form a polypeptide.

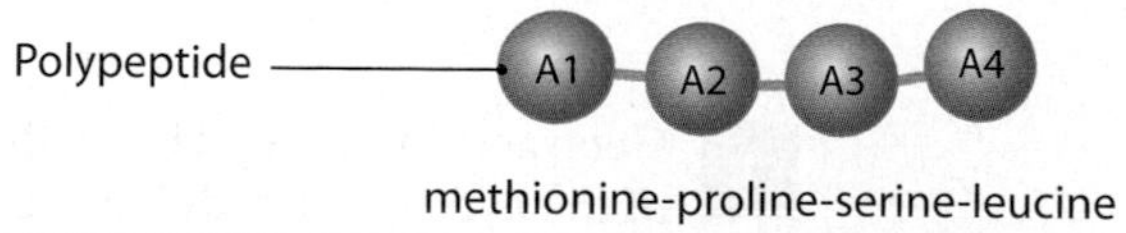

Polypeptide
A1
A2
A3
A4
methionine-proline-serine-leucine

So to summarize:

The DNA triplets determine the sequence of mRNA codons. → The mRNA codons determine the sequence of tRNAs. → The sequence of tRNAs determines the sequence of amino acids in the polypeptide or protein.

Module 3.10 Review

a. What is a gene?

b. Why is the genetic code described as a triplet code?

c. List the three types of RNA involved in protein synthesis.

3.10 Discuss the nature of the genetic code, and summarize the process of protein synthesis.

Transcription encodes genetic instructions on a strand of RNA

A gene not only contains the instructions for the synthesis of proteins and nucleic acids, it also contains triplets that say "do read this message" or "do not read this message," "message starts here," or "message ends here." **Transcription** is RNA production from a DNA template. The term *transcription* is appropriate, as it means "to copy" or "rewrite." All three types of RNA are formed through DNA transcription, but we focus here on the transcription of mRNA, which carries the information needed to synthesize proteins.

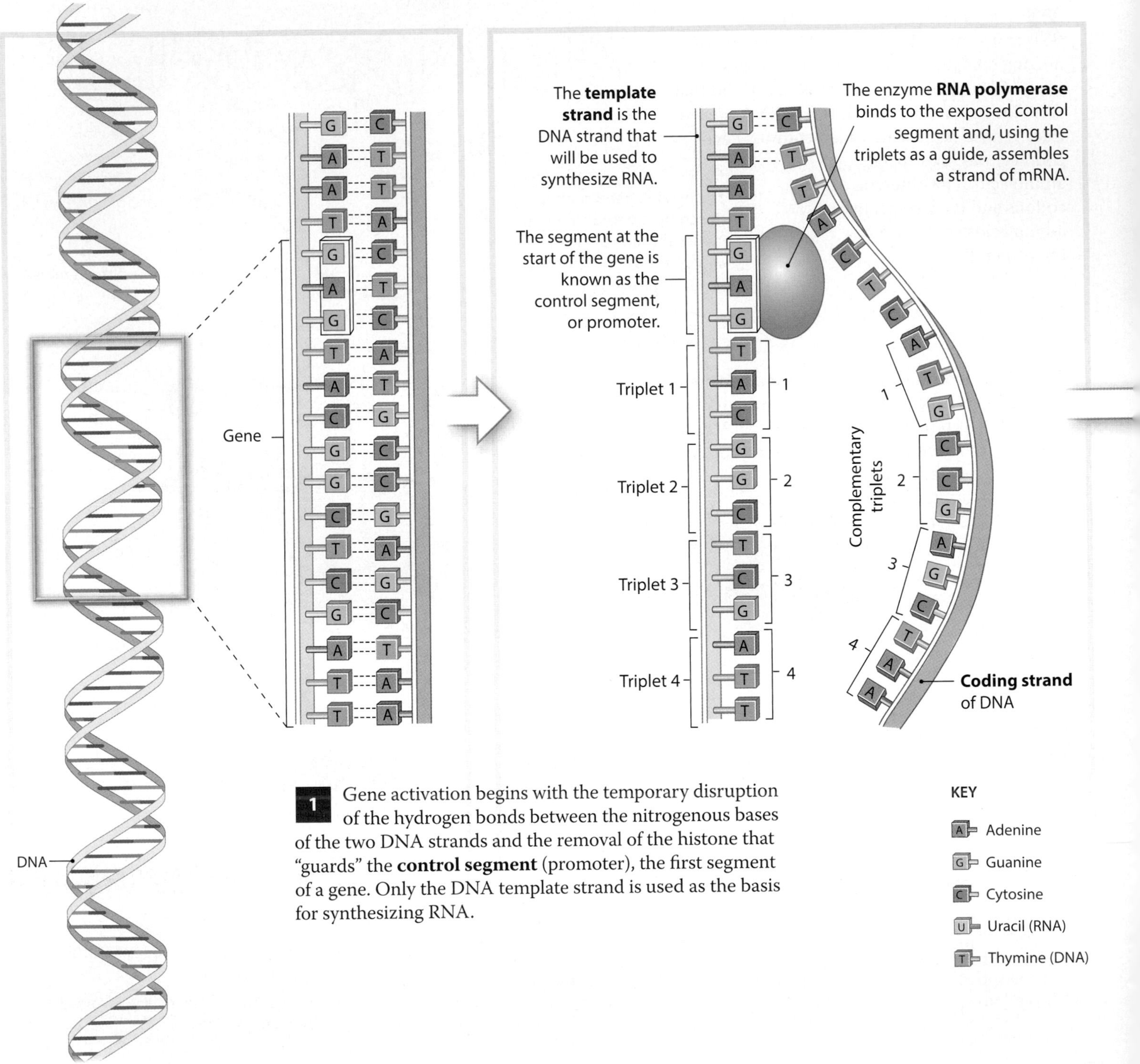

DNA
Gene
The **template strand** is the DNA strand that will be used to synthesize RNA.
The segment at the start of the gene is known as the control segment, or promoter.
Triplet 1
Triplet 2
Triplet 3
Triplet 4
The enzyme **RNA polymerase** binds to the exposed control segment and, using the triplets as a guide, assembles a strand of mRNA.
Complementary triplets
Coding strand of DNA
KEY
Adenine
Guanine
Cytosine
Uracil (RNA)
Thymine (DNA)

1 Gene activation begins with the temporary disruption of the hydrogen bonds between the nitrogenous bases of the two DNA strands and the removal of the histone that "guards" the **control segment** (promoter), the first segment of a gene. Only the DNA template strand is used as the basis for synthesizing RNA.

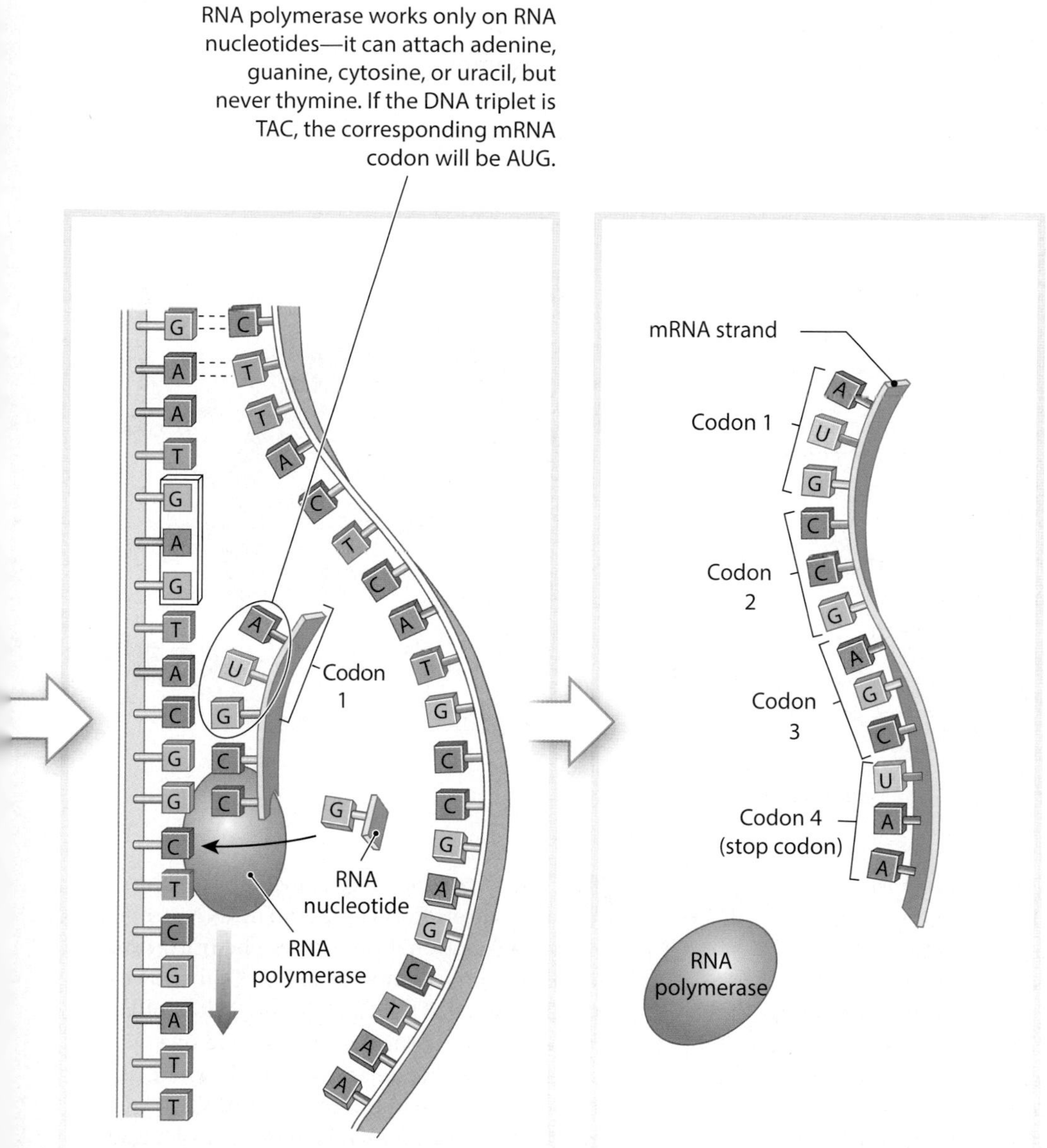

RNA polymerase works only on RNA nucleotides—it can attach adenine, guanine, cytosine, or uracil, but never thymine. If the DNA triplet is TAC, the corresponding mRNA codon will be AUG.
Codon 1
RNA nucleotide
RNA polymerase
mRNA strand
Codon 1
Codon 2
Codon 3
Codon 4 (stop codon)
RNA polymerase

2 RNA polymerase promotes hydrogen bonding between the nitrogenous bases of the DNA template strand and complementary RNA nucleotides in the nucleoplasm. The complementary nucleotides are then strung together by covalent bonding.

3 At the "stop" signal, the enzyme and the mRNA strand detach from the DNA strand and transcription ends. The complementary DNA strands now reassociate as hydrogen bonding occurs between complementary base pairs.

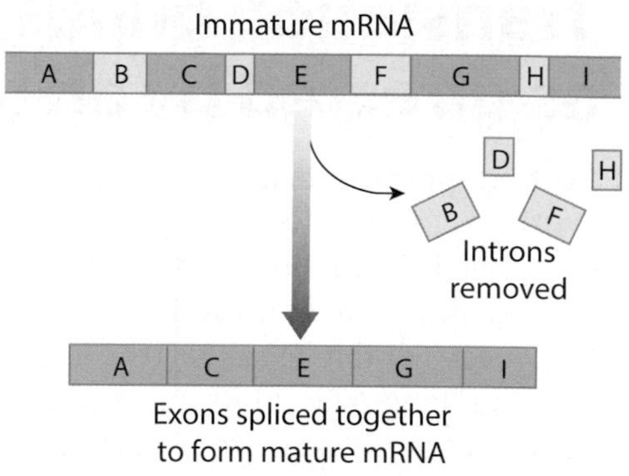

Immature mRNA
A B C D E F G H I
D H B F
Introns removed
A C E G I
Exons spliced together to form mature mRNA

4 Each gene includes a number of triplets that are not needed to build a functional protein. As a result, the mRNA strand assembled during transcription, sometimes called **immature mRNA** or pre-mRNA, must be "edited" before it leaves the nucleus to direct protein synthesis. In this RNA processing, noncoding intervening sequences, called **introns**, are snipped out, and the remaining coding segments, or **exons**, are spliced together.

The process creates a shorter, functional strand of mRNA that then enters the cytoplasm through a nuclear pore. Intron removal is extremely important and tightly regulated. By changing the editing instructions and removing different introns, a single gene can produce mRNAs that code for several different proteins. Some introns, however, act as enzymes to catalyze their own removal. How this variable editing is regulated is unknown.

Module 3.11 Review

a. What is transcription?
b. Define DNA template strand.
c. What process would be affected if a cell could not synthesize the enzyme RNA polymerase?

Translation builds polypeptides as directed by an mRNA strand

Protein synthesis is the assembling of functional polypeptides in the cytoplasm. Protein synthesis takes place through **translation**, the formation of a linear chain of amino acids, using the information from an mRNA strand. Again, the name is appropriate: To translate is to present the same information in a different language. In this case, a message written in the "language" of nucleic acids (the sequence of nitrogenous bases) is translated by ribosomes into the "language" of proteins (the sequence of amino acids in a polypeptide chain). Each mRNA codon designates a particular amino acid to be inserted into the polypeptide chain. There are three phases to translation: initiation, elongation, and termination.

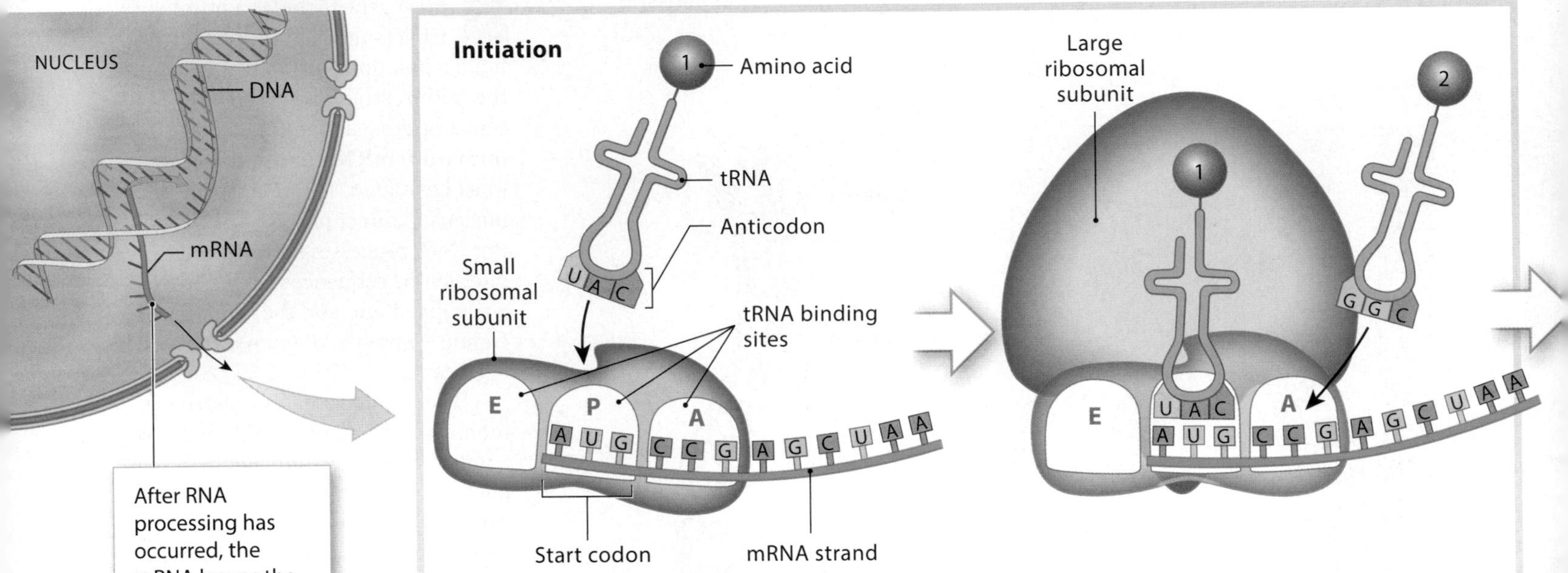
NUCLEUS
DNA
mRNA
Initiation
Amino acid
tRNA
Anticodon
Small ribosomal subunit
tRNA binding sites
E
P
A
Start codon
mRNA strand
Large ribosomal subunit
U A C
A U G C C G A G C U A A
G G C

After RNA processing has occurred, the mRNA leaves the nucleus through a nuclear pore. Translation then occurs in the cytoplasm.

1 The **initiation** phase of translation begins when the mRNA strand binds to a small ribosomal subunit near the P site, one of three adjacent tRNA binding sites. A tRNA then binds to the P site and to the start codon on the mRNA strand. Binding occurs between the nucleotides of the start codon and three complementary nucleotides in a segment of the tRNA strand known as the **anticodon**.

2 The small and large ribosomal subunits then interlock around the mRNA strand, forming a functional ribosome. The **initiation complex** is now complete and protein synthesis can proceed. The tRNA in the P site holds what will become the first amino acid of a peptide chain. The adjacent A site is where an additional tRNA can bind to the mRNA strand. More than 20 kinds of tRNA exist, each with a different nucleotide sequence in the anticodon. Each tRNA carries an amino acid, and there is at least one tRNA anticodon that corresponds to each of the amino acids used in protein synthesis.

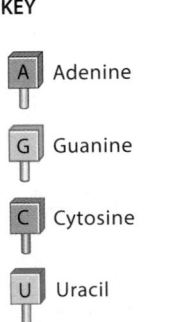

KEY

A Adenine

G Guanine

C Cytosine

U Uracil

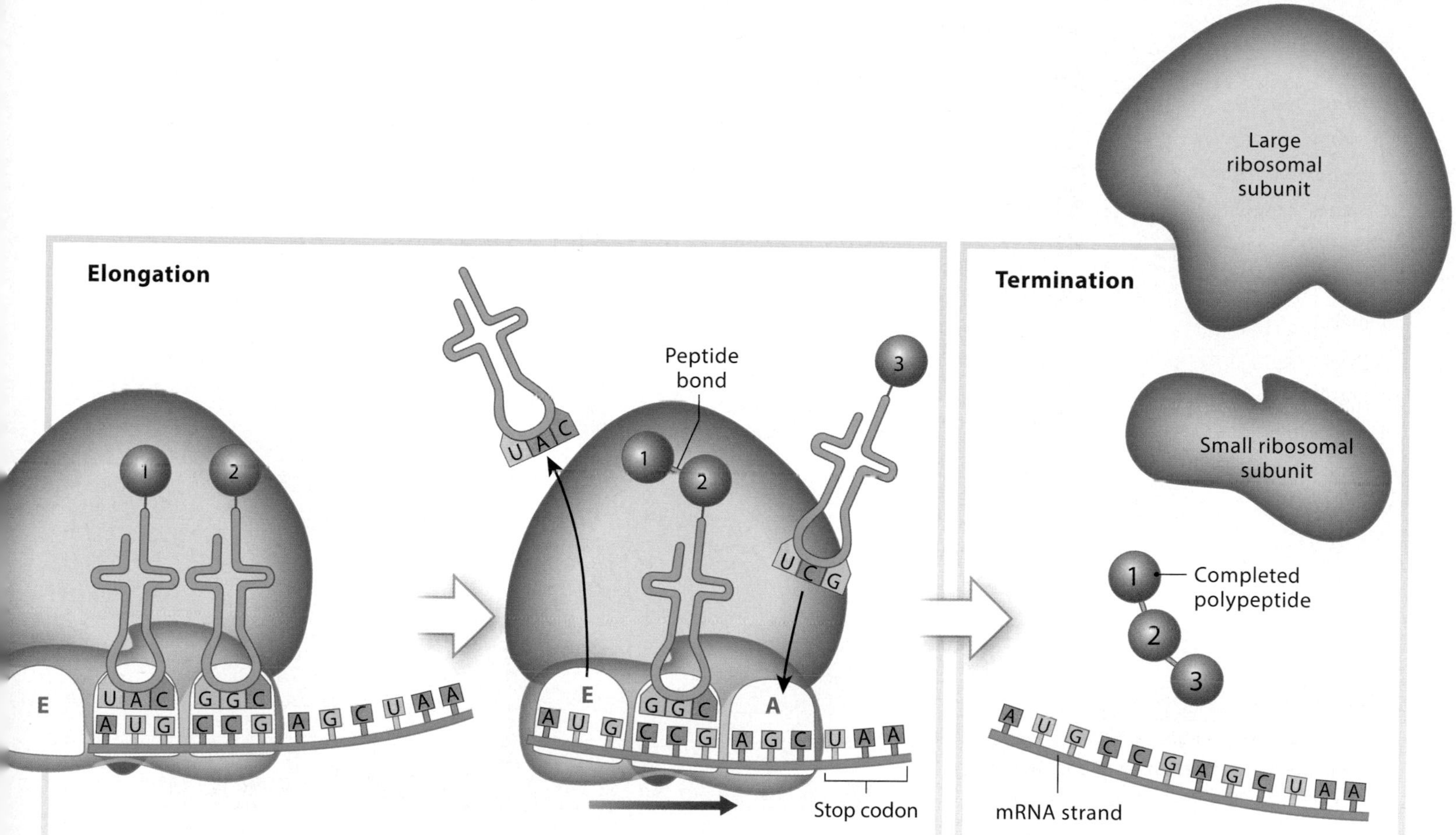
Elongation
Termination
Large ribosomal subunit
Small ribosomal subunit
Peptide bond
Completed polypeptide
Stop codon
mRNA strand
E
A

3 In **elongation**, amino acids are added one by one to the growing polypeptide chain. When a complementary tRNA binds to the A site, ribosomal enzymes remove the amino acid from the tRNA at the P site and attach it to the amino acid delivered to the A site. The ribosome then moves one codon farther along the mRNA strand, and the tRNA that moves from the P site to the E site is released into the cytoplasm. The released tRNA can now bind another amino acid of the same type and repeat the cycle. Elongation ends when the ribosome reaches the stop codon at the end of the mRNA strand.

4 **Termination** occurs as a protein releasing factor, not a tRNA molecule, recognizes the stop codon. A ribosomal enzyme then breaks the bond between the polypeptide and the tRNA in the P site, releasing the polypeptide. Other ribosomal enzymes separate the ribosomal subunits and free the intact strand of mRNA.

Translation proceeds swiftly, producing a typical protein in about 20 seconds. The mRNA strand remains intact, and it can interact with other ribosomes to create additional copies of the same polypeptide chain. The process does not continue indefinitely, however, because after a few minutes to a few hours, mRNA strands are broken down, and the nucleotides recycled. However, large numbers of protein chains can be produced during that time. Although only two mRNA codons are "read" by a ribosome at any one time, many ribosomes can bind to a single mRNA strand, and a multitude of identical proteins may be quickly and efficiently produced.

Module 3.12 Review

a. What is translation?

b. The nucleotide sequence of three mRNA codons is AUU-GCA-CUA. What is the complementary anticodon sequence for the second codon?

c. During the process of transcription, a nucleotide was deleted from an mRNA sequence that coded for a protein. What effect will this deletion have on the amino acid sequence of the protein?

LO **3.12** Summarize the process of translation.

Matching

Match each lettered term with the most closely related description.

a. introns
b. transcription
c. tRNA
d. chromosomes
e. exons
f. genetic information
g. nucleus
h. thymine
i. mRNA
j. gene
k. uracil
l. nuclear envelope
m. nuclear pore
n. nucleoli

1 DNA strands and histones
2 DNA nitrogenous base
3 Double membrane
4 mRNA noncoding regions
5 RNA nitrogenous base
6 Assemble ribosomal subunits
7 Passageway for functional mRNA
8 mRNA formation
9 Functional unit of heredity
10 mRNA coding regions
11 Codon
12 Anticodon
13 Cell control center
14 DNA nucleotide sequence

1 ______
2 ______
3 ______
4 ______
5 ______
6 ______
7 ______
8 ______
9 ______
10 ______
11 ______
12 ______
13 ______
14 ______

Short answer

The sequence of DNA bases at right are from a protein-coding gene. Use the sequence as the basis for answering the following questions.

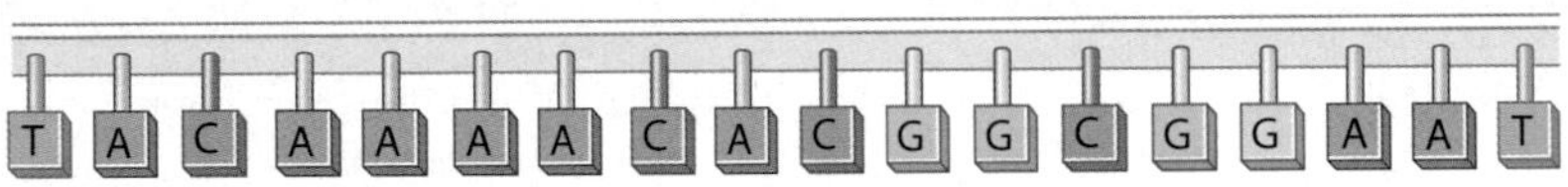
T A C A A A A C A C G G C G G A A T

15 Provide the corresponding mRNA base sequence, and insert a slash mark (/) between the codons.

16 Convert the mRNA codons you decoded above into tRNA anticodons.

15 ______

16 ______

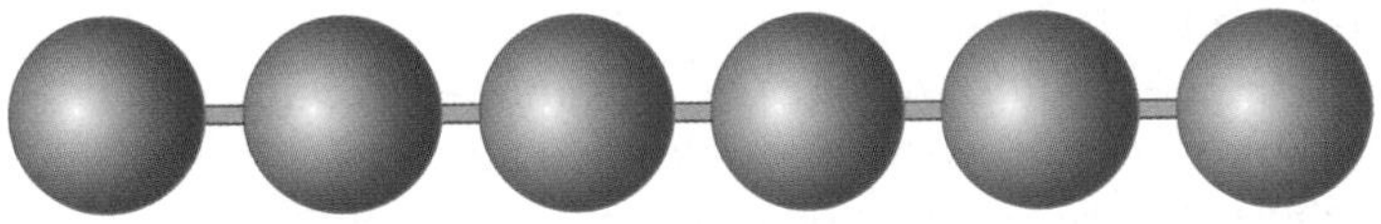

17 Using the genetic code table in Module 3.10, translate the anticodon sequence into the amino acid sequence of this polypeptide.

17 ______

Section integration

The nucleus is often described as the control center of the cell. Explain the role of the nucleus in maintaining homeostasis.

18 ______

The plasma membrane is a selectively permeable membrane

Watch
MasteringA&P®
A&PFlix™
Membrane Transport

Because the plasma membrane is an effective barrier, conditions inside the cell can be much different from conditions outside the cell. However, the barrier cannot be perfect, because cells are not self-sufficient, and their activities must be coordinated. In this section we consider how the plasma membrane selectively regulates the movement of materials into and out of the cell.

1 **Permeability** is the property of the plasma membrane that determines precisely which substances can enter or leave the cytoplasm.

Freely permeable membranes

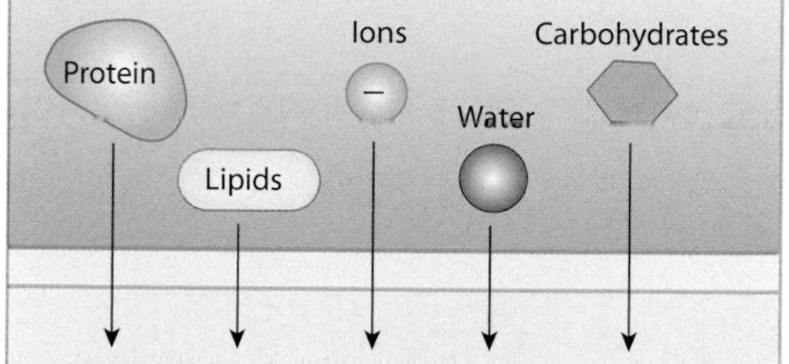
Protein
Ions
Carbohydrates
Water
Lipids
–

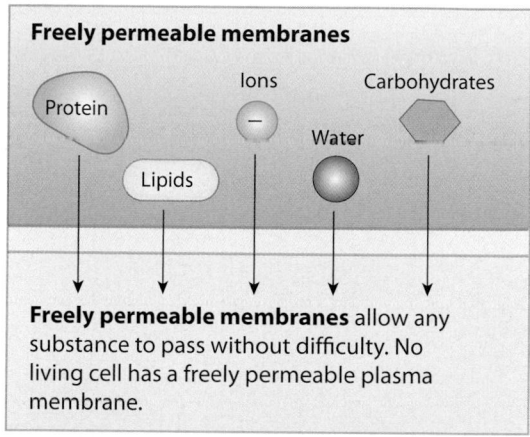

Freely permeable membranes allow any substance to pass without difficulty. No living cell has a freely permeable plasma membrane.

Selectively permeable membranes

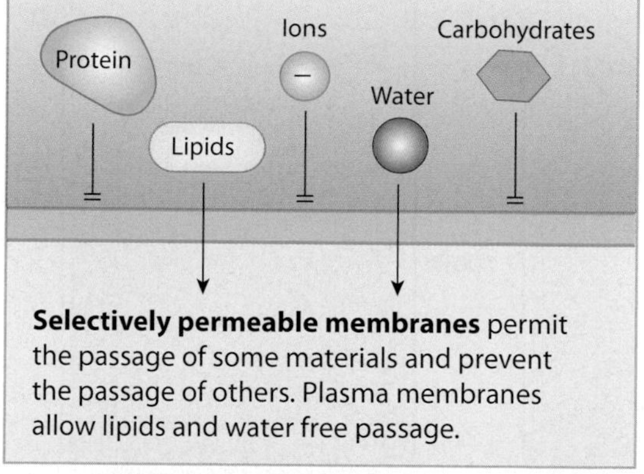
Protein
Ions
Carbohydrates
Water
Lipids
–

Selectively permeable membranes permit the passage of some materials and prevent the passage of others. Plasma membranes allow lipids and water free passage.

Impermeable membranes

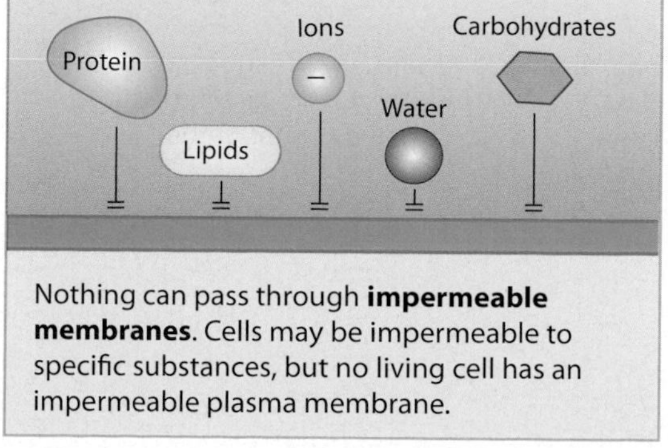
Protein
Ions
Carbohydrates
Water
Lipids
–

Nothing can pass through **impermeable membranes**. Cells may be impermeable to specific substances, but no living cell has an impermeable plasma membrane.

2 **Selectively permeable membranes** permit the free passage of some materials and restrict the passage of others. The distinction may be based on size, molecular shape, lipid solubility, electrical charge, or other factors. Cells differ in their permeabilities, depending on what lipids and proteins are present in the plasma membrane and how these components are arranged.

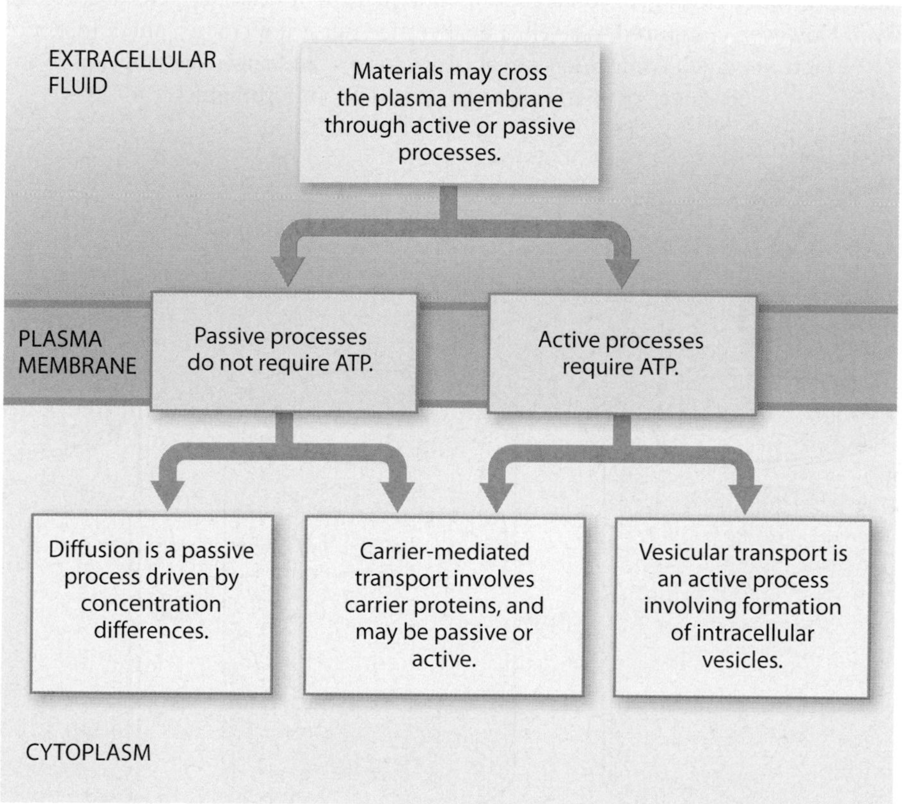
EXTRACELLULAR FLUID
Materials may cross the plasma membrane through active or passive processes.
PLASMA MEMBRANE
Passive processes do not require ATP.
Active processes require ATP.
Diffusion is a passive process driven by concentration differences.
Carrier-mediated transport involves carrier proteins, and may be passive or active.
Vesicular transport is an active process involving formation of intracellular vesicles.
CYTOPLASM

Module 3.13 Review

a. Define permeability.

b. Identify three different types of membranes based on permeability.

c. Distinguish between passive and active processes.

LO 3.13 Contrast permeable, selectively permeable, and impermeable membranes.

Module 3.14

Diffusion is passive movement driven by concentration differences

Ions and molecules in liquids and gases are constantly in motion, colliding and bouncing off one another and off obstacles in their paths. Driven by thermal energy, the movement is random: A molecule can bounce in any direction. One result of this continuous random motion is that, over time, the molecules in any given space will tend to become evenly distributed. This distribution process, the net movement of a substance from an area of higher concentration to an area of lower concentration, is called **diffusion**. The difference between the high and low concentrations is a **concentration gradient**. After the gradient has been eliminated, the molecular motion continues, but net movement no longer occurs in any particular direction.

1 Diffusion in air and water is slow, and it is most important over very short distances. A simple demonstration—observing a colored sugar cube placed in a beaker of water—can give you a good mental picture of what happens at the cellular level. However, compared to a cell, a beaker of water is enormous, and additional factors (which we will ignore) account for dye and sugar distribution over distances of centimeters as opposed to micrometers.

A colored sugar cube is placed in a large volume of clear water.

The dissolving cube establishes a steep concentration gradient for both the sugar and the dye. The sugar and dye concentrations are high near the cube and negligible elsewhere.

As time passes, the sugar and dye molecules spread through the solution.

Eventually the sugar and dye molecules are distributed evenly throughout the solution.

2 In extracellular fluids, water and dissolved solutes diffuse freely. A plasma membrane, however, acts as a barrier that selectively restricts diffusion: Some substances pass through easily, but others cannot penetrate the membrane. An ion or a molecule can diffuse across a plasma membrane only by (1) crossing the lipid portion of the membrane or (2) passing through a membrane channel.

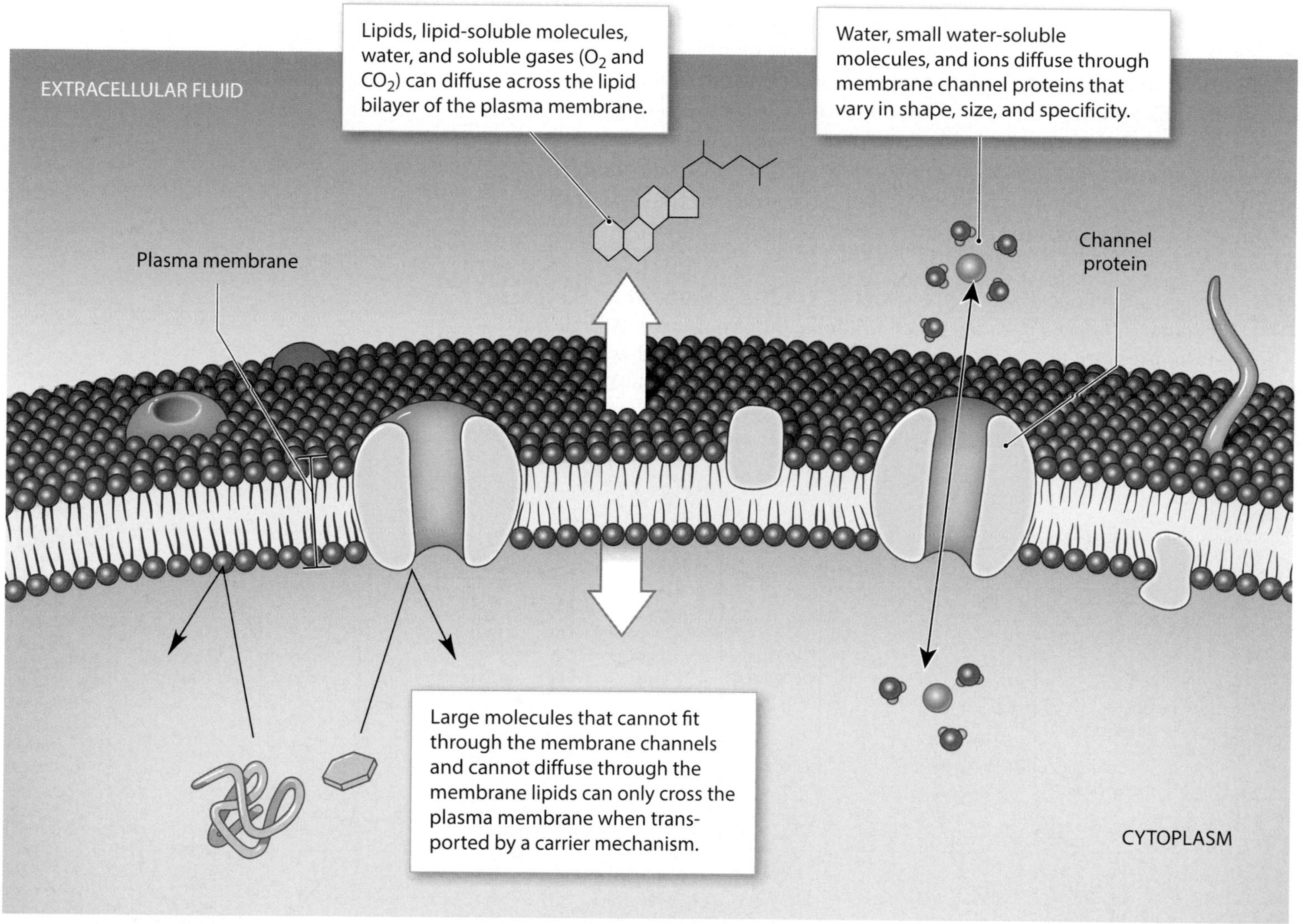

Lipids, lipid-soluble molecules, water, and soluble gases (O_2 and CO_2) can diffuse across the lipid bilayer of the plasma membrane.
Water, small water-soluble molecules, and ions diffuse through membrane channel proteins that vary in shape, size, and specificity.
EXTRACELLULAR FLUID
Plasma membrane
Channel protein
Large molecules that cannot fit through the membrane channels and cannot diffuse through the membrane lipids can only cross the plasma membrane when transported by a carrier mechanism.
CYTOPLASM

The diffusion of nutrients, waste products, and dissolved gases must keep pace with the demands of active cells. Important factors that influence diffusion rates are:

- Distance: The shorter the distance, the more quickly concentration gradients are eliminated. In the human body, few cells are farther than 25 μm from a blood vessel.
- Molecule size: The smaller the molecule size, the faster the diffusion rate. Ions and small organic molecules, such as glucose, diffuse more rapidly than do large proteins.
- Temperature: The higher the temperature, the faster the diffusion rate.
- Concentration gradient: The steeper the concentration gradient, the faster diffusion proceeds.
- Electrical forces: Opposite electrical charges (+ and –) attract each other; like charges (+ and + or – and –) repel each other. Electrical attraction or repulsion can accelerate or reduce the ion diffusion rate.

Module 3.14 Review

a. Define diffusion.

b. Identify factors that influence diffusion rates.

c. How would a decrease in the oxygen concentration in the lungs affect oxygen diffusion into the blood?

LO 3.14 Explain the process of diffusion, and identify its significance in the body.

Osmosis is the diffusion of water molecules across a selectively permeable membrane

Intracellular and extracellular fluids are solutions with a variety of dissolved materials. Each solute diffuses as though it were the only material in solution. If we ignore individual solute identities and simply count ions and molecules, we find that the total concentration of dissolved ions and molecules on either side of the plasma membrane stays the same. This state of equilibrium persists because a typical plasma membrane is freely permeable to water. The diffusion of water across a selectively permeable membrane is so important that it is given a special name: **osmosis** (oz-MŌ-sis; *osmos*, a push).

1 The movement of water driven by osmosis is called **osmotic flow**. The greater the initial difference in solute concentrations, the stronger the osmotic flow. The **osmotic pressure** of a solution is an indication of the force with which pure water moves into that solution as a result of its solute concentration. We can measure a solution's osmotic pressure in several ways. For example, an opposing pressure can prevent the osmotic flow of water into the solution.

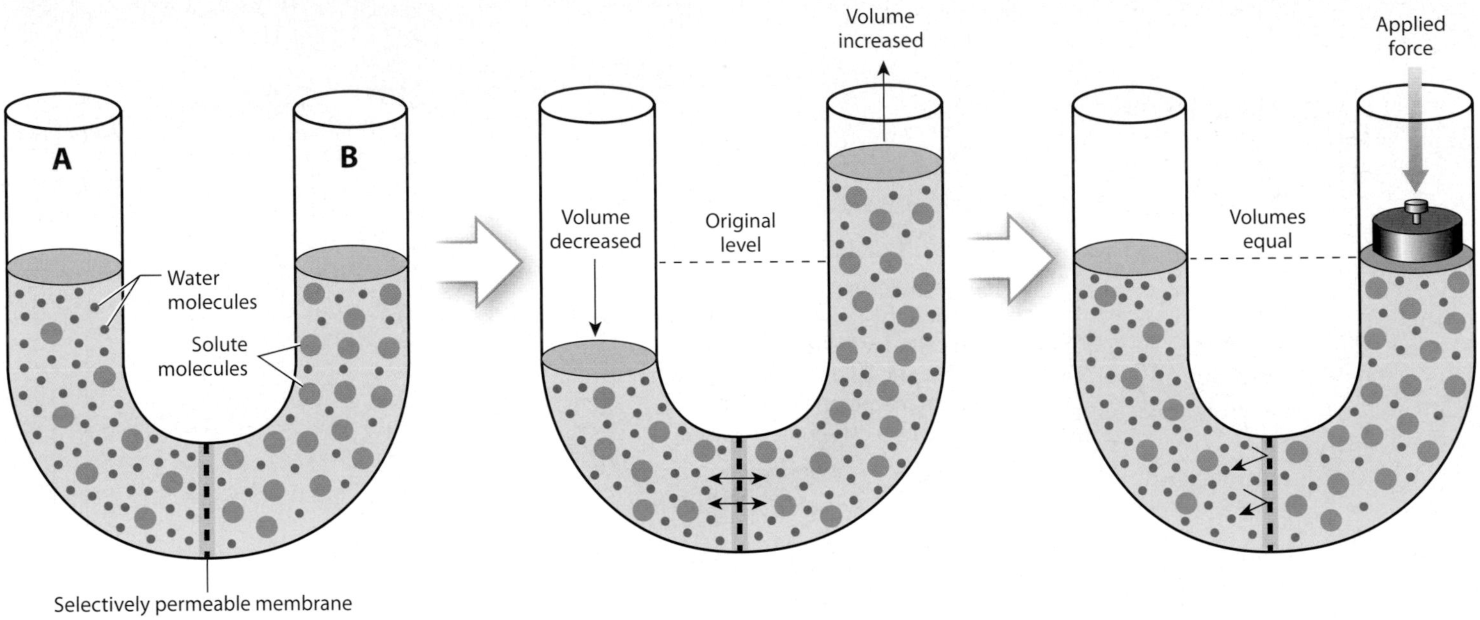
A
B
Water molecules
Solute molecules
Selectively permeable membrane
Volume decreased
Original level
Volume increased
Volumes equal
Applied force

A selectively permeable membrane separates these two solutions, which have different solute concentrations. Water molecules (small blue dots) begin to cross the membrane toward solution B, the solution with the higher concentration of solutes (larger pink circles). The larger solute molecules cannot diffuse across the membrane.

At equilibrium, the solute concentrations on the two sides of the membrane are equal. Note that the volume of solution B has increased at the expense of the volume of solution A.

Pushing against a fluid generates **hydrostatic pressure**. The osmotic pressure of solution B is equal to the amount of hydrostatic pressure, indicated by the weight, required to stop the osmotic flow.

2 The total solute concentration in an aqueous solution is the solution's **osmolarity**, or **osmotic concentration**. The nature of the solutes, however, is often as important as the total osmolarity. When we describe the effects of various osmotic solutions on cells, we usually use the term **tonicity** instead of osmolarity. Although often used interchangeably, the terms *osmolarity* and *tonicity* do not always mean the same thing. Osmolarity refers to the solute concentration of the solution, whereas tonicity is a description of how the solution affects a cell. Tonicity may have one of three effects, as illustrated here.

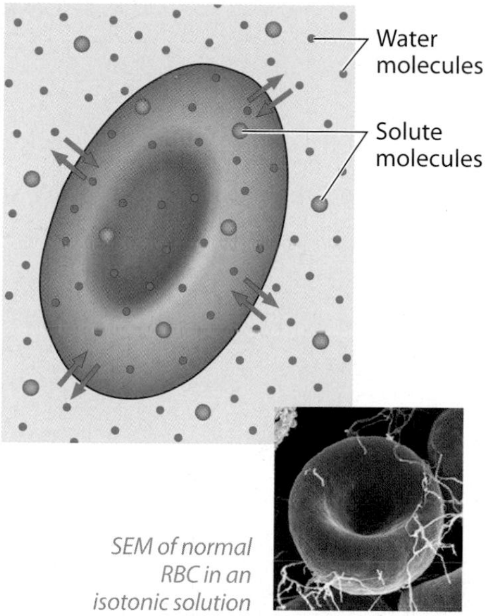
Water molecules
Solute molecules

SEM of normal RBC in an isotonic solution

A solution that does not cause an osmotic flow of water into or out of a cell is called **isotonic** (*iso-*, same + *tonos*, tension). The scanning electron micrograph (SEM) shows a red blood cell that has been stored in an isotonic solution.

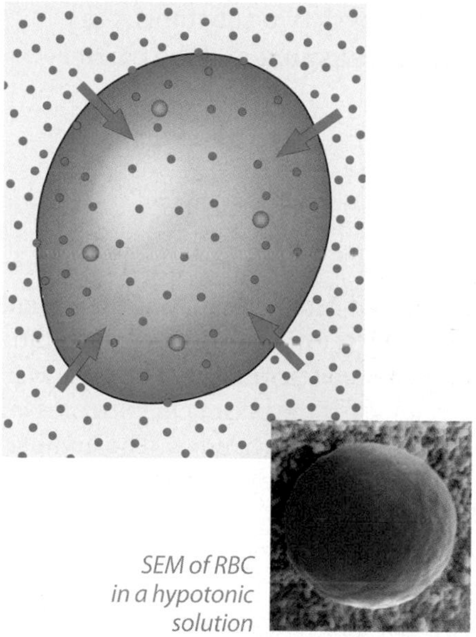

SEM of RBC in a hypotonic solution

A **hypotonic** solution is one that causes osmotic water flow into the cell. If a red blood cell is placed in a hypotonic solution, it will swell up like a balloon. The cell may eventually burst, releasing its contents. This event is **hemolysis** *(hemo-*, blood + *lysis*, a loosening).

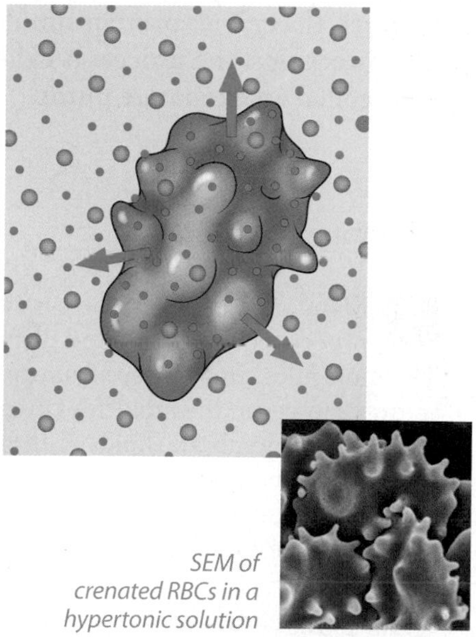

SEM of crenated RBCs in a hypertonic solution

Exposure to a **hypertonic** solution results in the osmotic movement of water out of the cell. As water moves out, the cell shrivels and dehydrates. The shrinking of red blood cells is called **crenation**, and the cells are said to be crenated.

It is often necessary to give patients large volumes of fluid to combat severe blood loss or dehydration. At such times the difference between osmolarity and tonicity is critically important. Consider a solution that has the same osmolarity as the intracellular fluid, but a higher concentration of one or more individual ions. If any of those ions can cross the plasma membrane and diffuse into the cell, the osmolarity of the intracellular fluid will gradually increase, and that of the extracellular solution will gradually decrease. Osmosis will then occur, moving water into the cell. If the process continues, the cell will gradually inflate like a water balloon. In this case, the extracellular solution and the intracellular fluid were initially equal in osmolarity, but the extracellular fluid was hypotonic rather than isotonic. In clinical emergencies, the fluid often administered is a 0.9 percent (0.9 g/dL) solution of sodium chloride (NaCl). This isotonic solution is called **normal saline**.

Module 3.15 Review

a. Describe osmosis.

b. Contrast the effects of a hypotonic solution and a hypertonic solution on a red blood cell.

c. Some pediatricians recommend using a 10 percent salt solution to relieve nasal congestion in infants. Explain the effects this treatment would have on the cells lining the nasal cavity. Would it be effective?

LO **3.15** Explain the process of osmosis, and identify its significance in the body.

In carrier-mediated transport, integral proteins facilitate membrane passage

Nutrients that are insoluble in lipids and too large to fit through membrane channels may be transported across the plasma membrane by **carrier proteins**. Many carrier proteins move a specific substance in one direction only, either into or out of the cell. Some carrier proteins simultaneously move more than one substance in the same direction. This process is called **cotransport**. If a carrier protein simultaneously moves two substances in opposite directions, the process is called **countertransport** and the carrier protein is called an **exchange pump**.

Facilitated Diffusion

1 Many substances can be passively transported across the plasma membrane in a process called **facilitated diffusion**. No ATP is expended in facilitated diffusion. The molecules simply move from an area of higher concentration to one of lower concentration. However, the molecules involved cannot cross the plasma membrane by diffusion through the membrane lipids or through open membrane channels. The number of suitable carrier proteins limits the rate of transport into the cell at any given moment. Once all the carrier proteins are saturated, the rate of transport cannot increase, regardless of further increases in the concentration of that substance in the ECF.

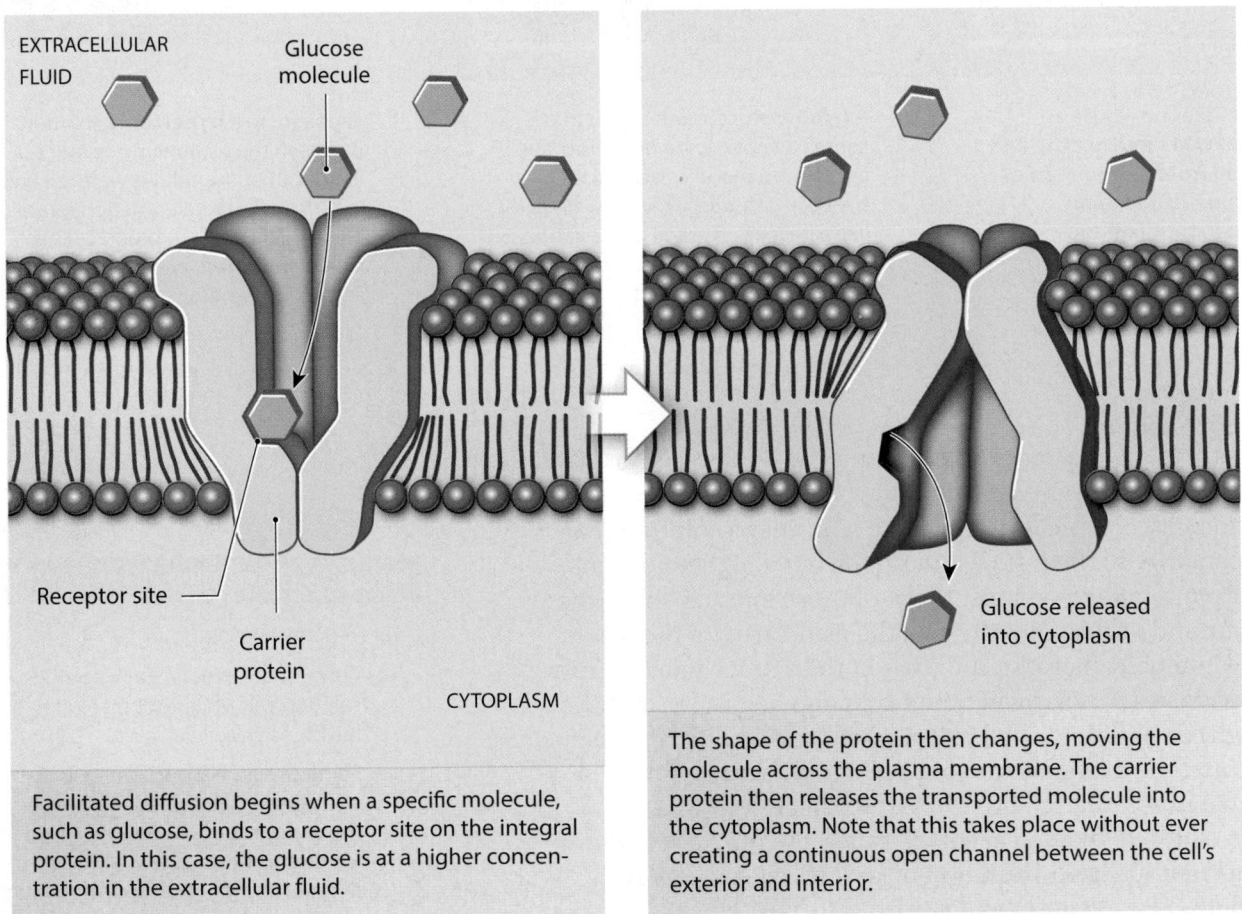

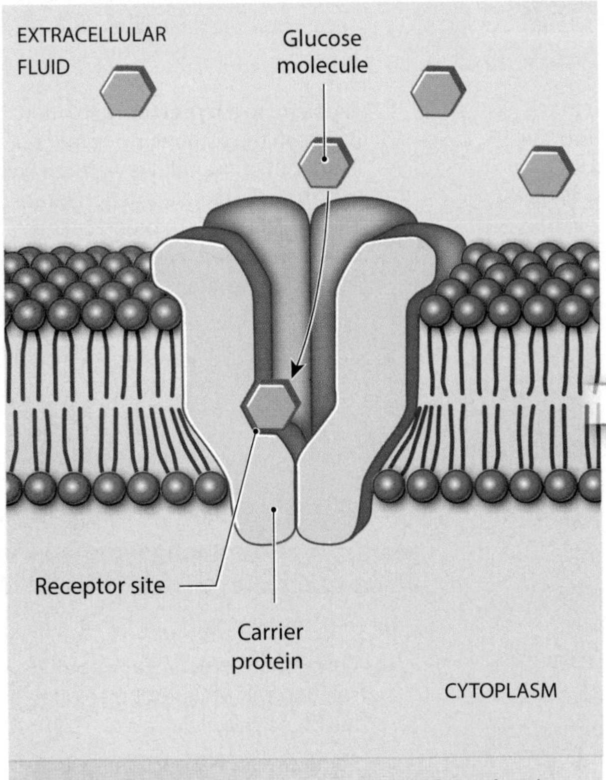
EXTRACELLULAR FLUID
Glucose molecule
Receptor site
Carrier protein
CYTOPLASM

Facilitated diffusion begins when a specific molecule, such as glucose, binds to a receptor site on the integral protein. In this case, the glucose is at a higher concentration in the extracellular fluid.

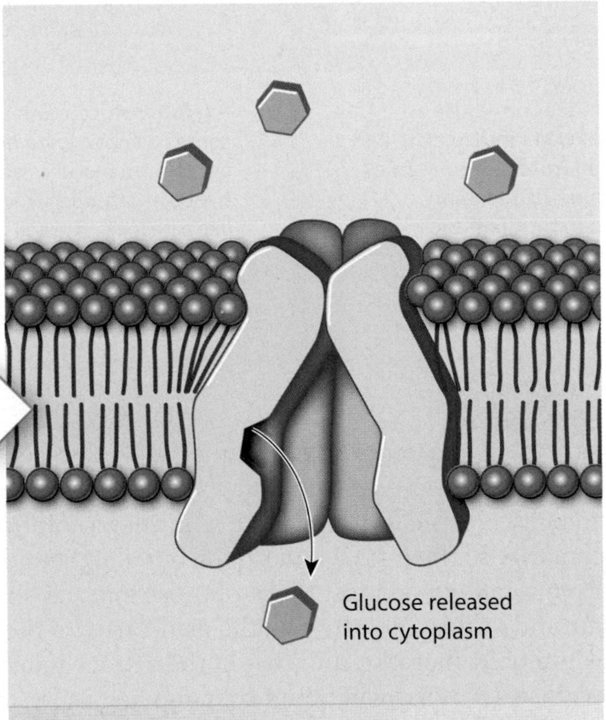
Glucose released into cytoplasm

The shape of the protein then changes, moving the molecule across the plasma membrane. The carrier protein then releases the transported molecule into the cytoplasm. Note that this takes place without ever creating a continuous open channel between the cell's exterior and interior.

Active Transport

2 In **active transport**, ATP (or another high-energy compound) provides the energy needed to move ions or molecules across the plasma membrane. Active transport offers one great advantage: It is not dependent on a concentration gradient. As a result, the cell can import or export specific substances, regardless of their intracellular or extracellular concentrations. All cells contain carrier proteins called **ion pumps**, which actively transport the cations sodium (Na^+), potassium (K^+), calcium (Ca^{2+}), and magnesium (Mg^{2+}) across their plasma membranes. Specialized cells can transport additional ions, such as iodide (I^-), chloride (Cl^-), and iron (Fe^{2+}).

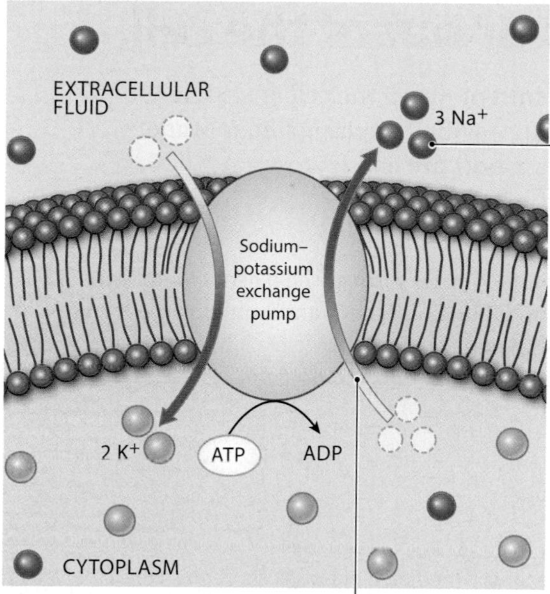
EXTRACELLULAR FLUID
3 Na+
Sodium–potassium exchange pump
2 K+
ATP
ADP
CYTOPLASM

1 Sodium ion concentrations are high in extracellular fluid, but low in the cytoplasm. Potassium ion distribution is just the opposite: low in extracellular fluid and high in the cytoplasm. As a result, sodium ions slowly diffuse into the cell, and potassium ions diffuse out through leak channels. Homeostasis within the cell depends on the ejection of sodium ions and the recapture of lost potassium ions. The sodium–potassium exchange pump is a carrier protein called **sodium–potassium ATPase**. It exchanges intracellular sodium for extracellular potassium.

2 On average, for each ATP molecule consumed, three sodium ions are ejected and the cell reclaims two potassium ions. The energy demands are impressive: Sodium–potassium ATPase may use up to 40 percent of the ATP produced by a resting cell!

Secondary Active Transport

3 In **secondary active transport**, the transport mechanism itself does not require energy from ATP, but the cell often needs to expend ATP at a later time to preserve homeostasis. Movement across the plasma membrane follows an existing concentration gradient for one of two substances transported, typically sodium ions, so transport does not require energy.

To preserve homeostasis, the cell must then expend ATP to pump the arriving sodium ions out of the cell by using the sodium–potassium exchange pump. It thus "costs" the cell one ATP for every three glucose molecules it transports into the cell.

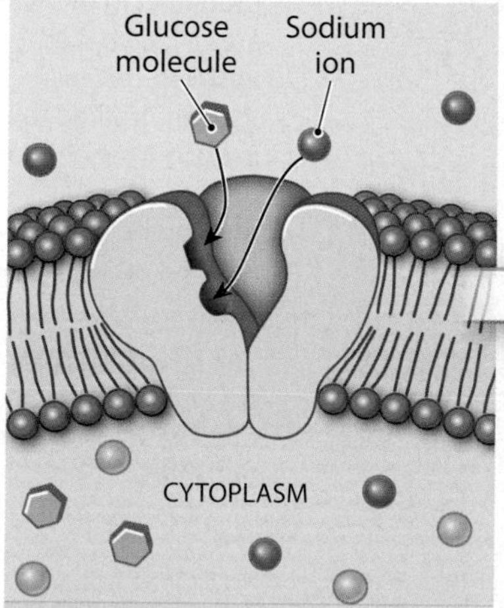
Glucose molecule
Sodium ion
CYTOPLASM

A sodium ion and a glucose molecule bind to receptor sites on the carrier protein.

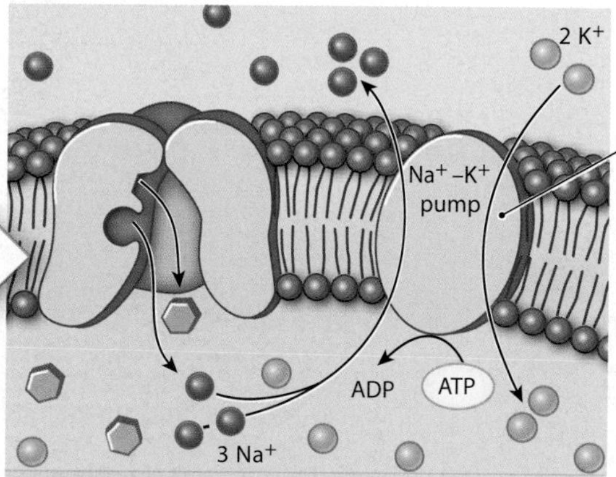
2 K+
Na+ –K+ pump
ADP
ATP
3 Na+

The carrier protein then changes shape, opening a path to the cytoplasm and releasing the transported materials. It then resumes its original shape and is ready to repeat the process.

Module 3.16 Review

a. Describe the process of carrier-mediated transport.

b. What do the transport processes of facilitated diffusion and active transport have in common?

c. During digestion, the concentration of hydrogen ions (H^+) in the stomach contents increases to many times that in cells lining the stomach. Which transport process could be responsible?

LO 3.16 Describe carrier-mediated transport and its role in the absorption and removal of specific substances.

In vesicular transport, vesicles selectively carry materials into or out of the cell

In **vesicular transport**, materials move into or out of the cell in **vesicles,** small membranous sacs that form at, or fuse with, the plasma membrane. The two major categories of vesicular transport are endocytosis and exocytosis, and both require energy in the form of ATP.

1 In **receptor-mediated endocytosis**, small vesicles form at the plasma membrane surface. Many important substances, including cholesterol and iron ions (Fe^{2+}), are distributed throughout the body attached to special transport proteins. These proteins are too large to pass through membrane pores, but they can and do enter cells by receptor-mediated endocytosis.

Endocytosis

Endocytosis is the importing of extracellular substances through the formation of vesicles at the cell surface. These vesicles are known as **endosomes**.

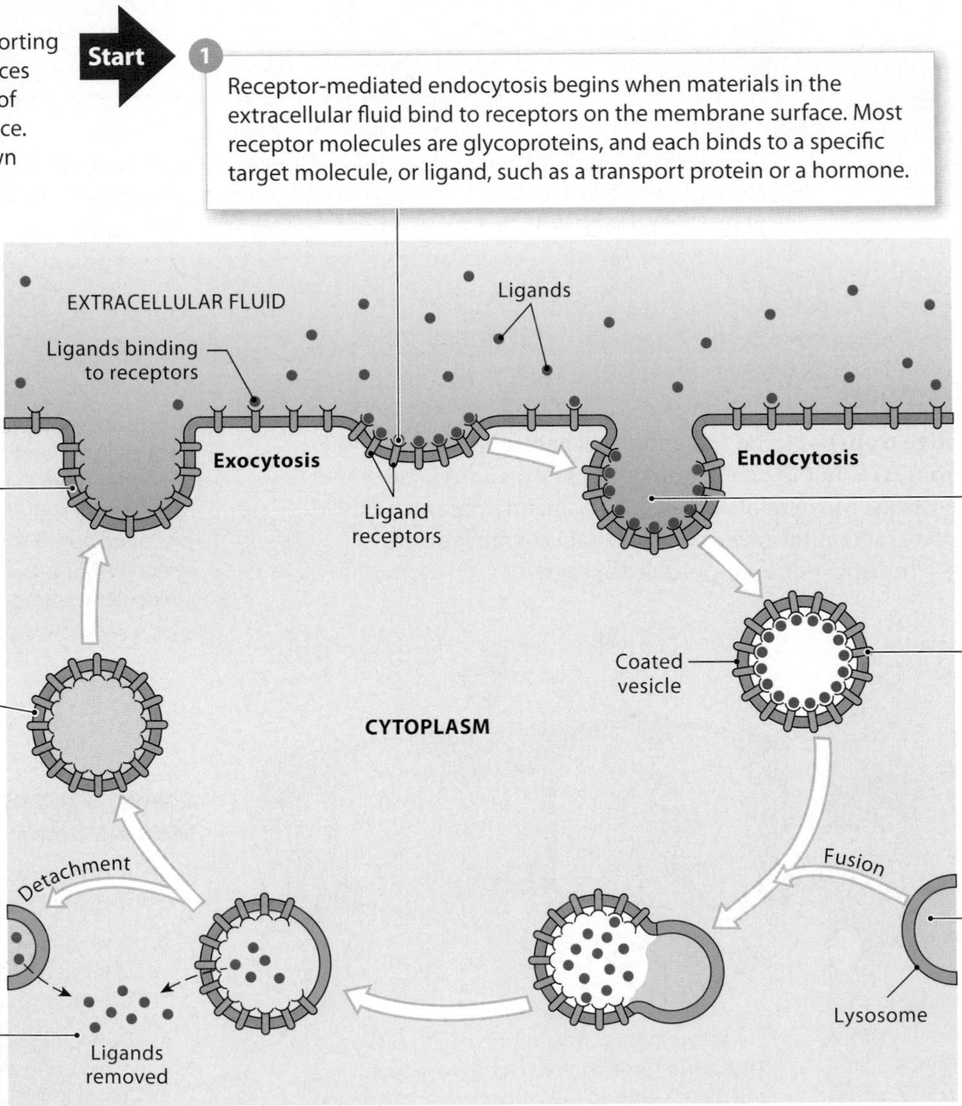

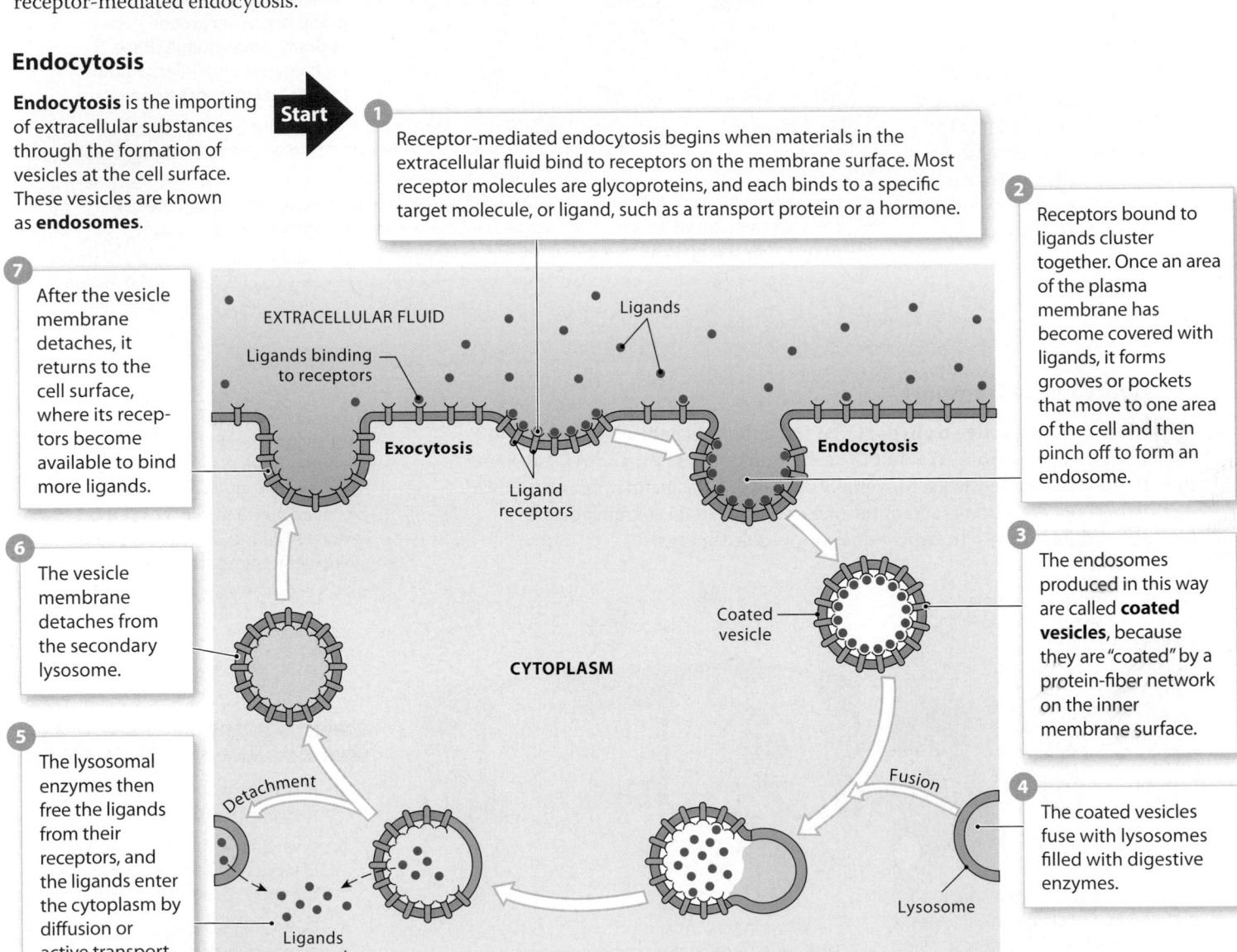

Start
EXTRACELLULAR FLUID
Ligands
Ligands binding to receptors
Exocytosis
Ligand receptors
Endocytosis
Coated vesicle
CYTOPLASM
Fusion
Lysosome
Detachment
Ligands removed

1 Receptor-mediated endocytosis begins when materials in the extracellular fluid bind to receptors on the membrane surface. Most receptor molecules are glycoproteins, and each binds to a specific target molecule, or ligand, such as a transport protein or a hormone.

2 Receptors bound to ligands cluster together. Once an area of the plasma membrane has become covered with ligands, it forms grooves or pockets that move to one area of the cell and then pinch off to form an endosome.

3 The endosomes produced in this way are called **coated vesicles**, because they are "coated" by a protein-fiber network on the inner membrane surface.

4 The coated vesicles fuse with lysosomes filled with digestive enzymes.

5 The lysosomal enzymes then free the ligands from their receptors, and the ligands enter the cytoplasm by diffusion or active transport.

6 The vesicle membrane detaches from the secondary lysosome.

7 After the vesicle membrane detaches, it returns to the cell surface, where its receptors become available to bind more ligands.

2 "Cell drinking," or **pinocytosis** (pi-nō-sī-TŌ-sis), is the formation of endosomes filled with extracellular fluid. This process is not as selective as receptor-mediated endocytosis, because no receptor proteins are involved. The target appears to be the fluid contents in general, rather than specific ligands. In a few specialized cells, pinocytosis produces vesicles on one side of the cell that are discharged on the other. This method of bulk transport is common in cells lining small blood vessels.

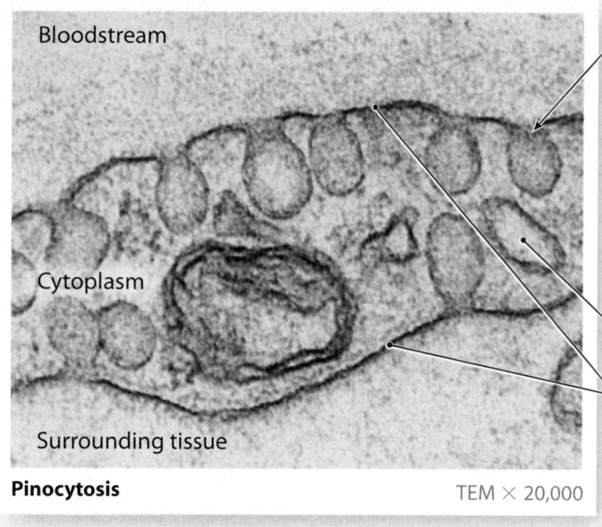

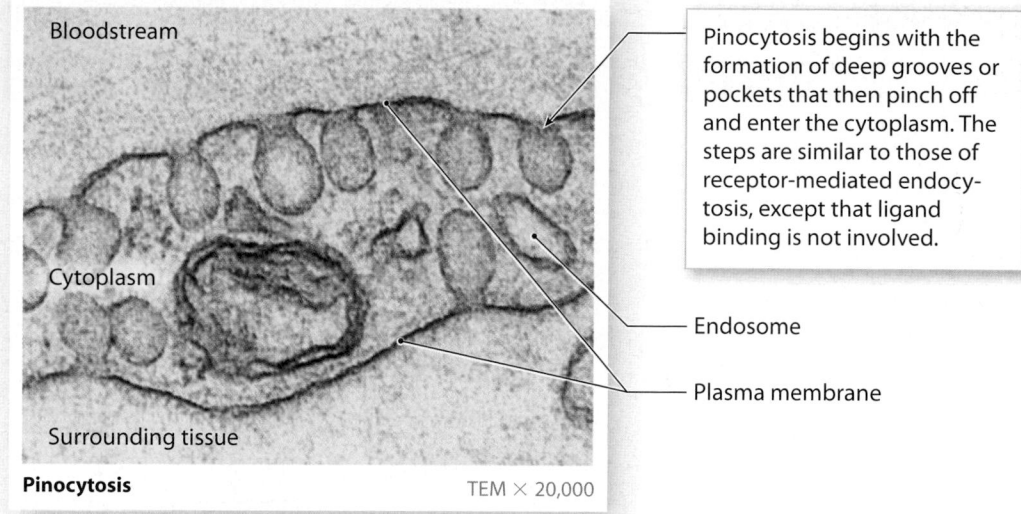

Bloodstream
Cytoplasm
Surrounding tissue
Pinocytosis begins with the formation of deep grooves or pockets that then pinch off and enter the cytoplasm. The steps are similar to those of receptor-mediated endocytosis, except that ligand binding is not involved.
Endosome
Plasma membrane

Pinocytosis TEM × 20,000

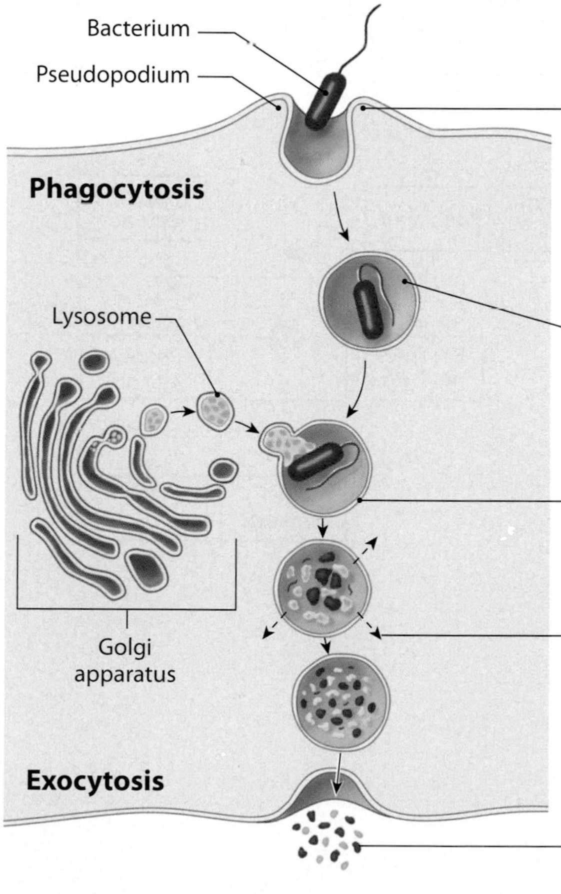

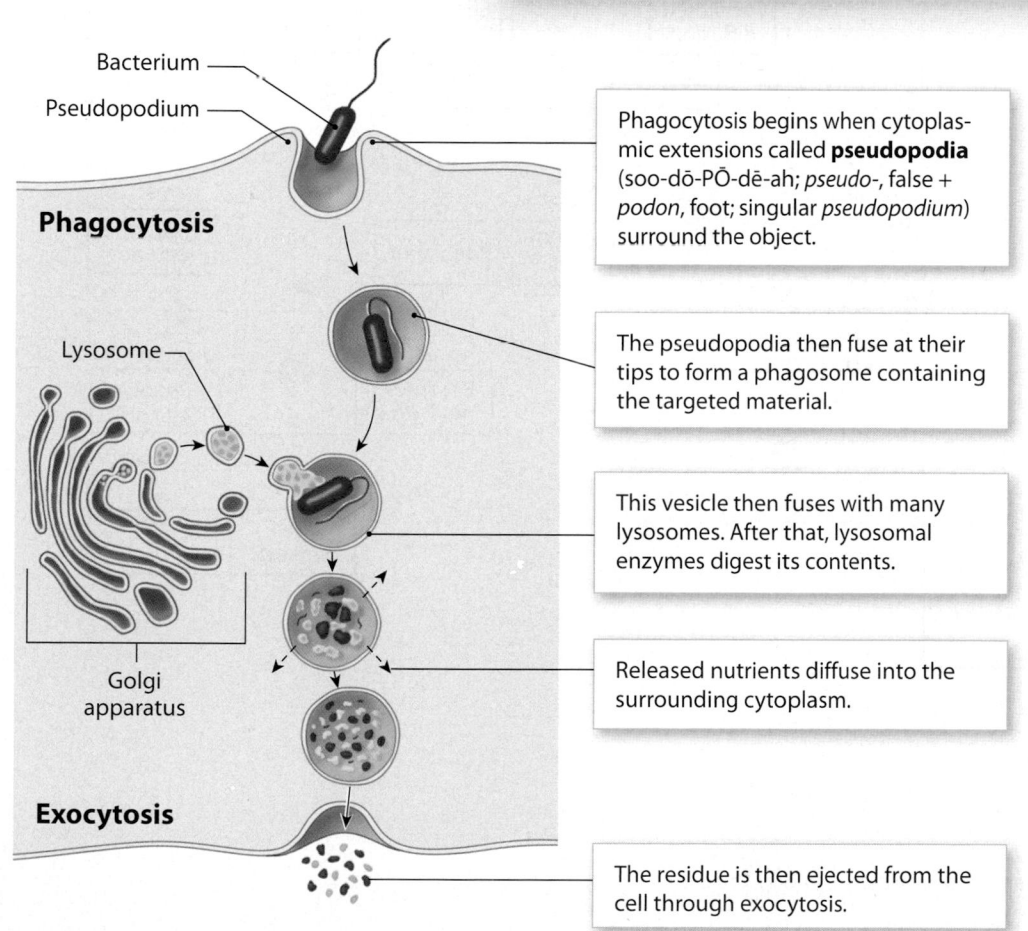

Bacterium
Pseudopodium
Phagocytosis
Lysosome
Golgi apparatus
Exocytosis
Phagocytosis begins when cytoplasmic extensions called **pseudopodia** (soo-dō-PŌ-dē-ah; *pseudo-*, false + *podon*, foot; singular *pseudopodium*) surround the object.
The pseudopodia then fuse at their tips to form a phagosome containing the targeted material.
This vesicle then fuses with many lysosomes. After that, lysosomal enzymes digest its contents.
Released nutrients diffuse into the surrounding cytoplasm.
The residue is then ejected from the cell through exocytosis.

3 "Cell eating," or **phagocytosis** (fag-ō-sī-TŌ-sis), produces **phagosomes** containing solid objects that may be as large as the cell itself. Although most cells perform pinocytosis, only specialized cells perform phagocytosis. These cells are called **phagocytes** or **macrophages**.

4 **Exocytosis** (ek-sō-sī-TŌ-sis) is the functional opposite of endocytosis. In exocytosis, a vesicle formed inside the cell fuses with, and becomes part of, the plasma membrane. When this takes place, the vesicle contents are released into the extracellular environment. The ejected material may be waste products (such as those accumulating in lysosomes) or secretory products (such as mucins or hormones).

Module 3.17 Review

a. Describe endocytosis.

b. When they encounter bacteria, certain types of white blood cells engulf the bacteria and bring them into the cell. What is this process called?

c. Describe exocytosis.

LO **3.17** Describe vesicular transport as a mechanism for facilitating the absorption or removal of specific substances from cells.

Concept map

Use each of the following terms once to fill in the blank boxes to correctly complete the map.

- exocytosis
- diffusion
- "cell eating"
- molecular size
- pinocytosis
- facilitated diffusion
- vesicular transport
- net diffusion of water
- active transport
- specificity

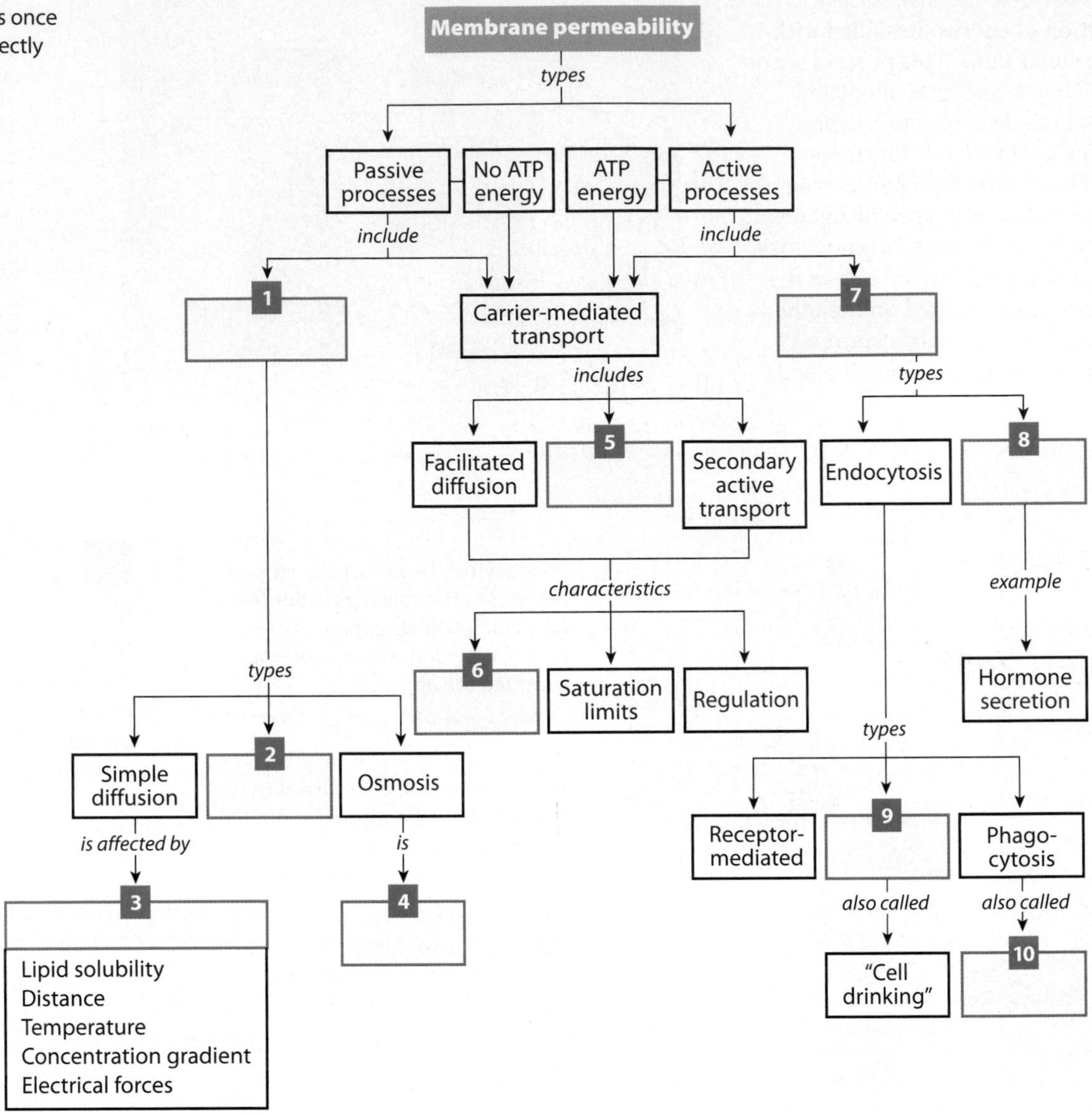

Membrane permeability
types
Passive processes
No ATP energy
ATP energy
Active processes
include
include
1
Carrier-mediated transport
7
includes
types
Facilitated diffusion
5
Secondary active transport
Endocytosis
8
characteristics
example
6
Saturation limits
Regulation
Hormone secretion
types
types
Simple diffusion
2
Osmosis
Receptor-mediated
9
Phago-cytosis
is affected by
is
also called
also called
3
4
"Cell drinking"
10
Lipid solubility
Distance
Temperature
Concentration gradient
Electrical forces

Short answer

Classify each of the following situations as an example of diffusion, osmosis, or neither.

11 You walk into a room and smell a balsam-scented candle.

12 Water flows through a garden hose.

13 Grass in the yard wilts after being exposed to excess chemical fertilizer.

14 A sugar cube placed in a cup of hot tea dissolves.

15 After soaking several hours in water containing sodium chloride, a stalk of celery weighs less than before it was placed in the salty water.

11 ______________________

12 ______________________

13 ______________________

14 ______________________

15 ______________________

Interphase and cell division make up the life cycle of a cell

The period between fertilization and physical maturity involves tremendous changes in organization and complexity. At fertilization, a single cell is all there is. At maturity, your body has about 75 trillion cells. **Cell division**, a form of cellular reproduction, makes this transformation possible.

Even when development is complete, cell division continues to be essential to survival. Physical wear and tear, toxic chemicals, temperature changes, and other environmental stresses damage cells. And, like individuals, cells age. The life span of a cell varies from hours to decades, depending on the type of cell and the stresses involved. Many cells apparently self-destruct after a certain period of time as a result of the activation of specific "suicide genes" in the nucleus. This genetically controlled cell death is called **apoptosis** (ap-op-TŌ-sis or ap-ō-TŌ-sis; *apo-*, separated from + *ptosis*, a falling).

There are two different forms of cell division. **Mitosis** (mī-TŌ-sis), the focus of this section, produces two daughter cells, each containing a complete set of 46 chromosomes. **Meiosis** (mī-Ō-sis), which we will examine in Chapter 26, produces sex cells (sperm or oocytes) containing only 23 chromosomes.

1 The division of a single cell produces a pair of **daughter cells**, each one half the size of the original. Before they themselves divide, each of the daughter cells will grow to the size of the original cell.

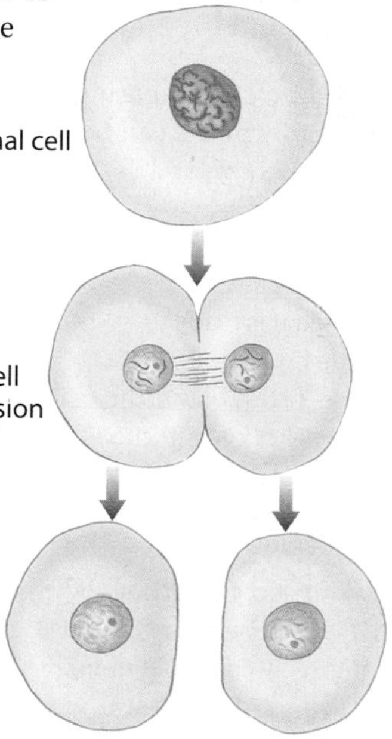
Original cell
Cell division
Daughter cells

2 The diagram at right details the life cycle of a typical cell. The cycle ends only when the cell dies.

Interphase is the period in which the cell is performing normal functions and not actively engaged in cell division. Some cells are in interphase indefinitely. Others are always either dividing or preparing to divide.

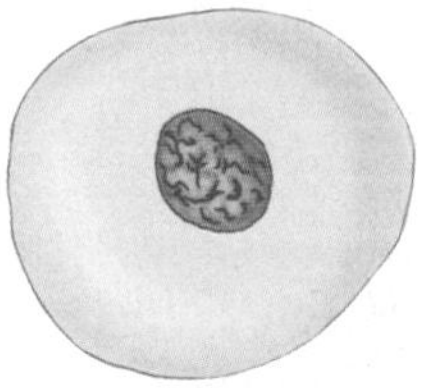

During interphase in a cell preparing to divide, the chromosomes of the cell are duplicated and associated proteins are synthesized.

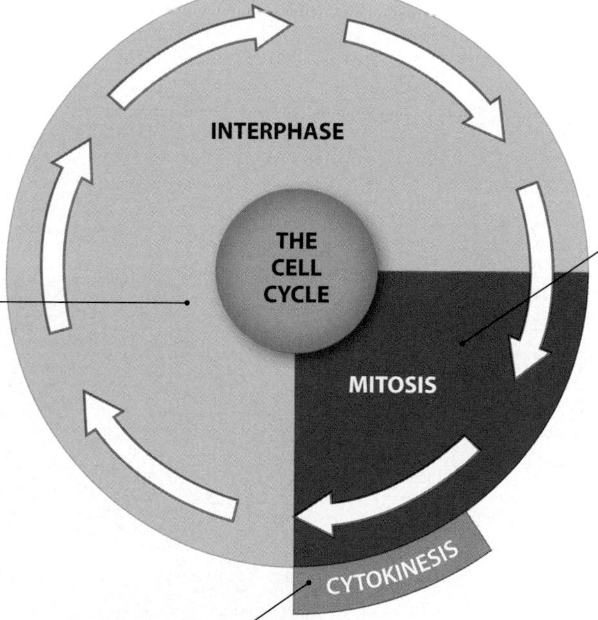
INTERPHASE
THE CELL CYCLE
MITOSIS
CYTOKINESIS

Cell division begins with mitosis. In mitosis, identical copies of the original chromosomes are distributed to each daughter cell.

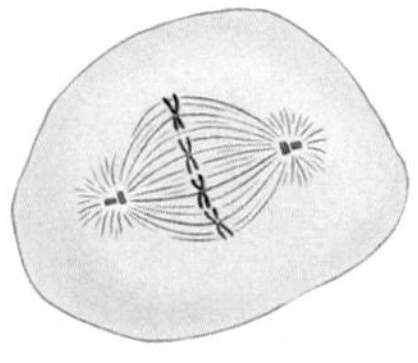

Cell division ends as **cytokinesis**, the division of the cytoplasm, physically separates the two daughter cells.

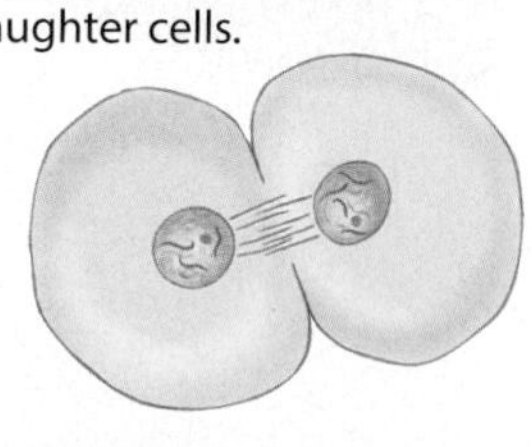

Module 3.18 Review

a. Explain why cell division is important.

b. Define apoptosis.

c. Distinguish between mitosis and meiosis.

LO 3.18 Distinguish between interphase and cell division in the cell cycle.

Module 3.19

During interphase, the cell prepares for cell division

Watch
MasteringA&P®
A&PFlix™
DNA Replication

Most cells spend only a small part of their time actively engaged in cell division. **Somatic** (*soma*, body) **cells** spend most of their functional lives in a state known as **interphase**. During interphase, a cell performs all its normal functions and, if necessary, prepares for cell division.

1 In a cell preparing to divide, interphase can be divided into the **G_1, S, and G_2 phases.**

When the G1 phase is complete, the cell enters the S phase. Over the next 6–8 hours, the cell duplicates its chromosomes. This involves DNA replication and the synthesis of histones and other proteins in the nucleus.

A cell that is ready to divide first enters the G_1 phase. In this phase, the cell makes enough mitochondria, cytoskeletal elements, endoplasmic reticula, ribosomes, Golgi membranes, and cytosol for two functional cells. Centriole replication begins in G_1 and commonly continues until G_2. In a cell dividing at top speed, G_1 may last just 8–12 hours. Such a cell pours all its energy into mitosis, and all other activities cease. If G_1 lasts for days, weeks, or months, preparation for mitosis occurs as the cell performs its normal functions.

When DNA replication ends, there is a brief (2- to 5-hour) G_2 phase devoted to last-minute protein synthesis and to completing centriole replication.

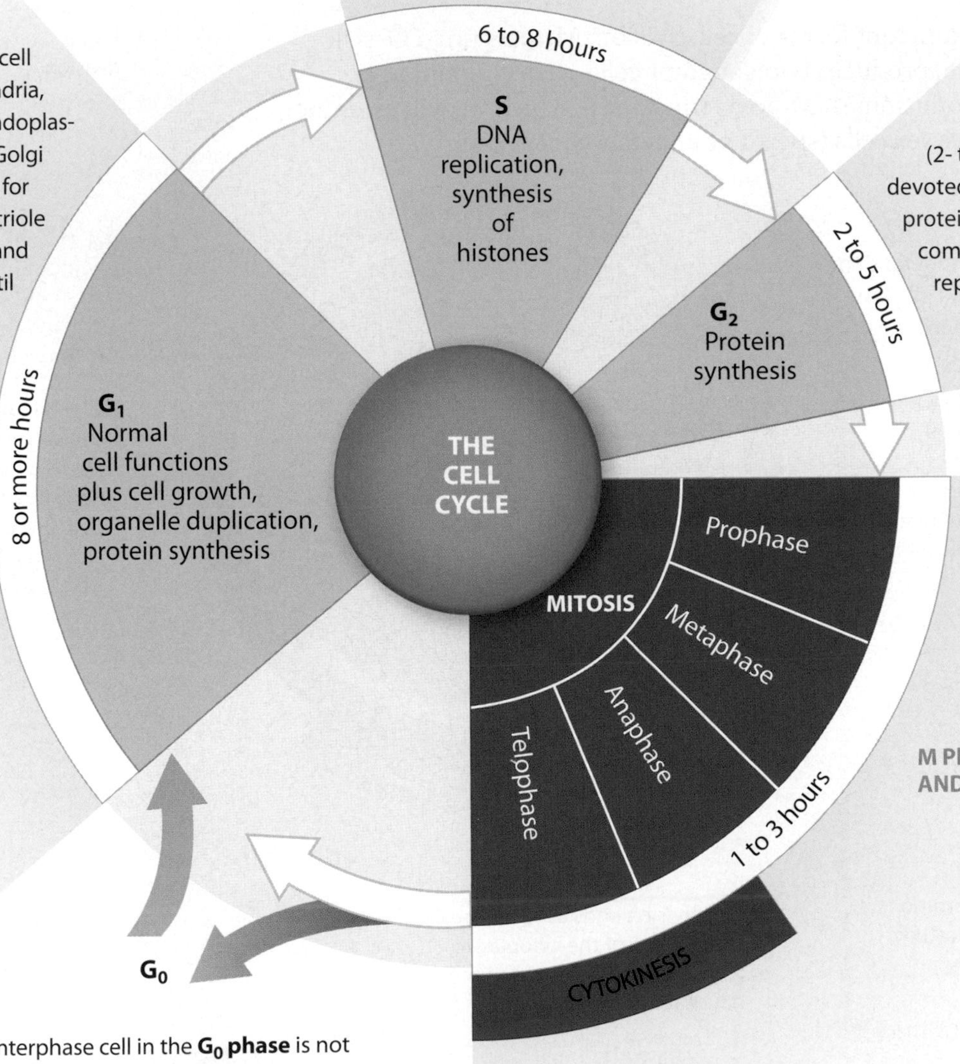

6 to 8 hours
S
DNA replication, synthesis of histones
2 to 5 hours
G2
Protein synthesis
8 or more hours
G1
Normal cell functions plus cell growth, organelle duplication, protein synthesis
THE CELL CYCLE
MITOSIS
Prophase
Metaphase
Anaphase
Telophase
1 to 3 hours
CYTOKINESIS
G0

M Phase – MITOSIS AND CYTOKINESIS

An interphase cell in the **G_0 phase** is not preparing for division, but is instead performing all of the other functions appropriate for that particular cell type. Some mature cells, such as skeletal muscle cells and most neurons, remain in G_0 indefinitely and never divide. In contrast, **stem cells**, which divide repeatedly with very brief interphase periods, never enter G_0.

2 During the S phase of the cell cycle, DNA is replicated. The goal of DNA replication is to copy the genetic information in the nucleus. The process occurs in cells preparing to undergo either mitosis or meiosis.

1 **DNA replication** begins when **DNA helicase** enzymes unwind the strands and disrupt the hydrogen bonds between the bases. As the strands unwind, molecules of **DNA polymerase** bind to the exposed nitrogenous bases. This enzyme (1) promotes bonding between the nitrogenous bases of the DNA strand and complementary DNA nucleotides in the nucleoplasm and (2) links the nucleotides by covalent bonds.

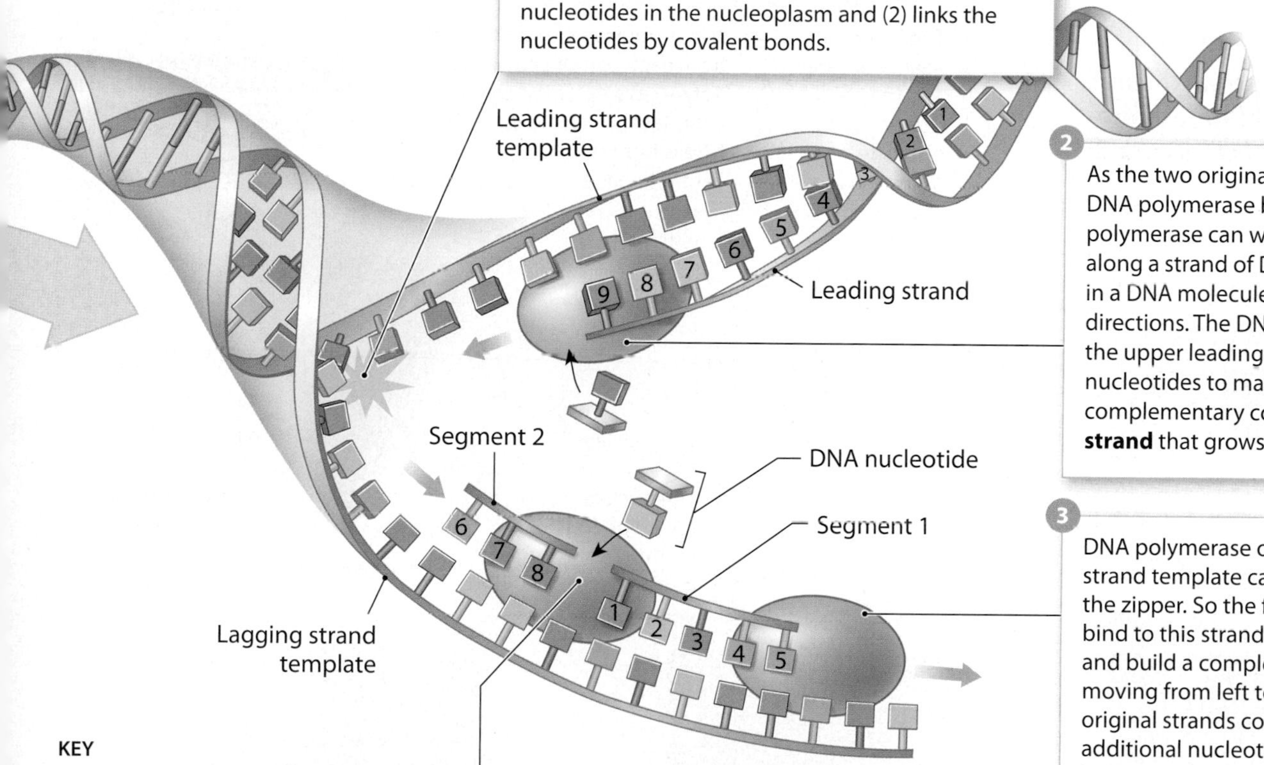

Leading strand template
Leading strand
Segment 2
DNA nucleotide
Segment 1
Lagging strand template

2 As the two original strands separate, DNA polymerase binds to them. DNA polymerase can work in only one direction along a strand of DNA, but the two strands in a DNA molecule are oriented in opposite directions. The DNA polymerase bound to the upper leading strand template adds nucleotides to make a single, continuous complementary copy called the **leading strand** that grows toward the "zipper."

3 DNA polymerase on the lower lagging strand template can work only away from the zipper. So the first DNA polymerase to bind to this strand must add nucleotides and build a complementary DNA strand moving from left to right. As the two original strands continue to unzip, additional nucleotides are continuously being exposed to the nucleoplasm. The first DNA polymerase on this strand cannot go into reverse. It can only continue to elongate the strand it already started.

4 Thus, a second DNA polymerase must bind closer to the point of unzipping and assemble a complementary copy (segment 2) that grows until it "bumps into" segment 1 created by the first DNA polymerase. Enzymes called **DNA ligases** (LĪ-gās-ez; *liga*, to tie) then splice together the two DNA segments into a strand called the **lagging strand**.

KEY
- Adenine
- Guanine
- Cytosine
- Thymine

3 Eventually, the unzipping completely separates the original strands. The copying ends, the last splicing is done, and two identical DNA molecules have formed. Once the DNA is replicated, the centrioles duplicated, and the necessary enzymes and proteins synthesized, the cell leaves interphase and is ready to proceed to mitosis.

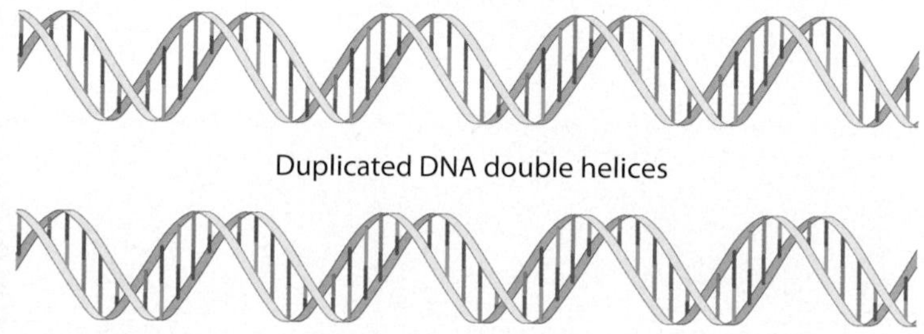

Duplicated DNA double helices

Module 3.19 Review

a. Describe interphase, and identify its stages.

b. What enzymes must be present for DNA replication to proceed normally?

c. A cell is actively manufacturing enough organelles to serve two functional cells. This cell is probably in what phase of interphase?

LO 3.19 Describe interphase, and explain its significance.

Module 3.20

Mitosis distributes chromosomes before cytokinesis separates the daughter cells

Watch
MasteringA&P®
A&PFlix™
Mitosis

The M phase of the cell cycle includes mitosis and cytokinesis. **Mitosis** is a series of events during which the duplicated chromosomes of a cell separate and migrate into two identical nuclei. The four stages of mitosis are prophase, metaphase, anaphase, and telophase.

MITOSIS

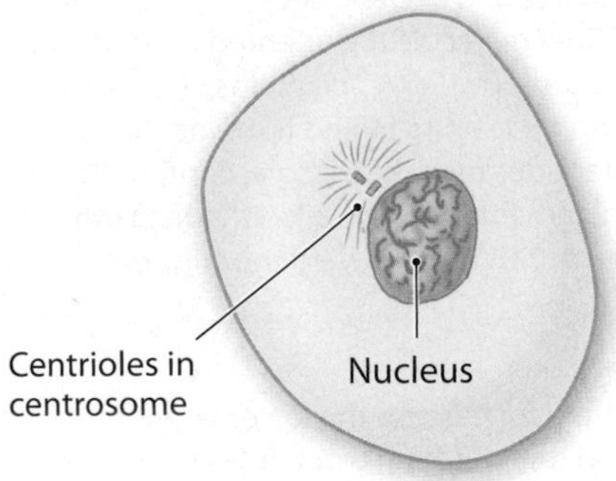

Centrioles in centrosome
Nucleus

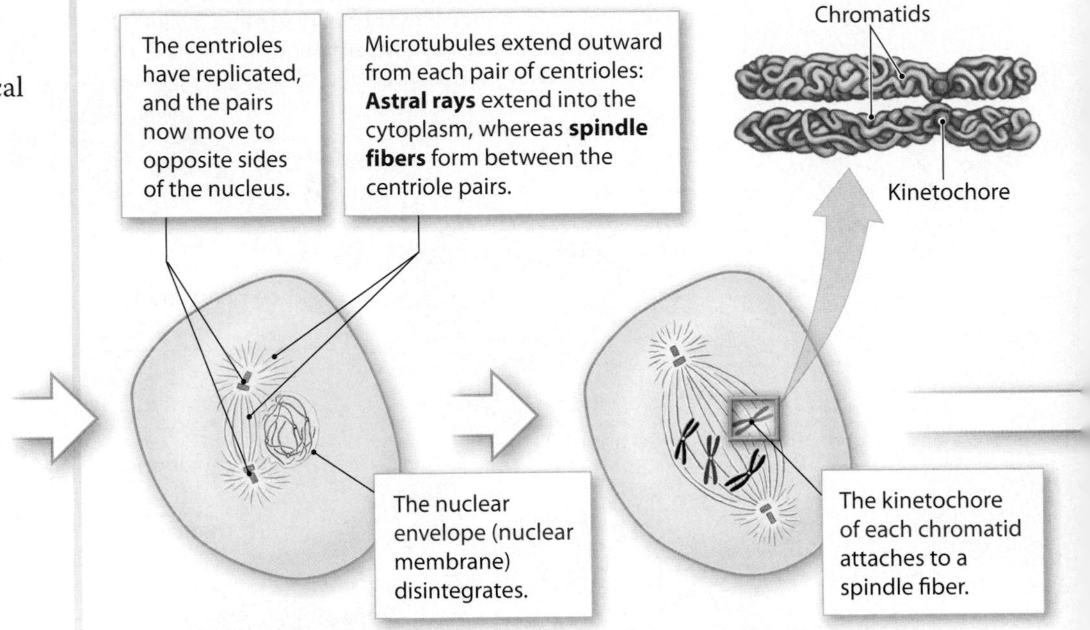

The centrioles have replicated, and the pairs now move to opposite sides of the nucleus.
Microtubules extend outward from each pair of centrioles: Astral rays extend into the cytoplasm, whereas spindle fibers form between the centriole pairs.
Chromatids
Kinetochore
The nuclear envelope (nuclear membrane) disintegrates.
The kinetochore of each chromatid attaches to a spindle fiber.

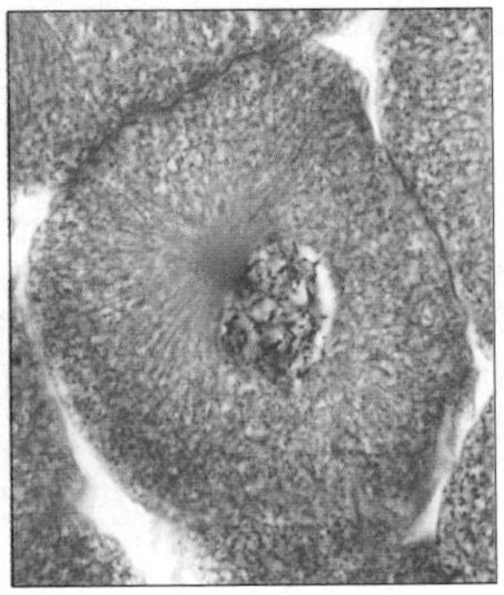

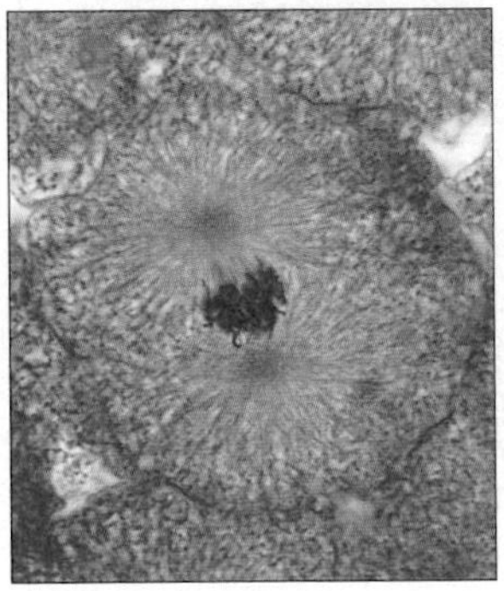

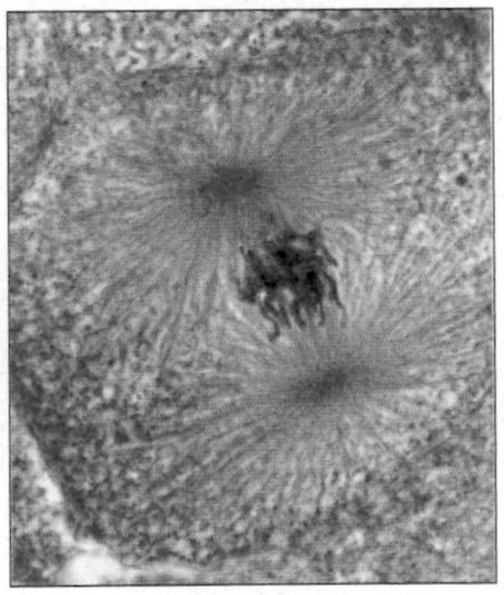

1 During interphase, the DNA strands are loosely coiled and chromosomes cannot be seen.

2 **Prophase** (PRŌ-fāz; *pro*, before) begins when the chromosomes coil so tightly that they become visible as individual structures under a light microscope. As a result of DNA replication during the S phase, two copies of each chromosome now exist. Each copy is called a **chromatid**, and the paired chromatids are connected at a region known as the centromere (p. 103). A raised region on the centromere, called the **kinetochore** (ki-NE-tō-kor), is where the spindle fibers attach to the chromatids.

The term mitosis specifically refers to division and duplication of the cell's nucleus alone. The cytoplasm—the rest of the cell—divides into two distinct new cells in a separate, but related, process known as **cytokinesis** (sī-tō-ki-NĒ-sis; *cyto-*, cell + *kinesis*, motion).

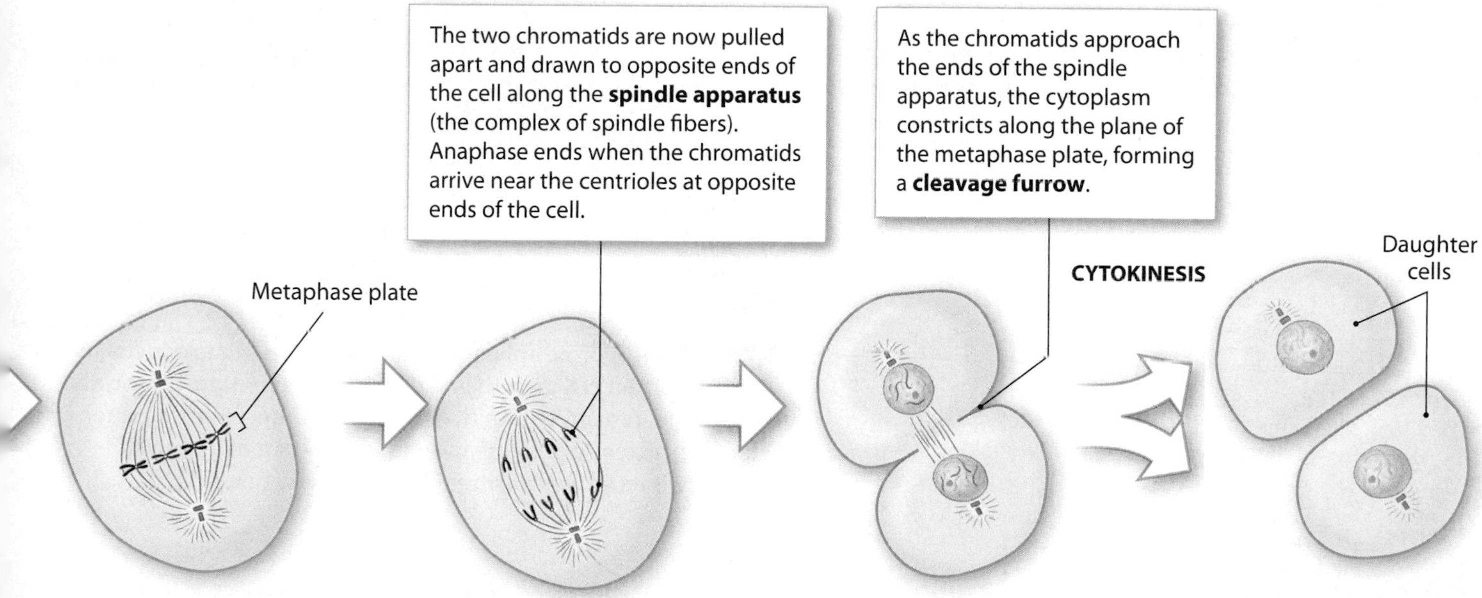
The two chromatids are now pulled apart and drawn to opposite ends of the cell along the **spindle apparatus** (the complex of spindle fibers). Anaphase ends when the chromatids arrive near the centrioles at opposite ends of the cell.
As the chromatids approach the ends of the spindle apparatus, the cytoplasm constricts along the plane of the metaphase plate, forming a **cleavage furrow**.
Metaphase plate
CYTOKINESIS
Daughter cells

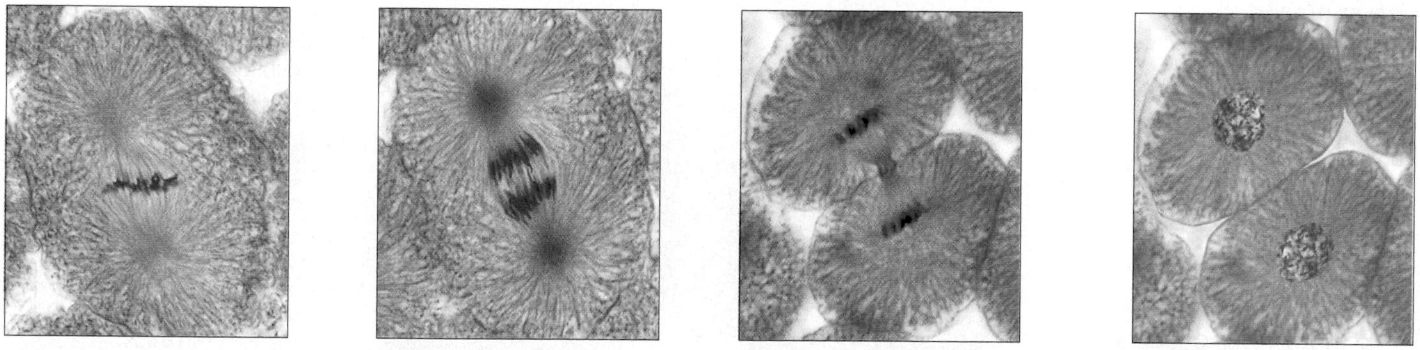

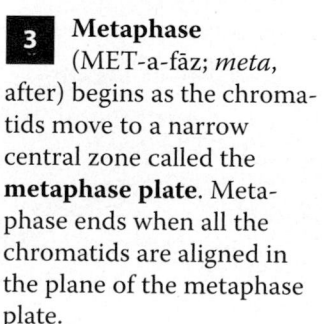

3 **Metaphase** (MET-a-fāz; *meta*, after) begins as the chromatids move to a narrow central zone called the **metaphase plate**. Metaphase ends when all the chromatids are aligned in the plane of the metaphase plate.

4 **Anaphase** (AN-a-fāz; *ana-*, apart) begins when the centromere of each chromatid pair splits and the chromatids separate. The chromatids are now pulled along the spindle fibers toward opposite sides of the dividing cell.

5 During **telophase** (TĒL-ō-fāz; *telo-*, end), each new cell prepares to return to interphase. The nuclear envelopes re-form, the nuclei enlarge, and the chromosomes gradually uncoil to the chromatin state. This stage marks the end of mitosis.

6 **Cytokinesis** usually begins with the formation of a cleavage furrow during anaphase and continues throughout telophase. The completion of cytokinesis marks the end of cell division.

Module 3.20 Review

a. Define mitosis, and list its four stages.

b. What is a chromatid, and how many would be present during normal mitosis in a human cell?

c. What would happen if spindle fibers failed to form in a cell during mitosis?

LO 3.20 Describe the process of mitosis, and its role in the cell life cycle.

Tumors and cancer are characterized by abnormal cell growth and division

When the rates of cell division and growth exceed the rate of cell death, a tissue begins to enlarge. The enlargement often results from the divisions of a single abnormal cell that no longer responds to the factors that regulate the rate of cell division. **Cancer** is an illness characterized by **mutations**—permanent changes in DNA nucleotide sequence and function—that disrupt normal control mechanisms that regulate the rates of cell division.

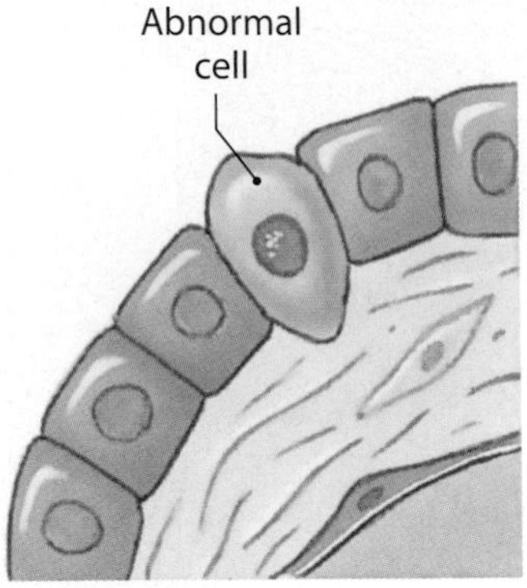
Abnormal cell

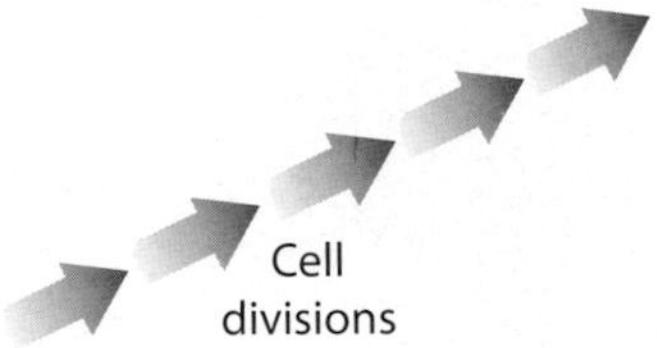
Cell divisions

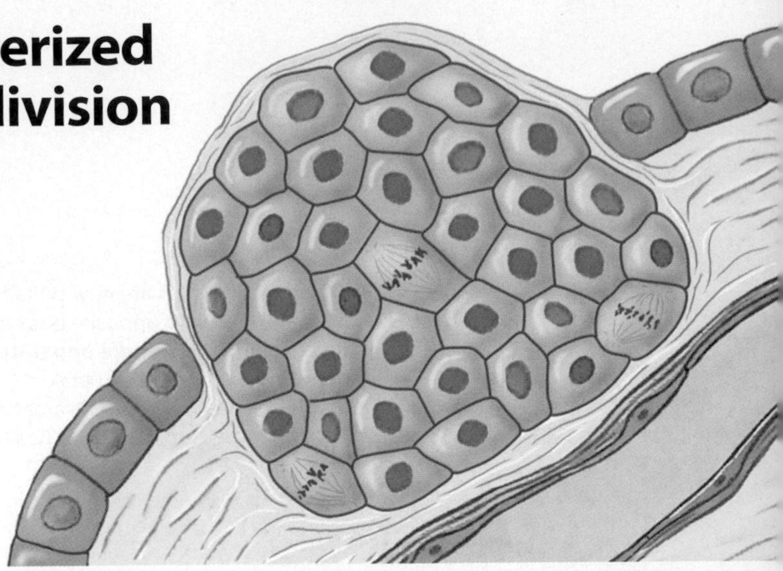

2 A **tumor**, or **neoplasm**, is a mass or swelling produced by abnormal cell growth and division. In a **benign tumor**, the cells usually remain within the originating tissue. Such a tumor seldom threatens a person's life and usually can be removed surgically if its size or position disturbs tissue function.

1 Cancer usually begins with a single abnormal cell. Every time a cell divides, there is a chance that something will go wrong with the control mechanism. As a result, cancers are most common in tissues where cells are dividing rapidly and continuously, such as the epithelium of the skin or the intestinal lining.

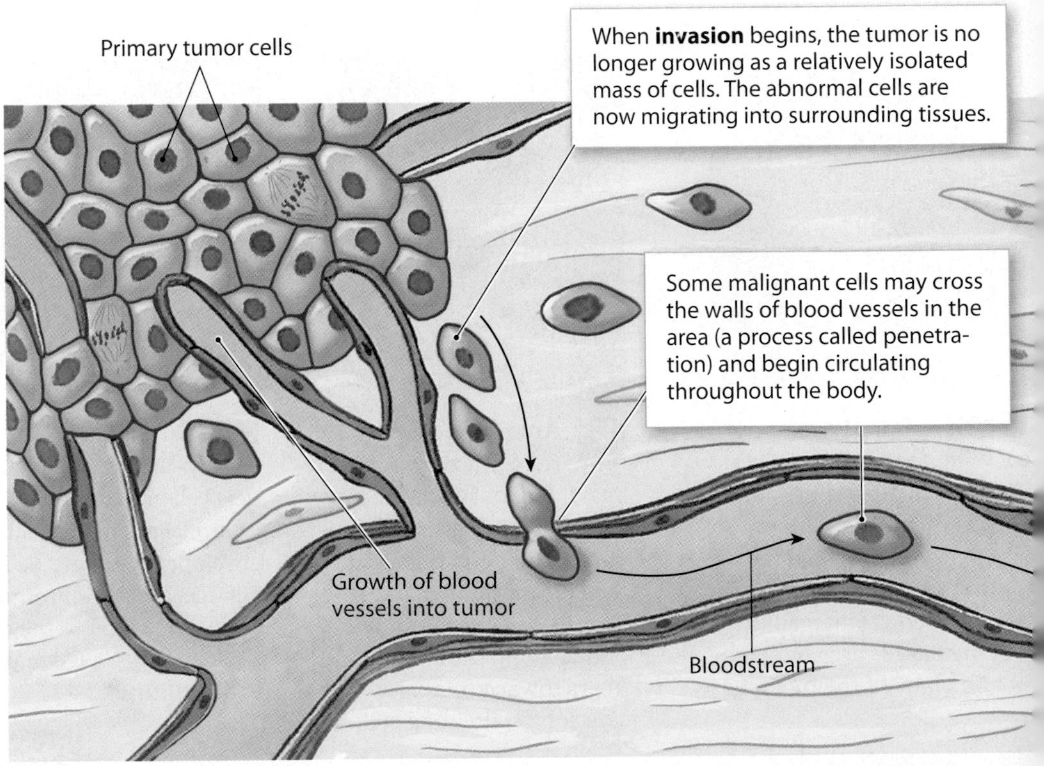
Primary tumor cells
When **invasion** begins, the tumor is no longer growing as a relatively isolated mass of cells. The abnormal cells are now migrating into surrounding tissues.
Some malignant cells may cross the walls of blood vessels in the area (a process called penetration) and begin circulating throughout the body.
Growth of blood vessels into tumor
Bloodstream

3 Cells in a **malignant tumor** divide very rapidly, releasing chemicals that stimulate the growth of blood vessels (**angiogenesis**) into the area. The availability of additional nutrients accelerates tumor growth, and malignant cells then migrate into surrounding tissues and nearby blood vessels. This process—**metastasis** (me-TAS-tuh-sis)—can produce secondary tumors in tissues remote from the primary tumor site.

4 Malignant cells may no longer perform their original functions, or they may perform normal functions in an abnormal way. For example, endocrine cancer cells may produce normal hormones, but in excessively large amounts. This photo shows a patient with a malignant tumor of the thyroid gland. Excessive amounts of thyroid hormone produced dramatic changes in metabolic activity throughout his body.

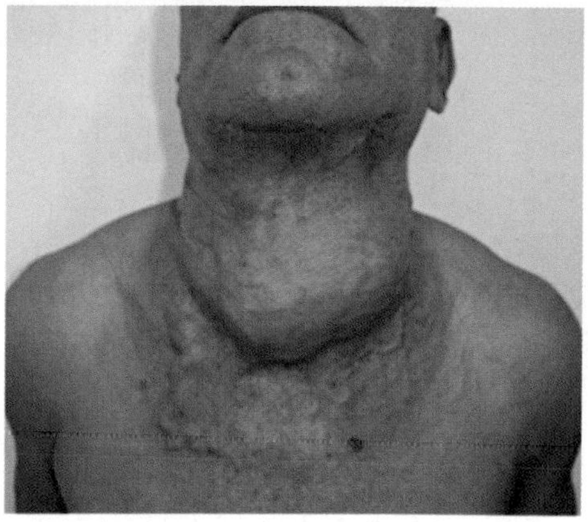

Responding to cues that are as yet unknown, cancer cells in the bloodstream ultimately escape out of blood vessels to establish secondary tumors at other sites. These tumors are extremely active metabolically, and their presence stimulates the growth of new blood vessels into the area. The increased blood supply provides additional nutrients to the cancer cells and further accelerates tumor growth and metastasis.

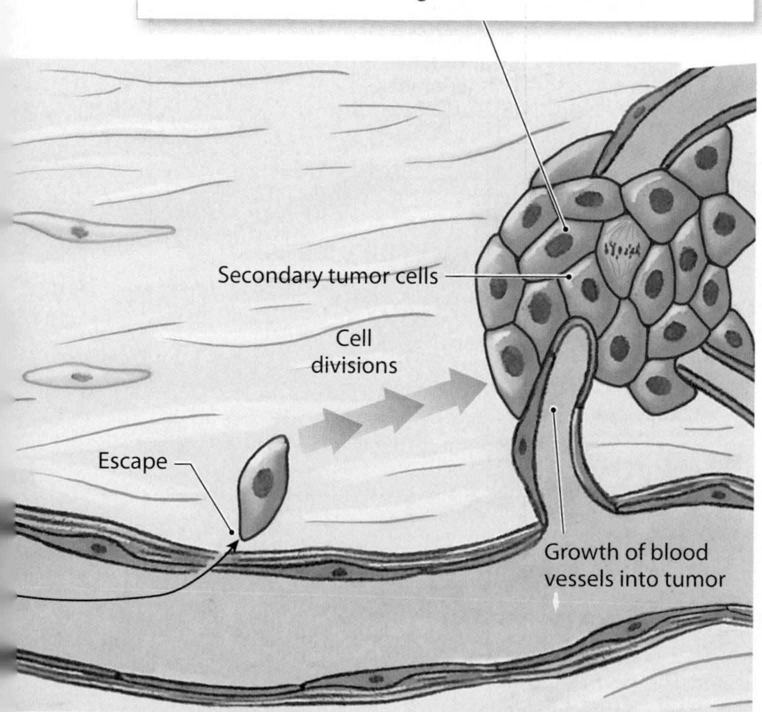
Secondary tumor cells
Cell divisions
Escape
Growth of blood vessels into tumor

Cancer cells do not use energy very efficiently. They grow and multiply at the expense of healthy tissues, competing with normal cells for space and nutrients. This competition contributes to the starved appearance of many patients in the late stages of cancer. Death may occur as a result of the compression of vital organs when nonfunctional cancer cells have killed or replaced the healthy cells in those organs, or when the cancer cells have starved normal tissues of essential nutrients. We will discuss cancer further in later chapters that deal with specific systems.

Module 3.21 Review

a. Define cancer.

b. What is a benign tumor?

c. Define metastasis.

LO **3.21** Discuss the relationship between cell division and cancer.

Concept map

Use each of the following terms once to fill in the blank boxes to correctly complete the map.

- metaphase
- DNA replication
- somatic cells
- G_2 phase
- telophase
- mitosis
- cytokinesis
- G_1 phase

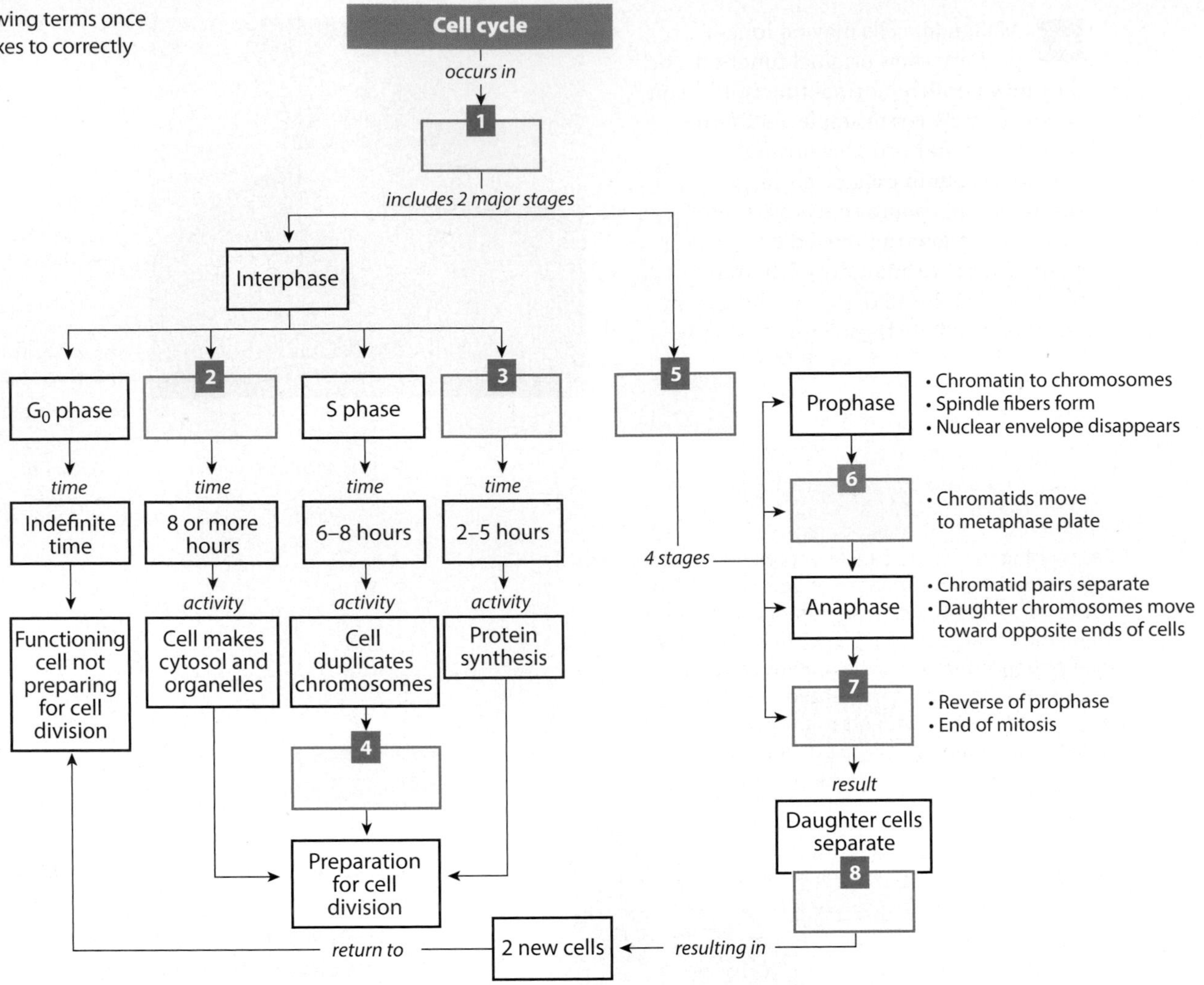

Cell cycle
occurs in
1
includes 2 major stages
Interphase
G_0 phase
2
S phase
3
5
time
time
time
time
Indefinite time
8 or more hours
6–8 hours
2–5 hours
activity
activity
activity
Functioning cell not preparing for cell division
Cell makes cytosol and organelles
Cell duplicates chromosomes
Protein synthesis
4
Preparation for cell division
return to
2 new cells
resulting in
4 stages
Prophase
• Chromatin to chromosomes
• Spindle fibers form
• Nuclear envelope disappears
6
• Chromatids move to metaphase plate
Anaphase
• Chromatid pairs separate
• Daughter chromosomes move toward opposite ends of cells
7
• Reverse of prophase
• End of mitosis
result
Daughter cells separate
8

Vocabulary

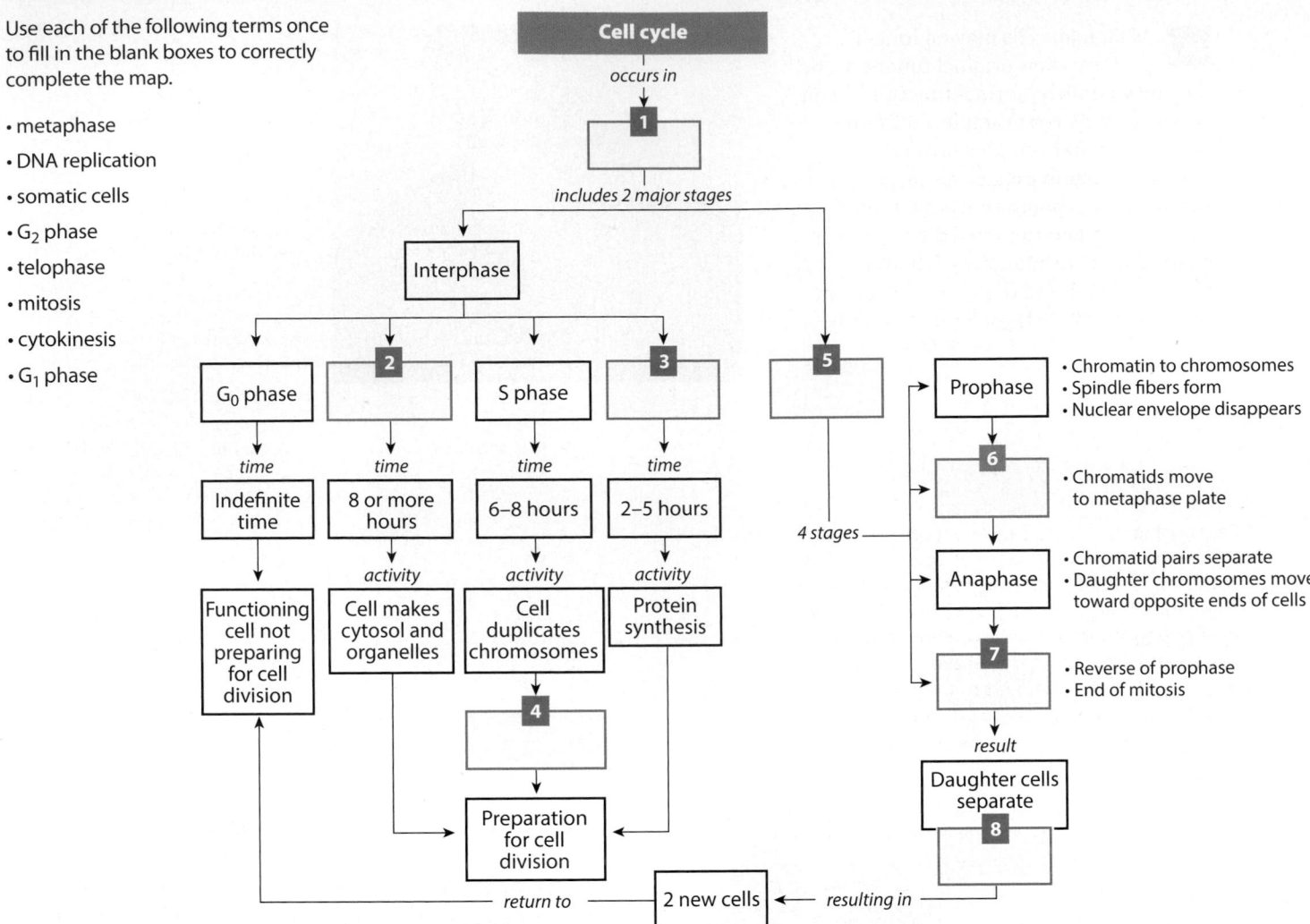

In the space provided, write the boldfaced terms introduced in this section that contain the indicated word part.

9 telo- *(end)*

10 pro- *(before)*

11 centro- *(in the middle)*

9 __________

10 __________

11 __________

Section integration

The muscle cells that that make up skeletal muscle tissue are large and multinucleated. They form early in development as groups of embryonic cells fuse together, contributing their nuclei and losing their individual plasma membranes. Describe an alternate mechanism that would also result in a large, multinucleated cell.

12 __________

Study Outline

SECTION 1 • Introduction to Cells

3.1 Cells differentiate for cellular specialization p. 87

1. **Cell theory** states that cells are the building blocks of all plants and animals, cells come from the division of preexisting cells, and cells are the smallest units that perform all vital physiological functions.
2. All cells are descendants of a single fertilized ovum.
3. Cells undergo a process of gradual specialization called **differentiation**.

3.2 Cells are the smallest living units of life p. 88

4. Cells are surrounded by **extracellular fluid** that in most cases is **interstitial fluid**.
5. The **cytoplasm** contains the fluid **cytosol** and the **organelles** suspended in the cytosol.
6. **Nonmembranous organelles** are not enclosed by membranes and their components are in contact with the cytosol. They include the **cytoskeleton**, **microvilli**, **centrioles**, **cilia**, **flagella**, and **ribosomes**.
7. **Membranous organelles** are isolated from the cytosol by phospholipid membranes. They include the **mitochondria**, **nucleus**, **endoplasmic reticulum**, **Golgi apparatus**, **lysosomes**, and **peroxisomes**.

3.3 The plasma membrane isolates the cell from its environment and performs varied functions p. 90

8. The **plasma membrane** is a **phospholipid bilayer** that separates the inside of the cell from the surrounding extracellular fluid.
9. The general functions of the plasma membrane include physical isolation, regulation of exchange with the environment, sensitivity to the environment, and structural support.
10. Membrane proteins are functionally classified as **anchoring proteins**, **recognition proteins**, **receptor proteins**, **carrier proteins**, and **channels**.

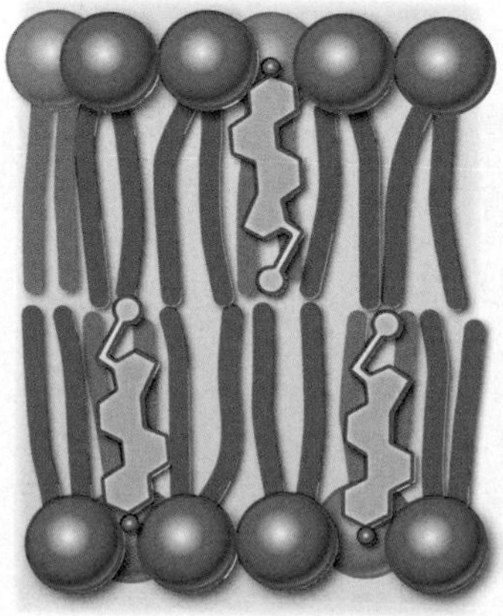

The plasma membrane is composed of two layers of phospholipid molecules with their hydrophilic heads at the membrane surface and their hydrophobic tails on the inside. The hydrophobic layer in the center of the membrane isolates the cytoplasm from the very different extracellular fluid.

3.4 The cytoskeleton plays both a structural and a functional role p. 92

11. The **cytoskeleton** provides an internal protein framework that gives the cytoplasm strength and flexibility.
12. Components of the cytoskeleton are **microfilaments**, **intermediate filaments**, **thick filaments**, **microtubules**, **centrioles**, and **cilia**.

3.5 Ribosomes are responsible for protein synthesis and are often associated with the endoplasmic reticulum p. 94

13. **Ribosomes** are organelles responsible for protein synthesis.
14. **Endoplasmic reticulum** (**ER**) is a network of intracellular membranes forming sheets or tubes called **cisternae. Smooth ER** is the site for the synthesis of phospholipids, cholesterol, steroid hormones, glycerides, and glycogen.

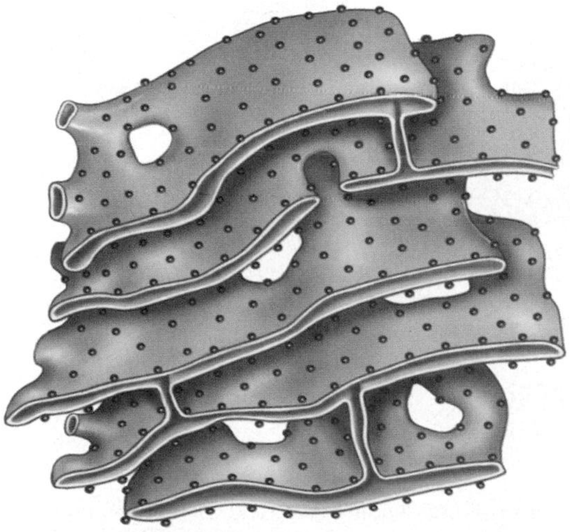

The rough endoplasmic reticulum (RER) functions as a combination workshop for chemical modification of proteins and shipping warehouse to send them to their next destination.

15. **Fixed ribosomes** are attached to **rough endoplasmic reticulum** (**RER**). Proteins are synthesized at the ribosomes and then modified and packaged in the RER.

3.6 The Golgi apparatus is a packaging center p. 96

16. The **Golgi apparatus** consists of five or six flattened membranous discs called cisternae.
17. The Golgi apparatus functions to renew or modify the plasma membrane; modify and package secretions for release through exocytosis; and package special enzymes within vesicles for use in the cytosol.

3.7 Mitochondria are the powerhouses of the cell p. 98

18. **Mitochondria** have a double membrane and contain their own DNA. The inner membrane contains numerous folds called **cristae**.
19. Mitochondria are responsible for ATP production through **aerobic metabolism**.

SECTION 2 • Structure and Function of the Nucleus

3.8 The nucleus is the control center for cellular homeostasis p. 101

20. The nucleus contains DNA and determines the structure and function of the cell by controlling the synthesis of proteins.

3.9 The nucleus contains DNA, RNA, organizing proteins, and enzymes p. 102

21. The **nucleus** serves as the control center for cellular processes. It is surrounded by a **nuclear envelope** (a double membrane with a **perinuclear space**), through which it communicates with the cytosol by way of **nuclear pores**.
22. The nucleus contains a supportive **nucleoplasm** and one or more **nucleoli**, which synthesize ribosomal RNA.
23. The DNA in the nucleus stores the instructions for protein synthesis.

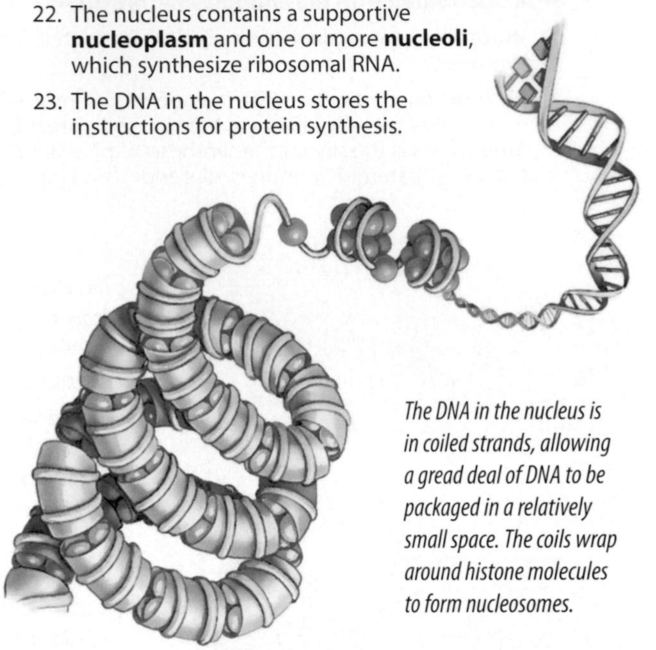

The DNA in the nucleus is in coiled strands, allowing a gread deal of DNA to be packaged in a relatively small space. The coils wrap around histone molecules to form nucleosomes.

24. The DNA in a cell not dividing is loosely coiled forming a tangle of fine filaments known as **chromatin**. The DNA in dividing cells is coiled tightly forming a complex of distinct structures called **chromosomes**.

3.10 Protein synthesis involves DNA, enzymes, and three types of RNA p. 104

25. DNA consists of long, parallel chains of nucleotides that are held together by hydrogen bonding between complementary base pairs. The nitrogenous bases involved are adenine (A), thymine (T), cytosine (C), and guanine (G). The information stored in the sequence of base pairs is called the **genetic code**.
26. The genetic code is called a **triplet code** because a sequence of three nitrogenous bases, or **triplet**, specifies a single amino acid.
27. A **gene** is the functional unit of heredity. It contains all the DNA nucleotides needed to produce specific proteins.
28. As **gene activation** begins, a portion of the DNA molecule containing that gene uncoils.
29. A strand of **messenger RNA** (**mRNA**) is assembled that contains information in triplets or **codons**.
30. **Transfer RNA** (**tRNA**) molecules bring amino acids to a ribosome. The amino acids form a polypeptide chain or protein.

3.11 Transcription encodes genetic instructions on a strand of RNA p. 106

31. **Transcription** is the production of RNA from a DNA template. After transcription a strand of mRNA carries instructions from the nucleus to the cytoplasm.

3.12 Translation builds polypeptides as directed by an mRNA strand p. 108

32. **Translation** is the formation of a linear chain of amino acids using the information provided by an mRNA strand. By complementary base pairing of tRNA **anticodons** to mRNA codons, amino acids are brought to the ribosome in the proper sequence. The three phases of translation are **initiation**, **elongation**, and **termination.**

SECTION 3 • How Substances Enter and Leave the Cell

3.13 The plasma membrane is a selectively permeable membrane p. 111

33. The plasma membrane selectively regulates the movement of materials into and out of the cell. **Selectively permeable membranes** permit the free passage of some materials and restrict the passage of others.
34. Materials may cross the plasma membrane through active processes (require ATP) or passive processes (do not require ATP).
35. Vesicular transport is an active process. Diffusion is a passive process. Carrier-mediated processes may be either passive or active.

3.14 Diffusion is passive movement driven by concentration differences p. 112

36. **Diffusion** is the net movement of material from an area of higher concentration to an area of lower concentration. This occurs until the **concentration gradient** is eliminated.
37. Important factors that influence diffusion rates include distance, molecule size, temperature, concentration gradient, and electrical forces.

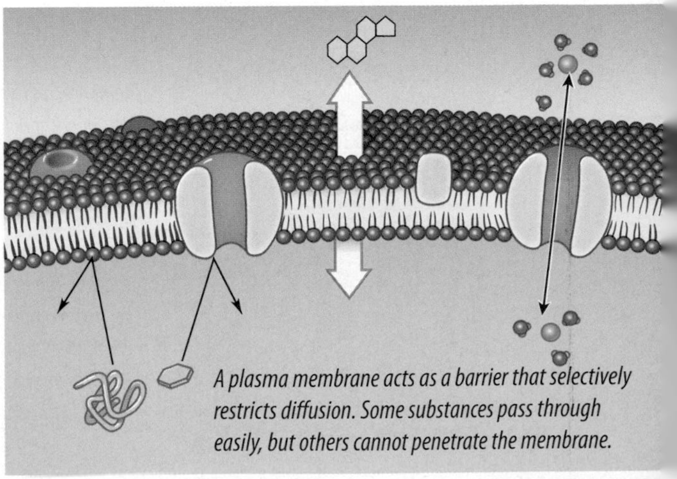

A plasma membrane acts as a barrier that selectively restricts diffusion. Some substances pass through easily, but others cannot penetrate the membrane.

3.15 Osmosis is the diffusion of water molecules across a selectively permeable membrane p. 114

38. **Osmosis** is the net diffusion of water across a membrane. The movement of water driven by osmosis is called **osmotic flow**. The **osmotic pressure** of a solution is the force of water movement into a solution resulting from its solute concentration.

39. **Tonicity** is the effects of osmotic solutions on cells. A solution that does not cause an osmotic flow is **isotonic**. A solution that causes water to flow into a cell is **hypotonic** and can lead to **hemolysis**. A solution that causes water to flow out of a cell is **hypertonic** and can lead to **crenation**.

3.16 In carrier-mediated transport, integral proteins facilitate membrane passage p. 116

40. Substances that are insoluble in lipids and too large to fit through membrane channels may be transported across the plasma membrane by **carrier proteins**.

41. **Facilitated diffusion** is a passive process, **active transport** is an active process, and **secondary active transport** does not require ATP (although the cell may require ATP later to maintain homeostasis).

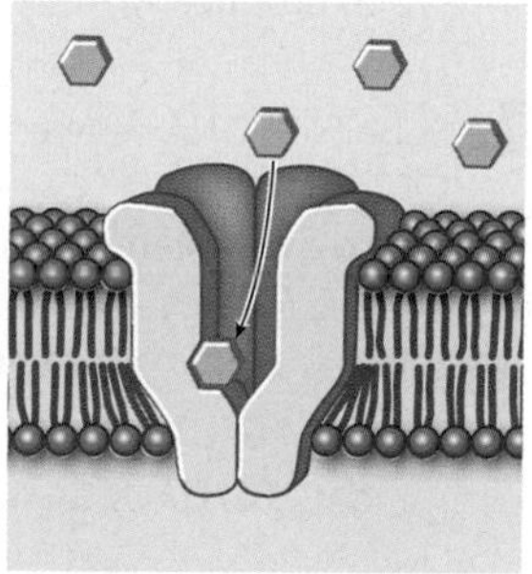

Facilitated diffusion allows many substances to be passively transported across the plasma membrane.

3.17 In vesicular transport, vesicles selectively carry materials into or out of the cell p. 118

42. In **vesicular transport**, materials move into or out of the cell in small membranous sacs called vesicles. The two major categories are **endocytosis** and **exocytosis**. Both require energy in the form of ATP.

43. **Pinocytosis** or "cell drinking" and **phagocytosis** or "cell eating" are two examples of endocytosis.

SECTION 4 • Cell Life Cycle

3.18 Interphase and cell division make up the life cycle of a cell p. 121

44. **Cell division** of a single cell produces a pair of **daughter cells**.

45. **Interphase** is the period in which a cell is performing normal functions and not actively engaged in cell division.

46. **Apoptosis** is the genetically controlled death of cells. **Mitosis** is the nuclear division of somatic cells. Sex cells (sperm or oocytes) are produced by **meiosis**.

3.19 During interphase, the cell prepares for cell division p. 122

47. **Somatic cells** spend the majority of their functional lives in interphase, which includes the G_1, **S**, and **G_2 phases**.

3.20 Mitosis distributes chromosomes before cytokinesis separates the daughter cells p. 124

48. **Mitosis** specifically refers to the division and duplication of the cell's nucleus. It proceeds in four stages: **prophase**, **metaphase**, **anaphase**, and **telophase**.

49. Division of the cytoplasm into two distinct new cells involves a separate process known as **cytokinesis**.

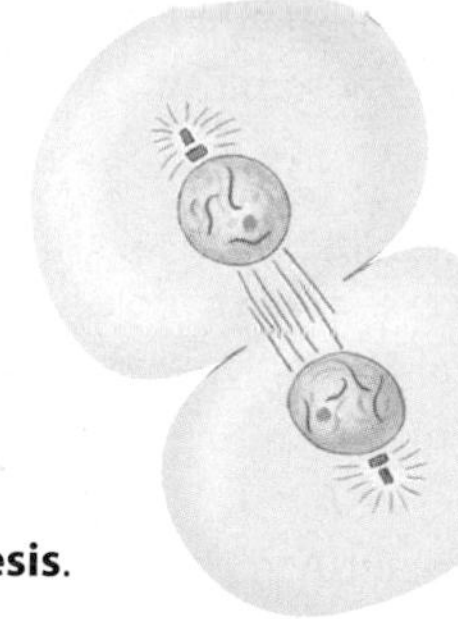

3.21 Tumors and cancer are characterized by abnormal cell growth and division p. 126

50. A **tumor**, or **neoplasm**, can be **benign** or **malignant**.

51. Malignant cells may spread locally by **invasion**, or to distant tissues and organs through **metastasis**.

Chapter Review Questions

Multiple choice

Select the correct answer from the list provided.

1 Which of the following is *not* a characteristic of the cell theory?
- ☐ a) Cells are the building blocks of life.
- ☐ b) Cells combine to form tissues.
- ☐ c) All new cells come from the division of pre-existing cells.
- ☐ d) Cells are the smallest units that perform all vital physiological functions.

2 The process of gradual cell specialization is called
- ☐ a) fertilization.
- ☐ b) meiosis.
- ☐ c) differentiation.
- ☐ d) metastasis.

3 The __________ is/are responsible for producing 95 percent of the ATP required by the cell.
- ☐ a) Golgi apparatus
- ☐ b) mitochondria
- ☐ c) rough endoplasmic reticulum
- ☐ d) smooth endoplasmic reticulum

4 Somatic cell nuclei contain __________ pairs of chromosomes.
- ☐ a) 8
- ☐ b) 23
- ☐ c) 46
- ☐ d) 92

5 The construction of a functional polypeptide by using the information in an mRNA strand is called
- a) translation.
- b) transcription.
- c) replication.
- d) differentiation.

6 Genetically controlled cell death is called
- a) mitosis.
- b) apoptosis.
- c) metastasis.
- d) mutation.

7 The __________ phase of protein synthesis encodes genetic instructions on a strand of mRNA.
- a) transcription
- b) translation
- c) transmigration
- d) transference

8 When substances pass through the plasma membrane by active processes, which molecule is required?
- a) RNA polymerase
- b) ATP
- c) DNA
- d) tRNA

9 The sodium–potassium exchange pump
- a) is an example of facilitated diffusion.
- b) does not require the input of cellular energy in the form of ATP.
- c) moves the sodium and potassium ions along their concentration gradients.
- d) is composed of a carrier protein located in the plasma membrane.

10 If a cell lacked ribosomes it would not be able to
- a) produce complex carbohydrates.
- b) synthesize proteins.
- c) divide.
- d) produce ATP.

11 Suppose a DNA segment has the following nucleotide sequence: CTC/ATA/CGA/TTC/AAG/TTA. Which nucleotide sequence would a complementary mRNA strand have?
- a) GAG/UAU/GAU/AAC/UUG/AAU
- b) GAG/TAT/GCT/AAG/TTC/AAT
- c) GAG/UAU/GCU/AAG/UUC/AAU
- d) GUG/UAU/GGA/UUG/AAG/GGU

12 How many amino acids are coded in the DNA segment in the previous question?
- a) 18
- b) 9
- c) 6
- d) 3

Short answer

13 Identify the five functional classes of membrane proteins.

14 Distinguish between mitosis and cytokinesis.

15 Differentiate between diffusion and osmosis.

16 Which organelle contains its own DNA, and what does it code for?

17 List the nonmembranous organelles and then list the membranous organelles.

18 List the stages of mitosis, and for each stage briefly describe the events that occur.

19 If a cell had microvilli on its plasma membrane, in which activity is it likely to be actively engaged?

20 If a red blood cell were immersed in a hypotonic solution, in which direction would water flow, and what effect would it have on the cell?

21 Order the following steps of protein synthesis into the proper sequence:

a) mRNA exits nucleus through nuclear pore;
b) introns are snipped from mRNA;
c) amino acids form peptide bonds;
d) ribosomal subunits detach from mRNA;
e) transcription of DNA forms mRNA;
f) tRNA anticodons bind to mRNA codons;
g) ribosomal subunits bind to mRNA

22 Describe the steps involved in phagocytosis.

23 Why does mitosis produce cells containing 46 chromosomes, whereas meiosis produces cells containing only 23 chromosomes?

24 Steroid hormones such as estrogen and testosterone are lipid molecules. Explain how these molecules cross the plasma membrane and enter the cell.

25 Describe the process used by malignant tumors to accelerate their growth.

MasteringA&P®

Access more chapter study tools online in the MasteringA&P Study Area:

- Chapter Quizzes, Chapter Practice Test, Art-labeling Activities, Animations, MP3 Tutor Sessions, and Clinical Case Studies

- Practice Anatomy Lab — PAL™
- Interactive Physiology — iP®
- A&P Flix — A&PFlix™
- PhysioEx — PhysioEx™

Chapter Integration • Applying what you have learned

Analyzing a stomach tumor

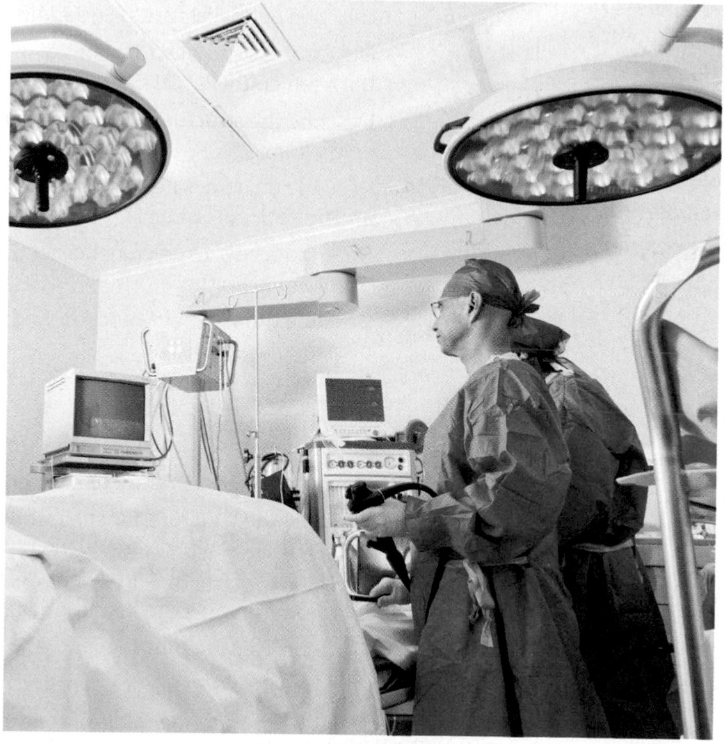

Vito has been experiencing persistent indigestion, heartburn, and loss of appetite. After many weeks of this discomfort he finally schedules an appointment with his physician. The doctor performs a diagnostic procedure in which he passes a tube through Vito's mouth into his stomach. Through this tube the doctor can use a scope to view the tissue lining Vito's stomach. Upon examination the doctor notices a tumor, or mass of abnormal cells, on the stomach wall. The doctor excises a small clump of tissue and sends it to the pathology laboratory for analysis. From what you have just learned about cells, answer the following questions about Vito's condition.

1 **Which word describing the tumor do Vito and his doctor hope to read in the lab report? Why? Which word do they not want to read in the report? Why?**

2 **Explain the cellular processes that likely caused the tumor.**

3 **Besides the obvious threat to Vito's stomach, does this tumor pose a threat to other organs?**

4 **Which cellular process would Vito and his doctor want to switch on in these tumor cells if they were able to do so?**

Exercising muscle

Corrine is a bodybuilder, certified personal trainer, and owner of her own fitness center. Five days per week she lifts weights for three hours per day. When working as a trainer, she often cites the benefits of electrolyte-laden sports drinks, claiming that they are good for fatigued cells.

5 **Corrine tells her clients that the muscle cells of people training for endurance sports, such as the marathon, have the potential for storing more energy and improving muscle performance, because those cells contain increased numbers of mitochondria. Is this true? Why or why not?**

6 **Using the principles of tonicity, explain the benefits of consuming energy drinks after strenuous exercise.**

4 Tissue Level of Organization

LEARNING OUTCOMES

These Learning Outcomes correspond by number to this chapter's modules and indicate what you should be able to do after completing the chapter.

SECTION 1 • Epithelial Tissue

- **4.1** Identify the four types of tissues in the body and describe their roles.
- **4.2** Describe three microscopy techniques.
- **4.3** Describe epithelial tissues, including cell shape, layers, and functions.
- **4.4** Discuss the types and functions of intercellular connections between epithelial cells.
- **4.5** Describe the structure and function of squamous epithelium.
- **4.6** Describe the structure, function, and locations of cuboidal and transitional epithelia.
- **4.7** Describe the structure, function, and locations of columnar epithelia.
- **4.8** Describe the structure, function, and locations of glandular epithelia.

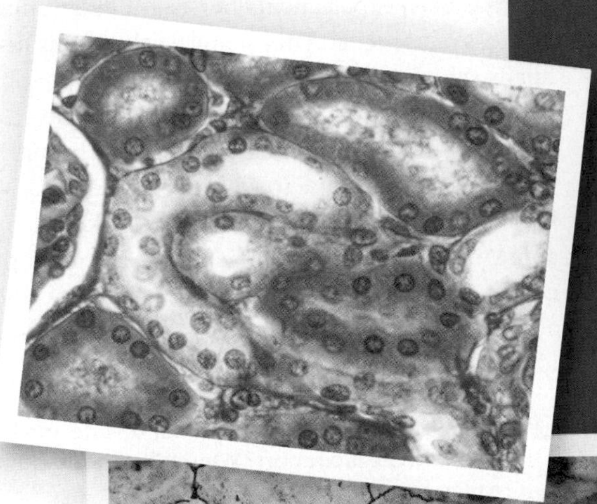

SECTION 2 • Connective Tissue

- **4.9** Describe the general structure of connective tissue.
- **4.10** Describe the structure, function, and locations of areolar tissue, adipose tissue, and reticular tissue.
- **4.11** Describe the structure, function, and locations of dense connective tissues and fluid connective tissues.
- **4.12** Describe the structure, function, and locations of cartilage.
- **4.13** Describe the structure and function of bone.
- **4.14** Describe the arrangements of epithelial and connective tissues in the four types of membranes, and the structures and locations of the three types of fasciae.

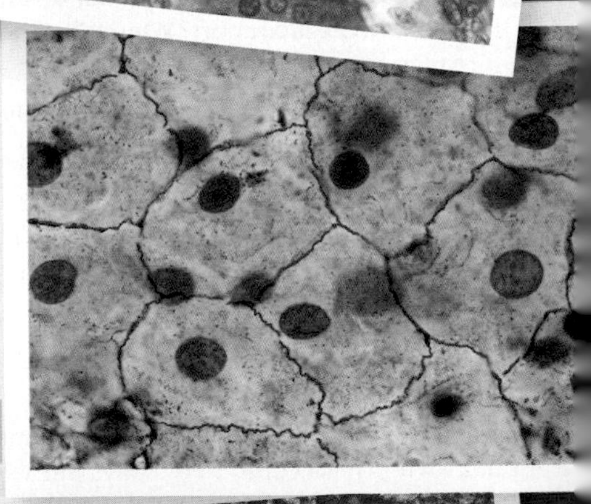

SECTION 3 • Muscle Tissue and Neural Tissue

- **4.15** Describe the relative proportions of muscle tissue and neural tissue in the body.
- **4.16** Specify the functions of muscle tissue and neural tissue.
- **4.17** **CLINICAL MODULE** Describe the roles of inflammation and regeneration in response to tissue injury.

Learning Outcomes are repeated at the bottom of each module.

Four types of tissue make up the body

Our perspective has gradually changed over the preceding chapters. Atoms and molecules can only be examined through special imaging techniques and experimental procedures. Cellular details often escape detection unless an electron microscope is used. Tissue structure, however, can be examined with a light microscope, and based on your experiences in the laboratory, you may already be able to identify some tissues without using a microscope.

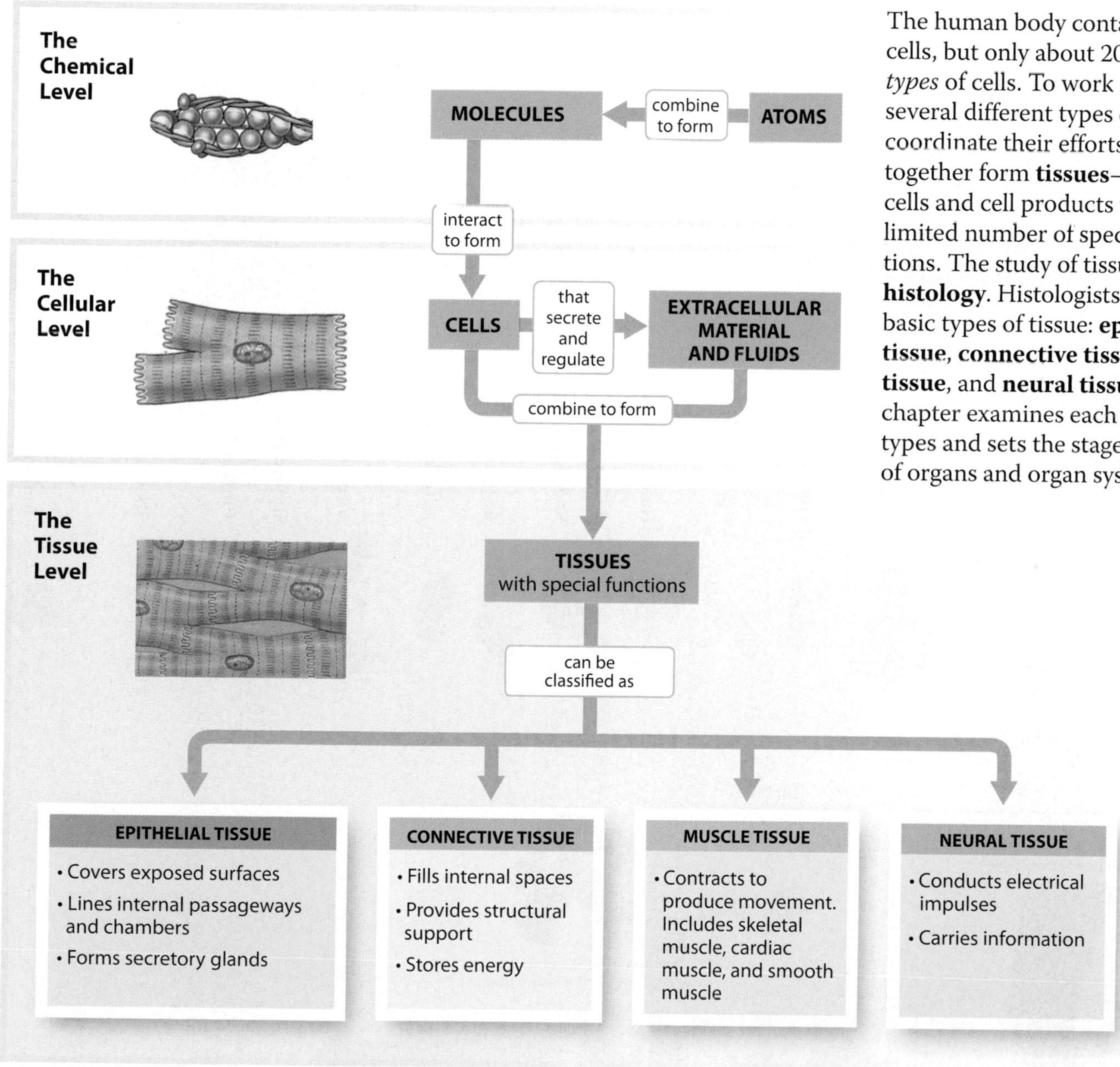
The Chemical Level
MOLECULES
combine to form
ATOMS
interact to form
The Cellular Level
CELLS
that secrete and regulate
EXTRACELLULAR MATERIAL AND FLUIDS
combine to form
The Tissue Level
TISSUES with special functions
can be classified as
EPITHELIAL TISSUE
• Covers exposed surfaces
• Lines internal passageways and chambers
• Forms secretory glands
CONNECTIVE TISSUE
• Fills internal spaces
• Provides structural support
• Stores energy
MUSCLE TISSUE
• Contracts to produce movement. Includes skeletal muscle, cardiac muscle, and smooth muscle
NEURAL TISSUE
• Conducts electrical impulses
• Carries information

The human body contains trillions of cells, but only about 200 different *types* of cells. To work efficiently, several different types of cells must coordinate their efforts. Cells working together form **tissues**—collections of cells and cell products that perform a limited number of specialized functions. The study of tissues is called **histology**. Histologists recognize four basic types of tissue: **epithelial tissue**, **connective tissue**, **muscle tissue**, and **neural tissue**. This chapter examines each of these tissue types and sets the stage for our study of organs and organ systems.

Module 4.1 Review

a. Name the term used to describe the study of tissues.

b. List the four basic tissue types.

c. Identify the functions of each tissue type.

LO **4.1** Identify the four types of tissues in the body and describe their roles.

Module 4.2

Microscopes are used to study cells and tissues

Much of what is studied in anatomy and physiology cannot be seen with the naked eye or a simple hand lens. **Microscopy** (the use of microscopes) began about 400 years ago with the inventions of **simple microscopes** with one lens and **compound microscopes** with more than one lens. However, these early light microscopes magnified objects only 10 to 20 times actual size. The light microscope in your lab likely magnifies up to 1000 times, the maximum when using light. Modern electron microscopes using beams of electrons instead of light can magnify over 1 million times!

1 Anatomy can be studied at different scales. The amount of fine detail that can be seen in an image, called resolution, varies with magnification and the type of microscope being used. The figure below shows the magnification ranges and degree of resolution achieved by different types of microscopes.

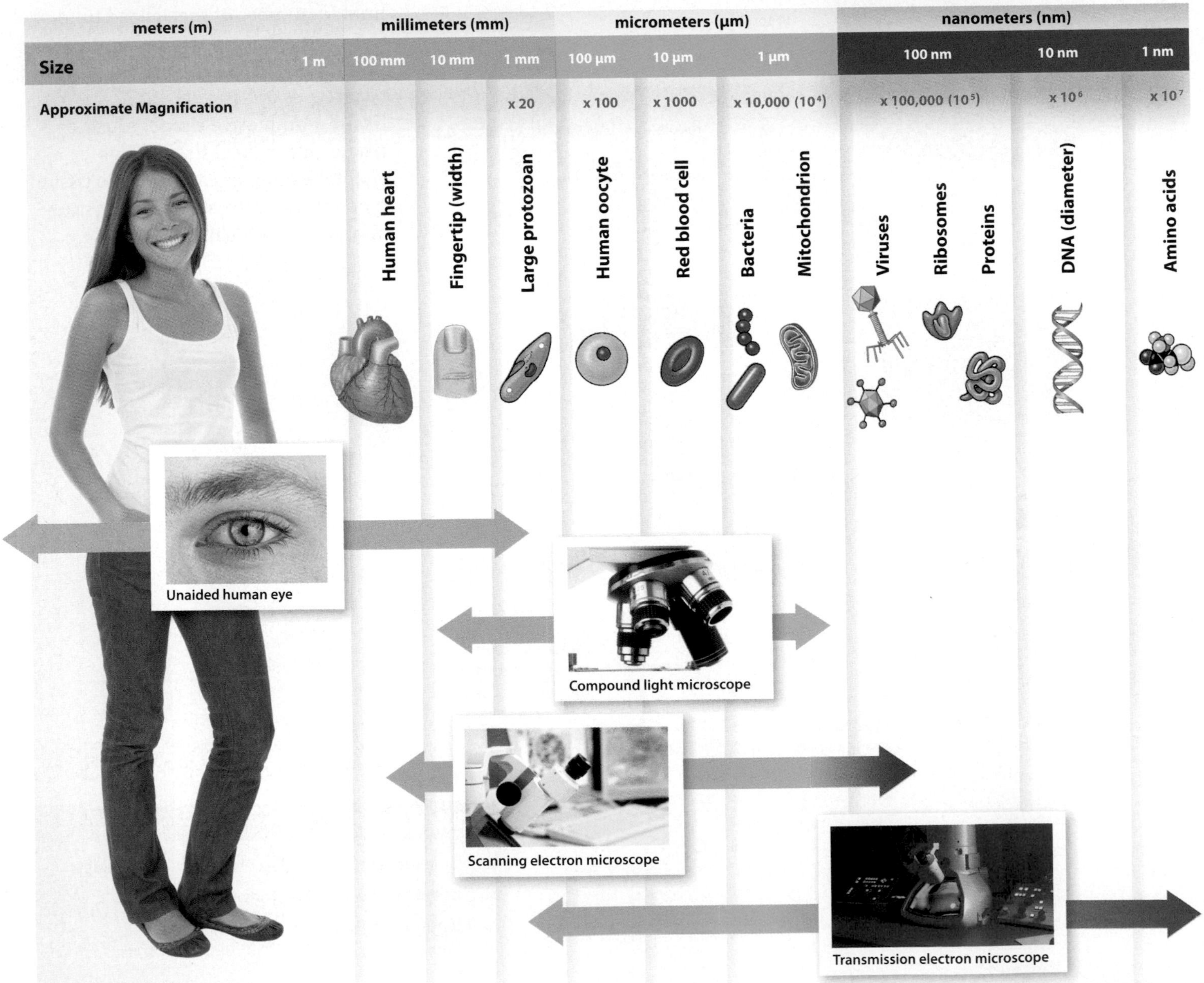

meters (m)
millimeters (mm)
micrometers (μm)
nanometers (nm)
Size
1 m
100 mm
10 mm
1 mm
100 μm
10 μm
1 μm
100 nm
10 nm
1 nm
Approximate Magnification
x 20
x 100
x 1000
x 10,000 (10⁴)
x 100,000 (10⁵)
x 10⁶
x 10⁷
Human heart
Fingertip (width)
Large protozoan
Human oocyte
Red blood cell
Bacteria
Mitochondrion
Viruses
Ribosomes
Proteins
DNA (diameter)
Amino acids
Unaided human eye
Compound light microscope
Scanning electron microscope
Transmission electron microscope

2 A **light microscope** is an optical instrument that directs visible light through a thin section of tissue. The specimen is first magnified with an objective lens and focused in the tube of the microscope. A **revolving nosepiece** holds several objective lenses of progressive magnifying power. A second lens in the eye piece, the ocular lens, magnifies the image further. The **total magnification** is calculated by multiplying the magnification of the objective lens times that of the ocular lens. The **resolution** of a microscope is the ability to distinguish between two separate points, or objects. The wavelengths of visible light limit the maximum resolution of a light microscope to about 200 nm (0.2 μm). Thus, objects closer together than 200 nm would appear as one object.

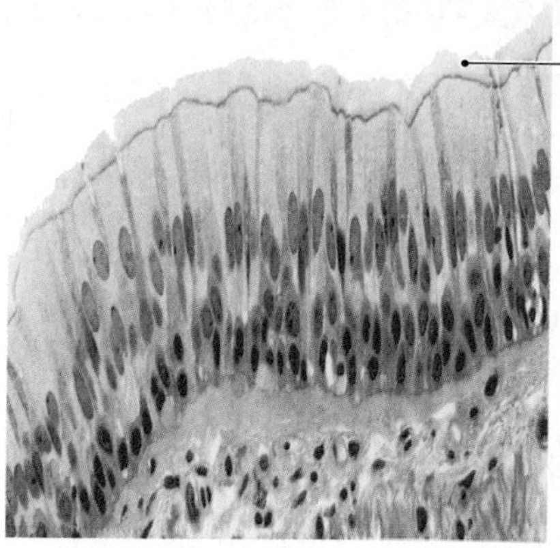
Cilia

LM × 400

This photographic image of cells from the lining of the respiratory tract was taken with a light microscope and is called a light micrograph (LM). It is magnified 400 times. In this text, this is indicated by the following notation: LM × 400.

3 A **transmission electron microscope** uses magnets to direct a beam of electrons through the surface of a finely sectioned object onto a photographic plate. The wavelength of the electron beam is about 0.00001 that of ordinary white light, which allows for far greater magnification and resolution. A transmission electron microscope has a maximum resolution of about 0.2 nm (0.0002 μm).

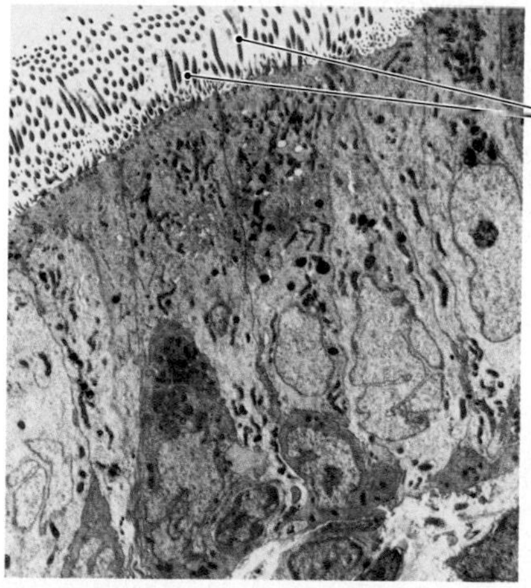
Cilia

TEM × 3000

This transmission electron microscope image is of cells from the lining of the respiratory tract. The cell and organelle details are much sharper. This type of image is called a transmission electron micrograph (TEM) and its magnification is indicated by the following notation: TEM × 3000.

4 The specimens prepared for **scanning electron microscopy** are not sectioned. The whole specimen is coated with an electron dense material, and then bombarded with electron beams. Because some of the electrons are reflected back, a three dimensional image of the specimen is created. A scanning electron microscope has a maximum resolution of about 10 nm (0.01 μm).

Cilia

SEM × 15,846

This dramatic image of the surface of the lining of the respiratory tract was taken with a scanning electron microscope and then color enhanced. The type of image is called a scanning electron micrograph (SEM), and its magnification is indicated by: SEM × 15,846.

5 You will probably be using a light microscope during your laboratory experiences. Begin by using the objective lens with the lowest magnifying power to focus on your specimen. Then carefully rotate objective lenses with greater magnification into position. For each tissue you examine, open your text to an image of the same tissue. The images will look most similar if you focus your microscope close to the magnification of the image in the book.

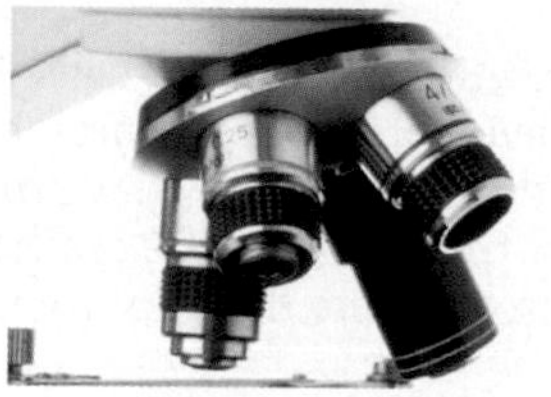

Module 4.2 Review

a. How do early microscopes compare with modern microscopes?

b. Differentiate between LM, TEM, and SEM.

c. Which kind of microscope is in your lab?

LO 4.2 Describe three microscopy techniques.

Epithelial tissue covers surfaces, lines structures, and forms secretory glands

1 Let's begin our discussion of tissue types with **epithelial tissue**, because it includes a very familiar feature: the surface of your skin. Epithelial tissue includes *epithelia* and *glands*.

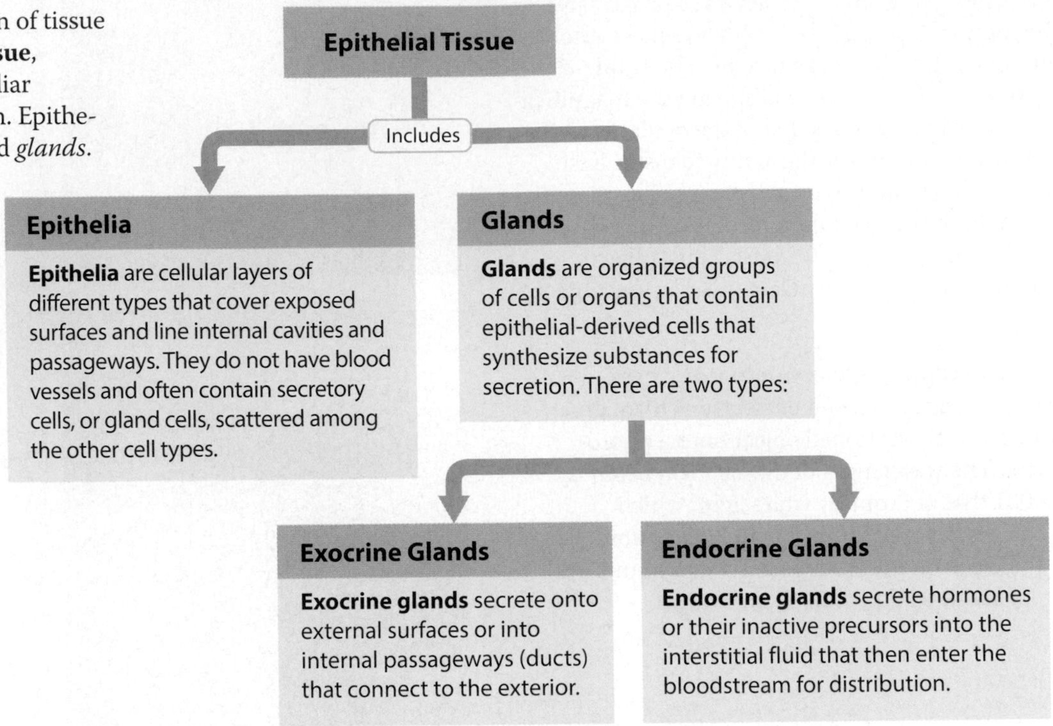
Epithelial Tissue
Includes
Epithelia
Epithelia are cellular layers of different types that cover exposed surfaces and line internal cavities and passageways. They do not have blood vessels and often contain secretory cells, or gland cells, scattered among the other cell types.
Glands
Glands are organized groups of cells or organs that contain epithelial-derived cells that synthesize substances for secretion. There are two types:
Exocrine Glands
Exocrine glands secrete onto external surfaces or into internal passageways (ducts) that connect to the exterior.
Endocrine Glands
Endocrine glands secrete hormones or their inactive precursors into the interstitial fluid that then enter the bloodstream for distribution.

2 The table at right describes four essential functions of epithelial tissue.

Functions of Epithelial Tissue

- **Provide physical protection**: Epithelia protect exposed and internal surfaces from abrasion, dehydration, and destruction by chemical or biological agents.
- **Control permeability**: Any substance that enters or leaves the body must cross an epithelium. Some epithelia are relatively impermeable, whereas others are permeable to compounds as large as proteins. Most are capable of selective absorption or secretion. The epithelial barrier can be regulated and modified in response to various stimuli. For example, think of the calluses that form on your hands when you do rough work for a period of time.
- **Provide sensation**: Sensory nerves extensively innervate most epithelia. Specialized epithelial cells can detect changes in the environment and convey information about such changes to the nervous system. For example, touch receptors respond to pressure by stimulating adjacent sensory nerves. A **neuroepithelium** is a sensory epithelium found in special sense organs that provide the sensations of smell, taste, sight, equilibrium, and hearing.
- **Produce specialized secretions**: Epithelial cells that produce secretions are called gland cells. Individual gland cells are often scattered among other cell types in an epithelium that may have many other functions. In a glandular epithelium, most or all of the epithelial cells produce secretions, and those secretions are the primary function of the tissue.

3 The cells of any epithelium share a number of basic features. An epithelium has an **apical** (Ā-pi-kal) **surface**, which faces the exterior of the body or some internal space, and a **base**, which is attached to adjacent tissues. The term **polarity** refers to the presence of structural and functional differences between the exposed and attached surfaces.

Microvilli are often found on the apical surfaces of epithelial cells that line internal passageways of the digestive, urinary, and reproductive tracts.

The apical surface is the region of the cell exposed to an internal or external environment. When the epithelium lines a tube, such as the intestinal tract, the apical surfaces of the epithelial cells are exposed to the space inside the tube, a passageway called the **lumen** (LOO-men).

The **basolateral surfaces** include both the base, where the cell attaches to underlying epithelial cells or deeper tissues, and the sides, where the cell contacts its neighbors.

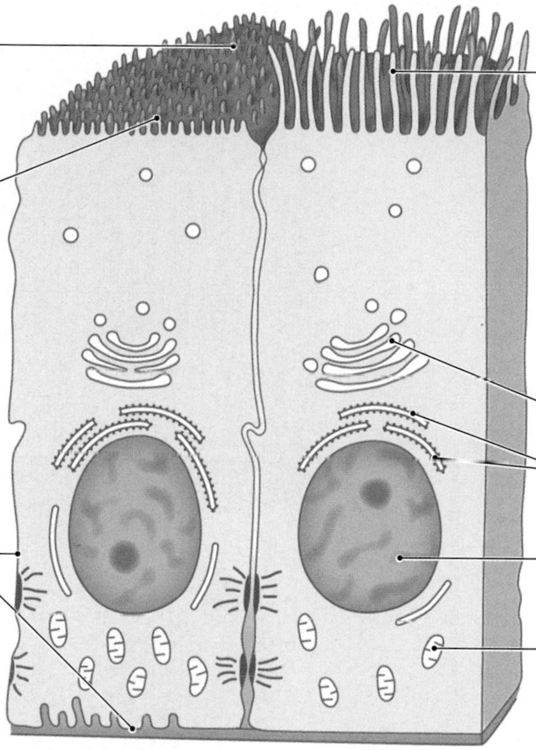

Motile cilia cover the apical surfaces of cells in portions of the respiratory and reproductive tracts. A typical ciliated cell contains about 250 cilia that beat in a coordinated manner (Module 3.4, p. 93).

Membranous Organelles
Most epithelial cells have membranous organelles comparable to those of other cell types.
Golgi apparatus (facing apical surface)
Endoplasmic reticulum (often extensive around the nucleus)
Nucleus (in a tall cell, located closer to the base than the apical surface)
Mitochondria (may be apical or basal, depending on cell functions)

Simple epithelia

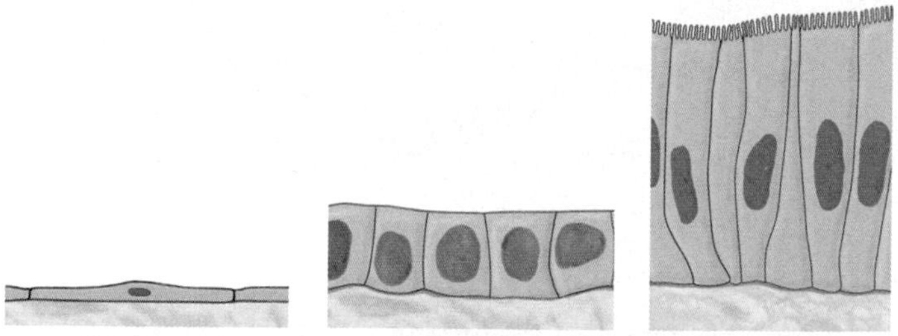

Squamous　Cuboidal　Columnar

Stratified epithelia

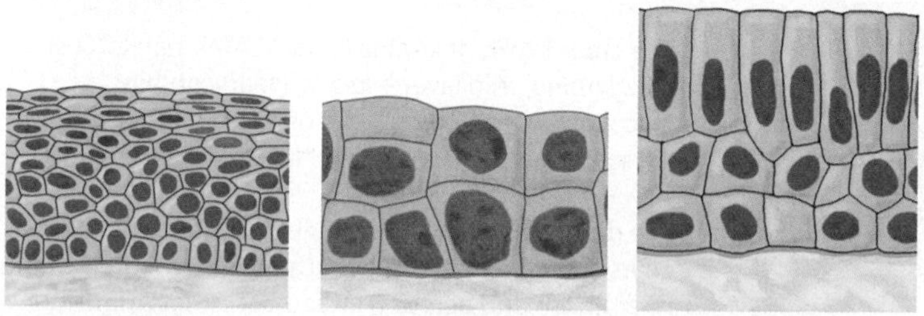

Squamous　Cuboidal　Columnar

4 Epithelial cells have three basic shapes: **squamous**, **cuboidal**, and **columnar**. For classification purposes, we look at the superficial cells in a section perpendicular to both the exposed surface and the basal surface. In sectional view, squamous cells appear thin and flat, cuboidal cells look like little boxes, and columnar cells look like tall, relatively slender rectangles. If only one layer of cells is present, that layer is a **simple epithelium**. In contrast, a **stratified epithelium** contains several layers of cells. Stratified epithelia are generally located in areas that need protection from mechanical or chemical stresses. The surface of the skin and the lining of the mouth are examples.

Module 4.3 Review

a. List four essential functions of epithelial tissue.

b. What is the probable function of an epithelial surface whose cells bear many cilia?

c. Summarize the classification of an epithelium based on cell shape and number of cell layers.

LO 4.3 Describe epithelial tissues, including cell shape, layers, and functions.

Epithelial cells are extensively interconnected, both structurally and functionally

To be effective as a barrier, an epithelium must form a complete cover or lining, and be able to replace lost or damaged cells through stem cell division. The physical integrity of an epithelium depends on intercellular connections and attachment to adjacent tissues.

1 The detailed structure of each form of intercellular attachment demonstrates the linkage between structure and function at all levels.

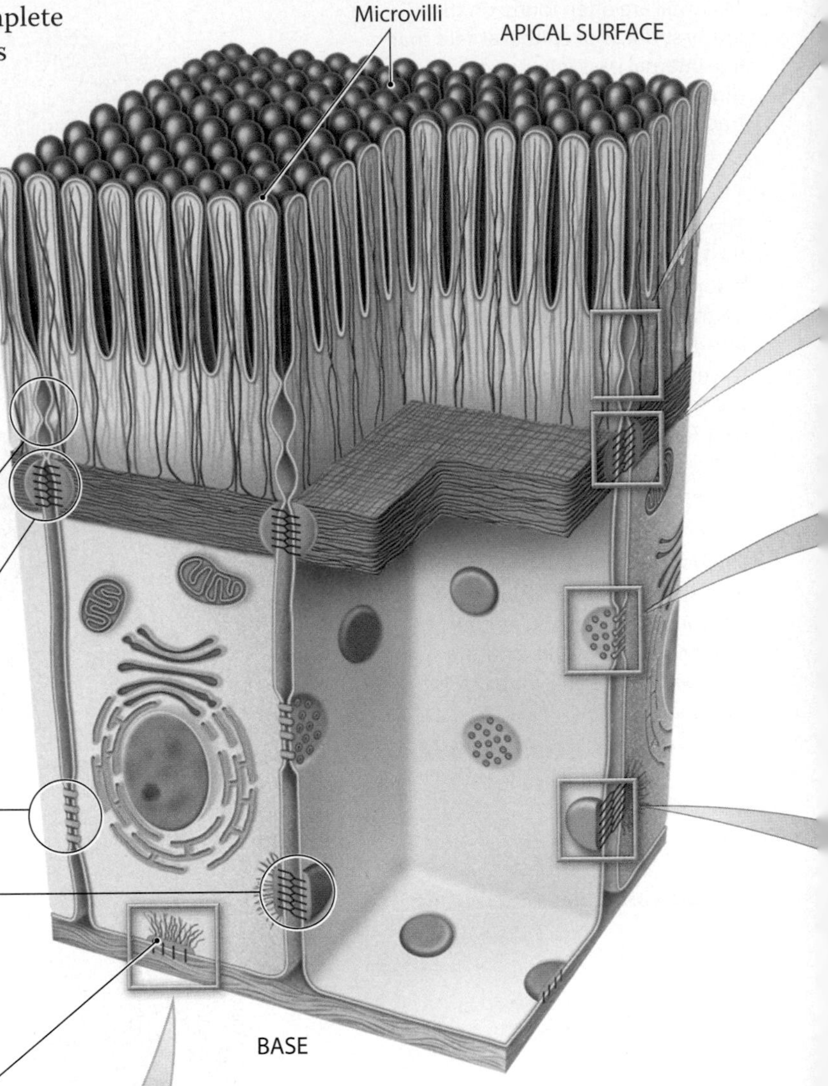
Microvilli
APICAL SURFACE
BASE

Intercellular Attachments

Tight (occluding) **junctions** are characteristic of epithelial cells lining the intestinal tract where they form a barrier that isolates the basolateral surfaces and deeper tissues from the contents of the lumen.

An **adhesion belt** locks together the terminal webs of neighboring cells, strengthening the apical region and preventing distortion and leakage at the occluding junctions.

Gap junctions permit chemical communication that coordinates the activities of adjacent cells.

Desmosomes (DEZ-mō-sōms; *desmos*, ligament + *soma*, body) provide firm attachment between neighboring cells by interlocking their cytoskeletons.

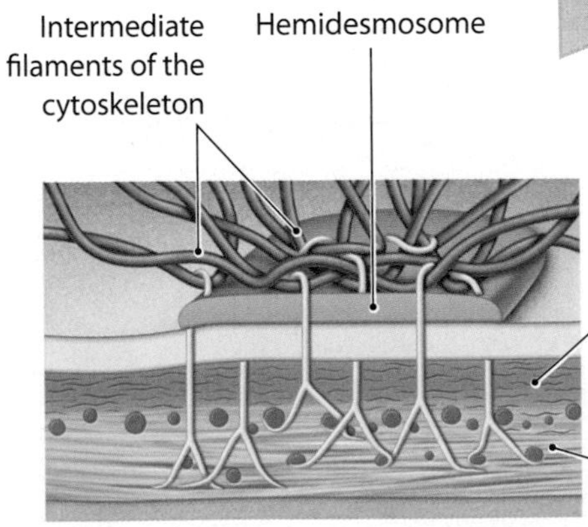
Intermediate filaments of the cytoskeleton
Hemidesmosome

2 **Hemidesmosomes** attach the deepest epithelial cells to the basement membrane. At a hemidesmosome, the basal cytoskeleton is locked to peripheral proteins and to transmembrane proteins that are firmly attached to a layer of extracellular protein filaments and fibers.

Basement Membrane

The **basement membrane**, which is also called the basal lamina, is a noncellular structure produced by the basal surface of the epithelium and the underlying connective tissue.

The **clear layer**, or lamina lucida (LAM-i-nah LOO-si-dah; *lamina*, thin layer + *lucida*, clear) contains glycoproteins and a network of fine protein filaments. It is produced by the adjacent layer of epithelial cells.

The **dense layer**, or lamina densa, contains bundles of coarse protein fibers. It gives the basement membrane its strength and acts as a filter that restricts diffusion between adjacent tissues and the epithelium.

3 At a tight junction, the attachment is so tight that it prevents water and solutes from passing between cells.

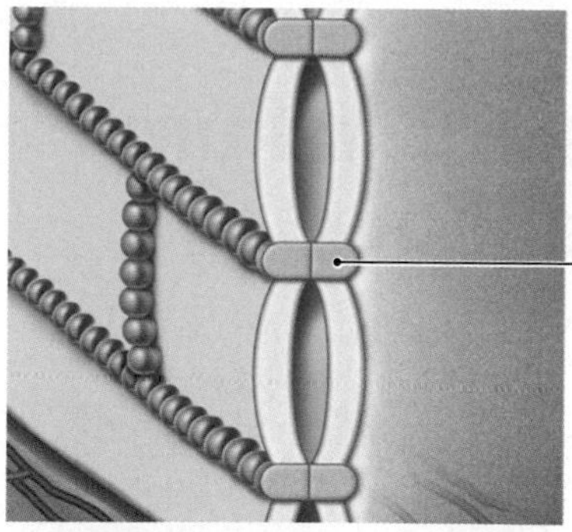

At a tight junction, the lipid portions of the two plasma membranes are tightly bound together by interlocking membrane proteins.

4 A continuous adhesion belt forms a band that encircles each cell and binds it to its neighbors. The bands are dense proteins that are attached to the microfilaments of the terminal web.

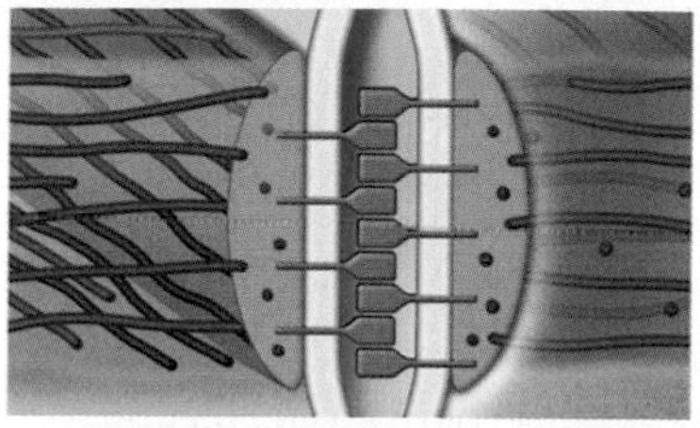

5 At a gap junction, two cells are held together by interlocking transmembrane proteins called **connexons**. Gap junctions between epithelial cells are common where the movement of ions helps coordinate functions such as secretion or cilia movement. Gap junctions are also common in other tissues. For example, gap junctions in cardiac muscle tissue and smooth muscle tissue are essential in coordinating muscle cell contractions.

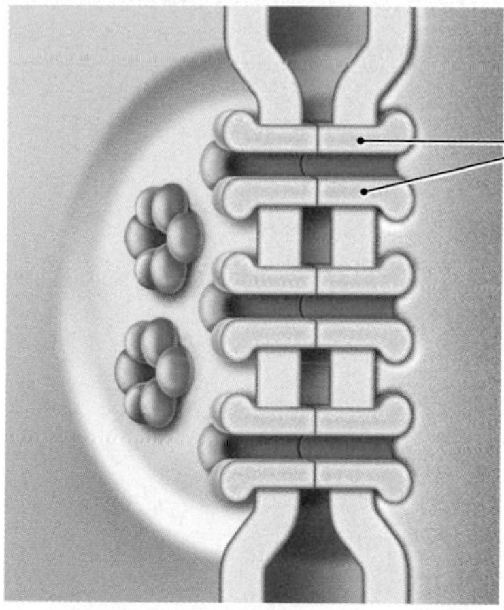

Connexons are channel proteins that form a narrow passageway and let small molecules and ions pass from cell to cell. Each connexon is composed of six connexin proteins that form a cylinder with a central channel. A number of human diseases are caused by mutations in the genes that code for different connexin proteins.

6 At a desmosome, the opposing plasma membranes are locked together. Desmosomes are very strong and durable, and help epithelial cells resist mechanical stresses such as stretching and twisting.

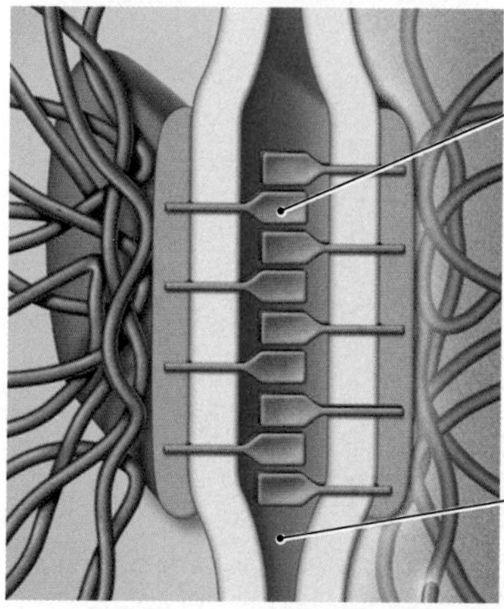

Cell adhesion molecules (CAMs) are transmembrane proteins that bind to each other and to extracellular materials. CAMs on the basolateral surface of an epithelium help bind the cell to the underlying basement membrane.

The membranes of adjacent cells may also be bonded by a thin layer of proteoglycans that contain polysaccharide derivatives, most notably **hyaluronan**.

Epithelia lack blood vessels, and for this reason they are said to be **avascular** (*a*, without + *vas*, vessel). Epithelial cells must obtain nutrients by diffusion or absorption across either the exposed or the attached epithelial surface. The cells forming the deepest layer of an epithelium must remain firmly attached to underlying connective tissues, because the blood vessels in those tissues nourish the entire epithelium.

Module 4.4 Review

a. Identify the various types of epithelial intercellular connections.

b. What is the functional significance of gap junctions?

c. How do epithelial tissues, which are avascular, obtain needed nutrients?

LO 4.4 Discuss the types and functions of intercellular connections between epithelial cells.

The cells in a squamous epithelium are flat and irregularly shaped

The cells in a **squamous epithelium** (SKWĀ-mus; *squama*, plate or scale) are thin, flat, and somewhat irregularly shaped, like pieces of a jigsaw puzzle. From the surface, the cells resemble fried eggs laid side by side. In sectional view, the disc-shaped nucleus occupies the thickest portion of each cell.

1 A **simple squamous epithelium** is the body's most delicate type of epithelium. This type of epithelium is located in protected regions where absorption or diffusion takes place, or where a slick, slippery surface reduces friction. Simple squamous epithelia are found along passageways in the kidneys, inside the eye, and at the gas exchange surfaces (alveoli) of the lungs. Simple squamous epithelia in certain locations have special names. A **mesothelium** lines the pericardial, pleural, and peritoneal body cavities. The simple squamous epithelium lining the inner surface of the heart and all blood vessels is called an **endothelium**.

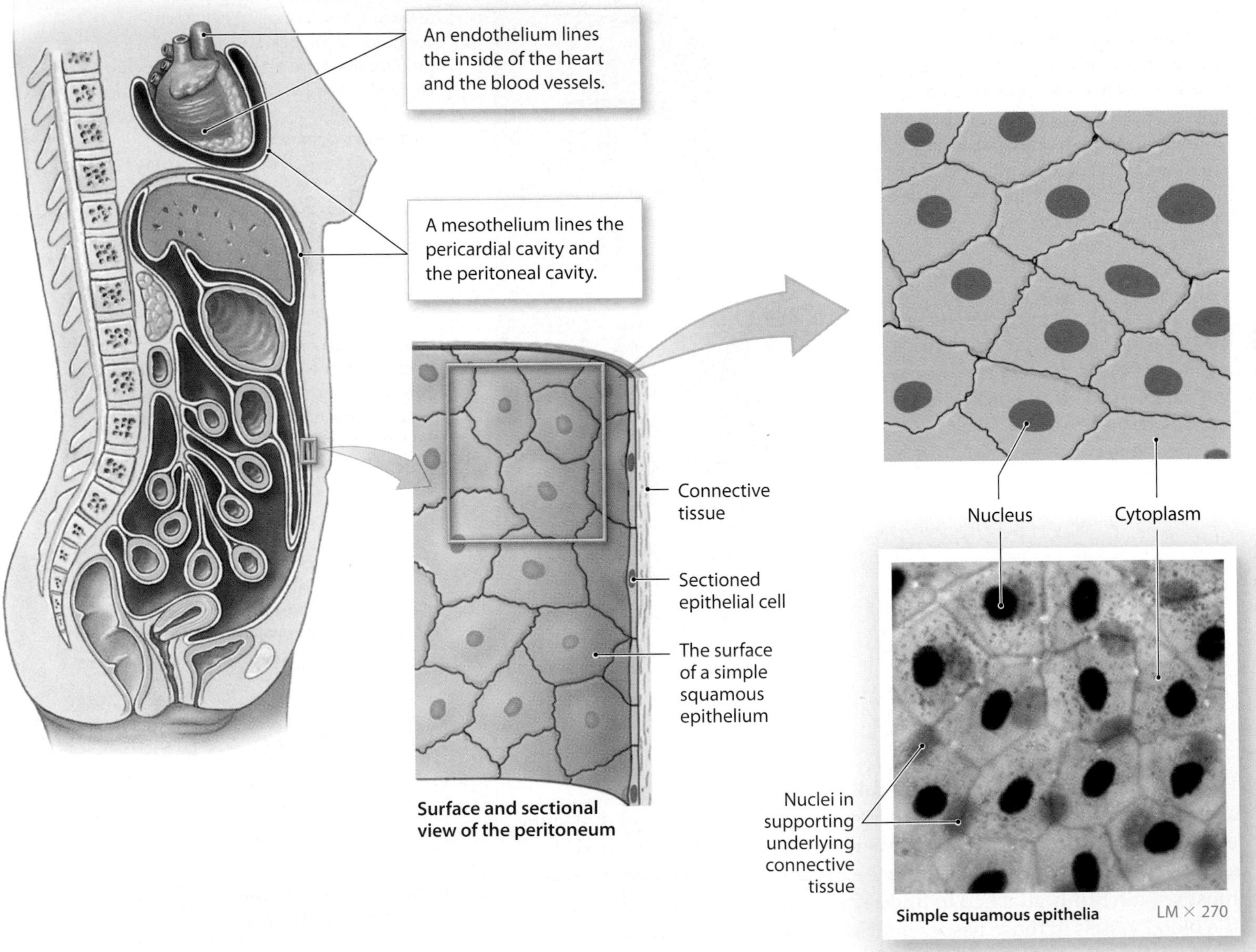
An endothelium lines the inside of the heart and the blood vessels.
A mesothelium lines the pericardial cavity and the peritoneal cavity.
Connective tissue
Sectioned epithelial cell
The surface of a simple squamous epithelium
Nucleus
Cytoplasm
Nuclei in supporting underlying connective tissue
LM × 270

Surface and sectional view of the peritoneum

Simple squamous epithelia

2 A **stratified squamous epithelium** is generally located where mechanical or chemical stresses are severe. The cells form a series of layers, like the layers in a sheet of plywood. Stratified squamous epithelia form the surface of the skin and line the mouth, throat, esophagus, rectum, anus, and vagina.

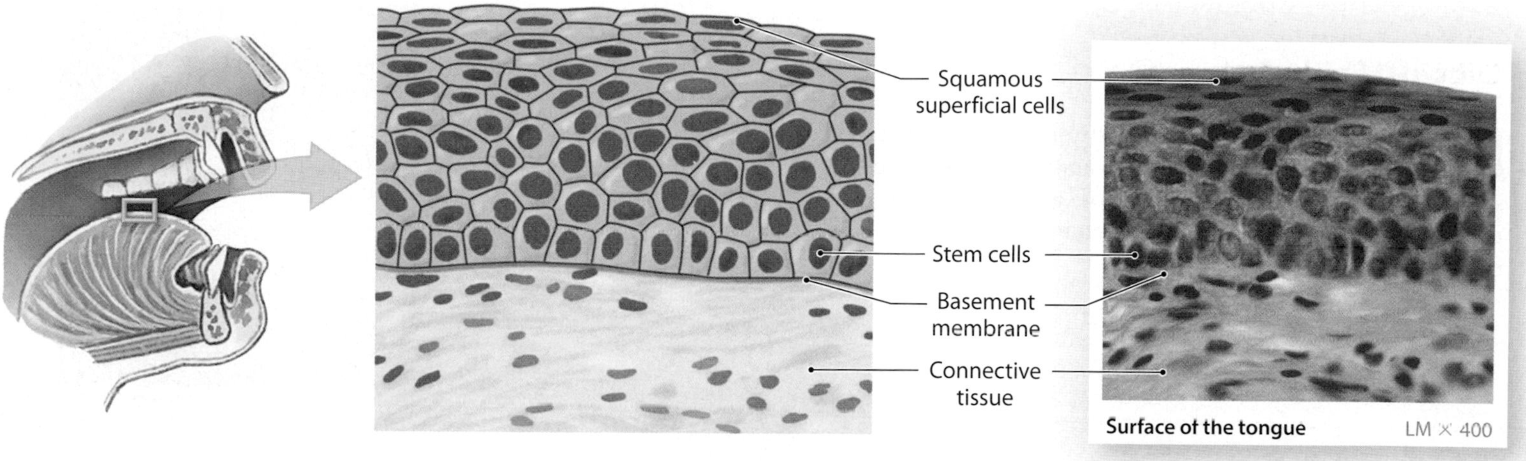

Squamous superficial cells
Stem cells
Basement membrane
Connective tissue

Surface of the tongue LM × 400

3 On exposed body surfaces, where mechanical stress and dehydration are potential problems, apical layers of epithelial cells are packed with filaments of the protein keratin. As a result, superficial layers are both tough and water resistant. Such an epithelium is said to be **keratinized**. A **nonkeratinized** stratified squamous epithelium resists abrasion but will dry out and deteriorate unless kept moist. Nonkeratinized stratified squamous epithelia are found in the mouth, pharynx, esophagus, anus, and vagina.

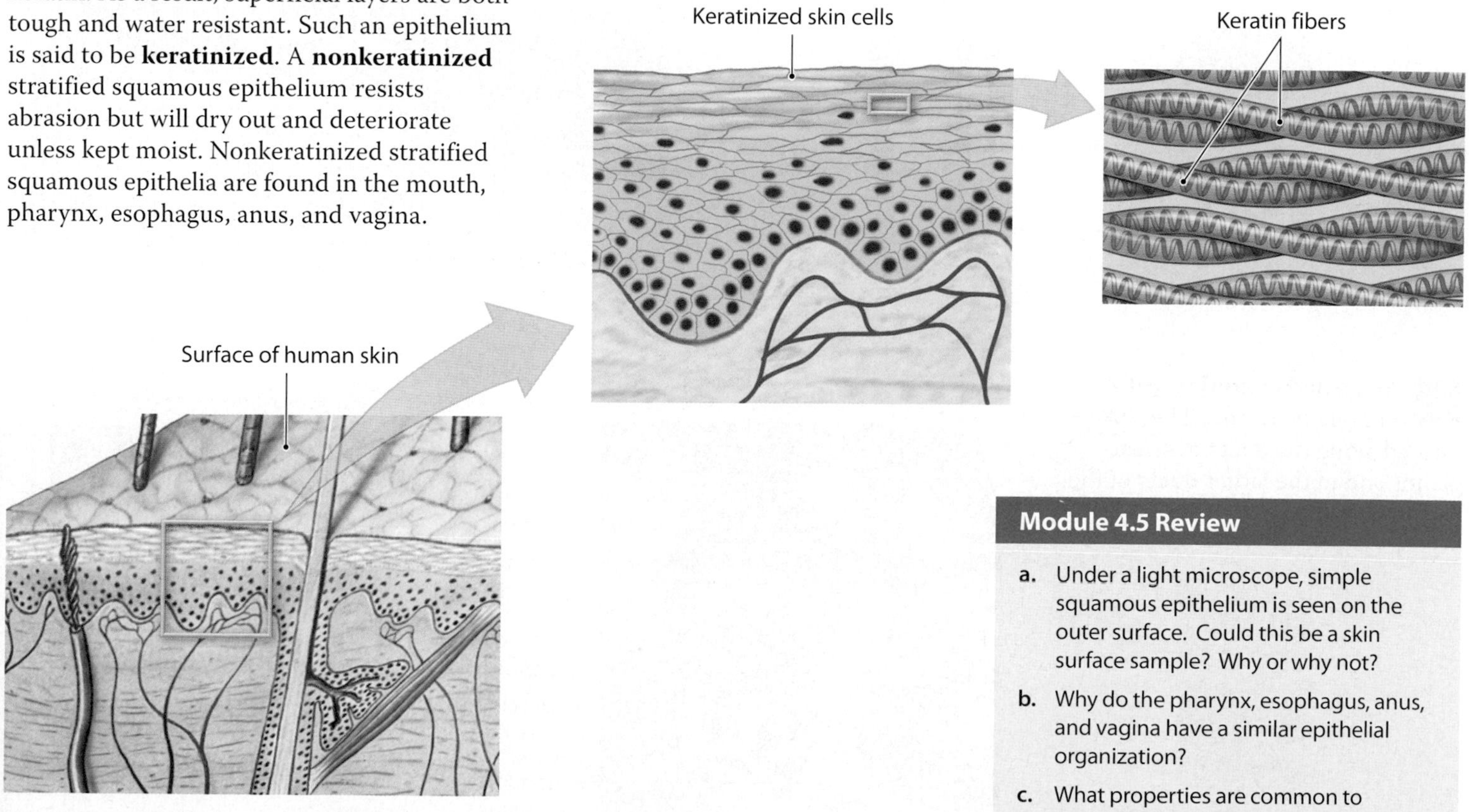

Keratinized skin cells
Keratin fibers
Surface of human skin

Module 4.5 Review

a. Under a light microscope, simple squamous epithelium is seen on the outer surface. Could this be a skin surface sample? Why or why not?

b. Why do the pharynx, esophagus, anus, and vagina have a similar epithelial organization?

c. What properties are common to keratinized epithelia?

LO 4.5 Describe the structure and function of squamous epithelium.

Cuboidal and transitional epithelia line several passageways and chambers connected to the exterior

Cuboidal Epithelium

1 The cells of a **cuboidal epithelium** resemble hexagonal boxes. (In sectional view, they appear square.) The spherical nuclei are near the center of each cell, and the distance between adjacent nuclei is roughly equal to the height of the epithelium.

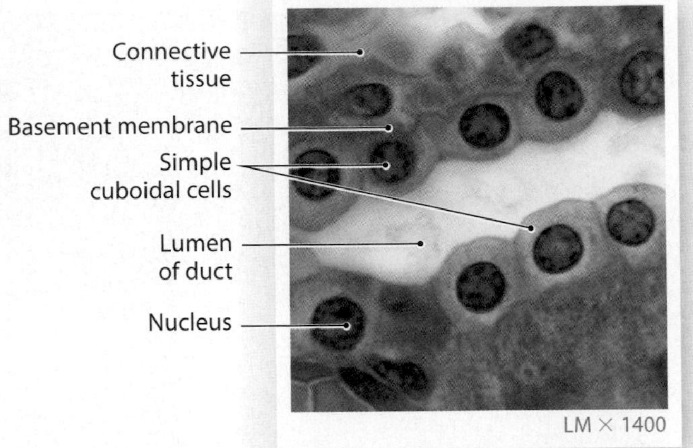

Connective tissue
Basement membrane
Simple cuboidal cells
Lumen of duct
Nucleus
LM × 1400

2 A **simple cuboidal epithelium** provides limited protection and occurs where secretion or absorption takes place. Such an epithelium lines portions of the kidney tubules. They also line secretory chambers in the thyroid gland.

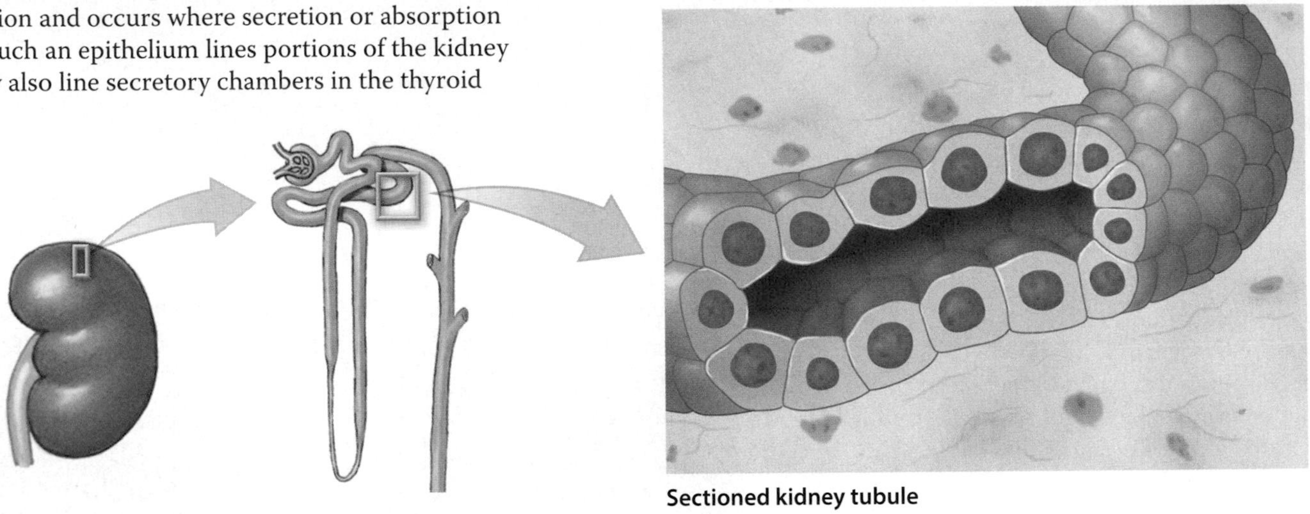

Sectioned kidney tubule

3 **Stratified cuboidal epithelia** are relatively rare. They are located along the ducts of sweat glands and in the larger ducts of the mammary glands.

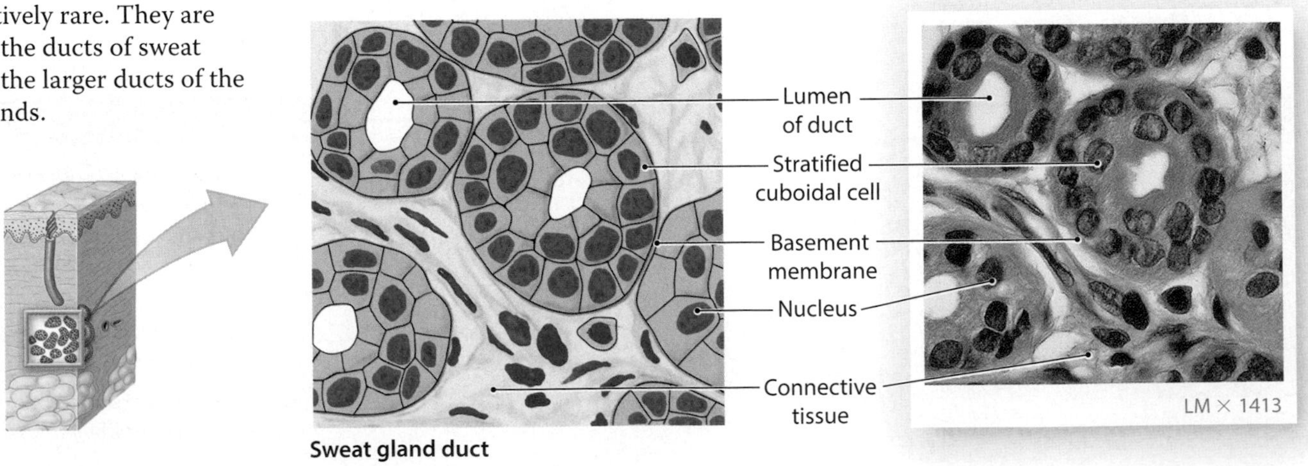

Lumen of duct
Stratified cuboidal cell
Basement membrane
Nucleus
Connective tissue
LM × 1413

Sweat gland duct

Transitional Epithelium

4 A **transitional epithelium** is an unusual stratified epithelium. Unlike most epithelia, it tolerates repeated cycles of stretching and recoiling (returning to its previous shape) without damage. It is called *transitional* because the appearance of the epithelium changes as stretching occurs. Transitional epithelia line the urinary bladder, the ureters, and the urine-collecting chambers within the kidneys, where large changes in volume occur.

Epithelium in a Relaxed Bladder

In an empty urinary bladder, the superficial cells are plump and cuboidal with a dome-shaped surface.

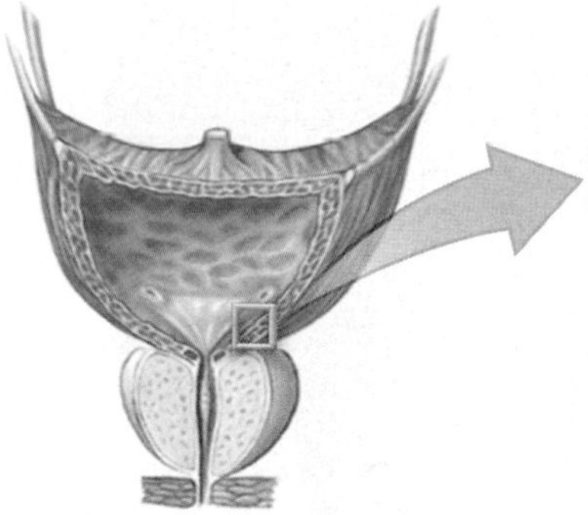

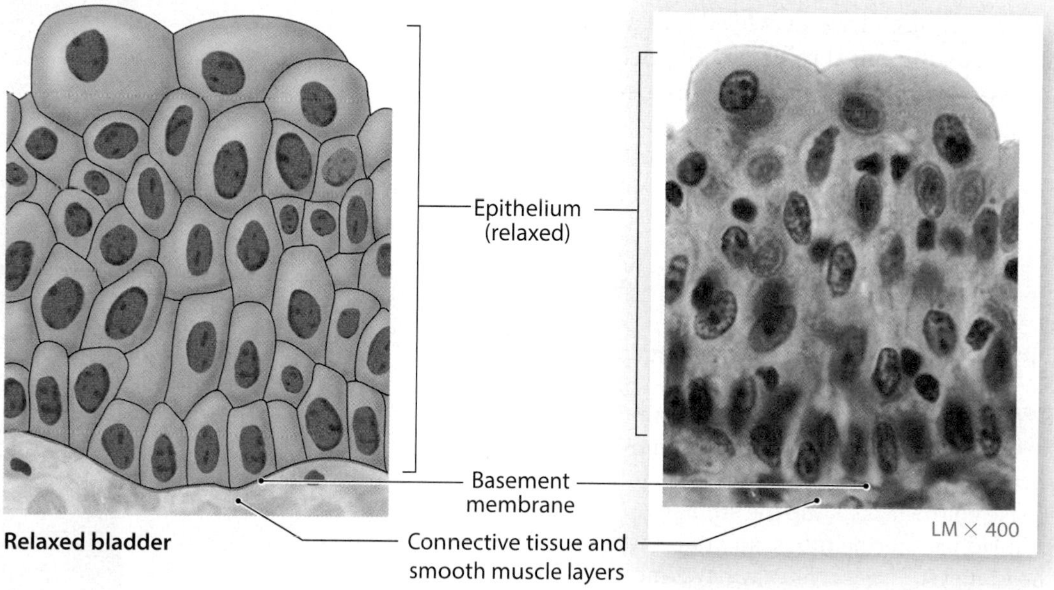
Epithelium (relaxed)
Basement membrane
Connective tissue and smooth muscle layers
LM × 400

Relaxed bladder

Epithelium in a Stretched Bladder

When the urinary bladder is full, the volume of urine has stretched the lining to its limits, and the epithelium appears flattened, and more like a stratified squamous epithelium.

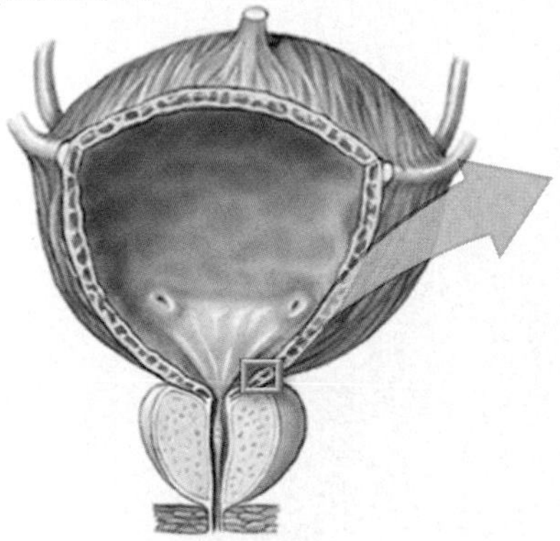

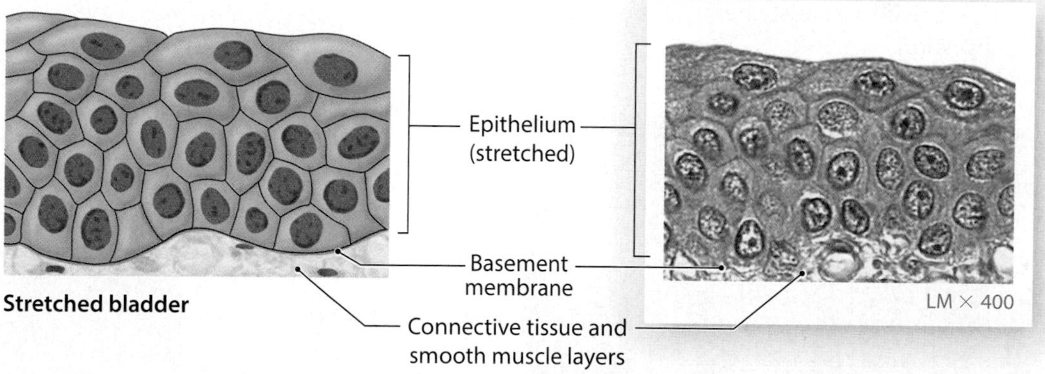
Epithelium (stretched)
Basement membrane
Connective tissue and smooth muscle layers
LM × 400

Stretched bladder

Module 4.6 Review

a. Describe the appearance of simple cuboidal epithelial cells in sectional view.

b. Stratified cuboidal epithelia are associated with what epithelial structures?

c. Identify the epithelium that lines the urinary bladder and changes in appearance as stretching occurs.

LO 4.6 Describe the structure, function, and locations of cuboidal and transitional epithelia.

Columnar epithelia absorb substances and protect the body from digestive chemicals

In a typical sectional view, the cells of a **columnar epithelium** appear rectangular. In reality, the densely packed cells are hexagonal, but they are taller and more slender than cells in a cuboidal epithelium. The elongated nuclei are crowded into a narrow band close to the basement membrane. The height of the epithelium is several times the distance between adjacent nuclei.

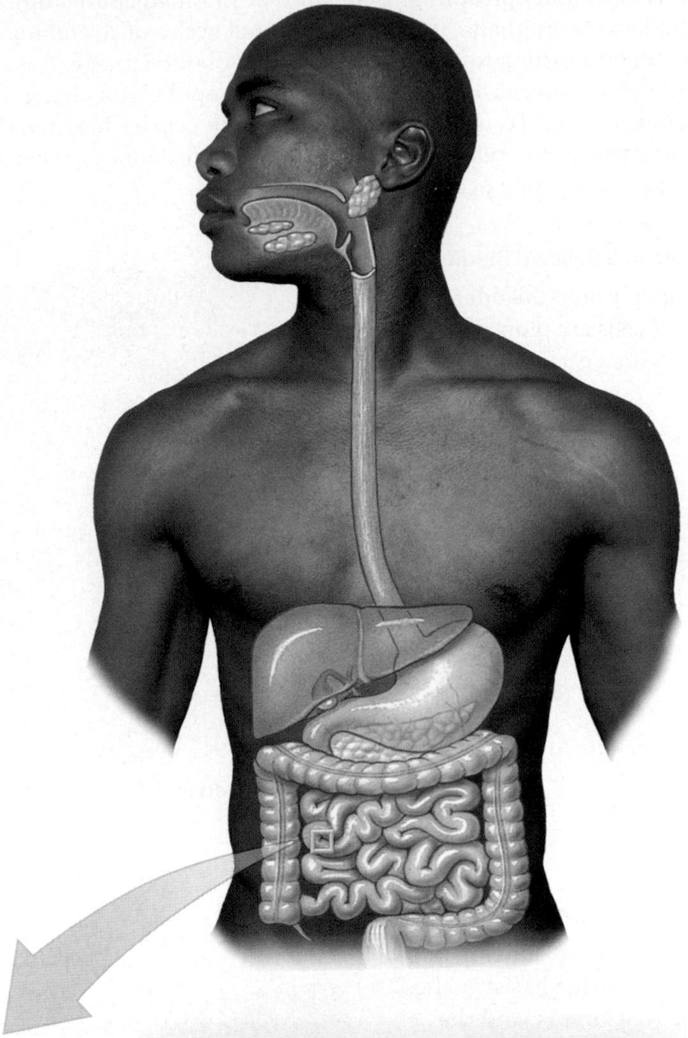

Simple Columnar Epithelium

1 A **simple columnar epithelium** is typically found where absorption or secretion takes place, as in the small intestine. These epithelia also line the stomach, gallbladder, uterine tubes, and ducts within the kidneys. These cells may have microvilli, which increase surface area for absorption, or cilia that move substances across the apical surface. In the stomach and intestinal tract, simple columnar epithelia secretions protect underlying tissues against chemicals involved with digestion.

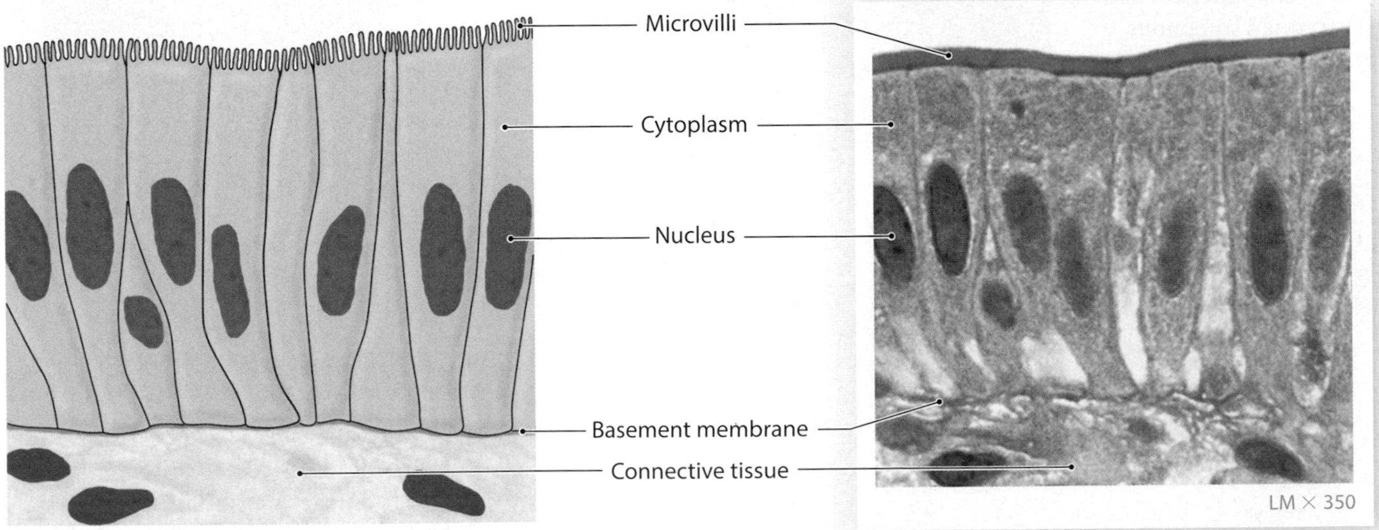
Microvilli
Cytoplasm
Nucleus
Basement membrane
Connective tissue
LM × 350

Intestinal lining

Pseudostratified Columnar Epithelium

Pseudostratified epithelia line the nasal cavities, the trachea, and larger airways of the lungs. They are also found along portions of the male reproductive tract.

2 A **pseudostratified columnar epithelium** includes several types of cells with varying shapes and functions. The distances between the cell nuclei and the exposed surface vary, so the epithelium appears to be layered, or stratified. It is not truly stratified, however, because every epithelial cell contacts the basement membrane. Pseudostratified columnar epithelial cells typically have motile cilia.

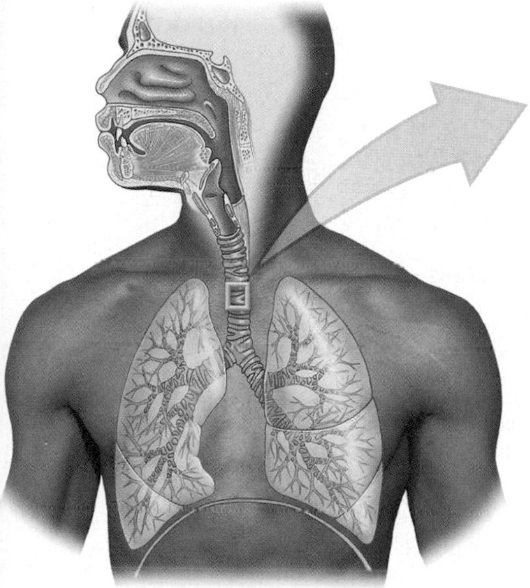

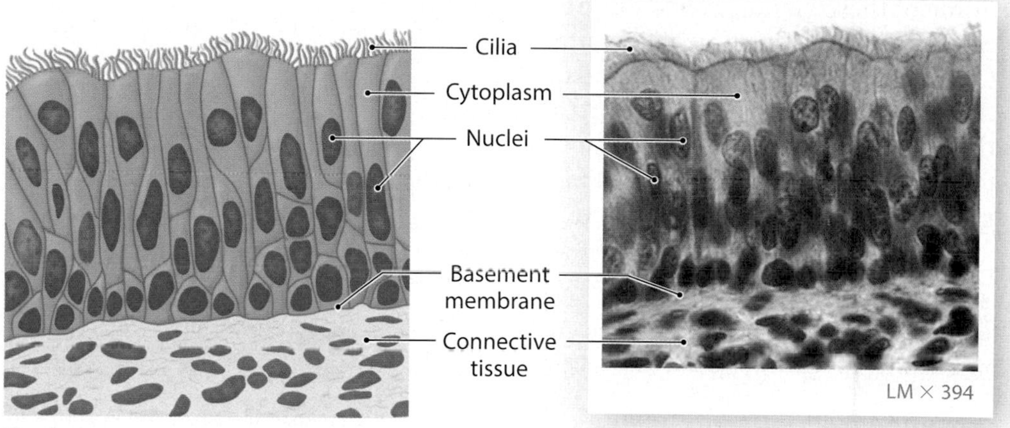
Cilia
Cytoplasm
Nuclei
Basement membrane
Connective tissue
LM × 394

Trachea

Stratified Columnar Epithelium

3 **Stratified columnar epithelia** are not widely distributed in the body. These epithelia may have two or more layers. In the latter case, only the superficial cells are columnar in shape. Stratified columnar epithelia are most often found lining large ducts such as those of the salivary glands or pancreas.

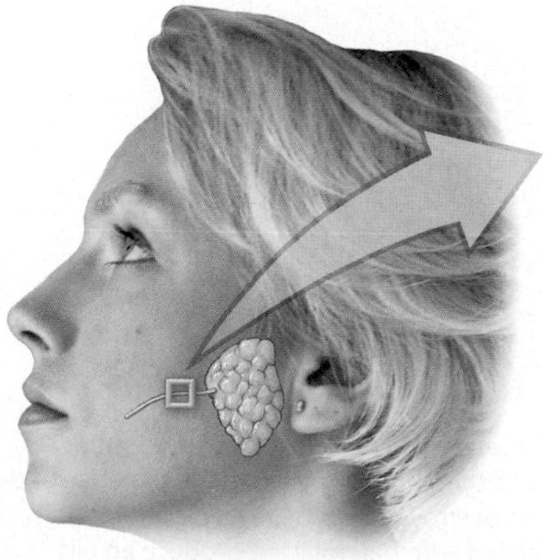

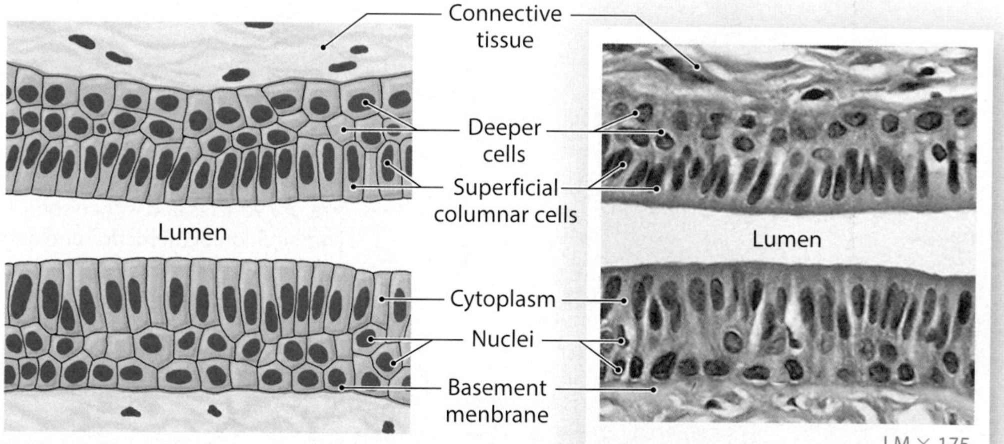
Connective tissue
Deeper cells
Superficial columnar cells
Lumen
Lumen
Cytoplasm
Nuclei
Basement menbrane
LM × 175

Salivary gland duct

Module 4.7 Review

a. Describe the appearance of simple columnar epithelial cells in a sectional view.

b. Explain why a pseudostratified columnar epithelium is not truly stratified.

c. The columnar epithelium lining the intestine typically has ________ on its apical surface.

LO 4.7 Describe the structure, function, and locations of columnar epithelia.

Module 4.8

Glandular epithelia are specialized for secretion

Secretion Methods

Many epithelia contain gland cells that are specialized for secretion. Collections of epithelial cells (or structures derived from epithelial cells) that produce secretions are called **glands**. They range from scattered cells to complex glandular organs. Some of these glands, called **endocrine glands**, release their secretions into the interstitial fluid. Others, known as **exocrine glands**, release their secretions into passageways called **ducts** that open onto an epithelial surface. We consider exocrine glands here. Endocrine glands are discussed in Chapter 16.

1 Glandular epithelial cells may release their secretions in one of three ways: merocrine, apocrine, or holocrine secretion.

Mucin is a merocrine secretion that mixes with water to form mucus. **Mucus** is an effective lubricant, a protective barrier, and a sticky trap for foreign particles and microorganisms.

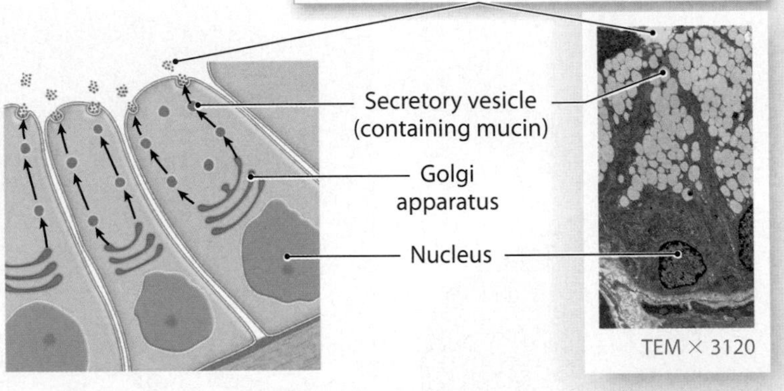
Secretory vesicle (containing mucin)
Golgi apparatus
Nucleus
TEM × 3120

In **merocrine secretion** (MER-u-krin; *meros*, part), the product is released from secretory vesicles by exocytosis. This is the most common mode of exocrine secretion.

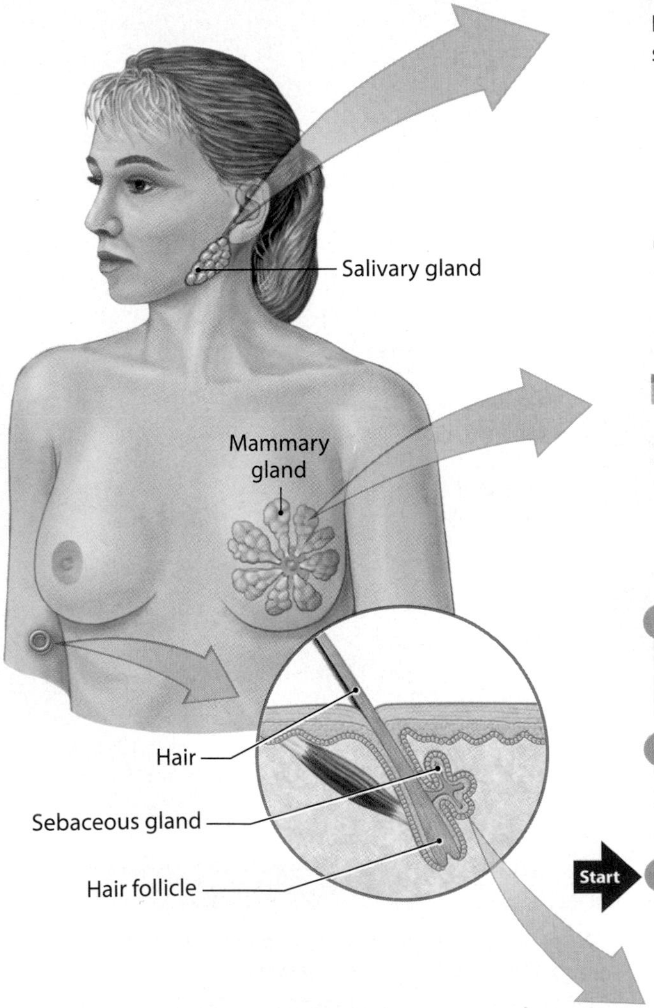
Salivary gland
Mammary gland
Hair
Sebaceous gland
Hair follicle

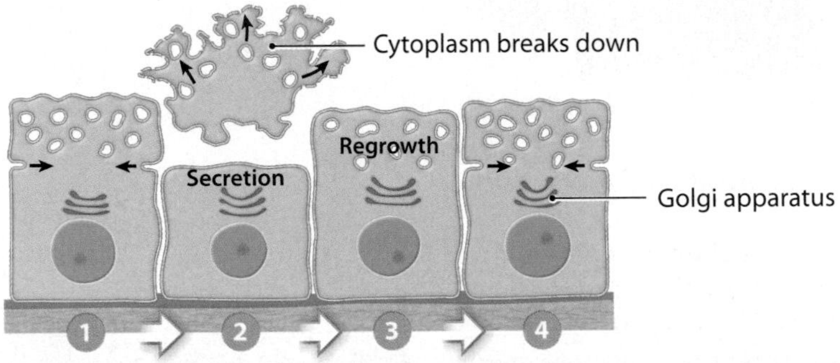
Cytoplasm breaks down
Secretion
Regrowth
Golgi apparatus
1
2
3
4

Apocrine secretion (AP-ō-krin; *apo-*, off) involves the loss of cytoplasm as well as the secretory product. The apical portion of the cytoplasm becomes packed with secretory vesicles and is then shed. Milk production in the mammary glands involves a combination of merocrine and apocrine secretions.

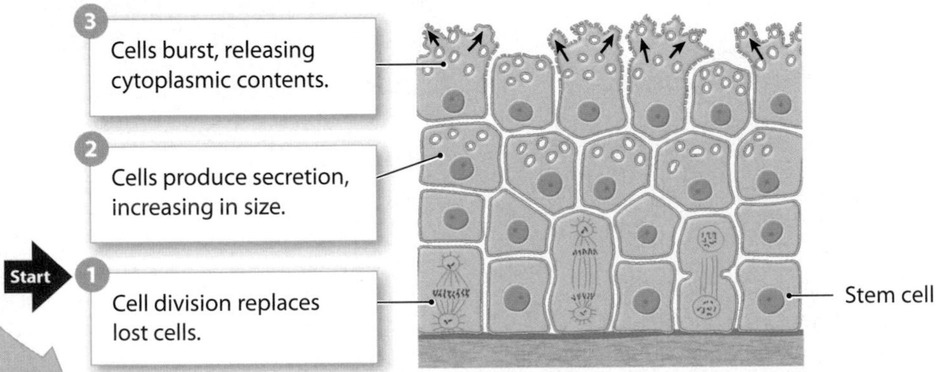
3 Cells burst, releasing cytoplasmic contents.
2 Cells produce secretion, increasing in size.
Start
1 Cell division replaces lost cells.
Stem cell

Holocrine secretion (HOL-ō-krin; *holos*, entire) destroys the gland cell. The entire cell becomes packed with secretory products and then bursts, releasing the secretion and killing the cell. Further secretion depends on destroyed cells being replaced through stem cell division. Sebaceous glands, associated with hair follicles, produce an oily hair coating by means of holocrine secretion.

Exocrine Gland Structure

Three characteristics are used to describe the structure of multicellular exocrine glands: the structure of the duct, the shape of the secretory area of the gland, and the relationship between the duct and the secretory areas.

In a branched gland, several secretory areas (tubular or acinar) share a duct. Note that "branched" refers to the glandular areas, not to the duct.

2 A **simple gland** has a single duct that does not divide on its way to the gland cells.

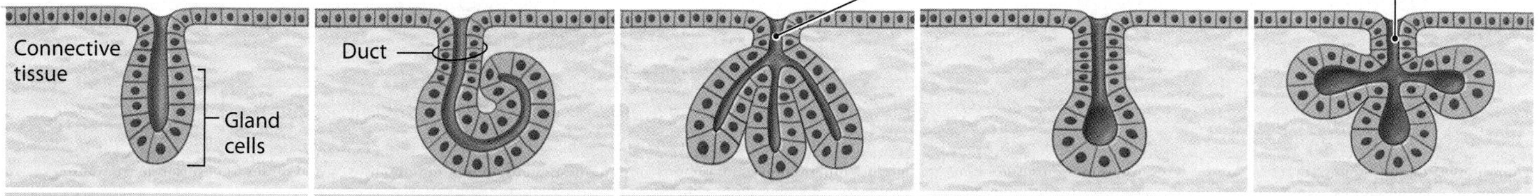

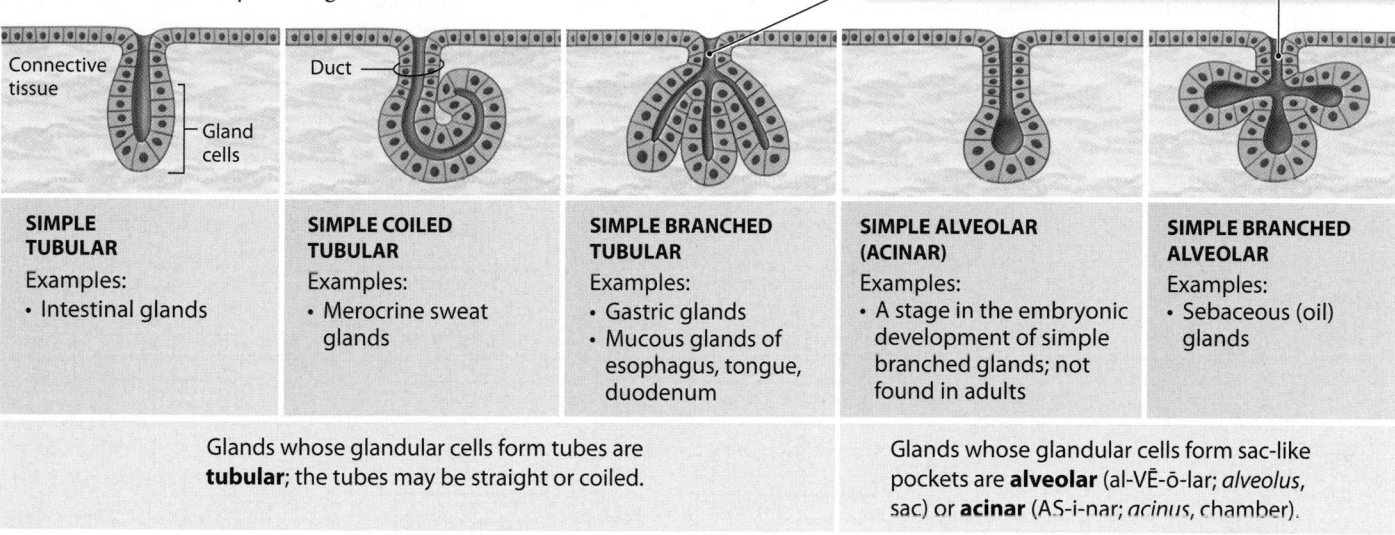

Connective tissue
Gland cells
Duct

SIMPLE TUBULAR	SIMPLE COILED TUBULAR	SIMPLE BRANCHED TUBULAR	SIMPLE ALVEOLAR (ACINAR)	SIMPLE BRANCHED ALVEOLAR
Examples: • Intestinal glands	Examples: • Merocrine sweat glands	Examples: • Gastric glands • Mucous glands of esophagus, tongue, duodenum	Examples: • A stage in the embryonic development of simple branched glands; not found in adults	Examples: • Sebaceous (oil) glands

Glands whose glandular cells form tubes are **tubular**; the tubes may be straight or coiled.

Glands whose glandular cells form sac-like pockets are **alveolar** (al-VĒ-ō-lar; *alveolus*, sac) or **acinar** (AS-i-nar; *acinus*, chamber).

3 In a **compound gland**, the duct divides one or more times on its way to the gland cells.

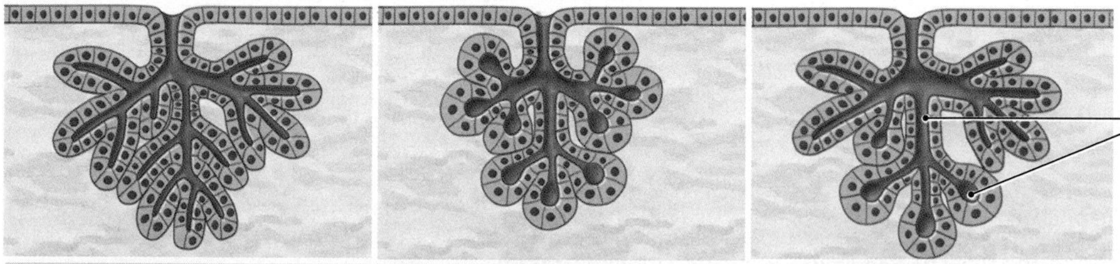

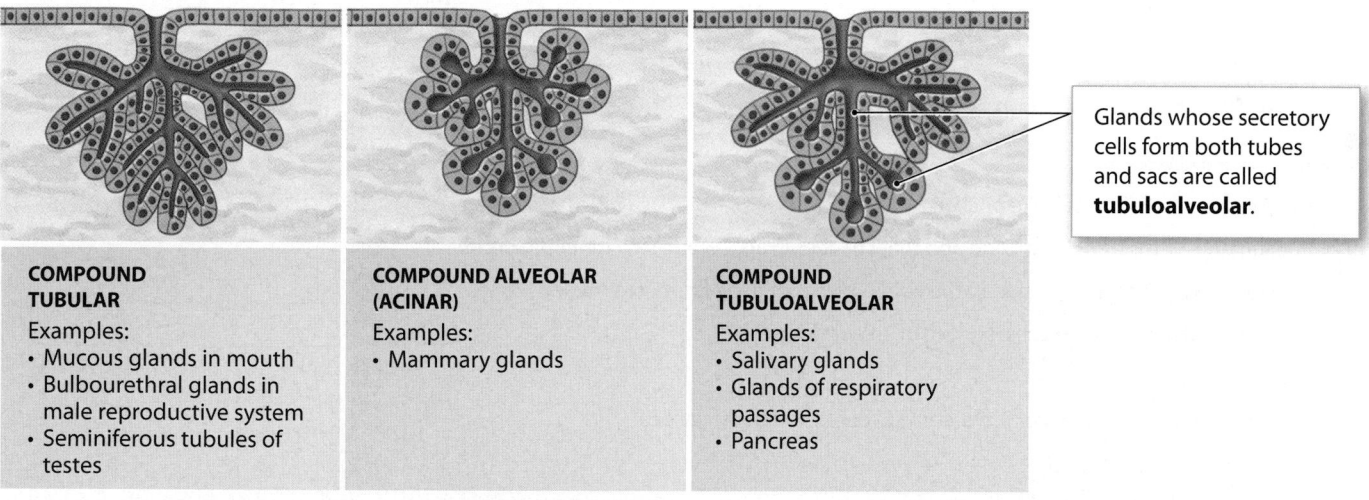

Glands whose secretory cells form both tubes and sacs are called **tubuloalveolar**.

COMPOUND TUBULAR	COMPOUND ALVEOLAR (ACINAR)	COMPOUND TUBULOALVEOLAR
Examples: • Mucous glands in mouth • Bulbourethral glands in male reproductive system • Seminiferous tubules of testes	Examples: • Mammary glands	Examples: • Salivary glands • Glands of respiratory passages • Pancreas

4 In epithelia that have independent, scattered gland cells, the individual secretory cells are called **mucous (goblet) cells**, and they secrete mucin. The apical cytoplasm is filled with large secretory vesicles that look clear or foamy in a light micrograph.

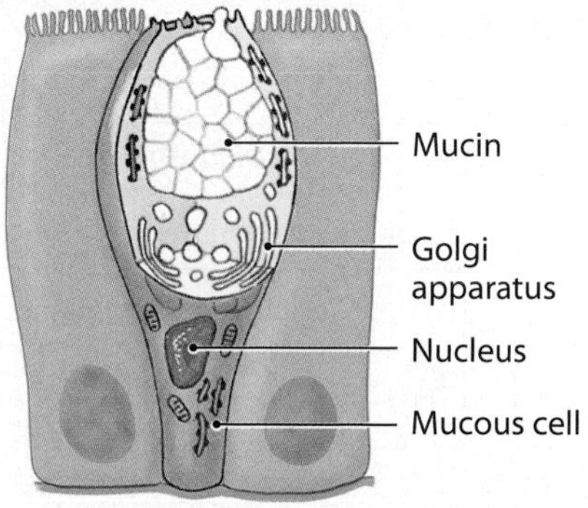

Mucin
Golgi apparatus
Nucleus
Mucous cell

Module 4.8 Review

a. Name the two primary types of glands.

b. Which type of gland has no ducts and its secretions are released directly into the interstitial fluid?

c. What mode of secretion occurs in the secretory cells of sebaceous glands, which fill with secretions and then rupture, releasing their contents?

LO **4.8** Describe the structure, function, and locations of glandular epithelia.

Labeling

Label the types of epithelial tissues shown in the drawing to the right.

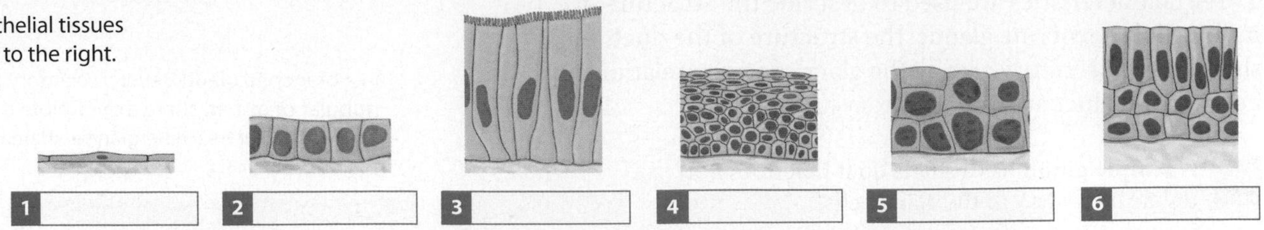

1
2
3
4
5
6

Concept map

Use the following terms once to fill in the blank boxes to correctly complete the map.

- endocrine glands
- ducts
- mucous cells
- apocrine secretion
- interstitial fluid
- mucus
- merocrine secretion
- epithelial surfaces
- mucin
- exocrine glands
- holocrine secretion

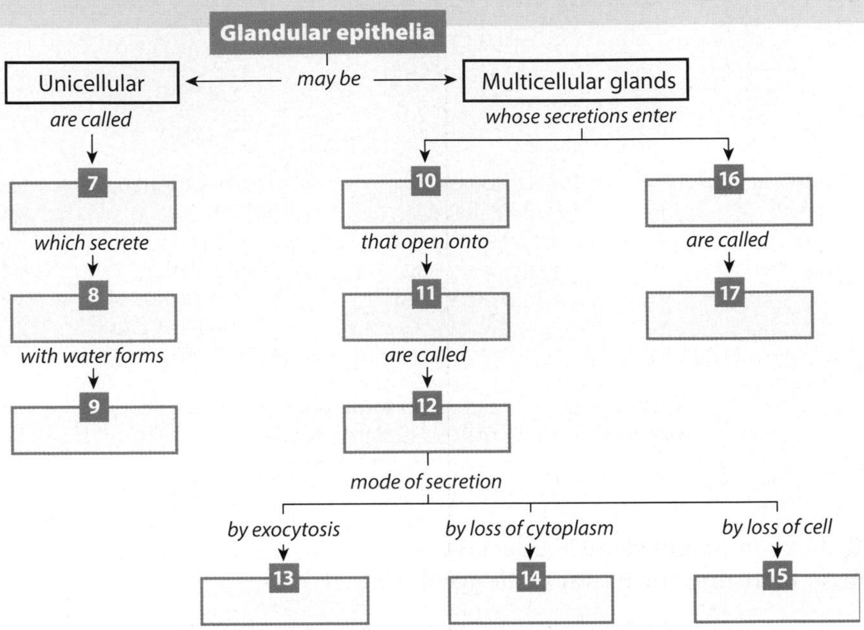

Glandular epithelia
may be
Unicellular
Multicellular glands
are called
7
which secrete
8
with water forms
9
whose secretions enter
10
that open onto
11
are called
12
mode of secretion
by exocytosis
13
by loss of cytoplasm
14
by loss of cell
15
16
are called
17

Vocabulary

Write the term for each of the following descriptions in the space provided.

18 A term meaning no blood vessels

19 A gland whose glandular cells form sac-like pockets

20 A type of epithelium that withstands stretching and that changes in appearance as stretching occurs

21 The cell junction formed by the partial fusion of the lipid portions of two plasma membranes

22 The complex structure attached to the basal surface of an epithelium

23 A gland that has a single duct

24 The type of epithelium lining the pericardial, pleural, and peritoneal body cavities

18 ______
19 ______
20 ______
21 ______
22 ______
23 ______
24 ______

Short answer

Fill in the missing epithelium type or structure.

Type of Epithelium	Structure (or Organ)
25	Lining of the trachea
Transitional epithelium	26
27	Surface of the skin
28	Lining of the small intestine

Type of Epithelium	Structure (or Organ)
Simple squamous epithelium	29
Cuboidal epithelium	30
31	Ducts of sweat glands

A matrix surrounds connective tissue cells

Connective tissue varies widely in appearance and function, but all forms share three basic components: (1) specialized cells, (2) extracellular protein fibers, and (3) a fluid known as **ground substance**. Together, the extracellular fibers and ground substance constitute the **matrix** that surrounds the cells. Whereas cells make up the bulk of epithelial tissue, the matrix typically accounts for most of the volume of connective tissue.

Functions of Connective Tissue

- Establish a structural framework for the body
- Transport fluids and dissolved materials
- Protect delicate organs
- Support, surround, and interconnect other types of tissue
- Store energy, especially in the form of triglycerides
- Defend the body from invading microorganisms

Connective Tissue

The various types of connective tissue occur throughout the body but they are never exposed to the outside environment. Many types of connective tissue are highly vascular (that is, they have many blood vessels) and contain sensory receptors that detect pain, pressure, temperature, and other stimuli.

Connective Tissue Proper

Connective tissue proper includes those connective tissues with many types of cells and extracellular fibers in a syrupy ground substance.

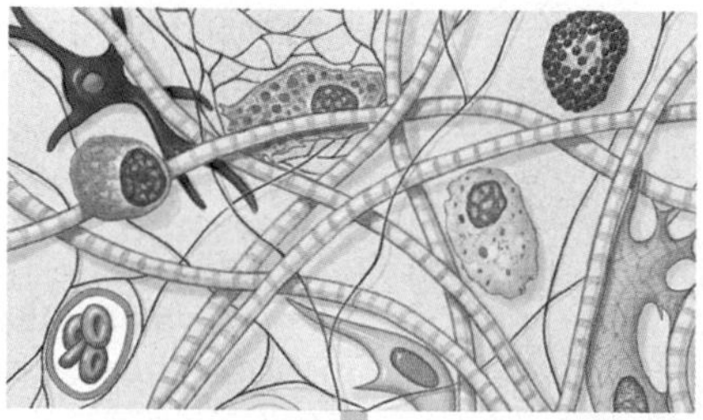

Loose

Fibers create loose, open framework
- areolar tissue
- adipose tissue
- reticular tissue

Dense

Fibers densely packed
- dense regular
- dense irregular
- elastic

Fluid Connective Tissues

Fluid connective tissues have distinctive populations of cells suspended in a watery matrix that contains dissolved proteins.

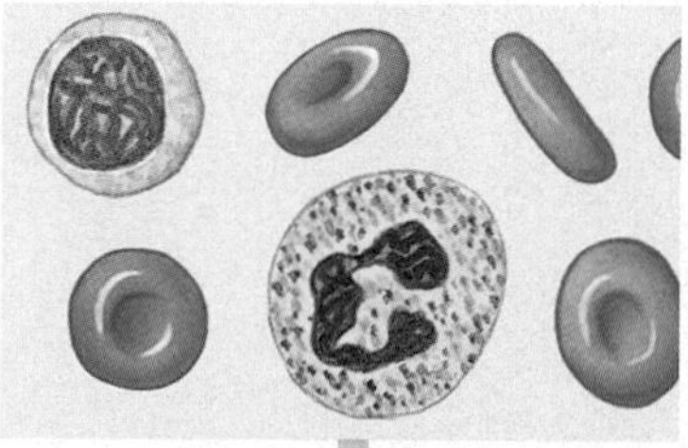

Blood

Flows within cardiovascular system

Lymph

Flows within lymphatic system

Supporting Connective Tissues

Supporting connective tissues differ from connective tissue proper in having a less diverse cell population and a matrix containing much more densely packed fibers. Supporting connective tissues protect soft tissues and support the weight of part or all of the body.

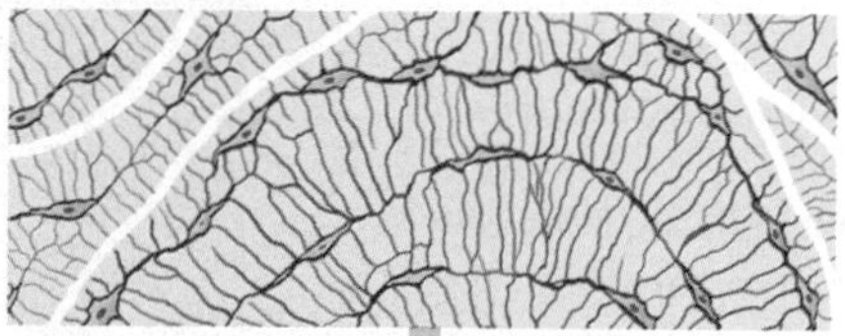

Cartilage

Solid, rubbery matrix
- hyaline cartilage
- elastic cartilage
- fibrocartilage

Bone

Solid, crystalline matrix

Module 4.9 Review

a. Identify the three basic components of connective tissue.

b. Summarize the functions of connective tissue.

c. Distinguish between connective tissue proper, fluid connective tissues, and supporting connective tissues.

LO 4.9 Describe the general structure of connective tissue.

Loose connective tissues support other tissue types

Connective tissue proper contains extracellular protein fibers, a viscous (syrupy) ground substance, and two classes of cells. Fixed cells are stationary and are involved primarily with local maintenance, repair, and energy storage. Wandering cells primarily defend and repair damaged tissues. The number of cells at any given moment varies depending on local conditions.

Loose Connective Tissues

The body contains three types of **loose connective tissues:** areolar tissue, adipose tissue, and reticular tissue.

1 **Areolar tissue** is the most common form of connective tissue proper in adults. It is the general packing material in the body. All of the cell types found in other forms of connective tissue proper can be found in areolar tissue.

Fibers

Reticular fibers are strong and form a branching network.	**Collagen fibers** are thick, straight or wavy, and often form bundles. They are very strong and resist stretching.	**Elastic fibers** are slender, branched, and very stretchy. They recoil to their original length after stretching or distortion.

Fixed Cells

Melanocytes are pigment cells that synthesize melanin, a brownish-yellow pigment.

Fixed macrophages are stationary phagocytic cells that engulf cell debris and pathogens.

Mast cells are fixed cells that stimulate local inflammation and mobilize tissue defenses.

Fibroblasts are fixed cells that synthesize the extracellular fibers of the connective tissue.

Adipocytes (fat cells) are fixed cells that store lipids in large intracellular vesicles.

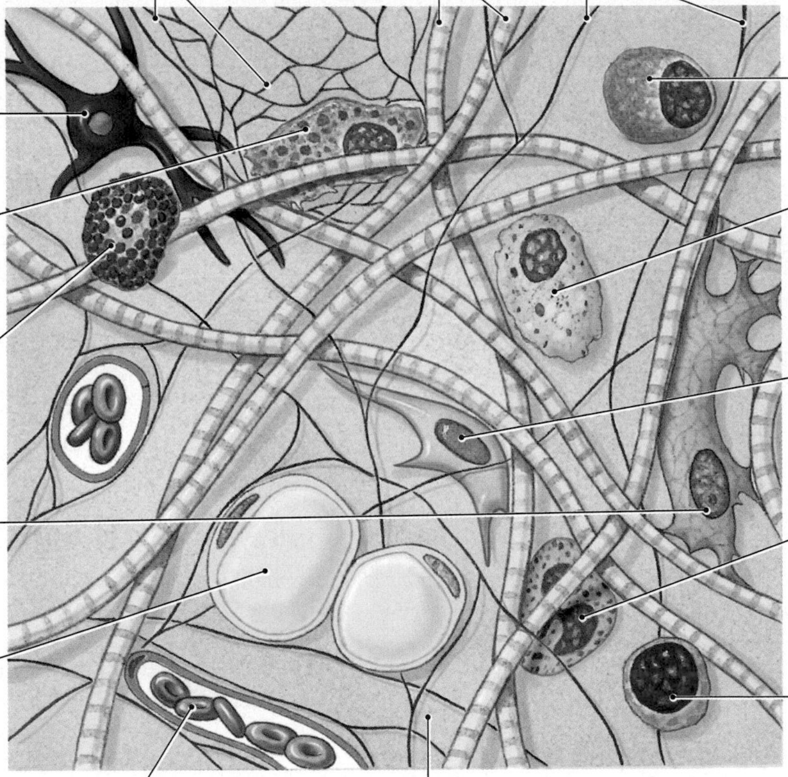
Red blood cell in vessel

Wandering Cells

Plasma cells are active, mobile immune cells that produce antibodies.

Free macrophages are wandering phagocytic cells that patrol the tissue, engulfing debris or pathogens.

Mesenchymal cells are mobile stem cells that repair damaged tissues.

Neutrophils and **eosinophils** are small, mobile, phagocytic blood cells that enter tissues during infection or injury.

Lymphocytes are mobile cells of the immune system.

Ground substance fills the spaces between cells and surrounds connective tissue fibers. In all forms of connective tissue proper, ground substance is clear, colorless, and viscous due to the presence of proteoglycans and glycoproteins.

2 **Adipose tissue** is found deep to the skin, especially at the flanks, buttocks, and breasts. It also forms a layer that provides padding within the orbit of the eyes, in the abdominopelvic cavity, and around the kidneys. The distinction between areolar tissue and adipose tissue is somewhat arbitrary. Adipocytes account for most of the volume of adipose tissue, but only a fraction of the volume of areolar tissue.

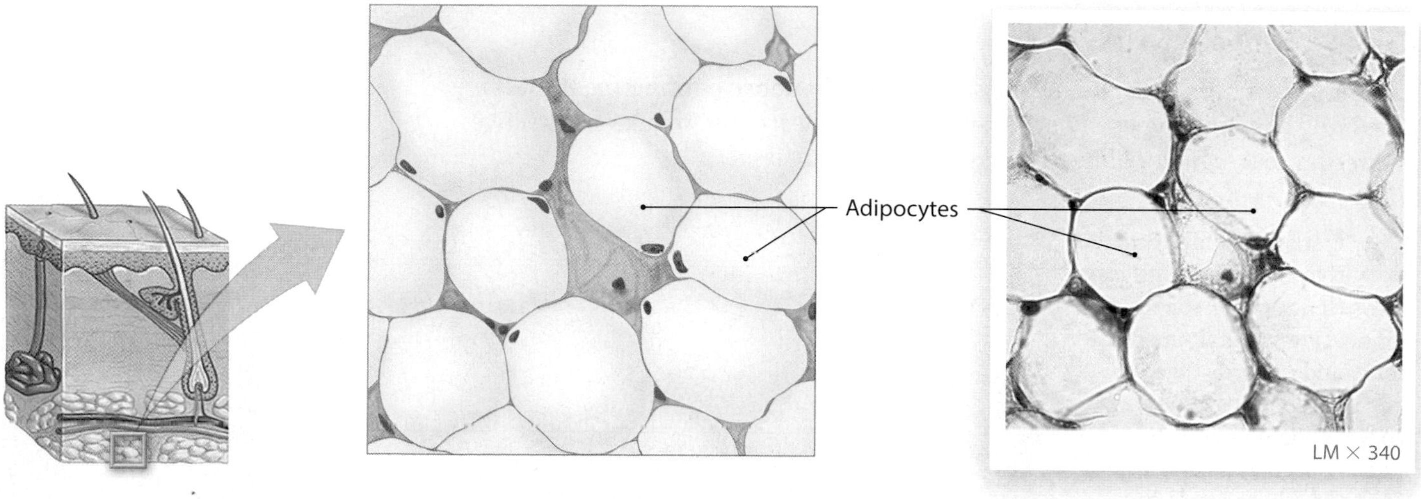

Adipocytes
LM × 340

3 **Reticular tissue** is found in the liver, kidney, spleen, lymph nodes, and bone marrow, where it forms a tough, flexible network that provides support and resists distortion. In reticular tissue, reticular fibers create a complex three-dimensional supporting network known as a **stroma**. Fixed macrophages and fibroblasts are present in reticular tissues, but these cells are seldom visible because the stroma is dominated by cells that are unique to the organ under consideration and have specialized functions.

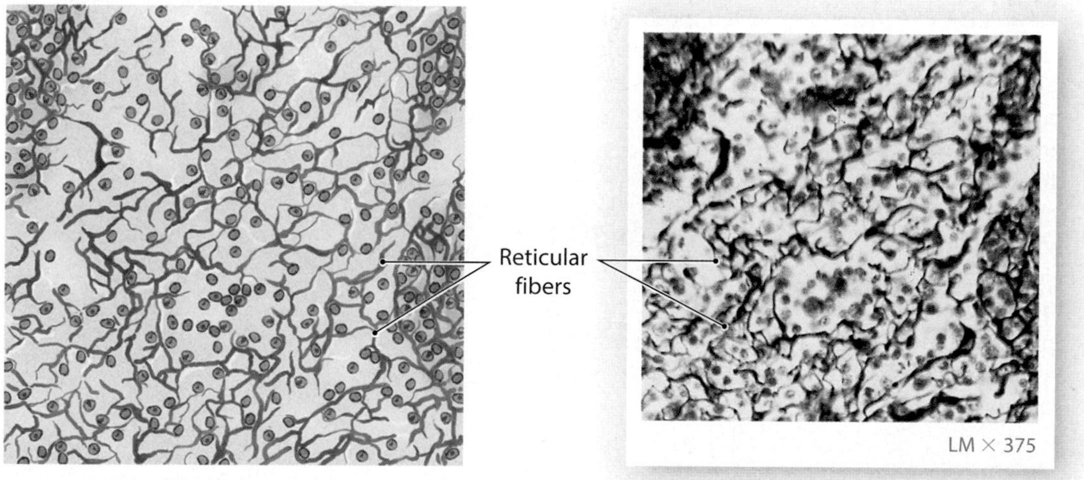

Reticular fibers
LM × 375

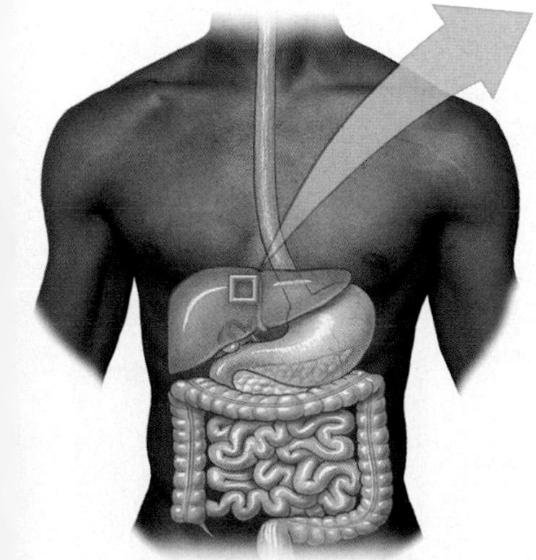

Module 4.10 Review

a. Identify the types of cells found in connective tissue proper.

b. Describe the role of fibroblasts in connective tissue.

c. Which type of connective tissue contains primarily lipids?

LO **4.10** Describe the structure, function, and locations of areolar tissue, adipose tissue, and reticular tissue.

Dense connective tissues are dominated by extracellular fibers, whereas fluid connective tissues have a liquid matrix

Dense Connective Tissues

Extracellular fibers make up most of the volume of **dense connective tissues**. The body has three types of dense connective tissues: dense regular connective tissue, dense irregular connective tissue, and elastic tissue.

1 **Dense regular connective tissue** is found in cords (tendons) or sheets connecting skeletal muscles to bone, and in cords (ligaments) that interconnect bones or stabilize the positions of internal organs. The forces applied to these cords and sheets arrive from a consistent direction: parallel to the long axis of the collagen fibers.

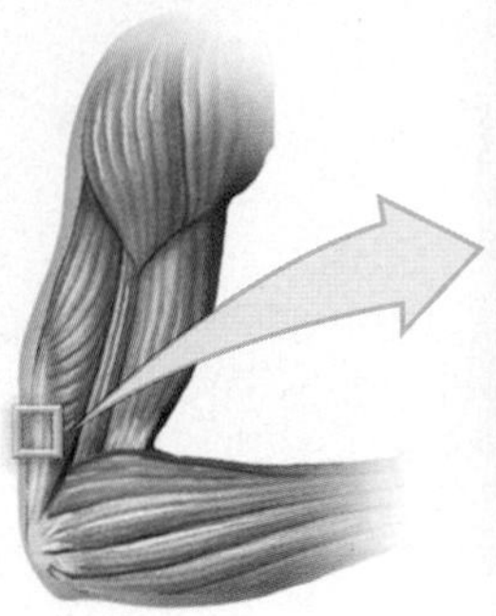

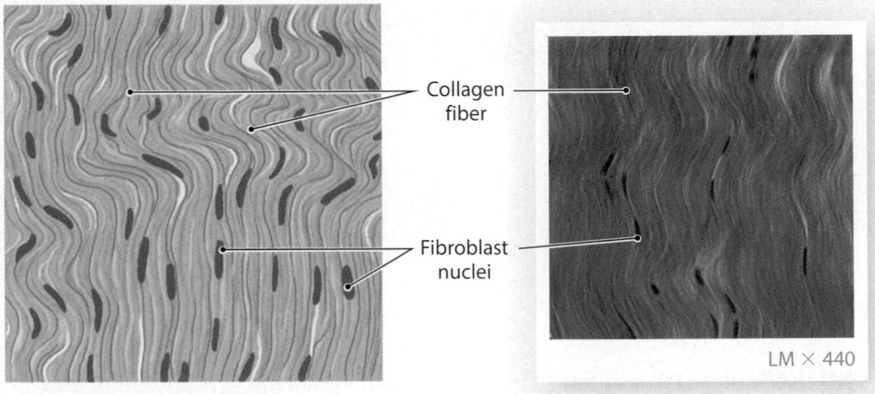
Collagen fiber
Fibroblast nuclei
LM × 440

Tendon from triceps muscle

2 In **dense irregular connective tissue**, the fibers form an interwoven meshwork in no consistent pattern. These tissues strengthen and support areas subjected to stresses from many directions. Dense irregular connective tissue forms (1) a covering, or capsule, that sheathes visceral organs; (2) a superficial layer covering bones, cartilages, and peripheral nerves; and (3) a thick supporting layer in the skin (the dermis).

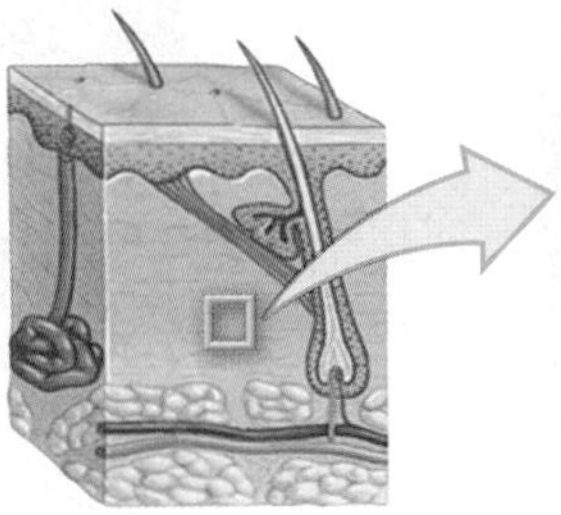

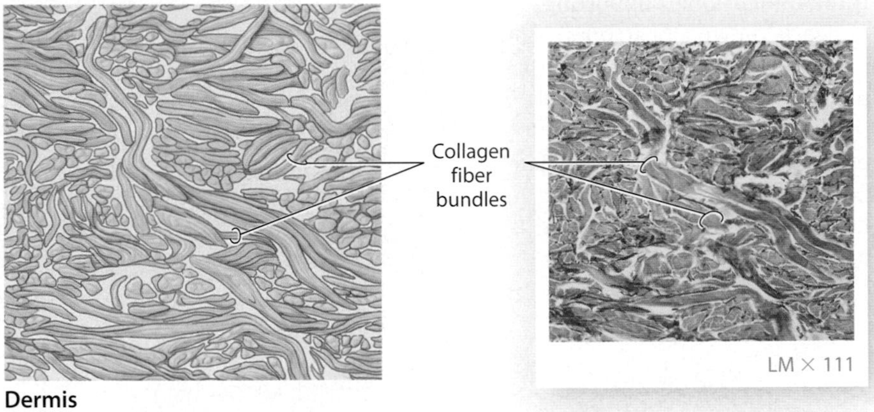
Collagen fiber bundles
LM × 111

Dermis

3 Dense regular and dense irregular connective tissues contain varying amounts of elastic fibers. When elastic fibers outnumber collagen fibers, the tissue is springy and resilient, allowing it to tolerate cycles of extension and recoil. This **elastic tissue** is found between vertebrae of the spinal column, in the walls of large blood vessels, in ligaments supporting transitional epithelia, and the erectile tissues of the penis.

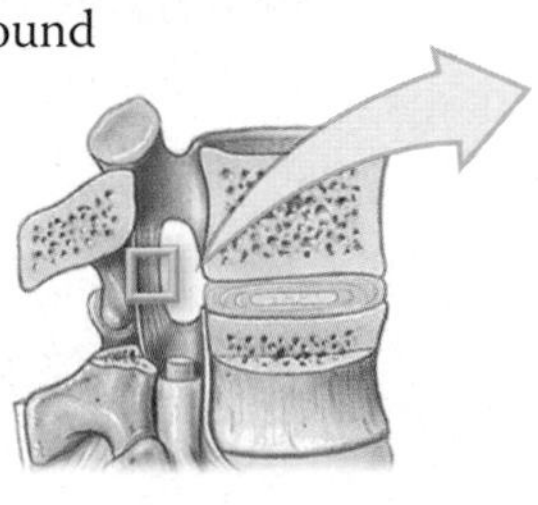

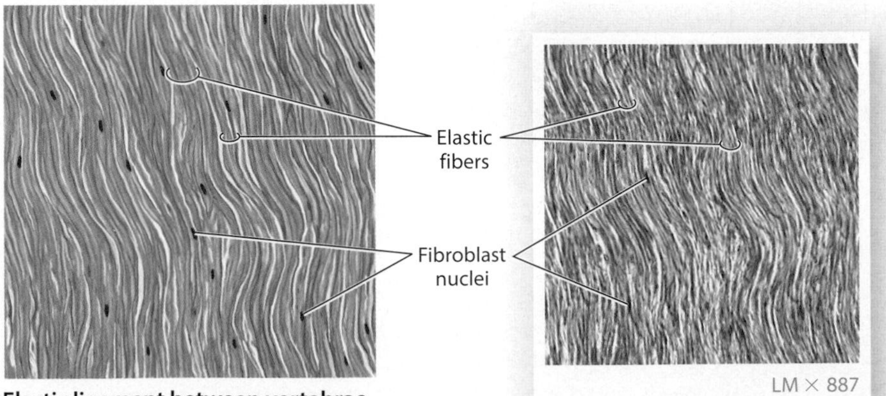
Elastic fibers
Fibroblast nuclei
LM × 887

Elastic ligament between vertebrae

Fluid Connective Tissues

4 **Fluid connective tissues** have a fluid matrix that includes many types of suspended proteins that under normal conditions do not form insoluble fibers. In blood, this watery matrix is called **plasma**. Blood cells and platelets, collectively known as **formed elements**, are suspended in the plasma.

Red Blood Cells

Red blood cells are formed elements that transport oxygen (and, to a lesser degree, carbon dioxide) in the blood.

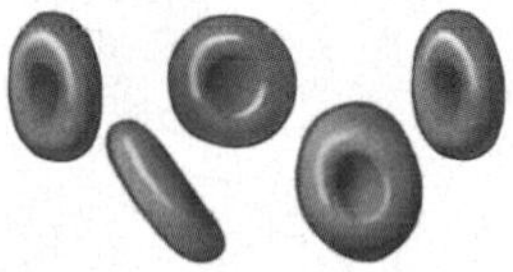

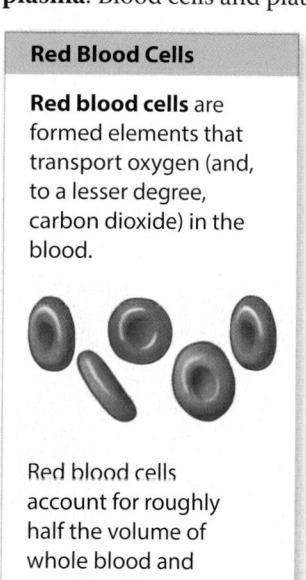

Red blood cells account for roughly half the volume of whole blood and give blood its color.

White Blood Cells

White blood cells are formed elements that help defend the body from infection and disease.

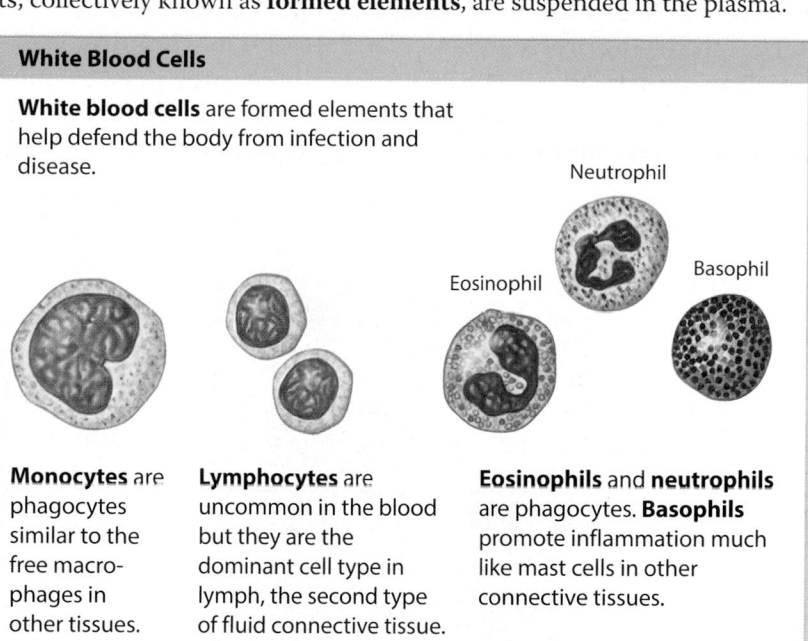

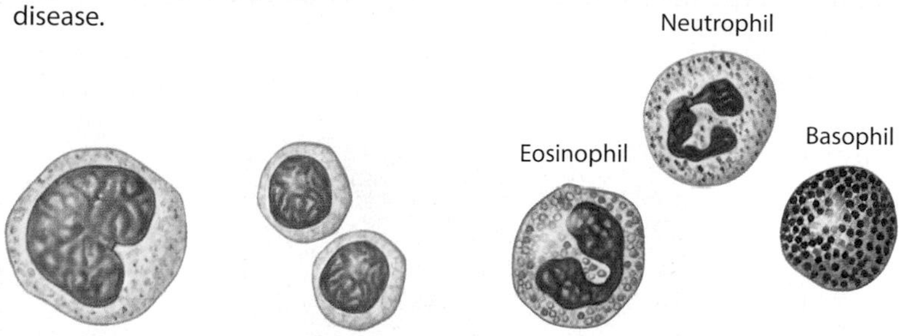
Neutrophil
Eosinophil
Basophil

Monocytes are phagocytes similar to the free macrophages in other tissues.

Lymphocytes are uncommon in the blood but they are the dominant cell type in lymph, the second type of fluid connective tissue.

Eosinophils and **neutrophils** are phagocytes. **Basophils** promote inflammation much like mast cells in other connective tissues.

Platelets

Platelets are formed elements consisting of membrane-enclosed packets of cytoplasm.

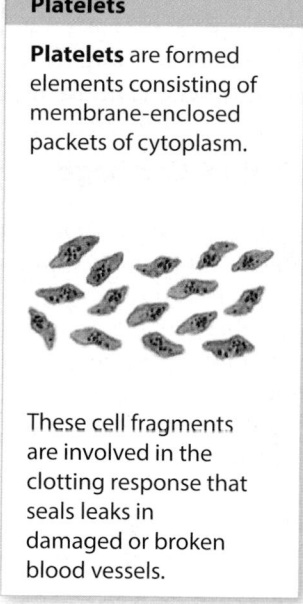

These cell fragments are involved in the clotting response that seals leaks in damaged or broken blood vessels.

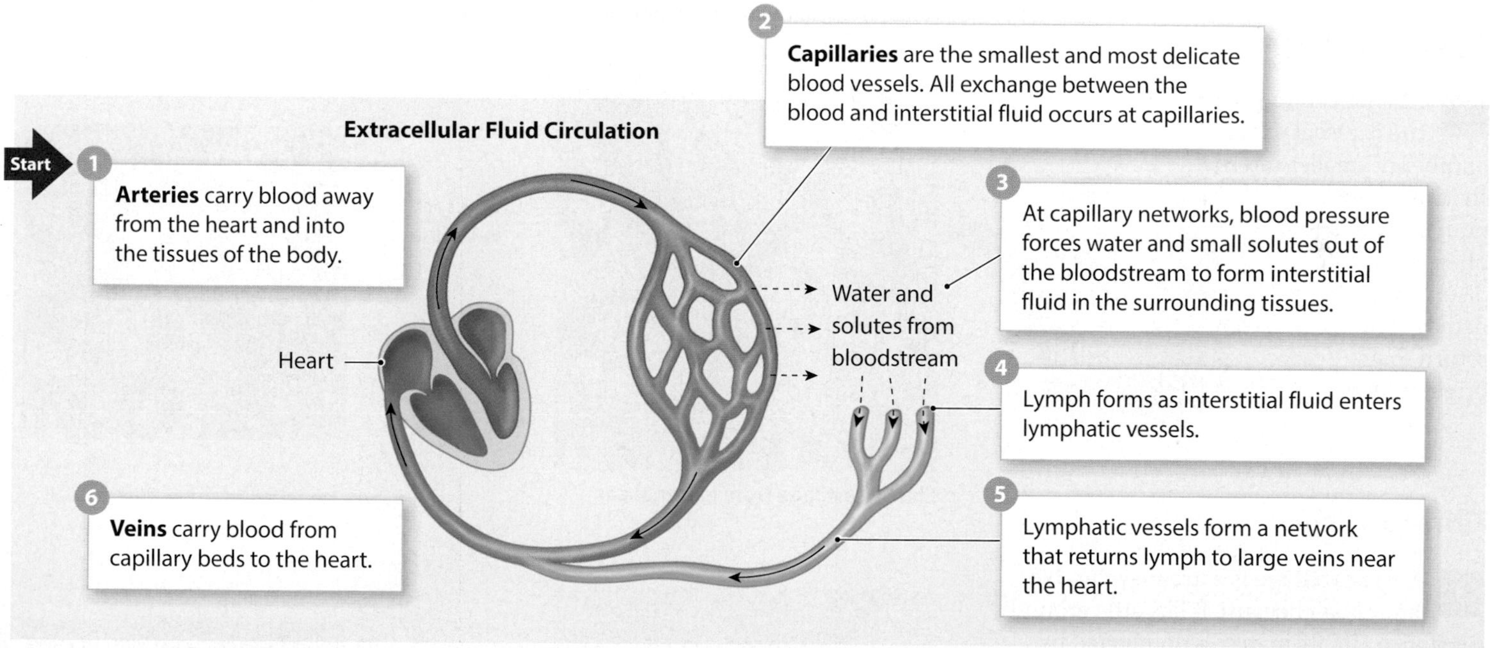
Extracellular Fluid Circulation
Start
1 Arteries carry blood away from the heart and into the tissues of the body.
2 Capillaries are the smallest and most delicate blood vessels. All exchange between the blood and interstitial fluid occurs at capillaries.
3 At capillary networks, blood pressure forces water and small solutes out of the bloodstream to form interstitial fluid in the surrounding tissues.
4 Lymph forms as interstitial fluid enters lymphatic vessels.
5 Lymphatic vessels form a network that returns lymph to large veins near the heart.
6 Veins carry blood from capillary beds to the heart.
Heart
Water and solutes from bloodstream

5 Extracellular fluid includes both plasma and interstitial fluid. **Blood** is normally confined to the vessels of the cardiovascular system, and contractions of the heart keep it in motion. As blood flows through body tissues, water and solutes move from the plasma into the surrounding interstitial fluid. **Lymph**, the second type of fluid connective tissue, is formed as interstitial fluid drains into lymphatic vessels that begin in peripheral tissues and empty into the venous system. The continuous recirculation of extracellular fluid is essential to homeostasis. It helps eliminate local differences in the levels of nutrients, wastes, or toxins; maintains blood volume; and alerts the immune system to infections that may be under way in peripheral tissues.

Module 4.11 Review

a. Lack of vitamin C in the diet interferes with the ability of fibroblasts to produce collagen. How might this affect connective tissue function?

b. Which two types of connective tissue have a liquid matrix?

c. Summarize the role of extracellular fluid in maintaining homeostasis.

LO **4.11** Describe the structure, function, and locations of dense connective tissues and fluid connective tissues.

Module 4.12

Cartilage provides a flexible support for body structures

In **cartilage**, the matrix is a firm gel that contains polysaccharide derivatives called **chondroitin sulfates** (kon-DROY-tin; *chondros*, cartilage). Chondroitin sulfates form complexes with proteins in the ground substance, producing proteoglycans. Cartilage cells, or **chondrocytes** (KON-drō-sīts), are the only cells in the avascular cartilage matrix. They occupy small chambers known as **lacunae** (la-KOO-nē; *lacus*, lake). The physical properties of cartilage depend on the proteoglycans in the matrix and on the type and abundance of extracellular fibers.

Hyaline Cartilage

1 **Hyaline cartilage** is found between the tips of the ribs and the bones of the sternum, covering bone surfaces at movable joints, supporting the respiratory passageways, and forming part of the nasal septum. It provides stiff but somewhat flexible support and reduces friction between bony surfaces.

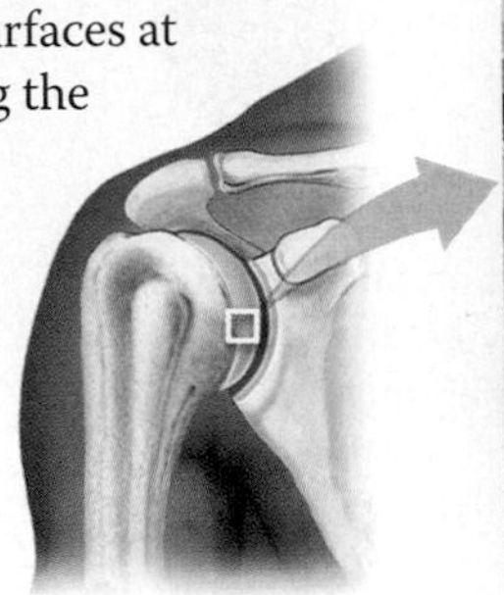

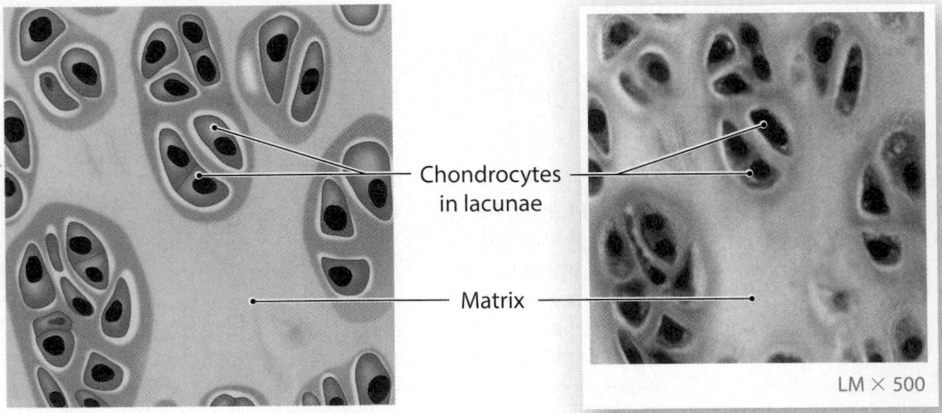
Chondrocytes in lacunae
Matrix
LM × 500

Hyaline cartilage from shoulder joint

Elastic Cartilage

2 **Elastic cartilage** supports the external ear and a number of smaller internal structures. Its numerous elastic fibers allow it to distort without damage and return to its original shape.

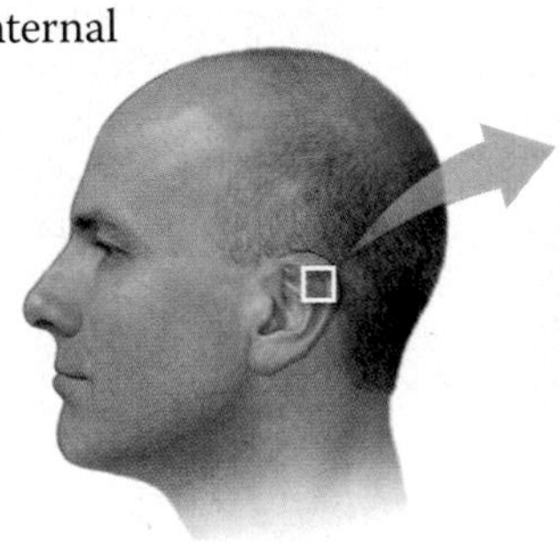

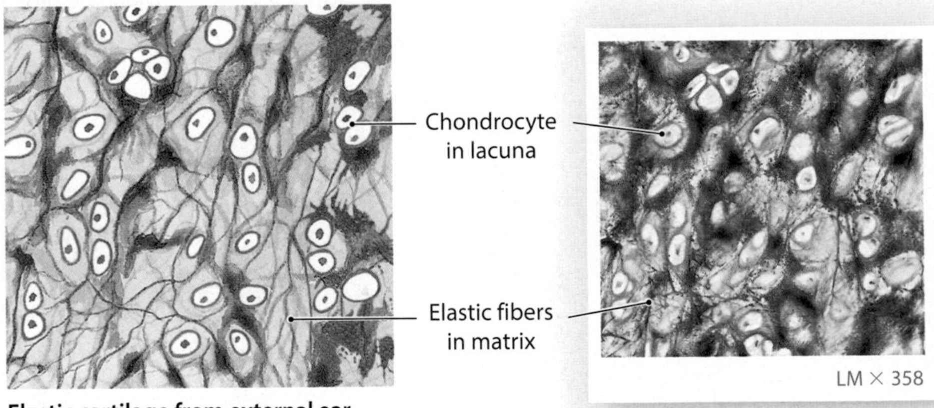
Chondrocyte in lacuna
Elastic fibers in matrix
LM × 358

Elastic cartilage from external ear

Fibrocartilage

3 **Fibrocartilage** is extremely durable and tough because it has little ground substance and its matrix is dominated by densely interwoven collagen fibers. Fibrocartilage pads are found within the knee joint, between the pubic bones of the pelvis, and in the intervertebral discs of the vertebral column. It resists compression, prevents bone-to-bone contact, and limits relative movement.

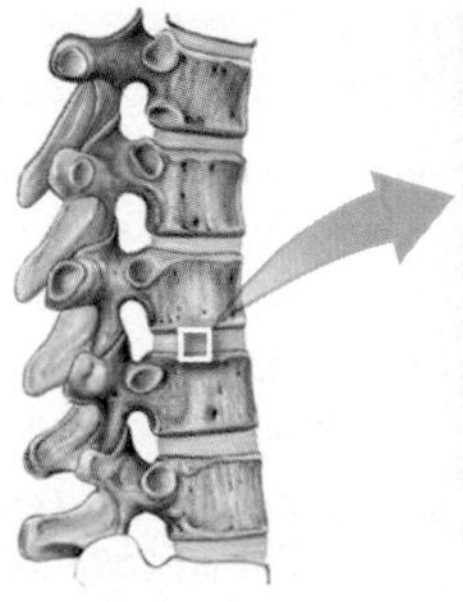

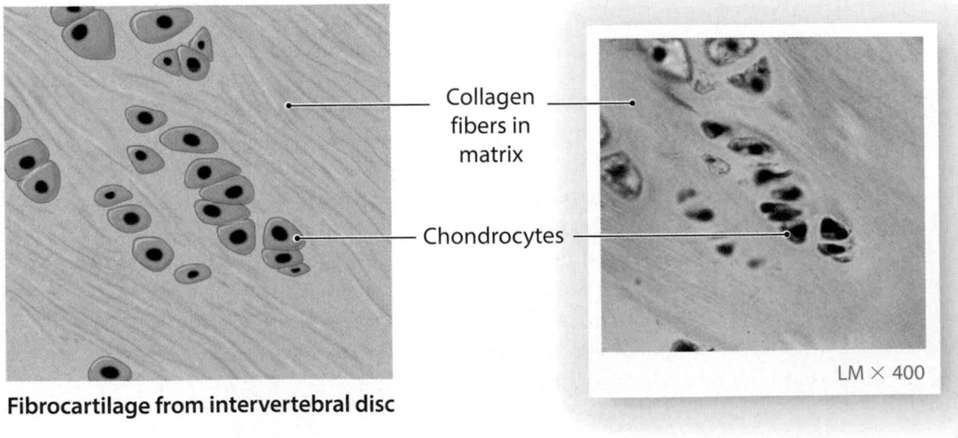
Collagen fibers in matrix
Chondrocytes
LM × 400

Fibrocartilage from intervertebral disc

4 Cartilage is generally set apart from surrounding tissues by a covering called a **perichondrium** (per-i-KON-drē-um; *peri-*, around). The perichondrium contains two distinct layers: an outer, fibrous layer of dense irregular connective tissue, and an inner, cellular layer. The fibrous layer gives mechanical support and protection and attaches the cartilage to other structures. The cellular layer is important to the growth and maintenance of the cartilage. Blood vessels in the perichondrium provide oxygen and nutrients to the underlying chondrocytes.

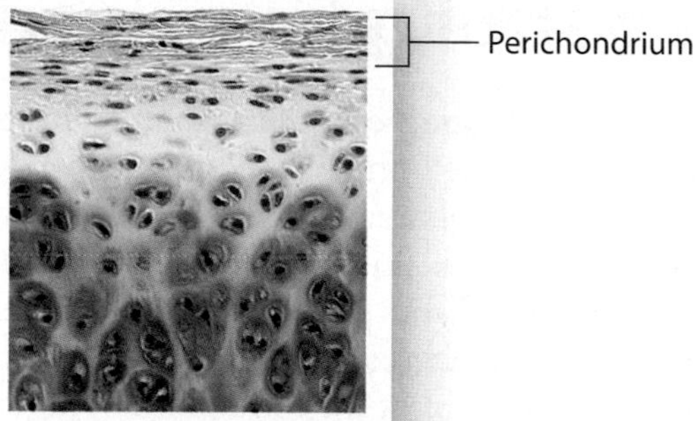

Perichondrium

Hyaline cartilage LM × 250

Appositional Growth

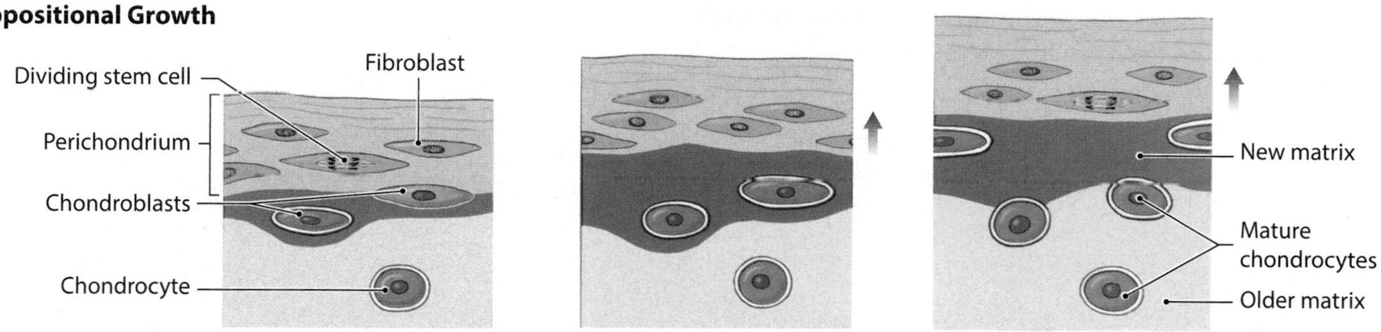

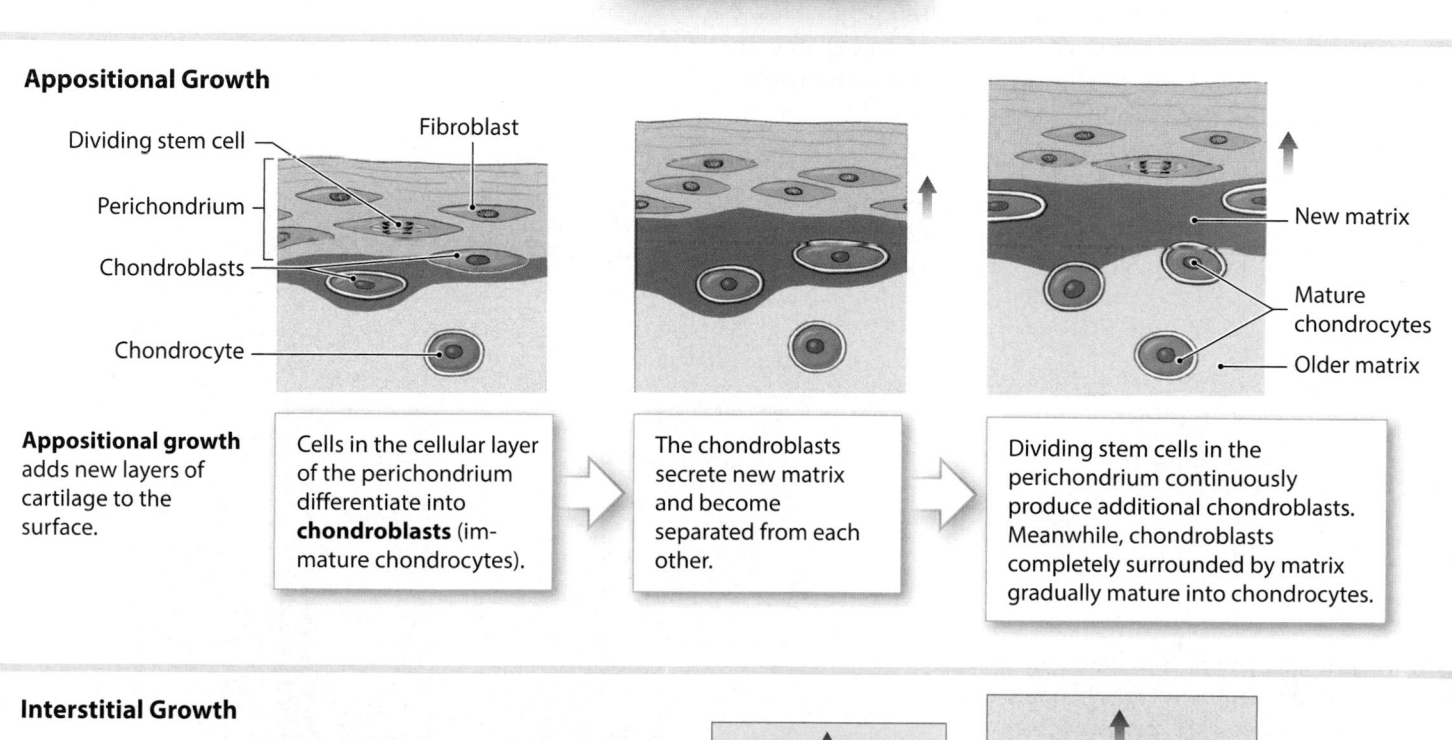

Dividing stem cell
Fibroblast
Perichondrium
Chondroblasts
Chondrocyte
New matrix
Mature chondrocytes
Older matrix

Appositional growth adds new layers of cartilage to the surface.

Cells in the cellular layer of the perichondrium differentiate into **chondroblasts** (immature chondrocytes).

The chondroblasts secrete new matrix and become separated from each other.

Dividing stem cells in the perichondrium continuously produce additional chondroblasts. Meanwhile, chondroblasts completely surrounded by matrix gradually mature into chondrocytes.

Interstitial Growth

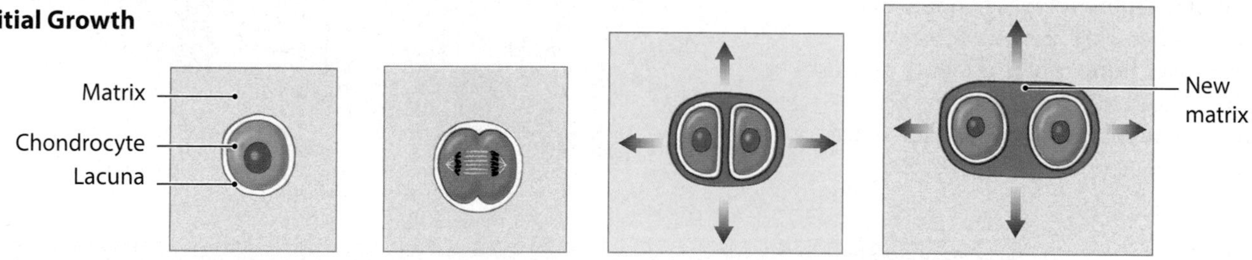

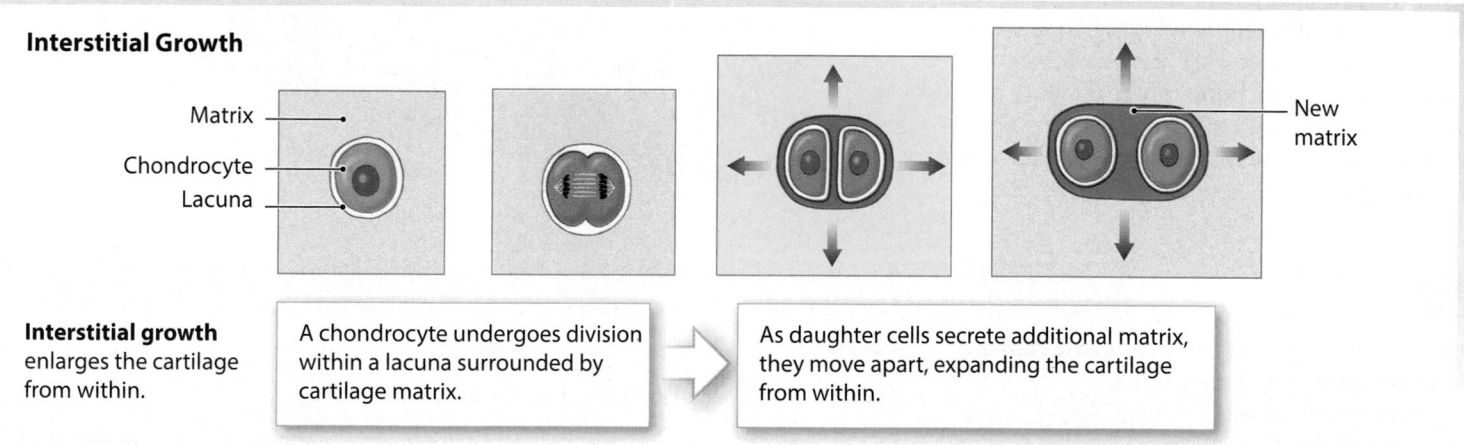

Matrix
Chondrocyte
Lacuna
New matrix

Interstitial growth enlarges the cartilage from within.

A chondrocyte undergoes division within a lacuna surrounded by cartilage matrix.

As daughter cells secrete additional matrix, they move apart, expanding the cartilage from within.

5 Both interstitial and appositional growth occur during development before birth and through adolescence, but interstitial growth predominates. In normal adults, cartilage growth no longer occurs. However, in unusual circumstances—such as after cartilage has been slightly damaged or subjected to excessive hormonal stimulation—some repair may occur by appositional growth. After severe damage, cartilage is not repaired, and a fibrous patch replaces the damaged area.

Module 4.12 Review

a. Mature cartilage cells are called ___________.

b. Which connective tissue fiber is characteristic of the cartilage supporting the ear?

c. If a person has a herniated intervertebral disc, which type of cartilage has been damaged?

LO **4.12** Describe the structure, function, and locations of cartilage.

Module 4.13

Bone provides a strong framework for the body

Bone, or **osseous** (OS-ē-us; *os*, bone) **tissue**, contains a small volume of ground substance. About two-thirds of the bone matrix consists of a mixture of calcium salts—primarily calcium phosphate, with lesser amounts of calcium carbonate. The rest of the matrix is dominated by collagen fibers. This combination gives bone truly remarkable properties. By themselves, calcium salts are hard but rather brittle, whereas collagen fibers are strong and relatively flexible. In bone, the presence of the minerals surrounding the collagen fibers produces a strong, somewhat flexible combination that is highly resistant to shattering. In its overall properties, bone can compete with the best steel-reinforced concrete. In essence, the collagen fibers in bone act like the steel reinforcing rods, and the mineralized matrix acts like the concrete.

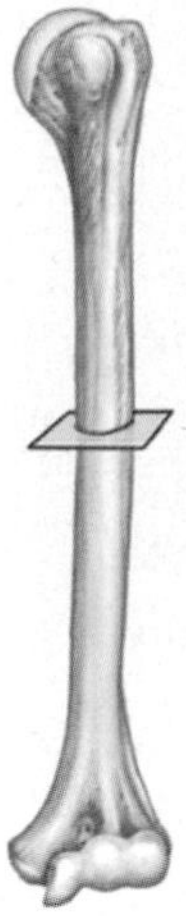

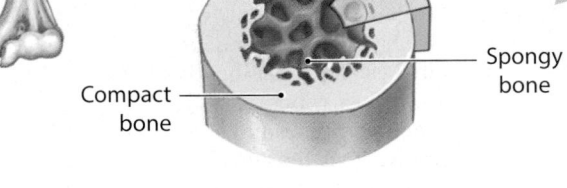

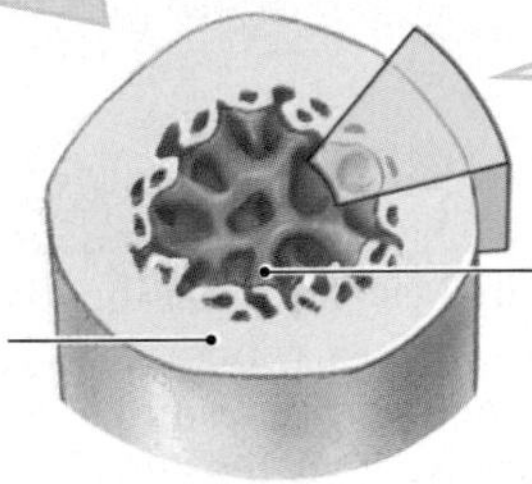
Compact bone
Spongy bone

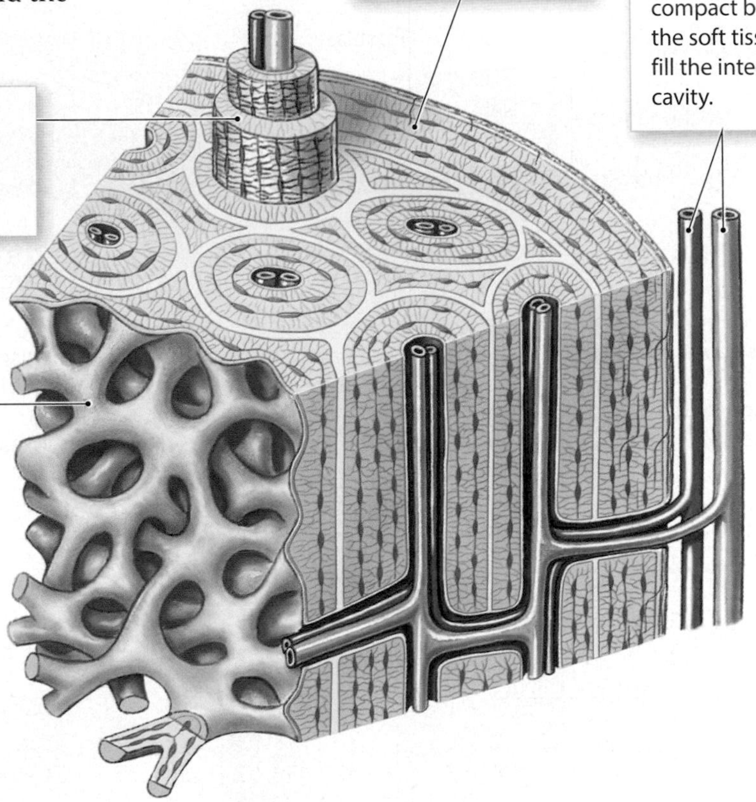
Compact bone also has a superficial layer of bone that was deposited during appositional growth of the bone. Because the matrix is solid and calcified, interstitial growth cannot occur in bone.
Unlike cartilage, bone is highly vascular. Large vessels outside the bone are connected to smaller vessels that supply areas of compact bone and the soft tissues that fill the interior cavity.
In compact bone the matrix is organized in concentric layers around branches of blood vessels within the bone.

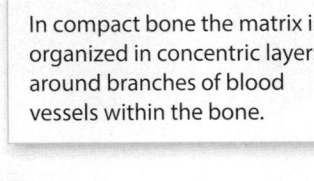

1 A typical long bone is hollow, and its walls contain two different types of bone. The weight-bearing outer layer consists of well-organized **compact bone**, whereas a finer network of **spongy bone** lines the internal cavity. We consider the structural and functional differences between the two bone types in Chapter 6.

2 Cartilage and bone share a number of functions, but these two supporting connective tissues are organized very differently.

A Comparison of Cartilage and Bone

Characteristic	Cartilage	Bone
Cells	Chondrocytes in lacunae	Osteocytes in lacunae
Ground substance	Chondroitin sulfate (in proteoglycan) and water	A small volume of liquid surrounding insoluble crystals of calcium salts
Fibers	Collagen, elastic, and reticular fibers in varying proportions	Collagen fibers predominate
Vascularity	None	Extensive
Covering	Perichondrium (two layers)	Periosteum (two layers)
Strength	Limited: bends easily, but hard to break	Strong: resists distortion until breaking point

3 Although the arrangement of the layers of bone may vary, the superficial and deeper layers share common structural features.

Except in joint cavities, where they are covered by a layer of hyaline cartilage, bone surfaces are sheathed by a **periosteum** (per-ē-OS-tē-um) composed of fibrous (outer) and cellular (inner) layers.

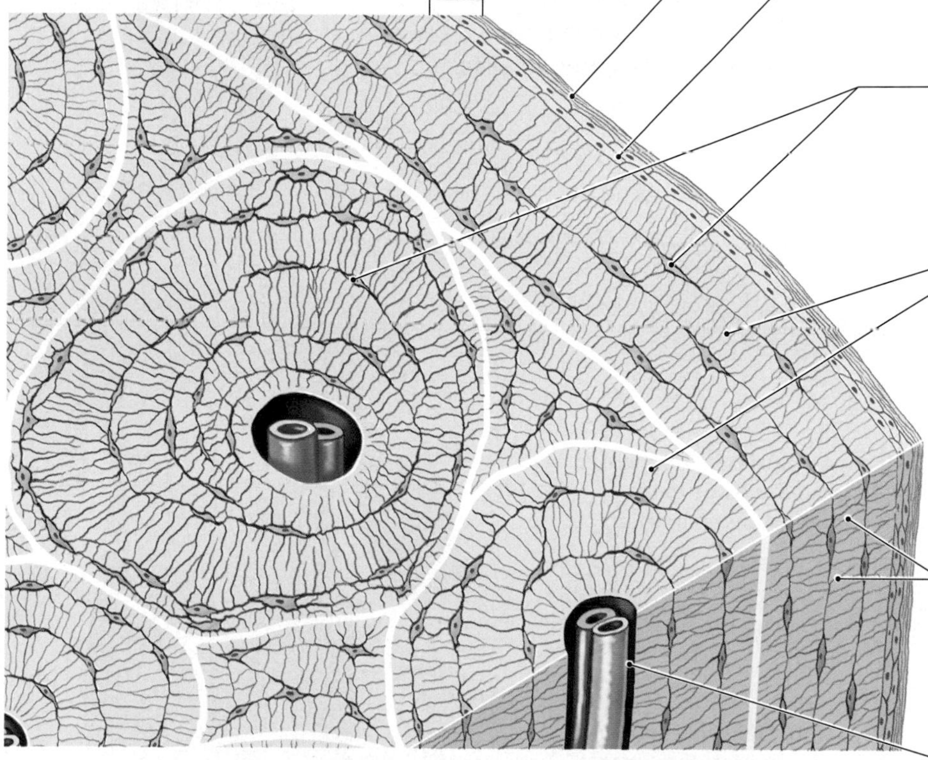

Layers of the Periosteum

The fibrous layer attaches a bone to surrounding tissues and to associated tendons and ligaments.

The cellular layer functions in appositional bone growth and participates in repairs after an injury.

Lacunae in the matrix contain **osteocytes** (OS-tē-ō-sīts), or bone cells. The lacunae are typically organized around blood vessels that branch through the bony matrix.

Layers of matrix separate the lacunae. These layers are oriented along the main axis of the bone.

Canaliculi (kan-a-LIK-ū-lē; little canals) are fine passageways that form a branching network for the exchange of materials between blood vessels and osteocytes. This is important because diffusion cannot occur through the calcified matrix of bone.

A **central canal** at the center of an osteon contains the blood vessels that provide oxygen and nutrients to the osteocytes.

4 The functional unit of compact bone is called an **osteon**.

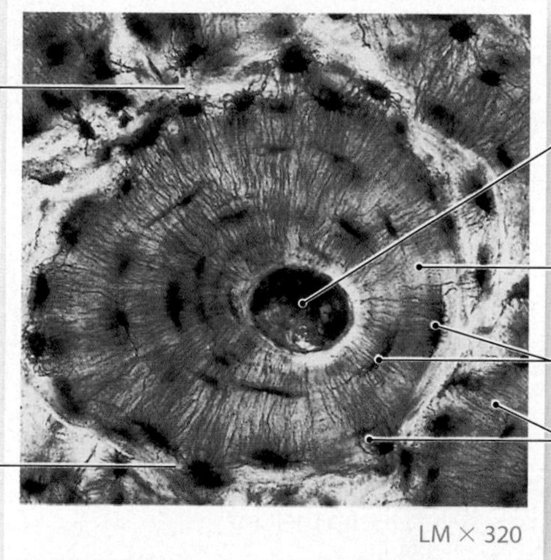
Osteon
Matrix
Lacunae
Canaliculi
LM × 320

Unlike cartilage, bone undergoes extensive remodeling throughout life, and complete repairs can be made even after severe damage has occurred. Bones also respond to the stresses placed on them, growing thicker and stronger with exercise and becoming thin and brittle with inactivity.

Module 4.13 Review

a. Distinguish between the two types of supporting connective tissues with respect to their characteristic fibers.

b. Mature bone cells in lacunae are called ________.

c. The functional unit of compact bone is called a(n) ________.

LO 4.13 Describe the structure and function of bone.

Membranes are physical barriers, and fasciae create internal compartments and divisions

Membranes

A membrane is a physical barrier. There are many different types of anatomical membranes—you encountered plasma membranes in Chapter 3, and you will find many other kinds of membranes in later chapters. Here we consider membranes that line or cover body surfaces. These membranes typically consist of an epithelium supported by connective tissue. Four of these membranes occur in the body: mucous membranes, serous membranes, the cutaneous membrane, and synovial membranes.

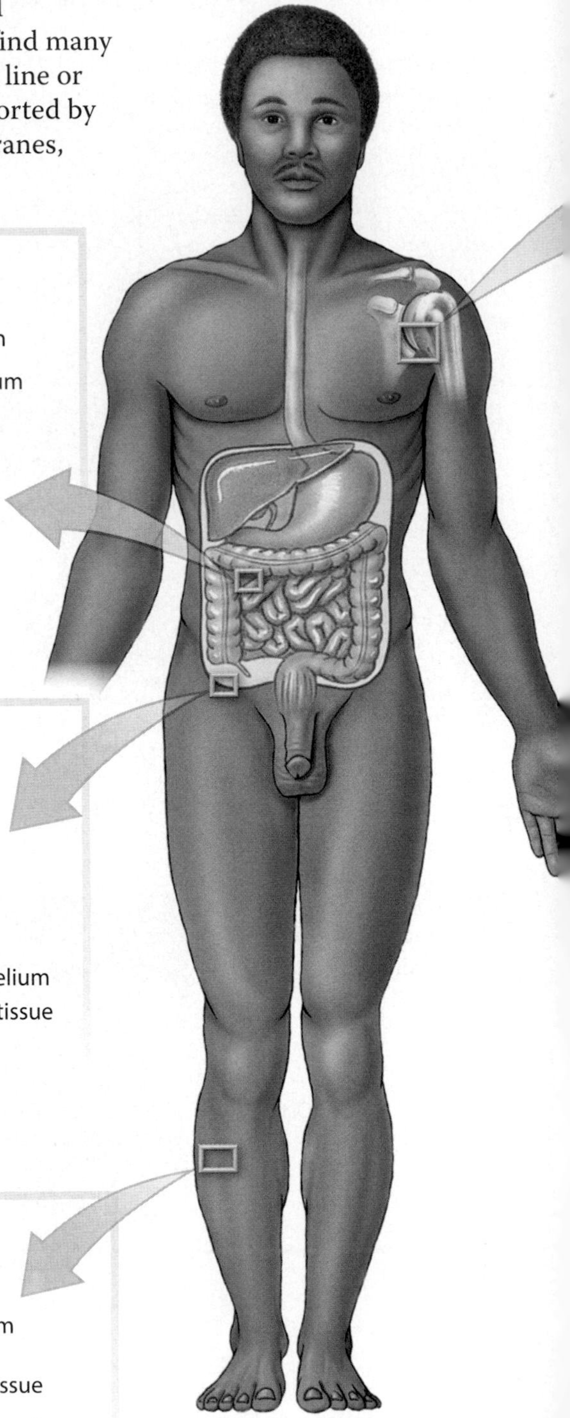

Mucous Membranes

Mucous membranes, or *mucosae* (mū-KŌ-sē), line passageways and chambers that communicate with the exterior, including those in the digestive, respiratory, reproductive, and urinary tracts. These epithelial surfaces must be kept moist to reduce friction and, in many cases, to facilitate absorption or secretion. The epithelial surfaces are lubricated either by mucus (produced by mucous cells or multicellular glands) or by fluids such as urine or semen.

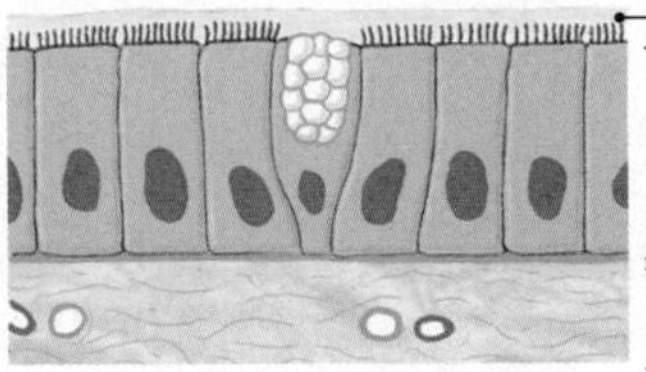
Mucous secretion
Epithelium

The **lamina propria** (PRŌ-prē-uh) is a layer of areolar tissue that supports the mucous epithelium.

Serous Membranes

Serous membranes, or *serosae* (se-RŌ-sē), consist of a mesothelium supported by areolar tissue. They are extremely delicate and never directly connected to the exterior. Three serous membranes line the body cavities of the trunk: (1) the **pleura**, which lines the pleural cavities and covers the lungs; (2) the **pericardium**, which lines the pericardial cavity and covers the heart; and (3) the **peritoneum**, which lines the peritoneal cavity and covers the surfaces of visceral organs.

A watery **serous fluid** diffusing from underlying tissues coats the surfaces of a serous membrane and prevents friction.

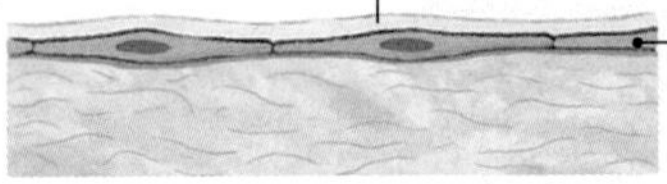
Mesothelium
Areolar tissue

The Cutaneous Membrane

The **cutaneous membrane** covers the surface of the body. It consists of a stratified squamous epithelium and a layer of areolar tissue reinforced by underlying dense irregular connective tissue. In contrast to serous and mucous membranes, the cutaneous membrane is thick, relatively waterproof, and usually dry. We take a closer look at the cutaneous membrane in Chapter 5.

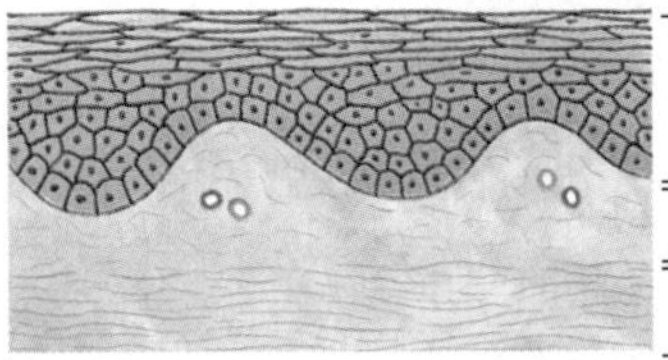
Epithelium
Areolar tissue
Dense irregular connective tissue

Synovial Membrane

A **synovial membrane** lines freely movable joint cavities but does not cover the opposing joint surfaces. Although the covering of the synovial membrane is often called an epithelium, it differs from true epithelia in four respects:
(1) It develops within a connective tissue, (2) no basement membrane is present, (3) gaps of up to 1 mm may separate adjacent cells, and (4) the synovial fluid and capillaries in the underlying connective tissue are continuously exchanging fluid and solutes.

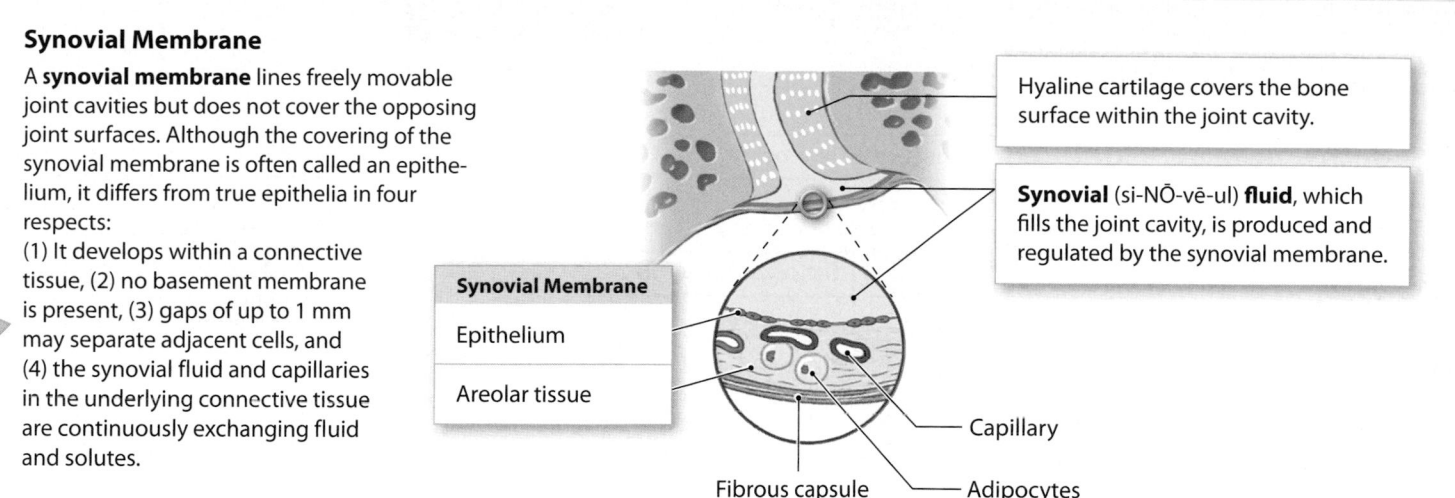

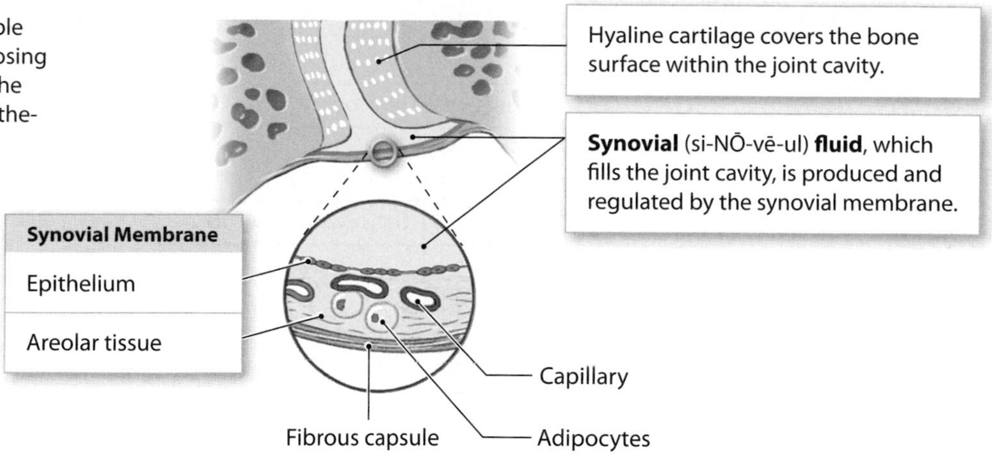
Hyaline cartilage covers the bone surface within the joint cavity.
Synovial (si-NŌ-vē-ul) fluid, which fills the joint cavity, is produced and regulated by the synovial membrane.
Synovial Membrane
Epithelium
Areolar tissue
Capillary
Fibrous capsule
Adipocytes

Fasciae

Fasciae (FASH-ē-ē; singular, fascia) are connective tissue layers that support and surround organs. The fasciae consist of three types of layers: the **superficial fascia**, the **deep fascia**, and the **subserous fascia**.

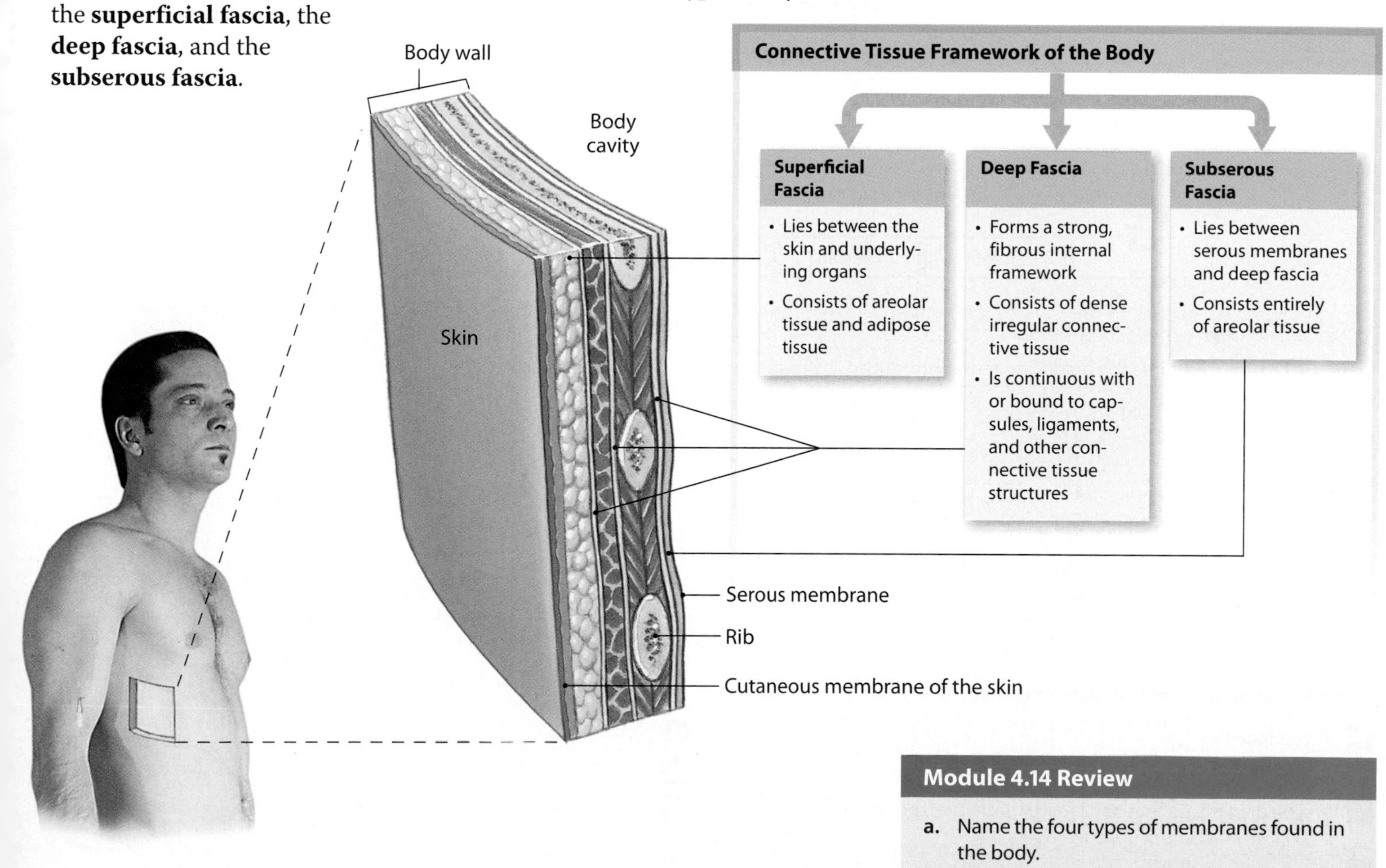
Body wall
Body cavity
Skin
Connective Tissue Framework of the Body
Superficial Fascia
• Lies between the skin and underlying organs
• Consists of areolar tissue and adipose tissue
Deep Fascia
• Forms a strong, fibrous internal framework
• Consists of dense irregular connective tissue
• Is continuous with or bound to capsules, ligaments, and other connective tissue structures
Subserous Fascia
• Lies between serous membranes and deep fascia
• Consists entirely of areolar tissue
Serous membrane
Rib
Cutaneous membrane of the skin

Module 4.14 Review

a. Name the four types of membranes found in the body.

b. Which cavities in the body are lined by serous membranes?

c. Name the three layers of fascia and their types of connective tissue.

4.14 Describe the arrangements of epithelial and connective tissues in the four types of membranes, and the structures and locations of the three types of fasciae.

Concept map

Use the following terms once to fill in the blank boxes to correctly complete the map.

- loose connective tissue
- chondrocytes in lacunae
- fluid connective tissue
- tendons
- ligaments
- regular
- hyaline
- blood
- adipose
- bone

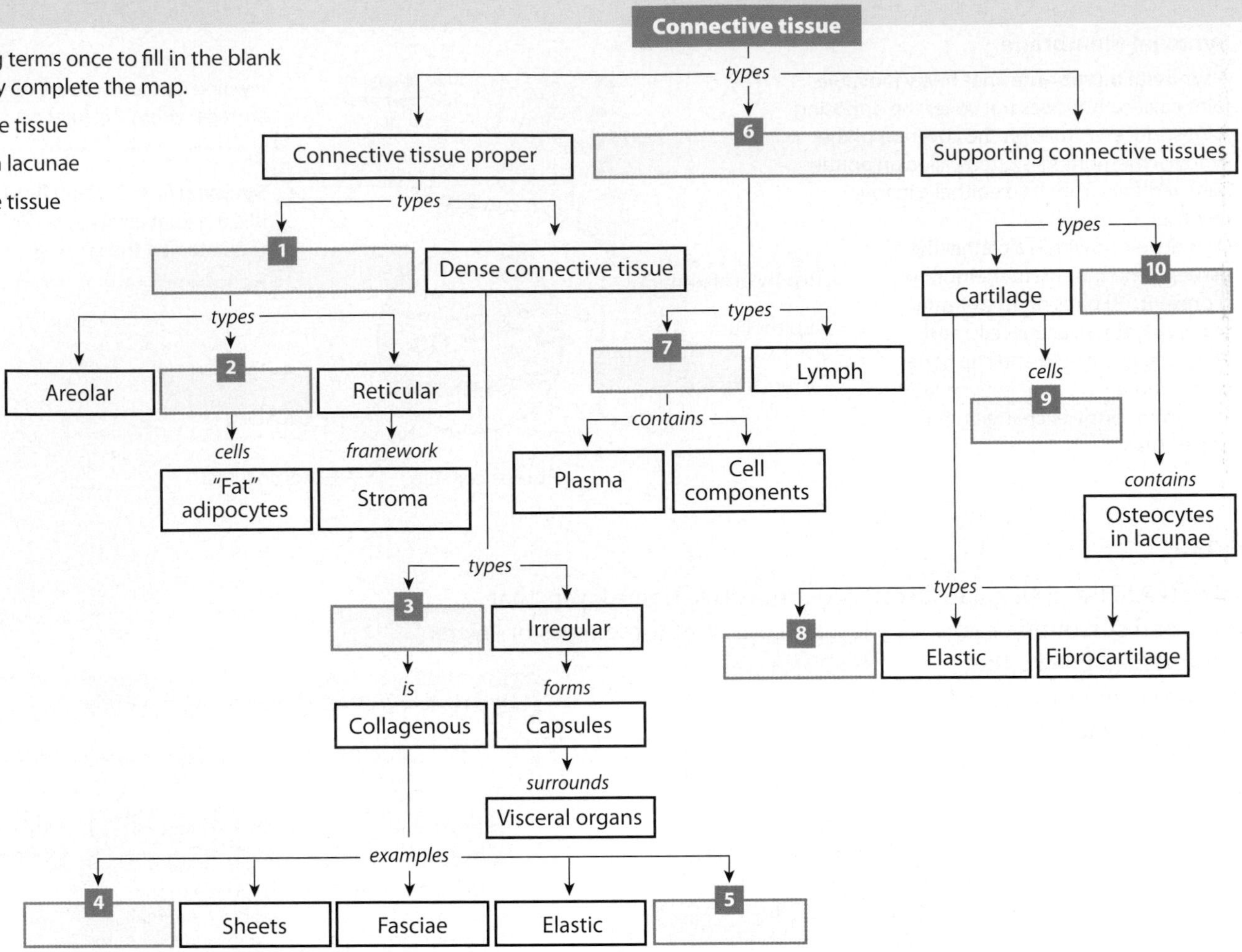

Connective tissue
types
Connective tissue proper
6
Supporting connective tissues
types
1
Dense connective tissue
types
Areolar
2
Reticular
cells
framework
"Fat" adipocytes
Stroma
types
7
Lymph
contains
Plasma
Cell components
types
Cartilage
10
cells
9
contains
Osteocytes in lacunae
types
3
Irregular
is
forms
Collagenous
Capsules
surrounds
Visceral organs
types
8
Elastic
Fibrocartilage
examples
4
Sheets
Fasciae
Elastic
5

Vocabulary

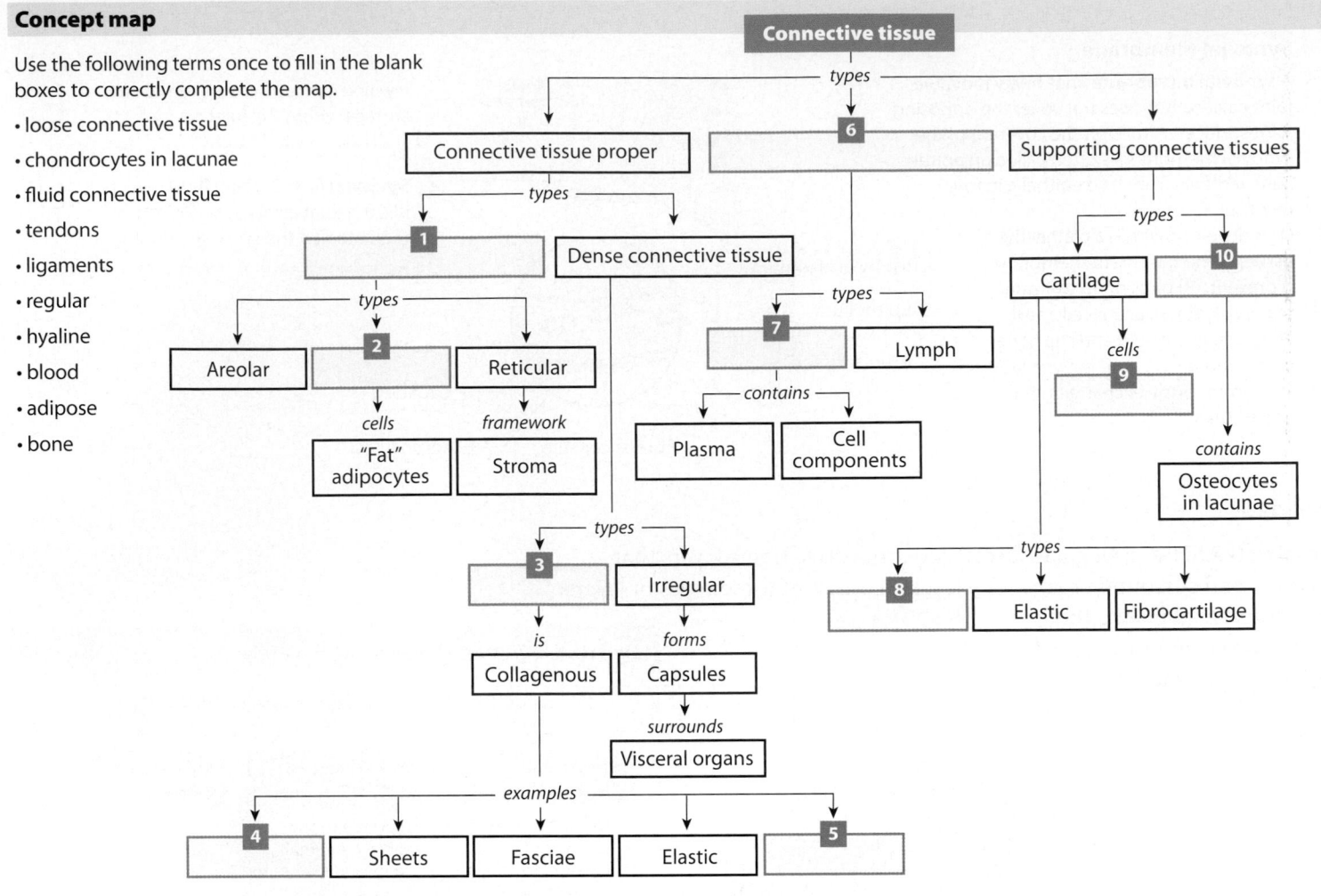

In the space provided, write the boldfaced terms introduced in this section that contain the indicated word part.

11 peri- *(around)*
12 os- *(bone)*
13 chondro- *(cartilage)*
14 inter- *(between)*
15 lacus- *(lake)*

11 ______
12 ______
13 ______
14 ______
15 ______

Enter the appropriate term for each description below.

16 A cartilage cell
17 Bone tissue
18 A type of cartilage that has a matrix with little ground substance and large amounts of collagen fibers
19 Cells that store lipid reserves
20 The membrane that lines freely movable joint cavities
21 The membrane that covers the surface of the body
22 Separates cartilage from surrounding tissues

16 ______
17 ______
18 ______
19 ______
20 ______
21 ______
22 ______

Muscle tissue outweighs neural tissue by 25:1

As we have seen, epithelia cover surfaces and line passageways, and connective tissues support and interconnect parts of the body. Together, these tissues provide a strong, interwoven framework within which the organs of the body can function. When we consider the collective weight of the four types of tissue in the body, connective tissues make up about 45 percent of the total body weight. Muscle tissue contributes the most, and neural tissue the least, to body weight.

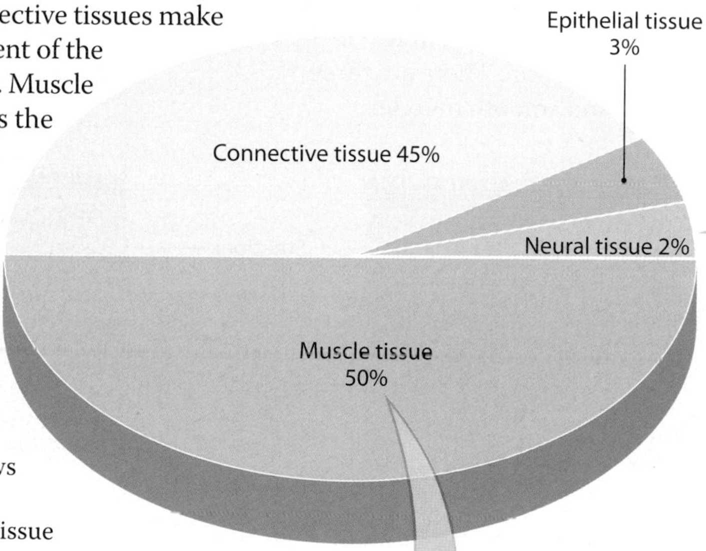
Epithelial tissue 3%
Connective tissue 45%
Neural tissue 2%
Muscle tissue 50%

This pie chart shows the percentages by weight of the four tissue types in the body.

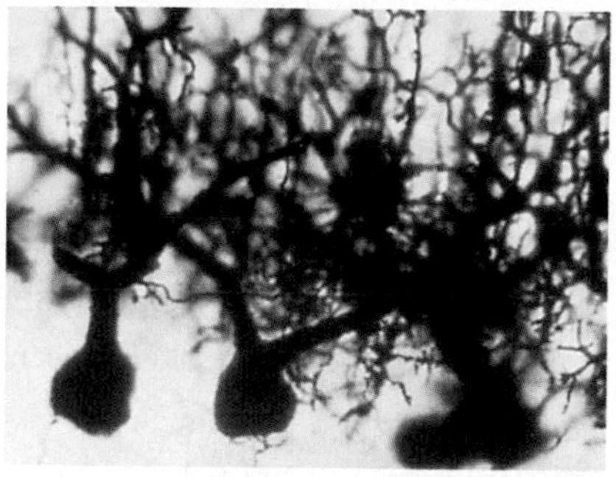

Specialized neurons in the brain LM × 350

Muscle Tissue

Skeletal Muscle Tissue

Skeletal muscle tissue moves the body by pulling on bones of the skeleton, making it possible for us to walk, dance, bite an apple, or play the guitar.

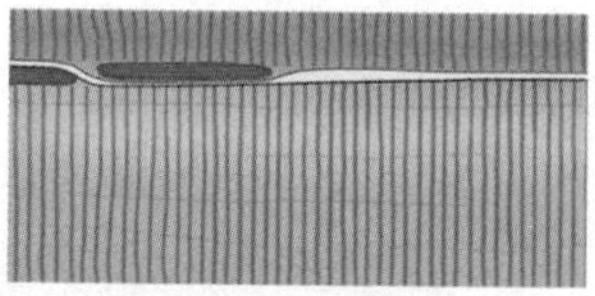

Cardiac Muscle Tissue

Cardiac muscle tissue contractions move blood within the heart and through the blood vessels.

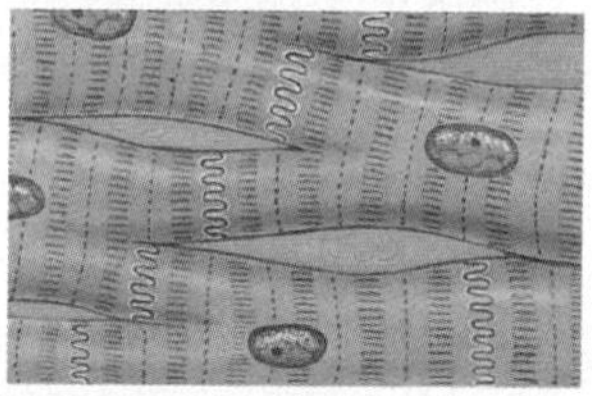

Smooth Muscle Tissue

Smooth muscle tissue contractions move fluids and solids along the digestive tract and regulate the diameters of small arteries, among other functions.

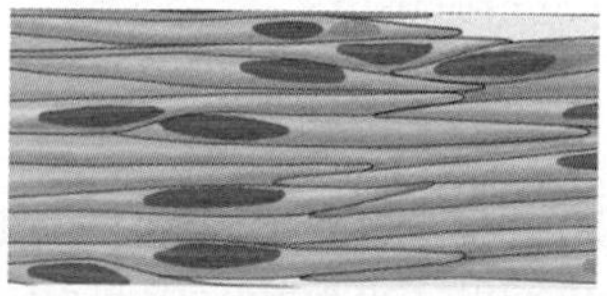

In this section we consider the histology of muscle tissue and neural tissue. We only provide an overview of these tissue types because additional details will be given in later chapters dealing with the muscular system, nervous system, cardiovascular system, and digestive system.

Module 4.15 Review

a. What is the relative percentage of body weight from each of the four tissue types?

b. List the three classifications of muscle tissue.

c. Describe a function for each type of muscle tissue.

LO 4.15 Describe the relative proportions of muscle tissue and neural tissue in the body.

Module 4.16

Muscle tissue is specialized for contraction and neural tissue is specialized for communication

Muscle Tissue

Many vital functions involve movement of one kind or another—movement of materials along the digestive tract, movement of blood around the cardiovascular system, or movement of the body from one place to another. Movement is produced by **muscle tissue**, which is specialized for contraction. There are three types of muscle tissue: skeletal muscle, cardiac muscle, and smooth muscle.

1 **Skeletal muscle tissue** is found in skeletal muscles, organs that also contain connective tissues and neural tissue. The cells are long, cylindrical, banded (**striated**), and have multiple nuclei (**multinucleate**).

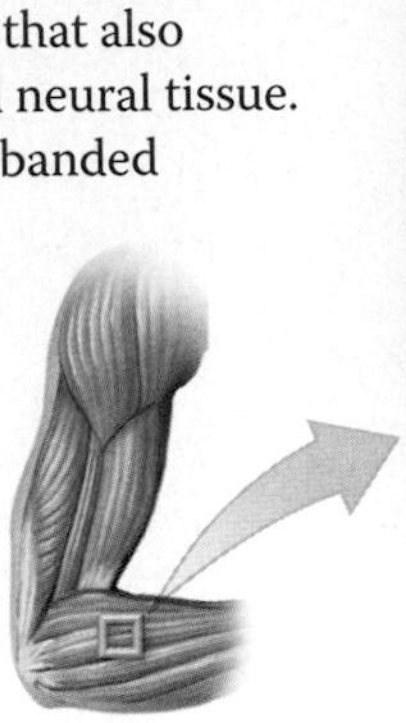

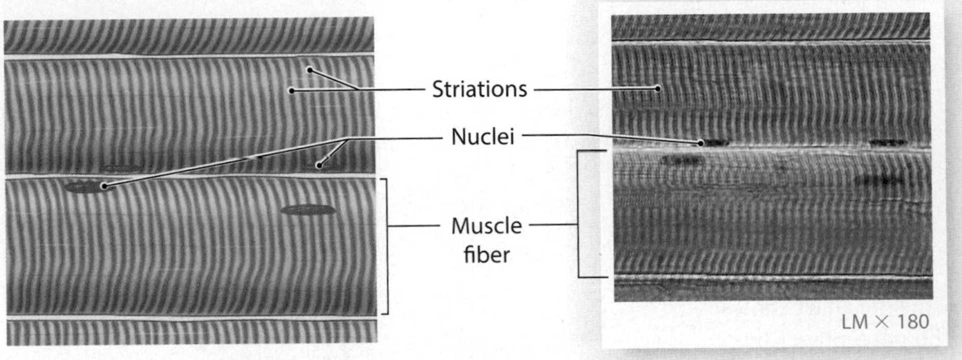
Striations
Nuclei
Muscle fiber
LM × 180

Skeletal muscles move or stabilize the position of the skeleton; guard entrances and exits to the digestive, respiratory, and urinary tracts; generate heat; and protect internal organs.

2 **Cardiac muscle tissue** is found in the heart. Cardiac muscle cells, also known as **cardiocytes**, are short, branched, and striated, usually with a single nucleus. These cells are interconnected at specialized intercellular junctions called **intercalated discs**. They help synchronize cardiocyte contractions.

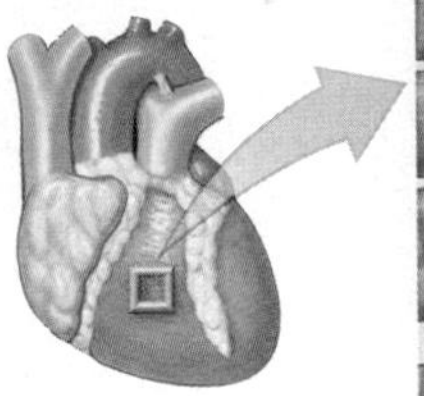

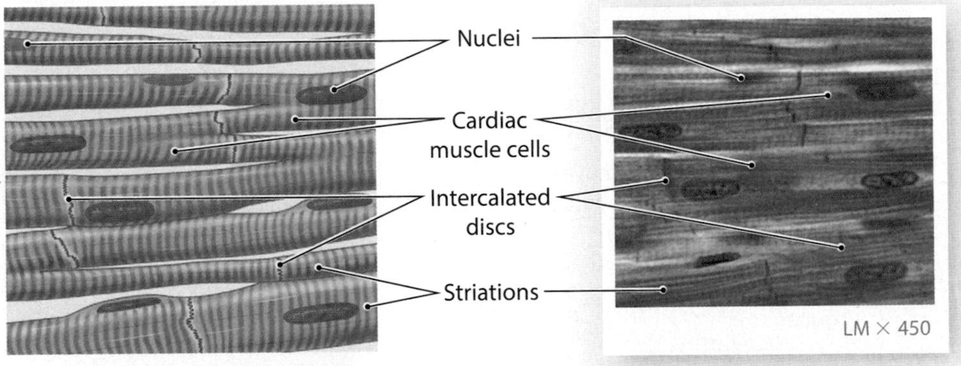
Nuclei
Cardiac muscle cells
Intercalated discs
Striations
LM × 450

Cardiac muscle moves blood and maintains blood pressure.

3 **Smooth muscle tissue** is found throughout the body. For example, smooth muscle is found in the skin, in the walls of blood vessels, and in many digestive, respiratory, urinary, and reproductive organs. The cells are short, spindle-shaped, and nonstriated, and have a single, central nucleus.

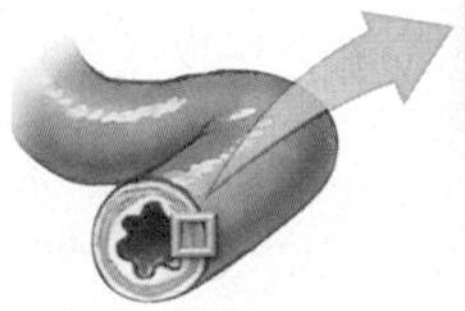

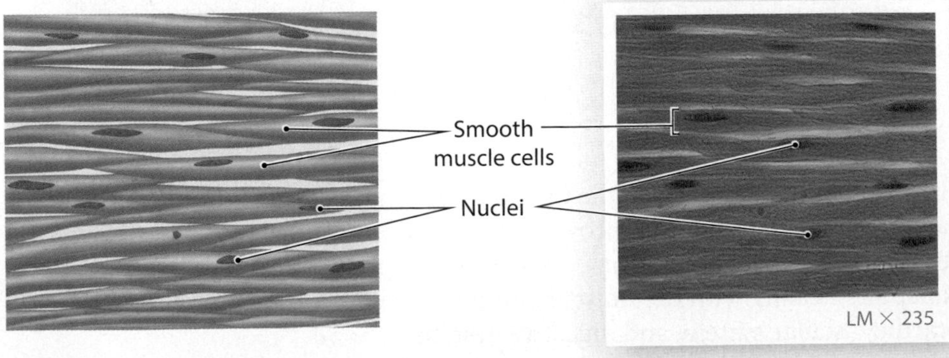
Smooth muscle cells
Nuclei
LM × 235

Smooth muscle moves food, urine, and reproductive tract secretions; controls diameter of respiratory passageways; and regulates diameter of blood vessels.

Neural Tissue

Neural tissue, which is also known as nervous tissue, is specialized to conduct electrical impulses from one region of the body to another. Ninety-eight percent of the neural tissue in the body is in the brain and spinal cord, which are the control centers of the nervous system. Neural tissue contains two types of cells: (1) **neurons** (NOOR-onz; *neuro*, nerve) and (2) several kinds of supporting cells, collectively called **neuroglia** (noo-ROG-lē-uh or noo-rō-GLĒ-uh), or **glial cells** (*glia*, glue).

4 Neurons transfer information from place to place and process information like living computer chips. Their sizes and shapes vary widely. The longest cells in your body are neurons. Many are as long as a meter (39 in.)!

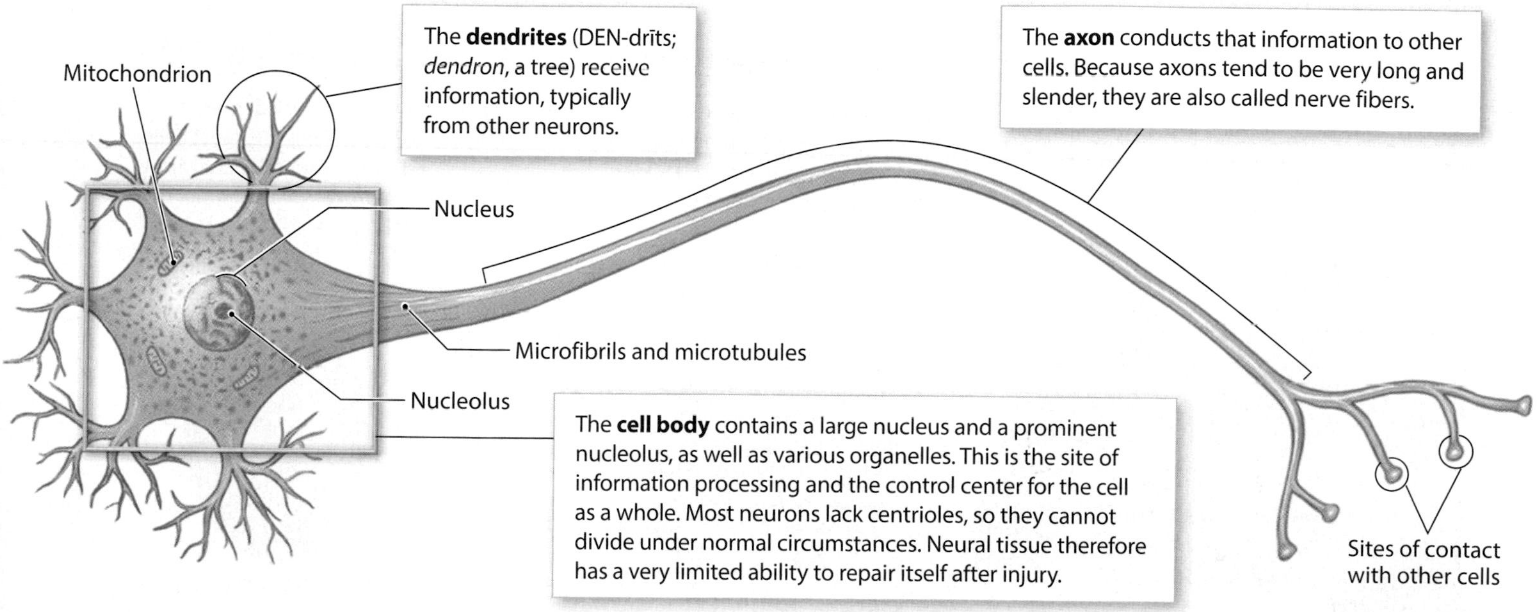
Mitochondrion
The **dendrites** (DEN-drīts; *dendron*, a tree) receive information, typically from other neurons.
The **axon** conducts that information to other cells. Because axons tend to be very long and slender, they are also called nerve fibers.
Nucleus
Microfibrils and microtubules
Nucleolus
The **cell body** contains a large nucleus and a prominent nucleolus, as well as various organelles. This is the site of information processing and the control center for the cell as a whole. Most neurons lack centrioles, so they cannot divide under normal circumstances. Neural tissue therefore has a very limited ability to repair itself after injury.
Sites of contact with other cells

5 There are several different types of neuroglia, each with specific functions. Each type has a distinctive appearance related to its primary function. In general, neuroglia protect, support, and repair neural tissue and maintain the nutrient supply to neurons.

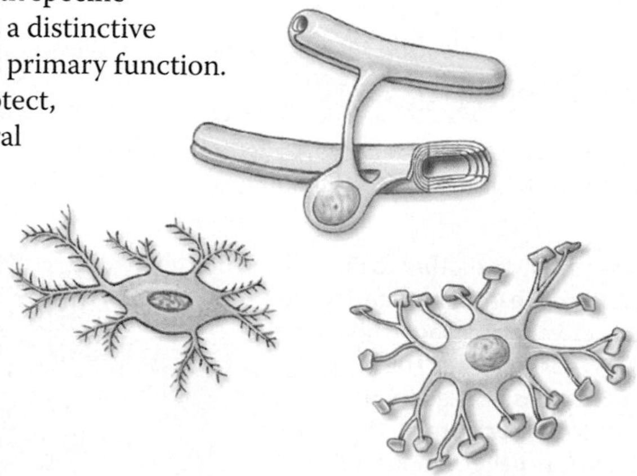

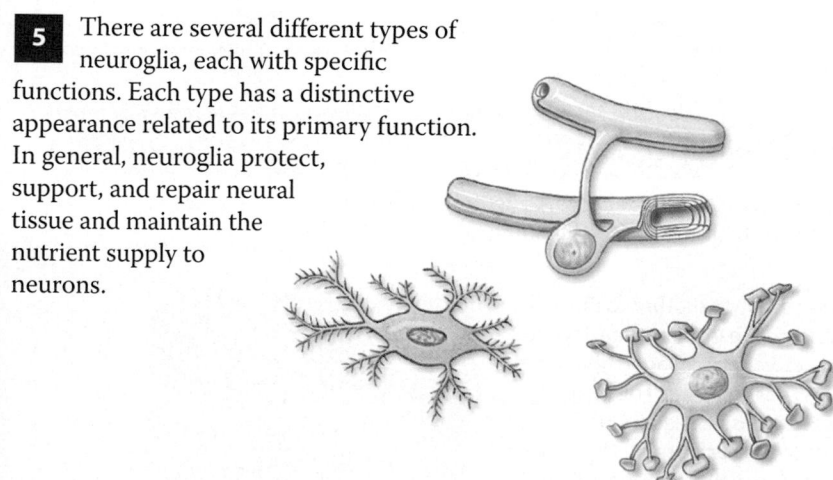

Functions of Neuroglia

- Maintain physical structure of neural tissue
- Repair neural tissue framework after injury
- Perform phagocytosis
- Provide nutrients to neurons
- Regulate the composition of the interstitial fluid surrounding neurons

Our conscious and unconscious thought processes reflect the communication among neurons in the brain. Such communication involves the propagation of electrical impulses that are generated at the cell body and travel along the axon to reach other cells.

Module 4.16 Review

a. Describe the three types of muscle tissue.

b. Which type of muscle tissue regulates blood vessel diameter?

c. Distinguish between neurons and neuroglia.

LO **4.16** Specify the functions of muscle tissue and neural tissue.

The response to tissue injury involves inflammation and regeneration

Let's consider what happens after an injury, focusing on the interaction among different tissues. Tissues are not isolated from each other; they combine to form organs with diverse functions. Therefore, any injury affects several types of tissue simultaneously. These tissues must respond in a coordinated way to preserve homeostasis. The body utilizes two related processes, inflammation and regeneration, to restore homeostasis after an injury.

1 Injury

When a tissue is injured, the body activates a general defense mechanism.

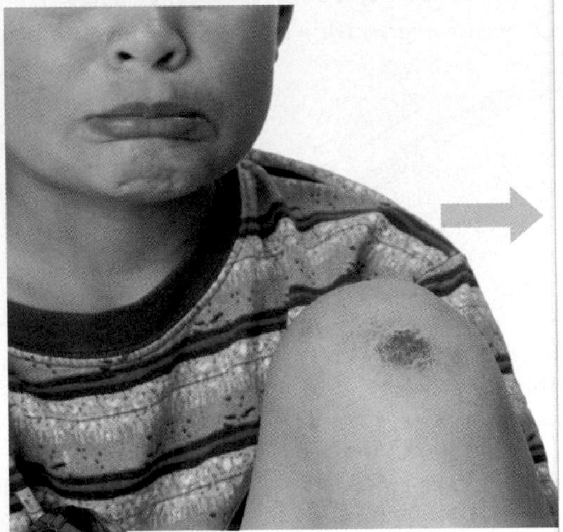

Exposure to Pathogens and Toxins

An injured tissue contains an abnormal concentration of pathogens, toxins, waste products, and the chemicals from injured cells.

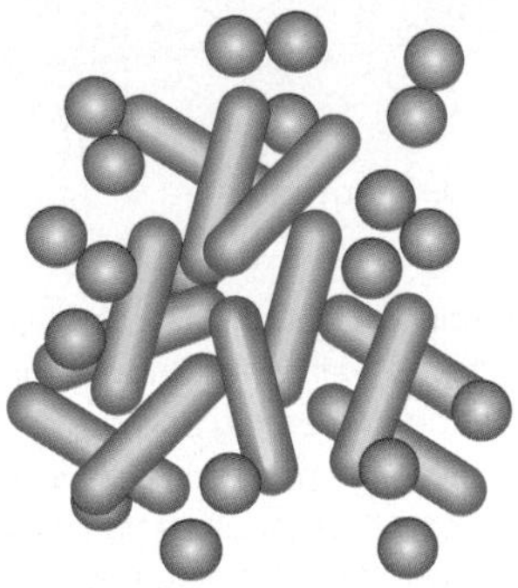

stimulates

Mast Cell Activation

When an injury damages connective tissue, mast cells release a variety of chemicals. This process, called **mast cell activation**, stimulates inflammation.

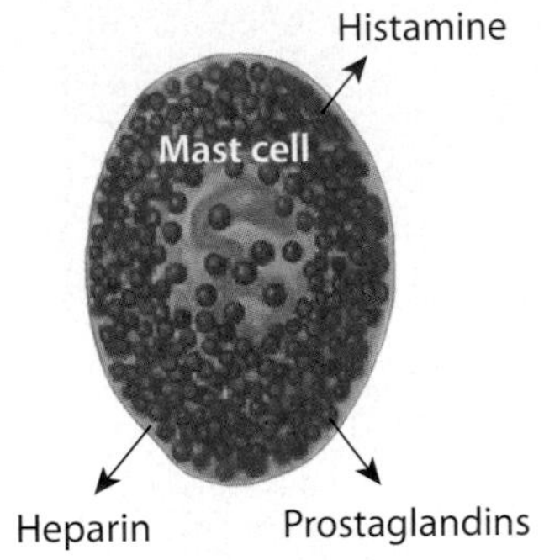
Histamine
Mast cell
Heparin
Prostaglandins

Regeneration inhibits further mast cell activation

3 Regeneration

Regeneration is the repair that occurs after the damaged tissue has been stabilized and the inflammation has subsided.

As tissue conditions return to normal, fibroblasts move into the area, laying down a network of collagen fibers that stabilizes the injury site. This process produces a dense, collagenous framework known as **scar tissue**. Over time, scar tissue is usually "remodeled" and gradually assumes a more normal appearance.

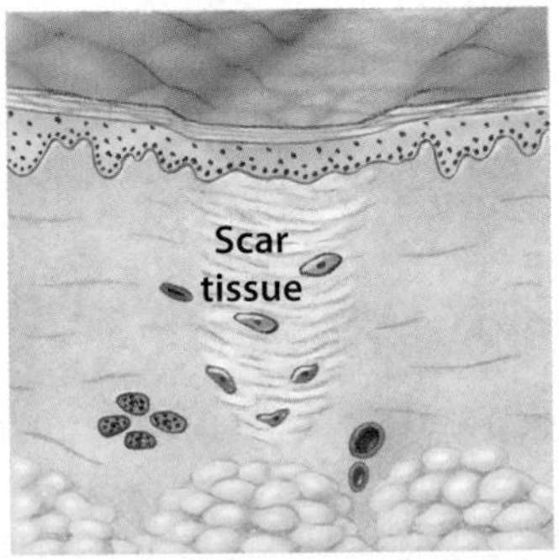
Scar tissue

Normal tissue conditions restored

2 Inflammation

Inflammation produces several familiar indications of injury: swelling, redness, heat (warmth), and pain. Inflammation may also result from the presence of pathogens, such as harmful bacteria, within the tissues. The presence of these pathogens constitutes an **infection**. Because all organs have connective tissues, inflammation can occur anywhere in the body.

Increased Blood Flow

In response to the released chemicals, the smooth muscle tissue that surrounds local blood vessels relaxes, and the vessels **dilate**, or enlarge in diameter. This dilation increases blood flow through the damaged tissue.

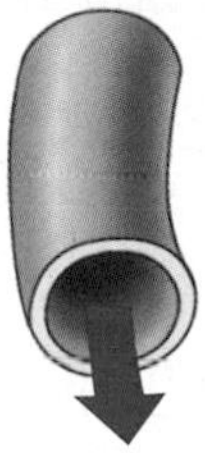

Increased Vessel Permeability

The dilation is accompanied by an increase in the permeability of the capillary walls. Plasma, including blood proteins, now diffuses into the injured tissue, so the area becomes swollen.

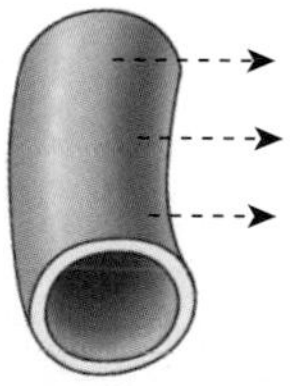

Pain

The combination of abnormal conditions within the tissue and the chemicals released by mast cells stimulates nerve endings that produce the sensation of pain.

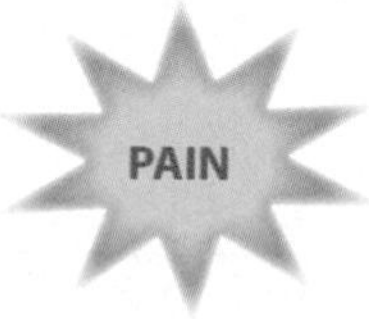
PAIN

Increased local temperature
The increased blood flow and permeability causes the tissue to become warm and red.

Increased oxygen and nutrients
Vessel dilation, increased blood flow, and increased vessel permeability result in enhanced delivery of oxygen and nutrients.

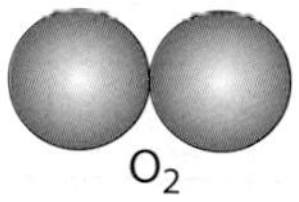
O_2

Increased phagocytosis
Phagocytes in the tissue are activated, and they begin engulfing tissue debris and pathogens. Additional phagocytes migrate into the tissue from the bloodstream, drawn to the site by the abnormal local conditions.

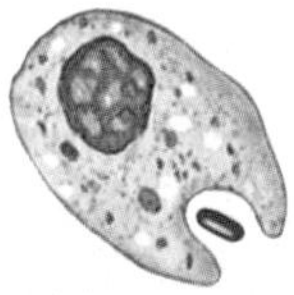

Removal of toxins and wastes
The enhanced circulation that increases the volume of interstitial fluid also carries away toxins and waste products, distributing them to the kidneys for excretion, or to the liver for inactivation.

Toxins
and wastes

Over a period of hours to days, the cleanup process generally succeeds in eliminating the inflammatory stimuli.

Each organ has a different ability to regenerate after injury. This ability is directly linked to the pattern of tissue organization in the injured organ. Epithelia, connective tissues (except cartilage), and smooth muscle tissue usually regenerate well, but skeletal and cardiac muscle tissues and neural tissue regenerate relatively poorly, if at all. Damaged areas that do regenerate are often replaced by scar tissue. The permanent replacement of normal tissue by scar tissue is called **fibrosis** (fī-BRŌ-sis). Fibrosis in muscle and other tissues may occur in response to injury, disease, or aging.

Module 4.17 Review

a. Identify the two processes in the response to tissue injury.
b. What are the four indications of inflammation that occur following an injury?
c. Why can inflammation occur in any organ in the body?

LO 4.17 Describe the roles of inflammation and regeneration in response to tissue injury.

Concept map

Complete the concept maps using key terms and concepts covered in this section.

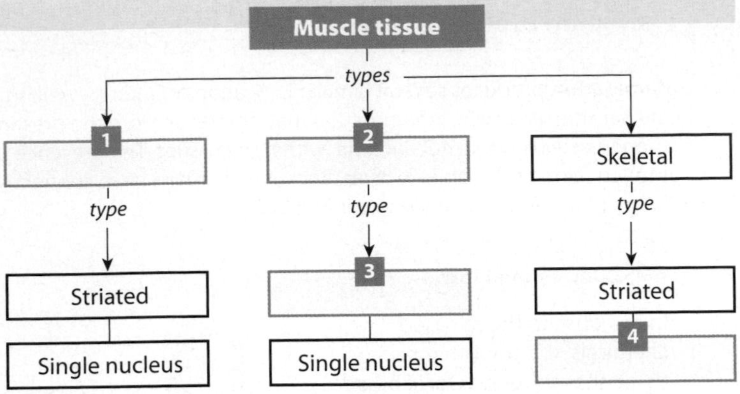
Muscle tissue
types
1
2
Skeletal
type
type
type
Striated
3
Striated
Single nucleus
Single nucleus
4

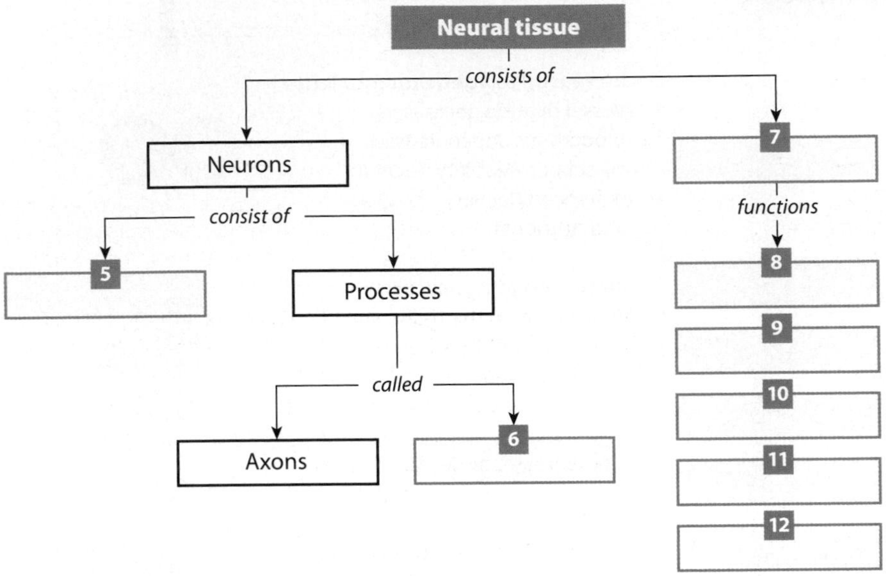
Neural tissue
consists of
Neurons
7
consist of
functions
5
Processes
8
9
called
10
Axons
6
11
12

Vocabulary

Write the term for each of the following descriptions in the space provided.

13 A single structure that extends from the cell body of a neuron and carries information to other cells

14 A specialized intercellular junction between cardiac muscle cells

15 The supporting cells found in neural tissue

16 Muscle tissue that contains large, multinucleate, striated cells

17 Muscle tissue that regulates the diameter of blood vessels and respiratory passageways

18 The repair process that occurs after inflammation has subsided

19 The first process in a tissue's response to injury

13 ______________________

14 ______________________

15 ______________________

16 ______________________

17 ______________________

18 ______________________

19 ______________________

Section integration

During inflammation, both blood flow and blood vessel permeability increase in the injured area.
Describe how these responses aid the cleanup process and eliminate the inflammatory stimuli in the injured area.

20 ______________________

____

Study Outline

SECTION 1 • Epithelial Tissue

4.1 Four types of tissue make up the body p. 135

1. **Tissues** are collections of cells and cell products that perform a limited number of specialized functions.
2. The study of tissues is called **histology**. The four basic tissue types recognized by histologists are **epithelial tissue**, **connective tissue**, **muscle tissue**, and **neural tissue**.

4.2 Microscopes are used to study cells and tissues p. 136

3. The earliest microscopes required a light source and could magnify only 10 or 20 times. Today's light microscopes can magnify up to 1000 times. **Simple microscopes** have one lens; **compound microscopes** feature more than one lens.
4. Modern microscopes use beams of electrons instead of light and can magnify over 1 million times. Common microscopes used in scientific study today are the **light microscope**, the **transmission electron microscope**, and the **scanning electron microscope.**

4.3 Epithelial tissue covers surfaces, lines structures, and forms secretory glands p. 138

5. **Epithelial tissue** includes **epithelia** and **glands**. Epithelia cover exposed surfaces and line internal passageways. Glands are secretory structures and there are two types of glands: **endocrine glands** and **exocrine glands**.
6. Endocrine glands secrete hormones or precursors into interstitial fluid usually for distribution into the bloodstream. Exocrine glands secrete onto external surfaces or into ducts that connect to the exterior. The functions of epithelial tissue are to provide physical protection, control permeability, provide sensation, and produce specialized secretions.
7. An epithelium has a **base** that is attached to adjacent tissues, and an **apical surface** that faces the body exterior or internal space. Epithelial cells have three basic shapes: **squamous**, **cuboidal**, and **columnar**. A **simple epithelium** is a single layer of cells; several layers compose a **stratified epithelium.**

4.4 Epithelial cells are extensively interconnected, both structurally and functionally p. 140

8. The physical integrity of an epithelium depends on intercellular connections and attachment to underlying tissues.

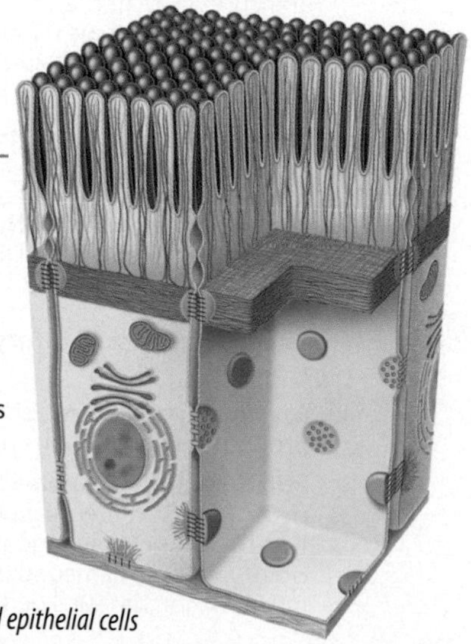

Interconnected epithelial cells

9. **Intercellular attachments** include: **tight** (occluding) **junctions**, **adhesion belts**, **gap junctions**, and **desmosomes**.
10. The **basement membrane** (basal lamina) of the cell consists of two layers: the **clear layer** (lamina lucida) and the **dense layer** (lamina densa). The basement membrane connects epithelium to underlying connective tissue by means of **hemidesmosomes**.
11. Epithelial tissue is **avascular**, that is, it lacks blood vessels.

4.5 The cells in a squamous epithelium are flat and irregularly shaped p. 142

12. **Simple squamous epithelium** is a single layer of irregularly shaped cells that fit together like a jigsaw puzzle. This tissue is delicate, reduces friction, and is found along passageways and in areas where absorption takes place.
13. Simple squamous epithelium has special names in certain body locations. **Mesothelium** lines the pericardial, pleural, and peritoneal body cavities, and **endothelium** lines the inner surface of the heart and blood vessels.
14. **Stratified squamous epithelium** has several layers and is located where mechanical or chemical stresses are severe. This tissue may be **keratinized** or **nonkeratinized**.

4.6 Cuboidal and transitional epithelia line passageways and chambers connected to the exterior p. 144

15. **Cuboidal epithelium** resembles hexagonal boxes. **Simple cuboidal epithelium** is found where absorption occurs, as in the kidney tubules, or where secretion takes place, as in the thyroid gland. **Stratified cuboidal epithelia** are rare and located along the ducts of certain exocrine glands.

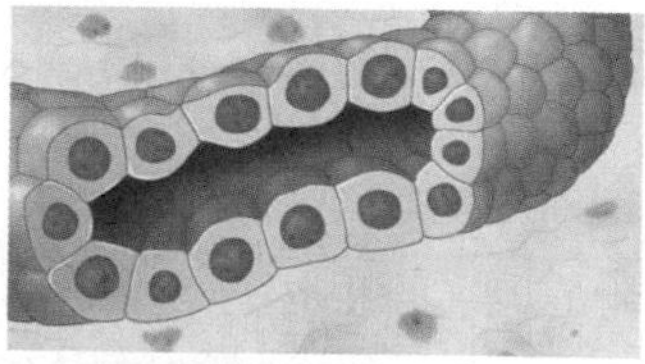

Cuboidal epithelia

16. **Transitional epithelium** is an unusual stratified epithelium. It allows for stretching without damage. It lines the urinary bladder and other regions of the urinary system where volume changes occur.

4.7 Columnar epithelia absorb substances and protect the body from digestive chemicals p. 146

17. **Columnar epithelium** appears rectangular. **Simple columnar epithelium** is found where absorption or secretion takes place, such as the small intestine, stomach, and gallbladder.
18. **Pseudostratified columnar epithelia** are often ciliated, and line the air passageways of the nasal cavities, trachea, and lungs. **Stratified columnar epithelia** are rare and line large ducts such as those in the salivary gland or pancreas.

4.8 Glandular epithelia are specialized for secretion p. 148

19. Collections of epithelial cells that produce secretions are called **glands. Endocrine glands** release their secretions directly into interstitial fluid. **Exocrine glands** release their secretions into **ducts** or onto an epithelial surface.

20. Glandular epithelial cells release their secretions by **merocrine**, **apocrine**, or **holocrine** modes. In merocrine secretion, the most common mode, the product is released by exocytosis. Apocrine secretion involves the loss of secretory product and cytoplasm. In holocrine secretion, gland cells become packed with secretions and then burst.

21. Exocrine glands are categorized as either **simple** or **compound**, based on the structure of the duct, the shape of the gland, and the relationship between the duct and the secretory areas. Forms of simple exocrine glands are simple tubular, simple coiled tubular, simple branched tubular, simple alveolar (acinar), or simple branched alveolar. Forms of compound exocrine glands are compound tubular, compound alveolar (acinar), or compound tubuloalveolar.

SECTION 2 • Connective Tissue

4.9 A matrix surrounds connective tissue cells p. 151

22. Connective tissue varies widely, but all forms contain specialized cells, extracellular protein fibers, and **ground substance** fluid. Connective tissues provide a structural framework, transport fluids and materials, protect organs, support and interconnect other tissues, store energy, and defend the body.

23. The three broad categories of connective tissue are **connective tissue proper**, which includes **loose** and **dense connective tissues**; **fluid connective tissues** (**blood** and **lymph**); and **supporting connective tissues** (**cartilage** and **bone**).

4.10 Loose connective tissues support other tissue types p. 152

24. Connective tissue proper contains extracellular protein fibers, a viscous ground substance, and cells that are either **fixed** or **wandering**.

25. The three types of loose connective tissues are **areolar tissue** (the most common), **adipose tissue**, and **reticular tissue**.

4.11 Dense connective tissues are dominated by extracellular fibers, whereas fluid connective tissues have a liquid matrix p. 154

26. Extracellular fibers make up most of the volume of **dense connective tissues**.

27. The body has three types of dense connective tissues: **dense regular connective tissue**, **dense irregular connective tissue**, and **elastic tissue**.

28. **Fluid connective tissues** have a fluid matrix containing many types of suspended proteins. In blood the watery matrix is called **plasma**. **Red blood cells**, **white blood cells**, and **platelets**, collectively known as **formed elements**, are suspended in the plasma. White blood cells in plasma include **monocytes**, **lymphocytes**, **eosinophils**, **neutrophils**, and **basophils**. **Lymph** is formed from interstitial fluid and has roles in maintaining blood volume and alerting the immune system to infections.

4.12 Cartilage provides a flexible support for body structures p. 156

29. In **cartilage** the matrix is a firm gel. Cartilage cells, or **chondrocytes**, occupy small chambers known as **lacunae**, and are the only cells in the avascular matrix.

30. There are three types of cartilage: **hyaline cartilage**, **elastic cartilage**, and **fibrocartilage**. Cartilage is separated from surrounding tissues by a structure known as the **perichondrium**.

31. Cartilage grows by two mechanisms. **Appositional growth** adds new layers of cartilage to the surface, while **interstitial growth** enlarges the cartilage from within.

4.13 Bone provides a strong framework for the body p. 158

32. Bone or **osseous tissue** contains a small volume of ground substance. The matrix of bone is mostly calcium salts (calcium phosphate and calcium carbonate) and collagen fibers.

33. A typical long bone is hollow and its walls contain either well-organized **compact bone**, or a network of **spongy bone**.

4.14 Membranes are physical barriers, and fasciae create internal compartments and divisions p. 160

34. Membranes line or cover body surfaces. **Mucous membranes** line passageways to the exterior, **serous membranes** line the cavities of the body trunk, the **cutaneous membrane** covers the surface of the body, and **synovial membranes** line freely movable joint cavities.

35. **Fasciae** are connective tissue layers that support and surround organs. Fasciae consist of three types of layers: the **superficial fascia**, the **deep fascia**, or the **subserous fascia**.

SECTION 3 • Muscle Tissue and Neural Tissue

4.15 Muscle tissue outweighs neural tissue by 25:1 p. 163

36. As a percentage of body weight, the four tissue types contribute the following: Muscle tissue 50 percent, connective tissue 45 percent, epithelial tissue 3 percent, and neural tissue 2 percent.

4.16 Muscle tissue is specialized for contraction and neural tissue is specialized for communication p. 164

37. Movement is produced by **muscle tissue** that is specialized for contraction. There are three types of muscle tissue: **skeletal muscle**, **cardiac muscle**, and **smooth muscle**.

38. Skeletal muscle tissue composes skeletal muscles, has long cylindrical cells, is **striated**, and has multiple nuclei. Cardiac muscle tissue forms the heart, has cells that are short and branched, is striated, and contains **intercalated discs**. Smooth muscle tissue is found throughout the body, and has cells that are spindle shaped, nonstriated, and with a single, central nucleus.

39. **Neural tissue**, or nervous tissue, is specialized to conduct electrical impulses. Neural tissue contains two basic cell types. **Neurons** are the cells that conduct impulses and transfer and process information. **Neuroglia**, or **glial cells**, protect, support, and repair neural tissue.

4.17 The response to tissue injury involves inflammation and regeneration p. 166

40. Restoration of homeostasis after an injury involves two processes: inflammation and regeneration.

41. **Inflammation** is indicated by swelling, redness, warmth, and pain. The inflammatory process removes damaged cells and tissue, and fights pathogens. **Regeneration** is the repair that occurs after the damaged tissue has been stabilized and the inflammation has subsided.

Chapter Review Questions

Labeling

Identify the three categories of cartilage shown below.

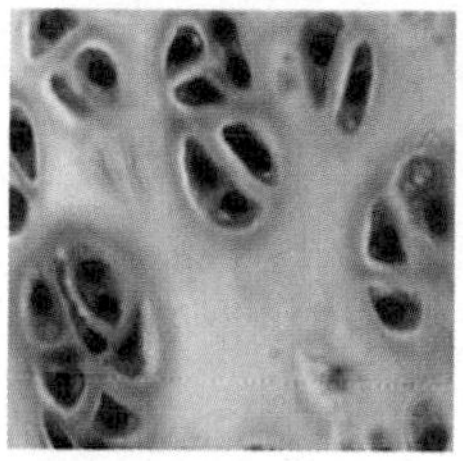

1

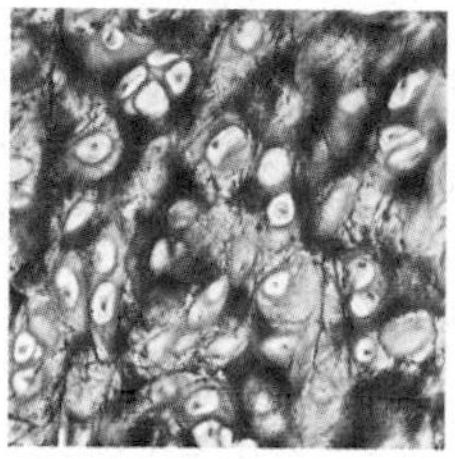

2

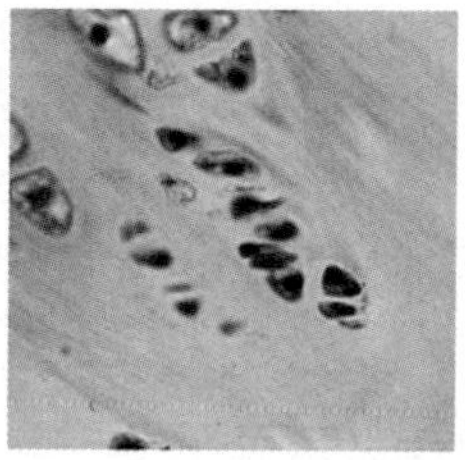

3

Identify the three categories of muscle shown below.

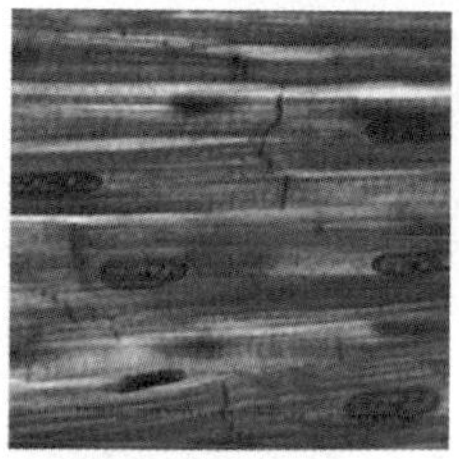

4

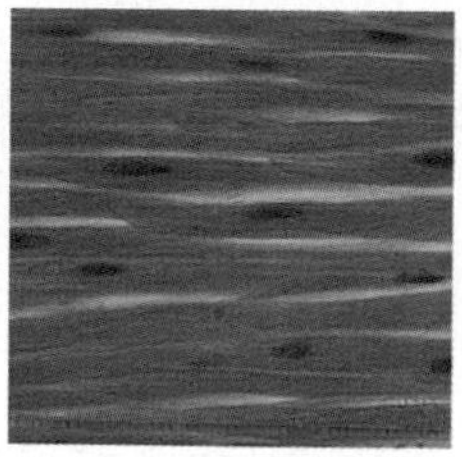

5

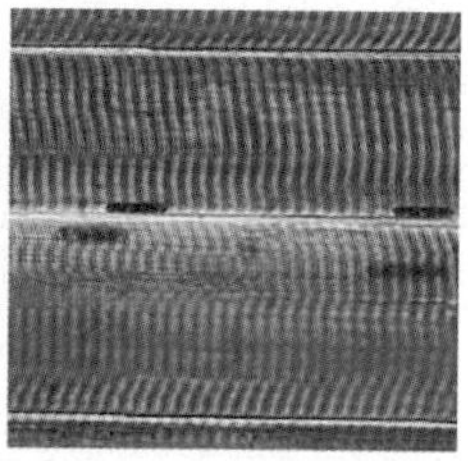

6

Identify the six connective tissue types shown below.

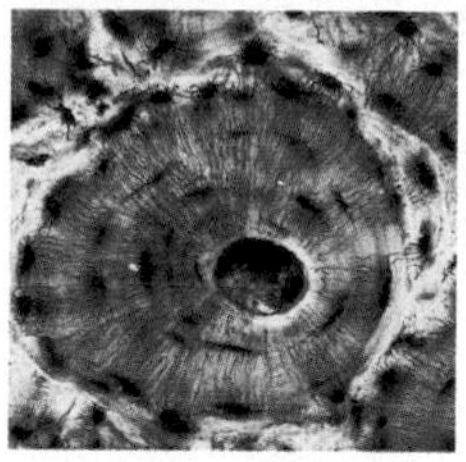

7

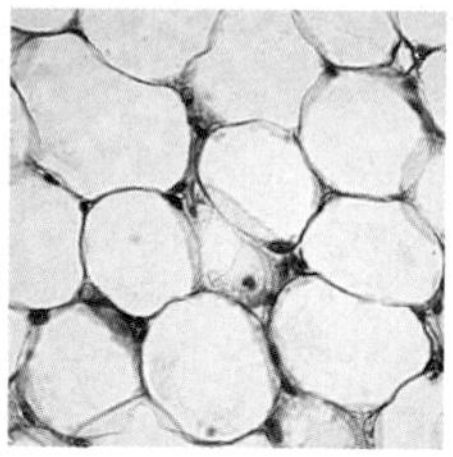

8

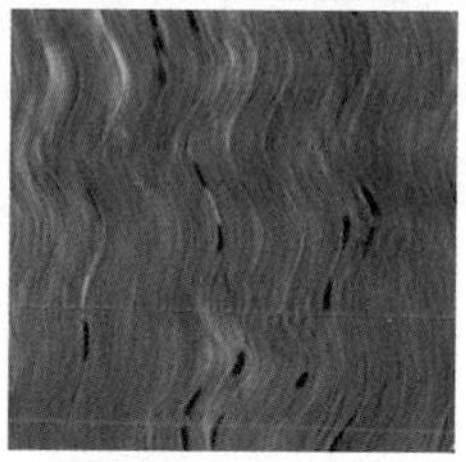

9

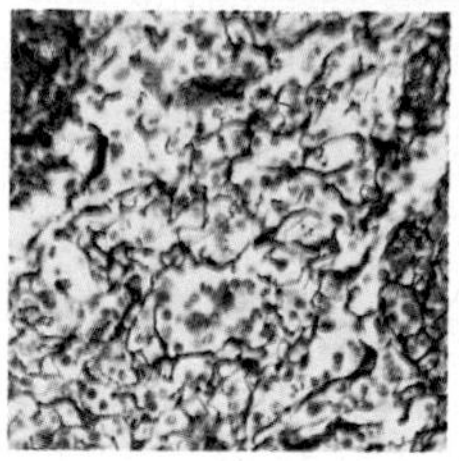

10

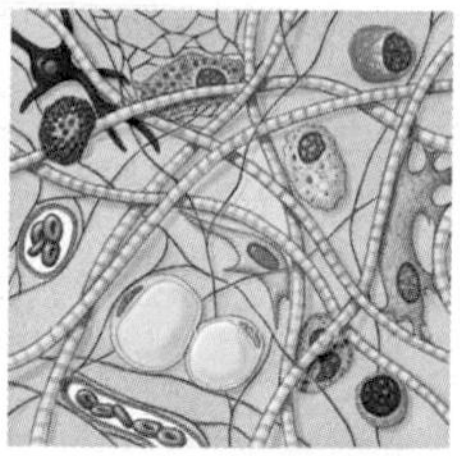

11

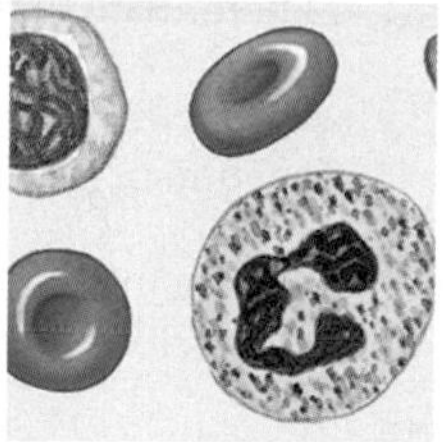

12

Multiple choice

Select the correct answer from the list provided.

13 Tissue that is specialized for contraction is
- ☐ a) epithelial tissue.
- ☐ b) connective tissue.
- ☐ c) muscle tissue.
- ☐ d) neural tissue.

14 The specialized intercellular junctions found in cardiac muscle are
- ☐ a) hemidesmosomes.
- ☐ b) intercalated discs.
- ☐ c) desmosomes.
- ☐ d) microvilli.

15 Collections of specialized cells and cell products that perform a limited number of functions are called
- ☐ a) cellular aggregates.
- ☐ b) organs.
- ☐ c) tissues.
- ☐ d) organ systems.

16 Axons, dendrites, and a cell body are characteristic of cells located in
- ☐ a) neural tissue.
- ☐ b) connective tissue.
- ☐ c) muscle tissue.
- ☐ d) epithelial tissue.

17 Specialized cells, extracellular protein fibers, and ground substance are characteristic of which type of tissue?
- ☐ a) muscle tissue
- ☐ b) epithelial tissue
- ☐ c) neural tissue
- ☐ d) connective tissue

18 Which of the following epithelia most easily permits diffusion?
- ☐ a) stratified squamous epithelium
- ☐ b) simple squamous epithelium
- ☐ c) transitional epithelium
- ☐ d) simple columnar epithelium

19 The tissue lining the peritoneal cavity is an example of a
- ☐ a) mucous membrane.
- ☐ b) serous membrane.
- ☐ c) cutaneous membrane.
- ☐ d) synovial membrane.

20 The addition of new layers of cartilage to the outer surface is which of the following types of growth?
- ☐ a) appositional
- ☐ b) interstitial

21 Which of the following is the most common form of connective tissue proper?
- ☐ a) reticular tissue
- ☐ b) areolar tissue
- ☐ c) adipose tissue
- ☐ d) elastic tissue

22 Which of the following membranes line movable joints?
- ☐ a) mucous membranes
- ☐ b) serous membranes
- ☐ c) cutaneous membranes
- ☐ d) synovial membranes

Short answer

23 Why is stratified squamous epithelium well suited to form the surface of the skin and the lining of the mouth and throat?

24 Differentiate between endocrine glands and exocrine glands.

25 Which two cell populations make up neural tissue? What is the function of each?

26 Why are infections always a serious threat after a severe skin burn or abrasion?

27 Describe the unique characteristics you would expect to see when examining cardiac muscle tissue under a light microscope.

28 During a lab practical, Jason examines a tissue that is composed of densely packed protein fibers that are running parallel. There are no striations, but small nuclei are visible. Jason identifies the tissue as skeletal muscle. Why is Jason's choice wrong, and which tissue is he probably observing?

29 While jogging through campus, Emily trips off a curb and falls onto the street. Her hands and exposed knees slide across the rough asphalt, resulting in deep abrasions. Outline the tissue injury response Emily's body will undergo.

Chapter Integration • Applying what you have learned

Some confusion in the histology lab

Alexis is a pre-nursing student enrolled in an anatomy and physiology class. Although she is familiar with basic microscope use, she is having difficulty identifying tissues in a histology lab. She has placed a slide labeled simple columnar epithelium onto the microscope stage, and has focused her microscope with the lowest objective. As a source of reference she opens her text, *Visual Anatomy & Physiology* to Module 4.7, and closely examines the light micrograph of simple columnar epithelium. To her dismay, the image in her microscope and the image in the text do not look at all alike.

1 **Explain the discrepancy.**

2 **If you were her lab partner, how would you help her?**

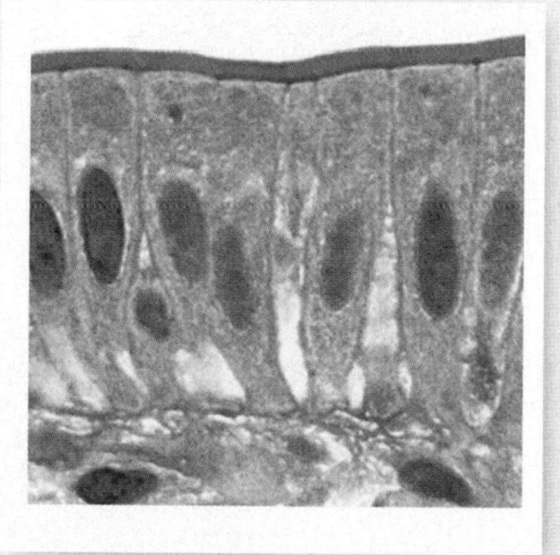

A 350× light micrograph of simple columnar epithelium as shown in Module 4.7.

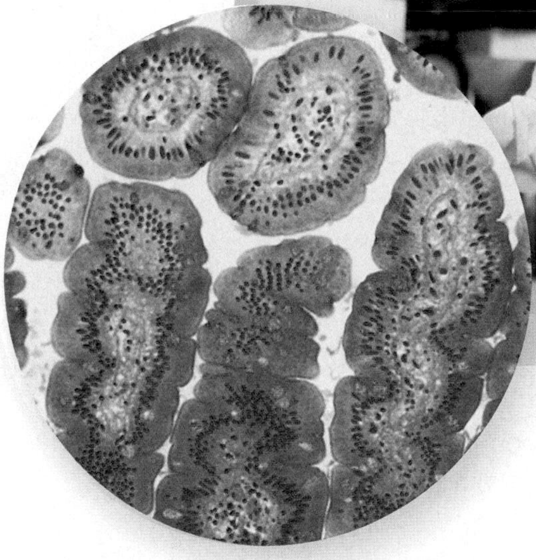

What Alexis is seeing through her microscope.

5 The Integumentary System

LEARNING OUTCOMES

These Learning Outcomes correspond by number to this chapter's modules and indicate what you should be able to do after completing the chapter.

SECTION 1 • Functional Anatomy of the Skin

5.1 Describe the tissue structure of the integument and the functions of the integumentary system.

5.2 Describe the main structural features of the epidermis, and explain the functional significance of each feature.

5.3 Explain what accounts for individual differences in skin color, and compare basal cell carcinoma with malignant melanoma.

5.4 Describe the structures and functions of the dermis and hypodermis.

5.5 Describe the classification of burns and the types of skin grafts.

SECTION 2 • Accessory Structures of the Skin

5.6 Describe the main functions of the accessory structures of the integumentary system.

5.7 Describe the mechanisms of hair production, and explain the structural basis for hair texture and color.

5.8 Describe the various kinds of exocrine glands in the skin, and discuss the secretions of each.

5.9 Describe the structure of a typical nail.

5.10 **CLINICAL MODULE** Summarize the effects of aging on the skin.

5.11 Describe the interaction between sunlight and endocrine functioning as they relate to the skin.

5.12 **CLINICAL MODULE** Explain how the skin responds to injury and is able to repair itself.

Learning Outcomes are repeated at the bottom of each module.

The integumentary system consists of the skin and various accessory structures

The **integumentary system** is the most accessible and often the least appreciated organ system. Often referred to simply as the skin or **integument** (in-TEG-ū-ment), this system makes up about 16 percent of your total body weight. Its surface, 1.5–2 m^2 in area, is constantly worn away, attacked by microorganisms, irradiated by sunlight, and exposed to environmental chemicals. The integumentary system is the place where you and the outside world meet—and your body's first line of defense against an often hostile environment.

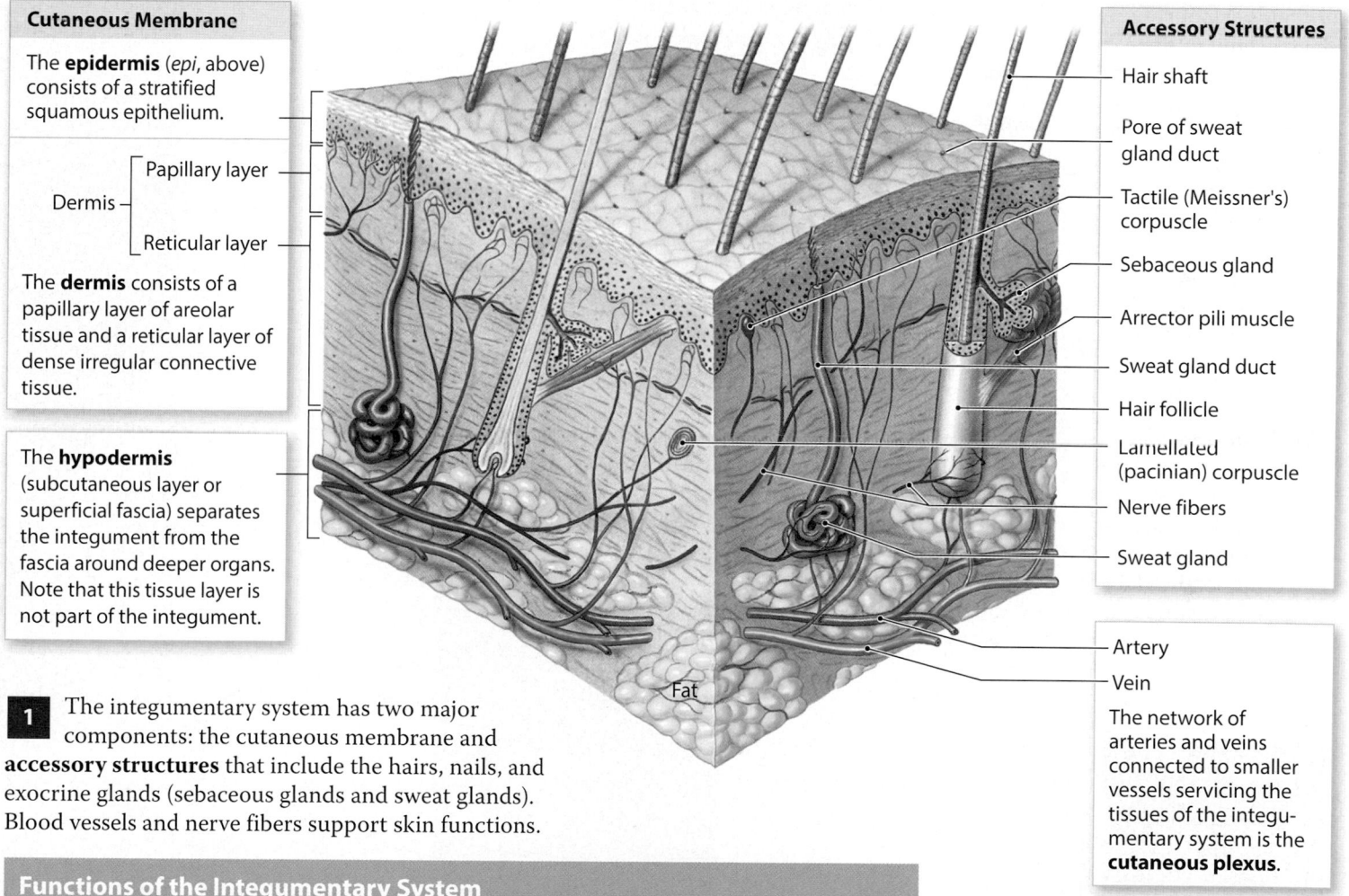

Cutaneous Membrane

The **epidermis** (*epi*, above) consists of a stratified squamous epithelium.

Dermis – Papillary layer; Reticular layer

The **dermis** consists of a papillary layer of areolar tissue and a reticular layer of dense irregular connective tissue.

The **hypodermis** (subcutaneous layer or superficial fascia) separates the integument from the fascia around deeper organs. Note that this tissue layer is not part of the integument.

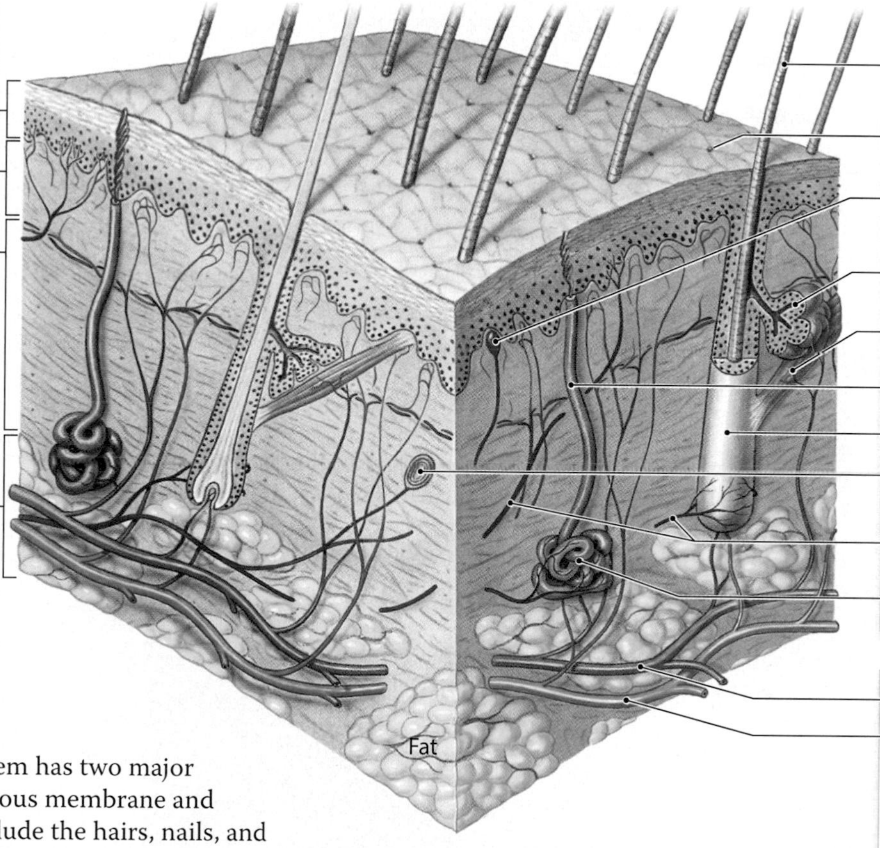
Fat

Accessory Structures

- Hair shaft
- Pore of sweat gland duct
- Tactile (Meissner's) corpuscle
- Sebaceous gland
- Arrector pili muscle
- Sweat gland duct
- Hair follicle
- Lamellated (pacinian) corpuscle
- Nerve fibers
- Sweat gland

- Artery
- Vein

The network of arteries and veins connected to smaller vessels servicing the tissues of the integumentary system is the **cutaneous plexus**.

1 The integumentary system has two major components: the cutaneous membrane and **accessory structures** that include the hairs, nails, and exocrine glands (sebaceous glands and sweat glands). Blood vessels and nerve fibers support skin functions.

Functions of the Integumentary System

- Protect underlying tissues and organs against impact, abrasion, fluid loss, and chemical attack
- Excrete salts, water, and organic wastes by integumentary glands
- Maintain normal body temperature through either insulation or evaporative cooling, as needed
- Produce melanin, which protects underlying tissue from ultraviolet radiation
- Produce keratin, which protects against abrasion and serves as a water repellent
- Synthesize vitamin D_3, a steroid that is subsequently converted to calcitriol, a hormone important to normal calcium metabolism
- Store lipids in adipocytes in the dermis and in adipose tissue in the hypodermis
- Detect touch, pressure, pain, and temperature stimuli, and relay that information to the nervous system

Module 5.1 Review

a. Identify the two major components of the cutaneous membrane.

b. List the various accessory structures of the integument.

c. Identify the major functions of the integumentary system.

LO **5.1** Describe the tissue structure of the integument and the functions of the integumentary system.

The epidermis is composed of strata (layers) that have various functions

The epidermis is dominated by **keratinocytes** (ke-RAT-i-nō-sīts), the body's most abundant epithelial cells. These cells form several layers, or **strata**. Keratinocytes are continuously produced by stem cell divisions in the deepest layers and are shed at the exposed surface.

1 The deeper layers of the epidermis form **epidermal ridges**, which extend into the dermis and are adjacent to dermal projections called **dermal papillae** (singular, *papilla*; small, nipple-like process) that project into the epidermis. These ridges and papillae are significant because they greatly increase the surface area for attachment, firmly binding the epidermis to the dermis.

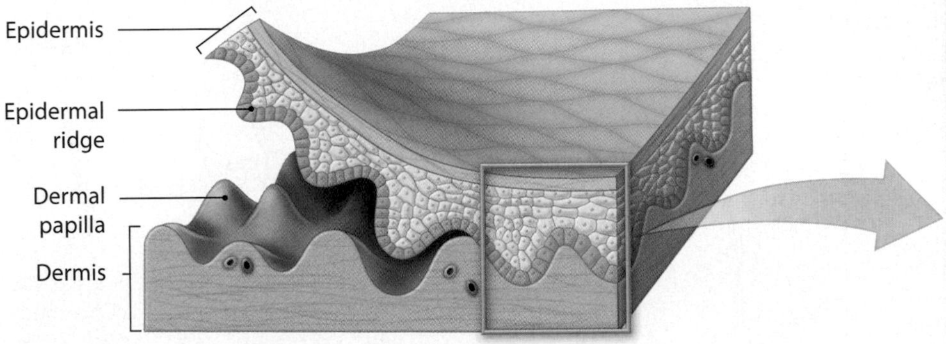
Epidermis
Epidermal ridge
Dermal papilla
Dermis

2 **Thin skin**, which covers most of the body surface, contains four strata. Thin skin is about as thick as the wall of a plastic sandwich bag (roughly 0.08 mm).

3 **Thick skin**, found on the palms of the hands and the soles of the feet, contains a fifth stratum. It is about as thick as a standard paper towel (roughly 0.5 mm). Note that the terms "thick" and "thin" refer to the relative thickness of just the epidermis, not to the cutaneous membrane as a whole.

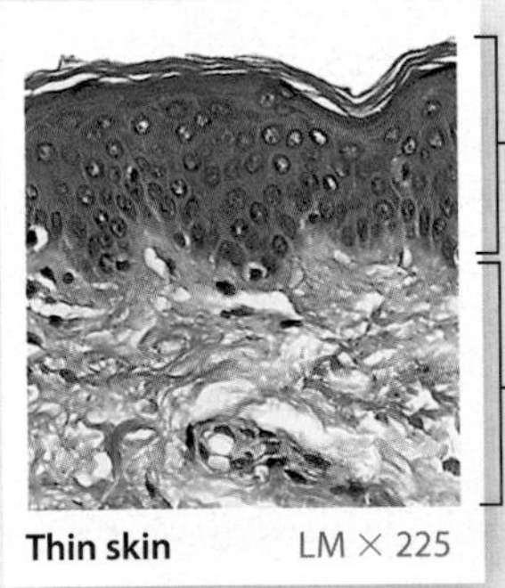
Epidermis
Dermis

Thin skin LM × 225

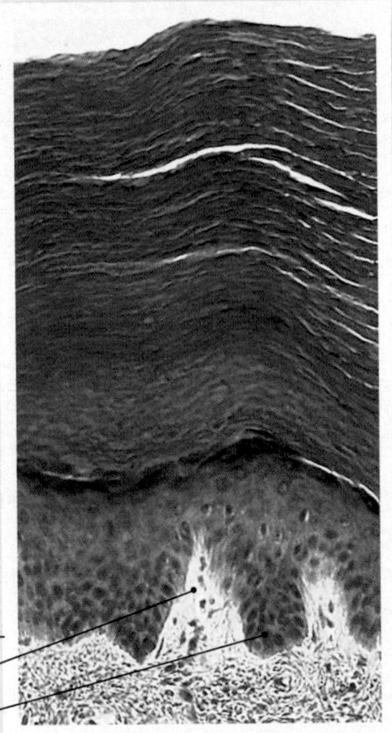
Dermal papilla
Epidermal ridge

LM × 225

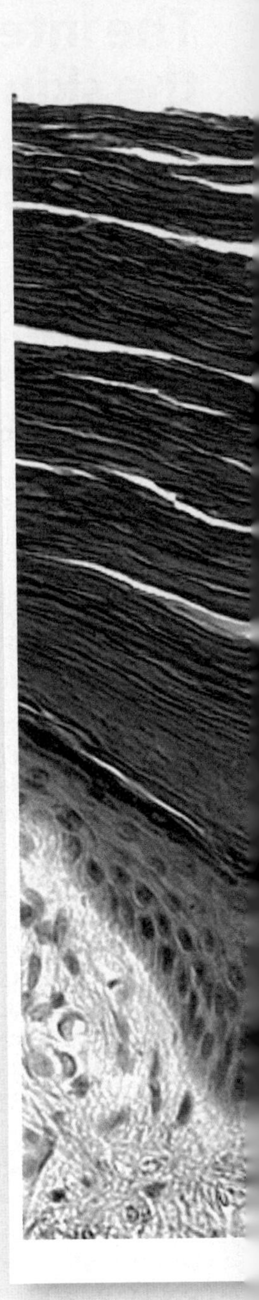

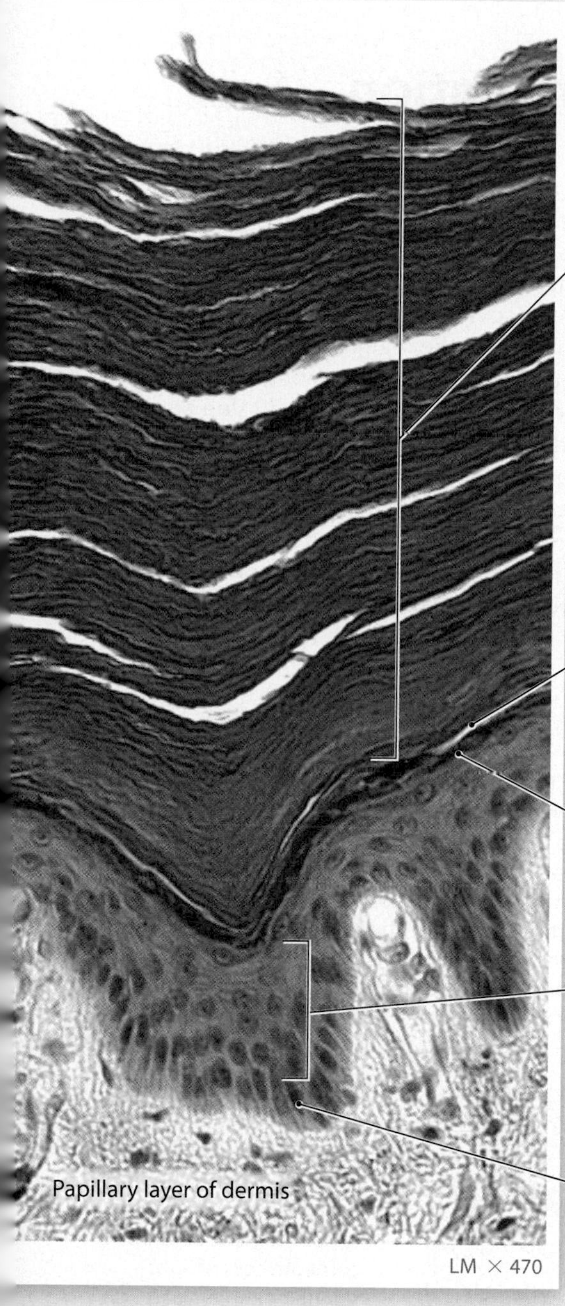

Papillary layer of dermis
LM × 470

4 Like all other epithelia, the epidermis lacks local blood vessels. Epidermal cells rely on the diffusion of nutrients and oxygen from capillaries within the dermis. As a result, the cells with the highest metabolic demand are closest to the underlying dermis.

Layers of the Epidermis

The **stratum corneum** (STRA-tum KOR-nē-um; *cornu*, horn) is at the exposed surface of both thick skin and thin skin. It normally contains 15 to 30 layers of keratinized cells. Keratinization is the formation of protective, superficial layers of cells filled with **keratin** (KER-a-tin; *keros*, horn). The dead cells in each layer of the stratum corneum remain tightly interconnected by desmosomes. It takes 7 to 10 days for a cell to move from the stratum basale to the stratum corneum. The dead cells generally remain in the exposed stratum corneum for an additional 2 weeks before they are shed or washed away. The stratum corneum is water resistant, but not waterproof. Water from interstitial fluid slowly penetrates to the surface and evaporates into the surrounding air. You lose about 500 mL (about 1 pt) of water in this way each day. The process is called **insensible perspiration**, because you are unable to see or feel (sense) the water loss. In contrast, you are usually very aware of the **sensible perspiration** produced by active sweat glands.

In the thick skin of the palms and soles, a **stratum lucidum** ("clear layer") separates the stratum corneum from deeper layers. The cells in this layer are flattened, densely packed, largely without organelles, and filled with the proteins keratin and **keratohyalin** (ker-a-tō-HĪ-a-lin). These proteins were synthesized while the cells were in the underlying stratum granulosum. By the time they reach the stratum lucidum, they are dead and undergoing dehydration.

The **stratum granulosum** ("grainy layer") consists of three to five layers of keratinocytes. By the time cells reach this layer, most have stopped dividing and have started making large amounts of keratin and keratohyalin. As these protein fibers develop, the cells grow thinner, and their membranes thicken and become less permeable.

The **stratum spinosum** ("spiny layer") consists of 8 to 10 layers of keratinocytes bound together by desmosomes. Its name refers to the fact that the cells look like miniature pincushions in standard histological sections. The stratum spinosum also contains **dendritic (Langerhans) cells**, which participate in the immune response by stimulating a defense against (1) microorganisms that manage to penetrate the superficial layers of the epidermis and (2) superficial skin cancers.

The **stratum basale** (STRA-tum buh-SAHL-āy) is the basal layer of the epidermis. Hemidesmosomes attach the cells of this layer to the basement membrane that separates the epidermis from the areolar tissue in the adjacent papillary layer of the dermis. **Basal cells** dominate the stratum basale. Basal cells are stem cells whose divisions replace the more superficial keratinocytes that are lost or shed at the epithelial surface. Merkel cells involved with touch sensations are scattered among the basal cells. A Merkel cell along with a sensory nerve terminal is called a tactile disc. We will discuss tactile discs and deeper-lying tactile corpuscles of the skin in Module 5.4.

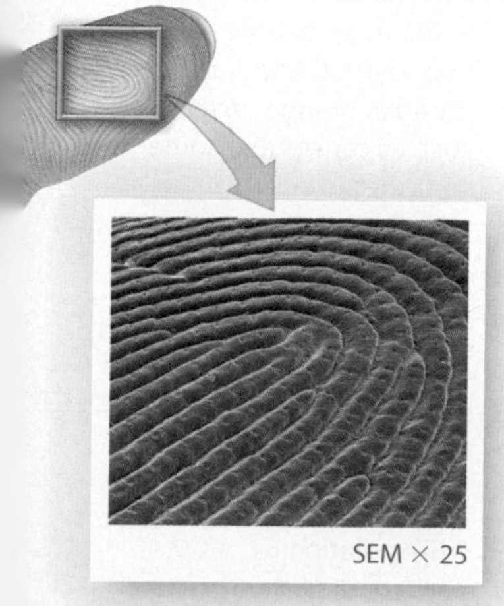

SEM × 25

5 This scanning electron micrograph of the tip of an index finger shows the ridge patterns in the thick skin that produce fingerprints. These ridge shapes are determined by genes and by the intrauterine environment during fetal development. Your epidermal ridge pattern is unique and does not change during your lifetime.

Module 5.2 Review

a. Identify the layers of the epidermis (from deep to superficial).

b. Dandruff is caused by excessive shedding of cells from the outer skin layers of the scalp. Thus, dandruff is composed of cells from which epidermal layer?

c. A splinter that penetrates to the third layer of the epidermis of the palm is lodged in which layer?

LO **5.2** Describe the main structural features of the epidermis, and explain the functional significance of each feature.

Module 5.3

Factors influencing skin color include epidermal pigmentation and dermal circulation

Skin color is influenced by the presence of pigments in the skin, the degree of dermal circulation, and the thickness and degree of keratinization in the epidermis.

1 The primary pigments involved in skin coloration are carotene and melanin. The micrograph at left shows a section through the thin skin of a light-skinned person. The ratio of melanocytes to basal cells ranges between 1:4 and 1:20, depending on the region of the body. The skin covering most areas of the body has about 1000 melanocytes per square millimeter. Differences in skin pigmentation among individuals do not reflect different numbers of melanocytes, but instead different levels of melanin synthesis. A deficiency or absence of melanin production leads to a disorder known as **albinism**. Individuals with this condition have a normal distribution of melanocytes, but the cells are incapable of producing melanin. The figure below shows a single melanocyte in the stratum basale.

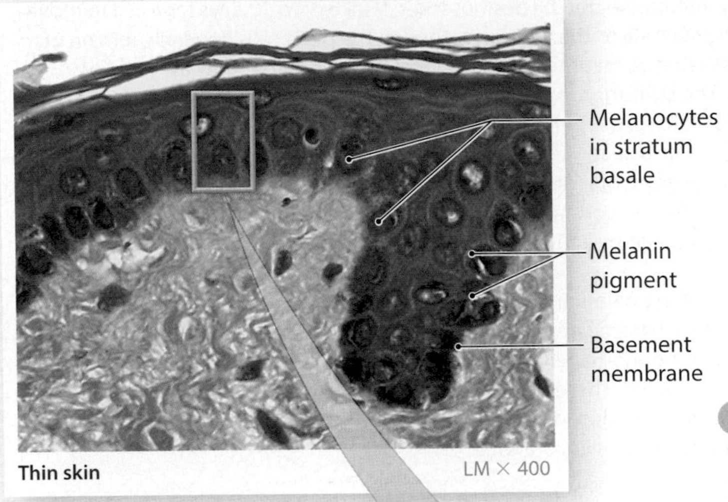
Melanocytes in stratum basale
Melanin pigment
Basement membrane
Thin skin
LM × 400

3 Melanosomes travel within the processes of melanocytes and are transferred intact to keratinocytes. The transfer of pigmentation colors the keratinocyte temporarily, until the melanosomes are destroyed by fusion with lysosomes. In people with pale skin, this transfer occurs in the stratum basale and stratum spinosum, and the cells of more superfical layers lose their pigmentation. In dark-skinned people, the melanosomes are larger, and the transfer may occur in the stratum granulosum as well. For this reason, skin pigmentation is darker and more persistent.

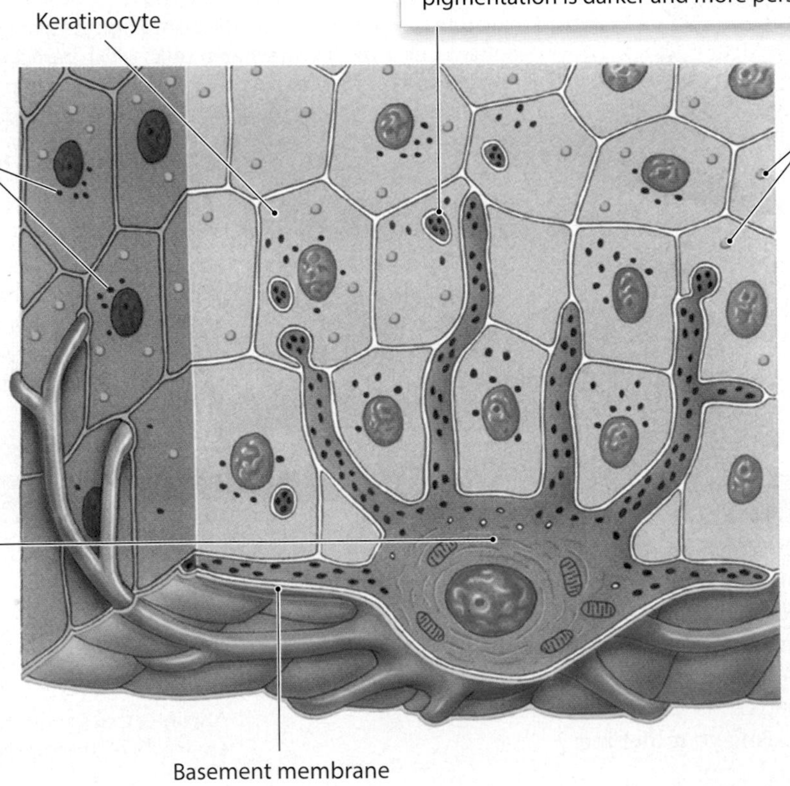
Keratinocyte
Basement membrane

1 **Melanin** is a brown, yellow-brown, or black pigment produced by **melanocytes**.

2 The melanocytes are located in the stratum basale, squeezed between or deep to the epithelial cells. Melanocytes manufacture melanin from the amino acid tyrosine and package it in intracellular vesicles called **melanosomes**.

4 **Carotene** (KAR-uh-tēn) is an orange-yellow pigment that normally accumulates in epidermal cells. It is most apparent in the stratum corneum cells of light-skinned people, but it also accumulates in fatty tissues in the deep dermis and hypodermis. Carotene is found in orange vegetables, such as carrots and some squashes.

5 The color of one's skin is genetically programmed. However, increased pigmentation, or tanning, can result in response to ultraviolet (UV) radiation.

2 The blood supply affects skin color because blood contains red blood cells filled with the red pigment hemoglobin. When bound to oxygen, hemoglobin is bright red, giving capillaries in the dermis a reddish tint that is most apparent in lightly pigmented people. If those vessels are dilated, the red tones become much more pronounced. For example, your skin becomes flushed and red when your body temperature rises because the superficial blood vessels dilate so that the skin can act like a radiator and lose heat. When its blood supply is temporarily reduced, the skin becomes pale. A light-skinned person who is frightened may "turn white" due to a sudden drop in blood supply to the skin. During a sustained reduction in blood flow, the oxygen levels in the tissues decline. Under these conditions, hemoglobin releases oxygen and turns a much darker red. Seen from the surface, the skin then takes on a bluish coloration called **cyanosis** (sī-uh-NŌ-sis; *kyanos*, blue). In people of any skin color, cyanosis is easiest to see in areas of very thin skin, such as the lips or beneath the nails.

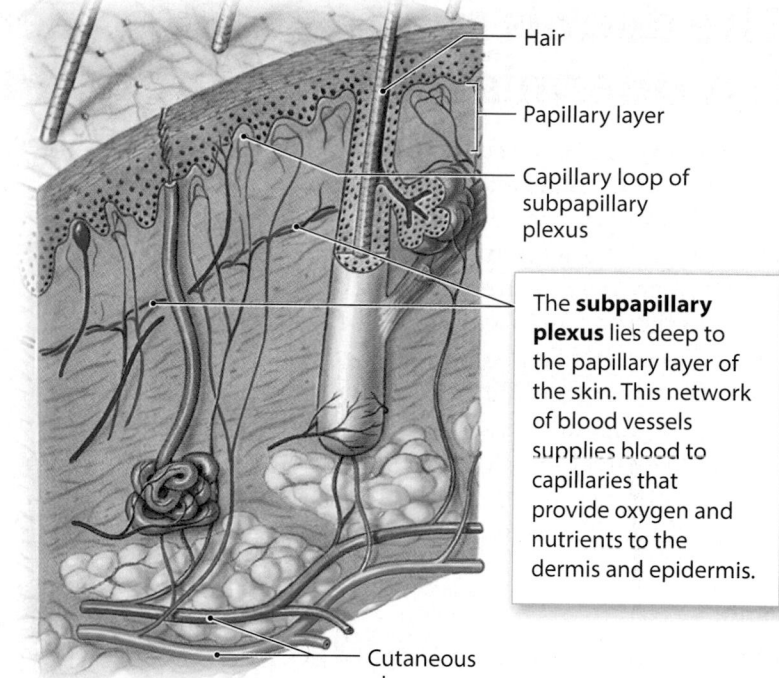

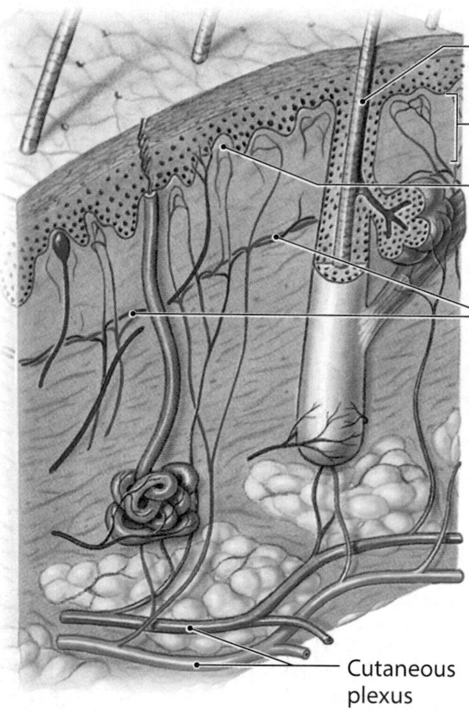
Hair
Papillary layer
Capillary loop of subpapillary plexus
The **subpapillary plexus** lies deep to the papillary layer of the skin. This network of blood vessels supplies blood to capillaries that provide oxygen and nutrients to the dermis and epidermis.
Cutaneous plexus

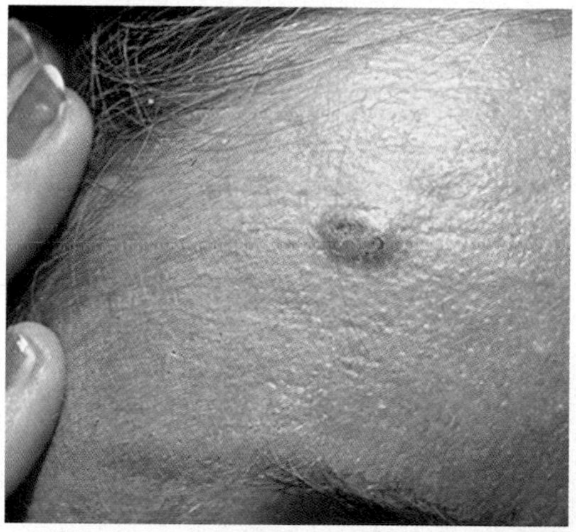

3 Skin cancers are the most common types of cancers. The most common form of skin cancer is a **basal cell carcinoma**. This cancer originates in the stratum basale, due to mutations caused by overexposure to the UV radiation in sunlight. Metastasis virtually never occurs in basal cell carcinomas, and most people survive these cancers. The melanin in keratinocytes provides some protection against the effects of UV radiation because the melanosomes are concentrated around the nucleus, where they act like sunshades for the enclosed DNA.

4 In contrast, a **malignant melanoma** (mel-a-NŌ-muh) is extremely dangerous. In this condition, cancerous melanocytes grow rapidly and metastasize through the lymphatic system. The outlook for long-term survival is in many cases determined by how early the condition is diagnosed. If the cancer is detected early, while it is still localized, the affected area can be surgically removed, and the 5-year survival rate is 99 percent. If the condition is not detected until extensive metastasis has occurred, the 5-year survival rate drops to 14 percent.

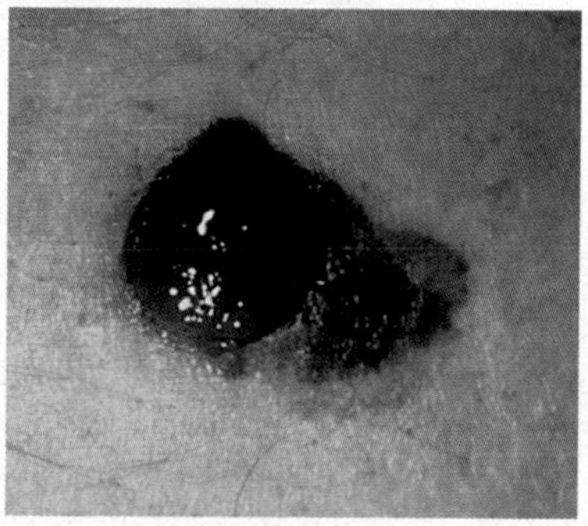

Module 5.3 Review

a. Name the two pigments contained in the epidermis.

b. Why does exposure to sunlight or sunlamps darken skin?

c. Why does the skin of a fair-skinned person appear red during exercise in hot weather?

LO **5.3** Explain what accounts for individual differences in skin color, and compare basal cell carcinoma with malignant melanoma.

The dermis supports the epidermis, and the hypodermis connects the dermis to the rest of the body

The **dermis** lies between the epidermis and the hypodermis. The presence of two types of fibers enables the dermis to tolerate limited stretching. Collagen fibers are very strong and resist stretching, but are easily bent or twisted. Elastic fibers permit stretching and then recoil to their original length. These elastic fibers allow flexibility, and the collagen fibers limit that flexibility to prevent tissue damage. Aging, hormonal changes, and the destructive effects of ultraviolet radiation permanently reduce the amount of elastin in the dermis. The results are wrinkles and sagging skin.

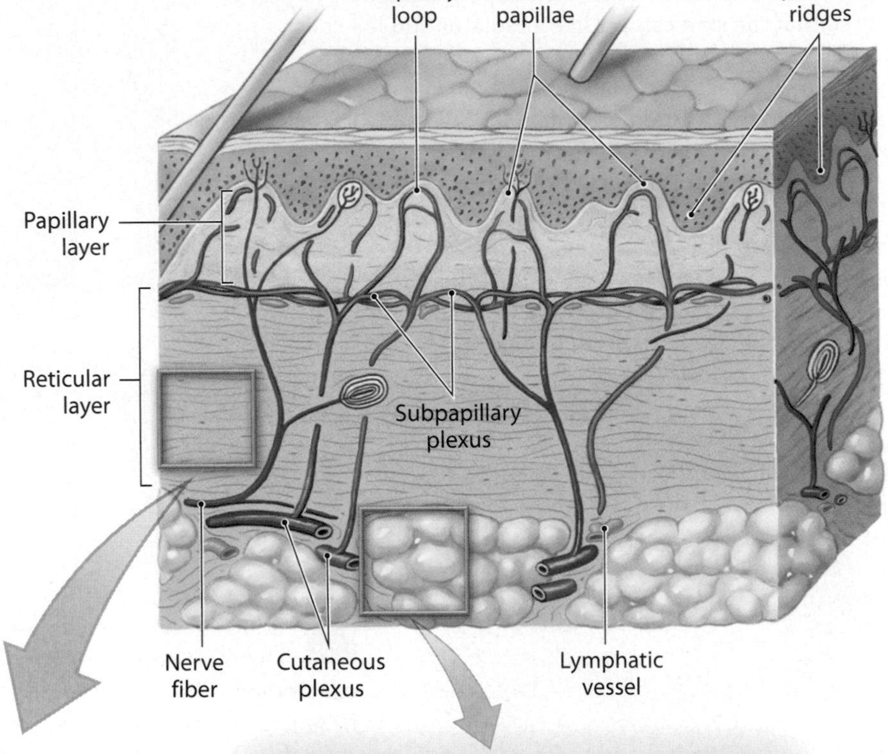
Capillary loop
Dermal papillae
Epidermal ridges
Papillary layer
Reticular layer
Subpapillary plexus
Nerve fiber
Cutaneous plexus
Lymphatic vessel

1 The **papillary layer** consists of a highly vascularized areolar tissue with all of the typical cell types within it. This layer also contains the capillaries, lymphatic vessels, and sensory neurons that supply the surface of the skin. The papillary layer gets its name from the dermal papillae that project between the epidermal ridges.

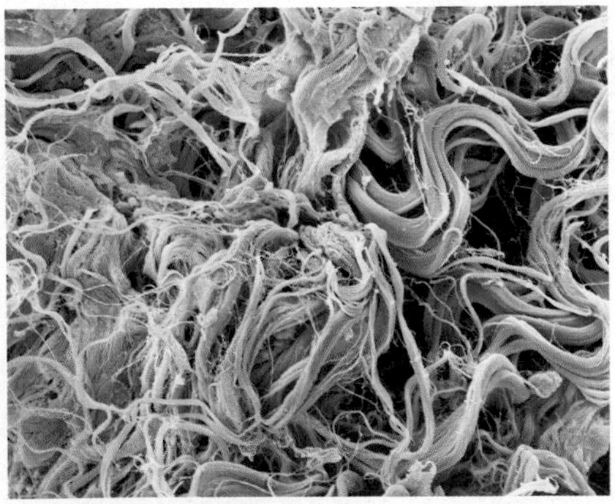

Reticular layer of dermis SEM × 1500

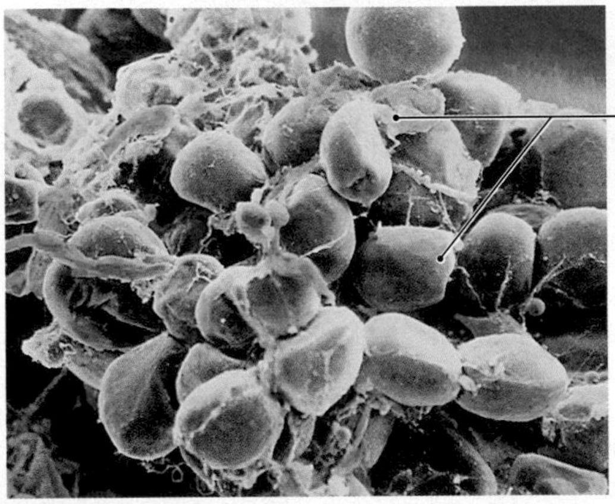
Adipocytes

Hypodermis SEM × 250

2 The **reticular layer** consists of an interwoven meshwork of dense irregular connective tissue containing both collagen and elastic fibers. Bundles of collagen fibers extend superficially beyond the reticular layer to blend into those of the papillary layer, so the boundary between the two layers is indistinct. Collagen fibers of the reticular layer also extend into the deeper hypodermis. In addition, the reticular and papillary layers of the dermis contain networks of blood vessels, lymphatic vessels, nerve fibers, and accessory organs such as hair follicles and sweat glands.

3 The **hypodermis** is not part of the skin. It separates the skin from deeper structures. The hypodermis stabilizes the skin's position relative to underlying tissues (such as skeletal muscles or other organs) while permitting independent movement. Because it is often dominated by adipose tissue, this layer is also an important energy storage site.

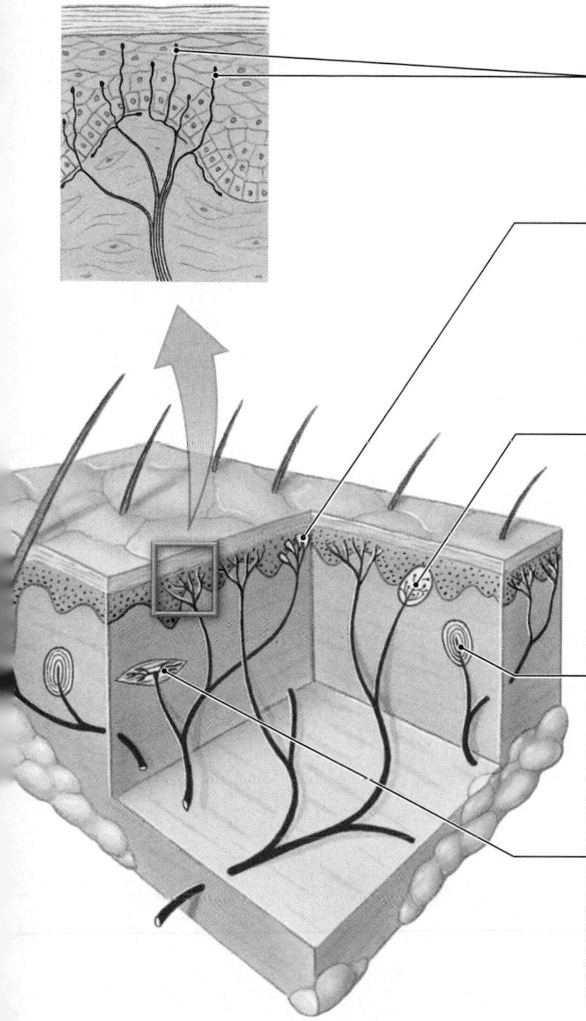

Touch Receptors of the Skin

Free nerve endings are sensitive to touch and pressure. They are found between epidermal cells.

Tactile discs detect sensations of texture and steady pressure. They are found in the deepest layer of the epidermis.

Meissner's (MĪS-nerz) **corpuscles** detect sensations of delicate touch ("fine touch"), pressure, and vibration. They are found in the papillary layer of the dermis.

Lamellated (pacinian) corpuscles are sensitive to deep pressure and vibration. They are found in both dermal layers and the hypodermis.

Ruffini corpuscles are sensitive to pressure and stretching of the skin. They are found in the reticular layer of the dermis.

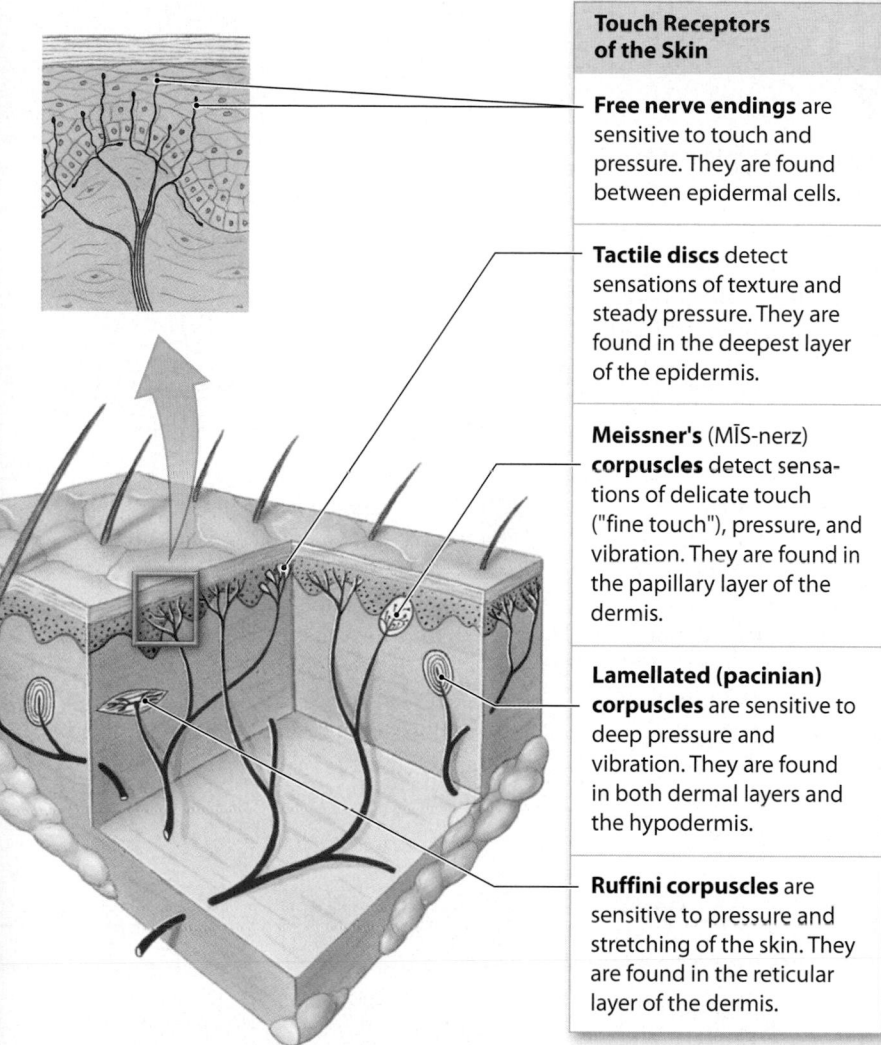

4 The integument contains many sensory receptors, and anything that comes in contact with the skin—from the lightest touch of a mosquito to the weight of a loaded backpack—initiates a nerve impulse that can reach our conscious awareness. Each square centimeter of skin contains approximately 400 centimeters of nerve fibers! We will focus on the main touch receptors found in the epidermis and dermis. Those found in the epidermis are free nerve endings and tactile discs. The dermis contains Meissner's corpuscles, lamellated (pacinian) corpuscles, and Ruffini corpuscles.

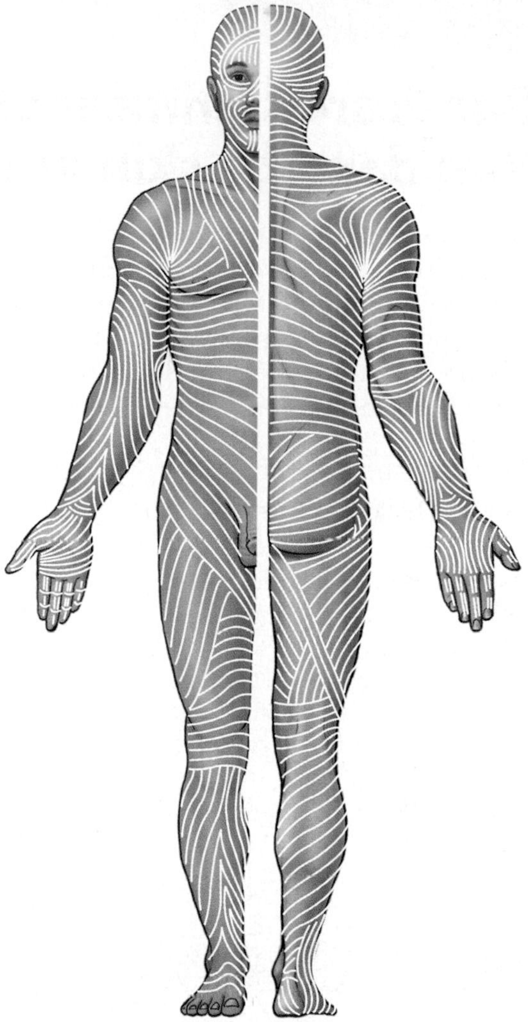
Anterior
Posterior

5 The banding shown here indicates the **lines of cleavage** in the skin. Cleavage lines are also called tension lines and Langer lines. Most of the collagen and elastic fibers at any location are arranged in parallel bundles oriented to resist the forces applied to the skin during normal movement. The resulting pattern of fiber bundles establishes lines of cleavage. These lines are clinically significant: A cut parallel to a cleavage line will usually remain closed and heal with little scarring. A cut at right angles to a cleavage line will be pulled open as severed elastic fibers recoil, resulting in greater scarring. For these reasons, surgeons choose to make neat incisions parallel to the lines of cleavage.

Beginning at puberty, men accumulate subcutaneous fat at the neck, on the arms, along the lower back, and over the buttocks. In contrast, women accumulate subcutaneous fat at the breasts, buttocks, hips, and thighs. In adults of either gender, the hypodermis of the backs of the hands and the upper surfaces of the feet contains few fat cells, whereas distressing amounts of adipose tissue can accumulate in the abdominal region, producing a prominent "potbelly."

Module 5.4 Review

a. Describe the location of the dermis.
b. What accounts for the ability of the dermis to undergo repeated stretching?
c. Where are the capillaries and sensory receptors that supply the epidermis located?

LO 5.4 Describe the structures and functions of the dermis and hypodermis.

Module 5.5

Burns are significant injuries that damage skin integrity

Burns are significant injuries because they can damage the integrity of large areas of the skin and compromise many essential functions. For example, dehydration and electrolyte imbalance can result, leading to kidney impairment and circulatory shock. Burns result from skin exposure to heat, friction, radiation, electrical shock, or strong chemical agents. The severity of the burn depends on the depth of penetration and the total area affected.

Partial-Thickness Burns

1 In a **first-degree burn**, only the surface of the epidermis is affected. In this type of burn, which includes most sunburns, the skin reddens and can be painful. The redness, a sign called **erythema** (er-i-THĒ-muh), results from inflammation of the sun-damaged tissues.

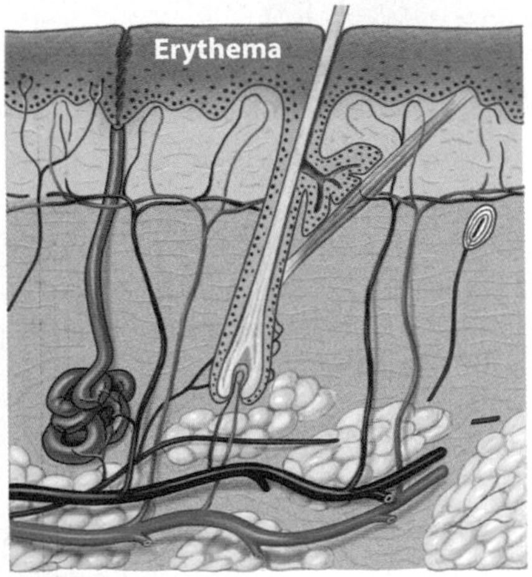
Erythema

2 In a **second-degree burn**, the entire epidermis and perhaps some of the dermis are damaged. Accessory structures such as hair follicles and glands are generally not affected, but blistering, pain, and swelling occur. If the blisters rupture at the surface, infection can easily develop. Healing typically takes 1 to 2 weeks, and some scar tissue may form. The gray shaded areas represent the depth of damaged skin.

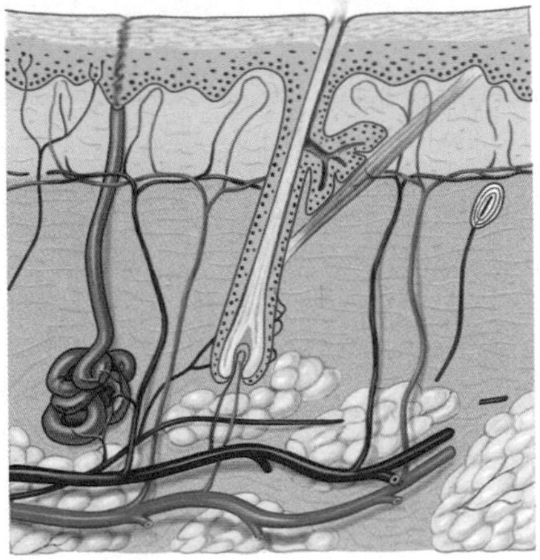

Full-Thickness Burns

3 **Third-degree burns**, or *full-thickness burns* destroy the epidermis and dermis, extending into the hypodermis. Despite swelling, these burns are less painful than second-degree burns, because sensory nerves are destroyed. Extensive third-degree burns cannot repair themselves, because granulation tissue cannot form and epithelial cells are unable to cover the injury. Skin grafting is usually necessary.

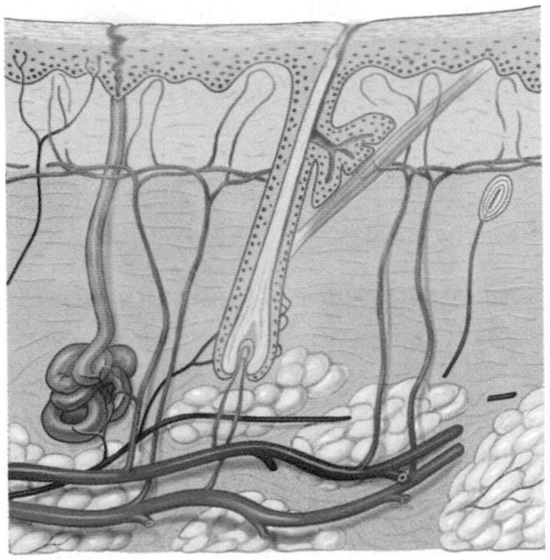

4 Burns that cover more than 20 percent of the skin surface threaten critical homeostatic functions of the skin.

Skin Functions Affected by Burns

- **Fluid and Electrolyte Balance**. Even areas with partial-thickness burns lose their effectiveness as barriers to fluid and electrolyte losses. In full-thickness burns, the rate of fluid loss through the skin may reach five times the normal level.
- **Thermoregulation**. Increased fluid loss means increased evaporative cooling. As a result, more energy must be expended to keep body temperature within acceptable limits.
- **Protection from Infection**. The dampness of the epidermal surface, resulting from uncontrolled fluid loss, encourages bacterial growth. If the skin is broken, infection is likely. Widespread bacterial infection, or **sepsis** (*sepsis*, rotting), is the leading cause of death in burn victims.

5 Each year in the United States, roughly 4000 people die in fires or die of burn injuries. In evaluating burns in a clinical setting, two key areas must be determined: the percentage of skin that has been burned and the depth of the burns.

The illustration to the right shows a simple method of estimating the percentage of the surface area affected by burns. This method is called the **rule of nines**, because the surface area in adults is divided into multiples of 9. The rule must be modified for children because their body proportions are different.

The depth of the burn can be quickly assessed with a pin. Because loss of sensation is characteristic of a full-thickness burn, the absence of a reaction to a pin prick indicates the presence of third-degree damage.

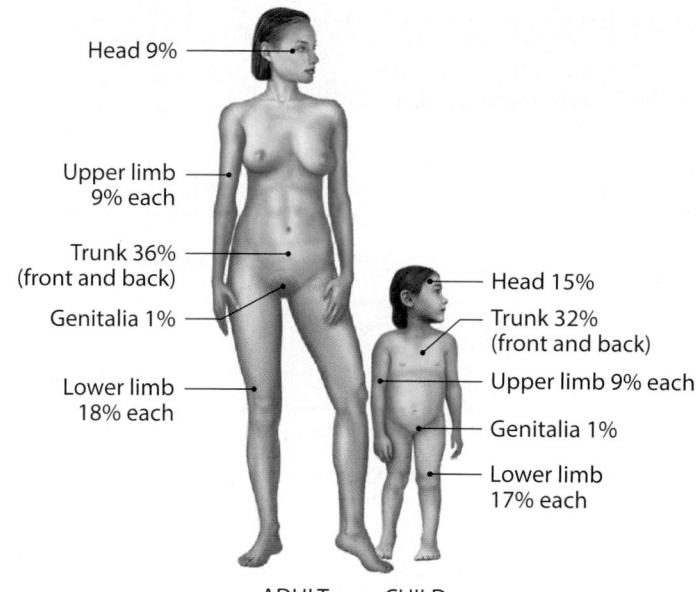

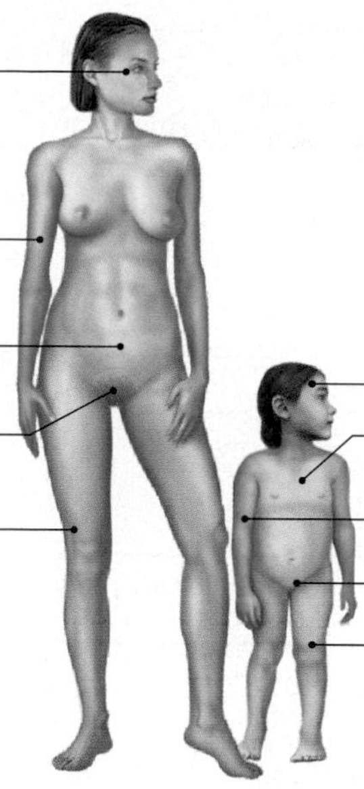
Head 9%
Upper limb 9% each
Trunk 36% (front and back)
Genitalia 1%
Lower limb 18% each
Head 15%
Trunk 32% (front and back)
Upper limb 9% each
Genitalia 1%
Lower limb 17% each
ADULT
CHILD

6 Because of the loss of critical homeostatic functions, emergency treatment of burns focuses on the following procedures:

- Replacing lost fluids and electrolytes.
- Providing sufficient nutrients to meet increased metabolic demands for thermoregulation and healing.
- Preventing infection by cleaning and covering the burn while administering antibiotic drugs.
- Assisting tissue repair. Because large full-thickness burns cannot heal unaided, surgical procedures are necessary to encourage healing. In a **skin graft**, areas of intact skin are transplanted to cover the site of the burn. A split-thickness graft involves a transfer of the epidermis and superficial portions of the dermis. A full-thickness graft involves the epidermis and both layers of the dermis. A skin graft made with a patient's own undamaged skin is called an **autograft**. If enough undamaged skin is lacking, then an **allograft** (frozen skin from a cadaver) or **xenograft** (animal skin) may be necessary. Due to rapid rejection by the immune system, both allografts and xenografts need to be replaced by an autograft.

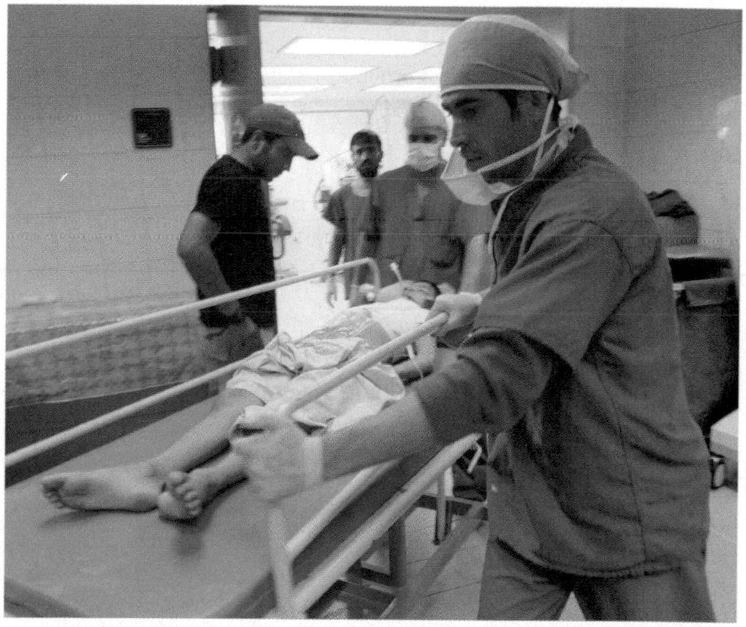

With fluid-replacement therapies, infection control methods, and grafting techniques, young patients with burns over 80 percent of the body have about a 50 percent chance of recovery. Recent advances in cell culturing may improve survival rates. After a square centimeter or so of undamaged epidermis is removed and grown in the laboratory, basal cell divisions produce large sheets of epidermal cells—up to several square meters in area—that can be transplanted to cover the burn area. Although questions remain about the strength and flexibility of the repairs, skin cultivation is a breakthrough in the treatment of serious burns.

Module 5.5 Review

a. Distinguish between a first-degree, second-degree, and third-degree burn.

b. Which type of burn usually requires skin grafting? Why?

c. Describe the three types of skin grafts. Which one is best? Why?

LO 5.5 Describe the classification of burns and the types of skin grafts.

Concept map

Use each of the following terms once to fill in the blank boxes to correctly complete the map.

- accessory structures
- collagen
- connective
- epidermis
- fat
- granulosum
- hypodermis
- nerves
- papillary layer
- reticular layer

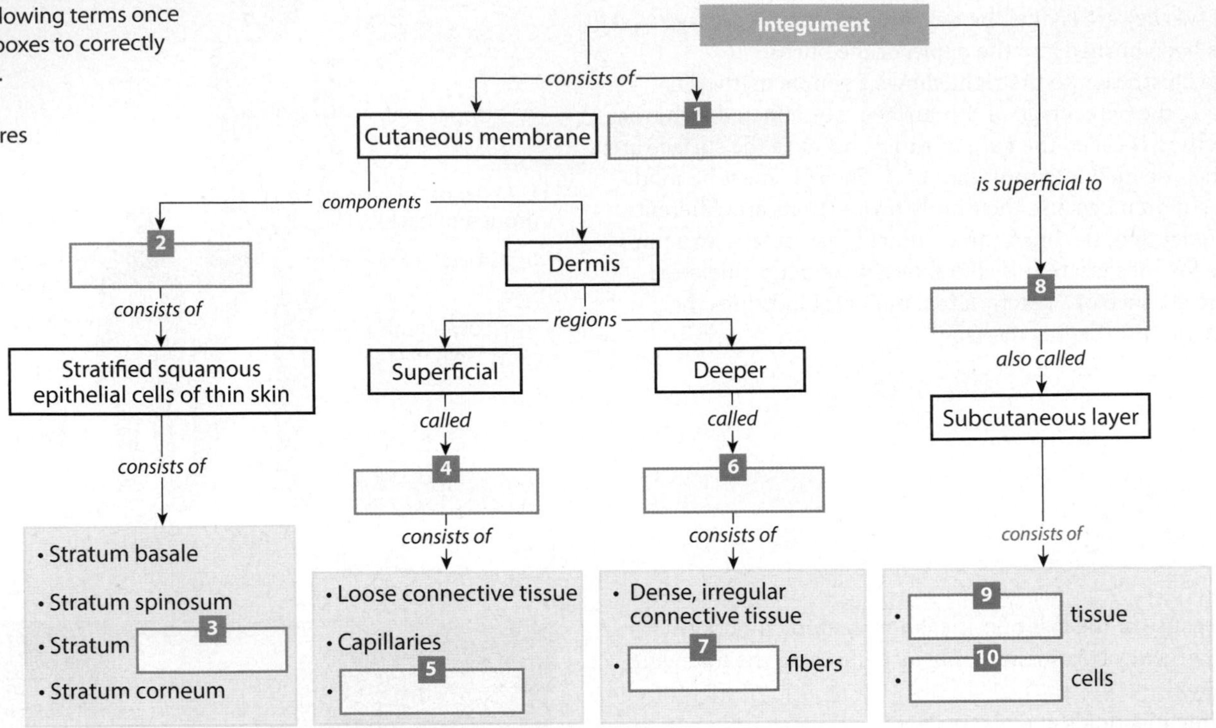

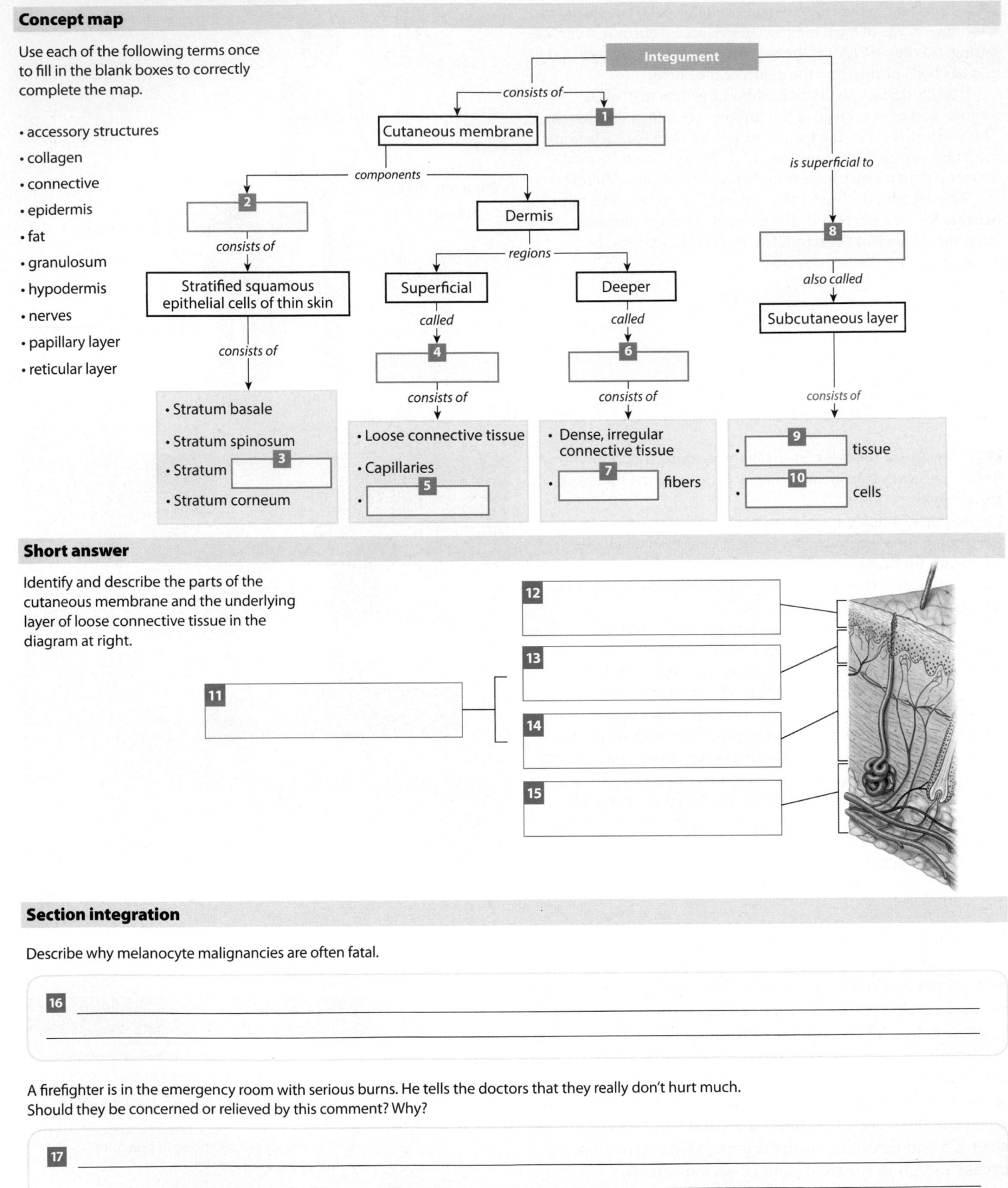

Integument
consists of
Cutaneous membrane
1
is superficial to
components
2
Dermis
8
consists of
regions
Stratified squamous epithelial cells of thin skin
Superficial
Deeper
also called
Subcutaneous layer
called
called
consists of
4
6
consists of
consists of
consists of
• Stratum basale
• Stratum spinosum
• Stratum 3
• Stratum corneum
• Loose connective tissue
• Capillaries
• 5
• Dense, irregular connective tissue
• 7 fibers
• 9 tissue
• 10 cells

Short answer

Identify and describe the parts of the cutaneous membrane and the underlying layer of loose connective tissue in the diagram at right.

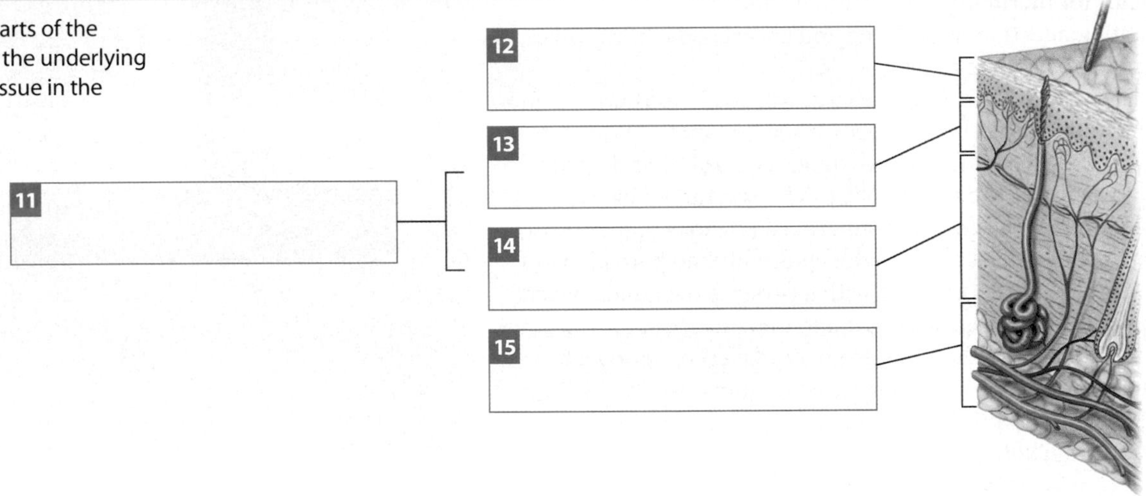

11
12
13
14
15

Section integration

Describe why melanocyte malignancies are often fatal.

16 ____

A firefighter is in the emergency room with serious burns. He tells the doctors that they really don't hurt much. Should they be concerned or relieved by this comment? Why?

17 ____

Hair follicles, exocrine glands, and nails are also components of the integumentary system

Hair and several other structures—hair follicles, exocrine glands (sebaceous and sweat glands), and nails—are considered **accessory structures** of the integument. During embryonic development, these structures originate from the epidermis, so they are also known as *epidermal derivatives*. Although located in the dermis, they project through the epidermis to the surface. The accessory structures of the integumentary system and their functions are shown below.

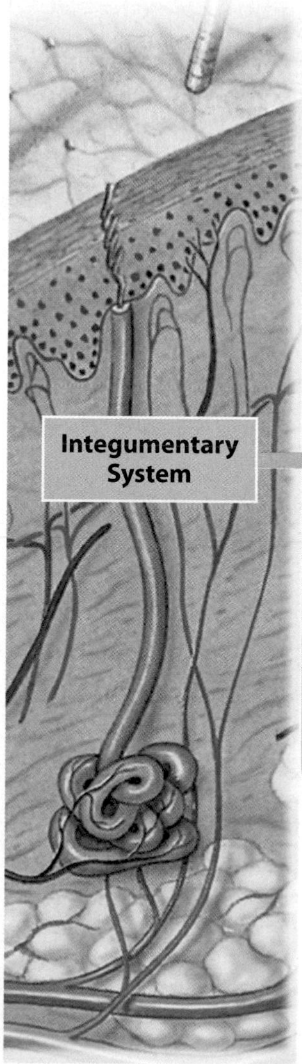

Integumentary System

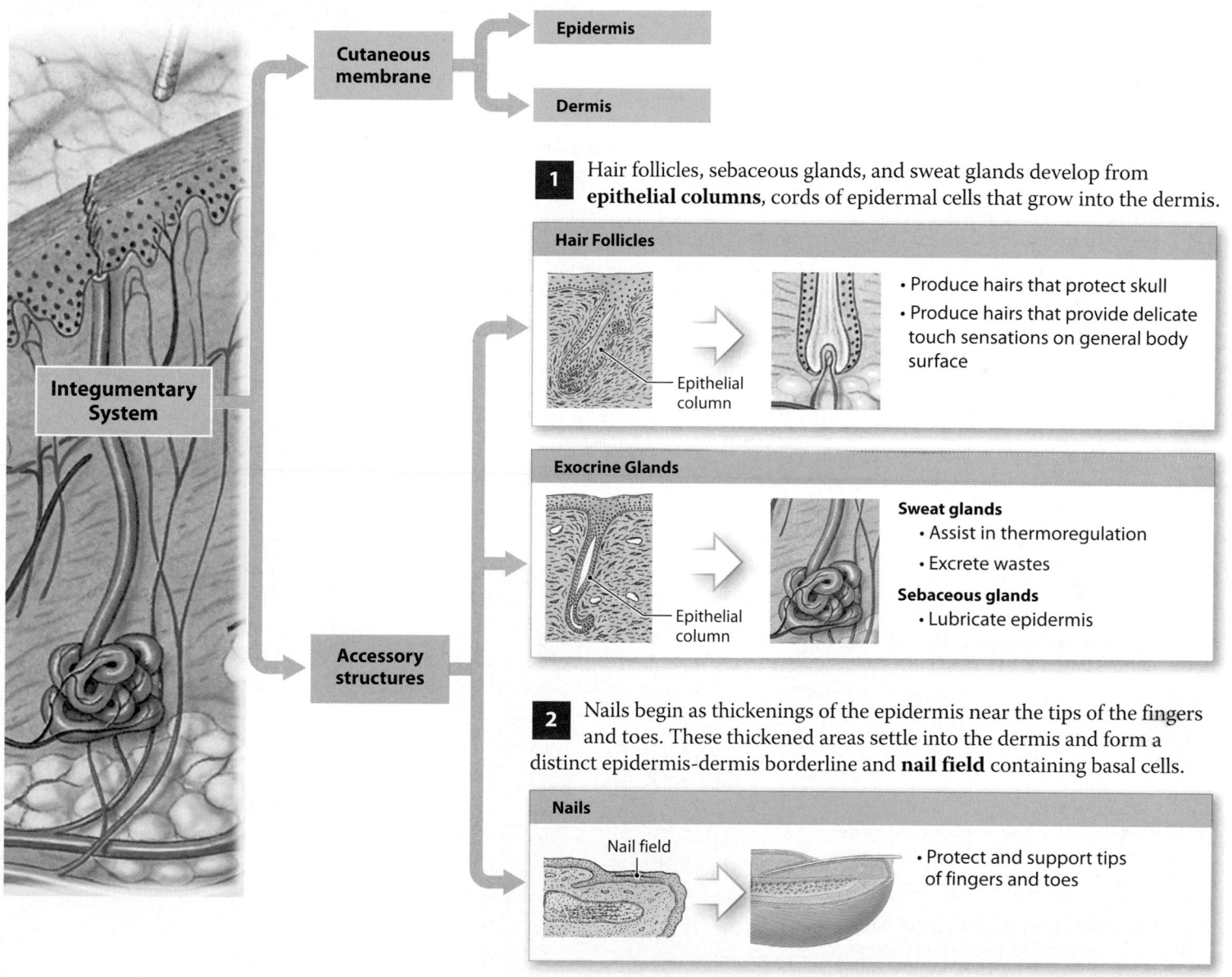

- **Cutaneous membrane**
  - **Epidermis**
  - **Dermis**
- **Accessory structures**

1 Hair follicles, sebaceous glands, and sweat glands develop from **epithelial columns**, cords of epidermal cells that grow into the dermis.

Hair Follicles

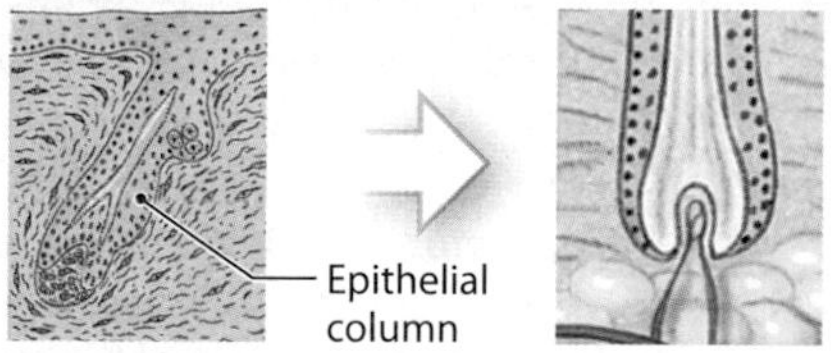
Epithelial column

- Produce hairs that protect skull
- Produce hairs that provide delicate touch sensations on general body surface

Exocrine Glands

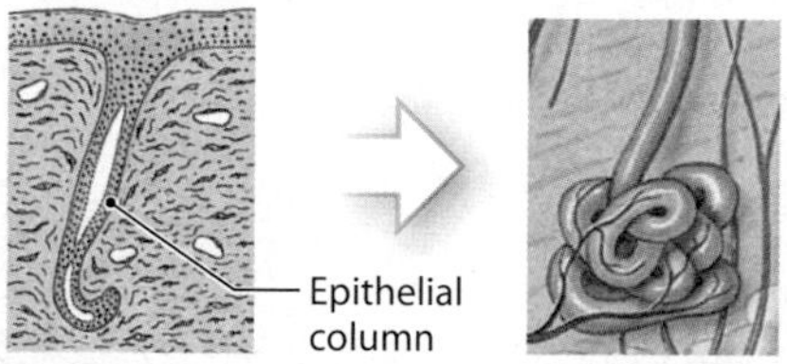
Epithelial column

Sweat glands
- Assist in thermoregulation
- Excrete wastes

Sebaceous glands
- Lubricate epidermis

2 Nails begin as thickenings of the epidermis near the tips of the fingers and toes. These thickened areas settle into the dermis and form a distinct epidermis-dermis borderline and **nail field** containing basal cells.

Nails

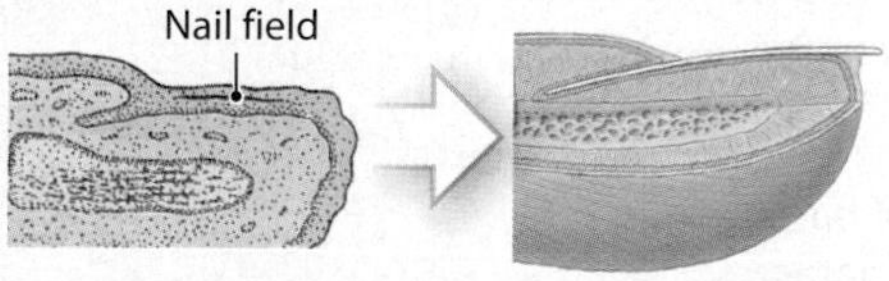
Nail field

- Protect and support tips of fingers and toes

Module 5.6 Review

a. What are epidermal derivatives?

b. What are the two examples of exocrine glands in the integument?

c. Describe the functions of the integument's accessory structures.

Hair is composed of dead, keratinized cells produced in a specialized hair follicle

Hairs project above the surface of the skin almost everywhere, except over the sides and soles of the feet, the palms of the hands, the sides of the fingers and toes, the lips, and portions of the external genitalia. The human body has about 2.5 million hairs, and 75 percent of them are on the general body surface, not on the head. Hairs are nonliving structures produced in organs called hair follicles.

1 A **hair follicle** is a complex structure composed of epithelial and connective tissues that forms a single hair. **Terminal hairs** are large, coarse, often darkly pigmented hairs such as those found on the scalp or the armpits. **Vellus hairs** are smaller, shorter, and more delicate hairs found on the general body surface.

A **sebaceous gland** produces secretions that coat the hair and the adjacent skin surface.

The **hair shaft**, which we see on the surface, begins deep within the hair follicle.

The **hair root**—the portion that anchors the hair into the skin—extends from the base of the hair follicle, where hair production begins, to the point where the hair shaft loses its connection with the walls of the follicle.

A connective tissue sheath surrounds the epithelial cells of the hair follicle.

A **root hair plexus** of sensory nerves surrounds the base of each hair follicle. As a result, you can feel the movement of the shaft of even a single hair.

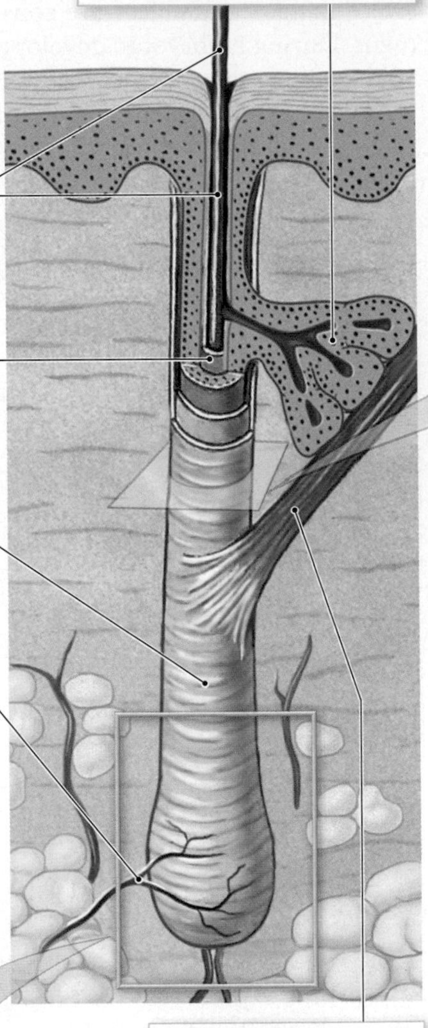

The **arrector pili** (a-REK-tor PĪ-lē; plural, arrectores pilorum) is a smooth muscle whose contraction pulls on the follicle, forcing the hair to stand erect.

2 Hair formation begins at the expanded base of a hair follicle called the **hair bulb**. The mass of epithelial cells making up the hair bulb forms a cap that surrounds a small hair papilla, a peg of connective tissue containing capillaries and nerves.

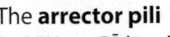

The **cuticle** of the hair consists of daughter cells produced at the edges of the hair matrix. The cuticle forms the surface of the hair.

The **cortex** of the hair is an intermediate layer of daughter cells deep to the cuticle.

The **medulla** of the hair consists of daughter cells formed at the center of the hair matrix.

The **hair matrix** consists of a layer of basal cells at the base of the hair bulb in contact with the hair papilla. These actively dividing cells produce the hair. As these cells divide, daughter cells are gradually pushed toward the surface. Melanocytes are also scattered among the cells of the hair matrix.

The **hair papilla** is a small connective tissue peg filled with blood vessels and nerves.

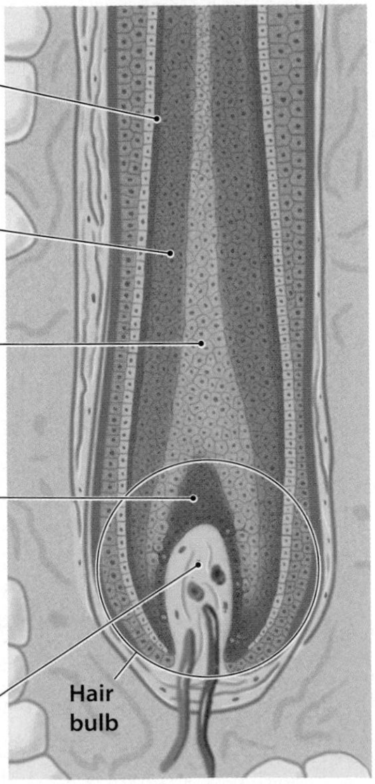
Hair bulb

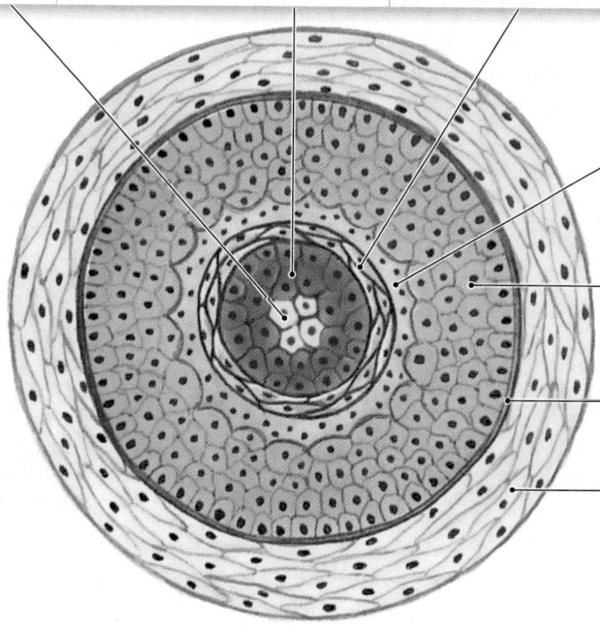
Hair Structure
The medulla, or core, of the hair contains a flexible **soft keratin**.
The cortex contains thick layers of **hard keratin**, which give the hair its stiffness.
The cuticle, although thin, is very tough and contains hard keratin.
Hair Follicle Structure
The **internal root sheath** surrounds the hair root and the deeper portion of the shaft. It is also produced from the hair matrix. The cells of this sheath disintegrate quickly, and this layer does not extend the entire length of the hair follicle.
The **external root sheath** consists of epithelial cells and extends from the skin surface to the hair matrix.
The **glassy membrane** is a thickened, clear basement membrane.
Connective tissue sheath

3 A cross section through a hair follicle reveals the internal structure of both the hair and the follicle.

4 Hairs grow and are shed according to a **hair growth cycle**. A hair in the scalp grows for 2 to 5 years, at a rate of about 0.33 mm per day. Variations in the growth rate and in the duration of the hair growth cycle account for individual differences in the length of uncut hair.

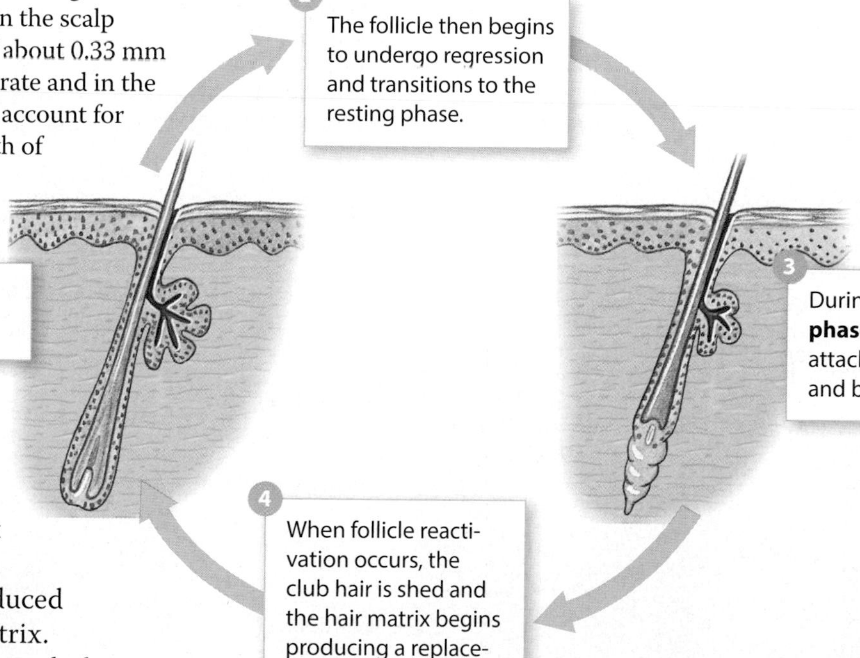
1
During the active phase the hair grows continuously.
2
The follicle then begins to undergo regression and transitions to the resting phase.
3
During the **resting phase** the hair loses its attachment to the follicle and becomes a **club hair**.
4
When follicle reactivation occurs, the club hair is shed and the hair matrix begins producing a replacement hair.

Variations in hair color reflect differences in structure and variations in the pigment produced by melanocytes at the hair matrix. Different forms of melanin give a dark brown, yellow-brown, or red color to the hair. As pigment production decreases with age, hair color lightens. Genes determine these structural and biochemical characteristics, but hormonal and environmental factors also influence your hair's condition. White hair results from the combination of a lack of pigment and the presence of air bubbles in the medulla of the hair shaft.

Module 5.7 Review

a. Describe a typical strand of hair.

b. What happens when an arrector pili muscle contracts?

c. Why is pulling a hair painful, yet cutting a hair is not?

LO **5.7** Describe the mechanisms of hair production, and explain the structural basis for hair texture and color.

Sebaceous glands and sweat glands are exocrine glands in the skin

In this module we examine the structure and function of the accessory structures that produce exocrine secretions, with particular attention to sebaceous glands and sweat glands. These exocrine glands assist in thermoregulation, excrete wastes, and lubricate the epidermis.

1 **Sebaceous** (se-BĀ-shus) **glands**, or oil glands, are holocrine glands that discharge an oily lipid secretion into hair follicles. Sebaceous glands that communicate with a single follicle share a duct and thus are classified as simple branched alveolar glands. The lipids released from gland cells enter the lumen (open passageway) of the gland. The arrector pili muscle contracts, squeezing the sebaceous gland and forcing its secretory product into the hair follicle and eventually onto the skin surface.

Sebaceous Glands

Sebaceous glands secrete **sebum** (SĒ-bum), which coats the hair shaft and surrounding epidermal surfaces. Sebum is a mixture of triglycerides, cholesterol, proteins, and electrolytes. Sebum lubricates and keeps the hair shaft from becoming dry and brittle. It also moisturizes the surrounding skin and inhibits the growth of bacteria.

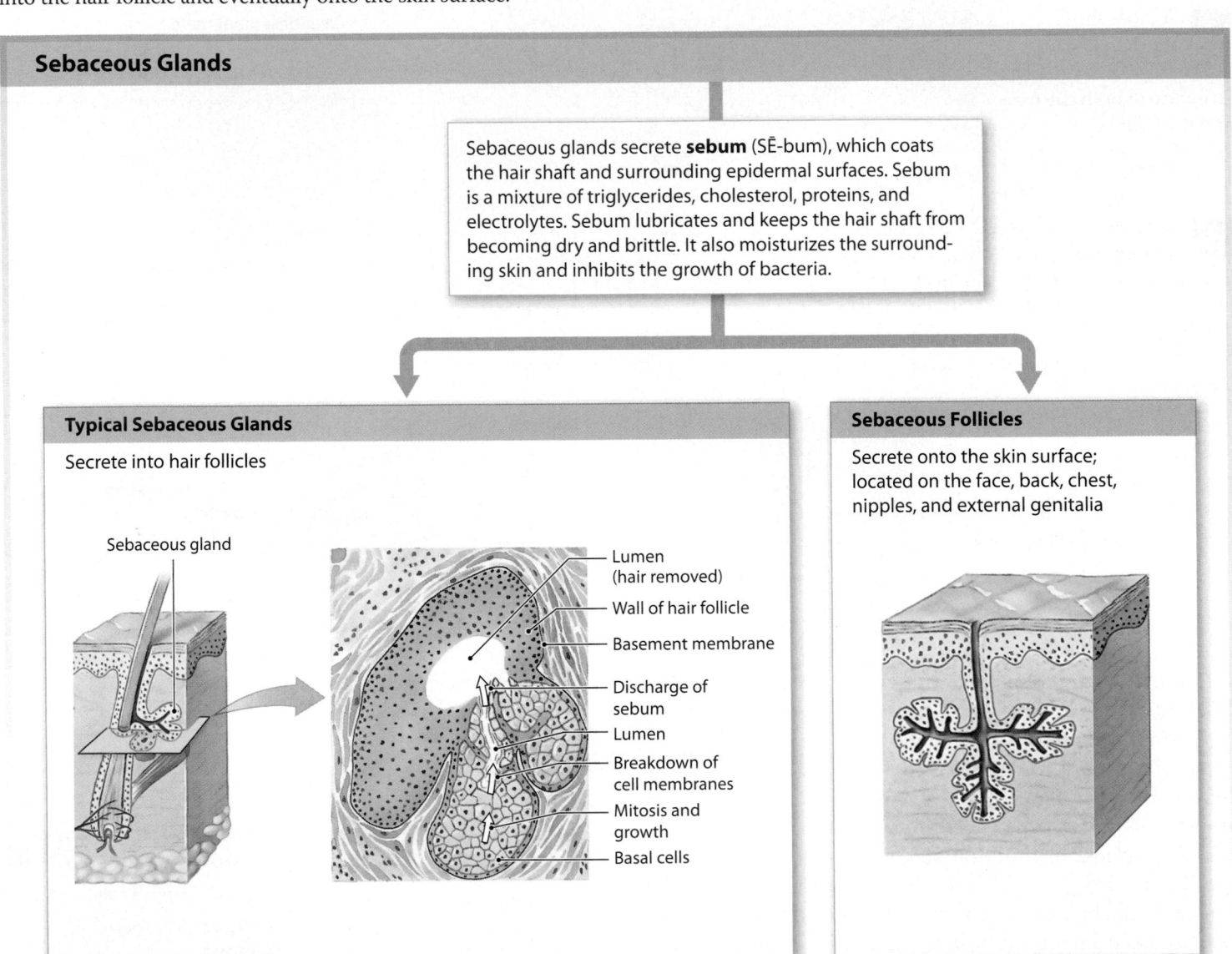

Typical Sebaceous Glands

Secrete into hair follicles

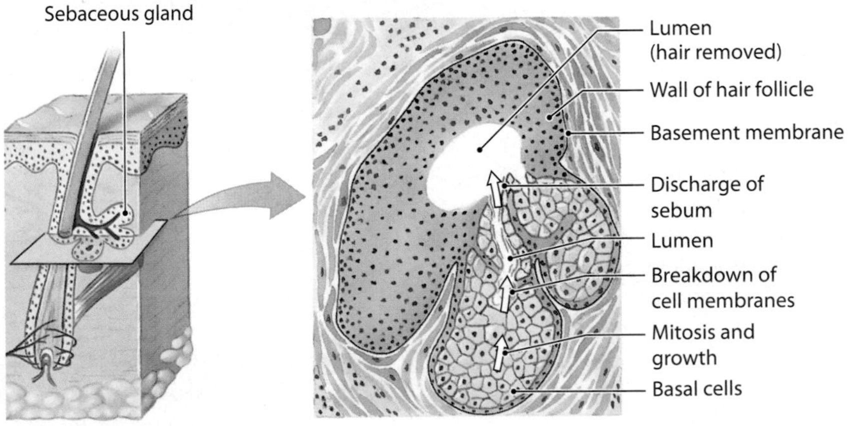
Sebaceous gland
Lumen (hair removed)
Wall of hair follicle
Basement membrane
Discharge of sebum
Lumen
Breakdown of cell membranes
Mitosis and growth
Basal cells

Sebaceous Follicles

Secrete onto the skin surface; located on the face, back, chest, nipples, and external genitalia

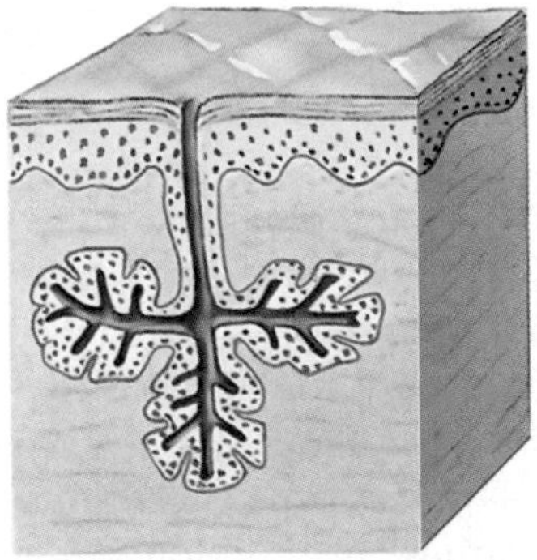

Sweat Glands

Sweat glands produce a watery solution by merocrine secretion, flush the epidermal surface, and perform other special functions.

Apocrine Sweat Glands

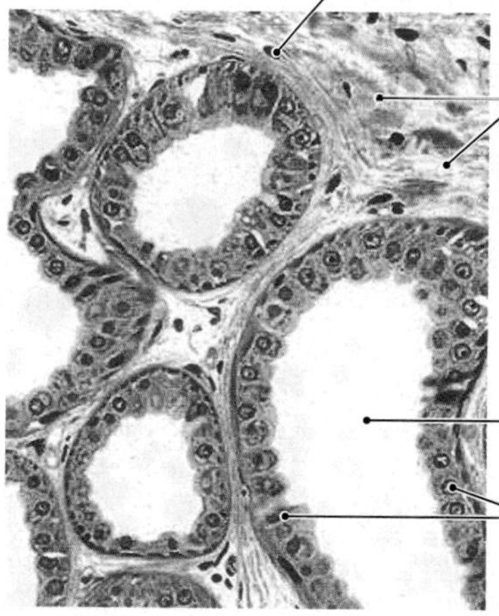

Connective tissue of dermis
Lumen
Apocrine gland cells
LM × 375

Section of apocrine sweat gland

- Limited distribution (axillae, groin, nipples)
- Produce a viscous secretion of complex composition
- Possible function in olfactory communication
- Strongly influenced by hormones
- Include ceruminous glands of the external ear and mammary glands that produce milk

Myoepithelial cells (*myo-*, muscle) are contractile cells that squeeze the gland and discharge the accumulated secretion. They are also found in mammary glands, lacrimal (tear) glands, and salivary glands.

Merocrine Sweat Glands

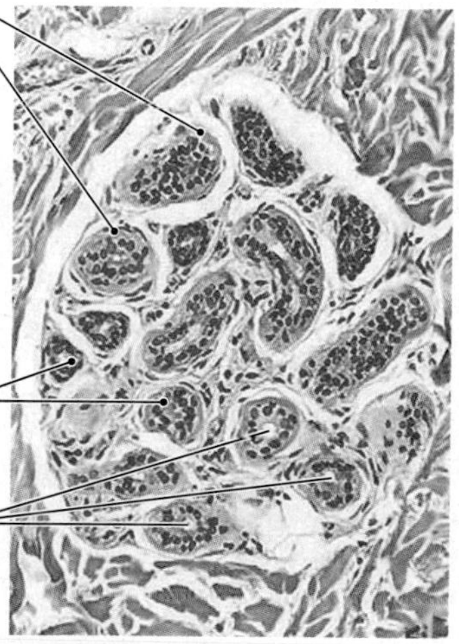

Merocrine gland cells
Lumen
LM × 210

Section of merocrine sweat gland

- Found in most areas of the skin
- Produce watery secretions containing electrolytes
- Merocrine secretion mechanism
- Controlled primarily by nervous system
- Important in thermoregulation and excretion
- Some antibacterial action

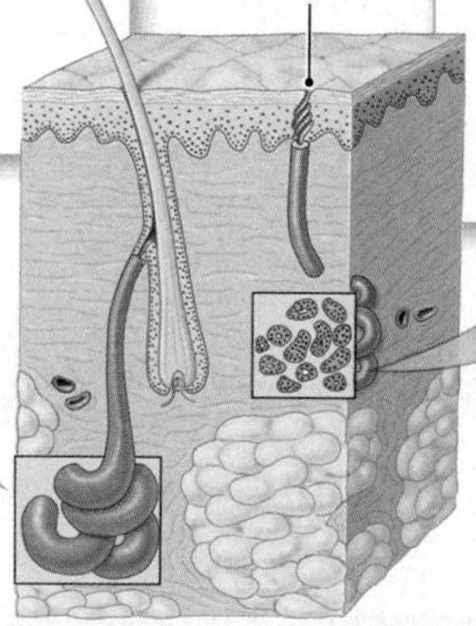

Sweat pore

2 **Apocrine sweat glands** are found in the armpits (axillae), around the nipples, and in the pubic region, where they secrete into hair follicles. These glands produce a sticky, cloudy, and potentially odorous secretion. Despite their name, these glands rely on merocrine secretion. The adult integument also contains 2–5 million **merocrine sweat glands** that discharge their secretions directly onto the surface of the skin. The palms and soles have the highest numbers, with the palm possessing an estimated 500 merocrine sweat glands per square centimeter (3000 per square inch).

Module 5.8 Review

a. Identify two types of exocrine glands found in the skin.

b. What are the functions of sebaceous secretions?

c. Deodorants are used to mask the effects of secretions from which type of skin gland?

LO **5.8** Describe the various kinds of exocrine glands in the skin, and discuss the secretions of each.

Nails are thick sheets of keratinized epidermal cells that protect the tips of fingers and toes

Nails protect the exposed dorsal surfaces of the tips of the fingers and toes. They also help limit distortion of the digits when they are subjected to mechanical stress—for example, when you run or grasp objects.

1 These superficial and sectional views illustrate common landmarks of nail structure. The **nail body**, the visible portion of the nail, is made up of dead, tightly compressed cells packed with keratin. The nail body is recessed deep to the level of the surrounding epithelium and is bounded on either side by **lateral nail grooves** (depressions) and **lateral nail folds** (ridges).

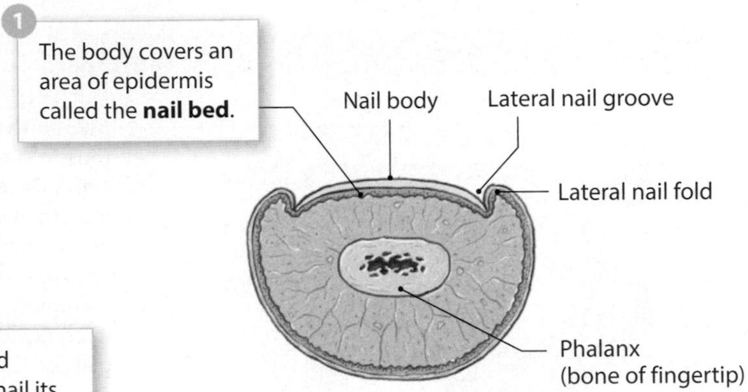
1
The body covers an area of epidermis called the **nail bed**.
Nail body
Lateral nail groove
Lateral nail fold
Phalanx (bone of fingertip)

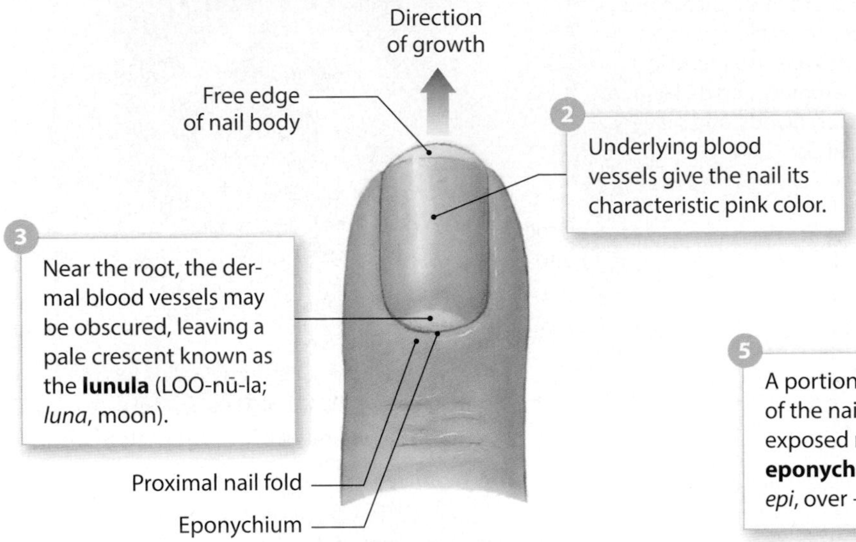
Direction of growth
Free edge of nail body
2
Underlying blood vessels give the nail its characteristic pink color.
3
Near the root, the dermal blood vessels may be obscured, leaving a pale crescent known as the **lunula** (LOO-nū-la; *luna*, moon).
Proximal nail fold
Eponychium

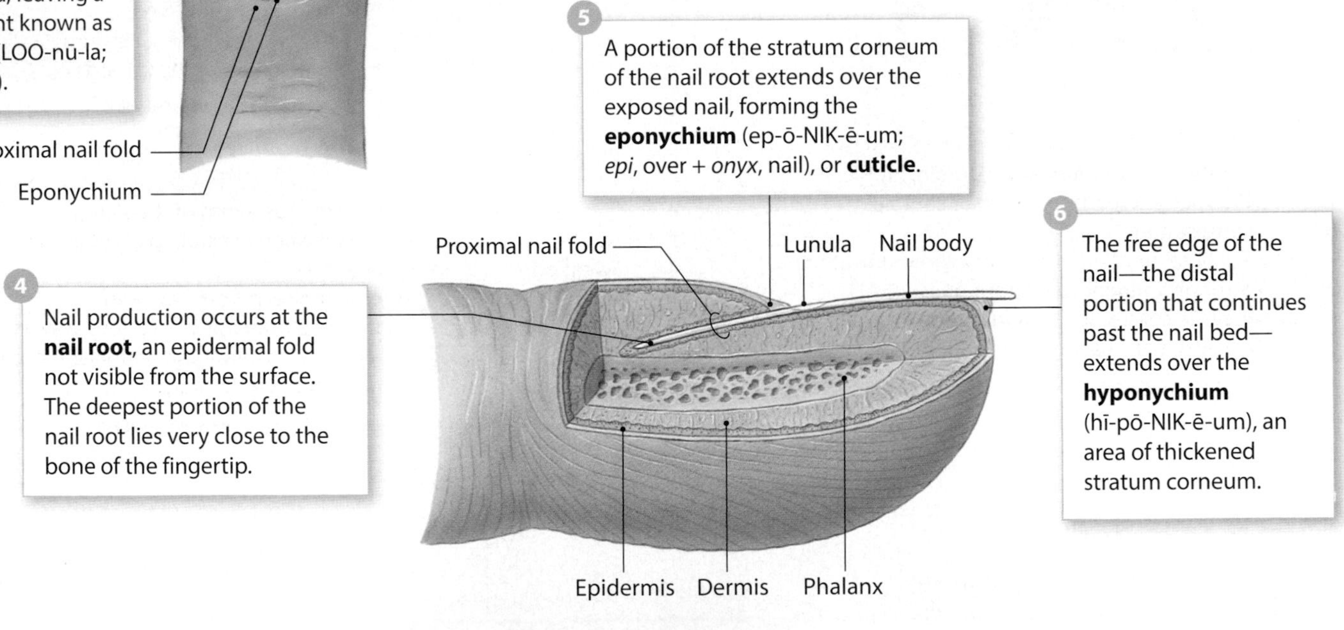
5
A portion of the stratum corneum of the nail root extends over the exposed nail, forming the **eponychium** (ep-ō-NIK-ē-um; *epi*, over + *onyx*, nail), or **cuticle**.
Proximal nail fold
Lunula
Nail body
6
The free edge of the nail—the distal portion that continues past the nail bed—extends over the **hyponychium** (hī-pō-NIK-ē-um), an area of thickened stratum corneum.
4
Nail production occurs at the **nail root**, an epidermal fold not visible from the surface. The deepest portion of the nail root lies very close to the bone of the fingertip.
Epidermis
Dermis
Phalanx

2 The cells producing the nails can be affected by conditions that alter body metabolism, so changes in the shape, structure, or appearance of the nails can provide useful diagnostic information. For example, the nails may become pitted and distorted as a result of **psoriasis** (a condition marked by rapid stem cell division in the stratum basale), or concave, as a result of some blood disorders.

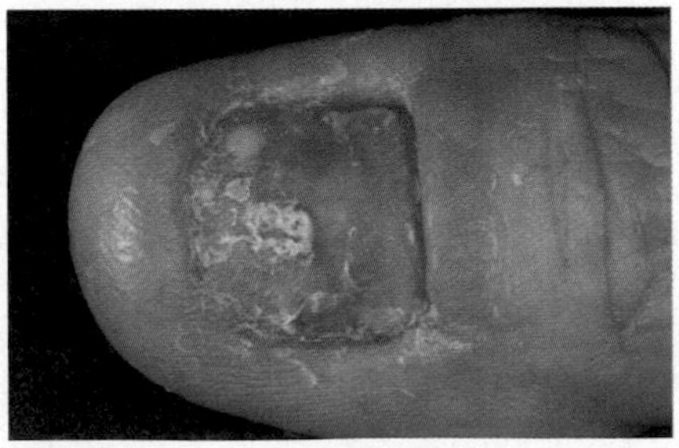

Module 5.9 Review

a. Describe a typical fingernail.

b. Where does nail production occur?

c. Define hyponychium.

LO 5.9 Describe the structure of a typical nail.

Module 5.10 CLINICAL MODULE

Age-related changes affect the integument

Fewer Melanocytes

Melanocyte activity declines, and in light-skinned people the skin becomes very pale. With less melanin in the skin, people become more sensitive to sun exposure and more likely to experience sunburn.

Drier Epidermis

Sebaceous gland secretion decreases, and the skin becomes dry and often scaly.

Thinning Epidermis

The epidermis thins as basal cell activity declines. Connections between the epidermis and dermis weaken, making older people more prone to injury, skin tears, and skin infections. The skin's metabolic activity also declines. Reduced vitamin D_3 production (see Module 5.11) leads to muscle weakness and brittle bones.

Diminished Immune Response

The number of dendritic (Langerhans) cells decreases to about half the levels seen at maturity (roughly age 21). This reduction in cells may decrease the sensitivity of the immune response and further encourage skin damage and infection.

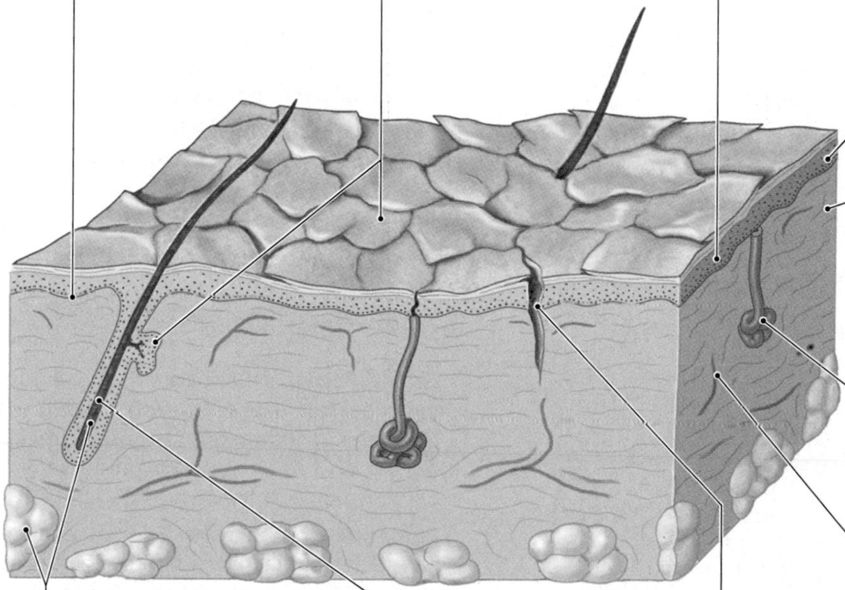

Thinning Dermis

The dermis becomes thinner and has fewer elastic fibers, making the integument weaker and less resilient. The results—sagging and wrinkling—are most pronounced in body regions with the most exposure to the sun.

Decreased Perspiration

Merocrine sweat glands become less active, and with impaired perspiration, older people cannot lose heat as fast as younger people. Thus, the elderly are at greater risk of overheating in warm environments.

Altered Hair and Fat Distribution

As sex hormone levels decrease with age, gender differences with respect to hair distribution and body-fat distribution begin to fade.

Fewer Active Follicles

Hair follicles stop functioning or produce thinner, finer hairs. Decreased melanocyte activity makes these hairs gray or white.

Slower Skin Repair

Skin repairs proceed more slowly. Thus, whereas repairs to an uninfected blister might take 3 to 4 weeks in a young adult, the same repairs could take 6 to 8 weeks at age 65–75.

Reduced Blood Supply

A reduction in dermal blood supply cools the skin, which can stimulate thermoreceptors and make a person feel cold even in a warm room. Reduced circulation and sweat gland function in the elderly lessens their ability to lose body heat, which can allow body temperatures to soar dangerously high with overexertion.

Aging affects all of the structures in the integument. The major structural and functional changes are summarized in this illustration. The gradual changes in the superficial appearance of the skin provide visual cues that we use unconsciously to estimate a person's age. This accounts for the popularity of cosmetic facial surgery.

Module 5.10 Review

a. Identify some common effects of the aging process on skin.

b. Why does hair turn white or gray with age?

c. Why do people tolerate summer heat less well and become more susceptible to heat-related illness when they become older?

LO 5.10 Summarize the effects of aging on the skin.

The integument responds to circulating hormones and has endocrine functions that are stimulated by ultraviolet radiation

Circulating Hormones

1 The integumentary system is physically and functionally tied to all other body systems. As a result, the state of the skin can be an indication of a person's health. Much of the communication between the skin and the rest of the body is chemical communication, as the specialized cells of the integument respond to levels of circulating hormones.

Glucocorticoids

Steroid hormones called **glucocorticoids** are released during times of stress. These hormones loosen the connections between keratinocytes and reduce the effectiveness of the epidermis as a barrier to infection.

Thyroid Hormones

Thyroid hormones maintain normal blood flow to the subpapillary plexus.

Sex Hormones

Sex hormones stimulate epidermal cell divisions, increasing epidermal thickness and accelerating wound repair. These hormones also increase the number of dendritic cells that defend against cancer cells and pathogens.

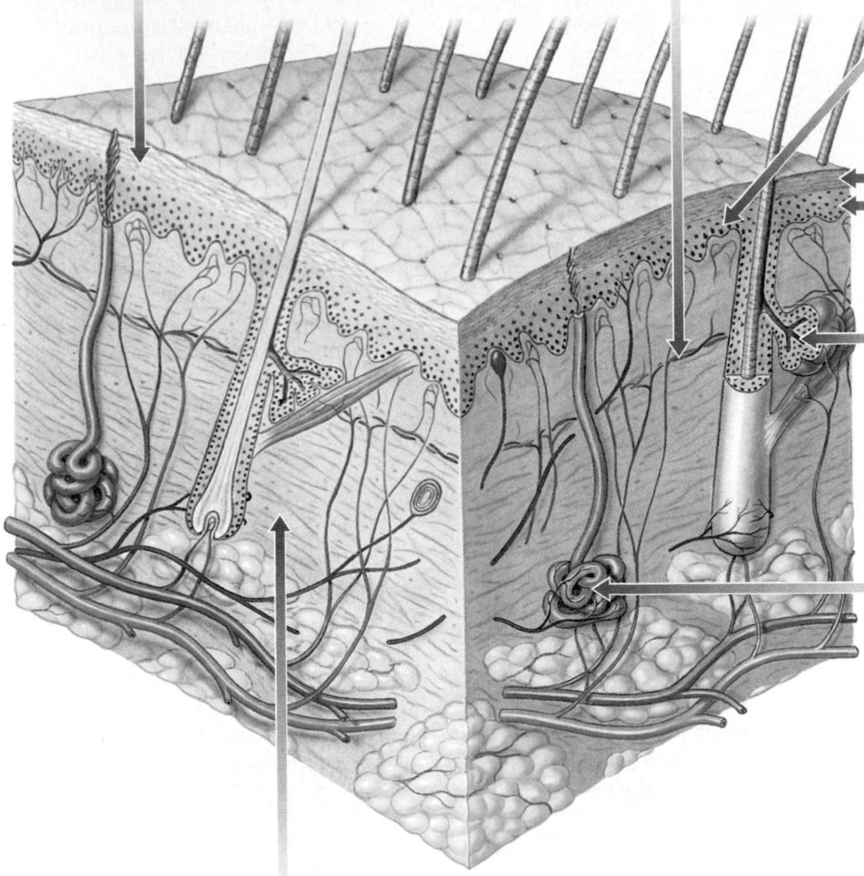

Growth Factors

Growth factors are compounds produced in the body that stimulate cell growth and cell division. Dozens of growth factors have been identified, but in many cases their origins and mechanisms of regulation are unknown.

Epidermal growth factor (EGF) is a peptide growth factor that has widespread effects on epithelia throughout the body. This peptide is produced by the salivary glands and glands of the duodenum, the initial segment of the small intestine.

EGF plays a role in:

- Promoting basal cell division in the stratum basale and stratum spinosum
- Accelerating keratin production in differentiating keratinocytes
- Stimulating epidermal development and epidermal repair after injury
- Stimulating synthetic activity and secretion by epithelial glands

Growth Hormone

Growth hormone (GH) stimulates fibroblast activity and collagen synthesis. Acting through intermediary compounds, GH also stimulates basal cell divisions, thickens the epidermis, and promotes wound repair.

Vitamin D_3 Production

2 Although too much sunlight can damage epithelial cells and deeper tissues, limited exposure to sunlight is beneficial because UV radiation plays a vital role in the synthesis of an important vitamin: **cholecalciferol** (kō-le-kal-SIF-er-ol), or **vitamin D_3**.

Sources of Vitamin D_3	
Sunlight: When exposed to ultraviolet radiation, epidermal cells in the stratum spinosum and stratum basale convert a cholesterol-related steroid into cholecalciferol. This vitamin then diffuses across the basement membrane and enters capillaries of the subpapillary plexus.	**Diet:** Cholecalciferol can be obtained from the diet, but few foods contain it other than fish, fish oils, and shellfish, and even there its presence and quantities vary. Today, many food products are "fortified with vitamin D"—most notably milk, soy milk, and orange juice.

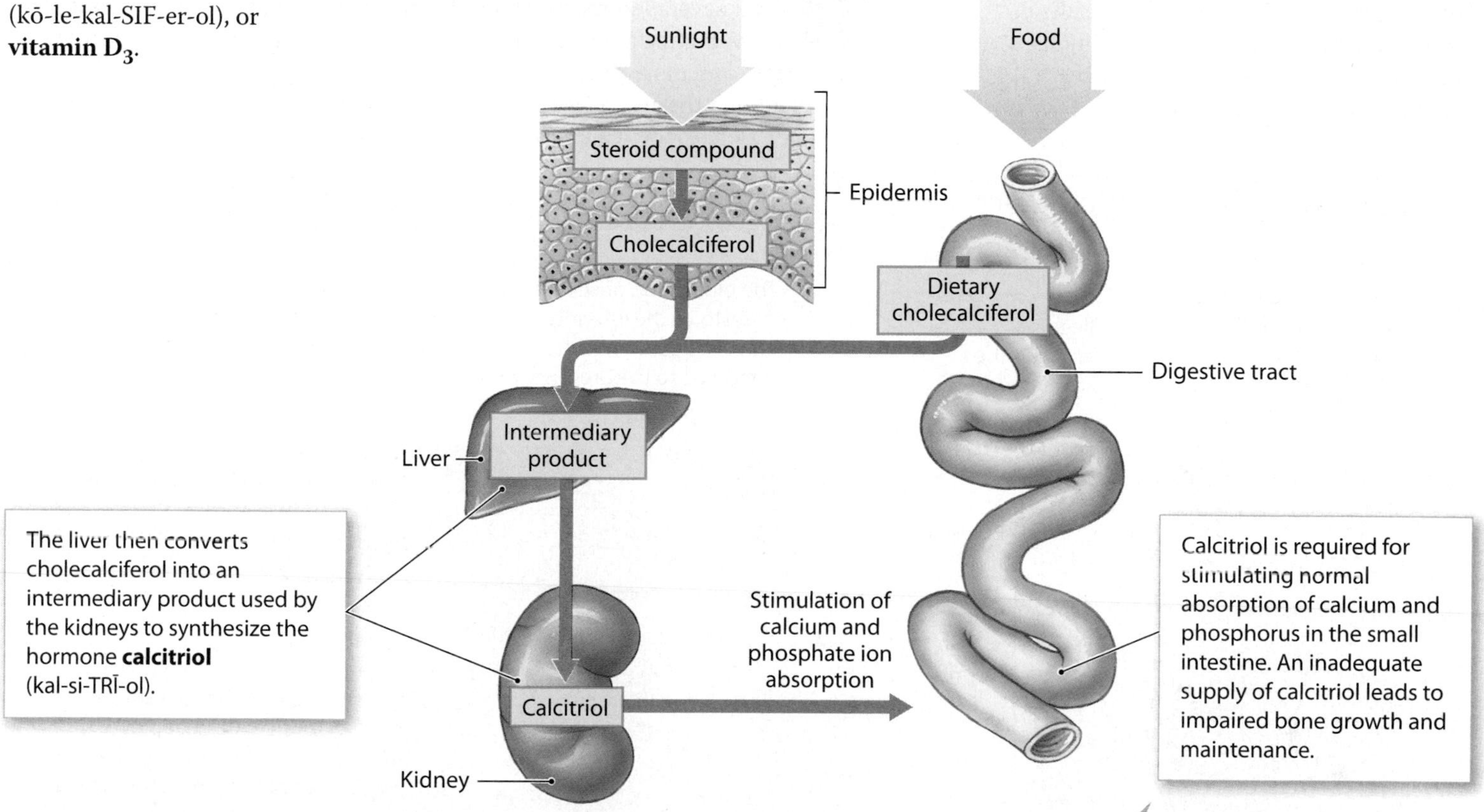
Sunlight
Food
Steroid compound
Cholecalciferol
Epidermis
Dietary cholecalciferol
Digestive tract
Liver
Intermediary product
The liver then converts cholecalciferol into an intermediary product used by the kidneys to synthesize the hormone **calcitriol** (kal-si-TRĪ-ol).
Stimulation of calcium and phosphate ion absorption
Calcitriol
Kidney
Calcitriol is required for stimulating normal absorption of calcium and phosphorus in the small intestine. An inadequate supply of calcitriol leads to impaired bone growth and maintenance.

3 The legs shown here are characteristic of **rickets**, a condition that results in flexible, poorly mineralized bones. Rickets develops in a growing child whose skin is not exposed to sunlight, and whose diet does not include a source of cholecalciferol (vitamin D_3). The bones have the proper shape, but they lack rigidity because the bone matrix contains inadequate amounts of calcium and phosphate. Rickets has largely been eliminated in the United States because dairy companies are required to add cholecalciferol, usually identified as "vitamin D," to the milk sold in grocery stores. However, cholecalciferol is required throughout life. Adults are less likely to develop rickets from cholecalciferol deficiency, but bone density will decrease, increasing the risk for fractures and slowing the healing process. The risks are especially acute for the elderly because skin production of cholecalciferol decreases by about 75 percent, even when exposed to sunlight.

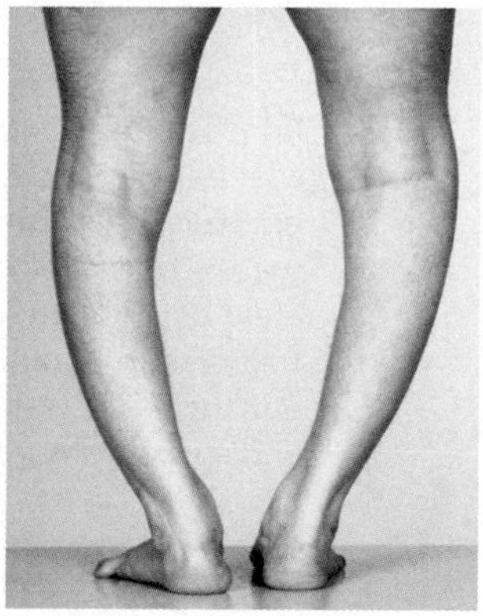

Module 5.11 Review

a. List some hormones that are necessary for maintaining a healthy integument.

b. Explain the relationship between sunlight exposure and vitamin D_3.

c. In some cultures, females must be covered from head to toe when they go outdoors. Explain why these women are at increased risk of developing bone problems later in life.

LO **5.11** Describe the interaction between sunlight and endocrine functioning as they relate to the skin.

The integument can often repair itself, even after extensive damage

The integumentary system often functions independently, responding directly and automatically to local influences without involving the nervous or endocrine systems. A dramatic display of local regulation can be seen after an injury to the skin. There are four phases in skin regeneration after injury: inflammatory, migratory, proliferation, and scarring.

1 An initial injury to the skin results in mast cell activation, which stimulates inflammation. Inflammation produces swelling, redness, heat (warmth), and pain (Module 4.17, p. 167).

2 After several hours, a scab has formed and cells of the stratum basale are migrating along the edges of the wound. Macrophages are removing debris, and more of these phagocytes are arriving via the enhanced circulation in the area. Clotting around the edges of the affected area partially isolates the region from adjacent undamaged tissues.

Inflammatory Phase

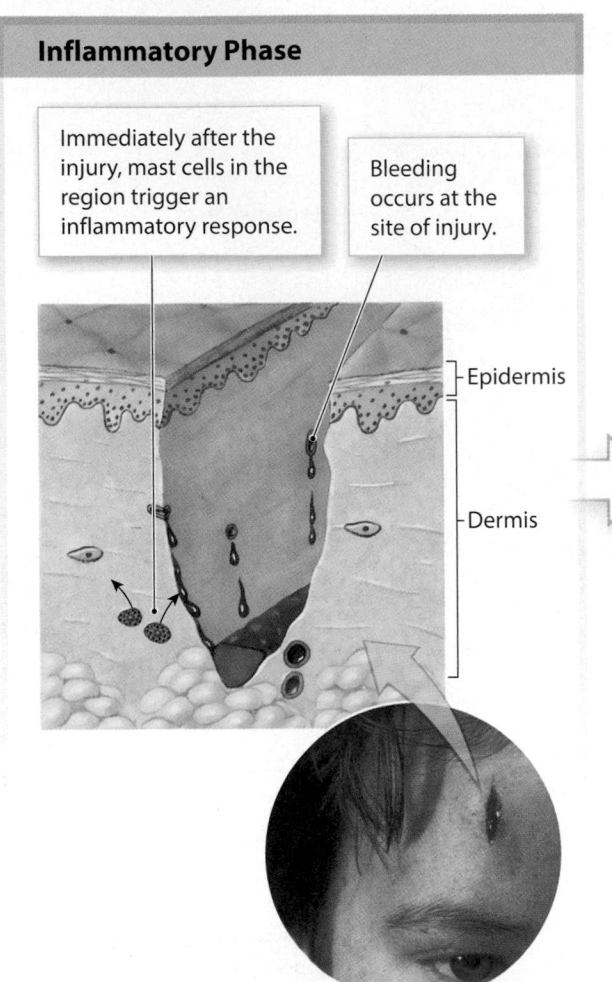

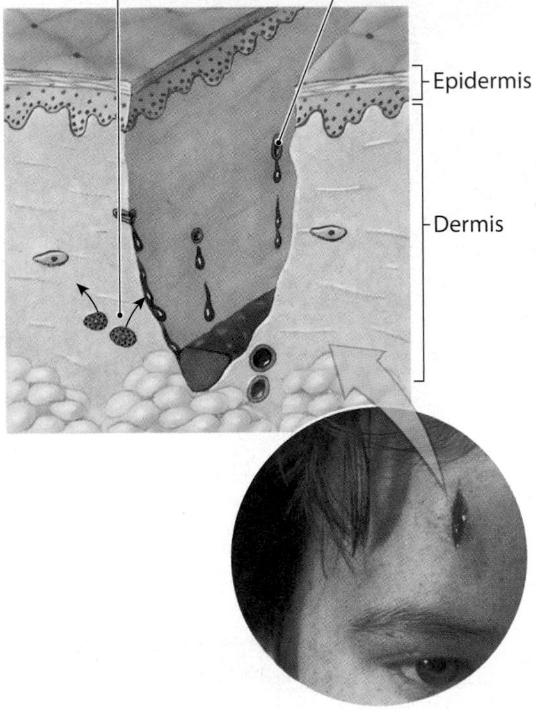
Immediately after the injury, mast cells in the region trigger an inflammatory response.
Bleeding occurs at the site of injury.
Epidermis
Dermis

Migratory Phase

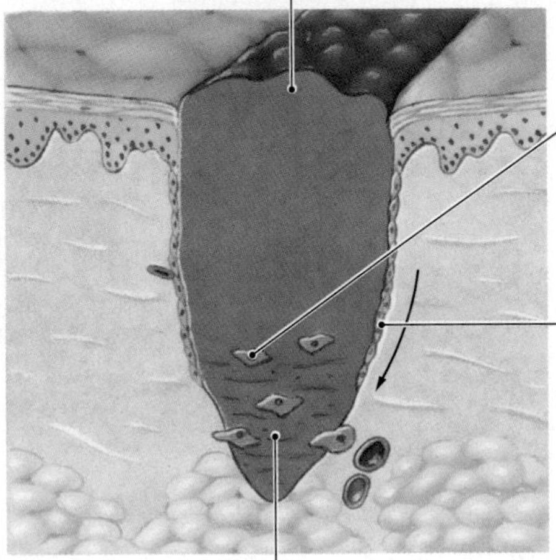
The blood clot, or **scab**, that forms at the surface temporarily restores the integrity of the epidermis and restricts the entry of additional microorganisms into the area. The scab is red due to the presence of trapped red blood cells.
Macrophages patrol the damaged area of the dermis, phagocytizing debris and pathogens.
Cells of the stratum basale divide rapidly and migrate along the edges of the wound to replace the missing epidermal cells.
If the wound occupies an extensive area or involves a region covered by thin skin, dermal repairs must be under way before epithelial cells can cover the surface. Divisions by fibroblasts and mesenchymal cells produce mobile cells that invade the deeper areas of injury. Endothelial cells of damaged blood vessels also begin to divide, and new capillaries grow in behind the fibroblasts, enhancing circulation. The combination of blood clot, fibroblasts, and an extensive capillary network is called **granulation tissue**.

3 One week after the injury, the scab has been undermined by epidermal cells migrating over a meshwork produced by fibroblast activity. Phagocytic activity around the site has almost ended, and the blood clot is disintegrating.

4 After several weeks, the scab has been shed, and the epidermis is complete. A shallow depression marks the injury site, but fibroblasts in the dermis continue to create scar tissue that will gradually elevate the overlying epidermis.

Proliferation Phase

Over time, deeper portions of the clot dissolve, and the number of capillaries declines. Continued proliferation and activity by the fibroblasts produces collagen fibers and ground substance. The repairs do not restore the integument to its original condition, however, because the dermis will contain an abnormally large number of collagen fibers and relatively few blood vessels.

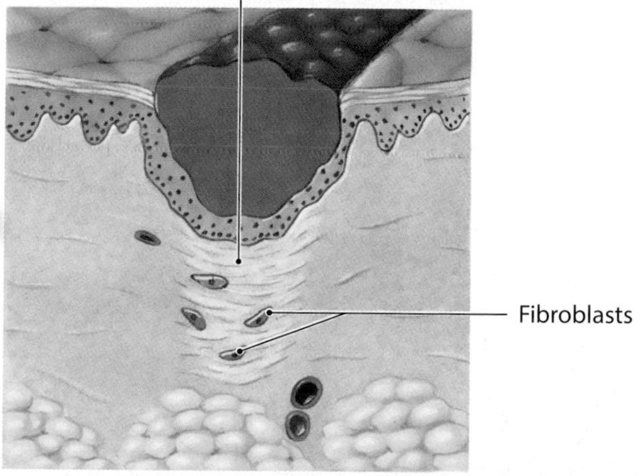
Fibroblasts

Scarring Phase

Severely damaged hair follicles, sebaceous or sweat glands, muscle cells, and nerves are seldom repaired, and they too are replaced by fibrous tissue. The formation of this rather inflexible, fibrous, noncellular **scar tissue** completes the repair process but fails to restore the tissue to its original condition.

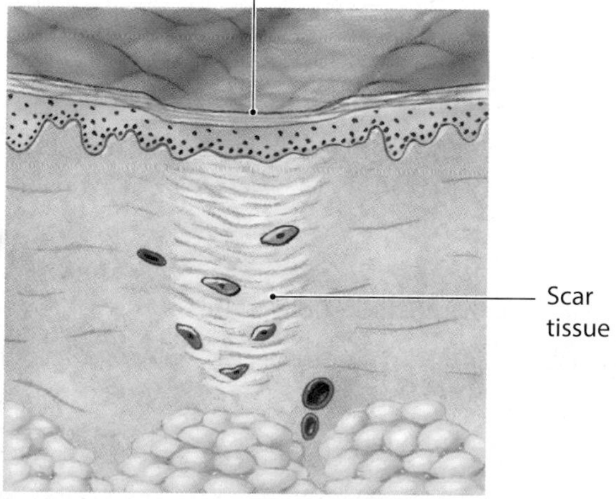
Scar tissue

5 In some adults, most often those with dark skin, scar tissue formation may continue beyond the requirements of tissue repair. The result is a **keloid** (KĒ-loyd), a raised, thickened mass of scar tissue that begins at the site of injury and grows into the surrounding dermis. Keloids are covered by a shiny, smooth epidermal surface. Keloids can develop anywhere on the body where the tissue has been injured, including as a result of surgeries. They are harmless and considered by some to be a cosmetic problem. In some cultures, however, keloids are produced intentionally as a form of body decoration.

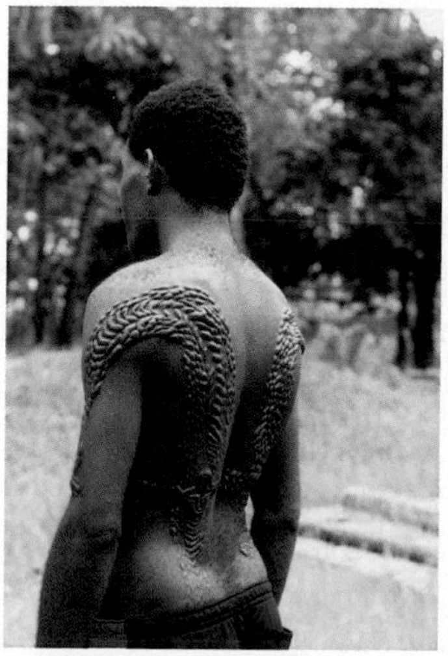

Module 5.12 Review

a. Identify the first step in skin repair.

b. Describe granulation tissue.

c. Why can skin regenerate effectively even after considerable damage?

LO 5.12 Explain how the skin responds to injury and is able to repair itself.

Labeling

Label the structures of a typical nail in the accompanying figures.

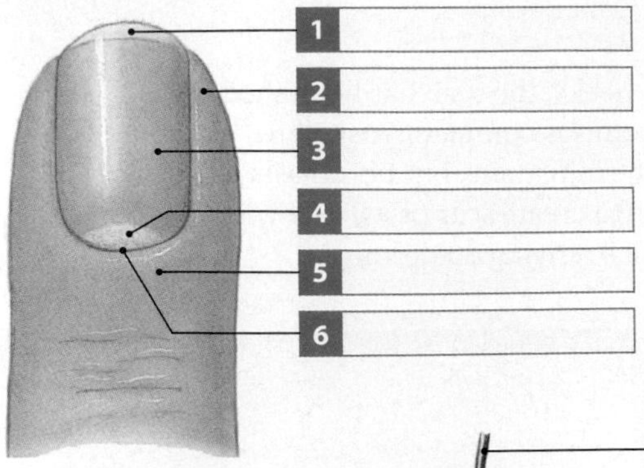
1
2
3
4
5
6

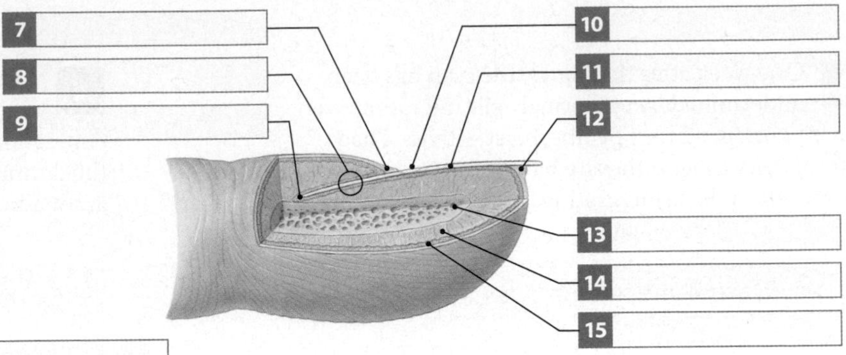
7
8
9
10
11
12
13
14
15

Label the structures of a hair follicle in the accompanying figure.

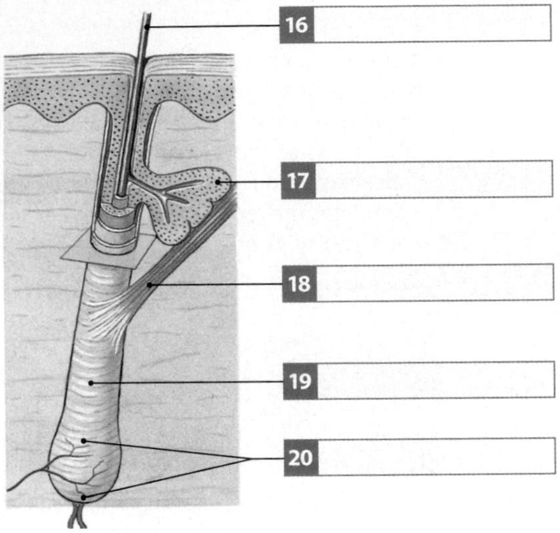
16
17
18
19
20

Matching

Match each lettered term with the most closely related description.

a. malignant melanoma
b. keloid
c. nail root
d. sebum
e. apocrine sweat glands
f. eponychium
g. EGF
h. vitamin D_3
i. reticular layer of dermis
j. merocrine sweat glands

21 Produced by epidermal cells stimulated by UV radiation
22 Epithelial fold not visible from the surface
23 Found in the armpit
24 Peptide produced by salivary glands
25 Site of hair production
26 Excessive scar tissue
27 Oily lipid secretion
28 Melanocytes metastasize through the lymphatic system
29 Abundant in the palms and soles
30 Cuticle

21 ________
22 ________
23 ________
24 ________
25 ________
26 ________
27 ________
28 ________
29 ________
30 ________

Section integration

Many people change the natural appearance of their hair, either by coloring it or by altering the degree of curl in it. Which layers of the hair do you suppose are affected by the chemicals added during these procedures? Why are the effects of the procedures not permanent?

31 ________

Study Outline

SECTION 1 • Functional Anatomy of the Skin

5.1 The integumentary system consists of the skin and various accessory structures p. 175

1. The **integument**, or **integumentary system**, is composed of the **cutaneous membrane** or skin and the **accessory structures**.
2. The cutaneous membrane includes the **epidermis** (made of stratified squamous epithelium) and **dermis** (containing papillary and reticular layers).
3. Beneath the dermis lies the **hypodermis** or subcutaneous layer. This tissue layer is not part of the integument.
4. Functions of the integument include protection, excretion, temperature regulation, vitamin D_3 synthesis, nutrient storage, and sensory detection.

5.2 The epidermis is composed of strata (layers) that have various functions p. 176

5. The epidermis contains **keratinocytes**, arranged in **strata** or layers. These cells are continuously produced in the deepest layers and shed at the exposed surface.
6. **Epidermal ridges** extend down into **dermal papillae** of the dermis. This arrangement firmly binds the epidermis to the dermis.

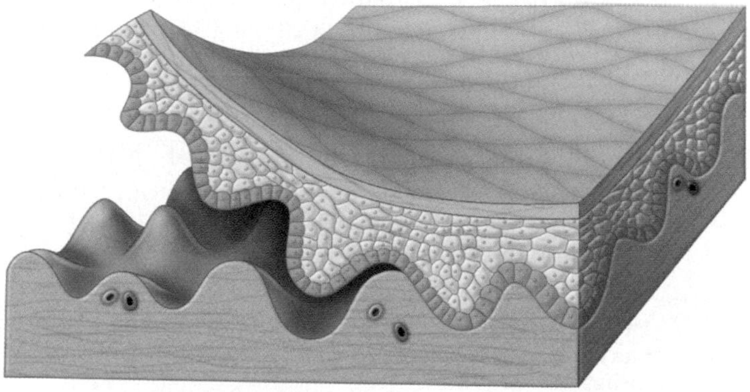

Epidermal ridges and dermal papillae greatly increase the surface area for attachment, firmly binding the epidermis to the dermis.

7. **Thin skin** contains four **strata**, whereas **thick skin**, which covers the palm of the hands and soles of the feet, contains a fifth stratum.
8. From deep to superficial, the layers of the epidermis are **stratum basale, stratum spinosum, stratum granulosum, stratum lucidum** (thick skin only), and **stratum corneum**.

5.3 Factors influencing skin color include epidermal pigmentation and dermal circulation p. 178

9. **Melanin** and **carotene** are the primary pigments involved in skin coloration.
10. **Melanocytes**, located in the stratum basale, produce melanin, a brown, yellow-brown, or black pigment. Carotene, an orange-yellow pigment found in vegetables such as carrots and some squashes, accumulates in the fatty tissues in the deep dermis and hypodermis.
11. Blood supply also affects skin color because it contains red blood cells filled with the red pigment hemoglobin. During exercise superficial blood vessels dilate to cool the body and the skin becomes flushed (red). During a sustained reduction in circulation, oxygen levels decline, giving the skin a bluish coloration called **cyanosis**.
12. The most common types of cancers are skin cancers. **Basal cell carcinoma** is the most common form of skin cancer, yet it is often survived. **Malignant melanoma** is extremely dangerous.

5.4 The dermis supports the epidermis, and the hypodermis connects the dermis to the rest of the body p. 180

13. Lying between the epidermis and hypodermis, the **dermis** contains collagen and elastic fibers. The collagen fibers provide strength; the elastic fibers allow for stretching.
14. The dermis consists of the superficial **papillary layer** and the deeper **reticular layer**. The papillary layer contains highly vascularized areolar tissue, capillaries, lymphatic vessels, and sensory neurons. The reticular layer is a meshwork of dense irregular connective tissue containing collagen and elastic fibers.
15. The **hypodermis**, which is not part of the integument, separates the skin from deeper structures.
16. The **touch receptors** of the integument are **free nerve endings, Meissner's corpuscles, tactile discs, lamellated (pacinian) corpuscles**, and **Ruffini corpuscles**.

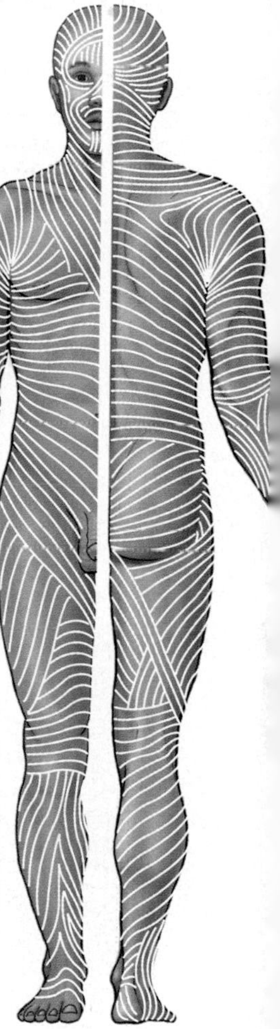

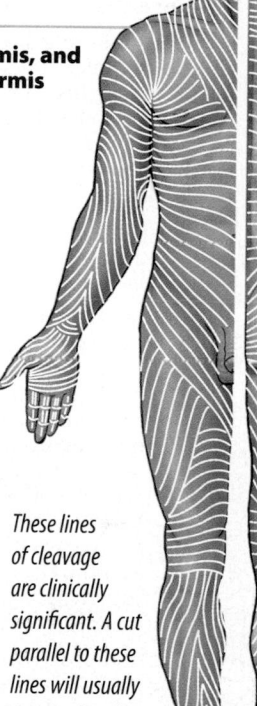

These lines of cleavage are clinically significant. A cut parallel to these lines will usually remain closed and heal with little scarring.

17. Most of the collagen and elastic fibers in any one area are arranged in parallel bundles to resist forces applied to the skin. Cuts parallel to the **lines of cleavage** will heal with little scarring, but a cut at right angles will pull open, as severed elastic fibers recoil.

5.5 Burns are significant injuries that damage skin integrity p. 182

18. **First-degree** and **second-degree burns** are called **partial-thickness burns**. First-degree burns damage only the surface of the epidermis, whereas second-degree burns damage the entire epidermis and perhaps some of the dermis.

19. **Third-degree burns** are called **full-thickness burns** and extend into the hypodermis. These burns cannot heal unaided and often require a **skin graft**.
20. The **rule of nines** is used to estimate the percentage of the surface area affected by burns.

SECTION 2 • Accessory Structures of the Skin

5.6 Hair follicles, exocrine glands, and nails are also components of the integumentary system p. 185

21. **Hair follicles** produce hairs that protect the skin and provide touch sensations to the body surface.
22. **Exocrine glands** assist in thermoregulation, excrete wastes, and lubricate the epidermis.
23. **Nails** protect and support the tips of the toes and fingers.

5.7 Hair is composed of dead, keratinized cells produced in a specialized hair follicle p. 186

24. A **hair follicle** is a complex structure made of connective and epithelial tissues that is responsible for the production of a single hair.
25. Hair formation begins at the base of the follicle at the **hair bulb**.
26. Each hair has a **root** and a **shaft**. At the base of the root is a **hair papilla**, surrounded by a hair bulb, and a **root hair plexus** of sensory nerves. Hairs have a central medulla of **soft keratin** surrounded by a **cortex** of **hard keratin**. The **cuticle** is a superficial layer of cells that protect the hair.
27. Composed of smooth muscle, each **arrector pili** muscle when contracted can make a single hair stand erect.

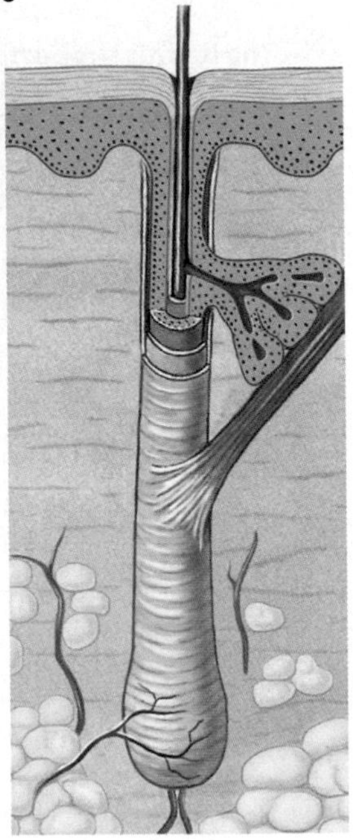

A hair follicle is a complex structure. The human body has about 2.5 million of them.

28. Hairs grow and shed according to the **hair growth cycle**. A typical scalp hair grows for 2–5 years and is then shed.

5.8 Sebaceous glands and sweat glands are exocrine glands in the skin p. 188

29. **Sebaceous glands** secrete **sebum**, which coats and protects the hair shaft.
30. The two types of sweat glands are **apocrine** and **merocrine sweat glands**. Apocrine sweat glands are found in the axillae, nipples, and pubic regions, where they secrete into hair follicles. Merocrine sweat glands discharge their secretion directly onto the skin surface.

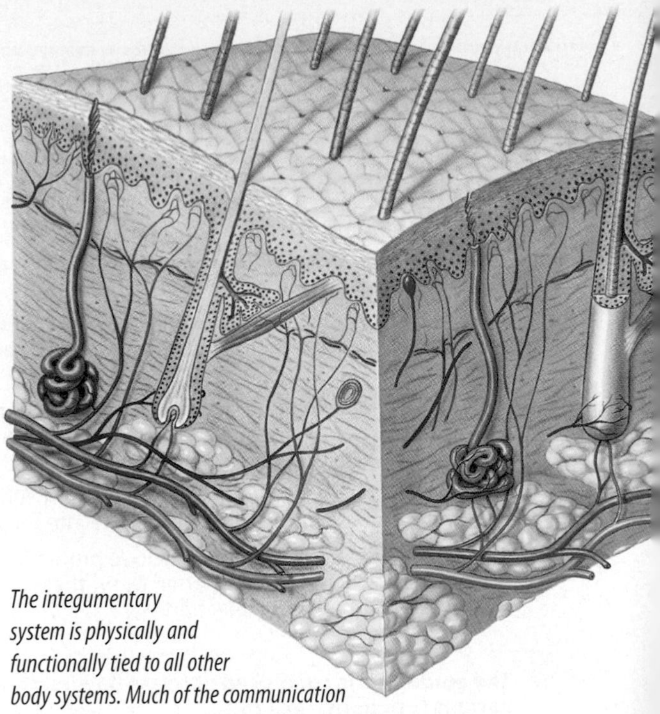

The integumentary system is physically and functionally tied to all other body systems. Much of the communication between the skin and the rest of the body is by chemical communication with hormones.

5.9 Nails are thick sheets of keratinized epidermal cells that protect the tips of fingers and toes p. 190

31. The **nail body** covers the **nail bed**. Nail production occurs at the **nail root**. A portion of the stratum corneum of the nail root extends over the nail, forming the **eponychium** or **cuticle**. The free edge of the nail extends over the **hyponychium**.

5.10 Age-related changes affect the integument p. 191

32. With aging, sebaceous and sweat gland activity decreases, the integument thins, blood flow decreases, cellular activity decreases, and repairs occur more slowly.

5.11 The integument responds to circulating hormones and has endocrine functions that are stimulated by ultraviolet radiation p. 192

33. Steroid, **thyroid, sex,** and **growth hormones** influence integument structure and function.
34. Limited exposure to sunlight is beneficial because UV radiation plays a role in the synthesis of **cholecalciferol** or **vitamin D_3**. A lack of vitamin D_3 in children can result in **rickets**.

5.12 The integument can often repair itself, even after extensive damage p. 194

35. The skin can regenerate itself after damage. The process begins with bleeding, and includes an **inflammatory phase, migratory phase, proliferation phase**, and **scarring phase**.
36. In some people, **scar tissue** formation may continue beyond the requirements of tissue repair, resulting in **keloid** formation.

Chapter Review Questions

Labeling

Identify the layers of the epidermis in this image.

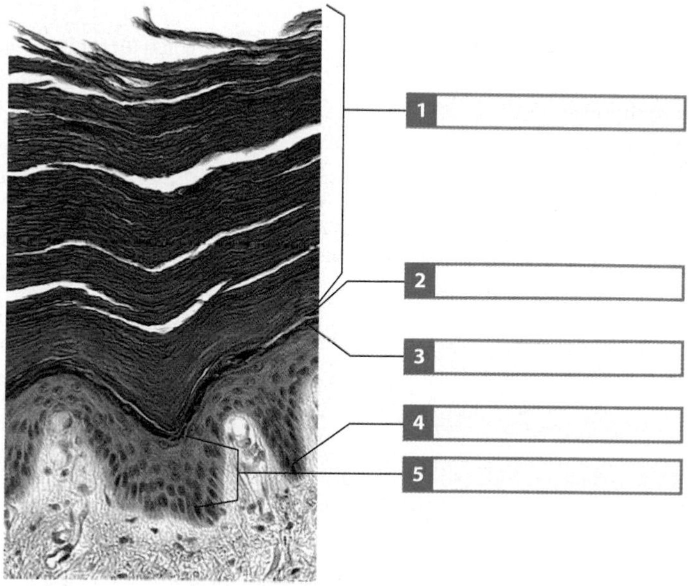
1
2
3
4
5

True/False

Indicate whether each statement is true or false.

6 Wrinkles and sagging skin are caused by damage to collagen fibers due to aging and exposure to ultraviolet radiation.

7 Another name for the hypodermis is the subcutaneous layer.

8 The perspiration you experience during intense exercise is called insensible perspiration.

9 The dermis is dominated by adipose tissue and therefore is an important site for energy storage.

10 Lamellated (pacinian) corpuscles are sensitive to deep pressure.

11 Bluish skin as a result of decreased oxygen in blood is called cyanosis.

12 The root hair plexus is a smooth muscle whose contraction pulls on the follicle, forcing the hair to stand erect.

13 The primary pigments involved in skin coloration are carotene and melanin.

6 ______________________

7 ______________________

8 ______________________

9 ______________________

10 ______________________

11 ______________________

12 ______________________

13 ______________________

Multiple choice

Select the correct answer from the list provided.

14 Which portion of the hair follicle produces the hair?

- a) hair shaft
- b) hair matrix
- c) hair root plexus
- d) arrector pili

15 Which of the following glands discharges an oily secretion that coats the hair and the adjacent surface of the skin?

- a) apocrine gland
- b) sebaceous gland
- c) merocrine gland
- d) sweat gland

16 Fine touch, pressure, and vibrations are perceived by which nerve endings?

- a) free nerve endings
- b) Meissner's corpuscles
- c) tactile discs
- d) Ruffini corpuscles

17 What is the primary function of sensible perspiration?

- a) to get rid of wastes
- b) to protect the skin from dryness
- c) to maintain electrolyte balance
- d) to reduce body temperature

18 In the elderly, blood supply to the dermis is reduced and sweat glands are less active. What is most affected by this combination of factors?

- a) ability to thermoregulate
- b) ability to heal injured skin
- c) ease with which skin is injured
- d) ability to grow hair

19 Which term is applied to the stratum corneum of the nail root that extends over the exposed nail?

- a) hyponychium
- b) eponychium
- c) lunula
- d) cerumen

20 Which factor is associated with darker skin color?

- a) more melanocytes present in skin
- b) more layers of epidermis
- c) more melanin produced by melanocytes
- d) more superficial blood vessels

21 Which anatomical feature is responsible for fingerprints?

- a) ridge patterns in the thick skin covering the fingertips
- b) dermal papillae
- c) the thickness of stratum lucidum
- d) the architectural arrangement of melanocytes

Fill-in

Fill in the following blanks in the spaces provided to the right.

22 From superficial to deep, the two layers of the cutaneous membrane are the __________ and the __________.

23 The __________ layer separates the integument from the fascia around deeper organs.

24 __________ sweat glands are found in the axillae, nipples, and pubic region.

25 The most common form of skin cancer is __________.

26 A deficiency or absence of __________ production leads to a disorder known as albinism.

22 __________, __________

23 __________

24 __________

25 __________

26 __________

Short answer

27 In which layer(s) of the epidermis does cell division occur?

28 What is the function of the arrector pili muscles?

29 Explain how hair color can vary from person to person, and over a person's lifetime.

30 What are the risks from having inadequate levels of vitamin D_3 in the body?

31 Explain why when warming your hands by a campfire, your face feels the heat more than your hands.

32 Why is it important for a surgeon to choose an incision pattern according to the cleavage lines of the skin?

33 A 32-year-old woman is admitted to the hospital with third-degree burns on her entire right leg, entire right arm, and the back of her trunk. Estimate the percentage of her surface area that is affected by these burns.

MasteringA&P®

Access more chapter study tools online in the MasteringA&P Study Area:

- **Chapter Quizzes, Chapter Practice Test, Art-labeling Activities, Animations, MP3 Tutor Sessions, and Clinical Case Studies**

- **Practice Anatomy Lab** PAL™
- **Interactive Physiology** iP®
- **A&P Flix** A&PFlix™
- **PhysioEx** PhysioEx™

Chapter Integration • Applying what you have learned

Thinking about the science of tattoos

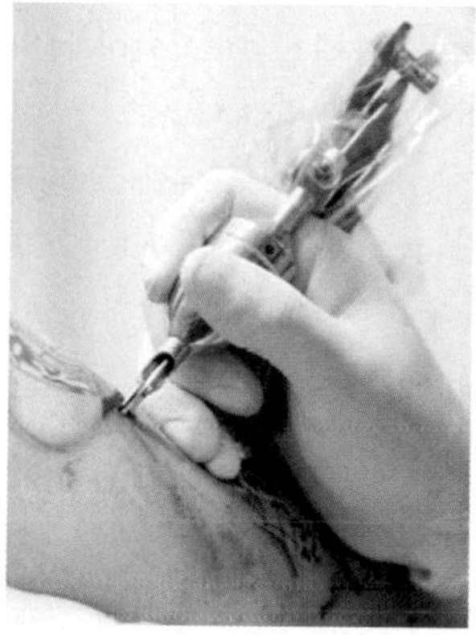

Tattoos are becoming increasingly visible in mainstream society. As tattoos become more fashionable, people of all ages and social classes are engaging in this age-old form of body art. Once considered a practice of only sailors, circus sideshow performers, or criminals, an increasingly diverse population is "sporting ink." Everyone from movie stars and athletes, to doctors and soccer moms have tattoos these days. It's safe to say that most of you reading this text are either tattooed or know someone who is tattooed.

The tattooing process has remained essentially the same for thousands of years. Originally, a sharpened bone tool was dipped in dye and then tapped into the skin with a small hammer to create the tattoo. Today's tattooing uses an electrical tool to deposit pigment through the epidermis and into the dermis. Similar to a sewing machine, a motor-driven cluster of needles punctures the skin many times each second. The artist progressively shades and colors the tattoo in a process that can take many hours.

In light of the popularity and prevalence of tattooing, use your knowledge of the integument to answer the following questions.

1 **Why does the tattoo ink stay in place?**

2 **Why do tattoos tend to fade over time?**

3 **Explain why it is somewhat painful to be tattooed.**

4 **Does tattooing carry a significant risk of infection?**

5 **Predict what is required for a tattoo to be successfully removed.**

6 Osseous Tissue and Bone Structure

LEARNING OUTCOMES

These Learning Outcomes correspond by number to this chapter's modules and indicate what you should be able to do after completing the chapter.

SECTION 1 • Introduction to the Structure and Growth of Bones

6.1 Describe the two main divisions of the skeleton, and list the major functions of the skeletal system.

6.2 Classify bones according to their shapes, identify the major types of surface markings, and explain the functional significance of surface markings.

6.3 Identify the parts of a typical long bone, and describe its internal structures.

6.4 Identify the cell types in bone, and list their major functions.

6.5 Compare the structures and functions of compact bone and spongy bone.

6.6 Describe the process of appositional bone growth.

6.7 Describe the mechanisms of endochondral ossification.

6.8 Describe the mechanisms of intramembranous ossification.

6.9 **CLINICAL MODULE** Discuss various abnormalities of bone formation and growth.

SECTION 2 • Physiology of Bones

6.10 List the minerals stored in the bones, and identify the organs involved in calcium homeostasis.

6.11 Discuss the effects of hormones on bone development, and explain the homeostatic mechanisms involved.

6.12 **CLINICAL MODULE** Describe the types of fractures, and explain how fractures heal.

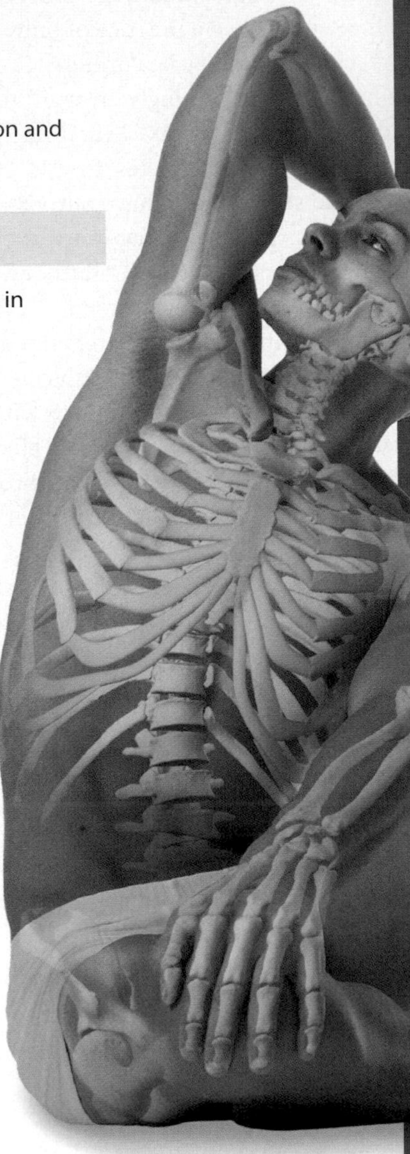

Learning Outcomes are repeated at the bottom of each module.

The skeletal system is made up of the axial and appendicular divisions

The skeletal system includes the varied bones of the skeleton and the cartilages, ligaments, and other connective tissues that stabilize or interconnect them. This chapter expands on the introduction to bone tissue presented in Chapter 4 by considering the gross anatomy of bones and the mechanisms involved in bone growth, remodeling, and repair.

1 The adult skeletal system includes about 206 separate bones and a number of associated cartilages. This body system is divided into the **axial skeleton** and the **appendicular skeleton**.

Axial Skeleton (80 Bones)

The axial skeleton consists of the bones of the skull, thorax, and vertebral column. These elements form the longitudinal axis of the body.

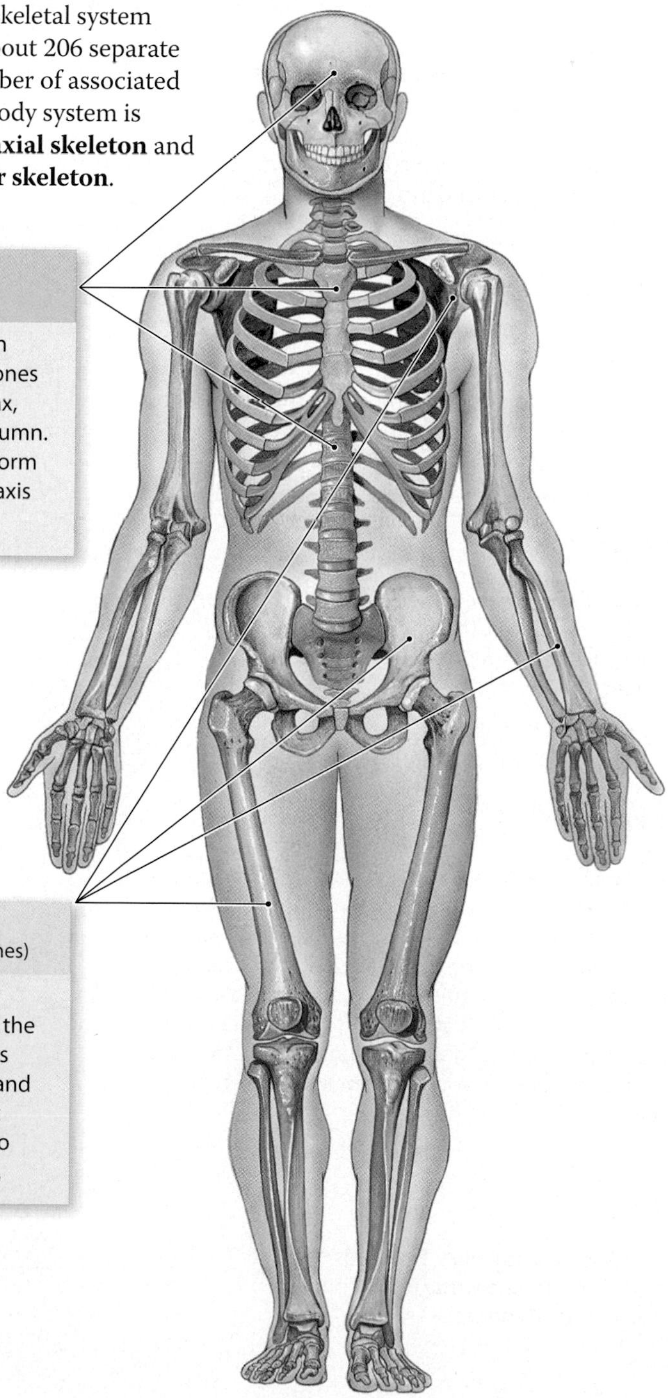

Appendicular Skeleton (126 Bones)

The appendicular skeleton includes the bones of the limbs and the pectoral and pelvic girdles that attach the limbs to the axial skeleton.

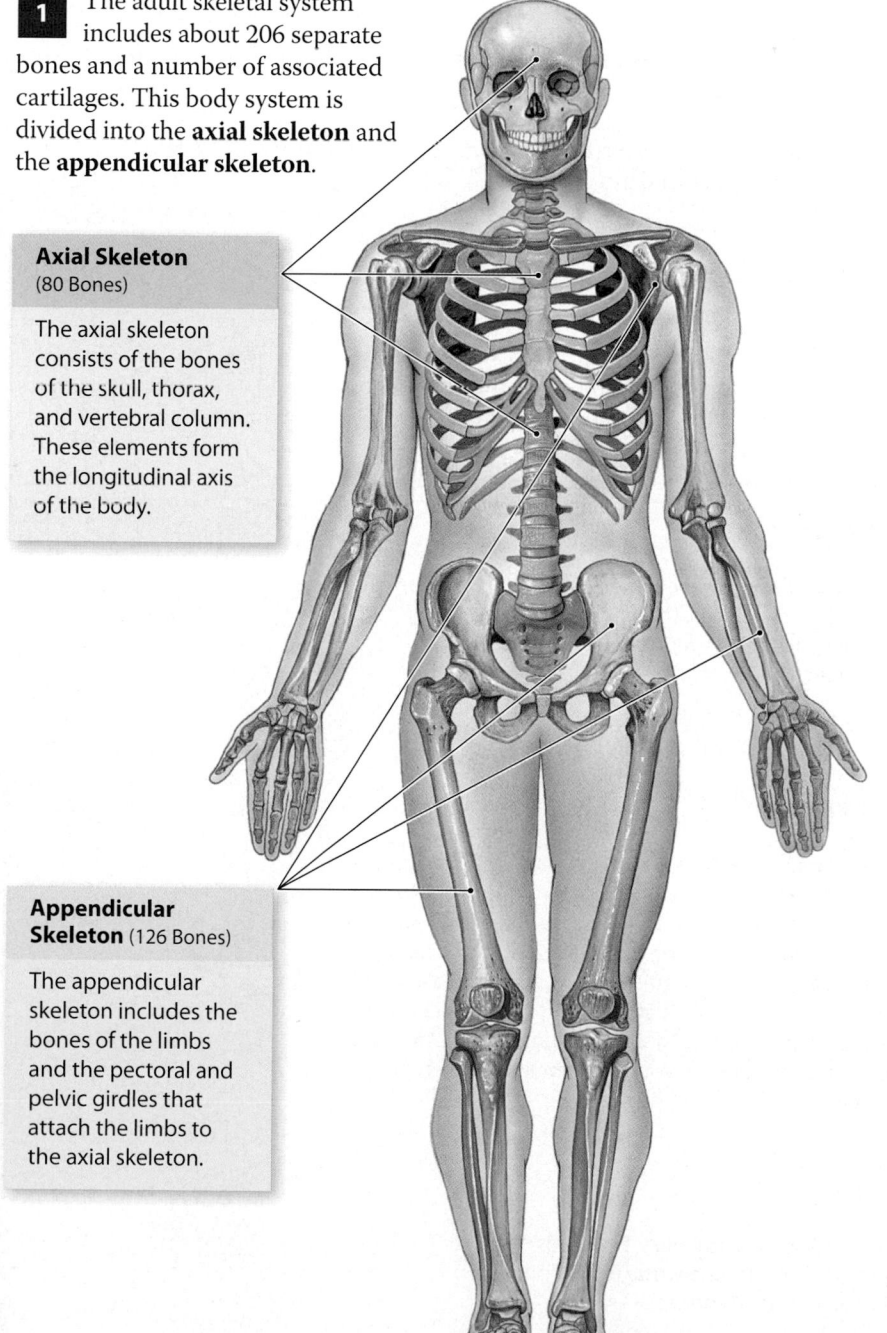

Functions of the Skeletal System

- **Support:** The skeletal system provides structural support for the entire body. Individual bones or groups of bones provide a framework for the attachment of soft tissues and organs.
- **Store minerals and lipids:** The calcium salts in bone are a valuable mineral reserve that maintains normal concentrations of calcium and phosphate ions in blood. Calcium is the most abundant mineral in the human body. Bones contain adipose tissue, which stores lipids as energy reserves.
- **Produce blood cells:** Red blood cells, white blood cells, and platelets are produced in the red bone marrow, which fills the internal cavities of many bones.
- **Protection:** The skeleton surrounds delicate tissues and organs. The ribs protect the heart and lungs, the skull encloses the brain, the vertebrae shield the spinal cord, and the pelvis cradles digestive and reproductive organs.
- **Leverage:** Many bones of the skeleton function as levers that can change the magnitude and direction of the forces skeletal muscles generate. The movements produced range from the fine motions of a fingertip to powerful changes in the position of the entire body.

2 The skeleton has many vital functions that are summarized in the table above. All of these functions ultimately depend on the unique and dynamic properties of bone tissue. The bone specimens that you study in lab or that you are familiar with from skeletons of dead animals are only the dry remains of this living tissue. They have the same relationship to the bone in a living organism as kiln-dried lumber does to a living tree.

Module 6.1 Review

a. Describe the axial skeleton.

b. Describe the appendicular skeleton.

c. Identify the functions of the skeletal system.

LO **6.1** Describe the two main divisions of the skeleton, and list the major functions of the skeletal system.

Bones are classified according to shape and structure and have varied surface markings

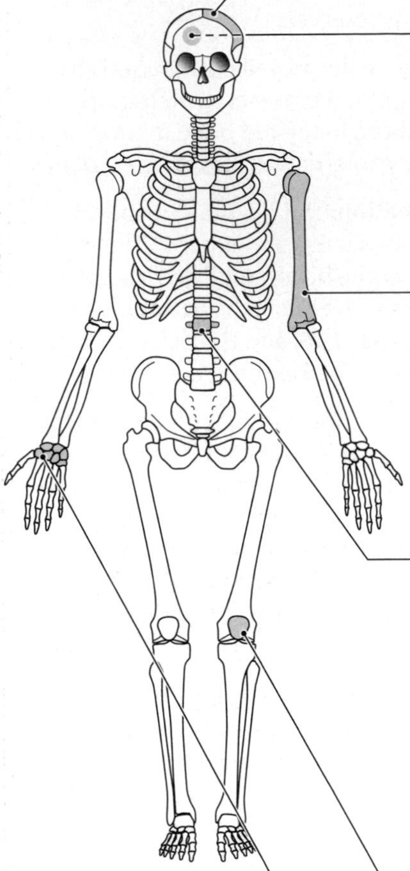

Flat Bones

Flat bones have thin, roughly parallel surfaces. Flat bones form the roof of the skull, the sternum, the ribs, and the scapulae. They protect underlying soft tissues and have an extensive surface area for the attachment of skeletal muscles.

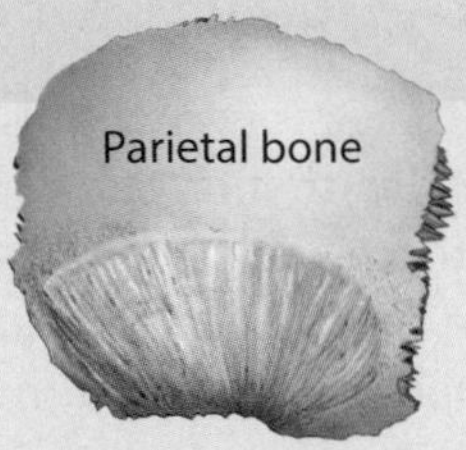
Parietal bone

Sutural Bones

Sutural bones, or Wormian bones, are small, flat, irregularly shaped bones between the flat bones of the skull. There are individual variations in the number, shape, and position of sutural bones. Their borders are like pieces of a jigsaw puzzle, and they range in size from a grain of sand to a quarter.

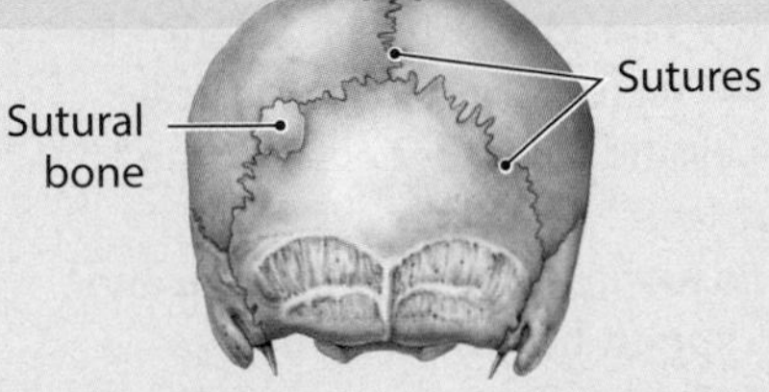
Sutural bone
Sutures

Long Bones

Long bones are elongated and slender. They are located in the arm and forearm, thigh and leg, palms, soles, fingers, and toes. The femur, the long bone of the thigh, is the largest and heaviest bone in the body.

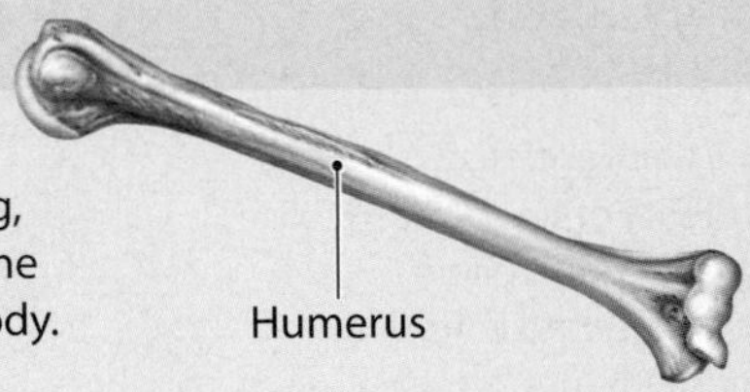
Humerus

Irregular Bones

Irregular bones have complex shapes with short, flat, notched, or ridged surfaces. The spinal vertebrae, the bones of the pelvis, and several skull bones are irregular bones.

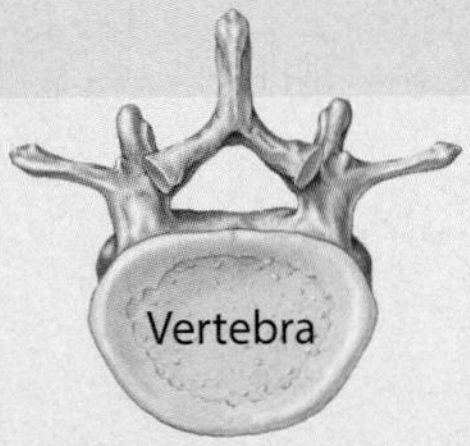
Vertebra

Sesamoid Bones

Sesamoid bones are generally small, flat, and shaped somewhat like a sesame seed. They develop inside tendons and are most commonly located near joints at the knees, the hands, and the feet. Sesamoid bones may form in at least 26 locations. Except for the patellae (pa-TEL-ē; singular, *patella*, a small shallow dish), or kneecaps, there are individual variations in the location and number of sesamoid bones. These variations, along with varying numbers of sutural bones, account for individual differences in the total number of bones in the skeleton.

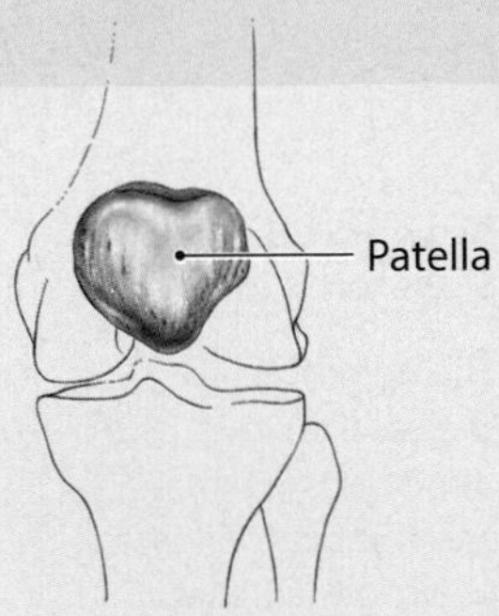
Patella

Short Bones

Short bones are small and boxy. Examples of short bones include bones in the wrists (carpal bones) and in the ankles (tarsal bones).

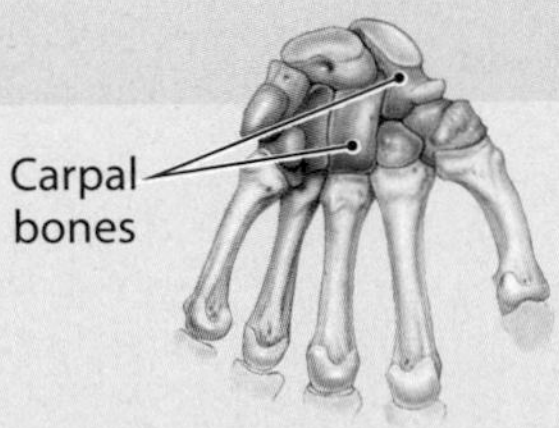
Carpal bones

1 The 206 major bones in a typical adult human skeleton can be divided into six broad categories based on their shape. The skeleton also contains a variable number of minor bones (usually sutural bones and sesamoid bones) whose numbers vary from individual to individual.

2 Illustrated here are the major types of **surface markings**, also known as bone features. Each bone in the body has characteristic surface markings related to its particular functions. Elevations or projections form where tendons and ligaments attach, and at joints, where adjacent bones articulate. Depressions, grooves, and tunnels in bone indicate sites where blood vessels or nerves lie alongside or penetrate the bone.

1 Surface Markings of the Skull

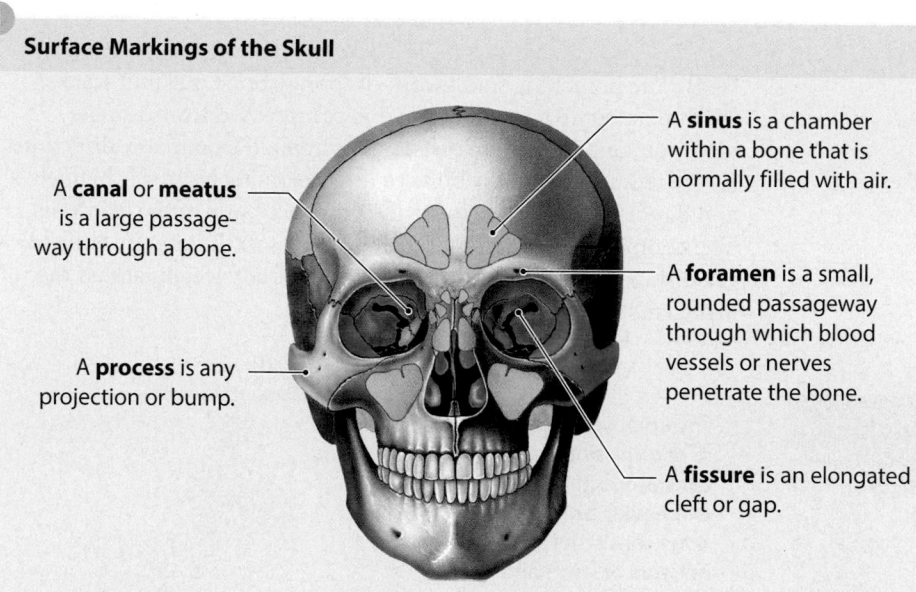

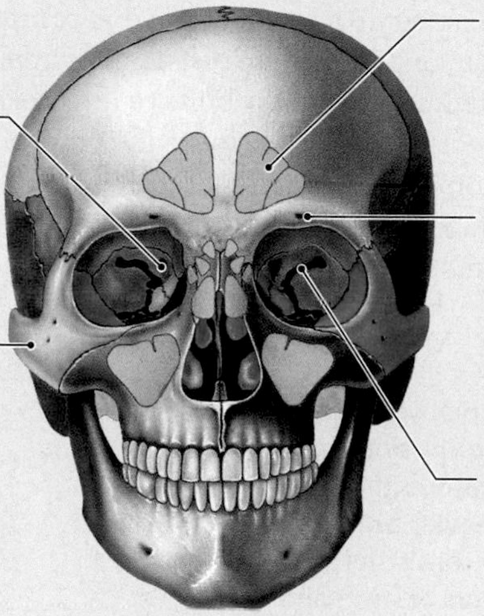
A **canal** or **meatus** is a large passageway through a bone.
A **process** is any projection or bump.
A **sinus** is a chamber within a bone that is normally filled with air.
A **foramen** is a small, rounded passageway through which blood vessels or nerves penetrate the bone.
A **fissure** is an elongated cleft or gap.

2 Surface Markings of the Humerus

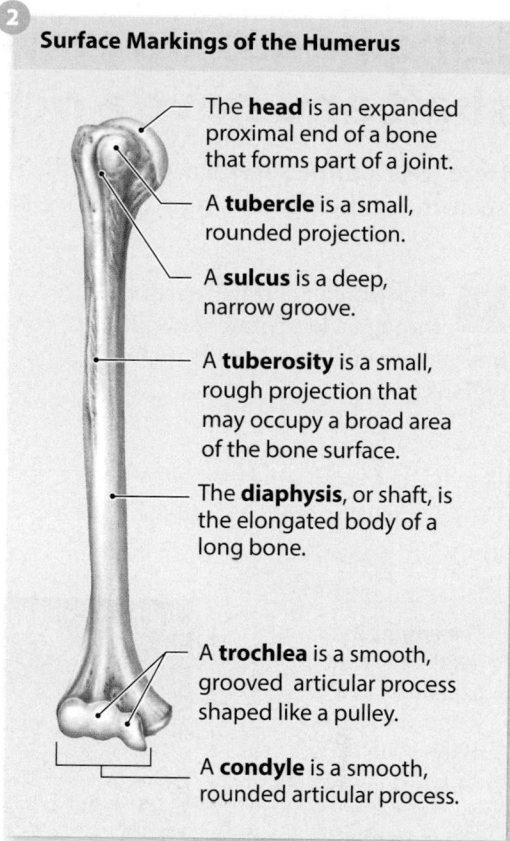

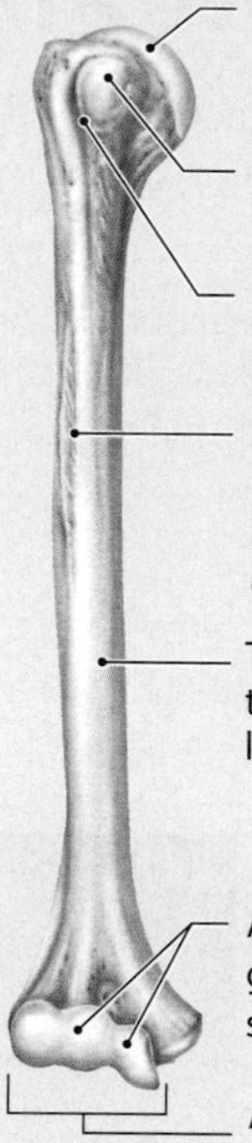
The **head** is an expanded proximal end of a bone that forms part of a joint.
A **tubercle** is a small, rounded projection.
A **sulcus** is a deep, narrow groove.
A **tuberosity** is a small, rough projection that may occupy a broad area of the bone surface.
The **diaphysis**, or shaft, is the elongated body of a long bone.
A **trochlea** is a smooth, grooved articular process shaped like a pulley.
A **condyle** is a smooth, rounded articular process.

3 Surface Markings of the Femur

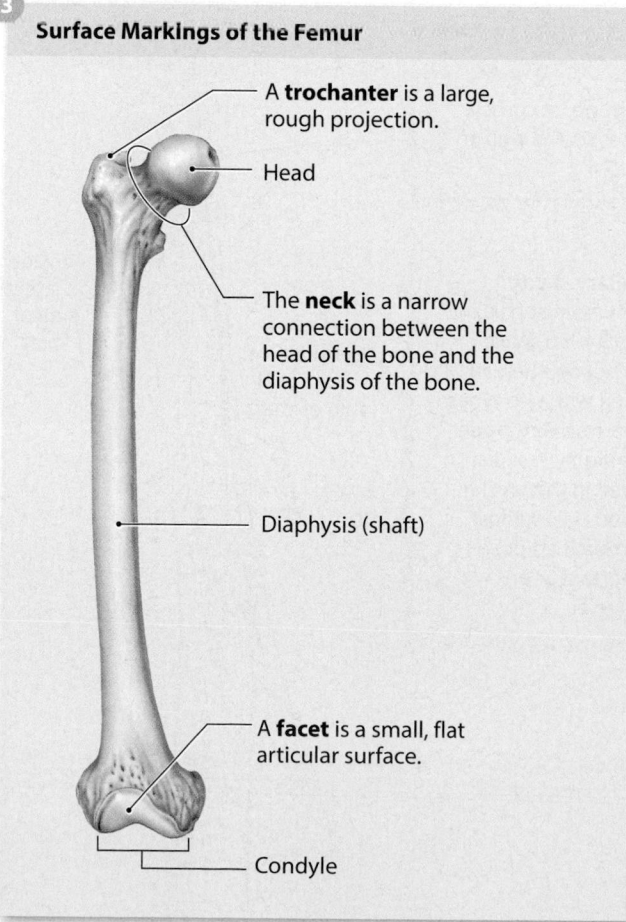

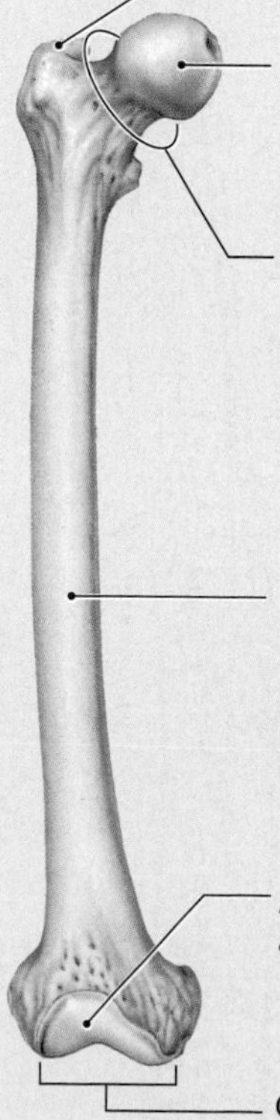
A **trochanter** is a large, rough projection.
Head
The **neck** is a narrow connection between the head of the bone and the diaphysis of the bone.
Diaphysis (shaft)
A **facet** is a small, flat articular surface.
Condyle

4 Surface Markings of the Pelvis

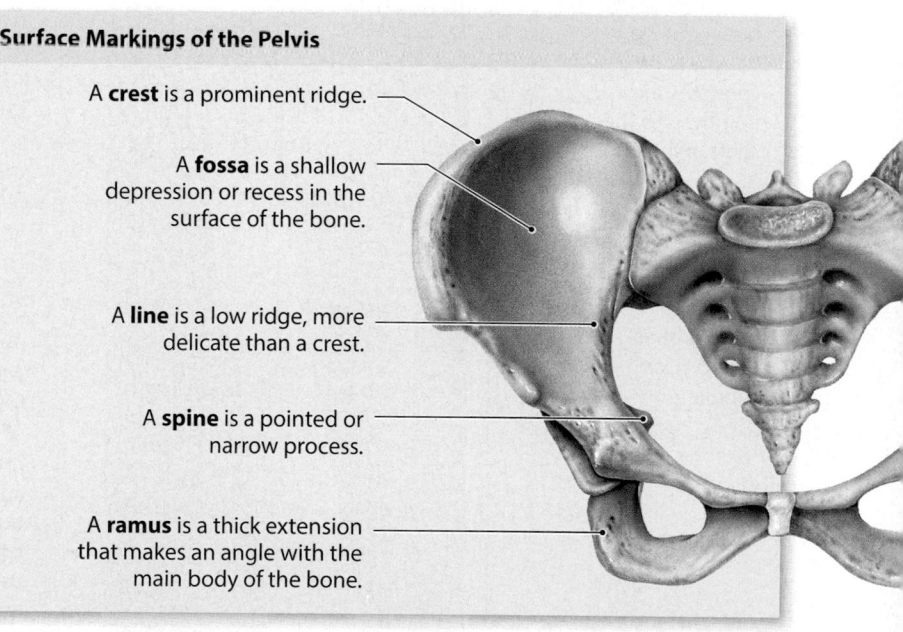

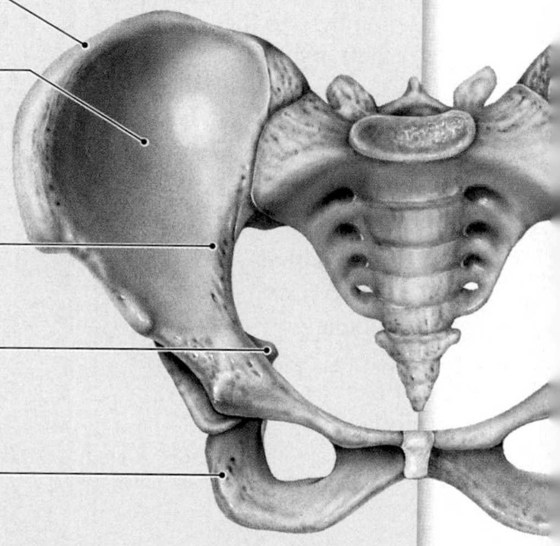
A **crest** is a prominent ridge.
A **fossa** is a shallow depression or recess in the surface of the bone.
A **line** is a low ridge, more delicate than a crest.
A **spine** is a pointed or narrow process.
A **ramus** is a thick extension that makes an angle with the main body of the bone.

Module 6.2 Review

a. Identify the six broad categories for classifying a bone according to shape.

b. Define surface markings.

c. Compare a tubercle with a tuberosity.

LO **6.2** Classify bones according to their shapes, identify the major types of surface markings, and explain the functional significance of surface markings.

Long bones transmit forces along the shaft and have a rich blood supply

Here we examine several aspects of the functional anatomy of the femur, a representative long bone.

1 This frontal (coronal) section through a left femur shows its internal organization and the major regions.

2 The branching framework of spongy bone makes it weaker than compact bone when it is compressed from a single direction, but it is able to resist forces applied from many directions. At the femur, the spongy bone of the proximal epiphysis channels and directs the body weight to the compact bone of the diaphysis. The compact bone then passes these forces to the spongy bone of the distal epiphysis, which distributes the body weight across the articular surface at the knee.

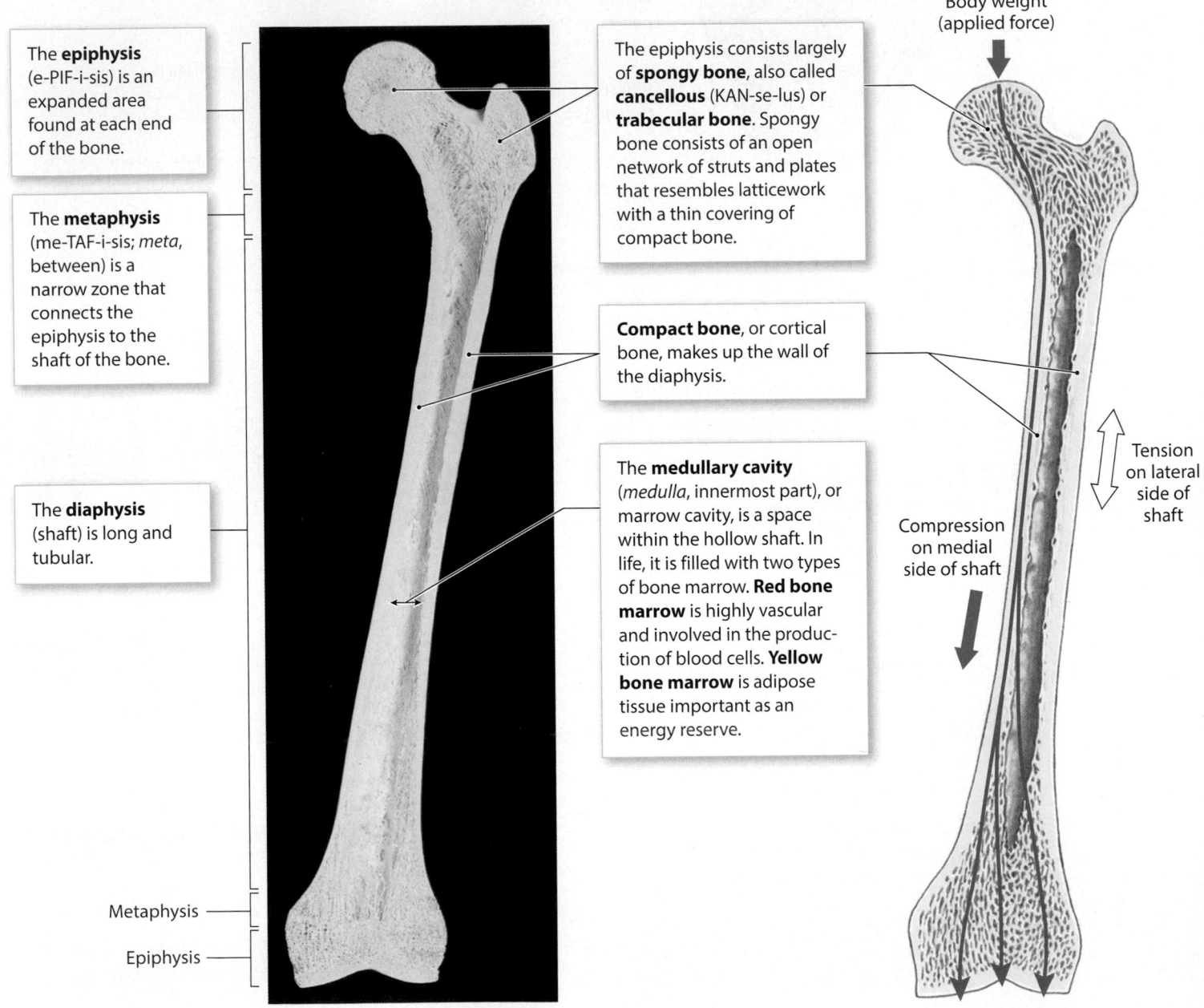

The **epiphysis** (e-PIF-i-sis) is an expanded area found at each end of the bone.

The **metaphysis** (me-TAF-i-sis; *meta*, between) is a narrow zone that connects the epiphysis to the shaft of the bone.

The **diaphysis** (shaft) is long and tubular.

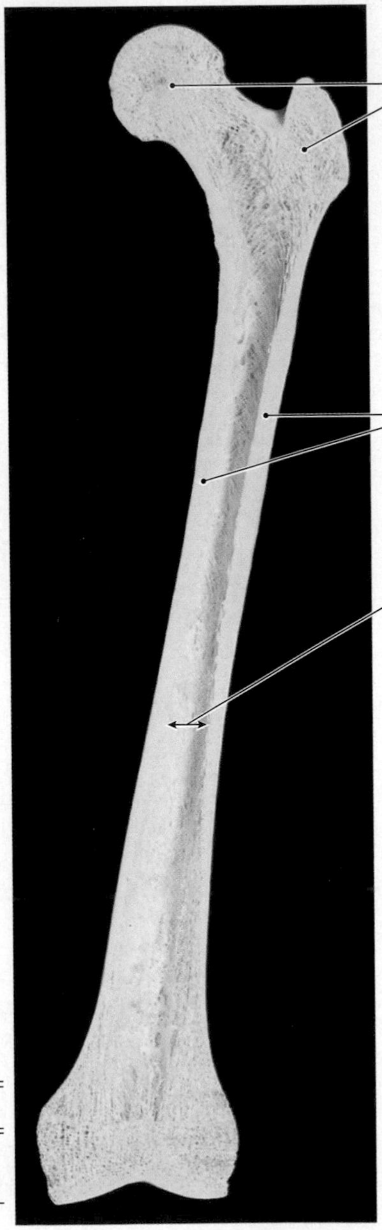
Metaphysis
Epiphysis

The epiphysis consists largely of **spongy bone**, also called **cancellous** (KAN-se-lus) or **trabecular bone**. Spongy bone consists of an open network of struts and plates that resembles latticework with a thin covering of compact bone.

Compact bone, or cortical bone, makes up the wall of the diaphysis.

The **medullary cavity** (*medulla*, innermost part), or marrow cavity, is a space within the hollow shaft. In life, it is filled with two types of bone marrow. **Red bone marrow** is highly vascular and involved in the production of blood cells. **Yellow bone marrow** is adipose tissue important as an energy reserve.

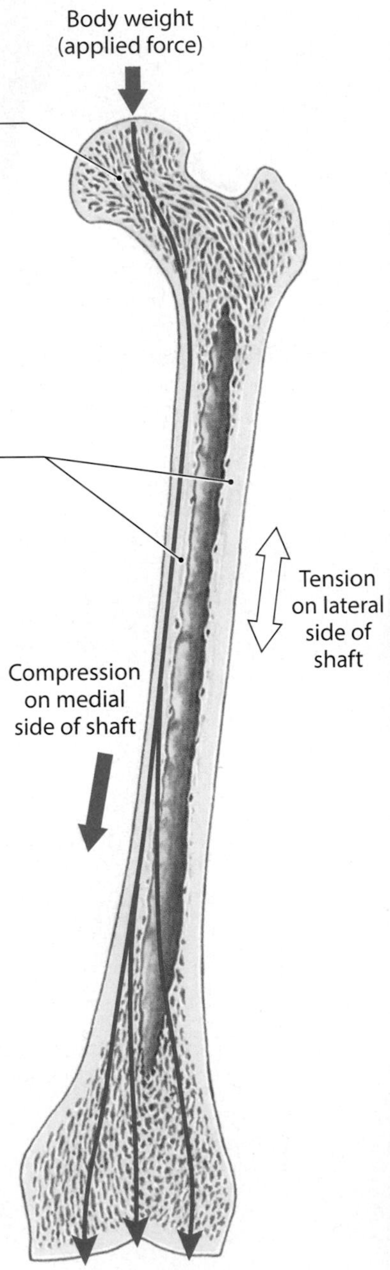
Body weight (applied force)
Tension on lateral side of shaft
Compression on medial side of shaft

3 In order for bones to grow and be maintained, they require an extensive blood supply. Therefore, osseous tissue is highly vascular. In a typical long bone such as the humerus, shown here, multiple sets of blood vessels develop. Extensive interconnections exist between them, so that if a blood vessel is compressed or cut, the vascular supply to any part of the bone will not be completely shut off.

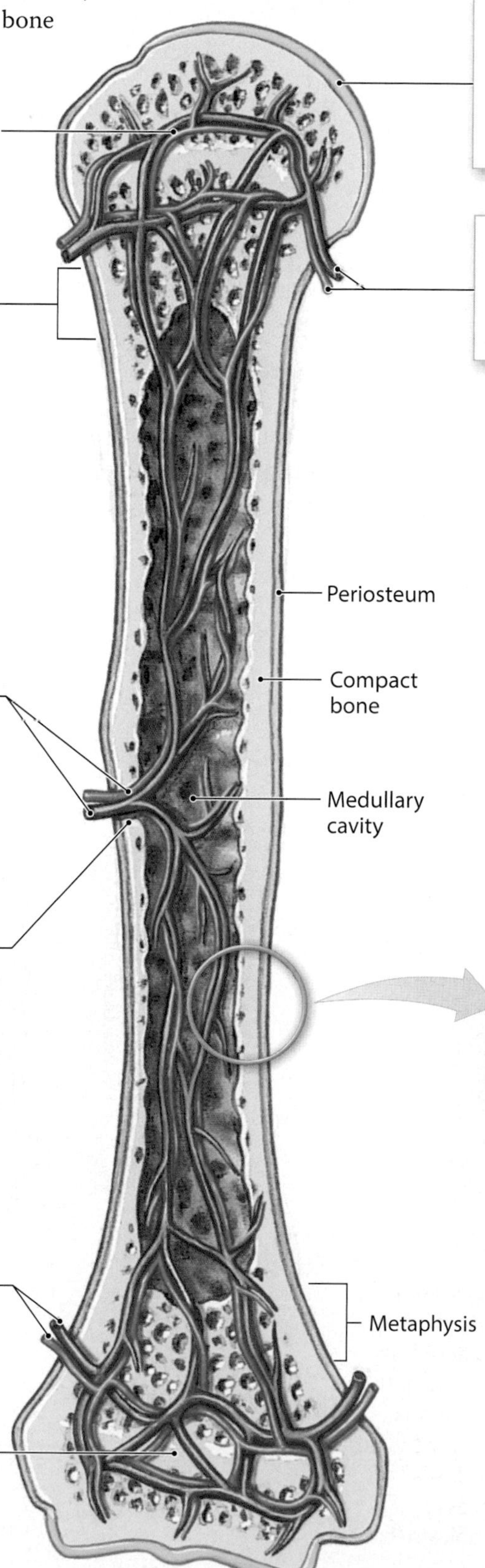
Epiphyseal artery and vein
Metaphysis
Periosteum
Compact bone
Medullary cavity
Metaphyseal artery and vein
Metaphysis
Epiphyseal line

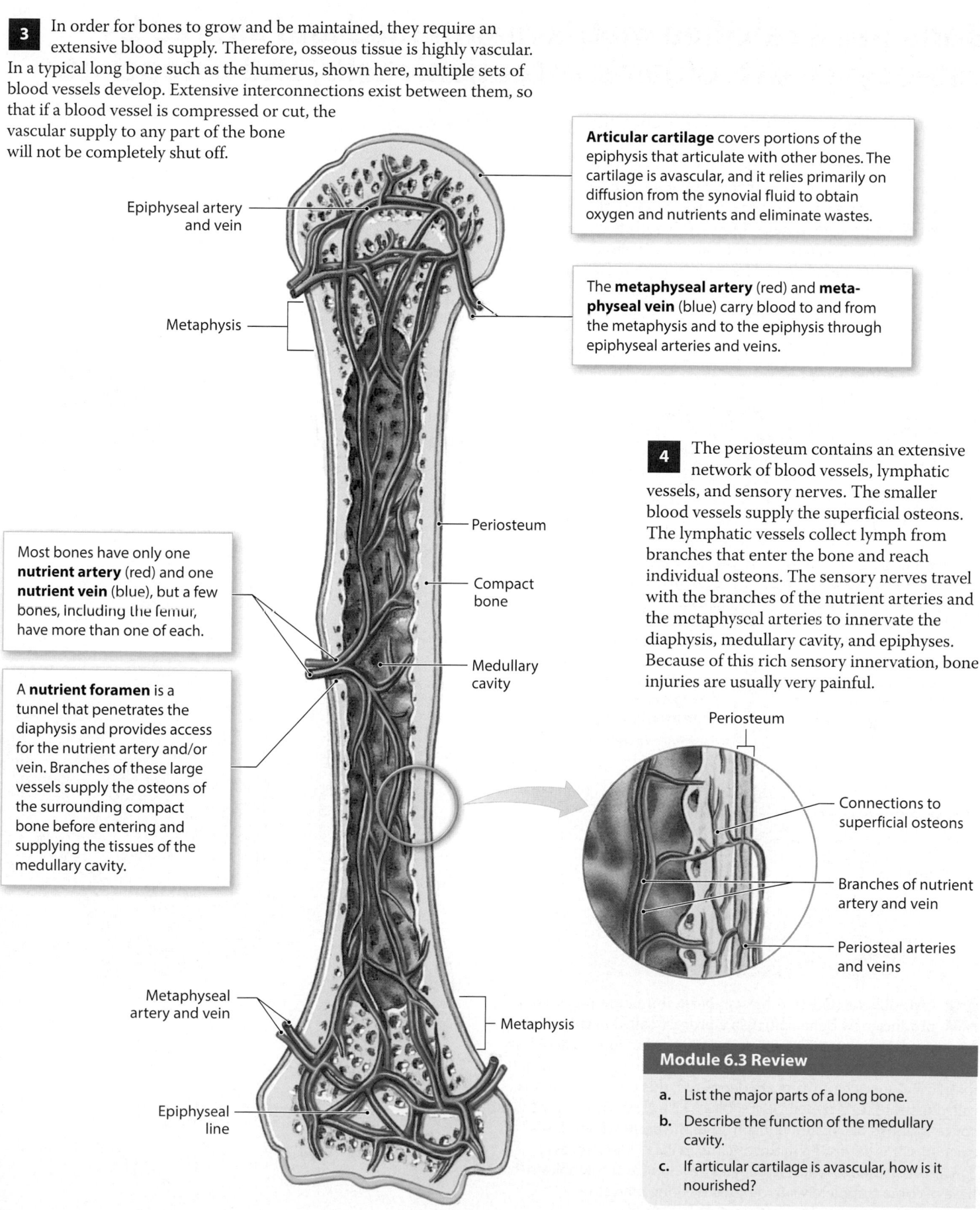

Articular cartilage covers portions of the epiphysis that articulate with other bones. The cartilage is avascular, and it relies primarily on diffusion from the synovial fluid to obtain oxygen and nutrients and eliminate wastes.

The **metaphyseal artery** (red) and **metaphyseal vein** (blue) carry blood to and from the metaphysis and to the epiphysis through epiphyseal arteries and veins.

Most bones have only one **nutrient artery** (red) and one **nutrient vein** (blue), but a few bones, including the femur, have more than one of each.

A **nutrient foramen** is a tunnel that penetrates the diaphysis and provides access for the nutrient artery and/or vein. Branches of these large vessels supply the osteons of the surrounding compact bone before entering and supplying the tissues of the medullary cavity.

4 The periosteum contains an extensive network of blood vessels, lymphatic vessels, and sensory nerves. The smaller blood vessels supply the superficial osteons. The lymphatic vessels collect lymph from branches that enter the bone and reach individual osteons. The sensory nerves travel with the branches of the nutrient arteries and the metaphyseal arteries to innervate the diaphysis, medullary cavity, and epiphyses. Because of this rich sensory innervation, bone injuries are usually very painful.

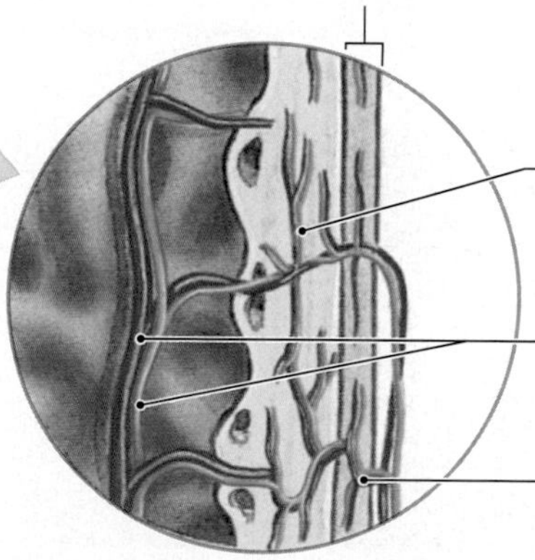
Periosteum
Connections to superficial osteons
Branches of nutrient artery and vein
Periosteal arteries and veins

Module 6.3 Review

a. List the major parts of a long bone.

b. Describe the function of the medullary cavity.

c. If articular cartilage is avascular, how is it nourished?

LO **6.3** Identify the parts of a typical long bone, and describe its internal structures.

Bone has a calcified matrix maintained and altered by osteocytes, osteoblasts, osteogenic cells, and osteoclasts

Both compact bone and spongy bone contain the same four cell types.

1 Mature bone cells called **osteocytes** (OS-tē-ō-sīts; (*osteo-*, bone + *cyte*, cell) maintain the protein and mineral content of the surrounding matrix through the turnover of matrix components. Osteocytes secrete chemicals that dissolve the adjacent matrix, and the released minerals enter the circulation. The osteocytes then rebuild the matrix, stimulating the deposition of mineral crystals. Osteocytes also participate in the repair of damaged bone.

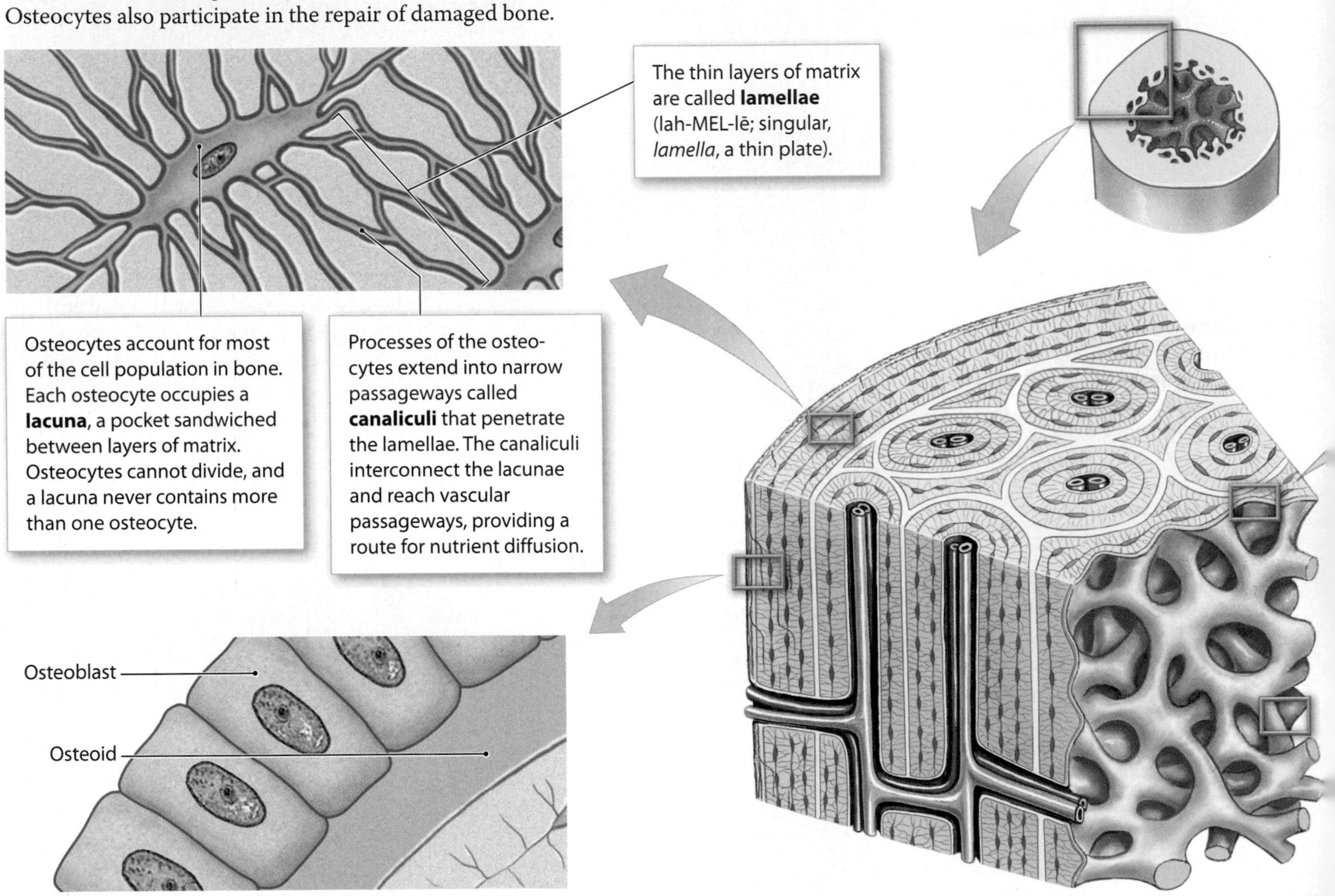
The thin layers of matrix are called **lamellae** (lah-MEL-lē; singular, *lamella*, a thin plate).
Osteocytes account for most of the cell population in bone. Each osteocyte occupies a **lacuna**, a pocket sandwiched between layers of matrix. Osteocytes cannot divide, and a lacuna never contains more than one osteocyte.
Processes of the osteocytes extend into narrow passageways called **canaliculi** that penetrate the lamellae. The canaliculi interconnect the lacunae and reach vascular passageways, providing a route for nutrient diffusion.
Osteoblast
Osteoid

2 **Osteoblasts** (OS-tē-ō-blasts; *-blast*, immature precursor cell) produce new bone matrix in a process called **ossification**, or **osteogenesis** (os-tē-ō-JEN-e-sis; *genesis*, production). Osteoblasts make and release the proteins and other organic components of the matrix. Before calcium salts are deposited, this organic matrix is called **osteoid** (OS-tē-oyd). Osteoblasts also assist in elevating local concentrations of calcium phosphate to the point where this calcium salt is deposited in the organic matrix. This process converts osteoid to bone. Osteocytes develop from osteoblasts that have become completely surrounded by bone matrix.

3 **Osteogenic** (os-tē-ō-JEN-ik) **cells,** also called osteoprogenitor cells, are mesenchymal cells present in bone (Module 4.10, p. 152). They are found in the inner, cellular layer of the periosteum; in an inner layer, or endosteum, that lines medullary cavities; and in the vascular passageways, which contain blood vessels that penetrate the matrix of compact bone. These stem cells divide to produce daughter cells that differentiate into osteoblasts, and are thus important in the repair of a fracture (a break or a crack in a bone).

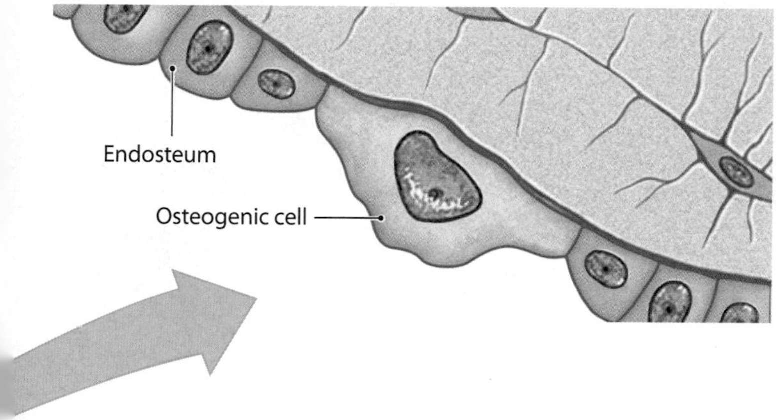
Endosteum
Osteogenic cell

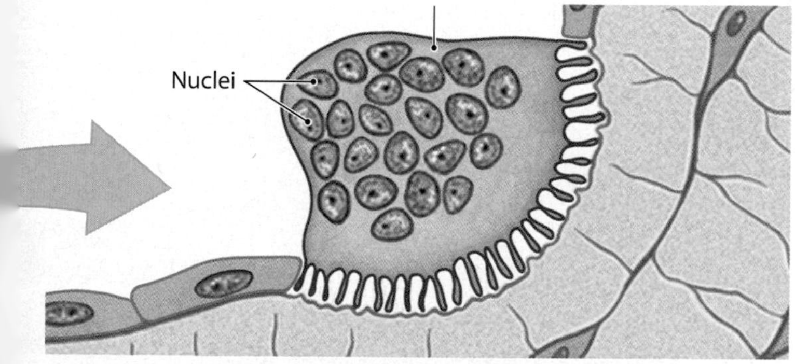
Osteoclast
Nuclei

4 **Osteoclasts** (OS-tē-ō-clasts; *klastos,* broken) are cells that remove and recycle bone matrix. These are large cells with 50 or more nuclei. Osteoclasts are not related to osteogenic cells or their descendants. Instead, they are derived from the same stem cells that produce monocytes and macrophages, cells involved in the body's defense mechanisms. Acids and proteolytic (protein-digesting) enzymes secreted by osteoclasts dissolve the matrix and release the stored minerals. This process, called **osteolysis** (os-tē-OL-i-sis; *lysis,* a loosening) or resorption, is important in regulating calcium and phosphate ion concentrations in blood and other body fluids.

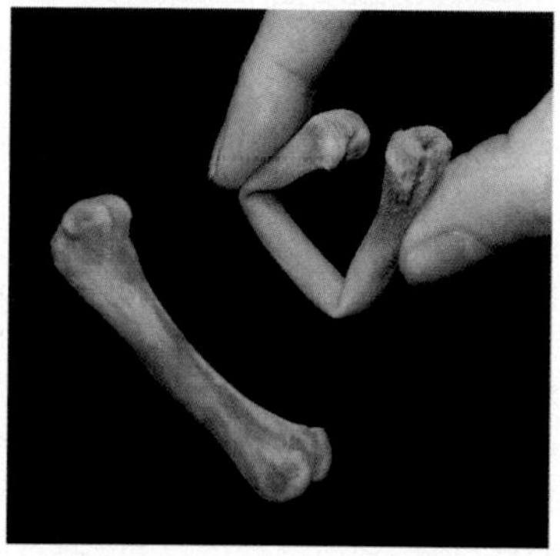

5 A bone without a calcified matrix looks normal, but is very flexible. Roughly one-third of the weight of bone is contributed by collagen fibers, and cells account for only 2 percent of the weight of a typical bone. A calcium salt, calcium phosphate, $Ca_3(PO_4)_2$, accounts for almost two-thirds of the weight of bone. Calcium phosphate interacts with calcium hydroxide, $Ca(OH)_2$, to form crystals of **hydroxyapatite,** $Ca_{10}(PO_4)_6(OH)_2$. As they form, these crystals incorporate other calcium salts, such as calcium carbonate ($CaCO_3$), and ions such as sodium, magnesium, and fluoride. Hydroxyapatite is very hard but inflexible and brittle. Collagen fibers are strong and flexible, but if they are compressed they bend. The protein–crystal combination in bone is strong, somewhat flexible, and highly resistant to shattering. In its overall properties, bone is on a par with the best steel-reinforced concrete. In fact, bone is far superior to concrete, because it can undergo remodeling (cycles of bone formation and resorption) as needed and can repair itself after injury.

Module 6.4 Review

a. Define osteocyte, osteoblast, osteogenic cell, and osteoclast.

b. If osteoclast activity exceeds osteoblast activity in a bone, what would be the effect on the bone?

c. How would the compressive strength of a bone be affected if the ratio of collagen to hydroxyapatite increased?

LO **6.4** Identify the cell types in bone, and list their major functions.

Compact bone consists of parallel osteons, and spongy bone consists of a network of trabeculae

The basic functional unit of mature compact bone is the **osteon** (OS-tē-on), or Haversian system. Many of the important features of osteons were introduced in Module 4.13 (p. 159).

1 The lamellae of each osteon form a series of nested cylinders around the central canal. In transverse section, these **concentric lamellae** resemble a target, with the **central canal** as the bull's-eye. You might think of a single osteon as a drinking straw with very thick walls: When you attempt to push the ends of the straw together or to pull them apart, the straw is quite strong. But if you hold the ends and push from the side, the straw will bend sharply with relative ease. The osteons in the diaphysis of a long bone are parallel to the long axis of the shaft. Thus, the shaft does not bend, even when extreme forces are applied to either end. The femur can withstand 10–15 times the body's weight without breaking. Yet a much smaller force applied to the side of the shaft can break any long bone, even the femur.

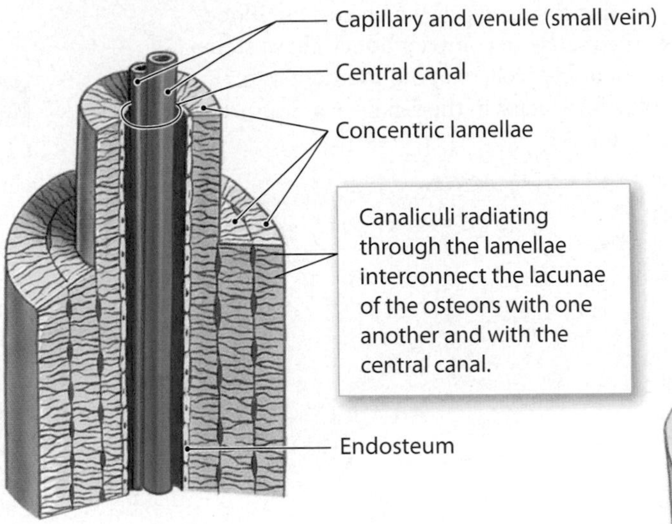
Capillary and venule (small vein)
Central canal
Concentric lamellae
Canaliculi radiating through the lamellae interconnect the lacunae of the osteons with one another and with the central canal.
Endosteum

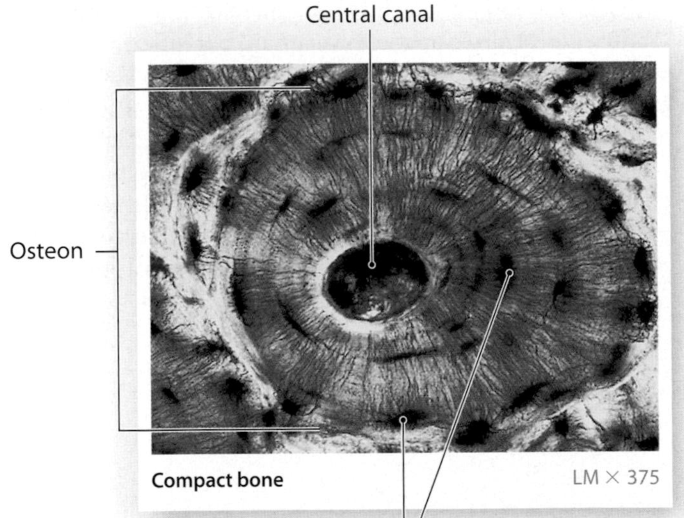
Central canal
Osteon

Compact bone LM × 375

The osteocytes occupy lacunae that lie between the lamellae. In preparing this micrograph, a small piece of bone was ground down until it was thin enough to transmit light. In this process, the lacunae and canaliculi are filled with bone dust, and thus appear black.

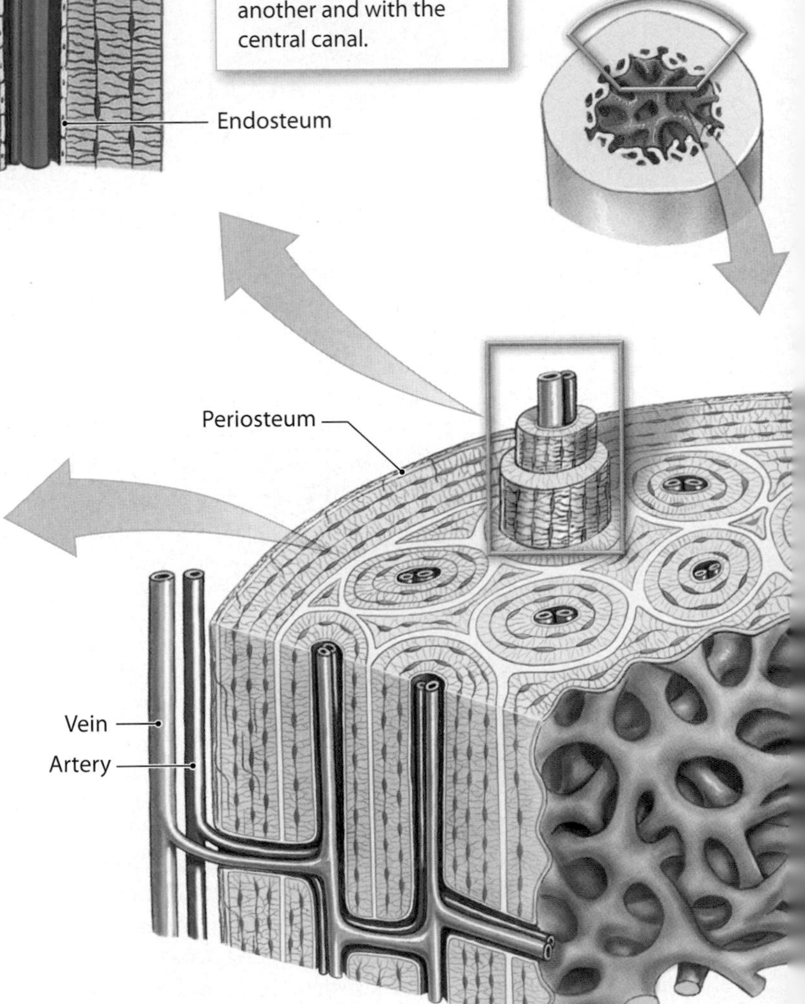
Periosteum
Vein
Artery

2 A section cut from the shaft of a long bone shows the organization of osteons and blood vessels, and allows a comparison between compact and spongy bone.

3 Spongy bone is found where bones are not heavily stressed or where stresses arrive from many directions. Below is the head of the femur, which transmits body weight to the shaft of that bone. The trabeculae are oriented along stress lines and are cross-braced extensively. In addition to being able to withstand stresses applied from many directions, spongy bone is much lighter than compact bone. Spongy bone thus reduces the weight of the skeleton, making it easier for muscles to move the bones.

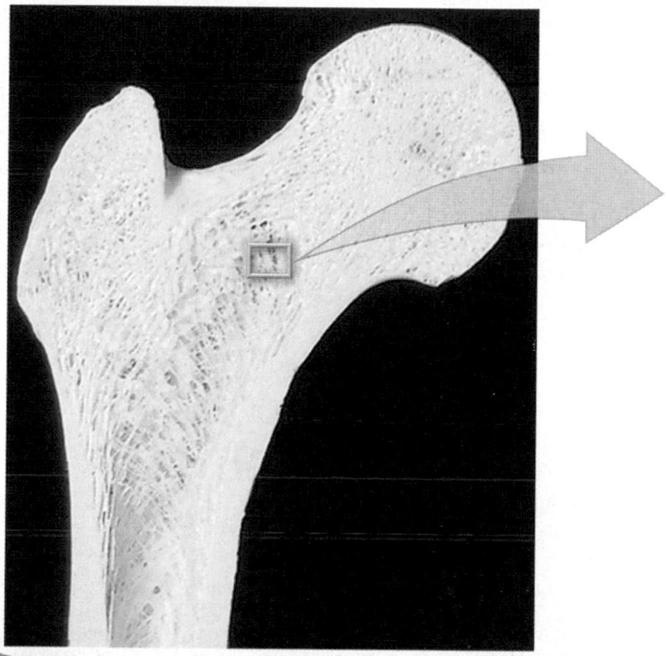

4 In spongy bone, lamellae are not arranged in osteons. The matrix in spongy bone forms struts and plates called **trabeculae**. The thin trabeculae branch, creating an open network. There are no capillaries or venules in the matrix of spongy bone. Nutrients reach the osteocytes by diffusion along canaliculi that open onto the surfaces of trabeculae. Red bone marrow is found between the trabeculae of spongy bone. Blood vessels within this tissue deliver nutrients to the trabeculae and remove wastes generated by the osteocytes.

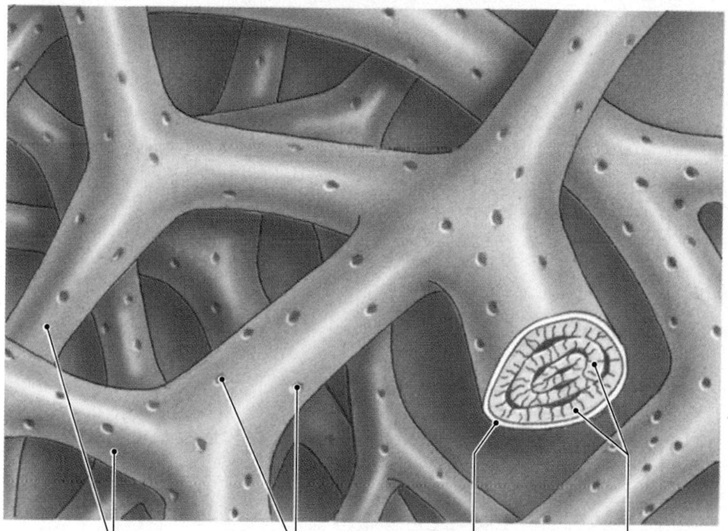
Trabeculae of spongy bone
Canaliculi opening on surface
Endosteum
Lamellae

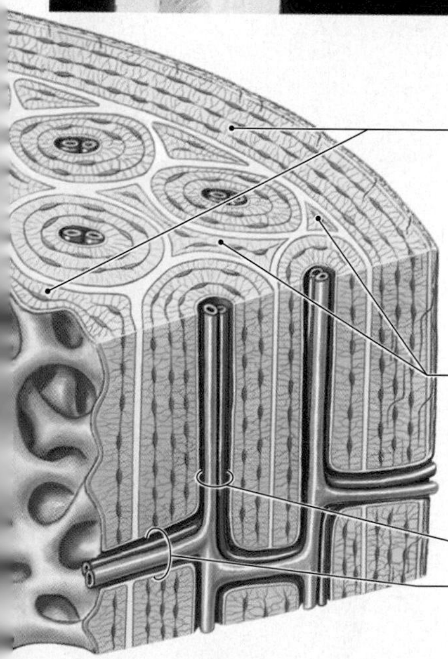

Circumferential lamellae (*circum-*, around + *ferre*, to bear) are found at the outer and inner surfaces of the bone, where they are covered by the periosteum and endosteum, respectively. These lamellae are produced during the growth and maintenance of the bone.

Interstitial lamellae fill in the spaces between the osteons in compact bone. These lamellae are remnants of osteons whose matrix components have been almost completely recycled by osteoclasts.

Central canal

Perforating canal

Central canals generally run parallel to the surface of the bone. Other passageways, known as **perforating canals**, extend perpendicular to the surface. Blood vessels in these canals supply blood to osteons deeper in the bone and to tissues of the medullary cavity.

Module 6.5 Review

a. Define osteon.

b. Compare the structures and functions of compact bone and spongy bone.

c. A sample of bone has lamellae that are not arranged in osteons. Is the sample more likely from the epiphysis or from the diaphysis?

LO 6.5 Compare the structures and functions of compact bone and spongy bone.

In appositional bone growth, layers of compact bone are added to the bone's outer surface

The appositional growth of cartilage was detailed in Module 4.12 (p. 157). As you will see in this module, appositional bone growth has many similarities to appositional cartilage growth.

1 The diameter of a bone enlarges through **appositional growth** at the outer surface. In appositional growth, osteogenic cells in the inner layer of the periosteum differentiate into osteoblasts and add bone matrix to the surface. This adds successive layers of circumferential lamellae to the outer surface of the bone. Osteoblasts trapped between these lamellae differentiate into osteocytes. Over time, the deeper lamellae are recycled and replaced with the osteons typical of compact bone.

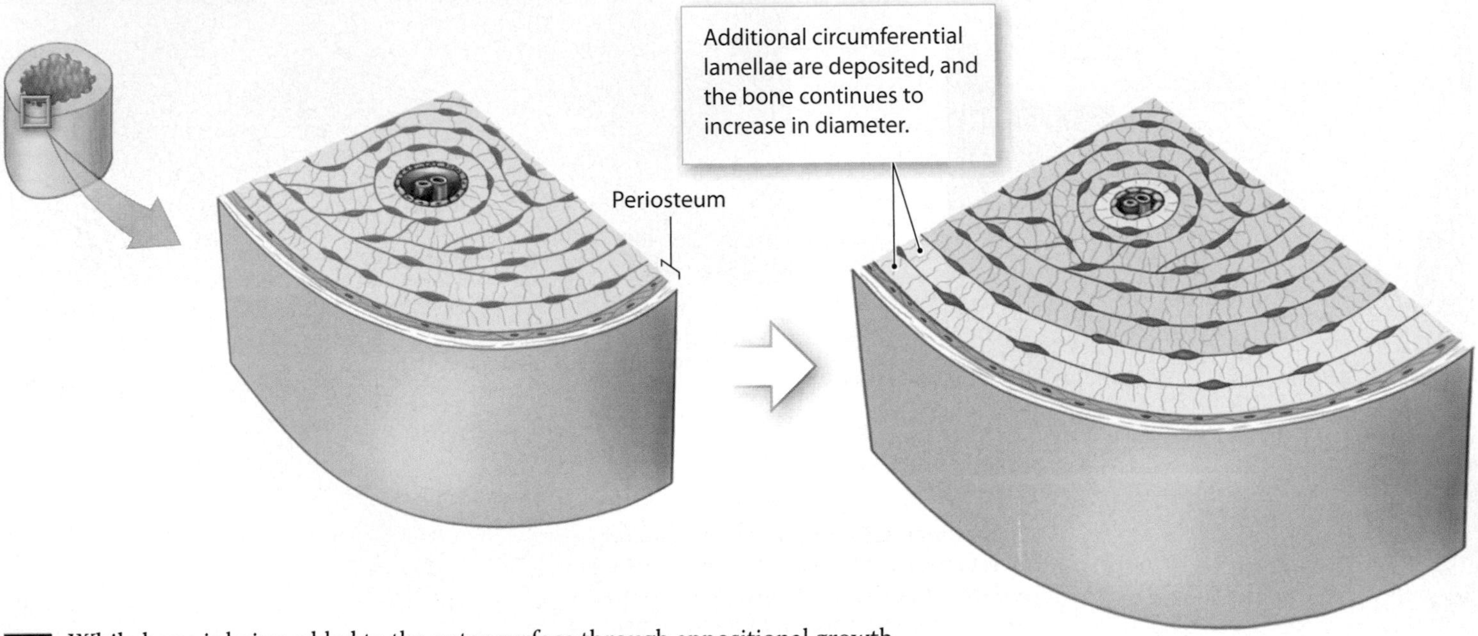
Additional circumferential lamellae are deposited, and the bone continues to increase in diameter.
Periosteum

2 While bone is being added to the outer surface through appositional growth, osteoclasts are removing and recycling lamellae at the inner surface. As a result, the medullary cavity gradually enlarges as the bone increases in diameter.

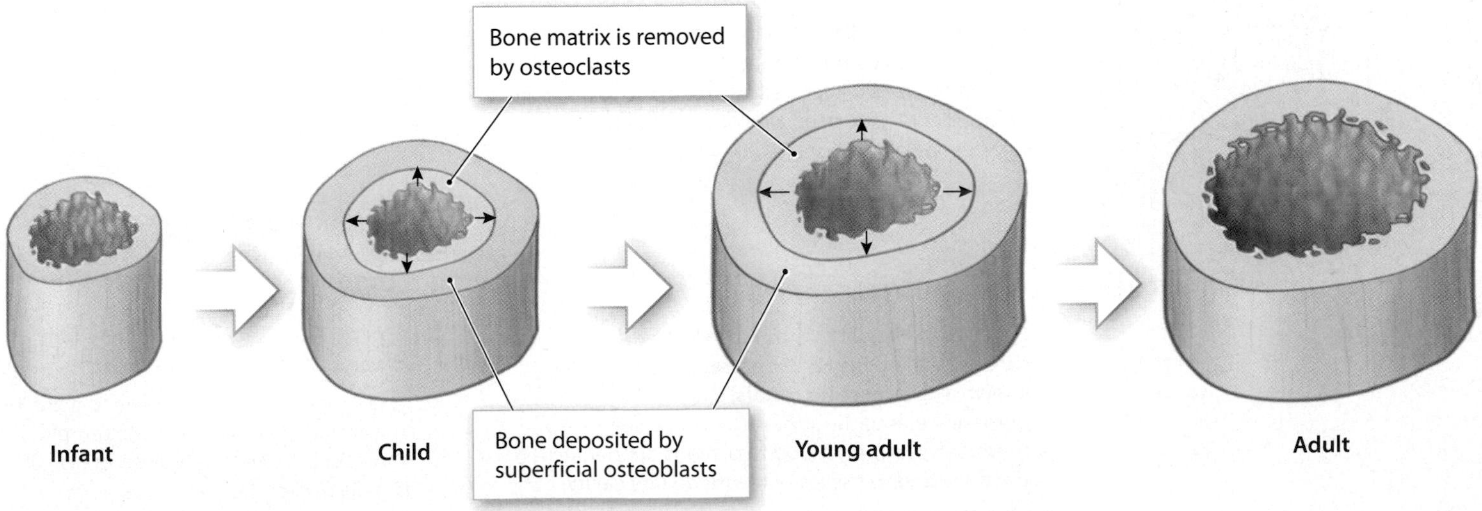
Bone matrix is removed by osteoclasts
Infant
Child
Bone deposited by superficial osteoblasts
Young adult
Adult

3 Except within joint cavities, the superficial layer of compact bone that covers all bones is wrapped by a **periosteum**, which has a fibrous outer layer and a cellular inner layer. The periosteum (1) isolates the bone from surrounding tissues, (2) provides a route for the blood and nervous supply, and (3) actively participates in bone growth and repair. As the bone grows, the collagen fibers from tendons, ligaments, and joint capsules are cemented into the circumferential lamellae by osteoblasts from the cellular layer of the periosteum. These are called **perforating fibers**, and this method of attachment is extremely strong. An excessive pull on a tendon or ligament will usually break a bone rather than snap the collagen fibers at the bone surface.

4 The **endosteum** is an incomplete cellular layer that lines the medullary cavity. This layer, which is active during bone growth, repair, and remodeling, covers the trabeculae of spongy bone and lines the inner surfaces of the central canals. It consists of a simple flattened layer of osteogenic cells that covers the bone matrix, generally without any intervening connective tissue fibers. Where the cellular layer is incomplete, the matrix is exposed. At these exposed sites, osteoclasts and osteoblasts can remove or deposit matrix components. The osteoclasts generally occur in shallow depressions, called **osteoclastic crypts** or Howship's lacunae, that they have eroded into the matrix. As a result of their activities, the inner circumferential lamellae are often incomplete or interrupted.

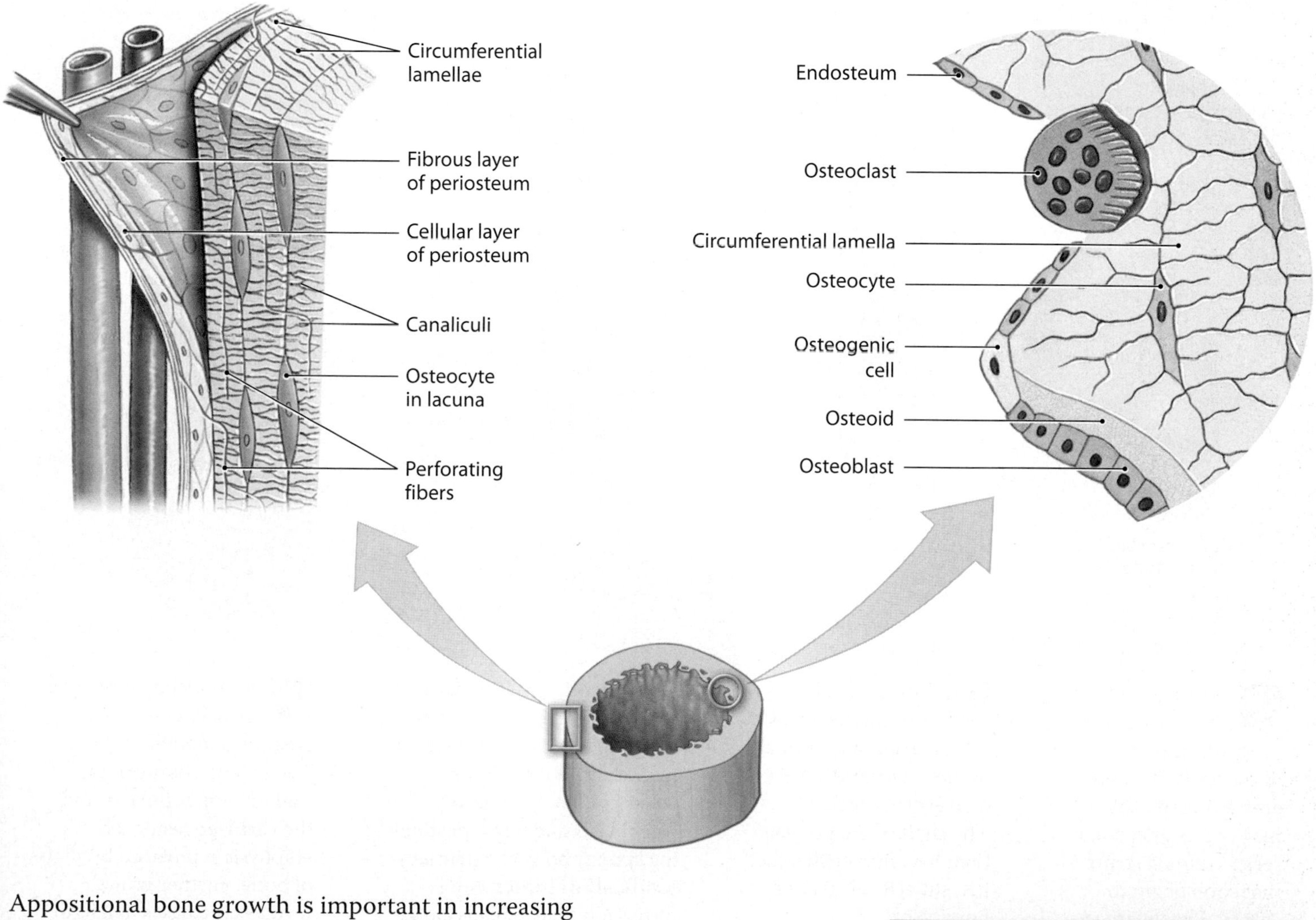

Circumferential lamellae
Fibrous layer of periosteum
Cellular layer of periosteum
Canaliculi
Osteocyte in lacuna
Perforating fibers
Endosteum
Osteoclast
Circumferential lamella
Osteocyte
Osteogenic cell
Osteoid
Osteoblast

Appositional bone growth is important in increasing the diameters of existing bones, but it does not form the original bones. Bone formation begins roughly 6 weeks after fertilization, when the embryo is 12 mm (0.5 in.) long. Two major processes are involved: endochondral ossification (Module 6.7) and intramembranous ossification (Module 6.8).

Module 6.6 Review

a. Define appositional growth.

b. As a bone increases in diameter, what happens to the medullary cavity?

c. Distinguish between the periosteum and the endosteum.

Module 6.7

Endochondral ossification replaces a cartilaginous model with bone

Watch
MasteringA&P®
A&P**Flix**™
Endochondral Ossification

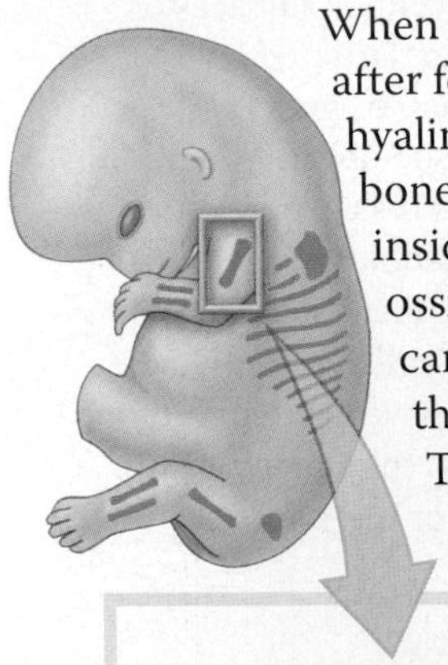

When bone formation begins in the embryo some 6 weeks after fertilization, all existing skeletal elements are made of hyaline cartilage. These cartilages are gradually replaced by bone through the process of **endochondral** (en-dō-KON-drul, *endo-*, inside + *chondros*, cartilage) **ossification**. The key steps in endochondral ossification are illustrated below. This process begins with a small cartilage that is basically a miniature model of the corresponding bone of the adult skeleton. As it forms, the bone grows in length and in diameter. The increase in diameter involves appositional bone deposition.

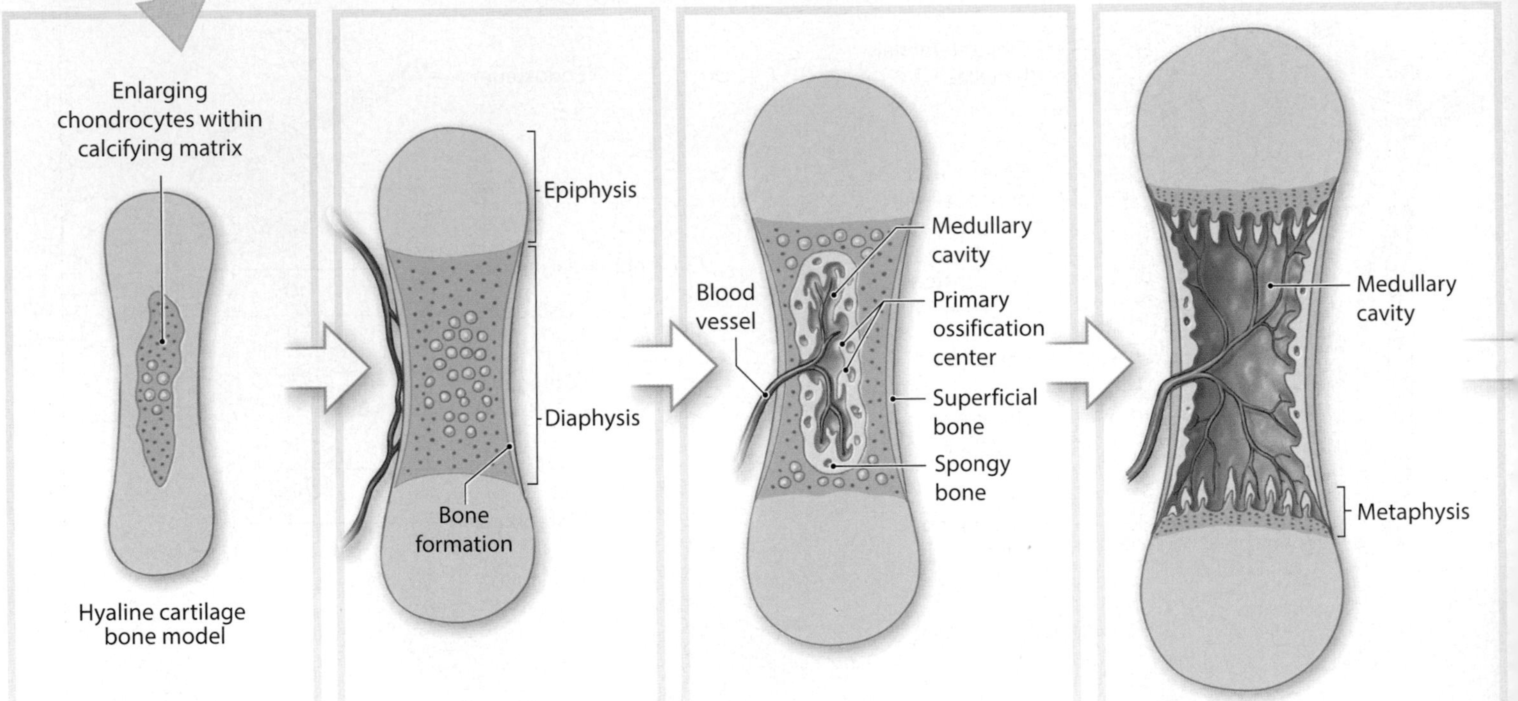

Enlarging chondrocytes within calcifying matrix
Hyaline cartilage bone model
Epiphysis
Diaphysis
Bone formation
Blood vessel
Medullary cavity
Primary ossification center
Superficial bone
Spongy bone
Medullary cavity
Metaphysis

1 As the cartilage enlarges, chondrocytes near the center of the shaft increase greatly in size. The matrix is reduced to a series of small struts that soon begin to calcify. The enlarged chondrocytes then die and disintegrate, leaving cavities within the cartilage.

2 Blood vessels grow around the edges of the cartilage, and the cells of the perichondrium convert to osteoblasts. The shaft of the cartilage then becomes ensheathed in a superficial layer of bone.

3 Blood vessels penetrate the cartilage and invade the central region. Fibroblasts migrating with the blood vessels differentiate into osteoblasts and begin producing spongy bone at a **primary ossification center**. Bone formation then spreads along the shaft toward both ends.

4 Remodeling occurs as growth continues, creating a medullary cavity. The osseous tissue of the shaft becomes thicker, and the cartilage near each epiphysis is replaced by shafts of bone. Further growth involves increases in length and diameter.

The chondrocytes within the epiphyseal cartilage are arranged into zones. On the epiphyseal side, chondrocytes continually add new cartilage. On the diaphyseal side, chondrocytes are degenerating. Osteoblasts are continuously invading this zone and replacing it with bone. The osteoblasts are therefore moving toward the epiphysis, which is being pushed away by the continued production of new cartilage.

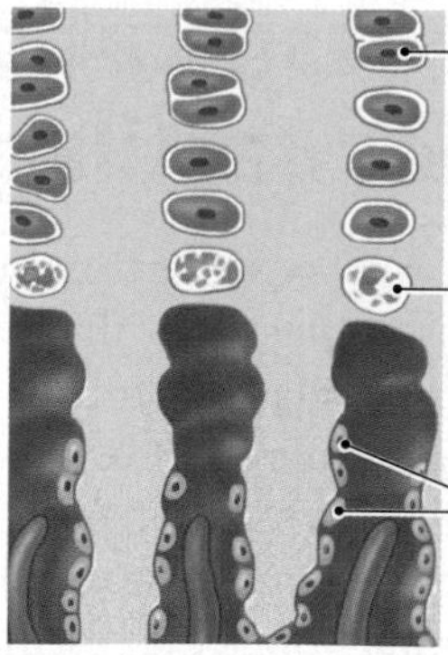
Chondrocytes at the epiphyseal side of the cartilage continue to divide and enlarge.
Chondrocytes degenerate at the diaphyseal side.
Osteoblasts migrate upward from the diaphysis and cartilage is gradually replaced by bone.

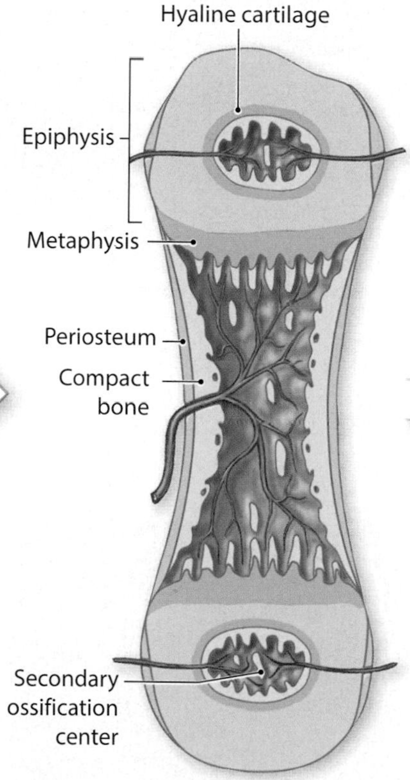
Hyaline cartilage
Epiphysis
Metaphysis
Periosteum
Compact bone
Secondary ossification center

5 Capillaries and osteoblasts migrate into the epiphyses, creating **secondary ossification centers**.

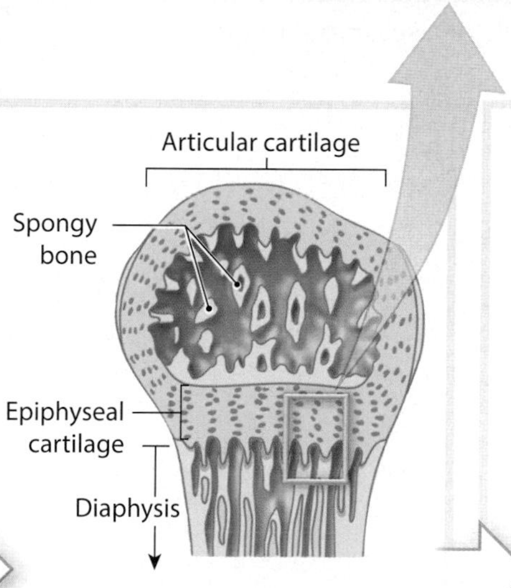
Articular cartilage
Spongy bone
Epiphyseal cartilage
Diaphysis

6 Soon the epiphyses are filled with spongy bone. Articular cartilage remains exposed to the joint cavity; over time it will be reduced to a thin superficial layer. At each metaphysis, an **epiphyseal cartilage**, or **epiphyseal plate**, separates the epiphysis from the diaphysis.

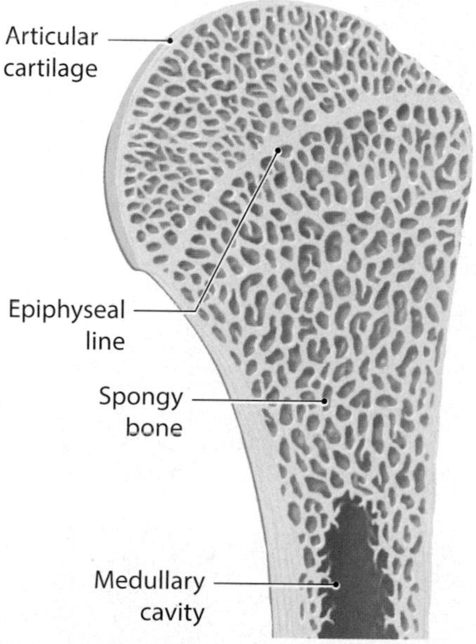
Articular cartilage
Epiphyseal line
Spongy bone
Medullary cavity

7 At puberty, the combination of rising levels of sex hormones, growth hormone, and thyroid hormones stimulates bone growth dramatically. Osteoblasts now begin producing bone faster than chondrocytes are producing new cartilage. As a result, the osteoblasts "catch up," and the epiphyseal cartilage gets narrower and narrower until it ultimately disappears. The completion of epiphyseal growth is called **epiphyseal closure**.

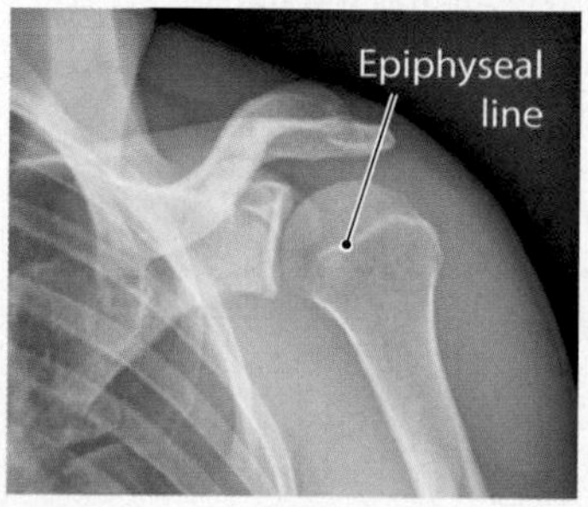
Epiphyseal line

In adults, the former location of the epiphyseal cartilage is often detectable in x-rays as a distinct **epiphyseal line**, which remains after epiphyseal growth has ended. In this x-ray of the left shoulder, the epiphyseal line of the head of the humerus is clearly visible.

Module 6.7 Review

a. Define endochondral ossification.

b. In endochondral ossification, what is the original source of osteoblasts?

c. How could x-rays of the femur be used to determine whether a person has reached full height?

Intramembranous ossification forms bone without a prior cartilaginous model

Intramembranous (in-tra-MEM-bra-nus) **ossification** begins when mesenchymal cells differentiate into osteoblasts within embryonic or fibrous connective tissue. This process normally occurs in the deeper layers of the dermis, and the bones that result are often called **dermal bones** or membrane bones. Examples of dermal bones include the roofing bones of the skull, the lower jaw (mandible), and the collarbone (clavicle). Sesamoid bones are membrane bones that form within tendons. The patella (kneecap) is an example of a sesamoid bone.

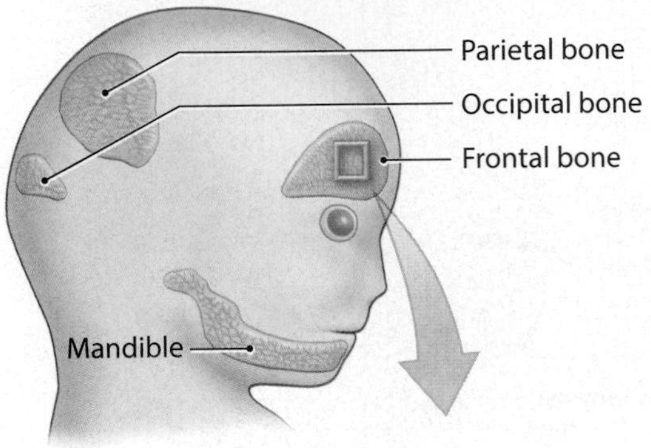

Parietal bone
Occipital bone
Frontal bone
Mandible

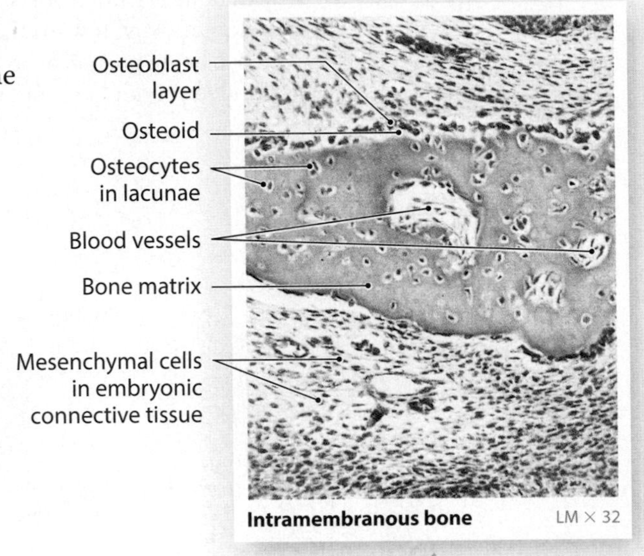

Osteoblast layer
Osteoid
Osteocytes in lacunae
Blood vessels
Bone matrix
Mesenchymal cells in embryonic connective tissue
LM × 32

Intramembranous bone

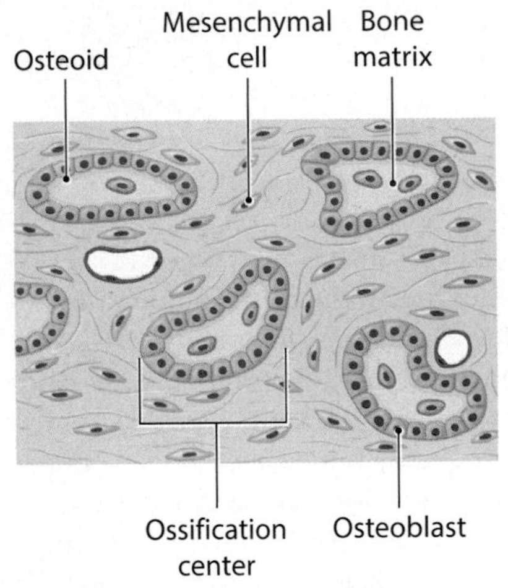

Osteoid
Mesenchymal cell
Bone matrix
Ossification center
Osteoblast

1 Mesenchymal cells first cluster together and start to secrete the organic components of the matrix. The resulting osteoid then becomes mineralized through the crystallization of calcium salts, and the mesenchymal cells differentiate into osteoblasts. The location in a tissue where ossification begins is called an **ossification center**.

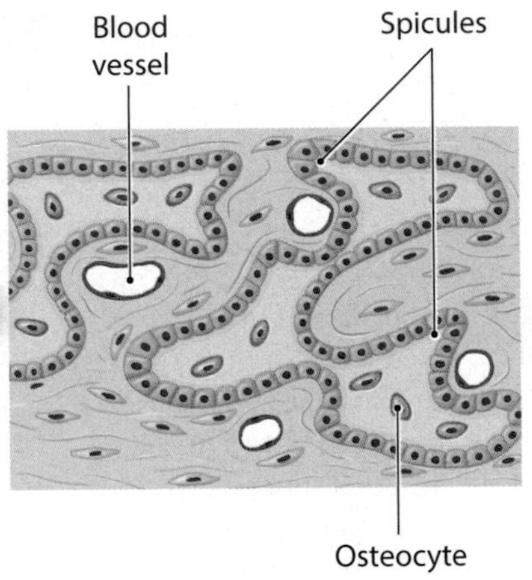

Blood vessel
Spicules
Osteocyte

2 The developing bone grows outward from the ossification center in small struts called **spicules**. As ossification proceeds, it traps some osteoblasts inside bony pockets; these cells differentiate into osteocytes. Meanwhile, mesenchymal cell divisions continue to produce additional osteoblasts.

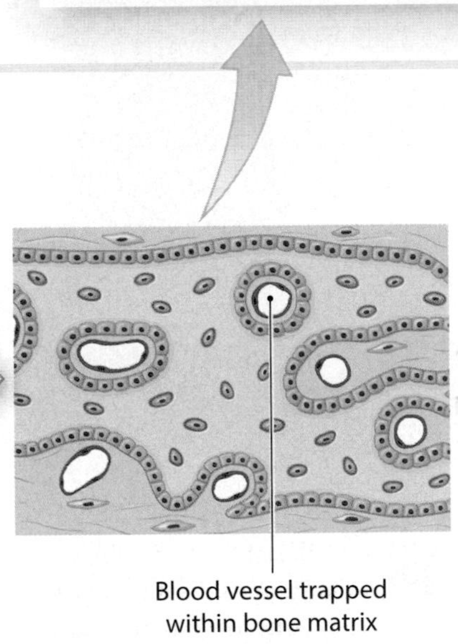

Blood vessel trapped within bone matrix

3 Bone growth is an active process, and osteoblasts require oxygen and a reliable supply of nutrients. Blood vessels begin to grow into the area. As spicules meet and fuse together, some of these blood vessels become trapped within the developing bone.

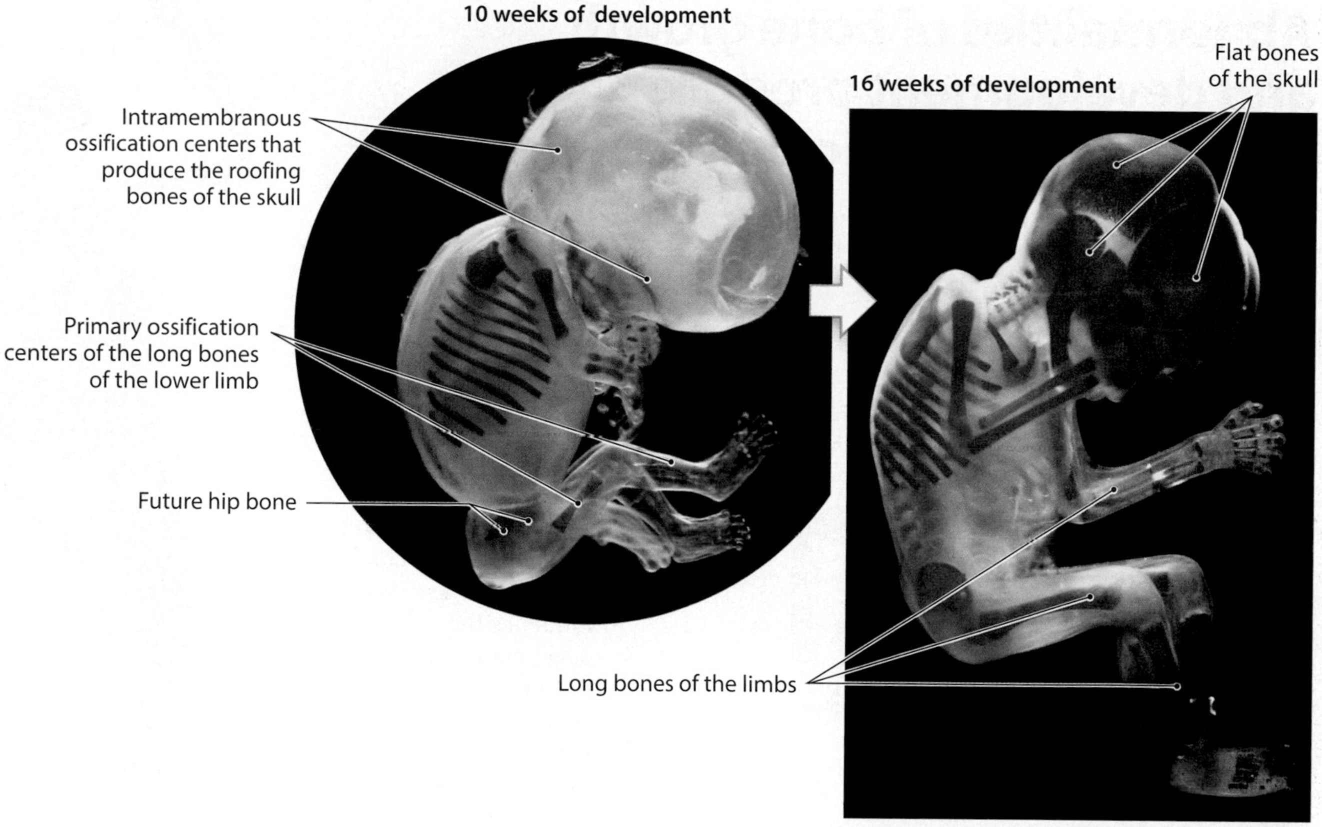
10 weeks of development
16 weeks of development
Intramembranous ossification centers that produce the roofing bones of the skull
Primary ossification centers of the long bones of the lower limb
Future hip bone
Flat bones of the skull
Long bones of the limbs

6 Intramembranous ossification starts approximately during the eighth week of embryonic development. These photos show the extent of intramembranous and endochondral ossification that occurs between 10 and 16 weeks of development. At 10 weeks, bone formation is under way but the skeleton is incomplete. At 16 weeks, most of the bones of the adult skeleton can be identified.

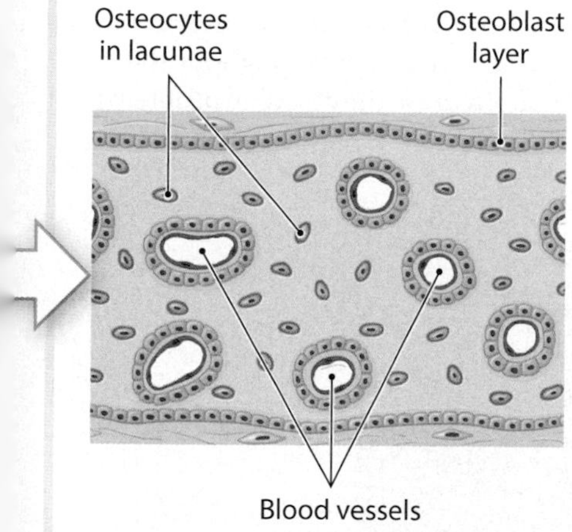
Osteocytes in lacunae
Osteoblast layer
Blood vessels

4 Continued deposition of bone by osteoblasts located close to blood vessels results in a plate of spongy bone with blood vessels weaving throughout.

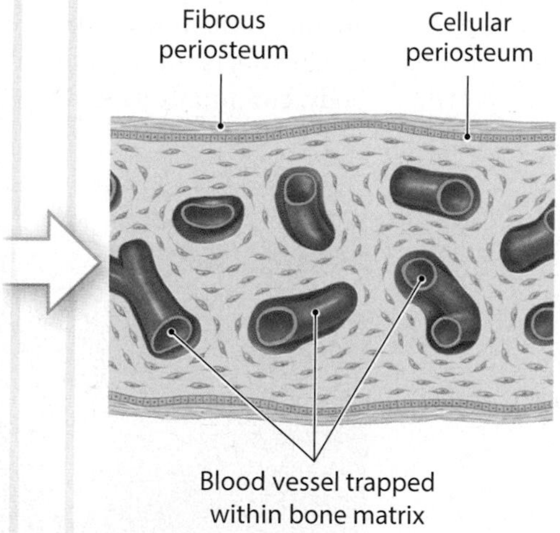
Fibrous periosteum
Cellular periosteum
Blood vessel trapped within bone matrix

5 Initially, the intramembranous bone consists of spongy bone only. Subsequent remodeling around trapped blood vessels can produce osteons typical of compact bone. As the rate of growth slows, the connective tissue around the bone organizes into the fibrous layer of the periosteum. The osteoblasts closest to the bone surface become less active but remain as the inner, cellular layer of the periosteum.

Module 6.8 Review

a. Define intramembranous ossification.

b. During intramembranous ossification, which type(s) of tissue is (are) replaced by bone?

c. Explain the primary difference between endochondral ossification and intramembranous ossification.

LO 6.8 Describe the mechanisms of intramembranous ossification.

Abnormalities of bone growth and development produce recognizable physical signs

A variety of endocrine or metabolic problems can result in atypical skeletal growth. Here we consider several conditions that affect the skeleton as a whole.

1 In **pituitary growth failure**, inadequate production of growth hormone (GH) leads to reduced epiphyseal cartilage activity and abnormally short bones. This condition is becoming increasingly rare in the United States, because children can be treated with synthetic human GH.

2 **Achondroplasia** (ā-kon-drō-PLĀ-zē-uh) results from abnormal epiphyseal activity. In this case the epiphyseal cartilages of the long bones grow unusually slowly and are replaced by bone early in life. As a result, the individual develops short, stocky limbs. Although other skeletal abnormalities occur, the trunk is normal in size, and sexual and mental development remain unaffected.

3 Several inherited metabolic conditions that affect many systems influence the growth and development of the skeletal system. These conditions produce characteristic variations in body proportions. For example, many people with **Marfan's syndrome** are very tall and have long, slender limbs, due to excessive cartilage formation at the epiphyseal cartilages. Although this is an obvious physical distinction, the characteristic body proportions are not in themselves dangerous. However, the underlying mutation, which affects the structure of connective tissue throughout the body, commonly causes life-threatening cardiovascular problems.

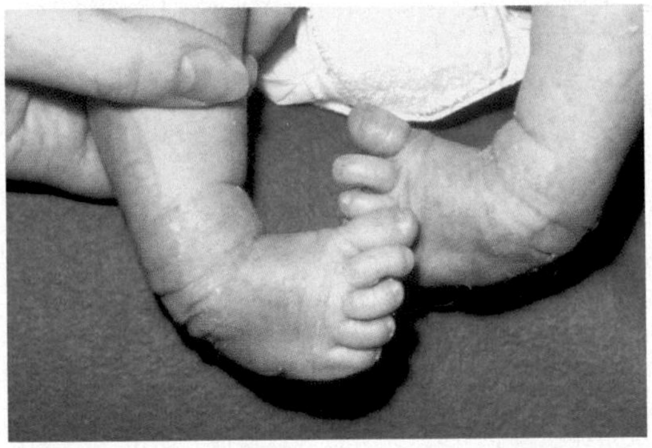

4 **Congenital talipes equinovarus** (TAL-i-pēz e-kwī-nō-VA-rus) (clubfoot) results from an inherited developmental abnormality that affects 2 in 1000 births. Boys are affected roughly twice as often as girls. One or both feet may be involved, and the condition may be mild, moderate, or severe. The underlying problem is abnormal muscle development that distorts growing bones and joints. In most cases the tibia, ankle, and foot are affected, and the feet are turned medially and inverted. If both feet are involved, the soles face one another. Prompt treatment with casts or other supports in infancy helps alleviate the problem, and fewer than half of the cases require surgery.

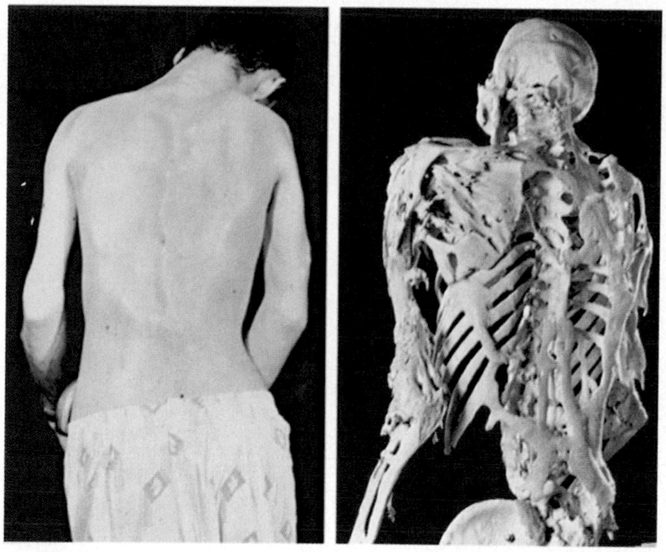

6 Under abnormal conditions, mesenchymal stem cells in any connective tissue can develop into osteoblasts that begin producing bone. The person at top left has **fibrodysplasia ossificans progressiva** (FOP), a rare single gene mutation disorder that involves bone deposits around skeletal muscles. The skeleton at top right shows the extent of abnormal ossification that can occur. Bones that develop in unusual places are called **heterotopic** (*hetero*, place), or **ectopic** (*ektos*, outside), **bones**. There is no effective treatment for this painful and debilitating condition, and patients seldom survive into their 40s.

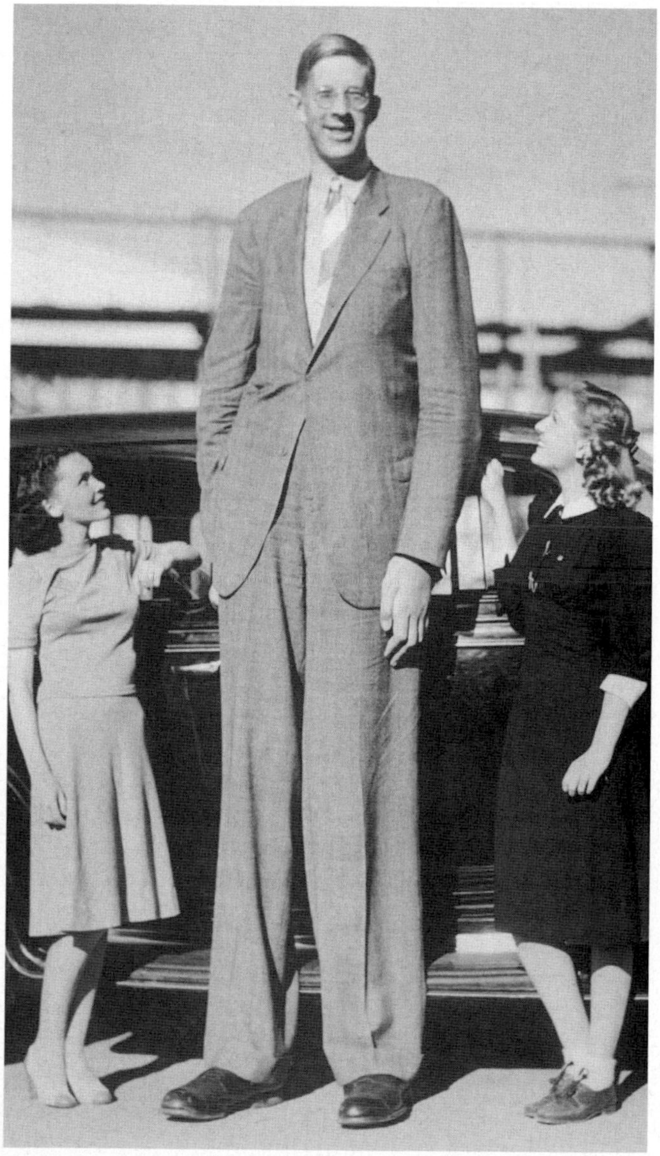

5 **Gigantism** (also called giantism) results from an overproduction of GH before puberty. Individuals can reach heights of over 2.7 m (8 ft 11 in.) and weights of over 200 kg (440 lb). Puberty is often delayed, and the facial features in adults resemble those of acromegaly (see **7**). The most common cause is a pituitary tumor, which may be treated by surgery, radiation, or drugs that suppress GH release.

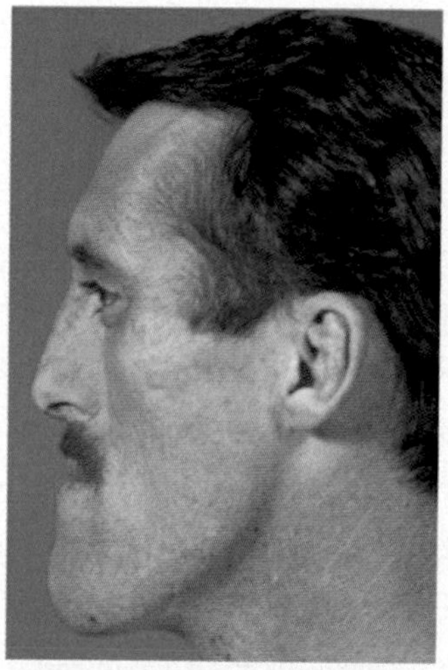

7 If GH levels rise abnormally after epiphyseal cartilages close, the skeleton does not grow longer, but bones get thicker, especially those in the face, jaw, and hands. Cartilage growth and alterations in soft-tissue structure lead to changes in physical features, such as the contours of the face. These physical changes occur in the disorder called **acromegaly**. Early diagnosis and treatment with pituitary surgery and/or drugs that reduce GH levels may prevent irreversible physical changes and extend life expectancy in both acromegaly and gigantism.

Module 6.9 Review

a. Why is pituitary dwarfism less common today in the United States?
b. Describe Marfan's syndrome.
c. Compare gigantism with acromegaly.

Concept map

Use each of the following terms once to fill in the blank boxes to correctly complete the map.

- lacunae
- osteocytes
- collagen
- intramembranous ossification
- compact bone
- periosteum
- hyaline cartilage

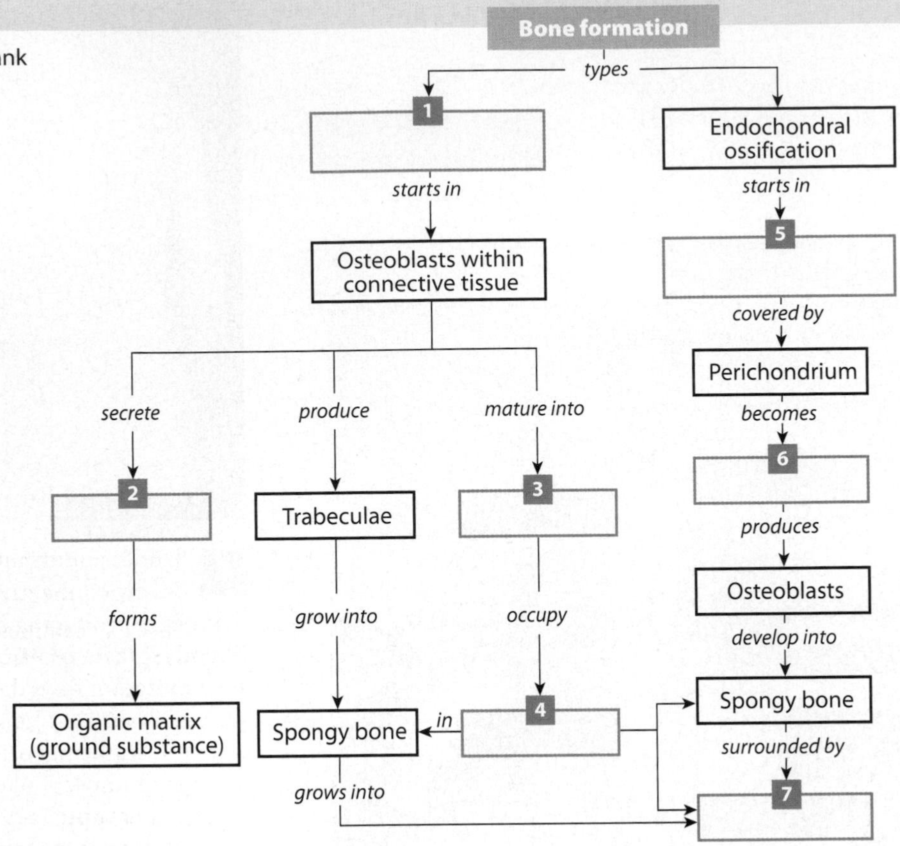

Bone formation
types
1
Endochondral ossification
starts in
starts in
Osteoblasts within connective tissue
5
covered by
Perichondrium
secrete
produce
mature into
becomes
2
Trabeculae
3
6
produces
Osteoblasts
forms
grow into
occupy
develop into
Organic matrix (ground substance)
Spongy bone
in
4
Spongy bone
surrounded by
grows into
7

Vocabulary

Write the term for each of the following descriptions in the space provided.

8 Bones with complex shapes

9 The expanded ends of a long bone

10 A shallow depression in the surface of a bone

11 The marrow-filled space within a bone

12 The strut- and plate-shaped matrix of spongy bone

13 Cells that remove and recycle bone matrix

14 Bones that develop in tendons

15 The process that forms new bone matrix

16 The basic functional unit of compact bone

17 Type of bone growth that increases bone diameter

18 Process by which cartilage is replaced by bone

8 ______________________

9 ______________________

10 ______________________

11 ______________________

12 ______________________

13 ______________________

14 ______________________

15 ______________________

16 ______________________

17 ______________________

18 ______________________

Section integration

While playing on her swing set, 10-year-old Rebecca falls and breaks her right leg. At the emergency room, the doctor tells her parents that the proximal end of the tibia where the epiphysis meets the diaphysis is fractured. The fracture is properly set and eventually heals. During a routine physical when she is 18, Rebecca learns that her right leg is 2.54 cm (1 in.) shorter than her left. What might account for this difference?

19 ______________________

Bones play an important role as mineral reservoirs

In this section we consider the dynamic relationship between calcium concentrations in the blood and calcium reserves in the skeletal system. We also examine the role of hormones in regulating calcium balance in the body.

1 This chemical analysis shows the importance of bones as mineral reservoirs. Minerals are inorganic ions that contribute to the osmotic balance of body fluids and are also vital in many physiological processes. In this section we focus on the homeostatic regulation of calcium ion concentrations in blood. We will consider other minerals in later chapters.

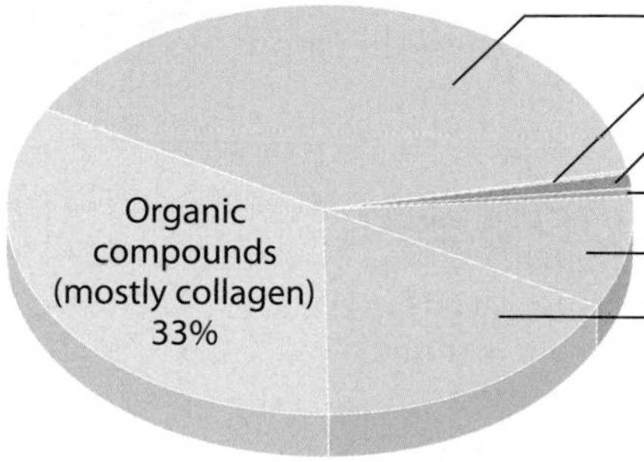
Organic compounds (mostly collagen) 33%

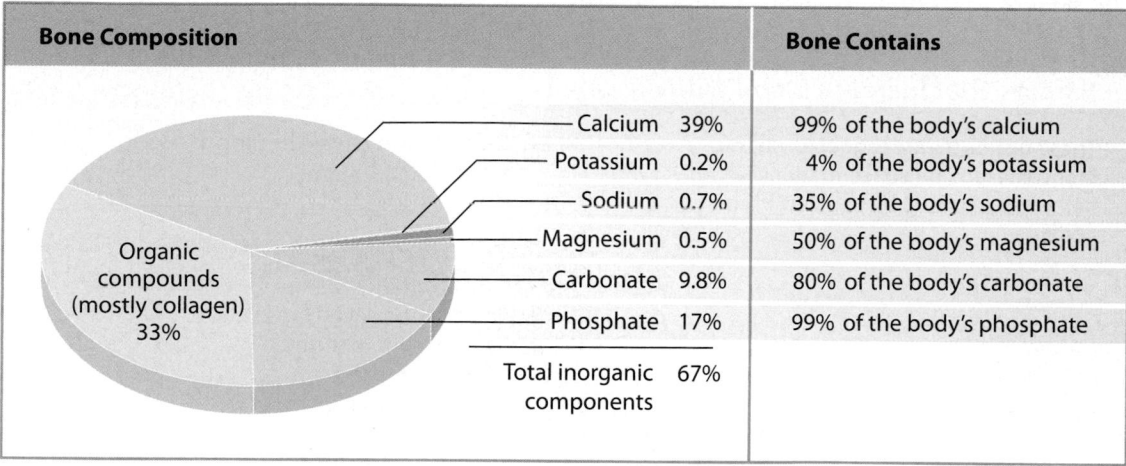

Bone Composition		Bone Contains
Calcium	39%	99% of the body's calcium
Potassium	0.2%	4% of the body's potassium
Sodium	0.7%	35% of the body's sodium
Magnesium	0.5%	50% of the body's magnesium
Carbonate	9.8%	80% of the body's carbonate
Phosphate	17%	99% of the body's phosphate
Total inorganic components	67%	

In the intestines, calcium and phosphate ions are absorbed from the diet. The absorption rate is hormonally regulated.

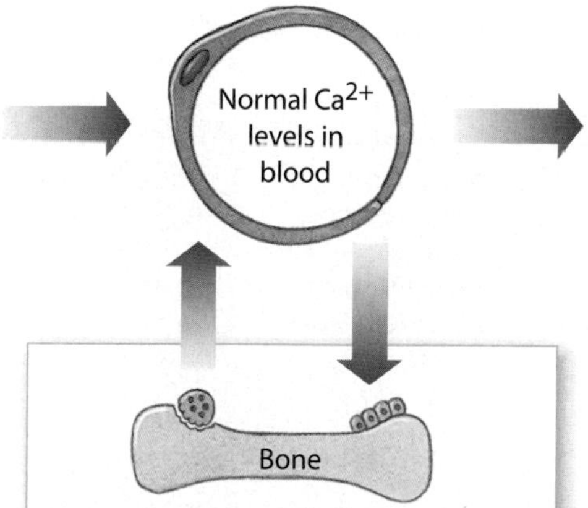
Normal Ca^{2+} levels in blood
Bone

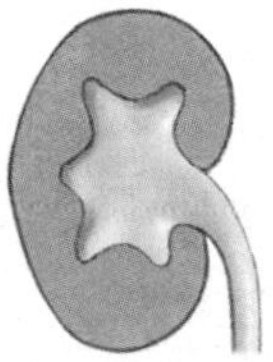

In the kidneys, the levels of calcium and phosphate ions lost in the urine are hormonally regulated.

Within the skeleton, osteoblasts continuously deposit new bone matrix. At the same time, osteoclasts erode existing matrix, releasing calcium and phosphate ions into the circulation. The balance between osteoblast and osteoclast activity is hormonally regulated.

2 **Calcium** is the most abundant mineral in the human body. The typical human body contains 1–2 kg (2.2–4.4 lb) of calcium, with roughly 99 percent of it deposited in the skeleton. Calcium ions play a role in a variety of physiological processes, such as muscle contraction, generating nerve impulses, and blood coagulation. The homeostatic regulation of calcium ion levels is a juggling act that balances activities under way in the intestines, bones, and kidneys.

Even small variations from the normal calcium concentration affect cellular operations. If the calcium concentration of blood increases or decreases by more than 30–35 percent, neuron and muscle cell function is disrupted, with potentially lethal results. Calcium ion concentration is so closely regulated, however, that daily fluctuations of more than 10 percent are highly unusual.

Module 6.10 Review

a. What is the ratio of organic compounds to inorganic components in the composition of bone?

b. Which three organ systems coordinate to maintain normal blood calcium levels?

c. If blood calcium levels are seriously decreased in a patient, what kind of symptoms would you expect to see?

LO **6.10** List the minerals stored in the bones, and identify the organs involved in calcium homeostasis.

Module 6.11

The primary hormones regulating calcium ion metabolism are parathyroid hormone, calcitonin, and calcitriol

Calcium ion homeostasis is maintained by hormones that target the skeletal system, the digestive tract, and the kidneys.

Factors That Increase Blood Calcium Levels

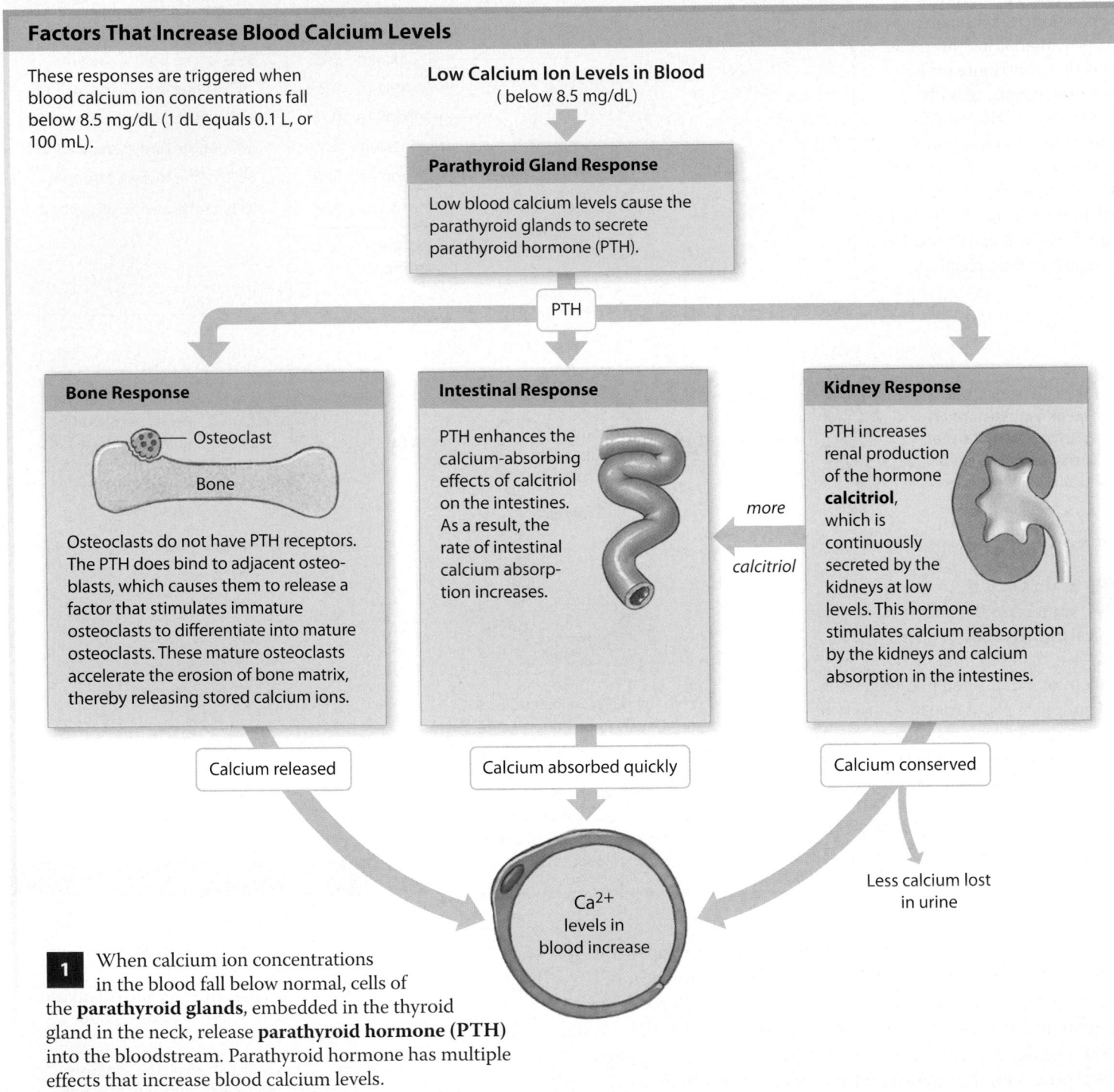

These responses are triggered when blood calcium ion concentrations fall below 8.5 mg/dL (1 dL equals 0.1 L, or 100 mL).

Low Calcium Ion Levels in Blood
(below 8.5 mg/dL)

Parathyroid Gland Response

Low blood calcium levels cause the parathyroid glands to secrete parathyroid hormone (PTH).

PTH

Bone Response

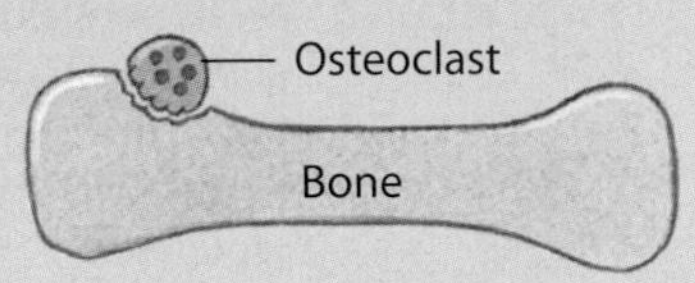
Osteoclast
Bone

Osteoclasts do not have PTH receptors. The PTH does bind to adjacent osteoblasts, which causes them to release a factor that stimulates immature osteoclasts to differentiate into mature osteoclasts. These mature osteoclasts accelerate the erosion of bone matrix, thereby releasing stored calcium ions.

Intestinal Response

PTH enhances the calcium-absorbing effects of calcitriol on the intestines. As a result, the rate of intestinal calcium absorption increases.

more calcitriol

Kidney Response

PTH increases renal production of the hormone **calcitriol**, which is continuously secreted by the kidneys at low levels. This hormone stimulates calcium reabsorption by the kidneys and calcium absorption in the intestines.

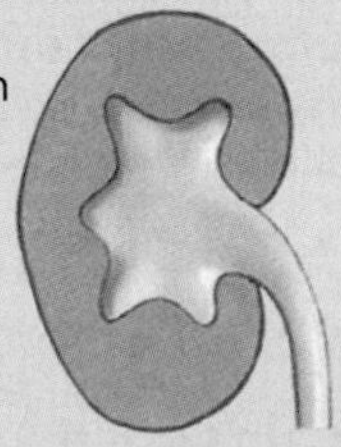

Calcium released

Calcium absorbed quickly

Calcium conserved

Less calcium lost in urine

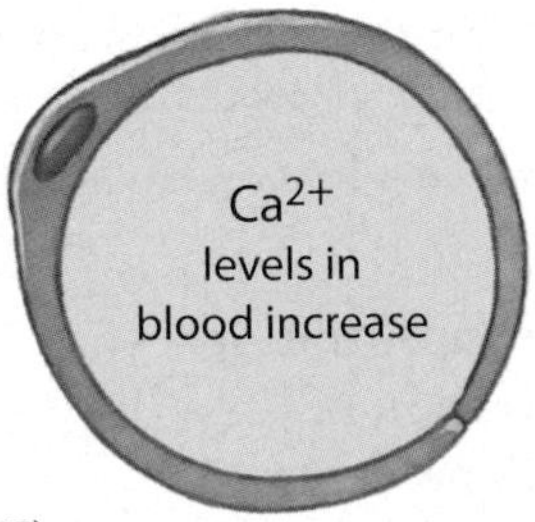
Ca^{2+}
levels in
blood increase

1 When calcium ion concentrations in the blood fall below normal, cells of the **parathyroid glands**, embedded in the thyroid gland in the neck, release **parathyroid hormone (PTH)** into the bloodstream. Parathyroid hormone has multiple effects that increase blood calcium levels.

Factors That Decrease Blood Calcium Levels

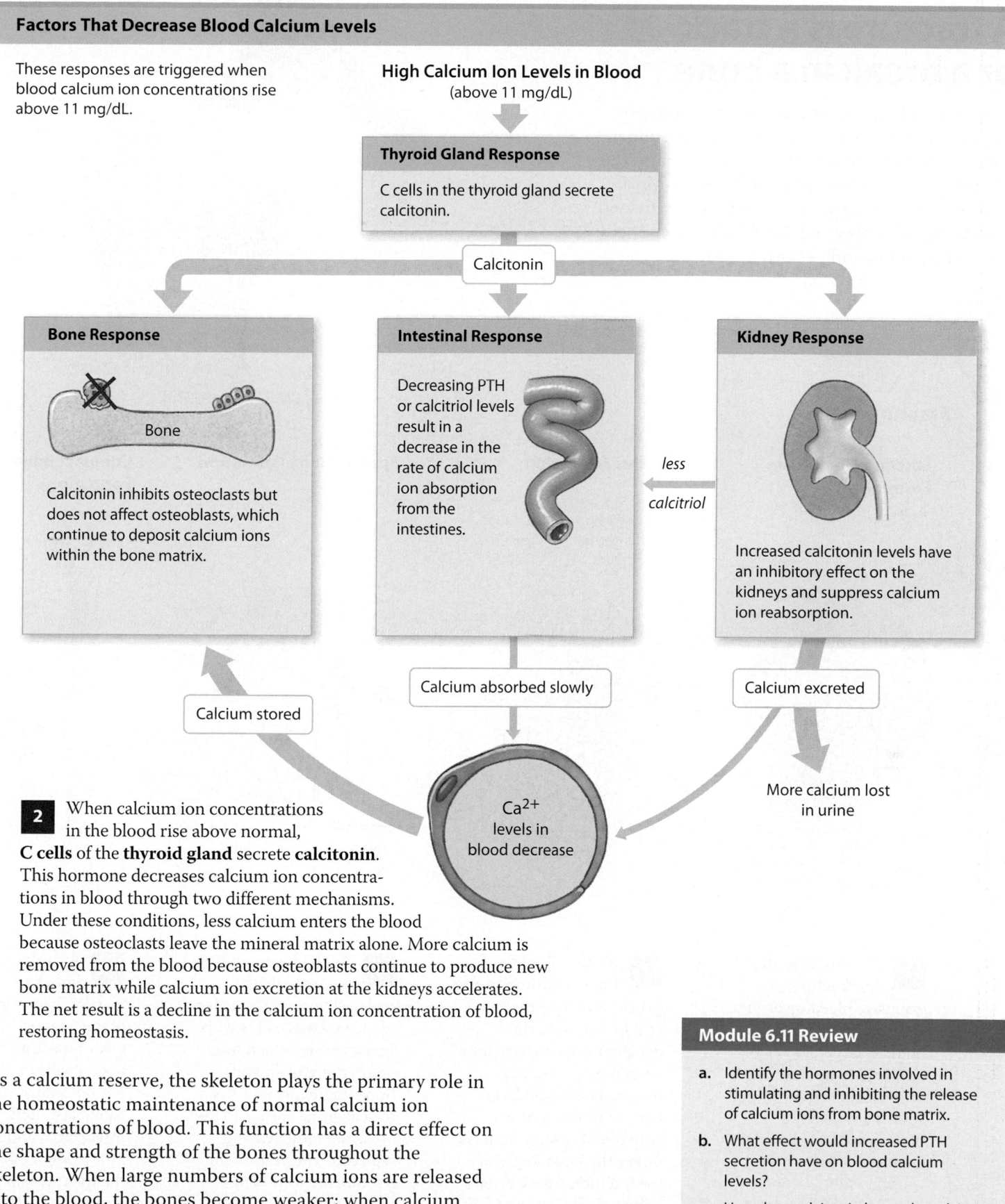

These responses are triggered when blood calcium ion concentrations rise above 11 mg/dL.

High Calcium Ion Levels in Blood
(above 11 mg/dL)

Thyroid Gland Response

C cells in the thyroid gland secrete calcitonin.

Calcitonin

Bone Response

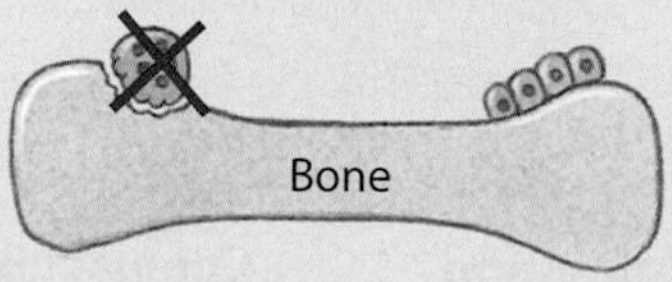
Bone

Calcitonin inhibits osteoclasts but does not affect osteoblasts, which continue to deposit calcium ions within the bone matrix.

Intestinal Response

Decreasing PTH or calcitriol levels result in a decrease in the rate of calcium ion absorption from the intestines.

less calcitriol

Kidney Response

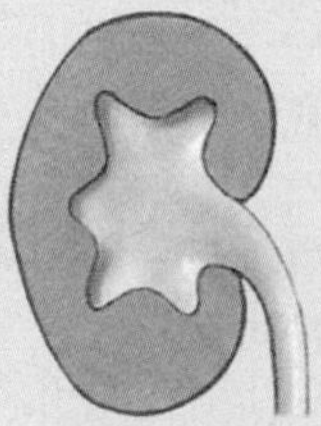

Increased calcitonin levels have an inhibitory effect on the kidneys and suppress calcium ion reabsorption.

Calcium stored

Calcium absorbed slowly

Calcium excreted

More calcium lost in urine

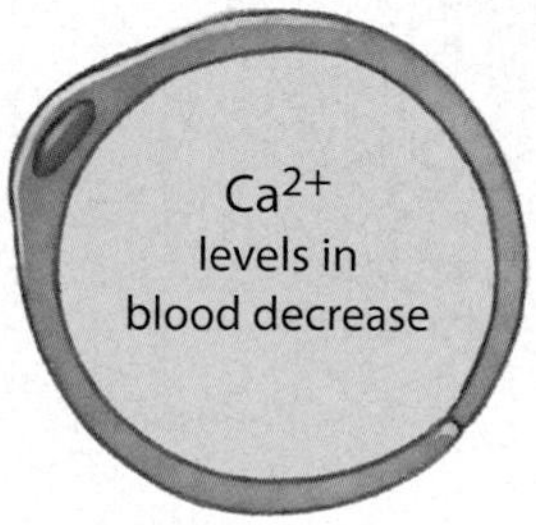
Ca^{2+}
levels in
blood decrease

2 When calcium ion concentrations in the blood rise above normal, **C cells** of the **thyroid gland** secrete **calcitonin**. This hormone decreases calcium ion concentrations in blood through two different mechanisms. Under these conditions, less calcium enters the blood because osteoclasts leave the mineral matrix alone. More calcium is removed from the blood because osteoblasts continue to produce new bone matrix while calcium ion excretion at the kidneys accelerates. The net result is a decline in the calcium ion concentration of blood, restoring homeostasis.

As a calcium reserve, the skeleton plays the primary role in the homeostatic maintenance of normal calcium ion concentrations of blood. This function has a direct effect on the shape and strength of the bones throughout the skeleton. When large numbers of calcium ions are released into the blood, the bones become weaker; when calcium salts are deposited, the bones become denser and stronger.

Module 6.11 Review

a. Identify the hormones involved in stimulating and inhibiting the release of calcium ions from bone matrix.

b. What effect would increased PTH secretion have on blood calcium levels?

c. How does calcitonin lower the calcium ion concentration of blood?

LO **6.11** Discuss the effects of hormones on bone development, and explain the homeostatic mechanisms involved.

A fracture is a crack or a break in a bone

Despite its mineral strength, bone can crack or even break if subjected to extreme loads, sudden impacts, or stresses from unusual directions. The damage produced constitutes a **fracture**. Most fractures heal even after severe damage, provided that the blood supply and the cellular components of the endosteum and periosteum survive.

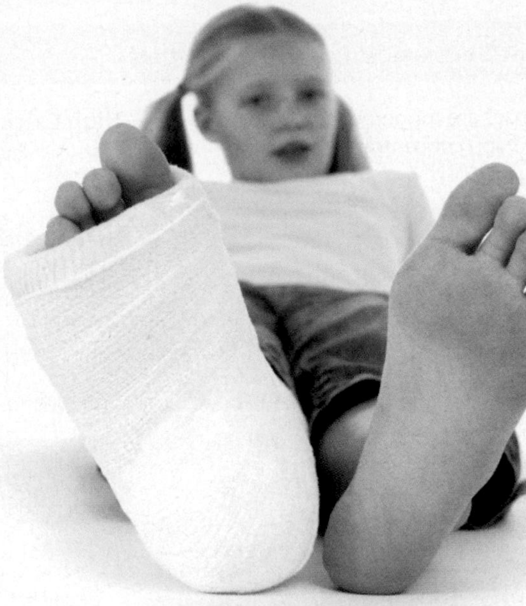

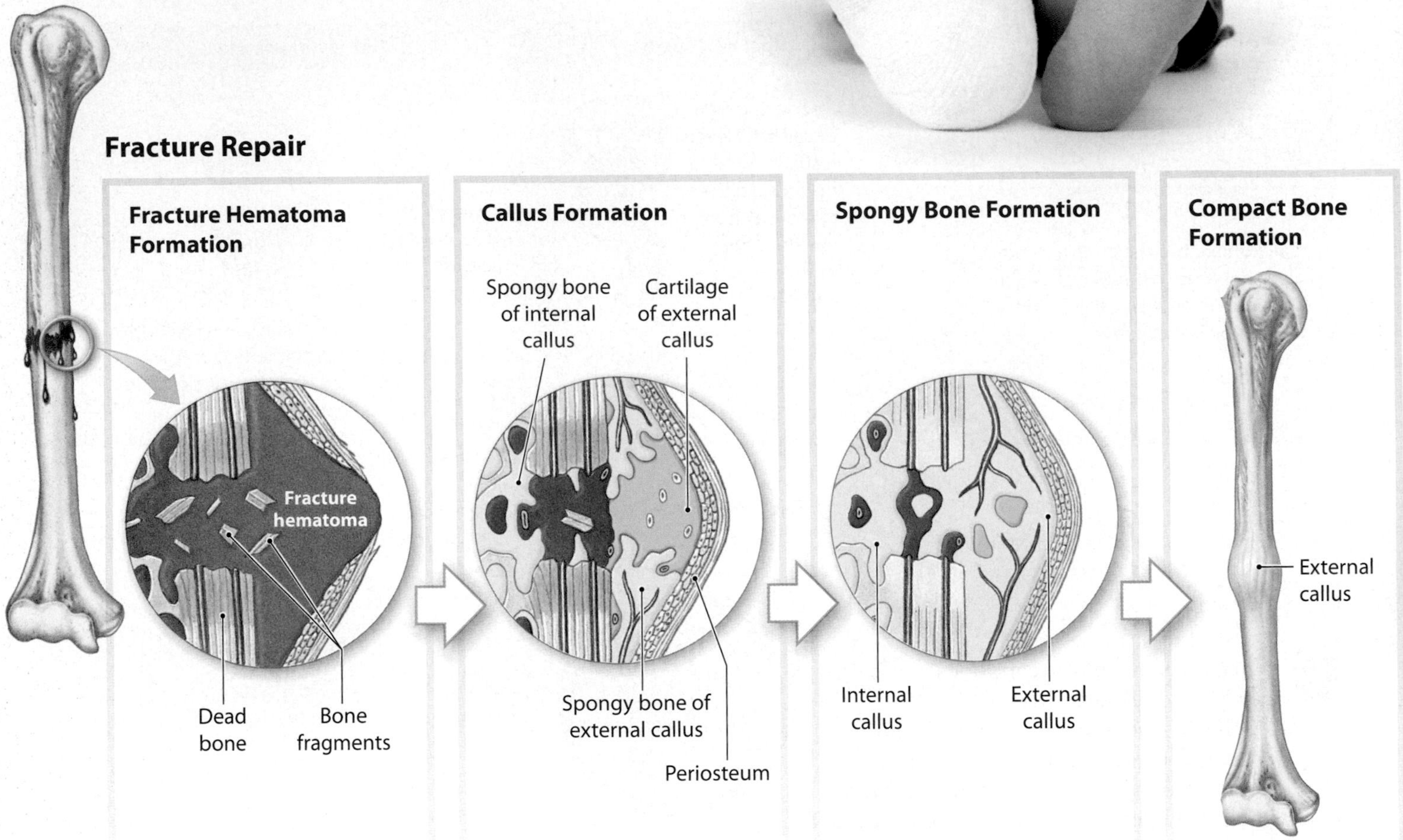
Fracture Repair
Fracture Hematoma Formation
Fracture hematoma
Dead bone
Bone fragments
Callus Formation
Spongy bone of internal callus
Cartilage of external callus
Spongy bone of external callus
Periosteum
Spongy Bone Formation
Internal callus
External callus
Compact Bone Formation
External callus

1 Immediately after the fracture, extensive bleeding occurs. Over a period of several hours, a large blood clot, or **fracture hematoma**, soon closes off the injured vessels and leaves a fibrous meshwork in the damaged area.

2 Next, cells of the intact endosteum and periosteum undergo rapid cell division, and the daughter cells migrate into the fracture zone. An **internal callus** (*callum*, hard skin) forms as a network of spongy bone unites the inner edges of the fracture. An **external callus** of cartilage and bone encircles and stabilizes the outer edges of the fracture.

3 As the repair continues, osteoblasts replace the central cartilage of the external callus with spongy bone, which then unites the broken ends. Bone fragments and areas of dead bone closest to the break are removed and replaced. The ends of the fracture are now held firmly in place and can withstand normal stresses from muscle contractions.

4 A swelling initially marks the location of the fracture. Over time, this region will be remodeled by osteoblasts and osteoclasts, and little evidence of the fracture will remain.

Types of Fractures

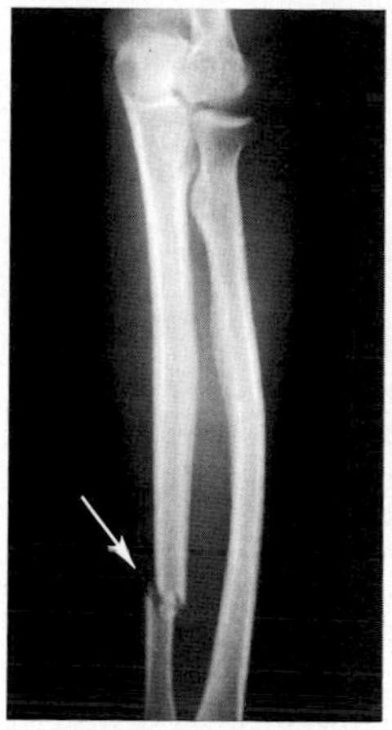

Transverse fractures, such as this fracture of the ulna, break a bone shaft across its long axis.

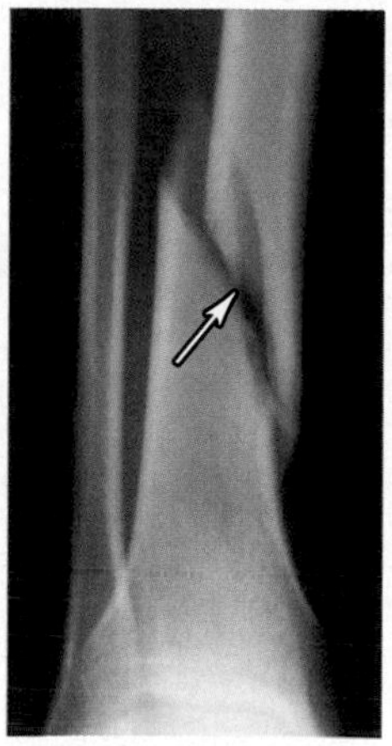

Spiral fractures, such as this fracture of the tibia, are produced by twisting stresses that spread along the length of the bone.

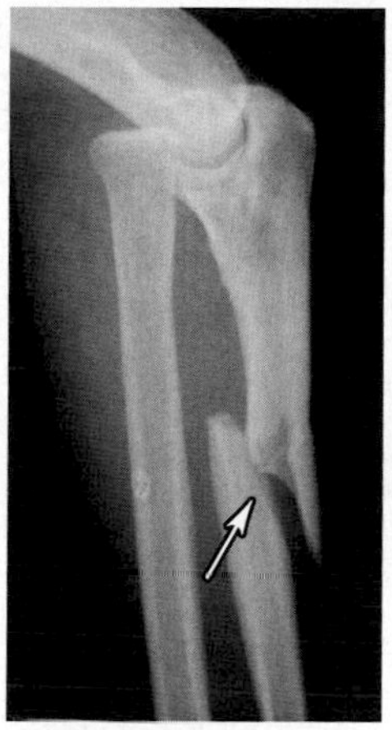

Displaced fractures produce new and abnormal bone alignments. **Nondisplaced fractures** retain the normal alignment of the bones or fragments.

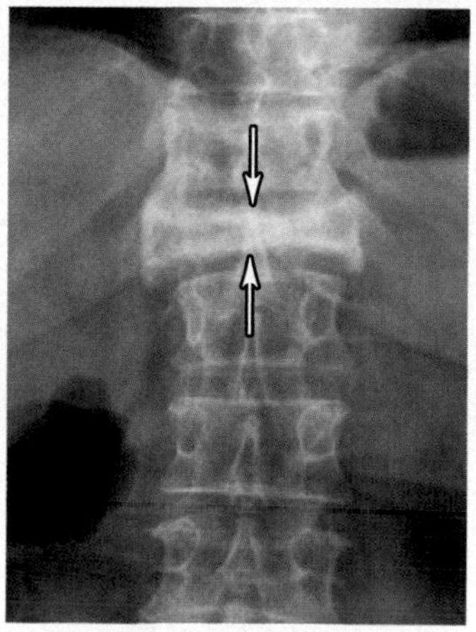

Compression fractures occur in vertebrae subjected to extreme stresses, such as the forces produced when you land on your buttocks in a fall. Compression fractures are often associated with osteoporosis.

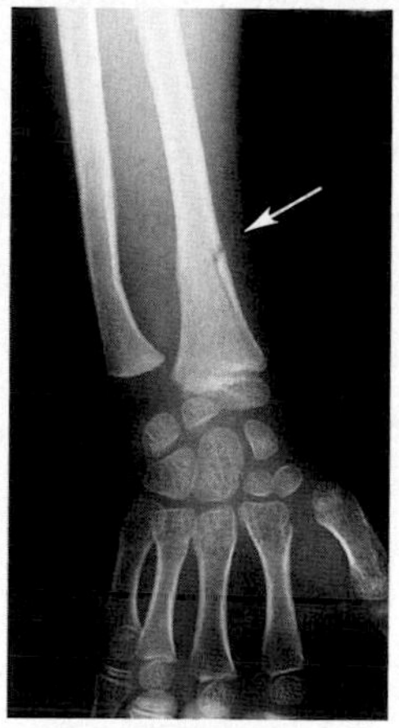

In a **greenstick fracture**, such as this fracture of the radius, only one side of the shaft is broken, and the other is bent. This type of fracture generally occurs in children, whose long bones have yet to ossify fully.

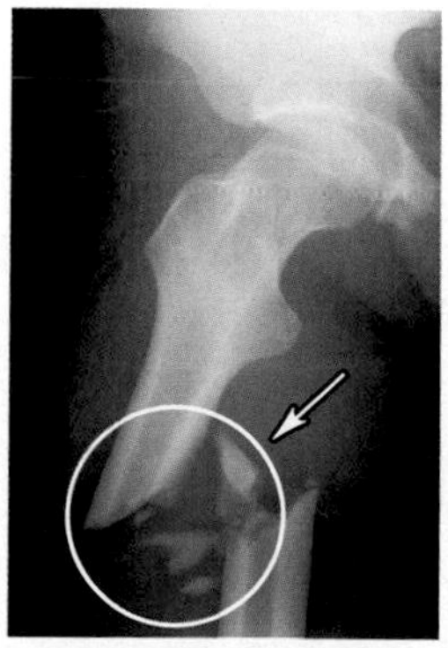

Comminuted fractures, such as this fracture of the femur, shatter the affected area into a multitude of bony fragments.

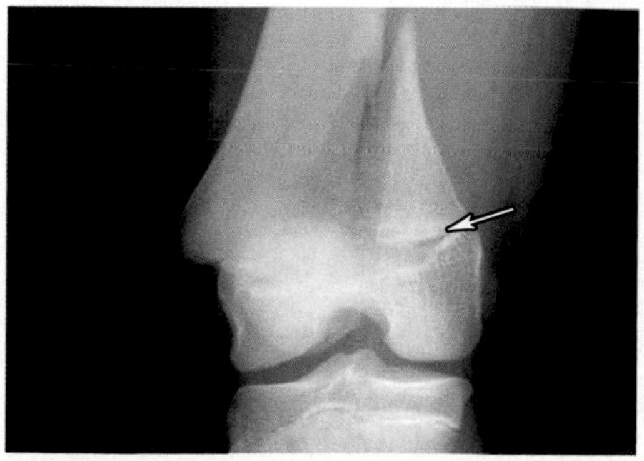

Epiphyseal fractures, such as this fracture of the femur, tend to occur where the bone matrix is undergoing calcification and chondrocytes are dying. A clean transverse fracture along this line generally heals well. Unless carefully treated, fractures between the epiphysis and the epiphyseal cartilage can permanently stop growth at this site.

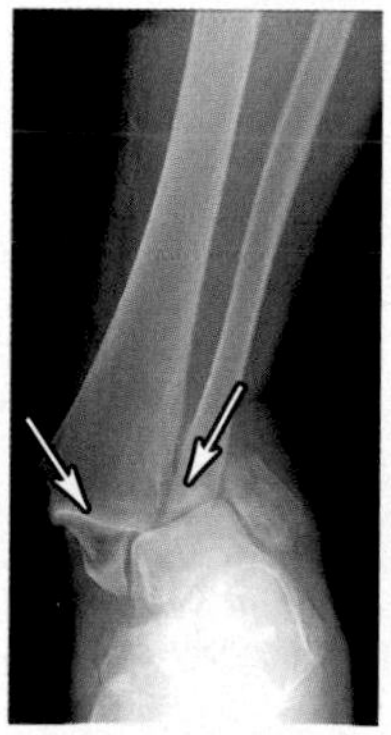

A **Pott's fracture**, also called a bimalleolar fracture, occurs at the ankle and affects both the medial malleolus of the distal tibia and the lateral malleolus of the distal fibula.

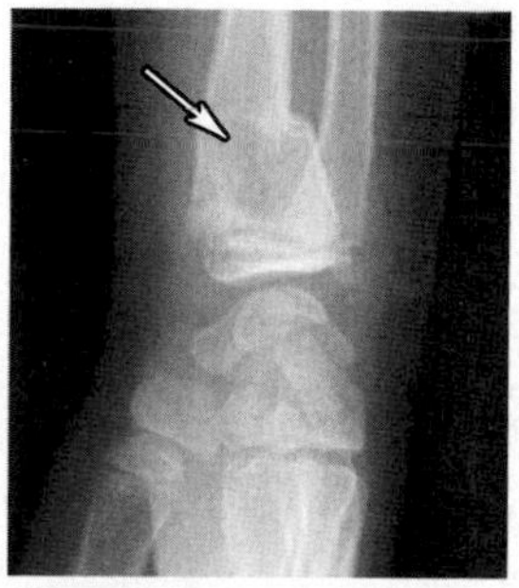

A **Colles fracture**, a break in the distal portion of the radius, is typically the result of reaching out to cushion a fall.

5 Fractures are named using various criteria, including their external appearance, their location, and the nature of the crack or break in the bone. Important types of fractures are illustrated here by representative x-rays. The broadest general categories are closed fractures and open fractures. **Closed**, or **simple, fractures** are completely internal. They can be seen only on x-rays, because they do not involve a break in the skin. **Open**, or **compound, fractures** project through the skin. These fractures, which are obvious on inspection, are more dangerous than closed fractures, due to the possibility of infection or uncontrolled bleeding. Many fractures fall into more than one category, because the terms overlap.

Module 6.12 Review

a. List the steps involved in fracture repair, beginning just after the fracture occurs.

b. When during fracture repair does an external callus form?

c. Define open fracture and closed fracture.

LO 6.12 Describe the types of fractures, and explain how fractures heal.

Concept map

Use each of the following terms once to fill in the blank boxes to correctly complete the map.

- ↓ Ca^{2+} level
- homeostasis
- release of stored Ca^{2+} from bone
- ↓ Ca^{2+} concentration in blood
- parathyroid glands
- calcitonin
- ↑ Ca^{2+} concentration in blood

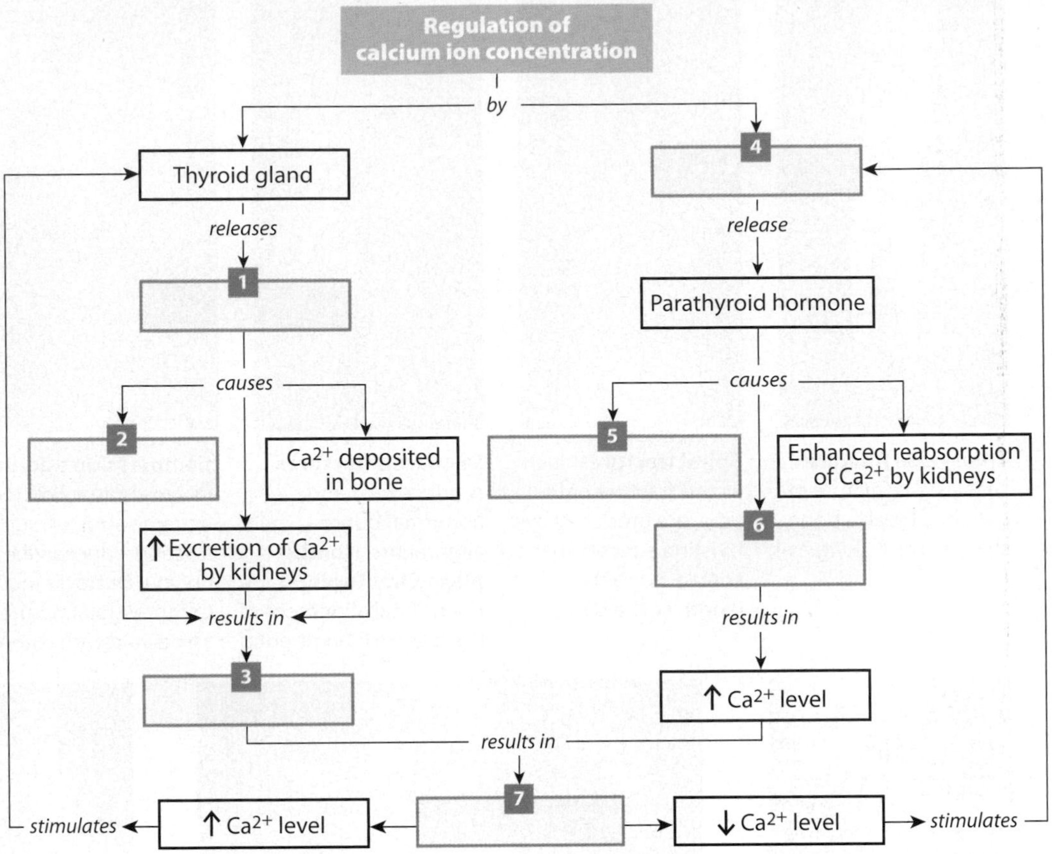
Regulation of calcium ion concentration
by
Thyroid gland
releases
1
causes
2
Ca2+ deposited in bone
↑ Excretion of Ca2+ by kidneys
results in
3
4
release
Parathyroid hormone
causes
5
Enhanced reabsorption of Ca2+ by kidneys
6
results in
↑ Ca2+ level
results in
7
stimulates
↑ Ca2+ level
↓ Ca2+ level
stimulates

Short answer

Identify the type of fracture and the bones involved in each of the following x-ray images.

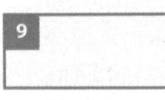

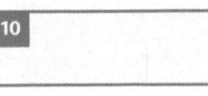

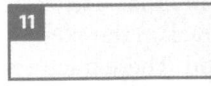

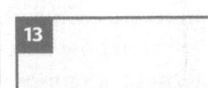

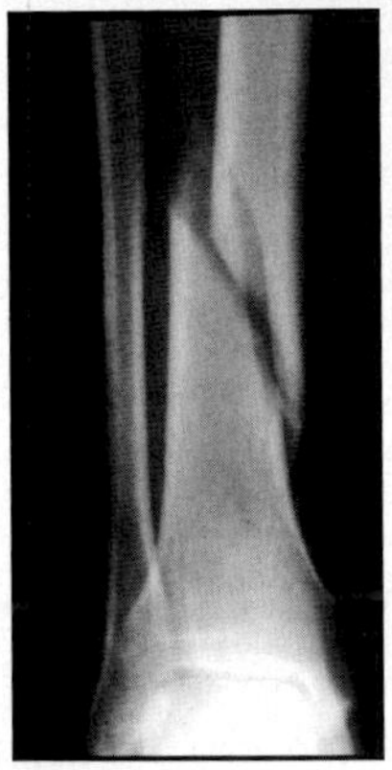
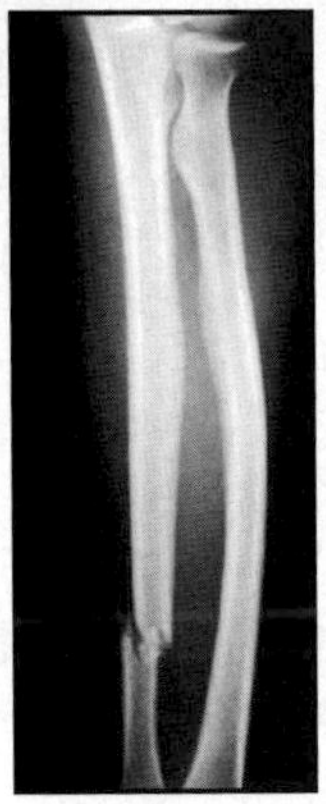
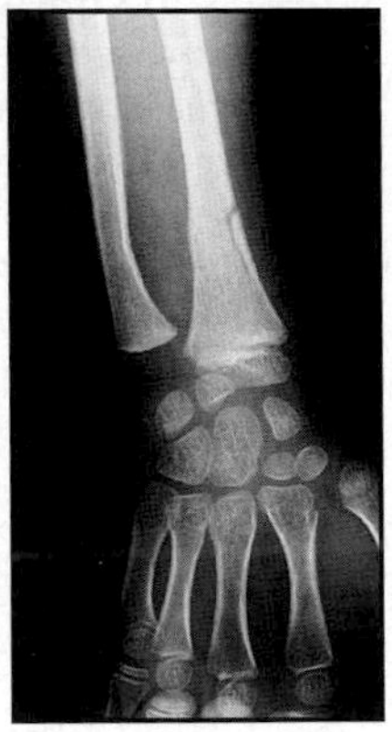
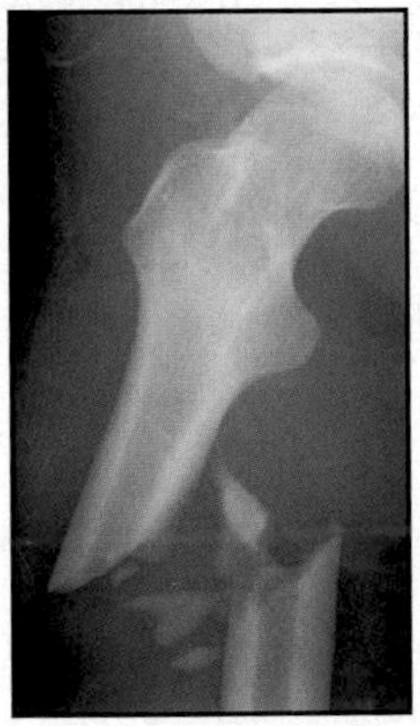
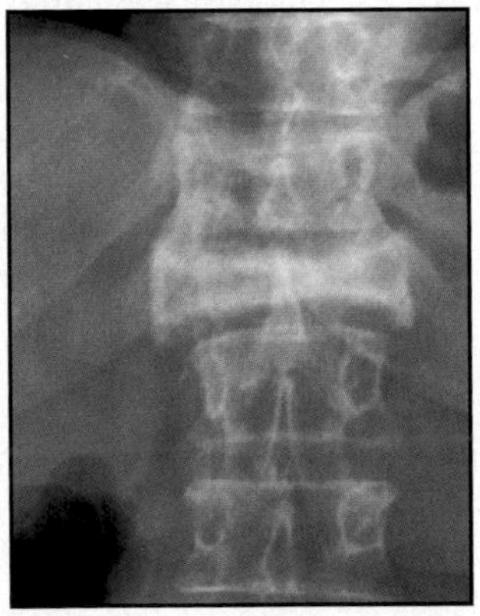
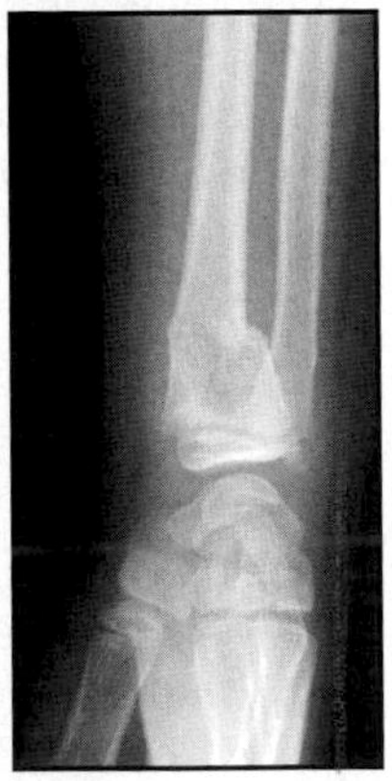

8

9

10

11

12

13

Study Outline

SECTION 1 • Introduction to the Structure and Growth of Bones

6.1 The skeletal system is made up of the axial and appendicular divisions p. 203

1. The **axial skeleton** has 80 bones and consists of the bones of the skull, thorax, and vertebral column.
2. The **appendicular skeleton** has 126 bones and consists of the limb bones, and the pectoral and pelvic girdles that attach the limbs to the axial skeleton.
3. The functions of the skeletal system are support, storage of minerals and lipids, blood cell production, protection, and leverage.

6.2 Bones are classified according to shape and structure and have varied surface markings p. 204

4. Bones can be divided into six broad categories based on their shapes: **flat bones**, **sutural bones**, **long bones**, **irregular bones**, **sesamoid bones**, and **short bones**.
5. Each bone has characteristic **surface markings**, including elevations or projections, depressions, grooves, and tunnels.

6.3 Long bones transmit forces along the shaft and have a rich blood supply p. 206

6. A representative long bone has an **epiphysis**, **metaphysis**, **diaphysis**, and a **medullary cavity**. The epiphyses consist mostly of **spongy bone** (also called **cancellous** or **trabecular bone**) while the diaphysis is made of **compact** (or cortical) **bone**.
7. The medullary cavity is filled with **red bone marrow** (for blood cell production) and **yellow bone marrow** (for lipid storage).
8. **Articular cartilage** covers the portion of the epiphysis that articulates with other bones. It is avascular and relies on the diffusion of synovial fluid for its metabolism.

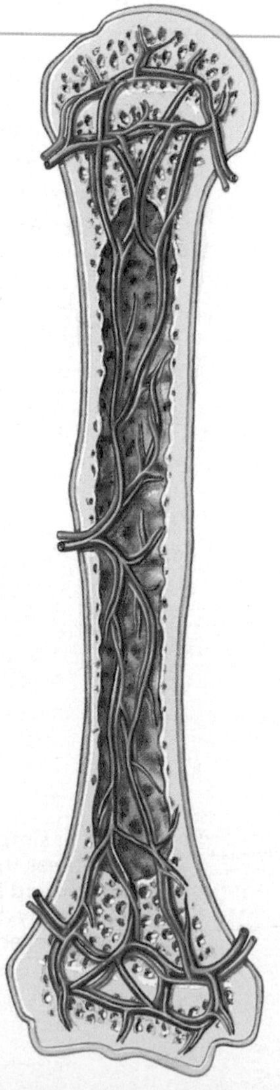

The growth and maintenance of bones require an extensive blood supply. Therefore, osseous tissue is highly vascular.

6.4 Bone has a calcified matrix maintained and altered by osteocytes, osteoblasts, osteogenic cells, and osteoclasts p. 208

9. **Osteocytes** are mature bone cells that account for most of the cell population in bone. These cells maintain the protein and mineral content of the bone matrix. They occupy a **lacuna**, and have processes that extend into **canaliculi**.
10. **Osteoblasts** produce new bone matrix in a process called **ossification** or **osteogenesis**. They develop into osteocytes once they have been surrounded by bone matrix.
11. **Osteogenic**, or osteoprogenitor, **cells** are mesenchymal cells present in bone. They reside in the inner layer of the periosteum, the endosteum, and in vascular passageways.
12. **Osteoclasts** are cells that remove and recycle bone matrix.

6.5 Compact bone consists of parallel osteons, and spongy bone consists of a network of trabeculae p. 210

13. The basic functional unit of compact bone is the **osteon** or Haversian system. Each osteon forms a series of **concentric lamellae** surrounding a **central canal**.

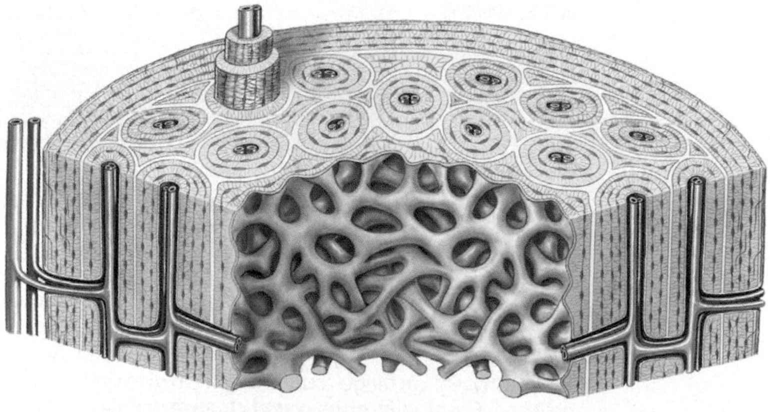

Compact bone consists of parallel osteons. You might think of a single osteon as a drinking straw with very thick walls. Osteons are parallel to the long axis of a bone and give it great strength even when extreme force is applied to either end.

14. Spongy bone is found in areas of less stress, as in the epiphyses of long bone. Spongy bone is formed by an open network of struts and plates called **trabeculae**.
15. **Circumferential lamellae** surround the outer and inner surfaces of bone. **Interstitial lamellae** fill in the spaces between the osteons of compact bone. **Perforating canals** run perpendicular to the long axis of bone.

6.6 In appositional bone growth, layers of compact bone are added to the bone's outer surface p. 212

16. The diameter of a long bone enlarges through **appositional growth** at the outer surface as layers of bone are added. During this process, osteoclasts remove and recycle lamellae at the inner surface, increasing the diameter of the medullary cavity.

17. **Periosteum** covers the superficial layer of compact bone except within the joint cavities. **Endosteum** lines the medullary cavity.

6.7 Endochondral ossification replaces a cartilaginous model with bone p. 214

18. **Endochondral ossification** begins with a hyaline cartilage model that is gradually replaced with bone. A **primary ossification center** forms at the diaphysis; **secondary ossification centers** form at the epiphyses.

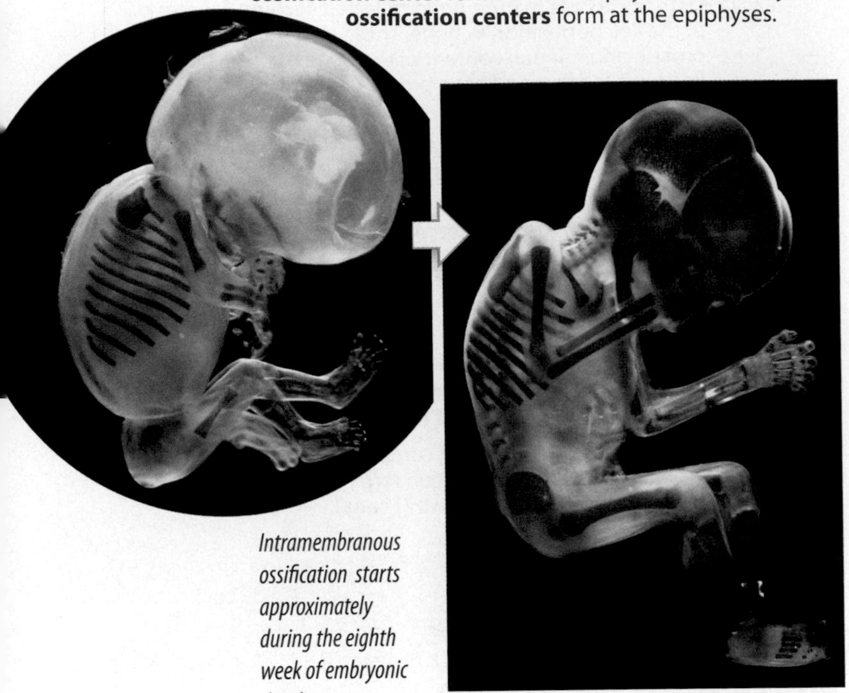

Intramembranous ossification starts approximately during the eighth week of embryonic development.
At 10 weeks (left), bone formation is under way but the skeleton is incomplete. At 16 weeks, most of the bones of the adult skeleton can be identified.

19. All the cartilage is replaced except at the **epiphyseal cartilage** of the metaphysis, and the articular cartilage. The epiphyseal cartilage regulates lengthwise growth of the bone. Eventually, **epiphyseal closure** will signal the end of epiphyseal growth.

6.8 Intramembranous ossification forms bone without a prior cartilaginous model p. 216

20. **Intramembranous ossification** begins when mesenchymal cells differentiate into osteoblasts. This process produces **dermal bones**, such as the roofing bones of the skull, the mandible, clavicle, and sesamoid bones such as the patella.

6.9 Abnormalities of bone growth and development produce recognizable physical signs p. 218

21. A variety of endocrine or metabolic problems can result in atypical skeletal growth. **Pituitary growth failure** and **achondroplasia** result in abnormally short bones. **Marfan's syndrome** and **gigantism** cause abnormal tallness. Other conditions include **congenital talipes equinovarus**, **acromegaly**, and **fibrodysplasia ossificans progressiva**.

SECTION 2 • Physiology of Bones

6.10 Bones play an important role as mineral reservoirs p. 221

22. **Calcium** is the most abundant mineral in the body; 99 percent of it is located in the skeleton.
23. The intestines, the skeleton, and the kidneys function to coordinate calcium homeostasis in blood.

6.11 The primary hormones regulating calcium ion metabolism are parathyroid hormone, calcitonin, and calcitriol p. 222

24. Low calcium ion levels in blood trigger **parathyroid hormone** (**PTH**) release from the **parathyroid glands**. PTH raises calcium ion levels in blood by indirectly stimulating osteoclasts, and enhancing calcium absorption in the intestines and calcium reabsorption by the kidneys. Calcitriol release from the kidneys is also stimulated by PTH and assists the intestines with calcium absorption.
25. High calcium ion levels in blood stimulate the **C cells** of the **thyroid gland** to release **calcitonin**. Calcitonin lowers blood calcium by inhibiting osteoclasts, and reducing calcitriol release and calcium reabsorption by the kidneys.

6.12 A fracture is a crack or a break in a bone p. 224

26. A break or crack in a bone is a **fracture**. The repair of a fracture involves the formation of a **fracture hematoma**, an **internal callus**, and an **external callus**.
27. There are many types of bone fractures depending on the forces causing the damage.

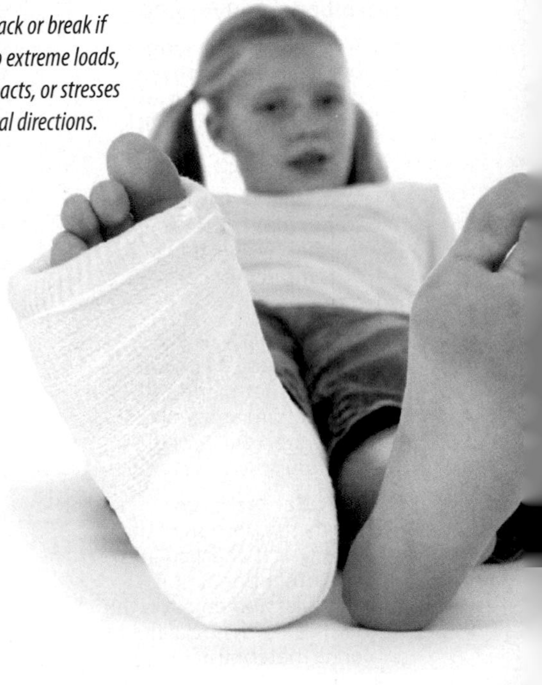

Bone can crack or break if subjected to extreme loads, sudden impacts, or stresses from unusual directions.

28. **Closed** or **simple fractures** are completely internal. They do not break the skin and can only be seen on x-ray. **Open** or **compound fractures** project through the skin. These are especially dangerous due to the possibility of infection or uncontrolled bleeding.

Chapter Review Questions

Labeling

Identify the anatomical regions and structures in this image of a long bone.

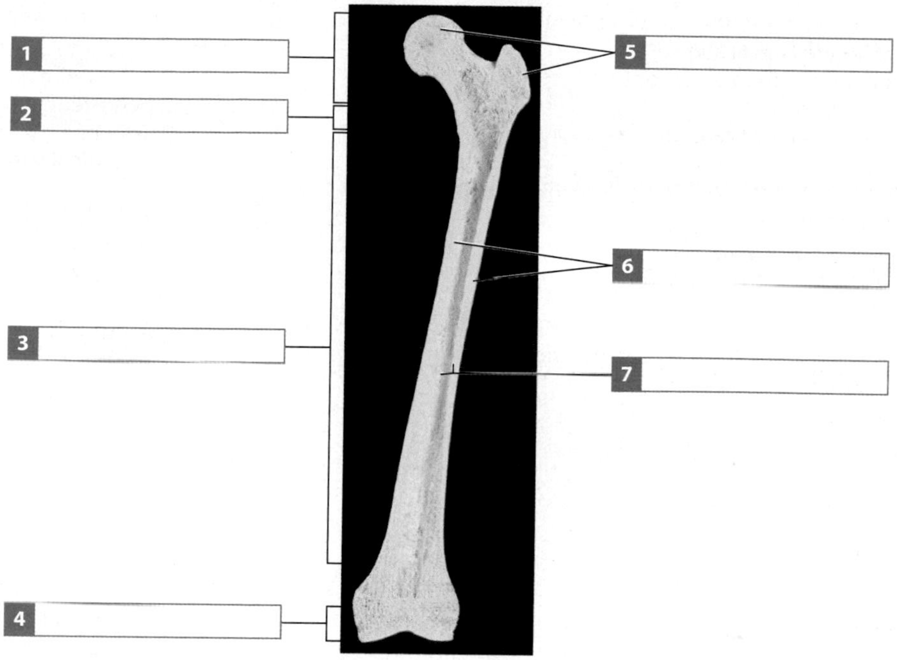

1
2
3
4
5
6
7

True/False

Indicate whether each statement is true or false.

8 A broken leg that results in a bone protruding through the skin is called a compound fracture.

9 The primary ossification centers are localized in the epiphyses.

10 Osteoblastic activity tends to deposit calcium into the bone.

11 Parathyroid hormone (PTH) is released in response to low blood calcium levels.

12 Cancellous bone and compact bone are synonymous terms.

13 A foramen is a small rounded passageway through which blood vessels or nerves penetrate.

14 Bone has a rich blood supply, but lacks sensory innervation.

15 Appositional growth increases bone diameter.

8 ____
9 ____
10 ____
11 ____
12 ____
13 ____
14 ____
15 ____

Multiple choice

Select the correct answer from the list provided.

16 Blood cell formation occurs in
- a) yellow bone marrow.
- b) red bone marrow.
- c) the matrix of bone tissue.
- d) the ground substance of bones.

17 Which of the following hormones is *not* involved with blood calcium ion homeostasis?
- a) calcitriol
- b) parathyroid hormone
- c) prolactin
- d) calcitonin

18 The presence of an epiphyseal line indicates
- a) epiphyseal growth has ended.
- b) the bone is fractured at that location.
- c) the bone will continue increasing in length for many years.
- d) vitamin D3 insufficiency.

19 The formation of bone without a prior cartilaginous model is
- a) endochondral ossification.
- b) intramembranous ossification.
- c) achondroplasia.
- d) fibrodysplasia ossificans progressiva.

20 Which of the following are examples of irregular bones?
- a) Wormian bones
- b) patellae
- c) carpals
- d) vertebrae

21 Which of the following terms is used to describe the architectural arrangement of spongy bone?
- a) osteon
- b) circumferential lamellae
- c) trabeculae
- d) Haversian system

22 Red bone marrow resides in the
- a) medullary cavity.
- b) sinus cavity.
- c) facets.
- d) fossa.

23 The membrane wrapping the bones, except at the joint cavity, is the
- a) endosteum.
- b) periosteum.
- c) osteon.
- d) interstitial lamellae.

24 Which of the following bones is *not* formed by endochondral ossification?
- a) mandible
- b) patella
- c) femur
- d) roofing bones of the skull

Fill-in

Fill in the following blanks in the spaces provided to the right.

25 The number of bones in the axial skeleton is _________ bones, while the appendicular skeleton contains _________ bones.

26 _________ bones develop inside tendons and are commonly located near the joints of the knees, hands, and feet.

27 An air-filled chamber within a bone is called a(an) _________.

28 The functional unit of mature compact bone is called a(an) _________.

29 _________ are cells that remove and recycle bone matrix.

25 ____________________ , ____________________

26 ____

27 ____

28 ____

29 ____

Short answer

30 What are the five primary functions of the skeletal system?

31 Which kind of fracture would you expect to occur in a patient whose shinbone broke when her foot stuck to the floor while she was dancing a pirouette?

32 Compare and contrast pituitary growth failure and achondroplasia.

33 What is the primary difference between endochondral ossification and intramembranous ossification?

34 Describe why epiphyseal fractures are of particular concern.

35 Why are impacts perpendicular to the shaft of a long bone more dangerous than stress applied parallel to the long axis of the bone?

36 Distinguish between acromegaly and gigantism.

Chapter Integration • Applying what you have learned

Assess the damage to a construction worker's spine from a one-story fall

A construction worker falls from the second floor of a building. He forcefully lands directly on his buttocks in a position as if he were sitting. He immediately feels intense pain in his lower back and is unable to get onto his feet. While waiting for the ambulance, his coworkers perform a quick assessment and notice there are no bones protruding through his skin. Upon arrival at the emergency room, a series of diagnostic tests are performed. He is informed that the trauma he has suffered is two severely damaged vertebrae. From what you have just learned about osseous tissue and bone structure, answer the following questions.

1 Taking into account the manner in which he fell and the forces his vertebral column experienced, which kind of fracture would you anticipate he has suffered? Is it likely to be a closed or open fracture?

2 An orthopedic surgeon was called to evaluate his case and she has recommended surgery. She informed the patient that the damage to the vertebrae was so severe that she would like to fuse them to the neighboring healthy vertebrae in order to stabilize his vertebral column. She mentions a technique called bone grafting, where she will shave bone from his iliac crest (hip), and lay these shavings between the damaged and healthy bones. Describe the cellular activity you would predict to occur that would result in neighboring bones fusing into a single bone.

3 What kind of processes would you expect to occur in the region of his hip where the bone for the graft was removed?

7 The Skeleton

LEARNING OUTCOMES

These Learning Outcomes correspond by number to this chapter's modules and indicate what you should be able to do after completing the chapter.

SECTION 1 • Axial Skeleton

- **7.1** List the four major components of the axial skeleton, and describe its major functions.
- **7.2** Identify the bones of the cranium and face, and locate and identify the cranial sutures.
- **7.3** Explain the significance of the markings and locations of the anterior and posterior aspects of the facial and cranial bones.
- **7.4** Explain the significance of the markings and locations of the lateral and medial aspects of the facial and cranial bones.
- **7.5** Explain the significance of the markings and locations of the inferior and interior aspects of the facial and cranial bones.
- **7.6** Describe and locate the surface features of the sphenoid, ethmoid, and palatine bones.
- **7.7** Describe the structure of the orbital complex and nasal complex and the functions of their individual bones.
- **7.8** Describe the mandible and the associated bones of the skull.
- **7.9** Describe key structural differences among the skulls of infants, children, and adults.
- **7.10** Identify and describe the curves of the spinal column and their functions, and identify the vertebral regions.
- **7.11** Describe the distinctive structural and functional characteristics of the cervical and thoracic vertebrae.
- **7.12** Describe the distinctive structural and functional characteristics of the lumbar vertebrae, sacrum, and coccyx.
- **7.13** Explain the significance of the articulations between the thoracic vertebrae and the ribs, and between the ribs and the sternum.

SECTION 2 • Appendicular Skeleton

- **7.14** List the four major components of the appendicular skeleton.
- **7.15** Identify the bones that form the pectoral girdles, their functions, and their superficial features.
- **7.16** Identify the bones of the arm and forearm, their functions, and their superficial features.
- **7.17** Identify the bones of the wrist and hand, and describe their locations using anatomical terminology.
- **7.18** Describe the hip bones that form the pelvic girdle, their functions, and their superficial features.
- **7.19** Identify the bones of the pelvis, and discuss the structural and functional differences between the pelvis in males and the pelvis in females.
- **7.20** Identify the bones of the thigh and leg, their functions, and their superficial features.
- **7.21** Identify the bones of the ankle and foot, and describe their locations using anatomical terminology.

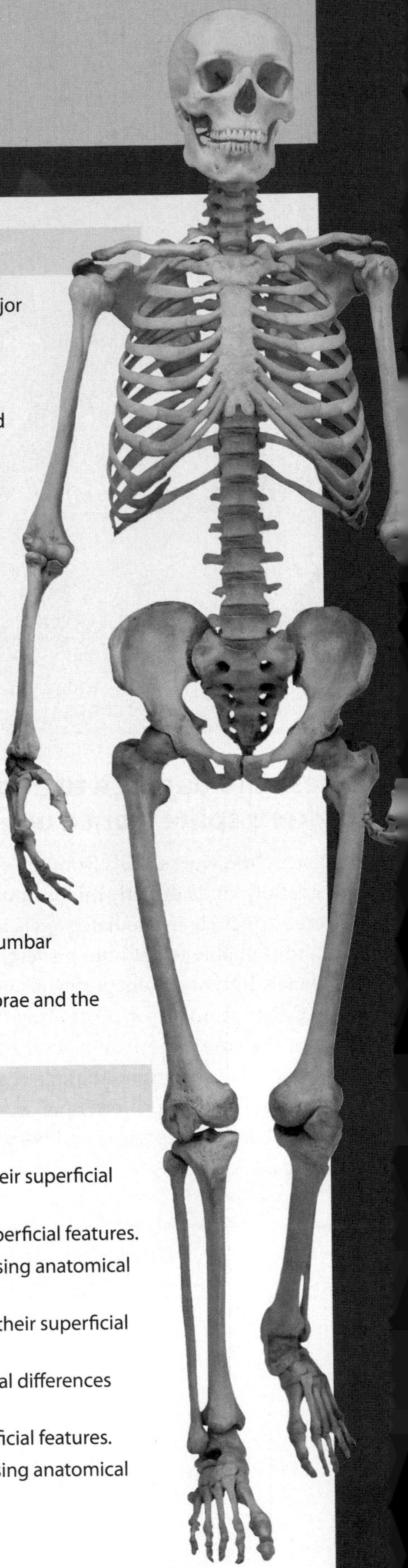

Learning Outcomes are repeated at the bottom of each module.

The axial skeleton includes bones of the head, vertebral column, and trunk

The **axial skeleton** forms the longitudinal axis of the body. This division of the skeletal system includes the skull and associated bones, the thoracic cage, the vertebral column, and various supplemental cartilages. There are typically 80 bones in the axial skeleton—roughly 40 percent of the bones in the human body.

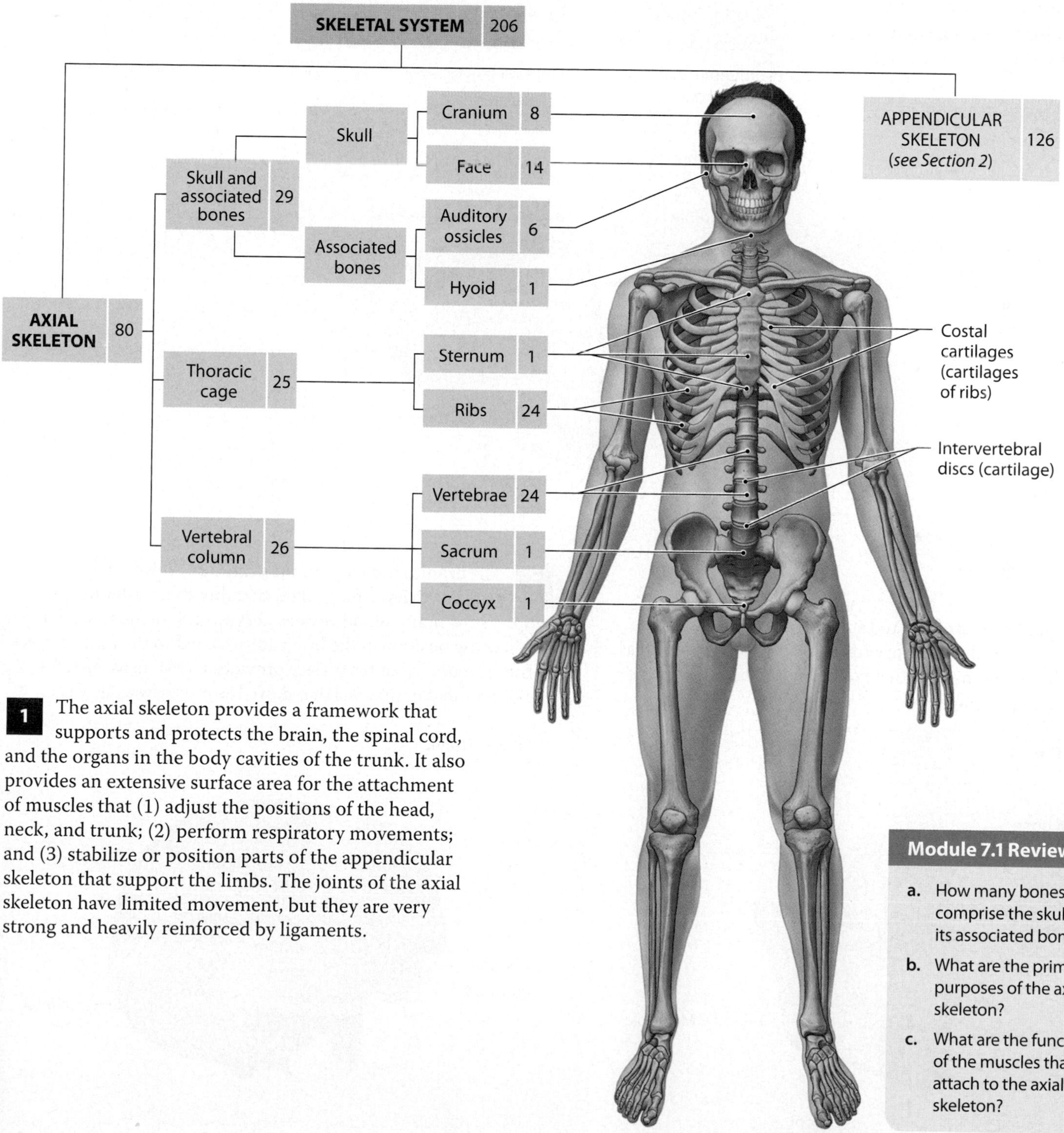
SKELETAL SYSTEM 206
AXIAL SKELETON 80
APPENDICULAR SKELETON 126 (see Section 2)
Skull and associated bones 29
Skull
Cranium 8
Face 14
Associated bones
Auditory ossicles 6
Hyoid 1
Thoracic cage 25
Sternum 1
Ribs 24
Vertebral column 26
Vertebrae 24
Sacrum 1
Coccyx 1
Costal cartilages (cartilages of ribs)
Intervertebral discs (cartilage)

1 The axial skeleton provides a framework that supports and protects the brain, the spinal cord, and the organs in the body cavities of the trunk. It also provides an extensive surface area for the attachment of muscles that (1) adjust the positions of the head, neck, and trunk; (2) perform respiratory movements; and (3) stabilize or position parts of the appendicular skeleton that support the limbs. The joints of the axial skeleton have limited movement, but they are very strong and heavily reinforced by ligaments.

Module 7.1 Review

a. How many bones comprise the skull and its associated bones?

b. What are the primary purposes of the axial skeleton?

c. What are the functions of the muscles that attach to the axial skeleton?

LO 7.1 List the four major components of the axial skeleton, and describe its major functions.

The skull has cranial and facial components that are usually bound together by sutures

The skull contains 22 bones: eight form the cranium, or braincase, and 14 form the face. Seven additional bones are associated with the skull: six auditory ossicles are located within the temporal bones of the cranium, and the hyoid bone is connected to the inferior surfaces of the temporal bones by a pair of ligaments.

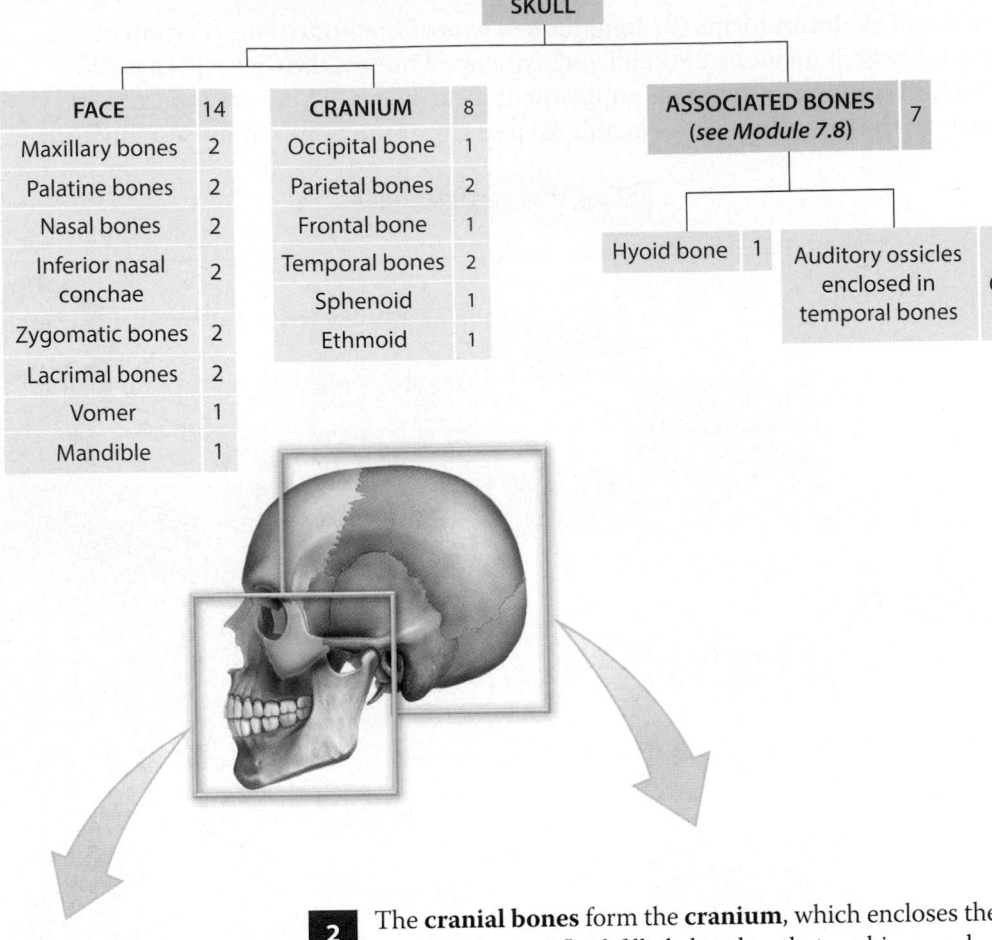

SKULL

FACE	14
Maxillary bones	2
Palatine bones	2
Nasal bones	2
Inferior nasal conchae	2
Zygomatic bones	2
Lacrimal bones	2
Vomer	1
Mandible	1

CRANIUM	8
Occipital bone	1
Parietal bones	2
Frontal bone	1
Temporal bones	2
Sphenoid	1
Ethmoid	1

ASSOCIATED BONES (*see Module 7.8*)	7
Hyoid bone	1
Auditory ossicles enclosed in temporal bones	6

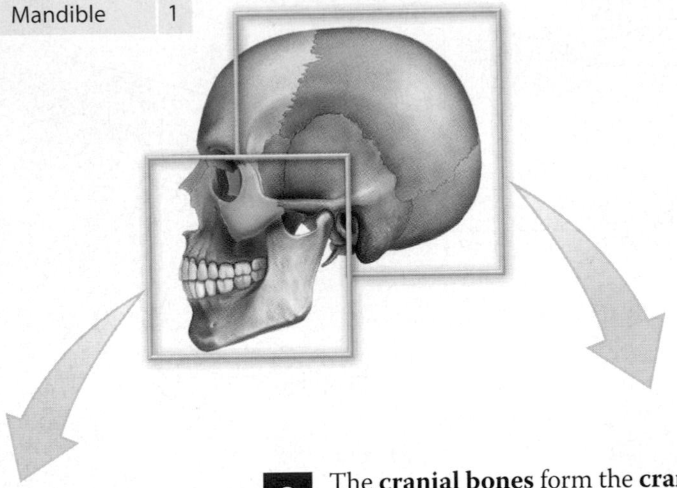

1 **Facial bones** protect and support the entrances to the digestive and respiratory tracts. They also provide areas for the attachment of muscles that control facial expressions and assist in manipulating food.

2 The **cranial bones** form the **cranium**, which encloses the **cranial cavity**, a fluid-filled chamber that cushions and supports the brain. Blood vessels, nerves, and membranes that stabilize the position of the brain are attached to the inner surface of the cranium. Its outer surface provides an extensive area for the attachment of muscles that move the eyes, jaws, and head.

Facial Bones

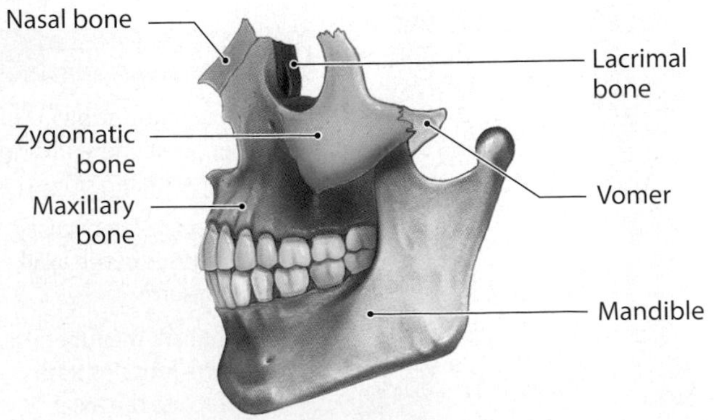
Nasal bone
Lacrimal bone
Zygomatic bone
Maxillary bone
Vomer
Mandible

Cranial Bones

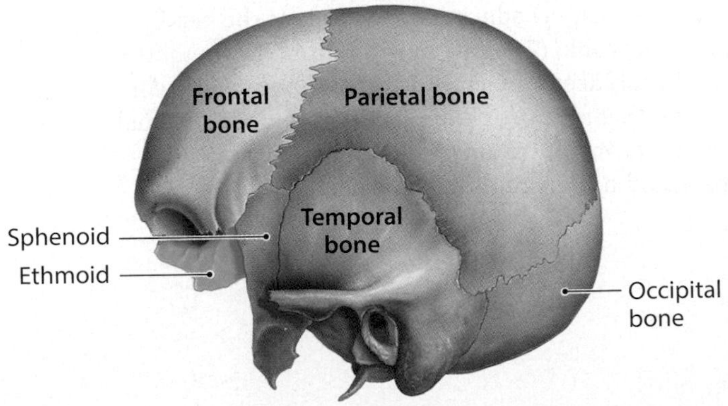
Frontal bone
Parietal bone
Sphenoid
Ethmoid
Temporal bone
Occipital bone

3 Joints, or articulations, form where two bones interconnect. Except where the mandible contacts the cranium, the connections between the skull bones of adults are immovable joints called **sutures**. At a suture, bones are tied firmly together with dense fibrous connective tissue. Each suture of the skull has a name, but at this point you need to know only four major sutures.

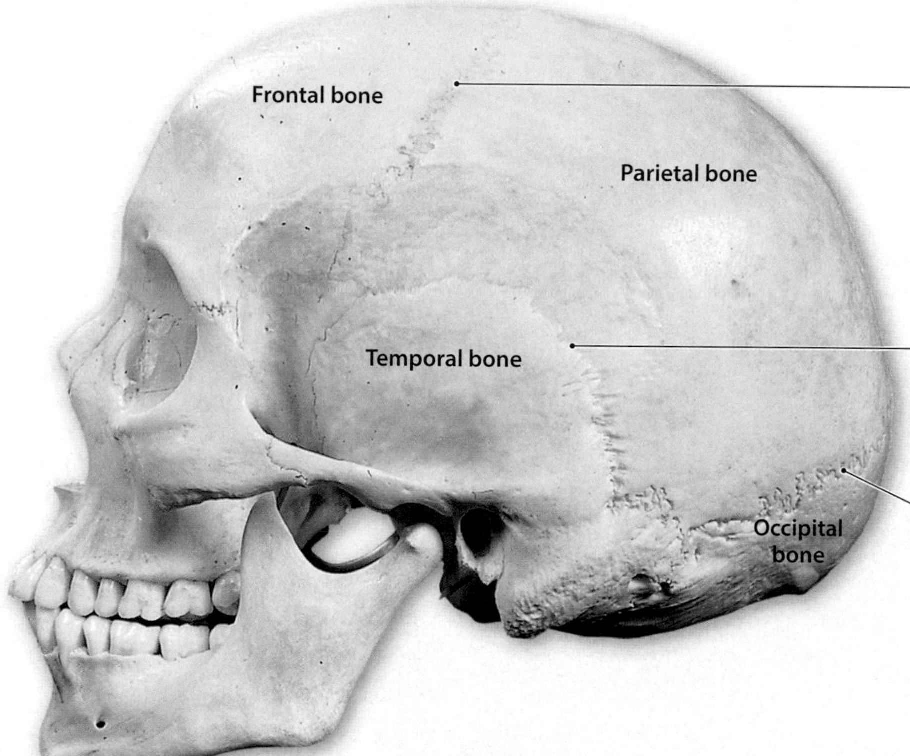

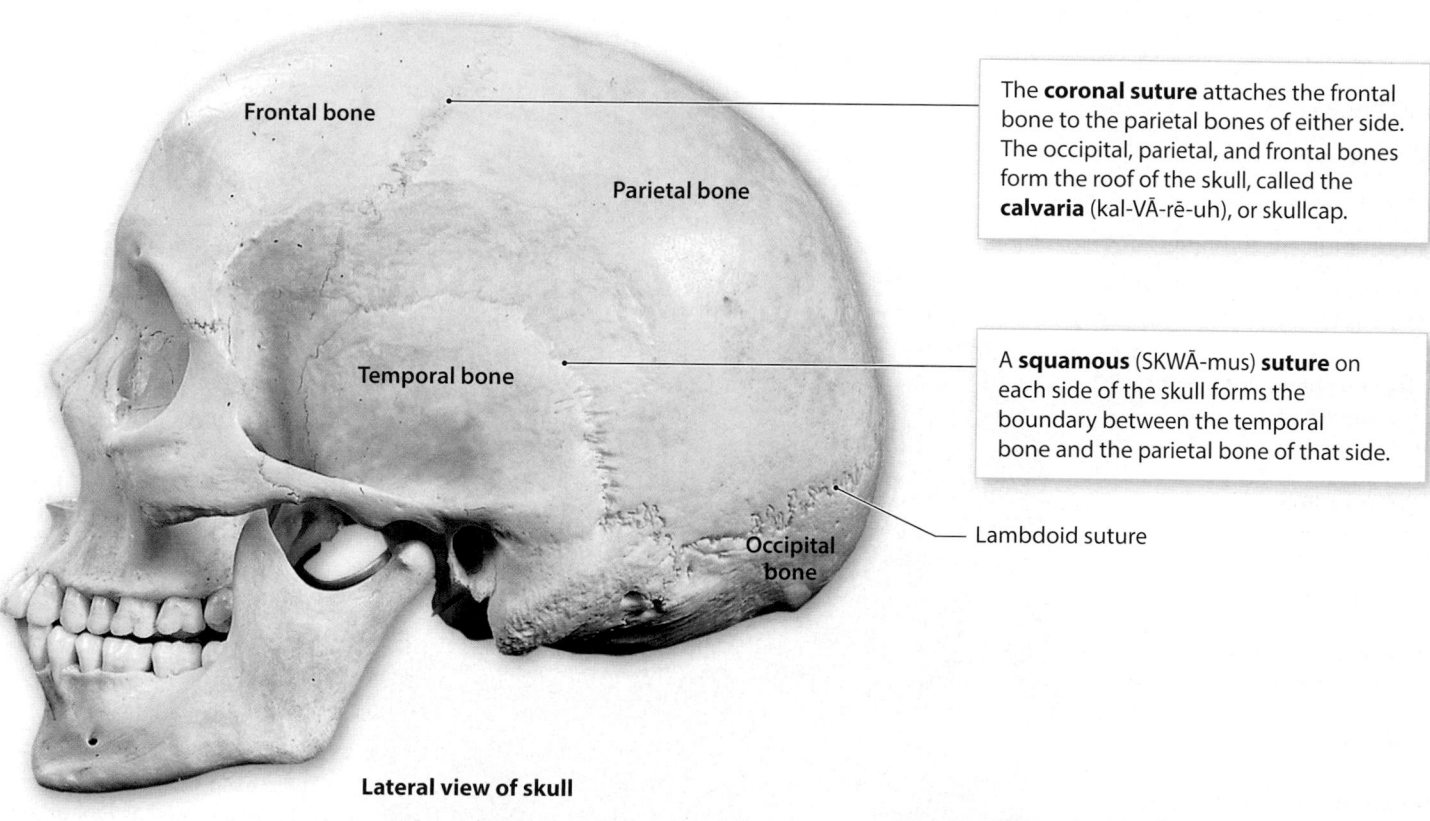

Frontal bone
Parietal bone
Temporal bone
Occipital bone
Lambdoid suture

The **coronal suture** attaches the frontal bone to the parietal bones of either side. The occipital, parietal, and frontal bones form the roof of the skull, called the **calvaria** (kal-VĀ-rē-uh), or skullcap.

A **squamous** (SKWĀ-mus) **suture** on each side of the skull forms the boundary between the temporal bone and the parietal bone of that side.

Lateral view of skull

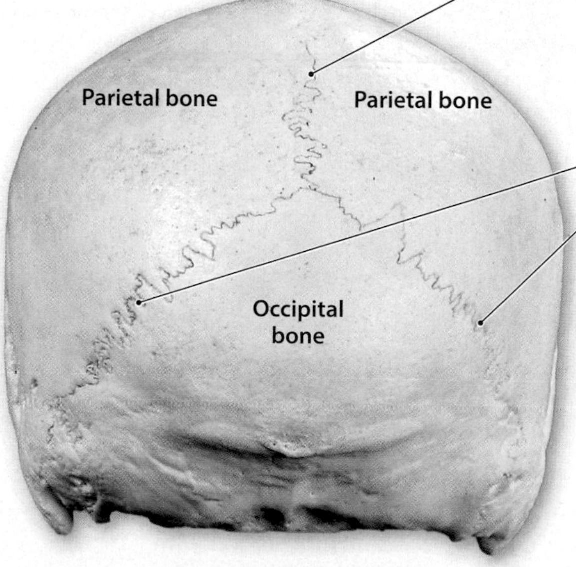

Parietal bone
Parietal bone
Occipital bone

The **sagittal suture** extends from the lambdoid suture to the coronal suture, between the parietal bones.

The **lambdoid** (LAM-doyd) **suture** arches across the posterior surface of the skull. This suture separates the occipital bone from the two parietal bones. Occasionally, one or more sutural bones (Wormian bones) may be present along the lambdoid suture (Module 6.2, p. 204).

Posterior view of skull

Module 7.2 Review

a. Identify the bones of the cranium.

b. Describe the functions of the facial bones.

c. Define suture.

LO **7.2** Identify the bones of the cranium and face, and locate and identify the cranial sutures.

Facial bones dominate the anterior aspect of the skull, and cranial bones dominate the posterior surface

1 We begin by examining the skull in anterior view. If you consider the cranium as the home of the brain, the facial bones form the front porch.

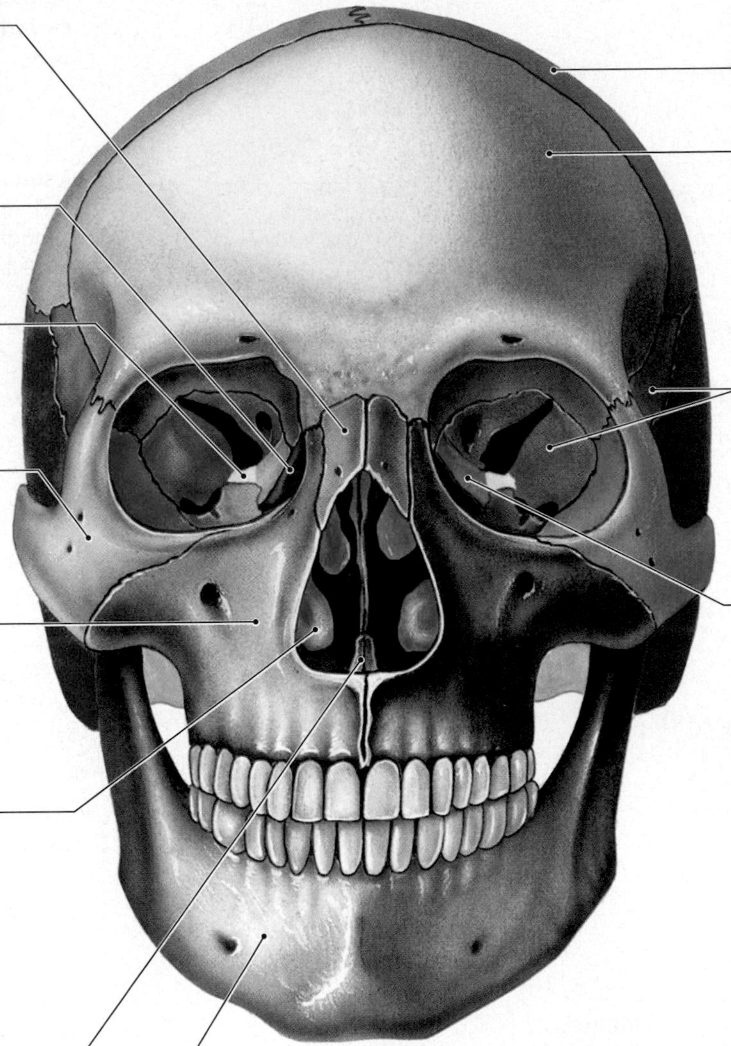

Facial Bones

Paired bones

The **nasal bones** support the superior portion of the bridge of the nose. They are connected to cartilages that support the distal portions of the nose.

The **lacrimal bones** form part of the medial wall of the orbits (eye sockets).

The **palatine bones** form the posterior portion of the hard palate and contribute to the floor of each orbit.

The **zygomatic bones** contribute to the rim and lateral wall of the orbit and form part of the cheekbone.

The **maxillae** support the upper teeth and form the inferior orbital rim, the lateral margins of the external nares, the upper jaw, and most of the hard palate.

The **inferior nasal conchae** create turbulence in air passing through the nasal cavity, and increase the epithelial surface area to warm and humidify inhaled air.

Single bones

The **vomer** forms the inferior portion of the bony nasal septum.

The **mandible** forms the lower jaw.

Cranial Bones

Parietal bone

The **frontal bone** forms the anterior portion of the cranium and the roof of the orbits. Mucus from the frontal sinuses within this bone help flush the nasal cavity surfaces.

The **sphenoid bone** forms part of the floor of the cranium, unites the cranial and facial bones, and acts as a cross-brace that strengthens the sides of the skull.

The **ethmoid bone** forms the anteromedial floor of the cranium, the roof of the nasal cavity, and part of the nasal septum and medial orbital wall.

2 Whereas the anterior view is dominated by facial bones, the posterior view is dominated by cranial bones. Several prominent landmarks on the occipital and temporal bones are identified here.

Cranial Bones

The **parietal bone** on each side forms part of the superior and lateral surfaces of the cranium.

The **occipital bone** contributes to the posterior, lateral, and inferior surfaces of the cranium.

The **temporal bone** on either side (1) forms part of the lateral wall of the cranium and articulates with facial bones, (2) forms an articulation with the mandible, (3) surrounds and protects the sense organs of the internal ear, and (4) is an attachment site for muscles that close the jaws and move the head.

- The **mastoid process** is an attachment site for muscles that rotate or extend the head.
- The **styloid** (STĪ-loyd; *stylos*, pillar) **process**, near the base of the mastoid process, is attached to ligaments that support the hyoid bone and to the tendons of several muscles.

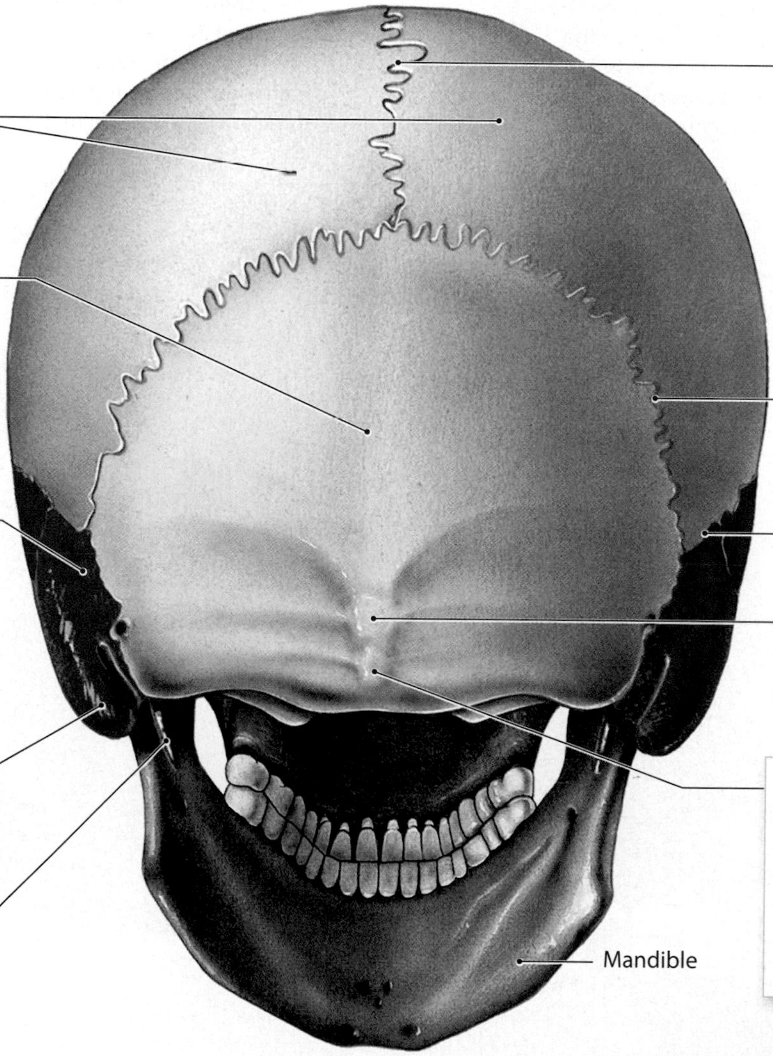

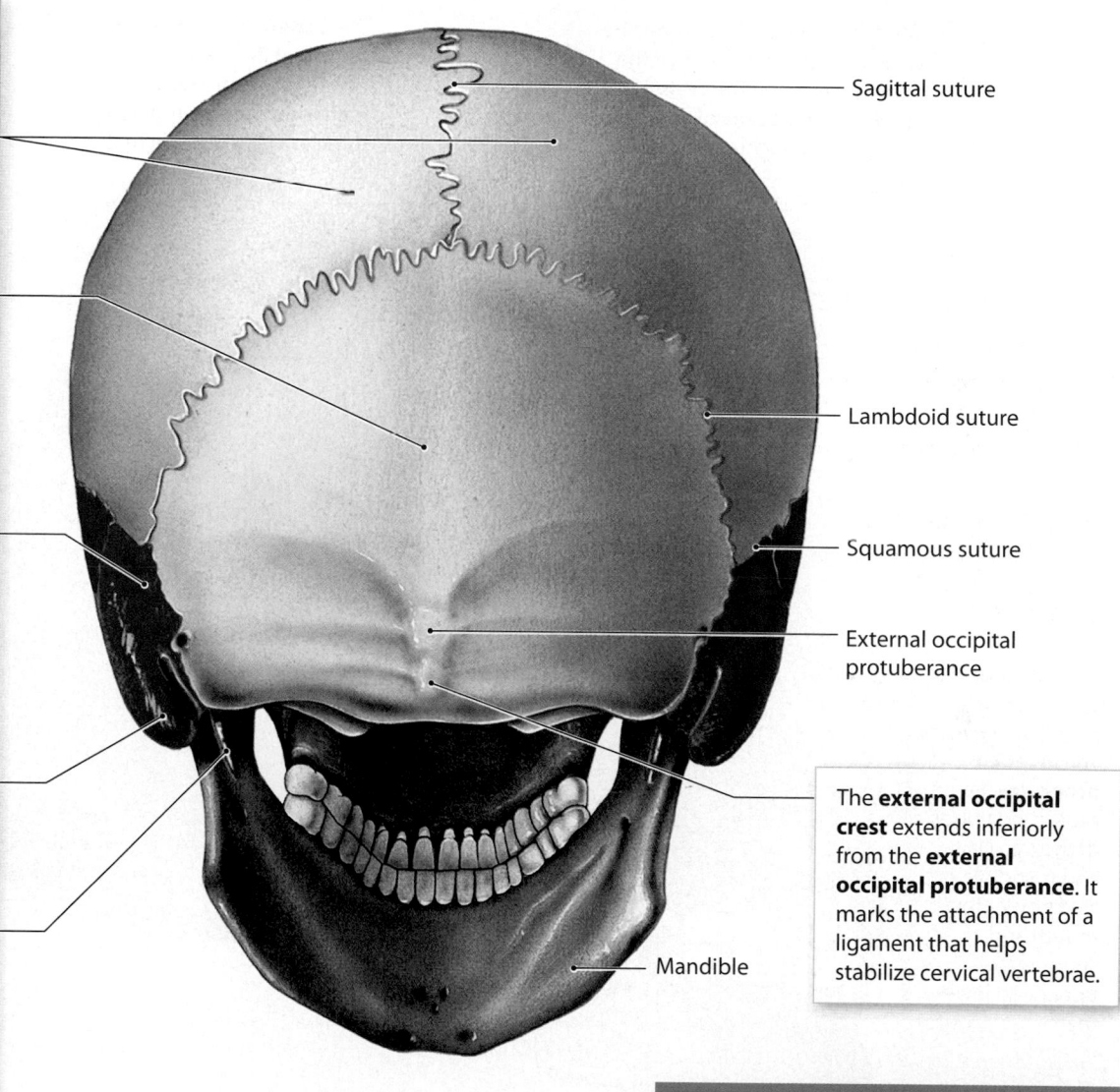

Sagittal suture
Lambdoid suture
Squamous suture
External occipital protuberance
Mandible

The **external occipital crest** extends inferiorly from the **external occipital protuberance**. It marks the attachment of a ligament that helps stabilize cervical vertebrae.

Module 7.3 Review

a. Identify the facial bones.

b. Quincy suffers a hit to the skull that fractures the right superior lateral surface of his cranium. Which bone is fractured?

c. Identify the following bones as either a facial bone or a cranial bone: vomer, ethmoid, sphenoid, temporal, and inferior nasal conchae.

LO **7.3** Explain the significance of the markings and locations of the anterior and posterior aspects of the facial and cranial bones.

The lateral and medial aspects of the skull share many surface markings

1 This lateral view of the skull shows how the large bones interconnect and reveals the surface markings of the individual cranial and facial bones.

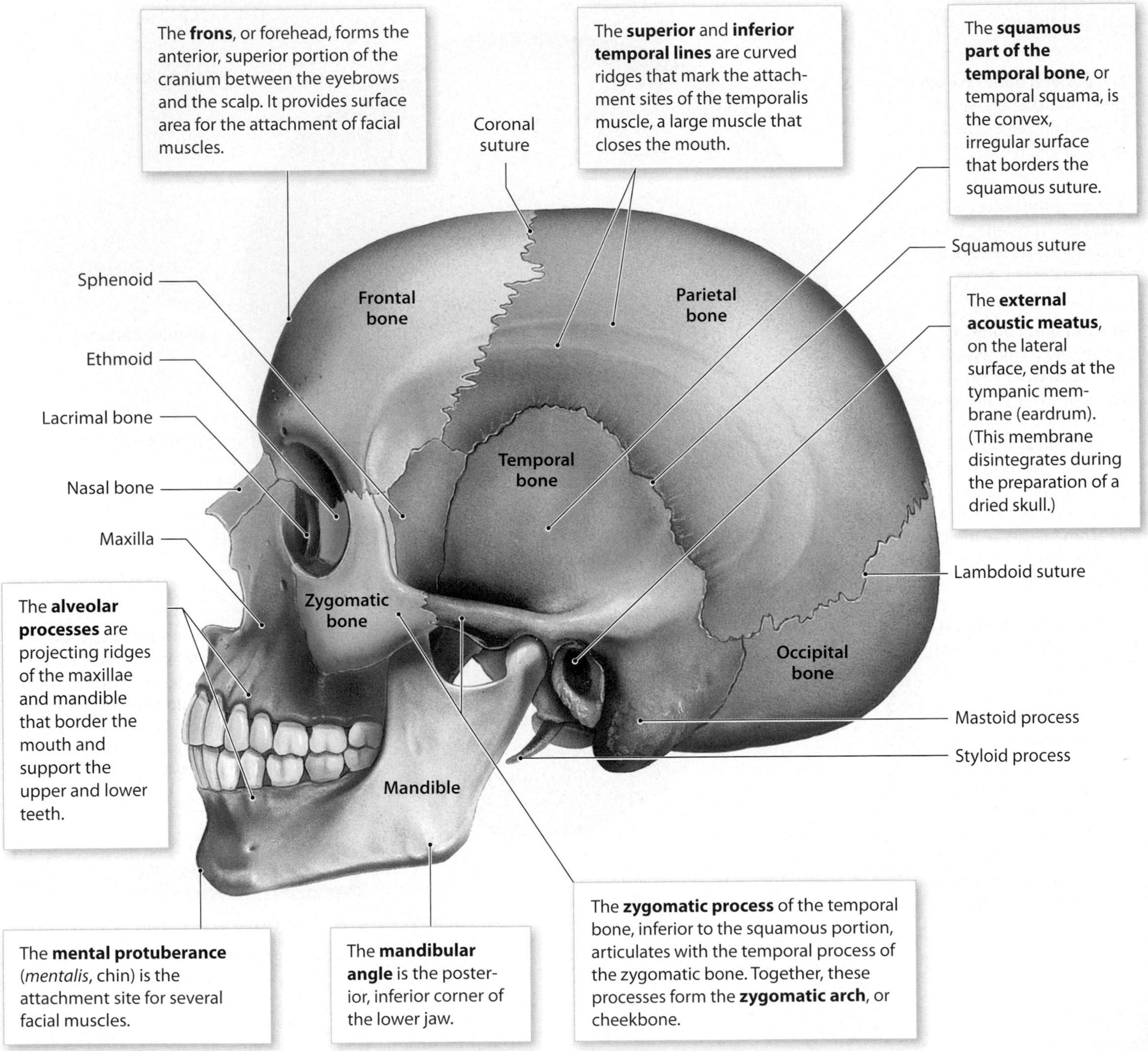
The **frons**, or forehead, forms the anterior, superior portion of the cranium between the eyebrows and the scalp. It provides surface area for the attachment of facial muscles.
Coronal suture
The **superior** and **inferior temporal lines** are curved ridges that mark the attachment sites of the temporalis muscle, a large muscle that closes the mouth.
The **squamous part of the temporal bone**, or temporal squama, is the convex, irregular surface that borders the squamous suture.
Squamous suture
Sphenoid
Ethmoid
Lacrimal bone
Nasal bone
Maxilla
Frontal bone
Parietal bone
Temporal bone
Zygomatic bone
Occipital bone
Mandible
The **external acoustic meatus**, on the lateral surface, ends at the tympanic membrane (eardrum). (This membrane disintegrates during the preparation of a dried skull.)
Lambdoid suture
Mastoid process
Styloid process
The **alveolar processes** are projecting ridges of the maxillae and mandible that border the mouth and support the upper and lower teeth.
The **mental protuberance** (*mentalis*, chin) is the attachment site for several facial muscles.
The **mandibular angle** is the posterior, inferior corner of the lower jaw.
The **zygomatic process** of the temporal bone, inferior to the squamous portion, articulates with the temporal process of the zygomatic bone. Together, these processes form the **zygomatic arch**, or cheekbone.

2 This sagittal section passes slightly to the left of the midline, leaving the vomer and the perpendicular plate of the ethmoid intact.

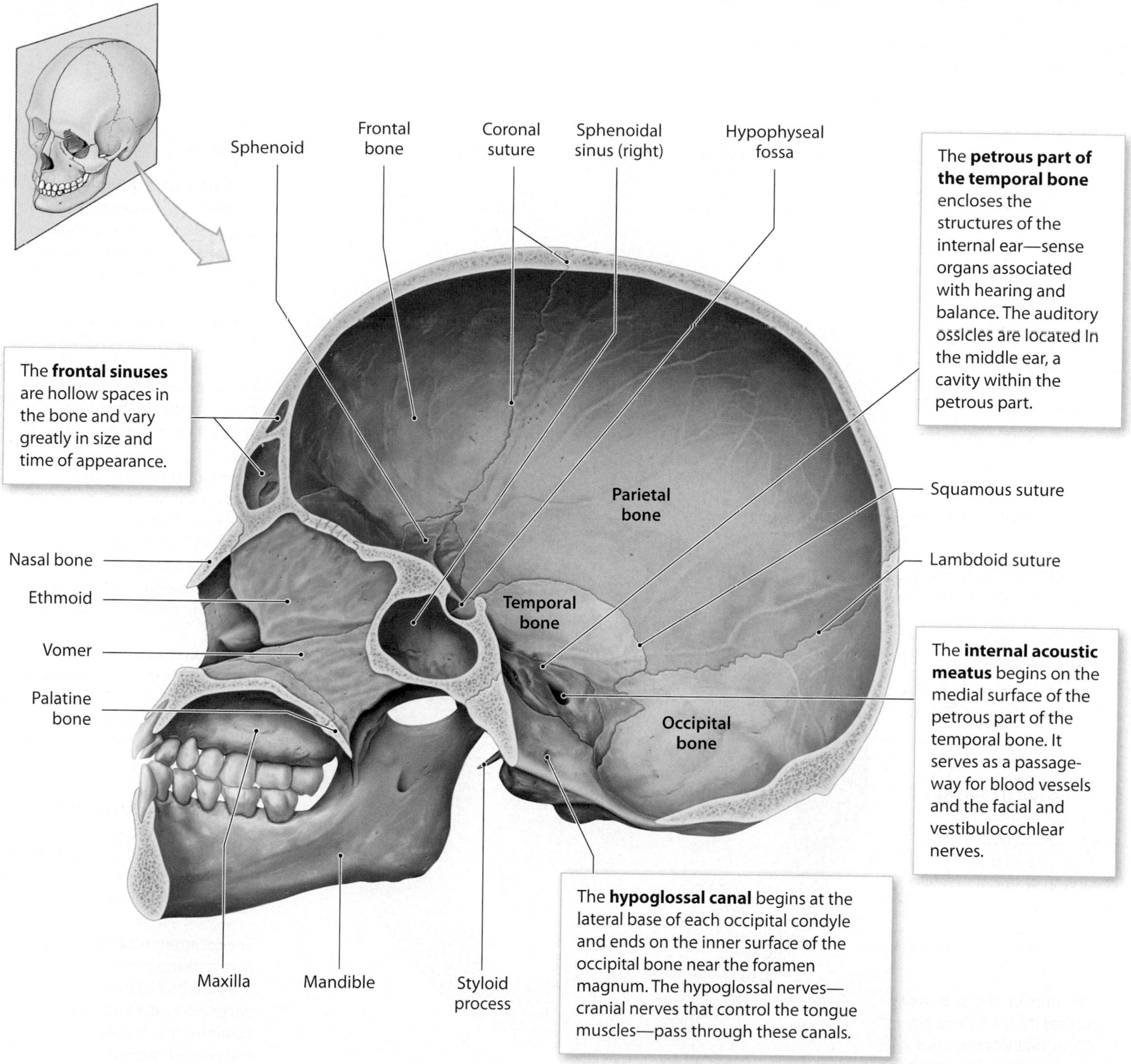

Sphenoid
Frontal bone
Coronal suture
Sphenoidal sinus (right)
Hypophyseal fossa
The **petrous part of the temporal bone** encloses the structures of the internal ear—sense organs associated with hearing and balance. The auditory ossicles are located in the middle ear, a cavity within the petrous part.
The **frontal sinuses** are hollow spaces in the bone and vary greatly in size and time of appearance.
Parietal bone
Squamous suture
Lambdoid suture
Nasal bone
Ethmoid
Vomer
Palatine bone
Temporal bone
Occipital bone
The **internal acoustic meatus** begins on the medial surface of the petrous part of the temporal bone. It serves as a passageway for blood vessels and the facial and vestibulocochlear nerves.
Maxilla
Mandible
Styloid process
The **hypoglossal canal** begins at the lateral base of each occipital condyle and ends on the inner surface of the occipital bone near the foramen magnum. The hypoglossal nerves—cranial nerves that control the tongue muscles—pass through these canals.

Module 7.4 Review

a. Name the meatuses found in the temporal bone.
b. The alveolar processes perform what functions in which bones?
c. What is the function of the internal acoustic meatus?

LO 7.4 Explain the significance of the markings and locations of the lateral and medial aspects of the facial and cranial bones.

The foramina on the inferior surface of the skull mark the passageways for nerves and blood vessels

1 In this inferior view you can see many of the important passageways for blood vessels and nerves, as well as the foramen used by the spinal cord.

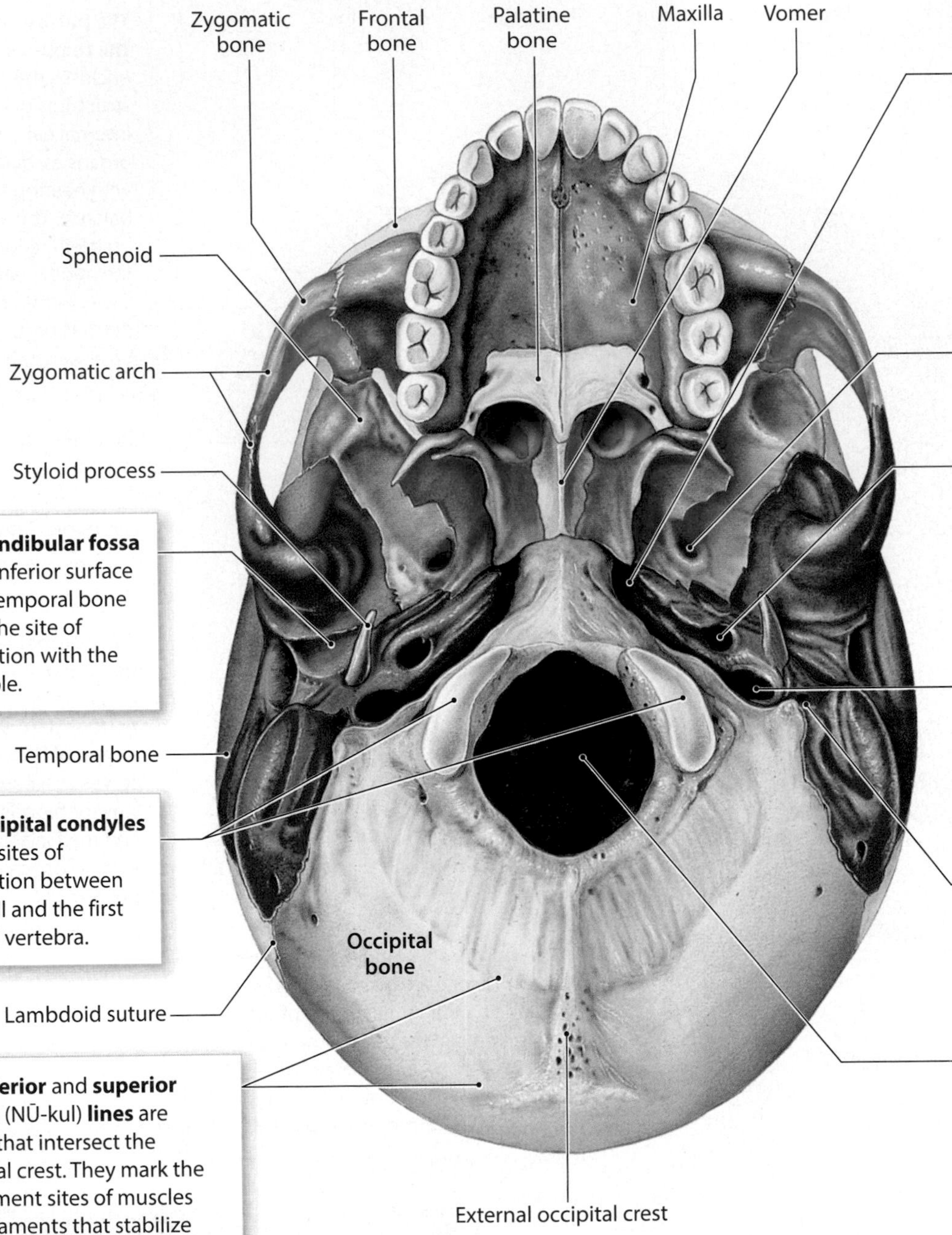
Zygomatic bone
Frontal bone
Palatine bone
Maxilla
Vomer
Sphenoid
Zygomatic arch
Styloid process
Temporal bone
Occipital bone
Lambdoid suture
External occipital crest

The **mandibular fossa** on the inferior surface of the temporal bone marks the site of articulation with the mandible.

The **occipital condyles** are the sites of articulation between the skull and the first cervical vertebra.

The **inferior** and **superior nuchal** (NŪ-kul) **lines** are ridges that intersect the occipital crest. They mark the attachment sites of muscles and ligaments that stabilize the articulation at the occipital condyles and balance the weight of the head over the cervical vertebrae.

Foramina

The **foramen lacerum** (LA-se-rum; *lacerare*, to tear) is a jagged slit extending between the sphenoid and the petrous portion of the temporal bone. It contains hyaline cartilage and small arteries that supply the inner surface of the cranium.

The **foramen ovale** (ō-VAH-lē) provides passage for nerves innervating the jaws.

The **carotid canal** provides passage for the internal carotid artery, a major artery to the brain. As it leaves the carotid canal, the internal carotid artery passes through the anterior portion of the foramen lacerum.

The **jugular foramen** lies between the occipital bone and the temporal bone. The internal jugular vein passes through this foramen, carrying venous blood from the brain.

The **stylomastoid foramen** lies posterior to the base of the styloid process. The facial nerve passes through this foramen to control the facial muscles.

The **foramen magnum** connects the cranial cavity with the vertebral canal, which is enclosed by the vertebral column. This foramen surrounds the connection between the brain and spinal cord.

2 Compare this superior view of the floor of the cranial cavity with the inferior view of the skull at left.

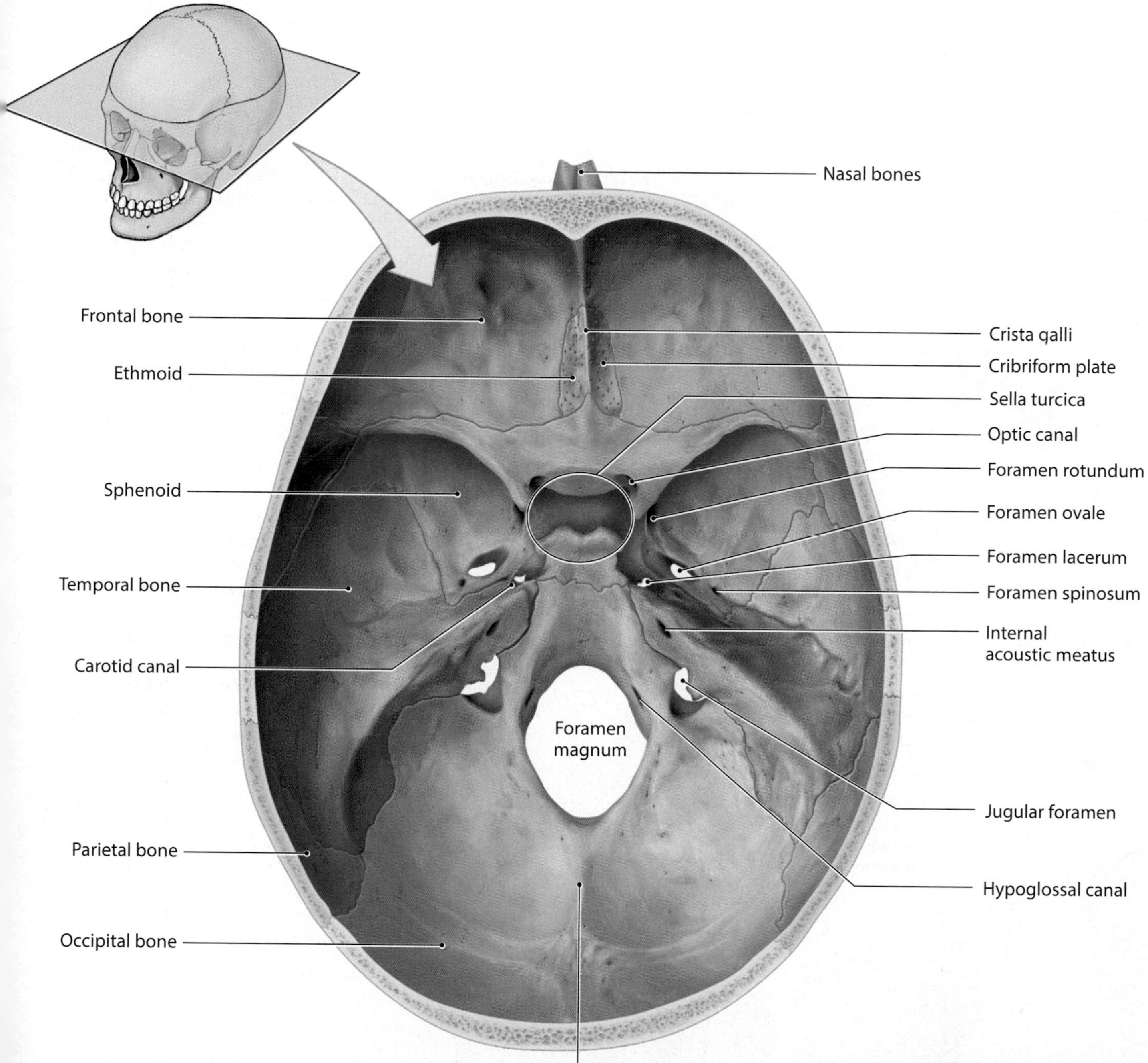

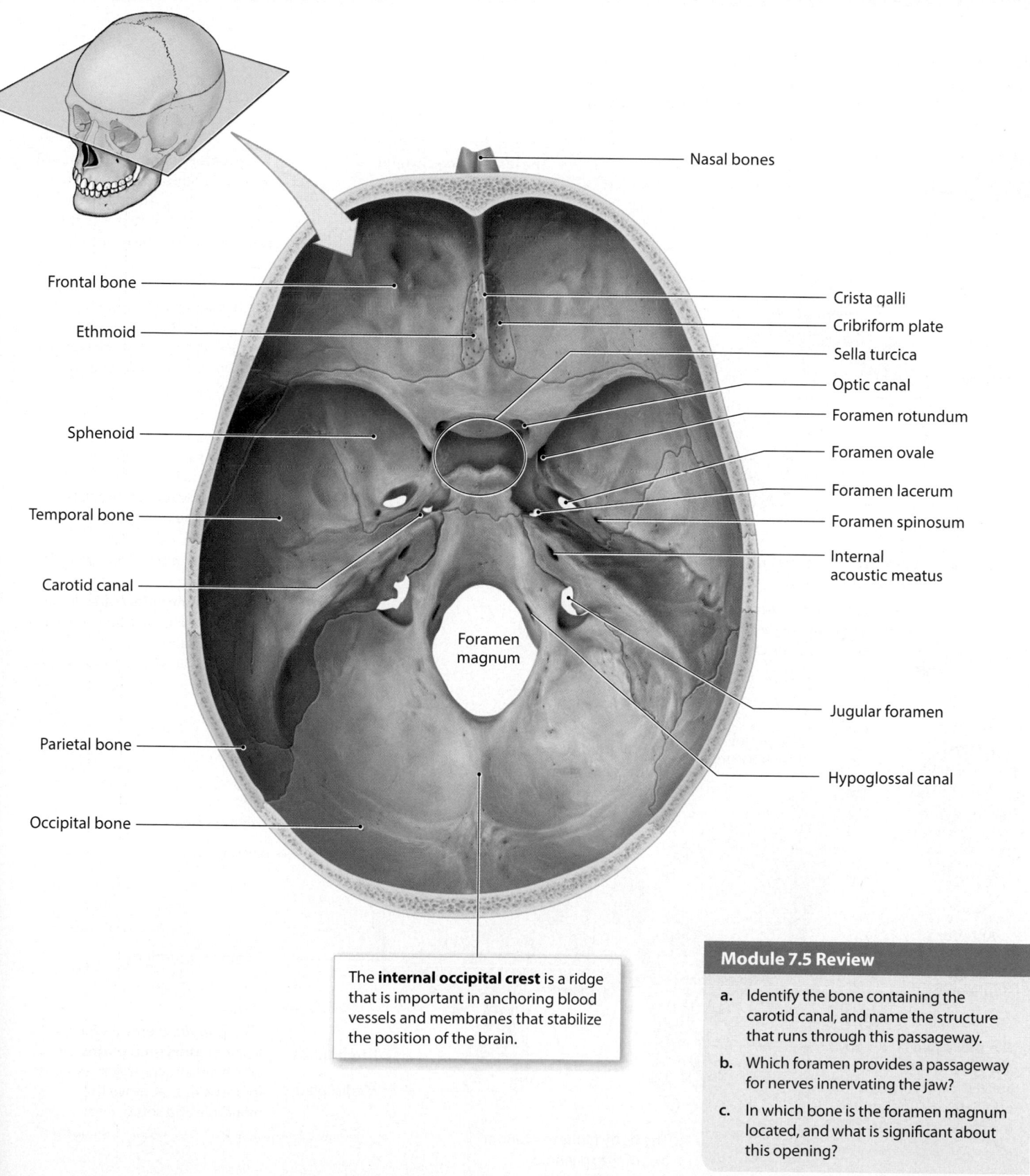

Nasal bones
Frontal bone
Crista galli
Ethmoid
Cribriform plate
Sella turcica
Optic canal
Foramen rotundum
Sphenoid
Foramen ovale
Foramen lacerum
Temporal bone
Foramen spinosum
Internal acoustic meatus
Carotid canal
Foramen magnum
Jugular foramen
Hypoglossal canal
Parietal bone
Occipital bone

The **internal occipital crest** is a ridge that is important in anchoring blood vessels and membranes that stabilize the position of the brain.

Module 7.5 Review

a. Identify the bone containing the carotid canal, and name the structure that runs through this passageway.

b. Which foramen provides a passageway for nerves innervating the jaw?

c. In which bone is the foramen magnum located, and what is significant about this opening?

LO **7.5** Explain the significance of the markings and locations of the inferior and interior aspects of the facial and cranial bones.

Module 7.6

The shapes and markings of the sphenoid, ethmoid, and palatine bones are best seen in the isolated bones

1 The **sphenoid** is an irregularly shaped bone that forms part of the floor of the cranium, unites the cranial and facial bones, and acts as a cross-bridge that strengthens the sides of the skull.

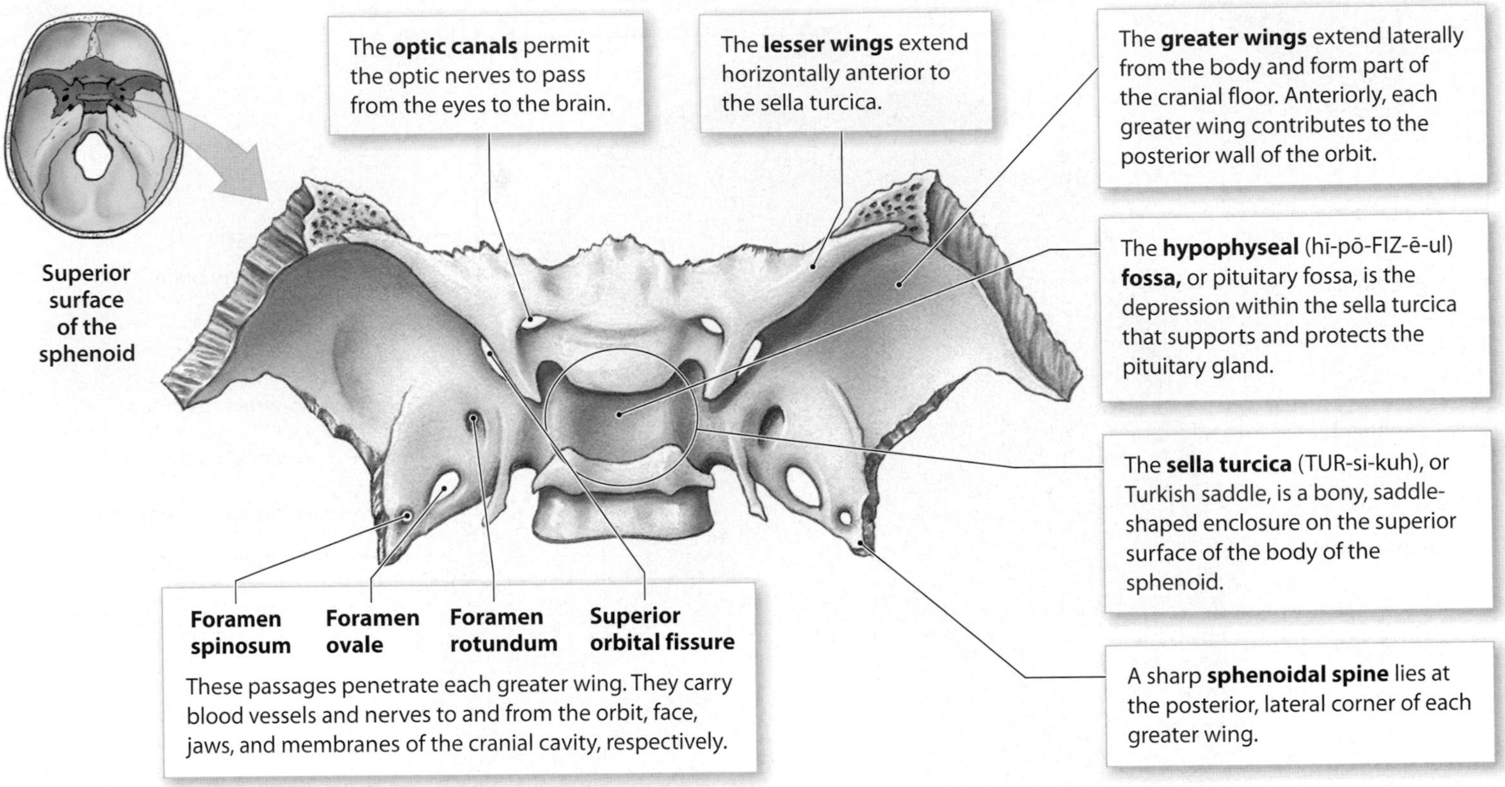
Superior surface of the sphenoid
The **optic canals** permit the optic nerves to pass from the eyes to the brain.
The **lesser wings** extend horizontally anterior to the sella turcica.
The **greater wings** extend laterally from the body and form part of the cranial floor. Anteriorly, each greater wing contributes to the posterior wall of the orbit.
The **hypophyseal** (hī-pō-FIZ-ē-ul) **fossa,** or pituitary fossa, is the depression within the sella turcica that supports and protects the pituitary gland.
The **sella turcica** (TUR-si-kuh), or Turkish saddle, is a bony, saddle-shaped enclosure on the superior surface of the body of the sphenoid.
A sharp **sphenoidal spine** lies at the posterior, lateral corner of each greater wing.
Foramen spinosum
Foramen ovale
Foramen rotundum
Superior orbital fissure
These passages penetrate each greater wing. They carry blood vessels and nerves to and from the orbit, face, jaws, and membranes of the cranial cavity, respectively.

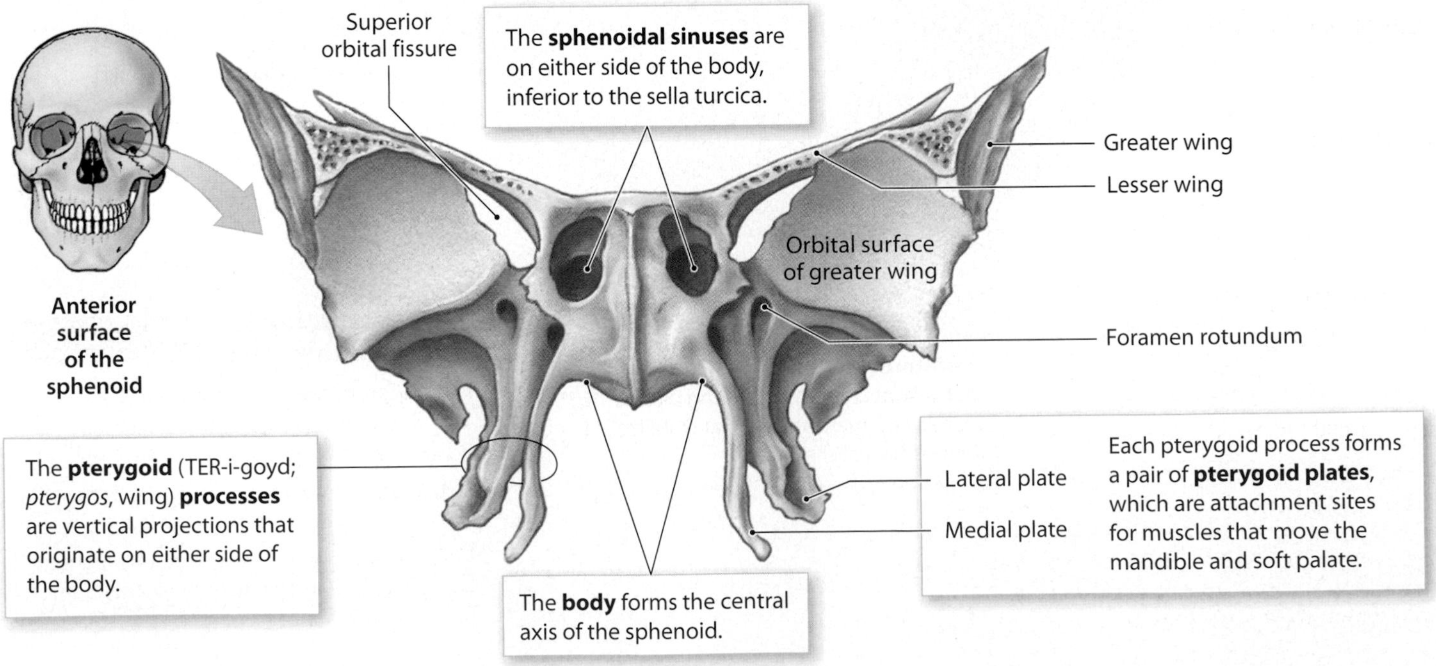
Anterior surface of the sphenoid
Superior orbital fissure
The **sphenoidal sinuses** are on either side of the body, inferior to the sella turcica.
Greater wing
Lesser wing
Orbital surface of greater wing
Foramen rotundum
The **pterygoid** (TER-i-goyd; *pterygos*, wing) **processes** are vertical projections that originate on either side of the body.
Lateral plate
Medial plate
Each pterygoid process forms a pair of **pterygoid plates,** which are attachment sites for muscles that move the mandible and soft palate.
The **body** forms the central axis of the sphenoid.

2 The **ethmoid** forms part of the floor of the anterior cranium, the roof of the nasal cavity, and part of the nasal septum and medial orbital wall. The ethmoid has three parts: (1) the cribriform plate, (2) the paired lateral masses, and (3) the perpendicular plate.

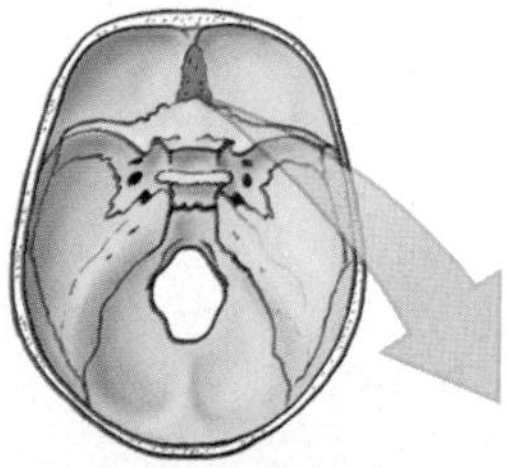

The **cribriform plate** (*cribrum*, sieve) forms the anteromedial floor of the cranium and the roof of the nasal cavity. The olfactory foramina in the cribriform plate permit passage of the olfactory nerves, which provide the sense of smell.

The **crista galli** (*crista*, crest + *gallus*, rooster, cock; cock's comb) is a bony ridge that projects superior to the cribriform plate. The falx cerebri, a membrane that stabilizes the position of the brain, attaches to this ridge.

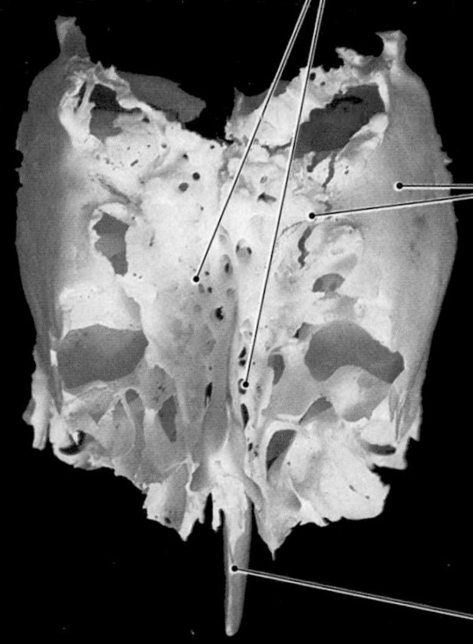

Superior surface

The **lateral masses** contain the **ethmoidal labyrinth**, which consists of the interconnected ethmoid air cells (ethmoid cells), small chambers that open into the nasal cavity on each side.

The **superior nasal conchae** (KONG-kē; singular, *concha*, a snail shell) and the **middle nasal conchae** are delicate projections of the lateral masses.

The **perpendicular plate** forms part of the nasal septum, along with the vomer and a piece of hyaline cartilage.

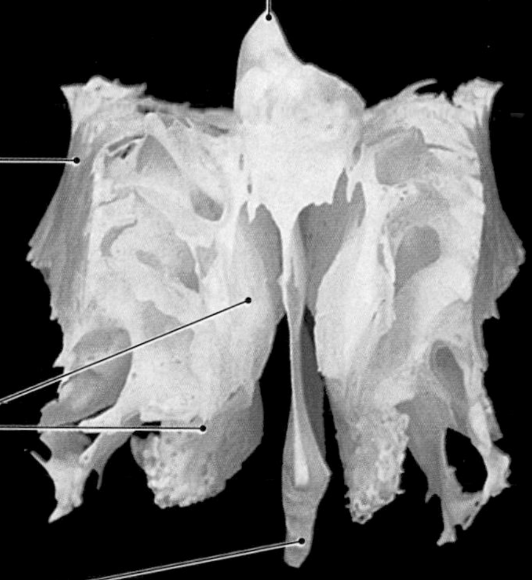

Posterior surface

3 The **palatine bones** form the posterior portion of the hard palate and contribute to the floor of each orbit.

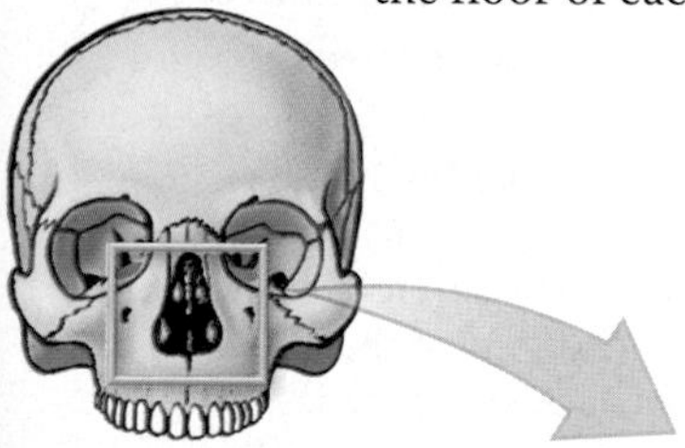

The **orbital process** forms part of the floor of the orbit. This process contains a small sinus that usually opens into the sphenoidal sinus.

The **perpendicular plate** of the palatine bone is a flat portion that extends from the horizontal plate to the orbital process.

The **horizontal plate** forms the posterior part of the hard palate.

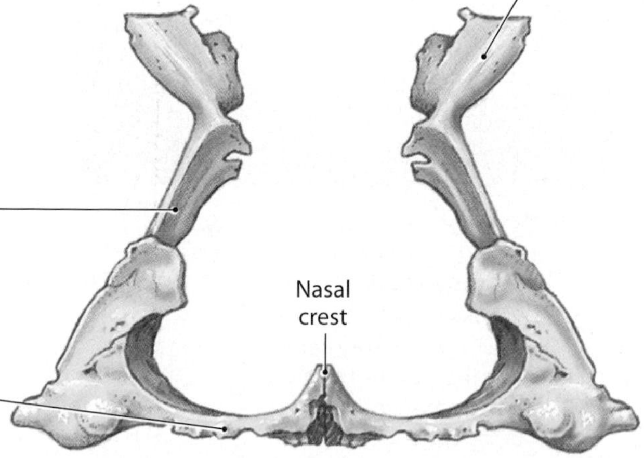

Nasal crest

Module 7.6 Review

a. Identify the bone containing the optic canal, and cite the structures using this passageway.

b. Which bone contains the sella turcica? What structure is in the depression (fossa) within the sella turcica?

c. Identify the bone containing the cribriform plate. What is significant about this structure?

LO 7.6 Describe and locate the surface features of the sphenoid, ethmoid, and palatine bones.

Each orbital complex contains one eye, and the nasal complex encloses the nasal cavities

The facial bones not only protect and support the openings of the digestive and respiratory systems, but also protect the delicate sense organs for vision and smell. Together, certain cranial bones and facial bones form an orbital complex containing one eye and the nasal complex that surrounds the nasal cavities.

1 The orbits are the bony recesses that contain the eyes. Each orbit is formed by the seven bones of the **orbital complex**. The frontal bone forms the roof, the zygomatic bone forms the lateral wall, and the maxilla provides most of the orbital floor. The orbital rim and the first portion of the medial wall are formed by the maxilla, the lacrimal bone, and the lateral mass of the ethmoid. The lateral mass articulates with the sphenoid and a small process of the palatine bone.

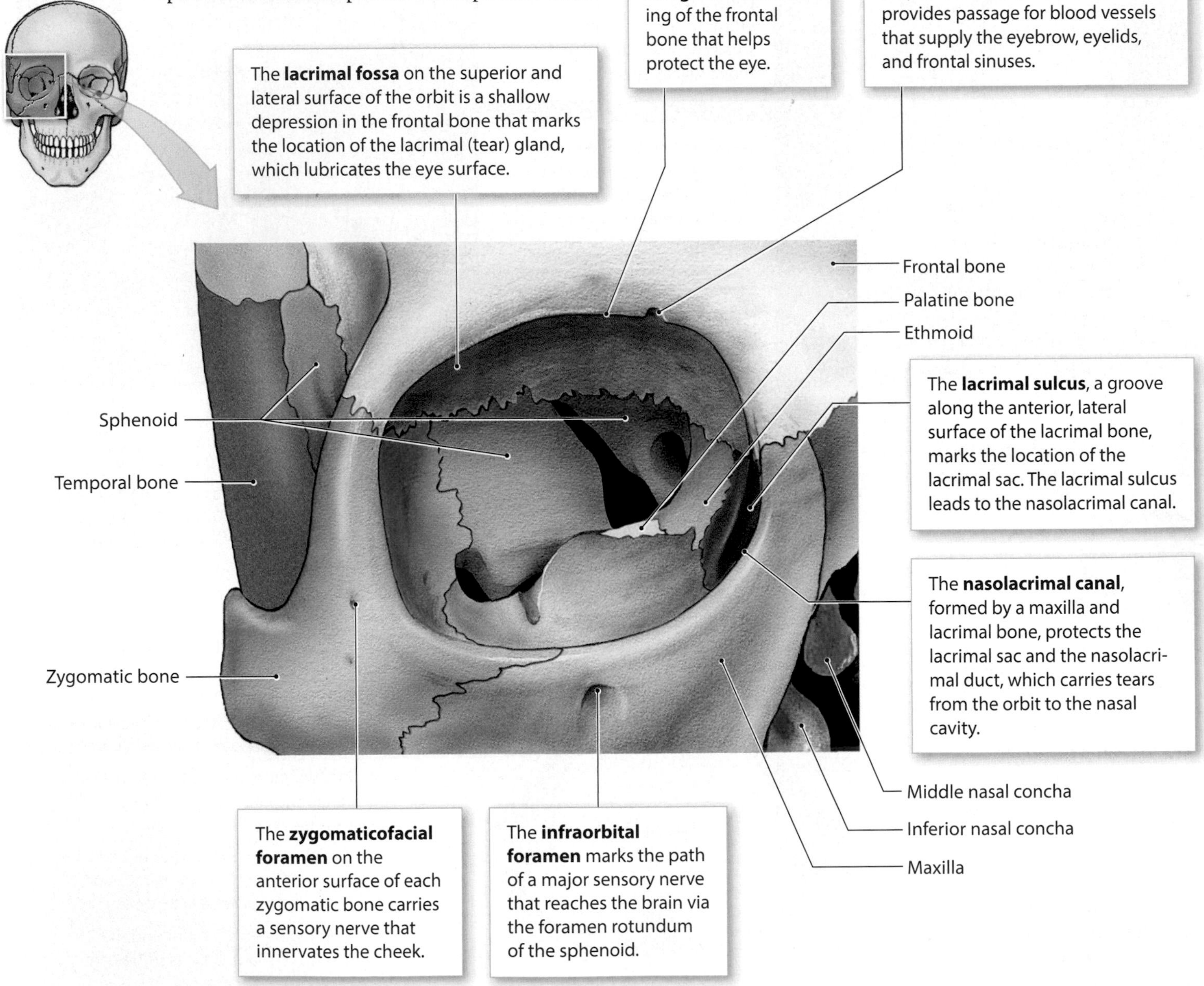
The **supraorbital margin** is a thickening of the frontal bone that helps protect the eye.
Either a **supraorbital notch** or a fully enclosed supraorbital foramen provides passage for blood vessels that supply the eyebrow, eyelids, and frontal sinuses.
The **lacrimal fossa** on the superior and lateral surface of the orbit is a shallow depression in the frontal bone that marks the location of the lacrimal (tear) gland, which lubricates the eye surface.
Frontal bone
Palatine bone
Ethmoid
The **lacrimal sulcus**, a groove along the anterior, lateral surface of the lacrimal bone, marks the location of the lacrimal sac. The lacrimal sulcus leads to the nasolacrimal canal.
Sphenoid
Temporal bone
The **nasolacrimal canal**, formed by a maxilla and lacrimal bone, protects the lacrimal sac and the nasolacrimal duct, which carries tears from the orbit to the nasal cavity.
Zygomatic bone
Middle nasal concha
Inferior nasal concha
Maxilla
The **zygomaticofacial foramen** on the anterior surface of each zygomatic bone carries a sensory nerve that innervates the cheek.
The **infraorbital foramen** marks the path of a major sensory nerve that reaches the brain via the foramen rotundum of the sphenoid.

2 The **nasal complex** includes the bones that enclose the nasal cavities and the paranasal sinuses, air-filled chambers connected to the nasal cavities. The sphenoid, ethmoid, frontal bone, palatine bone, and maxillae contain the **paranasal sinuses**. (The tiny palatine sinuses, not shown, generally open into the sphenoidal sinuses.) The paranasal sinuses lighten the skull bones, allow the voice to resonate, and provide an extensive area of mucous epithelium. Inflammation of the sinuses due to viral, bacterial, or fungal infection is called **sinusitis**.

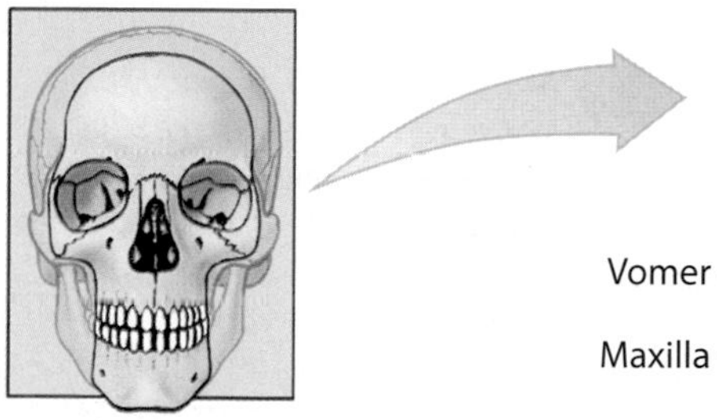
Frontal section

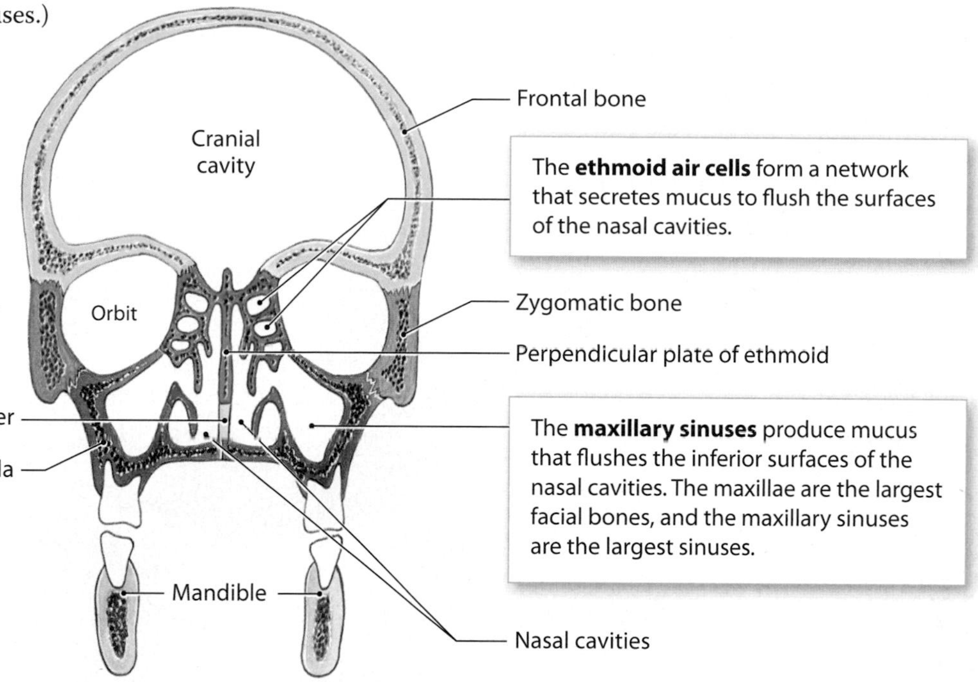
Frontal bone
Cranial cavity
The ethmoid air cells form a network that secretes mucus to flush the surfaces of the nasal cavities.
Orbit
Zygomatic bone
Perpendicular plate of ethmoid
Vomer
Maxilla
The maxillary sinuses produce mucus that flushes the inferior surfaces of the nasal cavities. The maxillae are the largest facial bones, and the maxillary sinuses are the largest sinuses.
Mandible
Nasal cavities

3 This is a sagittal section of the skull with the nasal septum removed. The frontal bone, sphenoid, and ethmoid form the superior wall of the nasal cavities. The lateral walls are formed by the maxillae and the lacrimal bones, the ethmoid (the superior and middle nasal conchae), and the inferior nasal conchae. Much of the anterior margin of the nasal cavity is formed by the soft tissues of the nose, but the bridge of the nose is supported by the maxillae and nasal bones.

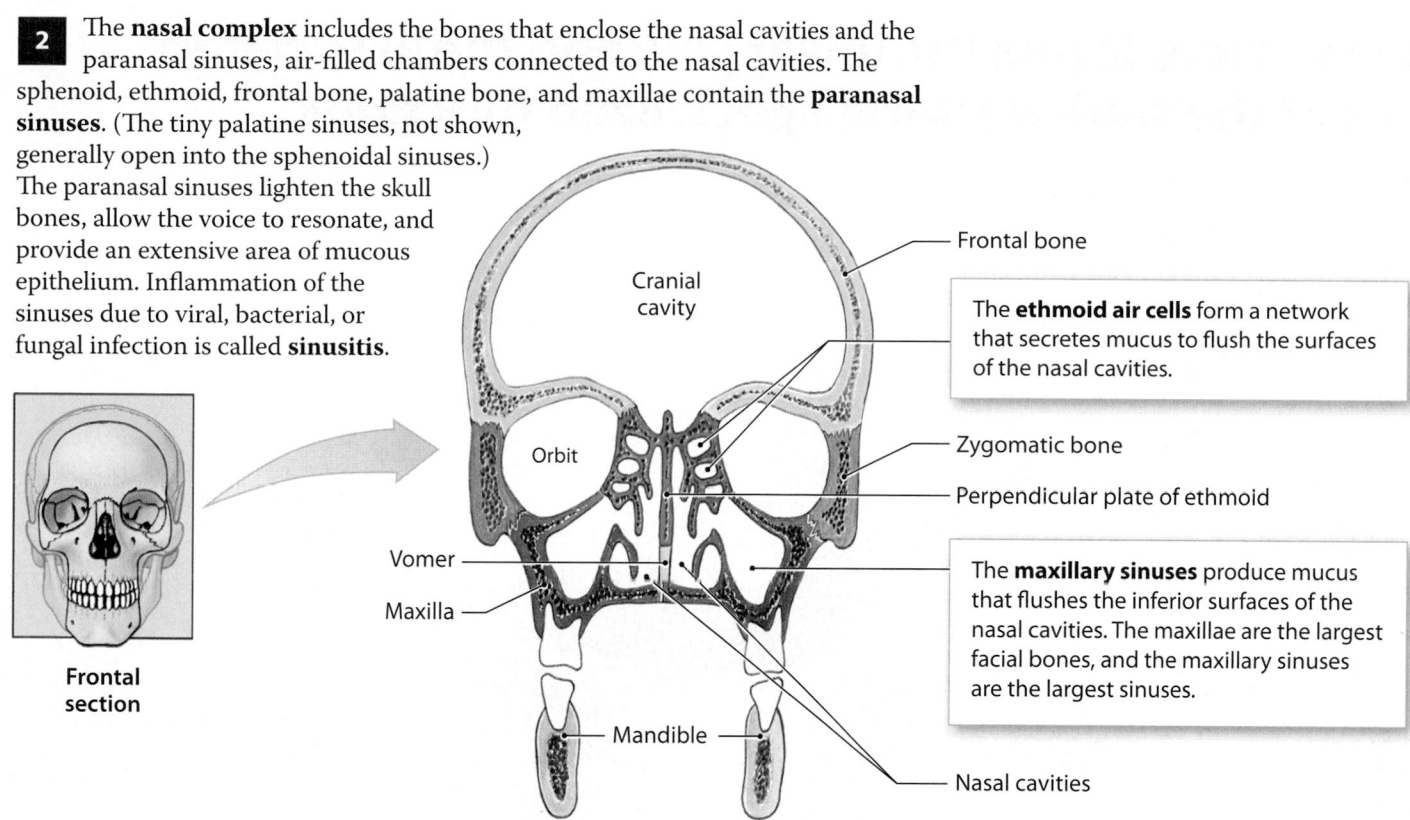

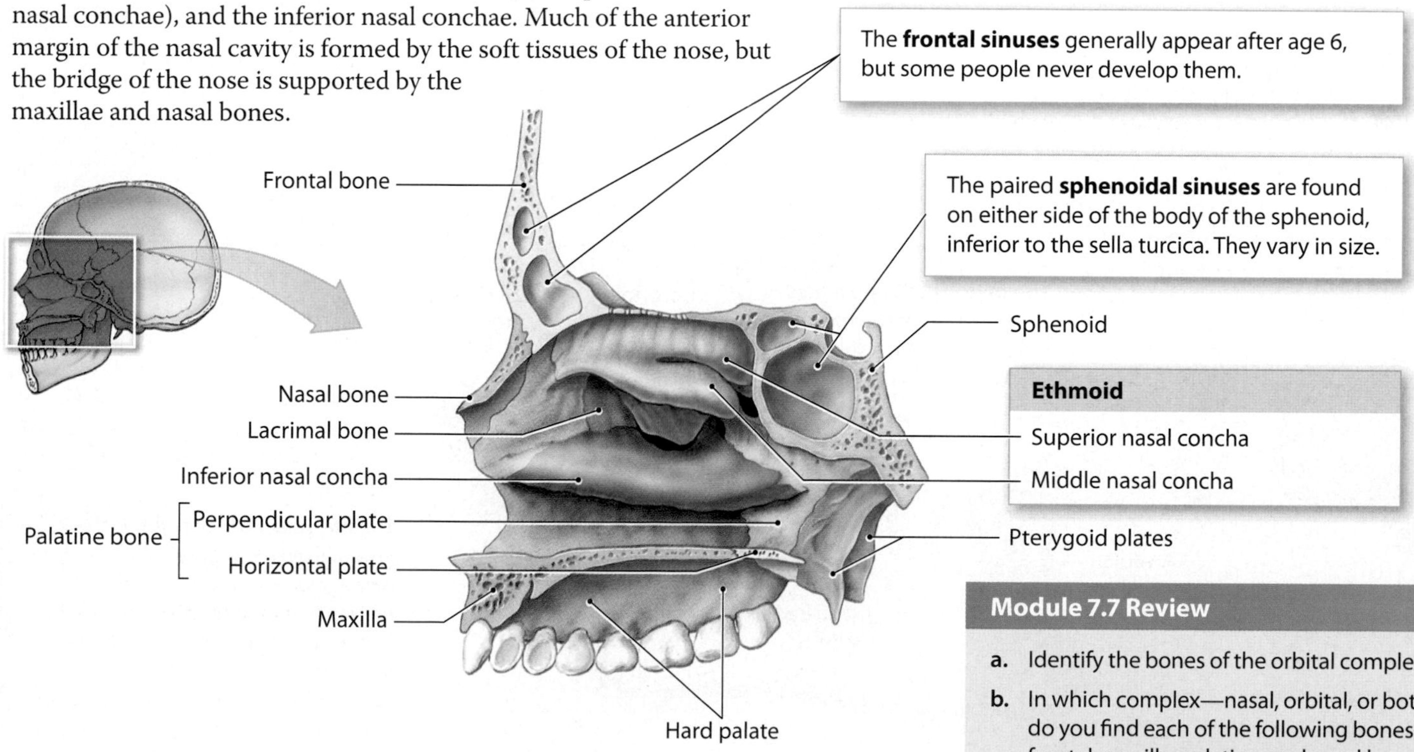
The frontal sinuses generally appear after age 6, but some people never develop them.
Frontal bone
The paired sphenoidal sinuses are found on either side of the body of the sphenoid, inferior to the sella turcica. They vary in size.
Sphenoid
Ethmoid
Superior nasal concha
Middle nasal concha
Nasal bone
Lacrimal bone
Inferior nasal concha
Palatine bone
Perpendicular plate
Horizontal plate
Maxilla
Pterygoid plates
Hard palate

Module 7.7 Review

a. Identify the bones of the orbital complex.

b. In which complex—nasal, orbital, or both—do you find each of the following bones? frontal, maxilla, palatine, and nasal bones

c. Describe the frontal sinuses.

LO **7.7** Describe the structure of the orbital complex and nasal complex and the functions of their individual bones.

The mandible forms the lower jaw, and the associated bones of the skull perform specialized functions

1 The mandible can be subdivided into the horizontal body and the ascending rami (singular, *ramus*) of the mandible. The lower teeth are supported by the mandibular body. The mandible forms the entire lower jaw and articulates with the mandibular fossae of the temporal bones.

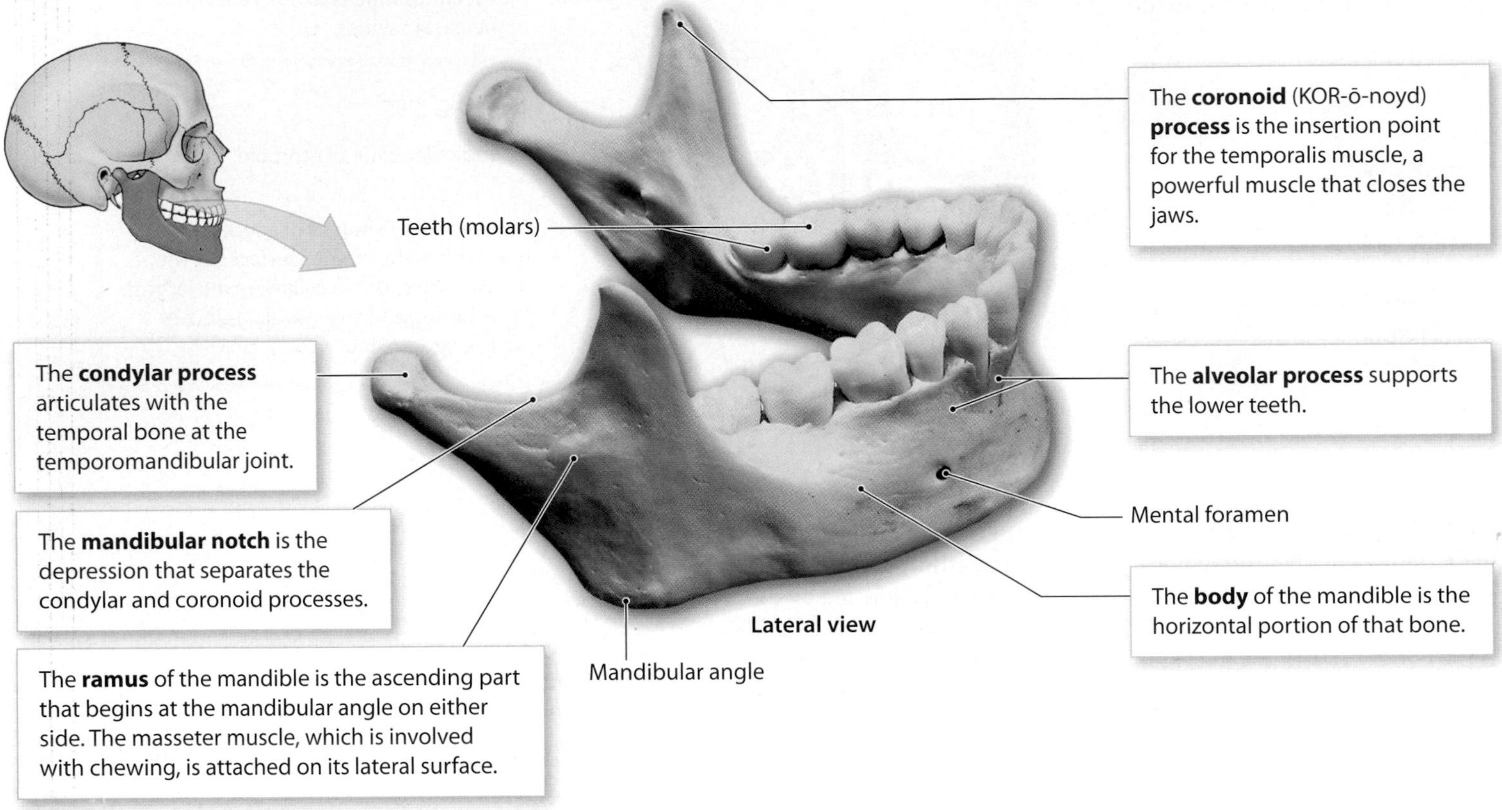
The **coronoid** (KOR-ō-noyd) **process** is the insertion point for the temporalis muscle, a powerful muscle that closes the jaws.
Teeth (molars)
The **condylar process** articulates with the temporal bone at the temporomandibular joint.
The **alveolar process** supports the lower teeth.
Mental foramen
The **mandibular notch** is the depression that separates the condylar and coronoid processes.
The **body** of the mandible is the horizontal portion of that bone.
Lateral view
The **ramus** of the mandible is the ascending part that begins at the mandibular angle on either side. The masseter muscle, which is involved with chewing, is attached on its lateral surface.
Mandibular angle

2 The medial surface of the mandible has prominent surface markings.

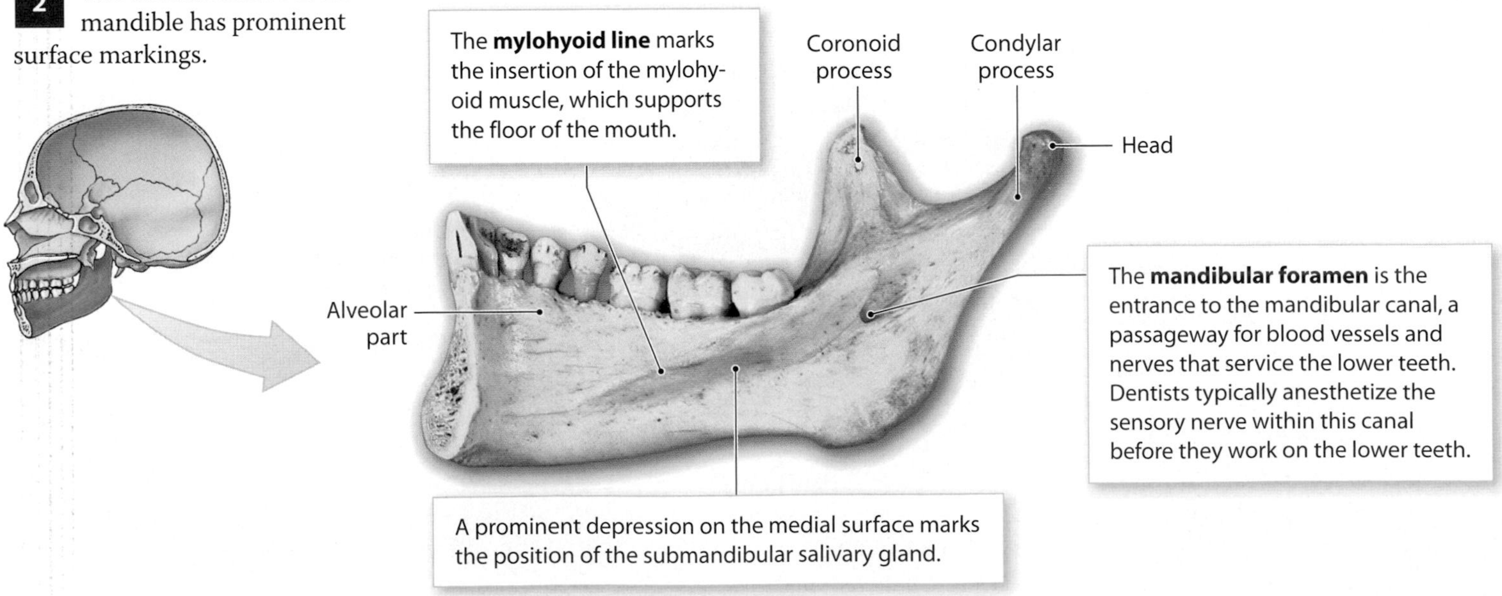
The **mylohyoid line** marks the insertion of the mylohyoid muscle, which supports the floor of the mouth.
Coronoid process
Condylar process
Head
Alveolar part
The **mandibular foramen** is the entrance to the mandibular canal, a passageway for blood vessels and nerves that service the lower teeth. Dentists typically anesthetize the sensory nerve within this canal before they work on the lower teeth.
A prominent depression on the medial surface marks the position of the submandibular salivary gland.

SKULL

FACE	14
Maxillary bones	2
Palatine bones	2
Nasal bones	2
Inferior nasal conchae	2
Zygomatic bones	2
Lacrimal bones	2
Vomer	1
Mandible	1

CRANIUM	8
Occipital bone	1
Parietal bones	2
Frontal bone	1
Temporal bones	2
Sphenoid	1
Ethmoid	1

ASSOCIATED BONES	7
Hyoid bone	1
Auditory ossicles enclosed in temporal bones (*see Chapter 15*)	6

4 Also associated with the skull are six tiny bones called **auditory ossicles**. Three bones are located within each middle ear cavity, enclosed in the petrous part of the temporal bone. They play a key role in hearing by conducting the vibrations produced by sound waves from the tympanic membrane to the receptor complex of the internal ear.

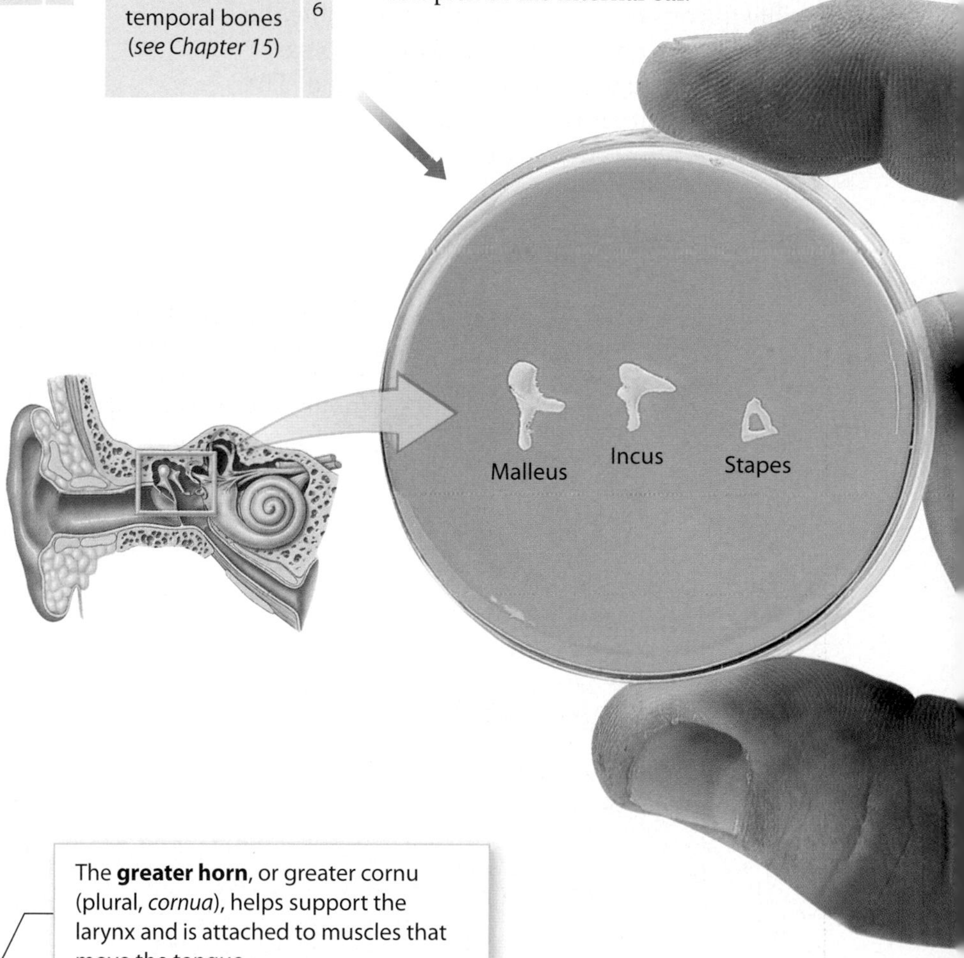
Malleus
Incus
Stapes

3 The **hyoid bone**—one of the seven associated bones of the skull—supports the larynx and is the attachment site for muscles of the larynx, pharynx, and tongue.

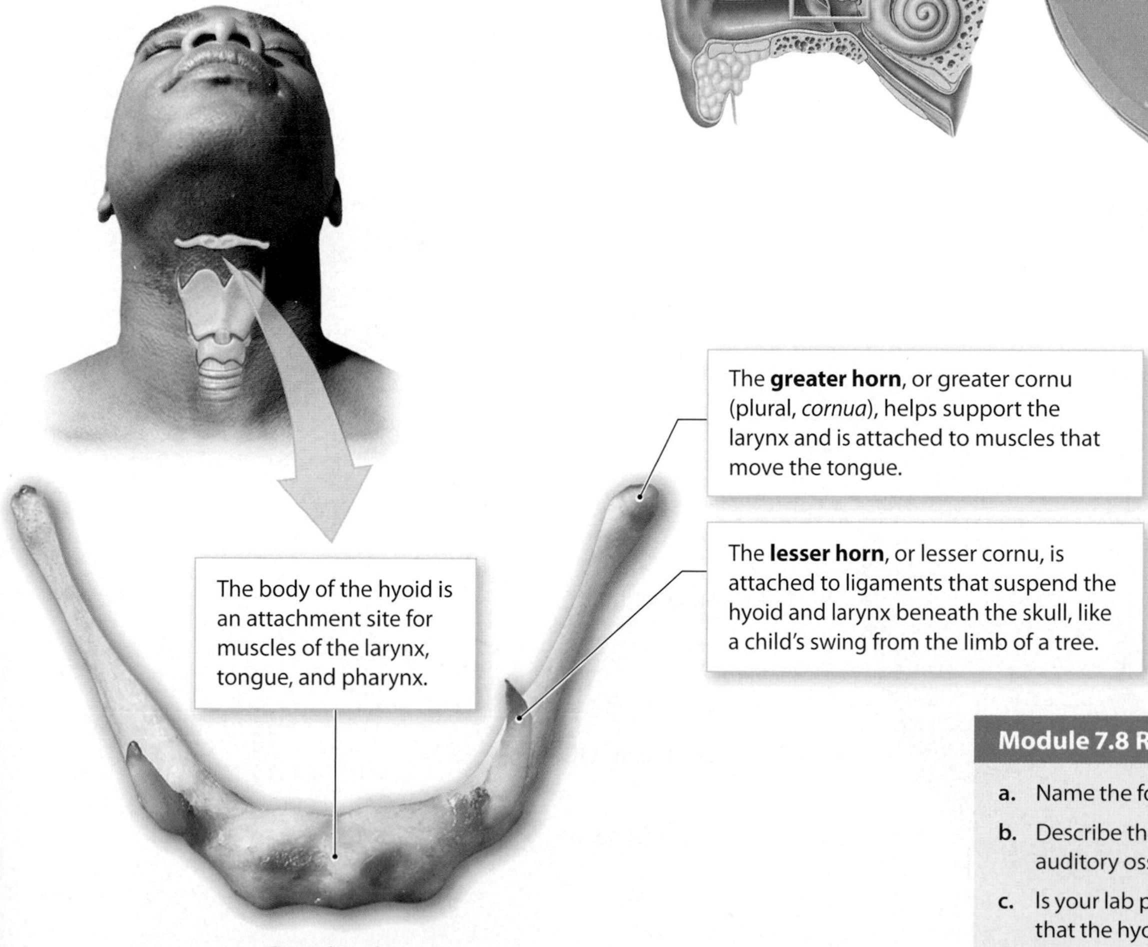
The body of the hyoid is an attachment site for muscles of the larynx, tongue, and pharynx.
The **greater horn**, or greater cornu (plural, *cornua*), helps support the larynx and is attached to muscles that move the tongue.
The **lesser horn**, or lesser cornu, is attached to ligaments that suspend the hyoid and larynx beneath the skull, like a child's swing from the limb of a tree.

Superior view

Module 7.8 Review

a. Name the foramina of the mandible.

b. Describe the location and function of the auditory ossicles.

c. Is your lab partner correct when she claims that the hyoid bone does not directly join (articulate with) any other bone?

LO **7.8** Describe the mandible and the associated bones of the skull.

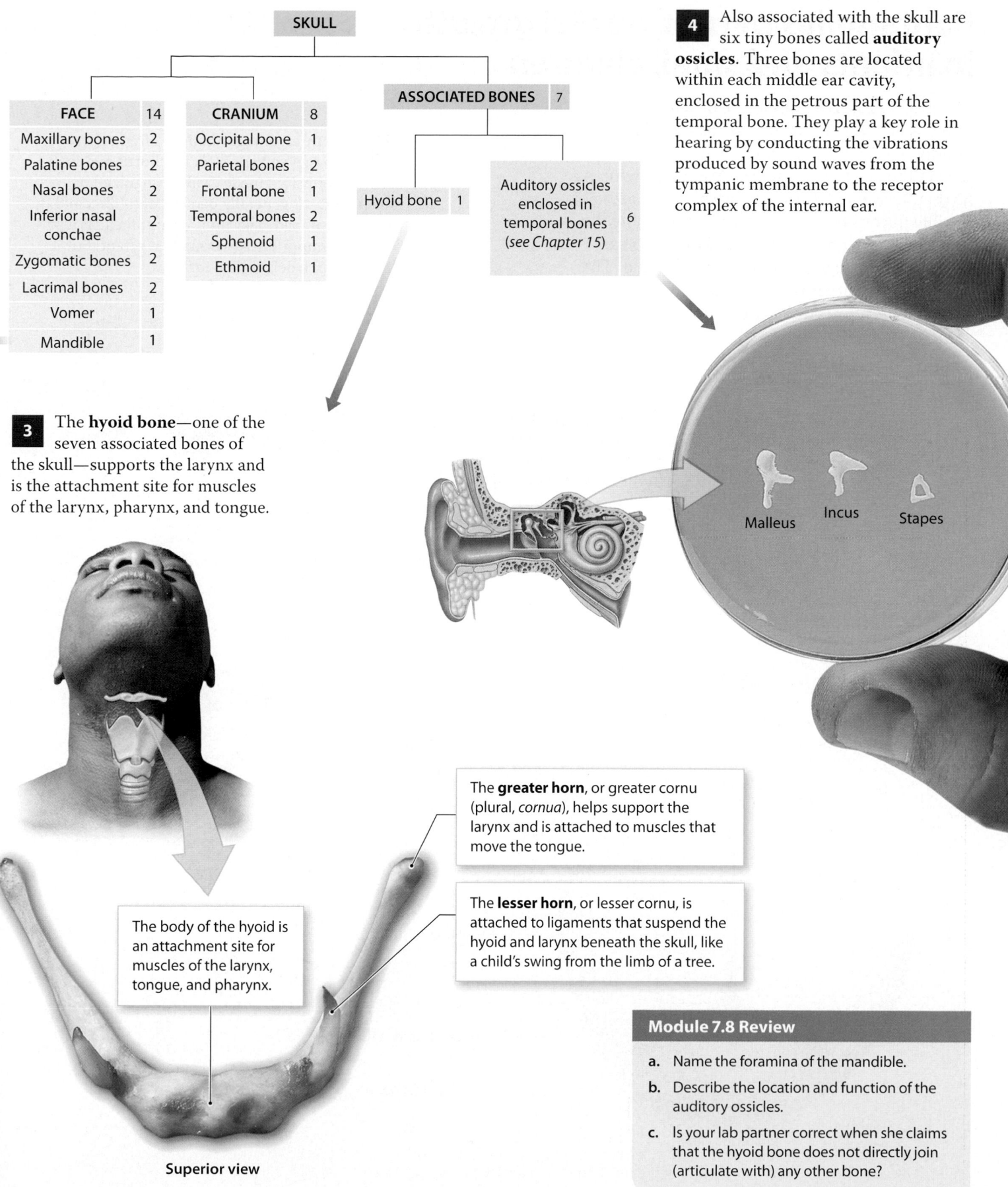

Fontanelles permit cranial growth in infants and small children

The skull organizes around the developing brain. As birth approaches, the brain enlarges rapidly. The skull bones are also growing, but they fail to keep pace with the brain's growth. At birth, the cranial bones are connected by areas of flexible fibrous connective tissue that enable the skull's shape to change without damage to ease the infant's passage through the birth canal. The largest fibrous areas between the cranial bones are known as **fontanelles** (fon-tuh-NELZ; sometimes spelled *fontanels*).

1 The **anterior fontanelle** is often referred to as the "soft spot" on newborns and is often the only fontanelle easily recognized by new parents. It generally persists until the child is nearly 2 years old. The anterior fontanelle is also of clinical importance. Normally it is taut, but it is shallow in a dehydrated infant.

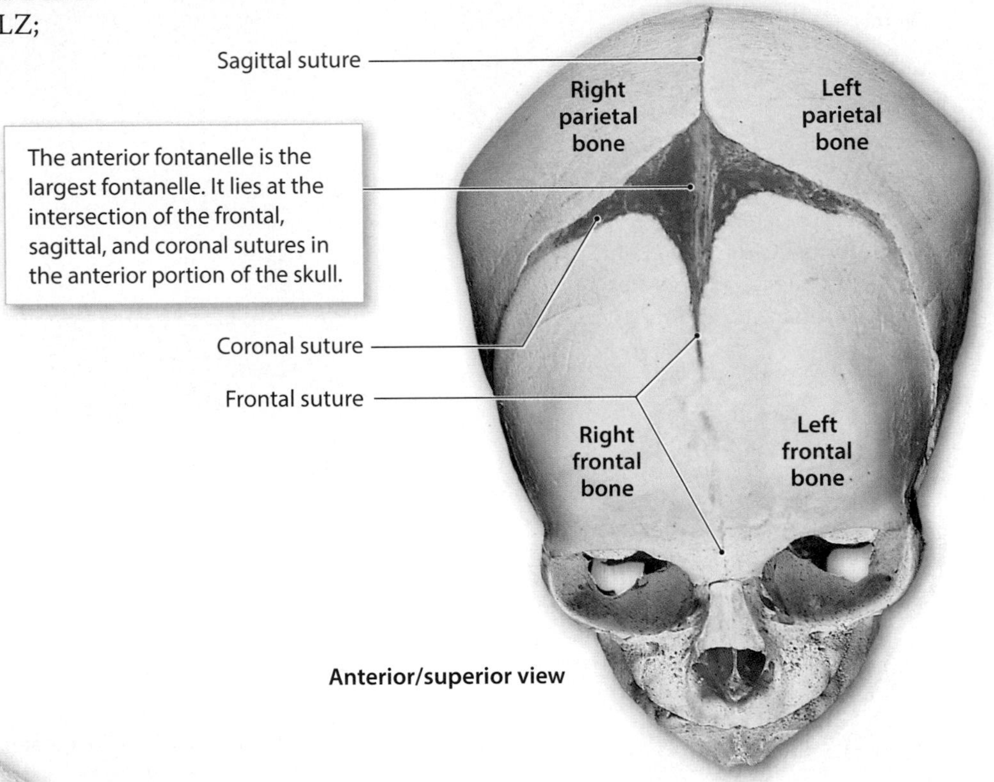
Sagittal suture
Right parietal bone
Left parietal bone
The anterior fontanelle is the largest fontanelle. It lies at the intersection of the frontal, sagittal, and coronal sutures in the anterior portion of the skull.
Coronal suture
Frontal suture
Right frontal bone
Left frontal bone

Anterior/superior view

2 Over time, the sutures gradually narrow and the fontanelles become smaller and smaller. In this posterior view of an infant skull, the occipital fontanelle has almost disappeared.

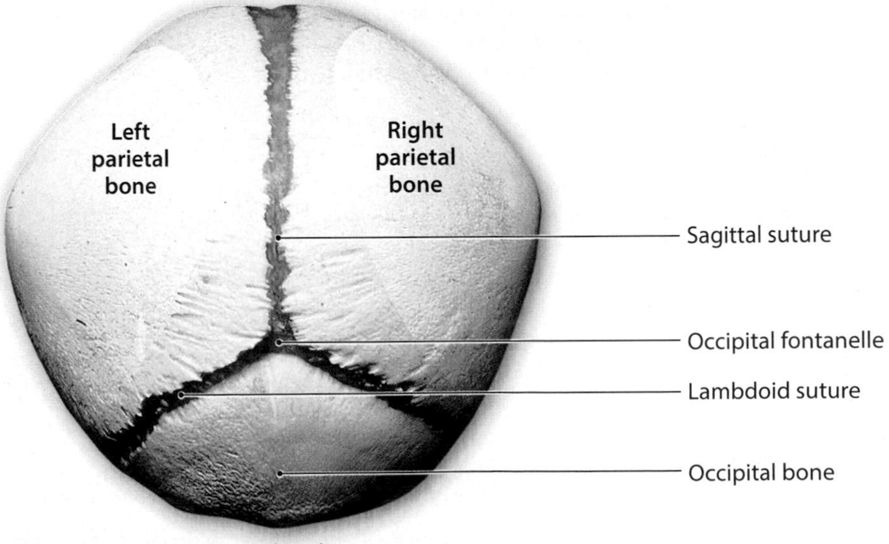
Left parietal bone
Right parietal bone
Sagittal suture
Occipital fontanelle
Lambdoid suture
Occipital bone

Posterior view

3 This lateral view shows how small an infant's facial bones are compared with the cranium.

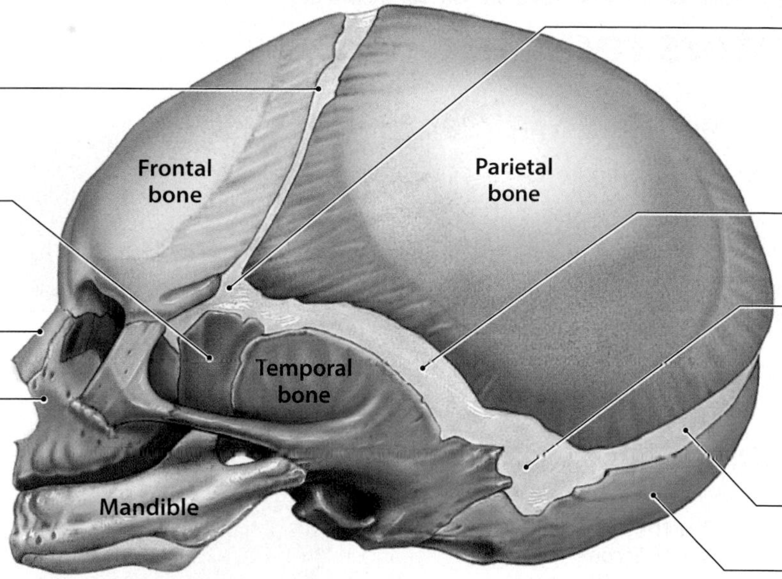

Coronal suture
Sphenoid
Nasal bone
Maxilla
Frontal bone
Parietal bone
Temporal bone
Mandible
The **sphenoidal fontanelle** on each side lies at the junction between the squamous suture and the coronal suture.
Squamous suture
The **mastoid fontanelle** on each side lies at the junction between the squamous suture and the lambdoid suture.
Lambdoid suture
Occipital bone

4 This superior view shows the size of the anterior fontanelle relative to the width of the fibrous areas destined to become the sutures of the adult skull. Because it is composed of fibrous connective tissue and covers a major blood vessel, the anterior fontanelle pulses as the heart beats.

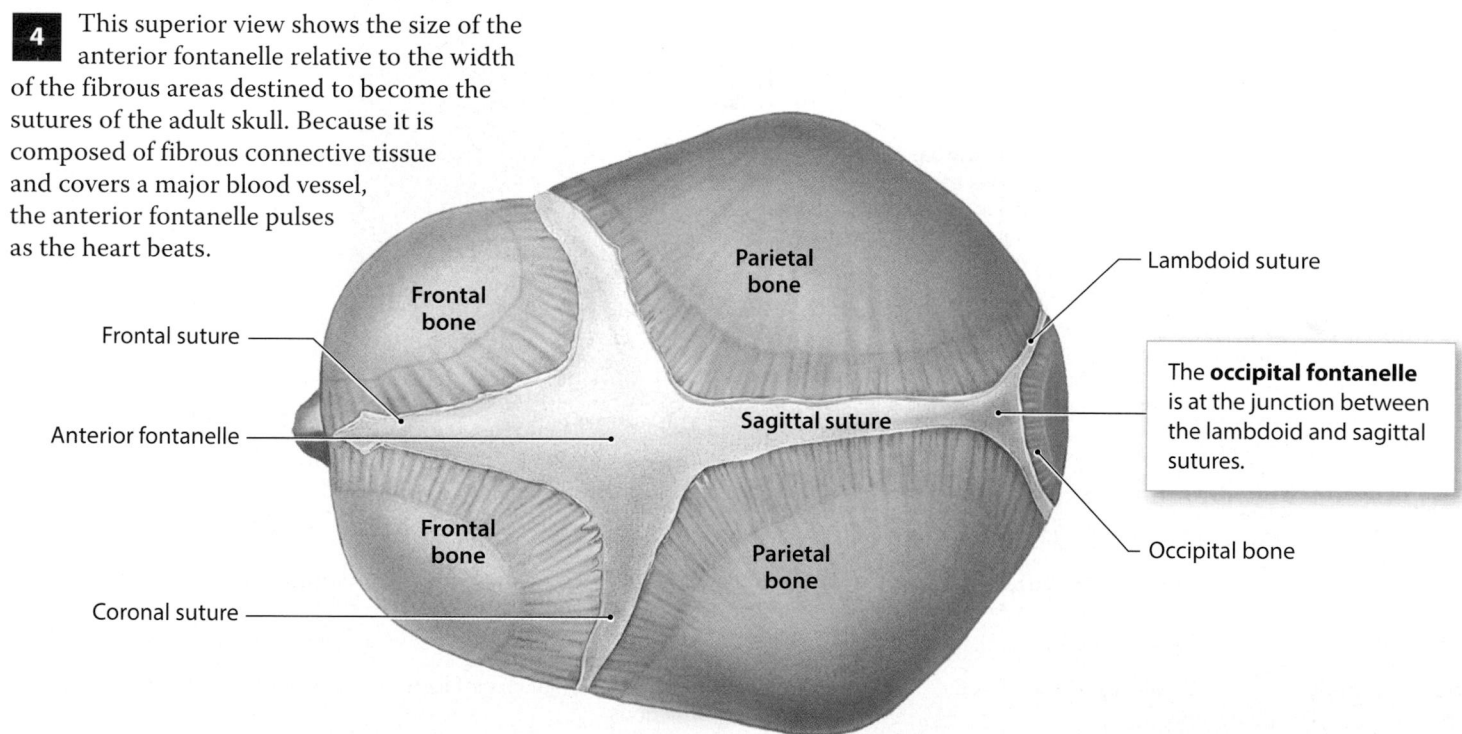

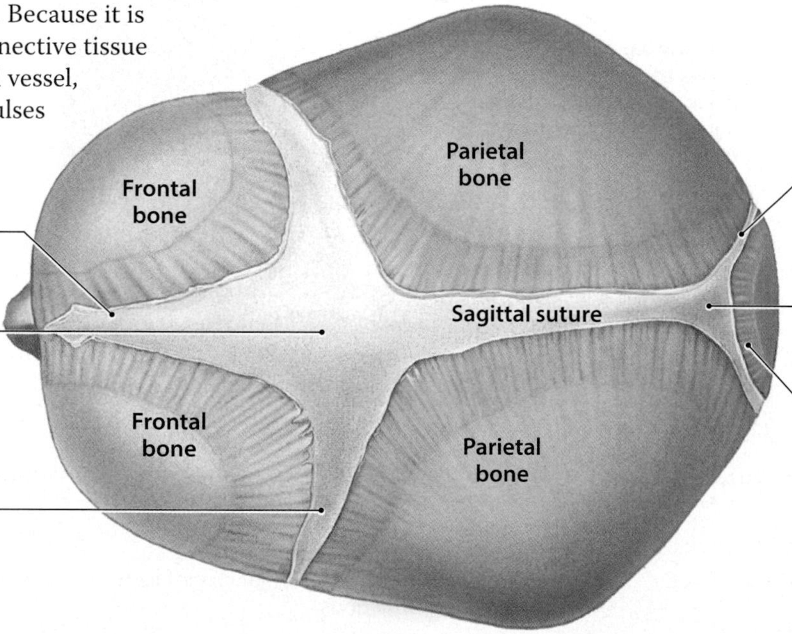

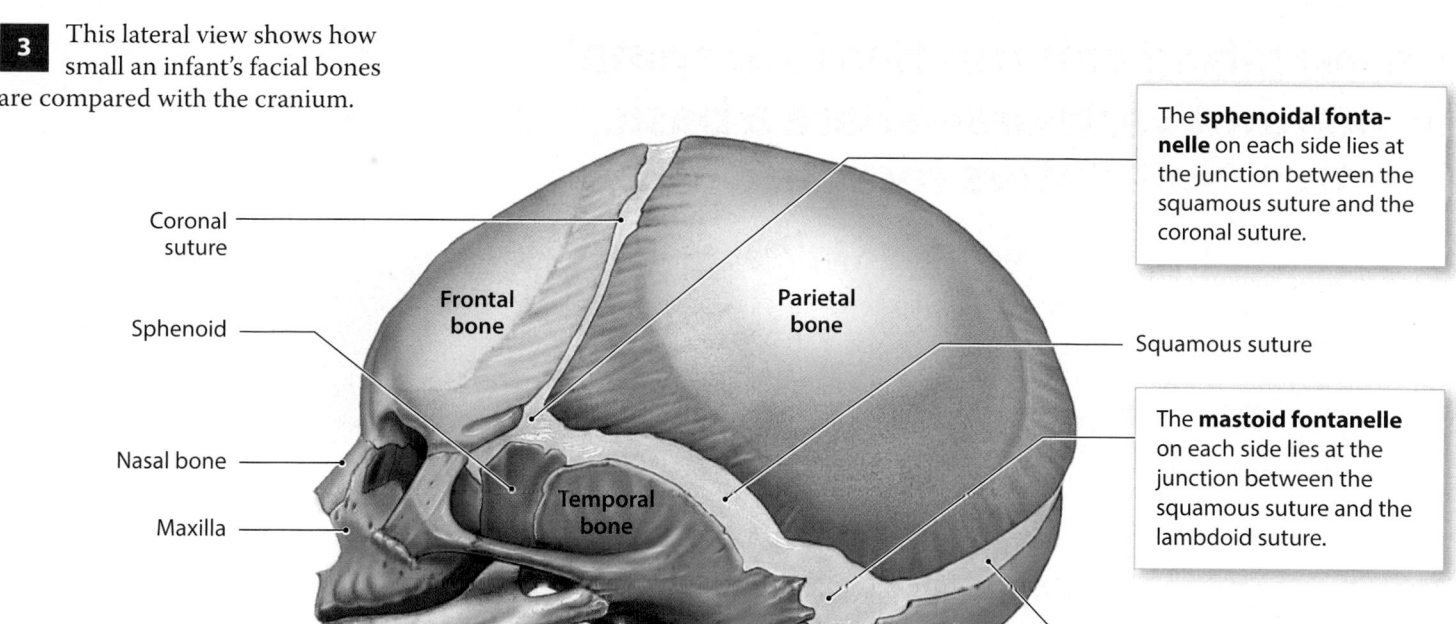

Frontal suture
Anterior fontanelle
Coronal suture
Frontal bone
Frontal bone
Parietal bone
Parietal bone
Sagittal suture
Lambdoid suture
The **occipital fontanelle** is at the junction between the lambdoid and sagittal sutures.
Occipital bone

The occipital, sphenoidal, and mastoid fontanelles disappear within a month or two after birth. Even after the fontanelles disappear, the bones of the skull remain separated by fibrous connections. The most significant growth in the skull occurs before age 5, because at that time the brain stops growing and the cranial sutures ossify.

Module 7.9 Review

a. Identify the major fontanelles.

b. What purposes do fontanelles serve?

c. Why do you think an infant's facial bones are so small compared with its cranium?

The vertebral column has four spinal curves, and vertebrae share a basic structure that differs regionally

1 The adult vertebral column (or spine) consists of 26 bones: the 24 vertebrae, the sacrum, and the coccyx (KOK-siks), or tailbone. The stacked vertebrae form a column of support that bears the weight of the head, neck, and trunk and ultimately transfers the body's weight to the appendicular skeleton of the lower limbs. The vertebrae also protect the spinal cord and help maintain an upright body position, as in sitting or standing. The length of the vertebral column of an adult averages 71 cm (28 in.).

Spinal Curves

Primary curves develop before birth, and secondary curves after birth.

The **cervical curve**, a secondary curve, develops as the infant learns to balance the weight of the head on the cervical vertebrae.

The **thoracic curve**, a primary curve, accommodates the thoracic organs.

The **lumbar curve**, a secondary curve, balances the weight of the trunk over the lower limbs. It develops with the ability to stand.

The **sacral curve,** a primary curve, accommodates the abdominopelvic organs.

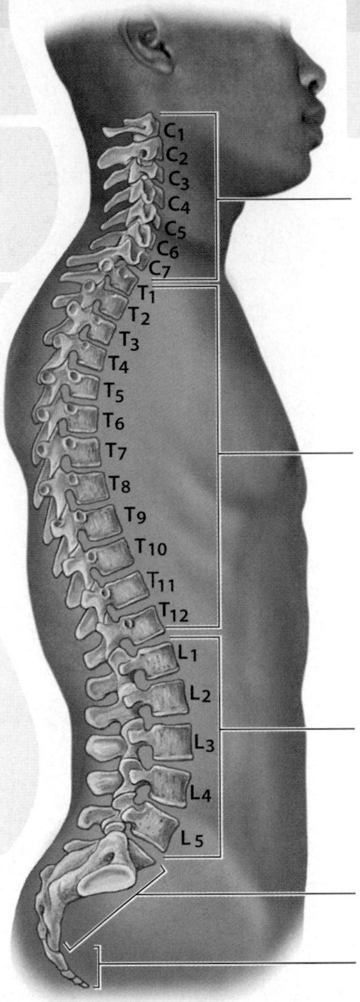
C1
C2
C3
C4
C5
C6
C7
T1
T2
T3
T4
T5
T6
T7
T8
T9
T10
T11
T12
L1
L2
L3
L4
L5
Cervical
(7 vertebrae)
Thoracic
(12 vertebrae)
Lumbar
(5 vertebrae)
Sacral
Coccygeal

Vertebral Regions

Regions are defined by anatomical characteristics of individual vertebrae.

Regional Comparison of Vertebral Structure and Function

	Cervical Vertebrae (7)	**Thoracic Vertebrae (12)**	**Lumbar Vertebrae (5)**
Location	Neck	Chest	Inferior portion of back
Vertebral Body	Small, oval, curved faces	Medium, heart-shaped, flat faces; facets for rib articulations	Massive, oval, flat faces
Vertebral Foramen	Large	Smaller	Smallest
Spinous Process	Long; split tip; points inferiorly	Long, slender; not split; points inferiorly	Blunt, broad; points posteriorly
Transverse Processes	Have transverse foramina	All but two have facets for rib articulations	Short; no articular facets or transverse foramina
Functions	Support skull, stabilize relative positions of brain and spinal cord, and allow controlled head movements	Support weight of head, neck, upper limbs, and chest; articulate with ribs to allow changes in volume of thoracic cage	Support weight of head, neck, upper limbs, and trunk

2 Each vertebra consists of three basic parts: (1) articular processes, (2) a vertebral arch, and (3) a vertebral body.

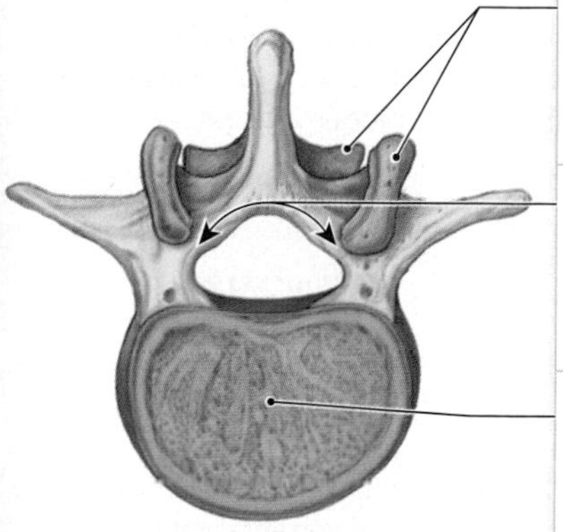

Superior view

Parts of a Vertebra

The **articular processes** extend superiorly and inferiorly to articulate with adjacent vertebrae.

The **vertebral arch** forms the posterior and lateral margins of the vertebral foramen.

The **vertebral body** is the part of a vertebra that transfers weight along the axis of the vertebral column.

3 The vertebral arch forms the posterior margin of each vertebral foramen. It consists of a pair each of laminae and pedicles, and supports various processes: articular, spinous, and transverse.

The **vertebral foramen** is framed by the vertebral body and the vertebral arch.

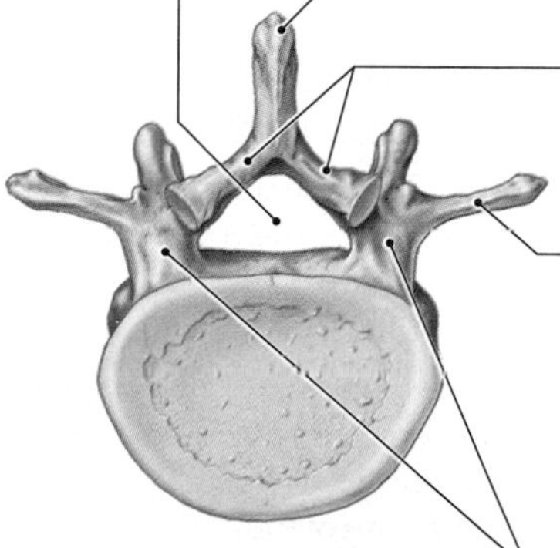

Inferior view

The Vertebral Arch

A **spinous process** projects posteriorly from the point where the vertebral laminae fuse to complete the vertebral arch.

The **laminae** (LAM-i-nē; singular, *lamina*, a thin plate) form the "roof" of the vertebral foramen.

Transverse processes project laterally on both sides from the point where the laminae join the pedicles. These processes are sites of muscle attachment and may also articulate with the ribs.

The **pedicles** (PED-i-kulz) form the sides or "walls" of the vertebral arch and connect it to the vertebral body.

4 This lateral view of three vertebrae shows how they combine to form the vertebral canal.

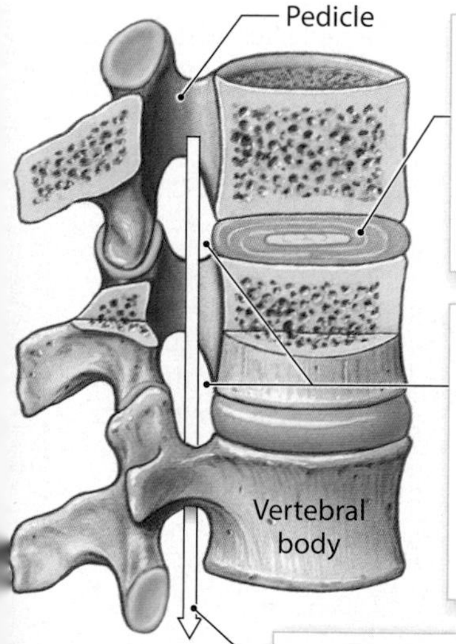
Pedicle
Vertebral body

The bodies of adjacent vertebrae are interconnected by ligaments but are separated by pads of fibrocartilage, the **intervertebral discs**.

The spaces that are formed between successive pedicles are called **intervertebral foramina**. Nerves and blood vessels running to or from the spinal cord pass through these foramina.

Together, the vertebral foramina of successive vertebrae form the **vertebral canal**, which encloses the spinal cord.

5 Each articular process has a smooth, concave surface called an **articular facet**. This facet is the portion of the articular process that forms the joint with the adjacent vertebra. This posterior view shows that the vertebral arches form the roof of the vertebral canal.

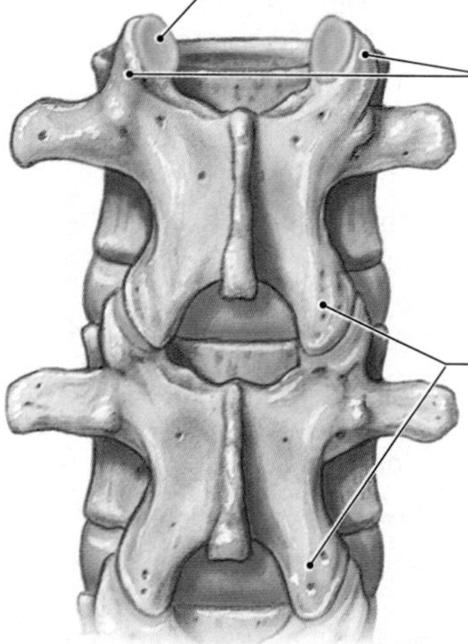
Articular facet

The **superior articular processes** articulate with the inferior articular processes of a superior vertebra.

The **inferior articular processes** articulate with the superior articular processes of an inferior vertebra (or the sacrum, in the case of the last lumbar vertebra).

Note that when referring to a specific vertebra, we use the capital letters C, T, L, S, and C_o to indicate the cervical, thoracic, lumbar, sacral, and coccygeal regions, respectively. In addition, we use a subscript number to indicate the relative position of the vertebra within that region, with 1 indicating the vertebra closest to the skull. For example, C_1 is in contact with the skull and C_3 is the third cervical vertebra. Similarly, T_{12} is in contact with L_1, and L_4 is the fourth lumbar vertebra.

Module 7.10 Review

a. What is the importance of the secondary curves of the spine?

b. Name the major components of a typical vertebra.

c. To which part of the vertebra do the intervertebral discs attach?

LO **7.10** Identify and describe the curves of the spinal column and their functions, and identify the vertebral regions.

Module 7.11

There are seven cervical vertebrae and twelve thoracic vertebrae

Cervical Vertebrae

The seven cervical vertebrae, the smallest in the vertebral column, extend from the occipital bone of the skull to the thorax.

1 The vertebral foramen of a cervical vertebra is very large. At this level, the spinal cord still contains most of the axons that connect the brain to the rest of the body. However, cervical vertebrae support only the weight of the head, so the vertebral body can be relatively small and light. As you continue toward the sacrum, the load increases and the vertebral bodies gradually enlarge.

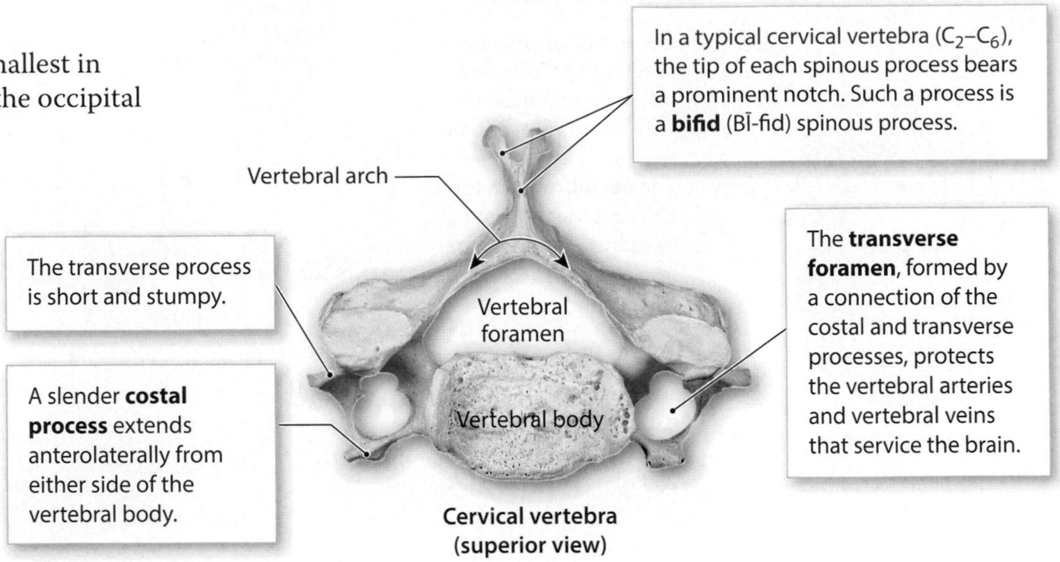

In a typical cervical vertebra (C_2–C_6), the tip of each spinous process bears a prominent notch. Such a process is a **bifid** (BĪ-fid) spinous process.
Vertebral arch
The transverse process is short and stumpy.
Vertebral foramen
A slender **costal process** extends anterolaterally from either side of the vertebral body.
Vertebral body
The **transverse foramen**, formed by a connection of the costal and transverse processes, protects the vertebral arteries and vertebral veins that service the brain.

Cervical vertebra (superior view)

2 The first two cervical vertebrae are specialized to support and stabilize the cranium while permitting head movement. The **atlas**, C_1, has no vertebral body and no spinous process, but it has a large, round vertebral foramen bounded by anterior and posterior arches. The **axis**, C_2, resembles inferior cervical vertebrae except for the presence of the prominent dens on the superior surface of the body.

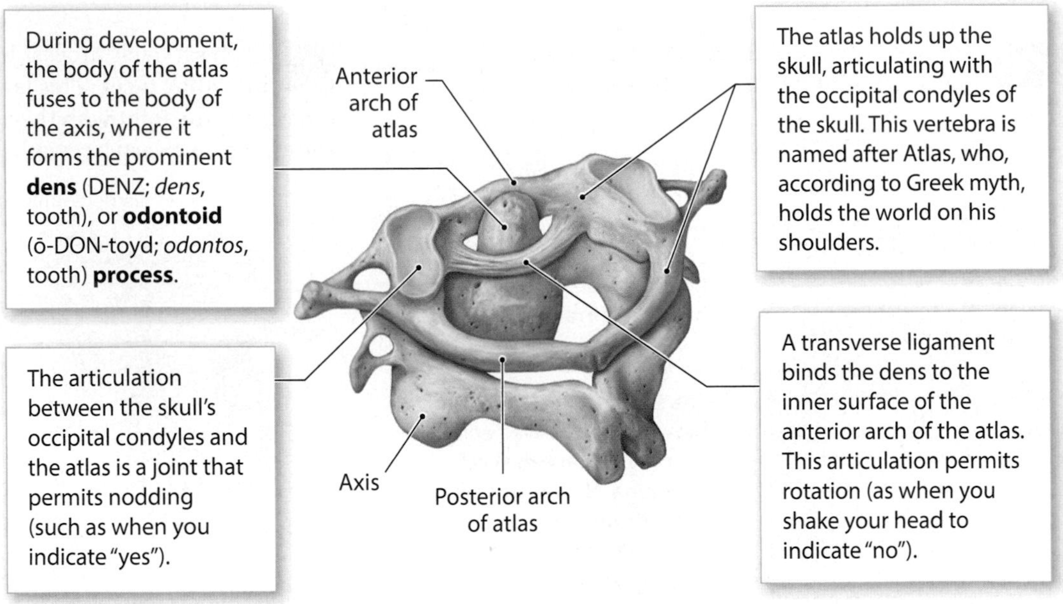

During development, the body of the atlas fuses to the body of the axis, where it forms the prominent **dens** (DENZ; *dens*, tooth), or **odontoid** (ō-DON-toyd; *odontos*, tooth) **process**.
Anterior arch of atlas
The atlas holds up the skull, articulating with the occipital condyles of the skull. This vertebra is named after Atlas, who, according to Greek myth, holds the world on his shoulders.
The articulation between the skull's occipital condyles and the atlas is a joint that permits nodding (such as when you indicate "yes").
Axis
Posterior arch of atlas
A transverse ligament binds the dens to the inner surface of the anterior arch of the atlas. This articulation permits rotation (as when you shake your head to indicate "no").

3 The last cervical vertebra has a robust spinous process that ends in a solid tubercle that can easily be felt through the skin. For this reason, C_7 is known as the **vertebra prominens** or prominent vertebra. The **ligamentum nuchae** (lig-uh-MEN-tum NŪ-kē; *nucha*, nape), a stout elastic ligament, begins at the vertebra prominens and extends to an insertion along the external occipital crest of the skull. When your head is upright, this ligament acts like the string on a bow to maintain the cervical curvature without muscular effort.

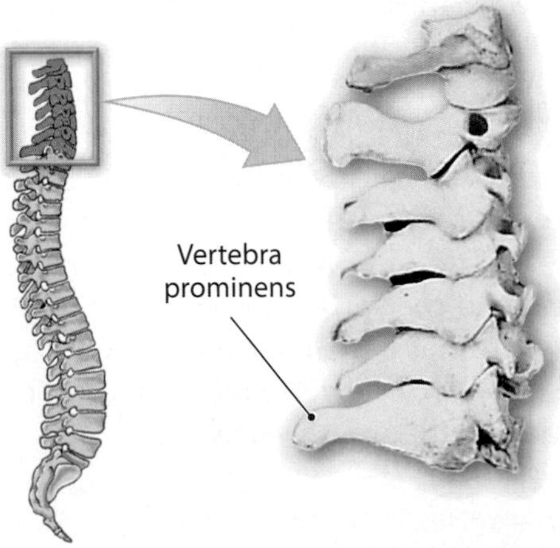

Vertebra prominens

Thoracic Vertebrae

4 There are 12 thoracic vertebrae. From T_1 to T_{12}, each vertebral body is slightly larger and more massive than the one superior to it, because the weight being transmitted along the vertebral column is steadily increasing.

5 A typical thoracic vertebra has a distinctive heart-shaped body that is much larger and more massive than that of a cervical vertebra, and the vertebral foramen is considerably smaller.

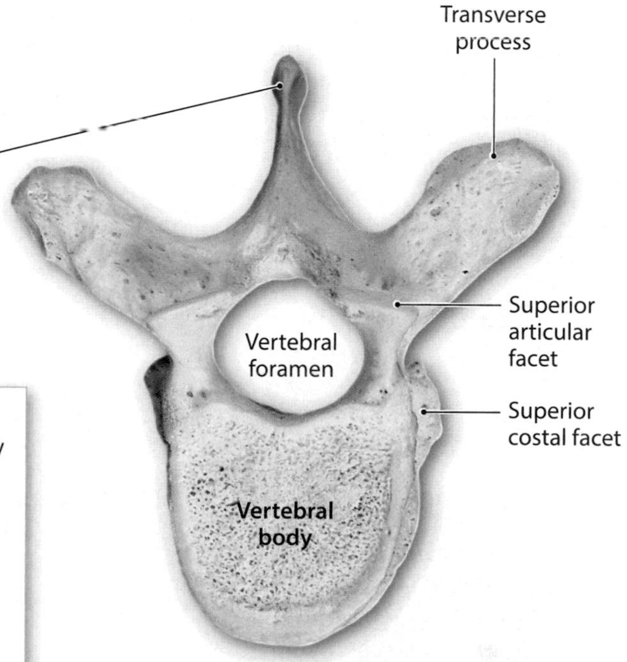

Transverse process
Superior articular facet
Vertebral foramen
Superior costal facet
Vertebral body

Thoracic vertebra (superior view)

The long, slender **spinous process** projects posteriorly and inferiorly. The spinous processes of T_{10}, T_{11}, and T_{12} increasingly resemble those of the lumbar region as the transition between the thoracic and lumbar curves approaches.

6 Each thoracic vertebra articulates with ribs at **costal facets** on the dorsolateral surfaces of the vertebral body. The transverse processes of vertebrae T_1–T_{10} also contain transverse costal facets for rib articulation. Thus, rib pairs 1 through 10 contact their vertebrae at two points: a costal facet and a transverse costal facet. Ribs 11 and 12 originate at costal facets on vertebrae T_{11} and T_{12}. Because these ribs do not contact transverse processes, they contact their vertebrae at only one point.

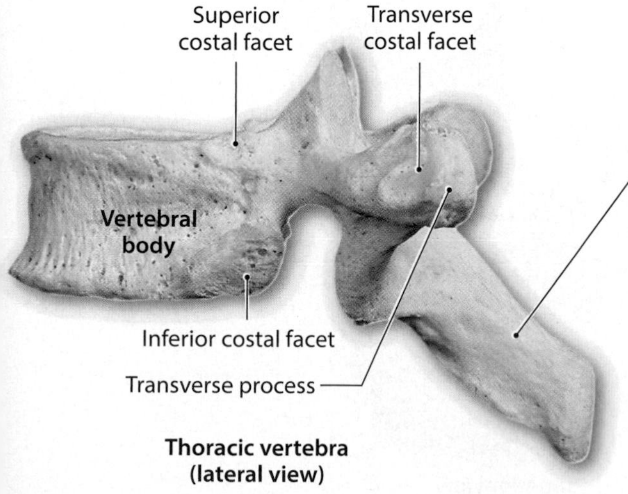

Superior costal facet
Transverse costal facet
Vertebral body
Inferior costal facet
Transverse process

Thoracic vertebra (lateral view)

Module 7.11 Review

a. Joe suffered a hairline fracture at the base of the dens. Which bone is fractured and where is the fractured bone located?

b. Examining a human vertebra, you notice that, in addition to the large foramen for the spinal cord, two smaller foramina are on either side of the bone in the region of the transverse processes. From which region of the vertebral column is this vertebra?

c. When you run your finger down the middle of a person's spine, what part of each vertebra are you feeling just beneath the skin?

LO **7.11** Describe the distinctive structural and functional characteristics of the cervical and thoracic vertebrae.

Module 7.12

There are five lumbar vertebrae; the sacrum and coccyx consist of fused vertebrae

Lumbar Vertebrae

1 The five lumbar vertebrae are the largest vertebrae, and they transmit the most weight. Compression fractures of these vertebrae may occur at any age, but they occur most often after aging and osteoporosis have reduced the strength of bones throughout the body.

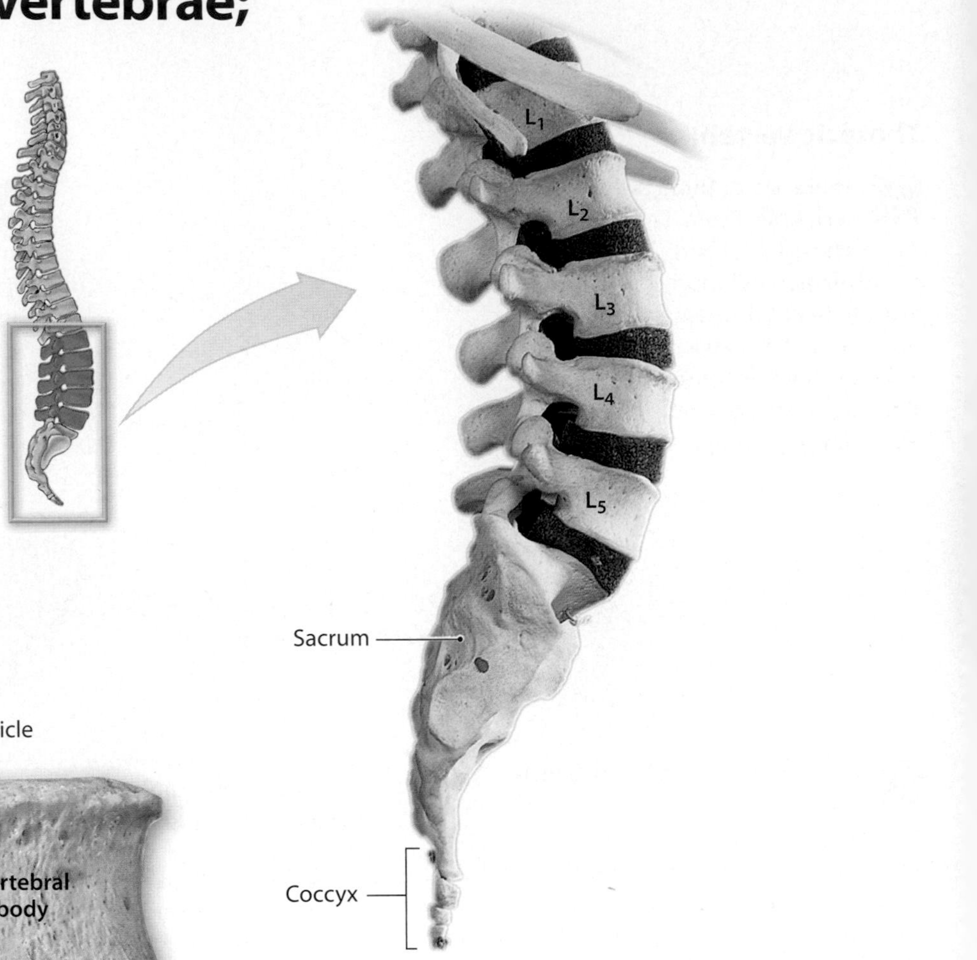

L_1
L_2
L_3
L_4
L_5
Sacrum
Coccyx

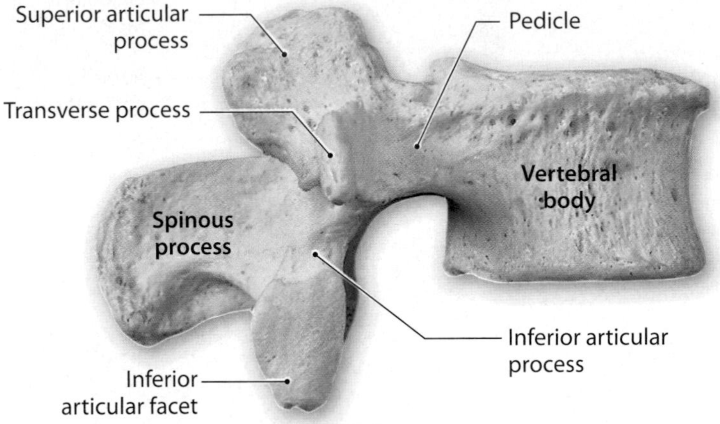

Superior articular process
Pedicle
Transverse process
Vertebral body
Spinous process
Inferior articular process
Inferior articular facet

Lateral view

2 The body of a typical lumbar vertebra is thicker than that of a thoracic vertebra, and the superior and inferior surfaces are oval rather than heart shaped. Other noteworthy features are that (1) lumbar vertebrae do not have costal facets; (2) the slender transverse processes, which lack transverse costal facets, project dorsolaterally; (3) the vertebral foramen is triangular; (4) the stumpy spinous processes project posteriorly; (5) the superior articular processes face medially ("up and in"); and (6) the inferior articular processes face laterally ("down and out").

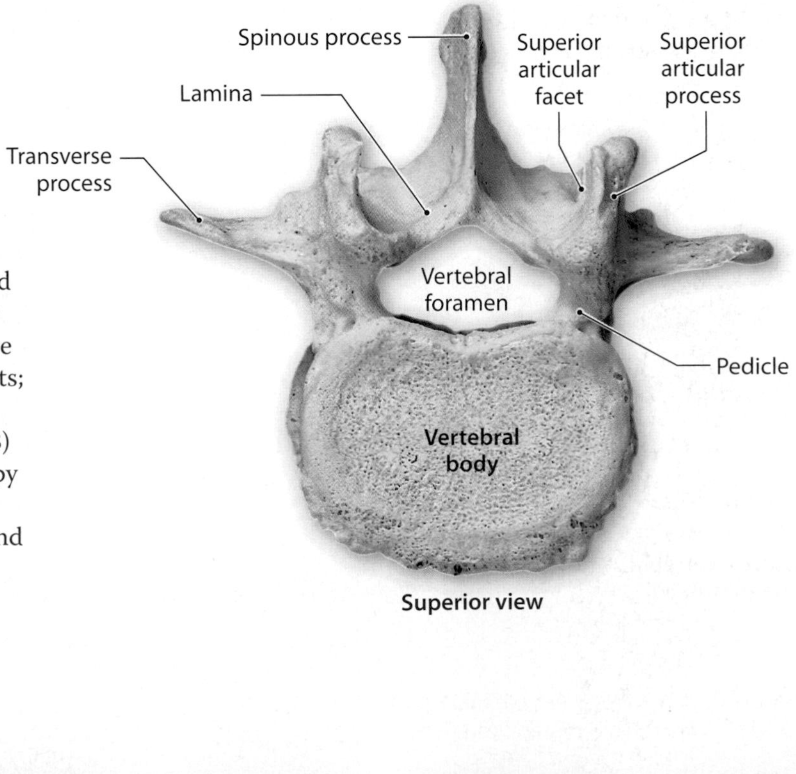

Spinous process
Superior articular facet
Superior articular process
Lamina
Transverse process
Vertebral foramen
Pedicle
Vertebral body

Superior view

Sacrum and Coccyx

The sacrum is a single bone that consists of the fused components of five sacral vertebrae. These vertebrae begin fusing shortly after puberty and, in general, are completely fused at age 25–30. The sacrum protects the reproductive, digestive, and urinary organs. It also attaches the axial skeleton to the appendicular skeleton by paired articulations with the hip bones of the pelvic girdle.

3 The anterior surface of the sacrum is concave, and the posterior surface is convex. The degree of curvature is more pronounced in males than in females.

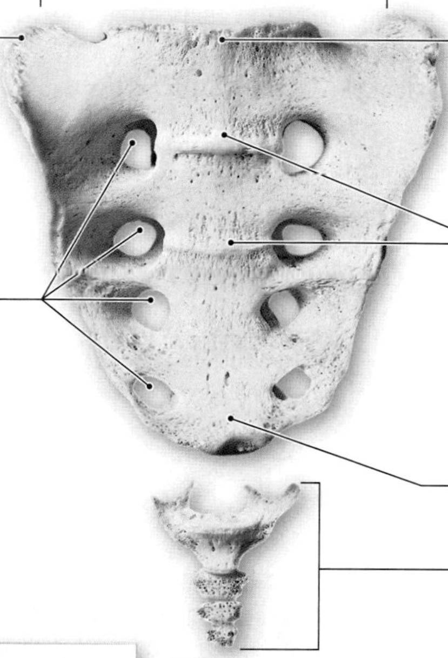
Coccyx

The **base** of the sacrum is the broad superior surface.

At the base of the sacrum, a broad sacral **ala**, or wing, extends on either side. The anterior and superior surfaces of each ala provide an extensive area for muscle attachment.

Four pairs of **sacral foramina** extend between the posterior and anterior surfaces. The intervertebral foramina of the fused sacral vertebrae open into these passageways.

The **sacral promontory** is an important surface marking in females during pelvic examinations and during labor and delivery.

Prominent transverse lines mark the former boundaries of individual vertebrae that fused during the formation of the sacrum.

The **apex** is the narrow, inferior portion of the sacrum.

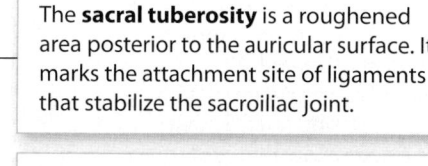

4 A posterior view of the sacrum shows the broad surface that provides an extensive area for the attachment of muscles.

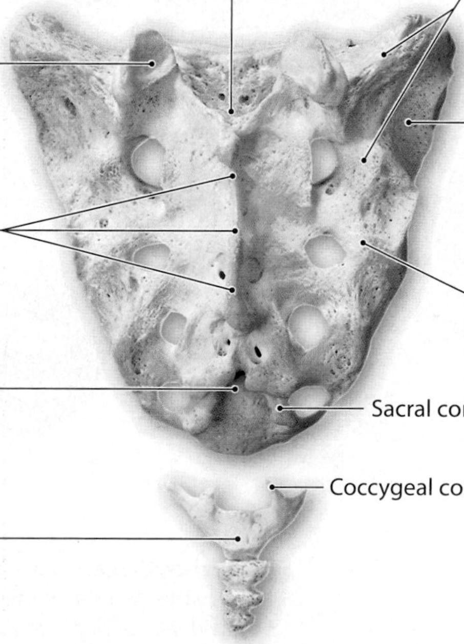
Sacral cornu
Coccygeal cornu

The **sacral canal** is a passageway that extends the length of the sacrum. Nerves and membranes that line the vertebral canal in the spinal cord continue into the sacral canal.

The **superior articular process** of the sacrum articulates with the last lumbar vertebra.

The **median sacral crest** is a ridge formed by the fused spinous processes of the sacral vertebrae.

The **sacral hiatus** (hī-Ā-tus) is the opening at the inferior end of the sacral canal.

The small **coccyx** is a single bone that consists of three to five coccygeal vertebrae that have generally begun fusing by age 26. Each **coccygeal cornu** (pl. *cornua*) curves to meet the sacral cornu superior to it.

The **sacral tuberosity** is a roughened area posterior to the auricular surface. It marks the attachment site of ligaments that stabilize the sacroiliac joint.

The **auricular surface** is a thickened, flattened area lateral and anterior to the superior portion of the lateral sacral crest. The auricular surface forms the sacroiliac joint with the hip bone of the pelvic girdle.

The **lateral sacral crest** is a ridge that represents the fused transverse processes of the sacral vertebrae.

Module 7.12 Review

a. How many vertebrae are present in the lumbar region? In the sacrum?

b. Why are the bodies of the lumbar vertebrae so large?

c. Which bone of the axial skeleton joins with the hip bones of the pelvic girdle?

LO **7.12** Describe the distinctive structural and functional characteristics of the lumbar vertebrae, sacrum, and coccyx.

Module 7.13

The thoracic cage protects organs in the chest and provides sites for muscle attachment

The skeleton of the chest, or **thoracic cage**, provides bony support to the walls of the thoracic cavity. It consists of the thoracic vertebrae, the ribs, and the sternum (breastbone). The ribs and the sternum form the rib cage, whose movements are important in breathing. The thoracic cage protects the heart, lungs, thymus, and other structures in the thoracic cavity, and serves as an attachment point for muscles involved in (1) breathing, (2) maintaining the position of the vertebral column, and (3) moving the pectoral girdles and upper limbs.

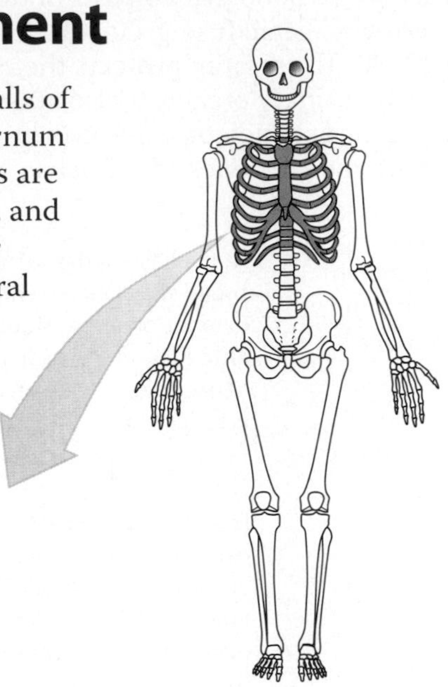

1 An anterior view of the thoracic cage shows the sternum, the costal cartilages, and the major classes of ribs.

The **jugular notch**, located between the clavicular articulations, is a shallow indentation on the superior surface of the manubrium.

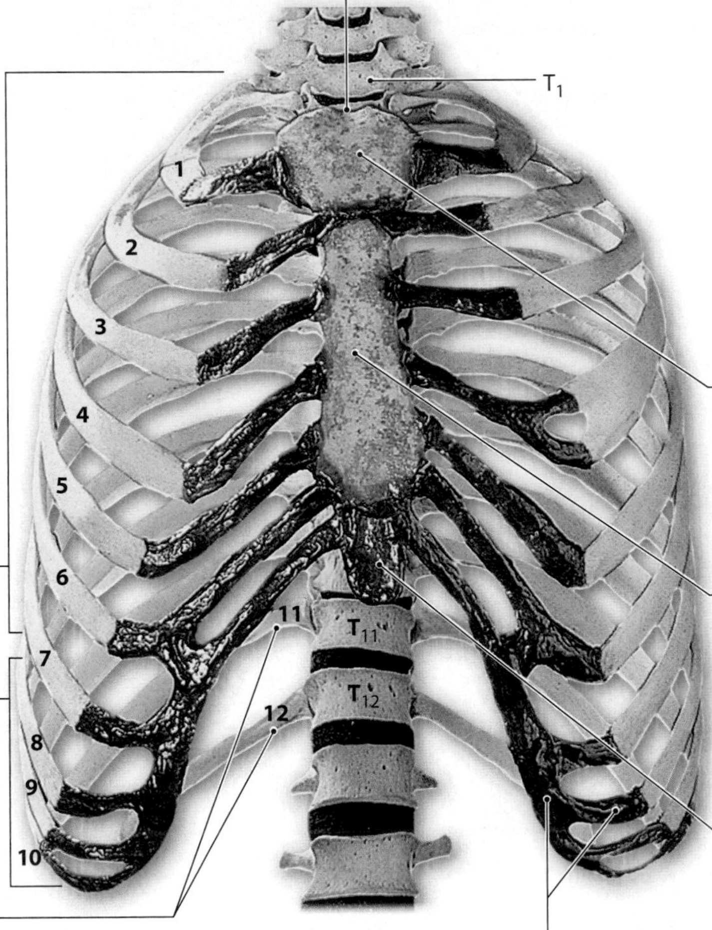

T1
1
2
3
4
5
6
7
8
9
10
11
12
T11
T12

Ribs

The **ribs** reinforce the posterior and lateral walls of the thoracic cavity. Ribs 1–7 gradually increase in length and curvature radius. Although the ribs are quite mobile and are among the most flexible of bones, they can be broken by a sharp blow or crushing impact. Fortunately, they are so stabilized by connective tissues and surrounding muscles that open fractures are rare, and splinting is unnecessary.

Vertebrosternal ribs (ribs 1–7), or **true ribs**, are connected to the sternum by individual **costal cartilages**.

The **vertebrochondral ribs** (ribs 8–10) are connected to the sternum by shared costal cartilages.

The last two pairs of ribs (11 and 12) are called **floating ribs**, because they have no connection with the sternum, or **vertebral ribs**, because they are attached only to the vertebrae and muscles of the body wall. Collectively, ribs 8–12 are also called **false ribs**.

Sternum

The adult **sternum**, or breastbone, is a flat bone that forms in the anterior midline of the thoracic wall. It has three distinct regions that usually fuse together during adulthood.

Manubrium: The broad, trapezoid-shaped manubrium (ma-NŪ-brē-um) articulates with the clavicles (collarbones) and the cartilages of the first pair of ribs.

Body: The body attaches to the inferior surface of the manubrium and extends inferiorly along the midline. Individual costal cartilages from rib pairs 2–7 are attached to this portion of the sternum.

Xiphoid process: The xiphoid (ZĪ-foyd) process, the smallest part of the sternum, is attached to the inferior surface of the body of the sternum.

Costal cartilages connect ribs to the sternum, either individually or in groups.

2 Ribs are long, curved, flattened bones that originate on or between the thoracic vertebrae and end in the wall of the thoracic cavity. Each of us, regardless of sex, has 12 pairs of ribs. Major surface markings on a representative rib (ribs 2–9) are shown in this posterior view.

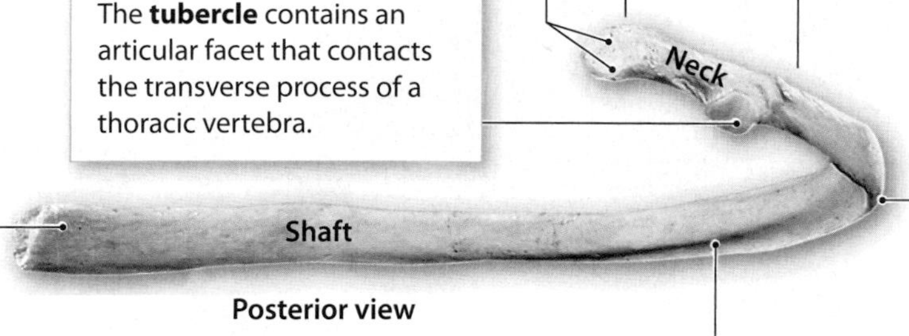
Articular facets on head
The vertebral end of the rib articulates with the vertebral column at the **head**, or **capitulum** (ka-PIT-ū-lum), where there are two **articular facets**.
The **tubercle** contains an articular facet that contacts the transverse process of a thoracic vertebra.
Neck
The bend, or **angle**, of the rib is the site where the tubular body, or **shaft**, begins curving toward the sternum.
The superficial surface is convex and provides an attachment site for muscles of the pectoral girdle and trunk. The intercostal muscles, which move the ribs, are attached to the superior and inferior surfaces.
Shaft
Posterior view
The **costal groove** along the inferior border marks the path of nerves and blood vessels.

3 This superior view of a representative rib shows sites of articulations with a thoracic vertebra. The heads of ribs 2–9 articulate with **costal facets** on two adjacent vertebrae, and their tubercular facets articulate with the transverse costal facets of the inferior vertebra. The heads of ribs 1, 10, 11, and 12 articulate with individual vertebrae at single costal facets. The tubercular facets of ribs 1 and 10 articulate with the costal facets of vertebrae T_1 and T_{10}, respectively. Ribs 11 and 12 articulate only at their heads, and there are no tubercular facets.

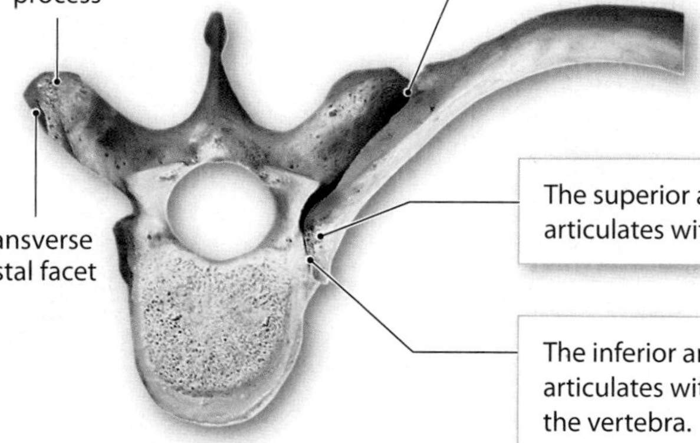
Transverse process
The tubercular facet of the rib articulates with the transverse costal facet of the inferior vertebra.
The superior articular facet of the rib articulates with the superior vertebra.
Transverse costal facet
The inferior articular facet of the rib articulates with the superior costal facet of the vertebra.

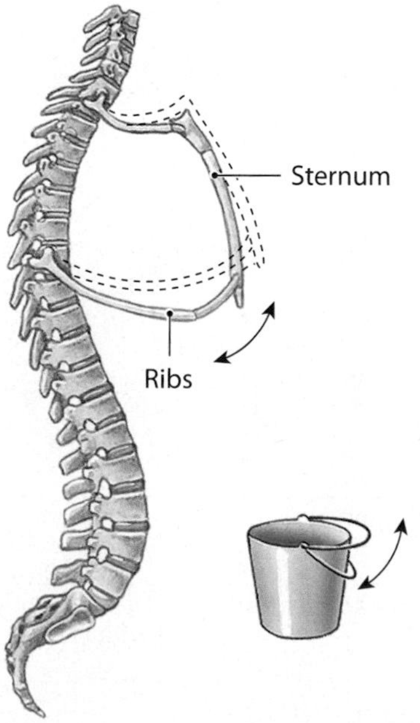
Sternum
Ribs

4 A typical rib acts like a bucket handle at a position just below horizontal. Pushing the handle down forces it inward; pulling it up swings it outward. Because of the curvature of the ribs, the same movements change the position of the sternum. Depression of the ribs pulls the sternum inward, whereas elevation moves it outward. As a result, movements of the ribs affect both the width and the depth of the thoracic cage, increasing or decreasing its volume accordingly.

Module 7.13 Review

a. How are vertebrosternal ribs distinguished from vertebrochondral ribs?

b. Improper chest compressions during cardiopulmonary resuscitation (CPR) can result in fractures of which bones?

c. In addition to the ribs and sternum, what other bones make up the thoracic cage?

LO **7.13** Explain the significance of the articulations between the thoracic vertebrae and the ribs, and between the ribs and the sternum.

Labeling

In the image below, label the bones of the orbital complex.

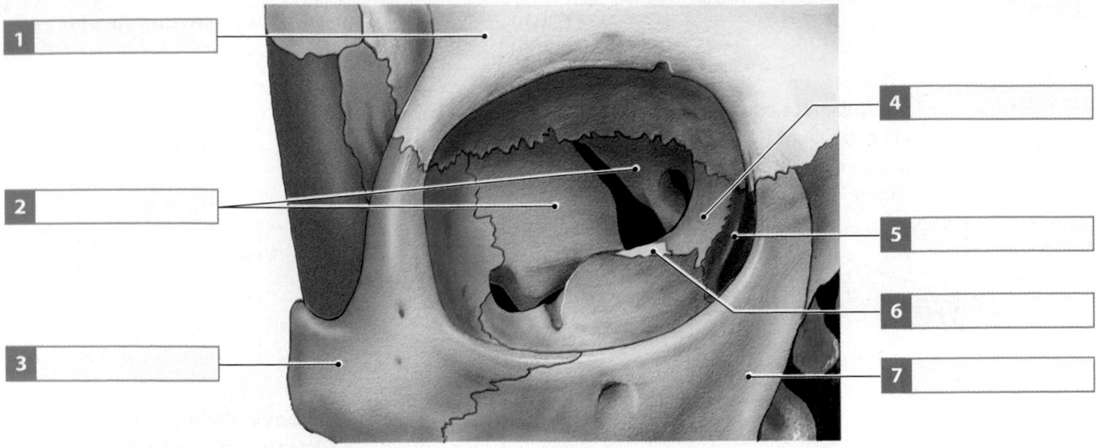
1
2
3
4
5
6
7

In the images below, label the sutures and fontanelles of the infant skull.

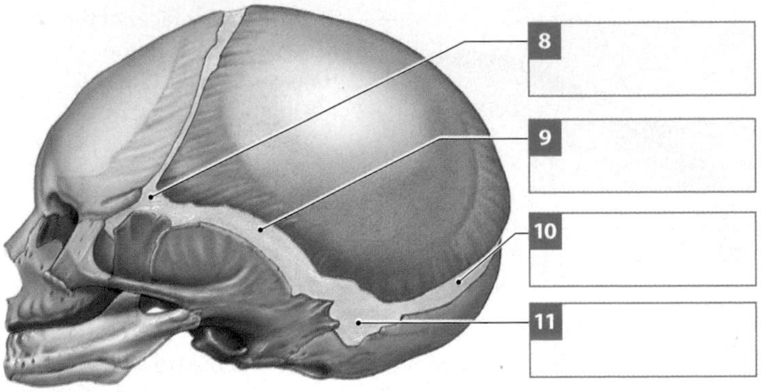
8
9
10
11

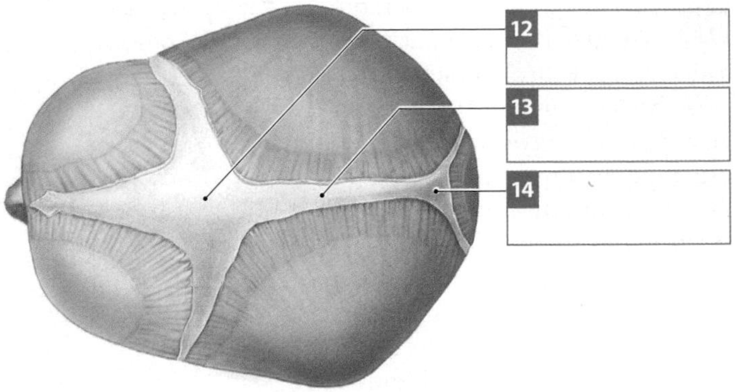
12
13
14

Short answer

Identify the vertebral region and regional characteristics for each vertebra.

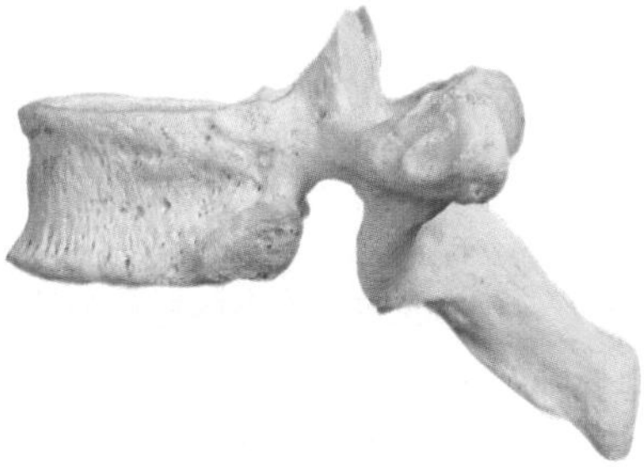

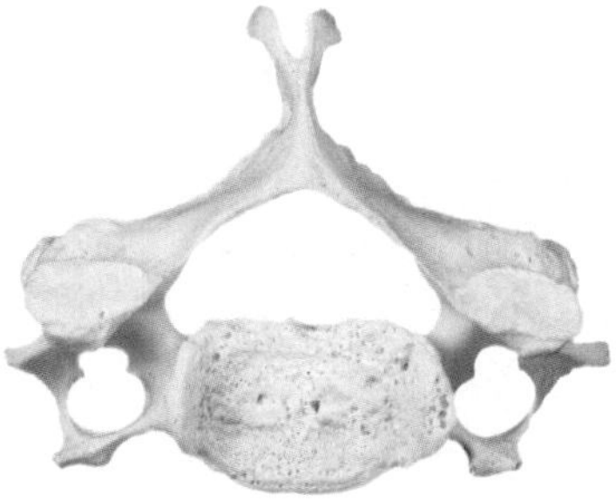

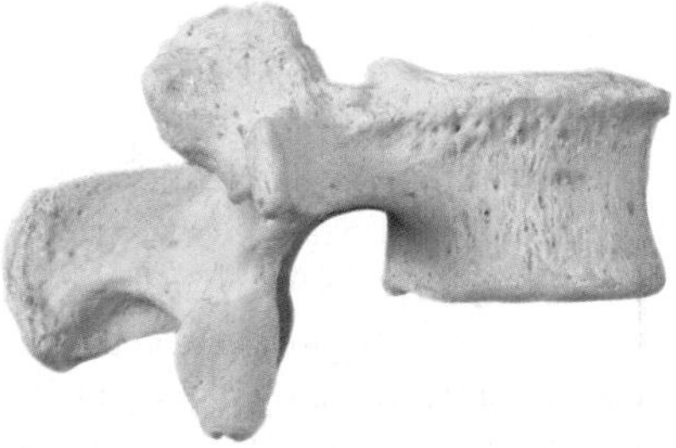

15 **Region:**

16 **Characteristics:**

17 **Region:**

18 **Characteristics:**

19 **Region:**

20 **Characteristics:**

The appendicular skeleton includes the limb bones and the pectoral and pelvic girdles

The **appendicular skeleton** includes the bones of the limbs and the supporting bone girdles that connect them to the trunk. The descriptions in this section emphasize surface markings that either have functional importance (such as the attachment sites for skeletal muscles and the paths of major nerves and blood vessels) or provide landmarks that define areas and locate structures of the body.

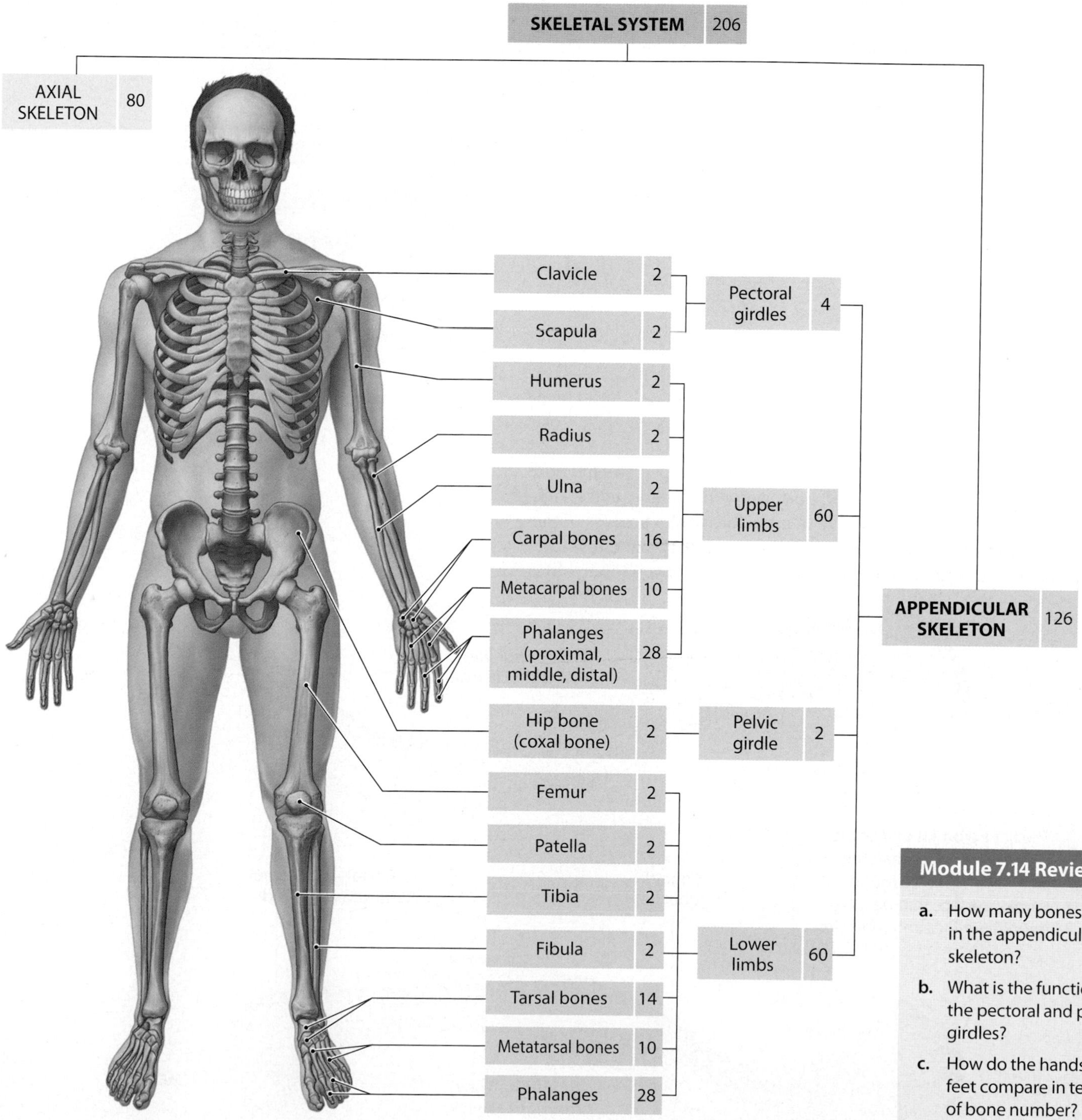
SKELETAL SYSTEM 206
AXIAL SKELETON 80
APPENDICULAR SKELETON 126
Clavicle 2
Scapula 2
Pectoral girdles 4
Humerus 2
Radius 2
Ulna 2
Carpal bones 16
Metacarpal bones 10
Phalanges (proximal, middle, distal) 28
Upper limbs 60
Hip bone (coxal bone) 2
Pelvic girdle 2
Femur 2
Patella 2
Tibia 2
Fibula 2
Tarsal bones 14
Metatarsal bones 10
Phalanges 28
Lower limbs 60

Module 7.14 Review

a. How many bones are in the appendicular skeleton?

b. What is the function of the pectoral and pelvic girdles?

c. How do the hands and feet compare in terms of bone number?

LO 7.14 List the four major components of the appendicular skeleton.

The pectoral girdles—the clavicles and scapulae—connect the upper limbs to the axial skeleton

Each arm articulates (forms a joint) with the trunk at the **pectoral girdle**, or shoulder girdle. The pectoral girdles consist of two S-shaped **clavicles** (KLAV-i-kulz; collarbones) and two broad, flat **scapulae** (SKAP-ū-lē; singular, *scapula*, SKAP-ū-luh; shoulder blades).

1 The clavicles originate at the superior, lateral border of the manubrium of the sternum, lateral to the jugular notch. It articulates with the sternum at the clavicular notch forming the **sternoclavicular joint.** These sternoclavicular joints are the only articulations between the pectoral girdles and the axial skeleton.

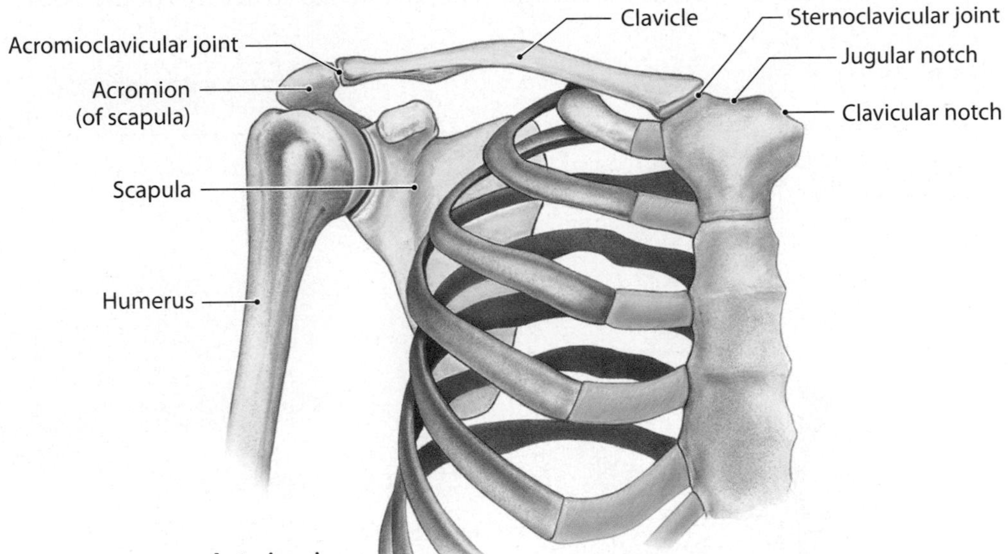

Acromioclavicular joint
Clavicle
Sternoclavicular joint
Jugular notch
Acromion
(of scapula)
Clavicular notch
Scapula
Humerus

Anterior view

2 From the sternoclavicular joint at the pyramid-shaped **sternal end**, each clavicle curves laterally and posteriorly for roughly half its length. It then forms a smooth posterior curve to articulate with a process of the scapula, the **acromion** (a-KRŌ-mē-on), at the **acromioclavicular joint**. Stabilizing ligaments attach to the conoid tubercle and costal tuberosity.

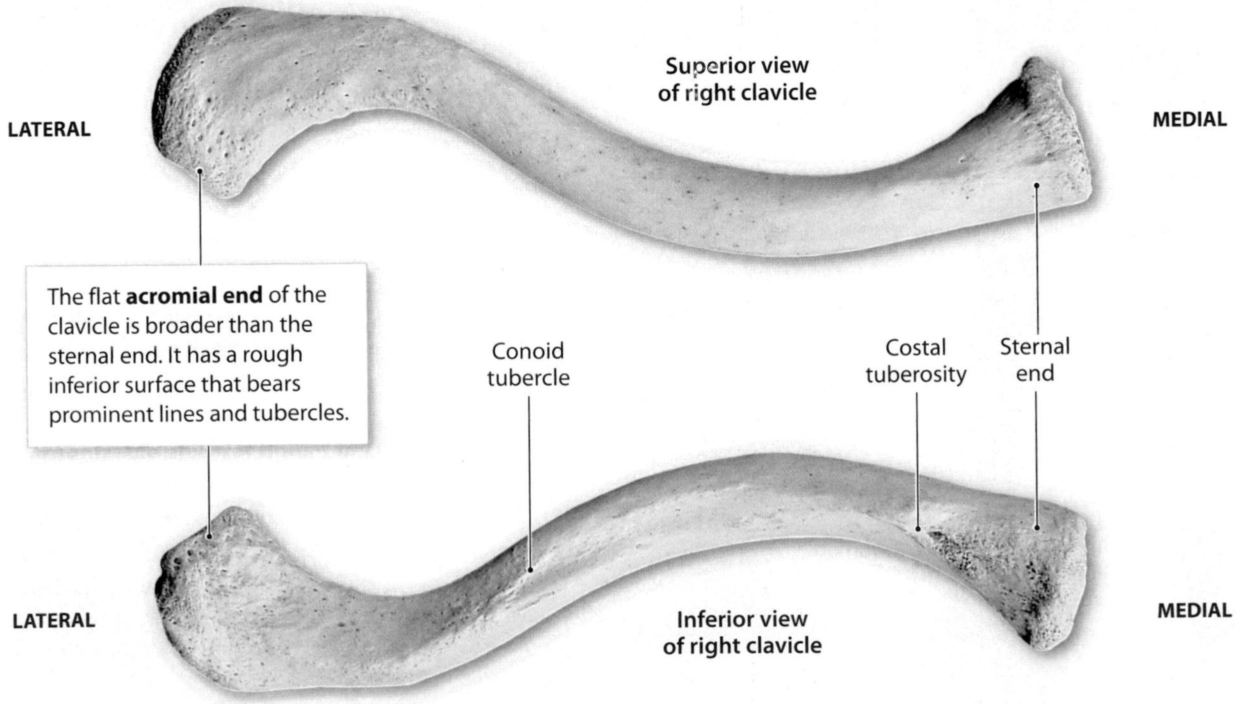

Superior view
of right clavicle
LATERAL
MEDIAL
The flat acromial end of the clavicle is broader than the sternal end. It has a rough inferior surface that bears prominent lines and tubercles.
Conoid
tubercle
Costal
tuberosity
Sternal
end
LATERAL
MEDIAL
Inferior view
of right clavicle

3 The anterior surface of the **body** of each scapula forms a broad, smooth triangle. The three sides of the triangle are the **superior border**; the **medial border**, or vertebral border; and the **lateral border**, or axillary (*axilla*, armpit) border. Muscles that position the scapula attach along these edges. The corners of the triangle are called the **superior angle**, the **inferior angle**, and the **lateral angle**. The depression in the anterior surface is called the **subscapular fossa**.

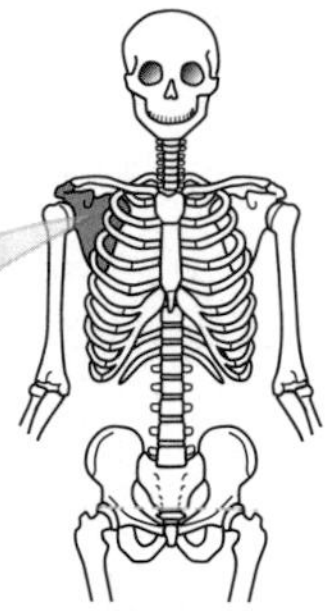

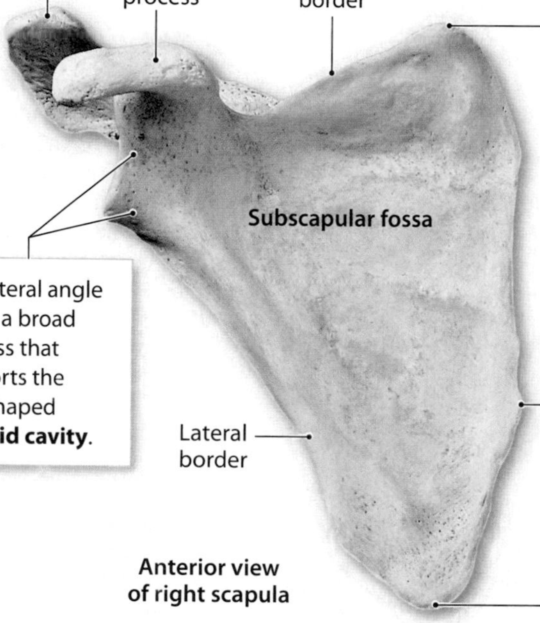

Acromion
Coracoid process
Superior border
Superior angle
Subscapular fossa
The lateral angle forms a broad process that supports the cup-shaped **glenoid cavity**.
Lateral border
Medial border
Inferior angle

Anterior view of right scapula

4 The posterior surface of the scapula is convex and has prominent ridges and processes for muscle attachment.

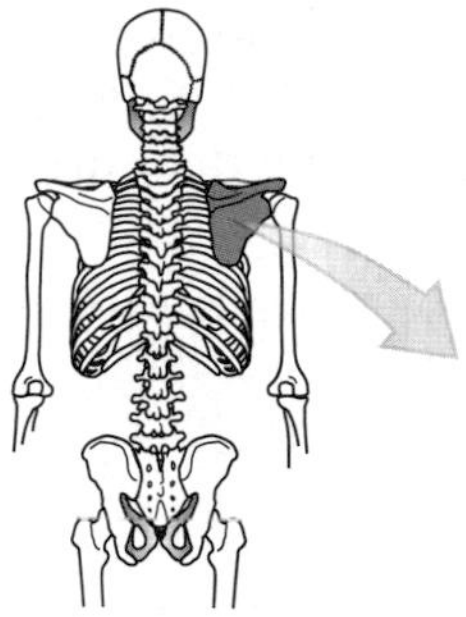

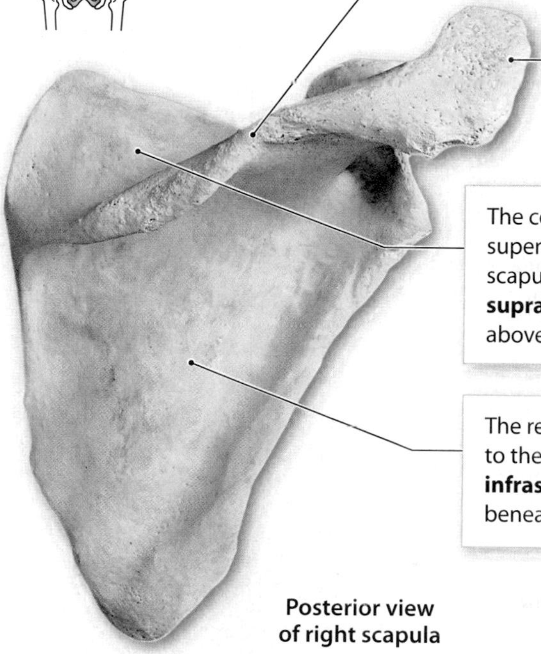

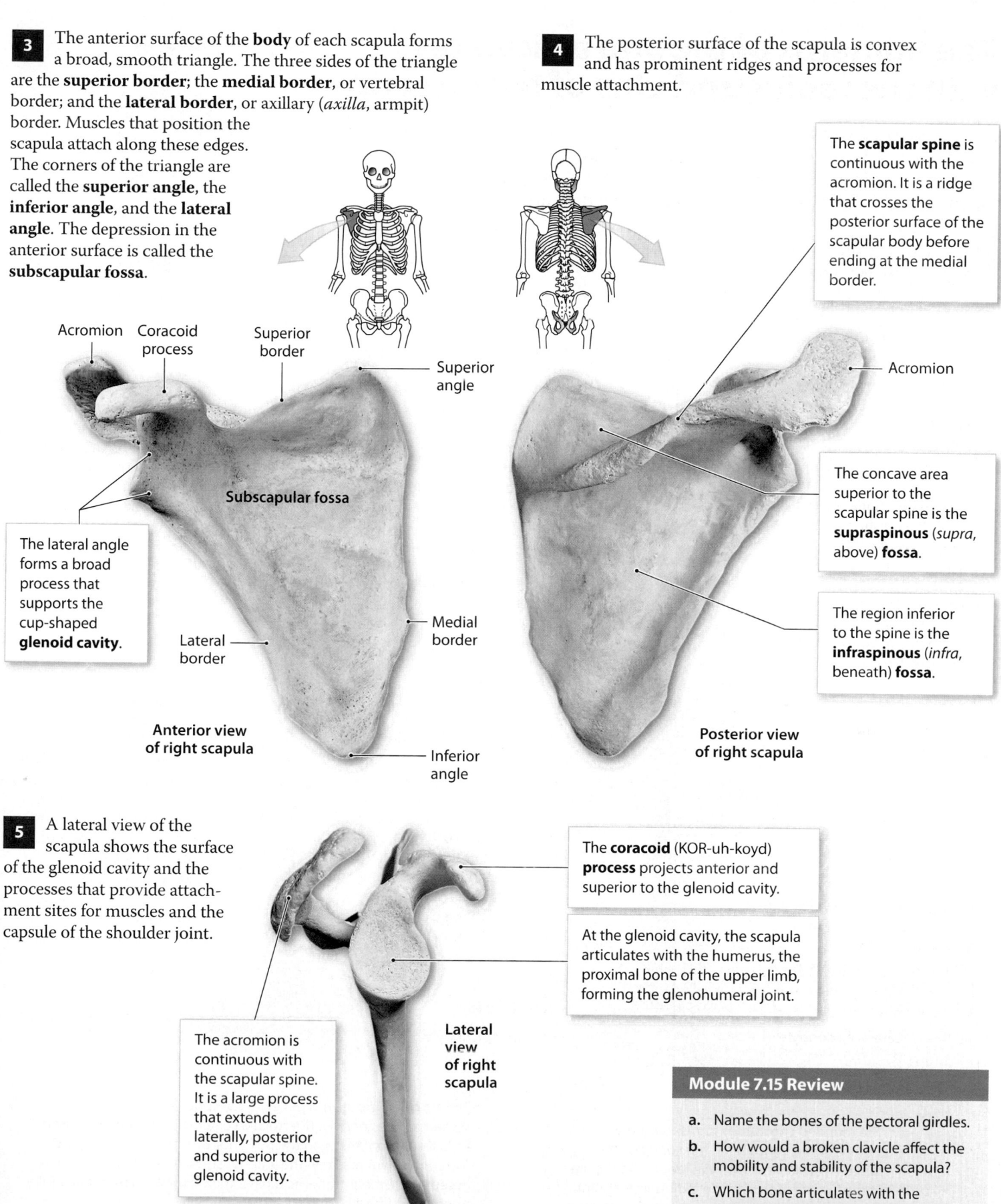

The **scapular spine** is continuous with the acromion. It is a ridge that crosses the posterior surface of the scapular body before ending at the medial border.
Acromion
The concave area superior to the scapular spine is the **supraspinous** (*supra*, above) **fossa**.
The region inferior to the spine is the **infraspinous** (*infra*, beneath) **fossa**.

Posterior view of right scapula

5 A lateral view of the scapula shows the surface of the glenoid cavity and the processes that provide attachment sites for muscles and the capsule of the shoulder joint.

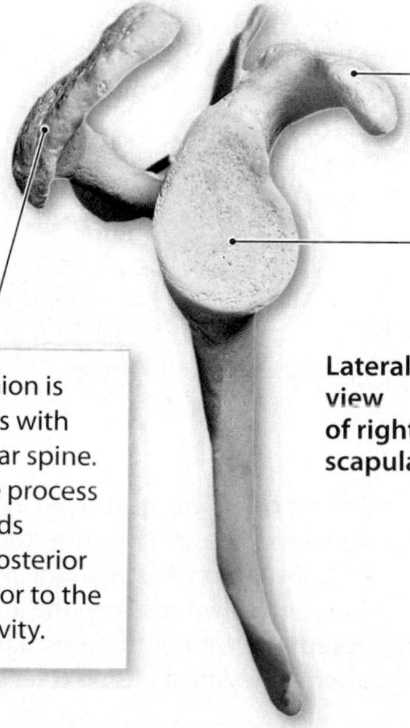

The **coracoid** (KOR-uh-koyd) **process** projects anterior and superior to the glenoid cavity.
At the glenoid cavity, the scapula articulates with the humerus, the proximal bone of the upper limb, forming the glenohumeral joint.
The acromion is continuous with the scapular spine. It is a large process that extends laterally, posterior and superior to the glenoid cavity.

Lateral view of right scapula

Module 7.15 Review

a. Name the bones of the pectoral girdles.

b. How would a broken clavicle affect the mobility and stability of the scapula?

c. Which bone articulates with the scapula at the glenoid cavity?

LO **7.15** Identify the bones that form the pectoral girdles, their functions, and their superficial features.

Module 7.16

The humerus of the arm articulates with the radius and ulna of the forearm

The skeleton of the upper limbs consists of the bones of the arms, forearms, wrists, and hands. Note that in anatomical descriptions, the term "arm" refers only to the proximal portion of the upper limb (from shoulder to elbow), not to the entire limb. We will examine the bones of the right upper limb.

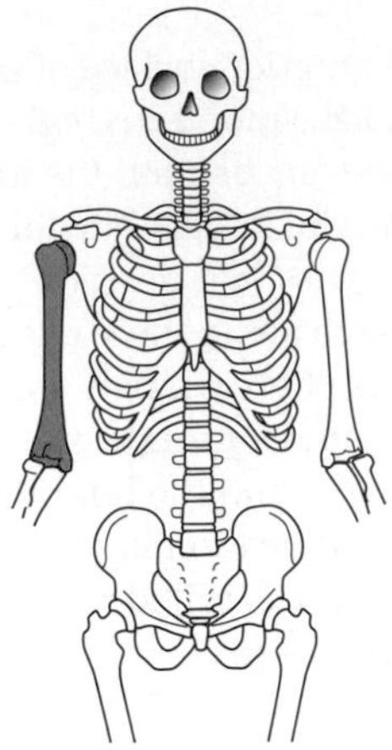

1 The arm, or **brachium**, contains one bone, the **humerus**, which extends from the scapula to the elbow. Near the distal articulation with the bones of the forearm, the shaft expands to either side at the **medial** and **lateral epicondyles**. Epicondyles are processes that develop proximal to an articulation. They provide additional surface area for muscle attachment.

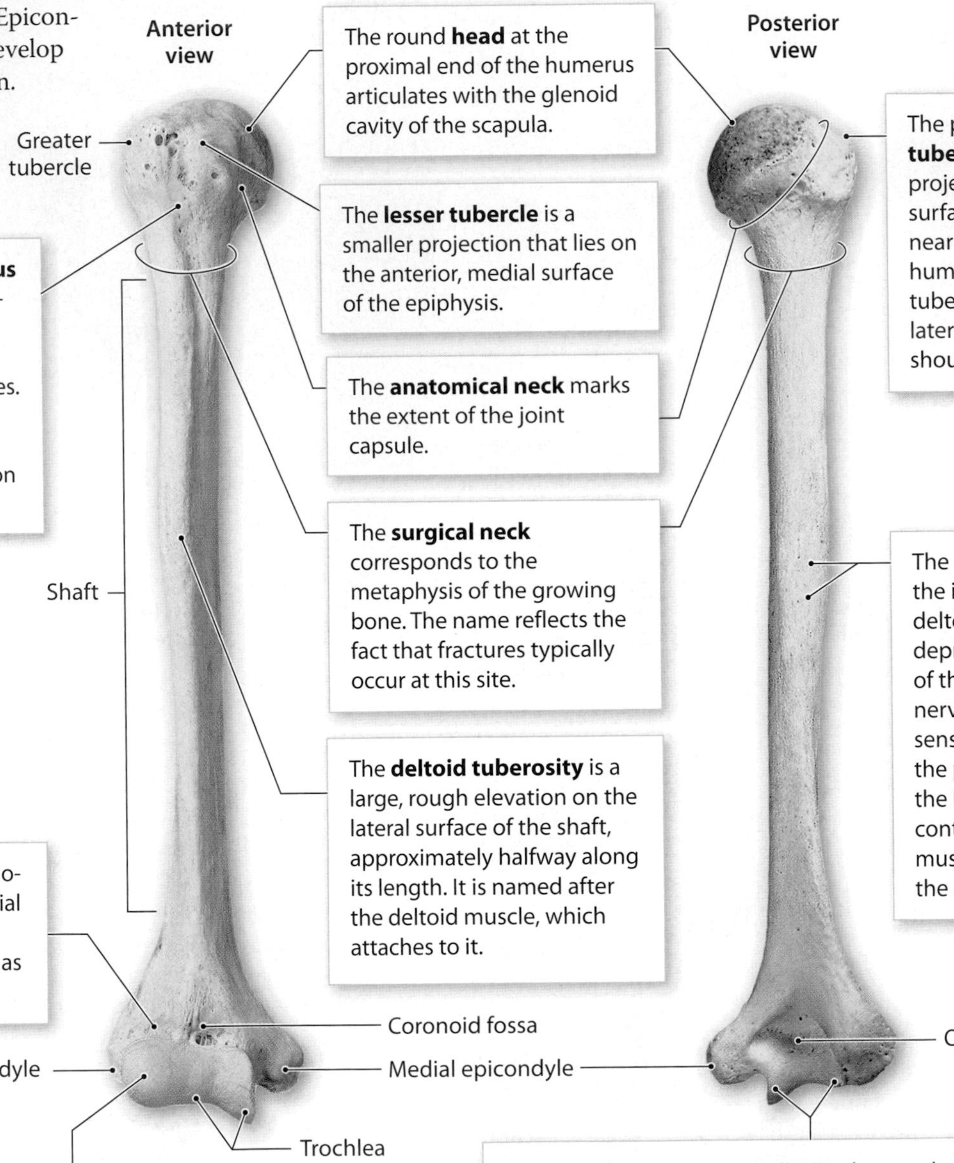

Anterior view
Posterior view
Greater tubercle
Shaft
Coronoid fossa
Lateral epicondyle
Medial epicondyle
Trochlea
Olecranon fossa
The round **head** at the proximal end of the humerus articulates with the glenoid cavity of the scapula.
The **lesser tubercle** is a smaller projection that lies on the anterior, medial surface of the epiphysis.
The **anatomical neck** marks the extent of the joint capsule.
The **surgical neck** corresponds to the metaphysis of the growing bone. The name reflects the fact that fractures typically occur at this site.
The **deltoid tuberosity** is a large, rough elevation on the lateral surface of the shaft, approximately halfway along its length. It is named after the deltoid muscle, which attaches to it.
The **intertubercular sulcus** (also called the intertubercular groove or bicipital groove) lies between the greater and lesser tubercles. Both tubercles are important sites for muscle attachment. A large tendon runs along the groove.
The prominent **greater tubercle** is a rounded projection on the lateral surface of the epiphysis, near the margin of the humeral head. The greater tubercle establishes the lateral contour of the shoulder.
The **radial groove** crosses the inferior end of the deltoid tuberosity. This depression marks the path of the radial nerve, a large nerve that provides both sensory information from the posterior surface of the limb and motor control over the large muscles that straighten the elbow.
The **radial fossa** accommodates a portion of the radial head when the forearm approaches the humerus as the elbow bends.
The rounded **capitulum** forms the lateral surface of the **condyle**. At the condyle, the humerus articulates with the radius and the ulna of the forearm, forming the radiohumeral joint. The condyle is divided into two articular regions.
The **trochlea** (*trochlea*, a pulley) is the spool-shaped medial portion of the condyle. The trochlea extends from the **olecranon** (ō-LEK-ruh-non) **fossa** on the posterior surface to the **coronoid** (*corona*, crown) **fossa** on the anterior surface. These depressions accept projections from the ulna as the elbow or humeroulnar joint approaches the limits of its range of motion.

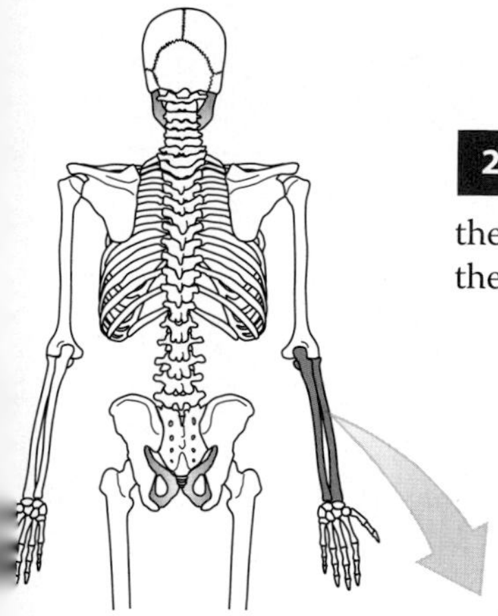

2 The **ulna** and **radius** are parallel bones that support the forearm, or **antebrachium**. In the anatomical position, the ulna lies medial to the radius.

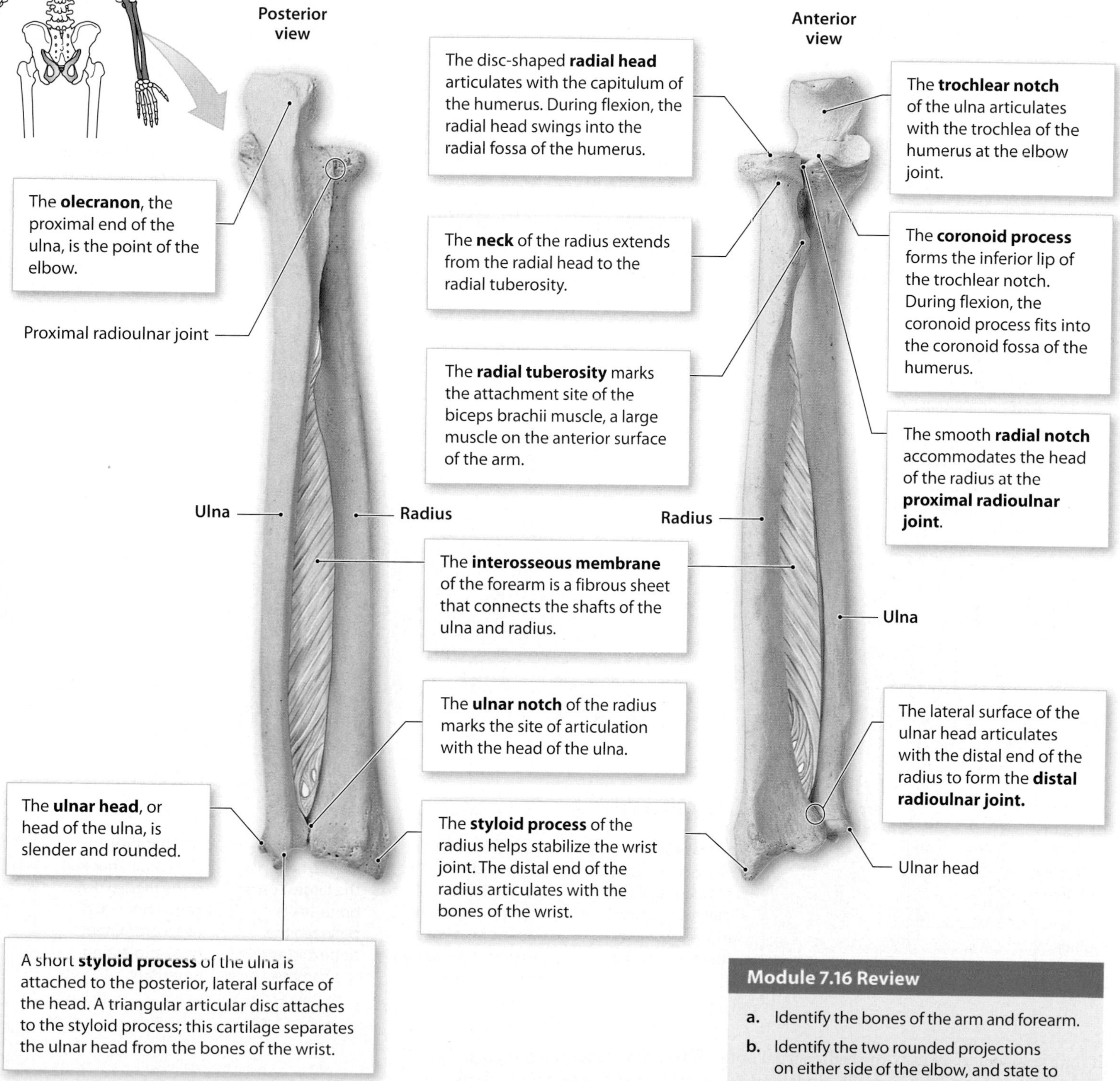

Posterior view
Anterior view
The **olecranon**, the proximal end of the ulna, is the point of the elbow.
Proximal radioulnar joint
Ulna
Radius
The disc-shaped **radial head** articulates with the capitulum of the humerus. During flexion, the radial head swings into the radial fossa of the humerus.
The **neck** of the radius extends from the radial head to the radial tuberosity.
The **radial tuberosity** marks the attachment site of the biceps brachii muscle, a large muscle on the anterior surface of the arm.
The **interosseous membrane** of the forearm is a fibrous sheet that connects the shafts of the ulna and radius.
The **ulnar notch** of the radius marks the site of articulation with the head of the ulna.
The **styloid process** of the radius helps stabilize the wrist joint. The distal end of the radius articulates with the bones of the wrist.
The **ulnar head**, or head of the ulna, is slender and rounded.
A short **styloid process** of the ulna is attached to the posterior, lateral surface of the head. A triangular articular disc attaches to the styloid process; this cartilage separates the ulnar head from the bones of the wrist.
The **trochlear notch** of the ulna articulates with the trochlea of the humerus at the elbow joint.
The **coronoid process** forms the inferior lip of the trochlear notch. During flexion, the coronoid process fits into the coronoid fossa of the humerus.
The smooth **radial notch** accommodates the head of the radius at the **proximal radioulnar joint**.
Radius
Ulna
The lateral surface of the ulnar head articulates with the distal end of the radius to form the **distal radioulnar joint.**
Ulnar head

Module 7.16 Review

a. Identify the bones of the arm and forearm.

b. Identify the two rounded projections on either side of the elbow, and state to which bone they belong.

c. Which bone of the forearm is positioned laterally while in anatomical position?

LO 7.16 Identify the bones of the arm and forearm, their functions, and their superficial features.

The wrist consists of carpal bones, and the hand consists of metacarpal bones and phalanges

1 The **carpus**, or wrist, contains eight carpal bones arranged in two rows: a row of four proximal carpal bones and a row of four distal carpal bones.

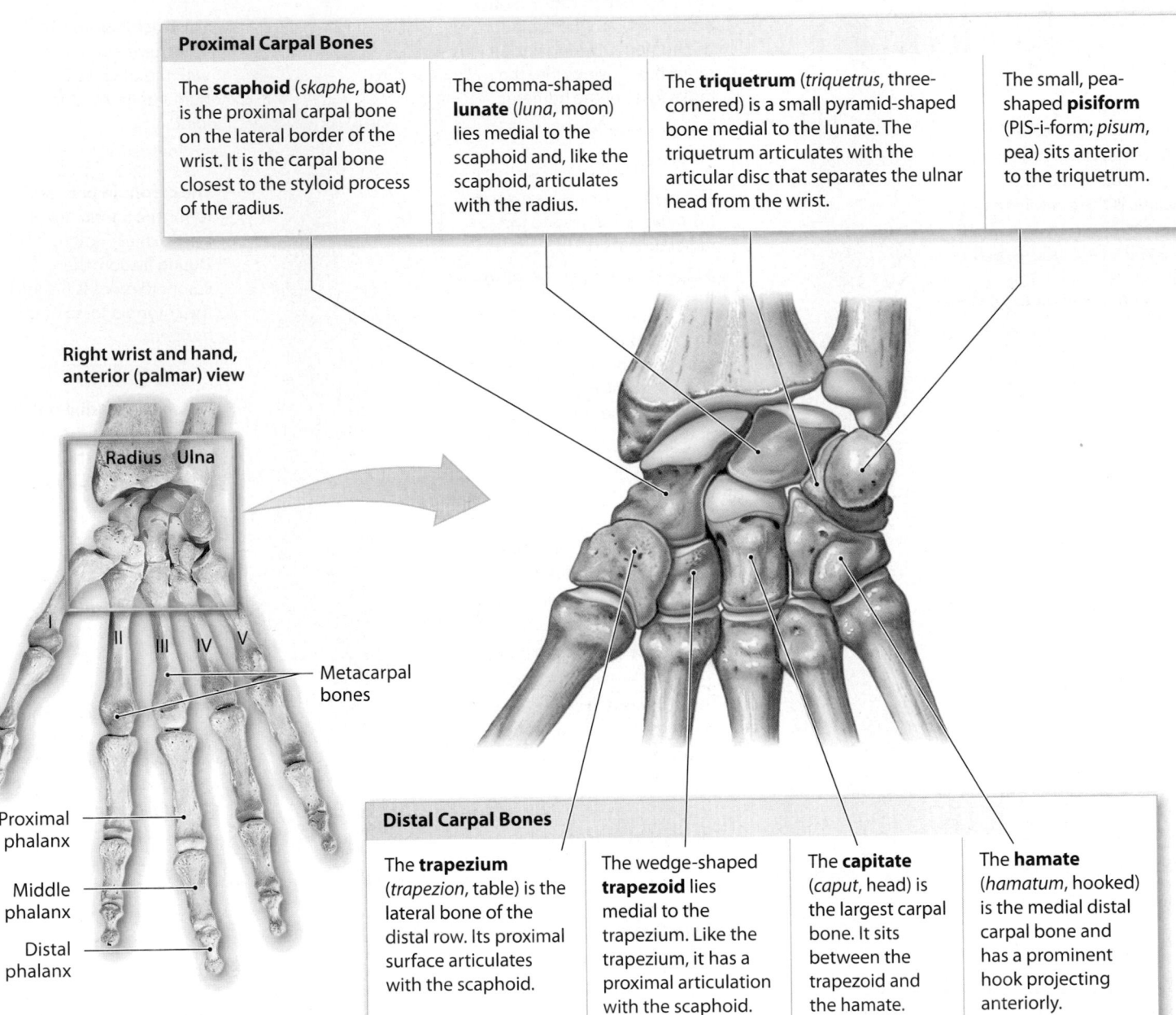

Proximal Carpal Bones

The **scaphoid** (*skaphe*, boat) is the proximal carpal bone on the lateral border of the wrist. It is the carpal bone closest to the styloid process of the radius.	The comma-shaped **lunate** (*luna*, moon) lies medial to the scaphoid and, like the scaphoid, articulates with the radius.	The **triquetrum** (*triquetrus*, three-cornered) is a small pyramid-shaped bone medial to the lunate. The triquetrum articulates with the articular disc that separates the ulnar head from the wrist.	The small, pea-shaped **pisiform** (PIS-i-form; *pisum*, pea) sits anterior to the triquetrum.

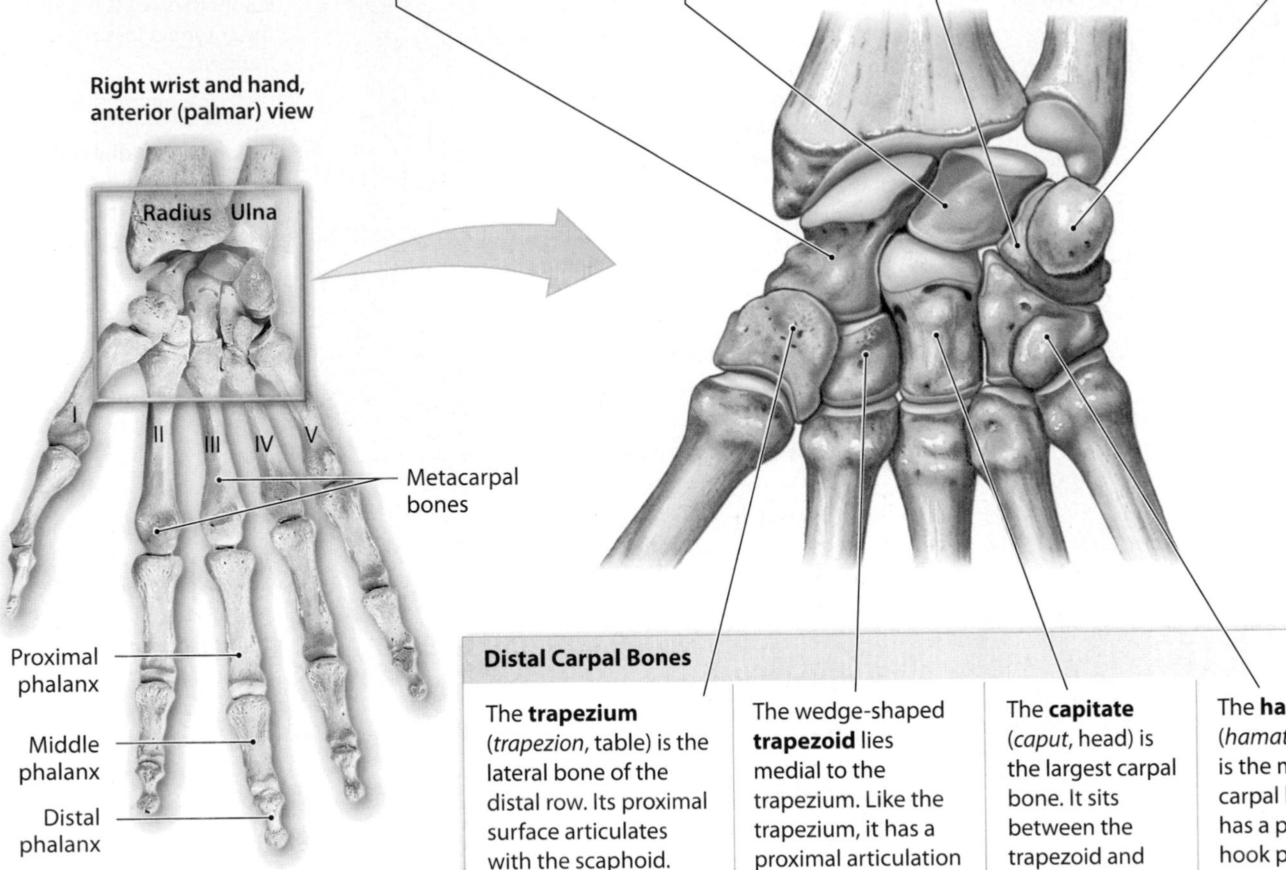

Right wrist and hand, anterior (palmar) view
Radius
Ulna
I
II
III
IV
V
Metacarpal bones
Proximal phalanx
Middle phalanx
Distal phalanx

Distal Carpal Bones

The **trapezium** (*trapezion*, table) is the lateral bone of the distal row. Its proximal surface articulates with the scaphoid.	The wedge-shaped **trapezoid** lies medial to the trapezium. Like the trapezium, it has a proximal articulation with the scaphoid.	The **capitate** (*caput*, head) is the largest carpal bone. It sits between the trapezoid and the hamate.	The **hamate** (*hamatum*, hooked) is the medial distal carpal bone and has a prominent hook projecting anteriorly.

It may help you to identify the eight carpal bones if you remember the sentence "Sam Likes To Push The Toy Car Hard." In lateral-to-medial order, the first four words stand for the proximal carpal bones (scaphoid, lunate, triquetrum, pisiform), and the last four stand for the distal carpal bones (trapezium, trapezoid, capitate, hamate).

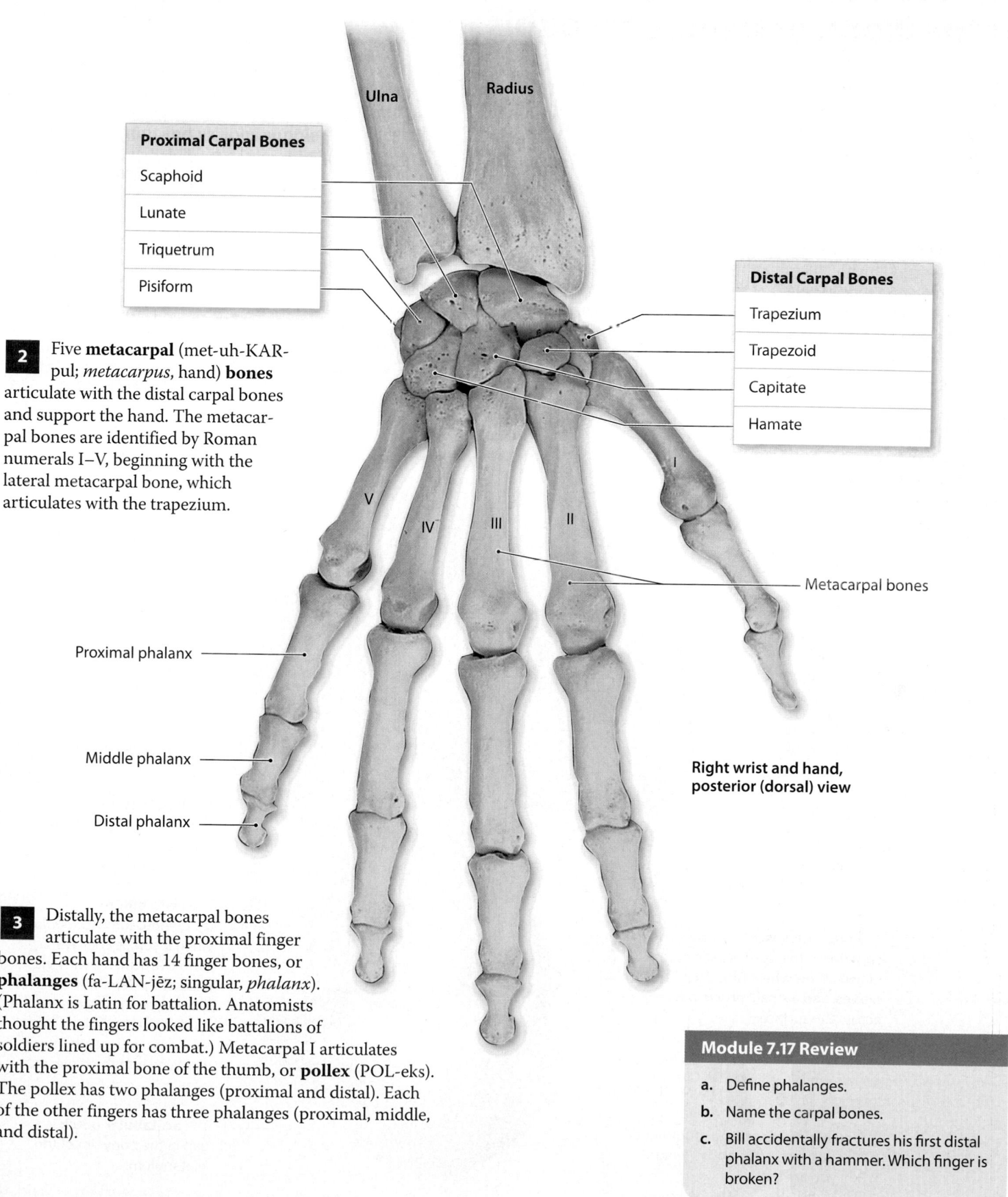

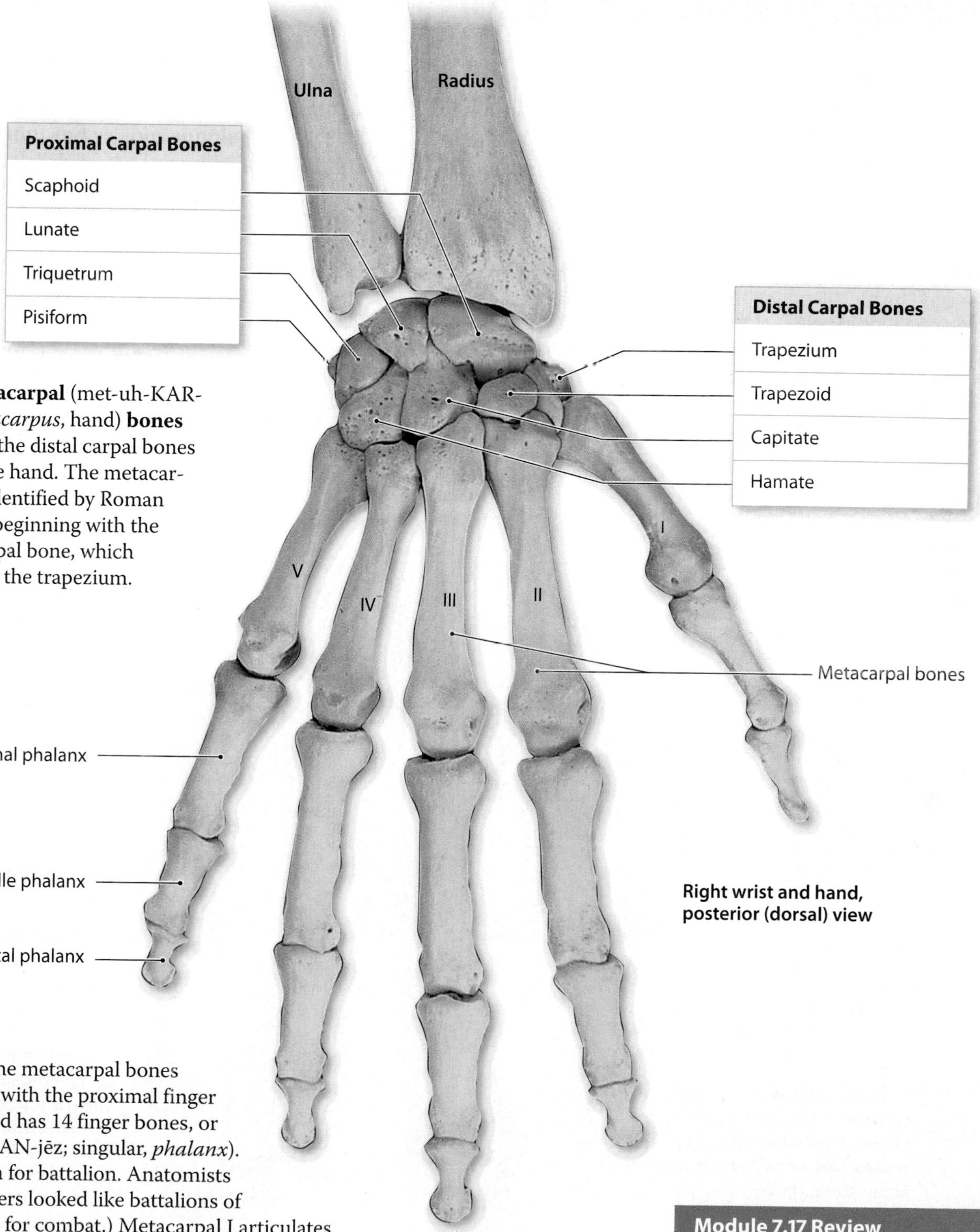
Ulna
Radius
Proximal Carpal Bones
Scaphoid
Lunate
Triquetrum
Pisiform
Distal Carpal Bones
Trapezium
Trapezoid
Capitate
Hamate
I
V
IV
III
II
Metacarpal bones
Proximal phalanx
Middle phalanx
Distal phalanx

Right wrist and hand, posterior (dorsal) view

2 Five **metacarpal** (met-uh-KAR-pul; *metacarpus*, hand) **bones** articulate with the distal carpal bones and support the hand. The metacarpal bones are identified by Roman numerals I–V, beginning with the lateral metacarpal bone, which articulates with the trapezium.

3 Distally, the metacarpal bones articulate with the proximal finger bones. Each hand has 14 finger bones, or **phalanges** (fa-LAN-jēz; singular, *phalanx*). (Phalanx is Latin for battalion. Anatomists thought the fingers looked like battalions of soldiers lined up for combat.) Metacarpal I articulates with the proximal bone of the thumb, or **pollex** (POL-eks). The pollex has two phalanges (proximal and distal). Each of the other fingers has three phalanges (proximal, middle, and distal).

Module 7.17 Review

a. Define phalanges.

b. Name the carpal bones.

c. Bill accidentally fractures his first distal phalanx with a hammer. Which finger is broken?

LO 7.17 Identify the bones of the wrist and hand, and describe their locations using anatomical terminology.

The hip bone forms by the fusion of the ilium, ischium, and pubis

1 The **pelvic girdle** consists of the paired **hip bones**, which are also called the **coxal bones** or innominate (no name) bones. Each hip bone forms by the fusion of three bones: an **ilium** (IL-ē-um; plural, *ilia*), an **ischium** (IS-kē-um; plural, *ischia*), and a **pubis** (PŪ-bis).

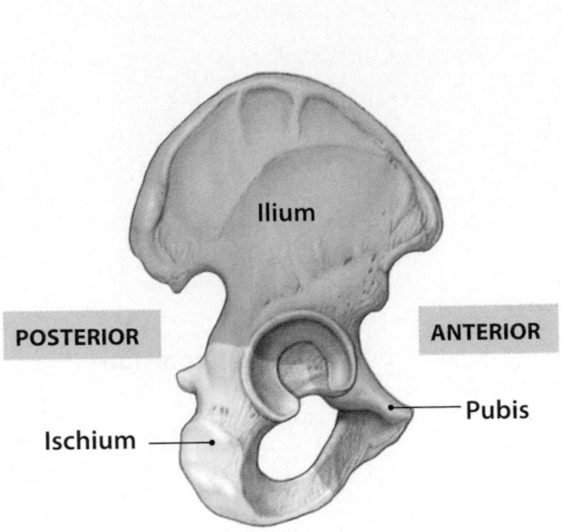
Ilium
POSTERIOR
ANTERIOR
Ischium
Pubis

2 In lateral view, the ilium dominates the hip bone. Surface markings along the margin of the ilium include the **iliac spines**, which mark the attachment sites of important muscles and ligaments; the **gluteal lines**, which mark the attachment of large hip muscles; and the **greater sciatic** (sī-AT-ik) **notch**, through which the large sciatic nerve reaches the lower limb.

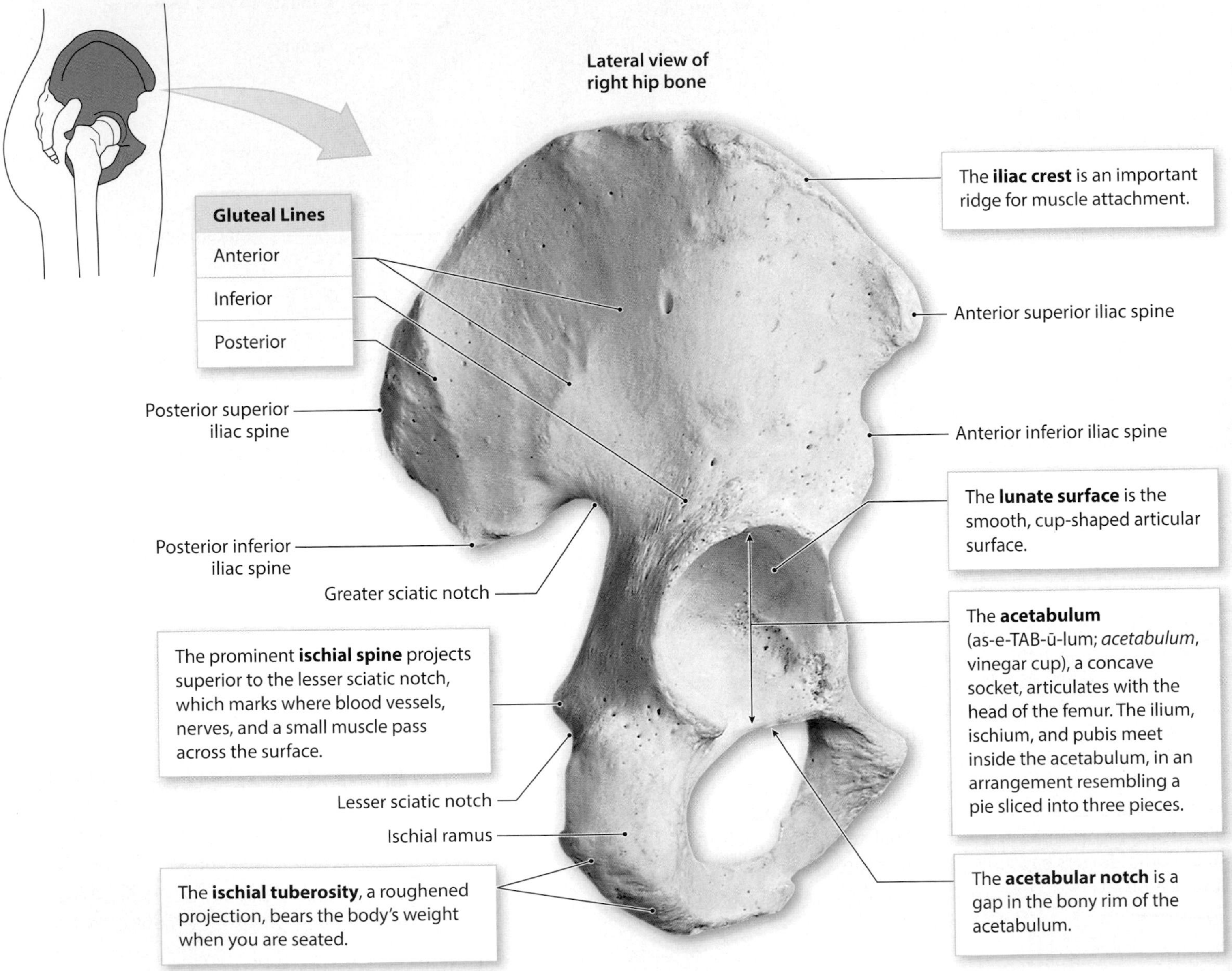
Lateral view of right hip bone
Gluteal Lines
Anterior
Inferior
Posterior
Posterior superior iliac spine
Posterior inferior iliac spine
Greater sciatic notch
The prominent ischial spine projects superior to the lesser sciatic notch, which marks where blood vessels, nerves, and a small muscle pass across the surface.
Lesser sciatic notch
Ischial ramus
The ischial tuberosity, a roughened projection, bears the body's weight when you are seated.
The iliac crest is an important ridge for muscle attachment.
Anterior superior iliac spine
Anterior inferior iliac spine
The lunate surface is the smooth, cup-shaped articular surface.
The acetabulum (as-e-TAB-ū-lum; acetabulum, vinegar cup), a concave socket, articulates with the head of the femur. The ilium, ischium, and pubis meet inside the acetabulum, in an arrangement resembling a pie sliced into three pieces.
The acetabular notch is a gap in the bony rim of the acetabulum.

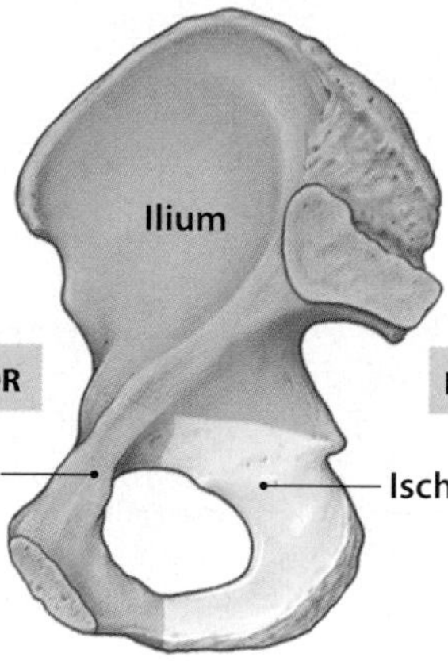

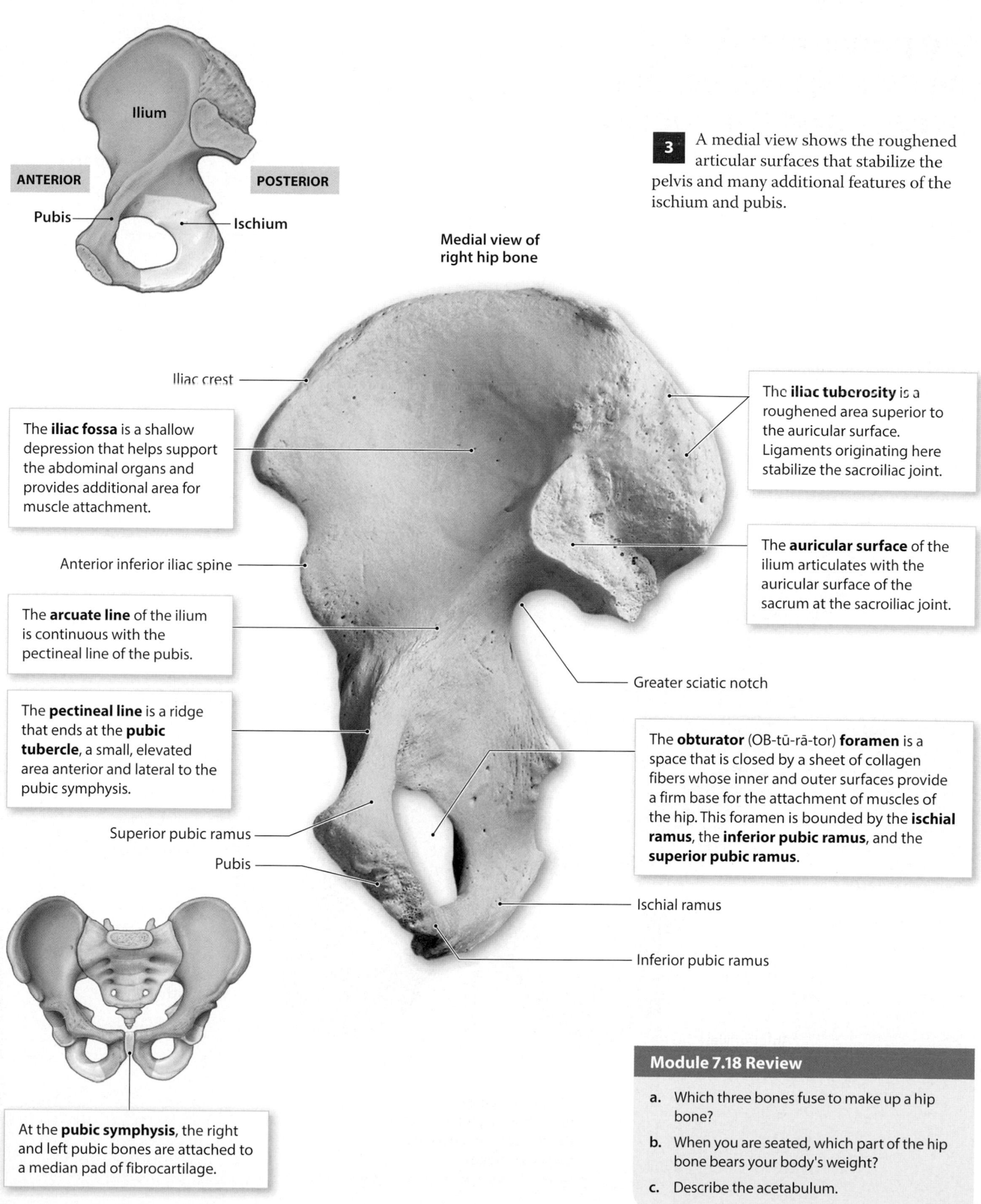

Ilium
ANTERIOR
POSTERIOR
Pubis
Ischium

3 A medial view shows the roughened articular surfaces that stabilize the pelvis and many additional features of the ischium and pubis.

Medial view of right hip bone

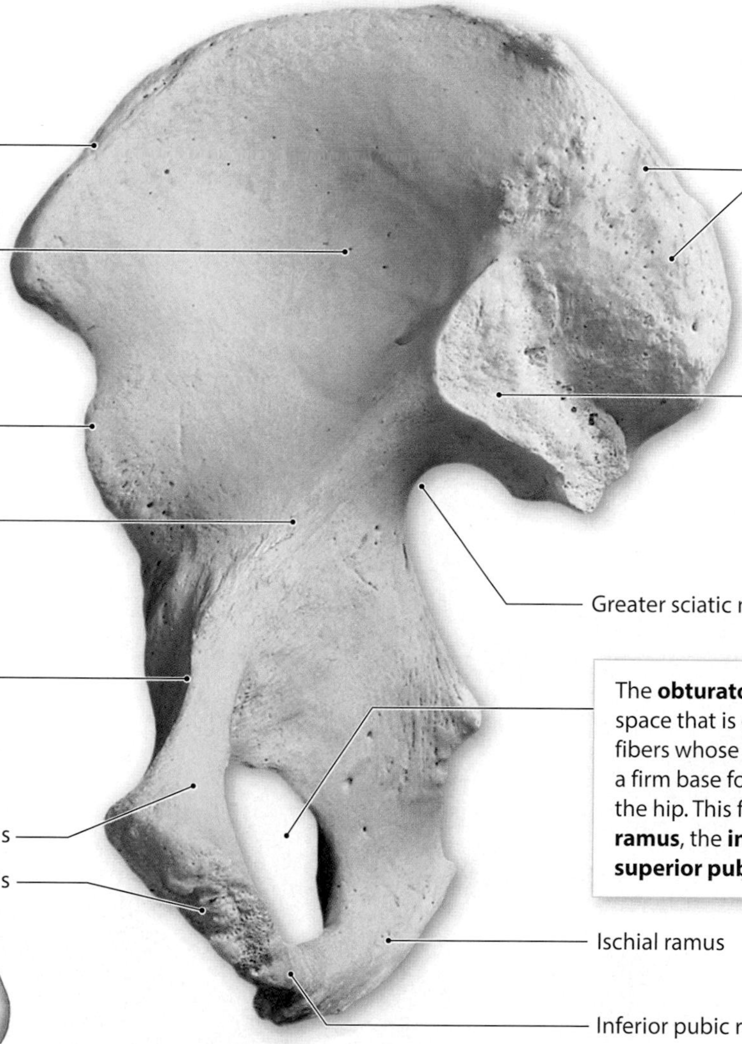

Iliac crest
Anterior inferior iliac spine
Superior pubic ramus
Pubis
Greater sciatic notch
Ischial ramus
Inferior pubic ramus

The **iliac fossa** is a shallow depression that helps support the abdominal organs and provides additional area for muscle attachment.

The **arcuate line** of the ilium is continuous with the pectineal line of the pubis.

The **pectineal line** is a ridge that ends at the **pubic tubercle**, a small, elevated area anterior and lateral to the pubic symphysis.

The **iliac tuberosity** is a roughened area superior to the auricular surface. Ligaments originating here stabilize the sacroiliac joint.

The **auricular surface** of the ilium articulates with the auricular surface of the sacrum at the sacroiliac joint.

The **obturator** (OB-tū-rā-tor) **foramen** is a space that is closed by a sheet of collagen fibers whose inner and outer surfaces provide a firm base for the attachment of muscles of the hip. This foramen is bounded by the **ischial ramus**, the **inferior pubic ramus**, and the **superior pubic ramus**.

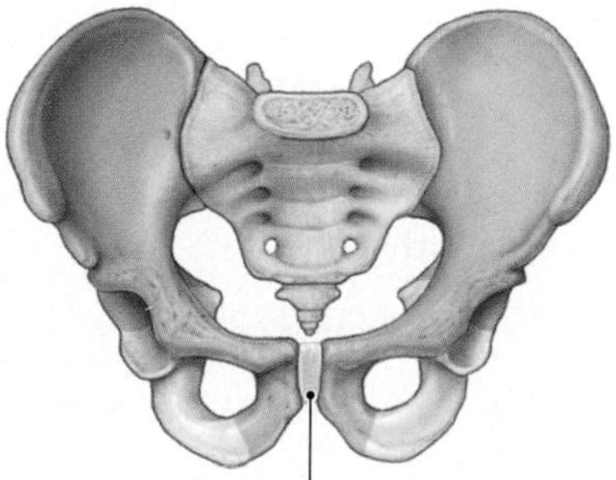

At the **pubic symphysis**, the right and left pubic bones are attached to a median pad of fibrocartilage.

Module 7.18 Review

a. Which three bones fuse to make up a hip bone?

b. When you are seated, which part of the hip bone bears your body's weight?

c. Describe the acetabulum.

LO **7.18** Describe the hip bones that form the pelvic girdle, their functions, and their superficial features.

The pelvis consists of the two hip bones, the sacrum, and the coccyx

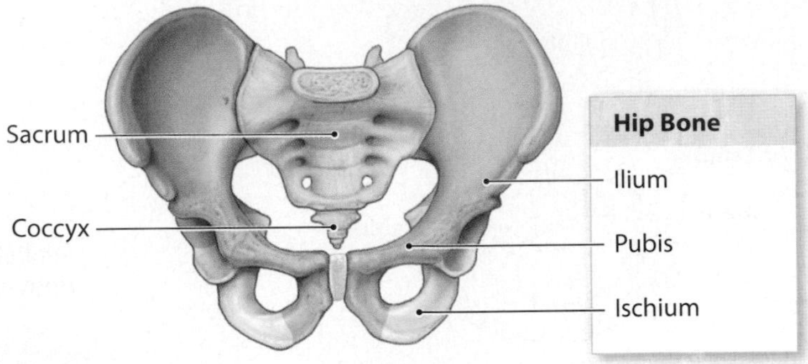
Sacrum
Coccyx
Hip Bone
Ilium
Pubis
Ischium

1 The **pelvis** contains bones of the axial skeleton (the sacrum and coccyx) and the appendicular skeleton (the hip bones). An extensive network of ligaments connects the lateral borders of the sacrum with the iliac crest, the ischial tuberosity, the ischial spine, and the arcuate line. Other ligaments tie the ilia to the posterior lumbar vertebrae. These interconnections increase the stability of the pelvis.

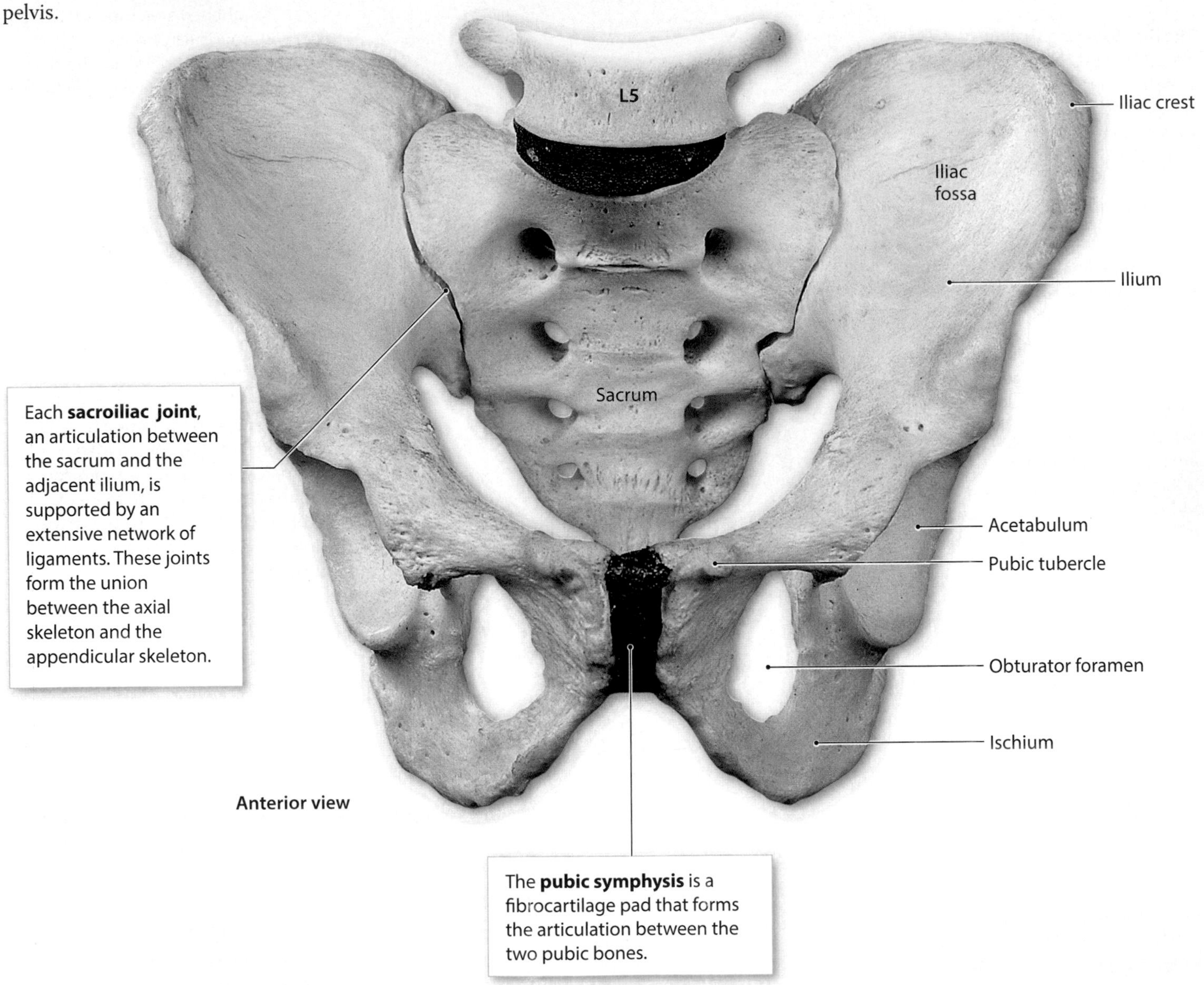
L5
Iliac crest
Iliac fossa
Ilium
Sacrum
Each **sacroiliac joint**, an articulation between the sacrum and the adjacent ilium, is supported by an extensive network of ligaments. These joints form the union between the axial skeleton and the appendicular skeleton.
Acetabulum
Pubic tubercle
Obturator foramen
Ischium
The **pubic symphysis** is a fibrocartilage pad that forms the articulation between the two pubic bones.

Anterior view

2 The pelvis may be divided into the **true** (lesser) **pelvis** (shown in purple) and the **false** (greater) **pelvis.** The true pelvis encloses the pelvic cavity, a subdivision of the abdominopelvic cavity. The superior limit of the true pelvis is a line that extends from either side of the base of the sacrum, along the arcuate line and pectineal line to the pubic symphysis.

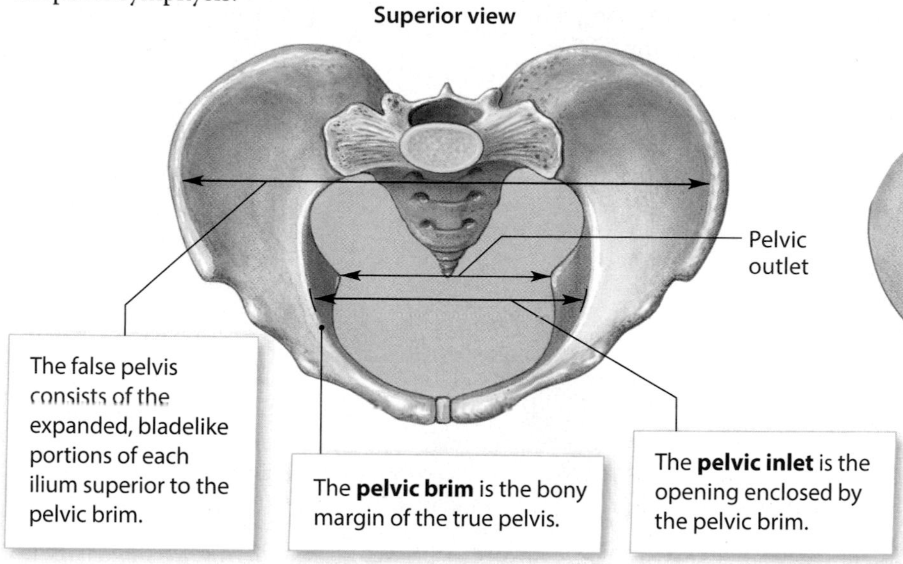
Superior view
Pelvic outlet
The false pelvis consists of the expanded, bladelike portions of each ilium superior to the pelvic brim.
The **pelvic brim** is the bony margin of the true pelvis.
The **pelvic inlet** is the opening enclosed by the pelvic brim.

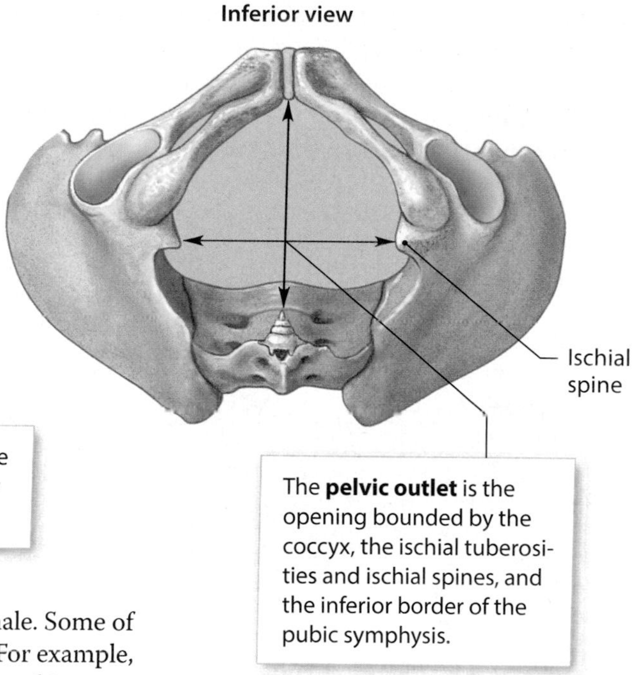
Inferior view
Ischial spine
The **pelvic outlet** is the opening bounded by the coccyx, the ischial tuberosities and ischial spines, and the inferior border of the pubic symphysis.

3 The shape of the pelvis of a female is somewhat different from that of a male. Some of the differences are the result of variations in body size and muscle mass. For example, in females, the pelvis is generally smoother and lighter and has less prominent markings. Other female skeletal variations appear to be adaptations for childbearing, including:

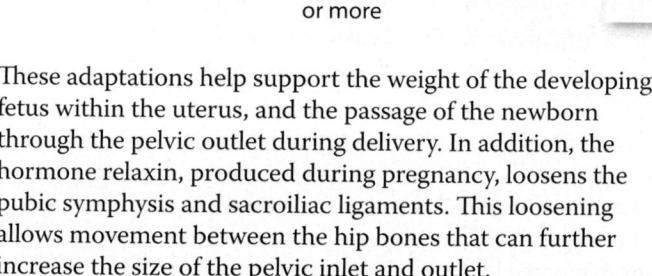

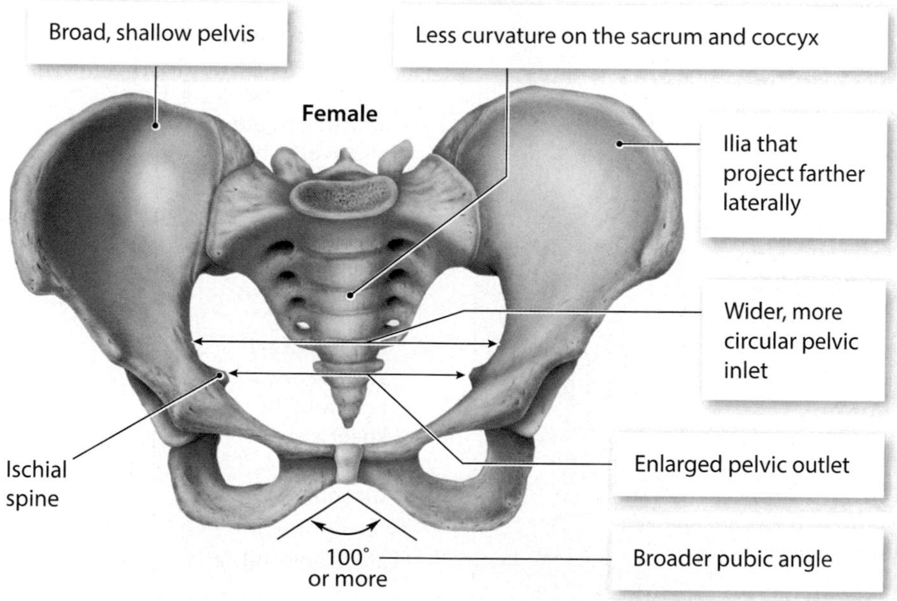
Broad, shallow pelvis
Less curvature on the sacrum and coccyx
Female
Ilia that project farther laterally
Wider, more circular pelvic inlet
Enlarged pelvic outlet
Ischial spine
100°
or more
Broader pubic angle

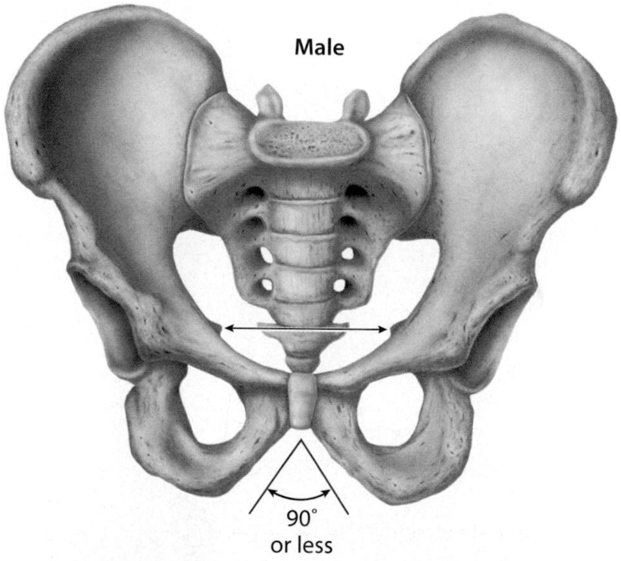
Male
90°
or less

These adaptations help support the weight of the developing fetus within the uterus, and the passage of the newborn through the pelvic outlet during delivery. In addition, the hormone relaxin, produced during pregnancy, loosens the pubic symphysis and sacroiliac ligaments. This loosening allows movement between the hip bones that can further increase the size of the pelvic inlet and outlet.

Module 7.19 Review

a. Name the bones of the pelvis.

b. The pubic bones are joined anteriorly by what structure?

c. How is the pelvis of females adapted for childbearing?

LO **7.19** Identify the bones of the pelvis, and discuss the structural and functional differences between the pelvis in males and the pelvis in females.

The femur, tibia, and patella meet at the knee

The skeleton of each lower limb consists of a **femur** (thigh), a **patella** (kneecap), a **tibia** and a **fibula** (leg), and the **tarsal bones**, **metatarsal bones**, and **phalanges** of the foot. Once again, anatomical terminology differs from common usage. In anatomical terms, "leg" refers only to the distal portion of the limb, not to the entire lower limb. Thus, we will use *thigh* and *leg*, rather than *upper leg* and *lower leg*. The functional anatomy of the lower limbs differs from that of the upper limbs, primarily because the lower limbs transfer the body weight to the ground.

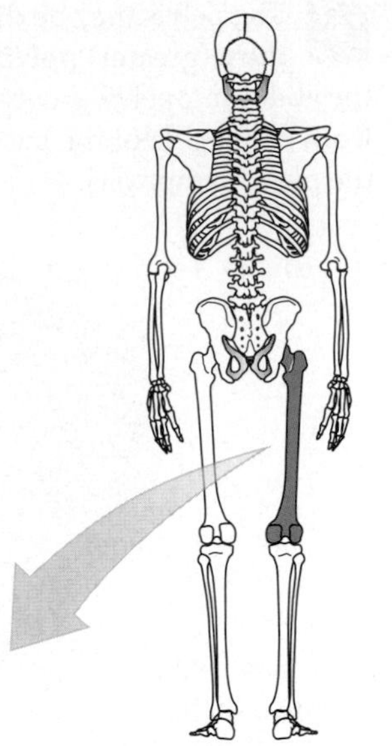

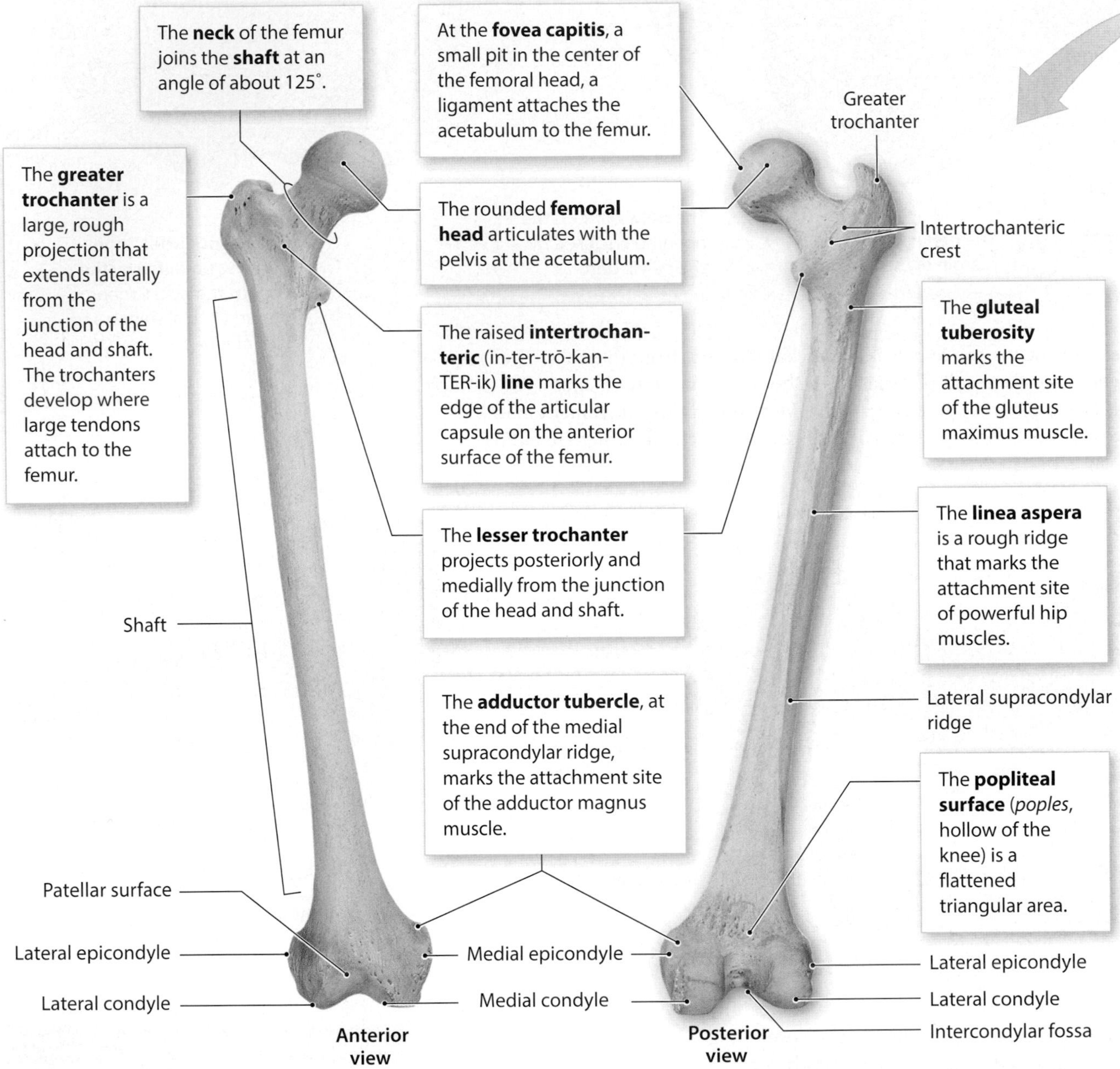
The neck of the femur joins the shaft at an angle of about 125°.
At the fovea capitis, a small pit in the center of the femoral head, a ligament attaches the acetabulum to the femur.
Greater trochanter
The greater trochanter is a large, rough projection that extends laterally from the junction of the head and shaft. The trochanters develop where large tendons attach to the femur.
The rounded femoral head articulates with the pelvis at the acetabulum.
Intertrochanteric crest
The raised intertrochanteric (in-ter-trō-kan-TER-ik) line marks the edge of the articular capsule on the anterior surface of the femur.
The gluteal tuberosity marks the attachment site of the gluteus maximus muscle.
The lesser trochanter projects posteriorly and medially from the junction of the head and shaft.
The linea aspera is a rough ridge that marks the attachment site of powerful hip muscles.
Shaft
The adductor tubercle, at the end of the medial supracondylar ridge, marks the attachment site of the adductor magnus muscle.
Lateral supracondylar ridge
The popliteal surface (poples, hollow of the knee) is a flattened triangular area.
Patellar surface
Lateral epicondyle
Medial epicondyle
Lateral epicondyle
Lateral condyle
Medial condyle
Lateral condyle
Intercondylar fossa
Anterior view
Posterior view

1 The **femur** is the longest and heaviest bone in the body. It articulates with the hip bone at the hip joint and with the tibia of the leg at the knee joint. Major surface markings are shown here on the anterior and posterior surfaces of the right femur.

2 At the distal end of the femur, the **medial** and **lateral condyles** are part of the knee joint. On the anterior and inferior surfaces, the two condyles are separated by the **patellar surface**, a smooth articular surface over which the patella glides. On the posterior surface, the medial and lateral condyles are separated by a deep **intercondylar fossa** that does not extend onto the anterior surface.

3 The patella is a large sesamoid bone that forms within the tendon of the quadriceps femoris, a group of muscles that extend (straighten) the knee. The right patella is shown here.

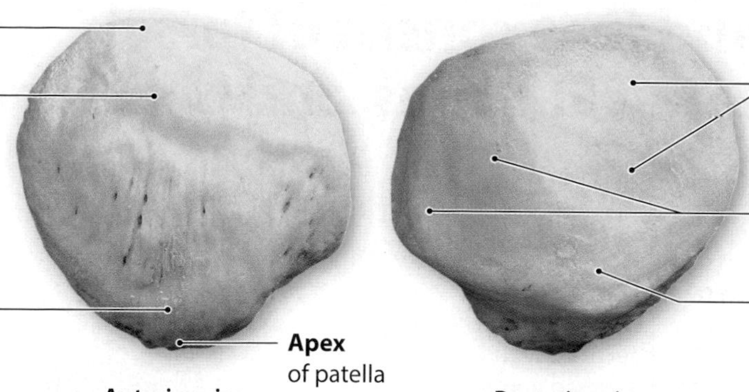

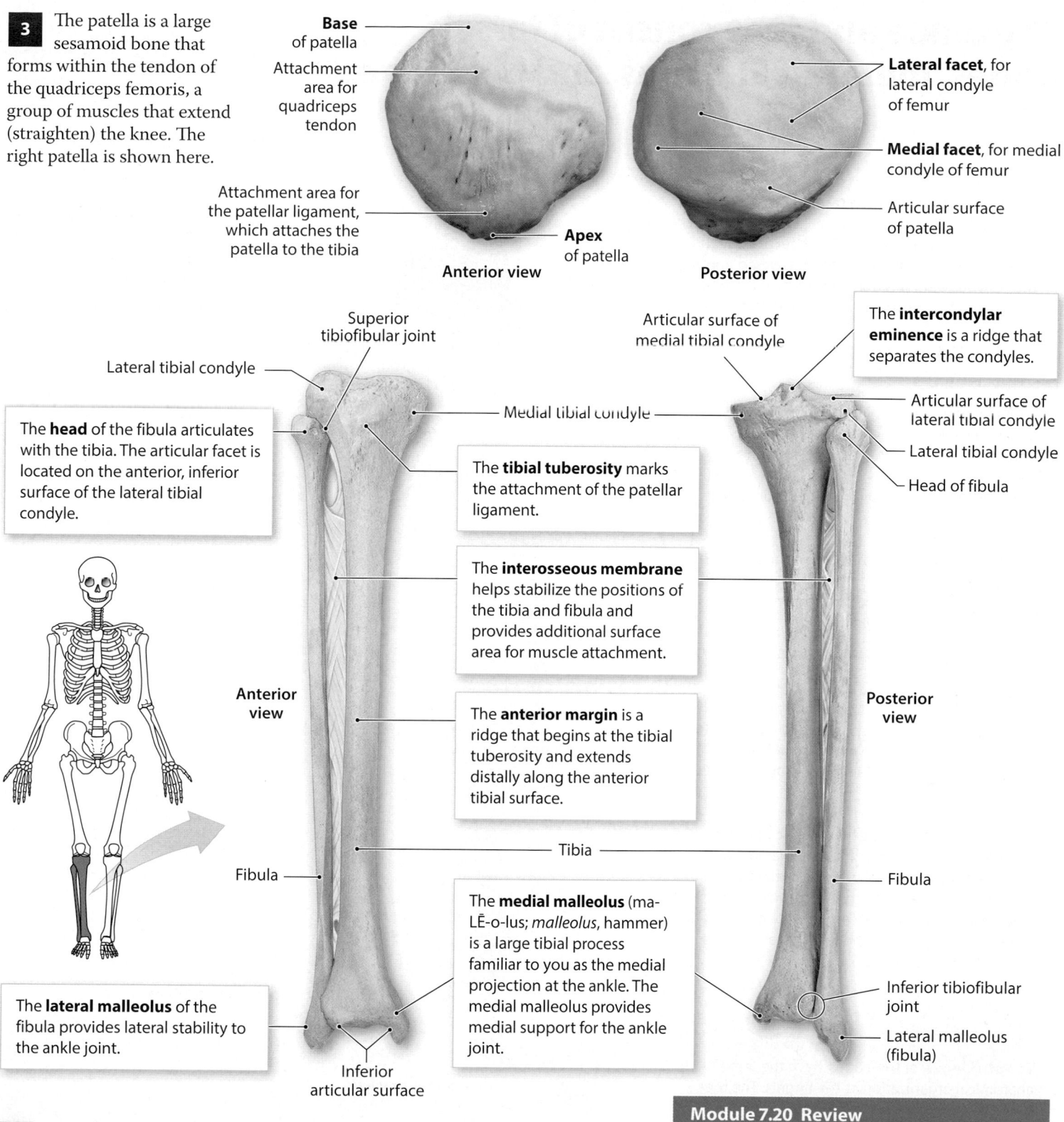

Base of patella
Attachment area for quadriceps tendon
Attachment area for the patellar ligament, which attaches the patella to the tibia
Apex of patella
Anterior view
Lateral facet, for lateral condyle of femur
Medial facet, for medial condyle of femur
Articular surface of patella
Posterior view

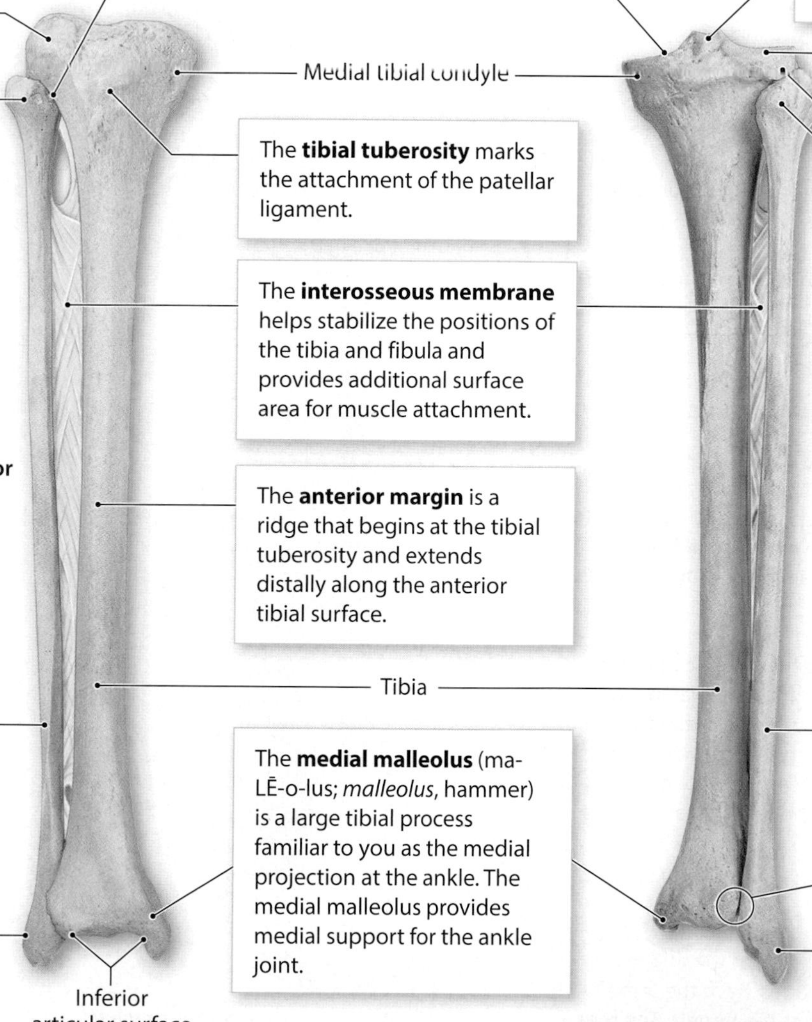

Superior tibiofibular joint
Lateral tibial condyle
The **head** of the fibula articulates with the tibia. The articular facet is located on the anterior, inferior surface of the lateral tibial condyle.
Medial tibial condyle
The **tibial tuberosity** marks the attachment of the patellar ligament.
The **interosseous membrane** helps stabilize the positions of the tibia and fibula and provides additional surface area for muscle attachment.
Anterior view
The **anterior margin** is a ridge that begins at the tibial tuberosity and extends distally along the anterior tibial surface.
Tibia
Fibula
The **medial malleolus** (ma-LĒ-o-lus; *malleolus*, hammer) is a large tibial process familiar to you as the medial projection at the ankle. The medial malleolus provides medial support for the ankle joint.
The **lateral malleolus** of the fibula provides lateral stability to the ankle joint.
Inferior articular surface
Articular surface of medial tibial condyle
The **intercondylar eminence** is a ridge that separates the condyles.
Articular surface of lateral tibial condyle
Lateral tibial condyle
Head of fibula
Posterior view
Fibula
Inferior tibiofibular joint
Lateral malleolus (fibula)

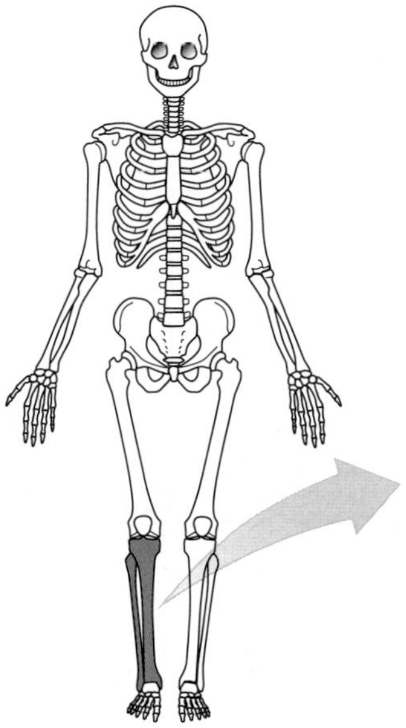

4 The **tibia** (TIB-ē-uh), or shinbone, is the large medial bone of the leg. At the proximal end of the tibia, the **medial** and **lateral tibial condyles** articulate with the medial and lateral condyles of the femur. The slender **fibula** (FIB-ū-luh) parallels the lateral border of the tibia but does not participate in the knee joint and bears no weight. However, the fibula is important as a site for the attachment of muscles that move the foot and toes. In addition, the distal tip of the fibula extends lateral to the ankle, providing important stability to that joint.

Module 7.20 Review

a. Identify the bones of the lower limb.
b. Which structure articulates with the acetabulum?
c. The fibula neither participates in the knee joint nor bears weight. Yet, when it is fractured, walking becomes difficult. Why?

LO 7.20 Identify the bones of the thigh and leg, their functions, and their superficial features.

The ankle and foot consist of tarsal bones, metatarsal bones, and phalanges

The Ankle (Tarsus)

The ankle consists of seven **tarsal bones**.

The **calcaneus** (kal-KĀ-nē-us), or heel bone, is the largest of the tarsal bones. When you stand normally, most of your weight is transmitted from the tibia, to the talus, to the calcaneus, and then to the ground.

The large **talus** transmits the weight of the body from the tibia toward the toes.

The **navicular** is anterior to the talus, on the medial side of the ankle. It articulates with the talus and with the three cuneiform (kū-NĒ-i-form) bones.

The **cuboid** articulates with the anterior surface of the calcaneus.

The three **cuneiform bones** are arranged in a row, with articulations between them. They are named according to their relative positions: medial, intermediate, and lateral.

Metatarsals

The distal surfaces of the cuboid and the cuneiform bones articulate with the **metatarsal bones** of the foot. The five long metatarsal bones form the distal portion of the foot.

The metatarsal bones are identified by Roman numerals I–V, proceeding from medial to lateral across the sole. Proximally, metatarsal bones I–III articulate with the three cuneiform bones, and metatarsal bones IV and V articulate with the cuboid. Distally, each metatarsal bone articulates with a different proximal phalanx.

Phalanges

The **phalanges**, or toe bones, have the same anatomical organization as the fingers. The toes contain 14 phalanges. The **hallux**, or great toe, has two phalanges (proximal and distal), and the other four toes each have three phalanges (proximal, middle, and distal).

1 The bones of the ankle accept the body weight from the leg and transfer it to the ground by distributing it through the bones of the foot. The combination of ankle and foot must be strong enough yet flexible enough to deal with the changes in loading that occur during walking, running, and jumping.

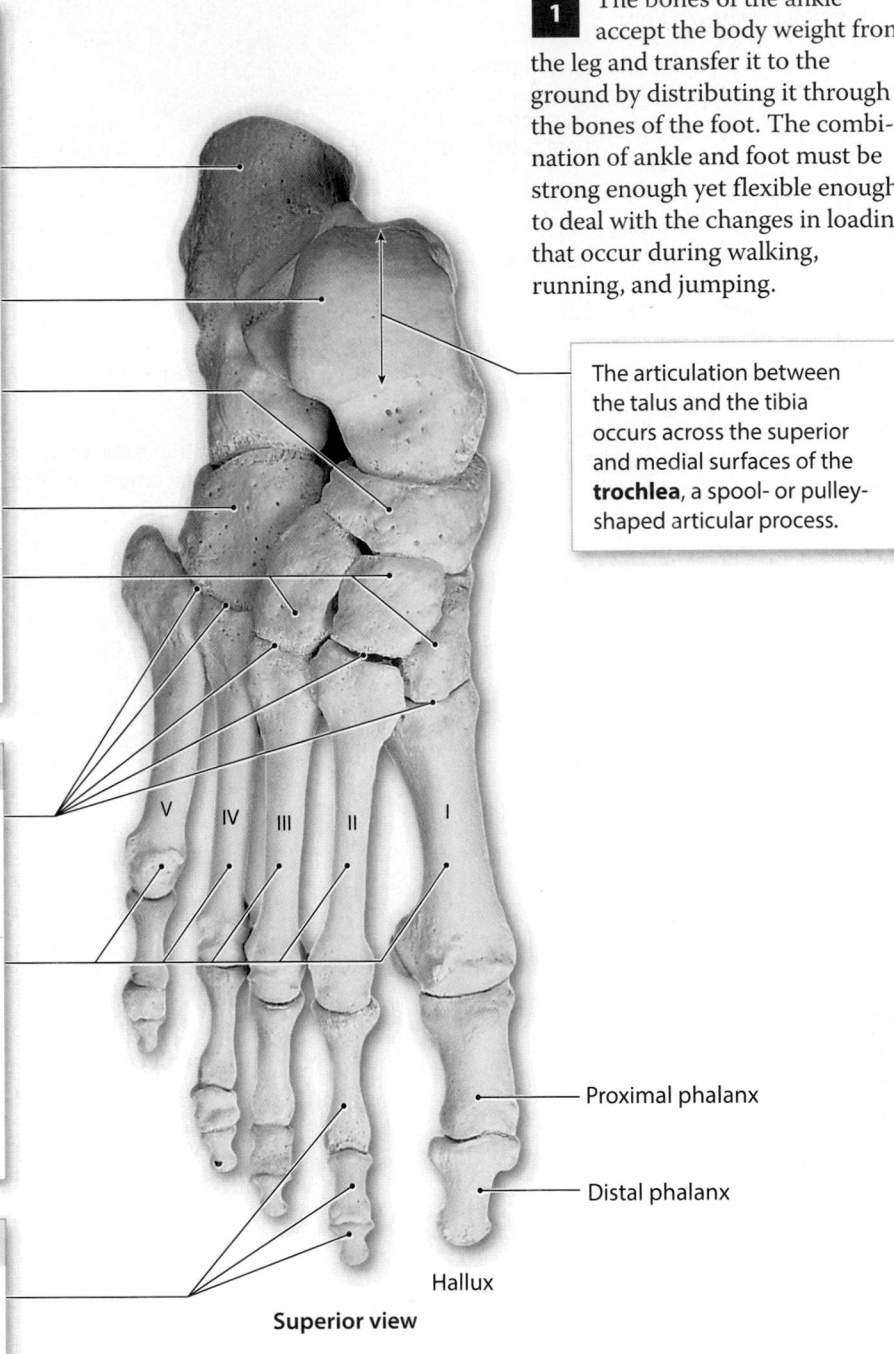

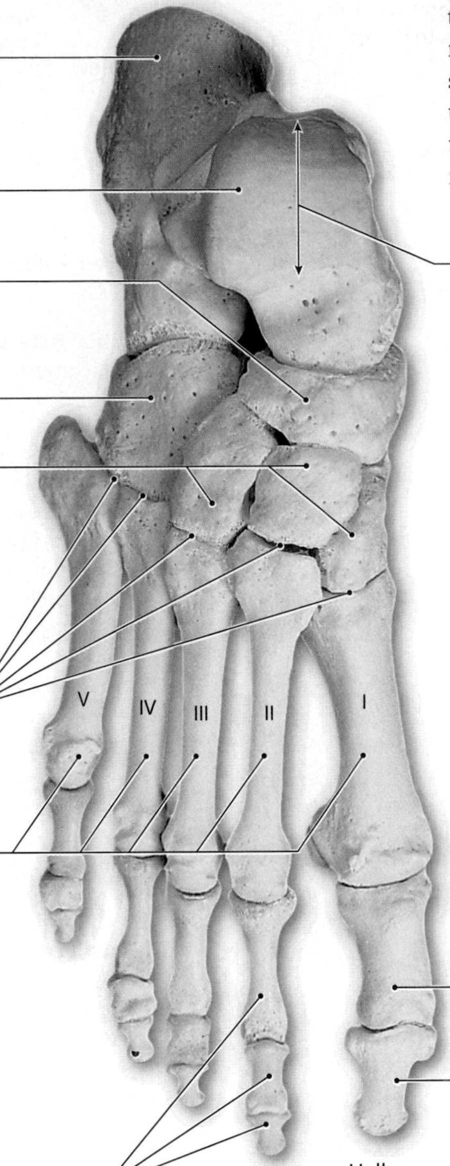
V
IV
III
II
I
Proximal phalanx
Distal phalanx
Hallux

The articulation between the talus and the tibia occurs across the superior and medial surfaces of the **trochlea**, a spool- or pulley-shaped articular process.

Superior view

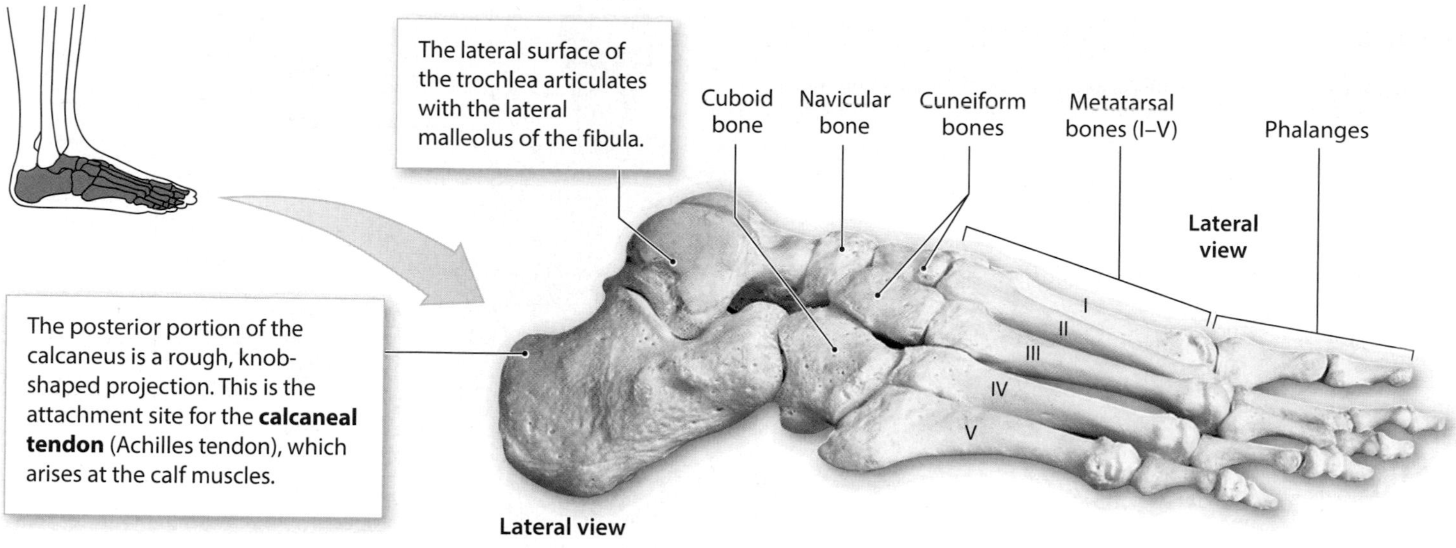
The lateral surface of the trochlea articulates with the lateral malleolus of the fibula.
Cuboid bone
Navicular bone
Cuneiform bones
Metatarsal bones (I–V)
Phalanges
Lateral view
I
II
III
IV
V
The posterior portion of the calcaneus is a rough, knob-shaped projection. This is the attachment site for the **calcaneal tendon** (Achilles tendon), which arises at the calf muscles.
Lateral view

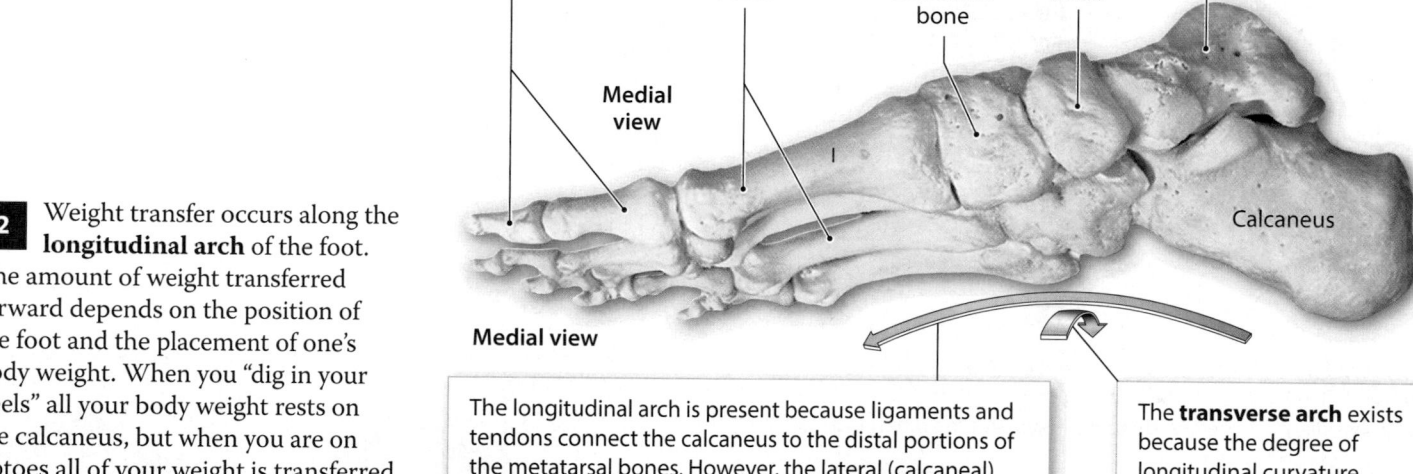

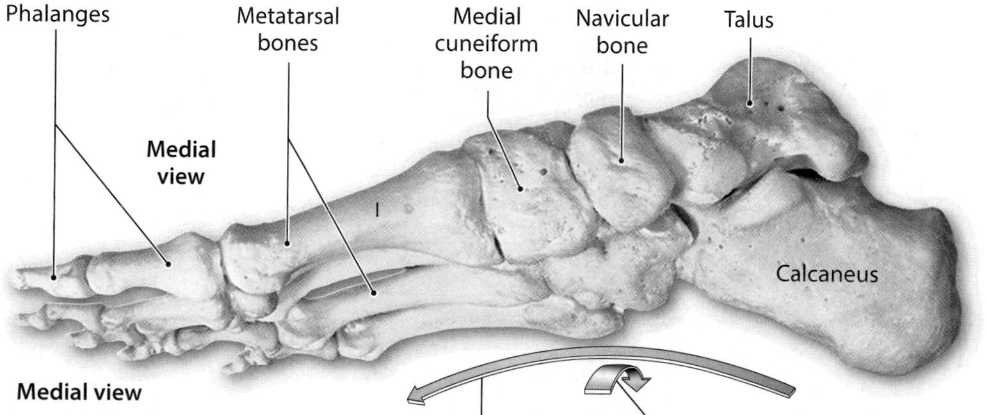
Phalanges
Metatarsal bones
Medial cuneiform bone
Navicular bone
Talus
Medial view
I
Calcaneus
Medial view

The longitudinal arch is present because ligaments and tendons connect the calcaneus to the distal portions of the metatarsal bones. However, the lateral (calcaneal) portion of the longitudinal arch has much less curvature than the medial (talar) portion, in part because the talar portion has considerably more elasticity. As a result, the medial plantar surface of the foot remains elevated, so that the muscles, nerves, and blood vessels that supply the inferior surface are not squeezed between the metatarsal bones and the ground.

The **transverse arch** exists because the degree of longitudinal curvature changes from the medial border to the lateral border of the foot.

2 Weight transfer occurs along the **longitudinal arch** of the foot. The amount of weight transferred forward depends on the position of the foot and the placement of one's body weight. When you "dig in your heels" all your body weight rests on the calcaneus, but when you are on tiptoes all of your weight is transferred to the metatarsal bones and phalanges. When you stand normally, your body weight is distributed evenly between the calcaneus and the distal ends of the metatarsal bones. In the condition known as flatfeet, normal arches are lost ("fall") or never form.

3 In a **dancer's fracture**, the diaphysis of the fifth metatarsal is broken. Most such cases occur while the body weight is being supported by the longitudinal arch of the foot (as in ballet dancing). A sudden shift in weight from the medial portion of the arch to the lateral, less elastic border breaks the fifth metatarsal close to the small toe.

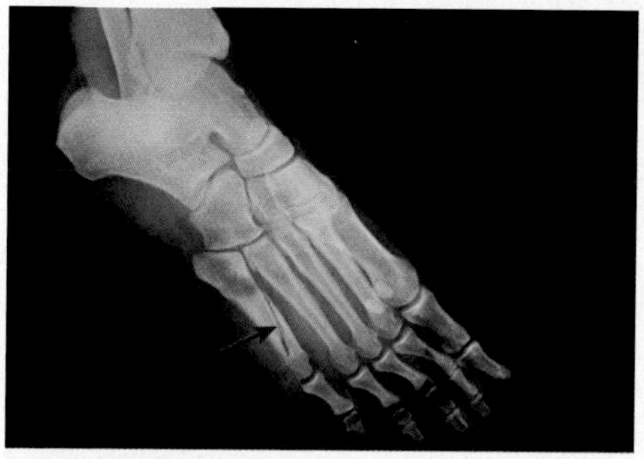

Module 7.21 Review

a. Identify the tarsal bones.

b. Which foot bone transmits the weight of the body from the tibia toward the toes?

c. Ten-year-old Joey jumps off the back porch, lands on his right heel, and breaks his foot. Which foot bone is most likely broken?

LO **7.21** Identify the bones of the ankle and foot, and describe their locations using anatomical terminology.

Labeling

Label the bones of the appendicular skeleton in the diagram at right.

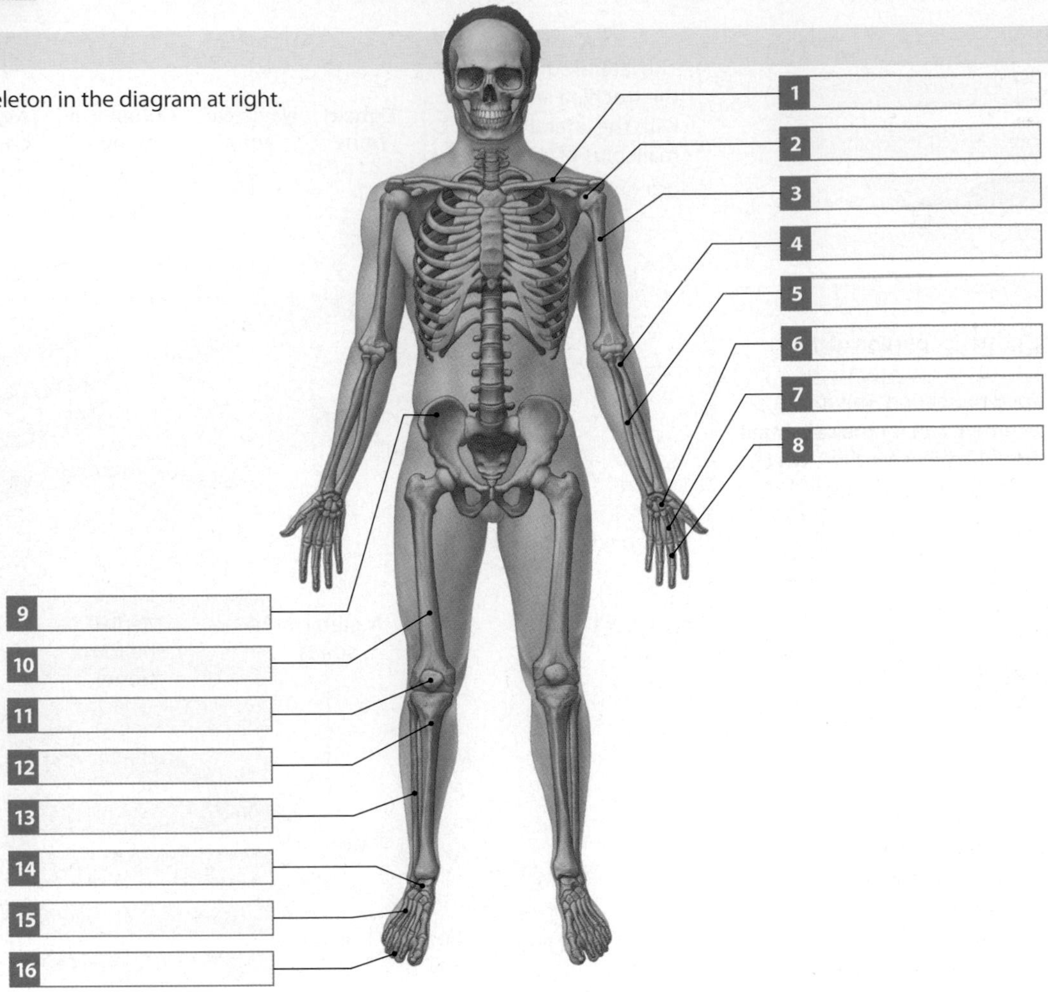
1
2
3
4
5
6
7
8
9
10
11
12
13
14
15
16

Short answer

In the pelvis diagrams below, identify the sex and the differences between them.

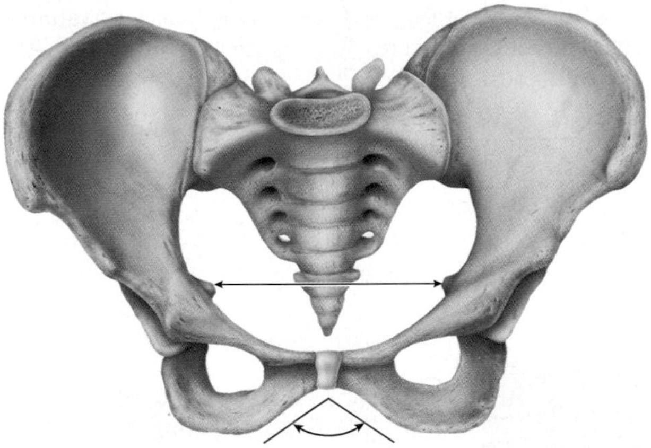

17 Sex:

18 Differences:

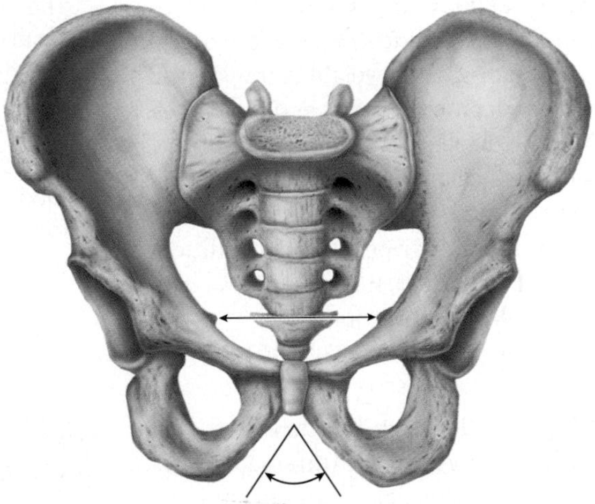

19 Sex:

20 Differences:

Study Outline

SECTION 1 • Axial Skeleton

7.1 The axial skeleton includes bones of the head, vertebral column, and trunk p. 233

1. The **axial skeleton** forms the longitudinal axis of the body and includes the skull and associated bones, the thoracic cage, the vertebral column, and various supplemental cartilages.
2. The axial skeleton supports and protects the brain, the spinal cord, and the organs in the body cavities of the trunk. It also provides an extensive surface area for the attachment of muscles.

7.2 The skull has cranial and facial components that are usually bound together by sutures p. 234

3. The skull contains 22 bones: eight form the cranium and 14 form the face. Seven other bones are associated with the skull: six auditory ossicles and the hyoid bone.
4. **Facial bones** protect and support the entrances to the digestive and respiratory tracts. They also provide areas for the attachment of facial muscles.
5. **Cranial bones** form the **cranium** and enclose the **cranial cavity**.
6. Except where the mandible contacts the cranium, the connections between the skull and bones of adults are immovable joints called **sutures**. The major sutures of the skull are the **coronal suture**, the **squamous suture**, the **sagittal suture**, and the **lambdoid suture**.

7.3 Facial bones dominate the anterior aspect of the skull, and cranial bones dominate the posterior surface p. 236

7. The facial bones visible on the anterior aspect of the skull are the **nasal**, **lacrimal**, **palatine**, **zygomatic**, **maxillae**, **inferior nasal conchae**, **vomer**, and **mandible**.
8. The cranial bones visible on the anterior aspect of the skull are the **parietal**, the **frontal**, **sphenoid**, and **ethmoid**.
9. The posterior view of the skull displays cranial bones only: **parietal**, **occipital**, and **temporal**.

7.4 The lateral and medial aspects of the skull share many surface markings p. 238

10. The cranial bones visible from a lateral view of the skull are the frontal bone, parietal bone, occipital bone, temporal bone, sphenoid, and ethmoid. Prominent features are the **superior** and **inferior temporal lines**, the **squamous part of the temporal bone**, and the **external acoustic meatus**.
11. The facial bones visible from a lateral view are the zygomatic bone, mandible, maxilla, nasal bone, and lacrimal bone. Important features are the **zygomatic arch**, the **mandibular angle**, the **mental protuberance**, and **alveolar processes**.

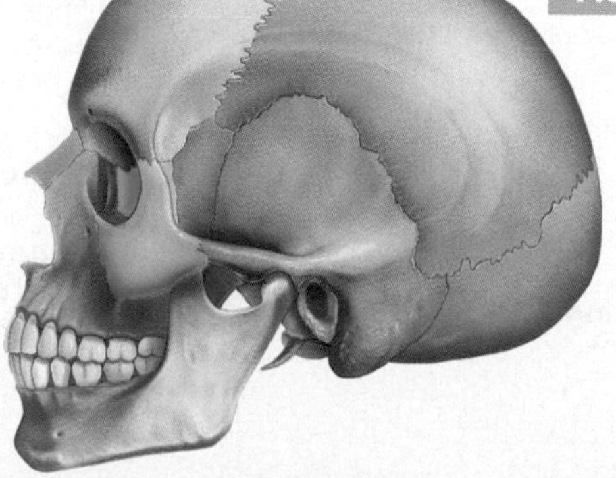

12. A sagittal section of a skull demonstrates the **frontal sinuses**, **petrous part of the temporal bone**, the **internal acoustic meatus**, and the **hypoglossal canal**.

7.5 The foramina on the inferior surface of the skull mark the passageways for nerves and blood vessels p. 240

13. Visible in an inferior view of a skull are passageways for blood vessels, nerves, and the spinal cord. Important foramina include the **foramen lacerum**, **foramen ovale**, **carotid canal**, **jugular foramen**, **stylomastoid foramen**, and the **foramen magnum**. Other landmarks include the **mandibular fossa**, **occipital condyles**, and the **inferior** and **superior nuchal lines**.
14. In horizontal section, many of the foramina are visible as they open into the cranial cavity. This section also demonstrates the crista galli, cribriform plate, sella turcica, and the **internal occipital crest**.

7.6 The shapes and markings of the sphenoid, ethmoid, and palatine bones are best seen in the isolated bones p. 242

15. The **sphenoid** forms part of the floor of the cranium, unites the cranial and facial bones, and acts as a cross-bridge that strengthens the sides of the skull.
16. The **ethmoid** forms part of the floor of the anterior cranium, the roof of the nasal cavity, a portion of the nasal septum and the medial orbital wall. It has three parts: the **cribriform plate**, the paired **lateral masses**, and the **perpendicular plate**. The **superior** and **middle nasal conchae** are projections of the lateral masses.
17. The **palatine bones** form the posterior portion of the hard palate, and contribute to the floor of each orbit.

7.7 Each orbital complex contains one eye, and the nasal complex encloses the nasal cavities p. 244

18. The orbits are bony recesses that contain the eyes and are formed by the seven bones of the **orbital complex**: frontal bone, zygomatic bone, maxilla, lacrimal, ethmoid, sphenoid, and palatine bone.
19. The **nasal complex** includes the bones of the nasal cavities, and the **paranasal sinuses**. Sphenoid, ethmoid, frontal bone, palatine bone, and maxillae contain the paranasal sinuses.

7.8 The mandible forms the lower jaw, and the associated bones of the skull perform specialized functions p. 246

20. The mandible can be subdivided into the **body** and the **rami**. Its most important features are the **coronoid process**, **alveolar process**, **condylar process**, **mandibular notch**, **mandibular foramen** and **mylohyoid line**.
21. The **hyoid bone** supports the larynx and is the attachment site for muscles. The **auditory ossicles** are enclosed in the petrous part of the temporal bone and are important in hearing.

7.9 Fontanelles permit cranial growth in infants and small children p. 248

22. At birth, cranial bones are connected by flexible fibrous connective tissue called **fontanelles**.
23. The important fontanelles are the **anterior fontanelle**, **sphenoidal fontanelle**, **mastoid fontanelle**, and **occipital fontanelle**.

7.10 The vertebral column has four spinal curves, and vertebrae share a basic structure that differs regionally p. 250

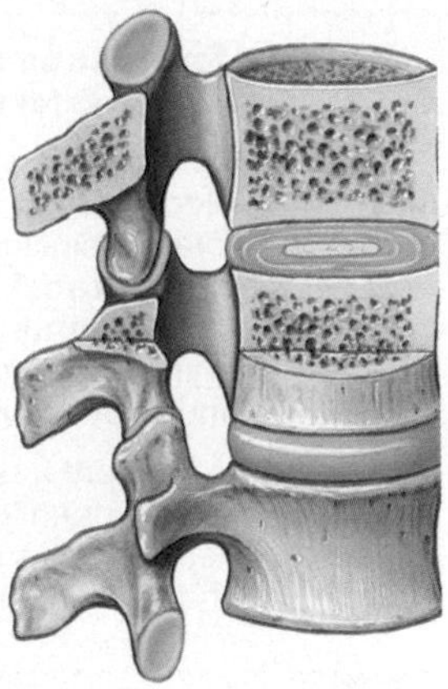

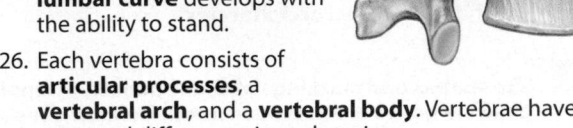

24. There are 26 bones in the adult vertebral column: the 24 vertebrae, the sacrum, and the coccyx.
25. The primary spinal curves develop before birth and remain in the adult as the **thoracic** and **sacral curves**. The secondary curves develop after birth. The **cervical curve** develops as the infant balances the weight of the head, and the **lumbar curve** develops with the ability to stand.
26. Each vertebra consists of **articular processes**, a **vertebral arch**, and a **vertebral body**. Vertebrae have structural differences in each region.

7.11 There are seven cervical vertebrae and twelve thoracic vertebrae p. 252

27. Regional characteristics of the seven cervical vertebrae are **transverse foramina**, and **bifid** spinous processes. Specialized cervical vertebrae are C_1 **(atlas)** and C_2 **(axis)** that allow for head movement. The **dens** or **odontoid process** of C_2 allows for head rotation.
28. Regional characteristics of the 12 thoracic vertebrae are **costal facets** for rib articulation, and inferiorly projecting **spinous processes**.

7.12 There are five lumbar vertebrae; the sacrum and coccyx consist of fused vertebrae p. 254

29. The five lumbar vertebrae are the largest vertebrae. They do not have costal facets, and have stumpier processes.
30. The sacrum is a single bone consisting of five fused sacral vertebrae. It begins fusing to the coccyx by age 26. Important features of the sacrum are the **base**, **promontory**, **foramina**, **apex**, **ala**, **canal**, **tuberosity**, **auricular surface**, **superior articular process**, **lateral** and **median crests**, and **hiatus**.

7.13 The thoracic cage protects organs in the chest and provides sites for muscle attachment p. 256

31. The **thoracic cage** consists of the thoracic vertebrae, the **ribs**, and the **sternum**.
32. Ribs 1–7 are called **vertebrosternal ribs** because they are connected to the sternum by **costal cartilages**. Ribs 8–10, or **vertebrochondral ribs**, connect to the sternum by shared costal cartilages. Ribs 11 and 12 are called **floating ribs** because they have no connection with the sternum.
33. The sternum has three distinct regions: the **manubrium**, **body**, and **xiphoid process**.
34. Ribs contain a **tubercle**, **capitulum**, **articular facets**, and a **costal groove**. Ribs assist with breathing by changing the volume of the thoracic cavity.

SECTION 2 • Appendicular Skeleton

7.14 The appendicular skeleton includes the limb bones and the pectoral and pelvic girdles p. 259

35. The **appendicular skeleton** contains 126 bones. It contains the pectoral girdle and upper limb, and the pelvic girdle and lower limb.

7.15 The pectoral girdles—the clavicles and the scapulae—connect the upper limbs to the axial skeleton p. 260

36. The **clavicles** originate on the superior, lateral border of the manubrium, and articulate laterally with the acromion of the **scapula**.
37. The important landmarks for muscle attachment to the scapula are the **borders**, **angles**, and **fossae**. A prominent spine on the posterior scapula ends laterally as the **acromion**. The **glenoid cavity** articulates with the head of the humerus to form the shoulder joint.

7.16 The humerus of the arm articulates with the radius and ulna of the forearm p. 262

38. The upper limbs consist of the bones of the arms, forearms, wrists, and hands.
39. The proximal **humerus** contains the **head**, **greater** and **lesser tubercles**, **intertubercular sulcus**, and the **anatomical** and **surgical necks**. The shaft has an elevation called the **deltoid tuberosity**, where the deltoid muscle attaches. The distal humerus has the **trochlea**, **capitulum**, **radial fossa**, and **coronoid fossa**.
40. The **ulna** and **radius** are the parallel bones of the forearm. The elbow is formed by the **olecranon** of the ulna. The **head** and **styloid process** are on the distal ulna. The **radial head** is on the proximal aspect of the bone and articulates with the capitulum of the humerus. It too has a **styloid process** on its distal aspect.

7.17 The wrist consists of carpal bones, and the hand consists of metacarpal bones and phalanges p. 264

41. Eight carpal bones in two rows of four form the wrist. The carpals of the proximal row are the **scaphoid**, **lunate**, **pisiform**, and **triquetrum**. The distal row of carpals is composed of the **trapezium**, **trapezoid**, **capitate**, and **hamate**.
42. Five **metacarpal bones** form the hand. Laterally to medially, they are identified by Roman numerals I–V.
43. Each hand has 14 finger bones or **phalanges**. Each finger contains a proximal, middle, and distal phalanx. The thumb or **pollex**, however, has only a proximal and distal phalanx.

7.18 The hip bone forms by the fusion of the ilium, ischium, and pubis p. 266

44. The **pelvic girdle** consists of the paired **hip** or **coxal bones**. Each hip bone is formed by the fusion of the **ilium**, **ischium**, and **pubis**.

45. In lateral view, the dominant features of a hip bone are the **iliac crest**, the **iliac** and **ischial spines**, **gluteal lines**, **ischial tuberosity**, and the socket or **acetabulum**. Visible from a medial view are the **iliac fossa**, **arcuate line**, **pectineal line**, **pubic tubercle**, **obturator foramen**, **ischial ramus**, and the **superior** and **inferior pubic rami**.

7.19 The pelvis consists of the two hip bones, the sacrum, and the coccyx p. 268

46. An extensive network of ligaments connects the sacrum with the posterior aspects of the two hip bones. The **pubic symphysis** is a pad of fibrocartilage that forms the anterior articulation between the two hip bones.
47. The pelvis can be divided into a **true** (lesser) and **false** (greater) **pelvis**.
48. Adaptations for childbearing cause the female pelvis to be shaped differently from a male pelvis.

7.20 The femur, tibia, and patella meet at the knee p. 270

49. Each lower limb consists of a **femur** (thigh), a **patella** (kneecap), a **tibia** and **fibula** (leg), and the **tarsal bones**, **metatarsal bones**, and **phalanges** of the foot.
50. The proximal femur has a **head**, **fovea capitis**, **neck**, **greater** and **lesser trochanters**, intertrochanteric crest, and **intertrochanteric line**. The distal femur contains **medial** and **lateral condyles**, medial and lateral epicondyles, a **patellar surface**, and an **intercondylar fossa**. The **linea aspera** runs the length of the posterior shaft.
51. The patella is a large sesamoid bone and is formed within the quadriceps femoris tendon.
52. The proximal tibia has **medial** and **lateral condyles** and an **intercondylar eminence**. The head of the fibula articulates with the proximal tibia. The projections of the ankle are formed by the **medial malleolus** of the tibia and the **lateral malleolus** of the fibula.

7.21 The ankle and foot consist of tarsal bones, metatarsal bones, and phalanges p. 272

53. The ankle consists of seven tarsal bones: the **calcaneus**, **talus**, **navicular**, **cuboid**, and the three **cuneiform bones**.
54. Five **metatarsals** form the foot. Medially to laterally, they are identified by Roman numerals I–V.
55. Each foot has 14 toe bones or **phalanges**. Each toe contains a proximal, middle, and distal phalanx. The great toe or **hallux**, however, has only a proximal and distal phalanx.
56. Arches assist with transferring the weight of the body to the feet. The **longitudinal arch** runs from the calcaneus to the distal metatarsals. The **transverse arch** runs perpendicular to the longitudinal arch.

Chapter Review Questions

Labeling

Identify the cranial and facial bones in the diagrams below.

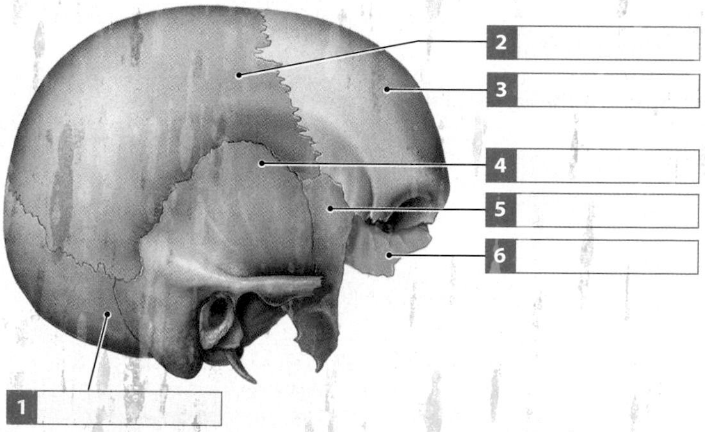

1
2
3
4
5
6

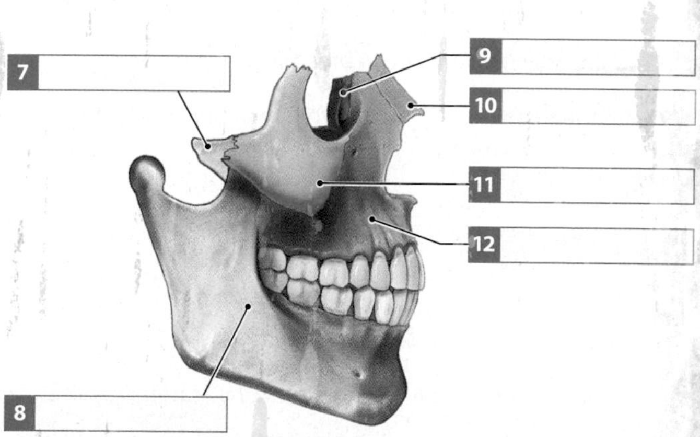

7
8
9
10
11
12

True/False

Indicate whether each statement is true or false.

13 The patella is a sesamoid bone.

14 The adductor tubercle is a small pit in the center of the femoral head.

15 The lateral border of the scapula is also called the vertebral border.

16 The last two pairs of ribs are also called floating ribs.

13 ____________________

14 ____________________

15 ____________________

16 ____________________

Multiple choice

Select the correct answer from the list provided.

17 When you move your head as if to say no,
- a) the atlas rotates on the occipital condyles.
- b) C_1 and C_2 rotate on the adjoining cervical vertebrae.
- c) the atlas rotates on the dens of the axis.
- d) the skull rotates with both C_1 and C_2.

18 Which of the following pairs of bones make up the bony nasal septum?
- a) inferior nasal conchae and vomer
- b) perpendicular plate of ethmoid and inferior nasal conchae
- c) vomer and perpendicular plate of ethmoid
- d) inferior nasal conchae and middle nasal conchae of ethmoid

19 The unpaired facial bones include the
- a) lacrimal and nasal.
- b) vomer and mandible.
- c) maxilla and mandible.
- d) zygomatic and palatine.

20 The joint between the frontal and parietal bones is correctly called the
- a) parietal suture.
- b) lambdoid suture.
- c) squamous suture.
- d) coronal suture.

21 Which part of the ulna forms the point of the elbow?
- a) styloid process
- b) olecranon
- c) coronoid process
- d) trochlear notch

22 Which of the following bones is *not* part of the orbital complex?
- a) lacrimal
- b) nasal
- c) ethmoid
- d) sphenoid

23 The head of the femur articulates with the pelvis at the
- a) true pelvis.
- b) acetabulum.
- c) obturator foramen.
- d) pubic symphysis.

24 What is the name of the flexible sheet that interconnects the radius and ulna (and the tibia and fibula)?
- a) interosseous membrane
- b) obturator foramen
- c) linea aspera
- d) intercondylar eminence

Short answer

25 What purpose do the fontanelles serve during birth?

26 Describe how ribs function in breathing.

27 Name the three bones that fuse to form a hip bone. Where do these bones meet?

28 Distinguish between the primary and secondary curves of the vertebral column.

29 Describe the arches of the feet.

Chapter Integration • Applying what you have learned

An illegal check from behind leaves a college hockey player with multiple injuries

Donny plays right wing on his college hockey team. He grew up in a hockey family in Peterborough, Canada. His dad played Juniors and had hopes of an NHL career, until marriage and children caused him to give up his dream. But he put skates on Donny as soon as Donny could walk, and taught him the game. After high school, Donny followed in his father's footsteps by choosing to play college hockey. He too hopes to have a career in professional hockey.

During a collegiate hockey game against his archrivals, Donny was handling the puck well and out-skating the defense, when he was checked from behind. Checking is a maneuver in which an opponent is neutralized through the proper use of one's body or stick; checking from behind is illegal. While lying on the ice being examined by the team's athletic trainer, Donny was experiencing chest and shoulder pain. His teammates helped him to the locker room, where a thorough physical exam revealed a badly bruised chest; dyspnea (difficulty breathing); and immobility of his left arm, especially in the anterior direction. X-rays subsequently taken at a nearby hospital revealed two fractured ribs.

1 **Why might Donny be experiencing difficulty breathing?**

2 **What is the probable reason for Donny's inability to move his left arm?**

3 **State whether Donny's injuries are related to the axial skeleton or to the appendicular skeleton.**

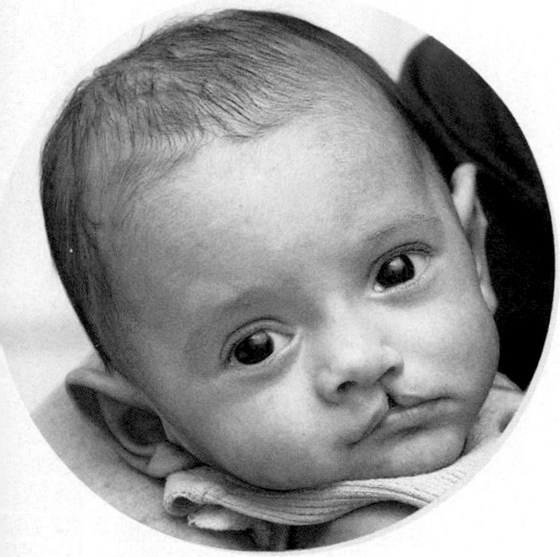

Examining a baby's cleft palate

There are occasions when certain sutures of the skull do not form properly in fetal development. An example of this is cleft palate, in which a permanent gap remains between the bones of the hard palate. From what you have just learned about the bones of the skull, answer the following questions.

4 **Which bones do not properly form into a suture in cleft palate? How many bones are involved?**

5 **If you were to look into the mouth of a baby with cleft palate, what would you expect to see?**

6 **What kinds of difficulties do you think a baby with cleft palate might have?**

8 Joints

LEARNING OUTCOMES

These Learning Outcomes correspond by number to this chapter's modules and indicate what you should be able to do after completing the chapter.

SECTION 1 • Joint Structure and Movement

8.1 Name and describe the three types of joints as classified by structure and movement.

8.2 Describe the basic structure of a synovial joint, and describe common accessory structures and their functions.

8.3 Explain the relationship between structure and function for each type of synovial joint.

8.4 Describe flexion/extension, abduction/adduction, and circumduction movements of the skeleton.

8.5 Describe rotational and special movements of the skeleton.

SECTION 2 • Axial and Appendicular Joints

8.6 Compare the general relationship between joint stability and range of motion for axial and appendicular joints.

8.7 Describe the joints between the vertebrae of the vertebral column.

8.8 Describe the structure and function of the shoulder and hip joints.

8.9 Describe the structure and function of the elbow and knee joints.

8.10 **CLINICAL MODULE** Explain arthritis, and describe its effects on joint structure and function.

Learning Outcomes are repeated at the bottom of each module.

Joints are classified according to structure and movement

The bones of the skeleton are relatively inflexible, so movements can occur only at **joints**, or **articulations**, where two bones meet.

Functional and Structural Classifications of Joints

Functional Category	Structural Category and Type	Description
Synarthrosis (no movement) At a synarthrosis (*syn-*, together + *arthrosis*, articulation), the bony edges are close together and may interlock. These extremely strong joints are located where movement between the bones must be prevented.	**Fibrous** **Suture** 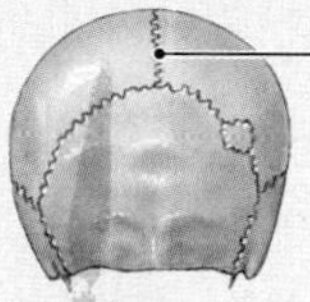	A **suture** (*sutura*, a sewing together) is a synarthrotic joint located only between the bones of the skull. The edges of the bones are interlocked and bound together at the suture by dense fibrous connective tissue.
	Gomphosis 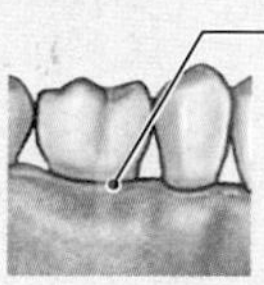	A **gomphosis** (gom-FŌ-sis; *gomphos*, bolt) is a synarthrosis that binds the teeth to bony sockets in the maxillae and mandible. The fibrous connection between a tooth and its socket is a periodontal (per-ē-ō-DON-tal) ligament (*peri*, around + *odontos*, tooth).
	Cartilaginous **Synchondrosis** 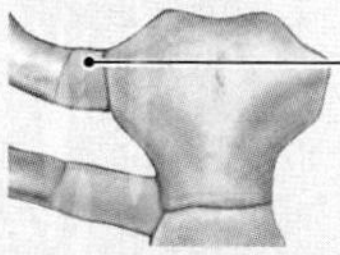	A **synchondrosis** (sin-kon-DRŌ-sis; *syn*, together + *chondros*, cartilage) is a rigid, cartilaginous bridge between two articulating bones. The cartilaginous connection between the ends of the first pair of ribs and the sternum is a synchondrosis.
	Bony **Synostosis** 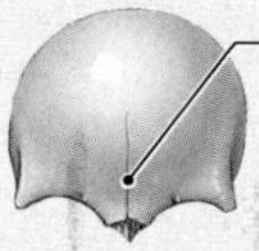	A **synostosis** (sin-os-TŌ-sis) is a totally rigid, immovable joint formed when two bones fuse and the boundary between them disappears. The frontal suture of the frontal bone and the epiphyseal lines of mature long bones are synostoses.
Amphiarthrosis (little movement) An amphiarthrosis (am-fē-ar-THRŌ-sis; *amphi-*, on both sides) permits more movement than a synarthrosis, but is much stronger than a freely movable joint. The articulating bones are connected by collagen fibers or cartilage.	**Fibrous** **Syndesmosis** 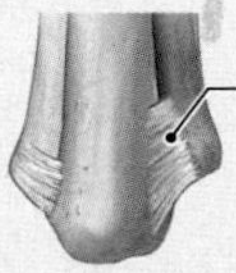	At a **syndesmosis** (sin-dez-MŌ-sis; *desmos*, a band or ligament), bones are connected by a ligament. One example is the distal joint between the tibia and fibula.
	Cartilaginous **Symphysis** 	At a **symphysis**, the articulating bones are separated by a pad of fibrocartilage. The joint between the two pubic bones (the pubic symphysis) is an example of a symphysis.
Diarthrosis (free movement) A diarthrosis (dī-ar-THRŌ-sis; *dia-*, through) permits the widest range of movement.	**Synovial** 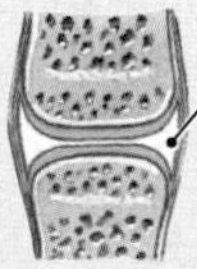	**Diarthroses**, or **synovial** (si-NŌ-vē-ul) **joints**, permit a wider range of motion than do other types of joints. They are typically located at the ends of long bones, such as those of the upper and lower limbs.

1 The anatomical structure of a joint determines the type and amount of movement the joint can execute. Each joint reflects a compromise between the need for strength and the need for mobility. The amount of movement at a joint is known as **range of motion (ROM)**. Joints are often categorized by their ROM, and are further divided into subgroups based on anatomical structure.

Module 8.1 Review

a. Define range of motion (ROM).

b. Distinguish between a synarthrosis and an amphiarthrosis.

c. Which structural category of joints allows for the greatest range of motion?

LO **8.1** Name and describe the three types of joints as classified by structure and movement.

Module 8.2

Synovial joints are freely movable diarthroses lined with a synovial membrane

In this module we review the general structure of a synovial joint, which was introduced in our earlier discussion of synovial membranes (Module 4.14, p. 161). We will focus on the key structures of synovial joints.

1 Under normal conditions, the opposing bony surfaces within a synovial joint cannot contact one another, because these surfaces are covered by special articular cartilages, which are slick and smooth. The articular cartilages alone can reduce friction during movement at the joint. However, even when pressure is applied across a joint, the smooth articular cartilages do not touch one another, because they are separated by a thin film of synovial fluid within the joint cavity.

Synovial Joint Components
Articular cartilage covers bones at a joint. It is like hyaline cartilage, but it has no perichondrium and its matrix contains more water than other cartilages.
The **joint capsule**, or **articular capsule**, is a sac that encloses the articulating ends of the bones in the joint. It is made up of an outer fibrous layer and an inner synovial membrane. It may be reinforced with accessory structures such as tendons or ligaments.
The **synovial membrane** lines the interior of the joint capsule and secretes synovial fluid into the joint cavity. This fluid lubricates, cushions shocks, prevents abrasion, and supports the chondrocytes of the articular cartilages. Even in a large joint such as the knee, the total quantity of synovial fluid in a joint is normally less than 3 mL.

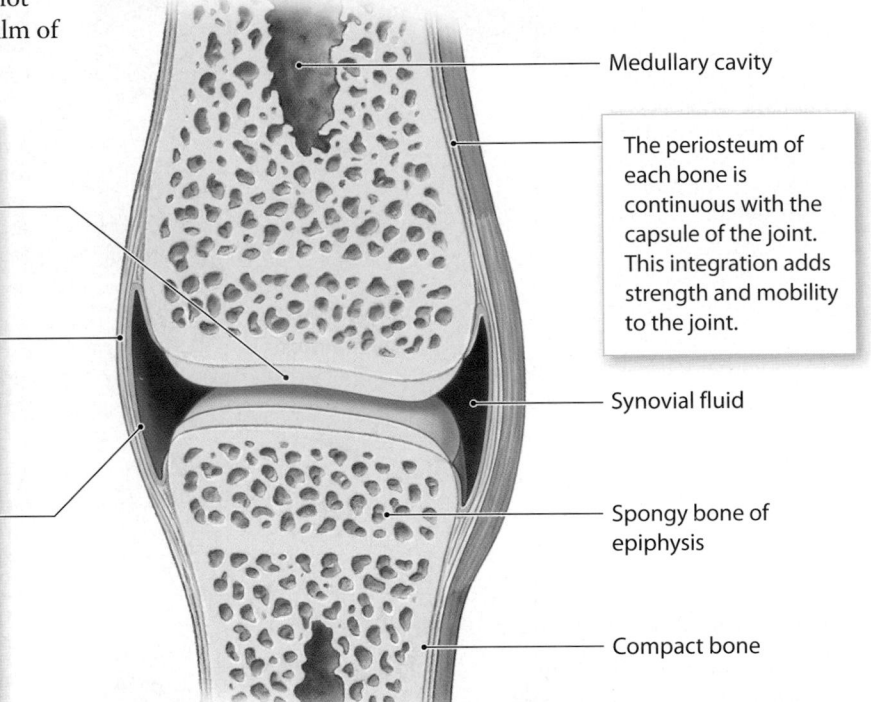

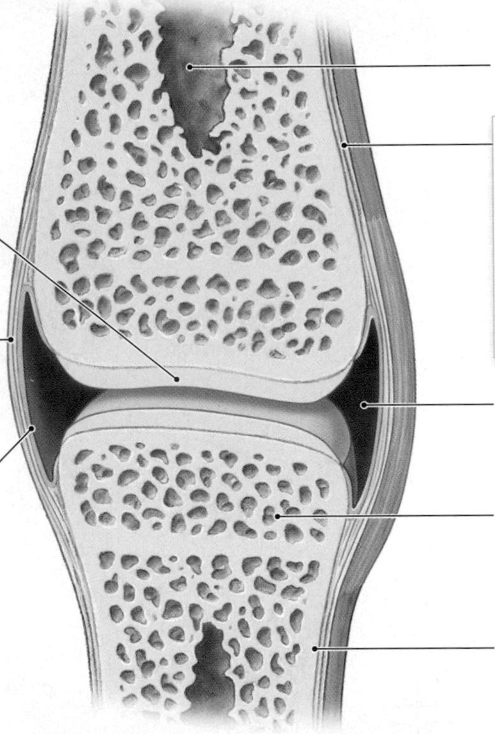
Medullary cavity
The periosteum of each bone is continuous with the capsule of the joint. This integration adds strength and mobility to the joint.
Synovial fluid
Spongy bone of epiphysis
Compact bone

2 This table details the major functions of synovial fluid. It is a clear, straw-colored, viscous fluid with the consistency of raw egg white. Its normal viscosity is due to a high concentration of hyaluronan (hyaluronic acid) (Module 4.4, p. 141). Synovial fluid is produced by the synovial membrane that lines the joint cavity. During normal movement, synovial fluid circulates from the areolar tissue into the joint cavity and percolates through the articular cartilage, providing oxygen and nutrients to the chondrocytes and carrying away their metabolic wastes.

Synovial Fluid Functions

- **Lubrication**. When part of an articular cartilage is compressed during movement, some of the synovial fluid squeezes out of the cartilage and into the space between the opposing surfaces. This thin layer of fluid markedly reduces friction between moving surfaces, just as a thin film of water reduces friction between a car's tires and a highway.
- **Nutrient Distribution**. Synovial fluid in a joint must circulate continuously to provide nutrients and dispose of waste for the chondrocytes in the articular cartilage. It circulates whenever the joint moves, and compression and reexpansion of the articular cartilage pump synovial fluid into and out of the cartilage matrix.
- **Shock Absorption**. Synovial fluid cushions shocks in joints that are subjected to sudden compression or impact. This cushioning effect occurs because the viscosity of synovial fluid increases with increasing pressure. For example, when you jog, your knees are severely compressed, and the viscosity of the synovial fluid increases. As the pressure lessens, viscosity decreases and it is again a lubricant.

3 In complex synovial joints, such as the knee, a variety of accessory structures provide support and additional stability. Several of these structures are shown in this diagram of a sagittal section of the knee.

The tendon of the quadriceps muscles attaches to the base of the patella. Although not part of the joint itself, tendons passing across or around a joint can limit the joint's range of motion and provide mechanical support for it.

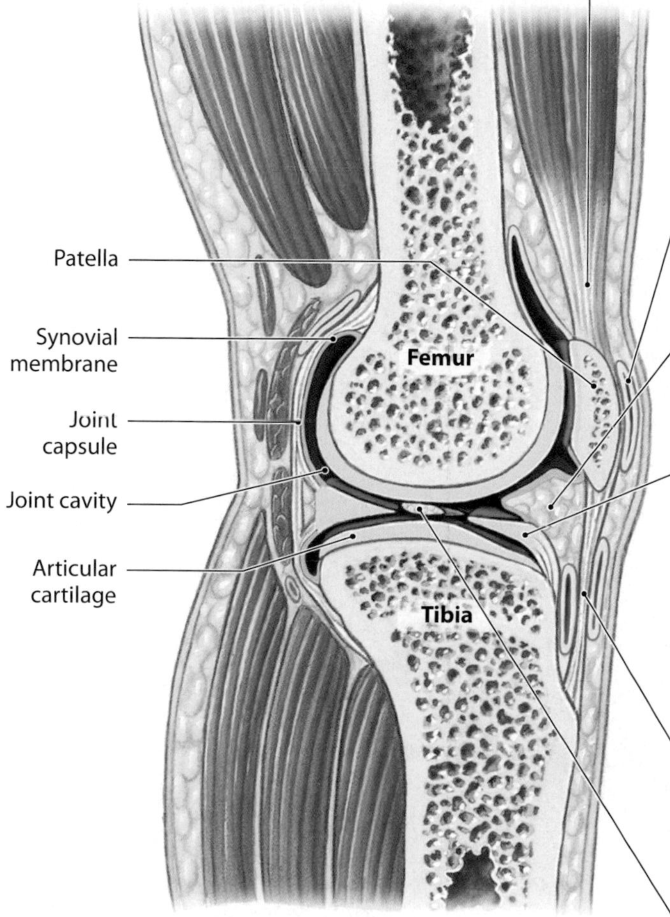

Patella
Synovial membrane
Joint capsule
Joint cavity
Articular cartilage
Femur
Tibia

Accessory Structures Supporting the Knee

A **bursa** (BUR-sa; a pouch; plural, *bursae*) is a small, thin, fluid-filled pocket that forms in connective tissue outside of a joint capsule. It contains synovial fluid and is lined by a synovial membrane. Bursae often form where a tendon or ligament rubs against other tissues. Located around most synovial joints, including the knee joint, bursae reduce friction and act as shock absorbers.

Fat pads are localized masses of adipose tissue covered by a layer of synovial membrane. They are commonly superficial to the joint capsule. Fat pads protect the articular cartilage and act as packing material for the joint. When the bones move, the fat pads fill in the spaces created as the joint cavity changes shape.

A **meniscus** (me-NIS-kus; a crescent; plural, *menisci*) is a pad of fibrocartilage between opposing bones within a synovial joint. Menisci, or articular discs, may subdivide a synovial cavity, channel the flow of synovial fluid, or allow for variations in the shapes of the articular surfaces.

Accessory ligaments support, strengthen, and reinforce synovial joints. **Capsular ligaments**, or intrinsic ligaments, are localized thickenings of the joint capsule. **Extrinsic ligaments** are separate from the joint capsule. Extrinsic ligaments may pass outside or inside the joint capsule, and are called extracapsular or intracapsular ligaments, respectively.

The **patellar ligament** extends from the apex of the patella to the tibial tuberosity. This is an example of an **extracapsular ligament**.

The **cruciate ligaments** that run through the interior of the knee joint are examples of **intracapsular ligaments**.

A joint cannot be both highly mobile and very strong. The greater the range of motion at a joint, the weaker it becomes. A synarthrosis, the strongest type of joint, has no movement, whereas a diarthrosis, such as the shoulder, is far weaker but has a broad range of motion. Any mobile diarthrosis will be damaged by movement beyond its normal range of motion. When reinforcing structures cannot protect a joint from extreme stresses, a **dislocation**, or **luxation** (luk-SĀ-shun), results. In a dislocation, the articulating surfaces are forced out of position. The displacement can damage the articular cartilage, tear ligaments, or distort the joint capsule. Although the inside of a joint has no pain receptors, sensitive nerves monitor the capsule, ligaments, and tendons, so dislocations are very painful.

Module 8.2 Review

a. Describe the components of a synovial joint, and identify the functions of each.

b. Why would improper circulation of synovial fluid lead to the degeneration of articular cartilages in the affected joint?

c. Define a joint dislocation (luxation).

LO **8.2** Describe the basic structure of a synovial joint, and describe common accessory structures and their functions.

Anatomical organization determines the motion at synovial joints

An accurate description of the functions of a joint includes terms that indicate the types of motion permitted. To show the general types of movement, we will use a simple model that you can try for yourself.

1 Take a pencil and stand it upright on the surface of a desk. The pencil represents a bone, and the desktop represents an articular surface. A little imagination and a lot of twisting, pushing, and pulling will demonstrate that there are only three ways to move the pencil. Considering them one at a time gives a frame of reference for us to analyze complex movements.

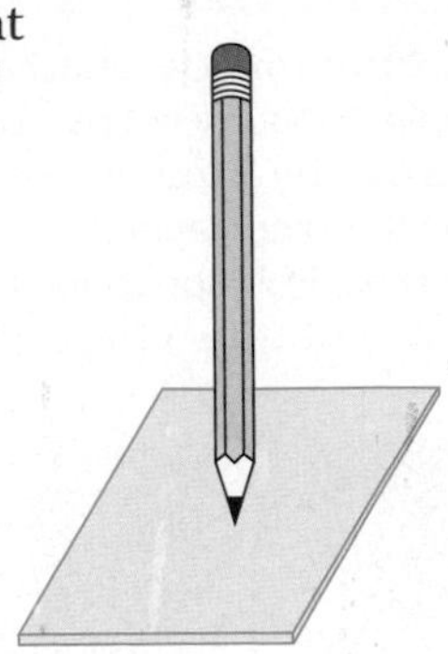

2 If you hold the pencil upright, without securing the point, you can push the pencil point across the surface. This kind of motion, **gliding**, is an example of linear motion. You could slide the point forward or backward, from side to side, or diagonally. However you move the pencil, the motion can be described by using two lines of reference (axes). One line represents forward–backward motion, the other left–right movement.

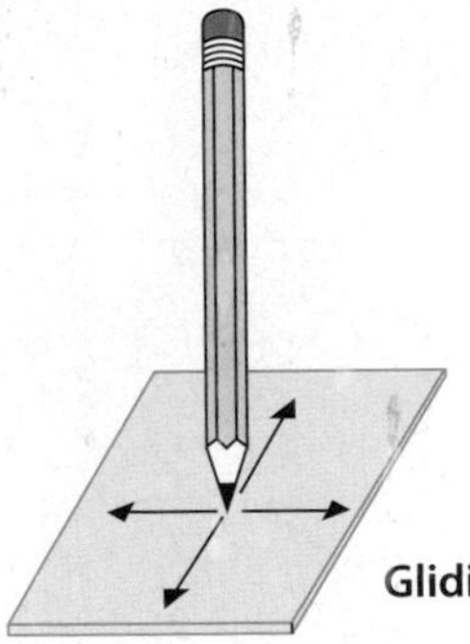

Gliding

3 If you hold the point in position, you can move the free (eraser) end of the pencil forward and backward, from side to side, or at some intermediate angle. These movements, which change the angle between the pencil shaft and the desktop, are examples of **angular motion**.

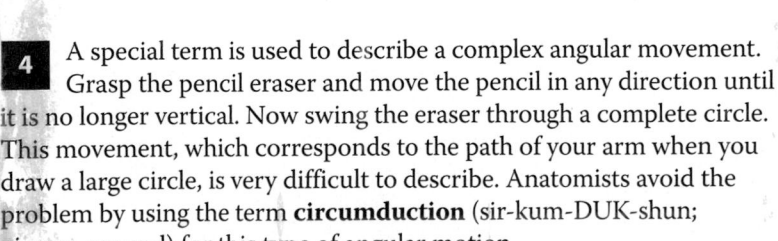

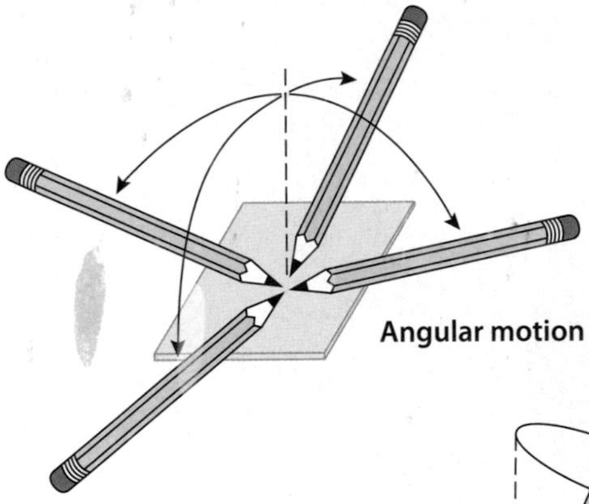

Angular motion

4 A special term is used to describe a complex angular movement. Grasp the pencil eraser and move the pencil in any direction until it is no longer vertical. Now swing the eraser through a complete circle. This movement, which corresponds to the path of your arm when you draw a large circle, is very difficult to describe. Anatomists avoid the problem by using the term **circumduction** (sir-kum-DUK-shun; *circum*, around) for this type of angular motion.

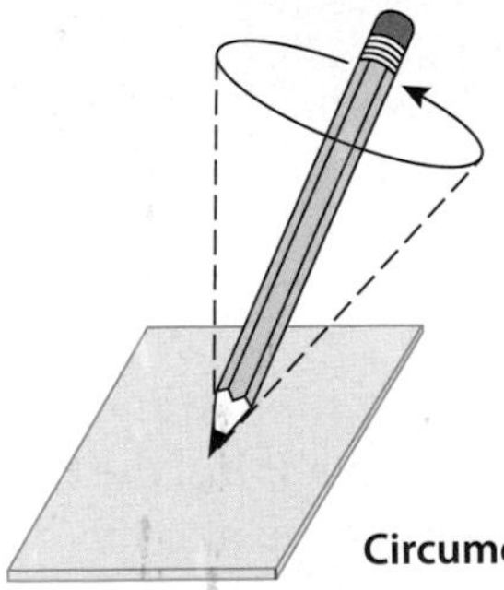

Circumduction

5 If you keep the shaft vertical and the point at one location, you can still spin the pencil around its longitudinal axis. This movement is called **rotation**. Several joints permit partial rotation, but none can rotate freely without limit because such a movement would hopelessly tangle the blood vessels, nerves, and muscles that cross the joint.

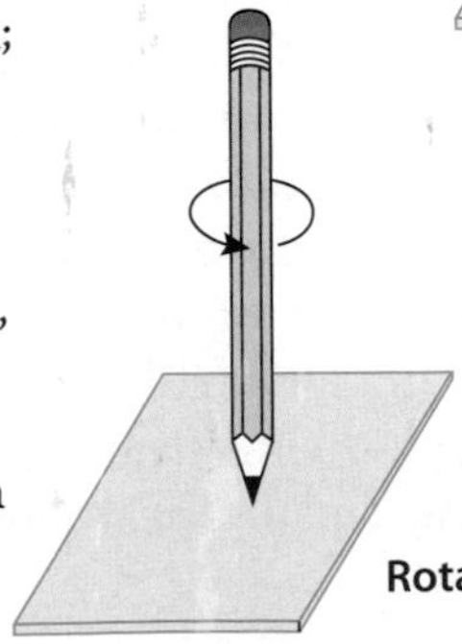

Rotation

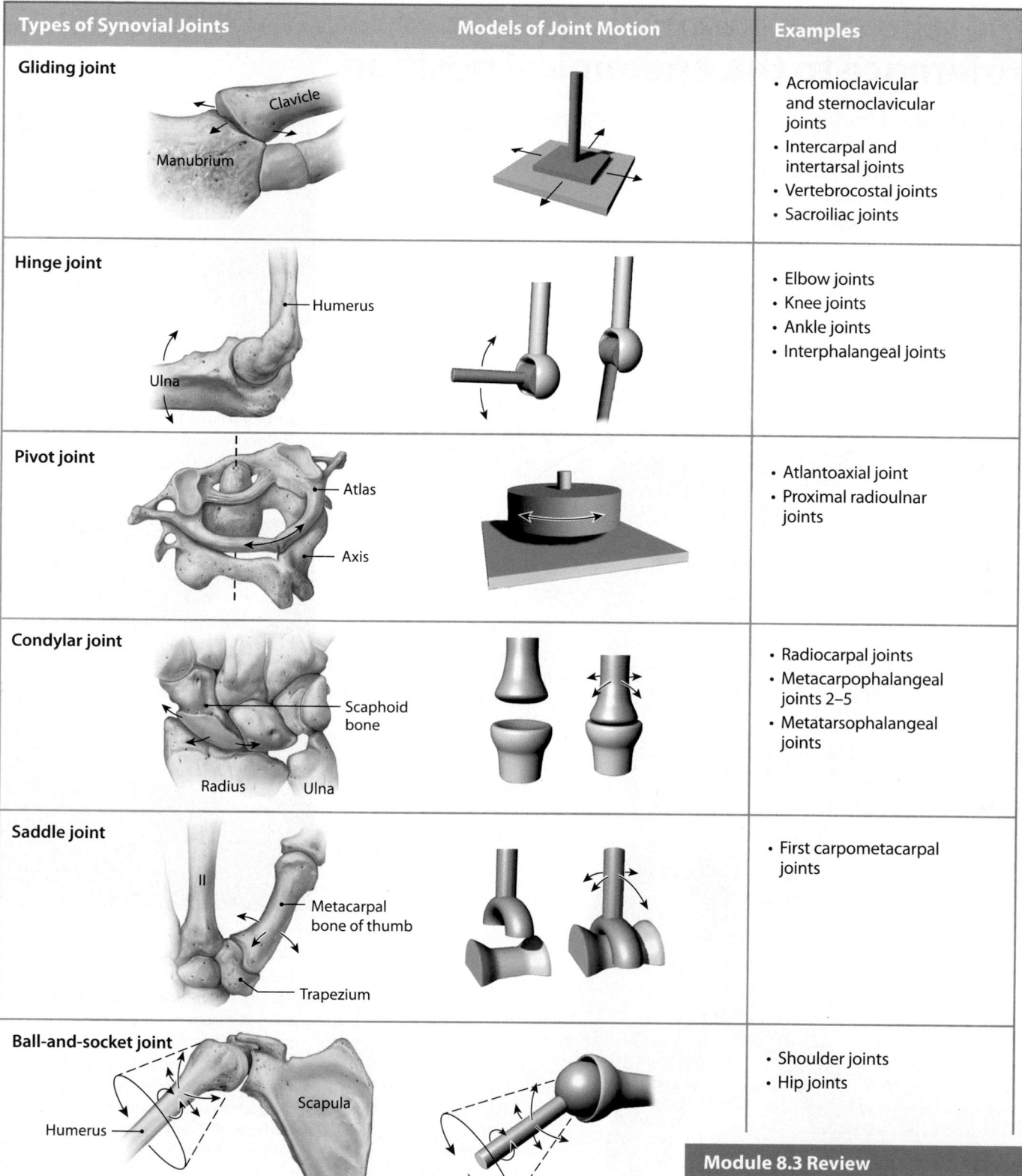

Types of Synovial Joints	Models of Joint Motion	Examples
Gliding joint 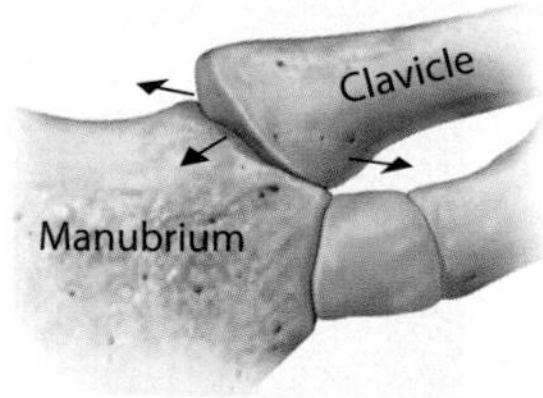 Clavicle Manubrium	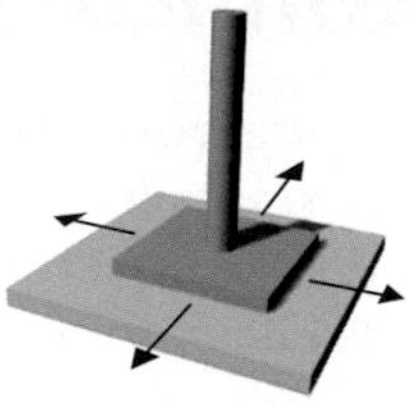	• Acromioclavicular and sternoclavicular joints • Intercarpal and intertarsal joints • Vertebrocostal joints • Sacroiliac joints
Hinge joint 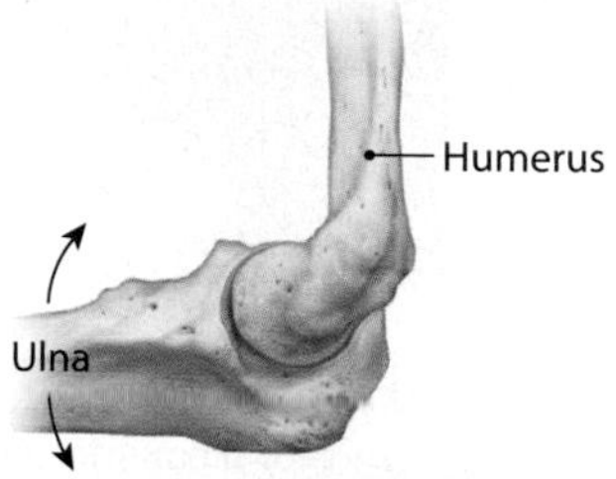 Humerus Ulna	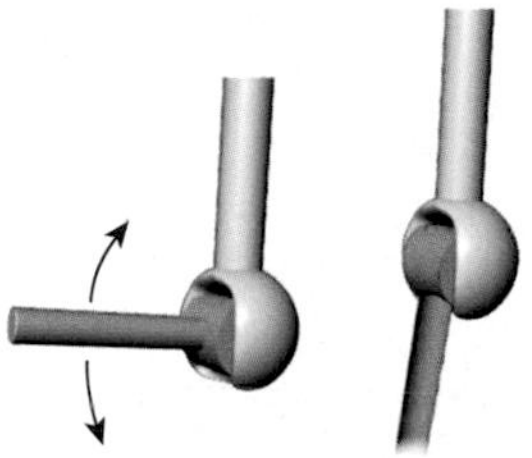	• Elbow joints • Knee joints • Ankle joints • Interphalangeal joints
Pivot joint 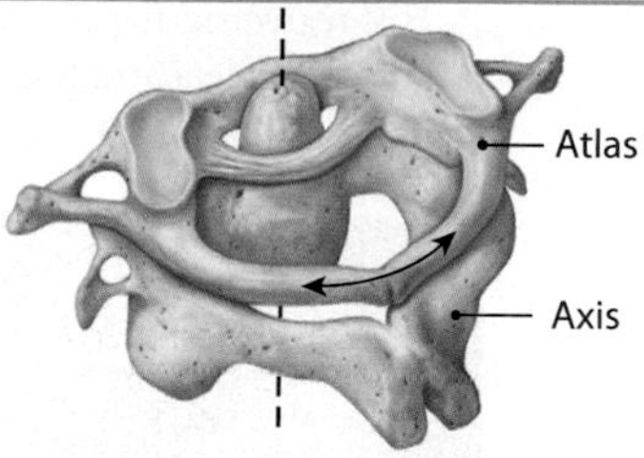 Atlas Axis	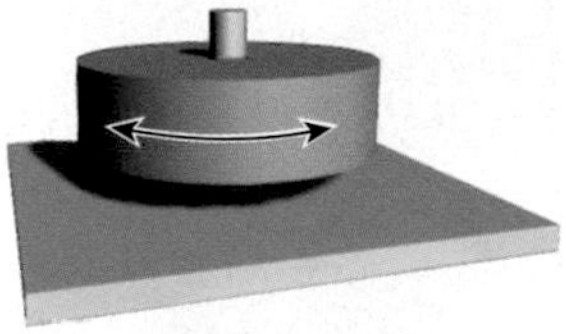	• Atlantoaxial joint • Proximal radioulnar joints
Condylar joint 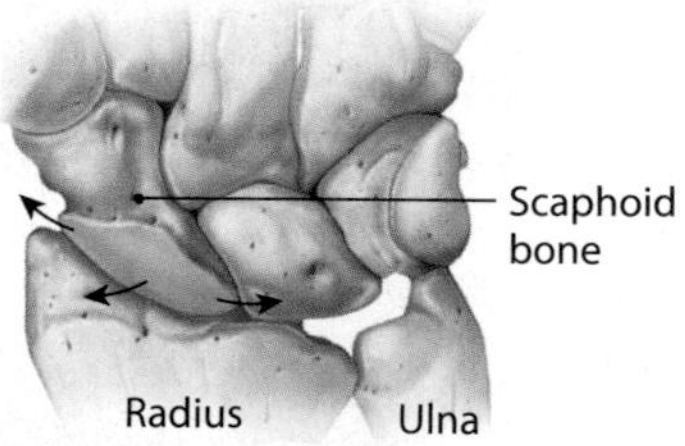 Scaphoid bone Radius Ulna	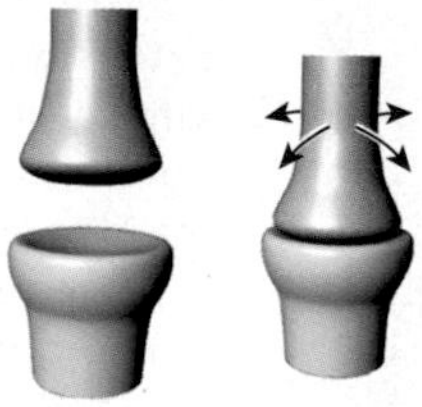	• Radiocarpal joints • Metacarpophalangeal joints 2–5 • Metatarsophalangeal joints
Saddle joint 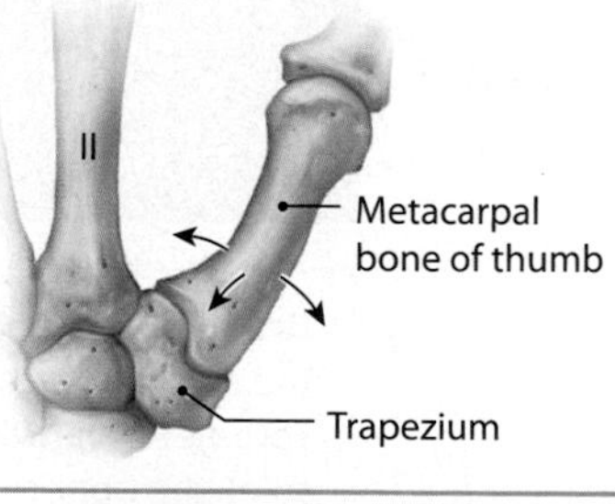 II Metacarpal bone of thumb Trapezium	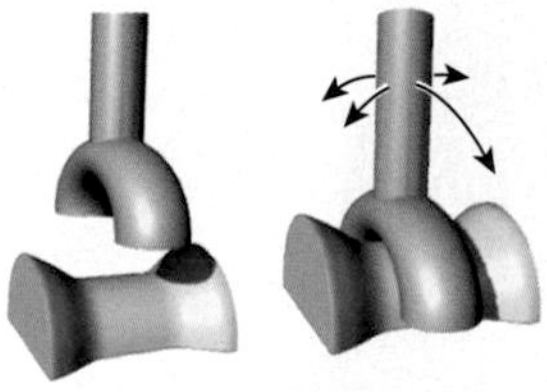	• First carpometacarpal joints
Ball-and-socket joint 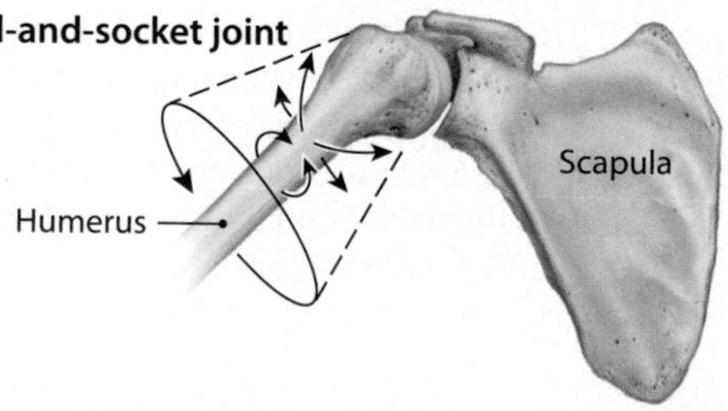 Humerus Scapula	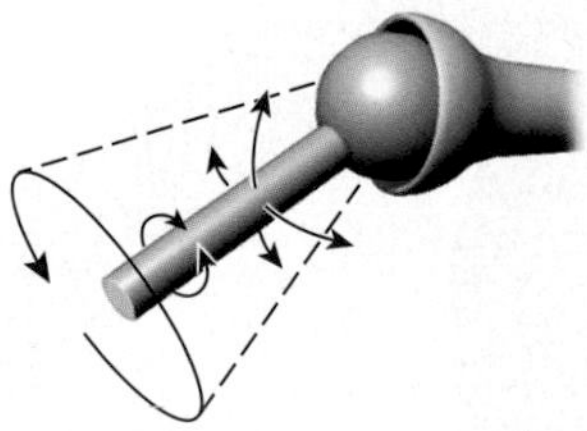	• Shoulder joints • Hip joints

6 This visual summary gives representative examples of the various anatomical classes of synovial joints based on the shapes of the articulating surfaces. It relates articular structure to simplified joint models and lists examples of each group of joints.

Module 8.3 Review

a. Identify the types of synovial joints based on the shapes of the articulating surfaces.

b. What type of synovial joint permits the greatest range of motion?

c. Indicate the type of synovial joint for each of the following: shoulder, elbow, ankle, and thumb.

LO 8.3 Explain the relationship between structure and function for each type of synovial joint.

Module 8.4

Specific terms are used to describe movements with reference to the anatomical position

It is easiest to understand the terms used to describe body movements when you see the actions under way. You should become very familiar with the descriptive terms presented here and in Module 8.5, because we will use them when we consider both the functions of specific joints and the actions of skeletal muscles.

Flexion and Extension

1 Flexion and extension describe the movements at the hinge joints of the long bones of the limbs, but they are also used to describe movements of the axial skeleton. For example, when you bring your head toward your chest, you flex the intervertebral joints of the neck.

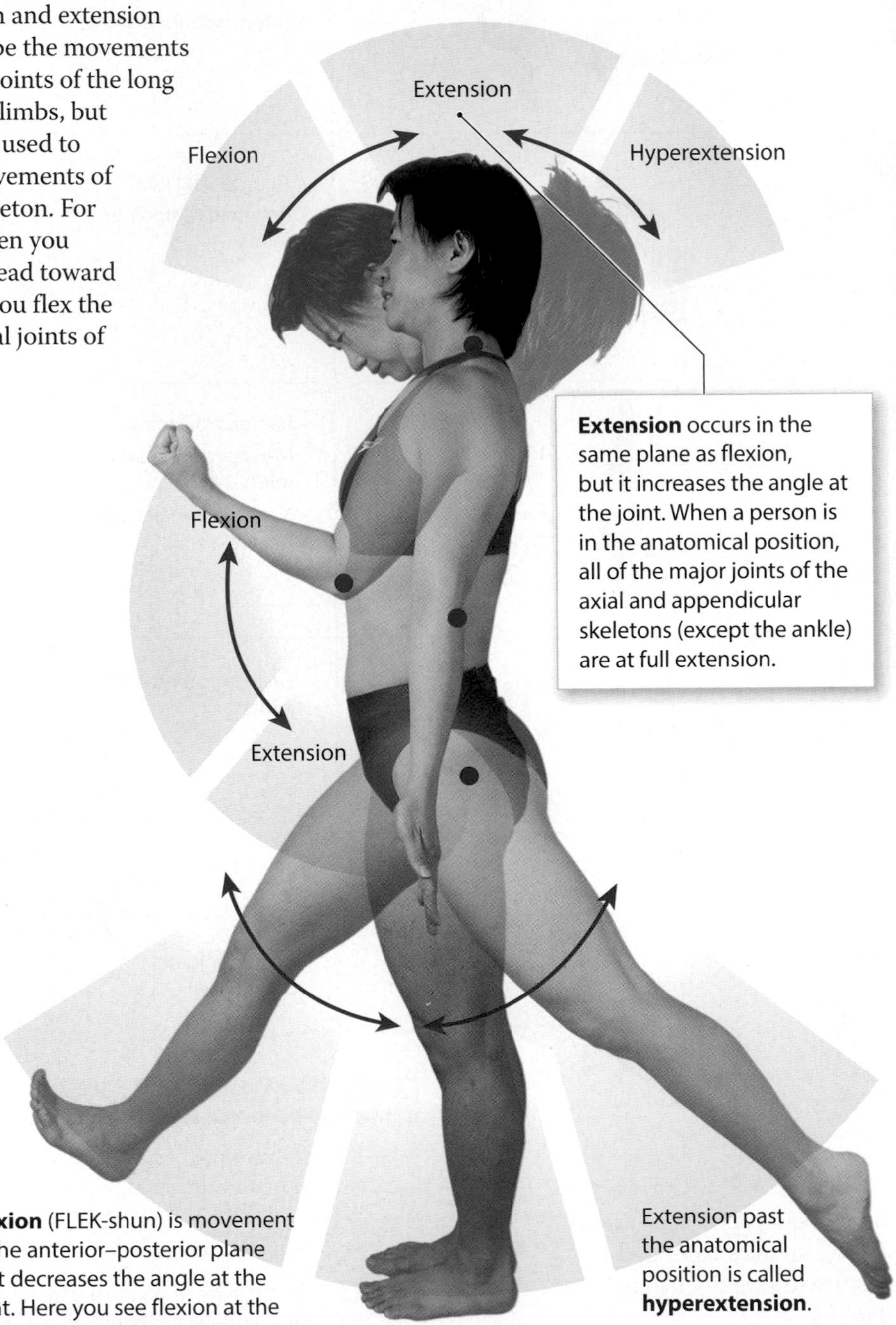

Extension
Flexion
Hyperextension
Flexion
Extension

Extension occurs in the same plane as flexion, but it increases the angle at the joint. When a person is in the anatomical position, all of the major joints of the axial and appendicular skeletons (except the ankle) are at full extension.

Flexion (FLEK-shun) is movement in the anterior–posterior plane that decreases the angle at the joint. Here you see flexion at the neck, the elbow, and the hip.

Extension past the anatomical position is called **hyperextension**.

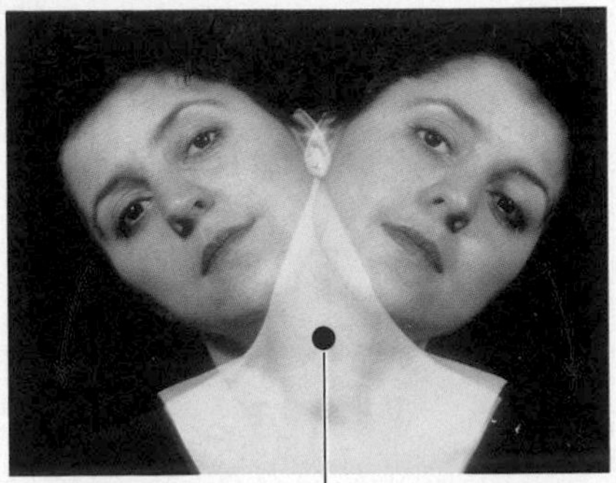

Lateral flexion occurs when your vertebral column bends to the side. This movement is most pronounced in the cervical and thoracic regions.

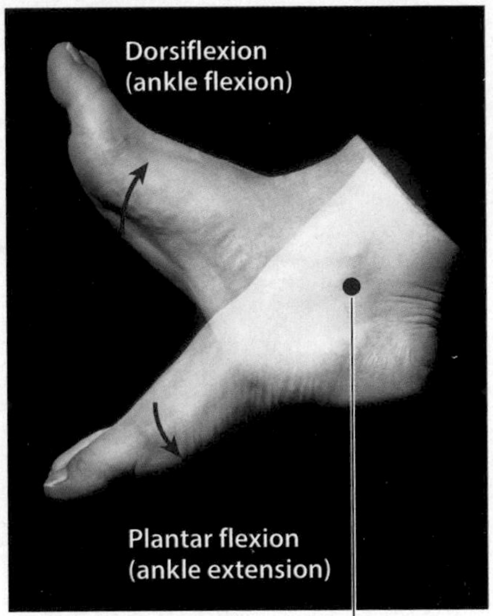

Dorsiflexion (ankle flexion)
Plantar flexion (ankle extension)

Dorsiflexion is upward movement of the foot or toes, as when you dig in your heel. **Plantar flexion** (*planta*, sole), the opposite movement, extends the ankle joint and bends the foot or toes, as when you stand on tiptoe. It is also acceptable (and simpler) to use "flexion and extension at the ankle," rather than "dorsiflexion and plantar flexion."

Spreading the fingers or toes apart abducts them, because they move away from a central digit. Bringing them together constitutes adduction. (Fingers move toward or away from the middle finger; toes move toward or away from the second toe.)

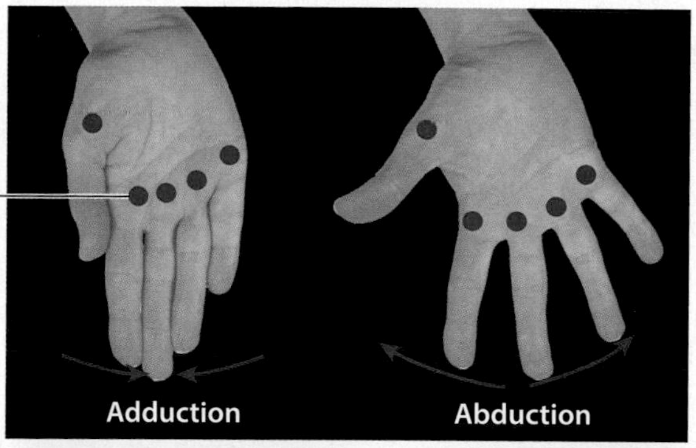

Adduction
Abduction

Abduction and Adduction

2 Abduction and adduction always refer to the movements of the appendicular skeleton, not to those of the axial skeleton.

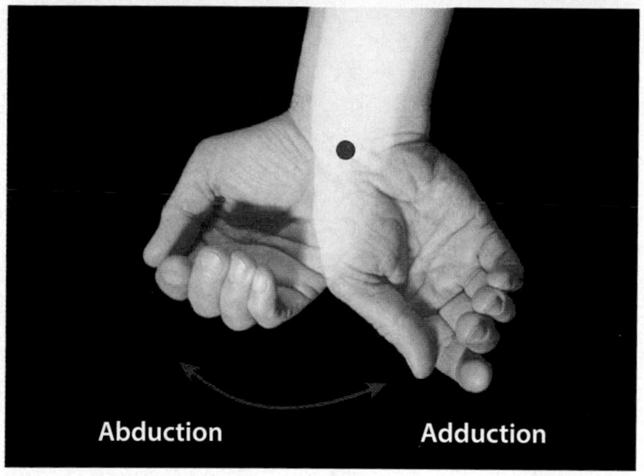

Abduction
Adduction

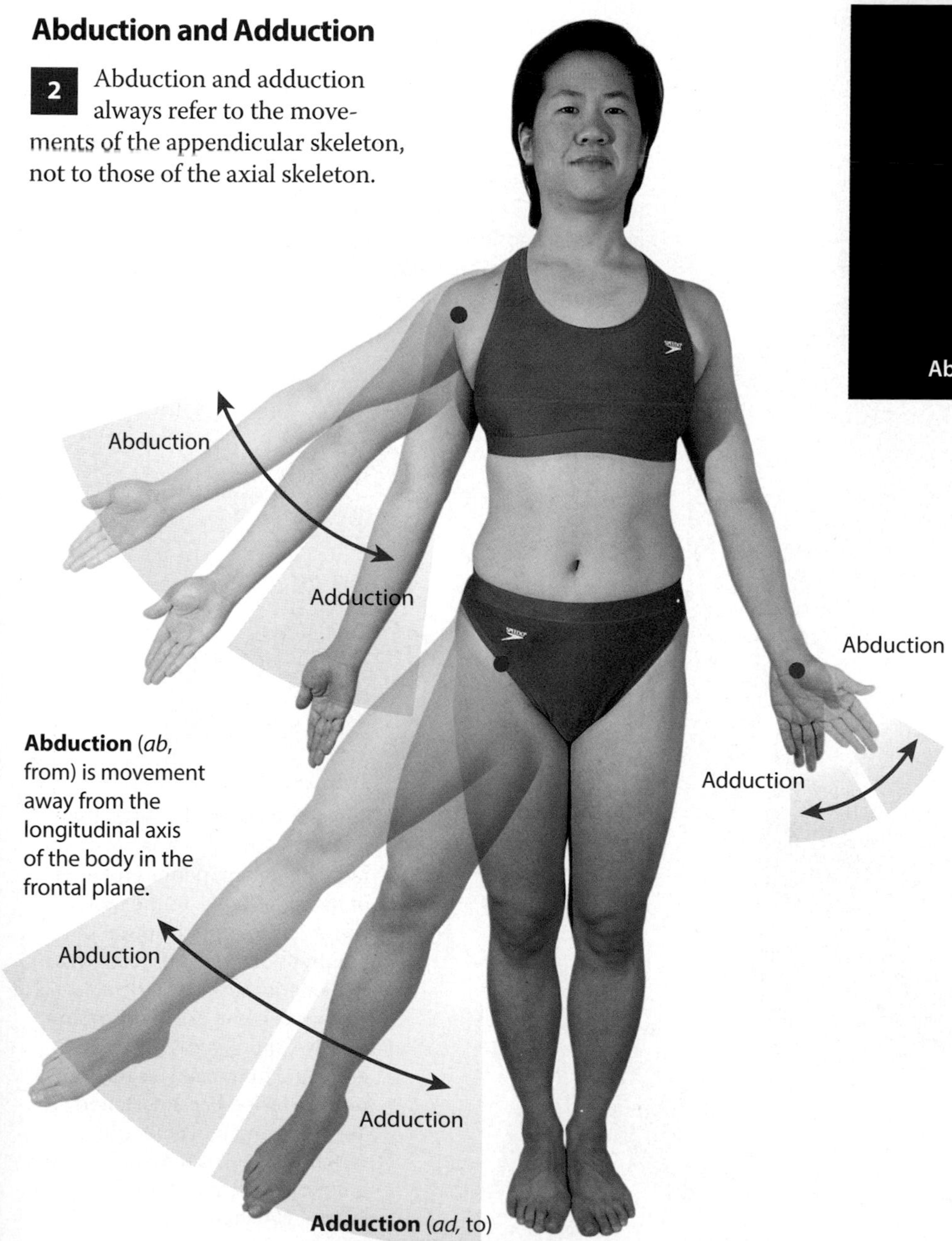

Abduction
Adduction
Abduction
Adduction
Abduction
Adduction

Abduction (*ab*, from) is movement away from the longitudinal axis of the body in the frontal plane.

Adduction (*ad*, to) is movement toward the longitudinal axis of the body in the frontal plane.

Circumduction

3 Moving your arm as if to draw a big circle on the wall is **circumduction**. In this movement your hand moves in a circle, but your arm does not rotate.

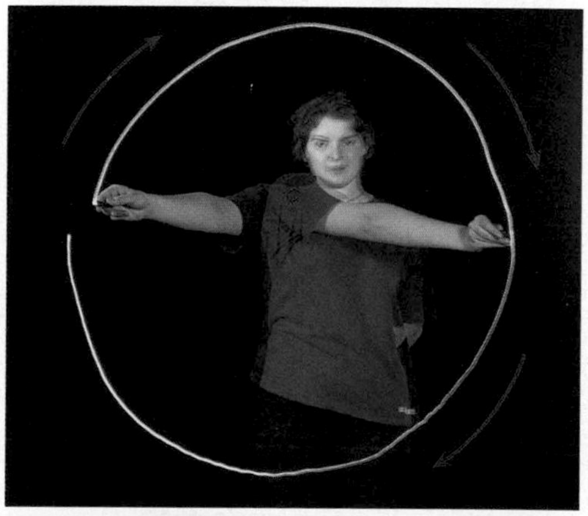

Module 8.4 Review

a. When doing jumping jacks, which limb movements are necessary?

b. Which movements are performed by hinge joints?

c. Compare dorsiflexion to plantar flexion.

LO **8.4** Describe flexion/extension, abduction/adduction, and circumduction movements of the skeleton.

Module 8.5

Specific terms describe rotation and special movements

Rotation

1 Trunk rotation is described as right or left rotation. Limb rotation can be described as medial or lateral rotation. Special terms are used to describe forearm rotation.

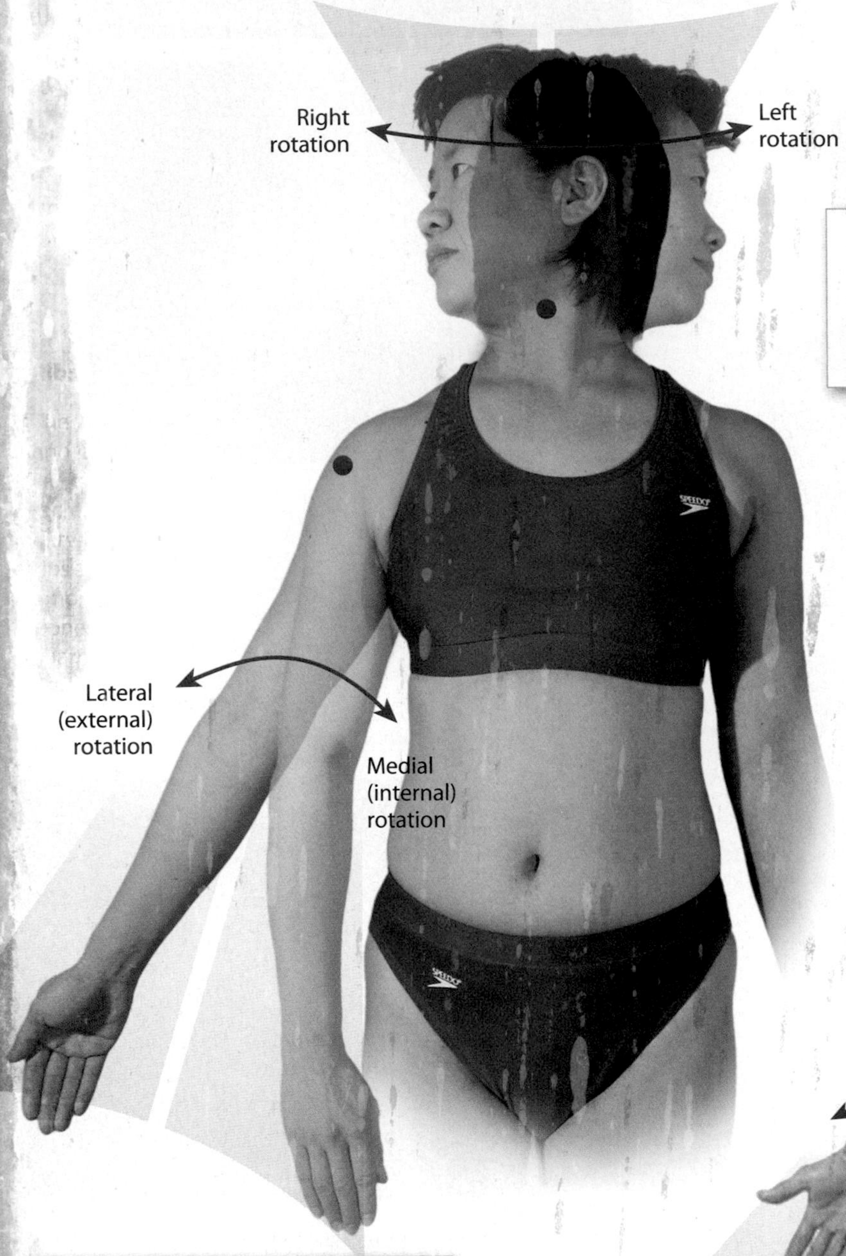

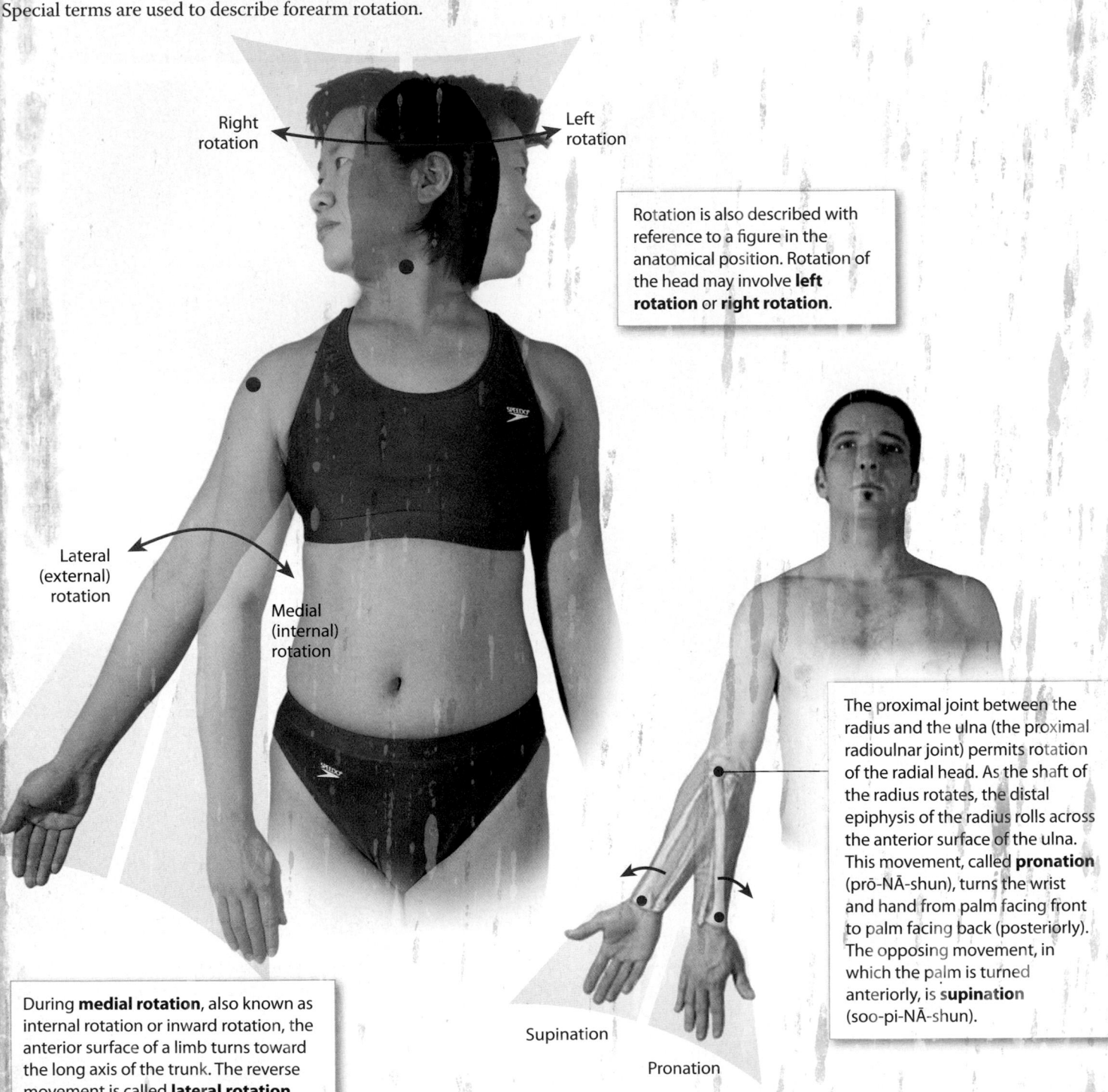

Right rotation
Left rotation
Lateral (external) rotation
Medial (internal) rotation

Rotation is also described with reference to a figure in the anatomical position. Rotation of the head may involve **left rotation** or **right rotation**.

During **medial rotation**, also known as internal rotation or inward rotation, the anterior surface of a limb turns toward the long axis of the trunk. The reverse movement is called **lateral rotation**, external rotation, or outward rotation.

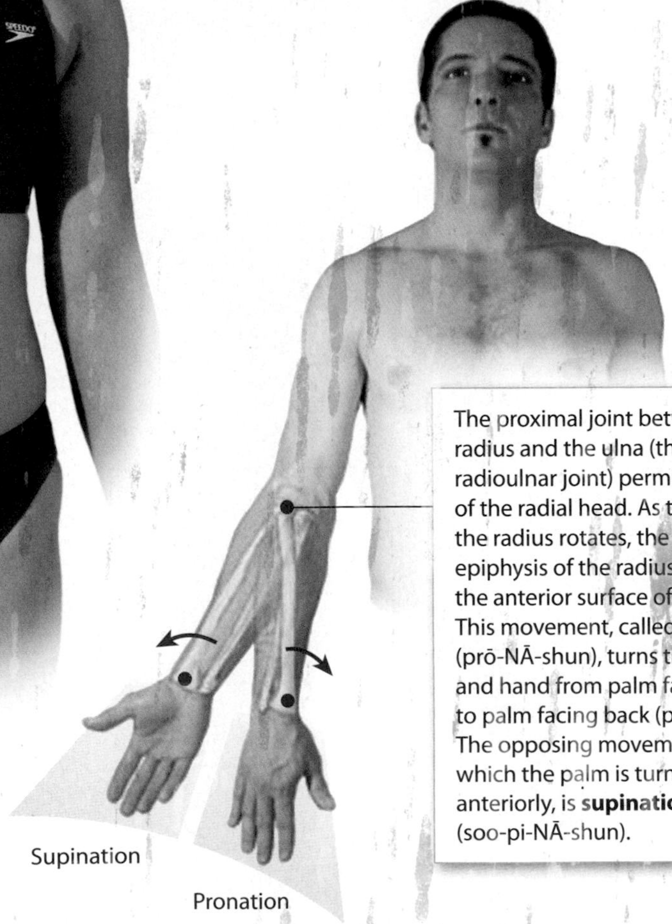

Supination
Pronation

The proximal joint between the radius and the ulna (the proximal radioulnar joint) permits rotation of the radial head. As the shaft of the radius rotates, the distal epiphysis of the radius rolls across the anterior surface of the ulna. This movement, called **pronation** (prō-NĀ-shun), turns the wrist and hand from palm facing front to palm facing back (posteriorly). The opposing movement, in which the palm is turned anteriorly, is **supination** (soo-pi-NĀ-shun).

Special Movements

2 Several specific terms apply to specific joints or unusual types of movement.

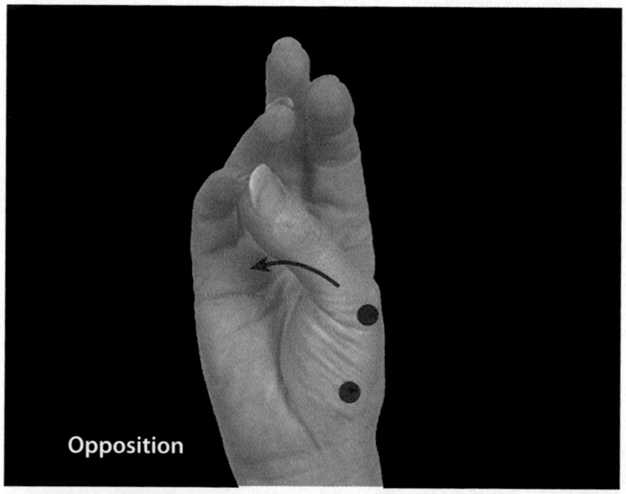
Opposition

Opposition is movement of the thumb toward the surface of the palm or the pads of other fingers. Opposition enables you to grasp and hold objects between your thumb and palm. It involves movement at the first carpometacarpal and metacarpophalangeal joints. Flexion at the fifth metacarpophalangeal joint can assist this movement.

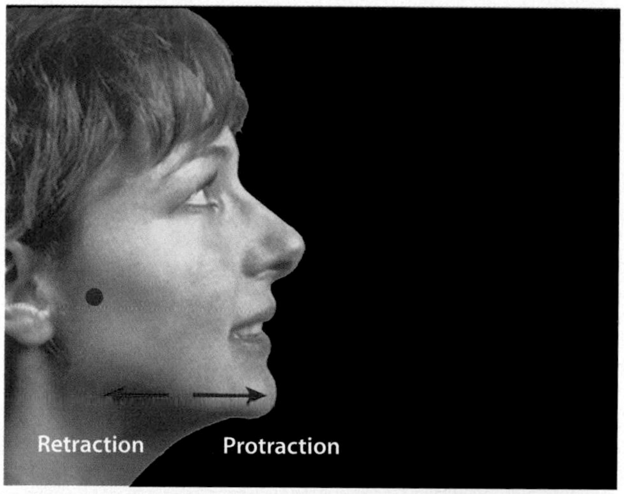
Retraction
Protraction

Protraction entails moving a part of the body anteriorly in the horizontal plane. **Retraction** is the reverse movement. You protract your jaw when you jut it forward, and you retract your jaw when you return it to its normal position. Although not shown here, you protract your clavicles when you cross your arms.

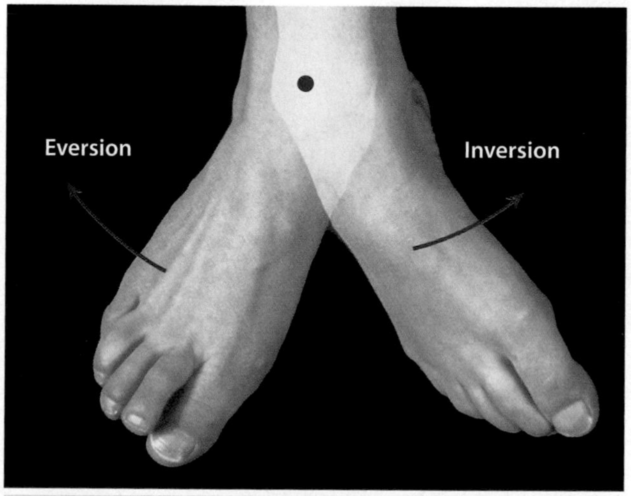
Eversion
Inversion

Inversion (*in*, into + *vertere*, to turn) is a twisting motion of the foot that turns the sole inward, elevating the medial edge of the sole. The opposite movement is called **eversion** (ē-VER-zhun).

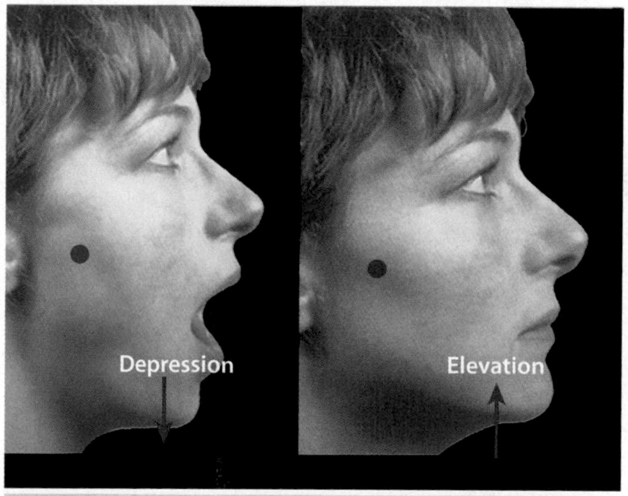
Depression
Elevation

Depression occurs when a body part moves inferiorly. **Elevation** is lifting a body structure superiorly.

Module 8.5 Review

a. What movements are made possible by the rotation of the head of the radius?

b. What hand movements occur when wriggling into tight-fitting gloves?

c. Snapping your fingers involves what movement of the thumb?

LO 8.5 Describe rotational and special movements of the skeleton.

Labeling

Label the structures in the synovial joint figure at right.

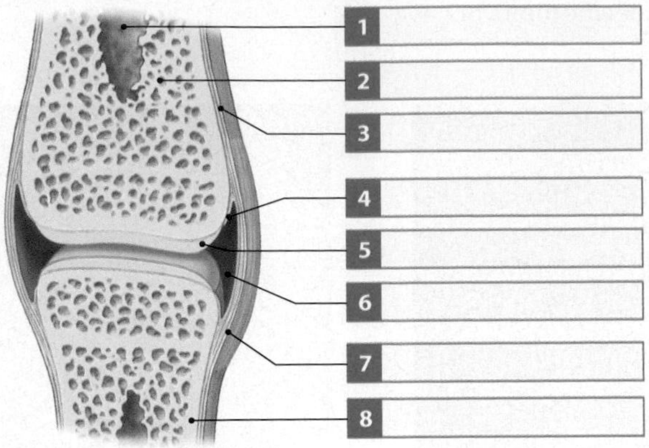
1
2
3
4
5
6
7
8

Identify each of the following movements.

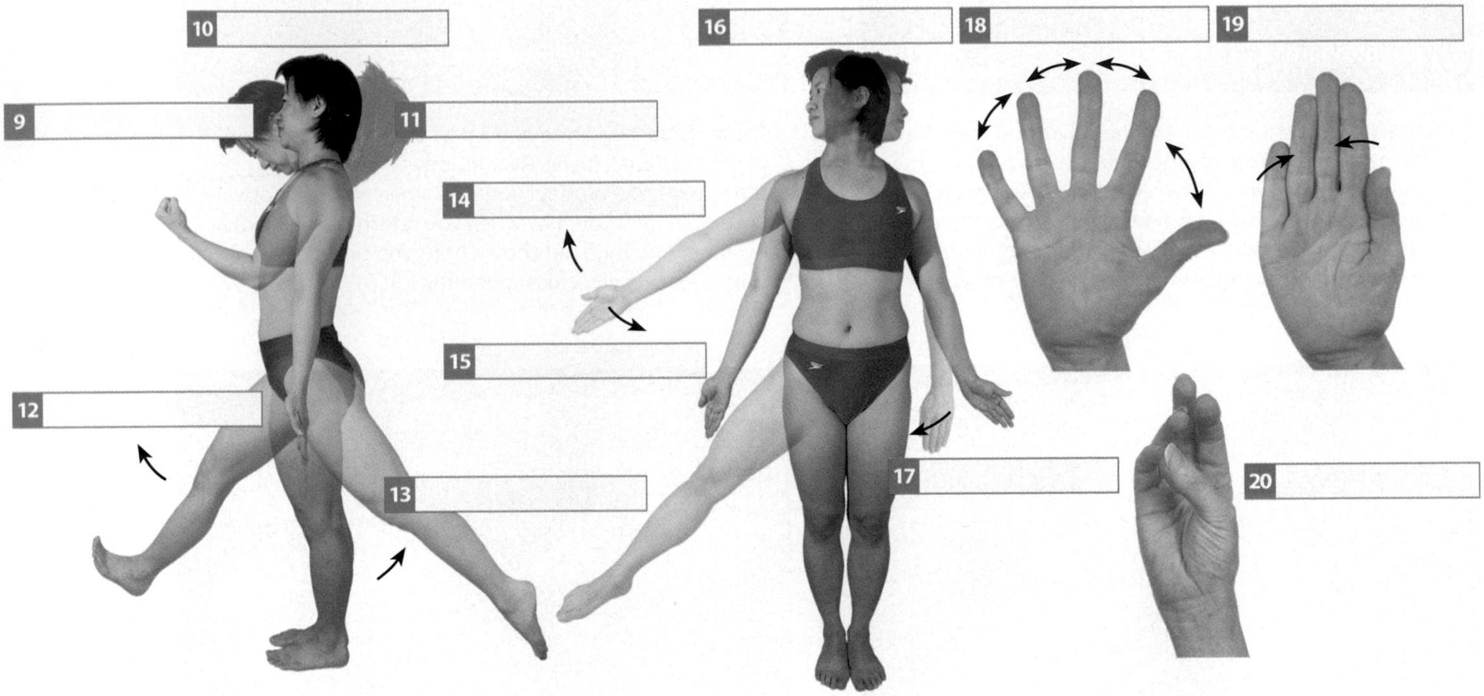
9
10
11
12
13
14
15
16
17
18
19
20

Matching

Match each lettered term with the most closely related description.

a. amphiarthrosis
b. synarthrosis
c. dislocation
d. pronation/supination
e diarthrosis
f. shoulder
g. articular discs
h. fluid-filled pouch

21 Freely movable joint
22 Movements of forearm bones
23 Ball-and-socket joint
24 Menisci
25 Immovable joint
26 Luxation
27 Bursa
28 Slightly movable joint

21 ______
22 ______
23 ______
24 ______
25 ______
26 ______
27 ______
28 ______

Axial joints have less range of motion than appendicular joints

1 A typical joint in the appendicular skeleton has an extensive range of motion. Because a joint cannot be both strong and highly mobile, these joints are often weaker than those of the axial skeleton. We will examine two joints in each limb: the shoulder and elbow in the upper limb, and the hip and knee in the lower limb.

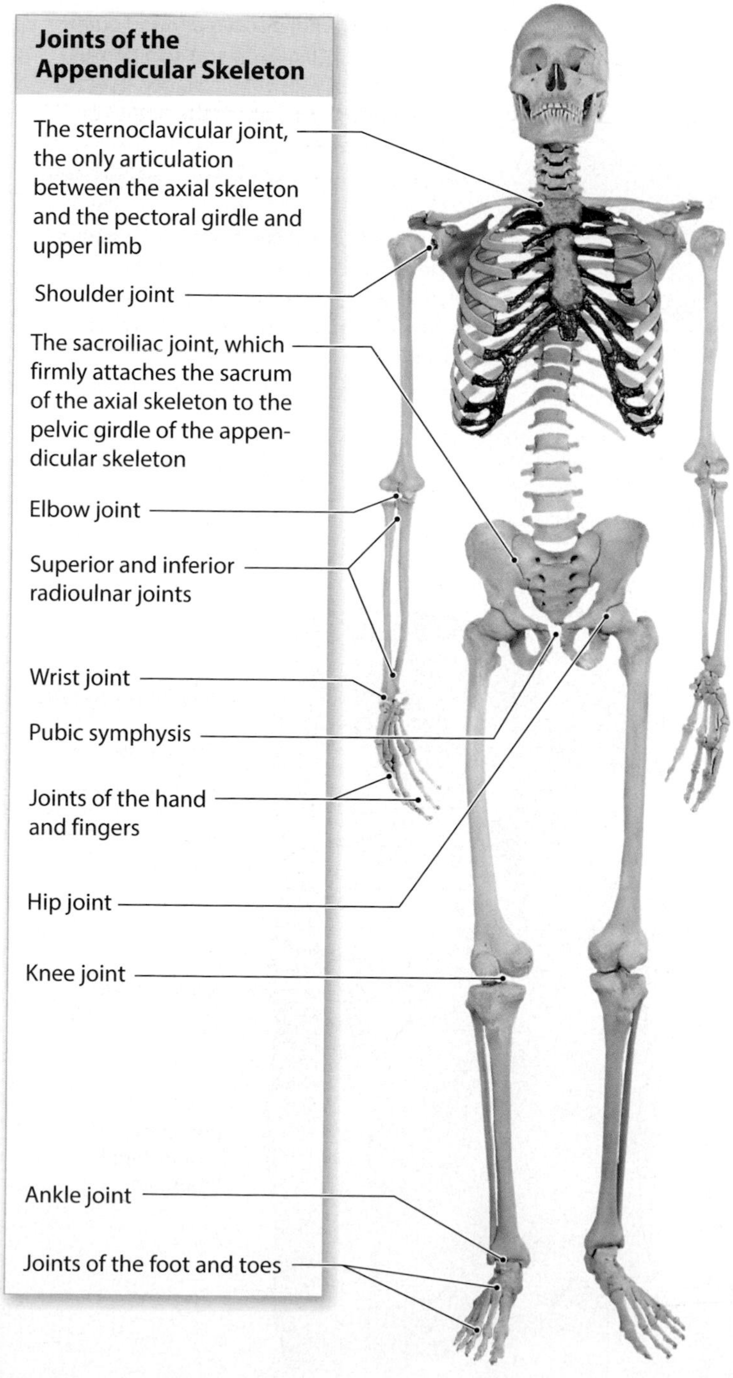
Joints of the Appendicular Skeleton
The sternoclavicular joint, the only articulation between the axial skeleton and the pectoral girdle and upper limb
Shoulder joint
The sacroiliac joint, which firmly attaches the sacrum of the axial skeleton to the pelvic girdle of the appendicular skeleton
Elbow joint
Superior and inferior radioulnar joints
Wrist joint
Pubic symphysis
Joints of the hand and fingers
Hip joint
Knee joint
Ankle joint
Joints of the foot and toes

2 Most joints in the axial skeleton are strong joints that permit very little movement. This section begins by examining the structure of the intervertebral joints, which permit limited but important movements of the vertebral column.

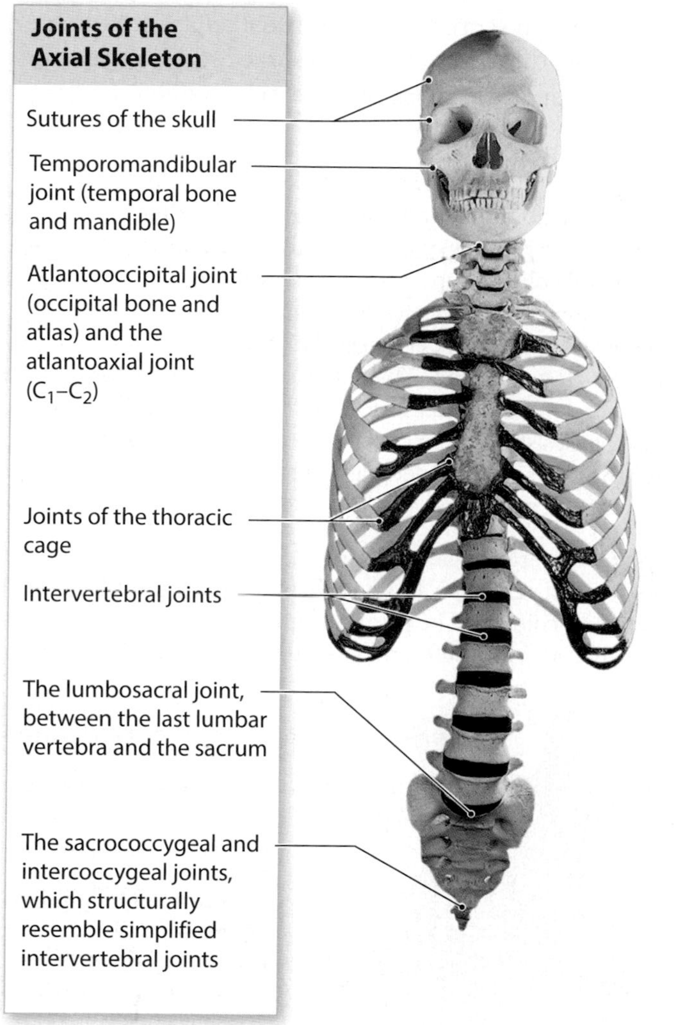
Joints of the Axial Skeleton
Sutures of the skull
Temporomandibular joint (temporal bone and mandible)
Atlantooccipital joint (occipital bone and atlas) and the atlantoaxial joint (C_1–C_2)
Joints of the thoracic cage
Intervertebral joints
The lumbosacral joint, between the last lumbar vertebra and the sacrum
The sacrococcygeal and intercoccygeal joints, which structurally resemble simplified intervertebral joints

Module 8.6 Review

a. Describe the relationship between joint strength and mobility.
b. Which division of the skeleton has the greatest range of motion?
c. Which joint attaches the upper limb to the axial skeleton?

LO **8.6** Compare the general relationship between joint stability and range of motion for axial and appendicular joints.

Adjacent vertebrae have gliding diarthroses between their articular processes and symphyseal joints between their vertebral bodies

The joints between the superior and inferior articular processes of adjacent vertebrae are gliding joints that permit flexion and rotation movements of the vertebral column. Little gliding occurs between adjacent vertebral bodies. From axis to sacrum, the bodies of adjacent vertebrae form symphyseal joints. At these joints the vertebrae are separated and cushioned by pads of fibrocartilage called **intervertebral discs.**

1 The intervertebral discs make a significant contribution to an individual's height: they account for roughly one-quarter the length of the vertebral column superior to the sacrum. As we grow older, the water content of the nucleus pulposus (see right) decreases and it becomes less effective as a cushion. This increases the chances for vertebral injury. Water loss from the intervertebral discs also causes shortening of the vertebral column, accounting for the characteristic decrease in height with advancing age.

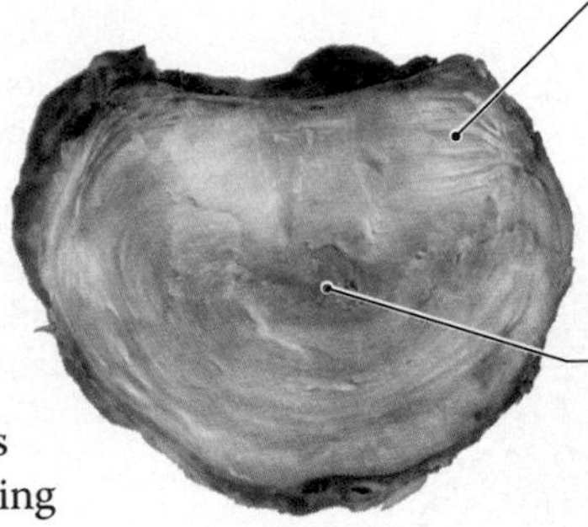

Superior view

Each intervertebral disc has a tough outer ring of fibrocartilage, the **anulus fibrosus** (AN-ū-lus fī-BRŌ-sus). The collagen fibers of this layer attach the disc to the bodies of adjacent vertebrae.

The anulus fibrosus surrounds the **nucleus pulposus** (pul-PŌ-sus), a soft, elastic, gelatinous core. The nucleus pulposus gives the disc resiliency and enables it to absorb shocks.

2 Numerous ligaments are attached to the bodies and processes of all vertebrae, binding them together and stabilizing the vertebral column. The primary ligaments have been identified in the anterior view (left) and lateral and sectional views (right).

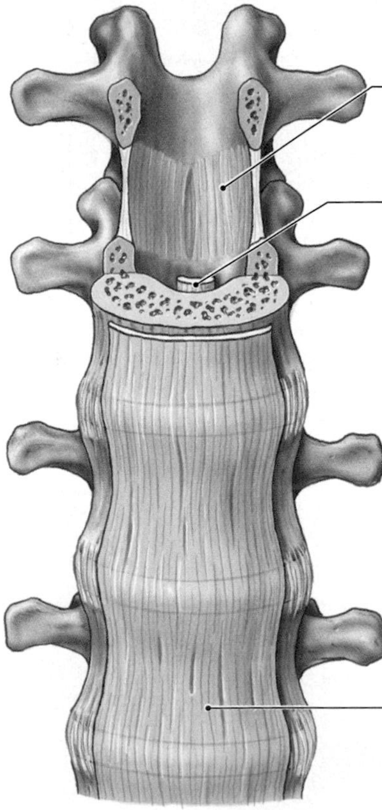

Primary Vertebral Ligaments
The **ligamentum flavum** (plural, *ligamenta flava*) connects the laminae of adjacent vertebrae.
The **posterior longitudinal ligament** parallels the anterior longitudinal ligament and connects the posterior surfaces of adjacent vertebral bodies.
The **interspinous ligament** connects the spinous processes of adjacent vertebrae.
The **supraspinous ligament** interconnects the tips of the spinous processes from the sacrum to vertebra C_7. The ligamentum nuchae extends from vertebra C_7 to the base of the skull.
The **anterior longitudinal ligament** connects the anterior surfaces of adjacent vertebral bodies.

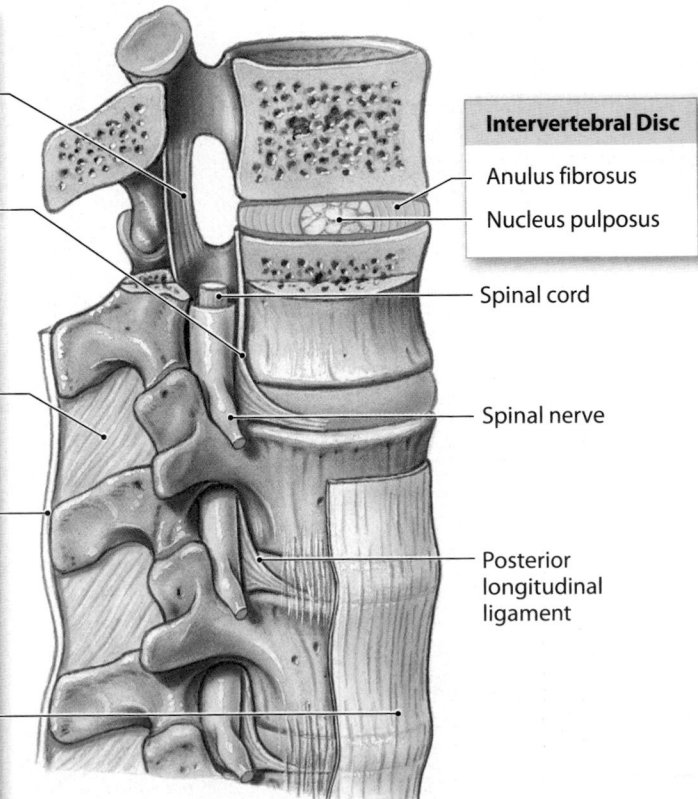

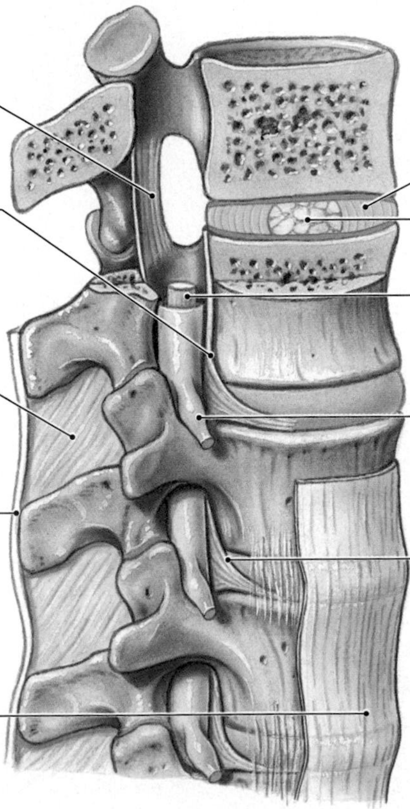
Intervertebral Disc
Anulus fibrosus
Nucleus pulposus
Spinal cord
Spinal nerve
Posterior longitudinal ligament

3 If the posterior longitudinal ligaments weaken, as often occurs with age, the compressed nucleus pulposus may distort the anulus fibrosus, forcing it partway into the vertebral canal. This condition, seen here in lateral view, is called a **bulging disc**. The tough outer layer of cartilage is actually bulging, and this is considered a normal part of aging.

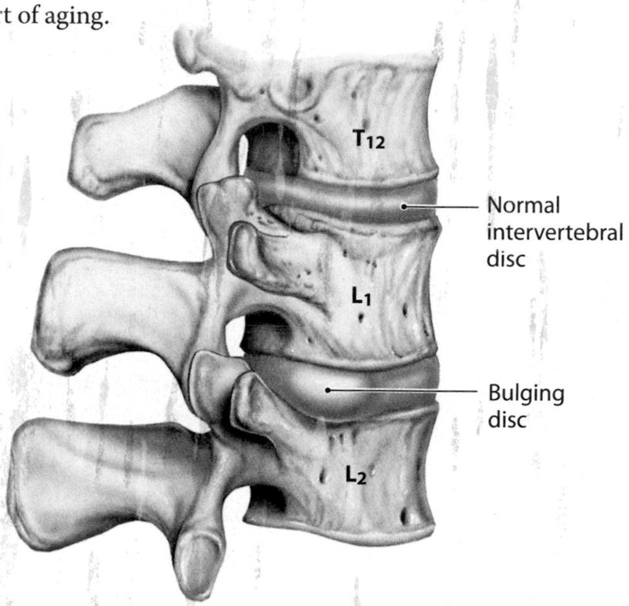

T12
Normal intervertebral disc
L1
Bulging disc
L2

4 If the nucleus pulposus breaks through the anulus fibrosus, it too may protrude into the vertebral canal. This condition, shown here in superior view, is called a **herniated disc**. When a disc herniates, it compresses spinal nerves.

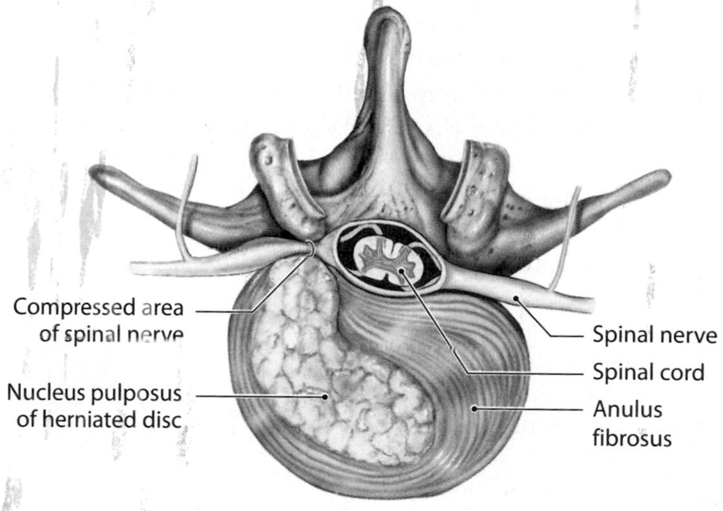

Compressed area of spinal nerve
Nucleus pulposus of herniated disc
Spinal nerve
Spinal cord
Anulus fibrosus

5 The bones of the skeleton become thinner and weaker as a normal part of the aging process. Inadequate ossification is called **osteopenia** (os-tē-ō-PĒ-nē-uh; *penia*, lacking). This reduction in bone mass begins between the ages of 30 and 40 as osteoblast activity begins to decline, while osteoclast activity continues at previous levels. Thereafter, women lose roughly 8 percent of their skeletal mass every decade; men lose roughly 3 percent per decade. When the reduction in bone mass is sufficient to compromise normal function, the condition is known as **osteoporosis** (os-tē-ō-po-RŌ-sis; *porosus*, porous). The combination of osteopenia or osteoporosis and the reduced cushioning properties of the intervertebral discs makes vertebral fractures a common problem among the elderly.

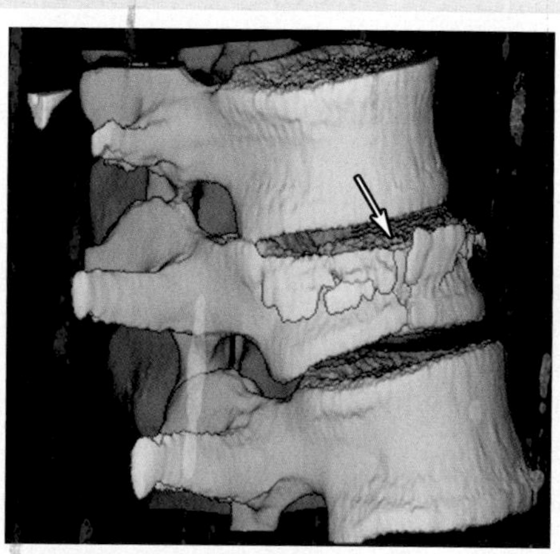

Clinical scan of a compression fracture in a lumbar vertebra

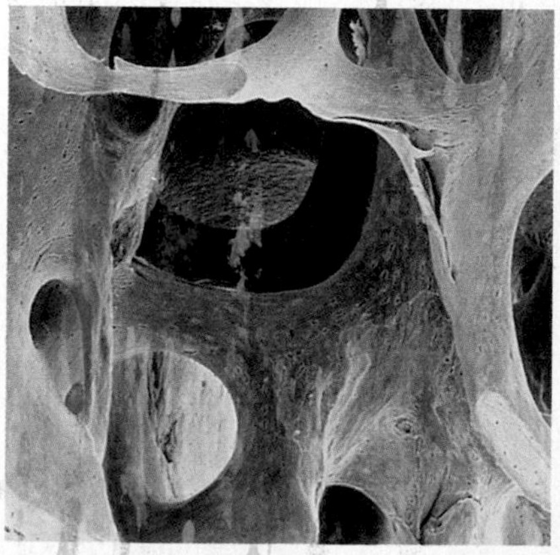

Normal spongy bone SEM × 25

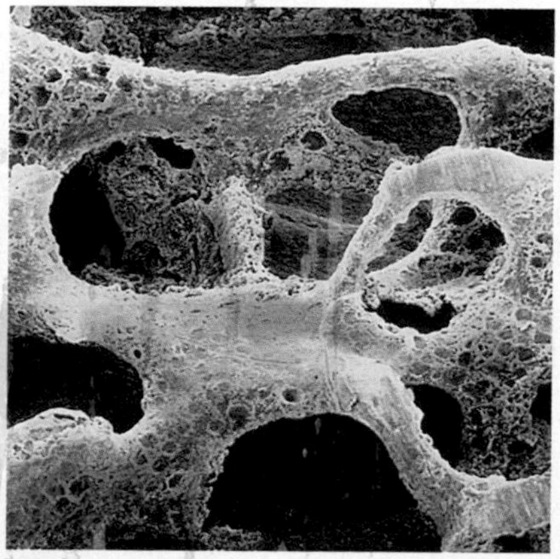

Spongy bone with osteoporosis SEM × 21

Module 8.7 Review

a. Identify the primary vertebral ligaments.

b. Describe the nucleus pulposus and anulus fibrosus of an intervertebral disc.

c. Compare a bulging disc with a herniated disc.

The shoulder and hip are ball-and-socket joints

Shoulder Joint

The shoulder joint, or **glenohumeral joint**, has the greatest range of motion of any joint. Because it is also the most frequently dislocated joint, it is an excellent example to show how stability is sacrificed for mobility. This joint is a ball-and-socket diarthrosis formed by the articulation of the head of the humerus with the glenoid cavity of the scapula. Another important joint in this area is the acromioclavicular joint, which is where the clavicle articulates with the scapula.

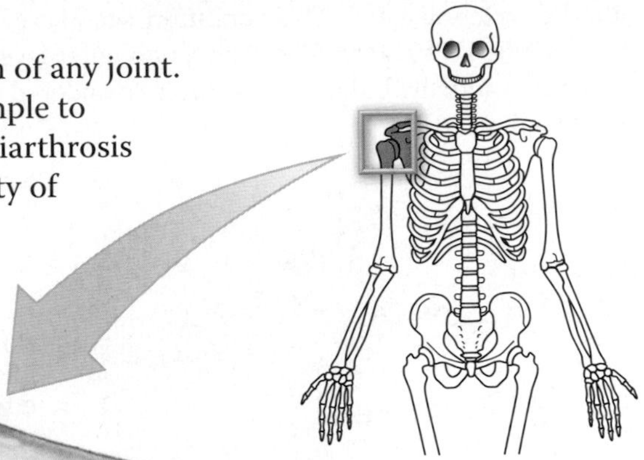

1 Five major ligaments help stabilize the shoulder joint. As at other joints, bursae at the shoulder reduce friction where large muscles and tendons pass across the joint capsule. All or some of these ligaments may be injured in a shoulder dislocation (affecting the glenohumeral joint), or a shoulder separation (involving the acromioclavicular joint).

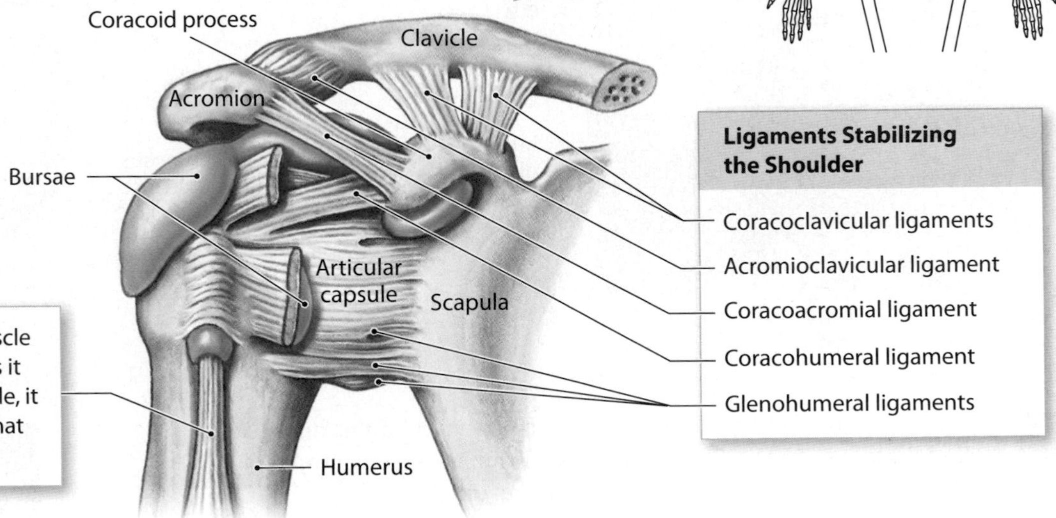
Coracoid process
Clavicle
Acromion
Bursae
Articular capsule
Scapula
Humerus
Ligaments Stabilizing the Shoulder
Coracoclavicular ligaments
Acromioclavicular ligament
Coracoacromial ligament
Coracohumeral ligament
Glenohumeral ligaments

A tendon of the biceps brachii muscle runs through the shoulder joint. As it passes through the articular capsule, it is surrounded by a tubular bursa that is continuous with the joint cavity.

2 This frontal section through the glenohumeral joint shows that the relatively loose articular capsule extends from the scapula to the anatomical neck of the humerus. The oversized articular capsule permits an extensive range of motion.

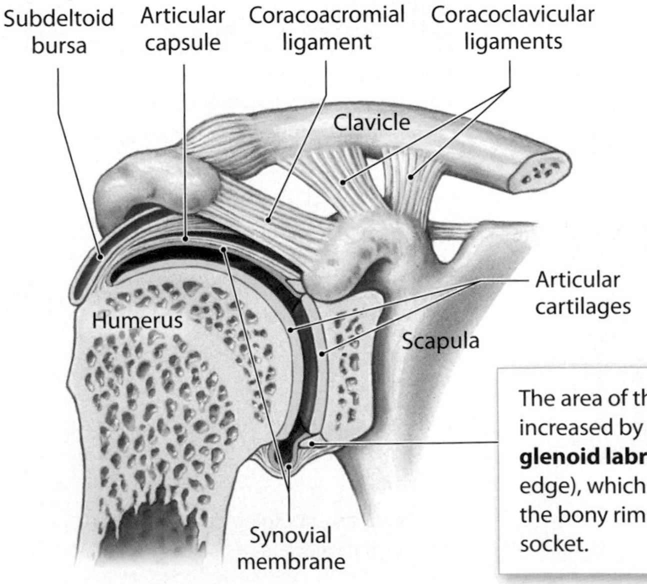
Subdeltoid bursa
Articular capsule
Coracoacromial ligament
Coracoclavicular ligaments
Clavicle
Articular cartilages
Humerus
Scapula
Synovial membrane

3 In lateral view, you can see how small the articular cartilage of the glenoid cavity is compared with the articular capsule. The small articular cartilage is compensated by the glenoid labrum, surrounding skeletal muscles, associated tendons, and various ligaments that stabilize the joint.

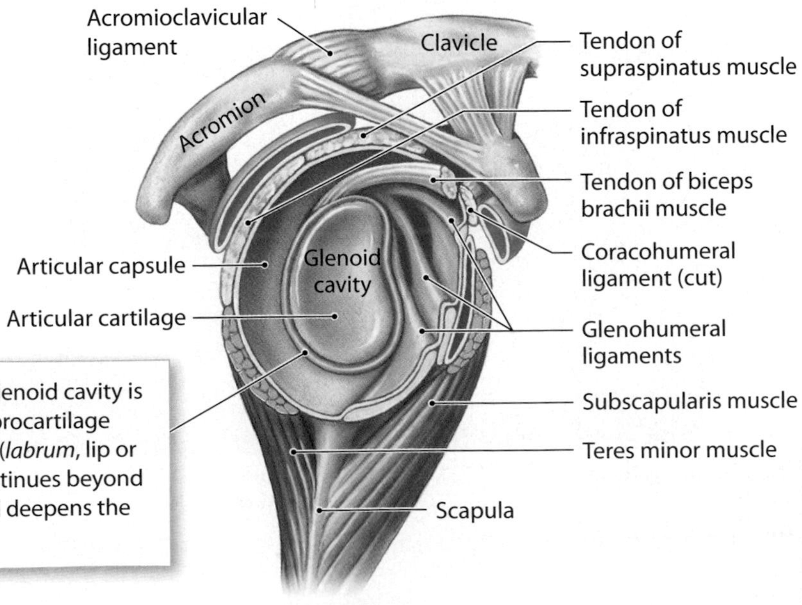
Acromioclavicular ligament
Clavicle
Acromion
Tendon of supraspinatus muscle
Tendon of infraspinatus muscle
Tendon of biceps brachii muscle
Coracohumeral ligament (cut)
Articular capsule
Glenoid cavity
Articular cartilage
Glenohumeral ligaments
Subscapularis muscle
Teres minor muscle
Scapula

The area of the glenoid cavity is increased by a fibrocartilage **glenoid labrum** (*labrum*, lip or edge), which continues beyond the bony rim and deepens the socket.

Hip Joint

The hip joint is a sturdy ball-and-socket diarthrosis that allows flexion and extension, adduction and abduction, circumduction, and rotation.

4 This lateral view shows the hip joint with the femur removed. Within the acetabulum, an articular cartilage pad extends like a horseshoe to either side of the acetabular notch. Two of the five ligaments that reinforce the articular capsule are shown here: the ligament of the femoral head and the transverse acetabular ligament.

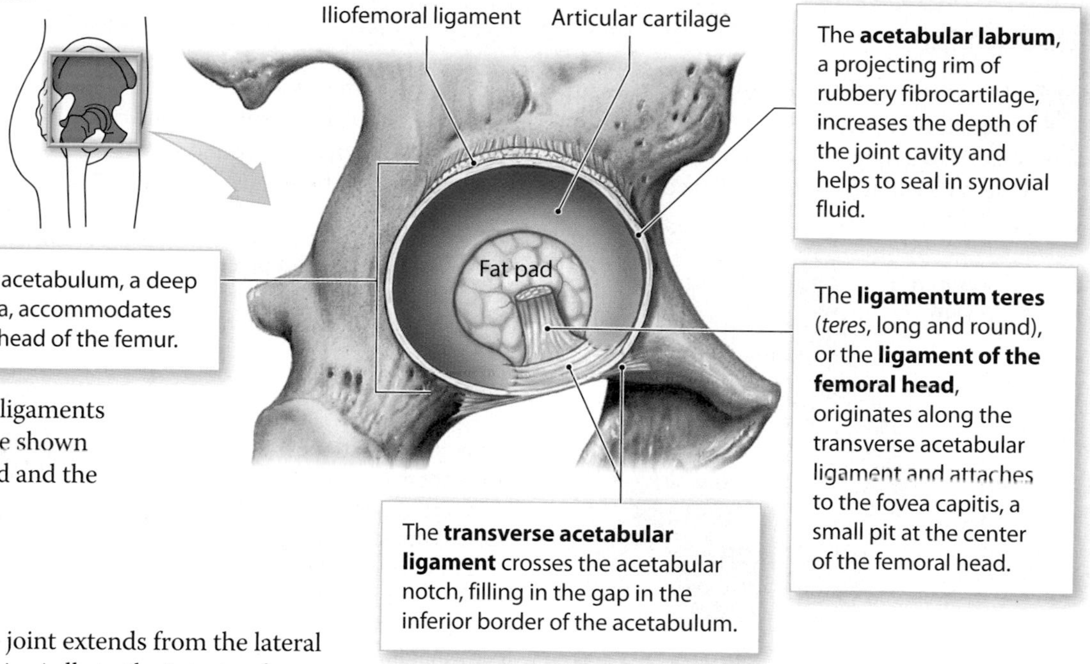

Iliofemoral ligament
Articular cartilage
The **acetabular labrum**, a projecting rim of rubbery fibrocartilage, increases the depth of the joint cavity and helps to seal in synovial fluid.
The acetabulum, a deep fossa, accommodates the head of the femur.
Fat pad
The **ligamentum teres** (*teres*, long and round), or the **ligament of the femoral head**, originates along the transverse acetabular ligament and attaches to the fovea capitis, a small pit at the center of the femoral head.
The **transverse acetabular ligament** crosses the acetabular notch, filling in the gap in the inferior border of the acetabulum.

5 The articular capsule of the hip joint extends from the lateral and inferior surfaces of the pelvic girdle to the intertrochanteric line and intertrochanteric crest of the femur, enclosing both the head and neck of the femur. The three remaining broad ligaments that reinforce the articular capsule are shown in these views.

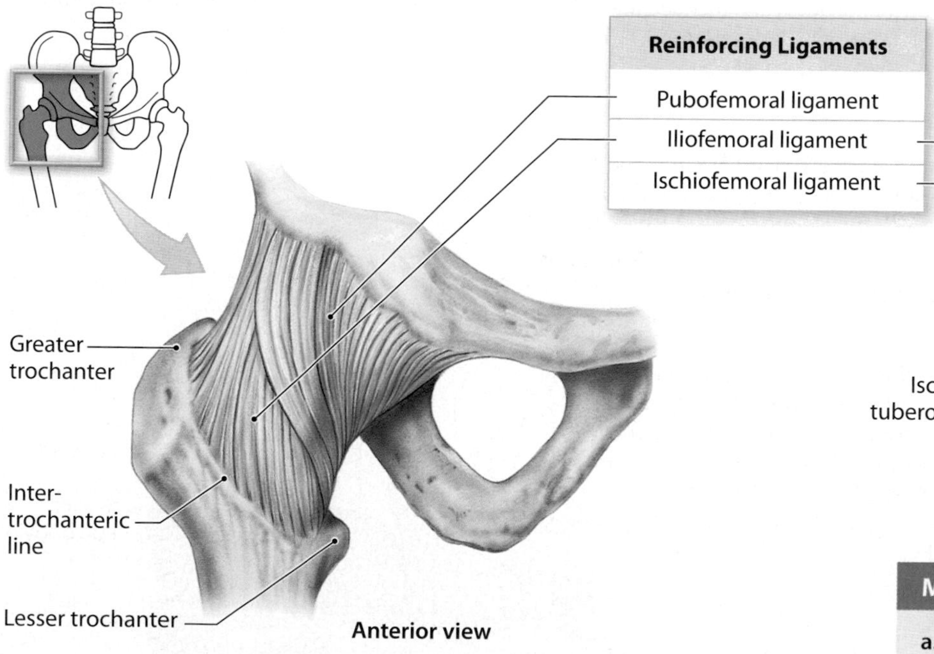

Reinforcing Ligaments
Pubofemoral ligament
Iliofemoral ligament
Ischiofemoral ligament
Greater trochanter
Inter-trochanteric line
Lesser trochanter
Anterior view

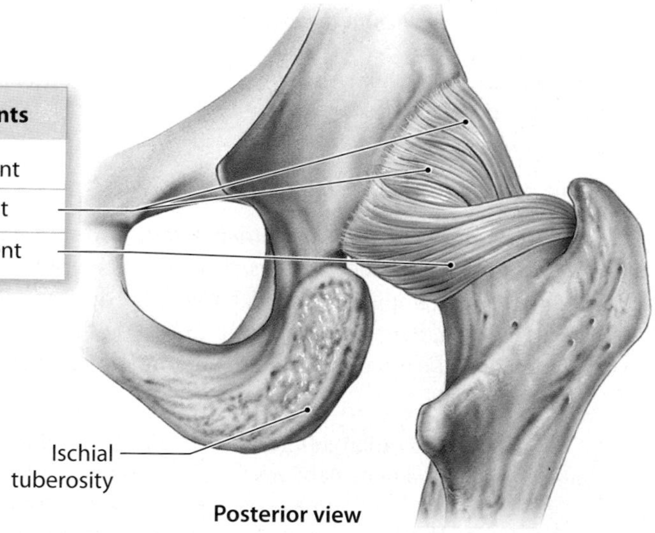

Ischial tuberosity
Posterior view

Although the head of the femur is well supported, the ball-and-socket joint is not directly aligned with the weight distribution along the shaft. As a result, hip fractures (fractures of the femoral neck or between the greater and lesser trochanters of the femur) are much more common than hip dislocations.

Module 8.8 Review

a. Which tissues or structures provide most of the stability for the shoulder joint?

b. At what site are the iliofemoral ligament, pubofemoral ligament, and ischiofemoral ligament located?

c. A football player is pushed out of bounds from behind. He falls onto his outstretched hand, pushing the humeral head forcefully upward. Which joints and ligaments are affected?

LO **8.8** Describe the structure and function of the shoulder and hip joints.

Module 8.9

The elbow and knee are hinge joints

Elbow Joint

1 The elbow joint is a complex hinge joint that involves the humerus, radius, and ulna. It is extremely stable because (1) the bony surfaces of the humerus and ulna interlock, (2) a single, thick articular capsule surrounds both the humeroulnar and proximal radioulnar joints, and (3) the articular capsule is reinforced by strong ligaments.

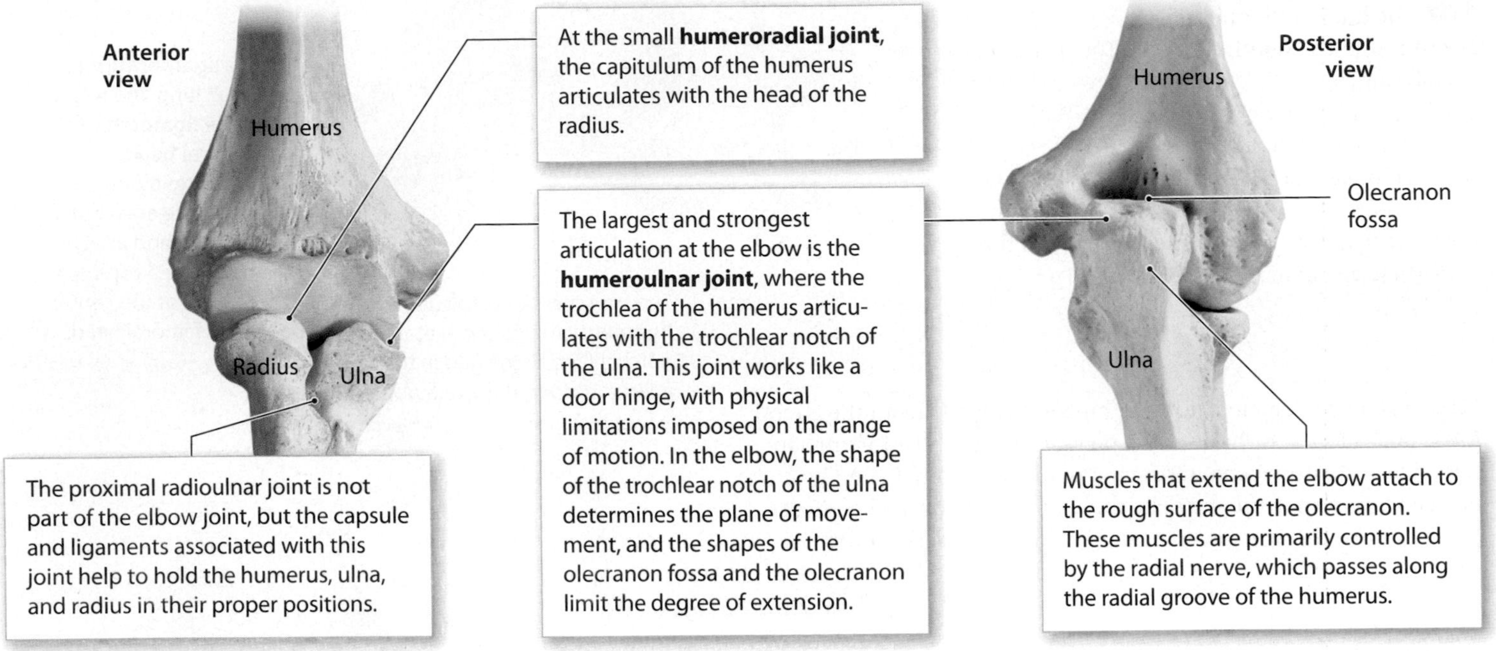

Anterior view
Humerus
Radius
Ulna
At the small **humeroradial joint**, the capitulum of the humerus articulates with the head of the radius.
The largest and strongest articulation at the elbow is the **humeroulnar joint**, where the trochlea of the humerus articulates with the trochlear notch of the ulna. This joint works like a door hinge, with physical limitations imposed on the range of motion. In the elbow, the shape of the trochlear notch of the ulna determines the plane of movement, and the shapes of the olecranon fossa and the olecranon limit the degree of extension.
The proximal radioulnar joint is not part of the elbow joint, but the capsule and ligaments associated with this joint help to hold the humerus, ulna, and radius in their proper positions.
Posterior view
Humerus
Olecranon fossa
Ulna
Muscles that extend the elbow attach to the rough surface of the olecranon. These muscles are primarily controlled by the radial nerve, which passes along the radial groove of the humerus.

2 Although the elbow is extremely strong and stable, severe stresses can produce dislocations or other injuries, especially if epiphyseal growth has not been completed. For example, parents in a rush may drag a toddler along behind them. This exerts an upward, twisting pull on the elbow joint that can result in a partial dislocation of the radial head from the annular ligament known as **nursemaid's elbow**.

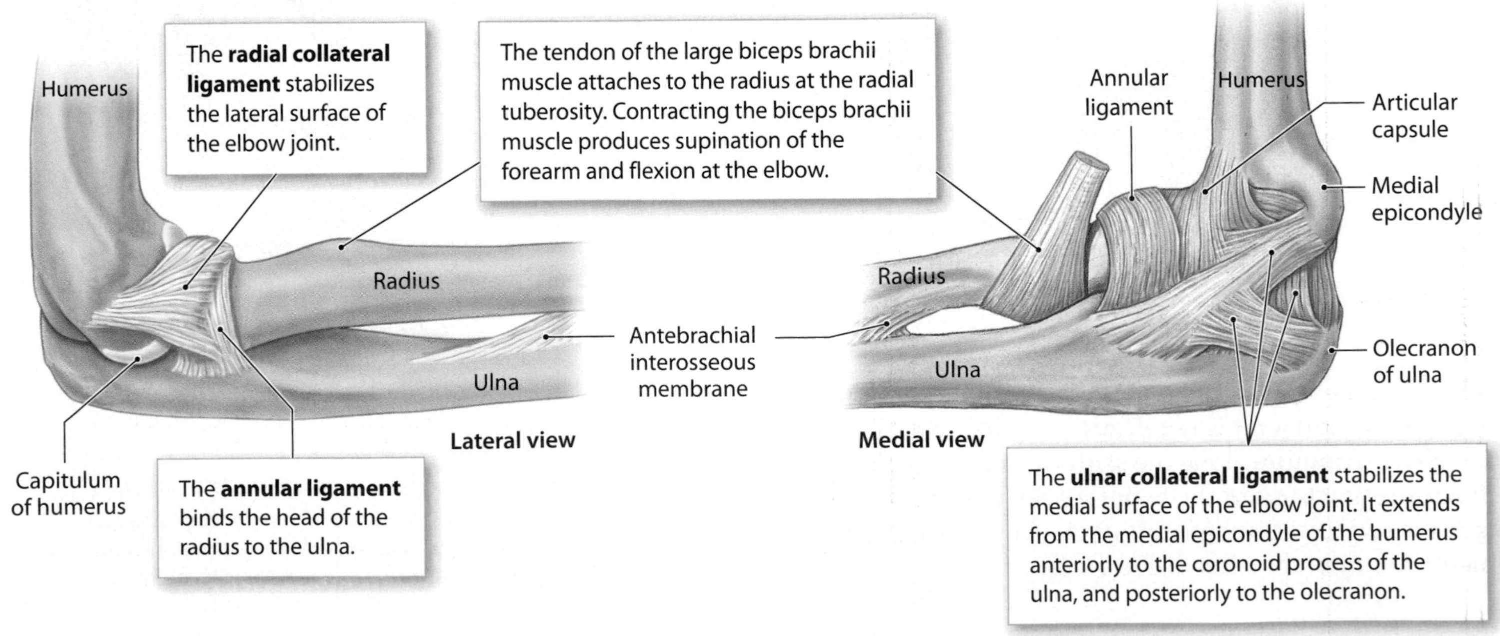

Humerus
The **radial collateral ligament** stabilizes the lateral surface of the elbow joint.
The tendon of the large biceps brachii muscle attaches to the radius at the radial tuberosity. Contracting the biceps brachii muscle produces supination of the forearm and flexion at the elbow.
Radius
Ulna
Antebrachial interosseous membrane
Lateral view
Capitulum of humerus
The **annular ligament** binds the head of the radius to the ulna.
Annular ligament
Humerus
Articular capsule
Medial epicondyle
Radius
Ulna
Olecranon of ulna
Medial view
The **ulnar collateral ligament** stabilizes the medial surface of the elbow joint. It extends from the medial epicondyle of the humerus anteriorly to the coronoid process of the ulna, and posteriorly to the olecranon.

Knee Joint

The knee joint actually consists of three separate joints: two between the femur and tibia (medial condyle to medial condyle, and lateral condyle to lateral condyle) and one between the patella and the patellar surface of the femur. Together, these joints permit flexion, extension, and very limited rotation. Note that the fibula does not form part of the knee joint.

3 The tendon from the quadriceps muscle passes over the anterior surface of the joint, embedding the patella, and the patellar ligament then continues to its attachment on the anterior surface of the tibia. The patellar ligament and adjacent ligamentous bands support the anterior surface of the knee joint.

4 The popliteal ligaments and several muscles that originate or insert on the femoral or tibial epiphyses reinforce the posterior surface of the joint.

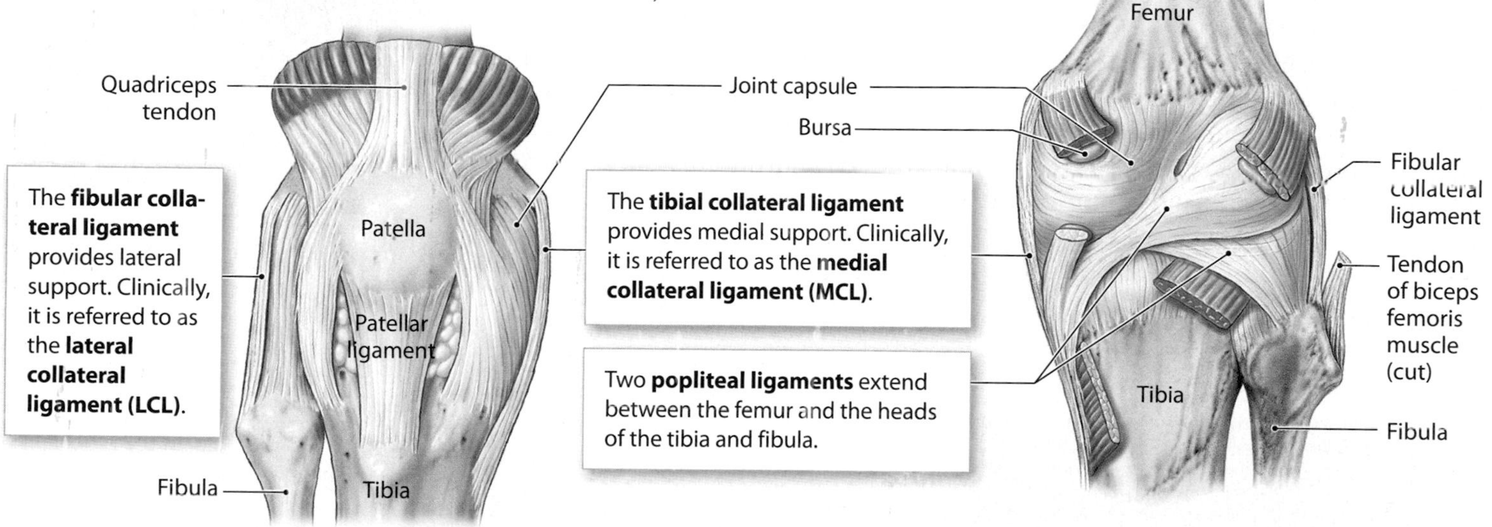
Quadriceps tendon
Patella
Patellar ligament
Fibula
Tibia
The **fibular collateral ligament** provides lateral support. Clinically, it is referred to as the **lateral collateral ligament (LCL)**.
Joint capsule
Bursa
The **tibial collateral ligament** provides medial support. Clinically, it is referred to as the **medial collateral ligament (MCL)**.
Two **popliteal ligaments** extend between the femur and the heads of the tibia and fibula.
Femur
Fibular collateral ligament
Tendon of biceps femoris muscle (cut)
Tibia
Fibula

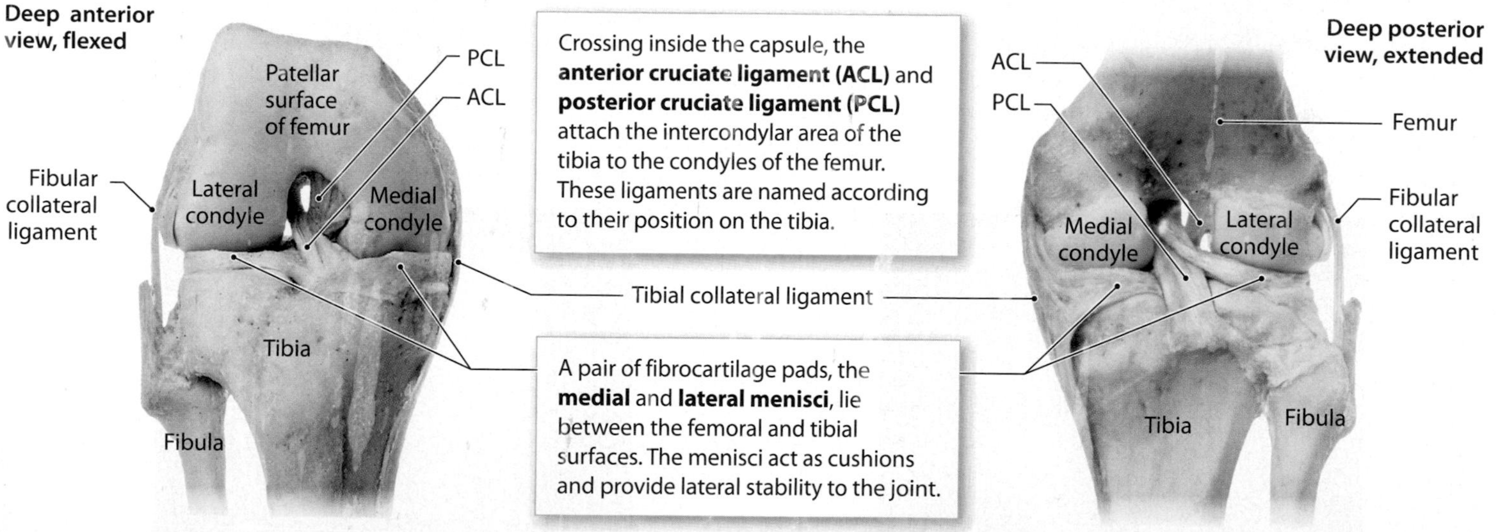
Deep anterior view, flexed
Patellar surface of femur
Fibular collateral ligament
Lateral condyle
Medial condyle
PCL
ACL
Tibia
Fibula
Crossing inside the capsule, the **anterior cruciate ligament (ACL)** and **posterior cruciate ligament (PCL)** attach the intercondylar area of the tibia to the condyles of the femur. These ligaments are named according to their position on the tibia.
Tibial collateral ligament
A pair of fibrocartilage pads, the **medial** and **lateral menisci**, lie between the femoral and tibial surfaces. The menisci act as cushions and provide lateral stability to the joint.
Deep posterior view, extended
ACL
PCL
Femur
Fibular collateral ligament
Medial condyle
Lateral condyle
Tibia
Fibula

Within the joint, the anterior cruciate (*cruciatus*, cross) ligament (ACL) and posterior cruciate ligament (PCL) limit the anterior and posterior movement of the tibia and maintain alignment of the femoral and tibial condyles. At full extension, a slight lateral rotation of the tibia tightens the ACL and forces the lateral meniscus between the tibia and femur. The knee joint is then locked in the extended position; unlocking requires medial rotation of the tibia or lateral rotation of the femur.

Module 8.9 Review

a. Of the elbow and knee joints, which has menisci?

b. What signs and symptoms would you expect in a person who has damaged the menisci of the knee joint?

c. Which ligament is a severely hyperextended knee more likely to damage: the ACL or the PCL?

Arthritis can disrupt normal joint structure and function

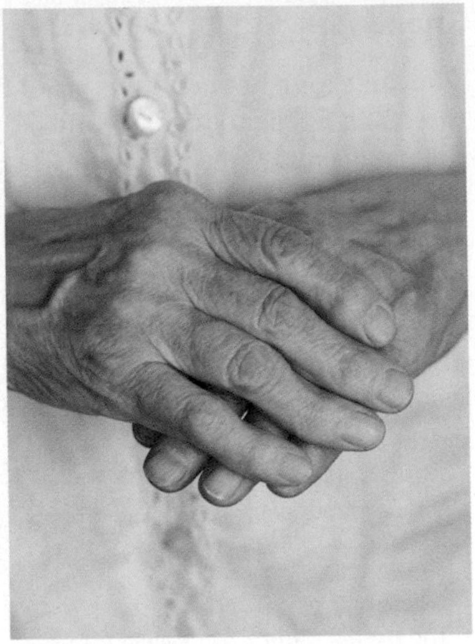

Joints are subjected to heavy use throughout our lifetime, and problems with joint function are relatively common, especially in older individuals. **Rheumatism** (RŪ-muh-tiz-um) is a general term that indicates pain and stiffness affecting the musculoskeletal system. **Arthritis** (ar-THRĪ-tis; *arthro,* joint + *itis,* inflammation) encompasses all the rheumatic diseases that affect synovial joints. Arthritis always involves damage to the articular cartilage, but the specific cause can vary. **Osteoarthritis** (os-tē-ō-ar-THRĪ-tis), also known as *degenerative arthritis* or *degenerative joint disease* (*DJD*) is the most common form of arthritis. It generally affects individuals age 60 or older. Osteoarthritis can result from the cumulative effects of wear and tear at joint surfaces or from genetic factors affecting collagen formation. In the U.S. population, 25 percent of women and 15 percent of men over age 60 show signs of this disease. Two other less common types of arthritis are **gouty arthritis** (Module 23.14, p. 909) and **rheumatoid arthritis** (Module 20.21, p. 778).

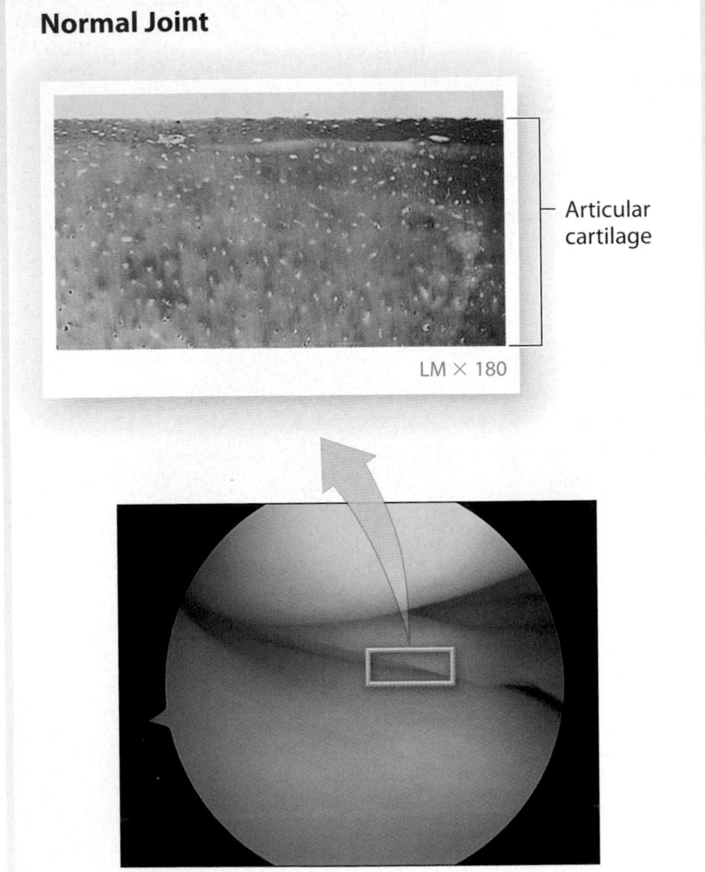
Normal Joint
Articular cartilage
LM × 180

Arthroscopic view of normal cartilage

1 This is a normal articular cartilage. Its surface is smooth and slick, and a sectional view reveals thick cartilage with a homogeneous matrix.

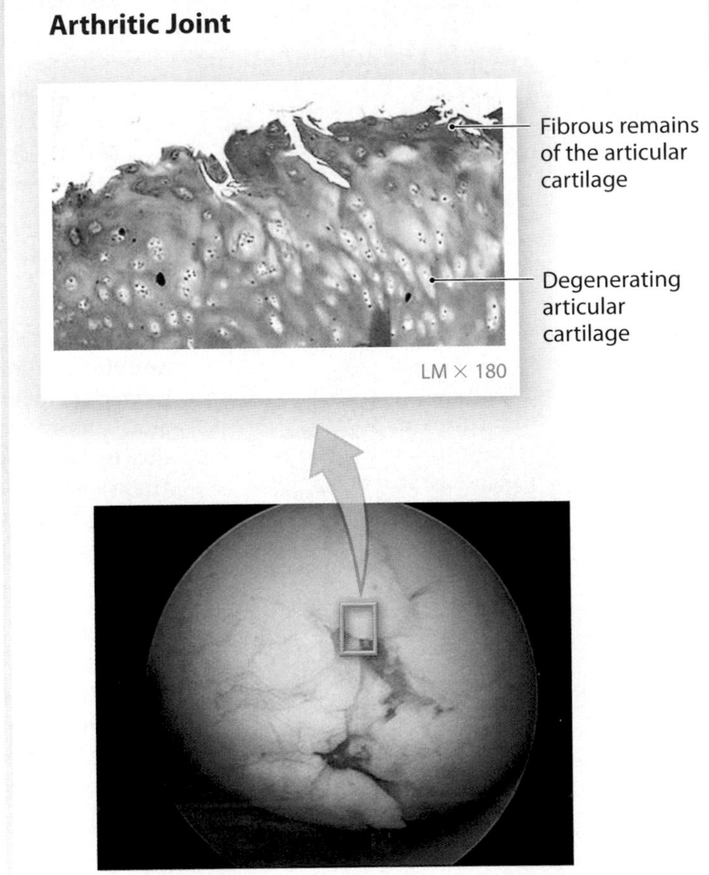
Arthritic Joint
Fibrous remains of the articular cartilage
Degenerating articular cartilage
LM × 180

Arthroscopic view of damaged cartilage

2 This is an articular cartilage damaged by osteoarthritis. The exposed surfaces change from a slick, smooth-gliding surface to a surface composed of a rough feltwork of bristly collagen fibers. Such a change drastically increases friction at the joint, which then promotes further degeneration of the articular cartilage.

3 There are several options available for examining the structure of problematic joints. This is an arthroscopic view of the interior of the left knee, showing injuries to the anterior and posterior cruciate ligaments. An **arthroscope** is a narrow, flexible fiberoptic scope containing a tiny camera that permits exploration of a joint without major surgery. Optical fibers are thin threads of glass or plastic that conduct light. The fibers can be bent around corners, so they can be introduced into a knee or other joint and moved around, enabling the physician to see inside the joint. If necessary, additional small incisions can be made to insert flexible instruments that permit surgery inside the joint, within view of the arthroscope. This procedure, called **arthroscopic surgery**, has greatly improved the treatment of knee and other joint injuries.

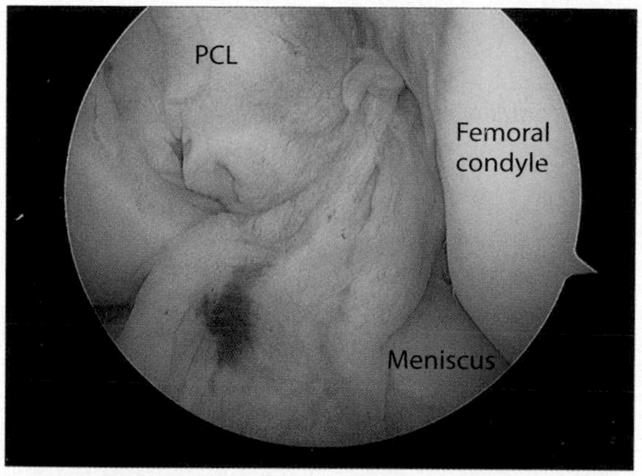
PCL
Femoral condyle
Meniscus

4 An arthroscope cannot show soft tissue details outside the joint cavity, and repeated arthroscopy eventually leads to the formation of scar tissue and other joint problems. Magnetic resonance imaging (MRI) is a cost-effective and noninvasive method of viewing, without injury, and examining soft tissues around the joint. These are superior and posterior MRI views of the left knee joint. Note the clarity of the soft tissue detail.

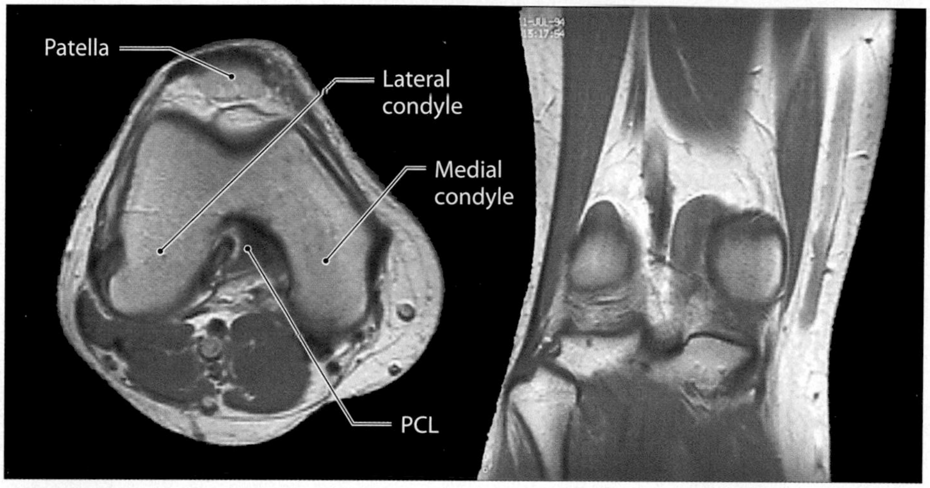
Patella
Lateral condyle
Medial condyle
PCL

5 Implantation of artificial joints such as these may be the method of last resort when regular exercise, physical therapy, and anti-inflammatory drugs (such as aspirin) fail to slow the progress of arthritis. Artificial joints can restore mobility and relieve pain. People who undergo joint replacement surgery will have lifelong restrictions. For example, high-impact activities such as jogging, running, and playing football are no longer allowed after a hip or knee replacement. New knees and hips can last for more than 15 years, especially if the patient follows restrictions and avoids high-impact joint stress and strain.

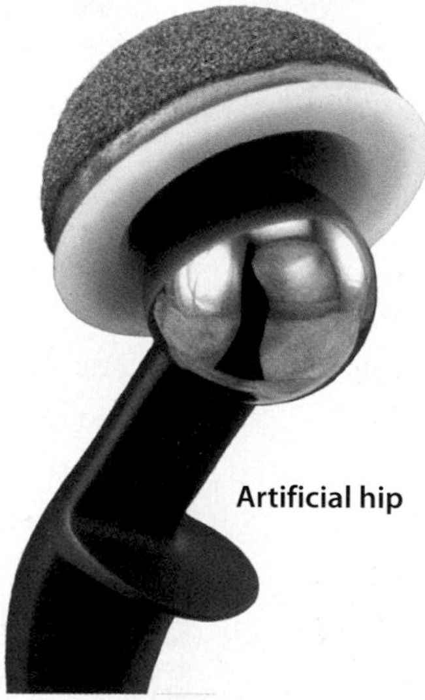

Artificial hip

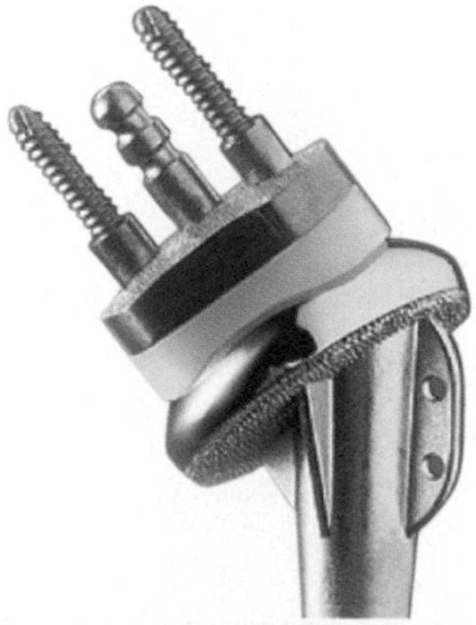

Artificial shoulder

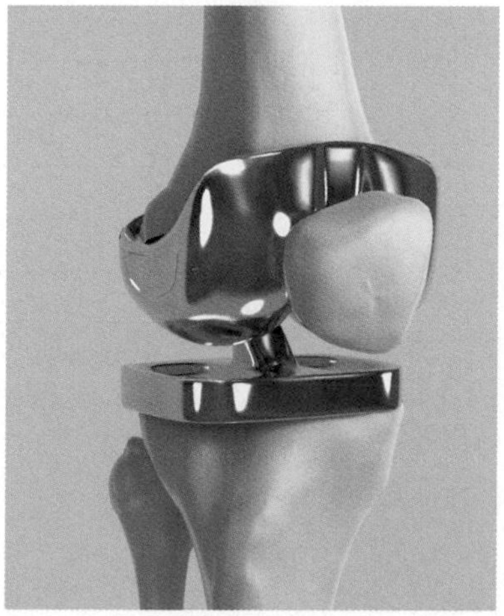

Artificial knee

Module 8.10 Review

- **a.** Compare rheumatism with osteoarthritis.
- **b.** Explain how an arthroscope is used.
- **c.** What can a person do to slow the progression of arthritis?

8.10 Explain arthritis, and describe its effects on joint structure and function.

Labeling

Label each of the structures in the accompanying diagram of the shoulder joint.

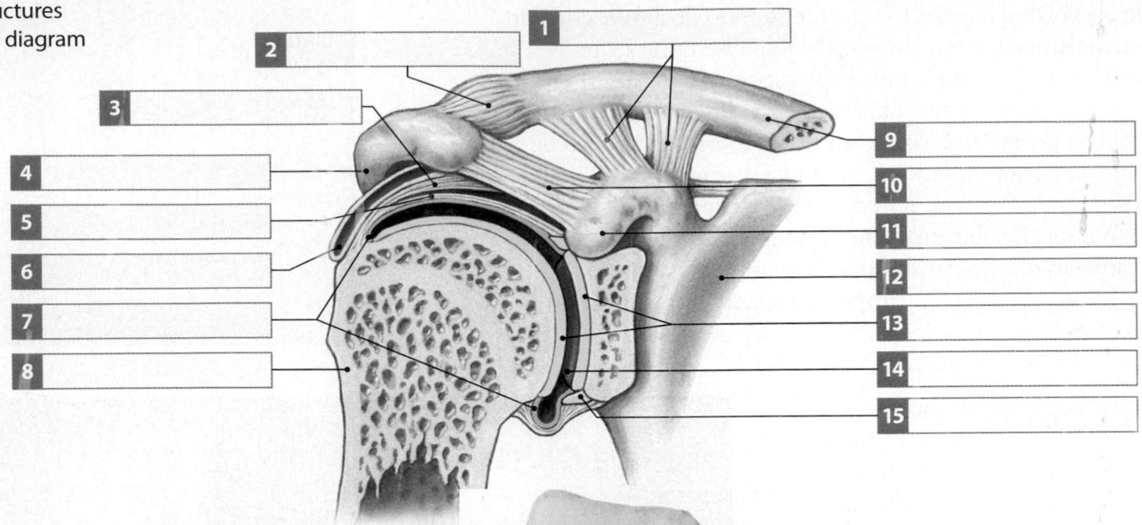
1
2
3
4
5
6
7
8
9
10
11
12
13
14
15

Label each of the structures in the accompanying photograph of the knee joint.

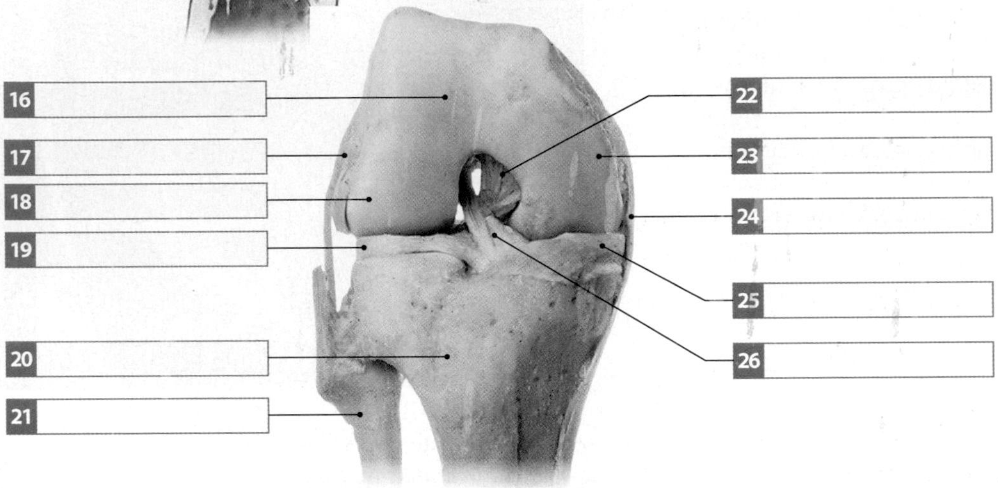
16
17
18
19
20
21
22
23
24
25
26

Matching

Match each lettered term with the most closely related description.

a. acetabulum
b. popliteal ligament
c. disc outer layer
d. dislocation
e. arthritis
f. disc inner layer
g. reinforce knee joint
h. osteoporosis

27 Knee joint posterior
28 Articular cartilage damage
29 Reduced bone mass
30 Cruciate ligaments
31 Anulus fibrosus
32 Deep fossa
33 Nucleus pulposus
34 Nursemaid's elbow

27 ______
28 ______
29 ______
30 ______
31 ______
32 ______
33 ______
34 ______

Short answer

35 Identify the joints that attach the pectoral and pelvic girdles to the axial skeleton.

______
______

Study Outline

SECTION 1 • Joint Structure and Movement

8.1 Joints are classified according to structure and movement p. 281

1. Movement of the skeleton occurs only at **joints** (**articulations**), which is where two bones meet.
2. A joint with no movement is called a **synarthrosis**, a joint with little movement is called an **amphiarthrosis**, and a freely movable joint is a **diarthrosis**.
3. The amount of movement at a joint is known as **range of motion (ROM)**.

8.2 Synovial joints are freely movable diarthroses lined with a synovial membrane p. 282

4. The components of a synovial joint are **articular cartilage**, a **joint capsule**, and a **synovial membrane**.
5. Synovial fluid lubricates the joint, provides nutrients for the chondrocytes of the articular cartilage, and helps to absorb shocks in the joints.
6. The accessory structures of a complex joint such as the knee include **bursae**, **fat pads**, **menisci**, and **accessory ligaments**.
7. Extreme stresses can cause joint **dislocation** or **luxation**.

8.3 Anatomical organization determines the motion at synovial joints p. 284

8. Joint movement is determined by the structural arrangement of the involved bones.
9. The types of synovial joints include **gliding**, **hinge**, **pivot**, **condylar**, **saddle**, and **ball-and-socket joints**.

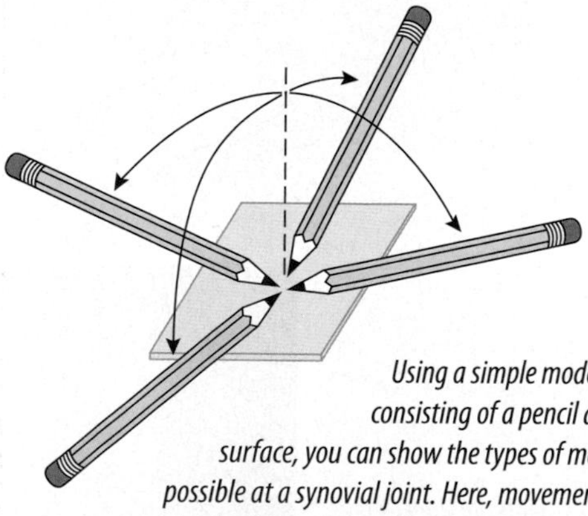

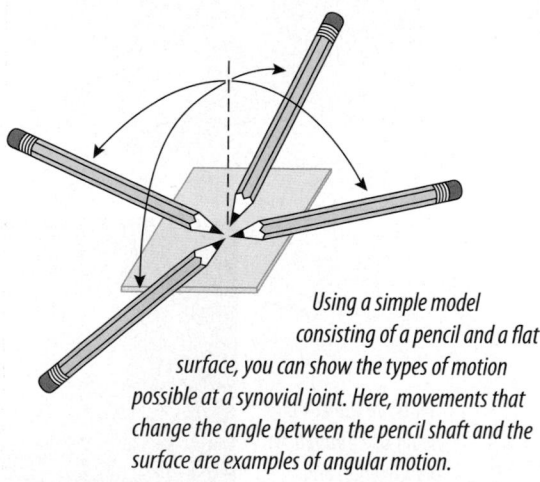

Using a simple model consisting of a pencil and a flat surface, you can show the types of motion possible at a synovial joint. Here, movements that change the angle between the pencil shaft and the surface are examples of angular motion.

8.4 Specific terms are used to describe movements with reference to the anatomical position p. 286

10. Terms used to describe joint movements include **flexion**, **extension**, **hyperextension**, **lateral flexion**, **dorsiflexion** (flexion at the ankle), **plantar flexion** (extension at the ankle), **abduction**, **adduction**, and **circumduction**.

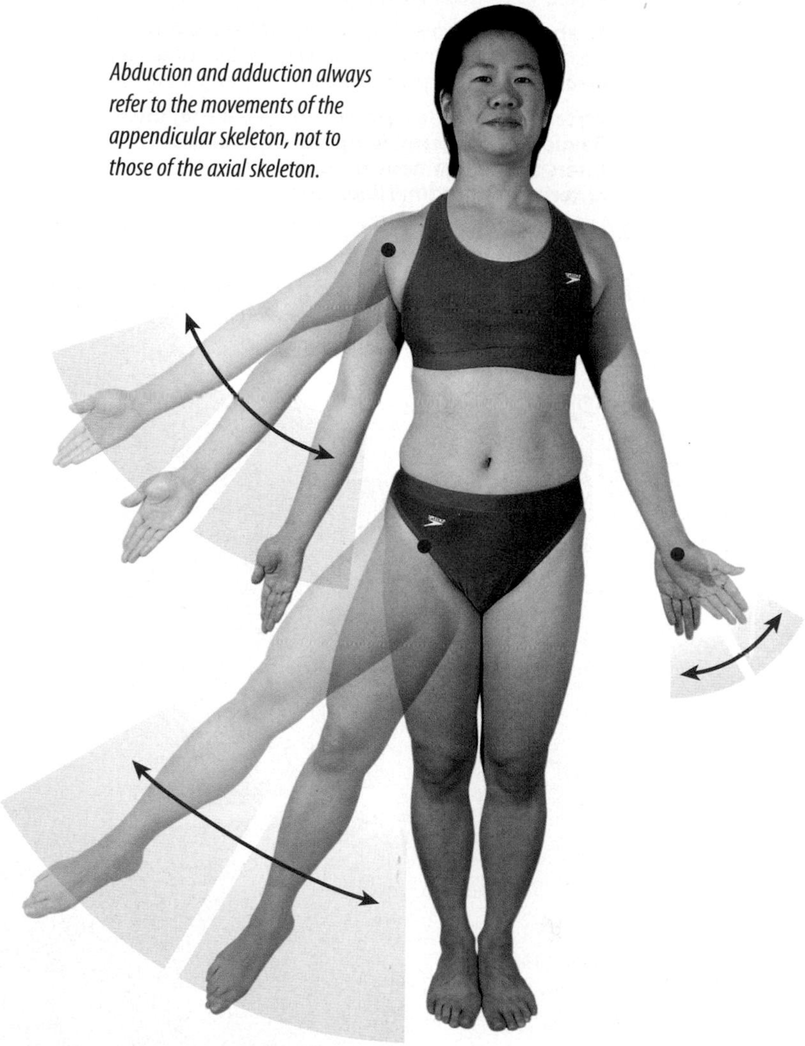

Abduction and adduction always refer to the movements of the appendicular skeleton, not to those of the axial skeleton.

8.5 Specific terms describe rotation and special movements p. 288

11. Rotations include **right rotation**, **left rotation**, **lateral** (external) **rotation**, **medial** (internal) **rotation**, **pronation**, and **supination**.
12. Special movements include **opposition**, **protraction**, **retraction**, **inversion**, **eversion**, **depression**, and **elevation**.

SECTION 2 • Axial and Appendicular Joints

8.6 Axial joints have less range of motion than appendicular joints p. 291

13. Most of the joints of the axial skeleton are strong and permit little movement.
14. Joints of the appendicular skeleton allow for extensive range of motion, but are often weaker or less stable.

8.7 Adjacent vertebrae have gliding diarthroses between their articular processes and symphyseal joints between their vertebral bodies p. 292

15. The superior and inferior articular processes of adjacent vertebrae form gliding joints.
16. The bodies of adjacent vertebrae are separated and cushioned by pads of fibrocartilage called **intervertebral discs** forming symphyseal joints.
17. Vertebral stabilizing ligaments include the **ligamentum flavum**, the **posterior longitudinal ligament**, the **interspinous ligament**, the **supraspinous ligament**, and the **anterior longitudinal ligament**.
18. An intervertebral disc can be distorted into a **bulging disc**, or a **herniated disc**.
19. Age-related loss of bone mass is **osteopenia**. **Osteoporosis** is a loss of bone mass that interferes with normal function.

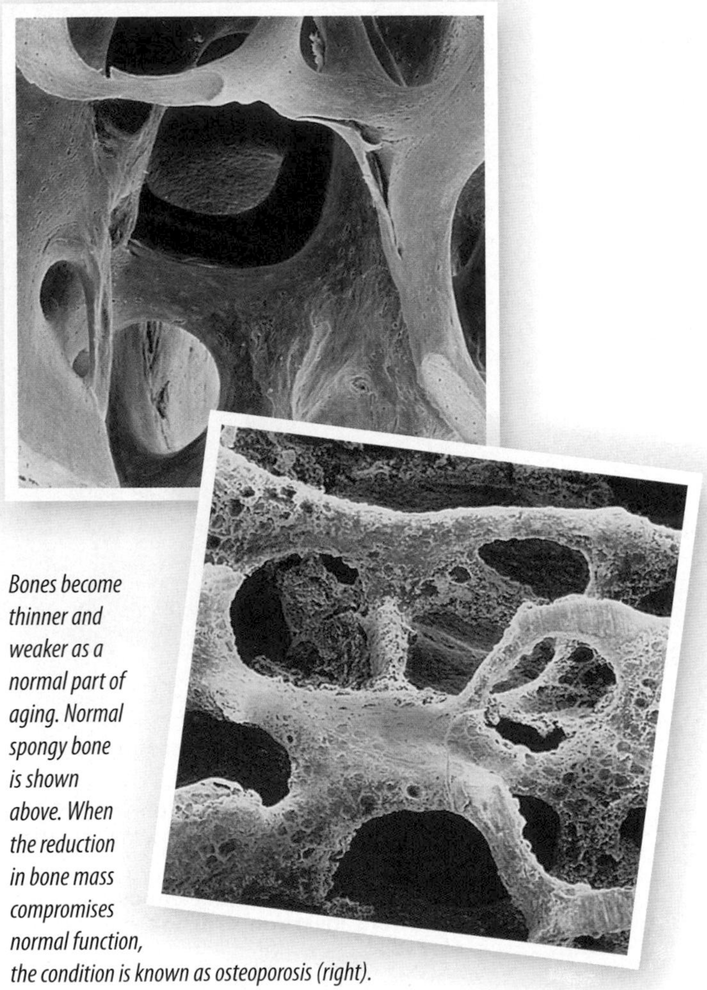

Bones become thinner and weaker as a normal part of aging. Normal spongy bone is shown above. When the reduction in bone mass compromises normal function, the condition is known as osteoporosis (right).

8.8 The shoulder and hip are ball-and-socket joints p. 294

20. The shoulder joint (**glenohumeral joint**) is a ball-and-socket joint with the greatest range of motion.
21. The ligaments stabilizing the shoulder joint are the coracoclavicular, acromioclavicular, coracoacromial, coracohumeral, and glenohumeral ligaments.
22. The depth of the glenoid cavity socket is increased by the **glenoid labrum**.
23. An **acetabular labrum**, the **ligamentum teres**, and the pubofemoral, the iliofemoral, and ischiofemoral ligaments stabilize the hip joint.

8.9 The elbow and knee are hinge joints p. 296

24. The elbow joint includes the **humeroradial joint** and the **humeroulnar joint**. The proximal radioulnar joint is not part of the elbow joint.
25. The knee joint includes three separate joints: two between the femur and tibia (medial condyle to medial condyle, and lateral condyle to lateral condyle) and one between the patella and the patellar surface of the femur.
26. The ligaments stabilizing the knee joint are the **fibular** and **tibial collateral ligaments**, the **anterior** and **posterior cruciate ligaments**, and the **medial** and **lateral menisci**.

8.10 Arthritis can disrupt normal joint structure and function p. 298

27. **Rheumatism** is a general term indicating musculoskeletal pain and stiffness.
28. **Arthritis** is joint inflammation. **Osteoarthritis** is the most common form of arthritis and is the result of age-related wear and tear. Other familiar types are **gouty arthritis** and **rheumatoid arthritis**.
29. Problematic joints can be treated with **arthroscopic surgery**, or joint replacement.

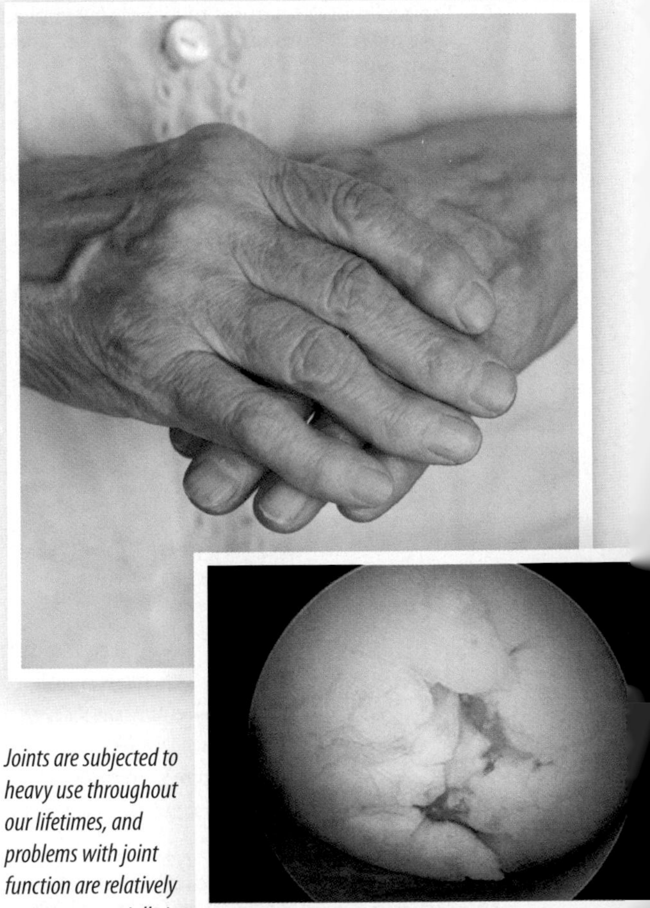

Joints are subjected to heavy use throughout our lifetimes, and problems with joint function are relatively common, especially in older individuals. This arthroscopic view of damaged cartilage shows how exposed surfaces change from a slick, smooth-gliding surface to a rough feltwork of bristly collagen fibers. This drastically increases friction at the joint and then promotes further degeneration.

Chapter Review Questions

True/False

Indicate whether each statement is true or false.

1. A bursa is a synovial fluid pocket outside of a joint capsule.
2. The acromioclavicular joint is an example of a condylar joint.
3. Standing up from a chair involves hip joint extension and knee joint flexion.
4. The ligamentum flavum connects the laminae of adjacent vertebrae.
5. The nucleus pulposus provides resiliency and shock absorption to the intervertebral disc.
6. The ligamentum teres attaches the head of the humerus into the glenoid cavity.

1 ____________

2 ____________

3 ____________

4 ____________

5 ____________

6 ____________

Multiple choice

Select the correct answer from the list provided.

7. The synarthrosis formed between the bones of the skull is a
   - ☐ a) gomphosis.
   - ☐ b) synostosis.
   - ☐ c) symphysis.
   - ☐ d) suture.

8. The joint between adjacent vertebral bodies is a
   - ☐ a) synostosis.
   - ☐ b) syndesmosis.
   - ☐ c) symphysis.
   - ☐ d) gomphosis.

9. The joints typically located at the end of adjacent long bones are
   - ☐ a) synarthroses.
   - ☐ b) amphiarthroses.
   - ☐ c) diarthroses.
   - ☐ d) symphyses.

10. The function of articular cartilage is to
    - ☐ a) reduce friction.
    - ☐ b) prevent bony surfaces from contacting one another.
    - ☐ c) provide lubrication.
    - ☐ d) both a and b

11. Which of the following is *not* a function of synovial fluid?
    - ☐ a) shock absorption
    - ☐ b) nutrient distribution
    - ☐ c) lubrication of articular surfaces
    - ☐ d) maintenance of ionic balance

12. A dislocation of an articulating surface is
    - ☐ a) circumduction.
    - ☐ b) hyperextension.
    - ☐ c) luxation.
    - ☐ d) supination.

13. Abduction and adduction always refer to movements of the
    - ☐ a) axial skeleton.
    - ☐ b) appendicular skeleton.
    - ☐ c) skull.
    - ☐ d) vertebral column.

14. Rotation of the radius that makes the palm face posteriorly is
    - ☐ a) supination.
    - ☐ b) pronation.
    - ☐ c) protraction.
    - ☐ d) inversion.

15. A ballerina standing on her toes is performing which action of the ankle joint?
    - ☐ a) dorsiflexion
    - ☐ b) plantar flexion
    - ☐ c) inversion
    - ☐ d) eversion

16. Increasing the angle between bones is
    - ☐ a) flexion.
    - ☐ b) extension.
    - ☐ c) hyperextension.
    - ☐ d) abduction.

17 The hip is an extremely stable joint because it has
- ☐ a) a complete bony socket.
- ☐ b) a strong articular capsule.
- ☐ c) supporting ligaments.
- ☐ d) all of the above

18 Movement of a limb away from the longitudinal axis of the body is
- ☐ a) abduction.
- ☐ b) adduction.
- ☐ c) hyperextension.
- ☐ d) protraction.

19 Although the knee is only considered to be one joint, how many separate joints does it actually contain?
- ☐ a) two
- ☐ b) three
- ☐ c) four
- ☐ d) five

20 The head of the radius is attached to the ulna by the
- ☐ a) radial collateral ligament.
- ☐ b) ulnar collateral ligament.
- ☐ c) annular ligament.
- ☐ d) medial cruciate ligament.

Short answer

21 Differentiate between a bulging disc and a herniated disc.

22 Explain how articular cartilage differs from other cartilage in the body.

23 List the six different types of diarthroses, and give an example of each.

24 Describe how a meniscus functions in a joint.

25 Distinguish between osteopenia and osteoporosis.

Chapter Integration • Applying what you have learned

A 65-year-old avid golfer considers total knee replacement surgery to address chronic pain

Jon loved his golf. He loved the feel of just being out on a golf course. Even into his early 60s he still played; his only concession to age—and to a chronically aching knee—was pushing his bag on a wheeled cart instead of carrying his bag over his shoulder. He chalked the bum knee up to aging and managed the pain with wraps, ibuprofen, ice, and hot baths. For years his family physician had been urging him to consider a knee replacement, because advances in the procedure made it almost miraculous in relieving pain and allowing near-normal activity. Finally realizing that the pain would only get worse and that he might be prevented from enjoying his golf, Jon did some research on knee replacement surgery, or knee arthroplasty.

Arthroplasty is used as a last resort when other treatments, such as physical therapy and anti-inflammatory drugs, do not relieve the signs and symptoms, and when the patient's discomfort becomes intense. In knee arthroplasty, a total knee dissection is performed, and damaged cartilage and bone, including the patella, are removed. The removed natural structures are replaced by synthetic structures made of metal and plastic—typically, a plastic patella rubs against the metal surfaces over the distal femur and proximal tibia. In the United States, the average age for a patient undergoing joint replacement is about 65–70 years. After reading about the surgery and speaking with several friends who had had the procedure, Jon decided to undergo knee arthroplasty. Here is an x-ray of Jon's knees taken 6 months after surgery. He can now walk without pain, and has returned to golf—although his game remains as bad as ever.

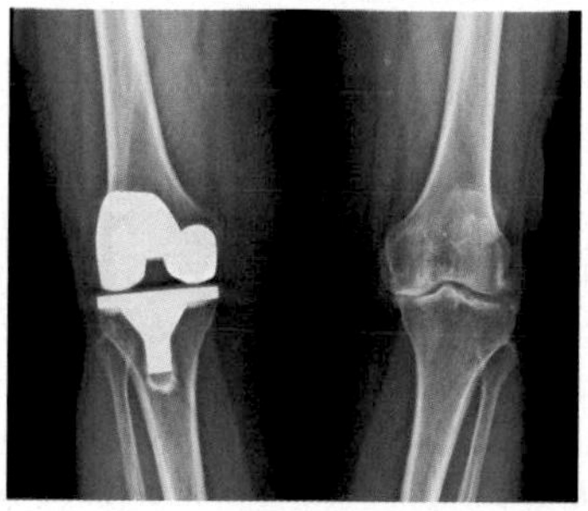

1 **Which bones compose the functioning knee joint?**

2 **What purpose would physical therapy serve for a person suffering from joint pain?**

3 **Classify the knee joint in terms of function and structure, and indicate the types of movements possible at this joint.**

4 **Predict which activities may be restricted following a total knee arthroplasty.**

Assess a high-school quarterback's shoulder injury

Tommy is the quarterback on his high school football team. Trailing by four points late in the game, he calls for a long pass play. After taking the snap, he drops back and looks downfield for an open receiver. His linemen block as best they can, but are not able to cover the blitzing linebacker. To avoid a sack, Tommy cocks back his arm and tries to rifle a pass to his tight end downfield. At the exact moment Tommy elevates the ball to throw, the linebacker is on him. The linebacker knocks the ball loose by chopping down onto Tommy's arm midway between his shoulder and elbow joints. When Tommy's arm is hit, he feels intense pain in his shoulder. As he runs to the sidelines in extreme discomfort, he is completely unable to move his shoulder joint.Using what you have just learned about joints and joint structure, answer the following questions.

5 **Do you think Tommy suffered a shoulder dislocation or a shoulder separation?**

6 **Which bones, ligaments, and joint structures would you expect to be affected in Tommy's injury?**

7 **Would you expect Tommy's injury to heal completely, or has he incurred a long-term joint instability?**

9 Skeletal Muscle Tissue

LEARNING OUTCOMES

These Learning Outcomes correspond by number to this chapter's modules and indicate what you should be able to do after completing the chapter.

SECTION 1 • Functional Anatomy of Skeletal Muscle Tissue

- **9.1** Describe the functions of skeletal muscle tissue.
- **9.2** Describe the organization of skeletal muscle at the tissue level.
- **9.3** Identify the structural components of a sarcomere.
- **9.4** Describe the structural components of a thin filament and a thick filament.
- **9.5** Describe a major characteristic of excitable membranes and its importance in generating an action potential.
- **9.6** Identify the components of the neuromuscular junction, and summarize the events involved in the control of skeletal muscles by motor neurons.
- **9.7** Describe the role of ATP in a muscle contraction, and explain the steps involved in the contraction of a skeletal muscle fiber.

SECTION 2 • Functional Properties of Skeletal Muscle

- **9.8** Describe how muscle tension develops with respect to neural control and excitation-contraction coupling.
- **9.9** Describe the mechanism responsible for tension production in a muscle fiber, and discuss the factors that determine the peak tension developed during a contraction.
- **9.10** Discuss the factors that affect peak tension production during the contraction of an entire skeletal muscle, and explain the significance of the motor unit in this process.
- **9.11** Compare the different types of muscle contractions.
- **9.12** Describe the mechanisms by which muscle fibers obtain the energy to power contractions.
- **9.13** Describe the factors that contribute to muscle fatigue, and discuss the stages and mechanisms involved in the muscle's subsequent recovery.
- **9.14** Relate the types of muscle fibers to muscle performance.
- **9.15** **CLINICAL MODULE** Explain the physiological factors responsible for muscle hypertrophy, atrophy, and paralysis.

LO Learning Outcomes are repeated at the bottom of each module.

Skeletal muscle tissue enables body movement and other vital functions

Muscle tissue, one of the four primary tissue types, consists chiefly of muscle cells that are highly specialized for contraction. Without the three types of muscle tissue—skeletal, cardiac, and smooth—nothing in the body would move, and no body movement could occur. In this chapter we consider the structure and function of skeletal muscle tissue. Each cell in skeletal muscle tissue is a single muscle fiber. **Skeletal muscles** are organs composed primarily of skeletal muscle tissue plus connective tissues, nerves, and blood vessels. Skeletal muscles are directly or indirectly attached to bones and have several functions, as detailed in the list of functions at right.

Muscle Tissue

Skeletal Muscle Tissue

Skeletal muscle tissue contractions are under voluntary control. They move the body by pulling on bones, thus making it possible for us to walk, dance, bite an apple, or play the piano.

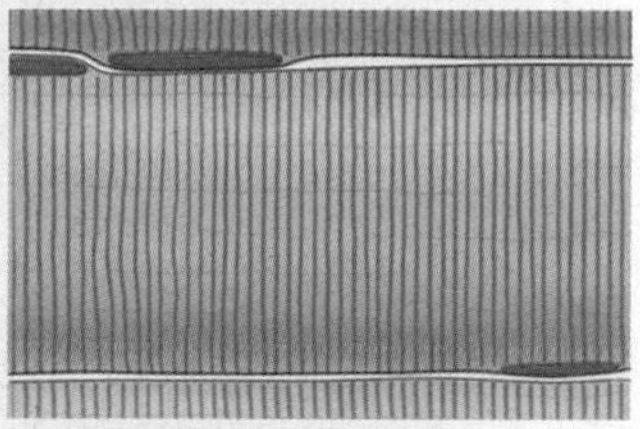

Cardiac Muscle Tissue

Cardiac muscle tissue is involuntary and is found in the heart. When it contracts, blood is pumped through the blood vessels.

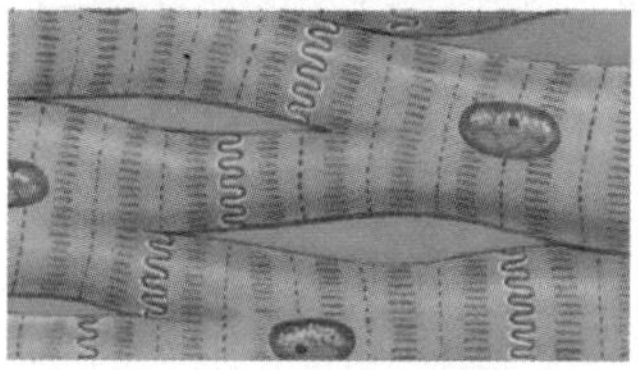

See Chapter 19.

Smooth Muscle Tissue

Smooth muscle tissue is involuntary and is found in the walls of hollow organs. In the digestive tract it moves fluids and solids. As a lining in small arteries, it regulates vessel diameter.

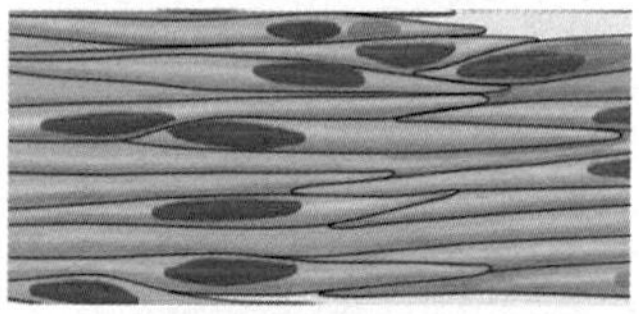

See Chapter 22.

In this section we examine the functional anatomy of a typical skeletal muscle. We will place particular emphasis on the microscopic structural features that make contractions possible.

Functions of Skeletal Muscle Tissue

- **Produce Skeletal Movement**. Skeletal muscle contractions pull on tendons and move the bones. The effects range from simple motions such as extending the arm or breathing, to the highly coordinated movements of swimming, skiing, or typing.
- **Maintain Posture and Body Position**. Tension in skeletal muscles stabilizes joints to help maintain body posture—for example, holding your head still when you read a book, or balancing your body weight above your feet when you walk. Without constant muscular activity, you could neither sit upright nor stand.
- **Support Soft Tissues**. The abdominal wall and the floor of the pelvic cavity consist of layers of skeletal muscle. These muscles support the weight of visceral organs and shield internal tissues from injury.
- **Guard Body Entrances and Exits**. Skeletal muscles (sphincters) encircle the openings of the digestive and urinary tracts. They provide voluntary control over swallowing, defecation, and urination.
- **Maintain Body Temperature**. Muscle contractions require energy; whenever energy is used in the body, some of it is converted to heat. The heat released by working muscles keeps body temperature in the range required for normal functioning.
- **Provide Nutrient Reserves**. When the diet does not contain sufficient protein or calories, the contractile proteins in skeletal muscles are broken down into amino acids, which are released into the circulation. Some of these amino acids can be used by the liver to synthesize glucose; others can be broken down to provide energy.

Module 9.1 Review

a. Name the three types of muscle tissue, identify where they are found, and list their functions.

b. Which muscle types are voluntary, and which are involuntary?

c. Describe the functions of skeletal muscle tissue.

LO **9.1** Describe the functions of skeletal muscle tissue.

A skeletal muscle contains skeletal muscle tissue, connective tissues, blood vessels, and nerves

Structure of a Skeletal Muscle

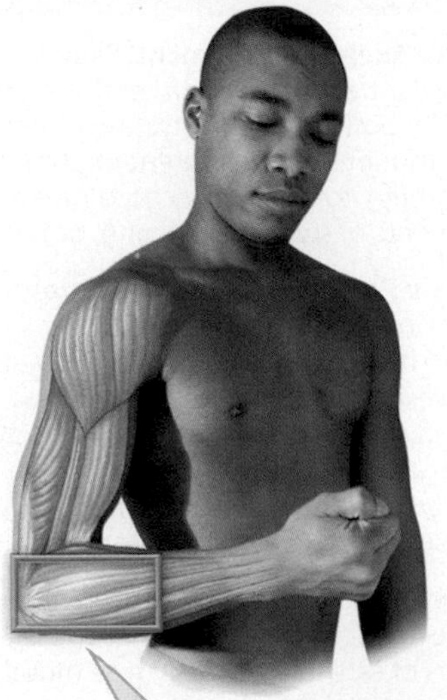

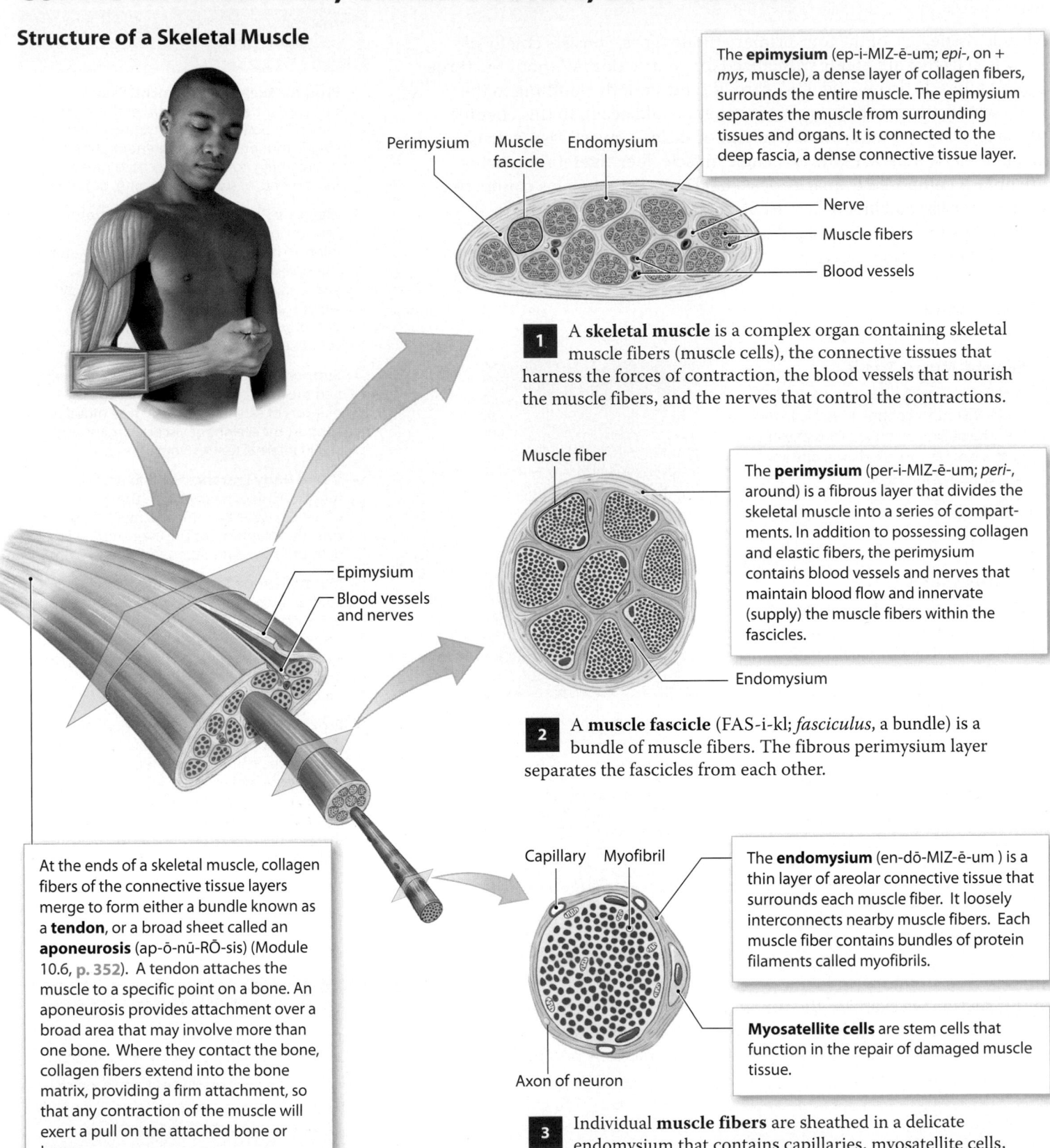

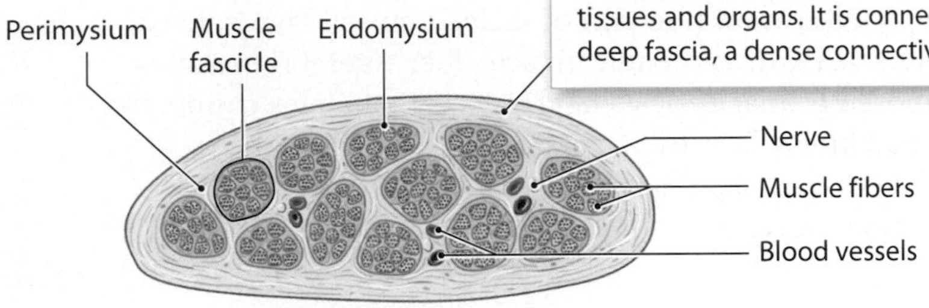
Perimysium
Muscle fascicle
Endomysium
Nerve
Muscle fibers
Blood vessels

The **epimysium** (ep-i-MIZ-ē-um; *epi-*, on + *mys*, muscle), a dense layer of collagen fibers, surrounds the entire muscle. The epimysium separates the muscle from surrounding tissues and organs. It is connected to the deep fascia, a dense connective tissue layer.

1 A **skeletal muscle** is a complex organ containing skeletal muscle fibers (muscle cells), the connective tissues that harness the forces of contraction, the blood vessels that nourish the muscle fibers, and the nerves that control the contractions.

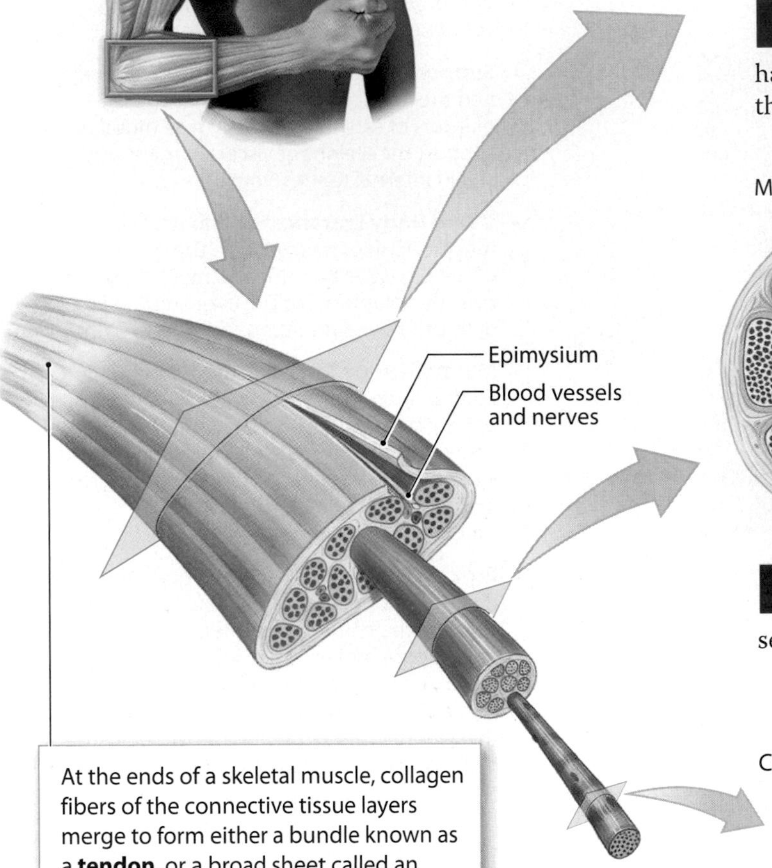
Epimysium
Blood vessels and nerves

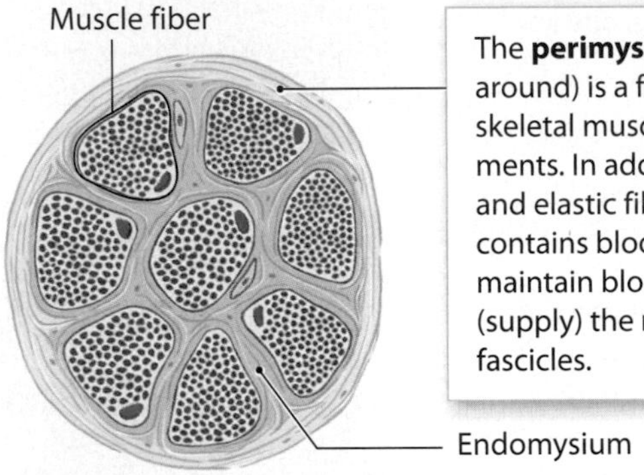
Muscle fiber
Endomysium

The **perimysium** (per-i-MIZ-ē-um; *peri-*, around) is a fibrous layer that divides the skeletal muscle into a series of compartments. In addition to possessing collagen and elastic fibers, the perimysium contains blood vessels and nerves that maintain blood flow and innervate (supply) the muscle fibers within the fascicles.

2 A **muscle fascicle** (FAS-i-kl; *fasciculus*, a bundle) is a bundle of muscle fibers. The fibrous perimysium layer separates the fascicles from each other.

At the ends of a skeletal muscle, collagen fibers of the connective tissue layers merge to form either a bundle known as a **tendon**, or a broad sheet called an **aponeurosis** (ap-ō-nū-RŌ-sis) (Module 10.6, p. 352). A tendon attaches the muscle to a specific point on a bone. An aponeurosis provides attachment over a broad area that may involve more than one bone. Where they contact the bone, collagen fibers extend into the bone matrix, providing a firm attachment, so that any contraction of the muscle will exert a pull on the attached bone or bones.

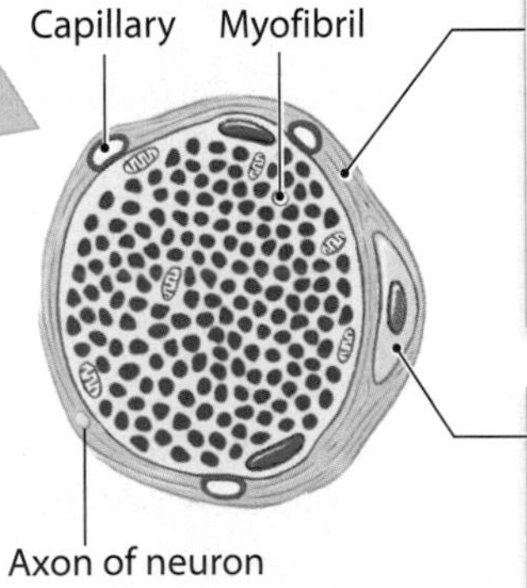
Capillary
Myofibril
Axon of neuron

The **endomysium** (en-dō-MIZ-ē-um) is a thin layer of areolar connective tissue that surrounds each muscle fiber. It loosely interconnects nearby muscle fibers. Each muscle fiber contains bundles of protein filaments called myofibrils.

Myosatellite cells are stem cells that function in the repair of damaged muscle tissue.

3 Individual **muscle fibers** are sheathed in a delicate endomysium that contains capillaries, myosatellite cells, and the axons of the neurons that control the muscle fibers.

Skeletal Muscle Development

4 During development, groups of embryonic cells called **myoblasts** (*myo-*, muscle + *blastos*, formative cell) fuse, forming multinucleate cells. These large cells then develop into distinctive skeletal muscle fibers. Each nucleus in a skeletal muscle fiber represents the contribution of a single myoblast. Some myoblasts, however, do not fuse with developing muscle fibers. These unfused cells remain in the endomysium of adult skeletal muscle tissue as myosatellite cells. After an injury, myosatellite cells may enlarge, divide, and fuse with damaged muscle fibers, helping to repair the tissue.

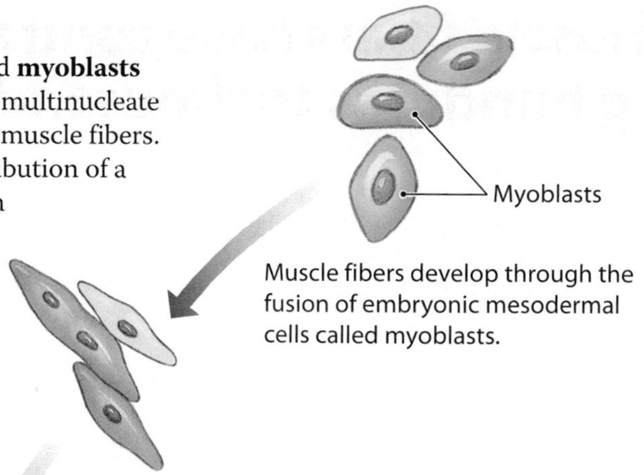
Myoblasts
Muscle fibers develop through the fusion of embryonic mesodermal cells called myoblasts.

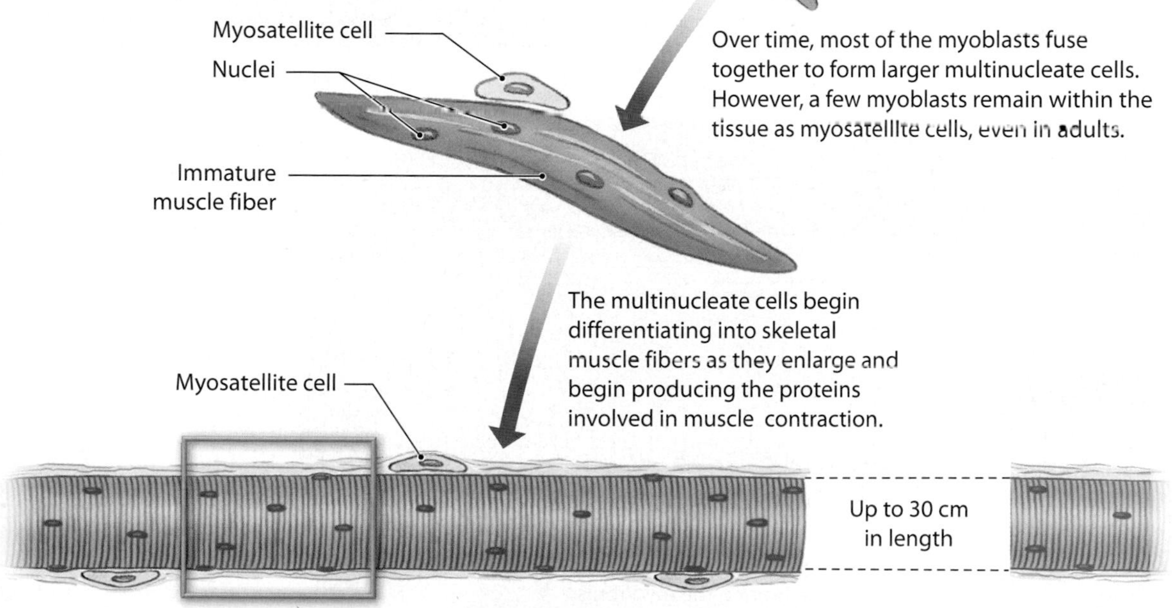
Myosatellite cell
Nuclei
Immature muscle fiber
Over time, most of the myoblasts fuse together to form larger multinucleate cells. However, a few myoblasts remain within the tissue as myosatellite cells, even in adults.
The multinucleate cells begin differentiating into skeletal muscle fibers as they enlarge and begin producing the proteins involved in muscle contraction.
Myosatellite cell
Up to 30 cm in length

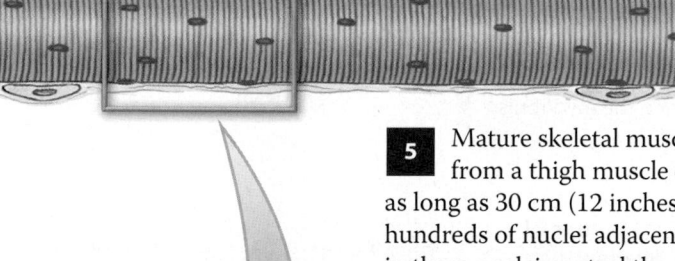

5 Mature skeletal muscle fibers are enormous. A muscle fiber from a thigh muscle could have a diameter of 100 μm and be as long as 30 cm (12 inches). Each skeletal muscle fiber contains hundreds of nuclei adjacent to the plasma membrane. The genes in these nuclei control the production of enzymes and structural proteins required for normal muscle contraction, and the more copies of these genes, the faster these proteins can be produced.

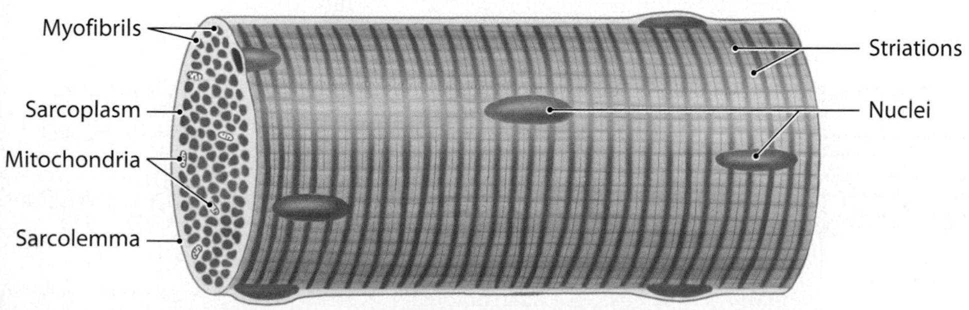
Myofibrils
Sarcoplasm
Mitochondria
Sarcolemma
Striations
Nuclei

6 Skeletal muscle fibers are so unusual that special terms are used to describe them. The plasma membrane is called the **sarcolemma** (sar-kō-LEM-uh; *sarkos*, flesh + *lemma*, husk), and the cytoplasm is called **sarcoplasm** (SAR-kō-plazm) (Module 3.2, p. 88). Most of the sarcoplasm consists of myofibrils.

Module 9.2 Review

a. Define tendon and aponeurosis.

b. Describe the connective tissue layers associated with skeletal muscle tissue.

c. How would severing the tendon attached to a muscle affect the muscle's ability to move a body part?

Skeletal muscle fibers have contractile myofibrils containing hundreds to thousands of sarcomeres

1 A **myofibril** is a cylindrical structure 1–2 μm in diameter and as long as the entire fiber. The sarcoplasm of a single skeletal muscle fiber may contain hundreds to thousands of myofibrils. Because each myofibril has a banded appearance and the fiber is jam-packed with myofibrils lying side-by-side, the entire muscle fiber appears to have light and dark bands, or **striations**.

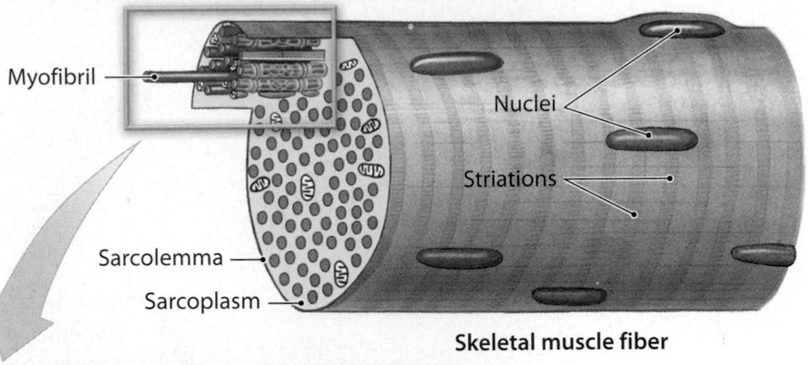
Myofibril
Nuclei
Striations
Sarcolemma
Sarcoplasm
Skeletal muscle fiber

2 Myofibrils consist of bundles of protein filaments called **myofilaments**. The most abundant myofilaments are **thin filaments** composed primarily of actin, and **thick filaments** composed primarily of myosin.

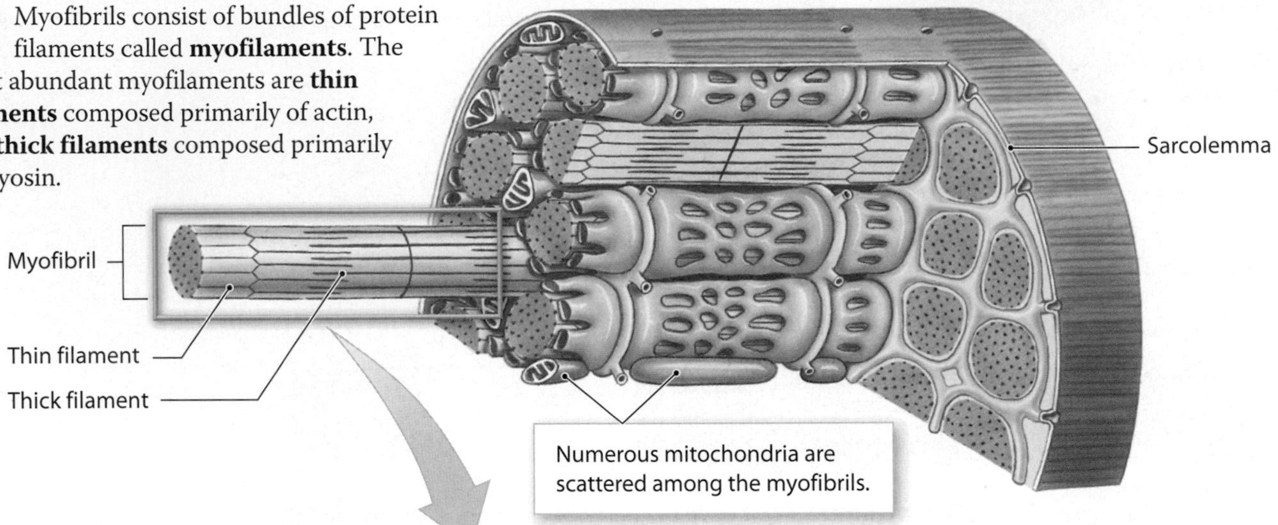
Sarcolemma
Myofibril
Thin filament
Thick filament
Numerous mitochondria are scattered among the myofibrils.

3 The myofilaments within each myofibril are arranged into repeating contractile units called **sarcomeres** (SAR-kō-mērz; *sarkos*, flesh + *meros*, part). Each myofibril consists of approximately 10,000 sarcomeres aligned end to end. Each sarcomere has a resting length of about 2 μm.

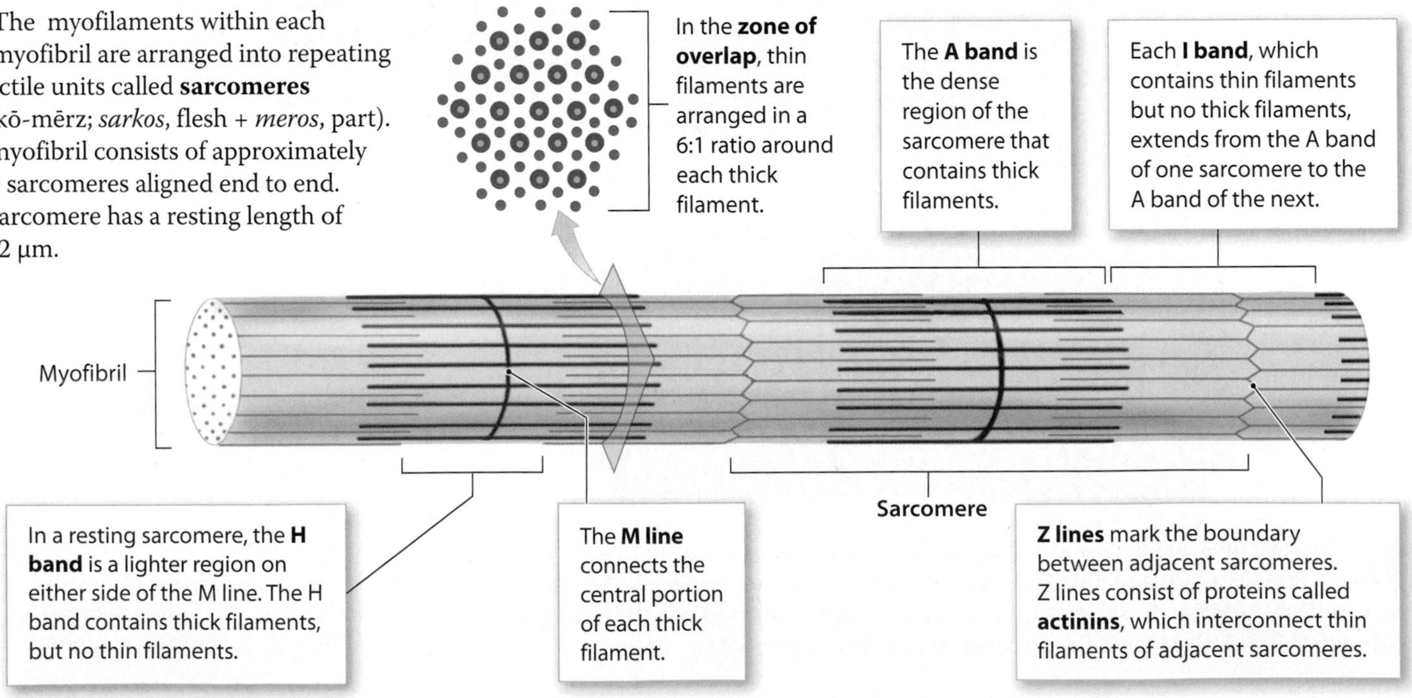
In the **zone of overlap**, thin filaments are arranged in a 6:1 ratio around each thick filament.
The **A band** is the dense region of the sarcomere that contains thick filaments.
Each **I band**, which contains thin filaments but no thick filaments, extends from the A band of one sarcomere to the A band of the next.
Myofibril
Sarcomere
In a resting sarcomere, the **H band** is a lighter region on either side of the M line. The H band contains thick filaments, but no thin filaments.
The **M line** connects the central portion of each thick filament.
Z lines mark the boundary between adjacent sarcomeres. Z lines consist of proteins called **actinins**, which interconnect thin filaments of adjacent sarcomeres.

4 The sarcolemma separates the sarcoplasm from the surrounding interstitial fluid. Negative electrical charges are on the internal surface and positive electrical charges are on the external surface. This uneven distribution of electrical charges depends on the selective permeability of the plasma membrane. In skeletal muscle fibers, the distribution of electrical charges can change markedly with changes in membrane permeability. In fact, a sudden reversal in charge distribution is the first step that leads to a muscle contraction. These changes, triggered by the activity of a nerve cell (neuron) at one location on the sarcolemma, are swiftly propagated (spread) across the entire sarcolemma of the muscle fiber.

Transverse tubules, or **T tubules**, are narrow tubes that are continuous with the sarcolemma and extend into the sarcoplasm at right angles to the cell surface. T tubules form passageways through the muscle fiber, like a network of tunnels through a mountain.

Inside the sarcoplasm, T tubules encircle each sarcomere at the zones of overlap. Wherever a transverse tubule encircles a myofibril, the tubule is tightly bound to the membranes of the sarcoplasmic reticulum.

The **sarcoplasmic reticulum (SR)**, similar to the smooth endoplasmic reticulum of other cells, forms a tubular network around each individual myofibril. On either side of a T tubule, the tubules of the SR enlarge, fuse, and form expanded chambers called **terminal cisternae** (sis-TUR-nē).

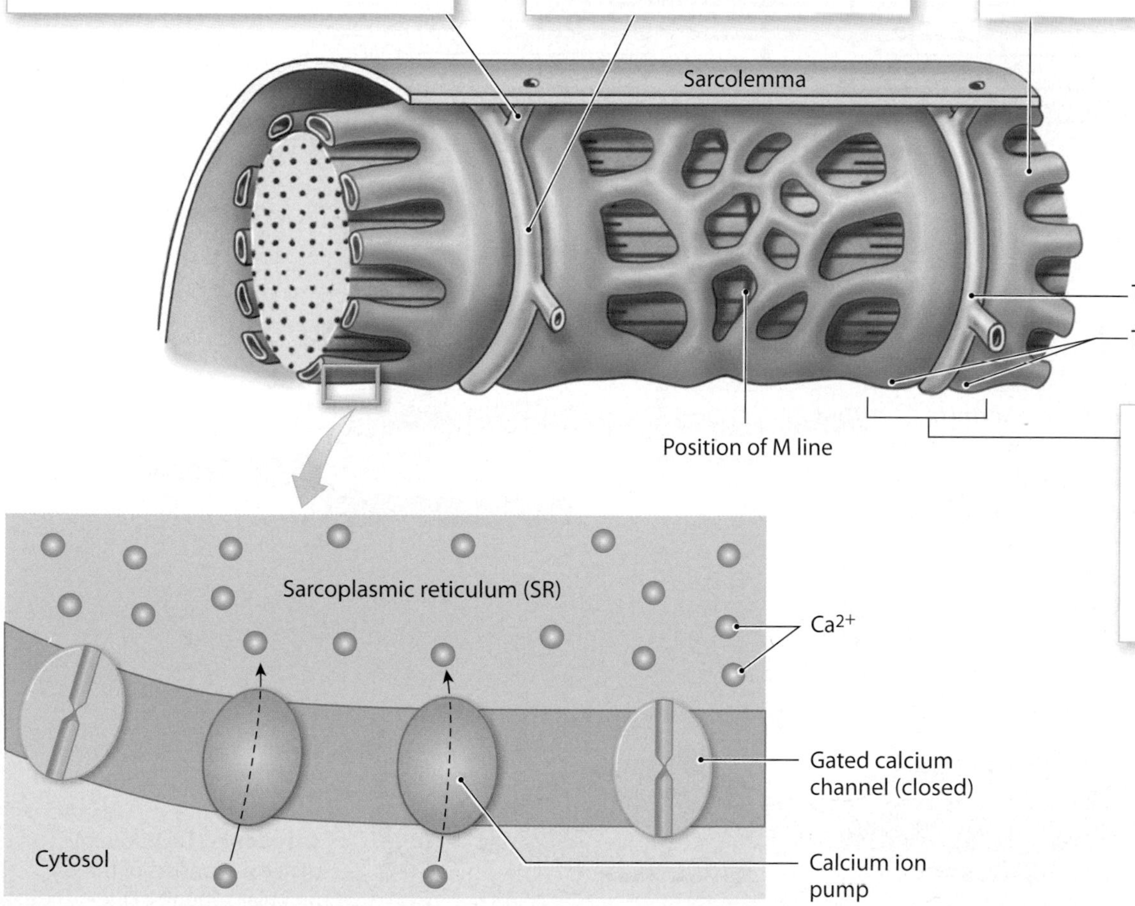
Sarcolemma
Transverse tubule
Terminal cisternae
Position of M line
Sarcoplasmic reticulum (SR)
Ca^{2+}
Gated calcium channel (closed)
Cytosol
Calcium ion pump

The combination of a pair of terminal cisternae plus a transverse tubule is known as a **triad**. Although the membranes of the triad are tightly bound together, their fluid contents are separate and distinct.

5 The membrane of the SR contains ion pumps that pump calcium ions from the cytosol into the SR. (To be precise, we use the term cytosol, instead of sarcoplasm, because sarcoplasm includes the cytosol and the organelles. Releasing Ca^{2+} into the sarcoplasm implies they are released into the entire sarcoplasm, including the SR.) In the SR, some of the calcium ions remain free in solution; the rest become reversibly bound to the protein **calsequestrin**. Including both the free calcium and the bound calcium, the total concentration of Ca^{2+} within terminal cisternae in a resting (inactive) skeletal muscle fiber can be 40,000 times that of the surrounding cytosol. A muscle contraction begins when stored calcium ions are released into the cytosol through gated calcium channels.

Module 9.3 Review

a. Describe the structural components of a sarcomere.

b. Define transverse tubules.

c. Where would you expect the greatest concentration of Ca^{2+} to be in a resting skeletal muscle fiber?

LO 9.3 Identify the structural components of a sarcomere.

The sliding filament theory of muscle contraction involves thin and thick filaments

A longitudinal section of a sarcomere enables us to examine the structure and organization of the thin and thick filaments in greater detail.

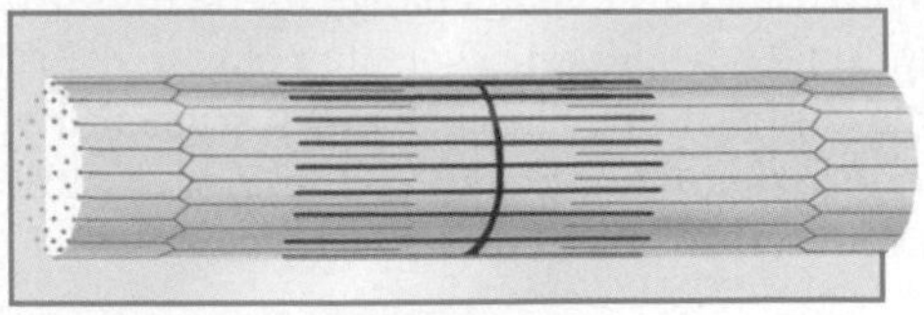

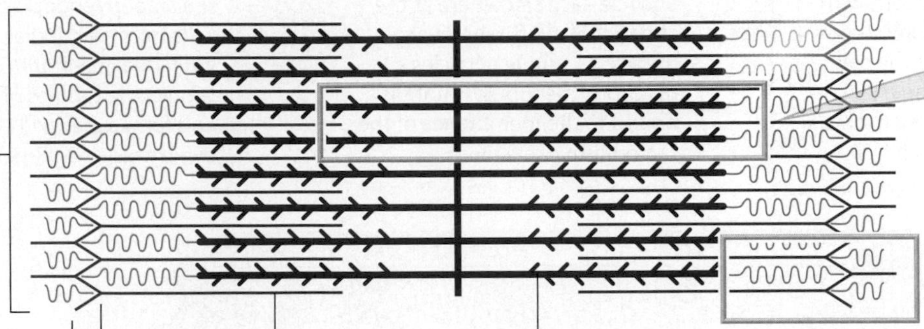
Myofibril
Z line
Thin filament
Thick filament

Structure of Thin Filaments

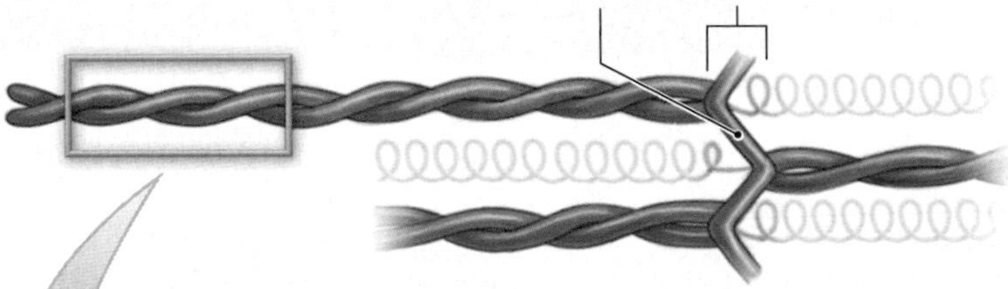
Actinin
Z line

1 At either end of the sarcomere, the thin filaments are attached to the Z line. **Actinin** interconnects the thin filaments at the Z line, creating an open meshwork at the end of the sarcomere. A typical thin filament is 5–6 nm in diameter and 1 μm long. As indicated in the magnified view below, each filament is primarily composed of actin, which is associated with other interacting proteins.

F-actin (filamentous actin) is a twisted strand composed of two rows of 300–400 individual molecules of G-actin (globular actin).

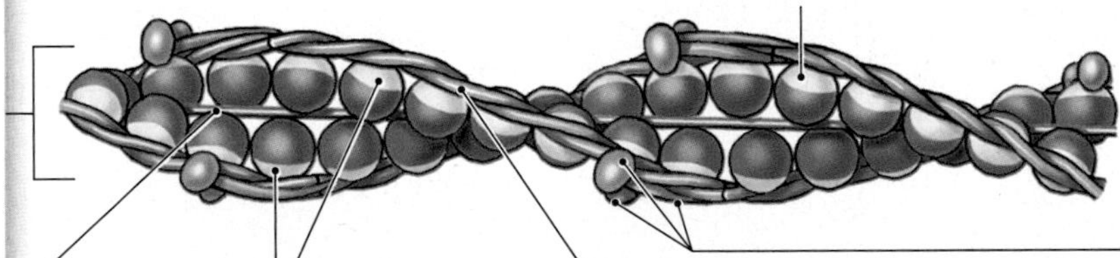
Active site

A **troponin** (TRŌ-pō-nin) molecule consists of three globular subunits. One subunit binds to tropomyosin, locking them together as a troponin–tropomyosin complex. A second subunit binds to one G-actin, holding the troponin–tropomyosin complex in position. And the third subunit has a receptor that binds two calcium ions.

Nebulin is a large protein that extends along the F-actin strand in the cleft between the rows of G-actin molecules. It holds the F-actin strand together.

Each **G-actin** molecule contains an **active site** to which myosin can bind. This is much like a substrate molecule binding to an enzyme's active site.

Strands of **tropomyosin** (trō-pō-MĪ-ō-sin; *tropos*, turning) cover the active sites on G-actin and prevent actin–myosin interaction. A tropomyosin molecule is a double-stranded protein that is bound to one molecule of troponin midway along its length.

Structure of Thick Filaments

2 Thick filaments are 10–12 nm in diameter and 1.6 μm long. A thick filament contains about 300 **myosin molecules**, each made up of a pair of myosin subunits twisted around one another. All the myosin molecules are arranged with their tails pointing toward the M line. The myosin heads are arranged in a spiral, each facing one of the surrounding thin filaments. There are no myosin heads in a small area on either side of the M line.

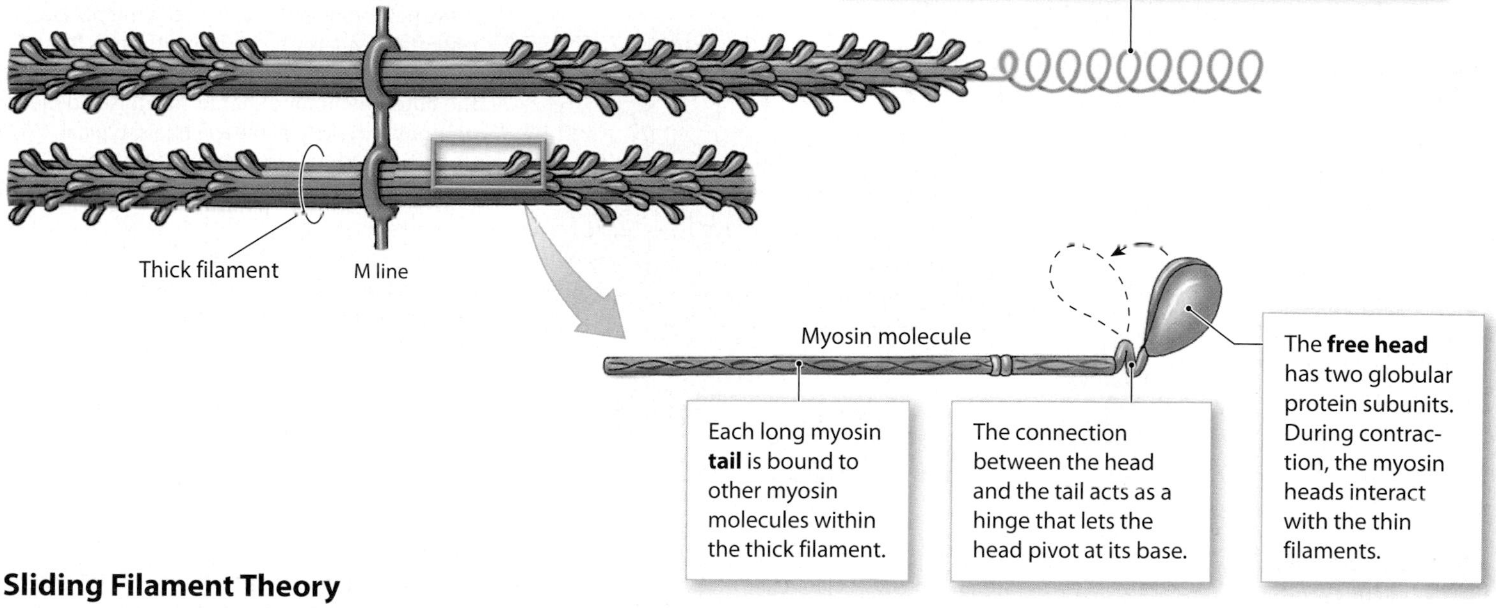

Each thick filament has a core of **titin**. From either end of the thick filament, a strand of titin continues across the I band to the Z line on that side. The portion of the titin strand exposed within the I band is elastic and recoils after stretching. In the normal resting sarcomere, the titin strands are completely relaxed; they become tense only when some external force stretches the sarcomere.
Thick filament
M line
Myosin molecule
Each long myosin **tail** is bound to other myosin molecules within the thick filament.
The connection between the head and the tail acts as a hinge that lets the head pivot at its base.
The **free head** has two globular protein subunits. During contraction, the myosin heads interact with the thin filaments.

Sliding Filament Theory

3 When a skeletal muscle fiber contracts, thin filaments slide past the thick filaments. In this process, (1) the H bands and I bands get smaller, (2) the zones of overlap get larger, (3) the Z lines move closer together, and (4) the width of the A band remains constant. This explanation is known as the **sliding filament theory**.

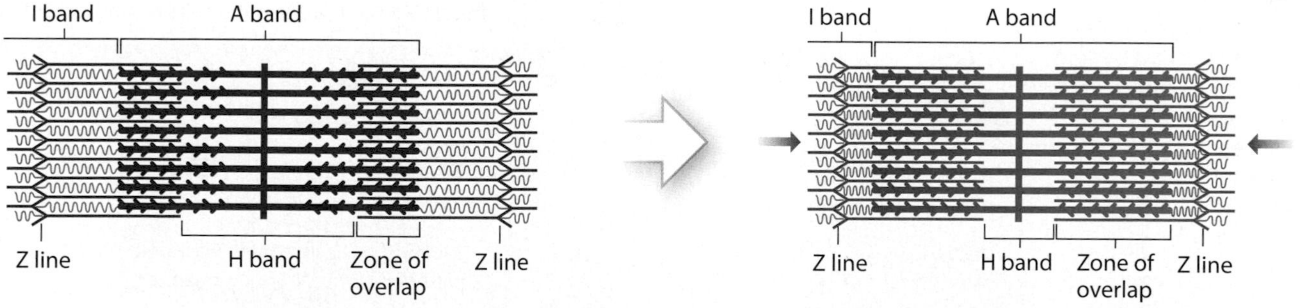

I band
A band
Z line
H band
Zone of overlap
Z line
I band
A band
Z line
H band
Zone of overlap
Z line

Sarcomere at rest

Sarcomere contraction and filament sliding

4 During a contraction, sliding occurs in every sarcomere along a myofibril, so the myofibril gets shorter. Because myofibrils are attached to the sarcolemma at each Z line and at either end of the muscle fiber, when myofibrils shorten, so does the muscle fiber.

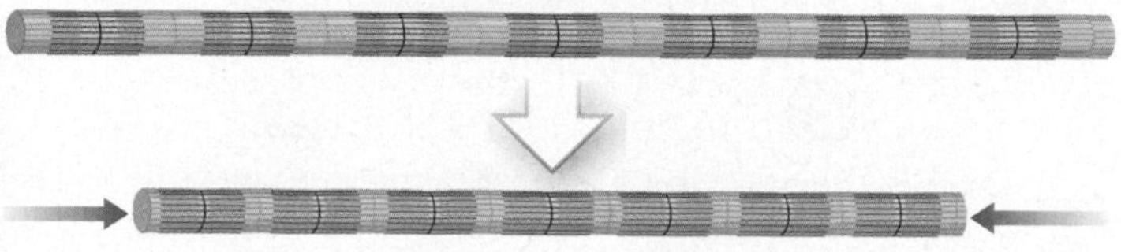

Module 9.4 Review

a. Describe the components of thin filaments and thick filaments.

b. Why do you think the zone of overlap is an important region of the sarcomere?

c. Summarize the sliding filament theory.

LO 9.4 Describe the structural components of a thin filament and a thick filament.

Skeletal muscle fibers and neurons have excitable plasma membranes that produce and carry electrical impulses called action potentials

Here we discuss the electrical nature of cells, what an action potential is, and how it is propagated along the axon of a neuron or sarcolemma of a skeletal muscle fiber.

1 Like miniature batteries with positive and negative electrical poles, all the cells of the body are polarized. A cell's polarization is due to an unequal distribution of positive and negative charges across its plasma membrane.

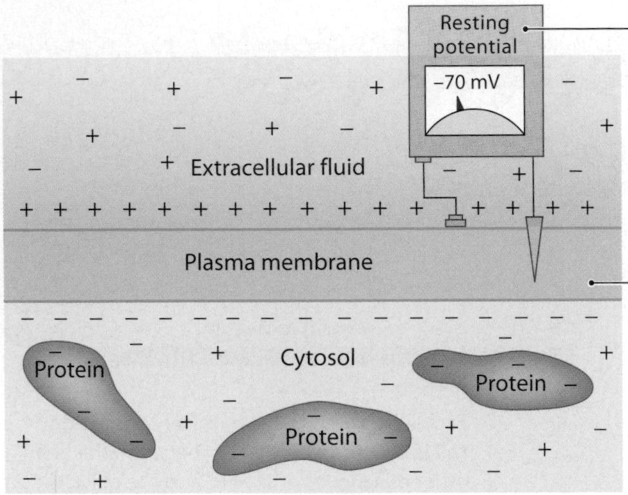

Resting potential
–70 mV
Extracellular fluid
Plasma membrane
Cytosol
Protein
Protein
Protein

The charge separation represents potential energy, or a potential difference, and the cell is said to be **polarized**. We refer to the potential difference across the plasma membrane as the **membrane potential**, or transmembrane potential. A cell's membrane potential is measured in millivolts (mV). It compares the inner membrane charge with the outer membrane charge. In undisturbed neurons and skeletal muscle fibers, typical **resting potentials** are about –70 mV and –85 mV, respectively.

For a resting cell, the inside surface of its plasma membrane has a slight negative charge with respect to its outside surface. Although the positive and negative charges are attracted to each other, the plasma membrane keeps them apart.

2 Plasma membranes are selectively permeable, and the cytosol and extracellular fluid (ECF) differ in composition.

The main contributors of positive electrical charges are sodium (Na^+) and potassium (K^+) ions. There is an excess of Na^+ outside of a cell (in the ECF) and an excess of K^+ inside a cell (in the cytosol).

There is a constant low rate of diffusion of Na^+ and K^+ down their concentration gradients across the membrane. Both require ion-specific "leak" membrane channels to cross the plasma membrane. Potassium ions cross the membrane more easily than sodium ions. However, sodium–potassium ion pumps constantly export 3 Na^+ from the cell and import 2 K^+ to maintain the cell's resting potential.

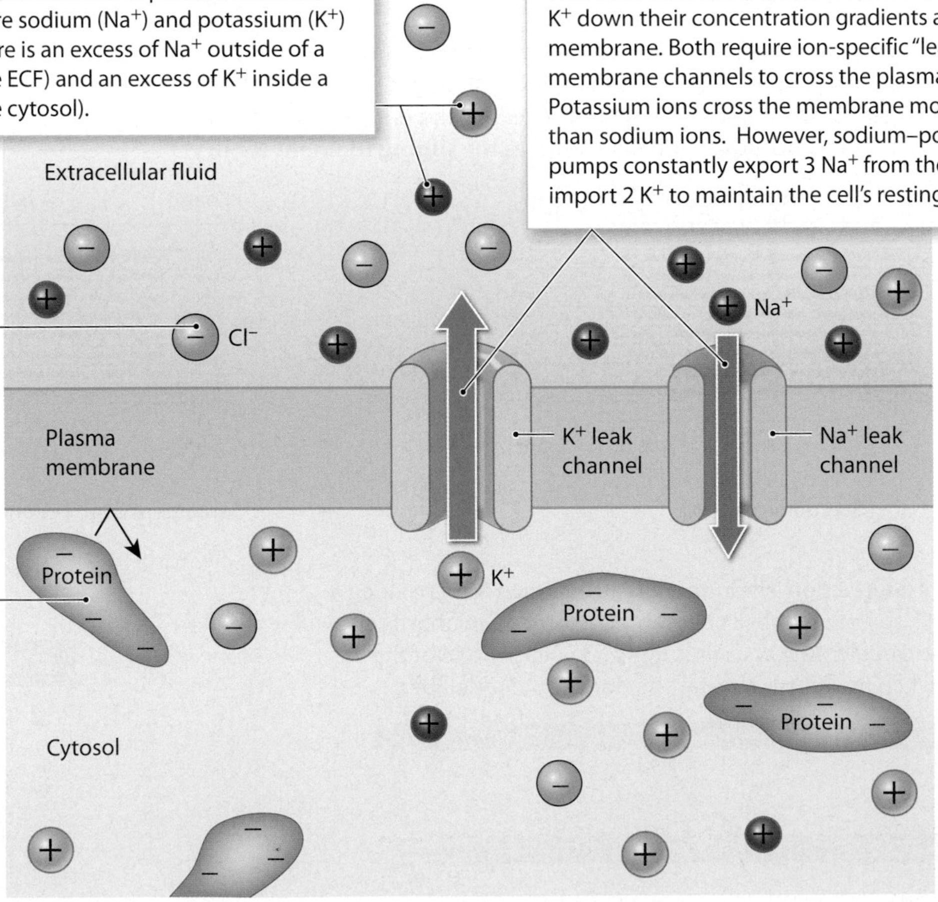

Extracellular fluid
Cl⁻
Na⁺
Plasma membrane
K⁺ leak channel
Na⁺ leak channel
K⁺
Protein
Protein
Protein
Cytosol

Although there is an excess of negatively charged chloride ions (Cl^-) in the ECF, little diffusion of these ions occurs across the plasma membrane.

The main contributors of negative charges within a cell are proteins that cannot cross the plasma membrane.

3 Neurons and skeletal muscle fibers have electrically excitable membranes. Such membranes contain **voltage-gated channels** that are activated (or inactivated) by changes in the membrane potential. Changes in the ion membrane permeability of a cell can temporarily reverse the distribution of electrical charges across the plasma membrane.

The events that produce an action potential begin at **1**.

1 A charge reversal begins with a small increase in sodium ion membrane permeability up to a set **threshold** (–55 mV in neurons).

2 Once the threshold is reached, voltage-gated Na^+ channels open and positively charged sodium ions rush into the cell. The membrane potential becomes positive and the cell is said to be **depolarized**.

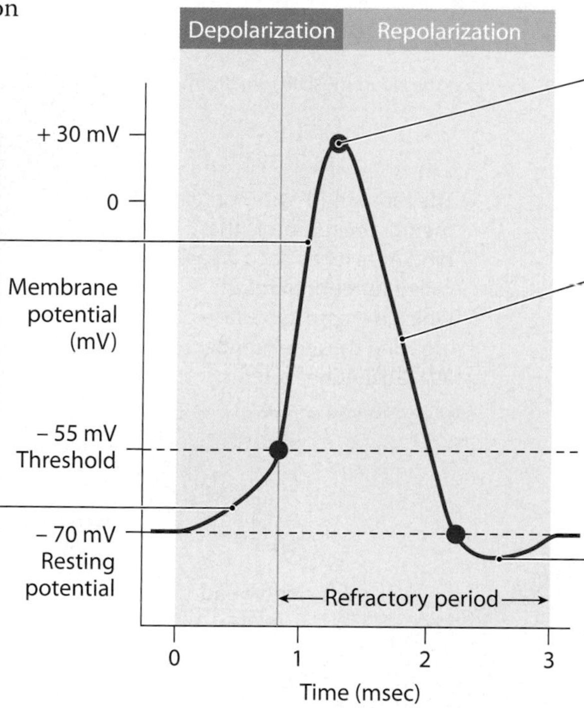

Depolarization
Repolarization
+ 30 mV
0
Membrane potential (mV)
– 55 mV Threshold
– 70 mV Resting potential
Refractory period
0
1
2
3
Time (msec)

3 The depolarization peaks at a membrane potential of +30 mV, at which point the voltage-gated Na^+ channels close and voltage-gated K^+ channels open. As potassium ions move out of the cell, **repolarization** (returning to the polarized state) begins.

4 Rapid repolarization continues until the resting potential is reached, when the voltage-gated K^+ channels begin closing.

5 As the voltage-gated K^+ channels close, the membrane potential stabilizes at resting levels and is once again negative. After the **refractory period** (a time when the membrane cannot respond to another stimulus), the former concentrations of sodium and potassium ions across the plasma membrane are restored. A second depolarization cannot occur until the refractory period is over.

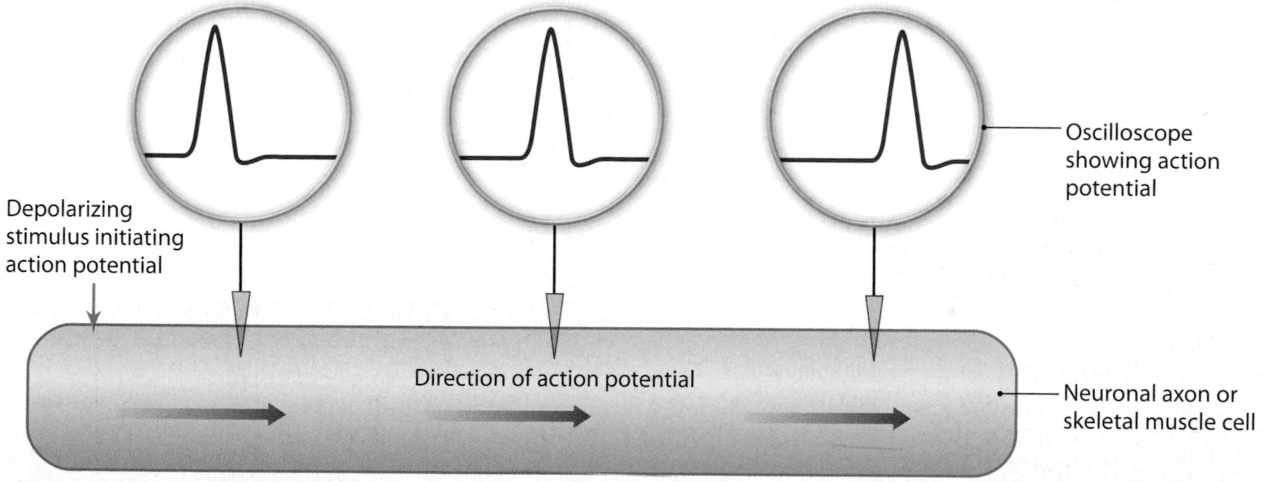

Oscilloscope showing action potential
Depolarizing stimulus initiating action potential
Direction of action potential
Neuronal axon or skeletal muscle cell

4 In most cells, the depolarization and repolarization of a plasma membrane is a localized change that does not spread across the entire plasma membrane. However, neurons and skeletal muscle fibers have electrically excitable membranes, and the initial depolarization and repolarization events produce an electrical impulse, or **action potential**, that is propagated along their plasma membranes. The opening and closing of voltage-gated Na^+ channels and the subsequent opening and closing of voltage-gated K^+ channels generate an action potential in less than 2 msec. An action potential travels in one direction because the refractory period prevents it from propagating back in the direction from where it was initiated. Excitable membranes permit rapid communication between different parts of a cell.

Module 9.5 Review

a. Describe the distribution of charges on either side of the plasma membrane for a resting (polarized) cell.

b. Explain the function of sodium–potassium ion pumps.

c. Define depolarization, and describe the events that follow it.

LO 9.5 Describe a major characteristic of excitable membranes and its importance in generating an action potential.

A skeletal muscle fiber contracts when stimulated by a motor neuron

Skeletal muscle fibers contract only under the control of the nervous system. Each skeletal muscle fiber is controlled by the nervous system at a specialized site known as a **neuromuscular junction (NMJ)**, located midway along the muscle fiber's length. The NMJ is made up of an axon terminal (synaptic terminal) of a motor neuron, a specialized region of the sarcolemma called the motor end plate and, in between, a space called the synaptic cleft. A **motor neuron** is a nerve cell that propagates an electrical impulse from the nervous system to skeletal muscle fibers.

Watch
MasteringA&P® ▸ ***A&PFlix***™
Events at the Neuromuscular Junction

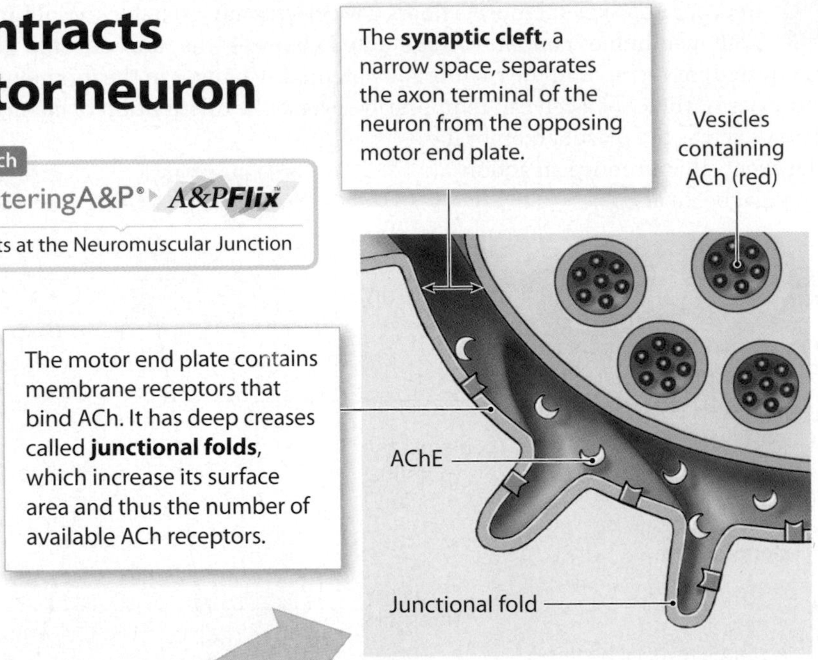

The **synaptic cleft**, a narrow space, separates the axon terminal of the neuron from the opposing motor end plate.
Vesicles containing ACh (red)
The motor end plate contains membrane receptors that bind ACh. It has deep creases called **junctional folds**, which increase its surface area and thus the number of available ACh receptors.
AChE
Junctional fold

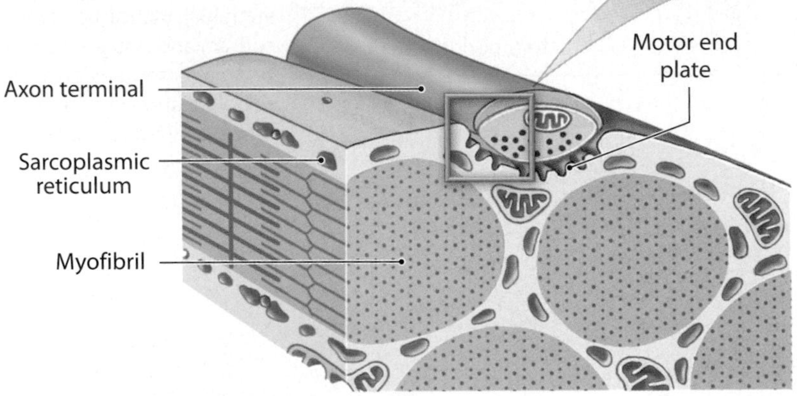

Axon terminal
Sarcoplasmic reticulum
Myofibril
Motor end plate

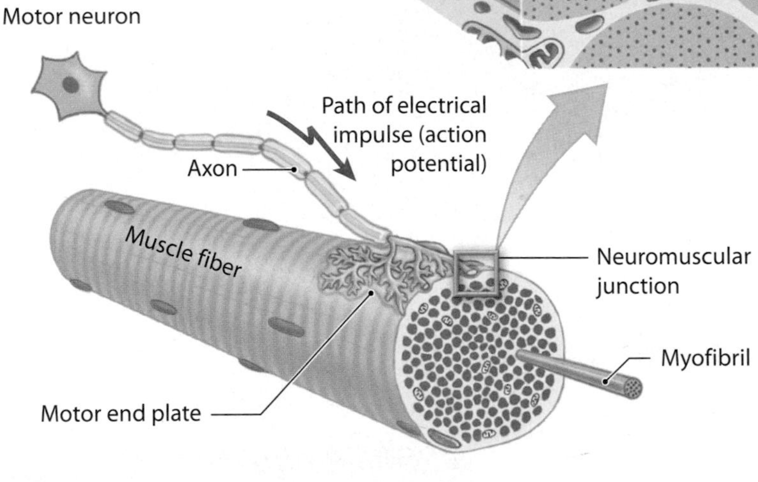

Motor neuron
Path of electrical impulse (action potential)
Axon
Muscle fiber
Neuromuscular junction
Myofibril
Motor end plate

1 A single axon may branch to control more than one skeletal muscle fiber, but each muscle fiber has only one NMJ. At the NMJ, the **axon terminal** of the neuron lies near the **motor end plate** of the muscle fiber.

2 The cytoplasm of the axon terminal contains vesicles filled with molecules of **acetylcholine** (as-e-til-KŌ-lēn), or **ACh**. Acetylcholine is a neurotransmitter, a chemical released by a neuron to change the permeability or other properties of another cell's plasma membrane. The synaptic cleft and the sarcolemma contain molecules of the enzyme **acetylcholinesterase** (**AChE**), which breaks down ACh.

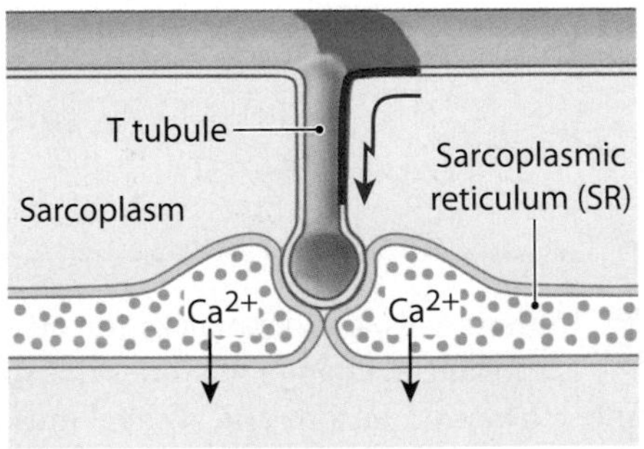

T tubule
Sarcoplasm
Sarcoplasmic reticulum (SR)
Ca2+
Ca2+

8 As the action potential sweeps down each T tubule and passes between the terminal cisternae, the SR's permeability changes, and calcium ions flood into the sarcomeres at the zones of overlap. This event, called **excitation-contraction coupling**, triggers the contraction of the muscle fiber (Module 9.7).

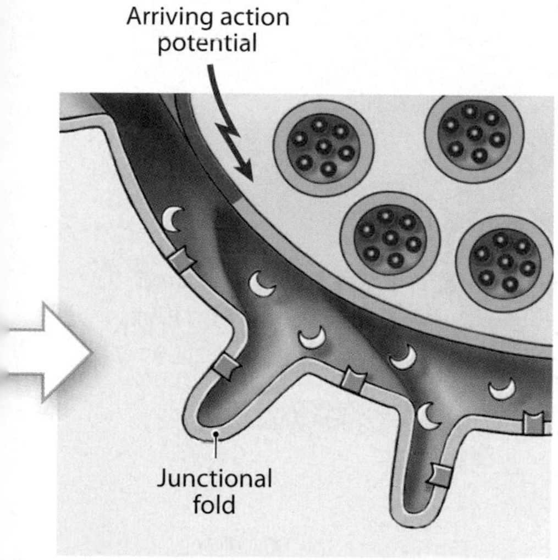
Arriving action potential
Junctional fold

3 The stimulus for ACh release is the arrival of an electrical impulse, or action potential, at the axon terminal. An action potential is a sudden change in the membrane potential that travels along the length of the axon.

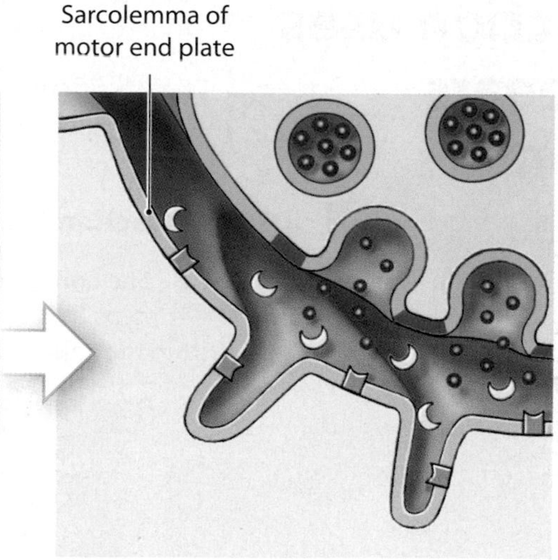
Sarcolemma of motor end plate

4 When the action potential reaches the neuron's axon terminal, changes in the membrane permeability trigger the exocytosis of ACh into the synaptic cleft. Exocytosis occurs as the synaptic vesicles fuse with the neuron's plasma membrane.

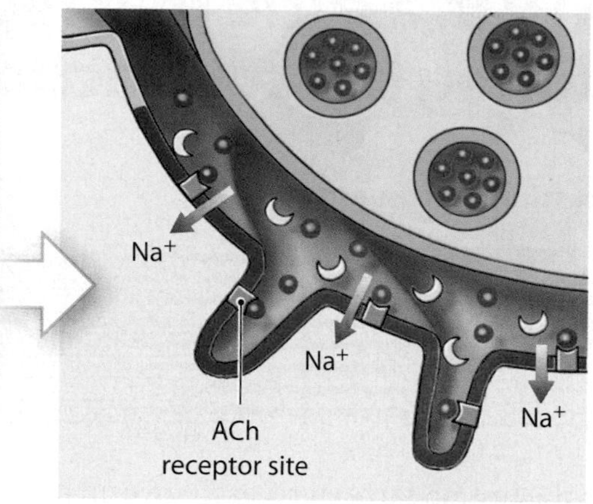
Na+
Na+
Na+
ACh receptor site

5 ACh molecules diffuse across the synaptic cleft and bind to ACh receptors on the surface of the motor end plate. ACh binding alters the membrane's permeability to Na^+. Because the extracellular fluid contains a high concentration of Na^+, and Na^+ concentration inside the cell is very low, sodium ions rush into the sarcoplasm.

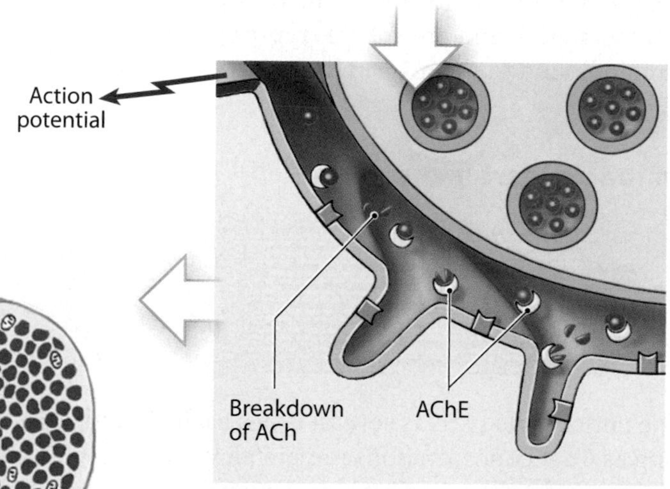
Action potential
Breakdown of ACh
AChE

6 The sudden inrush of sodium ions generates an action potential in the sarcolemma. ACh either diffuses away from the site or AChE quickly breaks it down in the synaptic cleft, thus inactivating the ACh receptor sites.

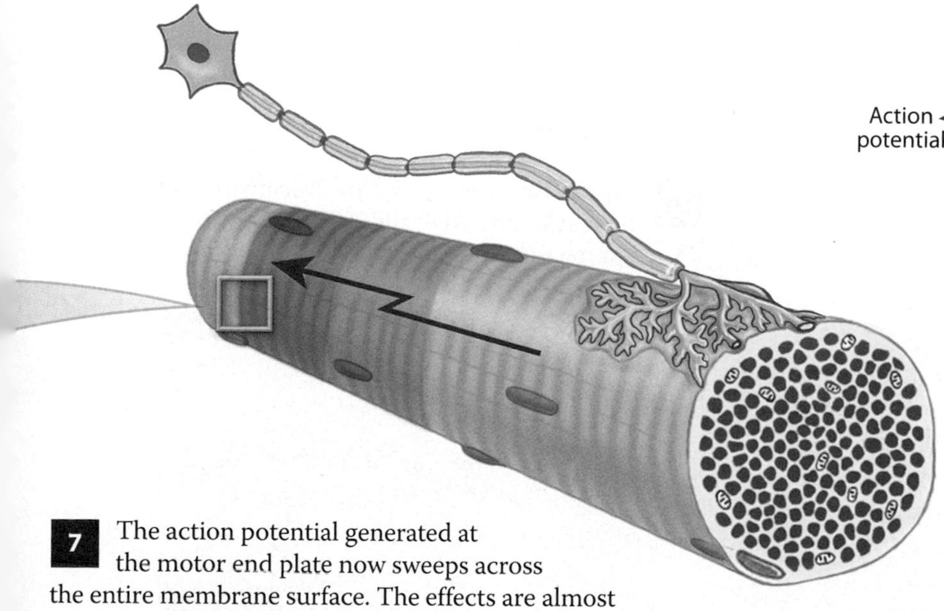

7 The action potential generated at the motor end plate now sweeps across the entire membrane surface. The effects are almost immediate because an action potential is an electrical event that flashes like a spark across the sarcolemmal surface. The effects are brief because the ACh has been removed, and no further stimulus acts upon the motor end plate until another action potential arrives at the axon terminal.

Module 9.6 Review

a. Describe the neuromuscular junction.

b. How would a drug that blocks acetylcholine release affect muscle contraction?

c. Predict what would happen if there were no AChE in the synaptic cleft.

LO **9.6** Identify the components of the neuromuscular junction, and summarize the events involved in the control of skeletal muscles by motor neurons.

Module 9.7

A muscle fiber contraction uses ATP in a cycle that repeats during the contraction

Watch
MasteringA&P®
A&P*Flix*™
The Cross-Bridge Cycle

Resting Sarcomere

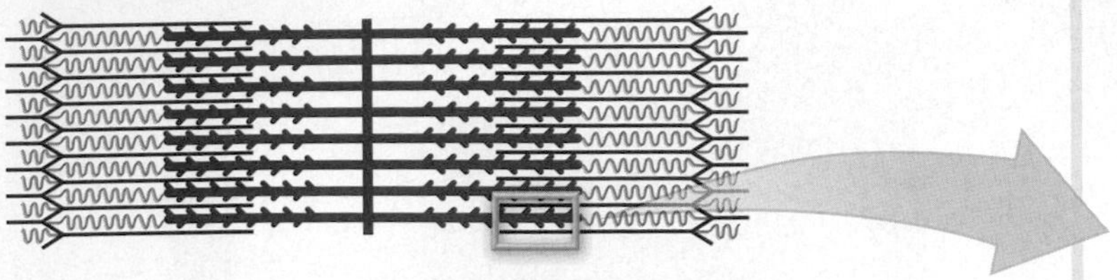

In the resting sarcomere, each myosin head is already "energized"—charged with the energy that will be used to power a contraction. Each myosin head points away from the M line. In this position, the myosin head is "cocked" like the spring in a mousetrap. Cocking the myosin head requires energy, which is obtained by breaking down ATP. In this process, the myosin head functions as ATPase, an enzyme that breaks down ATP. At the start of the contraction cycle, the breakdown products, ADP and phosphate (represented as P), remain bound to the myosin head.

Contracted Sarcomere

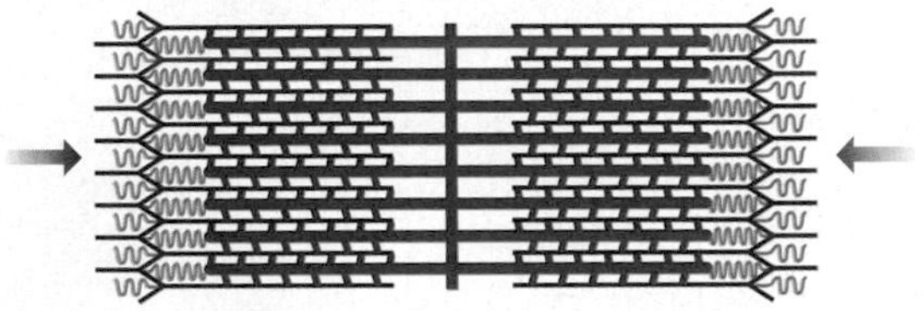

The entire cycle repeats several times each second, as long as Ca^{2+} concentrations remain elevated and ATP reserves are sufficient. Calcium ion levels will remain elevated only as long as action potentials continue to pass along the T tubules and stimulate the terminal cisternae. Once that stimulus is removed, the calcium channels in the SR close and calcium ion pumps pull Ca^{2+} from the cytosol and store it within the terminal cisternae. Troponin molecules then shift position, swinging the tropomyosin strands over the active sites and preventing further cross-bridge formation.

Contraction Cycle Begins

1 The **contraction cycle** is a series of interrelated steps that begins with the arrival of calcium ions within the zone of overlap in a sarcomere.

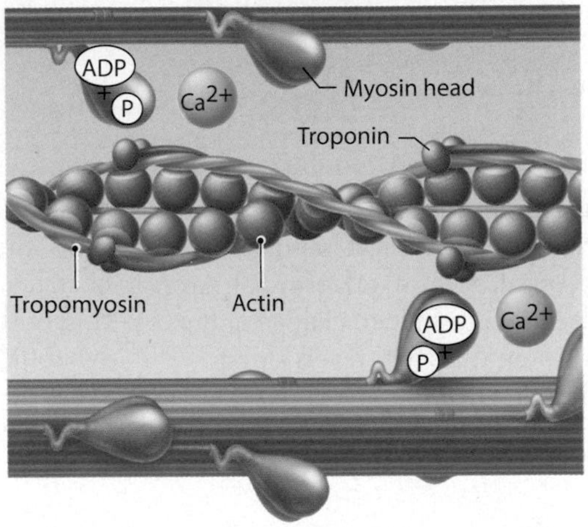
ADP
P
Ca2+
Myosin head
Troponin
Tropomyosin
Actin
ADP
P
Ca2+

Myosin Reactivates

6 Myosin reactivates when the free myosin head splits ATP into ADP and P. The energy released is used to "recock" the myosin head.

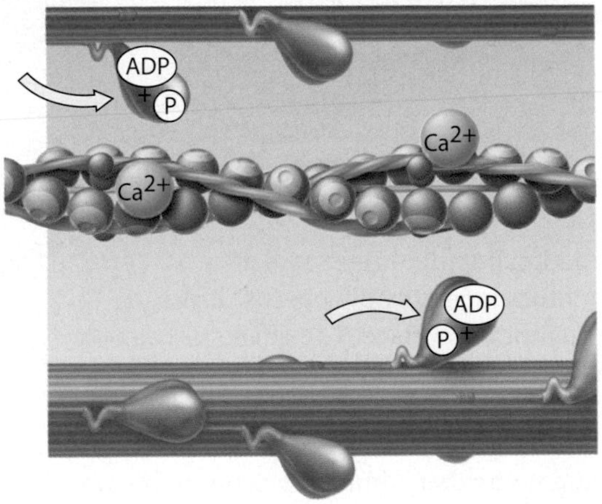
ADP
P
Ca2+
Ca2+
ADP
P

Active Sites Exposed

2 Calcium ions bind to troponin, weakening the bond between actin and the troponin–tropomyosin complex. The troponin molecule then changes position, rolling the tropomyosin molecule away from the active sites on actin and allowing interaction with the energized myosin heads.

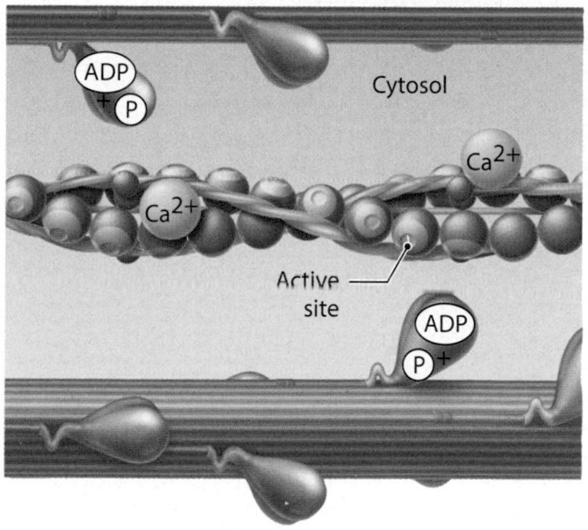

ADP
P
Cytosol
Ca2+
Ca2+
Active site
ADP
P

Cross-Bridges Form

3 Once the active sites are exposed, the energized myosin heads bind to them, forming **cross-bridges**.

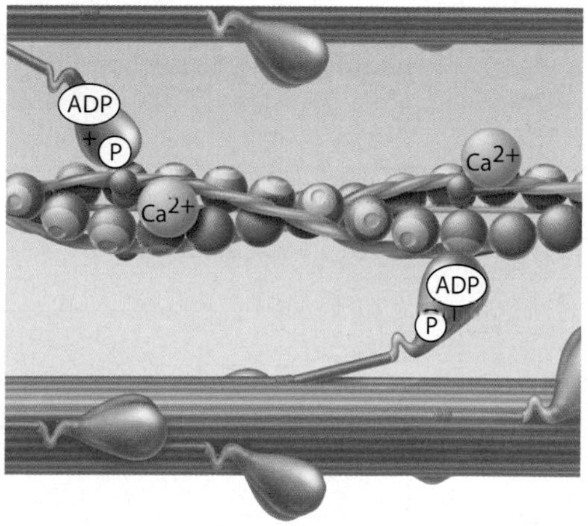

ADP
P
Ca2+
Ca2+
ADP
P

Myosin Heads Pivot

4 After the cross-bridges form, the energy that was stored in the resting state is released as the myosin heads pivot toward the M line. This action is called the power stroke; when it occurs, the bound ADP and phosphate group are released.

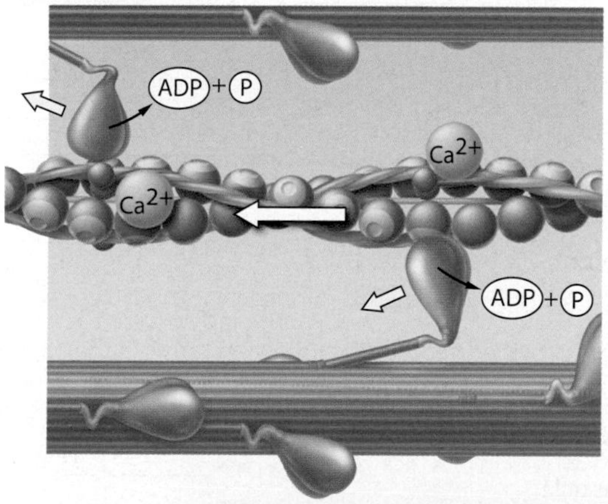

ADP + P
Ca2+
Ca2+
ADP + P

Cross-Bridges Detach

5 When another ATP binds to the myosin head, the link between the myosin head and the active site on the actin molecule is broken. The active site is now exposed and able to form another cross-bridge.

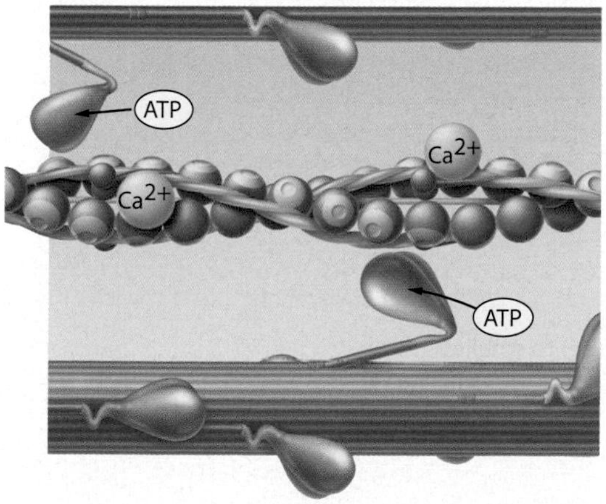

ATP
Ca2+
Ca2+
ATP

Module 9.7 Review

a. What molecule supplies the energy for a muscle fiber contraction?

b. List the interrelated steps that occur once the contraction cycle begins.

c. What triggers myosin reactivation?

LO **9.7** Describe the role of ATP in a muscle contraction, and explain the steps involved in the contraction of a skeletal muscle fiber.

Labeling

Label the structures in this figure of a skeletal muscle fiber.

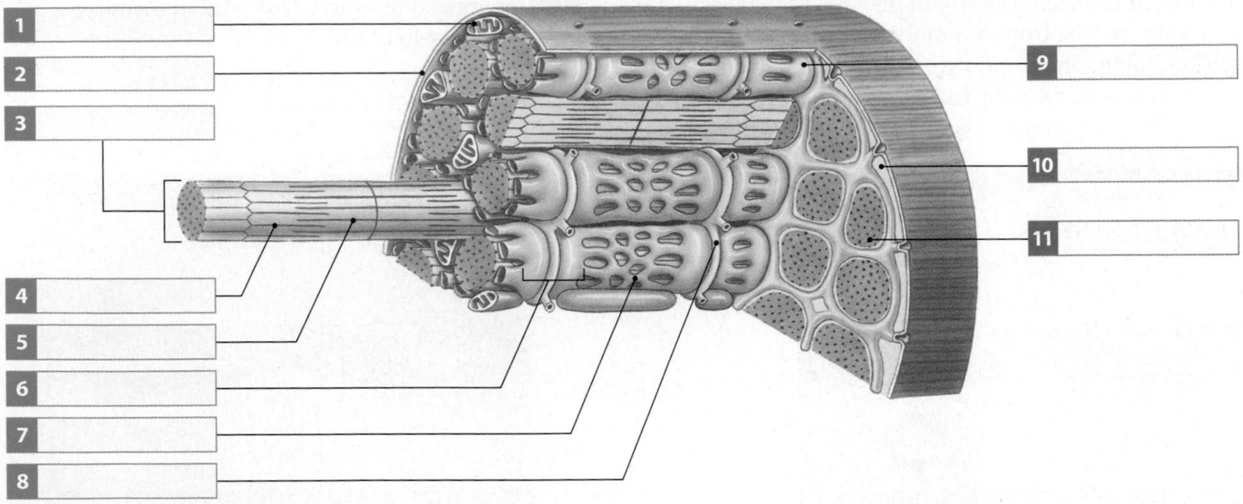
1
2
3
4
5
6
7
8
9
10
11

Label the structures in this diagram of adjacent sarcomeres.

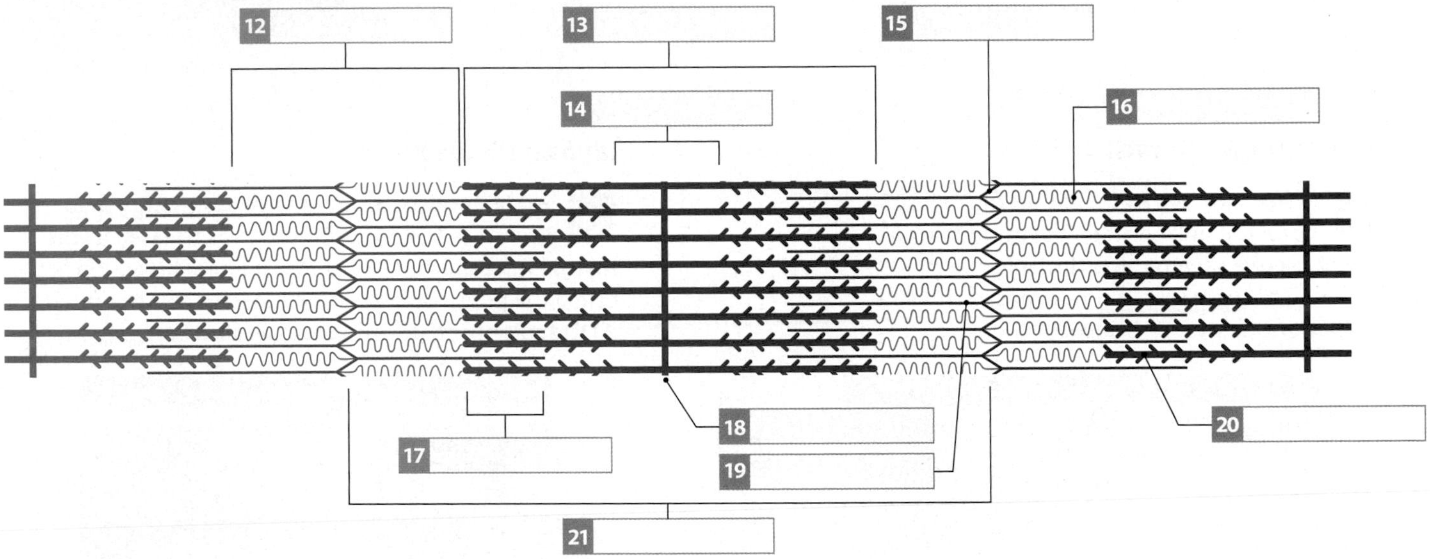
12
13
14
15
16
17
18
19
20
21

Fill-in

In the spaces provided, reorder the following terms in their proper sequence from largest (22) to smallest (27).

- muscle fiber
- muscle fascicle
- sarcomere
- myofibril
- skeletal muscle
- myofilament

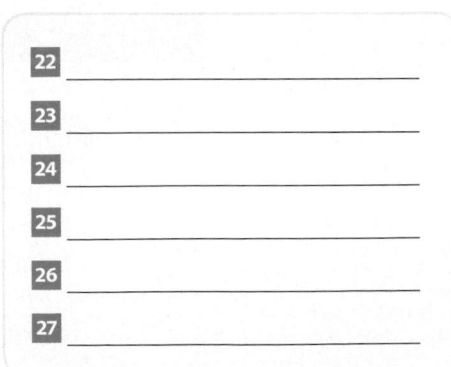

22 ____________________

23 ____________________

24 ____________________

25 ____________________

26 ____________________

27 ____________________

Muscle tension develops from the events that occur during excitation-contraction coupling

Watch
MasteringA&P®
A&P**Flix**™
Excitation-Contraction Coupling

1
Neural control

A skeletal muscle fiber contracts when stimulated by a motor neuron at a neuromuscular junction. The stimulus arrives in the form of an action potential at the axon terminal.

2
Excitation

The action potential causes the release of ACh into the synaptic cleft, which leads to excitation—the production of an action potential in the sarcolemma.

3
Calcium ion release

The muscle fiber action potential travels along the sarcolemma and into T tubules down to the triads. This triggers the release of Ca^{2+} from the terminal cisternae of the sarcoplasmic reticulum.

4
Contraction cycle begins

The contraction cycle begins when the Ca^{2+} bind to troponin, exposing the active sites on the thin filaments. This allows cross-bridge formation and will continue as long as ATP is available.

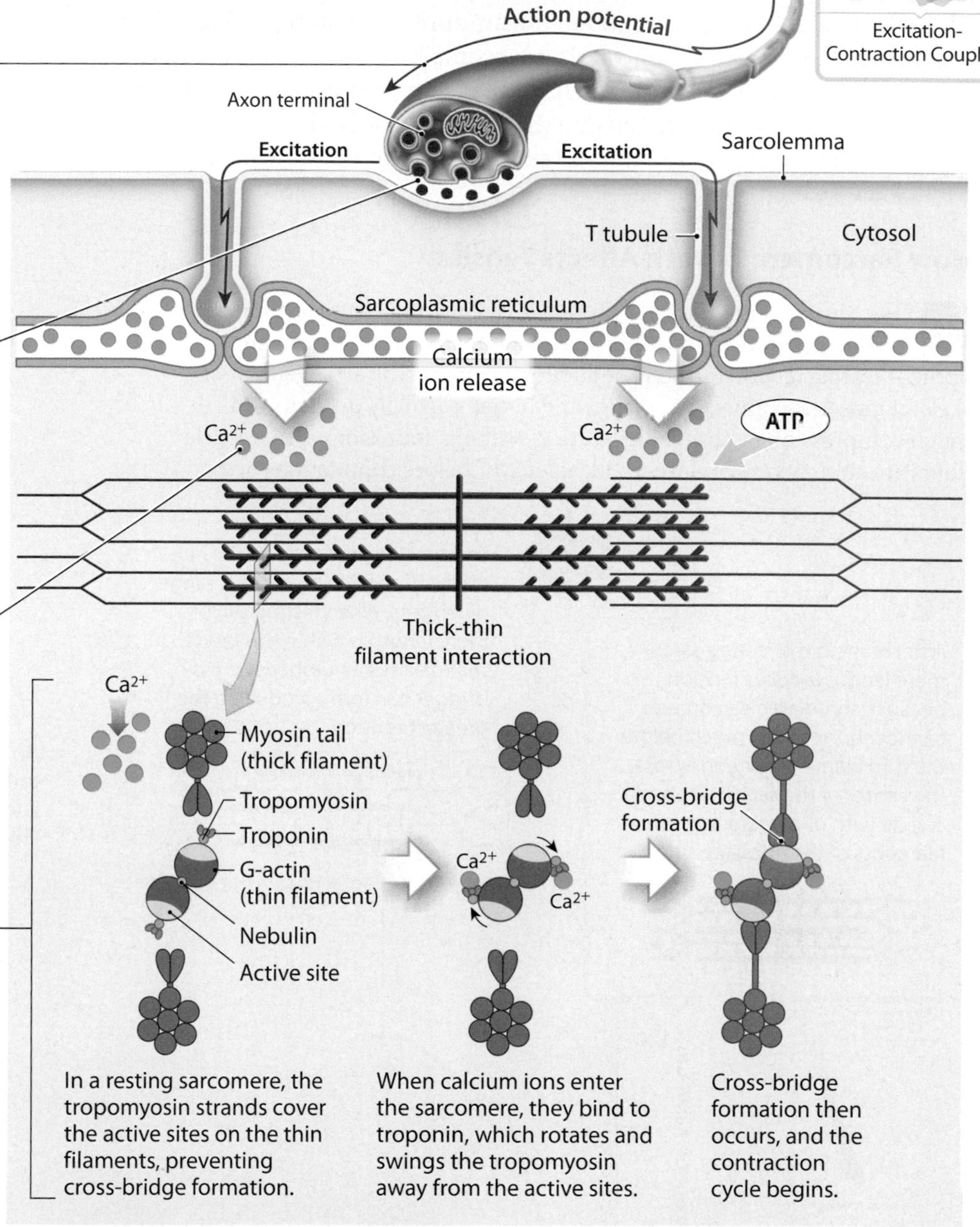
Action potential
Axon terminal
Excitation
Excitation
Sarcolemma
T tubule
Cytosol
Sarcoplasmic reticulum
Calcium ion release
Ca2+
Ca2+
ATP
Thick-thin filament interaction
Ca2+
Myosin tail (thick filament)
Tropomyosin
Troponin
G-actin (thin filament)
Nebulin
Active site
Ca2+
Ca2+
Cross-bridge formation
In a resting sarcomere, the tropomyosin strands cover the active sites on the thin filaments, preventing cross-bridge formation.
When calcium ions enter the sarcomere, they bind to troponin, which rotates and swings the tropomyosin away from the active sites.
Cross-bridge formation then occurs, and the contraction cycle begins.

5
Sarcomeres shorten

As the thick and thin filaments interact, the sarcomeres shorten, pulling the ends of the muscle fiber closer together.

6
Muscle tension produced

During the contraction, the entire skeletal muscle shortens and produces a pull, or **tension**, on the tendons at either end.

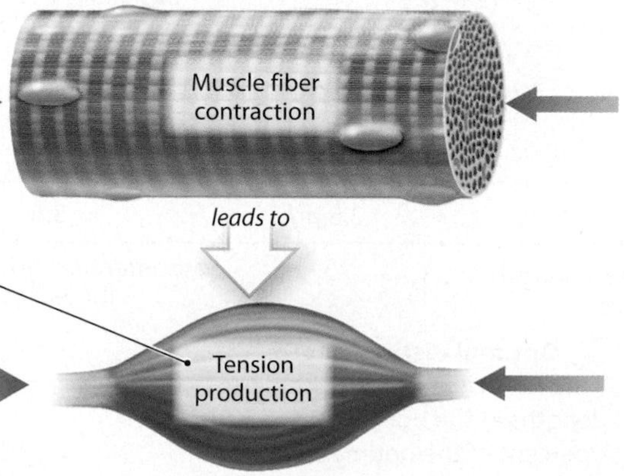
Muscle fiber contraction
leads to
Tension production

Module 9.8 Review

a. Summarize the events that occur at the neuromuscular junction.

b. What causes calcium to be released from the sarcoplasmic reticulum?

c. What is necessary for a contraction cycle to continue?

LO 9.8 Describe how muscle tension develops with respect to neural control and excitation-contraction coupling.

Tension production is greatest when a muscle is stimulated at its optimal length

There is no mechanism to regulate the amount of tension produced in a contraction by changing the number of contracting sarcomeres. When calcium ions are released, they are released from all triads in the muscle fiber. Thus, a muscle fiber is either "on" (producing tension) or "off" (relaxed). Tension produced at the individual muscle fiber does vary, however, depending on the fiber's resting length at the time of stimulation.

How Sarcomere Length Affects Tension

1 The tension a muscle fiber produces is related to sarcomere length. When sarcomeres are either stretched or compressed compared to optimal resting length, tension production declines. The arrangement of skeletal muscles, connective tissues, and bones normally prevents too much compression or stretching. During walking, for example, leg muscle fibers are stretched very close to "ideal length" before contractions occur.

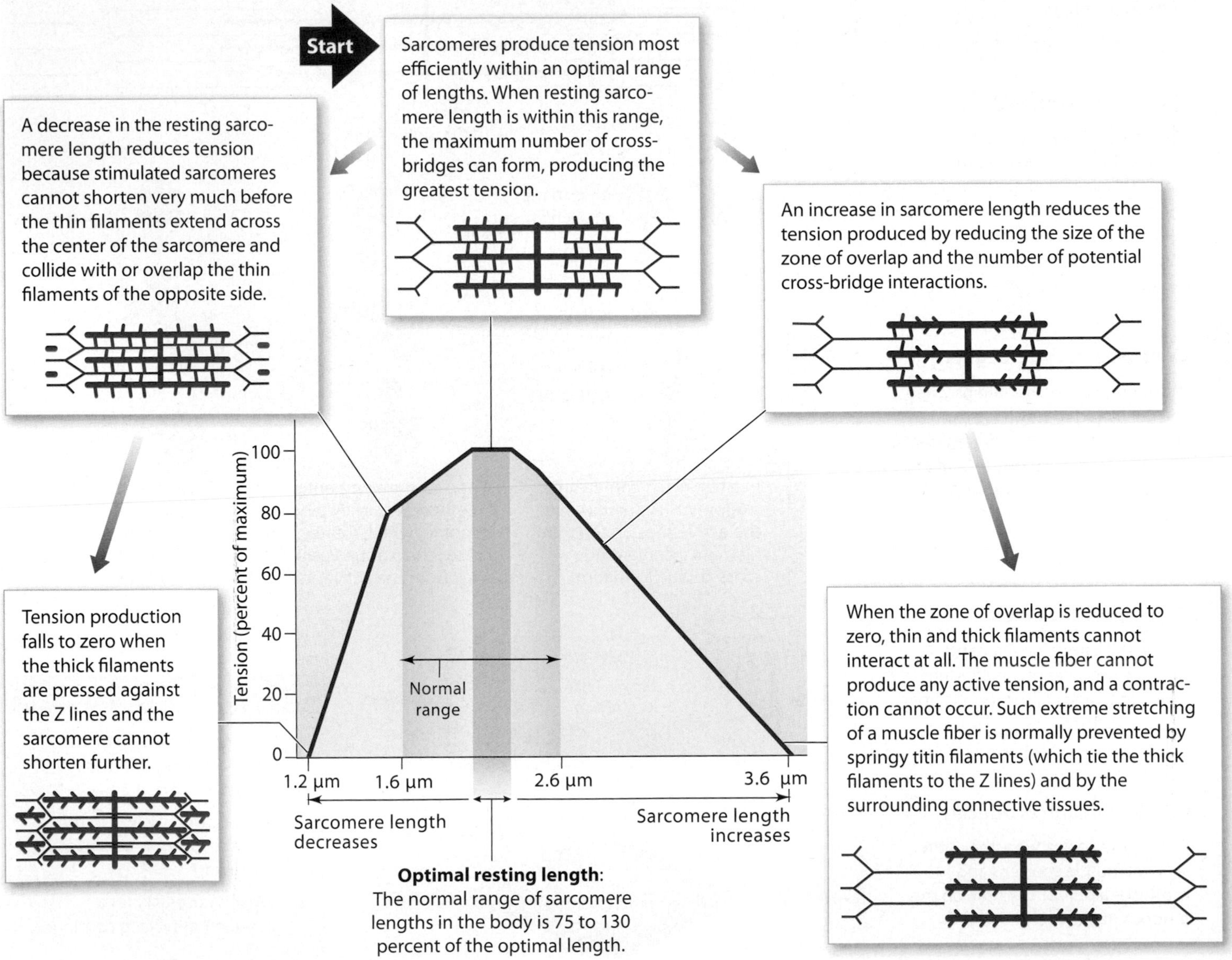
Start
Sarcomeres produce tension most efficiently within an optimal range of lengths. When resting sarcomere length is within this range, the maximum number of cross-bridges can form, producing the greatest tension.
A decrease in the resting sarcomere length reduces tension because stimulated sarcomeres cannot shorten very much before the thin filaments extend across the center of the sarcomere and collide with or overlap the thin filaments of the opposite side.
An increase in sarcomere length reduces the tension produced by reducing the size of the zone of overlap and the number of potential cross-bridge interactions.
Tension (percent of maximum)
100
80
60
40
20
0
Normal range
1.2 µm
1.6 µm
2.6 µm
3.6 µm
Sarcomere length decreases
Sarcomere length increases
Tension production falls to zero when the thick filaments are pressed against the Z lines and the sarcomere cannot shorten further.
When the zone of overlap is reduced to zero, thin and thick filaments cannot interact at all. The muscle fiber cannot produce any active tension, and a contraction cannot occur. Such extreme stretching of a muscle fiber is normally prevented by springy titin filaments (which tie the thick filaments to the Z lines) and by the surrounding connective tissues.
Optimal resting length: The normal range of sarcomere lengths in the body is 75 to 130 percent of the optimal length.

Measuring Muscle Fiber Tension

2 A **myogram** is a graphical representation of tension development in muscle fibers. This myogram shows three examples of a **twitch**, a single stimulus-contraction-relaxation sequence in a muscle fiber. Twitches vary in duration, depending on muscle type and location, internal and external environmental conditions, and other factors. We have all probably seen or felt an involuntary "muscle twitch" under the skin. Such a muscle twitch is called a **fasciculation**. A fasciculation involves more than one muscle fiber. It is the synchronous contraction of a motor unit, that is, a group of skeletal muscle fibers all controlled by a single motor neuron (Module 9.10).

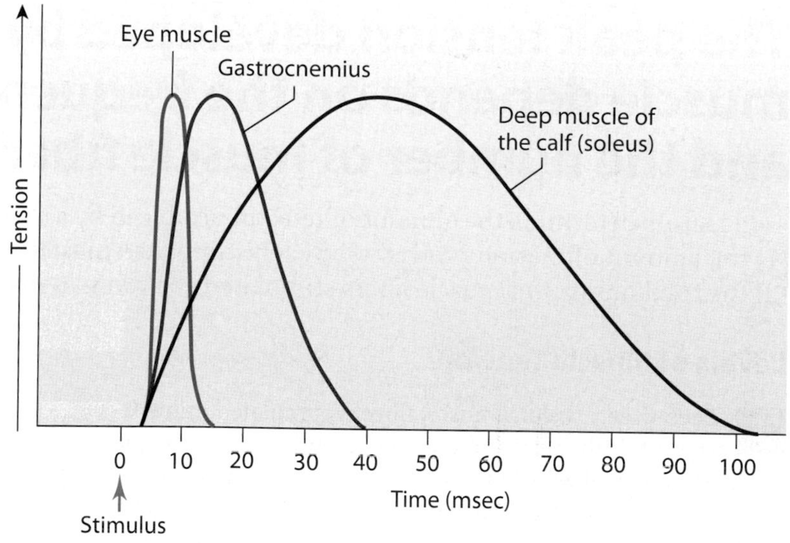

Eye muscle
Gastrocnemius
Deep muscle of the calf (soleus)
Tension
0
10
20
30
40
50
60
70
80
90
100
Time (msec)
Stimulus

3 This myogram shows the phases of a 40-msec twitch in a muscle fiber from the gastrocnemius muscle, a prominent superficial muscle of the calf. Remember that a millisecond is one thousandth of a second! So these events happen rapidly—in less time than a finger snap.

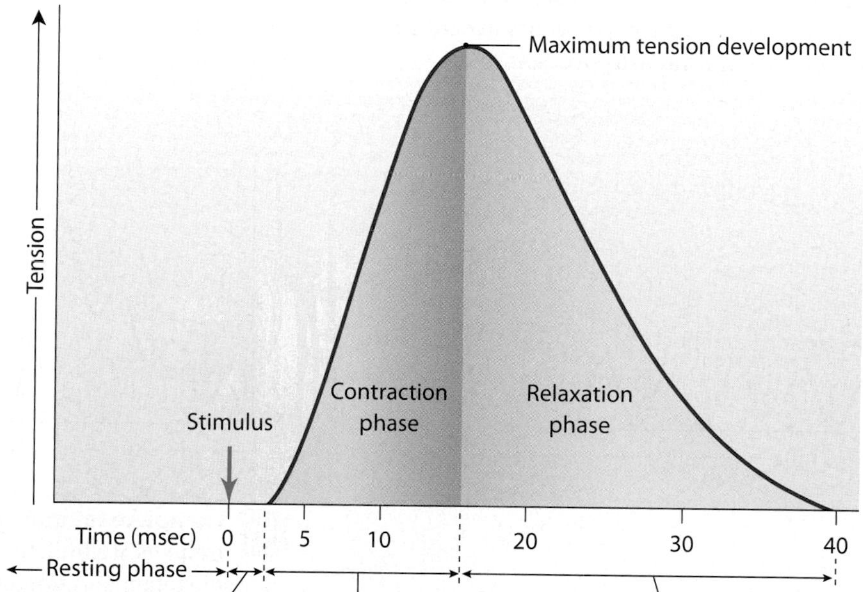

Maximum tension development
Tension
Stimulus
Contraction phase
Relaxation phase
Time (msec)
0
5
10
20
30
40
Resting phase

The **latent period** begins at stimulation and typically lasts about 2 msec. During this period, an action potential sweeps across the sarcolemma, and the sarcoplasmic reticulum releases calcium ions. The muscle fiber does not produce tension during the latent period, because the contraction cycle has yet to begin.

In the **contraction phase**, tension rises to a peak. As the tension rises, calcium ions are binding to troponin, active sites on thin filaments are being exposed, and cross-bridge interactions are occurring.

The **relaxation phase** lasts about 25 msec. During this period, calcium levels are falling, active sites are being covered by tropomyosin, and the number of active cross-bridges is declining as they detach. As a result, tension returns to resting levels.

Module 9.9 Review

a. Name a factor that affects the amount of tension produced when a skeletal muscle fiber contracts.

b. Explain two key concepts of the length–tension relationship.

c. Describe the events that occur during each phase of a twitch in a stimulated muscle fiber.

LO **9.9** Describe the mechanism responsible for tension production in a muscle fiber, and discuss the factors that determine the peak tension developed during a contraction.

The peak tension developed by a skeletal muscle depends on the frequency of stimulation and the number of muscle fibers stimulated

Two factors determine the amount of tension produced by a skeletal muscle:
(1) the amount of tension produced by each stimulated muscle fiber, and
(2) the total number of muscle fibers stimulated at a given moment.

Levels of Muscle Tension

1 Each time a skeletal muscle fiber is stimulated immediately after the relaxation phase has ended, the subsequent contraction will develop a slightly higher maximum tension than did the previous contraction. The increase in peak tension will continue over the first 30–50 stimulations. Because the tension rises like the steps in a staircase, this phenomenon is called **treppe** (TREP-eh, German for *staircase*). Most skeletal muscles do not demonstrate treppe. (However, treppe occurs in cardiac muscle tissue if stimuli of the same intensity are sent to the muscle fiber after a latent period.)

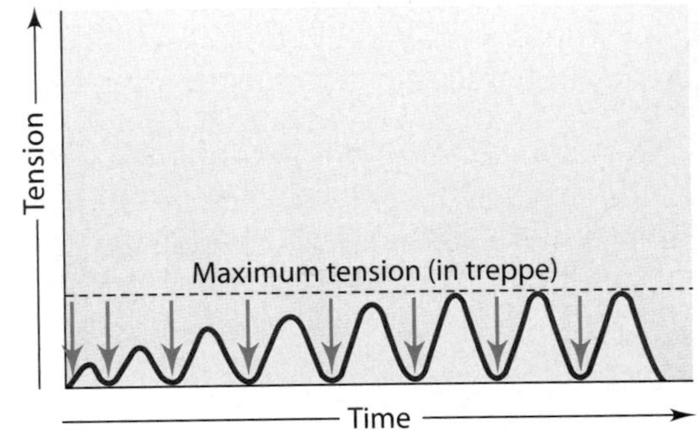
Tension
Maximum tension (in treppe)
Time

2 If a second stimulus arrives before the relaxation phase has ended, a second, more powerful contraction occurs. The addition of one twitch to another in this way is called **wave summation**. The duration of a single twitch determines the maximum time available to produce wave summation. For example, if a twitch lasts 20 msec (1/50 sec), a stimulus frequency of greater than 50 per second produces wave summation, whereas a stimulus frequency of less than 50 per second will produce individual twitches and treppe.

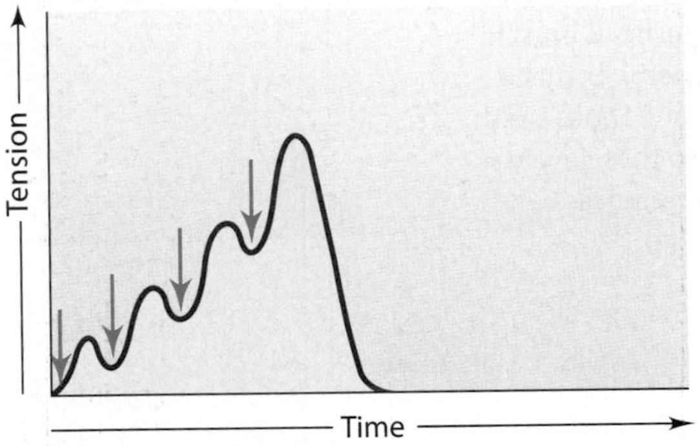
Tension
Time

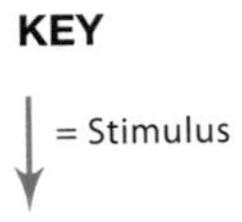
KEY
= Stimulus

3 A muscle producing almost peak tension during rapid cycles of contraction and relaxation is said to be in **incomplete tetanus** (*tetanos*, convulsive tension).

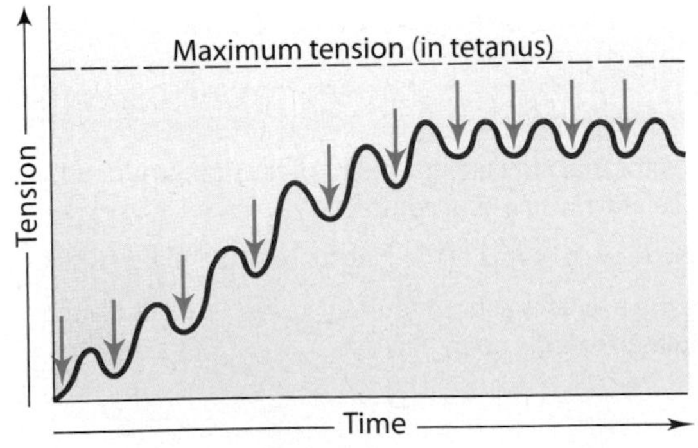
Tension
Maximum tension (in tetanus)
Time

4 **Complete tetanus** occurs when a higher stimulation frequency eliminates the relaxation phase. During complete tetanus, action potentials arrive so rapidly that the sarcoplasmic reticulum cannot reclaim calcium ions. The high cytosolic levels of Ca^{2+} prolong the contraction, making it continuous. Although muscles can be forced into tetanus in the laboratory, they seldom if ever develop peak tension in the course of normal activities. In part, this is because normal movements require precise control over, and continuous variation in, the amount of tension produced.

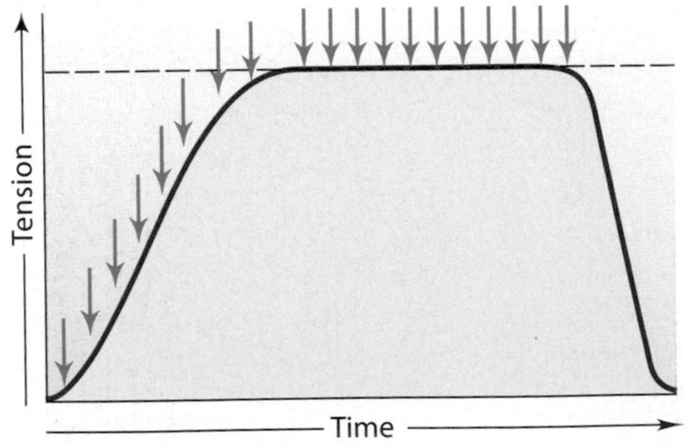
Tension
Time

Motor Units and Recruitment

5 A typical skeletal muscle contains thousands of muscle fibers. Although some motor neurons control just a few muscle fibers, most control hundreds of them. The amount of tension produced is controlled at the subconscious level through variations in the number of muscle fibers stimulated.

All the muscle fibers controlled by a single motor neuron constitute a **motor unit**. The size of a motor unit is related to the degree of control we have over our movements. In the muscles of the eye, where precise control is extremely important, a motor neuron may control 4–6 muscle fibers. We have much less precise control over our leg muscles, where a single motor neuron may control 1000–2000 muscle fibers.

The muscle fibers of each motor unit are intermingled with those of other motor units. As a result, the direction of pull exerted on the tendon does not change when the number of activated motor units changes. When you decide to perform a specific movement, the contraction begins with the activation of the smallest motor units in the stimulated muscle. As the movement continues, larger motor units containing faster and more powerful muscle fibers are activated, and tension rises steeply. The smooth but steady increase in muscular tension produced by increasing the number of active motor units is called **recruitment**.

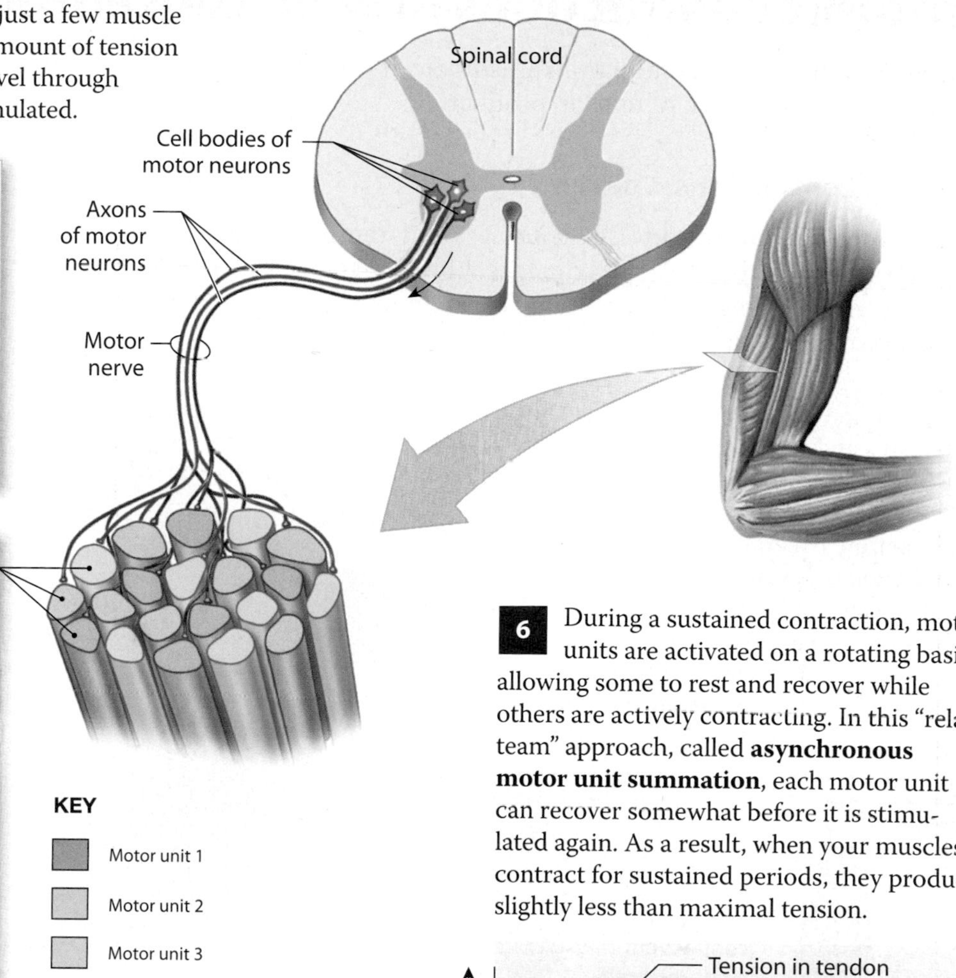

Spinal cord
Cell bodies of motor neurons
Axons of motor neurons
Motor nerve
KEY
Motor unit 1
Motor unit 2
Motor unit 3

6 During a sustained contraction, motor units are activated on a rotating basis, allowing some to rest and recover while others are actively contracting. In this "relay team" approach, called **asynchronous motor unit summation**, each motor unit can recover somewhat before it is stimulated again. As a result, when your muscles contract for sustained periods, they produce slightly less than maximal tension.

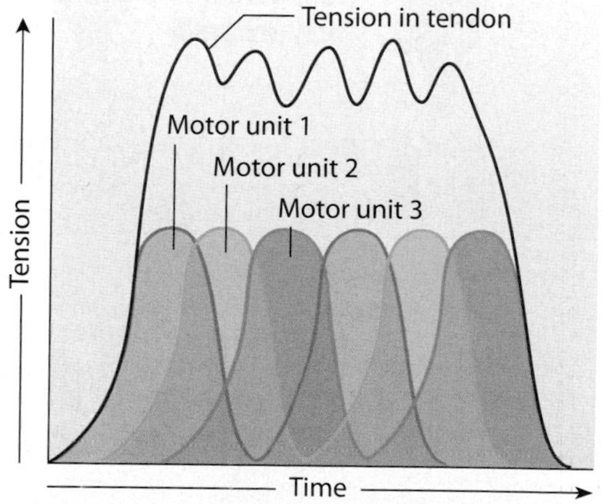

Tension in tendon
Motor unit 1
Motor unit 2
Motor unit 3
Tension
Time

7 A variable number of motor units is always active, even when the entire muscle is not contracting. Their contractions do not produce enough tension to cause movement, but they do tense and firm the muscle. This resting tension in a skeletal muscle is called **muscle tone**, and it is regulated at the subconscious level. The activity level of each motor neuron changes constantly, and individual muscle fibers can relax while a constant tension is maintained in the attached tendon. Activated muscle fibers use energy, so the greater the muscle tone, the higher the "resting" rate of metabolism. Elevated muscle tone increases resting energy consumption by a small amount, but the effects are cumulative, and they continue 24 hours per day.

Module 9.10 Review

a. Compare incomplete tetanus with wave summation.

b. Define motor unit.

c. Describe the relationship between the number of fibers in a motor unit and the precision of body movements.

9.10 Discuss the factors that affect peak tension production during the contraction of an entire skeletal muscle, and explain the significance of the motor unit in this process.

Module 9.11

Muscle contractions may be isotonic or isometric; isotonic contractions may be concentric or eccentric

We can classify muscle contractions as isotonic or isometric on the basis of their pattern of tension production.

Isotonic Contractions: Concentric

In an **isotonic contraction** (*iso-*, equal + *tonos*, tension), tension rises and the skeletal muscle's length changes. Lifting an object off a desk, walking, and running involve isotonic contractions. There are two types of isotonic contractions.

1 For one type, consider a skeletal muscle that is 1 cm^2 in cross-sectional area and can produce roughly 4 kg (8.8 lb) of tension in complete tetanus. If we hang a load of 2 kg (4.4 lb) from that muscle and stimulate it, the muscle will shorten. This is called a **concentric contraction**. An example of a concentric contraction is flexing the elbow joint from an extended position while holding a dumbbell. To lift the weight, the muscle must recruit enough motor units to overcome gravity and actively shorten.

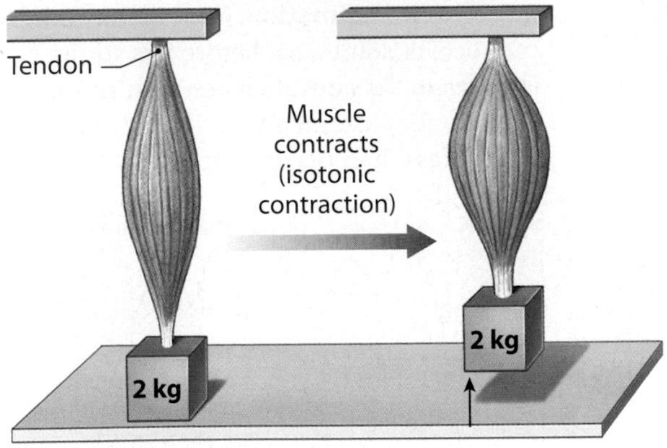

Tendon
Muscle contracts (isotonic contraction)
2 kg
2 kg

2 Before the muscle can shorten, the cross-bridges must produce enough tension to overcome the load—in this case, the 2-kg weight. During this initial period, tension in the muscle fibers rises until the tension in the tendon exceeds the load. As the muscle shortens, the tension in the skeletal muscle remains constant.

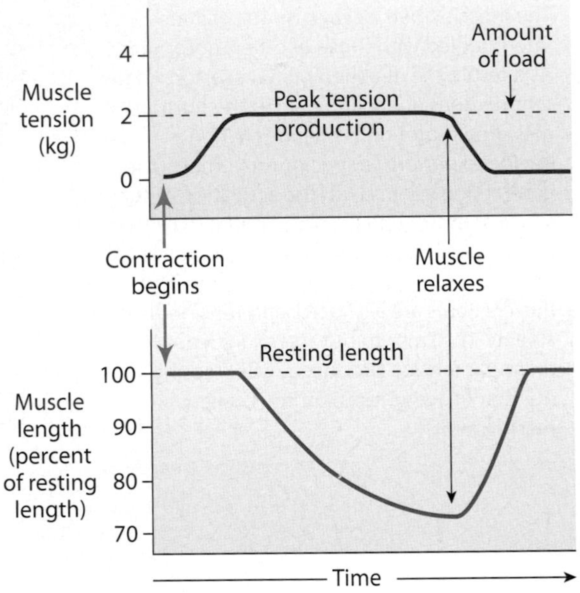

Muscle tension (kg)
4
2
0
Amount of load
Peak tension production
Contraction begins
Muscle relaxes
Resting length
Muscle length (percent of resting length)
100
90
80
70
Time

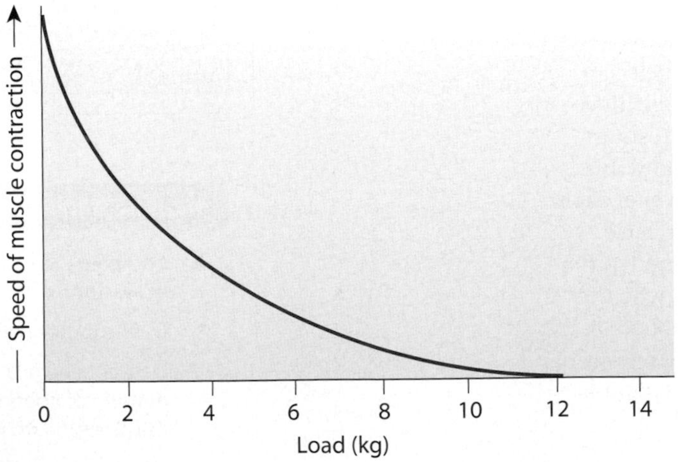

Speed of muscle contraction
0
2
4
6
8
10
12
14
Load (kg)

3 The speed, or rate, of muscle contraction varies inversely with the load on the muscle. If the load is relatively light, the muscle will contract quickly; if the load is heavy, the muscle will contract slowly or not at all. The speed of muscle contraction is fastest when the load equals zero.

Isotonic Contractions: Eccentric

4 For the second type of isotonic contraction, consider the opposite situation as in **1**. If the load is greater than the peak tension the muscle can develop, the muscle will elongate. This is called **eccentric contraction**. One example of eccentric contraction is returning the dumbbell in **1** from a flexed position to an extended one. To lower the weight, the muscle will decrease motor unit recruitment, the dumbbell will overcome the tension generated by the muscle, and the muscle will lengthen. In the example shown below, the muscle is still actively contracting, but not enough to lift the weight.

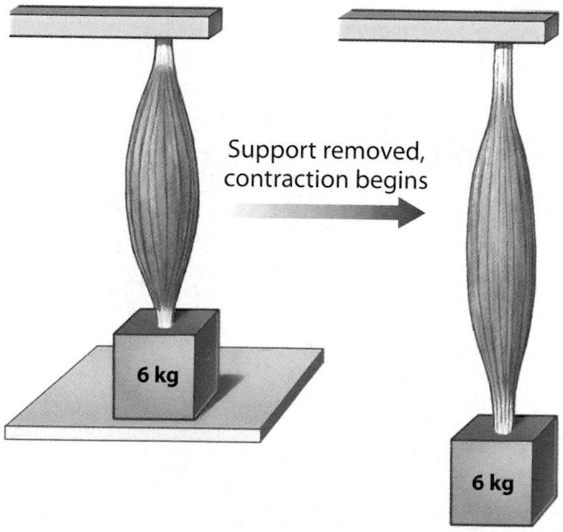

Support removed, contraction begins
6 kg
6 kg

5 The rate of elongation during an eccentric contraction depends on the difference between the tension developed by the active muscle fibers and the size of the load. By varying the tension in an eccentric contraction, you can control the rate of elongation.

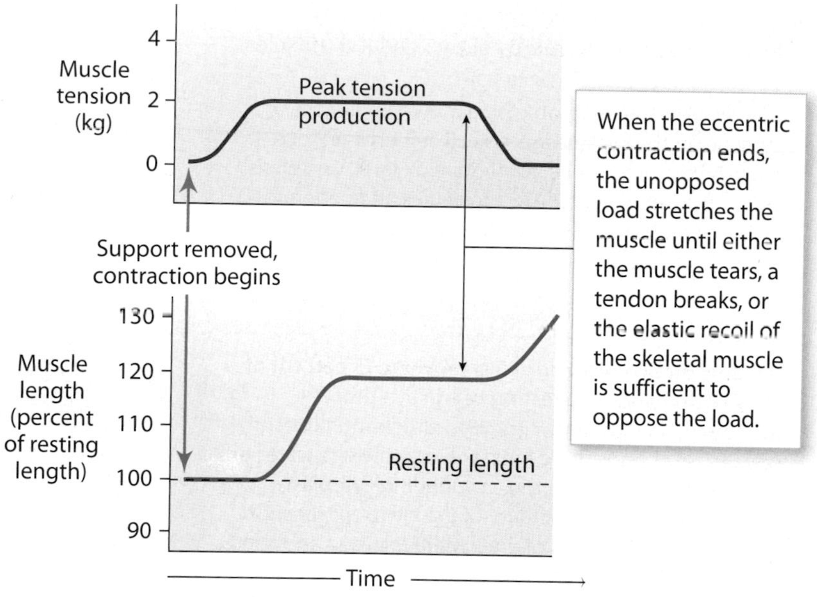

Muscle tension (kg)
4
2
0
Peak tension production
Support removed, contraction begins
Muscle length (percent of resting length)
130
120
110
100
90
Resting length
Time
When the eccentric contraction ends, the unopposed load stretches the muscle until either the muscle tears, a tendon breaks, or the elastic recoil of the skeletal muscle is sufficient to oppose the load.

Isometric Contractions

In an **isometric contraction** (*metric*, measure), the muscle as a whole does not change length, and the tension produced never exceeds the load. Many of the reflexive muscle contractions that keep your body upright when you stand or sit involve isometric contractions.

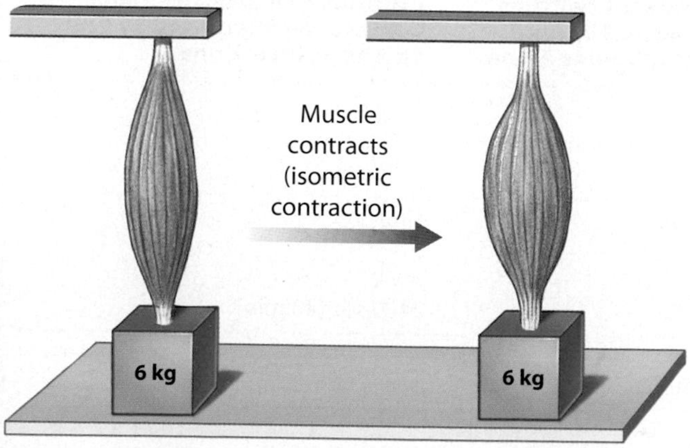

Muscle contracts (isometric contraction)
6 kg
6 kg

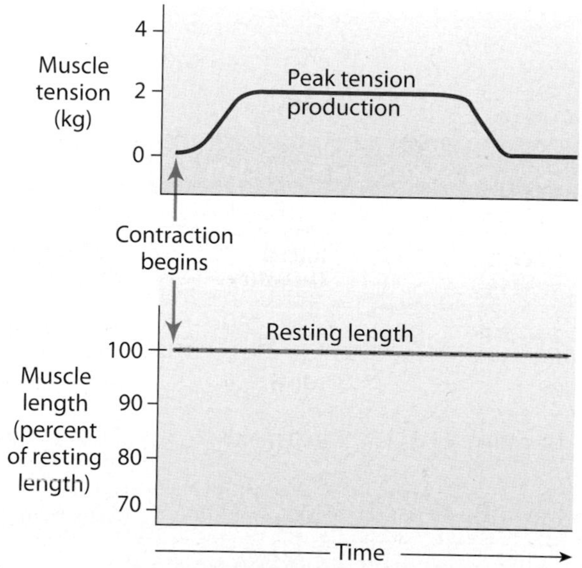

Muscle tension (kg)
4
2
0
Peak tension production
Contraction begins
Resting length
Muscle length (percent of resting length)
100
90
80
70
Time

6 If the load equals the peak tension, the load won't move when the muscle contracts. The contracting muscle bulges, but not as much as it does during an isotonic contraction. In an isometric contraction, although the muscle as a whole does not shorten, the individual muscle fibers shorten as connective tissues stretch. The muscle fibers cannot shorten further, because the tension does not exceed the load.

Module 9.11 Review

a. Compare isotonic and isometric contractions.

b. Explain the relationship between load and speed of muscle contraction.

c. Can a skeletal muscle contract without shortening? Why or why not?

LO **9.11** Compare the different types of muscle contractions.

Muscle contraction requires large amounts of ATP that may be produced anaerobically or aerobically

Energy Production

1 Mitochondrial activity is the ultimate source of the energy required by active skeletal muscles.

Glycolysis is the anaerobic breakdown of glucose to pyruvate in the cytoplasm of a cell. It is an anaerobic process because it does not require oxygen. Glycolysis provides a net gain of 2 ATP molecules and generates 2 pyruvate molecules from each glucose molecule.

Aerobic metabolism normally provides 95 percent of the ATP demands of a resting cell. In this process, introduced in Module 3.7 (p. 99), mitochondria absorb oxygen, ADP, phosphate ions, and organic substrates (such as pyruvate) from the surrounding cytoplasm. Primarily through the activity of the electron transport system, large amounts of energy are released and used to make ATP. The entire process is very efficient: For each molecule of pyruvate "fed" into the citric acid cycle, the cell gains 17 ATP molecules.

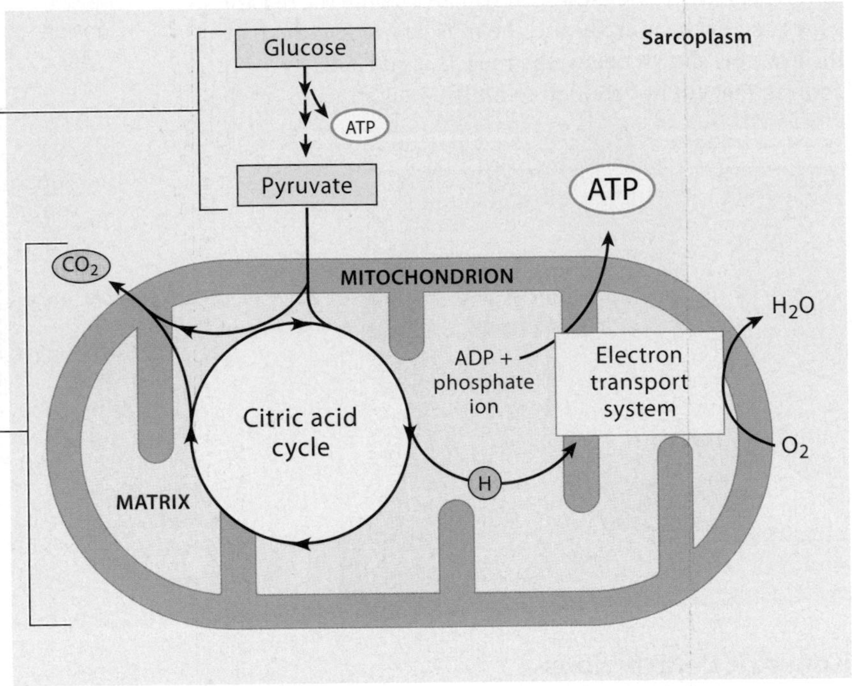
Glucose
ATP
Pyruvate
Sarcoplasm
ATP
CO_2
MITOCHONDRION
H_2O
ADP + phosphate ion
Electron transport system
Citric acid cycle
O_2
H
MATRIX

Sources of Energy Stored in a Typical Muscle Fiber

Energy Source	Initial Quantity	Utilization Process	Number of Twitches Supported by Each Energy Source Alone	Duration of Isometric Tetanic Contraction Supported by Each Energy Source Alone
Free ATP	3 mmol	ATP ⟶ ADP + P	10	2 sec
CP	20 mmol	ADP + CP ⟶ ATP + C	70	15 sec
Glycogen	100 mmol	Glycolysis (anaerobic metabolism)	670	130 sec
Glycogen	100 mmol	Aerobic metabolism	12,000	2400 sec (40 min)

2 This table characterizes the energy reserves of a typical skeletal muscle fiber. When energy demands are low and oxygen is abundant, a muscle fiber absorbs nutrients from the interstitial fluid and builds up its energy reserves. In addition to a small amount of free ATP in a cell, muscle fibers store a high-energy compound called **creatine phosphate** (**CP**). (Creatine is a compound assembled within skeletal muscles from catabolized amino acids.) However, muscle fibers store only a relatively small amount of energy in high-energy compounds. Most energy is stored as glycogen, which may account for 1.5 percent of total muscle weight and enables muscle contractions to continue for extended periods.

Energy Demands at Different Levels of Activity

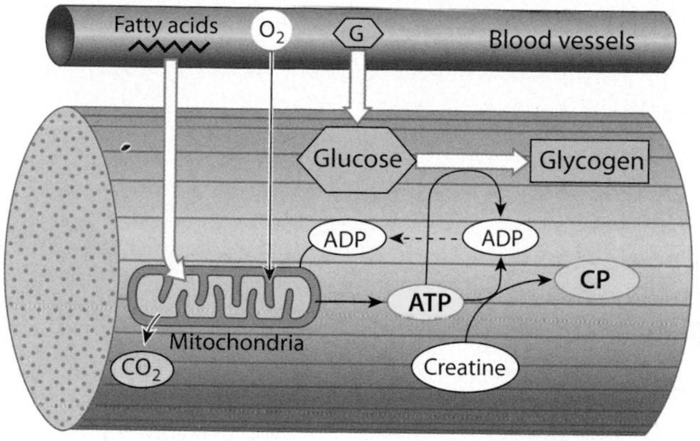
Fatty acids
O_2
G
Blood vessels
Glucose
Glycogen
ADP
ADP
CP
ATP
Mitochondria
CO_2
Creatine

Muscle at rest

3 In a resting skeletal muscle, the demand for ATP is low. More than enough oxygen is available for the mitochondria to meet that demand, and they produce a surplus of ATP. The extra ATP is used to build up reserves of CP and glycogen. Resting muscle fibers absorb fatty acids and glucose delivered by the bloodstream. The fatty acids are broken down in the mitochondria, generating ATP that is used to convert creatine to creatine phosphate and glucose to glycogen.

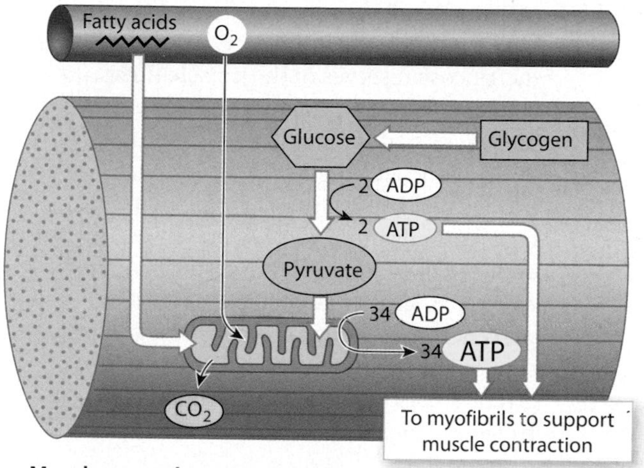
Fatty acids
O_2
Glucose
Glycogen
2 ADP
2 ATP
Pyruvate
34 ADP
34 ATP
CO_2
To myofibrils to support muscle contraction

Muscle at moderate activity levels

4 At moderate levels of activity, the demand for ATP increases. This demand is met by the mitochondria, and the rate of oxygen consumption increases as a result. The skeletal muscle now relies primarily on the aerobic metabolism of pyruvate to generate ATP, and the pyruvate is provided by glycolysis, using glucose obtained from stored glycogen reserves. As long as mitochondrial activity can meet the demand for ATP, the muscle will not become fatigued until glycogen, lipid, and amino acid reserves are exhausted. This type of fatigue affects the muscles of endurance athletes, such as marathon runners, after hours of exertion.

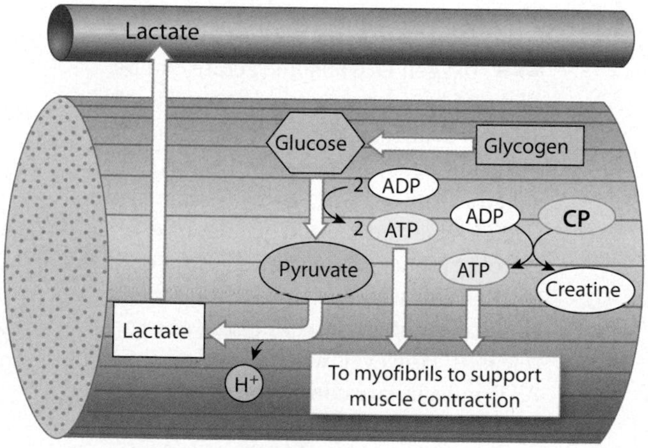
Lactate
Glucose
Glycogen
2 ADP
2 ATP
ADP
CP
ATP
Pyruvate
Creatine
Lactate
H^+
To myofibrils to support muscle contraction

Muscle at peak activity levels

5 At peak levels of activity, ATP demands are enormous and mitochondrial ATP production plateaus at a maximum rate determined by the availability of oxygen. At peak exertion, the mitochondria can provide only about one-third of the ATP needed. Glycolysis produces the rest. When glycolysis produces pyruvate faster than it can be utilized by the mitochondria, pyruvate levels rise in the sarcoplasm. Under these oxygen-limiting conditions, pyruvate is converted to **lactic acid**, which dissociates into a three-carbon **lactate** molecule and a hydrogen ion (H^+). The production of lactic acid during peak activity lowers the intracellular and extracellular pH. After only a few seconds of peak activity, changes in pH will alter the functional characteristics of key enzymes so that the muscle fiber can no longer contract. Sprinters usually experience this type of muscle fatigue.

Module 9.12 Review

a. Identify three sources of stored energy utilized by muscle fibers.

b. How do muscle cells continuously synthesize ATP?

c. When do muscle fibers produce lactic acid?

9.12 Describe the mechanisms by which muscle fibers obtain the energy to power contractions.

Module 9.13

Muscles fatigue and may need an extended recovery period

If a skeletal muscle can no longer perform at the required level of activity, it is **fatigued**. Many factors are involved in promoting muscle fatigue, but after peak activity a major factor is the decline in pH (from about 7.1 to 6.4) within the muscle fibers and the muscle as a whole, decreasing calcium ion binding to troponin and altering enzyme activities. In the **recovery period**, the conditions in muscle fibers are returned to normal, pre-exertion levels. After moderate activity, it may take several hours for muscle fibers to recover. After sustained activity at higher levels, complete recovery can take a week.

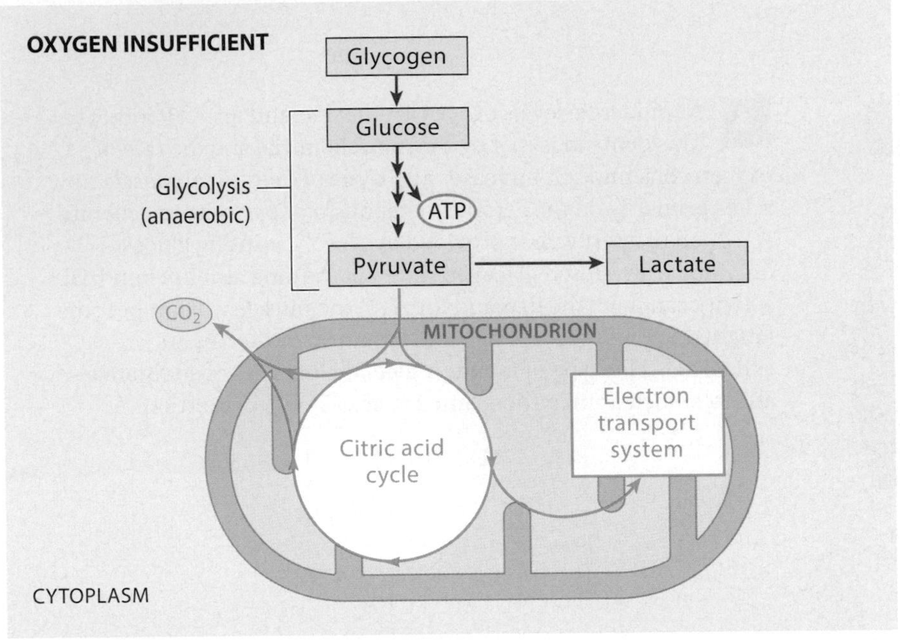

OXYGEN INSUFFICIENT
Glycogen
Glucose
Glycolysis (anaerobic)
ATP
Pyruvate
Lactate
CO_2
MITOCHONDRION
Electron transport system
Citric acid cycle
CYTOPLASM

1 Glycolysis enables a skeletal muscle to continue contracting even when insufficient oxygen limits mitochondrial activity. Recall that glycolysis is not the most efficient way to generate ATP. It squanders the glucose reserves of the muscle fibers, and it is potentially dangerous because the dissociation of lactic acid can lower the pH of the blood and tissues. Glycolysis can produce ATP faster than aerobic metabolism, but only until glycogen reserves are depleted (1–2 minutes). In terms of energy efficiency, glycolysis captures some 4–6 percent of the energy as ATP in the conversion of glucose to pyruvate. The reactions of glycolysis also elevate body temperature, triggering increased sweat gland activity.

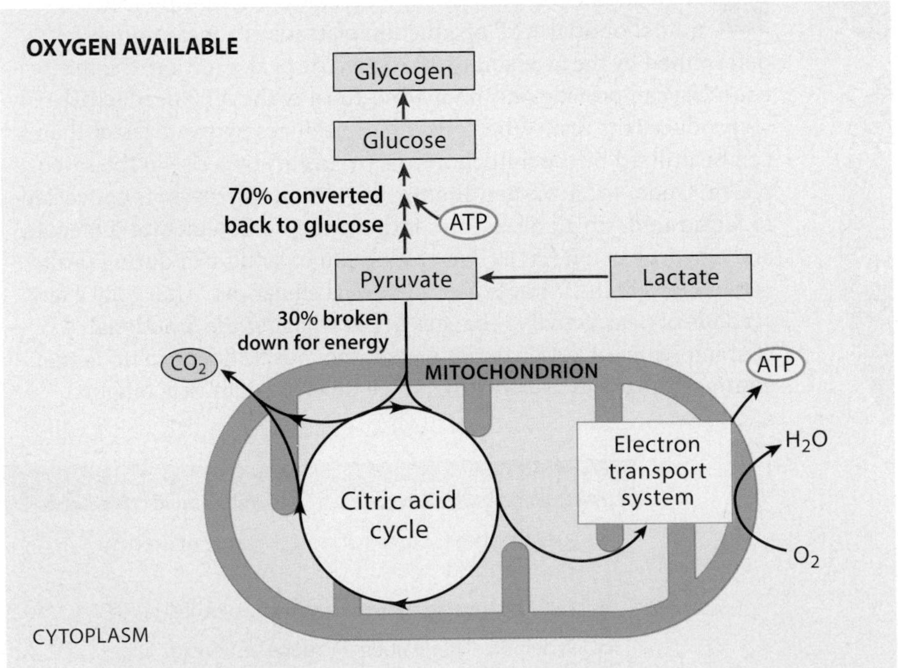

OXYGEN AVAILABLE
Glycogen
Glucose
70% converted back to glucose
ATP
Pyruvate
Lactate
30% broken down for energy
CO_2
MITOCHONDRION
ATP
Electron transport system
H_2O
Citric acid cycle
O_2
CYTOPLASM

2 During the recovery period, when oxygen is available, lactate can be recycled by conversion back to pyruvate. The pyruvate can then be used either by mitochondria to generate ATP or as a substrate for enzyme pathways that synthesize glucose and rebuild glycogen reserves. Aerobic metabolism is much more efficient than glycolysis—the muscle fiber can capture about 42 percent of the energy released. However, whenever they are using ATP, muscles are continuously releasing heat. About 70–80 percent of body heat is produced by resting skeletal muscles.

Peak Activity

Much of the lactate produced during peak exertion diffuses out of the muscle fibers and into the bloodstream. The liver absorbs this lactate and begins converting it into pyruvate.

Recovery

This process continues after exertion has ended, because lactate levels within muscle fibers remain relatively high, and lactate continues to diffuse into the bloodstream. After the absorbed lactate is converted to pyruvate in the liver, about 30 percent of the new pyruvate molecules are broken down in the mitochondria, providing the ATP needed to convert the remaining 70 percent of pyruvate molecules into glucose. The glucose molecules are then released into the circulation. Skeletal muscle fibers take up the molecules and use them to rebuild their glycogen reserves.

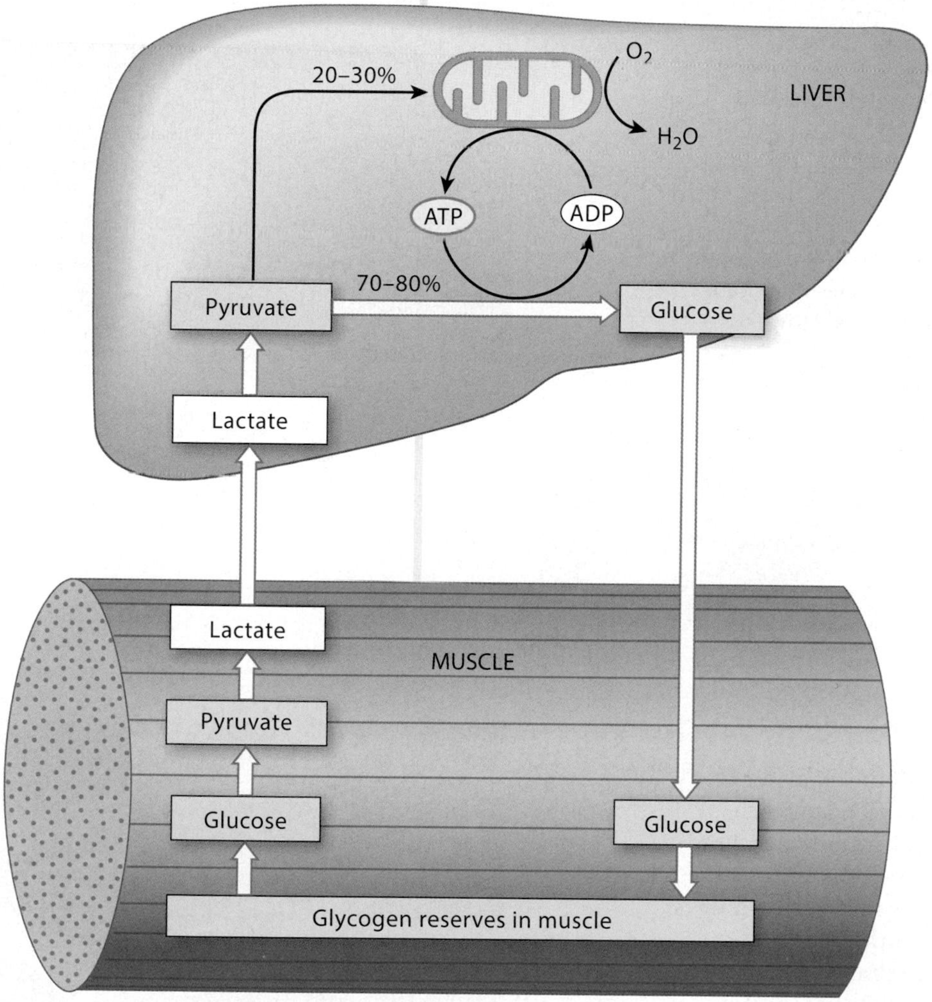

20–30%
O_2
LIVER
H_2O
ATP
ADP
70–80%
Pyruvate
Glucose
Lactate
Lactate
MUSCLE
Pyruvate
Glucose
Glucose
Glycogen reserves in muscle

3 Much of the lactate released by muscle fibers during strenuous activity diffuses from the muscle tissue into the bloodstream. The liver absorbs and recycles it, and releases glucose into the circulation. This shuffling of lactate to the liver and of glucose back to muscle cells is called the **Cori cycle**. Throughout the recovery period, the body's oxygen demand remains elevated above normal resting levels. The more ATP required, the more oxygen will be needed. The amount of oxygen required to restore normal, pre-exertion conditions is called the **oxygen debt**, or **excess postexercise oxygen consumption (EPOC)**. Most of the additional oxygen consumption occurs in skeletal muscle fibers, which must restore ATP, creatine phosphate, and glycogen concentrations to their former levels, and in liver cells, which generate the ATP needed to convert excess lactate to glucose.

Module 9.13 Review

a. How is skeletal muscle recovery different after moderate activity compared to sustained activity at higher levels?

b. What happens to the lactate produced inside skeletal muscle during peak activity?

c. Define oxygen debt (excess postexercise oxygen consumption).

LO **9.13** Describe the factors that contribute to muscle fatigue, and discuss the stages and mechanisms involved in the muscle's subsequent recovery.

Fast, slow, and intermediate skeletal muscle fibers differ in size, internal structure, metabolism, and resistance to fatigue

The human body has three major types of skeletal muscle fibers: fast fibers, slow fibers, and intermediate fibers.

Fast Fibers

1 Most of the skeletal muscle fibers in the body are called **fast fibers**, because they can reach peak twitch tension in 0.01 second or less after stimulation. Fast fibers have large diameters and contain densely packed myofibrils, large glycogen reserves, and relatively few mitochondria. The tension produced by a muscle fiber is directly proportional to the number of myofibrils, so muscles dominated by fast fibers produce powerful contractions. However, fast fibers fatigue rapidly because their contractions use ATP in massive amounts, and they have relatively few mitochondria to generate ATP. As a result, prolonged activity is supported primarily by anaerobic metabolism.

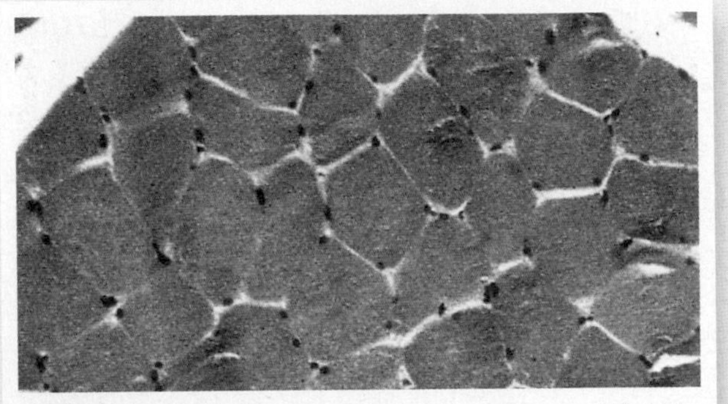

Fast fibers in cross section LM × 171

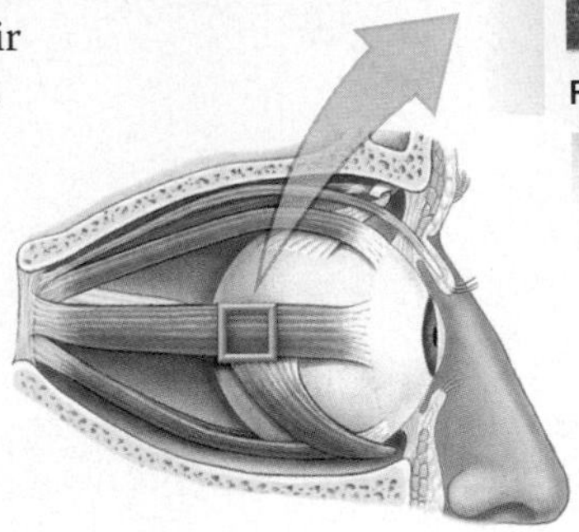

Slow Fibers

2 **Slow fibers** have only about half the diameter of fast fibers and take three times as long to reach peak tension after stimulation. Slow fibers are specialized to continue contracting for extended periods, long after a fast fiber would become fatigued. Slow muscle fibers are surrounded by a more extensive network of capillaries than fast muscle tissue, so they have a dramatically higher oxygen supply to support mitochondrial activity. Slow fibers also contain the red pigment **myoglobin** (MĪ-ō-glō-bin), a globular protein that is structurally related to hemoglobin, the red oxygen-carrying pigment in blood. Both myoglobin and hemoglobin reversibly bind oxygen molecules. Myoglobin is most abundant in slow fibers, so resting slow fibers contain substantial oxygen reserves that can be mobilized during a contraction. Because slow fibers have both an extensive capillary supply and a high concentration of myoglobin, skeletal muscles dominated by slow fibers appear dark red.

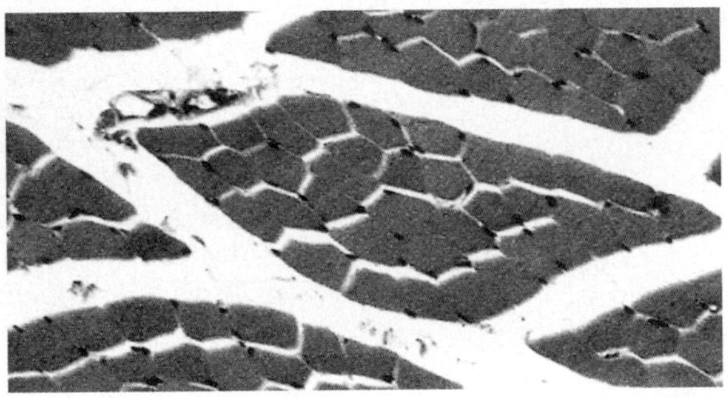

Slow fibers in cross section LM × 171

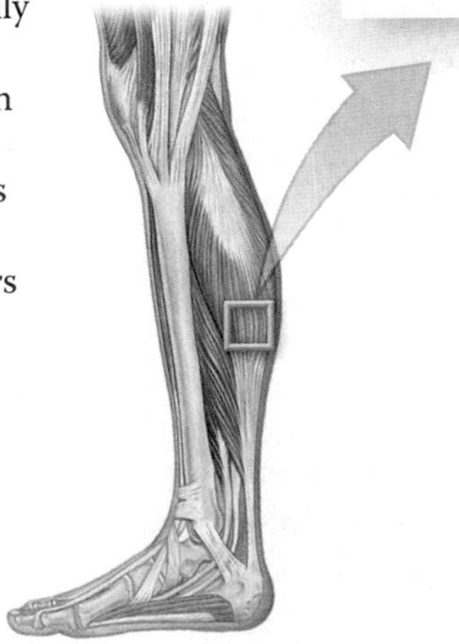

Intermediate Fibers

3 This light micrograph (a longitudinal section) shows that a fast muscle fiber (W, for white) and a slow muscle fiber (R, for red) have very different sizes and densities, and that slow fibers have more mitochondria (M) and a more extensive capillary supply. Our bodies also contain a third group of muscle fibers—**intermediate fibers**—that more closely resemble fast fibers, for they contain little myoglobin and are relatively pale. Intermediate fibers have a more extensive capillary network around them and are more resistant to fatigue than are fast fibers.

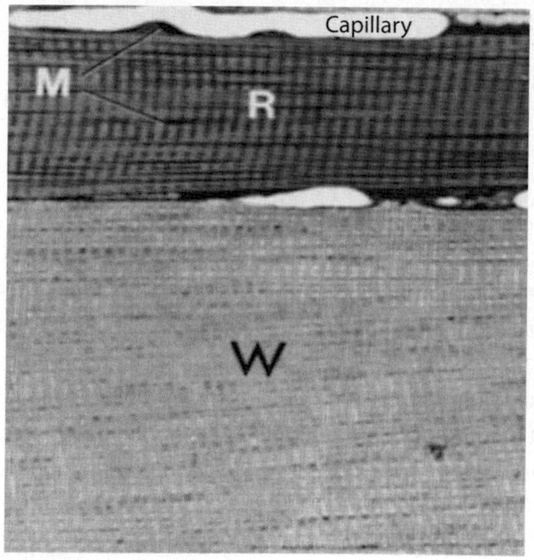

Capillary
M
R
W

Fast (W) and slow (R) fibers in longitudinal section LM × 783

Properties of Skeletal Muscle Fiber Types

Property	Fast Fibers	Slow Fibers	Intermediate Fibers
Cross-sectional diameter	Large	Small	Intermediate
Color	White	Red	Pink
Myoglobin content	Low	High	Low
Capillary supply	Scarce	Dense	Intermediate
Mitochondria	Few	Many	Intermediate
Time to peak tension	Rapid	Prolonged	Medium
Contraction speed	Fast	Slow	Fast
Fatigue resistance	Low	High	Intermediate
Glycolytic enzyme concentration in cytosol	High	Low	High
Substrates used for ATP generation during contraction (type of metabolism)	Carbohydrates (anaerobic)	Lipids, carbohydrates, amino acids (aerobic)	Primarily carbohydrates (anaerobic)
Alternative names	Type II-B, FF (fast fatigue), white, fast-twitch glycolytic	Type I, S (slow), red, SO (slow oxidative), slow-twitch oxidative	Type II-A, FR (fast resistant), fast-twitch oxidative

4 This table compares the properties of fast, slow, and intermediate muscle fibers. The percentage of each fiber type in a skeletal muscle can vary greatly. Most human muscles contain a mixture of fiber types and so appear pink. However, there are no slow fibers in muscles of the eye or hand, where swift but brief contractions are required. Many back and calf muscles are dominated by slow fibers; these muscles contract almost continuously to maintain an upright posture. The percentage of fast versus slow fibers in each muscle is genetically determined, but the ratio of intermediate fibers to fast fibers can increase as a result of athletic training. For example, if a muscle is used repeatedly for endurance events, some of the fast fibers will develop the appearance and functional capabilities of intermediate fibers.

Module 9.14 Review

a. Identify the three types of skeletal muscle fibers.

b. Why would a sprinter experience muscle fatigue before a marathon runner would?

c. Which type of muscle fiber would you expect to predominate in the leg muscles of an endurance athlete, such as a cyclist or a long-distance runner?

LO **9.14** Relate the types of muscle fibers to muscle performance.

Many factors can result in muscle hypertrophy, atrophy, or paralysis

Hypertrophy

1 As a result of repeated, exhaustive stimulation, muscle fibers develop more mitochondria, a higher concentration of glycolytic enzymes, and larger glycogen reserves. Such muscle fibers have more myofibrils than do fibers that are less used, and each myofibril contains more thick and thin filaments. The net effect is **hypertrophy**, or an enlargement of the stimulated muscle. The number of muscle fibers does not change significantly, but the muscle as a whole enlarges because each muscle fiber increases in diameter. The muscle also becomes stronger, since tension production is proportional to the cross-sectional area of the muscle. The muscles of a bodybuilder are excellent examples of muscular hypertrophy. Although hypertrophy can be promoted by administering steroid hormones such as androgens (male sex hormones), exhaustive training is still required, and the side effects of steroid use can be very dangerous.

Atrophy

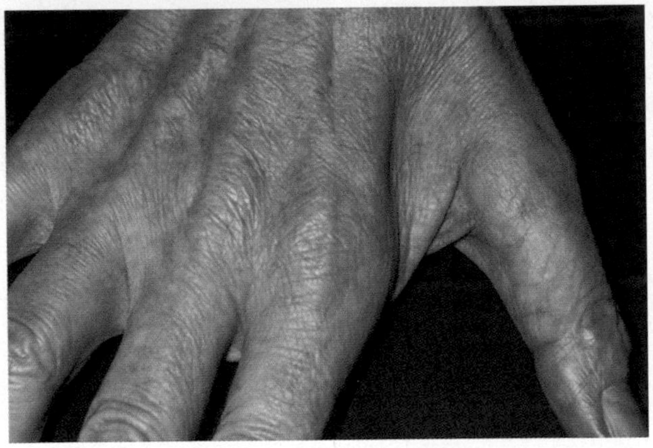

2 A skeletal muscle that is not regularly stimulated by a motor neuron loses muscle tone and mass. The muscle becomes flaccid, and the muscle fibers become smaller and weaker. This reduction in muscle size, tone, and power is called **atrophy**. A variable degree of muscle atrophy is a normal consequence of aging. Individuals of any age who are paralyzed by spinal injuries or other damage to the nervous system will gradually lose muscle tone and size in the areas affected. Even a temporary reduction in muscle use can lead to muscular atrophy; you can easily observe this effect by comparing "before and after" limb muscles in someone who has worn a cast. Muscle atrophy is initially reversible, but dying muscle fibers are not replaced. In extreme atrophy, the functional losses are permanent. That is why physical therapy is crucial for people who are temporarily unable to move normally.

Muscular Dystrophy

3 There are several different **muscular dystrophies** (DIS-trō-fēz), and all are inherited diseases that produce muscle weakness and deterioration. The most common and best understood is **Duchenne/Becker muscular dystrophy (DBMD)**. This form appears in childhood, and generally affects only males. It is estimated that 1 in 3500 to 5000 newborn males worldwide is born with the condition. A progressive muscular weakness develops, and most individuals die due to respiratory paralysis. Skeletal muscles are primarily affected, although for some reason the facial muscles continue to function normally. In later stages of the disease, the facial muscles and cardiac muscle tissue may also become involved. The inheritance of DBMD is sex-linked: Women carrying the defective genes are unaffected, but each of their male children will have a 50 percent chance of developing DBMD. Genetic tests are available to determine whether or not a woman is carrying the defective gene and prenatal tests can determine whether a fetus has the condition.

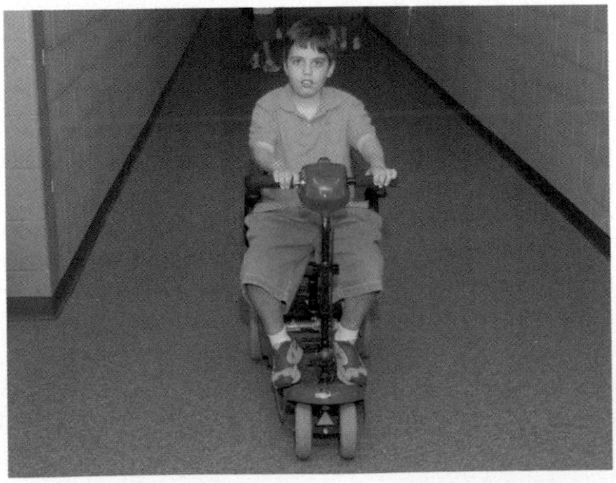

Clinical Conditions That Affect Skeletal Muscles

4 Because skeletal muscles depend on motor neurons for stimulation, disorders that impair the nervous system can indirectly affect the muscular system.

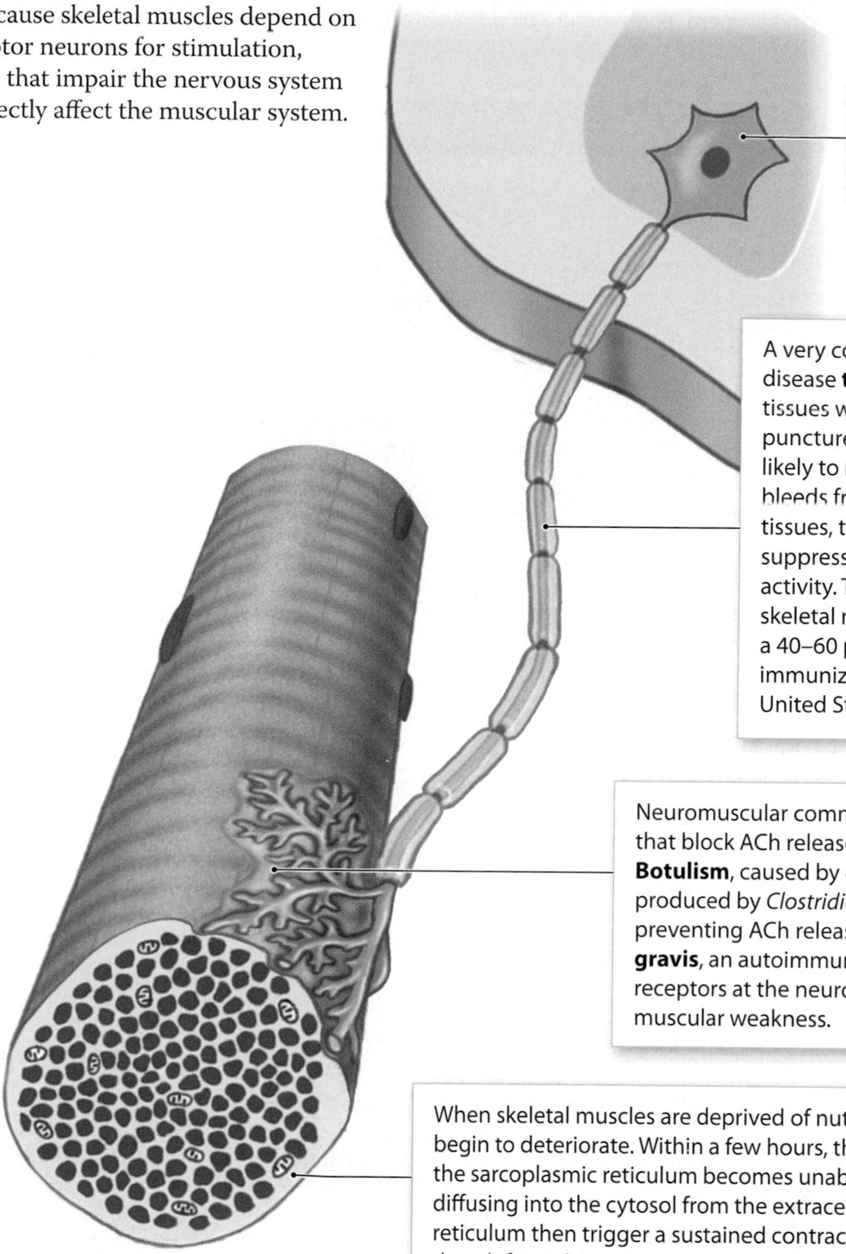

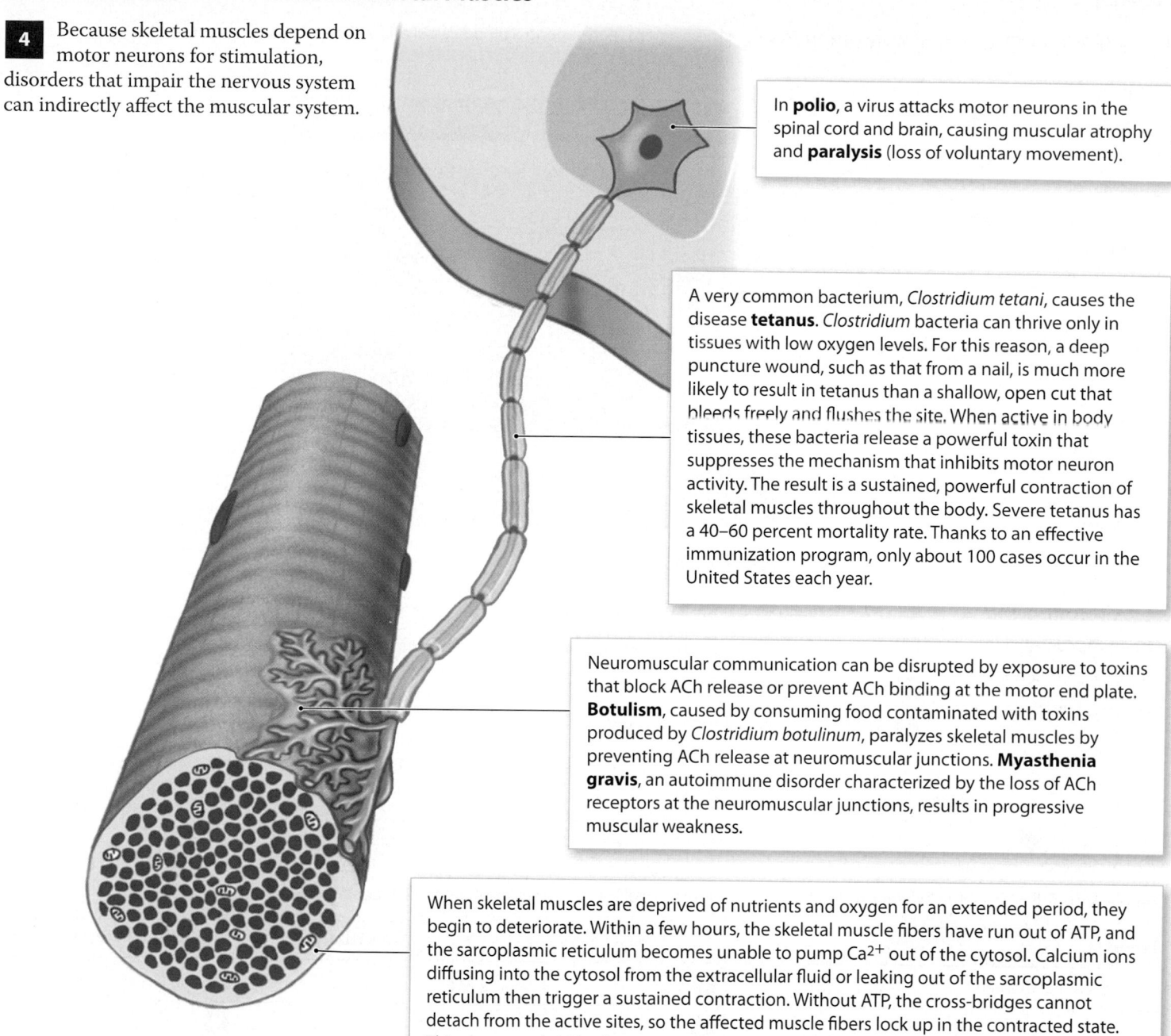

In **polio**, a virus attacks motor neurons in the spinal cord and brain, causing muscular atrophy and **paralysis** (loss of voluntary movement).

A very common bacterium, *Clostridium tetani*, causes the disease **tetanus**. *Clostridium* bacteria can thrive only in tissues with low oxygen levels. For this reason, a deep puncture wound, such as that from a nail, is much more likely to result in tetanus than a shallow, open cut that bleeds freely and flushes the site. When active in body tissues, these bacteria release a powerful toxin that suppresses the mechanism that inhibits motor neuron activity. The result is a sustained, powerful contraction of skeletal muscles throughout the body. Severe tetanus has a 40–60 percent mortality rate. Thanks to an effective immunization program, only about 100 cases occur in the United States each year.

Neuromuscular communication can be disrupted by exposure to toxins that block ACh release or prevent ACh binding at the motor end plate. **Botulism**, caused by consuming food contaminated with toxins produced by *Clostridium botulinum*, paralyzes skeletal muscles by preventing ACh release at neuromuscular junctions. **Myasthenia gravis**, an autoimmune disorder characterized by the loss of ACh receptors at the neuromuscular junctions, results in progressive muscular weakness.

When skeletal muscles are deprived of nutrients and oxygen for an extended period, they begin to deteriorate. Within a few hours, the skeletal muscle fibers have run out of ATP, and the sarcoplasmic reticulum becomes unable to pump Ca^{2+} out of the cytosol. Calcium ions diffusing into the cytosol from the extracellular fluid or leaking out of the sarcoplasmic reticulum then trigger a sustained contraction. Without ATP, the cross-bridges cannot detach from the active sites, so the affected muscle fibers lock up in the contracted state. This can occur locally after acute injury or after death.

5 Shortly after death, a generalized skeletal muscle contraction, called **rigor mortis**, occurs throughout the body, beginning with the smaller muscles of the face, neck, and arms. As the SR deteriorates, calcium ions are released and a sustained contraction begins. As ATP reserves are exhausted, the muscles become locked in the contracted state. Because all skeletal muscles are involved during rigor mortis, the individual becomes "stiff as a board." It typically begins 2–7 hours after death and disappears after 1–6 days or when decomposition begins, but the timing depends on environmental factors such as temperature. Forensic pathologists base time of death estimates on the degree of rigor mortis and environmental conditions.

Module 9.15 Review

a. Define muscle hypertrophy and muscle atrophy.

b. Six weeks after Fred broke his leg the cast is removed, and as he steps down from the exam table, his leg gives way and he falls. Propose a logical explanation.

c. Explain how the flexibility or rigidity of a dead body can provide a clue about a murder victim's time of death.

LO **9.15** Explain the physiological factors responsible for muscle hypertrophy, atrophy, and paralysis.

Labeling

Use the following terms to correctly label the structures in the diagram representing a blood vessel and a skeletal muscle at rest.

- creatine
- glycogen
- CP
- glucose
- O_2
- fatty acids

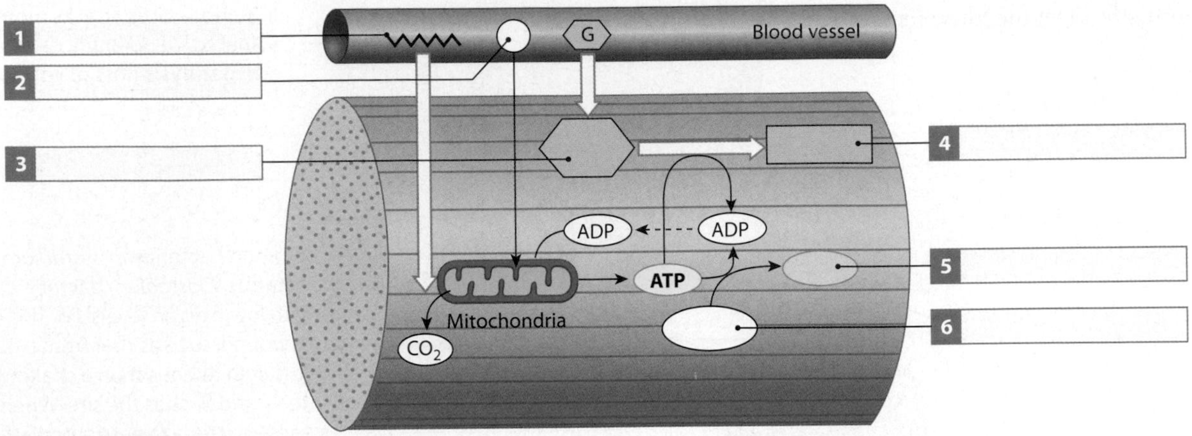
1
2
3
4
5
6
G
Blood vessel
ADP
ADP
ATP
Mitochondria
CO_2

Matching

Match each lettered term with the most closely related description.

a. eccentric contraction
b. isometric contraction
c. isotonic contraction
d. concentric contraction

7 Muscle does not change length during contraction
8 Muscle changes length during contraction
9 Peak tension less than load, and muscle elongates
10 Peak tension greater than load, and muscle shortens

7 ______
8 ______
9 ______
10 ______

Short answer

Complete the following table by writing in the anatomical and physiological properties of the three types of skeletal muscle fibers.

Property	Fast Fibers	Slow Fibers	Intermediate Fibers
Cross-sectional diameter	11	12	13
Color	14	RED	15
Myoglobin content	16	17	LOW
Capillary supply	SCARCE	18	19
Mitochondria	20	21	22
Time to peak tension	23	PROLONGED	24
Contraction speed	25	26	27
Fatigue resistance	28	29	INTERMEDIATE
Glycolytic enzyme concentration in cytosol	HIGH	30	31

Study Outline

SECTION 1 • Functional Anatomy of Skeletal Muscle Tissue

9.1 Skeletal muscle tissue enables body movement and other vital functions p. 307

1. The three types of muscle tissue (skeletal, cardiac and smooth) consist chiefly of muscle cells that are highly specialized for contraction.
2. **Skeletal muscles** are complex organs that are attached directly or indirectly to bones.
3. Skeletal muscle functions to produce skeletal movement, maintain posture and body position, support soft tissue, guard entrances and exits of the body, maintain body temperature, and provide nutrient reserves.

9.2 A skeletal muscle contains skeletal muscle tissue, connective tissues, blood vessels, and nerves p. 308

4. **Skeletal muscle** is surrounded by **epimysium**, a dense layer of collagen fibers.
5. A **muscle fascicle** is a bundle of muscle fibers surrounded by **perimysium**.
6. Individual muscle fibers are sheathed in **endomysium**.
7. Embryonic **myoblasts** fuse to form multinucleate cells that develop into distinct muscle fibers.
8. Skeletal muscle fibers have a plasma membrane called a **sarcolemma**, and a cytoplasm surrounding the myofibrils called **sarcoplasm**.

9.3 Skeletal muscle fibers have contractile myofibrils containing hundreds to thousands of sarcomeres p. 310

9. A single skeletal muscle fiber contains hundreds to thousands of banded, cylindrical structures called **myofibrils**.

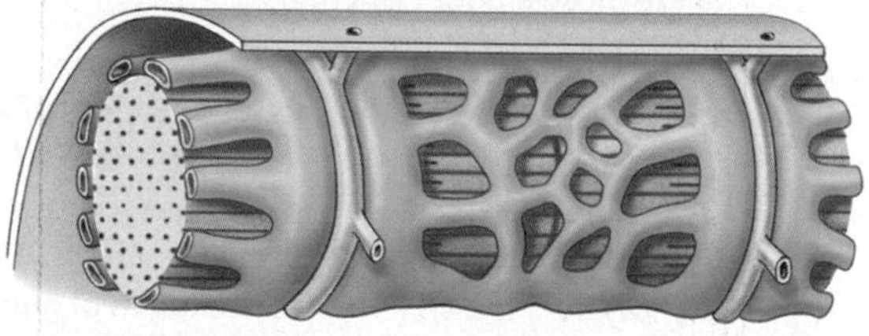

10. Myofibrils consist of bundles of protein **myofilaments**. **Thin filaments** are composed mostly of actin, and **thick filaments** are composed primarily of myosin.
11. Myofilaments have repeating, light and dark banded functional units called **sarcomeres**.
12. On either side of the sarcolemma is an uneven distribution of charges called a transmembrane potential.
13. **Transverse tubules** or **T tubules** are extensions of the sarcolemma into the sarcoplasm. When they are in contact with a pair of **terminal cisternae** of the **sarcoplasmic reticulum (SR)** it forms a **triad**.
14. Ca^{2+} ions are stored in the SR. A muscle contraction begins when the SR releases Ca^{2+} into the sarcoplasm through gated calcium channels.

9.4 The sliding filament theory of muscle contraction involves thin and thick filaments p. 312

15. The thin filaments are attached to the Z line by **actinin**.
16. **F-actin** (filamentous actin) is a twisted strand of individual **G-actin** (globular actin) molecules.
17. Strands of **tropomyosin** cover the active sites of G-actin. **Troponin** is bound to tropomyosin, and also has one site bound to G-actin, and two sites for Ca^{2+} ions.
18. Thick filaments contain **myosin molecules** arranged with their tails pointing toward the M line.

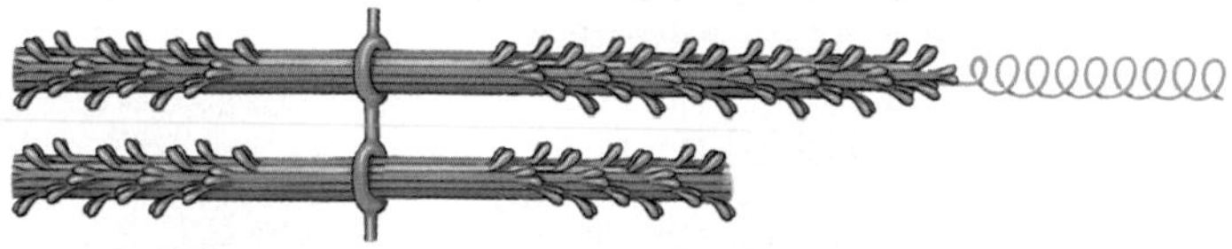

19. The myosin molecule has a **tail** bound to other myosin molecules, a **free head** that has two globular subunits, and a connection between the two that acts as a hinge.
20. During contraction myosin heads form cross-bridges with thin filaments.
21. When a skeletal muscle fiber contracts, thin filaments slide past thick filaments. This is known as the **sliding filament theory**. This occurs in every sarcomere along a myofibril, so the myofibril gets shorter. When myofibrils get shorter, so does the muscle fiber.

9.5 Skeletal muscle fibers and neurons have excitable plasma membranes that produce and carry electrical impulses called action potentials p. 314

22. At rest, a cell is **polarized**. The inside of its plasma membrane has a slight negative charge with respect to its outside surface.
23. In undisturbed neurons and skeletal muscle fibers, typical **resting potentials** are –70 mV and –85 mV, respectively.
24. Open leak channels in the plasma membrane of cells allow ions to move down their concentration gradients. Sodium–potassium ion pumps maintain the cell's resting potential by exporting three Na^+ from the cell, in exchange for two K^+.
25. A cell is **depolarized** when a threshold potential is reached (–55 mV for skeletal muscle) and Na^+ rush into the cell. The depolarization peaks at +30 mV. Na^+ inflow then abruptly stops and **repolarization** begins as K^+ rush out of the cell. This loss of positive ions causes the membrane potential to become negative again.
26. In neurons and skeletal muscle fibers, the depolarization produces an electrical impulse or **action potential** that is spread or propagated along their plasma membranes.

9.6 A skeletal muscle fiber contracts when stimulated by a motor neuron p. 316

27. Each skeletal muscle fiber is controlled by the nervous system at a specialized site called a **neuromuscular junction (NMJ)**.

28. The NMJ is made up of an **axon terminal**, a specialized region of the sarcolemma called the **motor end plate**, and a narrow space between the two called the **synaptic cleft**.
29. When the action potential reaches the synaptic terminal, it releases the neurotransmitter **acetylcholine (ACh)** into the synaptic cleft. The enzyme **acetylcholinesterase (AChE)** breaks down ACh.
30. ACh binds to receptors in the motor end plate changing the membrane permeability causing Na^+ to rush into the sarcoplasm. This generates an action potential in the skeletal muscle fiber.
31. The action potential sweeps down each T tubule to the terminal cisternae causing the SR to release Ca^{2+}. This event, called **excitation-contraction coupling**, triggers the contraction of the muscle fiber.

9.7 A muscle fiber contraction uses ATP in a cycle that repeats during the contraction p. 318

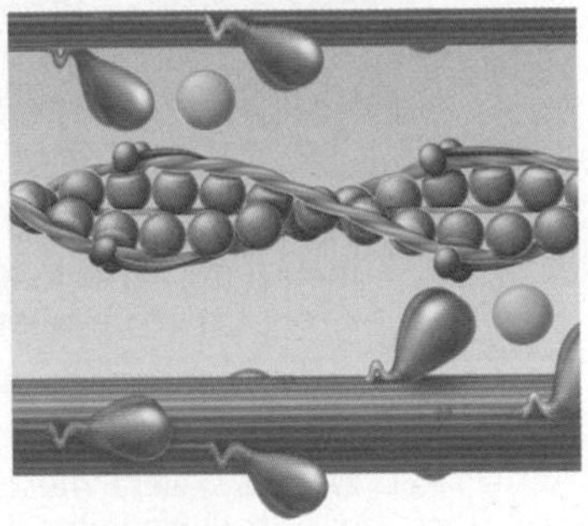

32. Ca^{2+} bind to troponin, which causes the tropomyosin to move away from the actin active site allowing interaction with the energized myosin heads, forming **cross-bridges**.
33. The myosin head pivots as it releases the bound ADP and P.
34. The cross-bridge is broken when another ATP binds to the myosin head.
35. Myosin reactivation occurs when the free myosin head splits ATP into ADP and P.
36. The entire cycle is repeated several times each second, as long as Ca^{2+} concentrations remain elevated and ATP reserves are sufficient.

SECTION 2 • Functional Properties of Skeletal Muscle

9.8 Muscle tension develops from the events that occur during excitation-contraction coupling p. 321

37. A skeletal muscle contracts when stimulated by a motor neuron at a neuromuscular junction.
38. Excitation-contraction coupling begins with the release of ACh at the axon terminal. This causes an action potential in the sarcolemma.
39. The action potential travels along the T tubules to the triads, where it triggers the release of Ca^{2+} from the terminal cisternae of the SR.
40. Ca^{2+} allow for thick and thin filaments to interact, causing the sarcomere to shorten, pulling the ends of the muscle fiber closer together.
41. The entire skeletal muscle shortens and produces **tension** on the tendons at either end.

9.9 Tension production is greatest when a muscle is stimulated at its optimal length p. 322

42. The tension a muscle fiber produces is related to sarcomere length.
43. Tension is produced most efficiently when the maximum number of cross-bridges can form. This is called **optimal resting length**.
44. A **twitch** is a single stimulus-contraction-relaxation sequence in a muscle fiber. These vary greatly depending on many factors.
45. The phases of a twitch include the **latent period**, **contraction phase**, and **relaxation phase**.

9.10 The peak tension developed by a skeletal muscle depends on the frequency of stimulation and the number of muscle fibers stimulated p. 324

46. The amount of tension produced by a skeletal muscle is determined by the amount of tension produced by each stimulated muscle fiber, and the total number of muscle fibers stimulated at a given moment.
47. **Treppe** is a progressive increase in twitch tension that is not demonstrated by most skeletal muscles.
48. Repeated stimulation before the relaxation phase ends may produce **wave summation**, in which one twitch is added to another.
49. A muscle producing almost peak tension during rapid cycles of contraction and relaxation is in **incomplete tetanus**. **Complete tetanus** occurs when the relaxation phase is eliminated.
50. All the muscle fibers controlled by a single motor neuron constitute a **motor unit**. Increasing the number of active motor units is called **recruitment**.

9.11 Muscle contractions may be isotonic or isometric; isotonic contractions may be concentric or eccentric p. 326

51. In **isotonic contractions**, the tension in the muscle rises and the length of the muscle changes. There are two types of isotonic contractions: concentric and eccentric contractions.
52. In **concentric contractions** the muscle shortens. In **eccentric contractions** the load is greater than peak tension, so the muscle lengthens.
53. In **isometric contractions** the tension rises, but the length of the muscle does not change.

9.12 Muscle contraction requires large amounts of ATP that may be produced anaerobically or aerobically p. 328

54. Mitochondrial activity is the ultimate source of the energy required by active skeletal muscles.
55. **Glycolysis** is the anaerobic breakdown of glucose to pyruvate in the cytoplasm of a cell. It is an anaerobic process, because it does not require oxygen.
56. **Aerobic metabolism** normally provides 95 percent of the ATP demands of a resting cell. In this process, mitochondria absorb oxygen, ADP, phosphate ions, and pyruvate. For each molecule of pyruvate the cell gains 17 ATP molecules.
57. In addition to a small amount of free ATP, muscle fibers store a high-energy compound called **creatine phosphate (CP)**.
58. In resting skeletal muscle the demand for ATP is low and mitochondria produce a surplus of ATP.
59. At moderate levels of activity, ATP demand increases. Skeletal muscle now relies primarily on aerobic metabolism of pyruvate.
60. At peak activity levels, skeletal muscle relies heavily on

glycolysis to generate ATP because the mitochondria cannot obtain enough oxygen to meet demand. The result is an accumulation of **lactic acid** in the muscle that lowers intracellular pH, and contributes to fatigue.

9.13 Muscles fatigue and may need an extended recovery period p. 330

61. An active skeletal muscle is **fatigued** when it can no longer continue to perform.
62. In the **recovery period** the conditions in muscle fibers are returned to normal, pre-exertion levels.
63. The **oxygen debt**, or **excess postexercise oxygen consumption (EPOC)**, resulting from muscle activity is the amount of oxygen required during the recovery period to restore the muscle to its normal condition.

9.14 Fast, slow, and intermediate skeletal muscle fibers differ in size, internal structure, metabolism, and resistance to fatigue p. 332

64. There are three major types of muscle fibers: fast, slow and intermediate.
65. **Fast fibers** are large in diameter, and contain densely packed myofibrils, large glycogen reserves, and relatively few mitochondria. They contract rapidly and powerfully, are supported by anaerobic metabolism, and fatigue easily.
66. **Slow fibers** are smaller in diameter, slower in contraction speed, contain more capillaries and mitochondria, and also contain the red oxygen-carrying pigment **myoglobin**.
67. **Intermediate fibers** are very similar to fast fibers, but have a greater resistance to fatigue.

9.15 Many factors can result in muscle hypertrophy, atrophy, or paralysis p. 334

68. **Hypertrophy** is the enlargement of stimulated muscle fiber as a result of more mitochondria, greater glycogen and enzyme reserves, and an increase in myofibrils.
69. **Atrophy** is the inverse effect from disuse or disease.
70. A variety of clinical conditions can affect skeletal muscles such as **Duchenne/Becker muscular dystrophy (DBMD)**, **polio**, **tetanus**, **botulism**, and **myasthenia gravis**.

Chapter Review Questions

True/False

Indicate whether each statement is true or false.

1. An increase in sarcomere length reduces the tension produced in a muscle fiber by reducing the size of the zone of overlap and the number of potential cross-bridge interactions.
2. Muscle hypertrophy is mostly explained by an increase in muscle fiber (cell) number as a result of exercise.
3. The Cori cycle describes the transfer of lactate to the liver and glucose back to the muscle cells during the recovery period.
4. In glycolysis, one molecule of glucose produces two molecules of pyruvate and 17 ATP.

1 ____________
2 ____________
3 ____________
4 ____________

Matching

Match each lettered term with the most closely related description.

a. myoblasts
b. sarcoplasmic reticulum
c. T tubules
d. tropomyosin
e. troponin
f. I band
g. A band
h. synaptic vesicles
i. motor end plate
j. triad

5. Contains Ca^{2+}
6. Contains thick filaments and thin filaments
7. Contain ACh receptors
8. Extensions of the sarcolemma into the sarcoplasm
9. Embryonic cells that fuse to form muscle fibers
10. Contains the neurotransmitter ACh
11. Formed by T tubules and a pair of terminal cisternae
12. Contains only thin filaments
13. Has receptor sites for Ca^{2+}
14. Cover the active sites of the G-actin

5 ____________
6 ____________
7 ____________
8 ____________
9 ____________
10 ____________
11 ____________
12 ____________
13 ____________
14 ____________

Multiple choice

Select the correct answer from the list provided.

15 The connective tissue coverings of a skeletal muscle, listed from superficial to deep, are

- ☐ a) endomysium, perimysium, and epimysium.
- ☐ b) endomysium, epimysium, and perimysium.
- ☐ c) epimysium endomysium, and perimysium.
- ☐ d) epimysium, perimysium, and endomysium.

16 The detachment of the myosin cross-bridges is directly triggered by

- ☐ a) the repolarization of T tubules.
- ☐ b) the attachment of ATP to the myosin heads.
- ☐ c) the hydrolysis of ATP.
- ☐ d) calcium ions.

17 A muscle producing almost peak tension during rapid cycles of contraction and relaxation is said to be in

- ☐ a) incomplete tetanus.
- ☐ b) treppe.
- ☐ c) complete tetanus.
- ☐ d) a twitch.

18 The type of contraction in which the tension rises, but the muscle does not change length is

- ☐ a) an isotonic contraction.
- ☐ b) an isometric contraction.
- ☐ c) a concentric contraction.
- ☐ d) an eccentric contraction.

19 Which of the following statements about myofibrils is *not* correct?

- ☐ a) Each skeletal muscle fiber contains hundreds to thousands of myofibrils.
- ☐ b) Myofibrils contain repeating units called sarcomeres.
- ☐ c) Myofibrils extend the length of a skeletal muscle fiber.
- ☐ d) Filaments consist of bundles of myofibrils.

20 Vesicles filled with acetylcholine (ACh) are found in the

- ☐ a) axon terminal.
- ☐ b) motor end plate.
- ☐ c) synaptic cleft.
- ☐ d) transverse tubule.

21 The properties of slow fibers include all of the following except

- ☐ a) red in color.
- ☐ b) low myoglobin content.
- ☐ c) many mitochondria.
- ☐ d) small cross-sectional diameter.

22 Which of the following activities involves eccentric muscle contractions?

- ☐ a) maintaining upright posture
- ☐ b) lifting a barbell over your head
- ☐ c) lowering yourself into a chair
- ☐ d) blinking your eyes

23 All the muscle fibers controlled by a single motor neuron constitute a

- ☐ a) motor unit.
- ☐ b) motor end plate.
- ☐ c) neuromuscular junction.
- ☐ d) cross-bridge.

24 The region of the A band in a sarcomere that contains only thick filaments is the

- ☐ a) Z line.
- ☐ b) I band.
- ☐ c) zone of overlap.
- ☐ d) H band.

Short answer

25 What structural feature of a skeletal muscle fiber is responsible for conducting action potentials into the interior of the cell?

26 Describe how it is possible that the muscles flexing the elbow can adjust their tension to lift a cup of coffee to your mouth or curl a dumbbell in the gym.

27 Research in exercise physiology suggests that for several days after a bout of intense aerobic exercise, people consume oxygen at a rate greater than when they are normally at rest. Explain these results, and cite the process involved.

28 What two mechanisms are used to generate ATP from glucose in muscle cells?

29 Explain the processes that cause a muscle to hypertrophy.

MasteringA&P®

Access more chapter study tools online in the MasteringA&P Study Area:

- **Chapter Quizzes, Chapter Practice Test, Art-labeling Activities, Animations, MP3 Tutor Sessions, and Clinical Case Studies**

Practice Anatomy Lab	PAL™
Interactive Physiology	iP®
A&P Flix	A&PFlix™
PhysioEx	PhysioEx™

Chapter Integration • Applying what you have learned

Two childhood friends have become very different types of athletes

Sean and Stuart are neighborhood friends who have played sports together and competed against each other since early childhood. Although they are both outstanding athletes, they have noticed they each have their own individual skills in which they excel. Sean is a star soccer player. He can run the length of a soccer field for an entire game and does not experience the fatigue that his friend Stuart does when they compete. Stuart, a top prospect baseball player, is much more explosive in shorter duration events. He always has much better times than his friend Sean when running from home plate to first base. From what you have just learned about skeletal muscle tissue, answer the following questions.

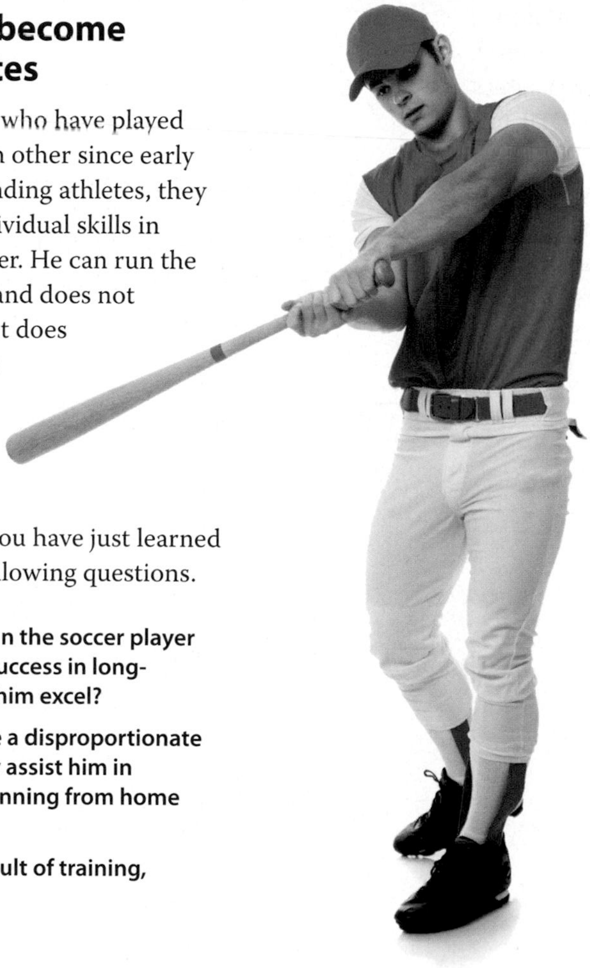

1. **Which muscle fiber type do you think Sean the soccer player has in greater abundance to explain his success in long-duration activities? How would this help him excel?**
2. **Stuart the baseball player is likely to have a disproportionate amount of which fiber type? How do they assist him in explosive, short duration activities like running from home plate to first base?**
3. **Do you think their athletic success is a result of training, genetics, or both?**

10 The Muscular System

LEARNING OUTCOMES

These Learning Outcomes correspond by number to this chapter's modules and indicate what you should be able to do after completing the chapter.

SECTION 1 • Functional Organization of the Muscular System

10.1 Describe the general function of the body's axial and appendicular muscles.

10.2 Describe fascicle organization and explain how levers affect muscle efficiency.

10.3 Explain how the name of a muscle can help identify its location, appearance, or function.

10.4 Describe the separation of muscles into axial and appendicular divisions.

SECTION 2 • Axial Muscles

10.5 Describe the four groups of axial muscles and their general functions.

10.6 Identify the facial expression muscles, and cite their origins, insertions, and actions.

10.7 Identify the eye and jaw muscles, and cite their origins, insertions, and actions.

10.8 Identify the tongue, pharynx, and neck muscles, and cite their origins, insertions, and actions.

10.9 Identify the vertebral column muscles, and cite their origins, insertions, and actions.

10.10 Identify the trunk muscles, and cite their origins, insertions, and actions.

10.11 Identify the pelvic floor muscles, and cite their origins, insertions, and actions.

SECTION 3 • Appendicular Muscles

10.12 Describe the general functions of the muscles of the upper and lower limbs.

10.13 Identify the principal appendicular muscles.

10.14 Identify the pectoral girdle muscles, and cite their origins, insertions, and actions.

10.15 Identify the muscles that move the arm, and cite their origins, insertions, and actions.

10.16 Identify the forearm muscles, and cite their origins, insertions, and actions.

10.17 Identify the muscles of the hand and fingers, and cite their origins, insertions, and actions.

10.18 Identify the intrinsic hand muscles, and cite their origins, insertions, and actions.

10.19 Identify the muscles that move the thigh, and cite their origins, insertions, and actions.

10.20 Identify the muscles that move the leg, and cite their origins, insertions, and actions.

10.21 Identify the muscles that move the foot and toes, and cite their origins, insertions, and actions.

10.22 Identify the intrinsic foot muscles, and cite their origins, insertions, and actions.

10.23 Describe the deep fascia and its relationship to the various limb muscle compartments.

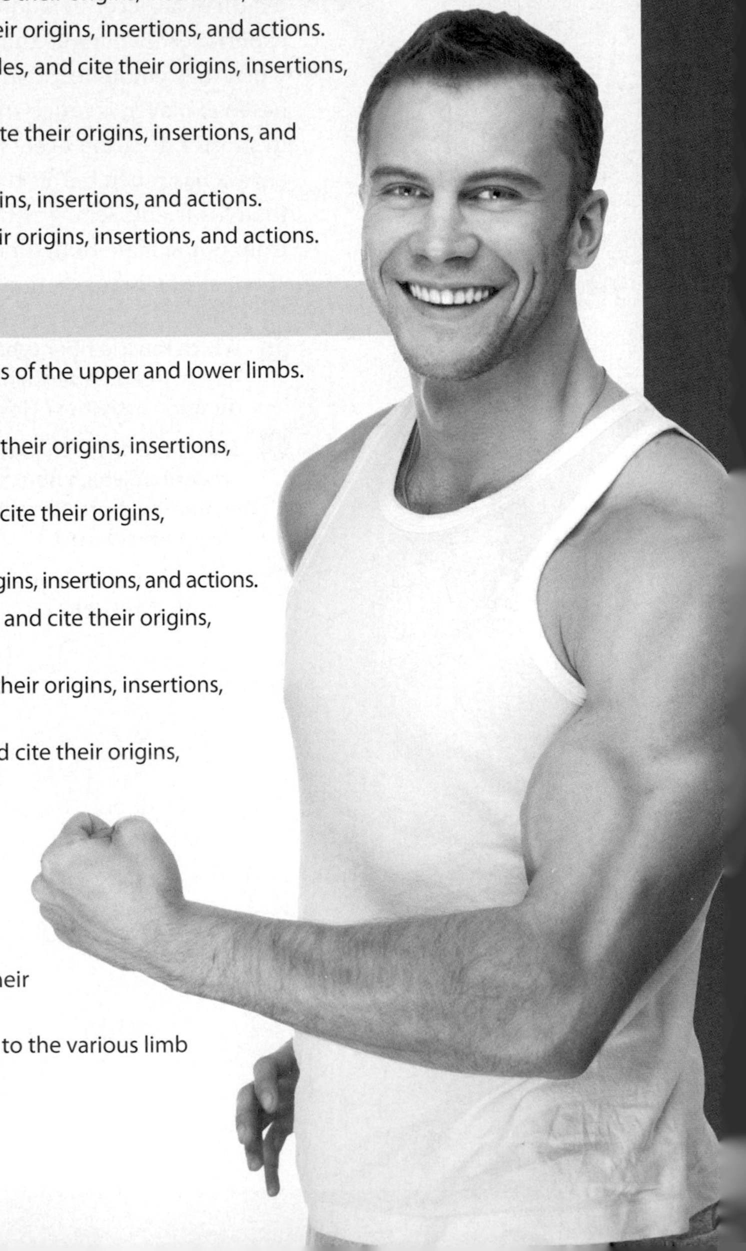

Learning Outcomes are repeated at the bottom of each module.

The axial and appendicular muscles have different functions

The skeletal muscles of the **muscular system** make up almost one-half the weight of your body. As we saw in Chapter 9, skeletal muscle tissue has multiple functions (Module 9.1, p. 307).

Watch MasteringA&P® ▸ A&P*Flix*™
- Origins, Insertions, Actions, and Innervations (Over 60 animations on the topic)
- Group Muscle Actions & Joints (Over 50 animations on the topic)

1 The muscular system contributes more to body weight than does any other organ system.

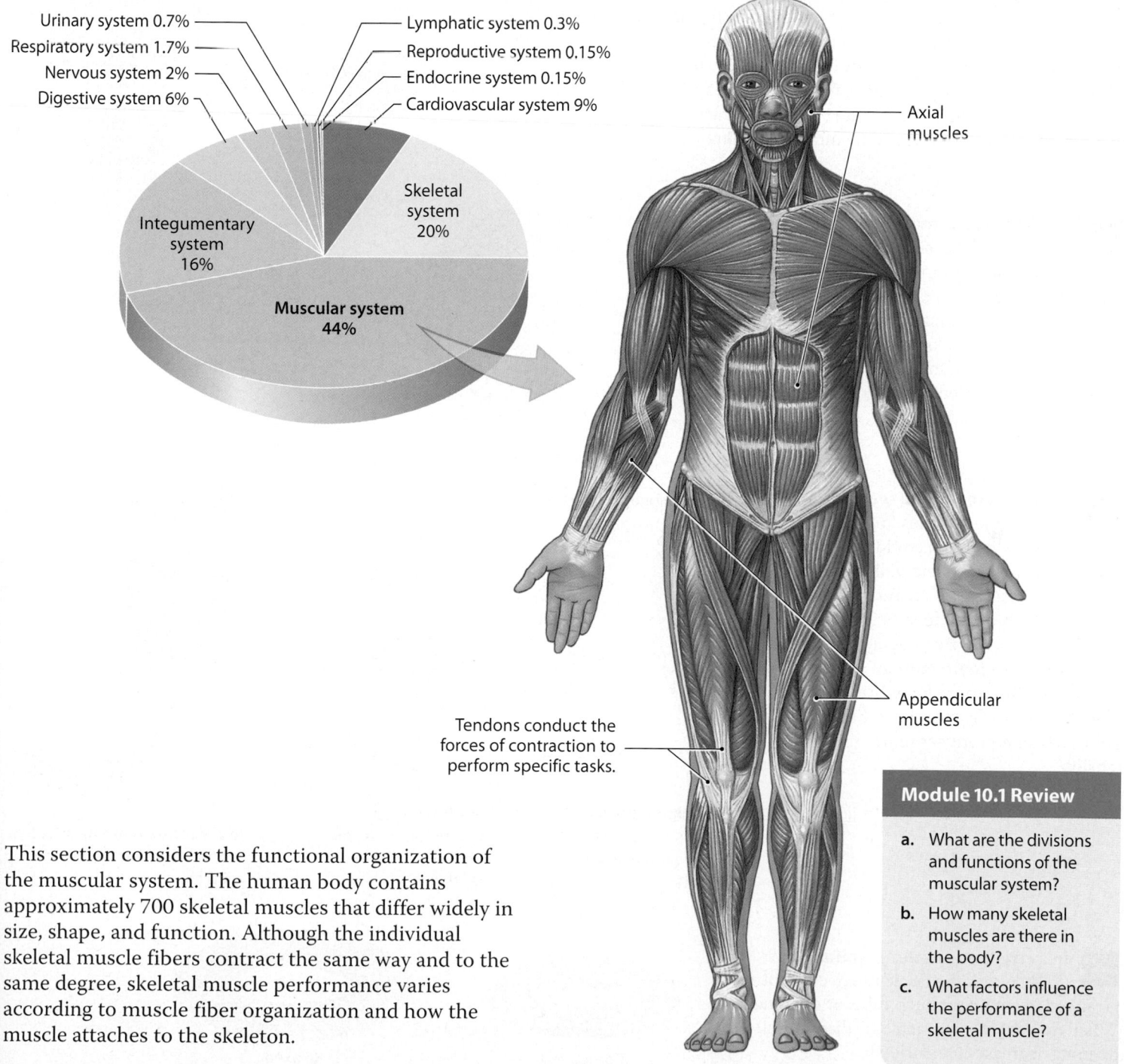

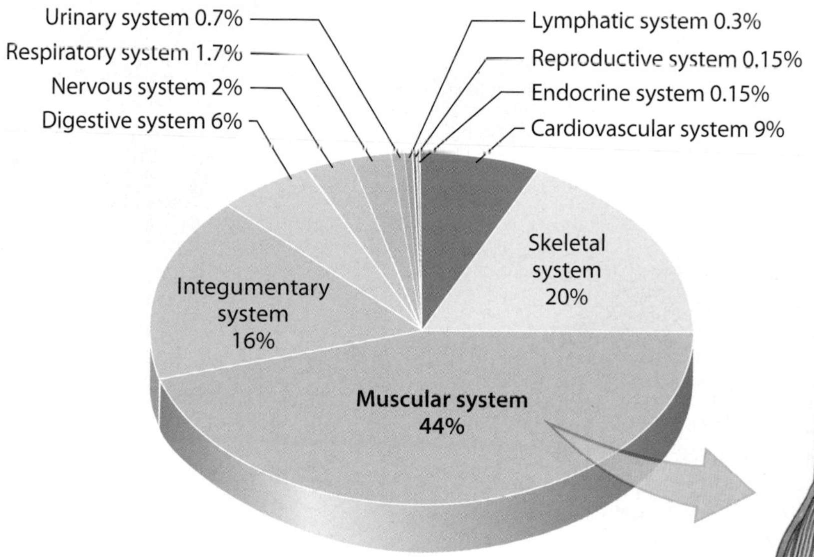
Urinary system 0.7%
Respiratory system 1.7%
Nervous system 2%
Digestive system 6%
Lymphatic system 0.3%
Reproductive system 0.15%
Endocrine system 0.15%
Cardiovascular system 9%
Skeletal system 20%
Integumentary system 16%
Muscular system 44%

2 The muscular system is divided into axial and appendicular divisions. **Axial muscles** support and position the axial skeleton. **Appendicular muscles** support, move, and brace the limbs.

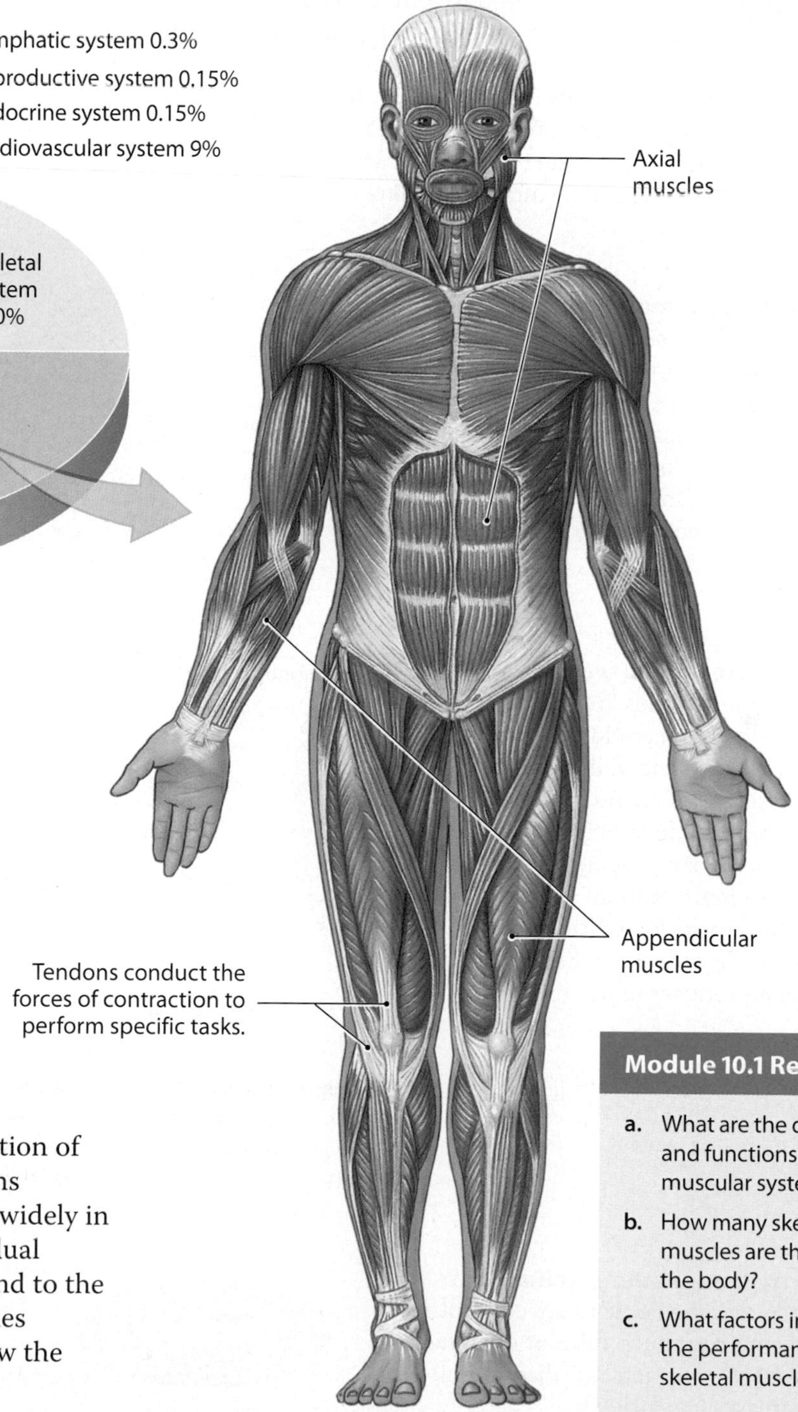
Axial muscles
Appendicular muscles
Tendons conduct the forces of contraction to perform specific tasks.

This section considers the functional organization of the muscular system. The human body contains approximately 700 skeletal muscles that differ widely in size, shape, and function. Although the individual skeletal muscle fibers contract the same way and to the same degree, skeletal muscle performance varies according to muscle fiber organization and how the muscle attaches to the skeleton.

Module 10.1 Review

a. What are the divisions and functions of the muscular system?

b. How many skeletal muscles are there in the body?

c. What factors influence the performance of a skeletal muscle?

LO 10.1 Describe the general function of the body's axial and appendicular muscles.

Muscular power and range of motion are influenced by fascicle organization and leverage

Fascicle Organization

1 In a **parallel muscle**, such as the biceps brachii, the fascicles are parallel to the long axis of the muscle. Most skeletal muscles in the body are parallel muscles. Some are flat bands with broad attachments (aponeuroses) at each end (Module 9.2, p. 308). Others are plump and cylindrical, with tendons at one or both ends. The muscle has a central **body**, also known as the belly. A skeletal muscle fiber can contract until it has shortened by about 30 percent. Because the muscle fibers in a parallel muscle are parallel to the long axis of the muscle, when those fibers contract together, the entire muscle shortens by about 30 percent. The tension developed during this contraction depends on the total number of myofibrils the muscle contains.

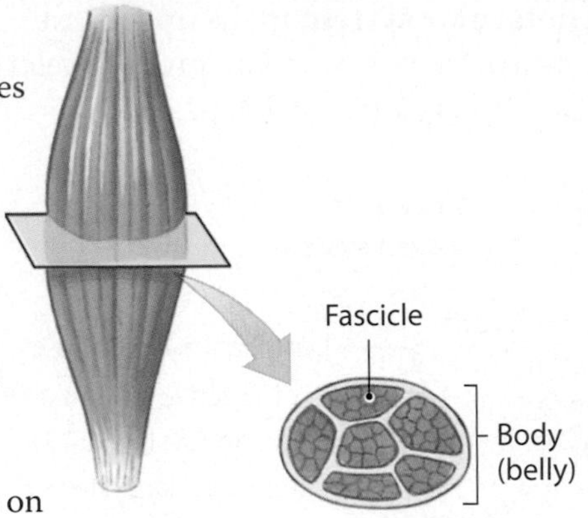

Fascicle
Body
(belly)

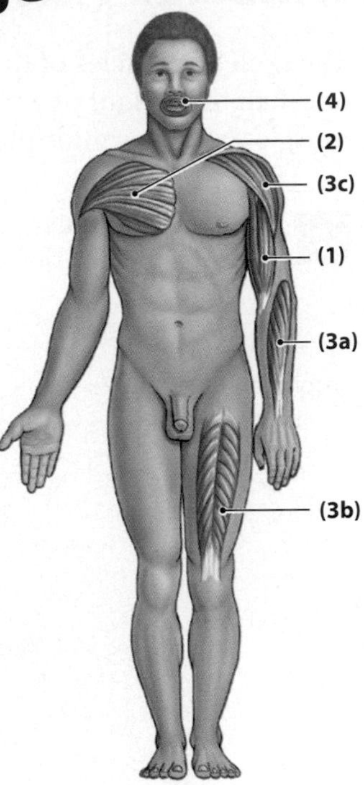

(4)
(2)
(3c)
(1)
(3a)
(3b)

2 In a **convergent muscle**, such as the pectoralis major, muscle fascicles extending over a broad area converge on a common attachment site. A convergent muscle is versatile, because the stimulation of different portions of the muscle can change the direction of pull. However, when the entire muscle contracts, the muscle fibers do not pull as hard on the attachment site as would a parallel muscle of the same size.

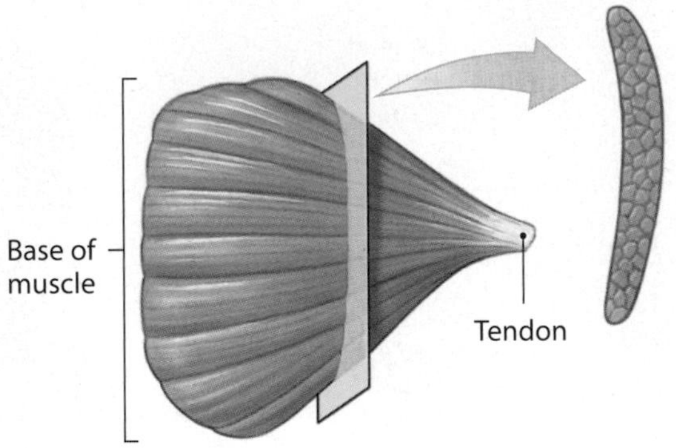

Base of
muscle
Tendon

3 In a **pennate muscle** (*penna*, feather), the fascicles form a common angle with the tendon. Because the muscle fibers pull at an angle, contracting pennate muscles do not move their tendons as far as parallel muscles do. But a pennate muscle contains more muscle fibers—and thus more myofibrils—than does a parallel muscle of the same size, so it produces more tension.

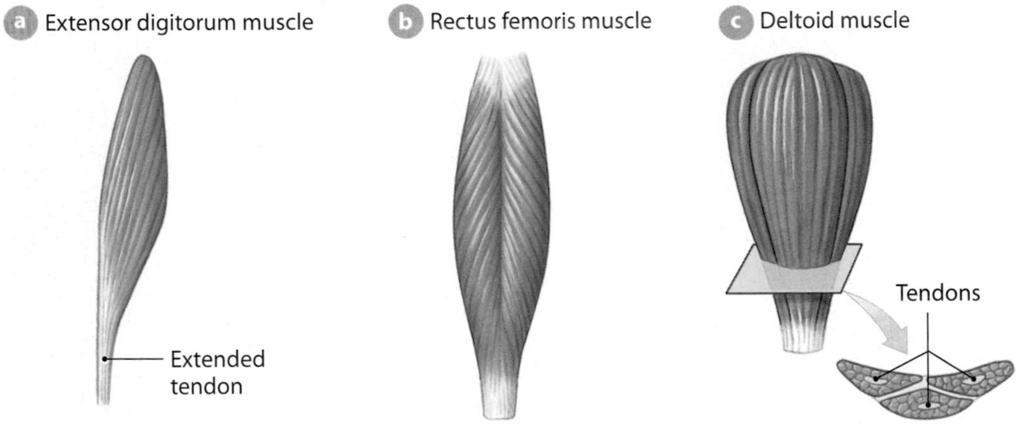

a Extensor digitorum muscle
Extended
tendon
b Rectus femoris muscle
c Deltoid muscle
Tendons

In a **unipennate** muscle, all the muscle fibers are on the same side of the tendon.

In a **bipennate** muscle, fibers are on both sides of the tendon.

In a **multipennate** muscle, the tendon branches within the muscle.

4 In a **circular muscle**, or **sphincter** (SFINK-ter), the fascicles are concentrically arranged to encircle a duct, tube, or opening. When the muscle contracts, the diameter of the opening decreases (constricts).

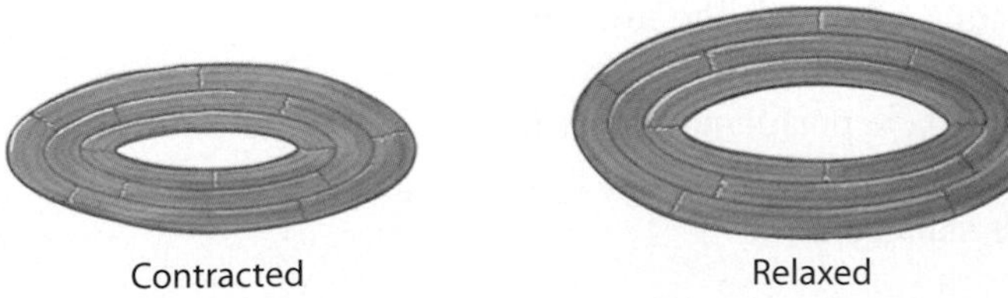

Contracted
Relaxed

Levers and Leverage

The force, speed, or direction of movement produced by muscle contraction can be modified by attaching the muscle to a lever. A **lever** is a rigid structure, such as a board or crowbar, used to lift or pry something that pivots on a fixed point called the **fulcrum**. A lever moves when an applied force is sufficient to overcome any load that would otherwise oppose or prevent such movement. In the body, each bone is a lever and each joint is a fulcrum, and muscles provide the applied force.

5 In a **first-class lever**, the fulcrum (F) lies between the applied force (AF) and the load (L). The body has few first-class levers, but this is an important example. First-class levers are like seesaws—the balance depends on the relative sizes of the force and the load, and how far each is from the fulcrum.

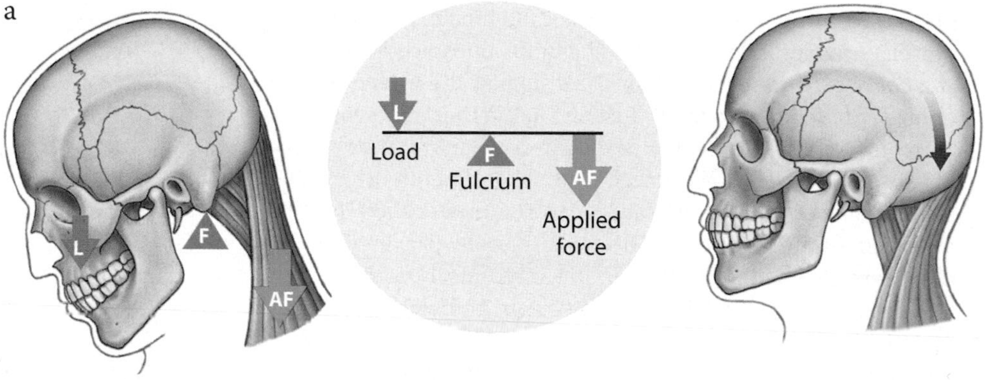

L
F
AF
Load
Fulcrum
Applied force

6 In a **second-class lever**, the load is located between the applied force and the fulcrum. A wheelbarrow is a familiar example of a second-class lever. Because the force is always farther from the fulcrum than the load is, a small force can move a larger weight, but the load moves more slowly and covers a shorter distance. Thus the effective force is increased at the expense of speed and distance traveled.

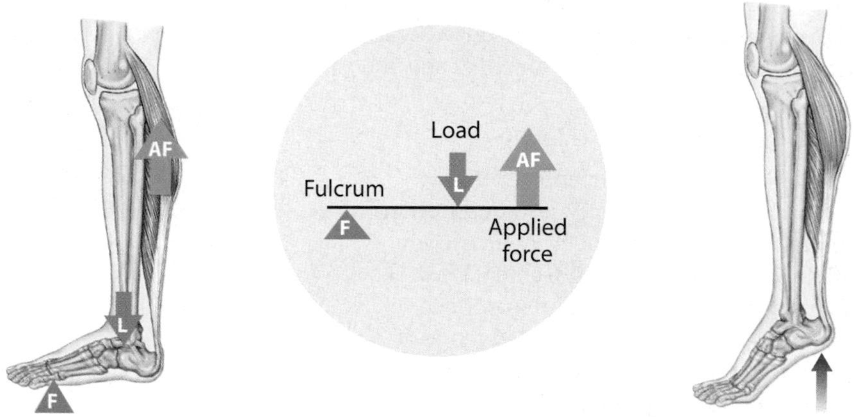

AF
L
F
Load
Fulcrum
Applied force

7 In **third-class levers** (the most common levers in the body), the force is applied between the load and the fulcrum. In contrast to second-class levers, speed and distance traveled are increased at the expense of effective force. In the example shown, the load is six times farther from the fulcrum than is the applied force. The effective force is reduced to the same degree. The muscle must generate 180 kg (396 lb) of tension at its attachment to the forearm to support 30 kg (66 lb) held in the hand. However, the distance traveled and the speed of movement are increased by that same 6:1 ratio: The load will travel 45 cm (18 in.) when the point of attachment moves 7.5 cm (2.9 in.).

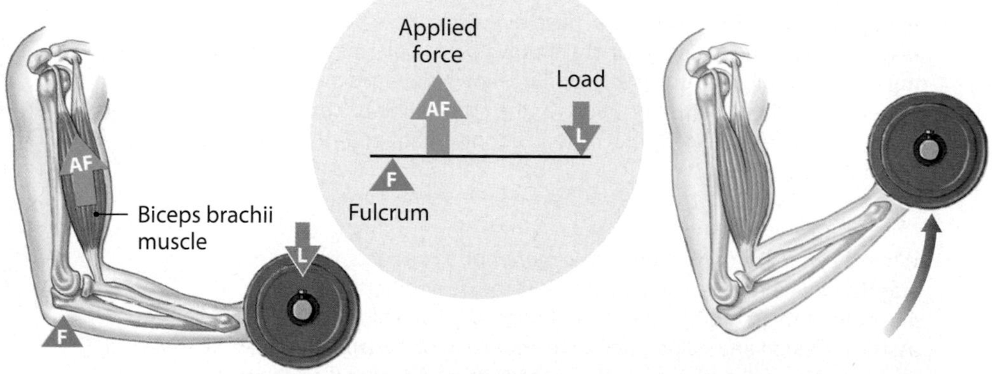

AF
Biceps brachii muscle
L
F
Applied force
Load
Fulcrum

Module 10.2 Review

a. Define a lever, and describe the three classes of levers.

b. The joint between the occipital bone of the skull and the first cervical vertebra (atlas) is which part of which class of lever system?

c. Why does a pennate muscle generate more tension than does a parallel muscle of the same size?

LO **10.2** Describe fascicle organization and explain how levers affect muscle efficiency.

The names of muscles can provide clues to their appearance and/or function

1 In most cases one end of a muscle is fixed in position, and the other end moves during a contraction. The place where the fixed end attaches is called the **origin** of the muscle. Most muscles originate at a bone, but some originate at a connective tissue sheath or band such as the **intermuscular septa** (components of the deep fascia that may separate adjacent skeletal muscles) or at the interosseous membranes of the forearm or leg (Module 7.16, p. 263). The site where the movable end attaches to another structure is called the **insertion** of the muscle. The origin is typically proximal to the insertion when the body is in the anatomical position. However, knowing which end is the origin and which is the insertion is ultimately less important than knowing where the two ends attach and what the muscle accomplishes when it contracts. When a muscle contracts, it produces a specific movement or **action**. In general, we will describe actions in terms of movement at specific joints.

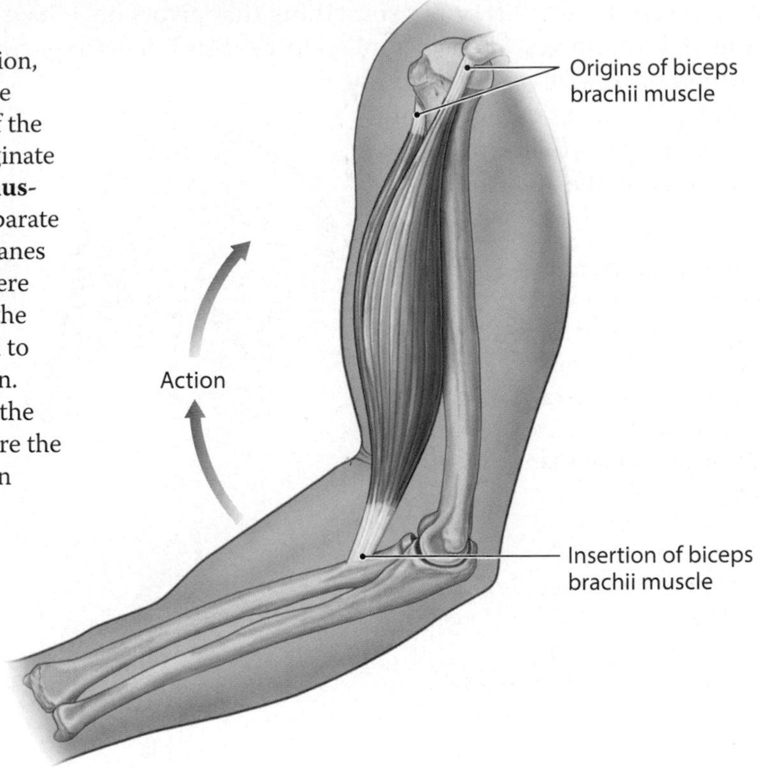
Origins of biceps brachii muscle
Action
Insertion of biceps brachii muscle

2 When complex movements occur, muscles commonly work in groups rather than individually. Their cooperation improves the efficiency of a particular movement. For example, large muscles of the limbs produce flexion or extension over an extended range of motion. Based on their functions, muscles may be described as agonists, antagonists, or synergists.

An **agonist**, or **prime mover**, is a muscle whose contraction is chiefly responsible for producing a particular movement. Determining which muscle in a group of muscles is the prime mover depends on the action under way and the relative positions of the articulating bones. In this simple example, the biceps brachii is an agonist that bends the elbow as when doing curls.

When a **synergist** (*syn-*, together + *ergon*, work) contracts, it helps a larger agonist work efficiently. Synergists may provide additional pull near the insertion or may stabilize the point of origin. The brachioradialis muscle assists in flexion and helps stabilize the elbow joint. Synergists that assist an agonist by preventing movement at another joint are called **fixators**.

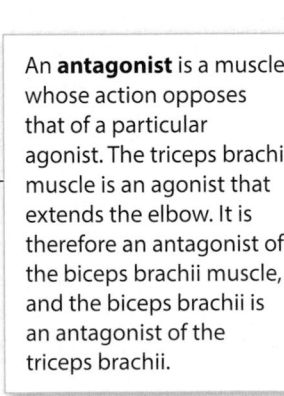

An **antagonist** is a muscle whose action opposes that of a particular agonist. The triceps brachii muscle is an agonist that extends the elbow. It is therefore an antagonist of the biceps brachii muscle, and the biceps brachii is an antagonist of the triceps brachii.

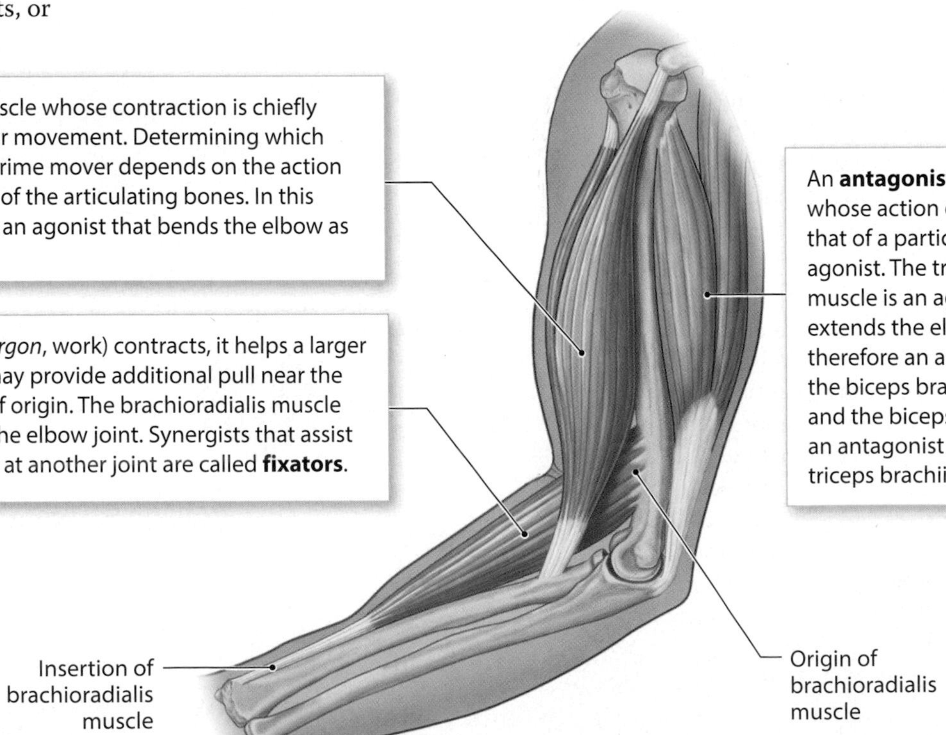
Insertion of brachioradialis muscle
Origin of brachioradialis muscle

Muscle Terminology

Terms Indicating Specific Regions of the Body*	Terms Indicating Position, Direction, or Fascicle Organization	Terms Indicating Structural Characteristics of the Muscle	Terms Indicating Actions
Abdominis (abdomen)	Anterior (front)	**Nature of Origin**	**General**
Ancon (elbow)	External (on the outside)	Biceps (two heads)	Abductor (movement away)
Auricular (auricle of ear)	Extrinsic (outside the structure)	Triceps (three heads)	Adductor (movement toward)
Brachial (brachium)	Inferior (below)	Quadriceps (four heads)	Depressor (lowering movement)
Capitis (head)	Internal (away from the surface)		Extensor (straightening movement)
Carpi (wrist)	Intrinsic (within the structure)	**Shape**	Flexor (bending movement)
Cervicis (neck)	Lateral (on the side)	Deltoid (triangle)	Levator (raising movement)
Coccygeal (coccyx)	Medial (middle)	Orbicularis (circle)	Pronator (turning into prone position)
Costal (rib)	Oblique (slanting)	Pectinate (comblike)	Supinator (turning into supine position)
Cutaneous (skin)	Posterior (back)	Piriformis (pear-shaped)	Tensor (tensing movement)
Femoris (femur)	Profundus (deep)	Platy- (flat)	
Glossal (tongue)	Rectus (straight)	Pyramidal (pyramid)	**Specific**
Hallux (great toe)	Superficial (toward the surface)	Rhomboid (parallelogram)	Buccinator (trumpeter)
Ilium (groin)	Superior (toward the head)	Serratus (serrated)	Risorius (laugher)
Inguinal (groin)	Transverse (crosswise)	Splenius (bandage)	Sartorius (like a tailor)
Lumbar (lumbar region)		Teres (long and round)	
Nasalis (nose)		Trapezius (trapezoid)	
Nuchal (back of neck)			
Ocular (eye)		**Other Striking Features**	
Oris (mouth)		Alba (white)	
Palpebra (eyelid)		Brevis (short)	
Pollex (thumb)		Gracilis (slender)	
Popliteal (posterior to knee)		Latae (wide)	
Psoas (loin)		Latissimus (widest)	
Radial (forearm)		Longissimus (longest)	
Scapular (scapula)		Longus (long)	
Temporal (temple)		Magnus (large)	
Thoracic (thorax)		Major (larger)	
Tibial (tibia; shin)		Maximus (largest)	
Ulnar (ulna)		Minimus (smallest)	
		Minor (smaller)	
		Vastus (great)	

* For other regional terms, refer to Module 1.15

3 This table includes a useful summary of the most important terms used in naming skeletal muscles. Familiarity with these terms will help you identify and remember specific muscles. Except for the platysma and the diaphragm, the complete names of all skeletal muscles include the term "muscle." Although the full name, such as the biceps brachii muscle, will usually appear in the text, for simplicity only the descriptive name (biceps brachii) will be used in figures and tables.

LO 10.3 Explain how the name of a muscle can help identify its location, appearance, or function.

Module 10.3 Review

a. Define the term *synergist* as it relates to muscle action.

b. Muscle A abducts the humerus, and muscle B adducts the humerus. What is the relationship between these two muscles?

c. What does the name *flexor carpi radialis longus* tell you about this muscle?

Module 10.4

The skeletal muscles can be assigned to the axial division or the appendicular division based on origins and functions

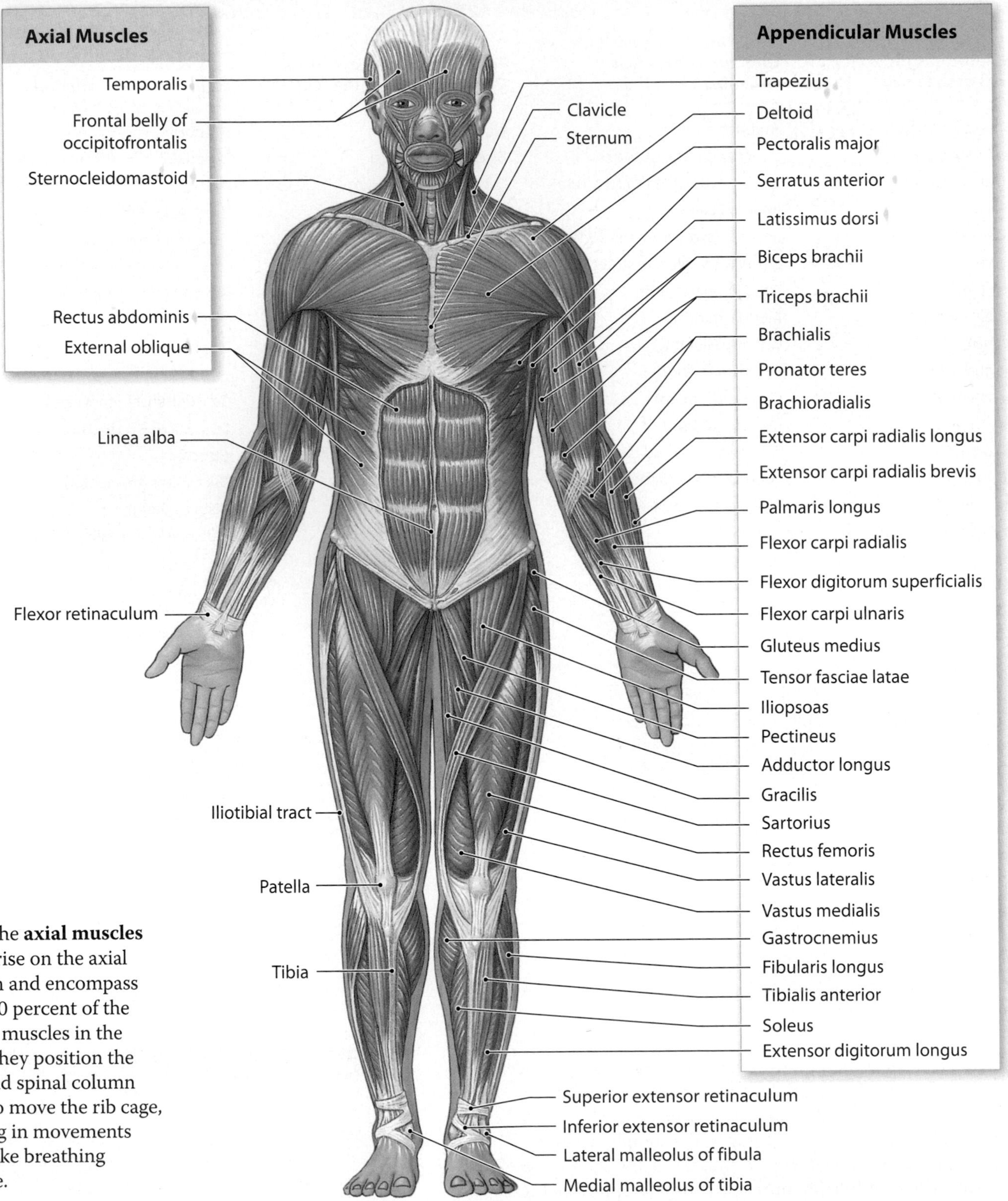

Axial Muscles
Temporalis
Frontal belly of occipitofrontalis
Sternocleidomastoid
Rectus abdominis
External oblique
Linea alba
Flexor retinaculum
Iliotibial tract
Patella
Tibia
Clavicle
Sternum
Appendicular Muscles
Trapezius
Deltoid
Pectoralis major
Serratus anterior
Latissimus dorsi
Biceps brachii
Triceps brachii
Brachialis
Pronator teres
Brachioradialis
Extensor carpi radialis longus
Extensor carpi radialis brevis
Palmaris longus
Flexor carpi radialis
Flexor digitorum superficialis
Flexor carpi ulnaris
Gluteus medius
Tensor fasciae latae
Iliopsoas
Pectineus
Adductor longus
Gracilis
Sartorius
Rectus femoris
Vastus lateralis
Vastus medialis
Gastrocnemius
Fibularis longus
Tibialis anterior
Soleus
Extensor digitorum longus
Superior extensor retinaculum
Inferior extensor retinaculum
Lateral malleolus of fibula
Medial malleolus of tibia

1 The **axial muscles** arise on the axial skeleton and encompass about 60 percent of the skeletal muscles in the body. They position the head and spinal column and also move the rib cage, assisting in movements that make breathing possible.

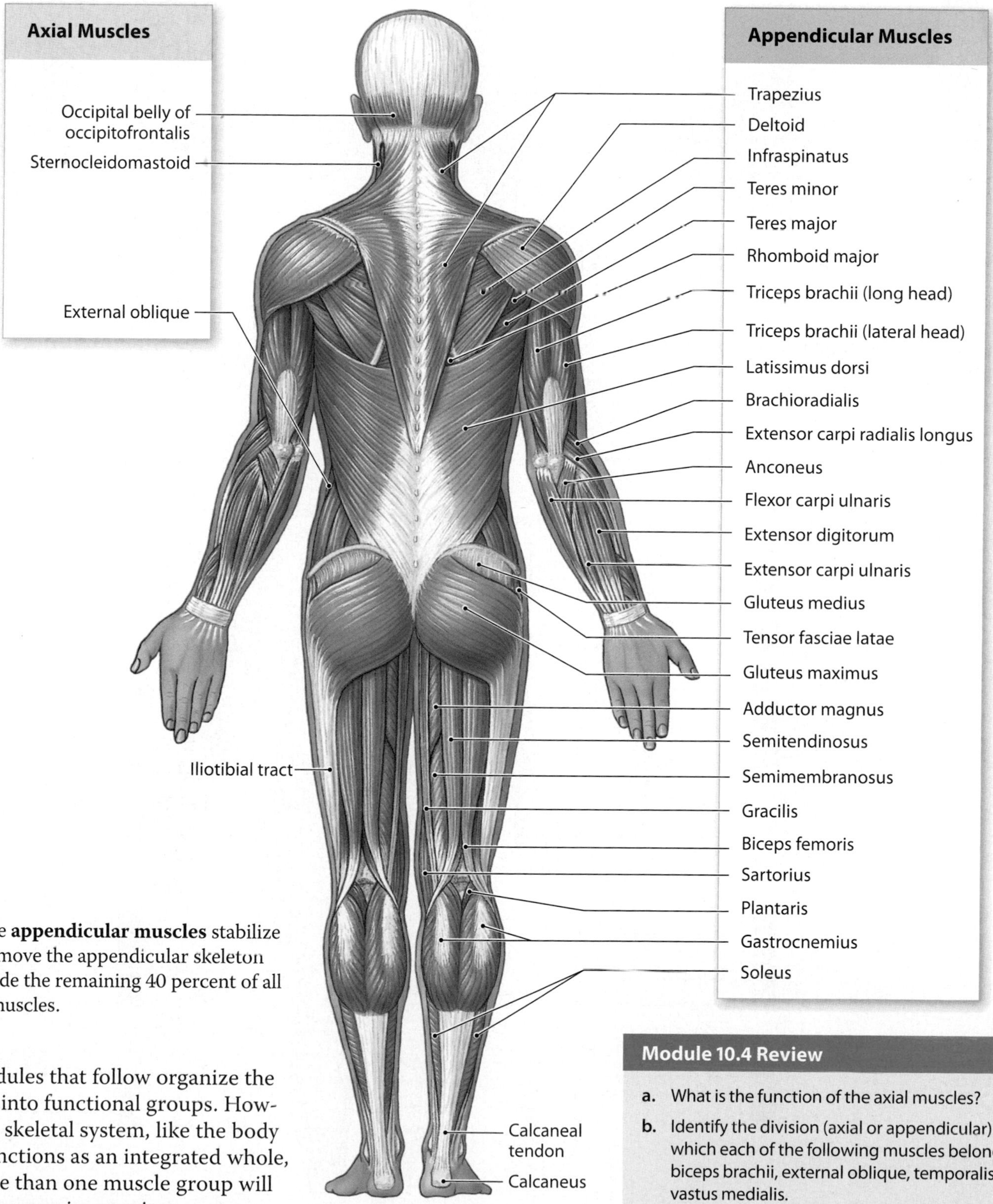

Axial Muscles
Occipital belly of occipitofrontalis
Sternocleidomastoid
External oblique
Iliotibial tract
Calcaneal tendon
Calcaneus
Appendicular Muscles
Trapezius
Deltoid
Infraspinatus
Teres minor
Teres major
Rhomboid major
Triceps brachii (long head)
Triceps brachii (lateral head)
Latissimus dorsi
Brachioradialis
Extensor carpi radialis longus
Anconeus
Flexor carpi ulnaris
Extensor digitorum
Extensor carpi ulnaris
Gluteus medius
Tensor fasciae latae
Gluteus maximus
Adductor magnus
Semitendinosus
Semimembranosus
Gracilis
Biceps femoris
Sartorius
Plantaris
Gastrocnemius
Soleus

2 The **appendicular muscles** stabilize or move the appendicular skeleton and include the remaining 40 percent of all skeletal muscles.

The modules that follow organize the muscles into functional groups. However, the skeletal system, like the body itself, functions as an integrated whole, and more than one muscle group will usually be seen in one view.

Module 10.4 Review

a. What is the function of the axial muscles?

b. Identify the division (axial or appendicular) to which each of the following muscles belongs: biceps brachii, external oblique, temporalis, and vastus medialis.

c. Which structures labeled in the figures in this module are not muscles?

LO **10.4** Describe the separation of muscles into axial and appendicular divisions.

Labeling

Label each of the muscle types below according to its fascicle organization.

1

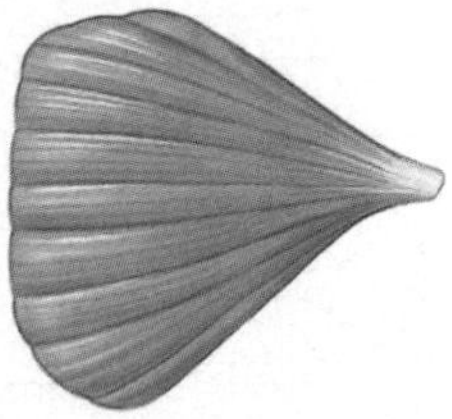

2

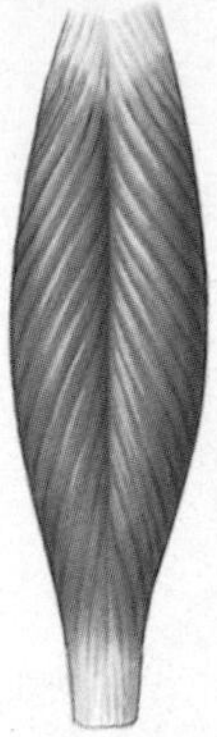

3

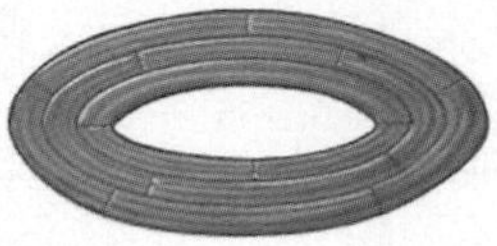

4

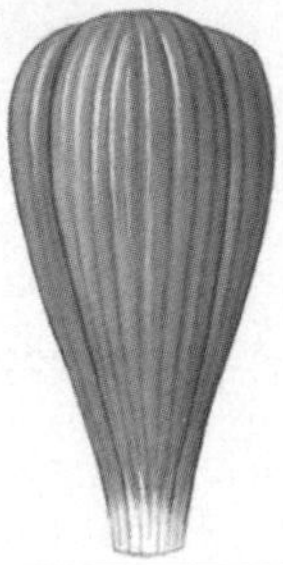

5

Label each of the indicated superficial muscles in the diagram to the right.

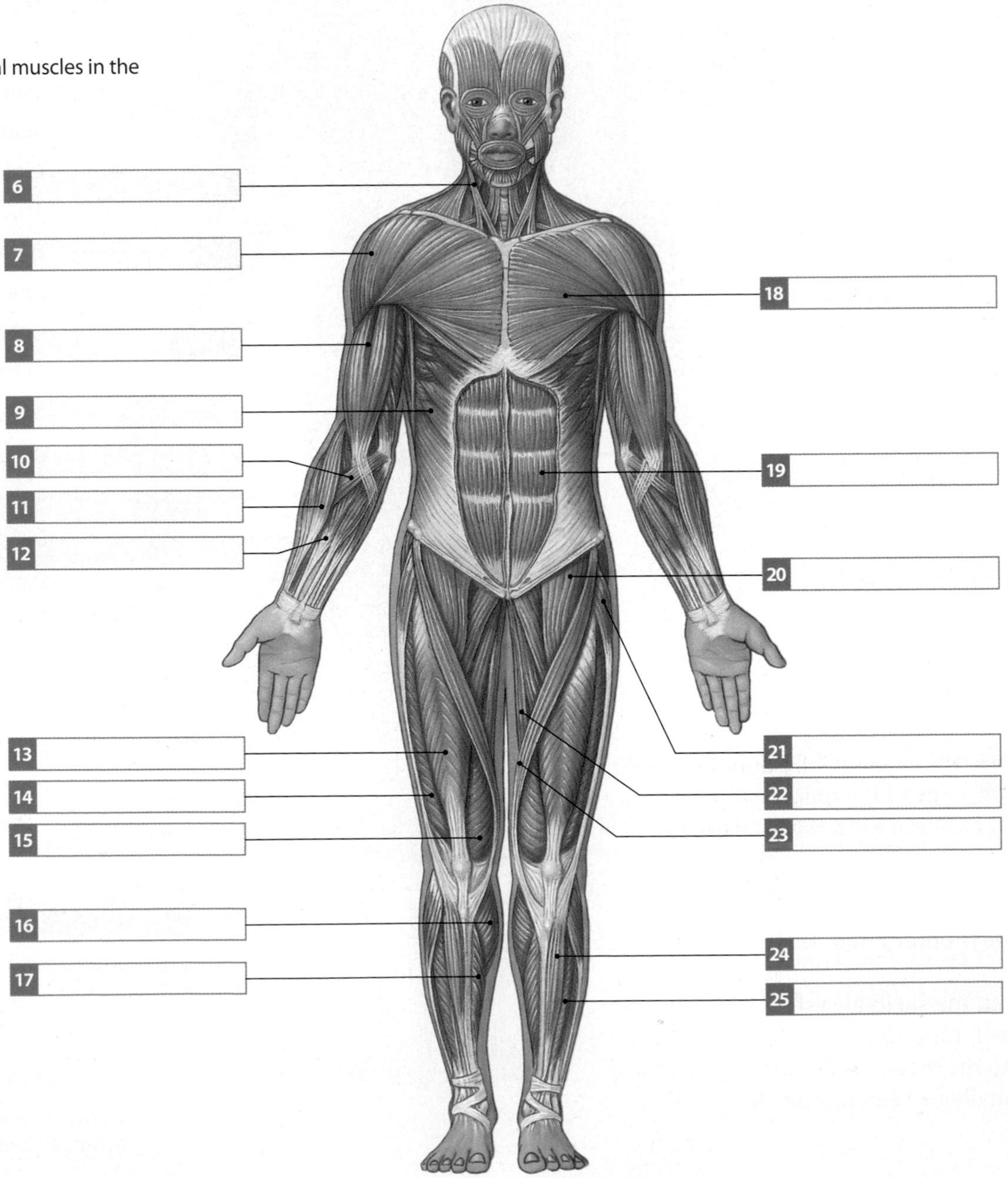
6
7
8
9
10
11
12
13
14
15
16
17
18
19
20
21
22
23
24
25

There are four groups of axial muscles

The axial musculature stabilizes and positions the head, neck, and trunk. Based on location and/or function, we can divide the axial muscles into the four groups shown here. The groups do not always have distinct anatomical boundaries. For example, a function such as the extension of the vertebral column involves muscles along its entire length.

1 The first group contains muscles of the head and neck that are not associated with the vertebral column. These muscles include the muscles of facial expression (Module 10.6), the extrinsic eye muscles (Module 10.7), and the muscles of the tongue, pharynx, and neck (Module 10.8).

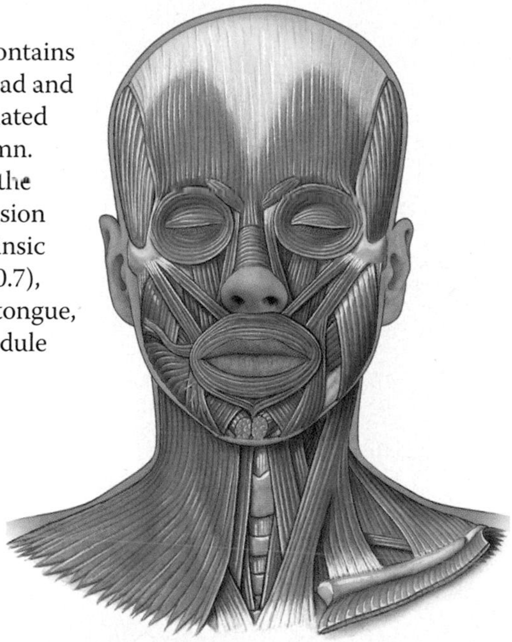

2 The second group—the muscles of the vertebral column—includes numerous muscles of varied size that stabilize, flex, extend, or rotate the vertebral column (Module 10.9).

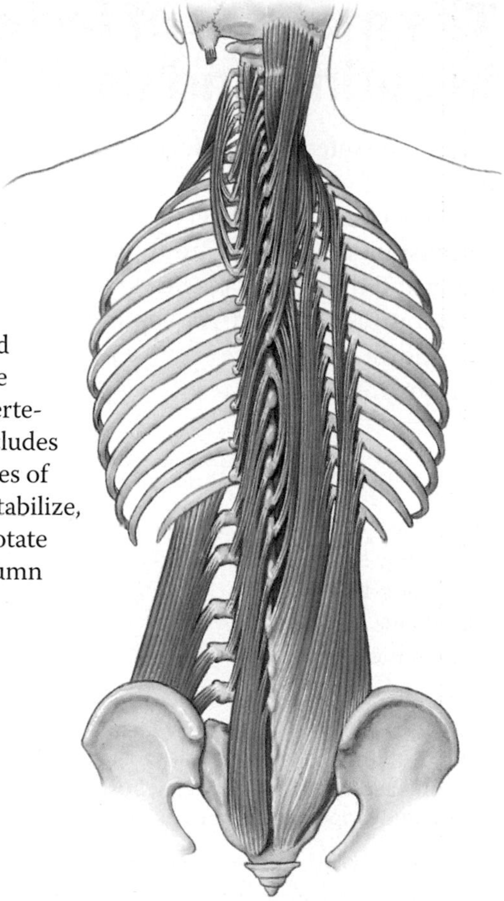

3 The third group consists of the oblique and rectus muscles of the trunk (Module 10.10). These muscles are broad sheets or bands that form the muscular walls of the thoracic and abdominopelvic cavities.

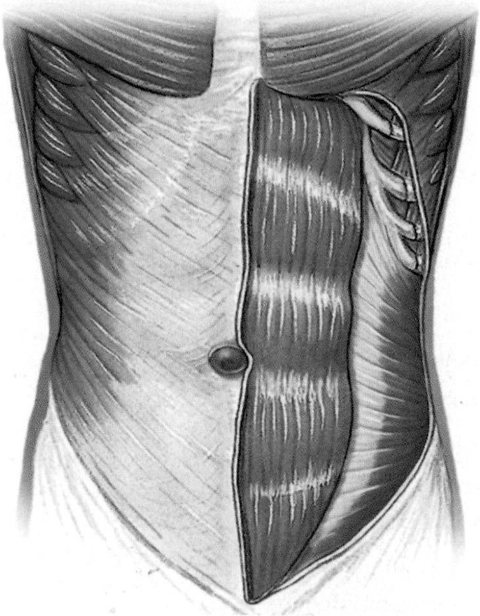

4 The fourth group—the muscles of the pelvic floor (Module 10.11)—spans the pelvic outlet and supports the organs of the pelvis.

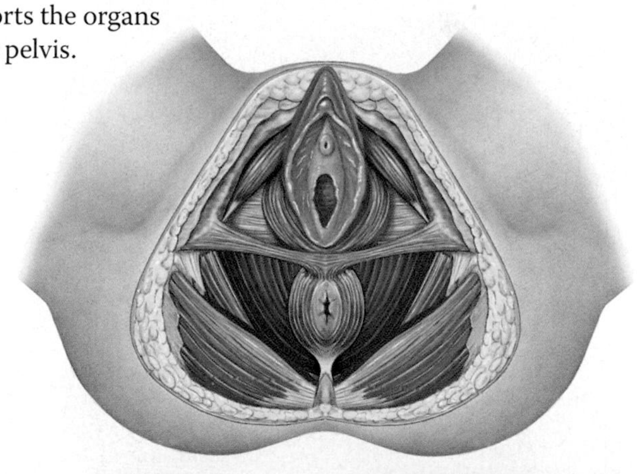

Module 10.5 Review

a. Which regions of the body are stabilized and positioned by the axial muscles?

b. The first and second axial muscle groups include which muscles?

c. The third and fourth axial muscle groups include which muscles?

LO 10.5 Describe the four groups of axial muscles and their general functions.

Module 10.6

The muscles of facial expression are important in eating and useful for communication

All the **muscles of facial expression** originate on the surface of the skull, except for the platysma of the neck. At their insertions, the fibers of the epimysium are woven into those of the superficial fascia and the dermis of the skin. Thus, when they contract, the skin moves and this allows you to have facial expressions.

1 This anterior view shows superficial muscles on the right side of the face and deeper muscles on the left side of the face.

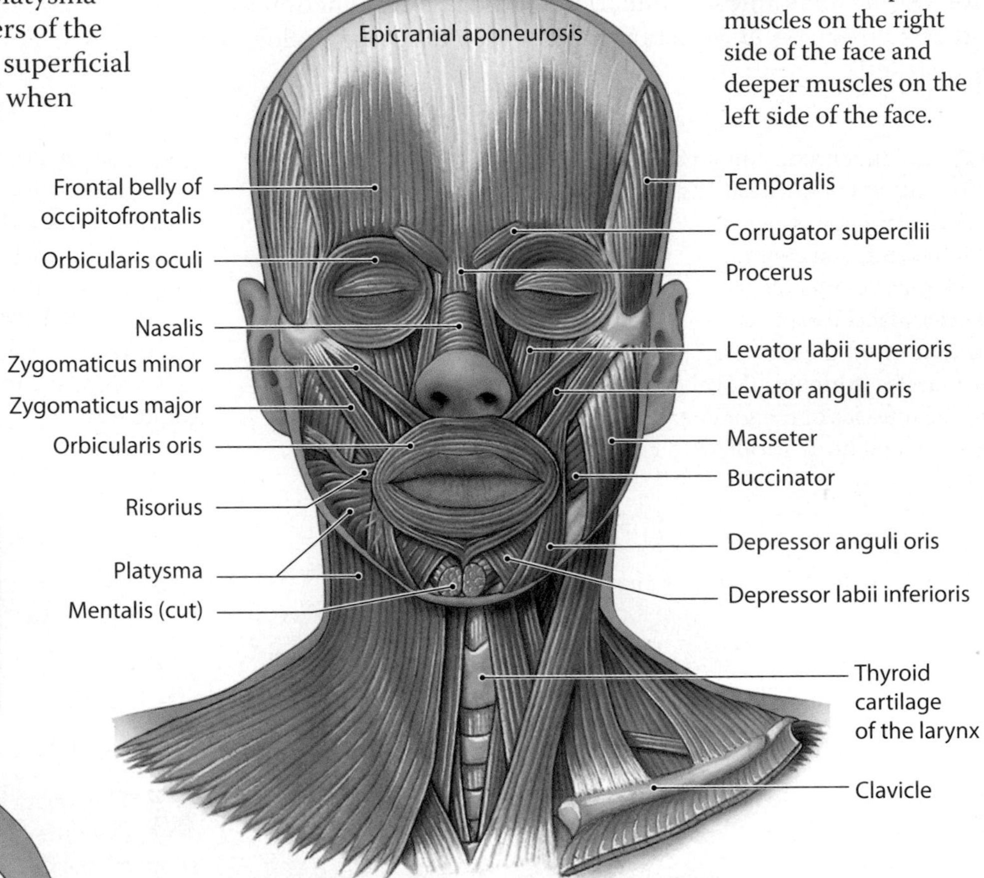

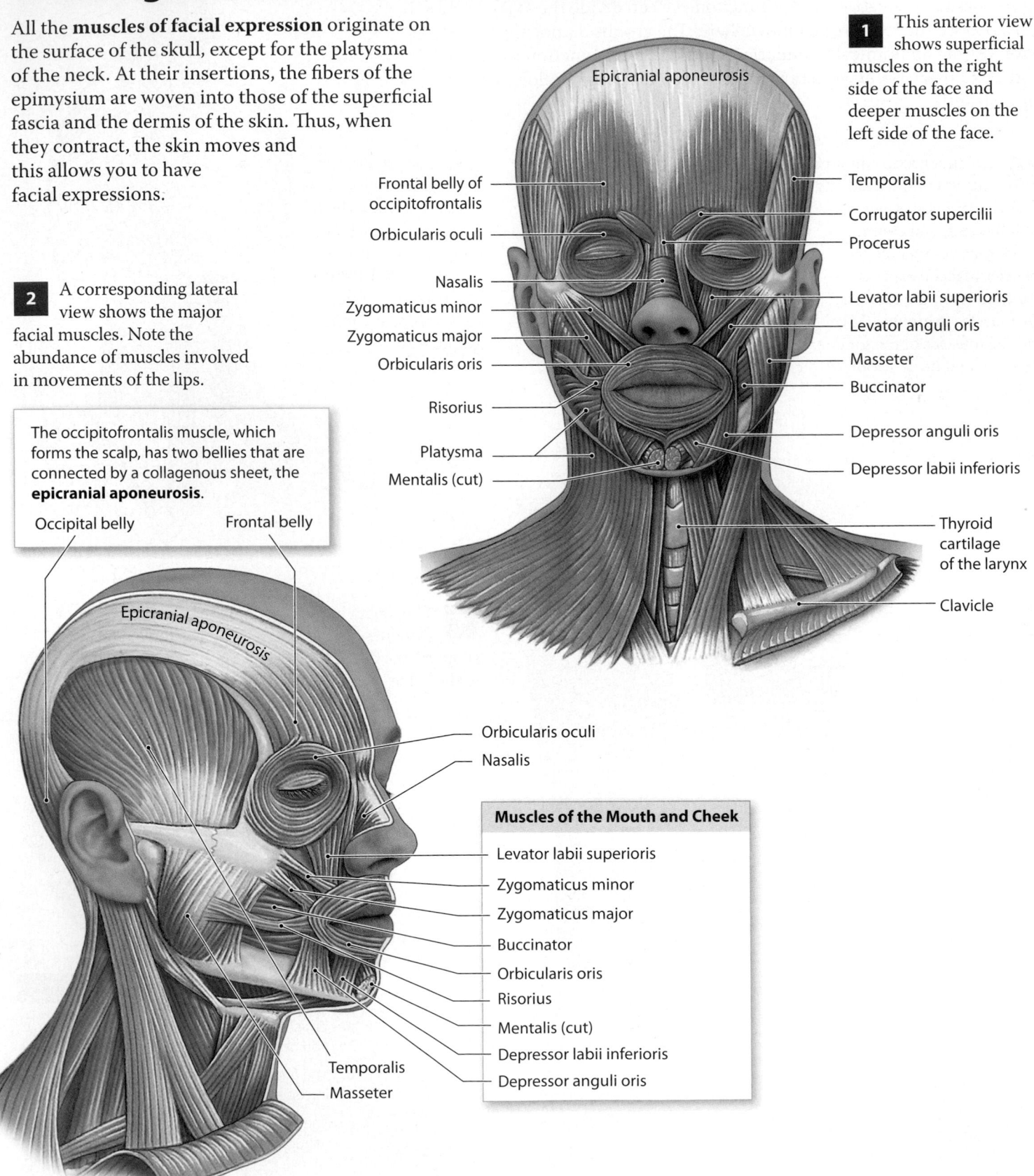

Epicranial aponeurosis
Frontal belly of occipitofrontalis
Orbicularis oculi
Nasalis
Zygomaticus minor
Zygomaticus major
Orbicularis oris
Risorius
Platysma
Mentalis (cut)
Temporalis
Corrugator supercilii
Procerus
Levator labii superioris
Levator anguli oris
Masseter
Buccinator
Depressor anguli oris
Depressor labii inferioris
Thyroid cartilage of the larynx
Clavicle

2 A corresponding lateral view shows the major facial muscles. Note the abundance of muscles involved in movements of the lips.

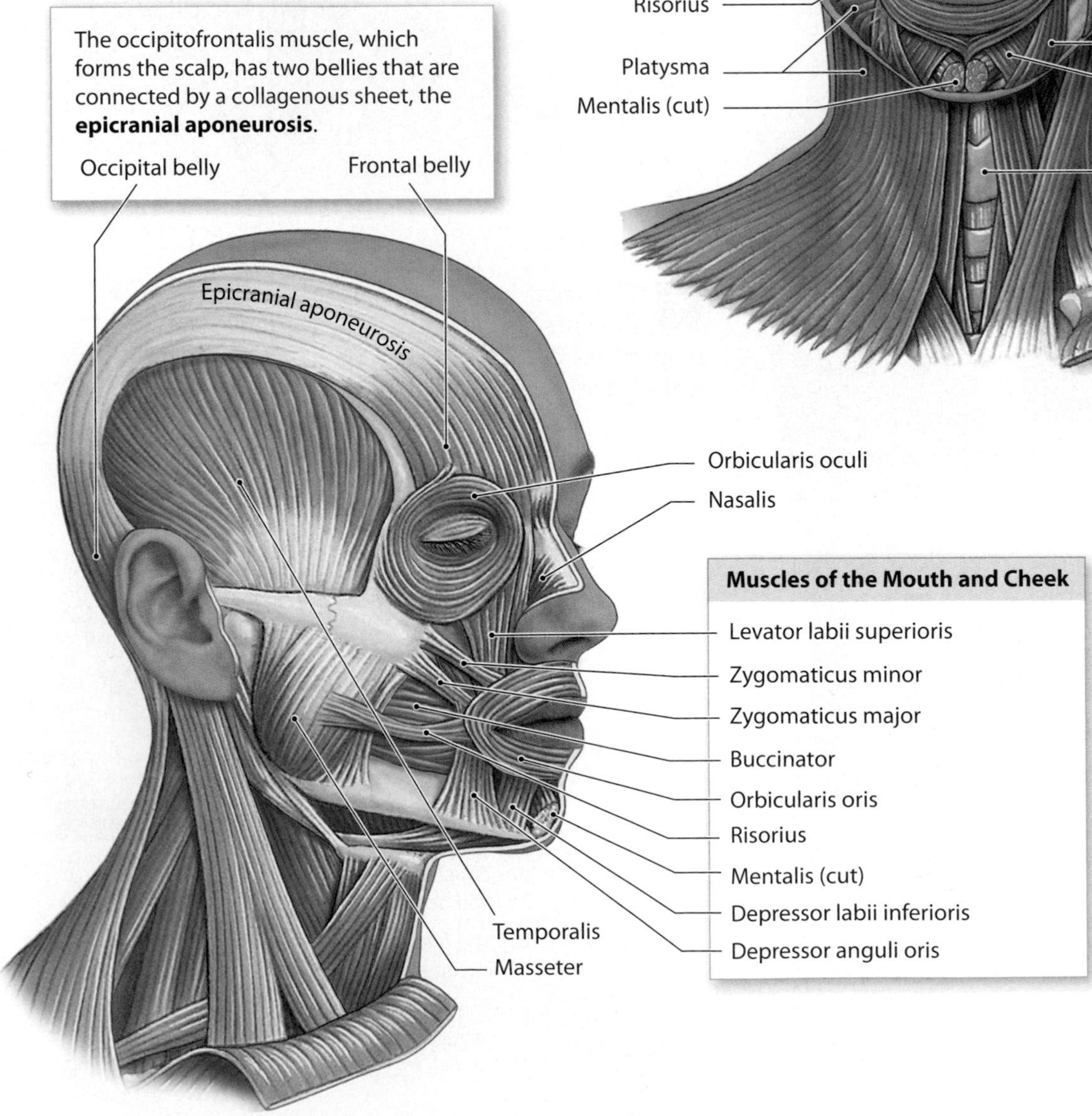

The occipitofrontalis muscle, which forms the scalp, has two bellies that are connected by a collagenous sheet, the **epicranial aponeurosis**.
Occipital belly
Frontal belly
Epicranial aponeurosis
Orbicularis oculi
Nasalis
Muscles of the Mouth and Cheek
Levator labii superioris
Zygomaticus minor
Zygomaticus major
Buccinator
Orbicularis oris
Risorius
Mentalis (cut)
Depressor labii inferioris
Depressor anguli oris
Temporalis
Masseter

Muscles of Facial Expression

Group and Muscle	Origin	Insertion	Action
Mouth			
Buccinator	Alveolar processes of maxillary bone and mandible	Blends into fibers of orbicularis oris	Compresses cheeks
Depressor labii inferioris	Mandible between the anterior midline and the mental foramen	Skin of lower lip	Depresses lower lip
Levator labii superioris	Inferior margin of orbit, superior to the infraorbital foramen	Orbicularis oris	Elevates upper lip
Levator anguli oris	Maxillary bone below the infraorbital foramen	Corner of mouth	Elevates the corner of the mouth
Mentalis	Incisive fossa of mandible	Skin of chin	Elevates and protrudes lower lip
Orbicularis oris	Maxillary bone and mandible	Lips	Compresses, purses lips
Risorius	Fascia surrounding parotid salivary gland	Angle of mouth	Draws corner of mouth to the side
Depressor anguli oris	Anterolateral surface of mandibular body	Skin at angle of mouth	Depresses corner of mouth
Zygomaticus major	Zygomatic bone near zygomaticomaxillary suture	Angle of mouth	Retracts and elevates corner of mouth
Zygomaticus minor	Zygomatic bone posterior to zygomaticotemporal suture	Upper lip	Retracts and elevates upper lip
Eye			
Corrugator supercilii	Orbital rim of frontal bone near nasal suture	Eyebrow	Pulls skin inferiorly and anteriorly; wrinkles brow
Levator palpebrae superioris	Tendinous band around optic foramen	Upper eyelid	Elevates upper eyelid
Orbicularis oculi	Medial margin of orbit	Skin around eyelids	Closes eye
Nose			
Procerus	Nasal bones and lateral nasal cartilages	Aponeurosis at bridge of nose and skin of forehead	Moves nose, changes position and shape of nostrils
Nasalis	Maxillary bone and alar cartilage of nose	Bridge of nose	Compresses bridge, depresses tip of nose; elevates corners of nostrils
Scalp			
Occipitofrontalis			
Frontal belly	Epicranial aponeurosis	Skin of eyebrow and bridge of nose	Raises eyebrows, wrinkles forehead
Occipital belly	Occipital and temporal bone	Epicranial aponeurosis	Tenses and retracts scalp
Neck			
Platysma	Superior thorax between cartilage of 2nd rib and acromion of scapula	Mandible and skin of cheek	Tenses skin of neck; depresses mandible and pulls lower lip inferiorly

3 This table groups the muscles of facial expression by region and summarizes their origins, insertions, and actions.

Module 10.6 Review

a. Name the muscles associated with the mouth , and identify the one involved in kissing or whistling.

b. State whether the following muscles involve the mouth, eye, nose, ear, scalp, or neck: buccinator, corrugator supercilii, mentalis, nasalis, platysma, procerus, and risorius.

c. Explain how a person is able to consciously move the skin on the scalp but is not able to consciously move the skin of the thigh.

LO 10.6 Identify the facial expression muscles, and cite their origins, insertions, and actions.

The extrinsic eye muscles position the eye, and the muscles of mastication move the lower jaw

1 Five of the six extrinsic eye muscles are visible in this lateral view of the right eye.

2 One additional muscle, the medial rectus, can be seen in this medial view of the right eye.

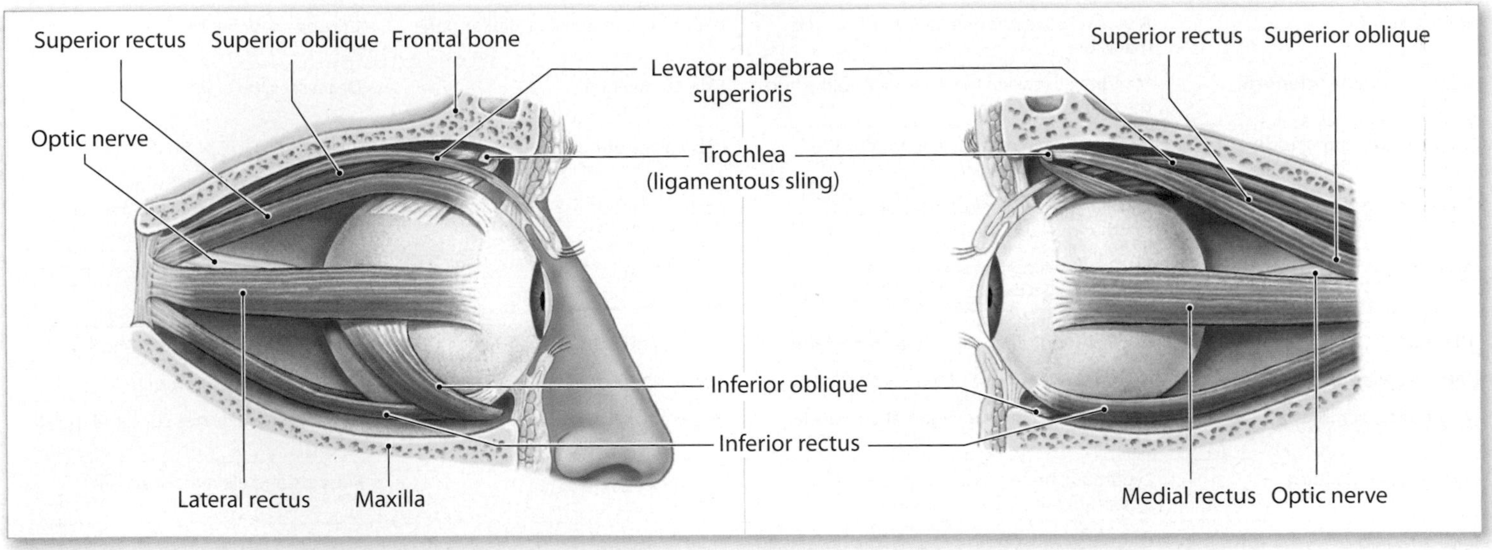
Superior rectus
Superior oblique
Frontal bone
Optic nerve
Levator palpebrae superioris
Trochlea (ligamentous sling)
Inferior oblique
Inferior rectus
Lateral rectus
Maxilla
Superior rectus
Superior oblique
Medial rectus
Optic nerve

3 This anterior view of the right eye shows the direction of eye movements produced by the contraction of each extrinsic eye muscle operating independently.

4 This anterior view of the right orbit shows the origins of the extrinsic eye muscles.

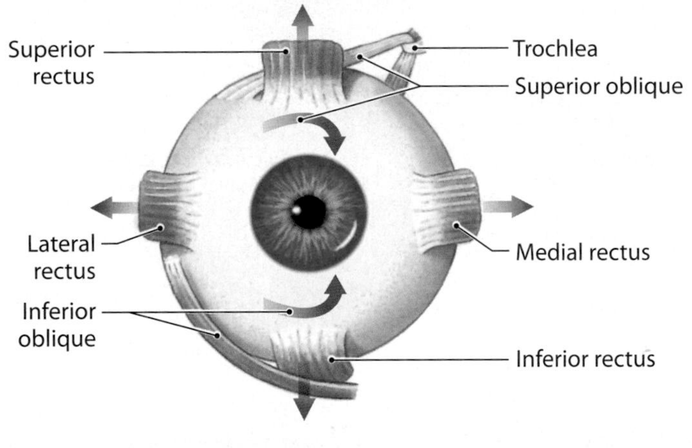
Superior rectus
Trochlea
Superior oblique
Lateral rectus
Medial rectus
Inferior oblique
Inferior rectus

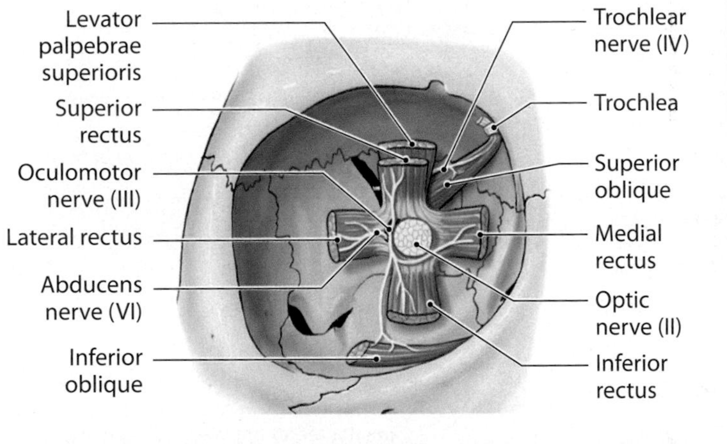
Levator palpebrae superioris
Superior rectus
Oculomotor nerve (III)
Lateral rectus
Abducens nerve (VI)
Inferior oblique
Trochlear nerve (IV)
Trochlea
Superior oblique
Medial rectus
Optic nerve (II)
Inferior rectus

Extrinsic Eye Muscles

Muscle	Origin	Insertion	Action
Inferior rectus	Sphenoid around optic canal	Inferior, medial surface of eyeball	Eye looks inferiorly
Medial rectus	Sphenoid around optic canal	Medial surface of eyeball	Eye looks medially
Superior rectus	Sphenoid around optic canal	Superior surface of eyeball	Eye looks superiorly
Lateral rectus	Sphenoid around optic canal	Lateral surface of eyeball	Eye looks laterally
Inferior oblique	Maxillary bone at anterior portion of orbit	Inferior, lateral surface of eyeball	Eye rolls, looks superiorly and laterally
Superior oblique	Sphenoid around optic canal	Superior, lateral surface of eyeball	Eye rolls, looks inferiorly and laterally

5 This lateral view shows the largest superficial muscles of mastication (chewing) of the right side of the head.

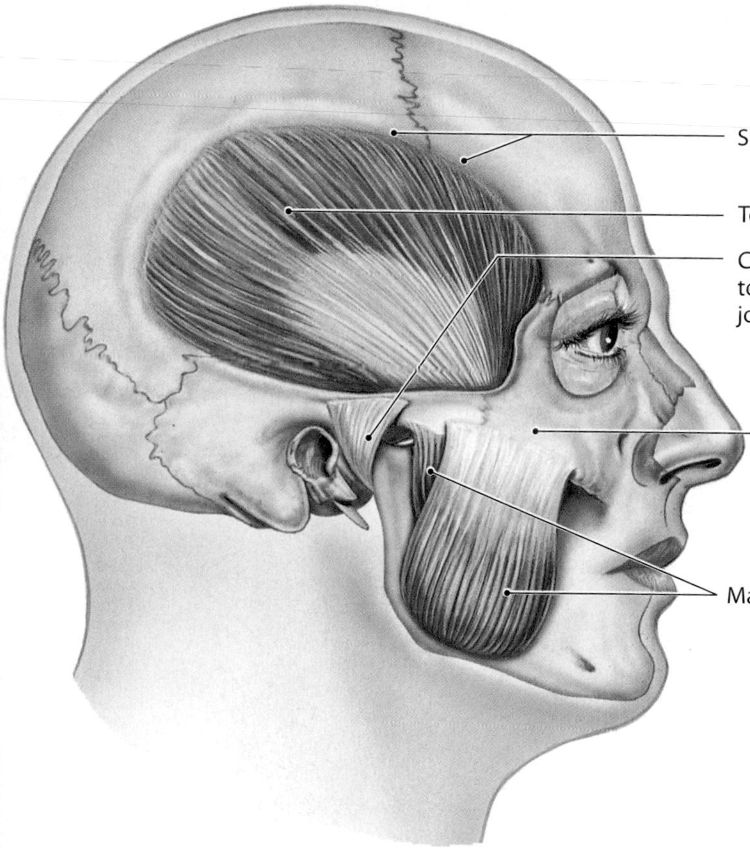

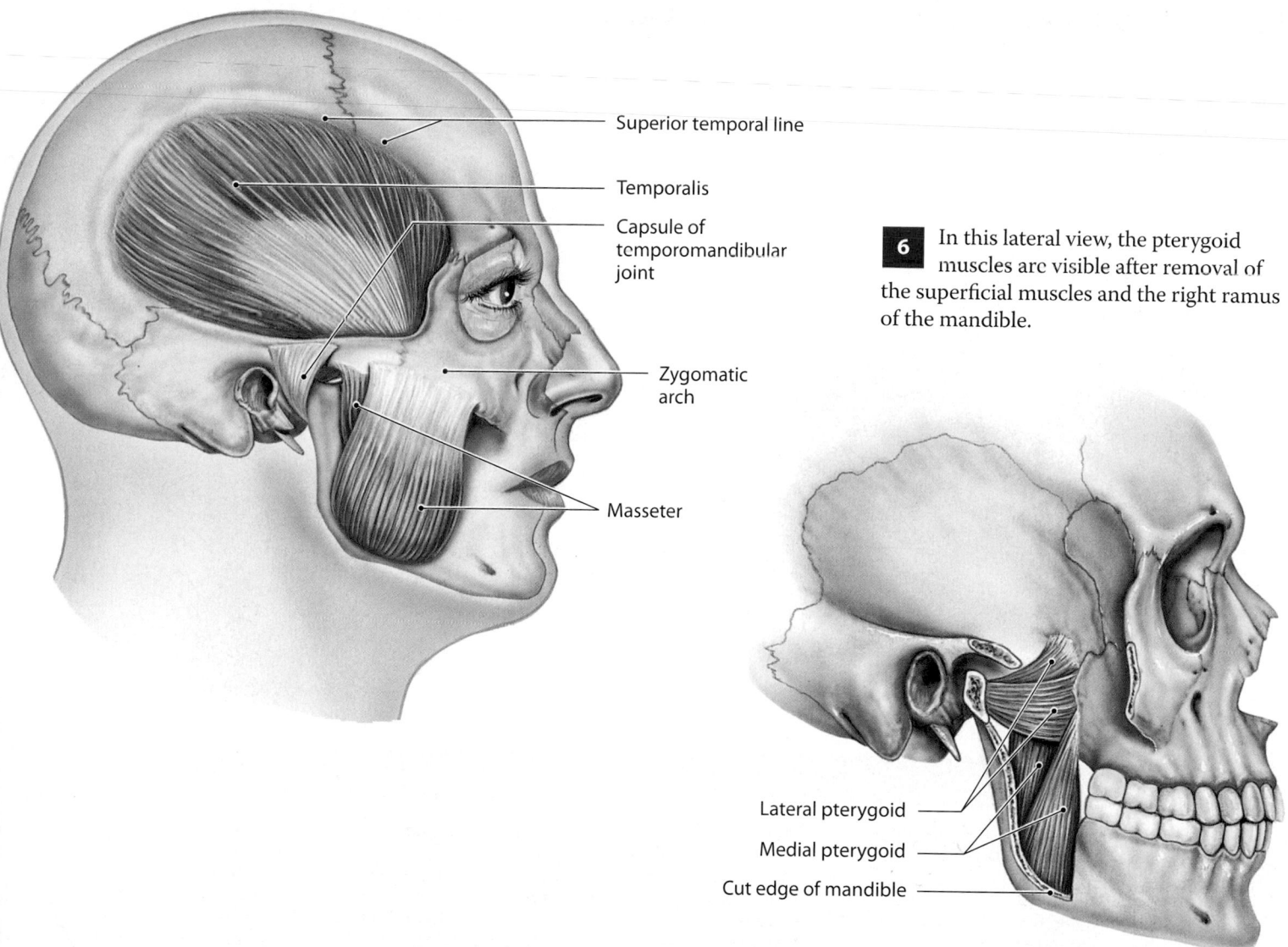

Superior temporal line
Temporalis
Capsule of temporomandibular joint
Zygomatic arch
Masseter

6 In this lateral view, the pterygoid muscles are visible after removal of the superficial muscles and the right ramus of the mandible.

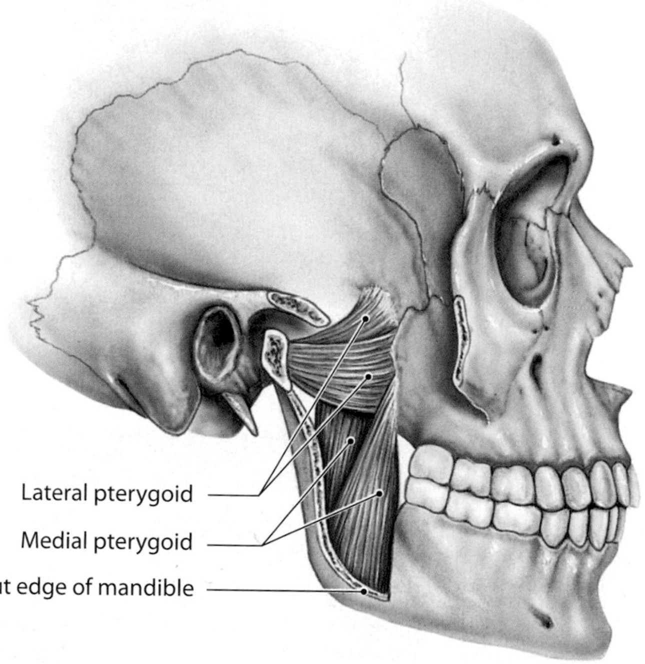

Lateral pterygoid
Medial pterygoid
Cut edge of mandible

Muscles of Mastication

Muscle	Origin	Insertion	Action
Masseter	Zygomatic arch	Lateral surface of mandibular ramus	Elevates mandible and closes the jaws
Temporalis	Along temporal lines of skull	Coronoid process of mandible	Elevates mandible
Pterygoids (medial and lateral)	Lateral pterygoid plate	Medial surface of mandibular ramus	Medial: Elevates the mandible and closes the jaws, or slides the mandible from side to side Lateral: Opens jaws, protrudes the mandible, or slides the mandible from side to side

Module 10.7 Review

a. Name the extrinsic eye muscles.

b. Which muscles have their origin on the lateral pterygoid plates and their insertion on the medial surface of the mandibular ramus?

c. If you were contracting and relaxing your masseter muscle, what would you probably be doing?

LO 10.7 Identify the eye and jaw muscles, and cite their origins, insertions, and actions.

The muscles of the tongue are closely associated with the muscles of the pharynx and neck

1 This lateral view shows the tongue muscles after removal of the left half of the mandible.

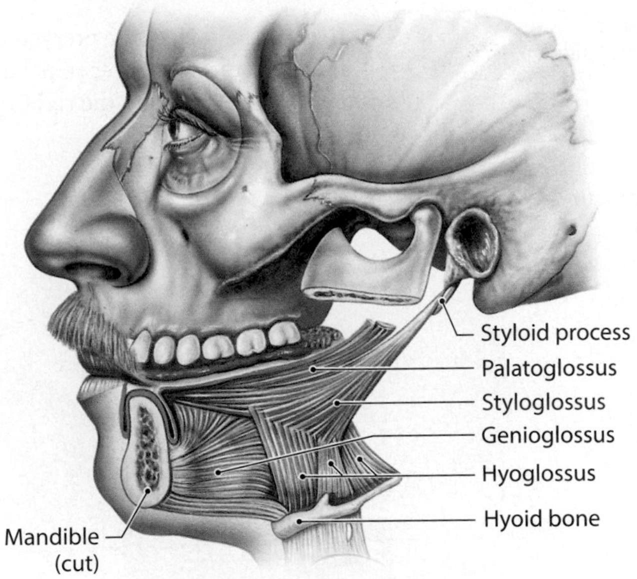
Styloid process
Palatoglossus
Styloglossus
Genioglossus
Hyoglossus
Hyoid bone
Mandible (cut)

2 This lateral view shows the major groups of the muscles of the pharynx. Their roles in swallowing are discussed in Module 22.8 (p. 843).

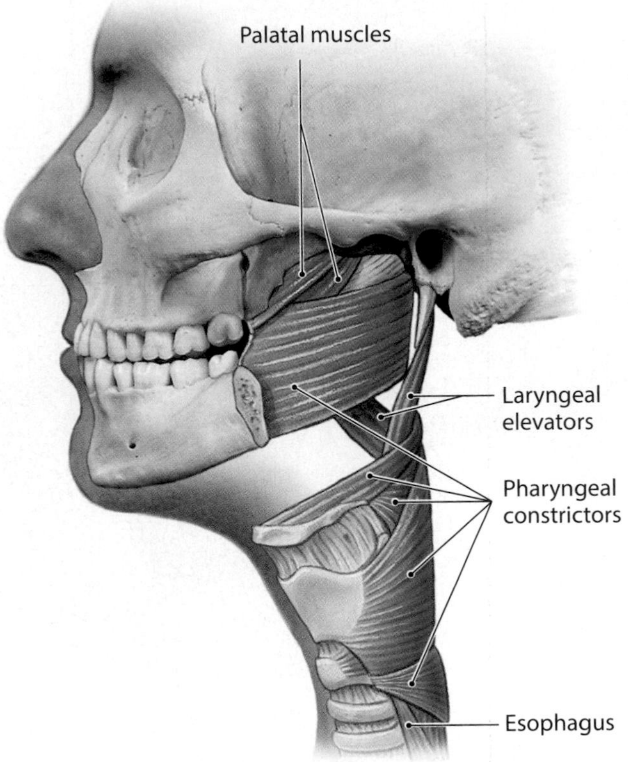
Palatal muscles
Laryngeal elevators
Pharyngeal constrictors
Esophagus

Muscles of the Tongue

Muscle	Origin	Insertion	Action
Genioglossus	Medial surface of mandible around chin	Body of tongue, hyoid bone	Depresses and protracts tongue
Hyoglossus	Body and greater horn of hyoid bone	Side of tongue	Depresses and retracts tongue
Palatoglossus	Anterior surface of soft palate	Side of tongue	Elevates tongue, depresses soft palate
Styloglossus	Styloid process of temporal bone	Along the side to tip and base of tongue	Retracts tongue, elevates side of tongue

Muscles of the Pharynx

Muscle	Origin	Insertion	Action
Pharyngeal constrictors	Pterygoid process of sphenoid, medial surfaces of mandible, horns of hyoid bone, cricoid and thyroid cartilages of larynx	Median raphe attached to occipital bone	Constrict pharynx to propel an ingested food mass into the esophagus
Laryngeal elevators	Soft palate, cartilage around inferior portion of auditory tube, styloid process	Thyroid cartilage	Elevate larynx
Palatal muscles	Petrous part of temporal bone and adjacent soft tissues, sphenoidal spine and adjacent soft tissues	Soft palate	Elevate soft palate

3 The anterior muscles of the neck are primarily involved in positioning the mandible, hyoid bone, and larynx.

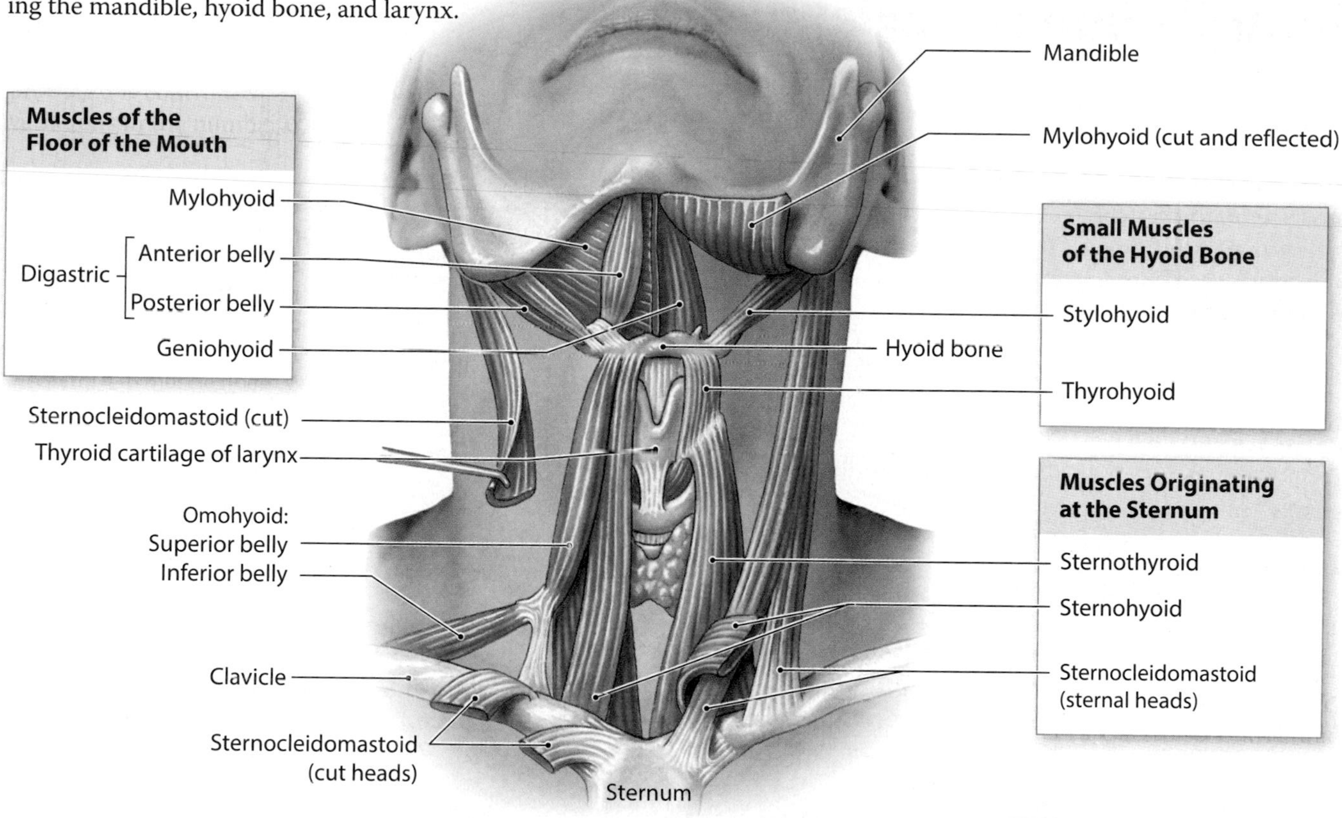
Mandible
Mylohyoid (cut and reflected)
Muscles of the Floor of the Mouth
Mylohyoid
Digastric
Anterior belly
Posterior belly
Geniohyoid
Small Muscles of the Hyoid Bone
Stylohyoid
Hyoid bone
Thyrohyoid
Sternocleidomastoid (cut)
Thyroid cartilage of larynx
Muscles Originating at the Sternum
Omohyoid:
Superior belly
Inferior belly
Sternothyroid
Sternohyoid
Clavicle
Sternocleidomastoid (sternal heads)
Sternocleidomastoid (cut heads)
Sternum

Anterior Muscles of the Neck

Muscle	Origin	Insertion	Action
Digastric	Inferior surface of mandible at chin and mastoid region	Hyoid bone	Depresses mandible or elevates larynx
Geniohyoid	Medial surface of mandible at chin	Hyoid bone	As above and pulls hyoid bone anteriorly
Mylohyoid	Mylohyoid line of mandible	Median raphe that runs to hyoid bone	Elevates hyoid bone or depresses mandible
Omohyoid	Superior border of scapula near scapular notch	Hyoid bone	Depresses hyoid bone and larynx
Sternohyoid	Clavicle and manubrium	Hyoid bone	Depresses hyoid bone and larynx
Sternothyroid	Manubrium and first costal cartilage	Thyroid cartilage of larynx	Depresses hyoid bone and larynx
Stylohyoid	Styloid process	Hyoid bone	Elevates larynx
Thyrohyoid	Thyroid cartilage of larynx	Hyoid bone	Elevates thyroid, depresses hyoid
Sternocleidomastoid	One head attaches to sternal end of clavicle; the other head attaches to manubrium	Mastoid region of skull and lateral portion of superior nuchal line	Flexes the neck; one alone bends head toward shoulder and rotates neck

4 This is a superior view of an isolated mandible. Several muscles extending from the hyoid bone to the mandible form the muscular floor of the mouth and support the tongue.

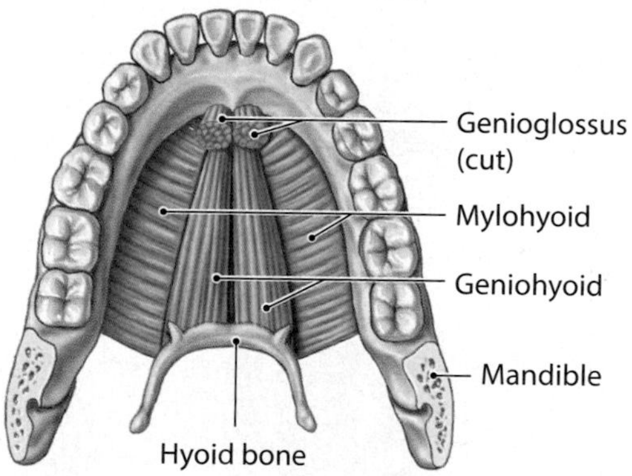
Genioglossus (cut)
Mylohyoid
Geniohyoid
Mandible
Hyoid bone

Module 10.8 Review

a. List the muscles of the tongue.

b. Which muscles elevate the soft palate?

c. Which muscles associated with the hyoid form the floor of the mouth?

LO 10.8 Identify the tongue, pharynx, and neck muscles, and cite their origins, insertions, and actions.

The muscles of the vertebral column support and align the axial skeleton

1 The **muscles of the vertebral column** are arranged in several layers. They include muscles originating or inserting on the ribs and the processes of the vertebrae. Although this mass of muscles extends from the sacrum to the skull, each muscle group is composed of numerous separate muscles of various lengths.

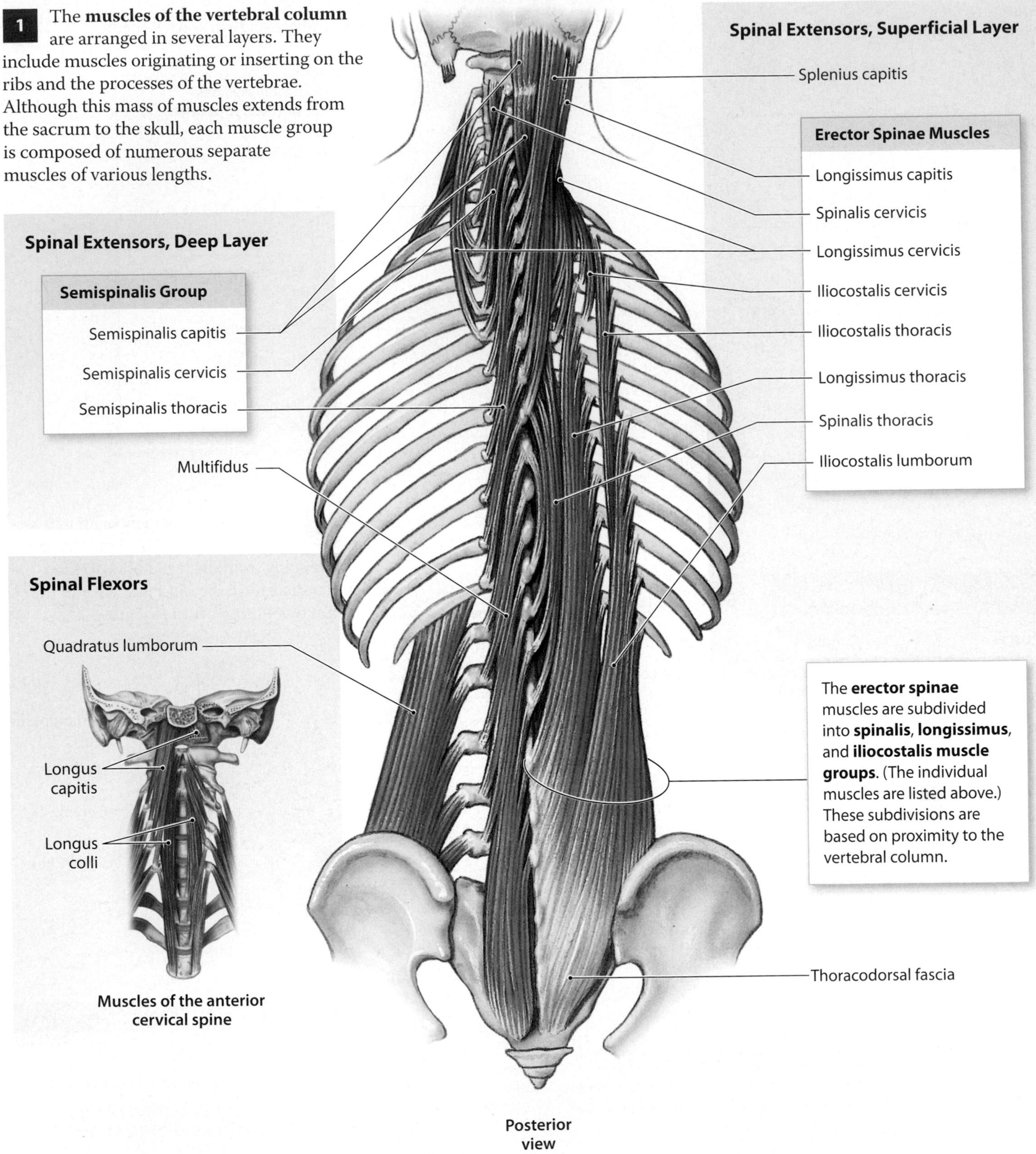

Spinal Extensors, Superficial Layer
Splenius capitis
Erector Spinae Muscles
Longissimus capitis
Spinalis cervicis
Longissimus cervicis
Iliocostalis cervicis
Iliocostalis thoracis
Longissimus thoracis
Spinalis thoracis
Iliocostalis lumborum
Spinal Extensors, Deep Layer
Semispinalis Group
Semispinalis capitis
Semispinalis cervicis
Semispinalis thoracis
Multifidus
Spinal Flexors
Quadratus lumborum
Longus capitis
Longus colli
Muscles of the anterior cervical spine
The erector spinae muscles are subdivided into spinalis, longissimus, and iliocostalis muscle groups. (The individual muscles are listed above.) These subdivisions are based on proximity to the vertebral column.
Thoracodorsal fascia
Posterior view

Muscles of the Vertebral Column

Group and Muscle		Origin	Insertion	Action
Spinal Extensors — Superficial Layer				
Splenius (splenius capitis, splenius cervicis)		Spinous processes and ligaments connecting inferior cervical and superior thoracic vertebrae	Mastoid process, occipital bone of skull, and superior cervical vertebrae	Together, the two sides extend neck; alone, each rotates and laterally flexes neck to that side
Erector spinae				
SPINALIS GROUP	**Spinalis cervicis**	Inferior portion of ligamentum nuchae and spinous process of C_7	Spinous process of axis	Extends neck
	Spinalis thoracis	Spinous processes of inferior thoracic and superior lumbar vertebrae	Spinous processes of superior thoracic vertebrae	Extends vertebral column
LONGISSIMUS GROUP	**Longissimus capitis**	Transverse processes of inferior cervical and superior thoracic vertebrae	Mastoid process of temporal bone	Together, the two sides extend head; alone, each rotates and laterally flexes neck to that side
	Longissimus cervicis	Transverse processes of superior thoracic vertebrae	Transverse processes of middle and superior cervical vertebrae	Together, the two sides extend head; alone, each rotates and laterally flexes neck to that side
	Longissimus thoracis	Broad aponeurosis and transverse processes of inferior thoracic and superior lumbar vertebrae; joins iliocostalis	Transverse processes of superior vertebrae and inferior surfaces of ribs	Extends vertebral column; alone, each produces lateral flexion to that side
ILIOCOSTALIS GROUP	**Iliocostalis cervicis**	Superior borders of vertebrosternal ribs near the angles	Transverse processes of middle and inferior cervical vertebrae	Extends or laterally flexes neck, elevates ribs
	Iliocostalis thoracis	Superior borders of inferior seven ribs medial to the angles	Upper ribs and transverse process of last cervical vertebra	Stabilizes thoracic vertebrae in extension
	Iliocostalis lumborum	Iliac crest, sacral crests, and spinous processes	Inferior surfaces of inferior seven ribs near their angles	Extends vertebral column, depresses ribs
Spinal Extensors — Deep Layer				
SEMISPINALIS GROUP	**Semispinalis capitis**	Articular processes of inferior cervical and transverse processes of superior thoracic vertebrae	Occipital bone, between nuchal lines	Together, the two sides extend head; alone, each extends and laterally flexes neck
	Semispinalis cervicis	Transverse processes of T_1–T_5 or T_6	Spinous processes of C_2–C_5	Extends vertebral column and rotates toward opposite side
	Semispinalis thoracis	Transverse processes of T_6–T_{10}	Spinous processes of C_5–T_4	Extends vertebral column and rotates toward opposite side
	Multifidus	Sacrum and transverse processes of each vertebra	Spinous processes of the third or fourth more superior vertebrae	Extends vertebral column and rotates toward opposite side
Spinal Flexors				
Longus capitis		Transverse processes of cervical vertebrae	Base of the occipital bone	Together, the two sides flex the neck; alone, each rotates head to that side
Longus colli		Anterior surfaces of cervical and superior thoracic vertebrae	Transverse processes of superior cervical vertebrae	Flexes or rotates neck; limits hyperextension
Quadratus lumborum		Iliac crest and iliolumbar ligament	Last rib and transverse processes of lumbar vertebrae	Together, they depress ribs; alone, each side laterally flexes vertebral column

2 Note that this table lists many extensors of the vertebral column, but few flexors. The vertebral column does not need a massive series of flexor muscles because (1) many of the large trunk muscles flex the vertebral column when they contract, and (2) most of the body weight lies anterior to the vertebral column, and gravity tends to flex the spine when unopposed by the extensor muscles.

Module 10.9 Review

a. List the spinal flexor muscles.

b. Which muscles enable you to extend your neck?

c. What might account for a lack of massive flexor muscles?

LO 10.9 Identify the vertebral column muscles, and cite their origins, insertions, and actions.

Module 10.10

The oblique and rectus muscles form the muscular walls of the trunk

1 The oblique and rectus muscle groups share embryological origins. The scalene muscles extend from the cervical vertebrae in the neck to the first two ribs.

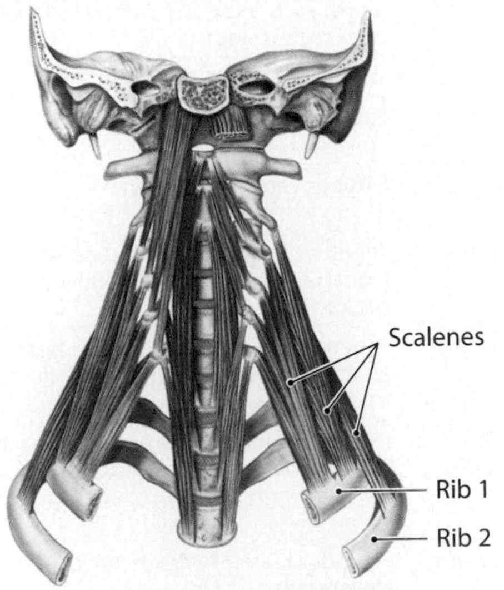

Scalenes
Rib 1
Rib 2

Anterior view, frontal section

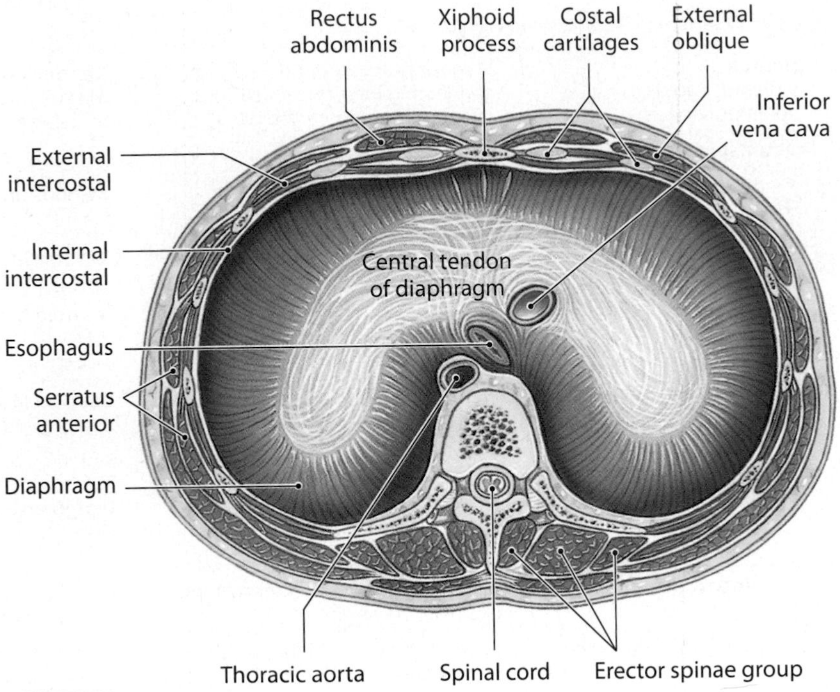

Rectus abdominis
Xiphoid process
Costal cartilages
External oblique
Inferior vena cava
External intercostal
Internal intercostal
Central tendon of diaphragm
Esophagus
Serratus anterior
Diaphragm
Thoracic aorta
Spinal cord
Erector spinae group

Superior view of the diaphragm

2 Superficial muscles are shown on the right side of the body, and deeper muscles of the oblique and rectus groups are shown on the left side.

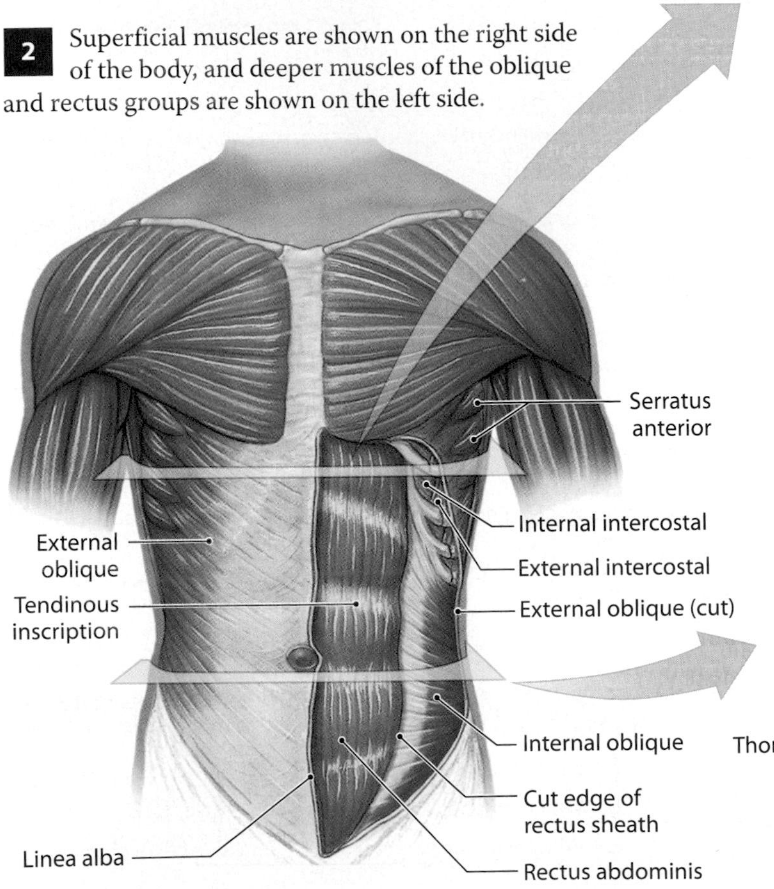

Serratus anterior
Internal intercostal
External intercostal
External oblique (cut)
External oblique
Tendinous inscription
Internal oblique
Cut edge of rectus sheath
Linea alba
Rectus abdominis

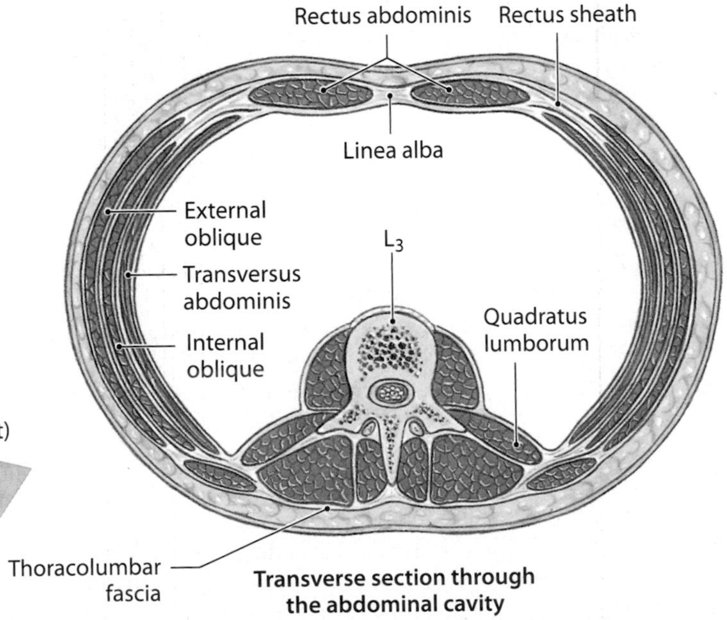

Rectus abdominis
Rectus sheath
Linea alba
External oblique
Transversus abdominis
Internal oblique
L3
Quadratus lumborum
Thoracolumbar fascia

Transverse section through the abdominal cavity

Oblique and Rectus Muscles

Region	Group and Muscle	Origin	Insertion	Action
Oblique Group				
CERVICAL REGION	**Scalenes**	Transverse and costal processes of cervical vertebrae	Superior surfaces of first two ribs	Elevate ribs or flex neck
THORACIC REGION	**External intercostals**	Inferior border of each rib	Superior border of more inferior rib	Elevate ribs
	Internal intercostals	Superior border of each rib	Inferior border of the preceding rib	Depress ribs
	Transversus thoracis	Posterior surface of sternum	Cartilages of ribs	Depress ribs
ABDOMINAL REGION	**External oblique**	External and inferior borders of ribs 5–12	Linea alba and iliac crest	Compresses abdomen, depresses ribs, flexes or bends spine
	Internal oblique	Lumbodorsal fascia and iliac crest	Inferior ribs, xiphoid process, and linea alba	Compresses abdomen, depresses ribs, flexes or bends spine
	Transversus abdominis	Cartilages of ribs 6–12, iliac crest, and lumbodorsal fascia	Linea alba and pubis	Compresses abdomen
Rectus Group				
THORACIC REGION	**Diaphragm**	Xiphoid process, cartilages of ribs 4–10, and anterior surfaces of lumbar vertebrae	Central tendinous sheet	Contraction expands thoracic cavity, compresses abdominopelvic cavity
ABDOMINAL REGION	**Rectus abdominis**	Superior surface of pubis around symphysis	Inferior surfaces of costal cartilages (ribs 5–7) and xiphoid process	Depresses ribs, flexes vertebral column, compresses abdomen

3 This is an inferior view of the diaphragm, showing its major landmarks.

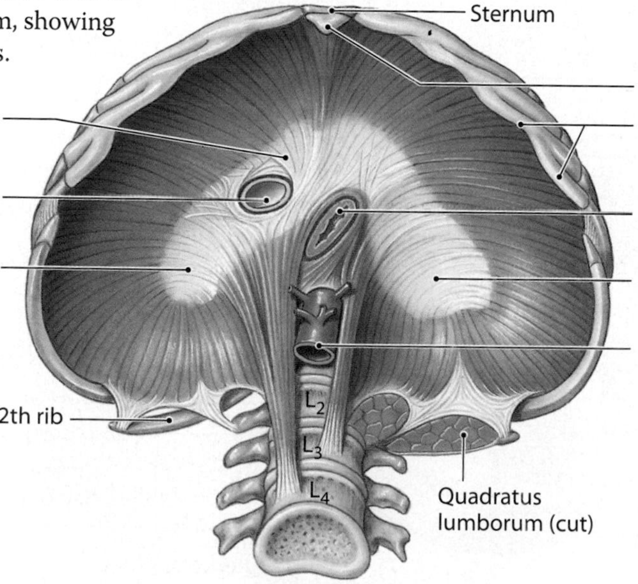

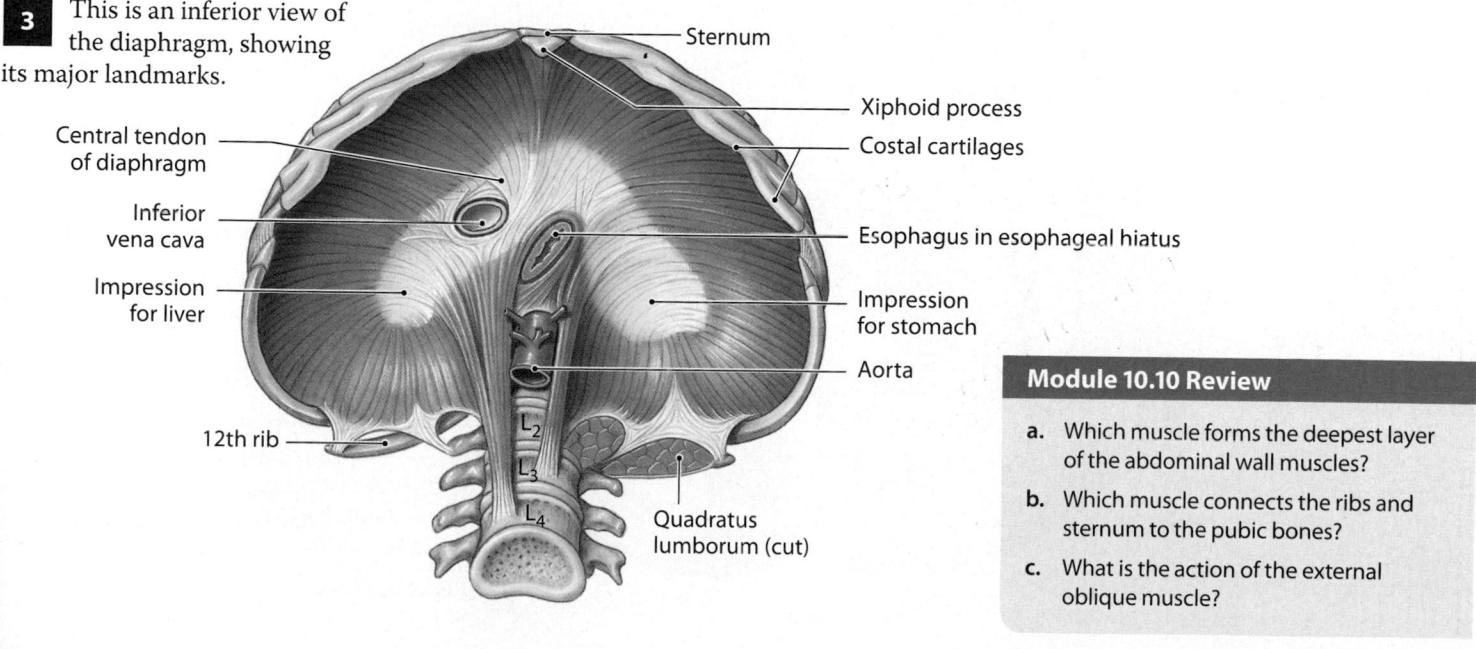

Sternum
Xiphoid process
Costal cartilages
Central tendon of diaphragm
Inferior vena cava
Esophagus in esophageal hiatus
Impression for liver
Impression for stomach
Aorta
12th rib
L_2
L_3
L_4
Quadratus lumborum (cut)

Module 10.10 Review

a. Which muscle forms the deepest layer of the abdominal wall muscles?

b. Which muscle connects the ribs and sternum to the pubic bones?

c. What is the action of the external oblique muscle?

LO **10.10** Identify the trunk muscles, and cite their origins, insertions, and actions.

The muscles of the pelvic floor support the organs of the abdominopelvic cavity

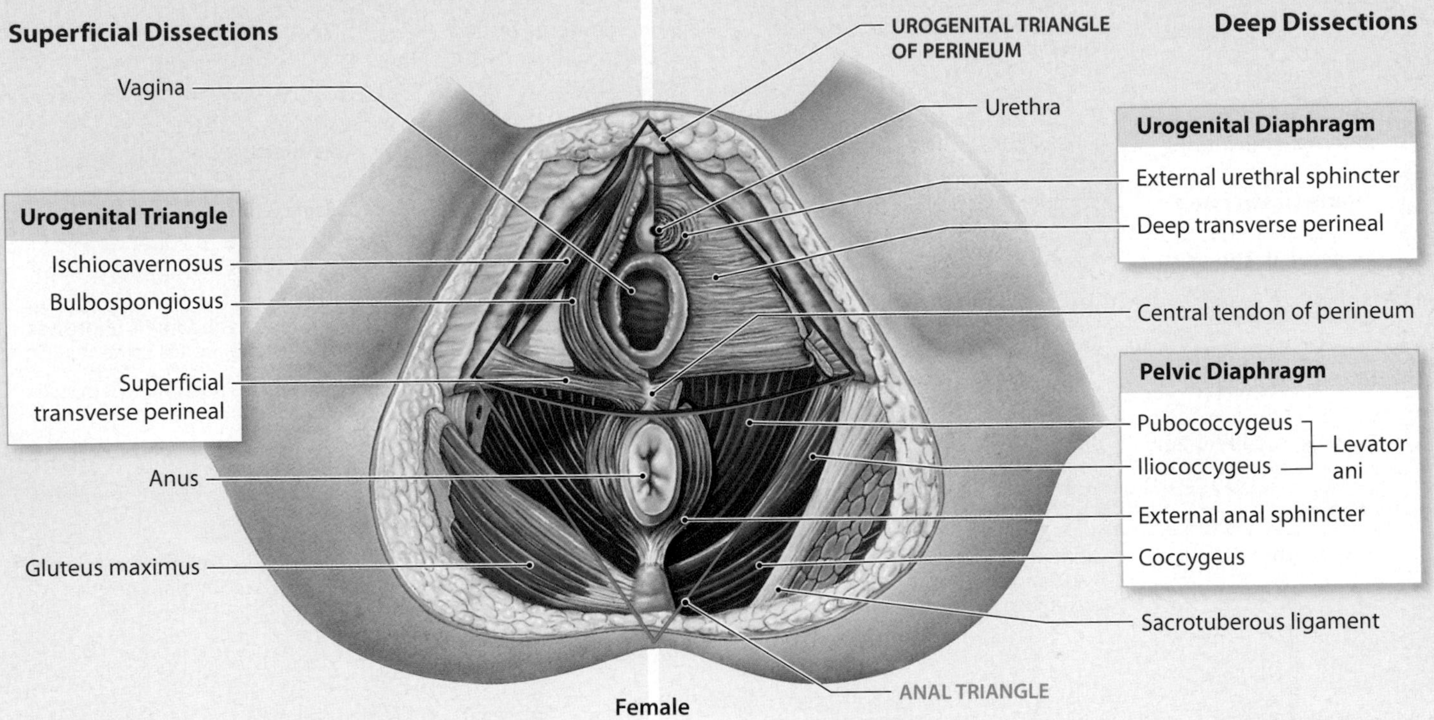
Superficial Dissections
Deep Dissections
Vagina
UROGENITAL TRIANGLE OF PERINEUM
Urethra
Urogenital Triangle
Ischiocavernosus
Bulbospongiosus
Superficial transverse perineal
Anus
Gluteus maximus
Urogenital Diaphragm
External urethral sphincter
Deep transverse perineal
Central tendon of perineum
Pelvic Diaphragm
Pubococcygeus
Iliococcygeus
Levator ani
External anal sphincter
Coccygeus
Sacrotuberous ligament
ANAL TRIANGLE

Female

1 The muscles of the pelvic floor form the **perineum** (per-i-NĒ-um), a muscular sheet that spans the pelvic outlet. Females and males have different superficial muscles associated with their external genitalia.

2 There are no differences between the deep perineal musculature of females and males.

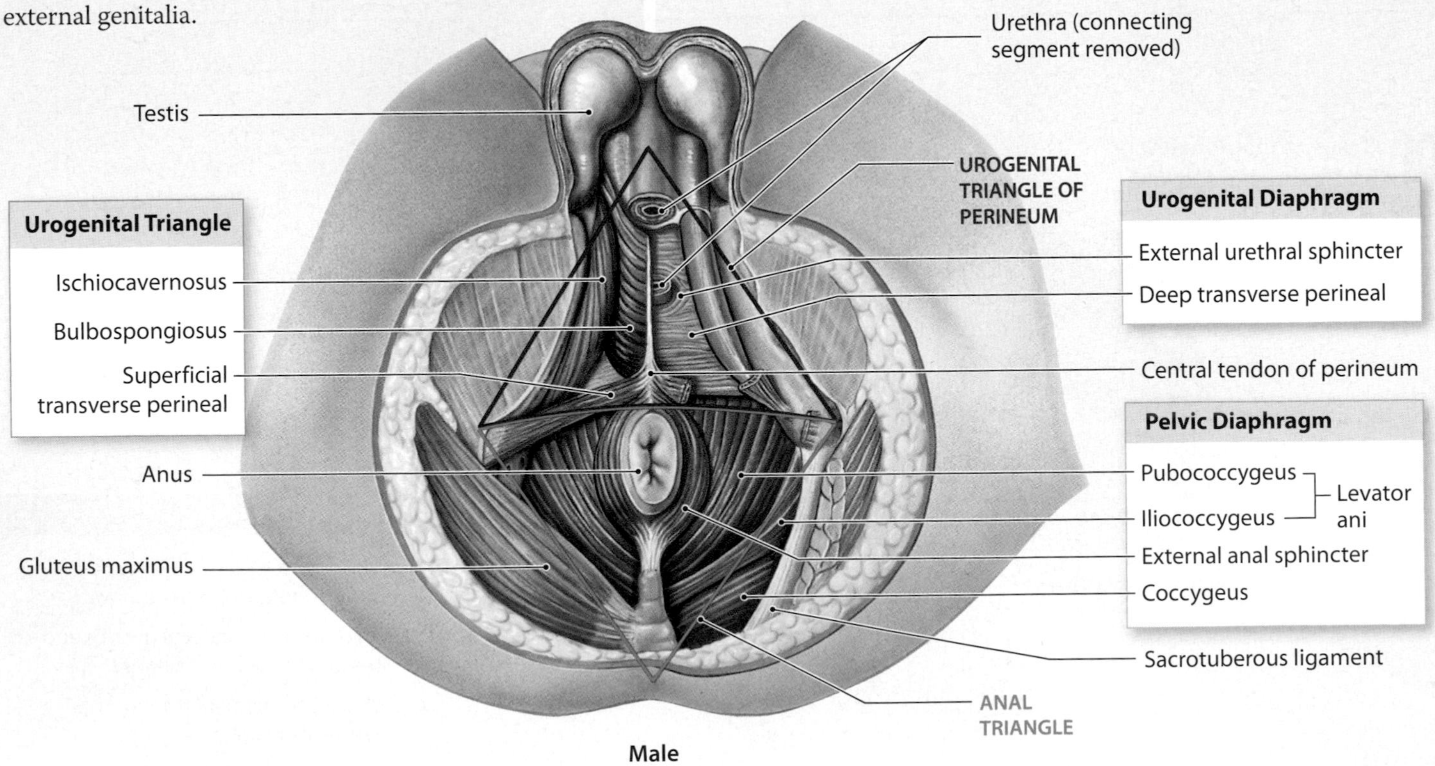
Testis
Urethra (connecting segment removed)
UROGENITAL TRIANGLE OF PERINEUM
Urogenital Triangle
Ischiocavernosus
Bulbospongiosus
Superficial transverse perineal
Anus
Gluteus maximus
Urogenital Diaphragm
External urethral sphincter
Deep transverse perineal
Central tendon of perineum
Pelvic Diaphragm
Pubococcygeus
Iliococcygeus
Levator ani
External anal sphincter
Coccygeus
Sacrotuberous ligament
ANAL TRIANGLE

Male

Muscles of the Pelvic Floor

Group and Muscle	Origin	Insertion	Action
Urogenital Triangle			
SUPERFICIAL MUSCLES			
Bulbospongiosus			
Females	Collagen sheath at base of clitoris; fibers run on either side of urethral and vaginal opening	Central tendon of perineum	Compresses and stiffens clitoris; narrows vaginal opening
Males	Collagen sheath at base of penis; fibers cross over urethra	Median raphe and central tendon of perineum	Compresses base and stiffens penis; ejects urine or semen
Ischiocavernosus	Ischial ramus and tuberosity	Pubic symphysis anterior to base of penis or clitoris	Compresses and stiffens penis or clitoris
Superficial transverse perineal	Ischial ramus	Central tendon of perineum	Stabilizes central tendon of perineum
DEEP MUSCLES			
Urogenital diaphragm			
Deep transverse perineal	Ischial ramus	Median raphe of urogenital diaphragm	Stabilizes central tendon of perineum
External urethral sphincter			
Females	Ischial and pubic rami	To median raphe; inner fibers encircle urethra	Closes urethra; compresses vagina and greater vestibular glands
Males	Ischial and pubic rami	To median raphe at base of penis; inner fibers encircle urethra	Closes urethra; compresses prostate and bulbourethral glands
Anal Triangle			
Pelvic diaphragm			
Coccygeus	Ischial spine	Lateral, inferior borders of sacrum and coccyx	Flexes coccygeal joints; tenses and supports pelvic floor
Levator ani			
Iliococcygeus	Ischial spine, pubis	Coccyx and median raphe	Tenses floor of pelvis; flexes coccygeal joints; elevates and retracts anus
Pubococcygeus	Inner margins of pubis	Coccyx and median raphe	Tenses floor of pelvis; flexes coccygeal joints; elevates and retracts anus
External anal sphincter	Coccyx	Encircles anal opening	Closes anal opening

The urogenital and pelvic diaphragms do not completely close the pelvic outlet, because the urethra, vagina, and anus pass through them to open at the exterior. Muscular sphincters surround their openings and permit voluntary control of urination and defecation. Muscles, nerves, and blood vessels also pass through the pelvic outlet as they travel to or from the lower limbs.

Module 10.11 Review

a. Which muscles make up the urogenital diaphragm?

b. In females, what is the action of the bulbospongiosus muscle?

c. The coccygeus muscle extends from the sacrum and coccyx to which structure?

LO **10.11** Identify the pelvic floor muscles, and cite their origins, insertions, and actions.

Labeling

Label each of the indicated muscles of the face in the following diagram.

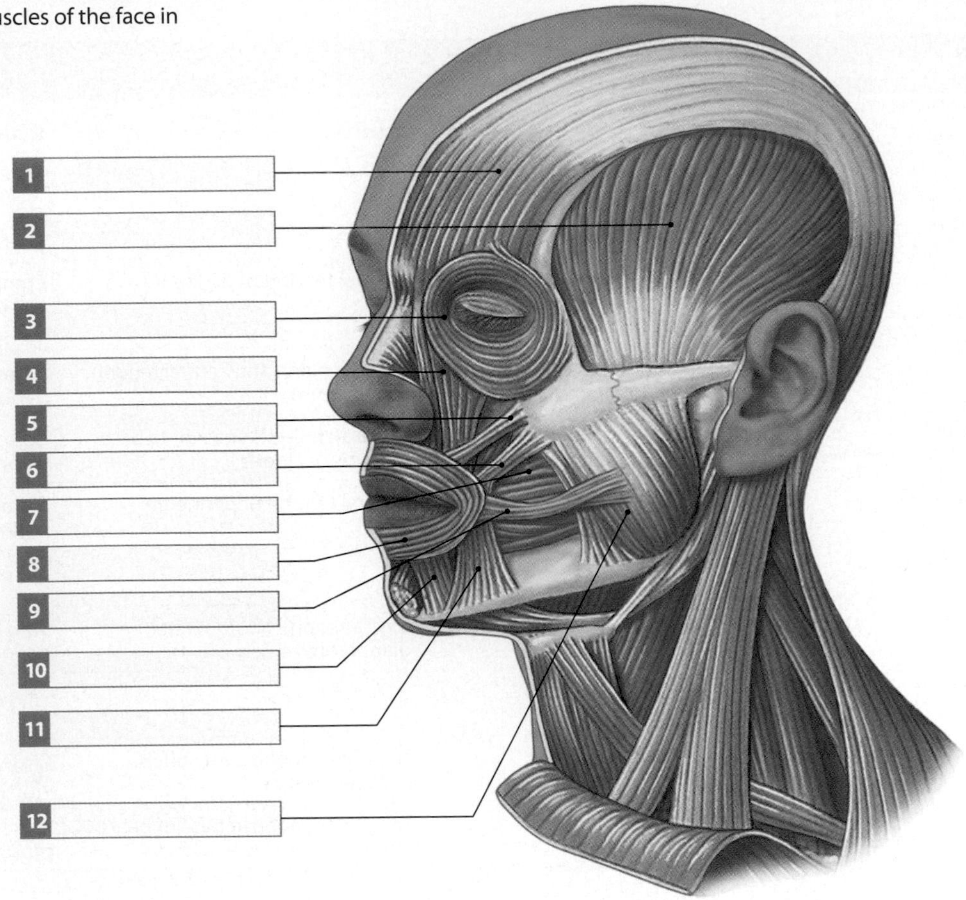
1
2
3
4
5
6
7
8
9
10
11
12

Label each of the indicated muscles of the neck in the following diagram.

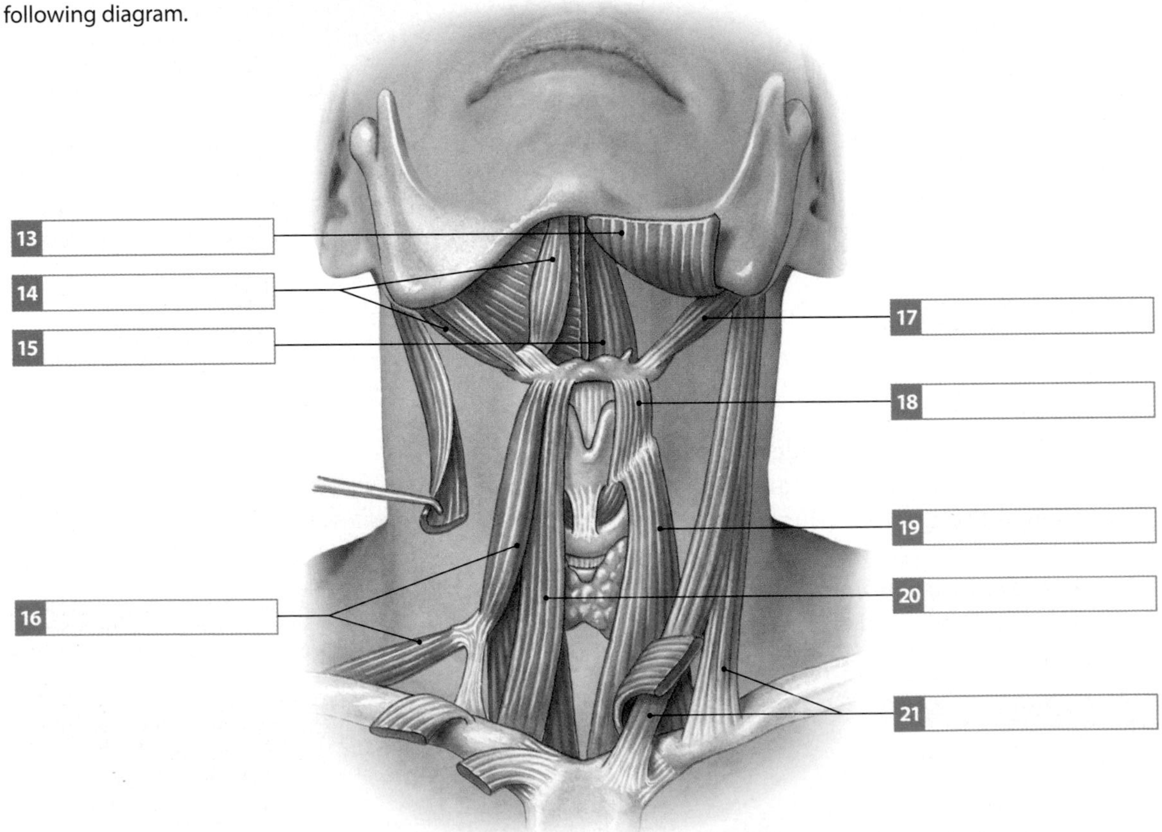
13
14
15
16
17
18
19
20
21

The appendicular muscles stabilize, position, and support the limbs

In the following modules, we will group the appendicular muscles by their actions and origins. Muscle actions can be described in two ways. The first describes actions in terms of the bone or region affected. For example, we say a muscle such as the biceps brachii performs "flexion of the forearm." Or, we list muscles as groups, such as "muscles that move the forearm."

Specialists, such as kinesiologists and physical therapists, use the alternate, second way by identifying the joint involved. In this approach, we say the action of the biceps brachii muscle is "flexion at (or of) the elbow." In general, we will use this second way of describing muscle actions.

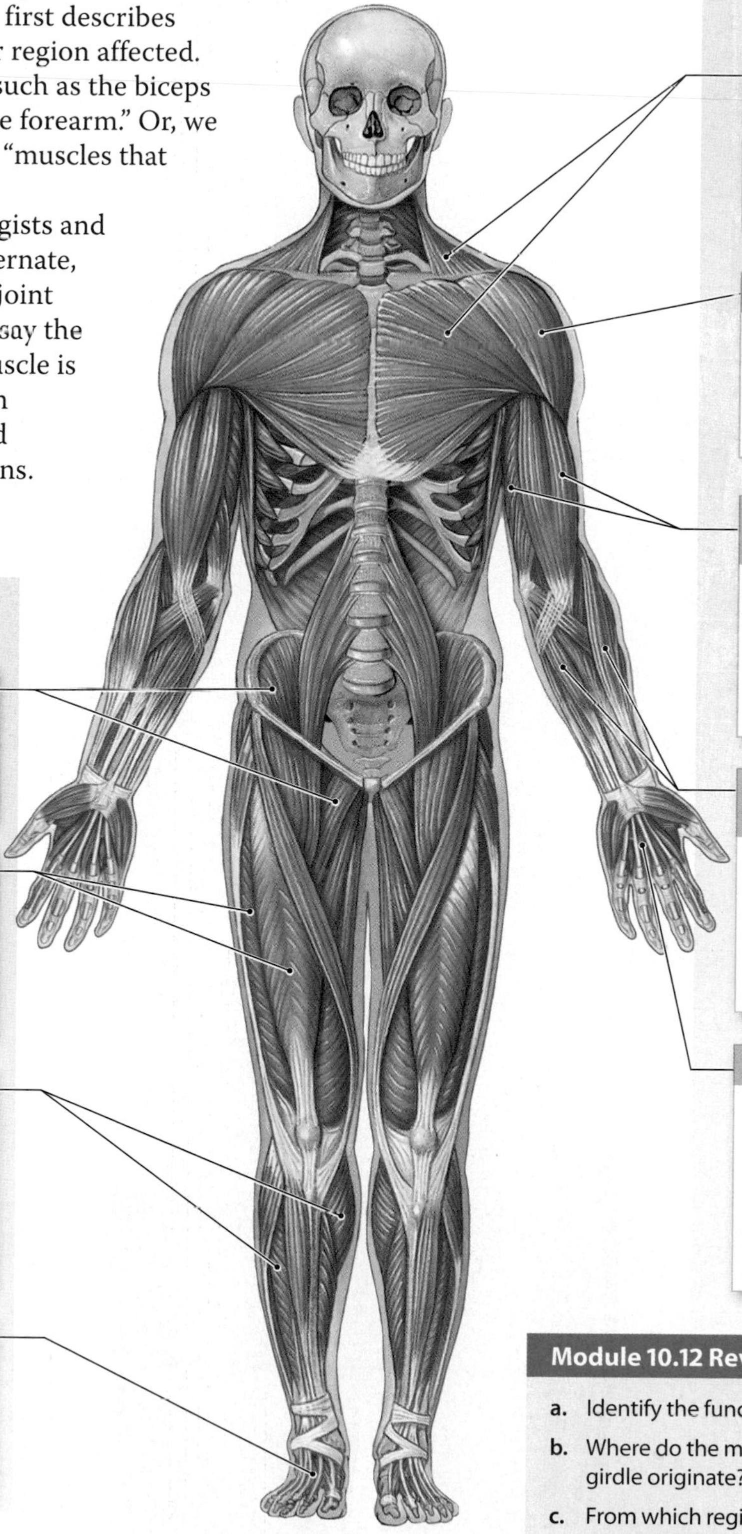
Upper Limb
Muscles That Position the Pectoral Girdle
These muscles originate on the axial skeleton and insert on the clavicle and scapula.
Muscles That Move the Arm
These muscles originate on the pectoral girdle and the thoracic cage and insert on the humerus.
Muscles That Move the Forearm and Hand
These muscles primarily originate on the pectoral girdle and arm, and insert on the radius, ulna, and/or carpals.
Extrinsic Muscles of the Hand and Fingers
These muscles primarily originate on the humerus, radius, and ulna, and insert on the metacarpals and phalanges.
Intrinsic Muscles of the Hand
These are the muscles that perform fine movements. They originate primarily on the carpal and metacarpal bones, and insert on the phalanges.
Lower Limb
Muscles That Move the Thigh
These muscles originate in the pelvic region, and typically insert on the femur.
Muscles That Move the Leg
These muscles originate on the pelvis and femur, and insert on the tibia and/or fibula.
Extrinsic Muscles That Move the Foot and Toes
These muscles originate on the tibia and fibula, and insert on the tarsals, metatarsals, and/or phalanges.
Intrinsic Muscles of the Foot
These muscles originate primarily on the tarsal and metatarsal bones, and insert on the phalanges.

Module 10.12 Review

a. Identify the function of the appendicular muscles.

b. Where do the muscles that position each pectoral girdle originate?

c. From which region is a muscle likely to originate if it inserts on the femur?

LO 10.12 Describe the general functions of the muscles of the upper and lower limbs.

Module 10.13

The largest appendicular muscles originate on the trunk

1 In general, muscles originating on the trunk control gross (large scale) movements of the limbs. These muscles are often large and powerful. Distally, the limb muscles get smaller and more numerous, and their movements become more precise.

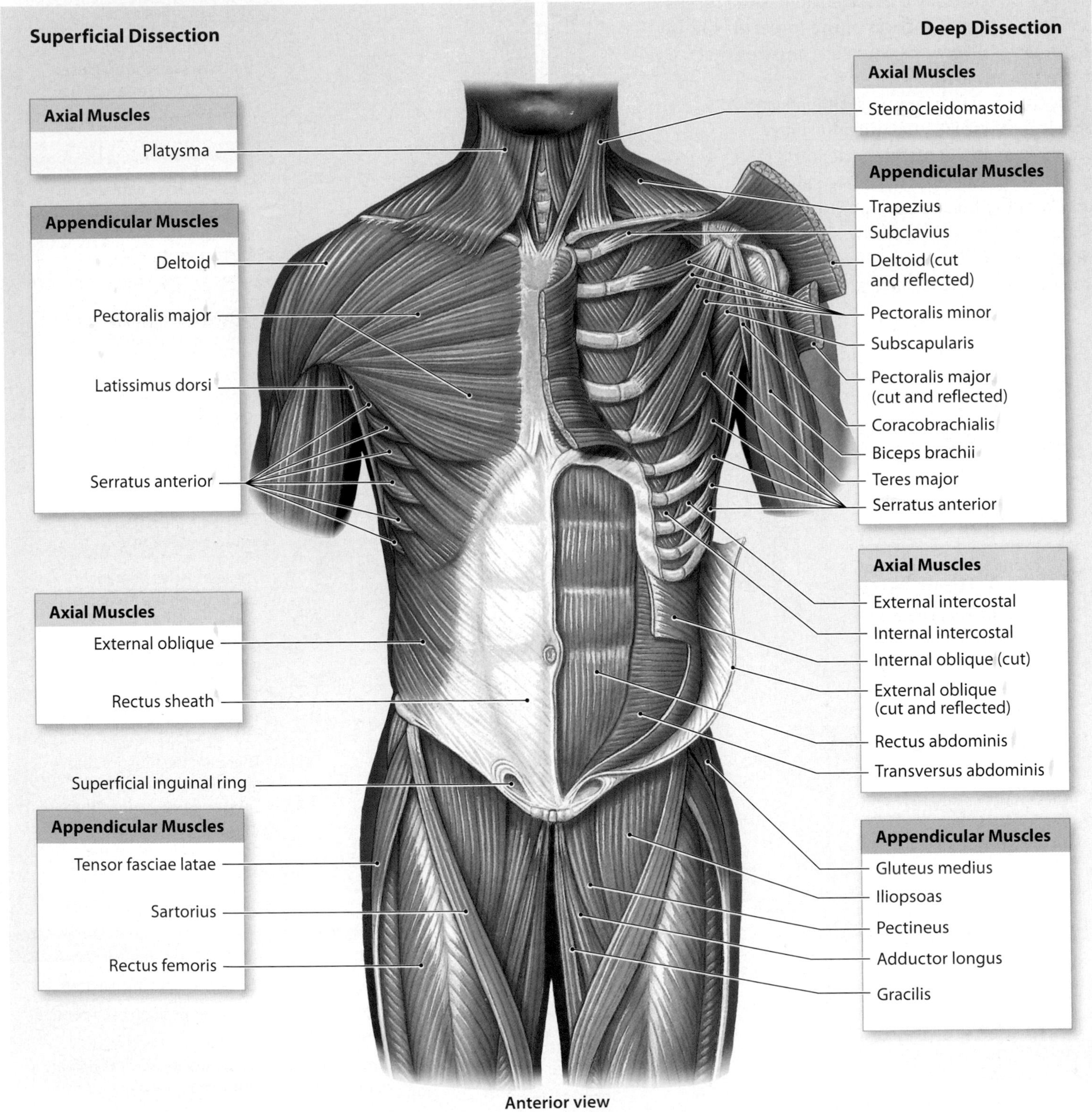

Superficial Dissection
Axial Muscles
Platysma
Appendicular Muscles
Deltoid
Pectoralis major
Latissimus dorsi
Serratus anterior
Axial Muscles
External oblique
Rectus sheath
Superficial inguinal ring
Appendicular Muscles
Tensor fasciae latae
Sartorius
Rectus femoris
Deep Dissection
Axial Muscles
Sternocleidomastoid
Appendicular Muscles
Trapezius
Subclavius
Deltoid (cut and reflected)
Pectoralis minor
Subscapularis
Pectoralis major (cut and reflected)
Coracobrachialis
Biceps brachii
Teres major
Serratus anterior
Axial Muscles
External intercostal
Internal intercostal
Internal oblique (cut)
External oblique (cut and reflected)
Rectus abdominis
Transversus abdominis
Appendicular Muscles
Gluteus medius
Iliopsoas
Pectineus
Adductor longus
Gracilis

Anterior view

2 Appendicular muscles that originate on the large bones of the limb girdles and the proximal bones of the limbs dominate the posterior trunk.

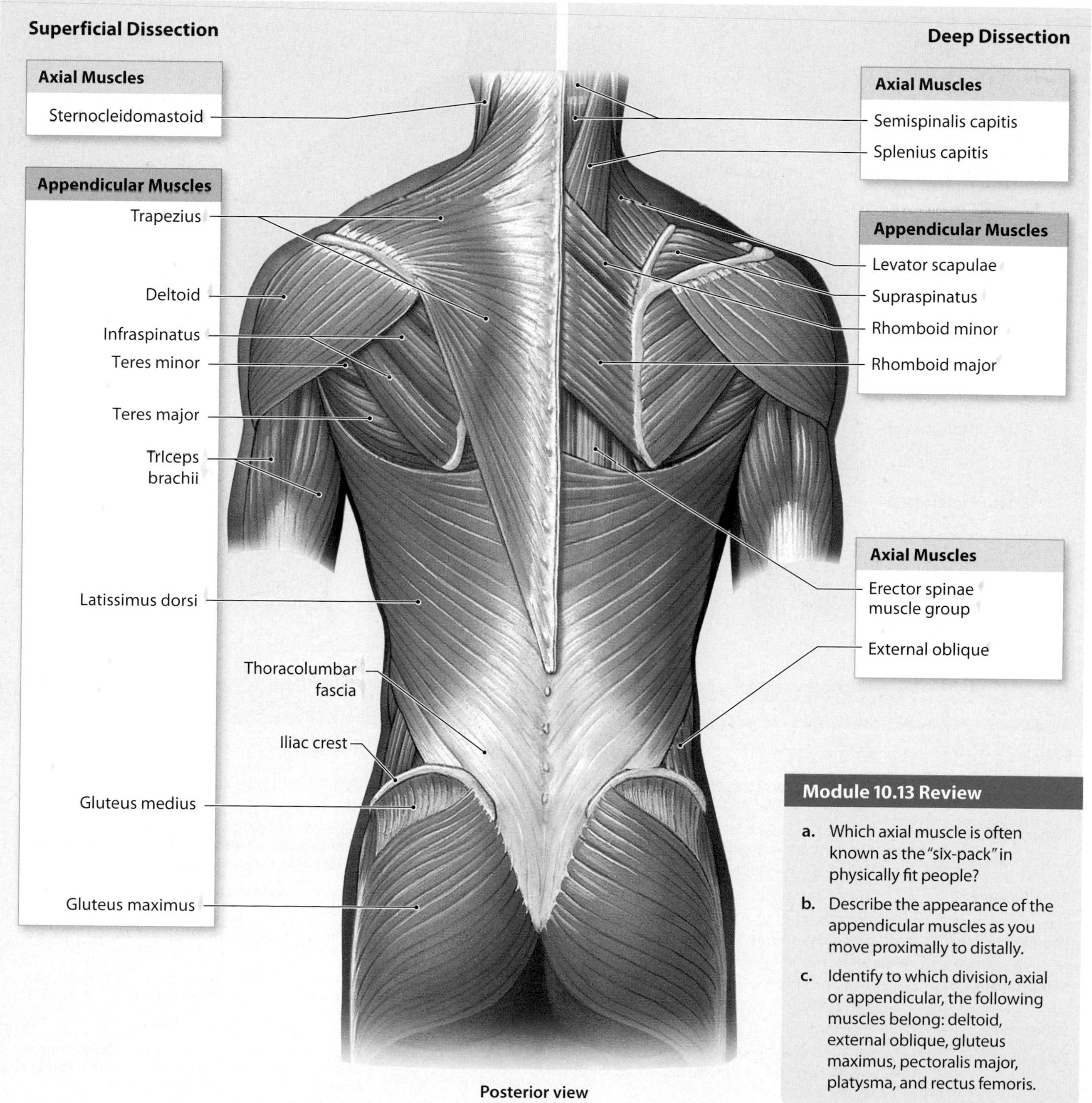

Superficial Dissection
Axial Muscles
Sternocleidomastoid
Appendicular Muscles
Trapezius
Deltoid
Infraspinatus
Teres minor
Teres major
Triceps brachii
Latissimus dorsi
Thoracolumbar fascia
Iliac crest
Gluteus medius
Gluteus maximus
Deep Dissection
Axial Muscles
Semispinalis capitis
Splenius capitis
Appendicular Muscles
Levator scapulae
Supraspinatus
Rhomboid minor
Rhomboid major
Axial Muscles
Erector spinae muscle group
External oblique

Posterior view

Module 10.13 Review

a. Which axial muscle is often known as the "six-pack" in physically fit people?

b. Describe the appearance of the appendicular muscles as you move proximally to distally.

c. Identify to which division, axial or appendicular, the following muscles belong: deltoid, external oblique, gluteus maximus, pectoralis major, platysma, and rectus femoris.

LO 10.13 Identify the principal appendicular muscles.

Module 10.14

Muscles that position each pectoral girdle originate on the occipital bone, superior vertebrae, and ribs

The muscles that position the pectoral girdles also anchor the pectoral girdles to the axial skeleton. Although these muscles have a smaller range of motion compared with other appendicular muscles, they help to increase upper limb mobility.

1 Several deep muscles of the chest, notably the pectoralis minor and serratus anterior, work with the trapezius and levator scapulae of the back to position and stabilize each pectoral girdle.

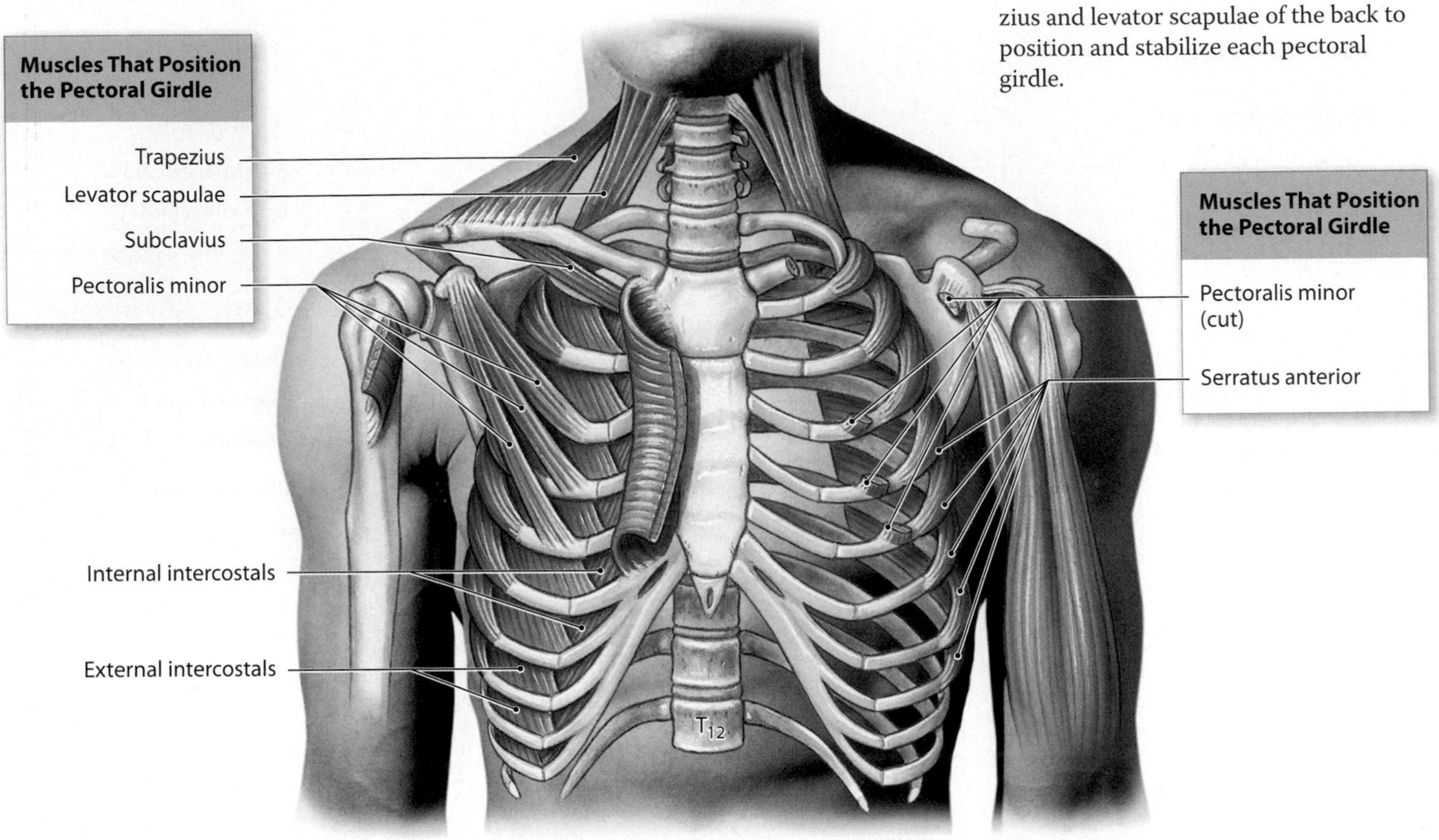

Muscles That Position the Pectoral Girdle
Trapezius
Levator scapulae
Subclavius
Pectoralis minor
Internal intercostals
External intercostals
T_{12}
Muscles That Position the Pectoral Girdle
Pectoralis minor (cut)
Serratus anterior

Anterior view

Muscles That Position the Pectoral Girdle

Muscle	Origin	Insertion	Action
Levator scapulae	Transverse processes of first four cervical vertebrae	Vertebral border of scapula near superior angle	Elevates scapula
Pectoralis minor	Anterior-superior surfaces of ribs 2–4, 2–5, or 3–5 depending on anatomical variation	Coracoid process of scapula	Depresses and protracts shoulder; rotates scapula so glenoid cavity moves inferiorly (downward rotation); elevates ribs if scapula is stationary
Rhomboid major	Spinous processes of superior thoracic vertebrae	Vertebral border of scapula from spine to inferior angle	Adducts scapula and performs downward rotation
Rhomboid minor	Spinous processes of vertebrae C_7–T_1	Vertebral border of scapula near spine	Adducts scapula and performs downward rotation
Serratus anterior	Anterior and superior margins of ribs 1–8 or 1–9	Anterior surface of vertebral border of scapula	Protracts shoulder; rotates scapula so glenoid cavity moves superiorly (upward rotation)
Subclavius	First rib	Clavicle (inferior border)	Depresses and protracts shoulder
Trapezius	Occipital bone, ligamentum nuchae, and spinous processes of thoracic vertebrae	Clavicle and scapula (acromion and scapular spine)	Depends on active region and state of other muscles; may (1) elevate, retract, depress, or rotate scapula upward, (2) elevate clavicle, or (3) extend neck

2 The broad trapezius is the largest muscle in this group. The rhomboid major and rhomboid minor muscles, and levator scapulae lie deep to the trapezius.

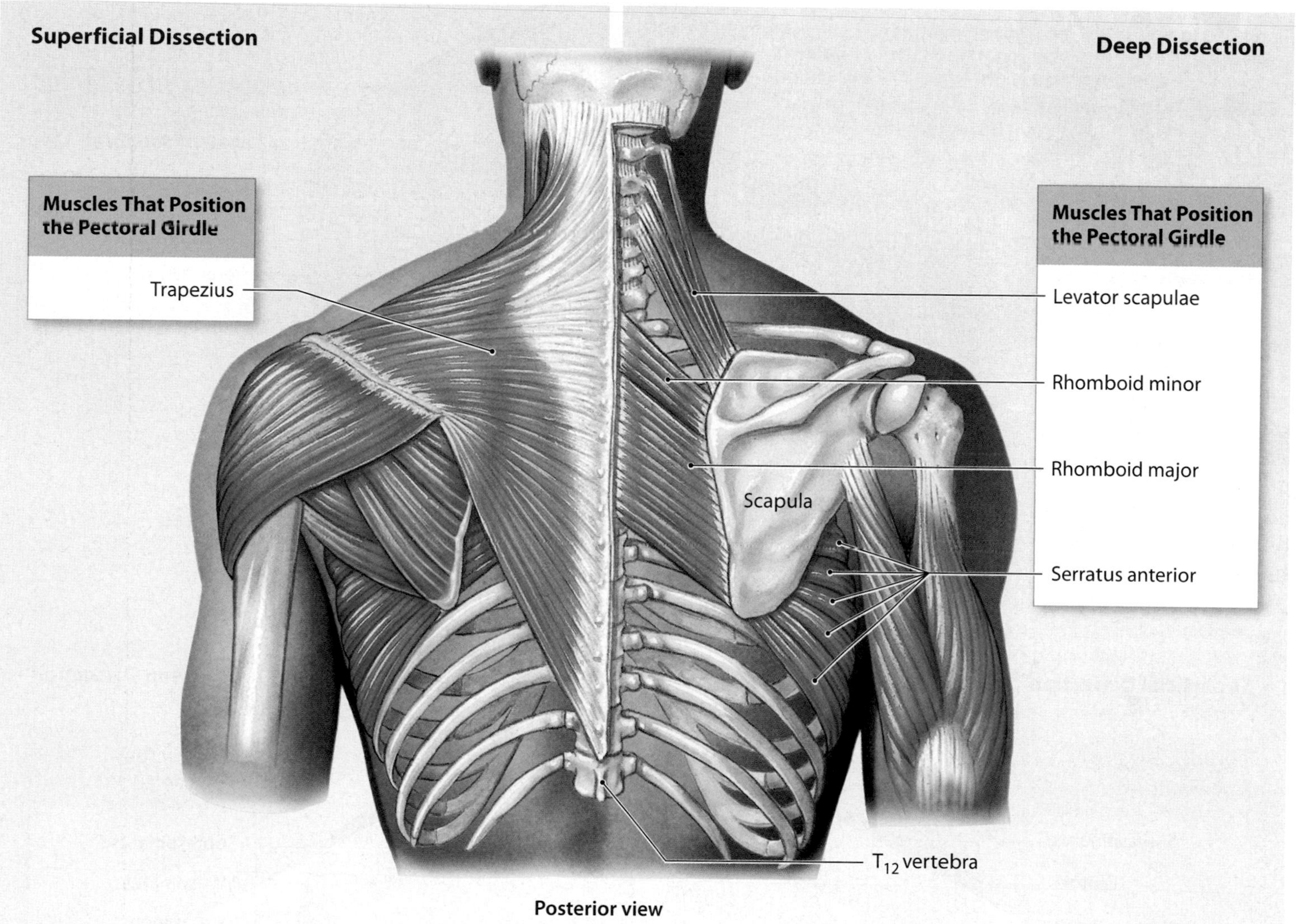

Superficial Dissection
Deep Dissection
Muscles That Position the Pectoral Girdle
Trapezius
Muscles That Position the Pectoral Girdle
Levator scapulae
Rhomboid minor
Rhomboid major
Scapula
Serratus anterior
T_{12} vertebra

Posterior view

The large, superficial trapezius muscles, commonly called the "traps," cover the back and portions of the neck, reaching to the base of the skull. These muscles are innervated by more than one nerve. For this reason, specific regions can be made to contract independently. As a result, their actions are quite varied.

Module 10.14 Review

a. Identify the largest of the superficial muscles that position the pectoral girdle.

b. Which muscles enable you to shrug your shoulders?

c. Which muscle originates on the first rib and inserts on the inferior border of the clavicle?

LO **10.14** Identify the pectoral girdle muscles, and cite their origins, insertions, and actions.

Muscles that move the arm originate on the clavicle, scapula, thoracic cage, and vertebral column

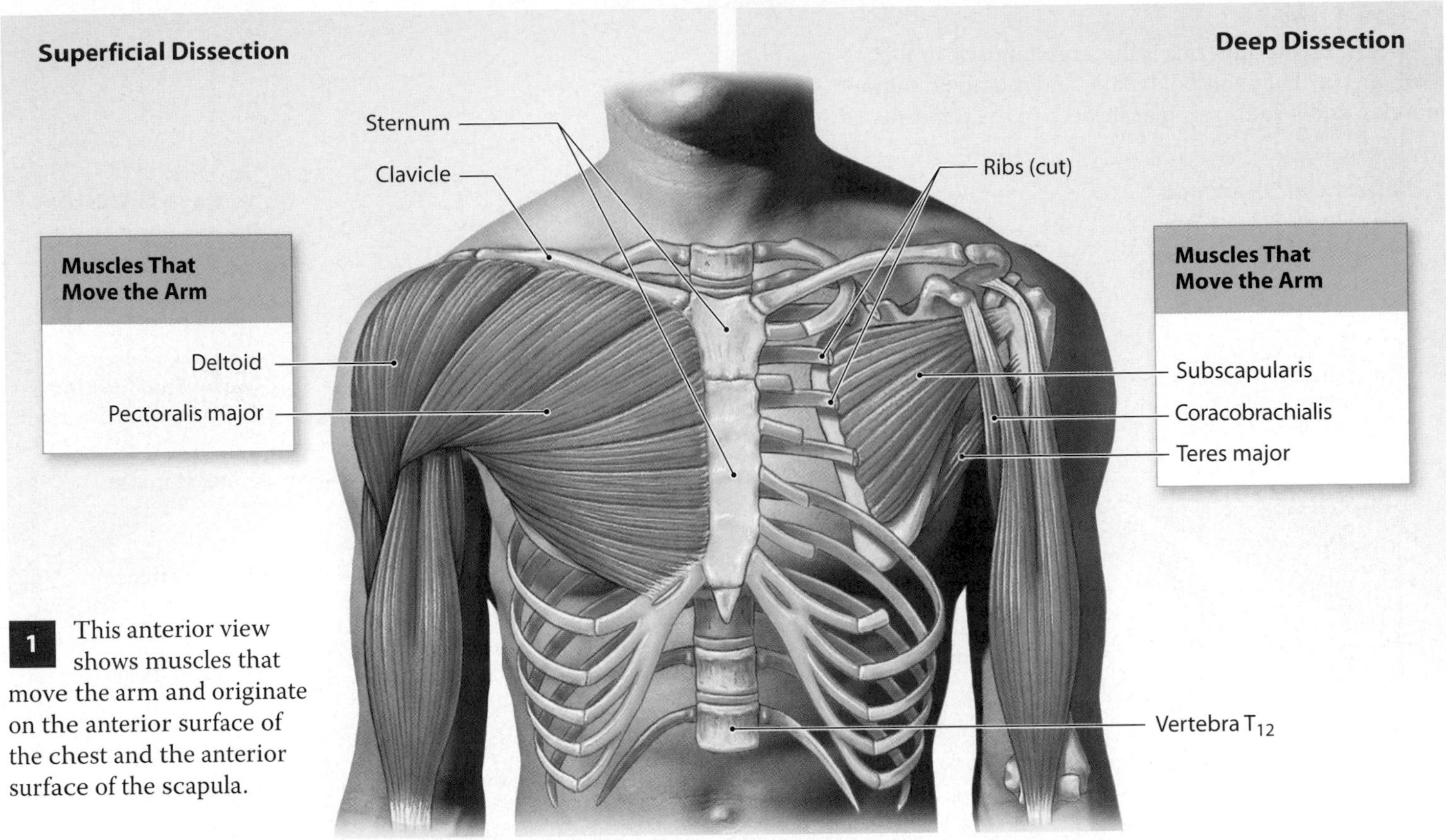
Superficial Dissection
Deep Dissection
Sternum
Clavicle
Ribs (cut)
Muscles That Move the Arm
Deltoid
Pectoralis major
Muscles That Move the Arm
Subscapularis
Coracobrachialis
Teres major
Vertebra T_{12}

1 This anterior view shows muscles that move the arm and originate on the anterior surface of the chest and the anterior surface of the scapula.

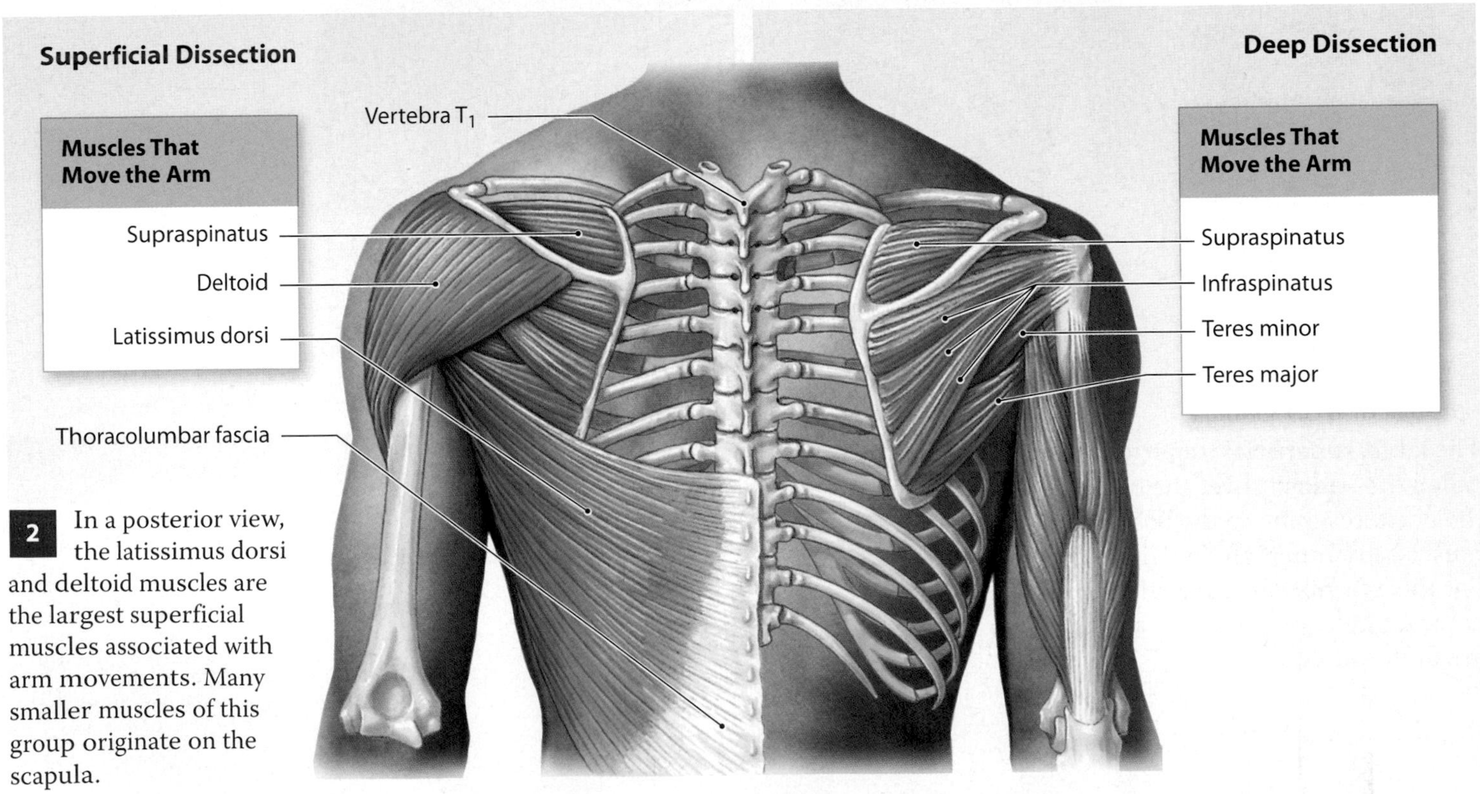
Superficial Dissection
Deep Dissection
Vertebra T_1
Muscles That Move the Arm
Supraspinatus
Deltoid
Latissimus dorsi
Thoracolumbar fascia
Muscles That Move the Arm
Supraspinatus
Infraspinatus
Teres minor
Teres major

2 In a posterior view, the latissimus dorsi and deltoid muscles are the largest superficial muscles associated with arm movements. Many smaller muscles of this group originate on the scapula.

Muscles That Move the Arm

Muscle	Origin	Insertion	Action
Deltoid	Clavicle and scapula (acromion and adjacent scapular spine)	Deltoid tuberosity of humerus	Whole muscle: abduction at shoulder; anterior part: flexion and medial rotation; posterior part: extension and lateral rotation
Supraspinatus	Supraspinous fossa of scapula	Greater tubercle of humerus	Abduction at shoulder
Subscapularis	Subscapular fossa of scapula	Lesser tubercle of humerus	Medial rotation at shoulder
Teres major	Inferior angle of scapula	Passes medially to reach the medial lip of intertubercular groove of humerus	Extension, adduction, and medial rotation at shoulder
Infraspinatus	Infraspinous fossa of scapula	Greater tubercle of humerus	Lateral rotation at shoulder
Teres minor	Lateral border of scapula	Passes laterally to reach the greater tubercle of humerus	Lateral rotation at shoulder
Coracobrachialis	Coracoid process	Medial margin of shaft of humerus	Adduction and flexion at shoulder
Pectoralis major	Cartilages of ribs 2–6, body of sternum, and inferior, medial portion of clavicle	Crest of greater tubercle and lateral lip of intertubercular groove of humerus	Flexion, adduction, and medial rotation at shoulder
Latissimus dorsi	Spinous processes of inferior thoracic and all lumbar vertebrae, ribs 8–12, and lumbodorsal fascia	Floor of intertubercular groove of the humerus	Extension, adduction, and medial rotation at shoulder

3 The action of muscles positioning the arm can be understood by considering their direction of pull relative to the center of the glenoid cavity. The arrows indicate the line of force, or **line of action**, produced when a muscle contracts.

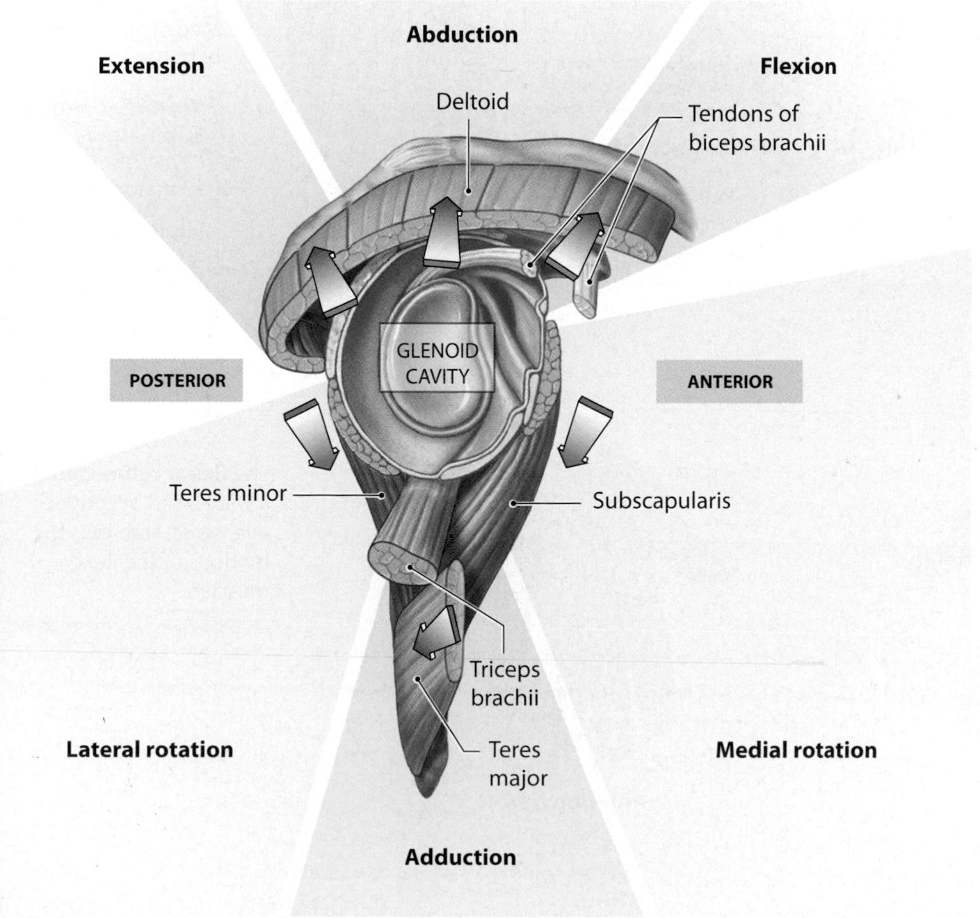

Abduction
Extension
Flexion
Deltoid
Tendons of biceps brachii
GLENOID CAVITY
POSTERIOR
ANTERIOR
Teres minor
Subscapularis
Triceps brachii
Teres major
Lateral rotation
Medial rotation
Adduction

Collectively, the supraspinatus, infraspinatus, teres minor, and subscapularis muscles and their associated tendons form the **rotator cuff**. The acronym SITS assists in remembering these four muscles. Sports that involve throwing a ball, such as baseball or football, place considerable strain on the rotator cuff, and rotator cuff injuries are relatively common.

Module 10.15 Review

a. Define line of action.

b. Name the muscle that abducts the upper arm.

c. Which muscle originates on the anterior surface of the scapula and inserts on the lesser tubercle of the humerus?

LO **10.15** Identify the muscles that move the arm, and cite their origins, insertions, and actions.

Muscles that move the forearm and hand originate on the scapula, humerus, radius, or ulna

1 This posterior view shows the superficial muscles involved in extension at the elbow and wrist.

2 This anterior view shows the superficial muscles involved in flexion at the elbow and wrist.

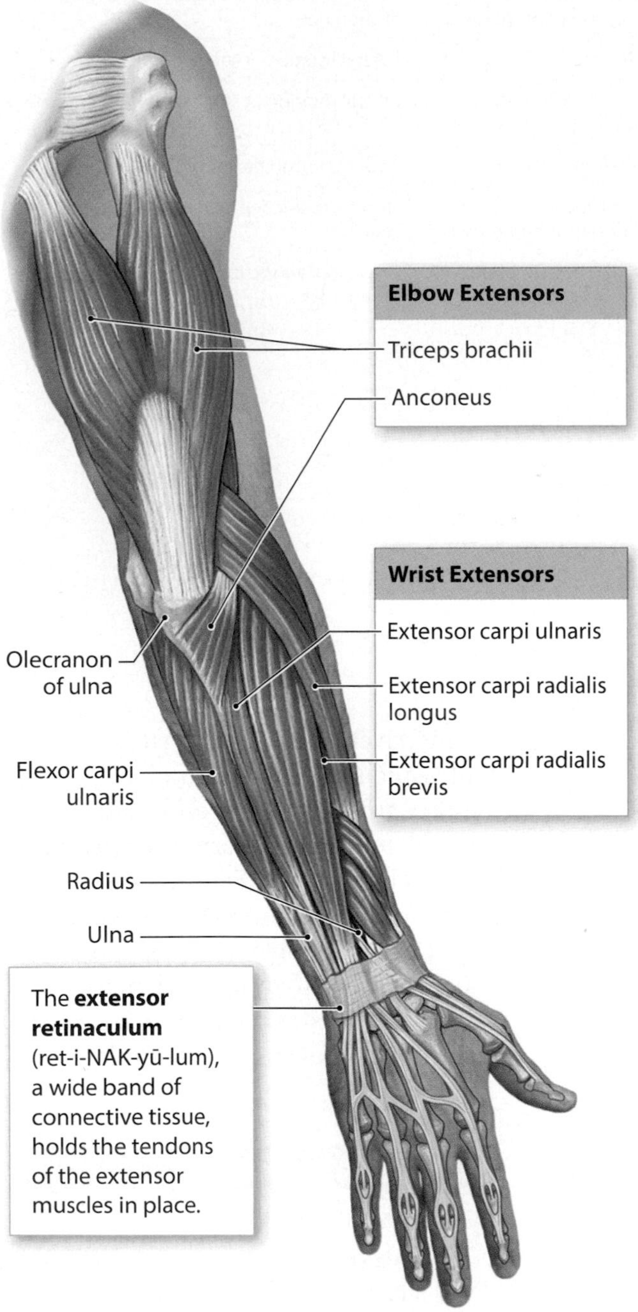
Elbow Extensors
Triceps brachii
Anconeus
Wrist Extensors
Extensor carpi ulnaris
Extensor carpi radialis longus
Extensor carpi radialis brevis
Olecranon of ulna
Flexor carpi ulnaris
Radius
Ulna
The **extensor retinaculum** (ret-i-NAK-yū-lum), a wide band of connective tissue, holds the tendons of the extensor muscles in place.

Posterior view

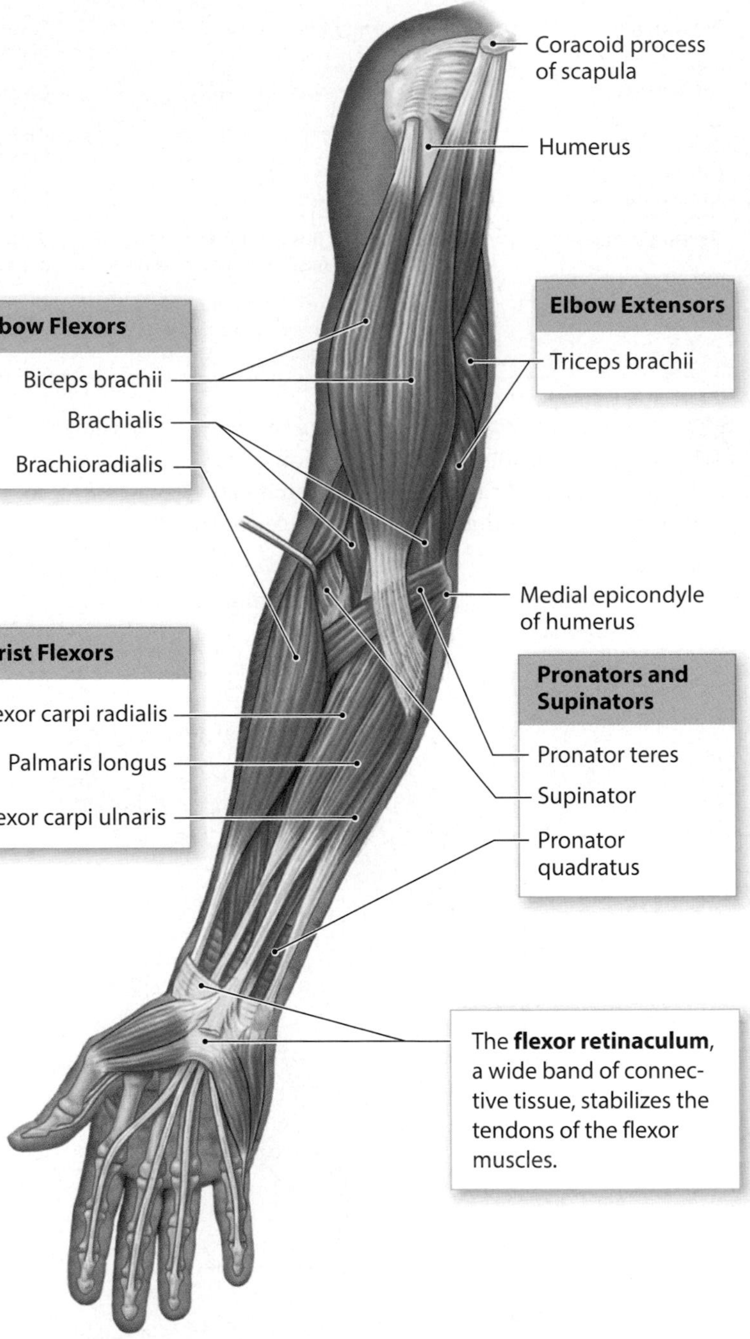
Coracoid process of scapula
Humerus
Elbow Flexors
Biceps brachii
Brachialis
Brachioradialis
Elbow Extensors
Triceps brachii
Medial epicondyle of humerus
Wrist Flexors
Flexor carpi radialis
Palmaris longus
Flexor carpi ulnaris
Pronators and Supinators
Pronator teres
Supinator
Pronator quadratus
The **flexor retinaculum**, a wide band of connective tissue, stabilizes the tendons of the flexor muscles.

Anterior view

Muscles That Move the Forearm and Hand

Group	Muscle	Origin	Insertion	Action
Action at the Elbow				
FLEXORS	**Biceps brachii**	One head from the coracoid process; the other from the supraglenoid tubercle (both on the scapula)	Tuberosity of radius	Flexion at elbow and shoulder; supination
FLEXORS	**Brachialis**	Anterior, distal surface of humerus	Tuberosity of ulna	Flexion at elbow
FLEXORS	**Brachioradialis**	Ridge superior to the lateral epicondyle of humerus	Lateral aspect of styloid process of radius	Flexion at elbow
EXTENSORS	**Anconeus**	Posterior, inferior surface of lateral epicondyle of humerus	Lateral margin of olecranon on ulna	Extension at elbow
EXTENSORS	**Triceps brachii**	One head from the superior, lateral margin of humerus, one from infraglenoid tubercle of scapula, and one from posterior surface of humerus inferior to radial groove	Olecranon of ulna	Extension at elbow, plus extension and adduction at the shoulder
PRONATORS / SUPINATORS	**Pronator quadratus**	Anterior and medial surfaces of distal portion of ulna	Anterolateral surface of distal portion of radius	Pronation
PRONATORS / SUPINATORS	**Pronator teres**	Medial epicondyle of humerus and coronoid process of ulna	Midlateral surface of radius	Pronation
PRONATORS / SUPINATORS	**Supinator**	Lateral epicondyle of humerus, annular ligament, and ridge near radial notch of ulna	Anterolateral surface of radius distal to the radial tuberosity	Supination
Action at the Hand				
FLEXORS	**Flexor carpi radialis**	Medial epicondyle of humerus	Bases of second and third metacarpal bones	Flexion and abduction at wrist
FLEXORS	**Flexor carpi ulnaris**	Medial epicondyle of humerus; adjacent medial surface of olecranon and anteromedial portion of ulna	Pisiform bone, hamate bone, and base of fifth metacarpal bone	Flexion and adduction at wrist
FLEXORS	**Palmaris longus**	Medial epicondyle of humerus	Palmar aponeurosis and flexor retinaculum	Flexion at wrist
EXTENSORS	**Extensor carpi radialis longus**	Lateral supracondylar ridge of humerus	Base of second metacarpal bone	Extension and abduction at wrist
EXTENSORS	**Extensor carpi radialis brevis**	Lateral epicondyle of humerus	Base of third metacarpal bone	Extension and abduction at wrist
EXTENSORS	**Extensor carpi ulnaris**	Lateral epicondyle of humerus; adjacent dorsal surface of ulna	Base of fifth metacarpal bone	Extension and adduction at wrist

3 **Synovial tendon sheaths** are tubular bursae that surround tendons where they cross bony surfaces. The tendons of the flexor muscles pass through such sheaths as they pass deep to the flexor retinaculum. Inflammation of the flexor retinaculum and synovial tendon sheaths can restrict movement and put pressure on the distal portions of the median nerve, a mixed (sensory and motor) nerve that innervates the hand. This condition, known as **carpal tunnel syndrome**, causes tingling, numbness, weakness, and chronic pain in the wrist and hand.

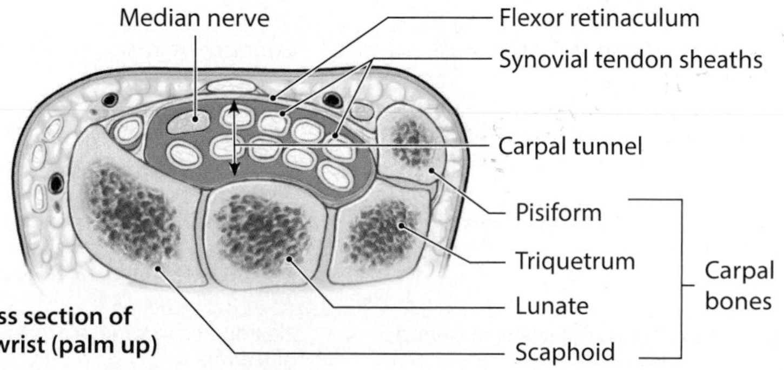

Median nerve
Flexor retinaculum
Synovial tendon sheaths
Carpal tunnel
Pisiform
Triquetrum
Lunate
Scaphoid
Carpal bones

Cross section of right wrist (palm up)

Module 10.16 Review

a. Define retinaculum.

b. Are the wrist extensors located on the anterior surface or the posterior surface of the forearm?

c. Which muscles are involved in turning a doorknob?

10.16 Identify the forearm muscles, and cite their origins, insertions, and actions.

Module 10.17

Muscles that move the hand and fingers originate on the humerus, radius, ulna, and interosseous membrane

1 In anterior view, the large flexor digitorum superficialis covers the smaller digital flexors.

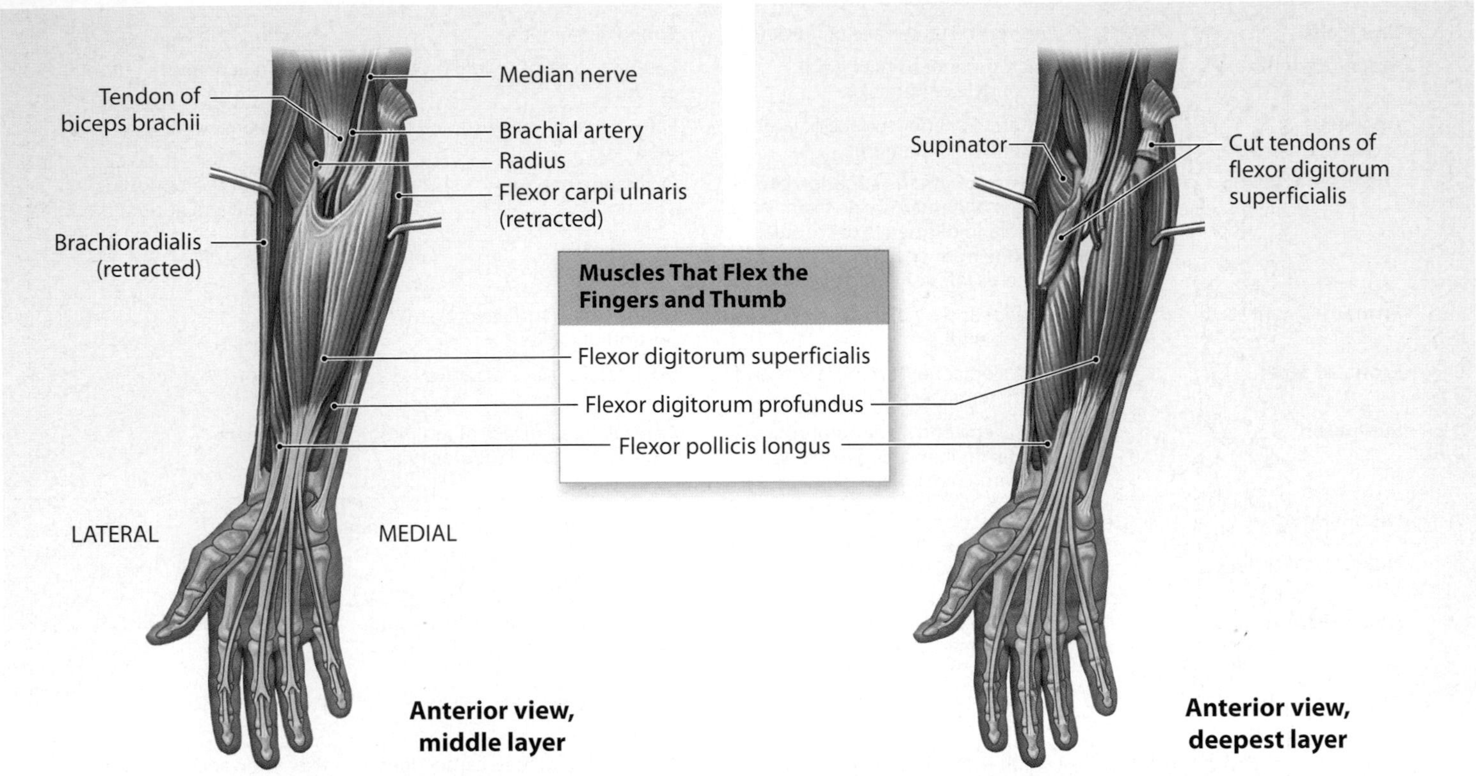

Tendon of biceps brachii
Median nerve
Brachial artery
Radius
Flexor carpi ulnaris (retracted)
Brachioradialis (retracted)
Supinator
Cut tendons of flexor digitorum superficialis
Muscles That Flex the Fingers and Thumb
Flexor digitorum superficialis
Flexor digitorum profundus
Flexor pollicis longus
LATERAL
MEDIAL
Anterior view, middle layer
Anterior view, deepest layer

Muscles That Move the Hand and Fingers

	Muscle	Origin	Insertion	Action
	Abductor pollicis longus	Proximal dorsal surfaces of ulna and radius	Lateral margin of first metacarpal bone	Abduction at carpometacarpal joint of thumb and wrist
EXTENSORS	**Extensor digitorum**	Lateral epicondyle of humerus	Posterior surfaces of the phalanges, fingers 2–5	Extension at finger joints and wrist
	Extensor pollicis brevis	Shaft of radius distal to origin of abductor pollicis longus	Base of proximal phalanx of thumb	Extension at carpometacarpal joint of thumb; abduction at wrist
	Extensor pollicis longus	Posterior and lateral surfaces of ulna and interosseous membrane	Base of distal phalanx of thumb	Extension at carpometacarpal joint of thumb; abduction at wrist
	Extensor indicis	Posterior surface of ulna and interosseous membrane	Posterior surface of phalanges of index finger (2), with tendon of extensor digitorum	Extension and adduction at joints of index finger
	Extensor digiti minimi	By extensor tendon to lateral epicondyle of humerus and from intermuscular septa	Posterior surface of proximal phalanx of little finger (5)	Extension at joints of little finger
FLEXORS	**Flexor digitorum superficialis**	Medial epicondyle of humerus; adjacent anterior surfaces of ulna and radius	Midlateral surfaces of middle phalanges of fingers 2–5	Flexion at proximal interphalangeal, metacarpophalangeal, and wrist joints
	Flexor digitorum profundus	Medial and posterior surfaces of ulna, medial surface of coronoid process, and interosseus membrane	Bases of distal phalanges of fingers 2–5	Flexion at distal interphalangeal joints and, to a lesser degree, proximal interphalangeal joints and wrist
	Flexor pollicis longus	Anterior shaft of radius, interosseous membrane	Base of distal phalanx of thumb	Flexion at carpometacarpal joint of thumb

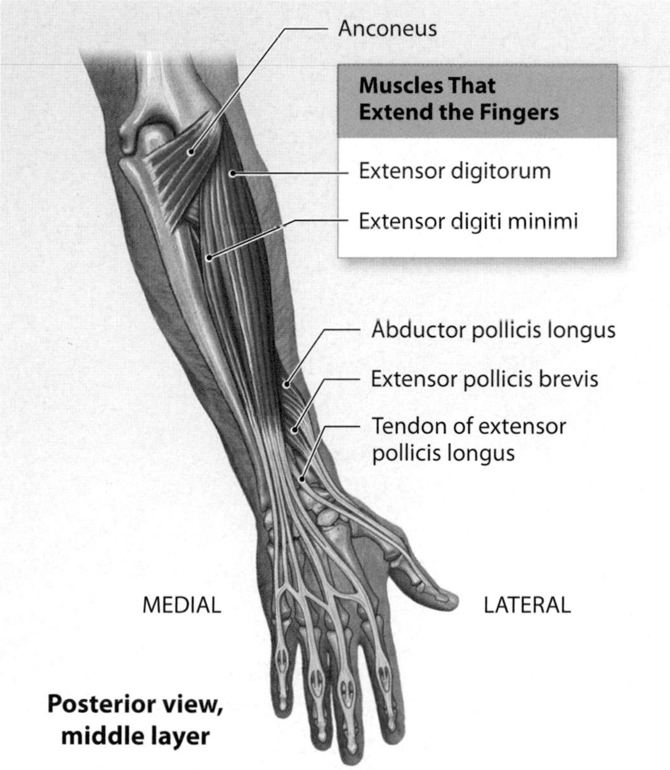
Anconeus
Muscles That Extend the Fingers
Extensor digitorum
Extensor digiti minimi
Abductor pollicis longus
Extensor pollicis brevis
Tendon of extensor pollicis longus
MEDIAL
LATERAL

Posterior view, middle layer

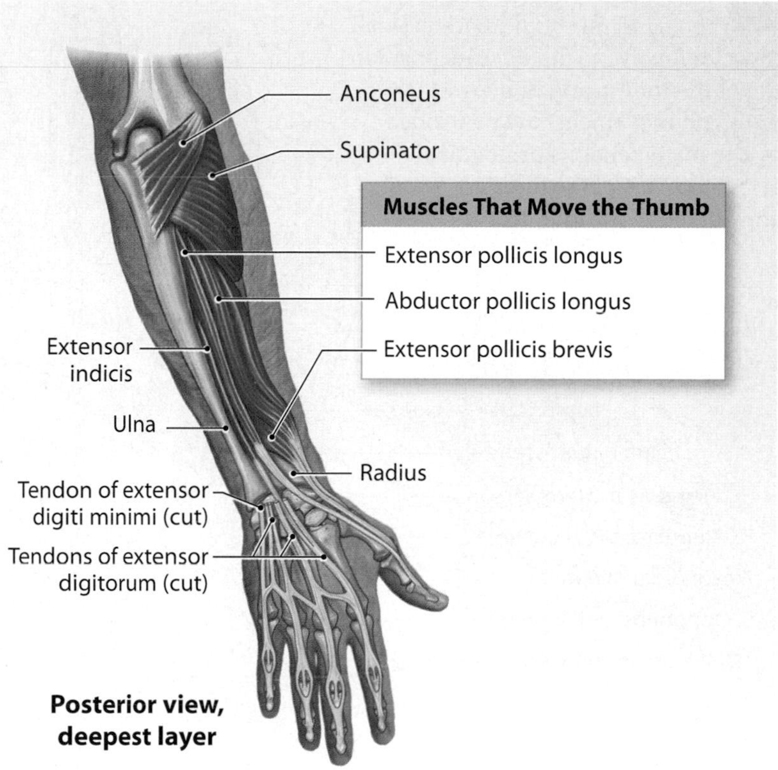
Anconeus
Supinator
Muscles That Move the Thumb
Extensor pollicis longus
Abductor pollicis longus
Extensor pollicis brevis
Extensor indicis
Ulna
Radius
Tendon of extensor digiti minimi (cut)
Tendons of extensor digitorum (cut)

Posterior view, deepest layer

2 In posterior view, the muscles that extend the fingers can only be seen after removal of the muscles involved in wrist movements. The deepest digital extensor muscles are those associated with movements of the thumb.

As you study these muscles you will notice that extensor muscles usually lie along the posterior and lateral surfaces of the forearm, whereas flexors are typically found on the anterior and medial surfaces. Remember, the limb must be in the anatomical position for this to be true. This information can be quite useful when you are trying to identify a particular muscle on a quiz or a lab practical.

Module 10.17 Review

a. List the muscles that extend the fingers.

b. Name the muscles that abduct the wrist.

c. The names of muscles associated with the thumb frequently include what term?

LO **10.17** Identify the muscles of the hand and fingers, and cite their origins, insertions, and actions.

The intrinsic muscles of the hand originate on the carpal and metacarpal bones and associated tendons and ligaments

1 Dissection of the palmar surface of the hand reveals numerous small muscles involved with delicate positioning of the thumb and fingers. More powerful movements are controlled by the many tendons originating at the muscles of the forearm considered in Module 10.17.

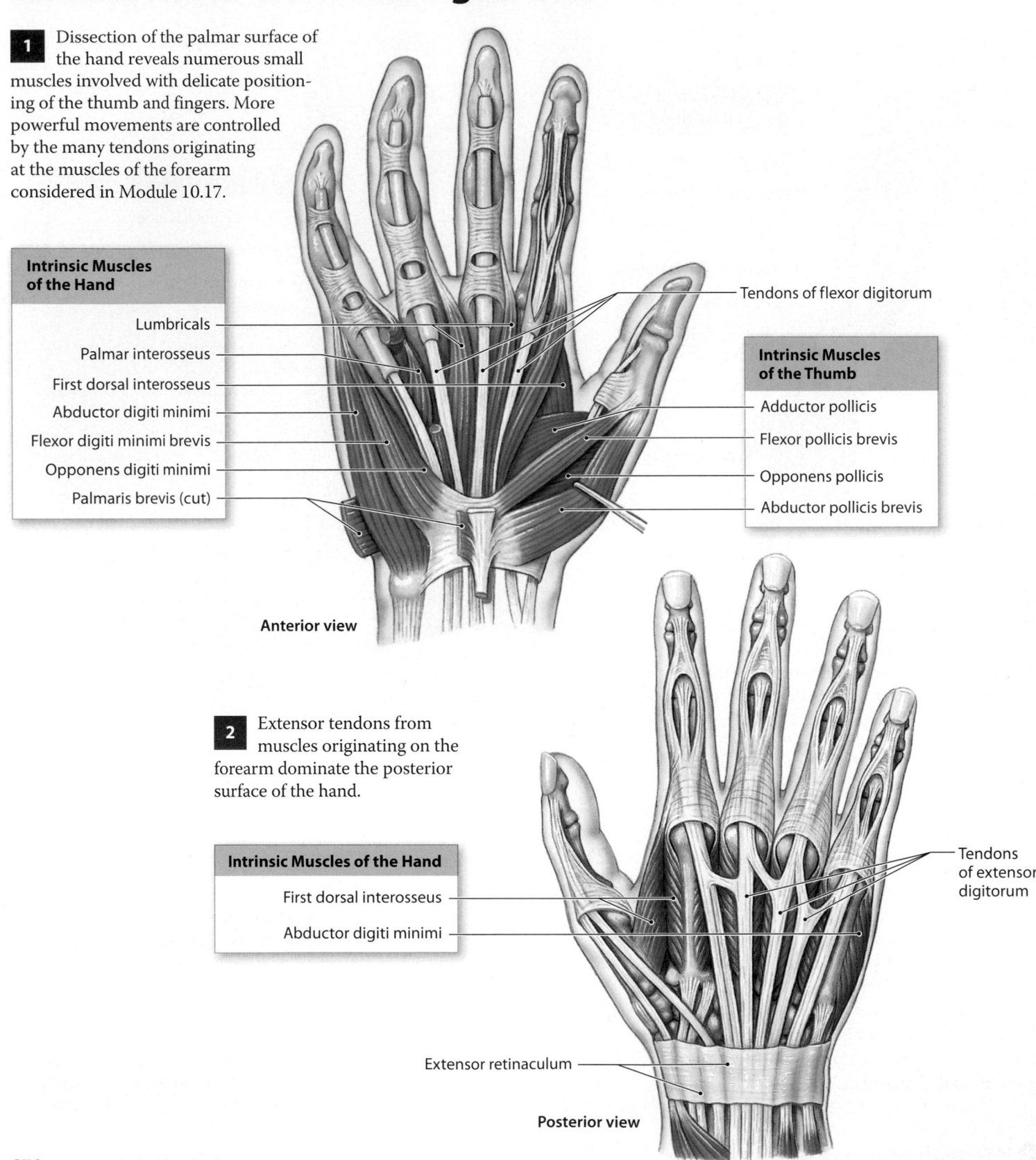

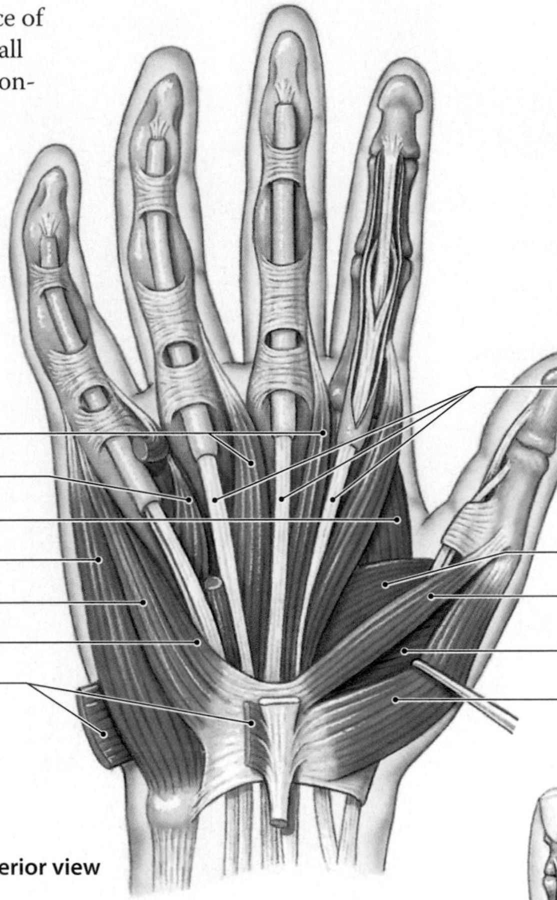
Intrinsic Muscles of the Hand
Lumbricals
Palmar interosseus
First dorsal interosseus
Abductor digiti minimi
Flexor digiti minimi brevis
Opponens digiti minimi
Palmaris brevis (cut)
Tendons of flexor digitorum
Intrinsic Muscles of the Thumb
Adductor pollicis
Flexor pollicis brevis
Opponens pollicis
Abductor pollicis brevis

Anterior view

2 Extensor tendons from muscles originating on the forearm dominate the posterior surface of the hand.

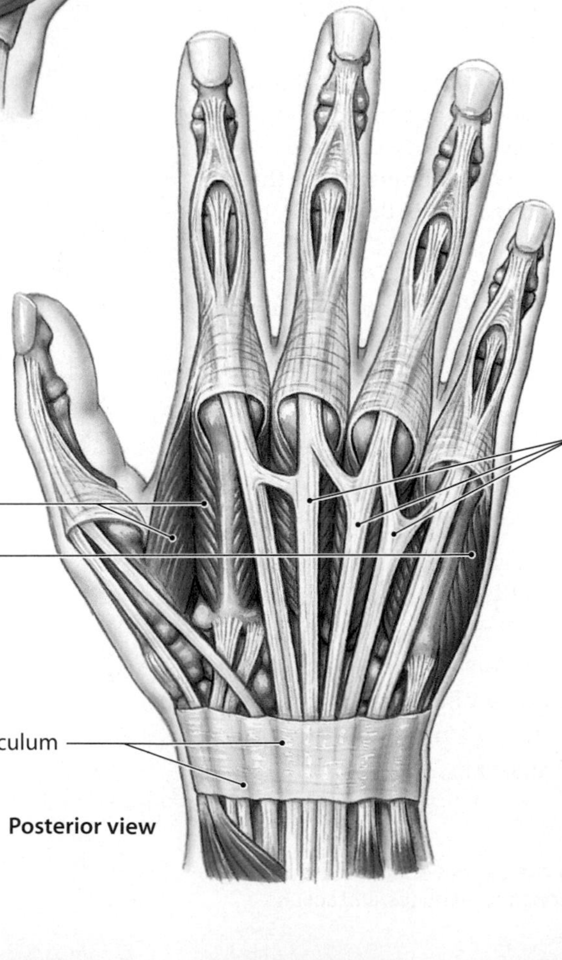
Intrinsic Muscles of the Hand
First dorsal interosseus
Abductor digiti minimi
Tendons of extensor digitorum
Extensor retinaculum

Posterior view

Intrinsic Muscles of the Hand

Group	Muscle	Origin	Insertion	Action
	Palmaris brevis	Palmar aponeurosis	Skin of medial border of hand	Moves skin on medial border toward midline of palm
ADDUCTION / ABDUCTION	**Adductor pollicis**	Metacarpal and carpal bones	Proximal phalanx of thumb	Adduction of thumb
ADDUCTION / ABDUCTION	**Palmar interosseus (3–4)**	Sides of metacarpal bones II, IV, and V	Bases of proximal phalanges of fingers 2, 4, and 5	Adduction at metacarpophalangeal joints of fingers 2, 4, and 5; flexion at metacarpophalangeal joints; extension at interphalangeal joints
ADDUCTION / ABDUCTION	**Abductor pollicis brevis**	Transverse carpal ligament, scaphoid bone, and trapezium	Radial side of base of proximal phalanx of thumb	Abduction of thumb
ADDUCTION / ABDUCTION	**Dorsal interosseus (4)**	Each originates from opposing faces of two metacarpal bones (I and II, II and III, III and IV, IV and V)	Bases of proximal phalanges of fingers 2–4	Abduction at metacarpophalangeal joints of fingers 2 and 4; flexion at metacarpophalangeal joints; extension at interphalangeal joints
ADDUCTION / ABDUCTION	**Abductor digiti minimi**	Pisiform bone	Proximal phalanx of little finger	Abduction of little finger and flexion at its metacarpophalangeal joint
FLEXION	**Flexor pollicis brevis**	Flexor retinaculum, trapezium, capitate bone, and ulnar side of first metacarpal bone	Radial and ulnar sides of proximal phalanx of thumb	Flexion and adduction of thumb
FLEXION	**Lumbrical (4)**	Tendons of flexor digitorum profundus	Tendons of extensor digitorum to digits 2–5	Flexion at metacarpophalangeal joints 2–5; extension at proximal and distal interphalangeal joints, digits 2–5
FLEXION	**Flexor digiti minimi brevis**	Hamate bone	Proximal phalanx of little finger	Flexion at joints of little finger
	Opponens pollicis	Trapezium and flexor retinaculum	First metacarpal bone	Opposition of thumb
	Opponens digiti minimi	Trapezium and flexor retinaculum	Fifth metacarpal bone	Opposition of fifth metacarpal bone

Fine control of the hand involves small intrinsic muscles that originate on the carpal and metacarpal bones. These intrinsic muscles are responsible for (1) flexion and extension of the fingers at the metacarpophalangeal joints, (2) abduction and adduction of the fingers at the metacarpophalangeal joints, and (3) opposition and reposition (relaxed position) of the thumb. No muscles originate on the phalanges, and only tendons extend across the distal joints of the fingers.

Module 10.18 Review

a. Name the intrinsic muscles of the thumb.

b. Which muscles originate on the phalanges?

c. If there are no muscles in the fingers, how are we able to move them?

LO **10.18** Identify the intrinsic hand muscles, and cite their origins, insertions, and actions.

The muscles that move the thigh originate on the pelvis and associated ligaments and fasciae

1 The **gluteal group** covers the posterior and lateral surfaces of the pelvis.

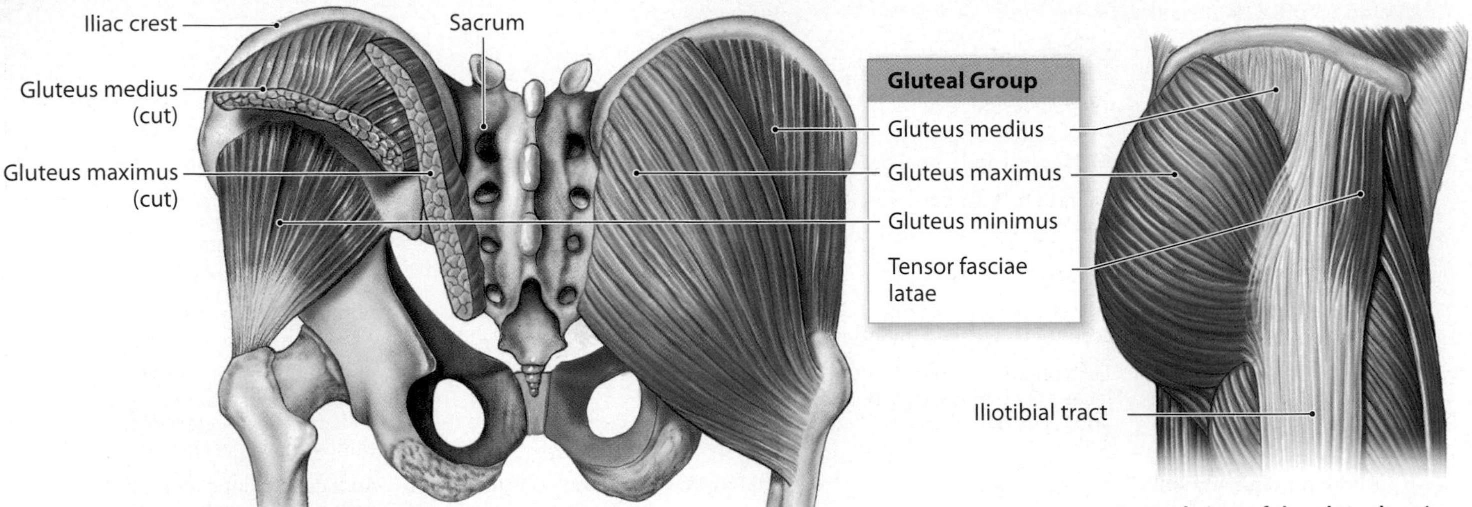
Iliac crest
Sacrum
Gluteus medius (cut)
Gluteus maximus (cut)
Gluteal Group
Gluteus medius
Gluteus maximus
Gluteus minimus
Tensor fasciae latae
Iliotibial tract

Gluteal muscles, posterior view

Lateral view of the gluteal region

2 This dissection of the gluteal region shows the orientation of the six lateral rotators.

3 This anterior view shows the isolated iliopsoas muscle group and the adductor group.

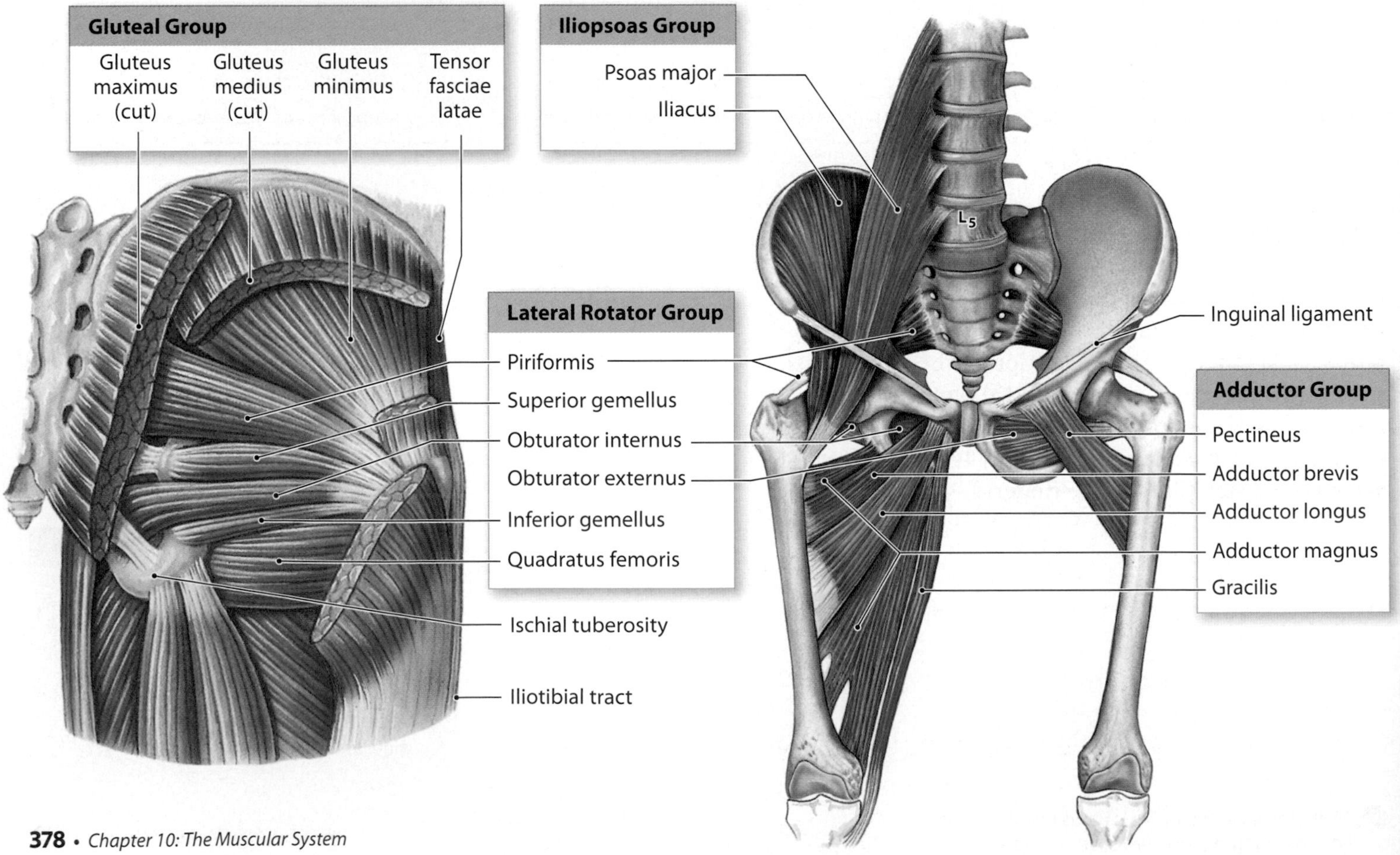
Gluteal Group
Gluteus maximus (cut)
Gluteus medius (cut)
Gluteus minimus
Tensor fasciae latae
Iliopsoas Group
Psoas major
Iliacus
L5
Lateral Rotator Group
Piriformis
Superior gemellus
Obturator internus
Obturator externus
Inferior gemellus
Quadratus femoris
Ischial tuberosity
Iliotibial tract
Inguinal ligament
Adductor Group
Pectineus
Adductor brevis
Adductor longus
Adductor magnus
Gracilis

Muscles That Move the Thigh

Group and Muscle	Origin	Insertion	Action
Gluteal Group			
Gluteus maximus	Iliac crest, posterior gluteal line, and lateral surface of ilium; sacrum, coccyx, and lumbodorsal fascia	Iliotibial tract and gluteal tuberosity of femur	Extension and lateral rotation at hip
Gluteus medius	Anterior iliac crest of ilium, lateral surface between posterior and anterior gluteal lines	Greater trochanter of femur	Abduction and medial rotation at hip
Gluteus minimus	Lateral surface of ilium between inferior and anterior gluteal lines	Greater trochanter of femur	Abduction and medial rotation at hip
Tensor fasciae latae	Iliac crest and lateral surface of anterior superior iliac spine	Iliotibial tract	Flexion and medial rotation at hip; tenses fascia lata, which laterally supports the knee
Lateral Rotator Group			
Obturator (externus and internus)	Lateral and medial margins of obturator foramen	Trochanteric fossa of femur (externus); medial surface of greater trochanter (internus)	Lateral rotation at hip
Piriformis	Anterolateral surface of sacrum	Greater trochanter of femur	Lateral rotation and abduction at hip
Gemellus (superior and inferior)	Ischial spine and tuberosity	Medial surface of greater trochanter with tendon of obturator internus	Lateral rotation at hip
Quadratus femoris	Lateral border of ischial tuberosity	Intertrochanteric crest of femur	Lateral rotation at hip
Adductor Group			
Adductor brevis	Inferior ramus of pubis	Linea aspera of femur	Adduction, flexion, and medial rotation at hip
Adductor longus	Inferior ramus of pubis anterior to adductor brevis	Linea aspera of femur	Adduction, flexion, and medial rotation at hip
Adductor magnus	Inferior ramus of pubis posterior to adductor brevis and ischial tuberosity	Linea aspera and adductor tubercle of femur	Adduction at hip; superior part produces flexion and medial rotation; inferior part produces extension and lateral rotation
Pectineus	Superior ramus of pubis	Pectineal line inferior to lesser trochanter of femur	Flexion, medial rotation, and adduction at hip
Gracilis	Inferior ramus of pubis	Medial surface of tibia inferior to medial condyle	Flexion at knee; adduction and medial rotation at hip
Iliopsoas Group*			
Iliacus	Iliac fossa of ilium	Femur distal to lesser trochanter; tendon fused with that of psoas major	Flexion at hip
Psoas major	Anterior surfaces and transverse processes of vertebrae (T_{12}–L_5)	Lesser trochanter in company with iliacus	Flexion at hip or lumbar intervertebral joints

* The psoas major and iliacus are often considered collectively as the iliopsoas

One method for understanding the actions of these diverse muscles is to consider their orientation around the hip joint. Muscles originating on the surface of the pelvis and inserting on the femur will produce characteristic movements determined by their position relative to the acetabulum. Many of the muscles that act on the hip are very large, and they have insertions that extend over a broad area. As a result, these muscles often have more than one action line, and therefore produce more than one action at the hip. For example, the action of the adductor magnus varies depending on what portion of the muscle is activated; when the entire muscle contracts, it produces a combination of flexion, extension, and adduction at the hip.

Module 10.19 Review

a. Name the muscles that compose the gluteal group.

b. Identify the muscle whose origin is the lateral border of the ischial tuberosity and whose insertion is the intertrochanteric crest of the femur.

c. Which leg movement would be impaired by injury to the obturator muscles?

LO 10.19 Identify the muscles that move the thigh, and cite their origins, insertions, and actions.

The muscles that move the leg originate on the pelvis and femur

1 **Flexors of the knee** originate on the pelvic girdle and extend along the posterior and medial surfaces of the thigh.

2 Most of the **extensors of the knee** originate on the femoral surface and extend along the anterior and lateral surfaces of the thigh. Collectively the knee extensors are called the **quadriceps muscles** or the **quadriceps femoris**.

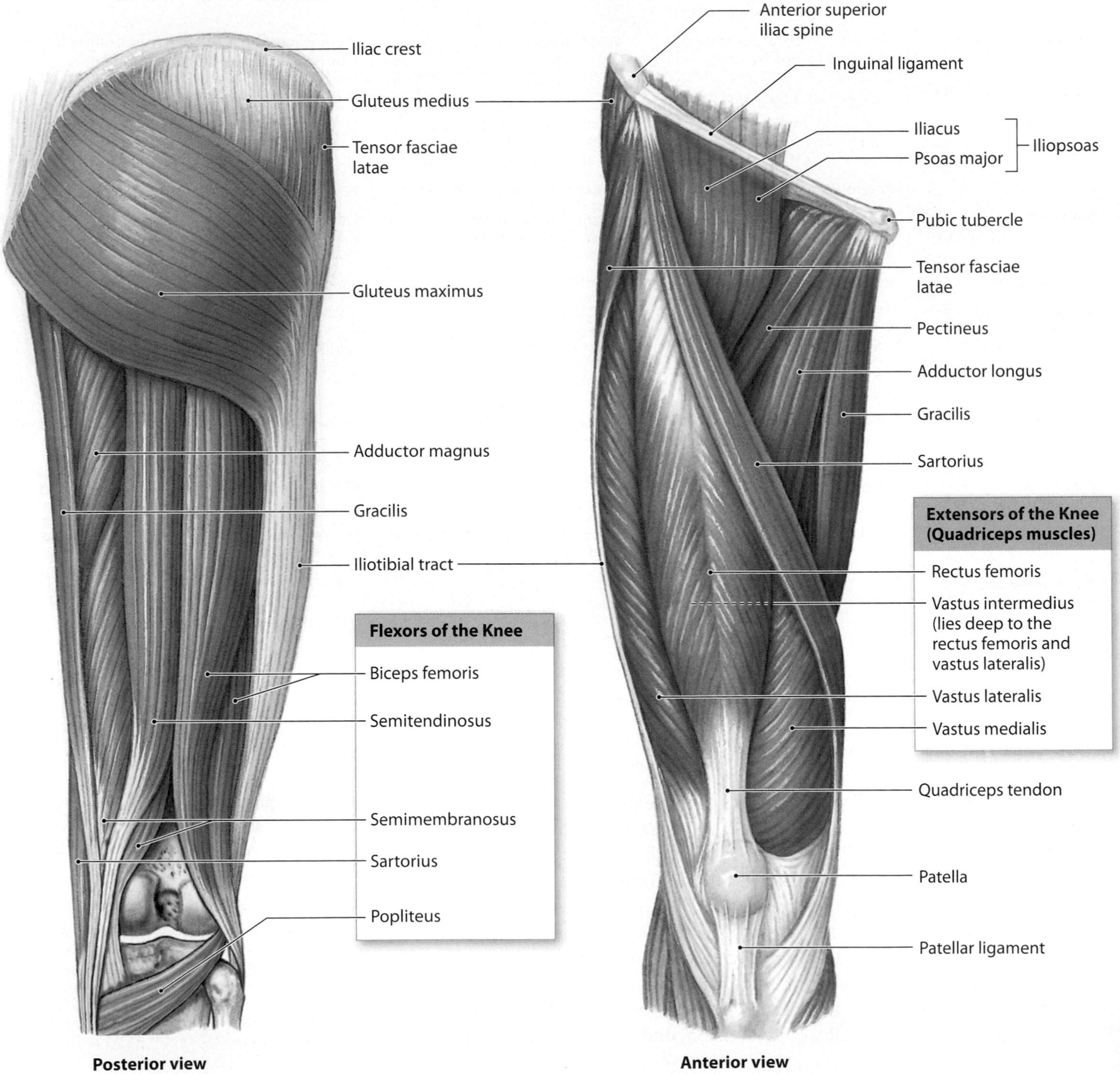

Iliac crest
Gluteus medius
Tensor fasciae latae
Gluteus maximus
Adductor magnus
Gracilis
Iliotibial tract
Flexors of the Knee
Biceps femoris
Semitendinosus
Semimembranosus
Sartorius
Popliteus
Anterior superior iliac spine
Inguinal ligament
Iliacus
Psoas major
Iliopsoas
Pubic tubercle
Tensor fasciae latae
Pectineus
Adductor longus
Gracilis
Sartorius
Extensors of the Knee (Quadriceps muscles)
Rectus femoris
Vastus intermedius (lies deep to the rectus femoris and vastus lateralis)
Vastus lateralis
Vastus medialis
Quadriceps tendon
Patella
Patellar ligament

Posterior view

Anterior view

Muscles That Move the Leg

Group and Muscle	Origin	Insertion	Action
Flexors of the Knee			
Biceps femoris	Ischial tuberosity and linea aspera of femur	Head of fibula, lateral condyle of tibia	Flexion at knee; extension and lateral rotation at hip
Semimembranosus	Ischial tuberosity	Posterior surface of medial condyle of tibia	Flexion at knee; extension and medial rotation at hip
Semitendinosus	Ischial tuberosity	Proximal, medial surface of tibia near insertion of gracilis	Flexion at knee; extension and medial rotation at hip
Sartorius	Anterior superior iliac spine	Medial surface of tibia near tibial tuberosity	Flexion at knee; flexion and lateral rotation at hip
Popliteus	Lateral condyle of femur	Posterior surface of proximal tibial shaft	Medial rotation of tibia (or lateral rotation of femur); flexion at knee
Extensors of the Knee			
Rectus femoris	Anterior inferior iliac spine and superior acetabular rim of ilium	Tibial tuberosity via patellar ligament	Extension at knee; flexion at hip
Vastus intermedius	Anterolateral surface of femur and linea aspera (distal half)	Tibial tuberosity via patellar ligament	Extension at knee
Vastus lateralis	Anterior and inferior to greater trochanter of femur and along linea aspera (proximal half)	Tibial tuberosity via patellar ligament	Extension at knee
Vastus medialis	Entire length of linea aspera of femur	Tibial tuberosity via patellar ligament	Extension at knee

3 This cross-sectional view shows the positions of the major thigh muscles relative to the femur. Together, the vastus muscles cradle the rectus femoris muscle the way a bun surrounds a hot dog. All four muscles insert on the patella via the quadriceps tendon.

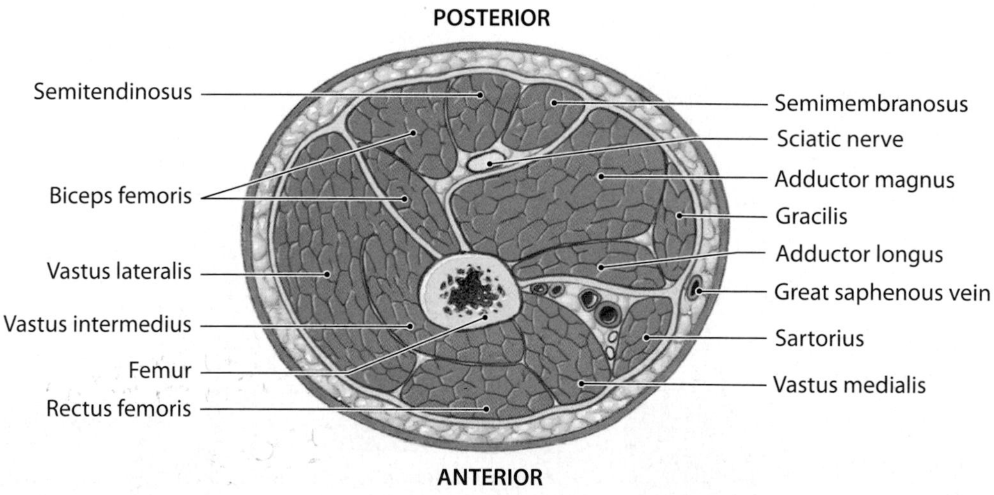
POSTERIOR
Semitendinosus
Biceps femoris
Vastus lateralis
Vastus intermedius
Femur
Rectus femoris
Semimembranosus
Sciatic nerve
Adductor magnus
Gracilis
Adductor longus
Great saphenous vein
Sartorius
Vastus medialis
ANTERIOR

Module 10.20 Review

a. Which muscles flex the knee?

b. Name the quadriceps muscles.

c. Identify the muscle whose origin is on the lateral condyle of the femur.

LO **10.20** Identify the muscles that move the leg, and cite their origins, insertions, and actions.

The extrinsic muscles that move the foot and toes originate on the tibia and fibula

1 These views show the multiple muscle layers in the posterior aspect of the leg.

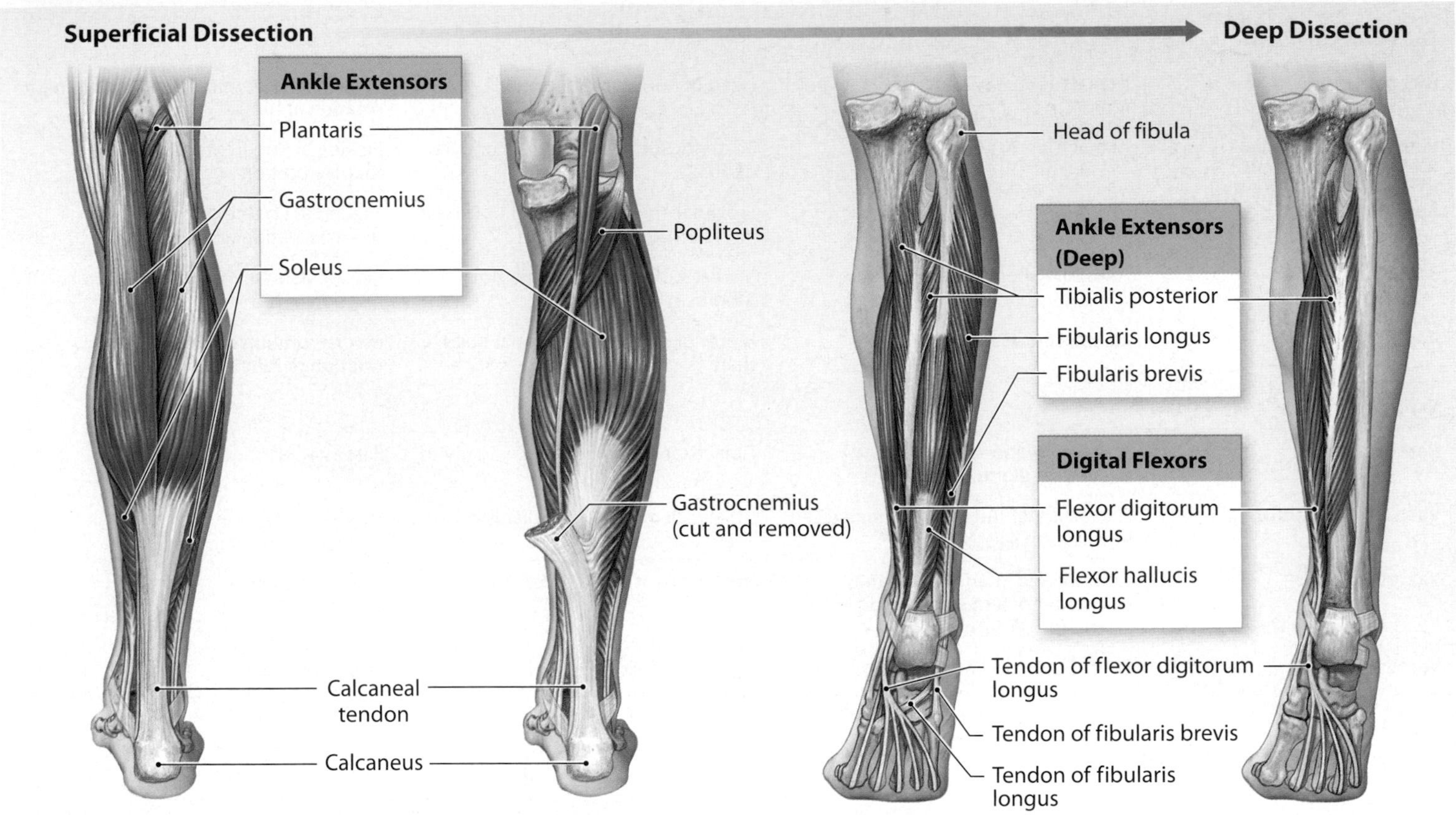
Superficial Dissection
Deep Dissection
Ankle Extensors
Plantaris
Gastrocnemius
Soleus
Popliteus
Gastrocnemius (cut and removed)
Calcaneal tendon
Calcaneus
Head of fibula
Ankle Extensors (Deep)
Tibialis posterior
Fibularis longus
Fibularis brevis
Digital Flexors
Flexor digitorum longus
Flexor hallucis longus
Tendon of flexor digitorum longus
Tendon of fibularis brevis
Tendon of fibularis longus

2 These lateral, medial, and anterior views show the arrangement of the major superficial muscles.

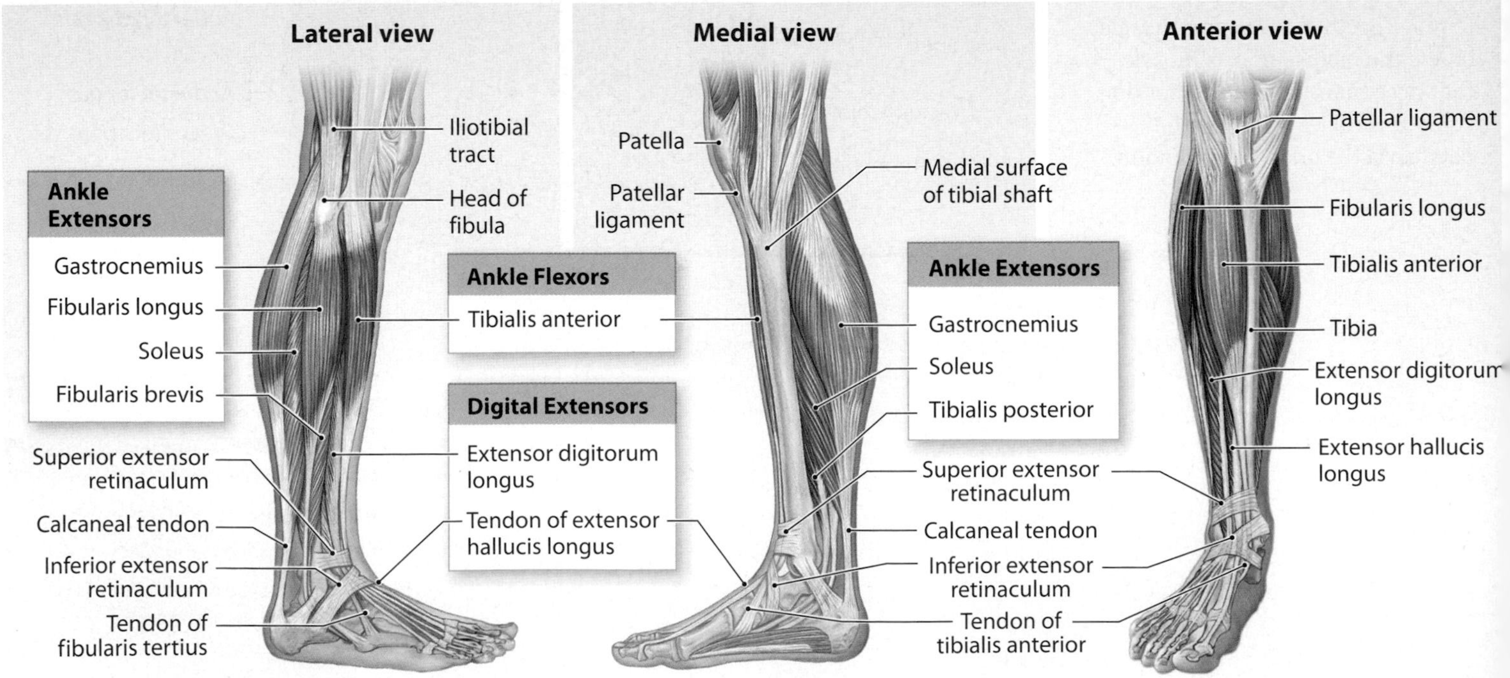
Lateral view
Medial view
Anterior view
Ankle Extensors
Gastrocnemius
Fibularis longus
Soleus
Fibularis brevis
Superior extensor retinaculum
Calcaneal tendon
Inferior extensor retinaculum
Tendon of fibularis tertius
Iliotibial tract
Head of fibula
Ankle Flexors
Tibialis anterior
Digital Extensors
Extensor digitorum longus
Tendon of extensor hallucis longus
Patella
Patellar ligament
Medial surface of tibial shaft
Ankle Extensors
Gastrocnemius
Soleus
Tibialis posterior
Superior extensor retinaculum
Calcaneal tendon
Inferior extensor retinaculum
Tendon of tibialis anterior
Patellar ligament
Fibularis longus
Tibialis anterior
Tibia
Extensor digitorum longus
Extensor hallucis longus

Extrinsic Muscles That Move the Foot and Toes

Group	Muscle	Origin	Insertion	Action
Action at the Ankle				
FLEXORS (Dorsiflexors)	**Tibialis anterior**	Lateral condyle and proximal shaft of tibia	Base of first metatarsal bone and medial cuneiform bone	Flexion (dorsiflexion) at ankle; inversion of foot
	Fibularis tertius	Distal anterior surface of fibula and interosseous membrane	Dorsal surface of fifth metatarsal bone	Flexion (dorsiflexion); eversion of foot
EXTENSORS (Plantar flexors)	**Gastrocnemius**	Femoral condyles	Calcaneus by calcaneal tendon	Extension (plantar flexion) at ankle; inversion of foot; flexion at knee
	Fibularis brevis	Midlateral margin of fibula	Base of fifth metatarsal bone	Eversion of foot and extension (plantar flexion) at ankle
	Fibularis longus	Lateral condyle of tibia, head and proximal shaft of fibula	Base of first metatarsal bone and medial cuneiform bone	Eversion of foot and extension (plantar flexion) at ankle; supports longitudinal arch
	Plantaris	Lateral supracondylar ridge	Posterior portion of calcaneus	Extension (plantar flexion) at ankle; flexion at knee
	Soleus	Head and proximal shaft of fibula and adjacent posteromedial shaft of tibia	Calcaneus by calcaneal tendon (with gastrocnemius)	Extension (plantar flexion) at ankle
	Tibialis posterior	Interosseous membrane and adjacent shafts of tibia and fibula	Tarsal and metatarsal bones	Adduction and inversion of foot; extension (plantar flexion) at ankle
Action at the Toes				
DIGITAL FLEXORS	**Flexor digitorum longus**	Posteromedial surface of tibia	Inferior surfaces of distal phalanges, toes 2–5	Flexion at joints of toes 2–5
	Flexor hallucis longus	Posterior surface of fibula	Inferior surface, distal phalanx of great toe	Flexion at joints of great toe
DIGITAL EXTENSORS	**Extensor digitorum longus**	Lateral condyle of tibia, anterior surface of fibula	Superior surfaces of phalanges, toes 2–5	Extension at joints of toes 2–5
	Extensor hallucis longus	Anterior surface of fibula	Superior surface, distal phalanx of great toe	Extension at joints of great toe

The largest muscles associated with ankle movement are the **gastrocnemius** and **soleus**. These muscles produce ankle extension (plantar flexion), a movement essential to walking and running (Module 8.4, p. 286). The muscles that move the toes are much smaller, and they originate on the surface of the tibia, fibula, or both. Large tendon sheaths surround the tendons of the tibialis anterior, extensor digitorum longus, and extensor hallucis longus muscles where they cross the ankle joint. The positions of these sheaths are stabilized by the **superior extensor retinaculum** and **inferior extensor retinaculum**, tough supporting bands of collagen fibers.

Module 10.21 Review

a. Name the muscles involved in extending the ankle.

b. How would a torn calcaneal tendon affect movement of the foot?

c. Name the muscles involved in flexing the toes.

LO **10.21** Identify the muscles that move the foot and toes, and cite their origins, insertions, and actions.

The intrinsic muscles of the foot originate on the tarsal and metatarsal bones and associated tendons and ligaments

1 This superior view introduces some of the **intrinsic muscles of the foot**. It also reveals the importance of the retinacula in stabilizing the positions of the tendons descending from the leg.

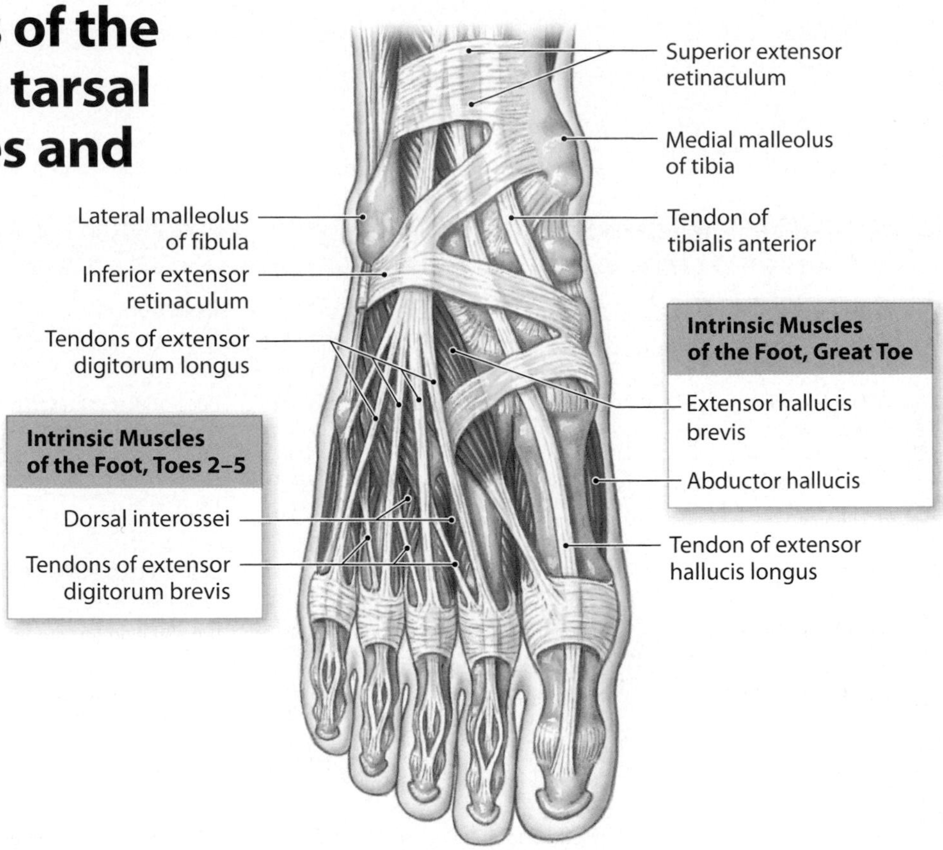

Superior extensor retinaculum
Medial malleolus of tibia
Tendon of tibialis anterior
Lateral malleolus of fibula
Inferior extensor retinaculum
Tendons of extensor digitorum longus
Intrinsic Muscles of the Foot, Great Toe
Extensor hallucis brevis
Abductor hallucis
Intrinsic Muscles of the Foot, Toes 2–5
Dorsal interossei
Tendons of extensor digitorum brevis
Tendon of extensor hallucis longus

2 Intrinsic muscles are more numerous on the inferior surface of the foot and occur in several layers.

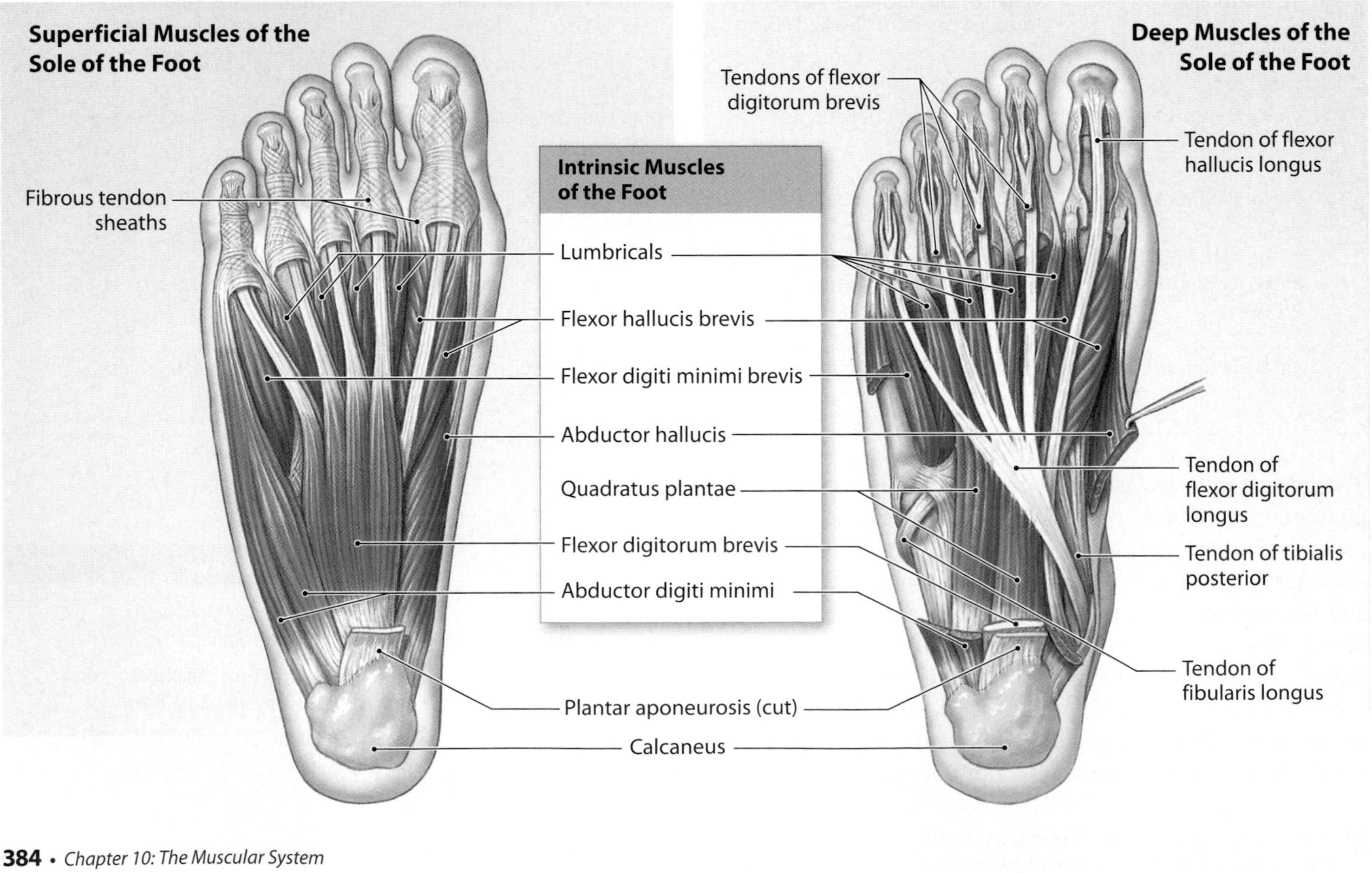

Superficial Muscles of the Sole of the Foot
Deep Muscles of the Sole of the Foot
Fibrous tendon sheaths
Tendons of flexor digitorum brevis
Tendon of flexor hallucis longus
Intrinsic Muscles of the Foot
Lumbricals
Flexor hallucis brevis
Flexor digiti minimi brevis
Abductor hallucis
Quadratus plantae
Flexor digitorum brevis
Abductor digiti minimi
Tendon of flexor digitorum longus
Tendon of tibialis posterior
Tendon of fibularis longus
Plantar aponeurosis (cut)
Calcaneus

Intrinsic Muscles of the Foot

	Muscle	Origin	Insertion	Action
FLEXION / EXTENSION	**Flexor hallucis brevis**	Cuboid and lateral cuneiform bones	Proximal phalanx of great toe	Flexion at metatarsophalangeal joint of great toe
	Flexor digitorum brevis	Inferior surface of calcaneus	Sides of middle phalanges, toes 2–5	Flexion at proximal interphalangeal joints of toes 2–5
	Quadratus plantae	Calcaneus (medial, inferior surfaces)	Tendon of flexor digitorum longus	Flexion at joints of toes 2–5
	Lumbrical (4)	Tendons of flexor digitorum longus	Tendons of extensor digitorum longus, toes 2 to 5	Flexion at metatarsophalangeal joints; extension at proximal interphalangeal joints of toes 2–5
	Flexor digiti minimi brevis	Base of metatarsal bone V	Lateral side of proximal phalanx of toe 5	Flexion at metatarsophalangeal joint of toe 5
	Extensor digitorum brevis	Calcaneus (superior and lateral surfaces)	Dorsal surfaces of toes 1–4	Extension at metatarsophalangeal joints of toes 1–4
	Extensor hallucis brevis	Superior surface of anterior calcaneus	Dorsal surface of the base of proximal phalanx of great toe	Extension of great toe
ADDUCTION / ABDUCTION	**Adductor hallucis**	Bases of metatarsal bones II–IV and plantar ligaments	Proximal phalanx of great toe	Adduction at metatarsophalangeal joint of great toe
	Abductor hallucis	Calcaneus (tuberosity on inferior surface)	Medial side of proximal phalanx of great toe	Abduction at metatarsophalangeal joint of great toe
	Plantar interosseus (3)	Bases and medial sides of metatarsal bones	Medial sides of toes 3–5	Adduction at metatarsophalangeal joints of toes 3–5
	Dorsal interosseus (4)	Sides of metatarsal bones	Medial and lateral sides of toe 2; lateral sides of toes 3 and 4	Abduction at metatarsophalangeal joints of toes 3 and 4
	Abductor digiti minimi	Inferior surface of calcaneus	Lateral side of proximal phalanx, toe 5	Abduction at metatarsophalangeal joint of toe 5

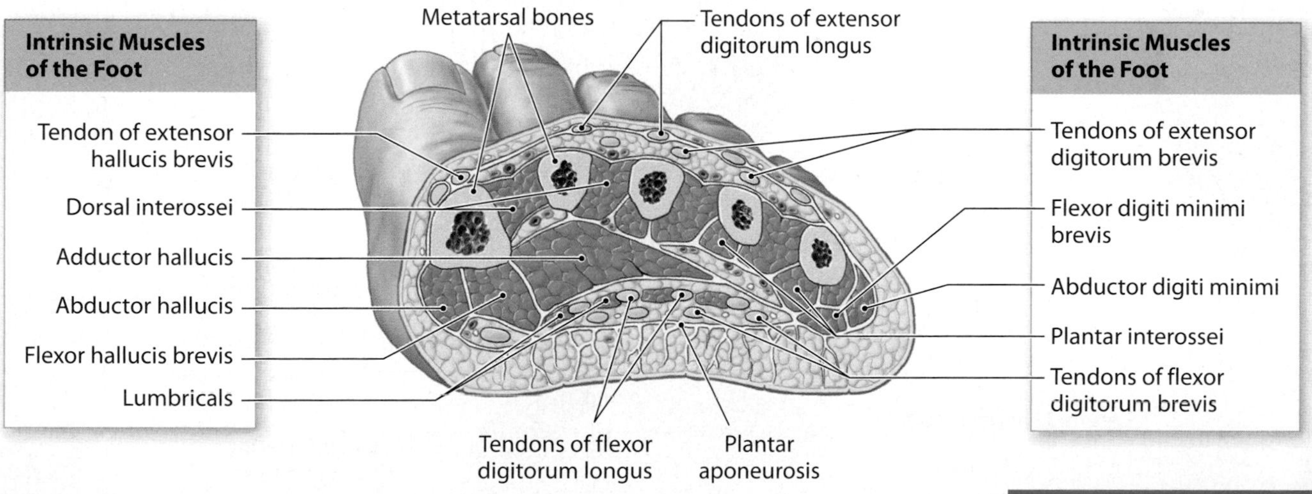

Intrinsic Muscles of the Foot
Tendon of extensor hallucis brevis
Dorsal interossei
Adductor hallucis
Abductor hallucis
Flexor hallucis brevis
Lumbricals
Metatarsal bones
Tendons of extensor digitorum longus
Tendons of flexor digitorum longus
Plantar aponeurosis
Intrinsic Muscles of the Foot
Tendons of extensor digitorum brevis
Flexor digiti minimi brevis
Abductor digiti minimi
Plantar interossei
Tendons of flexor digitorum brevis

3 As you see in this cross section, most of the muscle mass in the foot lies inferior to the metatarsal bones. Many of these muscles are flexors that tense during ankle extension and help you "push off" when walking. This anatomical arrangement provides padding and assists in maintaining the arches of the foot.

Module 10.22 Review

a. Identify the intrinsic muscle that flexes the great toe.

b. What are the functions of the superior and inferior retinacula of the foot?

c. Describe the origin, insertion, and action of the lumbrical muscles.

LO **10.22** Identify the intrinsic foot muscles, and cite their origins, insertions, and actions.

The deep fascia separates the limb muscles into separate compartments

Fibrous partitions of the deep fascia create partitioned-off sections called **compartments**. The muscles within each compartment have compatible functions, and each has a characteristic blood supply and innervation. Each compartment is generally not in communication with other compartments, and for this reason, infection or excess pressure usually remains confined there.

1 Here are sectional views at intervals along the length of an upper and lower limb. In each relevant section the compartments are labeled and representative muscles and important reference structures are indicated.

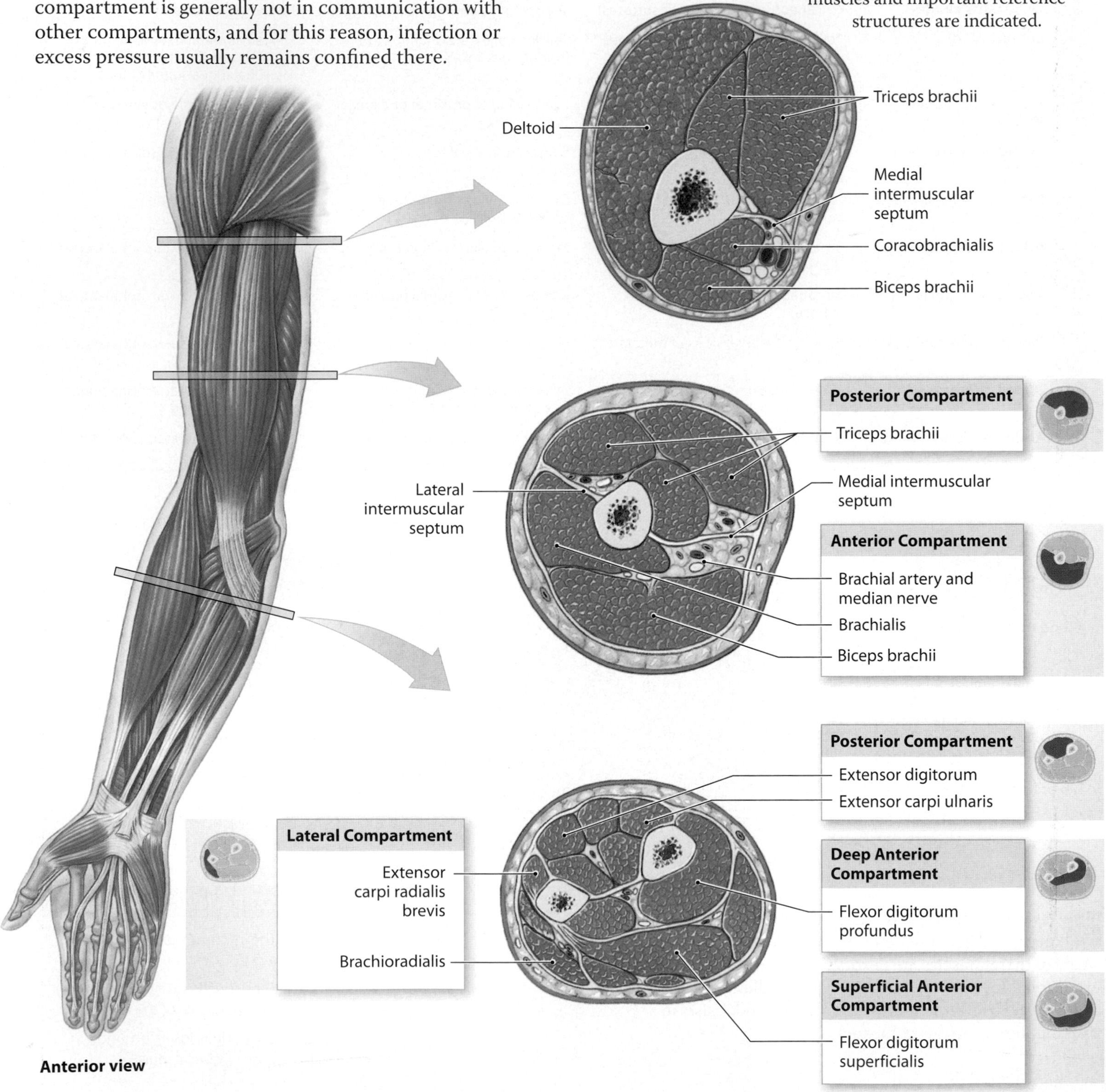

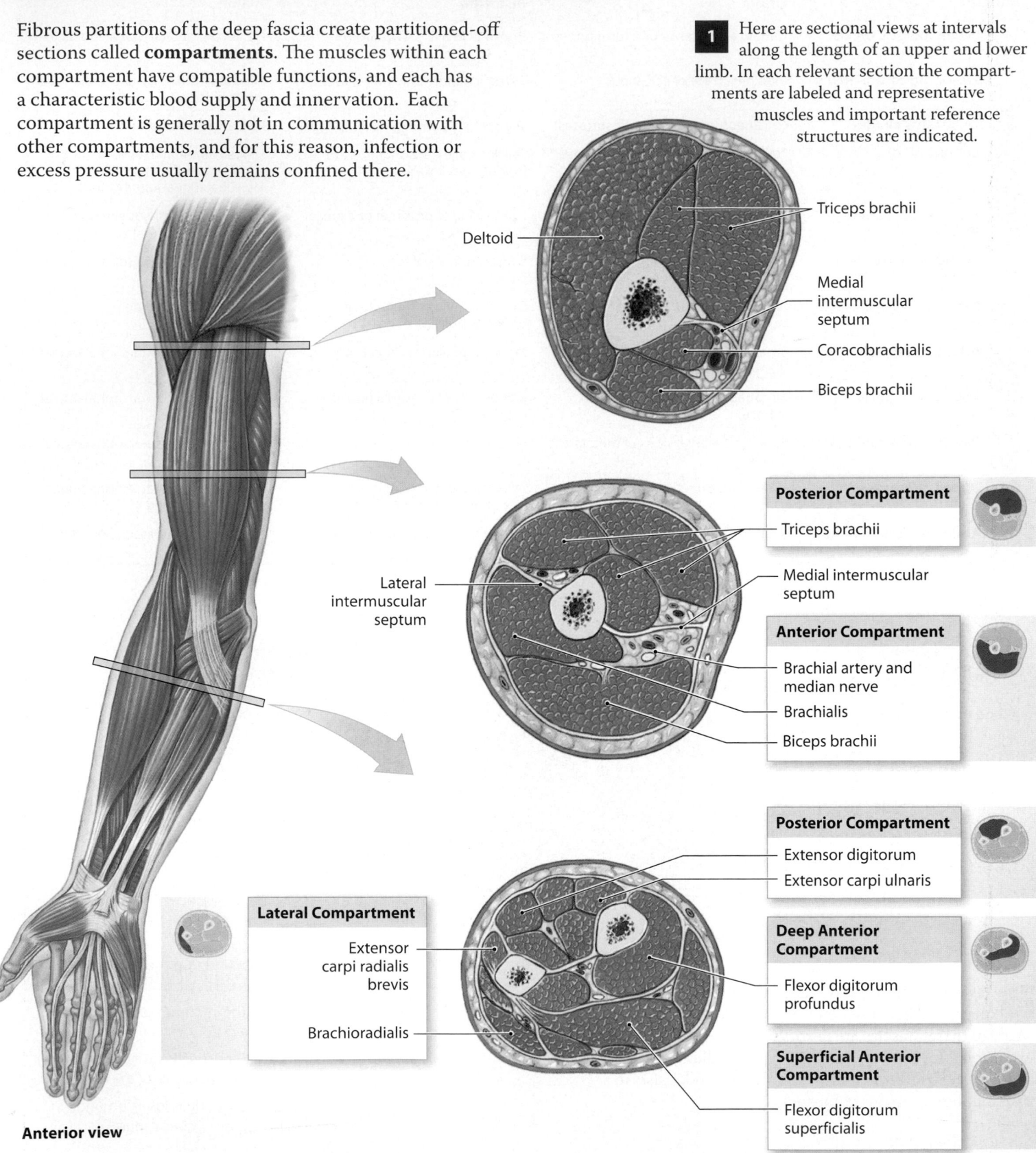

Deltoid
Triceps brachii
Medial intermuscular septum
Coracobrachialis
Biceps brachii
Posterior Compartment
Triceps brachii
Lateral intermuscular septum
Medial intermuscular septum
Anterior Compartment
Brachial artery and median nerve
Brachialis
Biceps brachii
Posterior Compartment
Extensor digitorum
Extensor carpi ulnaris
Lateral Compartment
Extensor carpi radialis brevis
Brachioradialis
Deep Anterior Compartment
Flexor digitorum profundus
Superficial Anterior Compartment
Flexor digitorum superficialis
Anterior view

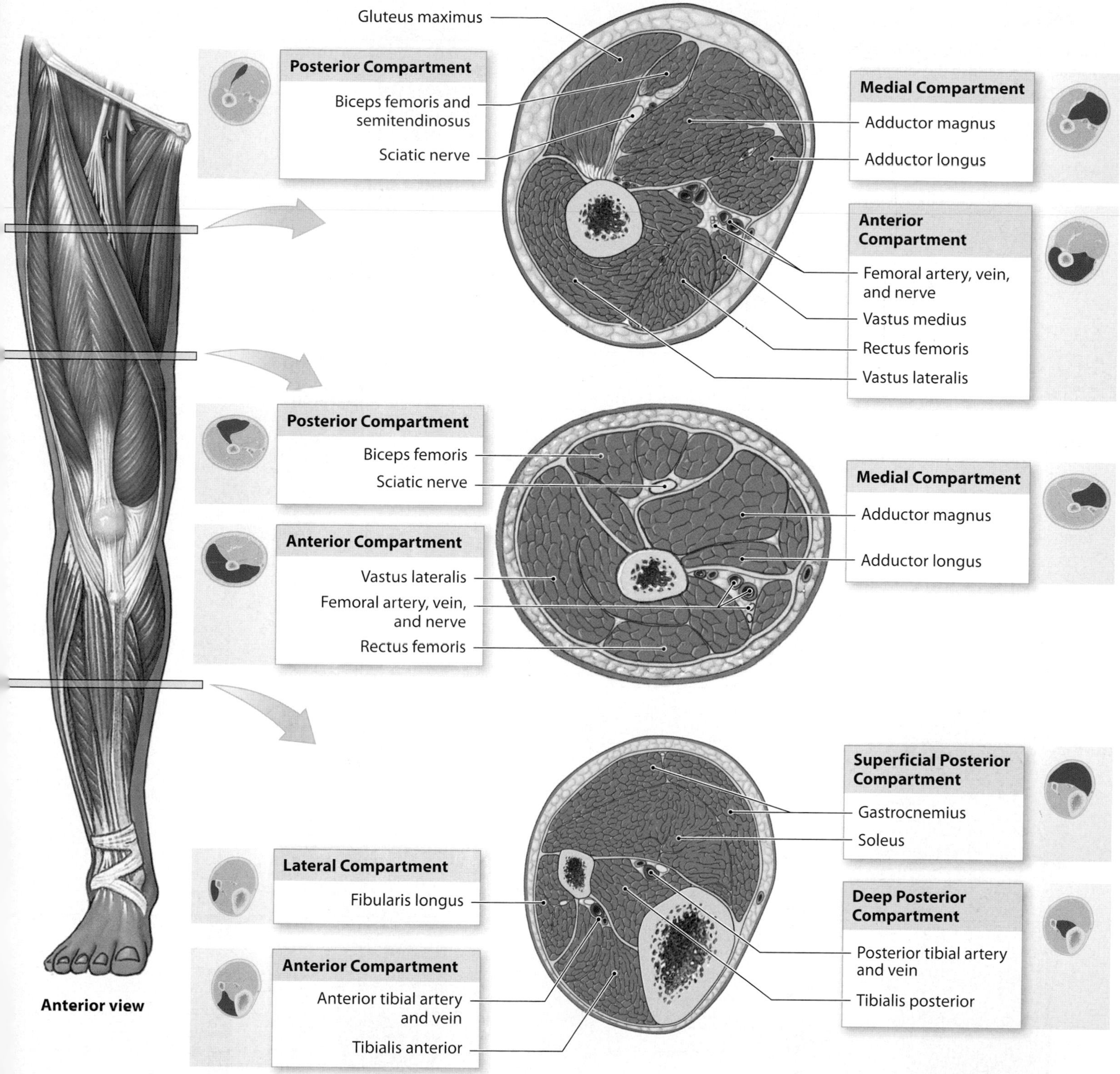

Gluteus maximus
Posterior Compartment
Biceps femoris and semitendinosus
Sciatic nerve
Medial Compartment
Adductor magnus
Adductor longus
Anterior Compartment
Femoral artery, vein, and nerve
Vastus medius
Rectus femoris
Vastus lateralis
Posterior Compartment
Biceps femoris
Sciatic nerve
Anterior Compartment
Vastus lateralis
Femoral artery, vein, and nerve
Rectus femoris
Medial Compartment
Adductor magnus
Adductor longus
Superficial Posterior Compartment
Gastrocnemius
Soleus
Lateral Compartment
Fibularis longus
Deep Posterior Compartment
Posterior tibial artery and vein
Tibialis posterior
Anterior Compartment
Anterior tibial artery and vein
Tibialis anterior
Anterior view

Compartments are clinically important. For example, trauma to a limb can cause bleeding, which elevates pressures and compresses blood vessels and nerves within the compartment. A lack of blood flow leads to "blood starvation," or ischemia. This condition, called **compartment syndrome**, can lead to the paralysis or death of the affected muscles if the pressure is not relieved within 2–4 hours.

Module 10.23 Review

a. Name the eight muscle compartments of the limbs.

b. Define compartment syndrome.

c. Propose a reason why compartment syndrome can be life threatening.

LO **10.23** Describe the deep fascia and its relationship to the various limb muscle compartments.

Labeling

Label each of the indicated muscles that move the forearm and hand in the diagram at right.

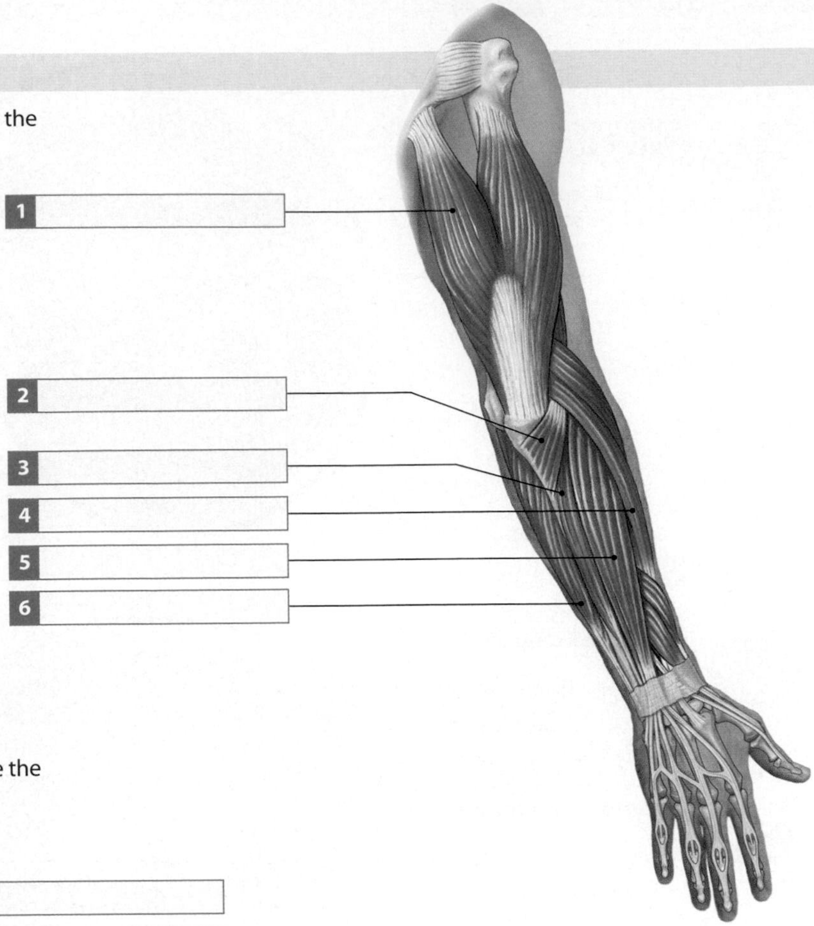
1
2
3
4
5
6

Label each of the indicated muscles that move the thigh and leg in the diagram below.

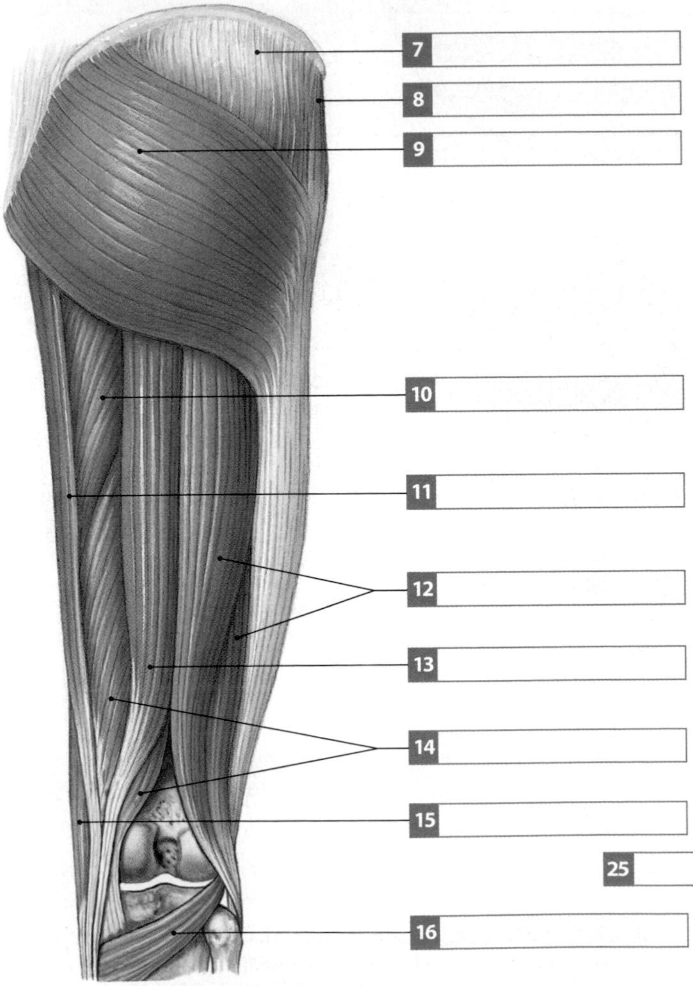
7
8
9
10
11
12
13
14
15
16

Label each of the indicated muscles that move the foot and toe in the diagram below.

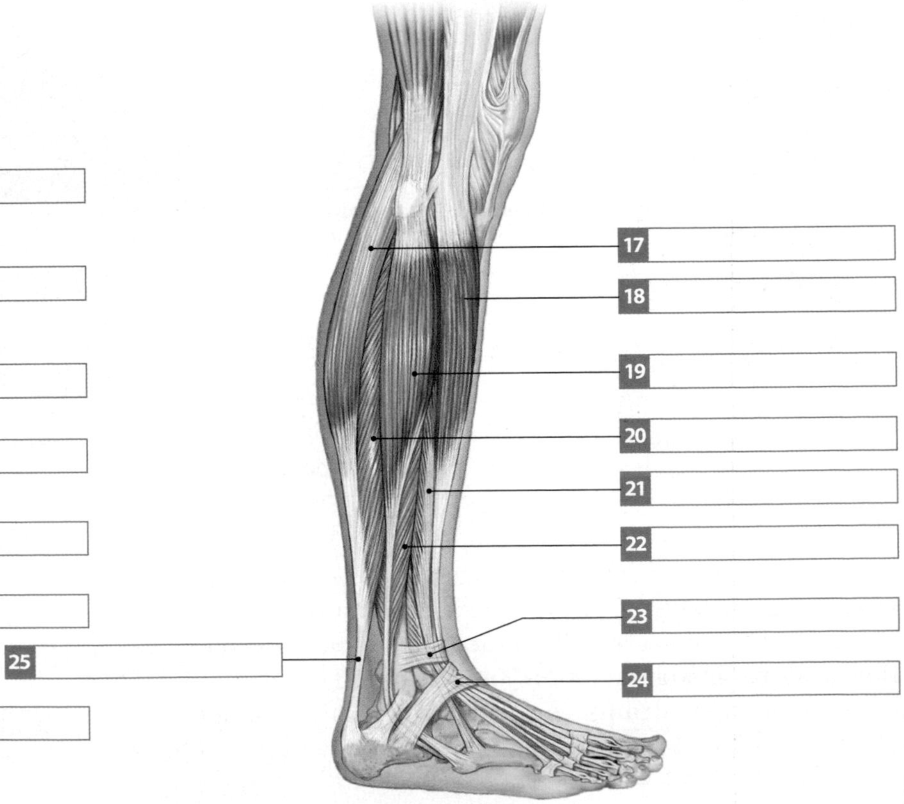
17
18
19
20
21
22
23
24
25

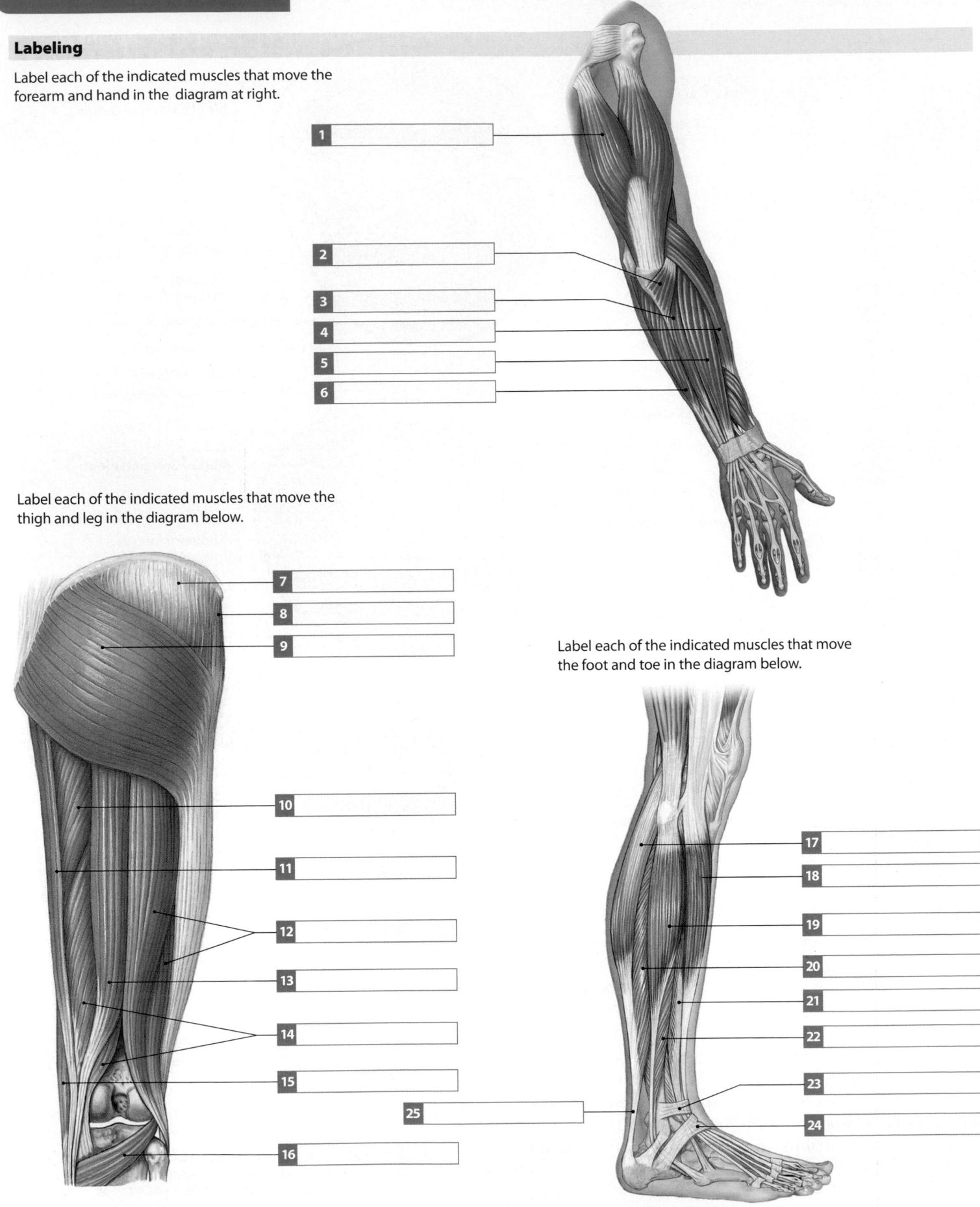

Study Outline

SECTION 1 • Functional Organization of the Muscular System

10.1 The axial and appendicular muscles have different functions p. 343

1. Skeletal muscle accounts for almost half the weight of your body.
2. The **muscular system** is divided into the **axial** and **appendicular muscles**.
3. The axial muscles support and position the axial skeleton while the appendicular muscles support and move the limbs.

10.2 Muscular power and range of motion are influenced by fascicle organization and leverage p. 344

4. Muscle fascicles can be organized as **parallel**, **convergent**, **pennate** (**unipennate**, **bipennate**, or **multipennate**) or **circular** (**sphincter**).

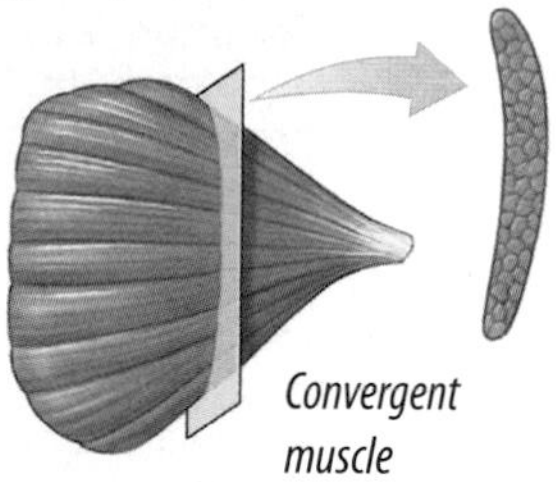

Convergent muscle

5. A **lever** is a rigid structure that pivots on a fixed point called a **fulcrum**. In the body, bones are levers, and joints are fulcrums.
6. Levers are classified as **first-class**, **second-class**, and **third-class levers**. Third-class levers are the most common levers in the body.

10.3 The names of muscles can provide clues to their appearance and/or function p. 346

7. Each muscle can be identified by its **origin**, **insertion**, and **action**.
8. The site of attachment at the fixed end of the muscle is the **origin**; the site where the movable end of the muscle attaches to another structure is called the **insertion**. The movement produced when a muscle contracts is the **action**.
9. A muscle can be classified as an **agonist** or **prime mover**, an **antagonist**, or a **synergist**.
10. Muscle terminology is associated with the location of the muscle, as well as its position, fascicle organization, structural characteristics, actions, and other features.

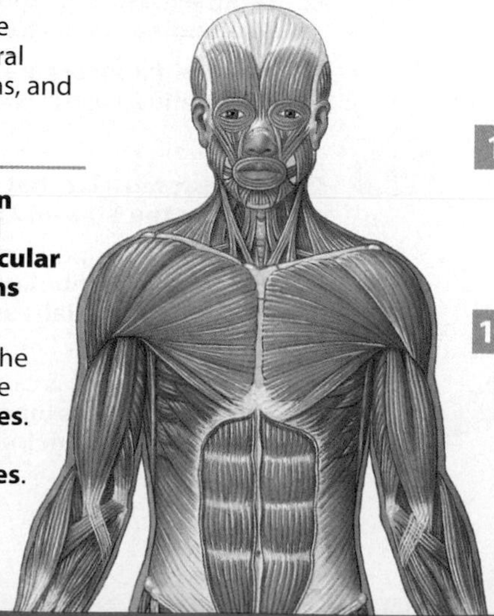

10.4 The skeletal muscles can be assigned to the axial division or the appendicular division based on origins and functions p. 348

11. About 60 percent of the skeletal muscles in the body are **axial muscles**. The remaining are **appendicular muscles**.

SECTION 2 • Axial Muscles

10.5 There are four groups of axial muscles p. 351

12. The first group of axial muscles is the muscles of the head and neck that are not associated with the vertebral column. These are the muscles of the face, extrinsic eye, tongue, pharynx, and neck.
13. The second group of axial muscles is the muscles of the vertebral column. The third group of muscles is the muscles of the trunk, and the muscles of the pelvic floor form the fourth group.

10.6 The muscles of facial expression are important in eating and useful for communication p. 352

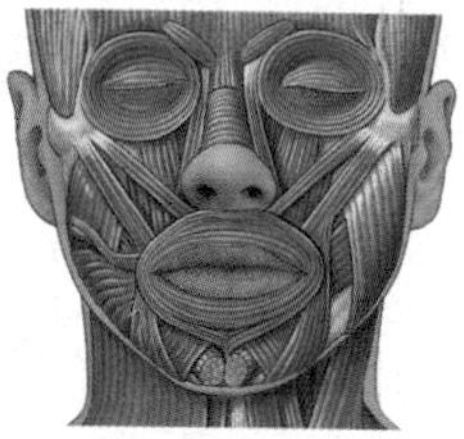

14. The **muscles of facial expression** originate on the surface of the skull. They insert on the superficial fascia and dermis of the skin.
15. The muscles of the mouth and cheek are **levator labii superioris**, **zygomaticus minor**, **zygomaticus major**, **buccinator**, **orbicularis oris**, **risorius**, **mentalis**, **depressor labii inferioris**, and **depressor anguli oris**.

10.7 The extrinsic eye muscles position the eye, and the muscles of mastication move the lower jaw p. 354

16. The extrinsic eye muscles are **inferior rectus**, **medial rectus**, **superior rectus**, **lateral rectus**, **inferior oblique**, and **superior oblique**.
17. The muscles of mastication are **masseter**, **temporalis**, **medial pterygoid**, and **lateral pterygoid**.

10.8 The muscles of the tongue are closely associated with the muscles of the pharynx and neck p. 356

18. The muscles of the tongue are **genioglossus**, **hyoglossus**, **palatoglossus**, and **styloglossus**.
19. The muscles of the pharynx are the **pharyngeal constrictors**, **laryngeal elevators**, and **palatal muscles**.
20. The anterior muscles of the neck are **digastric**, **geniohyoid**, **mylohyoid**, **omohyoid**, **sternohyoid**, **sternothyroid**, **stylohyoid**, **thyrohyoid**, and **sternocleidomastoid**.

10.9 The muscles of the vertebral column support and align the axial skeleton p. 358

21. The **erector spinae** muscles are subdivided into the **spinalis**, **longissimus**, and **iliocostalis muscle groups**.

10.10 The oblique and rectus muscles form the muscular walls of the trunk p. 360

22. The oblique muscles include the **scalenes**, **external** and **internal intercostals**, **transversus thoracis**, **external** and **internal obliques**, and **transversus abdominis**.

23. The rectus group includes the **diaphragm** and **rectus abdominis.**

10.11 **The muscles of the pelvic floor support the organs of the abdominopelvic cavity** p. 362

24. The muscles of the pelvic floor form the **perineum**, a muscular sheet that spans the pelvic outlet.
25. The muscles of the pelvic floor are divided into the muscles of the **urogenital triangle** and **anal triangle.**

SECTION 3 • Appendicular Muscles

10.12 **The appendicular muscles stabilize, position, and support the limbs** p. 365

26. The upper limb muscles include muscles of the pectoral girdle, arm, forearm, hand, and fingers.
27. The lower limb muscles include muscles of the thigh, leg, foot, and toes.

10.13 **The largest appendicular muscles originate on the trunk** p. 366

28. Gross movements of the limbs are controlled by muscles that originate on the trunk. Limb muscles get smaller, more numerous, and more precise as they are located more distally on the limb.

10.14 **Muscles that position each pectoral girdle originate on the occipital bone, superior vertebrae, and ribs** p. 368

29. The muscles of the pectoral girdle are **trapezius**, **levator scapulae**, **subclavius**, **pectoralis minor**, **serratus anterior**, **rhomboid major**, and **rhomboid minor**.

10.15 **Muscles that move the arm originate on the clavicle, scapula, thoracic cage, and vertebral column** p. 370

30. The muscles that move the arm are **deltoid**, **pectoralis major**, **coracobrachialis**, **teres major**, **latissimus dorsi**, and the four muscles of the rotator cuff: **supraspinatus**, **infraspinatus**, **teres minor**, and **subscapularis**.

10.16 **Muscles that move the forearm and hand originate on the scapula, humerus, radius, or ulna** p. 372

31. The elbow extensors are **triceps brachii** and **anconeus.** The elbow flexors are **biceps brachii**, **brachialis**, and **brachioradialis.**
32. The wrist extensors are **extensor carpi ulnaris**, **extensor carpi radialis longus**, and **extensor carpi radialis brevis.** The wrist flexors are **flexor carpi radialis**, **palmaris longus**, and **flexor carpi ulnaris.**
33. The muscles of pronation are **pronator teres** and **pronator quadratus. Supinator** causes supination.

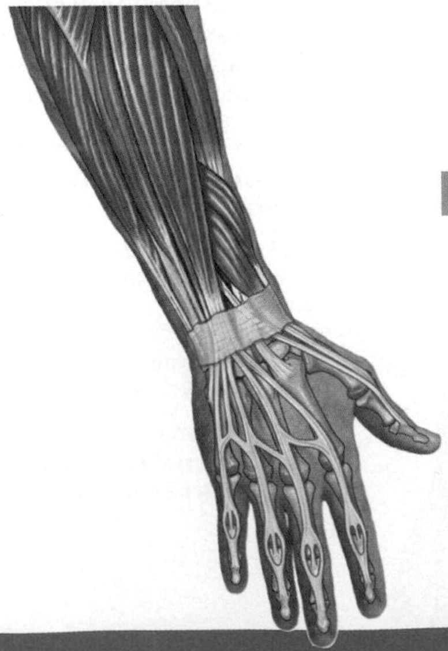

34. The tendons of the flexor muscles pass through **synovial tendon sheaths.** Inflammation of the flexor retinaculum and the synovial tendon sheaths can put pressure on the median nerve causing the pain known as **carpal tunnel syndrome.**

10.17 **Muscles that move the hand and fingers originate on the humerus, radius, ulna, and interosseous membrane** p. 374

35. The muscles that flex the fingers and the thumb are **flexor digitorum superficialis**, **flexor digitorum profundus**, and **flexor pollicis longus.** Finger extension is caused by **extensor digitorum** and **extensor digiti minimi.**
36. The muscles that move the thumb are **extensor pollicis longus**, **abductor pollicis longus**, and **extensor pollicis brevis.**

10.18 **The intrinsic muscles of the hand originate on the carpal and metacarpal bones and associated tendons and ligaments** p. 376

37. The intrinsic muscles of the hand perform flexion and extension of the fingers at the metacarpophalangeal joints; abduction and adduction of the fingers at the metacarpophalangeal joints; and opposition and reposition (relaxed position) of the thumb.

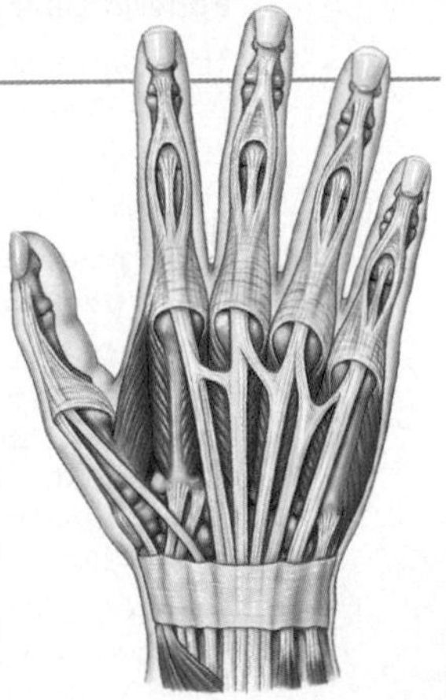

10.19 **The muscles that move the thigh originate on the pelvis and associated ligaments and fasciae** p. 378

38. The muscles of the **gluteal group** are **gluteus maximus**, **gluteus medius**, **gluteus minimus**, and **tensor fasciae latae.**
39. The iliopsoas group includes **psoas major** and **iliacus.**
40. The lateral rotator group muscles are **piriformis**, **superior** and **inferior gemellus**, **obturator internus** and **externus**, and **quadratus femoris.**
41. The adductor group muscles are **pectineus**, **adductor brevis**, **adductor longus**, **adductor magnus**, and **gracilis.**

10.20 **The muscles that move the leg originate on the pelvis and femur** p. 380

42. The **knee flexors** are **biceps femoris**, **semimembranosus**, **semitendinosus**, **sartorius**, and **popliteus.**
43. The **knee extensors** are **rectus femoris**, **vastus intermedius**, **vastus lateralis**, and **vastus medialis.**

10.21 **The extrinsic muscles that move the foot and toes originate on the tibia and fibula** p. 382

44. Plantar flexion is performed by **gastrocnemius**, **soleus**, **fibularis brevis**, **fibularis longus**, **plantaris**, and **tibialis posterior.** The **tibialis anterior** and **fibularis tertius** cause dorsiflexion.
45. Flexion of the toes is caused by **flexor digitorum longus** and **flexor hallucis longus.** Toe extension is caused by **extensor digitorum longus** and **extensor hallucis longus.**

10.22 The intrinsic muscles of the foot originate on the tarsal and metatarsal bones and associated tendons and ligaments p. 384

46. The **intrinsic muscles of the foot** stabilize the positions of the tendons descending from the leg. These muscles are more numerous on the inferior surface of the foot and occur in several layers.
47. These muscles are responsible for flexion and extension of the interphalangeal joints, as well as abduction and adduction of the metatarsophalangeal joints.

10.23 The **deep fascia separates the limb muscles into separate compartments** p. 386

48. Fibrous partitions create **compartments** containing muscles with compatible functions as well as blood supply and innervation.
49. Trauma to a limb may cause bleeding into a compartment. This may elevate pressures and compress blood vessels and nerves within a compartment, a condition called **compartment syndrome**.

Chapter Review Questions

Labeling

Identify the lever system in each of the images below.

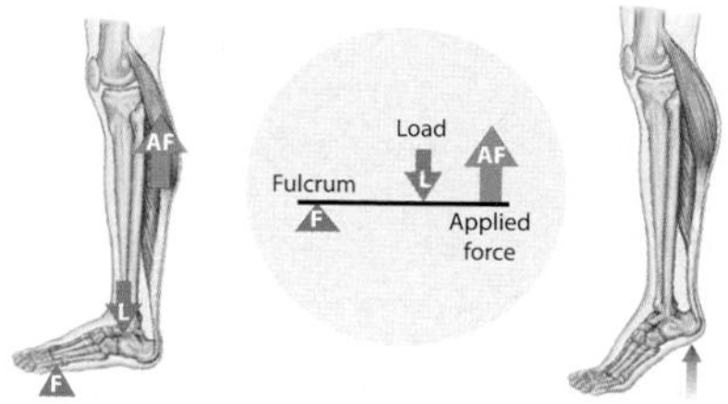

AF
Load
L
AF
Fulcrum
F
Applied force
L
F

1 ______________

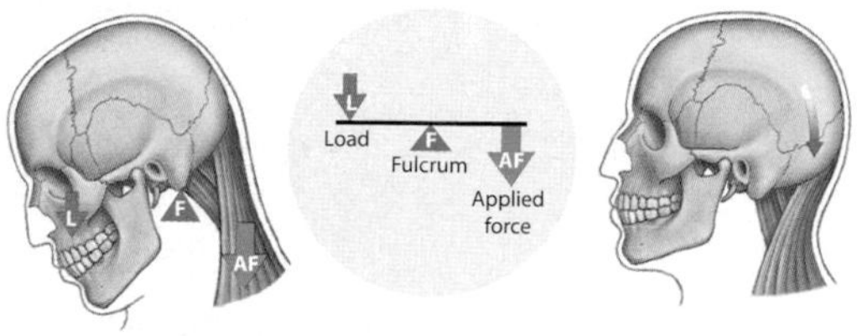

Load
L
F
Fulcrum
AF
Applied force
F
AF
L

2 ______________

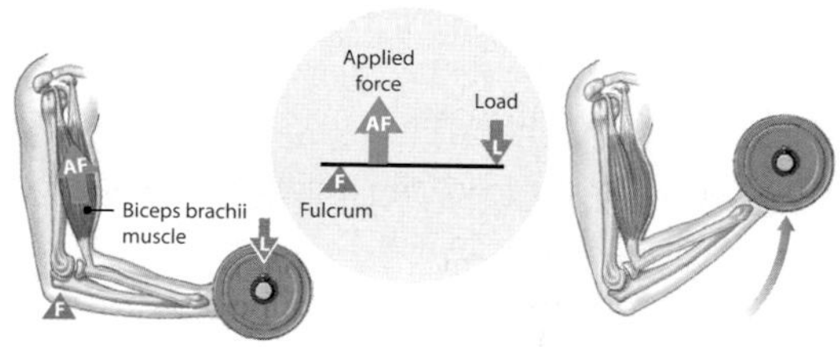

AF
Biceps brachii muscle
L
F
Applied force
AF
Load
L
F
Fulcrum

3 ______________

True/False

Indicate whether each statement is true or false.

4 Pectoralis major is an example of a convergent muscle.

5 A wheelbarrow is an example of a third-class lever.

6 A muscle agonist can also be referred to as a prime mover.

7 The inferior rectus and superior oblique are examples of extrinsic eye muscles.

8 The esophageal hiatus is an opening in the transversus abdominis muscle.

4 ______________
5 ______________
6 ______________
7 ______________
8 ______________

Matching

Match each lettered description with the most closely related muscle.

a. moves eye laterally
b. flexion at knee
c. extension at knee
d. abduction at shoulder
e. downward rotation of scapula
f. flexion at elbow
g. plantar flexion
h. mastication
i. medial rotation at shoulder
j. dorsiflexion

9 Tibialis anterior
10 Temporalis
11 Brachialis
12 Lateral rectus
13 Soleus
14 Biceps femoris
15 Rhomboid major
16 Vastus medialis
17 Deltoid
18 Subscapularis

9 ______________
10 ______________
11 ______________
12 ______________
13 ______________
14 ______________
15 ______________
16 ______________
17 ______________
18 ______________

Multiple choice

Select the correct answer from the list provided.

19 The site where the more movable end of a muscle attaches is the
- ☐ a) origin.
- ☐ b) insertion.
- ☐ c) belly.
- ☐ d) fascicle.

20 A muscle with a feather-shaped fascicle organization is called a
- ☐ a) parallel muscle.
- ☐ b) pennate muscle.
- ☐ c) convergent muscle.
- ☐ d) circular muscle.

21 The most common lever system in the body is
- ☐ a) first-class.
- ☐ b) second-class.
- ☐ c) third-class.
- ☐ d) fourth-class.

22 Which of the following muscles is an axial muscle?
- ☐ a) erector spinae
- ☐ b) trapezius
- ☐ c) deltoid
- ☐ d) flexor carpi radialis

23 The major extensor of the elbow is
- ☐ a) triceps brachii.
- ☐ b) biceps brachii.
- ☐ c) deltoid.
- ☐ d) subscapularis.

24 Inflammation of the retinaculum and synovial tendon sheaths resulting in pressure on the median nerve is called
- ☐ a) anterior compartment syndrome.
- ☐ b) rotator cuff syndrome.
- ☐ c) carpal tunnel syndrome.
- ☐ d) plantar fascitis.

25 When doing a pull-up exercise, which of the following muscles is responsible for adduction at the shoulder joint?
- ☐ a) levator scapulae
- ☐ b) deltoid
- ☐ c) supraspinatus
- ☐ d) latissimus dorsi

26 Which of the following muscles performs the hip and knee action required to kick a ball?
- ☐ a) rectus femoris
- ☐ b) pectineus
- ☐ c) gracilis
- ☐ d) biceps femoris

27 Which of the following locations is *not* an attachment site for biceps brachii?
- ☐ a) tuberosity of radius
- ☐ b) coracoid process of scapula
- ☐ c) acromion process of scapula
- ☐ d) supraglenoid tubercle of scapula

28 Which of the following is *not* a muscle of mastication?
- ☐ a) masseter
- ☐ b) buccinator
- ☐ c) temporalis
- ☐ d) lateral pterygoid

Short answer

29 What are the functions of the muscles of the pelvic floor?

30 Identify the muscles of the rotator cuff.

31 List the muscles of the quadriceps muscle group and describe their actions.

Chapter Integration • Applying what you have learned

Bodybuilding and lookin' good

Bodybuilders spend many hours in the gym lifting free weights to develop their muscles. Larger muscles and greater muscle definition are the goals, and looking "ripped" requires a lot of dedication. To sculpt their arms, for example, bodybuilders do a lot of biceps curls (flexion at the elbow holding weights in the anatomical position) and triceps curls (extension at the elbow).

As a 10-year-old, Jerry and his friends would go to the beach at Lion's Park, known locally as "Muscle Beach" because all the local bodybuilders would go there in the summer to work out and show off for the girls. Some female bodybuilders even started going there. Jerry was always amazed by the size, shape, and strength of these musclemen and vowed that someday he would become one of them. As he reached puberty, Jerry became a fitness fanatic who worked out many hours a day. As he learned more about bodybuilding, he eschewed the massive, heavily muscled look for one of a more athletic, lean, and well-defined musculature. Everyone came to admire his "six-pack abs," especially his girlfriend, DJ.

1 **Explain why doing both biceps curls and triceps curls helps achieve larger, well-toned arms.**

2 **Which exercises would be best for shaping your abdominal muscles into "six-pack abs"?**

Sports, muscles, and joints

Jennifer is a high school freshman and an up-and-coming volleyball player on the junior varsity team. It is her intent to join the varsity team as a sophomore. One element of her game that she knows she needs to improve upon is her vertical jump. She has committed her offseason workouts to spending more time in the weight room strengthening the muscles involved in jumping. From what you have just learned about the muscular system, answer the following questions and design a weightlifting program to help Jennifer achieve her goals.

3 **Identify the actions involved in the hip, knee, and ankle joints when jumping.**

4 **Which muscles are involved in each of these actions?**

5 **What kind of exercises would you suggest Jennifer perform in the weight room to increase the strength in these muscles?**

11 Neural Tissue

LEARNING OUTCOMES

These Learning Outcomes correspond by number to this chapter's modules and indicate what you should be able to do after completing the chapter.

SECTION 1 • Cellular Organization of the Nervous System

- **11.1** Describe the anatomical and functional divisions of the nervous system.
- **11.2** Sketch and label the structure of a typical neuron, and describe the functions of each component.
- **11.3** Classify and describe neurons on the basis of their structure and function.
- **11.4** Describe the locations and functions of neuroglia in the CNS.
- **11.5** Describe the locations and functions of Schwann cells and satellite cells.

SECTION 2 • Neurophysiology

- **11.6** Describe the general role of membrane potential changes in neuronal activity.
- **11.7** Explain how the resting potential is created and maintained.
- **11.8** Describe the functions of gated channels with respect to the permeability of the plasma membrane.
- **11.9** Describe graded potentials.
- **11.10** Describe the events involved in the generation and propagation of an action potential.
- **11.11** Describe continuous propagation and saltatory propagation, and discuss the factors that affect the speed with which action potentials are propagated.
- **11.12** Describe the general structure of synapses in the CNS and PNS, and discuss the events that occur at a chemical synapse.
- **11.13** Discuss the significance of postsynaptic potentials, including the roles of excitatory postsynaptic potentials and inhibitory postsynaptic potentials.
- **11.14** Discuss the interactions that make information processing in neural tissue possible.

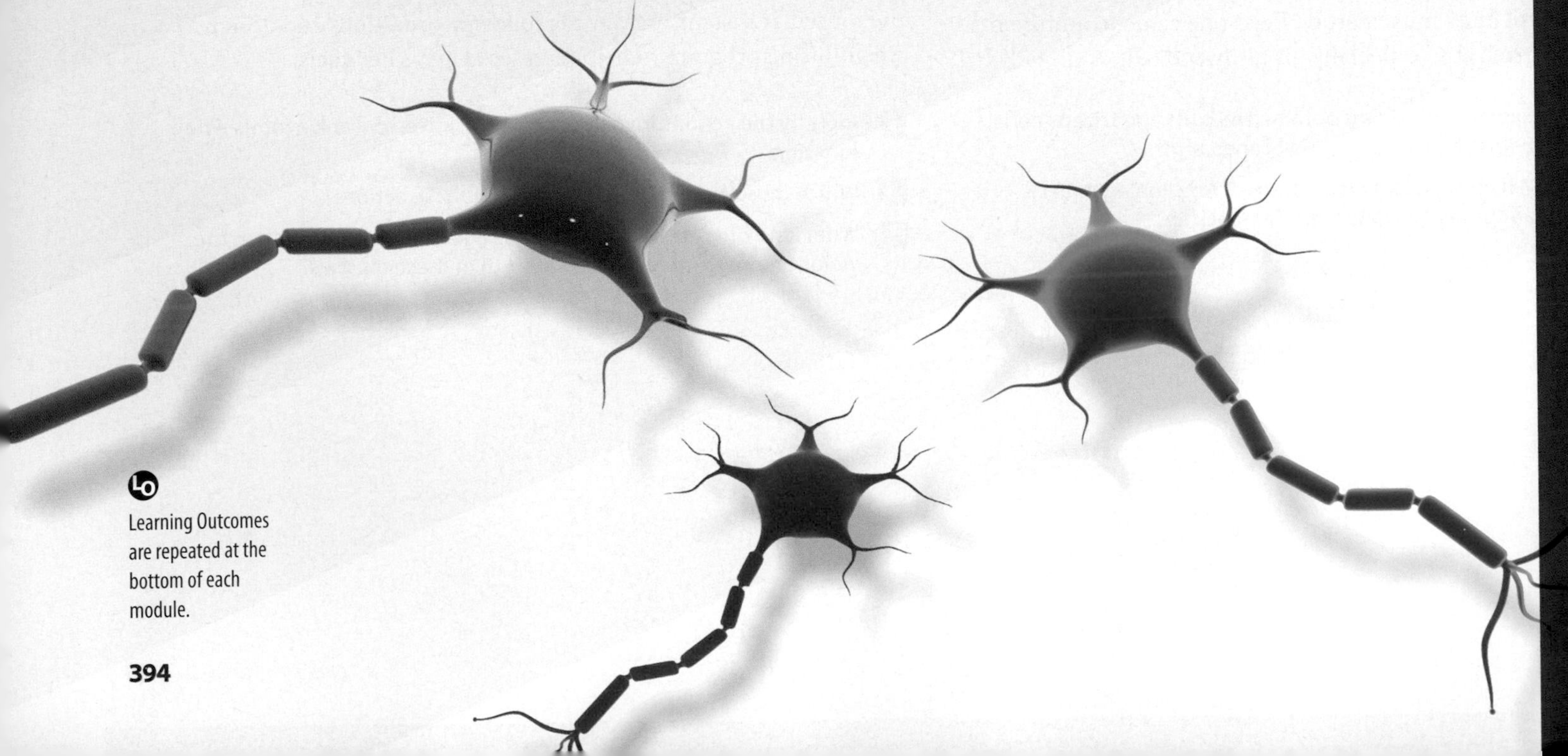

LO Learning Outcomes are repeated at the bottom of each module.

The nervous system has two divisions: the CNS and PNS

This is the first of three chapters on the nervous system. It considers the anatomical and functional divisions of the nervous system, the cellular organization of neural tissue, and the basic principles of neurophysiology. Anatomically, the nervous system is divided into the central nervous system (CNS) and peripheral nervous system (PNS). Their major components and general functions are described in this flowchart, beginning with the sensory receptors of the PNS ①.

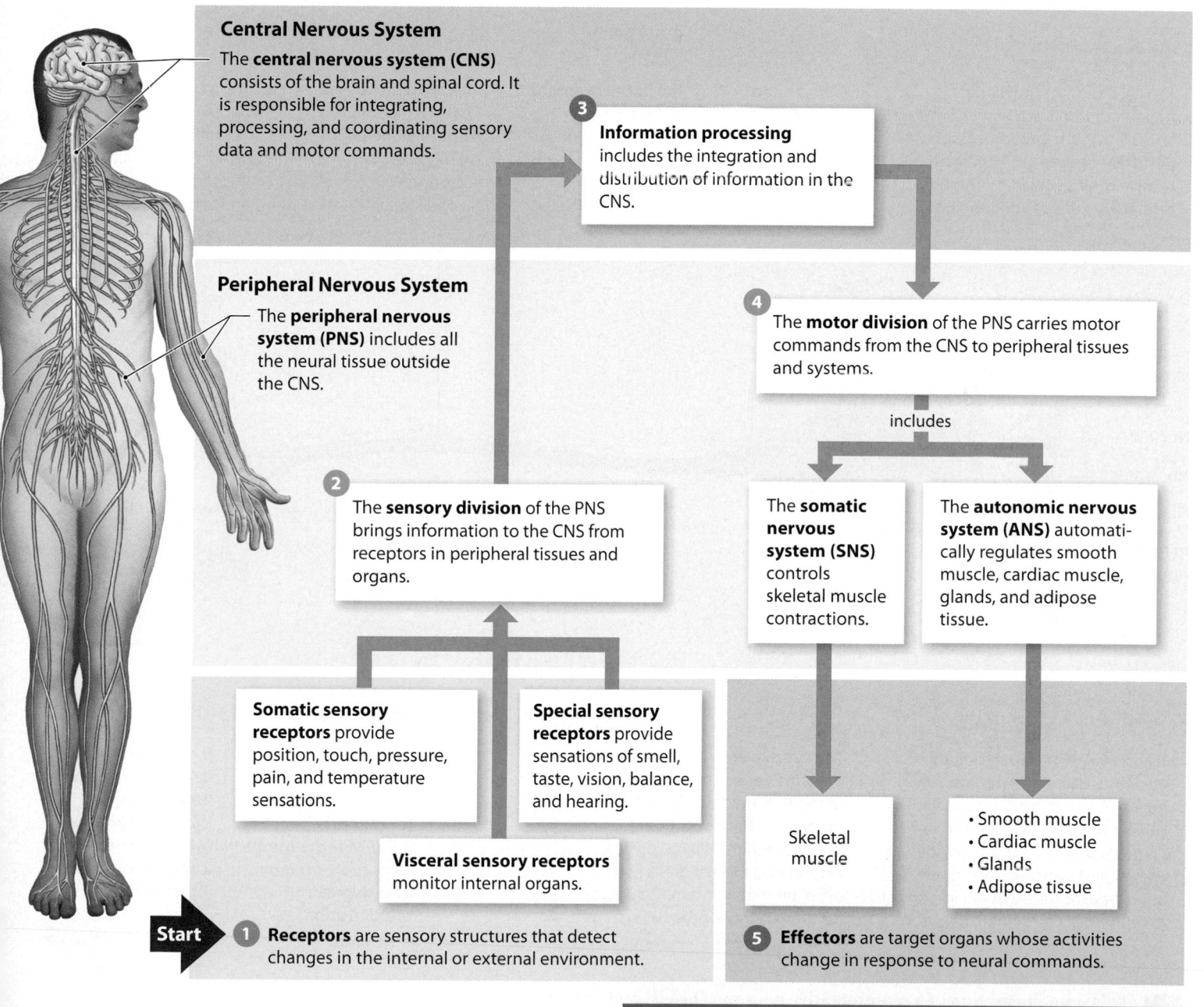

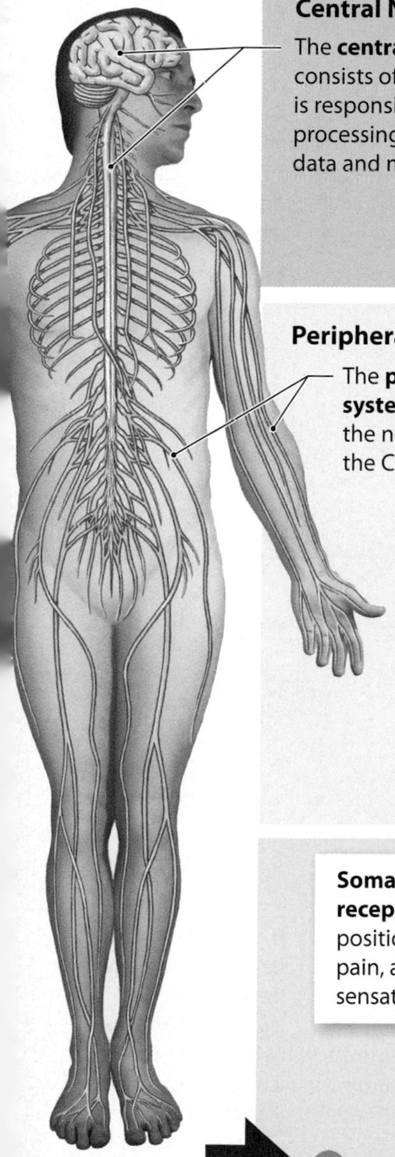

Central Nervous System

The **central nervous system (CNS)** consists of the brain and spinal cord. It is responsible for integrating, processing, and coordinating sensory data and motor commands.

Peripheral Nervous System

The **peripheral nervous system (PNS)** includes all the neural tissue outside the CNS.

Start ① **Receptors** are sensory structures that detect changes in the internal or external environment.

Somatic sensory receptors provide position, touch, pressure, pain, and temperature sensations.

Visceral sensory receptors monitor internal organs.

Special sensory receptors provide sensations of smell, taste, vision, balance, and hearing.

② The **sensory division** of the PNS brings information to the CNS from receptors in peripheral tissues and organs.

③ **Information processing** includes the integration and distribution of information in the CNS.

④ The **motor division** of the PNS carries motor commands from the CNS to peripheral tissues and systems.

includes

The **somatic nervous system (SNS)** controls skeletal muscle contractions.

The **autonomic nervous system (ANS)** automatically regulates smooth muscle, cardiac muscle, glands, and adipose tissue.

Skeletal muscle

- Smooth muscle
- Cardiac muscle
- Glands
- Adipose tissue

⑤ **Effectors** are target organs whose activities change in response to neural commands.

Module 11.1 Review

a. Compare the central and peripheral nervous systems.

b. Which division of the PNS brings information to the CNS?

c. Name the target organs of the ANS.

LO 11.1 Describe the anatomical and functional divisions of the nervous system.

Neurons are nerve cells specialized for intercellular communication

In this module we examine the structure of neurons. Recall from Chapter 4 (p. 165) that neurons have three general regions: **dendrites**, which receive stimuli from the environment or from other neurons; a **cell body**, which contains the nucleus and other organelles; and one **axon**, which carries information toward other cells. Specialized structures within each region contribute to intercellular communication.

Dendrites

Typical dendrites are highly branched, with each branch bearing fine 0.5 to 1 μm long processes called **dendritic spines**. CNS neurons receive most of their information primarily at the dendritic spines.

Axon

The **axon hillock** is the origin of the axon from the cell body.

The **initial segment** of the axon lies distally adjacent to the axon hillock. It is where an action potential is initiated.

The **axolemma** (*lemma*, husk) is a specialized portion of the plasma membrane that surrounds the cytoplasm (axoplasm) of the axon.

The **axoplasm** (AK-sō-plazm) contains neurofibrils, neurotubules, small vesicles, lysosomes, mitochondria, and various enzymes.

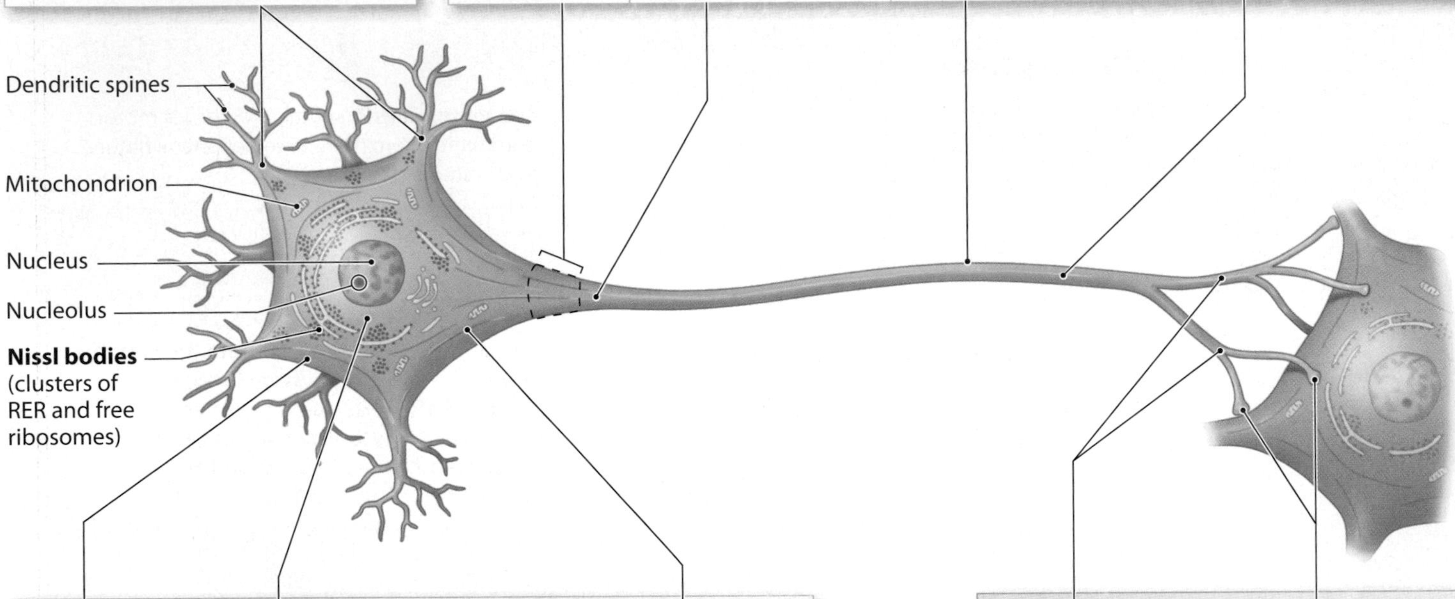
Dendritic spines
Mitochondrion
Nucleus
Nucleolus
Nissl bodies (clusters of RER and free ribosomes)

Cell Body

The cytoplasm surrounding the nucleus is called the **perikaryon** (per-i-KAR-ē-on; *peri*, around + *karyon*, nucleus). The perikaryon contains organelles that provide energy and synthesize the chemical neurotransmitters that are important in cell-to-cell communication.

The cytoskeleton of the perikaryon contains **neurofilaments** similar to the intermediate filaments in other cells. **Neurofibrils** are bundles of neurofilaments that extend into the dendrites and axon, providing internal support for these slender processes.

Telodendria

The main axon trunk ends in a series of fine extensions, or **telodendria** (tel-ō-DEN-drē-uh; *telo-*, end + *dendron*, tree; singular: **telodendrion**).

The telodendria of an axon end at **axon terminals**, or **synaptic terminals**, where the neuron communicates with other cells.

1 This is an illustration of a representative neuron. The cell body contains most of the organelles of a neuron. The axon is a long cytoplasmic process that extends away from the cell body. Many materials, including enzymes and lysosomes, travel the length of the axon along **neurotubules** (neuron microtubules) through a process called **axoplasmic transport**. Axoplasmic transport occurs in both directions. If debris or unusual chemicals appear in the axon terminal, **retrograde flow** soon delivers them to the cell body.

2 Each axon terminal is part of a **synapse**, a specialized site where the neuron communicates with another cell. Every synapse involves a **presynaptic cell** and a **postsynaptic cell**. Communication between those cells most commonly involves the release of chemicals called **neurotransmitters** into the **synaptic cleft**, a narrow space separating the two cells. Axon terminals receive (through axoplasmic transport) a continuous supply of neurotransmitters synthesized in the cell body. They also reabsorb and reassemble fragments of neurotransmitters broken down in the synaptic cleft.

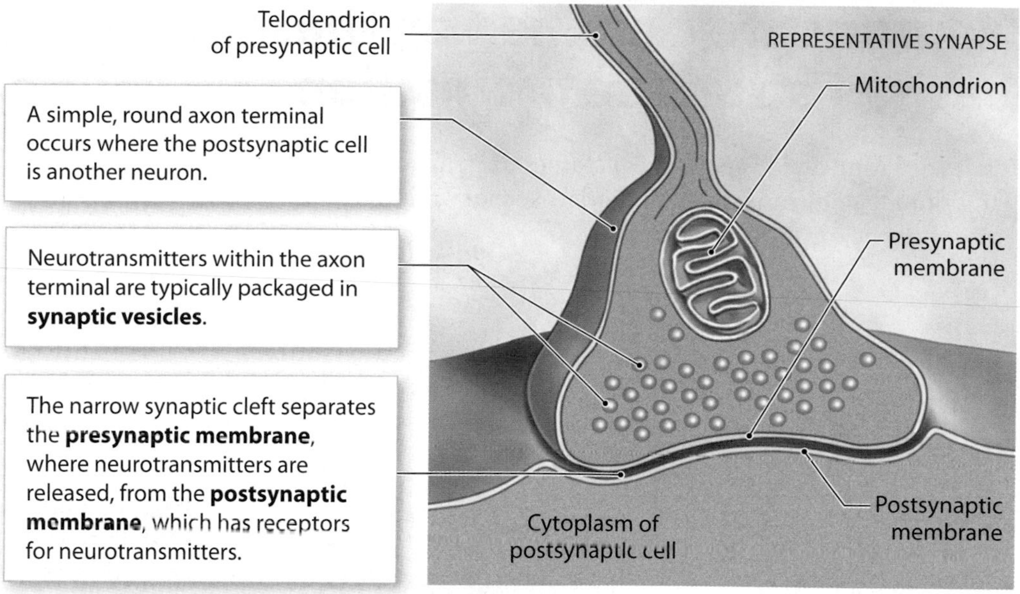

Telodendrion of presynaptic cell
A simple, round axon terminal occurs where the postsynaptic cell is another neuron.
Neurotransmitters within the axon terminal are typically packaged in **synaptic vesicles**.
The narrow synaptic cleft separates the **presynaptic membrane**, where neurotransmitters are released, from the **postsynaptic membrane**, which has receptors for neurotransmitters.
REPRESENTATIVE SYNAPSE
Mitochondrion
Presynaptic membrane
Cytoplasm of postsynaptic cell
Postsynaptic membrane

3 Except in a few special cases involving sensory receptors, a presynaptic cell is always a neuron. The postsynaptic cell can be either a neuron or another type of cell. Notice that axons may branch along their length, producing side branches known as **collateral branches**. Collateral branches enable a single neuron to communicate with several other cells.

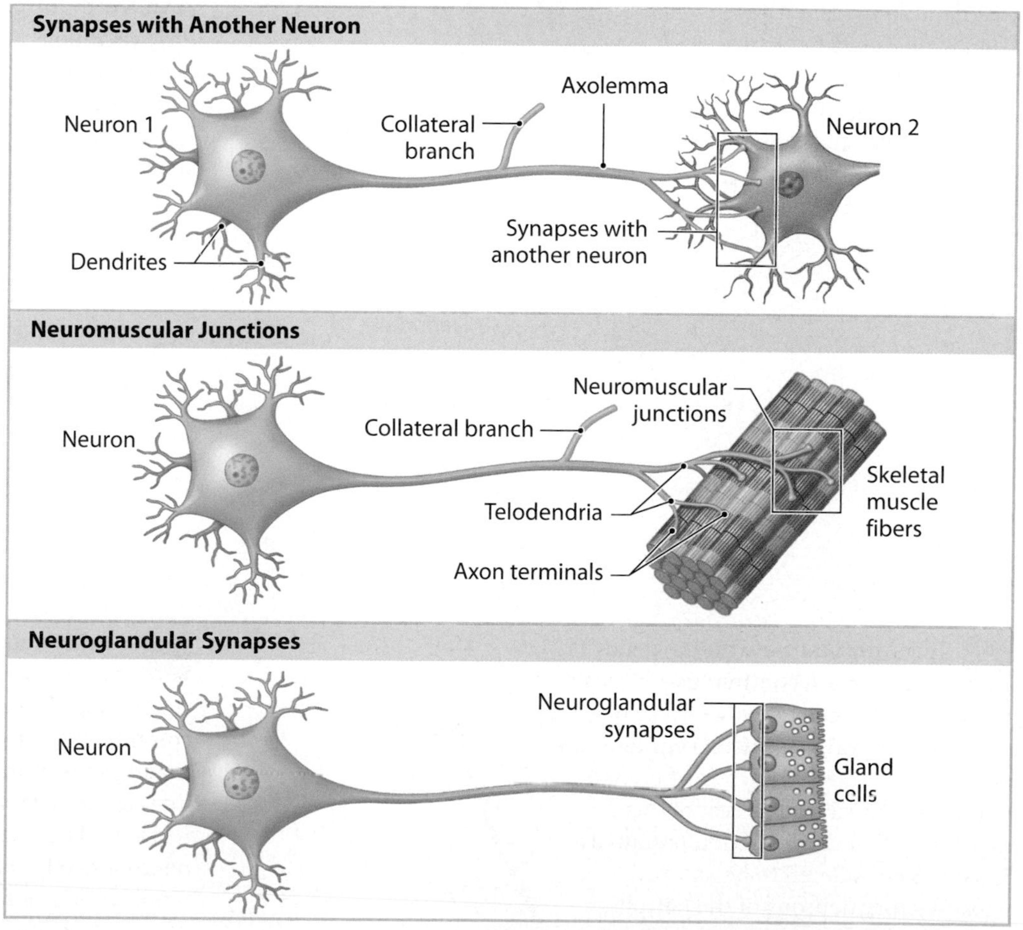

Synapses with Another Neuron
Neuron 1
Collateral branch
Axolemma
Neuron 2
Synapses with another neuron
Dendrites
Neuromuscular Junctions
Neuron
Collateral branch
Neuromuscular junctions
Skeletal muscle fibers
Telodendria
Axon terminals
Neuroglandular Synapses
Neuron
Neuroglandular synapses
Gland cells

Most CNS neurons lack centrioles and cannot divide. As a result, neurons lost to injury or disease are seldom replaced. Although neural stem cells persist in the adult nervous system, these cells are typically inactive (except in the epithelium responsible for our sense of smell, in the retina of the eye, and in the hippocampus, a portion of the brain involved with memory storage).

Module 11.2 Review

a. Name the structural components of a typical neuron.

b. Describe a synapse.

c. Why is a CNS neuron not usually replaced after it is injured?

LO **11.2** Sketch and label the structure of a typical neuron, and describe the functions of each component.

Neurons are classified on the basis of structure or function

There are four major anatomical classes of neurons. Functionally, neurons are classified as sensory neurons, interneurons, or motor neurons.

Neuron Structure

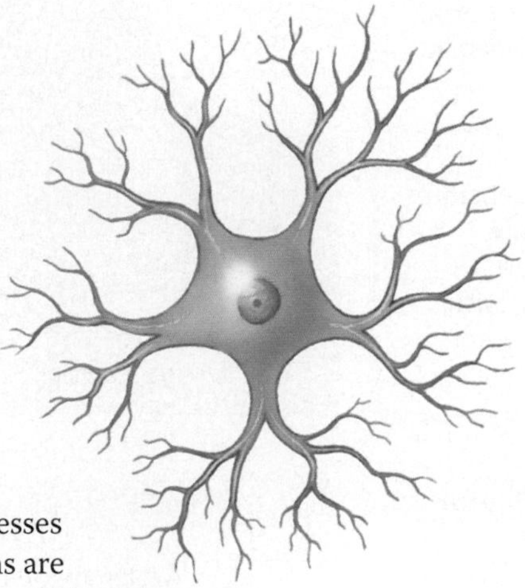

1 **Anaxonic** (an-AKS-on-ic) **neurons** are small and lack anatomical features that distinguish dendrites from axons; all the cell processes look alike. Anaxonic neurons are located in the brain and in special sense organs. Their functions are poorly understood.

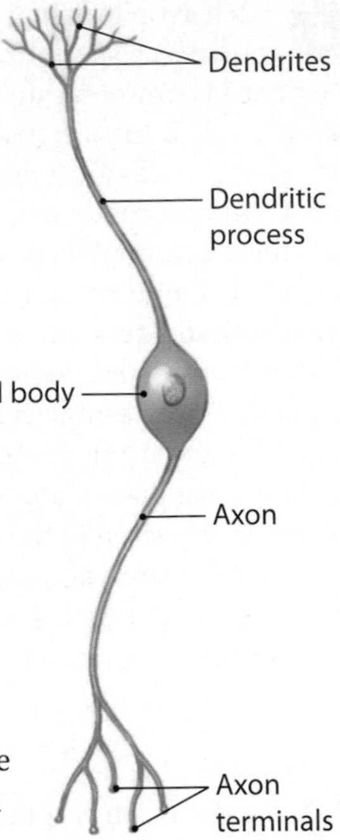
Dendrites
Dendritic process
Cell body
Axon
Axon terminals

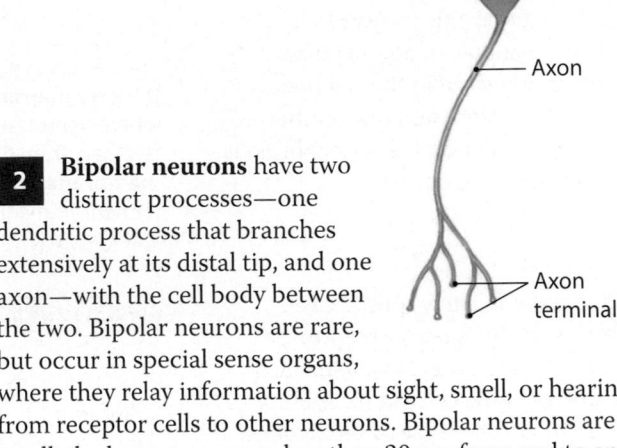

2 **Bipolar neurons** have two distinct processes—one dendritic process that branches extensively at its distal tip, and one axon—with the cell body between the two. Bipolar neurons are rare, but occur in special sense organs, where they relay information about sight, smell, or hearing from receptor cells to other neurons. Bipolar neurons are small; the largest measure less than 30 μm from end to end.

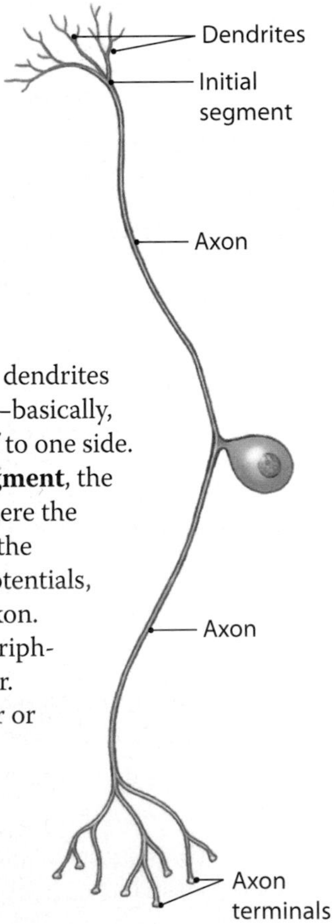
Dendrites
Initial segment
Axon
Axon
Axon terminals

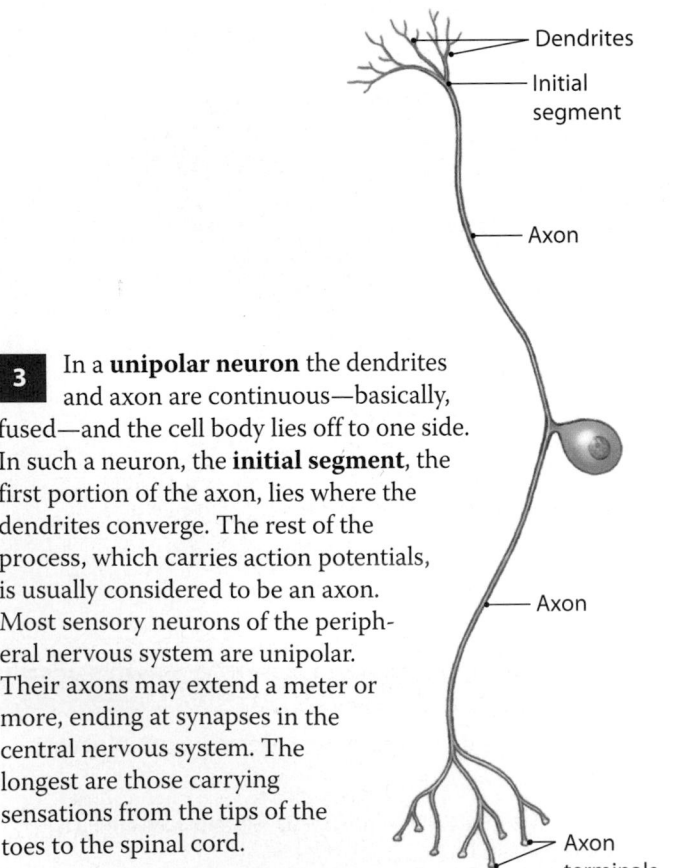

3 In a **unipolar neuron** the dendrites and axon are continuous—basically, fused—and the cell body lies off to one side. In such a neuron, the **initial segment**, the first portion of the axon, lies where the dendrites converge. The rest of the process, which carries action potentials, is usually considered to be an axon. Most sensory neurons of the peripheral nervous system are unipolar. Their axons may extend a meter or more, ending at synapses in the central nervous system. The longest are those carrying sensations from the tips of the toes to the spinal cord.

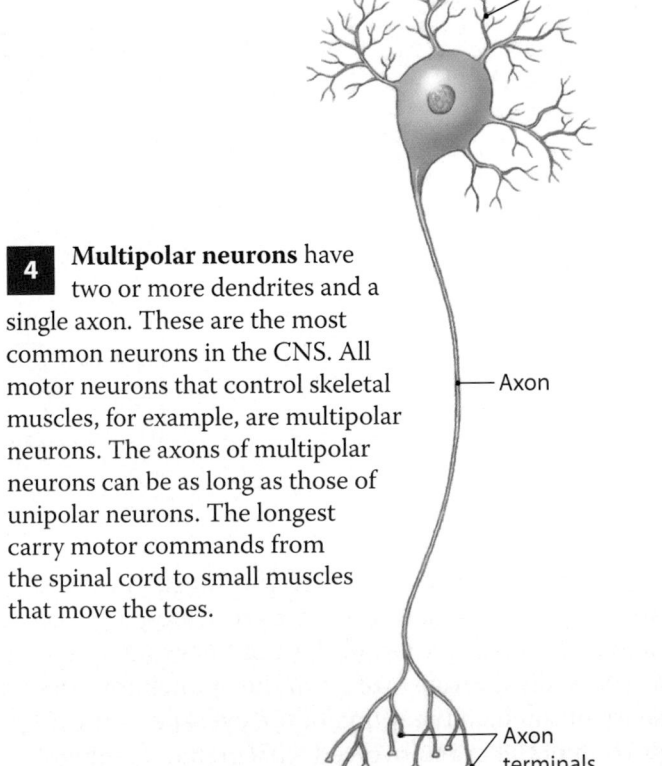

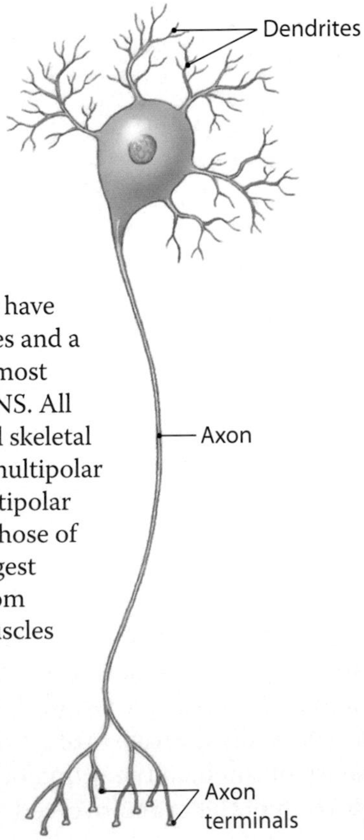
Dendrites
Axon
Axon terminals

4 **Multipolar neurons** have two or more dendrites and a single axon. These are the most common neurons in the CNS. All motor neurons that control skeletal muscles, for example, are multipolar neurons. The axons of multipolar neurons can be as long as those of unipolar neurons. The longest carry motor commands from the spinal cord to small muscles that move the toes.

Neuron Function within the CNS and PNS

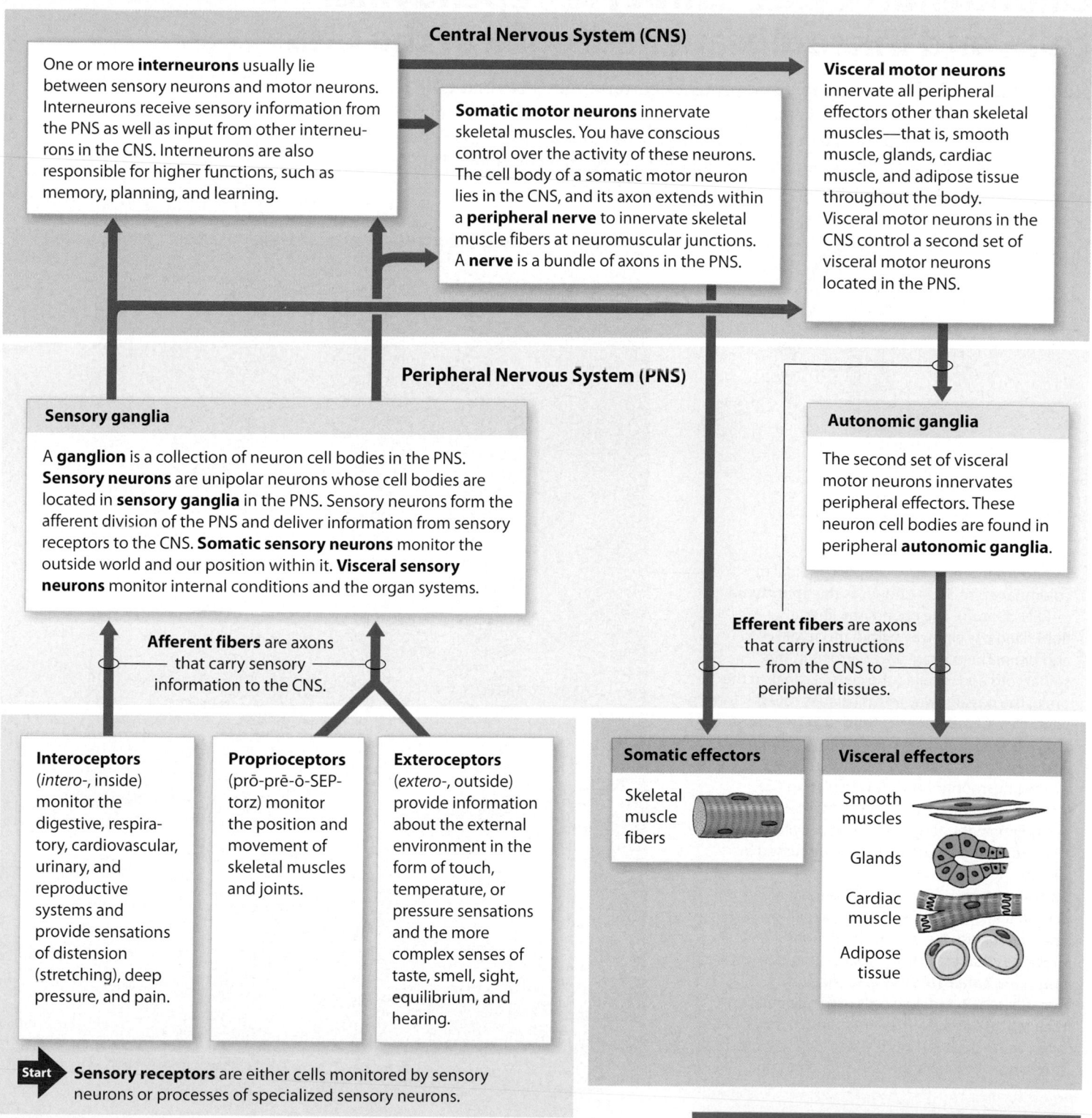

Central Nervous System (CNS)

One or more **interneurons** usually lie between sensory neurons and motor neurons. Interneurons receive sensory information from the PNS as well as input from other interneurons in the CNS. Interneurons are also responsible for higher functions, such as memory, planning, and learning.

Somatic motor neurons innervate skeletal muscles. You have conscious control over the activity of these neurons. The cell body of a somatic motor neuron lies in the CNS, and its axon extends within a **peripheral nerve** to innervate skeletal muscle fibers at neuromuscular junctions. A **nerve** is a bundle of axons in the PNS.

Visceral motor neurons innervate all peripheral effectors other than skeletal muscles—that is, smooth muscle, glands, cardiac muscle, and adipose tissue throughout the body. Visceral motor neurons in the CNS control a second set of visceral motor neurons located in the PNS.

Peripheral Nervous System (PNS)

Sensory ganglia

A **ganglion** is a collection of neuron cell bodies in the PNS. **Sensory neurons** are unipolar neurons whose cell bodies are located in **sensory ganglia** in the PNS. Sensory neurons form the afferent division of the PNS and deliver information from sensory receptors to the CNS. **Somatic sensory neurons** monitor the outside world and our position within it. **Visceral sensory neurons** monitor internal conditions and the organ systems.

Autonomic ganglia

The second set of visceral motor neurons innervates peripheral effectors. These neuron cell bodies are found in peripheral **autonomic ganglia**.

Afferent fibers are axons that carry sensory information to the CNS.

Efferent fibers are axons that carry instructions from the CNS to peripheral tissues.

Interoceptors (*intero-*, inside) monitor the digestive, respiratory, cardiovascular, urinary, and reproductive systems and provide sensations of distension (stretching), deep pressure, and pain.

Proprioceptors (prō-prē-ō-SEP-torz) monitor the position and movement of skeletal muscles and joints.

Exteroceptors (*extero-*, outside) provide information about the external environment in the form of touch, temperature, or pressure sensations and the more complex senses of taste, smell, sight, equilibrium, and hearing.

Somatic effectors

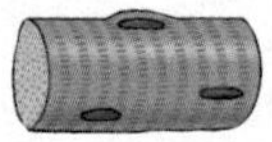

Skeletal muscle fibers

Visceral effectors

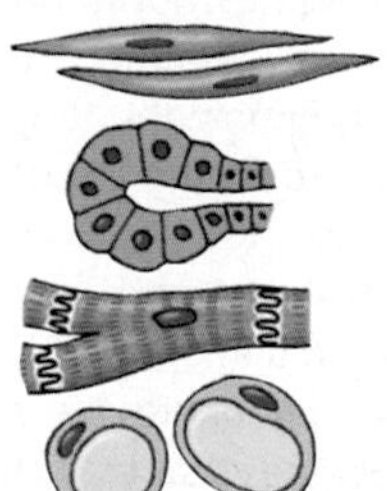

Smooth muscles
Glands
Cardiac muscle
Adipose tissue

Start **Sensory receptors** are either cells monitored by sensory neurons or processes of specialized sensory neurons.

5 This flowchart describes the basic relationships among the three functional classes of neurons: sensory neurons, interneurons, and motor neurons. The human body has about 10 million sensory neurons, half a million motor neurons, and an estimated 20 billion interneurons.

KEY

➡ = Somatic (sensory & motor)

➡ = Visceral (sensory & motor)

Module 11.3 Review

a. Classify neurons according to their structure.

b. Classify neurons according to their function.

c. Are unipolar neurons in a tissue sample of the PNS more likely to be sensory neurons or motor neurons?

LO **11.3** Classify and describe neurons on the basis of their structure and function.

Oligodendrocytes, astrocytes, ependymal cells, and microglia are neuroglia of the CNS

Neuroglia (or **glial cells**) support and protect neurons in the PNS and CNS. Neuroglia are abundant and diverse and make up about half the volume of the nervous system. Neural tissue in the CNS is organized differently than in the PNS, primarily because the CNS has a greater variety of glial cell types. This illustration summarizes information about neuroglia in the CNS.

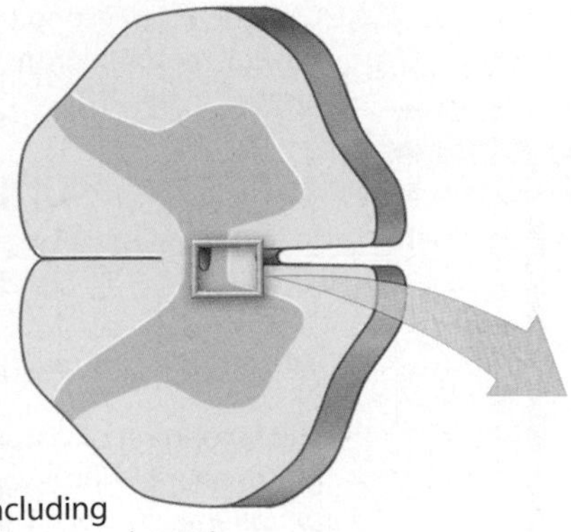

Section of spinal cord including a portion of the central canal

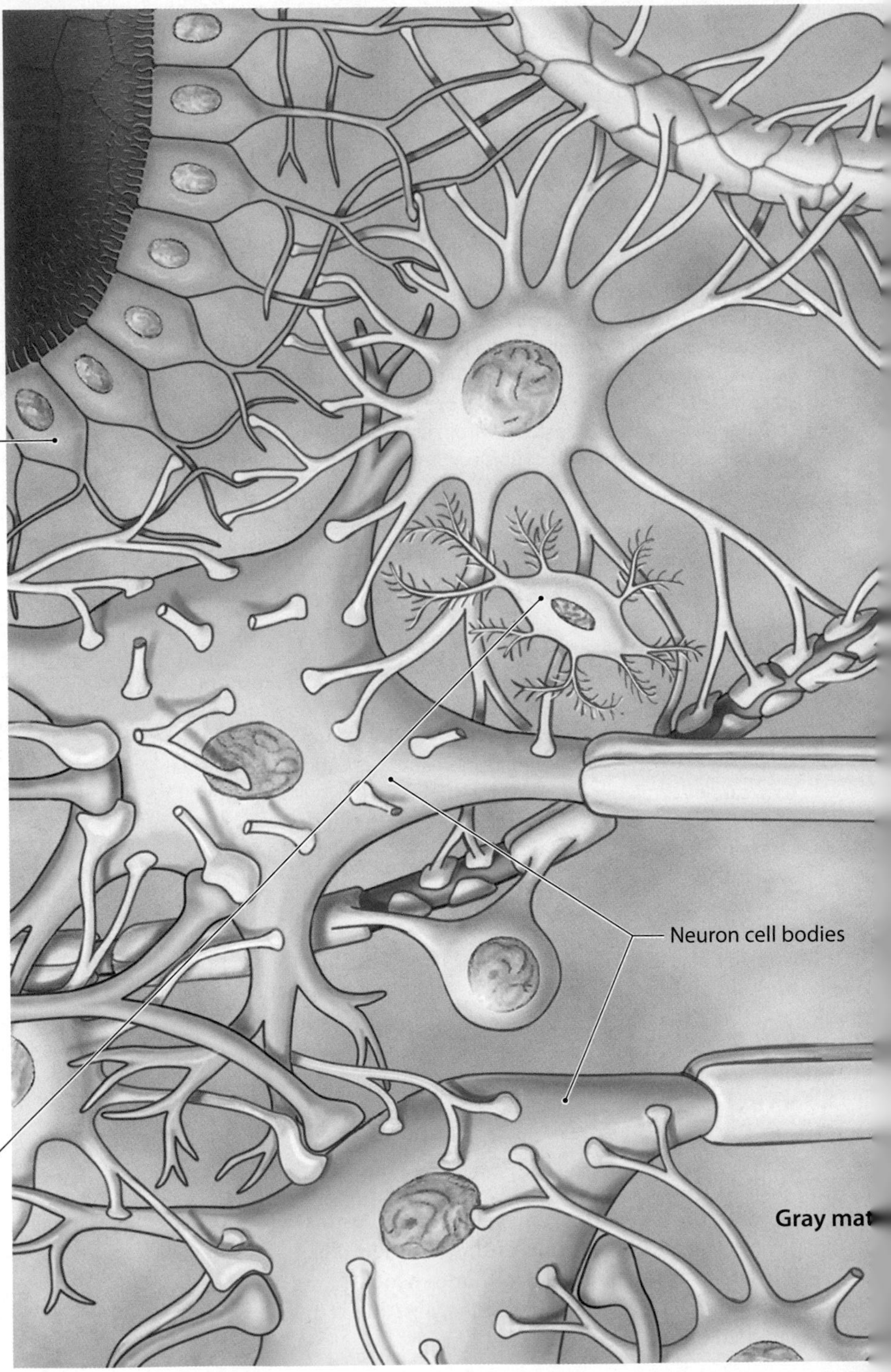
Neuron cell bodies
Gray mat

Ependymal cells form a simple cuboidal to columnar epithelium known as the **ependyma** (e-PEN-di-muh). The ependyma lines a fluid-filled passageway within the spinal cord and brain. The passageway narrows in the spinal cord and is called the central canal. In the brain, the passageway forms cavities called ventricles. **Cerebrospinal fluid** (**CSF**) fills these internal spaces and also surrounds the brain and spinal cord. Ependymal cells assist in producing, monitoring, and circulating CSF. There are three types of ependymal cells: ependymocytes, tanycytes, and specialized CSF-producing ependymal cells (discussed in Chapter 13).

Ependymocytes have motile cilia that aid in the circulation of CSF and also microvilli. Ependymocytes have long slender basal processes that branch and make contact with neuroglia. **Tanycytes** are specialized non-ciliated ependymal cells with microvilli on their apical surfaces. They are found in only one brain ventricle. It is thought that they transport substances between the CSF and the brain.

Microglia (mī-KRŌG-lē-uh) are embryologically related to monocytes and macrophages. Microglia migrate into the CNS as the nervous system forms and they persist as mobile cells, continuously moving through the neural tissue, removing cellular debris, wastes, and pathogens by phagocytosis.

Astrocytes maintain the **blood–brain barrier** that isolates the CNS from the chemicals and hormones circulating in the blood. They also provide structural support within neural tissue; regulate ion, nutrient, and dissolved gas concentrations in the interstitial fluid surrounding the neurons; absorb and recycle neurotransmitters that are not broken down or reabsorbed at synapses; and form scar tissue after CNS injury.

Oligodendrocytes (ol-i-gō-DEN-drō-sīts; *oligo-*, few) provide a structural framework within the CNS by stabilizing the positions of axons. They also produce **myelin** (MĪ-e-lin), a membranous wrapping that coats axons and increases the speed of nerve impulse transmission. When myelinating an axon, the tip of an oligodendrocyte process expands to form an enormous membranous pad containing very little cytoplasm. This flattened "pancake" somehow gets wound around the axon, forming concentric layers of plasma membrane. These layers constitute a **myelin sheath**.

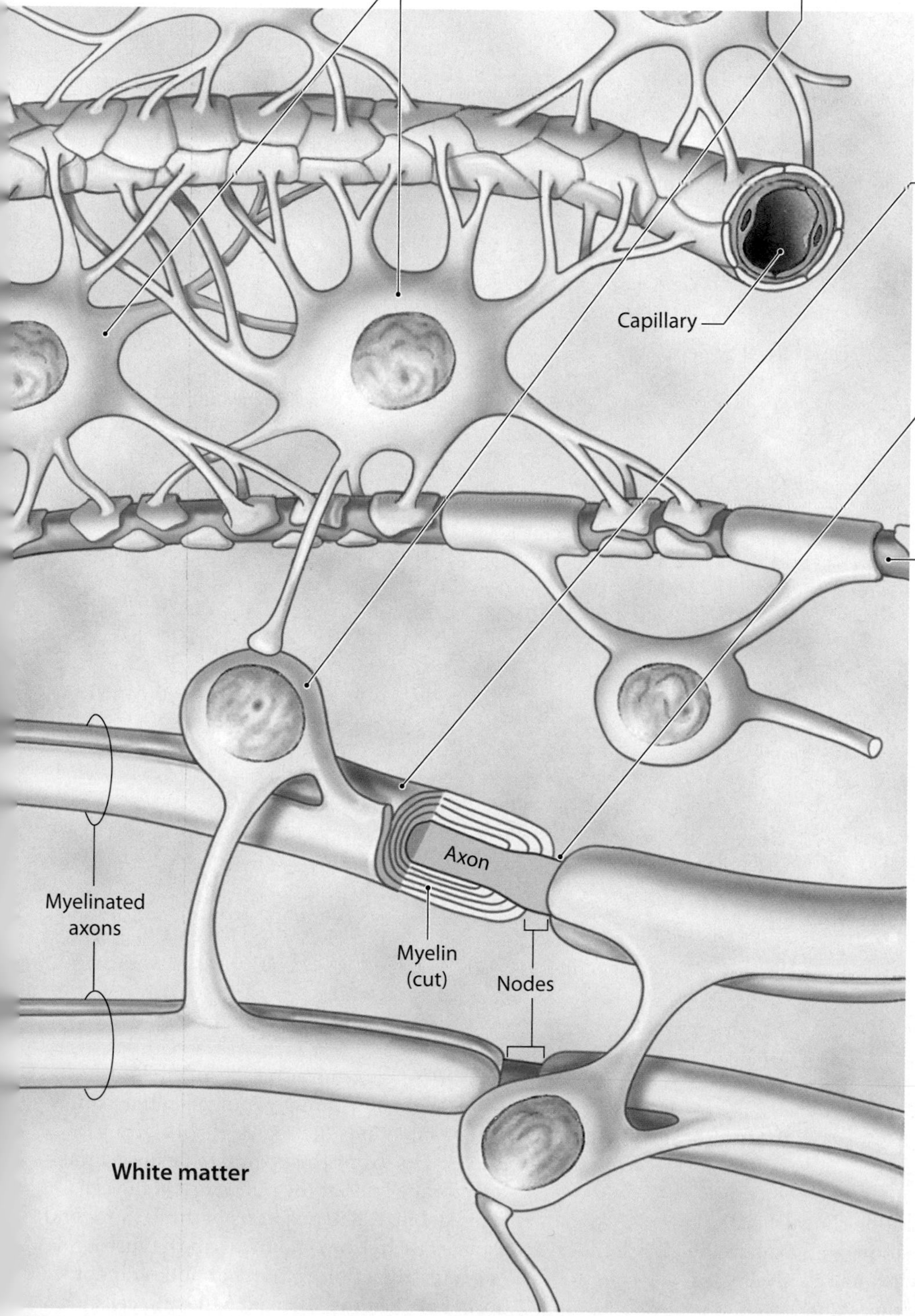

Capillary
Axon
Myelinated axons
Myelin (cut)
Nodes
White matter

Many oligodendrocytes cooperate in the formation of a myelin sheath along the length of an axon. Such an axon is said to be **myelinated**. Each oligodendrocyte myelinates segments of several axons. The relatively large areas of the axon that are thus wrapped in myelin are called **internodes** (*inter*, between).

The small unmyelinated gaps that separate adjacent internodes are called **nodes** (or nodes of Ranvier; RAHN-vē-ā). In dissection, myelinated axons appear glossy white, primarily because of the lipids within the myelin. As a result, regions dominated by myelinated axons make up the **white matter** of the CNS.

Not all axons in the CNS are myelinated. **Unmyelinated axons** may not be completely covered by the processes of neuroglia. Such axons are common where relatively short axons and collaterals form synapses with densely packed neuron cell bodies. Areas containing neuron cell bodies, dendrites, and unmyelinated axons are a dusky gray color, and they make up the **gray matter** of the CNS.

Module 11.4 Review

a. Identify the neuroglia of the central nervous system.

b. Which glial cell protects the CNS from chemicals and hormones circulating in the blood?

c. Which type of neuroglia would increase in the brain tissue of a person with a CNS infection?

LO 11.4 Describe the locations and functions of neuroglia in the CNS.

Schwann cells and satellite cells are the neuroglia of the PNS

Schwann cells cover peripheral axons in two different ways and participate in axon repair after injury. Satellite cells surround peripheral cell bodies.

Myelinated Axons

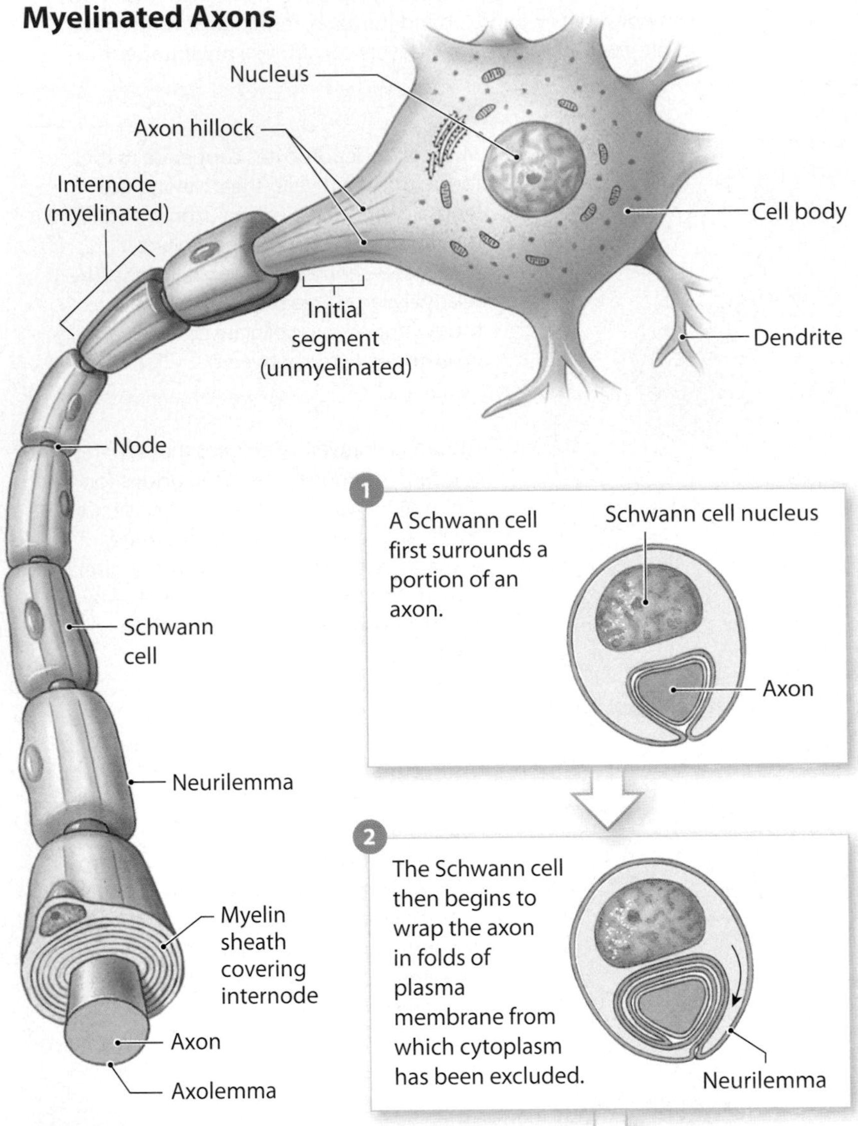
Nucleus
Axon hillock
Internode (myelinated)
Cell body
Initial segment (unmyelinated)
Dendrite
Node
Schwann cell
Neurilemma
Myelin sheath covering internode
Axon
Axolemma

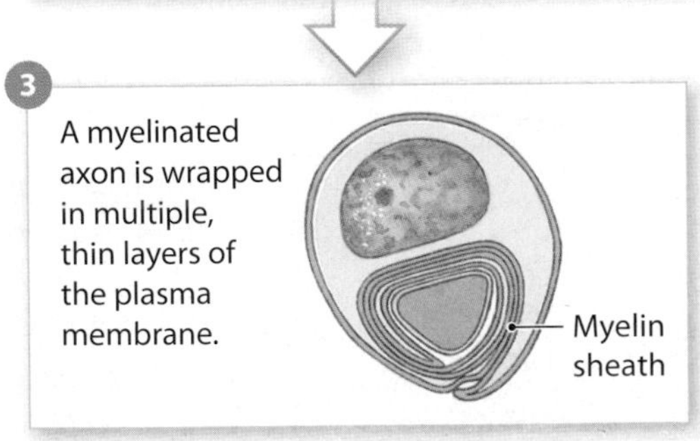
1
A Schwann cell first surrounds a portion of an axon.
Schwann cell nucleus
Axon
2
The Schwann cell then begins to wrap the axon in folds of plasma membrane from which cytoplasm has been excluded.
Neurilemma
3
A myelinated axon is wrapped in multiple, thin layers of the plasma membrane.
Myelin sheath

1 **Schwann cells** form a sheath around peripheral axons. Wherever a Schwann cell covers an axon, the outer surface of the Schwann cell is called the **neurilemma** (nū-ri-LEM-uh). Most axons in the PNS, whether myelinated or unmyelinated, are shielded from contact with interstitial fluid by Schwann cells.

2 This illustration shows the steps in the myelination of an axon in the PNS. A single Schwann cell myelinates one internode of one axon.

Unmyelinated Axons

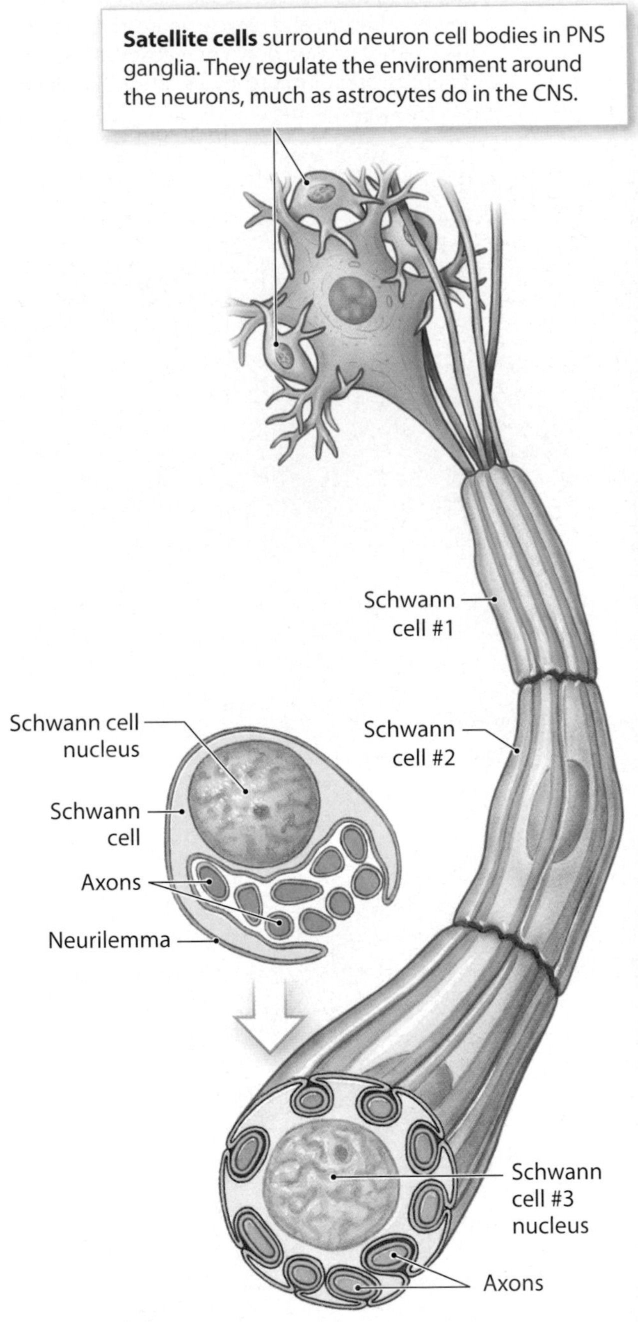
Satellite cells surround neuron cell bodies in PNS ganglia. They regulate the environment around the neurons, much as astrocytes do in the CNS.
Schwann cell #1
Schwann cell #2
Schwann cell nucleus
Schwann cell
Axons
Neurilemma
Schwann cell #3 nucleus
Axons

3 A single Schwann cell can enclose segments of many unmyelinated axons. It first surrounds the segments of a group of axons. The axons then come to lie in separate membrane folds at the periphery of the cell. This stabilizes the positions of these axons and isolates them from chemicals in the surrounding interstitial fluid. Note that nodes do not form between the adjacent Schwann cells.

Axon Injury and Repair

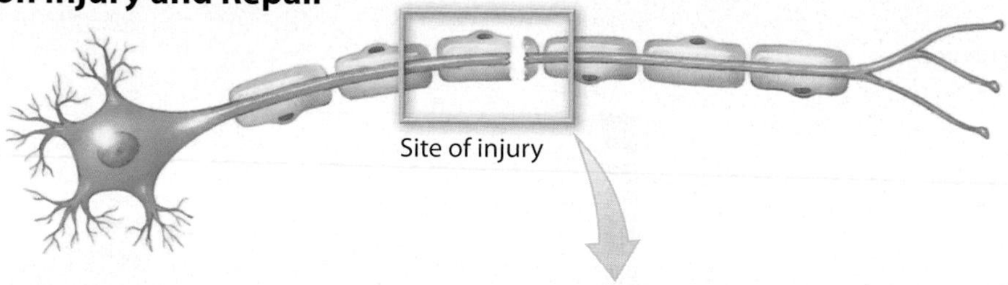
Site of injury

Step 1
Distal to the injury site, the axon and myelin degenerate and fragment.

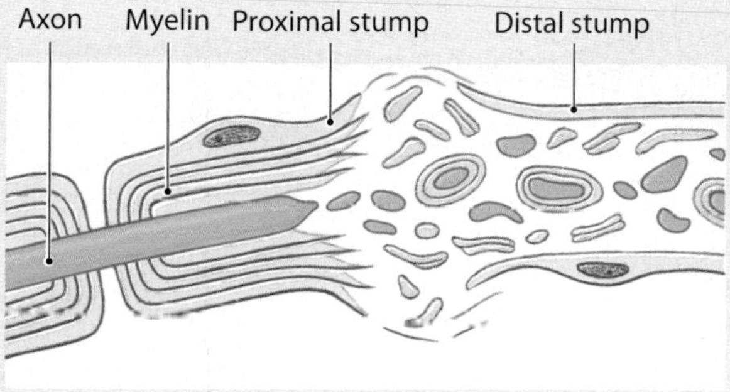
Axon
Myelin
Proximal stump
Distal stump

Step 2
The Schwann cells do not degenerate; instead, they proliferate along the path of the original axon. Over this period, macrophages move into the area and remove the degenerating debris distal to the injury site.

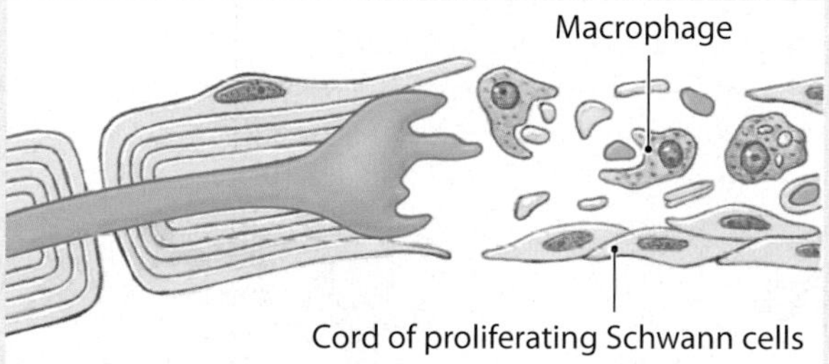
Macrophage
Cord of proliferating Schwann cells

Step 3
As the neuron recovers, its axon grows into the site of injury and then distally, along the path created by the newly divided Schwann cells.

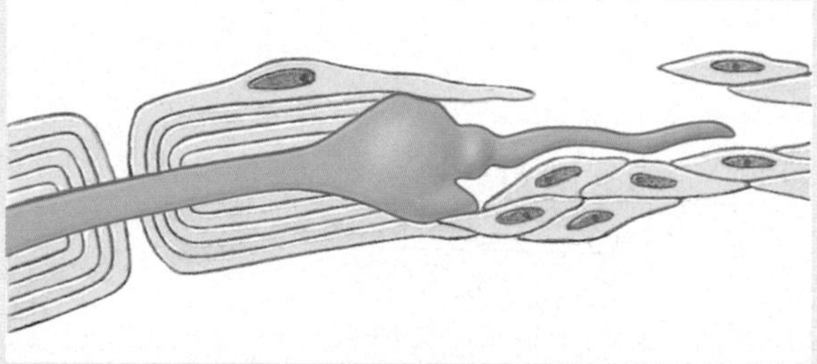

Step 4
As the axon elongates, the Schwann cells wrap around it. If the axon reestablishes its normal synaptic contacts, normal function may be regained. However, if it stops growing or grows in some new direction, normal function will not return.

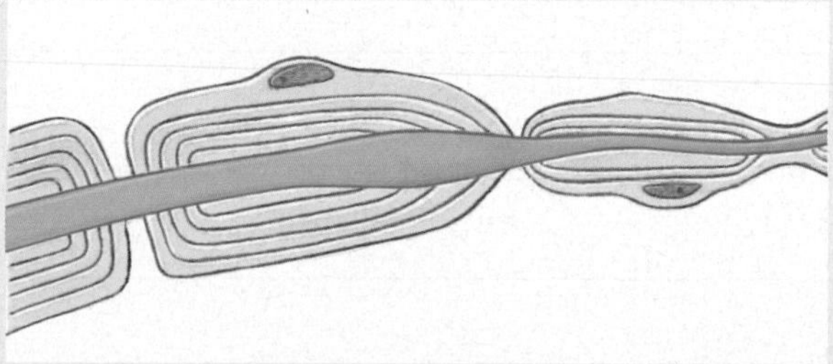

4 Schwann cells participate in the repair of damaged nerves in the PNS. The repair process, which often fails to restore full function, is known as **Wallerian degeneration**.

Limited regeneration can occur in the CNS, but there the situation is more complicated because (1) many more axons are likely to be involved, (2) astrocytes produce scar tissue that can prevent axon growth across the damaged area, and (3) astrocytes release chemicals that block the regrowth of axons.

Module 11.5 Review

a. Identify the neuroglia of the PNS.
b. Describe the neurilemma.
c. In which part of the nervous system does Wallerian degeneration occur?

LO 11.5 Describe the locations and functions of Schwann cells and satellite cells.

Labeling

Label each of the structures in the following diagram of a neuron.

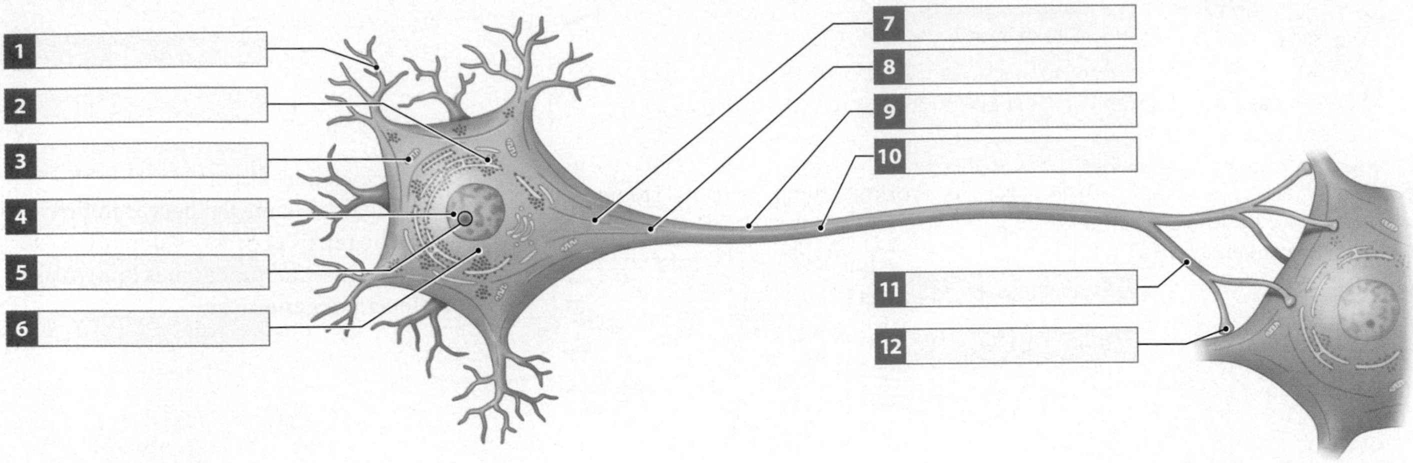
1
2
3
4
5
6
7
8
9
10
11
12

Label the anatomical classes of neurons shown below.

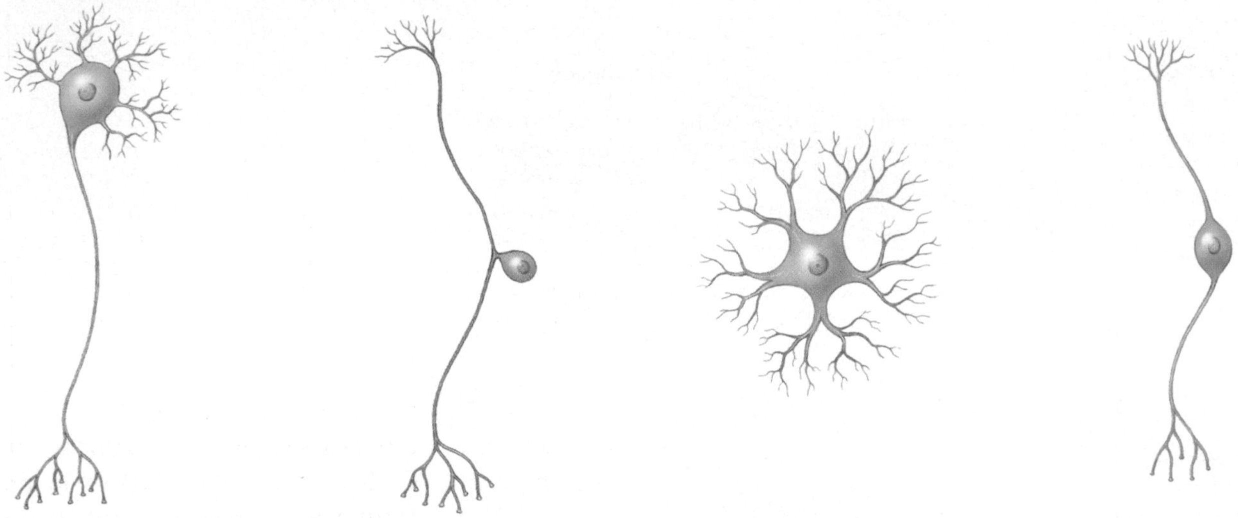

13 ____________ 14 ____________ 15 ____________ 16 ____________

Vocabulary

In the space provided, write the boldfaced terms introduced in this section that contain the indicated word part.

17 neur- *(nerve)* 17 ______________________________

18 dendr- *(tree)* 18 ______________________________

19 ef- *(away from)* 19 ______________________________

20 af- *(toward)* 20 ______________________________

Neuronal activity depends on changes in membrane potential

1 Plasma membranes are selectively permeable, and the cytosol and extracellular fluid differ in composition. Under normal circumstances, the inside of the plasma membrane has a slight negative charge with respect to the outside. The cause is a slight excess of positively charged ions outside the plasma membrane, and a slight excess of negatively charged ions and proteins inside the plasma membrane. Recall from Chapter 9 that this unequal charge distribution is called the **membrane potential**, or **transmembrane potential**, and it is a characteristic of all living cells. Membrane potential is the result of differences in the membrane's permeability to various ions and of active transport mechanisms.

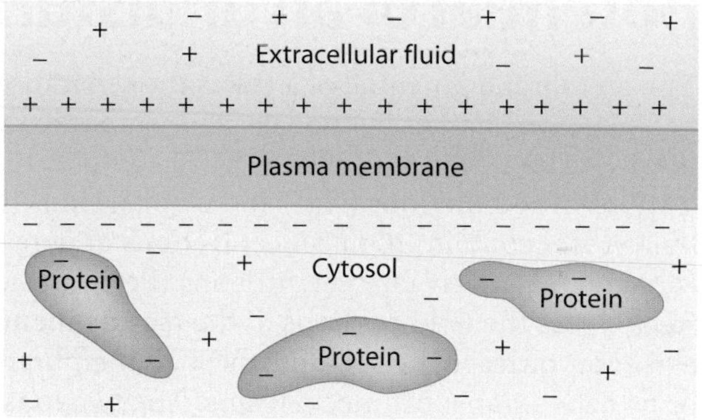
Extracellular fluid
Plasma membrane
Cytosol
Protein
Protein
Protein

1. The membrane potential of an undisturbed cell is called the **resting membrane potential**, or simply the **resting potential**. All neural activities begin with a change in the resting potential of a neuron.
See Modules 11.7 and 11.8

2. A typical stimulus produces a temporary, localized change in the resting potential. This change, which decreases with distance away from the stimulus, is called a **graded potential**.
See Module 11.9

3. If the graded potential is sufficiently large, it triggers an **action potential** in the excitable membrane of the axon. An action potential is an electrical event that involves one location on the membrane. Once an action potential develops in one location, it spreads along the surface of an axon toward the axon terminals.
See Modules 11.10 and 11.11

4. **Synaptic activity** then produces graded potentials in the plasma membrane of the postsynaptic cell. The process typically involves the release of neurotransmitters, such as ACh, by the presynaptic cell. These compounds bind to receptors on the postsynaptic plasma membrane, changing its permeability.
See Module 11.12

5. The response of the postsynaptic cell ultimately depends on what the stimulated receptors do and what other stimuli are influencing the cell at the same time. The integration of stimuli at the level of the individual cell is the simplest form of **information processing** in the nervous system.
See Modules 11.13 and 11.14

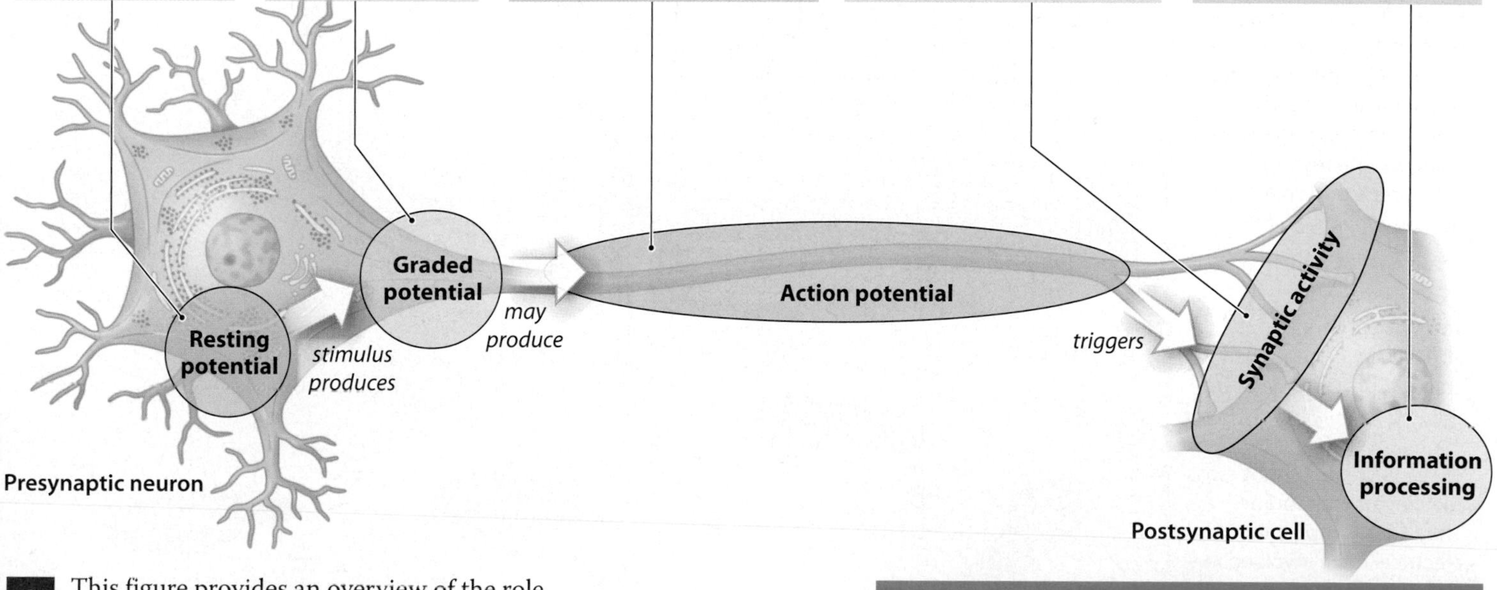
Resting potential
stimulus produces
Graded potential
may produce
Action potential
triggers
Synaptic activity
Information processing
Presynaptic neuron
Postsynaptic cell

2 This figure provides an overview of the role of the membrane potential in neural activity. Changes in the membrane potential have many important functions; for example, they can trigger muscle contraction and gland secretion and transfer information in the nervous system.

Module 11.6 Review

a. Define membrane potential.
b. Compare a graded potential with an action potential.
c. Define information processing.

LO 11.6 Describe the general role of membrane potential changes in neuronal activity.

Module 11.7

The resting potential is the membrane potential of an undisturbed cell

Watch
MasteringA&P®
A&P**Flix**
Resting Membrane Potential

The membrane potential of a cell is the separation of positive and negative charges by the plasma membrane (Module 9.5, p. 314). The extracellular fluid (ECF) contains high concentrations of sodium ions (Na^+) and chloride ions (Cl^-), whereas the cytosol contains high concentrations of potassium ions (K^+) and negatively charged proteins (Pr^-). The ions cannot freely cross the lipid portions of the plasma membrane. They can enter or leave the cell only through membrane channels or by active transport mechanisms. The two ions we are most concerned with are potassium and sodium, because they are the main factors that influence the membrane potential. This module discusses the roles of **1** the differing membrane permeabilities for K^+ and Na^+, **2** the sodium–potassium exchange pump, and **3** the chemical and electrical gradients for K^+ and Na^+.

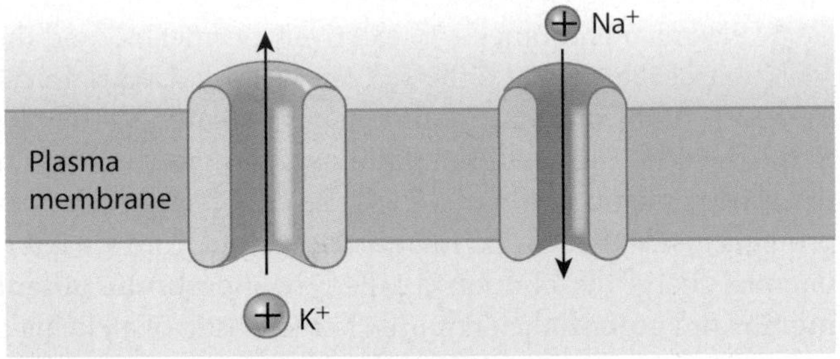
Na+
Plasma membrane
K+

1 The membrane potential exists primarily because plasma membranes contain passive **leak channels**, which are always open. Their size, shape, and structure determine what ions can pass through the membrane.

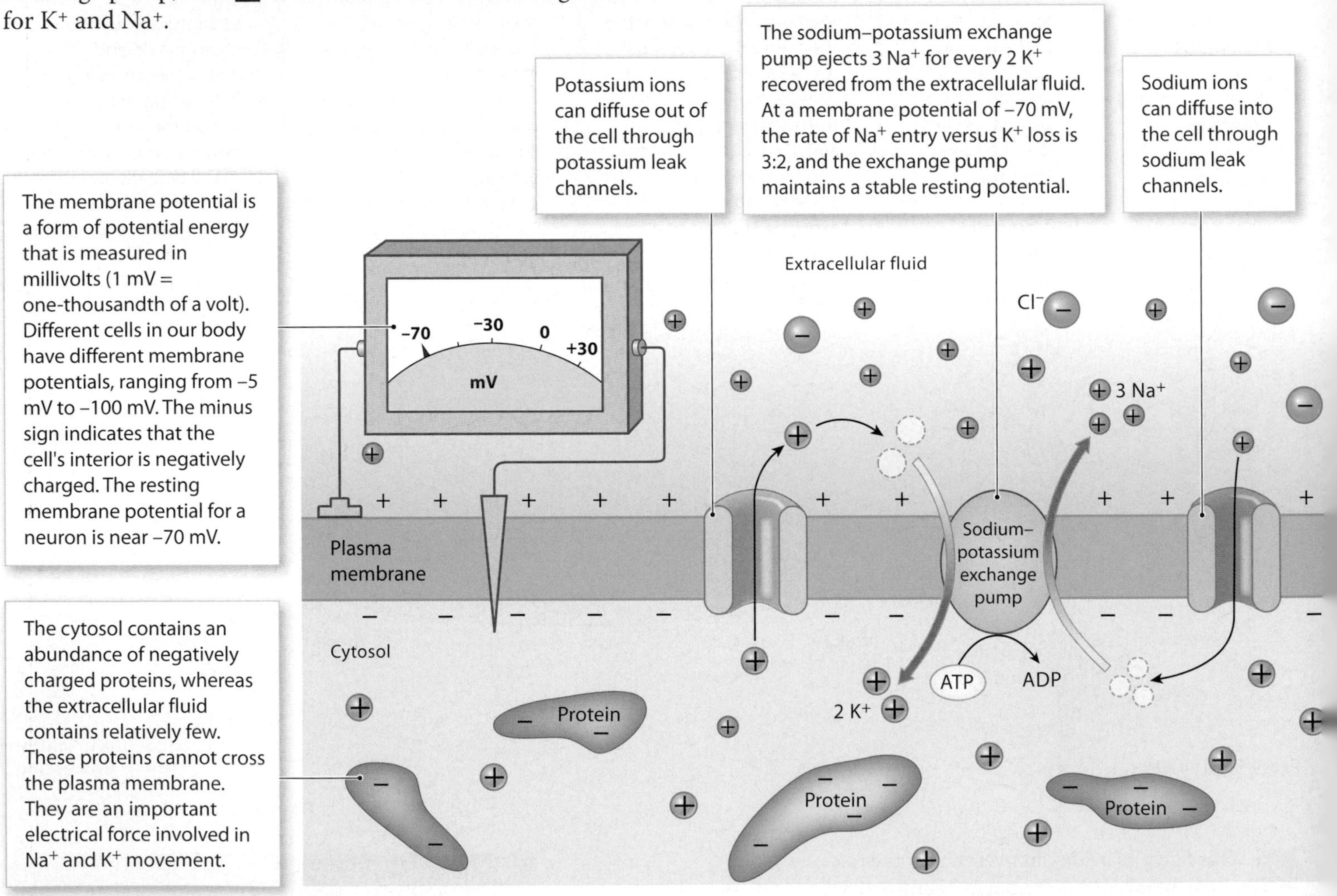
The membrane potential is a form of potential energy that is measured in millivolts (1 mV = one-thousandth of a volt). Different cells in our body have different membrane potentials, ranging from –5 mV to –100 mV. The minus sign indicates that the cell's interior is negatively charged. The resting membrane potential for a neuron is near –70 mV.
The cytosol contains an abundance of negatively charged proteins, whereas the extracellular fluid contains relatively few. These proteins cannot cross the plasma membrane. They are an important electrical force involved in Na+ and K+ movement.
Potassium ions can diffuse out of the cell through potassium leak channels.
The sodium–potassium exchange pump ejects 3 Na+ for every 2 K+ recovered from the extracellular fluid. At a membrane potential of –70 mV, the rate of Na+ entry versus K+ loss is 3:2, and the exchange pump maintains a stable resting potential.
Sodium ions can diffuse into the cell through sodium leak channels.
–70
–30
0
+30
mV
Extracellular fluid
Cl−
3 Na+
Plasma membrane
Sodium–potassium exchange pump
Cytosol
ATP
ADP
2 K+
Protein
Protein
Protein

2 An undisturbed (unstimulated) cell has a characteristic **resting membrane potential**, or **resting potential**. This is an overview of the events responsible for the resting potential of a neuron. Both passive forces and active processes act across the plasma membrane to produce and maintain the resting potential. The passive forces are driven by both chemical and electrical gradients. The Na^+–K^+ exchange pump is the active ATP-requiring process.

Potassium Ion Gradients
At normal resting potential, an electrical gradient opposes the chemical gradient for potassium ions (K^+). The net **electrochemical** gradient tends to force potassium ions out of the cell.

If the plasma membrane were freely permeable to potassium ions, the outflow of K^+ would continue until the **equilibrium potential** (–90 mV) was reached. Note how similar it is to the resting potential.

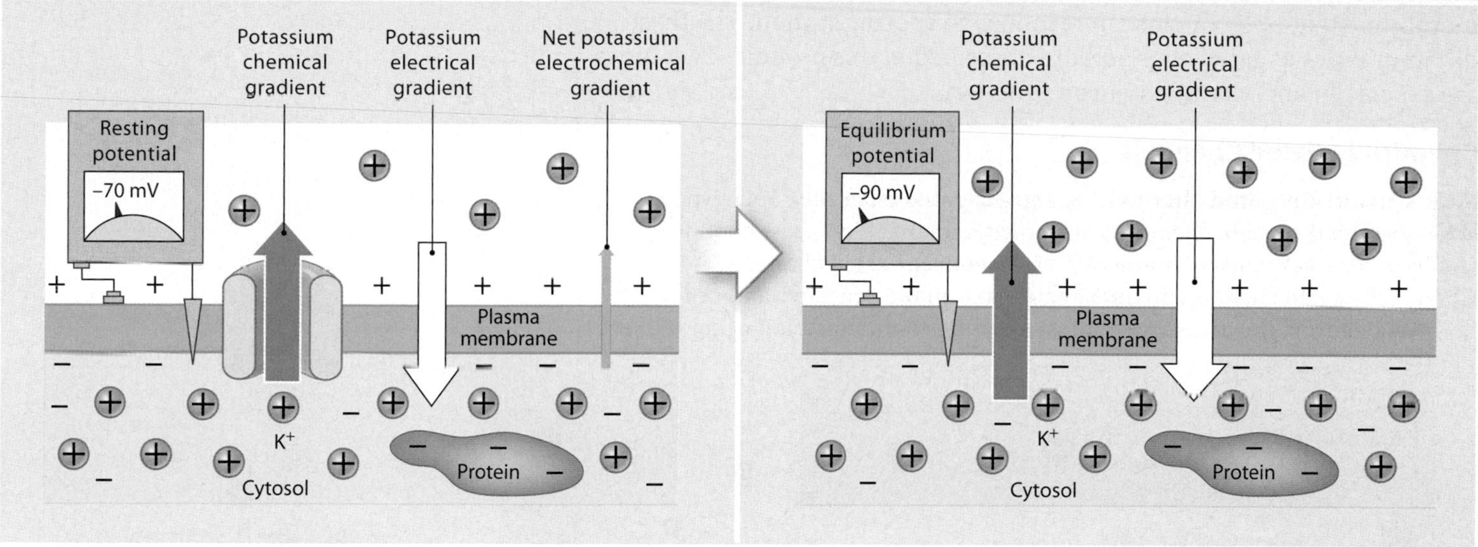

Potassium chemical gradient
Potassium electrical gradient
Net potassium electrochemical gradient
Resting potential
–70 mV
Plasma membrane
K+
Cytosol
Protein
Equilibrium potential
–90 mV

Sodium Ion Gradients
At the normal resting potential, chemical and electrical gradients combine to drive sodium ions (Na^+) into the cell.

If the plasma membrane were freely permeable to sodium ions, the influx of Na^+ would continue until the equilibrium potential (+66 mV) was reached. Note how different it is from the resting potential.

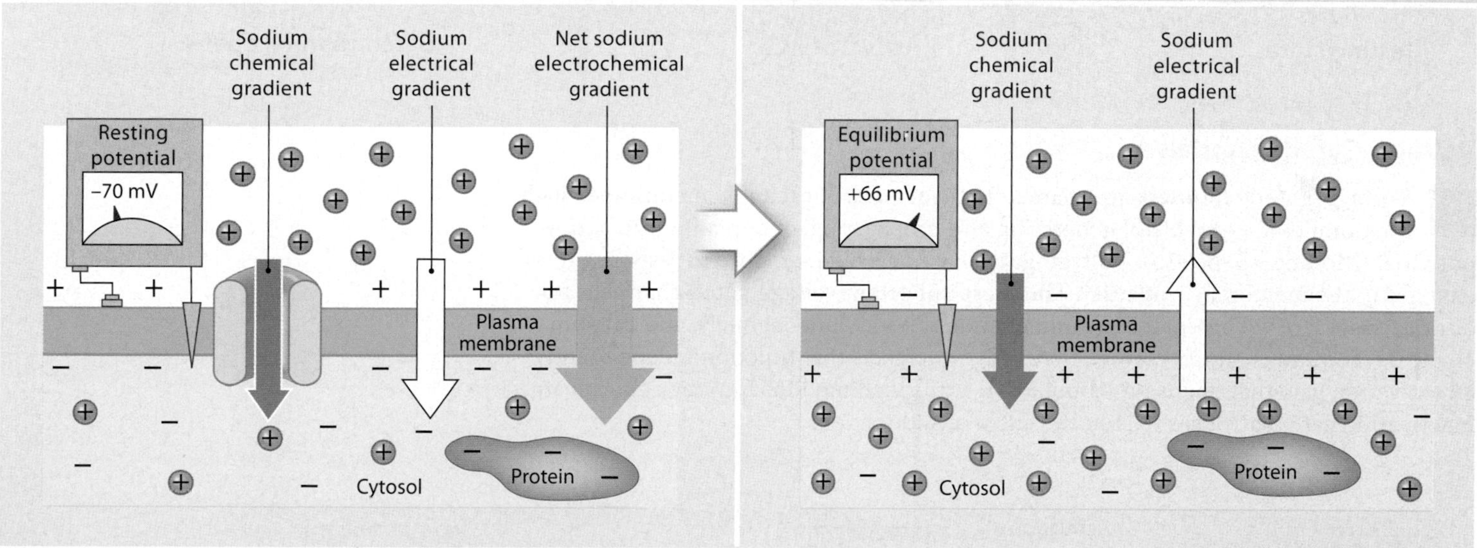

Sodium chemical gradient
Sodium electrical gradient
Net sodium electrochemical gradient
Resting potential
–70 mV
Plasma membrane
Cytosol
Protein
Equilibrium potential
+66 mV

3 The **chemical gradient** for any ion is its concentration gradient across the plasma membrane. The **electrical gradient** is created by the attraction between opposite charges, or the repulsion between like charges (+/+ or –/–). Together, these gradients determine an ion's **electrochemical gradient**, a form of potential energy. At a membrane potential known as the **equilibrium potential**, an ion's electrical and chemical gradients are equal and opposite, and there is no net movement of ions across the membrane. The equilibrium potential indicates an ion's contribution to the resting potential.

Module 11.7 Review

a. Define resting membrane potential.
b. What happens at the sodium–potassium exchange pump?
c. What effect would decreasing the concentration of extracellular potassium ions have on the membrane potential of a neuron?

LO 11.7 Explain how the resting potential is created and maintained.

Module 11.8

Three types of gated channels change the permeability of the plasma membrane

Permeability changes are due to the **gated channels** within the plasma membrane that open or close in response to specific stimuli. The three different types of gated channels are described in this module. Two types are especially important in neuron function.

Chemically Gated Channels

1 **Chemically gated channels**, or **ligand-gated channels**, open when they bind specific chemicals. The receptors that bind acetylcholine (ACh) at the neuromuscular junction are chemically gated channels. Chemically gated channels are most abundant on the dendrites and cell body of a neuron, the areas where most synaptic communication occurs.

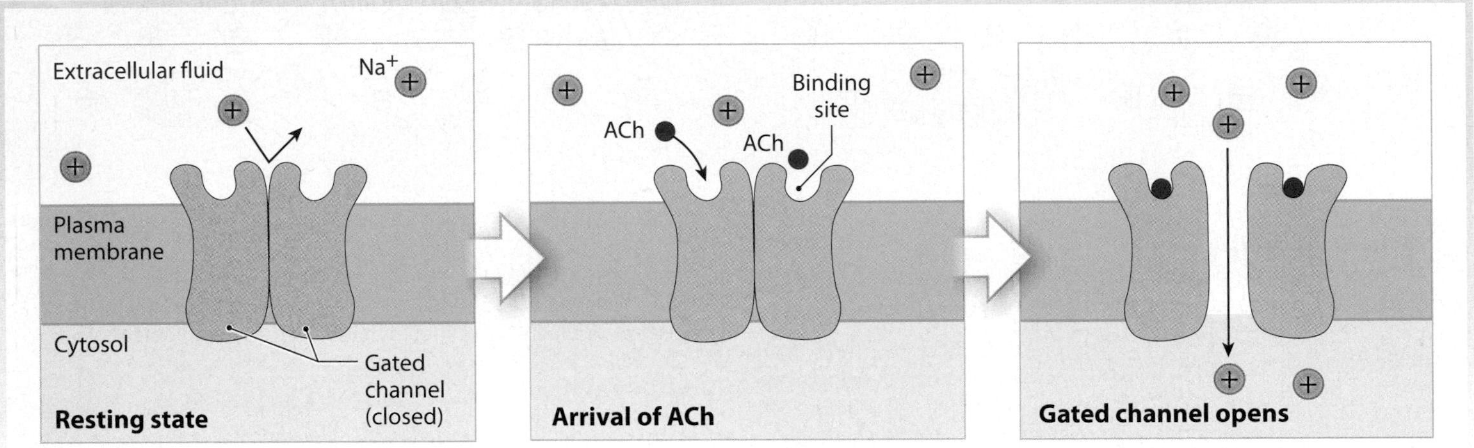

Extracellular fluid
Na^+
Plasma membrane
Cytosol
Gated channel (closed)
Resting state
Binding site
ACh
ACh
Arrival of ACh
Gated channel opens

Voltage-Gated Channels

2 **Voltage-gated channels** are characteristic of areas of excitable membrane. Such membranes are capable of generating and propagating, or spreading, an action potential (Module 9.5, p. 315). Voltage-gated channels open or close in response to changes in the membrane potential. The most important voltage-gated channels, for our purposes, are voltage-gated sodium channels, potassium channels, and calcium channels. Sodium channels (shown here) have two gates that function independently: an activation gate that opens on stimulation, letting sodium ions into the cell, and an inactivation gate that closes to stop the entry of sodium ions.

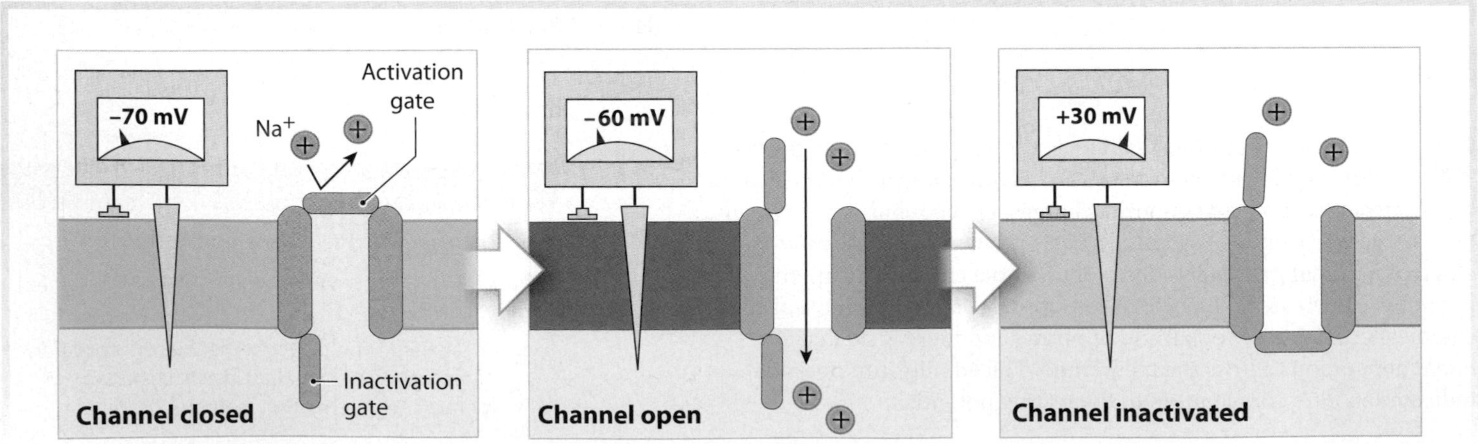

–70 mV
Na^+
Activation gate
Inactivation gate
Channel closed
–60 mV
Channel open
+30 mV
Channel inactivated

Mechanically Gated Channels

3 **Mechanically gated channels** open in response to physical distortion of the membrane surface. Such channels are important in sensory receptors that respond to touch, pressure, or vibration.

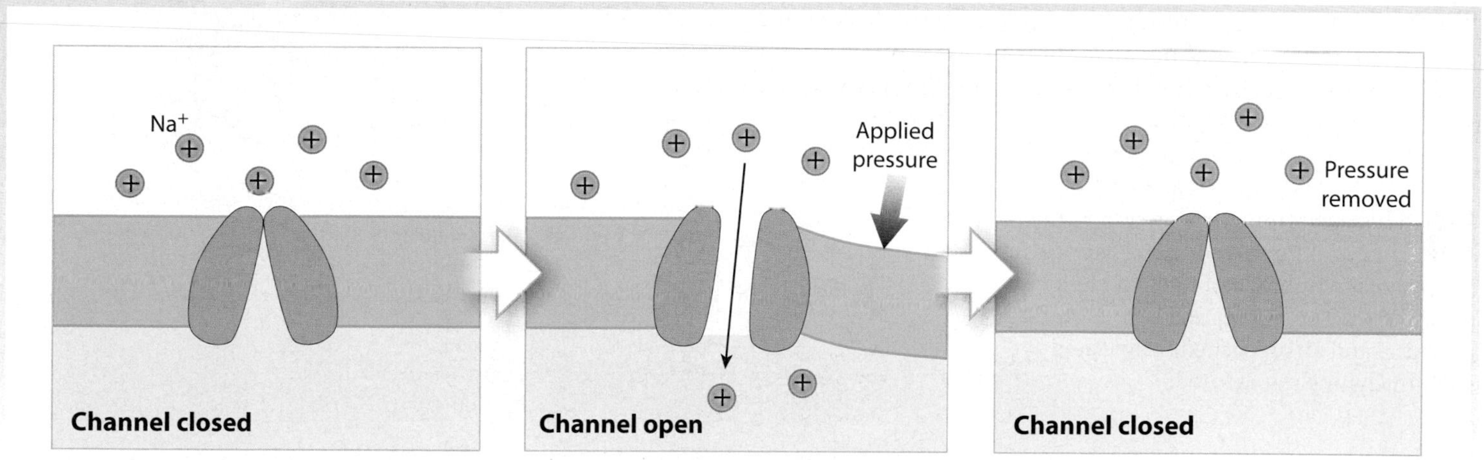

Na+
Channel closed
Applied pressure
Channel open
Pressure removed
Channel closed

Distribution of Gated Channels on a Neuron

4 Chemically gated channels are found on the neuron cell body and its dendrites. Voltage-gated Na^+ channels and voltage-gated K^+ channels are found along the axon of a neuron. As we will see, their distribution along an axon depends on whether it is myelinated or unmyelinated. Voltage-gated Ca^{2+} channels occur at axon terminals.

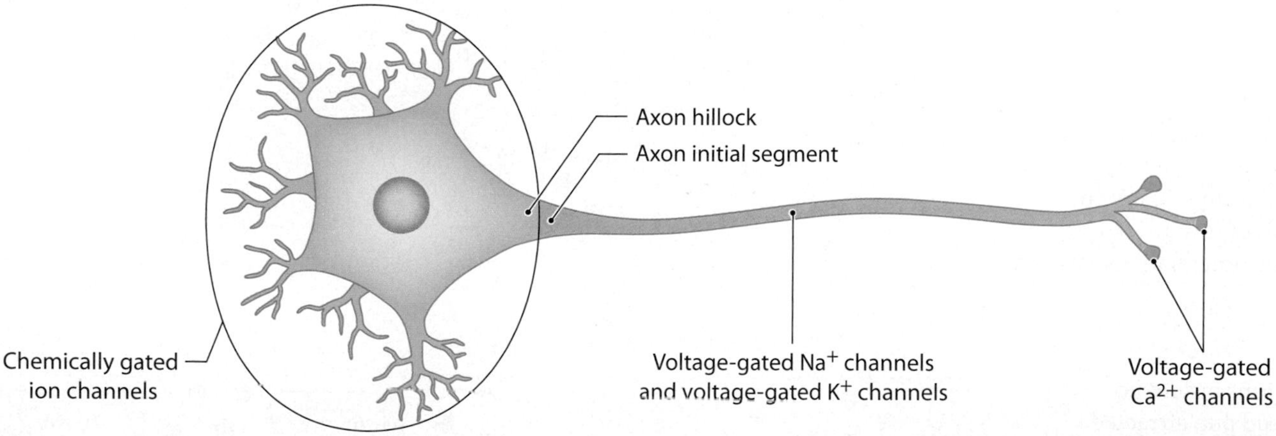

Axon hillock
Axon initial segment
Chemically gated ion channels
Voltage-gated Na+ channels and voltage-gated K+ channels
Voltage-gated Ca2+ channels

Most gated channels are closed at the resting potential. The opening of gated channels increases the rate of ion movement across the plasma membrane and this changes the membrane potential.

Module 11.8 Review

a. Define gated channels.

b. Identify the three types of gated channels, and state the conditions under which each operates.

c. What effect would a chemical that blocks voltage-gated sodium channels in a neuron's plasma membrane have on its membrane potential?

LO **11.8** Describe the functions of gated channels with respect to the permeability of the plasma membrane.

Graded potentials are localized changes in the membrane potential

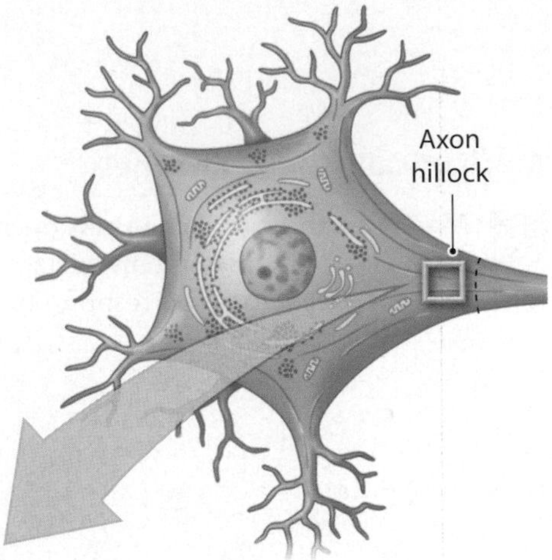
Axon
hillock

Graded potentials, or local potentials, are changes (currents) in the membrane potential that cannot spread far from the site of stimulation. Any stimulus that opens a gated channel will produce a graded potential. Note that when illustrating graded potentials, we can ignore the leak channels responsible for the resting potential because their properties do not change. In this module we primarily consider the gated channels for sodium ions.

1 This diagrammatic view shows the chemically gated sodium channels in the plasma membrane of the axon hillock of a neuron. The membrane is at its resting potential and all of its chemically gated sodium channels are closed.

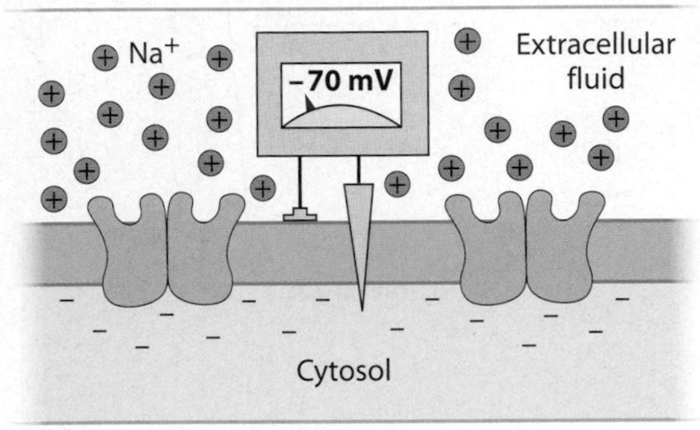
Na^+
–70 mV
Extracellular
fluid
Cytosol

2 When the membrane is exposed to a chemical that opens the chemically gated sodium channels, sodium ions enter the cell. This inrush of positive charges due to the large Na^+ electrochemical gradient reduces the membrane potential in this area. Any shift from the resting potential toward a more positive value is called a **depolarization**. Depolarization applies to changes in potential from –70 mV to smaller negative values (toward 0 mV) as well as to membrane potentials above 0 mV.

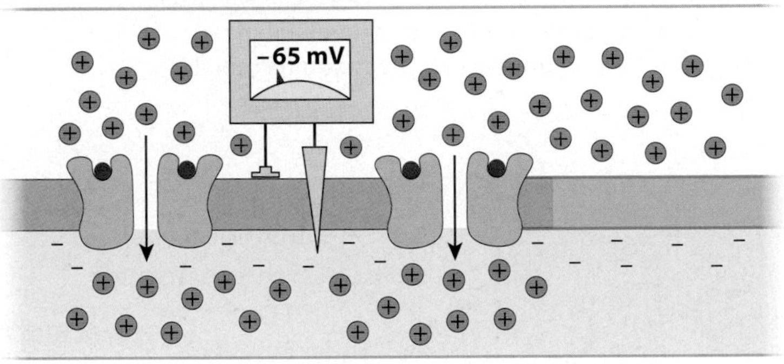
–65 mV

3 The sodium ions inside the cell now spread out, attracted by the negative charges along the inner surface of the membrane. The extent of depolarization spreads as well. As the plasma membrane depolarizes, extracellular sodium ions move toward the open channels, replacing ions that have entered the cell. This movement of positive charges parallel to the inner and outer surfaces of the membrane is called a **local current**.

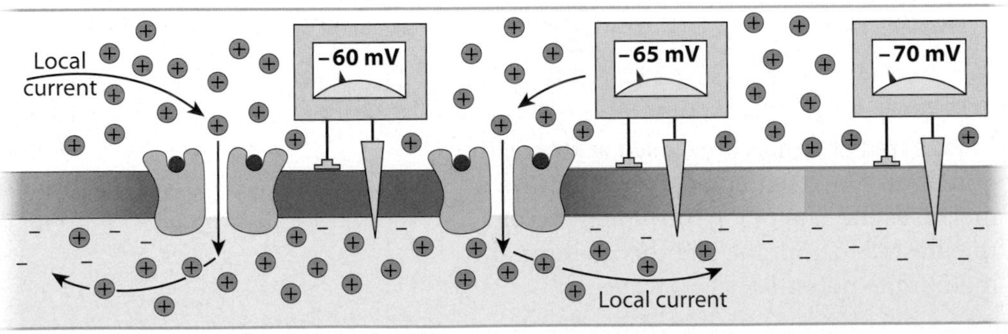
Local
current
–60 mV
–65 mV
–70 mV
Local current

4 The degree of depolarization decreases with distance away from the stimulation site, primarily because the ions are entering only in one location but spreading in all directions. The impact on the membrane potential is proportional to the size of the stimulus, because the intensity of the stimulus determines the number of open sodium channels. The more open channels there are, the more sodium ions enter the cell, the greater the membrane area affected, and the greater the degree of depolarization.

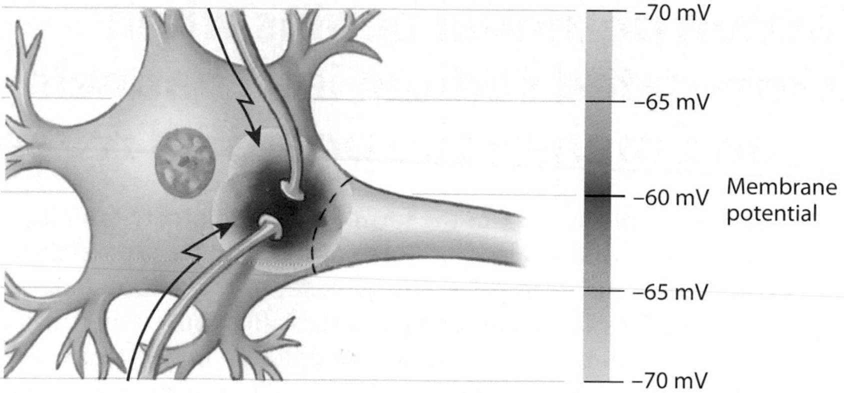

−70 mV
−65 mV
−60 mV
Membrane potential
−65 mV
−70 mV

5 This graph shows the changes in the membrane potential that occur over time when different chemical stimuli are applied to the axon hillock.

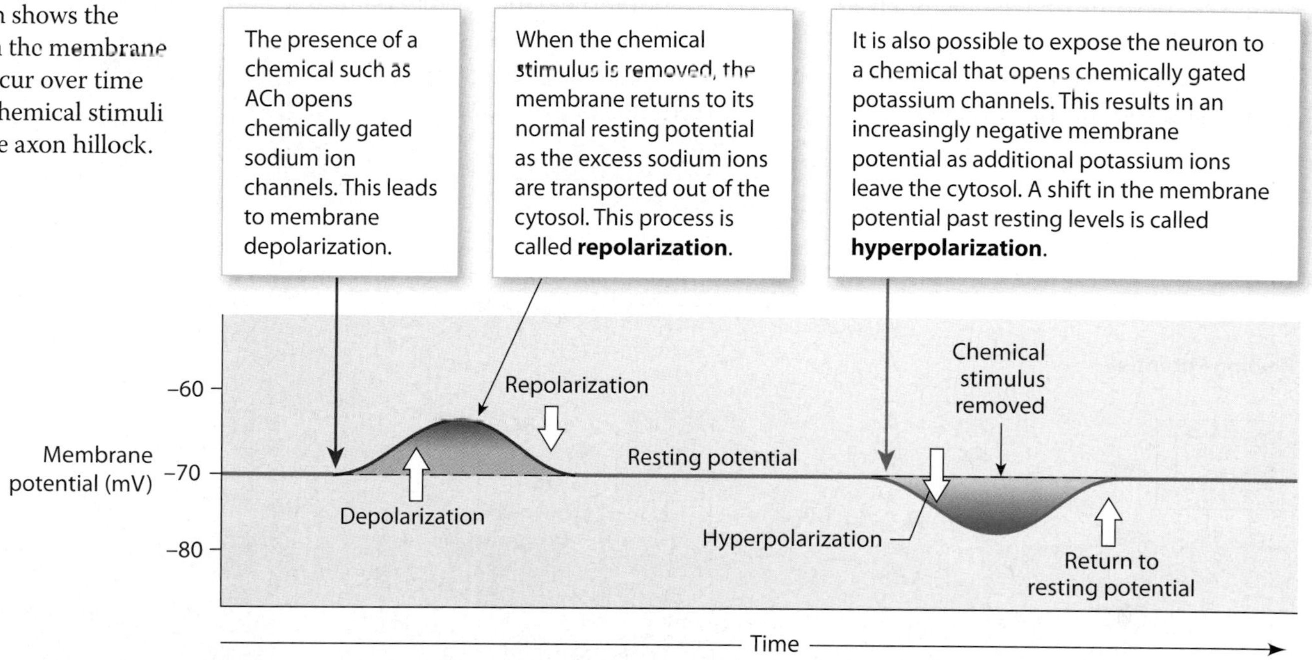

The presence of a chemical such as ACh opens chemically gated sodium ion channels. This leads to membrane depolarization.
When the chemical stimulus is removed, the membrane returns to its normal resting potential as the excess sodium ions are transported out of the cytosol. This process is called **repolarization**.
It is also possible to expose the neuron to a chemical that opens chemically gated potassium channels. This results in an increasingly negative membrane potential as additional potassium ions leave the cytosol. A shift in the membrane potential past resting levels is called **hyperpolarization**.
Membrane potential (mV)
−60
−70
−80
Repolarization
Depolarization
Resting potential
Chemical stimulus removed
Hyperpolarization
Return to resting potential
Time

Graded Potentials

Graded potentials, whether depolarizing or hyperpolarizing, share four basic characteristics:

1. The membrane potential is most affected at the site of stimulation, and the effect decreases with distance.
2. The effect spreads passively through local currents.
3. The graded change in membrane potential may involve either depolarization or hyperpolarization. The nature of the change is determined by the properties of the membrane channels involved. For example, in a resting membrane, the opening of sodium channels will cause depolarization, whereas the opening of potassium channels will cause hyperpolarization. Thus, the change in membrane potential reflects whether positive charges enter or leave the cell.
4. The stronger the stimulus, the greater the change in the membrane potential, and the larger the area affected.

6 This table summarizes the most important characteristics of graded potentials.

Module 11.9 Review

a. Define graded potential.

b. Describe depolarization, repolarization, and hyperpolarization.

c. What factors account for the local currents associated with graded potentials?

LO 11.9 Describe graded potentials.

An action potential begins when voltage-gated sodium ion channels open and membrane potential reverses

Watch
MasteringA&P®
A&PFlix
Generation of an Action Potential

Each neuron receives information in the form of graded potentials on its dendrites and cell body, and graded potentials at the axon terminals trigger the release of neurotransmitters. However, the two ends of the neuron may be a meter apart, and even the largest graded potentials affect only a tiny area. Such relatively long-range communication requires a different mechanism—the action potential. **Action potentials** are propagated changes in the membrane potential that, once initiated, affect an entire excitable membrane. Whereas the resting potential depends on leak channels and the graded potential depends on chemically gated channels, action potentials depend on voltage-gated channels. Recall that both neurons and skeletal muscle cells have areas of excitable membrane containing voltage-gated channels (Module 9.5, p. 314).

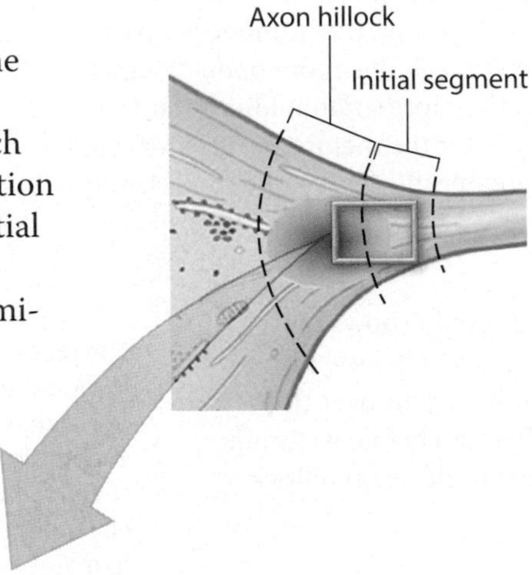
Axon hillock
Initial segment

Resting Potential

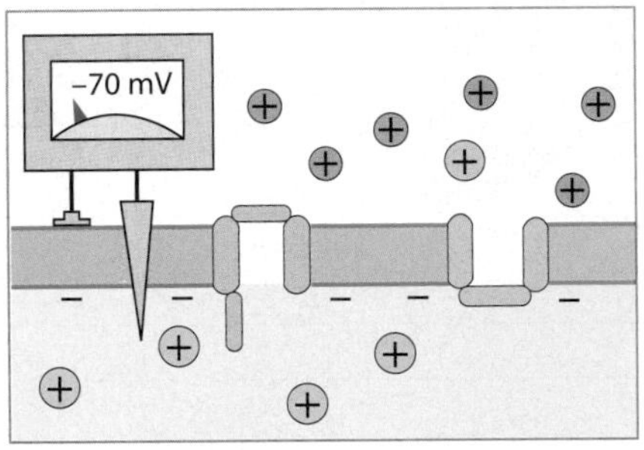
–70 mV

The axolemma contains both voltage-gated sodium channels and voltage-gated potassium channels. Both types are closed when the membrane is at the resting potential.

KEY

⊕ = Sodium ion
⊕ = Potassium ion

1 Depolarization to Threshold

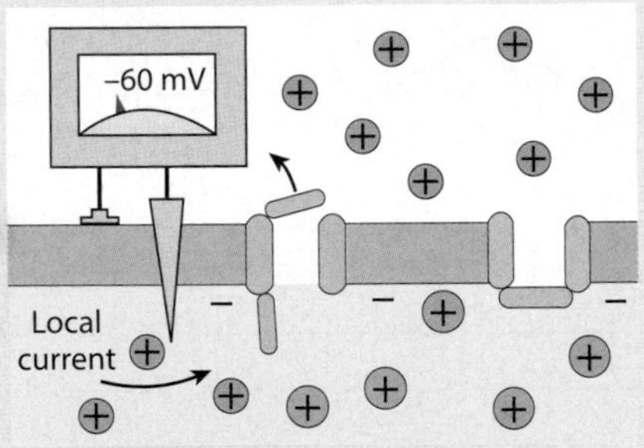
–60 mV
Local current

The stimulus that initiates an action potential is a graded depolarization large enough to open voltage-gated sodium channels. The opening of the channels occurs at a membrane potential known as the **threshold**.

2 Activation of Sodium Channels and Rapid Depolarization

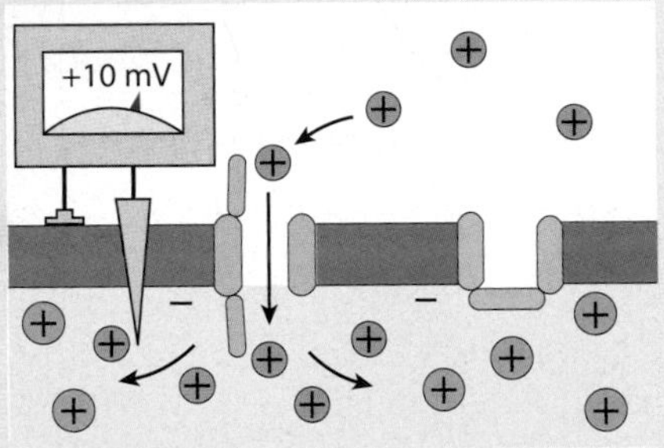
+10 mV

When the sodium channel activation gates open, the plasma membrane becomes much more permeable to Na^+. Driven by the large electrochemical gradient, sodium ions rush into the cytosol, and rapid depolarization occurs. The inner membrane surface now contains more positive ions than negative ones, and the membrane potential has changed from –60 mV to a positive value.

1 This figure illustrates steps in the formation of an action potential at the initial segment of an axon. The first step is a graded depolarization caused by the opening of chemically gated sodium ion channels, usually at the axon hillock. Note that when illustrating action potentials, we can ignore both the leak channels and the chemically gated channels, because their properties do not change.

2 This graph plots the changes in the membrane potential at one location during the generation of an action potential.

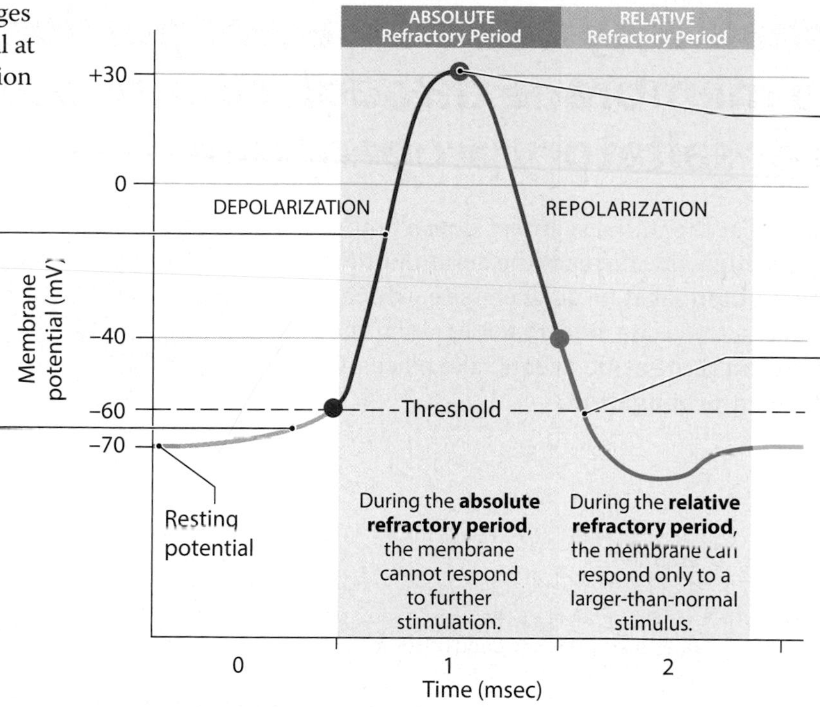
ABSOLUTE Refractory Period
RELATIVE Refractory Period
+30
0
–40
–60
–70
Membrane potential (mV)
DEPOLARIZATION
REPOLARIZATION
Threshold
Resting potential
During the **absolute refractory period**, the membrane cannot respond to further stimulation.
During the **relative refractory period**, the membrane can respond only to a larger-than-normal stimulus.
0
1
2
Time (msec)

1 A graded depolarization brings an area of excitable membrane to threshold (–60 mV).

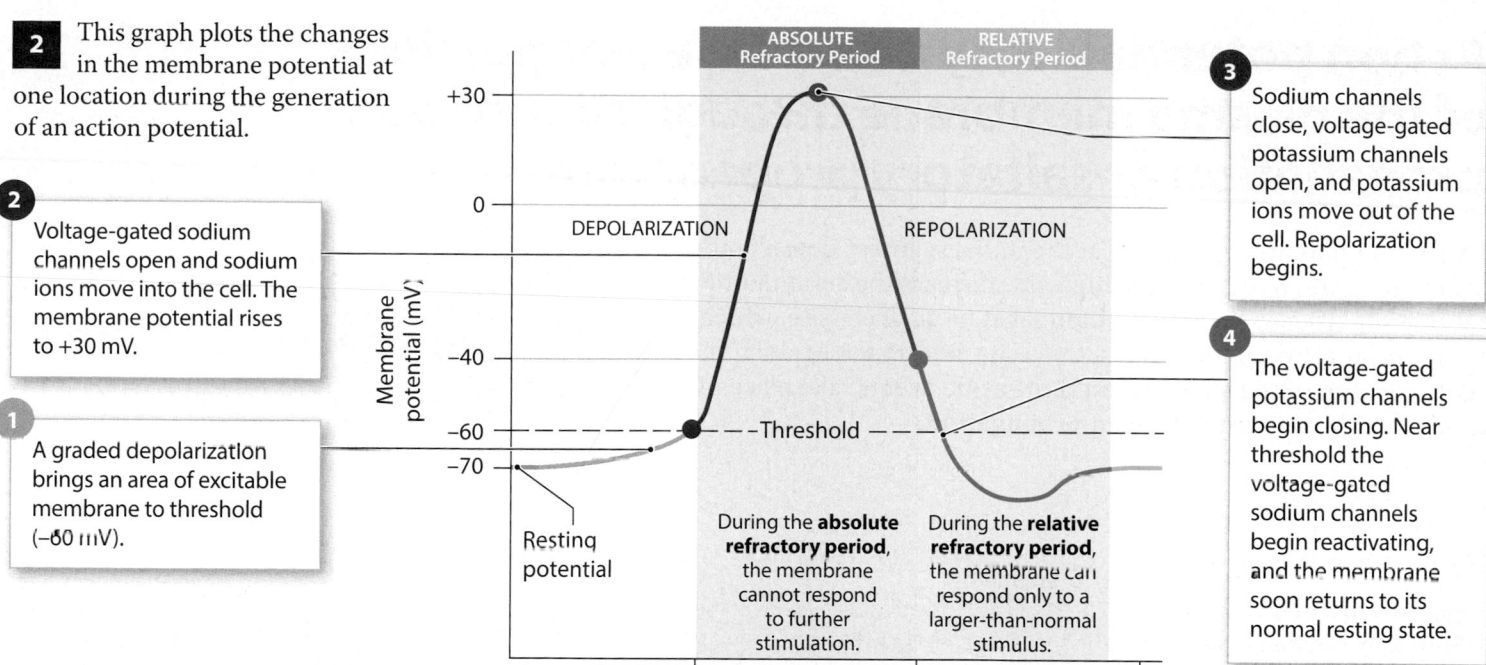

2 Voltage-gated sodium channels open and sodium ions move into the cell. The membrane potential rises to +30 mV.

3 Sodium channels close, voltage-gated potassium channels open, and potassium ions move out of the cell. Repolarization begins.

4 The voltage-gated potassium channels begin closing. Near threshold the voltage-gated sodium channels begin reactivating, and the membrane soon returns to its normal resting state.

3 **Inactivation of Sodium Channels and Activation of Potassium Channels**

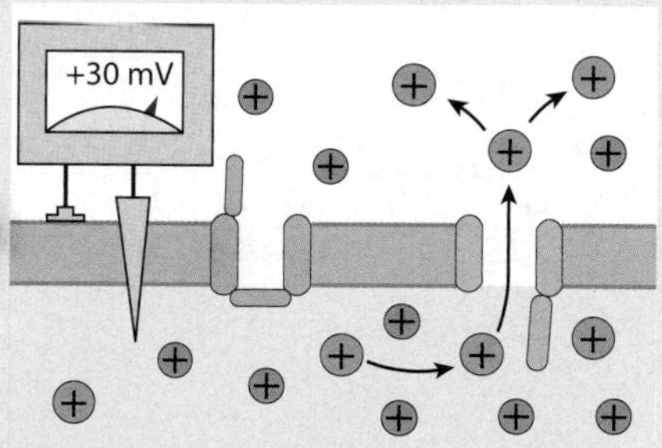
+30 mV

As the membrane potential approaches +30 mV, the inactivation gates of the voltage-gated sodium channels close. This step is known as **sodium channel inactivation**, and it coincides with the opening of voltage-gated potassium channels. Positively charged potassium ions move out of the cytosol, shifting the membrane potential back toward resting levels. Repolarization now begins.

4 **Potassium Channels Close**

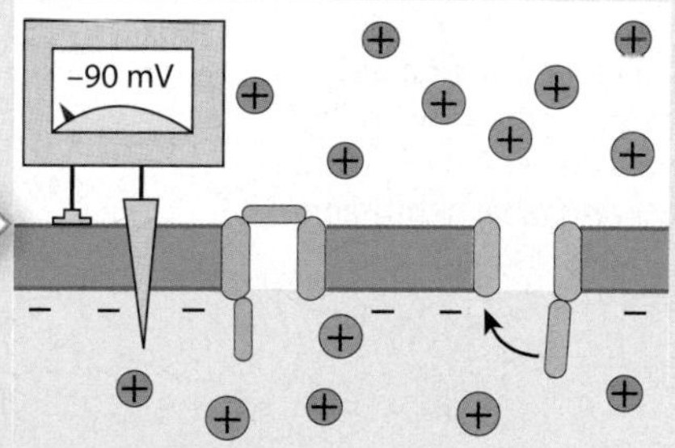
–90 mV

The voltage-gated sodium channels remain inactivated until the membrane has repolarized to near threshold levels. At this time, they regain their normal status: closed but capable of opening. The voltage-gated potassium channels begin closing as the membrane reaches the normal resting potential (about –70 mV). Until all of these potassium channels have closed, potassium ions continue to leave the cell. This produces a brief hyperpolarization.

Resting Potential

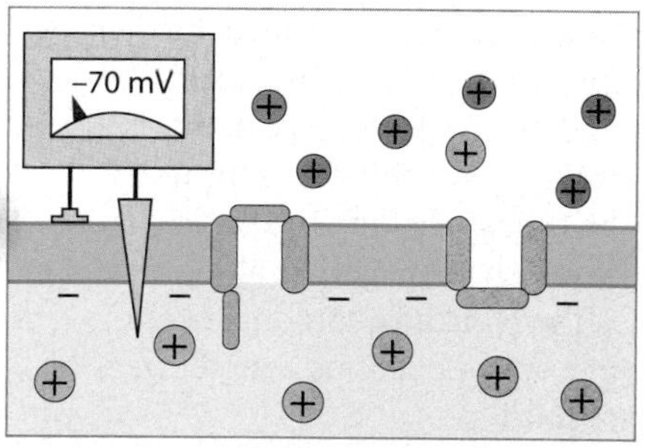
–70 mV

As the voltage-gated potassium channels close, the membrane potential returns to normal resting levels. The action potential is now over, and the membrane is once again at the resting potential.

In an excitable membrane, a graded depolarization is like the pressure on the trigger of a gun, and the action potential is like the firing of the gun. As long as the trigger is pulled hard enough, the gun fires, and it fires the same way every time. Similarly, all stimuli that bring the membrane to threshold generate identical action potentials. This concept is called the **all-or-none principle**, because a given stimulus either triggers a typical action potential, or it does not trigger one at all.

Module 11.10 Review

a. Define action potential.

b. List the events involved in the generation of an action potential.

c. Compare the absolute refractory period with the relative refractory period.

LO **11.10** Describe the events involved in the generation and propagation of an action potential.

Module 11.11

Action potentials may affect adjacent portions of the plasma membrane through continuous propagation or saltatory propagation

Watch
MasteringA&P®
A&PFlix™
Propagation of an Action Potential

An action potential generated at the initial segment doesn't move along the axon like a car on a highway. Instead, the action potential at one site triggers an action potential at an adjacent site, which triggers an action potential at a third site, and so forth, along the entire length of the axon. Because the same events take place over and over, this activity is termed **propagation**.

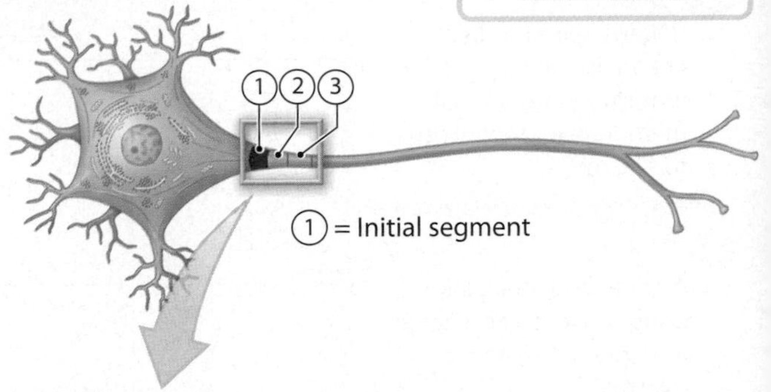

① ② ③
① = Initial segment

Continuous Propagation

1 In **continuous propagation**, the action potential appears to "move" along the axolemma of an axon in a series of tiny steps. Continuous propagation occurs along unmyelinated axons. This diagram shows how an action potential generated in the initial segment ① affects more distant portions of the axon ② and ③. Each step takes only a millisecond, but the steps must be repeated along the entire axon, so propagation along an unmyelinated axon only occurs at a speed of about 1 meter per second.

Step 1
As an action potential develops at the initial segment ①, the membrane potential at this site depolarizes to +30 mV.

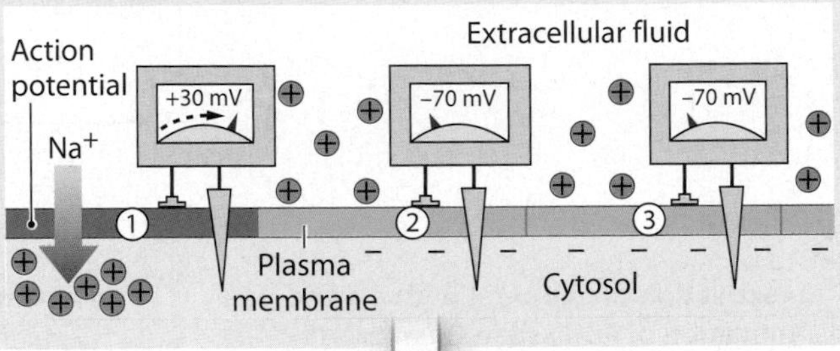

Action potential
Extracellular fluid
Na^+
+30 mV
–70 mV
–70 mV
Plasma membrane
Cytosol
① ② ③

Step 2
As the sodium ions entering at ① spread away from the open voltage-gated channels, a graded depolarization quickly brings the membrane in segment ② to threshold.

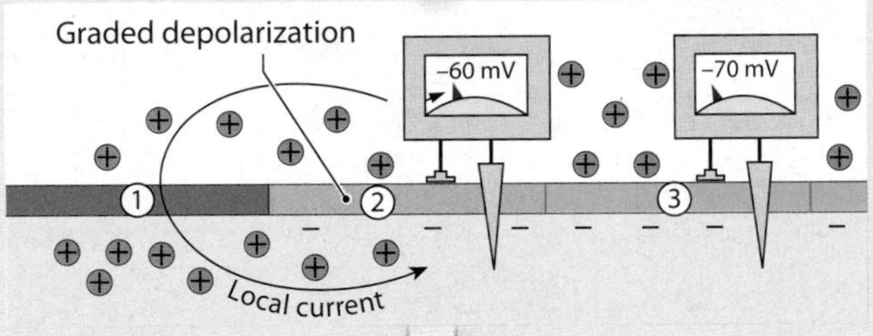

Graded depolarization
–60 mV
–70 mV
Local current
① ② ③

Step 3
An action potential now occurs in segment ② while segment ① begins repolarization.

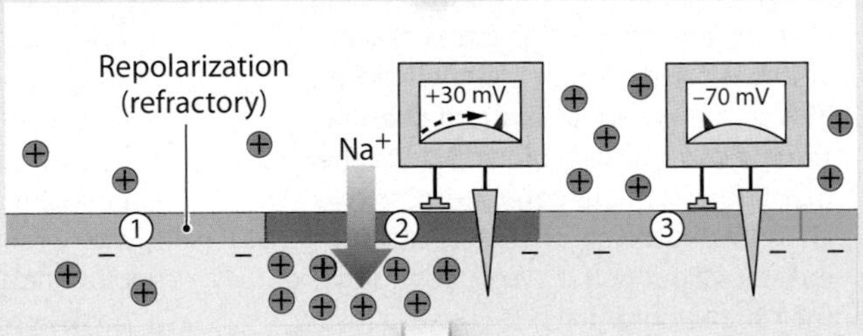

Repolarization (refractory)
Na^+
+30 mV
–70 mV
① ② ③

Step 4
As the sodium ions entering at segment ② spread laterally, a graded depolarization quickly brings the membrane in segment ③ to threshold. The action potential can only move forward, not backward, because the membrane at segment ① is in the absolute refractory period of repolarization.

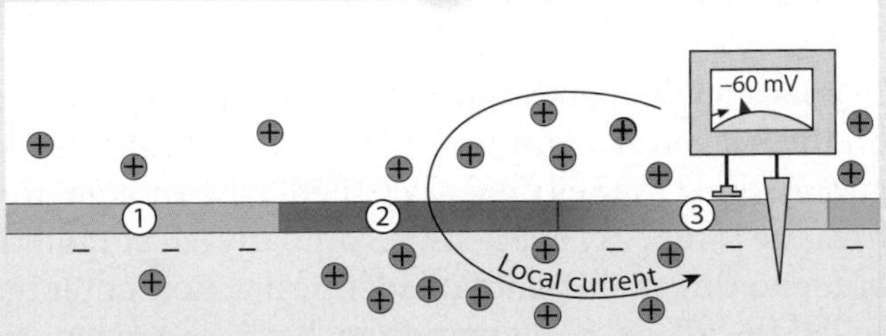

–60 mV
Local current
① ② ③

Saltatory Propagation

2 Continuous propagation cannot occur along a myelinated axon, because myelin blocks the flow of ions across the membrane. Ions can only cross the plasma membrane at the unmyelinated nodes (nodes of Ranvier). As a result, only the nodes can respond to a depolarizing stimulus. When an action potential appears at the initial segment of a myelinated axon, the local current skips the internode and depolarizes the closest node in a process called **saltatory propagation** (*saltare*, leaping).

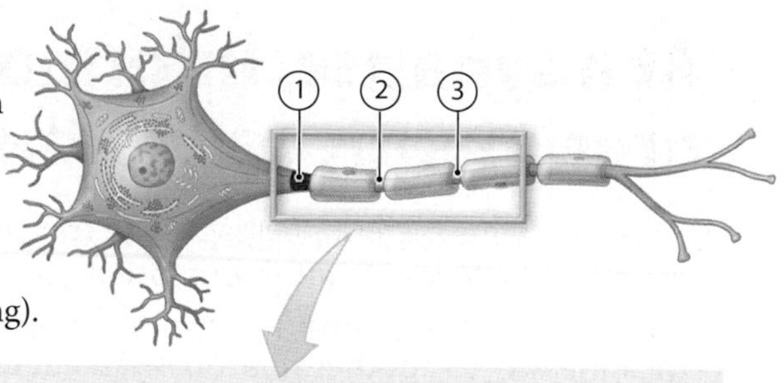

1
2
3

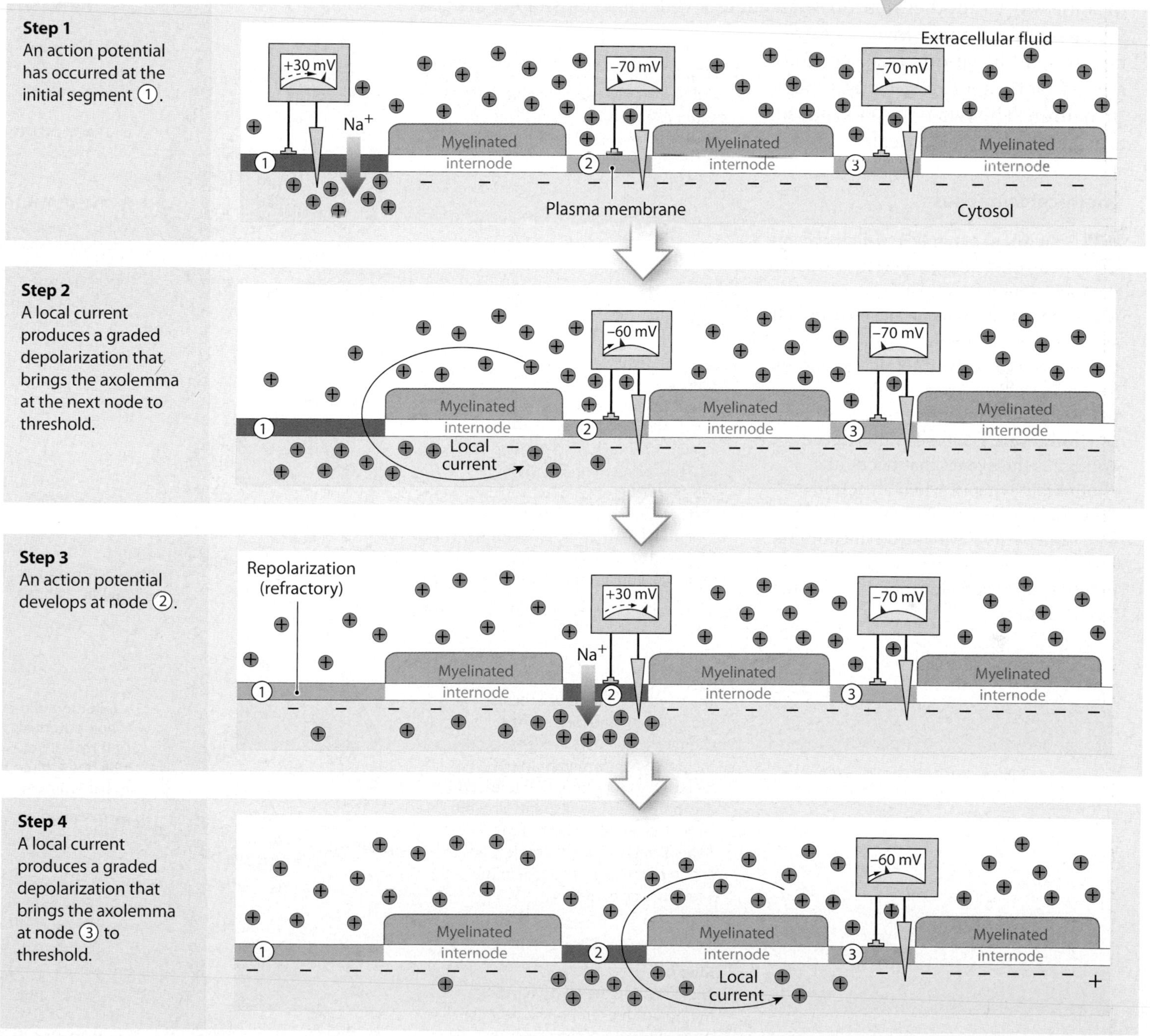

Step 1
An action potential has occurred at the initial segment ①.
+30 mV
−70 mV
−70 mV
Extracellular fluid
Na+
Myelinated internode
Plasma membrane
Cytosol
Step 2
A local current produces a graded depolarization that brings the axolemma at the next node to threshold.
−60 mV
−70 mV
Local current
Step 3
An action potential develops at node ②.
Repolarization (refractory)
+30 mV
−70 mV
Na+
Step 4
A local current produces a graded depolarization that brings the axolemma at node ③ to threshold.
−60 mV
Local current

Saltatory propagation is much faster than continuous propagation, but there are differences among myelinated axons. That is because the speed varies with axon diameter. The larger the diameter, the lower the resistance to ion movement, and the faster the action potential travels.

Module 11.11 Review

a. Define continuous propagation and saltatory propagation.

b. What is the relationship between myelin and the propagation speed of action potentials?

LO 11.11 Describe continuous propagation and saltatory propagation, and discuss the factors that affect the speed with which action potentials are propagated.

At a synapse, information travels from the presynaptic cell to the postsynaptic cell

In the nervous system, messages are transmitted from one location to another along axons in the form of action potentials, also known as "nerve impulses." To be effective, messages must be not only propagated along an axon but also transferred in some way to another neuron or an effector cell. That transfer occurs at a **synapse**. At a synapse involving two neurons, information is relayed from a presynaptic neuron to a postsynaptic neuron. The two types of synapses are chemical and electrical.

Chemical Synapses

1 **Chemical synapses**, which rely on neurotransmitter release, are by far the most abundant type of synapse. Most synapses between neurons, and all synapses between neurons and other types of cells, involve chemical synapses. Synapses that release acetylcholine (ACh) as a neurotransmitter are known as **cholinergic synapses**. This figure illustrates the events that occur at a cholinergic synapse when an action potential arrives at an axon terminal. The same steps are involved in communication across a neuromuscular junction, a process described in Module 9.6 (p. 316).

Step 1
The normal stimulus for neurotransmitter release is the depolarization of the axon terminal by the arrival of an action potential.

Step 2
The depolarization of the axon terminal opens voltage-gated calcium channels. Calcium ions rush into the axon terminal and trigger the exocytosis of synaptic vesicles and the release of ACh into the synaptic cleft. The calcium ions that triggered exocytosis are rapidly removed, ending the release of ACh.

Step 3
The released ACh diffuses across the synaptic cleft and binds to the chemically gated Na^+ receptors on the postsynaptic membrane. The greater the amount of ACh released, the more receptors respond, and the larger the depolarization. If the depolarization is great enough, an action potential will appear in the postsynaptic neuron.

Step 4
The effects on the postsynaptic membrane are temporary, because of the presence of the enzyme acetylcholinesterase (AChE) in the synaptic cleft. ACh molecules that bind to receptor sites are generally broken down within 20 msec of their arrival by AChE. Other ACh molecules diffuse away from the site.

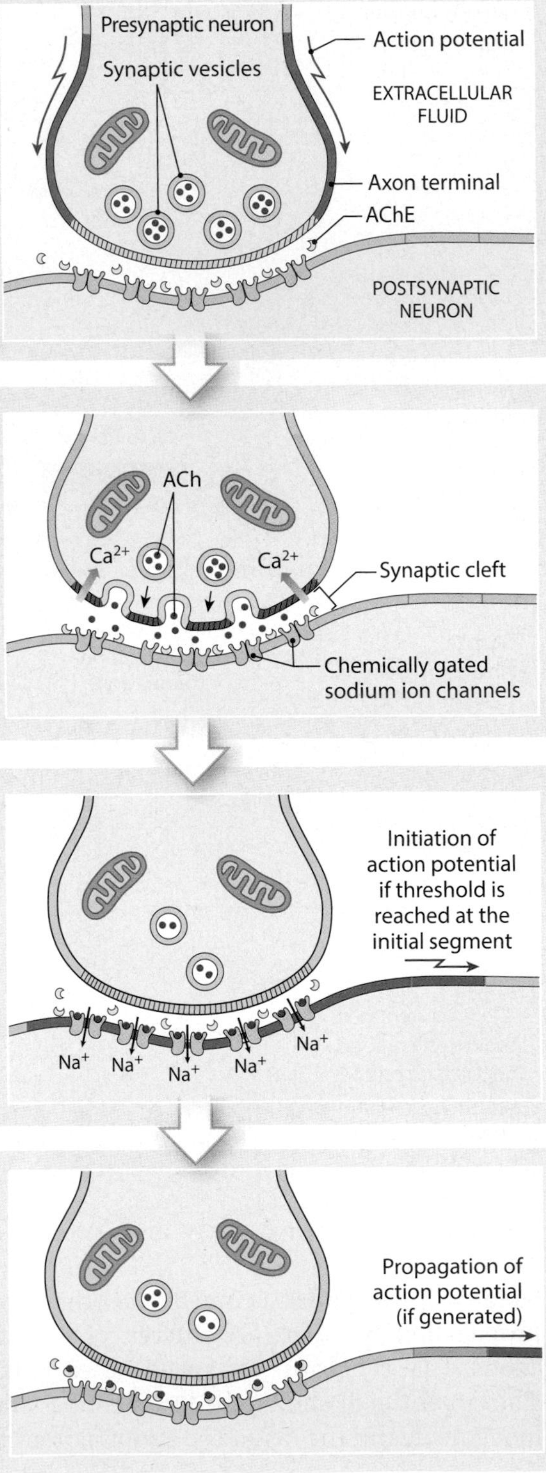

Presynaptic neuron
Synaptic vesicles
Action potential
EXTRACELLULAR FLUID
Axon terminal
AChE
POSTSYNAPTIC NEURON
ACh
Ca2+
Ca2+
Synaptic cleft
Chemically gated sodium ion channels
Initiation of action potential if threshold is reached at the initial segment
Na+
Na+
Na+
Na+
Na+
Propagation of action potential (if generated)

2 This figure summarizes the events that occur at a cholinergic synapse each time an action potential arrives at the **axon terminal**. Because neurotransmitters are reabsorbed and recycled, the **axon terminal** can continue to function for an extended period. However, after extended stimulation it may be unable to keep pace with the demand for neurotransmitter. **Synaptic fatigue** then occurs, and the synapse will be unable to function normally until its supply of ACh has been replenished.

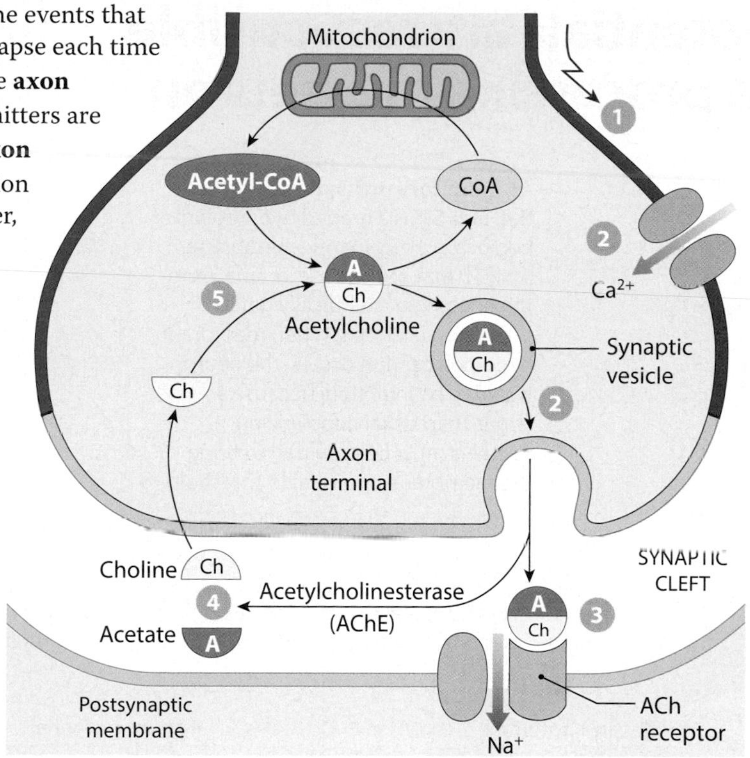

Mitochondrion
1
Acetyl-CoA
CoA
2
A
Ch
Ca2+
5
Acetylcholine
A
Ch
Synaptic vesicle
Ch
2
Axon terminal
SYNAPTIC CLEFT
Choline
Ch
Acetylcholinesterase (AChE)
4
A
Ch
3
Acetate
A
Postsynaptic membrane
Na+
ACh receptor

Events Occurring at a Cholinergic Synapse

1. An arriving action potential depolarizes the axon terminal.
2. Calcium ions enter the cytosol and, after a brief delay, ACh is released through the exocytosis of synaptic vesicles.
3. ACh binds to sodium channel receptors on the postsynaptic membrane, producing a graded depolarization.
4. Depolarization ends as ACh is broken down into acetate and choline by AChE.
5. The axon terminal reabsorbs choline from the synaptic cleft and uses it to synthesize new molecules of ACh.

3 A **synaptic delay** lasting 0.2–0.5 msec occurs between the arrival of the action potential at the axon terminal and the effect on the postsynaptic membrane. Although a delay of 0.5 msec is not very long, when information is being passed along a chain of interneurons the cumulative synaptic delay may be considerable. This is why reflexes are important for survival—they involve only a few synapses and thus provide rapid and automatic responses to stimuli. There is only one way to eliminate synaptic delay entirely—directly couple the presynaptic neuron to the postsynaptic cell at an electrical synapse.

Electrical Synapses

4 At an **electrical synapse**, the presynaptic and postsynaptic membranes are locked together by gap junctions. As a result, changes in the membrane potential of one cell will produce local currents that affect the other cell as if the two shared a common membrane. Electrical synapses are located in both the CNS and PNS, but they are extremely rare. They are present in some areas of the brain, in the eye, and in at least one pair of PNS ganglia (the ciliary ganglia). An action potential reaching an electrical synapse will always be propagated to the next cell, an arrangement that is efficient but not versatile. Much of the complexity and adaptability of the nervous system results from the fact that the responses of a postsynaptic cell can vary depending on the local chemical environment or the activities of synapses that release multiple neurotransmitters with varied effects.

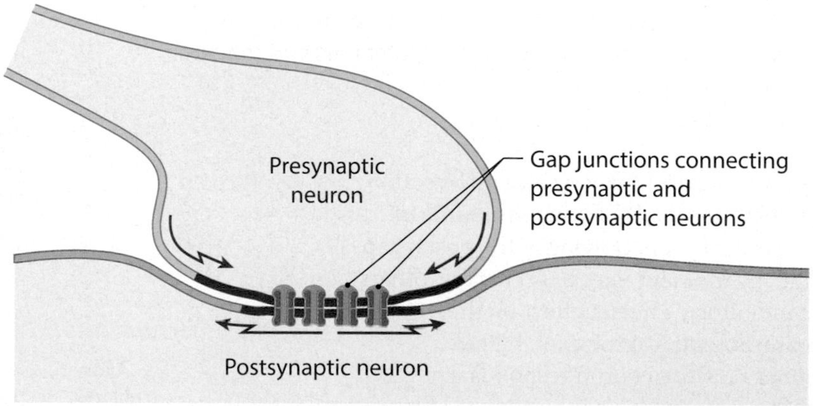

Presynaptic neuron
Gap junctions connecting presynaptic and postsynaptic neurons
Postsynaptic neuron

Module 11.12 Review

a. Describe the parts of a chemical synapse.
b. What is synaptic fatigue, and how does the synapse recover?
c. Contrast an electrical synapse with a chemical synapse.

LO **11.12** Describe the general structure of synapses in the CNS and PNS, and discuss the events that occur at a chemical synapse.

Module 11.13

Postsynaptic potentials are responsible for information processing in a neuron

An **excitatory postsynaptic potential**, or **EPSP**, is a graded depolarization caused by the arrival of a neurotransmitter at the postsynaptic membrane that shifts the membrane potential closer to the threshold. When that occurs, the membrane is said to be **facilitated**. The larger the degree of facilitation, the smaller the additional stimulus needed to trigger an action potential.

An **inhibitory postsynaptic potential**, or **IPSP**, is a graded hyperpolarization of the postsynaptic membrane. An IPSP may result, for example, from the opening of chemically gated potassium channels. When membrane hyperpolarization occurs, the neuron is said to be inhibited, because a larger-than-usual depolarizing stimulus must be provided to bring the membrane potential to threshold.

Summation is the integration of the effects of graded potentials on a segment of the plasma membrane. EPSPs and IPSPs result from the activation of different types of chemically gated channels. If two different neurotransmitters arrive simultaneously, and both sets of channels open, the net effect may be no change in the membrane potential.

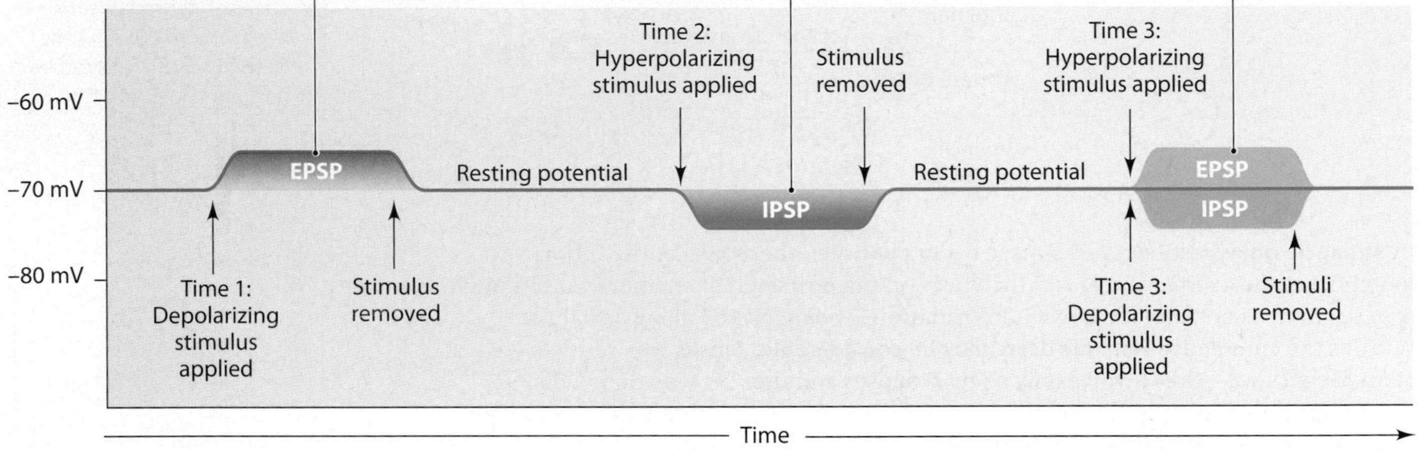

–60 mV
–70 mV
–80 mV
EPSP
Resting potential
Time 2: Hyperpolarizing stimulus applied
Stimulus removed
IPSP
Resting potential
Time 3: Hyperpolarizing stimulus applied
EPSP
IPSP
Time 1: Depolarizing stimulus applied
Stimulus removed
Time 3: Depolarizing stimulus applied
Stimuli removed
Time

1 **Postsynaptic potentials** are graded potentials that develop in the postsynaptic membrane in response to a neurotransmitter. Two major types of postsynaptic potentials develop at neuron-to-neuron synapses: excitatory postsynaptic potentials and inhibitory postsynaptic potentials.

2 A single neuron may receive information across thousands of synapses. Some of the neurotransmitters arriving at the postsynaptic cell at any moment may be excitatory, others may be inhibitory. The net effect on the membrane potential at the axon hillock determines how the neuron responds from moment to moment. This is because (1) the axon hillock is closest to the initial segment, and graded potentials are what trigger the action potential, and (2) the threshold at the boundary between the axon hillock and the initial segment is lower than it is elsewhere on the cell body. Thus it is the axon hillock that integrates the excitatory and inhibitory stimuli affecting the cell body and dendrites at any given moment and determines the rate of action potential generation at the initial segment. This integration process is the simplest level of information processing in the nervous system.

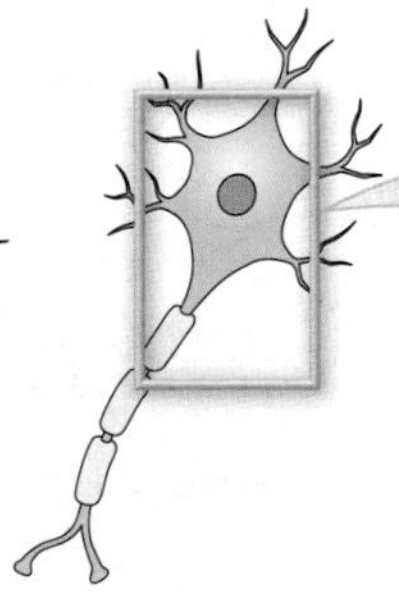

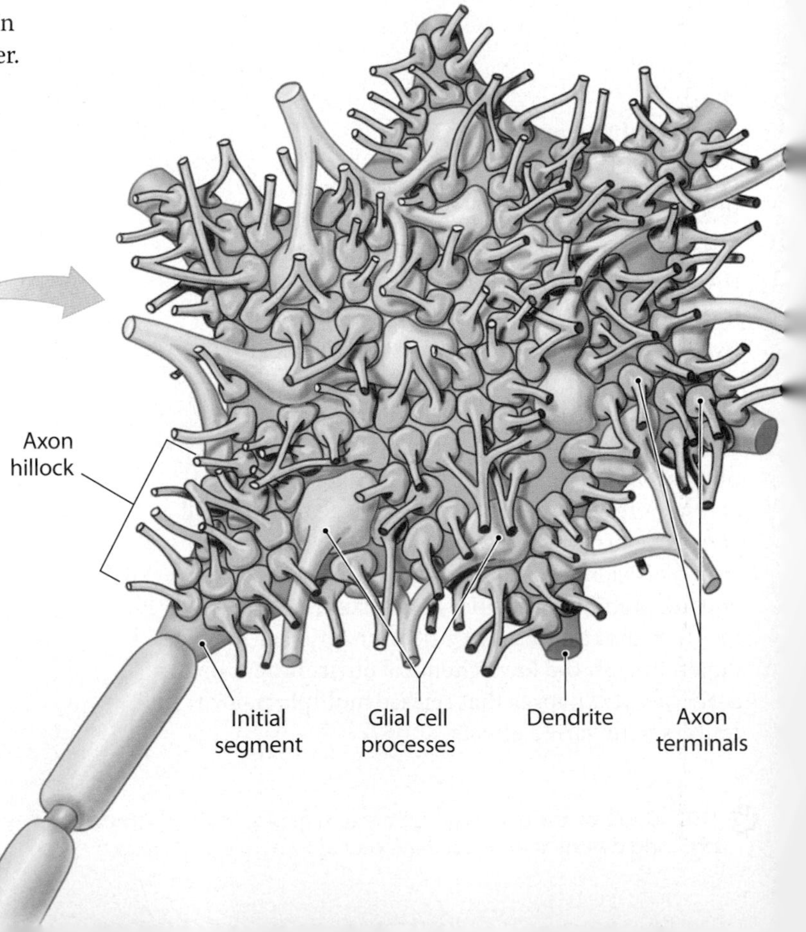

Axon hillock
Initial segment
Glial cell processes
Dendrite
Axon terminals

3 An individual EPSP or IPSP has a small effect on the membrane potential. A typical EPSP produces a depolarization of about 0.5 mV at the postsynaptic membrane. Before an action potential will arise in the initial segment, local currents must depolarize that region by at least 10 mV. Therefore, a single EPSP will not result in an action potential, even if the synapse is on the axon hillock. However, individual EPSPs combine through the process of summation. There are two forms of summation that can generate an action potential: temporal summation and spatial summation.

Temporal Summation

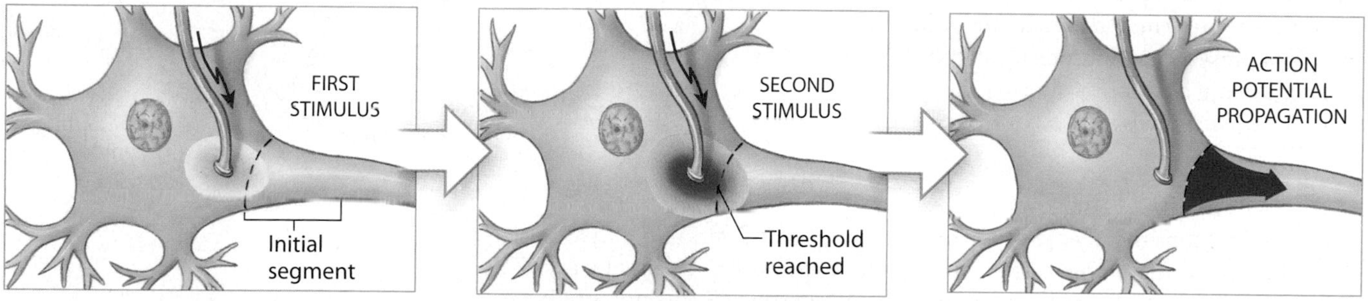
FIRST
STIMULUS
Initial
segment
SECOND
STIMULUS
Threshold
reached
ACTION
POTENTIAL
PROPAGATION

Temporal summation (*tempus*, time) occurs when a single synapse is active repeatedly. For example, a typical EPSP lasts about 20 msec, but under maximum stimulation an action potential can reach the axon terminal each millisecond. When a second EPSP arrives before the effects of the first EPSP have disappeared, another group of vesicles discharges ACh into the synaptic cleft, more ACh molecules arrive at the postsynaptic membrane, and the degree of depolarization increases. In this way, a series of small steps can eventually bring the initial segment to threshold.

Spatial Summation

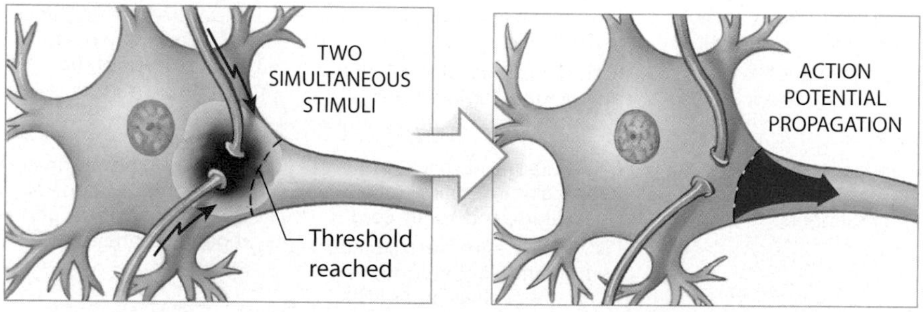
TWO
SIMULTANEOUS
STIMULI
Threshold
reached
ACTION
POTENTIAL
PROPAGATION

Spatial summation involves multiple synapses that are active simultaneously. The effects on the membrane potential are cumulative, because each active synapse opens gated channels that allow ion movement in or out of the cell. In this example, the active synapses are generating EPSPs and opening gated channels that let sodium ions enter the cell. The effects are cumulative, and the degree of depolarization at the initial segment depends on (1) how many excitatory synapses are active at any given moment, and (2) how far they are from the initial segment. As in temporal summation, an action potential results when the membrane potential at the initial segment reaches threshold.

Each neuron integrates the information arriving across synapses, and also monitors and responds to a variety of factors in the local environment. Neurons are dynamic, active cells, and their sensitivity to stimulation changes in response to variations in body temperature, oxygen or nutrient availability, or the presence of abnormal chemicals.

LO 11.13 Discuss the significance of postsynaptic potentials, including the roles of excitatory postsynaptic potentials and inhibitory postsynaptic potentials.

Module 11.13 Review

a. Define excitatory postsynaptic potential (EPSP) and inhibitory postsynaptic potential (IPSP).

b. If a single EPSP depolarizes the initial segment from a resting potential of –70 mV to –65 mV, and threshold is at –60 mV, will an action potential be generated? Explain your answer.

c. Compare temporal summation with spatial summation.

Module 11.14

Information processing involves interacting groups of neurons, and information is encoded in the frequency and pattern of action potentials

1 Cellular information processing occurs at the postsynaptic membrane, as the membrane potential at the axon hillock rises and falls. Higher levels of information processing involve **regulatory neurons** that facilitate or inhibit the activities of presynaptic neurons by affecting the membrane of the cell body or by altering the sensitivity of axon terminals.

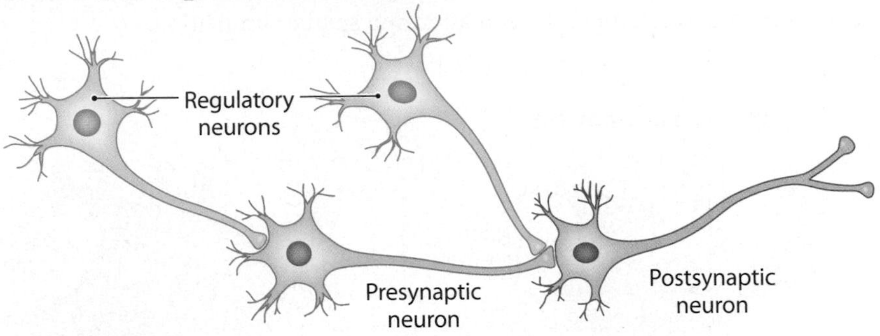
Regulatory
neurons
Presynaptic
neuron
Postsynaptic
neuron

2 One of the reasons the nervous system is so complex and versatile is that it uses more than 100 different neurotransmitters that each work in different ways. This table introduces a few of the important neurotransmitters you will encounter in later chapters. As you see, they may have indirect effects as well as direct effects on ion channels. The indirect effects usually involve binding to **G proteins**, a diverse family of enzyme complexes attached to membrane receptors. As we will see in Chapter 16, activated G proteins trigger the formation or release of other products called **second messengers** that alter conditions in the cytoplasm and change the activity of the postsynaptic cell.

Selected Neurotransmitters

Neurotransmitter	Chemical Structure	Mechanism of Action	Location	Comments
Acetylcholine (ACh)	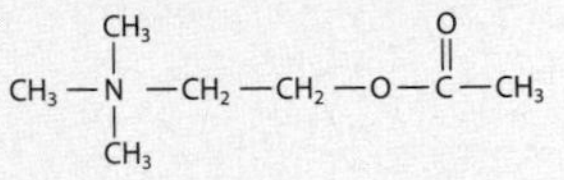	Primarily direct, through binding to chemically gated channels	CNS: Synapses throughout brain and spinal cord PNS: Neuromuscular junctions, neuroglandular junctions, and synapses in autonomic ganglia	Widespread in CNS and PNS; best known and most studied of the neurotransmitters
Norepinephrine (NE)	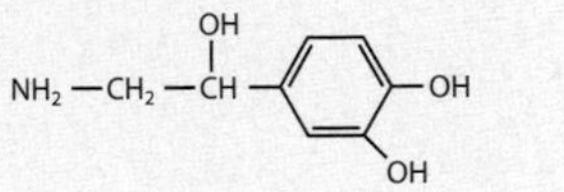	Indirect, through G proteins and second messengers	CNS: Cerebral cortex, hypothalamus, brain stem, cerebellum, and spinal cord PNS: Most neuromuscular and neuroglandular junctions of sympathetic division of ANS	Involved in attention and consciousness, control of body temperature, and regulation of pituitary gland secretion
Epinephrine (E)	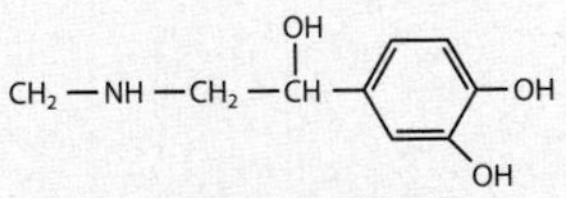	Indirect: G proteins and second messengers	CNS: Thalamus, hypothalamus, midbrain, and spinal cord	Generally excitatory effect along autonomic pathways
Serotonin	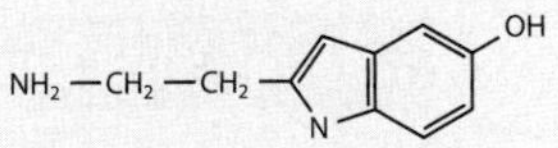	Primarily indirect: G proteins and second messengers	CNS: Hypothalamus, limbic system, cerebellum, spinal cord, and retina	Important in emotional states, moods, and body temperature; several illicit hallucinogenic drugs, such as Ecstasy, target serotonin receptors
Glutamate	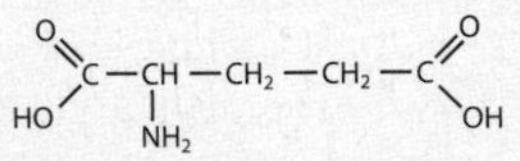	Indirect: G proteins and second messengers Direct: opens calcium/sodium channels	CNS: Cerebral cortex and brain stem	Important in memory and learning; most important excitatory neurotransmitter in the brain
Gamma-aminobutyric acid (GABA)	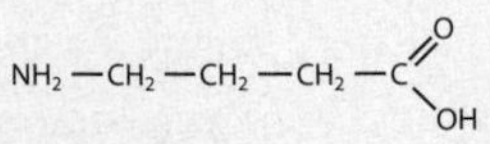	Direct or indirect (G proteins), depending on type of receptor	CNS: Cerebral cortex, cerebellum, interneurons throughout brain and spinal cord	Direct inhibitory effects: opens Cl^- channels; indirect effects: opens K^+ channels and blocks entry of Ca^{2+}

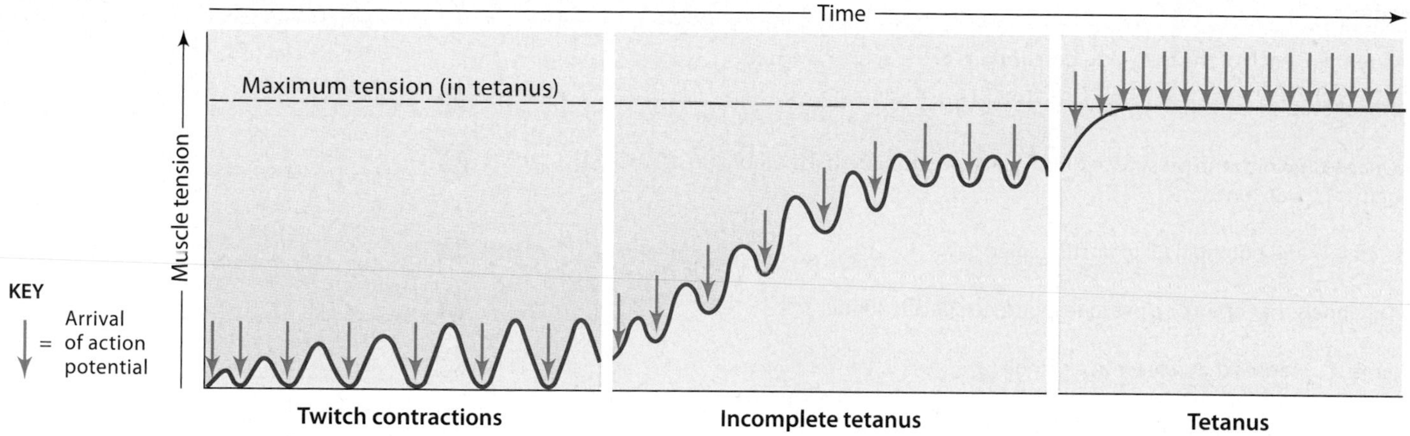
Time
Maximum tension (in tetanus)
Muscle tension
KEY
Arrival of action potential
Twitch contractions
Incomplete tetanus
Tetanus

3 In the nervous system, complex information is translated into action potentials that are propagated along axons. On arrival, the message is often interpreted solely on the basis of the frequency of action potentials. This figure shows how the rate of action potentials arriving at a neuromuscular junction determines the nature of the resulting contraction.

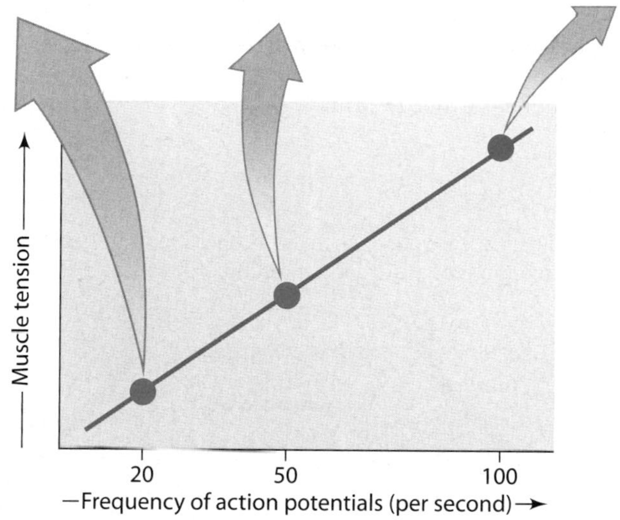
Muscle tension
20
50
100
Frequency of action potentials (per second)

4 This table summarizes the key concepts for understanding information processing in the nervous system.

Information Processing

- The neurotransmitters released at a synapse may have either excitatory or inhibitory effects. The effect on the axon's initial segment reflects a summation of the stimuli arriving at any moment. The frequency of action potentials generated is an indication of the degree of sustained depolarization at the axon hillock.
- Neurons may be facilitated or inhibited by extracellular chemicals other than neurotransmitters.
- The response of a postsynaptic neuron to the activation of a presynaptic neuron can be altered by (1) the presence of chemicals that cause facilitation or inhibition at the synapse, (2) activity under way at other synapses affecting the postsynaptic cell, and (3) modification of the rate of neurotransmitter release through facilitation or inhibition by regulatory neurons.
- Information is relayed in the form of action potentials. In general, the degree of sensory stimulation or the strength of the motor response is proportional to the frequency of action potentials.

Module 11.14 Review

a. Describe the role of regulatory neurons.

b. What determines the frequency of action potential generation?

c. The greater the degree of sustained depolarization at the axon hillock, the __________ (higher or lower) the frequency of action potentials generated.

LO 11.14 Discuss the interactions that make information processing in neural tissue possible.

SECTION 2 Review

Vocabulary

Write the term for each of the following descriptions in the space provided.

1 A propagated change in the membrane potential

2 A synapse in which the presynaptic and postsynaptic neuronal membranes are locked together by gap junctions

3 The membrane potential of an unstimulated cell

4 Ion channels that open or close in response to specific stimuli

5 Chemical synapses that release acetylcholine

6 A shift in the membrane potential from −70 mV to −85 mV

7 Movement of positive charges parallel to the inner and outer plasma membrane surfaces

8 A shift in the membrane potential from −70 mV to +30 mV

1 ____________________

2 ____________________

3 ____________________

4 ____________________

5 ____________________

6 ____________________

7 ____________________

8 ____________________

Short answer

For the following diagram of a cholinergic synapse, write the names of components 9–14 in the boxes at left, and then fill in the table at right with descriptions of the events represented by 15–20.

Components

9	
10	
11	
12	
13	
14	

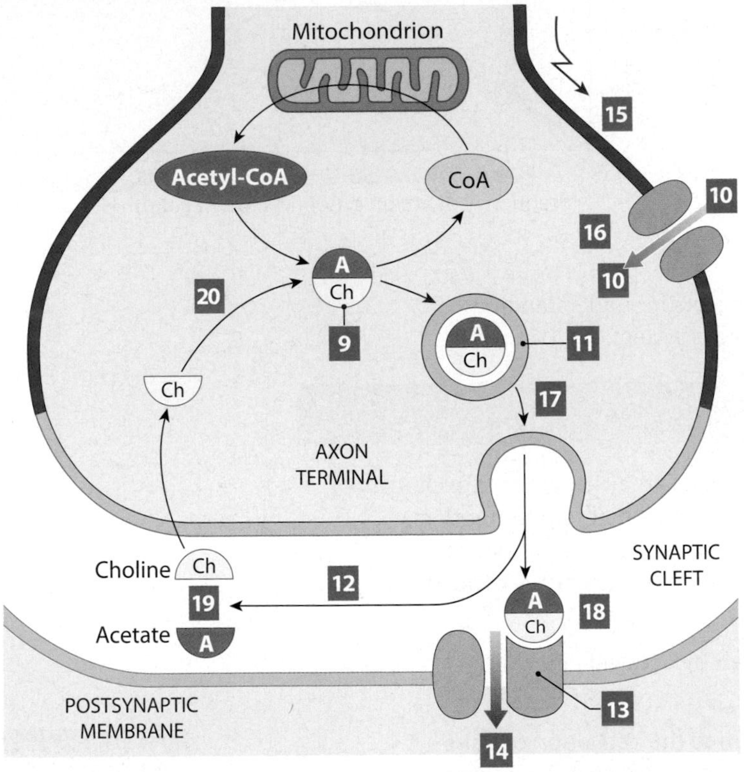

Mitochondrion
Acetyl-CoA
CoA
A
Ch
9
10
11
12
13
14
15
16
17
18
19
20
AXON TERMINAL
SYNAPTIC CLEFT
Choline
Acetate
POSTSYNAPTIC MEMBRANE

Events Occurring at Synapse
15
16
17
18
19
20

Section integration

Guillain-Barré *(GHEE-yan BAH-rā)* syndrome is a degeneration of myelin sheaths that ultimately may result in paralysis. Propose a mechanism by which myelin sheath degeneration can cause muscular paralysis.

21 ____

Study Outline

SECTION 1 • Cellular Organization of the Nervous System

11.1 The nervous system has two divisions: the CNS and PNS p. 395

1. The **central nervous system (CNS)** consists of the brain and spinal cord. It is responsible for integrating, processing, and coordinating sensory data and motor commands.
2. The **peripheral nervous system (PNS)** includes all the neural tissue outside the CNS.
3. **Receptors** detect changes in the internal and external environment. The **sensory division** of the PNS brings information from receptors to the CNS.
4. The **motor division** of the PNS carries motor commands from the CNS to the **effectors** or target organs.

11.2 Neurons are nerve cells specialized for intercellular communication p. 396

5. Neurons have three general regions: **dendrites** that receive stimuli; a **cell body** that contains the nucleus and other organelles; and an **axon** that carries information to other cells.
6. The **telodendria** of an axon end at **axon terminals**. Axon terminals are part of the **synapse** where the neuron communicates with another cell.
7. Axon terminals contain **synaptic vesicles** containing neurotransmitters.

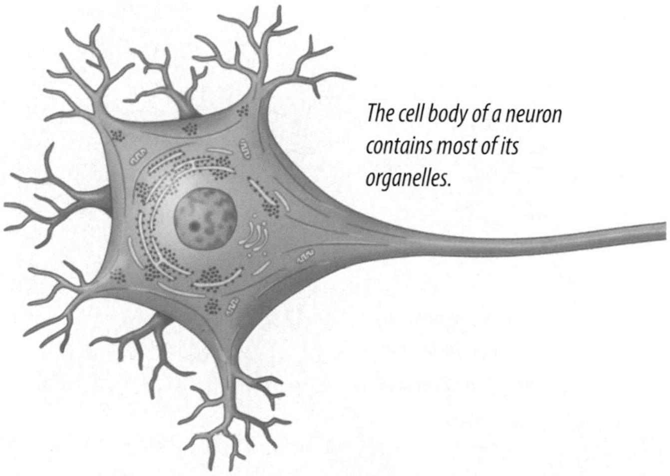

The cell body of a neuron contains most of its organelles.

11.3 Neurons are classified on the basis of structure or function p. 398

8. The four major anatomical classes of neurons are **anaxonic**, **bipolar**, **unipolar**, and **multipolar**.
9. Functionally, neurons are classified **sensory neurons**, **interneurons**, or **motor neurons**.

11.4 Oligodendrocytes, astrocytes, ependymal cells, and microglia are neuroglia of the CNS p. 400

10. **Neuroglia**, or **glial cells**, support and protect neurons.
11. **Ependymal cells** are associated with cerebrospinal fluid production and circulation. **Microglia** remove cellular debris and pathogens. **Astrocytes** maintain the **blood–brain barrier**.
12. **Oligodendrocytes** help form the myelin sheath that surrounds axons making up **white matter**. **Gray matter** is unmyelinated neuron cell bodies, dendrites, and unmyelinated cell axons.

11.5 Schwann cells and satellite cells are the neuroglia of the PNS p. 402

13. **Schwann cells** form a myelin sheath around or enclose segments of peripheral axons.
14. **Satellite cells** surround cell bodies in ganglia.
15. In the PNS, repair of nerves may follow **Wallerian degeneration**, a process that often fails to restore full function.

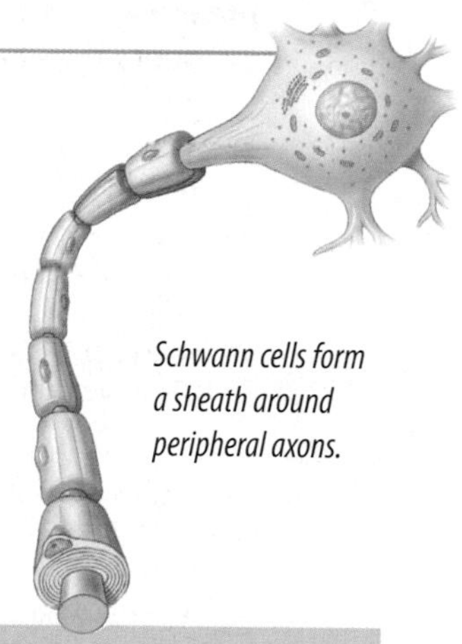

Schwann cells form a sheath around peripheral axons.

SECTION 2 • Neurophysiology

11.6 Neuronal activity depends on changes in membrane potential p. 405

16. **Membrane potential** is the unequal charge distribution between the inner and outer surfaces of the plasma membrane, where there is a slight negative charge inside the plasma membrane with respect to the outside.
17. All neural activities begin with a change in the **resting membrane potential**, or **resting potential**, of the neuron. If localized changes in resting potential, called **graded potentials**, are sufficient, they can trigger an **action potential**.
18. An action potential at the axon terminal causes release of neurotransmitter by the presynaptic cell and graded potentials in the postsynaptic cell. This entire process is called **synaptic activity**.

11.7 The resting potential is the membrane potential of an undisturbed cell p. 406

19. Passive **leak channels** allow the movement of Na^+ into the cell, and K^+ out of the cell.
20. The sodium–potassium exchange pump, an active ATP-requiring process, ejects 3 Na^+ for every 2 K^+.
21. Potassium ion gradients force K^+ out of the cell, and sodium ion gradients drive Na^+ into the cell
22. The resulting resting membrane potential for a neuron is approximately −70 mV.

11.8 Three types of gated channels change the permeability of the plasma membrane p. 408

23. Resting potential is stable until the cell is disturbed. When disturbed, ion permeability changes due to **gated channels** within the plasma membrane.

24. **Chemically gated channels**, or **ligand-gated channels**, open when specific chemicals such as ACh bind to them.
25. **Voltage-gated channels** open or close in response to changes in membrane potential and are characteristic of excitable membranes. These membranes are capable of generating and propagating action potentials.
26. **Mechanically gated channels** open in response to physical distortion of the membrane surface. They are found in sensory receptors that respond to touch, pressure, or vibration.

11.9 Graded potentials are localized changes in the membrane potential p. 410

27. **Graded potentials** cannot spread far from the site of stimulation.
28. Any shift from resting potential toward a more positive value is called **depolarization**.
29. A return to normal resting potential is called **repolarization**, and results when excess Na^+ leaves the cell.
30. **Hyperpolarization** is a shift in membrane potential more negative than resting potential.

11.10 An action potential begins when voltage-gated sodium ion channels open and membrane potential reverses p. 412

31. A graded potential that reaches **threshold** (−60 mV) initiates an **action potential**.
32. Once threshold is reached, voltage-gated Na^+ channels open, Na^+ enters the cell, and membrane potential rises to +30 mV.
33. Repolarization begins when Na^+ channels close, and then voltage-gated K^+ channels open, allowing K^+ to move out of the cell.
34. Repolarization continues as K^+ channels close, and both Na^+ and K^+ channels return to their normal state.
35. During the **absolute refractory period**, the membrane cannot respond to further stimulation. During the **relative refractory period**, the membrane can respond only to a larger-than-normal stimulus.

11.11 Action potentials may affect adjacent portions of the plasma membrane through continuous propagation or saltatory propagation p. 414

36. In **continuous propagation** the action potential moves in a series of tiny steps along the axon.
37. In **saltatory propagation**, depolarization occurs only at the nodes (nodes of Ranvier) of a myelinated axon. Action potentials travel much faster in saltatory propagation than in continuous propagation.
38. Propagation speed along myelinated axons also depends on axon diameter. The larger the diameter, the lower the resistance to ion movement, and the faster the action potential travels.

11.12 At a synapse, information travels from the presynaptic cell to the postsynaptic cell p. 416

39. **Chemical synapses**, the most abundant type of synapse, release neurotransmitter into the synaptic cleft. **Cholinergic synapses** release ACh.
40. When an action potential depolarizes an axon terminal, calcium ions enter the cell, and ACh is released through exocytosis of synaptic vesicles from the presynaptic cell.
41. ACh binds to sodium ion channel receptors on the postsynaptic membrane, producing a graded depolarization.
42. Depolarization ends as ACh is broken down into acetate and choline by AChE.
43. The axon terminal reabsorbs choline from the synaptic cleft and uses it to synthesize new ACh molecules.
44. In **electrical synapses**, the presynaptic and postsynaptic membranes are locked together by gap junctions. These structures allow for changes in membrane potential to be transferred from one cell to the other. These rare synapses are located in both the PNS and CNS.

11.13 Postsynaptic potentials are responsible for information processing in a neuron p. 418

45. An **excitatory postsynaptic potential**, or **EPSP**, is a graded depolarization that shifts membrane potential toward threshold.
46. When the membrane potential shifts toward threshold, the membrane is **facilitated**. The larger the facilitation, the smaller the additional stimulus needed to trigger action potential.
47. An **inhibitory postsynaptic potential**, or **IPSP**, is a graded depolarization that opens chemically gated K^+ channels and causes hyperpolarization.
48. **Summation** is the integration of EPSP and IPSP. **Temporal summation** occurs when a single synapse is active repeatedly. **Spatial summation** involves multiple synapses that are active simultaneously.

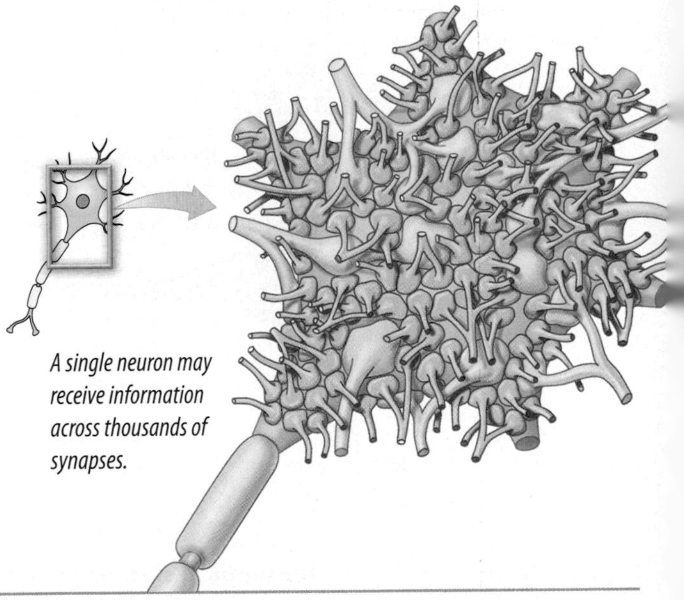

A single neuron may receive information across thousands of synapses.

11.14 Information processing involves interacting groups of neurons, and information is encoded in the frequency and pattern of action potentials p. 420

49. Neurotransmitter release at a synapse may be either excitatory or inhibitory, depending on the summation of the stimuli, extracellular chemicals other than neurotransmitters, and other complex factors.
50. There are over 100 different neurotransmitters that each work in different ways.

Chapter Review Questions

Labeling

Label the structures in the following diagram.

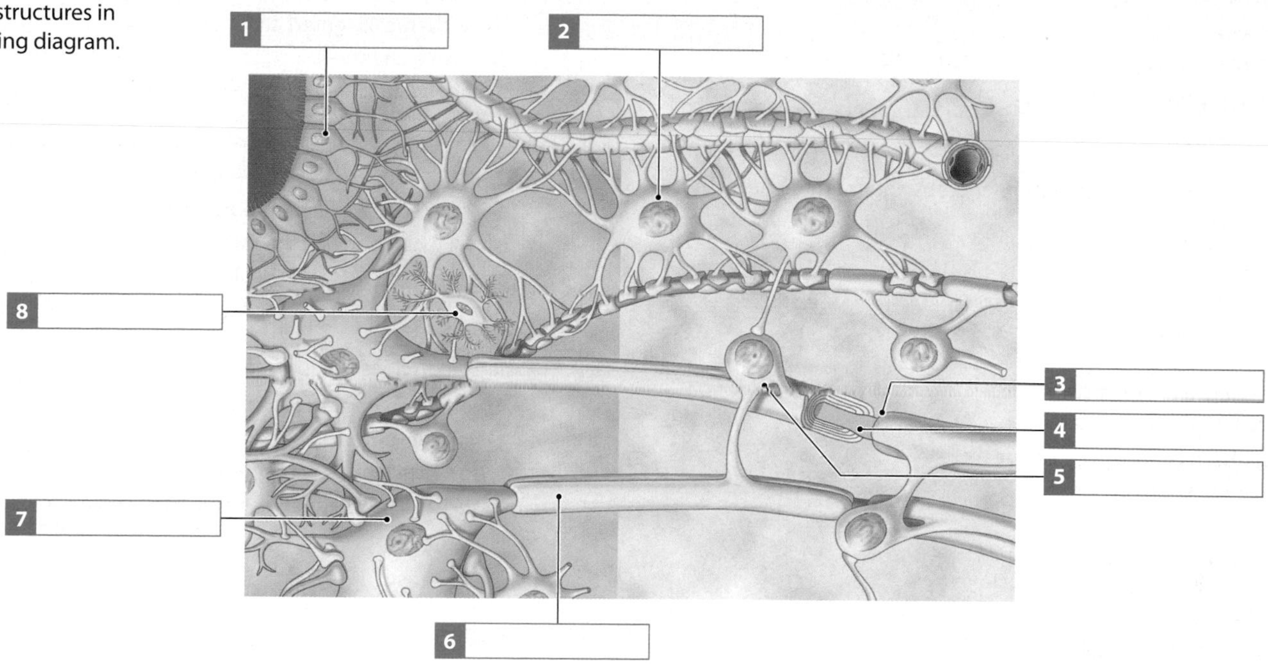

1
2
3
4
5
6
7
8

True/False

Indicate whether each statement is true or false.

9 Somatic sensory receptors monitor internal organs.

10 Synaptic vesicles contain neurotransmitters.

11 Microglia maintain the blood–brain barrier.

12 Schwann cells form the neurilemma.

13 The resting membrane potential for a neuron is near −70 mV.

9 ______

10 ______

11 ______

12 ______

13 ______

Matching

Match each lettered term with the most closely related description.

a. relative refractory period
b. voltage-gated channel
c. oligodendrocyte
d. chemically gated channel
e. mechanically gated channel
f. Schwann cell
g. absolute refractory period
h. astrocyte

14 Produces myelin in the CNS

15 Opens in response to physical distortion

16 A time when a membrane can respond only to a larger-than-normal stimulus

17 Opens or closes in response to changes in membrane potential

18 Produces myelin in the PNS

19 A time when a membrane cannot respond to further stimulation

20 Maintains the blood–brain barrier

21 Opens in response to neurotransmitters

14 ______

15 ______

16 ______

17 ______

18 ______

19 ______

20 ______

21 ______

Multiple choice

Select the correct answer from the list provided.

22 In the CNS, a neuron typically receives information from other neurons at its

- [ ] a) axon.
- [ ] b) Nissl bodies.
- [ ] c) dendrites.
- [ ] d) nucleus.

23 The neural cells between sensory neurons and motor neurons are

- [ ] a) neuroglia.
- [ ] b) interneurons.
- [ ] c) sensory ganglia.
- [ ] d) autonomic ganglia.

24 Any shift from resting potential toward a more positive value is called

- [ ] a) repolarization.
- [ ] b) depolarization.
- [ ] c) membrane potential.
- [ ] d) hyperpolarization.

25 Which of the following statements is true regarding plasma membrane leak channels?

- [ ] a) Na^+ passively moves through leak channels to exit cell.
- [ ] b) Na^+ actively moves through leak channels to exit cell.
- [ ] c) K^+ passively moves through leak channels to exit cell.
- [ ] d) K^+ actively moves through leak channels to exit cell.

26 Receptors that bind to ACh at the postsynaptic membrane are

- [ ] a) voltage-gated channels.
- [ ] b) mechanically gated channels.
- [ ] c) passive channels.
- [ ] d) chemically gated channels.

27 If the resting membrane potential of a neuron is −70 mV, and threshold is −60 mV, a membrane potential of −55 mV

- [ ] a) will produce an action potential.
- [ ] b) will hyperpolarize the membrane.
- [ ] c) is referred to as the absolute refractory period.
- [ ] d) will cause an IPSP.

28 Repolarization does not include

- [ ] a) voltage-gated Na^+ channels closing.
- [ ] b) voltage-gated Na^+ channels opening.
- [ ] c) voltage-gated K^+ channels closing.
- [ ] d) Na^+ and K^+ channels returning to their normal states.

Short answer

29 What are the major components of the central nervous system (CNS) and the peripheral nervous system (PNS)?

30 What three functional classes of neurons are found in the nervous system? What is the function of each type of neuron?

31 Distinguish between continuous propagation and saltatory propagation.

32 How does a graded potential differ from an action potential?

33 What is the difference between temporal summation and spatial summation?

MasteringA&P®

Access more chapter study tools online in the MasteringA&P Study Area:

- Chapter Quizzes, Chapter Practice Test, Art-labeling Activities, Animations, MP3 Tutor Sessions, and Clinical Case Studies

- Practice Anatomy Lab — PAL™
- Interactive Physiology — iP®
- A&P Flix — A&PFlix™
- PhysioEx — PhysioEx™

Chapter Integration • Applying what you have learned

Multiple sclerosis is a progressive, debilitating, demyelinating disease

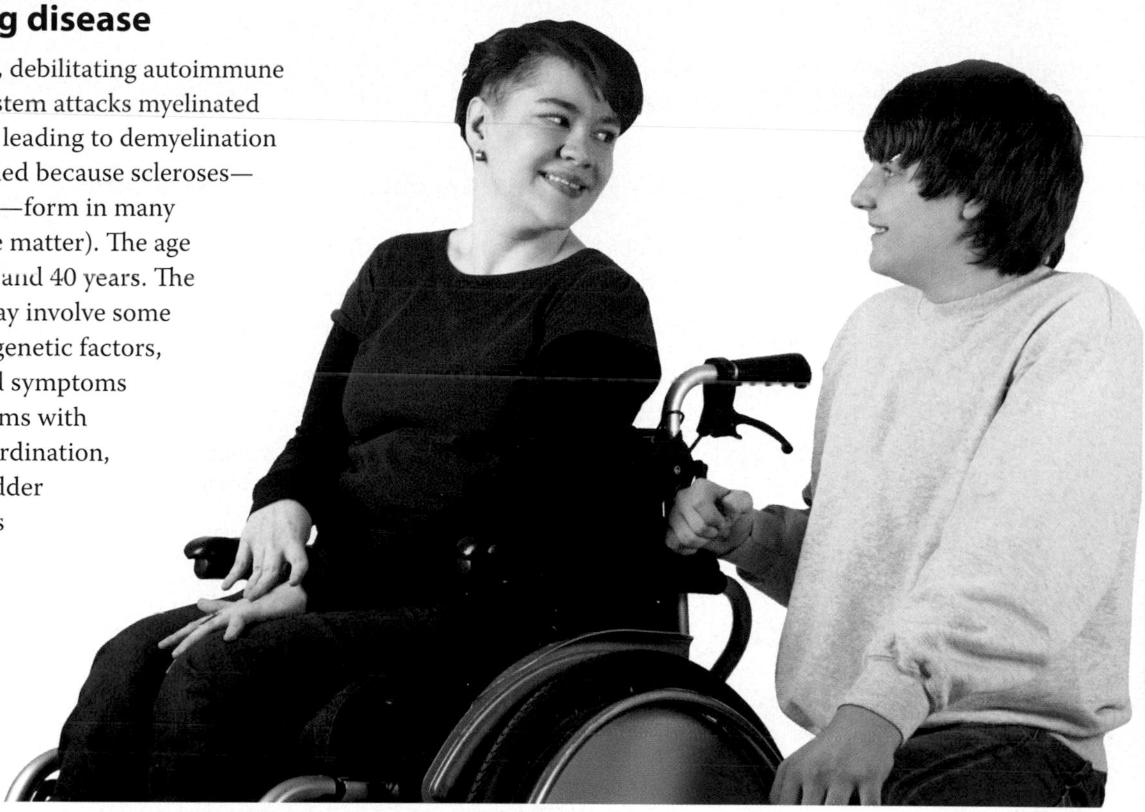

Multiple sclerosis (MS) is a progressive, debilitating autoimmune disease in which the body's immune system attacks myelinated portions of the central nervous system, leading to demyelination of affected axons. The disease is so named because scleroses—also known as scars, plaques, or lesions—form in many places within myelinated regions (white matter). The age at onset is most commonly between 20 and 40 years. The cause of the disease is unknown, but may involve some combination of environmental agents, genetic factors, and viral infections. Common signs and symptoms include partial loss of vision and problems with speech, balance, and general motor coordination, including loss of bowel and urinary bladder control. The incidence among women is about twice that of men. Individuals with MS experience unpredictable, recurrent cycles of deterioration, remission, and relapse. There is no cure for MS, although drugs that alter the sensitivity or responses of the immune system can slow the progression of the disease.

1 **Define demyelination.**

2 **Why would individuals with MS experience generalized motor coordination dysfunction?**

3 **Which glial cells would be affected in MS?**

12 The Spinal Cord, Spinal Nerves, and Spinal Reflexes

LEARNING OUTCOMES

These Learning Outcomes correspond by number to this chapter's modules and indicate what you should be able to do after completing the chapter.

SECTION 1 • Functional Organization of the Spinal Cord

12.1 Describe how the spinal cord can function without input from the brain.

12.2 Discuss the anatomical features of the spinal cord.

12.3 Describe the three meningeal layers that surround the spinal cord.

12.4 Explain the roles of white matter and gray matter in processing and relaying sensory information and motor commands.

12.5 Describe the major components of a spinal nerve.

12.6 Describe the rami associated with spinal nerves.

12.7 Relate the distribution pattern of spinal nerves to the region they innervate, and describe the cervical plexus.

12.8 Relate the distribution pattern of the brachial plexus to its function.

12.9 Relate the distribution patterns of the lumbar plexus and sacral plexus to their functions.

SECTION 2 • Introduction to Reflexes

12.10 Discuss the significance of neuronal pools, and describe their major patterns of neuron interaction.

12.11 Describe the steps in a reflex.

12.12 Describe the steps in the stretch reflex.

12.13 Explain withdrawal reflexes and crossed extensor reflexes and the responses produced by each.

12.14 **CLINICAL MODULE** Explain the value of reflex testing and how the brain may control and modify reflex responses.

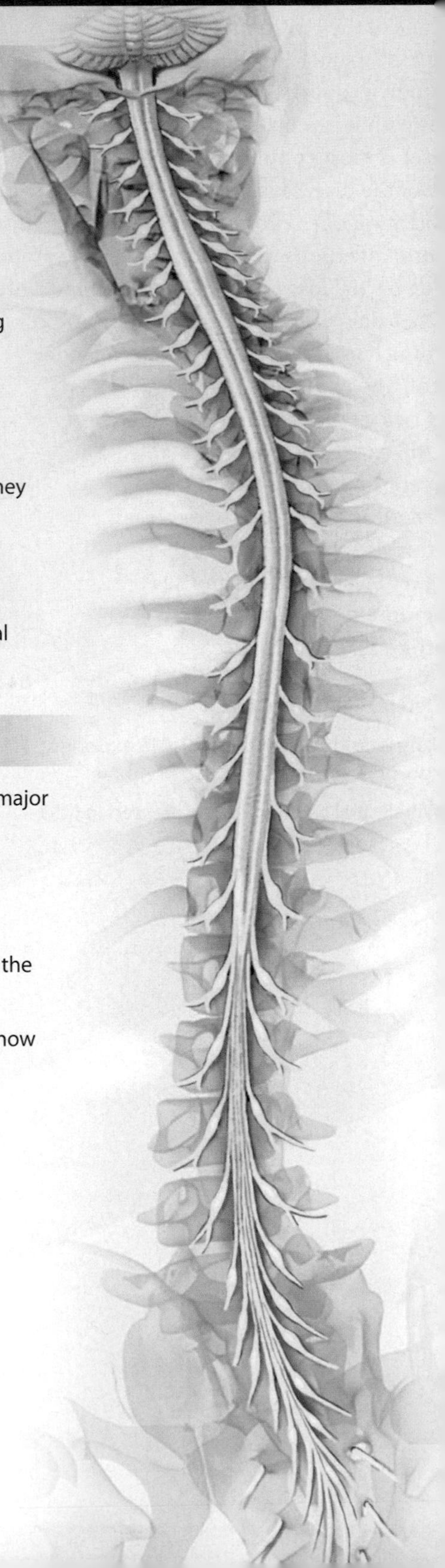

Learning Outcomes are repeated at the bottom of each module.

The spinal cord can function independently from the brain

Organization is usually the key to success in any complex environment. In a large corporation, for example, only the most important problems reach the desk of the president. The nervous system works in much the same way. We are consciously aware of only a small fraction of the day-to-day activities monitored by sensory receptors and controlled by motor neurons. Throughout our lives, input pathways are routing sensations, processing centers are prioritizing and distributing information, and motor centers are directing responses to stimuli, most often outside of our awareness. This is possible only because the nervous system is so highly organized. Because our primary interest here is how the nervous system functions, we will consider the system from a functional perspective. Our approach is represented in the diagram below.

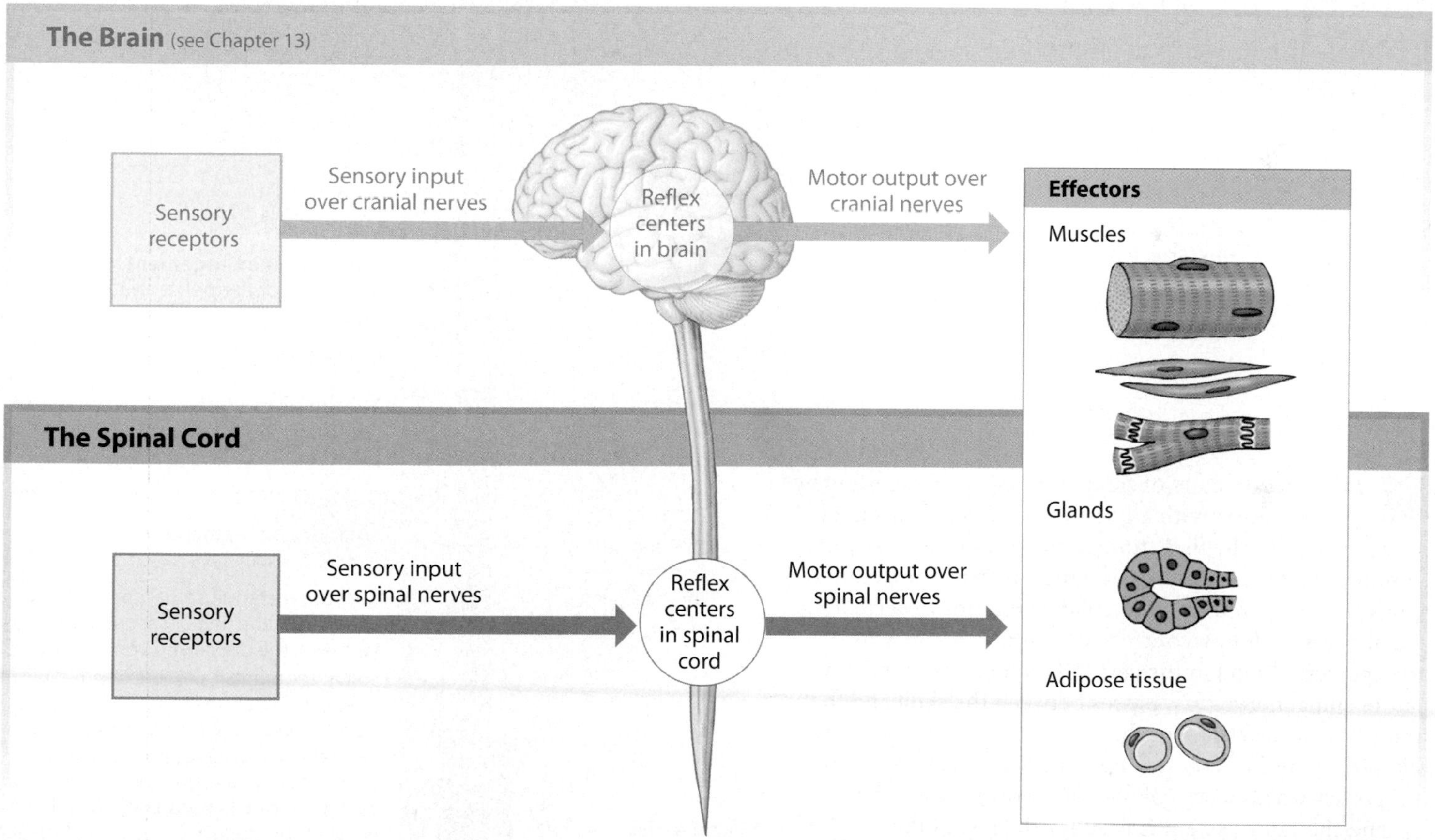
The Brain (see Chapter 13)
Sensory receptors
Sensory input over cranial nerves
Reflex centers in brain
Motor output over cranial nerves
Effectors
Muscles
Glands
Adipose tissue
The Spinal Cord
Sensory receptors
Sensory input over spinal nerves
Reflex centers in spinal cord
Motor output over spinal nerves

This chapter considers the spinal cord, spinal nerves, and spinal reflexes. We will begin with the **spinal cord**, the simplest part of the CNS, because the functional and structural relationships are relatively easy to understand. A **reflex** is a rapid, automatic response triggered by specific stimuli. **Spinal reflexes** are controlled in the spinal cord. They can function without any input from the brain. Chapter 13 examines the brain, cranial nerves, cranial reflexes, and the major pathways that interconnect the brain and spinal cord.

Module 12.1 Review

a. Describe the direction of sensory input and motor commands relative to the spinal cord.

b. What is a reflex?

c. Define spinal reflex.

LO 12.1 Describe how the spinal cord can function without input from the brain.

Module 12.2

The spinal cord contains gray matter and white matter

The adult spinal cord is approximately 45 cm (18 in.) long with a maximum width of roughly 14 mm (0.55 in.). Note that the cord itself is not as long as the vertebral column—instead, the adult spinal cord ends between vertebrae L_1 and L_2. In sectional view, it has an outer layer of white matter and an inner layer of gray matter surrounding a small central canal. The amount of gray matter is greatest in segments of the spinal cord dedicated to the sensory and motor control of the limbs. These segments are expanded, forming the cervical and thoracic enlargements of the spinal cord.

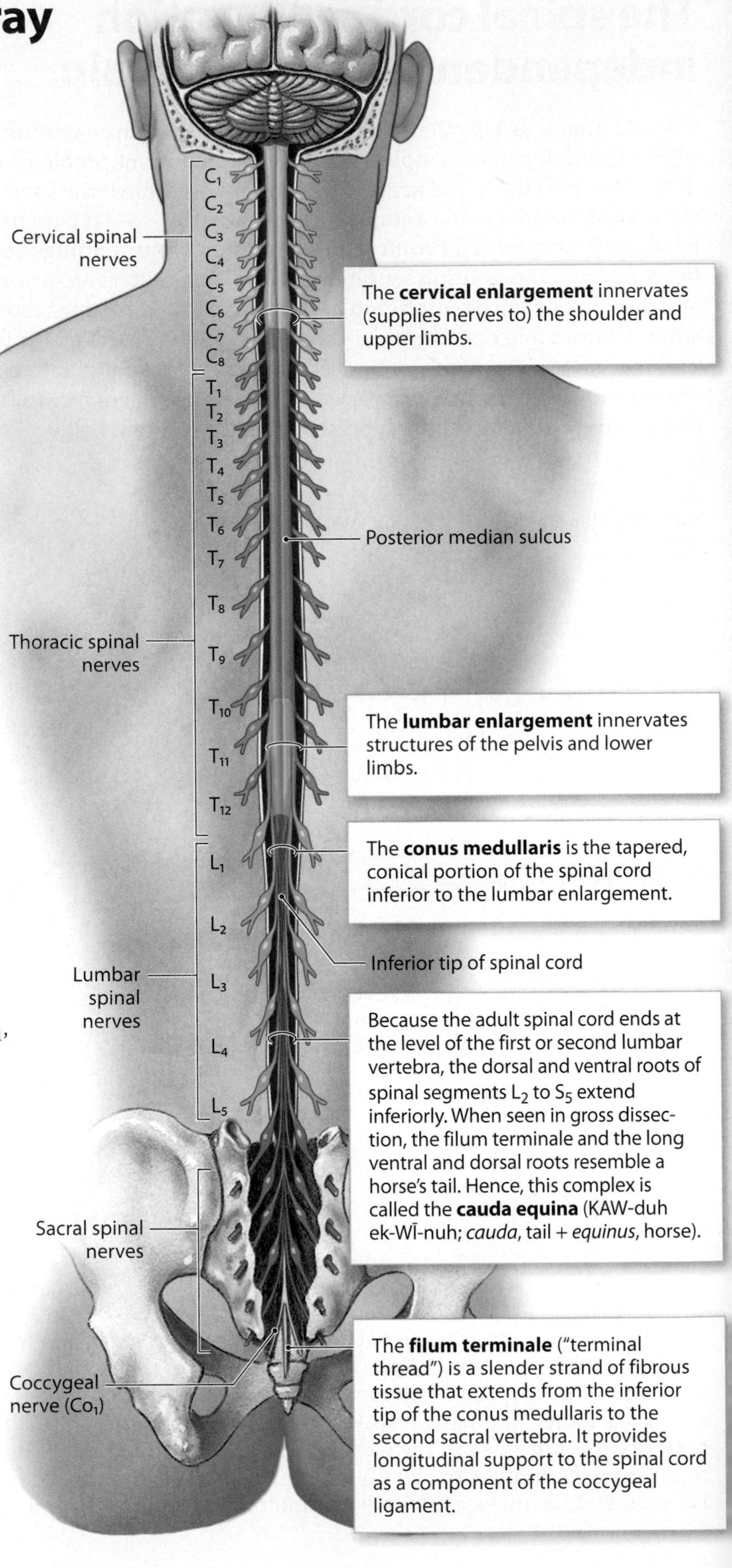

Cervical spinal nerves
C_1
C_2
C_3
C_4
C_5
C_6
C_7
C_8
The **cervical enlargement** innervates (supplies nerves to) the shoulder and upper limbs.
T_1
T_2
T_3
T_4
T_5
T_6
T_7
Posterior median sulcus
T_8
Thoracic spinal nerves
T_9
T_{10}
T_{11}
T_{12}
The **lumbar enlargement** innervates structures of the pelvis and lower limbs.
The **conus medullaris** is the tapered, conical portion of the spinal cord inferior to the lumbar enlargement.
L_1
L_2
Inferior tip of spinal cord
Lumbar spinal nerves
L_3
L_4
L_5
Because the adult spinal cord ends at the level of the first or second lumbar vertebra, the dorsal and ventral roots of spinal segments L_2 to S_5 extend inferiorly. When seen in gross dissection, the filum terminale and the long ventral and dorsal roots resemble a horse's tail. Hence, this complex is called the **cauda equina** (KAW-duh ek-WĪ-nuh; *cauda*, tail + *equinus*, horse).
Sacral spinal nerves
Coccygeal nerve (Co_1)
The **filum terminale** ("terminal thread") is a slender strand of fibrous tissue that extends from the inferior tip of the conus medullaris to the second sacral vertebra. It provides longitudinal support to the spinal cord as a component of the coccygeal ligament.

1 There are 31 pairs of spinal nerves, each identified by its association with adjacent vertebrae. Each spinal nerve inferior to the first thoracic vertebra takes its name from the vertebra immediately superior to it. For example, spinal nerve T_1 emerges immediately inferior to vertebra T_1, spinal nerve T_2 follows vertebra T_2, and so forth. This arrangement differs in the cervical region. There the first pair of spinal nerves, C_1, passes between the skull and the first cervical vertebra. For this reason, each cervical nerve takes its name from the vertebra immediately inferior to it. In other words, cervical nerve C_2 precedes vertebra C_2, and the same system is used for the rest of the cervical series. The transition from one numbering system to another occurs between the last cervical vertebra and the first thoracic vertebra. The spinal nerve found at this location is designated C_8. Therefore, although there are only seven cervical vertebrae, there are eight cervical nerves.

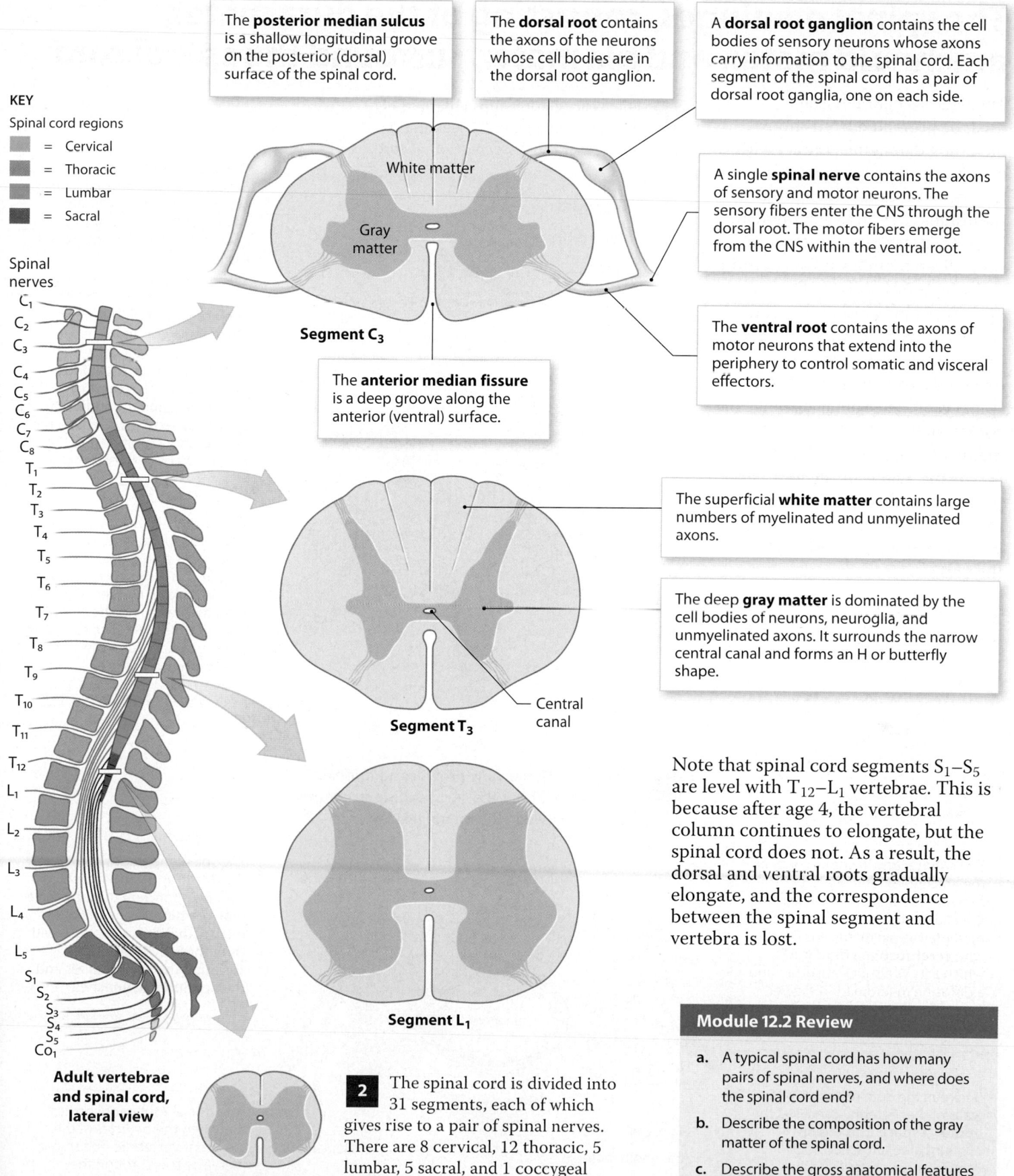

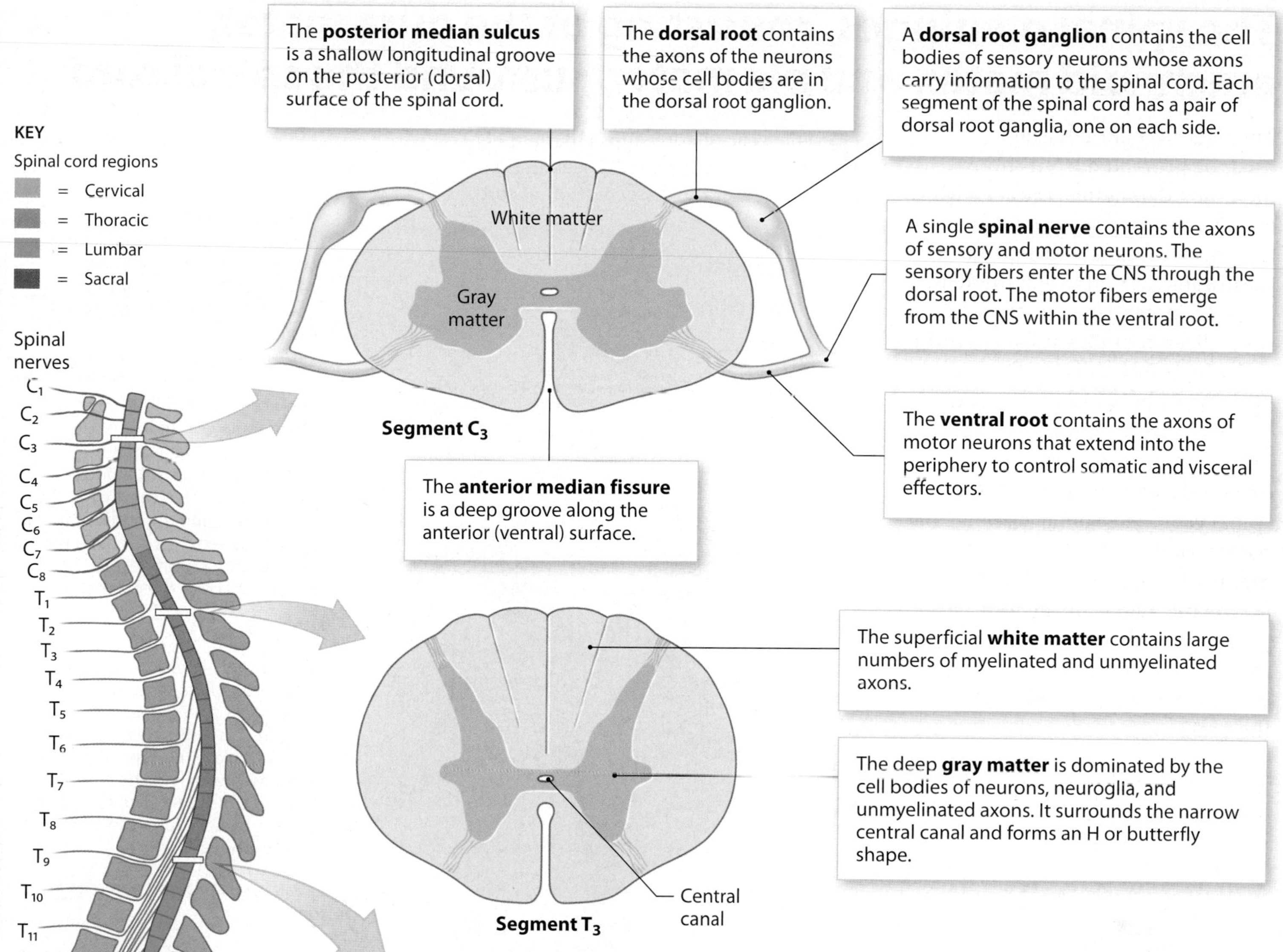
KEY
Spinal cord regions
= Cervical
= Thoracic
= Lumbar
= Sacral
Spinal nerves
The **posterior median sulcus** is a shallow longitudinal groove on the posterior (dorsal) surface of the spinal cord.
The **dorsal root** contains the axons of the neurons whose cell bodies are in the dorsal root ganglion.
A **dorsal root ganglion** contains the cell bodies of sensory neurons whose axons carry information to the spinal cord. Each segment of the spinal cord has a pair of dorsal root ganglia, one on each side.
White matter
Gray matter
A single **spinal nerve** contains the axons of sensory and motor neurons. The sensory fibers enter the CNS through the dorsal root. The motor fibers emerge from the CNS within the ventral root.
Segment C_3
The **ventral root** contains the axons of motor neurons that extend into the periphery to control somatic and visceral effectors.
The **anterior median fissure** is a deep groove along the anterior (ventral) surface.
The superficial **white matter** contains large numbers of myelinated and unmyelinated axons.
The deep **gray matter** is dominated by the cell bodies of neurons, neuroglia, and unmyelinated axons. It surrounds the narrow central canal and forms an H or butterfly shape.
Central canal
Segment T_3
C_1 C_2 C_3 C_4 C_5 C_6 C_7 C_8
T_1 T_2 T_3 T_4 T_5 T_6 T_7 T_8 T_9 T_{10} T_{11} T_{12}
L_1 L_2 L_3 L_4 L_5
S_1 S_2 S_3 S_4 S_5 Co_1
Segment L_1

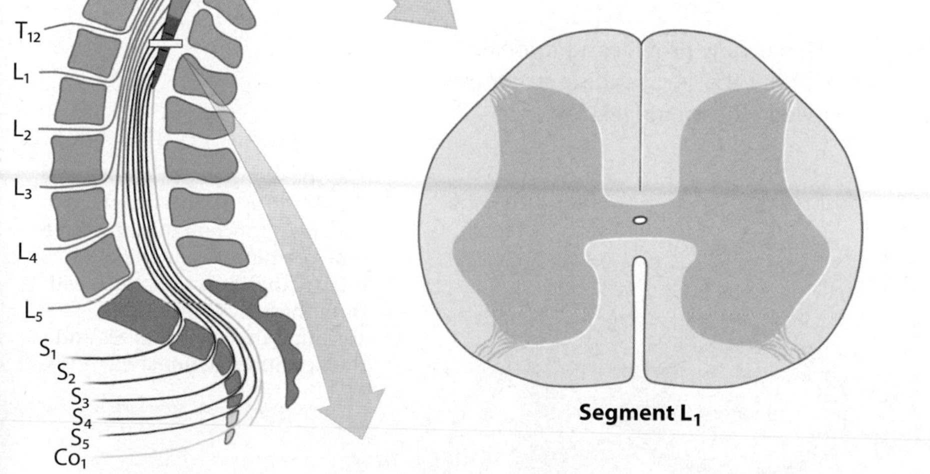

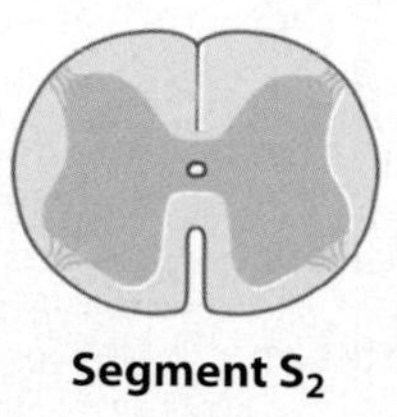
Segment S_2

Adult vertebrae and spinal cord, lateral view

Note that spinal cord segments S_1–S_5 are level with T_{12}–L_1 vertebrae. This is because after age 4, the vertebral column continues to elongate, but the spinal cord does not. As a result, the dorsal and ventral roots gradually elongate, and the correspondence between the spinal segment and vertebra is lost.

2 The spinal cord is divided into 31 segments, each of which gives rise to a pair of spinal nerves. There are 8 cervical, 12 thoracic, 5 lumbar, 5 sacral, and 1 coccygeal spinal segments.

Module 12.2 Review

a. A typical spinal cord has how many pairs of spinal nerves, and where does the spinal cord end?

b. Describe the composition of the gray matter of the spinal cord.

c. Describe the gross anatomical features of a cross section of spinal cord.

LO 12.2 Discuss the anatomical features of the spinal cord.

The spinal meninges, consisting of the dura mater, arachnoid mater, and pia mater, surround the spinal cord

The delicate neural tissues must be protected from shocks, including damaging contact with the surrounding bony walls of the vertebral canal. The **spinal meninges** (me-NIN-jēz; singular, *meninx*, membrane), a series of specialized membranes surrounding the spinal cord, provide the necessary physical stability and shock absorption. Blood vessels branching within these layers deliver oxygen and nutrients to the spinal cord. The spinal meninges consist of three layers: (1) the dura mater, (2) the arachnoid mater, and (3) the pia mater. At the foramen magnum of the skull, the spinal meninges are continuous with the cranial meninges, which surround the brain.

1 The basic relationships among the spinal meninges can be seen in this posterior view of the dissected spinal cord.

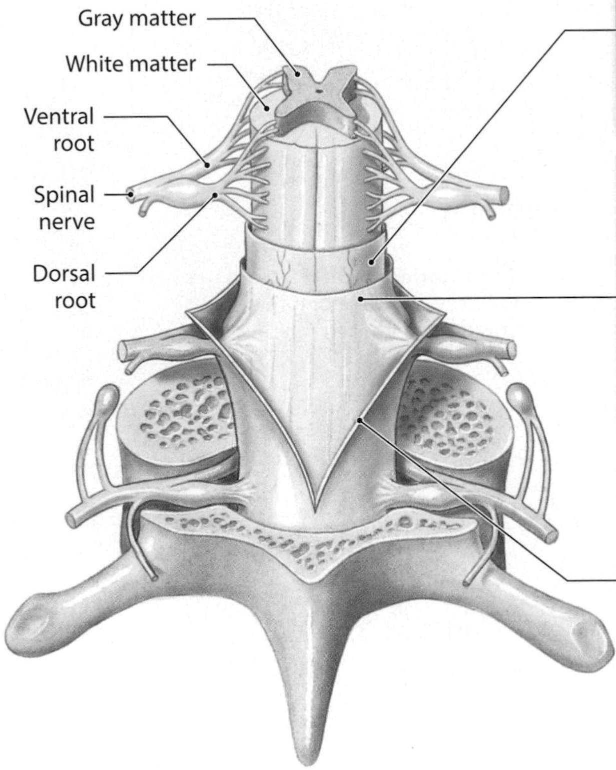
Gray matter
White matter
Ventral root
Spinal nerve
Dorsal root

The **pia mater** (PĒ-uh MĀ-ter; *pia*, delicate + *mater*, mother) consists of a meshwork of elastic and collagen fibers that is firmly bound to the underlying neural tissue.

The **arachnoid** (a-RAK-noyd; *arachne*, spider) **mater** is the middle meningeal layer. It includes a simple squamous epithelium, called the arachnoid membrane, and the subarachnoid space that extends between the arachnoid membrane and the outer surface of the pia mater.

The tough, fibrous **dura mater** (DOO-ruh; *dura*, hard) is the outermost covering of the spinal cord. It contains dense collagen fibers that are oriented along the longitudinal axis of the cord. A narrow subdural space separates the dura mater from the arachnoid mater.

2 A cross-sectional view provides additional information about the surrounding structures and the spaces between the meningeal layers.

The **subarachnoid space** contains the arachnoid trabeculae, a network of collagen and elastic fibers that attaches the arachnoid mater to the pia mater. It is filled with **cerebrospinal fluid (CSF)**, which acts as a shock absorber and a diffusion medium for dissolved gases, nutrients, chemical messengers, and wastes.

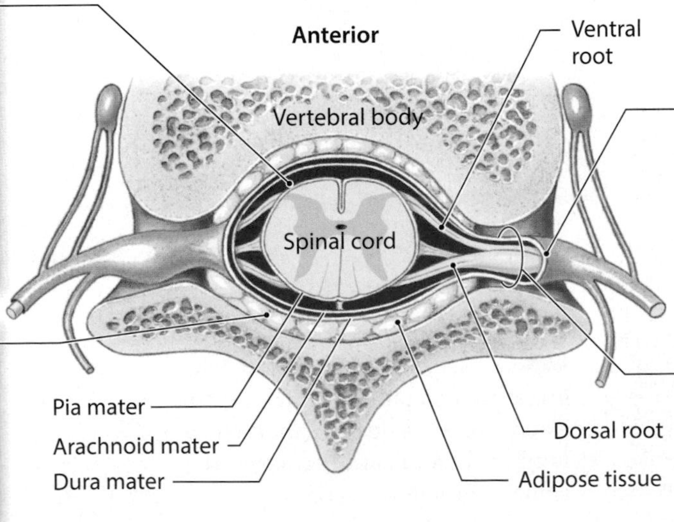
Anterior
Vertebral body
Spinal cord
Ventral root
Pia mater
Arachnoid mater
Dura mater
Dorsal root
Adipose tissue

Between the dura mater and the walls of the vertebral canal lies the **epidural space**. This region contains areolar tissue, blood vessels, and a protective padding of adipose tissue.

The spinal meninges accompany the dorsal and ventral roots as they pass through the intervertebral foramina. The meningeal membranes are continuous with the connective tissues that surround the spinal nerves and their peripheral branches.

The dorsal root ganglion lies between the pedicles of the adjacent vertebrae. The spinal nerve passes through the intervertebral foramen.

3 This anterior view of the cervical spinal cord shows the actual appearance of the meninges, supporting ligaments, and the roots of the spinal nerves.

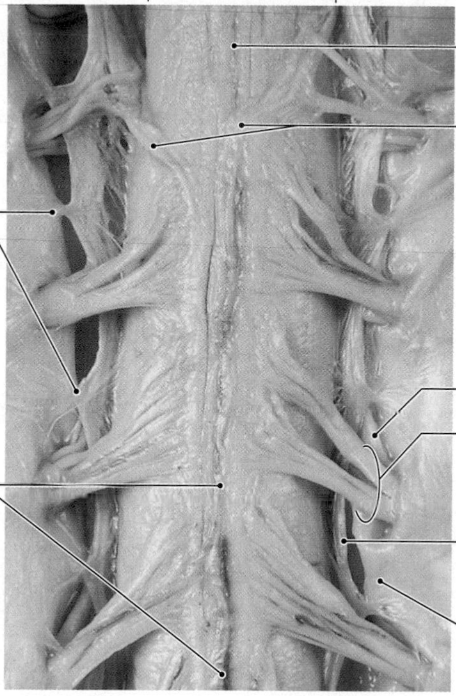

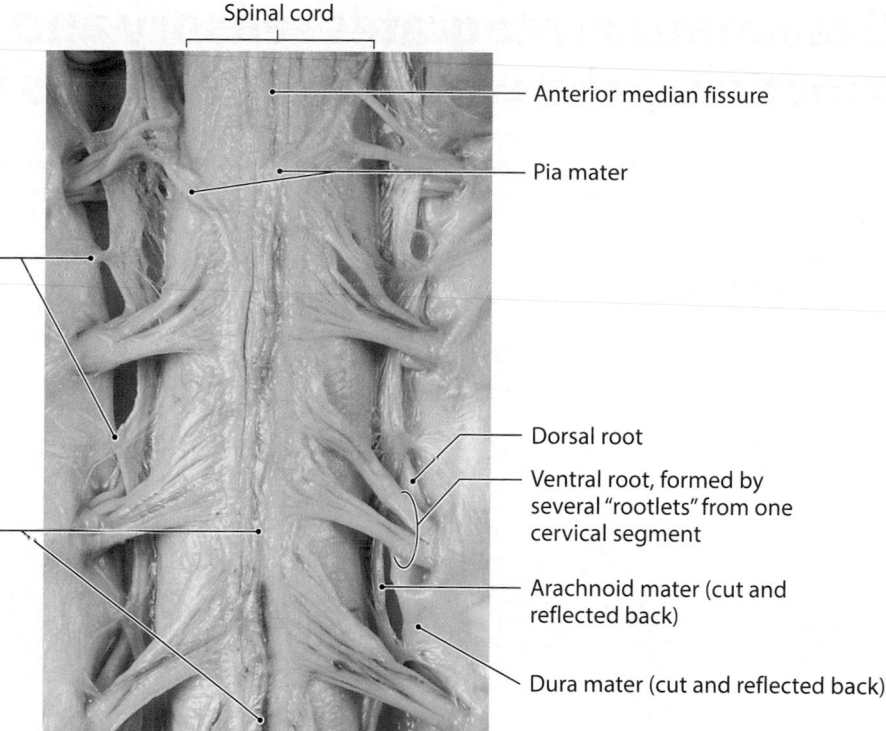

Spinal cord
Anterior median fissure
Pia mater
Dorsal root
Ventral root, formed by several "rootlets" from one cervical segment
Arachnoid mater (cut and reflected back)
Dura mater (cut and reflected back)

Along the length of the spinal cord, paired **denticulate ligaments** extend from the pia mater through the arachnoid mater to the dura mater. Denticulate ligaments prevent lateral movement of the spinal cord. Together, the dural connections at the foramen magnum and the coccygeal ligament at the sacrum prevent superior–inferior movement of the spinal cord.

The blood vessels servicing the spinal cord run along the surface of the spinal pia mater, within the subarachnoid space.

4 This is an x-ray of the inferior lumbar vertebrae. The vertebral canal inferior to vertebra L_1 contains the cauda equina bathed in the CSF of the subarachnoid space.

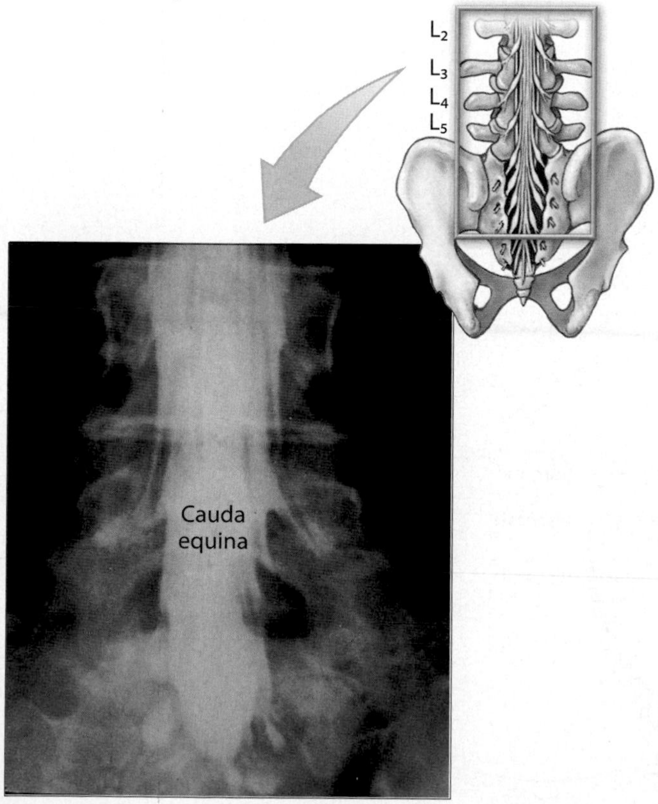

L_2
L_3
L_4
L_5
Cauda equina

5 In adults, cerebrospinal fluid can be safely withdrawn in a procedure known as a **lumbar puncture**, or spinal tap. A needle is inserted into the subarachnoid space in the lumbar region inferior to the tip of the conus medullaris.

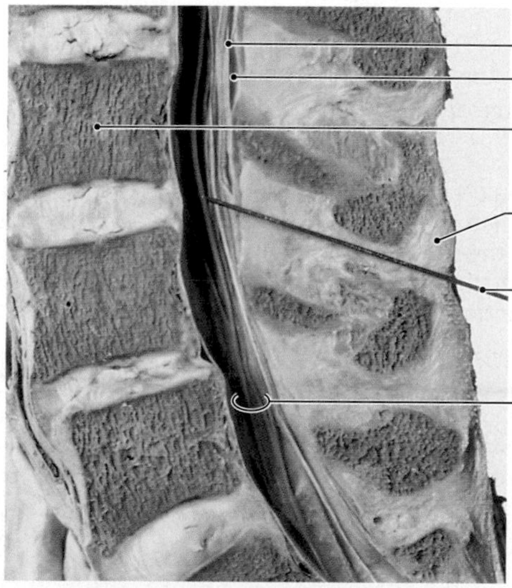

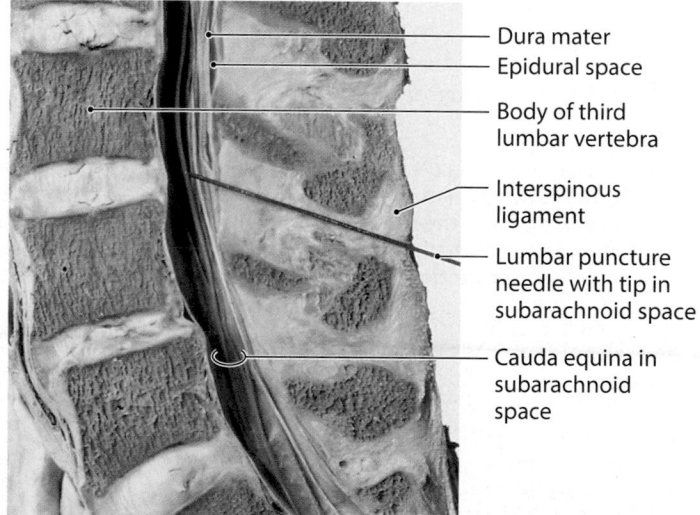

Dura mater
Epidural space
Body of third lumbar vertebra
Interspinous ligament
Lumbar puncture needle with tip in subarachnoid space
Cauda equina in subarachnoid space

Module 12.3 Review

a. Identify and describe the three spinal meninges.

b. Where is the cerebrospinal fluid that surrounds the spinal cord located?

c. Name the structures and spinal coverings that are penetrated during a lumbar puncture procedure.

LO 12.3 Describe the three meningeal layers that surround the spinal cord.

Gray matter integrates sensory and motor functions, and white matter carries information

1 This micrograph shows the major landmarks in and around the spinal cord. Compare this cross section with the diagrammatic views in this module.

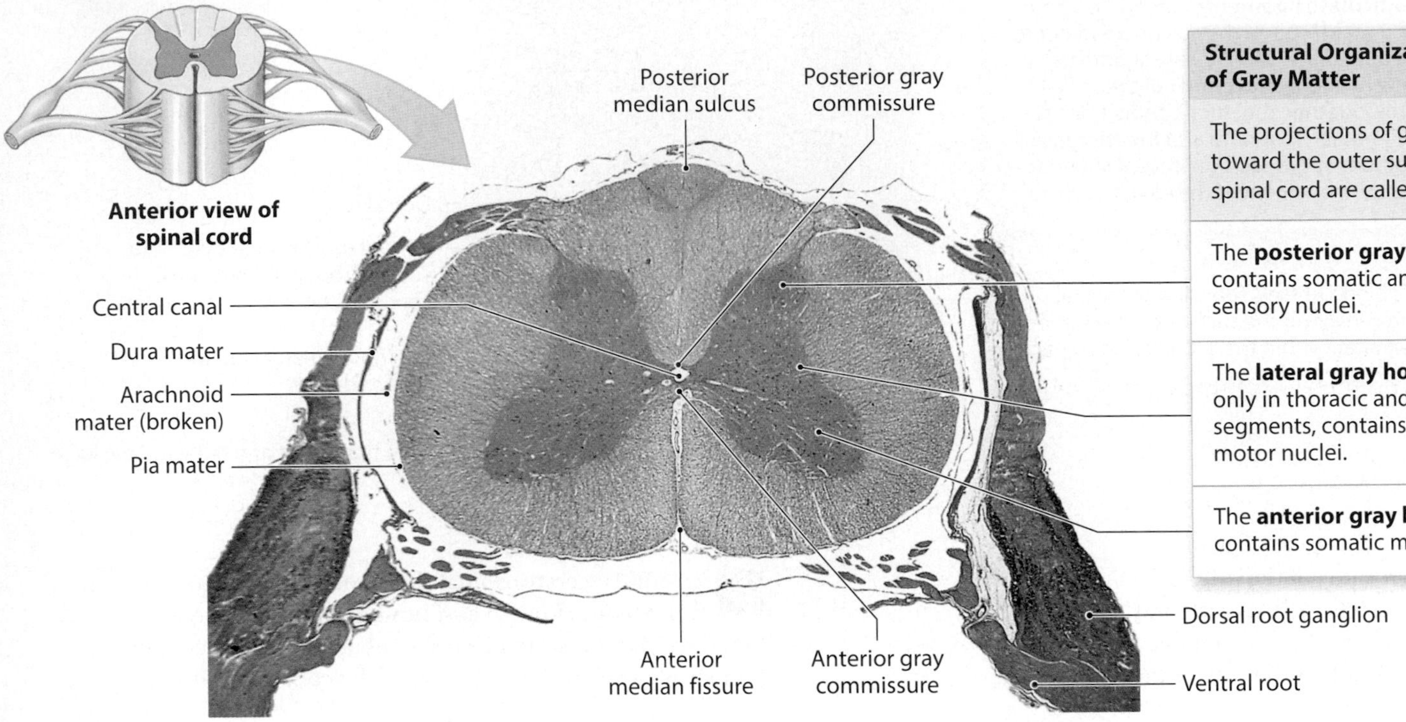
Anterior view of spinal cord
Posterior median sulcus
Posterior gray commissure
Central canal
Dura mater
Arachnoid mater (broken)
Pia mater
Anterior median fissure
Anterior gray commissure
Dorsal root ganglion
Ventral root

Structural Organization of Gray Matter

The projections of gray matter toward the outer surface of the spinal cord are called **horns**.

The **posterior gray horn** contains somatic and visceral sensory nuclei.

The **lateral gray horn**, located only in thoracic and lumbar segments, contains visceral motor nuclei.

The **anterior gray horn** contains somatic motor nuclei.

2 This diagrammatic view focuses on the organization of the gray matter of the spinal cord.

A frontal section along the length of the central canal (dashed line) of the spinal cord separates the sensory (posterior, or dorsal) nuclei from the motor (anterior, or ventral) nuclei.

The **gray commissures** (*commissura*, a joining together) are posterior and anterior to the central canal. They contain axons that cross from one side of the cord to the other before they reach a destination in the gray matter.

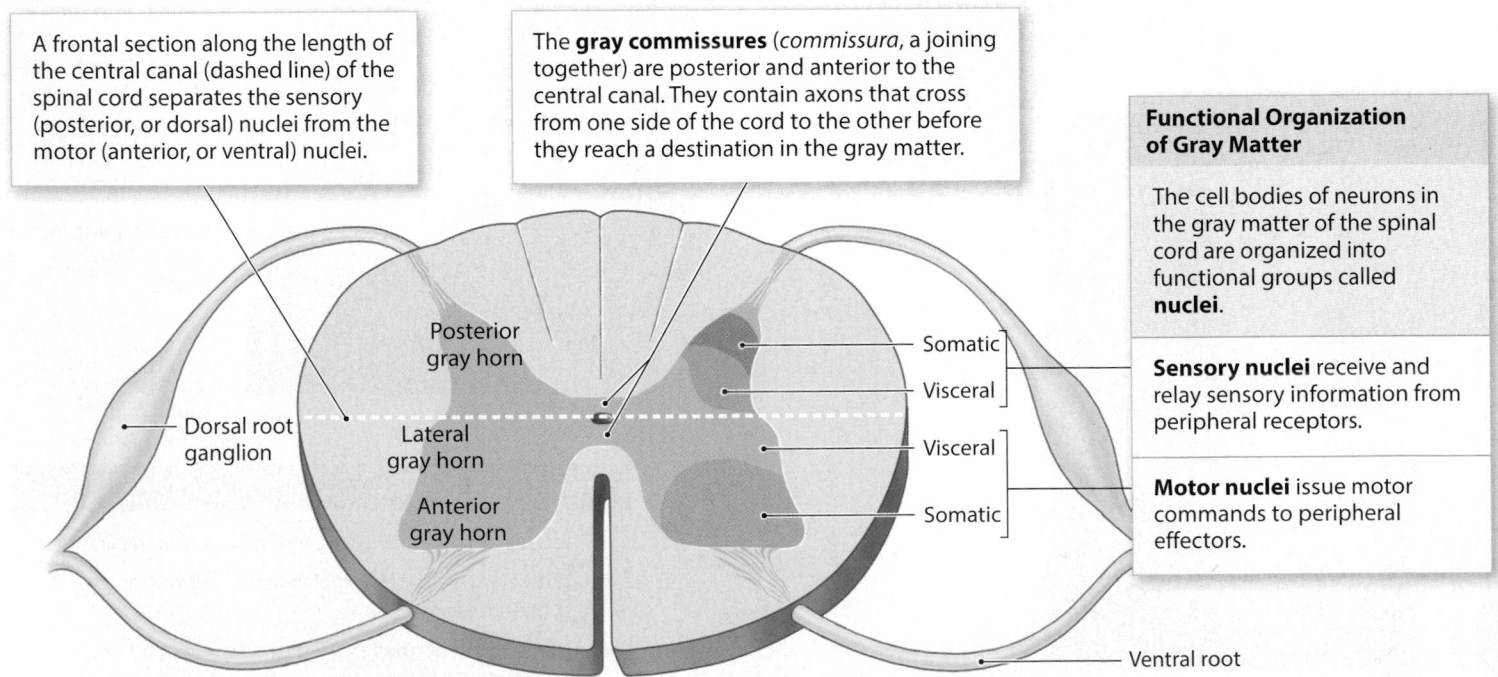

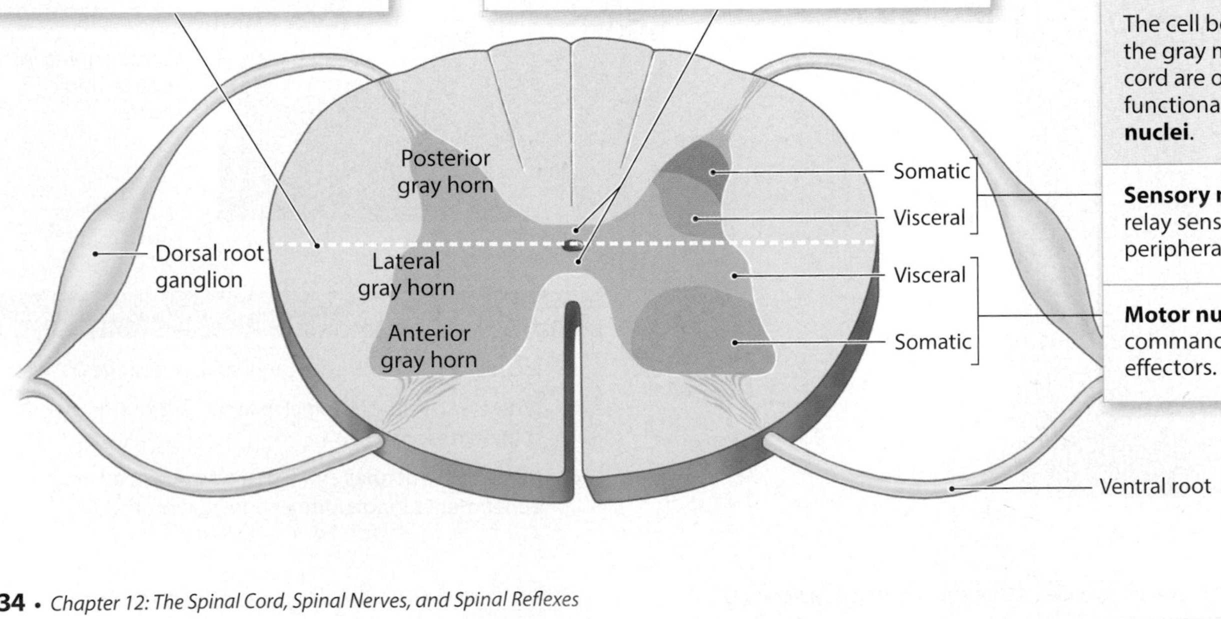
Posterior gray horn
Lateral gray horn
Anterior gray horn
Dorsal root ganglion
Somatic
Visceral
Visceral
Somatic
Ventral root

Functional Organization of Gray Matter

The cell bodies of neurons in the gray matter of the spinal cord are organized into functional groups called **nuclei**.

Sensory nuclei receive and relay sensory information from peripheral receptors.

Motor nuclei issue motor commands to peripheral effectors.

3 Like the gray horns, the white matter is organized according to the region of the body innervated. The white matter on each side of the spinal cord can be divided into three regions called **columns**. These columns contain tracts. A **tract** is a bundle of axons in the CNS that is relatively uniform with respect to diameter, myelination, and propagation speed. All the axons within a tract relay the same type of information (sensory or motor) in the same direction. **Ascending tracts** carry sensory information toward the brain, and **descending tracts** convey motor commands to the spinal cord.

Organization of Tracts in the Posterior White Column

The posterior white column contains ascending tracts providing sensations from the trunk and limbs.

Leg
Hip
Trunk
Arm

Structural and Functional Organization of White Matter

The **posterior white column** lies between the posterior gray horns and the posterior median sulcus.

The **lateral white column** includes the white matter on either side of the spinal cord, between the anterior and posterior columns.

The **anterior white column** lies between the anterior gray horns and the anterior median fissure.

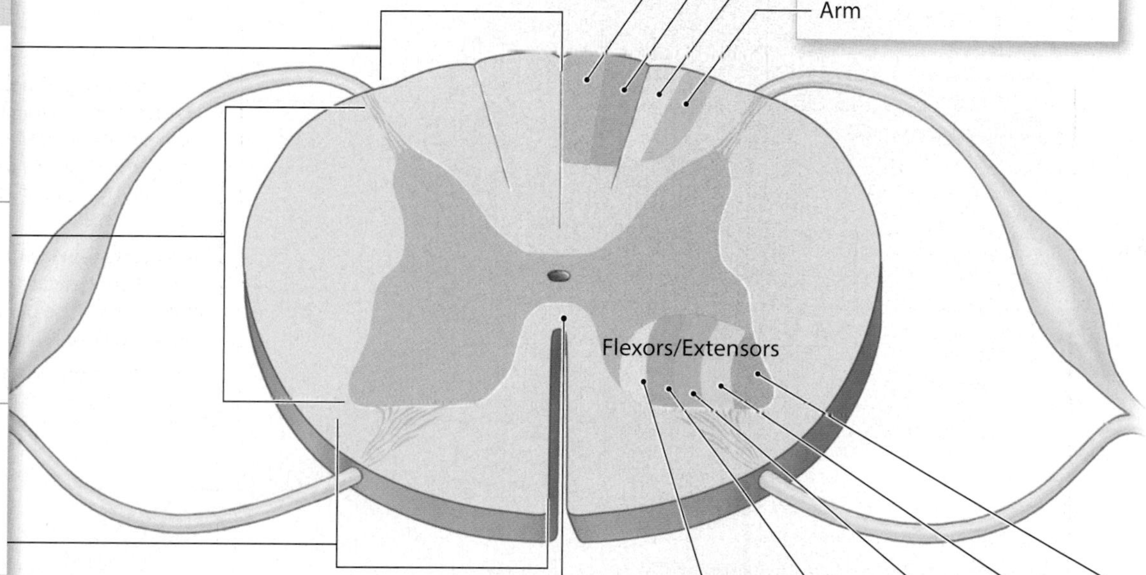
Flexors/Extensors

The **anterior white commissure** interconnects the anterior white columns. This is where axons cross from one side of the spinal cord to the other.

Trunk Shoulder Arm Forearm Hand

In the cervical enlargement, which contains neurons involved with sensations and motor control of the upper limbs, the motor nuclei of the anterior gray horn are grouped by region. The motor neurons controlling flexor muscles are medial to those controlling extensor muscles.

The high degree of organization of the spinal cord is clinically important. For example, damage to a specific area of gray matter allows clinicians to predict which muscles will be affected. Similarly, because spinal tracts have very specific functions, damage to one produces a characteristic loss of sensation or motor control.

Module 12.4 Review

a. Differentiate between sensory nuclei and motor nuclei.

b. A person with polio has lost the use of his leg muscles. In which area of his spinal cord would you expect the virus-infected motor neurons to be?

c. A disease that damages myelin sheaths would affect which portion of the spinal cord?

LO **12.4** Explain the roles of white matter and gray matter in processing and relaying sensory information and motor commands.

Spinal nerves have a similar anatomical structure and distribution pattern

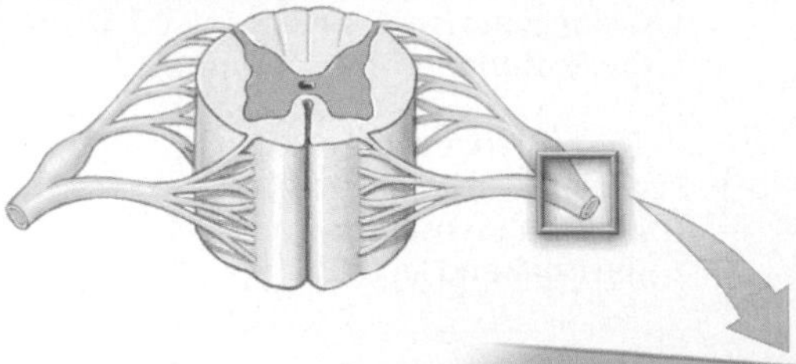

1 Every segment of the spinal cord is connected to a pair of spinal nerves. Surrounding each spinal nerve is a series of connective tissue layers continuous with those of their associated peripheral nerves. These layers, best seen in sectional view, are comparable with those associated with skeletal muscles.

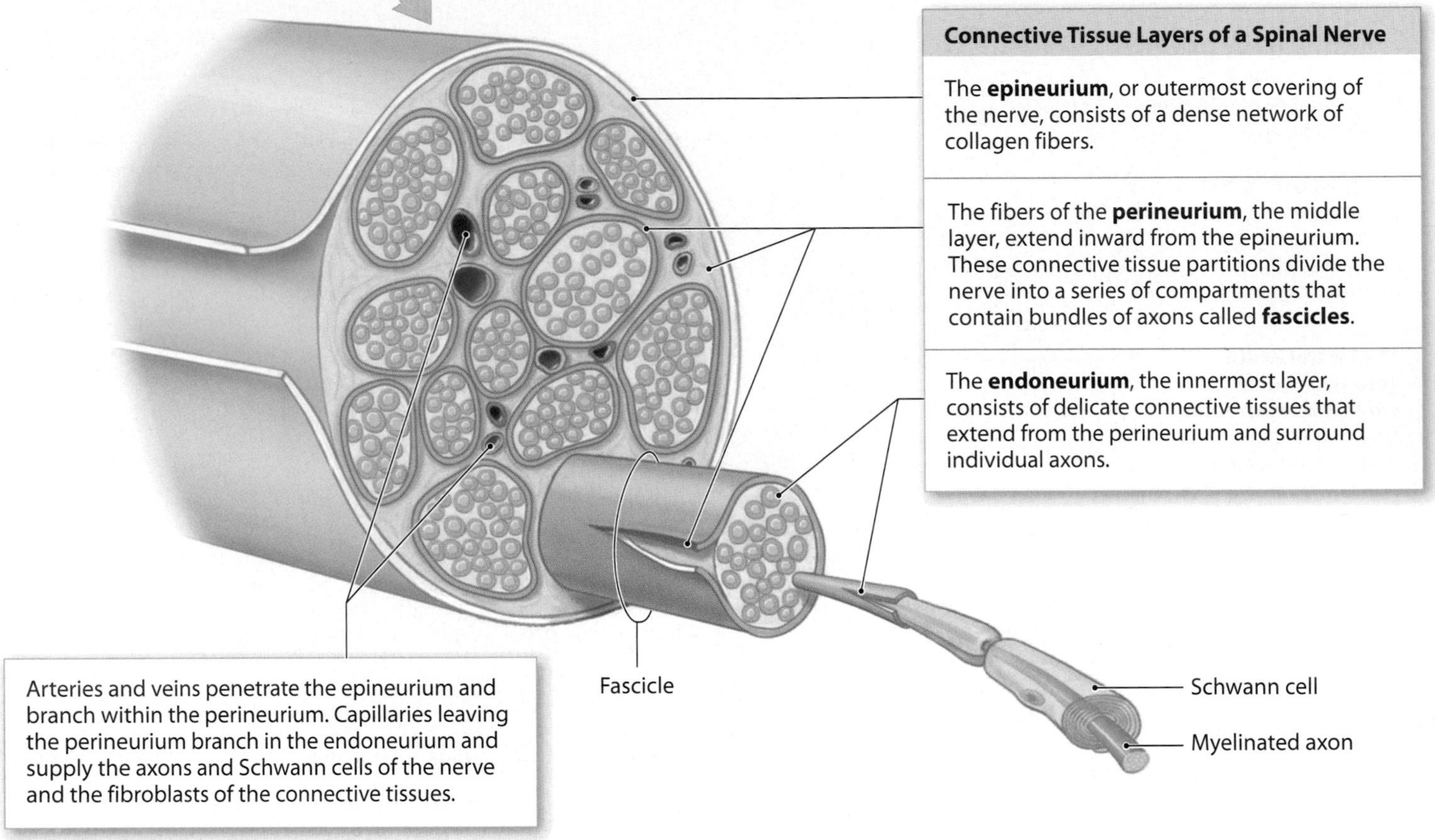
Connective Tissue Layers of a Spinal Nerve
The **epineurium**, or outermost covering of the nerve, consists of a dense network of collagen fibers.
The fibers of the **perineurium**, the middle layer, extend inward from the epineurium. These connective tissue partitions divide the nerve into a series of compartments that contain bundles of axons called **fascicles**.
The **endoneurium**, the innermost layer, consists of delicate connective tissues that extend from the perineurium and surround individual axons.
Arteries and veins penetrate the epineurium and branch within the perineurium. Capillaries leaving the perineurium branch in the endoneurium and supply the axons and Schwann cells of the nerve and the fibroblasts of the connective tissues.
Fascicle
Schwann cell
Myelinated axon

2 Each spinal nerve divides to form **rami** (RĀ-mī; singular *ramus*, a branch). Some of these rami carry visceral motor fibers of the autonomic nervous system (ANS). Spinal nerves in the thoracic and upper lumbar segments of the spinal cord carry the motor output of the **sympathetic division** that is responsible for the "fight or flight" response.

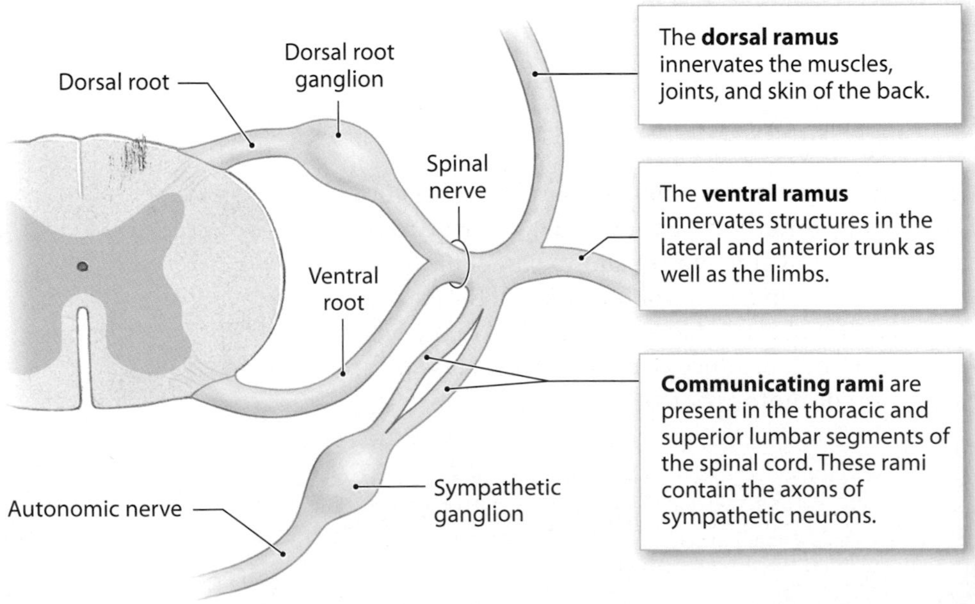
Dorsal root
Dorsal root ganglion
Spinal nerve
Ventral root
Autonomic nerve
Sympathetic ganglion
The **dorsal ramus** innervates the muscles, joints, and skin of the back.
The **ventral ramus** innervates structures in the lateral and anterior trunk as well as the limbs.
Communicating rami are present in the thoracic and superior lumbar segments of the spinal cord. These rami contain the axons of sympathetic neurons.

3 The specific bilateral region of the skin surface monitored by a single pair of spinal nerves is known as a **dermatome**. Each pair of spinal nerves supplies its own dermatome, but the boundaries of adjacent dermatomes overlap to some degree. Spinal nerve C_1 typically lacks a sensory branch to the skin; when present, it innervates the scalp with C_2 and C_3. The face is monitored by the fifth pair of cranial nerves (N V).

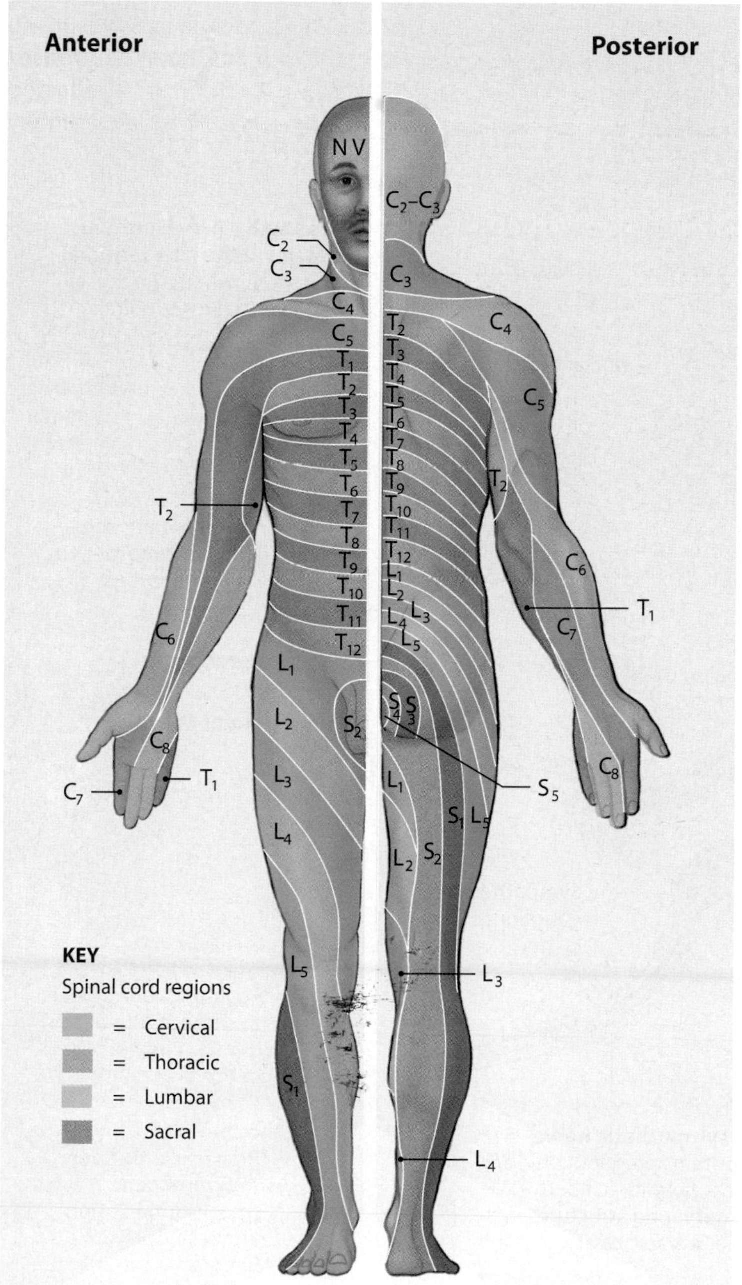
Anterior
Posterior
NV
C2–C3
C2
C3
C4
C5
T1
T2
T3
T4
T5
T6
T7
T8
T9
T10
T11
T12
L1
L2
L3
L4
L5
S1
S2
S3
S4
S5
C6
C7
C8
KEY
Spinal cord regions
= Cervical
= Thoracic
= Lumbar
= Sacral

4 Dermatomes are clinically important because damage or infection of a spinal nerve or dorsal root ganglion produces a loss of sensation in the corresponding region of the skin. Additionally, characteristic signs may appear on the skin supplied by that specific nerve. The skin eruptions shown here are characteristic of **shingles**, a viral infection of dorsal root ganglia. Shingles (derived from the Latin *cingulum*, girdle) is caused by the varicella-zoster virus (VZV), the same herpes virus that causes chickenpox. This herpes virus attacks neurons within the dorsal roots of spinal nerves and sensory ganglia of cranial nerves. This disorder produces a painful rash and blisters whose distribution corresponds to that of the affected sensory nerve and its associated dermatome. Any person who has had chickenpox is at risk of developing shingles, because the virus can remain dormant within the anterior gray horns of the spinal cord. It is not known what triggers the reactivation of the virus. In 2006, the U.S. Food and Drug Administration approved a VZV vaccine (Zostavax) for use in people ages 60 and above who have had chickenpox.

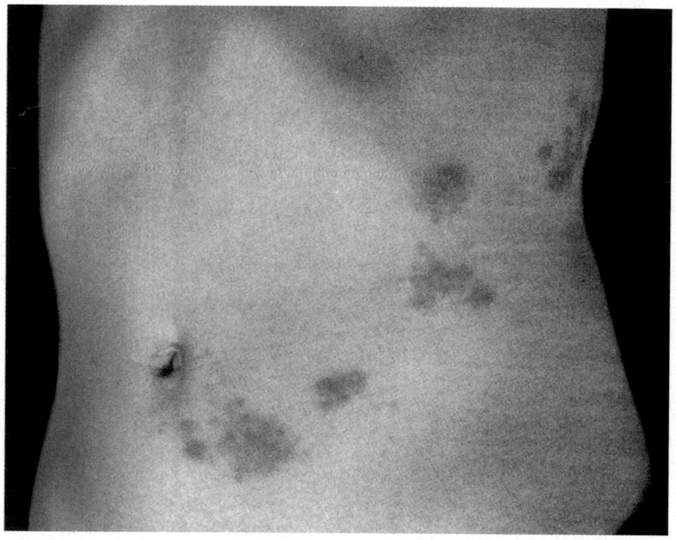

Module 12.5 Review

a. Identify the three layers of connective tissue of a spinal nerve, and identify the major peripheral branches of a spinal nerve.

b. Describe a dermatome.

c. Explain the cause of shingles.

Module 12.6

Each ramus of a spinal nerve provides motor and sensory innervation to a specific region

1 This diagram illustrates how a spinal nerve distributes motor commands that originate in motor nuclei of the thoracic or superior lumbar segments of the spinal cord.

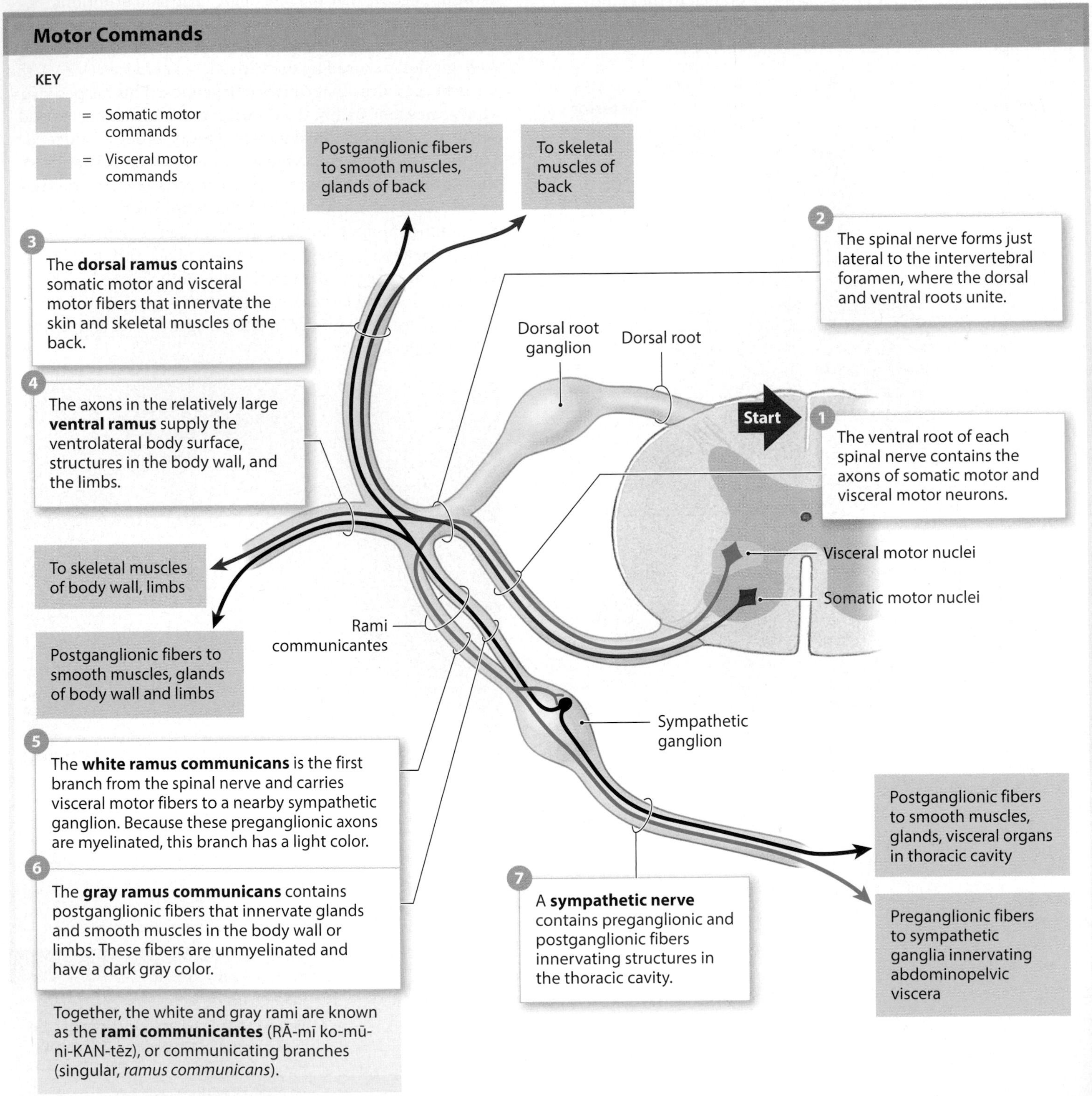
Motor Commands
KEY
= Somatic motor commands
= Visceral motor commands
Postganglionic fibers to smooth muscles, glands of back
To skeletal muscles of back
2 The spinal nerve forms just lateral to the intervertebral foramen, where the dorsal and ventral roots unite.
3 The **dorsal ramus** contains somatic motor and visceral motor fibers that innervate the skin and skeletal muscles of the back.
Dorsal root ganglion
Dorsal root
4 The axons in the relatively large **ventral ramus** supply the ventrolateral body surface, structures in the body wall, and the limbs.
Start
1 The ventral root of each spinal nerve contains the axons of somatic motor and visceral motor neurons.
Visceral motor nuclei
Somatic motor nuclei
To skeletal muscles of body wall, limbs
Postganglionic fibers to smooth muscles, glands of body wall and limbs
Rami communicantes
Sympathetic ganglion
5 The **white ramus communicans** is the first branch from the spinal nerve and carries visceral motor fibers to a nearby sympathetic ganglion. Because these preganglionic axons are myelinated, this branch has a light color.
6 The **gray ramus communicans** contains postganglionic fibers that innervate glands and smooth muscles in the body wall or limbs. These fibers are unmyelinated and have a dark gray color.
Together, the white and gray rami are known as the **rami communicantes** (RĀ-mī ko-mū-ni-KAN-tēz), or communicating branches (singular, *ramus communicans*).
7 A **sympathetic nerve** contains preganglionic and postganglionic fibers innervating structures in the thoracic cavity.
Postganglionic fibers to smooth muscles, glands, visceral organs in thoracic cavity
Preganglionic fibers to sympathetic ganglia innervating abdominopelvic viscera

2 This diagram illustrates how a spinal nerve collects sensory information from peripheral structures and delivers it to sensory nuclei in the thoracic or superior lumbar segments of the spinal cord. The dorsal, ventral, and white rami also contain sensory fibers. Somatic sensory information arrives over the dorsal and ventral rami. Visceral sensory information reaches the dorsal root through the dorsal, ventral, and white rami.

Sensory Information

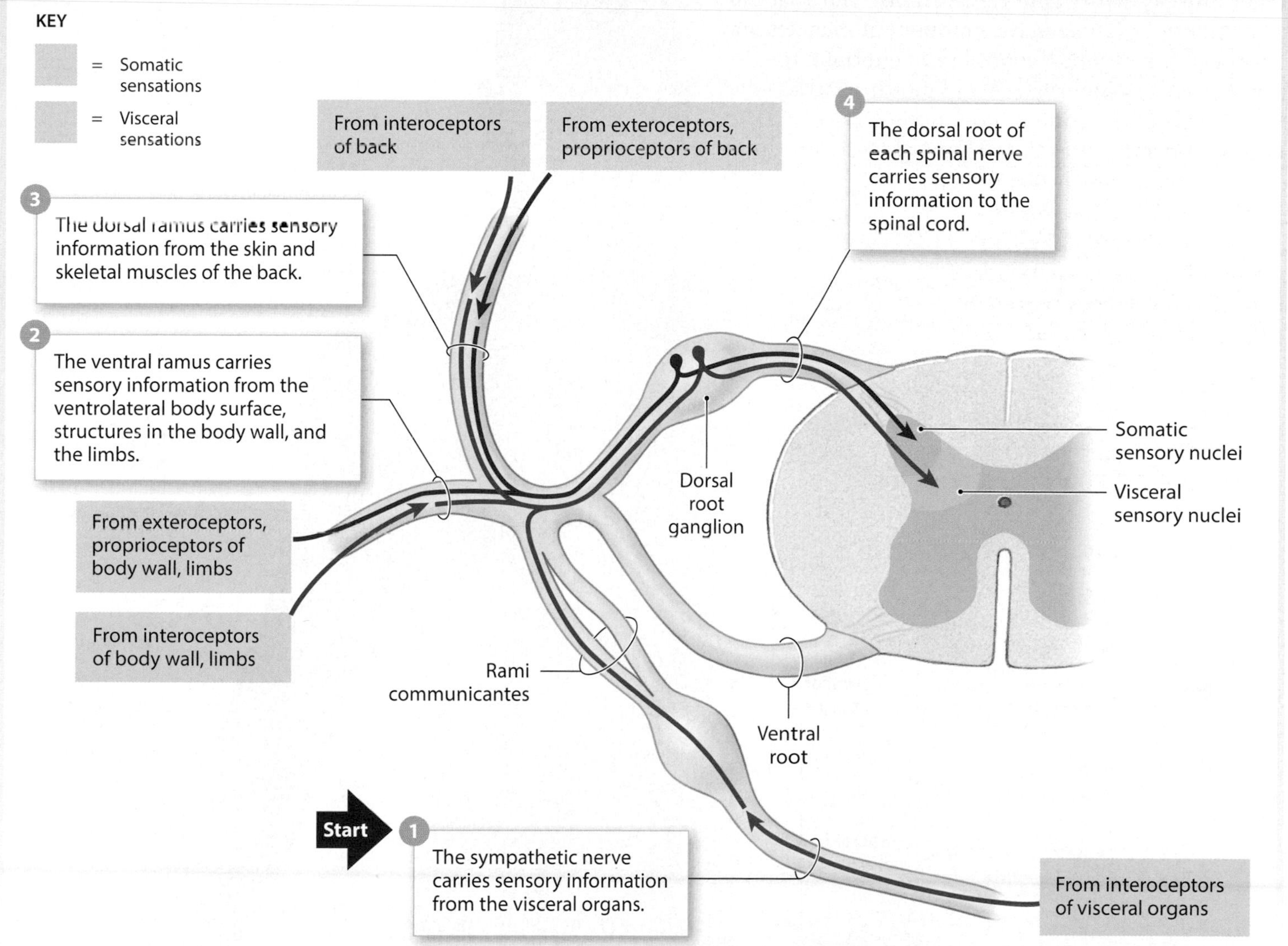

KEY
= Somatic sensations
= Visceral sensations
From interoceptors of back
From exteroceptors, proprioceptors of back
4
The dorsal root of each spinal nerve carries sensory information to the spinal cord.
3
The dorsal ramus carries sensory information from the skin and skeletal muscles of the back.
2
The ventral ramus carries sensory information from the ventrolateral body surface, structures in the body wall, and the limbs.
Somatic sensory nuclei
Visceral sensory nuclei
Dorsal root ganglion
From exteroceptors, proprioceptors of body wall, limbs
From interoceptors of body wall, limbs
Rami communicantes
Ventral root
Start
1
The sympathetic nerve carries sensory information from the visceral organs.
From interoceptors of visceral organs

Module 12.6 Review

a. Describe gray ramus and white ramus.

b. Indicate whether the following fibers make up the white rami or gray rami:
1) preganglionic fibers connecting a spinal nerve with a sympathetic ganglion in the thoracic and lumbar region of the spinal cord
2) postganglionic fibers connecting a sympathetic ganglion in the thoracic or lumbar region with the spinal nerve

c. Which ramus innervates the skin and skeletal muscles of the back?

LO **12.6** Describe the rami associated with spinal nerves.

Spinal nerves form nerve plexuses that innervate the skin and skeletal muscles; the cervical plexus is the smallest of these nerve plexuses

During development, small skeletal muscles innervated by different ventral rami typically fuse to form larger muscles with compound origins. The anatomical distinctions between the component muscles may disappear, but separate ventral rami continue to provide sensory innervation and motor control to each part of the compound muscle.

As they converge, the ventral rami of adjacent spinal nerves blend their fibers, producing a series of compound nerve trunks. This complex interwoven network of nerves is called a **nerve plexus** (PLEK-sus; *plexus*, braid).

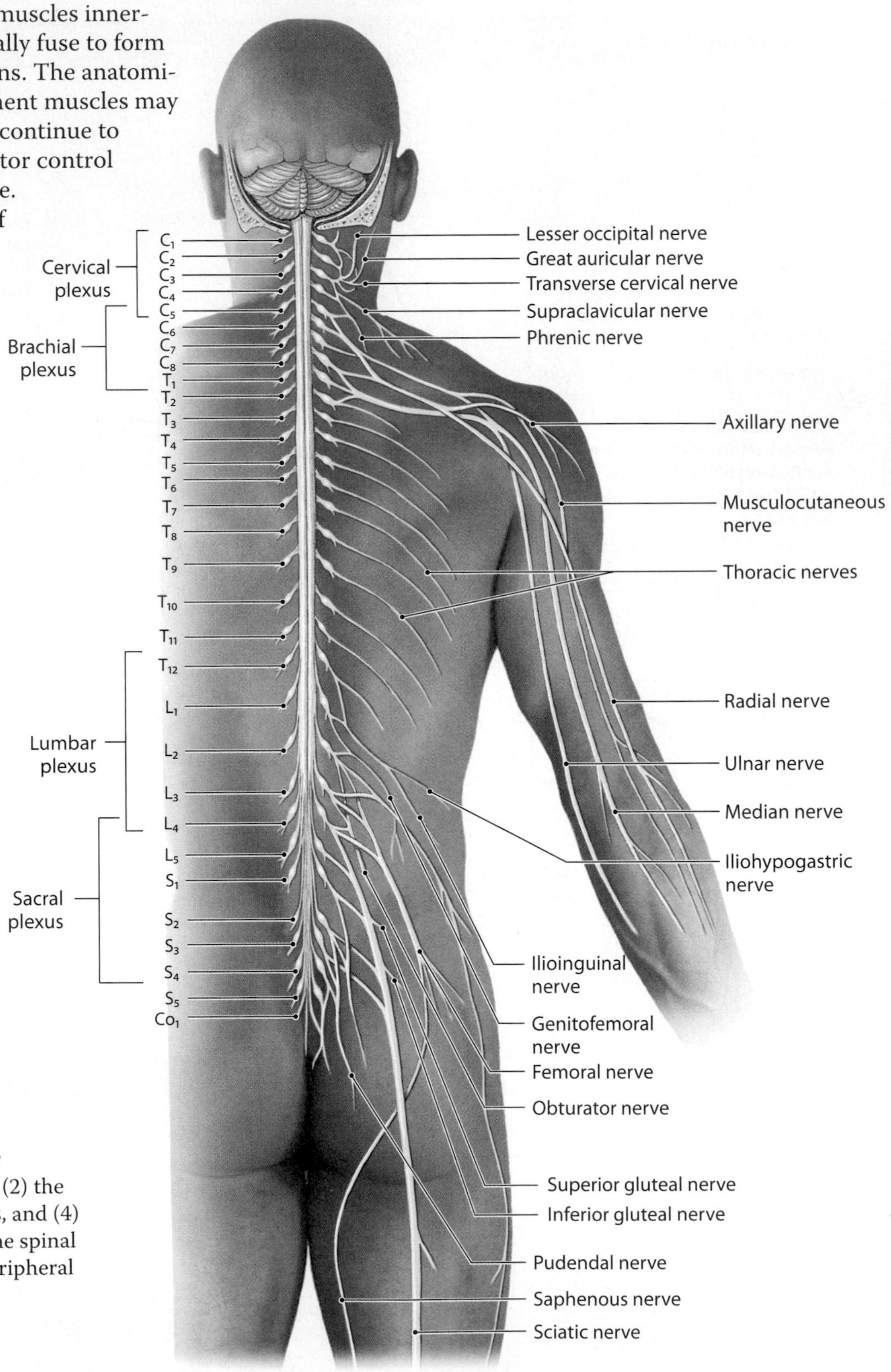

Cervical plexus
Brachial plexus
Lumbar plexus
Sacral plexus
C1
C2
C3
C4
C5
C6
C7
C8
T1
T2
T3
T4
T5
T6
T7
T8
T9
T10
T11
T12
L1
L2
L3
L4
L5
S1
S2
S3
S4
S5
Co1
Lesser occipital nerve
Great auricular nerve
Transverse cervical nerve
Supraclavicular nerve
Phrenic nerve
Axillary nerve
Musculocutaneous nerve
Thoracic nerves
Radial nerve
Ulnar nerve
Median nerve
Iliohypogastric nerve
Ilioinguinal nerve
Genitofemoral nerve
Femoral nerve
Obturator nerve
Superior gluteal nerve
Inferior gluteal nerve
Pudendal nerve
Saphenous nerve
Sciatic nerve

1 The ventral rami form four major plexuses: (1) the **cervical plexus**, (2) the **brachial plexus**, (3) the **lumbar plexus**, and (4) the **sacral plexus**. In this illustration, the spinal nerves are on the left, and the major peripheral nerves are on the right.

2 The cervical plexus consists of the ventral rami of spinal nerves C_1–C_5. The branches of the cervical plexus innervate the muscles of the neck and extend into the thoracic cavity, where they control the diaphragm, a key muscle of breathing. The **phrenic nerve**, the major nerve of the cervical plexus, provides the entire nerve supply to the diaphragm. Other branches of this nerve plexus are distributed to the skin of the neck and the superior part of the chest.

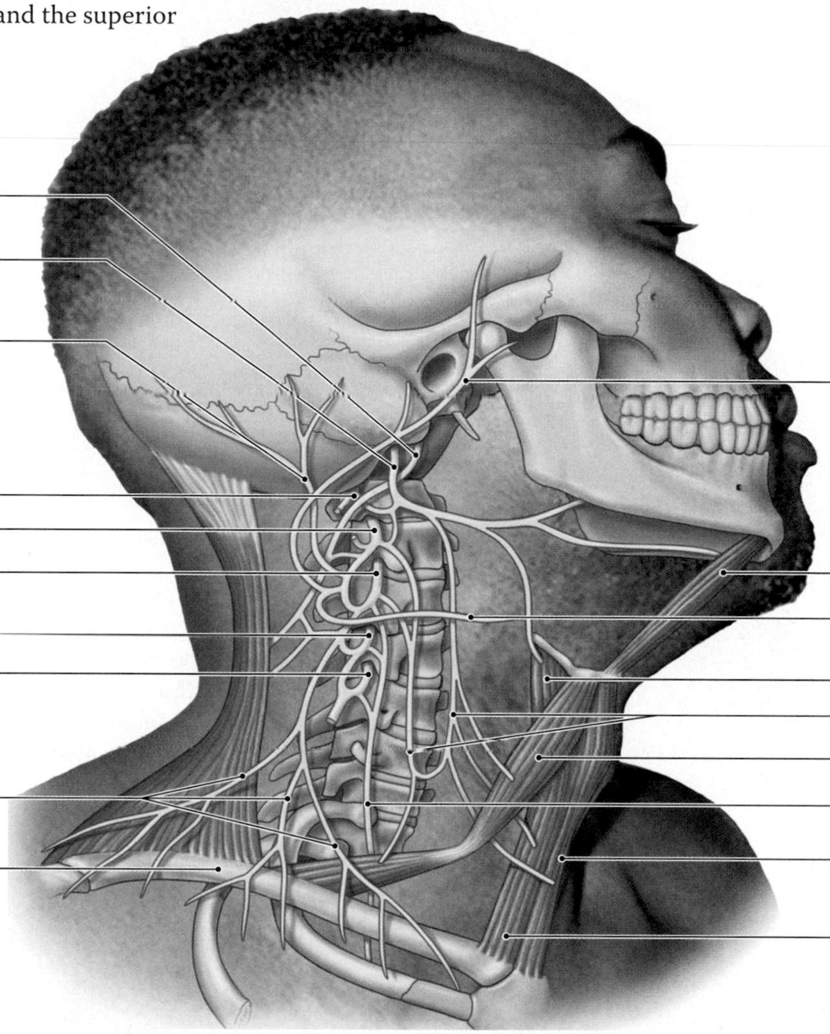

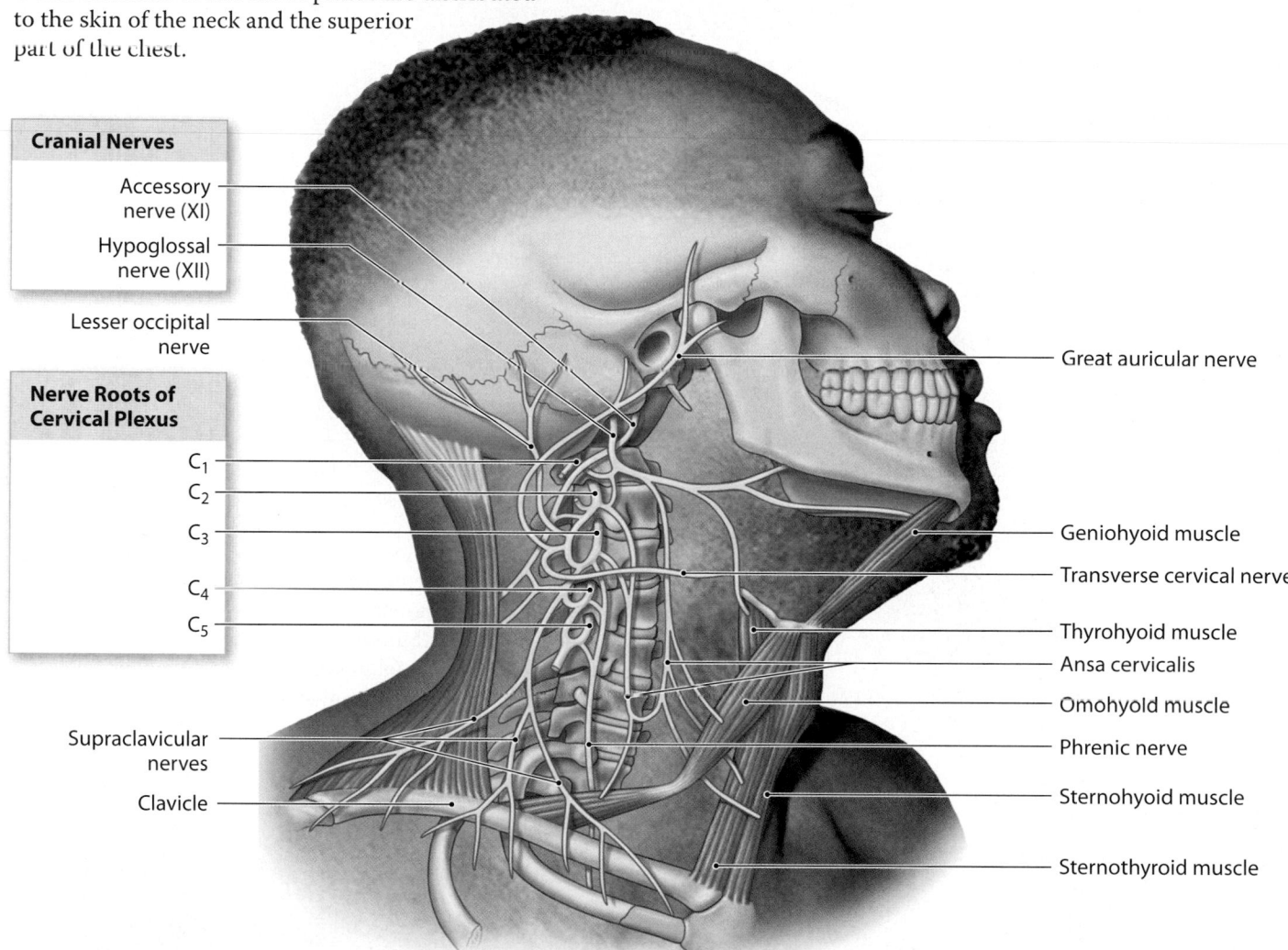

Cranial Nerves
Accessory nerve (XI)
Hypoglossal nerve (XII)
Lesser occipital nerve
Nerve Roots of Cervical Plexus
C1
C2
C3
C4
C5
Supraclavicular nerves
Clavicle
Great auricular nerve
Geniohyoid muscle
Transverse cervical nerve
Thyrohyoid muscle
Ansa cervicalis
Omohyoid muscle
Phrenic nerve
Sternohyoid muscle
Sternothyroid muscle

The Cervical Plexus

Nerve	Spinal Segments	Distribution
Ansa cervicalis (superior and inferior branches)	C_1–C_4	Five of the extrinsic laryngeal muscles: sternothyroid, sternohyoid, omohyoid, geniohyoid, and thyrohyoid muscles (by cranial nerve XII)
Lesser occipital, transverse cervical, supraclavicular, and great auricular nerves	C_2–C_3	Skin of upper chest, shoulder, neck, and ear
Phrenic nerve	C_3–C_5	Diaphragm
Cervical nerves	C_1–C_5	Levator scapulae, scalene, sternocleidomastoid, and trapezius muscles (with cranial nerve XI)

Module 12.7 Review

a. Define nerve plexus, and list the major nerve plexuses.

b. When an anesthetic blocks the function of the ventral rami of the cervical spinal nerves, which areas of the body will be affected?

c. Injury to which of the nerve plexuses would interfere with the ability to breathe?

LO **12.7** Relate the distribution pattern of spinal nerves to the region they innervate, and describe the cervical plexus.

Module 12.8

The brachial plexus innervates the pectoral girdle and upper limbs

1 The brachial plexus innervates the pectoral girdle and upper limbs, with contributions from the ventral rami of spinal nerves C_4–T_1.

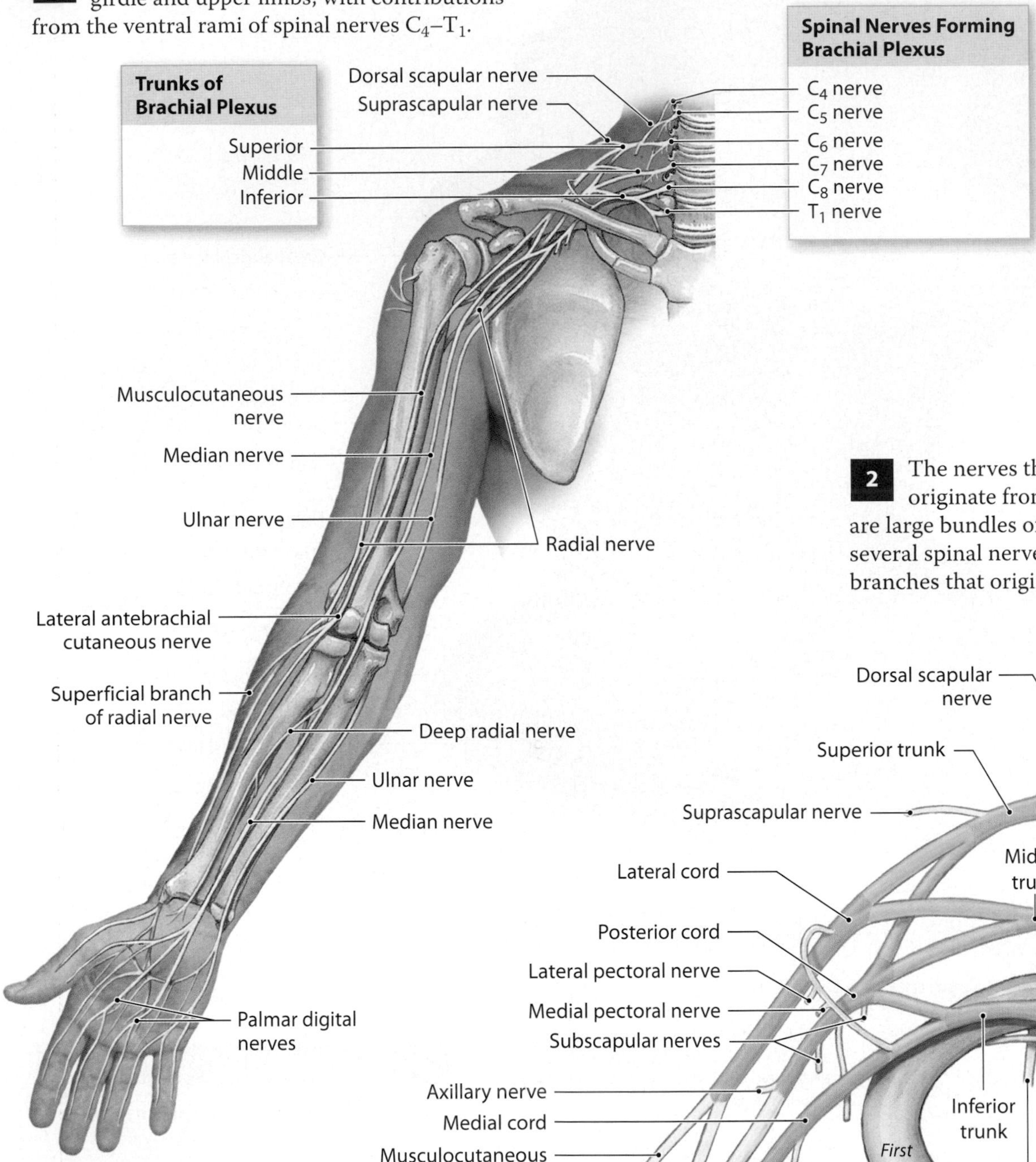

Trunks of Brachial Plexus
Superior
Middle
Inferior
Dorsal scapular nerve
Suprascapular nerve
Spinal Nerves Forming Brachial Plexus
C_4 nerve
C_5 nerve
C_6 nerve
C_7 nerve
C_8 nerve
T_1 nerve
Musculocutaneous nerve
Median nerve
Ulnar nerve
Radial nerve
Lateral antebrachial cutaneous nerve
Superficial branch of radial nerve
Deep radial nerve
Ulnar nerve
Median nerve
Palmar digital nerves

2 The nerves that form the brachial plexus originate from trunks and cords. **Trunks** are large bundles of axons contributed by several spinal nerves. **Cords** are smaller branches that originate at trunks.

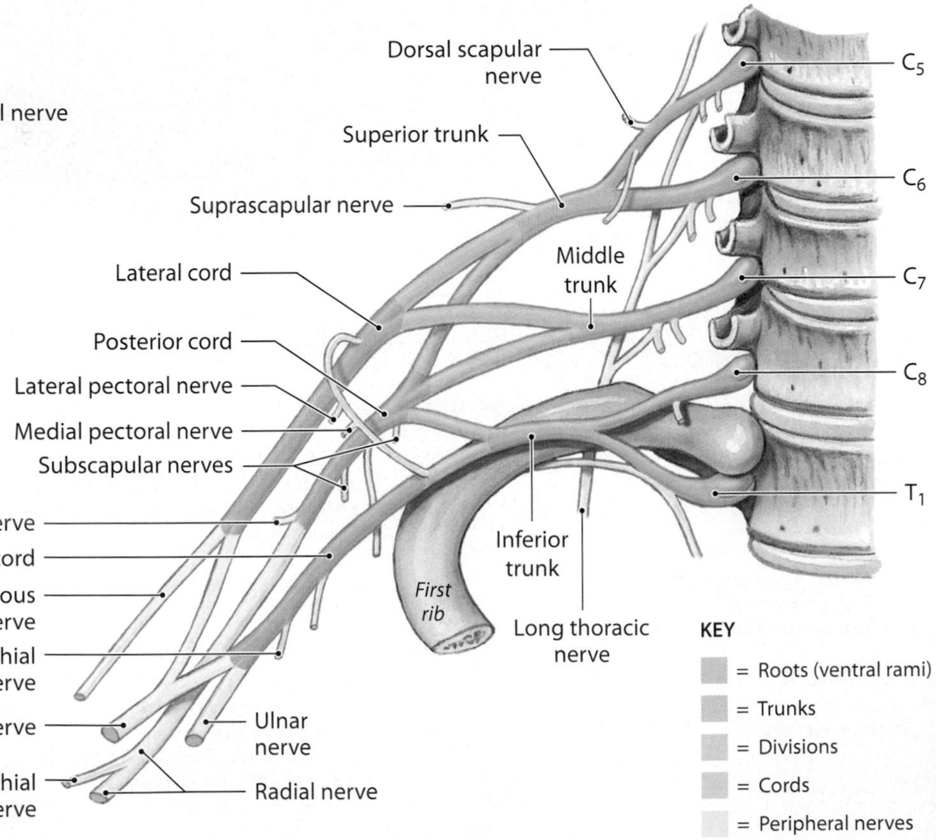

Dorsal scapular nerve
Superior trunk
Suprascapular nerve
Lateral cord
Posterior cord
Lateral pectoral nerve
Medial pectoral nerve
Subscapular nerves
Axillary nerve
Medial cord
Musculocutaneous nerve
Medial antebrachial cutaneous nerve
Median nerve
Posterior brachial cutaneous nerve
Ulnar nerve
Radial nerve
Middle trunk
Inferior trunk
First rib
Long thoracic nerve
C_5
C_6
C_7
C_8
T_1
KEY
= Roots (ventral rami)
= Trunks
= Divisions
= Cords
= Peripheral nerves

3 The distribution of the cutaneous nerves of the wrist and hand is very important in clinical medicine. Nerve damage or injury in this region can be precisely localized by carefully testing the sensory function of the hand.

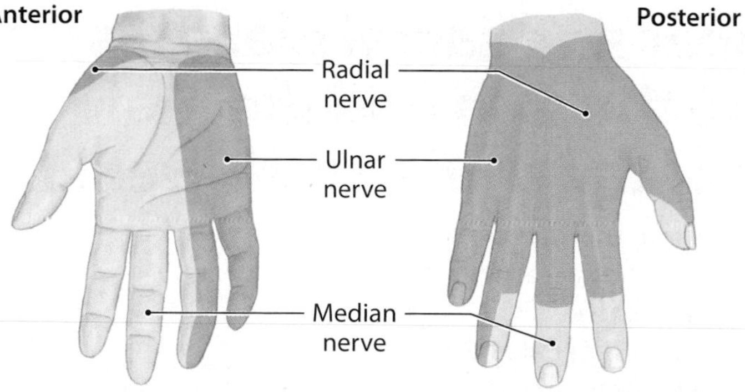

Anterior
Posterior
Radial nerve
Ulnar nerve
Median nerve

The Brachial Plexus

Nerve	Spinal Segments	Distribution
Nerve to subclavius	C_4–C_6	Subclavius muscle
Dorsal scapular nerve	C_5	Rhomboid and levator scapulae muscles
Long thoracic nerve	C_5–C_7	Serratus anterior muscle
Suprascapular nerve	C_5, C_6	Supraspinatus and infraspinatus muscles; sensory from shoulder joint and scapula
Pectoral nerves (medial and lateral)	C_5–T_1	Pectoralis muscles
Subscapular nerves	C_5, C_6	Subscapularis and teres major muscles
Thoracodorsal nerve	C_6–C_8	Latissimus dorsi muscle
Axillary nerve	C_5, C_6	Deltoid and teres minor muscles; sensory from the skin of the shoulder
Medial antebrachial cutaneous nerve	C_8, T_1	Sensory from skin over anterior, medial surface of arm and forearm
Radial nerve	C_5–T_1	Many extensor muscles on the arm and forearm (triceps brachii, anconeus, extensor carpi radialis, extensor carpi ulnaris, and brachioradialis muscles); supinator muscle, digital extensor muscles, and abductor pollicis muscle by the deep branch; sensory from skin over the posterolateral surface of the limb through the posterior brachial cutaneous nerve (arm), posterior antebrachial cutaneous nerve (forearm), and the superficial branch (radial half of hand)
Musculocutaneous nerve	C_5–T_1	Flexor muscles on the arm (biceps brachii, brachialis, and coracobrachialis muscles); sensory from skin over lateral surface of the forearm through the lateral antebrachial cutaneous nerve
Median nerve	C_6–T_1	Flexor muscles on the forearm (flexor carpi radialis and palmaris longus muscles); pronator quadratus and pronator teres muscles; digital flexors (through the anterior interosseous nerve); sensory from skin over anterolateral surface of the hand
Ulnar nerve	C_8, T_1	Flexor carpi ulnaris muscle, flexor digitorum profundus muscle, adductor pollicis muscle, and small digital muscles by the deep branch; sensory from skin over medial surface of the hand through the superficial branch

Module 12.8 Review

a. Describe the brachial plexus.

b. Name the major nerves associated with the brachial plexus.

c. Define a nerve plexus trunk and cord.

LO 12.8 Relate the distribution pattern of the brachial plexus to its function.

The lumbar and sacral plexuses innervate the skin and skeletal muscles of the trunk and lower limbs

The **lumbar plexus** and the **sacral plexus** arise from the lumbar and sacral segments of the spinal cord, respectively. The nerves arising at these plexuses innervate the pelvic girdle and lower limbs.

1 These illustrations show the origins of the spinal nerves of the lumbar and sacral plexuses. Notice that the lumbosacral trunk provides some nerve input from the L_4 spinal nerve to the sacral plexus.

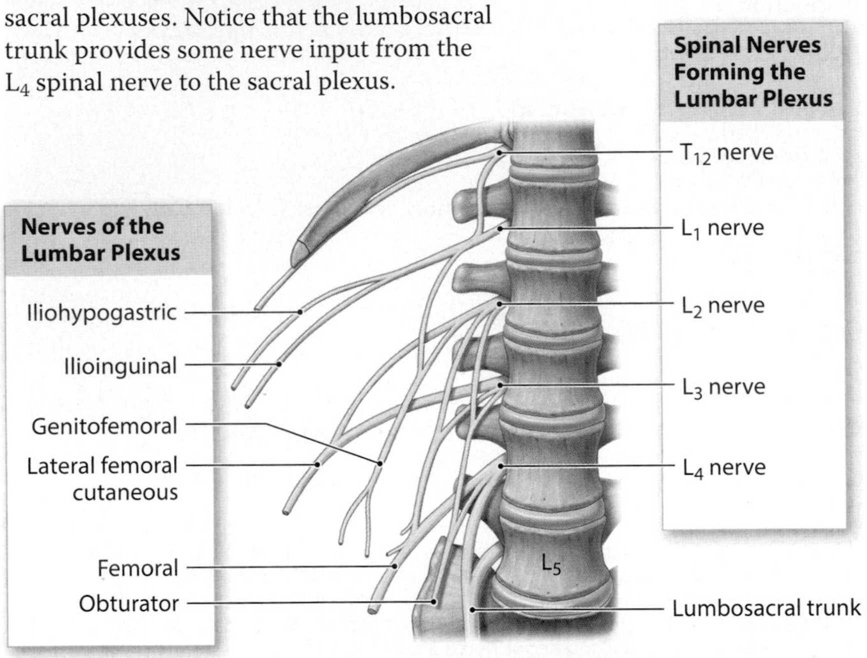
Spinal Nerves Forming the Lumbar Plexus
T_{12} nerve
L_1 nerve
L_2 nerve
L_3 nerve
L_4 nerve
L_5
Lumbosacral trunk
Nerves of the Lumbar Plexus
Iliohypogastric
Ilioinguinal
Genitofemoral
Lateral femoral cutaneous
Femoral
Obturator

Lumbar plexus, anterior view

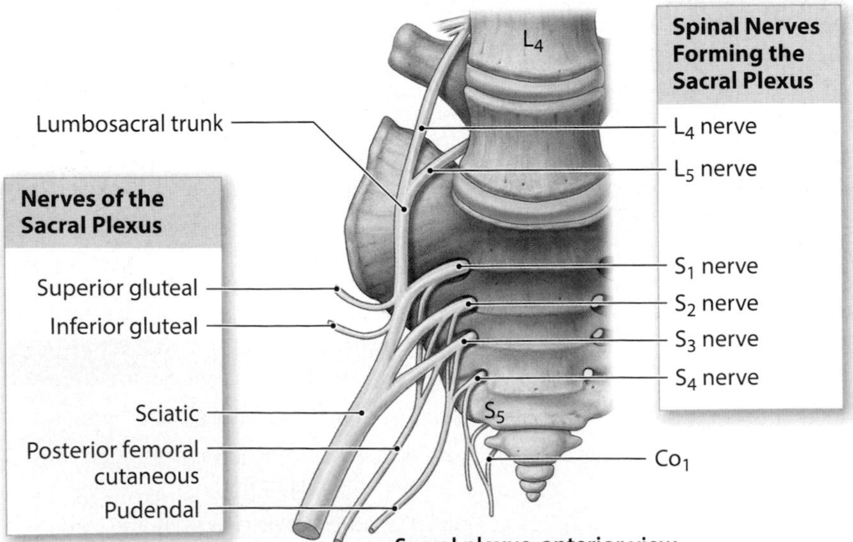
L_4
Spinal Nerves Forming the Sacral Plexus
L_4 nerve
L_5 nerve
S_1 nerve
S_2 nerve
S_3 nerve
S_4 nerve
S_5
Co_1
Lumbosacral trunk
Nerves of the Sacral Plexus
Superior gluteal
Inferior gluteal
Sciatic
Posterior femoral cutaneous
Pudendal

Sacral plexus, anterior view

2 This posterior view of the lower limb shows the distribution of the nerves of the sacral plexus. The **sciatic nerve** is the largest and longest nerve in the body.

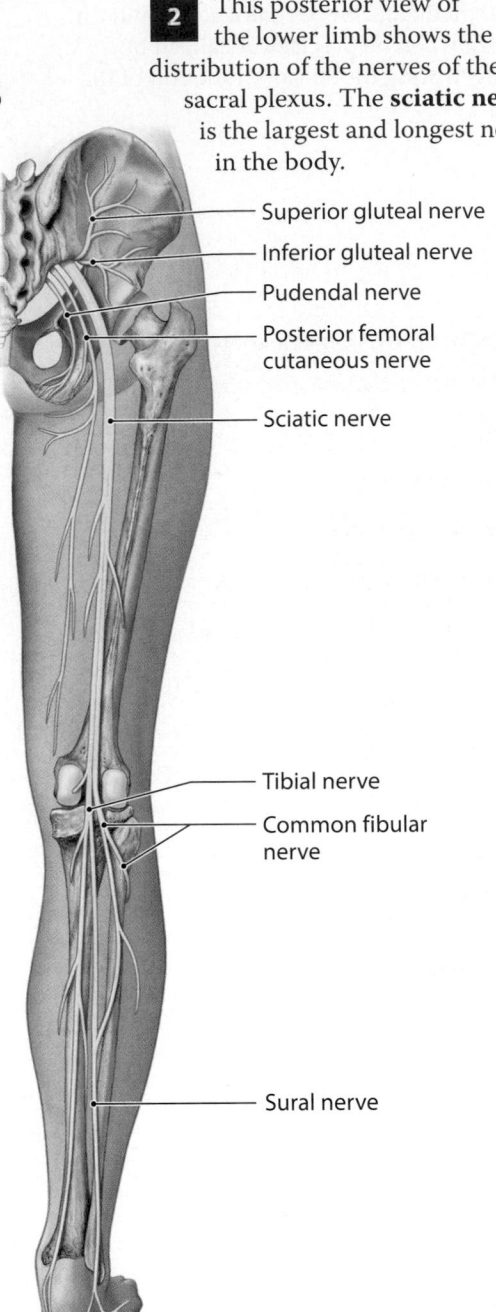

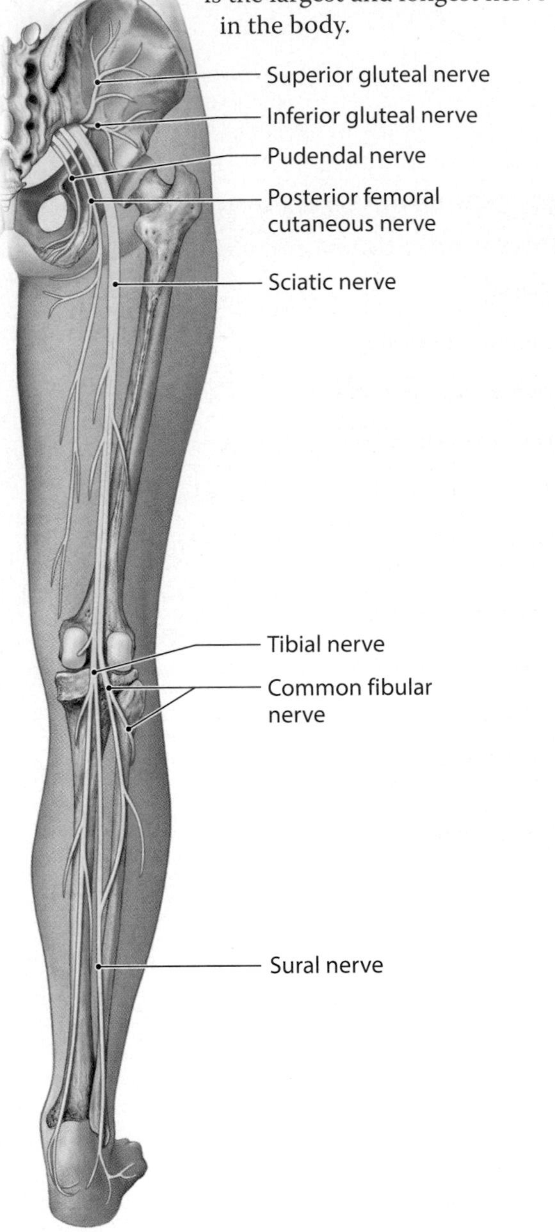
Superior gluteal nerve
Inferior gluteal nerve
Pudendal nerve
Posterior femoral cutaneous nerve
Sciatic nerve
Tibial nerve
Common fibular nerve
Sural nerve

3 These illustrations show the dermatomes of the sensory nerves innervating the ankle and foot.

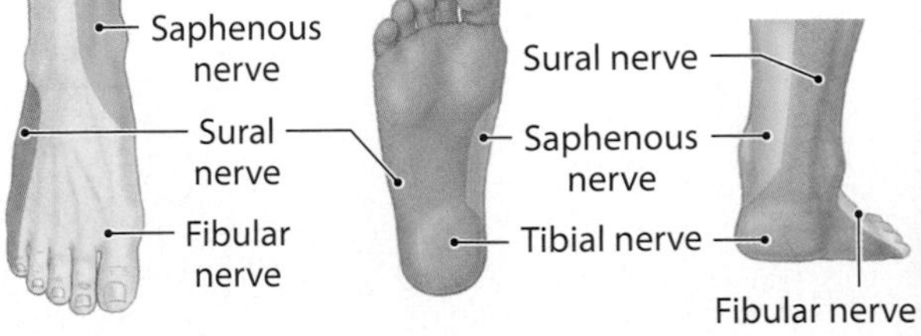
Saphenous nerve
Sural nerve
Fibular nerve
Sural nerve
Saphenous nerve
Tibial nerve
Fibular nerve

4 In this anterior view of the lower trunk and lower limb, you can see the distribution of the nerves of both the lumbar and sacral plexuses.

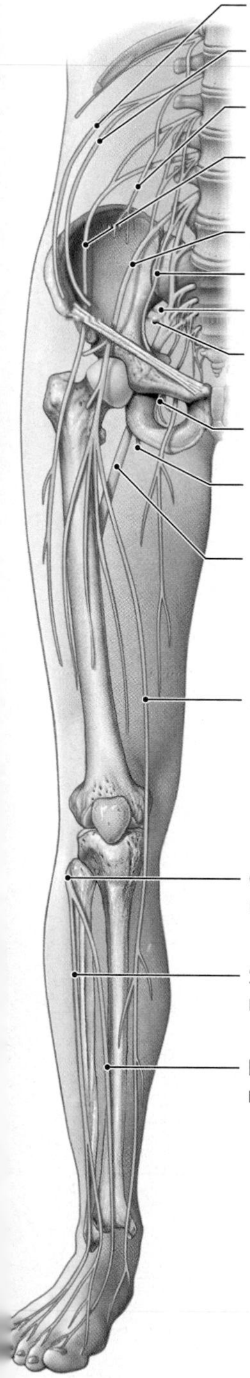

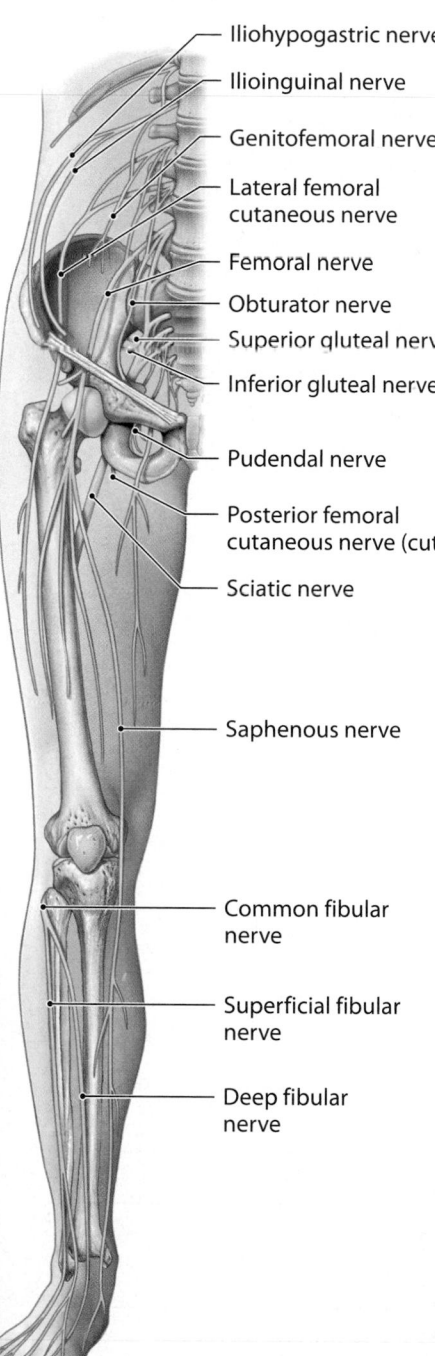

Iliohypogastric nerve
Ilioinguinal nerve
Genitofemoral nerve
Lateral femoral cutaneous nerve
Femoral nerve
Obturator nerve
Superior gluteal nerve
Inferior gluteal nerve
Pudendal nerve
Posterior femoral cutaneous nerve (cut)
Sciatic nerve
Saphenous nerve
Common fibular nerve
Superficial fibular nerve
Deep fibular nerve

The Lumbar and Sacral Plexuses

Nerve	Spinal Segment(s)	Distribution
Lumbar Plexus		
Iliohypogastric nerve	T_{12}, L_1	Abdominal muscles (external and internal oblique muscles, transversus abdominis muscle); skin over inferior abdomen and buttocks
Ilioinguinal nerve	L_1	Abdominal muscles (with iliohypogastric nerve); skin over superior, medial thigh and portions of external genitalia
Genitofemoral nerve	L_1, L_2	Skin over anteromedial surface of thigh and portions of external genitalia
Lateral femoral cutaneous nerve	L_2, L_3	Skin over anterior, lateral, and posterior surfaces of thigh
Femoral nerve	L_2–L_4	Anterior muscles of thigh (sartorius muscle and quadriceps group); flexors and adductors of hip (pectineus and iliopsoas muscles); skin over anteromedial surface of thigh, medial surface of leg and foot
Obturator nerve	L_2–L_4	Adductors of hip (adductors magnus, brevis, and longus muscles); gracilis muscle; skin over medial surface of thigh
Saphenous nerve	L_2–L_4	Skin over medial surface of leg
Sacral Plexus		
Gluteal nerves	L_4–S_2	
Superior		Abductors of hip (gluteus minimus, gluteus medius, and tensor fasciae latae muscles)
Inferior		Extensor of hip (gluteus maximus muscle)
Posterior femoral cutaneous nerve	S_1–S_3	Skin of perineum and posterior surfaces of thigh and leg
Sciatic nerve	L_4–S_3	Two of the hamstrings (semimembranosus and semitendinosus muscles); adductor magnus muscle (with obturator nerve)
Tibial nerve		Flexors of knee and extensors (plantar flexors) of ankle (popliteus, gastrocnemius, soleus, and tibialis posterior muscles, and the long head of the biceps femoris muscle); flexors of toes; skin over posterior surface of leg, plantar surface of foot
Common fibular nerve		Biceps femoris muscle (short head); fibularis muscles (brevis and longus) and tibialis anterior muscle; extensors of toes; skin over anterior surface of leg and dorsal surface of foot; skin over lateral portion of foot (through the sural nerve)
Pudendal nerve	S_2–S_4	Muscles of perineum, including urogenital diaphragm and external anal and urethral sphincter muscles; skin of external genitalia and related skeletal muscles (bulbospongiosus and ischiocavernosus muscles)

Module 12.9 Review

a. Describe the lumbar plexus and sacral plexus.

b. List the major nerves of the sacral plexus.

c. Which nerve divides into the tibial nerve and common fibular nerve?

LO 12.9 Relate the distribution patterns of the lumbar plexus and sacral plexus to their functions.

Labeling

Label each of the structures in the following cross-sectional diagram of the spinal cord.

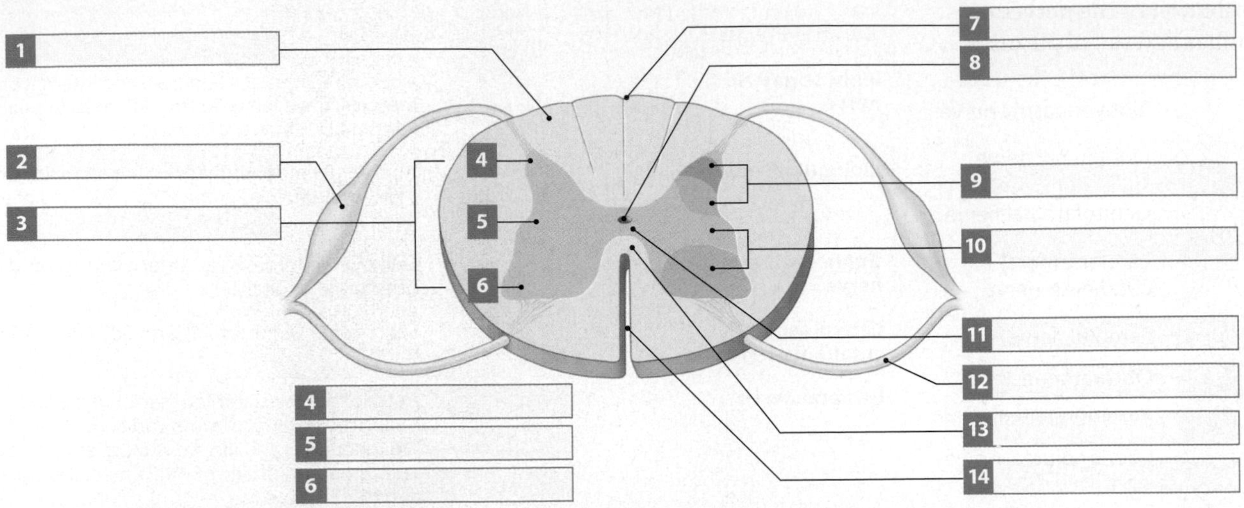
1
2
3
4
5
6
4
5
6
7
8
9
10
11
12
13
14

Label the views (15, 19) and the nerves that innervate the indicated regions of the hands (16–18).

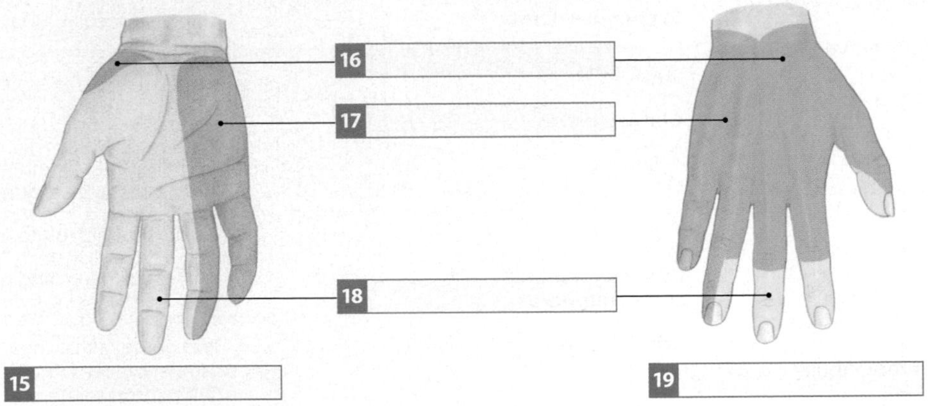
16
17
18
15
19

Vocabulary

Write the term for each of the following descriptions in the space provided.

20 The regions of white matter in the spinal cord

21 The tapered, conical portion of the spinal cord inferior to the lumbar enlargement

22 Bundles of axons in the PNS plus their associated blood vessels and connective tissues

23 Specialized membranes that provide stability and support for the spinal cord and brain

24 The complex made up of the filum terminale and the long dorsal and ventral spinal nerve roots inferior to the spinal cord

25 The plexus from which the radial, median, and ulnar nerves originate

26 The outermost covering of the spinal cord

27 The connective tissue partition that separates adjacent bundles of nerve fibers in a spinal nerve or a peripheral nerve

28 A bundle of unmyelinated, postganglionic fibers that innervates glands and smooth muscles in the body wall or limbs

20 ____________________

21 ____________________

22 ____________________

23 ____________________

24 ____________________

25 ____________________

26 ____________________

27 ____________________

28 ____________________

CNS neurons are grouped into neuronal pools, which form neural circuits

Your body has about 10 million sensory neurons, one-half million motor neurons, and 20 billion interneurons. The interneurons of the CNS are organized into a much smaller number of **neuronal pools**—functional groups of interconnected neurons. A neuronal pool may involve neurons in several regions of the brain, or the neurons may be in one specific location in the brain or spinal cord. Estimates of the number of neuronal pools range between a few hundred and a few thousand. The pattern of interaction among neurons provides clues to the function of a neuronal pool. We refer to the "wiring diagram" as a **neural circuit**, like electrical circuits in the wiring of a house. Here are a few common circuit patterns.

Divergence

Divergence is the spread of information from one neuron to several neurons, or from one pool to multiple pools. Divergence permits the broad distribution of a specific input. Considerable divergence occurs when sensory neurons bring information into the CNS: The information is distributed to neuronal pools throughout the spinal cord and brain.

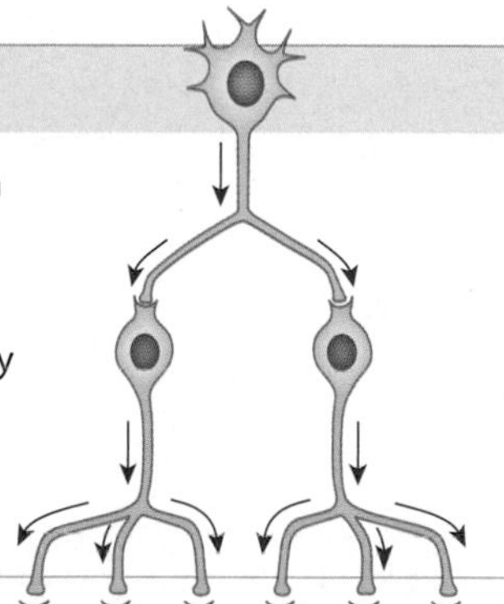

Parallel Processing

Parallel processing occurs when several neurons or neuronal pools process the same information simultaneously. Divergence must take place before parallel processing can occur. As a result, many responses can occur simultaneously.

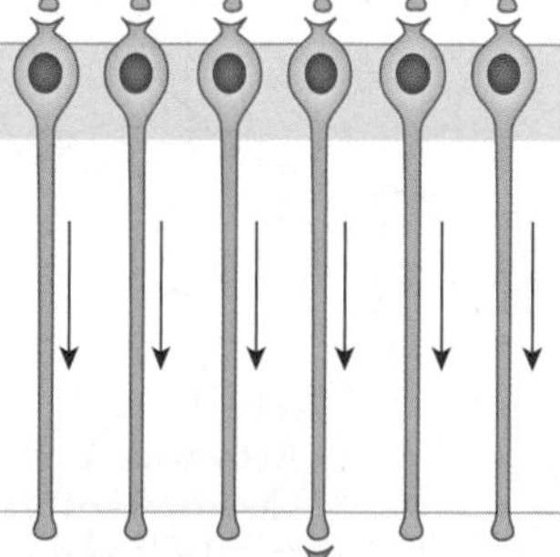

Serial Processing

In **serial processing**, information is relayed in a stepwise fashion, from one neuron to another or from one neuronal pool to the next. This pattern occurs as sensory information is relayed from one part of the brain to another.

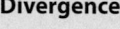

Convergence

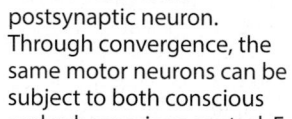

In **convergence**, several neurons synapse on a single postsynaptic neuron. Several patterns of activity in the presynaptic neurons can therefore have the same effect on the postsynaptic neuron. Through convergence, the same motor neurons can be subject to both conscious and subconscious control. For example, the movements of your diaphragm and ribs are now being controlled by your brain at the subconscious level. But you can also consciously control the same motor neurons, as when you take a deep breath and hold it.

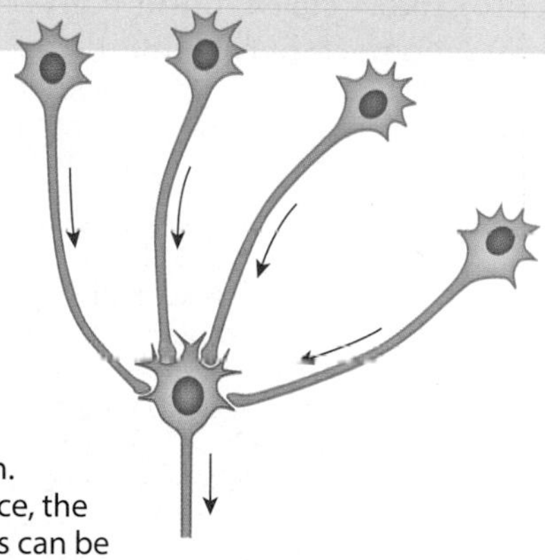

Reverberation

In **reverberation**, collateral branches of axons somewhere along the circuit extend back toward the source of an impulse and further stimulate the presynaptic neurons. Reverberation is like a positive feedback loop involving neurons: Once a reverberating circuit has been activated, it will continue to function until synaptic fatigue or inhibitory stimuli break the cycle.

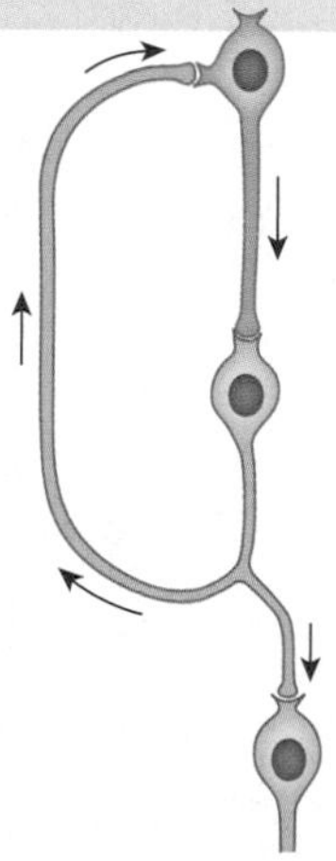

The most complex neural processing occurs in the brain. The simplest circuits, which occur within the PNS and the spinal cord, control the automatic responses of reflexes. Reflexes are a bit like LEGO® bricks: Individually, they are quite simple, but they can be combined in a great variety of ways to create very complex motor responses.

Module 12.10 Review

a. Differentiate between divergent and convergent neural circuits.

b. Which kind of neural circuit processes information in a stepwise fashion, one neuron to another?

c. Where does the most complex neural processing occur?

LO 12.10 Discuss the significance of neuronal pools, and describe their major patterns of neuron interaction.

Reflexes are vital to homeostasis

Reflexes are rapid, automatic responses to specific stimuli. Reflexes preserve homeostasis by making rapid adjustments in the function of organs or organ systems. The response shows little variability: Each time a particular reflex is activated, it usually produces the same motor response. In a reflex, sensory fibers deliver information from peripheral receptors to an integration center in the CNS, and motor fibers carry motor commands to peripheral effectors.

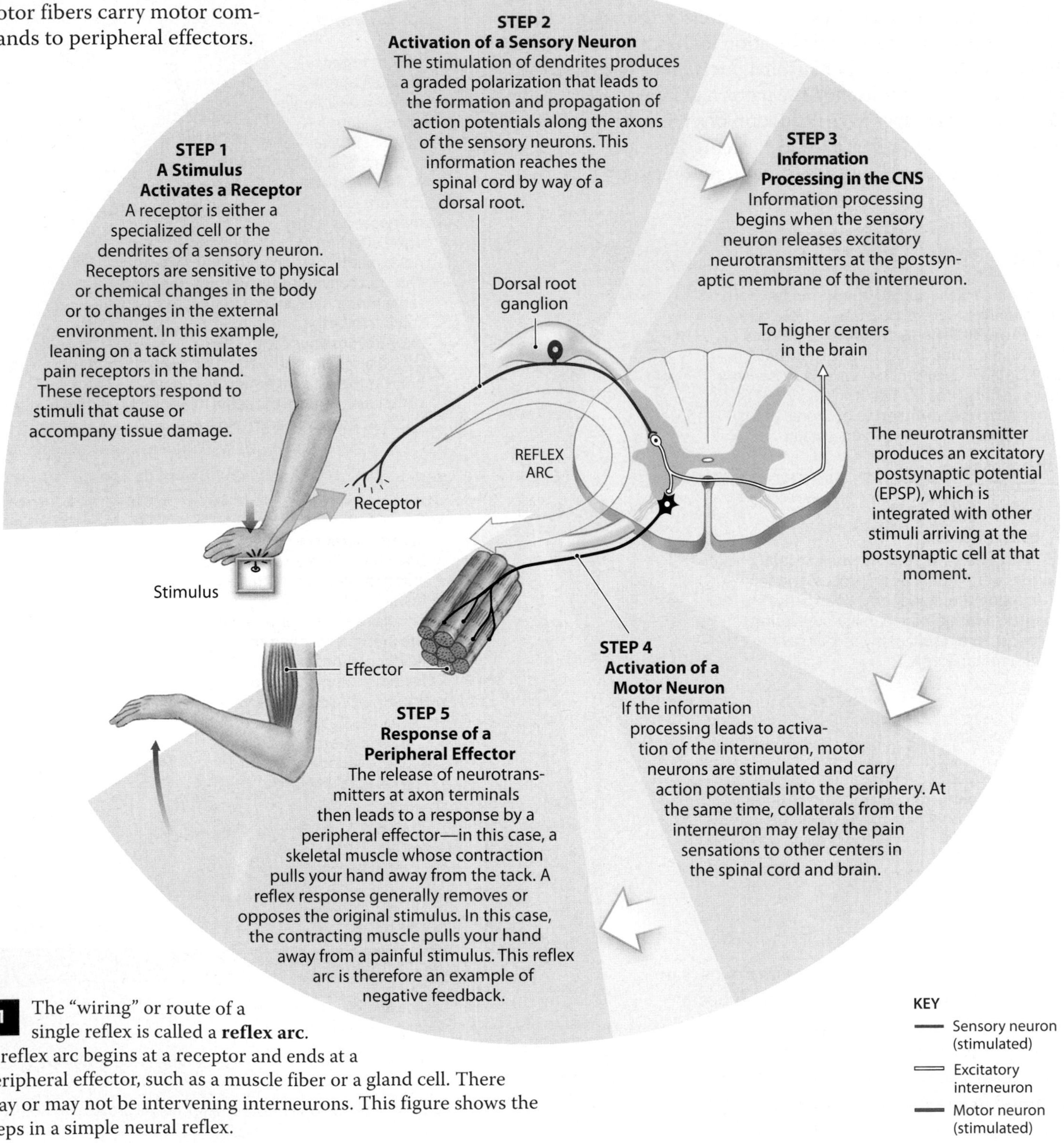
STEP 1
A Stimulus Activates a Receptor
A receptor is either a specialized cell or the dendrites of a sensory neuron. Receptors are sensitive to physical or chemical changes in the body or to changes in the external environment. In this example, leaning on a tack stimulates pain receptors in the hand. These receptors respond to stimuli that cause or accompany tissue damage.
STEP 2
Activation of a Sensory Neuron
The stimulation of dendrites produces a graded polarization that leads to the formation and propagation of action potentials along the axons of the sensory neurons. This information reaches the spinal cord by way of a dorsal root.
STEP 3
Information Processing in the CNS
Information processing begins when the sensory neuron releases excitatory neurotransmitters at the postsynaptic membrane of the interneuron.
Dorsal root ganglion
To higher centers in the brain
The neurotransmitter produces an excitatory postsynaptic potential (EPSP), which is integrated with other stimuli arriving at the postsynaptic cell at that moment.
REFLEX ARC
Receptor
Stimulus
Effector
STEP 4
Activation of a Motor Neuron
If the information processing leads to activation of the interneuron, motor neurons are stimulated and carry action potentials into the periphery. At the same time, collaterals from the interneuron may relay the pain sensations to other centers in the spinal cord and brain.
STEP 5
Response of a Peripheral Effector
The release of neurotransmitters at axon terminals then leads to a response by a peripheral effector—in this case, a skeletal muscle whose contraction pulls your hand away from the tack. A reflex response generally removes or opposes the original stimulus. In this case, the contracting muscle pulls your hand away from a painful stimulus. This reflex arc is therefore an example of negative feedback.
KEY
Sensory neuron (stimulated)
Excitatory interneuron
Motor neuron (stimulated)

1 The "wiring" or route of a single reflex is called a **reflex arc**. A reflex arc begins at a receptor and ends at a peripheral effector, such as a muscle fiber or a gland cell. There may or may not be intervening interneurons. This figure shows the steps in a simple neural reflex.

Reflex Classifications

Development

Innate reflexes result from the connections that form between neurons during development and are genetically or developmentally programmed. Such reflexes generally appear in a predictable sequence, from the simplest reflex responses (withdrawal from pain) to more complex motor patterns (chewing, suckling, or tracking objects with the eyes).

Acquired reflexes, or conditioned reflexes, are rapid and automatic, but they were learned rather than preestablished. Such reflexes are enhanced by repetition.

The Nature of the Response

Somatic reflexes provide for the involuntary control of skeletal muscles. (The withdrawal reflex at left is one example.) Somatic reflexes are essential because they are immediate. Somatic reflexes provide a rapid response that can later be supplemented by voluntary motor commands.

Visceral reflexes, or autonomic reflexes, control or adjust the activities of smooth muscle, cardiac muscle, glands, and adipose tissues.

The Complexity of the Circuit

Polysynaptic reflexes involve at least one interneuron in addition to a sensory neuron and a motor neuron. These reflexes have a longer delay between stimulus and response, with the length of the delay proportional to the number of synapses. Polysynaptic reflexes can produce far more complicated responses than monosynaptic reflexes, because the interneurons can control motor neurons that activate several muscle groups simultaneously.

A **monosynaptic reflex** is the simplest reflex arc, involving only one synapse in the CNS. The sensory neuron synapses directly on a motor neuron, which serves as the processing center. Transmission across a chemical synapse always involves a synaptic delay, but with only one synapse, the delay between the stimulus and the response is minimized.

The Processing Site

In **spinal reflexes**, the important interconnections and processing events occur in nuclei of the spinal cord. In a simple segmental reflex, all of the processing occurs in a single spinal segment. **Intersegmental reflexes** involve multiple segments of the spinal cord, and they can produce very complex and coordinated responses.

In **cranial reflexes**, the important interconnections and processing events occur in nuclei of the brain.

2 Neural reflexes can be classified according to (1) their development, (2) the nature of the resulting motor response, (3) the complexity of the neural circuit involved, or (4) the site of information processing. These categories are not mutually exclusive. They represent different ways of describing a single reflex.

Module 12.11 Review

a. List the components of a reflex arc.

b. What are common characteristics of reflexes?

c. Describe the various classifications of neural reflexes.

LO 12.11 Describe the steps in a reflex.

The stretch reflex is a monosynaptic reflex involving muscle spindles

Watch
MasteringA&P®
A&PFlix™
The Stretch Reflex

1 The best-known monosynaptic reflex is the **stretch reflex**, or myotatic reflex. It automatically regulates skeletal muscle length. The patellar reflex shown here is a well-known example.

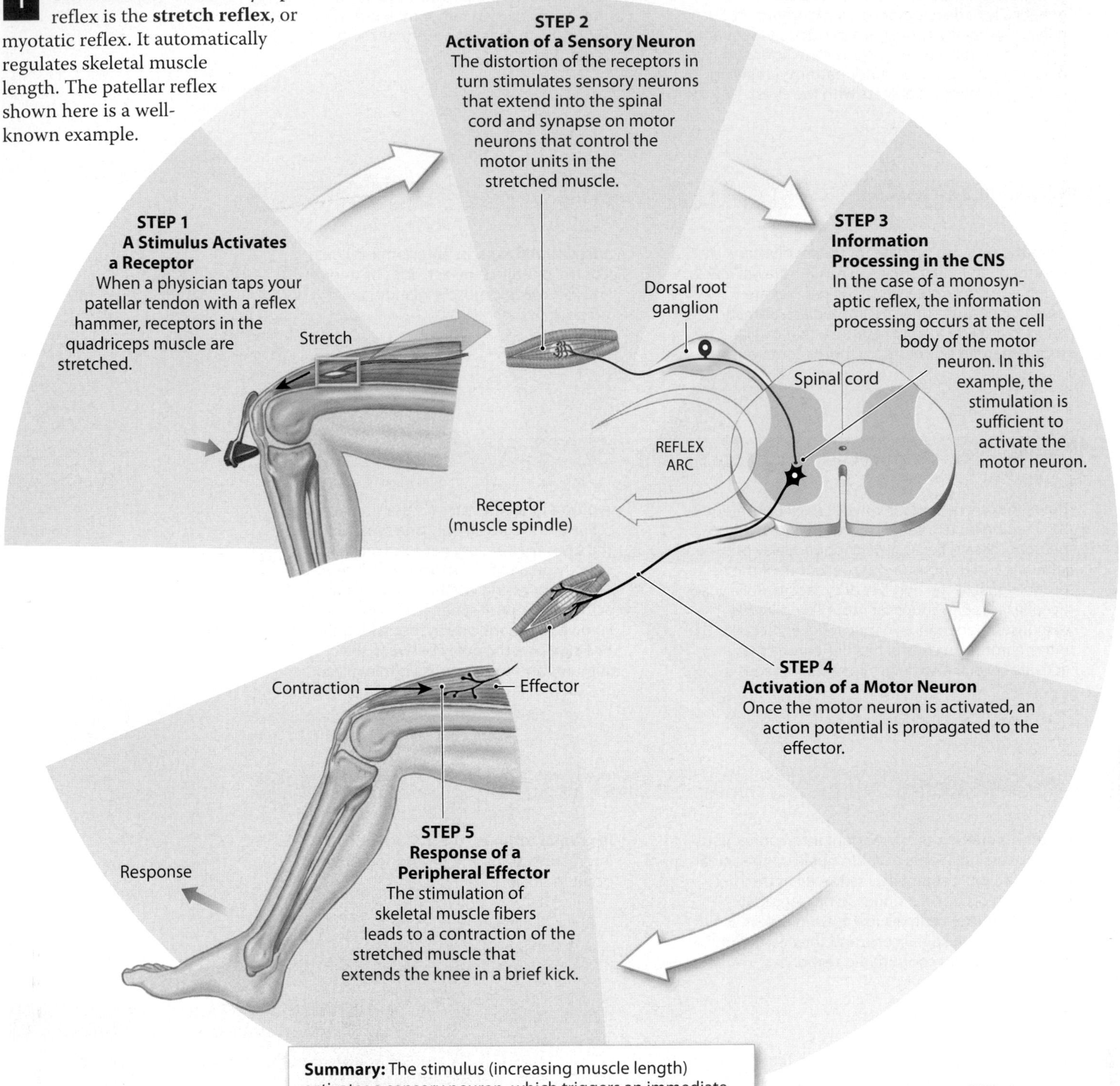
STEP 1
A Stimulus Activates a Receptor
When a physician taps your patellar tendon with a reflex hammer, receptors in the quadriceps muscle are stretched.
Stretch
STEP 2
Activation of a Sensory Neuron
The distortion of the receptors in turn stimulates sensory neurons that extend into the spinal cord and synapse on motor neurons that control the motor units in the stretched muscle.
Dorsal root ganglion
Spinal cord
REFLEX ARC
Receptor (muscle spindle)
STEP 3
Information Processing in the CNS
In the case of a monosynaptic reflex, the information processing occurs at the cell body of the motor neuron. In this example, the stimulation is sufficient to activate the motor neuron.
STEP 4
Activation of a Motor Neuron
Once the motor neuron is activated, an action potential is propagated to the effector.
Contraction
Effector
STEP 5
Response of a Peripheral Effector
The stimulation of skeletal muscle fibers leads to a contraction of the stretched muscle that extends the knee in a brief kick.
Response

Summary: The stimulus (increasing muscle length) activates a sensory neuron, which triggers an immediate motor response (contraction of the stretched muscle) that counteracts the stimulus. The entire reflex is completed within 20–40 msec because the action potentials traveling toward and away from the spinal cord are conducted along large-diameter myelinated fibers.

KEY
- Sensory neuron (stimulated)
- Motor neuron (stimulated)

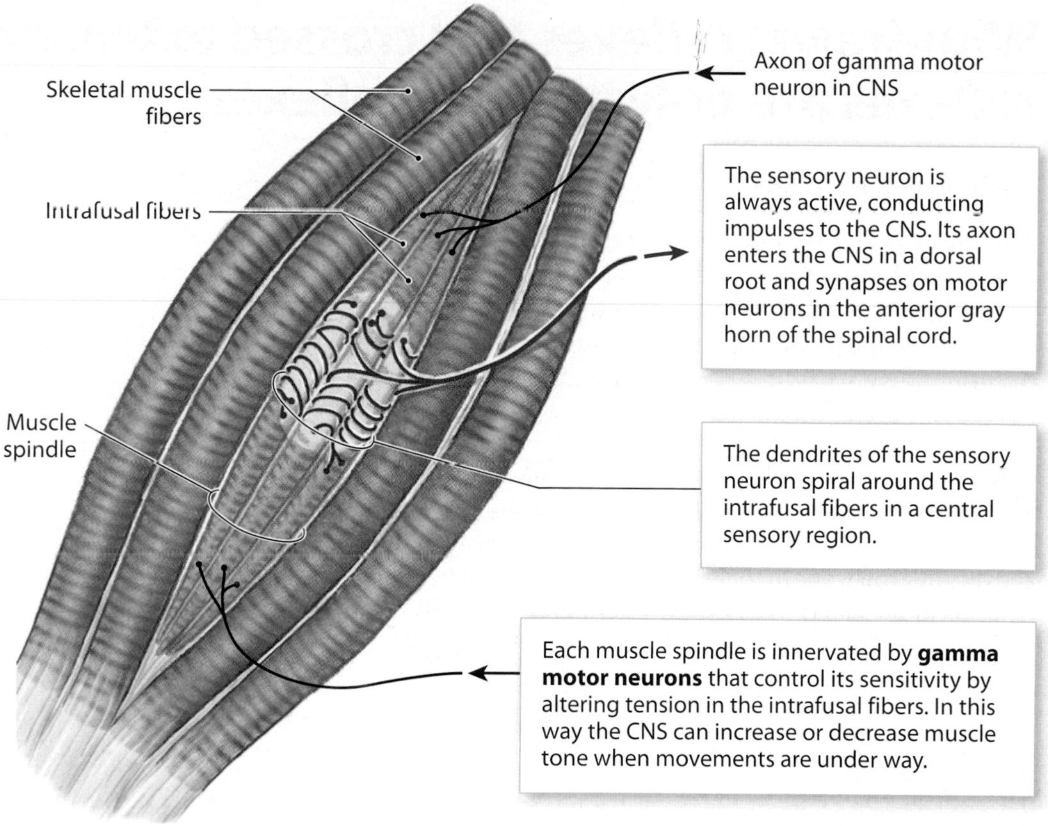

Skeletal muscle fibers
Intrafusal fibers
Muscle spindle
Axon of gamma motor neuron in CNS
The sensory neuron is always active, conducting impulses to the CNS. Its axon enters the CNS in a dorsal root and synapses on motor neurons in the anterior gray horn of the spinal cord.
The dendrites of the sensory neuron spiral around the intrafusal fibers in a central sensory region.
Each muscle spindle is innervated by **gamma motor neurons** that control its sensitivity by altering tension in the intrafusal fibers. In this way the CNS can increase or decrease muscle tone when movements are under way.

2 The sensory receptors involved in the stretch reflex are **muscle spindles**. Each muscle spindle consists of a bundle of small, specialized skeletal muscle fibers called **intrafusal muscle fibers**. The muscle spindle is surrounded by larger skeletal muscle fibers responsible for resting muscle tone and, at greater levels of stimulation, for the contraction of the entire muscle. Both sensory and motor neurons innervate each intrafusal fiber.

3 Stretching the central portion of the intrafusal fiber distorts the dendrites and stimulates the sensory neuron. This in turn stimulates the motor neurons, increasing muscle tone. Compressing the central portion inhibits the sensory neuron, which reduces stimulation of the motor neurons, and muscle tone decreases.

Sensory Region	Action Potential Frequency in Sensory Neuron	Effect on Skeletal Muscle
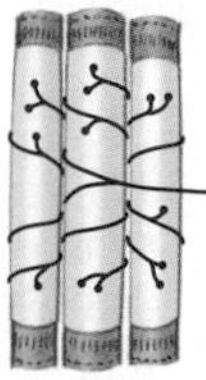 Resting length	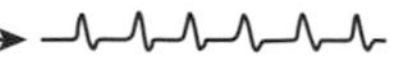	Normal muscle tone persists
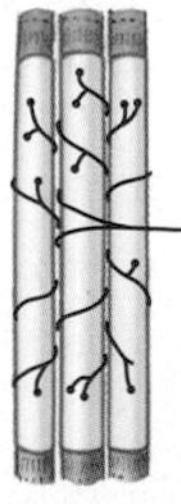 Stretched	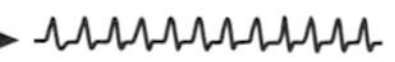	Muscle tone increases
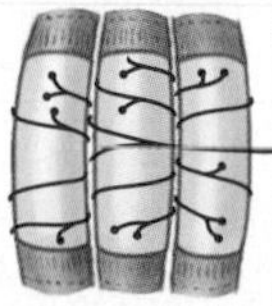 Compressed	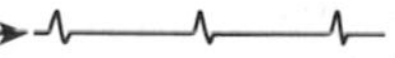	Muscle tone decreases

Many stretch reflexes are **postural reflexes**—reflexes that help us maintain a normal upright posture. Standing, for example, requires cooperation among many muscle groups. Some of these muscles work in opposition to one another, exerting forces that keep the body's weight balanced over the feet. If your body leans forward, stretch receptors in your calf muscles are stimulated. Those muscles then respond by increasing muscle tone, returning your body to an upright position. Postural muscles generally have a firm muscle tone and extremely sensitive stretch receptors. As a result, very fine adjustments are continually being made, and you are not aware of the cycles of contraction and relaxation that occur.

Module 12.12 Review

a. Define stretch reflex.

b. In the patellar reflex, identify the response observed and the effectors involved.

c. In the patellar reflex, how does stimulation of the muscle spindle by gamma motor neurons affect sensitivity and reaction time?

LO 12.12 Describe the steps in the stretch reflex.

Module 12.13

Withdrawal reflexes and crossed extensor reflexes are polysynaptic reflexes

Withdrawal Reflexes

Withdrawal reflexes move affected parts of the body away from a stimulus. The strongest withdrawal reflexes are triggered by painful stimuli, but these reflexes are sometimes initiated by the stimulation of touch receptors or pressure receptors. The **flexor reflex**, a representative withdrawal reflex, affects the muscles of a limb.

1 If you accidentally grab a hot pan on the stove, a dramatic flexor reflex will occur. When the pain receptors in your hand are stimulated, the sensory neurons activate interneurons in the spinal cord that stimulate motor neurons in the anterior gray horns. The result is a contraction of flexor muscles that yanks your hand away from the stove.

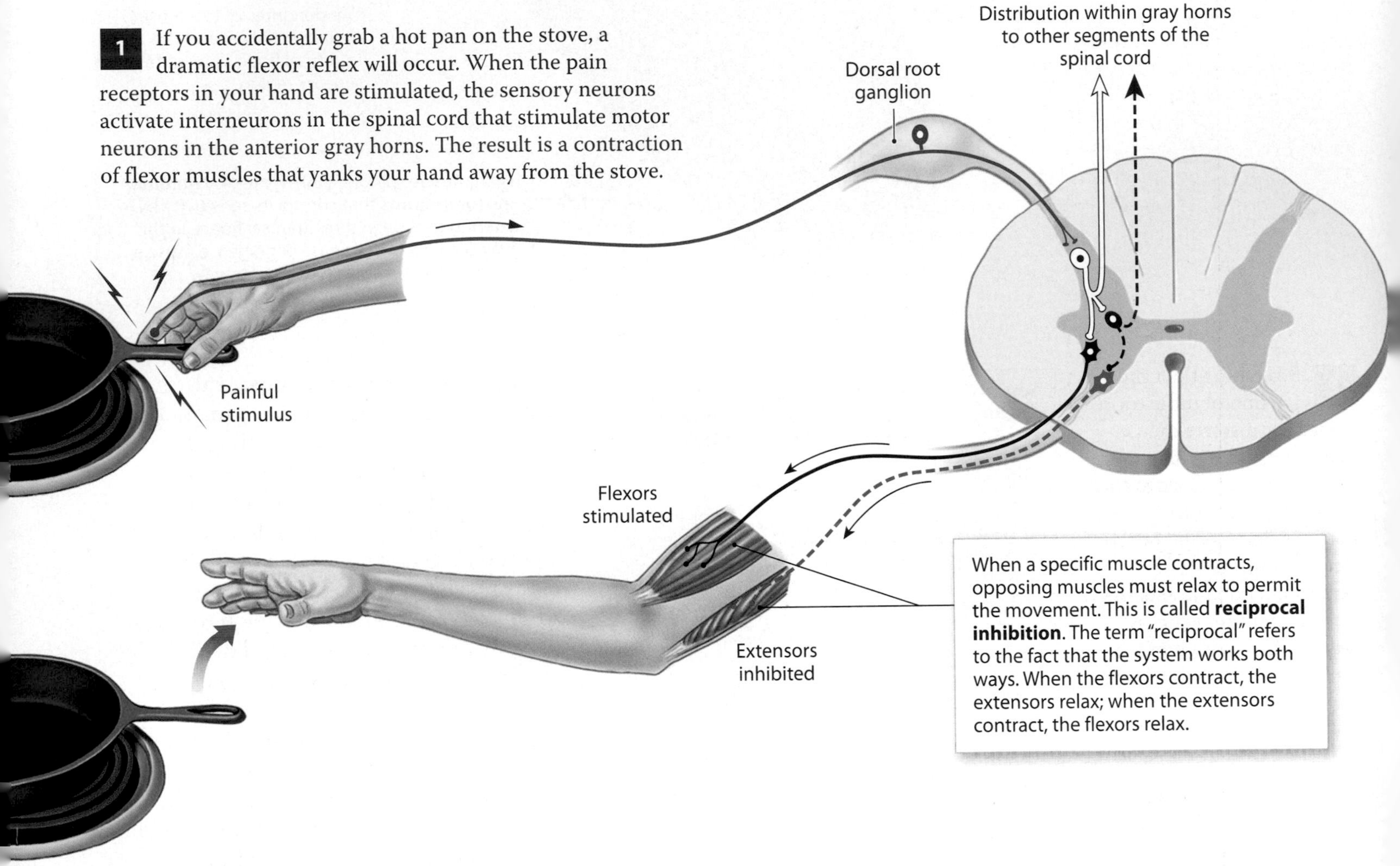

Distribution within gray horns to other segments of the spinal cord
Dorsal root ganglion
Painful stimulus
Flexors stimulated
Extensors inhibited
When a specific muscle contracts, opposing muscles must relax to permit the movement. This is called **reciprocal inhibition**. The term "reciprocal" refers to the fact that the system works both ways. When the flexors contract, the extensors relax; when the extensors contract, the flexors relax.

Withdrawal reflexes show tremendous versatility, because the sensory neurons activate many pools of interneurons. The distribution of the effects and the strength and character of the motor responses depend on the intensity and location of the stimulus. Mild discomfort might provoke a brief contraction in muscles of your hand and wrist. More powerful stimuli would produce coordinated muscular contractions affecting the positions of your hand, wrist, forearm, and arm. Severe pain would also stimulate contractions of your shoulder, trunk, and arm muscles. These contractions could persist for several seconds, due to the activation of reverberating circuits.

KEY

- Sensory neuron (stimulated)
- Excitatory interneuron
- Motor neuron (stimulated)
- Motor neuron (inhibited)
- Inhibitory interneuron

Crossed Extensor Reflexes

The stretch reflexes and withdrawal reflexes involve **ipsilateral reflex arcs** (*ipsi*, same + *lateral*, side): The sensory stimulus and the motor response occur on the same side of the body. The **crossed extensor reflex** involves a **contralateral reflex arc** (*contra*, opposite), because an additional motor response occurs on the side opposite the stimulus.

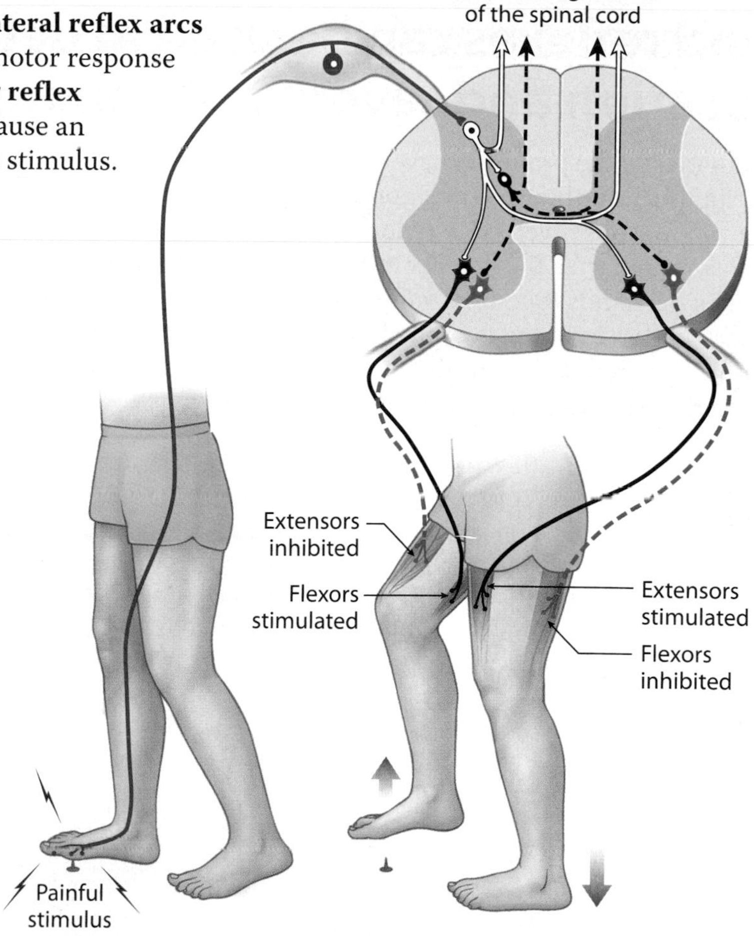

To motor neurons in other segments of the spinal cord
Extensors inhibited
Flexors stimulated
Extensors stimulated
Flexors inhibited
Painful stimulus

2 When you step on a tack, the flexor reflex pulls the affected foot away from the ground as excitatory interneurons stimulate the flexor muscles of that limb and inhibitory interneurons relax the extensor muscles that would otherwise oppose that movement. The crossed extensor reflex occurs simultaneously because collaterals of the excitatory and inhibitory interneurons cross to the other side of the spinal cord to motor neurons controlling muscles in the uninjured leg. The excitatory interneurons stimulate motor neurons controlling extensor muscles while the inhibitory interneurons relax the flexor muscles. As a result, your opposite leg straightens to support the shifting weight.

3 **Polysynaptic reflexes** are responsible for the automatic actions involved in complex movements such as walking and running. All polysynaptic reflexes share the basic characteristics summarized in this table.

Properties of Polysynaptic Reflexes

- **They Involve Pools of Interneurons.** Processing takes place in pools of interneurons before motor neurons are activated. The result may be excitation or inhibition.
- **They Are Intersegmental in Distribution.** The interneuron pools extend across spinal segments and may activate muscle groups in many parts of the body.
- **They Involve Reciprocal Inhibition.** Reciprocal inhibition coordinates muscular contractions and reduces resistance to movement.
- **They Have Reverberating Circuits, Which Prolong the Reflexive Motor Response.** Positive feedback between interneurons that innervate motor neurons and the processing pool maintains the stimulation even after the initial stimulus has faded.
- **Several Reflexes May Cooperate to Produce a Coordinated, Controlled Response.** As a reflex movement gets under way, antagonistic reflexes are inhibited. In complex polysynaptic reflexes, commands may be distributed along the length of the spinal cord, producing a well-coordinated response.

Module 12.13 Review

a. Identify the basic characteristics of polysynaptic reflexes.

b. Describe the flexor reflex.

c. During a withdrawal reflex of the foot, what happens to the limb on the side opposite the stimulus? What is this response called?

LO 12.13 Explain withdrawal reflexes and crossed extensor reflexes and the responses produced by each.

The brain can inhibit or facilitate spinal reflexes, and reflexes can be used to determine the location and severity of damage to the CNS

Activities in the brain can have a profound effect on the performance of a reflex by facilitating or inhibiting the motor neurons or interneurons involved. The facilitation of motor neurons involved in reflexes is called **reinforcement**. For example, a method used to overemphasize the patellar reflex is the Jendrassik maneuver. To do this, the person hooks the hands together by interlocking the fingers and then tries to pull the hands apart while a light tap is applied to the patella. This reinforcement produces a big kick rather than a twitch. This distractive technique still produces a larger reflex response even if the person realizes it is just a distraction.

1 The **biceps reflex**, **triceps reflex**, and **ankle-jerk reflex** are stretch reflexes often tested during a physical exam. Each reflex is controlled by specific segments of the spinal cord. As a result, testing these reflexes provides information about the status of the corresponding spinal segments. The table on the right presents information on several reflexes commonly used in physical exams.

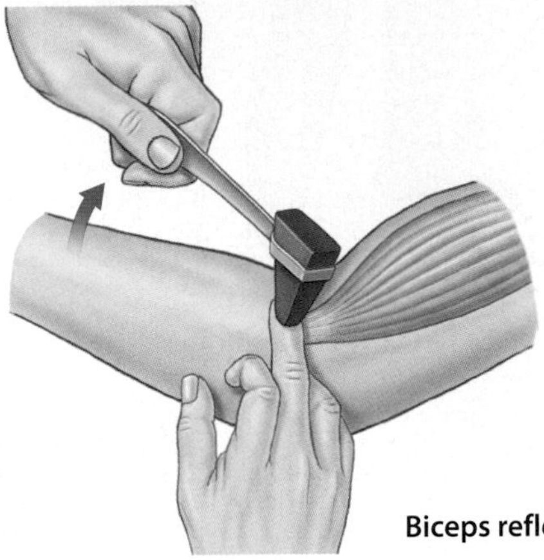

Biceps reflex

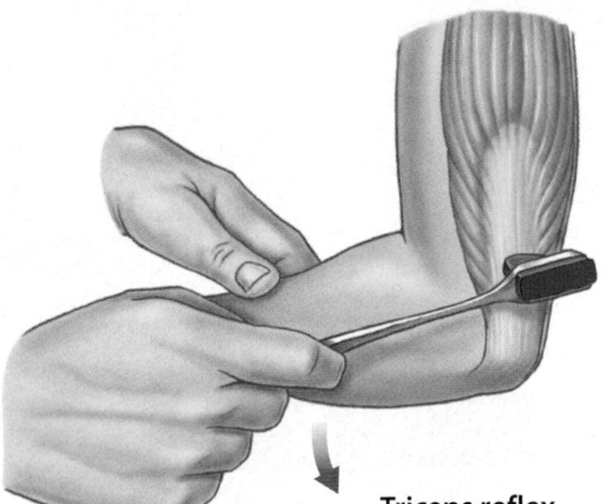

Triceps reflex

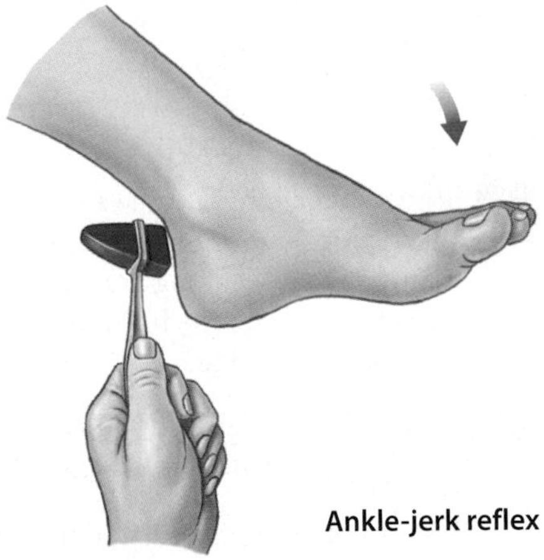

Ankle-jerk reflex

2 Descending fibers may also have an inhibitory effect on spinal reflexes. Stroking an infant's foot on the lateral side of the sole produces an extension of the hallux and a fanning out of the other toes known as the **Babinski sign**, or positive Babinski reflex. This response disappears as descending motor pathways develop, because those pathways inhibit this reflex response.

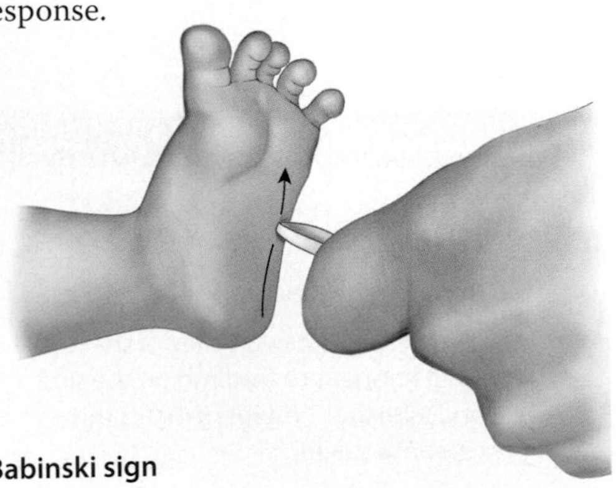

Babinski sign

3 In normal adults, stroking the lateral side of the sole produces a curling of the toes (plantar flexion), called a **plantar reflex** (or negative Babinski reflex). If either the higher centers or the descending tracts are damaged, the Babinski sign will reappear in an adult. As a result, this reflex is often tested if CNS injury is suspected.

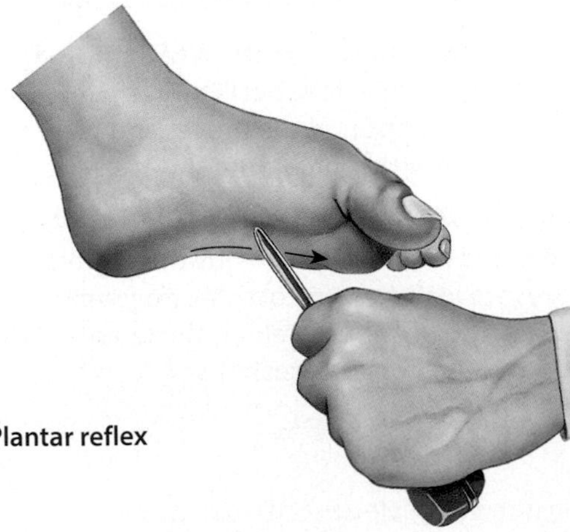

Plantar reflex

Reflexes Used in Diagnostic Testing

Reflex	Stimulus	Afferent Nerve(s)	Spinal Segment	Efferent Nerve(s)	Normal Response
Superficial Reflexes					
Abdominal reflex	Light stroking of skin of abdomen	T_7–T_{12} depending on region stroked	T_7–T_{12} at level of arrival	Same as afferent	Contractions of abdominal muscles that pull navel toward the stimulus
Cremasteric reflex	Stroking of skin of upper thigh	Femoral nerve	L_1	Genitofemoral nerve	Contraction of cremaster, elevation of scrotum
Plantar reflex	Longitudinal stroking of lateral side of sole of foot	Tibial nerve	S_1, S_2	Tibial nerve	Flexion at toe joints
Anal reflex	Stroking of region around the anus	Pudendal nerve	S_4, S_5	Pudendal nerve	Constriction of external anal sphincter
Stretch Reflexes					
Biceps reflex	Tap to tendon of biceps brachii muscle near its insertion	Musculocutaneous nerve	C_5, C_6	Musculocutaneous nerve	Flexion at elbow
Triceps reflex	Tap to tendon of triceps brachii muscle near its insertion	Radial nerve	C_6, C_7	Radial nerve	Extension at elbow
Brachioradialis reflex	Tap to forearm near styloid process of the radius	Radial nerve	C_5, C_6	Radial nerve	Flexion at elbow, supination, and flexion at finger joints
Patellar reflex	Tap to patellar tendon	Femoral nerve	L_2–L_4	Femoral nerve	Extension at knee
Ankle-jerk reflex	Tap to calcaneal tendon	Tibial nerve	S_1, S_2	Tibial nerve	Extension (plantar flexion) at ankle

4 The **abdominal reflex** depends on facilitation, rather than inhibition, of the descending tracts. In this reflex, seen in normal adults, a light stroking of the skin produces a reflexive twitch in the abdominal muscles that moves the navel toward the stimulus. The absence of an abdominal reflex may indicate damage to the descending tracts.

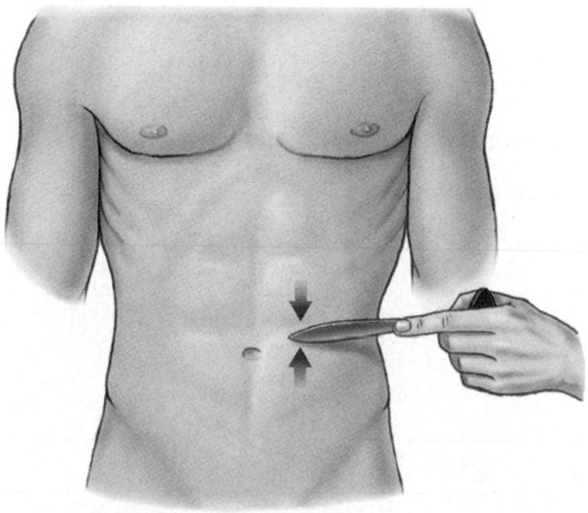

Abdominal reflex

Module 12.14 Review

a. Define reinforcement as it pertains to spinal reflexes.

b. What purpose does reflex testing serve?

c. After injuring her back, 22-year-old Tina exhibits a positive Babinski reflex. What does this imply about her injury?

LO **12.14** Explain the value of reflex testing and how the brain may control and modify reflex responses.

Labeling

Label the neural circuit patterns in the following diagrams.

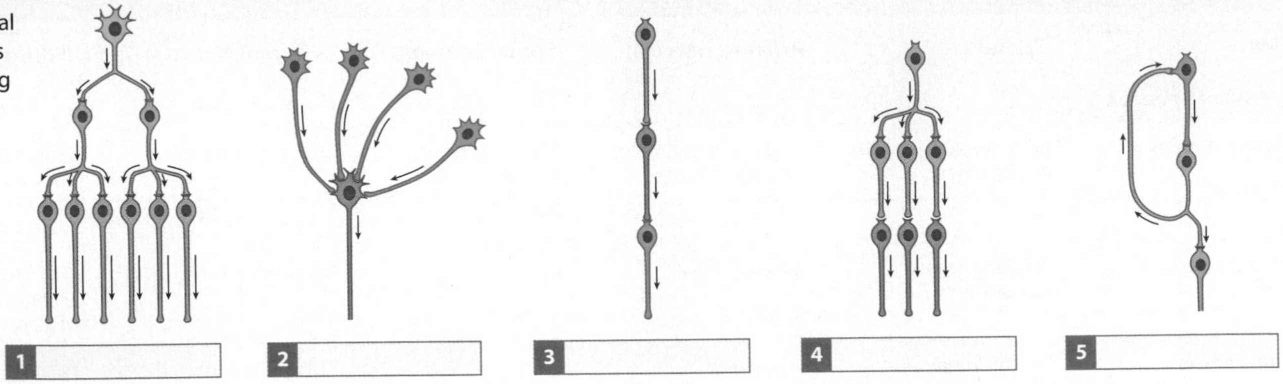
1
2
3
4
5

Label the indicated structures in the accompanying diagram of a reflex arc.

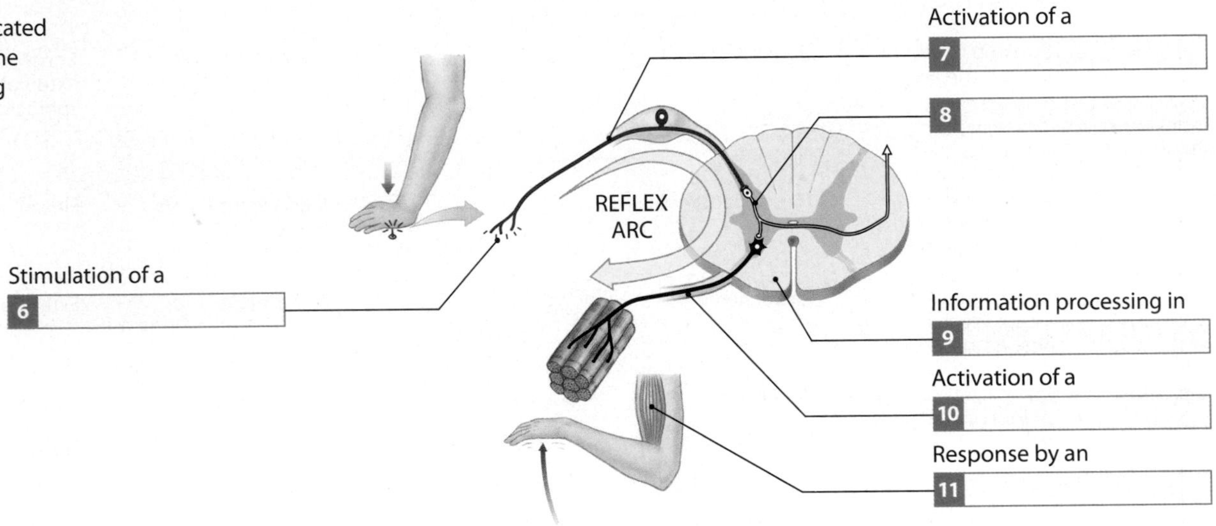
Activation of a
7
8
REFLEX ARC
Stimulation of a
6
Information processing in
9
Activation of a
10
Response by an
11

Vocabulary

Write the term for each of the following descriptions in the space provided.

12 A reflex in which the sensory stimulus and motor response occur on the same side of the body

13 Reflexes that move affected parts of the body away from a stimulus

14 Controls the sensitivity of a muscle spindle

15 Withdrawal reflex that affects the muscles of a limb

16 Autonomic reflexes that adjust nonskeletal muscles, glands, and adipose tissue

17 Type of reflexes enhanced by repetition

18 A reflex in which the motor response occurs on the side opposite to the stimulus

19 Process involved in preventing opposing muscles from contracting during a reflex

20 The enhancement of reflexes due to facilitation of motor neurons

12 ______________________

13 ______________________

14 ______________________

15 ______________________

16 ______________________

17 ______________________

18 ______________________

19 ______________________

20 ______________________

Short answer

Classify the withdrawal reflex illustrated in the reflex arc labeling exercise above according to development, response, complexity of circuit, and processing site.

21 ____

____

____

Study Outline

SECTION 1 • Functional Organization of the Spinal Cord

12.1 The spinal cord can function independently from the brain p. 429

1. The **spinal cord** receives sensory input and sends motor commands on spinal nerves.
2. A **reflex** is a rapid, automatic response triggered by specific stimuli.
3. **Spinal reflexes** are controlled in the spinal cord.

12.2 The spinal cord contains gray matter and white matter p. 430

4. There are 31 spinal segments, each giving rise to a pair of spinal nerves. There are 8 cervical, 12 thoracic, 5 lumbar, 5 sacral, and 1 coccygeal spinal segments.
5. The **cervical enlargement** supplies nerves to the shoulder and upper limb. The **lumbar enlargement** innervates the structures of the pelvis and lower limb.
6. The **conus medullaris** is the inferior tip of the spinal cord. Spinal segments L_2–S_5 extend as the **cauda equina**. The **filum terminale** extends from the conus medullaris to vertebra S_2.
7. Each spinal nerve connects to the spinal cord by a **dorsal root** and a **ventral root**.
8. The dorsal root contains axons of the **dorsal root ganglion**. The dorsal root ganglion contains the bodies of sensory neurons whose axons bring information to the spinal cord.
9. The ventral root contains somatic and visceral motor neuron axons.
10. The **white matter** of the spinal cord contains myelinated and unmyelinated axons. The **gray matter** contains cell bodies, neuroglia, and unmyelinated axons and surrounds a central canal.

12.3 The spinal meninges, consisting of the dura mater, arachnoid, and pia mater, surround the spinal cord p. 432

11. The **spinal meninges** are a series of specialized membranes that surround the spinal cord, providing stability and shock absorption.
12. From superficial to deep, the meninges are: (1) **dura mater**, (2) **arachnoid**, and (3) **pia mater**.
13. Between the dura mater and the walls of the vertebral canal is the **epidural space** containing protective adipose tissue padding. The **subarachnoid space** is filled with **cerebrospinal fluid (CSF)** that acts as a shock absorber and a medium for the diffusion of gases, nutrients, and wastes.

The dissected spinal cord

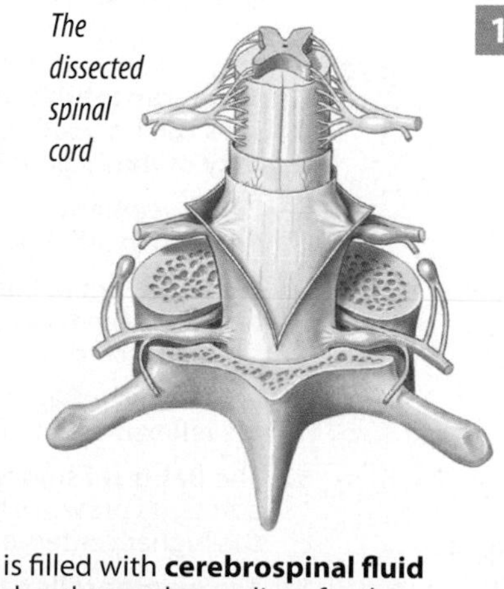

14. CSF is withdrawn in a procedure known as a **lumbar puncture**, or spinal tap.

12.4 Gray matter integrates sensory and motor functions, and white matter carries information p. 434

15. Projections of gray matter are called **horns.** The **posterior gray horn** contains somatic and visceral sensory nuclei. The **lateral gray horn** (only in thoracic and lumbar segments) contains visceral motor nuclei. The **anterior gray horn** contains somatic motor nuclei.
16. The **gray commissures** surround the central canal and contain axons that cross the spinal cord.
17. White matter is divided into **columns**: the **posterior white column**, the **lateral white column**, and the **anterior white column**.
18. Columns contain **tracts**, which are bundles of axons uniform in structure and function.
19. **Ascending tracts** carry sensory information to the brain. **Descending tracts** convey motor commands to the spinal cord.

12.5 Spinal nerves have a similar anatomical structure and distribution pattern p. 436

20. **Epineurium**, a dense network of collagen fibers, surrounds spinal nerves. **Perineurium** divides the axons into bundles called **fascicles**. **Endoneurium** surrounds individual axons.
21. Each spinal nerve divides to form **rami** that carry visceral and motor fibers of the autonomic nervous system (ANS). Thoracic and lumbar spinal segments carry motor output for the **sympathetic division** of the ANS.

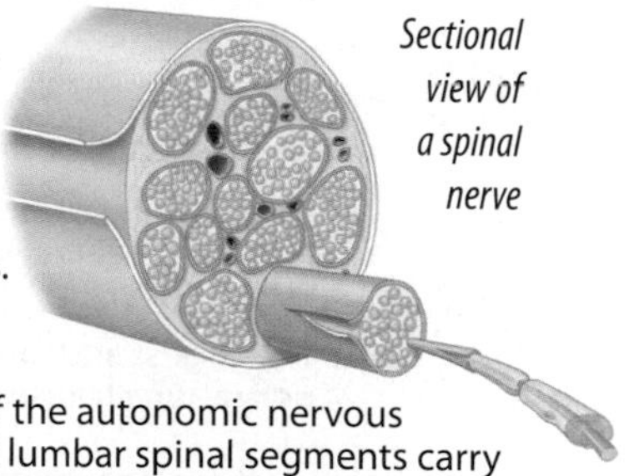

Sectional view of a spinal nerve

22. The specific region of the skin surface monitored by a single pair of spinal nerves is a **dermatome**. The varicella-zoster virus, the virus that causes chickenpox, can cause skin eruptions called **shingles** along a dermatome.

12.6 Each ramus of a spinal nerve provides motor and sensory innervation to a specific region p. 438

23. The ventral root of each spinal nerve contains axons of somatic motor and visceral motor neurons. The dorsal root of each spinal nerve carries sensory information to the spinal cord.
24. The **dorsal ramus** contains somatic motor and visceral motor fibers that innervate the skin and skeletal muscles of the back. The dorsal ramus also carries sensory information from the skin and skeletal muscles of the back
25. The **ventral ramus** axons supply the ventrolateral body surface, structures in the body wall, and the limbs. The ventral ramus also carries sensory information from the ventrolateral body surface, structures in the body wall, and the limbs.
26. The white and gray rami are together known as the **rami communicantes**, and they contain fibers of the ANS.

12.7 Spinal nerves form nerve plexuses that innervate the skin and skeletal muscles; the cervical plexus is the smallest of these nerve plexuses p. 440

27. The ventral rami form four major plexuses: (1) the **cervical plexus**, (2) the **brachial plexus**, (3) the **lumbar plexus**, and (4) the **sacral plexus**.
28. The cervical plexus consists of the ventral rami of spinal nerves C_1–C_5.
29. The cervical plexus innervates the muscles of the neck, the diaphragm, and the skin of the neck and superior chest.
30. The **phrenic nerve**, a major nerve of the cervical plexus, supplies the diaphragm.

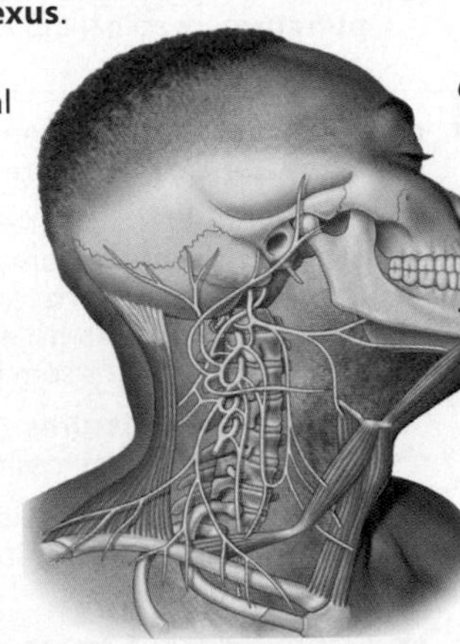

The cervical plexus

12.8 The brachial plexus innervates the pectoral girdle and upper limbs p. 442

31. The ventral rami of spinal nerves C_4–T_1 form the brachial plexus.
32. **Trunks** (large bundles of axons from several spinal nerves) and **cords** (smaller branches that originate at trunks) form the brachial plexus.

12.9 The lumbar and sacral plexuses innervate the skin and skeletal muscles of the trunk and lower limbs p. 444

33. The **lumbar plexus** arises from spinal cord segments T_{12}–L_4. The **sacral plexus** is formed from spinal cord segments L_4–S_4.
34. The largest and longest nerve in the body is the **sciatic nerve**, which arises from the sacral plexus.

SECTION 2 • Introduction to Reflexes

12.10 CNS neurons are grouped into neuronal pools, which form neural circuits p. 447

35. The interneurons of the CNS are organized into functional groups of interconnected neurons called **neuronal pools**.
36. The neuronal circuits have varying patterns, termed **divergence**, **parallel processing**, **serial processing**, **convergence**, and **reverberation**.

12.11 Reflexes are vital to homeostasis p. 448

37. **Reflexes** are rapid, automatic responses to specific stimuli. The "wiring" of a single reflex is called a **reflex arc**.
38. The five steps in a simple neural reflex are: (1) **a stimulus activates a receptor**, (2) **activation of a sensory neuron**, (3) **information processing**, (4) **activation of a motor neuron**, and (5) **response of a peripheral effector**.
39. Neural reflexes are classified according to their development, the nature of the resulting motor response, the complexity of the neural circuit, and the information processing site.

12.12 The stretch reflex is a monosynaptic reflex involving muscle spindles p. 450

40. The **stretch reflex** is a monosynaptic reflex regulating skeletal muscle length. The patellar reflex is an example.
41. **Muscle spindles** are the sensory receptors involved in a stretch reflex. Each muscle spindle consists of specialized skeletal muscle fibers called **intrafusal muscle fibers**.
42. Stretching the intrafusal fiber stimulates a sensory neuron, which in turn stimulates the motor neurons.
43. Many stretch reflexes are **postural reflexes** that help us to maintain normal upright posture.

12.13 Withdrawal reflexes and crossed extensor reflexes are polysynaptic reflexes p. 452

44. **Withdrawal reflexes** move affected parts of the body away from a stimulus. Painful stimuli trigger the strongest withdrawal reflexes.
45. The **flexor reflex** affects the muscles of the limb. When a muscle contracts in a flexor reflex, opposing muscles relax to permit the movement. This is called **reciprocal inhibition**.

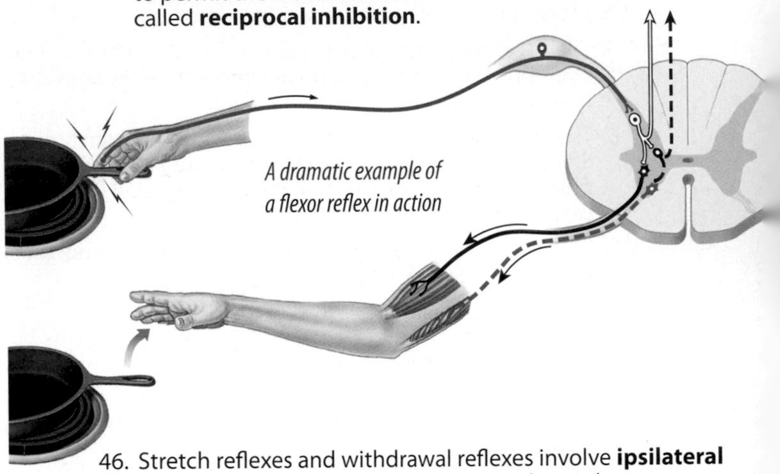

A dramatic example of a flexor reflex in action

46. Stretch reflexes and withdrawal reflexes involve **ipsilateral reflex arcs**, meaning the sensory stimulus and motor response occur on the same side of the body.
47. The **crossed extensor reflex** involves a **contralateral reflex arc** because an additional motor response occurs on the side opposite the stimulus.
48. **Polysynaptic reflexes** are responsible for the automatic actions involved in complex movements such as walking and running.

12.14 The brain can inhibit or facilitate spinal reflexes, and reflexes can be used to determine the location and severity of damage to the CNS p. 454

49. **Reinforcement** is the facilitation of motor neurons involved in reflexes.
50. Physical examinations test the status of spinal segments because each reflex is controlled by specific segments of the spinal cord.
51. The **biceps reflex**, **triceps reflex**, and **ankle-jerk reflex** are reflexes commonly tested in physical exams.
52. The **Babinski sign** is present in infants but disappears in adults. Its presence in adults is a sign of damage to either the higher centers of the brain or descending tracts.
53. The **plantar reflex** and the **abdominal reflex** are normally present in adults.

Chapter Review Questions

Labeling

Label the structures in the following diagram.

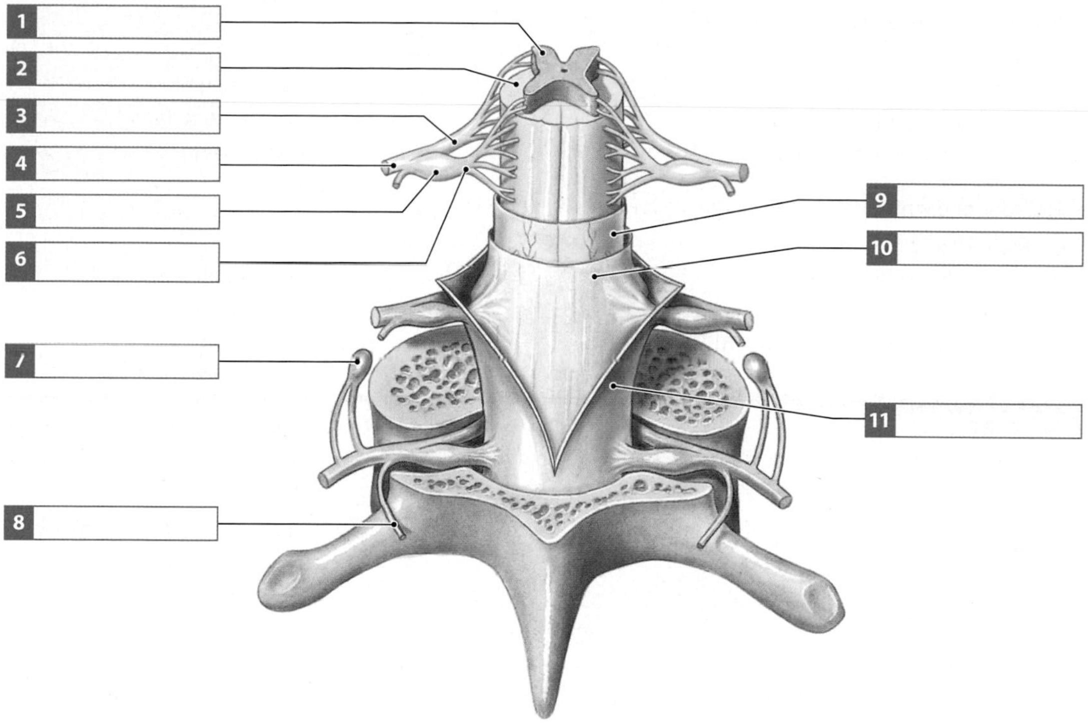
1
2
3
4
5
6
7
8
9
10
11

True/False

Indicate whether each statement is true or false.

12 Sensory input travels toward the spinal cord and motor commands travel away from the spinal cord.

13 Endoneurium is the outermost covering of a spinal nerve.

14 The dorsal root of each spinal nerve carries motor information away from the spinal cord.

15 There are eight cervical spinal nerves.

16 Motor nuclei are located in the posterior gray horn of the spinal cord.

12 ____________________

13 ____________________

14 ____________________

15 ____________________

16 ____________________

Multiple choice

Select the correct answer from the list provided.

17 The following steps are involved in a simple neural reflex: (1) activation of a sensory neuron; (2) activation of a motor neuron; (3) response of a peripheral effector; (4) a stimulus activates a receptor; (5) information processing. The proper sequence of these steps is

- ☐ a) 1, 3, 4, 5, 2.
- ☐ b) 4, 5, 3, 1, 2.
- ☐ c) 4, 1, 5, 2, 3.
- ☐ d) 4, 3, 1, 5, 2.

18 The layers of the spinal meninges from superficial to deep are

- ☐ a) pia mater, arachnoid, dura mater.
- ☐ b) arachnoid, pia mater, dura mater.
- ☐ c) dura mater, arachnoid, pia mater.
- ☐ d) dura mater, pia mater, arachnoid.

19 Fascicles containing bundles of axons are surrounded by

- [ ] a) epineurium.
- [ ] b) perineurium.
- [ ] c) endoneurium.
- [ ] d) dermatome.

20 The adult spinal cord extends only to the

- [ ] a) coccyx.
- [ ] b) sacrum.
- [ ] c) third or fourth lumbar vertebra.
- [ ] d) first or second lumbar vertebra.

21 The ventral root of each spinal nerve contains axons of

- [ ] a) visceral sensory neurons.
- [ ] b) somatic motor neurons only.
- [ ] c) visceral motor neurons only.
- [ ] d) both somatic motor and visceral motor neurons.

22 A neuronal circuit that spreads information from one neuron to several neurons is

- [ ] a) divergence.
- [ ] b) convergence.
- [ ] c) serial processing.
- [ ] d) parallel processing.

23 Reflex arcs in which the sensory stimulus and the motor response occur on the same side of the body are

- [ ] a) contralateral.
- [ ] b) ipsilateral.
- [ ] c) monosynaptic.
- [ ] d) crossed extensor.

24 The tapered conical portion of the spinal cord is the

- [ ] a) cauda equina.
- [ ] b) conus medullaris.
- [ ] c) filum terminale.
- [ ] d) coccygeal nerve.

25 The median nerve is a component of the

- [ ] a) cervical plexus.
- [ ] b) brachial plexus.
- [ ] c) lumbar plexus.
- [ ] d) sacral plexus.

26 Stroking an infant's foot on the lateral side of the sole produces extension of the hallux and a fanning of the toes known as the

- [ ] a) Babinski reflex.
- [ ] b) plantar reflex.
- [ ] c) cremasteric reflex.
- [ ] d) ankle jerk reflex.

Short answer

27 Describe the cause and symptoms of shingles.

28 A student forcefully bumps the posterior medial aspect of her elbow on her desktop and feels pain in her hand. Which nerve is involved, and where in her hand is she experiencing pain?

29 Why do cervical nerves outnumber cervical vertebrae?

30 Where is cerebrospinal fluid located, and what are its functions?

Chapter Integration • Applying what you have learned

A helmet-to-helmet collision causes a "stinger"

Dominic is a defenseman on his high school's junior varsity lacrosse team. He plays an aggressive style of defense and initiates extensive physical contact. As the final seconds of a recent game counted down, an opposing player was about to take a shot on goal that would have tied the game. Dominic ran full speed into the shooter. Their helmets collided dead-on, and Dominic's head snapped hard to the left. As his opponent fell backward onto the ground, Dominic felt an intense pain from the right side of his neck that radiated down his entire right upper limb to the tips of his fingers. The referee blew the whistle, indicating the end of the game. As the winning players and coaches congregated, the team's athletic trainer asked Dominic how he was feeling after delivering such a significant hit. Dominic replied that upon impact he felt a severe pain on the right side of his neck and entire right upper limb that was both numbing and burning, but that it lasted only a few moments. The trainer informed him that he likely experienced a *stinger* and that although the pain was quite intense, he would not likely have a lasting injury.

From what you have just learned about spinal nerves, answer the following questions.

1 **Why did Dominic's pain radiate down only the right upper limb and not both upper limbs? Why was the pain also in his neck?**

2 **Which spinal nerves would you suspect to be involved?**

3 **Why do you think he experienced the numbing, burning pain that he did, and why do you think the pain extended all the way to his fingertips?**

A really bad day

Karen had just finished her A&P midterm exam and was hurrying out of the classroom to get home and finish packing for her first-ever spring break trip to Cancun, Mexico. She was excited about her upcoming trip, and equally excited about how well she was doing in her first year of nursing school. She thought back to how she had struggled with every high school science class she'd had, but her experience in her A&P class was completely different. Her professor was wonderful at presenting the information in an understandable manner and in guiding students to numerous ancillary tools to help them comprehend a seemingly overwhelming subject.

Seeing the crowd of students waiting for the elevator, Karen chose to take the stairs. In her haste to get home, she slipped on the edge of a step and fell down a flight of stairs. During the fall she suffered lumbar and sacral spinal cord damage due to a hyperextension of her back. The injury resulted in edema around the central canal that compressed the anterior horn of the lumbar region.

An additional consequence of her injury was the loss of the ability to control her bowels and urinary bladder, because control of these functions involves spinal reflex arcs located in the sacral region of the spinal cord. In both instances, two sphincter muscles—an inner sphincter of smooth muscle and an outer sphincter of skeletal muscle—control the passage of wastes (whether feces or urine) out of the body.

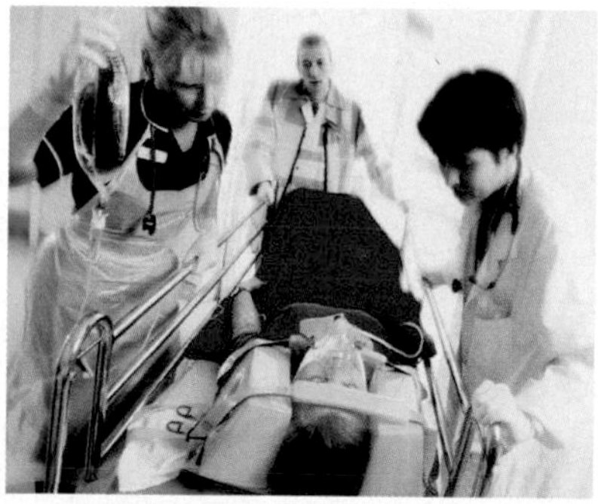

4 **What signs and symptoms would you expect Karen to exhibit during her neurological evaluation?**

5 **How would a transection of the spinal cord at the L_1 level affect a person's bowel and urinary bladder control?**

6 **Compare the effects of a transection of the spinal cord at L_1 in an adult with a newborn's inability to control urination.**

13 The Brain, Cranial Nerves, and Sensory and Motor Pathways

LEARNING OUTCOMES

These Learning Outcomes correspond by number to this chapter's modules and indicate what you should be able to do after completing the chapter.

SECTION 1 • Functional Anatomy of the Brain and Cranial Nerves

13.1 Describe the origins of the different regions of the brain from the embryonic neural tube.

13.2 Name the six major regions of the brain, and describe their functions.

13.3 Explain how the brain is protected and supported, and how cerebrospinal fluid forms and circulates.

13.4 List the main components of the medulla oblongata and pons, and specify the functions of each.

13.5 List the main components of the cerebellum, and specify the functions of each.

13.6 List the main components of the midbrain, and specify the functions of each.

13.7 List the main components of the diencephalon, and specify the functions of each.

13.8 Identify the main components of the limbic system, and specify the locations and functions of each.

13.9 Describe the structure and function of the basal nuclei of the cerebrum.

13.10 Identify the major superficial landmarks of the cerebrum, and cite the location of each.

13.11 Identify the locations of the motor, sensory, and association areas of the cerebral cortex, and discuss the functions of each.

13.12 Discuss the significance of the white matter of the cerebral hemispheres.

13.13 **CLINICAL MODULE** Discuss the origin and significance of the major categories of brain waves seen in an electroencephalogram (EEG).

13.14 Identify the cranial nerves by name and number, and cite the functions of each.

SECTION 2 • Sensory and Motor Pathways

13.15 Describe the basic events that occur along a sensory pathway, and explain the difference between a sensation and a perception.

13.16 Explain the ways in which receptors can be classified.

13.17 List the types of tactile receptors, and specify the functions of each.

13.18 Identify and describe the major sensory pathways.

13.19 Describe the components, processes, and functions of the somatic motor pathways.

13.20 Describe the levels of information processing involved in motor control.

13.21 **CLINICAL MODULE** Describe the roles of the nervous system in referred pain, Parkinson's disease, rabies, cerebral palsy, amyotrophic lateral sclerosis, Alzheimer's disease, and multiple sclerosis.

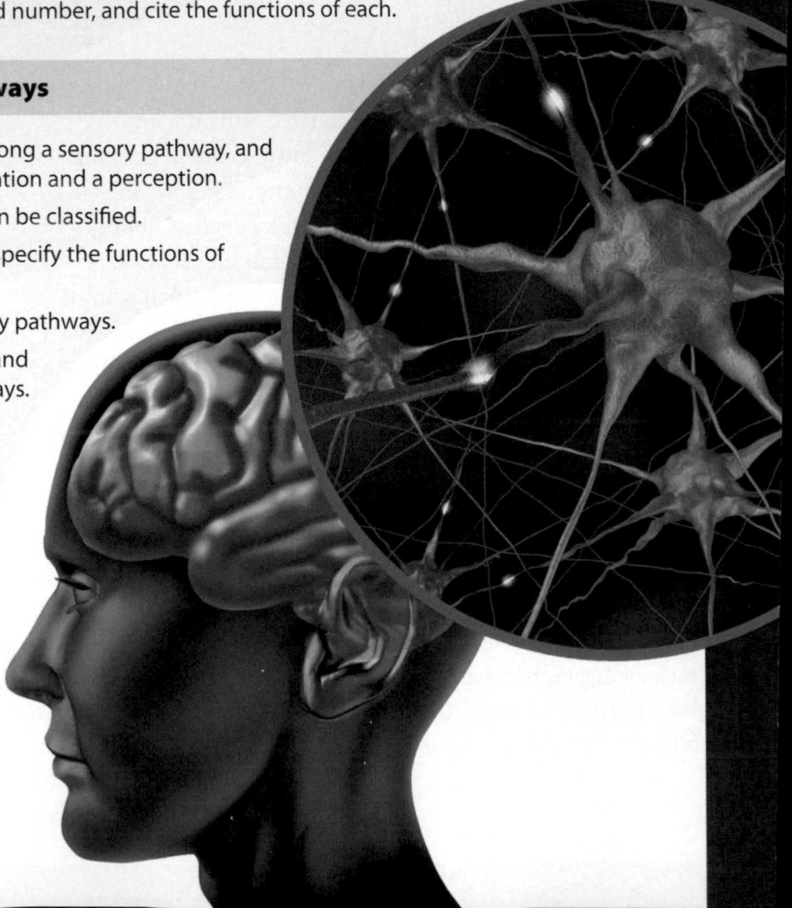

LO Learning Outcomes are repeated at the bottom of each module.

The brain develops from a hollow neural tube

The first section of this chapter introduces the functional anatomy of the brain and cranial nerves. In adult humans, the brain contains almost 97 percent of the body's neural tissue. A "typical" brain weighs 1.4 kg (3 lb) and has a volume of 1200 mL (71 in.^{3}). Brain size varies considerably among individuals. On average, the brains of males are about 10 percent larger than those of females, due to differences in average body size. No correlation exists between brain size and intelligence.

1 This lateral view of the brain of an embryo after 4 weeks of development shows the **neural tube**, the hollow cylinder that is the beginning of the central nervous system (CNS). The central cavity is called the **neurocoel** (NŪ-rō-sēl). In the cephalic portion of the neural tube, three areas enlarge rapidly through expansion of the neurocoel. This creates three prominent divisions called **primary brain vesicles**. The primary brain vesicles are named for their relative positions along the neural tube.

The **mesencephalon** (mez-en-SEF-a-lon), or "midbrain," is an expansion caudal to the prosencephalon.

The **prosencephalon** (prōz-en-SEF-a-lon; *proso*, forward + *encephalos*, brain), or "forebrain," is at the anterior tip of the neural tube.

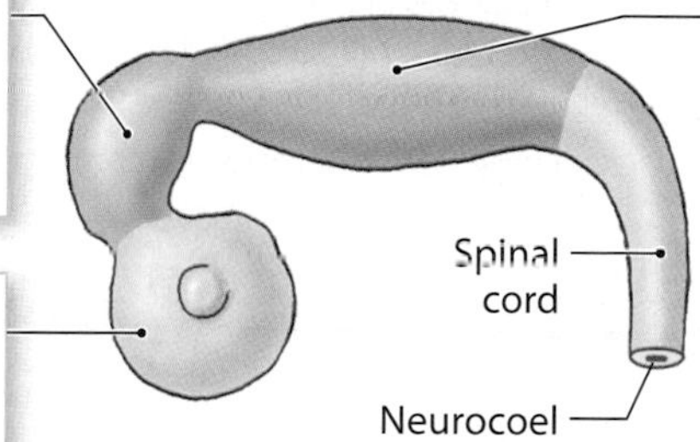
Spinal cord
Neurocoel

The **rhombencephalon** (rom-ben-SEF-a-lon), or "hindbrain," is the most caudal of the primary brain vesicles. It is continuous with the spinal cord.

2 By week 5 of development, the primary brain vesicles have changed position and the prosencephalon and rhombencephalon have subdivided, forming **secondary brain vesicles**.

Prosencephalon

The **diencephalon** (dī-en-SEF-a-lon; *dia*, through + *encephalos*, brain) becomes the major relay and processing center for information headed to and from the cerebrum.

The **telencephalon** (tel-en-SEF-a-lon; *telos*, end) begins to expand rapidly, eventually becoming the **cerebrum**, the largest part of the adult brain.

Rhombencephalon

The **metencephalon** (met-en-SEF-a-lon; *meta*, after) is caudal to the **midbrain** (mesencephalon). This region will form the **cerebellum** and the **pons** of the adult brain.

The **myelencephalon** (mī-el-en-SEF-a-lon; *myelon*, spinal cord) will become the **medulla oblongata**.

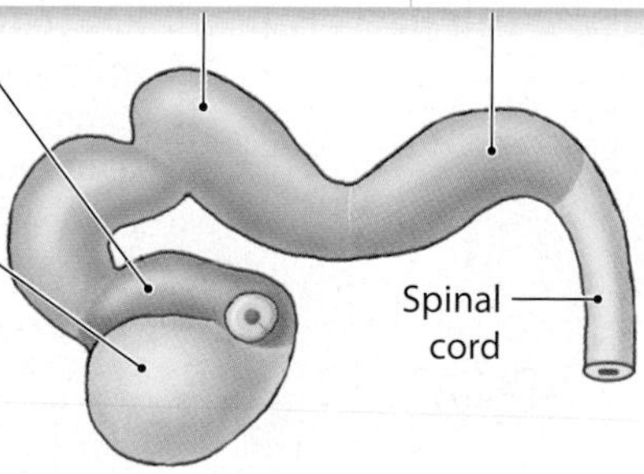
Spinal cord

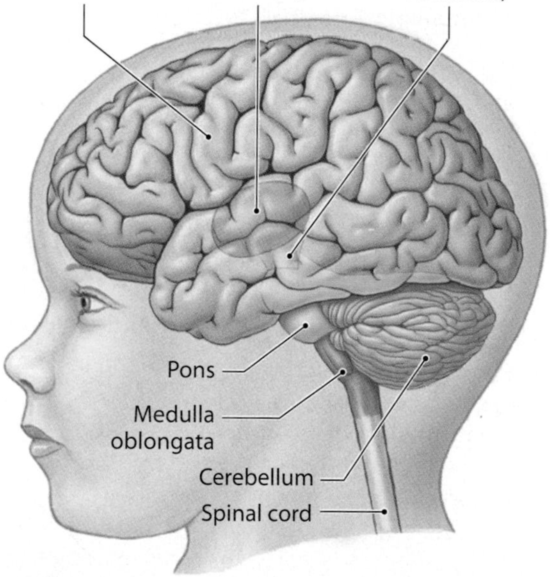
Cerebrum
Diencephalon (covered by cerebrum)
Midbrain (covered by cerebrum)
Pons
Medulla oblongata
Cerebellum
Spinal cord

3 As development continues, the cerebrum enlarges to the point where it covers other regions of the brain.

Module 13.1 Review

a. Name the three primary brain vesicles.
b. Which structures give rise to the secondary brain vesicles?
c. Which region of the adult brain is largest?

LO 13.1 Describe the origins of the different regions of the brain from the embryonic neural tube.

Module 13.2

Each region of the brain has distinct structural and functional characteristics

1 This diagrammatic view of the brain introduces the six major regions of the brain and their general functions. The regions are color coded, and these colors will be used as appropriate in illustrations throughout the chapter.

Cerebrum

The **cerebrum** (se-RĒ-brum) of the adult brain is divided into a pair of large **cerebral hemispheres.** Their surfaces are highly folded and covered by a 1.5 to 4.5 mm thick superficial layer of gray matter called the **cerebral cortex** (*cortex*, rind or bark). Functions include conscious thought, memory storage and processing, sensory processing, and regulating skeletal muscle contractions.

Fissures are deep grooves that subdivide the cerebral hemisphere.	**Gyri** (JĪ-rī; singular, *gyrus*) are folds in the cerebral hemispheres that increase its surface area.	**Sulci** (SUL-sī; singular, *sulcus*) are shallow depressions in the cerebral hemispheres that separate adjacent gyri.

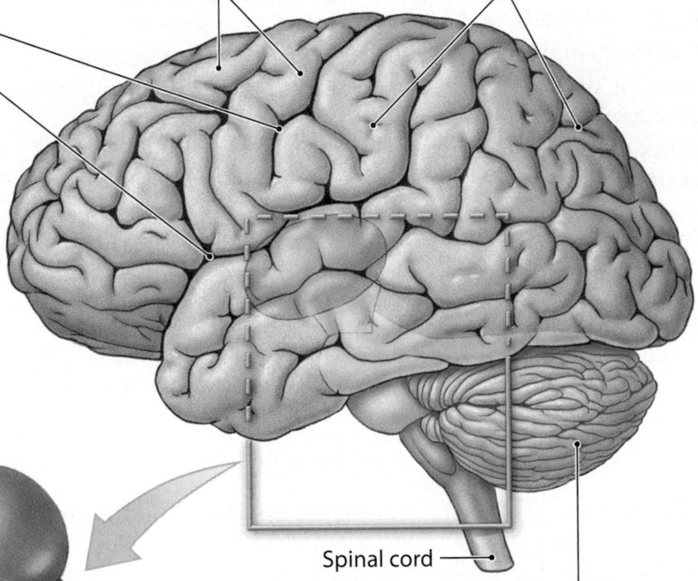

Spinal cord

Diencephalon

The **diencephalon** is the structural and functional link between the cerebral hemispheres and the rest of the CNS.

The **thalamus** (THAL-a-mus) contains relay and processing centers for sensory information.

The **hypothalamus** (*hypo-*, below), or floor of the diencephalon, contains centers involved with emotions, autonomic function, and hormone production.

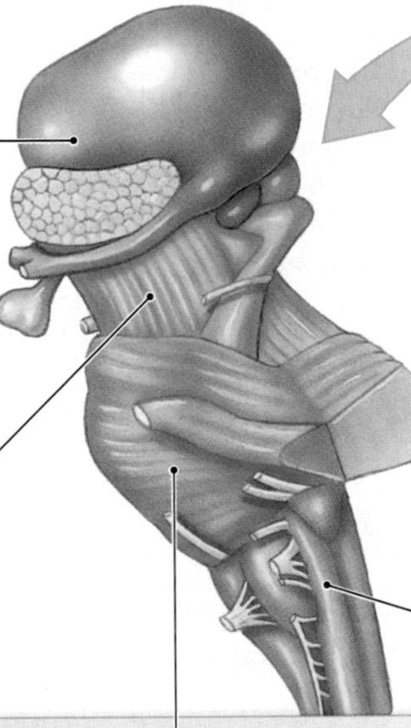

Cerebellum

The **cerebellum** (ser-e-BEL-um), partially hidden by the cerebral hemispheres, is the second-largest structure in the brain. The functions of the cerebellum include coordinating and modulating motor commands from the cerebral cortex. The cerebellum has only 10 percent of the brain's volume, but over 50 percent of the brain's neurons.

Brain Stem

The **brain stem** includes the midbrain, pons, and medulla oblongata.	The **midbrain** contains nuclei that process visual and auditory information and control reflexes triggered by these stimuli. It also contains centers that help maintain consciousness.	The **pons** (*pons*, bridge) connects the cerebellum to the brain stem. In addition to tracts and relay centers, the pons also contains nuclei that function in somatic and visceral motor control.	The **medulla oblongata** relays sensory information to other portions of the brain stem and to the thalamus. The medulla oblongata also contains major centers that regulate autonomic function, such as heart rate and blood pressure.

2 During development, the neurocoel within the cerebral hemispheres, diencephalon, metencephalon, and medulla oblongata expands to form chambers called **ventricles** (VEN-tri-kls). The ventricles are filled with cerebrospinal fluid and lined by ependymal cells (Module 11.4, p. 400).

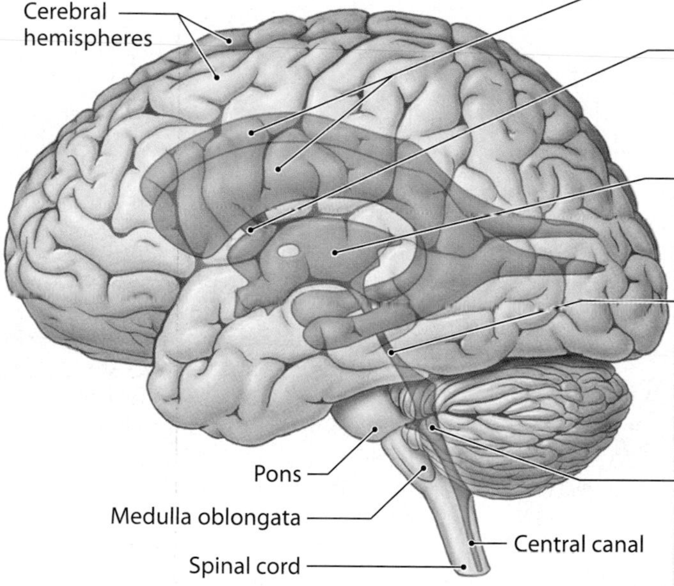

Cerebral hemispheres
Pons
Medulla oblongata
Spinal cord
Central canal

Ventricular system, lateral view

Ventricles of the Brain

Each cerebral hemisphere contains a large **lateral ventricle**.

Each lateral ventricle communicates with the third ventricle through an **interventricular foramen**.

The **third ventricle** is located in the diencephalon.

The **cerebral aqueduct** is a slender canal within the midbrain that connects the third ventricle to the fourth ventricle.

The **fourth ventricle** begins in the metencephalon and extends into the superior portion of the medulla oblongata. It then narrows and becomes the central canal of the spinal cord.

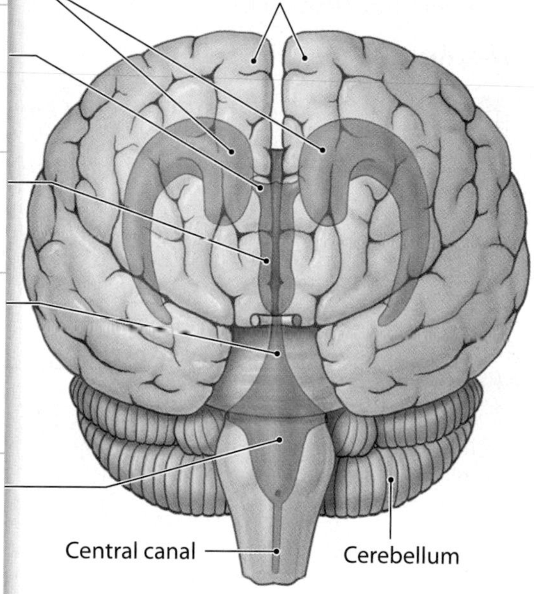

Cerebral hemispheres
Central canal
Cerebellum

Ventricular system, anterior view

3 The interconnections between the ventricles are seen in this frontal section of the brain.

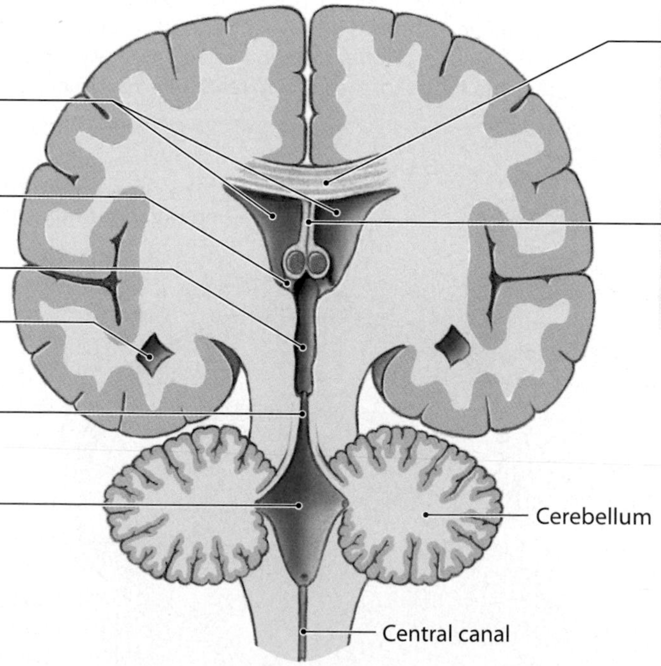

Lateral ventricles
Interventricular foramen
Third ventricle
Inferior tip of lateral ventricle
Cerebral aqueduct
Fourth ventricle
Cerebellum
Central canal

The **corpus callosum** is a thick tract of white matter that interconnects the two cerebral hemispheres.

The **septum pellucidum** is a thin partition that separates the two lateral ventricles.

Module 13.2 Review

a. Name the six major regions of the brain and the distinct structures of each.

b. Describe the role of the medulla oblongata.

c. Compare the corpus callosum with the septum pellucidum.

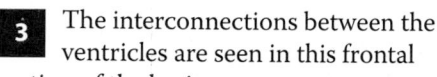

LO **13.2** Name the six major regions of the brain, and describe their functions.

The cranial meninges and cerebrospinal fluid protect and support the brain

The cranial bones, **cranial meninges** (membranous brain coverings), and the cerebrospinal fluid protect the delicate brain tissues. In addition, the neural tissue of the brain is biochemically isolated from the general circulation by the blood–brain barrier.

Meninges

1 The three layers that make up the cranial meninges—the cranial **dura mater**, **arachnoid mater**, and **pia mater**—are continuous with those of the spinal meninges (Module 12.3, p. 432).

1 Dura mater

The cranial dura mater consists of outer and inner fibrous layers. The outer layer is fused to the periosteum of the cranial bones. As a result, there is no epidural space. The outer (periosteal) and inner (meningeal) layers of the cranial dura mater are typically separated by a slender gap that contains tissue fluids and blood vessels, including several large dural sinuses. The dural sinuses collect blood from the veins of the brain.

2 Arachnoid mater

The cranial arachnoid mater consists of the arachnoid membrane and the arachnoid trabeculae (which connect to the pia mater). The arachnoid membrane is a smooth covering that does not follow the brain's underlying folds. The subarachnoid space lies between the arachnoid membrane and the pia mater.

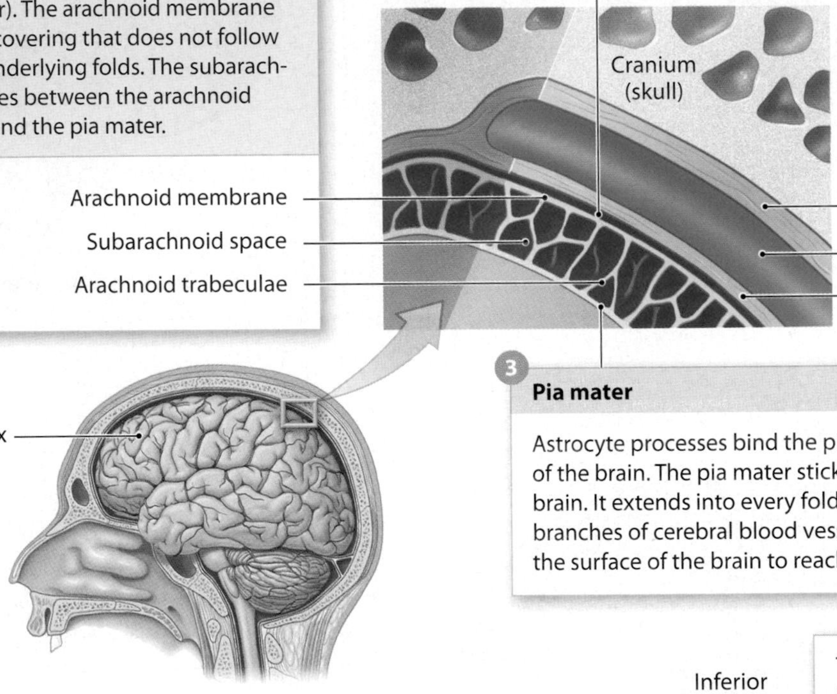
Subdural space
Cranium (skull)
Arachnoid membrane
Subarachnoid space
Arachnoid trabeculae
Dura mater (periosteal layer)
Dural sinus
Dura mater (meningeal layer)
Cerebral cortex

3 Pia mater

Astrocyte processes bind the pia mater to the surface of the brain. The pia mater sticks to the surface of the brain. It extends into every fold and accompanies the branches of cerebral blood vessels as they penetrate the surface of the brain to reach internal structures.

2 At several sites, the inner layer of the dura mater extends into the cranial cavity, forming **dural folds**—sheets that dip inward and then return. These stabilize and support the brain. **Dural sinuses** are large collecting veins located within the dural folds. There are three large dural folds.

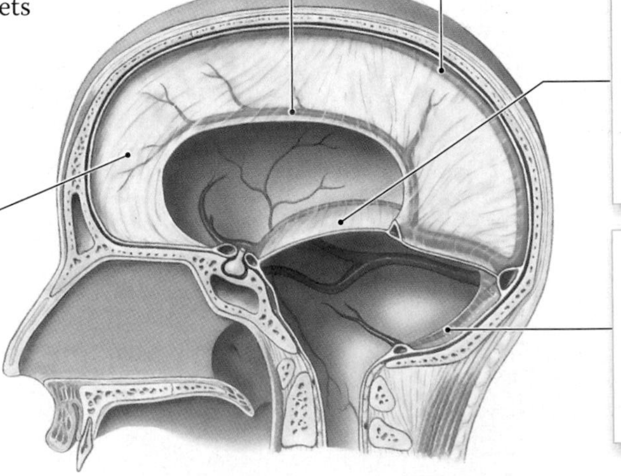
Inferior sagittal sinus

The **superior sagittal sinus** is the largest dural sinus.

The **falx cerebri** (FALKS SER-e-brī; *falx*, sickle shaped) is a fold of dura mater between the cerebral hemispheres. Its inferior portions attach anteriorly to the crista galli of the ethmoid bone and posteriorly to the internal occipital crest of the occipital bone. The superior and inferior sagittal sinuses lie within this dural fold.

The **tentorium cerebelli** (ten-TŌ-rē-um ser-e-BEL-ī; *tentorium*, a tent) separates the cerebral hemispheres from the cerebellum.

The **falx cerebelli** separates the two cerebellar hemispheres along the midsagittal line inferior to the tentorium cerebelli.

Cerebrospinal Fluid

3 **Cerebrospinal fluid (CSF)** completely surrounds and bathes the exposed surfaces of the CNS. Each of the ventricles contains an area of **choroid plexus** (*choroid*, vascular coat; *plexus*, network), which is involved in producing and maintaining CSF. Each choroid plexus consists of a combination of specialized ependymal cells with tight junctions and capillaries.

The CSF circulates from the choroid plexuses through the ventricles and fills the central canal of the spinal cord. As it circulates, materials diffuse between the CSF and the interstitial fluid of the CNS across the ependymal cells.

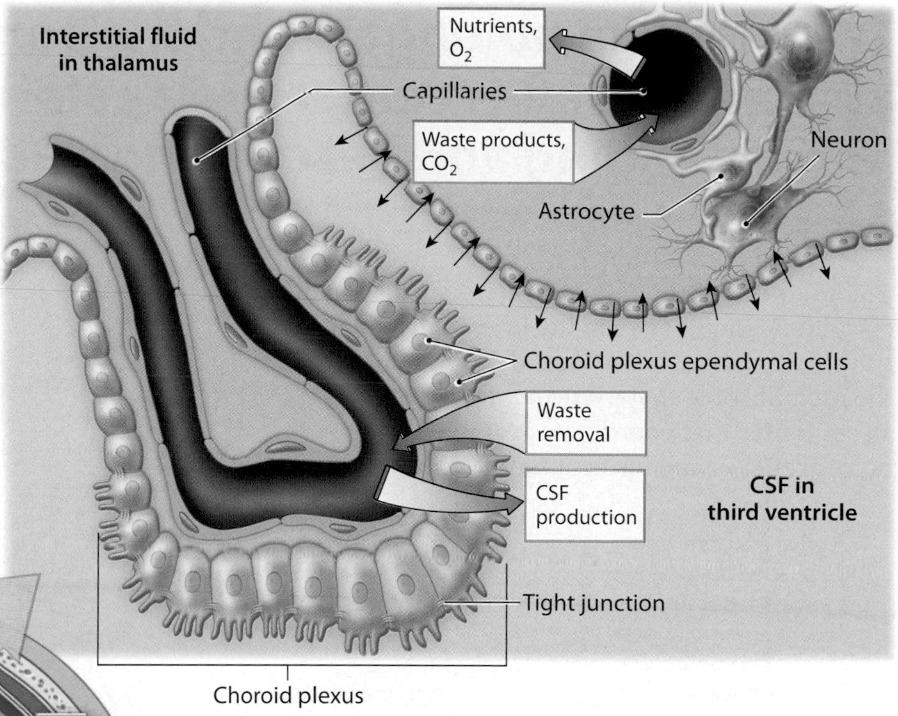

Interstitial fluid in thalamus
Nutrients, O_2
Capillaries
Waste products, CO_2
Neuron
Astrocyte
Choroid plexus ependymal cells
Waste removal
CSF production
CSF in third ventricle
Tight junction
Choroid plexus

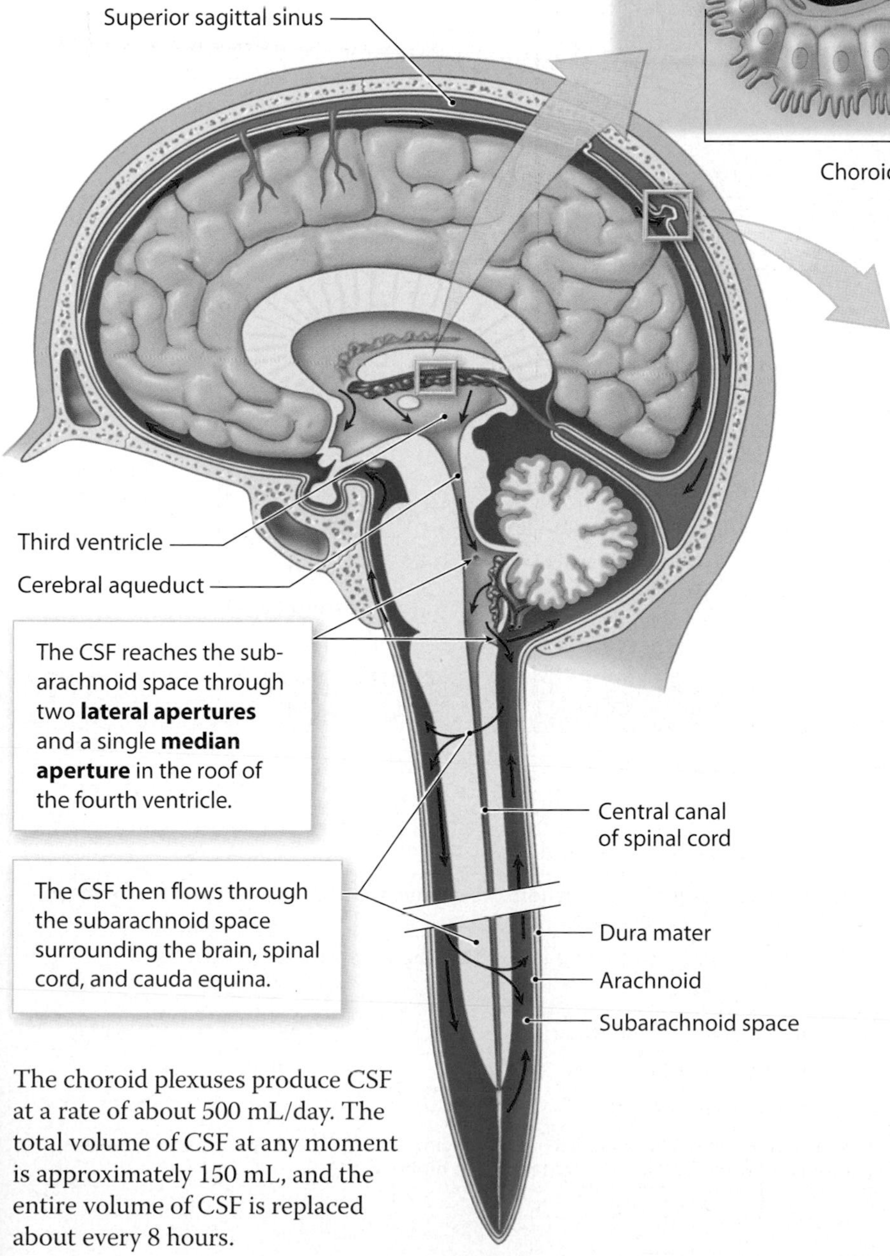

Superior sagittal sinus
Third ventricle
Cerebral aqueduct
The CSF reaches the subarachnoid space through two **lateral apertures** and a single **median aperture** in the roof of the fourth ventricle.
The CSF then flows through the subarachnoid space surrounding the brain, spinal cord, and cauda equina.
Central canal of spinal cord
Dura mater
Arachnoid
Subarachnoid space

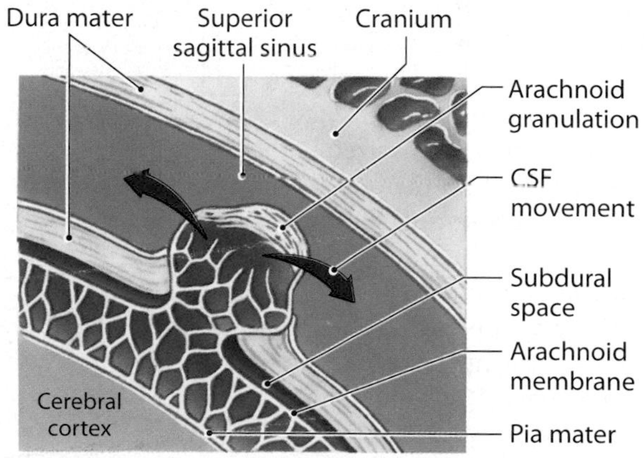

Dura mater
Superior sagittal sinus
Cranium
Arachnoid granulation
CSF movement
Subdural space
Arachnoid membrane
Pia mater
Cerebral cortex

4 Fingerlike extensions of the arachnoid membrane penetrate the meningeal layer of the dura mater and extend into the superior sagittal sinus. In adults, these extensions form large **arachnoid granulations**, sites where CSF is absorbed into the venous circulation.

The choroid plexuses produce CSF at a rate of about 500 mL/day. The total volume of CSF at any moment is approximately 150 mL, and the entire volume of CSF is replaced about every 8 hours.

Module 13.3 Review

a. From superficial to deep, name the layers that constitute the cranial meninges.

b. What would happen if the normal circulation or absorption of CSF became blocked?

c. How would decreased diffusion across the arachnoid granulations affect the volume of cerebrospinal fluid in the ventricles?

LO **13.3** Explain how the brain is protected and supported, and how cerebrospinal fluid forms and circulates.

Module 13.4

The medulla oblongata and the pons contain autonomic reflex centers, relay stations, and ascending and descending tracts

Medulla Oblongata

The **medulla oblongata** is a very busy place: All communication between the brain and spinal cord travels along tracts that ascend or descend through the medulla oblongata. The medulla oblongata is also a center that coordinates complex autonomic reflexes and visceral functions.

1 Descending tracts cover the anterior surface of the medulla oblongata.

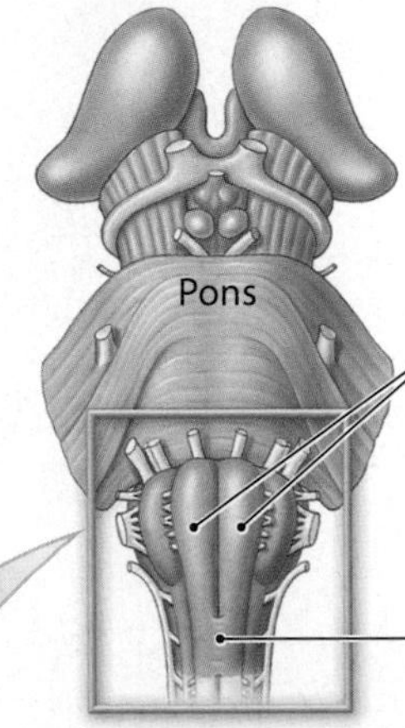

Pons

The **pyramids** contain tracts of motor fibers that originate at the cerebral cortex.

Some of the pyramidal fibers cross to the opposite side of the medulla oblongata as they descend into the spinal cord. That crossing is called a **decussation** (de-kuh-SĀ-shun; *decussation*, crossing over).

The **olive** is a prominent olive-shaped bulge along the anterolateral surface of the medulla oblongata. It follows the contours of the olivary nucleus.

2 Landmarks and structures of the medulla oblongata are shown in these illustrations, and major components and functions are summarized in the table below. The medulla oblongata contains **autonomic centers** controlling vital functions, and **relay stations** along sensory and motor pathways. The medulla oblongata also contains the nuclei associated with five cranial nerves, although those nuclei are not illustrated here.

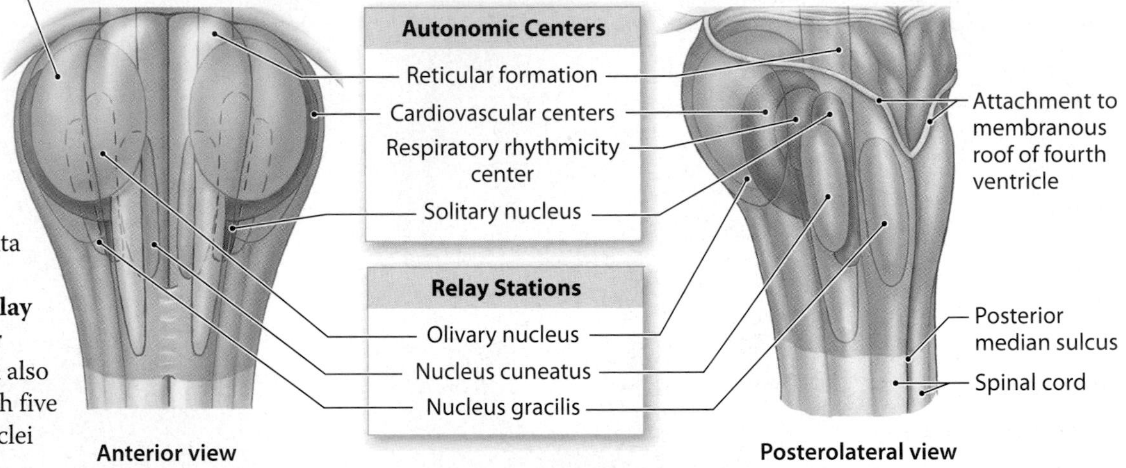

Autonomic Centers
Reticular formation
Cardiovascular centers
Respiratory rhythmicity center
Solitary nucleus
Relay Stations
Olivary nucleus
Nucleus cuneatus
Nucleus gracilis
Attachment to membranous roof of fourth ventricle
Posterior median sulcus
Spinal cord
Anterior view
Posterolateral view

Components of the Medulla Oblongata and Their Functions

Component	Function
Gray Matter	
Nucleus gracilis, nucleus cuneatus	Relay somatic sensory information to the thalamus
Olivary nuclei	Located within the olives; relay information from the red nucleus and other nuclei of the midbrain, and the cerebral cortex to the cerebellum
Solitary nucleus	Integrates and relays visceral sensory information to autonomic processing centers
Autonomic reflex centers	
Cardiac centers	Regulate heart rate and force of contraction
Vasomotor centers	Regulate distribution of blood flow
Respiratory rhythmicity centers	Set the pace of respiratory movements
Other nuclei/centers	Contain sensory and motor nuclei of cranial nerves VIII (in part), IX, X, XI (in part), and XII; relay ascending sensory information from the spinal cord to higher centers
White Matter	
Ascending and descending tracts	Link the brain with the spinal cord

Pons

3 The **pons** links the cerebellum with the midbrain, diencephalon, cerebrum, medulla oblongata, and spinal cord.

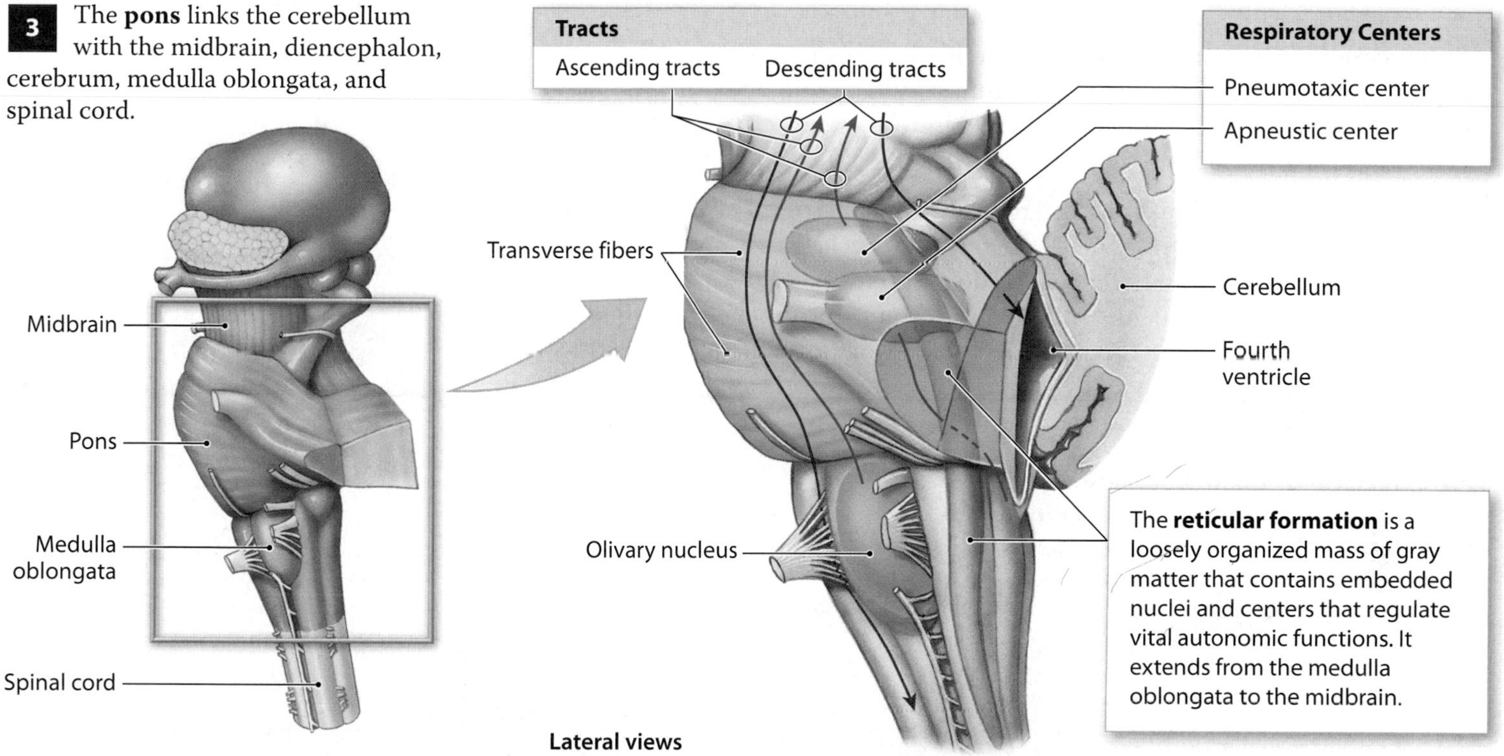

Midbrain
Pons
Medulla oblongata
Spinal cord
Tracts
Ascending tracts
Descending tracts
Respiratory Centers
Pneumotaxic center
Apneustic center
Transverse fibers
Cerebellum
Fourth ventricle
Olivary nucleus
The **reticular formation** is a loosely organized mass of gray matter that contains embedded nuclei and centers that regulate vital autonomic functions. It extends from the medulla oblongata to the midbrain.

Lateral views

Components of the Pons and Their Functions

Component	Function
Gray Matter	
Nuclei associated with cranial nerves V, VI, VII, and VIII (in part)	Relay sensory information and issue somatic motor commands
Apneustic and pneumotaxic centers	Adjust activities of the respiratory rhythmicity centers in the medulla oblongata
Relay centers	Relay sensory and motor information to the cerebellum
White Matter	
Ascending tracts	Carry sensory information from the nucleus cuneatus and nucleus gracilis to the thalamus
Descending tracts	Carry motor commands from higher centers to motor nuclei of cranial or spinal nerves
Transverse fibers	Interconnect processing centers in the cerebellar hemispheres

Module 13.4 Review

a. What is the function of the ascending and descending tracts in the medulla oblongata?

b. Which medulla oblongata components relay somatic sensory information to the thalamus?

c. Describe the pyramids of the medulla oblongata and a decussation.

LO 13.4 List the main components of the medulla oblongata and pons, and specify the functions of each.

The cerebellum coordinates learned and reflexive patterns of muscular activity at the subconscious level

The **cerebellum** is an automatic processing center that monitors proprioceptive, visual, tactile, balance, and auditory sensations. It has two primary functions:

- **Adjusting the Postural Muscles of the Body.** The cerebellum coordinates rapid, automatic adjustments that maintain balance and equilibrium. The cerebellum makes these adjustments by modifying the activities of motor centers in the brain stem.
- **Programming and Fine-Tuning Movements Controlled at the Conscious and Subconscious Levels.** The cerebellum refines learned movement patterns indirectly by regulating activity along motor pathways at the cerebral cortex, basal nuclei, and motor centers in the brain stem. The cerebellum compares the motor commands with proprioceptive information (perception of muscle and joint position) and makes any necessary adjustments so the movement you intended to do is actually accomplished.

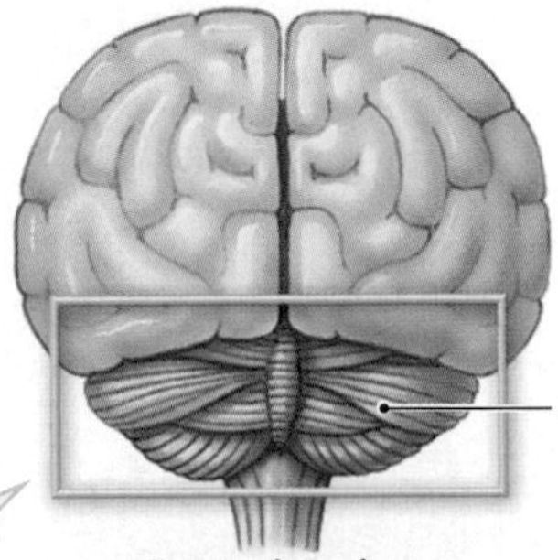
Cerebellum

Posterior view

1 As seen in this view of the posterior superior surface, the cerebellum has large **anterior** and **posterior lobes**. Like the cerebrum, the cerebellum has two hemispheres and the surface is covered with a thin layer of gray matter called the **cerebellar cortex**.

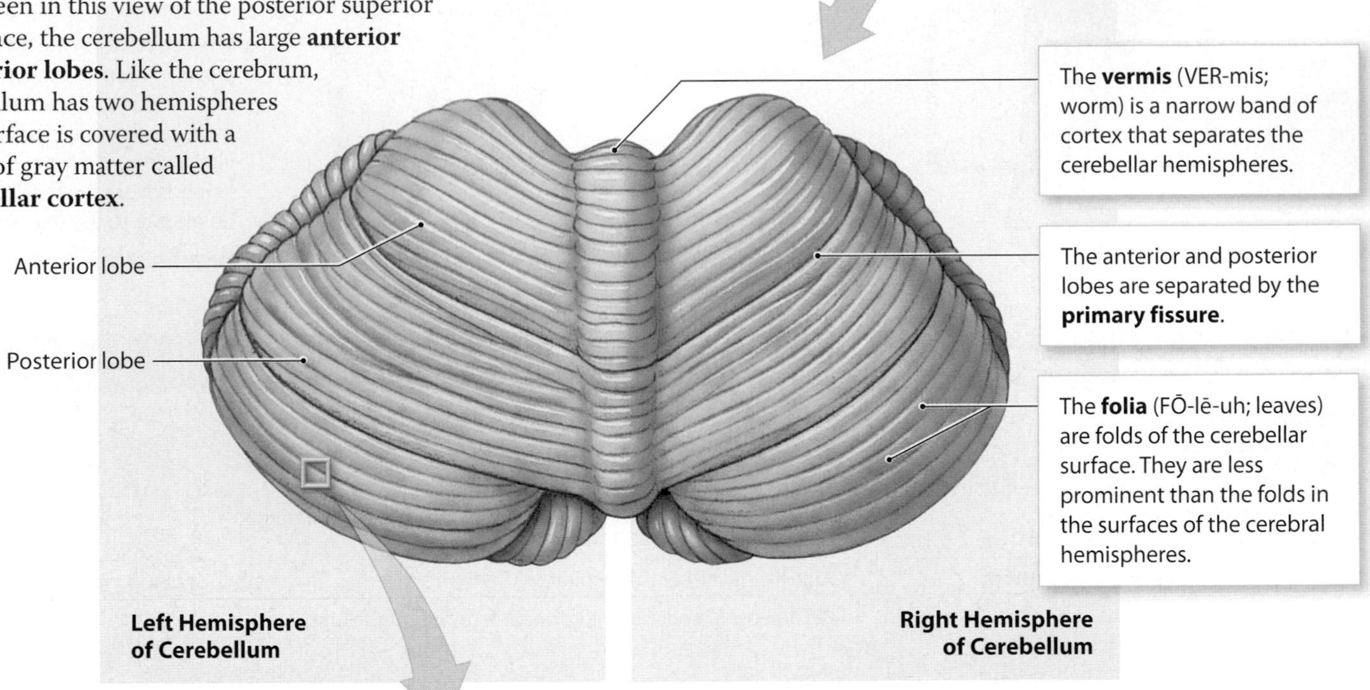
Anterior lobe
Posterior lobe
The **vermis** (VER-mis; worm) is a narrow band of cortex that separates the cerebellar hemispheres.
The anterior and posterior lobes are separated by the **primary fissure**.
The **folia** (FŌ-lē-uh; leaves) are folds of the cerebellar surface. They are less prominent than the folds in the surfaces of the cerebral hemispheres.
Left Hemisphere of Cerebellum
Right Hemisphere of Cerebellum

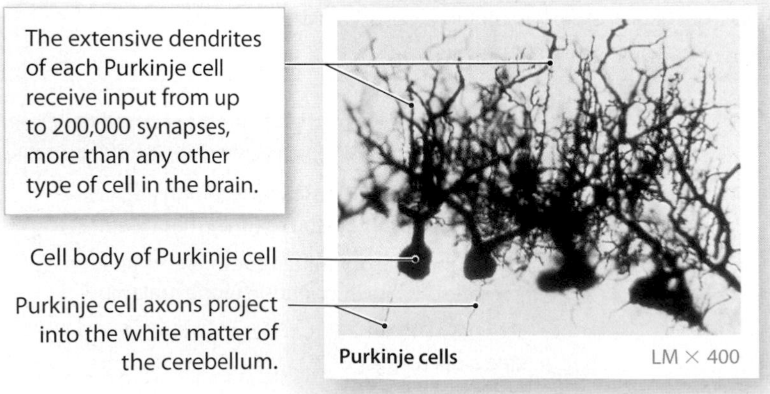
The extensive dendrites of each Purkinje cell receive input from up to 200,000 synapses, more than any other type of cell in the brain.
Cell body of Purkinje cell
Purkinje cell axons project into the white matter of the cerebellum.

Purkinje cells LM × 400

2 The cerebellar cortex has an outer molecular layer, a single layer of Purkinje cells, and an inner granular layer. This cortex is about one-fourth the thickness of the cerebral cortex. The **Purkinje** (pur-KIN-jē) **cells** are highly branched and create a layer of neuron cell bodies each with a single axon. Most axons carrying sensory information to the cerebellum do not synapse in the cerebellar nuclei but pass through the deeper layers of the cerebellum on their way to the dendrites of Purkinje cells. Information about the motor commands issued at the conscious and subconscious levels reaches the Purkinje cells, after being relayed by nuclei in the pons or by the cerebellar nuclei. The only axons that leave the cerebellar cortex are those of the Purkinje cells.

3 This sagittal section through the vermis shows the internal organization of the cerebellum. The locations of the cerebellar peduncles are also shown.

Cerebellar Peduncles

The three **cerebellar peduncles** (*pedunculus*, foot) on each side contain tracts that link the cerebellum with the brain stem, cerebrum, and spinal cord. The middle and inferior peduncles carry most of the input to the cerebellum, and the superior peduncle carries most of the cerebellar output.

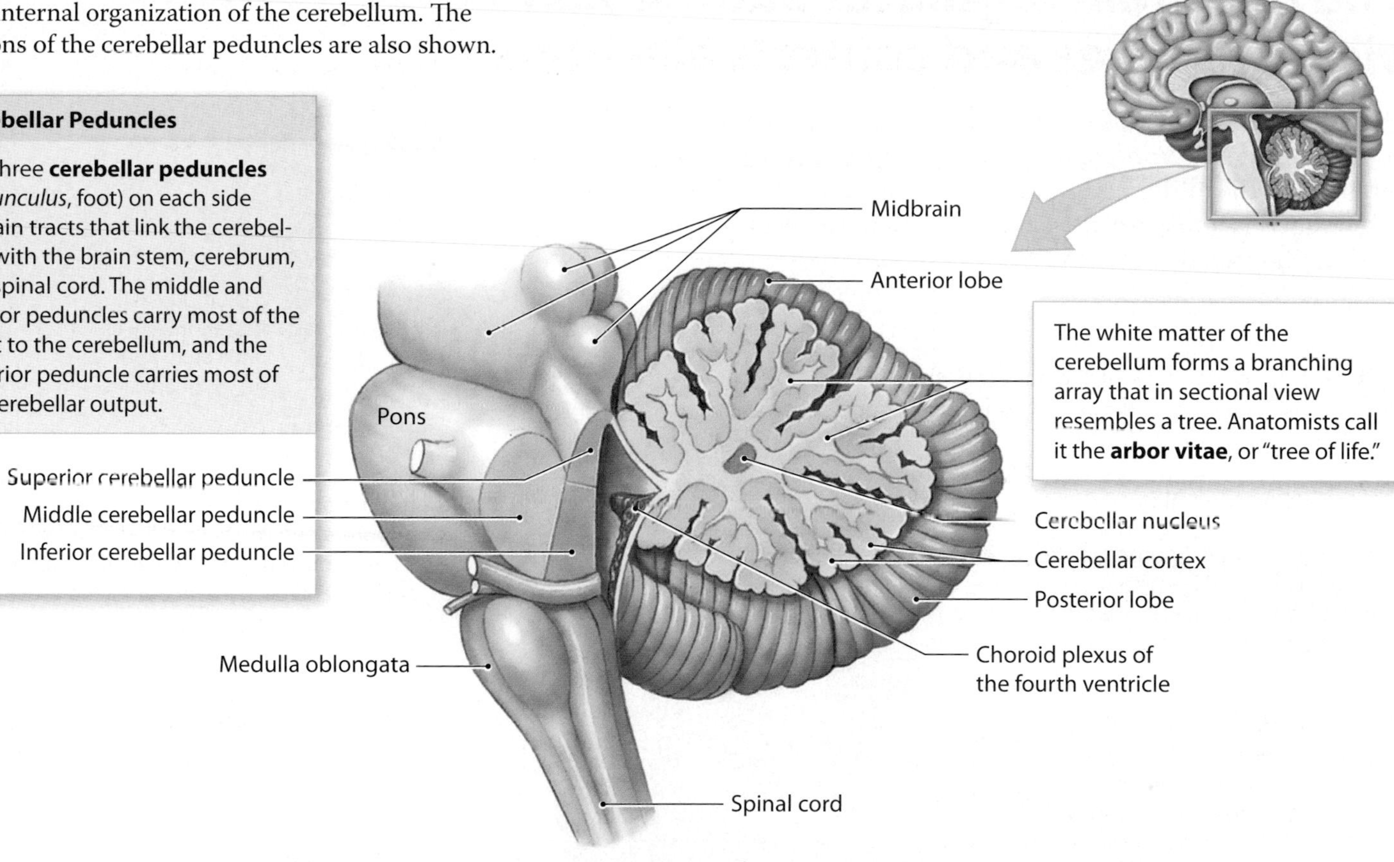

Superior cerebellar peduncle
Middle cerebellar peduncle
Inferior cerebellar peduncle
Pons
Midbrain
Anterior lobe
Cerebellar nucleus
Cerebellar cortex
Posterior lobe
Choroid plexus of the fourth ventricle
Medulla oblongata
Spinal cord

Lateral view

The white matter of the cerebellum forms a branching array that in sectional view resembles a tree. Anatomists call it the **arbor vitae**, or "tree of life."

Components of the Cerebellum and Their Functions

Component	Function
Gray Matter	
Cerebellar cortex	Involuntary coordination and control of ongoing body movements
Cerebellar nuclei	Involuntary coordination and control of ongoing body movements
White Matter	
Arbor vitae	Connects cerebellar cortex and nuclei with cerebellar peduncles
Cerebellar peduncles	
Superior	Link the cerebellum with midbrain, diencephalon, and cerebrum
Middle	Contain transverse fibers and carry communications between the cerebellum and pons
Inferior	Link the cerebellum with the medulla oblongata and spinal cord
Transverse fibers	Interconnect nuclei in the pons with the cerebellar hemisphere on the opposite side

The cerebellum can be permanently damaged by trauma or stroke, or temporarily affected by drugs such as alcohol. The result is **ataxia** (a-TAK-sē-uh; *ataxia*, lack of order), a disturbance in muscular coordination. In severe ataxia, the individual cannot sit or stand without assistance.

Module 13.5 Review

a. Identify the components of the cerebellar gray matter.

b. Describe the arbor vitae, including its makeup, location, and function.

c. Describe ataxia.

LO 13.5 List the main components of the cerebellum, and specify the functions of each.

The midbrain regulates auditory and visual reflexes and controls alertness

The **midbrain** is the most complex and integrative portion of the brain stem. Working independently or with the cerebellum, the midbrain can direct complex motor patterns at the subconscious level. It also influences the level of activity in the entire nervous system.

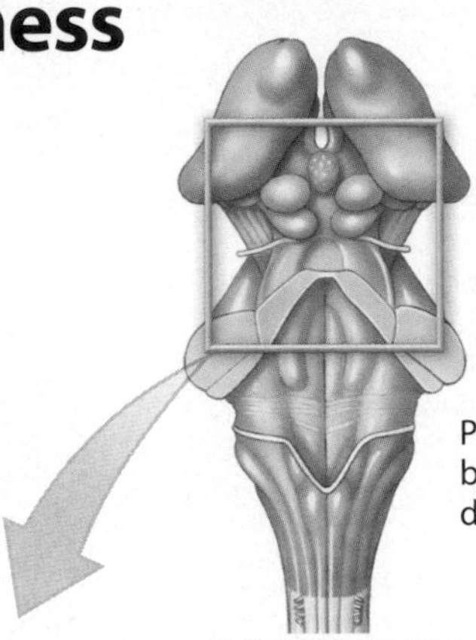

Posterior view of brain stem and diencephalon

1 This posterior view of the midbrain shows the major superficial landmarks as well as underlying nuclei.

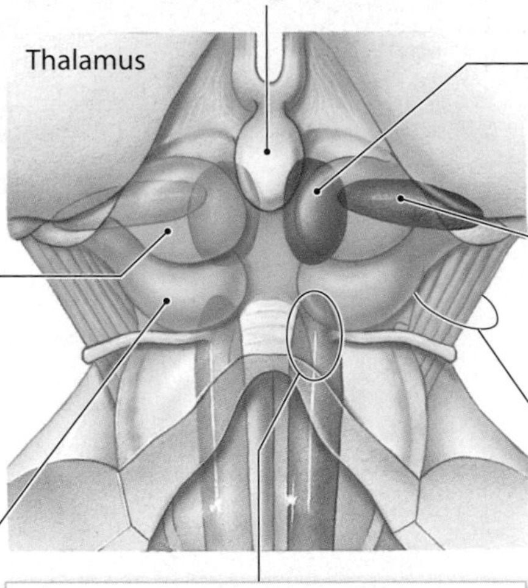
Pineal gland
Thalamus

Corpora Quadrigemina

The **corpora quadrigemina** (KOR-por-uh qua-dri-JEM-i-nuh) are two pairs of sensory nuclei located in the roof of the midbrain.

Each **superior colliculus** (ko-LIK-ū-lus; *colliculus*, hill) receives visual inputs from the thalamus and controls the reflex movements of the eyes, head, and neck in response to these visual stimuli (see **2** below).

Each **inferior colliculus** receives auditory input from nuclei in the medulla oblongata and pons and controls reflex movements of the head, neck, and trunk in response to these auditory stimuli.

The **reticular activating system (RAS)** is a specialized part of the reticular formation. Stimulation of the RAS makes you more alert and attentive; damage to the RAS produces unconsciousness.

The **red nucleus** receives information from the cerebrum and cerebellum and issues subconscious motor commands that affect upper limb position and background muscle tone.

The **substantia nigra** (NĪ-gruh; *nigra*, black) contains darkly pigmented cells that adjust activity in the basal nuclei of the cerebrum.

The **cerebral peduncles** are nerve fiber bundles on the ventrolateral surfaces of the midbrain. They contain (1) descending fibers that reach the cerebellum by way of the pons and (2) descending fibers that carry voluntary motor commands issued by the cerebral hemispheres.

2 This superior view of a horizontal section through the midbrain shows the internal subdivisions relative to the cerebral aqueduct.

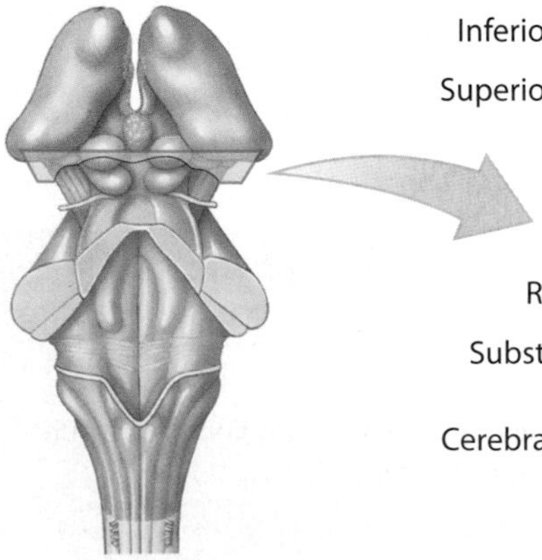

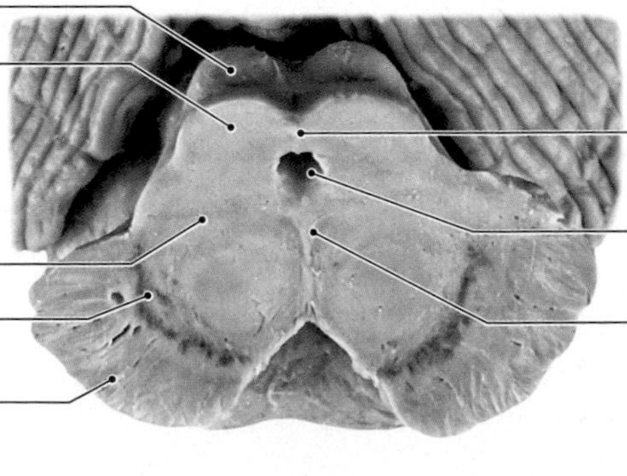
Posterior
Inferior colliculus
Superior colliculus
Red nucleus
Substantia nigra
Cerebral peduncle
Cerebral aqueduct
Anterior

The **tectum**, or roof of the midbrain, is the region posterior to the cerebral aqueduct.

The **tegmentum** is the area anterior to the cerebral aqueduct.

Components of the Midbrain and Their Functions

Component	Region/Nuclei	Function
Gray Matter		
Tectum (roof)	**Superior colliculi**	Integrate visual information with other sensory inputs; initiate reflex responses to visual stimuli
	Inferior colliculi	Relay auditory information to medial geniculate nuclei; initiate reflex responses to auditory stimuli
Walls and floor	**Red nuclei**	Provide subconscious control of upper limb position and background muscle tone
	Substantia nigra	Regulates activity in the basal nuclei
	Reticular formation (RAS headquarters)	Processes incoming sensations and outgoing motor commands automatically; can initiate involuntary motor responses to stimuli; helps maintain consciousness (RAS)
	Other nuclei/centers	Are associated with cranial nerves III and IV
White Matter		
	Cerebral peduncles	Connect primary motor cortex with motor neurons in brain and spinal cord; carry ascending sensory information to thalamus

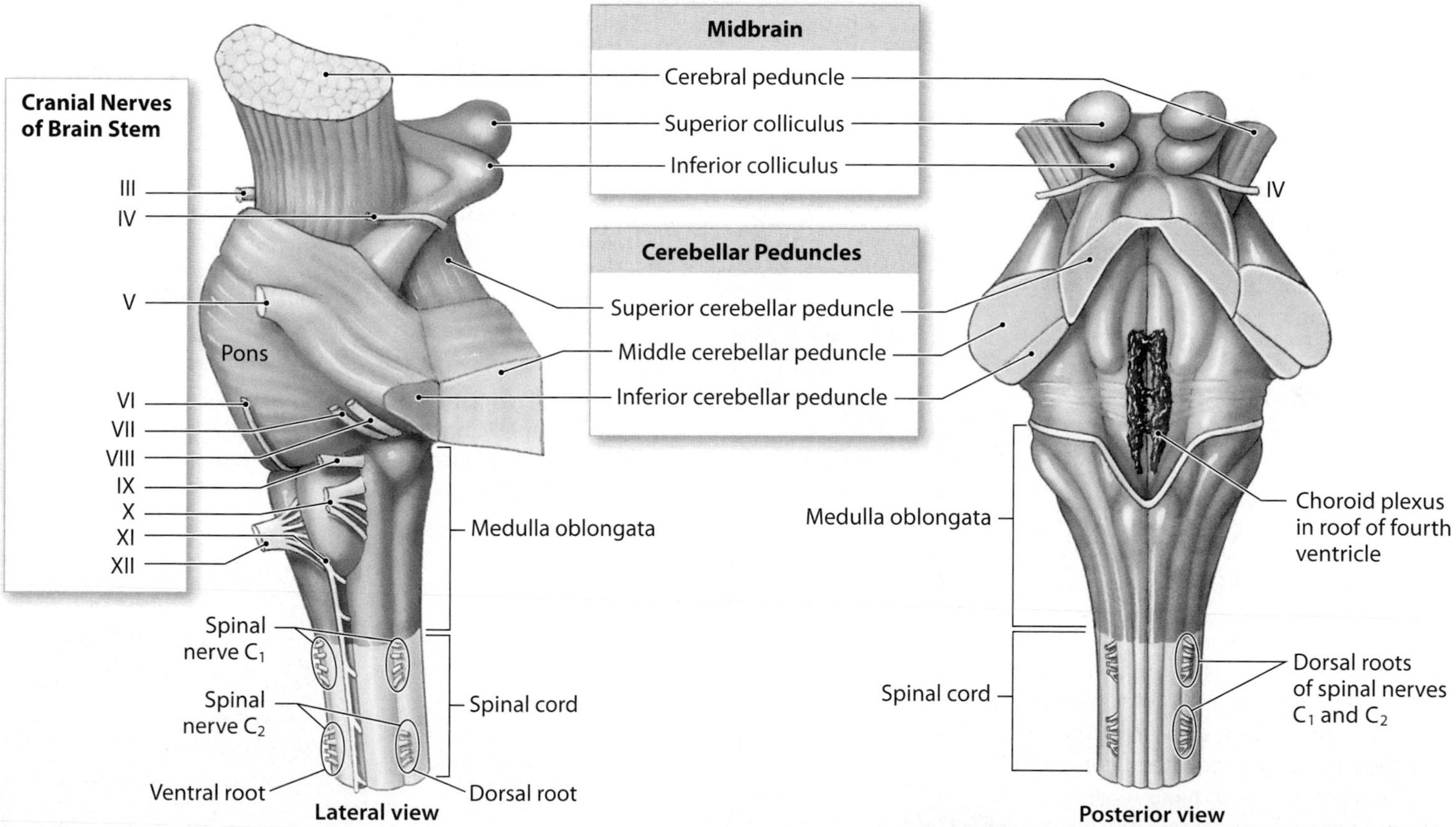

Cranial Nerves of Brain Stem
III
IV
V
VI
VII
VIII
IX
X
XI
XII
Pons
Midbrain
Cerebral peduncle
Superior colliculus
Inferior colliculus
Cerebellar Peduncles
Superior cerebellar peduncle
Middle cerebellar peduncle
Inferior cerebellar peduncle
Medulla oblongata
Spinal nerve C_1
Spinal nerve C_2
Spinal cord
Ventral root
Dorsal root
Lateral view
IV
Medulla oblongata
Choroid plexus in roof of fourth ventricle
Spinal cord
Dorsal roots of spinal nerves C_1 and C_2
Posterior view

3 These views of the brain stem show the anatomy of the midbrain in relation to the brain stem as a whole.

Module 13.6 Review

a. Identify the sensory nuclei contained within the corpora quadrigemina.

b. Which area(s) of the midbrain control(s) reflexive movements of the eyes, head, and neck?

c. Cranial nerves III to XII arise from which structure?

LO 13.6 List the main components of the midbrain, and specify the functions of each.

The diencephalon consists of the epithalamus, thalamus (left and right), and hypothalamus

Epithalamus

1 As seen in this sagittal section of the brain, the **epithalamus** is the roof of the diencephalon superior to the third ventricle. The anterior portion of the epithalamus contains an extensive area of choroid plexus that extends through the interventricular foramina.

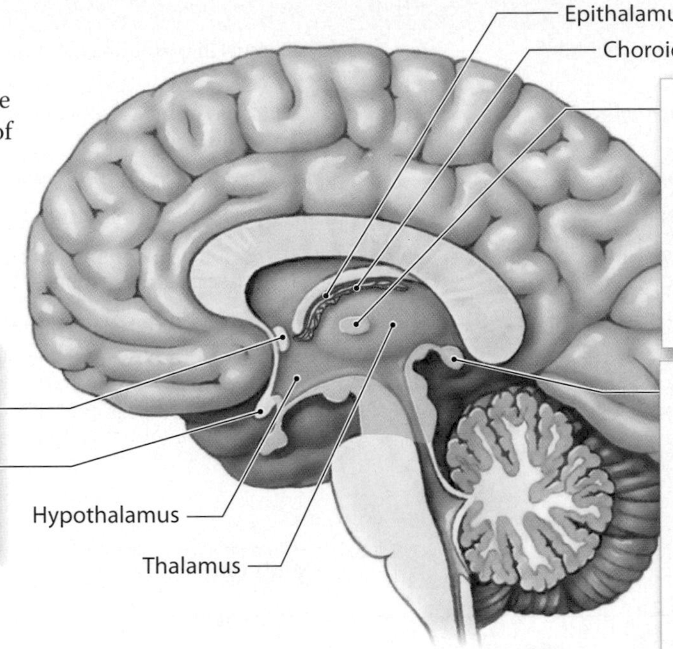
Epithalamus
Choroid plexus
Hypothalamus
Thalamus

The anterior limit of the diencephalon is marked by the **anterior commissure**, a tract that interconnects the cerebral hemispheres, and the **optic chiasm**, where the optic nerves connect to the brain.

A projection of gray matter called an **interthalamic adhesion** extends into the third ventricle from the thalamus on either side, although no fibers cross the midline. It is absent in about 20 percent of human brains.

The **pineal gland** lies in the posterior, inferior portion of the epithalamus. It is an endocrine structure that secretes the hormone **melatonin**. Melatonin is important in the regulation of day–night cycles and also in the regulation of reproductive functions.

Thalamus

2 On each side of the brain, a **thalamus** sits superior to the midbrain. The important landmarks in this lateral view can be seen only after the cerebral hemispheres and cerebral peduncles have been removed.

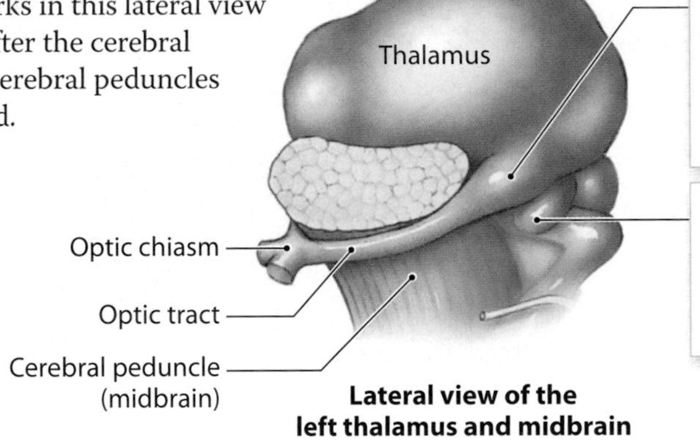
Thalamus
Optic chiasm
Optic tract
Cerebral peduncle (midbrain)

Lateral view of the left thalamus and midbrain

The **lateral geniculate** (je-NIK-ū-lāt; *genicula*, little knee) **nucleus** of each thalamus receives visual information over the **optic tract** and sends signals to both the midbrain and the occipital lobe of the cerebral hemisphere on that side.

The **medial geniculate nucleus** of each thalamus relays auditory information from specialized receptors of the inner ear to the appropriate area of the cerebral cortex.

3 The thalamus is the final relay point for ascending sensory information that will be relayed, or **projected**, to the cerebral cortex. The thalamus acts as a filter, passing on only a small portion of the arriving sensory information to the cerebral hemispheres. Each region of the thalamus contains nuclei or groups of nuclei that connect to specific regions of the cerebral cortex (see table at the upper right).

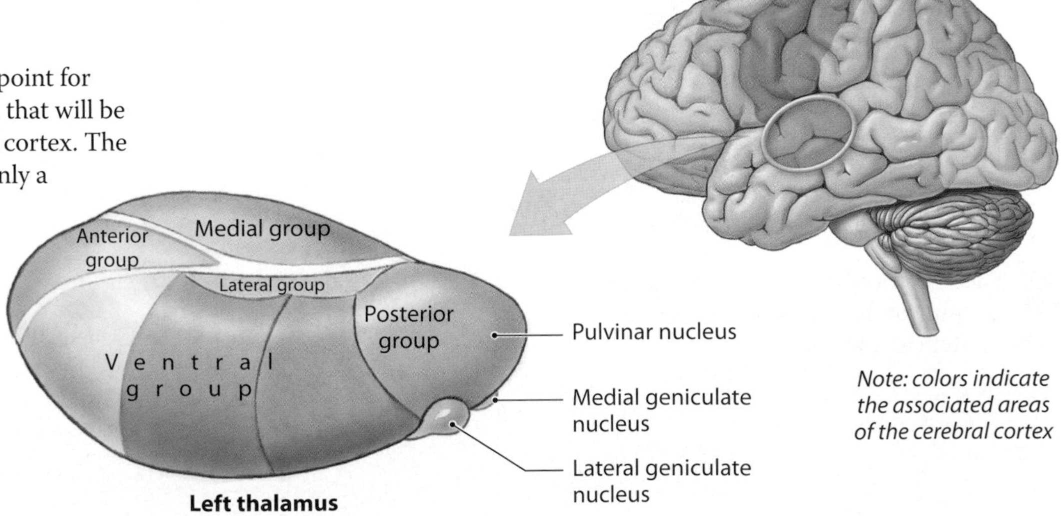
Anterior group
Medial group
Lateral group
Posterior group
Ventral group
Pulvinar nucleus
Medial geniculate nucleus
Lateral geniculate nucleus
Note: colors indicate the associated areas of the cerebral cortex

Left thalamus

Components of the Thalamus and Their Functions	
Component	**Function**
Anterior group	Part of the limbic system (Module 13.8)
Medial group	Integrates sensory information for projection to the frontal lobes of the cerebral hemispheres
Ventral group	Projects sensory information to the primary sensory cortex; relays information from the cerebellum and basal nuclei to the motor area of the cerebral cortex
Posterior group	
Pulvinar nuclei	Integrate sensory information for projection to association areas of the cerebral cortex
Lateral geniculate nuclei	Project visual information to the visual cortex
Medial geniculate nuclei	Project auditory information to the auditory cortex
Lateral group	Integrates sensory information and influences emotional states

Hypothalamus

4 The **hypothalamus** contains important control and integrative centers that are shown in this sagittal section. Hypothalamic centers may be stimulated by (1) sensory information from the cerebrum, brain stem, and spinal cord; (2) changes in the composition of the CSF and interstitial fluid; or (3) chemicals in the circulating blood that rapidly enter the hypothalamus because this region lacks a blood–brain barrier.

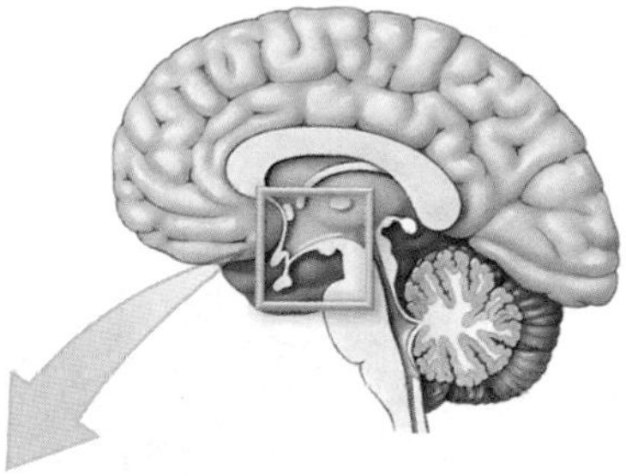

Hypothalamic Nuclei

Autonomic centers control the cardiovascular and vasomotor centers of the medulla oblongata.

The **preoptic area** is the body's thermostat. It regulates body temperature by coordinated adjustments in blood flow and sweat gland activity.

The **suprachiasmatic nucleus** coordinates day–night cycles of activity/inactivity.

Hormonal centers secrete chemical messengers that control endocrine cells of the anterior pituitary gland and secrete two hormones into the circulation at the posterior pituitary gland.

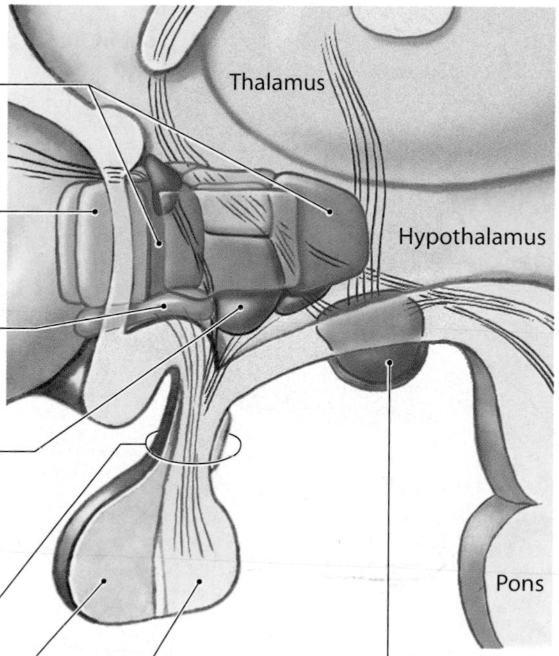
Thalamus
Hypothalamus
Pons
Anterior pituitary gland
Posterior pituitary gland

A narrow stalk called the **infundibulum** (in-fun-DIB-ū-lum; *infundibulum*, funnel) extends inferiorly, connecting the floor of the hypothalamus to the pituitary gland.

The **mammillary bodies** control feeding reflexes such as licking and swallowing.

Module 13.7 Review

a. Name the main components of the diencephalon.

b. Damage to the lateral geniculate nuclei of the thalamus would interfere with what particular function?

c. Which component of the diencephalon is stimulated by changes in body temperature?

LO **13.7** List the main components of the diencephalon, and specify the functions of each.

The limbic system is a functional group of tracts and nuclei located in the cerebrum and diencephalon

The **limbic system** (*limbus*, border) includes nuclei and tracts along the border between the cerebrum and diencephalon. This system is a functional grouping rather than an anatomical one. Functions of the limbic system include (1) establishing emotional states; (2) linking the conscious, intellectual functions of the cerebral cortex with the unconscious and autonomic functions of the brain stem; (3) facilitating memory storage and retrieval; and (4) affecting motivation.

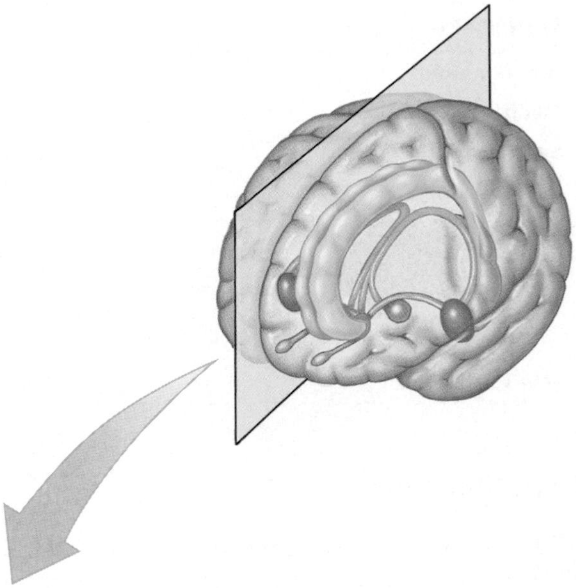

1 This diagrammatic sagittal section shows the position and orientation of the major components of the limbic system.

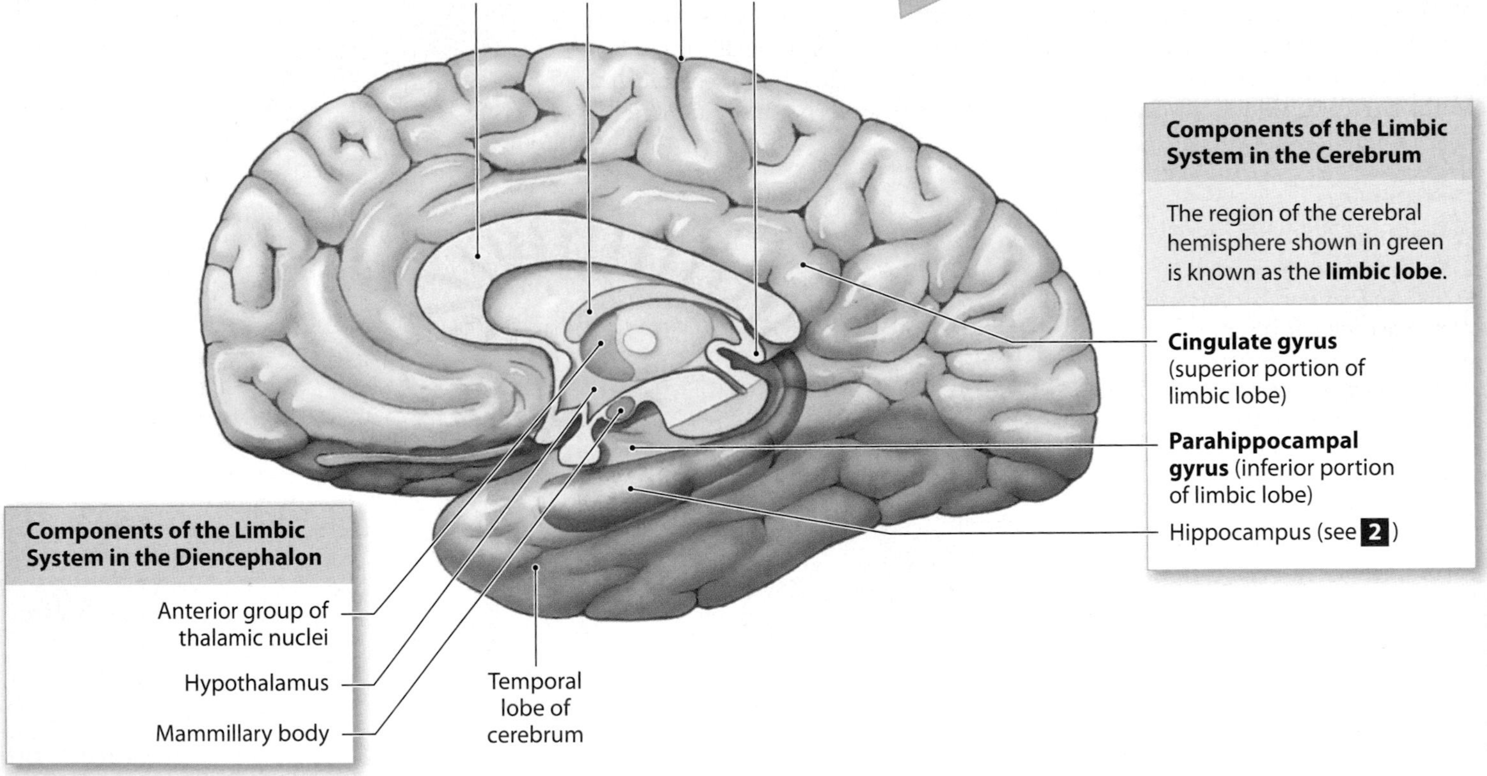
Central sulcus
Corpus callosum
Fornix
Pineal gland
Components of the Limbic System in the Cerebrum
The region of the cerebral hemisphere shown in green is known as the **limbic lobe**.
Cingulate gyrus (superior portion of limbic lobe)
Parahippocampal gyrus (inferior portion of limbic lobe)
Hippocampus (see 2)
Components of the Limbic System in the Diencephalon
Anterior group of thalamic nuclei
Hypothalamus
Mammillary body
Temporal lobe of cerebrum

2 The specific functions of important limbic system components and nuclei are indicated in this sectional view.

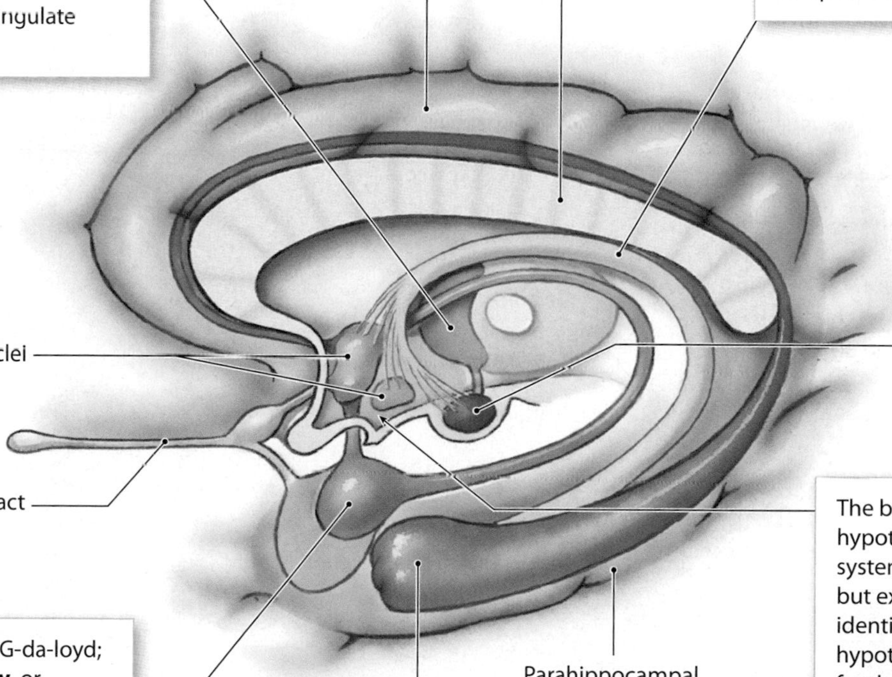

Cingulate gyrus
Corpus callosum
Hypothalamic nuclei
Mammillary body
Olfactory tract
Parahippocampal gyrus

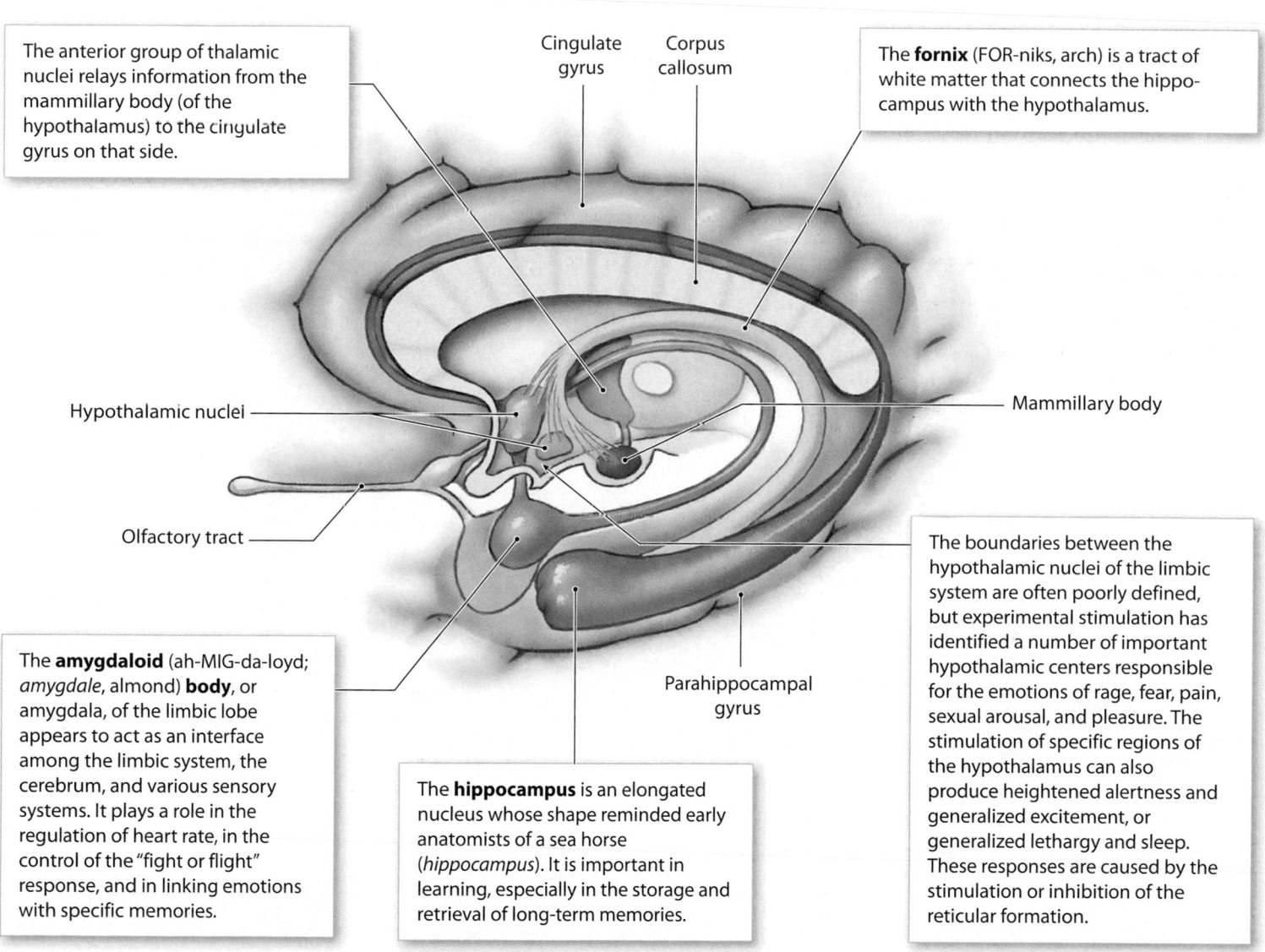

The anterior group of thalamic nuclei relays information from the mammillary body (of the hypothalamus) to the cingulate gyrus on that side.

The **fornix** (FOR-niks, arch) is a tract of white matter that connects the hippocampus with the hypothalamus.

The **amygdaloid** (ah-MIG-da-loyd; *amygdale*, almond) **body**, or amygdala, of the limbic lobe appears to act as an interface among the limbic system, the cerebrum, and various sensory systems. It plays a role in the regulation of heart rate, in the control of the "fight or flight" response, and in linking emotions with specific memories.

The **hippocampus** is an elongated nucleus whose shape reminded early anatomists of a sea horse (*hippocampus*). It is important in learning, especially in the storage and retrieval of long-term memories.

The boundaries between the hypothalamic nuclei of the limbic system are often poorly defined, but experimental stimulation has identified a number of important hypothalamic centers responsible for the emotions of rage, fear, pain, sexual arousal, and pleasure. The stimulation of specific regions of the hypothalamus can also produce heightened alertness and generalized excitement, or generalized lethargy and sleep. These responses are caused by the stimulation or inhibition of the reticular formation.

Whereas the sensory cortex, motor cortex, and association areas of the cerebral cortex enable you to perform complex tasks, it is largely the limbic system that makes you want to do them. For this reason, the limbic system is also known as the motivational system.

Module 13.8 Review

a. List the primary functions of the limbic system.

b. Which region of the limbic system is particularly important for the storage and retrieval of long-term memories?

c. What are some functions of the amygdaloid body?

LO **13.8** Identify the main components of the limbic system, and specify the locations and functions of each.

Module 13.9

The basal nuclei of the cerebrum adjust and refine ongoing voluntary movements

The **basal nuclei** are masses of gray matter that lie within each cerebral hemisphere deep to the floor of the lateral ventricle. The basal nuclei provide subconscious control of skeletal muscle tone and help coordinate learned movement patterns. Under normal conditions, these nuclei do not initiate particular movements. But once a movement is under way, the basal nuclei provide the general pattern and rhythm, especially for movements of the trunk and proximal limb muscles.

1 The basal nuclei consist of the **caudate nucleus** and the **lentiform** (lens-shaped) **nucleus**. The lentiform nucleus is subdivided into a medial **globus pallidus** (GLŌ-bus PAL-i-dus; pale globe) and a lateral **putamen** (pū-TĀ-men). The axon bundles that link the cerebral cortex to the diencephalon and brain stem pass between and around the basal nuclei. Together these fibers form the **internal capsule** of the cerebrum.

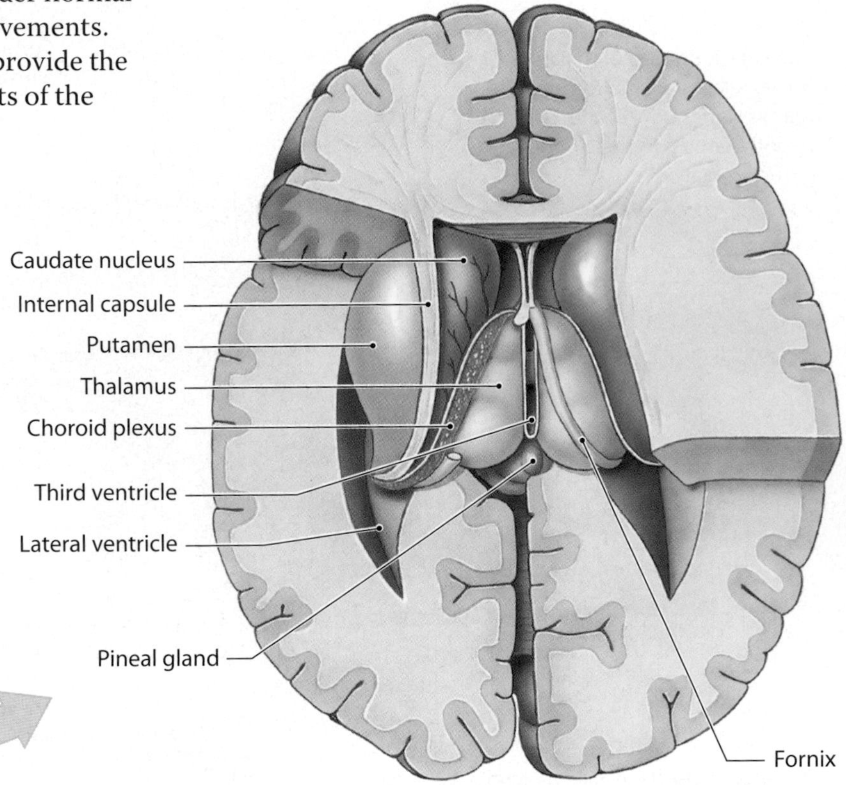

Caudate nucleus
Internal capsule
Putamen
Thalamus
Choroid plexus
Third ventricle
Lateral ventricle
Pineal gland
Fornix

Superior view of two transverse sections at different levels, dissected

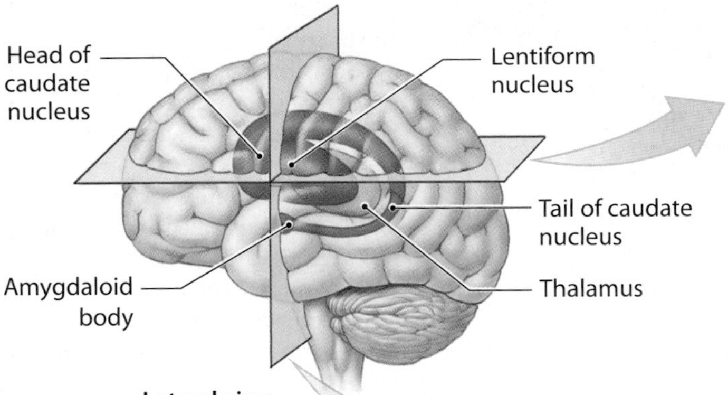

Head of caudate nucleus
Lentiform nucleus
Tail of caudate nucleus
Amygdaloid body
Thalamus

Lateral view

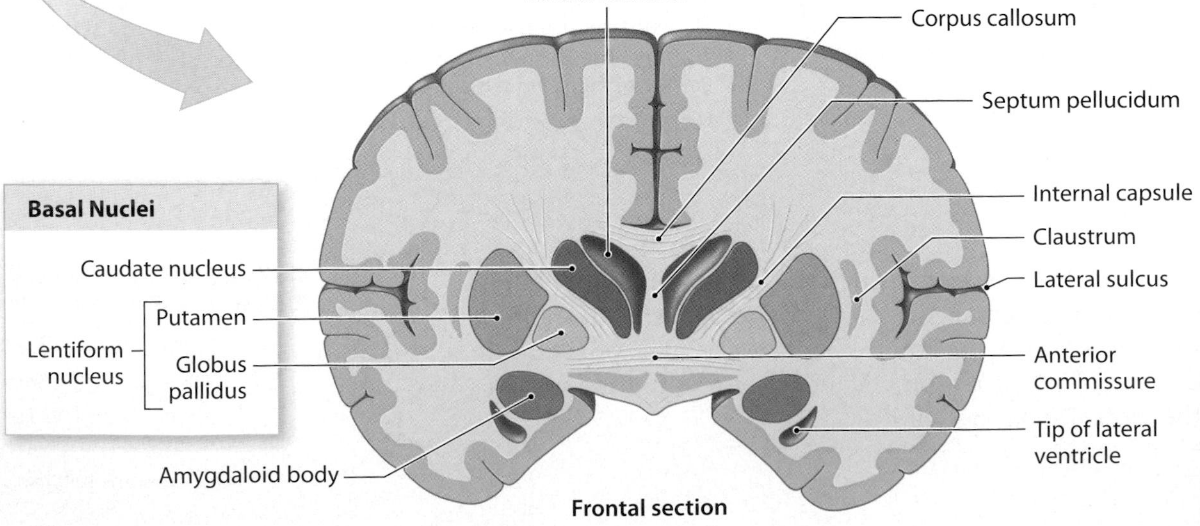

Basal Nuclei
Caudate nucleus
Lentiform nucleus
Putamen
Globus pallidus
Amygdaloid body
Lateral ventricle
Corpus callosum
Septum pellucidum
Internal capsule
Claustrum
Lateral sulcus
Anterior commissure
Tip of lateral ventricle

Frontal section

2 This diagram indicates the roles of the basal nuclei in modifying ongoing movements.

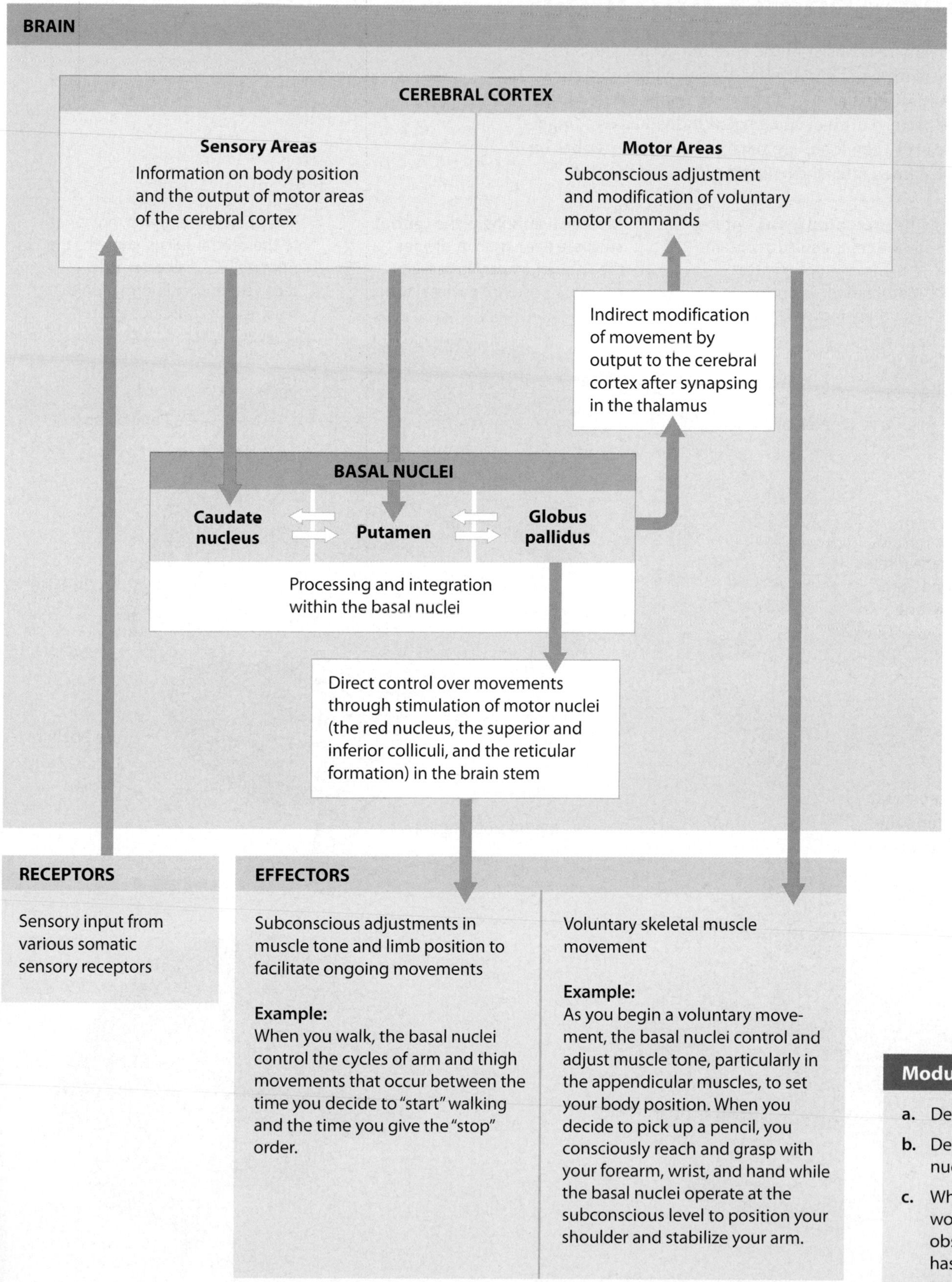
BRAIN
CEREBRAL CORTEX
Sensory Areas
Information on body position and the output of motor areas of the cerebral cortex
Motor Areas
Subconscious adjustment and modification of voluntary motor commands
Indirect modification of movement by output to the cerebral cortex after synapsing in the thalamus
BASAL NUCLEI
Caudate nucleus
Putamen
Globus pallidus
Processing and integration within the basal nuclei
Direct control over movements through stimulation of motor nuclei (the red nucleus, the superior and inferior colliculi, and the reticular formation) in the brain stem
RECEPTORS
Sensory input from various somatic sensory receptors
EFFECTORS
Subconscious adjustments in muscle tone and limb position to facilitate ongoing movements
Example:
When you walk, the basal nuclei control the cycles of arm and thigh movements that occur between the time you decide to "start" walking and the time you give the "stop" order.
Voluntary skeletal muscle movement
Example:
As you begin a voluntary movement, the basal nuclei control and adjust muscle tone, particularly in the appendicular muscles, to set your body position. When you decide to pick up a pencil, you consciously reach and grasp with your forearm, wrist, and hand while the basal nuclei operate at the subconscious level to position your shoulder and stabilize your arm.

Module 13.9 Review

a. Define the basal nuclei.

b. Describe the caudate nucleus.

c. What clinical signs would you expect to observe in a person who has damage to the basal nuclei?

LO 13.9 Describe the structure and function of the basal nuclei of the cerebrum.

Superficial landmarks divide the cerebral hemispheres into lobes

1 The cerebral hemispheres consist of the cerebral cortex and its associated fiber system together with the deeper-lying subcortical nuclei. Each cerebral hemisphere can be divided into regions called **lobes**. Your brain has a unique pattern of sulci and gyri, as individual as a fingerprint, but the boundaries between lobes are reliable landmarks. Lobes on the external surfaces are named after the overlying bones of the skull.

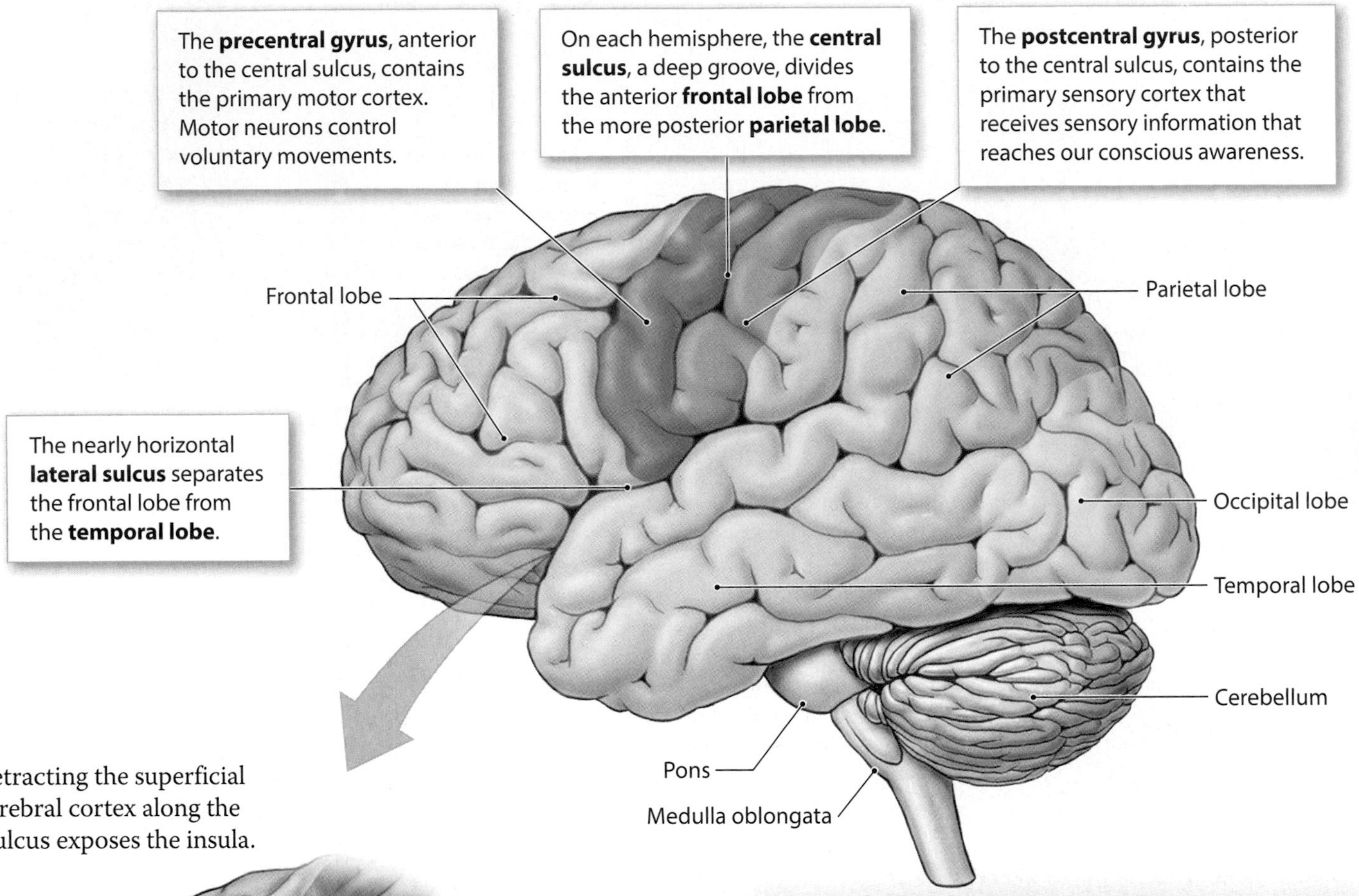
The **precentral gyrus**, anterior to the central sulcus, contains the primary motor cortex. Motor neurons control voluntary movements.
On each hemisphere, the **central sulcus**, a deep groove, divides the anterior **frontal lobe** from the more posterior **parietal lobe**.
The **postcentral gyrus**, posterior to the central sulcus, contains the primary sensory cortex that receives sensory information that reaches our conscious awareness.
Frontal lobe
Parietal lobe
The nearly horizontal **lateral sulcus** separates the frontal lobe from the **temporal lobe**.
Occipital lobe
Temporal lobe
Cerebellum
Pons
Medulla oblongata

2 Retracting the superficial cerebral cortex along the lateral sulcus exposes the insula.

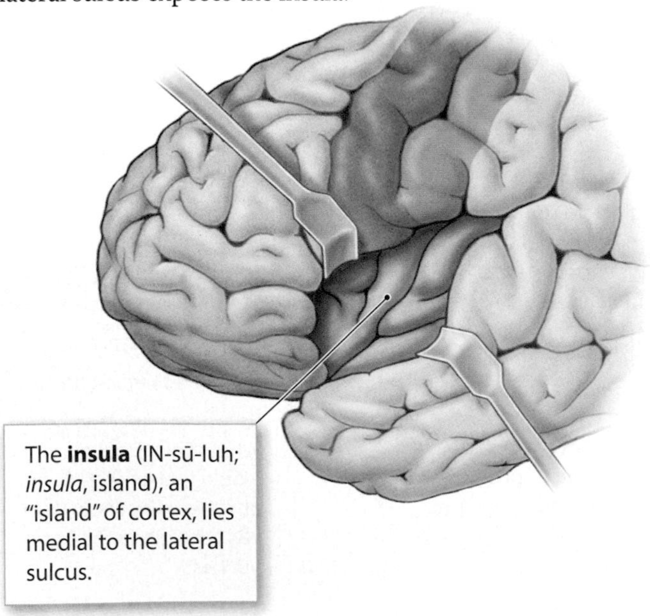
The **insula** (IN-sū-luh; *insula*, island), an "island" of cortex, lies medial to the lateral sulcus.

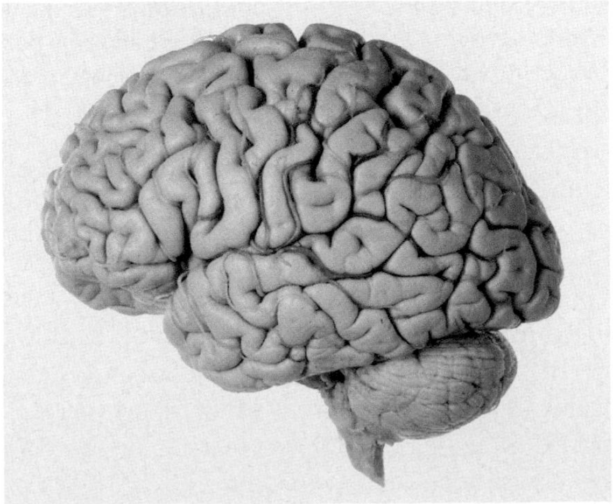

Lateral view, cadaver brain

3 This midsagittal view indicates the inner boundaries of the lobes and highlights the way the cerebral hemispheres cover the rest of the brain. For clarity, structures and regions outside of the cerebrum are labeled with gray lettering.

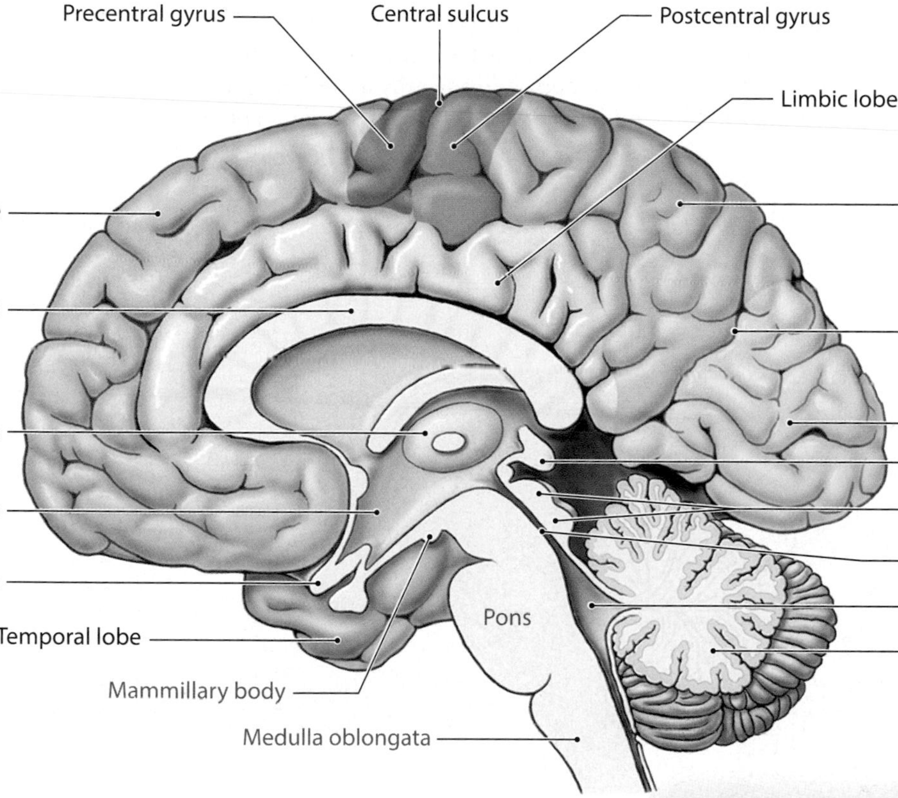

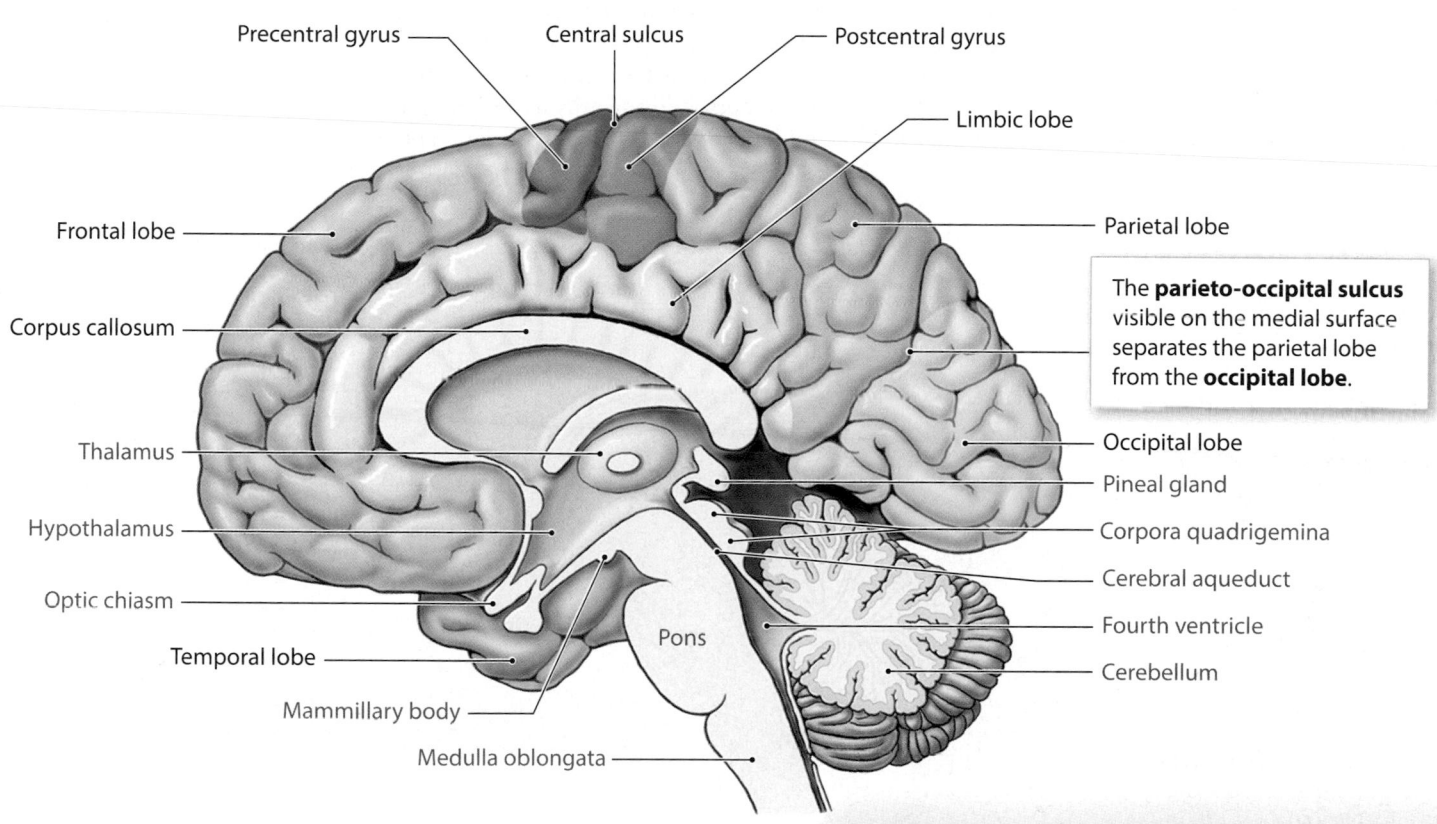

Precentral gyrus
Central sulcus
Postcentral gyrus
Limbic lobe
Frontal lobe
Parietal lobe
Corpus callosum
The **parieto-occipital sulcus** visible on the medial surface separates the parietal lobe from the **occipital lobe**.
Thalamus
Occipital lobe
Pineal gland
Hypothalamus
Corpora quadrigemina
Cerebral aqueduct
Optic chiasm
Fourth ventricle
Pons
Temporal lobe
Cerebellum
Mammillary body
Medulla oblongata

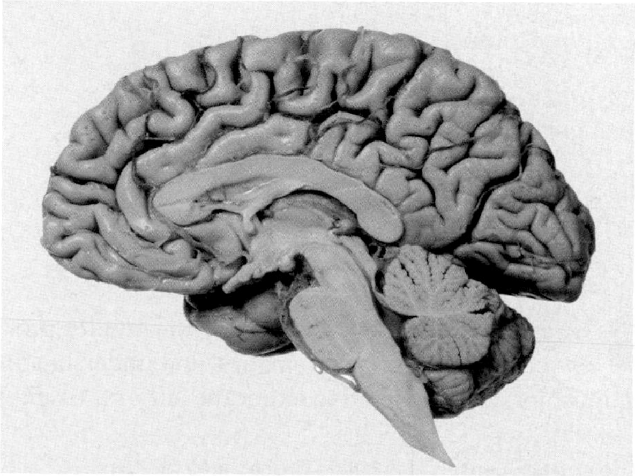

Midsagittal section, cadaver brain

It's important to remember the following general facts about the cerebral hemispheres:

- Each cerebral hemisphere receives sensory information from, and sends motor commands to, the opposite side of the body. Thus the motor areas of the left cerebral hemisphere control muscles on the right side, and motor areas of the right cerebral hemisphere control muscles on the left side. This crossing over, which occurs in the brain stem and spinal cord, has no known functional significance.
- Even though the two hemispheres may look identical and have many similar functions, important differences exist.
- The correspondence between a specific function and a specific region of the cerebral hemisphere is imprecise. The boundaries are not distinct and have considerable overlap, and some cortical functions, such as consciousness, cannot easily be assigned to any single region. However, we know that normal individuals use all portions of the brain.

Module 13.10 Review

a. Identify the lobes of the cerebrum, and indicate the basis for their names.

b. Describe the insula.

c. What effect would damage to the left postcentral gyrus produce?

LO **13.10** Identify the major superficial landmarks of the cerebrum, and cite the location of each.

Module 13.11

The lobes of the cerebral cortex have regions with specific functions

1 The cerebral cortex is divided into six functional categories. The **primary motor cortex** issues voluntary commands to skeletal muscles, and the **primary sensory cortex** receives general somatic sensory information. The special senses of sight, sound, smell, and taste reach other portions of the cerebral cortex. Each sensory and motor region of the cortex is connected to a nearby **association area**. Association areas are regions of the cortex that interpret incoming data or coordinate a motor response.

Motor Cortex

Neurons of the primary motor cortex are called **pyramidal cells**, because their cell bodies resemble little pyramids.

The **somatic motor association area** is responsible for the coordination of learned movements.

Gustatory Cortex

The **gustatory cortex** of the insula receives information from taste receptors.

Olfactory Cortex

The **olfactory cortex** receives sensory information from the olfactory receptors.

Auditory Cortex

The **primary auditory cortex** is responsible for monitoring auditory (sound) information.

The **auditory association area** monitors sensory activity in the auditory cortex and recognizes sounds, such as spoken words.

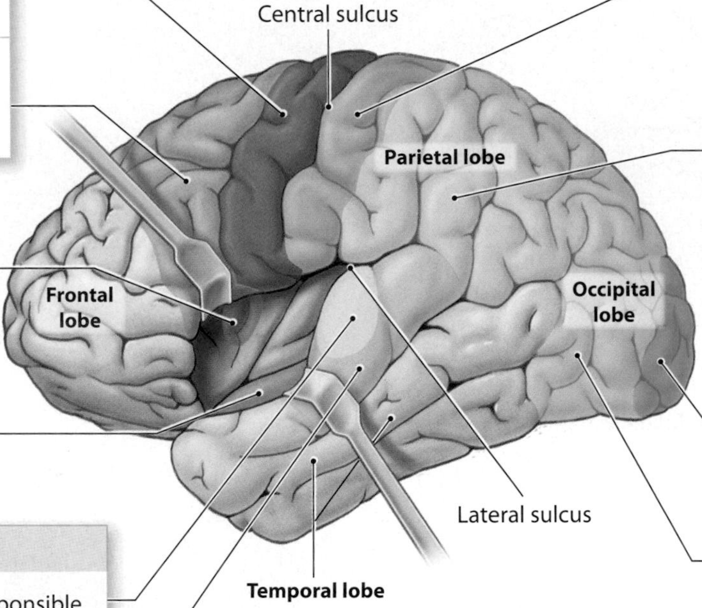

Central sulcus
Parietal lobe
Frontal lobe
Occipital lobe
Lateral sulcus
Temporal lobe

Sensory Cortex

Neurons in the primary sensory cortex receive somatic sensory information from receptors for touch, pressure, pain, vibration, or temperature.

The **somatic sensory association area** monitors activity in the primary sensory cortex. It allows you to recognize a light touch, such as a mosquito landing on your arm.

Visual Cortex

The **primary visual cortex** receives information from the lateral geniculate nuclei.

The **visual association area** monitors the patterns of activity in the visual cortex and interprets the results. When you see the symbols c, a, and r, your visual association area recognizes that they form the word "car."

2 **Integrative centers** concerned with complex processes, such as speech, writing, mathematics, and understanding spatial relationships, are restricted to either the left or the right hemisphere.

The **speech center,** also called Broca's area or the motor speech area, lies in the same hemisphere as the general interpretive area. The speech center regulates the patterns of breathing and vocalization needed for normal speech.

The **prefrontal cortex** coordinates information relayed from the association areas of the cortex. In the process, it performs abstract intellectual functions such as predicting the consequences of events or actions.

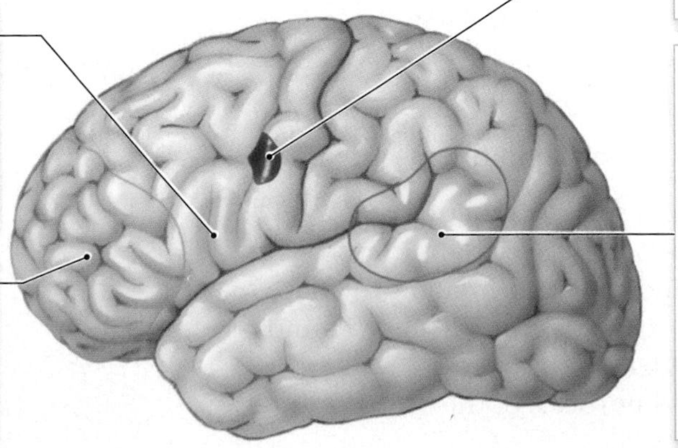

The **frontal eye field** controls learned eye movements, such as when you scan these lines of text.

The **general interpretive area**, or Wernicke's area, receives information from all the sensory association areas. This analytical center is present in only one hemisphere (typically the left). This region plays an essential role in your personality by integrating sensory information and coordinating access to complex visual and auditory memories.

3 Each of the two cerebral hemispheres is responsible for specific functions that are not ordinarily performed by the opposite hemisphere. This regional specialization is called **hemispheric lateralization**.

Left Cerebral Hemisphere

In most people, the left hemisphere contains the general interpretive and speech centers and is responsible for language-based skills. Reading, writing, and speaking, for example, depend on processing done in the left cerebral hemisphere. In addition, the premotor cortex that controls hand movements is larger on the left side for right-handed people than for left-handed people. The left hemisphere is also important for analytical tasks, such as mathematics and logic.

Right Cerebral Hemisphere

The right cerebral hemisphere analyzes sensory information and relates the body to the sensory environment. Interpretive centers in this hemisphere enable you to identify familiar objects by touch, smell, sight, taste, or feel. For example, the right hemisphere plays a dominant role in recognizing faces and in understanding three-dimensional relationships. It is also important in analyzing the emotional context of a conversation—for instance, distinguishing between the threat "Get lost!" and the question "Get lost?"

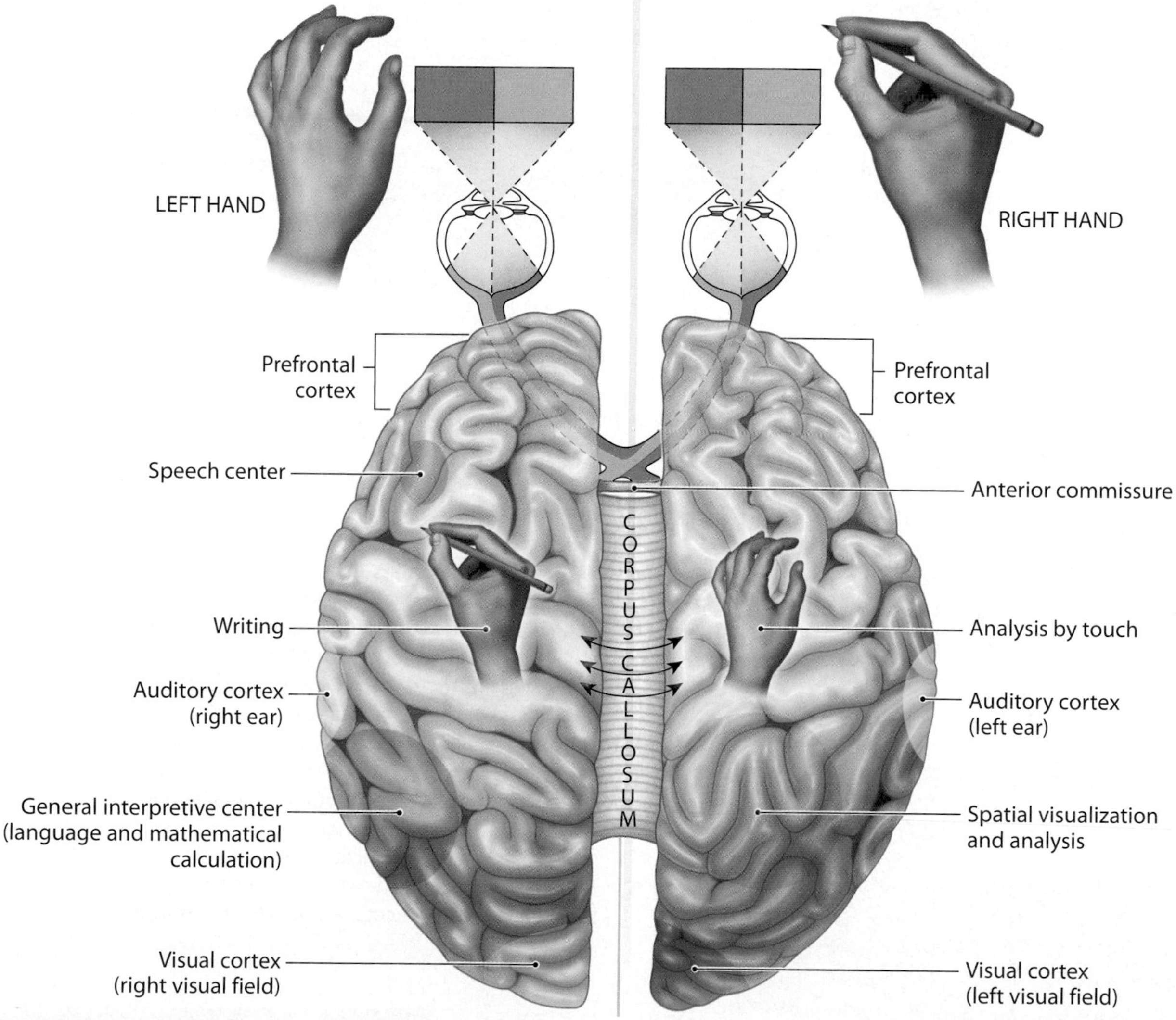

LEFT HAND
RIGHT HAND
Prefrontal cortex
Prefrontal cortex
Speech center
Anterior commissure
CORPUS CALLOSUM
Writing
Analysis by touch
Auditory cortex (right ear)
Auditory cortex (left ear)
General interpretive center (language and mathematical calculation)
Spatial visualization and analysis
Visual cortex (right visual field)
Visual cortex (left visual field)

Left-handed people represent about 9 percent of the human population. Although in most cases the primary motor cortex of the right hemisphere controls motor function for the dominant left hand, the centers involved with speech and analytical function are in the left hemisphere. Interestingly, an unusually high percentage of musicians and artists are left-handed, and the primary motor cortex and association areas on the right cerebral hemisphere are near the association areas involved with spatial visualization and emotions.

Module 13.11 Review

a. Where is the primary motor cortex located?

b. Which senses are affected by damage to the temporal lobes?

c. Which brain region has been affected in a stroke victim who is unable to speak?

LO **13.11** Identify the locations of the motor, sensory, and association areas of the cerebral cortex, and discuss the functions of each.

Module 13.12

White matter connects the cerebral hemispheres and the lobes of each hemisphere, and links the cerebrum to the rest of the brain

The interior of the cerebral hemispheres consists primarily of white matter (myelinated axons or fibers) organized into groups that share common functions.

1 **Association fibers** interconnect areas of neural cortex within a single cerebral hemisphere.

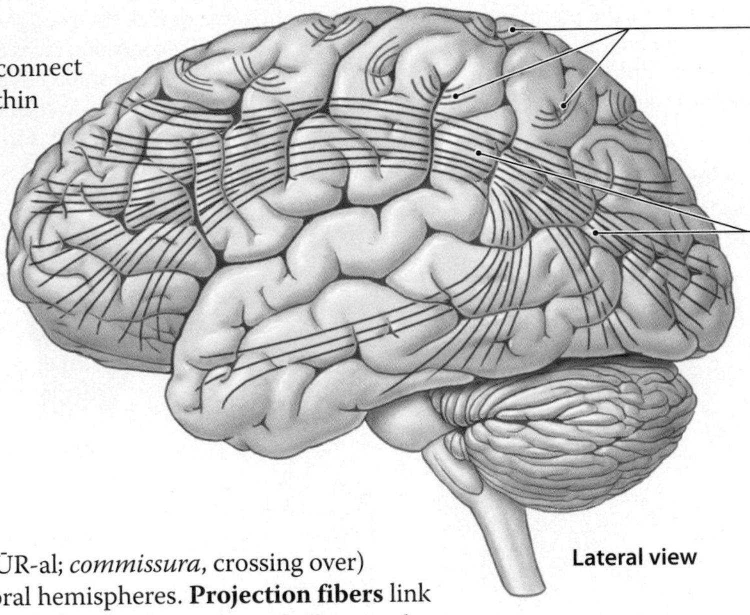

The shortest association fibers are called **arcuate** (AR-kū-āt) **fibers**, because they curve in an arc to pass from one gyrus to another.
Longer association fibers are organized into discrete bundles, or fasciculi. The **longitudinal fasciculi** connect the frontal lobe to the other lobes of the same cerebral hemisphere.

Lateral view

2 **Commissural** (kom-i-SŪR-al; *commissura*, crossing over) **fibers** connect the cerebral hemispheres. **Projection fibers** link the cerebral cortex to the diencephalon, brain stem, cerebellum, and spinal cord. All projection fibers must pass through the diencephalon, where axons heading to sensory areas of the cerebral cortex pass among the axons descending from motor areas of the cortex. In gross dissection, the ascending fibers and descending fibers look alike, and the entire mass is known as the **internal capsule**.

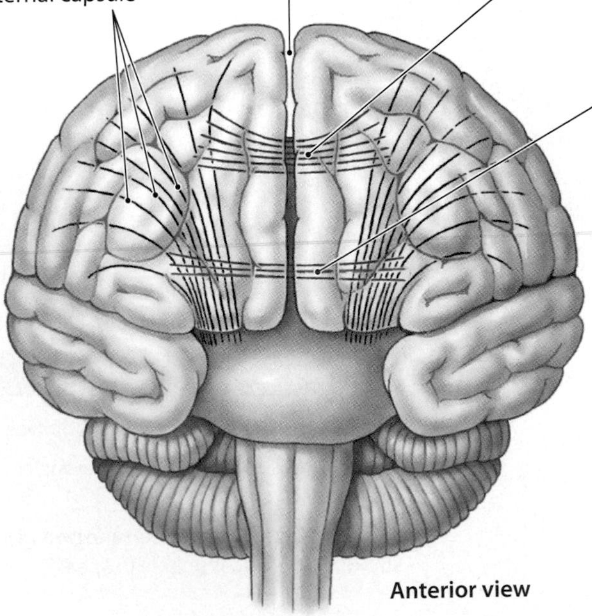

Projection fibers of internal capsule
Longitudinal fissure
The **corpus callosum** is the most important band of commissural fibers because it allows communication and coordination between the left and right cerebral hemispheres. It contains more than 200 million axons carrying some 4 billion impulses per second.
The **anterior commissure** is a smaller tract of commissural fibers that provides another route for communication between the cerebral hemispheres. Its importance increases if the corpus callosum is damaged.

Anterior view

Module 13.12 Review

a. What special names are given to axons in the white matter of the cerebral hemispheres?

b. What is the function of the longitudinal fasciculi?

c. What are fibers carrying information between the brain and spinal cord called, and through which brain regions do they pass?

LO 13.12 Discuss the significance of the white matter of the cerebral hemispheres.

Module 13.13 CLINICAL MODULE

Brain activity can be monitored using external electrodes; the record is called an electroencephalogram, or EEG

1 Neural function depends on electrical impulses, and the brain contains billions of neurons and the axons of the central white matter. The activity under way at any given moment generates an electrical field that can be measured by placing electrodes on the scalp. The electrical activity changes constantly, as nuclei and cortical areas are stimulated or quieted down. A printed report of the electrical activity of the brain is called an **electroencephalogram (EEG)**. The electrical patterns observed are called **brain waves**.

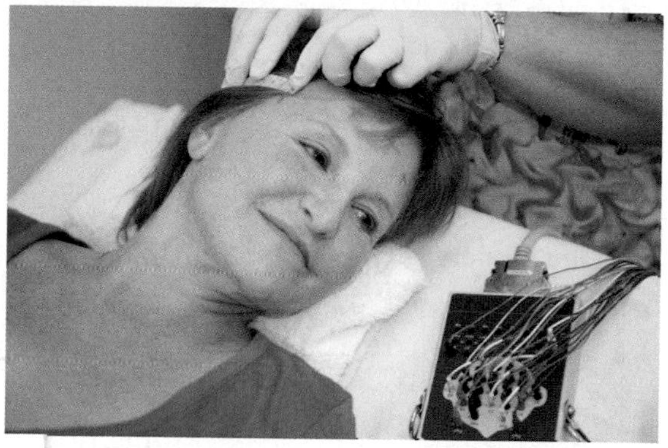

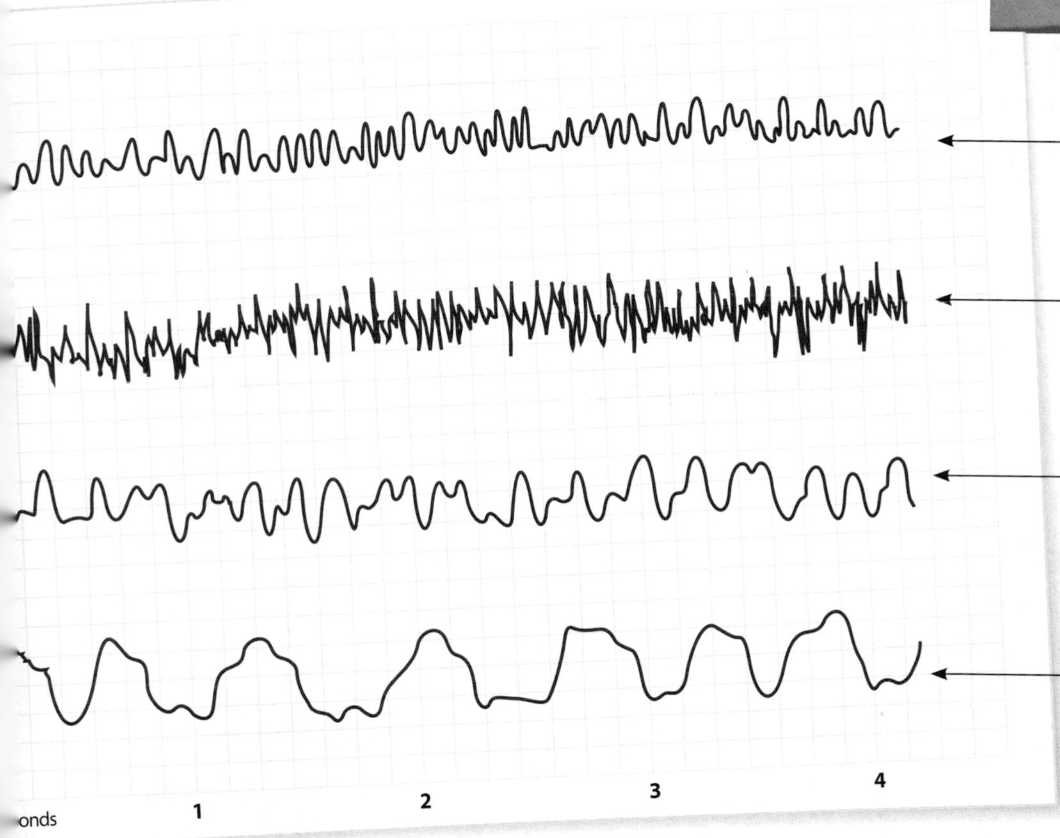
onds
1
2
3
4

Alpha waves occur in the brains of healthy, awake adults who are resting with their eyes closed. Alpha waves disappear during sleep, but they also vanish when the individual begins to concentrate on some specific task.

Beta waves are higher-frequency waves that appear in people who are either concentrating on a task, under stress, or in a state of psychological tension.

Theta waves may appear transiently during sleep in normal adults but are most often observed in children and in intensely frustrated adults. The presence of theta waves under other circumstances may indicate the presence of a brain disorder, such as a tumor.

Delta waves are large-amplitude, low-frequency waves. They are normally seen during deep sleep in people of all ages. Delta waves are also seen in the brains of infants (in whom cortical development is still incomplete) and in awake adults when a tumor, vascular blockage, or inflammation has damaged portions of the brain.

Electrical activity in the two hemispheres is generally synchronized by a "pacemaker" mechanism that seems to involve the thalamus. Asynchrony between the hemispheres can therefore indicate localized damage or other cerebral abnormalities. A tumor or injury affecting one hemisphere, for example, typically changes the pattern in that hemisphere, and the patterns of the two hemispheres are no longer aligned. A **seizure** is a temporary cerebral disorder accompanied by abnormal movements, unusual sensations, inappropriate behavior, or some combination of these signs and symptoms. Clinical conditions characterized by seizures are known as seizure disorders, or **epilepsies**. Seizures of all kinds are accompanied by a marked change in the pattern of electrical activity recorded in an EEG. The change begins in one portion of the cerebral cortex but may spread across the entire cortical surface, like a wave on the surface of a pond.

Module 13.13 Review

a. Define electroencephalogram (EEG).

b. Name and describe the four wave types associated with an EEG.

c. Differentiate between a seizure and epilepsy.

LO **13.13** Discuss the origin and significance of the major categories of brain waves seen in an electroencephalogram (EEG).

The twelve pairs of cranial nerves are classified as sensory, special sensory, motor, or mixed nerves

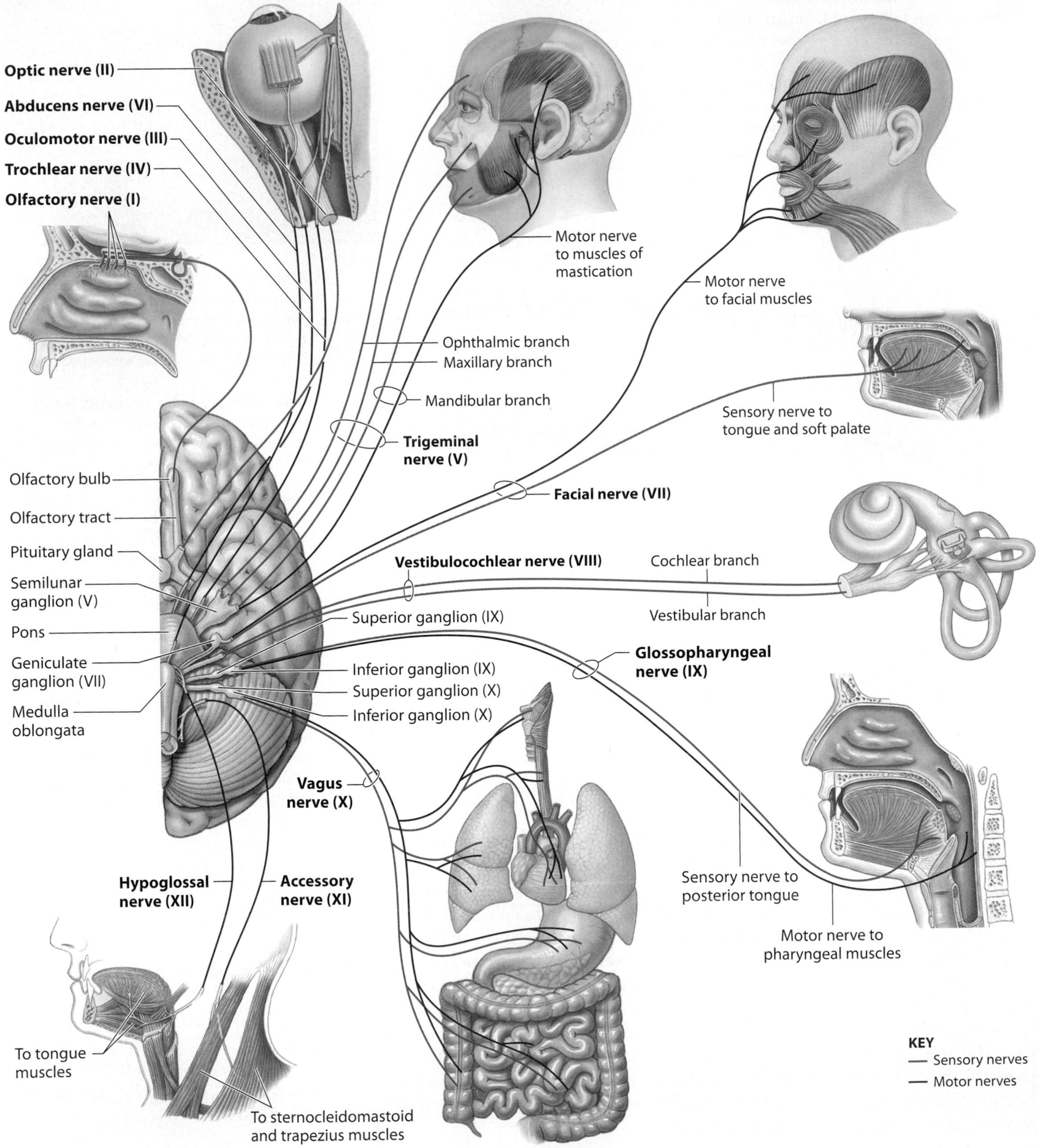
Optic nerve (II)
Abducens nerve (VI)
Oculomotor nerve (III)
Trochlear nerve (IV)
Olfactory nerve (I)
Motor nerve to muscles of mastication
Motor nerve to facial muscles
Ophthalmic branch
Maxillary branch
Mandibular branch
Trigeminal nerve (V)
Sensory nerve to tongue and soft palate
Facial nerve (VII)
Olfactory bulb
Olfactory tract
Pituitary gland
Semilunar ganglion (V)
Pons
Geniculate ganglion (VII)
Medulla oblongata
Vestibulocochlear nerve (VIII)
Cochlear branch
Vestibular branch
Superior ganglion (IX)
Glossopharyngeal nerve (IX)
Inferior ganglion (IX)
Superior ganglion (X)
Inferior ganglion (X)
Vagus nerve (X)
Hypoglossal nerve (XII)
Accessory nerve (XI)
Sensory nerve to posterior tongue
Motor nerve to pharyngeal muscles
To tongue muscles
To sternocleidomastoid and trapezius muscles
KEY
Sensory nerves
Motor nerves

Branches and Functions of Cranial Nerves

Cranial Nerve (Number)	Sensory Ganglion	Branch	Primary Function	Foramen	Innervation
Olfactory (I)			Special sensory	Olfactory foramina of ethmoid	Olfactory epithelium
Optic (II)			Special sensory	Optic canal	Retina of eye
Oculomotor (III)			Motor	Superior orbital fissure	Inferior, medial, superior rectus, inferior oblique, and levator palpebrae superioris muscles; intrinsic eye muscles
Trochlear (IV)			Motor	Superior orbital fissure	Superior oblique muscle
Trigeminal (V)	Semilunar		Mixed	Superior orbital fissure	Areas associated with the jaws
		Ophthalmic	Sensory	Superior orbital fissure	Orbital structures, nasal cavity, skin of forehead, upper eyelid, eyebrows, and part of nose
		Maxillary	Sensory	Foramen rotundum	Lower eyelid; superior lip, gums, and teeth; cheek, part of nose, palate, and part of pharynx
		Mandibular	Mixed	Foramen ovale	*Sensory:* inferior gums, teeth, lips, part of palate, and part of tongue *Motor:* muscles of mastication
Abducens (VI)			Motor	Superior orbital fissure	Lateral rectus muscle
Facial (VII)	Geniculate		Mixed	Internal acoustic meatus to facial canal; exits at stylomastoid foramen	*Sensory:* taste receptors on anterior two-thirds of tongue *Motor:* muscles of facial expression, lacrimal gland, submandibular gland, and sublingual salivary glands
Vestibulocochlear (Acoustic) (VIII)		Cochlear Vestibular	Special sensory	Internal acoustic meatus	Cochlea (receptors for hearing) Vestibule (receptors for motion and balance)
Glossopharyngeal (IX)	Superior and inferior		Mixed	Jugular foramen	*Sensory:* posterior third of tongue; pharynx and part of palate; receptors for blood pressure, pH, oxygen, and carbon dioxide concentrations *Motor:* pharyngeal muscles and parotid salivary gland
Vagus (X)	Superior and inferior		Mixed	Jugular foramen	*Sensory:* pharynx; auricle and external acoustic canal; diaphragm; visceral organs in thoracic and abdominopelvic cavities *Motor:* palatal and pharyngeal muscles and visceral organs in thoracic and abdominopelvic cavities
Accessory (XI)		Internal	Motor	Jugular foramen	Skeletal muscles of palate, pharynx, and larynx (with vagus nerve)
		External	Motor	Jugular foramen	Sternocleidomastoid and trapezius muscles
Hypoglossal (XII)			Motor	Hypoglossal canal	Tongue musculature

Two useful mnemonics for remembering the names of the cranial nerves in order are "**O**h **O**h **O**h, **T**o **T**ouch **A**nd **F**eel **V**ery **G**reen **V**egetables, **A**h **H**eaven!" and "**O**h, **O**nce **O**ne **T**akes **T**he **A**natomy **F**inal, **V**ery **G**ood **V**acations **A**re **H**eavenly!".

Module 13.14 Review

a. Identify the cranial nerves by name and number.

b. Which cranial nerves have motor functions only?

c. Which cranial nerves are mixed nerves?

LO **13.14** Identify the cranial nerves by name and number, and cite the functions of each.

Labeling

Label the structures in the accompanying figure of a lateral view of the human brain.

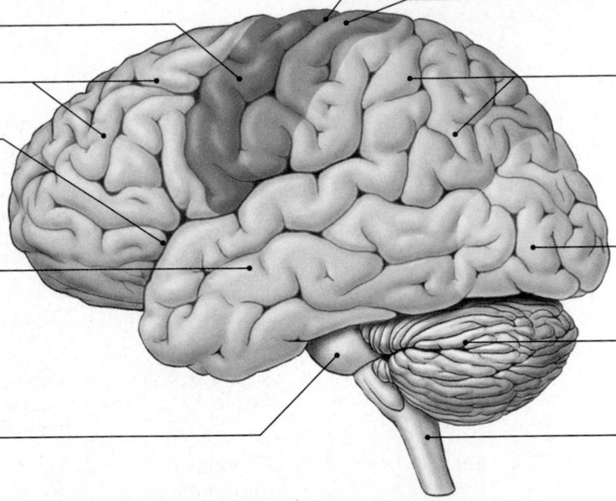

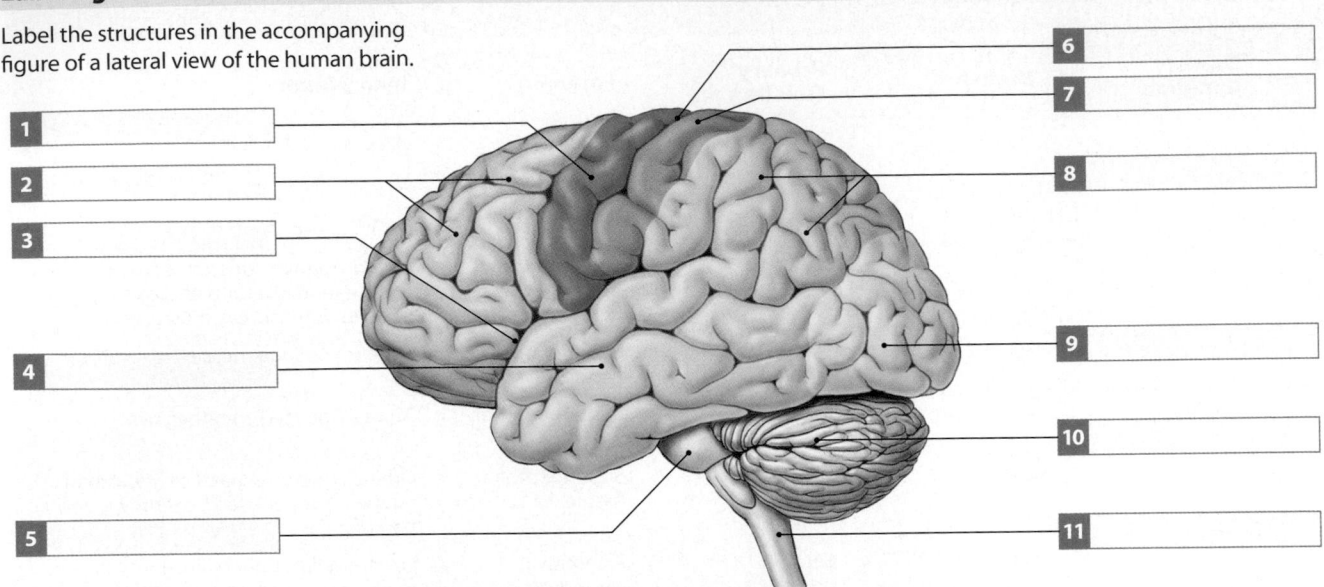

1 ______
2 ______
3 ______
4 ______
5 ______
6 ______
7 ______
8 ______
9 ______
10 ______
11 ______

Identify the cranial nerves in the accompanying figure, and indicate the function of each: M = motor, S = sensory, or B = both motor and sensory.

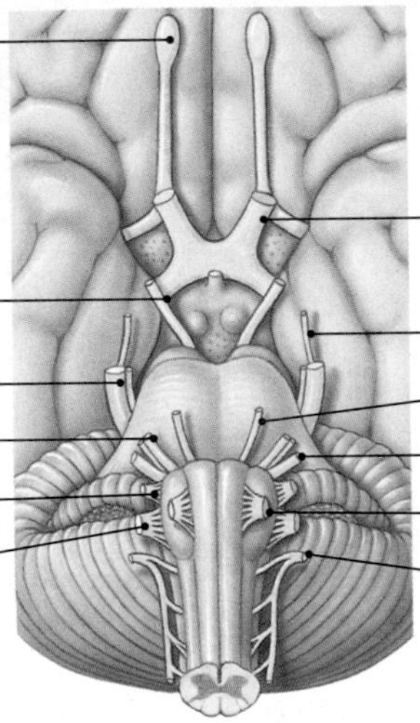

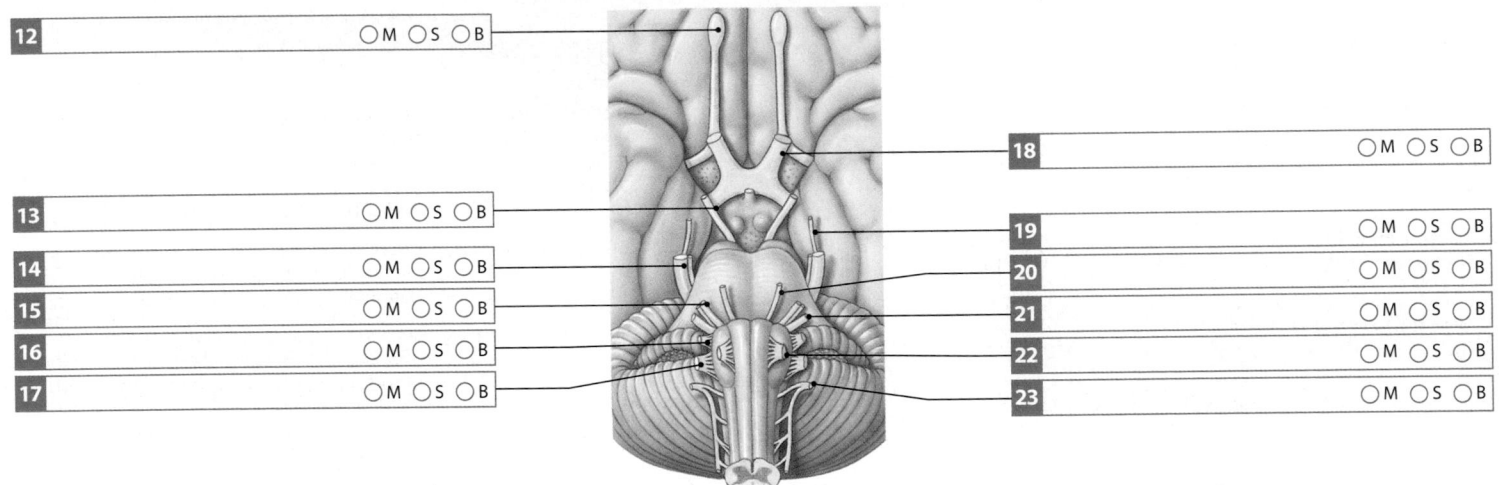

12 ______ ○M ○S ○B
13 ______ ○M ○S ○B
14 ______ ○M ○S ○B
15 ______ ○M ○S ○B
16 ______ ○M ○S ○B
17 ______ ○M ○S ○B
18 ______ ○M ○S ○B
19 ______ ○M ○S ○B
20 ______ ○M ○S ○B
21 ______ ○M ○S ○B
22 ______ ○M ○S ○B
23 ______ ○M ○S ○B

Vocabulary

Write the term for each of the following descriptions in the space provided.

24 Forms the walls of the diencephalon
25 The shortest association fibers in the CNS white matter
26 The tract of white matter that connects the hippocampus with the hypothalamus
27 The fibers that permit communication between the two cerebral hemispheres
28 The nuclei made up of the caudate nucleus and the lentiform nucleus

24 ______
25 ______
26 ______
27 ______
28 ______

Section integration

29 Smelling salts may restore consciousness after a person has fainted. The active ingredient of smelling salts is ammonia, and it acts by irritating the lining of the nasal cavity. Propose a mechanism by which smelling salts would raise a person from the unconscious state to the conscious state.

29 ______

Sensations carried by sensory pathways to the CNS begin with transduction at a sensory receptor

Sensory receptors are specialized cells or cell processes that inform your central nervous system about conditions inside or outside the body. The term **general senses** is used to describe our sensitivity to temperature, pain, touch, pressure, vibration, and proprioception. General sensory receptors are distributed throughout the body, and they are relatively simple in structure. Sensory pathways begin at peripheral receptors and end within the CNS, often at the diencephalon and/or cerebral hemispheres. Much of the information carried by a sensory pathway never reaches the primary sensory cortex and our awareness. The information carried by a sensory pathway is called a **sensation**, and the conscious awareness of a sensation is called a **perception**.

1 The area monitored by a single receptor cell is its **receptive field**. The larger the receptive field, the poorer your ability to localize a stimulus. A touch receptor on the general body surface, for example, may have a receptive field 7 cm (2.5 in.) in diameter. As a result, you can describe a light touch there as affecting only a general area, not an exact spot. On the tongue or fingertips, where the receptive fields are less than a millimeter in diameter, you can be very precise about the location of a stimulus.

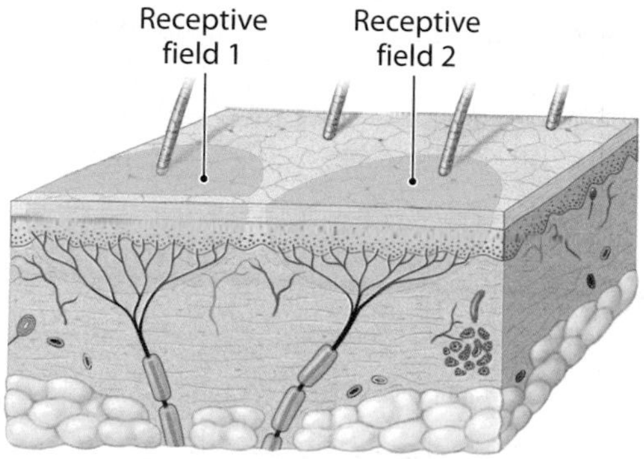
Receptive field 1
Receptive field 2

2 This diagram presents the basic events that occur along sensory and motor pathways.

Arriving stimulus

Sensory Pathway

Depolarization of Receptor

The process begins when a physical or chemical stimulus results in a graded change in the membrane potential of a receptor cell. The conversion of a stimulus to a change in membrane potential is called **transduction**.*

Action Potential Generation

If the stimulus depolarizes the receptor cell to threshold, action potentials develop in the initial segment. The greater the degree of sustained depolarization, the higher the frequency of action potentials.

Propagation over Labeled Line

A **labeled line** consists of axons carrying information about one type of stimulus (touch, pressure, temperature). The CNS interprets the stimulus according to the nature of the axon over which it arrives. That's why you see lights when you bump your eyes—the stimulated receptor sends information over axons that normally carry visual information.

CNS Processing

Information processing occurs at every synapse along the labeled line. The line may branch repeatedly, distributing the sensory information to multiple nuclei and centers in the spinal cord and brain.

Motor Pathway (involuntary)

Immediate Involuntary Response

Processing centers in the spinal cord or brain stem may direct an immediate reflex response even before sensations reach the cerebral cortex.

Motor Pathway (voluntary)

Perception

Only about 1 percent of arriving sensations are relayed to the primary sensory cortex, and often those are sensations that merit a voluntary response.

Voluntary Response

The voluntary response, which isn't immediate, can moderate, enhance, or supplement the relatively simple reflexive response.

*Note: A transducer is a device that converts a physical signal into an electrical signal, or the reverse. For example, a digital scale has a transducer that converts a force into an electrical signal proportional to the applied force.

Module 13.15 Review

a. Define the term general senses.
b. Relate receptive field size to localization.
c. Outline the sensory pathway.

LO 13.15 Describe the basic events that occur along a sensory pathway, and explain the difference between a sensation and a perception.

Module 13.16

Receptors for the general senses are classified by function

1 The simplest receptors are the dendrites of sensory neurons. As shown at the right, the branching tips of these dendrites, called free nerve endings, are not protected by accessory structures. Free nerve endings extend through a tissue the way grass roots extend into the soil. They can be stimulated by many different stimuli and therefore exhibit little receptor specificity. For example, free nerve endings that respond to tissue damage by providing pain sensations may be stimulated by chemical stimulation, pressure, temperature changes, or trauma. The sensitivity and specificity of a free nerve ending may be altered by its location, the presence of accessory structures, or both.

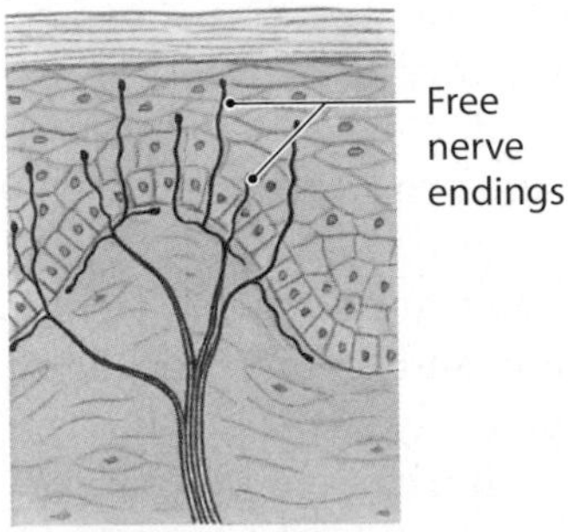
Free nerve endings

2 As shown in this concept map, general sensory receptors can be classified according to the nature of the primary stimulus.

A Functional Classification of General Sensory Receptors

Nociceptors

Nociceptors are pain receptors. They are free nerve endings with large receptive fields and broad sensitivity. Two types of axons—Type A and Type C fibers—carry pain sensations. Type A fibers propagate action potentials faster and respond to weaker intensity stimuli than do C fibers.

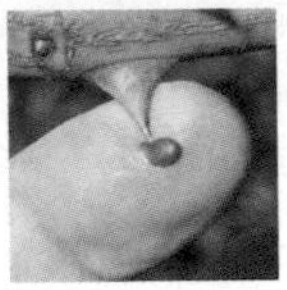

Myelinated **Type A fibers** carry sensations of **fast pain**, or prickling pain. An injection or a deep cut produces this type of pain. These sensations quickly reach the CNS, where they often trigger somatic reflexes. They are also relayed to the primary sensory cortex and so receive conscious attention. In most cases, the arriving information permits the stimulus to be localized to an area several centimeters in diameter.

Slower, unmyelinated **Type C fibers** carry sensations of **slow pain**, or burning and aching pain. These sensations cause a generalized activation of the reticular formation and thalamus. The person becomes aware of the pain but has only a general idea of the area affected.

Thermoreceptors

Thermoreceptors, or temperature receptors, are free nerve endings located in the dermis, in skeletal muscles, in the liver, and in the hypothalamus. Cold receptors are three or four times more numerous than warm receptors. No structural differences between warm and cold thermoreceptors have been identified.

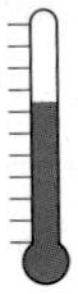

Chemoreceptors

Chemoreceptors respond to water-soluble and lipid-soluble substances that are dissolved in body fluids (interstitial fluid, blood, and CSF).

Mechanoreceptors

Mechanoreceptors are sensitive to stimuli that distort their plasma membranes. These membranes contain mechanically gated ion channels whose gates open or close in response to stretching, compression, twisting, or other distortions of the membrane.

Proprioceptors monitor the positions of joints and muscles. They are the most structurally and functionally complex of the general sensory receptors. One example is the muscle spindle, discussed in Module 12.12.

Baroreceptors (bar-ō-rē-SEP-torz; *baro-*, pressure) detect pressure changes in the walls of blood vessels and in portions of the digestive, respiratory, and urinary tracts.

Tactile receptors provide the sensations of touch, pressure, and vibration. Touch sensations provide information about shape or texture, whereas pressure sensations indicate the degree and frequency of mechanical distortion. Extremely sensitive **fine touch and pressure receptors** provide detailed information about a stimulation, whereas **crude touch and pressure receptors** provide poor localization and give little information.

3 Receptors can also be categorized based on the nature of their response to stimulation.

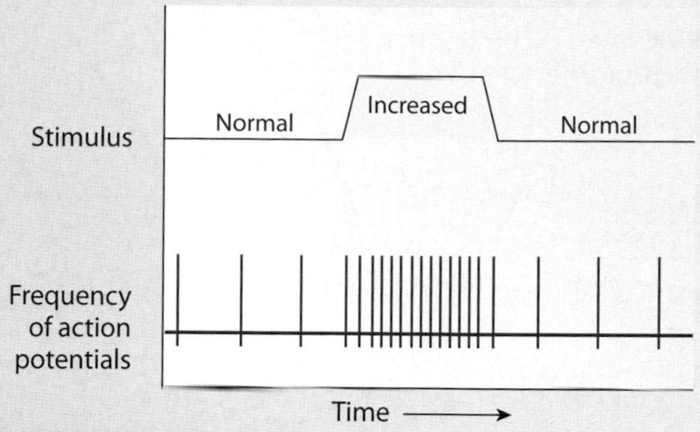
Stimulus
Normal
Increased
Normal
Frequency of action potentials
Time

Tonic receptors are always active and generate action potentials at a frequency that reflects the background level of stimulation. When the stimulus increases or decreases, the rate of action potential generation changes accordingly.

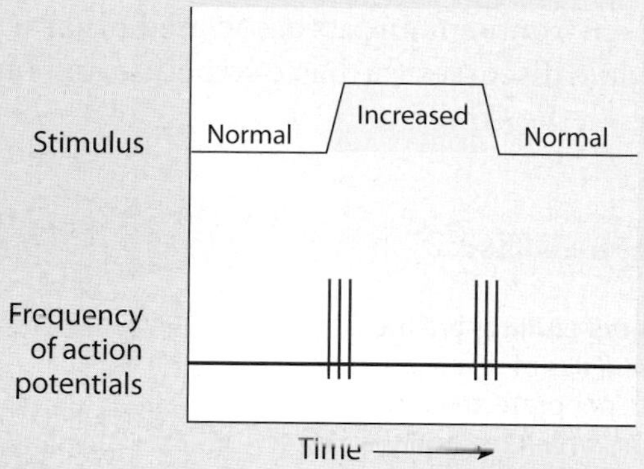
Stimulus
Normal
Increased
Normal
Frequency of action potentials
Time

Phasic receptors are normally inactive, but become active for a short time in response to a change in the conditions they are monitoring.

4 A reduction in sensitivity in the presence of a constant stimulus is called **adaptation**. You seldom notice the rumble of the tires when you ride in a car, or the background noise of the air conditioner, because your nervous system quickly adapts to stimuli that are painless and constant. Adaptation may be peripheral or central.

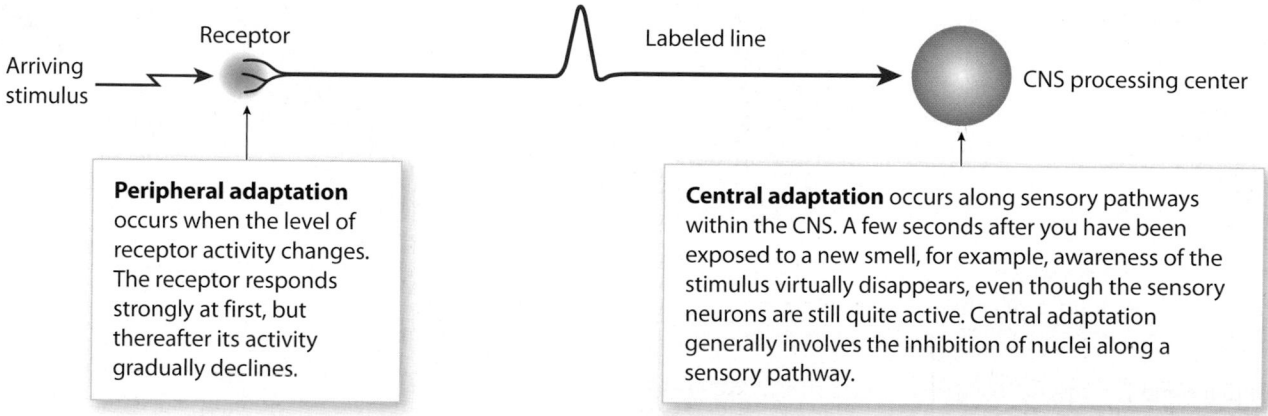

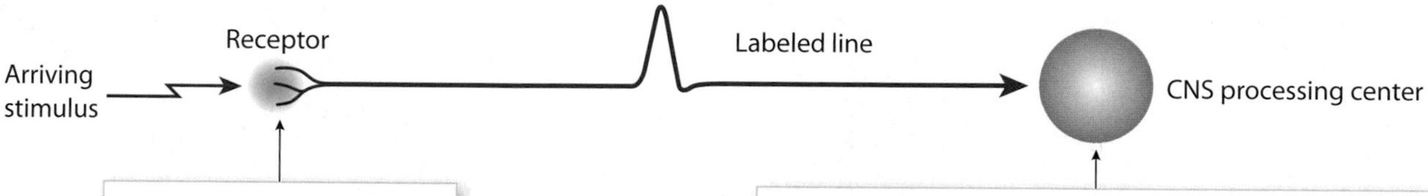
Arriving stimulus
Receptor
Labeled line
CNS processing center

Peripheral adaptation occurs when the level of receptor activity changes. The receptor responds strongly at first, but thereafter its activity gradually declines.

Central adaptation occurs along sensory pathways within the CNS. A few seconds after you have been exposed to a new smell, for example, awareness of the stimulus virtually disappears, even though the sensory neurons are still quite active. Central adaptation generally involves the inhibition of nuclei along a sensory pathway.

Module 13.16 Review

a. List the four types of general sensory receptors based on function, and identify the type of stimulus that excites each type.

b. Describe the three classes of mechanoreceptors.

c. Explain adaptation, and differentiate between peripheral adaptation and central adaptation.

LO **13.16** Explain the ways in which receptors can be classified.

Module 13.17

General sensory receptors have a simple structure and are widely distributed in the body

There are millions of general sensory receptors in the body. Not surprisingly, the greatest diversity is found in the skin, which is in constant contact with the external environment and its associated hazards and threats to homeostasis. This module discusses the basic structure and function of the important tactile receptors in the skin.

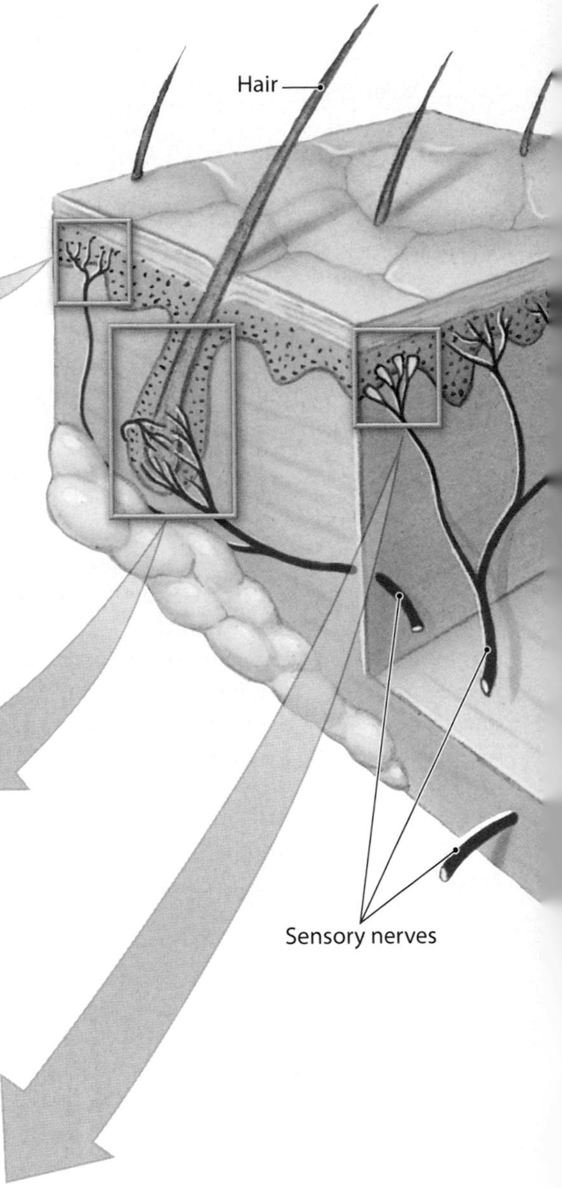

Hair
Sensory nerves

Free Nerve Endings

Free nerve endings are the branching tips of sensory neurons. They are not protected by any accessory structures, and they are nonspecific: They can respond to touch and pressure, pain, and temperature stimuli. Free nerve endings are the most common receptors in the skin.

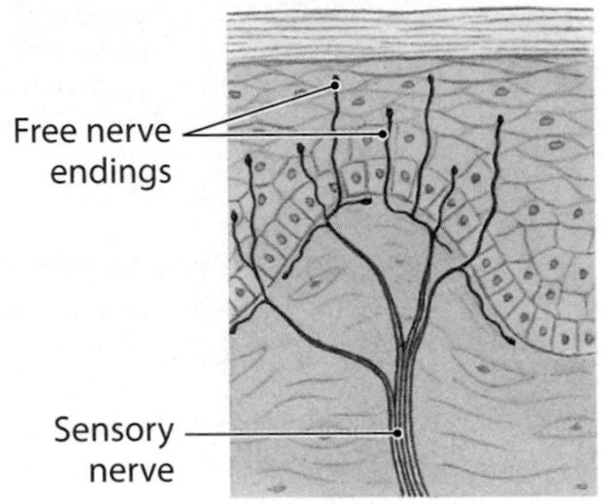

Free nerve endings
Sensory nerve

Root Hair Plexuses

Wherever hairs are located, the nerve endings of the **root hair plexus** monitor distortions and movements across the body surface (Module 5.7, p. 186). When a hair is displaced, the movement of the follicle distorts the sensory dendrites and produces action potentials. These receptors adapt rapidly, so they are best at detecting initial contact and subsequent movements.

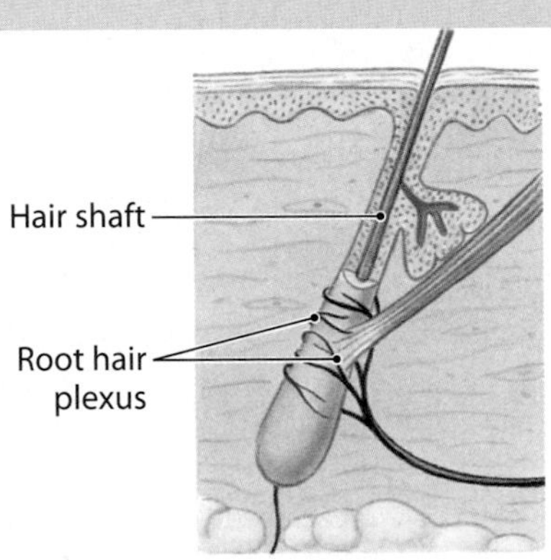

Hair shaft
Root hair plexus

Tactile Discs

Tactile discs are fine touch and pressure receptors. They are extremely sensitive tonic receptors with very small receptive fields. The dendritic processes of many nerve terminals branching from a single myelinated afferent fiber make close contact with **Merkel cells**, unusually large epithelial cells in the stratum basale, the deepest layer of the epidermis. Each Merkel cell and its nerve terminal make up a tactile disc.

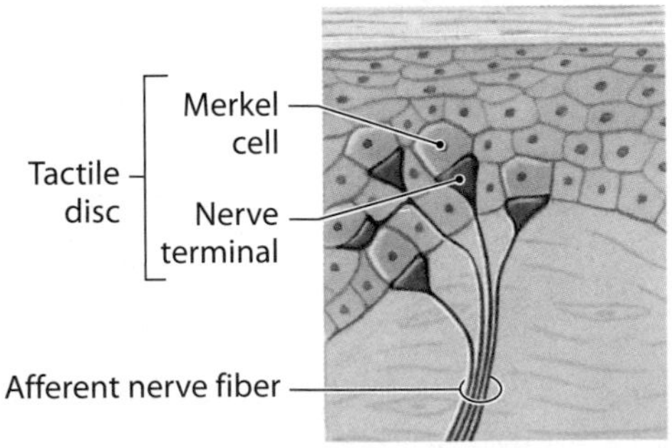

Tactile disc
Merkel cell
Nerve terminal
Afferent nerve fiber

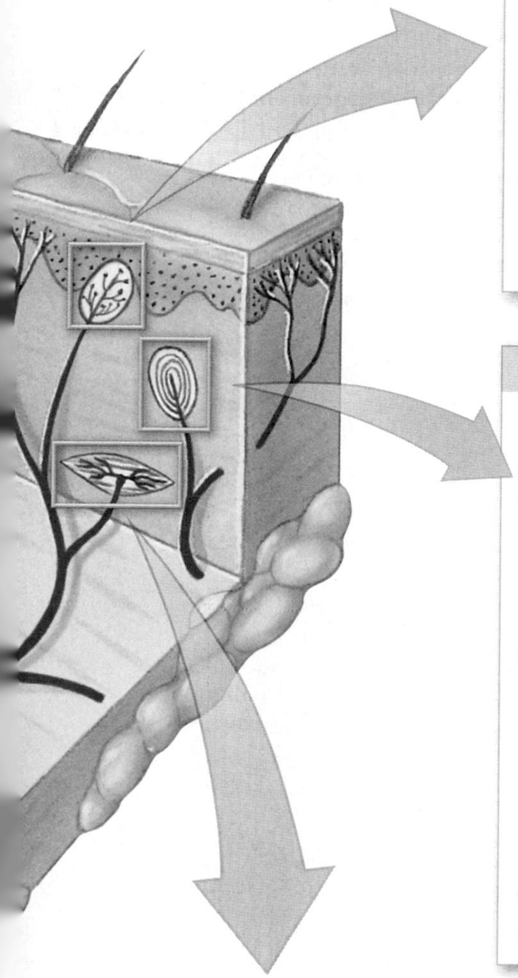

Tactile Corpuscles

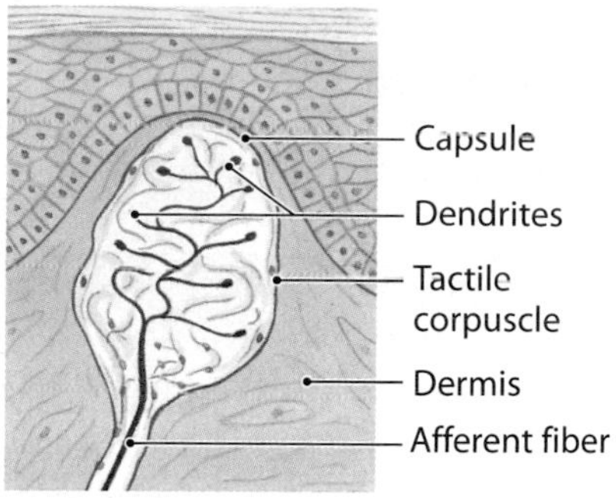

Capsule
Dendrites
Tactile corpuscle
Dermis
Afferent fiber

Tactile corpuscles, or **Meissner's** (MĪS-nerz) **corpuscles**, provide sensations of fine touch and pressure and low-frequency vibration. They adapt to stimulation within a second after contact. Tactile corpuscles are fairly large structures, measuring roughly 100 μm long and 50 μm wide. These receptors are most abundant in the eyelids, lips, fingertips, nipples, and external genitalia. The dendrites are highly coiled and interwoven, and they are surrounded by modified Schwann cells. A fibrous capsule surrounds the entire complex and anchors it within the papillary dermis.

Lamellated Corpuscles

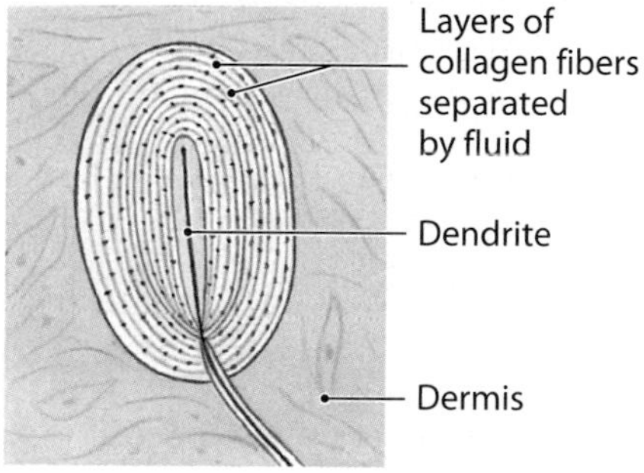

Layers of collagen fibers separated by fluid
Dendrite
Dermis

Lamellated (LAM-e-lāt-ed; *lamella*, thin plate) **corpuscles**, or **pacinian** (pa-SIN-ē-an) **corpuscles**, are sensitive to deep pressure. Because they are fast-adapting receptors, they are most sensitive to pulsing or high-frequency vibrating stimuli. A single dendrite lies within concentric layers of collagen fibers and specialized fibroblasts. The entire corpuscle may reach 4 mm long and 1 mm in diameter. The concentric layers, separated by interstitial fluid, shield the dendrite from virtually every source of stimulation other than direct pressure. Somatic sensory information is provided by lamellated corpuscles located throughout the dermis, notably in the fingers, mammary glands, and external genitalia; in the superficial and deep fasciae; and in joint capsules. Visceral sensory information is provided by lamellated corpuscles in mesenteries, in the pancreas, and in the walls of the urethra and urinary bladder.

Ruffini Corpuscles

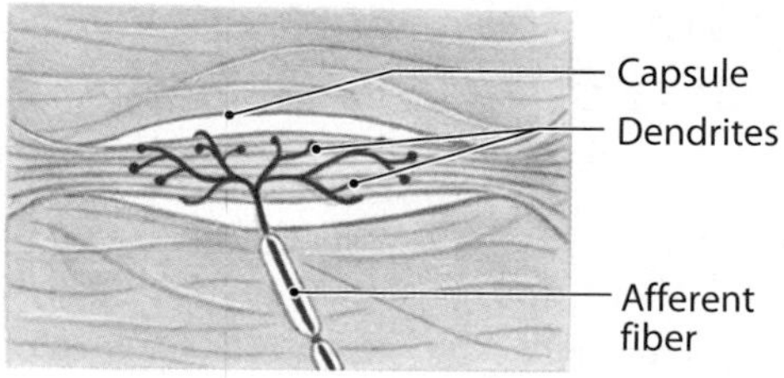

Capsule
Dendrites
Afferent fiber

Ruffini (roo-FĒ-nē) **corpuscles** are sensitive to pressure and distortion of the reticular (deep) dermis as when the skin is stretched. These receptors are tonic and show little if any adaptation. A capsule surrounds a core of collagen fibers that are continuous with those of the surrounding dermis. Within the capsule, a network of dendrites is intertwined with the collagen fibers. Any tension or distortion of the dermis tugs or twists the capsular fibers, stretching or compressing the attached dendrites and altering the activity in the myelinated afferent fiber.

Module 13.17 Review

a. Identify the six types of tactile receptors located in the skin, and describe their sensitivities.

b. Which types of tactile receptors are located only in the dermis?

c. Which is likely to be more sensitive to continuous deep pressure: a lamellated corpuscle or a Ruffini corpuscle?

LO 13.17 List the types of tactile receptors, and specify the functions of each.

Three major somatic sensory pathways carry information from the skin and muscles to the CNS

1 Two tracts within the **spinothalamic pathway** provide conscious sensations of poorly localized ("crude") touch, pressure, pain, and temperature. In this pathway, axons of **first-order neurons** enter the spinal cord and synapse on **second-order neurons** within the posterior gray horns. The axons of these interneurons cross to the opposite side of the spinal cord before ascending to the thalamus. **Third-order neurons** synapse in the primary sensory cortex.

Spinothalamic Pathway

The **anterior spinothalamic tracts** of the spinothalamic pathway carry crude touch and pressure sensations.

The **lateral spinothalamic tracts** of the spinothalamic pathway carry pain and temperature sensations.

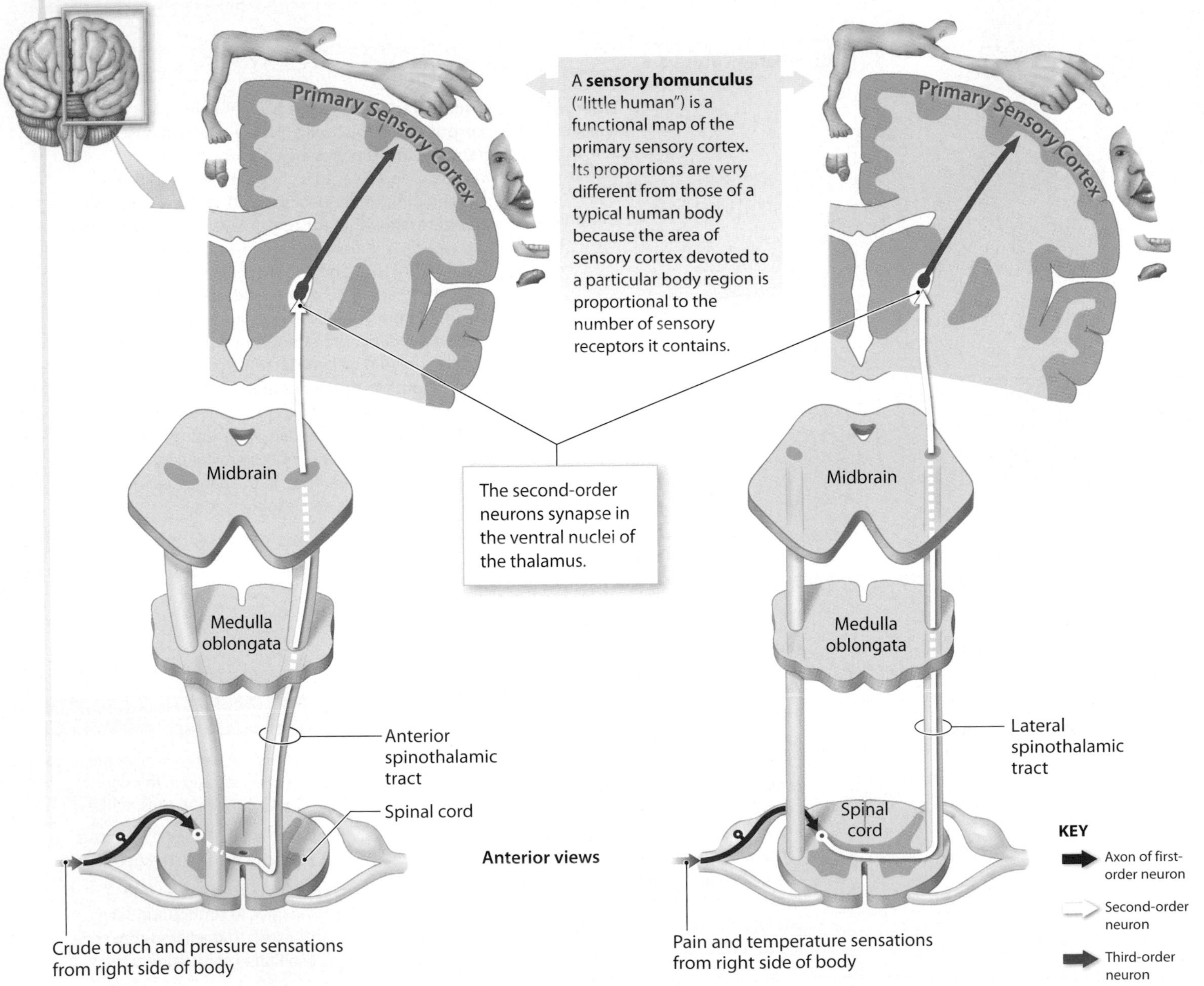

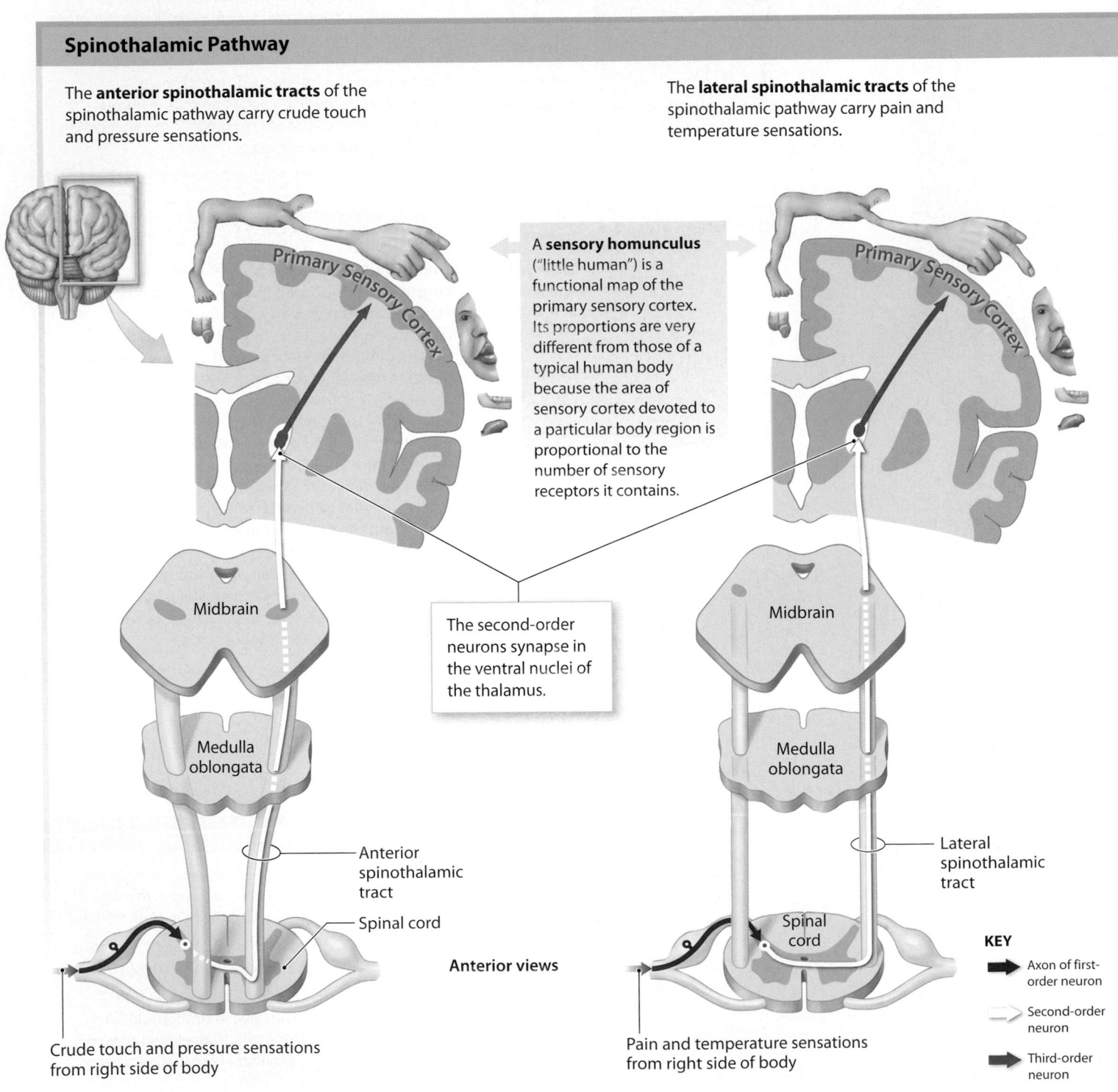

Primary Sensory Cortex
Primary Sensory Cortex
A **sensory homunculus** ("little human") is a functional map of the primary sensory cortex. Its proportions are very different from those of a typical human body because the area of sensory cortex devoted to a particular body region is proportional to the number of sensory receptors it contains.
The second-order neurons synapse in the ventral nuclei of the thalamus.
Midbrain
Midbrain
Medulla oblongata
Medulla oblongata
Anterior spinothalamic tract
Spinal cord
Lateral spinothalamic tract
Spinal cord
Anterior views
Crude touch and pressure sensations from right side of body
Pain and temperature sensations from right side of body
KEY
Axon of first-order neuron
Second-order neuron
Third-order neuron

2 The **posterior column pathway** carries sensations of highly localized ("fine") touch, pressure, vibration, and proprioception. This pathway begins at a peripheral receptor and ends at the primary sensory cortex of the cerebral hemispheres.

Posterior Column Pathway

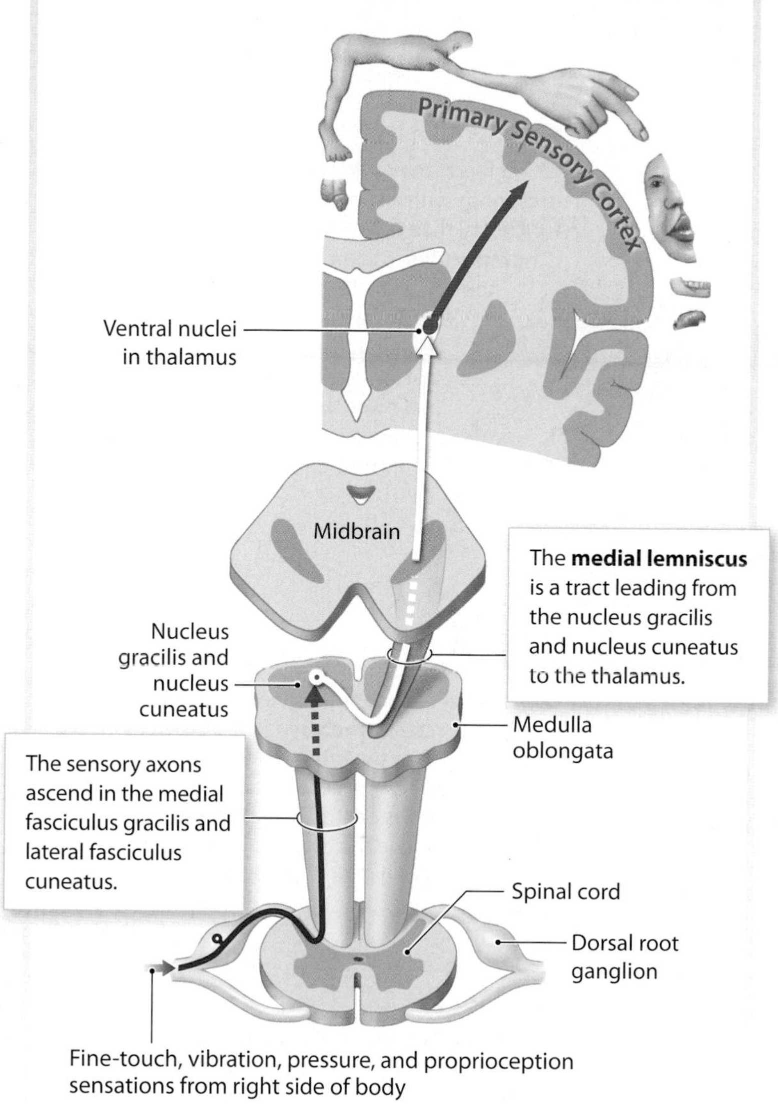

Primary Sensory Cortex
Ventral nuclei in thalamus
Midbrain
The **medial lemniscus** is a tract leading from the nucleus gracilis and nucleus cuneatus to the thalamus.
Nucleus gracilis and nucleus cuneatus
Medulla oblongata
The sensory axons ascend in the medial fasciculus gracilis and lateral fasciculus cuneatus.
Spinal cord
Dorsal root ganglion
Fine-touch, vibration, pressure, and proprioception sensations from right side of body

3 The cerebellum receives proprioceptive information about the position of skeletal muscles, tendons, and joints along the **spinocerebellar pathway**. The posterior spinocerebellar tracts contain axons that do not cross over to the opposite side of the spinal cord. These axons reach the cerebellar cortex by the inferior cerebellar peduncle of that side. The anterior spinocerebellar tracts are dominated by axons that have crossed to the opposite side of the spinal cord.

Spinocerebellar Pathway

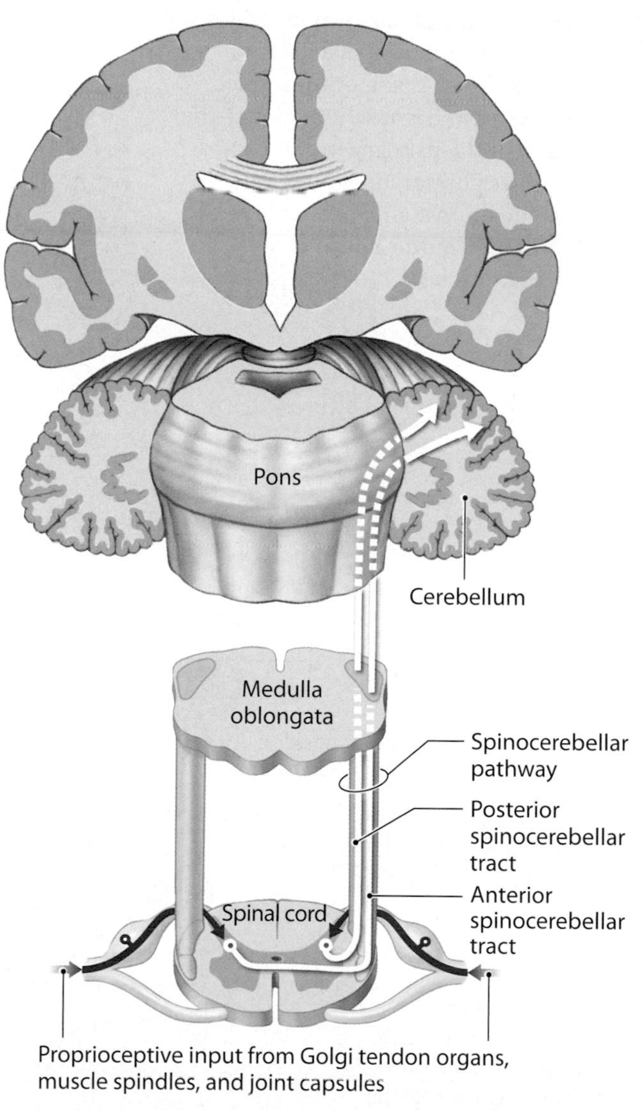

Pons
Cerebellum
Medulla oblongata
Spinocerebellar pathway
Posterior spinocerebellar tract
Anterior spinocerebellar tract
Spinal cord
Proprioceptive input from Golgi tendon organs, muscle spindles, and joint capsules

4 This cross section through the spinal cord shows the locations of the somatic sensory pathways.

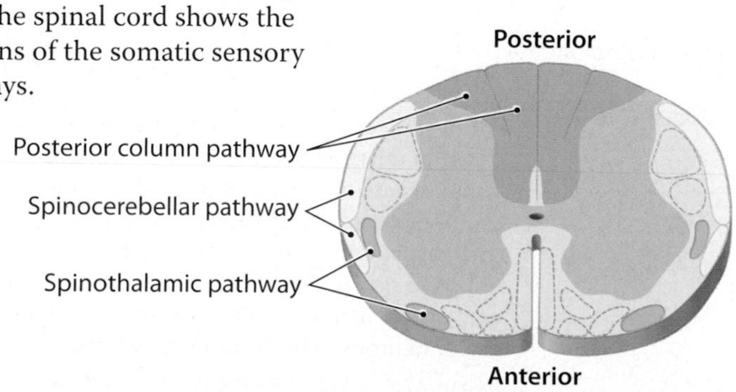

Posterior
Posterior column pathway
Spinocerebellar pathway
Spinothalamic pathway
Anterior

Module 13.18 Review

a. Define sensory homunculus.

b. Which spinal tracts carry action potentials generated by nociceptors?

c. Which cerebral hemisphere receives impulses conducted by the right fasciculus gracilis of the spinal cord?

LO 13.18 Identify and describe the major sensory pathways.

The somatic nervous system controls skeletal muscles through upper and lower motor neurons

Somatic motor pathways always involve at least two motor neurons: an **upper motor neuron**, whose cell body lies in a CNS processing center, and a **lower motor neuron**, whose cell body lies in a nucleus of the brain stem or spinal cord. The upper motor neuron synapses on the lower motor neuron, which in turn innervates a single motor unit in a skeletal muscle.

1 The **corticospinal pathway** provides voluntary control over skeletal muscles. This pathway is sometimes called the pyramidal system because it begins at the pyramidal cells of the primary motor cortex. The axons of these upper motor neurons descend into the brain stem and spinal cord to synapse on lower motor neurons that control skeletal muscles.

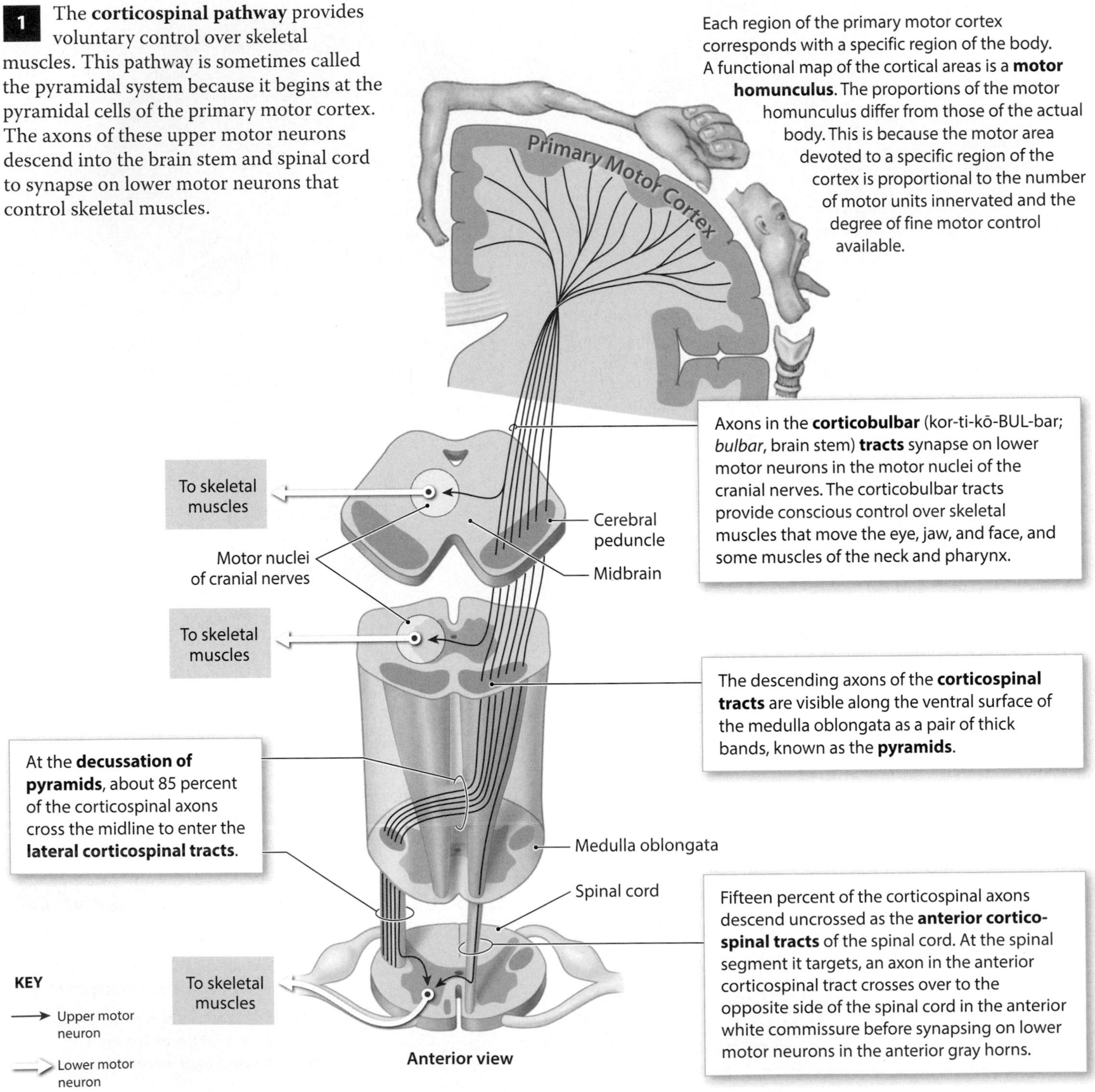

Each region of the primary motor cortex corresponds with a specific region of the body. A functional map of the cortical areas is a **motor homunculus**. The proportions of the motor homunculus differ from those of the actual body. This is because the motor area devoted to a specific region of the cortex is proportional to the number of motor units innervated and the degree of fine motor control available.
Primary Motor Cortex
Axons in the **corticobulbar** (kor-ti-kō-BUL-bar; *bulbar*, brain stem) **tracts** synapse on lower motor neurons in the motor nuclei of the cranial nerves. The corticobulbar tracts provide conscious control over skeletal muscles that move the eye, jaw, and face, and some muscles of the neck and pharynx.
To skeletal muscles
Motor nuclei of cranial nerves
Cerebral peduncle
Midbrain
To skeletal muscles
The descending axons of the **corticospinal tracts** are visible along the ventral surface of the medulla oblongata as a pair of thick bands, known as the **pyramids**.
At the **decussation of pyramids**, about 85 percent of the corticospinal axons cross the midline to enter the **lateral corticospinal tracts**.
Medulla oblongata
Spinal cord
Fifteen percent of the corticospinal axons descend uncrossed as the **anterior corticospinal tracts** of the spinal cord. At the spinal segment it targets, an axon in the anterior corticospinal tract crosses over to the opposite side of the spinal cord in the anterior white commissure before synapsing on lower motor neurons in the anterior gray horns.
KEY
Upper motor neuron
Lower motor neuron
To skeletal muscles
Anterior view

2 Several centers in the cerebrum, diencephalon, and brain stem may issue somatic motor commands as a result of processing performed at a subconscious level. The locations of those centers are indicated in the diagram to the right. The components of the **medial pathway** help control gross movements of the trunk and proximal limb muscles, whereas those of the **lateral pathway** help control the distal limb muscles that perform more precise movements.

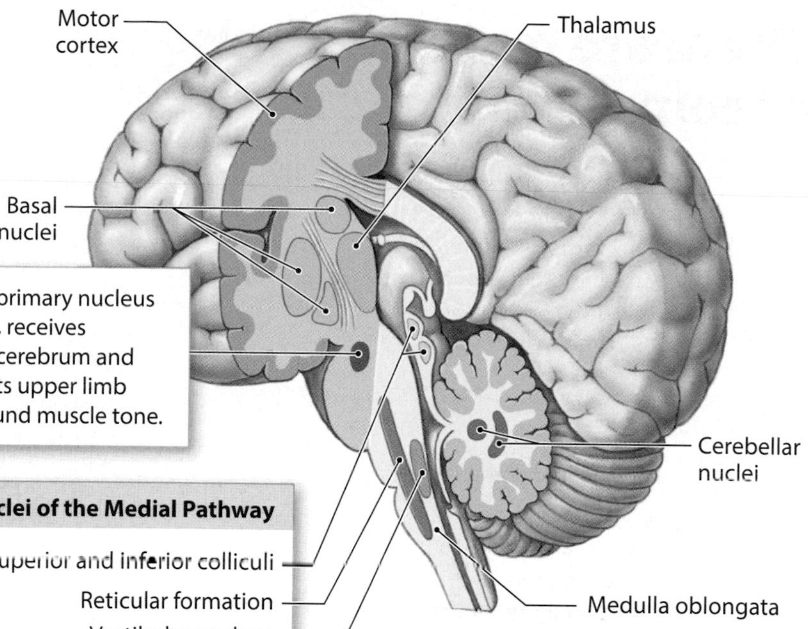

Motor cortex
Thalamus
Basal nuclei
Cerebellar nuclei
Medulla oblongata

The **red nucleus**, the primary nucleus of the lateral pathway, receives information from the cerebrum and cerebellum and adjusts upper limb position and background muscle tone.

Nuclei of the Medial Pathway

- Superior and inferior colliculi
- Reticular formation
- Vestibular nucleus

3 The locations of the medial and lateral pathways are indicated in this cross section of the spinal cord.

Medial Pathway

The medial pathway is primarily concerned with the control of muscle tone and gross movements of the neck, trunk, and proximal limb muscles. The upper motor neurons of the medial pathway are located in the vestibular nuclei, the superior and inferior colliculi, and the reticular formation.

The **reticulospinal tracts** contain the axons of upper motor neurons in the reticular formation. The reticular formation receives input from almost every ascending and descending pathway. It also has extensive connections with the cerebrum, the cerebellum, and brain stem nuclei.

The **vestibulospinal tracts** begin at the vestibular nuclei of cranial nerve VIII. These nuclei receive sensory information from the inner ear about the position and movement of the head. These nuclei respond to changes in the orientation of the head by issuing motor commands that alter the muscle tone and position of the neck, eyes, head, and limbs.

The **tectospinal tracts** contain the axons of upper motor neurons in the superior and inferior colliculi of the midbrain. Axons in the tectospinal tracts direct reflexive changes in the position of the head, neck, and upper limbs in response to bright lights, sudden movements, or loud noises.

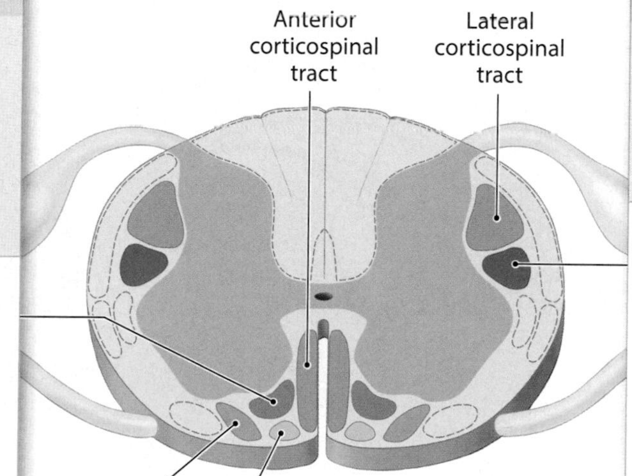

Anterior corticospinal tract
Lateral corticospinal tract

Lateral Pathway

The lateral pathway is primarily concerned with the control of muscle tone and the more precise movements of the distal parts of the limbs.

The upper motor neurons of the lateral pathway lie within the red nuclei of the midbrain. Axons of these motor neurons cross to the opposite side of the brain and descend into the spinal cord in the **rubrospinal tracts** (*ruber*, red). In humans, the rubrospinal tracts are small and extend only to the cervical spinal cord, where they provide motor control over distal muscles of the upper limbs.

Module 13.19 Review

a. Define corticospinal tracts.

b. Describe the role of the corticobulbar tracts.

c. What effect would increased stimulation of the motor neurons of the red nucleus have on muscle tone?

LO **13.19** Describe the components, processes, and functions of the somatic motor pathways.

There are multiple levels of somatic motor control

1 In the last two chapters you encountered many nuclei in the spinal cord and brain that play roles in skeletal muscle contractions. In general, the closer a motor center is to the cerebral cortex, the more complex and variable the motor activities will be. The cerebellum is involved in coordinating the motor activities at multiple levels.

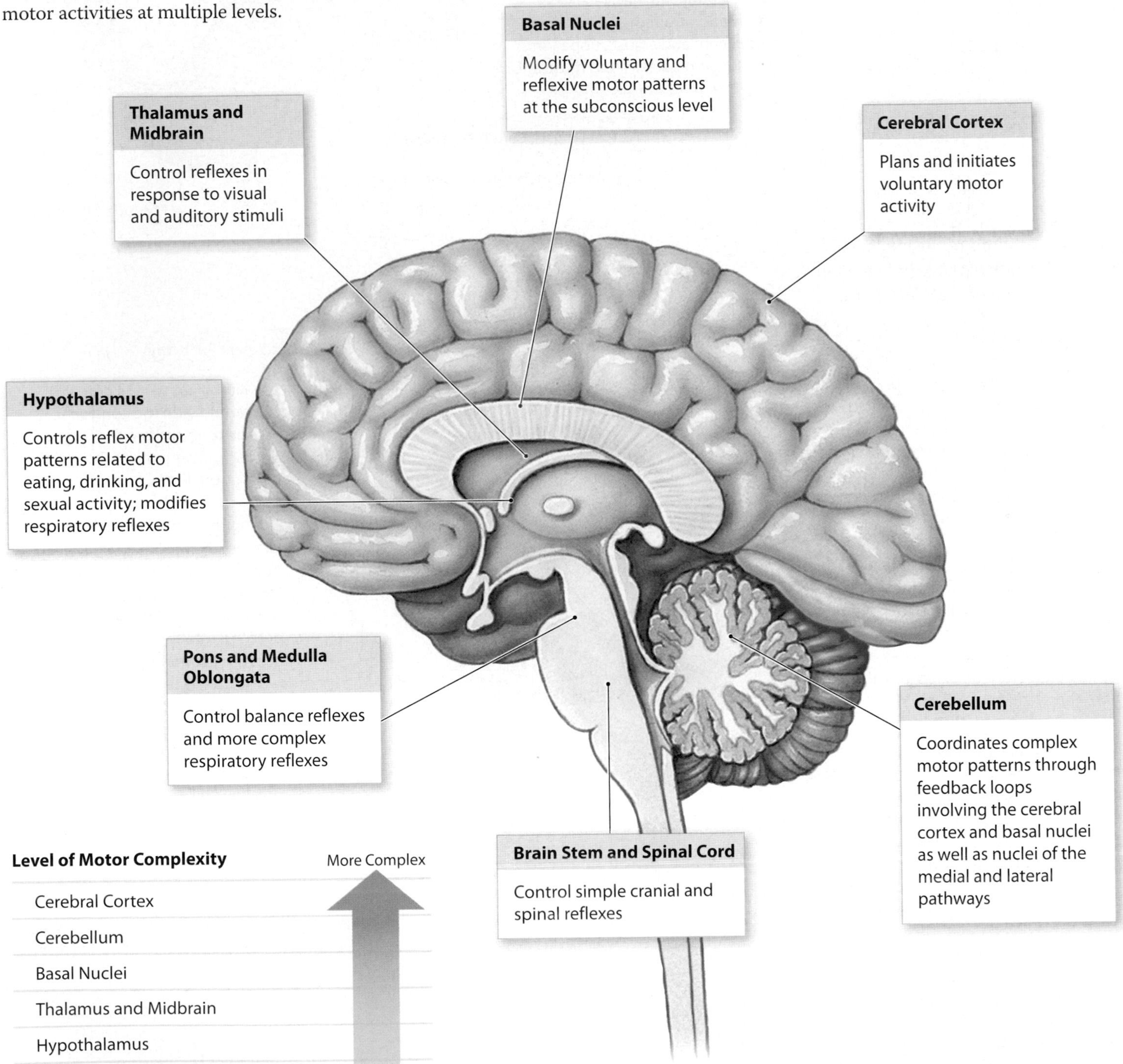
Basal Nuclei
Modify voluntary and reflexive motor patterns at the subconscious level
Thalamus and Midbrain
Control reflexes in response to visual and auditory stimuli
Cerebral Cortex
Plans and initiates voluntary motor activity
Hypothalamus
Controls reflex motor patterns related to eating, drinking, and sexual activity; modifies respiratory reflexes
Pons and Medulla Oblongata
Control balance reflexes and more complex respiratory reflexes
Cerebellum
Coordinates complex motor patterns through feedback loops involving the cerebral cortex and basal nuclei as well as nuclei of the medial and lateral pathways
Brain Stem and Spinal Cord
Control simple cranial and spinal reflexes

Level of Motor Complexity	More Complex
Cerebral Cortex	
Cerebellum	
Basal Nuclei	
Thalamus and Midbrain	
Hypothalamus	
Pons and Medulla Oblongata	
Brain Stem and Spinal Cord	Less Complex

Preparing for Movement

2 When you make a conscious decision to do a specific movement, information is relayed from the frontal lobes to motor association areas. These areas in turn relay the information to the cerebellum and basal nuclei.

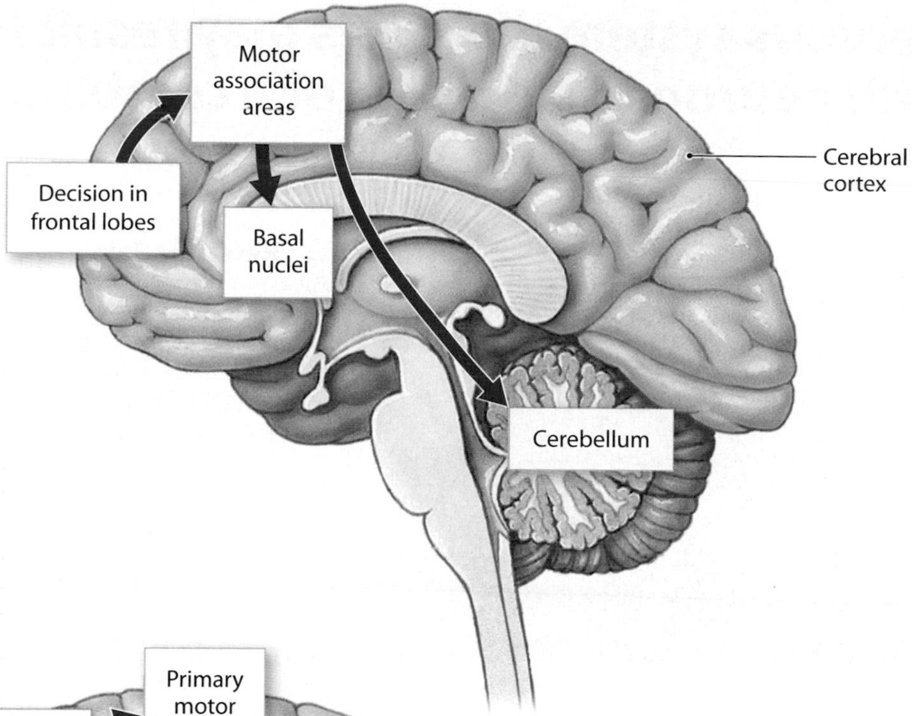

Motor association areas
Decision in frontal lobes
Basal nuclei
Cerebral cortex
Cerebellum

Performing a Movement

3 As the movement begins, the motor association areas send instructions to the primary motor cortex. Feedback from the basal nuclei and cerebellum modifies those commands, and output along the medial and lateral pathways directs involuntary adjustments in position and muscle tone.

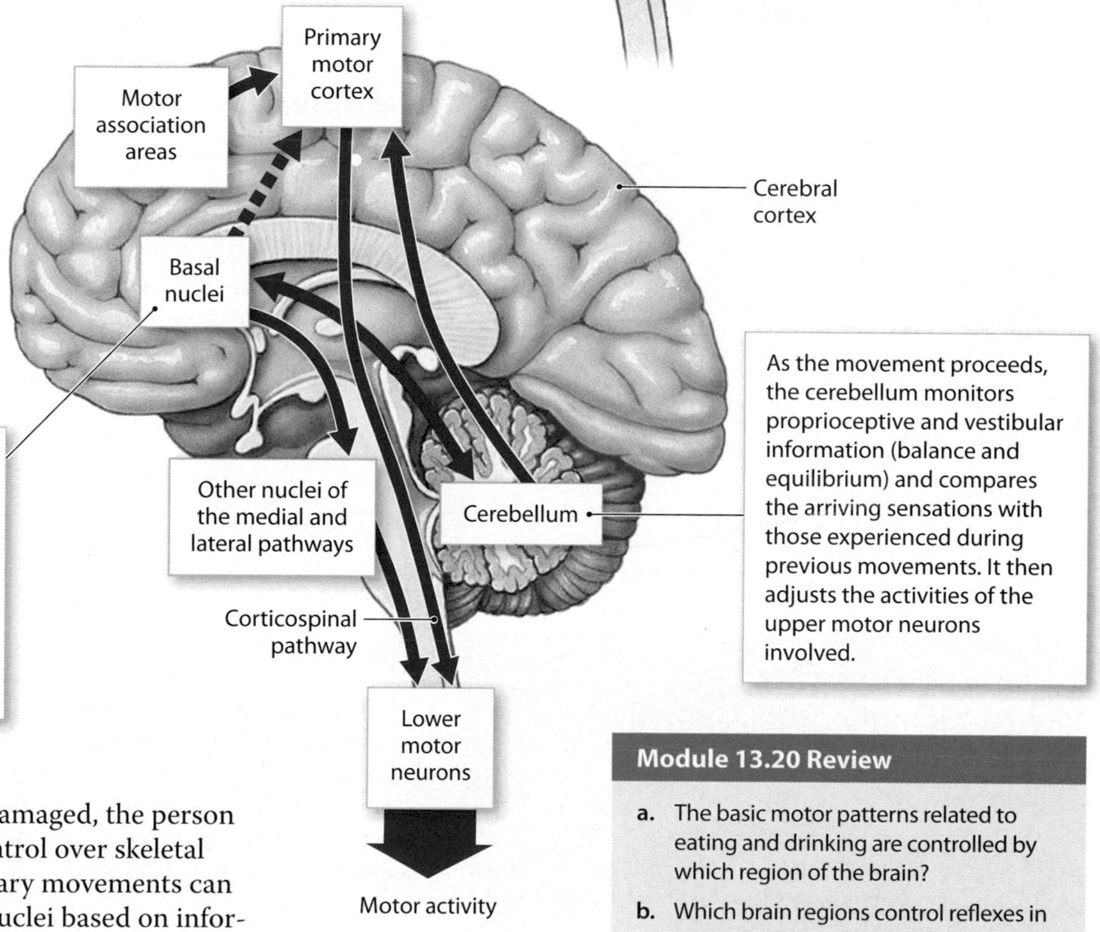

Primary motor cortex
Motor association areas
Cerebral cortex
Basal nuclei
Other nuclei of the medial and lateral pathways
Cerebellum
Corticospinal pathway
Lower motor neurons
Motor activity
As the movement proceeds, the cerebellum monitors proprioceptive and vestibular information (balance and equilibrium) and compares the arriving sensations with those experienced during previous movements. It then adjusts the activities of the upper motor neurons involved.

The basal nuclei adjust patterns of movement in two ways:

1. They alter the sensitivity of the pyramidal cells to adjust the output along the corticospinal tract.
2. They change the excitatory or inhibitory output of the medial and lateral pathways.

If the primary motor cortex is damaged, the person loses the ability to exert fine control over skeletal muscles. However, some voluntary movements can still be controlled by the basal nuclei based on information from the prefrontal cortex concerning planned movements. But because the corticospinal pathway is inoperative, the cerebellar feedback cannot fine-tune the ongoing movements. A person in this condition can stand, maintain balance, and even walk, but all movements are hesitant, awkward, and poorly controlled.

Module 13.20 Review

a. The basic motor patterns related to eating and drinking are controlled by which region of the brain?

b. Which brain regions control reflexes in response to visual and auditory stimuli that are experienced while viewing a movie?

c. During a tennis match, you decide how and where to hit the ball. Explain how the motor association areas are involved in your decisions.

LO 13.20 Describe the levels of information processing involved in motor control.

Nervous system disorders may result from problems with neurons, pathways, or a combination of the two

Heart

Referred Pain

1 **Referred pain** is the sensation of pain in a part of the body other than its actual source. A familiar example is the pain of a heart attack, which is frequently felt in the left arm. Strong visceral pain sensations arriving at a segment of the spinal cord can stimulate interneurons that are part of the spinothalamic pathway. Activity in these interneurons leads to the stimulation of the primary sensory cortex, so the person feels pain in a specific part of the body surface.

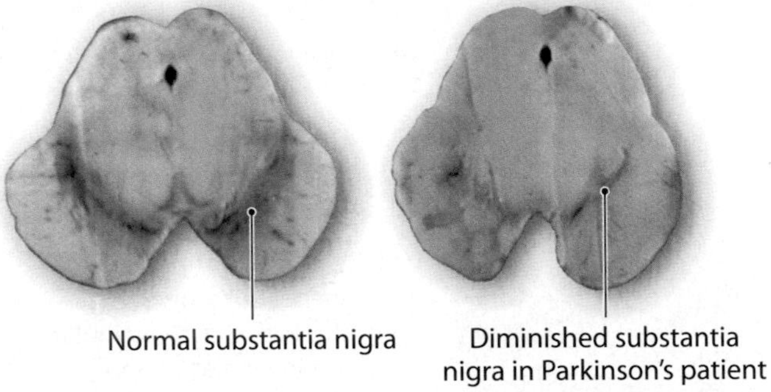
Normal substantia nigra
Diminished substantia nigra in Parkinson's patient

Parkinson's Disease

2 **Parkinson's disease** results when neurons of the substantia nigra are damaged or secrete less dopamine. The basal nuclei become more active, which raises skeletal muscle tone and produces rigidity and stiffness. People who have Parkinson's disease have difficulty starting voluntary movements, because opposing muscle groups do not relax; they must be overpowered. Once a movement is under way, every aspect must be voluntarily controlled through intense effort and concentration.

Rabies

3 **Rabies** is a dramatic example of a clinical condition directly related to retrograde flow in peripheral axons. A bite from a rabid animal injects the rabies virus into peripheral tissues, where virus particles quickly enter axon terminals. Retrograde flow then carries the virus into the CNS, with potentially fatal results. Many toxins (including heavy metals), some pathogenic bacteria, and other viruses also bypass CNS defenses by exploiting axoplasmic transport (Module 11.2, p. 396).

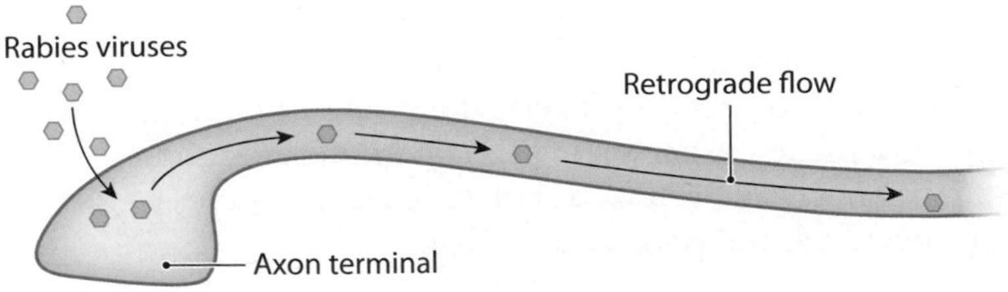
Rabies viruses
Retrograde flow
Axon terminal

Cerebral Palsy

4 **Cerebral palsy (CP)** refers to a number of disorders that affect voluntary motor movement. The motor dysfunction is nonprogressive, appears during infancy or childhood, and persists throughout the person's lifetime. The cause may be trauma associated with premature or unusually stressful birth, maternal exposure to drugs (including alcohol), or a genetic defect that causes improper development of motor pathways.

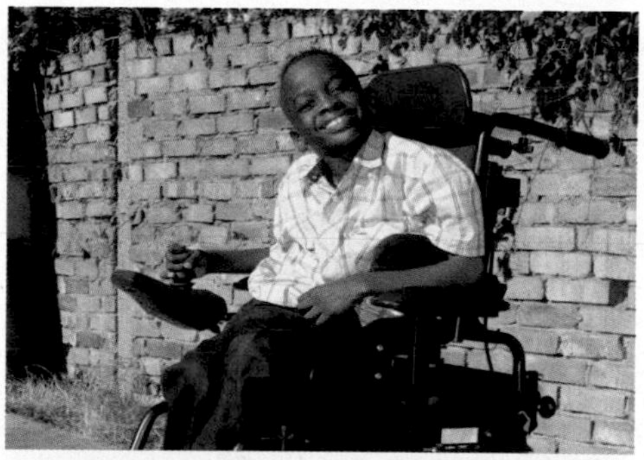

Amyotrophic Lateral Sclerosis

5 **Amyotrophic lateral sclerosis (ALS)** is a progressive, degenerative disorder that affects motor neurons in the spinal cord, brain stem, and cerebral hemispheres. The degeneration affects both upper and lower motor neurons. A defect in axonal transport is thought to underlie the disease. Because a motor neuron and its dependent muscle fibers are so intimately related, the destruction of CNS neurons causes atrophy of the associated skeletal muscles. ALS is commonly known as *Lou Gehrig's disease*, named after the famous New York Yankees player who died of the disorder. Noted physicist Stephen Hawking is also afflicted with this condition.

Alzheimer's Disease

6 **Alzheimer's disease (AD)** is a progressive disorder characterized by the loss of higher-order cerebral functions. It is the most common cause of **senile dementia**, or senility. Signs and symptoms may appear at 50–60 years of age or later, although the disease occasionally affects younger people. An estimated 2 million people in the United States—including roughly 15 percent of those over age 65, and nearly half of those over age 85—have some form of the condition, and it causes approximately 100,000 deaths each year. Microscopic examination of the brains of AD patients reveals intracellular and extracellular abnormalities in brain regions that are specifically associated with memory processing, such as the hippocampus.

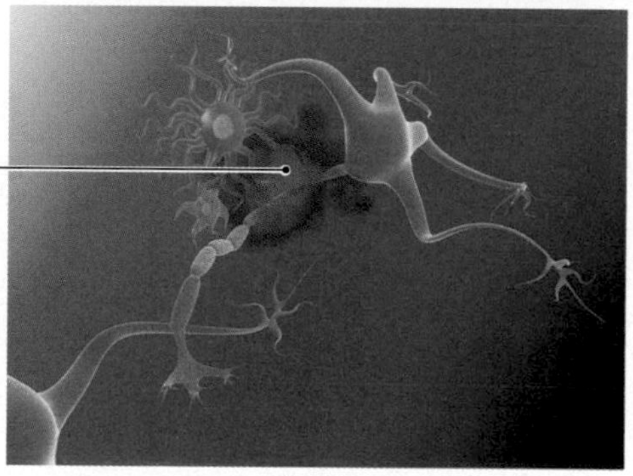
Abnormal dendrites, axons, and extracellular proteins form complexes known as Alzheimer's plaques.

Multiple Sclerosis

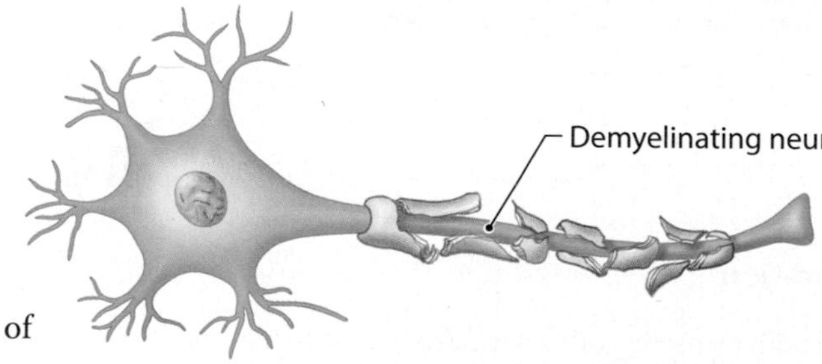
Demyelinating neuron

7 **Multiple sclerosis** (skler-Ō-sis; *sklerosis*; hardness), or **MS**, is a disease characterized by recurrent incidents of demyelination that affects axons in the optic nerve, brain, and spinal cord. Common signs and symptoms include partial loss of vision and problems with speech, balance, and general motor coordination, including bowel and urinary bladder control. The time between incidents and the degree of recovery vary from case to case. In about one-third of all cases, the disorder is progressive, and functional impairment increases following each new incident. The first attack typically occurs at 30–40 years of age; the incidence among women is 1.5 times that among men.

Module 13.21 Review

a. Define referred pain.
b. Describe how rabies is contracted.
c. Describe amyotrophic lateral sclerosis (ALS).

LO **13.21** Describe the roles of the nervous system in referred pain, Parkinson's disease, rabies, cerebral palsy, amyotrophic lateral sclerosis, Alzheimer's disease, and multiple sclerosis.

Labeling

Label each type of tactile receptor found in the skin.

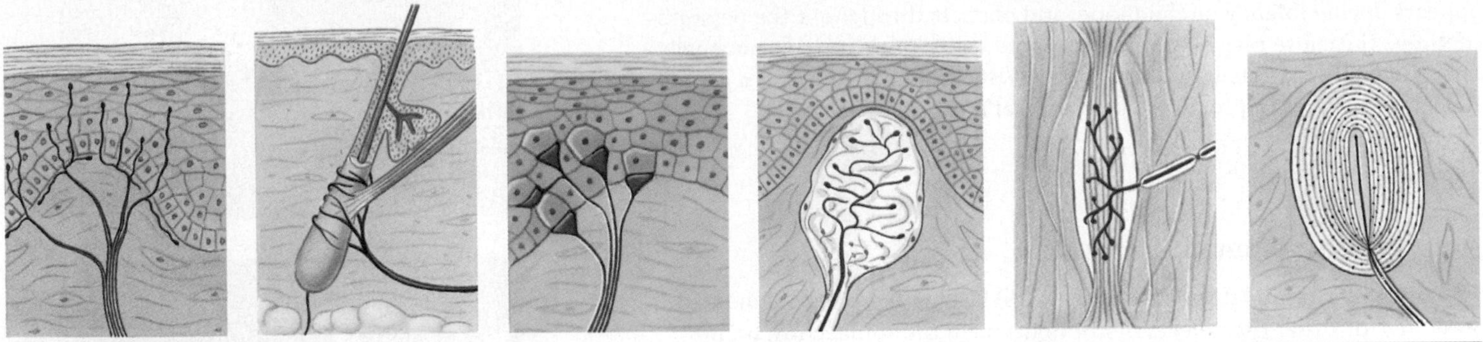

1

2

3

4

5

6

Fill-in

The general organization of the spinal cord is such that motor tracts are 7 ______ (anterior or posterior), and sensory tracts are 8 ______ (anterior or posterior).

Short answer

Identify the descending and ascending tracts and pathways in the accompanying sectional diagram of the spinal cord, and then describe the general functions of the tracts of each pathway.

9

10

11

12

13

14

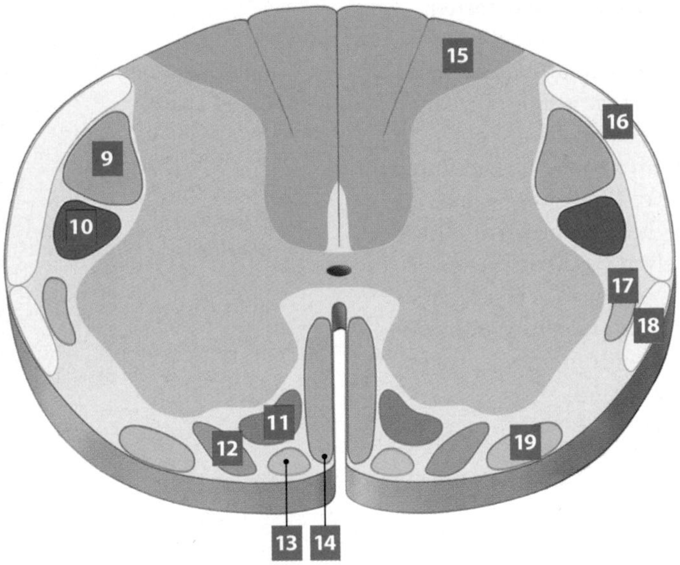
15
16
9
10
17
18
11
12
19
13
14

15

16

17

18

19

Section integration

A person whose primary motor cortex has been injured retains the ability to walk, maintain balance, and perform other voluntary and involuntary movements. Even though the movements lack precision and are awkward and poorly controlled, why is the ability to walk and maintain balance possible?

20 ______

Study Outline

SECTION 1 • Functional Anatomy of the Brain and Cranial Nerves

13.1 The brain develops from a hollow neural tube p. 463

1. The neural tube contains a hollow internal passageway called a **neurocoel**.
2. The cephalic portion of the neural tube enlarges into three divisions called the **primary brain vesicles**: the **prosencephalon**, **mesencephalon**, and **rhombencephalon**.
3. **Secondary brain vesicles** develop from the prosencephalon (**diencephalon** and **telencephalon**) and rhombencephalon (**metencephalon** and **myelencephalon**).

13.2 Each region of the brain has distinct structural and functional characteristics p. 464

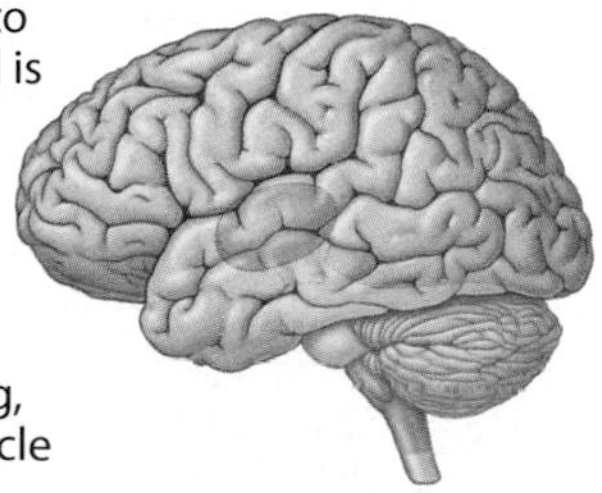

4. The **cerebrum** is divided into **cerebral hemispheres**, and is covered by a superficial layer of gray matter called **cerebral cortex**.
5. The functions of the cerebrum include thought, memory, sensory processing, and regulating skeletal muscle contractions.
6. Visible on the cerebrum are **fissures** (deep grooves), **gyri** (folds), and **sulci** (shallow depressions).
7. The **diencephalon** links cerebral hemispheres with the rest of the CNS. It is composed of the **thalamus** and **hypothalamus**.
8. The **brain stem** includes the **midbrain**, **pons**, and **medulla oblongata**.
9. The **cerebellum** controls motor commands from the cerebral cortex.
10. Chambers within the brain called **ventricles** are filled with cerebrospinal fluid and lined by ependymal cells. The ventricles of the brain contain two **lateral ventricles**, a **third ventricle**, and a **fourth ventricle**. Each lateral ventricle is connected to the third ventricle through an **interventricular foramen**, and the **cerebral aqueduct** connects the third ventricle to the fourth ventricle.
11. The **corpus callosum** interconnects the two cerebral hemispheres. The **septum pellucidum** is a thin partition separating the two lateral ventricles.

13.3 The cranial meninges and cerebrospinal fluid protect and support the brain p. 466

12. The cranial **dura mater**, **arachnoid mater**, and **pia mater** are continuous with the layers of the spinal meninges.
13. The cranial dura mater is made of an outer periosteal layer that is fused to the periosteum of the cranial bones, and typically separated from the inner meningeal layer by the **dural sinus**.

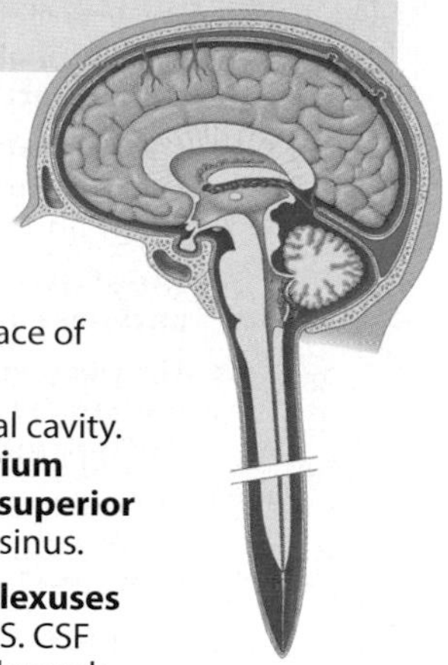

14. The cranial arachnoid mater consists of the arachnoid membrane and the arachnoid trabeculae that connect it to the pia mater.
15. The pia mater adheres to the surface of the brain by astrocyte processes.
16. **Dural folds** extend into the cranial cavity. These are the **falx cerebri**, **tentorium cerebelli**, and **falx cerebelli**. The **superior sagittal sinus** is the largest dural sinus.
17. CSF is produced by the **choroid plexuses** and bathes the surfaces of the CNS. CSF reaches the subarachnoid space through two **lateral apertures** and a single **median aperture** in the roof of the fourth ventricle.
18. CSF is absorbed into the venous circulation at **arachnoid granulations**.

13.4 The medulla oblongata and the pons contain autonomic reflex centers, relay stations, and ascending and descending tracts p. 468

19. All communication between the brain and spinal cord involves tracts that ascend or descend through the **medulla oblongata**. It is also a center for autonomic reflexes and visceral functions.
20. In the medulla oblongata, **pyramids** contain tracts of motor fibers. Some of these fibers cross over to the opposite side of the medulla. That crossing is called a **decussation**.
21. The medulla also contains **autonomic centers** and **relay stations**.
22. The **pons** links the cerebellum with the midbrain, diencephalon, cerebrum, medulla oblongata, and spinal cord.

13.5 The cerebellum coordinates learned and reflexive patterns of muscular activity at the subconscious level p. 470

23. The **cerebellum** functions to (1) adjust the postural muscles of the body and (2) fine-tune movements controlled at the conscious and subconscious levels.
24. The cerebellar hemispheres are each made of an **anterior** and **posterior lobe**, and separated by the **vermis**. The **folia** are folds on the cerebellar surface.
25. The three **cerebellar peduncles** on each side are tracts that link the cerebellum with the brain stem, cerebrum, and spinal cord.

13.6 The midbrain regulates auditory and visual reflexes and controls alertness p. 472

26. The **midbrain** is the most complex and integrative portion of the brain stem.
27. Major landmarks of the midbrain are the **corpora quadrigemina** (formed by the paired **superior** and **inferior colliculi**), the **red nucleus**, the **substantia nigra**, and the **cerebral peduncles**.

28. The **tectum** and **tegmentum** surround the cerebral aqueduct.

13.7 The diencephalon consists of the epithalamus, thalamus (left and right), and hypothalamus p. 474

29. The **epithalamus** is the roof of the diencephalon. Its anterior portion contains an extensive choroid plexus that extends through the interventricular foramina.
30. The anterior aspect of the diencephalon is marked by the **anterior commissure** and the **optic chiasm**.
31. The **pineal gland** is an endocrine structure that secretes **melatonin**, an important regulator of day–night cycles, and reproductive functions.
32. Important structures of the **thalamus** are the **lateral geniculate nuclei**, **medial geniculate nuclei**, and the **optic tract**.
33. The **hypothalamus** contains important control and integrative centers. **Hypothalamic nuclei** include the **preoptic area** and **suprachiasmatic nucleus**.
34. The **infundibulum** connects the floor of the hypothalamus to the pituitary gland.

13.8 The limbic system is a functional group of tracts and nuclei located in the cerebrum and diencephalon p. 476

35. The **limbic system** is a functional grouping rather than an anatomical one. It functions to establish emotional states, link conscious and unconscious functions, and facilitate memory.
36. The components of the limbic system in the diencephalon are the anterior group of thalamic nuclei, hypothalamus, and mammillary body.
37. The **cingulate gyrus** and **parahippocampal gyrus** form the **limbic lobe**, the region of the limbic system in the cerebrum.
38. The **amygdaloid body** of the limbic lobe plays a role in the regulation of heart rate, the "fight or flight" response, emotions, and memory.
39. The **hippocampus** is important in learning and long-term memory, and is connected to the hypothalamus by the **fornix**.

13.9 The basal nuclei of the cerebrum adjust and refine ongoing voluntary movements p. 478

40. The **basal nuclei** are masses of gray matter in the floor of the lateral ventricles. They provide subconscious control of skeletal muscle tone and help coordinate learned movement patterns.
41. The basal nuclei consist of the **caudate nucleus** and **lentiform nucleus**. The lentiform nucleus is subdivided into the **globus pallidus** and **putamen**.

13.10 Superficial landmarks divide the cerebral hemispheres into lobes p. 480

42. Each cerebral hemisphere can be divided into **lobes** that are named after the overlying bones of the skull.
43. Important landmarks of the cerebrum are the **precentral gyrus**, **central sulcus**, **postcentral gyrus**, and the **lateral sulcus**.
44. Retraction of the cerebral cortex along the lateral sulcus exposes the **insula**.
45. Each cerebral hemisphere receives information from and sends commands to the opposite side of the body. Although each hemisphere has similar appearances, they have different functions. The correspondence between function and region in the cerebral cortex is imprecise.

13.11 The lobes of the cerebral cortex have regions with specific functions p. 482

46. The **primary motor cortex** of the precentral gyrus sends voluntary commands to skeletal muscles. **Pyramidal cells** are the neurons of this area. The **somatic motor association area** coordinates learned movements.
47. The **primary sensory cortex** receives general somatic sensory information.
48. **Association areas** interpret incoming data or coordinate a motor response
49. The **gustatory cortex** of the insula receives information from taste receptors, and the **olfactory cortex** of the insula receives sensory information from the olfactory receptors.
50. The **auditory cortex**, located in the temporal lobe, contains the **primary auditory cortex** and the **auditory association area**.
51. The **visual cortex** in the occipital lobe contains the **primary visual cortex** and the **visual association area**.
52. **Integrative centers** concerned with complex processes such as speech, writing, math, and spatial relationships, are restricted to either the left or right hemisphere. These include the **speech center** (Broca's area), **prefrontal cortex**, **frontal eye field**, and the **general interpretive area**.
53. Each cerebral hemisphere is responsible for specific functions not performed by the opposite hemisphere. This regional specialization is called **hemispheric lateralization**.

13.12 White matter connects the cerebral hemispheres and the lobes of each hemisphere, and links the cerebrum to the rest of the brain p. 484

54. **Association fibers** such as the **arcuate fibers** and the **longitudinal fasciculi** interconnect areas of the neural cortex within a single cerebral hemisphere.
55. **Commissural fibers** such as the **corpus callosum** and **anterior commissure** interconnect the cerebral hemispheres.

13.13 Brain activity can be monitored using external electrodes; the record is called an electroencephalogram, or EEG p. 485

56. **EEG** electrical patterns are called **brain waves**.
57. **Alpha waves** appear in awake adults, **beta waves** are visible when a person is concentrating or under stress, **theta waves** appear mostly in children, and **delta waves** are normally seen during sleep.

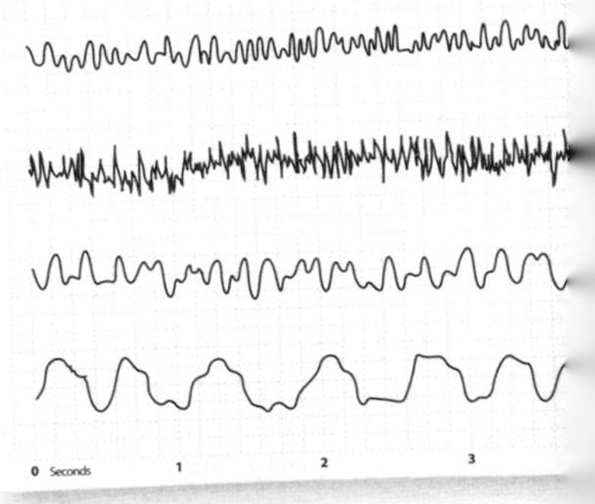
0 Seconds
1
2
3

13.14 **The twelve pairs of cranial nerves are classified as sensory, special sensory, motor, or mixed nerves** p. 486

58. The sensory cranial nerves are the **ophthalmic** and **maxillary** branches of the trigeminal nerve (V).
59. The special sensory cranial nerves are the **olfactory (I)**, **optic (II)**, and **vestibulocochlear (VIII) nerves**.
60. The motor cranial nerves are the **oculomotor (III)**, **trochlear (IV)**, **abducens (VI)**, **accessory (XI)**, and **hypoglossal (XII) nerves**.
61. The mixed cranial nerves are the **trigeminal (V)**, **facial (VII)**, **glossopharyngeal (IX)**, and **vagus (X) nerves**.

SECTION 2 • Sensory and Motor Pathways

13.15 **Sensations carried by sensory pathways to the CNS begin with transduction at a sensory receptor** p. 489

62. The **general senses** detect temperature, pain, touch, pressure, vibration, and proprioception. The information carried by a sensory pathway is a **sensation**, and the conscious awareness of a sensation is a **perception**.

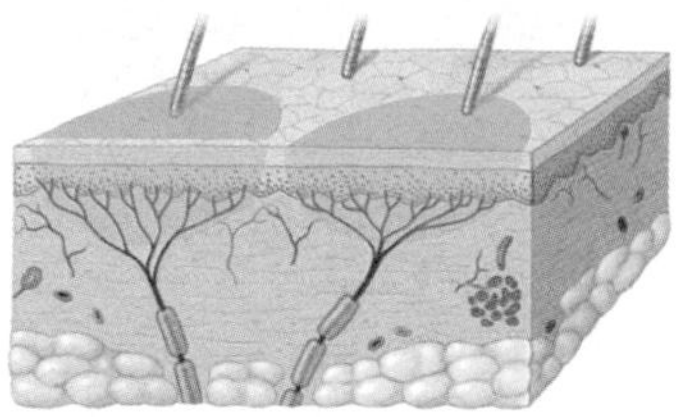

63. The area monitored by a single receptor cell is its **receptive field**. The larger a receptive field is, the poorer your ability to localize a stimulus.

13.16 **Receptors for the general senses are classified by function** p. 490

64. The simplest receptors are the dendrites of sensory neurons. These free nerve endings are not protected by accessory structures.
65. The general sensory receptors are **nociceptors**, **thermoreceptors**, **chemoreceptors**, and **mechanoreceptors**.

13.17 **General sensory receptors have a simple structure and are widely distributed in the body** p. 492

66. The greatest diversity of sensory receptors is found in the skin.
67. The important receptors of the skin are **free nerve endings**, **root hair plexuses**, **tactile discs**, **tactile corpuscles**, **lamellated corpuscles**, and **Ruffini corpuscles**.

13.18 **Three major somatic sensory pathways carry information from the skin and muscles to the CNS** p. 494

68. The **spinothalamic pathway** has two tracts. The **anterior spinothalamic tracts** carry crude touch and pressure sensations. The **lateral spinothalamic tracts** carry pain and temperature sensations.
69. In the spinothalamic pathway, axons of **first-order neurons** enter the spinal cord and synapse with **second-order neurons** within the posterior gray horns. The axons of these interneurons cross to the opposite side of the spinal cord before ascending to the thalamus. **Third-order neurons** synapse in the primary sensory cortex.
70. A **sensory homunculus** is a functional map illustrating the area of primary sensory cortex devoted to a particular body region.
71. The **posterior column pathway** carries sensations of fine touch, pressure, vibration, and proprioception.
72. The **spinocerebellar pathway** sends proprioceptive information to the cerebellum.

13.19 **The somatic nervous system controls skeletal muscles through upper and lower motor neurons** p. 496

73. Somatic motor pathways involve at least two motor neurons: an **upper motor neuron**, whose cell body lies in a CNS processing center, and a **lower motor neuron**, whose cell body lies in a nucleus of the brain stem or spinal cord.
74. The **corticospinal pathway** provides voluntary control over skeletal muscles.
75. A **motor homunculus** is functional map illustrating the primary motor cortex area devoted to a particular body region.
76. The **medial pathway** controls gross movements and proximal limb muscles. The **lateral pathway** controls the distal limb muscles that perform precise movements.

13.20 **There are multiple levels of somatic motor control** p. 498

77. The closer a motor center is to the cerebral cortex, the more complex the motor activities will be.
78. Increasing levels of complexity begin at the brain stem and spinal cord, and continue to the pons and medulla oblongata, hypothalamus, thalamus and midbrain, basal nuclei, and cerebral cortex. The cerebellum is involved at multiple levels.
79. Voluntary movement begins with a conscious decision in the frontal lobes, and then information is relayed to the motor association areas.
80. As the movement begins, the motor association areas send instructions to the primary motor cortex.

13.21 **Nervous system disorders may result from problems with neurons, pathways, or a combination of the two** p. 500

81. **Referred pain** is pain in a part of the body other than its actual source. **Parkinson's disease** is caused by a decrease in dopamine secretion by the substantia nigra. **Rabies** damages the CNS through retrograde flow beginning at the axon terminals.
82. **Cerebral palsy** refers to a number of disorders that affect voluntary motor movement. **Amyotrophic lateral sclerosis (ALS)** is a progressive degeneration of CNS motor neurons.
83. **Alzheimer's disease** is a progressive disorder characterized by the loss of higher-order cerebral functions. **Multiple sclerosis (MS)** results from demyelination of the axons of the optic nerve, brain, and spinal cord.

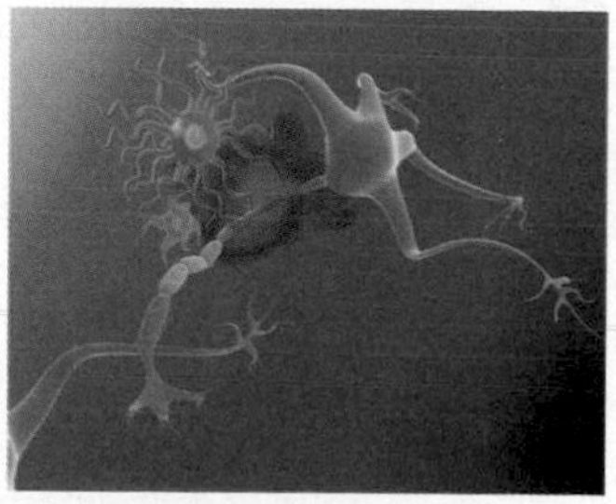

An Alzheimer's plaque

Chapter Review Questions

Labeling

Label the structures of the brain indicated in the following diagram.

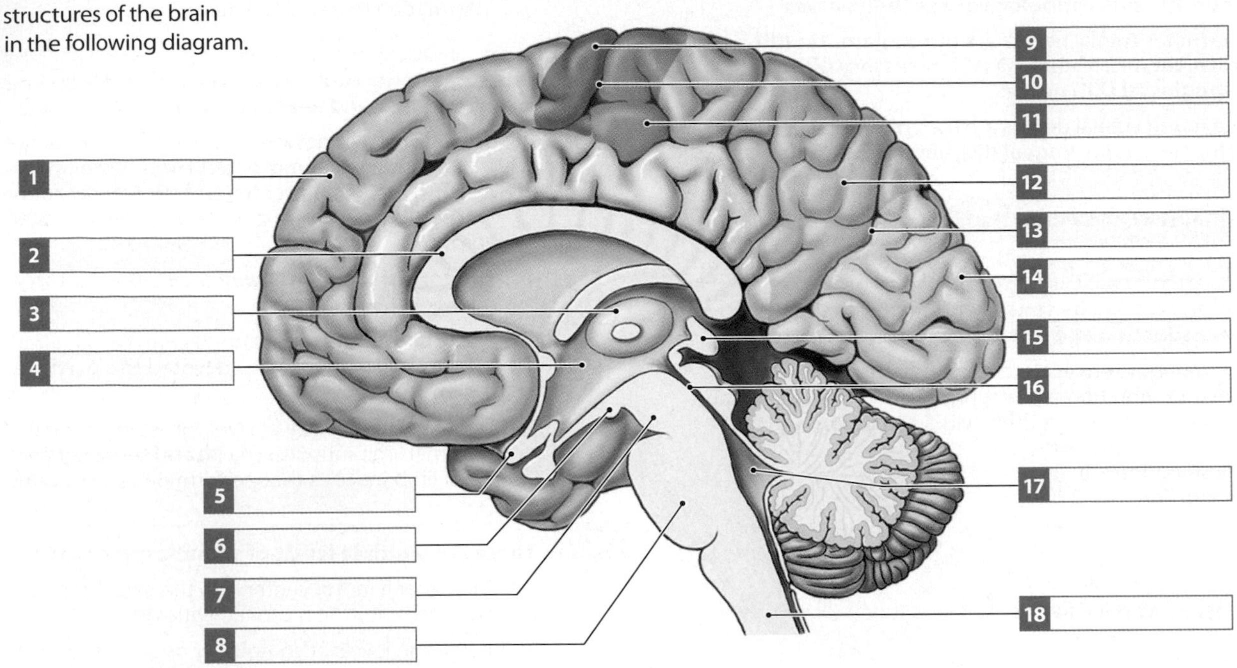
1
2
3
4
5
6
7
8
9
10
11
12
13
14
15
16
17
18

True/False

Indicate whether each statement is true or false.

19 The special sensory cranial nerves are I, II, VIII, and X.

20 Proprioceptors are the general sensory receptors that monitor the positions of joints and muscles.

21 Parkinson's disease is characterized by demyelination that affects axons of the optic nerve, brain, and spinal cord.

22 Ruffini corpuscles are often referred to as free nerve endings.

23 The posterior column pathway carries sensations of touch, pressure, vibration, and proprioception.

19 ____

20 ____

21 ____

22 ____

23 ____

Multiple choice

Select the correct answer from the list provided.

24 The adult cerebrum develops from the
- ☐ a) diencephalon.
- ☐ b) telencephalon.
- ☐ c) mesencephalon.
- ☐ d) metencephalon.

25 The cerebral aqueduct connects the
- ☐ a) lateral ventricles to the third ventricle.
- ☐ b) third ventricle to the fourth ventricle.
- ☐ c) fourth ventricle to the central canal.
- ☐ d) lateral ventricles to the interventricular foramen.

26 The fold of dura mater that projects between the cerebral hemispheres is the
- ☐ a) falx cerebri.
- ☐ b) falx cerebelli.
- ☐ c) tentorium cerebelli.
- ☐ d) superior sagittal sinus.

27 The leading site where CSF is absorbed into venous circulation is the
- ☐ a) lateral apertures.
- ☐ b) median aperture.
- ☐ c) choroid plexuses.
- ☐ d) arachnoid granulations.

28 The superior and inferior colliculi collectively form the

- a) thalamus.
- b) corpora quadrigemina.
- c) limbic system.
- d) reticular activating system.

29 The centers in the pons that adjust the respiratory rhythmicity centers in the medulla oblongata are the

- a) apneustic and pneumotaxic centers.
- b) inferior and superior peduncles.
- c) cardiac and vasomotor centers
- d) nucleus gracilis and nucleus cuneatus.

30 The final relay point for ascending sensory information that will be projected to the primary sensory cortex is the

- a) hypothalamus.
- b) thalamus.
- c) spinal cord.
- d) pons.

31 The brain waves normally seen during deep sleep in people of all ages are called

- a) alpha waves.
- b) beta waves.
- c) theta waves.
- d) delta waves.

Short answer

32 Cerebral meningitis is a condition in which the meninges of the brain become inflamed as the result of viral or bacterial infection. This condition can be life threatening. Why?

33 Julia is about to have a cavity in her tooth filled. The tooth in need of attention is on the right side of her bottom jaw in the far back. Which nerve do you think the dentist will block when he gives her an injection to control the pain? Is this a sensory nerve, motor nerve, or mixed nerve? Besides stopping the pain from the tooth, what other sensory or motor activities would you expect to be affected by the injection?

Chapter Integration • Applying what you have learned

A frightening awakening

Upon awakening one morning, John feels an unusual sensation on the left side of his face. He looks in the mirror and sees that half of his face is drooping. He tries to make a variety of facial expressions such as smiling and frowning, but can only do so on his right side. He is completely unable to control the facial muscles on his left side. Extremely alarmed at this condition, he drives himself to his physician's office right away.

John's doctor asks him a few questions about how his symptoms developed, and then gives him a physical and neurological examination. After asking a few more questions, the doctor tells John that his condition is called Bell's palsy. He assures John that although his symptoms are alarming, he has not had a stroke or a heart attack, and the Bell's palsy is likely to last only a few weeks.

Using what you have learned about cranial nerves, answer the following questions.

1 **Which cranial nerve do you think is most likely to be involved in John's Bell's palsy?**

2 **Besides the inability to move his facial muscles, which other symptoms would you expect John to be experiencing?**

3 **What advice would you give John to help him manage his condition?**

14 The Autonomic Nervous System

LEARNING OUTCOMES

These Learning Outcomes correspond by number to this chapter's modules and indicate what you should be able to do after completing the chapter.

SECTION 1 • Functional Anatomy of the Autonomic Nervous System (ANS)

- **14.1** Describe the control of skeletal muscles by the SNS and the control of visceral effectors by the ANS.
- **14.2** List the divisions of the ANS and the general functions of each.
- **14.3** Describe the structures and functions of the sympathetic and parasympathetic divisions of the ANS.
- **14.4** Describe the innervation patterns of the sympathetic and parasympathetic divisions of the ANS.
- **14.5** Describe the types of sympathetic and parasympathetic receptors and their associated neurotransmitters.
- **14.6** Describe the mechanisms of neurotransmitter release in the ANS, and explain the effects of neurotransmitters on target organs and tissues.

SECTION 2 • Autonomic Regulation and Control Mechanisms

- **14.7** Describe the role of the ANS in maintaining homeostasis during unconsciousness.
- **14.8** Discuss the relationship between the two divisions of the ANS and the significance of dual innervation.
- **14.9** Define a visceral reflex, and explain the significance of such reflexes.
- **14.10** Explain the roles of baroreceptors and chemoreceptors in homeostasis.
- **14.11** Describe the hierarchy of interacting levels of control in the ANS beginning with the hypothalamus.

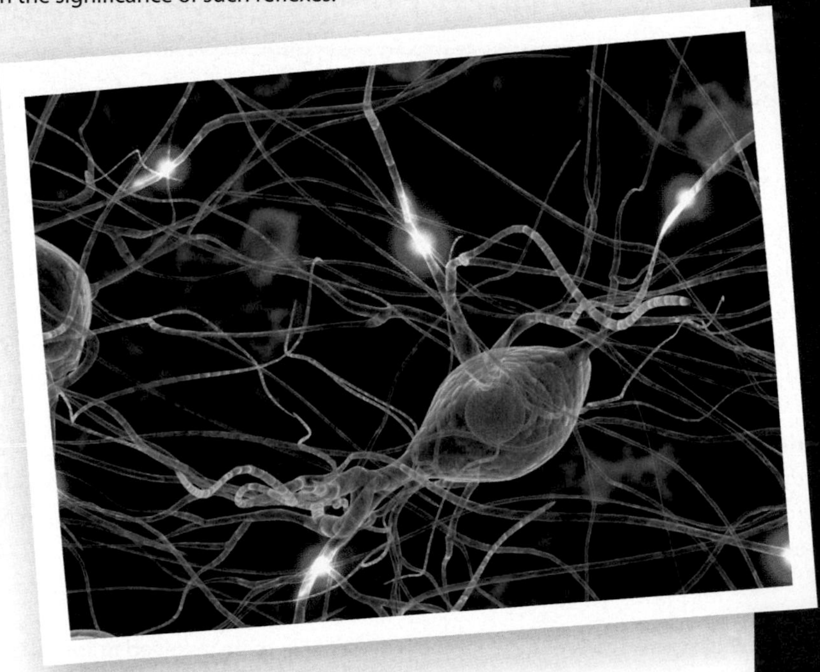

LO Learning Outcomes are repeated at the bottom of each module.

Ganglionic neurons of the ANS control visceral effectors

In Chapters 12 and 13 we considered the organization of the **somatic nervous system (SNS)**. The SNS provides conscious and subconscious control over the skeletal muscles of the body. In this section we consider the **autonomic nervous system (ANS)**, which controls visceral function mostly outside of our awareness.

1 In the SNS, motor neurons of the central nervous system exert direct control over skeletal muscles. The lower motor neurons may be controlled by reflexes based in the spinal cord or brain, or by upper motor neurons whose cell bodies lie within nuclei of the brain or at the primary motor cortex.

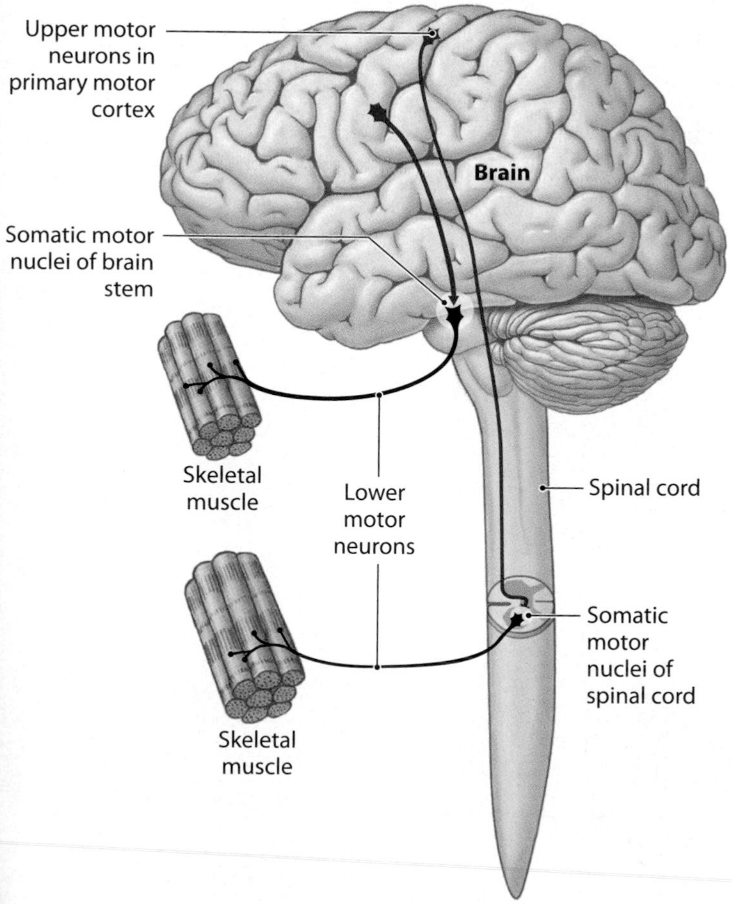

Upper motor neurons in primary motor cortex
Brain
Somatic motor nuclei of brain stem
Skeletal muscle
Lower motor neurons
Spinal cord
Somatic motor nuclei of spinal cord
Skeletal muscle

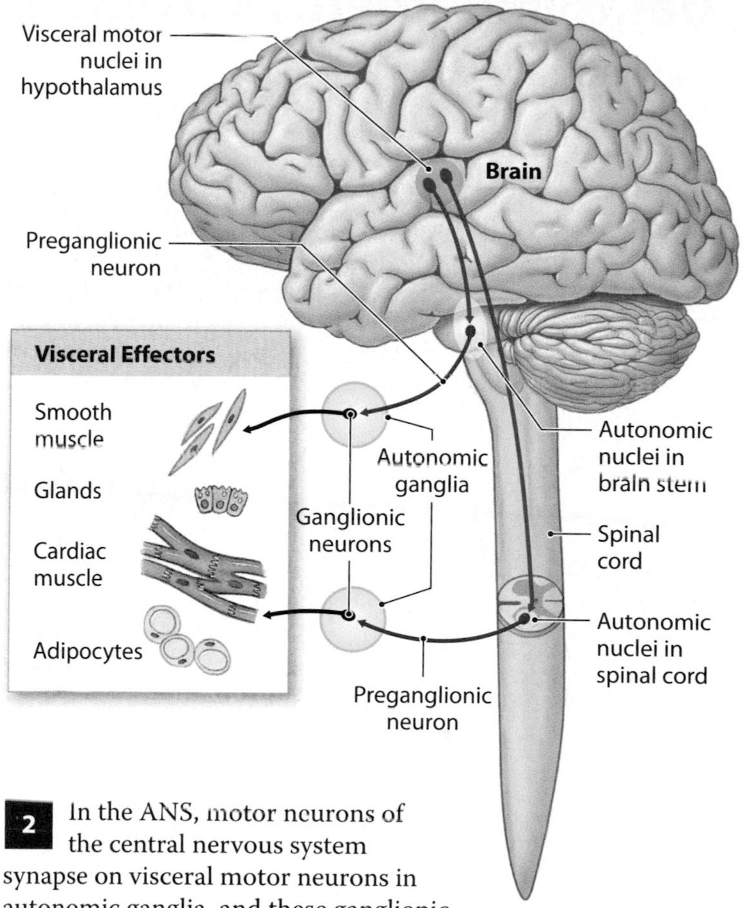

Visceral motor nuclei in hypothalamus
Brain
Preganglionic neuron
Visceral Effectors
Smooth muscle
Glands
Cardiac muscle
Adipocytes
Autonomic ganglia
Ganglionic neurons
Autonomic nuclei in brain stem
Spinal cord
Autonomic nuclei in spinal cord
Preganglionic neuron

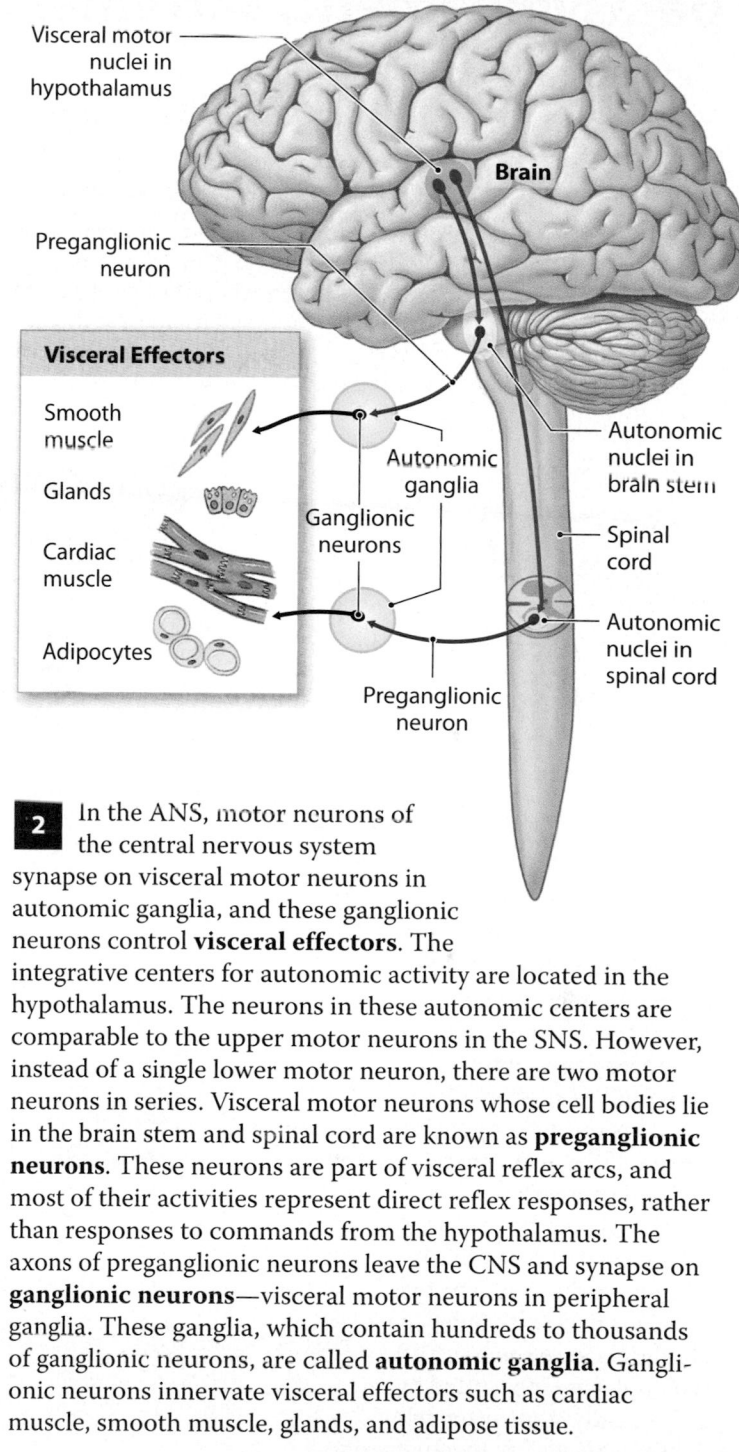

2 In the ANS, motor neurons of the central nervous system synapse on visceral motor neurons in autonomic ganglia, and these ganglionic neurons control **visceral effectors**. The integrative centers for autonomic activity are located in the hypothalamus. The neurons in these autonomic centers are comparable to the upper motor neurons in the SNS. However, instead of a single lower motor neuron, there are two motor neurons in series. Visceral motor neurons whose cell bodies lie in the brain stem and spinal cord are known as **preganglionic neurons**. These neurons are part of visceral reflex arcs, and most of their activities represent direct reflex responses, rather than responses to commands from the hypothalamus. The axons of preganglionic neurons leave the CNS and synapse on **ganglionic neurons**—visceral motor neurons in peripheral ganglia. These ganglia, which contain hundreds to thousands of ganglionic neurons, are called **autonomic ganglia**. Ganglionic neurons innervate visceral effectors such as cardiac muscle, smooth muscle, glands, and adipose tissue.

Module 14.1 Review

a. Compare the SNS with the ANS.

b. Describe the role of preganglionic neurons.

c. Explain the function of autonomic ganglia.

LO 14.1 Describe the control of skeletal muscles by the SNS and the control of visceral effectors by the ANS.

Module 14.2

The ANS consists of sympathetic, parasympathetic, and enteric divisions

Autonomic Nervous System

Sympathetic Division

In the sympathetic division, or **thoracolumbar** (thor-a-kō-LUM-bar) **division**, axons emerge from the cell bodies within the lateral gray horns of thoracic and superior lumbar segments of the spinal cord (T_1–L_2). These axons innervate ganglia relatively close to the spinal cord.

Parasympathetic Division

In the parasympathetic division, or **craniosacral** (krā-nē-ō-SĀ-krul) **division**, axons emerge from the brain stem nuclei and the lateral gray horns of the sacral segments of the spinal cord. These axons innervate ganglia very close to (or within) target organs.

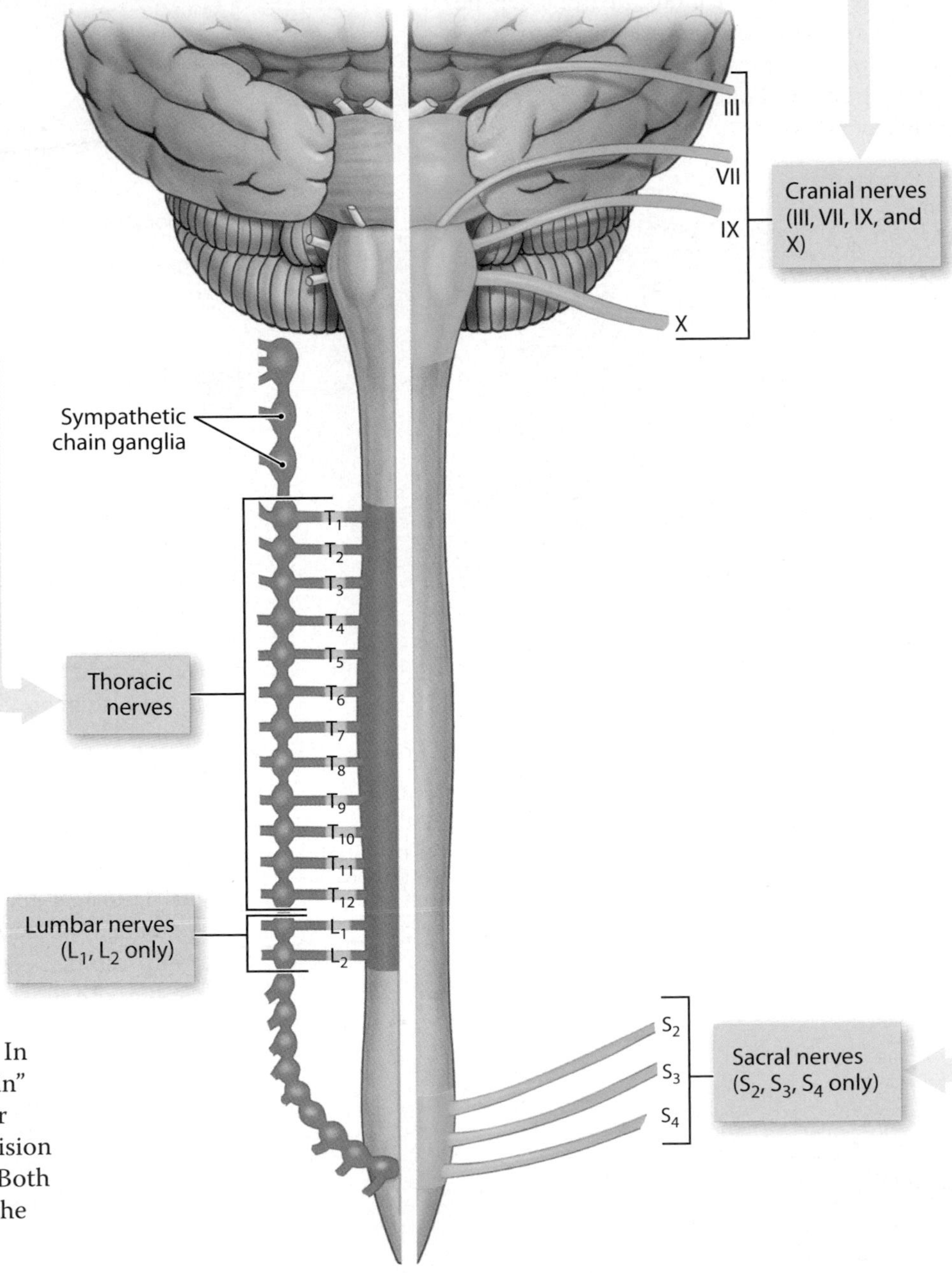

III
VII
IX
X
Cranial nerves (III, VII, IX, and X)
Sympathetic chain ganglia
T1
T2
T3
T4
T5
T6
T7
T8
T9
T10
T11
T12
L1
L2
Thoracic nerves
Lumbar nerves (L1, L2 only)
S2
S3
S4
Sacral nerves (S2, S3, S4 only)

1 The ANS contains two well-known divisions whose names are probably already familiar to you: the **sympathetic division** and the **parasympathetic division**. Most often, these two divisions have opposing effects: If the sympathetic division causes excitation, the parasympathetic causes inhibition. However, this is not always the case. The two divisions may work independently, because some structures are innervated by only one division; or the two divisions may work together, each controlling a stage of a complex process. In general, the sympathetic division "kicks in" only during periods of exertion, stress, or emergency, and the parasympathetic division predominates under resting conditions. Both primary divisions of the ANS influence the third autonomic division, known as the enteric nervous system.

2 The **enteric nervous system (ENS)** is an extensive network of neurons and nerve networks located in the walls of the digestive tract. Although the sympathetic and parasympathetic divisions influence the activities of the ENS, many complex visceral reflexes are initiated and coordinated locally, without instructions from the CNS. Altogether, the ENS has about 100 million neurons—at least as many as the spinal cord. It also uses the same neurotransmitters found in the brain. The ENS will be discussed in our examination of the digestive system in Chapter 22.

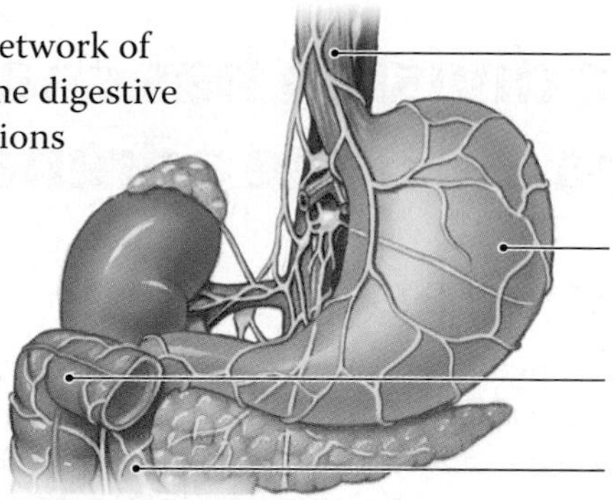

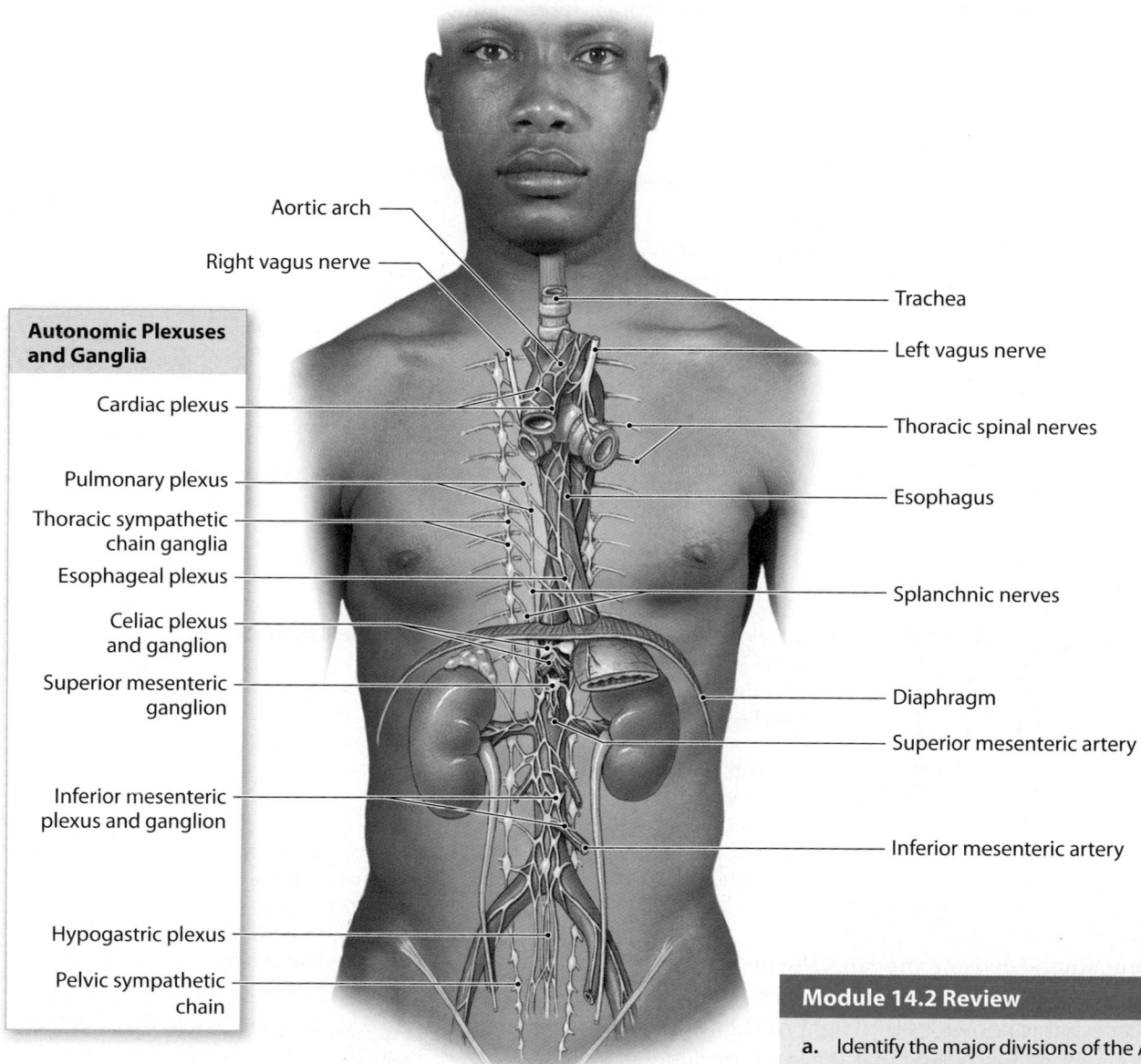

Esophagus
Stomach
Large intestine
Small intestine

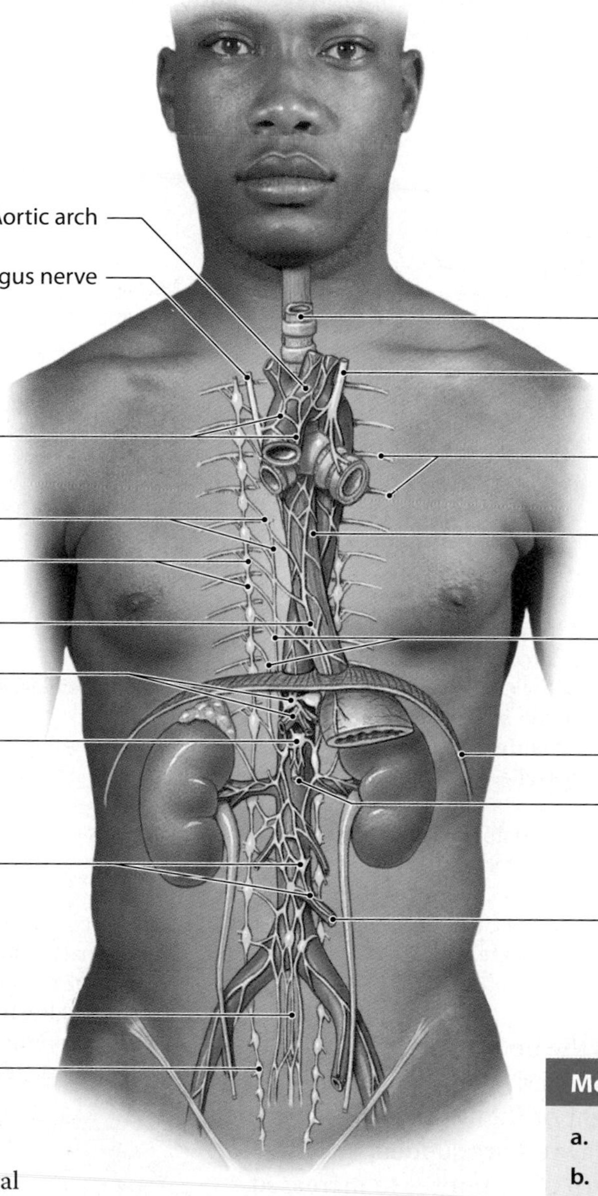

Aortic arch
Right vagus nerve
Autonomic Plexuses and Ganglia
Cardiac plexus
Pulmonary plexus
Thoracic sympathetic chain ganglia
Esophageal plexus
Celiac plexus and ganglion
Superior mesenteric ganglion
Inferior mesenteric plexus and ganglion
Hypogastric plexus
Pelvic sympathetic chain
Trachea
Left vagus nerve
Thoracic spinal nerves
Esophagus
Splanchnic nerves
Diaphragm
Superior mesenteric artery
Inferior mesenteric artery

3 In this chapter we discuss the general organization of the sympathetic and parasympathetic divisions. In dissection, the sympathetic and parasympathetic nerves are not clearly segregated. Instead, they form a complex of intertwining nerves and plexuses. To clarify the structure of these divisions, we will rely on diagrams.

Module 14.2 Review

a. Identify the major divisions of the ANS.
b. Which division of the ANS is responsible for the physiological changes you experience when startled by a loud noise?
c. Compare the anatomy of the sympathetic division with that of the parasympathetic division.

LO **14.2** List the divisions of the ANS and the general functions of each.

The sympathetic division has chain ganglia, collateral ganglia, and the adrenal medullae ...

Organization of the Sympathetic Division

1 In the sympathetic division, preganglionic neurons from the thoracic and superior lumbar segments of the spinal cord synapse on ganglionic neurons that may be located (1) within the sympathetic chain of ganglia near the spinal cord, (2) in collateral ganglia that lie within the thoracic or abdominopelvic cavities, or (3) within modified ganglion cells in the adrenal medullae. Because the ganglionic neurons are relatively close to the vertebral column, the axons of the preganglionic neurons (**preganglionic fibers**) are short compared to axons of the ganglionic neurons (**postganglionic fibers**).

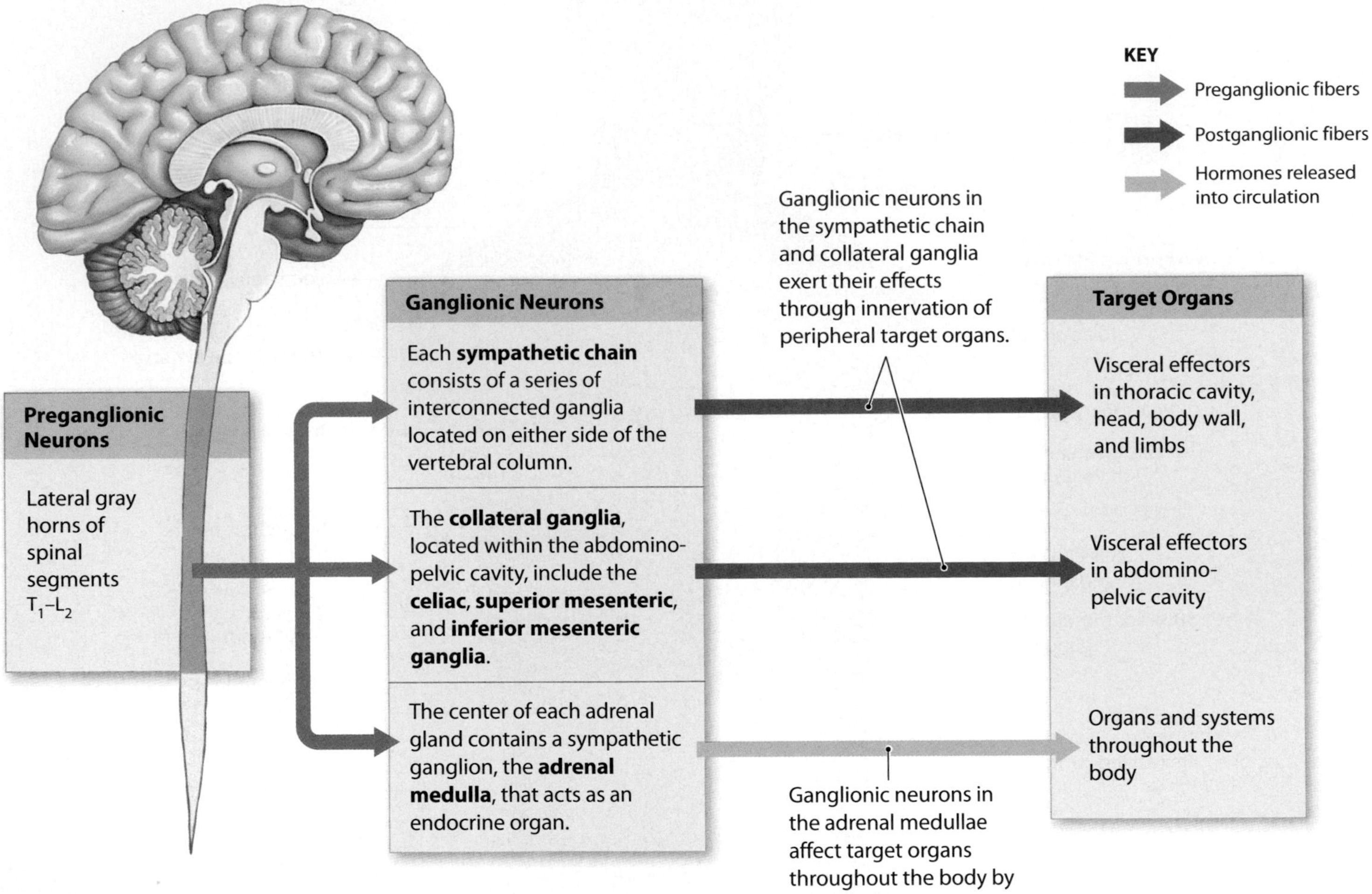
KEY
Preganglionic fibers
Postganglionic fibers
Hormones released into circulation
Preganglionic Neurons
Lateral gray horns of spinal segments T_1–L_2
Ganglionic Neurons
Each sympathetic chain consists of a series of interconnected ganglia located on either side of the vertebral column.
The collateral ganglia, located within the abdomino-pelvic cavity, include the celiac, superior mesenteric, and inferior mesenteric ganglia.
The center of each adrenal gland contains a sympathetic ganglion, the adrenal medulla, that acts as an endocrine organ.
Ganglionic neurons in the sympathetic chain and collateral ganglia exert their effects through innervation of peripheral target organs.
Ganglionic neurons in the adrenal medullae affect target organs throughout the body by releasing hormones into the general circulation.
Target Organs
Visceral effectors in thoracic cavity, head, body wall, and limbs
Visceral effectors in abdomino-pelvic cavity
Organs and systems throughout the body

The sympathetic division prepares the body for heightened levels of somatic activity. When fully activated, this division produces what is known as the "fight or flight" response, which readies the body for a crisis that may require sudden, intense physical activity. The general pattern of responses to increased levels of sympathetic activity includes the following changes: (1) heightened mental alertness, (2) increased metabolic rate, (3) decreased digestive and urinary functions, (4) mobilization of energy reserves, (5) dilation of respiratory passageways and increased respiratory rate, (6) increased heart rate and blood pressure, and (7) activation of sweat glands.

... whereas the parasympathetic division has terminal or intramural ganglia

Organization of the Parasympathetic Division

2 In the parasympathetic division, a typical preganglionic fiber synapses on six to eight ganglionic neurons. These neurons may be located in **terminal ganglia**, located near the target organ, or in **intramural ganglia** (*murus*, wall), which are embedded in the tissues of the target organ. Terminal ganglia are usually paired. Examples of terminal ganglia include the parasympathetic ganglia associated with the cranial nerves. Intramural ganglia typically consist of interconnected masses and clusters of ganglion cells.

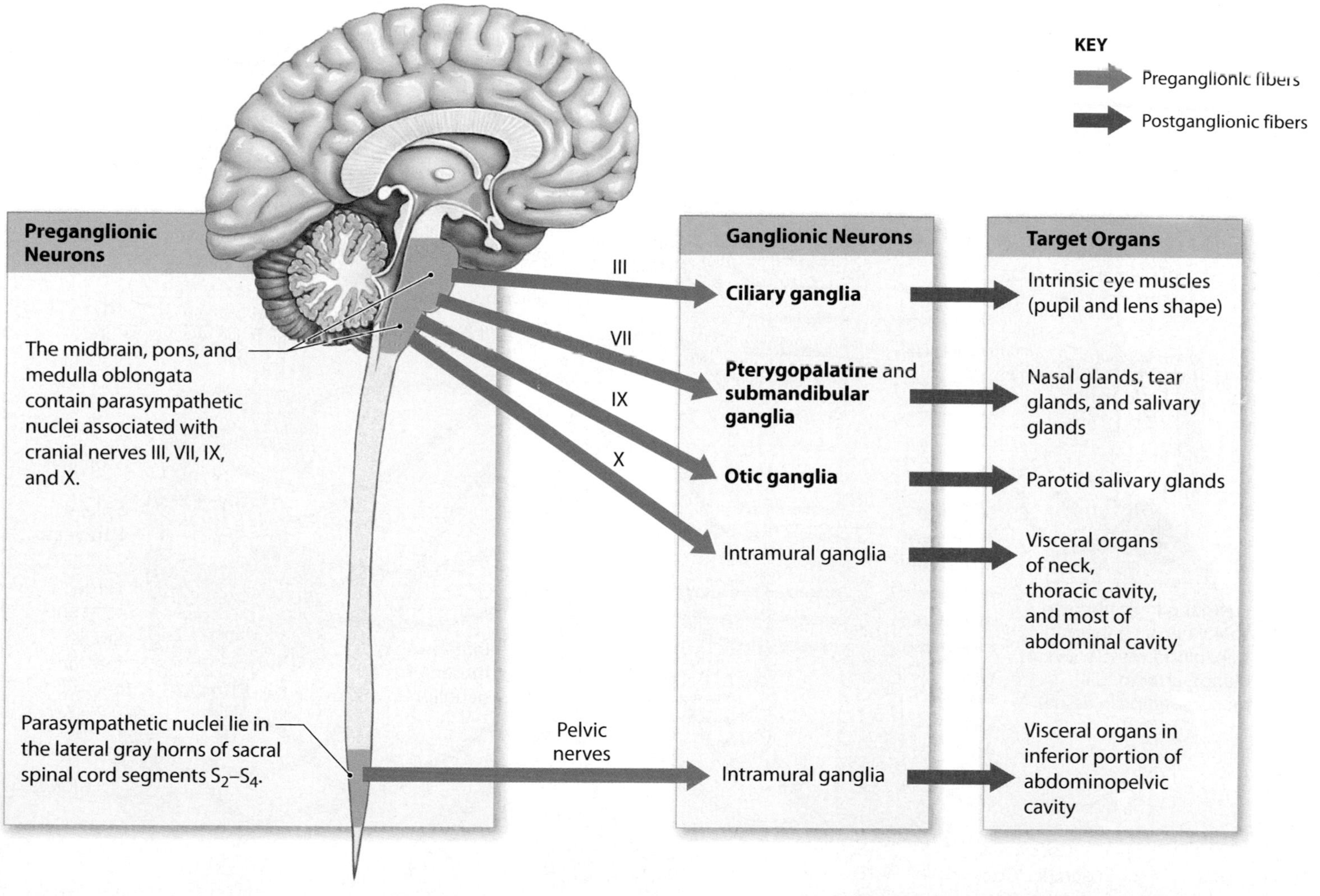
KEY
Preganglionic fibers
Postganglionic fibers
Preganglionic Neurons
The midbrain, pons, and medulla oblongata contain parasympathetic nuclei associated with cranial nerves III, VII, IX, and X.
Parasympathetic nuclei lie in the lateral gray horns of sacral spinal cord segments S_2–S_4.
III
VII
IX
X
Pelvic nerves
Ganglionic Neurons
Ciliary ganglia
Pterygopalatine and submandibular ganglia
Otic ganglia
Intramural ganglia
Intramural ganglia
Target Organs
Intrinsic eye muscles (pupil and lens shape)
Nasal glands, tear glands, and salivary glands
Parotid salivary glands
Visceral organs of neck, thoracic cavity, and most of abdominal cavity
Visceral organs in inferior portion of abdominopelvic cavity

The parasympathetic division is concerned with conserving energy and regulating visceral function. It is known as the "rest and digest" system. The overall pattern of responses to increased levels of parasympathetic activity includes the following changes: (1) decreased metabolic rate, (2) decreased heart rate and blood pressure, (3) increased secretion by salivary and digestive glands, (4) increased motility and blood flow in the digestive tract, and (5) stimulation of urination and defecation.

Module 14.3 Review

a. List general responses to increased sympathetic activity and to parasympathetic activity.

b. Starting in the spinal cord, trace the path of a nerve impulse through the sympathetic ANS to its target organ in the abdominopelvic cavity.

c. Describe an intramural ganglion.

LO **14.3** Describe the structures and functions of the sympathetic and parasympathetic divisions of the ANS.

The two ANS divisions innervate many of the same structures, but the innervation patterns are different

Innervation in the Sympathetic Division

1 The distribution of sympathetic fibers is the same on both sides of the body. For simplicity, the left side of this image shows the distribution to the skin (and to skeletal muscles and other tissues of the body wall), whereas the right side depicts the innervation of visceral organs.

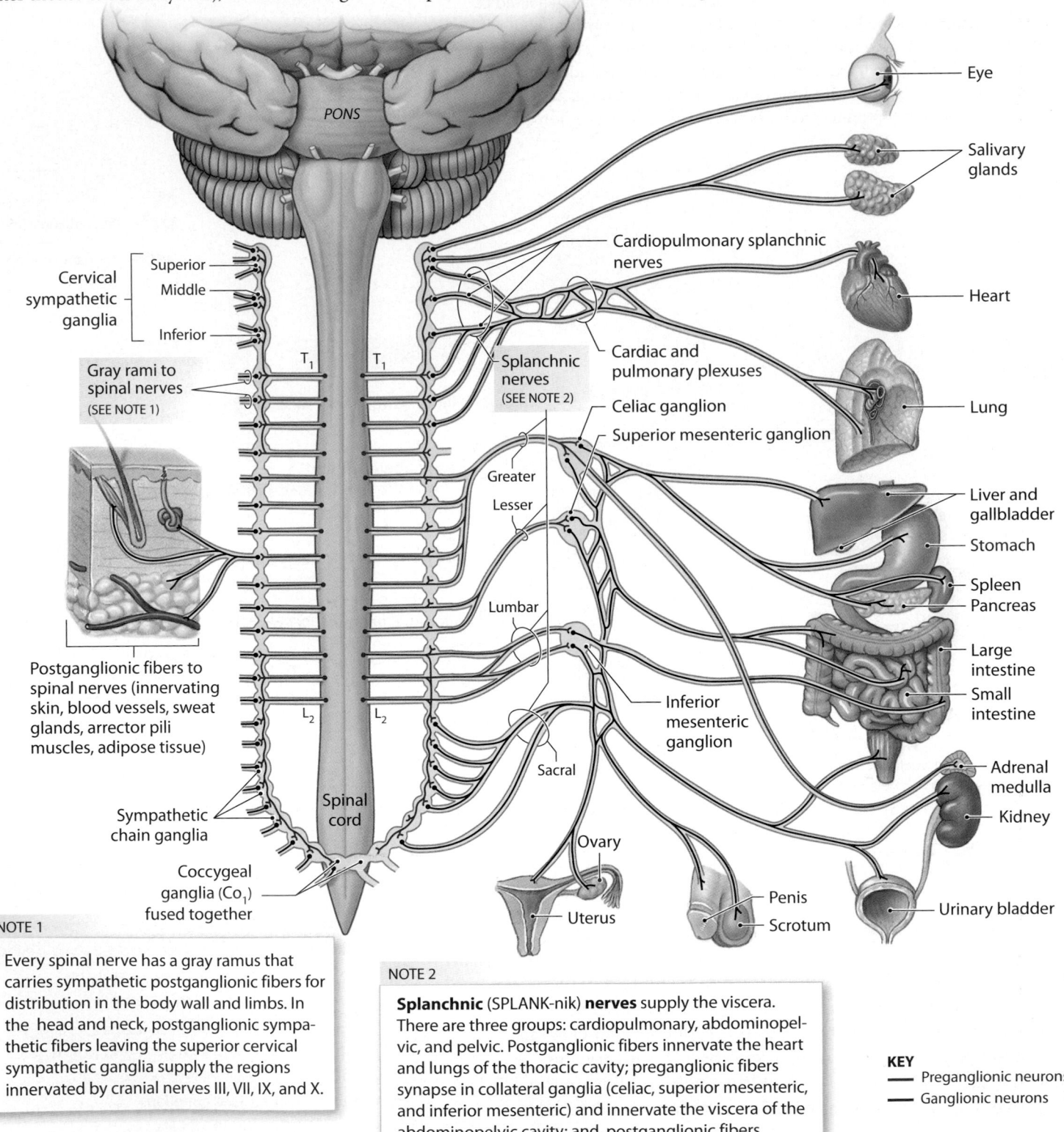

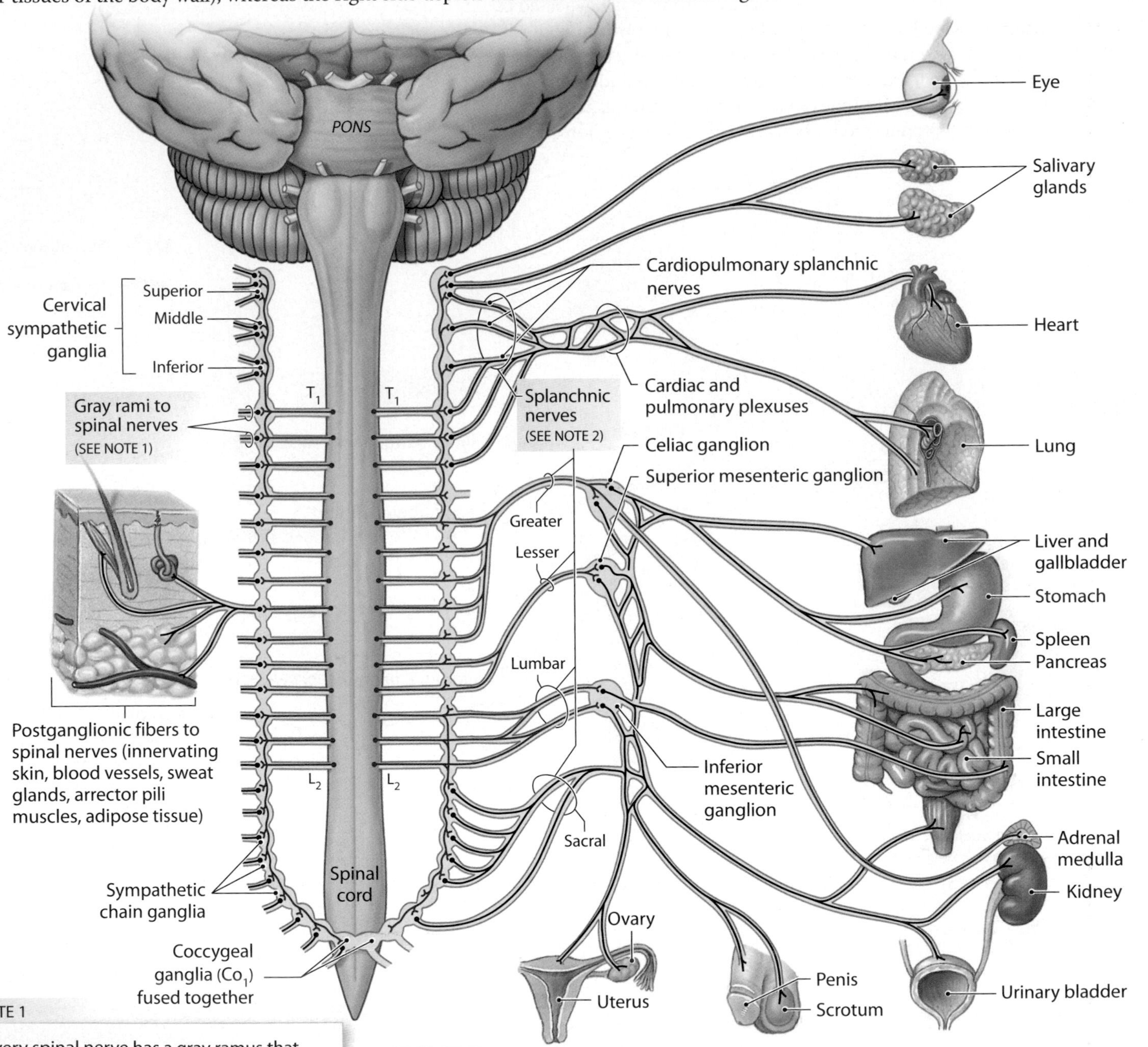
PONS
Eye
Salivary glands
Cardiopulmonary splanchnic nerves
Heart
Cervical sympathetic ganglia
Superior
Middle
Inferior
Cardiac and pulmonary plexuses
Splanchnic nerves (SEE NOTE 2)
Gray rami to spinal nerves (SEE NOTE 1)
T_1
T_1
Celiac ganglion
Lung
Superior mesenteric ganglion
Greater
Lesser
Liver and gallbladder
Stomach
Spleen
Pancreas
Lumbar
Large intestine
Small intestine
Postganglionic fibers to spinal nerves (innervating skin, blood vessels, sweat glands, arrector pili muscles, adipose tissue)
L_2
L_2
Inferior mesenteric ganglion
Sacral
Adrenal medulla
Kidney
Sympathetic chain ganglia
Spinal cord
Ovary
Coccygeal ganglia (Co_1) fused together
Uterus
Penis
Scrotum
Urinary bladder
KEY
Preganglionic neurons
Ganglionic neurons

NOTE 1

Every spinal nerve has a gray ramus that carries sympathetic postganglionic fibers for distribution in the body wall and limbs. In the head and neck, postganglionic sympathetic fibers leaving the superior cervical sympathetic ganglia supply the regions innervated by cranial nerves III, VII, IX, and X.

NOTE 2

Splanchnic (SPLANK-nik) **nerves** supply the viscera. There are three groups: cardiopulmonary, abdominopelvic, and pelvic. Postganglionic fibers innervate the heart and lungs of the thoracic cavity; preganglionic fibers synapse in collateral ganglia (celiac, superior mesenteric, and inferior mesenteric) and innervate the viscera of the abdominopelvic cavity; and, postganglionic fibers inferior to L_2 innervate the pelvic cavity.

Innervation in the Parasympathetic Division

2 This image shows the parasympathetic innervation on one side of the body; the innervation on the opposite side (not shown) follows the same pattern.

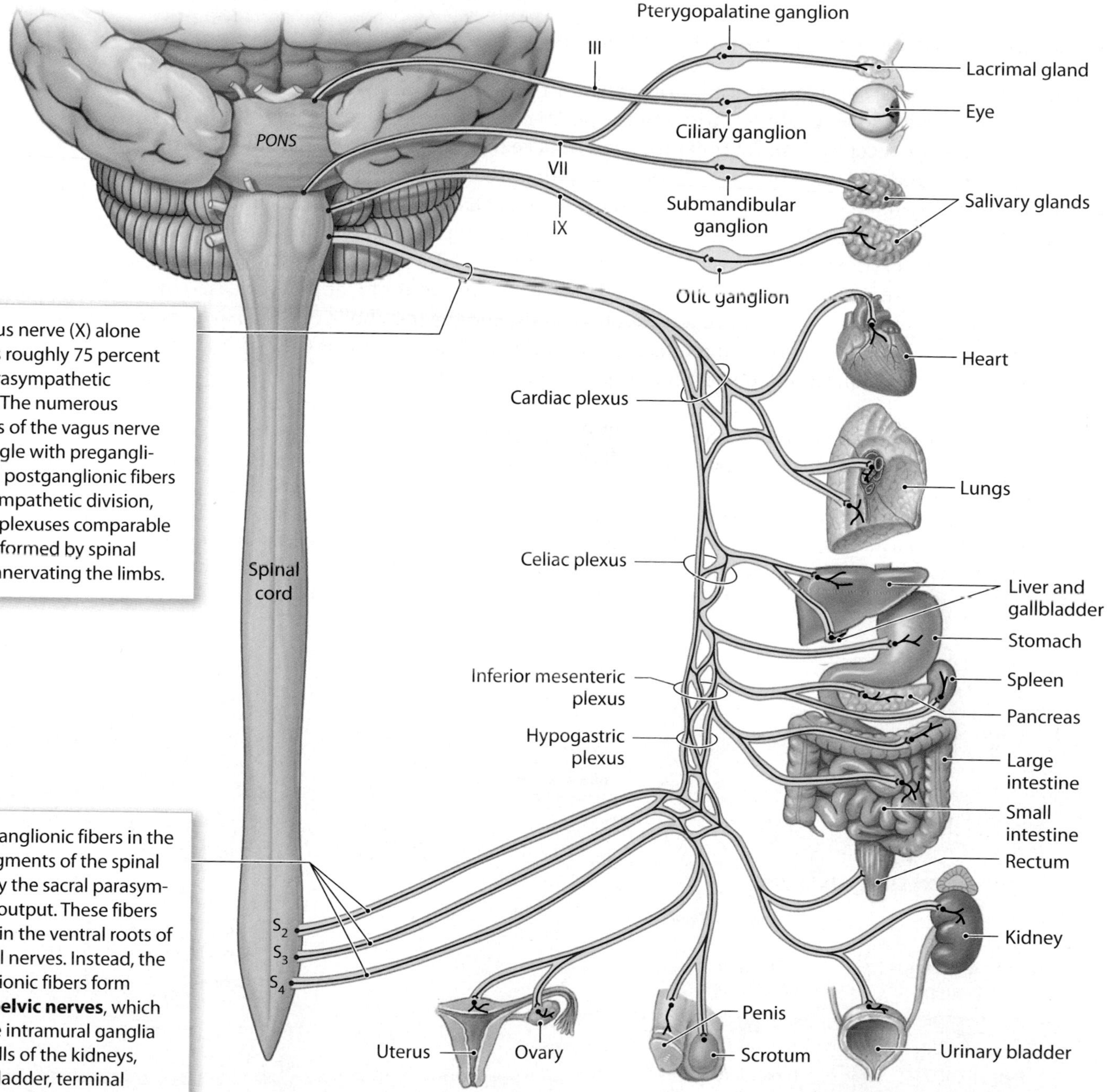
Pterygopalatine ganglion
III
Lacrimal gland
Eye
Ciliary ganglion
PONS
VII
Submandibular ganglion
Salivary glands
IX
Otic ganglion
Heart
Cardiac plexus
Lungs
Celiac plexus
Liver and gallbladder
Stomach
Spinal cord
Inferior mesenteric plexus
Spleen
Pancreas
Hypogastric plexus
Large intestine
Small intestine
Rectum
S_2
S_3
S_4
Kidney
Penis
Uterus
Ovary
Scrotum
Urinary bladder

The vagus nerve (X) alone provides roughly 75 percent of all parasympathetic outflow. The numerous branches of the vagus nerve intermingle with preganglionic and postganglionic fibers of the sympathetic division, forming plexuses comparable to those formed by spinal nerves innervating the limbs.

The preganglionic fibers in the sacral segments of the spinal cord carry the sacral parasympathetic output. These fibers do not join the ventral roots of the spinal nerves. Instead, the preganglionic fibers form distinct **pelvic nerves**, which innervate intramural ganglia in the walls of the kidneys, urinary bladder, terminal portions of the large intestine, and sex organs.

Module 14.4 Review

a. Define splanchnic nerves.

b. Name the plexuses innervated by the vagus nerve.

c. Which nerve carries most of the parasympathetic outflow?

LO **14.4** Describe the innervation patterns of the sympathetic and parasympathetic divisions of the ANS.

Membrane receptors at target organs mediate the effects of sympathetic and parasympathetic stimulation

Neurotransmitter Release in the Sympathetic Division

The effects of sympathetic stimulation result primarily from the interactions of norepinephrine (NE) and epinephrine (E) with **adrenergic receptors** in the plasma membrane. There are two classes of sympathetic adrenergic receptors: **alpha receptors** and **beta receptors**. In general, NE stimulates alpha receptors to a greater degree than it does beta receptors, whereas E stimulates both classes of receptors. Localized sympathetic activity involving the release of NE at sympathetic terminals primarily affects alpha receptors located near those terminals. The effects typically persist for a few seconds. By contrast, generalized sympathetic activation and release of E and NE by the adrenal medullae affect alpha and beta receptors throughout the body. Tissue concentrations of E and NE released by the adrenal medullae remain elevated for as long as 30 seconds, and the metabolic effects may persist for several minutes. Because the adrenal medullae release three times as much E as they do NE, during sympathetic activation the effects of beta receptors predominate.

1 Stimulation of alpha (α) receptors activates their associated G proteins on the cytoplasmic side of the plasma membrane. Depending on receptor type, their activation triggers different activities in the cell. Stimulation of alpha-1 receptors generally excites the target cell, whereas stimulation of alpha-2 receptors generally inhibits the target cell.

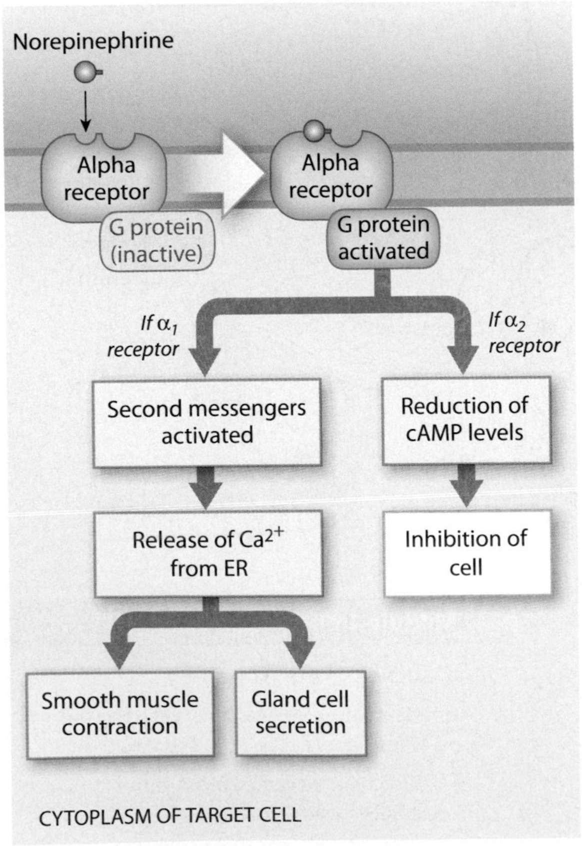
Norepinephrine
Alpha receptor
G protein (inactive)
Alpha receptor
G protein activated
If α_1 receptor
If α_2 receptor
Second messengers activated
Reduction of cAMP levels
Release of Ca^{2+} from ER
Inhibition of cell
Smooth muscle contraction
Gland cell secretion
CYTOPLASM OF TARGET CELL

2 Beta (β) receptors are located on the plasma membranes of cells in many organs, including skeletal muscles, the lungs, the heart, and the liver. Stimulation of beta receptors and G protein activation triggers changes in the metabolic activity of the target cell. The three types of beta receptors are beta-1 (β_1), beta-2 (β_2), and beta-3 (β_3).

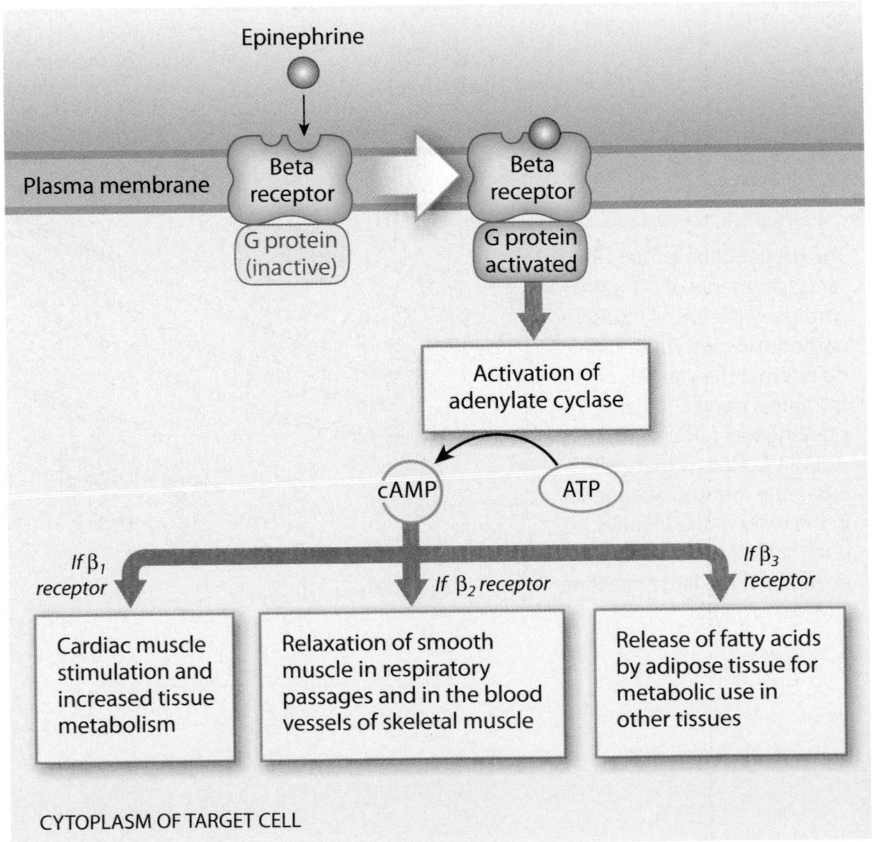
Epinephrine
Plasma membrane
Beta receptor
G protein (inactive)
Beta receptor
G protein activated
Activation of adenylate cyclase
cAMP
ATP
If β_1 receptor
If β_2 receptor
If β_3 receptor
Cardiac muscle stimulation and increased tissue metabolism
Relaxation of smooth muscle in respiratory passages and in the blood vessels of skeletal muscle
Release of fatty acids by adipose tissue for metabolic use in other tissues
CYTOPLASM OF TARGET CELL

Neurotransmitter Release in the Parasympathetic Division

Although all the synapses and neuromuscular or neuroglandular junctions of the parasympathetic division use the same transmitter—acetylcholine (ACh)—two types of **cholinergic receptors** occur on the postsynaptic membranes: nicotinic receptors and muscarinic receptors.

3 **Nicotinic** (nik-ō-TIN-ik) **receptors** are found on all postganglionic neurons, adrenal medullae cells, and at the neuromuscular junctions of skeletal muscle fibers. ACh always excites the ganglionic neuron or muscle fiber by opening chemically gated channels in the postsynaptic membrane. Nicotinic receptors are also stimulated by nicotine, a powerful toxin that can be obtained from a variety of sources, including tobacco.

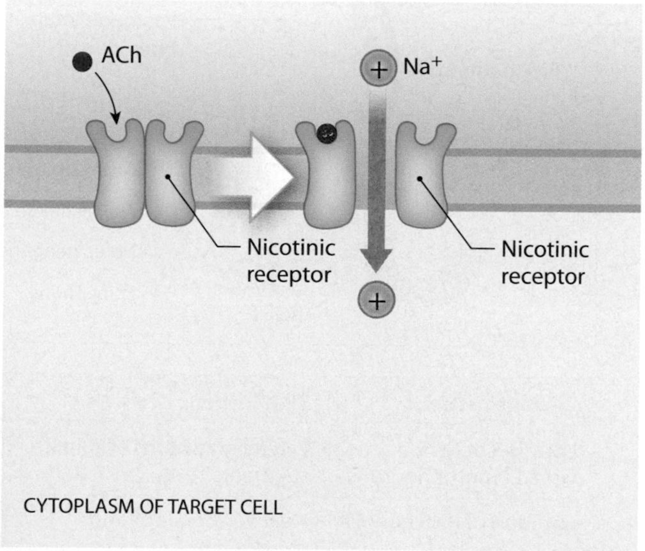
ACh
Na^+
Nicotinic receptor
Nicotinic receptor
CYTOPLASM OF TARGET CELL

4 **Muscarinic** (mus-ka-RIN-ik) **receptors** are G protein–coupled receptors. They occur at cholinergic neuromuscular or neuroglandular junctions in the parasympathetic division, and at the few cholinergic junctions in the sympathetic division. Muscarinic receptor stimulation and G protein activation produces longer-lasting effects than does the stimulation of nicotinic receptors. The response, which reflects the activation or inactivation of specific enzymes, can be either excitatory or inhibitory. Muscarinic receptors are stimulated by muscarine, a toxin produced by some poisonous mushrooms.

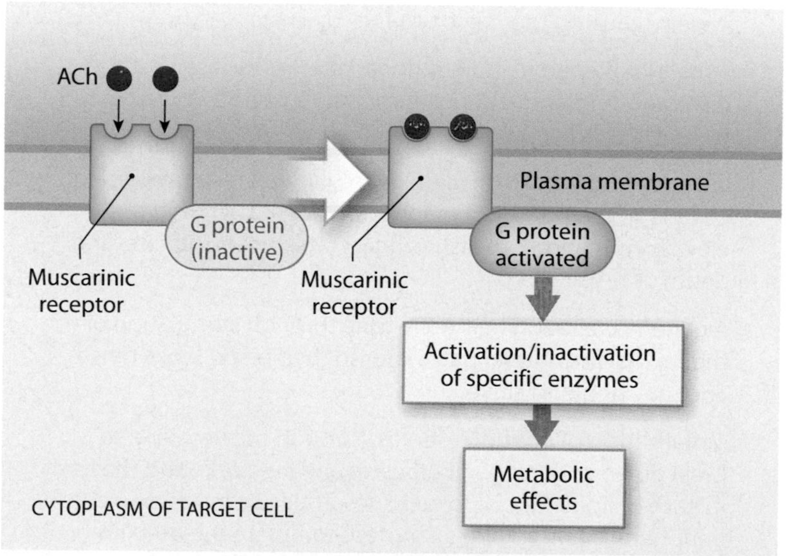
ACh
Plasma membrane
G protein (inactive)
Muscarinic receptor
Muscarinic receptor
G protein activated
Activation/inactivation of specific enzymes
Metabolic effects
CYTOPLASM OF TARGET CELL

Module 14.5 Review

a. Compare and contrast alpha and beta receptors.
b. A person with high blood pressure (hypertension) is prescribed a drug that blocks beta receptors. How could this medication alleviate hypertension?
c. Compare nicotinic receptors with muscarinic receptors.

LO **14.5** Describe the types of sympathetic and parasympathetic receptors and their associated neurotransmitters.

The functional differences between the two ANS divisions reflect their divergent anatomical and physiological characteristics

Functional Characteristics of the Sympathetic Division

1 Unlike the parasympathetic division, the entire sympathetic division can be quickly activated. This table summarizes the effects of sympathetic activation.

Effects of Sympathetic Activation

The sympathetic division can change the activities of tissues and organs by releasing NE at peripheral synapses, and by distributing E and NE throughout the body in the bloodstream. The visceral motor fibers that target specific effectors, such as smooth muscle fibers in blood vessels of the skin, can be activated in reflexes that do not involve other visceral effectors.

In a crisis, however, the entire division responds. This event, called **sympathetic activation**, is controlled by sympathetic centers in the hypothalamus. The effects are not limited to peripheral tissues; sympathetic activation also alters CNS activity. During sympathetic activation, the following changes occur:

- Increased alertness by stimulation of the reticular activating system, causing the person to feel "on edge."
- A feeling of energy and euphoria, often associated with a disregard for danger and a temporary insensitivity to painful stimuli.
- Increased activity in the cardiovascular and respiratory centers of the pons and medulla oblongata, leading to elevations in blood pressure, heart rate, breathing rate, and depth of respiration.
- A general elevation in muscle tone through stimulation of the medial and lateral pathways, so the person looks tense and may begin to shiver.
- Mobilization of energy reserves, through the accelerated breakdown of glycogen in muscle and liver cells and the release of lipids by adipose tissues. These changes, plus the peripheral changes already noted, complete the preparations necessary for the person to cope with a stressful situation.

2 This illustration and the table below it summarize the anatomical characteristics of the sympathetic division of the ANS.

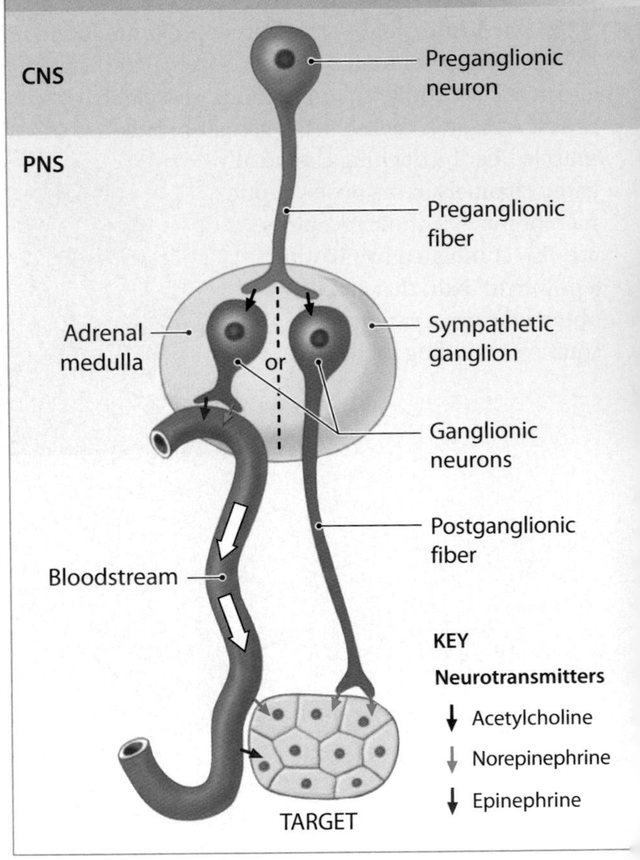
Sympathetic
CNS
PNS
Preganglionic neuron
Preganglionic fiber
Adrenal medulla
or
Sympathetic ganglion
Ganglionic neurons
Postganglionic fiber
Bloodstream
TARGET
KEY
Neurotransmitters
Acetylcholine
Norepinephrine
Epinephrine

Characteristic	Sympathetic Division
Location of CNS visceral motor neurons	Lateral gray horns of spinal segments T_1–L_2
Location of PNS ganglia	Near vertebral column
Preganglionic fibers Neurotransmitter	Short ACh
Postganglionic fibers Neurotransmitter	Long Normally NE, sometimes NO or ACh
General functions	Stimulates metabolism; increases alertness; prepares for emergency ("fight or flight")

Functional Characteristics of the Parasympathetic Division

3 This illustration and the table below it summarize the anatomical characteristics of the parasympathetic division of the ANS.

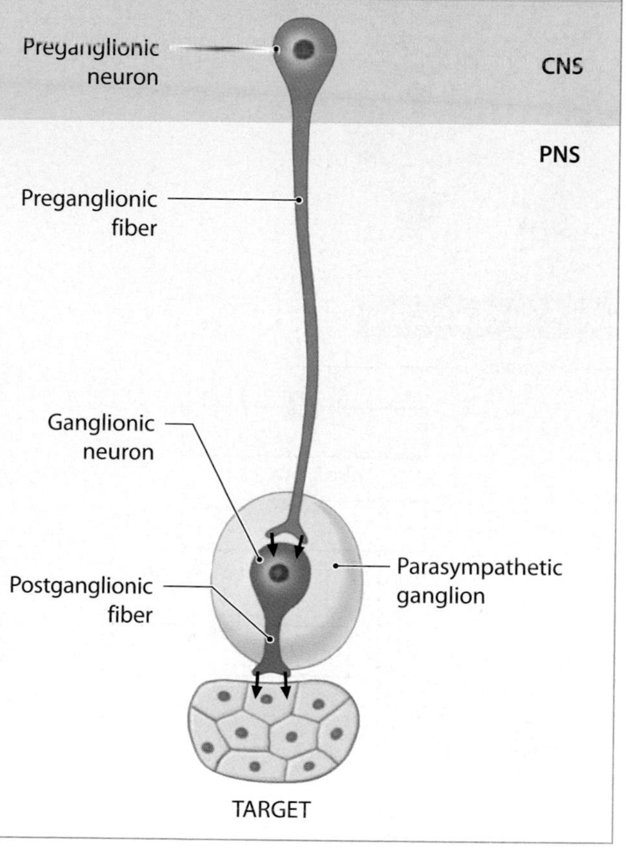
Parasympathetic
Preganglionic neuron
CNS
PNS
Preganglionic fiber
Ganglionic neuron
Postganglionic fiber
Parasympathetic ganglion
TARGET

Characteristic	Parasympathetic Division
Location of CNS visceral motor neurons	Brain stem and spinal segments S_2–S_4
Location of PNS ganglia	Typically intramural
Preganglionic fibers Neurotransmitter	Relatively long ACh
Postganglionic fibers Neurotransmitter	Relatively short ACh
General functions	Promotes relaxation, nutrient uptake, energy storage ("rest and digest")

4 The parasympathetic division does not release neurotransmitters directly into the bloodstream, so it does not undergo a division-wide activation. This table summarizes the effects of parasympathetic stimulation.

Effects of Parasympathetic Activation

Under normal conditions, the entire parasympathetic division—unlike the sympathetic division—is neither controlled nor activated as a whole. Although it is active continuously, the activities are reflex responses to conditions within specific structures or regions. Examples of the major effects produced by the parasympathetic division include the following:

- Constriction of the pupils (to restrict the amount of light that enters the eyes) and focusing of the lenses of the eyes on nearby objects.
- Secretion by digestive glands, including salivary glands, gastric glands, duodenal glands, intestinal glands, the pancreas (exocrine and endocrine), and the liver.
- Secretion of hormones that promote the absorption and utilization of nutrients by peripheral cells.
- Changes in blood flow and glandular activity associated with sexual arousal.
- Increased smooth muscle activity along the digestive tract.
- Stimulation and coordination of defecation.
- Contraction of the urinary bladder during urination.
- Constriction of the respiratory passageways.
- Reduction in heart rate and in the force of contraction.

These functions center on relaxation, food processing, and energy absorption. The parasympathetic division has been called the **anabolic system** (*anabole*, a raising up) because its stimulation leads to a general increase in the nutrient content of the blood. Cells throughout the body respond to this increase by absorbing nutrients and using them to support growth, cell division, and the creation of energy reserves in the form of lipids or glycogen.

Module 14.6 Review

a. What physiological changes are typical in a tense (or anxious) person?

b. What neurotransmitter is released by all parasympathetic neurons?

c. Why is the parasympathetic division called the anabolic system?

LO 14.6 Describe the mechanisms of neurotransmitter release in the ANS, and explain the effects of neurotransmitters on target organs and tissues.

Labeling

Fill in the missing labels in this diagram of the sympathetic division of the ANS.

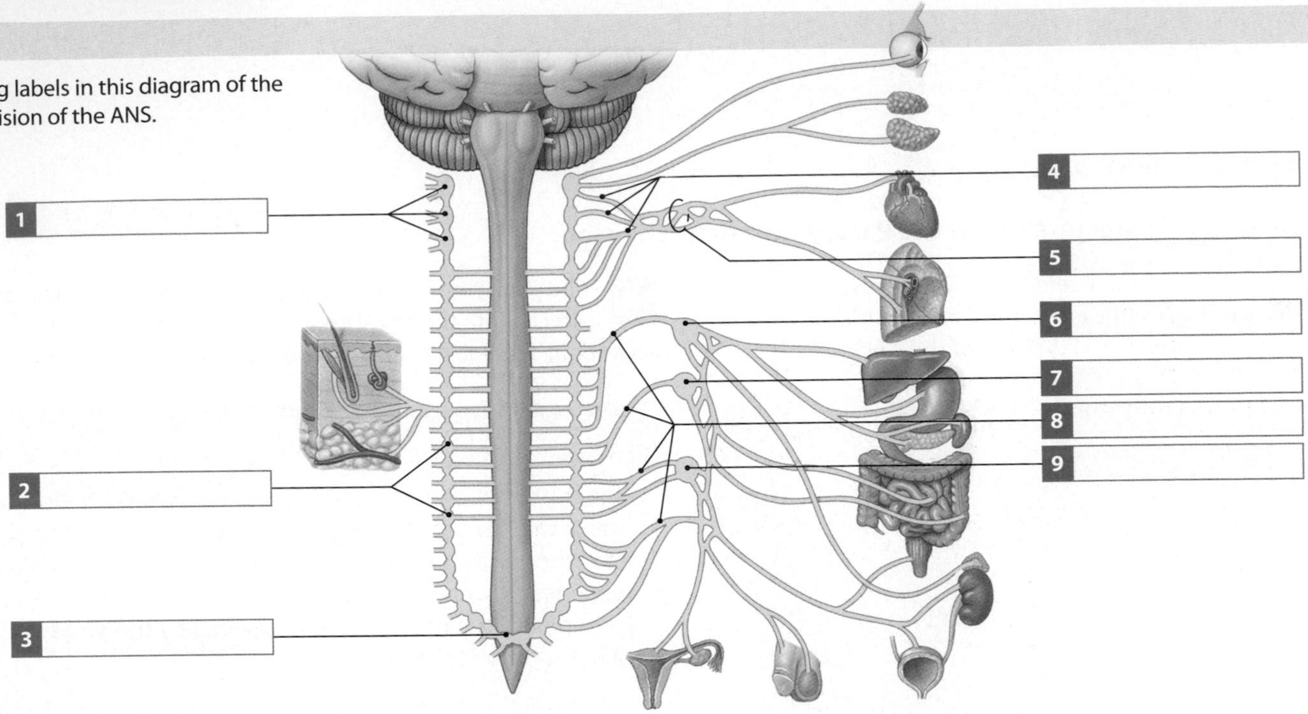
1
2
3
4
5
6
7
8
9

Concept map

Use each of the following terms once to fill in the blank boxes to correctly complete the map.

- sympathetic division
- craniosacral division
- cranial nerves III, VII, IX, X
- thoracolumbar division
- parasympathetic division
- sacral nerves
- thoracic nerves
- enteric nervous system
- lumbar nerves

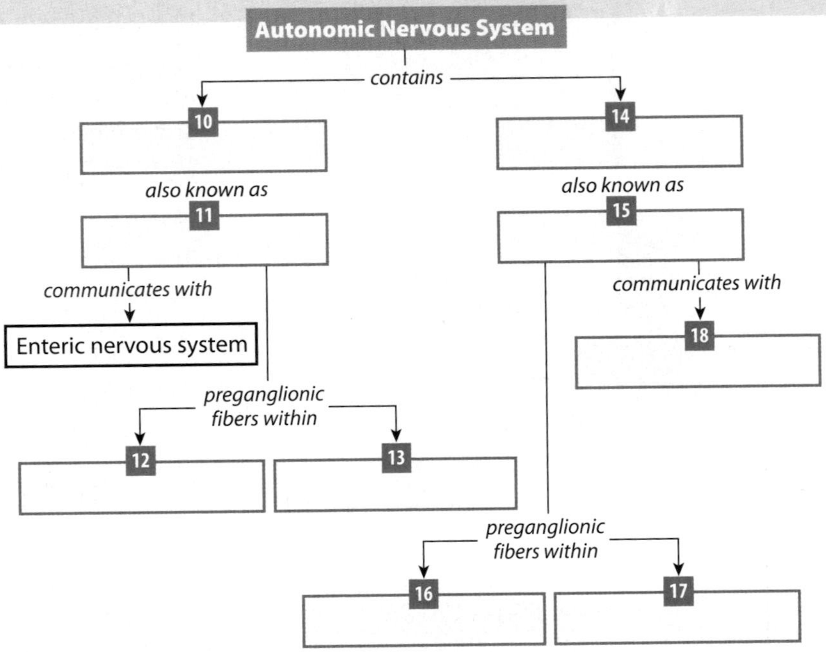
Autonomic Nervous System
contains
10
14
also known as
11
also known as
15
communicates with
Enteric nervous system
communicates with
18
preganglionic fibers within
12
13
preganglionic fibers within
16
17

Matching

Match each lettered term with the most closely related description.

a. nicotinic, muscarinic
b. secrete norepinephrine
c. receptors
d. alpha, beta
e. cholinergic
f. splanchnic nerves
g. acetylcholine
h. parasympathetic activation

19 Collateral ganglia
20 Parasympathetic neurotransmitter
21 Sexual arousal
22 Adrenal medullae
23 Cholinergic receptors
24 Determines neurotransmitter effects
25 All parasympathetic neurons
26 Adrenergic receptors

19 ______
20 ______
21 ______
22 ______
23 ______
24 ______
25 ______
26 ______

The ANS adjusts visceral motor responses to maintain homeostasis

If all consciousness were eliminated, vital physiological processes would continue virtually unchanged; after all, a night's sleep is not a life-threatening event. Longer, deeper states of unconsciousness are not necessarily more dangerous, as long as nourishment and other basic care are provided. People who have suffered severe brain injuries can survive in a coma for decades. Over that period, the ANS continues to adjust activities of the digestive, cardiovascular, respiratory, and reproductive systems to maintain homeostasis without instructions or interference from the conscious mind.

1 This flowchart gives a general overview of the way sensory information is distributed and motor commands are issued by the nervous system.

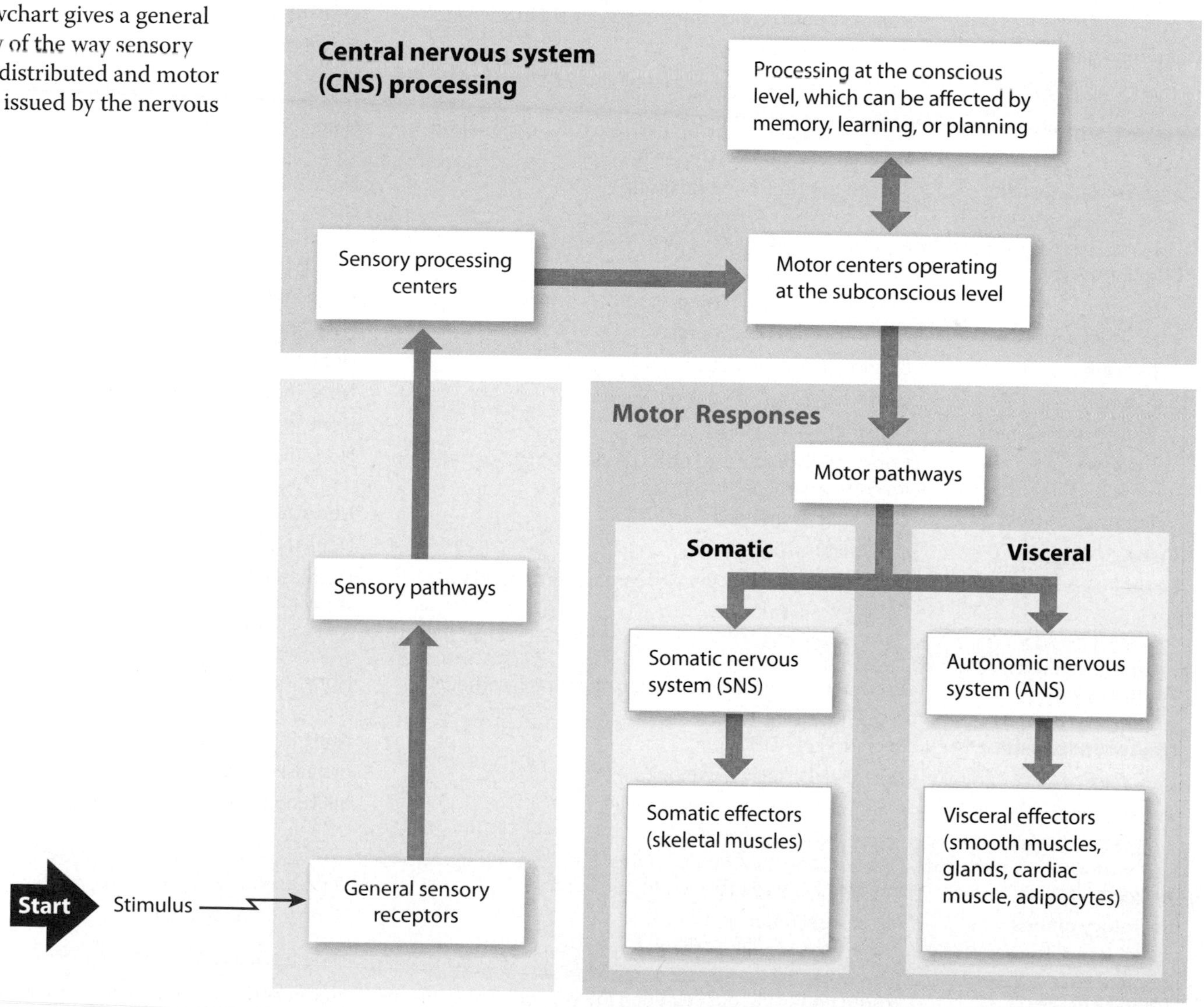

Central nervous system (CNS) processing
Processing at the conscious level, which can be affected by memory, learning, or planning
Sensory processing centers
Motor centers operating at the subconscious level
Motor Responses
Motor pathways
Somatic
Visceral
Sensory pathways
Somatic nervous system (SNS)
Autonomic nervous system (ANS)
Somatic effectors (skeletal muscles)
Visceral effectors (smooth muscles, glands, cardiac muscle, adipocytes)
Start
Stimulus
General sensory receptors

The output of the ANS has an impact on virtually every body system. This section considers those effects and how autonomic output is coordinated and directed.

Module 14.7 Review

a. Explain the significance of the ANS to homeostasis.
b. Name the two types of motor pathways.
c. Identify somatic effectors and visceral effectors.

LO 14.7 Describe the role of the ANS in maintaining homeostasis during unconsciousness.

Module 14.8

The ANS provides precise control over visceral functions

1 This table lists the effects of sympathetic and parasympathetic innervation on major organs and systems of the body. Even in the absence of stimuli, autonomic motor neurons maintain a continuous level of spontaneous activity called **autonomic tone**. Many vital organs receive **dual innervation**, receiving instructions from both the sympathetic and parasympathetic divisions. The effects may be opposing or complementary. In cases where only the sympathetic division provides innervation, the response may still vary widely depending on the type of receptor stimulated.

A Functional Comparison of the Sympathetic and Parasympathetic Divisions of the ANS

Structure	Sympathetic Effects (receptor type)	Parasympathetic Effects (all cholinergic)
Eye		
	Dilation of pupil (α_1); accommodation for distance vision (β_2)	Constriction of pupil; accommodation for close vision
Lacrimal glands	None (not innervated)	Secretion
Skin		
Sweat glands	Increased secretion, palms and soles (α_1); generalized increase in secretion (cholinergic)	None (not innervated)
Arrector pili muscles	Contraction; erection of hairs (α_1)	None (not innervated)
Cardiovascular System		
Blood vessels		
To skin	Dilation (β_2 and cholinergic); constriction (α_1)	None (not innervated)
To skeletal muscles	Dilation (β_2 and cholinergic)	None (not innervated)
To heart	Dilation (β_2); constriction (α_1, α_2)	None (not innervated)
To lungs	Dilation (β_2); constriction (α_2)	None (not innervated)
To digestive viscera	Constriction (α_1); dilation (α_2)	None (not innervated)
To kidneys	Constriction, decreased urine production (α_1, α_2); dilation, increased urine production (β_1, β_2)	None (not innervated)
To brain	Dilation (cholinergic)	None (not innervated)
Veins	Constriction (α_1, β_2)	None (not innervated)
Heart	Increased heart rate, force of contraction, and blood pressure (α_1, β_1)	Decreased heart rate, force of contraction, and blood pressure
Endocrine System		
Adrenal gland	Secretion of epinephrine, norepinephrine by adrenal medulla	None (not innervated)
Neurohypophysis	Secretion of ADH (β_1)	None (not innervated)
Pancreas	Decreased insulin secretion (α_2)	Increased insulin secretion
Pineal gland	Increased melatonin secretion (β_1, β_2)	Inhibition of melatonin synthesis
Respiratory System		
Airways	Increased airway diameter (β_2)	Decreased airway diameter
Secretory glands	Mucous secretion (α_1)	None (not innervated)
Digestive System		
Salivary glands	Production of viscous secretion (α_1, β_1) containing mucins and enzymes	Production of copious, watery secretion
Sphincters	Constriction (α_1)	Dilation
General level of activity	Decreased (α_2, β_2)	Increased
Secretory glands	Inhibition (α_2)	Stimulation
Liver	Glycogen breakdown, glucose synthesis and release (α_1, β_2)	Glycogen synthesis
Pancreas	Decreased exocrine secretion (α_1)	Increased exocrine secretion

A Functional Comparison of the Sympathetic and Parasympathetic Divisions of the ANS (Continued)		
Structure	**Sympathetic Effects** (receptor type)	**Parasympathetic Effects** (all cholinergic)
Muscular System		
	Increased force of contraction, glycogen breakdown (β_2)	None (not innervated)
	Facilitation of ACh release at neuromuscular junction (α_2)	None (not innervated)
Adipose Tissue		
	Lipolysis, fatty acid release (α_1, β_1, β_3)	None (not innervated)
Urinary System		
Kidneys	Secretion of renin (β_1)	Uncertain effects on urine production
Urinary bladder	Constriction of internal sphincter; relaxation of urinary bladder (α_1, β_2)	Tensing of urinary bladder, relaxation of internal sphincter to eliminate urine
Male Reproductive System		
	Increased glandular secretion and ejaculation (α_1)	Erection
Female Reproductive System		
	Increased glandular secretion; contraction of pregnant uterus (α_1)	Variable (depending on hormones present)
	Relaxation of nonpregnant uterus (β_1)	Variable (depending on hormones present)

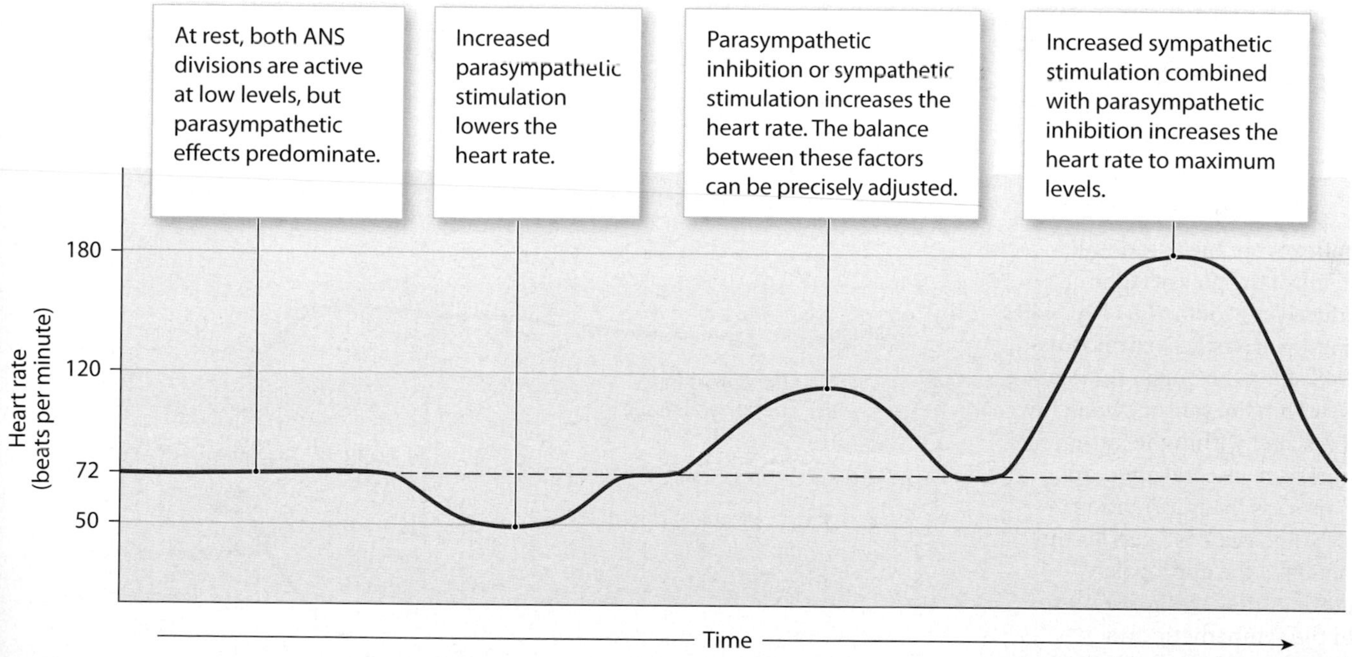

At rest, both ANS divisions are active at low levels, but parasympathetic effects predominate.
Increased parasympathetic stimulation lowers the heart rate.
Parasympathetic inhibition or sympathetic stimulation increases the heart rate. The balance between these factors can be precisely adjusted.
Increased sympathetic stimulation combined with parasympathetic inhibition increases the heart rate to maximum levels.
180
120
72
50
Heart rate (beats per minute)
Time

2 This graph shows the effects of dual innervation on the heart. The heart consists of cardiac muscle tissue, and its contractions are triggered by specialized pacemaker cells. The two autonomic divisions have opposing effects on pacemaker function. ACh released by the parasympathetic division reduces the heart rate, whereas NE released by the sympathetic division accelerates the heart rate. Because autonomic tone exists, small amounts of both neurotransmitters are released continuously. However, under resting conditions, the effects of parasympathetic innervation predominate.

Module 14.8 Review

a. Define dual innervation.

b. Explain autonomic tone and its significance in controlling visceral function.

c. You go outside on a cold day and blood flow to your skin is reduced, conserving body heat. You become angry, and your face turns red. Explain these changes.

LO **14.8** Discuss the relationship between the two divisions of the ANS and the significance of dual innervation.

Module 14.9

Most visceral functions are controlled by visceral reflexes

Visceral reflexes provide automatic motor responses that can be modified, facilitated, or inhibited by higher centers, especially those of the hypothalamus. All visceral reflexes are polysynaptic. Each visceral reflex arc consists of a receptor, a sensory neuron, a processing center (one or more interneurons), and one or two visceral motor neurons.

1 **Short reflexes** bypass the CNS entirely. They involve sensory neurons and interneurons whose cell bodies are located within autonomic ganglia. The interneurons synapse on ganglionic neurons, and the motor commands are then distributed to effectors by postganglionic fibers. Short reflexes control very simple motor responses with localized effects. In general, short reflexes control activity in one small part of a target organ, whereas long reflexes coordinate the activities of an entire organ. Short reflexes predominate in the enteric nervous system, which operates largely outside of the awareness and control of the CNS.

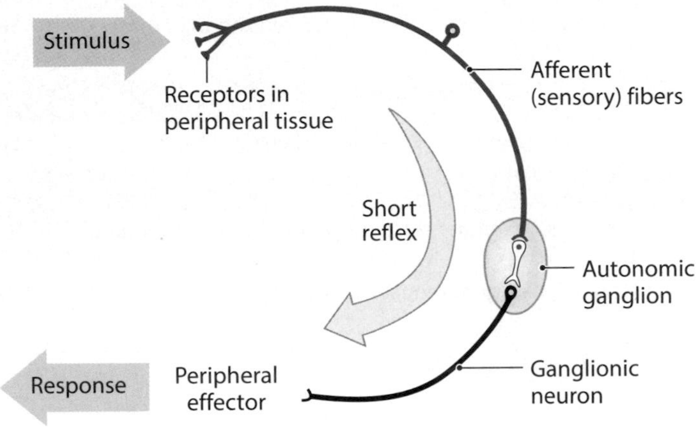

Stimulus
Receptors in peripheral tissue
Afferent (sensory) fibers
Short reflex
Autonomic ganglion
Response
Peripheral effector
Ganglionic neuron

2 **Long reflexes** are the autonomic equivalents of the polysynaptic reflexes introduced in Module 12.11 (p. 449). Visceral sensory neurons deliver information to the CNS along the dorsal roots of spinal nerves, within the sensory branches of cranial nerves, and within the autonomic nerves that innervate visceral effectors. Interneurons process the information within the CNS, and the ANS carries the motor commands to the appropriate visceral effectors. Although short reflexes are present in the sympathetic and parasympathetic divisions, long reflexes predominate, and they are responsible for coordinating responses involving multiple organ systems.

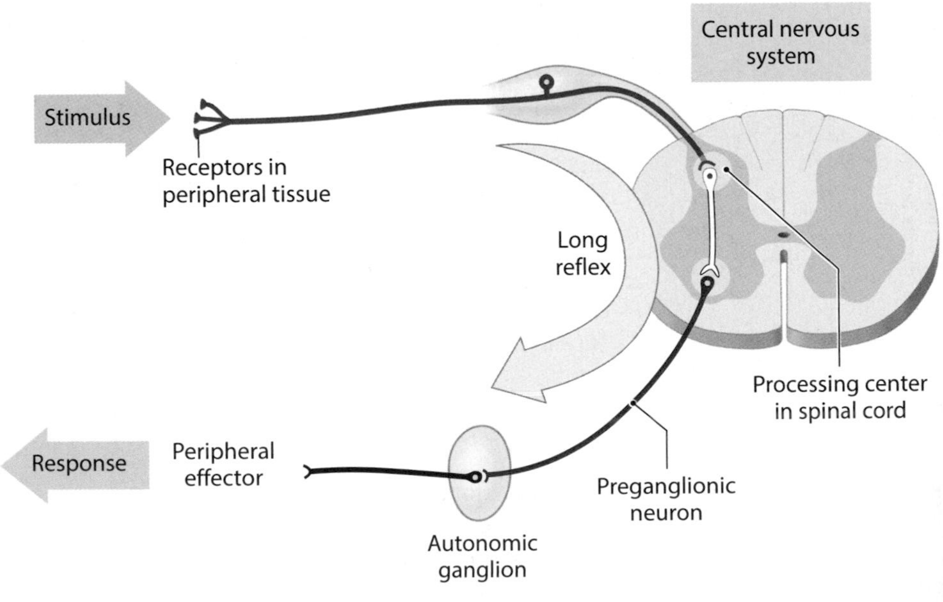

Central nervous system
Stimulus
Receptors in peripheral tissue
Long reflex
Processing center in spinal cord
Response
Peripheral effector
Preganglionic neuron
Autonomic ganglion

Common Visceral Reflexes

	Reflex	Stimulus	Response	Comments
SYMPATHETIC	**Cardioacceleratory reflex**	Sudden decrease in carotid blood pressure	Increase in heart rate and force of contraction	Coordinated in cardiac centers of medulla oblongata
SYMPATHETIC	**Vasomotor reflexes**	Changes in blood pressure in major arteries	Changes in diameter of peripheral vessels	Coordinated in vasomotor center in medulla oblongata
SYMPATHETIC	**Pupillary reflex**	Low light level reaching visual receptors	Dilation of pupil	
SYMPATHETIC	**Ejaculation (in males)**	Erotic stimuli (primarily tactile)	Skeletal muscle contractions ejecting semen	
PARASYMPATHETIC	**Gastric and intestinal reflexes**	Pressure and physical contact	Smooth muscle contractions that propel food materials and mix them with secretions	By vagus nerve
PARASYMPATHETIC	**Defecation**	Distention of rectum	Relaxation of internal anal sphincter	Requires voluntary relaxation of external anal sphincter
PARASYMPATHETIC	**Urination**	Distention of urinary bladder	Contraction of walls of urinary bladder; relaxation of internal urethral sphincter	Requires voluntary relaxation of external urethral sphincter
PARASYMPATHETIC	**Direct light and consensual light reflexes**	Bright light shining in eye(s)	Constriction of pupils of both eyes	
PARASYMPATHETIC	**Swallowing reflex**	Movement of food and liquids into pharynx	Smooth muscle and skeletal muscle contractions	Coordinated by medullary swallowing center
PARASYMPATHETIC	**Coughing reflex**	Irritation of respiratory tract	Sudden explosive ejection of air	Coordinated by medullary coughing center
PARASYMPATHETIC	**Baroreceptor reflex**	Sudden increase in carotid blood pressure	Reduction in heart rate and force of contraction	Coordinated in cardiac centers of medulla oblongata
PARASYMPATHETIC	**Sexual arousal**	Erotic stimuli (visual or tactile)	Increased glandular secretions, sensitivity, erection	

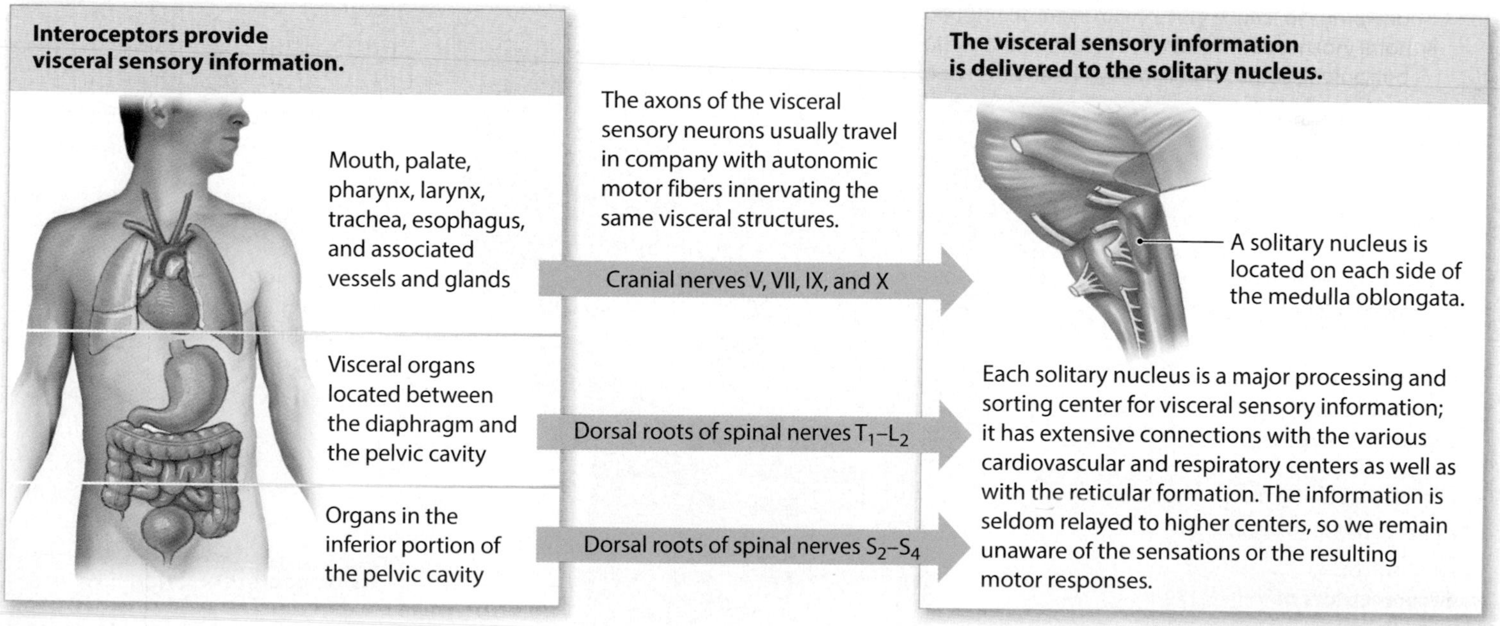

Interoceptors provide visceral sensory information.
Mouth, palate, pharynx, larynx, trachea, esophagus, and associated vessels and glands
Visceral organs located between the diaphragm and the pelvic cavity
Organs in the inferior portion of the pelvic cavity
The axons of the visceral sensory neurons usually travel in company with autonomic motor fibers innervating the same visceral structures.
Cranial nerves V, VII, IX, and X
Dorsal roots of spinal nerves T_1–L_2
Dorsal roots of spinal nerves S_2–S_4
The visceral sensory information is delivered to the solitary nucleus.
A solitary nucleus is located on each side of the medulla oblongata.
Each solitary nucleus is a major processing and sorting center for visceral sensory information; it has extensive connections with the various cardiovascular and respiratory centers as well as with the reticular formation. The information is seldom relayed to higher centers, so we remain unaware of the sensations or the resulting motor responses.

3 Visceral sensory information is collected by interoceptors monitoring visceral tissues and organs, primarily within the thoracic and abdominopelvic cavities. These interoceptors include nociceptors, thermoreceptors, tactile receptors, baroreceptors, and chemoreceptors, although none of them are as numerous as they are in somatic tissues. Almost all of the processing of this sensory information is done at the subconscious level, in nuclei in the spinal cord and the **solitary nuclei** in the brain stem.

LO 14.9 Define a visceral reflex, and explain the significance of such reflexes.

Module 14.9 Review

a. Define visceral reflex.

b. Compare short reflexes with long reflexes.

c. Describe the solitary nucleus.

Baroreceptors and chemoreceptors initiate important autonomic reflexes involving visceral sensory pathways

Baroreceptors and chemoreceptors play key roles in the control of visceral function by the autonomic nervous system.

1 **Baroreceptors** are stretch receptors that monitor changes in pressure. The receptor consists of free nerve endings that branch within the elastic tissues in the walls of hollow organs, blood vessels, and tubes in the respiratory, digestive, and urinary tract. When the pressure changes, the elastic walls of these structures stretch or recoil. These changes in shape distort the receptor's dendritic branches and alter the rate of action potential generation. Baroreceptors monitor blood pressure in the walls of major vessels, including the carotid artery (at the **carotid sinus**) and the aorta (at the **aortic sinus**). The information provided by these baroreceptors plays a major role in regulating cardiac function and adjusting blood flow to vital tissues. Baroreceptors in the lungs monitor the degree of lung expansion. This information is relayed to the respiratory rhythmicity centers in the medulla oblongata, which sets the pace of respiration. Baroreceptors in the urinary and digestive tracts trigger a variety of visceral reflexes, such as urination.

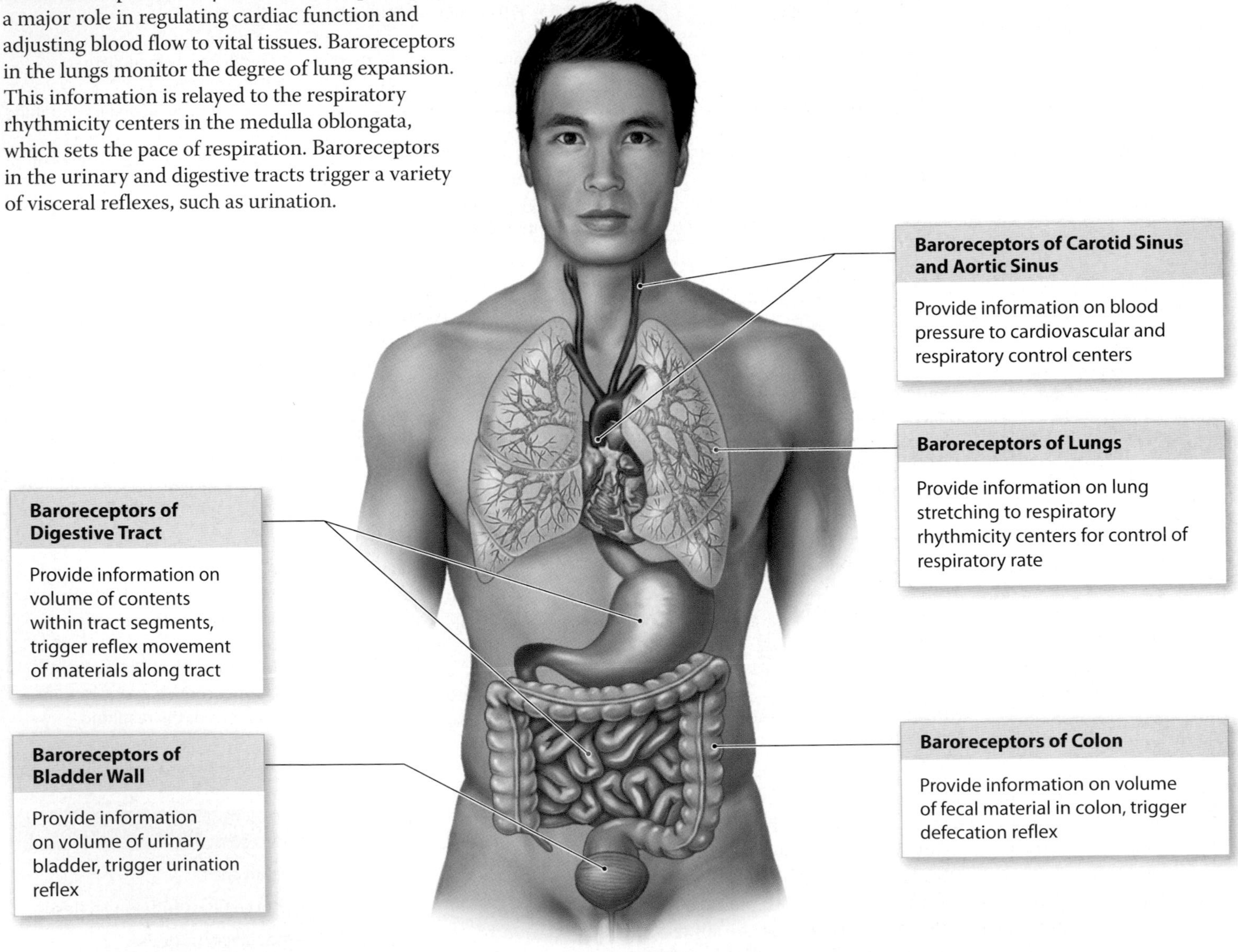

Baroreceptors of Carotid Sinus and Aortic Sinus
Provide information on blood pressure to cardiovascular and respiratory control centers
Baroreceptors of Lungs
Provide information on lung stretching to respiratory rhythmicity centers for control of respiratory rate
Baroreceptors of Digestive Tract
Provide information on volume of contents within tract segments, trigger reflex movement of materials along tract
Baroreceptors of Colon
Provide information on volume of fecal material in colon, trigger defecation reflex
Baroreceptors of Bladder Wall
Provide information on volume of urinary bladder, trigger urination reflex

2 **Chemoreceptors** are specialized neurons that detect small changes in the concentrations of specific chemicals or compounds. Chemoreceptors are found (1) within the medulla oblongata and elsewhere in the brain, (2) in the **carotid bodies**, near the origin of the internal carotid arteries on each side of the neck, and (3) in the **aortic bodies** between the major branches of the aortic arch. (The carotid bodies and aortic bodies are small tissue masses of receptors and supporting cells.) The neurons in the respiratory centers in the medulla oblongata monitor the pH and P_{CO_2} in cerebrospinal fluid. The carotid and aortic bodies monitor the pH and levels of carbon dioxide (P_{CO_2}) and oxygen (P_{O_2}) in arterial blood. These chemoreceptors play an important role in the reflexive control of respiration and cardiovascular function. (The symbol P refers to the partial pressure of a single gas in a mixture of gases, such as air. As we will discuss in Chapter 21, differences in partial pressure drive the diffusion of gases in the lungs and body tissues.)

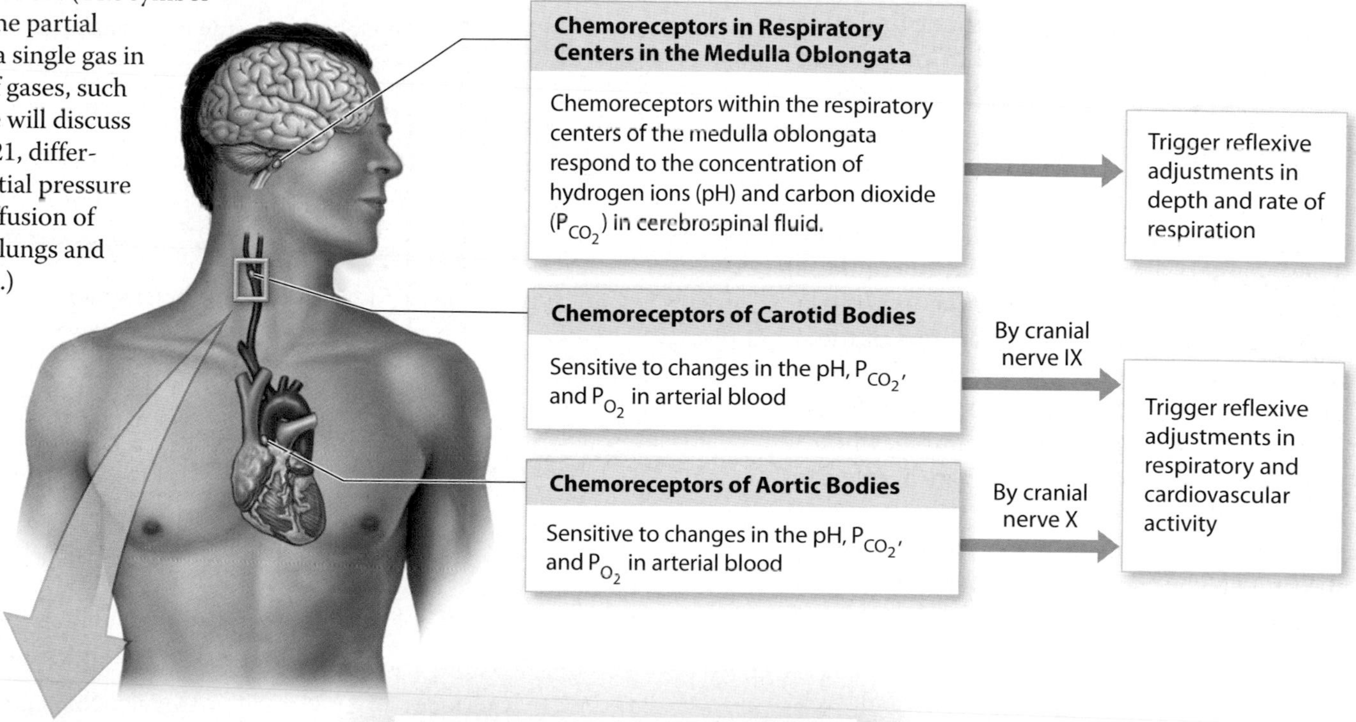

Chemoreceptors in Respiratory Centers in the Medulla Oblongata
Chemoreceptors within the respiratory centers of the medulla oblongata respond to the concentration of hydrogen ions (pH) and carbon dioxide (P_{CO_2}) in cerebrospinal fluid.
Trigger reflexive adjustments in depth and rate of respiration
Chemoreceptors of Carotid Bodies
Sensitive to changes in the pH, P_{CO_2}, and P_{O_2} in arterial blood
By cranial nerve IX
Chemoreceptors of Aortic Bodies
Sensitive to changes in the pH, P_{CO_2}, and P_{O_2} in arterial blood
By cranial nerve X
Trigger reflexive adjustments in respiratory and cardiovascular activity

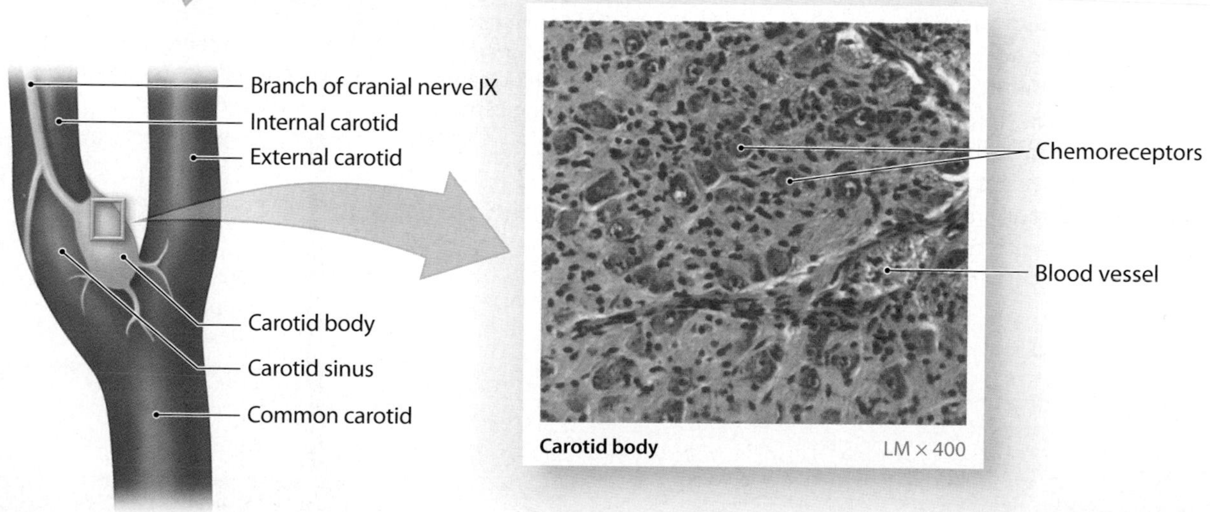

Branch of cranial nerve IX
Internal carotid
External carotid
Carotid body
Carotid sinus
Common carotid
Chemoreceptors
Blood vessel
Carotid body
LM × 400

3 Enlarged view of the carotid sinus and the location of the carotid body.

Module 14.10 Review

a. Define baroreceptor and chemoreceptor.

b. Where are baroreceptors located within the body?

c. Which type of receptor is sensitive to changes in blood pH?

LO **14.10** Explain the roles of baroreceptors and chemoreceptors in homeostasis.

Module 14.11

The autonomic nervous system has multiple levels of motor control

The levels of activity in the sympathetic and parasympathetic divisions of the ANS are controlled by centers in the brain stem that regulate specific visceral functions. As in the SNS, simple ANS reflexes based in the autonomic ganglia and the spinal cord provide relatively rapid and automatic responses to stimuli. More complex sympathetic and parasympathetic reflexes are coordinated by processing centers in the medulla oblongata. In addition to the cardiovascular and respiratory centers, the medulla oblongata contains centers and nuclei involved with salivating, swallowing, digestive secretions, peristalsis, and urinary function. These centers are in turn subject to regulation by the hypothalamus. Because the hypothalamus interacts with all other portions of the brain, activity in the limbic system, thalamus, or cerebral cortex can have dramatic effects on autonomic function. For example, when you become angry, your heart rate accelerates, your blood pressure rises, and your respiratory rate increases; when you consider your next meal, your stomach "growls" and your mouth waters.

1 This diagram shows the pathways and levels of ANS control. The hypothalamus acts as the "headquarters" of both the sympathetic and parasympathetic divisions, and hypothalamic output is indicated by solid red arrows. Nuclei in the spinal cord and brain stem control basic visceral reflex patterns. There is also continual feedback between higher brain centers and the hypothalamus and brain stem; this subconscious communication is indicated by the dashed red arrows.

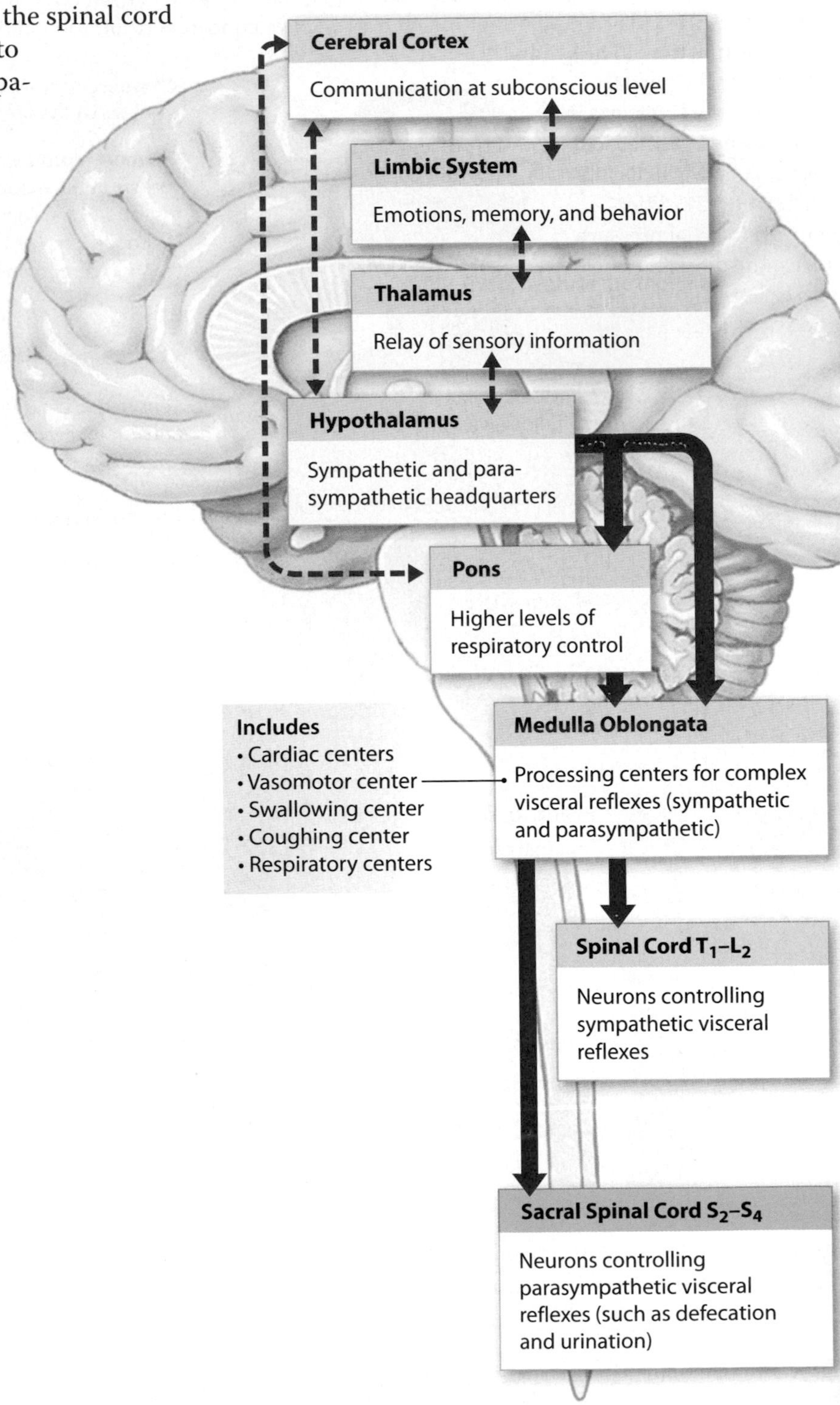

Cerebral Cortex
Communication at subconscious level
Limbic System
Emotions, memory, and behavior
Thalamus
Relay of sensory information
Hypothalamus
Sympathetic and parasympathetic headquarters
Pons
Higher levels of respiratory control
Includes
• Cardiac centers
• Vasomotor center
• Swallowing center
• Coughing center
• Respiratory centers
Medulla Oblongata
Processing centers for complex visceral reflexes (sympathetic and parasympathetic)
Spinal Cord T_1–L_2
Neurons controlling sympathetic visceral reflexes
Sacral Spinal Cord S_2–S_4
Neurons controlling parasympathetic visceral reflexes (such as defecation and urination)

2 The SNS and the ANS have parallel organization and are integrated at the level of the brain stem. This diagram shows the basic patterns of organization and integration. Blue arrows indicate ascending sensory information; red arrows, descending motor commands; dashed lines, pathways of communication and feedback among higher centers. The activities of the SNS and those of the ANS are integrated at many levels. Although we have considered somatic and visceral motor pathways separately, the two have many parallels in terms of both organization and function. Sensory pathways may carry information distributed to both the SNS and the ANS, triggering integrated and compatible reflexes. Higher levels of integration involve the brain stem, and both systems are influenced, if not controlled, by higher brain centers.

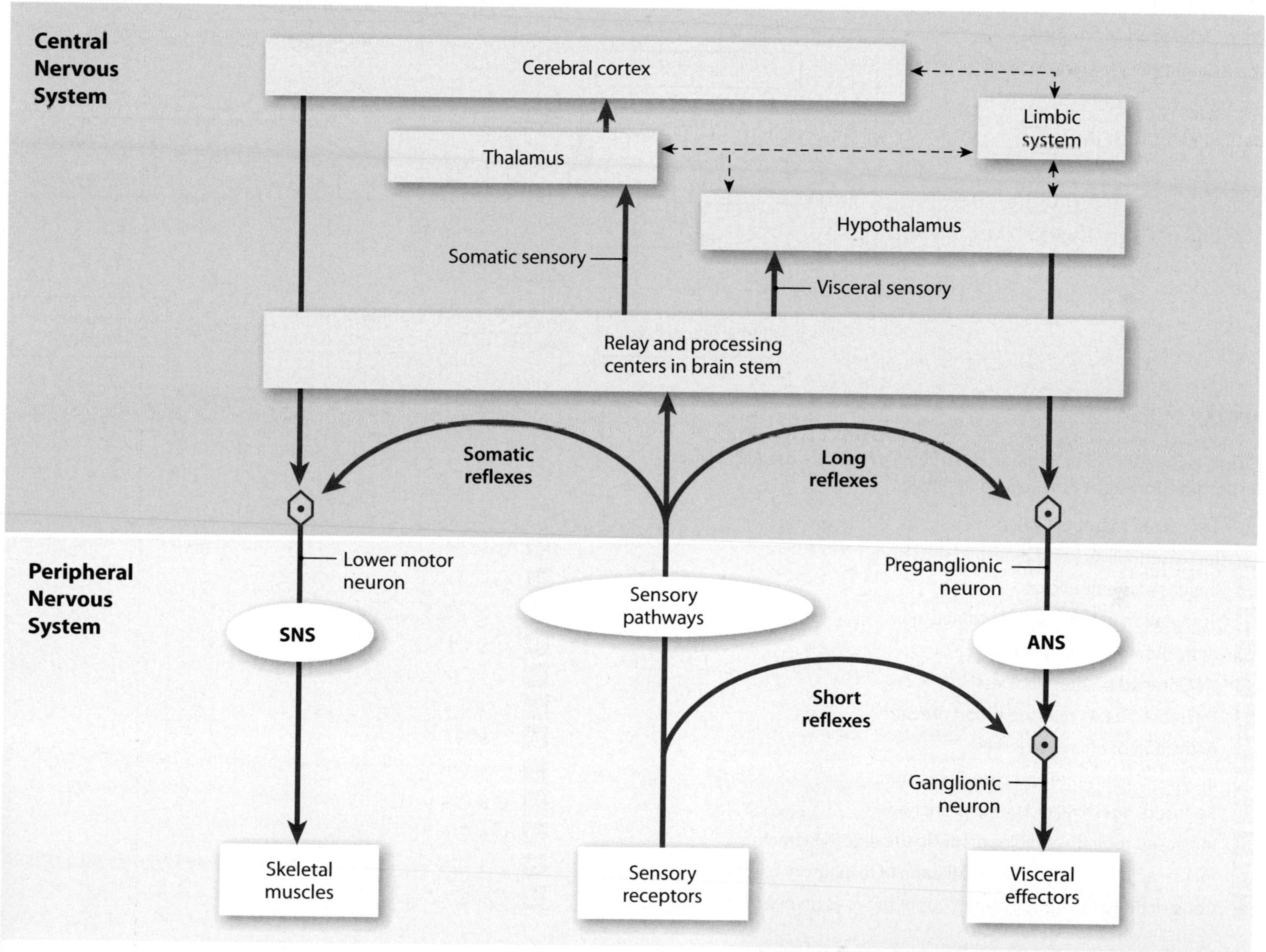

Central Nervous System
Cerebral cortex
Limbic system
Thalamus
Hypothalamus
Somatic sensory
Visceral sensory
Relay and processing centers in brain stem
Somatic reflexes
Long reflexes
Peripheral Nervous System
Lower motor neuron
Preganglionic neuron
Sensory pathways
SNS
ANS
Short reflexes
Ganglionic neuron
Skeletal muscles
Sensory receptors
Visceral effectors

Module 14.11 Review

a. What brain structure is considered to be the headquarters for the ANS?

b. Harry has a brain tumor that is pressing against his hypothalamus. Would you expect this tumor to interfere with autonomic function? Why or why not?

c. What brain structure relays somatic sensory information?

LO 14.11 Describe the hierarchy of interacting levels of control in the ANS beginning with the hypothalamus.

Concept map

Use each of the following terms once to fill in the blank boxes to correctly complete the map.

- respiratory
- pons
- spinal cord T_1–L_2
- vasomotor
- coughing
- hypothalamus
- sympathetic visceral reflexes
- parasympathetic visceral reflexes
- complex visceral reflexes
- limbic system and thalamus

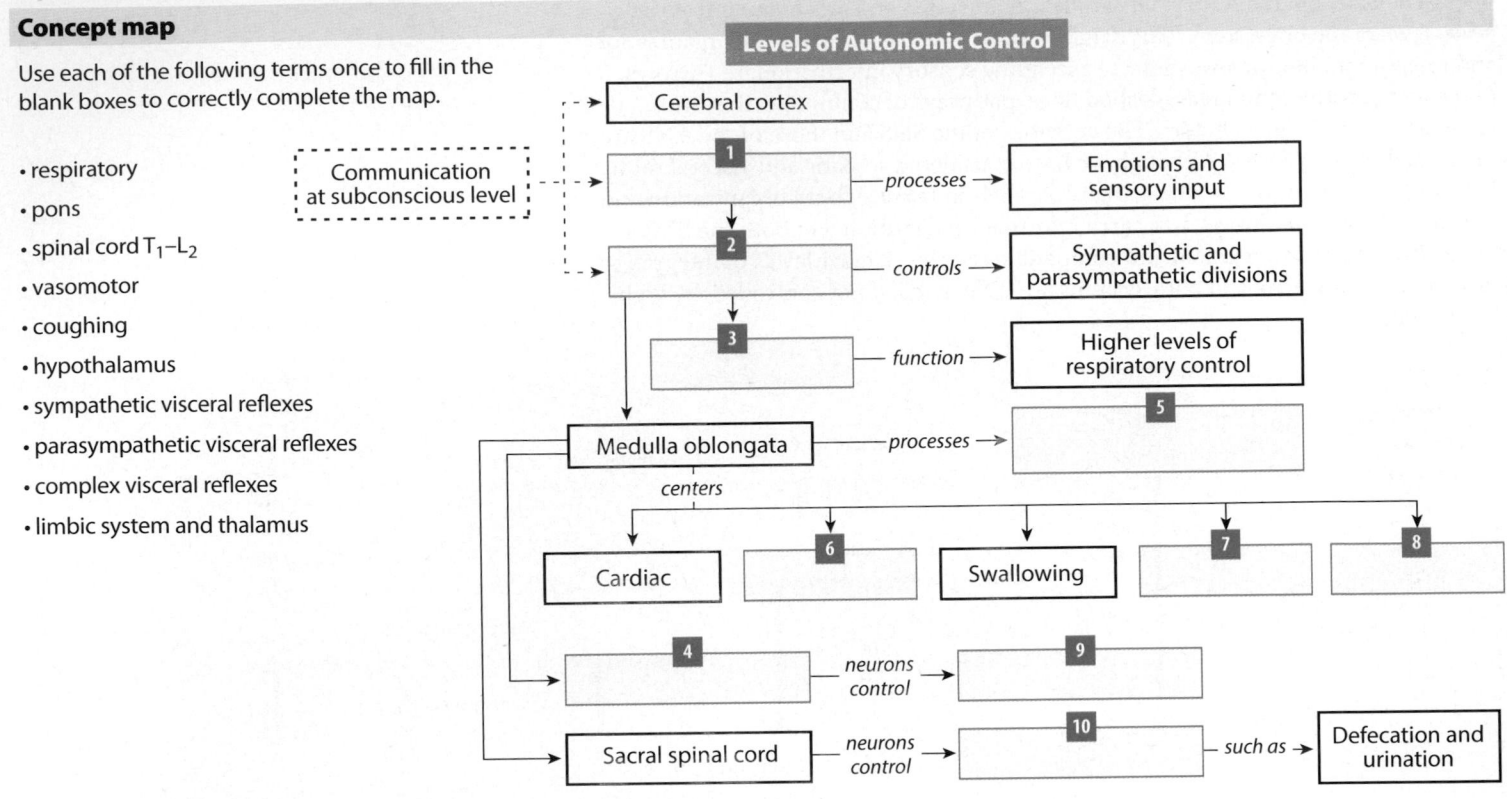

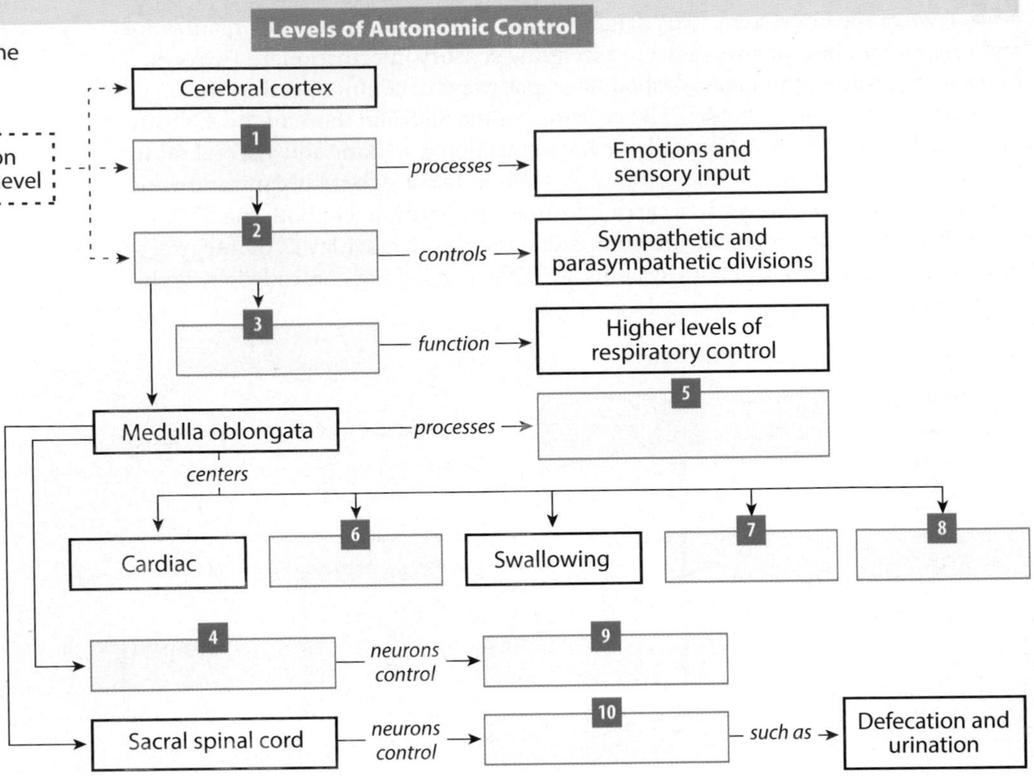
Levels of Autonomic Control
Communication at subconscious level
Cerebral cortex
1
processes
Emotions and sensory input
2
controls
Sympathetic and parasympathetic divisions
3
function
Higher levels of respiratory control
5
Medulla oblongata
processes
centers
Cardiac
6
Swallowing
7
8
4
neurons control
9
Sacral spinal cord
neurons control
10
such as
Defecation and urination

Fill-in

Fill in S (sympathetic) or P (parasympathetic) to indicate the ANS division responsible for each of the following effects.

11 Decreased metabolic rate — ○ S ○ P
12 Increased salivary and digestive secretions — ○ S ○ P
13 Increased metabolic rate — ○ S ○ P
14 Stimulation of urination and defecation — ○ S ○ P
15 Activation of sweat glands — ○ S ○ P
16 Heightened mental alertness — ○ S ○ P
17 Decreased heart rate and blood pressure — ○ S ○ P
18 Activation of energy reserves — ○ S ○ P
19 Increased heart rate and blood pressure — ○ S ○ P
20 Reduced digestive and urinary functions — ○ S ○ P
21 Increased motility and blood flow in the digestive tract — ○ S ○ P
22 Increased respiratory rate and dilation of respiratory passages — ○ S ○ P
23 Constriction of the pupils and focus of the eyes on nearby objects — ○ S ○ P

Short answer

Recent surveys show that about one-third of the American adult population is involved in some type of exercise program. What contributions does sympathetic activation make to help the body adjust to changes that occur during exercise and still maintain homeostasis?

24 ______________________________

____

Study Outline

SECTION 1 • Functional Anatomy of the Autonomic Nervous System (ANS)

14.1 Ganglionic neurons of the ANS control visceral effectors p. 509

1. The **autonomic nervous system (ANS)** controls visceral functions outside of our conscious awareness.
2. In the **somatic nervous system (SNS)** neurons control skeletal muscles. The ANS controls the **visceral effectors**: smooth muscle, glands, cardiac muscle, and adipocytes.
3. ANS motor neurons synapse with visceral motor neurons in autonomic ganglia. These **ganglionic neurons** control visceral effectors.
4. The visceral motor neurons whose cell bodies lie in the brain stem and spinal cord are known as **preganglionic neurons**. The axons of these neurons leave the CNS and synapse with ganglionic neurons at **autonomic ganglia**.

14.2 The ANS consists of sympathetic, parasympathetic, and enteric divisions p. 510

5. The most familiar divisions of the ANS are the **sympathetic division** and the **parasympathetic division**.
6. While it is not always the case, the sympathetic division causes excitation and the parasympathetic division causes inhibition.
7. The two divisions may work independently because some structures are innervated by only one division, or together during complex processes.
8. Both divisions influence the third ANS division, the **enteric nervous system**.
9. The enteric nervous system is an extensive network of neurons in the walls of the digestive tract.

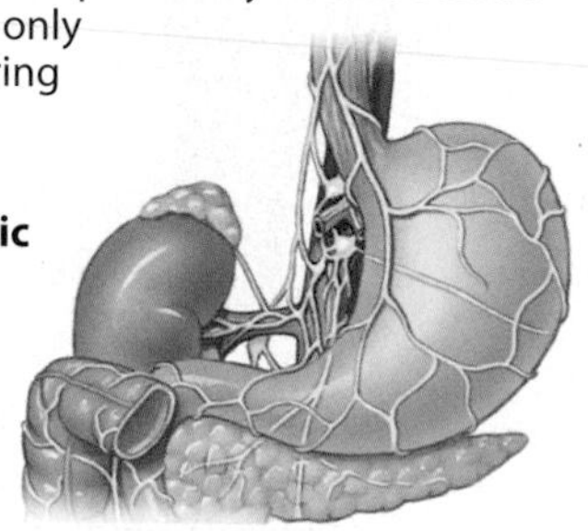

14.3 The sympathetic division has chain ganglia, collateral ganglia, and the adrenal medullae . . . whereas the parasympathetic division has terminal or intramural ganglia p. 512

10. The preganglionic neurons of the sympathetic division have their cell bodies in the lateral gray horns of spinal segments T_1–L_2.
11. They synapse with ganglionic neurons located at the **sympathetic chain**, **collateral ganglia**, or modified ganglion cells in the **adrenal medullae**.
12. The axons of preganglionic neurons **(preganglionic fibers)** are relatively short compared with the axons of ganglionic neurons **(postganglionic fibers)**.
13. The sympathetic division prepares the body for heightened somatic activity such as the "fight or flight" response.
14. The preganglionic neurons of the parasympathetic division have their cell bodies in cranial nerves III, VII, IX, and X and in the lateral gray horns of sacral spinal segments S_2–S_4.
15. They synapse with ganglionic neurons in **terminal ganglia** (located near the target organ) or in **intramural ganglia** (embedded in the wall of the target organ).
16. The parasympathetic division is known as the "rest and digest" system because it regulates visceral function and energy conservation.

14.4 The two ANS divisions innervate many of the same structures, but the innervation patterns are different p. 514

17. Every spinal nerve has a gray ramus that carries sympathetic postganglionic fibers for distribution in the body wall and limbs.
18. Postganglionic sympathetic fibers of the cervical sympathetic ganglia supply the regions innervated by cranial nerves III, VII, IX, and X.
19. **Splanchnic nerves** supply the viscera. There are three groups: cardiopulmonary, abdominopelvic (greater, lesser, and lumbar), and pelvic (sacral).
20. The vagus nerve (X) carries 75 percent of parasympathetic outflow.
21. The preganglionic fibers of the sacral segments carry sacral parasympathetic output and form distinct **pelvic nerves**. These nerves innervate intramural ganglia in the kidneys, urinary bladder, terminal portions of the large intestine, and sex organs.

14.5 Membrane receptors at target organs mediate the effects of sympathetic and parasympathetic stimulation p. 516

22. Sympathetic stimulation is a result of the interactions of the neurotransmitters norepinephrine (NE) and epinephrine (E) with **adrenergic receptors** in the plasma membrane.
23. There are two classes of adrenergic receptors: **alpha receptors** and **beta receptors**.
24. There are two types of alpha receptors: alpha-1 and alpha-2.
25. Stimulation of alpha-1 receptors generally excites the target cell. Stimulation of alpha-2 receptors generally inhibits the target cell.
26. There are three types of beta receptors: beta-1, beta-2, and beta-3.
27. Beta receptor stimulation changes the metabolic activity of the target cell.
28. All parasympathetic stimulation uses the neurotransmitter ACh that binds to **cholinergic receptors**.
29. There are two types of cholinergic receptors: **nicotinic receptors** and **muscarinic receptors**.
30. ACh always excites nicotinic receptors. ACh can excite or inhibit muscarinic receptors.

Muscarinic receptors are stimulated by muscarine, a toxin produced by some poisonous mushrooms.

14.6 The functional differences between the two ANS divisions reflect their divergent anatomical and physiological characteristics p. 518

31. In crisis, the entire sympathetic division responds. This is called **sympathetic activation**.
32. During sympathetic activation, the following changes occur: increased alertness, feelings of energy and euphoria, temporary insensitivity to painful stimuli, increased heart rate, increased blood pressure, increased breathing rate and depth of respiration, elevated muscle tone, and a mobilization of energy reserves.
33. The major effects produced by the parasympathetic division include constriction of the pupils, focusing of the lens on near objects, secretion of digestive glands, secretion of hormones promoting the absorption of nutrients, blood flow and glandular activity in response to sexual arousal, control of defecation, control of urination, constriction of the respiratory passageways, and reduction of heart rate and force of contraction.
34. The parasympathetic division has been called the **anabolic system** because its stimulation increases nutrient content in blood.

SECTION 2 • Autonomic Regulation and Control Mechanisms

14.7 The ANS adjusts visceral motor responses to maintain homeostasis p. 521

35. The ANS adjusts activities of the digestive, cardiovascular, respiratory, and reproductive systems to maintain homeostasis without instructions or interference from the conscious mind.
36. General sensory receptors are stimulated, sending sensory information toward the CNS where it is processed.
37. Motor commands are issued from the CNS toward either somatic or visceral effectors.

14.8 The ANS provides precise control over visceral functions p. 522

38. Even in the absence of stimuli, autonomic motor neurons show a continuous level of spontaneous activity called **autonomic tone**.
39. Many organs receive instructions from both ANS divisions. This is called **dual innervation**, and the effects may be opposing or complementary.

14.9 Most visceral functions are controlled by visceral reflexes p. 524

40. **Visceral reflexes** provide automatic motor responses that can be influenced by higher centers in the brain.
41. All visceral reflexes are polysynaptic reflex arcs that consist of a receptor, a sensory neuron, a processing center, and one or two visceral motor neurons.
42. **Short reflexes** control simple motor responses with localized effects, and operate outside the awareness and control of the CNS.
43. **Long reflexes** are the autonomic equivalents of the polysynaptic reflexes. Visceral sensory neurons send information to the CNS along the dorsal roots of spinal nerves, sensory branches of cranial nerves, and autonomic nerves that innervate visceral effectors.
44. Long reflexes are more common than short reflexes, and they are responsible for controlling multiple organ systems.
45. Visceral sensory information is collected by interoceptors that monitor visceral tissues and organs. These include nociceptors, thermoreceptors, tactile receptors, baroreceptors, and chemoreceptors.
46. The processing of sensory information is performed at the subconscious level in the nuclei in the spinal cord and the **solitary nuclei** in the brain stem.

14.10 Baroreceptors and chemoreceptors initiate important autonomic reflexes involving visceral sensory pathways p. 526

47. **Baroreceptors** are stretch receptors that monitor changes in pressure.
48. Baroreceptors monitor blood pressure in the walls of major vessels such as the **carotid sinus** and **aortic sinus**. This allows for the regulation of cardiac function and blood flow to vital tissues.
49. Baroreceptors in the lung monitor lung expansion to control respiratory rhythmicity, and those in the bladder and colon control urination and defecation.
50. **Chemoreceptors** detect changes in the concentration of specific chemicals or compounds.
51. Chemoreceptors are located in the medulla oblongata, the **carotid bodies**, and the **aortic bodies**.

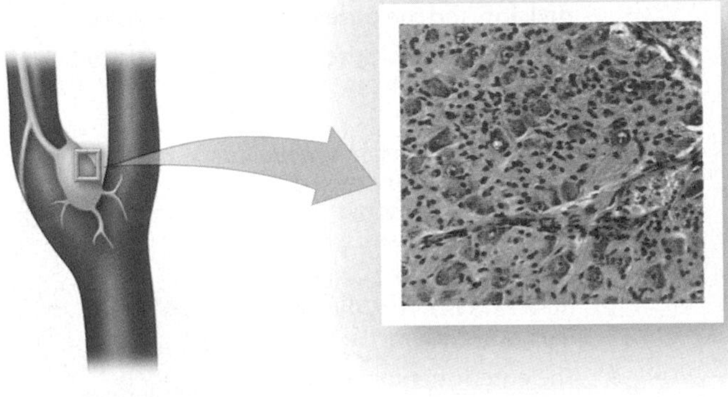

52. The chemoreceptors of the medulla oblongata respond to changes in pH and P_{CO_2} in CSF to regulate depth and rate of respiration.
53. The chemoreceptors of the carotid bodies and aortic bodies respond to changes in pH, P_{CO_2}, and P_{O_2} in arterial blood to adjust respiratory and cardiovascular activity.

14.11 The autonomic nervous system has multiple levels of motor control p. 528

54. The hypothalamus acts as the headquarters of both ANS divisions.
55. Nuclei in the spinal cord and brain stem control basic visceral reflex patterns.
56. There is also continued feedback between higher brain centers and the hypothalamus and brain stem.
57. The somatic nervous system and autonomic nervous system have parallel organization and are integrated at the level of the brain stem.
58. Sensory pathways may distribute information to both the somatic nervous system and autonomic nervous system, triggering integrated and compatible reflexes.

Chapter Review Questions

Labeling

Label the anatomical characteristics of the ANS shown in the diagram below.

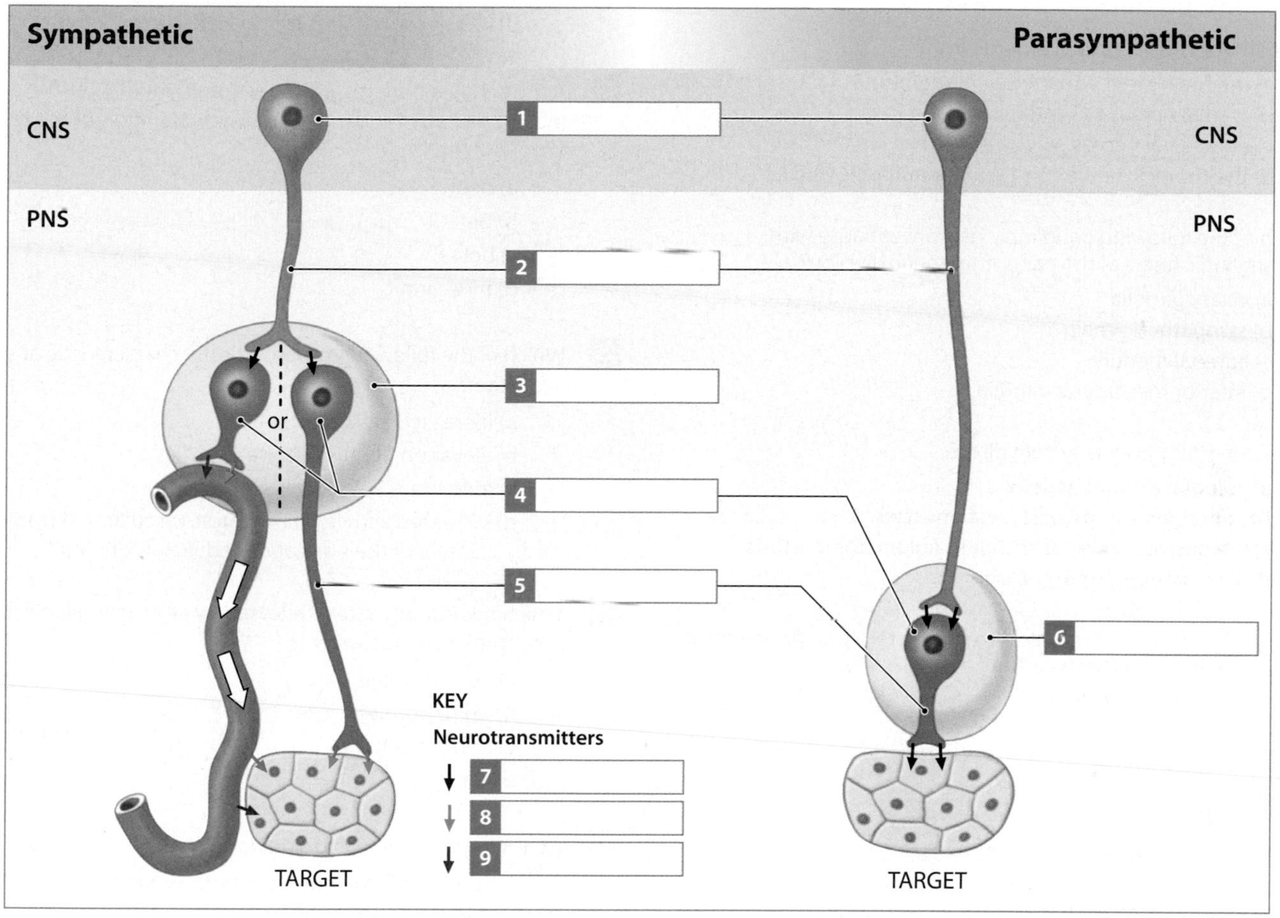

Sympathetic
Parasympathetic
CNS
PNS
CNS
PNS
1
2
3
4
5
6
or
KEY
Neurotransmitters
7
8
9
TARGET
TARGET

True/False

Indicate whether each statement is true or false.

10 Ganglionic neurons have their cell bodies in the CNS.

11 The enteric nervous system is influenced by both the sympathetic and parasympathetic divisions.

12 The celiac ganglion is associated with sympathetic regulation of the cardiovascular and respiratory systems.

13 All effector organs receive dual innervation.

14 Chemoreceptors of the carotid and aortic bodies are sensitive to changes in blood pressure.

15 Nicotinic receptors can be found at the neuromuscular junctions of skeletal muscle fibers.

10 ____________________

11 ____________________

12 ____________________

13 ____________________

14 ____________________

15 ____________________

Multiple choice

Select the correct answer from the list provided.

16 Which of the following is *not* a visceral effector?
- ☐ a) smooth muscle
- ☐ b) glands
- ☐ c) skeletal muscle
- ☐ d) adipocytes

17 Sympathetic division axons emerge from
- ☐ a) the brain stem and the sacral segments of the spinal cord.
- ☐ b) cranial nerves III, VIII, IX, and X.
- ☐ c) only sacral nerves.
- ☐ d) the thoracic nerves and lumbar nerves L_1 and L_2.

18 Which of the following ganglionic neurons synapse with preganglionic fibers of the parasympathetic division?
- ☐ a) ciliary ganglia
- ☐ b) sympathetic chain
- ☐ c) adrenal medulla
- ☐ d) inferior mesenteric ganglia

19 The autonomic nervous system directs
- ☐ a) voluntary motor activity.
- ☐ b) conscious control of skeletal muscles.
- ☐ c) unconscious processes that maintain homeostasis.
- ☐ d) sensory input from the skin.

20 The division of the autonomic nervous system that prepares the body for activity and stress is the
- ☐ a) sympathetic division.
- ☐ b) parasympathetic division.
- ☐ c) enteric division.
- ☐ d) somatomotor division.

21 The two types of cholinergic receptors are
- ☐ a) alpha-1 and alpha-2 receptors.
- ☐ b) beta-1 and beta-2 receptors.
- ☐ c) nicotinic and muscarinic receptors.
- ☐ d) alpha-1 and beta-1 receptors.

22 Which receptors would you block to cause a decreased heart rate?
- ☐ a) alpha-1 receptors
- ☐ b) beta-1 receptors
- ☐ c) nicotinic receptors
- ☐ d) muscarinic receptors

23 Where are the chemoreceptors found that are sensitive to changes in pH and P_{CO_2} in arterial blood?
- ☐ a) carotid and aortic bodies
- ☐ b) carotid and aortic sinuses
- ☐ c) baroreceptors of the lungs
- ☐ d) respiratory centers of the medulla oblongata

24 A drug that stimulates which receptor would relax the smooth muscle passages and increase airway diameter of the respiratory tract?
- ☐ a) beta-1
- ☐ b) beta-2
- ☐ c) beta-3
- ☐ d) muscarinic

25 Which of the following responses is *not* characteristic of sympathetic activation?
- ☐ a) increased alertness
- ☐ b) elevation of muscle tone
- ☐ c) erection
- ☐ d) increased activity in the cardiovascular and respiratory centers of the pons and medulla oblongata

26 Which neurotransmitter is released by all preganglionic fibers of the autonomic nervous system?
- ☐ a) acetylcholine
- ☐ b) epinephrine
- ☐ c) norepinephrine
- ☐ d) serotonin

27 Dual innervation refers to situations in which vital organs
- ☐ a) receive information from both sympathetic and parasympathetic fibers.
- ☐ b) receive information from both the somatic and autonomic nervous systems.
- ☐ c) are receptive to multiple neurotransmitters.
- ☐ d) are under both conscious and unconscious control.

Short answer

28 Identify the components of a visceral reflex arc, and distinguish between short reflexes and long reflexes.

29 Identify at least three differences between the somatic and autonomic nervous systems.

30 Which cranial and sacral nerves are associated with the parasympathetic division?

MasteringA&P®

Access more chapter study tools online in the MasteringA&P Study Area:

- **Chapter Quizzes, Chapter Practice Test, Art-labeling Activities, Animations, MP3 Tutor Sessions, and Clinical Case Studies**

- **Practice Anatomy Lab** — PAL™
- **Interactive Physiology** — iP®
- **A&P Flix** — A&PFlix™
- **PhysioEx** — PhysioEx™

Chapter Integration • Applying what you have learned

A patch for the ocean

Andrew, a recent high school graduate who has spent his entire life in his hometown of Tulsa, Oklahoma, is very excited to have been accepted to a university in San Diego. One of the first social events organized by his dormitory is an all-day whale-watching tour. Although he had fished from a boat in lakes as a young boy, Andrew has never been out on the ocean. Despite being apprehensive, he signed up.

Prior to the tour, Andrew visited the campus health center to ask about seasickness medicine. The doctor recommended that Andrew use a patch that he can place behind one of his ears. The doctor said that even though it's not fully understood how the patch's active ingredient, scopolamine, works to prevent seasickness, his patients have had good results with it. Andrew decided to give it a try.

As he applied the patch on the morning of the whale-watching tour, he read the package label. It described scopolamine as antimuscarinic and anticholinergic. Andrew had no idea what that meant. Using what you have just learned about the autonomic nervous system, answer the following questions.

1 **Which division of the autonomic nervous system would be affected if muscarinic receptors were blocked?**

2 **Blocking muscarinic receptors would affect which neurotransmitter?**

3 **Why does the label also describe scopolamine as anticholinergic?**

4 **Predict some side effects Andrew might experience while wearing the patch.**

Bee careful in the garden!

Gregor is outside gardening when he is stung on his finger by a bee. Shortly afterwards, he experiences pain, swelling, redness, and itching in the immediate area of the sting; his throat begins to swell and breathing becomes more difficult.

An ambulance is called, and Gregor is taken to the emergency room. Drugs are administered to prevent anaphylaxis, a severe, life-threatening allergic response. Anaphylaxis is characterized by the contraction of smooth muscles along the respiratory passageways and a peripheral vasodilation that produces a fall in blood pressure, which would affect multiple organ systems.

5 **Would acetylcholine or epinephrine be more helpful in relieving his condition? Why?**

6 **If the condition progressed to the point where there is a fall in blood pressure and inadequate blood flow to body tissues, describe what ANS responses would attempt to restore normal blood flow and blood pressure.**

15 The Special Senses

LEARNING OUTCOMES

These Learning Outcomes correspond by number to this chapter's modules and indicate what you should be able to do after completing the chapter.

SECTION 1 • Olfaction and Gustation

15.1 Explain the roles of a generator potential and receptor depolarization in olfaction and gustation.

15.2 Describe the sensory organs of smell, trace the olfactory pathways to their destinations in the cerebrum, and explain how olfactory perception occurs.

15.3 Describe the sensory organs of gustation.

15.4 Describe gustatory reception, briefly describe the physiological processes involved in taste, and trace the gustatory pathway.

SECTION 2 • Equilibrium and Hearing

15.5 Describe the sensory receptors of the internal ear.

15.6 Describe the structures of the external, middle, and internal ear, and explain how they function.

15.7 Describe the structures and functions of the bony labyrinth and membranous labyrinth.

15.8 Describe the functions of hair cells in the semicircular ducts, utricle, and saccule.

15.9 Describe the structures and functions of the organ of Corti.

15.10 Explain the anatomical and physiological basis for pitch and volume sensations for hearing.

15.11 Trace the pathways for the sensations of equilibrium and hearing to their respective destinations in the brain.

SECTION 3 • Vision

15.12 Outline the embryonic development of the eye.

15.13 Identify the accessory structures of the eye and explain their functions.

15.14 Describe the layers of the wall of the eye and the anterior and posterior cavities of the eye.

15.15 Explain how light is directed to the fovea of the retina.

15.16 Describe the process by which images are focused on the retina.

15.17 Describe the structure and function of the retina's layers of cells, and the distribution of rods and cones and their relation to visual acuity.

15.18 Describe the structure of the photoreceptors and how we are able to distinguish colors.

15.19 Explain photoreception and how visual pigments are activated.

15.20 Explain how the visual pathways distribute information to their destinations in the brain.

15.21 **CLINICAL MODULE** Describe various refraction problems associated with the cornea, lens, or shape of the eye.

15.22 **CLINICAL MODULE** Describe age-related disorders of olfaction, gustation, vision, equilibrium, and hearing.

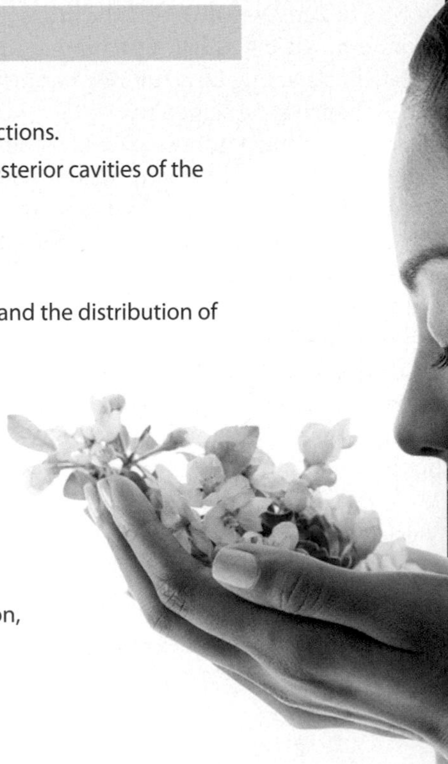

LO Learning Outcomes are repeated at the bottom of each module.

Olfaction is the sense of smell and gustation is the sense of taste

The special senses give us important information about our environment. Although the sensory information provided is diverse and complex, each special sense originates at receptor cells that may be neurons or specialized receptor cells that communicate with sensory neurons.

1 Olfactory receptors are the dendrites of specialized neurons involved with the sense of smell. When dissolved chemicals contact the dendritic processes, there is a depolarization, called a **generator potential**. This graph shows the action potentials produced by a generator potential.

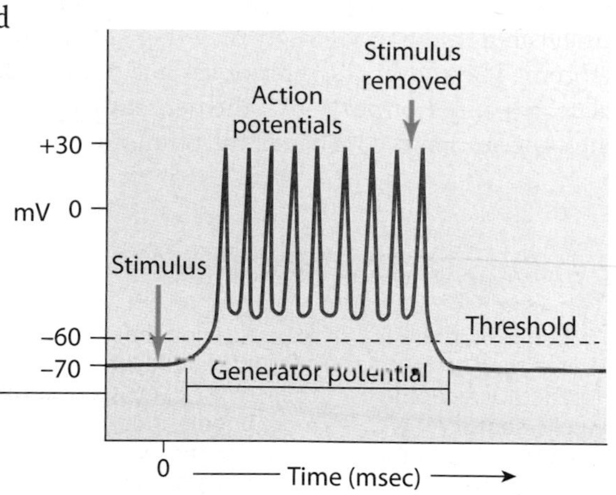
Stimulus
removed
Action
potentials
+30
mV 0
Stimulus
Threshold
–60
–70
Generator potential
0
Time (msec)

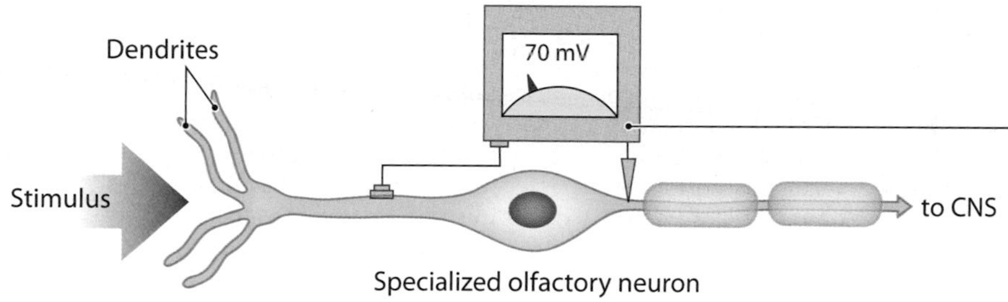
Dendrites
70 mV
Stimulus
to CNS
Specialized olfactory neuron

2 The receptors for the senses of taste, vision, equilibrium, and hearing are specialized cells that have inexcitable membranes and form synapses with the processes of sensory neurons. When stimulated, the membrane of the receptor cell undergoes a graded depolarization that triggers the release of chemical transmitters at the synapse. These transmitters then depolarize the sensory neuron, inducing a generator potential capable of producing action potentials that are propagated to the CNS. Because a synapse is involved, there is a slight synaptic delay. However, this arrangement permits modification of the sensitivity of the receptor cell by presynaptic facilitation or inhibition.

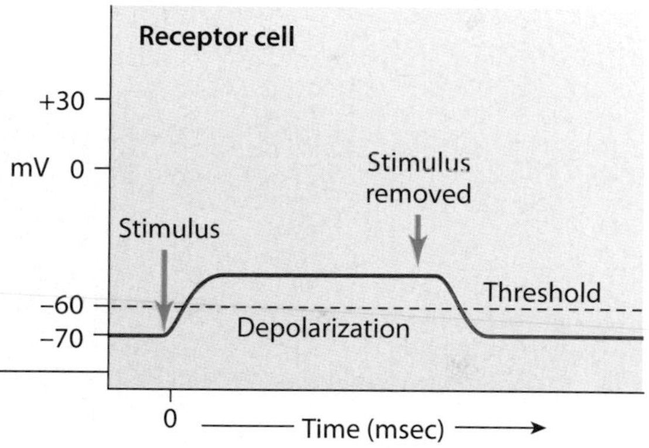
Receptor cell
+30
mV 0
Stimulus
removed
Stimulus
Threshold
–60
–70
Depolarization
0
Time (msec)

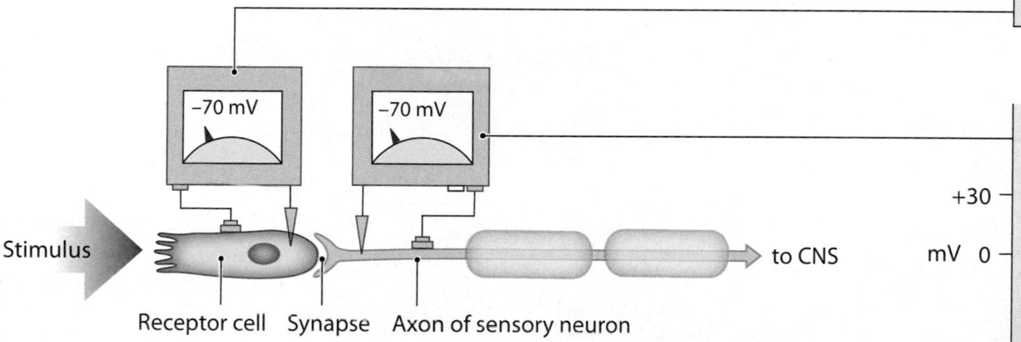
–70 mV
–70 mV
Stimulus
to CNS
Receptor cell
Synapse
Axon of sensory neuron

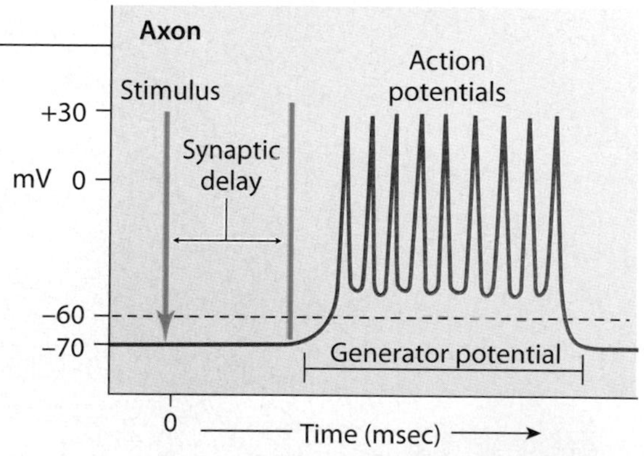
Axon
Stimulus
Action
potentials
+30
Synaptic
delay
mV 0
–60
–70
Generator potential
0
Time (msec)

In the following modules we will consider the organization of receptors and the sensory pathways for each of the special senses. We begin with the special senses of olfaction and gustation.

Module 15.1 Review

a. Where do the special senses originate?
b. What is a generator potential?
c. How does the origin of the other senses differ from olfaction?

LO **15.1** Explain the roles of a generator potential and receptor depolarization in olfaction and gustation.

Module 15.2

Olfaction involves specialized chemoreceptive neurons and delivers sensations directly to the cerebrum

1 **Olfaction** is the sense of smell. **Olfactory organs** are paired structures that provide olfaction. These organs are located in the nasal cavity on either side of the nasal septum. They cover the inferior surface of the cribriform plate, the superior portion of the perpendicular plate, and the superior nasal conchae of the ethmoid.

Olfactory Pathway to the Cerebrum

Chemicals in the air stimulate the olfactory receptor cells, sensory neurons within the olfactory organ.	Axons leaving the olfactory epithelium collect into 20 or more bundles that penetrate the cribriform plate of the ethmoid bone.	The first synapse occurs in the **olfactory bulb**, which is located just superior to the cribriform plate.	Axons leaving the olfactory bulb travel along the **olfactory tract** to reach the olfactory cortex, the hypothalamus, and portions of the limbic system.	The distribution of olfactory information to the limbic system and hypothalamus explains the profound emotional and behavioral responses, as well as the memories, that can be triggered by certain smells.

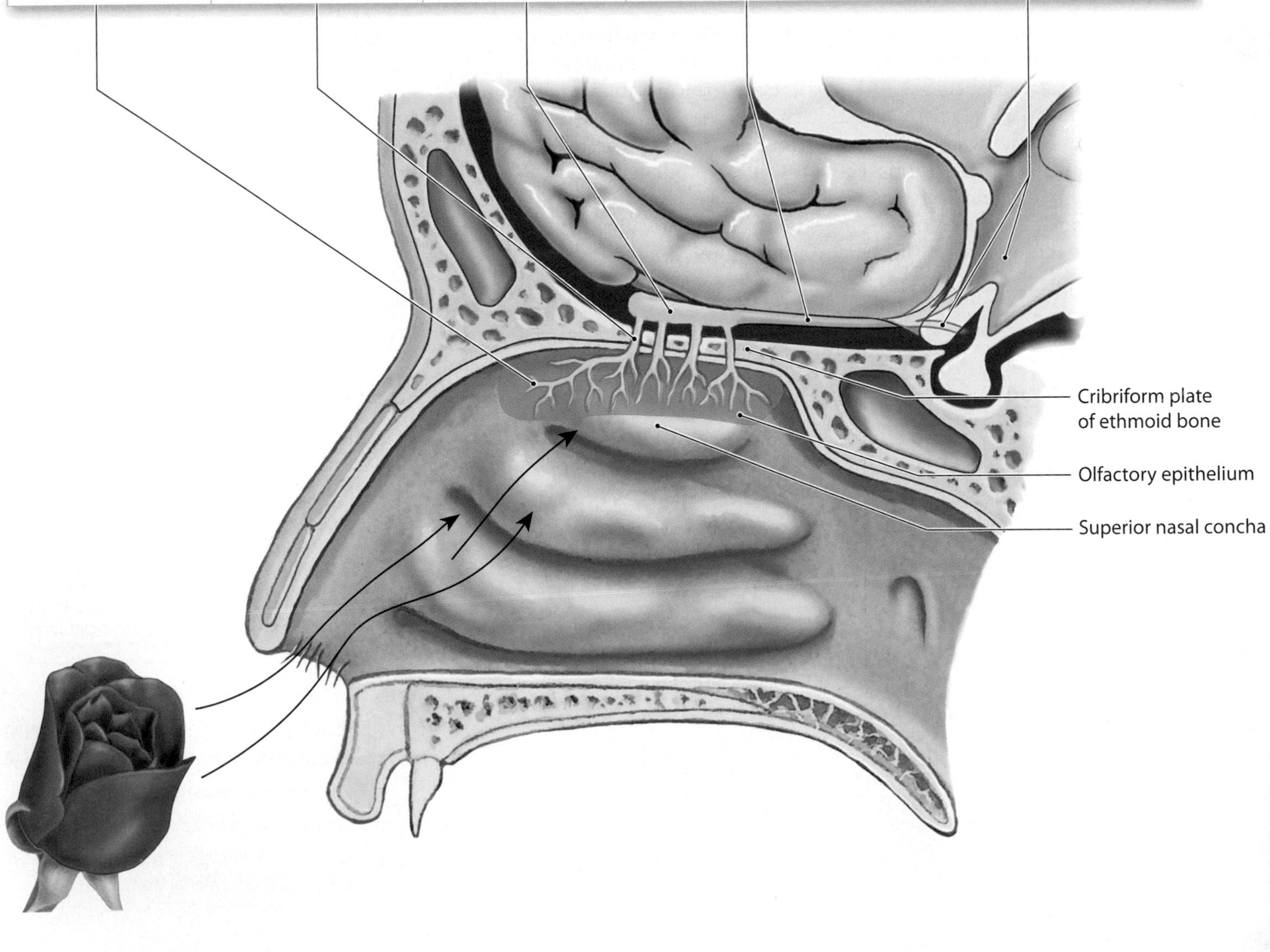

Cribriform plate of ethmoid bone
Olfactory epithelium
Superior nasal concha

2 The olfactory organs are made up of two layers: the **olfactory epithelium** and the **lamina propria**. The olfactory receptors are modified neurons within the epithelium. The exposed tip of each receptor cell forms a knob that projects beyond the epithelial surface. The knob provides a base for up to 20 cilia-shaped dendrites that extend into the surrounding mucus and lie parallel to the epithelial surface, exposing their considerable surface area to dissolved chemicals. Between 10 and 20 million olfactory receptors are packed into an area of roughly 5 cm^2. If we take into account the exposed dendritic surfaces, the actual sensory area probably approaches that of the entire body surface.

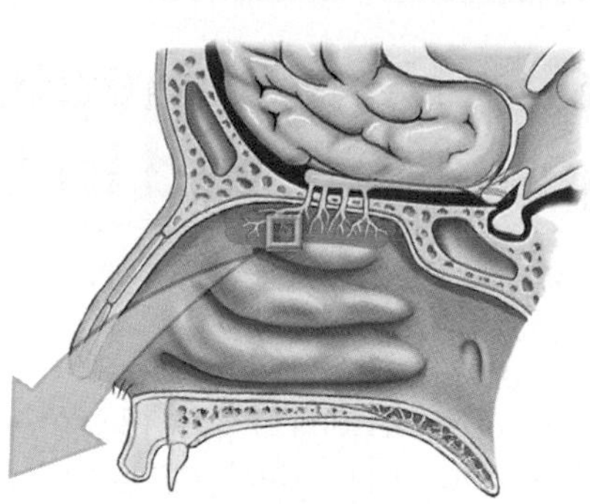

The underlying lamina propria consists of areolar tissue, numerous blood vessels, and nerves. This layer also contains **olfactory glands**, or Bowman's glands, whose secretions absorb water and form a thick, pigmented mucus.

The olfactory epithelium contains the **olfactory receptor cells**, **supporting cells**, and regenerative **basal cells** (stem cells). The olfactory receptor population undergoes considerable turnover; new receptor cells are produced by the division and differentiation of basal cells. This turnover is one of the few examples of neuronal replacement in adult humans.

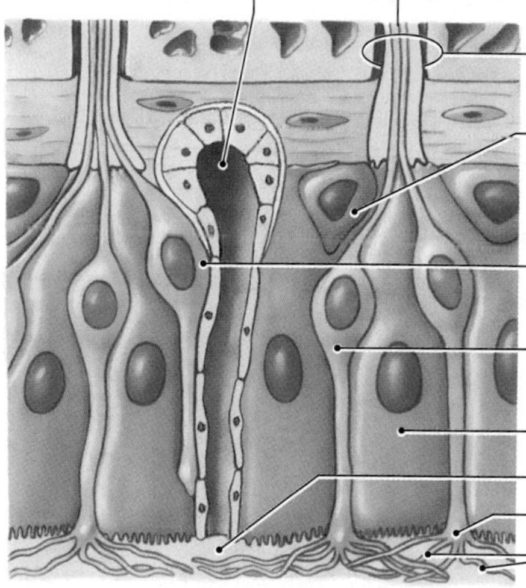

Olfactory gland
To olfactory bulb
Olfactory nerve fibers
Basal cell: divides to replace worn-out olfactory receptor cells
Developing olfactory receptor cell
Olfactory receptor cell
Supporting cell
Mucous layer
Knob
Olfactory dendrites: surfaces contain receptor proteins

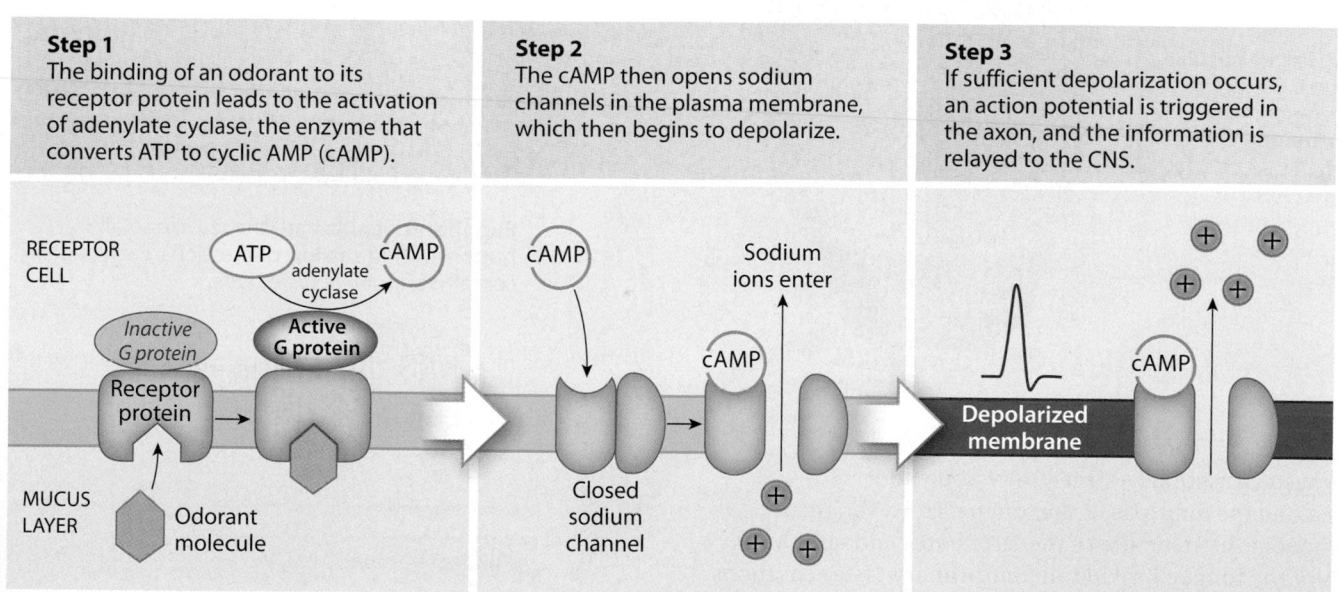

Step 1
The binding of an odorant to its receptor protein leads to the activation of adenylate cyclase, the enzyme that converts ATP to cyclic AMP (cAMP).

Step 2
The cAMP then opens sodium channels in the plasma membrane, which then begins to depolarize.

Step 3
If sufficient depolarization occurs, an action potential is triggered in the axon, and the information is relayed to the CNS.

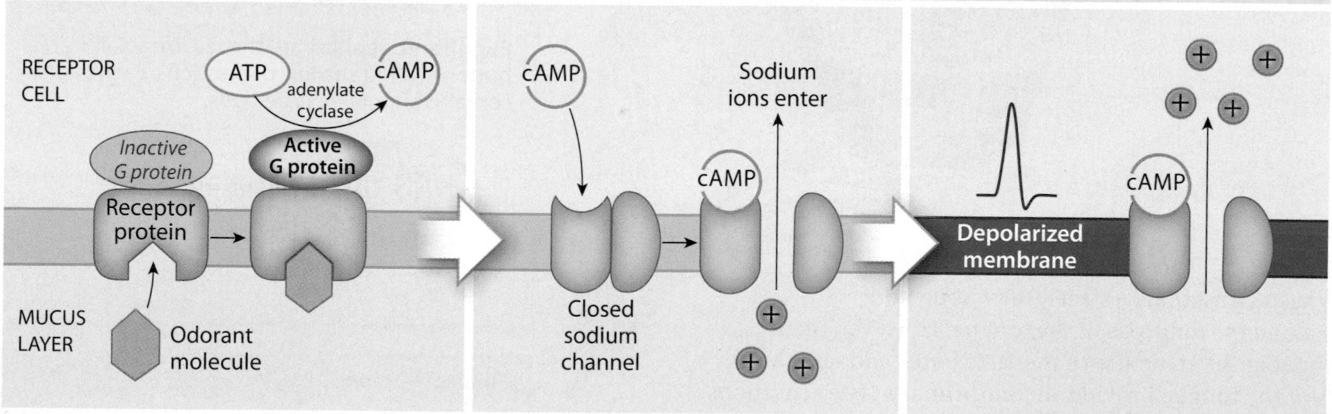

RECEPTOR CELL
ATP
adenylate cyclase
cAMP
Inactive G protein
Active G protein
Receptor protein
MUCUS LAYER
Odorant molecule
cAMP
Sodium ions enter
cAMP
Closed sodium channel
Depolarized membrane
cAMP

3 Olfactory reception occurs on the dendrites of the olfactory receptor cells. **Odorants**—dissolved chemicals that stimulate olfactory neurons—interact with membrane receptors called odorant binding proteins on the membrane surface. In general, odorants are small organic molecules; the strongest smells are associated with molecules of either high water or high lipid solubilities. As few as four odorant molecules can activate an olfactory receptor cell.

Module 15.2 Review

a. Describe olfaction.

b. Which neurons associated with olfaction are continually regenerated?

c. Trace the olfactory pathway, beginning at the olfactory epithelium.

LO 15.2 Describe the sensory organs of smell, trace the olfactory pathways to their destinations in the cerebrum, and explain how olfactory perception occurs.

Gustation involves epithelial chemoreceptor cells located in taste buds

Gustation, the sense of taste, provides information about the foods and liquids we eat and drink. **Taste receptors**, or gustatory (GUS-ta-tor-ē) receptors, are distributed over the superior surface of the tongue and adjacent portions of the pharynx and larynx. The most important taste receptors are on the tongue. By the time we reach adulthood, the taste receptors on the pharynx, larynx, and epiglottis have decreased in number.

1 The superior surface of the tongue has numerous variously shaped epithelial projections called **lingual papillae** (pa-PIL-ē; *papilla*, a nipple-shaped mound). The human tongue has four types of lingual papillae: (1) vallate (VAL-āt; *vallum*, wall) papillae, (2) foliate (FŌ-lē-āt) papillae, (3) fungiform (*fungus*, mushroom) papillae, and (4) filiform (*filum*, thread) papillae. The vallate, foliate, and fungiform papillae contain taste receptors and specialized epithelial cells in sensory structures called **taste buds**. An adult has about 5000 taste buds.

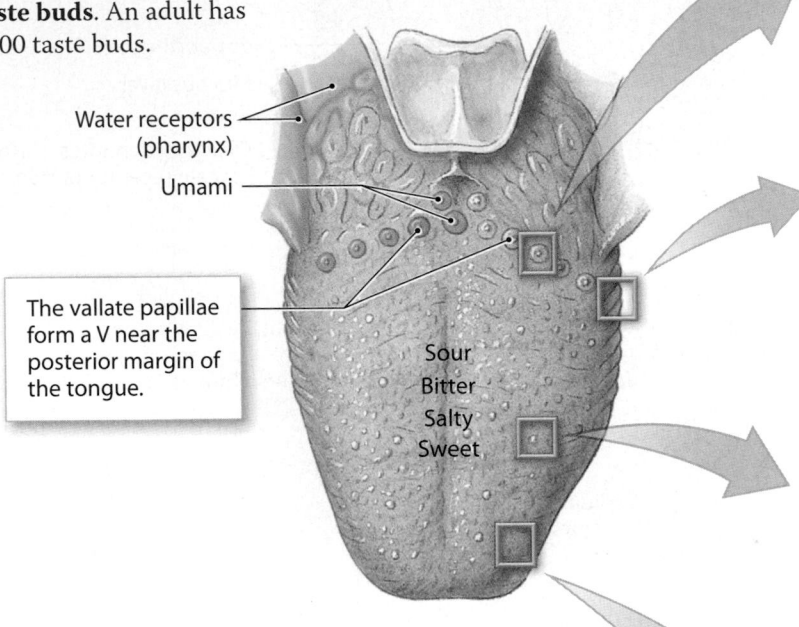

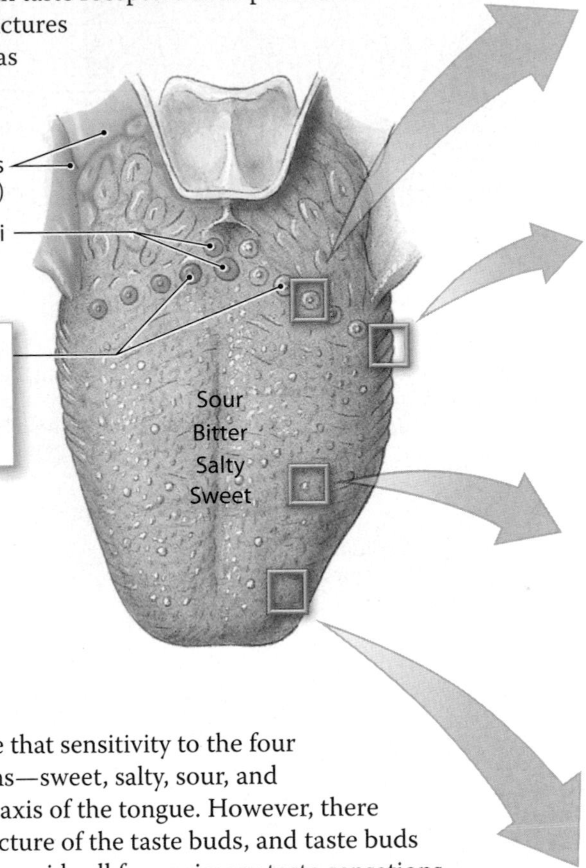
Water receptors (pharynx)
Umami
Sour
Bitter
Salty
Sweet

The vallate papillae form a V near the posterior margin of the tongue.

Vallate Papillae

Vallate papillae are relatively large, shaped like the tip of a pencil eraser, and surrounded by deep epithelial folds. Each papilla contains as many as 100 taste buds.

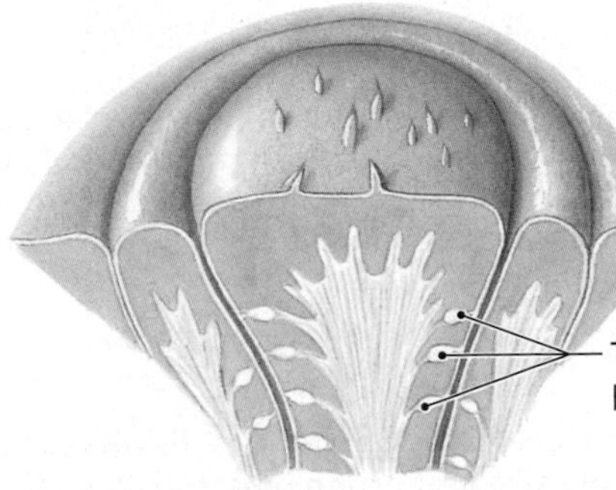
Taste buds

Foliate Papillae

Foliate papillae are found on the lateral margins of the posterior region of the tongue.

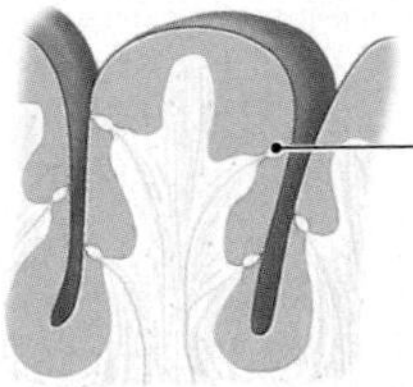
Taste bud

Fungiform Papillae

Fungiform papillae are shaped like small buttons within shallow depressions. Each papilla contains about five taste buds.

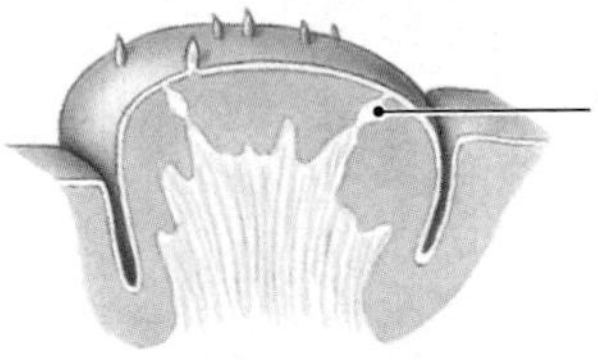
Taste bud

Filiform Papillae

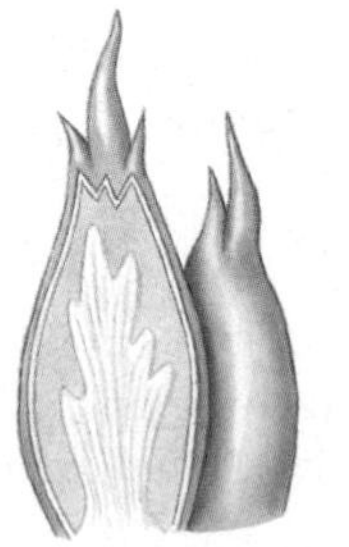

Filiform papillae provide friction that helps the tongue move objects around in the mouth but do not contain taste buds. Filiform and fungiform papillae are found on the anterior two-thirds of the superior surface of the tongue.

2 There is some evidence that sensitivity to the four primary taste sensations—sweet, salty, sour, and bitter—varies along the long axis of the tongue. However, there are no differences in the structure of the taste buds, and taste buds in all portions of the tongue provide all four primary taste sensations. There are also two other taste sensations. **Umami** (oo-MAH-mē) is a pleasant, savory taste that is characteristic of beef broth, chicken broth, and Parmesan cheese. This taste is detected by receptors sensitive to amino acids, small peptides, and nucleotides. These receptors are present in taste buds of the vallate papillae. **Water receptors** have been demonstrated in humans, and they appear to be especially concentrated in the pharynx. The sensory output of these receptors is processed in the hypothalamus and affects water balance and the regulation of blood volume. For example, drinking stimulates water receptors, that in turn stimulate the hypothalamus to secrete antidiuretic hormone, which ultimately prevents over-ingestion of water.

3 Taste buds are recessed into the surrounding epithelium, isolated from the relatively unprocessed contents of the mouth. Each taste bud contains 40–100 receptor cells and many small stem cells.

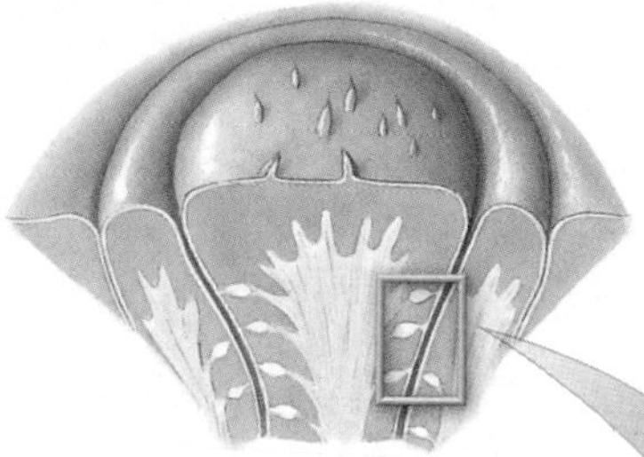

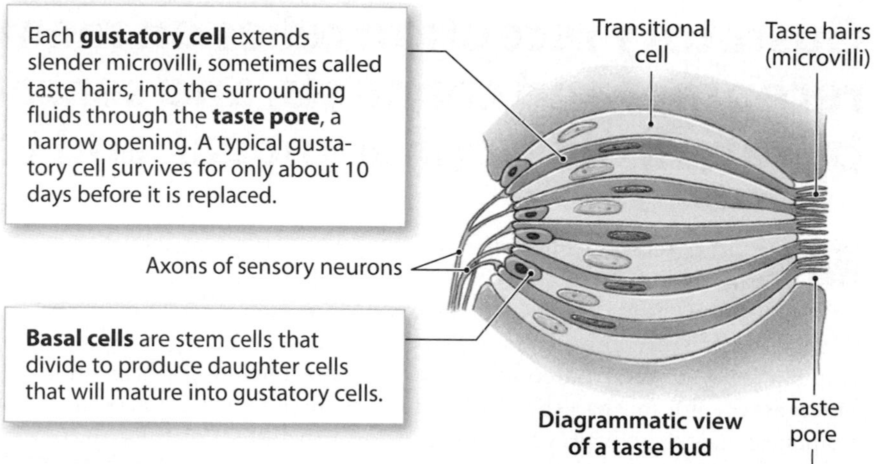
Each **gustatory cell** extends slender microvilli, sometimes called taste hairs, into the surrounding fluids through the **taste pore**, a narrow opening. A typical gustatory cell survives for only about 10 days before it is replaced.
Transitional cell
Taste hairs (microvilli)
Axons of sensory neurons
Basal cells are stem cells that divide to produce daughter cells that will mature into gustatory cells.
Taste pore

Diagrammatic view of a taste bud

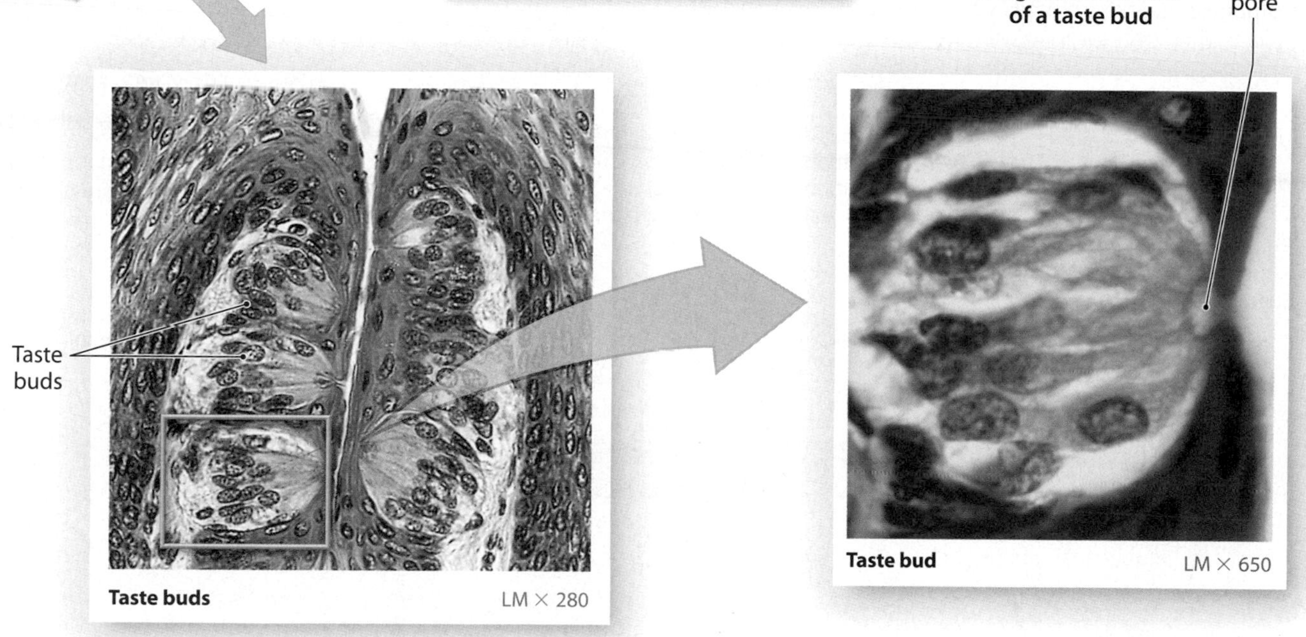
Taste buds
Taste buds
LM × 280
Taste bud
LM × 650

4 The threshold for taste receptor stimulation varies for each of the primary taste sensations, and taste receptors respond more readily to unpleasant than to pleasant stimuli. For example, we are 100,000 times more sensitive to bitter substances and almost 1000 times more sensitive to acids (sour) than to either sweet or salty chemicals. This sensitivity has survival value, because acids can damage the mucous membranes of the mouth and pharynx, and many potent biological toxins have an extremely bitter taste. Our tasting abilities change with age. We begin life with more than 10,000 taste buds, but the number begins declining dramatically by age 50. The sensory loss becomes especially significant because aging people also experience a decline in the number of olfactory receptors. As a result, many elderly people find that their food tastes bland and unappetizing, whereas children tend to find the same foods too spicy.

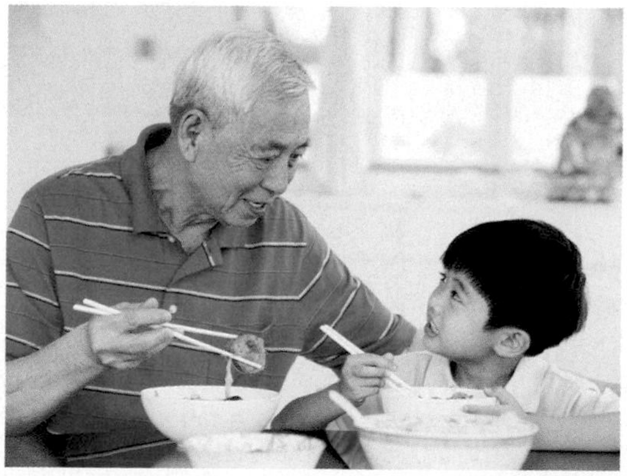

Module 15.3 Review

a. Define gustation.

b. Describe filiform papillae.

c. Which taste receptors offer a survival advantage when tasting something for the first time?

LO 15.3 Describe the sensory organs of gustation.

Module 15.4

Gustatory reception relies on membrane receptors and channels, and sensations are carried by facial, glossopharyngeal, and vagus nerves

1 Gustatory reception, like olfactory reception, is stimulated by dissolved chemicals. Dissolved chemicals contacting the taste hairs may either diffuse through plasma membrane leak channels or bind to receptor proteins of the gustatory receptor cell. Some 90 percent of the gustatory receptor cells respond to two or more different taste stimuli. The different tastes involve different receptor mechanisms. Taste receptors adapt slowly, but central adaptation quickly reduces your sensitivity to a new taste.

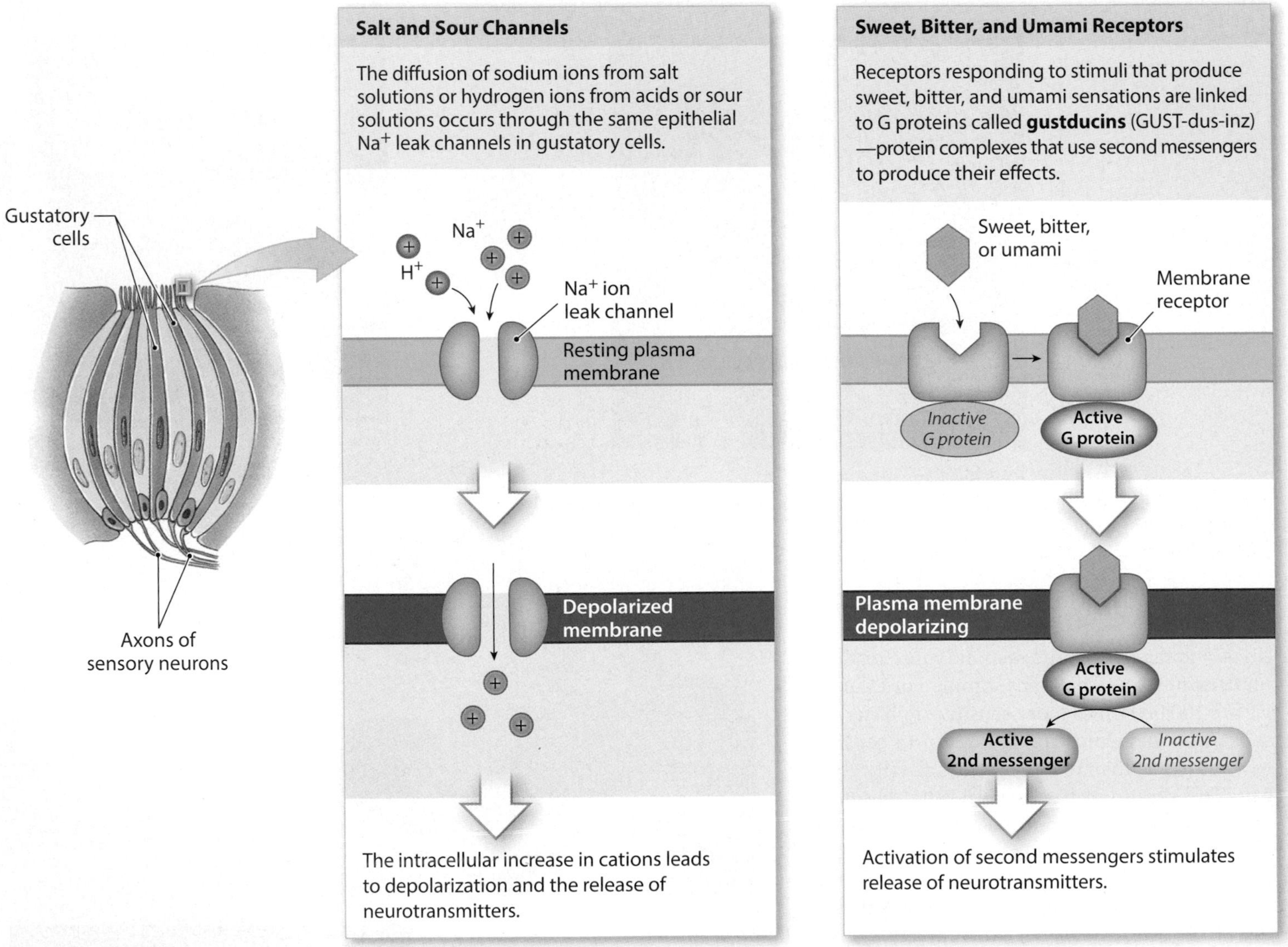

Gustatory cells
Axons of sensory neurons
Salt and Sour Channels
The diffusion of sodium ions from salt solutions or hydrogen ions from acids or sour solutions occurs through the same epithelial Na^+ leak channels in gustatory cells.
Na^+
H^+
Na^+ ion leak channel
Resting plasma membrane
Depolarized membrane
The intracellular increase in cations leads to depolarization and the release of neurotransmitters.
Sweet, Bitter, and Umami Receptors
Receptors responding to stimuli that produce sweet, bitter, and umami sensations are linked to G proteins called **gustducins** (GUST-dus-inz)—protein complexes that use second messengers to produce their effects.
Sweet, bitter, or umami
Membrane receptor
Inactive G protein
Active G protein
Plasma membrane depolarizing
Active G protein
Active 2nd messenger
Inactive 2nd messenger
Activation of second messengers stimulates release of neurotransmitters.

The released neurotransmitters enter synapses with the dendrites of adjacent sensory neurons. Depolarization of sensory neurons leads to a generator potential and the propagation of action potentials along the gustatory pathway to the CNS.

2 This illustration follows the **gustatory pathway** from the receptors to the cerebral cortex. A conscious perception of taste is produced by processing at the primary sensory cortex, as the information received from the taste buds is correlated with other sensory data. Information about the texture of food, along with taste-related sensations such as "peppery" or "burning hot," comes from sensory afferents in the trigeminal nerve (V). In addition, the level of stimulation from the olfactory receptors plays an overwhelming role in taste perception. Thus, you are several thousand times more sensitive to "tastes" when your olfactory organs are fully functional. By contrast, when you have a cold and your nose is stuffed up, airborne molecules cannot reach your olfactory receptors, so meals taste dull and unappealing. This reduction in taste perception occurs even though the taste buds are responding normally.

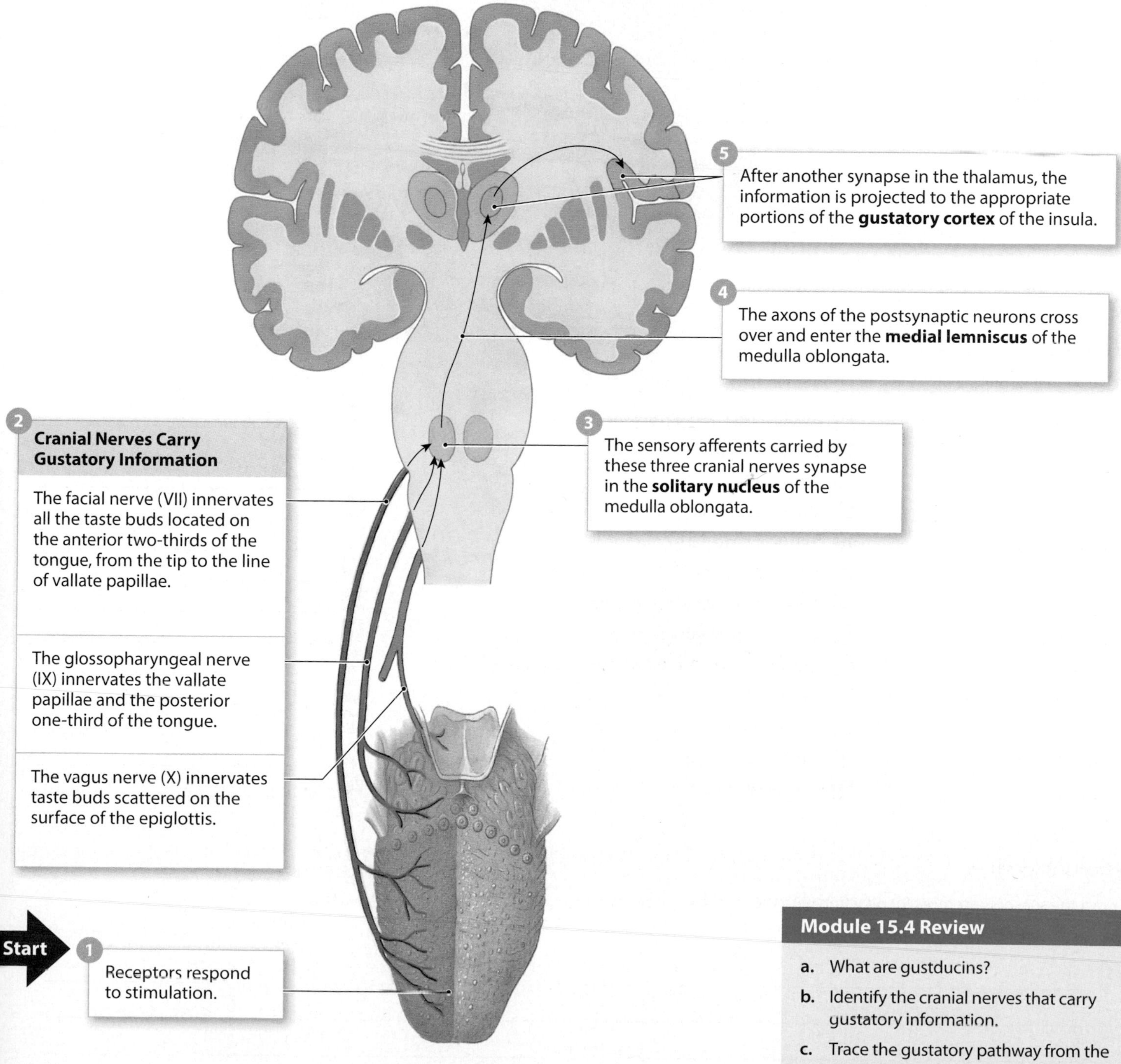

5
After another synapse in the thalamus, the information is projected to the appropriate portions of the **gustatory cortex** of the insula.
4
The axons of the postsynaptic neurons cross over and enter the **medial lemniscus** of the medulla oblongata.
3
The sensory afferents carried by these three cranial nerves synapse in the **solitary nucleus** of the medulla oblongata.
2
Cranial Nerves Carry Gustatory Information
The facial nerve (VII) innervates all the taste buds located on the anterior two-thirds of the tongue, from the tip to the line of vallate papillae.
The glossopharyngeal nerve (IX) innervates the vallate papillae and the posterior one-third of the tongue.
The vagus nerve (X) innervates taste buds scattered on the surface of the epiglottis.
Start
1
Receptors respond to stimulation.

Module 15.4 Review

a. What are gustducins?

b. Identify the cranial nerves that carry gustatory information.

c. Trace the gustatory pathway from the taste receptors to the cerebral cortex.

LO 15.4 Describe gustatory reception, briefly describe the physiological processes involved in taste, and trace the gustatory pathway.

Labeling

Label taste sensations associated with areas on the tongue (1–5) and the four types of lingual papillae (6–9).

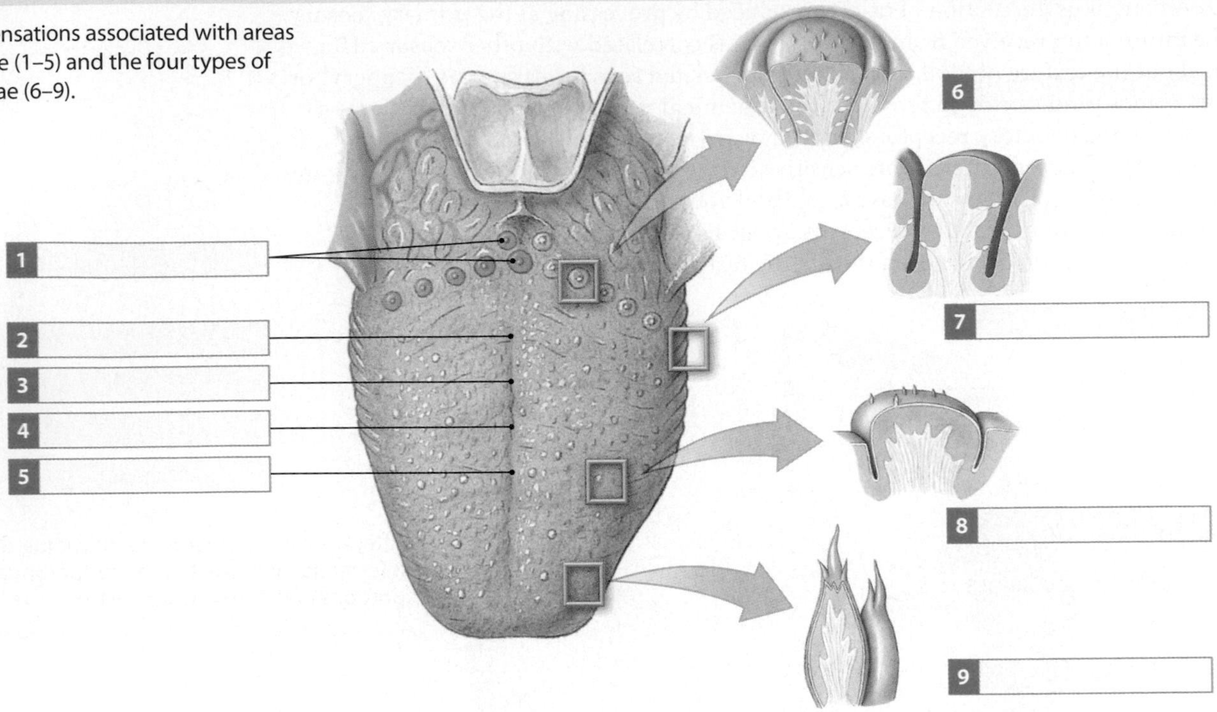
1
2
3
4
5
6
7
8
9

Matching

Match each lettered term with the most closely related description.

a. gustation
b. depolarization
c. Bowman's glands
d. lingual papillae
e. G proteins
f. stem cells
g. bitter
h. olfactory dendrites
i. olfaction
j. taste bud
k. cerebral cortex
l. olfactory bulb
m. odorant

10 Sweet, bitter, and umami sensations
11 Sense of smell
12 Basal cells
13 Chemical stimulus
14 Sense of taste
15 Olfactory glands
16 Receives all special senses stimuli
17 Cluster of gustatory receptors
18 Site of first synapse by olfactory receptors
19 Contain olfactory receptor proteins
20 Most sensitive taste sensation
21 Produces generator potential
22 Epithelial projections of tongue

10 ____
11 ____
12 ____
13 ____
14 ____
15 ____
16 ____
17 ____
18 ____
19 ____
20 ____
21 ____
22 ____

Section integration

Contrast the sensory receptors for olfaction with the sensory receptors for taste, vision, equilibrium, and hearing.

23 ____

Equilibrium and hearing involve the internal ear

1 In olfaction, the sensory receptors are modifed sensory neurons. In gustation, sensory receptor cells communicate with sensory neurons. In each case, the sensory receptor cells are located within epithelia exposed to the external environment, and the information is routed directly to the CNS for processing.

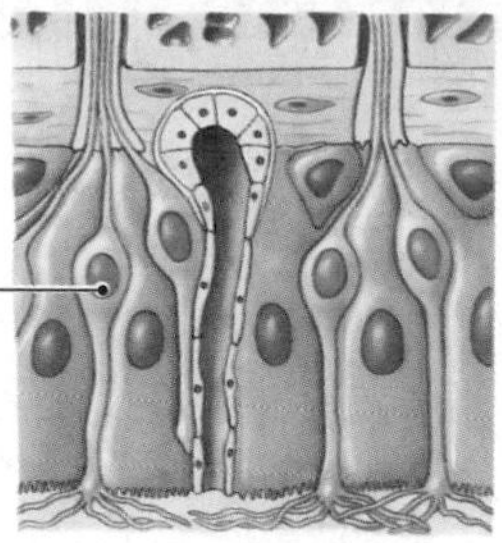
Olfactory receptor cell

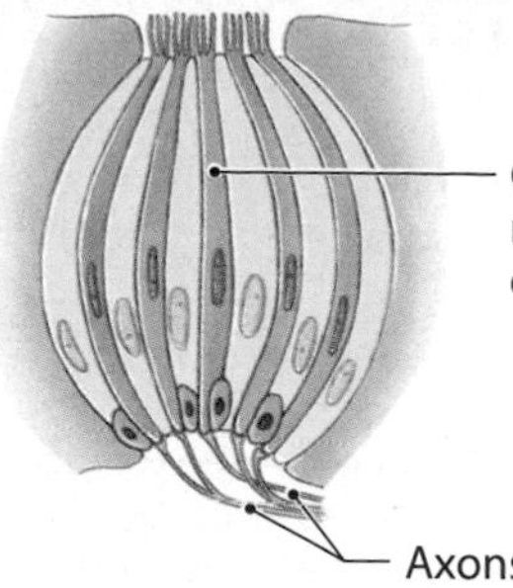
Gustatory receptor cell
Axons of sensory neurons

2 In contrast, the receptor cells for the sensations of equilibrium and hearing are isolated and protected from the external environment. These sensory receptors are located within the **internal ear**, a complex sense organ, where the sensory information is integrated and organized before it is forwarded to the CNS.

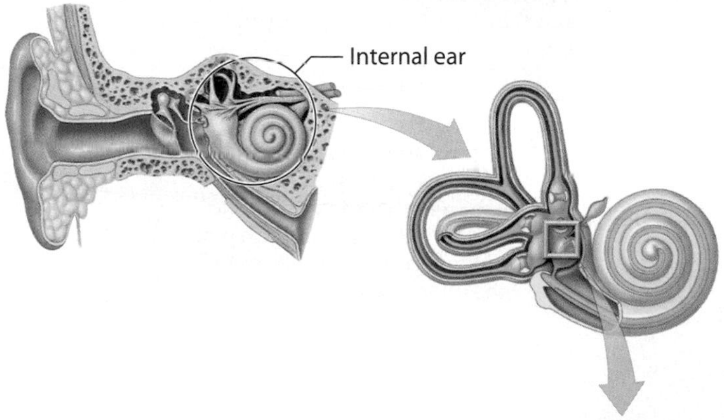
Internal ear

3 The receptor cells of the internal ear are called **hair cells** because their free surfaces are covered with specialized processes similar to the cilia and microvilli of other cells. They are basically mechanoreceptors sensitive to contact or movement, and they are always surrounded by supporting cells and monitored by the dendrites of sensory neurons. When an external force pushes against the processes of a hair cell, the distortion of the plasma membrane alters the rate at which the hair cell releases neurotransmitters. In this way, hair cells provide information about the direction and strength of mechanical stimuli.

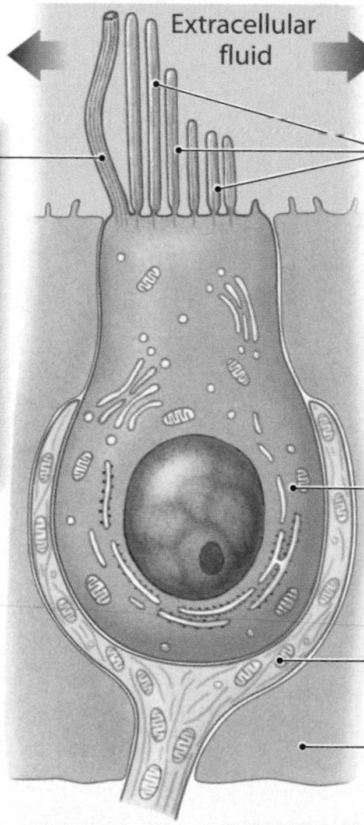
Displacement in this direction stimulates hair cell
Extracellular fluid
Displacement in this direction inhibits hair cell
Certain hair cells involved with balance contain a **kinocilium** (ki-nō-SIL-ē-um), a single large cilium. Hair cells do not actively move their kinocilia (or stereocilia).
The free surface of every hair cell supports 80–100 long **stereocilia**, which resemble very long microvilli.
Hair cell
Dendrite of sensory neuron
Supporting cell

The internal ear has a complex three-dimensional structure that determines what stimuli can reach the hair cells in each region. For example, hair cells in one region can respond only to gravity or acceleration, whereas those in other regions respond only to rotation or only to sound. In this section you will learn how the anatomy of the internal ear enables a single receptor cell type to provide such a diversity of information.

Module 15.5 Review

a. Contrast the olfactory and gustatory receptors with those of equilibrium and hearing.

b. Describe the internal ear receptors.

c. What kind of stimuli can the internal ear sense?

LO 15.5 Describe the sensory receptors of the internal ear.

Module 15.6

The ear is divided into the external ear, the middle ear, and the internal ear

1 The ear is divided into three anatomical regions: the external ear, the middle ear, and the internal ear.

External Ear	Middle Ear	Internal Ear
The **external ear**—the visible portion of the ear—collects and directs sound waves toward the middle ear.	The **middle ear**, or tympanic cavity, is an air-filled chamber separated from the external acoustic meatus by the tympanic membrane. It is connected to the pharynx by the auditory tube.	The **internal ear** contains the sensory organs for hearing and equilibrium. It receives amplified sound waves from the middle ear.

Elastic cartilage keeps the external acoustic meatus open and makes the **auricle** flexible.

The **auditory ossicles** are three tiny bones within the middle ear. These bones connect the tympanic membrane with one of the receptor complexes of the internal ear.

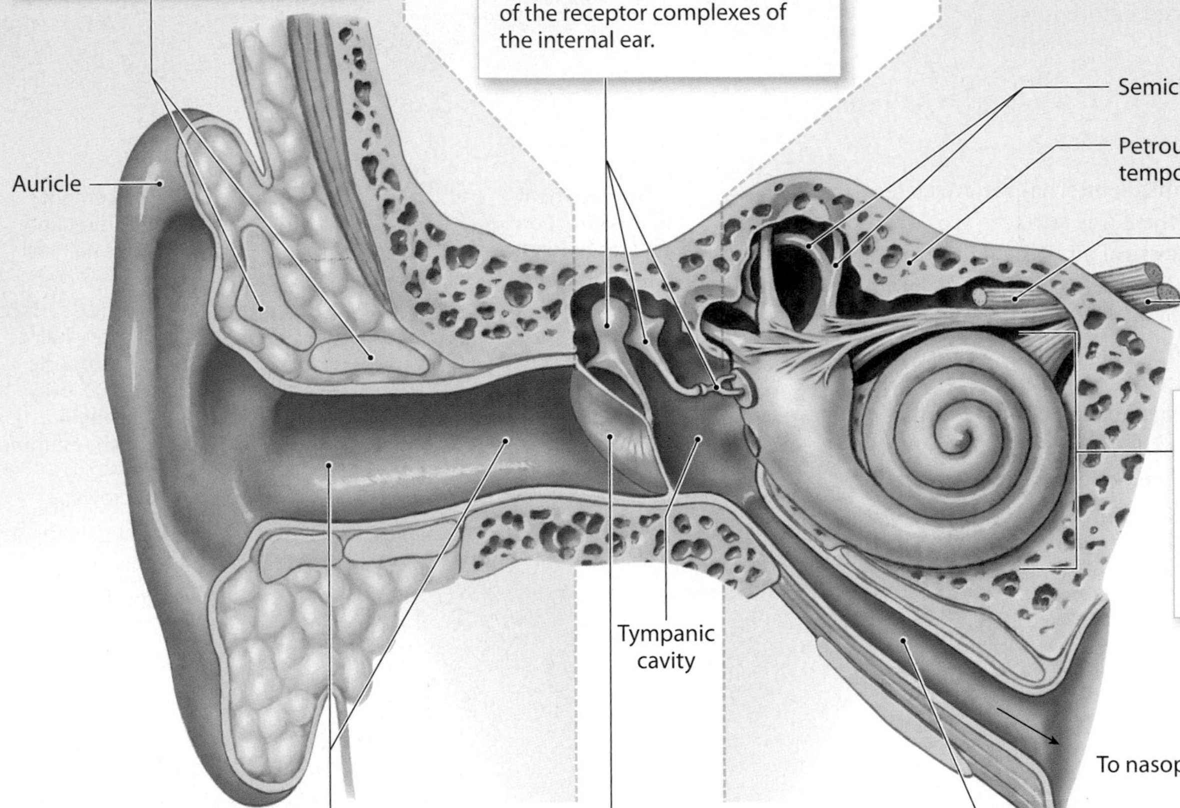
Auricle
Semicircular canals
Petrous part of temporal bone
Facial nerve (VII)
Vestibulocochlear nerve (VIII)
Tympanic cavity
To nasopharynx

The superficial contours of the internal ear are established by a layer of dense bone known as the **bony labyrinth** (*labyrinthos*, network of canals).

The **external acoustic meatus** is a passageway within the temporal bone. The skin lining this passage contains **ceruminous glands** that secrete a waxy material, **cerumen**, and it has many small, outwardly projecting hairs. Together, cerumen and the hairs help keep foreign objects and insects from reaching more delicate internal structures. Cerumen also slows the growth of microorganisms in the external acoustic meatus, reducing the chances for an external ear infection.

The **tympanic membrane**, also called the tympanum or eardrum, lies at the end of the external acoustic meatus. The tympanic membrane is a thin, semitransparent sheet that separates the external ear from the middle ear.

The **auditory tube** is also called the pharyngotympanic tube or the eustachian tube. The auditory tube permits pressure equalization on either side of the tympanic membrane, but can also allow microorganisms to travel from the nasopharynx into the middle ear. Invasion by microorganisms can lead to an unpleasant middle ear infection known as **otitis media**.

2 The middle ear contains the three auditory ossicles and communicates with both the superior portion of the pharynx (the nasopharynx), through the auditory tube, and the mastoid air cells, through a number of small connections. The articulations between the auditory ossicles are the smallest synovial joints in the body. Each ossicle has a tiny capsule and supporting extracapsular ligaments.

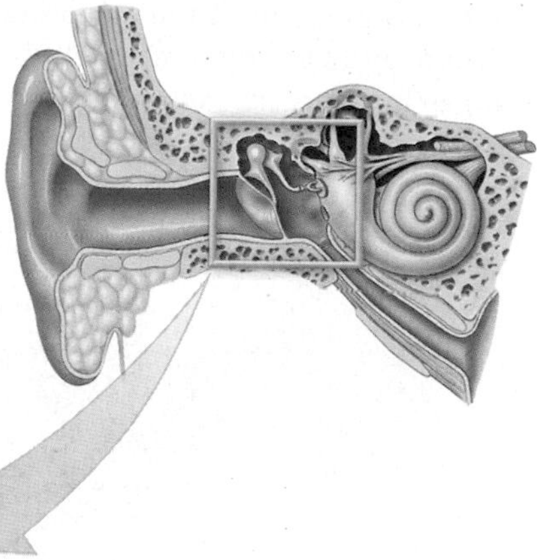

Auditory Ossicles

The **malleus** (*malleus*, hammer) attaches at three points to the interior surface of the tympanic membrane.

The **incus** (*incus*, anvil), the middle ossicle, attaches the malleus to the stapes.

The edges of the base of the **stapes** (STĀ-pēz; *stapes*, stirrup) are bound to the edges of the oval window, an opening in the bone that surrounds the internal ear.

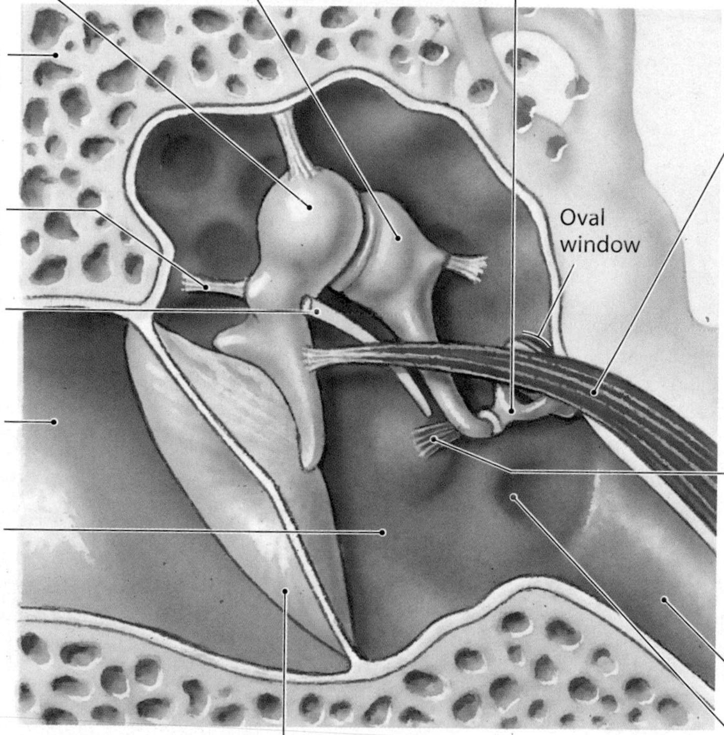

Temporal bone (petrous part)
Stabilizing ligament
Branch of facial nerve VII (cut)
External acoustic meatus
Tympanic cavity (middle ear)
Oval window
Auditory tube
Round window

Muscles of the Middle Ear

The **tensor tympani** (TEN-sor tim-PAN-ē) **muscle** is a short ribbon of muscle that originates on the petrous part of the temporal bone and the auditory tube, and inserts on the "handle" of the malleus. When the tensor tympani contracts, the malleus is pulled medially, stiffening the tympanic membrane and reducing the amount it can vibrate in response to a sound. The tensor tympani muscle is innervated by motor fibers of the mandibular branch of the trigeminal nerve (V).

The **stapedius** (sta-PĒ-dē-us) **muscle**, innervated by the facial nerve (VII), originates from the posterior wall of the middle ear and inserts on the stapes. Contraction of the stapedius pulls the stapes, reducing movement of the stapes at the oval window.

Arriving sound waves vibrate the tympanic membrane, thereby converting the waves into mechanical movements. The auditory ossicles conduct those vibrations to the internal ear, because they are connected in such a way that an in–out movement of the tympanic membrane produces a rocking motion of the stapes. The ossicles thus function as a lever system that collects the force applied to the tympanic membrane and focuses it on the oval window. Because the tympanic membrane is 22 times larger and heavier than the oval window, considerable amplification occurs, so we can hear very faint sounds. But that degree of amplification can be a problem when we are exposed to very loud noises. Contractions of the tensor tympani and the stapedius muscles protect the tympanic membrane and ossicles from violent movements under very noisy conditions.

Module 15.6 Review

a. Why are external ear infections relatively uncommon?

b. Name the three tiny bones located in the middle ear.

c. What is the function of the auditory tube?

LO 15.6 Describe the structures of the external, middle, and internal ear, and explain how they function.

The bony labyrinth protects the membranous labyrinth

1 The internal ear contains the receptors for the special senses of equilibrium and hearing. It can be subdivided into an outer bony labyrinth and an inner membranous labyrinth.

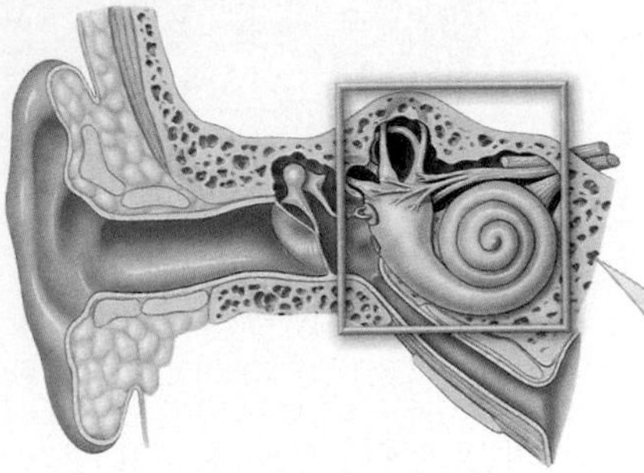

Bony Labyrinth

The **bony labyrinth** is a shell of dense bone that surrounds and protects the membranous labyrinth. Between the bony and membranous labyrinths flows the **perilymph** (PER-i-limf), a liquid that closely resembles cerebrospinal fluid. The bony labyrinth consists of three parts: the **semicircular canals**, the **vestibule** (VES-ti-būl), and the **cochlea** (KOK-lē-a; *cochlea*, snail shell).

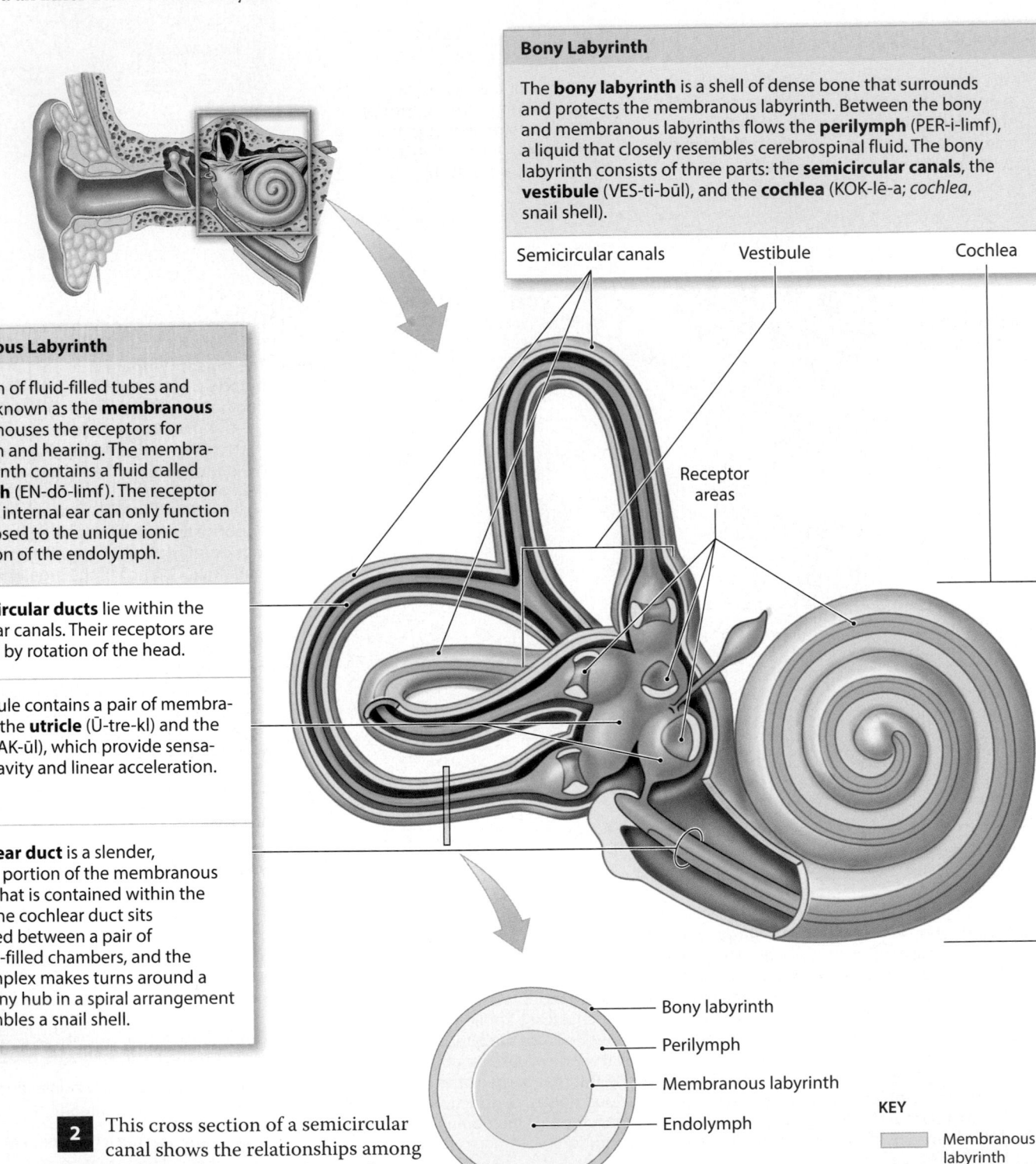

Semicircular canals | Vestibule | Cochlea

Membranous Labyrinth

A collection of fluid-filled tubes and chambers known as the **membranous labyrinth** houses the receptors for equilibrium and hearing. The membranous labyrinth contains a fluid called **endolymph** (EN-dō-limf). The receptor cells of the internal ear can only function when exposed to the unique ionic composition of the endolymph.

The **semicircular ducts** lie within the semicircular canals. Their receptors are stimulated by rotation of the head.

The vestibule contains a pair of membranous sacs, the **utricle** (Ū-tre-kl) and the **saccule** (SAK-ūl), which provide sensations of gravity and linear acceleration.

The **cochlear duct** is a slender, elongated portion of the membranous labyrinth that is contained within the cochlea. The cochlear duct sits sandwiched between a pair of perilymph-filled chambers, and the entire complex makes turns around a central bony hub in a spiral arrangement that resembles a snail shell.

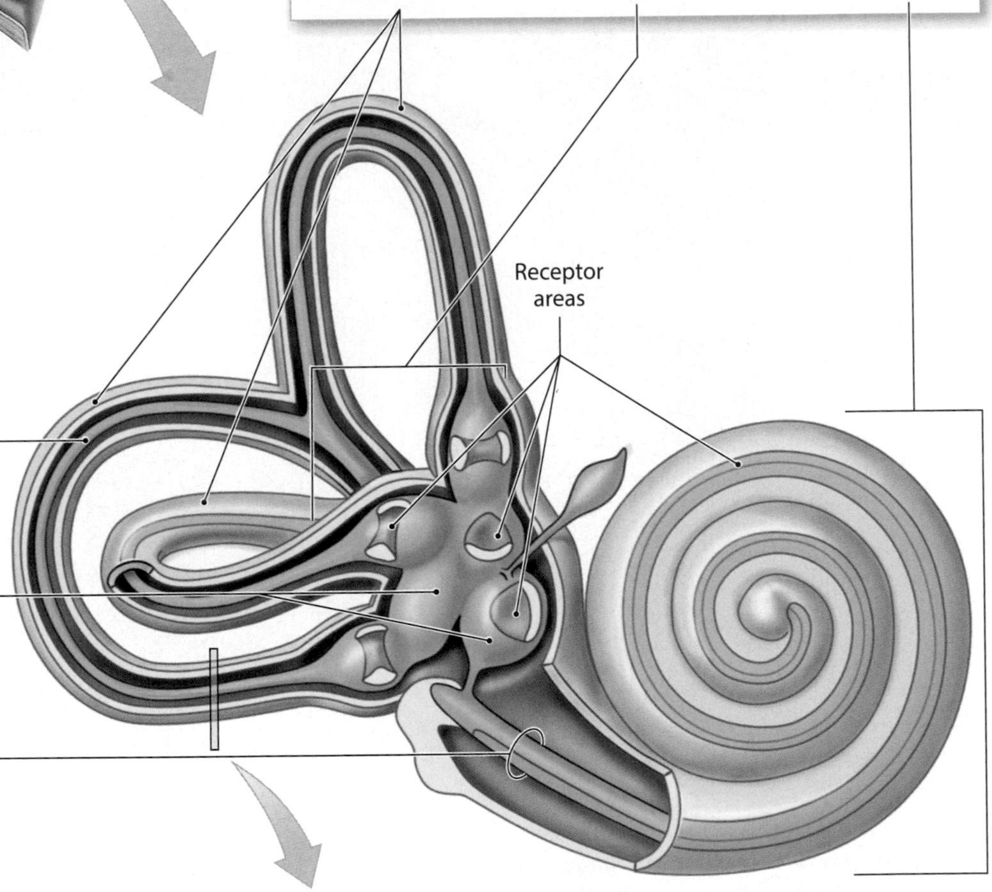
Receptor areas

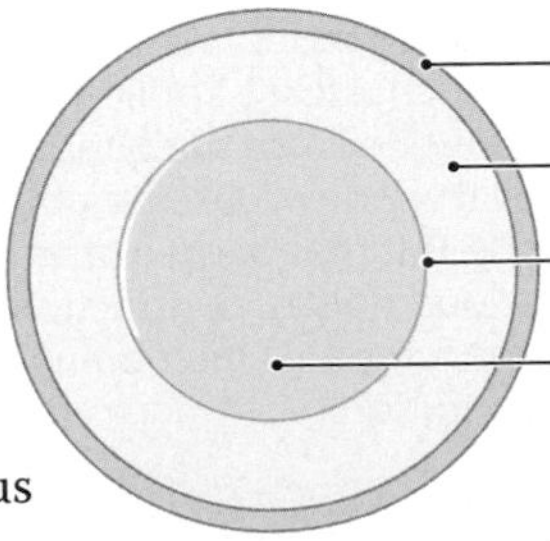
Bony labyrinth
Perilymph
Membranous labyrinth
Endolymph

2 This cross section of a semicircular canal shows the relationships among the bony labyrinth, perilymph, membranous labyrinth, and endolymph.

KEY

- Membranous labyrinth
- Bony labyrinth

3 This concept map summarizes the regions and functions of the membranous labyrinth.

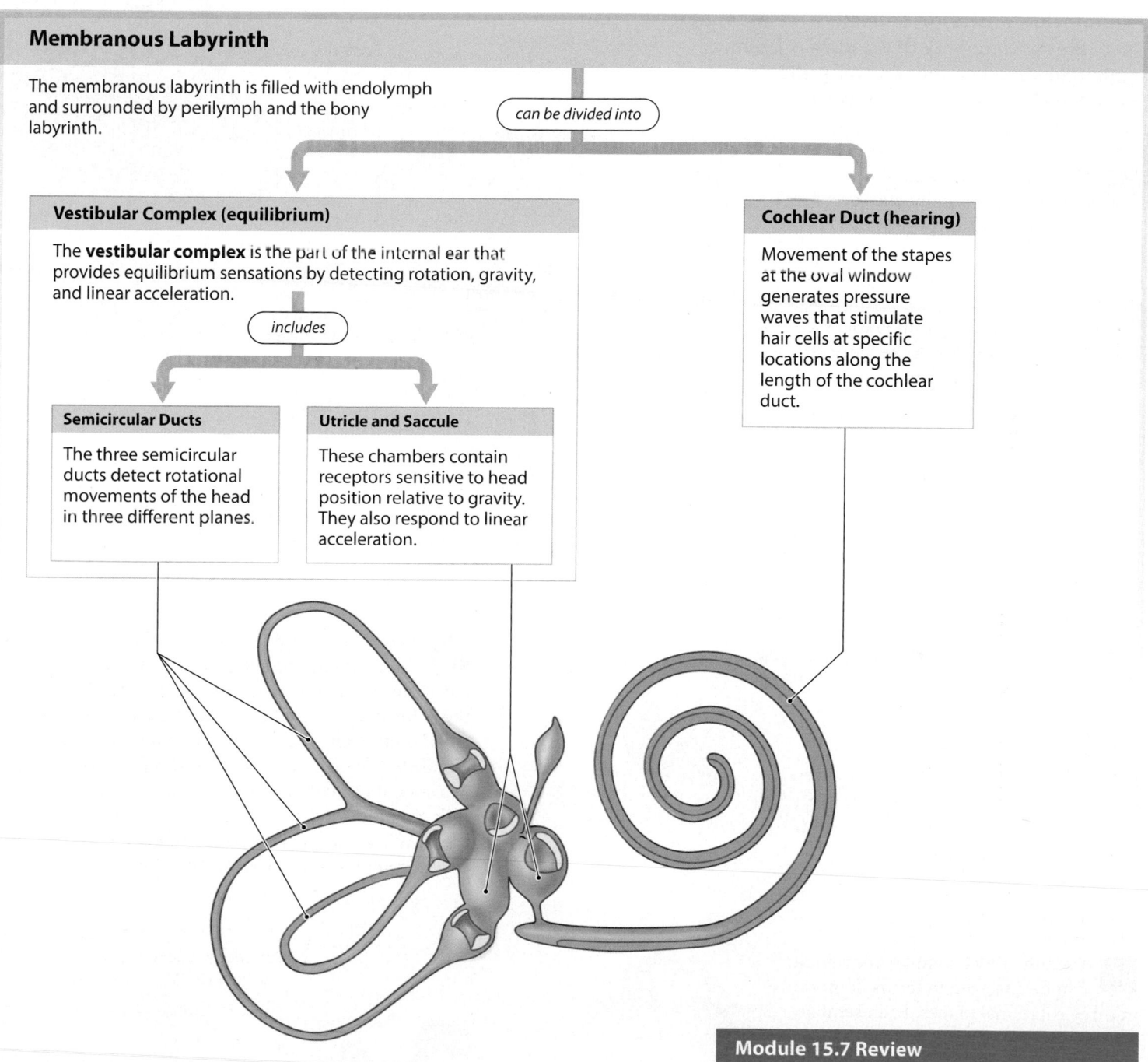

Membranous Labyrinth

The membranous labyrinth is filled with endolymph and surrounded by perilymph and the bony labyrinth.

can be divided into

Vestibular Complex (equilibrium)

The **vestibular complex** is the part of the internal ear that provides equilibrium sensations by detecting rotation, gravity, and linear acceleration.

includes

Semicircular Ducts

The three semicircular ducts detect rotational movements of the head in three different planes.

Utricle and Saccule

These chambers contain receptors sensitive to head position relative to gravity. They also respond to linear acceleration.

Cochlear Duct (hearing)

Movement of the stapes at the oval window generates pressure waves that stimulate hair cells at specific locations along the length of the cochlear duct.

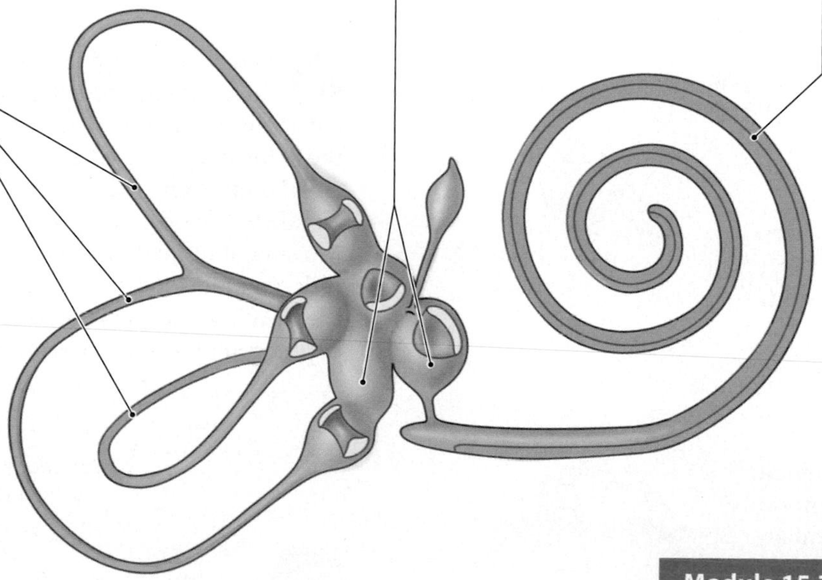

The structure and orientation of the receptor complex vary from one part of the membranous labyrinth to another. As a result, each region has a different sensitivity. Receptors in the vestibule respond to gravity or acceleration, those in the semicircular ducts respond only to rotation, and hair cells in the cochlear duct are sensitive only to sound.

Module 15.7 Review

a. Identify the structures of the bony labyrinth.

b. How do the semicircular canals and the semicircular ducts differ?

c. Describe the regional differences among the receptor complexes in the membranous labyrinth.

LO 15.7 Describe the structures and functions of the bony labyrinth and membranous labyrinth.

Module 15.8

Hair cells in the semicircular ducts respond to rotation; hair cells in the utricle and saccule respond to gravity and linear acceleration

The **anterior**, **posterior**, and **lateral semicircular ducts** are continuous with the utricle. Each semicircular duct contains an **ampulla**, an expanded region that contains the receptors.

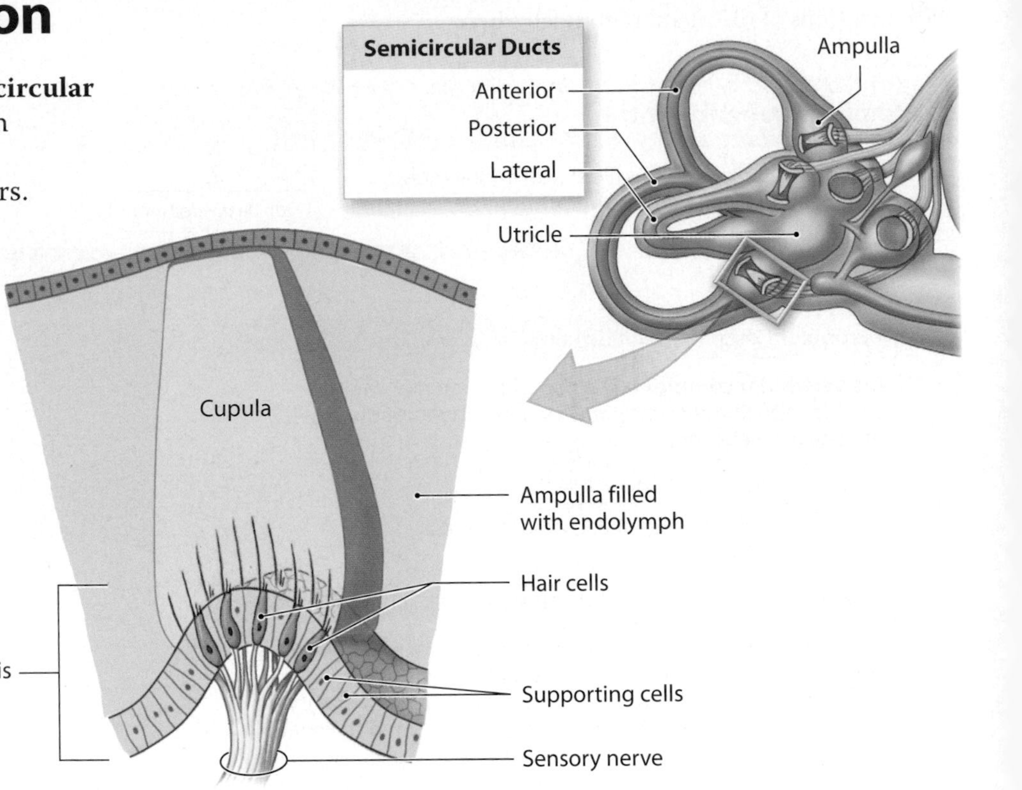

Semicircular Ducts
Anterior
Posterior
Lateral
Utricle
Ampulla
Cupula
Ampulla filled with endolymph
Hair cells
Crista ampullaris
Supporting cells
Sensory nerve

1 The region in the wall of the ampulla that contains the receptors is known as a **crista ampullaris**. Each crista ampullaris is bound to a **cupula** (KŪ-pū-luh), a flexible, elastic, gelatinous structure that extends the full width of the ampulla. At a crista ampullaris, the kinocilia and stereocilia of the hair cells are embedded in the cupula.

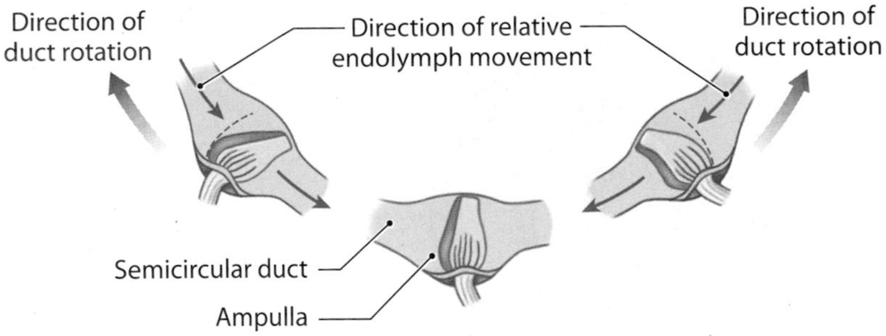

Direction of duct rotation
Direction of relative endolymph movement
Direction of duct rotation
Semicircular duct
Ampulla

2 The cupula has a density very close to that of the surrounding endolymph, so it essentially floats above the receptor surface. When your head rotates in the plane of a semicircular duct, the movement of endolymph along the length of the duct pushes the cupula to the side, distorting the receptor processes. Movement of fluid in one direction stimulates the hair cells, and movement in the opposite direction inhibits them. When the endolymph stops moving, the elastic cupula rebounds to its normal position.

3 Even the most complex movement can be analyzed in terms of motion in three rotational planes. Each semicircular duct responds to one of these rotational movements. A horizontal rotation, as in shaking your head "no," stimulates the hair cells of the lateral semicircular duct. Nodding "yes" excites the anterior duct, and tilting your head from side to side activates receptors in the posterior duct.

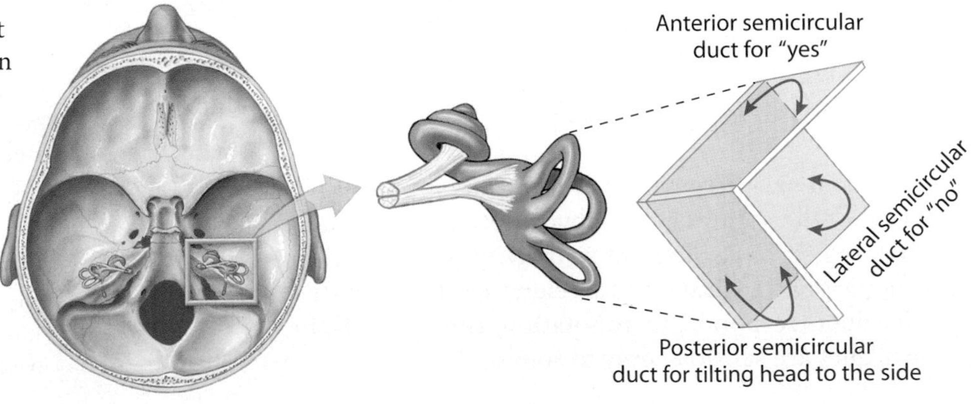

Anterior semicircular duct for "yes"
Lateral semicircular duct for "no"
Posterior semicircular duct for tilting head to the side

4 The utricle and saccule provide equilibrium sensations, whether the body is moving or stationary. These two chambers are connected by a slender passageway that is continuous with the narrow **endolymphatic duct**, which ends in a blind pouch called the **endolymphatic sac**. This sac projects into the subarachnoid space, where it is surrounded by a capillary network. Portions of the cochlear duct secrete endolymph continuously, and endolymph returns to the general circulation at the endolymphatic sac.

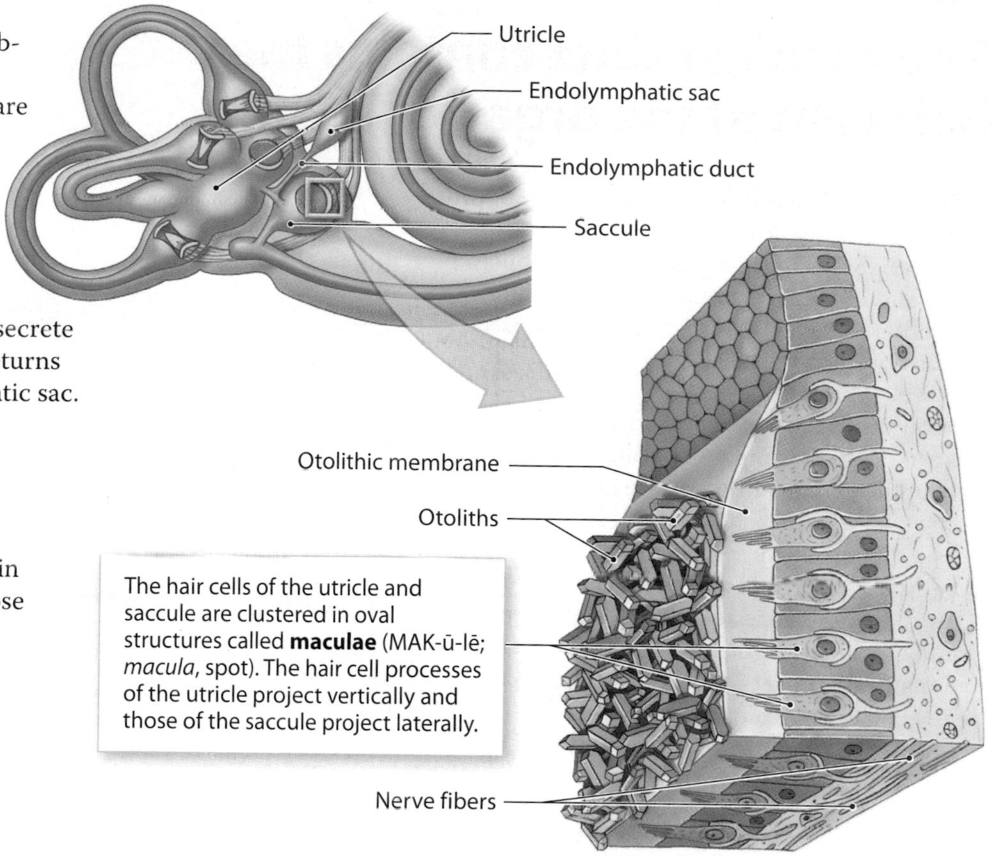

Utricle
Endolymphatic sac
Endolymphatic duct
Saccule
Otolithic membrane
Otoliths
Nerve fibers

The hair cells of the utricle and saccule are clustered in oval structures called **maculae** (MAK-ū-lē; *macula*, spot). The hair cell processes of the utricle project vertically and those of the saccule project laterally.

5 The hair cell processes are embedded in a gelatinous **otolithic membrane** whose surface contains densely packed calcium carbonate crystals. These crystals are called **otoliths** ("ear stones").

6 Changes in the position of the head cause distortion of the hair cell processes in the maculae and send signals to the brain.

When your head is in the normal, upright position, the otoliths sit atop the otolithic membrane of the macula in the utricle. Their weight presses on the macular surface, pushing the hair cell processes down rather than to one side or another.

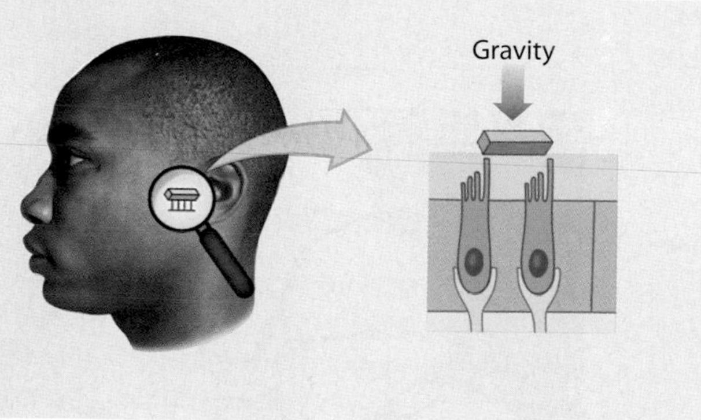

Gravity

When your head is tilted, the pull of gravity on the otoliths shifts them to the side, thereby distorting the hair cell processes and stimulating the macular receptors. This mechanism accounts for your perception of linear acceleration, as when your car speeds up suddenly. The otoliths lag behind, and the effect on the hair cells is comparable to tilting your head back.

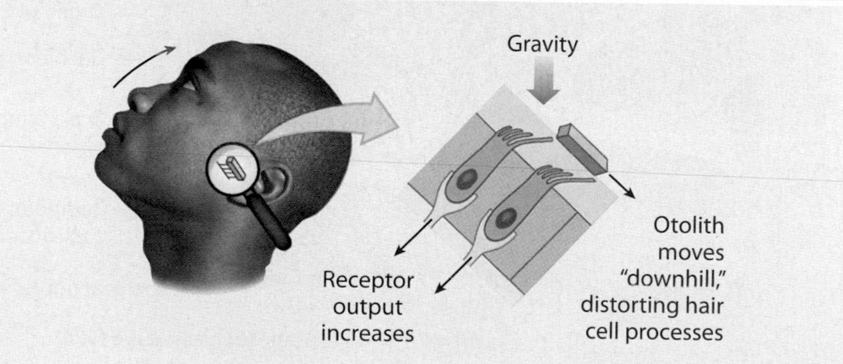

Gravity
Receptor output increases
Otolith moves "downhill," distorting hair cell processes

Module 15.8 Review

a. Damage to the cupula of the lateral semicircular duct would interfere with what perception?

b. Define otoliths.

c. Cite the functions of receptors in the saccule and utricle.

LO **15.8** Describe the functions of hair cells in the semicircular ducts, utricle, and saccule.

The cochlear duct contains the hair cells of the organ of Corti

1 The **cochlear duct** lies between a pair of perilymphatic chambers: the **scala vestibuli** (SKĀ-luh ve-STIB-yuh-lī; vestibular duct) and the **scala tympani** (tympanic duct). The outer surfaces of these ducts are encased by the bony labyrinth everywhere except at the **oval window** (the base of the scala vestibuli) and the **round window** (the base of the scala tympani). Because the vestibuli and tympani scalae are interconnected at the tip of the cochlear spiral, they really form one long and continuous perilymphatic chamber. This chamber begins at the oval window, extends through the scala tympani, and proceeds around the top of the cochlea and along the scala tympani; it ends at the round window.

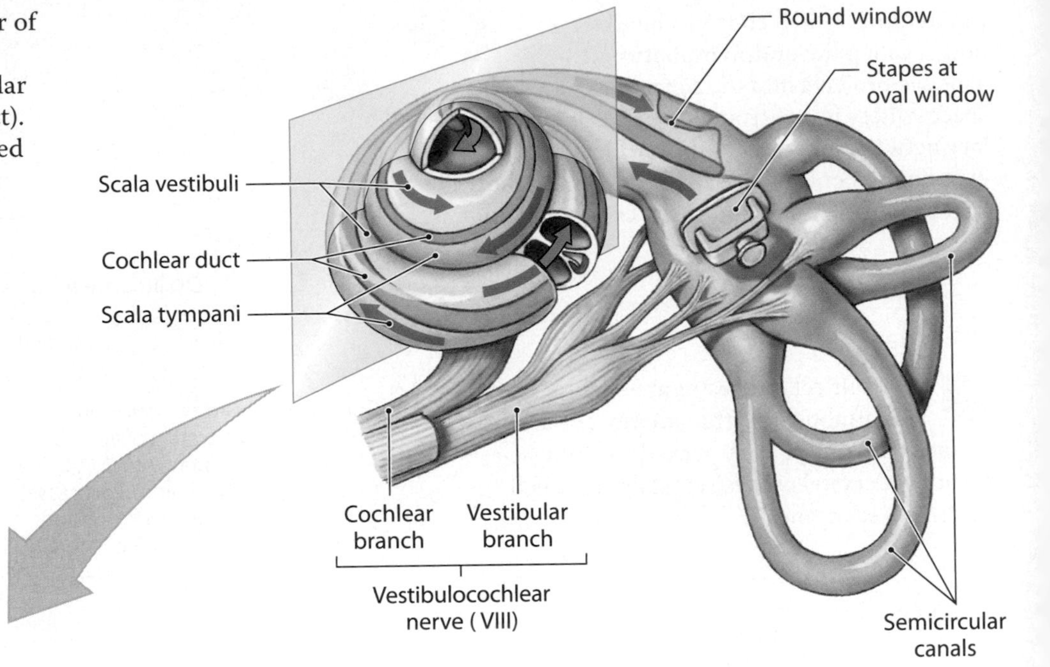
Round window
Stapes at oval window
Scala vestibuli
Cochlear duct
Scala tympani
Cochlear branch
Vestibular branch
Vestibulocochlear nerve (VIII)
Semicircular canals

KEY

From oval window to tip of cochlear spira

From tip of cochlear spiral to round windo

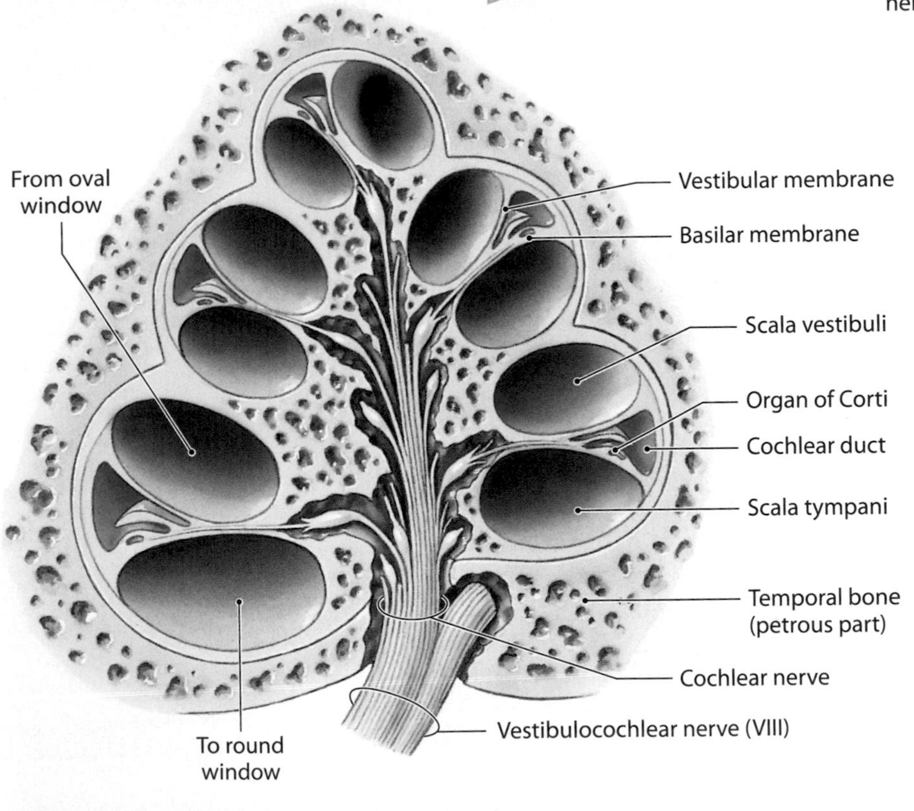
From oval window
Vestibular membrane
Basilar membrane
Scala vestibuli
Organ of Corti
Cochlear duct
Scala tympani
Temporal bone (petrous part)
Cochlear nerve
Vestibulocochlear nerve (VIII)
To round window

2 The cochlear duct is a long, coiled tube suspended between the scala vestibuli and the scala tympani. The **vestibular membrane** separates the cochlear duct from the scala vestibuli, and the **basilar membrane** separates the cochlear duct from the scala tympani. The hair cells of the cochlear duct are located in a structure called the **organ of Corti**, or spiral organ, which is located on the basilar membrane.

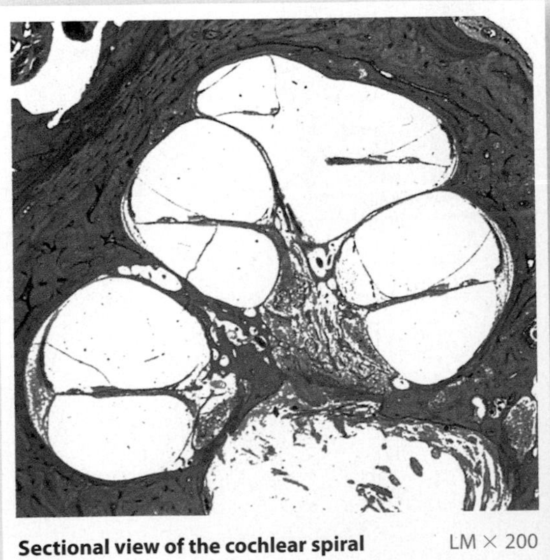

Sectional view of the cochlear spiral LM × 200

3 This sectional view shows a single turn of the cochlea. The scala vestibuli and scala tympani are filled with perilymph, whereas the cochlear duct, which contains the organ of Corti, is filled with endolymph.

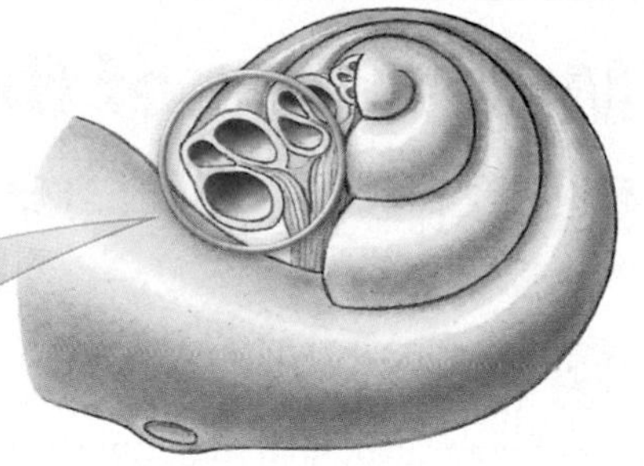

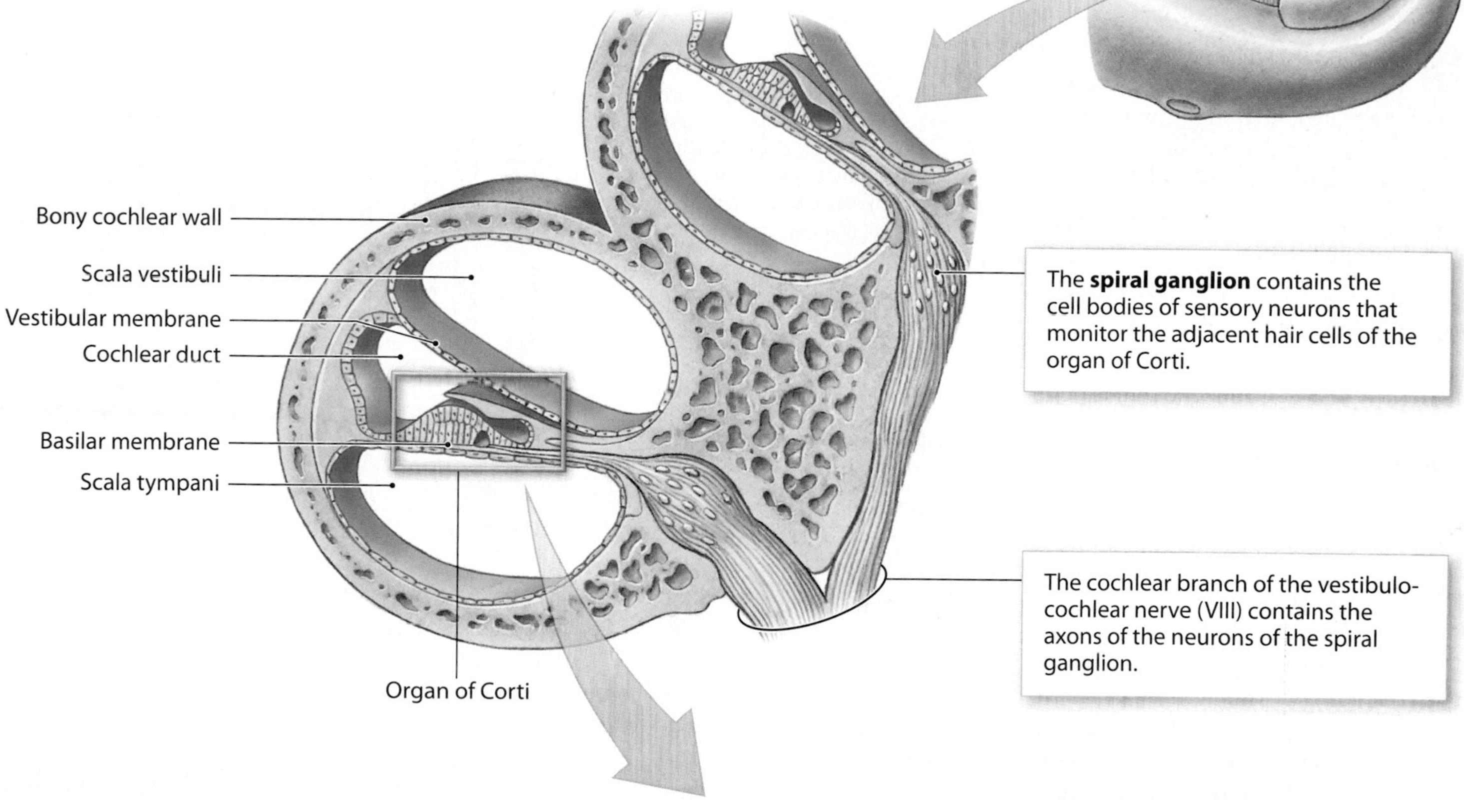

Bony cochlear wall
Scala vestibuli
Vestibular membrane
Cochlear duct
Basilar membrane
Scala tympani
Organ of Corti
The **spiral ganglion** contains the cell bodies of sensory neurons that monitor the adjacent hair cells of the organ of Corti.
The cochlear branch of the vestibulo-cochlear nerve (VIII) contains the axons of the neurons of the spiral ganglion.

4 The hair cells of the organ of Corti are arranged in a series of longitudinal rows. They lack kinocilia, and their stereocilia are in contact with the overlying **tectorial** (tek-TOR-ē-al; *tectum*, roof) **membrane**. This membrane is firmly attached to the inner wall of the cochlear duct. When a portion of the basilar membrane bounces up and down in response to pressure fluctuations within the perilymph, the stereocilia of the hair cells are pressed against the tectorial membrane and distorted. If the amount of movement increases, more hair cells—and more rows of hair cells—are stimulated. These pressure changes within the perilymph are triggered by sound waves arriving at the tympanic membrane.

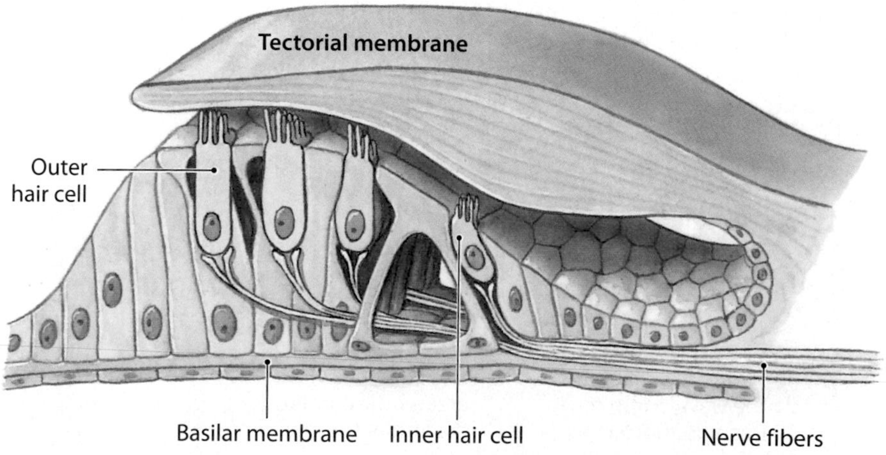

Tectorial membrane
Outer hair cell
Basilar membrane
Inner hair cell
Nerve fibers

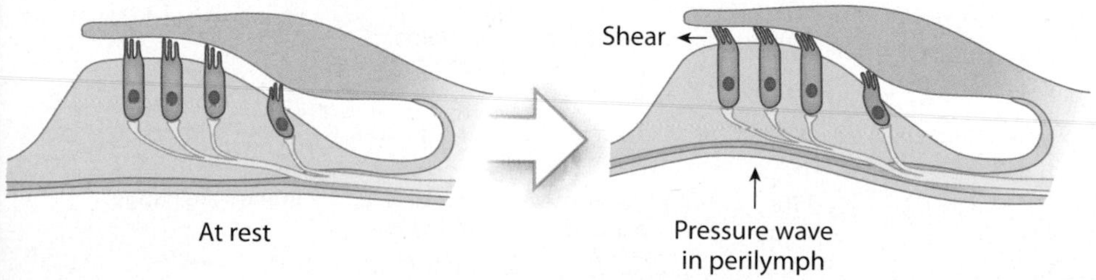

At rest
Shear
Pressure wave in perilymph

Module 15.9 Review

a. Where is the organ of Corti located?

b. Name the fluids found within the scala vestibuli, scala tympani, and cochlear duct.

c. When the basilar membrane moves, what happens to the hair cells of the organ of Corti?

LO **15.9** Describe the structures and functions of the organ of Corti.

Movement of the basilar membrane is the stimulus for the sensations of pitch and volume

1 Hearing is the perception of sound, which consists of waves of pressure conducted through a medium such as air or water. In air, each pressure wave has a region where the air molecules are crowded together and one where they are farther apart.

The **wavelength** of sound is the distance between two adjacent wave crests (peaks) or, equivalently, the distance between two adjacent wave troughs.

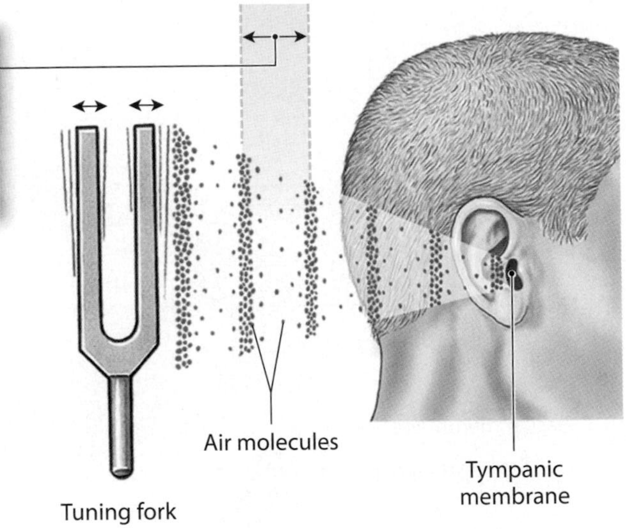

Air molecules
Tuning fork
Tympanic membrane

2 Sound waves can be graphed as S-shaped curves that repeat in a regular pattern. At sea level, sound waves travel through the air at about 1235 km/h (768 mph). The **frequency** is the number of waves that pass a fixed reference point—such as the tympanic membrane—in a given time. Physicists use the term "cycles" rather than waves. Hence, the frequency of a sound is measured in terms of the number of cycles per second (cps), a unit called **hertz (Hz)**. What we perceive as the pitch of a sound is our sensory response to its frequency. A high-frequency sound (high pitch, short wavelength) might have a frequency of 15,000 Hz or more; a very low frequency sound (low pitch, long wavelength) could have a frequency of 100 Hz or less.

Because all sound waves travel at the same speed, as the frequency increases, the wavelength must become shorter.

The **amplitude** is determined by the amount of energy carried by the wave. The greater the amplitude, the louder the sound.

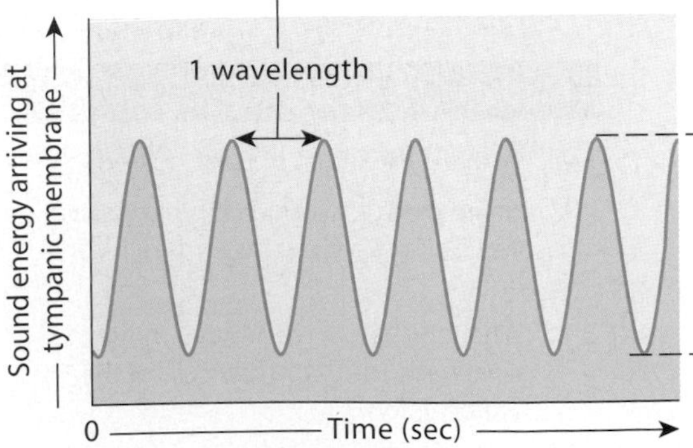

1 wavelength
Sound energy arriving at tympanic membrane
0
Time (sec)

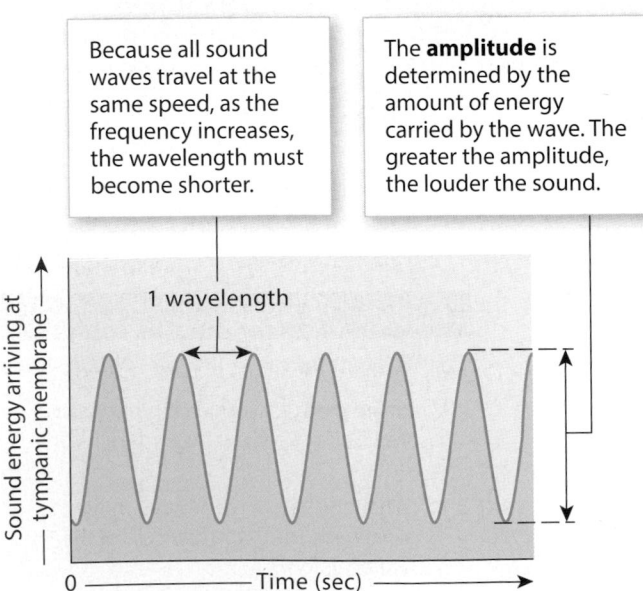

3 This table lists the **intensity**—or energy in sound waves—of familiar sounds. Intensity determines how loud it seems. The greater the energy content, the larger the amplitude, and the louder the sound. Sound energy is reported in **decibels** (DES-i-belz) (**dB**).

Intensity of Representative Sounds

Typical Decibel Level	Example	Dangerous Time Exposure
0	Lowest audible sound	
30	Quiet library; soft whisper	
40	Quiet office; living room; bedroom away from traffic	
50	Light traffic at a distance; refrigerator; gentle breeze	
60	Air conditioner from 20 feet; conversation; sewing machine in operation	
70	Busy traffic; noisy restaurant	Some damage if continuous
80	Subway; heavy city traffic; alarm clock at 2 feet; factory noise	More than 8 hours
90	Truck traffic; noisy home appliances; shop tools; gas lawn mower	Less than 8 hours
100	Chain saw; boiler shop; pneumatic drill	2 hours
120	"Heavy metal" rock concert; sandblasting; thunderclap nearby	Immediate danger
140	Gunshot; jet plane	Immediate danger
160	Rocket launching pad	Hearing loss inevitable

4 The energy of sound waves is a physical pressure. When sound waves strike a flexible object, the object responds to that pressure. Given the right combination of frequencies and amplitudes, the object will begin to vibrate at the same frequency as the sound, a phenomenon called **resonance**. The higher the amplitude, the greater the amount of vibration. For you to be able to hear a sound, your tympanic membrane must vibrate in resonance with the sound waves. Pressure waves at the tympanic membrane generate movement of the stapes at the oval window. The flexibility of the basilar membrane varies along its length, so pressure waves of different frequencies affect different parts of the membrane. This diagram depicts the movement produced by sound waves with a frequency of 6000 Hz. When the stapes moves inward, the basilar membrane distorts toward the round window, which bulges into the middle ear cavity. When the stapes moves outward, the basilar membrane rebounds and distorts toward the oval window. The location of the vibration is interpreted as **pitch**. The number of stimulated hair cells is interpreted as **volume**.

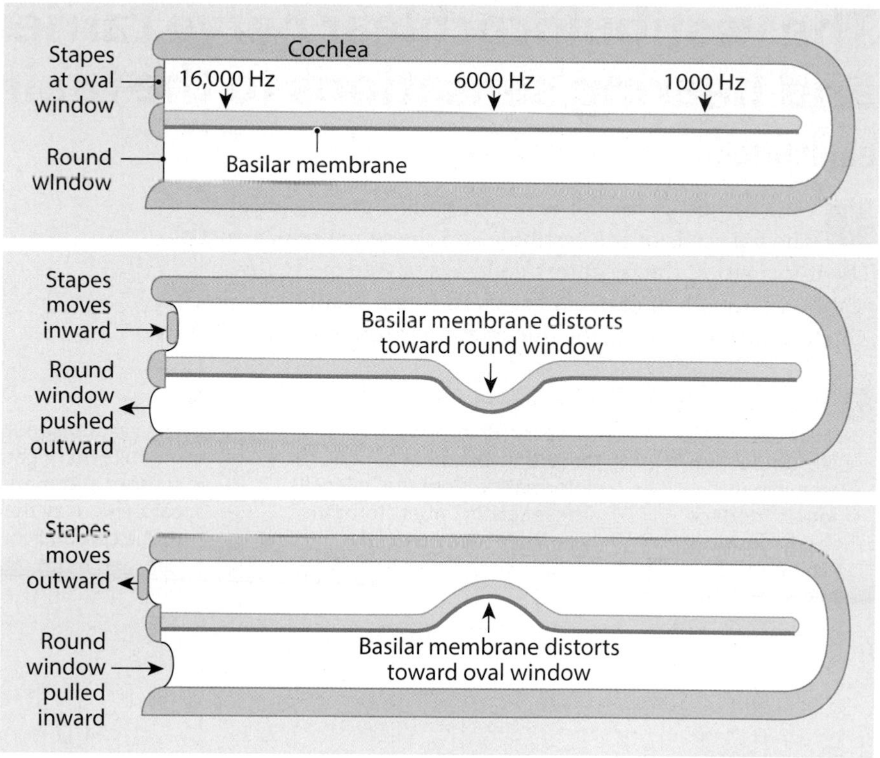

Stapes at oval window
Cochlea
16,000 Hz
6000 Hz
1000 Hz
Round window
Basilar membrane
Stapes moves inward
Basilar membrane distorts toward round window
Round window pushed outward
Stapes moves outward
Round window pulled inward
Basilar membrane distorts toward oval window

5 This illustration summarizes the events involved in hearing.

Events Involved in Hearing

1	2	3	4	5	6
Sound waves arrive at the tympanic membrane.	Movement of the tympanic membrane causes displacement of the auditory ossicles.	Movement of the stapes at the oval window establishes pressure waves in the perilymph of the scala vestibuli.	The pressure waves distort the basilar membrane on their way to the round window of the scala tympani.	Vibration of the basilar membrane causes vibration of hair cells against the tectorial membrane.	Information about the region and the intensity of stimulation is relayed to the CNS over the cochlear branch of cranial nerve VIII.

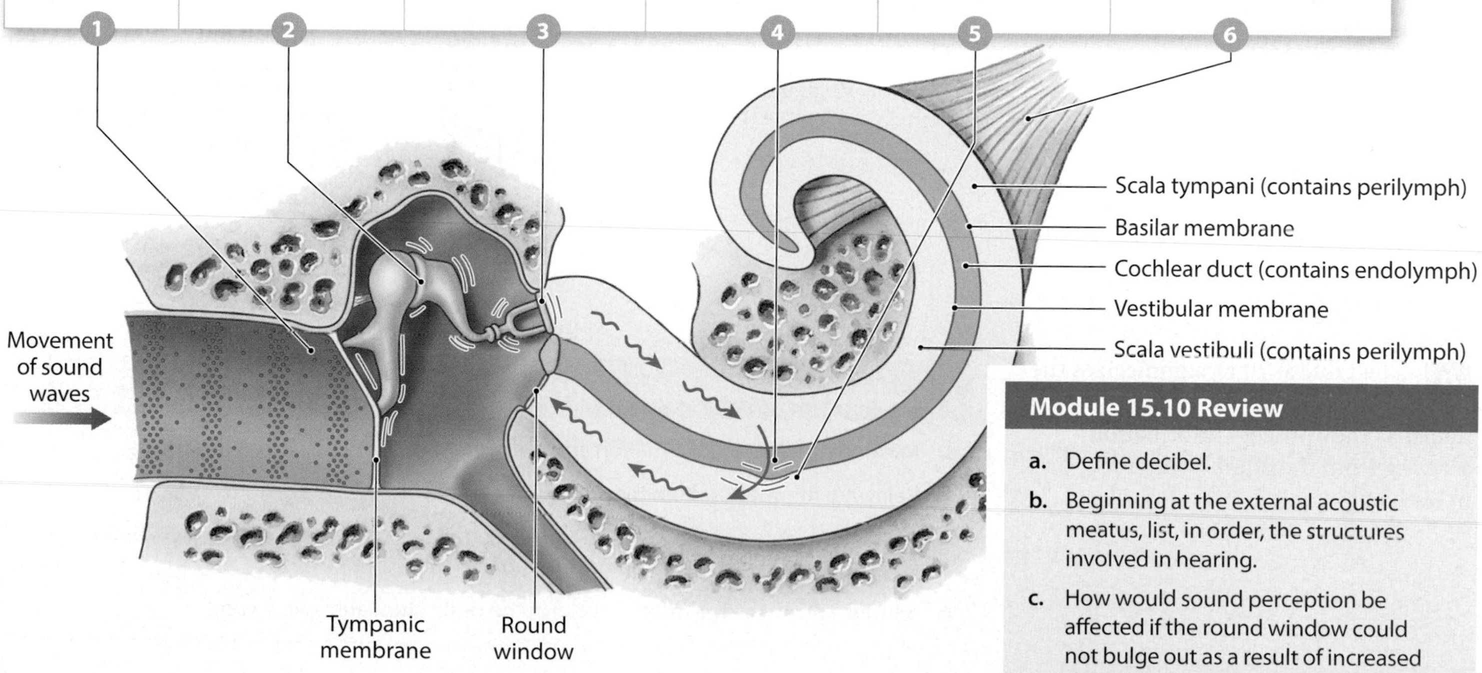

Scala tympani (contains perilymph)
Basilar membrane
Cochlear duct (contains endolymph)
Vestibular membrane
Scala vestibuli (contains perilymph)
Movement of sound waves
Tympanic membrane
Round window

Module 15.10 Review

a. Define decibel.

b. Beginning at the external acoustic meatus, list, in order, the structures involved in hearing.

c. How would sound perception be affected if the round window could not bulge out as a result of increased perilymph pressure?

LO 15.10 Explain the anatomical and physiological basis for pitch and volume sensations for hearing.

The vestibulocochlear nerve carries equilibrium and hearing sensations to the brain stem

Equilibrium

1 The receptors for the sense of equilibrium (balance) are the hair cells of the vestibule and the semicircular ducts. The information the receptors collect is passed along the vestibular branch of cranial nerve VIII to the vestibular nuclei, which in turn distribute the information throughout the CNS.

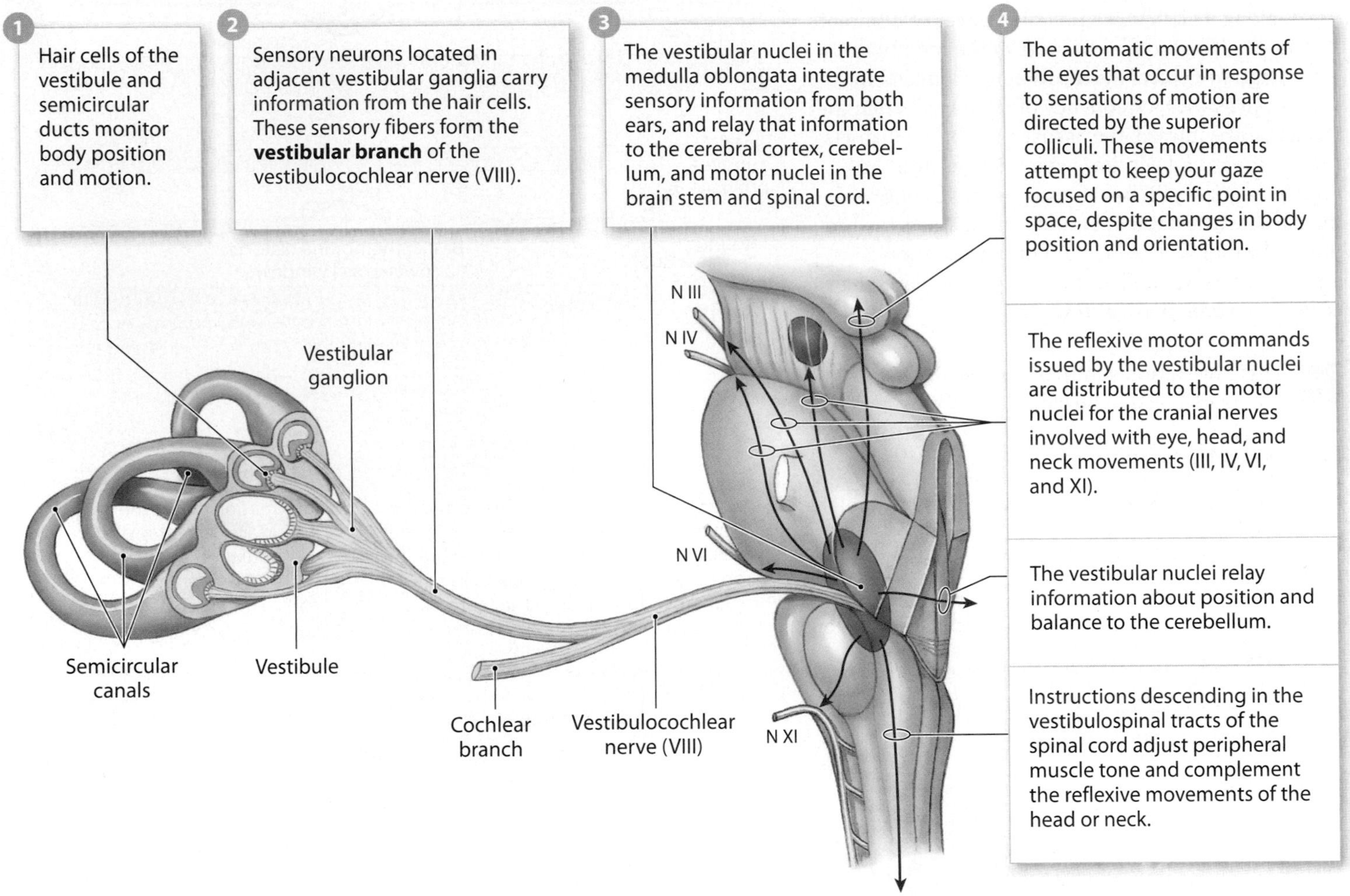

1
Hair cells of the vestibule and semicircular ducts monitor body position and motion.
2
Sensory neurons located in adjacent vestibular ganglia carry information from the hair cells. These sensory fibers form the **vestibular branch** of the vestibulocochlear nerve (VIII).
3
The vestibular nuclei in the medulla oblongata integrate sensory information from both ears, and relay that information to the cerebral cortex, cerebellum, and motor nuclei in the brain stem and spinal cord.
4
The automatic movements of the eyes that occur in response to sensations of motion are directed by the superior colliculi. These movements attempt to keep your gaze focused on a specific point in space, despite changes in body position and orientation.
The reflexive motor commands issued by the vestibular nuclei are distributed to the motor nuclei for the cranial nerves involved with eye, head, and neck movements (III, IV, VI, and XI).
The vestibular nuclei relay information about position and balance to the cerebellum.
Instructions descending in the vestibulospinal tracts of the spinal cord adjust peripheral muscle tone and complement the reflexive movements of the head or neck.
Vestibular ganglion
Semicircular canals
Vestibule
Cochlear branch
Vestibulocochlear nerve (VIII)
N III
N IV
N VI
N XI

2 The table at right summarizes the primary functions of the vestibular nuclei as they process information arriving from the equilibrium receptors in both ears.

Functions of the Vestibular Nuclei

1. Integrating sensory information about equilibrium that arrives from both ears.
2. Relaying information from the vestibular complex to the cerebellum.
3. Relaying information from the vestibular complex to the cerebral cortex, providing a conscious sense of head position and movement.
4. Sending commands to motor nuclei in the brain stem and spinal cord.

Hearing

3 Hair cells along the basilar membrane are the receptors for the sense of hearing. Information they collect is passed along the cochlear branch of the vestibulocochlear nerve (VIII) to the cochlear nuclei, which in turn distribute that information to the superior olivary nuclei, inferior colliculi, thalamus, and auditory cortex.

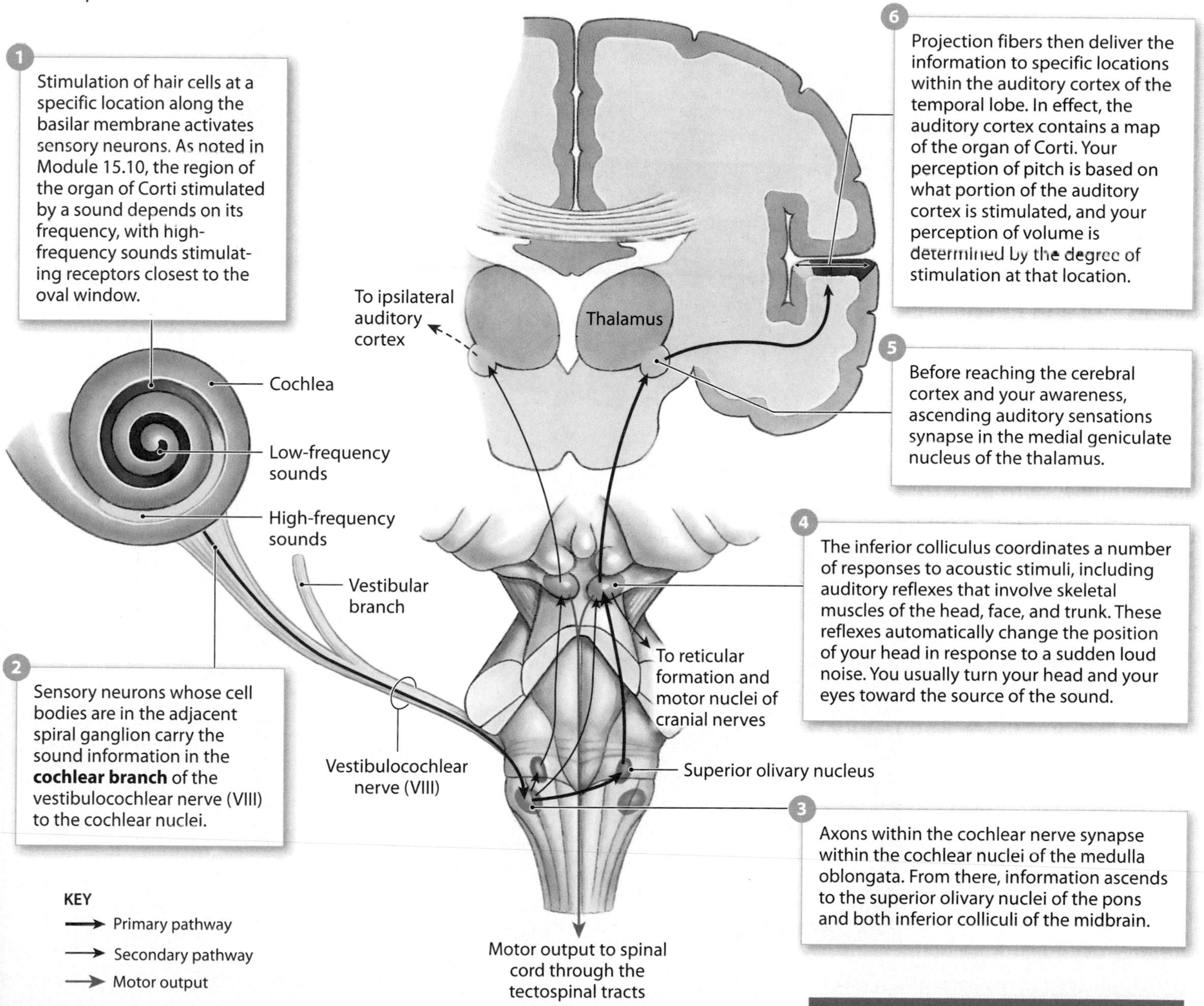

1
Stimulation of hair cells at a specific location along the basilar membrane activates sensory neurons. As noted in Module 15.10, the region of the organ of Corti stimulated by a sound depends on its frequency, with high-frequency sounds stimulating receptors closest to the oval window.
6
Projection fibers then deliver the information to specific locations within the auditory cortex of the temporal lobe. In effect, the auditory cortex contains a map of the organ of Corti. Your perception of pitch is based on what portion of the auditory cortex is stimulated, and your perception of volume is determined by the degree of stimulation at that location.
To ipsilateral auditory cortex
Thalamus
Cochlea
Low-frequency sounds
High-frequency sounds
5
Before reaching the cerebral cortex and your awareness, ascending auditory sensations synapse in the medial geniculate nucleus of the thalamus.
Vestibular branch
4
The inferior colliculus coordinates a number of responses to acoustic stimuli, including auditory reflexes that involve skeletal muscles of the head, face, and trunk. These reflexes automatically change the position of your head in response to a sudden loud noise. You usually turn your head and your eyes toward the source of the sound.
To reticular formation and motor nuclei of cranial nerves
2
Sensory neurons whose cell bodies are in the adjacent spiral ganglion carry the sound information in the cochlear branch of the vestibulocochlear nerve (VIII) to the cochlear nuclei.
Vestibulocochlear nerve (VIII)
Superior olivary nucleus
3
Axons within the cochlear nerve synapse within the cochlear nuclei of the medulla oblongata. From there, information ascends to the superior olivary nuclei of the pons and both inferior colliculi of the midbrain.
KEY
Primary pathway
Secondary pathway
Motor output
Motor output to spinal cord through the tectospinal tracts

Most of the auditory information from one cochlea is projected to the auditory cortex of the cerebral hemisphere on the opposite side of the brain. However, each auditory cortex also receives information from the cochlea on that side. These interconnections play a role in localizing sounds (left/right). They can also reduce the functional impact of damage to a cochlea or ascending pathway.

Module 15.11 Review

a. Where are the hair cell receptors for equilibrium located?

b. Which cranial nerves are involved with eye, head, and neck movements?

c. What is your reflexive response to hearing a loud noise, such as a firecracker?

LO 15.11 Trace the pathways for the sensations of equilibrium and hearing to their respective destinations in the brain.

Labeling

Label the structures in the following diagram of the right ear.

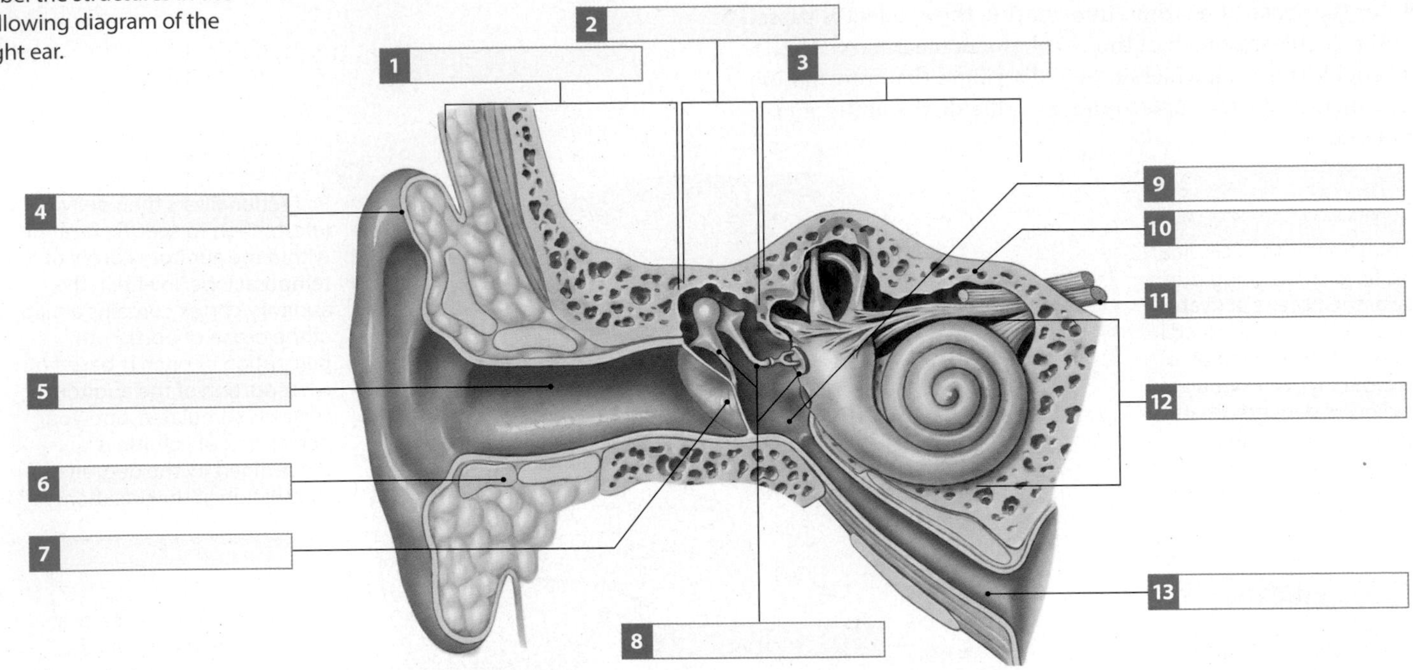
1
2
3
4
5
6
7
8
9
10
11
12
13

Label the structures in the following micrograph of the internal ear.

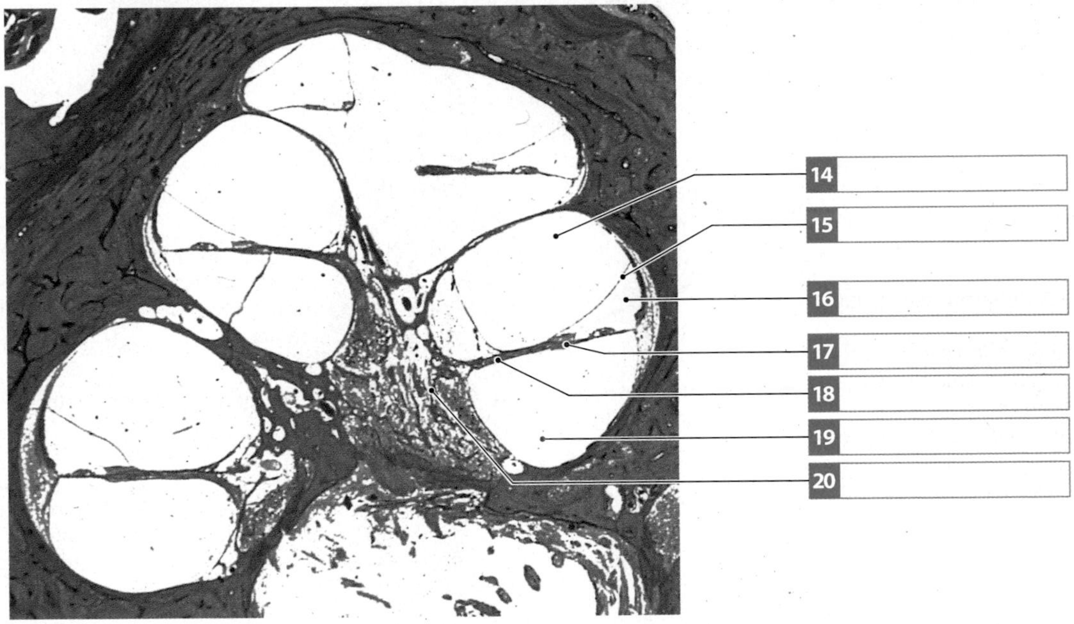
14
15
16
17
18
19
20

Section integration

For a few seconds after you ride an express elevator from the 25th floor to the ground floor, you still feel as if you are descending, even though you have come to a stop. Why?

21 ______________________________

____

____

The eyes form early in embryonic development

This section will examine the anatomy and physiology of the adult eyes. The eyes, which give us the sense of vision, are our most complex sense organs. A tremendous amount of sensory integration, feedback, and processing occurs within the eye before any sensations are relayed to the CNS. This functional complexity far exceeds that of the receptor complexes responsible for olfaction, gustation, equilibrium, and hearing. That complexity is directly related to the way the eyes form during embryonic development. In many ways the eyes resemble detached CNS nuclei that are totally devoted to processing visual stimuli.

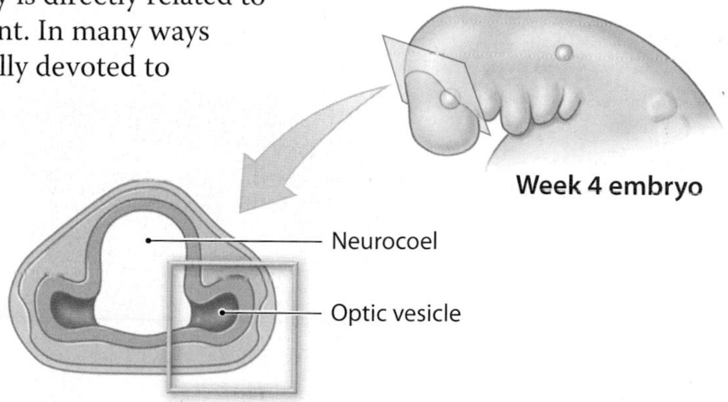
Week 4 embryo
Neurocoel
Optic vesicle

1 The first indication of eye development appears as a pair of bulges called **optic vesicles** form in the lateral walls of the prosencephalon. These bulges extend to either side like a pair of dumbbells, each containing a cavity continuous with the neurocoel (Module 13.1, p. 463).

2 The lateral bulges become indented, forming a pair of **optic cups** that remain connected to the diencephalon by slender stalks. The epidermis overlying the optic cup responds by forming a pocket that later pinches off and develops into the lens of the eye. The inner and outer layers of the optic cup develop into the retina. The retina will house the light-sensing photoreceptor cells.

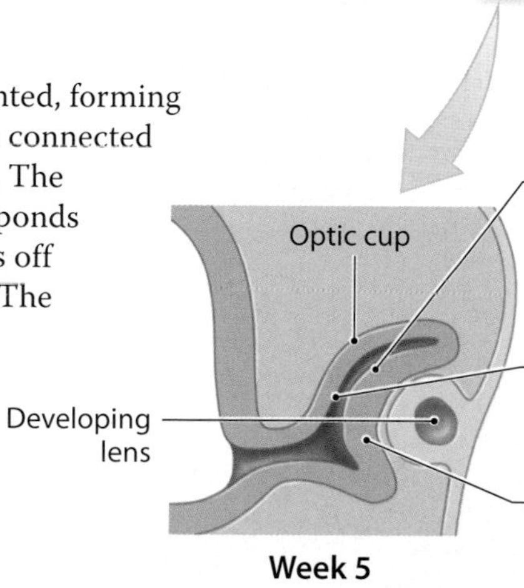
Optic cup
Developing lens
Week 5

Layers of the Developing Retina

The ependymal cells in this region of the outer layer of the optic cup develop into the photoreceptors.

The ependymal cells on the inner layer of the optic cup develop into pigment cells that absorb light that has passed through the photoreceptor layer. The separation gradually decreases until the photoreceptors and pigment cell layers are in contact.

This neural tissue of the outer layer of the optic cup forms layers of neurons, ganglion cells, and specialized glial cells that process and integrate visual information.

3 Embryonic cells aggregating around the optic cup form supporting layers of connective tissue that isolate the neural tissues from the rest of the body. The eye develops interior chambers filled with fluid that is continuously generated and reabsorbed. Modules within this section will examine each of these components of the eye in greater detail.

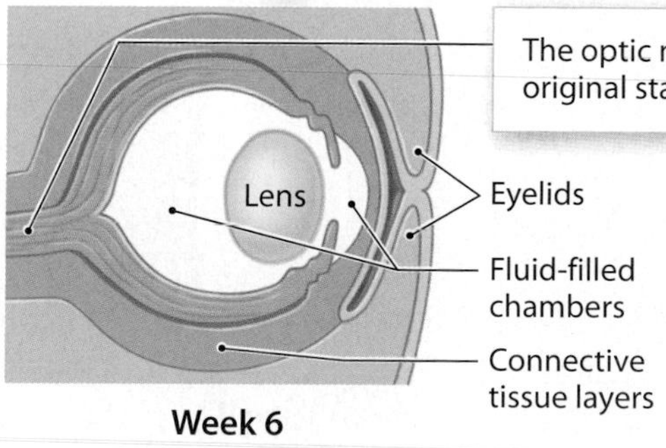
Lens
Eyelids
Fluid-filled chambers
Connective tissue layers
Week 6

The optic nerve (II) follows the path of the original stalk connected to the optic cup.

Module 15.12 Review

a. What are the first structures that form during eye development?

b. Which structures develop into the retina?

c. Which cells develop into the photoreceptors?

LO 15.12 Outline the embryonic development of the eye.

Module 15.13

Accessory structures of the eye provide protection while allowing light to reach the interior of the eye

1 **Accessory structures** of the eye include the eyelids, eyelashes, the superficial epithelium of the eye, and the structures associated with the production, secretion, and removal of tears.

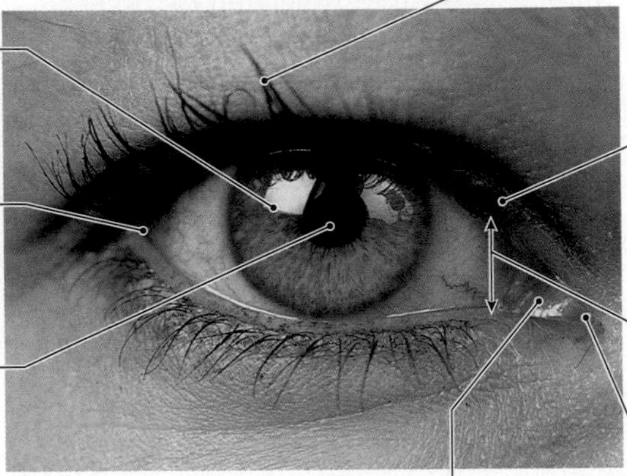

The **cornea** is a transparent area on the anterior surface of the eye.

Laterally, the two eyelids are connected at the **lateral canthus**.

Light enters the eye by passing through the cornea and then through the **pupil**, an opening at the center of the colored **iris**.

The **lacrimal caruncle** (KAR-ung-kul) is a small, reddish body at the medial angle of the eye. It produces the thick secretions that cause the gritty deposits that sometimes appear after a good night's sleep.

Eyelids and Eyelashes

The **eyelashes**, along the margins of the eyelids, are very robust hairs that help prevent foreign matter from reaching the surface of the eye.

The eyelid, or **palpebra** (PAL-peh-bruh), is a continuation of the skin. The continual blinking of the palpebrae keeps the surface of the eye lubricated, and removes dust and debris. The eyelids can also close firmly to protect the delicate surface of the eye.

The **palpebral fissure** is the gap that separates the free margins of the upper and lower eyelids.

Medially, the two eyelids are connected at the **medial canthus** (KAN-thus).

2 The epithelium covering the inner surfaces of the eyelids and the outer surface of the eye is called the **conjunctiva** (kon-junk-TĪ-vuh). It is a mucous membrane covered by a specialized stratified squamous epithelium.

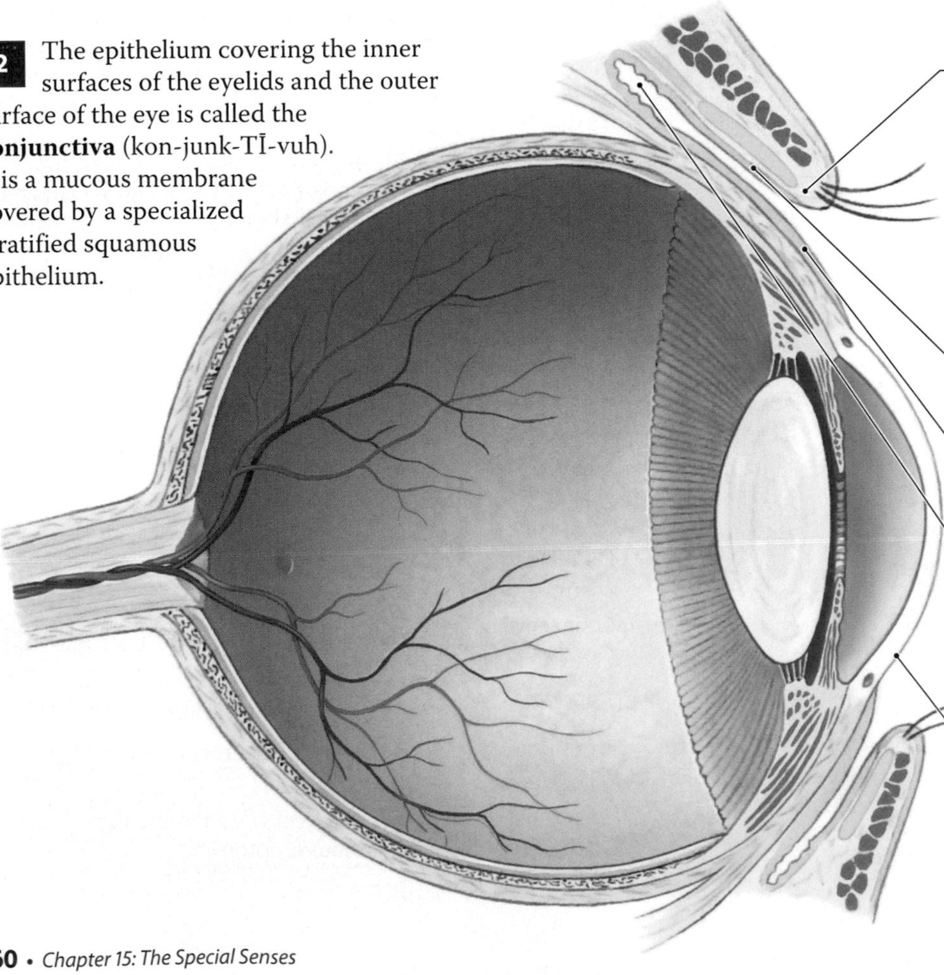

Modified sebaceous glands called **tarsal glands**, or Meibomian (mī-BŌ-mē-an) glands, are along the inner margin of the lid. These glands secrete a lipid-rich product that helps keep the eyelids from sticking together.

Conjunctiva

The **palpebral conjunctiva** covers the inner surface of the eyelids.

The **ocular conjunctiva** covers the anterior surface of the eye.

The pocket created where the palpebral conjunctiva becomes continuous with the ocular conjunctiva is known as the **fornix** of the eye.

The ocular conjunctiva is continuous with the very delicate corneal epithelium that covers the surface of the cornea. The cornea is a transparent portion of the eye's anterior surface.

3 A constant flow of tears keeps conjunctival surfaces moist and clean. Tears reduce friction, remove debris, prevent bacterial infection, and provide nutrients and oxygen to portions of the conjunctival epithelium. The **lacrimal apparatus** produces, distributes, and removes tears. The lacrimal apparatus of each eye consists of (1) a lacrimal gland (or tear gland) with associated ducts, (2) paired lacrimal canaliculi, (3) a lacrimal sac, and (4) a nasolacrimal duct.

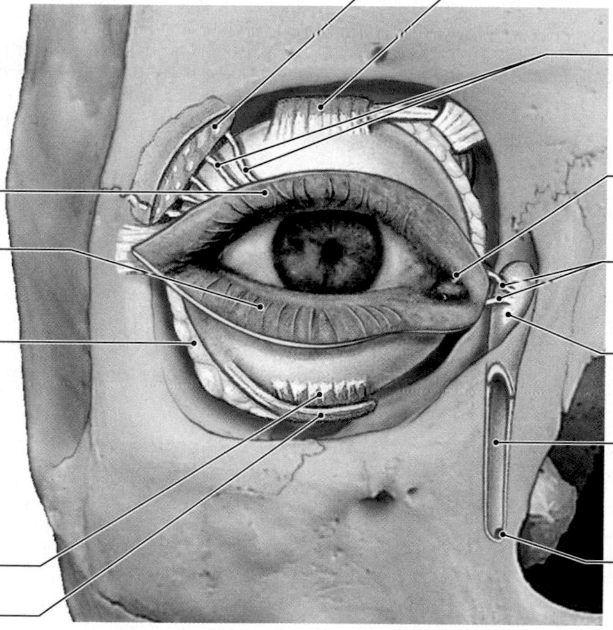

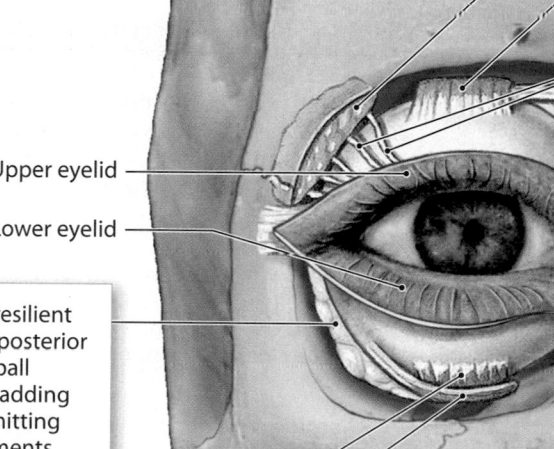

Superior rectus muscle
Upper eyelid
Lower eyelid
Inferior rectus muscle
Inferior oblique muscle

A layer of resilient orbital fat posterior to the eyeball provides padding while permitting eye movements.

Components of the Lacrimal Apparatus

The almond-shaped **lacrimal gland** is about 12–20 mm (0.5–0.75 in.) long. Each day it produces about 1 mL of watery, slightly alkaline tears. The tears lubricate, nourish, and oxygenate the corneal cells. Lacrimal gland secretions contain the antibacterial enzyme **lysozyme** and antibodies that attack pathogens before they enter the body.

There are 10–12 **tear ducts** that deliver tears from the lacrimal gland to the space behind the upper eyelid.

The **lacrimal puncta** (singular, punctum) are two small pores that drain the lacrimal lake.

The **lacrimal canaliculi** are small canals that connect the lacrimal puncta to the lacrimal sac.

The **lacrimal sac** is a small chamber that nestles within the lacrimal sulcus of the orbit.

The **nasolacrimal duct** originates at the inferior tip of the lacrimal sac. It passes through the nasolacrimal canal to deliver tears to the nasal cavity.

The nasolacrimal duct empties into the inferior meatus, a narrow passageway inferior and lateral to the inferior nasal concha.

4 **Conjunctivitis**, or pinkeye, is an inflammation of the conjunctiva. The most obvious sign, redness, is due to the dilation of blood vessels deep to the conjunctival epithelium. This condition may be caused by pathogenic infection or by physical, allergic, or chemical irritation of the conjunctival surface.

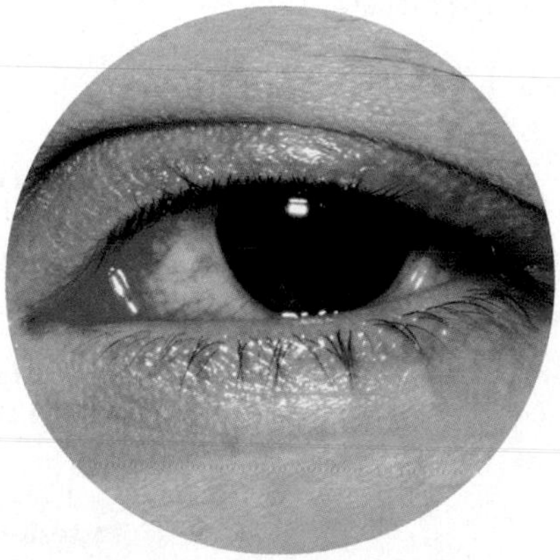

Module 15.13 Review

a. List the accessory structures associated with the eye.

b. Which layer of the eye would be the first affected by inadequate tear production?

c. Describe conjunctivitis.

LO 15.13 Identify the accessory structures of the eye and explain their functions.

Module 15.14

The eye has a layered wall; it is hollow, with fluid-filled anterior and posterior cavities

1 The wall of the eye has three **layers**, formerly called tunics.

Fibrous Layer

The **fibrous layer**, the outermost layer of the eye, consists of the cornea and the white **sclera** (SKLER-uh). These two components are continuous, and the border between the two is called the **corneal limbus**. The fibrous layer (1) supports and protects, (2) is an attachment site for the extrinsic eye muscles, and (3) contains the transparent cornea. Light first enters the eye through the cornea; its curvature aids in the focusing process.

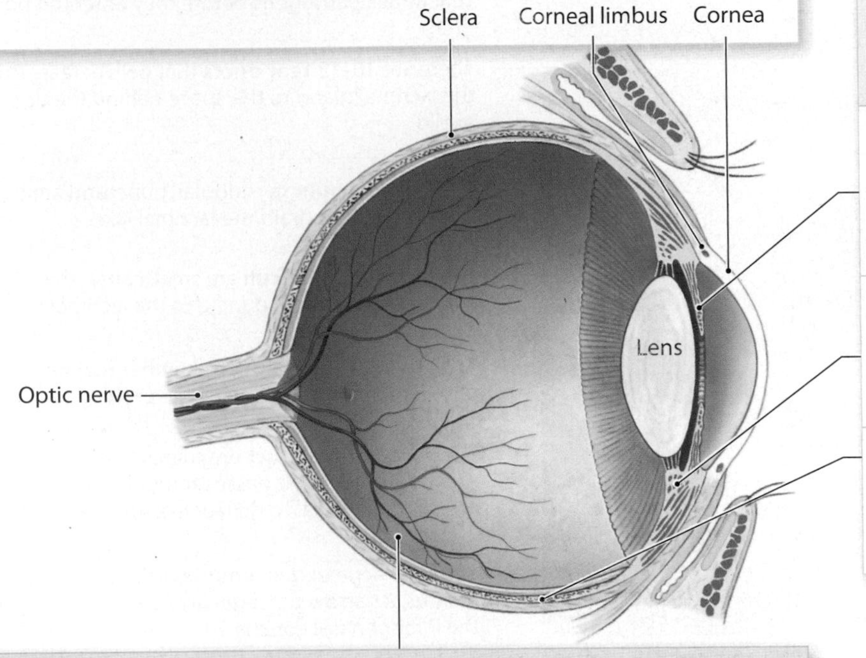

Sclera
Corneal limbus
Cornea
Lens
Optic nerve

Vascular Layer

The **vascular layer**, or **uvea** (Ū-vē-uh), contains numerous blood vessels, lymphatic vessels, and the intrinsic (smooth) muscles of the eye. The functions of this middle layer include (1) providing a route for blood vessels and lymphatics that supply tissues of the eye; (2) regulating the amount of light that enters the eye; (3) secreting and reabsorbing the fluid called aqueous humor that circulates within the chambers of the eye; and (4) controlling the shape of the lens, an essential part of the focusing process.

The colored **iris** is visible through the transparent corneal surface. It contains blood vessels, pigment cells, and layers of smooth muscle fibers. When these muscles contract, they change the diameter of the pupil.

The **ciliary body** is a thickened region that bulges into the interior of the eye. Suspensory ligaments extend from the ciliary body to the lens, holding it in position posterior to the pupil.

The **choroid** is a vascular layer that is covered by the sclera. The choroid has an extensive capillary network that delivers oxygen and nutrients to the neural tissue within the neural layer.

Inner Layer

The **inner layer**, or **retina**, is the innermost layer of the eye. The retina consists of a thin outer layer (the pigmented part) that absorbs light, and a thick inner layer (the neural part) that contains the **photoreceptors**—the cells that are sensitive to light.

2 This sectional view shows that the ciliary body and the lens divide the interior of the eye into a small **anterior cavity** and a large **posterior cavity**.

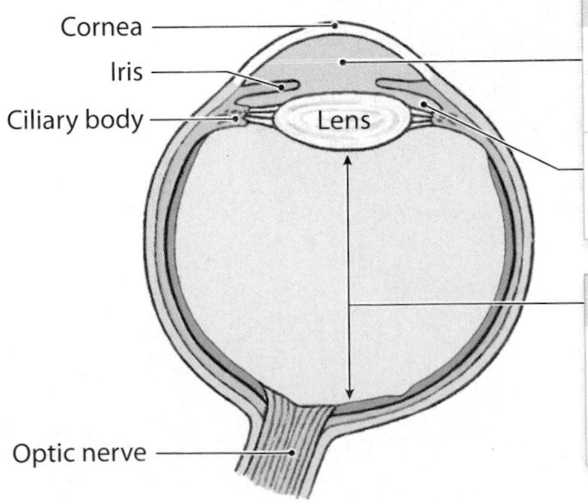

Cornea
Iris
Ciliary body
Lens
Optic nerve

Anterior Cavity

The **anterior chamber** extends from the cornea to the iris.

The **posterior chamber** extends between the iris and the ciliary body and lens.

Posterior Cavity

Most of the posterior cavity's volume is taken up by a gelatinous substance known as the **vitreous body**. **Vitreous humor** is the fluid part of the vitreous body.

3 A view of the anterior cavity at greater magnification reveals additional details about the structure of the iris and ciliary body. Eye color is determined by (1) genes that influence the density and distribution of melanocytes on the anterior surface and interior of the iris, and (2) the density of the pigmented epithelium. When the connective tissue of the iris contains few melanocytes, light passes through it and bounces off the pigmented epithelium; the eye then appears blue. Individuals with green, brown, or black eyes have increasing numbers of melanocytes in the body and surface of the iris. The eyes of people with albinism appear a very pale gray or blue-gray.

4 **Aqueous humor** is a fluid that circulates within the anterior cavity, passing from the posterior to the anterior chamber through the pupil. It also freely diffuses through the vitreous body and across the surface of the retina, where it is called vitreous humor. Circulation of the aqueous humor provides an important route for nutrient and waste transport, in addition to forming a fluid cushion. The pressure exerted by the aqueous humor helps retain the eye's shape; it also stabilizes the position of the retina, pressing the neural part against the pigmented part. A procedure called **tonometry** measures the eye's intraocular pressure—the fluid pressure within the anterior chamber. Normal intraocular pressure ranges from 12 to 21 mm Hg.

The body of the iris consists of a highly vascular, pigmented, loose connective tissue. The anterior surface has no epithelial covering; instead, it has an incomplete layer of fibroblasts and melanocytes. The posterior surface is covered by a pigmented epithelium that is part of the neural tunic.

Aqueous humor forms through active secretion by epithelial cells of the ciliary body's ciliary processes at a rate of 1–2 μL per minute. The epithelial cells regulate its composition, which is similar to cerebrospinal fluid.

Aqueous humor leaves the eye through the **scleral venous sinus** (canal of Schlemm), a passageway that extends completely around the eye at the level of the corneal limbus. Collecting channels deliver the aqueous humor from this canal to veins in the sclera. The rate of removal normally keeps pace with the rate of secretion, and aqueous humor is removed and recycled within a few hours of its formation.

The bulk of the ciliary body consists of the **ciliary muscle**, a smooth muscular ring that projects into the interior of the eye.

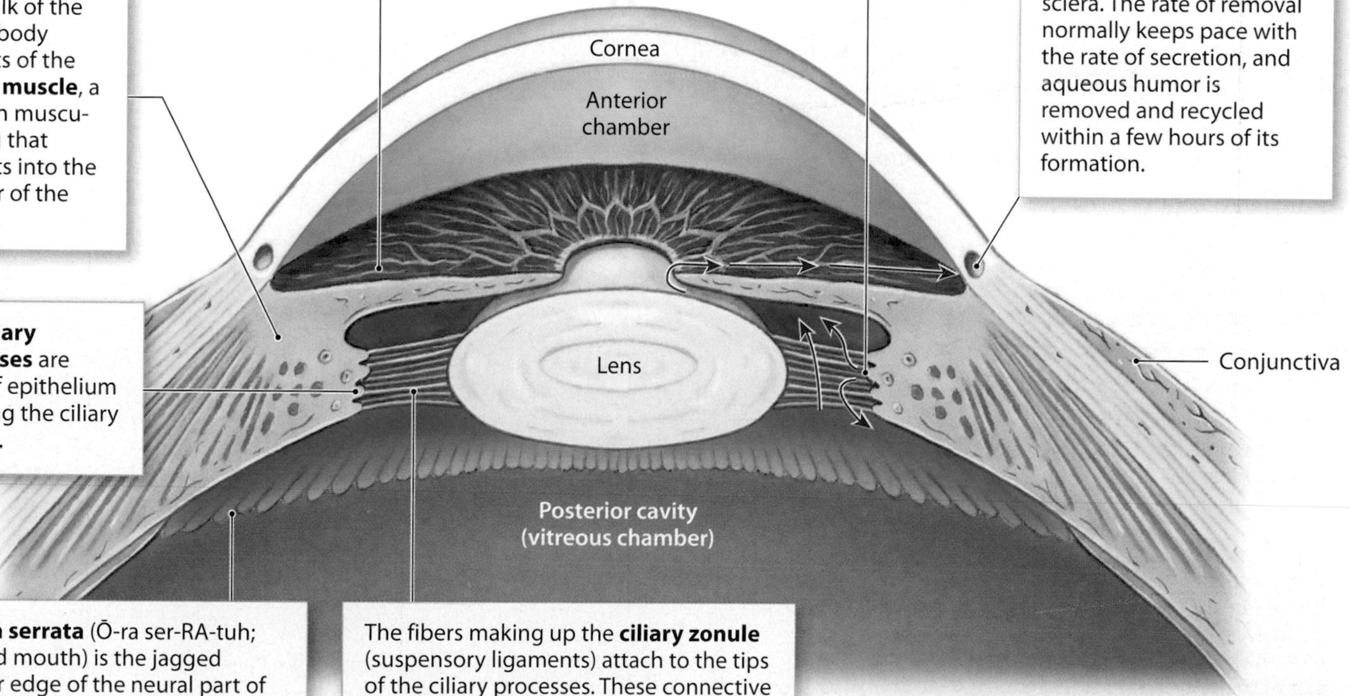

Cornea
Anterior chamber
Lens
Conjunctiva
Posterior cavity (vitreous chamber)

The **ciliary processes** are folds of epithelium covering the ciliary muscle.

The **ora serrata** (Ō-ra ser-RA-tuh; serrated mouth) is the jagged anterior edge of the neural part of the retina. The outer pigmented part of the retina continues anteriorly across the posterior surface of the iris.

The fibers making up the **ciliary zonule** (suspensory ligaments) attach to the tips of the ciliary processes. These connective tissue fibers hold the lens posterior to the iris and centered on the pupil. As a result, any light passing through the pupil will also pass through the lens.

Module 15.14 Review

a. Name the three layers of the eye.

b. What give eyes their characteristic color?

c. Where in the eye is aqueous humor located?

Module 15.15

The eye is highly organized and has a consistent visual axis that directs light to the fovea of the retina

1 The sectional view below presents key aspects of eye anatomy that are associated with positioning the eye and allowing light to reach the photoreceptors of the retina.

The cornea allows light to enter the eye, so its transparency and clarity are vital to eye function. The cornea consists primarily of a dense matrix containing multiple layers of collagen fibers, organized so as not to interfere with the passage of light. The cornea is avascular; the superficial epithelial cells must obtain oxygen and nutrients from the tears that flow across their free surfaces.

The lens lies posterior to the cornea, held in place by the ciliary zonule that originates on the ciliary body. The lens consists of concentric layers of cells surrounded by a dense fibrous capsule. The cells are slender, long, and filled with transparent proteins called **crystallins**, which are responsible for both the clarity and the focusing power of the lens. Around the edges of the lens, its capsular fibers intermingle with those of the ciliary zonule. The primary function of the lens is to focus the visual image on the photoreceptors. The lens does this by changing shape.

Tension in the series of fibers making up the ciliary zonule resists the tendency of the lens to assume a spherical shape.

The ciliary body supports the lens and controls its shape.

The retina contains the photoreceptors, pigment cells, supporting cells, and neurons.

The blood vessels of the choroid directly or indirectly provide nutrients to all structures within the eye.

The sclera, or "white of the eye," consists of dense fibrous connective tissue containing both collagen and elastic fibers. This layer is thickest over the posterior surface of the eye, near the exit of the optic nerve, and thinnest over the anterior surface. The sclera stabilizes the shape of the eye during eye movements. The six extrinsic eye muscles insert on the sclera, blending their collagen fibers with those of the sclera.

The **optic nerve** (II) carries visual information to the brain.

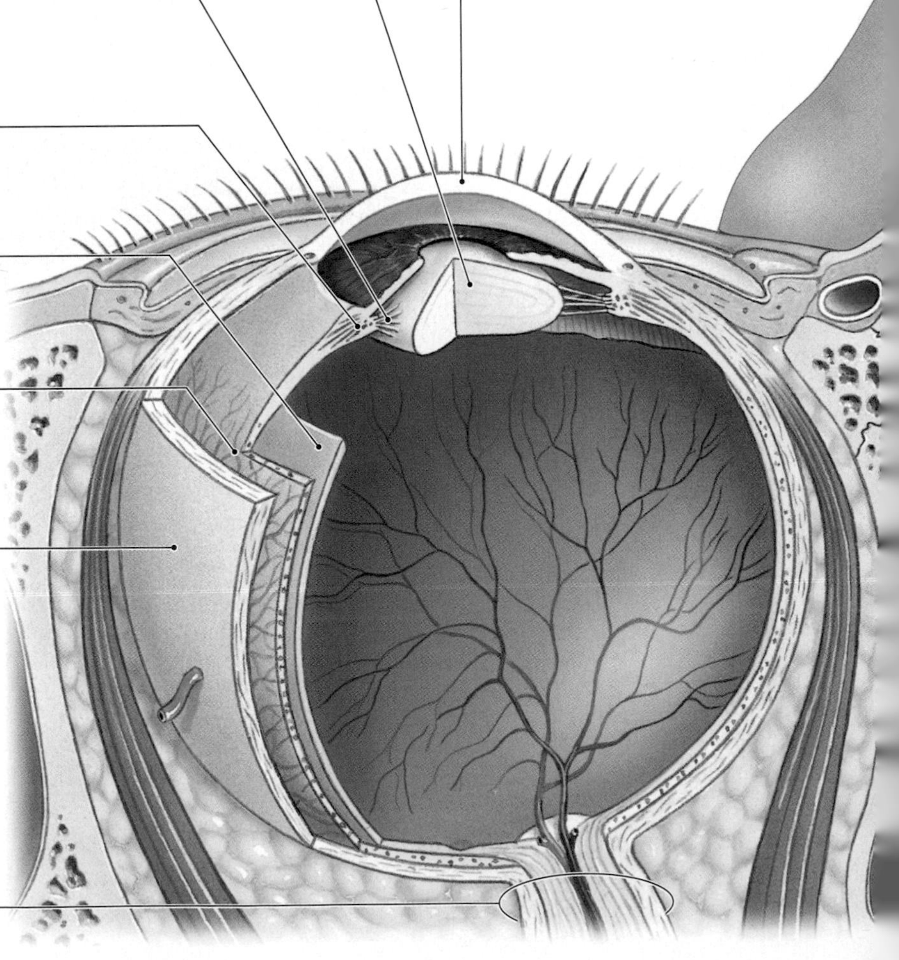

2 The amount of light entering the eye and passing through the lens is controlled by the two layers of the pupillary muscles of the iris. When these smooth muscles contract, they change the diameter of the pupil. Both muscle layers are controlled by the autonomic nervous system. Parasympathetic activation in response to bright light causes the pupils to constrict, whereas sympathetic activation in response to dim light causes the pupils to dilate.

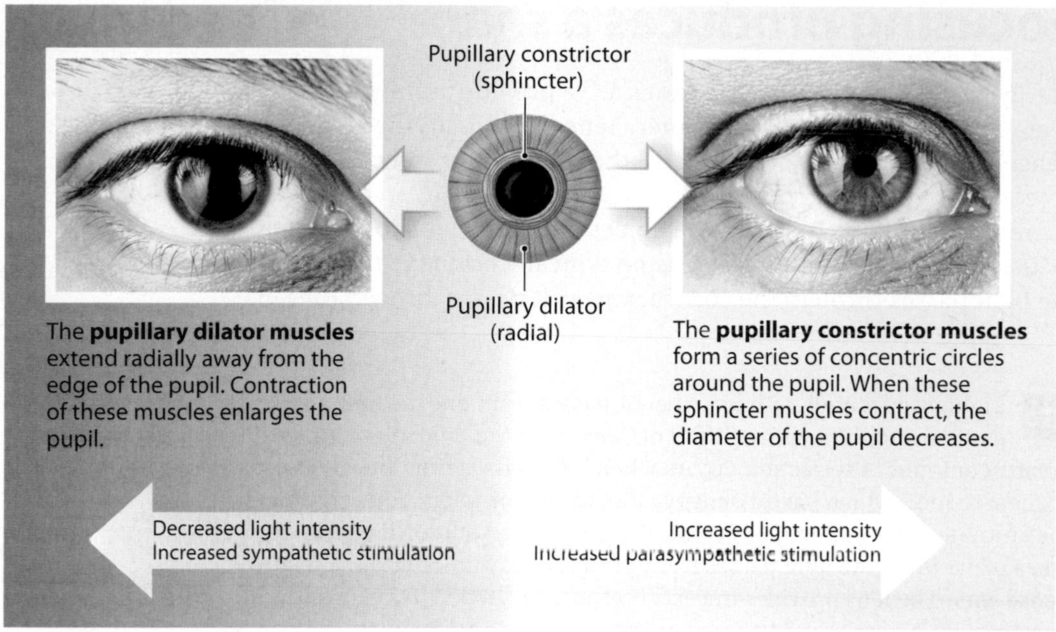

Pupillary constrictor (sphincter)
Pupillary dilator (radial)
The **pupillary dilator muscles** extend radially away from the edge of the pupil. Contraction of these muscles enlarges the pupil.
The **pupillary constrictor muscles** form a series of concentric circles around the pupil. When these sphincter muscles contract, the diameter of the pupil decreases.
Decreased light intensity
Increased sympathetic stimulation
Increased light intensity
Increased parasympathetic stimulation

3 Light passing through the center of the cornea and the center of the lens strikes a specific location that contains the highest density of photoreceptors anywhere in the eye.

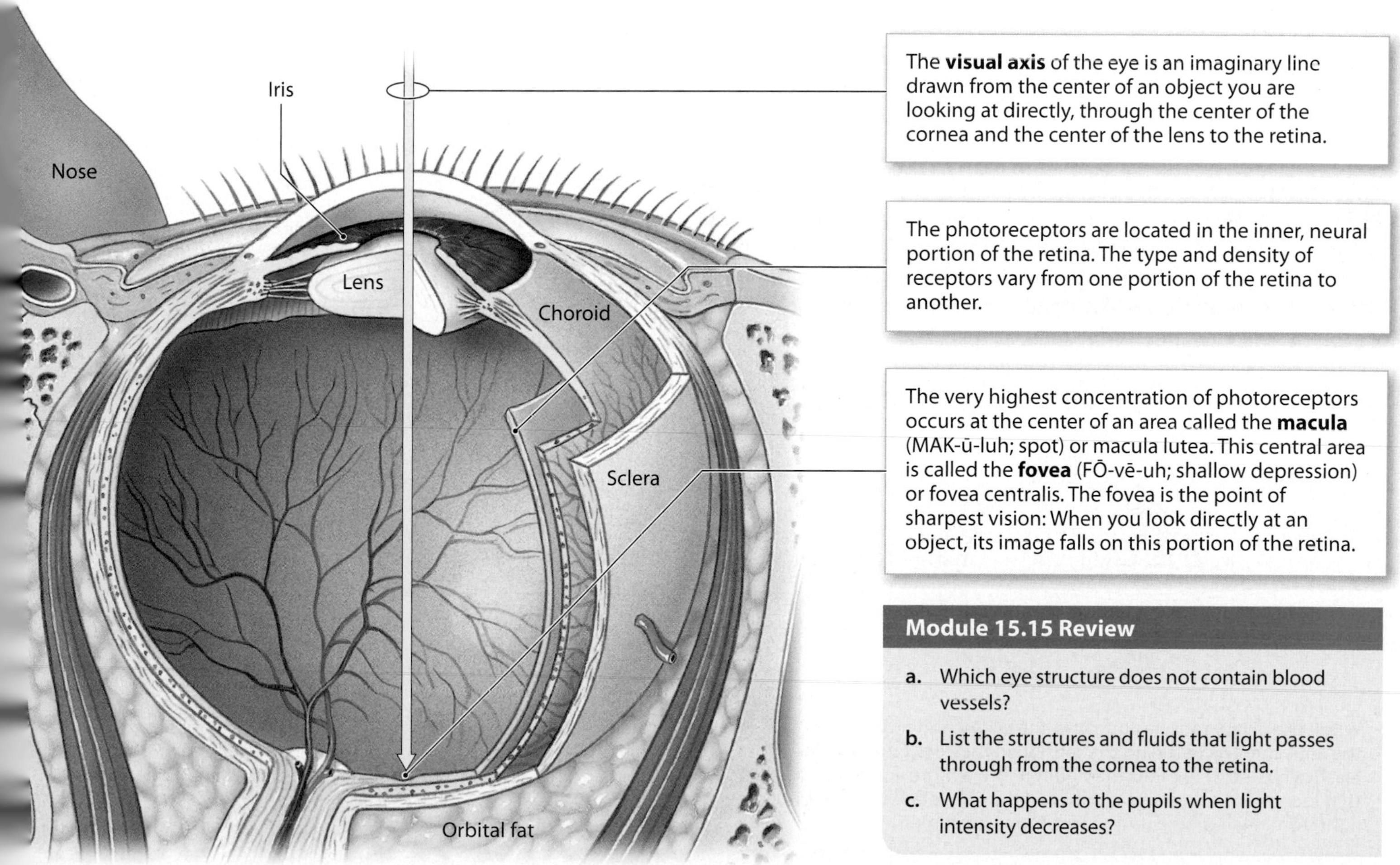

Iris
Nose
Lens
Choroid
Sclera
Orbital fat
The **visual axis** of the eye is an imaginary line drawn from the center of an object you are looking at directly, through the center of the cornea and the center of the lens to the retina.
The photoreceptors are located in the inner, neural portion of the retina. The type and density of receptors vary from one portion of the retina to another.
The very highest concentration of photoreceptors occurs at the center of an area called the **macula** (MAK-ū-luh; spot) or macula lutea. This central area is called the **fovea** (FŌ-vē-uh; shallow depression) or fovea centralis. The fovea is the point of sharpest vision: When you look directly at an object, its image falls on this portion of the retina.

Module 15.15 Review

a. Which eye structure does not contain blood vessels?

b. List the structures and fluids that light passes through from the cornea to the retina.

c. What happens to the pupils when light intensity decreases?

LO 15.15 Explain how light is directed to the fovea of the retina.

Focusing produces a sharply defined image on the retina

The eye is often compared to a camera. To provide useful information, the lens of the eye, like a camera lens, must focus the arriving image. For an image to be "in focus," the rays of light arriving from an object must strike the sensitive surface of the retina in precise order, forming a miniature image of the object. If the rays are not perfectly focused, the image is blurry. Focusing typically occurs in two steps: as the light passes through the cornea, and as it passes through the lens.

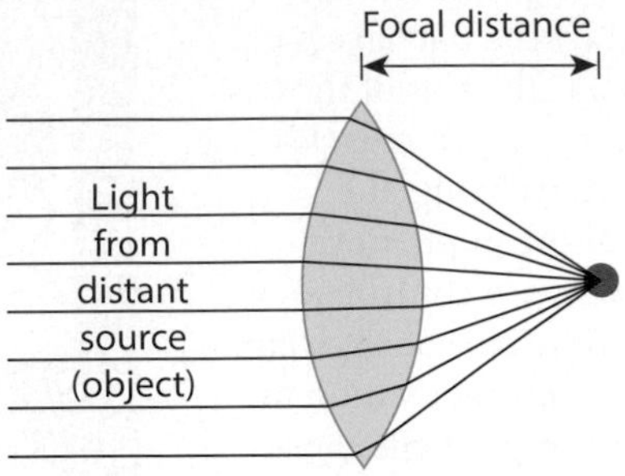
Focal distance
Light from distant source (object)

1 Light is **refracted**, or bent, when it passes from one medium to another medium with a different density. In the human eye, the greatest amount of refraction occurs when light passes from the air into the corneal tissues, which have a density close to that of water. You cannot vary the amount of refraction that occurs at the cornea. Additional refraction takes place when the light passes from the aqueous humor into the relatively dense lens. The lens provides the extra refraction needed to focus the light rays from an object toward a **focal point**—a specific point of intersection on the retina. The distance between the center of the lens and its focal point is the **focal distance** of the lens. Whether in the eye or in a camera, the focal distance is determined by the distance from the object to the lens, and the shape of the lens.

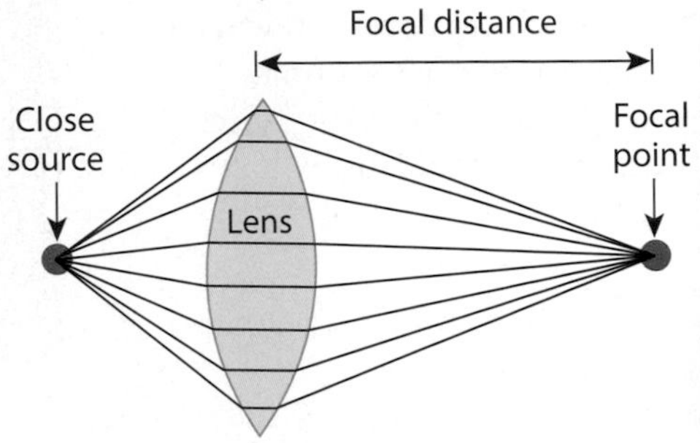
Focal distance
Close source
Lens
Focal point

The closer the light source, the longer the focal distance

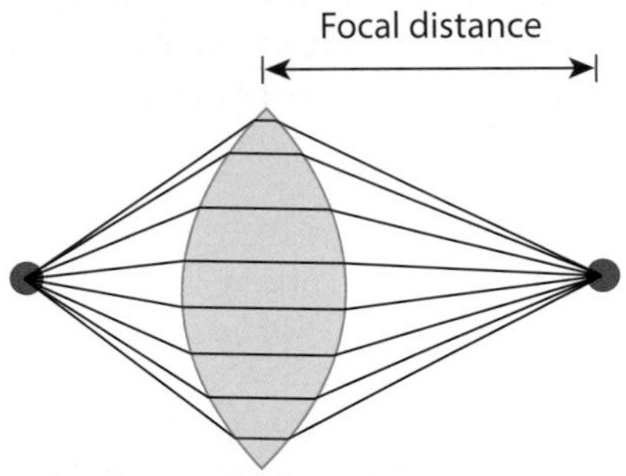
Focal distance

The rounder the lens, the shorter the focal distance

2 A camera focuses an image by moving the lens toward or away from the digital image sensor. This method of focusing cannot work in our eyes, because the distance from the lens to the retina cannot change. We focus images on the retina by changing the shape of the lens to keep the focal distance constant and give us clear vision, a process called **accommodation**.

For Close Vision: Ciliary Muscle Contracted, Lens Rounded

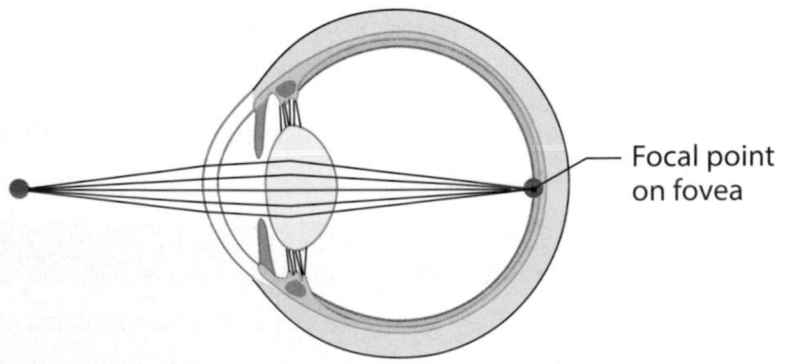
Focal point on fovea

When the ciliary muscle contracts, the ciliary body moves toward the lens, thereby reducing the tension in the ciliary zonule. The elastic capsule of the lens then pulls it into a more spherical shape that increases the refractive power of the lens, enabling it to bring light from nearby objects into focus on the retina.

For Distant Vision: Ciliary Muscle Relaxed, Lens Flattened

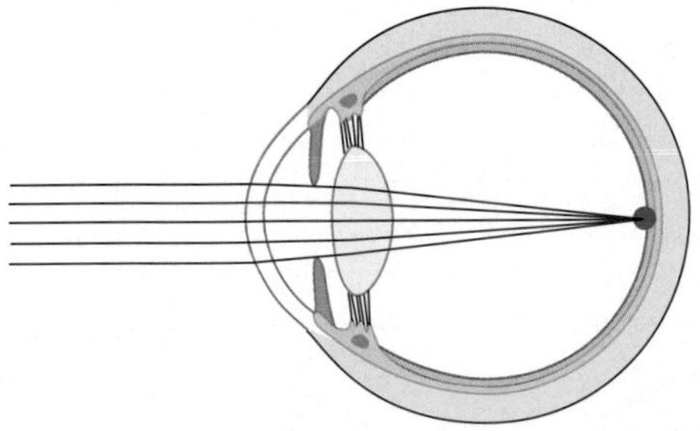

When the ciliary muscle relaxes, the ciliary zonule pulls at the circumference of the lens, making the lens flatter and bringing the image of a distant object into focus on the retina.

3 An object in view is not a single point. The image consists of a large number of individual points, like the pixels on a computer screen. In the eye, light from each point is focused on the retina, creating a miniature image of the original. However, the image is inverted and reversed. The brain compensates for this image reversal, and we are not aware of any difference between the orientation of the image on the retina and that of the object. The compensation is learned by experience—a person wearing glasses that reverse and invert the visual image can adapt to the change relatively quickly.

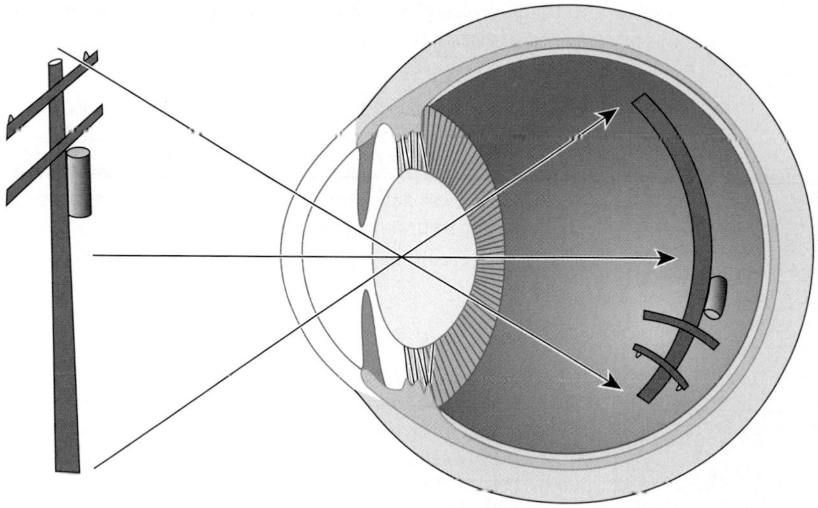

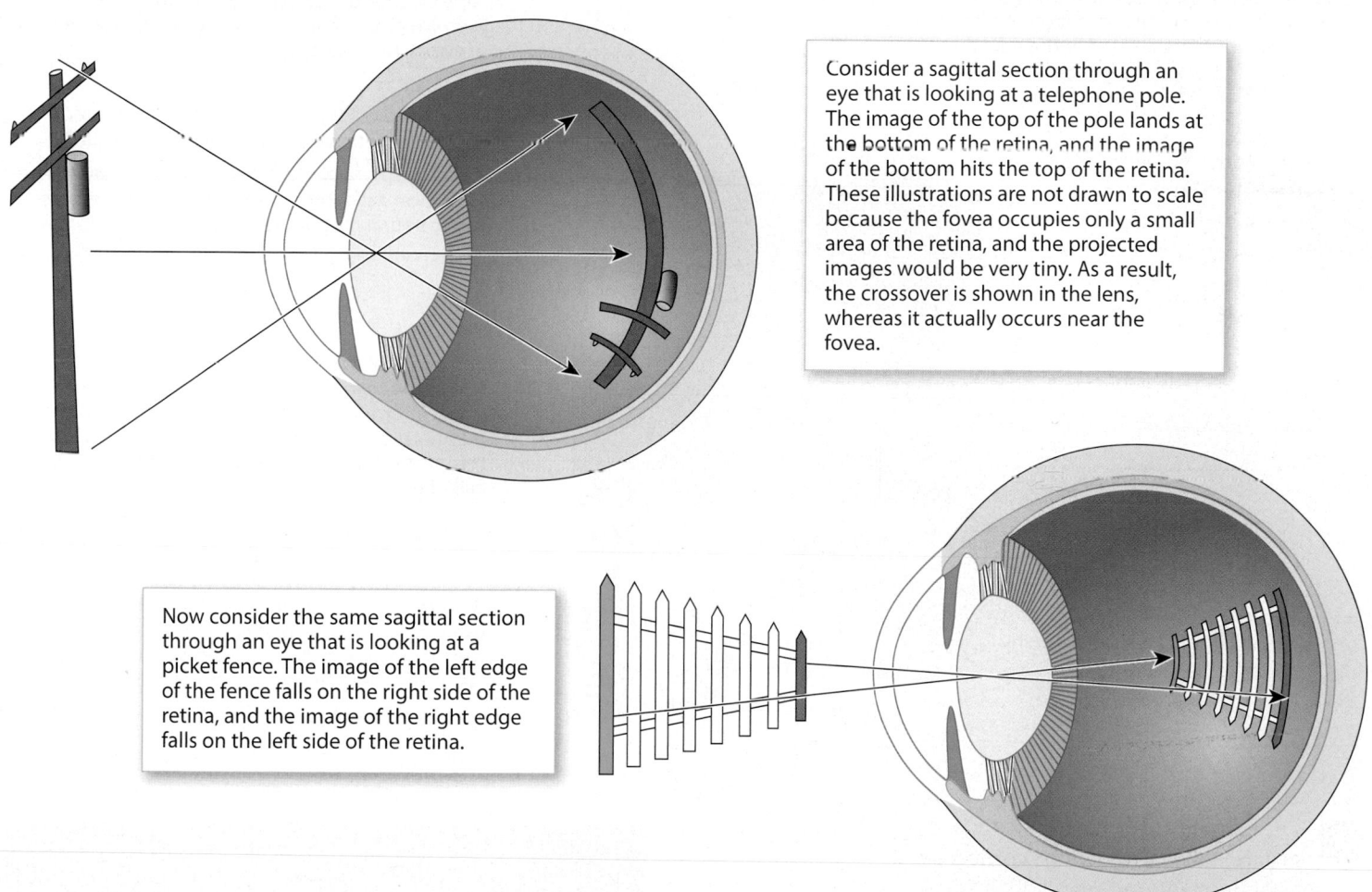

Consider a sagittal section through an eye that is looking at a telephone pole. The image of the top of the pole lands at the bottom of the retina, and the image of the bottom hits the top of the retina. These illustrations are not drawn to scale because the fovea occupies only a small area of the retina, and the projected images would be very tiny. As a result, the crossover is shown in the lens, whereas it actually occurs near the fovea.

Now consider the same sagittal section through an eye that is looking at a picket fence. The image of the left edge of the fence falls on the right side of the retina, and the image of the right edge falls on the left side of the retina.

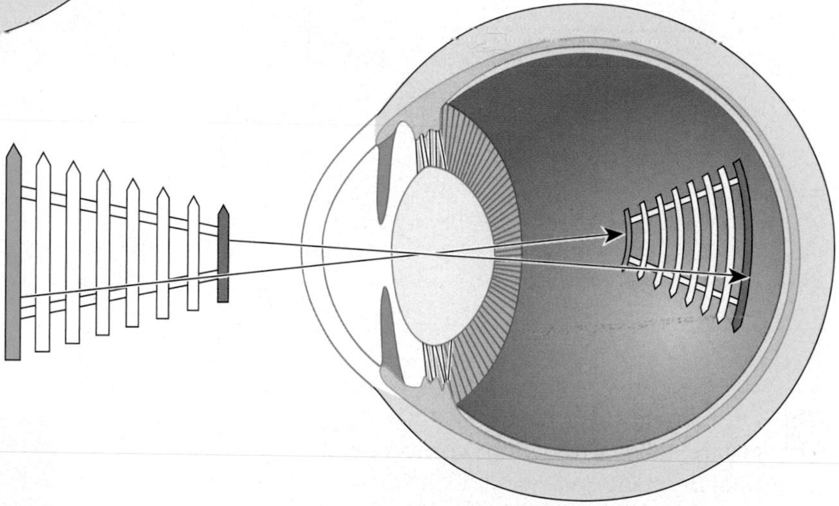

The greatest amount of refraction is required to view objects that are very close to the lens. The inner limit of clear vision, known as the **near point of vision**, is determined by the degree of elasticity in the lens. Children can usually focus on something 7–9 cm (3–4 in.) from the eye, but over time the lens tends to become stiffer and less responsive. A young adult can usually focus on objects 15–20 cm (6–8 in.) away. As we age, this distance gradually increases. The near point at age 60 is typically about 83 cm (33 in.).

Module 15.16 Review

a. Define focal point.

b. When the ciliary muscles are relaxed, are you viewing something close up or something in the distance?

c. Why does the near point of vision typically increase with age?

Module 15.17

The neural part of the retina contains multiple layers of specialized photoreceptors, neurons, and supporting cells

1 This diagrammatic sectional view through the eye shows the retina near the origin of the optic nerve.

Pigmented Part of the Retina

The pigmented part of the retina absorbs light that passes through the neural part, preventing light from bouncing back and producing visual "echoes." The pigment cells also have important biochemical interactions with the retina's light receptors, which are located in the neural part of the retina.

Neural Part of the Retina

The neural part of the retina contains photoreceptors, supporting cells, and neurons that perform preliminary processing and integration of visual information.

The outermost layer, closest to the pigmented part of the retina, contains the photoreceptors.

Ganglion cells form the innermost layer of cells in the neural part of the retina.

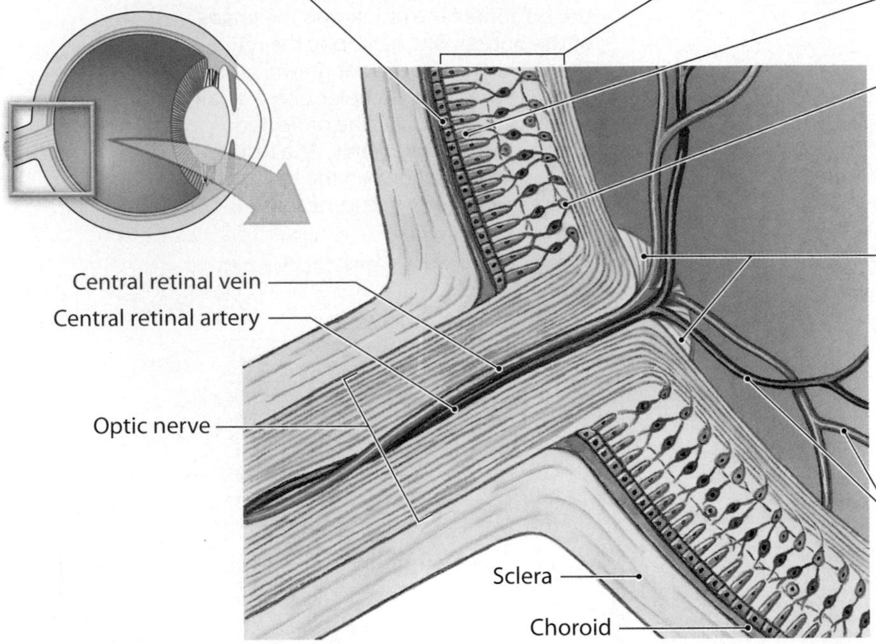
Central retinal vein
Central retinal artery
Optic nerve
Sclera
Choroid

The axons of the ganglion cells converge at the **optic disc** to form the optic nerve, which carries visual information to the brain. The optic disc has no photoreceptors. Because an image falling on this portion of the retina cannot be detected, it is called the **blind spot**.

Blood vessels enter and leave the interior of the eye within the optic nerve. They radiate across the inner surface of the eye, servicing the inner layers of cells in the neural part of the retina.

2 This is a photograph of the retinal surface, taken through the cornea, pupil, and lens of the right eye.

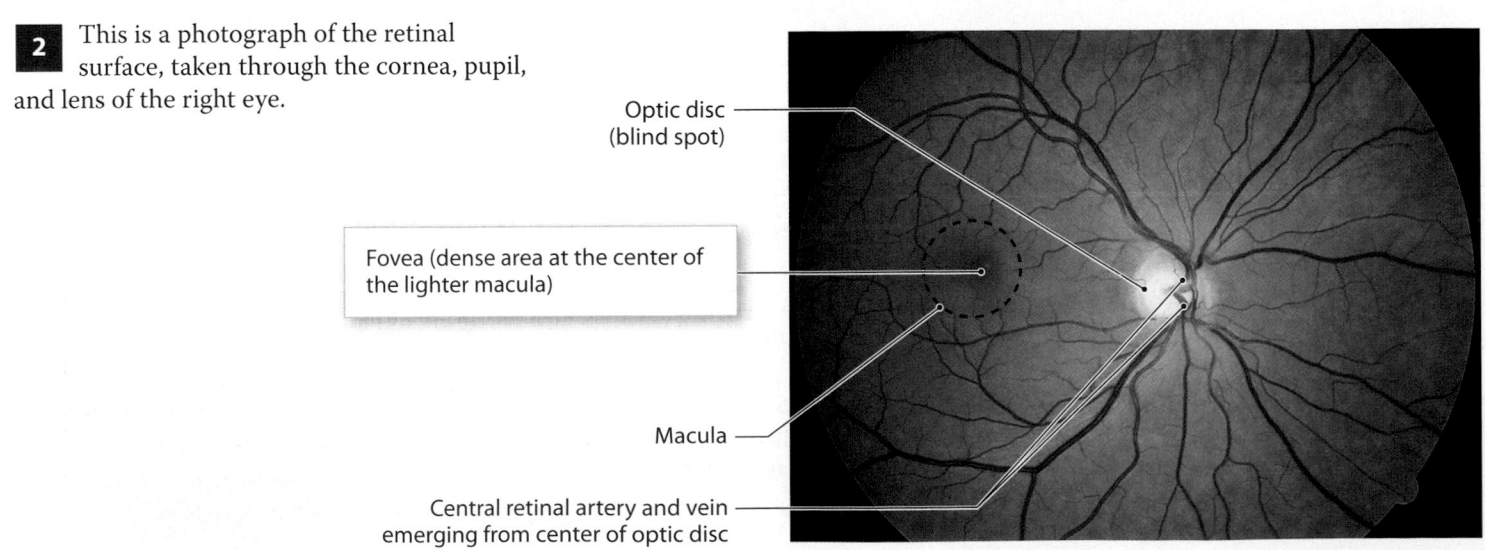

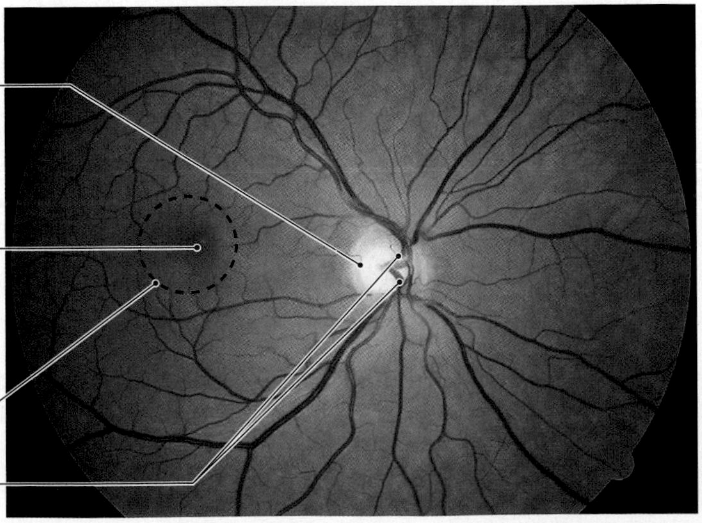
Optic disc
(blind spot)
Fovea (dense area at the center of the lighter macula)
Macula
Central retinal artery and vein
emerging from center of optic disc

3 This sectional view shows that the retina contains multiple layers of specialized cells, including two types of photoreceptors: rods and cones.

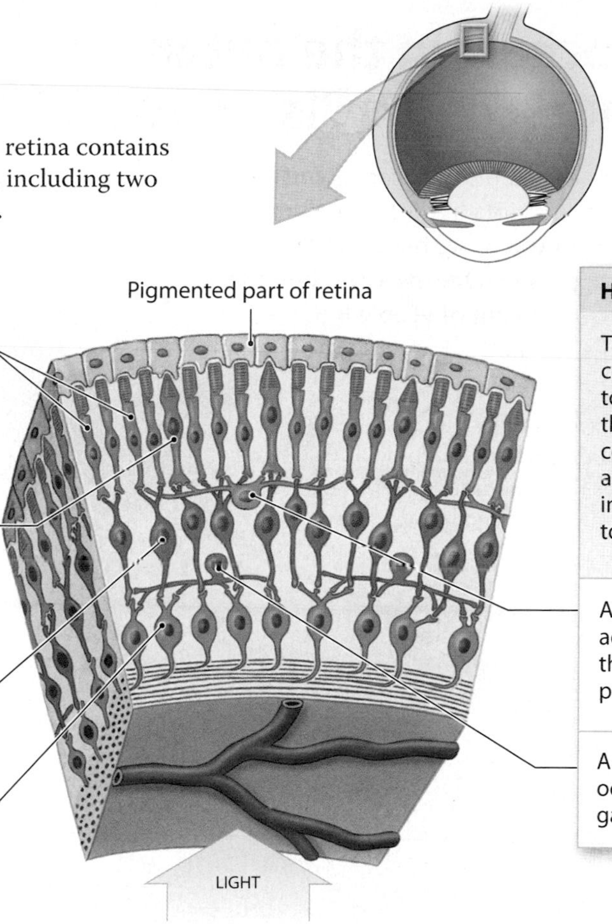

Pigmented part of retina
LIGHT

Photoreceptors of the Retina

Rods do not discriminate among colors of light. Highly sensitive, they enable us to see in dimly lit rooms, at twilight, and in pale moonlight.

Cones provide us with color vision. They give us sharper, clearer images than rods do, but cones require more intense light.

Rods and cones synapse with neurons called **bipolar cells**.

Bipolar cells synapse on ganglion cells.

Horizontal and Amacrine Cells

These cells can facilitate or inhibit communication between photoreceptors and ganglion cells, thereby altering the sensitivity of the retina. The effect is comparable to adjusting the contrast on a television set. These cells play an important role in the eye's adjustment to dim or brightly lit environments.

A network of **horizontal cells** extends across the outer portion of the retina at the level of the synapses between photoreceptors and bipolar cells.

A layer of **amacrine** (AM-a-krin) **cells** occurs where bipolar cells synapse with ganglion cells.

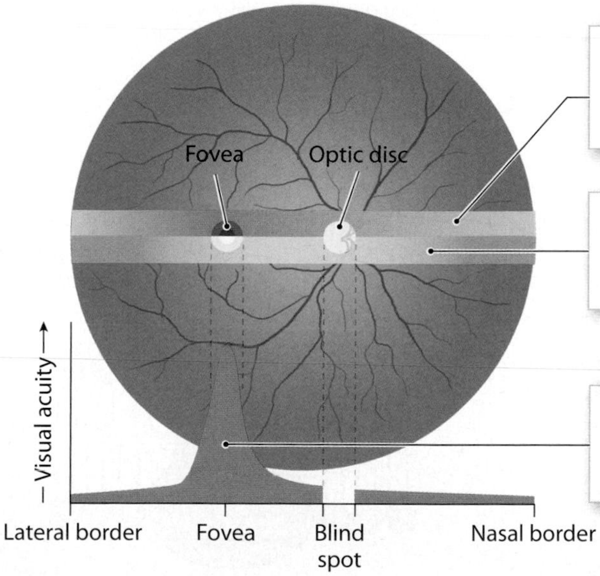

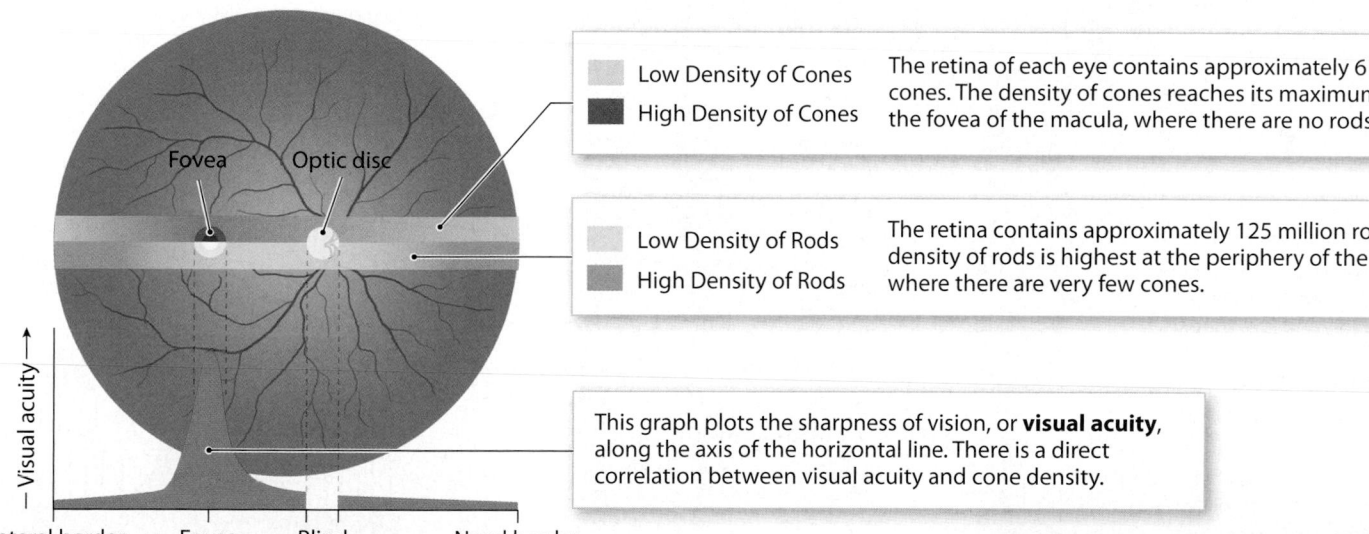

Fovea
Optic disc
Visual acuity
Lateral border
Fovea
Blind spot
Nasal border

Low Density of Cones
High Density of Cones

The retina of each eye contains approximately 6 million cones. The density of cones reaches its maximum at the fovea of the macula, where there are no rods.

Low Density of Rods
High Density of Rods

The retina contains approximately 125 million rods. The density of rods is highest at the periphery of the retina, where there are very few cones.

This graph plots the sharpness of vision, or **visual acuity**, along the axis of the horizontal line. There is a direct correlation between visual acuity and cone density.

4 The two color bands in this illustration of the retinal surface indicate the relative densities of cones and rods on either side of a horizontal line passing through the fovea and optic disc of the right eye. When you look directly at an object, the image falls on the fovea, the center of color vision and image sharpness. However, in very dim light, cones cannot function. That is why you cannot see a dim star if you stare directly at it, but you can see it if you shift your gaze to one side or the other so that the image falls on the more sensitive rods.

Module 15.17 Review

a. Compare rods with cones.

b. If you enter a dimly lit room, will you be able to see clearly? Why or why not?

c. If you had been born without cones in your eyes, explain why you would or would not be able to see.

LO 15.17 Describe the structure and function of the retina's layers of cells, and the distribution of rods and cones and their relation to visual acuity.

Module 15.18

Photoreception occurs in the outer segment of rod and cone cells

The rods and cones of the retina are called photoreceptors because they detect photons, the basic units of visible light. Light energy is a form of radiant energy that travels in waves with a characteristic wavelength (distance between wave peaks). Our eyes are sensitive to wavelengths of 400–700 nm, the spectrum of visible light. A nanometer (nm) is one billionth of a meter.

1 This illustration shows the major structural features of rods and cones, and the adjacent pigment epithelium and bipolar cells. The outer segments of both rods and cones have membranous plates, or discs, that contain special organic compounds called **visual pigments**.

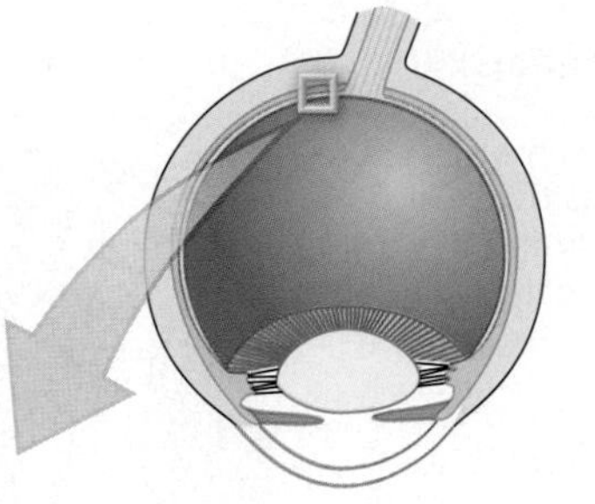

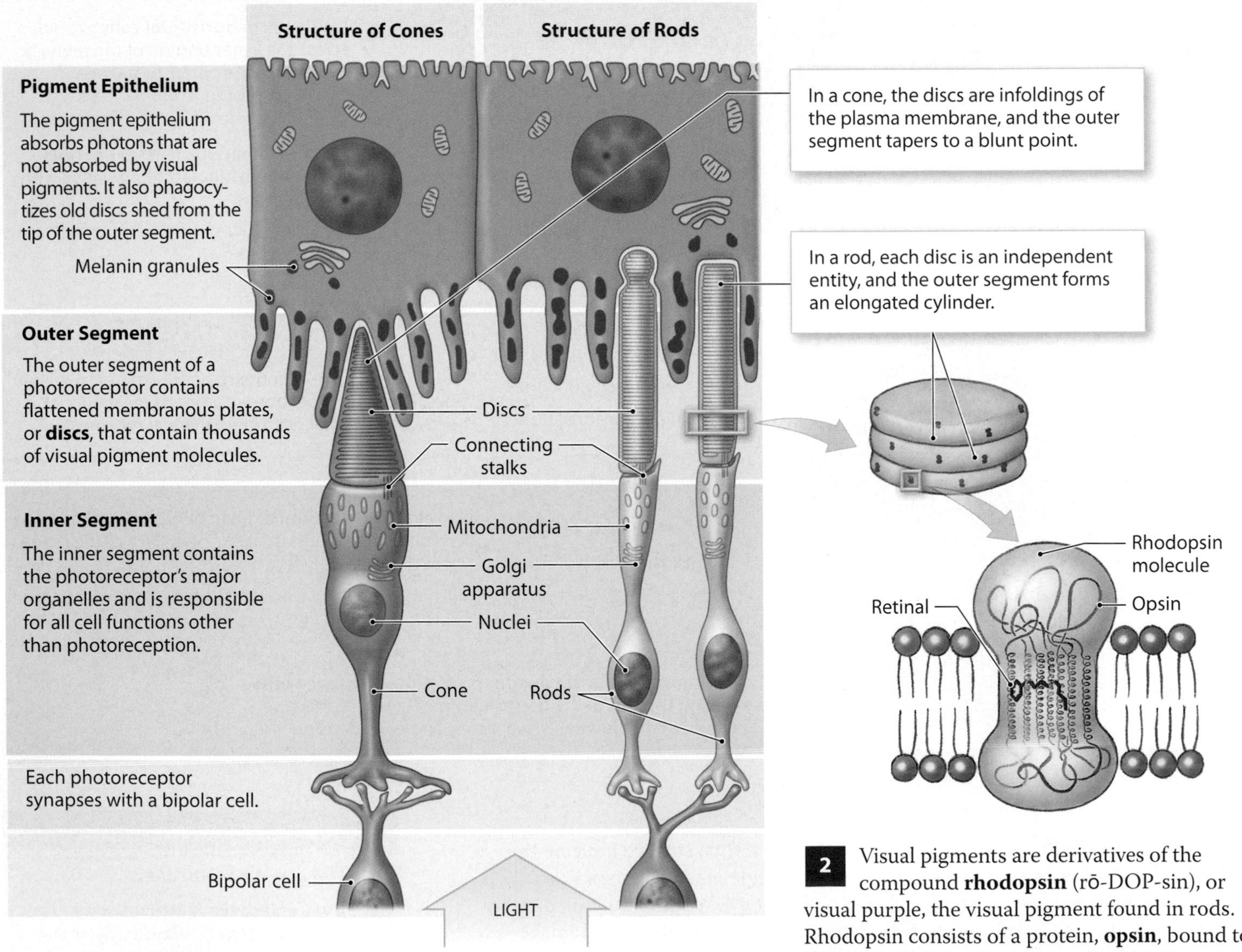

Structure of Cones
Structure of Rods
Pigment Epithelium
The pigment epithelium absorbs photons that are not absorbed by visual pigments. It also phagocytizes old discs shed from the tip of the outer segment.
Melanin granules
Outer Segment
The outer segment of a photoreceptor contains flattened membranous plates, or **discs**, that contain thousands of visual pigment molecules.
Inner Segment
The inner segment contains the photoreceptor's major organelles and is responsible for all cell functions other than photoreception.
Each photoreceptor synapses with a bipolar cell.
Bipolar cell
Discs
Connecting stalks
Mitochondria
Golgi apparatus
Nuclei
Cone
Rods
LIGHT
In a cone, the discs are infoldings of the plasma membrane, and the outer segment tapers to a blunt point.
In a rod, each disc is an independent entity, and the outer segment forms an elongated cylinder.
Rhodopsin molecule
Retinal
Opsin

2 Visual pigments are derivatives of the compound **rhodopsin** (rō-DOP-sin), or visual purple, the visual pigment found in rods. Rhodopsin consists of a protein, **opsin**, bound to the pigment **retinal** (RET-i-nal), which is synthesized from vitamin A. The type of opsin present determines the wavelength of light that can be absorbed by retinal.

3 All rods contain the same type of opsin. It is most sensitive to blue-green wavelengths of light. There are three types of cones: **blue cones**, **green cones**, and **red cones**. Each type has a different form of opsin that is sensitive to a different range of wavelengths. Their stimulation in various combinations is the basis for color vision. In a person with normal color vision, the cone population consists of 16 percent blue cones, 10 percent green cones, and 74 percent red cones. Although their wavelength sensitivities overlap, each type is most sensitive to a specific portion of the visual spectrum. If all three cone populations are stimulated, we perceive the color as white. We also perceive white if rods (but not cones) are stimulated, which is why everything appears black-and-white when we enter dimly lit surroundings or walk by starlight.

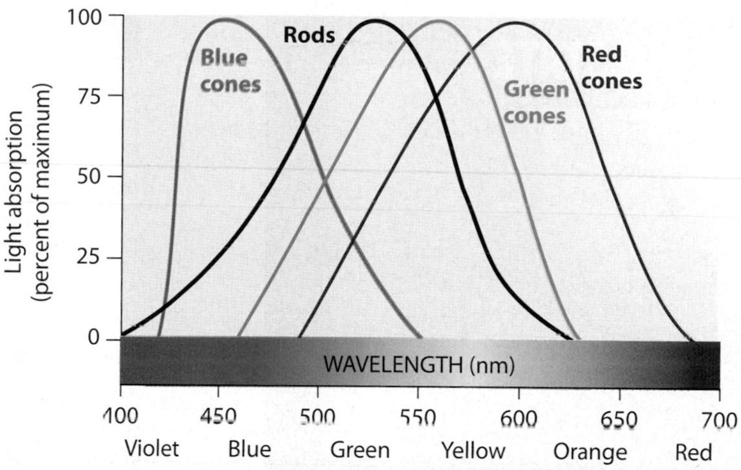

Light absorption (percent of maximum)
100
75
50
25
0
Rods
Blue cones
Green cones
Red cones
WAVELENGTH (nm)
400
450
500
550
600
650
700
Violet
Blue
Green
Yellow
Orange
Red

4 People who cannot distinguish certain colors have a form of **color blindness**. The standard tests for color vision involve picking numbers or letters out of a complex colored picture such as the one to the right. Color blindness occurs when one or more types of cones are nonfunctional. The cones may be absent, or they may be present but unable to manufacture the necessary visual pigments. In the most common type of color blindness (red–green color blindness), the red cones are missing, so the person cannot distinguish red light from green light. Inherited color blindness involving one or two cone pigments is not unusual. Ten percent of all males show some color blindness, but only about 0.67 percent of all females are color blind. Total color blindness is extremely rare. Only 1 person in 300,000 does not manufacture any cone pigments.

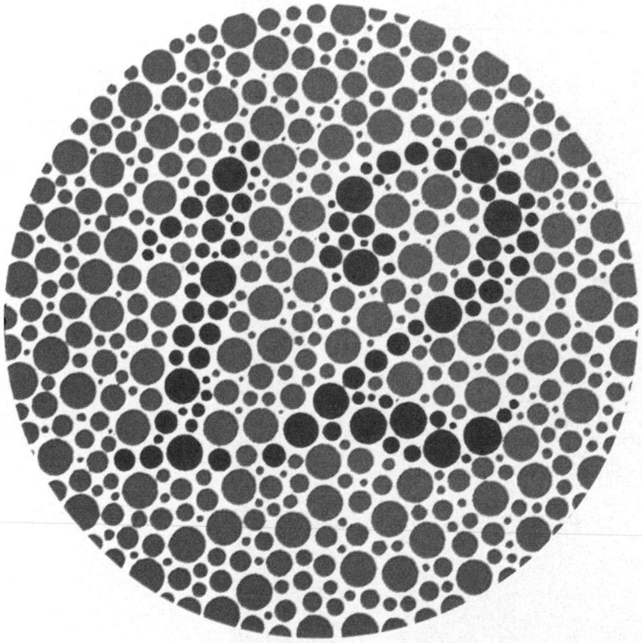

Module 15.18 Review

a. Describe the structure of a photoreceptor.

b. How could a diet deficient in vitamin A affect vision?

c. Identify the three types of cones.

LO 15.18 Describe the structure of the photoreceptors and how we are able to distinguish colors.

Photoreception involves activation, bleaching, and reassembly of visual pigments

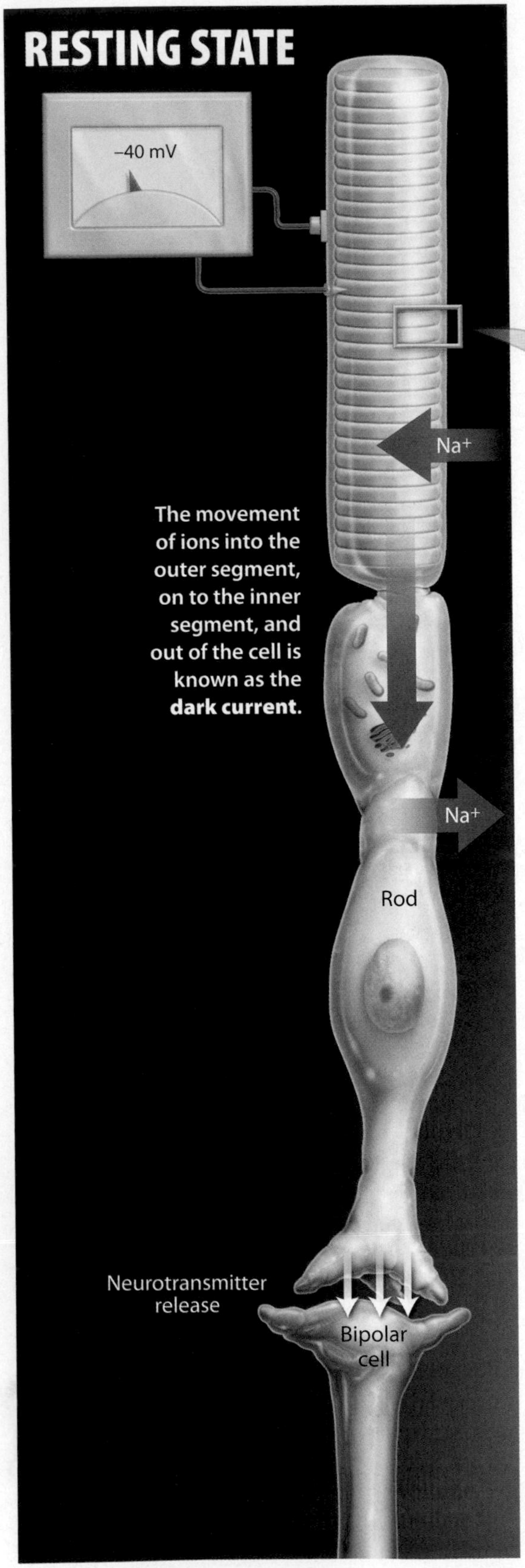
RESTING STATE
−40 mV
Na+
The movement of ions into the outer segment, on to the inner segment, and out of the cell is known as the **dark current.**
Na+
Rod
Neurotransmitter release
Bipolar cell

1 The plasma membrane in the outer segment of the photoreceptor contains chemically gated sodium ion channels. In darkness, these gated channels are kept open in the presence of cGMP (cyclic guanosine monophosphate), a derivative of the high-energy compound guanosine triphosphate (GTP). Because the channels are open, the membrane potential is approximately −40 mV, rather than the −70 mV typical of resting neurons. At the −40 mV membrane potential, the photoreceptor is continuously releasing neurotransmitters across synapses to bipolar cells. The inner segment also continuously pumps sodium ions (Na^+) out of the cytosol.

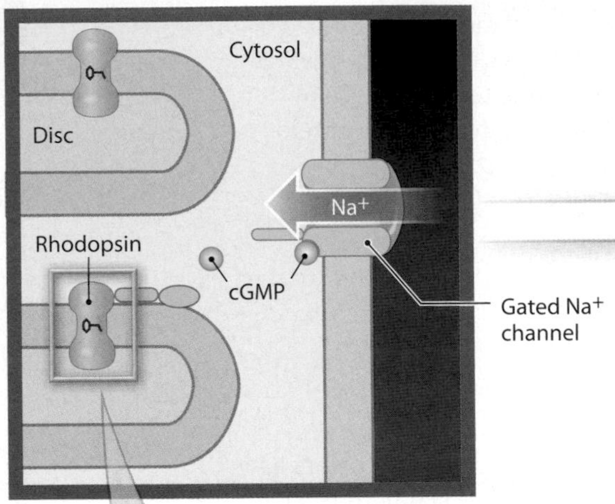
Cytosol
Disc
Na+
Rhodopsin
cGMP
Gated Na+ channel

2 The bound retinal molecule in rhodopsin has two possible configurations: the bent or curved 11-*cis* form and the more linear 11-*trans* form. Normally, in the dark, the molecule is in the 11-*cis* form. On absorbing light it changes to the 11-*trans* form. This change activates the opsin molecule.

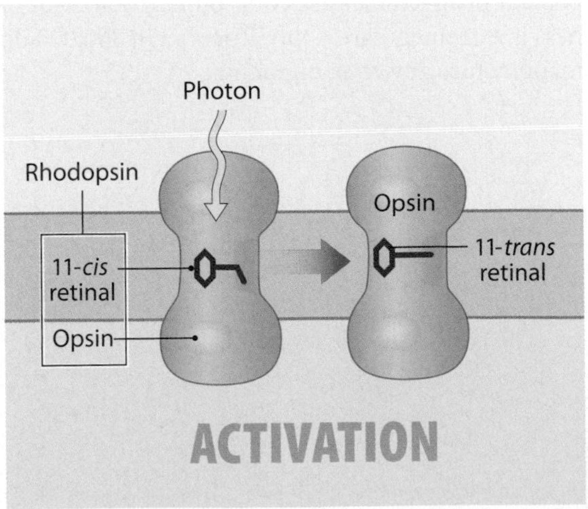
Photon
Rhodopsin
Opsin
11-*cis* retinal
11-*trans* retinal
Opsin
ACTIVATION

3 Opsin then activates **transducin**, a G protein bound to the disc membrane. The transducin in turn activates the enzyme **phosphodiesterase (PDE)**.

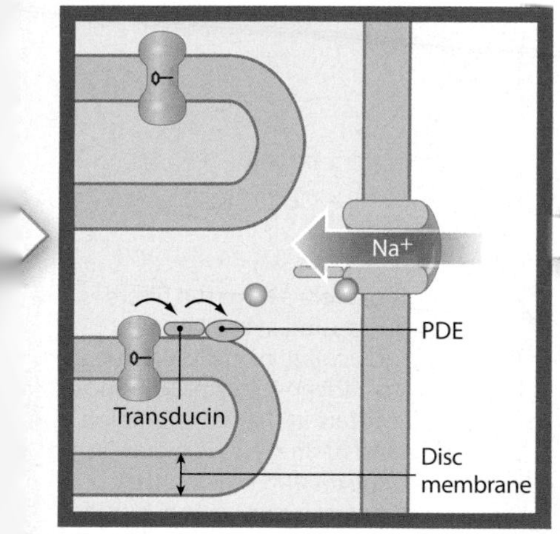
Na+
PDE
Transducin
Disc membrane

4 Phosphodiesterase is an enzyme that breaks down cGMP. The removal of cGMP from the gated sodium channels results in their inactivation. The rate of Na^+ entry into the cytosol then decreases.

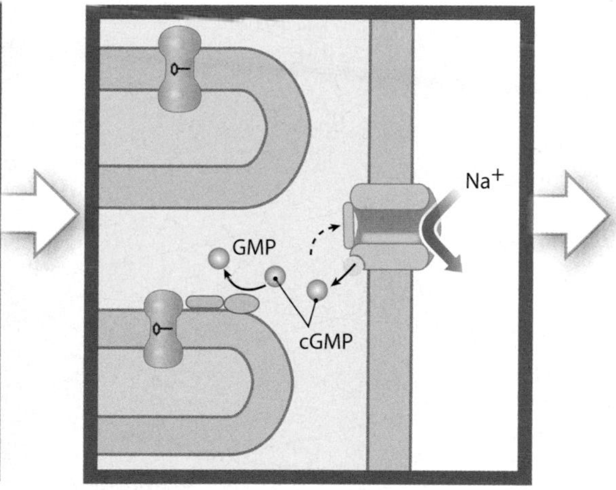
Na+
GMP
cGMP

ACTIVE STATE

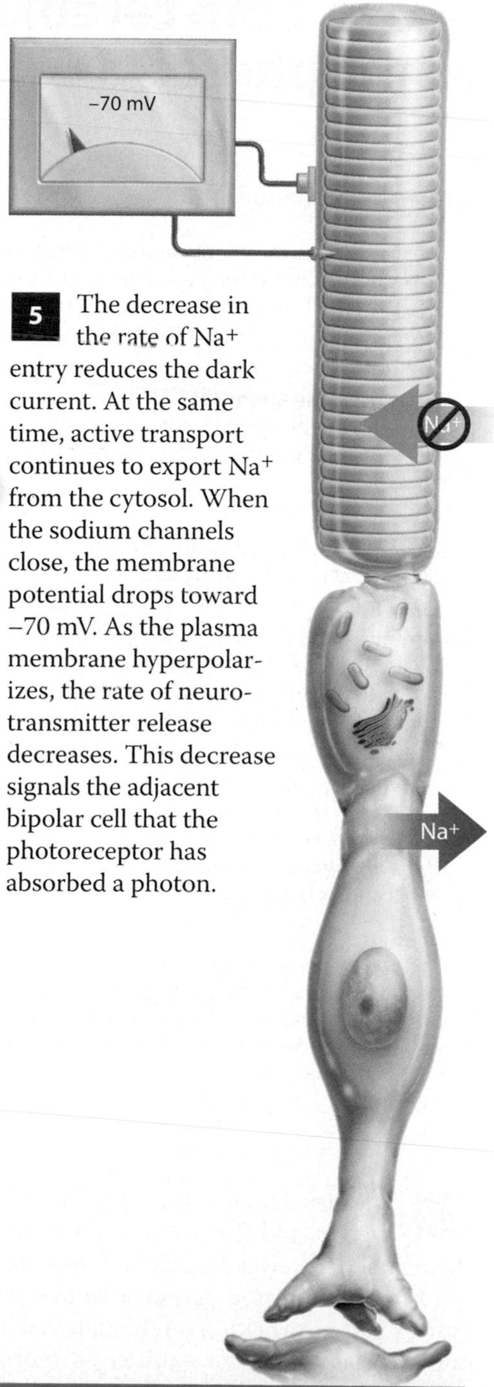
–70 mV
Na+
Na+

5 The decrease in the rate of Na^+ entry reduces the dark current. At the same time, active transport continues to export Na^+ from the cytosol. When the sodium channels close, the membrane potential drops toward –70 mV. As the plasma membrane hyperpolarizes, the rate of neurotransmitter release decreases. This decrease signals the adjacent bipolar cell that the photoreceptor has absorbed a photon.

6 Rhodopsin cannot respond to additional photons until its retinal component regains its original shape. It does not spontaneously revert to the 11-*cis* form. Instead, the entire rhodopsin molecule must be broken down into retinal and opsin in a process called **bleaching**. The retinal is then converted to its original *cis* shape. This conversion requires energy in the form of ATP. Opsin and 11-*cis* retinal are reassembled and the rhodopsin molecule is ready to repeat the cycle.

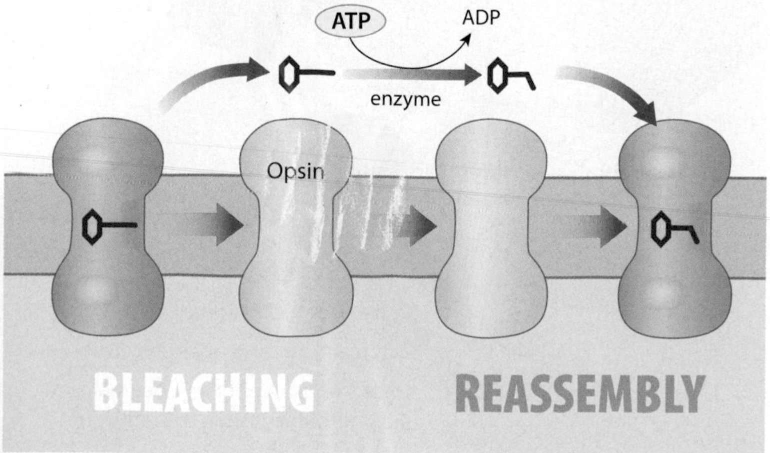
ATP
ADP
enzyme
Opsin
BLEACHING
REASSEMBLY

Module 15.19 Review

a. Visual pigments undergo which three changes during photoreception?

b. What are the two configurations of retinal?

c. When during photoreception is ATP required?

LO 15.19 Explain photoreception and how visual pigments are activated.

Module 15.20

The visual pathways distribute visual information from each eye to both cerebral hemispheres

The Visual Pathways

The visual pathways begin at the photoreceptors in the retina. Each photoreceptor monitors a specific receptive field, and when stimulated, passes the information through a bipolar cell to a ganglion cell.

Axons from the approximately 1 million ganglion cells converge on the optic disc, penetrate the wall of the eye, and proceed toward the diencephalon as the optic nerve (II).

The two optic nerves, one from each eye, reach the diencephalon at the **optic chiasm**.

From the optic chiasm, about half the fibers proceed toward the lateral geniculate nucleus of the same side of the brain, whereas the other half cross over to reach the lateral geniculate nucleus of the opposite side.

From each lateral geniculate nucleus, visual information travels to the occipital cortex of the cerebral hemisphere on that side. The bundle of projection fibers linking each lateral geniculate nucleus with the visual cortex is known as the **optic radiation**.

The perception of a visual image reflects the integration of information that arrives at the visual cortex of the occipital lobes. Each eye receives a slightly different visual image, because (1) the foveae are 5–7.5 cm (2–3.0 in.) apart, and (2) the nose and eye socket block the view of the opposite side.

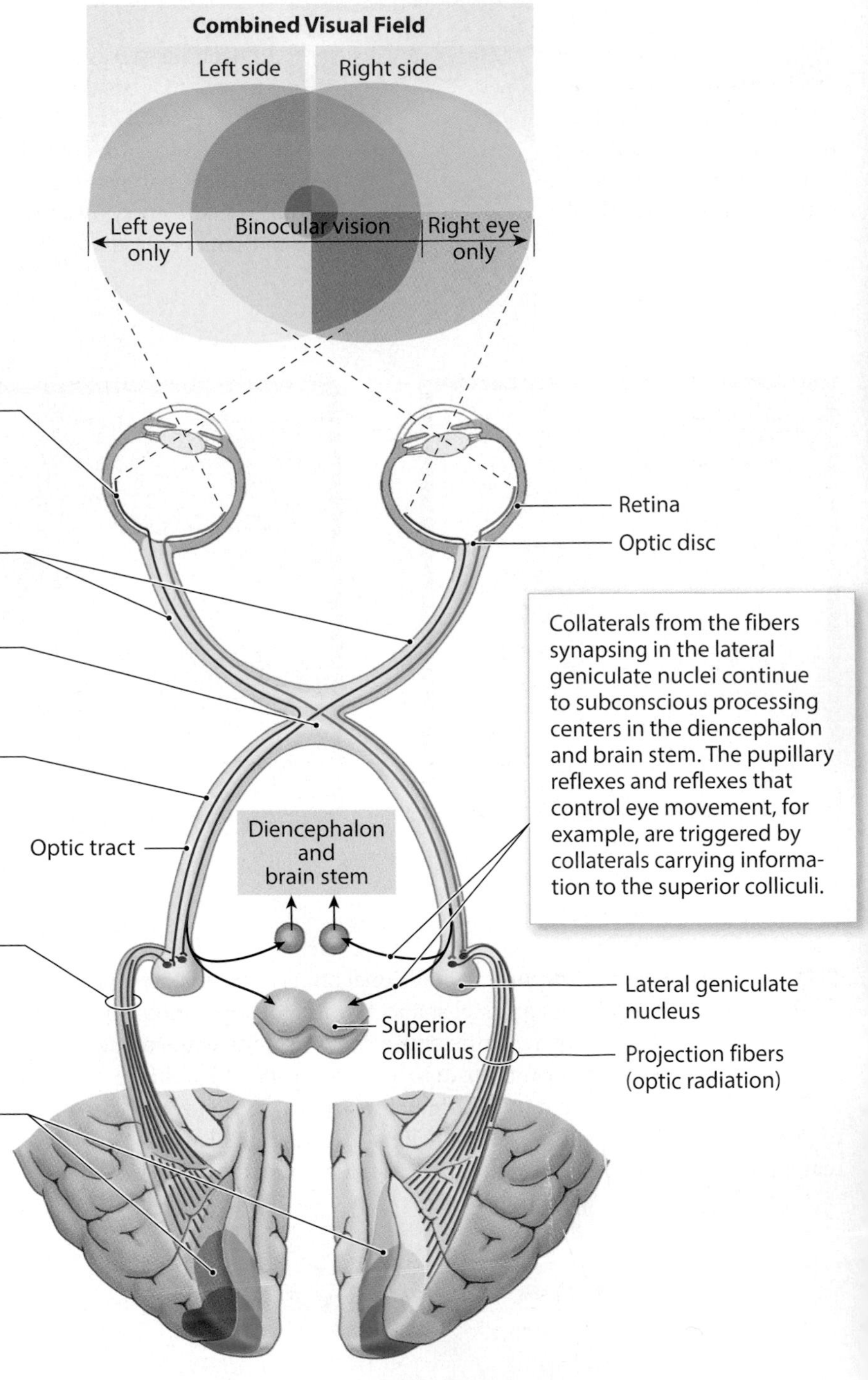

Combined Visual Field
Left side
Right side
Left eye only
Binocular vision
Right eye only
Retina
Optic disc
Collaterals from the fibers synapsing in the lateral geniculate nuclei continue to subconscious processing centers in the diencephalon and brain stem. The pupillary reflexes and reflexes that control eye movement, for example, are triggered by collaterals carrying information to the superior colliculi.
Optic tract
Diencephalon and brain stem
Lateral geniculate nucleus
Superior colliculus
Projection fibers (optic radiation)
Left cerebral hemisphere
Right cerebral hemisphere

1 The visual images from the left and right eyes overlap, and the visual cortex of each cerebral hemisphere receives information from both eyes. The information is sorted, however, so that the left visual cortex gets information on the right half of the visual field, and the right visual cortex receives information on the left half of the visual field. **Depth perception** is the ability to judge depth or distance by interpreting the three-dimensional relationships among objects in view. Your brain perceives depth by comparing the relative positions of objects within the images received by both eyes. The map in the visual cortex is upside down and backward, duplicating the orientation of the visual image at the retina.

Module 15.20 Review

a. Define optic radiation.

b. Where are visual images perceived?

c. Trace the visual pathway, beginning at the photoreceptors in the retina.

LO 15.20 Explain how the visual pathways distribute information to their destinations in the brain.

Refraction problems result from abnormalities in the cornea or lens, or in the shape of the eye

Emmetropia

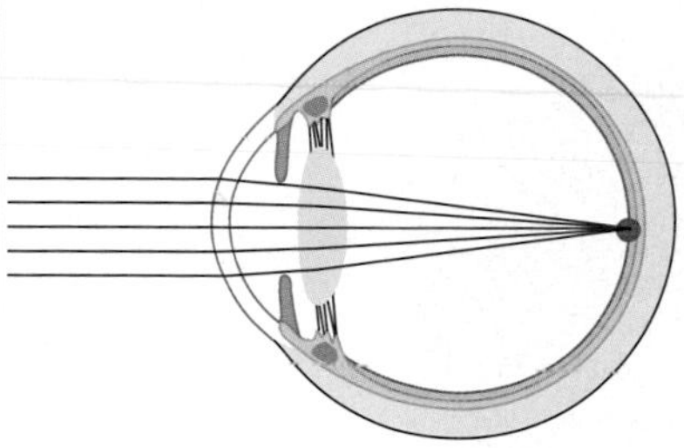

In the normal healthy eye, when the ciliary muscle is relaxed and the lens is flattened, the image of a distant object will be focused on the retina's surface. This condition is called **emmetropia** (*emmetro-*, proper + *opia*, vision), or normal vision.

Myopia

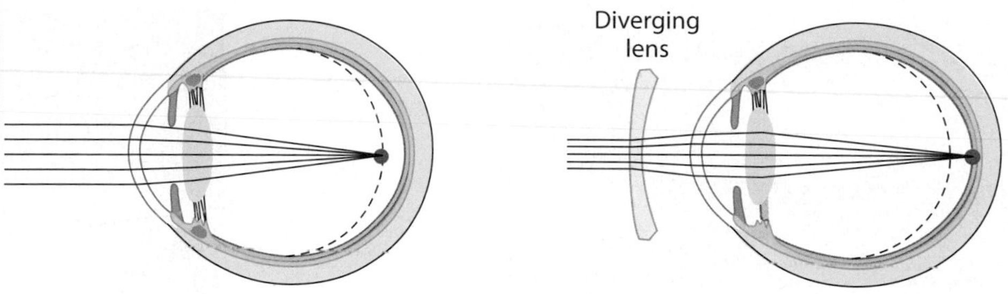
Diverging lens

If the eyeball is too deep or the resting curvature of the lens is too great, the image of a distant object is projected in front of the retina. Such individuals are said to be nearsighted because vision at close range is clear but distant objects are blurry and out of focus. Their condition is more formally termed **myopia** (*myein*, to shut + *ops*, eye).

Myopia can be treated by placing a diverging lens in front of the eye. Diverging lenses have at least one concave surface and spread the light rays apart as if the object were closer to the viewer.

1 Small variations in the performance of the lens or the structure of the eye can be corrected with external lenses (glasses or contact lenses).

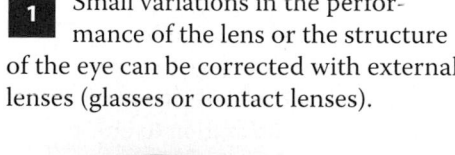

Hyperopia

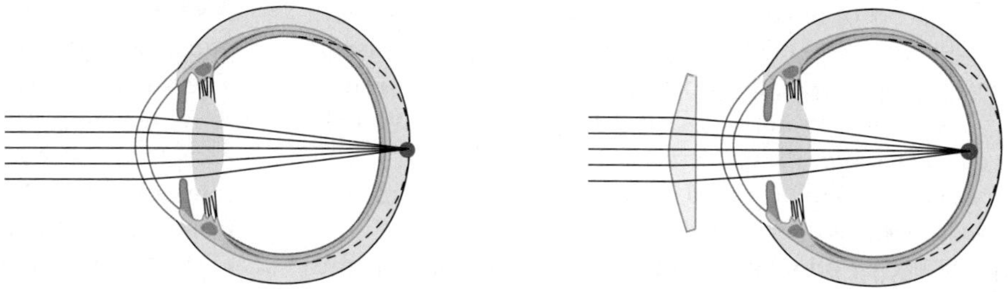

If the eyeball is too shallow or the lens is too flat, **hyperopia** results. The ciliary muscle must contract to focus even a distant object on the retina, and at close range the lens cannot provide enough refraction to focus an image on the retina. Hyperopic people are said to be farsighted, because they can see distant objects most clearly.

Hyperopia can be corrected by placing a converging lens in front of the eye. Converging lenses have at least one convex surface and provide the additional refraction needed to bring nearby objects into focus.

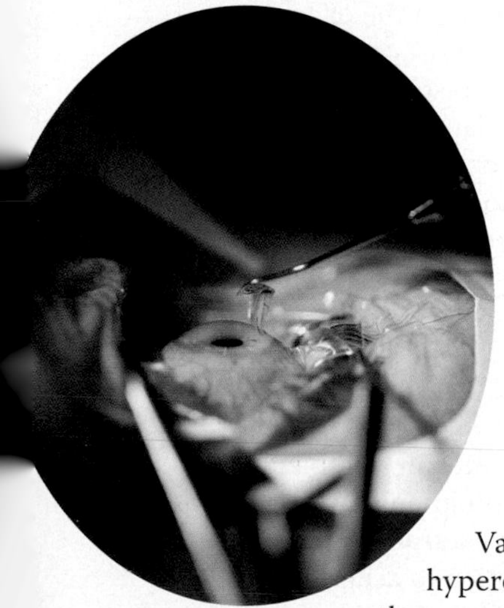

Variable success at correcting myopia and hyperopia has been achieved by surgery that reshapes the cornea. In **photorefractive keratectomy (PRK)** a computer-guided laser shapes the cornea to exact specifications. Tissue is removed only to a depth of 10–20 μm—no more than about 10 percent of the cornea's thickness. The entire procedure can be done in less than a minute. A variation on PRK is called **laser-assisted in-situ keratomileusis (LASIK)**. In this procedure the interior layers of the cornea are reshaped and covered by a flap of the normal corneal epithelium. Roughly 70 percent of LASIK patients achieve normal vision; it is now the most common form of refractive surgery. Each year, an estimated 100,000 people undergo PRK therapy in the United States. Corneal scarring is rare, and approximately 10 million Americans have had corneal refractive surgery. However, many still need reading glasses, and both immediate and long-term visual problems can occur.

Module 15.21 Review

a. Define emmetropia.
b. Which type of lens would correct hyperopia?
c. Discuss two surgical procedures for correcting myopia and hyperopia.

LO **15.21** Describe various refraction problems associated with the cornea, lens, or shape of the eye.

Aging is associated with many disorders of the special senses; trauma, infection, and abnormal stimuli may cause problems at any age

Olfaction

1 Disorders of the sense of smell may result from a head injury or from normal, age-related changes. If an injury to the head damages the olfactory nerves (I), then the sense of smell may be impaired. Unlike other populations of neurons, olfactory receptor cells are regularly replaced by the division of stem cells. Despite this process, the total number of receptors declines with age, and the remaining receptors become less sensitive. As a result, elderly people have difficulty detecting odors in low concentrations. This explains why your grandmother may overdo her perfume, and your grandfather's aftershave may seem so strong—they must use more of the odorous solution to be able to smell it themselves.

Gustation

2 Disorders of the sense of taste can be caused by problems with olfactory receptors, damage to taste buds, damage to cranial nerves, and age-related changes. The sense of smell also makes a large contribution to our sense of taste, so conditions that affect the olfactory receptors—such as the common cold—can also dull your sense of taste. A reduced sense of taste may also result from damage to taste buds by inflammation or infections of the mouth. Alternatively, the cranial nerves carrying taste sensations—facial (VII), glossopharyngeal (IX), and vagus (X)—may be damaged through trauma or compression by a tumor.

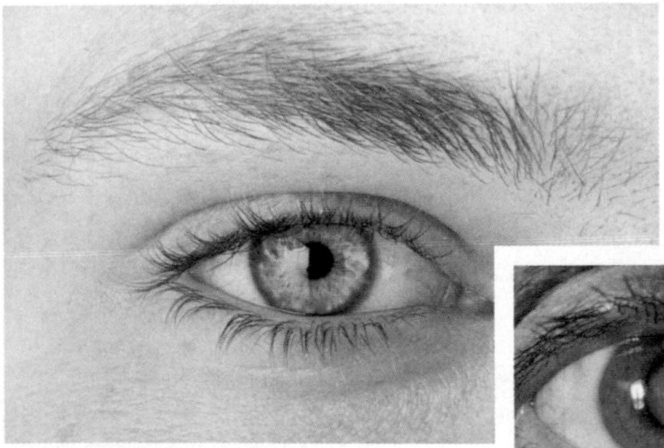

Normal eye

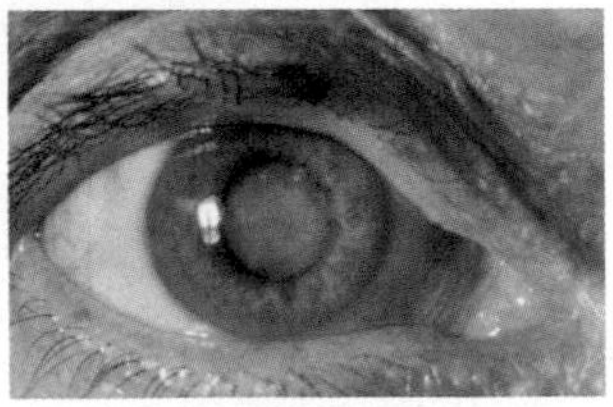

Eye with cataract

Vision

3 We discussed several common eye problems earlier in the chapter. A condition in which the lens loses its transparency—called a **cataract**—can result from injuries, radiation, or reaction to drugs. **Senile cataracts**, however, are a natural consequence of aging and are the most common form. As the condition advances, the person needs brighter and brighter light for reading, and visual acuity may eventually decrease to the point of blindness. Cataracts may be corrected by surgery. Cataract surgery involves removing the lens, either intact or after it has been shattered with high-frequency sound waves. The missing lens is then replaced by an artificial substitute. Vision is then refined with glasses or contact lenses.

Equilibrium

4 **Vertigo** is a feeling that you are dizzily spinning or that things are dizzily turning about you. Vertigo is usually caused by conditions that alter the function of the internal ear receptor complex, the vestibular branch of the vestibulocochlear nerve, or sensory nuclei and pathways in the central nervous system. Any event that sets endolymph into motion can stimulate the equilibrium receptors and produce vertigo; flushing the external auditory canal with cold water may chill the endolymph in the outermost portions of the labyrinth and establish a temperature-related circulation of fluid that produces mild and temporary vertigo. Although vertigo is usually caused by a problem with the internal ear, it can also be due to vision problems or drug intake. Excessive consumption of alcohol or exposure to certain drugs can also produce vertigo by changing endolymph composition or disturbing hair cells. Perhaps the most common cause of vertigo is **motion sickness**. Its unpleasant signs and symptoms include headache, sweating, flushing of the face, nausea, and vomiting. The drugs commonly administered to prevent motion sickness appear to depress activity in the brain stem.

Hearing

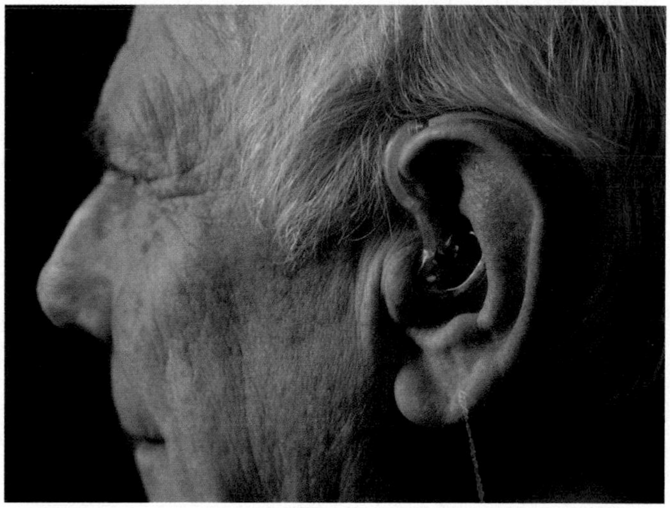

5 An estimated 6 million Americans have at least a partial hearing deficit, or deafness. **Conductive deafness** results from interference with the normal transfer of vibrations from the tympanic membrane to the oval window. Causes include excess ear wax or trapped water in the external acoustic meatus, scarring or perforation of the tympanic membrane, or immobilization of one or more auditory ossicles by fluid or a tumor. In **nerve deafness**, the problem lies within the cochlea or somewhere along the auditory pathway. Vibrations reach the oval window, but the receptors either cannot respond or their response cannot reach the central nervous system. Very loud noises can cause nerve deafness by damaging the sensory cilia on the receptor cells. Bacterial or viral infections may also kill receptor cells and damage sensory nerves. Young children have the greatest hearing range: They can detect sounds ranging from a 20-Hz buzz to a 20,000-Hz whine. With age, damage due to loud noises or other injuries accumulates: The tympanic membrane gets less flexible, the articulations between the auditory ossicles stiffen, and the round window may begin to ossify. As a result, older people typically show some degree of hearing loss.

Module 15.22 Review

a. Which cranial nerves provide taste sensations from the tongue?

b. What causes vertigo?

c. Identify two common classes of hearing-related disorders.

LO 15.22 Describe age-related disorders of olfaction, gustation, vision, equilibrium, and hearing.

Labeling

Label the structures in this diagram of a sagittal section of the left eye.

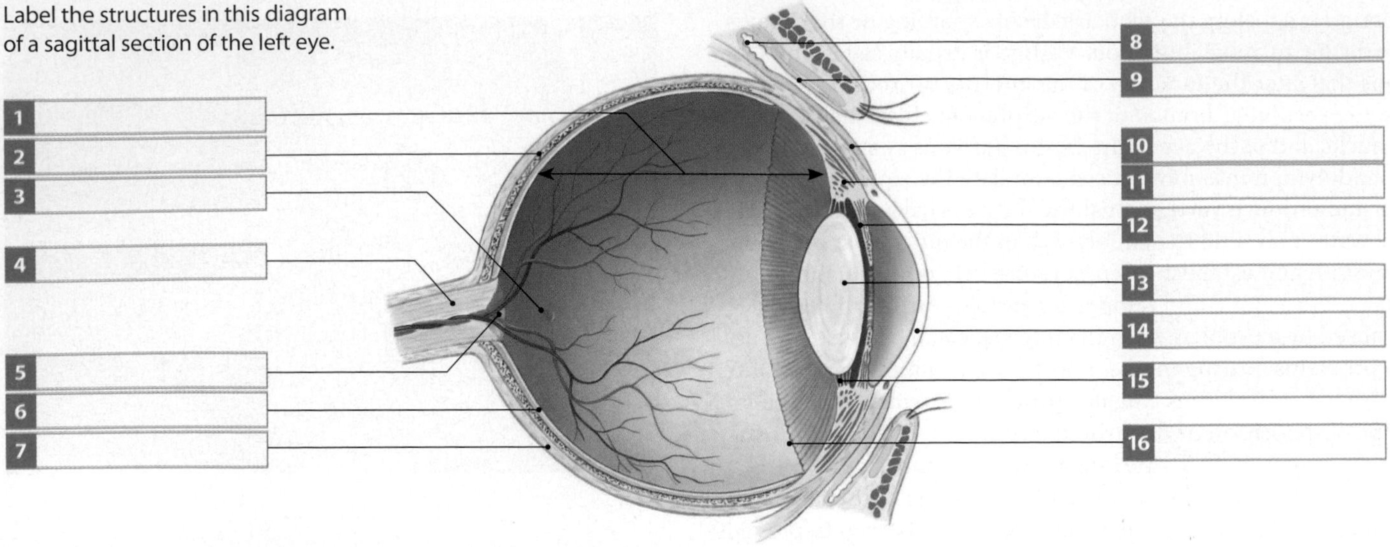
1
2
3
4
5
6
7
8
9
10
11
12
13
14
15
16

Matching

Match each lettered term with the most closely related description.

a. ganglion cells
b. cones
c. vascular layer
d. rods
e. optic disc
f. posterior chamber
g. crystallins
h. sclera
i. retina
j. occipital lobe
k. palpebrae
l. rhodopsin
m. fovea
n. posterior cavity
o. pupil

17 Visual pigment
18 Eyelids
19 Transparent proteins in the cells of a lens
20 White of the eye
21 Opening surrounded by the iris
22 Inner layer
23 Site of vitreous body
24 Photoreceptors that enable vision in dim light
25 Extends between the iris and the ciliary body and lens
26 Sharpest vision
27 Visual cortex
28 Photoreceptors that provide the perception of color
29 Their axons form the optic nerves
30 Iris, ciliary body, and choroid
31 Region of retina called the "blind spot"

17 _____
18 _____
19 _____
20 _____
21 _____
22 _____
23 _____
24 _____
25 _____
26 _____
27 _____
28 _____
29 _____
30 _____
31 _____

Section integration

A bright flash of light from nearby exploding fireworks blinds Rachel's eyes. The result is a "ghost" image that temporarily remains on her retinas. What might account for the images and their subsequent disappearance?

32 _____

Study Outline

SECTION 1 • Olfaction and Gustation

15.1 Olfaction is the sense of smell and gustation is the sense of taste p. 537

1. Receptors for olfaction, or the sense of smell, are the dendrites of specialized neurons. Dissolved chemicals contact these dendrites, causing a **generator potential**.
2. Receptors for taste, vision, equilibrium, and hearing are specialized cells with inexcitable membranes. When stimulated, graded depolarization triggers chemical transmitter release at the synapse with a sensory neuron.
3. Depolarization of the sensory neuron creates a generator potential and action potentials that are propagated to the CNS.

15.2 Olfaction involves specialized chemoreceptive neurons and delivers sensations directly to the cerebrum p. 538

4. The sense of smell, or **olfaction**, is provided by paired **olfactory organs**.
5. The olfactory pathway to the cerebrum is as follows: sensory neurons within the olfactory organ are stimulated by chemicals in the air; axons leave the olfactory epithelium and penetrate the cribriform plate of the ethmoid bone; synapses occur within the **olfactory bulb**; axons travel along the **olfactory tract** to the olfactory cortex, hypothalamus, and limbic system; distribution of olfactory information triggers not only the sensation of smell, but emotions, behaviors, and memories.
6. The olfactory organs are made up of the **olfactory epithelium** and the **lamina propria**.
7. The olfactory epithelium contains **olfactory receptor cells**, **supporting cells**, and **basal cells**. The exposed tip of each receptor cell forms a knob that serves as a base for dendrites that are exposed to dissolved compounds.
8. The lamina propria consists of areolar tissue, blood vessels, nerves, and **olfactory glands**.
9. Dissolved chemicals called **odorants** stimulate olfactory receptors at their dendrites by binding to specific receptors.

15.3 Gustation involves epithelial chemoreceptor cells located in taste buds p. 540

10. **Gustation**, or the sense of taste, provides information about the foods and liquids we eat and drink.
11. **Taste receptors** are distributed along the superficial tongue and adjacent portions of the pharynx and larynx.
12. The superior surface of the tongue has four types of **lingual papillae** that contain taste receptors and specialized epithelial cells in sensory structures called **taste buds**.
13. There are four primary taste sensations: sweet, salty, sour, and bitter. There are also receptors for **umami**, a savory taste characteristic of beef and chicken broth and Parmesan cheese. **Water receptors** are concentrated in the pharynx.

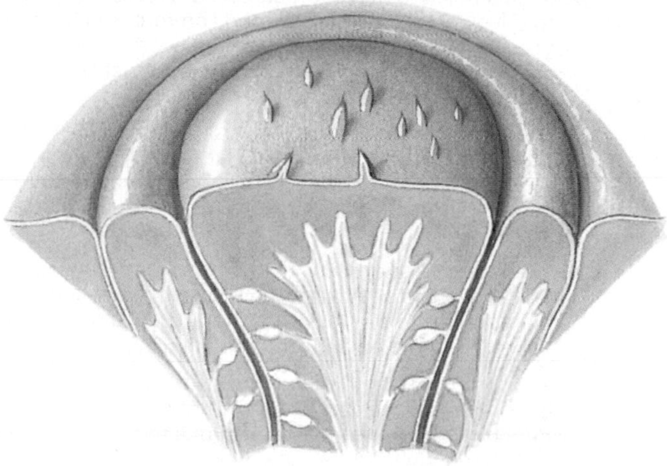

Vallate papillae are projections on the tongue that contain taste buds.

14. Taste buds are recessed deep into the surrounding epithelium, and contain 40–100 receptor cells called **gustatory cells**.
15. Each gustatory cell extends microvilli (taste hairs) through **taste pores**.
16. **Basal cells** are stem cells within a taste bud that divide to produce daughter cells that mature into gustatory receptor cells.
17. Taste reception stimulation threshold varies by primary taste sensation. Receptors also respond more readily to unpleasant stimuli. A decline in the number of taste buds occurs as we age.

15.4 Gustatory reception relies on membrane receptors and channels, and sensations are carried by facial, glossopharyngeal, and vagus nerves p. 542

18. Gustatory reception is stimulated by dissolved chemicals that contact the taste hairs and either diffuse through membrane channels or bind to receptor proteins of the gustatory cell.
19. Taste hairs stimulated by the entry of Na^+ or H^+ through leak channels are responsible for salt and sour sensations. Membrane receptors provide the sensations of sweet, bitter, and umami.
20. The **gustatory pathway** to the cerebral cortex begins with receptor stimulation. The cranial nerve involved is dependent on which region of the tongue the receptors are stimulated. The facial nerve (VII) innervates taste buds on the anterior two-thirds of the tongue. The glossopharyngeal nerve (IX) innervates taste buds on the posterior one-third of the tongue, and the vagus nerve (X) innervates the taste buds on the epiglottis.
21. Sensory afferents from the involved cranial nerve synapse in the **solitary nucleus** of the medulla oblongata. Axons then enter the **medial lemniscus** of the medulla oblongata and synapse in the thalamus. Information is then projected into the **gustatory cortex** of the insula.

SECTION 2 • Equilibrium and Hearing

15.5 Equilibrium and hearing involve the internal ear p. 545

22. In olfaction and gustation the receptor cells are exposed to the external environment and communicate with the CNS. By contrast, in equilibrium and hearing the receptors are isolated and protected from the external environment within the **internal ear**.
23. The receptors for hearing and equilibrium are called **hair cells** because their free surfaces are covered with processes similar to cilia and microvilli. Certain hair cells contain a **kinocilium**, a single large cilium, and many **stereocilia**.
24. These processes are sensitive to contact or movement. Forces pushing against the hair cell processes alter the rate of neurotransmitter release, thereby providing information about the direction and strength of the stimuli.
25. The complex three-dimensional structure of the internal ear can determine acceleration, rotation, or sound depending on which region is stimulated.

15.6 The ear is divided into the external ear, the middle ear, and the internal ear p. 546

26. The ear is divided into three anatomical regions: the **external ear**, the **middle ear**, and the **internal ear**.
27. The external ear consists of the **auricle**, the **external acoustic meatus**, and the **ceruminous glands**. **Cerumen** is a waxy material secreted by the ceruminous glands that, along with hairs projecting from the skin, helps keep foreign objects from reaching deeper delicate structures.
28. The middle ear, or tympanic cavity, is an air-filled cavity that contains the three **auditory ossicles**: **malleus**, **incus**, and **stapes**.
29. The internal ear contains the sensory organs for hearing and equilibrium. It contains a layer of dense bone called the **bony labyrinth**.
30. The **auditory tube** connects the internal ear with the nasopharynx and allows pressure equalization on either side of the tympanic membrane.
31. An infection of the internal ear is known as **otitis media**.

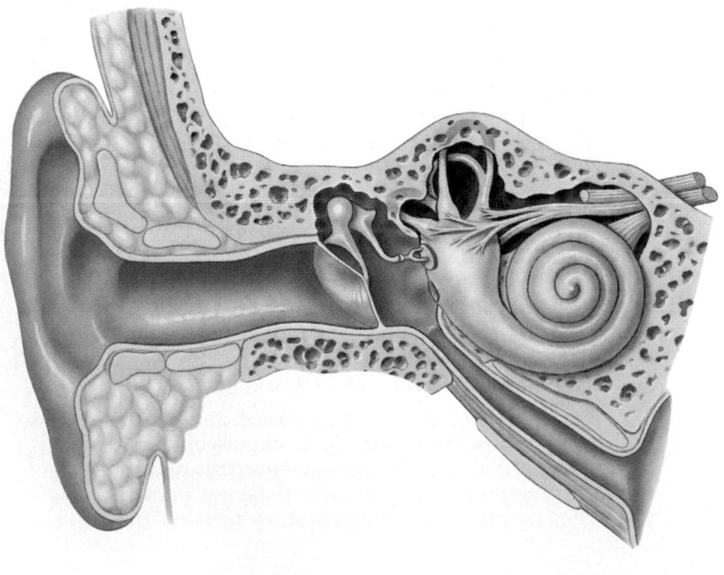

15.7 The bony labyrinth protects the membranous labyrinth p. 548

32. The internal ear can be subdivided into a **bony labyrinth** and **membranous labyrinth**. It contains receptors for the special senses of equilibrium and hearing.
33. The membranous labyrinth is a collection of tubes and chambers that contain **endolymph**.
34. The **semicircular ducts** of the membranous labyrinth lie within the semicircular canals of the bony labyrinth and are stimulated by the rotation of the head.

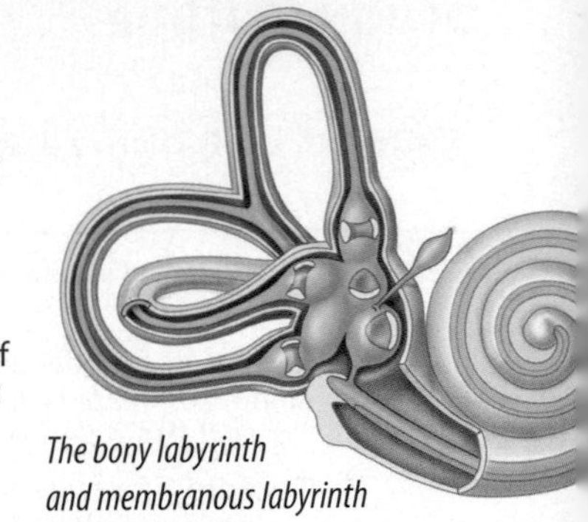

The bony labyrinth and membranous labyrinth of the internal ear

35. The **cochlear duct** is an extension of the membranous labyrinth within the cochlea and is surrounded by a pair of perilymph-filled chambers. It is responsible for hearing.
36. The bony labyrinth surrounds and protects the membranous labyrinth. Between the bony and membranous labyrinths flows **perilymph**.
37. The bony labyrinth consists of the **semicircular canals**, the **vestibule**, and the **cochlea**.
38. The membranous labyrinth, filled with endolymph and surrounded by perilymph, contains the **vestibular complex**, which is responsible for equilibrium. The three semicircular ducts detect rotational movements of the head, and the utricle and saccule monitor head position relative to gravity and linear acceleration.

15.8 Hair cells in the semicircular ducts respond to rotation; hair cells in the utricle and saccule respond to gravity and linear acceleration p. 550

39. The **anterior**, **posterior**, and **lateral semicircular ducts** are continuous with the utricle. Each duct contains an **ampulla**, an extended region that contains receptors.
40. The **crista ampullaris** is a region in the ampulla wall that contains receptors. Here the kinocilia and stereocilia are embedded in the **cupula**, a flexible, elastic, gelatinous structure that extends the full width of the ampulla.
41. Head rotation in the plane of a semicircular duct causes the endolymph in that duct to push the cupula to the side and distort the receptor processes.
42. Equilibrium sensations are provided by **maculae** in the utricle and saccule. Hair cell processes of the utricle and saccule are embedded in a gelatinous **otolithic membrane** containing **otoliths**.
43. Gravity holds the otoliths in place when the head is in the normal upright position. Changes in head position cause the otoliths to move and distort the hair cells. A similar mechanism can also detect linear acceleration.

15.9 The cochlear duct contains the hair cells of the organ of Corti p. 552

44. The **cochlear duct** lies between two perilymph-filled chambers: the **scala vestibuli** and **scala tympani**. The scalae vestibuli and tympani are one long continuous duct interconnected at the tip of the cochlear spiral.

45. The outer surfaces of these three ducts are encased by the bony labyrinth except at the **oval window** (the base of the scala vestibuli) and the **round window** (base of the scala tympani).
46. The **vestibular membrane** separates the cochlear duct from the scala vestibuli, and the **basilar membrane** separates the cochlear duct from the scala tympani.
47. The hair cells of the cochlear duct are within the **organ of Corti**, a structure atop the basilar membrane.
48. Movement of the basilar membrane in response to perilymph pressure changes causes the stereocilia of the hair cells to press against the **tectorial membrane**.
49. The **spiral ganglion** contains the cell bodies of sensory neurons that monitor the organ of Corti hair cells.

15.10 Movement of the basilar membrane is the stimulus for the sensations of pitch and volume p. 554

50. Air pressure at various **frequencies** and **amplitudes** vibrate the tympanic membrane.
51. The location of the vibration along the basilar membrane is interpreted as **pitch**; the number of hair cells stimulated is interpreted as **volume**.
52. The events involved in hearing are as follows: sound waves arrive at the tympanic membrane; auditory ossicles vibrate; the stapes at the oval window causes pressure waves in perilymph of scala vestibuli; pressure waves on their way to the round window distort the basilar membrane; hair cells vibrate against tectorial membrane; region and intensity of stimulation are relayed to CNS over the cochlear branch of cranial nerve VIII.

15.11 The vestibulocochlear nerve carries equilibrium and hearing sensations to the brain stem p. 556

53. The receptors for equilibrium are the hair cells of the vestibule and the semicircular ducts. Sensory information is passed along the **vestibular branch** of cranial nerve VIII to the vestibular nuclei, and then throughout the CNS.
54. Hair cells of the basilar membrane are the receptors for hearing. Information is sent along the **cochlear branch** of cranial nerve VIII to the cochlear nuclei, superior olivary nuclei, inferior colliculi, thalamus, and auditory cortex.

SECTION 3 • Vision

15.12 The eyes form early in embryonic development p. 559

55. The first indication of eye development is a pair of bulges called **optic vesicles** that appear in the 4th week of embryonic development.
56. **Optic cups** form from the optic vesicles. Ependymal cells on the outer layer of the optic cups develop into photoreceptors. Ependymal cells on the inner layer develop into pigment cells.
57. The lens develops from cells superficial to the optic cups. The eye develops interior chambers filled with fluid.

15.13 Accessory structures of the eye provide protection while allowing light to reach the interior of the eye p. 560

58. The **accessory structures** of the eye include the eyelids, eyelashes, the superficial epithelium of the eye, and the structures associated with tears.
59. The two eyelids (**palpebrae**) are connected laterally by the **lateral canthus**, and medially by the **medial canthus**. The **palpebral fissure** is the gap between the two eyelids.
60. The **palpebral conjunctiva** covers the inner surface of the eyelids and the **ocular conjunctiva** covers the anterior surface of the eye. **Conjunctivitis** is an inflammation of the **conjunctiva**.
61. The **lacrimal apparatus** of each eye produces, distributes, and removes tears. It consists of a **lacrimal gland** with associated ducts, paired **lacrimal canaliculi**, a **lacrimal sac**, and a **nasolacrimal duct**.

15.14 The eye has a layered wall; it is hollow, with fluid-filled anterior and posterior cavities p. 562

62. The **layers** (tunics) of the wall of the eye are the **fibrous layer**, **vascular layer**, and **inner layer**.
63. The fibrous layer, the outermost layer of the eye, consists of the cornea and **sclera**. The fibrous layer supports and protects the eye, serves as an attachment site for extrinsic eye muscles, and through the cornea light passes to begin the focusing process.

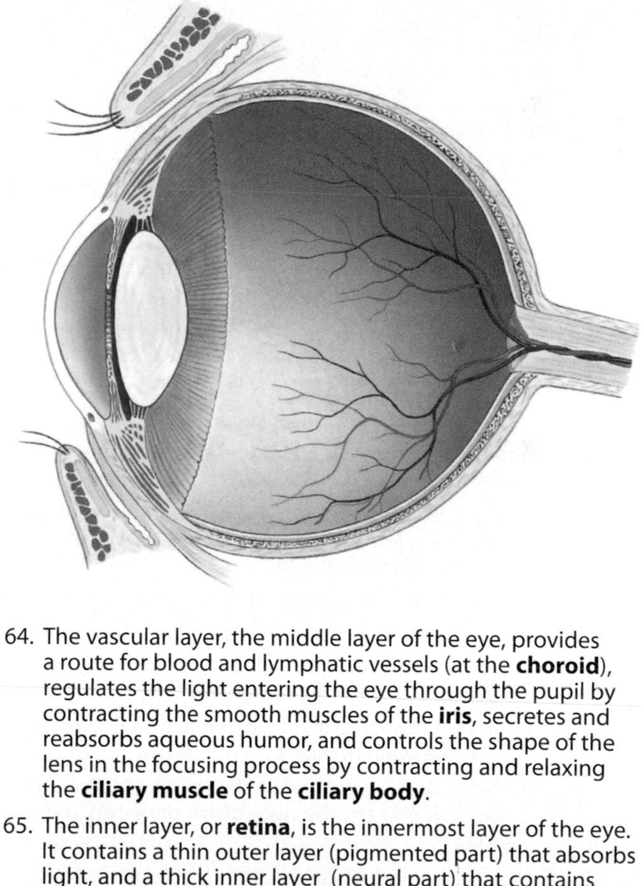

64. The vascular layer, the middle layer of the eye, provides a route for blood and lymphatic vessels (at the **choroid**), regulates the light entering the eye through the pupil by contracting the smooth muscles of the **iris**, secretes and reabsorbs aqueous humor, and controls the shape of the lens in the focusing process by contracting and relaxing the **ciliary muscle** of the **ciliary body**.
65. The inner layer, or **retina**, is the innermost layer of the eye. It contains a thin outer layer (pigmented part) that absorbs light, and a thick inner layer (neural part) that contains light-sensitive **photoreceptors**.
66. The eye is divided by the ciliary body and lens into an **anterior cavity** and a larger **posterior cavity**. The anterior cavity contains an **anterior chamber** and a **posterior chamber**.
67. **Aqueous humor** circulates within the anterior cavity. It is secreted by the ciliary body and reabsorbed at the **scleral venous sinus**. The gelatinous **vitreous body** occupies most of the posterior cavity. The **vitreous humor** is the fluid part of the vitreous body.

15.15 The eye is highly organized and has a consistent visual axis that directs light to the fovea of the retina p. 564

68. The cornea is avascular and permits light to enter the eye.
69. The lens focuses the visual image on the photoreceptors. The ciliary body supports the lens and controls its shape.
70. Blood vessels in the choroid provide nutrients to the eye.
71. The thickest layer of the eye is the sclera. It stabilizes the shape of the eye and is the site of insertion for the six extrinsic eye muscles.
72. The two layers of pupillary muscles control the amount of light entering the eye by changing the diameter of the pupil. These muscles are the **pupillary dilator muscles** and **pupillary constrictor muscles**.

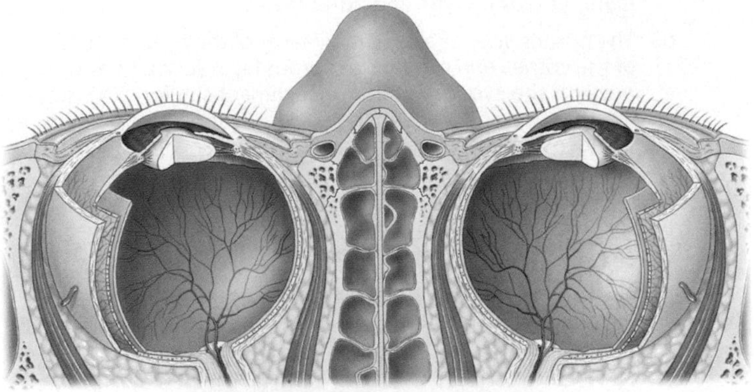

73. The **visual axis** of the eye is an imaginary line from the object of direct sight through the center of the cornea, the center of the lens, and to the retina.
74. The highest concentration of photoreceptors is at the **fovea** of the **macula**. This is the area of sharpest vision.

15.16 Focusing produces a sharply defined image on the retina p. 566

75. Light rays must be focused on the retina, or the image will be blurry. Focusing occurs first as light passes through the cornea and then the lens.
76. Light is bent or **refracted** by the cornea, aqueous humor, and then the lens toward a **focal point**.
77. **Accommodation** is the process of the lens changing shape to focus an image on the retina.
78. Light is refracted in a way that results in an inverted image that the brain reverses.
79. Refraction is greatest for closer objects. The ability to focus on near objects, the **near point of vision**, decreases as we age.

15.17 The neural part of the retina contains multiple layers of specialized photoreceptors, neurons, and supporting cells p. 568

80. The pigmented part of the retina absorbs light that passes through the neural part, preventing the light from bouncing back and causing "echoes."
81. The neural part of the retina contains photoreceptors, supporting cells, and neurons that process and integrate visual information. **Ganglion cells** form the innermost layer of the neural part of the retina.
82. Axons of the ganglion cells converge to form the optic nerve. This region—the **optic disc**—cannot detect an image because it has no photoreceptors, and is therefore called the **blind spot**.
83. The **photoreceptors of the retina** are rods, cones, and bipolar cells.
84. **Rods** are not color sensitive, but due to their high sensitivity allow for vision in dim light. **Cones** allow for color vision as well as sharp, clear vision. Rods and cones synapse with **bipolar cells**.
85. **Horizontal** and **amacrine cells** facilitate or inhibit communication between photoreceptor and ganglion cells.
86. **Visual acuity** is best at the fovea where cone density is greatest.

15.18 Photoreception occurs in the outer segment of rod and cone cells p. 570

87. The outer segments of rods and cones have membranous **discs** containing **visual pigments**. These pigments are derivatives of the compound **rhodopsin**.
88. Rhodopsin consists of the protein **opsin** bound to the pigment **retinal**. Retinal is synthesized from vitamin A.
89. There are three types of cones: **blue cones**, **green cones**, and **red cones**. Each is sensitive to a different range of light wavelength. Their stimulation in various combinations is the basis for color vision. A person with **color blindness** lacks one or more cone pigments.

15.19 Photoreception involves activation, bleaching, and reassembly of visual pigments p. 572

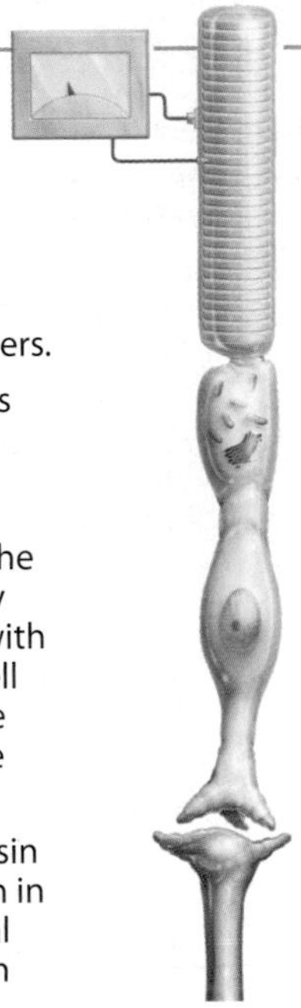

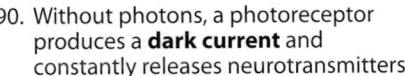

90. Without photons, a photoreceptor produces a **dark current** and constantly releases neurotransmitters.
91. On absorbing light, retinal changes shape and activates the opsin molecule. This activation leads to changes in the permeability of the outer segment to Na^+ that alters the rate of neurotransmitter release by the inner segment at its synapse with a bipolar cell. Change in bipolar cell activity is detected by one or more ganglion cells. This determines the region of retina stimulation.
92. After absorbing a photon, rhodopsin breaks down into retinal and opsin in a process called **bleaching**. Retinal is converted to its original shape in a process that requires ATP. Once retinal is converted, it can recombine with opsin, and repeat the cycle.

15.20 The visual pathways distribute visual information from each eye to both cerebral hemispheres p. 574

93. The visual pathway begins at the photoreceptor in the retina. Each photoreceptor monitors a specific receptive field. Information is passed to bipolar and ganglion cells.
94. Ganglion cell axons converge at the optic disc and proceed to the diencephalon as the optic nerve (II). The two optic nerves meet at the diencephalon at the **optic chiasm**. At

that point, about half the fibers continue to the lateral geniculate nucleus of the same side of the brain and the other half cross to the opposite side.

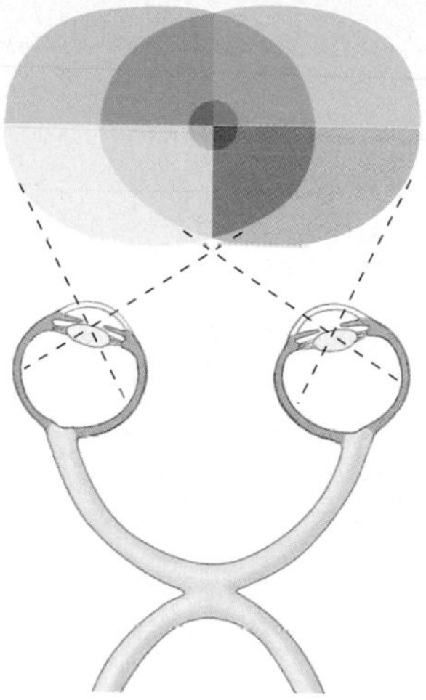

95. The **optic radiation** links each lateral geniculate nucleus with the visual cortex of the occipital lobe.
96. The visual images from the left and right eyes overlap, and the visual cortex of each cerebral hemisphere receives information from both eyes. The sorting of this information allows for **depth perception**.

15.21 Refraction problems result from abnormalities in the cornea or lens, or in the shape of the eye p. 575

97. In a normal relaxed eye, the image will be focused on the retina. Normal vision is called **emmetropia**.
98. In **myopia**, or nearsightedness, the image is projected in front of the retina. This is corrected with a diverging lens.
99. In **hyperopia**, or farsightedness, the image cannot focus on the retina. This condition is corrected with a converging lens.

15.22 Aging is associated with many disorders of the special senses; trauma, infection, and abnormal stimuli may cause problems at any age p. 576

100. Aging and a variety of conditions can decrease the number of receptor cells, resulting in diminished senses. Both the number of olfactory receptors and taste buds decline with age. Lens transparency lessens with age, leading to **senile cataracts**. The tympanic membrane loses some flexibility with age, leading to hearing loss. **Vertigo** and **nerve deafness** are examples of disorders that may result from damage to the cranial nerves and brain.

Chapter Review Questions

Labeling

Label the structures in this sectional view through the back of the eye.

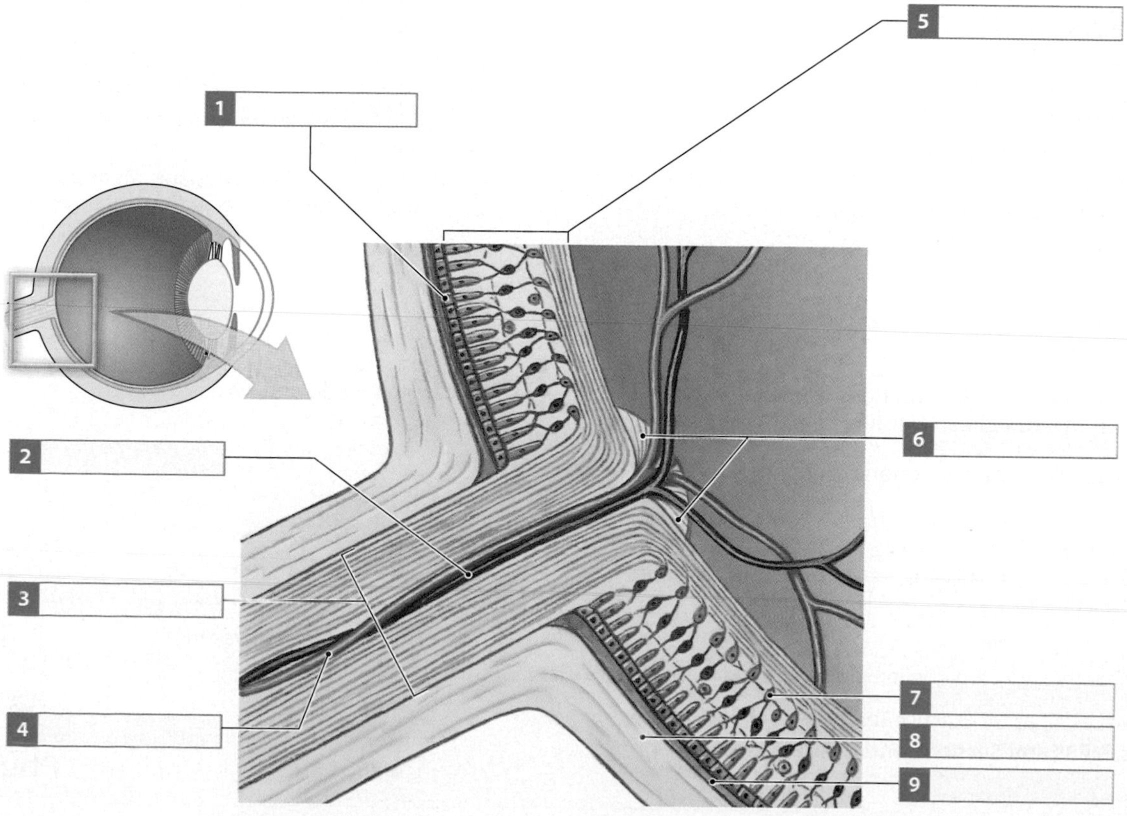
1
2
3
4
5
6
7
8
9

True/False

Indicate whether each statement is true or false.

10 Humans are most sensitive to the sweet taste.

11 The cochlear duct and scala tympani both contain perilymph, and the scala vestibuli contains endolymph.

12 Vibration of the basilar membrane causes vibration of hair cells against the tectorial membrane.

13 The accessory structures of the eye include the fibrous layer, vascular layer, and inner layer.

14 Rods are the photoreceptors that enable vision in dim light.

15 The retinal area of greatest visual acuity is the fovea.

10 ____________________

11 ____________________

12 ____________________

13 ____________________

14 ____________________

15 ____________________

Multiple choice

Select the correct answer from the list provided.

16 A blind spot occurs in the retina where
- ☐ a) the fovea is located.
- ☐ b) ganglion cells synapse with bipolar cells.
- ☐ c) axons of ganglion cells converge at the optic nerve.
- ☐ d) rod cells are clustered to form the macula.

17 The choroid is found in the
- ☐ a) fibrous layer.
- ☐ b) vascular layer.
- ☐ c) inner layer.
- ☐ d) vitreous body.

18 Receptors in the saccule and utricle provide the sensation of
- ☐ a) gravity and linear acceleration.
- ☐ b) hearing.
- ☐ c) rotational movements of the head.
- ☐ d) vibration.

19 The equalization of pressure on either side of the tympanic membrane is achieved by the
- ☐ a) round window.
- ☐ b) oval window.
- ☐ c) auditory tube.
- ☐ d) scala tympani.

20 What accounts for the feeling you experience when you accelerate down rollercoaster tracks?
- ☐ a) Pressure waves in the perilymph distort the basilar membrane.
- ☐ b) The otoliths move horizontally and distort the hair cell processes, stimulating the macular receptors.
- ☐ c) The hair cells in the organ of Corti press against the tectorial membrane.
- ☐ d) Movement of the tympanic membrane causes displacement of the auditory ossicles.

Short answer

21 Identify the four primary taste sensations. Name the two additional taste sensations that have been identified.

22 What is a cataract? What causes a cataract and how is it treated?

23 What is otitis media and what causes it?

Chapter Integration • Applying what you have learned

Curious complications from the common cold

Earl Drummond, 75 years old, visits his physician for a routine examination. During the history and physical, Mr. Drummond states that he has been experiencing difficulty seeing at a distance while driving and that at times he feels dizzy, especially when he closes his eyes. Additionally, Mr. Drummond tells his physician that he has a slight "cold." The physician asks Mr. Drummond to stand with his feet together and arms extended forward. The doctor discovers that as long as Mr. Drummond keeps his eyes open, he exhibits very little swaying movement. However, when he closes his eyes, his body begins to sway a great deal, and his arms tend to drift together toward the left side of his body. After a complete workup, which includes a neurological evaluation and a basic vision-screening test, the preliminary diagnosis is a slight fever, myopia, and vertigo.

It is estimated that 20–25 percent of all adults in the United States have myopia, while 40 percent of people in the United States experience vertigo at least once during their lifetime. The prevalence for both conditions increases with age.

1 **Explain the diagnosis of myopia.**

2 **What treatment is available for people with myopia?**

3 **Based on the information given, provide an explanation for Mr. Drummond's vertigo.**

4 **Why is Mr. Drummond's vertigo worse when his eyes are closed?**

5 **Why might Mr. Drummond's arms move toward his left side when he has his eyes closed?**

Flight deck hearing loss

Daniel is a veteran of the U.S. Navy who spent several years working on the flight deck of an aircraft carrier regularly working 12-hour shifts. He consistently wore earplugs as well as a helmet designed to provide ear protection. There were, however, occasions when he was exposed to the sound of jet engines that approached 140 dB with no ear protection at all. Such events were inevitable despite his best efforts to minimize his exposure. At the time of his discharge from the navy it was noted that he had suffered appreciable hearing loss. His physician informed Daniel that he is now partially deaf. Using what you have just learned about hearing, answer the following questions.

6 **What kind of deafness do you think Daniel has experienced from his prolonged exposure to the sound of jet engines?**

7 **What has occurred at the cellular level that has resulted in his deafness?**

8 **Do you think it is possible for Daniel to regain his hearing? Why or why not?**

FLY
107

16 The Endocrine System

LEARNING OUTCOMES

These Learning Outcomes correspond by number to this chapter's modules and indicate what you should be able to do after completing the chapter.

SECTION 1 • Hormones and Intercellular Communication

16.1 Describe the similarities between the endocrine and nervous systems and their specific modes of intercellular communication.

16.2 Explain the classification of hormones, and identify key functions of hormones secreted by organs and tissues of the endocrine system.

16.3 Explain the general mechanisms of hormonal action.

16.4 Describe how the hypothalamus controls endocrine organs.

16.5 Describe the location and structure of the pituitary gland, and identify pituitary hormones and their functions.

16.6 Describe the role of negative feedback in the functional relationship between the hypothalamus and the pituitary gland.

16.7 Describe the location and structure of the thyroid gland, identify the hormones it produces, and specify the functions of those hormones.

16.8 Describe the location of the parathyroid glands, and identify the functions of the hormone they produce.

16.9 Describe the location, structure, and functions of the adrenal glands, identify the hormones produced, and specify the functions of each hormone.

16.10 Describe the location and structure of the pancreas, identify the hormones it produces, and specify the functions of those hormones.

16.11 Describe the location of the pineal gland, and identify the functions of the hormone that it produces.

16.12 **CLINICAL MODULE** Explain diabetes mellitus: its types, clinical manifestations, and treatments.

SECTION 2 • Hormones and System Integration

16.13 Explain how hormones interact to produce coordinated physiological responses.

16.14 Describe the functions of the hormones produced by the kidneys and the heart, and explain the roles of other endocrine organs and hormones in normal growth and development.

16.15 Define the general adaptation syndrome, and compare homeostatic responses with stress responses.

16.16 **CLINICAL MODULE** Describe key endocrine disorders, citing their characteristic signs and symptoms.

LO Learning Outcomes are repeated at the bottom of each module.

The nervous and endocrine systems release chemical messengers that bind to target cells

To preserve homeostasis, cellular activities must be coordinated throughout the body. The table below introduces the various ways cells communicate and coordinate their activities using chemical messengers.

Mechanisms of Intercellular Communication

Mechanism	Transmission	Chemical Mediators	Distribution of Effects
Direct communication 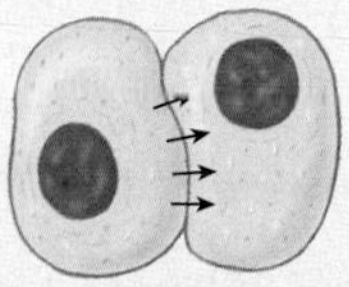	Through gap junctions	Ions, small solutes, lipid-soluble materials	Usually limited to adjacent cells of the same type that are interconnected by connexons
Paracrine communication 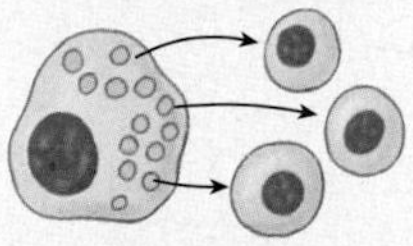	Through extracellular fluid	**Paracrine factors**	Primarily limited to the local area, where paracrine factor concentrations are relatively high; target cells must have appropriate receptors
Endocrine communication 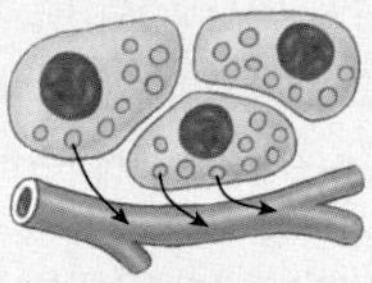	Through the bloodstream	**Hormones**	Target cells are mainly in other distant tissues and organs and must have appropriate receptors
Synaptic communication 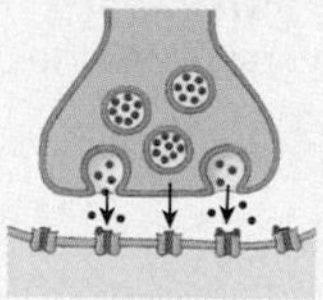	Across synapses	Neurotransmitters	Limited to very specific area; target cells must have appropriate receptors

Viewed from a general perspective, the differences between the nervous and endocrine systems seem relatively clear. In fact, these broad organizational and functional differences are the basis for treating them as two separate systems. Yet when we consider them in detail, we see similarities in the ways they are organized:

- Both systems rely on the release of chemicals that bind to specific receptors on their target cells.
- The two systems share many chemical messengers; for example, norepinephrine and epinephrine are called hormones when released into the bloodstream, but they are called neurotransmitters when released across synapses.
- Both systems are regulated primarily by negative feedback control mechanisms.
- The two systems share a common goal: to preserve homeostasis by coordinating and regulating the activities of other cells, tissues, organs, and systems.

In this section we will consider the structure and function of the endocrine system, hormones, and the integration of neural and endocrine activities.

Module 16.1 Review

a. How does paracrine communication differ from endocrine communication?

b. What are the chemical means of cellular communication in the nervous and endocrine systems?

c. What is the common goal of the nervous and endocrine systems?

LO **16.1** Describe the similarities between the endocrine and nervous systems and their specific modes of intercellular communication.

Module 16.2

Hormones may be amino acid derivatives, peptides, or lipid derivatives

1 The hormones and paracrine factors of the body can be divided into three groups on the basis of their chemical structure: (1) **amino acid derivatives**, (2) **peptide hormones**, and (3) **lipid derivatives**.

Amino Acid Derivatives

This group of hormones includes (1) thyroid hormones, produced by the thyroid gland; (2) the compounds epinephrine (E), norepinephrine (NE), and dopamine, which are sometimes called **catecholamines** (kat-e-KŌ-la-mēnz); and (3) melatonin (mel-a-TŌ-nin), a derivative of tryptophan that is secreted by the pineal gland.

Thyroid Hormones

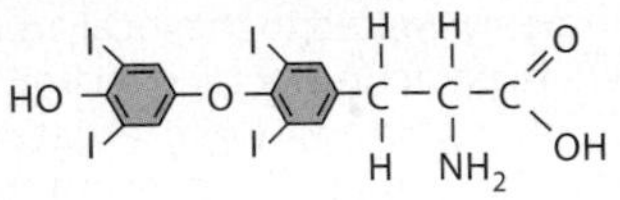

Example: Thyroxine (T_4)

Catecholamines

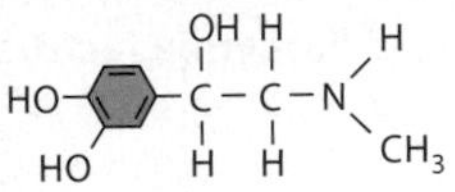

Example: Epinephrine

Tryptophan Derivatives

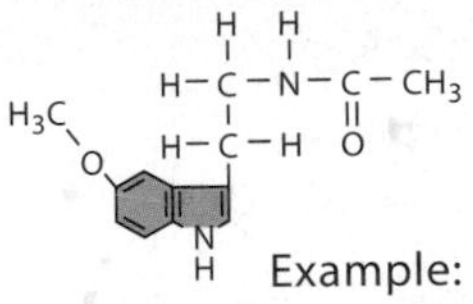

Example: Melatonin

Peptide Hormones

In general, peptide hormones are synthesized as **prohormones**—inactive precursor molecules that are converted to active hormones. Peptide hormones range from short polypeptide chains of amino acids, such as antidiuretic hormone (ADH) and oxytocin (9 amino acids apiece), to small proteins, such as insulin (51 amino acids), growth hormone (GH; 191 amino acids) and prolactin (PRL; 198 amino acids). This group includes all the hormones secreted by the hypothalamus, heart, thymus, digestive tract, and pancreas, and most of the hormones of the pituitary gland. Glycoproteins—polypeptides that have carbohydrate side chains—may also function as hormones. Examples include thyroid-stimulating hormone (TSH), luteinizing hormone (LH), and follicle-stimulating hormone (FSH) from the pituitary gland, and several hormones produced in other organs.

Lipid Derivatives

These hormones consist of carbon rings and side chains built either from fatty acids (eicosanoids) or cholesterol (steroid hormones).

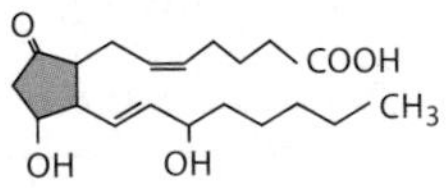

Eicosanoids

Example: Prostaglandin E

Eicosanoids (ī-kō-sa-noydz) are important paracrine factors that coordinate cellular activities and affect enzymatic processes (such as blood clotting) in extracellular fluids. Some eicosanoids, such as **leukotrienes** (loo-kō-TRĪ-ēns), have secondary roles as hormones. A second group of eicosanoids—**prostaglandins**—is involved primarily in coordinating local cellular activities.

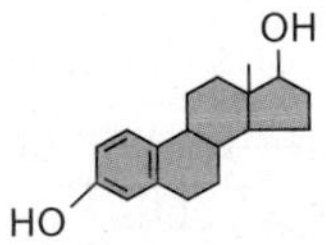

Steroid Hormones

Example: Estrogen

Steroid hormones are released by the reproductive organs (androgens by the testes in males, estrogens and progesterone by the ovaries in females), by the cortex of the adrenal glands (corticosteroids), and by the kidneys (calcitriol). Because circulating steroid hormones are bound to specific transport proteins in the blood, they remain in circulation longer than do secreted peptide hormones.

2 The **endocrine system** includes those organs (indicated in purple in this figure) whose primary function is the production of hormones or paracrine factors. Many other organs contain tissues that secrete hormones, but their endocrine functions are secondary. Examples include the heart, kidneys, intestines, thymus, and reproductive organs. The endocrine functions of these organs will be considered primarily in later chapters.

Hypothalamus

The **hypothalamus** secretes hormones involved with fluid balance, smooth muscle contraction, and the control of hormone secretion by the anterior lobe of the pituitary gland.

Pituitary Gland

The **pituitary gland** secretes multiple hormones that regulate the endocrine activities of the adrenal cortex, thyroid gland, and reproductive organs, and a hormone that stimulates melanin production.

Thyroid Gland

The **thyroid gland** secretes hormones that affect metabolic rate and calcium ion levels in body fluids.

Adrenal Glands

The two **adrenal glands** secrete hormones involved with mineral balance, metabolic control, and resistance to stress. The adrenal medullae release epinephrine and norepinephrine during sympathetic activation.

Pancreas (Pancreatic Islets)

The **pancreatic islets** secrete hormones regulating the rate of glucose uptake and utilization by body tissues.

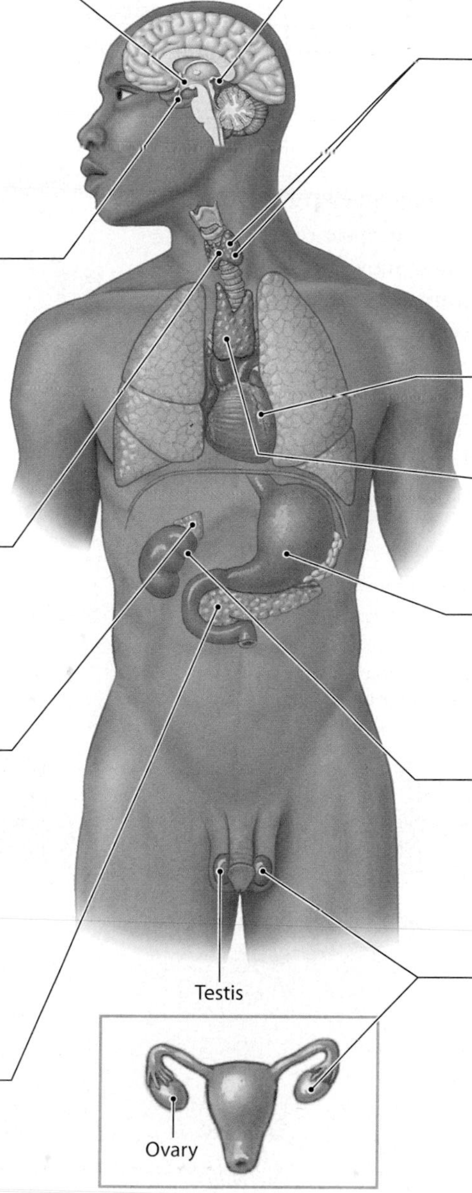

Testis
Ovary

Pineal Gland

The **pineal gland** secretes melatonin, which affects reproductive function and helps establish circadian (day/night) rhythms.

Parathyroid Glands

The **parathyroid glands** secrete a hormone important to the regulation of calcium ion levels in body fluids.

Organs with Secondary Endocrine Functions

Heart: Secretes hormones involved in regulating blood volume	See Chapter 19
Thymus: Secretes hormones involved in stimulating and coordinating the immune response	See Chapter 20
Digestive Tract: Secretes numerous hormones involved in coordinating system functions, glucose metabolism, and appetite	See Chapter 22
Kidneys: Secrete hormones that regulate blood cell production and the rates of calcium and phosphate absorption by the intestinal tract	See Chapter 24
Gonads: Secrete hormones affecting growth, metabolism, and sexual characteristics, and other hormones that coordinate the activities of organs in the reproductive system	See Chapter 26

Module 16.2 Review

a. Describe the structural classification of hormones.

b. Define endocrine system.

c. Name the organs of the endocrine system.

LO 16.2 Explain the classification of hormones, and identify key functions of hormones secreted by organs and tissues of the endocrine system.

Module 16.3

Hormones affect target cells after binding to receptors in the plasma membrane, cytoplasm, or nucleus

To affect a target cell, a hormone must first interact with an appropriate receptor—a protein molecule to which a particular molecule binds strongly. Each cell has receptors for responding to several different hormones, but cells in different tissues have different combinations of receptors. This arrangement is one reason hormones have differential effects on specific tissues. For every cell, the presence or absence of a specific receptor determines the cell's hormonal sensitivities. If a cell has a receptor that can bind a particular hormone, that cell will respond to the hormone's presence. If a cell lacks the proper receptor for that hormone, the hormone will have no effect on that cell. Hormone receptors are located either in the plasma membrane or inside the cell.

Watch
MasteringA&P® ▸ A&P**Flix**
Mechanism of Hormone Action: Second Messenger cAMP

1 The receptors for catecholamines (E, NE, and dopamine), peptide hormones, and eicosanoids are in the plasma membranes of their target cells. However, binding of the hormone with its receptor on a target cell does not produce a direct effect on its intracellular activities. Instead, the hormone, or **first messenger**, uses an intracellular intermediary, or **second messenger**, to exert the hormone's effects in the cell. Thus hormone binding gives rise to a second messenger, which may act as an enzyme activator, inhibitor, or cofactor. Regardless, the net result is a change in the rates of various metabolic reactions. The two most important second messengers are **cyclic AMP (cAMP)**, a derivative of ATP, and calcium ions (Ca^{2+}).

The link between the first messenger and the second messenger generally involves a **G protein**, an enzyme complex coupled to a membrane receptor.

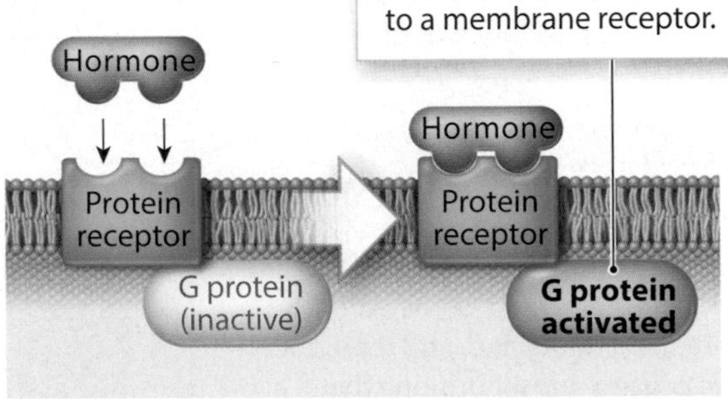

Hormone
Protein receptor
G protein (inactive)
Hormone
Protein receptor
G protein activated

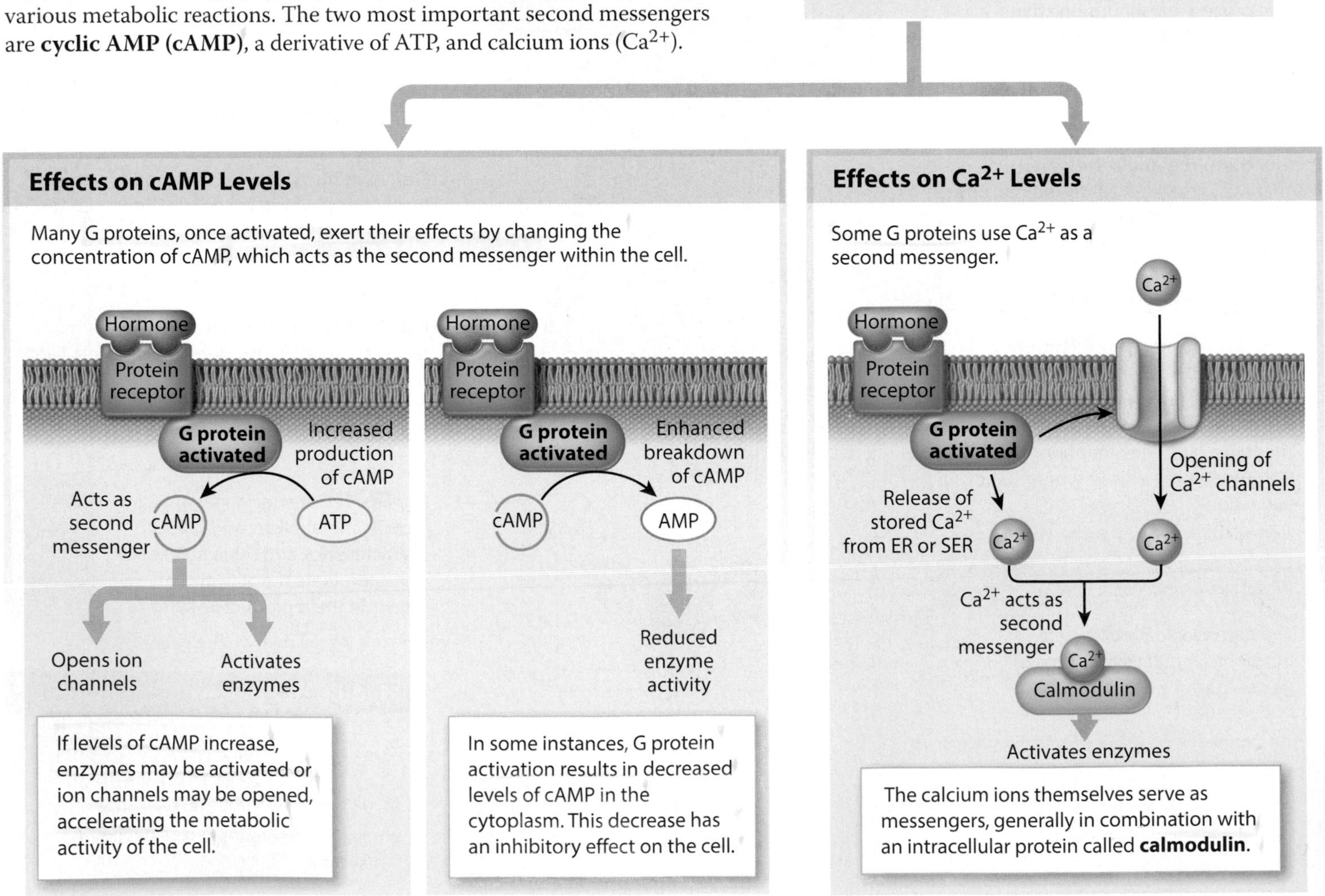

Effects on cAMP Levels
Many G proteins, once activated, exert their effects by changing the concentration of cAMP, which acts as the second messenger within the cell.
Hormone
Protein receptor
G protein activated
Increased production of cAMP
Acts as second messenger
cAMP
ATP
Opens ion channels
Activates enzymes
If levels of cAMP increase, enzymes may be activated or ion channels may be opened, accelerating the metabolic activity of the cell.
Hormone
Protein receptor
G protein activated
Enhanced breakdown of cAMP
cAMP
AMP
Reduced enzyme activity
In some instances, G protein activation results in decreased levels of cAMP in the cytoplasm. This decrease has an inhibitory effect on the cell.
Effects on Ca2+ Levels
Some G proteins use Ca2+ as a second messenger.
Ca2+
Hormone
Protein receptor
G protein activated
Opening of Ca2+ channels
Release of stored Ca2+ from ER or SER
Ca2+
Ca2+
Ca2+ acts as second messenger
Ca2+
Calmodulin
Activates enzymes
The calcium ions themselves serve as messengers, generally in combination with an intracellular protein called calmodulin.

2 Steroid hormones diffuse across the phospholipid bilayer of the plasma membrane and bind to receptors in the cytoplasm or nucleus. The hormone–receptor complexes then alter the activity of specific genes. By this mechanism, steroid hormones can alter the rate of DNA transcription in the nucleus, changing the pattern of protein synthesis. The resulting changes in the synthesis of enzymes or structural proteins directly affect the target cell's metabolic activity and structure. For example, in response to the sex hormone testosterone, skeletal muscle fibers increase production of enzymes and structural proteins, causing increases in muscle size and strength.

Steroid hormone

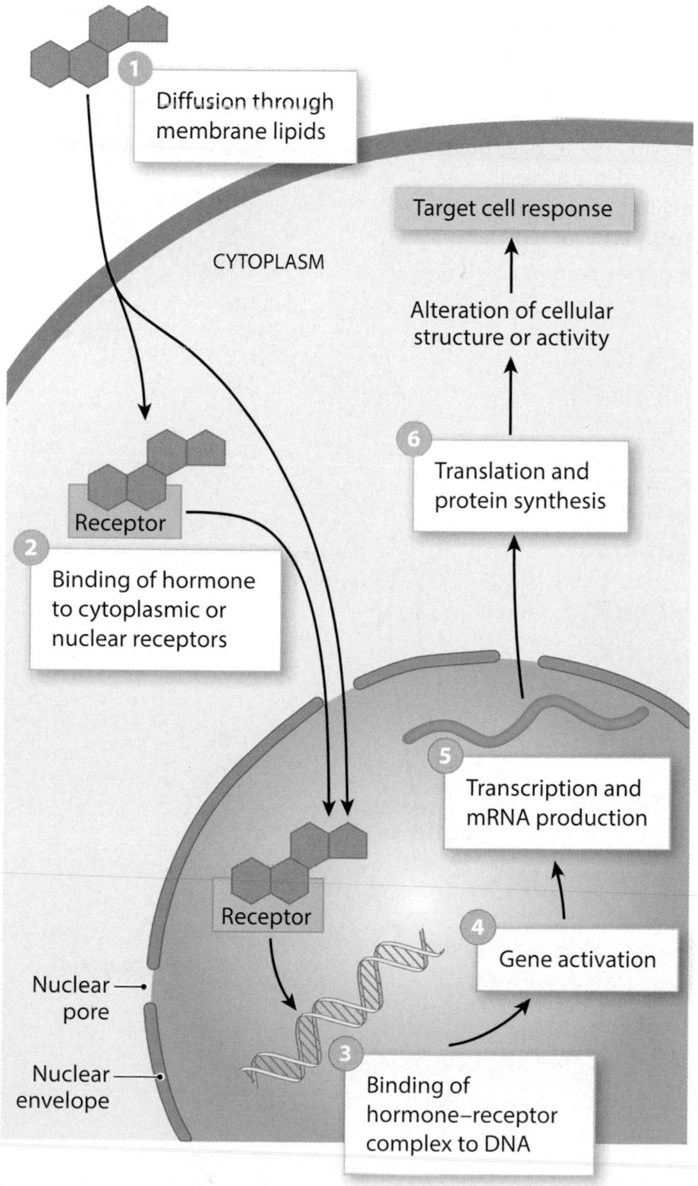

1 Diffusion through membrane lipids
CYTOPLASM
Target cell response
Alteration of cellular structure or activity
6 Translation and protein synthesis
Receptor
2 Binding of hormone to cytoplasmic or nuclear receptors
5 Transcription and mRNA production
Receptor
4 Gene activation
Nuclear pore
Nuclear envelope
3 Binding of hormone–receptor complex to DNA

3 Thyroid hormones are primarily transported across the plasma membrane by carrier-mediated processes. Once in the cells, these hormones bind to receptors on mitochondria and within the nucleus. Thyroid hormones bound to mitochondria increase the rate of ATP synthesis in the mitochondria. Hormone–receptor complexes in the nucleus activate specific genes or change the rate of transcription, which affects the metabolic activities of the cell by increasing or decreasing the concentrations of specific enzymes.

Thyroid hormone

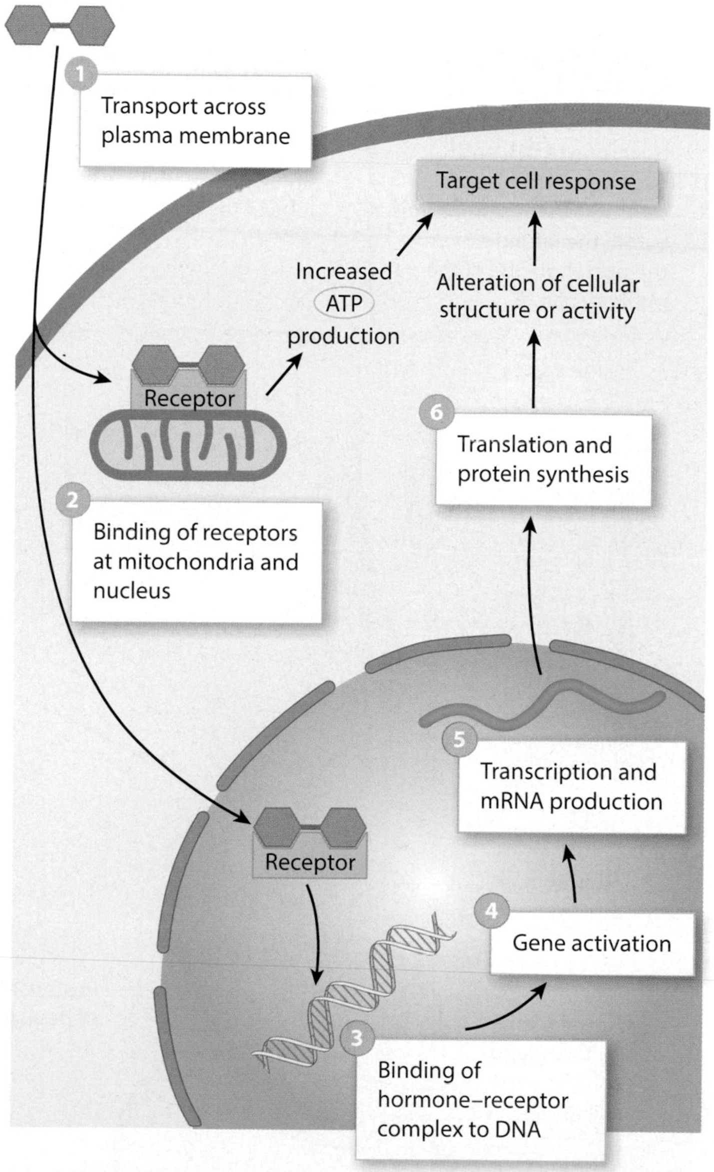

1 Transport across plasma membrane
Target cell response
Increased ATP production
Alteration of cellular structure or activity
Receptor
6 Translation and protein synthesis
2 Binding of receptors at mitochondria and nucleus
5 Transcription and mRNA production
Receptor
4 Gene activation
3 Binding of hormone–receptor complex to DNA

Module 16.3 Review

a. Define hormone receptor.

b. Differentiate between a first messenger and a second messenger.

c. Which type of hormone diffuses across the plasma membrane and binds to receptors in the cytoplasm?

LO 16.3 Explain the general mechanisms of hormonal action.

Module 16.4

The hypothalamus exerts direct or indirect control over the activities of many different endocrine organs

1 The **hypothalamus** provides the highest level of endocrine control by integrating the activities of the nervous and endocrine systems. The hypothalamus does this through three mechanisms.

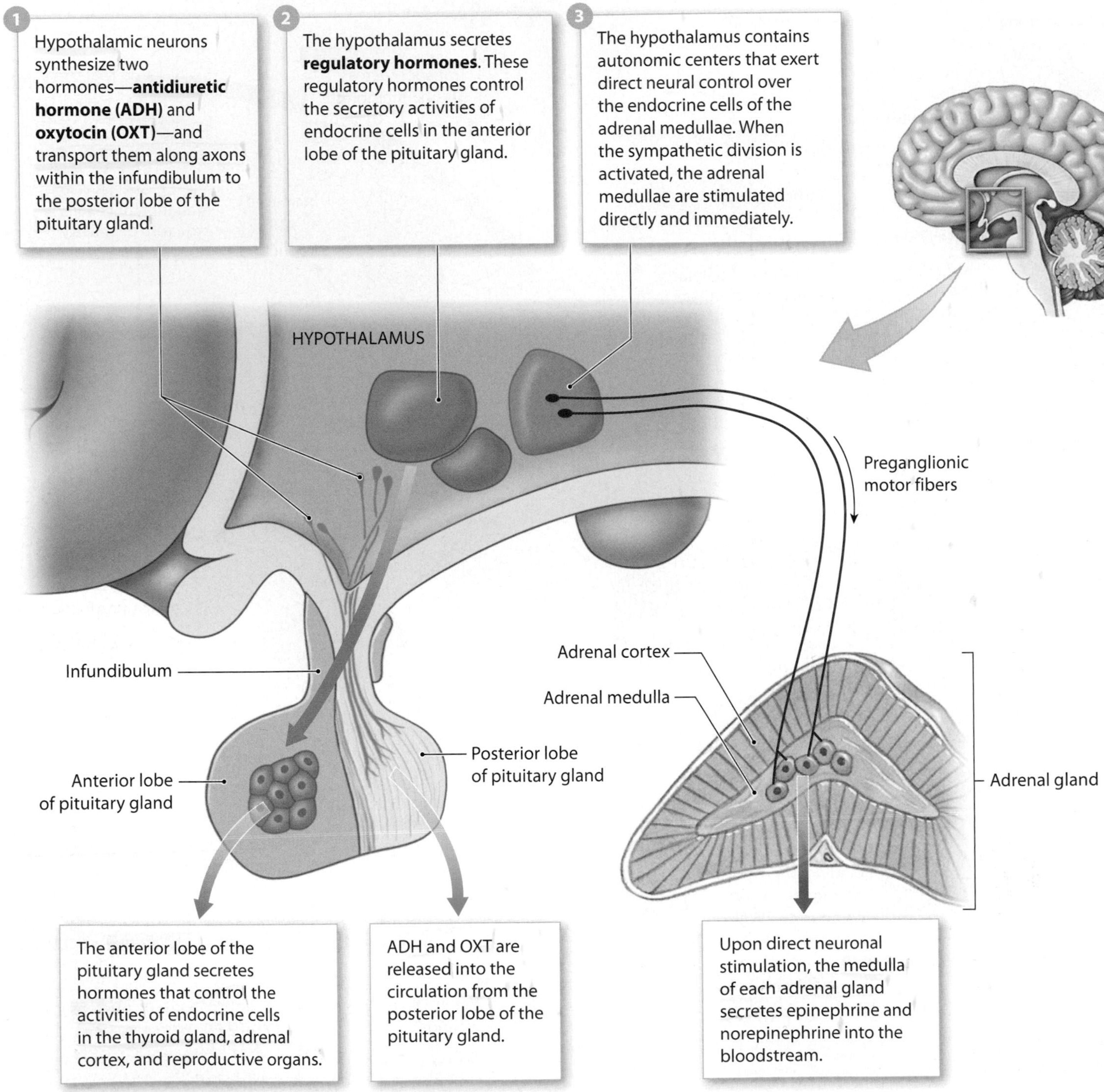

1
Hypothalamic neurons synthesize two hormones—**antidiuretic hormone (ADH)** and **oxytocin (OXT)**—and transport them along axons within the infundibulum to the posterior lobe of the pituitary gland.
2
The hypothalamus secretes **regulatory hormones**. These regulatory hormones control the secretory activities of endocrine cells in the anterior lobe of the pituitary gland.
3
The hypothalamus contains autonomic centers that exert direct neural control over the endocrine cells of the adrenal medullae. When the sympathetic division is activated, the adrenal medullae are stimulated directly and immediately.
HYPOTHALAMUS
Preganglionic motor fibers
Infundibulum
Adrenal cortex
Adrenal medulla
Posterior lobe of pituitary gland
Anterior lobe of pituitary gland
Adrenal gland
The anterior lobe of the pituitary gland secretes hormones that control the activities of endocrine cells in the thyroid gland, adrenal cortex, and reproductive organs.
ADH and OXT are released into the circulation from the posterior lobe of the pituitary gland.
Upon direct neuronal stimulation, the medulla of each adrenal gland secretes epinephrine and norepinephrine into the bloodstream.

2 By secreting specific regulatory hormones, the hypothalamus controls the production of hormones in the anterior lobe of the pituitary gland. At the **median eminence**, a swelling near the attachment of the infundibulum, hypothalamic neurons release regulatory factors into the surrounding interstitial fluids. These secretions enter the bloodstream quite easily, because the endothelial cells lining the capillaries in this region are unusually permeable. These **fenestrated** (FEN-es-trā-ted; *fenestra*, window) **capillaries** allow relatively large molecules to enter or leave the bloodstream.

Hypophyseal Portal System

The capillary networks and the interconnecting vessels make up a portal system. Portal systems are named after their destinations. This particular system is known as the **hypophyseal** (hī-po-FIZ-ē-al) **portal system** (*hypophysis* is the Latin name for the pituitary gland). Portal systems are an efficient means of chemical communication. This one ensures that all the hypothalamic hormones entering the portal vessels will reach their target cells in the anterior lobe before being diluted through mixing with the general circulation. The communication is strictly one way, however. Any chemicals released by the cells "downstream" must do a complete circuit of the cardiovascular system before they reach the capillaries of the portal system.

The capillary networks in the median eminence are supplied by the superior hypophyseal artery. Before leaving the hypothalamus, the capillary networks unite to form a series of larger vessels that spiral around the infundibulum to reach the anterior lobe.

The vessels between the median eminence and the anterior lobe carry blood from one capillary network to another. Blood vessels that link two capillary networks are called **portal vessels**. In this case, they have the histological structure of veins, so they are called portal veins.

Once within the anterior lobe, these vessels form a second capillary network that branches among the endocrine cells.

Neurons of the supraoptic and paraventricular nuclei manufacture ADH and OXT, respectively. These hormones are released by axon terminals at fenestrated capillaries in the posterior lobe of the pituitary gland.

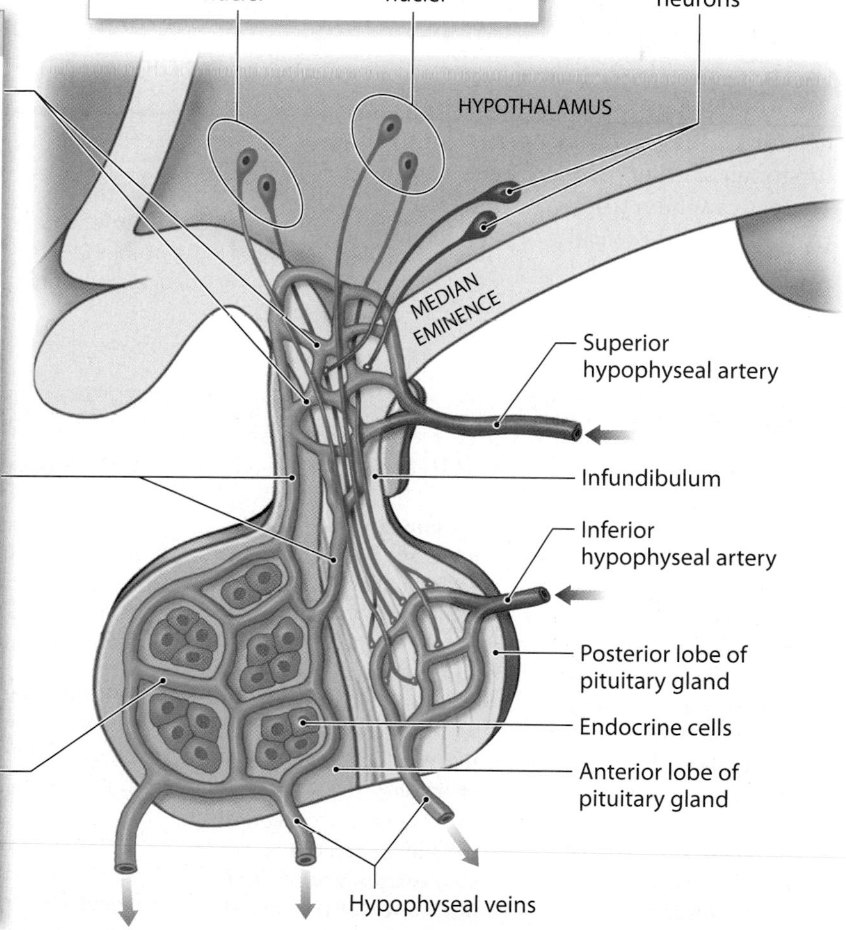

Supraoptic nuclei
Paraventricular nuclei
Neurosecretory neurons
HYPOTHALAMUS
MEDIAN EMINENCE
Superior hypophyseal artery
Infundibulum
Inferior hypophyseal artery
Posterior lobe of pituitary gland
Endocrine cells
Anterior lobe of pituitary gland
Hypophyseal veins

The regulatory hormones secreted by the hypothalamus are transported directly to the anterior lobe by the hypophyseal portal system. Two classes of hypothalamic regulatory hormones exist: releasing hormones and inhibiting hormones. A **releasing hormone (RH)** stimulates the synthesis and secretion of one or more hormones at the anterior lobe. In contrast, an **inhibiting hormone (IH)** prevents the synthesis and secretion of hormones from the anterior lobe. Releasing hormones, inhibiting hormones, or some combination of the two may control an endocrine cell in the anterior lobe.

Module 16.4 Review

a. Define regulatory hormone.
b. Identify the three mechanisms by which the hypothalamus integrates neural and endocrine function.
c. Name and describe the characteristics and functions of the blood vessels that link the hypothalamus with the anterior lobe of the pituitary gland.

LO **16.4** Describe how the hypothalamus controls endocrine organs.

Module 16.5

The pituitary gland consists of an anterior lobe and a posterior lobe

1 The **pituitary gland**, or **hypophysis** (hī-POF-i-sis), is a small, oval gland that lies nestled within the sella turcica, a depression in the sphenoid bone. Nine important peptide hormones are released by the pituitary gland—seven by the anterior lobe and two by the posterior lobe. All nine hormones bind to membrane receptors, and all nine use cAMP as a second messenger. The hormones of the anterior lobe are also called **tropic hormones** (*trope*, a turning), because they "turn on" endocrine glands or support the functions of other organs. (Some sources call them trophic hormones [*trophe*, nourishment].)

The **infundibulum** (in-fun-DIB-ū-lum; funnel) is a funnel-shaped stalk that connects the pituitary gland to the inferior surface of the hypothalamus.

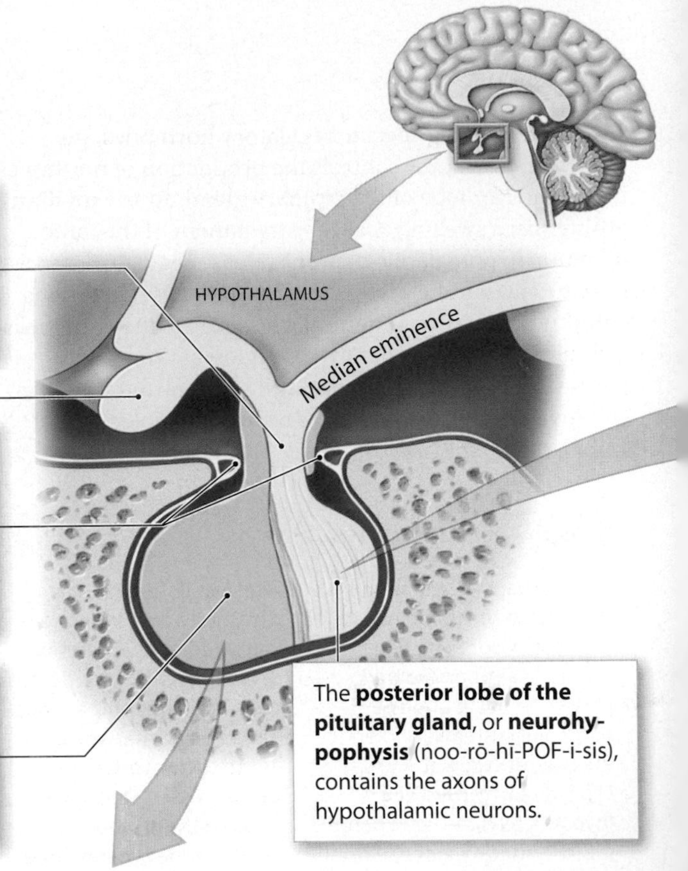

HYPOTHALAMUS
Median eminence
Optic chiasm

A fold of the dura mater encircles the base of the infundibulum, locking the pituitary gland in position and isolating it from the cranial cavity.

The **anterior lobe of the pituitary gland**, or **adenohypophysis** (ad-e-nō-hī-POF-i-sis), contains a variety of endocrine cells.

The **posterior lobe of the pituitary gland**, or **neurohypophysis** (noo-rō-hī-POF-i-sis), contains the axons of hypothalamic neurons.

Hormones of the Anterior Lobe

TSH

Thyroid-stimulating hormone (TSH) targets the thyroid gland, where it triggers the release of thyroid hormones. TSH is released in response to **thyrotropin-releasing hormone (TRH)** from the hypothalamus. As circulating concentrations of thyroid hormones rise, the rate of TRH and TSH production falls—another example of negative feedback.

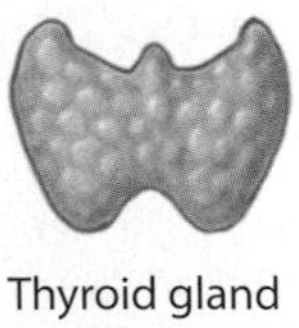

Thyroid gland

ACTH

Adrenocorticotropic hormone (ACTH), also known as corticotropin, stimulates the release of steroid hormones by the adrenal cortex, the outer portion of the adrenal gland. ACTH specifically targets cells that produce hormones that affect glucose metabolism. ACTH release occurs under the stimulation of **corticotropin-releasing hormone (CRH)** from the hypothalamus.

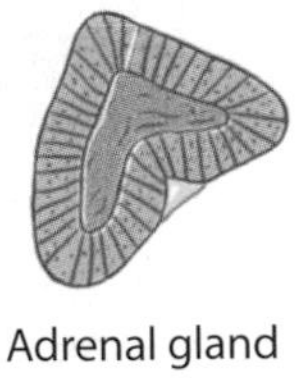

Adrenal gland

Gonadotropins (FSH and LH)

The hormones called **gonadotropins** (gō-nad-ō-TRŌ-pinz) regulate the activities of the gonads. (These organs—the testes in males and ovaries in females—produce reproductive cells as well as hormones.) The production of gonadotropins occurs under stimulation by **gonadotropin-releasing hormone (GnRH)** from the hypothalamus.

Follicle-stimulating hormone (FSH) promotes ovarian follicle development in females and, in combination with luteinizing hormone, stimulates the secretion of **estrogens** (ES-trō-jenz) by ovarian cells. In males, FSH promotes the physical maturation of developing sperm. FSH production is inhibited by **inhibin**, a peptide hormone released by cells in the testes and ovaries.

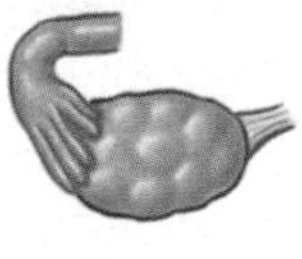

Ovary

Luteinizing (LŪ-tē-in-ī-zing) **hormone (LH)** induces ovulation, the release of reproductive cells in females. It also promotes the secretion, by the ovaries, of estrogens and progesterone, which prepare the body for possible pregnancy. In males, this gonadotropin stimulates the production of sex hormones by the interstitial cells of the testes. These sex hormones are called **androgens** (AN-drō-jenz; *andros*, man); the most important one is testosterone.

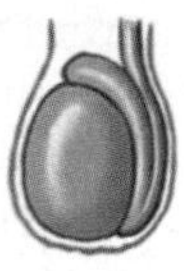

Testis

Hormones of the Posterior Lobe

ADH

Antidiuretic hormone (ADH), also known as **vasopressin (VP)**, is released in response to a variety of stimuli, most notably an increase in the solute concentration in the blood or a decrease in blood volume or blood pressure. An increase in the solute concentration stimulates specialized hypothalamic neurons. Because they respond to a change in the osmotic concentration of body fluids, these neurons are called **osmoreceptors**. The osmoreceptors then stimulate the neurosecretory neurons that release ADH. The primary function of ADH is to act on the kidneys to retain water and decrease urination. With losses minimized, any water absorbed from the digestive tract will be retained, reducing the concentrations of electrolytes in the extracellular fluid. In high concentrations, ADH also causes vasoconstriction, a narrowing of peripheral blood vessels that helps increase blood pressure. Alcohol inhibits ADH release, which explains why people find themselves making frequent trips to the bathroom after consuming alcoholic beverages.

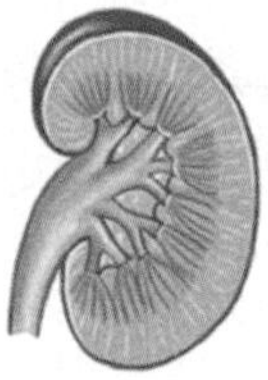

Kidney

OXT

In women, **oxytocin** (*okytokos*, swift birth), or **OXT**, stimulates smooth muscle contraction in the wall of the uterus, promoting labor and delivery. After delivery, OXT promotes the ejection of milk by stimulating the contraction of myoepithelial cells around the secretory alveoli and the ducts of the mammary glands. Although the functions of OXT in sexual activity remain unclear, it is known that circulating concentrations of oxytocin rise during sexual arousal and peak at orgasm in both sexes. Oxytocin release is triggered by sensory input, so it is an example of a **neuroendocrine reflex**.

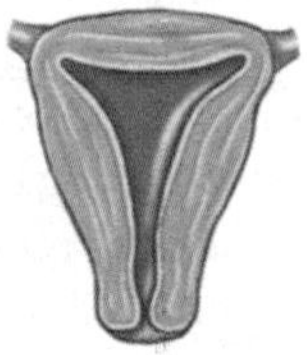

Uterus

GH

Growth hormone (GH) stimulates cell growth and reproduction by accelerating the rate of protein synthesis. Skeletal muscle cells and chondrocytes are particularly sensitive to GH. The production of GH is regulated by **growth hormone–releasing hormone (GH–RH)** and **growth hormone–inhibiting hormone (GH–IH)** from the hypothalamus. The stimulation of growth by GH involves two different mechanisms of control:

Musculo-skeletal system

The primary control mechanism is indirect. Liver cells respond to GH by synthesizing and releasing **somatomedins**, known also as **insulin-like growth factors (IGFs)**. These peptide hormones stimulate tissue growth by binding to receptors on a variety of plasma membranes. In skeletal muscle fibers, cartilage cells, and other target cells, somatomedins increase the uptake of amino acids and their incorporation into new proteins.

The direct actions of GH are more selective:

- In epithelia and connective tissues, GH stimulates stem cell divisions and the differentiation of daughter cells.
- In adipose tissue, GH stimulates the breakdown of stored triglycerides by adipocytes (fat cells), which then release fatty acids into the blood. Many tissues then stop breaking down glucose and use fatty acids instead to generate ATP. This is termed a **glucose-sparing effect**.
- In the liver, GH stimulates the breakdown of glycogen reserves, leading to the release of glucose into the bloodstream.

PRL

Prolactin (*pro-*, before + *lac*, milk) **(PRL)** works with other hormones to stimulate mammary gland development. In pregnancy and during the nursing period that follows delivery, PRL also stimulates milk production by the mammary glands. Prolactin production is inhibited by **prolactin-inhibiting hormone (PIH)** and stimulated by several prolactin-releasing factors.

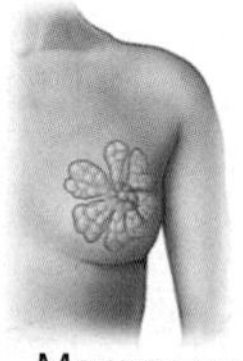

Mammary gland

MSH

The pars intermedia, a narrow portion of the anterior lobe closest to the posterior lobe, may secrete **melanocyte-stimulating hormone (MSH)**. MSH stimulates the melanocytes of the skin to increase their production of melanin. In adults, this portion of the anterior lobe is virtually nonfunctional, and the circulating blood usually does not contain MSH.

Module 16.5 Review

a. Name the two lobes of the pituitary gland.

b. Identify the nine pituitary hormones and their target tissues.

c. In a dehydrated person, how would the amount of ADH released by the posterior lobe of the pituitary change?

LO **16.5** Describe the location and structure of the pituitary gland, and identify pituitary hormones and their functions.

Negative feedback mechanisms control the secretion rates of the hypothalamus and the pituitary gland

1 Hormone secretion is typically controlled through negative feedback; the relationships among the hypothalamus, the pituitary, and the endocrine target organ provide many useful examples. In the common arrangement shown here, the hypothalamus produces a releasing hormone (or factor) that triggers the release of a hormone by the anterior lobe of the pituitary gland. The pituitary hormone stimulates release of a second hormone by the target organ. This second hormone suppresses secretion of both the hypothalamic releasing hormone and the pituitary hormone. The table at right lists major hormones whose secretion is controlled in this way.

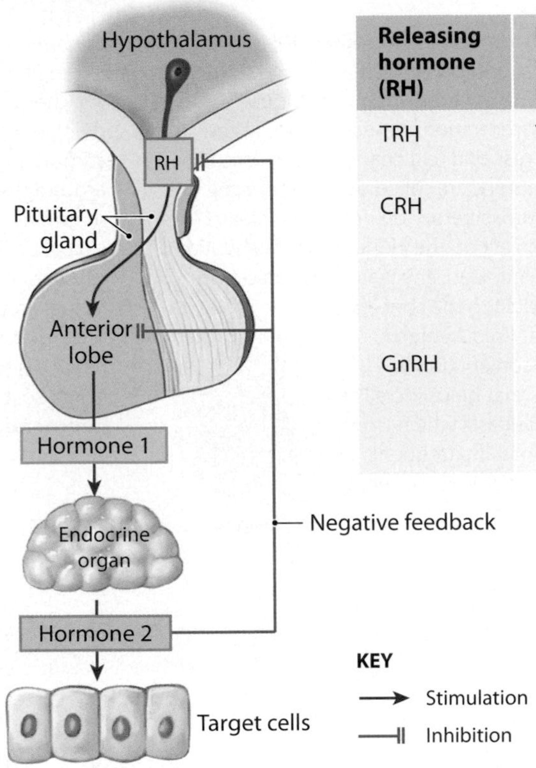
Hypothalamus
RH
Pituitary gland
Anterior lobe
Hormone 1
Endocrine organ
Hormone 2
Target cells
Negative feedback
KEY
Stimulation
Inhibition

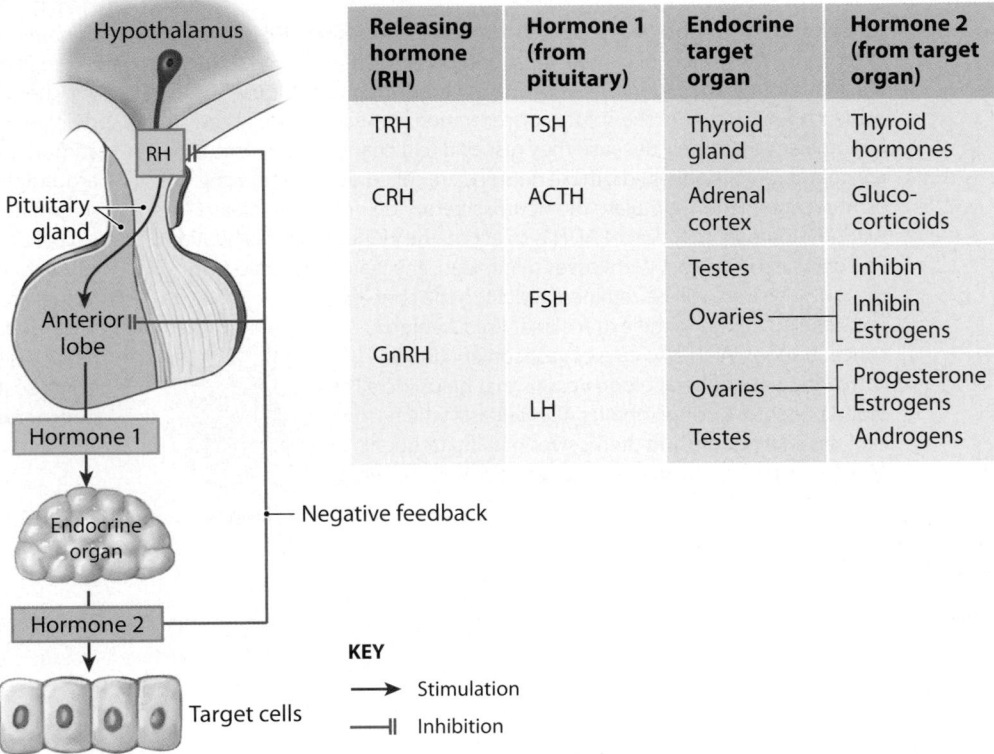

Releasing hormone (RH)	Hormone 1 (from pituitary)	Endocrine target organ	Hormone 2 (from target organ)
TRH	TSH	Thyroid gland	Thyroid hormones
CRH	ACTH	Adrenal cortex	Gluco-corticoids
GnRH	FSH	Testes	Inhibin
		Ovaries	Inhibin, Estrogens
	LH	Ovaries	Progesterone, Estrogens
		Testes	Androgens

2 Growth hormone secretion involves both releasing and inhibiting hormones. The hypothalamic neurons releasing these hormones are sensitive to levels of **somatomedins**, which function like Hormone 2 in the general pattern shown above. However, the somatomedins not only suppress secretion of GH–RH but stimulate secretion of GH–IH. This combination provides more rapid and precise regulation of GH levels.

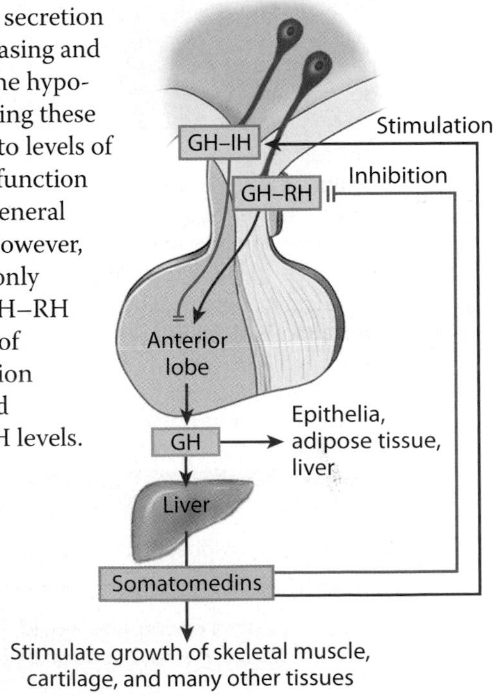
GH–IH
GH–RH
Stimulation
Inhibition
Anterior lobe
GH
Epithelia, adipose tissue, liver
Liver
Somatomedins
Stimulate growth of skeletal muscle, cartilage, and many other tissues

3 Prolactin secretion is also controlled by a pair of regulatory hormones. Secretion of prolactin-releasing factor (PRF) is inhibited by PRL, whereas secretion of prolactin-inhibiting hormone (PIH) is stimulated by PRL.

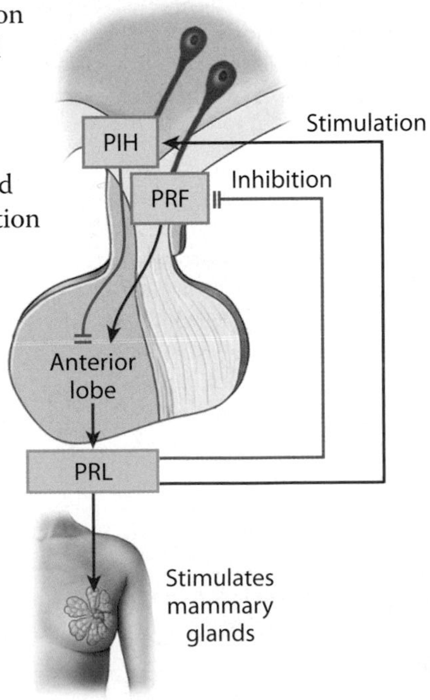
PIH
PRF
Stimulation
Inhibition
Anterior lobe
PRL
Stimulates mammary glands

4 This figure gives an overview of the hormonal relationships we have considered in Modules 16.3 through 16.5.

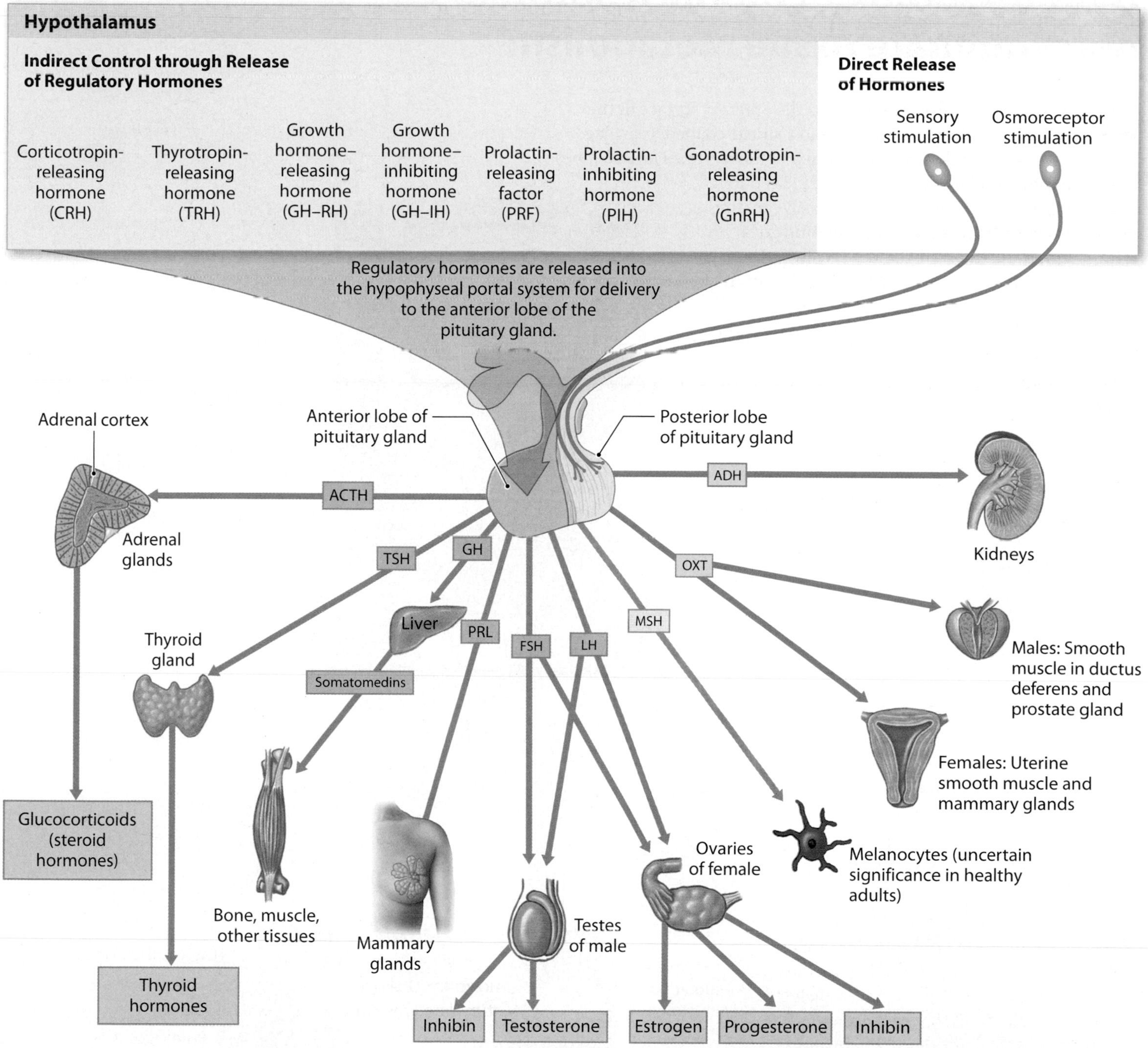
Hypothalamus
Indirect Control through Release of Regulatory Hormones
Corticotropin-releasing hormone (CRH)
Thyrotropin-releasing hormone (TRH)
Growth hormone–releasing hormone (GH–RH)
Growth hormone–inhibiting hormone (GH–IH)
Prolactin-releasing factor (PRF)
Prolactin-inhibiting hormone (PIH)
Gonadotropin-releasing hormone (GnRH)
Direct Release of Hormones
Sensory stimulation
Osmoreceptor stimulation
Regulatory hormones are released into the hypophyseal portal system for delivery to the anterior lobe of the pituitary gland.
Adrenal cortex
Anterior lobe of pituitary gland
Posterior lobe of pituitary gland
ACTH
ADH
Adrenal glands
Kidneys
TSH
GH
OXT
Liver
PRL
FSH
LH
MSH
Thyroid gland
Somatomedins
Males: Smooth muscle in ductus deferens and prostate gland
Females: Uterine smooth muscle and mammary glands
Glucocorticoids (steroid hormones)
Ovaries of female
Melanocytes (uncertain significance in healthy adults)
Bone, muscle, other tissues
Testes of male
Mammary glands
Thyroid hormones
Inhibin
Testosterone
Estrogen
Progesterone
Inhibin

Module 16.6 Review

a. List the hypothalamic releasing hormones.

b. The release of which pituitary hormone would lead to an increased level of somatomedins in the blood?

c. What effects would increased circulating levels of glucocorticoids have on the pituitary secretion of ACTH?

LO 16.6 Describe the role of negative feedback in the functional relationship between the hypothalamus and the pituitary gland.

The thyroid gland contains follicles and requires iodine to produce hormones that stimulate tissue metabolism

1 The **thyroid gland** curves across the anterior surface of the trachea just inferior to the thyroid ("shield-shaped") cartilage, which forms most of the anterior surface of the larynx. The two lobes of the thyroid gland are united by a slender connection, the **isthmus** (IS-mus). You can easily feel the gland with your fingers. When something goes wrong with it, the thyroid gland typically becomes visible as it enlarges and distorts the surface of the neck. The size of the gland varies, depending on heredity and environmental and nutritional factors, but its average weight is about 34 g (1.2 oz). An extensive blood supply gives the thyroid gland a deep red color.

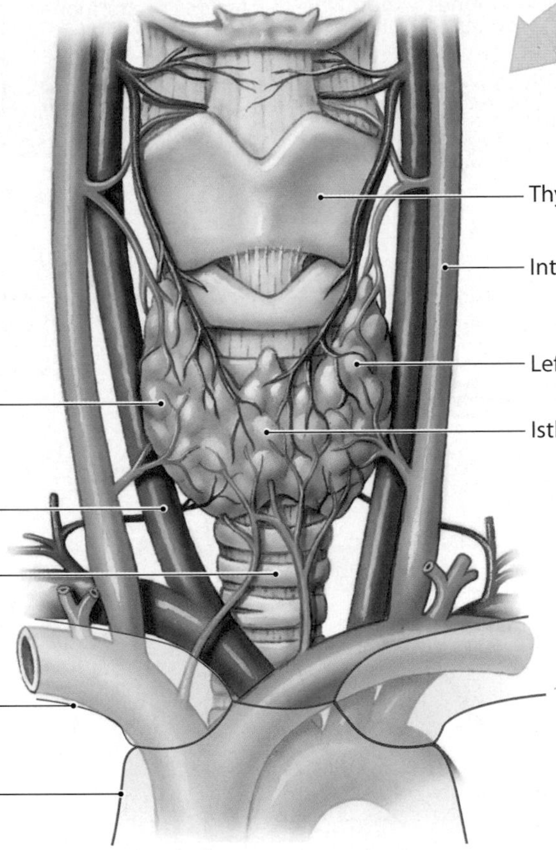
Thyroid cartilage
Internal jugular vein
Left lobe of thyroid gland
Right lobe of thyroid gland
Isthmus of thyroid gland
Common carotid artery
Trachea
Outline of clavicle
Outline of sternum

2 The thyroid gland contains large numbers of **thyroid follicles**, hollow spheres lined by a simple cuboidal epithelium. The follicle cells surround a follicle cavity that holds a viscous colloid, a fluid containing large quantities of dissolved proteins. A network of capillaries surrounds each follicle, delivering nutrients and regulatory hormones to the glandular cells and accepting their secretory products and metabolic wastes. The follicle cells synthesize a globular protein called **thyroglobulin** (thī-rō-GLOB-ū-lin) and secrete it into the colloid of the thyroid follicles. Thyroglobulin molecules contain the amino acid tyrosine, the building block of thyroid hormones.

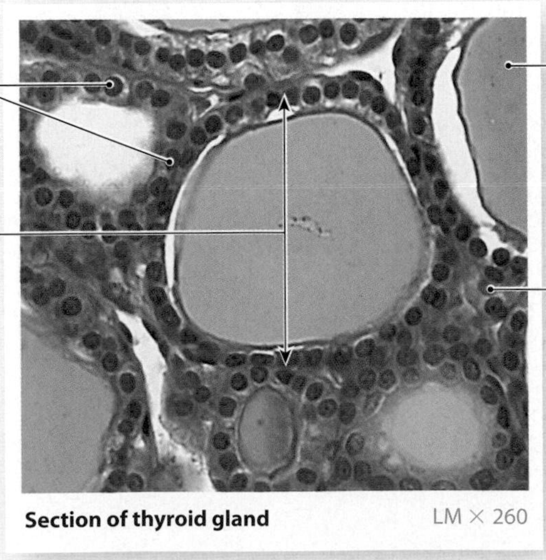
Simple cuboidal epithelium of follicle
Thyroid follicle
Thyroglobulin in colloid

Section of thyroid gland LM × 260

A second population of endocrine cells lies sandwiched between the basement membrane of the follicle cells. These large, pale cells—called **C (clear) cells**—produce the hormone **calcitonin (CT)**, which helps to regulate Ca^{2+} concentrations in body fluids.

3 This figure shows the continuous process by which thyroid hormones are produced and then stored within thyroglobulin in thyroid follicles. Thyroid hormones are released into and removed from the circulation as needed.

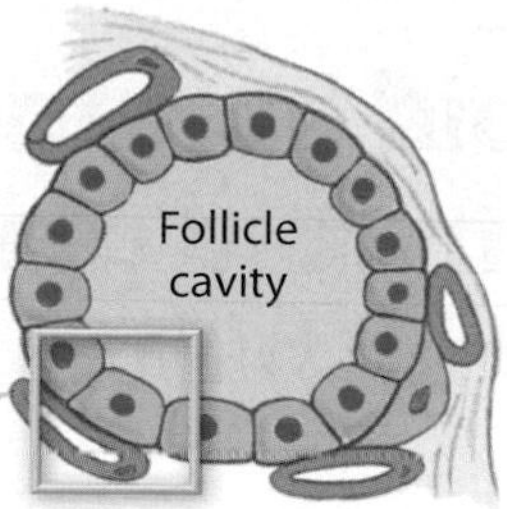

Follicle cavity

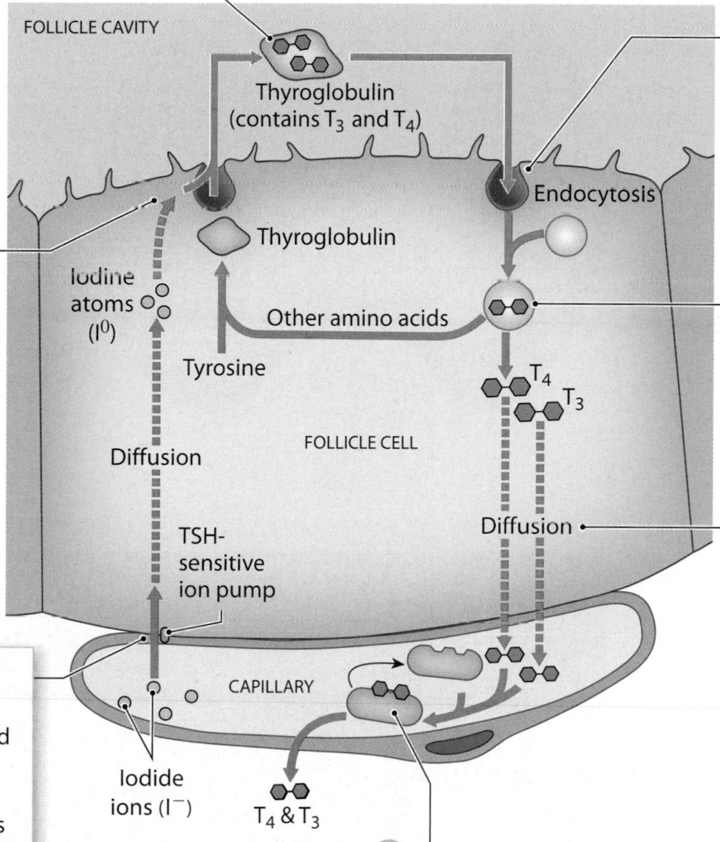

FOLLICLE CAVITY
Thyroglobulin (contains T_3 and T_4)
Endocytosis
Thyroglobulin
Iodine atoms (I^0)
Other amino acids
Tyrosine
T_4
T_3
FOLLICLE CELL
Diffusion
Diffusion
TSH-sensitive ion pump
CAPILLARY
Iodide ions (I^-)
T_4 & T_3

Start

1 Iodide ions are absorbed from the diet and are delivered to the thyroid gland by the bloodstream. Carrier proteins in the basement membrane of the follicle cells actively transport iodide ions (I^-) into the cytoplasm.

2 The iodide ions diffuse to the apical surface of each follicle cell, where they lose an electron and are converted to an iodine atom (I^0) by the enzyme thyroid peroxidase. This reaction also attaches one or two iodine atoms to the tyrosine portions of a thyroglobulin molecule within the follicle lumen.

3 The hormone **thyroxine** (thī-ROK-sēn), or $\mathbf{T_4}$, contains four iodine atoms. A related molecule called **triiodothyronine (T_3)** contains three iodine atoms. Eventually, each molecule of thyroglobulin contains four to eight molecules of T_3, T_4, or both.

4 Follicle cells remove thyroglobulin from the follicle cavity by endocytosis.

5 Lysosomal enzymes break the thyroglobulin down, and the released amino acids and thyroid hormones enter the cytoplasm. The amino acids are then recycled and used to synthesize more thyroglobulin.

6 The released molecules of T_3 and T_4 diffuse across the basement membrane and enter the bloodstream. About 90 percent of all thyroid secretions is T_4. T_3 is secreted in comparatively small amounts although its metabolic effects are much stronger than those of T_4.

7 Roughly 75 percent of the T_4 molecules and 70 percent of the T_3 molecules entering the bloodstream become attached to transport proteins called **thyroid-binding globulins (TBGs)**. The transport proteins release thyroid hormones only gradually, and the bound thyroid hormones represent a substantial reserve: The bloodstream normally contains more than a week's supply of thyroid hormones.

Effects of Thyroid Hormones on Peripheral Tissues

- Increased rates of oxygen consumption and energy consumption; in children, may cause a rise in body temperature
- Increased heart rate and force of contraction; generally results in a rise in blood pressure
- Increased sensitivity to sympathetic stimulation
- Maintenance of normal sensitivity of respiratory centers to changes in oxygen and carbon dioxide concentrations
- Stimulation of red blood cell formation and thus enhanced oxygen delivery
- Stimulation of activity in other endocrine tissues
- Accelerated turnover of minerals in bone

Module 16.7 Review

a. Name the hormones of the thyroid gland.

b. What thyroid hormone aids in calcium regulation?

c. Why do signs and symptoms of decreased thyroxine concentrations not appear until about a week after a thyroidectomy (surgical removal of the thyroid gland)?

LO 16.7 Describe the location and structure of the thyroid gland, identify the hormones it produces, and specify the functions of those hormones.

Parathyroid hormone, produced by the parathyroid glands, is the primary regulator of blood calcium ion levels

1 Two pairs of **parathyroid glands** are embedded in the posterior surface of the thyroid gland. Altogether, the four parathyroid glands weigh a mere 1.6 g (0.06 oz).

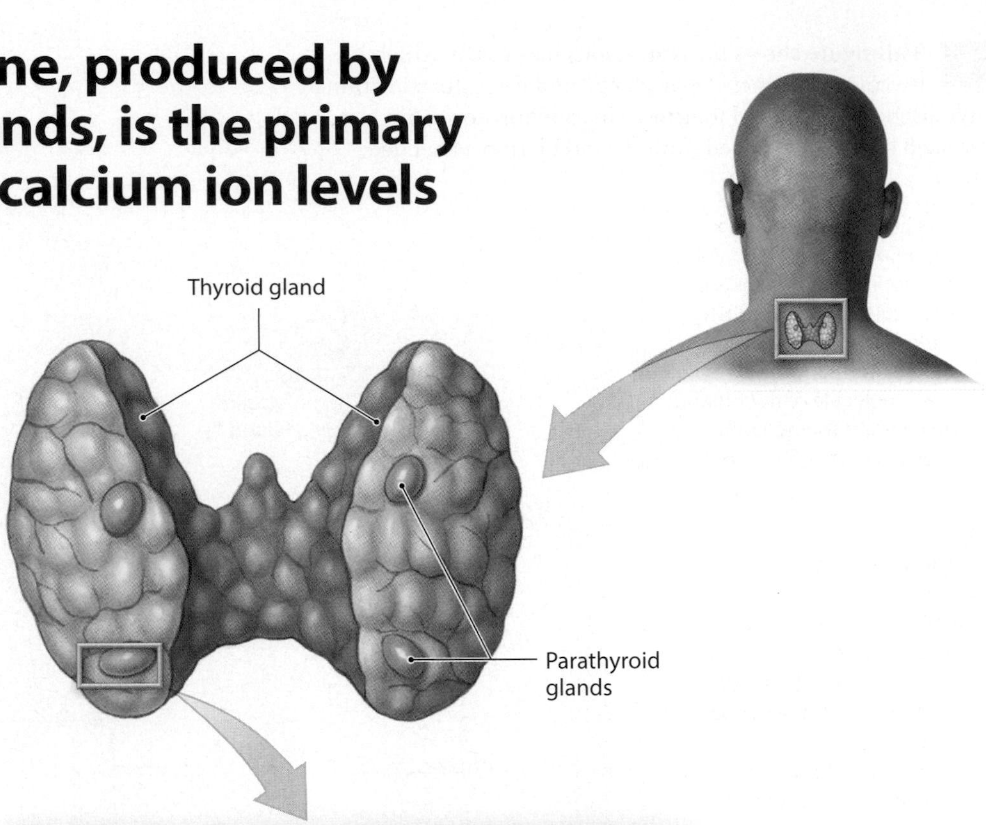
Thyroid gland
Parathyroid glands

2 The parathyroid glands have at least two cell populations: parathyroid cells and oxyphil cells. **Parathyroid** (chief) **cells** produce parathyroid hormone. The functions of the oxyphil cells are unknown. Like the C cells of the thyroid gland, the parathyroid cells monitor the circulating concentration of calcium ions. When the Ca^{2+} concentration of the blood falls below normal, the parathyroid cells secrete **parathyroid hormone (PTH)**. The net result of PTH secretion is an increase in Ca^{2+} concentration in body fluids.

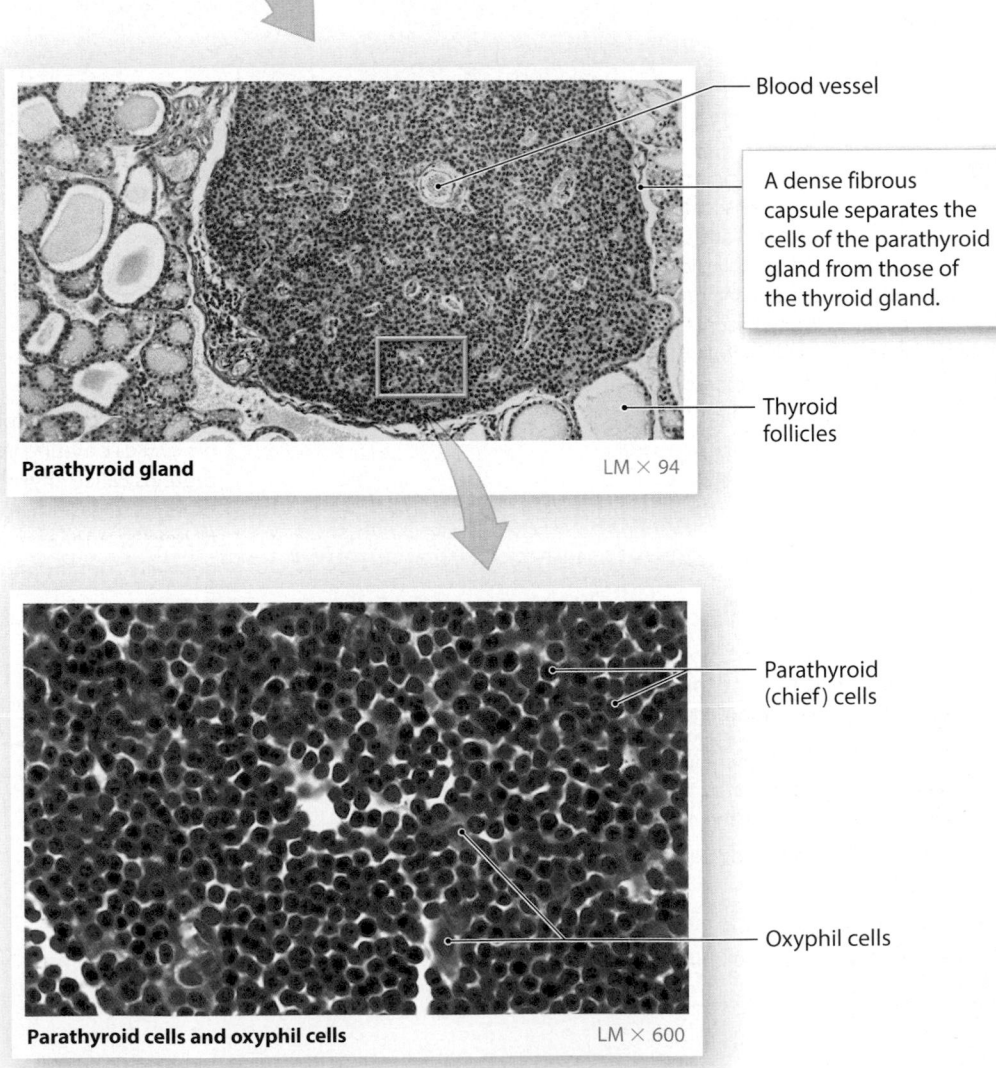

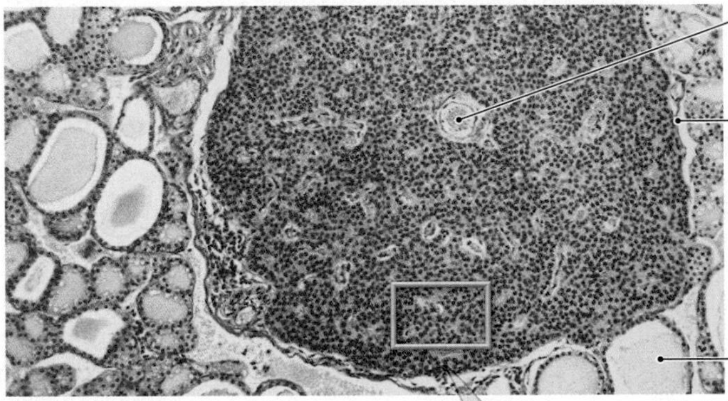
Blood vessel
A dense fibrous capsule separates the cells of the parathyroid gland from those of the thyroid gland.
Thyroid follicles

Parathyroid gland LM × 94

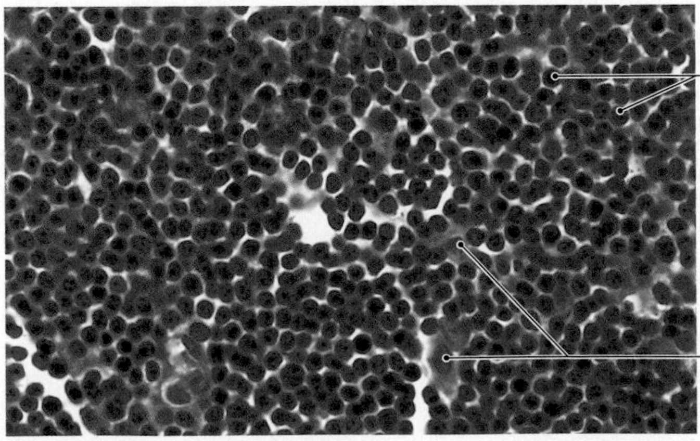
Parathyroid (chief) cells
Oxyphil cells

Parathyroid cells and oxyphil cells LM × 600

3 Parathyroid hormone and calcitonin (from the thyroid) have opposing effects on levels of calcium ions in body fluids. However, in healthy adults PTH, aided by calcitriol secreted by the kidneys, is the primary regulator of circulating calcium ion concentrations. Removal of the thyroid gland seldom affects calcium ion homeostasis because dietary intake and metabolic demand are so closely balanced that increased blood calcium levels are very rare. However, calcitonin can be administered clinically to treat several metabolic disorders that cause increased calcium levels and excessive bone formation.

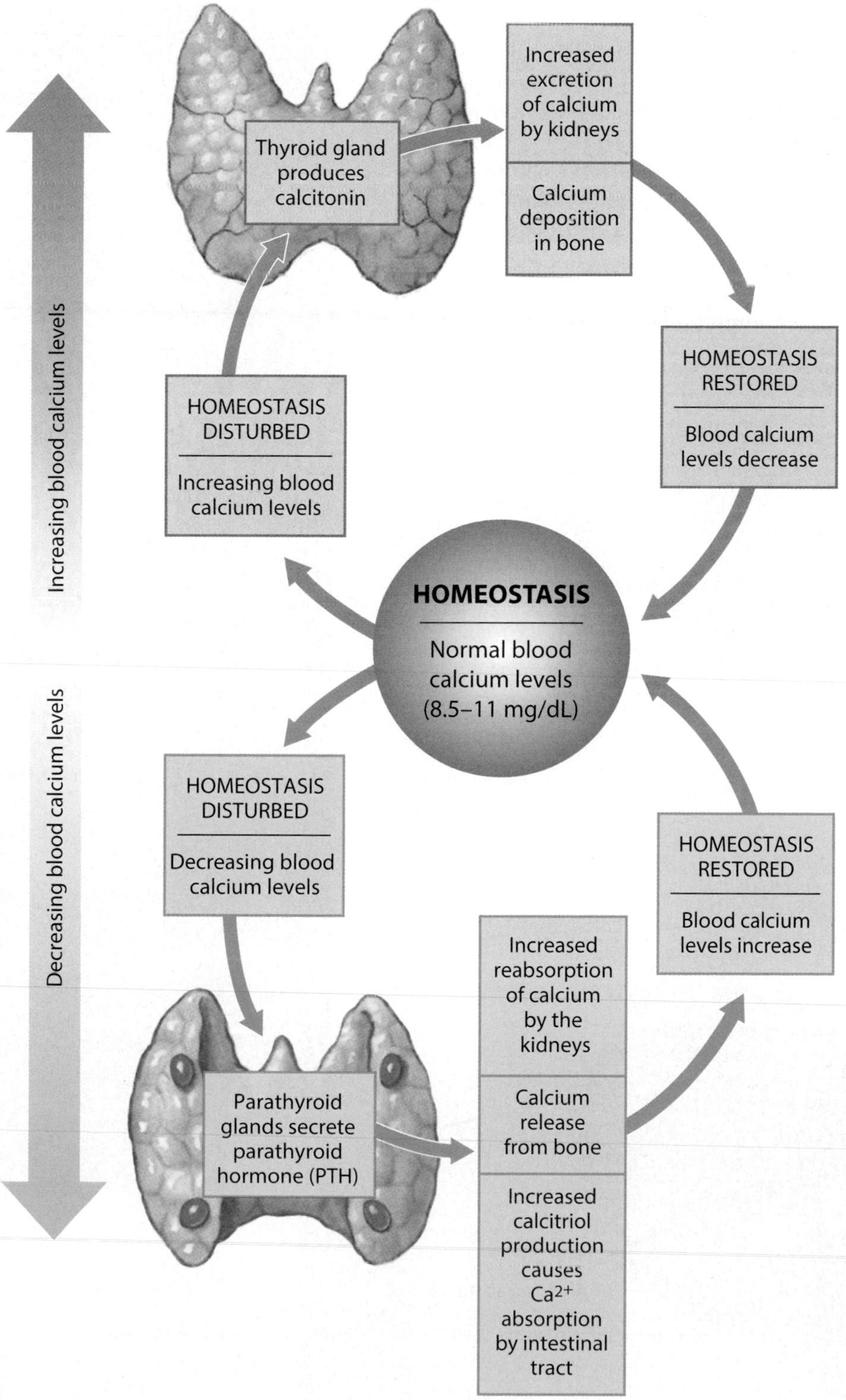

Increasing blood calcium levels
Thyroid gland produces calcitonin
Increased excretion of calcium by kidneys
Calcium deposition in bone
HOMEOSTASIS RESTORED
Blood calcium levels decrease
HOMEOSTASIS DISTURBED
Increasing blood calcium levels
HOMEOSTASIS
Normal blood calcium levels (8.5–11 mg/dL)
Decreasing blood calcium levels
HOMEOSTASIS DISTURBED
Decreasing blood calcium levels
HOMEOSTASIS RESTORED
Blood calcium levels increase
Increased reabsorption of calcium by the kidneys
Calcium release from bone
Increased calcitriol production causes Ca2+ absorption by intestinal tract
Parathyroid glands secrete parathyroid hormone (PTH)

Effects of Parathyroid Hormone on Peripheral Tissues

- PTH mobilizes calcium from bone by affecting osteoblast and osteoclast activity. PTH stimulates osteoblasts to secrete a growth factor known as RANKL. Osteoclasts have no PTH receptors, but both precursor and mature osteoclasts have RANKL receptors. This growth factor results in an increase in osteoclasts and osteoclast activity. With more osteoclasts, the rates of mineral turnover and Ca^{2+} release accelerate. As bone matrix erodes, blood Ca^{2+} rises.
- PTH enhances the reabsorption of Ca^{2+} by the kidneys, reducing urinary losses.
- PTH stimulates the formation and secretion of calcitriol by the kidneys. In general, the effects of calcitriol complement or enhance those of PTH, but calcitriol also enhances Ca^{2+} and PO_4^{3-} absorption by the instestinal tract.

Module 16.8 Review

a. Describe the locations of the parathyroid glands.

b. Explain how parathyroid hormone increases blood calcium levels.

c. Decreased blood calcium levels would result in increased secretion of which hormone?

LO 16.8 Describe the location of the parathyroid glands, and identify the functions of the hormone they produce.

The adrenal glands produce hormones involved in metabolic regulation

1 A yellow, pyramid-shaped **adrenal gland**, or suprarenal (sū-pra-RĒ-nal; *supra-*, above + *ren*, kidney) gland, sits on the superior border of each kidney. The adrenal glands are retroperitoneal, as are the kidneys, and only their anterior surfaces are covered by a layer of parietal peritoneum. Like other endocrine glands, the adrenal glands are richly supplied with blood vessels.

2 The adrenal cortex is yellowish due to stored lipids, especially cholesterol and various fatty acids. The adrenal cortex produces more than two dozen steroid hormones, collectively called **corticosteroids**. Like other steroid hormones, corticosteroids exert their effects by determining which genes in the nuclei of their target cells are transcribed, and at what rate. The resulting changes in the nature and concentration of enzymes in the cytoplasm affect cellular metabolism. Corticosteroids are vital: If the adrenal glands are destroyed or removed, the person will die unless corticosteroids are administered.

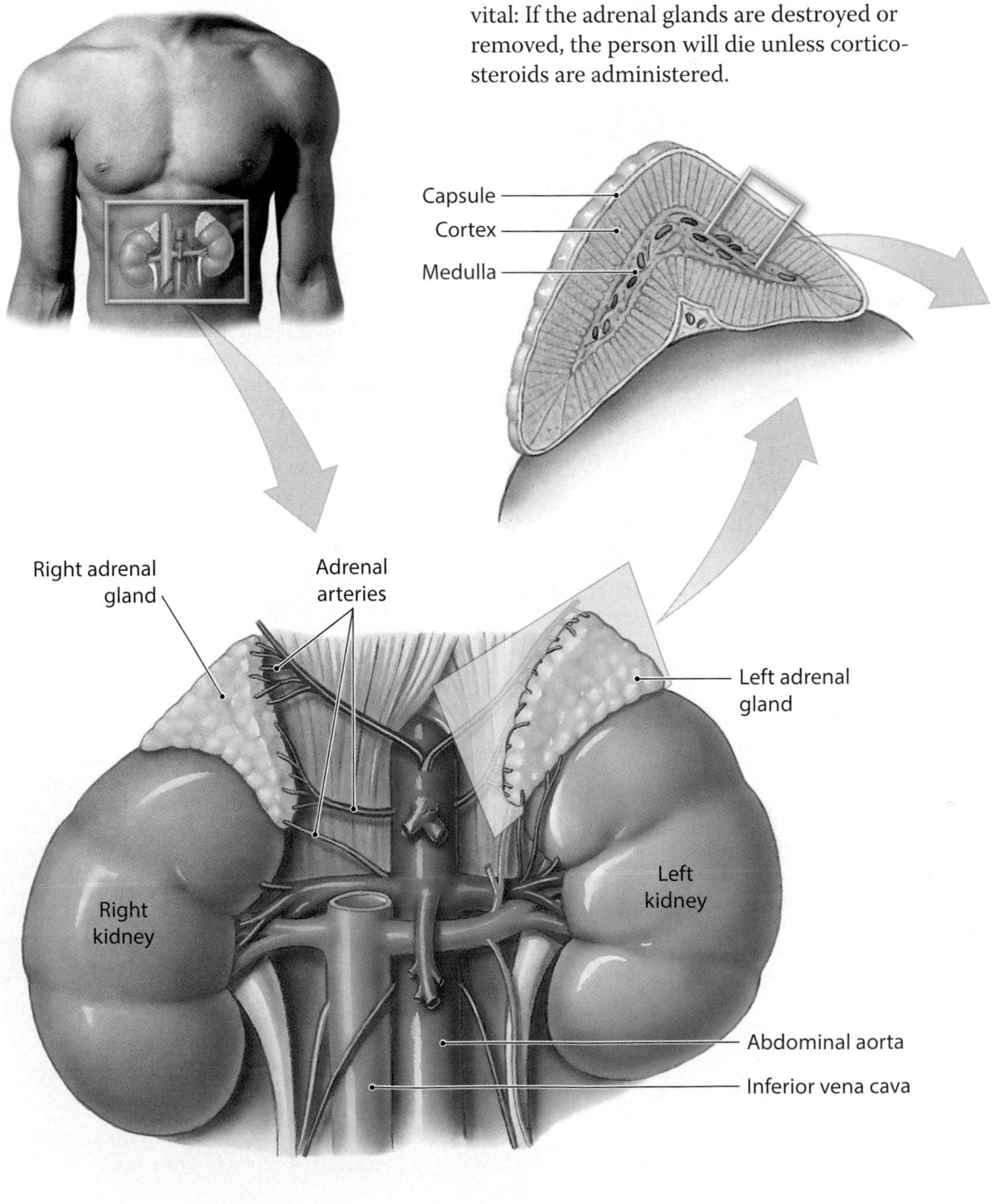
Capsule
Cortex
Medulla
Right adrenal gland
Adrenal arteries
Left adrenal gland
Right kidney
Left kidney
Abdominal aorta
Inferior vena cava

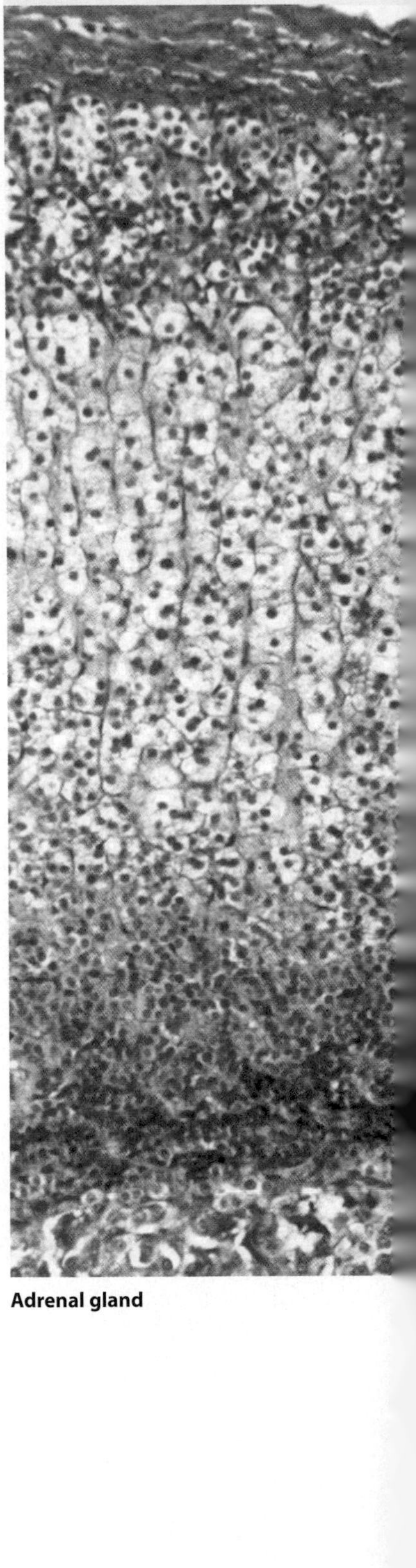

Adrenal gland

3 Deep to the adrenal capsule are three distinct regions, or zones, in the adrenal cortex. Each zone synthesizes specific steroid hormones. Deep to the cortex lies the adrenal medulla, which synthesizes epinephrine and norepinephrine.

The Adrenal Hormones

Region/Zone	Hormones	Primary Targets	Hormonal Effects	Regulatory Control
ADRENAL CAPSULE				
ADRENAL CORTEX				
The **zona glomerulosa** (glō-mer-ū-LŌ-suh) is the outer region of the adrenal cortex.	**Mineralocorticoids**, primarily **aldosterone**	Kidneys	Aldosterone increases renal reabsorption of Na^+ and water, especially in the presence of ADH. It also accelerates urinary loss of K^+.	Mineralocorticoid secretion is stimulated by activating the renin-angiotensin-aldosterone system (Module 16.14) and inhibited by hormones opposing that system.
The **zona fasciculata** (fa-sik-ū-LA-tuh; *fasciculus*, little bundle) is the large, central portion of the adrenal cortex.	**Glucocorticoids** (glū-kō-KOR-ti-koydz) are steroid hormones that affect glucose metabolism. The primary hormones are **cortisol** (KOR-ti-sol), also called hydrocortisone, and smaller amounts of the related steroid **corticosterone** (kor-ti-KOS-te-rōn). The liver converts some of the circulating cortisol to **cortisone**, another metabolically active glucocorticoid.	Most cells	Glucocorticoids increase rates of glucose and glycogen formation by the liver. They also stimulate the release of amino acids from skeletal muscles, and lipids from adipose tissues, and they promote lipid catabolism within peripheral cells. These actions supplement the glucose-sparing effect of growth hormone (noted in Module 16.5). Cortisol also reduces inflammation (an **anti-inflammatory effect**).	Glucocorticoid secretion is stimulated by ACTH from the anterior lobe of the pituitary gland.
The **zona reticularis** (re-tik-ū-LAR-is; *reticulum*, network) forms a narrow band bordering each adrenal medulla.	Small quantities of androgens (male sex hormones) that may be converted to estrogens in the bloodstream	Skin, bones, and other tissues, but minimal effects in normal adults	Adrenal androgens stimulate the development of pubic hair in boys and girls before puberty.	Androgen secretion is stimulated by ACTH.
ADRENAL MEDULLA	Epinephrine (E), norepinephrine (NE)	Most cells	Epinephrine and norepinephrine increase cardiac activity, blood pressure, glycogen breakdown, and blood glucose levels.	Epinephrine and norepinephrine secretion is stimulated by sympathetic preganglionic fibers during sympathetic activation.

Module 16.9 Review

a. Identify the two regions of an adrenal gland, and cite the hormones secreted by each.

b. List the three zones of the adrenal cortex.

c. What effect would increased cortisol levels have on blood glucose levels?

LO **16.9** Describe the location, structure, and functions of the adrenal glands, identify the hormones produced, and specify the functions of each hormone.

Module 16.10

The pancreatic islets secrete insulin and glucagon and regulate glucose use by most cells

1 The **pancreas** is mostly retroperitoneal and lies in the loop formed between the inferior border of the stomach and the proximal portion of the small intestine. It is a slender, pale organ with a nodular (lumpy) texture. In adults, the pancreas is 20–25 cm (8–10 in.) long and weighs about 80 g (2.8 oz). The **exocrine pancreas**, roughly 99 percent of the organ's volume, consists of clusters of gland cells and their attached ducts. Together, the gland and duct cells secrete large quantities of an alkaline, enzyme-rich fluid that reaches the lumen of the intestinal tract through one or two **pancreatic ducts**.

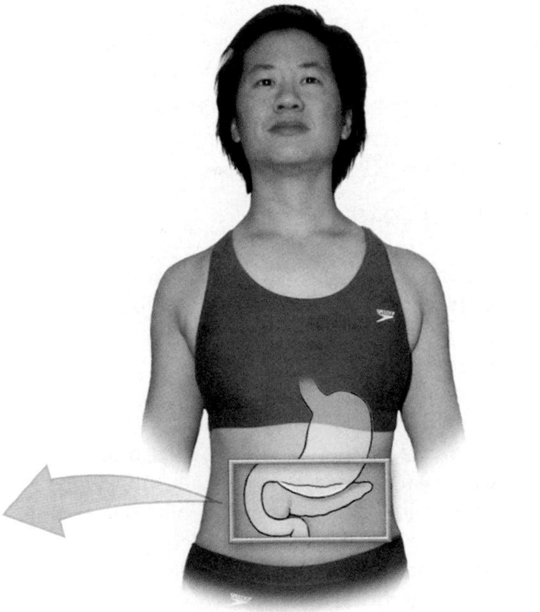

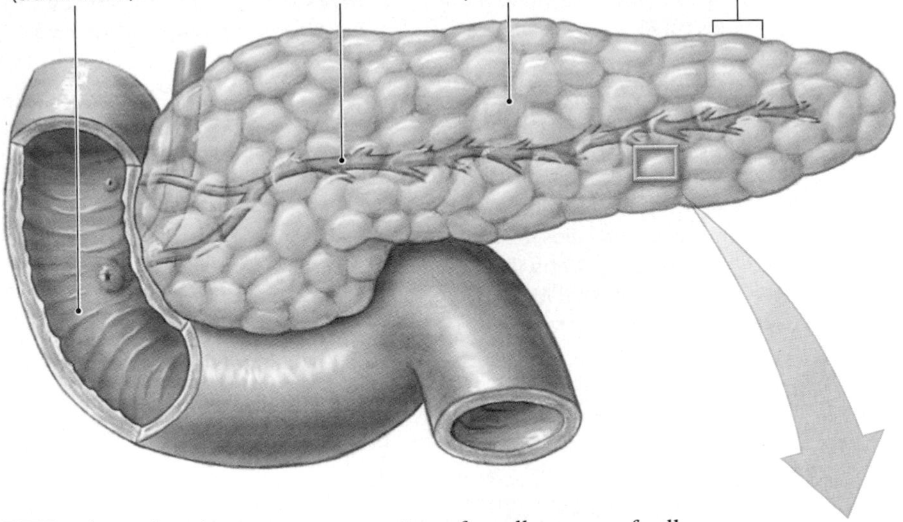

Small intestine (duodenum)
Pancreatic duct
Body of pancreas
Lobule

2 The **endocrine pancreas** consists of small groups of cells scattered among the exocrine cells. The endocrine clusters are known as **pancreatic islets**, or the islets of Langerhans (LAN-ger-hanz). Pancreatic islets account for only about 1 percent of all cells in the pancreas. Nevertheless, a typical pancreas contains about 2 million pancreatic islets, and their secretions are vital to our survival.

The exocrine cells form small clusters that secrete into a lumen continuous with a pancreatic duct. The glandular clusters are called **pancreatic acini** (singular, *acinus*).

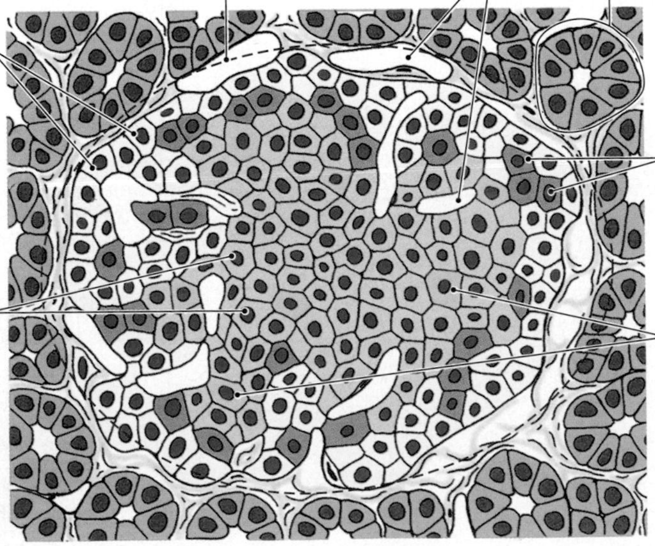

Capillaries
Pancreatic islet

Alpha cells produce the hormone **glucagon** (GLŪ-ka-gon). Glucagon raises blood glucose levels by increasing the rates of glycogen breakdown and glucose release by the liver.

Beta cells produce the hormone **insulin** (IN-suh-lin). Insulin lowers blood glucose levels by increasing the rate of glucose uptake and utilization by cells, and by increasing glycogen synthesis in skeletal muscles and the liver.

Delta cells produce a peptide hormone identical to growth hormone–inhibiting hormone (GH–IH). GH–IH suppresses the release of glucagon and insulin by other islet cells and slows the rates of food absorption and enzyme secretion along the intestinal tract.

F cells produce the hormone **pancreatic polypeptide (PP)**. PP inhibits gallbladder contractions and regulates the production of some pancreatic enzymes. It may also help control the rate of nutrient absorption by the intestinal tract.

3 Insulin and glucagon are the primary hormones responsible for the regulation of blood glucose levels. When blood glucose levels increase, beta cells secrete insulin, which then stimulates the transport of glucose across plasma membranes and into target cells. When blood glucose levels decrease, alpha cells secrete glucagon, which stimulates glycogen breakdown and glucose release by the liver.

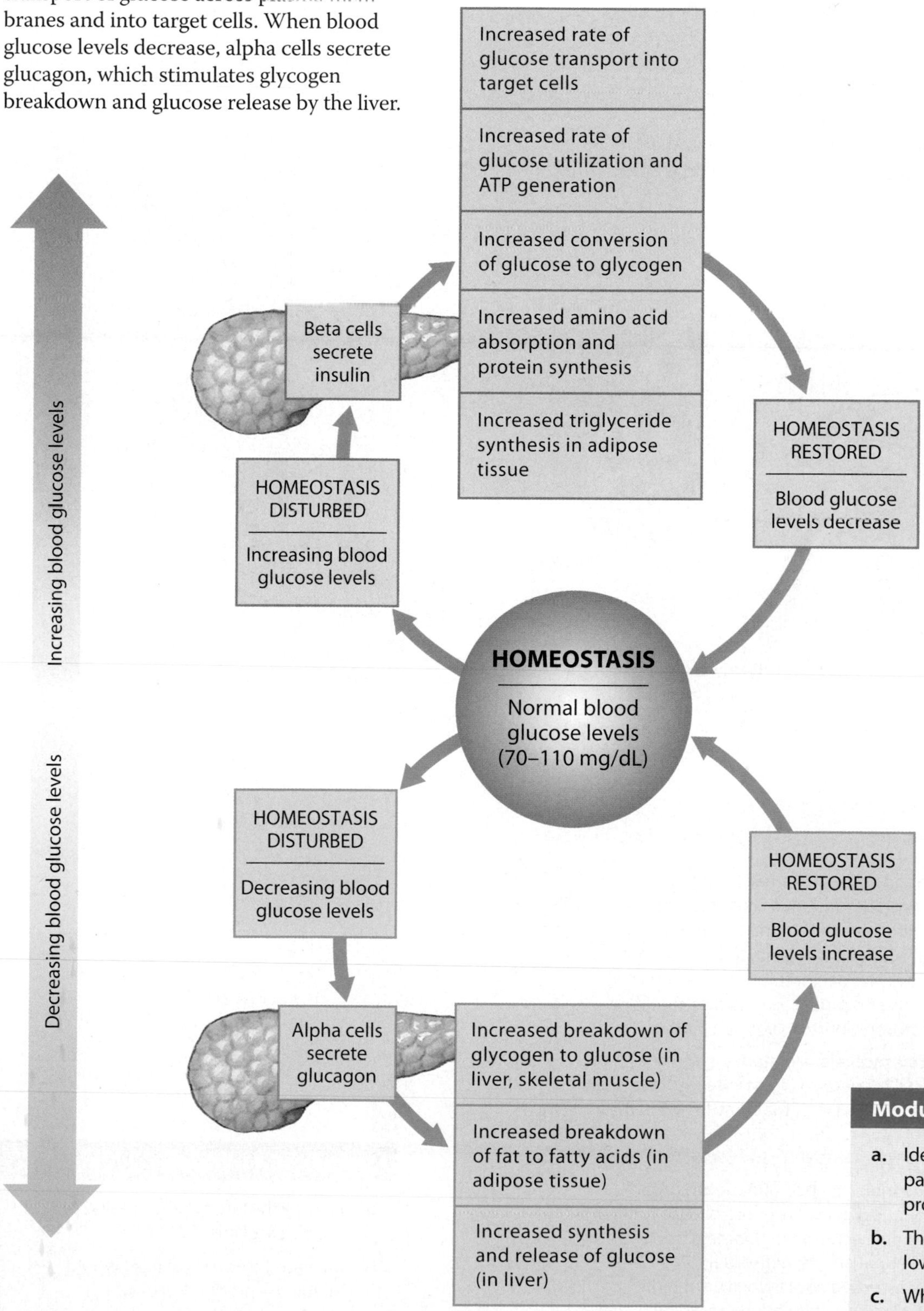

Increasing blood glucose levels
Decreasing blood glucose levels
Increased rate of glucose transport into target cells
Increased rate of glucose utilization and ATP generation
Increased conversion of glucose to glycogen
Increased amino acid absorption and protein synthesis
Increased triglyceride synthesis in adipose tissue
Beta cells secrete insulin
HOMEOSTASIS DISTURBED
Increasing blood glucose levels
HOMEOSTASIS RESTORED
Blood glucose levels decrease
HOMEOSTASIS
Normal blood glucose levels (70–110 mg/dL)
HOMEOSTASIS DISTURBED
Decreasing blood glucose levels
HOMEOSTASIS RESTORED
Blood glucose levels increase
Alpha cells secrete glucagon
Increased breakdown of glycogen to glucose (in liver, skeletal muscle)
Increased breakdown of fat to fatty acids (in adipose tissue)
Increased synthesis and release of glucose (in liver)

Module 16.10 Review

a. Identify the types of cells in the pancreatic islets and the hormones produced by each.

b. The secretion of which hormone lowers blood glucose concentrations?

c. What is the effect of increased glucagon levels on the amount of glycogen stored in the liver?

LO 16.10 Describe the location and structure of the pancreas, identify the hormones it produces, and specify the functions of those hormones.

Module 16.11

The pineal gland of the epithalamus secretes melatonin

1 The **pineal gland**, part of the epithalamus, lies in the posterior portion of the roof of the third ventricle in the brain. The pineal gland contains neurons, neuroglia, and special secretory cells called **pinealocytes** (pin-Ē-al-ō-sīts). These cells synthesize the hormone **melatonin** from molecules of the neurotransmitter serotonin. Collaterals from the visual pathways enter the pineal gland and affect the rate of melatonin production. This rate is lowest during daylight hours and highest at night.

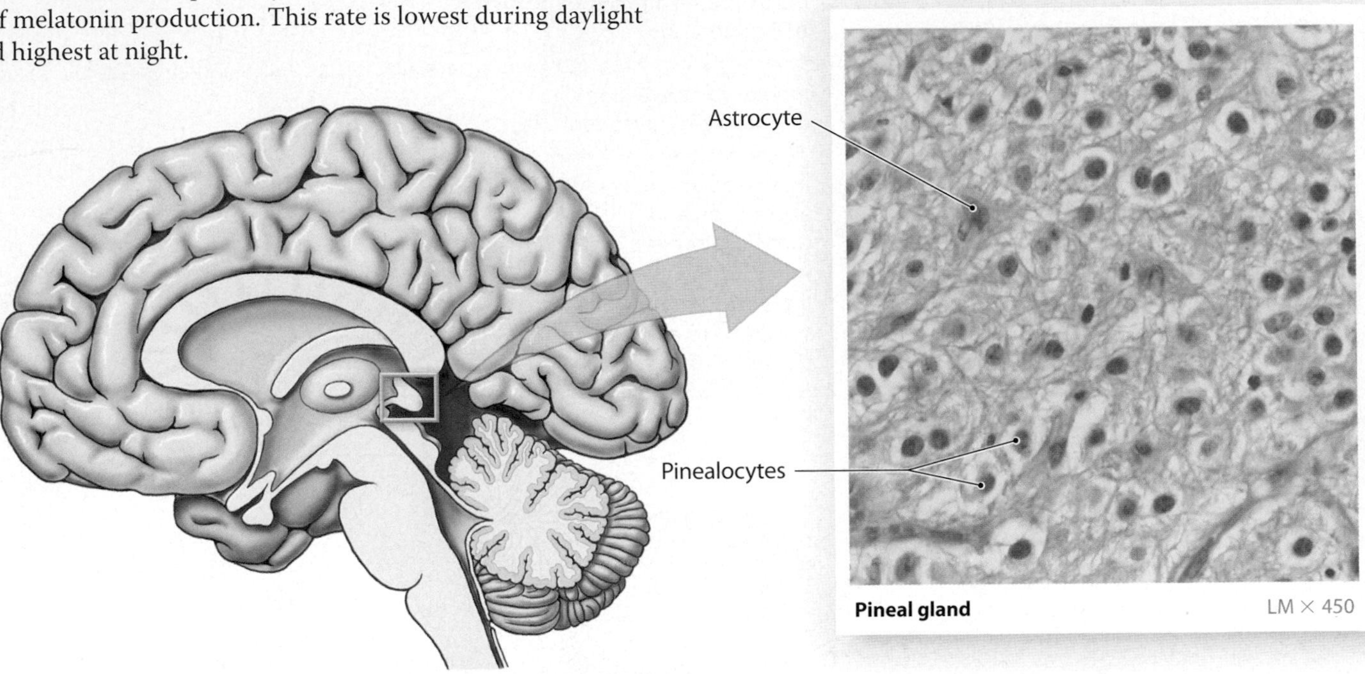
Astrocyte
Pinealocytes

Pineal gland LM × 450

Functions of Melatonin in Humans

- **Inhibiting reproductive functions**. In some mammals, melatonin slows the maturation of sperm, oocytes, and reproductive organs by reducing the rate of GnRH secretion. The significance of this effect in humans remains unclear, but circumstantial evidence suggests that melatonin may play a role in the timing of human sexual maturation. Melatonin levels in the blood decline at puberty, and pineal tumors that eliminate melatonin production cause premature puberty in young children.
- **Protecting against damage by free radicals**. Melatonin is a very effective antioxidant. It may protect CNS neurons from **free radicals**, such as nitric oxide (NO) or hydrogen peroxide (H_2O_2), that may be formed in active neural tissue.
- **Setting circadian rhythms**. Because pineal activity is cyclical, the pineal gland may also be involved with the maintenance of basic **circadian rhythms**—daily changes in physiological processes that follow a regular day/night pattern. Increased melatonin secretion in darkness has been suggested as a primary cause of seasonal affective disorder (SAD). This condition can develop during the winter in people who live at high latitudes, where sunlight is scarce or lacking. It is characterized by changes in mood, eating habits, and sleeping patterns.

Module 16.11 Review

a. Identify the hormone-secreting cells of the pineal gland.

b. Increased amounts of light would inhibit the production of which hormone?

c. List three functions of melatonin.

LO 16.11 Describe the location of the pineal gland, and identify the functions of the hormone that it produces.

Diabetes mellitus is an endocrine disorder characterized by excessively high blood glucose levels

Diabetes Mellitus

Diabetes mellitus (mel-Ī-tus; *mellitum*, honey) is characterized by glucose concentrations that are high enough to overwhelm the reabsorption capabilities of the kidneys. The presence of abnormally high blood glucose levels is called **hyperglycemia** (hī-per-glī-SĒ-mē-ah). In diabetes mellitus, glucose appears in the urine (**glycosuria**; glī-kō-SOO-rē-a), and urine volume generally becomes excessive (**polyuria**). Diabetes mellitus can be caused by genetic abnormalities or mutations that result in inadequate insulin production, the synthesis of abnormal insulin molecules, or the production of defective insulin-receptor proteins.

Type 1 Diabetes

Type 1 diabetes is characterized by inadequate insulin production by the pancreatic beta cells. People with Type 1 diabetes must receive insulin to live—typically multiple injections daily, or continuous infusion through an insulin pump or other device. Type 1 diabetes accounts for only about 5–10 percent of diabetes cases and often develops in children and young adults.

Type 2 Diabetes

Type 2 diabetes is the most common form of diabetes mellitus. Most people with this form of diabetes produce normal amounts of insulin, at least initially, but their tissues do not respond properly—a condition known as insulin resistance. Type 2 diabetes is associated with obesity, and weight loss through diet and exercise can be an effective treatment, especially when coupled with drugs that alter rates of glucose synthesis and release by the liver.

1 Untreated diabetes mellitus disrupts metabolic activities throughout the body. Clinical problems arise because the tissues involved are experiencing an energy crisis—in essence, most of the tissues are responding as they would during chronic starvation, breaking down lipids and even proteins because they are unable to absorb glucose from their surroundings. Problems involving abnormal changes in blood vessel structure are particularly dangerous. An estimated 25.8 million people in the United States have some form of diabetes.

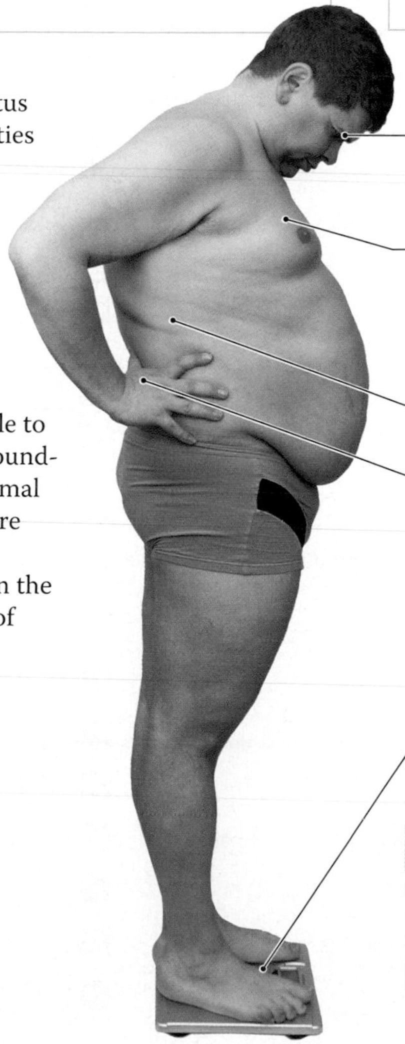

Clinical Problems Caused by Diabetes Mellitus

The proliferation of capillaries and hemorrhaging at the retina may cause partial or complete blindness. This condition is called **diabetic retinopathy**.

Degenerative blockages in cardiac circulation can lead to early heart attacks. For a given age group, heart attacks are three to five times more likely in people with diabetes than in people who do not have the condition.

Degenerative changes in the kidneys, a condition called **diabetic nephropathy**, can lead to kidney failure.

Abnormal blood flow to neural tissues is probably responsible for a variety of problems with peripheral nerves, including abnormal autonomic function. As a group, these disorders are termed **diabetic neuropathy**.

Blood flow to the distal portions of the limbs is reduced, and peripheral tissues may be damaged as a result. A reduction in blood flow to the feet, for example, can lead to tissue death, ulceration, infection, and the loss of toes or a major portion of one or both feet.

Module 16.12 Review

a. Define diabetes mellitus.
b. Identify and describe the two types of diabetes mellitus.
c. Identify some clinical problems associated with diabetes mellitus.

LO 16.12 Explain diabetes mellitus: its types, clinical manifestations, and treatments.

Concept map

Use each of the following terms once to fill in the blank boxes to correctly complete the map.

- steroid hormones
- tryptophan derivatives
- glycoproteins
- short polypeptides
- catecholamines
- peptide hormones
- thyroid hormones
- transport proteins
- lipid derivatives
- small proteins
- eicosanoids

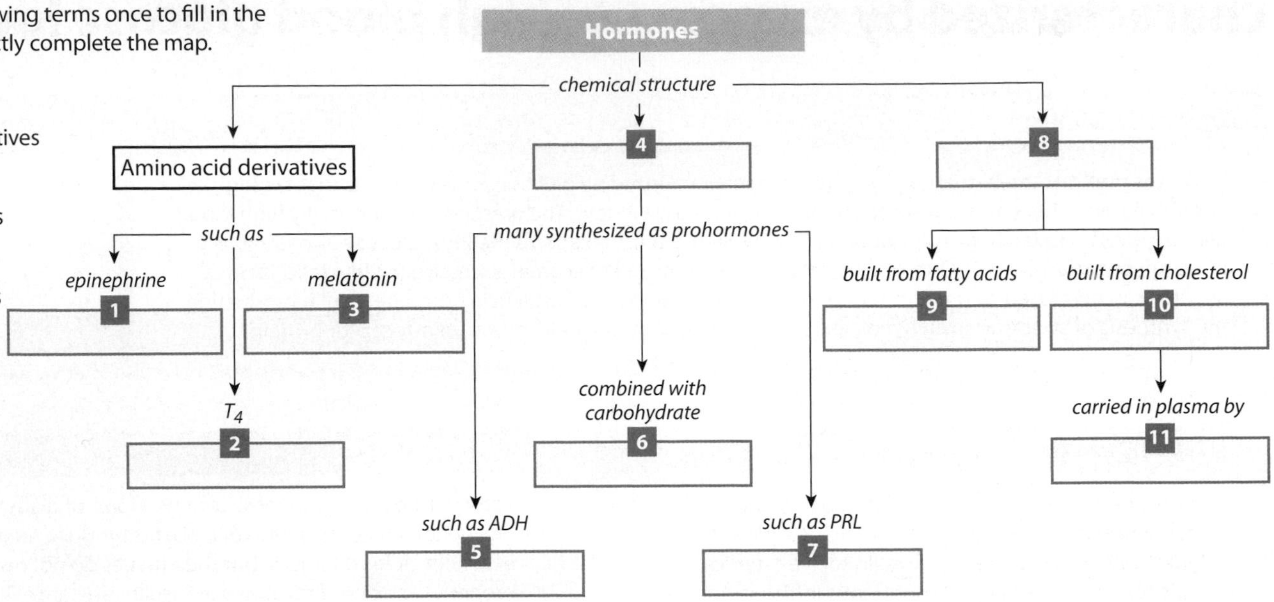
Hormones
chemical structure
Amino acid derivatives
such as
epinephrine
1
melatonin
3
T_4
2
4
many synthesized as prohormones
combined with carbohydrate
6
such as ADH
5
such as PRL
7
8
built from fatty acids
9
built from cholesterol
10
carried in plasma by
11

Matching

Match each lettered term with the most closely related description.

a. FSH
b. androgens
c. F cells
d. parathyroid glands
e. epinephrine
f. direct communication
g. tropic hormones
h. secretes releasing hormones
i. prostaglandins
j. cyclic AMP

12 Pancreatic polypeptide
13 Adrenal medulla
14 Gap junctions
15 Pituitary gland
16 Second messenger
17 Hypothalamus
18 Zona reticularis
19 Eicosanoids
20 Gonadotropins
21 Chief cells

12 __________
13 __________
14 __________
15 __________
16 __________
17 __________
18 __________
19 __________
20 __________
21 __________

Short answer

Identify the endocrine gland—or endocrine cells—based on the major effects produced by its/their secreted hormone(s).

22 Stimulates and coordinates the immune response
23 Establishes circadian rhythms
24 Secretes insulin and regulates glucose uptake and utilization
25 Controls hormone secretion of the pituitary gland
26 Regulates RBC production and the absorption of calcium and phosphate by the intestinal tract
27 Regulates mineral balance, metabolic control, and resistance to stress
28 Regulates secretions of adrenal cortex, thyroid gland, and reproductive organs
29 Affects growth, metabolism, and sexual characteristics

22 __________
23 __________
24 __________
25 __________
26 __________
27 __________
28 __________
29 __________

Hormones interact to produce coordinated physiological responses

Extracellular fluids contain a mixture of hormones whose concentrations change daily or even hourly. Because cells have more than one type of receptor, they respond to multiple hormones simultaneously. When a cell receives instructions from two hormones at the same time, there are four possible outcomes.

Antagonistic Effects

The two hormones may have **antagonistic** (opposing) **effects**, as in the negative feedback cases of PTH and calcitonin, or insulin and glucagon. The net result depends on the balance between the two hormones. In general, when antagonistic hormones are present, the observed effects are weaker than those produced by either hormone acting unopposed.

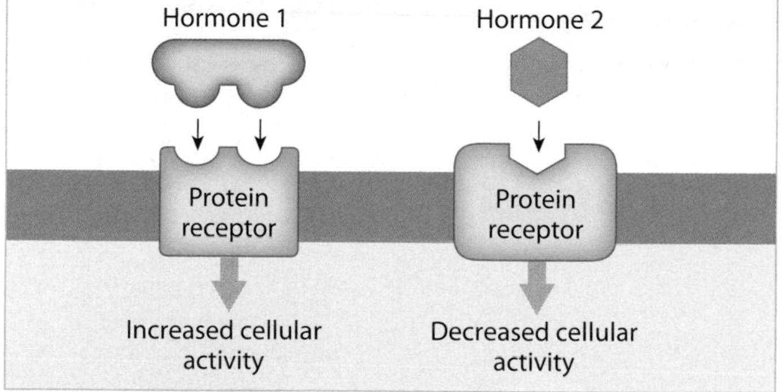
Hormone 1
Hormone 2
Protein receptor
Protein receptor
Increased cellular activity
Decreased cellular activity

Additive Effects

The two hormones may have **additive effects**, so that the net result is greater than the effect that each would produce acting alone. In some cases, the net result is greater than the sum of the hormones' individual effects. This interaction is a **synergistic effect** (sin-er-JIS-tik; *synairesis*, a drawing together). An example is the enhancement of the glucose-sparing action of GH in the presence of glucocorticoids.

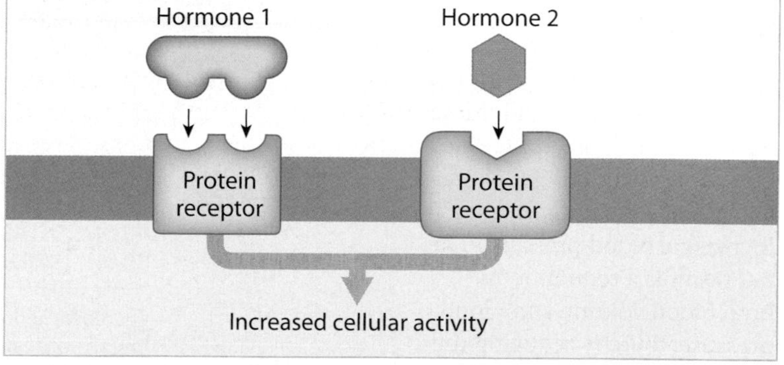
Hormone 1
Hormone 2
Protein receptor
Protein receptor
Increased cellular activity

Permissive Effects

One hormone can have a **permissive effect** on another. In such cases, the first hormone is needed for the second to produce its effect. For example, epinephrine does not change the rate of energy consumption in a tissue unless thyroid hormones are also present in normal concentrations.

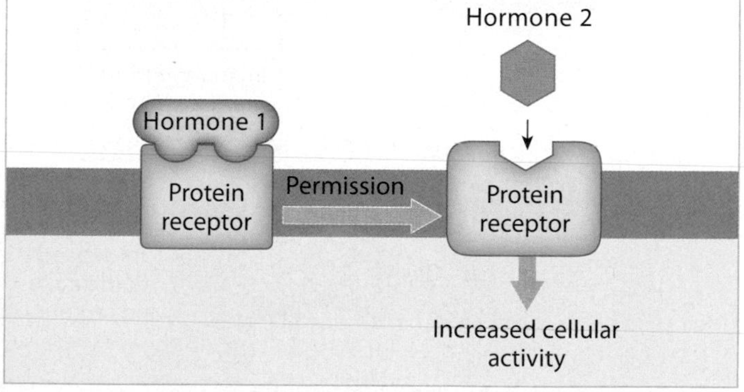
Hormone 2
Hormone 1
Protein receptor
Permission
Protein receptor
Increased cellular activity

Integrative Effects

Hormones may produce different, but complementary, effects in specific tissues and organs. These **integrative effects** are important in coordinating the activities of diverse physiological systems. The differing effects of calcitriol and parathyroid hormone on tissues involved in calcium metabolism are an example.

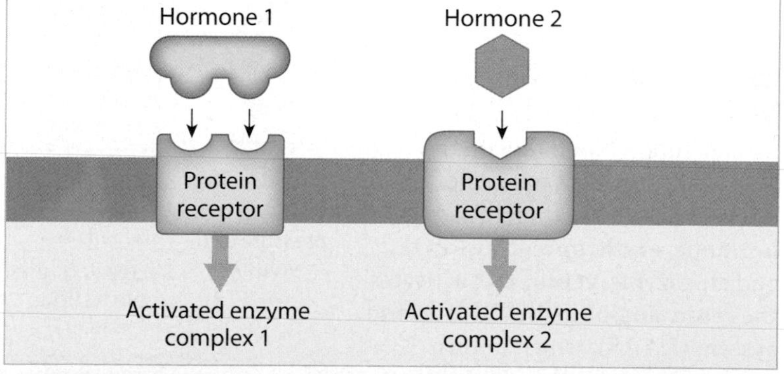
Hormone 1
Hormone 2
Protein receptor
Protein receptor
Activated enzyme complex 1
Activated enzyme complex 2

In this section, we will consider the ways hormones interact to preserve homeostasis and control short-term or long-term processes.

Module 16.13 Review

a. Which of the four hormonal effects are involved in a negative feedback response?

b. Define synergistic effect.

c. What kind of effect do hormones have if they produce different but complementary effects?

Long-term regulation of blood pressure, blood volume, and growth involves hormones produced by the endocrine system and by endocrine tissues in other systems

The long-term regulation of blood pressure and blood volume involves not only the pituitary and adrenal glands, but also endocrine cells in the heart and kidneys.

1 The endocrine cells of the heart are located in the heart walls. If blood volume becomes too great, these cells are stretched excessively and begin to secrete **natriuretic peptides** (nā-trē-ū-RET-ik; *natrium*, sodium + *ouresis*, urination). Natriuretic peptides promote the loss of Na^+ and water by the kidneys, and inhibit renin release and the secretion of ADH and aldosterone. They also suppress thirst and prevent antagonistic hormones from increasing blood pressure. The net result is a reduction in both blood volume and blood pressure, thereby reducing the stretching of the heart walls.

2 If blood volume or blood pressure falls below the normal range, blood flow to the kidneys decreases. Endocrine cells in the kidneys then release a hormone, **erythropoietin (EPO)**, and an enzyme, **renin**, that activates the **renin-angiotensin-aldosterone system (RAAS),** which in turn creates a cascade of responses that leads to increased fluid intake and fluid retention. These processes will be further described in Chapters 19 and 23.

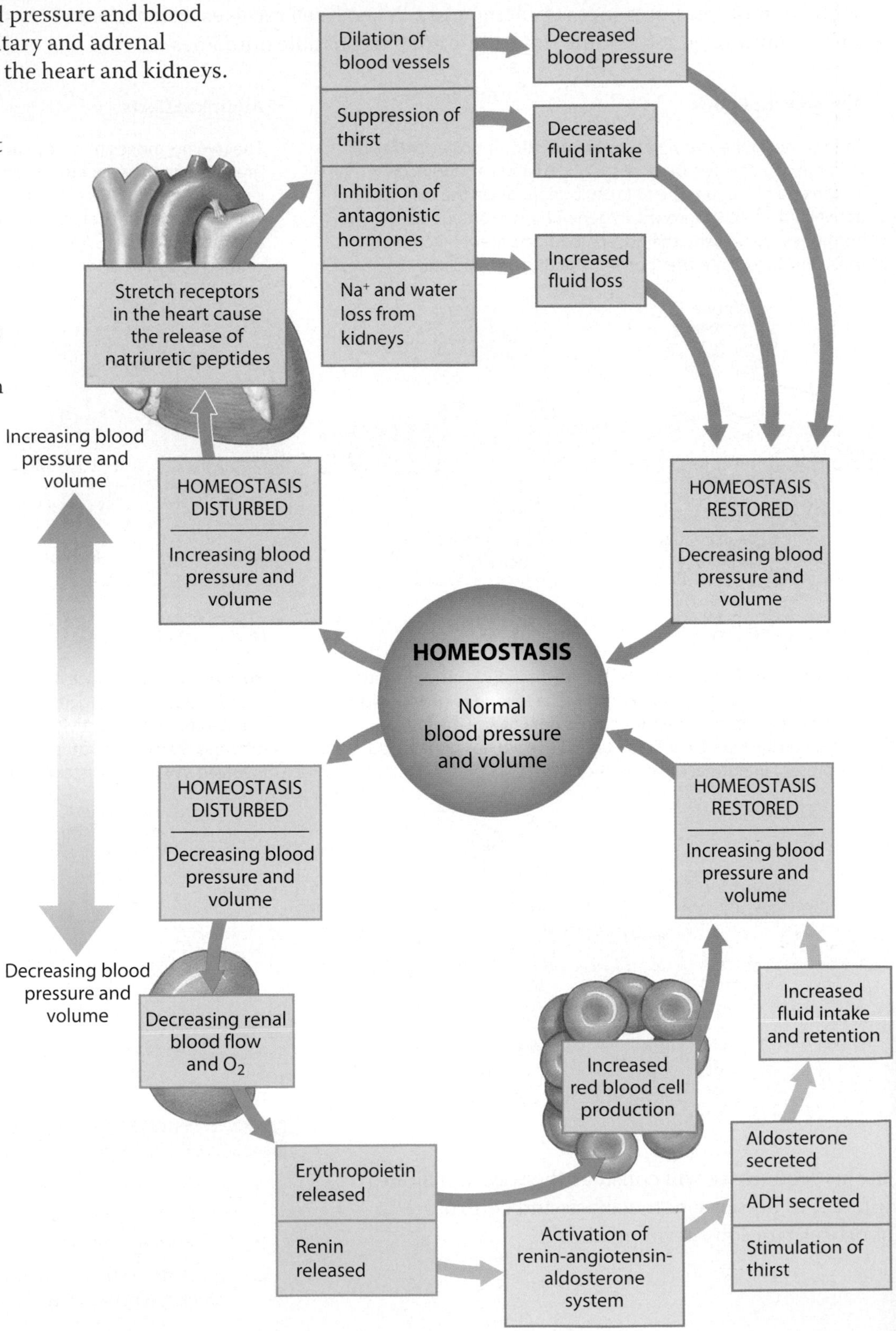

Dilation of blood vessels
Decreased blood pressure
Suppression of thirst
Decreased fluid intake
Inhibition of antagonistic hormones
Na⁺ and water loss from kidneys
Increased fluid loss
Stretch receptors in the heart cause the release of natriuretic peptides
Increasing blood pressure and volume
HOMEOSTASIS DISTURBED
Increasing blood pressure and volume
HOMEOSTASIS RESTORED
Decreasing blood pressure and volume
HOMEOSTASIS
Normal blood pressure and volume
HOMEOSTASIS DISTURBED
Decreasing blood pressure and volume
HOMEOSTASIS RESTORED
Increasing blood pressure and volume
Decreasing blood pressure and volume
Decreasing renal blood flow and O2
Increased red blood cell production
Increased fluid intake and retention
Erythropoietin released
Renin released
Activation of renin-angiotensin-aldosterone system
Aldosterone secreted
ADH secreted
Stimulation of thirst

3 Normal growth requires the cooperation of many endocrine organs. Several hormones—GH, thyroid hormones, insulin, PTH, calcitriol, and reproductive hormones—are especially important, although many others have secondary effects on growth. The circulating levels of these hormones are regulated independently. Every time the hormonal mixture changes, metabolic operations are modified to some degree. The modifications vary in duration and intensity, producing unique growth patterns in different people.

Insulin

Growing cells need adequate supplies of energy and nutrients. Without insulin, the passage of glucose and amino acids across plasma membranes is drastically reduced or eliminated.

Parathyroid Hormone, Calcitriol, and Calcitonin

Parathyroid hormone (PTH) and calcitriol promote the absorption of calcium salts for later deposition in bone; calcitonin accelerates the rate of deposition. Without adequate levels of both hormones, bones can still enlarge, but they will be poorly mineralized, weak, and flexible.

Thyroid Hormones

Normal growth requires appropriate levels of thyroid hormones. If these hormones are absent during fetal development or the first year of life, the nervous system will not develop normally, and mental retardation will result. If thyroid hormone levels decline before puberty, normal skeletal development will not continue.

Reproductive Hormones

The activity of osteoblasts in key locations and the growth of specific cell populations are affected by the presence or absence of reproductive hormones (androgens in males, estrogens in females). These sex hormones stimulate cell growth and differentiation in their target tissues. The targets differ for androgens and estrogens, and the differential growth induced by each accounts for sex-related differences in skeletal proportions and secondary sex characteristics.

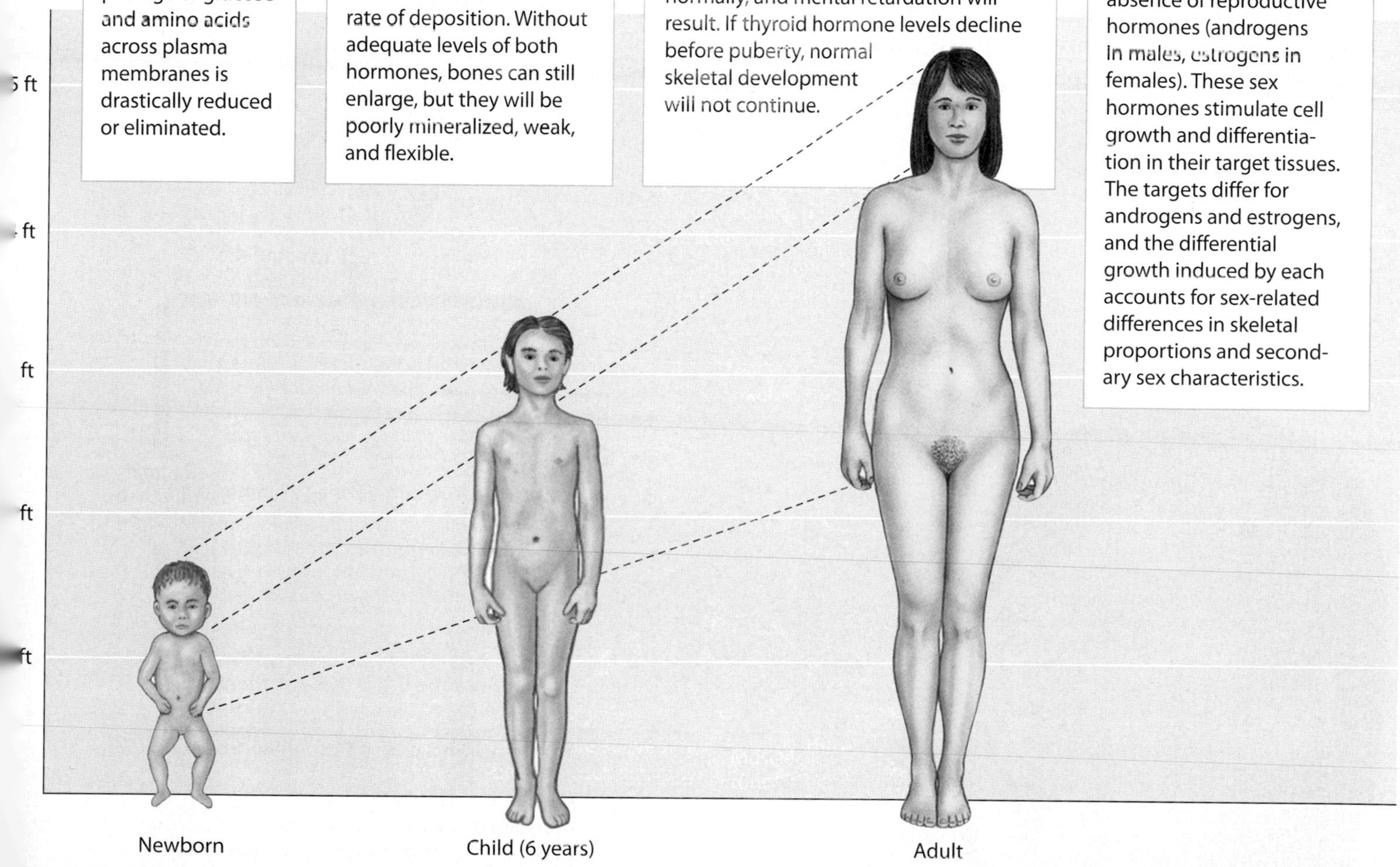

5 ft
ft
ft
ft
ft
Newborn
Child (6 years)
Adult

Growth Hormone

In Children	In Adults
The effects of GH on protein synthesis and cellular growth are most apparent in children. GH supports their muscular and skeletal development.	In adults, GH helps to maintain normal blood glucose concentrations and to mobilize lipid reserves stored in adipose tissue. It is not the primary hormone involved, however, and an adult with a GH deficiency but normal levels of thyroid hormones, insulin, and glucocorticoids will have no physiological problems.

Module 16.14 Review

a. Name the hormones secreted by the heart and a hormone released by the kidneys.

b. Explain the action of renin in the bloodstream.

c. Identify several hormones necessary for normal growth and development.

LO 16.14 Describe the functions of the hormones produced by the kidneys and the heart, and explain the roles of other endocrine organs and hormones in normal growth and development.

Module 16.15

The stress response is a predictable response to any significant threat to homeostasis

Any condition—whether it is physical or emotional—that threatens homeostasis is a form of **stress**. Many stresses are opposed by specific homeostatic adjustments. For example, a drop in body temperature leads to shivering or changes in the pattern of blood flow, which can restore body temperature to normal. In addition, the body has a general response to stress that can occur while other, more specific responses are under way. Exposure to a wide variety of stress-causing factors will produce the same general pattern of hormonal and physiological adjustments. These responses are part of the **stress response**, also known as the **general adaptation syndrome (GAS)**. The stress response is divided into three phases.

1 Alarm Phase ("Fight or Flight")

During the **alarm phase**, an immediate response to the stress occurs. This response is directed by the sympathetic division of the autonomic nervous system. In the alarm phase, (1) energy reserves are mobilized, mainly in the form of glucose, and (2) the body prepares to deal with the stress-causing factor through "fight or flight" responses. Epinephrine is the dominant hormone of the alarm phase. Its secretion is part of a generalized sympathetic activation.

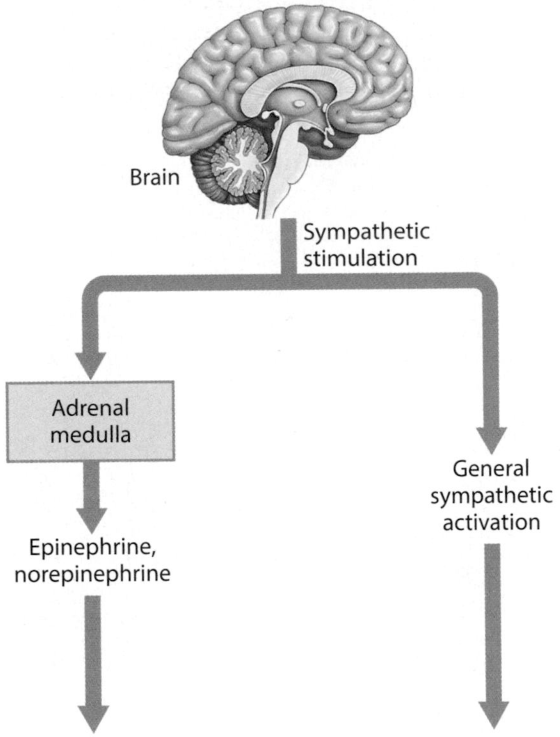
Brain
Sympathetic stimulation
Adrenal medulla
Epinephrine, norepinephrine
General sympathetic activation

Immediate Short-Term Responses to Crises

- Increased mental alertness
- Increased energy use by all cells
- Mobilization of glycogen and lipid reserves
- Changes in circulation
- Reduction in digestive activity and urine production
- Increased sweat gland secretion
- Increased heart rate and respiratory rate

2 Resistance Phase

If a stress lasts longer than a few hours, the person enters the **resistance phase** of the stress response. Glucocorticoids are the dominant hormones of the resistance phase. Epinephrine, GH, and thyroid hormones are also involved. Energy demands in the resistance phase remain higher than normal, due to the combined effects of these hormones. Neural tissue has a high demand for energy, and neurons must have a reliable supply of glucose. Glycogen reserves are adequate to maintain normal glucose concentrations during the alarm phase but are nearly exhausted after several hours. The hormones of the resistance phase mobilize lipids and amino acids, thus shifting tissue metabolism away from glucose, so that whatever glucose becomes available can be used by neural tissues.

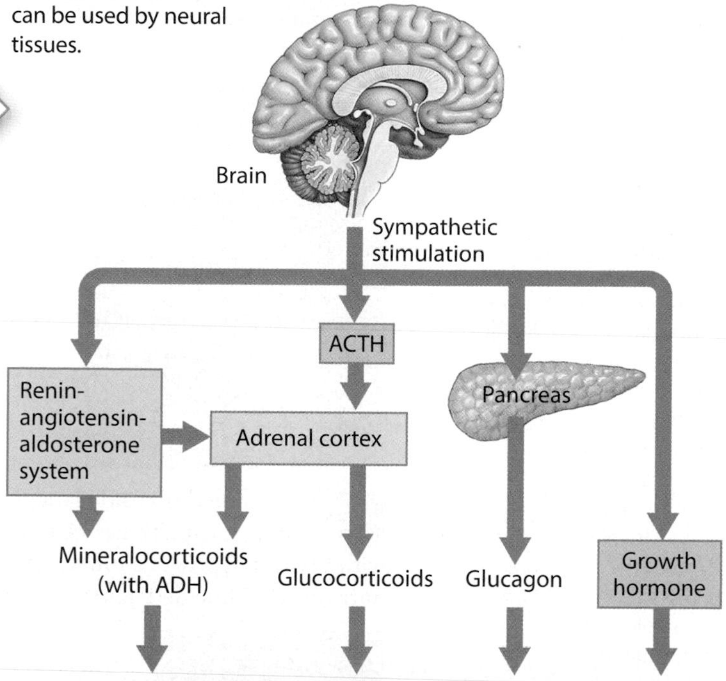

Brain
Sympathetic stimulation
ACTH
Renin-angiotensin-aldosterone system
Adrenal cortex
Pancreas
Mineralocorticoids (with ADH)
Glucocorticoids
Glucagon
Growth hormone

Long-Term Metabolic Adjustments

- Mobilization of remaining energy reserves: Lipids are released by adipose tissue; amino acids are released by skeletal muscle
- Conservation of glucose: Peripheral tissues (except neural) break down lipids to obtain energy
- Increased blood glucose concentrations: Liver synthesizes glucose from other carbohydrates, amino acids, and lipids
- Conservation of salts and water, loss of K^+ and H^+

3 Exhaustion Phase

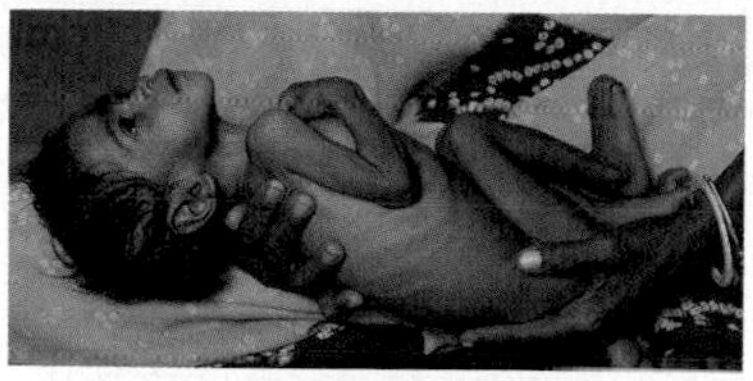

The body's lipid reserves are sufficient to maintain the resistance phase for a period of weeks or even months. But the resistance phase cannot be sustained indefinitely. When the resistance phase ends, homeostatic regulation breaks down and the **exhaustion phase** begins. Unless corrective actions are taken almost immediately, the failure of one or more organ systems will prove fatal. Mineral imbalances contribute to the existing problems with major systems. The production of aldosterone throughout the resistance phase results in a conservation of Na^+ at the expense of K^+. As the body's K^+ content declines, a variety of cells—notably neurons and muscle fibers—begin to malfunction. Although a single cause (such as heart failure) may be listed as the cause of death, the underlying problem is the body's inability to sustain the endocrine and metabolic adjustments of the resistance phase.

Factors That Can Trigger the Exhaustion Phase

- Exhaustion of lipid reserves and the breakdown of structural proteins as the body's primary energy source, damaging vital organs
- Infections that develop due to suppression of inflammation and of the immune response, a secondary effect of the glucocorticoids that are essential to the metabolic activities of the resistance phase
- Cardiovascular damage and complications that are related to the ADH and aldosterone-related elevations in blood pressure and blood volume
- Inability of the adrenal cortex to continue producing glucocorticoids, which results in a failure to maintain acceptable blood glucose concentrations
- Failure to maintain adequate fluid and electrolyte balance

Module 16.15 Review

a. List the three phases of the stress response.

b. Describe the resistance phase.

c. During which phase of the general adaptation syndrome is there a collapse of vital systems?

LO **16.15** Define the general adaptation syndrome, and compare homeostatic responses with stress responses.

Overproduction or underproduction of hormones can cause endocrine disorders

Endocrine disorders may develop for a variety of reasons, including abnormalities in the endocrine gland, the endocrine or neural regulatory mechanisms, or the target tissues. For example, a hormone level may rise because its target organs are becoming less responsive, because a tumor has formed among the gland cells, or because something has interfered with the normal feedback control mechanism. When naming endocrine disorders, clinicians use the prefix *hyper-* when referring to excessive hormone production and *hypo-* when referring to inadequate hormone production. Endocrine tumors can result in **hypersecretion**, but its incidence is relatively rare. Most endocrine disorders are the result of problems within the endocrine gland that result in **hyposecretion**, the production of inadequate levels of a particular hormone.

Common Causes of Hormone Hyposecretion

Metabolic Factors

Hyposecretion may result from a deficiency in some key substance needed to synthesize the hormone. For example, hypothyroidism can be caused by inadequate levels of iodine in the diet.

Physical Damage

Any condition that interrupts the normal circulatory supply to endocrine cells or that physically damages those cells may cause them to become inactive immediately or after an initial surge of hormone release.

Congenital Disorders

A person may be unable to produce normal amounts of a particular hormone because (1) the gland is too small, (2) the required enzymes are abnormal, (3) the receptors that trigger secretion are relatively insensitive, or (4) the gland cells lack the receptors normally involved in stimulating secretory activity.

Endocrine abnormalities can also be caused by the presence of abnormal hormonal receptors in target tissues. In such cases, the gland and the regulatory mechanisms involved are normal, but the peripheral cells are unable to respond to the circulating hormone. The best example of this type of abnormality is Type 2 diabetes, in which peripheral cells do not respond normally to insulin.

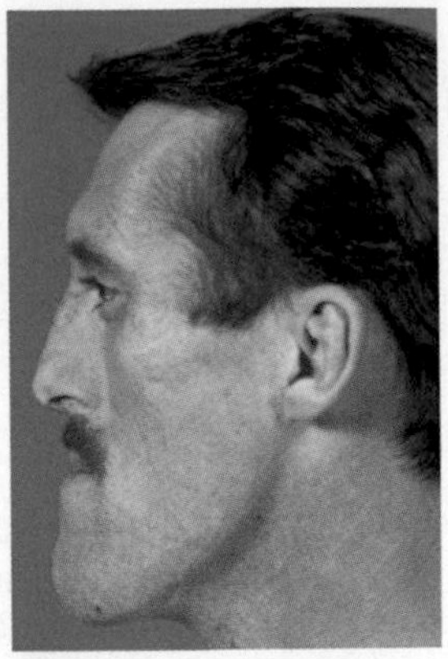

1 **Acromegaly** results from the overproduction of growth hormone after the epiphyseal plates have fused. Bone shapes change and cartilaginous areas of the skeleton enlarge. Signs of acromegaly include broad facial features and enlarged lower jaw.

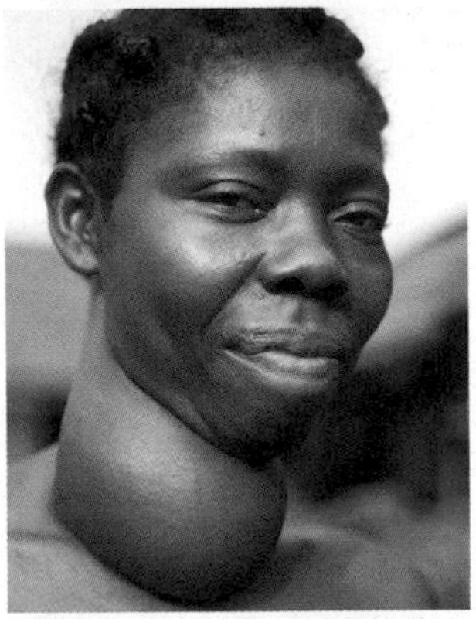

2 An enlarged thyroid gland, or **goiter**, is usually associated with thyroid hyposecretion due to nutritional iodine deficiency.

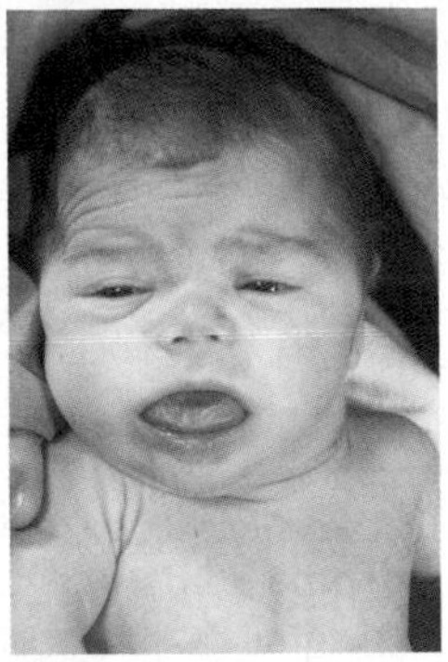

3 **Infantile hypothyroidism** is a congential disorder due to thyroid hormone deficiency. It is characterized by mental disability, puffy face, and a thick tongue.

An Overview of Endocrine Disorders

Hormone	Results of Under-production or Tissue Insensitivity	Principal Signs and Symptoms	Results of Over-production or Tissue Hypersensitivity	Principal Signs and Symptoms
Growth hormone (GH)	Pituitary growth failure	Delayed growth, abnormal fat distribution, low blood glucose hours after a meal	Gigantism, acromegaly	Excessive growth
Antidiuretic hormone (ADH)	Diabetes insipidus	Polyuria, dehydration, thirst	Syndrome of inappropriate ADH secretion (SIADH)	Increased body weight and water content
Thyroxine (T_4) and **triiodothyronine (T_3)**	Hypothyroidism, infantile hypothyroidism, myxedema	Low metabolic rate, low body temperature, impaired physical and mental development	Hyperthyroidism, Graves disease	High metabolic rate and body temperature
Parathyroid hormone (PTH)	Hypoparathyroidism	Muscular weakness, neurological problems, formation of dense bones, tetany due to low blood Ca^{2+} concentrations	Hyperparathyroidism	Neurological, mental, and muscular problems due to high blood Ca^{2+} concentrations; weak and brittle bones
Insulin	Diabetes mellitus (Type 1)	High blood glucose, impaired glucose utilization, dependence on lipids for energy, glycosuria	Excess insulin production (also caused by adminis-tering too much insulin)	Low blood glucose levels, possibly causing coma
Mineralocorticoids (MCs) Example: aldosterone	Hypoaldosteronism	Polyuria, low blood volume, high blood K^+, and low blood Na^+ concentrations	Aldosteronism	Increased body weight due to Na^+ and water retention, low blood K^+ concentration
Glucocorticoids (GCs) Example: cortisol	Addison's disease	Inability to tolerate stress, mobilize energy reserves, or maintain normal blood glucose concentrations	Cushing's disease	Excessive breakdown of tissue proteins and lipid reserves, impaired glucose metabolism
Epinephrine (E) and **norepinephrine (NE)**	None identified	None identified	Pheochromocytoma	High metabolic rate, body temperature, and heart rate; increased blood glucose levels
Estrogens (females)	Hypogonadism	Sterility, lack of secondary sex characteristics	Adrenogenital syndrome	Overproduction of androgens by zona reticularis of adrenal cortex; leads to masculinization
			Precocious puberty	Premature sexual maturation and related behavioral changes
Androgens (males)	Hypogonadism	Sterility, lack of secondary sex characteristics	Adrenogenital syndrome (gynecomastia)	Abnormal production of estrogens, sometimes due to adrenal or interstitial cell tumors; leads to breast enlargement
			Precocious puberty	Premature sexual maturation and related behavioral changes

4 **Addison's disease** is caused by hyposecretion of corticosteroids, especially gluco-corticoids. Pigment changes result from stimulation of melanocytes by ACTH, which is similar in structure to MSH.

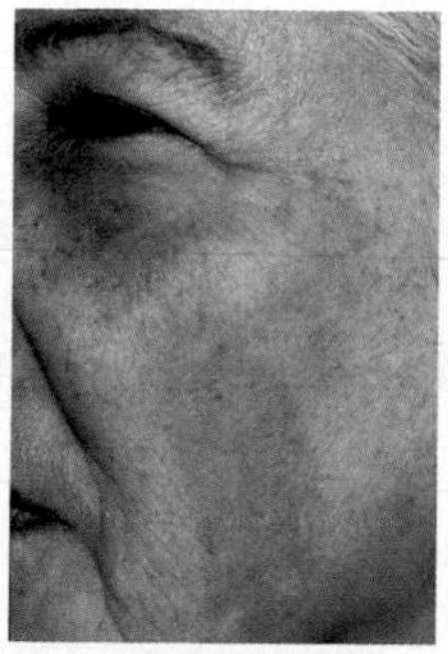

5 **Cushing's disease** is caused by hypersecretion of glucocorticoids. Lipid reserves are mobilized, and adipose tissue accumulates in the cheeks and at the base of the neck.

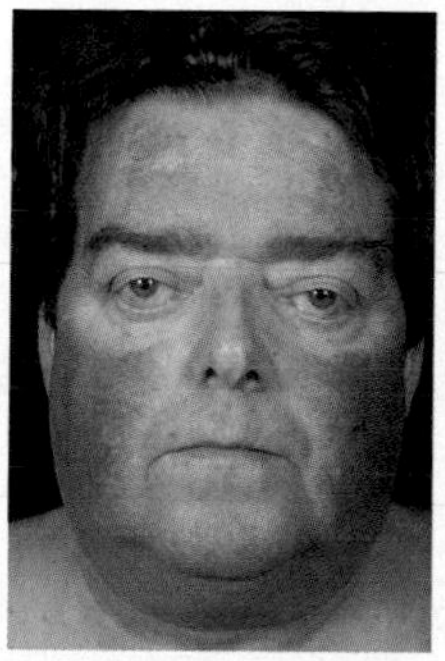

Module 16.16 Review

a. Define the prefixes *hyper-* and *hypo-* in the context of endocrine disorders.

b. Identify three common causes of hormone hyposecretion.

c. What condition is characterized by increased body weight due to Na^+ and water retention and a low blood K^+ concentration?

LO **16.16** Describe key endocrine disorders, citing their characteristic signs and symptoms.

Labeling

Write the phrases at left in the boxes to complete the diagram of the homeostatic regulation of blood pressure and volume.

- increased fluid loss
- erythropoietin released
- decreased blood pressure
- aldosterone secreted
- suppression of thirst
- decreasing blood pressure and volume
- increasing blood pressure and volume
- renin released
- ADH secreted
- release of natriuretic peptides
- Na^+ and H_2O loss from kidneys
- increased red blood cell production

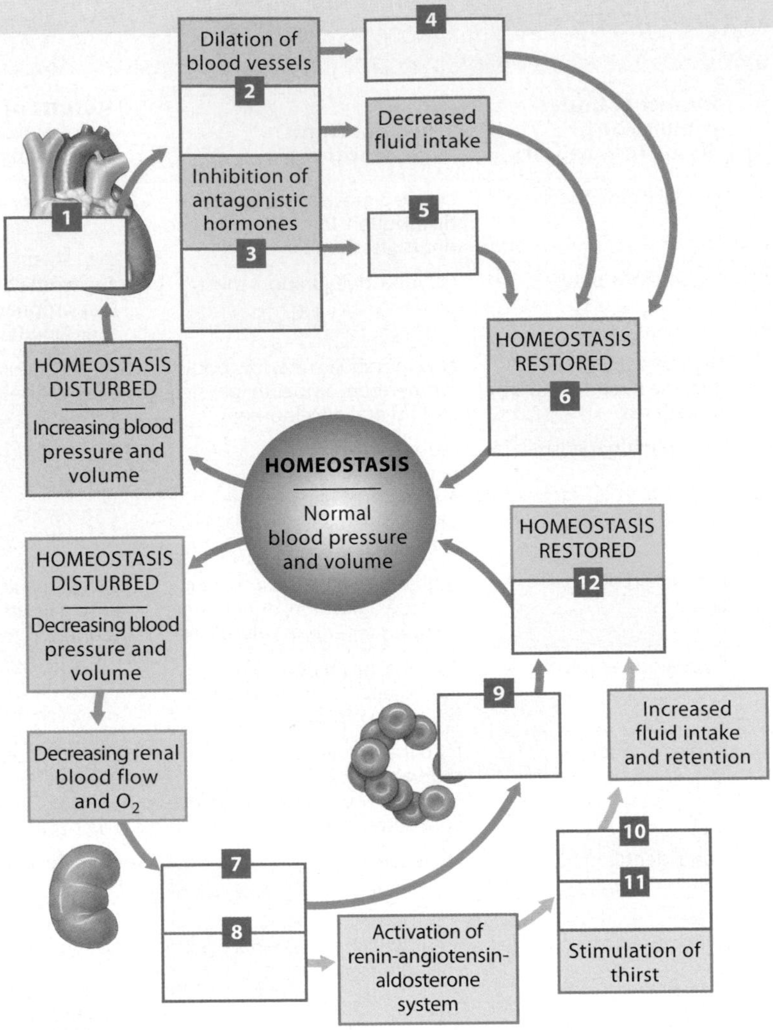

Dilation of blood vessels
2
4
Decreased fluid intake
Inhibition of antagonistic hormones
3
5
1
HOMEOSTASIS DISTURBED
Increasing blood pressure and volume
HOMEOSTASIS RESTORED
6
HOMEOSTASIS
Normal blood pressure and volume
HOMEOSTASIS DISTURBED
Decreasing blood pressure and volume
HOMEOSTASIS RESTORED
12
9
Increased fluid intake and retention
Decreasing renal blood flow and O_2
10
11
7
8
Activation of renin-angiotensin-aldosterone system
Stimulation of thirst

Matching

Match each lettered term with the most closely related description.

a. sympathetic activation
b. decrease blood pressure and volume
c. PTH and calcitonin
d. GH and glucocorticoids
e. increase blood pressure and volume
f. homeostasis threat
g. glucocorticoids
h. PTH and calcitriol
i. gigantism
j. protein synthesis

13 Antagonistic effect
14 Resistance phase
15 Alarm phase
16 Renin and EPO effect
17 Growth hormone (GH)
18 Additive effect
19 Excessive GH in children
20 Natriuretic peptides effect
21 Stress
22 Integrative effect

13 ______
14 ______
15 ______
16 ______
17 ______
18 ______
19 ______
20 ______
21 ______
22 ______

Section integration

Describe, and give an example of, the four possible effects that may occur when a cell receives instructions from two different hormones.

23 ______

Study Outline

SECTION 1 • Hormones and Intercellular Communication

16.1 The nervous and endocrine systems release chemical messengers that bind to target cells p. 587

1. Cells communicate in a variety of ways including through gap junctions, extracellular fluid (by **paracrine factors**), the bloodstream (by **hormones**), or across synapses (by neurotransmitters).
2. The nervous and endocrine systems both rely on the release of chemicals that bind to receptors on target cells, share many chemical messengers, are regulated by negative feedback mechanisms, and preserve homeostasis.

16.2 Hormones may be amino acid derivatives, peptides, or lipid derivatives p. 588

3. Hormones and paracrine factors can be grouped as **amino acid derivatives**, **peptide hormones**, and **lipid derivatives**.
4. Amino acid derivatives include thyroid hormones and **catecholamines**.
5. Peptide hormones are synthesized as **prohormones** that are converted to active hormones.
6. Lipid derivatives are built from either fatty acids **(eicosanoids)** or cholesterol **(steroid hormones)**.

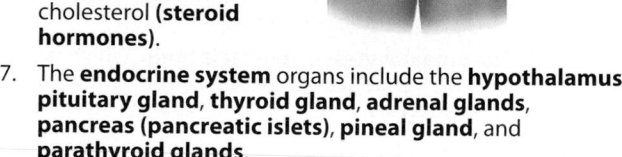

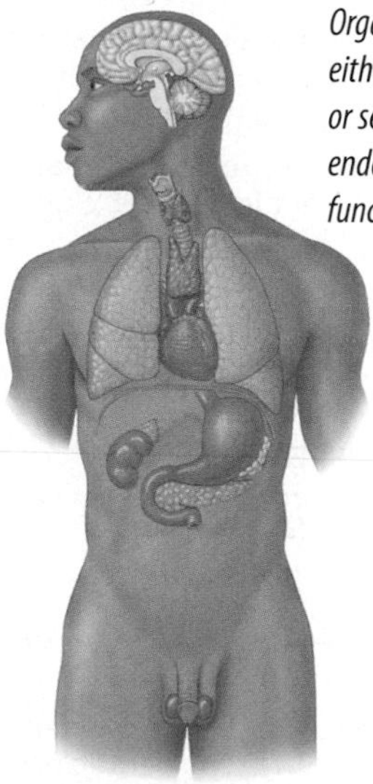
Organs with either primary or secondary endocrine functions

7. The **endocrine system** organs include the **hypothalamus**, **pituitary gland**, **thyroid gland**, **adrenal glands**, **pancreas (pancreatic islets)**, **pineal gland**, and **parathyroid glands**.
8. Organs with secondary endocrine functions are the heart, thymus, digestive tract, kidneys, and gonads.

16.3 Hormones affect target cells after binding to receptors in the plasma membrane, cytoplasm, or nucleus p. 590

9. Each cell has receptors for different hormones. Cells in different tissues have different combinations of receptors.
10. A hormone binding to plasma membrane receptors does not have a direct effect on cellular activities. This hormone, or **first messenger**, requires a **second messenger**.
11. The two most important second messengers are **cyclic AMP (cAMP)**, and calcium ions (Ca^{2+}).
12. Steroid hormones diffuse across the plasma membrane and bind to receptors in the cytoplasm or nucleus.
13. Thyroid hormones are transported across the plasma membrane by carrier-mediated processes.

16.4 The hypothalamus exerts direct or indirect control over the activities of many different endocrine organs p. 592

14. The **hypothalamus** regulates the nervous and endocrine systems through three mechanisms: synthesizing **antidiuretic hormone (ADH)** and **oxytocin (OXT)** and transporting them to the posterior lobe of the pituitary; secreting **regulatory hormones** that control endocrine cells in the anterior lobe of the pituitary gland; and controlling the endocrine cells of the adrenal medullae.

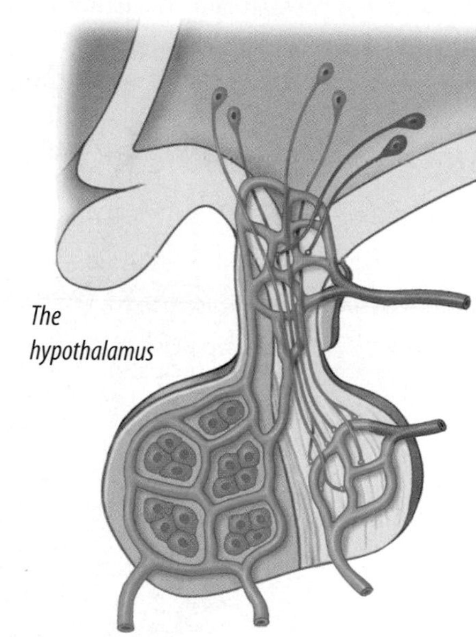
The hypothalamus

15. The **hypophyseal portal system** ensures that hypothalamic hormones reach their target cells in the anterior lobe of the pituitary gland before being diluted by the general circulation.
16. Hypothalamic regulatory hormones can be either a **releasing hormone** or an **inhibiting hormone**. Individually or in combination, these hormones control endocrine cells in the anterior lobe of the pituitary gland.

16.5 The pituitary gland consists of an anterior lobe and a posterior lobe p. 594

17. The **pituitary gland** is connected to the hypothalamus by the **infundibulum** and lies within the sella turcica of the sphenoid bone.
18. The pituitary gland releases nine hormones: seven from the **anterior lobe of the pituitary (adenohypophysis)**, two from the **posterior lobe of the pituitary (neurohypophysis)**.
19. The hormones of the anterior lobe of the pituitary are **thyroid-stimulating hormone (TSH)**, **adrenocorticotropic hormone (ACTH)**, **follicle-stimulating hormone (FSH)**, **luteinizing hormone (LH)**, **growth hormone (GH)**, **prolactin (PRL)**, and **melanocyte-stimulating hormone (MSH)**.
20. The hormones of the posterior lobe of the pituitary are **antidiuretic hormone (ADH)**, or **vasopressin**, and **oxytocin (OXT)**.

16.6 Negative feedback mechanisms control the secretion rates of the hypothalamus and the pituitary gland p. 596

21. Hormone secretion is controlled through negative feedback.

22. The hypothalamus produces a releasing hormone (or factor) that triggers the release of a hormone from the anterior lobe of the pituitary gland.
23. The pituitary hormone stimulates the release of a second hormone by the target. This second hormone suppresses secretion of the hypothalamic and pituitary hormones.

16.7 The thyroid gland contains follicles and requires iodine to produce hormones that stimulate tissue metabolism p. 598

24. The **thyroid gland** is located inferior to the thyroid cartilage. The **isthmus** connects its two lobes.

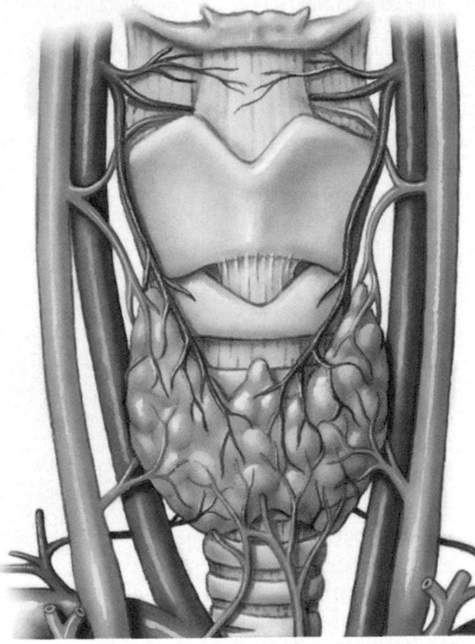

The thyroid gland

25. The thyroid gland contains **thyroid follicles** that produce **thyroglobulin**, which is used to make thyroid hormones.
26. Thyroid hormones are produced by the following process: iodine ions are absorbed from the diet and delivered to the thyroid gland by the bloodstream, iodine ions diffuse to the apical surface of each follicle cell and are converted to iodine atoms, thyroglobulin produces the thyroid hormones **thyroxine (T_4) and triiodothyronine (T_3)**, thyroglobulin is removed from a follicle cavity and recycled, T_3 and T_4 diffuse into the bloodstream, where the majority of thyroid hormones are bound to **thyroid-binding globulins (TBGs)**.
27. The effects of thyroid hormones on peripheral tissues include increased oxygen and energy consumption, increased heart rate and force of contraction, increased blood pressure, increased sensitivity to sympathetic stimulation, stimulation of red blood cell formation, increased activity of other endocrine tissues, and accelerated turnover of bone minerals.

16.8 Parathyroid hormone, produced by the parathyroid glands, is the primary regulator of blood calcium ion levels p. 600

28. Two pairs of **parathyroid glands** are located on the posterior surface of the thyroid gland.
29. Parathyroid cells produce parathyroid hormone and monitor the circulating concentration of calcium ions.
30. When Ca^{2+} decreases below normal, the parathyroid cells secrete **parathyroid hormone (PTH)**, resulting in increased Ca^{2+} concentration in body fluids.
31. PTH functions by inhibiting osteoblasts, increasing osteoclast number, enhancing Ca^{2+} reabsorption in the kidneys, and stimulating formation and release of calcitriol from the kidney that helps with calcium reabsorption by the intestinal tract.

16.9 The adrenal glands produce hormones involved in metabolic regulation p. 602

32. An **adrenal gland**, or suprarenal gland, sits atop each kidney.
33. The adrenal cortex releases more than two dozen steroid hormones called **corticosteroids**. These hormones determine which genes are transcribed in the nuclei of their target cells, and affect cellular metabolism.
34. The adrenal medulla synthesizes epinephrine and norepinephrine.

16.10 The pancreatic islets secrete insulin and glucagon and regulate glucose use by most cells p. 604

35. The **pancreas** is retroperitoneal and lies in the loop between the inferior border of the stomach and the proximal small intestine.
36. The **exocrine pancreas** is approximately 99 percent of the organ's volume. It secretes an alkaline, enzyme-rich fluid into the intestinal tract through **pancreatic ducts**.
37. The **endocrine pancreas** consists of small groups of endocrine cell clusters called **pancreatic islets**, or islets of Langerhans.
38. Pancreatic islets contain **alpha cells** that produce the hormone **glucagon** (which raises blood glucose), and **beta cells** that produce the hormone **insulin** (which lowers blood glucose).
39. **Delta cells** produce growth hormone–inhibiting hormone, and **F cells** produce the hormone **pancreatic polypeptide (PP)**.

16.11 The pineal gland of the epithalamus secretes melatonin p. 606

40. The **pineal gland** lies in the posterior portion of the roof of the third ventricle in the brain.
41. Its **pinealocytes** secrete **melatonin**, which functions to inhibit reproductive functions, protect against damage by **free radicals**, and set basic **circadian rhythms**.

16.12 Diabetes mellitus is an endocrine disorder characterized by excessively high blood glucose levels p. 607

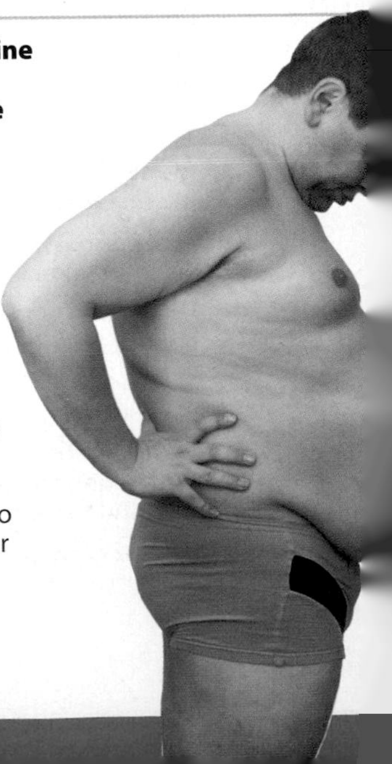

42. **Diabetes mellitus** is characterized by high blood glucose concentration **(hyperglycemia)**, glucose in the urine **(glycosuria)**, and excessive urine volume **(polyuria)**.
43. **Type 1 diabetes** is characterized by inadequate insulin production by the pancreatic beta cells, and requires those who have it to receive insulin by injection or pump.

44. **Type 2 diabetes** is the most common form of diabetes mellitus, and is the result of insulin resistance. It is associated with obesity.

45. Clinical problems caused by diabetes include blindness from **diabetic retinopathy**, heart attack from cardiac circulation blockage, kidney damage from **diabetic nephropathy**, problems with peripheral nerves from **diabetic neuropathy**, and a reduction in peripheral blood flow that can lead to foot amputation.

SECTION 2 • Hormones and System Integration

16.13 Hormones interact to produce coordinated physiological responses p. 609

46. Cells never respond to only one hormone; they respond to multiple hormones simultaneously.

47. When a cell receives instructions from multiple hormones, there are four possible outcomes: **antagonistic effects**, **permissive effects**, **additive (synergistic) effects**, or **integrative effects**.

16.14 Long-term regulation of blood pressure, blood volume, and growth involves hormones produced by the endocrine system and by endocrine tissues in other systems p. 610

48. Blood pressure and volume are regulated not only by the pituitary gland and adrenal glands, but also by endocrine cells in the heart and kidney.

49. If blood volume is too high, the endocrine cells located in the heart walls secrete **natriuretic peptides**. The result is a decrease in blood volume and pressure by the loss of water and Na^+ at the kidneys, an inhibition of renin release, and the secretion of ADH and aldosterone.

50. If blood pressure or volume is too low, blood flow to the kidney decreases, endocrine cells in the kidney release the hormone **erythropoietin**, and the enzyme **renin** that activates the **renin-angiotensin-aldosterone system (RAAS)**. This action results in a series of responses that increase fluid intake and retention.

51. Normal growth is regulated by hormones including insulin, parathyroid hormone, calcitriol, calcitonin, thyroid hormones, reproductive hormones, and growth hormone.

16.15 The stress response is a predictable response to any significant threat to homeostasis p. 612

52. **Stress** is a physical or emotional condition that threatens homeostasis.

53. Stress-causing factors produce a pattern of responses called the **stress response**, or the **general adaptation syndrome (GAS)**.

54. The three phases of the stress response are the **alarm phase** ("fight or flight"), the **resistance phase**, and the **exhaustion phase**.

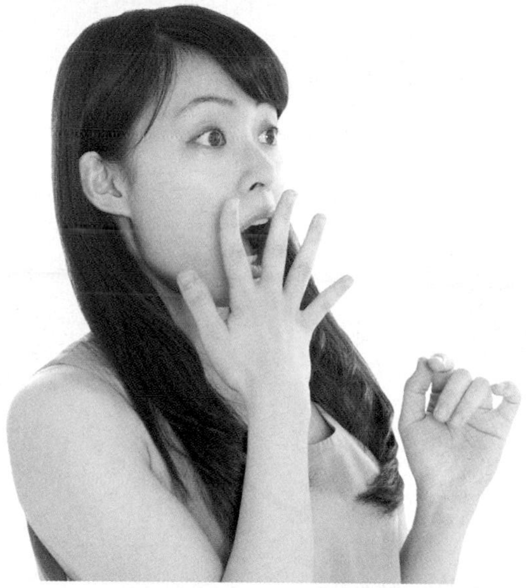

16.16 Overproduction or underproduction of hormones can cause endocrine disorders p. 614

55. Endocrine disorders may develop for a variety of reasons, including abnormalities of the endocrine gland, regulatory mechanisms, or the target tissues.

56. Endocrine disorders named with the prefix hyper- refer to excessive hormone production, or **hypersecretion**.

57. Endocrine disorders named with the prefix hypo- refer to inadequate hormone production, or **hyposecretion**.

Chapter Review Questions

True/False

Indicate whether each statement is true or false.

1. Intercellular communication through extracellular fluid is achieved by paracrine factors.
2. Catecholamines are lipid derivative hormones.
3. Hormones that bind to receptors in the plasma membrane require a second messenger.
4. Rising levels of blood calcium cause the parathyroid glands to secrete PTH.
5. People with Type 2 diabetes must receive insulin daily by injection or through an insulin pump.

1 ____________________

2 ____________________

3 ____________________

4 ____________________

5 ____________________

Fill-in

Hypothalamus

Identify the seven regulatory hormones of the hypothalamus.

6	7	8	9	10	11	12

Fill in the boxes with the names of the missing hormones.

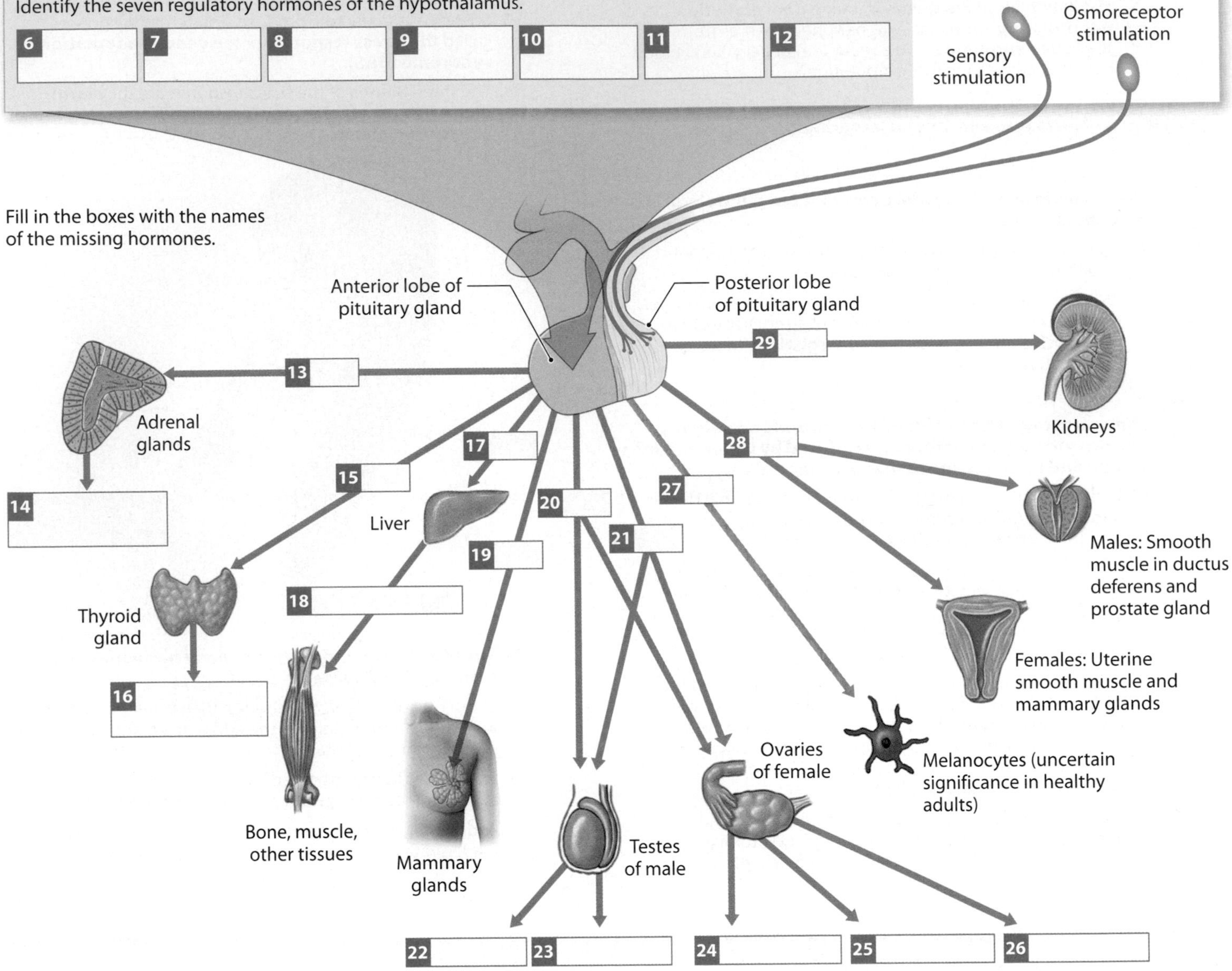

Sensory stimulation
Osmoreceptor stimulation
Anterior lobe of pituitary gland
Posterior lobe of pituitary gland
29
Kidneys
13
Adrenal glands
14
15
17
Liver
20
27
28
Males: Smooth muscle in ductus deferens and prostate gland
21
19
18
Thyroid gland
16
Females: Uterine smooth muscle and mammary glands
Melanocytes (uncertain significance in healthy adults)
Ovaries of female
Bone, muscle, other tissues
Mammary glands
Testes of male
22
23
24
25
26

Multiple choice

Select the correct answer from the list provided.

30 Secretions from the pancreatic islets are
- ☐ a) glucagon from delta cells and insulin from beta cells.
- ☐ b) glucagon from alpha cells and insulin from beta cells.
- ☐ c) glucagon from beta cells and insulin from alpha cells.
- ☐ d) glucagon from beta cells and insulin from delta cells.

31 The pineal gland secretes
- ☐ a) corticosteroids.
- ☐ b) thyroxine.
- ☐ c) melanocyte-stimulating hormone.
- ☐ d) melatonin.

32 The two most important second messengers are
- ☐ a) cAMP and G proteins.
- ☐ b) G proteins and calcium ions.
- ☐ c) catecholamines and eicosanoids.
- ☐ d) cAMP and calcium ions.

33 FSH production in males supports
- ☐ a) physical maturation of developing sperm.
- ☐ b) development of muscle size.
- ☐ c) the release of androgens.
- ☐ d) an increased desire for sexual activity.

34 Which of the following is *not* an effect of thyroid hormones on peripheral tissues?

- [ ] a) Increased rates of oxygen consumption and energy
- [ ] b) Decreased heart rate
- [ ] c) Accelerated turnover of minerals in bone
- [ ] d) Increased sensitivity to sympathetic stimulation

35 Steroid hormones bind to receptors in the

- [ ] a) plasma membrane only.
- [ ] b) plasma membrane or cytoplasm.
- [ ] c) plasma membrane or nucleus.
- [ ] d) cytoplasm or nucleus.

Short answer

36 What are the functions of the kidney hormones?

37 What effects do calcitonin and parathyroid hormone have on blood calcium levels?

Chapter Integration • Applying what you have learned

What's wrong with me?

Sherry, a 43-year-old mother of two, has been experiencing some slightly troublesome physical and mental problems that began about 8 months ago. She had been a physically fit 120-lb person with a bright, happy, and cheerful disposition all her adult life. Within the past few months, however, her husband has repeatedly commented about her overreactions to otherwise normal family situations, and that she seemed irritated by the slightest problems. She has also lost a considerable amount of weight, even though her diet and exercise levels haven't changed. Because Sherry just wasn't feeling "right," she decided to make an appointment with her family physician.

Sherry tells her physician that she has been restless, anxious, and quite irritable lately. She is having a hard time sleeping, complains of diarrhea and weight loss, and has an increased appetite without weight gain. Visual examination of her neck reveals an enlargement of the right anterior neck. During the examination, her physician notices a higher-than-normal heart rate and a fine tremor in her outstretched fingers. Although further clinical tests are necessary, the doctor suspects hyperthyroidism.

1 **Why did the physician suspect hyperthyroidism?**

2 **Explain hyperthyroidism.**

3 **What tests could the physician order to make a positive diagnosis of Sherry's condition?**

4 **Which of Sherry's signs and symptoms indicate nervous system involvement?**

17 Blood

LEARNING OUTCOMES

These Learning Outcomes correspond by number to this chapter's modules and indicate what you should be able to do after completing the chapter.

SECTION 1 • Plasma and Formed Elements

17.1 List the components of the cardiovascular system, and describe several important functions of blood.

17.2 Describe the important components and major properties of blood.

17.3 Explain the origins and differentiation of the formed elements.

SECTION 2 • Structure and Function of Formed Elements

17.4 Define hematology, describe the elements of a complete blood count (CBC), and give examples of red blood cell lab tests.

17.5 List the characteristics and functions of red blood cells, and describe the structure and functions of hemoglobin.

17.6 Describe how the components of aged or damaged red blood cells are recycled.

17.7 Explain the importance of blood typing and the basis for ABO and Rh incompatibilities.

17.8 **CLINICAL MODULE** Describe hemolytic disease of the newborn, explain the clinical significance of the cross-reaction between fetal and maternal blood types, and cite preventive measures.

17.9 Categorize the various types of white blood cells on the basis of their structures and functions.

17.10 Discuss the mechanisms that control blood loss after an injury, and describe the reaction sequences responsible for blood clotting.

17.11 **CLINICAL MODULE** Explain how blood disorders are detected, and describe examples of the various categories of blood disorders.

Learning Outcomes are repeated at the bottom of each module.

Blood is the fluid portion of the cardiovascular system

1 The **cardiovascular system** includes a fluid (blood), a series of conducting tubes (the blood vessels) that distribute the fluid throughout the body, and a pump (the heart) that keeps the fluid in motion.

Components of the Cardiovascular System

The **HEART** propels blood and maintains blood pressure

BLOOD VESSELS distribute blood around the body

Arteries	carry blood away from the heart to the capillaries
Capillaries	permit diffusion between blood and interstitial fluids
Veins	return blood from capillaries to the heart

BLOOD distributes oxygen, carbon dioxide, and blood cells; delivers nutrients and hormones; transports waste products; and assists in temperature regulation and defense against disease.

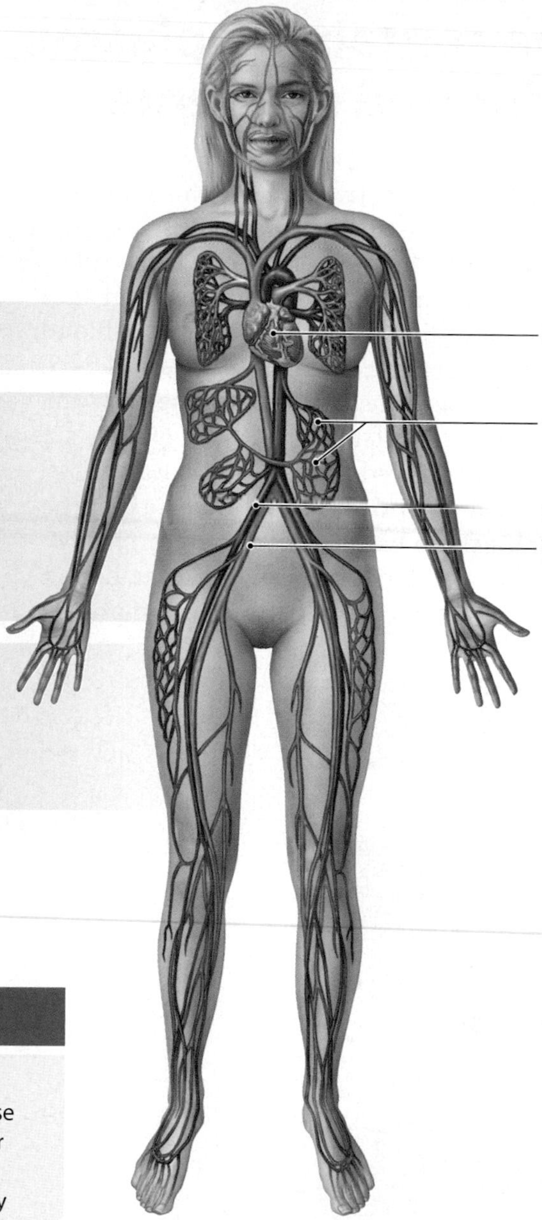
Heart
Capillaries
Artery
Vein

Functions of Blood

- **Transport Dissolved Gases, Nutrients, Hormones, and Metabolic Wastes.** Blood carries oxygen from the lungs to peripheral tissues, and carbon dioxide from those tissues to the lungs. Blood distributes nutrients absorbed by the digestive tract or released from storage in adipose tissue or in the liver. It carries hormones from endocrine glands toward their target cells. It also absorbs the wastes produced by tissue cells and carries them to the kidneys for excretion.
- **Regulate the pH and Ion Composition of Interstitial Fluids.** Diffusion between interstitial fluids and blood eliminates local deficiencies or excesses of ions such as calcium or potassium. Blood also absorbs and neutralizes acids generated by active tissues, such as lactic acid produced by skeletal muscles.
- **Restrict Fluid Losses at Injury Sites.** Blood contains enzymes and other substances that respond to breaks in vessel walls by initiating the clotting process. A blood clot acts as a temporary patch that prevents further blood loss.
- **Defend against Toxins and Pathogens.** Blood transports white blood cells, specialized cells that migrate into peripheral tissues to fight infections or remove debris. Blood also transports antibodies, proteins that specifically attack invading organisms or foreign substances.
- **Stabilize Body Temperature.** Blood absorbs the heat generated by active skeletal muscles and redistributes it to other tissues. If body temperature is already high, that heat will be lost across the surface of the skin. If body temperature is too low, the warm blood is directed to the brain and to other temperature-sensitive organs.

This section describes the vital roles of blood and introduces its two main components: plasma and formed elements. In Section 2, we will examine the structure and function of the formed elements.

Module 17.1 Review

a. Identify the components of the cardiovascular system.

b. List three types of blood vessels, and identify their functions.

c. What are the functions of blood?

LO 17.1 List the components of the cardiovascular system, and describe several important functions of blood.

Blood is a fluid connective tissue containing plasma and formed elements

Blood is a fluid connective tissue with a unique composition. It consists of **plasma** (PLAZ-muh), a liquid matrix, and **formed elements** (cells and cell fragments). The cardiovascular system of an adult male contains 5–6 liters (5.3–6.4 quarts) of blood; that of an adult female contains 4–5 liters (4.2–5.3 quarts). The difference in blood volume between the sexes is due to differences in average body size. After blood is removed for analysis or storage, the term **whole blood** is used to indicate that the blood composition has not been altered. The components of whole blood can, however, be separated, or fractionated, if only one component is of interest.

1 Plasma forms 55 percent of the volume of whole blood. In many respects, the composition of plasma resembles that of interstitial fluid. This similarity exists because water, ions, and small solutes are continuously exchanged between plasma and interstitial fluids across the walls of capillaries. The primary differences between plasma and interstitial fluid involve (1) the levels of respiratory gases (oxygen and carbon dioxide), due to the respiratory activities of tissue cells, and (2) the concentrations and types of dissolved proteins (because plasma proteins cannot cross capillary walls).

PLASMA COMPOSITION	
Plasma proteins	7%
Other solutes	1%
Water	92%
Transports organic and inorganic molecules, formed elements, and heat	

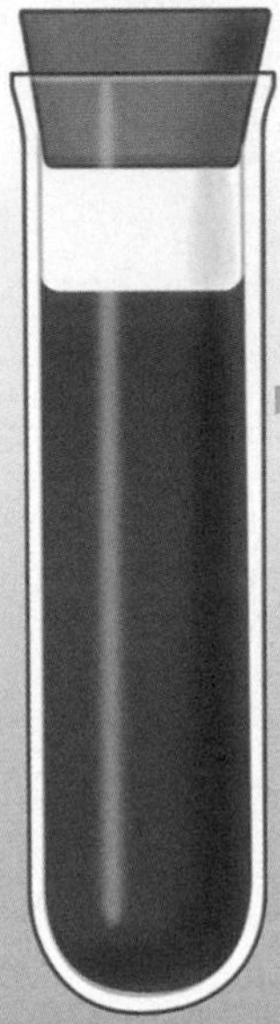

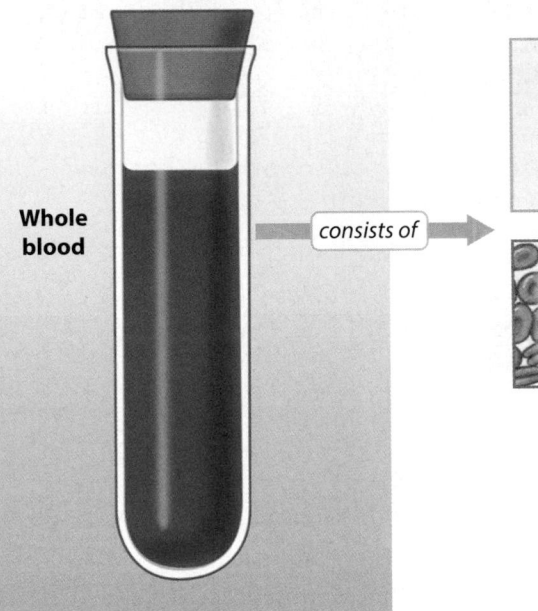

Whole blood *consists of*

Plasma
55%
(range: 46–63%)

+

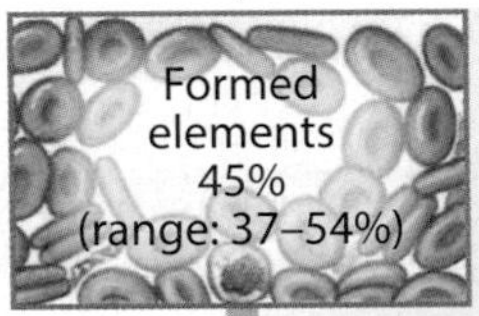
Formed elements
45%
(range: 37–54%)

The **hematocrit** (he-MAT-ō-krit) is the percentage of formed elements in a sample of whole blood. Red blood cells make up 99.9% of these formed elements. In adult males, the normal hematocrit, or **packed cell volume (PCV)**, averages 47 (range: 40–54); the average for adult females is 42 (range: 37–47). The difference in hematocrit between the sexes primarily reflects the fact that androgens (male hormones) stimulate red blood cell production, whereas estrogens (female hormones) do not.

FORMED ELEMENTS	
Platelets	< .1%
White blood cells	< .1%
Red blood cells	99.9%

2 Formed elements are blood cells and cell fragments suspended in plasma. They make up 45 percent of the volume of whole blood.

Properties of Whole Blood

- Blood temperature is about 38°C (100.4°F), slightly above normal body temperature.
- Blood is five times as viscous as water—that is, five times as resistant to flow. Blood's high viscosity results from interactions among dissolved proteins, formed elements, and water molecules in the plasma.
- Blood is slightly alkaline, with a pH between 7.35 and 7.45 (average: 7.4).

Plasma Proteins

Plasma proteins are in solution rather than forming insoluble fibers like those in other connective tissues, such as loose connective tissue or cartilage. On average, each 100 mL of plasma contains 7.6 g of protein, almost five times the concentration in interstitial fluid. The large size and globular shapes of most blood proteins usually prevent them from leaving the bloodstream. The liver synthesizes and releases more than 90 percent of all plasma proteins.

Albumins (al-BŪ-minz) make up about 60 percent of the plasma proteins. As the most abundant plasma proteins, they are major contributors to the osmotic pressure of plasma.

Globulins (GLOB-ū-linz) make up approximately 35 percent of the proteins in plasma. Important plasma globulins include antibodies and transport globulins. Antibodies, also called **immunoglobulins** (i-mū-nō-GLOB-ū-linz), attack foreign proteins and pathogens. **Transport globulins** bind small ions, hormones, lipids, and other compounds.

Fibrinogen (fī-BRIN-ō-jen) functions in clotting and normally makes up about 4 percent of plasma proteins. Under certain conditions, fibrinogen molecules interact to form large, insoluble strands of **fibrin** (FĪ-brin) that form the basic framework for a blood clot.

The plasma also contains active and inactive enzymes and hormones whose concentrations vary widely.

Other Solutes

Other solutes are generally present in plasma in concentrations similar to interstitial fluid. However, differences in the concentrations of nutrients and wastes can exist between arterial blood and venous blood.

Electrolytes: Normal extracellular ion composition is essential for vital cellular activities. The major plasma electrolytes are Na^+, K^+, Ca^{2+}, Mg^{2+}, Cl^-, HCO_3^-, HPO_4^-, and SO_4^{2-}.

Organic nutrients: Organic nutrients are used for ATP production, growth, and cell maintenance. This category includes lipids (fatty acids, cholesterol, and glycerides), carbohydrates (primarily glucose), and amino acids.

Organic wastes: Wastes are carried to sites for breakdown or excretion. Examples of organic wastes include urea, uric acid, creatinine, bilirubin, and ammonium ions.

Platelets

Platelets (PLĀT-lets) are small, membrane-bound cell fragments that contain enzymes and other substances important to blood clotting.

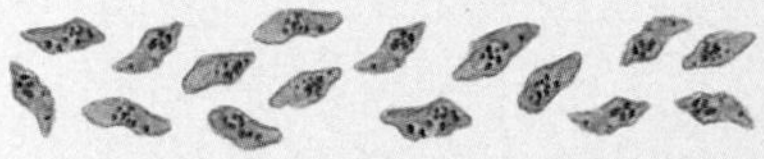

White Blood Cells

White blood cells (WBCs), or leukocytes (LŪ-kō-sīts; *leukos*, white + *-cyte*, cell), play a role in the body's defense mechanisms. There are five classes of leukocytes, each with slightly different functions that will be explored later in the chapter.

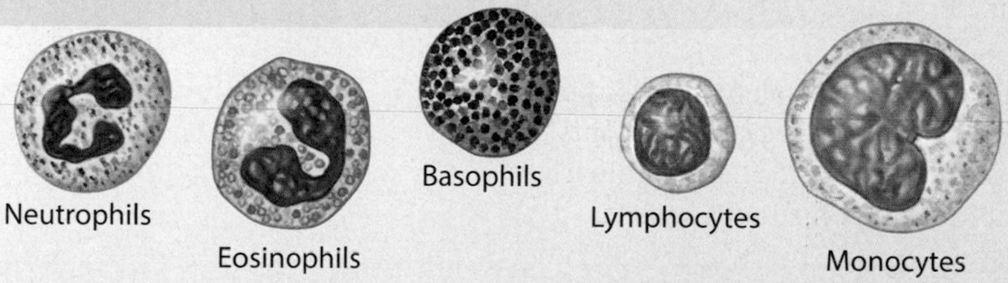
Neutrophils
Eosinophils
Basophils
Lymphocytes
Monocytes

Red Blood Cells

Red blood cells (RBCs), or erythrocytes (e-RITH-rō-sits; *erythros*, red + *-cyte*, cell), are the most abundant blood cells. These specialized cells are essential for oxygen transport in the blood.

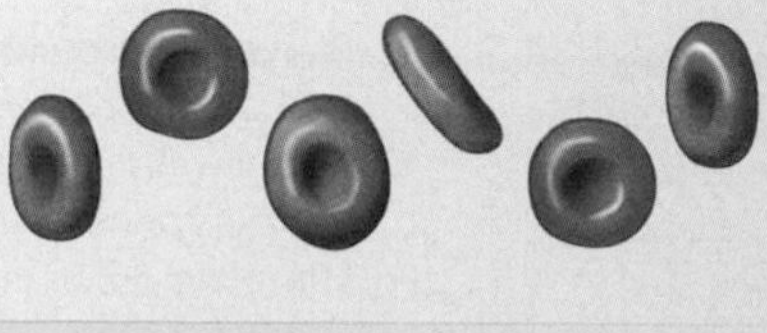

Module 17.2 Review

a. Identify the two components making up whole blood, and list the composition of each.

b. Define hematocrit.

c. Which specific plasma proteins would you expect to be elevated during an infection?

LO 17.2 Describe the important components and major properties of blood.

Module 17.3

Formed elements are produced by stem cells in red bone marrow

1 The formed elements develop in red bone marrow in a process called **hemopoiesis** (hē-mō-poy-Ē-sis (*hemo-*, blood + *poiesis*, a making), or **hematopoiesis**.

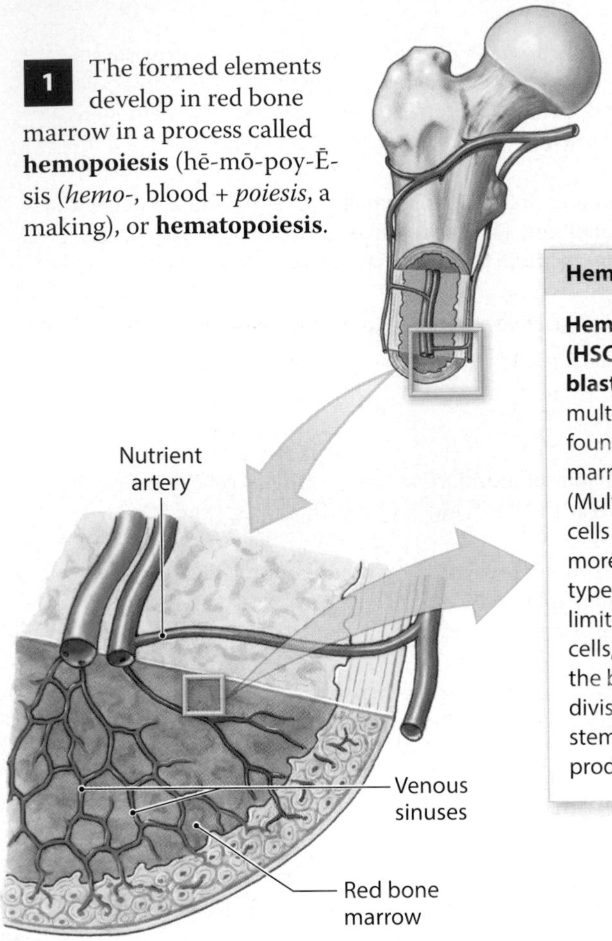
Nutrient artery
Venous sinuses
Red bone marrow

Hematopoietic stem cells (HSCs)

Hematopoietic stem cells (HSCs), also called **hemocytoblasts**, are self-renewing, multipotent stem cells found in the red bone marrow of adults. (Multipotent stem cells can give rise to more than one cell type. They are more limited than pluripotent stem cells, which can give rise to all of the body's cell types.) HSC divisions produce two types of stem cells responsible for producing all formed elements.

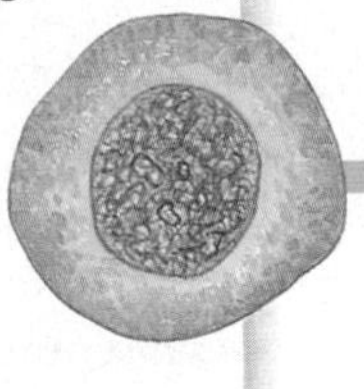

Lymphoid Stem Cells

Lymphoid stem cells, which are responsible for the production of lymphocytes, originate in the red bone marrow. Some remain there, while others migrate to **lymphoid tissues**, including the thymus, spleen, and lymph nodes. As a result, lymphocytes are produced in these organs as well as in the red bone marrow.

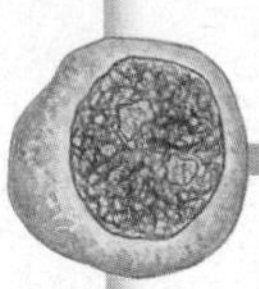

Myeloid Stem Cells

Myeloid stem cells are stem cells in red bone marrow that divide to give rise to all types of formed elements other than lymphocytes.

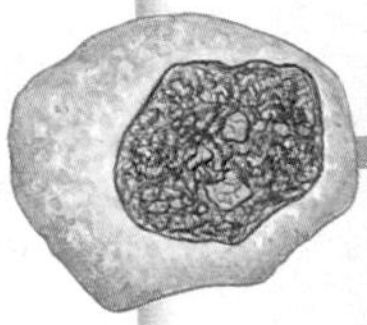

2 The term "formed elements" is appropriate because platelets are cell fragments, rather than specialized cells. This table summarizes important information about platelets.

Structure and Function of Platelets

Appearance in a Stained Blood Smear	Abundance (Average Number per µL)	Function	Remarks
Platelets (PLĀT-lets) are flattened discs that appear round when viewed from above, and spindle-shaped in section or in a blood smear. 	350,000 (range: 150,000–500,000)	Platelets clump together and stick to damaged vessel walls, and they release chemicals that stimulate blood clotting.	Platelets are continuously replaced. Each platelet circulates for 9–12 days before being removed by phagocytes, mainly in the spleen.

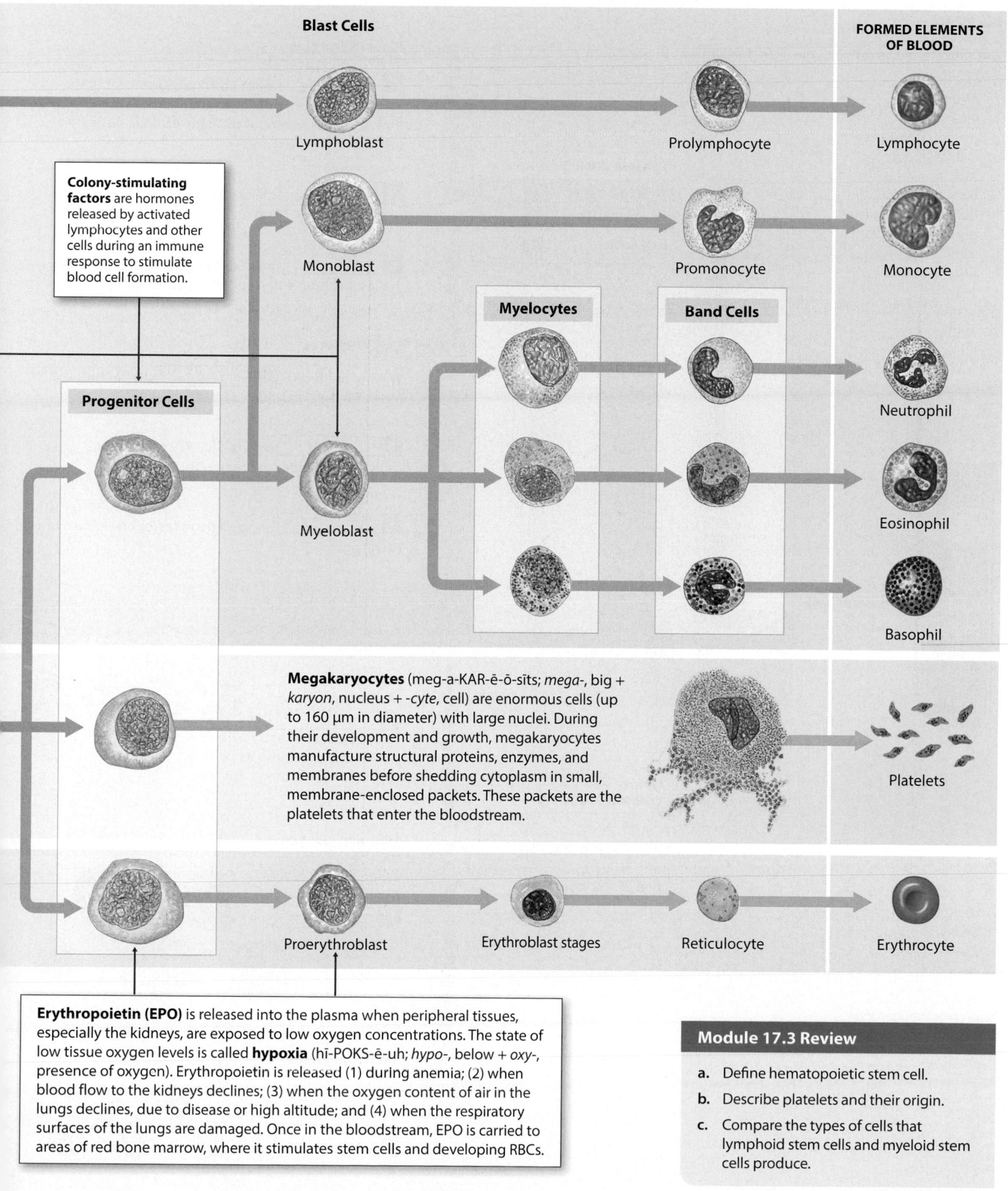

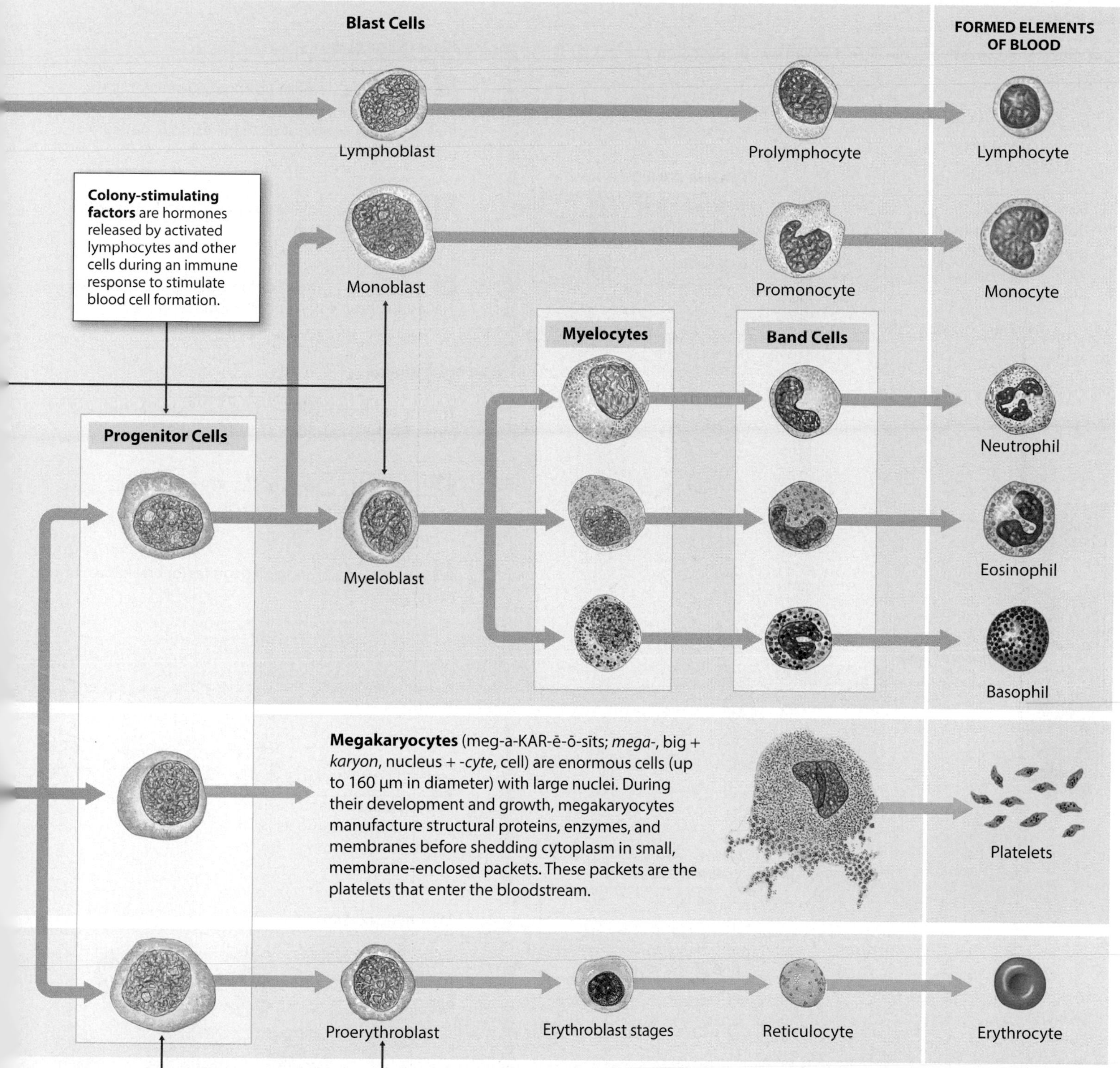
Blast Cells
FORMED ELEMENTS OF BLOOD
Lymphoblast
Prolymphocyte
Lymphocyte
Colony-stimulating factors are hormones released by activated lymphocytes and other cells during an immune response to stimulate blood cell formation.
Monoblast
Promonocyte
Monocyte
Myelocytes
Band Cells
Progenitor Cells
Neutrophil
Myeloblast
Eosinophil
Basophil
Megakaryocytes (meg-a-KAR-ē-ō-sīts; *mega-*, big + *karyon*, nucleus + *-cyte*, cell) are enormous cells (up to 160 μm in diameter) with large nuclei. During their development and growth, megakaryocytes manufacture structural proteins, enzymes, and membranes before shedding cytoplasm in small, membrane-enclosed packets. These packets are the platelets that enter the bloodstream.
Platelets
Proerythroblast
Erythroblast stages
Reticulocyte
Erythrocyte

Erythropoietin (EPO) is released into the plasma when peripheral tissues, especially the kidneys, are exposed to low oxygen concentrations. The state of low tissue oxygen levels is called **hypoxia** (hī-POKS-ē-uh; *hypo-*, below + *oxy-*, presence of oxygen). Erythropoietin is released (1) during anemia; (2) when blood flow to the kidneys declines; (3) when the oxygen content of air in the lungs declines, due to disease or high altitude; and (4) when the respiratory surfaces of the lungs are damaged. Once in the bloodstream, EPO is carried to areas of red bone marrow, where it stimulates stem cells and developing RBCs.

Module 17.3 Review

a. Define hematopoietic stem cell.

b. Describe platelets and their origin.

c. Compare the types of cells that lymphoid stem cells and myeloid stem cells produce.

LO 17.3 Explain the origins and differentiation of the formed elements.

Matching

Use the lettered terms below to fill in the blanks in the diagram.

a. electrolytes
b. globulins
c. 99.9%
d. oxygen transport
e. cell fragments
f. 7%
g. fibrinogen
h. defense mechanisms
i. organic nutrients
j. 92%
k. organic wastes
l. 1%
m. <.1%
n. <.1%
o. albumins

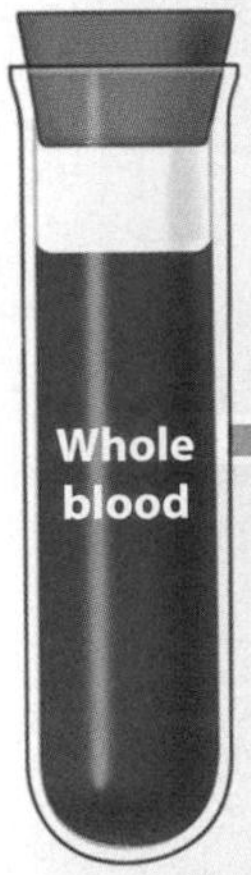
Whole blood

consists of

Plasma

+

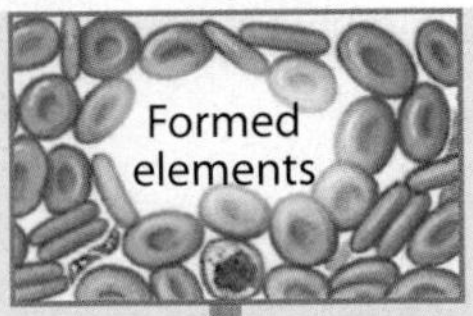
Formed elements

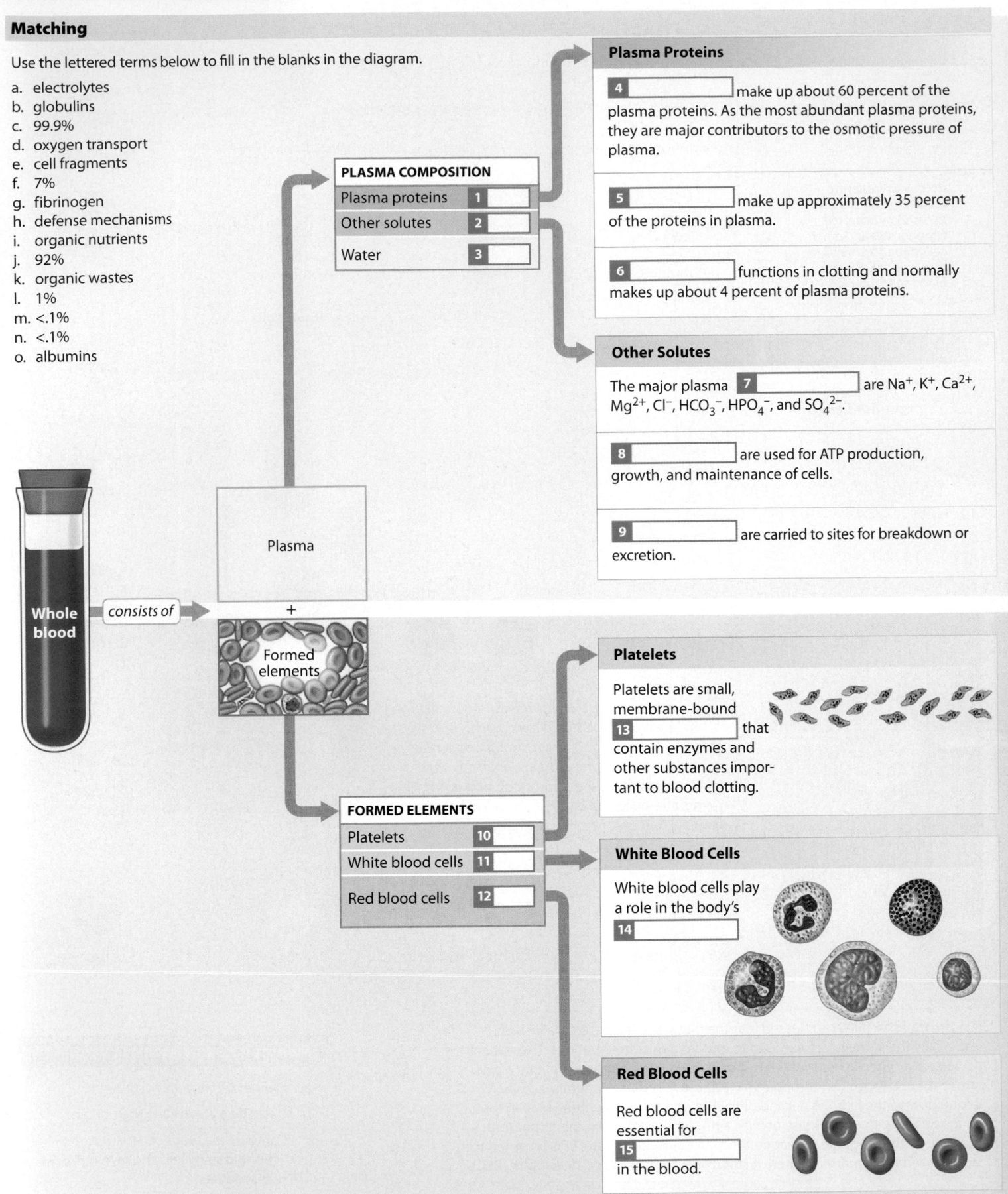

PLASMA COMPOSITION

Plasma proteins 1 ______
Other solutes 2 ______
Water 3 ______

Plasma Proteins

4 ______ make up about 60 percent of the plasma proteins. As the most abundant plasma proteins, they are major contributors to the osmotic pressure of plasma.

5 ______ make up approximately 35 percent of the proteins in plasma.

6 ______ functions in clotting and normally makes up about 4 percent of plasma proteins.

Other Solutes

The major plasma 7 ______ are Na^+, K^+, Ca^{2+}, Mg^{2+}, Cl^-, HCO_3^-, HPO_4^-, and SO_4^{2-}.

8 ______ are used for ATP production, growth, and maintenance of cells.

9 ______ are carried to sites for breakdown or excretion.

FORMED ELEMENTS

Platelets 10 ______
White blood cells 11 ______
Red blood cells 12 ______

Platelets

Platelets are small, membrane-bound 13 ______ that contain enzymes and other substances important to blood clotting.

White Blood Cells

White blood cells play a role in the body's 14 ______

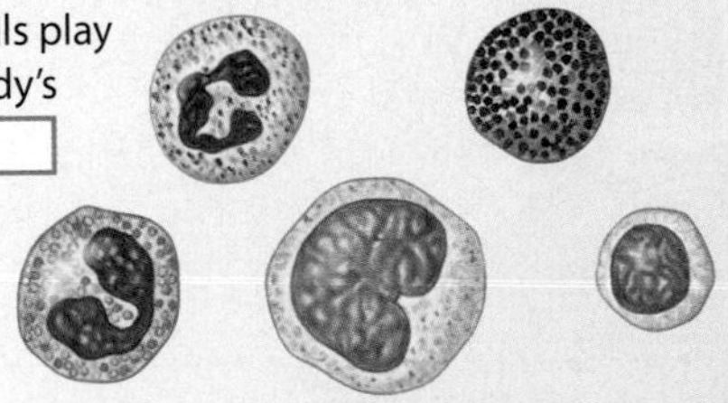

Red Blood Cells

Red blood cells are essential for 15 ______ in the blood.

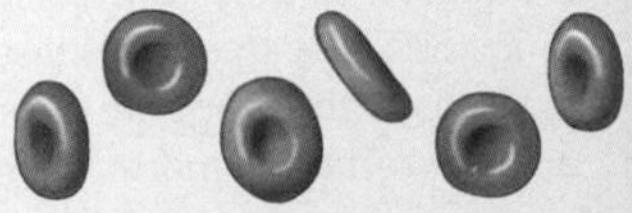

Hematology is the study of blood and blood-forming tissues

Hematology (hē-muh-TOL-ō-jē) is the study of blood, blood-forming tissues, and blood disorders. It is important as a branch of science because much can be learned about a person's health by evaluating blood. Blood provides a readily accessible medium for diagnosing and treating disease. Blood tests can detect disorders such as anemia, infection, and clotting disorders.

1 Since blood courses through nearly every tissue of the body, **dyscrasias** (dis-KRĀ-zē-uz) or blood disorders, can have systemic effects. Blood tests are used for many reasons. Some are simple, such as determining blood type; others are done to evaluate the types and numbers of RBCs, WBCs, and platelets. One test commonly performed is a **complete blood count (CBC)**. A CBC determines the RBC count, WBC count, erythrocyte indices (such as hemoglobin content), hematocrit, and the platelet count in 1 cubic millimeter of blood (1 microliter, μL). A differential count identifies the types and numbers for each white blood cell. Blood clotting tests and bleeding time give information on blood coagulation (Module 17.10). Abnormal values may indicate underlying medical conditions. At right is a vial of whole blood and a representative medical report showing a complete blood count.

Complete Blood Count (CBC)

Hematocrit: 42%
Hemoglobin: 14.9 g/dL
MCH: 30 μg/RBC
MCV: 90 μm³/cell
Platelet count: 350,000/mm³
RBC: 5.2 million/μL
WBC: 7000/μL
WBC differential count:

Neutrophils:	67%
Lymphocytes:	25%
Monocytes:	3%
Eosinophils:	4%
Basophils:	1%

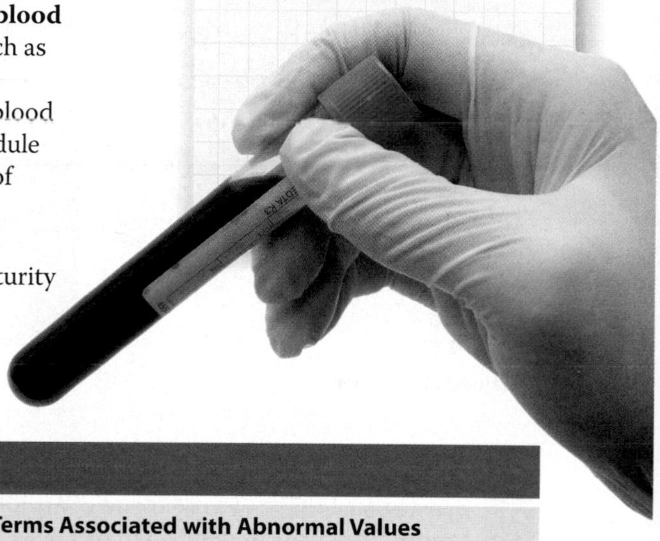
EDTA K2

2 Several common tests focus on RBCs and assess the number, size, shape, and maturity of circulating RBCs, indicating the erythropoietic activities under way (Module 17.6). The tests can also be useful in detecting problems, such as internal bleeding, that may not produce obvious signs or symptoms. The table below lists examples of important RBC tests and terms associated with abnormal values.

RBC Tests and Related Terminology

Test	Determines	Terms Associated with Abnormal Values	
		Elevated	**Depressed**
Hematocrit (Hct)	Percentage of formed elements in whole blood Normal = 37–54%	Polycythemia	Anemia
Hemoglobin concentration (Hb, Hgb)	Concentration of hemoglobin in blood Normal = 12–18 g/dL	Polycythemia	Anemia
Mean corpuscular hemoglobin concentration (MCH)	Average weight of Hb in one RBC Normal = 27–34 μg/RBC (normochromic)	Hyperchromic	Hypochromic
Mean corpuscular volume (MCV)	Average volume of one RBC Normal = 82–101 μm³/cell (normocytic)	Macrocytic	Microcytic
RBC count	Number of RBCs per μL of whole blood Normal = 4.2–6.3 million cells/μL	Erythrocytosis/Polycythemia	Anemia
Reticulocyte count (Retic.)	Percentage of circulating reticulocytes Normal = 0.8%	Reticulocytosis	Diminished erythropoiesis

This section examines the structure and function of the formed elements of blood, hemostasis (how bleeding is stopped), and various blood disorders.

Module 17.4 Review

a. What is hematology?
b. Describe a complete blood count (CBC).
c. Which condition would a patient have if she had a depressed hematocrit level?

LO **17.4** Define hematology, describe the elements of a complete blood count (CBC), and give examples of red blood cell lab tests.

Module 17.5

Red blood cells, the most common formed elements, contain hemoglobin

1 The human body contains an enormous number of red blood cells (RBCs). A standard blood test—the **red blood cell count**—reports the number of RBCs per microliter (μL) of whole blood. In adult males, 1 μL, or 1 cubic millimeter (mm^3), of whole blood contains 4.5–6.3 million RBCs; in adult females, 1 μL contains 4.2–5.5 million. A single drop of whole blood contains approximately 260 million RBCs, and the blood of an average adult has 25 trillion RBCs. RBCs thus account for roughly one-third of all cells in the human body.

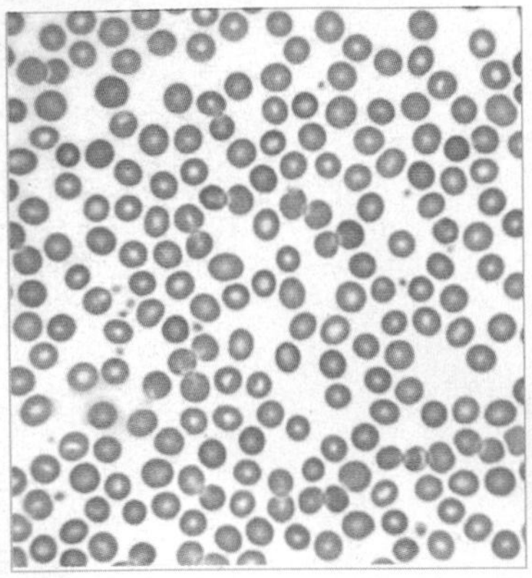

Stained blood smear LM × 450

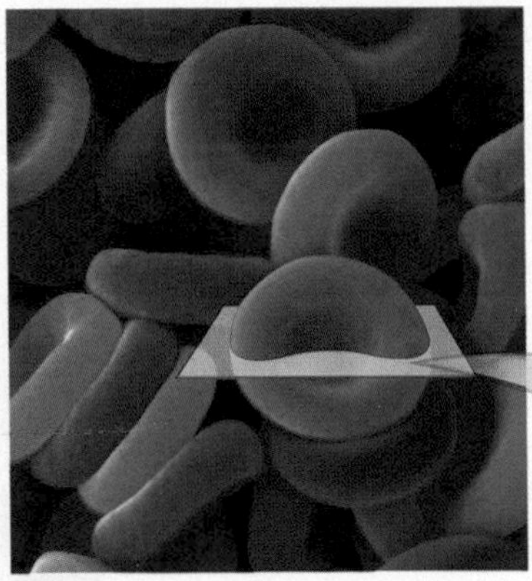

RBCs Colorized SEM × 2100

2 Each RBC is a biconcave disc with a thin central region and a thicker outer margin.

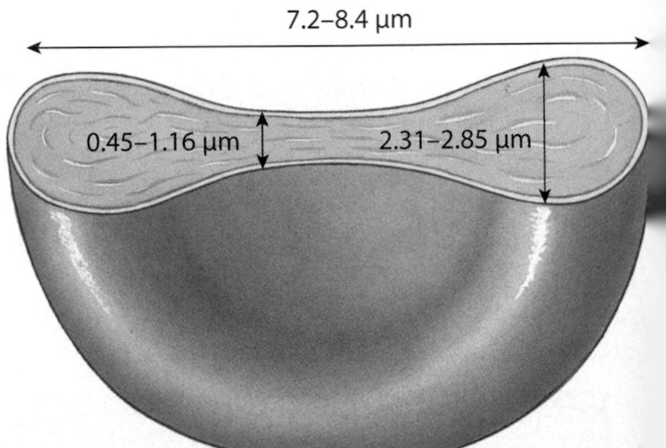
7.2–8.4 μm
0.45–1.16 μm
2.31–2.85 μm

3 The unusual shape of a red blood cell is important for the three reasons detailed in this table.

Functional Aspects of Red Blood Cells

- **Large surface area-to-volume ratio**. Each RBC carries oxygen bound to hemoglobin, an intracellular protein, and that oxygen must be absorbed or released quickly as the RBC passes through the capillaries. The greater the surface area per unit volume, the faster the exchange between the RBC's interior and the surrounding plasma. The total surface area of all the RBCs in the blood of a typical adult is about 3800 square meters, roughly 2000 times the total surface area of the body.
- **RBCs can form stacks**. Like dinner plates, RBCs can form stacks that ease the flow through narrow blood vessels. An entire stack can pass along a blood vessel only slightly larger than the diameter of a single RBC, whereas individual cells would bump the walls, bang together, and form logjams that could restrict or prevent blood flow.
- **Flexibility**. Red blood cells are very flexible and can bend and flex when entering small capillaries and branches. By changing shape, individual RBCs can squeeze through capillaries as narrow as 4 μm.

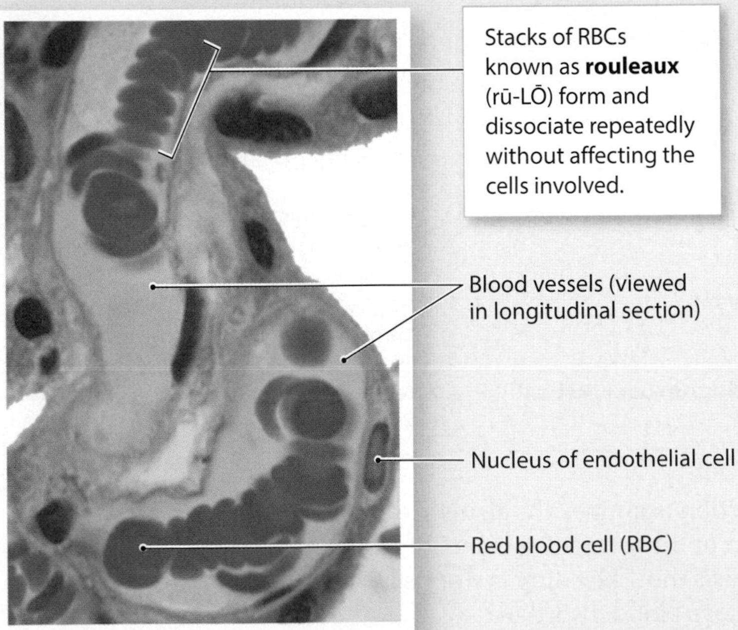
Stacks of RBCs known as **rouleaux** (rū-LŌ) form and dissociate repeatedly without affecting the cells involved.
Blood vessels (viewed in longitudinal section)
Nucleus of endothelial cell
Red blood cell (RBC)

Sectional view of capillaries LM × 1430

4 A red blood cell is very different from the "typical cell" we discussed in Chapter 3. As our RBCs develop, they lose most of their organelles, including nuclei; they retain only the cytoskeleton. Because mature RBCs lack nuclei, a condition called **anucleate**, and also lack ribosomes, they cannot divide or synthesize structural proteins or enzymes. As a result, RBCs cannot repair themselves, and their life span is normally less than 120 days. In effect, a developing RBC loses any organelle not directly associated with this primary function: the transport of respiratory gases. That function is performed by molecules of **hemoglobin (Hb, Hgb)**, which account for more than 95 percent of an RBC's intracellular proteins. The hemoglobin content of whole blood is reported in grams of Hb per deciliter (100 mL) of whole blood (g/dL). Normal ranges are 14–18 g/dL in males and 12–16 g/dL in females.

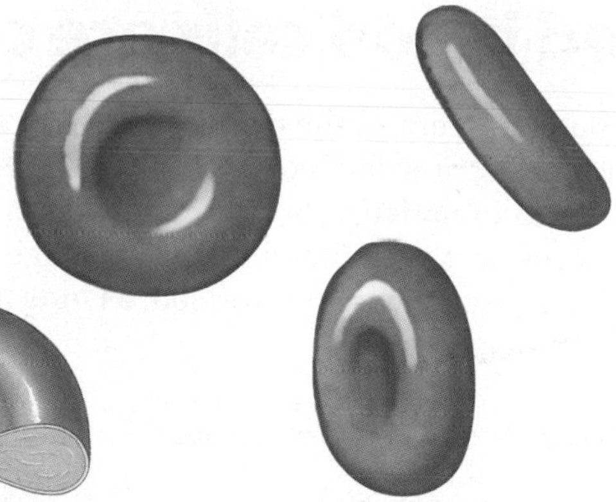

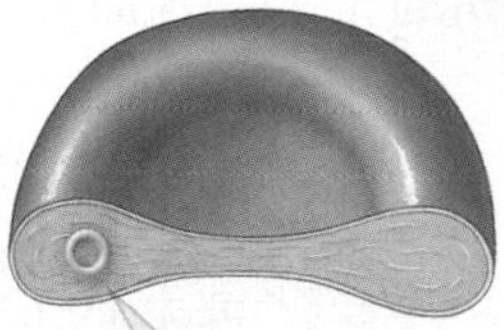

5 Hemoglobin has a complex quaternary structure. Each Hb molecule has two **alpha (α) chains** and two **beta (β) chains** of polypeptides. Each chain is a globular protein subunit that resembles the myoglobin in skeletal and cardiac muscle cells. Like myoglobin, each Hb chain contains a single molecule of **heme**, a nonprotein pigment complex.

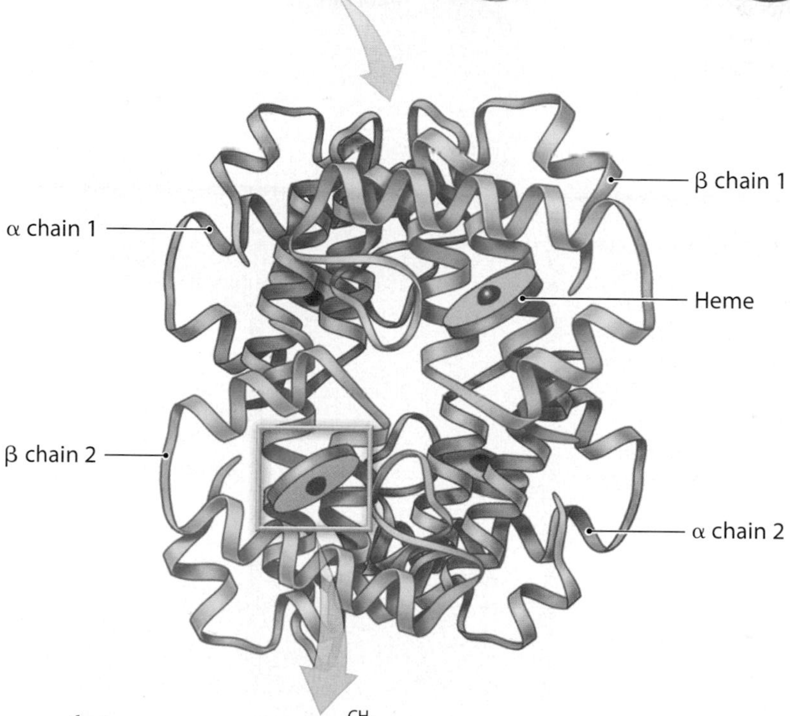
β chain 1
α chain 1
Heme
β chain 2
α chain 2

6 Each heme unit holds an iron ion in such a way that the iron can interact with an oxygen molecule, forming **oxyhemoglobin, HbO_2**. Blood containing RBCs filled with oxyhemoglobin is bright red. The iron–oxygen interaction is very weak, and the two can easily dissociate without damaging the heme unit or the oxygen molecule. The binding of an oxygen molecule to the iron in a heme unit is therefore completely reversible. A hemoglobin molecule whose iron is not bound to oxygen is called **deoxyhemoglobin** or reduced hemoglobin. Blood containing RBCs filled with deoxyhemoglobin is dark red—almost burgundy.

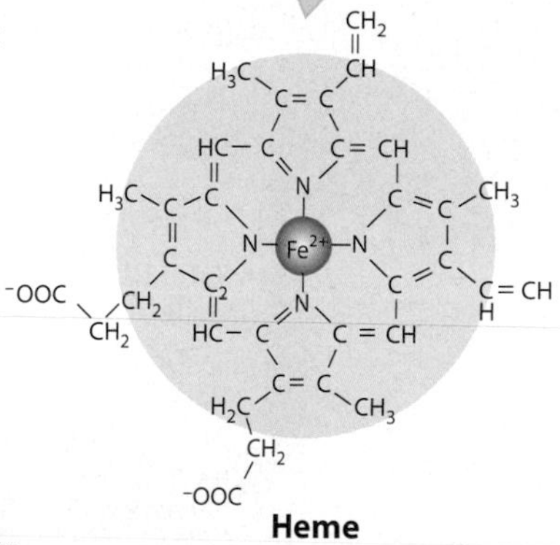
CH_2
CH
H_3C
C= C
HC– C
C= CH
N
H_3C
C
C =
CH_3
N
Fe^{2+}
N
C
-OOC
CH_2
CH_2
N
C= CH
H
HC– C
C = CH
C= C
H_2C
CH_3
CH_2
-OOC
Heme

Each red blood cell contains about 280 million Hb molecules. Because an Hb molecule contains four heme units, each RBC can potentially carry more than a billion molecules of oxygen at a time. Roughly 98.5 percent of the oxygen carried by the blood travels through the bloodstream bound to Hb molecules inside RBCs. The rest is dissolved in the plasma.

Module 17.5 Review

a. Define rouleaux.
b. Describe hemoglobin.
c. Compare oxyhemoglobin with deoxyhemoglobin.

LO 17.5 List the characteristics and functions of red blood cells, and describe the structure and functions of hemoglobin.

Module 17.6

Red blood cells are continually produced and recycled

About 1 percent of the circulating RBCs are replaced each day, and in the process approximately 3 million new RBCs enter the bloodstream each second! Such a rapid rate of replacement is necessary because a typical RBC has a relatively short life span. After it travels about 700 miles in 120 days, either its plasma membrane ruptures or it is engulfed by macrophages in the spleen, liver, or bone marrow. The continual elimination of RBCs usually goes unnoticed, as long as new ones enter the bloodstream at a comparable rate.

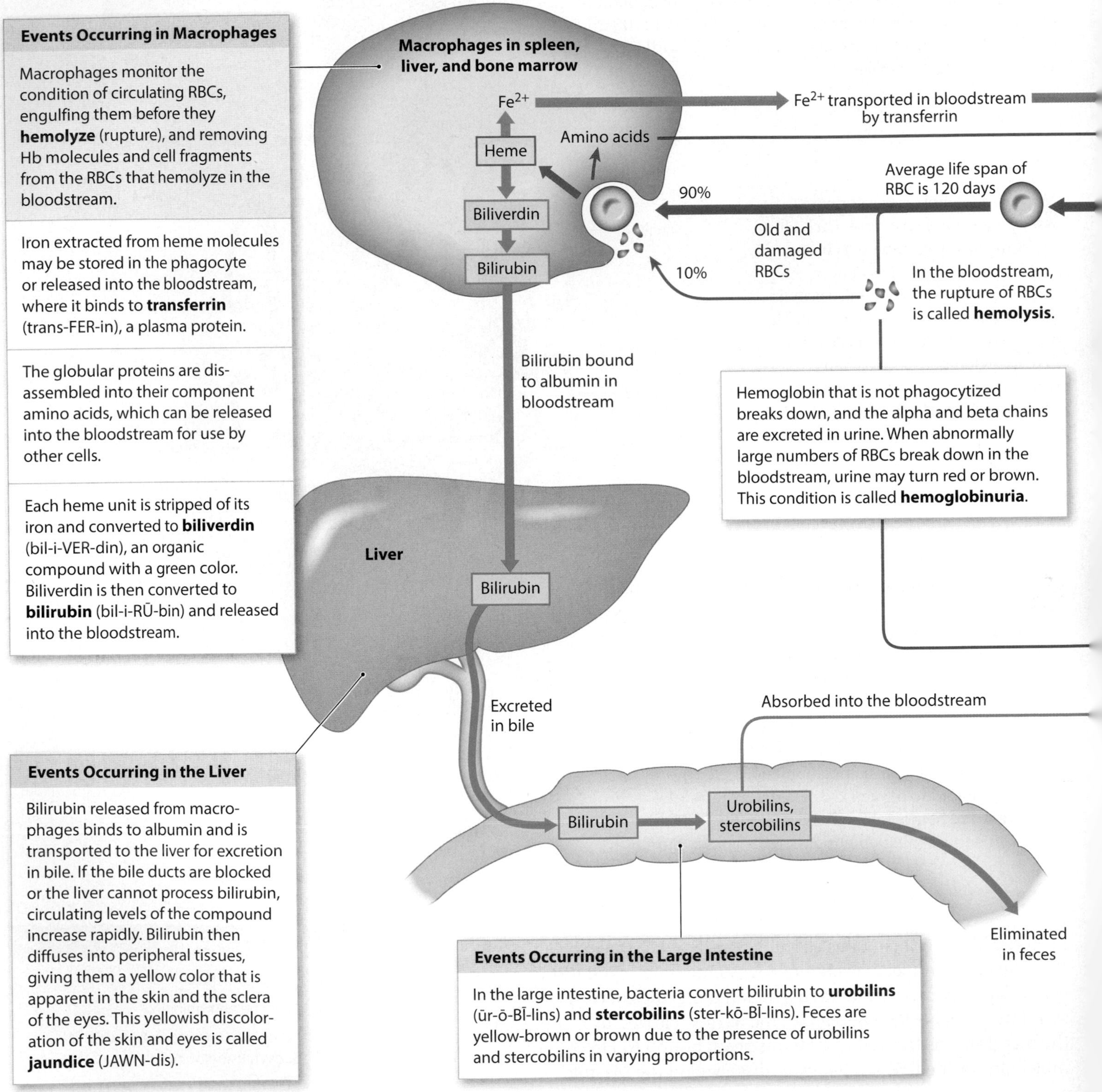

Macrophages in spleen, liver, and bone marrow
Fe^{2+}
Heme
Biliverdin
Bilirubin
Amino acids
Fe^{2+} transported in bloodstream by transferrin
Average life span of RBC is 120 days
90%
Old and damaged RBCs
10%
In the bloodstream, the rupture of RBCs is called **hemolysis**.
Bilirubin bound to albumin in bloodstream
Liver
Bilirubin
Excreted in bile
Absorbed into the bloodstream
Bilirubin
Urobilins, stercobilins
Eliminated in feces

Events Occurring in Macrophages

Macrophages monitor the condition of circulating RBCs, engulfing them before they **hemolyze** (rupture), and removing Hb molecules and cell fragments from the RBCs that hemolyze in the bloodstream.

Iron extracted from heme molecules may be stored in the phagocyte or released into the bloodstream, where it binds to **transferrin** (trans-FER-in), a plasma protein.

The globular proteins are disassembled into their component amino acids, which can be released into the bloodstream for use by other cells.

Each heme unit is stripped of its iron and converted to **biliverdin** (bil-i-VER-din), an organic compound with a green color. Biliverdin is then converted to **bilirubin** (bil-i-RŪ-bin) and released into the bloodstream.

Hemoglobin that is not phagocytized breaks down, and the alpha and beta chains are excreted in urine. When abnormally large numbers of RBCs break down in the bloodstream, urine may turn red or brown. This condition is called **hemoglobinuria**.

Events Occurring in the Liver

Bilirubin released from macrophages binds to albumin and is transported to the liver for excretion in bile. If the bile ducts are blocked or the liver cannot process bilirubin, circulating levels of the compound increase rapidly. Bilirubin then diffuses into peripheral tissues, giving them a yellow color that is apparent in the skin and the sclera of the eyes. This yellowish discoloration of the skin and eyes is called **jaundice** (JAWN-dis).

Events Occurring in the Large Intestine

In the large intestine, bacteria convert bilirubin to **urobilins** (ūr-ō-BĪ-lins) and **stercobilins** (ster-kō-BĪ-lins). Feces are yellow-brown or brown due to the presence of urobilins and stercobilins in varying proportions.

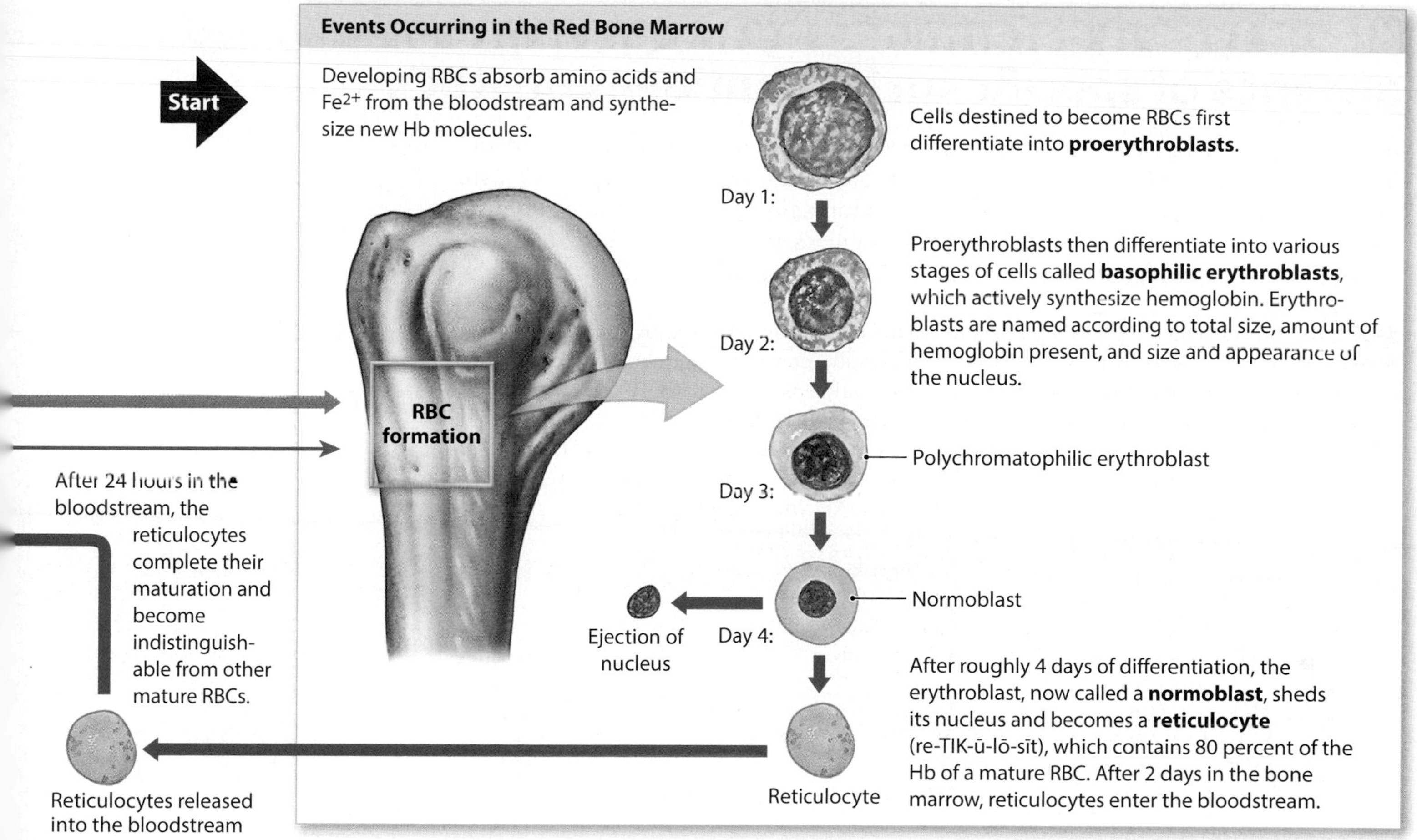
Events Occurring in the Red Bone Marrow
Start
Developing RBCs absorb amino acids and Fe^{2+} from the bloodstream and synthesize new Hb molecules.
Cells destined to become RBCs first differentiate into **proerythroblasts**.
Day 1:
Proerythroblasts then differentiate into various stages of cells called **basophilic erythroblasts**, which actively synthesize hemoglobin. Erythroblasts are named according to total size, amount of hemoglobin present, and size and appearance of the nucleus.
Day 2:
RBC formation
Polychromatophilic erythroblast
Day 3:
After 24 hours in the bloodstream, the reticulocytes complete their maturation and become indistinguishable from other mature RBCs.
Normoblast
Ejection of nucleus
Day 4:
After roughly 4 days of differentiation, the erythroblast, now called a **normoblast**, sheds its nucleus and becomes a **reticulocyte** (re-TIK-ū-lō-sīt), which contains 80 percent of the Hb of a mature RBC. After 2 days in the bone marrow, reticulocytes enter the bloodstream.
Reticulocyte
Reticulocytes released into the bloodstream

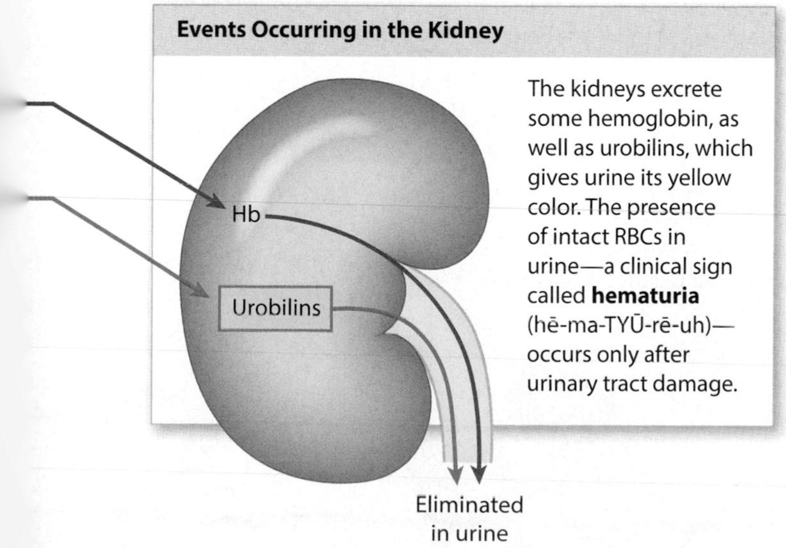
Events Occurring in the Kidney
The kidneys excrete some hemoglobin, as well as urobilins, which gives urine its yellow color. The presence of intact RBCs in urine—a clinical sign called **hematuria** (hē-ma-TYŪ-rē-uh)—occurs only after urinary tract damage.
Hb
Urobilins
Eliminated in urine

In adults, red blood cell formation, or **erythropoiesis** (e-rith-rō-poy-Ē-sis), occurs only in red bone marrow, or **myeloid** (MĪ-e-loyd) **tissue** (*myelos*, marrow). This tissue is located in portions of the vertebrae, sternum, ribs, skull, scapulae, pelvis, and proximal limb bones. Other marrow areas contain a fatty tissue known as yellow bone marrow. Under extreme stimulation, such as severe and sustained blood loss, areas of yellow marrow can convert to red marrow, increasing the rate of RBC formation.

Module 17.6 Review

a. Define hemolysis.

b. Identify the products formed during the breakdown of heme.

c. In what way would a liver disease affect the level of bilirubin in the blood?

LO 17.6 Describe how the components of aged or damaged red blood cells are recycled.

Blood type is determined by the presence or absence of specific surface antigens on RBCs

Antigens are substances that can trigger a protective defense mechanism called an **immune response**. Most antigens are proteins, although some other types of organic molecules are antigens as well. The plasma membranes of your cells contain **surface antigens**, substances your immune system recognizes as "normal" or "self." In other words, your immune system ignores these substances rather than attacking them as "foreign."

1 Your **blood type** is a classification determined by specific surface antigens in RBC plasma membranes. These surface antigens are genetically determined membrane glycoproteins or glycolipids. Your immune system ignores the surface antigens on your own RBCs. However, your plasma may contain antibodies that will attack the antigens on "foreign" RBCs. Although red blood cells have at least 50 kinds of surface antigens, three surface antigens are of particular importance: A, B, and Rh (also called D). There are four blood types based on the presence or absence of the A and B surface antigens. These surface antigens are also known as **agglutinogens**.

Type A	Type B	Type AB	Type O
Type A blood has RBCs with surface antigen A only.	**Type B** blood has RBCs with surface antigen B only.	**Type AB** blood has RBCs with both A and B surface antigens.	**Type O** blood has RBCs without both A and B surface antigens.
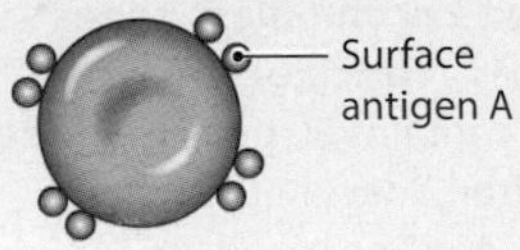Surface antigen A	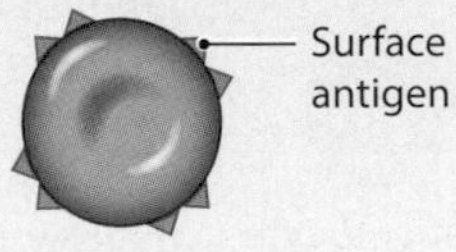Surface antigen B	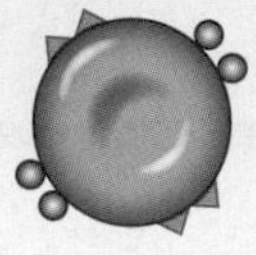	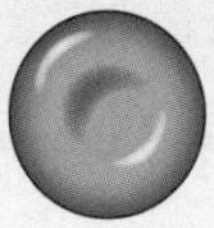
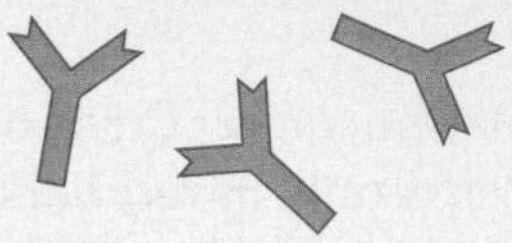			
If you have type A blood, your plasma contains anti-B antibodies, which will attack type B surface antigens.	If you have type B blood, your plasma contains anti-A antibodies.	If you have type AB blood, your plasma has neither anti-A nor anti-B antibodies.	If you have type O blood, your plasma contains both anti-A and anti-B antibodies.

2 If surface antigens on RBCs of one blood type are exposed to the corresponding antibodies also known as **agglutinins**, from another blood type, the RBCs will **agglutinate**, or clump together. This process is called **agglutination**. The cells may also undergo hemolysis. Such a **cross-reaction** is very dangerous because clumps and fragments of RBCs can plug small blood vessels in the kidneys, lungs, heart, or brain, damaging or destroying affected tissues. Accidental cross-reactions may occur if a person being treated for severe blood loss is accidentally given a transfusion of the wrong blood type.

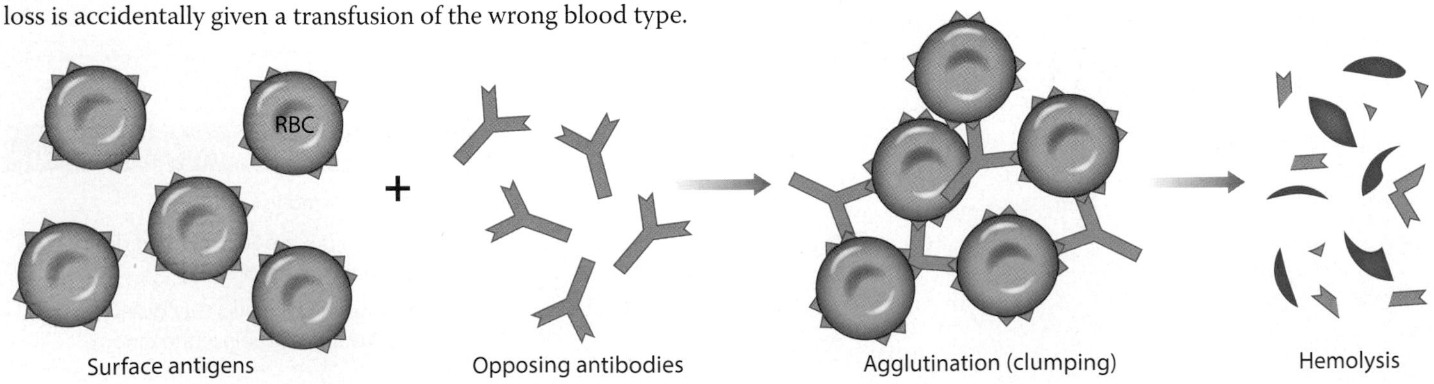
RBC
+
Surface antigens
Opposing antibodies
Agglutination (clumping)
Hemolysis

3 As this table indicates, the various blood types are not evenly distributed throughout the world. The term **Rh positive (Rh⁺)** indicates the presence of the **Rh** surface antigen. It is referred to as "Rh" because this antigen was first discovered in *Rh*esus monkeys before it was found in humans. The absence of this antigen is indicated as **Rh negative (Rh⁻)**. When the complete blood type is recorded, "Rh" is usually omitted and blood type is reported as O negative (O⁻), A positive (A⁺), and so on.

Differences in Blood Type Distribution

Population	Percentage with Each Blood Type				
	O	A	B	AB	Rh⁺
U.S. (Average)	46	40	10	4	85
African American	49	27	20	4	95
Caucasian	45	40	11	4	85
Chinese American	42	27	25	6	100
Filipino American	44	22	29	6	100
Hawaiian	46	46	5	3	100
Japanese American	31	39	21	10	100
Korean American	32	28	30	10	100
Native North American	79	16	4	<1	100
Native South American	100	0	0	0	100
Australian Aborigine	44	56	0	0	100

4 Shown here are the results of blood typing tests on blood samples from four people. Drops of each person's blood are mixed with solutions containing antibodies to the surface antigens A, B, and D (Rh). Clumping (agglutination) occurs when the sample contains the corresponding surface antigen(s). The blood type for each person is shown at right. Because cross-reactions, or **transfusion reactions**, are so dangerous, care must be taken to ensure that the blood types of a blood donor and the recipient are **compatible**—that is, the donor's blood cells and the recipient's plasma will not cross-react.

Anti-A	Anti-B	Anti-D	Blood type	Can receive blood from
			A⁺	A⁺, O⁺, O⁻
		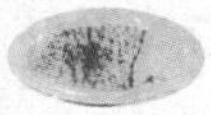	B⁺	B⁺, O⁺, O⁻
		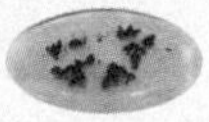	AB⁺	A⁺, B⁺, AB⁺, O⁺, O⁻ (universal recipient)
			O⁻	O⁻ (universal donor)

The presence of anti-A and/or anti-B antibodies is genetically determined and remains constant throughout life, regardless of whether the person has ever been exposed to foreign RBCs. In contrast, the plasma of an Rh-negative person does not necessarily contain anti-Rh antibodies. These antibodies are present only if the person has been sensitized by previous exposure to Rh-positive RBCs.

Module 17.7 Review

a. What is determined by the surface antigens on RBCs?

b. Which blood type(s) can be safely transfused into a person with type O⁻ blood?

c. Why can't a person with type A blood safely receive blood from a person with type B blood?

LO **17.7** Explain the importance of blood typing and the basis for ABO and Rh incompatibilities.

Hemolytic disease of the newborn is an RBC-related disorder caused by a cross-reaction between fetal and maternal blood types

Genes controlling the presence or absence of any surface antigen in the plasma membrane of a red blood cell are provided by both parents, so a child can have a blood type different from that of either parent. During pregnancy, when fetal and maternal vascular systems are closely intertwined, the mother's antibodies against RBC surface antigens may cross the placenta, attacking and destroying fetal RBCs. The resulting condition, called **hemolytic disease of the newborn (HDN)**, has many forms, some so mild as to remain undetected. Those involving the Rh surface antigen are quite dangerous, because unlike anti-A and anti-B antibodies, anti-Rh antibodies are able to cross the placenta and enter the fetal bloodstream.

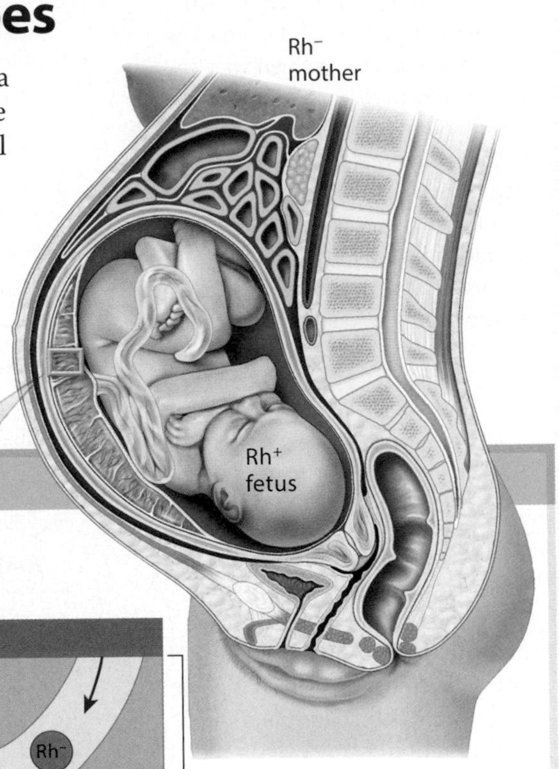

Rh⁻
mother
Rh⁺
fetus

First Pregnancy of an Rh⁻ Mother with an Rh⁺ Infant

The most common form of hemolytic disease of the newborn develops after an Rh⁻ woman has carried an Rh⁺ fetus.

1 Problems seldom develop during a first pregnancy, because very few fetal cells enter the maternal bloodstream then, and thus the mother's immune system is not stimulated to produce anti-Rh antibodies.

2 Exposure to fetal red blood cell antigens generally occurs during delivery, when bleeding takes place at the placenta and uterus. Such mixing of fetal and maternal blood can stimulate the mother's immune system to produce anti-Rh antibodies, leading to **sensitization**.

3 About 20 percent of Rh⁻ mothers who carried Rh⁺ children become sensitized within 6 months of delivery. Because the anti-Rh antibodies are not produced in significant amounts until after delivery, a woman's first infant is not affected.

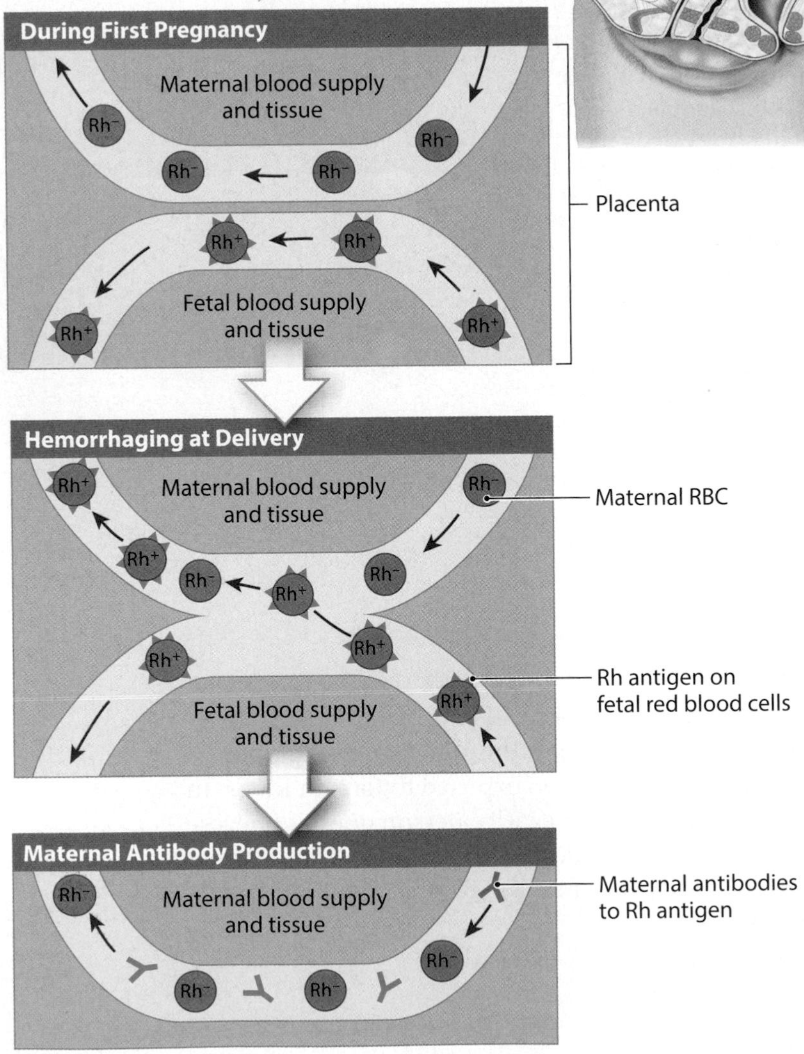

During First Pregnancy
Maternal blood supply and tissue
Rh⁻
Rh⁺
Fetal blood supply and tissue
Placenta
Hemorrhaging at Delivery
Maternal RBC
Rh antigen on fetal red blood cells
Maternal Antibody Production
Maternal antibodies to Rh antigen

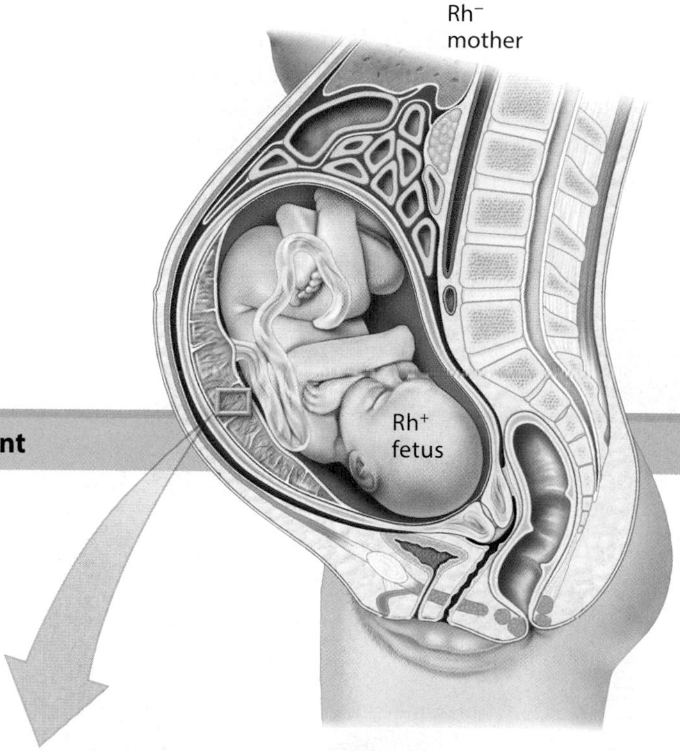

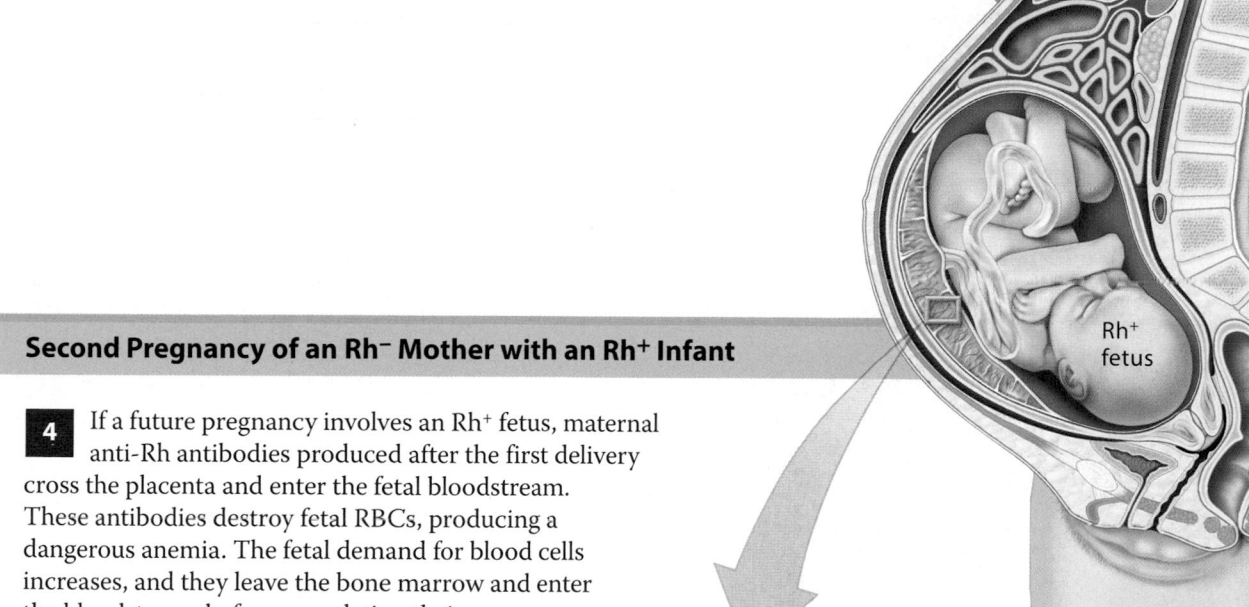

Rh⁻
mother
Rh⁺
fetus

Second Pregnancy of an Rh⁻ Mother with an Rh⁺ Infant

4 If a future pregnancy involves an Rh⁺ fetus, maternal anti-Rh antibodies produced after the first delivery cross the placenta and enter the fetal bloodstream. These antibodies destroy fetal RBCs, producing a dangerous anemia. The fetal demand for blood cells increases, and they leave the bone marrow and enter the bloodstream before completing their development. Because these immature RBCs are erythroblasts, HDN is also known as **erythroblastosis fetalis** (e-rith-rō-blas-TŌ-sis fē-TAL-is). Without treatment, the fetus may die before delivery or shortly thereafter. A newborn with severe HDN is anemic, and the high concentration of circulating bilirubin produces jaundice. Because the maternal antibodies remain active in the newborn for 1–2 months after delivery, the infant's entire blood volume may require replacement to remove the maternal anti-Rh antibodies, as well as the damaged RBCs. Fortunately, the mother's anti-Rh antibody production can be prevented if anti-Rh antibodies (available under the name RhoGAM) are administered to the mother in weeks 26–28 of pregnancy and during and after delivery. These antibodies destroy any fetal RBCs that cross the placenta before they can stimulate a maternal immune response. Because maternal sensitization does not occur, no anti-Rh antibodies are produced. In the United States, this relatively simple procedure has almost entirely eliminated HDN mortality caused by Rh incompatibilities.

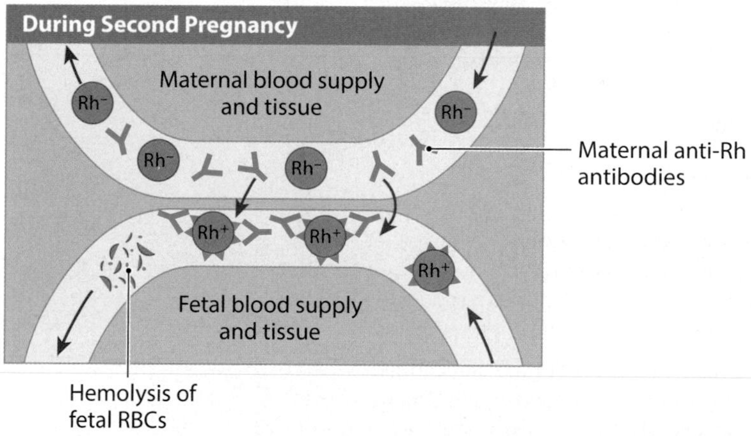

During Second Pregnancy
Maternal blood supply and tissue
Rh⁻
Rh⁻
Rh⁻
Rh⁻
Maternal anti-Rh antibodies
Rh⁺
Rh⁺
Rh⁺
Fetal blood supply and tissue
Hemolysis of fetal RBCs

Module 17.8 Review

a. Define hemolytic disease of the newborn (HDN).

b. Why is RhoGAM administered to pregnant Rh⁻ women?

c. Does an Rh⁺ mother carrying an Rh⁻ fetus require a RhoGAM injection? Explain your answer.

17.8 Describe hemolytic disease of the newborn, explain the clinical significance of the cross-reaction between fetal and maternal blood types, and cite preventive measures.

Module 17.9

White blood cells defend the body against pathogens, toxins, cellular debris, and abnormal or damaged cells

White blood cells (WBCs), or **leukocytes**, share the properties described below. A common phrase to help you remember the types of WBCs is "Never let monkeys eat bananas" where "N" stands for neutrophils, "L" for lymphocytes, "M" for monocytes, "E" for eosinophils, and "B" for basophils.

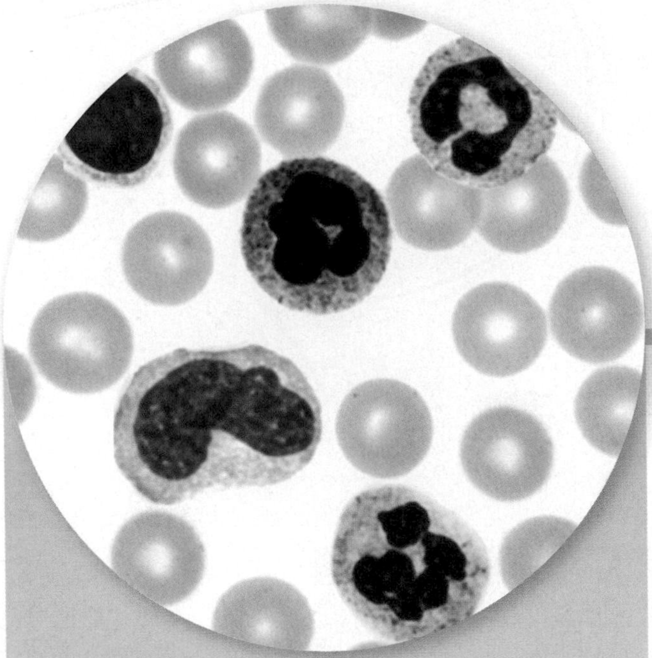

Shared Properties of WBCs

- WBCs circulate for only a short portion of their life span, using the bloodstream primarily to travel between organs and to rapidly reach areas of infection or injury. WBCs spend most of their time migrating through loose and dense connective tissues throughout the body.
- All WBCs can migrate out of the bloodstream. When circulating WBCs in the bloodstream become activated, they contact and adhere to the vessel walls and squeeze between adjacent endothelial cells to enter the surrounding tissue. This process is called **emigration**, or **diapedesis** (*dia*, through + *pedesis*, a leaping).
- All WBCs are attracted to specific chemical stimuli. This characteristic, called **positive chemotaxis** (kē-mō-TAK-sis), guides WBCs to invading pathogens, damaged tissues, and other active WBCs.
- Neutrophils, eosinophils, and monocytes are capable of phagocytosis. These phagocytes can engulf pathogens, cell debris, or other materials. Macrophages are monocytes that have moved out of the bloodstream and have become actively phagocytic.

WBCs can be divided into two classes

1 **Granular leukocytes**, or granulocytes, have abundant cytoplasmic granules (secretory vesicles and lysosomes) that absorb histological stains, such as Wright stain or Giemsa stain.

GRANULAR LEUKOCYTES
Neutrophils
Eosinophils
Basophils

AGRANULAR LEUKOCYTES
Monocytes
Lymphocytes

2 **Agranular leukocytes**, or agranulocytes, contain secretory vesicles and lysosomes smaller than those in granulocytes. However, few, if any, of these cytoplasmic granules absorb histological stain.

White Blood Cells

Cell	Quantity (Average Number per µL)	Appearance in a Stained Blood Smear	Functions	Remarks
Neutrophils 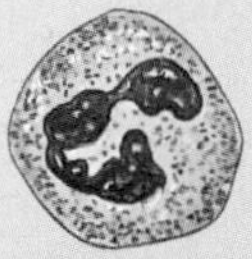	4150 (range: 1800–7300) Differential count: 50–70%	Round cell; nucleus lobed and may resemble a string of beads; cytoplasm contains large, pale inclusions	Phagocytic: engulf pathogens or debris in injured or infected tissues; release cytotoxic enzymes and chemicals	Move into tissues after several hours; may survive for minutes to days, depending on tissue activity; produced in red bone marrow
Eosinophils 	165 (range: 0–700) Differential count: 2–4%	Round cell; nucleus generally has two lobes; cytoplasm contains large granules that generally stain bright red	Phagocytic: engulf antibody-labeled materials; release cytotoxic enzymes; reduce inflammation; increase in abundance in allergies and parasitic infections	Move into tissues after several hours; may survive for minutes to days, depending on activity in tissues; produced in red bone marrow
Basophils 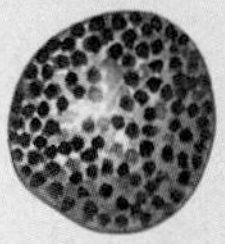	44 (range: 0–150) Differential count: <1%	Round cell; nucleus generally cannot be seen through dense, blue-stained granules in cytoplasm	Enter damaged tissues and release histamine and other chemicals that promote inflammation	Survival time unknown; assist mast cells of tissues in producing inflammation; produced in red bone marrow
Monocytes 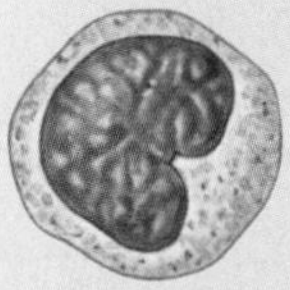	456 (range: 200–950) Differential count: 2–8%	Very large cell; nucleus kidney bean–shaped; abundant cytoplasm	Enter tissues and become macrophages; engulf pathogens or debris	Move into tissues after 1–2 days; survive for months or longer; produced primarily in red bone marrow
Lymphocytes 	2185 (range: 1500–4000) Differential count: 20–40%	Generally round cell, slightly larger than RBC; round nucleus; very little cytoplasm	Cells of lymphatic system; provide defense against specific pathogens or toxins	Survive for months to decades; circulate from blood to tissues and back; produced in red bone marrow and lymphoid tissues

A variety of conditions, including infections, inflammation, and allergic reactions, cause characteristic changes in the populations of circulating WBCs. Examining a stained blood smear can provide a window on such changes. A **differential count** identifies the types and numbers of white blood cells in a sample of blood. The values reported indicate the number, as a percentage, of each type of cell in a sample of 100 WBCs (Module 17.4).

Module 17.9 Review

a. Identify the five types of white blood cells.

b. Which type of white blood cell would you find in the greatest numbers in an infected cut?

c. How do basophils respond during inflammation?

LO **17.9** Categorize the various types of white blood cells on the basis of their structures and functions.

Module 17.10

The clotting response is a complex cascade of events that reduces blood loss

The process of **hemostasis** (*haima*, blood + *stasis*, halt) is responsible for stopping the loss of blood through the walls of damaged vessels. At the same time, it establishes a framework for tissue repairs. Although usually divided into three phases, hemostasis is a complex cascade in which many things happen at once, and all of them interact to some degree.

Vascular Phase

The **vascular phase** of hemostasis lasts for roughly 30 minutes after the injury occurs. It is dominated by the response of the endothelial cells and the contraction, or **vascular spasm**, of smooth muscle of the vessel walls.

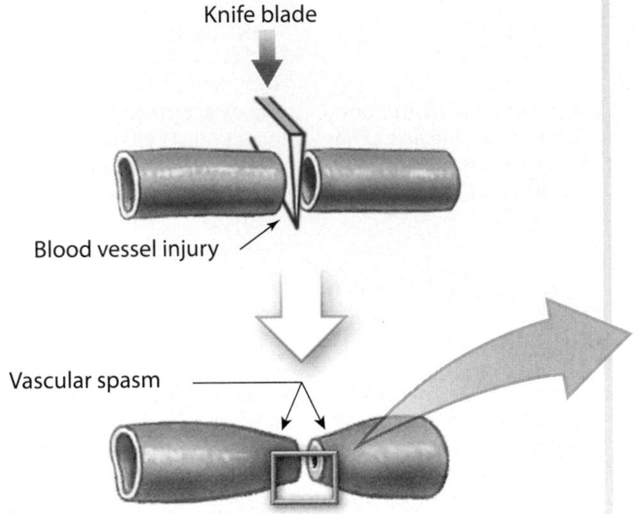

Knife blade
Blood vessel injury
Vascular spasm

Events of the Vascular Phase

- The endothelial cells contract and expose the underlying basement membrane to the bloodstream.
- The endothelial cells begin releasing chemical factors and local hormones. Endothelial cells also release **endothelins**, peptide hormones that (1) stimulate smooth muscle contraction and promote vascular spasms and (2) stimulate the division of endothelial cells, smooth muscle cells, and fibroblasts to accelerate the repair process.
- The endothelial plasma membranes become "sticky." A tear in the wall of a small artery or vein may be partially sealed off by the attachment of endothelial cells on either side of the break. In small capillaries, endothelial cells on opposite sides of the vessel may stick together and prevent blood flow along the damaged vessel. The stickiness also facilitates the attachment of platelets as the platelet phase gets under way.

Platelet Phase

The **platelet phase** of hemostasis begins with the attachment of platelets to sticky endothelial surfaces, to the basement membrane, to exposed collagen fibers, and to each other.

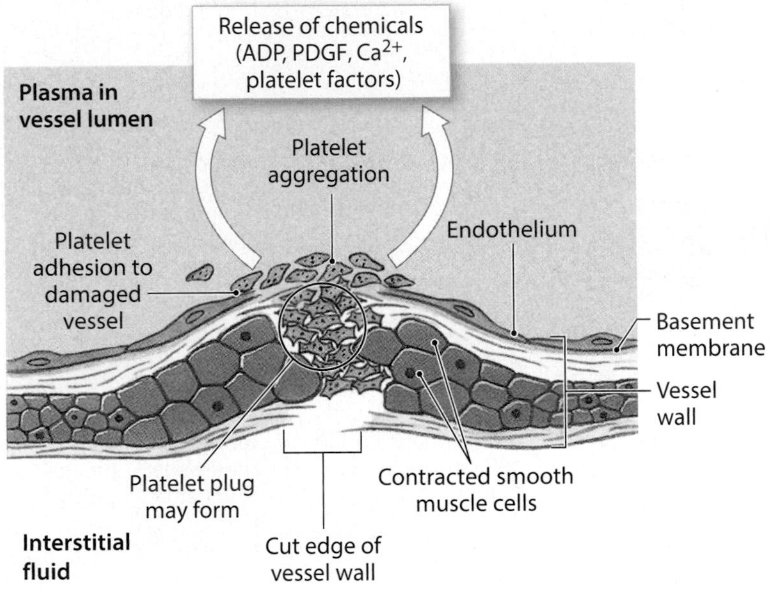

Release of chemicals (ADP, PDGF, Ca^{2+}, platelet factors)
Plasma in vessel lumen
Platelet aggregation
Endothelium
Platelet adhesion to damaged vessel
Basement membrane
Vessel wall
Platelet plug may form
Contracted smooth muscle cells
Interstitial fluid
Cut edge of vessel wall

Chemicals Released by Activated Platelets

- Adenosine diphosphate (ADP), which stimulates platelet aggregation and secretion
- Several chemicals that stimulate vascular spasms
- **Platelet factors**, proteins that play a role in blood clotting
- **Platelet-derived growth factor (PDGF)**, a peptide that promotes vessel repair
- Calcium ions (Ca^{2+}), which are required for platelet aggregation and in several steps in the clotting process

Coagulation Phase

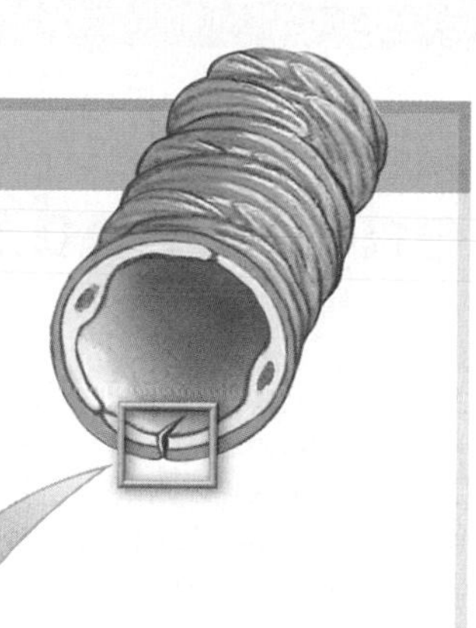

The **coagulation** (cō-ag-ū-LĀ-shun) **phase** of hemostasis does not start until 30 seconds or more after the vessel has been damaged. **Coagulation**, or blood clotting, involves a complex sequence of steps leading to the conversion of circulating fibrinogen (a soluble protein) into **fibrin** (an insoluble protein). As the fibrin network grows, blood cells and additional platelets are trapped in the fibrous tangle, forming a blood clot that seals off the damaged portion of the vessel. **Procoagulants** (clotting factors) in the plasma play a key role in this phase. Important clotting factors include Ca^{2+} and 11 different proteins (identified by Roman numerals). Many clotting factors are proenzymes, which, when converted to active enzymes, direct essential reactions in the clotting response. The activation of one proenzyme commonly creates a chain reaction, or **cascade**.

Common Pathway

The **common pathway** begins when enzymes from either the extrinsic or intrinsic pathway activate Factor X. Activated Factor X activates a complex called prothrombin activator. Prothrombin activator converts the proenzyme **prothrombin** into the enzyme **thrombin** (THROM-bin). Thrombin then completes the clotting process by converting fibrinogen to fibrin.

Extrinsic Pathway

The **extrinsic pathway** begins with the release of **tissue factor** (Factor III) by damaged endothelial cells or peripheral tissues. The greater the damage, the more tissue factor is released and the faster clotting occurs. Tissue factor then combines with Ca^{2+} and another clotting factor to form an enzyme complex capable of activating Factor X, the first step in the common pathway.

Intrinsic Pathway

The **intrinsic pathway** begins with the activation of proenzymes exposed to collagen fibers at the injury site. This pathway proceeds with the assistance of **PF-3**, a platelet factor released by aggregating platelets. After a series of linked reactions, activated clotting factors combine to form an enzyme complex capable of activating Factor X.

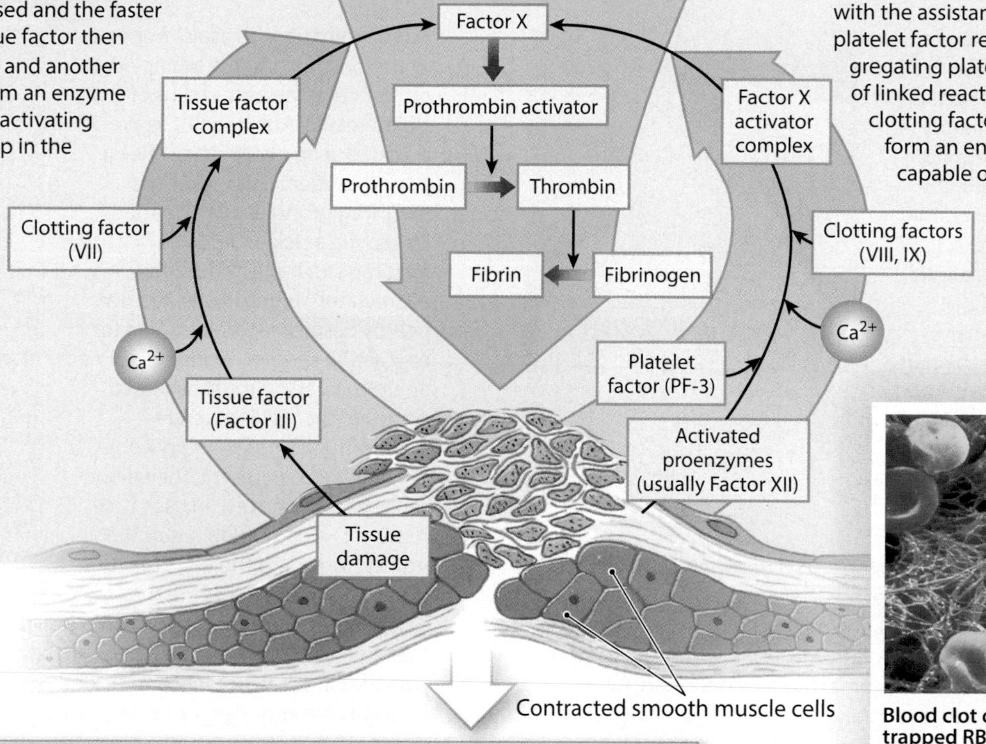

Factor X
Prothrombin activator
Prothrombin
Thrombin
Fibrin
Fibrinogen
Tissue factor complex
Clotting factor (VII)
Ca2+
Tissue factor (Factor III)
Tissue damage
Factor X activator complex
Clotting factors (VIII, IX)
Ca2+
Platelet factor (PF-3)
Activated proenzymes (usually Factor XII)
Contracted smooth muscle cells

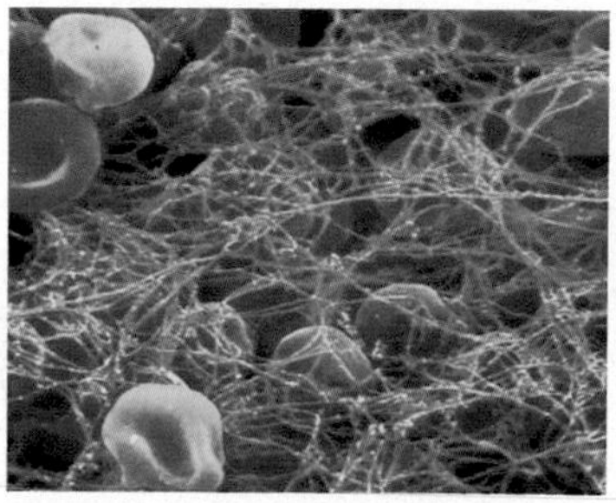

Blood clot containing trapped RBCs SEM × 1200

Clot Retraction

Once the fibrin meshwork has formed, platelets and red blood cells stick to the fibrin strands. The platelets then contract, and the entire clot begins to undergo **clot retraction**, a process that continues over a period of 30–60 minutes and pulls the cut edges together.

As repairs proceed, the clot gradually dissolves. This process, called **fibrinolysis** (fī-bri-NOL-i-sis), begins with the activation of the proenzyme **plasminogen** by thrombin, produced by the common pathway, and **tissue plasminogen activator (t-PA)**, released by damaged tissues. The activation of plasminogen produces the enzyme **plasmin** (PLAZ-min), which erodes the foundation of the clot.

Module 17.10 Review

a. Define hemostasis.

b. Briefly describe the vascular, platelet, and coagulation phases of hemostasis.

c. List the following events in the process of hemostasis and clot dissolution in the proper order of their occurrence: coagulation, fibrinolysis, vascular spasm, retraction, platelet phase.

17.10 Discuss the mechanisms that control blood loss after an injury, and describe the reaction sequences responsible for blood clotting.

Blood disorders can be classified by their origins and the changes in blood characteristics

1 The procedure called **venipuncture** (VĒN-i-punk-chur; *vena*, vein + *punctura*, a piercing) is crucial in diagnosing blood disorders. Fresh whole blood is generally collected from a superficial vein, such as the median cubital vein on the anterior surface of the elbow. Venipuncture is commonly used because (1) superficial veins are easy to locate, (2) the walls of veins are thinner than those of comparably sized arteries, and (3) blood pressure in the venous system is relatively low, so the puncture wound seals quickly. The most common clinical procedures examine venous blood.

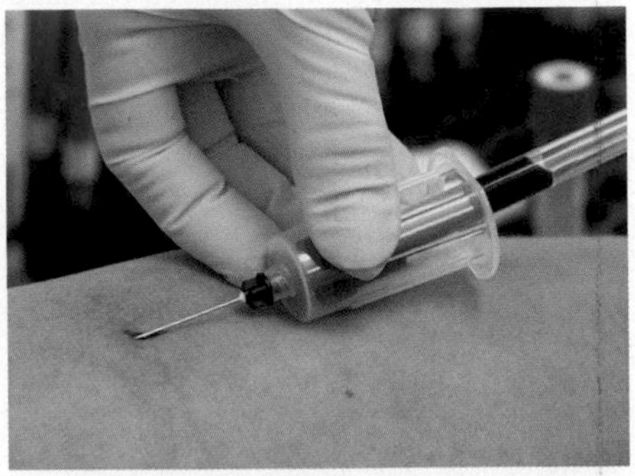

2 Blood disorders can have a variety of causes. The major categories of blood disorders are introduced here.

Nutritional Blood Disorders

In **iron deficiency anemia**, normal hemoglobin synthesis cannot occur because either iron reserves or the dietary intake of iron is inadequate. Because developing RBCs cannot synthesize functional hemoglobin, they are unusually small **(microcytic)**. Women are especially dependent on a normal dietary supply of iron because their iron reserves are about one-half that of a typical man.

A deficiency in **vitamin B_{12}** prevents normal stem cell divisions in the bone marrow, which can result in **pernicious** (per-NISH-us) **anemia**. Fewer red blood cells are produced, and those that are produced are abnormally large **(macrocytic)** and may develop bizarre shapes. Pernicious anemia can also result from a lack of intrinsic factor, a mucoprotein secreted by the stomach that is necessary for adequate vitamin B_{12} absorption.

Calcium ions and **vitamin K** affect almost every aspect of the clotting process. All three pathways (intrinsic, extrinsic, and common) require Ca^{2+}, so any disorder that lowers blood Ca^{2+} levels can impair blood clotting. Additionally, the liver requires adequate amounts of vitamin K for synthesizing four of the clotting factors, including prothrombin.

Congenital Blood Disorders

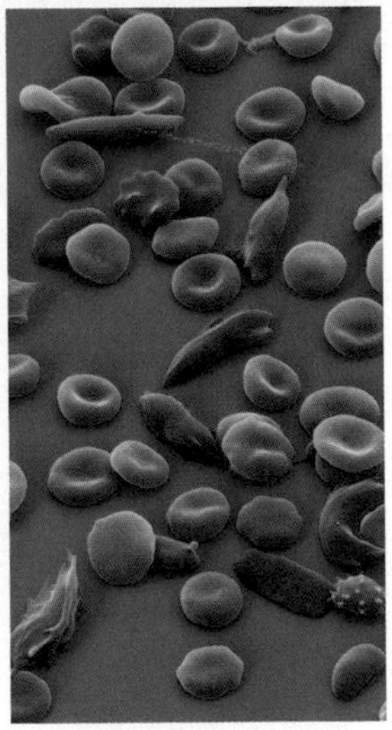

Normal and sickled RBCs

Sickle cell anemia results from a mutation affecting the amino acid sequence of the beta chains of the Hb molecule. Affected RBCs take on a sickled shape when they release bound oxygen. This makes the RBCs fragile and easily damaged. Moreover, a sickled RBC can become stuck in a narrow capillary. A circulatory blockage results, and nearby tissues become starved for oxygen. To develop sickle cell anemia, a person must have two copies of the sickling gene—one from each parent. If only one sickling gene is present, the person has the **sickling trait**. In such cases, most of the hemoglobin is of the normal form, and the RBCs function normally. However, having the sickling trait gives the person some resistance to malaria, a deadly mosquito-borne parasitic disease. Infection of RBCs by the parasites induces sickling, and sickled cells containing the parasites are engulfed and destroyed by macrophages.

Hemophilia (hē-mō-FĒ-lē-a) is an inherited bleeding disorder. About 1 person in 10,000 is a hemophiliac, and of those, 80–90 percent are males. In most cases, hemophilia is caused by the reduced production of a single clotting factor. The severity of hemophilia varies, depending on how little clotting factor is produced. In severe cases, extensive bleeding occurs with relatively minor contact, and bleeding occurs at joints and around muscles.

Thalassemias (thal-ah-SĒ-mē-uhs) are a diverse group of inherited blood disorders caused by an inability to produce adequate amounts of normal protein subunits of hemoglobin. The severity of different types of thalassemias depends on which and how many protein subunits are abnormal.

Blood Infections

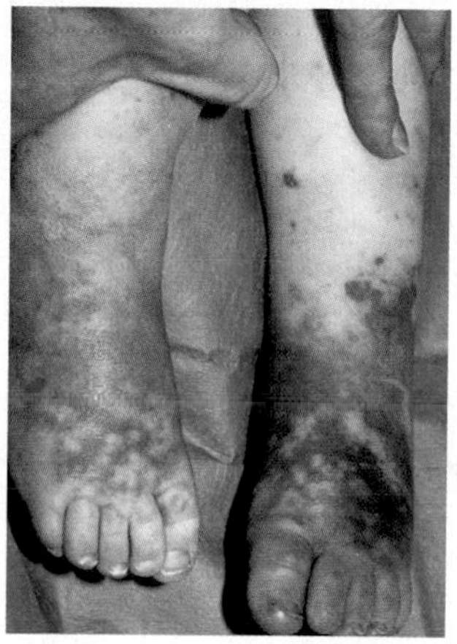
Infant with septicemia from a meningococcal bacteria

Blood is normally free of microorganisms, but they can enter the blood through a wound or infection. **Bacteremia** is a condition in which bacteria circulate in blood, but do not multiply there. **Viremia** is a similar condition associated with viruses. **Sepsis** (SEP-sis) is a widespread pathogenic infection of body tissues. Sepsis of the blood, or **septicemia** (formerly known as "blood poisoning"), results if pathogens are present and multiplying in the blood, and spreading throughout the body.

Mosquito biting a human

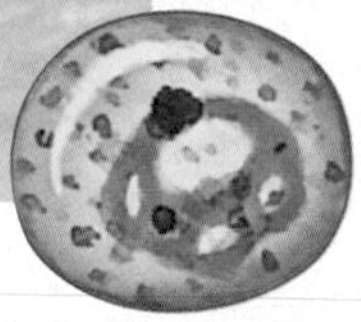
Red blood cell infected with *Plasmodium*

Malaria is a parasitic disease caused by several species of the protozoan *Plasmodium*. It is one of the most severe diseases in tropical countries, killing 1.5–3 million people per year, of which up to half are children under the age of 5. Malaria is transmitted from person to person by a mosquito. The parasite initially infects liver cells, then enlarges and fragments into smaller forms that infect red blood cells. Periodically, at intervals of 2–3 days, all of the infected RBCs rupture simultaneously and release more parasites that infect additional RBCs. The release and reinfection of RBCs corresponds to the cycles of fever and chills that characterize malaria. Dead RBCs can block blood vessels leading to vital organs, such as the kidney and brain, resulting in tissue death.

Blood Cell Cancers

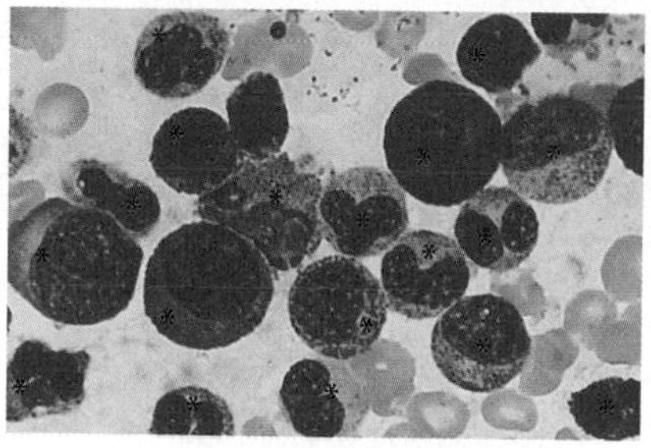
Abnormal WBCs (*) seen in a blood smear of a patient with myeloid leukemia

Leukemias are cancers of blood-forming tissues. The cancerous cells of leukemia do not form a compact tumor, but instead are spread throughout the body from their origin in red bone marrow. Both of the two types of leukemia—myeloid and lymphoid—are characterized by elevated levels of circulating WBCs. **Myeloid leukemia** is characterized by the presence of abnormal granulocytes (neutrophils, eosinophils, and basophils) or other cells of the bone marrow. **Lymphoid leukemia** involves lymphocytes and their stem cells. The first symptoms appear when immature and abnormal white blood cells appear in the bloodstream. Untreated leukemia is invariably fatal.

Degenerative Blood Disorders

In **disseminated intravascular coagulation (DIC)**, bacterial toxins activate several steps in the coagulation process that converts fibrinogen to fibrin within the circulating blood. Although much of the fibrin is removed by phagocytes or is dissolved by plasmin, small clots may block small vessels and damage nearby tissues. If the liver cannot produce enough circulating fibrinogen to keep pace with the rate at which fibrinogen is being removed, clotting abilities decline and uncontrolled bleeding may occur.

Module 17.11 Review

a. Define venipuncture.

b. Compare pernicious anemia with iron deficiency anemia.

c. Identify the two types of leukemia.

17.11 Explain how blood disorders are detected, and describe examples of the various categories of blood disorders.

Concept map

Use each of the following terms once to fill in the blank boxes to correctly complete the map.

- globulins
- electrolytes, glucose, urea
- proteins
- albumins
- leukocytes
- fibrinogen
- solutes
- basophils
- erythrocytes
- formed elements
- plasma
- eosinophils
- lymphocytes
- water
- platelets
- monocytes
- neutrophils

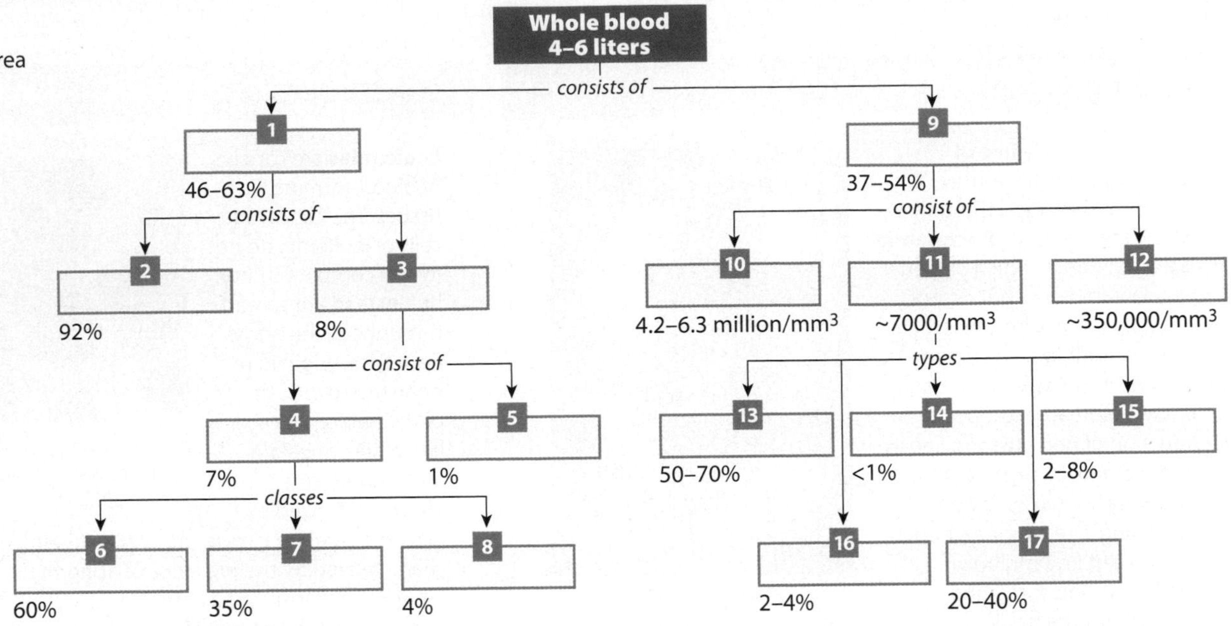
Whole blood
4–6 liters
consists of
1
46–63%
consists of
2
92%
3
8%
consist of
4
7%
5
1%
classes
6
60%
7
35%
8
4%
9
37–54%
consist of
10
4.2–6.3 million/mm3
11
~7000/mm3
12
~350,000/mm3
types
13
50–70%
14
<1%
15
2–8%
16
2–4%
17
20–40%

Matching

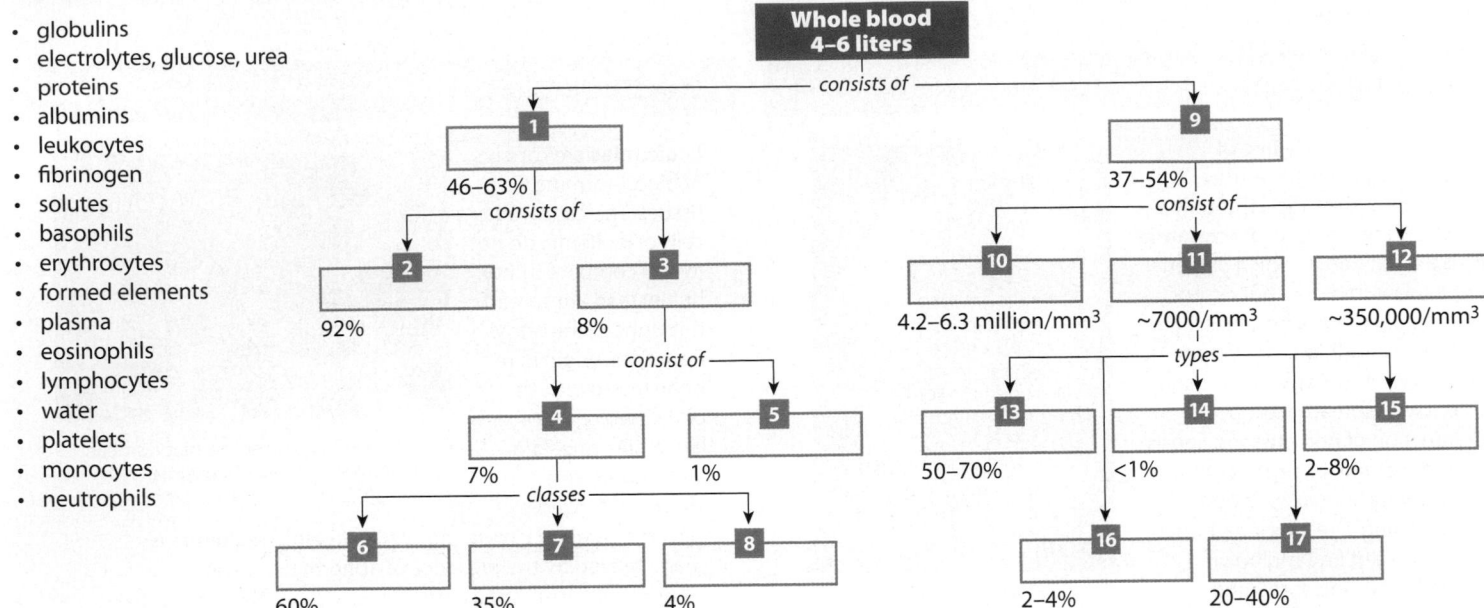

Match each lettered description with the most closely related term.

a. matrix
b. transport protein
c. jaundice
d. venipuncture
e. red bone marrow
f. mature RBCs
g. pigment complex
h. platelets
i. erythropoietin
j. cross-reaction
k. monocytes
l. lymphocytes

18 Myeloid tissue
19 Anucleated
20 Plasma
21 Macrophages
22 Globulin
23 Agglutination
24 Specific immunity
25 Bilirubin
26 Median cubital vein
27 Heme
28 Hormone
29 Blood clotting

18 ______
19 ______
20 ______
21 ______
22 ______
23 ______
24 ______
25 ______
26 ______
27 ______
28 ______
29 ______

Section integration

Why are mature red blood cells in humans incapable of protein synthesis and mitosis?

30 ______

Study Outline

SECTION 1 • Plasma and Formed Elements

17.1 Blood is the fluid portion of the cardiovascular system p. 623

1. The **cardiovascular system** includes the heart, blood vessels, and blood.
2. The **heart** propels blood and maintains blood pressure, and the **blood vessels** distribute blood around the body.
3. **Blood** functions to transport gases, nutrients, hormones, and waste products; regulate pH and ion composition of interstitial fluid; restrict fluid loss at injury sites; provide defense against disease; and assist in temperature regulation.

17.2 Blood is a fluid connective tissue containing plasma and formed elements p. 624

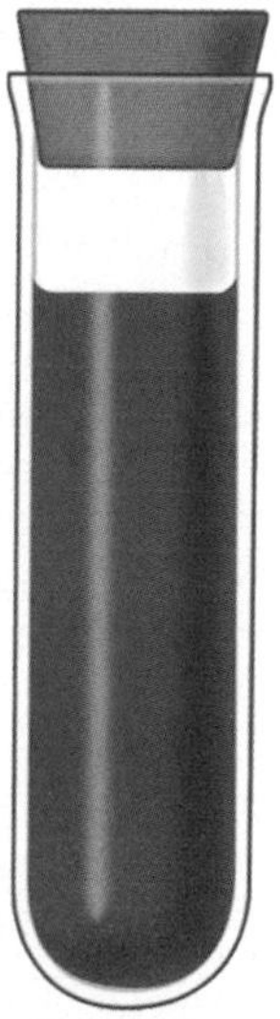

4. Blood is composed of a liquid matrix called **plasma**. Suspended in the plasma are blood cells and cell fragments called **formed elements**.
5. Plasma forms 55 percent of blood volume; formed elements are 45 percent of blood volume.
6. Plasma is primarily composed of water (92 percent), which functions in the transportation of organic and inorganic molecules, formed elements, and heat.
7. The remaining composition of plasma is 7 percent **plasma proteins** (**albumins**, **globulins**, **fibrinogen**, enzymes and hormones), and 1 percent other solutes (**electrolytes**, **organic nutrients**, and **organic wastes**).
8. Formed elements are 99.9 percent **red blood cells (RBCs)**, or erythrocytes, which transport oxygen in blood.
9. The remaining formed elements are **platelets** (cell fragments important in blood clotting) and **white blood cells (WBCs)**, or leukocytes. There are five classes of leukocytes that function in body defense mechanisms.

17.3 Formed elements are produced by stem cells in red bone marrow p. 626

10. **Hemopoiesis** is the process of formed element development.
11. All formed elements are produced by multipotent **hematopoietic stem cells (HSCs)**, or **hemocytoblasts**.
12. HSCs give rise to **lymphoid stem cells** and **myeloid stem cells**. Lymphoid stem cells give rise to lymphocytes. Myeloid stem cells give rise to all other types of formed elements.
13. **Erythropoietin (EPO)** stimulates RBC development. EPO is released into the plasma by the kidneys during **hypoxia** (low tissue oxygen).

SECTION 2 • Structure and Function of Formed Elements

17.4 Hematology is the study of blood and blood-forming tissues p. 629

14. **Hematology** is the study of blood, blood-forming tissues, and blood disorders.
15. Blood circulates through nearly every tissue of the body, so **dyscrasias**, or blood disorders, can have systemic effects.
16. Various diagnostic blood tests measure changes in the proportions and quantities of formed elements.
17. A **complete blood count (CBC)** is a common blood test that determines the RBC count, WBC count, erythrocyte indices (hemoglobin content), hematocrit, platelet count, and WBC differential count.
18. Several common RBC tests assess the number, size, shape, and maturity of red blood cells. Various terms are used to describe abnormal measured values.

17.5 Red blood cells, the most common formed elements, contain hemoglobin p. 630

19. The **red blood cell count** is a blood test that reports the number of RBCs per microliter of whole blood.
20. Mature RBCs do not have a nucleus and are **anucleate**. They have also lost most of their other organelles (including ribosomes) and have a life span of less than 120 days.
21. Ninety-five percent of RBC intracellular protein is **hemoglobin (Hb)**, which is responsible for transporting respiratory gases.
22. Each Hb molecule contains two **alpha** (α) **chains** and two **beta** (β) **chains** of polypeptides. Each chain contains a single **heme** molecule.

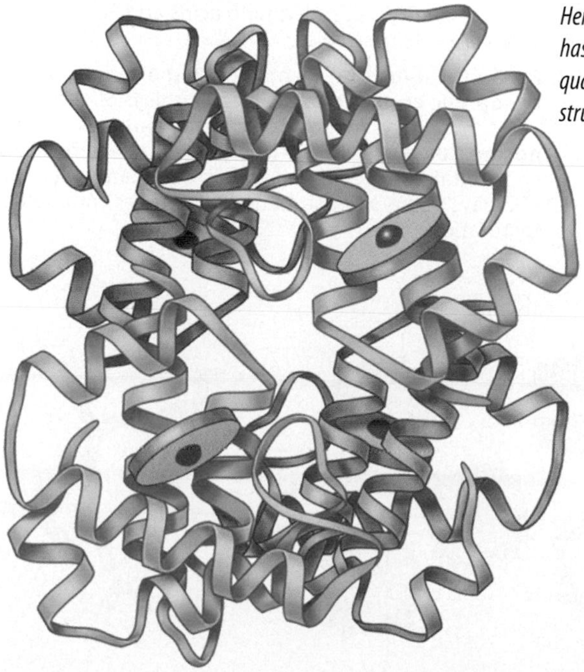

Hemoglobin has a complex quaternary structure.

23. Each heme holds an iron ion and interacts with an oxygen molecule to form **oxyhemoglobin** or **HbO_2**. This is a weak interaction that is easily reversed.
24. If a hemoglobin molecule is not bound to oxygen, it is called **deoxyhemoglobin**, or reduced hemoglobin.

17.6 Red blood cells are continually produced and recycled

p. 632

25. About 1 percent of the circulating RBCs are replaced each day.
26. Red blood cell formation or **erythropoiesis** occurs only in red bone marrow, or **myeloid tissue**, within the vertebrae, sternum, ribs, skull, scapulae, pelvis, and proximal limb bones.

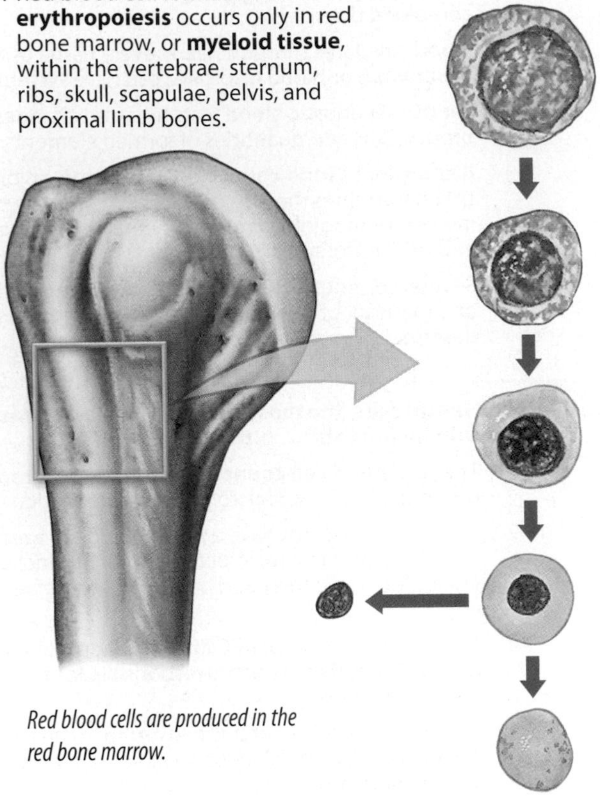

Red blood cells are produced in the red bone marrow.

27. Developing RBCs absorb amino acids and Fe^{2+} and synthesize Hb molecules.
28. **Proerythroblasts** differentiate into the various stages of **basophilic erythroblasts**. After four days of differentiation and synthesizing hemoglobin, the erythroblast becomes a **normoblast** and sheds its nucleus. It is now a **reticulocyte** and contains 80 percent of the Hb of a mature RBC. After two days in the bloodstream, reticulocytes complete their maturation.
29. After an average life span of 120 days, RBCs are engulfed by macrophages. Ten percent of RBCs will **hemolyze** (rupture) before this, and macrophages will engulf their cell fragments and Hb molecules.
30. The iron extracted from the heme molecules is either stored in the macrophage, or transported back to bone marrow by **transferrin**. The globular proteins are broken down into their amino acids and used by other cells. The heme unit is stripped of iron, converted to **biliverdin**, which is converted to **bilirubin**, and released into the bloodstream.
31. Bilirubin binds to albumin and is transported to the liver for excretion in bile. In the large intestine, bacteria convert bilirubin to **urobilins** and **stercobilins**, which are eliminated in feces. The kidneys excrete some hemoglobin and urobilins.
32. Hemoglobin that is not phagocytized breaks down, and the alpha and beta chains are eliminated in urine. The resulting red or brown urine is called **hemoglobinuria**. The presence of intact RBCs in urine is called **hematuria**.

17.7 Blood type is determined by the presence or absence of specific surface antigens on RBCs

p. 634

33. **Blood type** is a classification determined by specific genetically determined **surface antigens** in RBC plasma membranes. These surface antigens are recognized as "self" and ignored by your immune system, but your plasma may contain antibodies that will attack the antigens on "foreign" RBCs.
34. There are four blood types based on the presence or absence of the A and B surface antigens (agglutinogens).
35. **Type A** blood has surface antigen A and contains the anti-B antibodies, which will attack type B surface antigens.
36. **Type B** blood has surface antigen B and contains the anti-A antibodies, which will attack type A surface antigens.
37. **Type AB** blood has both A and B surface antigens, and neither anti-A nor anti-B antibodies.
38. **Type O** blood has neither A nor B surface antigens, and has both anti-A and anti-B antibodies.
39. RBCs will **agglutinate**, or clump, if the surface antigens on RBCs of one blood type are exposed to the corresponding antibodies **(agglutinins)** from another blood type. Such a **cross-reaction** is dangerous because the RBC clumps and fragments can plug small blood vessels and damage many organs and tissues. The RBCs may undergo hemolysis.
40. **Rh positive (Rh^+)** indicates the presence of the Rh surface antigen. The absence of this surface antigen is indicated as **Rh negative (Rh^-)**.

RBCs will agglutinate if the surface antigens on RBCs of one blood type are exposed to the corresponding antibodies of another blood type.

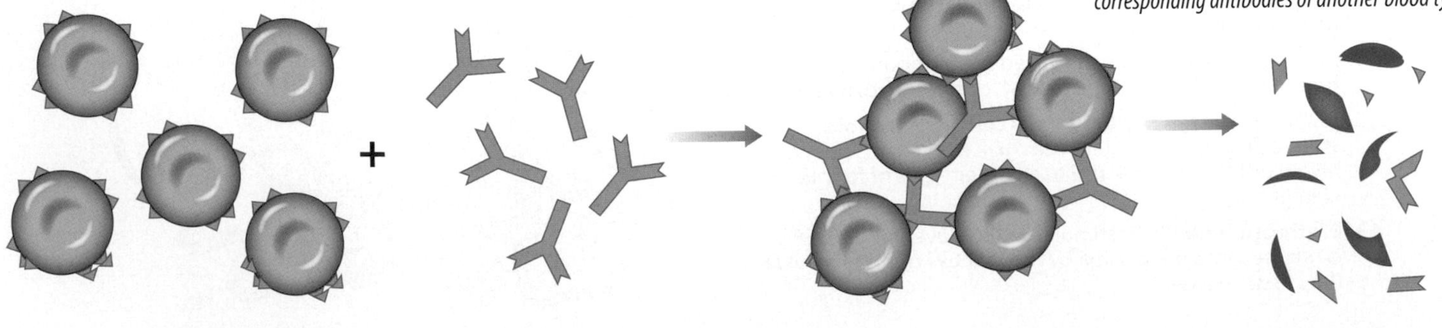

41. A **transfusion reaction** is so dangerous that blood from the donor and blood from the recipient must be **compatible**. A blood type test can determine if the donor's RBCs are compatible with the recipient's blood plasma.

17.8 Hemolytic disease of the newborn is an RBC-related disorder caused by a cross-reaction between fetal and maternal blood types p. 636

42. Blood type is genetically controlled, so a child can have a blood type different from that of either parent.
43. **Hemolytic disease of the newborn (HDN)**, or **erythroblastosis fetalis**, can result if the mother's antibodies cross the placenta and attack and destroy fetal RBCs.

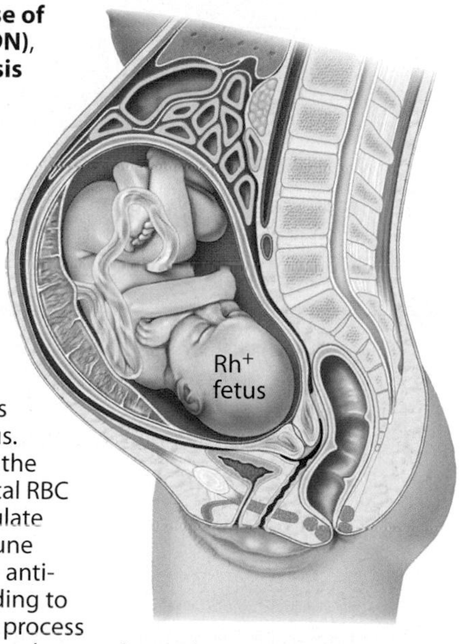

Rh+
fetus

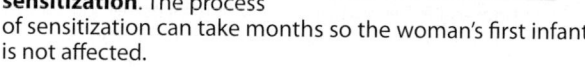

44. The most common form of HDN develops after an Rh^- woman has carried an Rh^+ fetus. During delivery of the first pregnancy, fetal RBC antigens will stimulate the mother's immune system to produce anti-Rh antibodies, leading to **sensitization**. The process of sensitization can take months so the woman's first infant is not affected.
45. If a future pregnancy involves an Rh^+ fetus, maternal anti-Rh antibodies can cross the placenta and cause HDN. Without treatment the fetus will die before delivery or shortly thereafter.
46. Injections of RhoGAM during and after a first pregnancy can prevent the mother from producing anti-Rh antibodies.

17.9 White blood cells defend the body against pathogens, toxins, cellular debris, and abnormal or damaged cells p. 638

47. The shared properties of **white blood cells (WBCs)**, or **leukocytes**, are circulation in bloodstream for only a short period of time; **emigration**, or **diapedesis**, out of the bloodstream and into areas of tissue or organ infection or injury; attraction to pathogens or tissue damage through **positive chemotaxis**; and phagocytosis (neutrophils, eosinophils, and monocytes).
48. The granular leukocytes are neutrophils, eosinophils, and basophils.
49. The agranular leukocytes are monocytes and lymphocytes.
50. A **differential count** of the WBC population can indicate a variety of pathogenic conditions.

17.10 The clotting response is a complex cascade of events that reduces blood loss p. 640

51. **Hemostasis** is the process of stopping blood loss through damaged blood vessels.
52. The **vascular phase** of hemostasis is dominated by endothelial cells and contraction, or **vascular spasm**, of smooth muscle of the vessel walls. The endothelial cells release **endothelins** that stimulate cell division, and their plasma membranes become sticky, which helps close the vessel and aid platelet attachment.

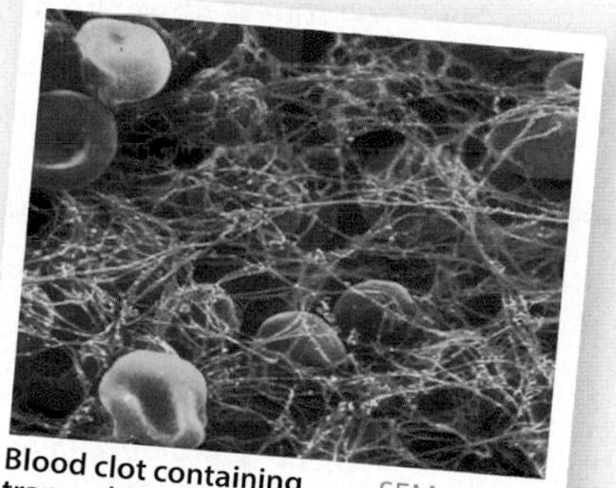

Blood clot containing trapped RBCs SEM × 1200

53. The **platelet phase** of hemostasis begins when platelets attach to the sticky endothelial surfaces, to exposed collagen fibers, and to each other.
54. The **coagulation phase** of hemostasis includes **coagulation** (blood clotting), a complex process leading to the conversion of fibrinogen to **fibrin**, forming a fibrin meshwork of platelets and RBCs stuck to the fibrin strands. The platelets contract in **clot retraction** to pull the cut edges together.
55. The clot gradually dissolves in a process called **fibrinolysis**.

17.11 Blood disorders can be classified by their origins and the changes in blood characteristics p. 642

56. Blood disorder diagnosis requires **venipuncture**, usually from the median cubital vein, to collect fresh whole blood.
57. Blood disorders can be classified as nutritional, congenital, infectious, cancerous, or degenerative.

Chapter Review Questions

Labeling

Label the blood cells shown here.

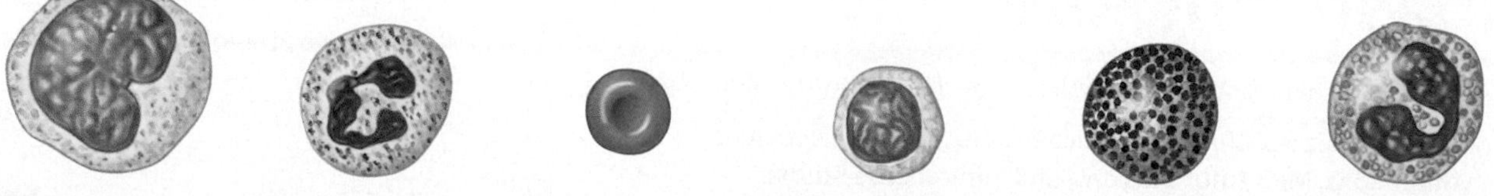

1 ________ 2 ________ 3 ________ 4 ________ 5 ________ 6 ________

True/False

Indicate whether each statement is true or false.

7 Globulins make up 60 percent of the plasma proteins.

8 Monocytes enter damaged tissues and release histamines that promote inflammation.

9 Red blood cells have an average life span of 120 days.

10 The first phase of hemostasis is the coagulation phase.

11 Neutrophils are the most abundant white blood cells.

7 ____________

8 ____________

9 ____________

10 ____________

11 ____________

Multiple choice

Select the correct answer from the list provided.

12 Plasma percentage of the whole blood volume; formed elements percentage of whole blood volume.
- ☐ a) 55, 45
- ☐ b) 45, 55
- ☐ c) 92, 8
- ☐ d) 63, 37

13 A hemoglobin molecule is composed of
- ☐ a) two protein chains and two heme units.
- ☐ b) four protein chains and two heme units.
- ☐ c) four protein chains and four heme units.
- ☐ d) one protein chain and one heme unit.

14 The correct order of erythrocyte formation is
- ☐ a) hematopoietic stem cell, myeloid stem cell, progenitor cell, proerythroblast, erythroblast stages, reticulocyte, erythrocyte.
- ☐ b) hematopoietic stem cell, progenitor cell, myeloid stem cell, proerythroblast, reticulocyte, erythroblast stages, erythrocyte.
- ☐ c) hematopoietic stem cell, myeloid stem cell, progenitor cell, reticulocyte, proerythroblast, erythroblast stages, erythrocyte.
- ☐ d) myeloid stem cell, hematopoietic stem cell, progenitor cell, proerythroblast, reticulocyte, erythroblast stages, erythrocyte.

15 Megakaryocytes shed cytoplasm in small, membrane-enclosed packets that enter the bloodstream as
- ☐ a) neutrophils.
- ☐ b) lymphocytes.
- ☐ c) monocytes.
- ☐ d) platelets.

16 Which of the following blood types is the "universal donor"?
- ☐ a) O^+
- ☐ b) O^-
- ☐ c) AB^+
- ☐ d) AB^-

17 When an RBC is engulfed by a macrophage, the hemoglobin heme units are stripped of their iron and converted to
- ☐ a) bilirubin.
- ☐ b) biliverdin.
- ☐ c) urobilins.
- ☐ d) stercobilins.

18 The percentage of formed elements in a sample of whole blood is called
- ☐ a) hemoglobinuria.
- ☐ b) hematocrit.
- ☐ c) hematuria.
- ☐ d) oxyhemoglobin.

Short answer

19 Describe the various types of leukemias.

20 What is the role of blood in stabilizing and maintaining body temperature?

21 Identify the RBC surface antigens and plasma antibodies for blood types A, B, AB, and O.

Chapter Integration • Applying what you have learned

The dangerous search for better cycling

Piero is an elite competitive cyclist who aspires to compete on the world stage. Despite his intense training in the last year along the coast in his hometown of Miami, he has been unable to achieve "the next level" of performance. He has heard about a drug that endurance athletes can take that could improve his athletic performance, but he has concerns because he also heard there could be serious side effects from using this drug.

Using what you have just learned about blood, answer the following questions.

1 **Which hormone do you think this drug is performing like?**

2 **How is it that this drug could help him in his cycling?**

3 **Which serious side effects do you think Piero would be at risk for if he were to use this drug?**

4 **Are there training techniques other than drugs that Piero could try instead?**

An unwelcome change in blood cells

Ursula is a 27-year-old, single female with no known medical history. A normally active person, with a well-balanced diet, she rarely was ill. One weekend, Ursula felt as though she was "coming down with something." On Saturday, she was fatigued, but had no trouble eating, but by Sunday, she was extremely tired and simply could not eat. Monday afternoon, her lethargy was so profound, she was unable to get out of bed, so she called a family member to take her to the physician's office.

While her physician was performing a physical exam, her blood pressure plummeted to 60/20, her heart rate increased to 190 beats per minute, and her body temperature was three degrees below normal (95.6°F). She was rushed to the hospital, where blood was drawn and tested. Obtaining blood using a venipuncture was nearly impossible, because Ursula was in the beginning stages of vascular collapse. Her gums and skin were pale.

When viewed through a microscope, Ursula's red blood cells were smaller than normal and spherically shaped, and thus were characterized as "spherocytes." Antibodies in her plasma were attacking her red blood cells, causing them to hemolyze and transform into "spherocytes." Additionally, her blood's hemoglobin levels were well below normal. Ultimately, Ursula was diagnosed with immune-mediated hemolytic anemia, and crisis treatment—a blood transfusion and the administration of corticosteroids and chemotherapeutic agents—was begun.

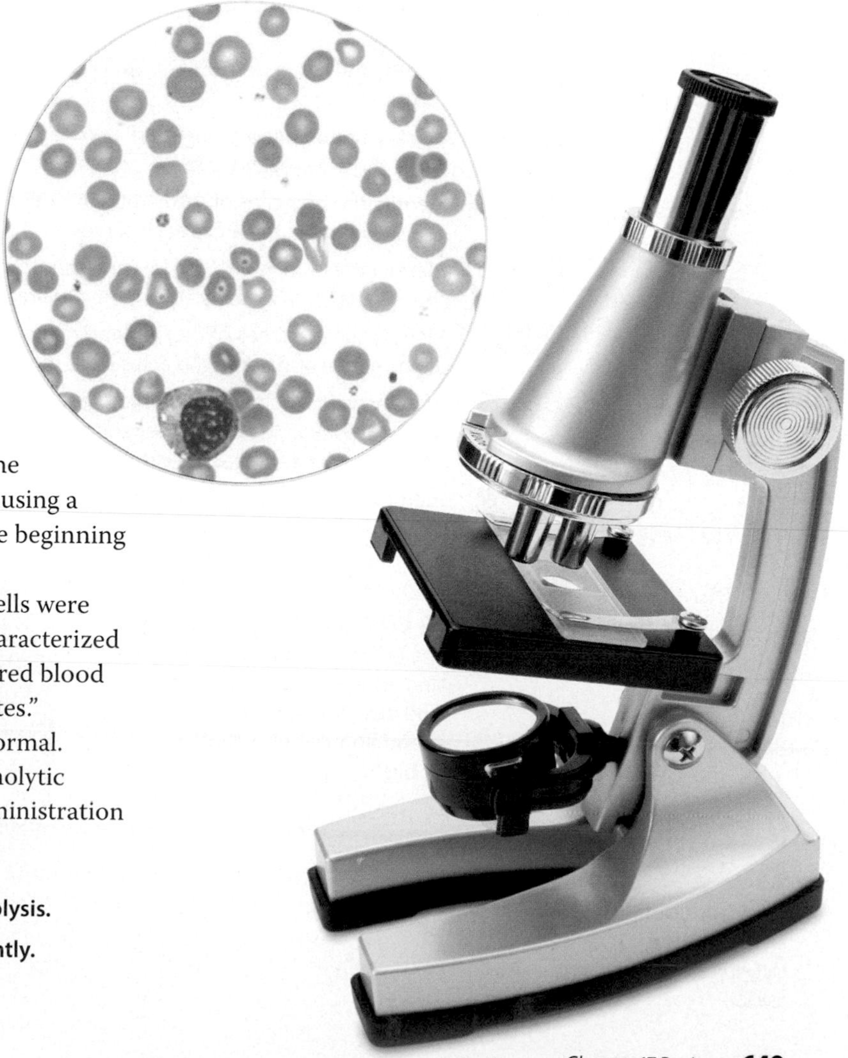

5 **Describe the shape of normal red blood cells, and define hemolysis.**

6 **Explain why Ursula's body temperature had dropped significantly.**

7 **Provide a plausible explanation for Ursula's fatigue.**

18 Blood Vessels and Circulation

LEARNING OUTCOMES

These Learning Outcomes correspond by number to this chapter's modules and indicate what you should be able to do after completing the chapter.

SECTION 1 • Functional Anatomy of Blood Vessels

18.1 Distinguish between the pulmonary and systemic circuits, and identify afferent and efferent blood vessels.

18.2 Distinguish among the types of blood vessels on the basis of their structure and function.

18.3 Describe the structures of capillaries and their functions in the exchange of dissolved materials between blood and interstitial fluid.

18.4 Describe the venous system, and indicate the distribution of blood within the cardiovascular system.

SECTION 2 • Patterns of Blood Flow

18.5 Distinguish between vasculogenesis and angiogenesis.

18.6 Identify the major arteries and veins of the pulmonary circuit, and name the areas each serves.

18.7 Identify the major arteries and veins of the systemic circuit, and name the areas each serves.

18.8 Identify the branches of the aortic arch and the branches of the superior vena cava, and name the areas each serves.

18.9 Identify the branches of the carotid arteries and the branches of the external jugular veins, and name the areas each serves.

18.10 Identify the branches of the internal carotid and vertebral arteries and the branches of the internal jugular veins, and name the areas each serves.

18.11 Identify the branches of the descending aorta and the branches of the venae cavae, and name the areas each serves.

18.12 Identify the branches of the visceral arterial vessels and the venous branches of the hepatic portal system, and name the areas each serves.

18.13 Identify the branches of the common iliac arteries and the branches of the common iliac veins, and name the areas each serves.

18.14 **CLINICAL MODULE** Identify the differences between fetal and adult circulation patterns, and describe the changes in blood flow patterns that occur at birth.

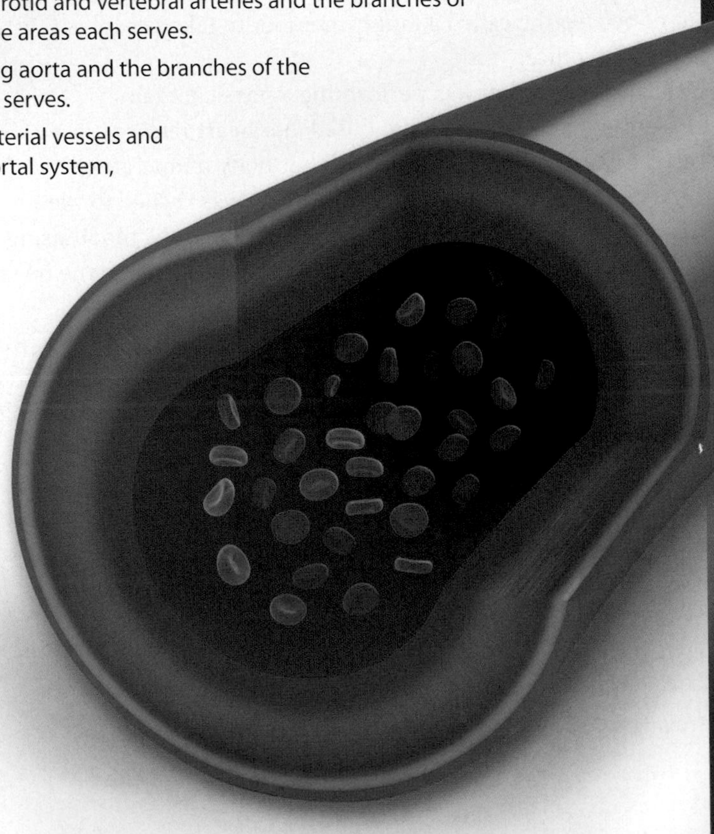

Learning Outcomes are repeated at the bottom of each module.

The heart pumps blood, in sequence, through the arteries, capillaries, and veins of the pulmonary and systemic circuits

Blood flows through a network of blood vessels that extend between the heart and peripheral tissues. Those blood vessels are organized into a **pulmonary circuit**, which carries blood to and from the gas exchange surfaces of the lungs, and a **systemic circuit**, which transports blood to and from the rest of the body. Each circuit begins and ends at the heart, and blood travels through these circuits in sequence. Thus, blood returning to the heart from the systemic circuit must complete the pulmonary circuit before reentering the systemic circuit. Blood is carried away from the heart by **arteries** and returns to the heart by way of **veins**. Microscopic, thin-walled vessels called **capillaries** interconnect the smallest arteries and the smallest veins. Capillaries are called exchange vessels, because their thin walls allow the exchange of nutrients, dissolved gases, and wastes between blood and the surrounding interstitial fluid.

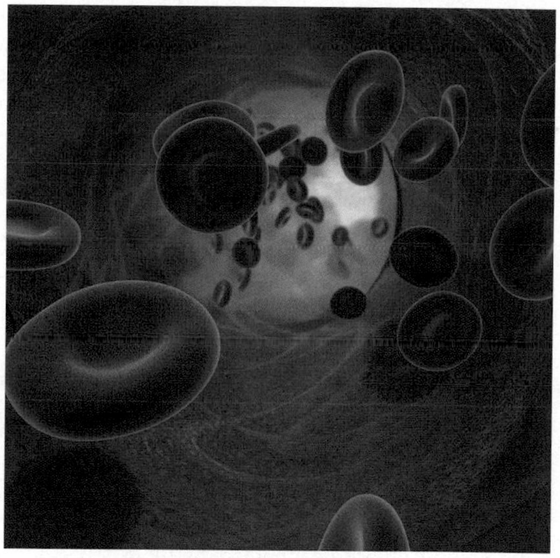

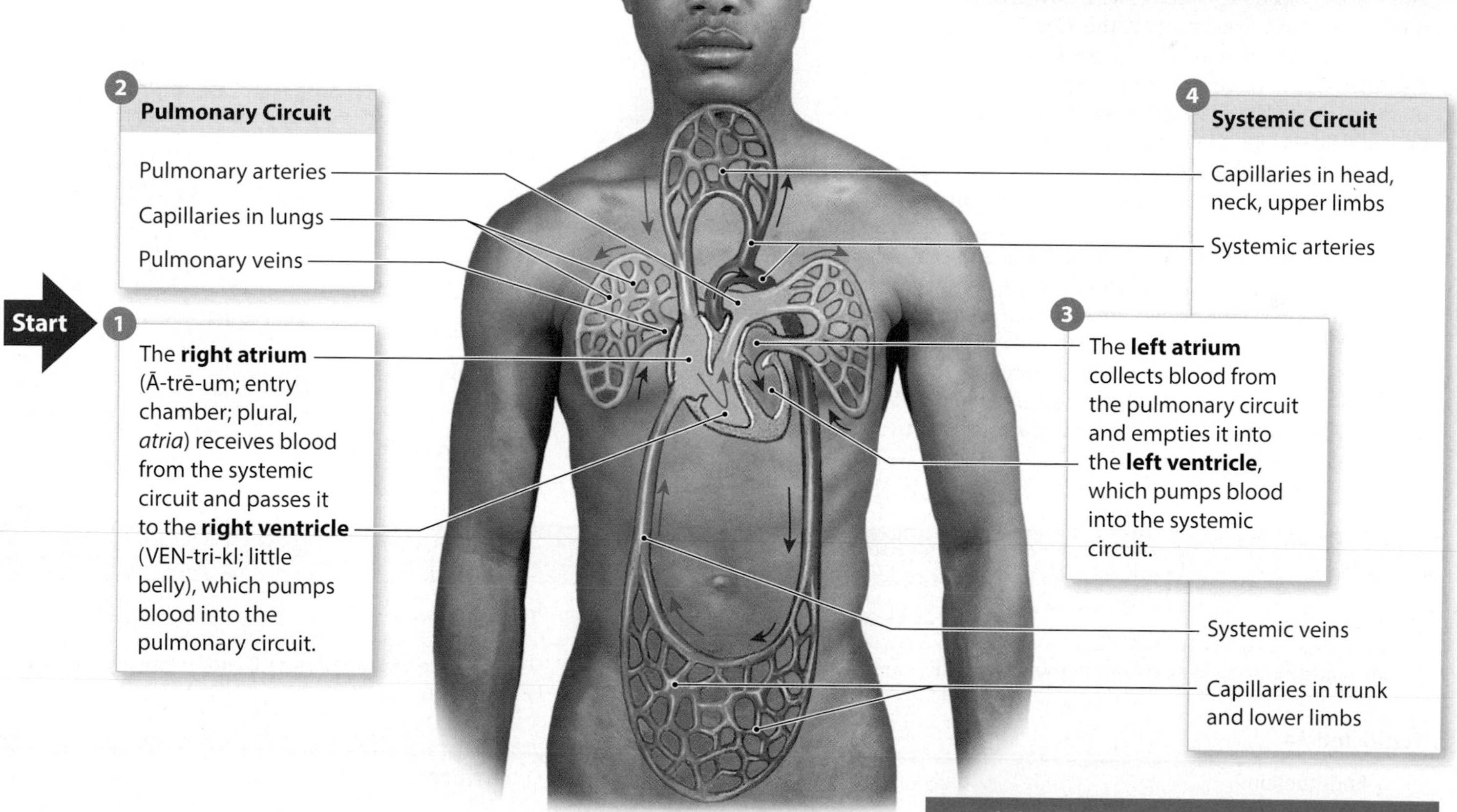
2
Pulmonary Circuit
Pulmonary arteries
Capillaries in lungs
Pulmonary veins
Start
1
The **right atrium** (Ā-trē-um; entry chamber; plural, *atria*) receives blood from the systemic circuit and passes it to the **right ventricle** (VEN-tri-kl; little belly), which pumps blood into the pulmonary circuit.
4
Systemic Circuit
Capillaries in head, neck, upper limbs
Systemic arteries
3
The **left atrium** collects blood from the pulmonary circuit and empties it into the **left ventricle**, which pumps blood into the systemic circuit.
Systemic veins
Capillaries in trunk and lower limbs

In this section we will take a close look at the structure of the major blood vessels in the body.

Module 18.1 Review

a. Describe the pulmonary circuit.

b. Describe the systemic circuit.

c. Which chamber of the heart receives blood from the systemic circuit?

LO **18.1** Distinguish between the pulmonary and systemic circuits, and identify afferent and efferent blood vessels.

Arteries and veins differ in the structure and thickness of their walls

1 The walls of arteries and veins contain three distinct layers: the tunica intima, tunica media, and tunica externa. These vessel walls are too thick to allow diffusion between the blood and tissues or even between blood and the vessel itself. Thus, large vessel walls contain small arteries and veins that supply the smooth muscle cells and fibroblasts of the tunica media and tunica externa. These blood vessels are called the **vasa vasorum** ("vessels of vessels").

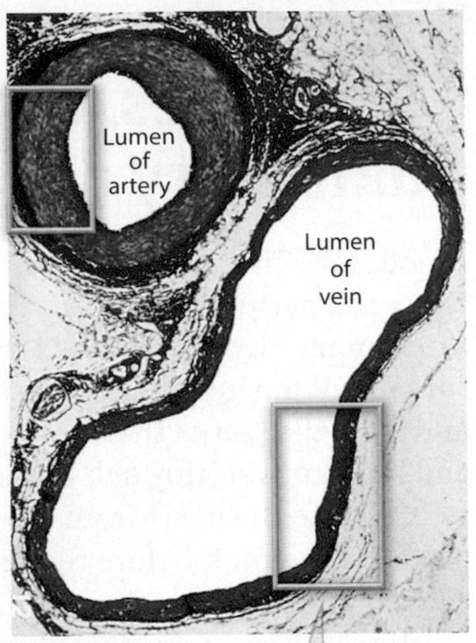

Lumen of artery
Lumen of vein

Artery and vein LM × 60

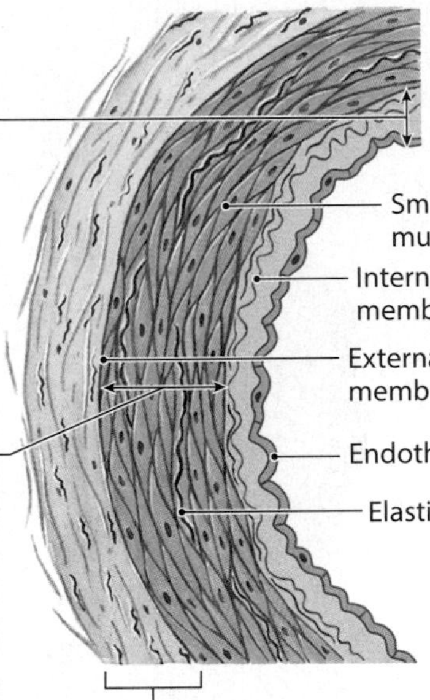

Artery
Smooth muscle
Internal elastic membrane
External elastic membrane
Endothelium
Elastic fiber

The **tunica intima** (IN-ti-muh), or tunica interna, is the innermost layer of a blood vessel. This layer includes the endothelial lining and an underlying layer of connective tissue containing elastic fibers. In arteries, the outer margin of the tunica intima contains a thick layer of elastic fibers called the **internal elastic membrane**.

The **tunica media**, the middle layer, contains concentric sheets of smooth muscle tissue in a framework of loose connective tissue. When these smooth muscles contract, the vessel decreases in diameter; this is called **vasoconstriction**. When the smooth muscles relax, the diameter increases; this is called **vasodilation**. Collagen fibers bind the tunica media to the tunica intima and tunica externa.

The **tunica externa** (eks-TER-nuh), or tunica adventitia, the outermost layer of a blood vessel, is a connective tissue sheath. In arteries, this layer contains collagen fibers with scattered bands of elastic fibers. In veins, it is generally thicker than the tunica media and contains networks of elastic fibers and bundles of smooth muscle cells. The connective tissue fibers of the tunica externa typically blend into those of adjacent tissues, stabilizing and anchoring the blood vessel.

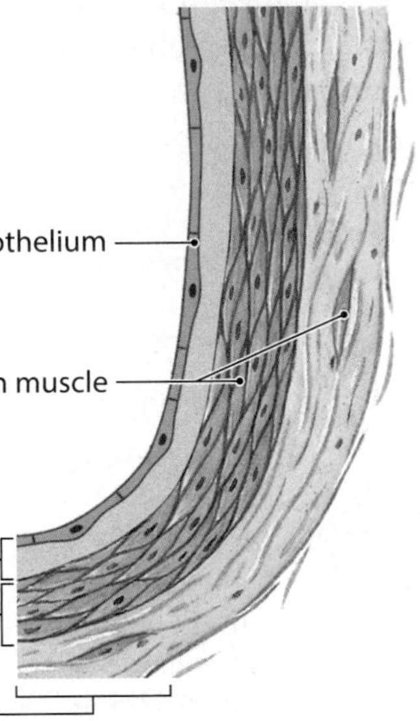

Vein
Endothelium
Smooth muscle
Tunica intima
Tunica media
Tunica externa

Features of Typical (Medium-Sized) Arteries and Veins

Feature	Typical Artery	Typical Vein
General appearance in sectional view	Usually round, with relatively thick wall Small lumen	Usually flattened or collapsed, with relatively thin wall Large lumen
Tunica Intima		
Endothelium	Usually rippled, due to vessel constriction	Often smooth
Internal elastic membrane	Present	Absent
Tunica Media	Thick, dominated by smooth muscle cells and elastic fibers	Thin, dominated by smooth muscle cells and collagen fibers
External elastic membrane	Present	Absent
Tunica Externa	Collagen and elastic fibers	Collagen, elastic fibers, and smooth muscle cells

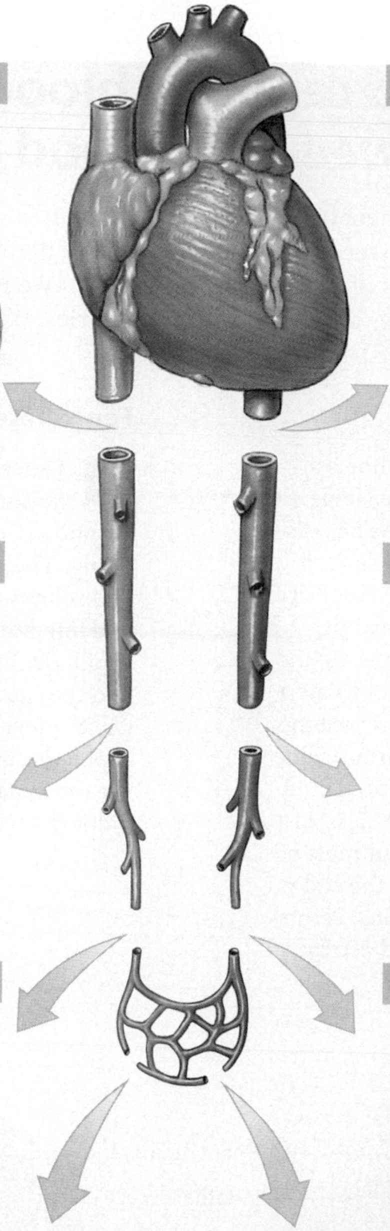

Large Vein

Large veins include the superior and inferior venae cavae and their branches. All three vessel wall layers are present in all large veins. The slender tunica media is surrounded by a thick tunica externa composed of a mixture of elastic and collagen fibers.

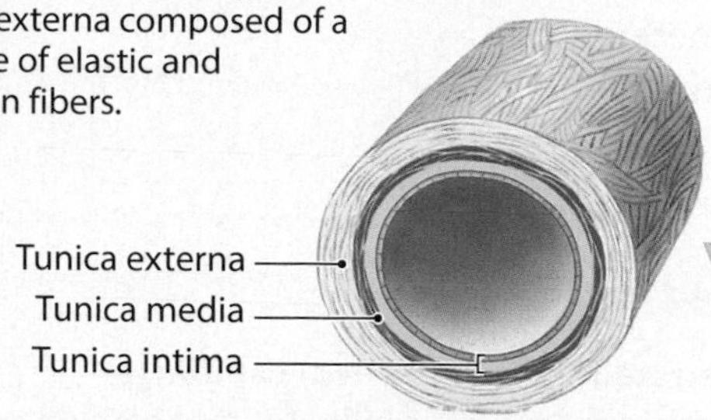

Tunica externa
Tunica media
Tunica intima

Elastic Artery

Elastic arteries are large vessels that transport blood away from the heart. The pulmonary trunk and aorta, as well as their major arterial branches, are elastic arteries. These are resilient, elastic vessels capable of stretching and recoiling as the heart beats and arterial pressures change.

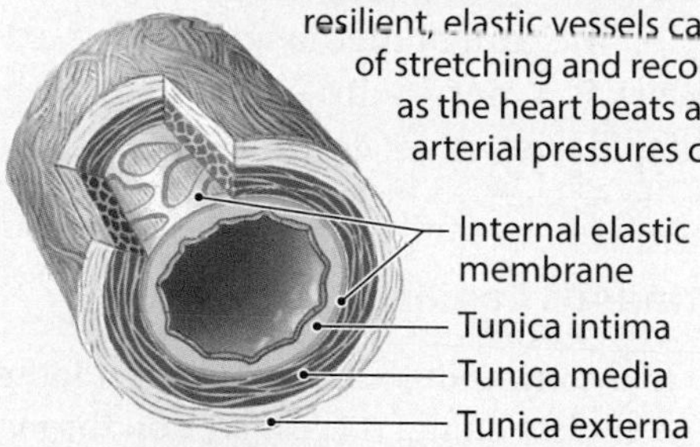

Internal elastic membrane
Tunica intima
Tunica media
Tunica externa

Medium-Sized Vein

Medium-sized veins range from 2 to 9 mm in internal diameter. In these veins, the tunica media is thin and contains smooth muscle cells and collagen fibers. The thickest layer is the tunica externa, which contains smooth muscle cells and longitudinal bundles of elastic and collagen fibers.

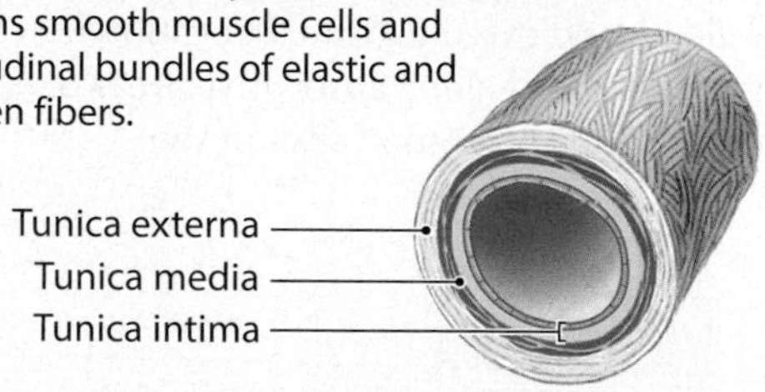

Tunica externa
Tunica media
Tunica intima

Muscular Artery

Muscular arteries, or medium-sized arteries, distribute blood to the body's skeletal muscles and internal organs. They are characterized by a thick tunica media. It contains more smooth muscle cells than does the tunica media of elastic arteries.

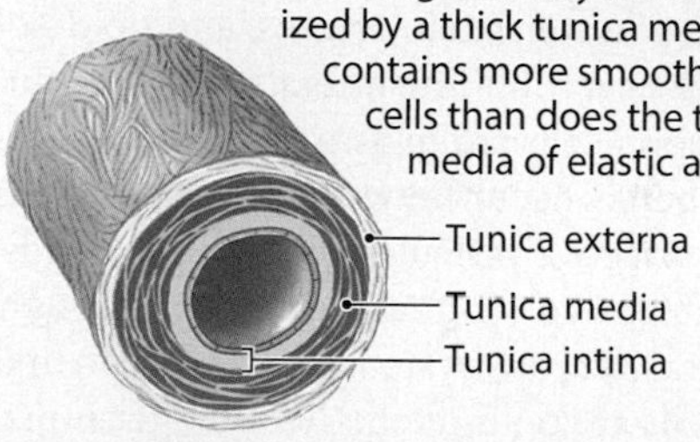

Tunica externa
Tunica media
Tunica intima

Venule

Venules collect blood from capillary beds and are the smallest venous vessels. Venules smaller than 50 μm lack a tunica media and resemble expanded capillaries.

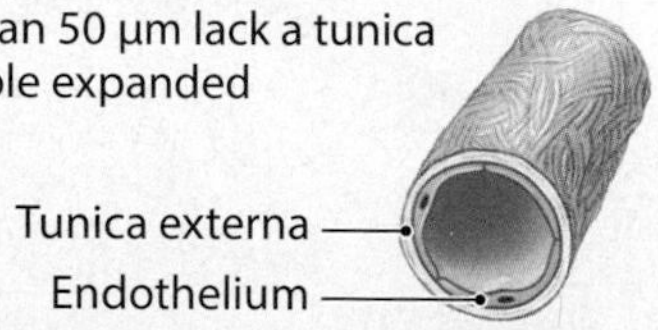

Tunica externa
Endothelium

Arteriole

Arterioles have a poorly defined tunica externa, and the tunica media consists of only one or two layers of smooth muscle cells.

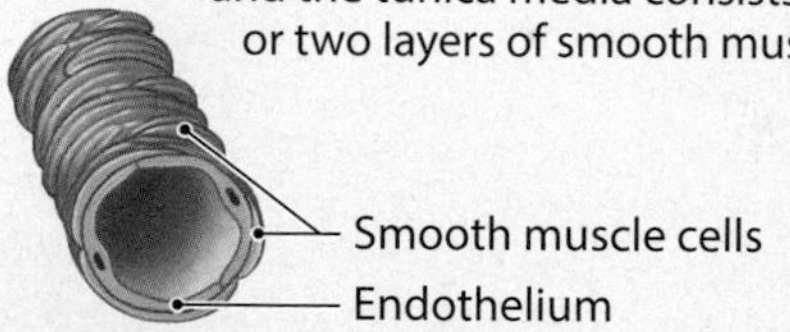

Smooth muscle cells
Endothelium

Capillaries

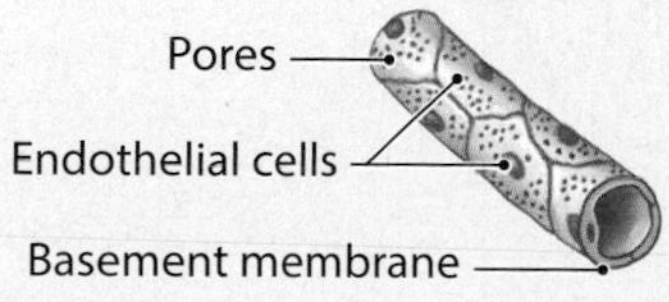

Pores
Endothelial cells
Basement membrane

Capillaries are the only blood vessels whose walls permit exchange between blood and the surrounding interstitial fluids. Because capillary walls are thin, diffusion distances are short, so exchange can occur quickly. Capillary structure is examined in greater detail in Module 18.3.

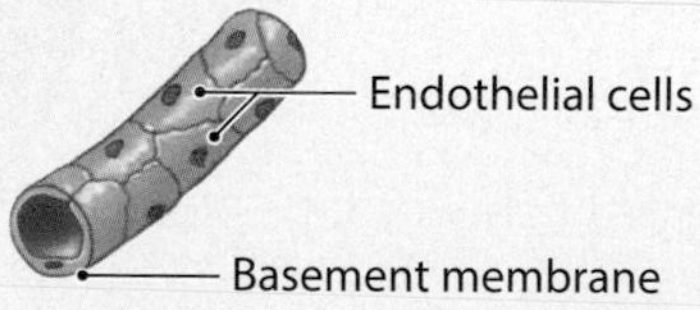

Endothelial cells
Basement membrane

2 There are five general classes of blood vessels in the cardiovascular system. **Arteries** carry blood away from the heart. As arteries enter peripheral tissues they branch repeatedly, and the branches decrease in diameter. The smallest arterial branches are called **arterioles** (ar-TĒR-ē-ōls). From the arterioles, blood moves into **capillaries**, where diffusion occurs between blood and interstitial fluid. From the capillaries, blood enters small **venules** (VEN-yūls), which unite to form larger **veins** that return blood to the heart.

LO 18.2 Distinguish among the types of blood vessels on the basis of their structure and function.

Module 18.2 Review

a. List the five general classes of blood vessels.

b. Describe a capillary.

c. A cross section of tissue shows several small, thin-walled vessels with very little smooth muscle tissue in the tunica media. Which type of vessels are these?

Module 18.3

Capillary structure and capillary blood flow affect the rates of exchange between the blood and interstitial fluid

A typical capillary consists of a tube of endothelial cells within a basement membrane. The tunica media and tunica externa are absent. The average diameter of a capillary is a mere 8 μm, very close to that of a single red blood cell. Two major types of capillaries exist: continuous capillaries and fenestrated capillaries.

Continuous Capillary

1 In a **continuous capillary**, the endothelium is a complete lining. A cross section through a large continuous capillary cuts across several endothelial cells. In a small continuous capillary, a single endothelial cell may completely encircle the lumen. Continuous capillaries are located throughout the body, in all tissues except epithelia and cartilage. Continuous capillaries permit the diffusion of water, small solutes, and lipid-soluble materials into the surrounding interstitial fluid, but prevent the loss of blood cells and plasma proteins. In addition, some exchange may occur between blood and interstitial fluid through selective vesicular transport (Module 3.17, p. 118). In specialized continuous capillaries throughout most of the central nervous system and in the thymus, the endothelial cells are bound together by tight junctions. Permeability in these capillaries is both restricted and precisely regulated.

Fenestrated Capillary

2 **Fenestrated** (FEN-es-trā-ted) **capillaries** (*fenestra*, window) are capillaries that contain "windows," or pores, that penetrate the endothelial lining. The pores permit the rapid exchange of water and solutes as large as small peptides between blood and interstitial fluid. Examples of fenestrated capillaries include the choroid plexus of the brain and the capillaries of the hypothalamus, pituitary gland, pineal gland, and thyroid gland. Fenestrated capillaries are also located along absorptive areas of the intestinal tract and at filtration sites in the kidneys.

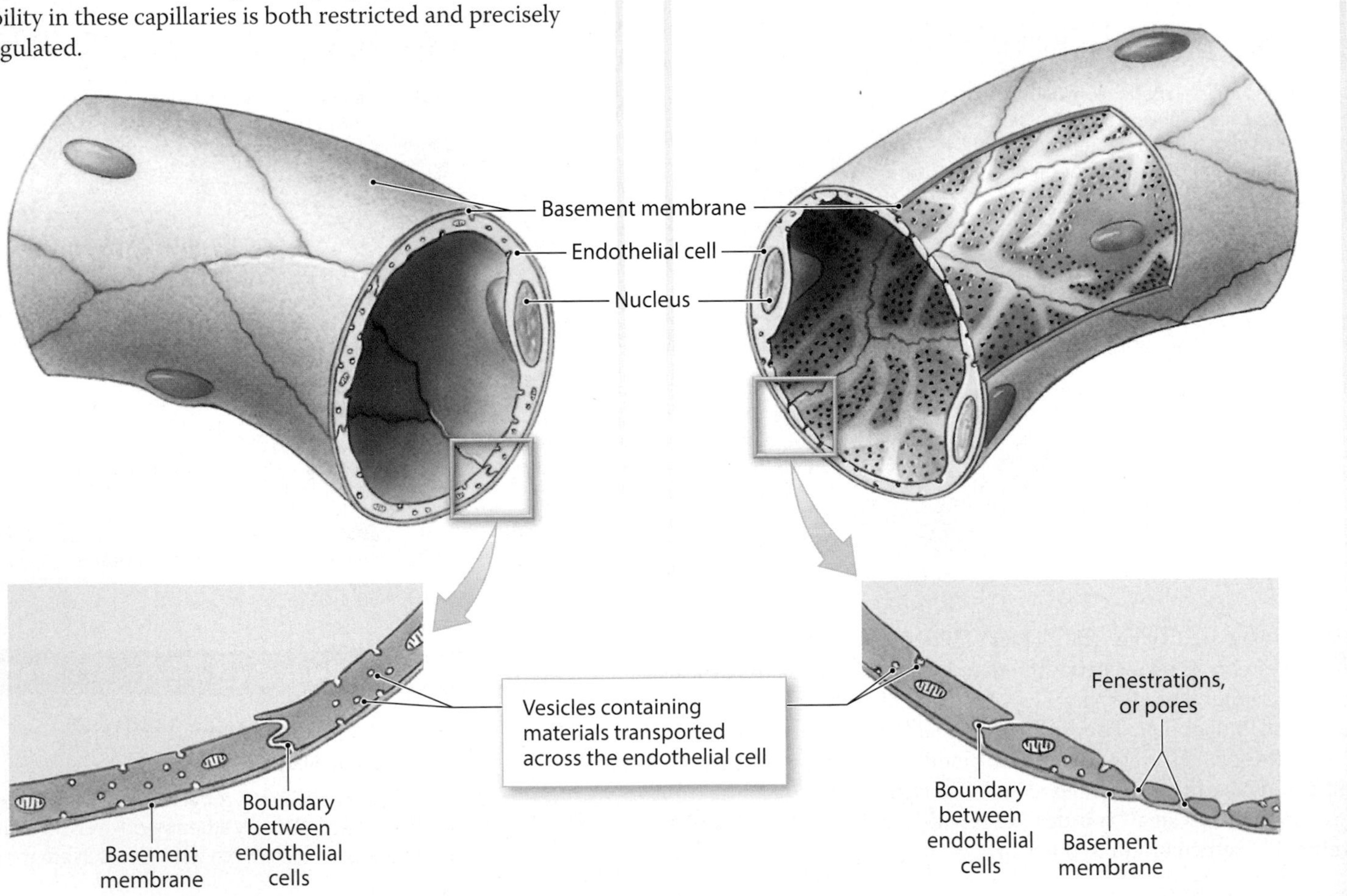

Basement membrane
Endothelial cell
Nucleus
Vesicles containing materials transported across the endothelial cell
Boundary between endothelial cells
Basement membrane
Fenestrations, or pores
Boundary between endothelial cells
Basement membrane

3 **Sinusoids** (SĪ-nuh-soydz) resemble fenestrated capillaries that are flattened and irregularly shaped. In contrast to fenestrated capillaries, sinusoids commonly have gaps between adjacent endothelial cells, and the basement membrane is either thinner or absent. As a result, sinusoids permit the free exchange of water and solutes as large as plasma proteins between the slow-moving blood and interstitial fluid. Sinusoids occur in the liver, bone marrow, spleen, and many endocrine organs, including the pituitary and adrenal glands.

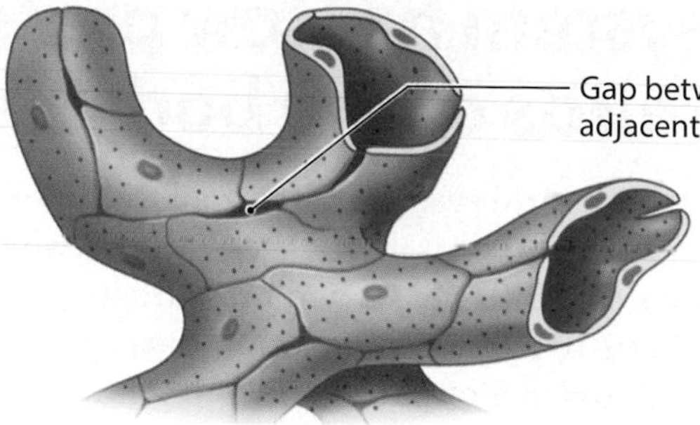

Gap between adjacent cells

More than one artery may supply blood to a capillary bed. The multiple arteries are called **collaterals**. They fuse before giving rise to arterioles. The fusion of two collateral arteries that supply a capillary bed is an example of an **arterial anastomosis**. (An *anastomosis* is the joining of blood vessels.) An arterial anastomosis acts like an insurance policy: If one artery is compressed or blocked, capillary circulation will continue.

A single arteriole generally gives rise to dozens of capillaries that empty into several venules.

A **precapillary sphincter** guards the entrance to each capillary. Contraction or relaxation of the smooth muscle cells changes the diameter of the capillary entrance, thereby controlling the flow of blood through the sphincter.

An **arteriovenous** (ar-tēr-ē-ō-VĒ-nus) **anastomosis** is a direct connection between an arteriole and a venule. When this anastomosis is dilated, blood will bypass the capillary bed and flow directly into the venous circulation. The pattern of blood flow through these anastomoses is regulated primarily by sympathetic innervation under the control of the cardiovascular centers of the medulla oblongata.

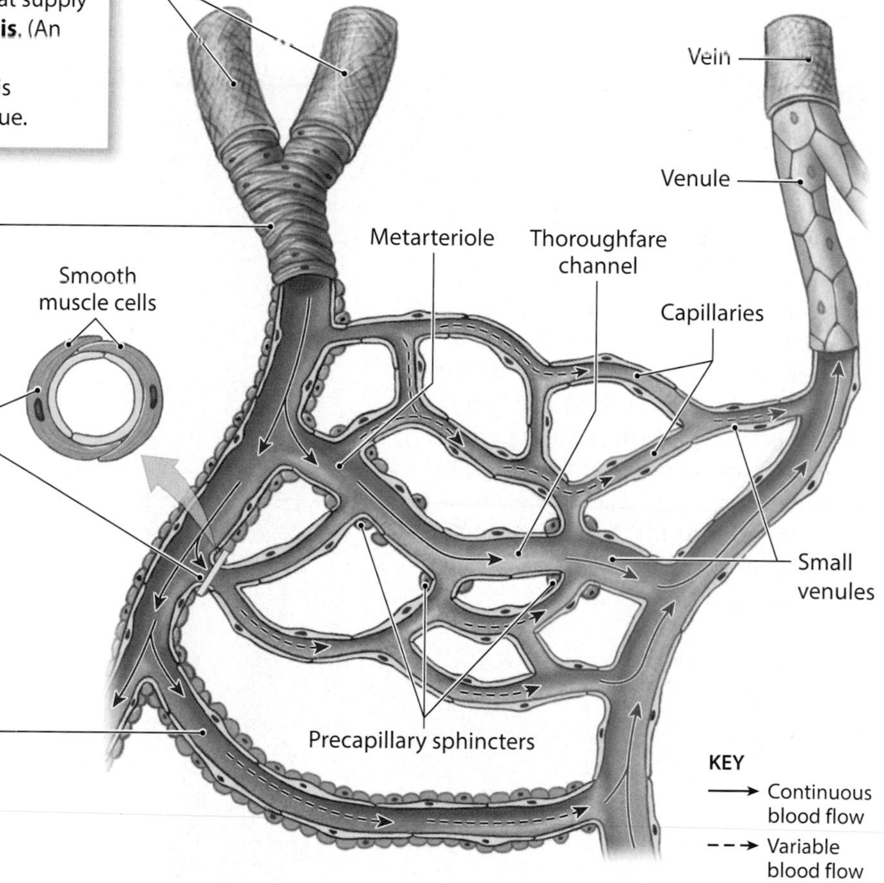

Vein
Venule
Metarteriole
Thoroughfare channel
Smooth muscle cells
Capillaries
Small venules
Precapillary sphincters
KEY
Continuous blood flow
Variable blood flow

4 Capillaries function as part of an interconnected network called a **capillary bed**. A capillary bed contains several relatively direct connections between arterioles and venules. The wall in the initial part of such a passageway contains smooth muscle that can change its diameter. This segment is called a **metarteriole** (met-ar-TĒR-ē-ōl) or **precapillary arteriole**. The rest of the passageway, which resembles a typical capillary in structure, is called a **thoroughfare channel**. Although blood normally flows at a constant rate from the arteriole to the venule across the capillary bed, the flow within each capillary is quite variable. Bands of smooth muscle at the entrance to each capillary—called **precapillary sphincters**—alternately contract and relax, perhaps a dozen times per minute. As a result, the blood flow within the associated capillary occurs in pulses rather than as a steady and constant stream. The cycling of contraction and relaxation of smooth muscle cells that change blood flow through capillary beds is called **vasomotion**.

Module 18.3 Review

a. Identify the two types of capillaries.

b. At what sites in the body are fenestrated capillaries located?

c. Why do capillaries permit the diffusion of materials, whereas arteries and veins do not?

LO **18.3** Describe the structures of capillaries and their functions in the exchange of dissolved materials between blood and interstitial fluid.

Module 18.4

The venous system has low pressures and contains almost two-thirds of the body's blood volume

The arterial system is a high-pressure system: Almost all the force developed by the heart is required to push blood along the network of arteries and through miles of capillaries. Blood pressure in a peripheral venule is only about 10 percent of that in the ascending aorta, and pressures continue to fall along the venous system.

1 The blood pressure in venules and medium-sized veins is so low that it cannot overcome the force of gravity. In the limbs, veins of this size contain **valves**, folds of the tunica intima that project from the vessel wall and point in the direction of blood flow. Venous valves (like valves in the heart) permit blood flow in one direction only, thereby preventing the backflow of blood toward the capillaries. If the walls of the veins near the valves weaken or become stretched and distorted, the valves may no longer work properly. Blood then pools in the veins, which become grossly distended. The effects range from mild discomfort and a cosmetic problem, as in superficial **varicose veins** in the thighs and legs, to painful distortion of adjacent tissues, as in the **hemorrhoids** that form in venous networks of the anal canal.

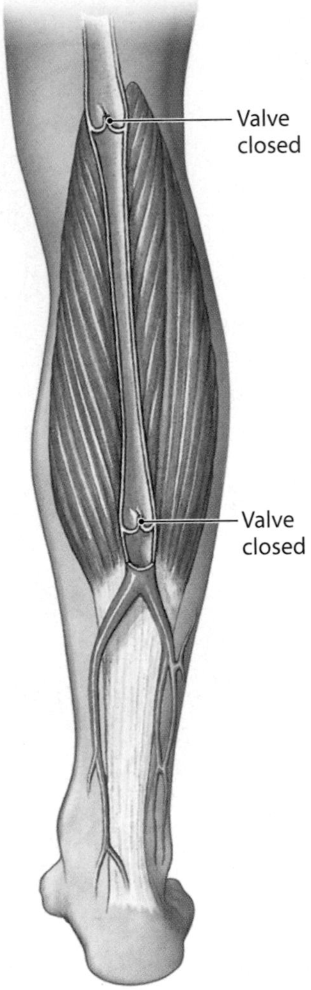

Valve closed
Valve closed

When you are standing, venous blood from your feet must overcome the pull of gravity to ascend to the heart. Valves compartmentalize the blood within the veins, thereby dividing the weight of the blood between the compartments.

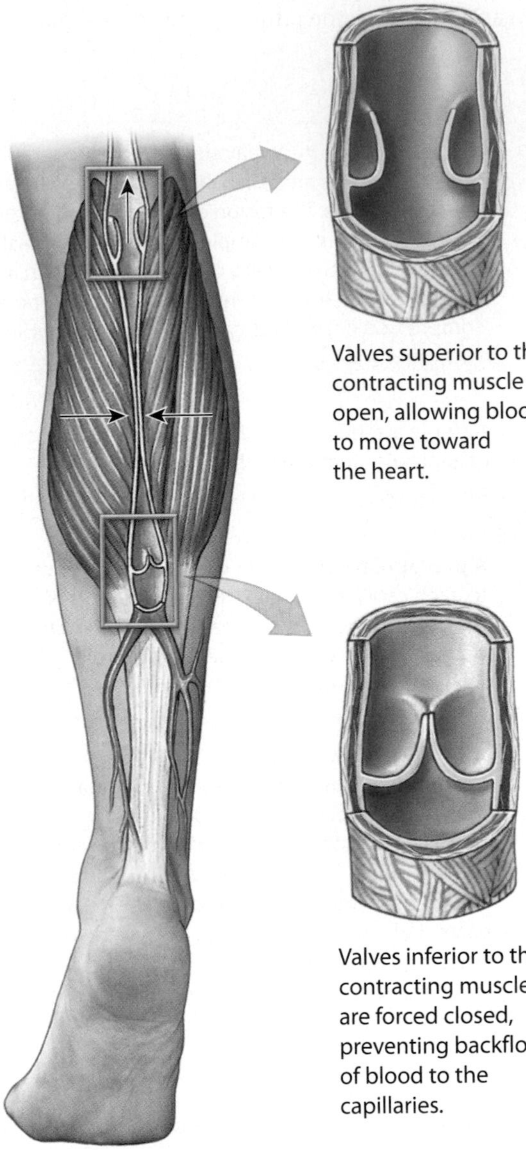

Valves superior to the contracting muscle open, allowing blood to move toward the heart.
Valves inferior to the contracting muscle are forced closed, preventing backflow of blood to the capillaries.

Any contraction of the surrounding skeletal muscles squeezes the blood toward the heart. Although you are probably not aware of it, when you stand, rapid cycles of contraction and relaxation are occurring within your leg muscles, helping to push blood toward the trunk.

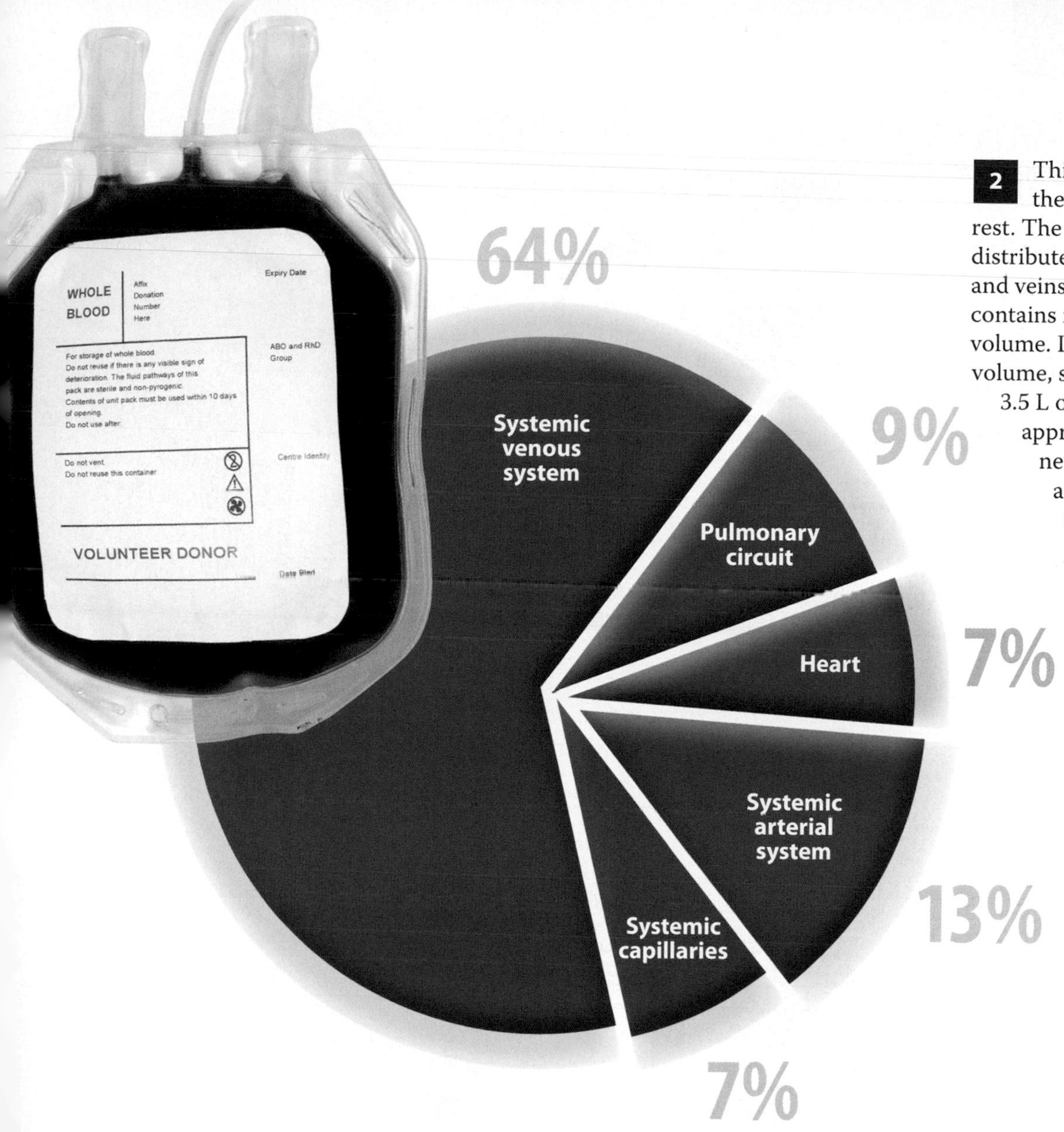
WHOLE
BLOOD
VOLUNTEER DONOR
64%
Systemic venous system
9%
Pulmonary circuit
7%
Heart
13%
Systemic arterial system
7%
Systemic capillaries

2 This chart shows the distribution of the body's total blood volume at rest. The total blood volume is unevenly distributed among arteries, capillaries, and veins. The systemic venous system contains nearly two-thirds of total blood volume. In a man with a 6 L total blood volume, systemic veins contain roughly 3.5 L of whole blood. Of that amount, approximately 1 L is found in venous networks in the liver, bone marrow, and skin. The pulmonary circuit, the heart, and the systemic arteries and capillaries contain the rest of the blood volume (about 1.5 L).

3 If serious hemorrhaging occurs, the body maintains blood volume within the arterial system at near-normal levels by reducing the volume of blood in the venous system. In the mechanism involved, the vasomotor center in the medulla oblongata stimulates sympathetic nerves innervating smooth muscle cells in the tunica media of medium-sized veins. The resulting contraction produces **venoconstriction**, reducing the diameter of the veins and the amount of blood contained in the venous system. In addition, blood enters the general circulation from venous networks in the liver, bone marrow, and skin. Reducing the amount of blood in the venous system can maintain the volume within the arterial system at near-normal levels despite a significant blood loss.

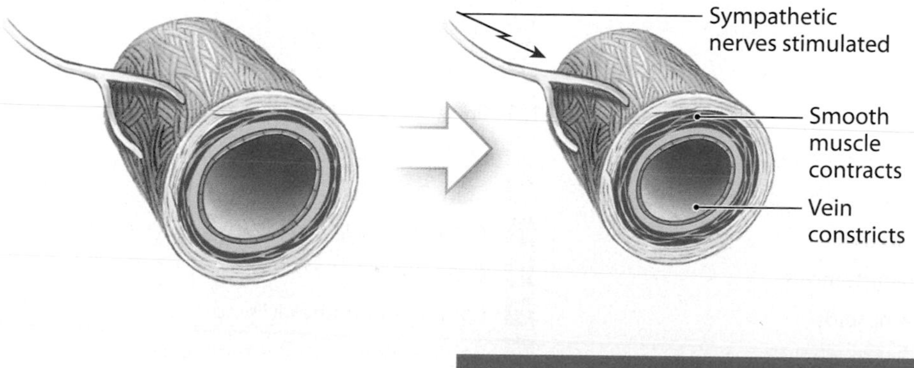
Sympathetic nerves stimulated
Smooth muscle contracts
Vein constricts

Module 18.4 Review

a. Why are valves located in veins, but not in arteries?

b. How is blood pressure maintained in veins to counter the force of gravity?

c. Define varicose veins.

LO **18.4** Describe the venous system, and indicate the distribution of blood within the cardiovascular system.

Labeling

Label the blood vessels and their components in the images below.

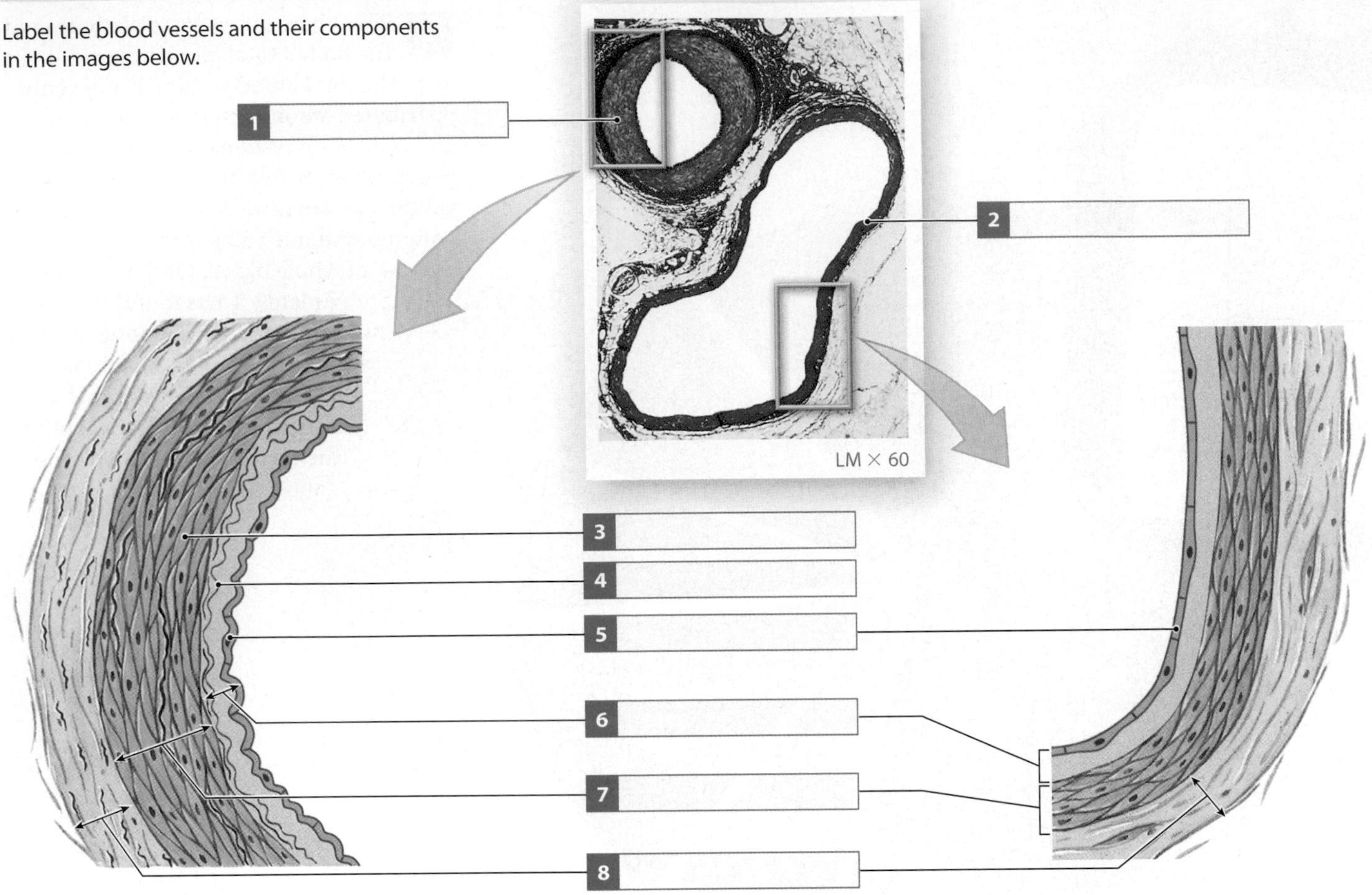
1
2
LM × 60
3
4
5
6
7
8

Matching

Match each lettered term with the most closely related description.

a. pulmonary circuit
b. vein
c. exchange vessels
d. continuous capillary
e. systemic circuit
f. right atrium
g. venous valves
h. fenestrated capillaries
i. artery
j. anastomosis
k. sinusoids
l. tunica media

9 Left ventricle pumps blood into this
10 Relatively thick wall
11 Joining of blood vessels
12 Present only in veins
13 Blood to and from the lungs
14 Present in liver, bone marrow, and spleen
15 Relatively thin wall
16 Contain pores
17 Smooth muscle tissue
18 Capillaries
19 Complete endothelial lining
20 Receives blood from systemic circuit

9 ______
10 ______
11 ______
12 ______
13 ______
14 ______
15 ______
16 ______
17 ______
18 ______
19 ______
20 ______

Section integration

What is the function of precapillary sphincters, and what role would you expect them to play during exercise and in cold temperatures?

21 ______

New blood vessels form through vasculogenesis and angiogenesis

1 Blood vessels begin as blood islands in the yolk sac around the seventh day of embryonic development. This marks the beginning of hemopoiesis (Module 17.3, p. 626). Two complementary processes are involved in forming blood vessels: vasculogenesis and angiogenesis. **Vasculogenesis** is the formation of the first vessels by precursor endothelial cells called hemangioblasts. **Angiogenesis** is the growth of new blood vessels from pre-existing vessels. Angiogenic remodeling occurs with the migration of endothelial cells to form vascular networks.

Vasculogenesis

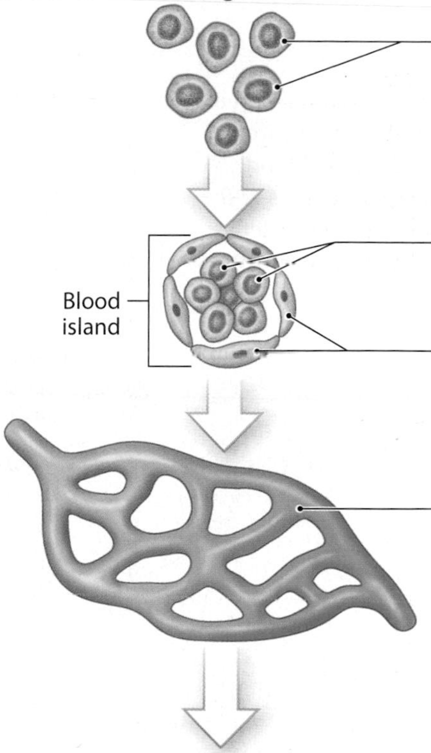
Blood island

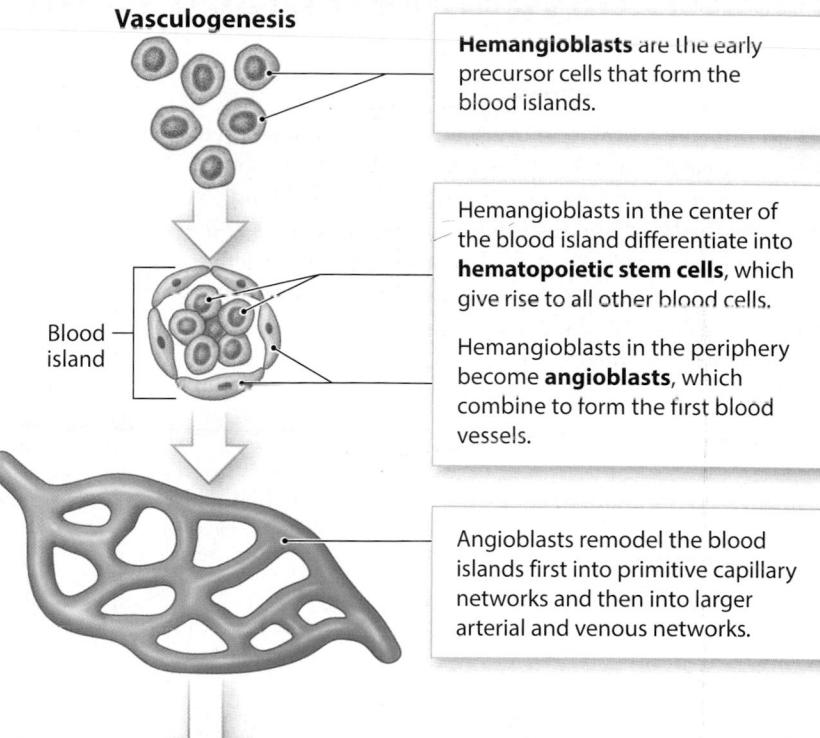

Hemangioblasts are the early precursor cells that form the blood islands.

Hemangioblasts in the center of the blood island differentiate into **hematopoietic stem cells**, which give rise to all other blood cells.

Hemangioblasts in the periphery become **angioblasts**, which combine to form the first blood vessels.

Angioblasts remodel the blood islands first into primitive capillary networks and then into larger arterial and venous networks.

Angiogenesis

Formation of the Aortic Arch and Its Branches

The **aortic arches** are a series of arterial channels that ultimately form the carotid arteries, the aortic arch, and part of the pulmonary arteries. The **dorsal aorta** becomes the descending aorta.

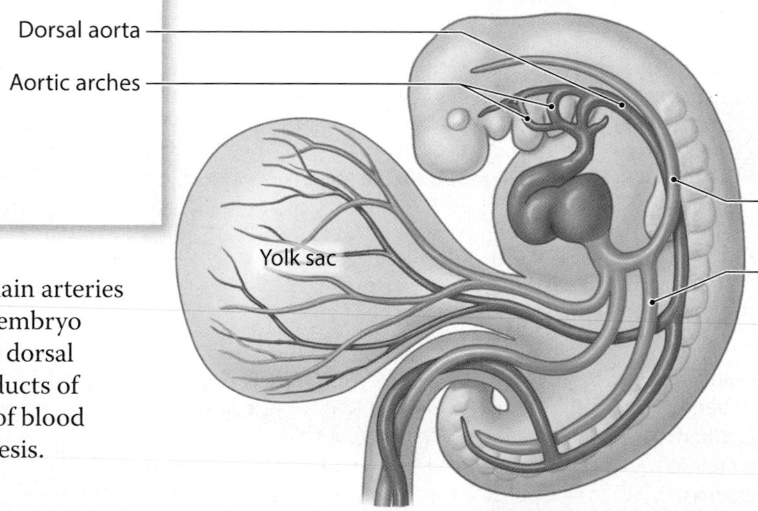
Dorsal aorta
Aortic arches
Yolk sac

4 WEEKS

Formation of the Venae Cavae

The **cardinal veins** are a series of venous channels that ultimately form the superior and inferior venae cavae and vessels returning to the heart.

Anterior cardinal vein (from cephalic region)

Posterior cardinal vein (from caudal region)

2 This diagram shows the main arteries and veins in a 4-week-old embryo when it is about 4 mm long. The dorsal aorta and cardinal veins are products of vasculogenesis. Further growth of blood vessels occurs through angiogenesis.

In this section we will follow the distribution of the major blood vessels in the body.

Module 18.5 Review

a. Distinguish between vasculogenesis and angiogenesis.

b. What are blood islands, and from which cells do they form?

c. What is the function of angioblasts?

LO 18.5 Distinguish between vasculogenesis and angiogenesis.

Module 18.6

The pulmonary circuit carries deoxygenated blood from the right ventricle to the lungs and returns oxygenated blood to the left atrium

1 This flowchart is an overview of the organization of the cardiovascular system. The **pulmonary circuit** is composed of arteries and veins that transport blood between the heart and the lungs. This circuit begins at the right ventricle and ends at the left atrium. From the left ventricle, the arteries of the **systemic circuit** transport oxygenated blood and nutrients to all organs and tissues, ultimately returning deoxygenated blood to the right atrium.

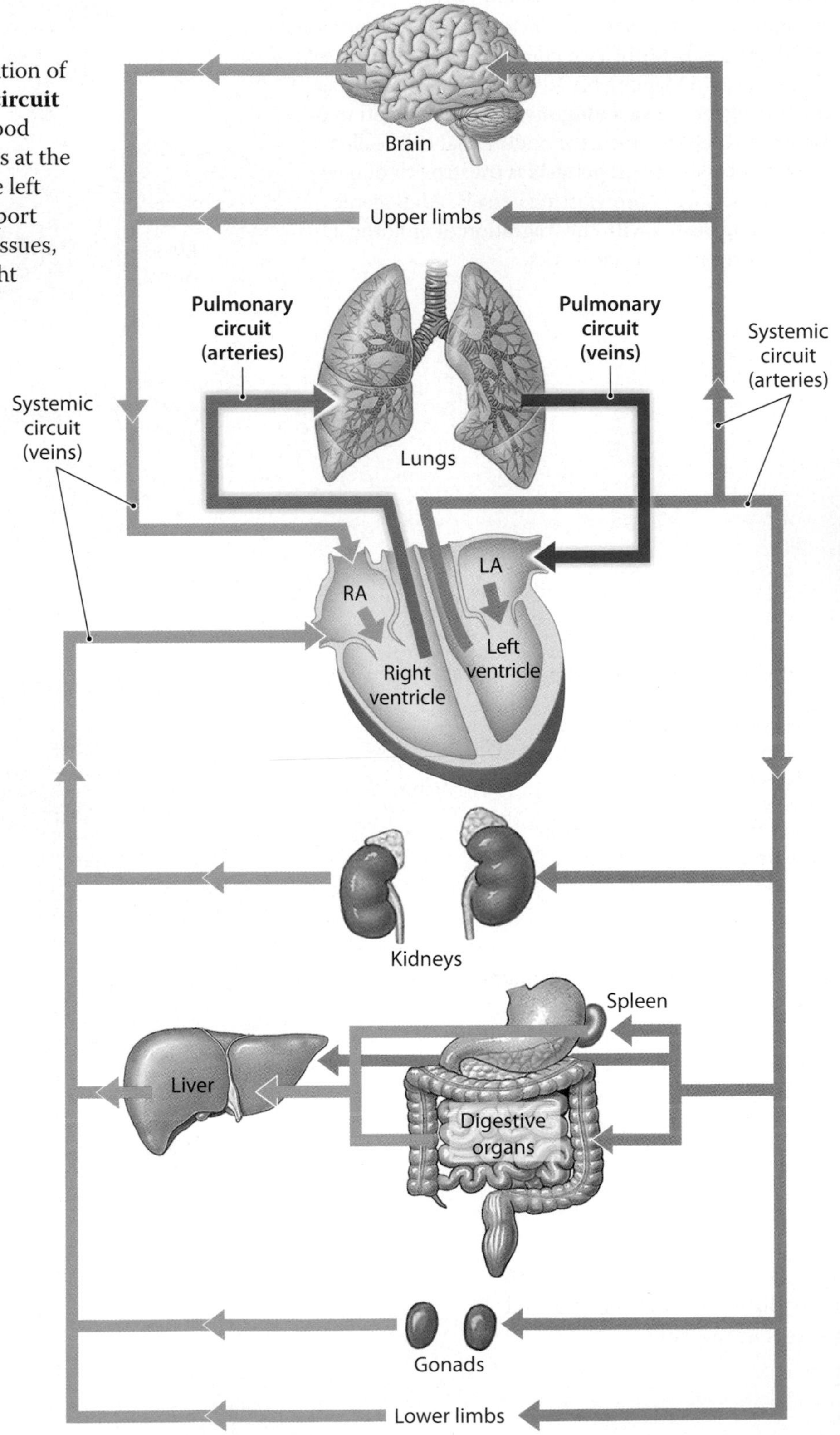
Brain
Upper limbs
Pulmonary circuit (arteries)
Pulmonary circuit (veins)
Systemic circuit (arteries)
Systemic circuit (veins)
Lungs
LA
RA
Left ventricle
Right ventricle
Kidneys
Spleen
Liver
Digestive organs
Gonads
Lower limbs

2 The table below summarizes the general organization and naming of the body's blood vessels.

General Patterns of Blood Vessel Organization

1. The peripheral distributions of arteries and veins on the body's left and right sides are generally identical, except near the heart, where the largest vessels connect to the atria or ventricles.
2. A single vessel may have several names as it crosses specific anatomical boundaries, making accurate anatomical descriptions possible when the vessel extends far into the periphery. For example, the external iliac artery becomes the femoral artery as it leaves the trunk and enters the lower limb.
3. Tissues and organs are usually serviced by several arteries and veins. Often, anastomoses between adjacent arteries or veins reduce the impact of a temporary or even permanent **occlusion** (blockage) of a single blood vessel.

3 Arteries of the pulmonary circuit differ from those of the systemic circuit in that they carry deoxygenated blood. (This is why most color-coded diagrams show the pulmonary arteries in blue, the same color as systemic veins.) By convention, several large arteries are called **trunks**; the **pulmonary trunk** is one important example. As the pulmonary trunk curves over the superior border of the heart, it gives rise to the left and right **pulmonary arteries**. These large arteries enter the lungs before branching repeatedly, giving rise to smaller and smaller arteries. The smallest branches, the **pulmonary arterioles**, provide blood to **alveolar capillaries** that surround small air pockets called **alveoli** (al-VĒ-ō-lī; singular, *alveolus*). The walls of the alveoli are thin enough for gas exchange between the capillary blood and inspired air; the blood absorbs oxygen and eliminates carbon dioxide. Oxygenated blood leaving the alveolar capillaries enters venules that in turn unite to form larger vessels carrying blood toward the **pulmonary veins**. These four veins, two from each lung, deliver oxygenated blood into the left atrium, completing the pulmonary circuit.

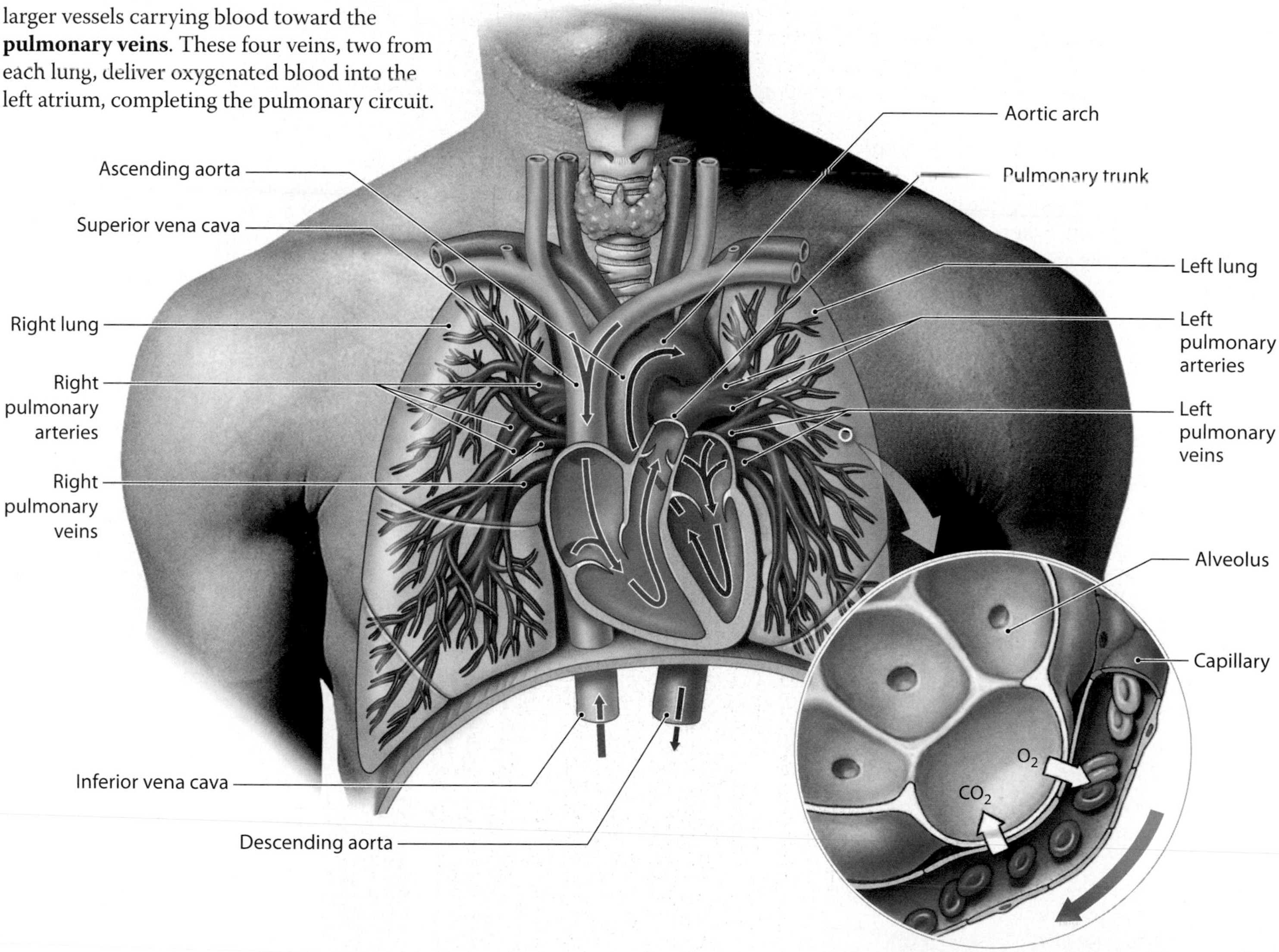

Aortic arch
Pulmonary trunk
Ascending aorta
Superior vena cava
Left lung
Right lung
Left pulmonary arteries
Right pulmonary arteries
Left pulmonary veins
Right pulmonary veins
Alveolus
Capillary
O_2
CO_2
Inferior vena cava
Descending aorta

Module 18.6 Review

a. Compare the oxygen content in the two circulatory circuits.

b. Briefly describe the three general patterns of blood vessel organization.

c. Trace a drop of blood through the lungs, beginning at the right ventricle and ending at the left atrium.

LO **18.6** Identify the major arteries and veins of the pulmonary circuit, and name the areas each serves.

The systemic arterial and venous systems operate in parallel, and the major vessels often have similar names

1 This figure is an overview of the systemic **arterial system**. Note that all the vessels of the systemic arterial system originate from the **aorta**, the large elastic artery extending from the left ventricle of the heart. In each of the six modules that follow, the left-hand page will illustrate and discuss the systemic arterial vessels and their branches. To reduce clutter, "artery" will not be repeated in every label. Because most of the major arteries are paired, with one artery of each pair on either side of the body, the terms "right" and "left" will appear in figure labels only when the arteries on both sides are labeled.

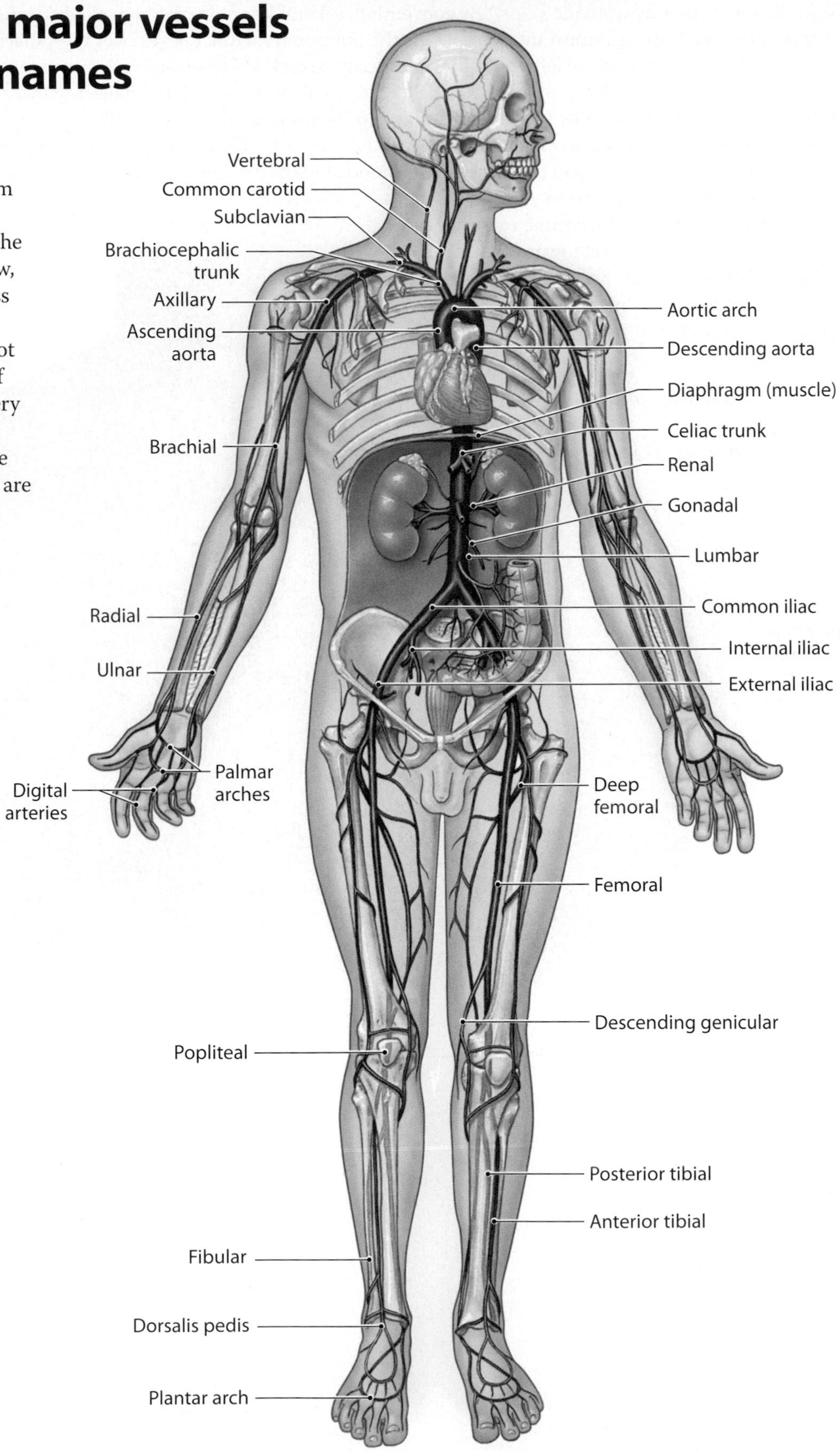
Vertebral
Common carotid
Subclavian
Brachiocephalic trunk
Axillary
Ascending aorta
Brachial
Radial
Ulnar
Digital arteries
Palmar arches
Popliteal
Fibular
Dorsalis pedis
Plantar arch
Aortic arch
Descending aorta
Diaphragm (muscle)
Celiac trunk
Renal
Gonadal
Lumbar
Common iliac
Internal iliac
External iliac
Deep femoral
Femoral
Descending genicular
Posterior tibial
Anterior tibial

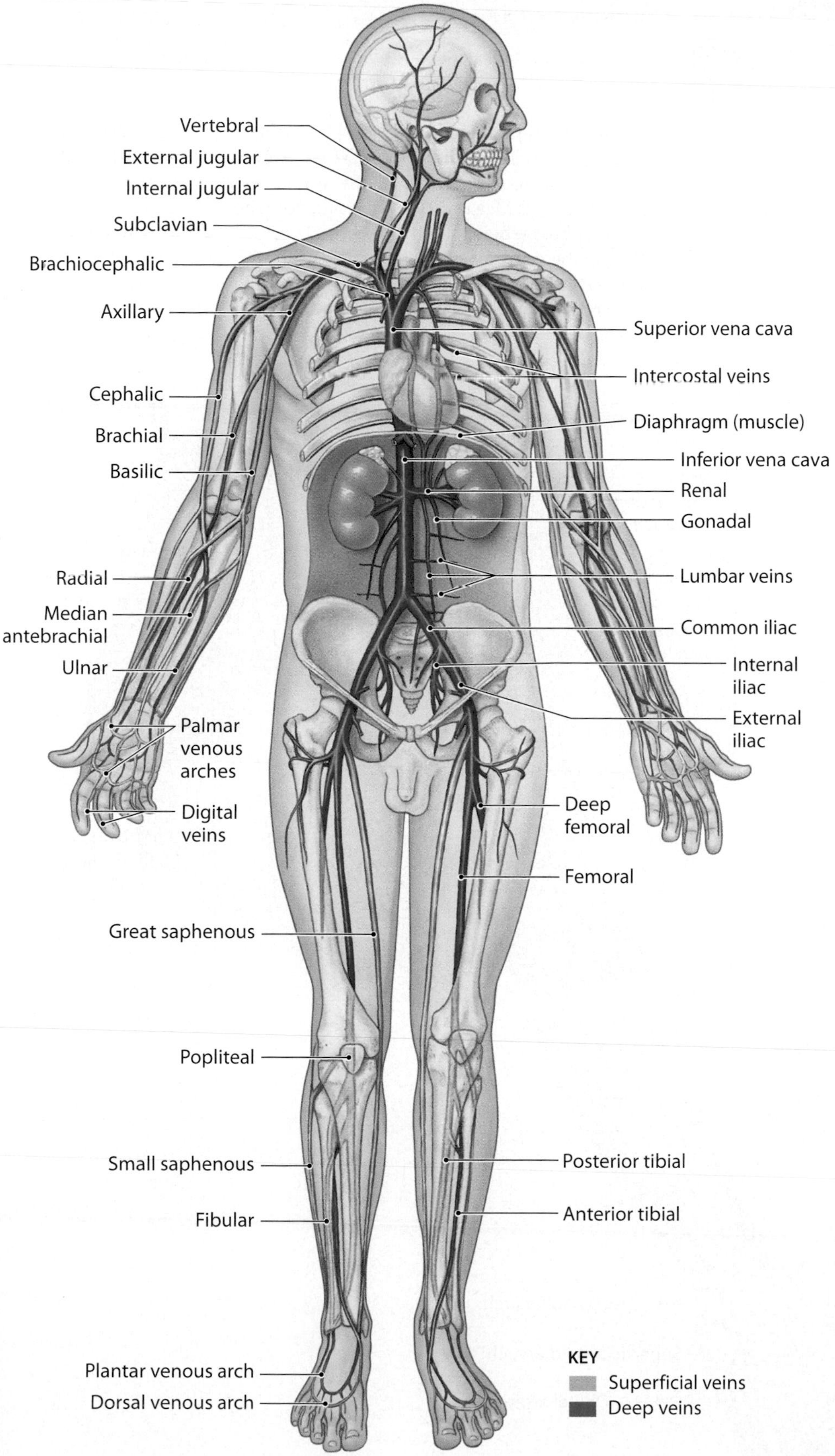
Vertebral
External jugular
Internal jugular
Subclavian
Brachiocephalic
Axillary
Cephalic
Brachial
Basilic
Radial
Median antebrachial
Ulnar
Palmar venous arches
Digital veins
Great saphenous
Popliteal
Small saphenous
Fibular
Plantar venous arch
Dorsal venous arch
Superior vena cava
Intercostal veins
Diaphragm (muscle)
Inferior vena cava
Renal
Gonadal
Lumbar veins
Common iliac
Internal iliac
External iliac
Deep femoral
Femoral
Posterior tibial
Anterior tibial
KEY
Superficial veins
Deep veins

2 This figure is an overview of the systemic **venous system**. Note that all of the vessels of the systemic venous system merge into two large veins: the **superior vena cava**, which collects systemic blood from the head, chest, and upper limbs, and the **inferior vena cava**, which collects systemic blood from all structures inferior to the diaphragm. In each of the six modules that follow, the right-hand page will illustrate and discuss the systemic venous vessels and their branches. To reduce clutter, "vein" will not be repeated in every label; the same name often applies to both the artery and the vein servicing a particular structure or region. Additionally, "right" and "left" will appear in figure labels only when the veins on both sides are shown.

One significant difference between the arterial and venous systems concerns the distribution of major veins in the neck and limbs. Arteries in these areas are located deep beneath the skin, protected by bones and surrounding soft tissues. In contrast, the neck and limbs generally have two sets of peripheral veins, one superficial and the other deep. This dual venous drainage is important for controlling body temperature. In hot weather, venous blood flows through superficial veins, where heat loss can occur; in cold weather, blood is routed to the deep veins to minimize heat loss.

Module 18.7 Review

a. Identify the largest artery in the body.
b. Name the two large veins that collect blood from the systemic circuit.
c. Besides containing valves, cite another major difference between the arterial and venous systems.

LO **18.7** Identify the major arteries and veins of the systemic circuit, and name the areas each serves.

The branches of the aortic arch supply structures . . .

Start

Branches of the Aortic Arch

Three elastic arteries originate along the aortic arch and deliver blood to the head, neck, shoulders, and upper limbs:

The **brachiocephalic** (brā-kē-ō-se-FAL-ik) **trunk** is the first branch off the aortic arch. It ascends for a short distance before branching to form the **right subclavian artery** and the **right common carotid artery**.	The **left common carotid artery**, the second vessel arising from the aortic arch, supplies the left side of the head and neck.	The **left subclavian artery**, the third artery arising from the aortic arch, has the same distribution pattern as the right subclavian artery.

The Right Subclavian Artery

Two major branches arise before a subclavian artery leaves the thoracic cavity: (1) the **internal thoracic artery** (internal mammary**)**, supplying the pericardium and anterior wall of the chest; and (2) the **vertebral artery**, which provides blood to the brain and spinal cord.

Arteries of the Arm

After leaving the thoracic cavity and passing across the superior border of the first rib, the subclavian is called the **axillary artery**. This artery crosses the axilla to enter the arm, where it becomes the **brachial artery**, which supplies blood to the upper limb. The brachial artery gives rise to the deep brachial artery, which supplies deep structures on the posterior aspect of the arm, and the ulnar collateral arteries, which supply the area around the elbow.

Arteries of the Forearm

As it approaches the coronoid fossa of the humerus, the brachial artery divides into the **radial artery**, which follows the radius, and the **ulnar artery**, which follows the ulna to the wrist. At the wrist, the radial and ulnar arteries fuse to form the superficial and deep **palmar arches**, which supply blood to the hand and to the **digital arteries** of the thumb and fingers.

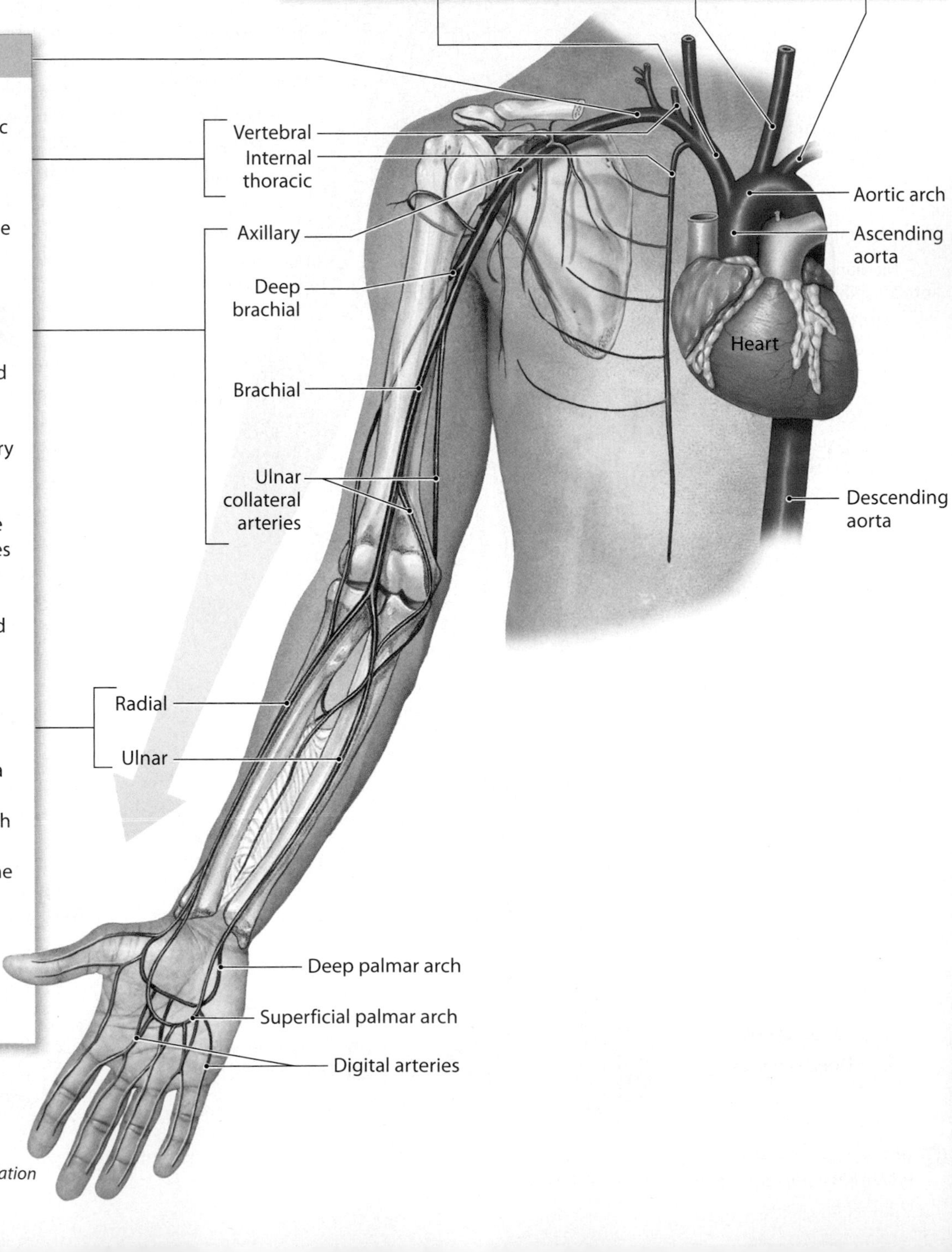
Vertebral
Internal thoracic
Axillary
Deep brachial
Brachial
Ulnar collateral arteries
Radial
Ulnar
Aortic arch
Ascending aorta
Heart
Descending aorta
Deep palmar arch
Superficial palmar arch
Digital arteries

. . . that are drained by the superior vena cava

Veins of the Neck

The **external jugular vein** drains superficial structures of the head and neck.	The **vertebral vein** drains the cervical spinal cord and the posterior surface of the skull.	The **internal jugular vein** drains deep structures of the head and neck.

The Right Subclavian Vein

The axillary vein is joined by the cephalic vein on the lateral surface of the first rib, forming the **subclavian vein**, which continues into the chest.

Veins of the Arm

As the **brachial vein** merges with the **basilic vein** it becomes the **axillary vein**, which enters the axilla.

The **cephalic vein** extends along the lateral side of the arm.

Veins of the Forearm

The **median cubital vein** interconnects the cephalic and basilic veins. (The median cubital is the vein from which venous blood samples are typically collected.)

The **ulnar vein** and the **radial vein** drain the **deep palmar arch**. Before crossing the elbow, these veins fuse to form the brachial vein.

The cephalic vein, the **median antebrachial vein**, and the basilic vein drain the **superficial palmar arch**.

The **digital veins** empty into superficial and deep veins of the hand, which are interconnected to form the palmar venous arches.

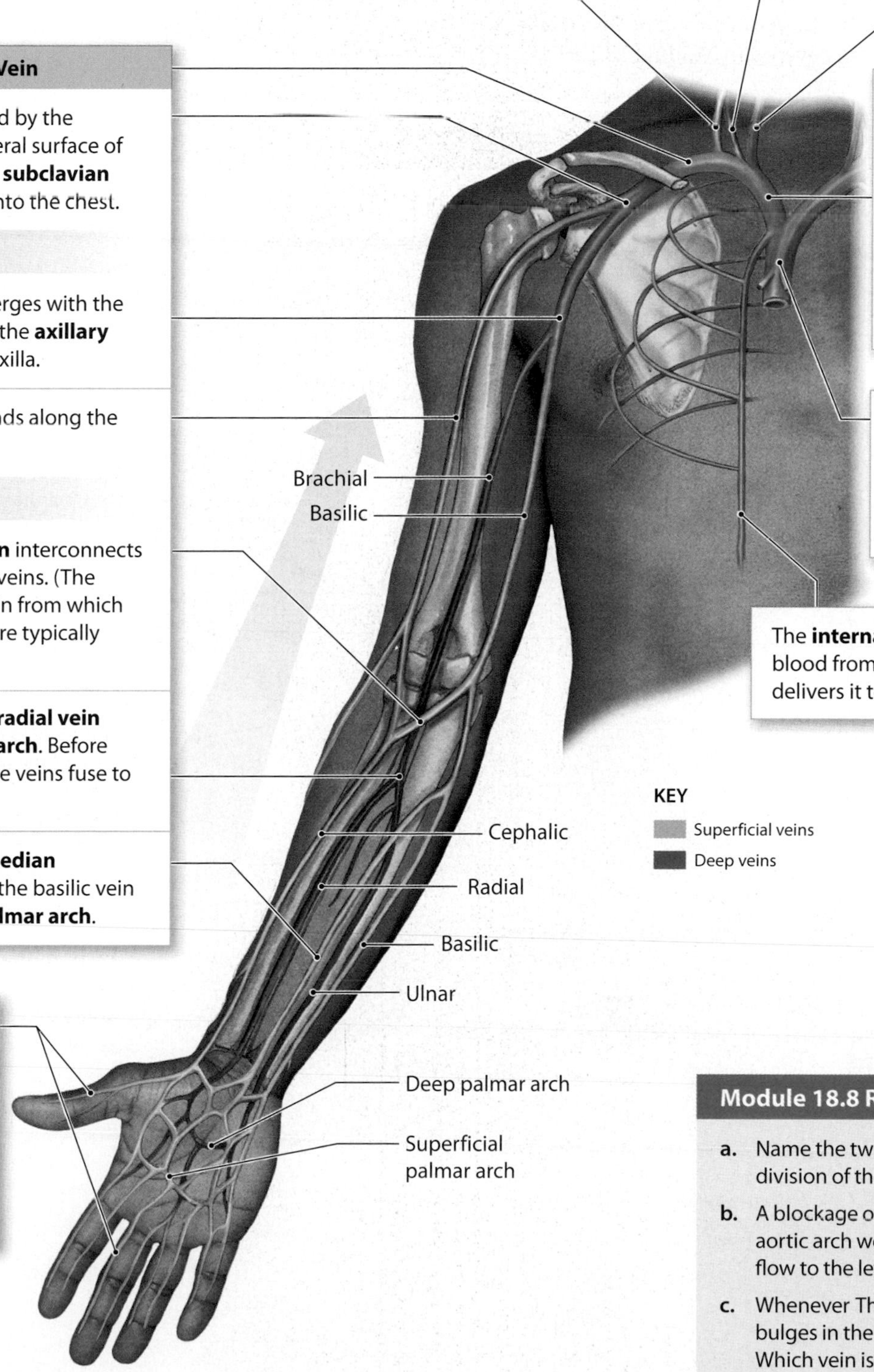
Brachial
Basilic
Cephalic
Radial
Basilic
Ulnar
Deep palmar arch
Superficial palmar arch
KEY
Superficial veins
Deep veins

The **brachiocephalic vein** forms as the jugular veins empty into the subclavian vein. It receives blood from the vertebral vein and the internal thoracic vein draining the anterior chest wall.

The **superior vena cava (SVC)** carries blood from the two brachiocephalic veins to the right atrium of the heart.

The **internal thoracic vein** collects blood from the intercostal veins and delivers it to the brachiocephalic vein.

Module 18.8 Review

a. Name the two arteries formed by the division of the brachiocephalic trunk.

b. A blockage of which branch of the aortic arch would interfere with blood flow to the left arm?

c. Whenever Thor gets angry, a large vein bulges in the lateral region of his neck. Which vein is this?

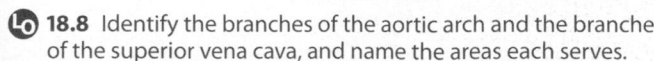

LO **18.8** Identify the branches of the aortic arch and the branches of the superior vena cava, and name the areas each serves.

The external carotid arteries supply the neck, lower jaw, and face, and the internal carotid and vertebral arteries supply the brain . . .

1 The **common carotid arteries** ascend deep in the tissues of the neck and supply blood to the structures of the face, neck, and brain. To locate the carotid artery, gently press a finger along either side of the windpipe (trachea) until you feel a strong pulse. Details of the arterial supply of the brain are covered in Module 18.10.

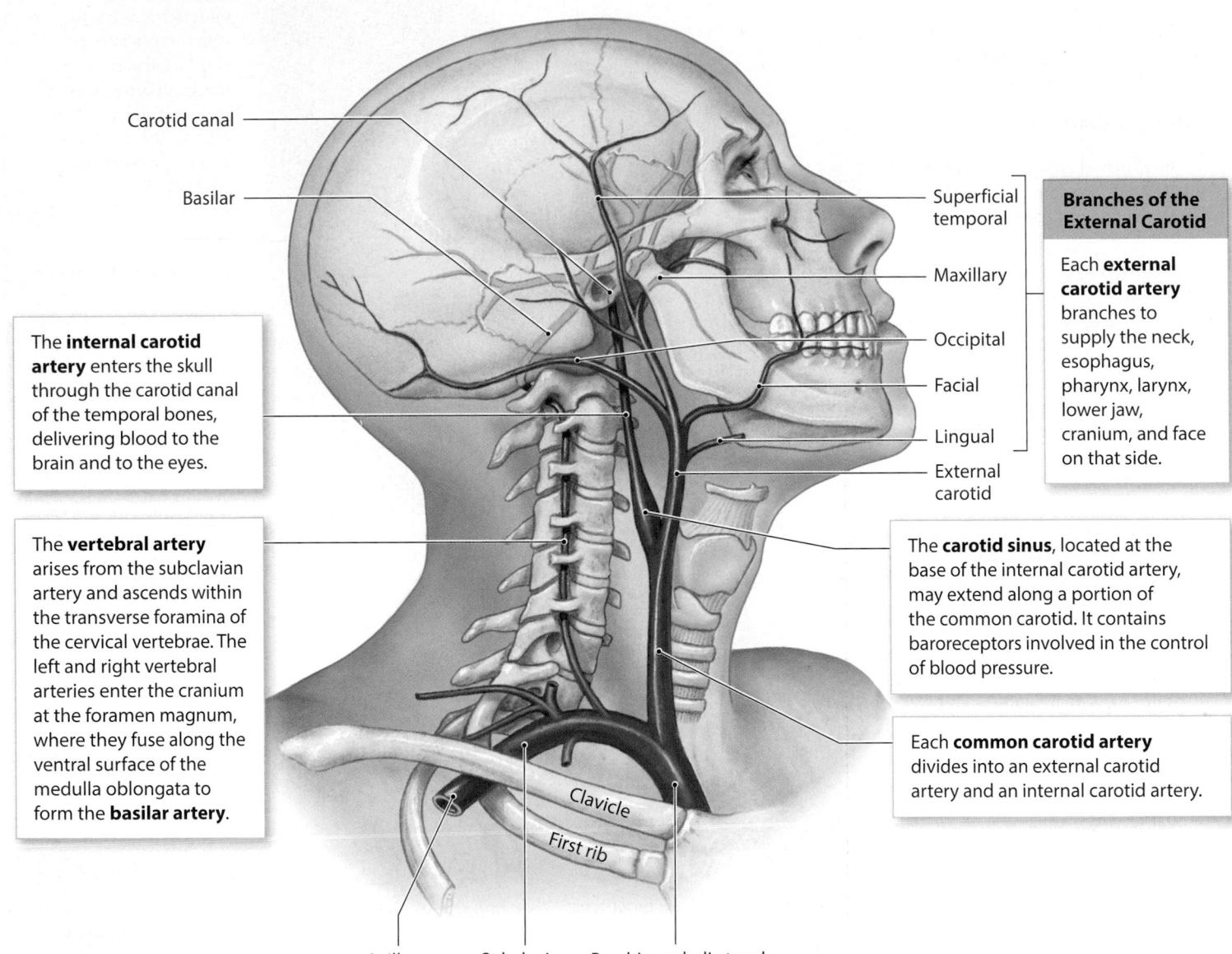

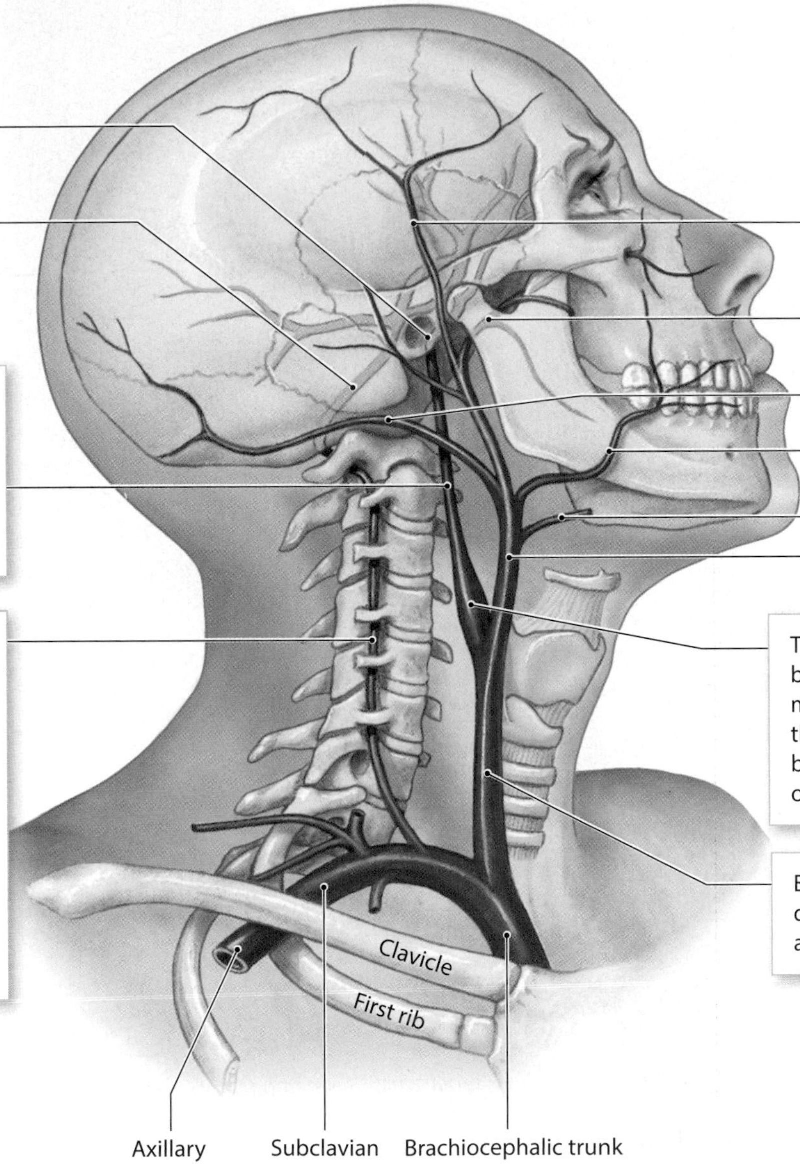
Carotid canal
Basilar
Superficial temporal
Maxillary
Occipital
Facial
Lingual
External carotid
Clavicle
First rib
Axillary
Subclavian
Brachiocephalic trunk

The **internal carotid artery** enters the skull through the carotid canal of the temporal bones, delivering blood to the brain and to the eyes.

The **vertebral artery** arises from the subclavian artery and ascends within the transverse foramina of the cervical vertebrae. The left and right vertebral arteries enter the cranium at the foramen magnum, where they fuse along the ventral surface of the medulla oblongata to form the **basilar artery**.

Branches of the External Carotid

Each **external carotid artery** branches to supply the neck, esophagus, pharynx, larynx, lower jaw, cranium, and face on that side.

The **carotid sinus**, located at the base of the internal carotid artery, may extend along a portion of the common carotid. It contains baroreceptors involved in the control of blood pressure.

Each **common carotid artery** divides into an external carotid artery and an internal carotid artery.

. . . while the external jugular veins drain the regions supplied by the external carotid arteries, and the internal jugular veins drain the brain

2 The **external jugular veins** are formed by the maxillary and temporal veins, while the **internal jugular veins** drain the blood from the various venous sinuses within the cranium (see Module 18.10 for additional details). The external and internal jugular veins combine with the vertebral and subclavian to form the **brachiocephalic vein**.

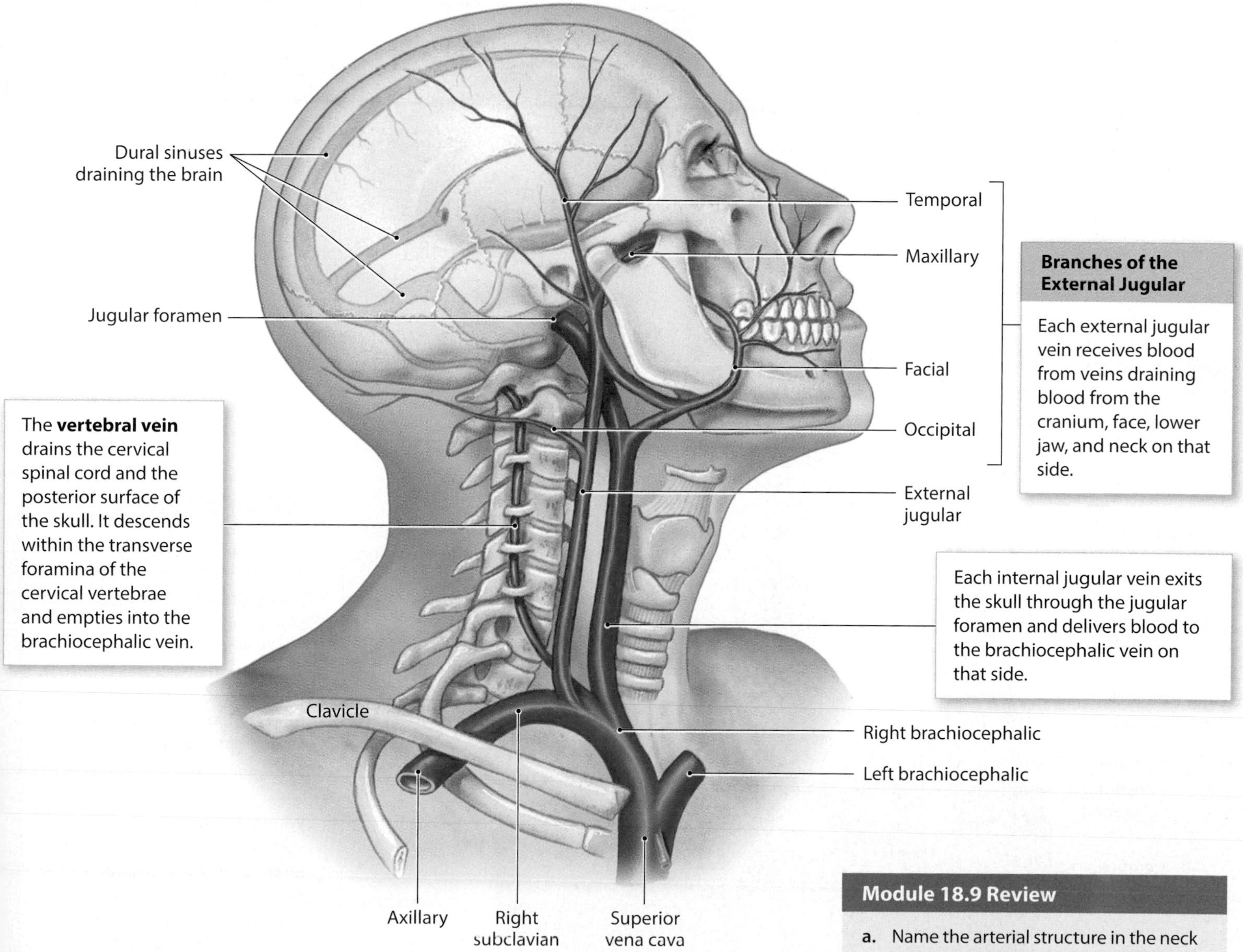

Dural sinuses draining the brain
Jugular foramen
The **vertebral vein** drains the cervical spinal cord and the posterior surface of the skull. It descends within the transverse foramina of the cervical vertebrae and empties into the brachiocephalic vein.
Clavicle
Axillary
Right subclavian
Superior vena cava
Temporal
Maxillary
Facial
Occipital
Branches of the External Jugular
Each external jugular vein receives blood from veins draining blood from the cranium, face, lower jaw, and neck on that side.
External jugular
Each internal jugular vein exits the skull through the jugular foramen and delivers blood to the brachiocephalic vein on that side.
Right brachiocephalic
Left brachiocephalic

Module 18.9 Review

a. Name the arterial structure in the neck region that contains baroreceptors.

b. Identify the branches of the external carotid artery.

c. Identify the veins that combine to form the brachiocephalic vein.

LO **18.9** Identify the branches of the carotid arteries and the branches of the external jugular veins, and name the areas each serves.

The internal carotid arteries and the vertebral arteries supply the brain . . .

1 This lateral view shows the major arteries supplying the brain. The internal carotid arteries normally supply the arteries of the anterior half of the cerebrum, and the rest of the brain receives blood from the vertebral and basilar arteries. The internal carotid artery ascends to the level of the optic nerves, where each artery divides into three branches: (1) an **ophthalmic artery**, which supplies the eyes; (2) an **anterior cerebral artery**, which supplies the frontal and parietal lobes of the brain; and (3) a **middle cerebral artery**, which supplies the midbrain and the lateral surfaces of the cerebral hemispheres.

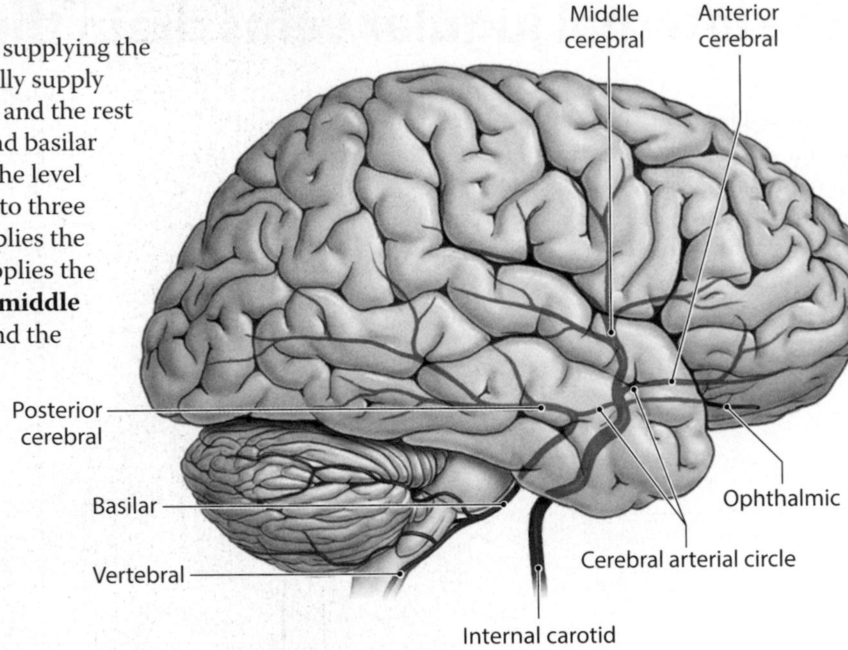
Middle cerebral
Anterior cerebral
Posterior cerebral
Basilar
Vertebral
Ophthalmic
Cerebral arterial circle
Internal carotid

2 The internal carotid arteries and the basilar artery are interconnected in a ring-shaped anastomosis called the **cerebral arterial circle**.

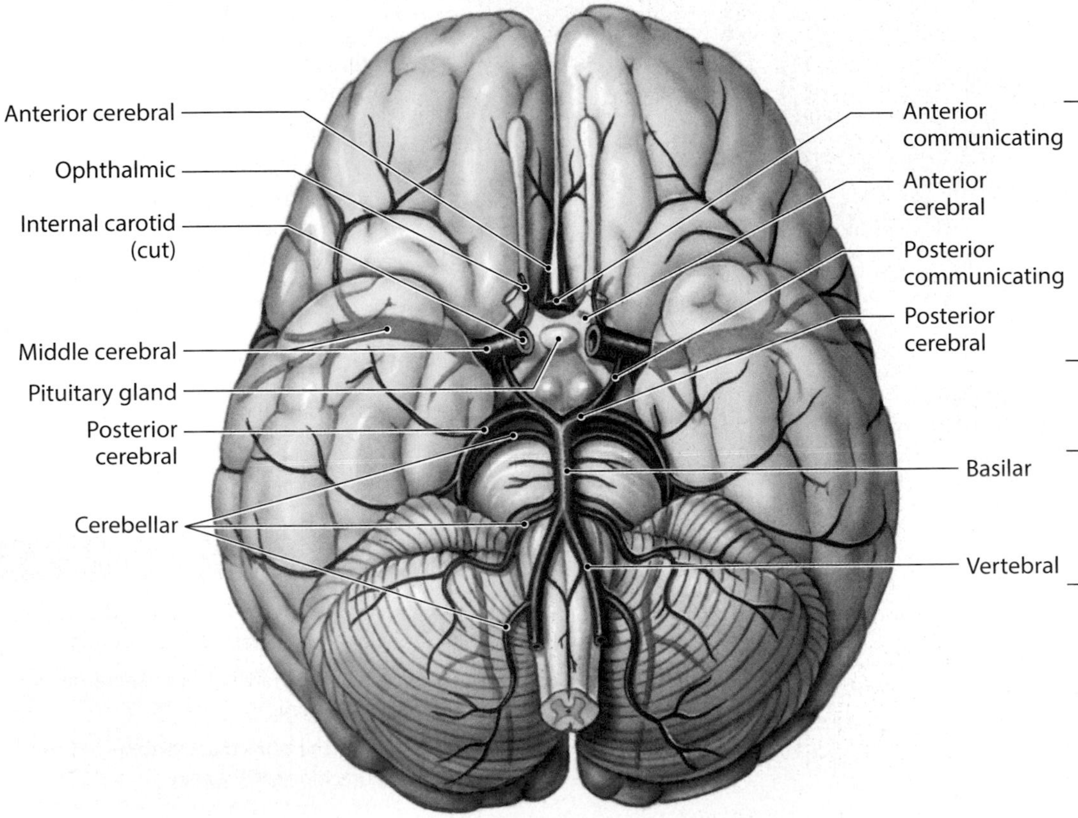
Anterior cerebral
Ophthalmic
Internal carotid (cut)
Middle cerebral
Pituitary gland
Posterior cerebral
Cerebellar
Anterior communicating
Anterior cerebral
Posterior communicating
Posterior cerebral
Basilar
Vertebral

Cerebral Arterial Circle

The cerebral arterial circle, or circle of Willis, encircles the infundibulum of the pituitary gland. This arrangement reduces the likelihood of a serious interruption of cerebral blood flow, because the brain can receive blood from either the carotid or the vertebral arteries.

Within the cranium, the vertebral arteries and the basilar artery supply blood to the spinal cord, medulla oblongata, pons, and cerebellum before dividing into the posterior cerebral arteries, which in turn branch off into the posterior communicating arteries.

. . . which is drained by the dural sinuses and the internal jugular veins

3 The superficial cerebral veins and small veins of the brain stem empty into a network of dural sinuses. Most of the deep cerebral veins converge within the brain to form the **great cerebral vein**, which delivers blood from the interior of the cerebral hemispheres and the choroid plexus to the **straight sinus**. Numerous small veins from the orbit and other cerebral veins drain into the **cavernous sinus**.

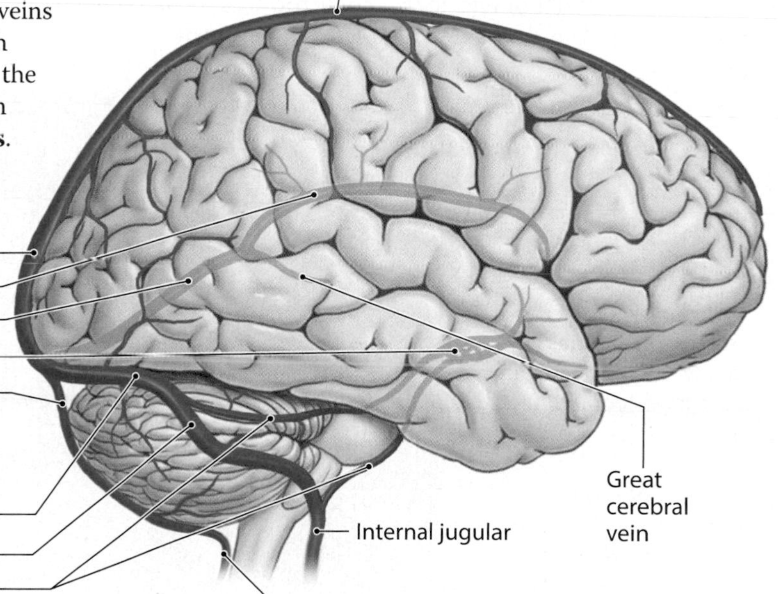

The **superior sagittal sinus**, in the falx cerebri, is the largest dural sinus.
Unpaired Median Sinuses
Superior sagittal sinus
Inferior sagittal sinus
Straight sinus
Cavernous sinus
Occipital sinus
Paired Lateral Sinuses
Right transverse sinus
Right sigmoid sinus
Petrosal sinuses
Internal jugular
Great cerebral vein
The **vertebral vein** on each side receives blood from the transverse sinus and occipital sinus as well as superficial veins of the skull and veins draining the cervical vertebrae.

4 The cavernous sinus empties into the two **petrosal sinuses**, which in turn drain into the **transverse sinuses**. The transverse sinuses, the straight sinus, and the superior sagittal sinus converge to form the **sigmoid sinuses**, which penetrate the jugular foramina and leave the skull as the internal jugular veins.

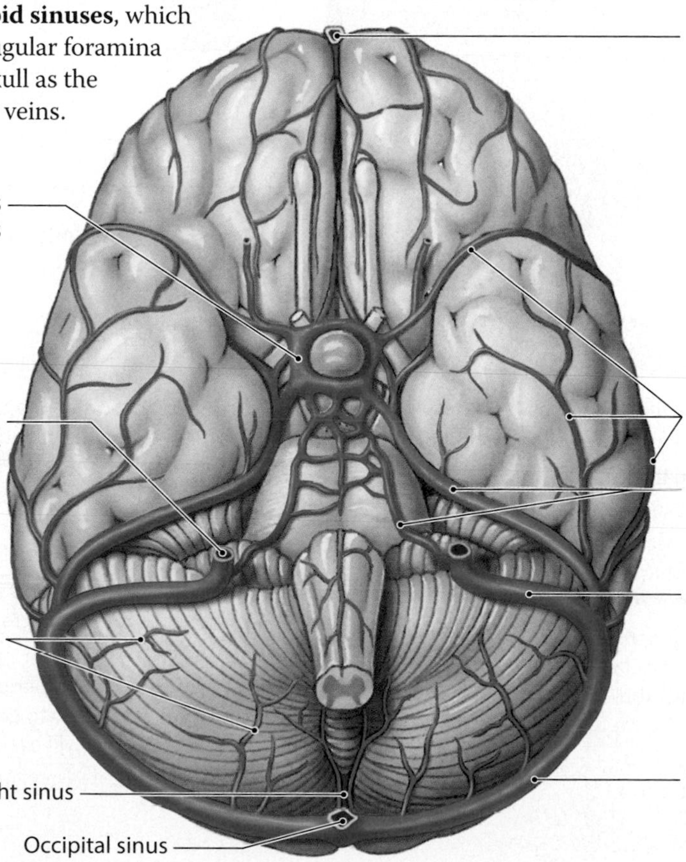

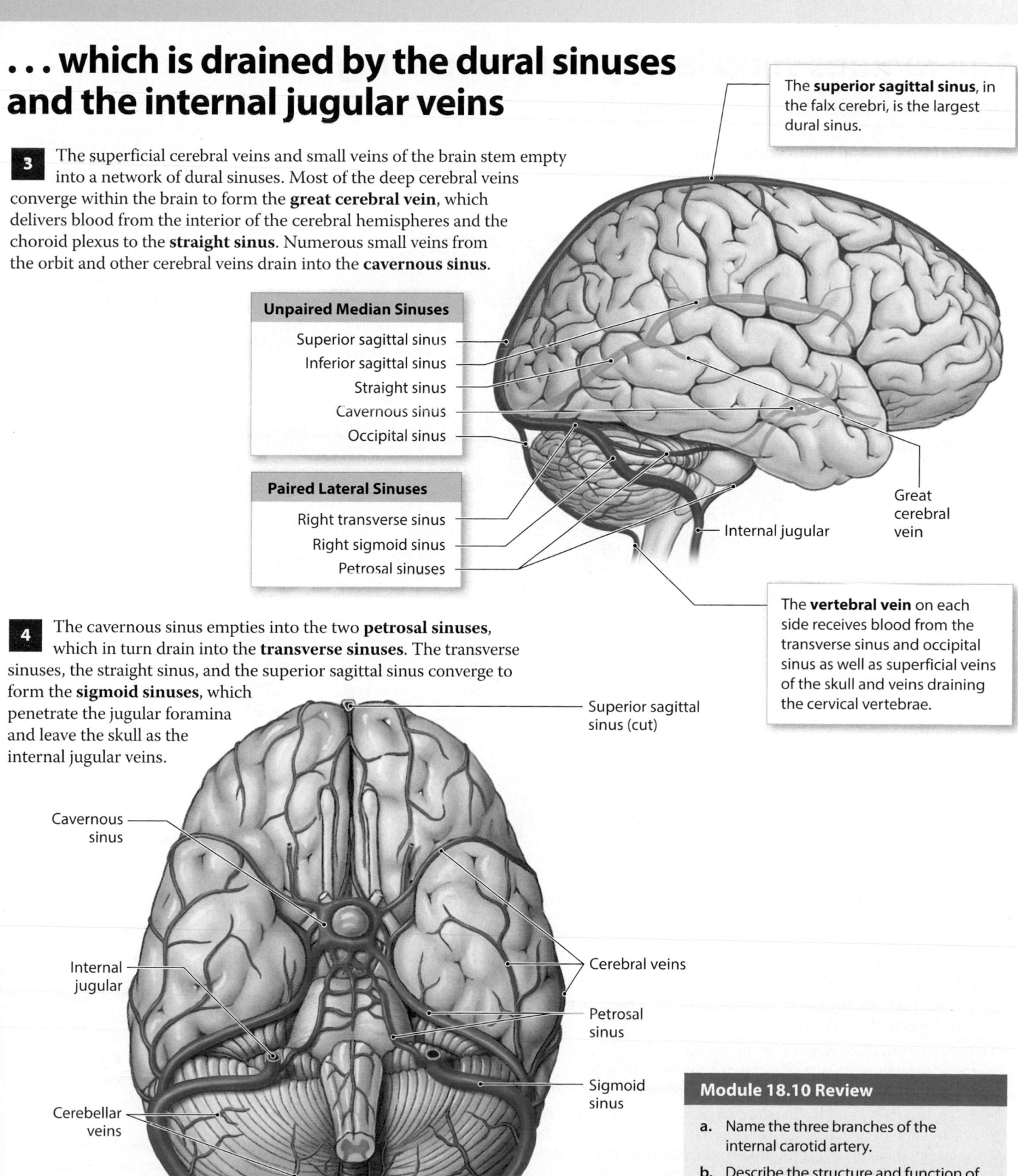

Superior sagittal sinus (cut)
Cavernous sinus
Internal jugular
Cerebral veins
Petrosal sinus
Sigmoid sinus
Cerebellar veins
Straight sinus
Transverse sinus
Occipital sinus

Module 18.10 Review

a. Name the three branches of the internal carotid artery.

b. Describe the structure and function of the cerebral arterial circle.

c. Name the veins that drain the dural sinuses of the brain.

LO **18.10** Identify the branches of the internal carotid and vertebral arteries and the branches of the internal jugular veins, and name the areas each serves.

The regions supplied by the descending aorta . . .

1 The descending aorta is continuous with the aortic arch. The diaphragm divides the descending aorta into a superior **thoracic aorta** and an inferior **abdominal aorta**. The figure details the branches of the thoracic aorta and introduces the branches of the abdominal aorta that are detailed further in the table below.

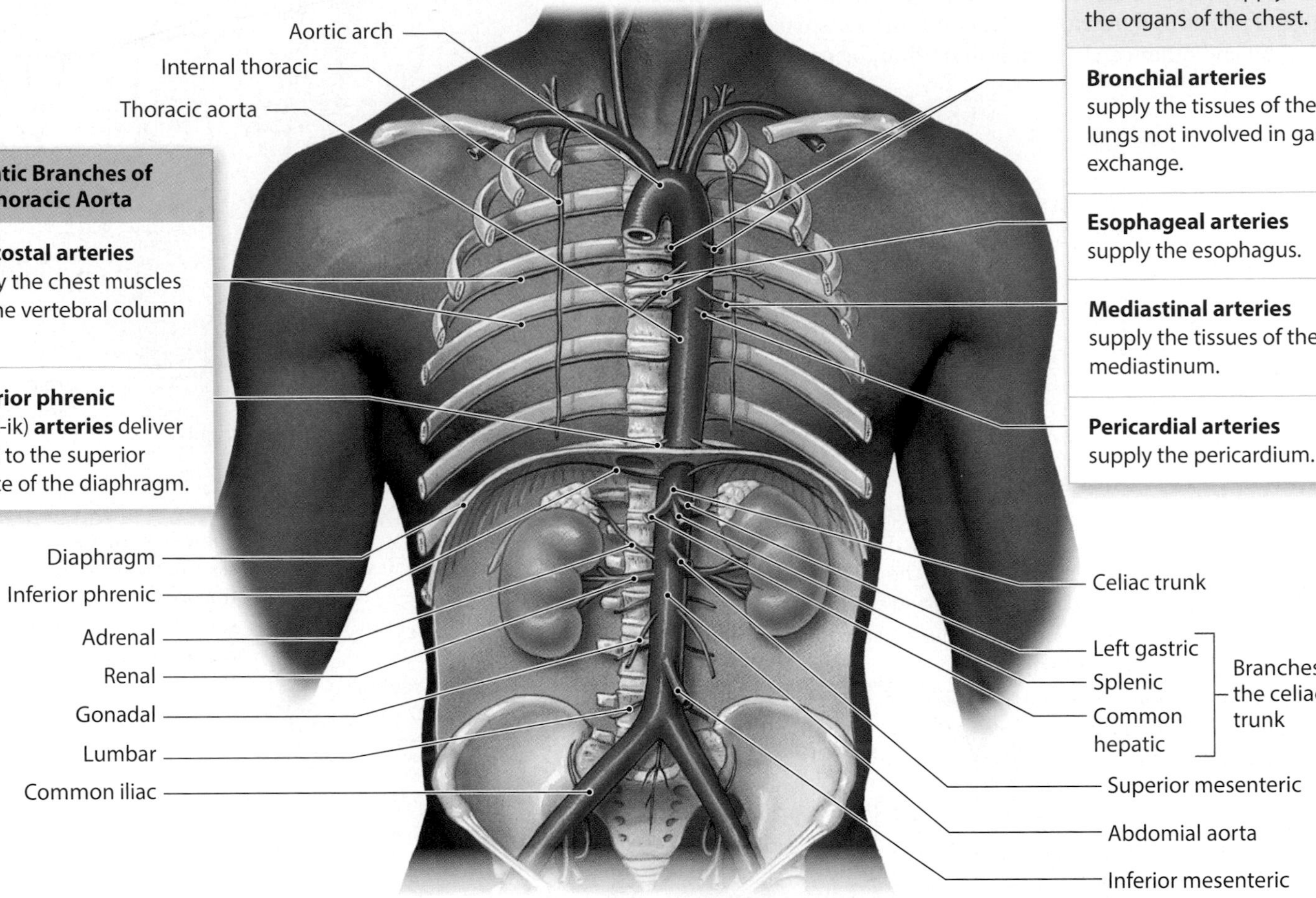
Aortic arch
Internal thoracic
Thoracic aorta
Diaphragm
Inferior phrenic
Adrenal
Renal
Gonadal
Lumbar
Common iliac
Celiac trunk
Left gastric
Splenic
Common hepatic
Branches of the celiac trunk
Superior mesenteric
Abdomial aorta
Inferior mesenteric

Somatic Branches of the Thoracic Aorta

Intercostal arteries supply the chest muscles and the vertebral column area.

Superior phrenic (FREN-ik) **arteries** deliver blood to the superior surface of the diaphragm.

Visceral Branches of the Thoracic Aorta

Visceral branches of the thoracic aorta supply the organs of the chest.

Bronchial arteries supply the tissues of the lungs not involved in gas exchange.

Esophageal arteries supply the esophagus.

Mediastinal arteries supply the tissues of the mediastinum.

Pericardial arteries supply the pericardium.

Major Paired Branches of the Abdominal Aorta

- The **inferior phrenic arteries** supply the inferior surface of the diaphragm and the inferior portion of the esophagus.
- The **adrenal arteries** supply the adrenal glands, which cap the superior part of each kidney.
- The short **renal arteries** arise along the posterolateral surface of the abdominal aorta, just inferior to the superior mesenteric artery. We will consider the branches of the renal arteries in Chapter 24.
- The **gonadal** (gō-NAD-al) **arteries** originate between the superior and inferior mesenteric arteries. In males, they are called testicular arteries; in females, they are termed ovarian arteries. The distribution of gonadal vessels (both arteries and veins) differs by sex; we will examine the differences in Chapter 26.
- Small **lumbar arteries** arise on the posterior surface of the aorta and supply the vertebrae, spinal cord, and abdominal wall.

Major Unpaired Branches of the Abdominal Aorta

- The **celiac** (SĒ-lē-ak) **trunk** divides into three branches: (1) the left gastric artery, which supplies the stomach and the inferior portion of the esophagus; (2) the splenic artery, which supplies the spleen and arteries to the stomach; and (3) the common hepatic artery, which supplies arteries to the liver, stomach, gallbladder, and the proximal portion of the small intestine.
- The **superior mesenteric** (mez-en-TER-ik) **artery** arises inferior to the celiac trunk to supply arteries to the pancreas and duodenum, and to most of the large intestine.
- The **inferior mesenteric artery** arises just superior to the bifurcation of the aorta, and it delivers blood to the terminal portions of the colon and the rectum. We will examine the branches of these vessels in Module 18.12.

. . . are drained by the superior and inferior venae cavae

2 The superior vena cava drains blood from the head, neck, shoulders, chest, and upper limbs. The chief collecting vessels of the thorax are the **azygos** (AZ-i-gos) **vein** and the **hemiazygos vein**. These veins receive blood from (1) **intercostal veins**, which in turn receive blood from the chest muscles; (2) **esophageal veins**, which drain blood from the inferior portion of the esophagus; (3) **bronchial veins** draining the passageways of the lungs; and (4) **mediastinal veins** draining other mediastinal structures. The inferior vena cava collects most of the blood inferior to the diaphragm.

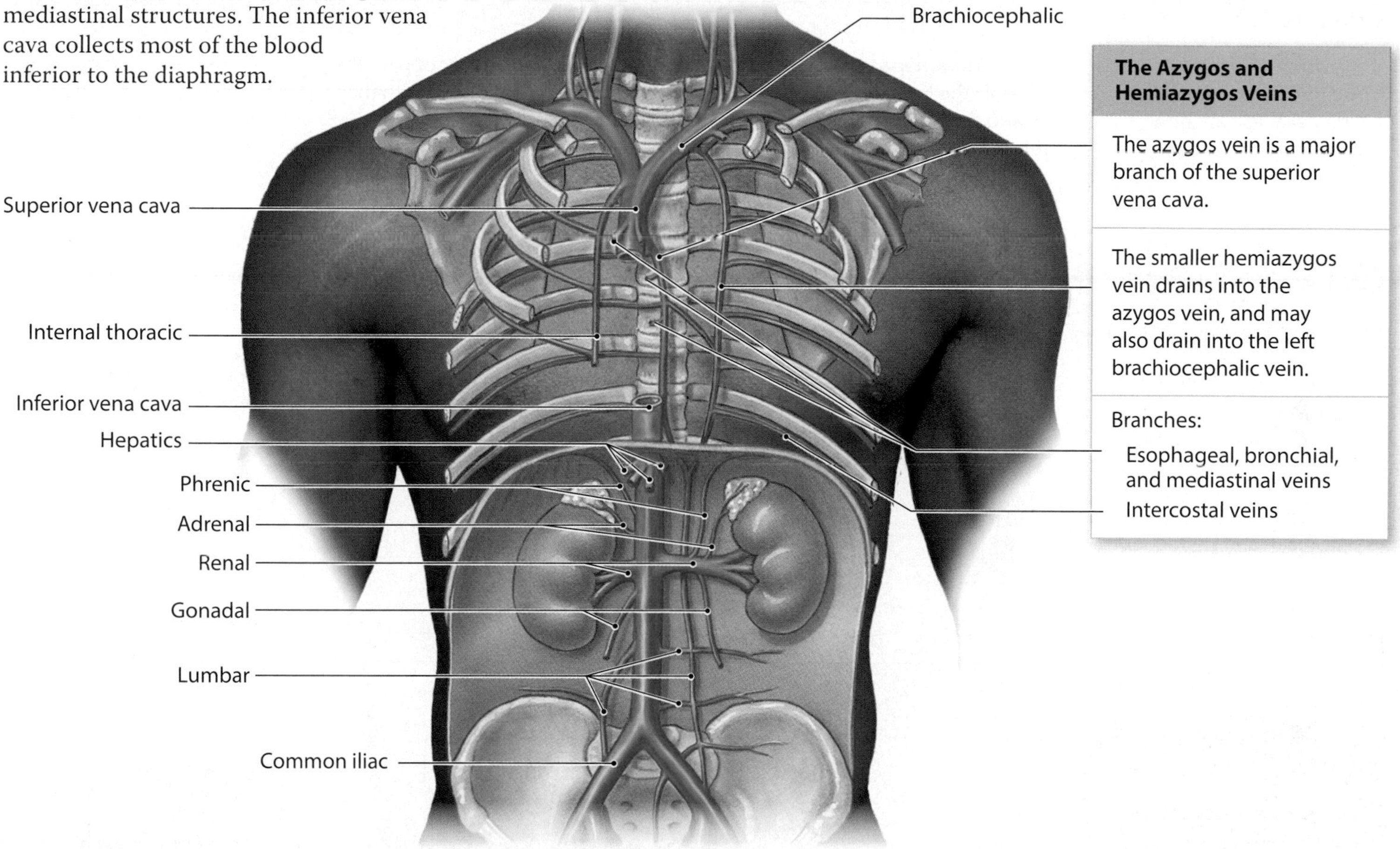
Brachiocephalic
Superior vena cava
Internal thoracic
Inferior vena cava
Hepatics
Phrenic
Adrenal
Renal
Gonadal
Lumbar
Common iliac
The Azygos and Hemiazygos Veins
The azygos vein is a major branch of the superior vena cava.
The smaller hemiazygos vein drains into the azygos vein, and may also drain into the left brachiocephalic vein.
Branches:
Esophageal, bronchial, and mediastinal veins
Intercostal veins

Major Branches of the Inferior Vena Cava

- **Lumbar veins** drain the lumbar portion of the abdomen, including the spinal cord and muscles of the body wall.
- **Gonadal** (ovarian or testicular) **veins** drain the ovaries or testes. The right gonadal vein empties into the inferior vena cava; the left gonadal vein generally drains into the left renal vein.
- **Hepatic veins** drain the sinusoids of the liver.
- **Renal veins**, the largest branches of the inferior vena cava, collect blood from the kidneys.
- **Adrenal veins** drain the adrenal glands. In most individuals, only the right adrenal vein drains into the inferior vena cava; the left adrenal vein drains into the left renal vein.
- **Phrenic veins** drain the diaphragm. Only the right phrenic vein drains into the inferior vena cava; the left drains into the left renal vein.

Module 18.11 Review

a. Grace is in an automobile accident, and her celiac trunk is ruptured. Which organs will be affected most directly by this injury?

b. Which vessel collects most of the venous blood inferior to the diaphragm?

c. Identify the major branches of the inferior vena cava.

LO 18.11 Identify the branches of the descending aorta and the branches of the venae cavae, and name the areas each serves.

The viscera supplied by the celiac trunk and mesenteric arteries . . .

1 The abdominal aorta begins immediately inferior to the diaphragm. Three unpaired branches supply the abdominal viscera: the celiac trunk, the superior mesenteric artery, and the inferior mesenteric artery.

The Celiac Trunk

The **celiac trunk** divides into the common hepatic artery, the left gastric artery, and the splenic artery.	The **common hepatic artery** branches to supply the liver, stomach, gallbladder, and the duodenum (the proximal segment of the small intestine).	The **left gastric artery** supplies the stomach. Its anastomosis with the right gastric artery ensures the stomach a continuous blood supply.	The **splenic artery** supplies the spleen and sends branches to the stomach and pancreas.

Branches of the Common Hepatic Artery

- Hepatic artery proper (liver)
- Cystic (gallbladder)
- Gastroduodenal (stomach and duodenum)
- Right gastric (stomach)
- Right gastroepiploic (stomach and duodenum)
- Superior pancreaticoduodenal (duodenum)

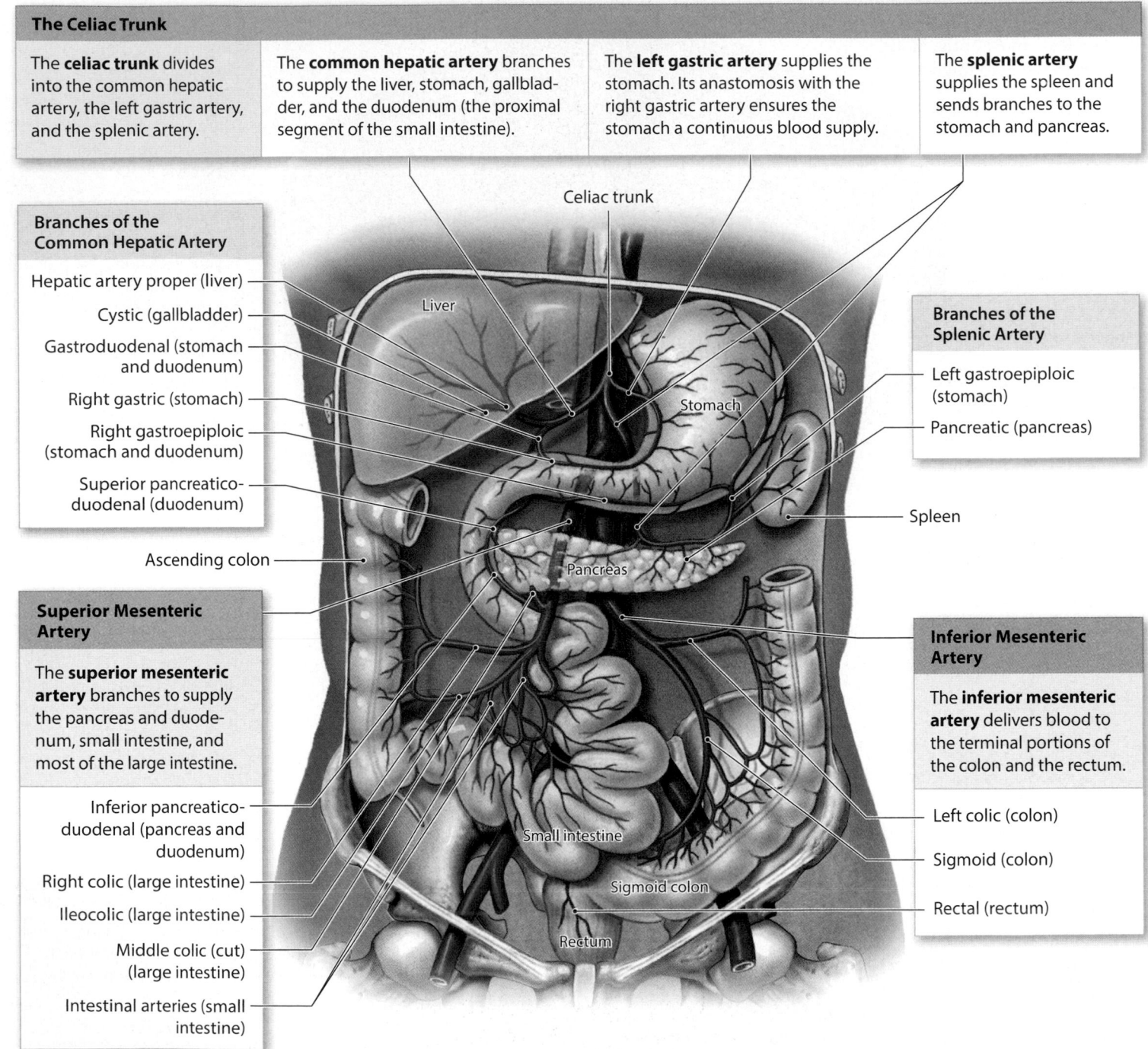

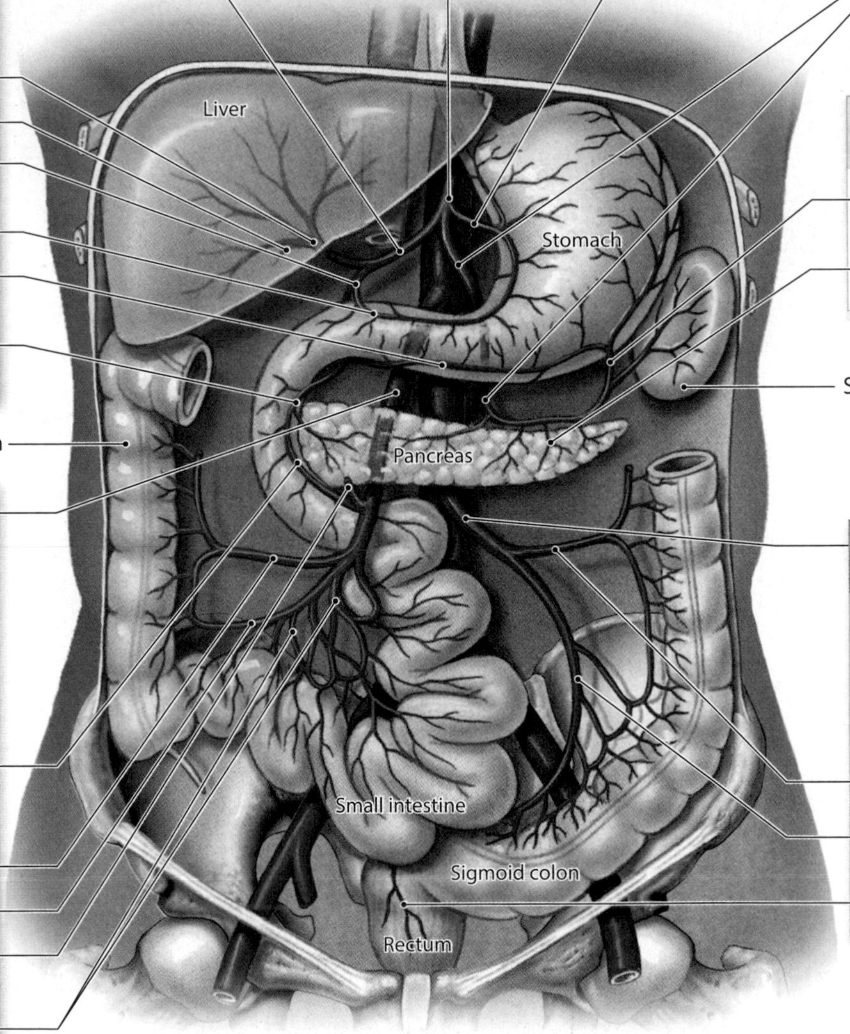
Celiac trunk
Liver
Stomach
Pancreas
Small intestine
Sigmoid colon
Rectum
Ascending colon
Spleen

Branches of the Splenic Artery

- Left gastroepiploic (stomach)
- Pancreatic (pancreas)

Superior Mesenteric Artery

The **superior mesenteric artery** branches to supply the pancreas and duodenum, small intestine, and most of the large intestine.

- Inferior pancreaticoduodenal (pancreas and duodenum)
- Right colic (large intestine)
- Ileocolic (large intestine)
- Middle colic (cut) (large intestine)
- Intestinal arteries (small intestine)

Inferior Mesenteric Artery

The **inferior mesenteric artery** delivers blood to the terminal portions of the colon and the rectum.

- Left colic (colon)
- Sigmoid (colon)
- Rectal (rectum)

. . . are drained by the branches of the hepatic portal vein

2 The **hepatic portal vein** forms through the fusion of the **superior mesenteric**, **inferior mesenteric**, and **splenic veins**. The largest volume of blood (and most of the nutrients) flows through the superior mesenteric vein. The hepatic portal vein receives blood from the left and right **gastric veins**, which drain the medial border of the stomach, and from the **cystic vein**, coming from the gallbladder. This circulatory pattern is called the **hepatic portal system**. It directs blood with absorbed nutrients from the digestive system to the liver for processing.

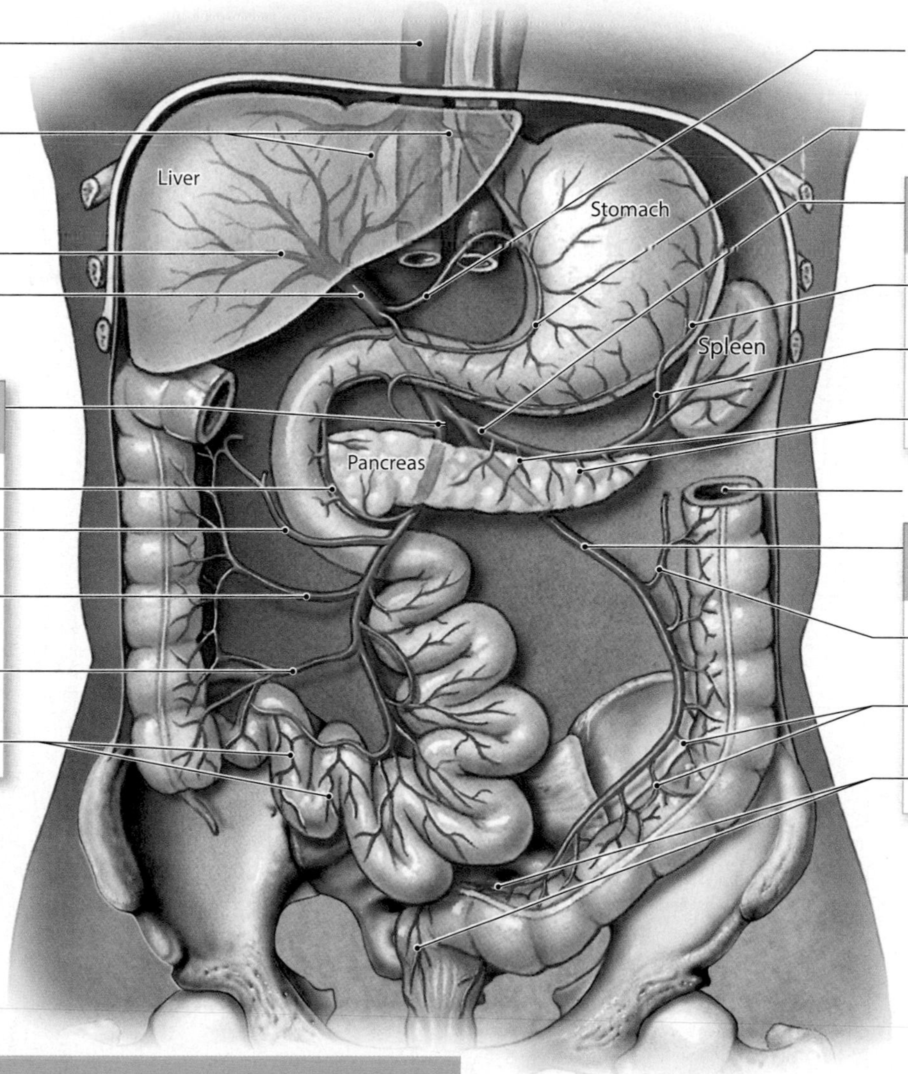

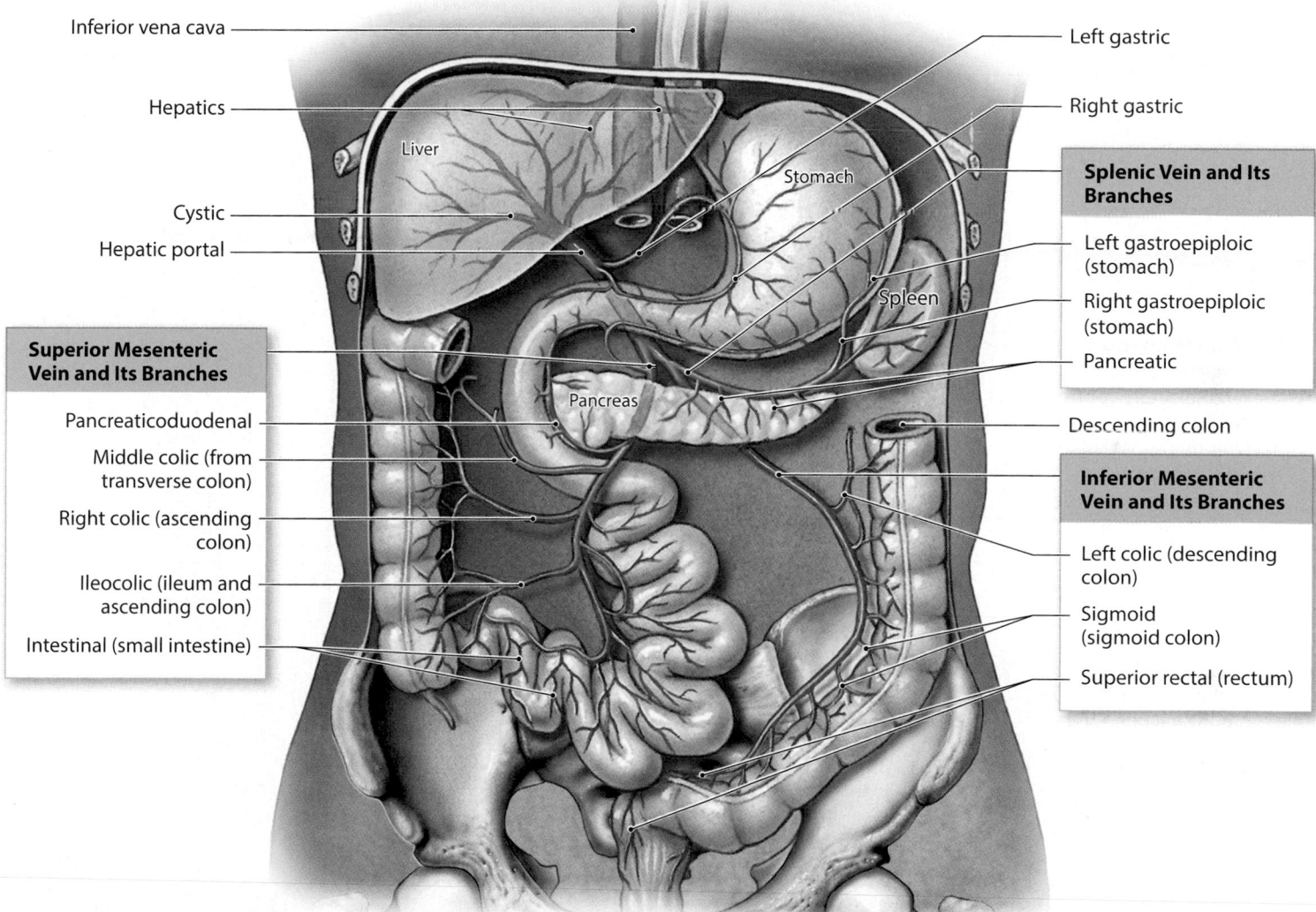

Inferior vena cava
Hepatics
Liver
Cystic
Hepatic portal
Stomach
Spleen
Pancreas
Left gastric
Right gastric
Splenic Vein and Its Branches
Left gastroepiploic (stomach)
Right gastroepiploic (stomach)
Pancreatic
Descending colon
Superior Mesenteric Vein and Its Branches
Pancreaticoduodenal
Middle colic (from transverse colon)
Right colic (ascending colon)
Ileocolic (ileum and ascending colon)
Intestinal (small intestine)
Inferior Mesenteric Vein and Its Branches
Left colic (descending colon)
Sigmoid (sigmoid colon)
Superior rectal (rectum)

Branches of the Hepatic Portal Vein

- The inferior mesenteric vein collects blood from capillaries along the inferior portion of the large intestine. It drains the left colic vein and the superior rectal veins, which collect venous blood from the descending colon, sigmoid colon, and rectum.
- The splenic vein is formed by the union of the inferior mesenteric vein and veins from the spleen, the lateral border of the stomach (left gastroepiploic vein), and the pancreas (pancreatic veins).
- The superior mesenteric vein collects blood from veins draining the stomach (right gastroepiploic vein), the small intestine (intestinal and pancreaticoduodenal veins), and two-thirds of the large intestine (ileocolic, right colic, and middle colic veins).

Module 18.12 Review

a. List the unpaired branches of the abdominal aorta that supply blood to the visceral organs.

b. Identify the three veins that merge to form the hepatic portal vein.

c. Identify two veins that carry blood away from the stomach.

18.12 Identify the branches of the visceral arterial vessels and the venous branches of the hepatic portal system, and name the areas each serves.

The pelvis and lower limbs are supplied by branches of the common iliac arteries . . .

1 Near the level of vertebra L_4, the abdominal aorta divides to form a pair of elastic arteries: the **right** and **left common iliac** (IL-ē-ak) **arteries**. At the level of the lumbosacral joint, each common iliac divides to form an **internal iliac artery** and an **external iliac artery**.

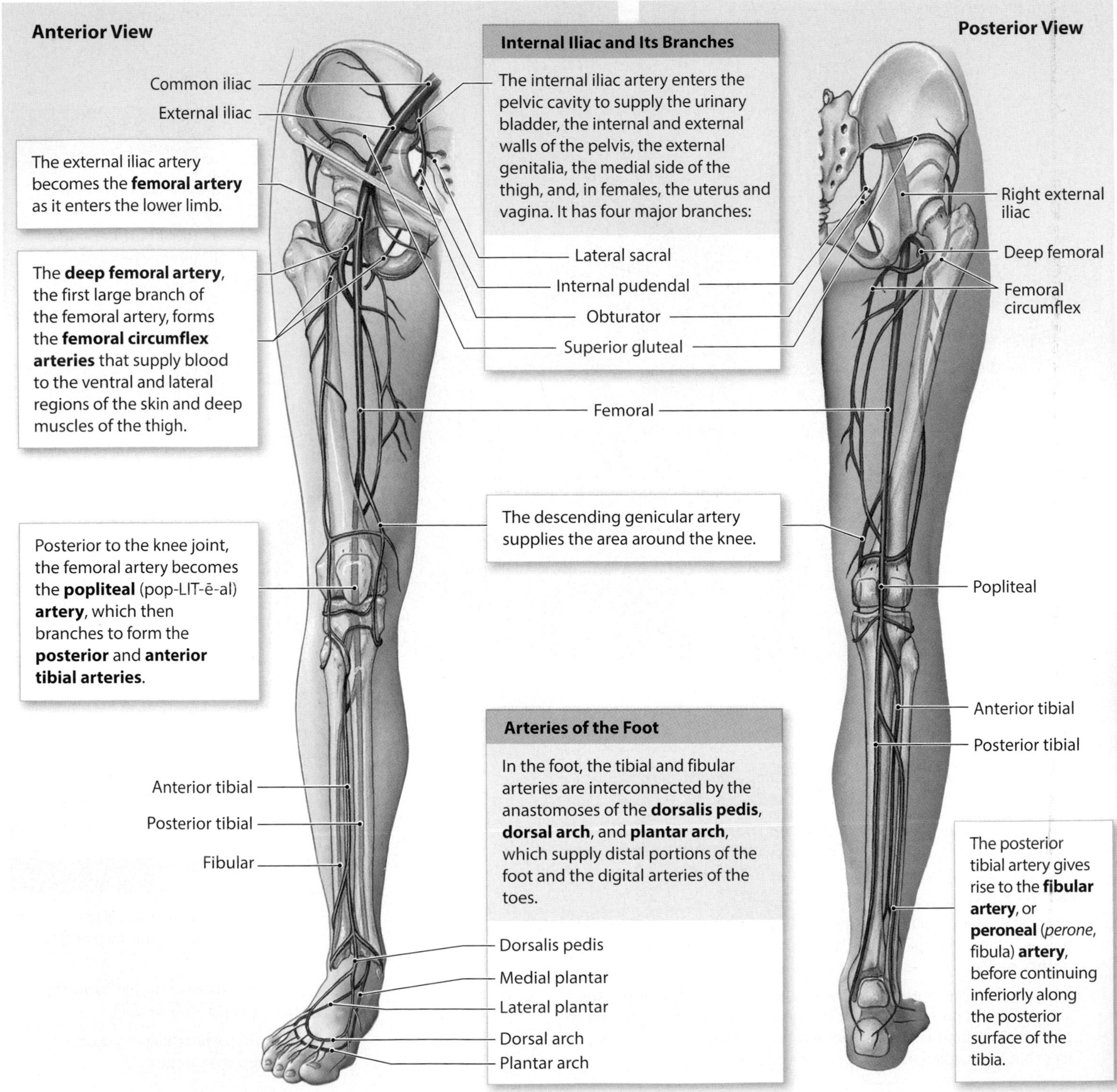
Anterior View
Common iliac
External iliac
The external iliac artery becomes the femoral artery as it enters the lower limb.
The deep femoral artery, the first large branch of the femoral artery, forms the femoral circumflex arteries that supply blood to the ventral and lateral regions of the skin and deep muscles of the thigh.
Posterior to the knee joint, the femoral artery becomes the popliteal (pop-LIT-ē-al) artery, which then branches to form the posterior and anterior tibial arteries.
Anterior tibial
Posterior tibial
Fibular
Internal Iliac and Its Branches
The internal iliac artery enters the pelvic cavity to supply the urinary bladder, the internal and external walls of the pelvis, the external genitalia, the medial side of the thigh, and, in females, the uterus and vagina. It has four major branches:
Lateral sacral
Internal pudendal
Obturator
Superior gluteal
Femoral
The descending genicular artery supplies the area around the knee.
Arteries of the Foot
In the foot, the tibial and fibular arteries are interconnected by the anastomoses of the dorsalis pedis, dorsal arch, and plantar arch, which supply distal portions of the foot and the digital arteries of the toes.
Dorsalis pedis
Medial plantar
Lateral plantar
Dorsal arch
Plantar arch
Posterior View
Right external iliac
Deep femoral
Femoral circumflex
Popliteal
Anterior tibial
Posterior tibial
The posterior tibial artery gives rise to the fibular artery, or peroneal (perone, fibula) artery, before continuing inferiorly along the posterior surface of the tibia.

. . . and drained by branches of the common iliac veins

2 The external iliac veins receive blood from the lower limbs, the pelvis, and the lower abdomen. As the left and right external iliac veins cross the inner surface of the ilium, they are joined by the internal iliac veins, which drain the pelvic organs. The internal iliac veins are formed by the fusion of the gluteal, internal pudendal, obturator, and lateral sacral veins. The union of external and internal iliac veins forms the **common iliac vein**.

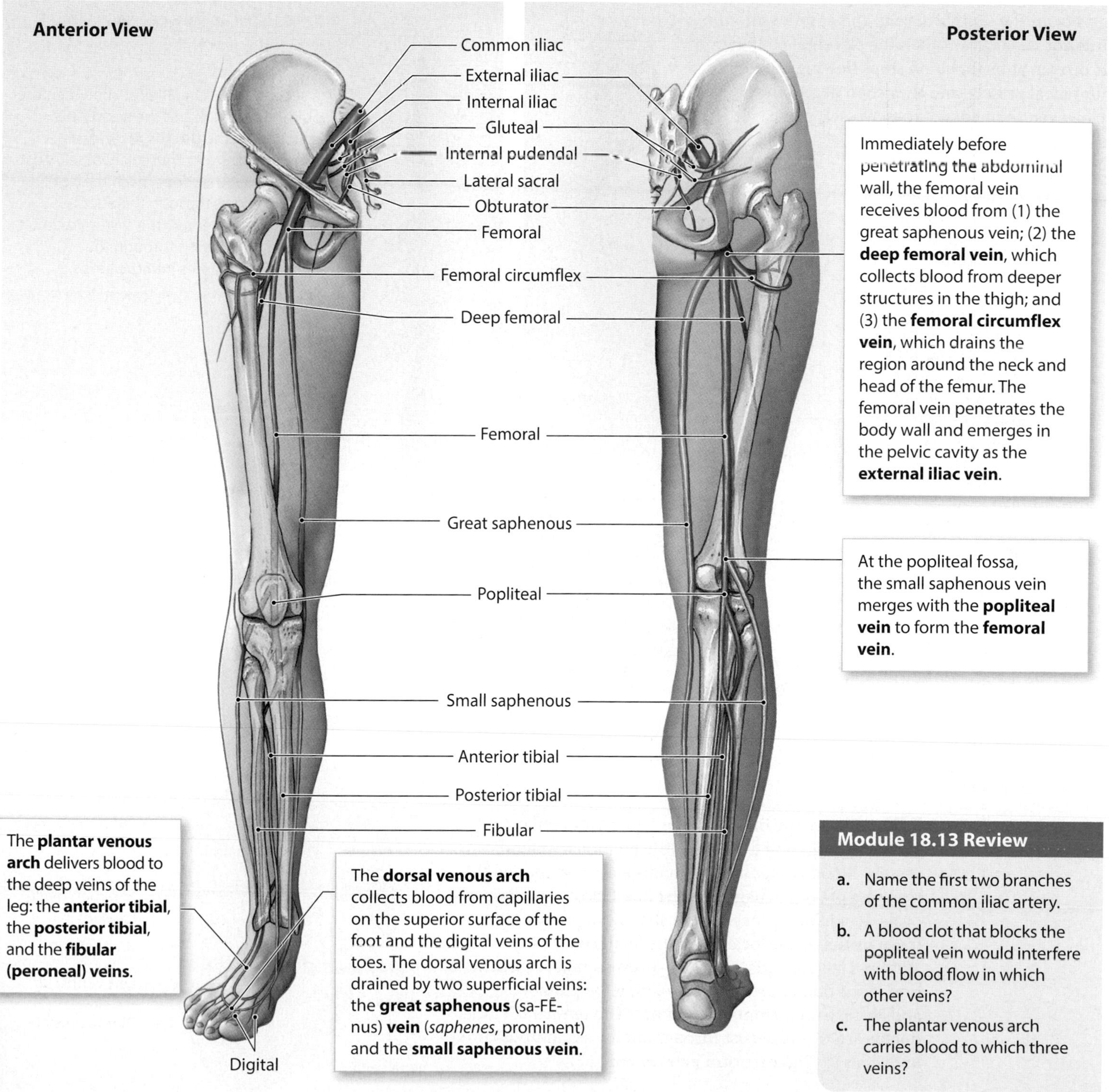

Anterior View
Posterior View
Common iliac
External iliac
Internal iliac
Gluteal
Internal pudendal
Lateral sacral
Obturator
Femoral
Femoral circumflex
Deep femoral
Femoral
Great saphenous
Popliteal
Small saphenous
Anterior tibial
Posterior tibial
Fibular
Digital
Immediately before penetrating the abdominal wall, the femoral vein receives blood from (1) the great saphenous vein; (2) the **deep femoral vein**, which collects blood from deeper structures in the thigh; and (3) the **femoral circumflex vein**, which drains the region around the neck and head of the femur. The femoral vein penetrates the body wall and emerges in the pelvic cavity as the **external iliac vein**.
At the popliteal fossa, the small saphenous vein merges with the **popliteal vein** to form the **femoral vein**.
The **plantar venous arch** delivers blood to the deep veins of the leg: the **anterior tibial**, the **posterior tibial**, and the **fibular (peroneal) veins**.
The **dorsal venous arch** collects blood from capillaries on the superior surface of the foot and the digital veins of the toes. The dorsal venous arch is drained by two superficial veins: the **great saphenous** (sa-FĒ-nus) **vein** (*saphenes*, prominent) and the **small saphenous vein**.

Module 18.13 Review

a. Name the first two branches of the common iliac artery.

b. A blood clot that blocks the popliteal vein would interfere with blood flow in which other veins?

c. The plantar venous arch carries blood to which three veins?

18.13 Identify the branches of the common iliac arteries and the branches of the common iliac veins, and name the areas each serves.

The pattern of blood flow through the fetal heart and the systemic circuit must change at birth

1 Fetal blood flows to the placenta through a pair of **umbilical arteries**, which arise from the internal iliac arteries and enter the umbilical cord. Blood returns from the placenta in the single **umbilical vein**, bringing oxygen and nutrients to the developing fetus. The umbilical vein drains into the **ductus venosus**, a vascular connection to an intricate network of veins within the developing liver. The ductus venosus collects blood from the veins of the liver and from the umbilical vein, and empties into the inferior vena cava. When the placental connection is broken at birth, blood stops flowing in the umbilical vessels, and they soon degenerate. However, remnants of these vessels persist throughout life as fibrous cords.

Prior to birth, the lungs are collapsed and most blood bypasses the pulmonary circuit completely. The **foramen ovale**, or interatrial opening, allows blood to pass from the right atrium to the left atrium, but any backflow is prevented by a flap that acts like a one-way valve.

A second short circuit exists between the pulmonary trunk and the aorta. This connection, the **ductus arteriosus**, consists of a short, muscular vessel. Most of the blood that does reach the right ventricle flows through the ductus arteriosus and enters the systemic circuit rather than flowing through the pulmonary arteries into the lungs.

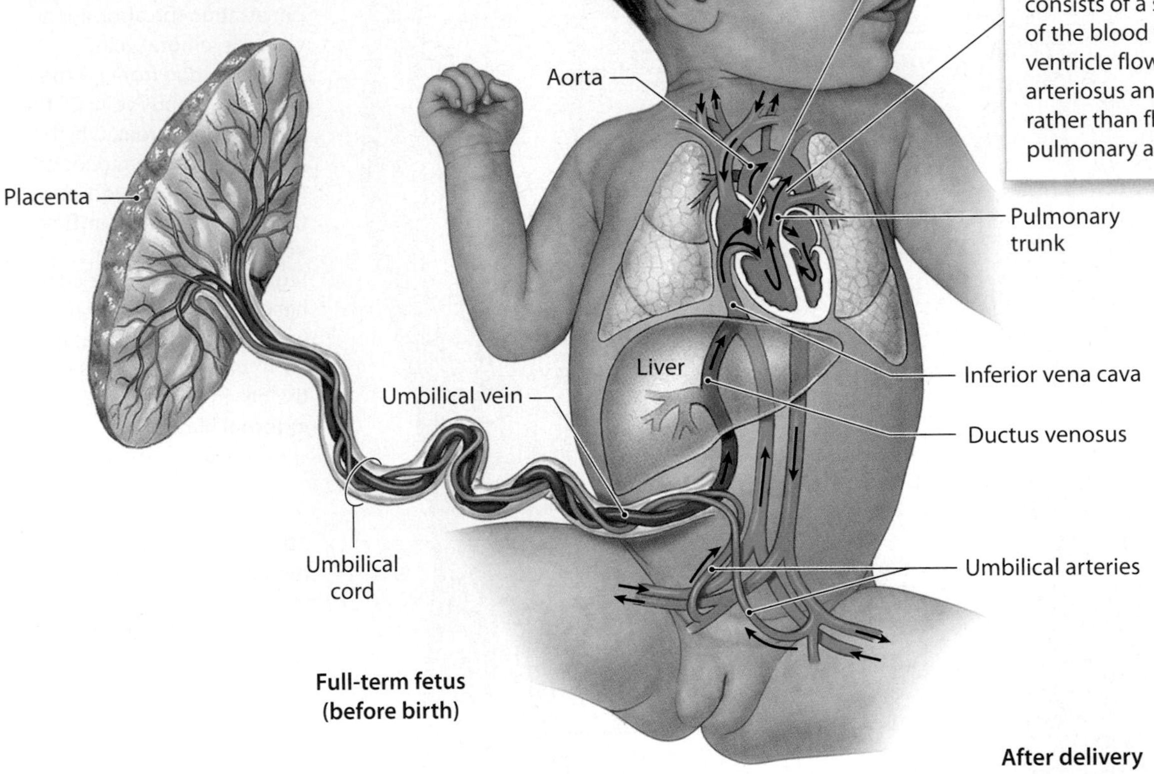
Aorta
Placenta
Pulmonary trunk
Liver
Inferior vena cava
Umbilical vein
Ductus venosus
Umbilical cord
Umbilical arteries

Full-term fetus (before birth)

2 At birth, the infant takes a first breath, inflating the lungs and expanding the pulmonary blood vessels. Blood rushes into the pulmonary vessels, and the resulting pressure changes at the heart close the foramen ovale. In adults, the heart has a shallow depression, the **fossa ovalis**, at the location of the fetal passageway. Within a few seconds, rising O_2 levels stimulate constriction of the ductus arteriosus, isolating the pulmonary and aortic trunks from one another. The remnants of the ductus arteriosus persist throughout life as a fibrous cord known as the **ligamentum arteriosum**.

After delivery

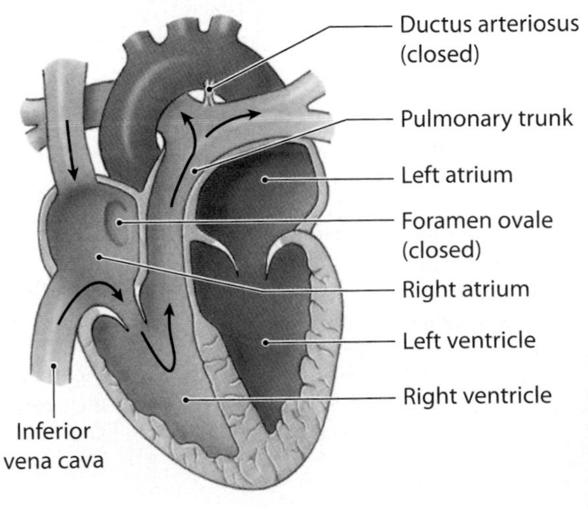
Ductus arteriosus (closed)
Pulmonary trunk
Left atrium
Foramen ovale (closed)
Right atrium
Left ventricle
Right ventricle
Inferior vena cava

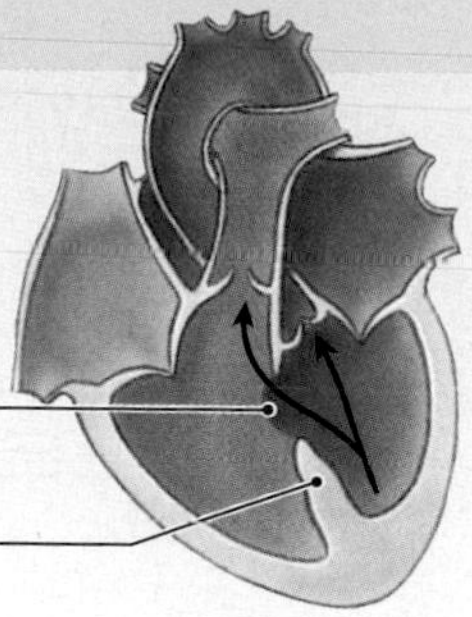
Ventricular septal defect
Ventricular septum

Ventricular Septal Defect

Ventricular septal defects are openings in the interventricular septum that separate the right and left ventricles. These defects are the most common congenital heart problems, affecting 0.12 percent of newborns. The opening between the two ventricles has an effect similar to a connection between the atria: When the more powerful left ventricle beats, it ejects blood into the right ventricle and pulmonary circuit.

3 Although minor individual variations in the vascular network are quite common, congenital cardiovascular problems serious enough to threaten homeostasis are relatively rare. Most congenital heart problems result from abnormal formation of the heart or problems with the connections between the heart and the great vessels. If diagnosed early, most can be surgically corrected—sometimes prior to delivery.

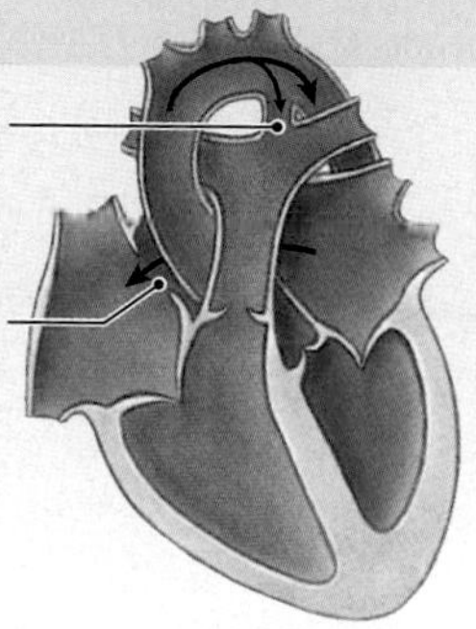
Patent ductus arteriosus
Patent foramen ovale

Patent Foramen Ovale and Patent Ductus Arteriosus

If the foramen ovale remains open, or **patent**, blood recirculates through the pulmonary circuit instead of entering the left ventricle. The movement, driven by the relatively high systemic pressure, is called a "left-to-right shunt." Arterial oxygen content is normal, but the left ventricle must work much harder than usual to provide adequate blood flow through the systemic circuit. Hence, pressures rise in the pulmonary circuit. If the pulmonary pressures rise enough, they may force blood into the systemic circuit through the ductus arteriosus. This condition—a **patent ductus arteriosus**—creates a "right-to-left shunt." Because the circulating blood is not adequately oxygenated, it develops a deep red color. The skin then develops the blue tones typical of cyanosis and the infant is known as a "blue baby."

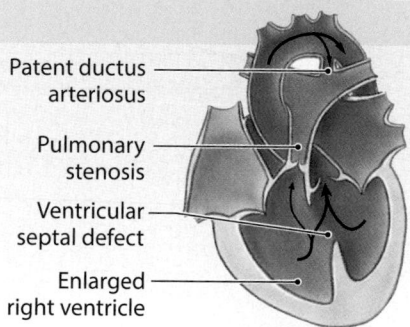

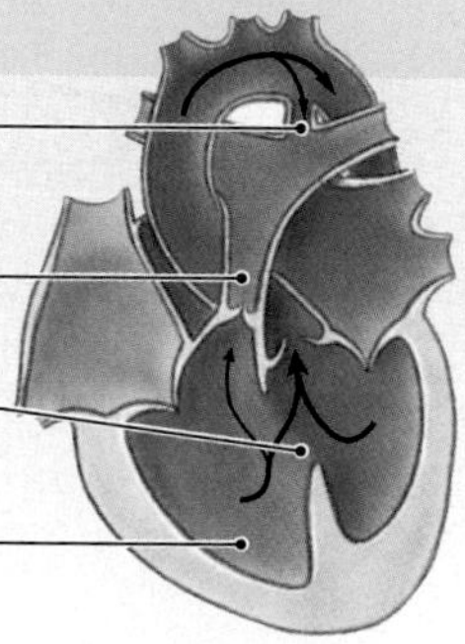
Patent ductus arteriosus
Pulmonary stenosis
Ventricular septal defect
Enlarged right ventricle

Tetralogy of Fallot

The **tetralogy of Fallot** (fa-LŌ) is a complex group of heart and circulatory defects that affect 0.10 percent of newborn infants. In this condition, (1) the pulmonary trunk is abnormally narrow (pulmonary stenosis), (2) the interventricular septum is incomplete, (3) the aorta originates where the interventricular septum normally ends, and (4) the right ventricle is enlarged and both ventricles thicken in response to the increased workload.

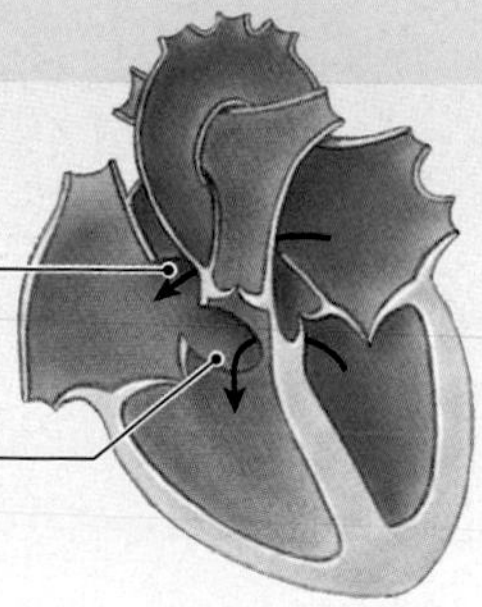
Atrial defect
Ventricular defect

Atrioventricular Septal Defect

In an **atrioventricular septal defect**, both the atria and ventricles are incompletely separated. The results are quite variable, depending on the extent of the defect and the effects on the atrioventricular valves. This type of defect most commonly affects infants with Down's syndrome, a disorder caused by the presence of an extra copy of chromosome 21.

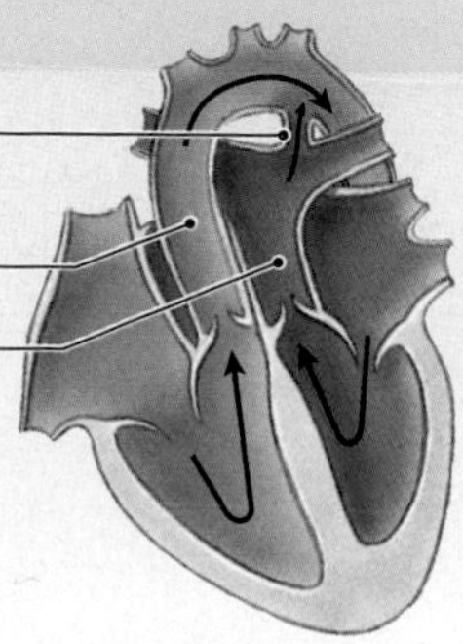
Patent ductus arteriosus
Aorta
Pulmonary trunk

Transposition of the Great Vessels

In the **transposition of the great vessels**, the aorta is connected to the right ventricle instead of to the left ventricle, and the pulmonary artery is connected to the left ventricle instead of to the right ventricle. This malformation affects 0.05 percent of newborn infants.

Module 18.14 Review

a. Describe the pattern of fetal blood flow to and from the placenta.

b. Identify the six structures that are necessary in the fetal circulation but cease to function at birth, and describe what becomes of these structures.

c. Compare a ventricular septal defect with tetralogy of Fallot.

LO 18.14 Identify the differences between fetal and adult circulation patterns, and describe the changes in blood flow patterns that occur at birth.

Labeling

Label the major arteries in the diagram at right.

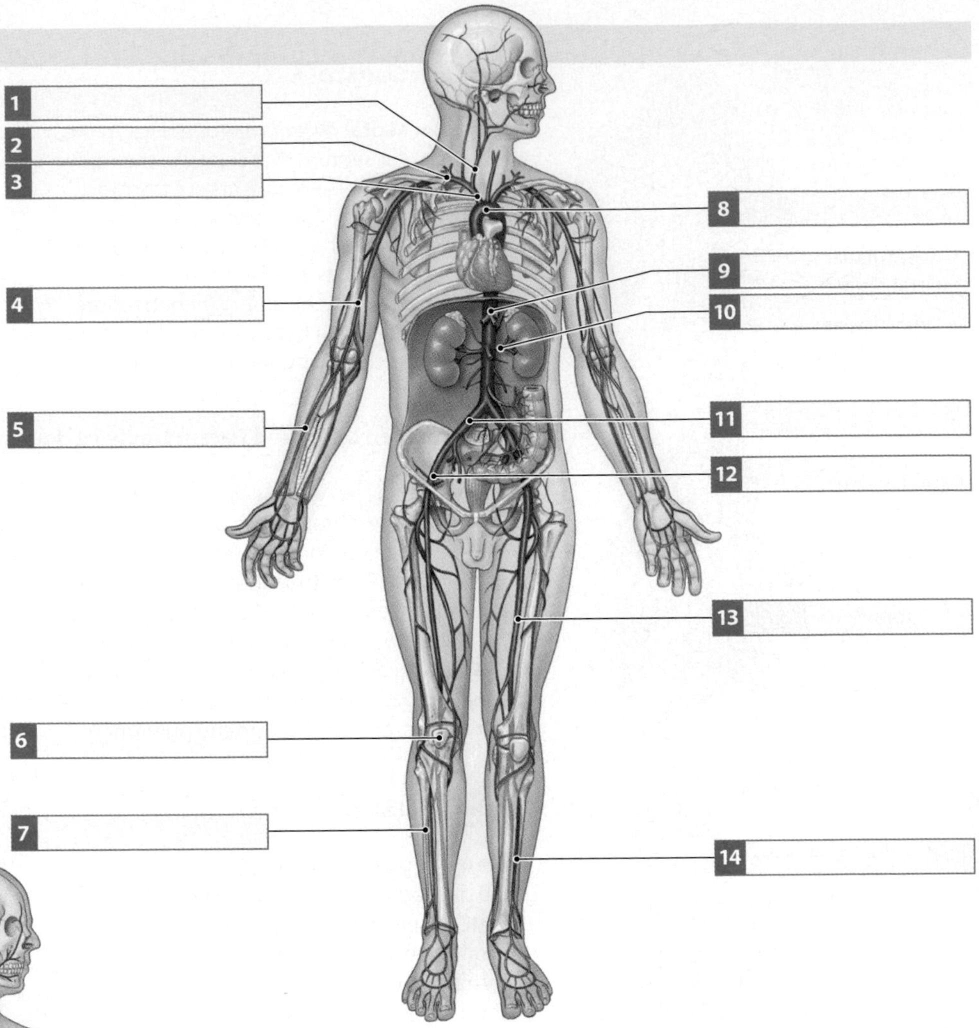
1
2
3
4
5
6
7
8
9
10
11
12
13
14

Label the major veins in the diagram below.

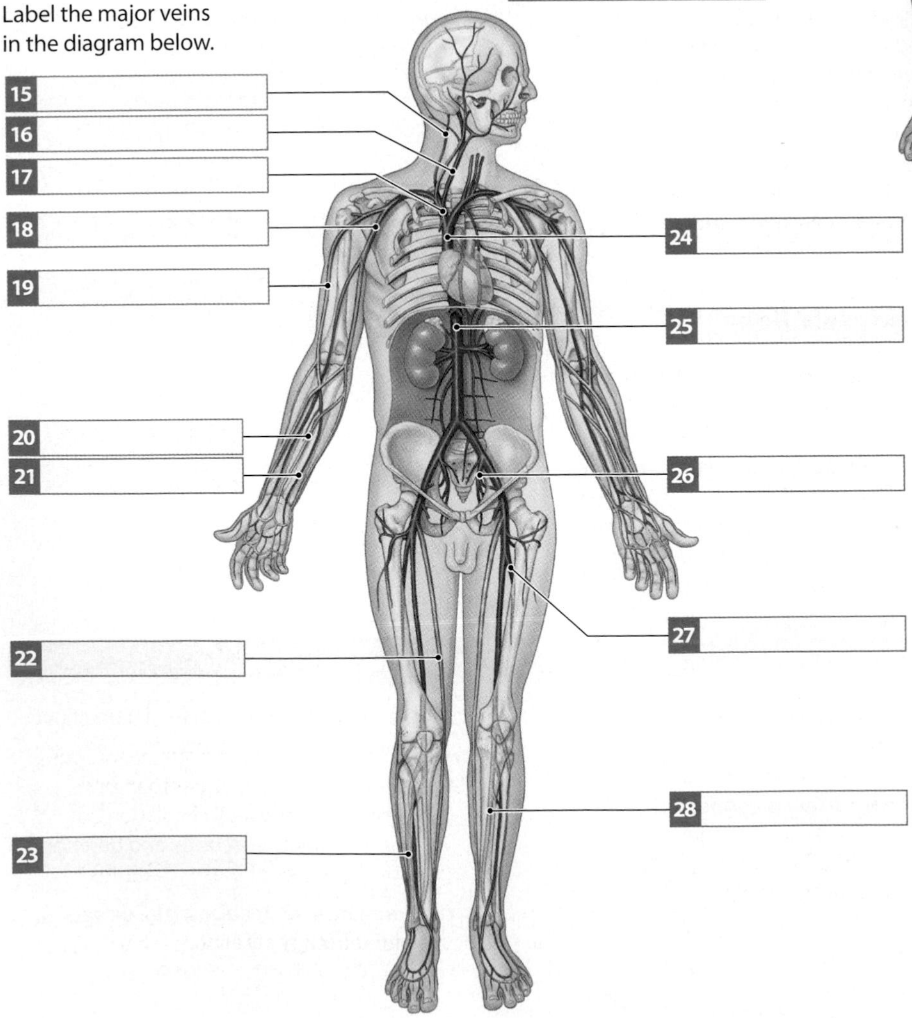
15
16
17
18
19
20
21
22
23
24
25
26
27
28

Study Outline

SECTION 1 • Functional Anatomy of Blood Vessels

18.1 The heart pumps blood, in sequence, through the arteries, capillaries, and veins of the pulmonary and systemic circuits p. 651

1. The **pulmonary circuit** carries blood to and from the lungs. The **systemic circuit** transports blood to the rest of the body.
2. Blood is carried away from the heart by **arteries**, and returns to the heart by **veins**.
3. Thin walled **capillaries** interconnect the smallest arteries and the smallest veins. The thin walls of capillaries allow for the exchange of nutrients, dissolved gases, and wastes between the blood and tissues.

18.2 Arteries and veins differ in the structure and thickness of their walls p. 652

4. The walls of arteries and veins have three distinct layers: the tunica intima, tunica media, and tunica externa.
5. The **tunica intima** is the innermost layer, and it includes an endothelial lining and underlying layer of connective tissue containing elastic fibers. Arteries also have an **internal elastic membrane**.
6. The **tunica media** is the middle layer, and it contains concentric sheets of smooth muscle tissue in a framework of loose connective tissue. This layer allows for **vasoconstriction** and **vasodilation**.
7. The **tunica externa**, or tunica adventitia, is the outermost layer and is a connective tissue sheath.
8. The classes of blood vessels as they carry blood away from the heart are **arteries**, **arterioles**, and **capillaries**. Blood then returns to the heart by **venules** and **veins**.

18.3 Capillary structure and capillary blood flow affect the rates of exchange between the blood and interstitial fluid p. 654

9. A typical capillary is a tube of endothelial cells with a delicate basement membrane. The tunica media and tunica externa are absent.
10. In a **continuous capillary**, the endothelium is a complete lining. These capillaries are located throughout the body.
11. In **fenestrated capillaries**, there are pores that penetrate the endothelial lining that permit rapid exchange of water and solutes as large as small peptides. These are located in the choroid plexus of the brain, and the capillaries of the hypothalamus, pituitary gland, pineal gland, and thyroid gland.
12. **Sinusoids** resemble fenestrated capillaries that are flat and irregularly shaped. They are present in the liver, bone marrow, spleen, and many endocrine organs.
13. Capillaries function as part of an interconnected **capillary bed**. Capillary beds are often supplied with blood as **collateral** arteries fuse in **arterial anastomoses**.
14. **Precapillary sphincters** at the entrance to capillaries continually contract and relax in a process called **vasomotion**.

18.4 The venous system has low pressures and contains almost two-thirds of the body's blood volume p. 656

15. Blood pressure in a peripheral vein is about 10 percent of that in the ascending aorta.
16. Folds of the tunica intima in some veins form **valves** that permit blood flow in only one direction.
17. Veins distended from weakness in their walls near valves can form **varicose veins** or **hemorrhoids**.

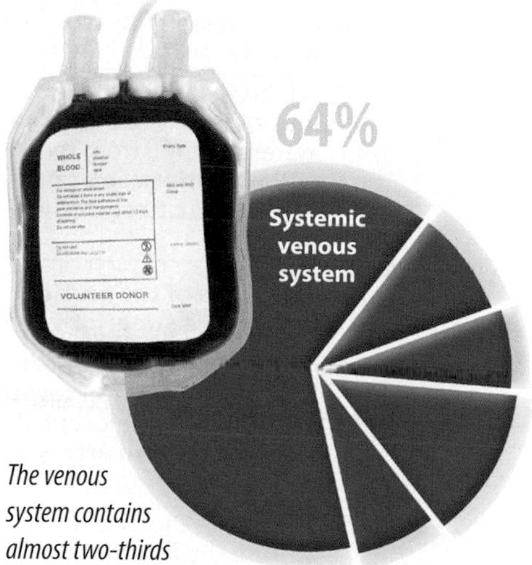
64%
Systemic venous system
WHOLE BLOOD
VOLUNTEER DONOR

The venous system contains almost two-thirds of the body's blood volume.

18. Blood volume is unevenly distributed among blood vessels in the following pattern: 9 percent, pulmonary circuit; 7 percent, heart; 13 percent, systemic arterial system; 7 percent, systemic capillaries; 64 percent, systemic venous system.
19. During hemorrhaging, **venoconstriction** (contraction of smooth muscle cells in medium-sized veins) maintains blood volume in the arterial system at near-normal levels.

SECTION 2 • Patterns of Blood Flow

18.5 New blood vessels form through vasculogenesis and angiogenesis p. 659

20. Blood vessels arise from blood islands in the embryonic yolk sac. Blood islands form from **hemangioblasts**.
21. Hemangioblasts within the central regions of the blood islands give rise to **hematopoietic stem cells** and those in the periphery become **angioblasts**.
22. Angioblasts remodel the blood islands into capillaries, and then into arterial and venous networks in a process called **vasculogenesis**. **Angiogenesis** is the formation of new blood vessels from pre-existing vessels.

18.6 The pulmonary circuit carries deoxygenated blood from the right ventricle to the lungs and returns oxygenated blood to the left atrium p. 660

23. The **pulmonary circuit** is composed of arteries and veins that transport blood between the heart and the lungs. This circuit begins at the right ventricle and ends at the left atrium.
24. Blood from the right ventricle enters the **pulmonary trunk**, then the left and right **pulmonary arteries**, **pulmonary arterioles**, and then the **alveolar capillaries** that surround the **alveoli.** This is where gas exchange occurs between the capillary blood and inspired air.

25. Oxygenated blood leaves the alveolar capillaries and enters venules that unite to form four **pulmonary veins** (two from each lung). These veins deliver blood to the left atrium, completing the pulmonary circuit.

18.7 The systemic arterial and venous systems operate in parallel, and the major vessels often have similar names p. 662

26. All the vessels from the systemic **arterial system** originate from the **aorta**.
27. All the vessels from the systemic **venous system** merge into either the **superior vena cava** (which collects systemic blood from the head, chest, and upper limbs), or the **inferior vena cava** (which collects systemic blood from all structures inferior to the diaphragm).

18.8 The branches of the aortic arch supply structures that are drained by the superior vena cava p. 664

28. The **branches of the aortic arch** are the **brachiocephalic trunk**, **left common carotid artery**, and **left subclavian artery**. The brachiocephalic trunk branches to form the **right subclavian artery** and **right common carotid artery**.
29. Within the thoracic cavity, the subclavian artery branches to form the **internal thoracic artery** (which supplies the pericardium and anterior chest wall) and **vertebral artery** (which supplies the brain and spinal cord).

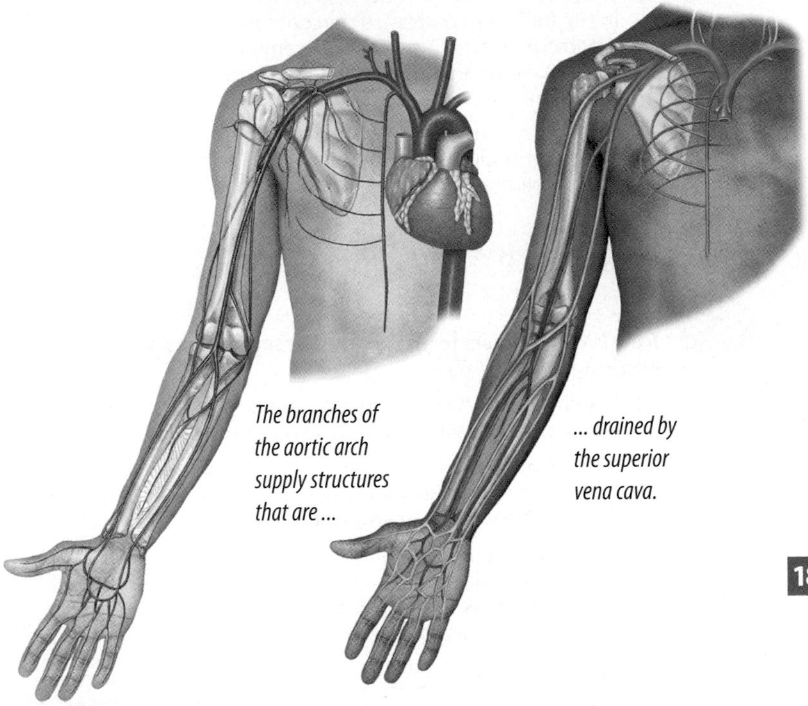
The branches of the aortic arch supply structures that are ...
... drained by the superior vena cava.

30. After leaving the thoracic cavity, the subclavian artery forms the **axillary artery** and then the **brachial artery**. The brachial artery divides into the **radial artery** and **ulnar artery**. These two arteries fuse at the wrist to form the superficial and deep **palmar arches**. These supply blood to the hand and to the **digital arteries** that supply the fingers.
31. The **digital veins** collect blood from the palmar venous arches. The cephalic vein, **median antebrachial vein**, and the basilic vein drain the **superficial palmar arch**.
32. The **ulnar vein** and **radial vein** drain the **deep palmar arch**. These two veins join to form the brachial vein.
33. The **median cubital vein** interconnects the cephalic and basilic vein. This vein is often the site from which venous blood samples are collected.
34. The **cephalic vein** extends along the forearm. The **brachial vein** and **basilic vein** merge to become the **axillary vein**.
35. The **external jugular vein** drains the superficial head and neck. The **internal jugular vein** drains the deep head and neck. The **vertebral vein** drains the cervical spinal cord and the posterior surface of the skull.
36. The **brachiocephalic vein** forms as the jugular veins join the axillary vein. The **superior vena cava** carries blood from the two brachiocephalic veins to the right atrium.

18.9 The external carotid arteries supply the neck, lower jaw, and face, and the internal carotid and vertebral arteries supply the brain, while the external jugular veins drain the regions supplied by the external carotid arteries, and the internal jugular veins drain the brain. p. 666

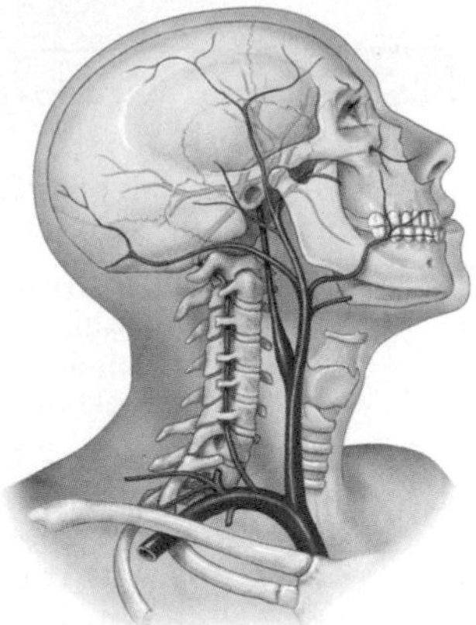

37. The **common carotid arteries** supply the face, neck, and brain. Each common carotid artery branches to form an external and internal carotid artery.
38. The **internal carotid artery** enters the skull through the carotid canal of the temporal bones to deliver blood to the brain and eyes. The **carotid sinus** is located at the base of the internal carotid artery.
39. The **vertebral artery** travels within the vertebral foramina of the cervical vertebrae. The left and right vertebral arteries fuse along the ventral surface of the medulla oblongata to form the **basilar artery**.
40. The **external carotid artery** branches to supply the neck, esophagus, pharynx, larynx, cranium, and face.
41. The maxillary and temporal veins form the **external jugular veins**. The **internal jugular veins** drain blood from the venous sinuses within the cranium. The external and internal jugular veins combine with the **vertebral** and subclavian veins to form the brachiocephalic vein.

18.10 The internal carotid arteries and the vertebral arteries supply the brain, which is drained by the dural sinuses and the internal jugular veins p. 668

42. The internal carotid arteries normally supply the arteries of the anterior half of the cerebrum, and the rest of the brain receives blood from the vertebral and basilar arteries.
43. The internal carotid divides into an **ophthalmic artery** (which supplies the eyes), **anterior cerebral artery** (which supplies frontal and parietal lobes of the brain), and **middle cerebral artery** (which supplies the midbrain and lateral surfaces of cerebral hemispheres).
44. The internal carotid arteries and basilar artery are interconnected in a ring-shaped anastomosis called the **cerebral arterial circle** (circle of Willis). This arrangement reduces the likelihood of a serious interruption of cerebral blood flow.

45. The superficial cerebral veins and small veins of the brain stem empty into a network of dural sinuses. Most of the deep cerebral veins converge into the **great cerebral vein**, which delivers blood to the **straight sinus**. Veins from the orbit and other cerebral veins drain into the **cavernous sinus**.

46. The **superior sagittal sinus**, in the falx cerebri, is the largest dural sinus. Other sinuses are the **petrosal sinuses**, **transverse sinuses**, and **sigmoid sinuses**.

18.11 The regions supplied by the descending aorta are drained by the superior and inferior venae cavae p. 670

47. The descending aorta is continuous with the aortic arch. The diaphragm divides the descending aorta into a **thoracic aorta** and an **abdominal aorta**.

48. The visceral branches of the thoracic aorta are the **bronchial arteries**, **esophageal arteries**, **mediastinal arteries**, and **pericardial arteries**.

49. The somatic branches of the thoracic aorta are the **intercostal** and **superior phrenic arteries**.

50. The major paired branches of the abdominal aorta are the **inferior phrenic**, **adrenal**, **renal**, **gonadal**, and **lumbar arteries**.

51. The major unpaired branches of the abdominal aorta are the **celiac trunk**, **superior mesenteric**, and **inferior mesenteric arteries**.

52. The major collecting vessels of the thorax are the **azygos** and **hemiazygos veins**. These veins receive blood from the **intercostal**, **esophageal**, **bronchial**, and **mediastinal veins**.

53. The major branches of the inferior vena cava are the **lumbar**, **gonadal** (ovarian or testicular), **hepatic**, **renal**, **adrenal**, and **phrenic veins**.

18.12 The viscera supplied by the celiac trunk and mesenteric arteries are drained by the branches of the hepatic portal vein p. 672

54. The **celiac trunk** divides into the **common hepatic artery** (which supplies the liver, stomach, gallbladder, and duodenum), **left gastric artery** (which supplies the stomach), and **splenic artery** (which supplies the spleen, stomach, and pancreas).

55. The **superior mesenteric artery** branches to supply the pancreas and duodenum, small intestine, and most of the large intestine.

56. The **inferior mesenteric artery** delivers blood to the terminal portions of the colon and the rectum.

57. The **hepatic portal vein** forms through the fusion of the **superior mesenteric**, **inferior mesenteric**, and **splenic veins**.

58. The hepatic portal vein receives blood from the left and right **gastric veins** (medial border of stomach) and the **cystic vein** (from the gallbladder). This circulation directs blood with absorbed nutrients from the digestive system and is called the **hepatic portal system**.

18.13 The pelvis and lower limbs are supplied by branches of the common iliac arteries and drained by branches of the common iliac veins p. 674

59. Near vertebra L_4, the abdominal aorta divides to form the **right** and **left common iliac arteries**. Near the lumbosacral joint, each common iliac divides to form an **internal** and **external iliac artery**.

60. The external iliac artery becomes the **femoral artery** when it enters the lower limb. The first branch is the **deep femoral artery**, which branches into the **femoral circumflex arteries**.

61. Posterior to the knee joint the femoral artery becomes the **popliteal artery**. This branches into the **posterior** and **anterior tibial arteries**.

62. The internal iliac artery supplies the organs of the pelvic cavity.

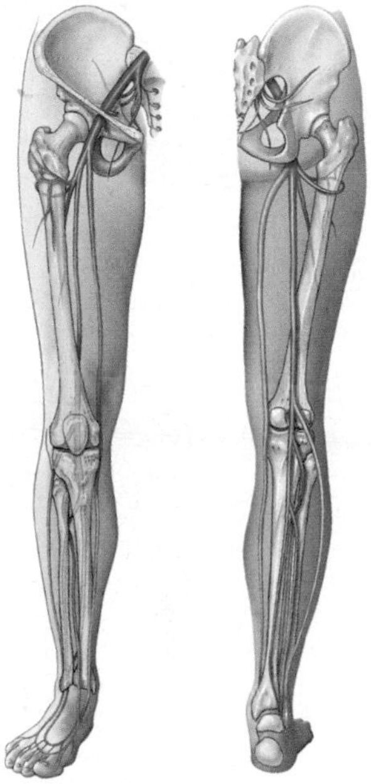

63. The tibial and fibular (peroneal) arteries are interconnected by the anastomoses of the **dorsalis pedis**, **dorsal arch**, and **plantar arch**. These arteries supply the distal portions of the foot and digital arteries of the toes.

64. The external iliac veins receive blood from the lower limbs, pelvis, and lower abdomen. The internal iliac veins drain the pelvic organs. The union of the external and internal iliac veins forms the **common iliac vein**.

65. In the foot, the **dorsal venous arch** is drained by the **great saphenous vein** and **small saphenous vein**. The small saphenous vein merges with the **popliteal vein** to form the **femoral vein**, which also receives blood from the **deep femoral vein** and the **femoral circumflex vein**.

18.14 The pattern of blood flow through the fetal heart and the systemic circuit must change at birth p. 676

66. Fetal blood flows into the placenta through a pair of **umbilical arteries** that arise from the internal iliac arteries. Blood returns from the placenta through a single **umbilical vein**.

67. The umbilical vein drains into the **ductus venosus**, which connects to a collection of veins in the developing liver, and empties into the inferior vena cava. At birth, blood stops flowing into the umbilical vessels, and they soon degenerate.

68. Prior to birth the lungs are collapsed, and most blood bypasses the pulmonary circuit. Blood passes from the right atrium to the left atrium through the **foramen ovale**. Blood also bypasses the lungs through the **ductus arteriosus**, a short vessel between the pulmonary trunk and aorta.

69. At birth the foramen ovale closes, forming the **fossa ovalis**. The ductus arteriosus forms the **ligamentum arteriosum**.

70. Most congenital heart problems result from abnormal formation of the heart or problems with the connections between the heart and the great vessels.

71. Common congenital heart problems are **ventricular septal defect**, **patent foramen ovale** and **patent ductus arteriosus**, **tetralogy of Fallot**, **atrioventricular septal defect**, and **transposition of the great vessels**.

Chapter Review Questions

Labeling

Label the arteries of the brain shown in the image below.

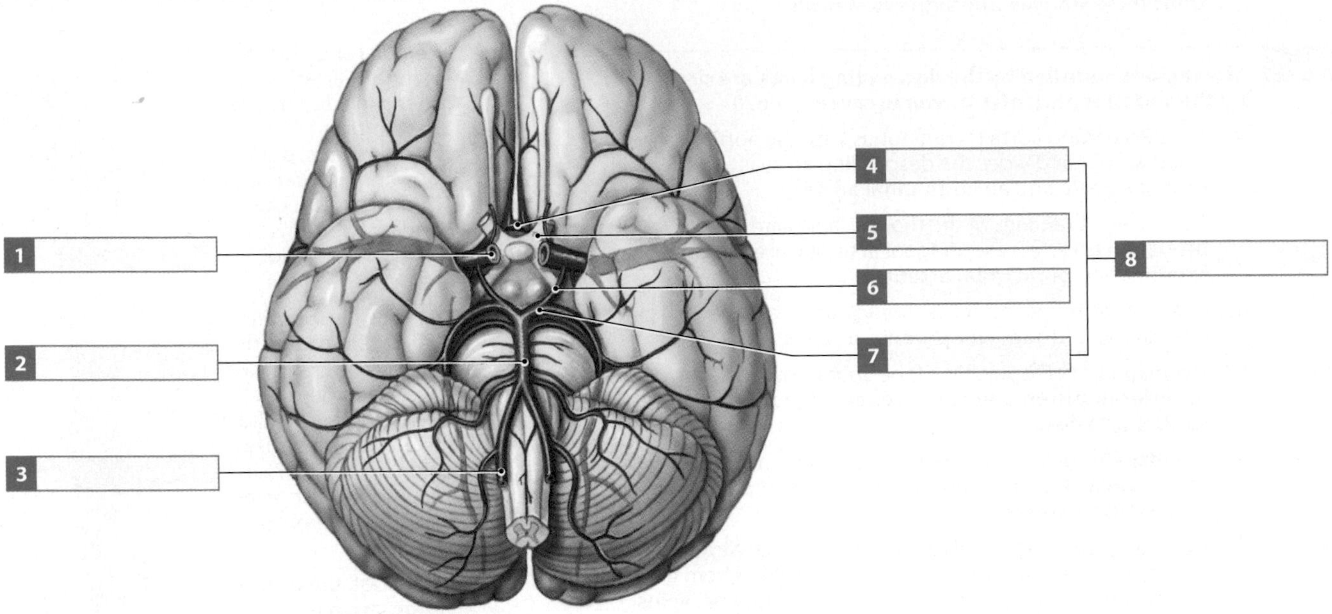
1
2
3
4
5
6
7
8

Multiple choice

Select the correct answer from the list provided.

9 The heart pumps blood into the systemic circuit from the

- ☐ a) right atrium.
- ☐ b) right ventricle.
- ☐ c) left atrium.
- ☐ d) left ventricle.

10 The blood vessel layer containing smooth muscle fibers allowing for the processes of vasoconstriction and vasodilation is the

- ☐ a) tunica intima.
- ☐ b) tunica media.
- ☐ c) tunica adventitia.
- ☐ d) tunica externa.

11 Capillaries containing pores that permit the passage of large molecules are

- ☐ a) continous capillaries.
- ☐ b) precapillary sphincters.
- ☐ c) collaterals.
- ☐ d) fenestrated capillaries.

12 Most of the body's total blood volume is distributed within the

- ☐ a) pulmonary circuit.
- ☐ b) systemic arterial system.
- ☐ c) systemic venous system.
- ☐ d) systemic capillaries.

13 The process of generating blood vessels from pre-existing vessels is called

- ☐ a) angiogenesis.
- ☐ b) vasculogenesis.
- ☐ c) fenestration.
- ☐ d) arteriosus.

14 Which of the following is the correct combination of celiac trunk branches?

- ☐ a) common hepatic artery, superior mesenteric artery, inferior mesenteric artery
- ☐ b) common hepatic artery, splenic artery, superior mesenteric artery
- ☐ c) common hepatic artery, left gastric artery, phrenic artery
- ☐ d) common hepatic artery, left gastric artery, splenic artery

15 The vein that drains the venous sinuses of the brain is the

- ☐ a) internal jugular vein.
- ☐ b) external jugular vein.
- ☐ c) vertebral vein.
- ☐ d) great cerebral vein.

16 The two branches from the brachiocephalic trunk are the

- ☐ a) left subclavian and left common carotid arteries.
- ☐ b) right subclavian and right common carotid arteries.
- ☐ c) left subclavian and left common carotid veins.
- ☐ d) right subclavian and right common carotid veins.

Short answer

17 The brachiocephalic trunk branches from the right side of the aortic arch. Why is there no such vessel on the left side?

18 List the distribution of the body's total blood volume.

19 Describe the function of the hepatic portal system.

MasteringA&P®

Access more chapter study tools online in the MasteringA&P Study Area:

- **Chapter Quizzes, Chapter Practice Test, Art-labeling Activities, Animations, MP3 Tutor Sessions, and Clinical Case Studies**

- **Practice Anatomy Lab** PAL™
- **Interactive Physiology** iP®
- **A&P Flix** A&PFlix™
- **PhysioEx** PhysioEx™

Chapter Integration • Applying what you have learned

Mapping cerebral circulation with angiography

Mr. Samuel was having bouts of light-headedness and slurred speech that alarmed his family, so they rushed him to the hospital. The attending physician was concerned that Mr. Samuel may be experiencing a stroke (damaged brain cells from lack of blood supply) due to a narrowed artery or blood clot in his brain. The physician ordered a cerebral angiogram, an x-ray image that would show the circulation in Mr. Samuel's head and neck. The physician explained to the family that the doctor performing the procedure would insert a tube called a catheter into Mr. Samuel's thigh, and guide it through his arteries to his neck. There, a contrasting agent, or dye, would be released to generate the image.

The image obtained from the procedure showed that Mr. Samuel had suffered a minor stroke. The report said that he had an 80 percent blockage of the vertebral artery on his left side. The doctor told the family that Mr. Samuel was fortunate to have had the blockage where he did because he would have suffered a more serious stroke if the blockage had occurred elsewhere.

1 **The angiography required a catheter to be guided from arteries in Mr. Samuel's thigh, all the way to the base of his neck. Name the vessels through which this catheter would travel.**

2 **Where do you think the dye would be released? Why is that the best place for it to be released?**

3 **Can you explain what the doctor meant when he said that Mr. Samuel was lucky to have had the blockage where he did? Where do you think a more serious blockage would be?**

19 The Heart and Cardiovascular Function

LEARNING OUTCOMES

These Learning Outcomes correspond by number to this chapter's modules and indicate what you should be able to do after completing the chapter.

SECTION 1 • Structure of the Heart

- **19.1** Describe the heart's location, shape, and borders.
- **19.2** Describe the structure of the pericardium and explain its functions, identify the layers of the heart wall, and describe the structures and functions of cardiac muscle.
- **19.3** Describe the location and general features of the heart.
- **19.4** Describe the cardiac chambers and the heart's external anatomy.
- **19.5** Describe the major vessels supplying the heart, and cite their locations.
- **19.6** Trace blood flow through the heart, identifying the major blood vessels, chambers, and heart valves.
- **19.7** Describe the relationship between the AV and semilunar valves during a heartbeat.
- **19.8** + **CLINICAL MODULE** Define arteriosclerosis, and explain its significance to health.

SECTION 2 • Cardiac Cycle

- **19.9** Explain the complete round of cardiac systole and diastole.
- **19.10** Explain the events of the cardiac cycle, and relate the heart sounds to specific events.
- **19.11** Describe the components and functions of the conducting system of the heart.
- **19.12** Describe an action potential in cardiac muscle, and explain the role of calcium ions.
- **19.13** Describe the factors affecting the heart rate.
- **19.14** Describe the variables that influence stroke volume.
- **19.15** Explain how stroke volume and cardiac output are coordinated.
- **19.16** + **CLINICAL MODULE** Identify the electrical events shown on an electrocardiogram.

SECTION 3 • Coordination of Cardiac Output and Blood Flow

- **19.17** Explain the effects of pressure, resistance, and venous return on cardiac output.
- **19.18** Describe the factors that influence total peripheral resistance.
- **19.19** Describe the factors that determine blood flow.
- **19.20** Describe the movement of fluids between capillaries and interstitial spaces.
- **19.21** Explain central regulation, autoregulation, and baroreceptor reflexes in response to changes in blood pressure and blood composition.
- **19.22** Explain the hormonal regulation of blood pressure and blood volume.
- **19.23** Describe the role of chemoreceptor reflexes in adjusting cardiovascular activity.
- **19.24** Explain how the cardiovascular system responds to the demands of exercise.
- **19.25** + **CLINICAL MODULE** Explain the body's response to blood loss.

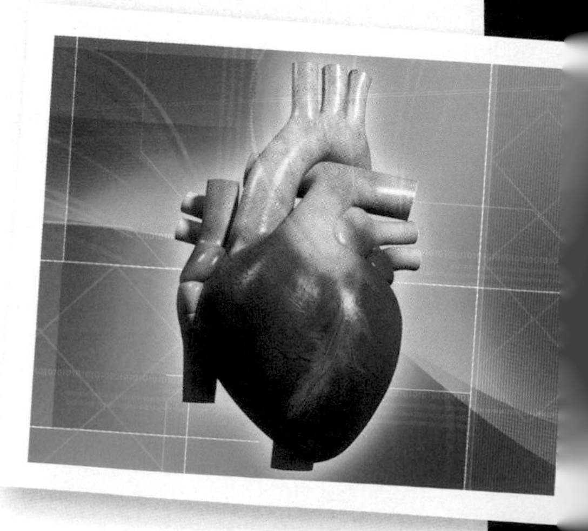

LO Learning Outcomes are repeated at the bottom of each module.

The heart has a superior base, an inferior apex, and four borders

1 The heart is located near the anterior chest wall, directly posterior to the sternum. A midsagittal section through the trunk does not divide the heart into equal halves. Note that (1) the center of the base lies slightly to the left of the midline, (2) a line drawn between the center of the base and the apex points further to the left, and (3) the entire heart is rotated to the left around this line, so that the right atrium and right ventricle dominate an anterior view of the heart (**2**).

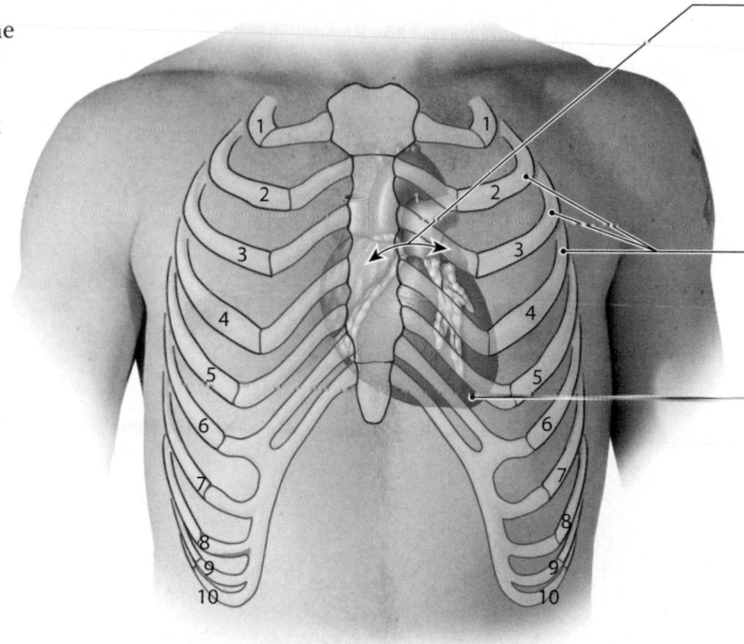
The **base** of the heart is at its superior border, where the great veins and arteries are attached. The base sits posterior to the sternum at the level of the third costal cartilage, centered about 1.2 cm (0.5 in.) to the left side.
Ribs
The inferior, pointed tip of the heart is the free **apex** (Ā-peks). A typical adult heart measures approximately 12.5 cm (5 in.) from the base to the apex, which reaches the fifth intercostal space approximately 7.5 cm (3 in.) to the left of the midline.

2 This anterior view illustrates the borders of the heart. The base forms the **superior border**. The **right border** of the heart is formed by the right atrium. The **left border** is formed by the left ventricle and a small portion of the left atrium. The left border extends to the apex, where it meets the inferior border. The **inferior border** is formed mainly by the inferior wall of the right ventricle.

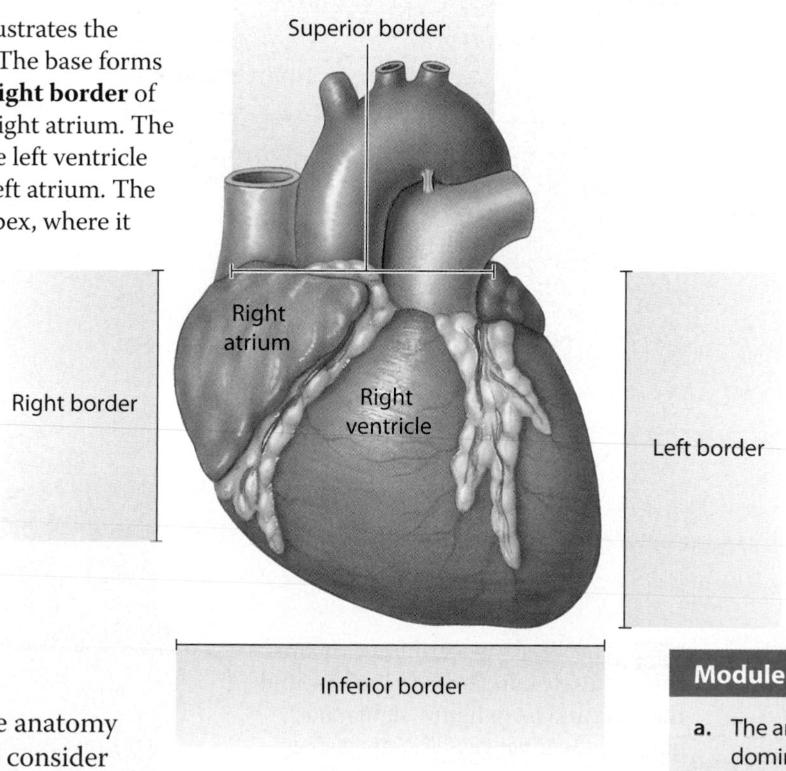
Superior border
Right atrium
Right border
Right ventricle
Left border
Inferior border

This section examines the anatomy of the heart. We will then consider the regulation of cardiac function (Section 2) before considering how cardiac and vasomotor activities are coordinated.

Module 19.1 Review

a. The anterior view of the heart is dominated by which structures?

b. The great veins and arteries are attached to which aspect of the heart?

c. Is the apex located on the superior or inferior aspect of the heart?

LO 19.1 Describe the heart's location, shape, and borders.

Module 19.2

The heart wall contains concentric layers of cardiac muscle tissue

1 This is a view of a section taken from the wall of the heart and the surrounding pericardium. The heart wall contains three layers: epicardium, myocardium, and endocardium.

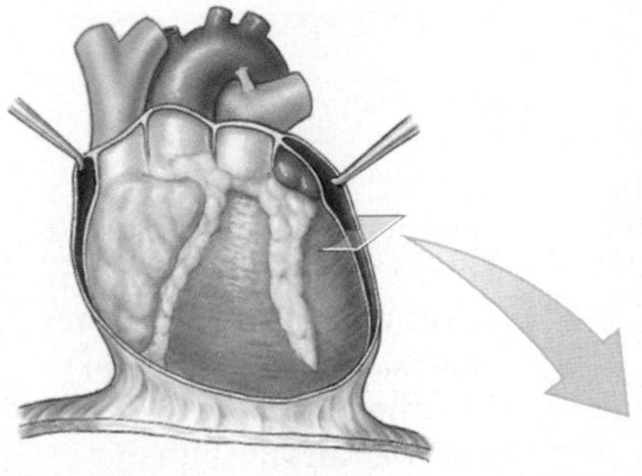

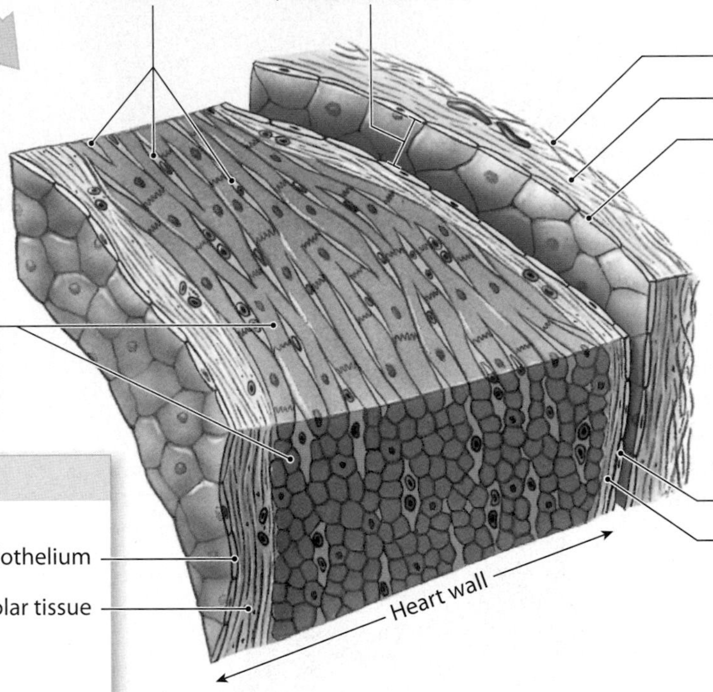

Connective tissues
Pericardial cavity (contains serous fluid)
Heart wall

The Pericardium

The **pericardium** is the serous membrane that lines the pericardial cavity and covers the heart (Module 4.14, p. 160).

Parietal Pericardium

The **parietal pericardium** is the portion of the serous membrane that lines the outer wall of the pericardial cavity. The parietal pericardium is reinforced by a dense fibrous layer; together they form the **pericardial sac** that surrounds the heart.

- Dense fibrous layer
- Areolar tissue
- Mesothelium

Epicardium

The **epicardium**, or visceral pericardium, covers the outer surface of the heart. This portion of the serous membrane consists of an exposed mesothelium and an underlying layer of areolar tissue that is attached to the myocardium.

- Mesothelium
- Areolar tissue

Myocardium

The **myocardium**, or muscular wall of the heart, forms both atria and ventricles. This middle layer contains cardiac muscle tissue, blood vessels, and nerves. The myocardium consists of concentric layers of cardiac muscle tissue.

Endocardium

The inner surfaces of the heart, including those of the heart valves, are covered by the **endocardium**, a simple squamous epithelium and underlying areolar tissue. The squamous epithelial lining of the cardiovascular system is called an endothelium. The endothelium of the heart is continuous with the endothelium of the attached great vessels.

- Endothelium
- Areolar tissue

2 The atrial myocardium contains muscle bundles that wrap around the atria and form figure-eights that encircle the great vessels. Superficial ventricular muscles wrap around both ventricles; deeper muscle layers spiral around and between the ventricles toward the apex in a figure-eight pattern.

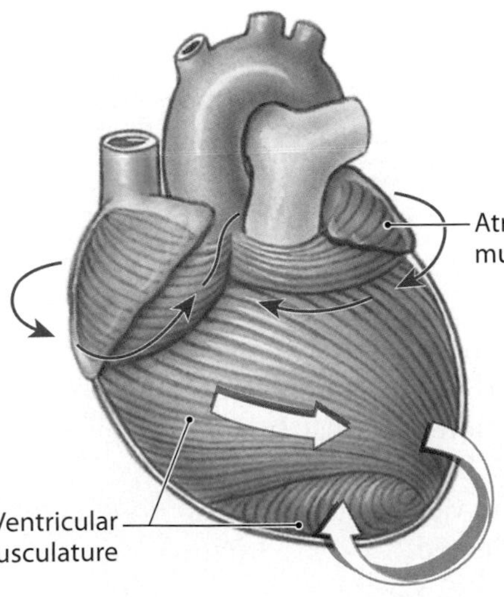

Atrial musculature
Ventricular musculature

3 This light micrograph shows the histological characteristics that distinguish cardiac muscle tissue from skeletal muscle tissue: (1) small cell size, (2) a single, centrally located nucleus, (3) branching interconnections between cells, and (4) specialized intercellular connections.

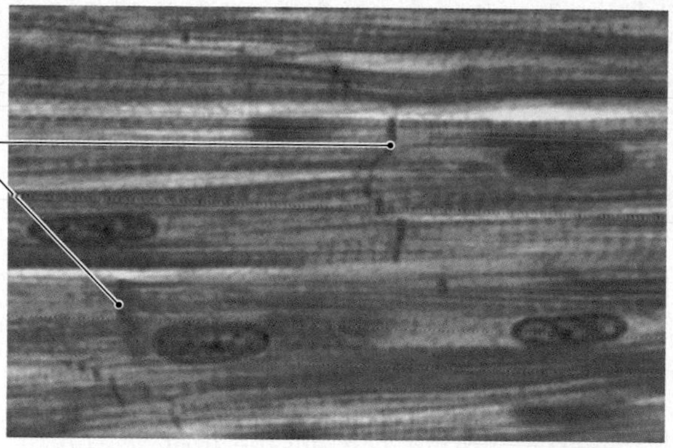

Each cardiac muscle cell is connected to several others at specialized sites known as **intercalated** (in-TER-ka-lā-ted) **discs**.

Cardiac muscle tissue LM × 575

4 Cardiac muscle cells are found only in the heart. Like skeletal muscle fibers, cardiac muscle cells contain organized myofibrils, and the presence of many aligned sarcomeres gives the cells a striated appearance. They are almost totally dependent on aerobic metabolism to obtain the energy they need to continue contracting. The sarcoplasm of a cardiac muscle cell contains many mitochondria and abundant reserves of myoglobin that store oxygen. Because these cells are metabolically very active and have a high demand for oxygen and nutrients, cardiac tissues are richly supplied with capillaries.

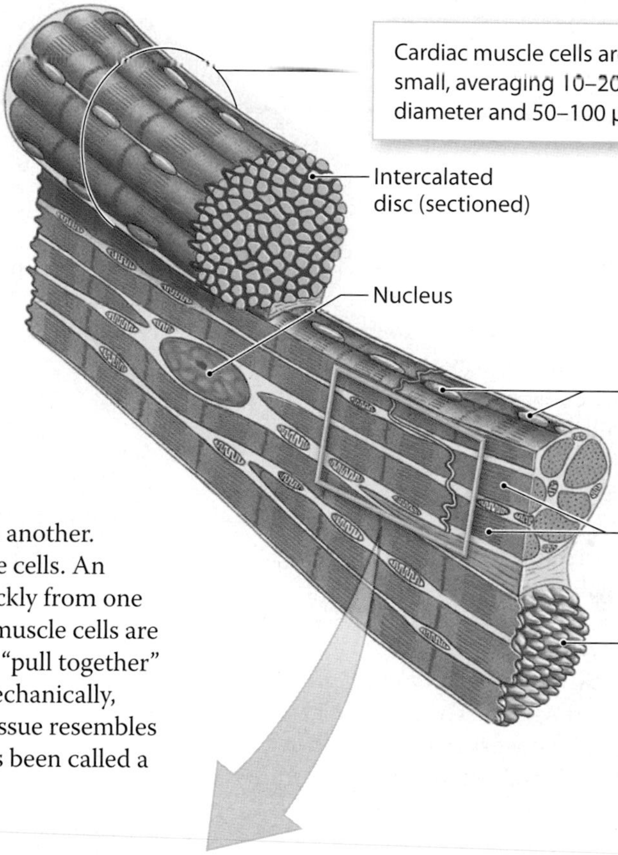

Cardiac muscle cells are relatively small, averaging 10–20 μm in diameter and 50–100 μm in length.
Intercalated disc (sectioned)
Nucleus
Mitochondria
Bundles of myofibrils
Intercalated disc

5 At an intercalated disc, the plasma membranes of two adjacent cardiac muscle cells are extensively intertwined and bound together by gap junctions and desmosomes. These connections help stabilize the positions of adjacent cells. The gap junctions allow ions and small molecules to move from one cell to another. This creates a direct electrical connection between the two muscle cells. An action potential can travel across an intercalated disc, moving quickly from one cardiac muscle cell to another. Myofibrils in the two interlocking muscle cells are firmly anchored to the membrane at the intercalated disc and can "pull together" with maximum efficiency. Because the cardiac muscle cells are mechanically, chemically, and electrically connected to one another, the entire tissue resembles a single, enormous muscle cell. For this reason, cardiac muscle has been called a **functional syncytium** (sin-SISH-ē-um; a fused mass of cells).

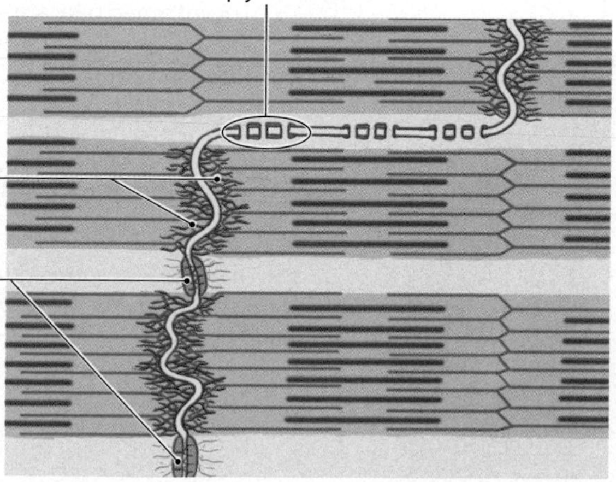

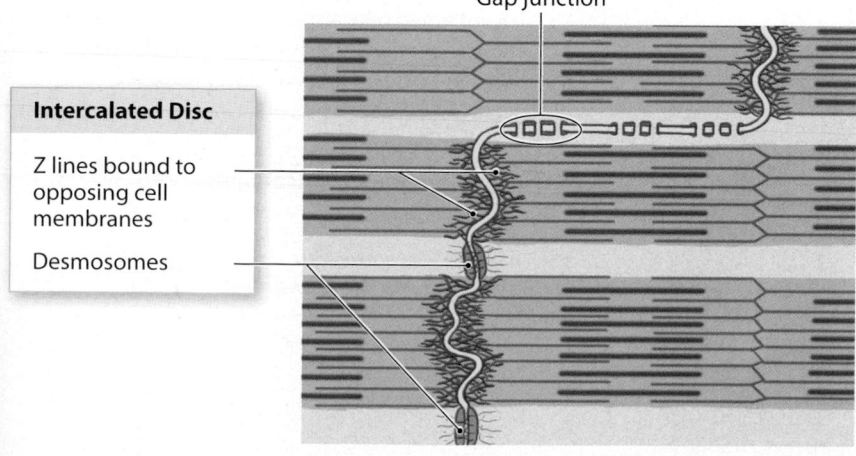

Gap junction
Intercalated Disc
Z lines bound to opposing cell membranes
Desmosomes

Module 19.2 Review

a. From superficial to deep, name the layers of the heart wall.

b. Describe the tissue layers of the epicardium.

c. Why is it important that cardiac tissue contain many mitochondria and capillaries?

LO **19.2** Describe the structure of the pericardium and explain its functions, identify the layers of the heart wall, and describe the structures and functions of cardiac muscle.

The heart is located in the mediastinum, and enclosed by the pericardial cavity

1 The position and orientation of the heart relative to the major vessels and the ribs, sternum, and lungs can be seen in this anterior view. The heart, surrounded by the pericardial sac, sits in the anterior portion of the **mediastinum** (mē-dē-as-TĪ-num), the region between the two pleural cavities. The mediastinum also contains the great vessels, thymus, esophagus, and trachea.

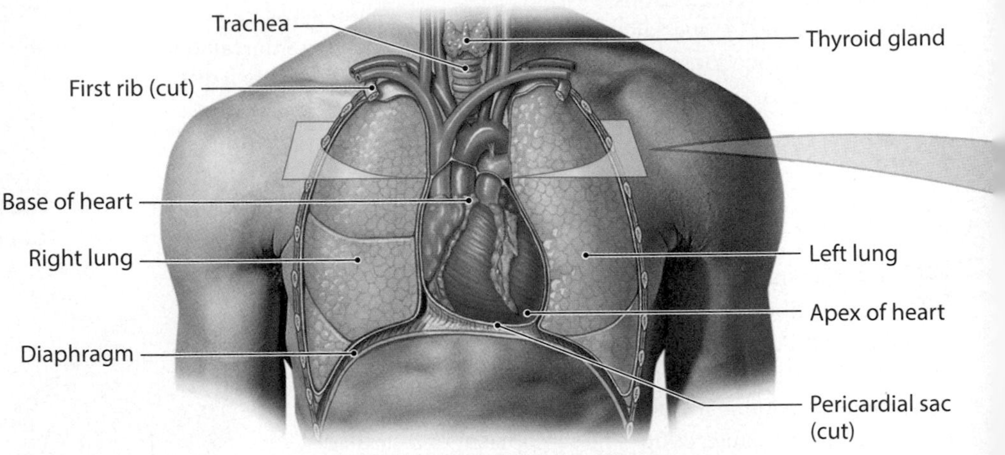
Trachea
Thyroid gland
First rib (cut)
Base of heart
Right lung
Left lung
Apex of heart
Diaphragm
Pericardial sac (cut)

Anterior view of chest cavity

2 To visualize the relationship between the heart, pericardium, and the pericardial cavity, imagine pushing your fist toward the center of a large, partially inflated balloon. The balloon represents the pericardium, and your fist is the heart. Your wrist, where the balloon folds back on itself, corresponds to the base of the heart, to which the **great vessels**, the largest veins and arteries in the body, are attached. The air space inside the balloon corresponds to the pericardial cavity.

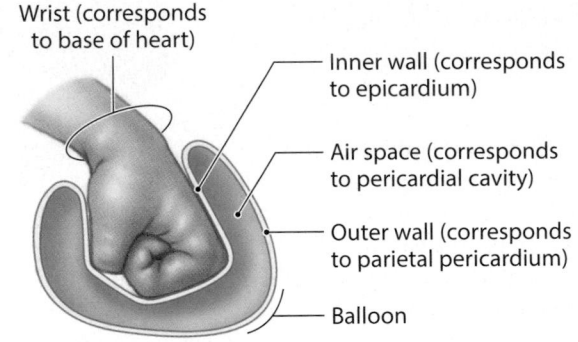

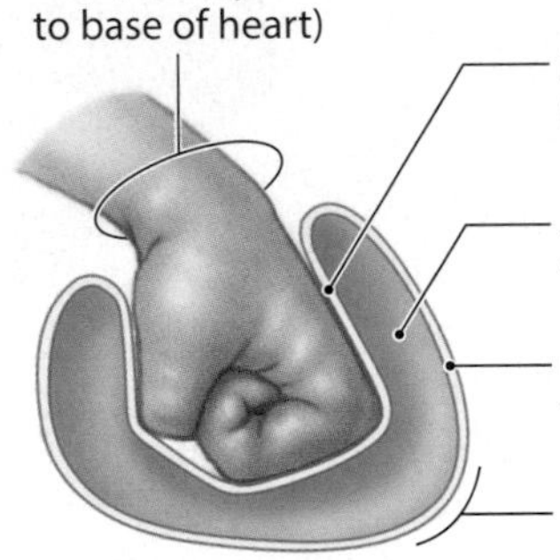
Wrist (corresponds to base of heart)
Inner wall (corresponds to epicardium)
Air space (corresponds to pericardial cavity)
Outer wall (corresponds to parietal pericardium)
Balloon

3 The **pericardial sac**, or *fibrous pericardium*, surrounds the heart. The pericardial sac consists of a dense network of collagen fibers. It attaches to the central tendon of the diaphragm and sternum, and stabilizes the position of the heart and associated vessels within the mediastinum. The parietal pericardium lines its inner surface.

The pericardial cavity contains 15–50 mL of pericardial fluid, secreted by the pericardial membranes. This fluid acts as a lubricant that reduces friction between the opposing surfaces as the heart beats. Pathogens can infect the pericardium, producing the condition **pericarditis**. The inflamed pericardial surfaces rub against one another, producing a distinctive scratching sound that can be heard through a stethoscope.

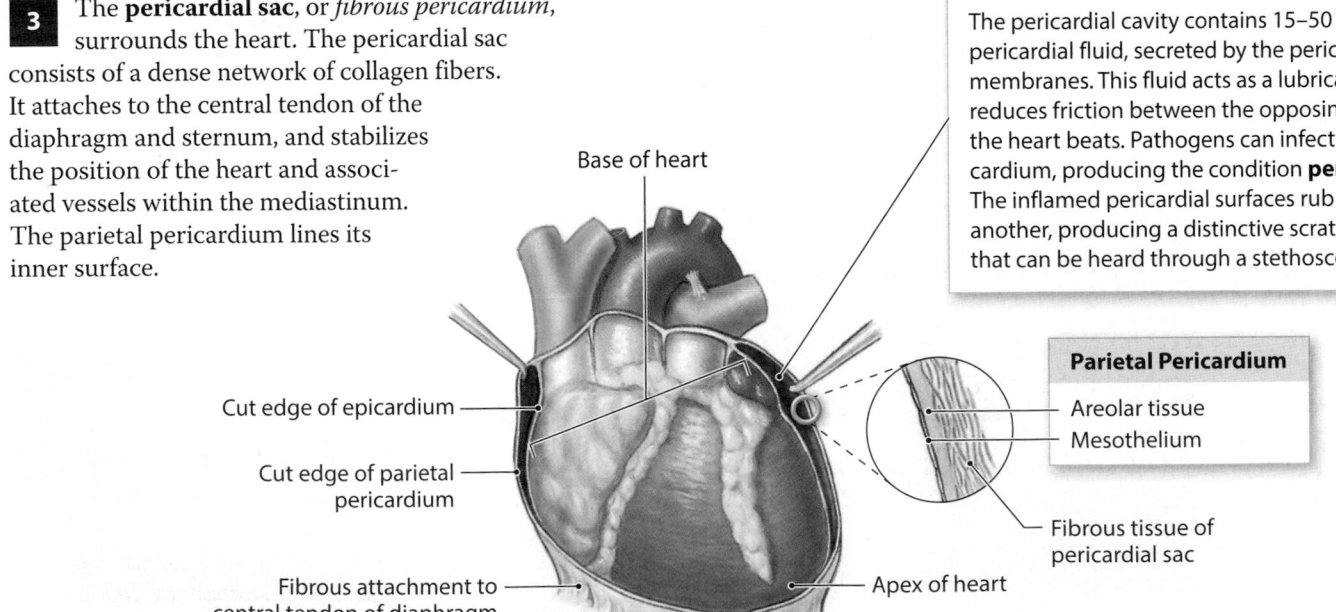

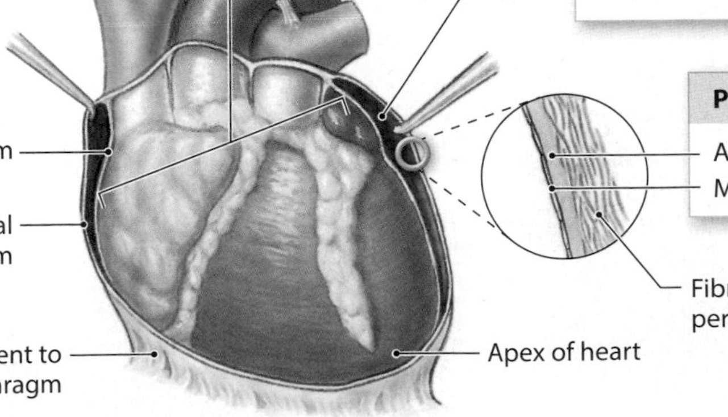
Base of heart
Cut edge of epicardium
Cut edge of parietal pericardium
Fibrous attachment to central tendon of diaphragm
Apex of heart
Parietal Pericardium
Areolar tissue
Mesothelium
Fibrous tissue of pericardial sac

4 This is a diagrammatic superior view of a partial dissection of the thoracic cavity. The image illustrates the position of the pericardial cavity and the physical relationships among the components in the mediastinum.

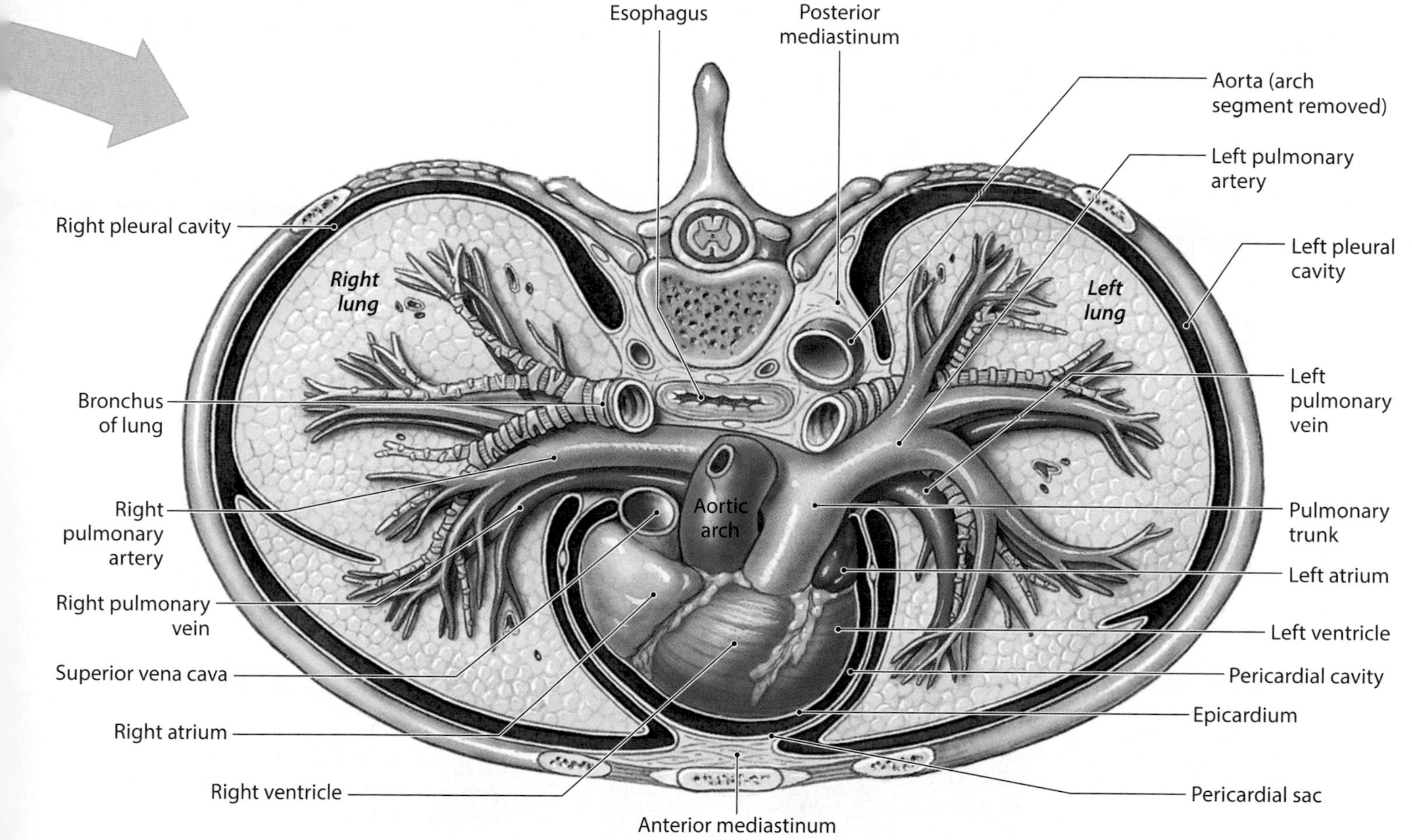

Esophagus
Posterior mediastinum
Aorta (arch segment removed)
Left pulmonary artery
Right pleural cavity
Left pleural cavity
Right lung
Left lung
Bronchus of lung
Left pulmonary vein
Right pulmonary artery
Aortic arch
Pulmonary trunk
Left atrium
Right pulmonary vein
Left ventricle
Superior vena cava
Pericardial cavity
Epicardium
Right atrium
Right ventricle
Pericardial sac
Anterior mediastinum

As you can see in this image, the heart does not have a lot of empty space around it—the thoracic cavity is very crowded. Traumatic injuries that damage the pericardium or chest wall can result in fluid accumulation within the pericardial cavity, which can restrict the movement of the heart. This condition, called **cardiac tamponade** (tam-po-NĀD; *tampon*, plug), can also be caused by acute pericarditis.

Module 19.3 Review

a. Define mediastinum.
b. Describe the heart's location in the body.
c. Why can cardiac tamponade be a life-threatening condition?

Module 19.4

The boundaries between the chambers of the heart can be identified on its external surface

1 The four cardiac chambers can easily be identified in a superficial view of the **anterior surface** of the heart. The two atria have relatively thin muscular walls and are highly expandable. When not filled with blood, the outer portion of each atrium deflates and becomes a lumpy, wrinkled flap. Shallow grooves, or **sulci** (singular, *sulcus*), mark the boundaries between the atria and ventricles and between the left and right ventricles. The connective tissue of the epicardium generally contains substantial amounts of fat, especially along the sulci. In fresh or preserved hearts, this fat must be stripped away to expose the underlying grooves. These sulci also contain the arteries and veins that carry blood to and from the cardiac muscle.

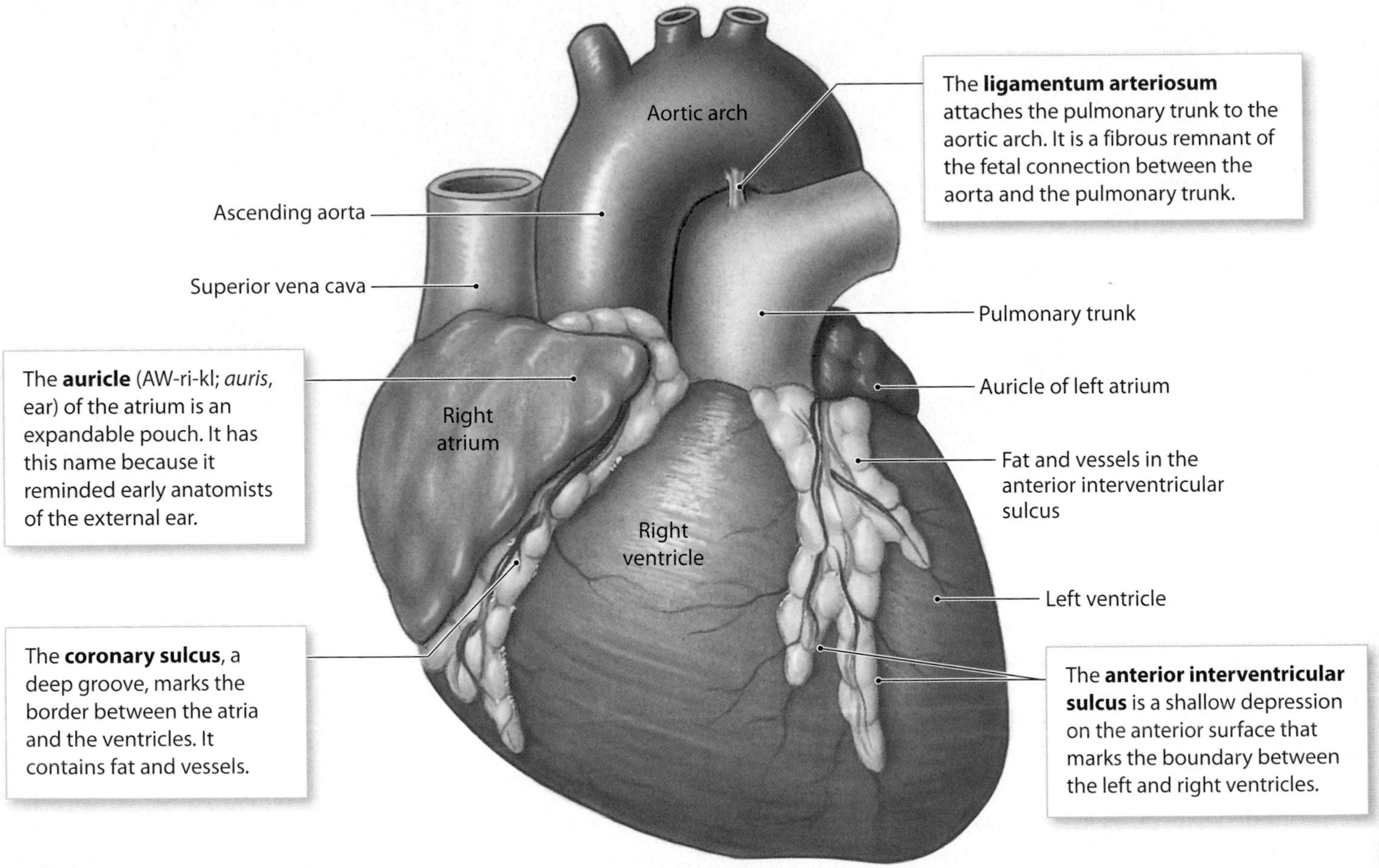

Aortic arch
The **ligamentum arteriosum** attaches the pulmonary trunk to the aortic arch. It is a fibrous remnant of the fetal connection between the aorta and the pulmonary trunk.
Ascending aorta
Superior vena cava
Pulmonary trunk
The **auricle** (AW-ri-kl; *auris*, ear) of the atrium is an expandable pouch. It has this name because it reminded early anatomists of the external ear.
Right atrium
Auricle of left atrium
Fat and vessels in the anterior interventricular sulcus
Right ventricle
Left ventricle
The **coronary sulcus**, a deep groove, marks the border between the atria and the ventricles. It contains fat and vessels.
The **anterior interventricular sulcus** is a shallow depression on the anterior surface that marks the boundary between the left and right ventricles.

Anterior surface

2 This view of the **posterior surface** of the heart shows the left atrium and its connection to the pulmonary veins. It also shows the right atrium and its connection to the coronary veins and the venae cavae.

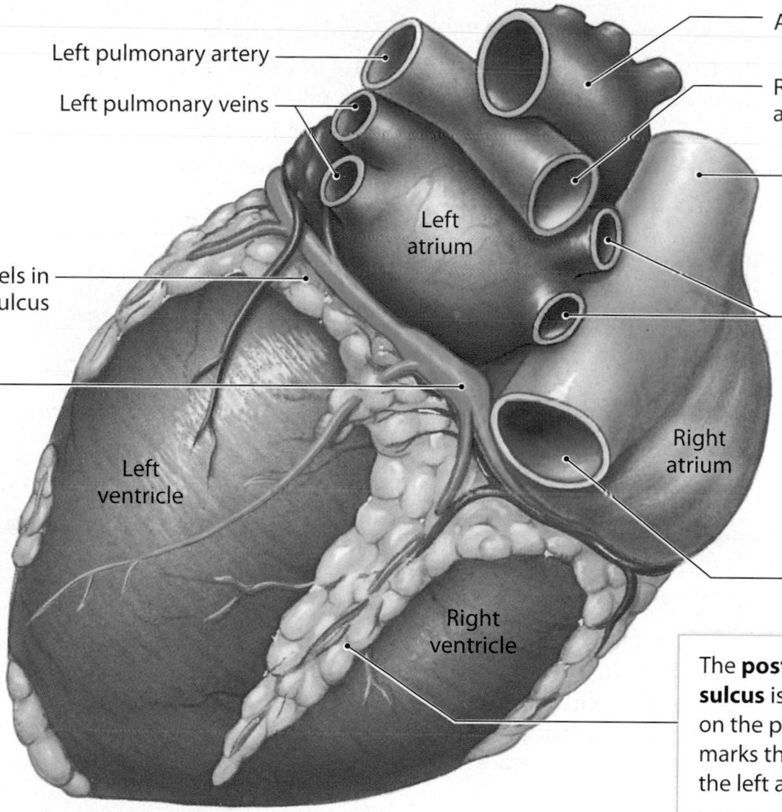

Left pulmonary artery
Left pulmonary veins
Aortic arch
Right pulmonary artery
Superior vena cava
Left atrium
Fat and vessels in the coronary sulcus
Right pulmonary veins (superior and inferior)
The **coronary sinus** carries blood collected from the myocardium by numerous coronary veins and conveys the blood to the right atrium.
Right atrium
Left ventricle
Inferior vena cava
Right ventricle
The **posterior interventricular sulcus** is a shallow depression on the posterior surface that marks the boundary between the left and right ventricles.

Posterior surface

3 As you see in this anterior view, a dissected heart from a preserved cadaver is not conveniently color-coded.

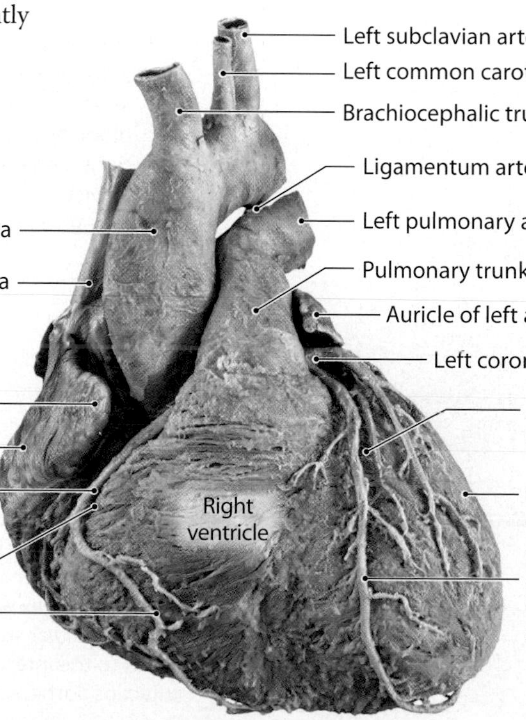

Left subclavian artery
Left common carotid artery
Brachiocephalic trunk
Ligamentum arteriosum
Ascending aorta
Left pulmonary artery
Superior vena cava
Pulmonary trunk
Auricle of left atrium
Left coronary artery
Auricle of right atrium
Anterior interventricular sulcus
Right atrium
Right coronary artery
Left ventricle
Right ventricle
Coronary sulcus
Anterior interventricular artery
Marginal branch of right coronary artery

Cadaver dissection, anterior view

Module 19.4 Review

a. Name the four cardiac chambers.

b. Name and describe the shallow depressions and grooves found on the heart's external surface.

c. Which structures collect blood from the myocardium, and into which heart chamber does this blood flow?

LO 19.4 Describe the cardiac chambers and the heart's external anatomy.

The heart has an extensive blood supply

The heart works continuously, so cardiac muscle cells require reliable supplies of oxygen and nutrients. Although a great volume of blood flows through the chambers of the heart, the myocardium needs its own, separate blood supply. The coronary circulation supplies that blood to the muscle tissue of the heart. During maximum exertion, blood flow to the myocardium may increase to nine times that of resting levels.

1 The left and right **coronary arteries** originate at the base of the ascending aorta, where blood pressure is the highest in the systemic circuit. However, myocardial blood flow is not steady: It peaks while the heart muscle is relaxed, and almost ceases while it contracts.

Right Coronary Artery

The **right coronary artery**, which follows the coronary sulcus around the heart, supplies blood to the right atrium, portions of both ventricles, and portions of the conducting system of the heart, which contols and coordinates the heartbeat.

Marginal arteries from the right coronary artery supply the surface of the right ventricle.

Left Coronary Artery

The **left coronary artery** supplies blood to the left ventricle, left atrium, and interventricular septum.

Circumflex artery

The large **anterior interventricular** (left anterior descending) **artery** runs along the surface within the anterior interventricular sulcus.

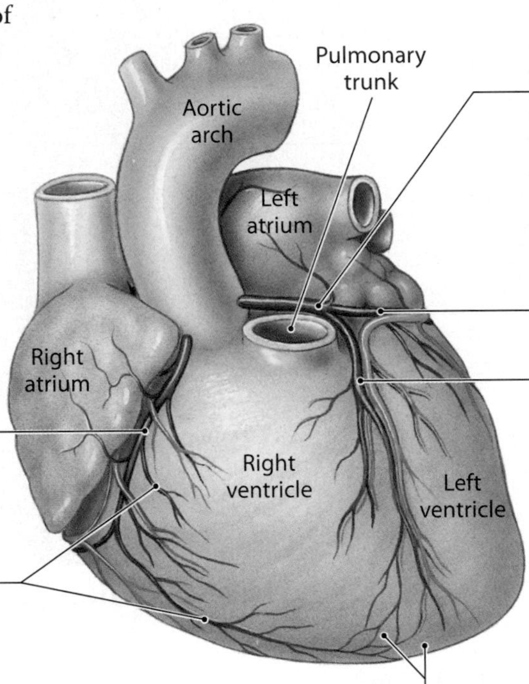
Pulmonary trunk
Aortic arch
Left atrium
Right atrium
Right ventricle
Left ventricle

Anterior view

Arterial anastomoses between the anterior and posterior interventricular arteries maintain a fairly continuous blood flow despite pressure fluctuations in the left and right coronary arteries.

2 Branches of the left and right coronary arteries continue onto the posterior surface of the heart.

The **circumflex artery** is a branch of the left coronary artery that curves to the left around the coronary sulcus, eventually meeting and fusing with small branches of the right coronary artery. A marginal artery branches from the circumflex artery to supply the posterior surface of the left ventricle.

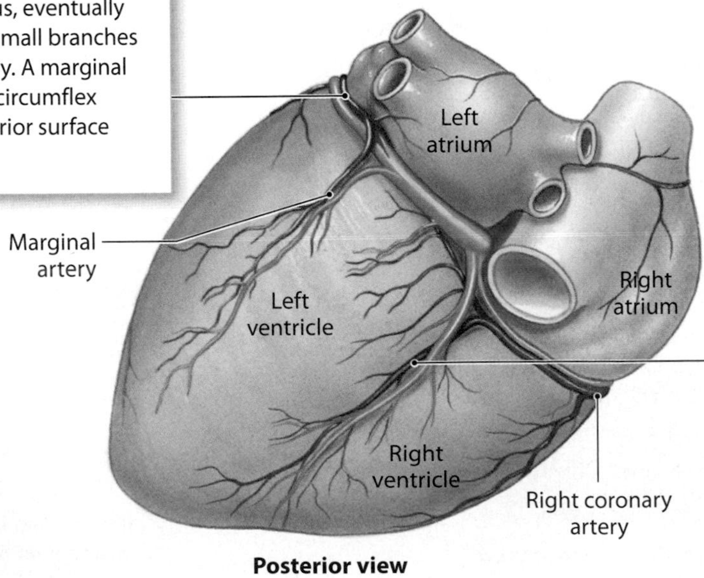
Marginal artery
Left atrium
Left ventricle
Right atrium
Right ventricle
Right coronary artery

Posterior view

The right coronary artery continues across the posterior surface of the heart, supplying the **posterior interventricular** (posterior descending) **artery**, which runs toward the apex within the posterior interventricular sulcus. This vessel supplies blood to the interventricular septum and adjacent portions of the ventricles.

3 This view identifies the major collecting vessels on the anterior surface of the heart.

The **anterior cardiac veins**, which drain the anterior surface of the right ventricle, empty directly into the right atrium.

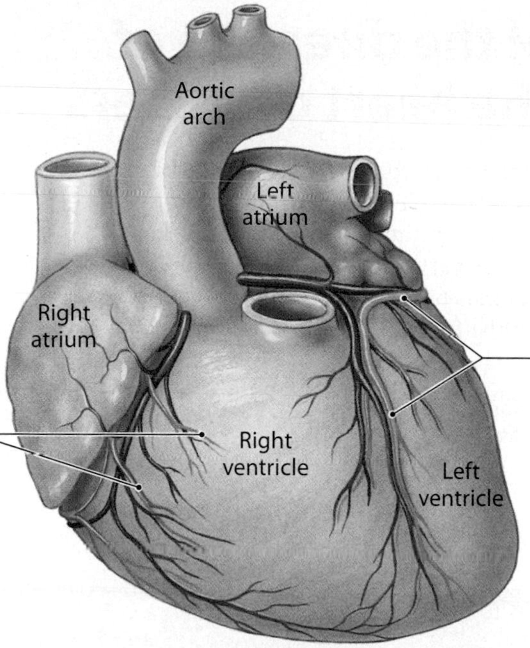
Aortic arch
Left atrium
Right atrium
Right ventricle
Left ventricle

The **great cardiac vein** begins on the anterior surface of the ventricles, along the interventricular sulcus. This vein drains blood from the region supplied by the anterior interventricular artery. The great cardiac vein reaches the level of the atria and then curves around the left side of the heart within the coronary sulcus to empty into the coronary sinus.

Anterior view

4 This view identifies the major collecting vessels on the posterior surface of the heart.

The **coronary sinus** is an expanded vein that opens into the right atrium near the base of the inferior vena cava.

The **posterior cardiac vein** drains the area supplied by the circumflex artery.

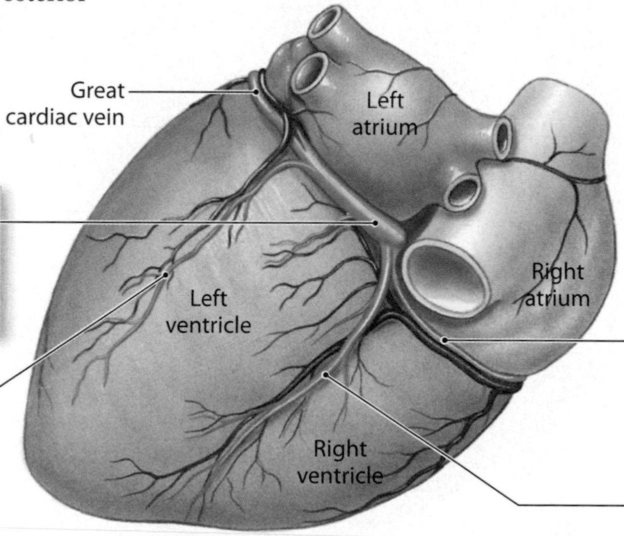
Great cardiac vein
Left atrium
Left ventricle
Right atrium
Right ventricle

The **small cardiac vein** receives blood from the posterior surfaces of the right atrium and ventricle. It empties into the coronary sinus along with the middle cardiac vein.

The **middle cardiac vein**, draining the area supplied by the posterior interventricular artery, empties into the coronary sinus.

Posterior view

Each time the left ventricle contracts, it forces blood into the aorta. The arrival of additional blood at elevated pressures stretches the elastic walls of the aorta. When the left ventricle relaxes, pressure decreases, and the walls of the aorta recoil. This recoil, called **elastic rebound**, pushes blood both forward, into the systemic circuit, and backward, into the coronary arteries. Thus, the combination of blood pressure and elastic rebound ensures a continuous flow of blood to meet the demands of active cardiac muscle tissue.

Module 19.5 Review

a. List the arteries and veins of the heart.

b. Compare the anterior cardiac veins to the posterior cardiac vein.

c. Describe what happens to blood flow during elastic rebound.

Internal valves control the direction of blood flow between the heart chambers

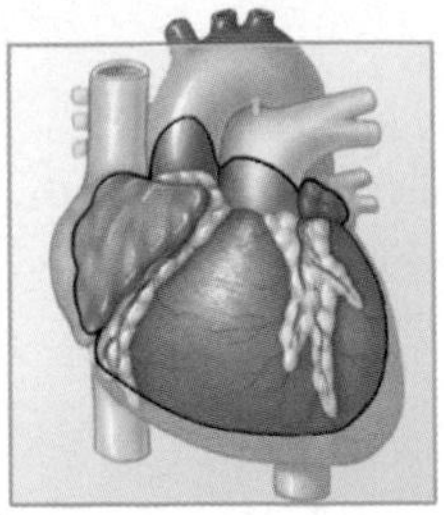

1 In a sectional view, you can see that the right atrium communicates with the right ventricle, and the left atrium with the left ventricle. The atria are separated by the **interatrial septum** (*septum*, wall). The ventricles are separated by the much thicker **interventricular septum**. Each septum is a muscular partition. **Atrioventricular (AV) valves**, folds of fibrous tissue, extend into the openings between the atria and ventricles. These valves permit blood flow in one direction only: from the atria to the ventricles.

Right Atrium

The right atrium receives blood from the superior and inferior venae cavae. It also receives blood from the cardiac veins through the coronary sinus.

The **fossa ovalis** is an oval depression and is a remnant of the foramen ovale. The foramen ovale closes at birth, and this opening between the atria is permanently sealed off during the next 3 months.

The anterior atrial wall and the inner surface of the auricle contain prominent muscular ridges called the **pectinate muscles** (*pectin*, comb).

The opening of the coronary sinus carries blood from the cardiac veins.

Right Ventricle

Blood travels from the right atrium into the right ventricle through a broad opening bordered by the **right atrioventricular (AV) valve**, also known as the **tricuspid** (trī- KUS-pid; *tri*, three) **valve**.

The free edge of each valve consists of three flaps, or **cusps**, attached to tendinous connective tissue fibers called the **chordae tendineae** (KOR-dē TEN-di-nē-ē; tendinous cords).

The fibers of the chordae tendineae originate at conical muscular projections called the **papillary** (PAP-i-ler-ē) **muscles**.

The superior portion of the right ventricle tapers toward the **pulmonary valve**, or pulmonary semilunar valve. Blood leaving the right ventricle passes through this valve to enter the pulmonary trunk.

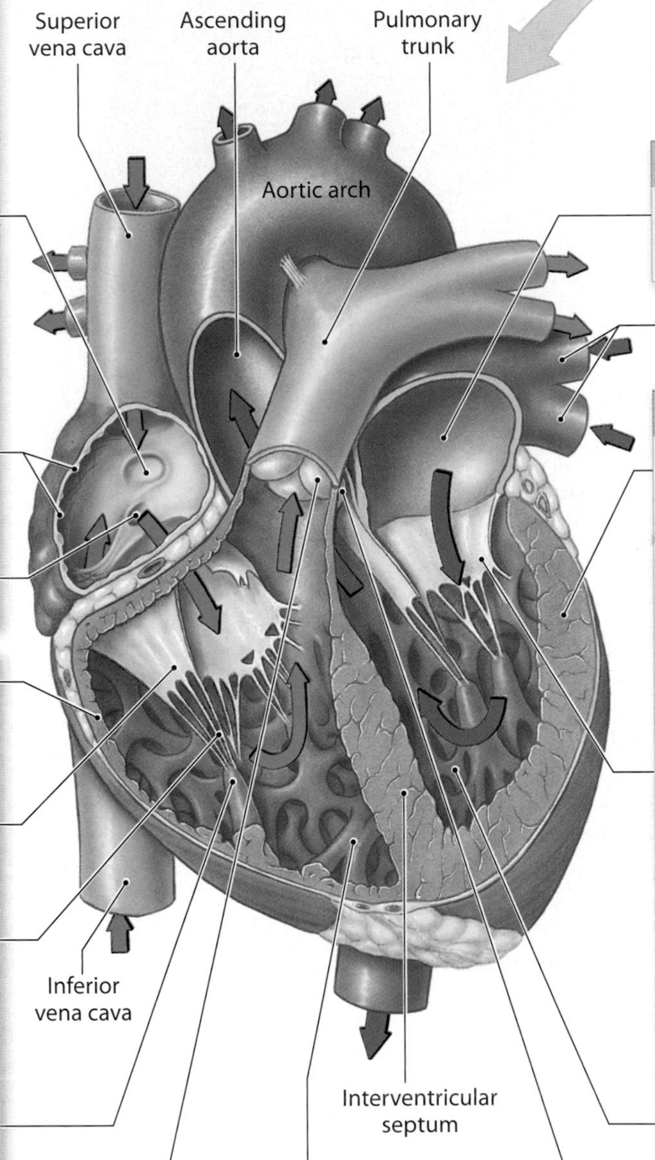
Superior vena cava
Ascending aorta
Pulmonary trunk
Aortic arch
Left pulmonary veins
Inferior vena cava
Interventricular septum

The **moderator band** provides a rapid-conduction path that tenses the papillary muscles before the ventricular myocardium contracts. This prevents "slamming" of the right AV cusps.

Left Atrium

The left atrium receives blood from the pulmonary veins.

Left Ventricle

The left ventricle is much larger than the right ventricle. Its thick, muscular wall enables the left ventricle to develop pressure sufficient to push blood through the large systemic circuit, whereas the right ventricle needs to pump blood, at lower pressure, through the nearby lungs.

The **left atrioventricular** (AV) **valve**, or **bicuspid** (bī-KUS-pid) **valve**, permits the flow of blood from the left atrium into the left ventricle but prevents backflow during ventricular contraction. As the name bicuspid implies, the left AV valve contains a pair, not a trio, of cusps. Clinicians often call this valve the **mitral** (MĪ-tral; *mitre*, a bishop's hat) **valve**.

The **trabeculae carneae** (tra-BEK-ū-lē KAR-nē-ē; *carneus*, fleshy) are a series of muscular ridges on the inner surfaces of the right and left ventricles.

Blood leaves the left ventricle by passing through the **aortic valve** (aortic semilunar valve) and into the ascending aorta.

2 The function of an atrium is to collect blood that is returning to the heart and convey it to the attached ventricle. The functional demands on the right and left atria are similar, and the two chambers look almost identical. The demands on the right and left ventricles, however, are very different, and the two have significant structural differences. These differences are best seen in sectional view.

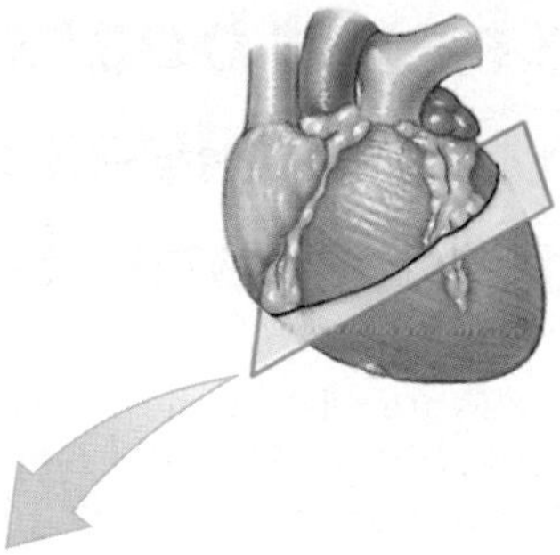

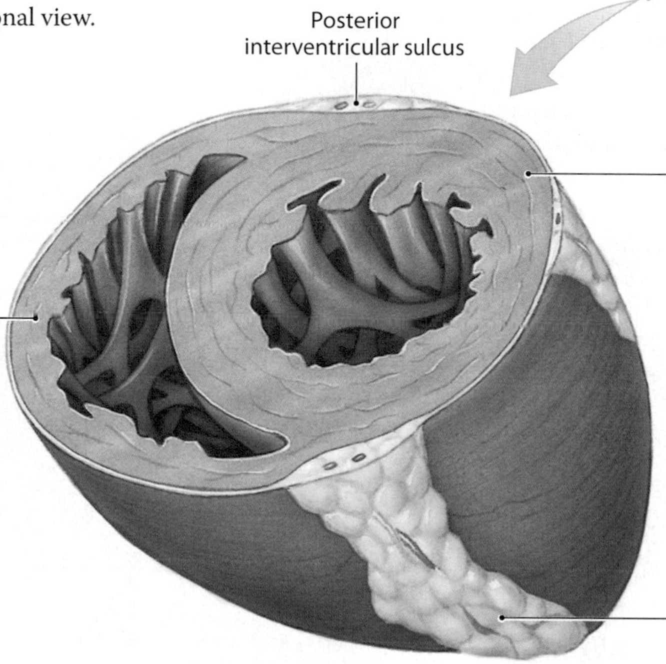

Posterior interventricular sulcus
Fat in anterior interventricular sulcus

In sectional view, the relatively thin wall of the right ventricle resembles a pouch attached to the massive wall of the left ventricle. Such a thin wall is adequate because the right ventricle normally does not need to work very hard to push blood through the pulmonary circuit—the lungs are close to the heart, and the pulmonary vessels are relatively short and wide.

The left ventricle has an extremely thick muscular wall and is round in cross section. It must develop four to six times as much pressure to push blood around the systemic circuit as the right ventricle develops to push blood around the pulmonary circuit.

3 These sections indicate the changes in ventricular shape that occur when the ventricles contract.

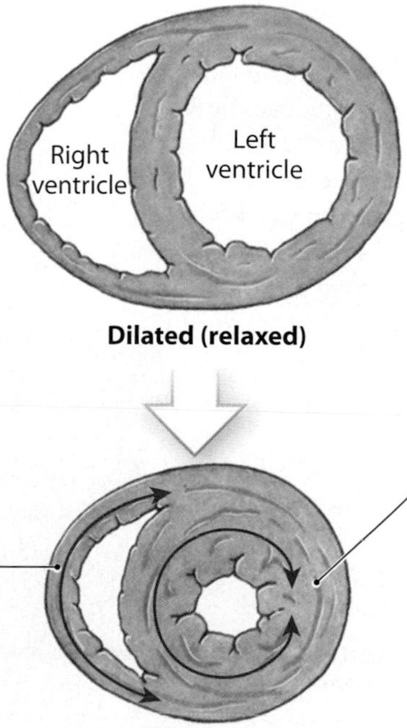

Right ventricle
Left ventricle
Dilated (relaxed)
Contracted

When the right ventricle contracts, it acts like a bellows, squeezing the blood against the thick wall of the left ventricle. This mechanism moves blood very efficiently with minimal effort, but it develops relatively low pressures.

When the left ventricle contracts, (1) the diameter of the ventricular chamber decreases, and (2) the distance between the base and apex decreases. The effect is similar to simultaneously squeezing and rolling up the end of a toothpaste tube. This motion also reduces the volume of the right ventricle, and a person whose right ventricular musculature has been severely damaged may still survive, because the contraction of the left ventricle helps push blood into the pulmonary circuit.

Module 19.6 Review

a. Damage to the semilunar valve on the right side of the heart would affect blood flow to which vessel?

b. What prevents the AV valves from swinging into the atria?

c. Why is the left ventricle more muscular than the right ventricle?

LO **19.6** Trace blood flow through the heart, identifying the major blood vessels, chambers, and heart valves.

When the heart beats, the AV valves close before the semilunar valves open, and the semilunar valves close before the AV valves open

1 When the ventricles are relaxed, the chordae tendineae are loose, and the AV valves offer no resistance to the flow of blood from the atria into the ventricles. Blood pressure in the pulmonary and systemic circuits keeps the aortic and pulmonary valves closed until the ventricles contract.

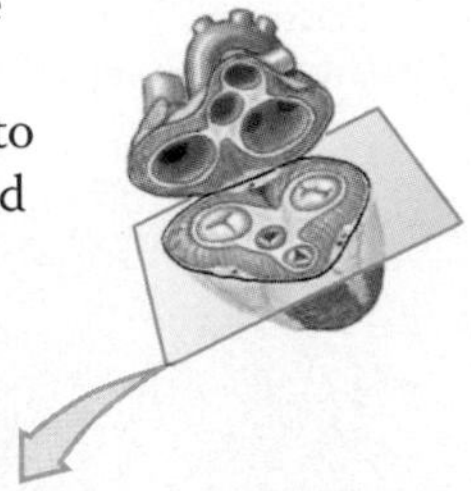

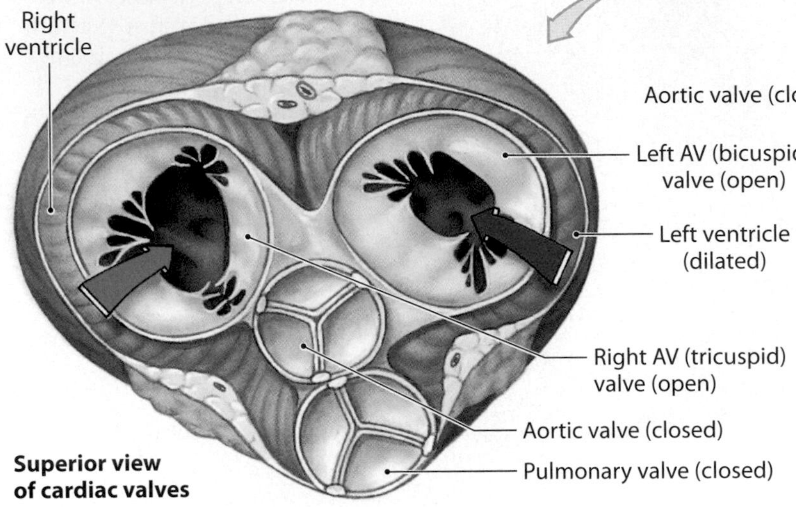
Right ventricle
Left AV (bicuspid) valve (open)
Left ventricle (dilated)
Right AV (tricuspid) valve (open)
Aortic valve (closed)
Pulmonary valve (closed)

Superior view of cardiac valves

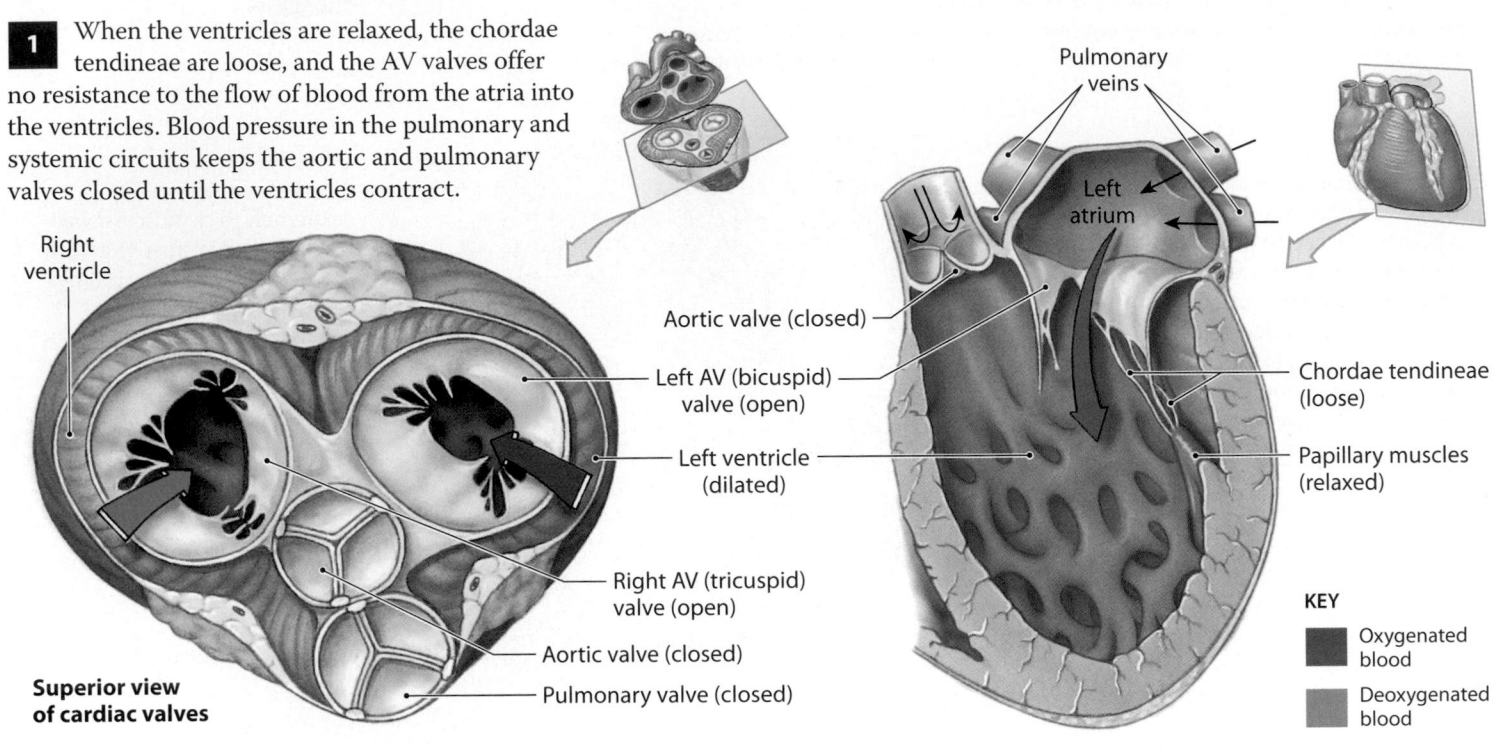

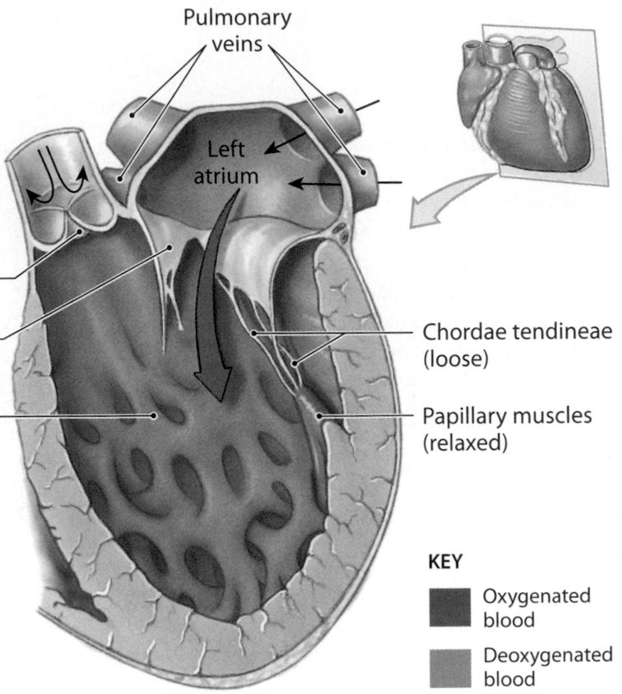
Pulmonary veins
Left atrium
Aortic valve (closed)
Chordae tendineae (loose)
Papillary muscles (relaxed)
KEY
Oxygenated blood
Deoxygenated blood

2 When the ventricles contract, blood moving back toward the atria pushes the cusps of the AV valves together, closing them and preventing backflow. At the same time, the contraction of the papillary muscles tenses the chordae tendineae, stopping the cusps before they swing into the atria. If the chordae tendineae are cut or the papillary muscles are damaged, **regurgitation** (backflow) of blood into the atria occurs each time the ventricles contract. As the ventricular pressures rise above those in the pulmonary and systemic circuits, the aortic and pulmonary valves open and blood flows out of the ventricles.

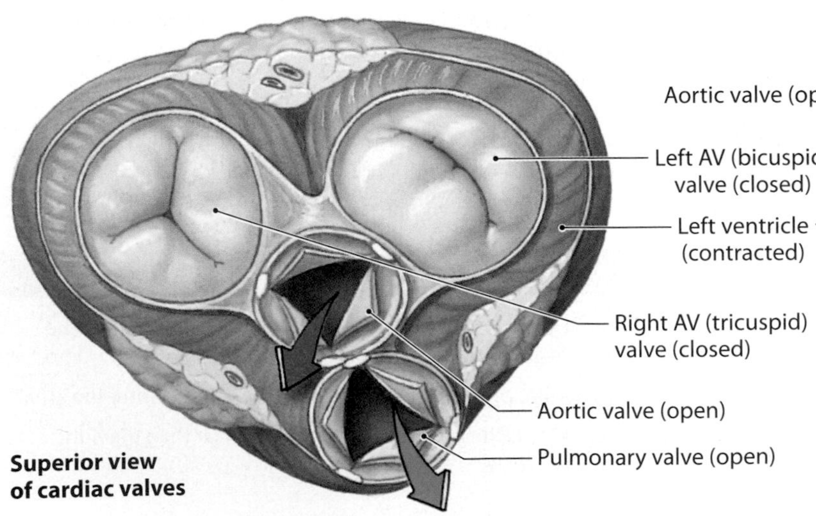
Left AV (bicuspid) valve (closed)
Left ventricle (contracted)
Right AV (tricuspid) valve (closed)
Aortic valve (open)
Pulmonary valve (open)

Superior view of cardiac valves

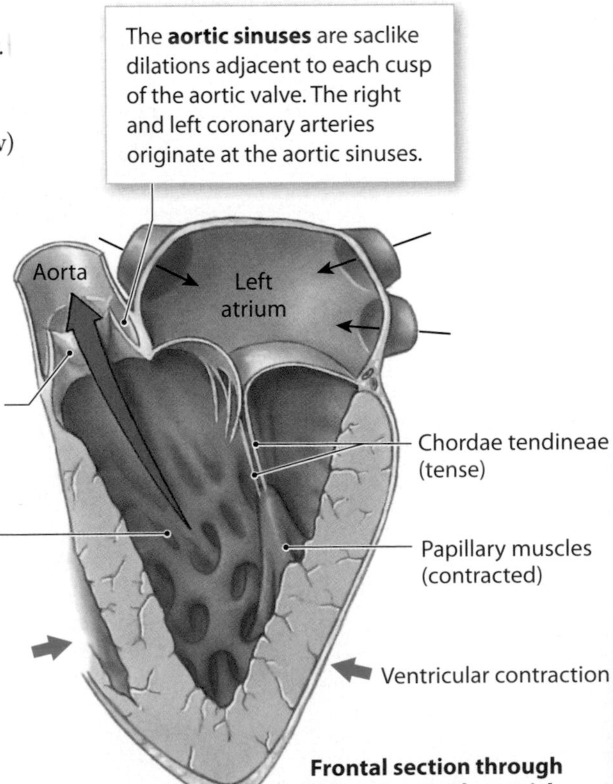
The aortic sinuses are saclike dilations adjacent to each cusp of the aortic valve. The right and left coronary arteries originate at the aortic sinuses.
Aorta
Left atrium
Aortic valve (open)
Chordae tendineae (tense)
Papillary muscles (contracted)
Ventricular contraction

Frontal section through left atrium and ventricle

3 The heart valves are encircled and supported by a flexible connective tissue frame known as the **cardiac skeleton** (fibrous skeleton). The cardiac skeleton consists of interconnected bands of dense connective tissue that encircle the heart valves and the bases of the pulmonary trunk and aorta. The cardiac skeleton stabilizes the positions of the heart valves and ventricular muscle cells and electrically isolates the ventricular myocardium from the atrial myocardium.

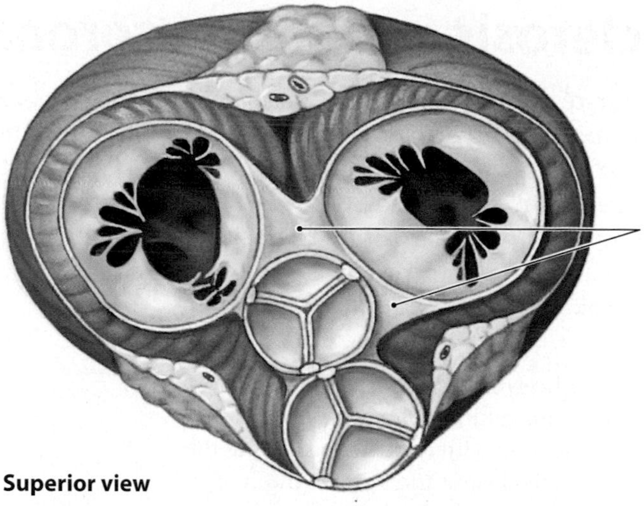
Cardiac skeleton

Superior view

4 The pulmonary and aortic valves each consist of three semilunar (half-moon shaped) cusps of thick connective tissue. Unlike the AV valves, the **semilunar valves** do not require muscular braces, because the cusps are stable. When the semilunar valves close, the three symmetrical cusps support one another like the legs of a tripod.

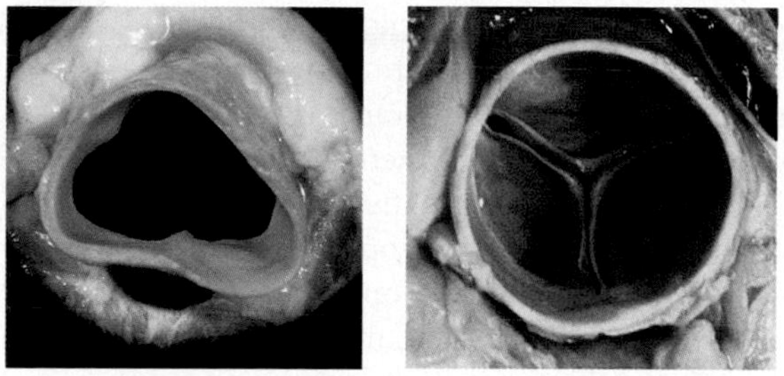

Open Closed

Superior views

5 Serious valve problems can interfere with cardiac function. If valve function deteriorates to the point at which the heart cannot maintain adequate blood flow, symptoms of **valvular heart disease (VHD)** appear. Congenital malformations may be responsible, but in many cases the condition develops as a consequence of **carditis**, an inflammation of the heart. In severe cases, the only option may be to replace the damaged valve with a prosthetic valve.

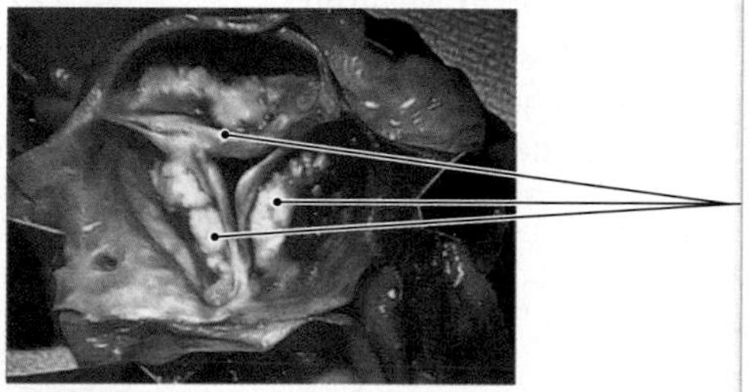

This is a superior view of a damaged aortic valve. The cusps are irregular in shape; they are also stiff and relatively inflexible. Such a valve would not open properly and cannot close completely.

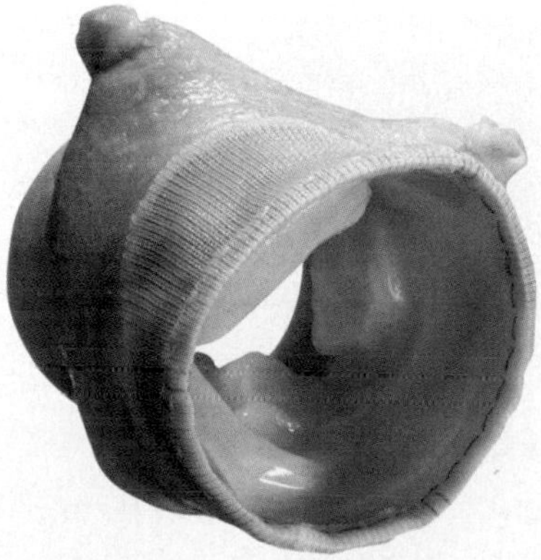

This bioprosthetic valve is an example of an artificial valve that uses the cusps from a pig's heart. Pig or cow valves do not stimulate the clotting system, but they may wear out after about 10 years.

Module 19.7 Review

a. Define cardiac regurgitation.

b. Compare the structure of the tricuspid valve with that of the pulmonary valve.

c. What do semilunar valves prevent?

LO 19.7 Describe the relationship between the AV and semilunar valves during a heartbeat.

Arteriosclerosis can lead to coronary artery disease

Arteriosclerosis (ar-tē-rē-ō-skle-RŌ-sis; *arterio-*, artery + *sklerosis,* hardness) is a thickening and toughening of arterial walls. This condition may not sound life-threatening, but complications related to arteriosclerosis account for roughly half of all deaths in the United States. The effects of arteriosclerosis are varied; for example, arteriosclerosis of coronary vessels is responsible for coronary artery disease (CAD), and arteriosclerosis of arteries supplying the brain can lead to strokes.

1 **Atherosclerosis** (ath-er-ō-skler-Ō-sis; *athero-*, fatty degeneration) is the formation of lipid deposits in the arterial tunica media associated with damage to the endothelial lining. Here is a histological view and a photograph of a dissected atherosclerotic artery. Atherosclerosis, the most common form of arteriosclerosis, tends to develop in people whose blood contains elevated levels of plasma lipids—specifically, cholesterol. The circulating lipids are removed and deposited within arterial walls, which become increasingly abnormal in structure. The result is an atherosclerotic **plaque**, a fatty mass of tissue that projects into the lumen of the vessel and restricts blood flow. Elderly people—especially elderly men—are most likely to develop atherosclerotic plaques. In addition to being older and male, other important risk factors for atherosclerosis include high blood cholesterol levels, high blood pressure, and cigarette smoking. About 20 percent of middle-aged men have all three of these risk factors; these men are four times as likely to experience a heart attack as other men in their age group. Although fewer women develop atherosclerotic plaques, elderly female smokers with high blood cholesterol and high blood pressure are at much greater risk than other women.

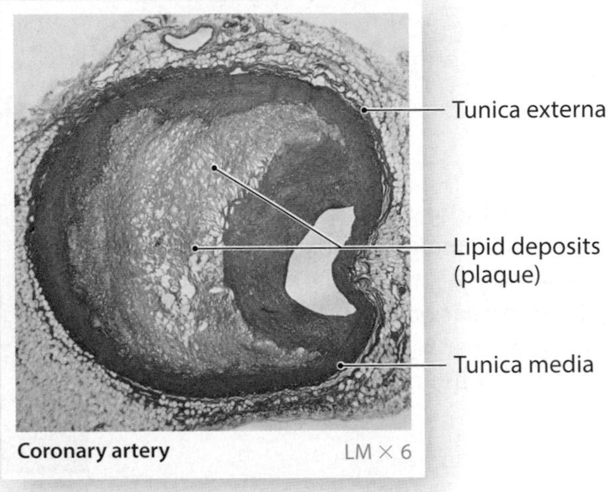
Tunica externa
Lipid deposits (plaque)
Tunica media
Coronary artery
LM × 6

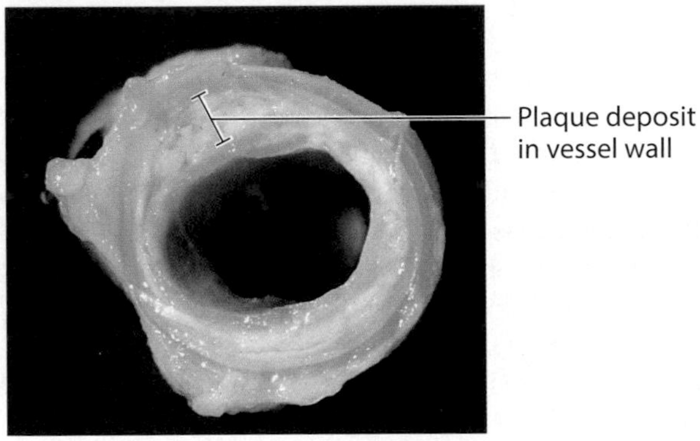
Plaque deposit in vessel wall

2 Plaques can be treated by removing the damaged segment of the vessel and replacing it (often with a superficial vein removed from the leg), but such surgery can be difficult and dangerous. In **balloon angioplasty** (AN-jē-ō-plas-tē; *angeion*, vessel), the tip of a catheter contains an inflatable balloon. Once in position, the balloon is inflated, pressing the plaque against the vessel walls. This photo shows the catheter within an artery of a cadaver. The artery has been opened to show the orientation and relative sizes of the catheter and artery. Balloon angioplasty is most effective in treating small, soft plaques. Several factors make this a highly attractive treatment: (1) The mortality rate during surgery is only about 1 percent; (2) the success rate is over 90 percent; and (3) the procedure can be performed on an outpatient basis.

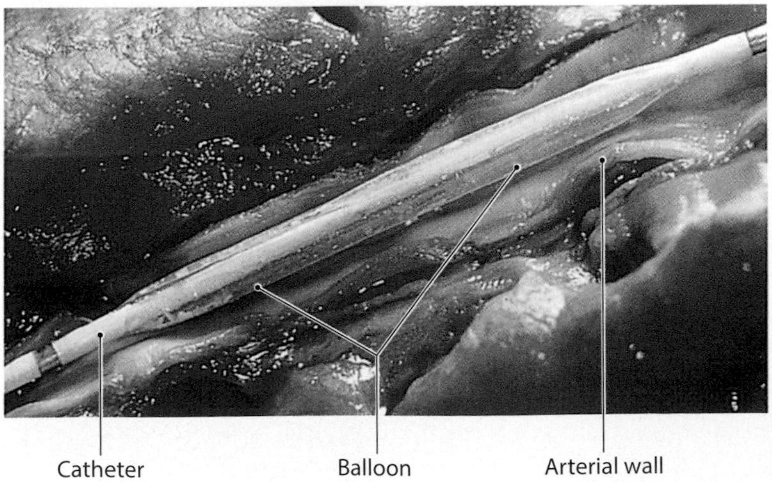
Catheter
Balloon
Arterial wall

3 The term **coronary artery disease (CAD)** refers to areas of partial or complete blockage of coronary circulation. Cardiac muscle cells need a constant supply of oxygen and nutrients, so any reduction in blood flow to the heart muscle produces a corresponding reduction in cardiac performance. Such reduced circulatory supply, known as **coronary ischemia** (is-KĒ-mē-uh), generally results from partial or complete blockage of the coronary arteries. The usual cause is the formation of an atherosclerotic plaque in a coronary artery. The plaque, or an associated **thrombus** (blood clot), then narrows the passageway and reduces blood flow. Spasms in the smooth muscles of the vessel wall can further decrease or even stop blood flow. Plaques may be visible in clinical scans or high-resolution ultrasound images.

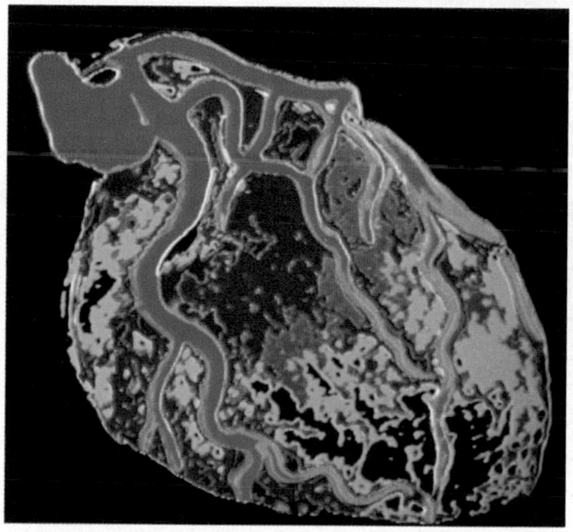

This is a color-enhanced **digital subtraction angiography (DSA)** scan of a normal heart. The major branches of the left and right coronary arteries are clearly visible.

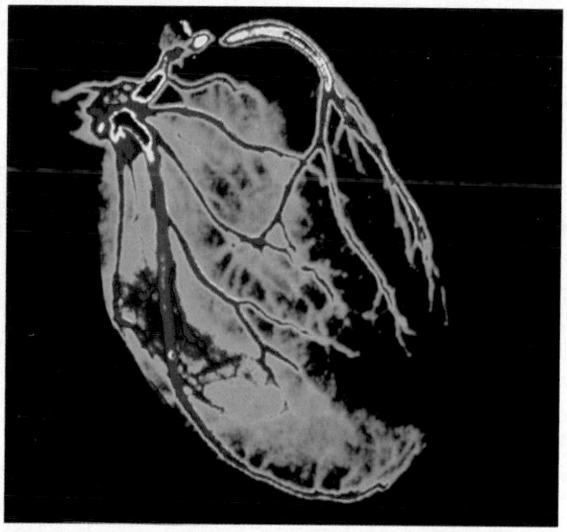

This is a color-enhanced DSA scan of the heart of a person with advanced CAD. Blood flow to the ventricular myocardium is severely restricted.

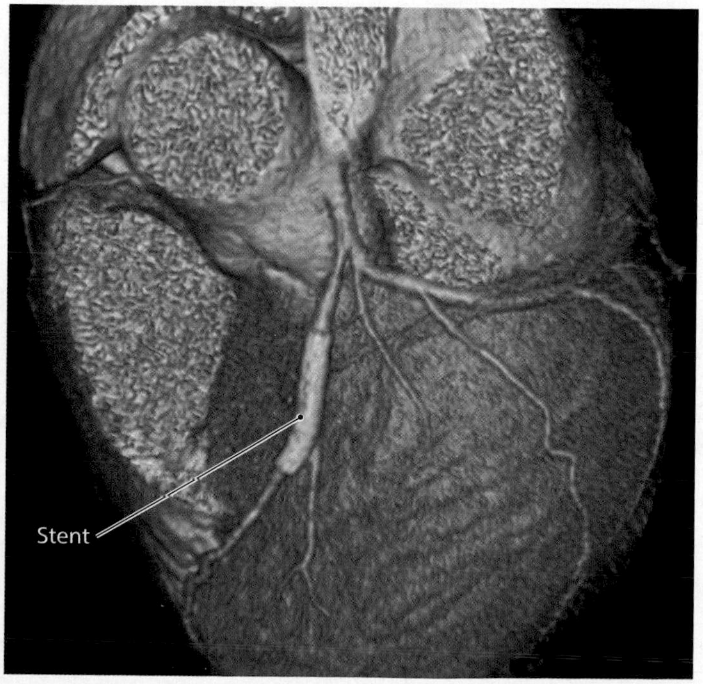
Stent

4 Because plaques commonly redevelop after angioplasty, a fine wire-mesh tube called a **stent** may be inserted into the vessel. The stent pushes against the vessel wall, holding it open. This is a scan of a stent that has been placed in the anterior interventricular artery. (This imaging technique reveals the lumen of the blood vessels, rather than the vessel, so the stent appears to surround the vessel.) Stents are routinely used by many cardiac specialists because their long-term success rate and incidence of complications are significantly lower than those for balloon angioplasty alone. If the circulatory blockage is extensive, multiple stents can be inserted along the length of the vessel.

Module 19.8 Review

a. Compare arteriosclerosis with atherosclerosis.

b. What is coronary ischemia?

c. Describe the purpose of a stent.

LO **19.8** Define arteriosclerosis, and explain its significance to health.

Labeling

Label each of the structures in this figure.

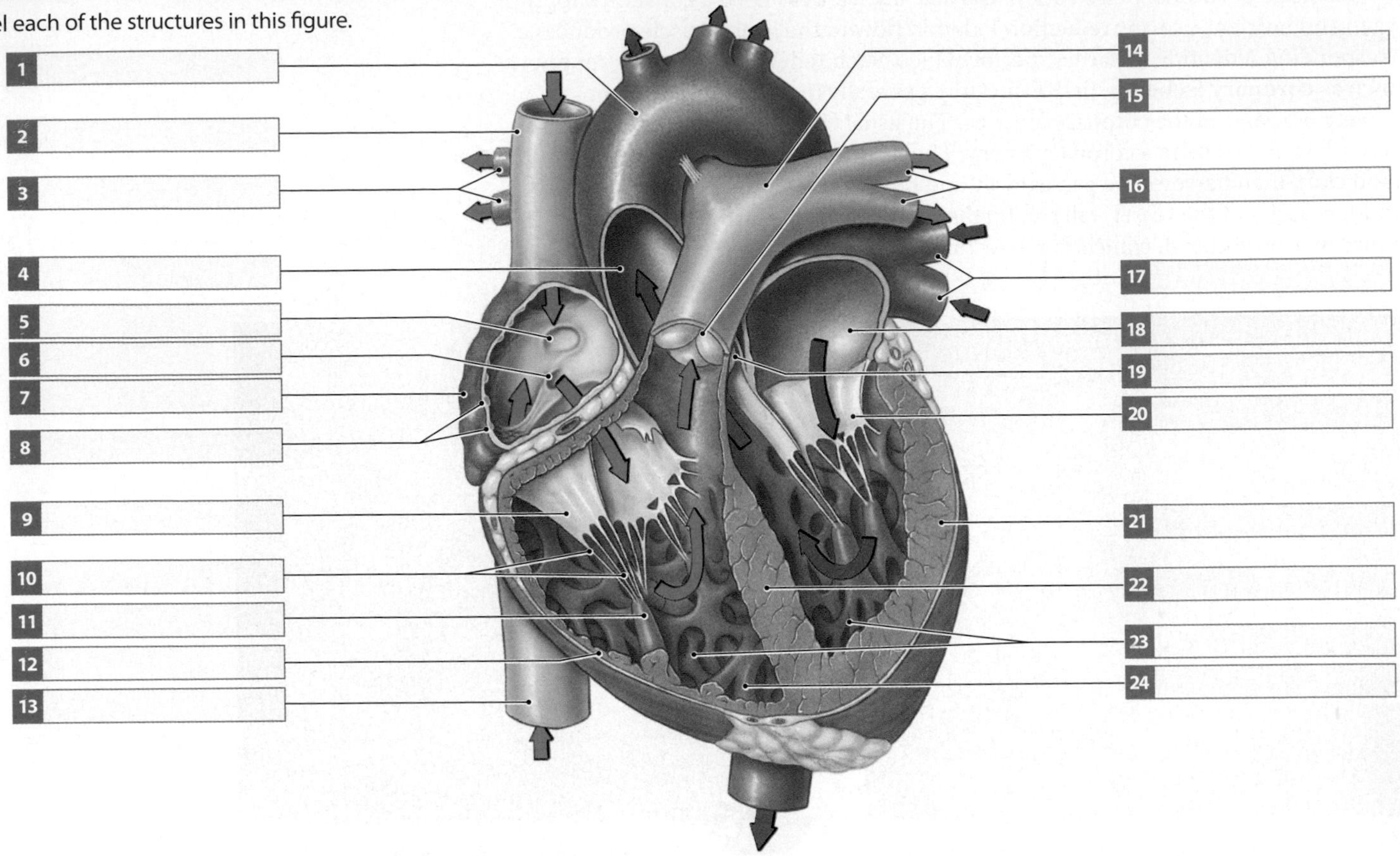
1
2
3
4
5
6
7
8
9
10
11
12
13
14
15
16
17
18
19
20
21
22
23
24

Matching

Match each lettered term with the most closely related description.

a. fossa ovalis
b. intercalated discs
c. serous membrane
d. tricuspid valve
e. aortic valve
f. endocardium
g. aorta
h. myocardium
i. mitral valve
j. coronary sinus

25 Blood to systemic arteries
26 Muscular wall of heart
27 Carries blood from coronary veins to right atrium
28 Depression in interatrial septum
29 Right atrioventricular valve
30 Alternate name for bicuspid valve
31 Cardiac muscle fiber connections
32 Pericardium
33 Inner surface of heart
34 Semilunar valve

25 ____________________
26 ____________________
27 ____________________
28 ____________________
29 ____________________
30 ____________________
31 ____________________
32 ____________________
33 ____________________
34 ____________________

Short answer

Beginning with the right atrium, list in order the heart chambers and valves through which (a) deoxygenated blood and (b) oxygenated blood flows.

35 ____________________

The cardiac cycle is a complete round of systole and diastole

1 Each heartbeat is followed by a brief resting phase, which allows time for the chambers to relax and prepare for the next heartbeat. The period between the start of one heartbeat and the beginning of the next is a single **cardiac cycle**. The cardiac cycle, therefore, includes alternating periods of contraction and relaxation.

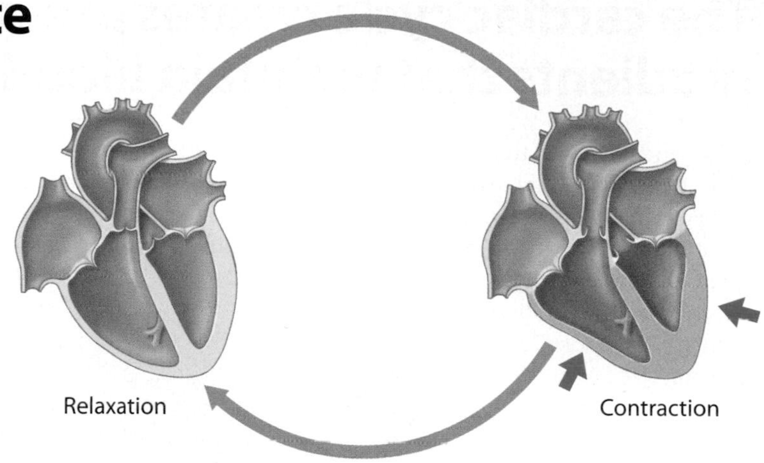
Relaxation
Contraction

2 Although we think of the heart as a pump, it is really four pumps that work in pairs, and thus a heartbeat is a complicated event. If all four chambers contracted at once, normal blood flow couldn't occur. Instead, the two atria contract first, pushing blood into the ventricles, and then the two ventricles contract, pushing blood through the pulmonary and systemic circuits and into the atria. The elaborate pacemaking and conducting systems within the heart normally provide the required spacing between atrial and ventricular contractions.

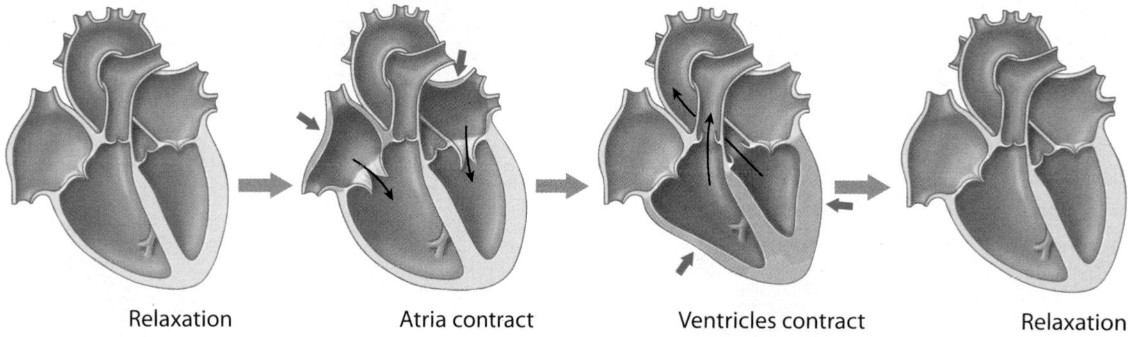
Relaxation
Atria contract
Ventricles contract
Relaxation

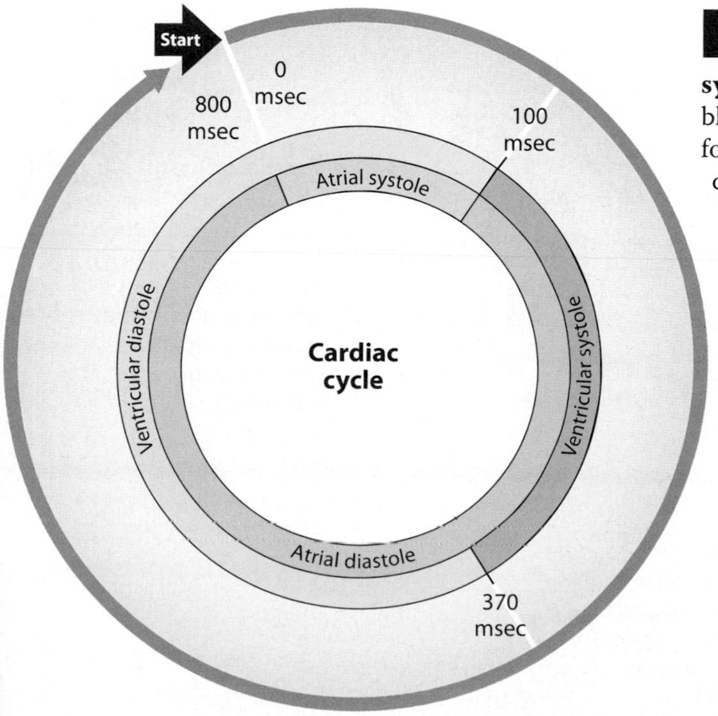
Start
0 msec
100 msec
370 msec
800 msec
Atrial systole
Ventricular systole
Atrial diastole
Ventricular diastole
Cardiac cycle

3 For any one chamber in the heart, the cardiac cycle can be divided into two phases: (1) systole and (2) diastole. During **systole** (SIS-tō-lē), or contraction, the chamber contracts and pushes blood into an adjacent chamber or into an arterial trunk. Systole is followed by **diastole** (dī-AS-tō-lē), or relaxation. During diastole, the chamber fills with blood and prepares for the next contraction. At a representative heart rate of 75 beats per minute (bpm), a sequence of systole and diastole in either the atria or the ventricles lasts 800 msec. For convenience, we will assume that the cardiac cycle is determined by the atria, and that it includes one cycle of atrial systole and atrial diastole. This section will examine the details of a representative cardiac cycle—how it is initiated and coordinated and how the pressures generated by the contracting chambers result in directional blood flow.

Module 19.9 Review

a. Define the cardiac cycle.

b. Give the alternate terms for heart contraction and heart relaxation.

c. When a chamber is relaxed, which phase is it in?

LO **19.9** Explain the complete round of cardiac systole and diastole.

The cardiac cycle creates pressure gradients that maintain blood flow

1 The phases of the cardiac cycle are diagrammed here for a heart rate of 75 beats per minute (bpm). When the heart rate increases, all the phases of the cardiac cycle are shortened. The greatest reduction occurs in the length of time spent in diastole. When the heart rate climbs from 75 bpm to 200 bpm, the time spent in systole drops by less than 40 percent, but the duration of diastole is reduced by almost 75 percent.

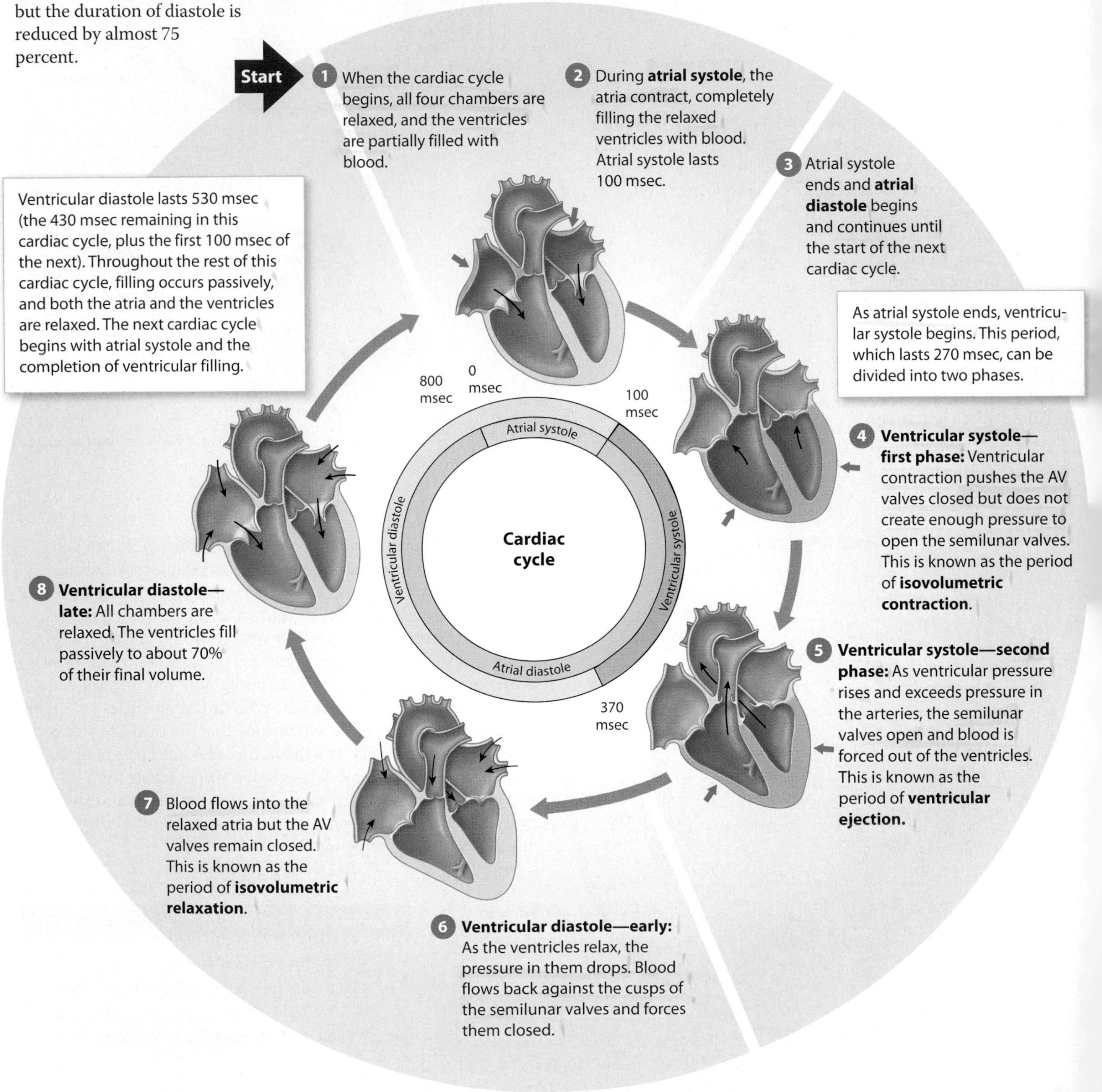
Start
1 When the cardiac cycle begins, all four chambers are relaxed, and the ventricles are partially filled with blood.
2 During **atrial systole**, the atria contract, completely filling the relaxed ventricles with blood. Atrial systole lasts 100 msec.
3 Atrial systole ends and **atrial diastole** begins and continues until the start of the next cardiac cycle.
As atrial systole ends, ventricular systole begins. This period, which lasts 270 msec, can be divided into two phases.
4 **Ventricular systole—first phase:** Ventricular contraction pushes the AV valves closed but does not create enough pressure to open the semilunar valves. This is known as the period of **isovolumetric contraction**.
5 **Ventricular systole—second phase:** As ventricular pressure rises and exceeds pressure in the arteries, the semilunar valves open and blood is forced out of the ventricles. This is known as the period of **ventricular ejection.**
6 **Ventricular diastole—early:** As the ventricles relax, the pressure in them drops. Blood flows back against the cusps of the semilunar valves and forces them closed.
7 Blood flows into the relaxed atria but the AV valves remain closed. This is known as the period of **isovolumetric relaxation**.
8 **Ventricular diastole—late:** All chambers are relaxed. The ventricles fill passively to about 70% of their final volume.
Ventricular diastole lasts 530 msec (the 430 msec remaining in this cardiac cycle, plus the first 100 msec of the next). Throughout the rest of this cardiac cycle, filling occurs passively, and both the atria and the ventricles are relaxed. The next cardiac cycle begins with atrial systole and the completion of ventricular filling.
0 msec
100 msec
370 msec
800 msec
Atrial systole
Ventricular systole
Atrial diastole
Ventricular diastole
Cardiac cycle

2 This graph plots the pressure changes within the aorta, the left atrium, and the left ventricle during the cardiac cycle. Although pressures are lower in the right atrium and right ventricle, the same principles apply; both sides of the heart contract at the same time, and they eject equal volumes of blood.

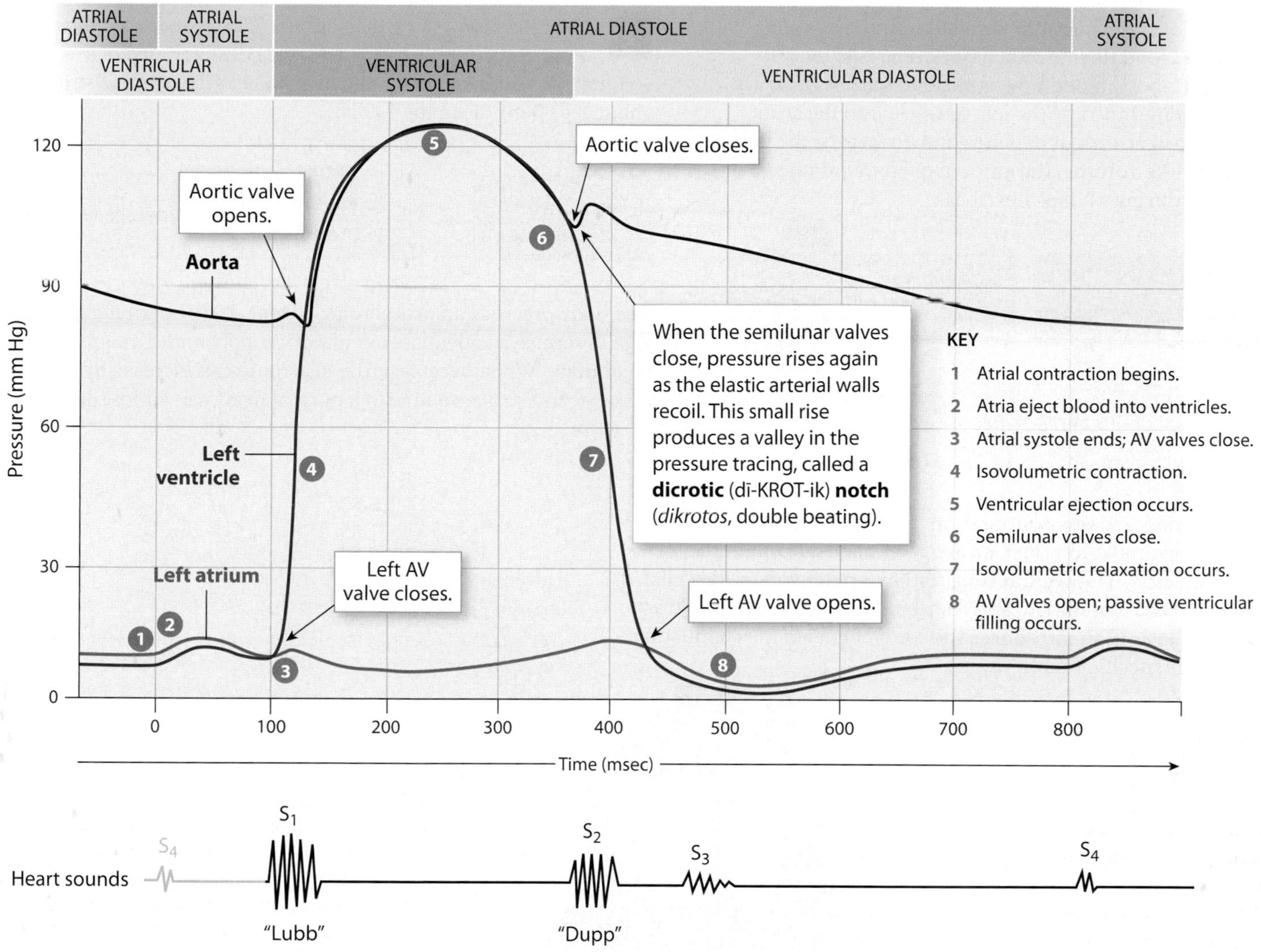

ATRIAL DIASTOLE
ATRIAL SYSTOLE
ATRIAL DIASTOLE
ATRIAL SYSTOLE
VENTRICULAR DIASTOLE
VENTRICULAR SYSTOLE
VENTRICULAR DIASTOLE
120
90
60
30
0
Pressure (mm Hg)
Aortic valve opens.
Aorta
Aortic valve closes.
Left ventricle
Left atrium
Left AV valve closes.
Left AV valve opens.
When the semilunar valves close, pressure rises again as the elastic arterial walls recoil. This small rise produces a valley in the pressure tracing, called a **dicrotic** (dī-KROT-ik) **notch** (*dikrotos*, double beating).
KEY
1 Atrial contraction begins.
2 Atria eject blood into ventricles.
3 Atrial systole ends; AV valves close.
4 Isovolumetric contraction.
5 Ventricular ejection occurs.
6 Semilunar valves close.
7 Isovolumetric relaxation occurs.
8 AV valves open; passive ventricular filling occurs.
0
100
200
300
400
500
600
700
800
Time (msec)
Heart sounds
S4
S1
S2
S3
S4
"Lubb"
"Dupp"

3 There are four heart sounds, designated as S_1 through S_4. If you listen to your own heart with a stethoscope, you will clearly hear the first and second heart sounds. The first heart sound, known as "lubb" (S_1), lasts a little longer than the second, called "dupp" (S_2). S_1, which marks the start of ventricular contraction, is produced as the AV valves close. S_2 occurs when the semilunar valves close. Third and fourth heart sounds are usually very faint and seldom are audible in healthy adults. These sounds are associated with blood flowing into the ventricles (S_3) and atrial contraction (S_4), rather than with valve action.

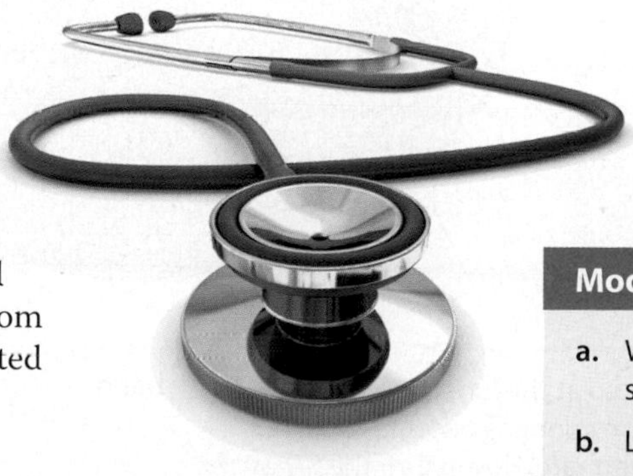

Module 19.10 Review

a. What are the two phases of ventricular systole?

b. List the phases of the cardiac cycle.

c. Is the heart always pumping blood when pressure in the left ventricle is rising? Explain.

LO 19.10 Explain the events of the cardiac cycle, and relate the heart sounds to specific events.

Module 19.11

The heart rate is established by the sinoatrial node and distributed by the conducting system

1 The goal of cardiovascular regulation is to maintain adequate blood flow to vital tissues. The best overall indicator of peripheral blood flow is **cardiac output (CO)**, the amount of blood pumped by the left ventricle into the aorta each minute. Cardiac output depends on two factors: the heart rate and the **stroke volume**, the amount of blood pumped out of the ventricle during a single heartbeat.

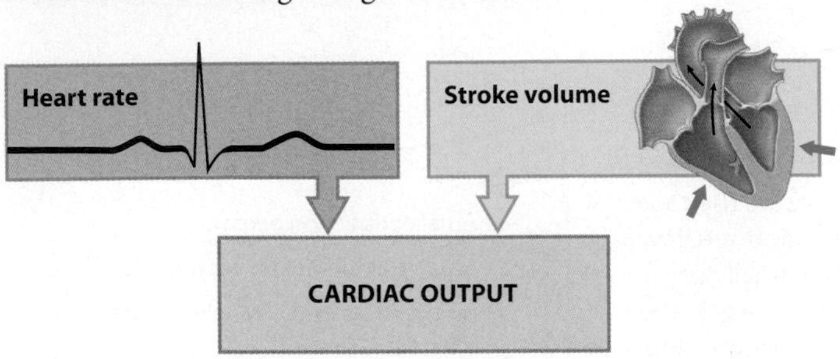
Heart rate
Stroke volume
CARDIAC OUTPUT

2 The calculation of cardiac output (CO) is very straightforward: you multiply the heart rate (HR) by the average stroke volume (SV). For example, if the heart rate is 75 bpm and stroke volume is 80 mL/beat, the CO is:

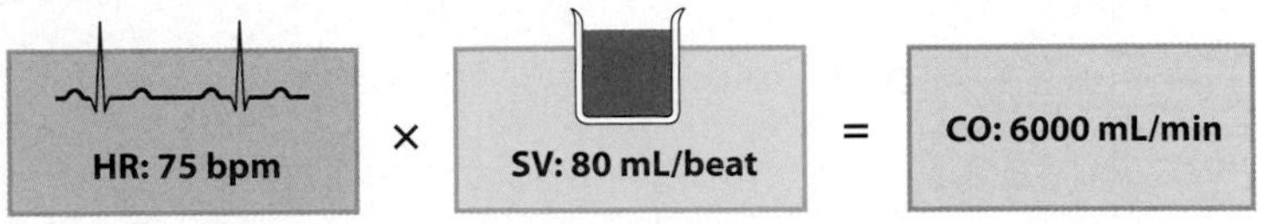
HR: 75 bpm
×
SV: 80 mL/beat
=
CO: 6000 mL/min

The body precisely adjusts cardiac output so that peripheral tissues receive an adequate circulatory supply under a variety of conditions. When necessary, the heart rate can increase by 250 percent, and stroke volume in a normal heart can almost double.

3 Cardiac muscle tissue contracts on its own, without neural or hormonal stimulation. This property is called **automaticity** (or autorhythmicity). The **conducting system** is a network of specialized cardiac muscle cells that initiate and distribute this stimulus to contract. The following illustration introduces the components of the conducting system and their specific functions.

1. Each heartbeat begins with an action potential generated at the **sinoatrial** (sī-nō-Ā-trē-al) **node**, or simply the **SA node**. The SA node is embedded in the posterior wall of the right atrium, near the entrance of the superior vena cava. The electrical impulse generated by this cardiac pacemaker is then distributed by other cells of the conducting system.

2. In the atria, conducting cells are found in **internodal pathways**, which distribute the contractile stimulus to atrial muscle cells as the impulse travels toward the ventricles.

3. The **atrioventricular (AV) node** is located at the junction between the atria and ventricles. The AV node also contains pacemaker cells, but they do not ordinarily affect the heart rate. However, if the SA node or internodal pathways are damaged, the heart will continue to beat because without commands from the SA node, the AV node will generate impulses at a rate of 40–60 bpm.

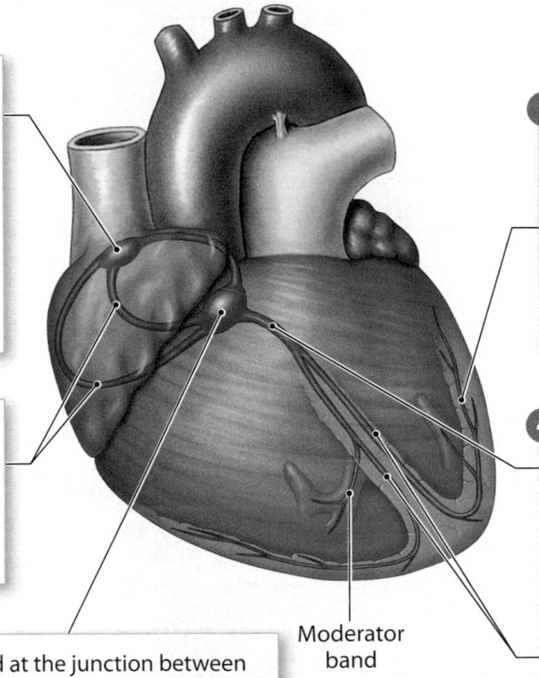
Moderator band

4. The AV node delivers the stimulus to the **AV bundle**, located within the interventricular septum. The AV bundle is normally the only electrical connection between the atria and the ventricles.

   The AV bundle leads to the right and left **bundle branches**. The left bundle branch, which supplies the massive left ventricle, is much larger than the right bundle branch. Both branches extend toward the apex of the heart, turn, and fan out deep to the endocardial surface.

5. **Purkinje fibers** are large-diameter conducting cells that propagate action potentials very rapidly—as fast as small myelinated axons. Purkinje cells are the final link in the distribution network and are responsible for the depolarization of the ventricular myocardial cells that triggers ventricular systole.

1 An action potential is generated at the SA node, and atrial activation begins.

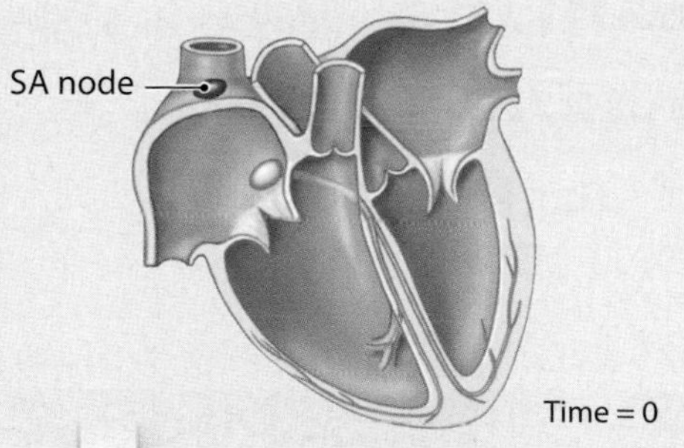
SA node
Time = 0

2 The stimulus spreads across the atrial surfaces by cell-to-cell contact within the internodal pathways and soon reaches the AV node.

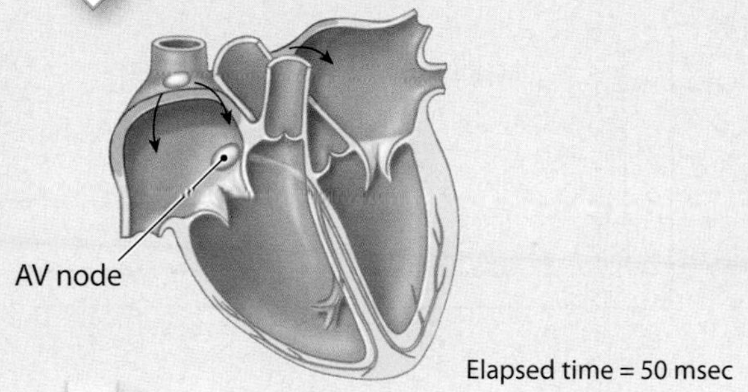
AV node
Elapsed time = 50 msec

3 A 100-msec delay occurs at the AV node. During this delay, atrial contraction occurs.

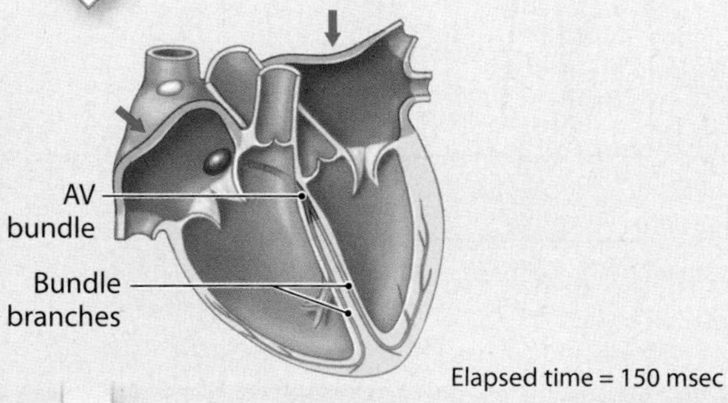
AV bundle
Bundle branches
Elapsed time = 150 msec

4 As atrial contraction is completed, the impulse travels along the interventricular septum within the AV bundle and the bundle branches to the Purkinje fibers and, by the moderator band, to the papillary muscles of the right ventricle. Ventricular contraction begins.

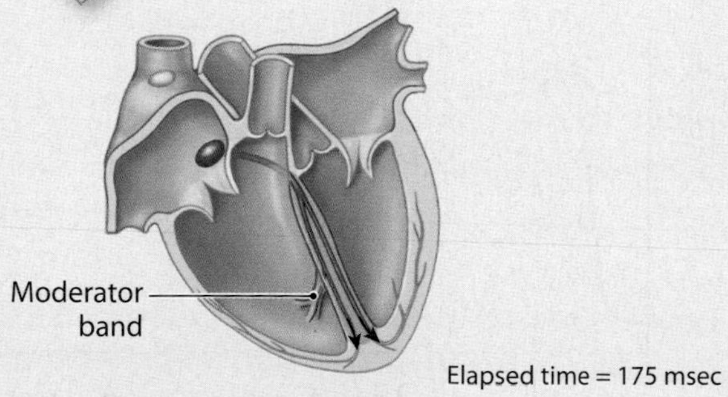
Moderator band
Elapsed time = 175 msec

5 The impulse is distributed by Purkinje fibers and relayed throughout the ventricular myocardium. Ventricular contraction reaches full force and proceeds to completion.

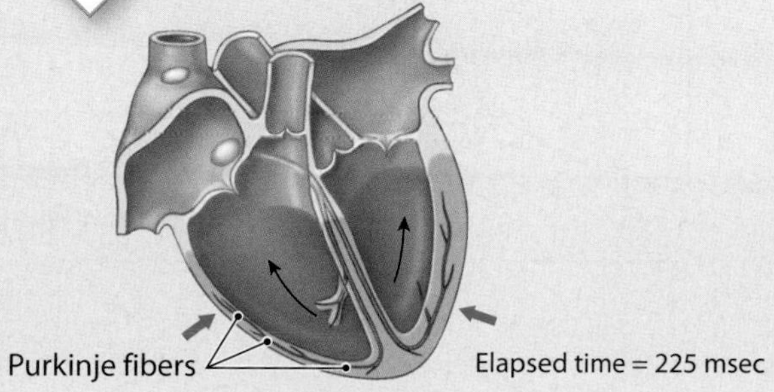
Purkinje fibers
Elapsed time = 225 msec

4 This sequence illustrates the distribution of the contractile stimulus and shows how the conducting system coordinates the contractions of the cardiac cycle. This conducting system allows the atria to contract before the ventricles. Although the cardiac cycle is continuous, we conventionally begin with atrial contraction, about 50 msec after an action potential is generated at the SA node.

The cells of the AV node can conduct impulses at a maximum rate of 230 per minute. Because each impulse results in a ventricular contraction, this value is the maximum normal heart rate. Even if the SA node generates impulses at a faster rate, the ventricles will still contract at 230 bpm. Higher heart rates occur only when the heart or the conducting system has been damaged or stimulated by drugs.

Module 19.11 Review

a. Define automaticity.

b. If the cells of the SA node failed to function, how would the heart rate be affected?

c. Why is it important for impulses from the atria to be delayed at the AV node before they pass into the ventricles?

LO **19.11** Describe the components and functions of the conducting system of the heart.

Module 19.12

Cardiac muscle cell contractions last longer than skeletal muscle fiber contractions primarily due to differences in membrane permeability

Muscle Cell Contractions

Skeletal Muscle

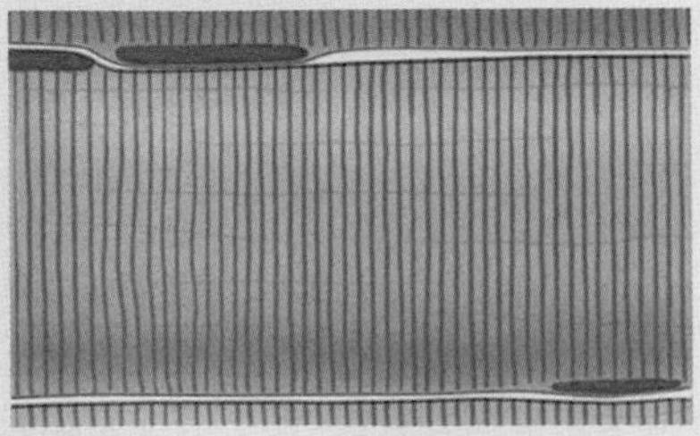

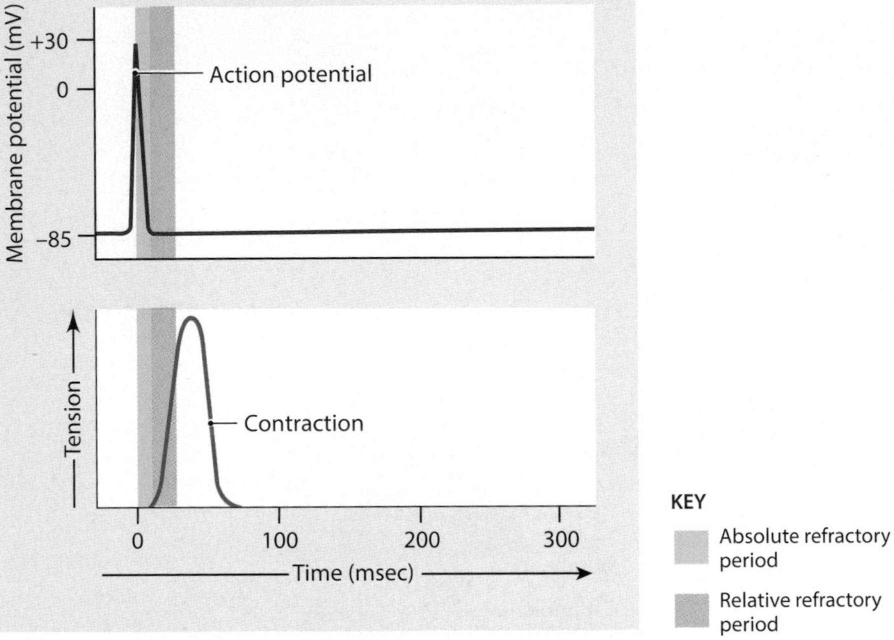
Membrane potential (mV)
+30
0
–85
Action potential
Tension
Contraction
0
100
200
300
Time (msec)
KEY
Absolute refractory period
Relative refractory period

1 In a skeletal muscle fiber, the action potential is relatively brief and ends as the related twitch contraction begins. The twitch contraction is short and ends as the sarcoplasmic reticulum reclaims the Ca^{2+} it released. Note that the refractory period ends before peak tension develops. As a result, twitches can summate and tetanus can occur.

Cardiac Muscle

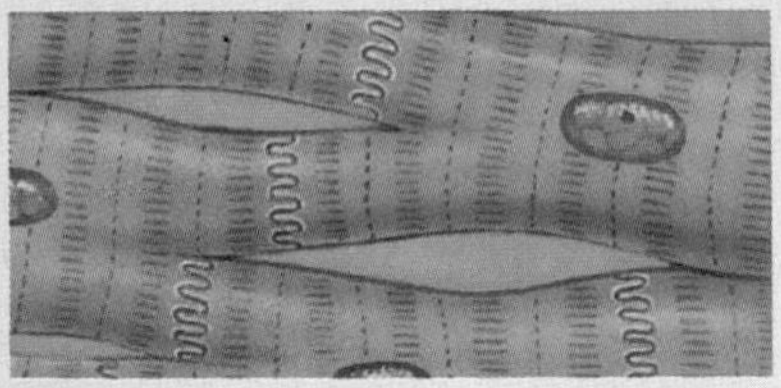

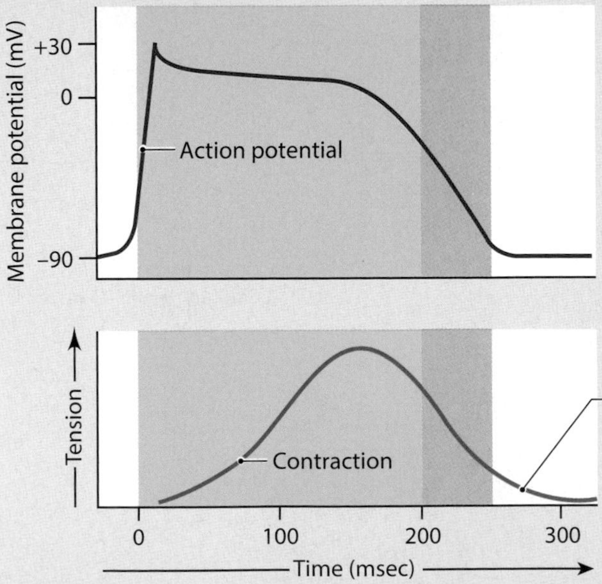
Membrane potential (mV)
+30
0
–90
Action potential
Tension
Contraction
0
100
200
300
Time (msec)

2 In a cardiac muscle cell, the action potential is prolonged because calcium ions continue to enter the cell for an extended period. As a result, the period of active muscle cell contraction is also extended. Note that the refractory period continues until relaxation is well under way. Thus, summation cannot occur, and tetanic contractions do not occur in normal cardiac muscle tissue. This feature is vital: A heart in tetany could not pump blood.

With a single twitch lasting 250 msec or longer, a normal cardiac muscle cell could reach 300–400 contractions per minute under maximum stimulation. The normal heart rate never gets that high because the stimulus for contraction cannot spread that quickly through the heart muscle.

Cardiac Action Potentials

3 The action potential in a cardiac muscle cell can be divided into three stages.

Rapid Depolarization

The stage of rapid depolarization in a cardiac muscle cell resembles that in a skeletal muscle fiber. At threshold, voltage-gated sodium channels open, and the membrane suddenly becomes permeable to Na^+. The result is a massive influx of sodium ions and the rapid depolarization of the sarcolemma. The channels involved are called **fast sodium channels**, because they open quickly and remain open for only a few milliseconds.

Plateau

During the plateau, the membrane potential remains near 0 mV. Two opposing factors are involved. As the membrane potential approaches +30 mV, the voltage-gated sodium channels close and the cell begins actively pumping Na^+ out of the cell. However, as the sodium channels are closing, voltage-gated calcium channels are opening. These channels are called **slow calcium channels**, because they open slowly and remain open for a relatively long period—roughly 175 msec. While the slow calcium channels are open, the entry of Ca^{2+} nearly balances the loss of Na^+, and the membrane potential hovers near 0 mV.

Repolarization

After approximately 175 msec, slow calcium channels begin closing, and slow potassium channels begin opening. As the channels open, potassium ions (K^+) rush out of the cell, and the net result is a period of rapid repolarization that restores the resting potential.

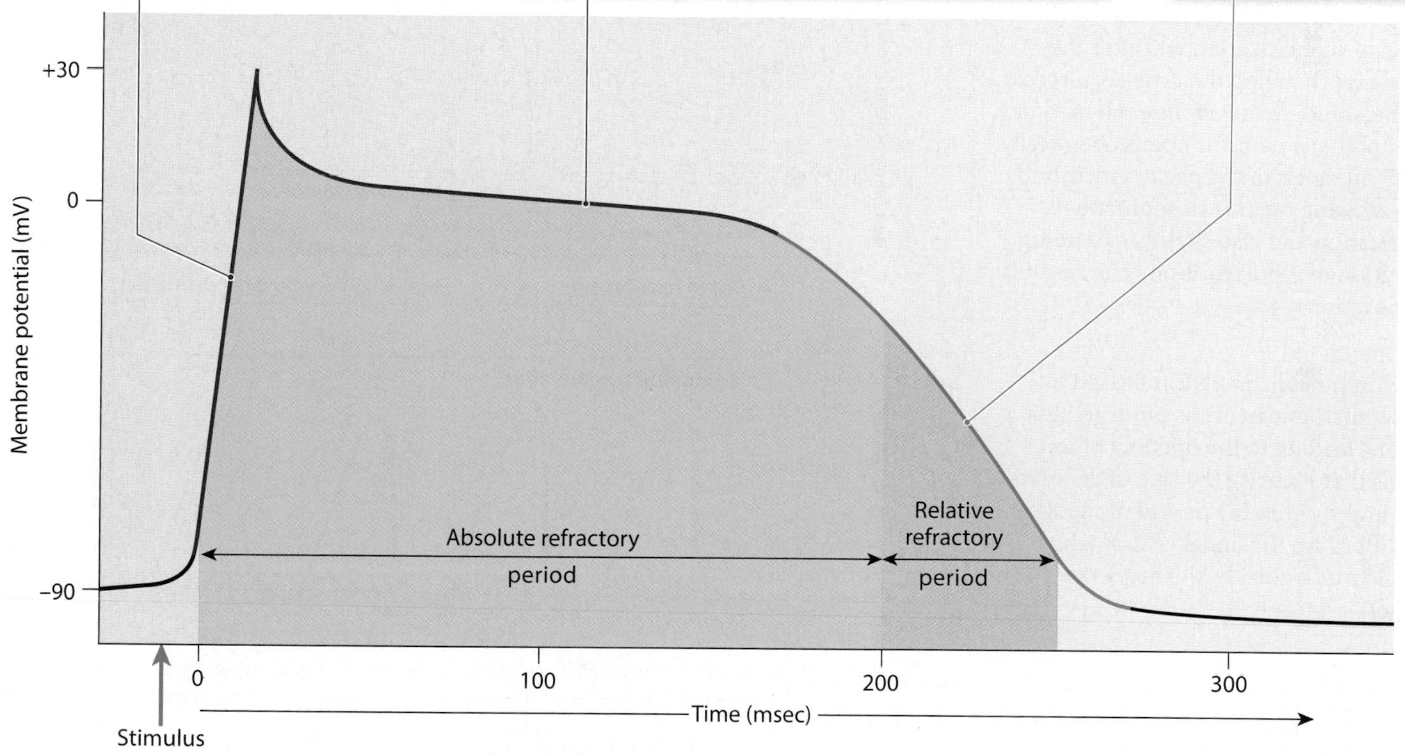
+30
0
−90
Membrane potential (mV)
Absolute refractory
period
Relative
refractory
period
0
100
200
300
Time (msec)
Stimulus

Module 19.12 Review

a. Why does tetany not occur in cardiac muscle?

b. List the three stages of an action potential in a cardiac muscle cell.

c. Describe slow calcium channels and the significance of their activity.

LO 19.12 Describe an action potential in cardiac muscle, and explain the role of calcium ions.

Module 19.13

The intrinsic heart rate can be altered by autonomic activity

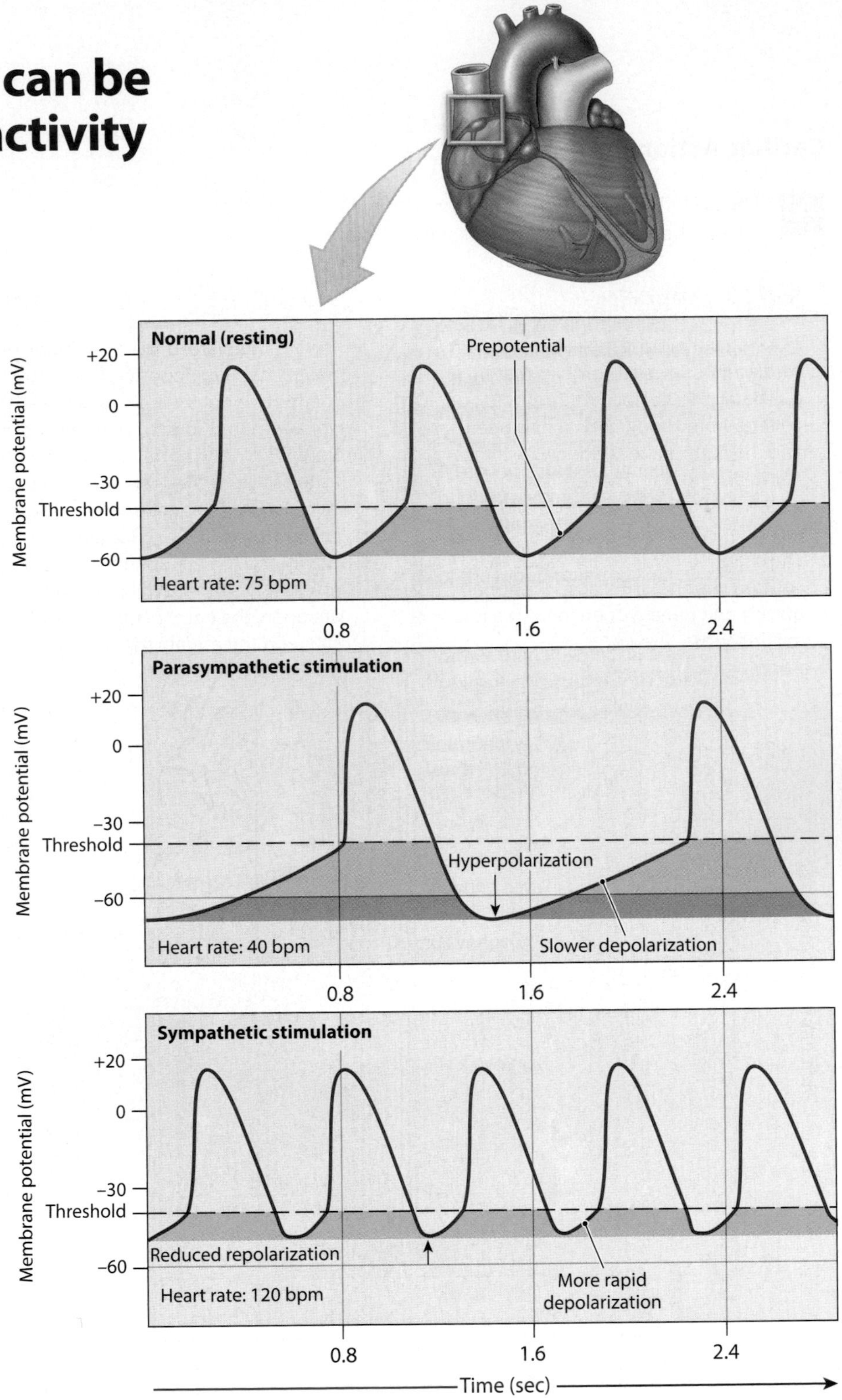

Normal (resting)
Prepotential
+20
0
−30
Threshold
−60
Heart rate: 75 bpm
0.8
1.6
2.4
Membrane potential (mV)
Parasympathetic stimulation
Hyperpolarization
Slower depolarization
Heart rate: 40 bpm
Sympathetic stimulation
Reduced repolarization
More rapid depolarization
Heart rate: 120 bpm
Time (sec)

1 Cells of the SA and AV nodes cannot maintain a stable resting potential. After each repolarization, the membrane gradually drifts toward threshold. This gradual spontaneous depolarization is called a **prepotential** or **pacemaker potential**. The rate of spontaneous depolarization is fastest at the SA node, which without neural or hormonal stimulation generates action potentials at a rate of 80–100 per minute. Because the SA node reaches threshold first, it establishes the heart rate—the impulse generated by the SA node brings the AV nodal cells to threshold before the prepotential of the AV nodal cells can do so.

2 Any factor that changes the rate of spontaneous depolarization or the duration of repolarization will alter the heart rate by changing the time required to reach threshold. Acetylcholine released by parasympathetic neurons opens chemically gated K^+ channels in the plasma membrane, thereby slowing the rate of spontaneous depolarization and also slightly extending the duration of repolarization. The result is a decline in heart rate.

3 Norepinephrine (NE) released by sympathetic neurons binds to beta-1 receptors, leading to the opening of ion channels that increase the rate of depolarization and shorten the period of repolarization. Because the nodal cells reach threshold more quickly, the heart rate increases.

4 Every person has a characteristic resting heart rate that varies with age, general health, and physical conditioning. However, there is a normal range of heart rates. The American Heart Association considers 60–100 bpm to be the normal range of resting heart rates.

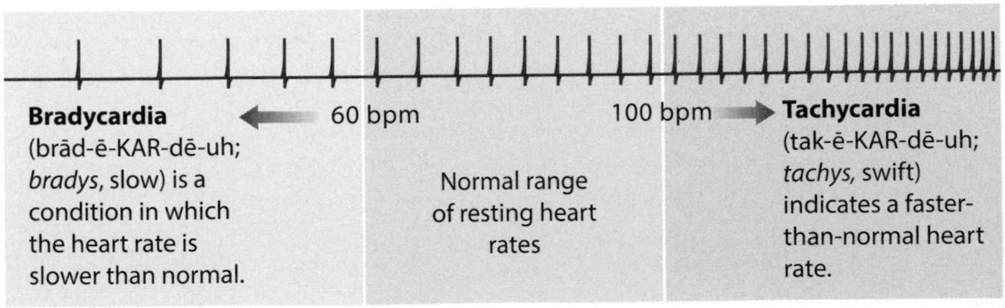

Bradycardia (brād-ē-KAR-dē-uh; *bradys*, slow) is a condition in which the heart rate is slower than normal.
60 bpm
Normal range of resting heart rates
100 bpm
Tachycardia (tak-ē-KAR-dē-uh; *tachys*, swift) indicates a faster-than-normal heart rate.

5 The cardiac centers of the medulla oblongata contain the autonomic headquarters for cardiac control. These centers innervate the heart by means of the cardiac plexus.

The **cardioinhibitory center** controls the parasympathetic neurons that slow the heart rate.

The **cardioacceleratory center** controls sympathetic neurons that increase the heart rate.

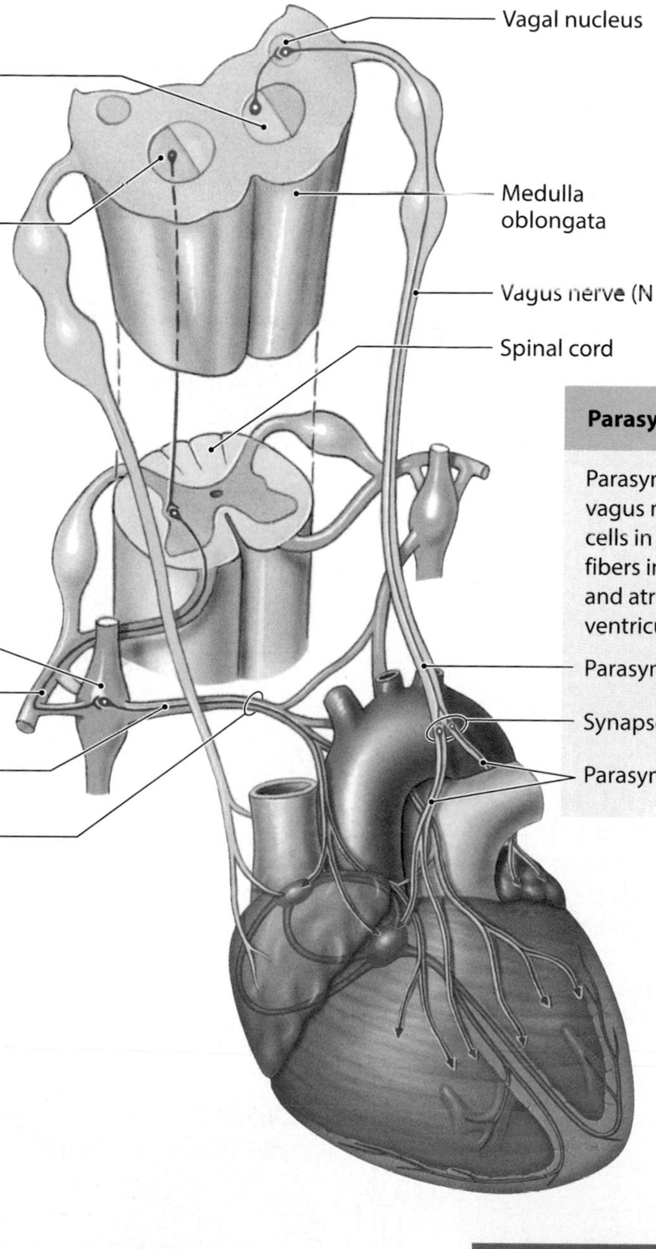

Vagal nucleus
Medulla oblongata
Vagus nerve (N X)
Spinal cord
Cervical sympathetic ganglion
Sympathetic preganglionic fiber
Sympathetic postganglionic fiber
Cardiac nerve
Parasympathetic preganglionic fiber
Synapses in cardiac plexus
Parasympathetic postganglionic fibers

Sympathetic

Sympathetic innervation arrives in postganglionic fibers within the cardiac nerves. These fibers innervate the nodes, the conducting system, and the atrial and ventricular myocardium.

Parasympathetic

Parasympathetic innervation arrives in the vagus nerve and synapses with ganglion cells in the cardiac plexus. Postganglionic fibers innervate the SA node, AV node, and atrial musculature. Innervation of the ventricular musculature is very limited.

Reflex pathways and input from higher centers, especially the sympathetic and parasympathetic headquarters in the hypothalamus, regulate the cardiac centers. Autonomic tone exists at the heart, as both divisions are chronically active. The normal resting heart rate is somewhat slower than the intrinsic SA nodal stimulation rate of 80–100 bpm, because at rest the effects of parasympathetic innervation predominate.

Module 19.13 Review

a. Caffeine has effects on conducting cells and contractile cells that are similar to those of NE. What effect would drinking large amounts of caffeinated beverages have on the heart rate?

b. Compare bradycardia with tachycardia.

c. Describe the sites and actions of the cardioinhibitory and cardioacceleratory centers.

LO **19.13** Describe the factors affecting the heart rate.

Stroke volume depends on the relationship between end-diastolic volume and end-systolic volume

1 The **stroke volume** of the heart can be compared to pumping water with a manual pump. The amount pumped varies with the amount of movement of the pump handle. Although the heart contains two large pumps—the right and left ventricles—we can use a single pump as a model because the two work in the same way and pump equal amounts of blood.

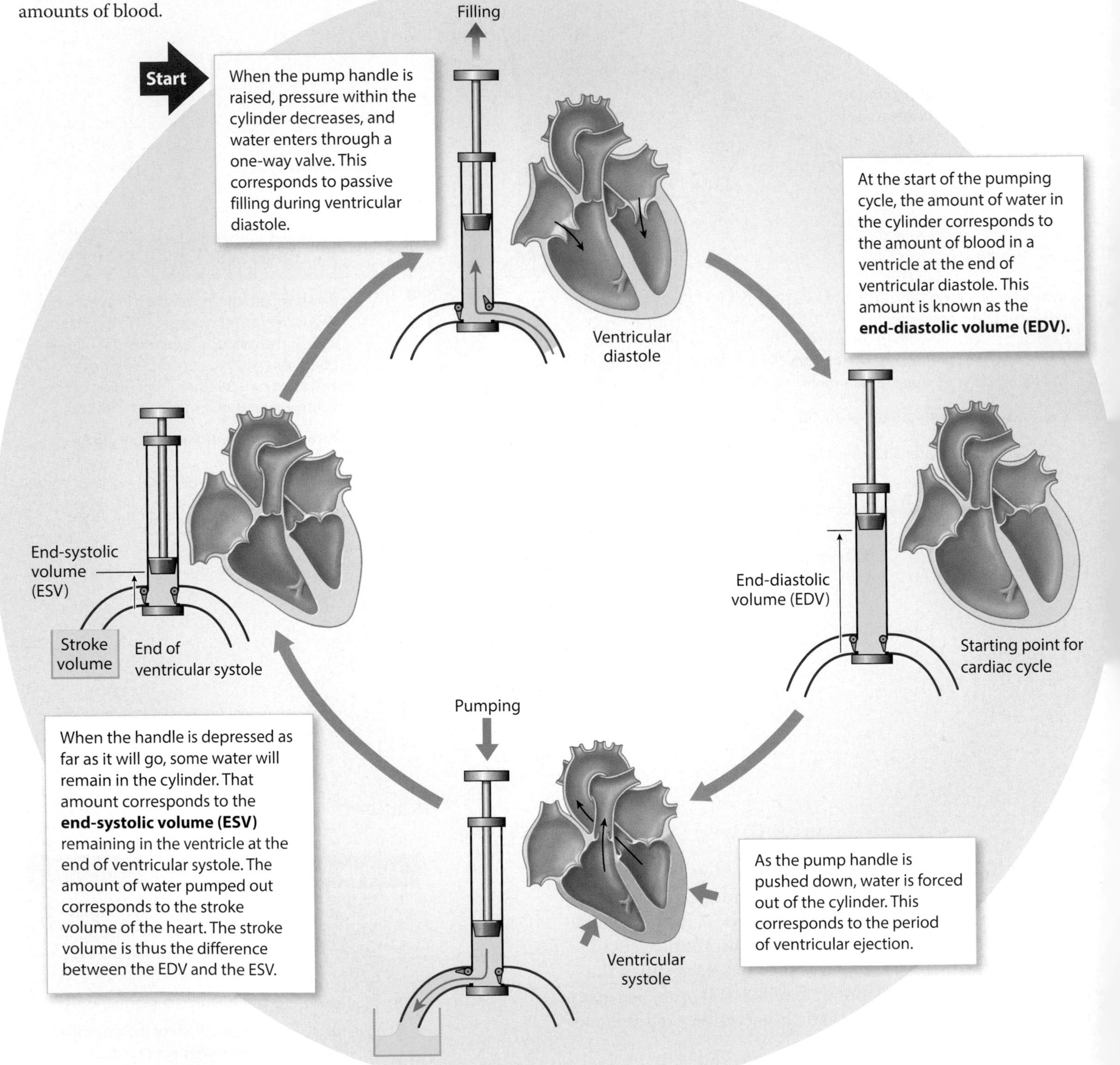
Filling
Start
When the pump handle is raised, pressure within the cylinder decreases, and water enters through a one-way valve. This corresponds to passive filling during ventricular diastole.
Ventricular diastole
At the start of the pumping cycle, the amount of water in the cylinder corresponds to the amount of blood in a ventricle at the end of ventricular diastole. This amount is known as the **end-diastolic volume (EDV).**
End-diastolic volume (EDV)
Starting point for cardiac cycle
End-systolic volume (ESV)
Stroke volume
End of ventricular systole
Pumping
When the handle is depressed as far as it will go, some water will remain in the cylinder. That amount corresponds to the **end-systolic volume (ESV)** remaining in the ventricle at the end of ventricular systole. The amount of water pumped out corresponds to the stroke volume of the heart. The stroke volume is thus the difference between the EDV and the ESV.
Ventricular systole
As the pump handle is pushed down, water is forced out of the cylinder. This corresponds to the period of ventricular ejection.

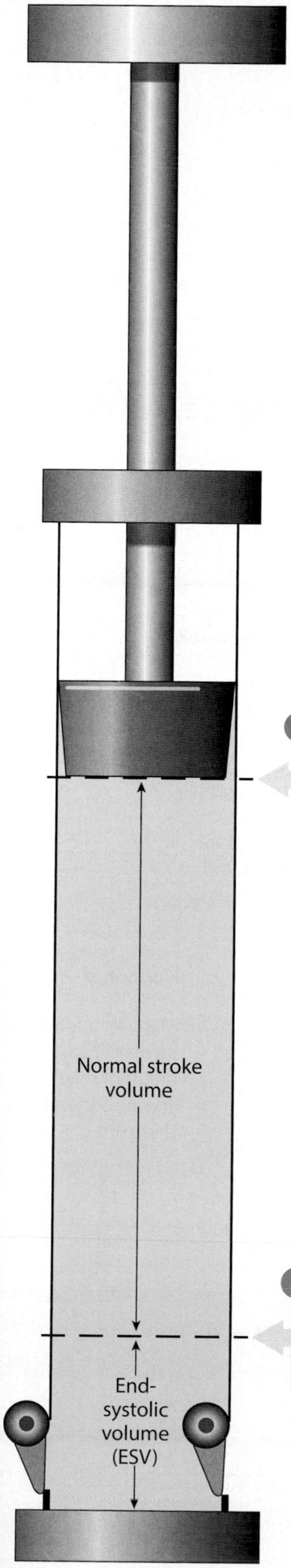
Normal stroke volume
End-systolic volume (ESV)

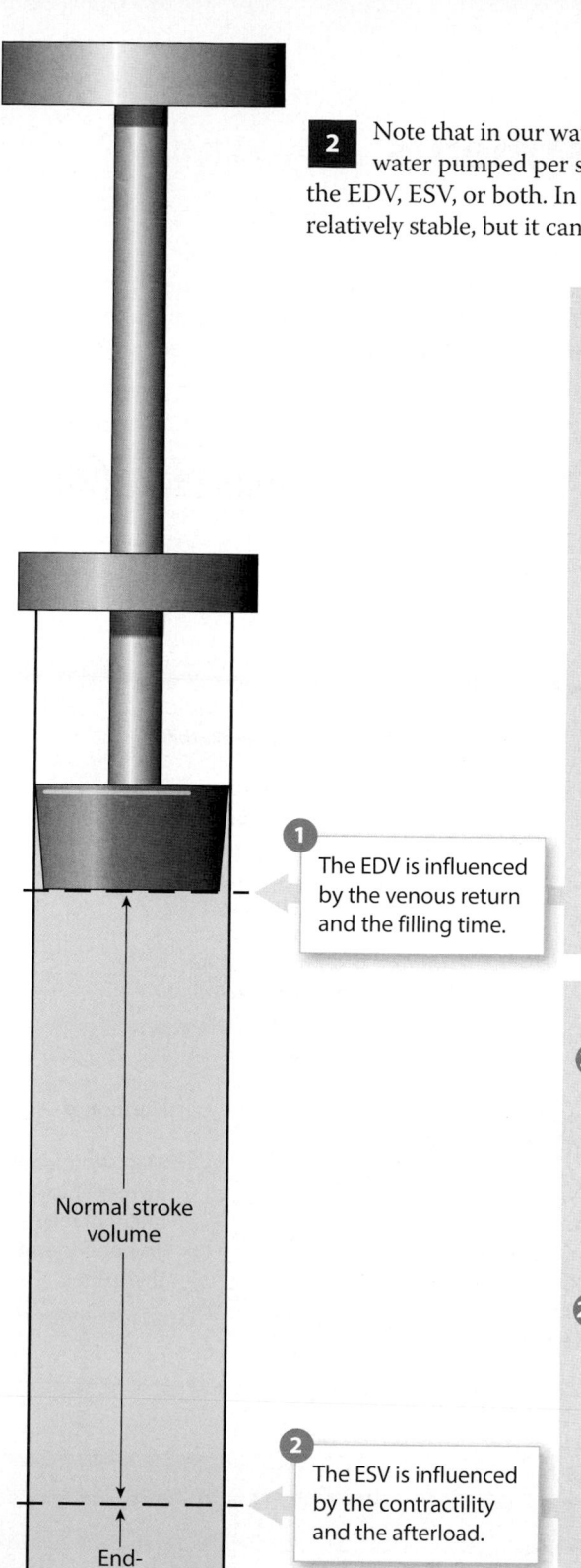

2 Note that in our water pump example, the amount of water pumped per stroke could be changed by altering the EDV, ESV, or both. In the resting heart, stroke volume is relatively stable, but it can increase substantially if necessary.

1 The EDV is influenced by the venous return and the filling time.

1a **Venous return** is the amount of blood delivered to the right atrium by the venae cavae and coronary sinus. Blood volume, muscular activity, and the rate of blood flow through peripheral capillaries are the major factors that determine venous return. When blood volume goes down significantly, venous return is reduced. When skeletal muscles contract and compress adjacent veins, or when peripheral tissue activity increases (which accelerates regional blood flow), venous return increases.

The **filling time** is the duration of ventricular diastole. Assuming that venous return remains constant, slowing the heart rate increases the EDV, and increasing the heart rate decreases the EDV.

1b The amount of myocardial stretching is called the **preload**. The greater the EDV, the larger the preload. When the EDV is very small, the sarcomeres in the cardiac muscle cells are too short to develop much power. As the EDV increases and the sarcomeres approach optimal length, the ventricular muscle cells can contract more efficiently and produce more powerful contractions and eject more blood. This relationship—greater EDV = greater stroke volume—is called the *Frank-Starling law of the heart*.

2 The ESV is influenced by the contractility and the afterload.

2a The **contractility** is the amount of force produced during a contraction at a given amount of preload. Sympathetic stimulation increases contractility. Many hormones—including epinephrine, norepinephrine, thyroid hormones, and glucagon—increase contractility to some degree. Many drugs used in clinical practice are intended to reduce contractility; examples include the "beta-blockers" such as propranolol or timolol and calcium channel blockers such as nifedipine or verapamil.

2b The **afterload** is the amount of tension the contracting ventricle must produce to force open the semilunar valve and eject blood. The greater the afterload, the longer the period of isovolumetric contraction, the shorter the duration of ventricular ejection, and the larger the ESV. In other words, as the afterload increases, the stroke volume decreases. Afterload is increased by any factor that restricts blood flow through the arterial system. For example, vasoconstriction (the narrowing of peripheral blood vessels) increases the afterload, whereas vasodilation (the relaxation and widening of peripheral blood vessels) decreases the afterload.

Module 19.14 Review

a. Define end-diastolic volume (EDV) and end-systolic volume (ESV).

b. What effect would an increase in venous return have on the stroke volume?

c. What effect would an increase in sympathetic stimulation of the heart have on the end-systolic volume (ESV)?

LO **19.14** Describe the variables that influence stroke volume.

Module 19.15

Cardiac output is regulated by adjustments in heart rate and stroke volume

1 As this figure indicates, cardiac output varies widely to meet metabolic demands. The body must closely monitor and tightly control those variations.

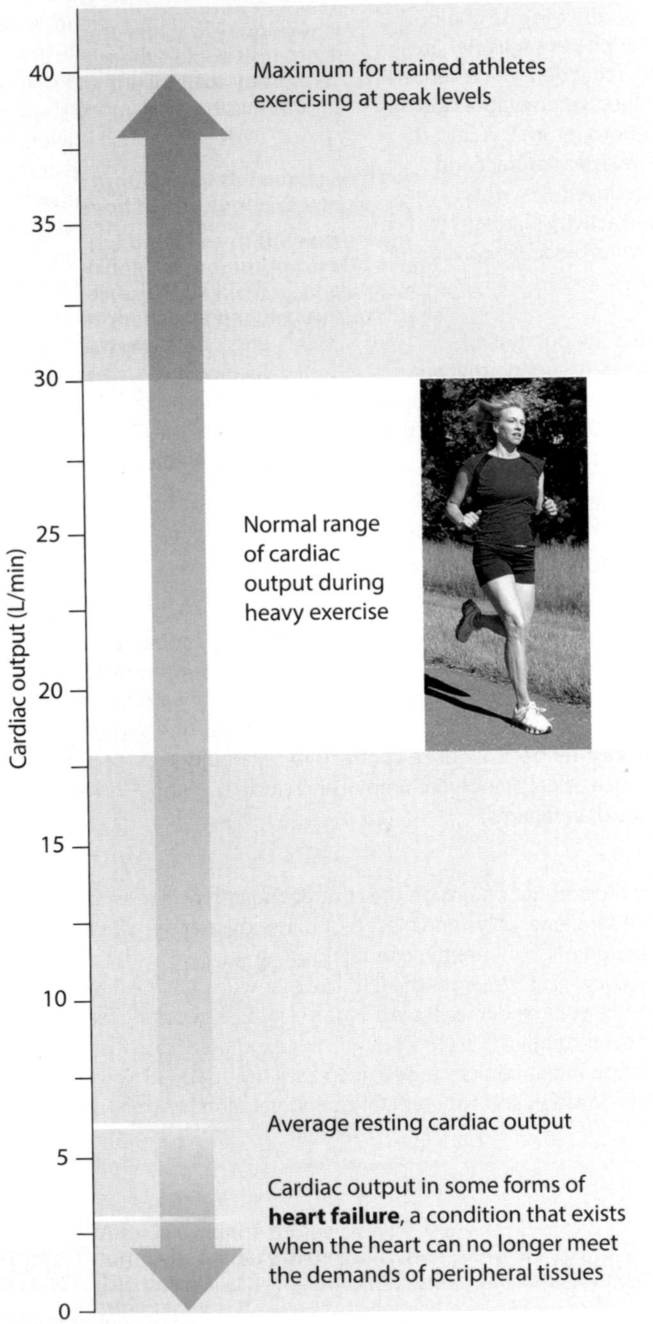

Cardiac output (L/min)
0
5
10
15
20
25
30
35
40
Maximum for trained athletes exercising at peak levels
Normal range of cardiac output during heavy exercise
Average resting cardiac output
Cardiac output in some forms of **heart failure**, a condition that exists when the heart can no longer meet the demands of peripheral tissues

2 This diagram summarizes the important information presented thus far concerning the factors that affect cardiac output. Review this carefully before proceeding to the next section.

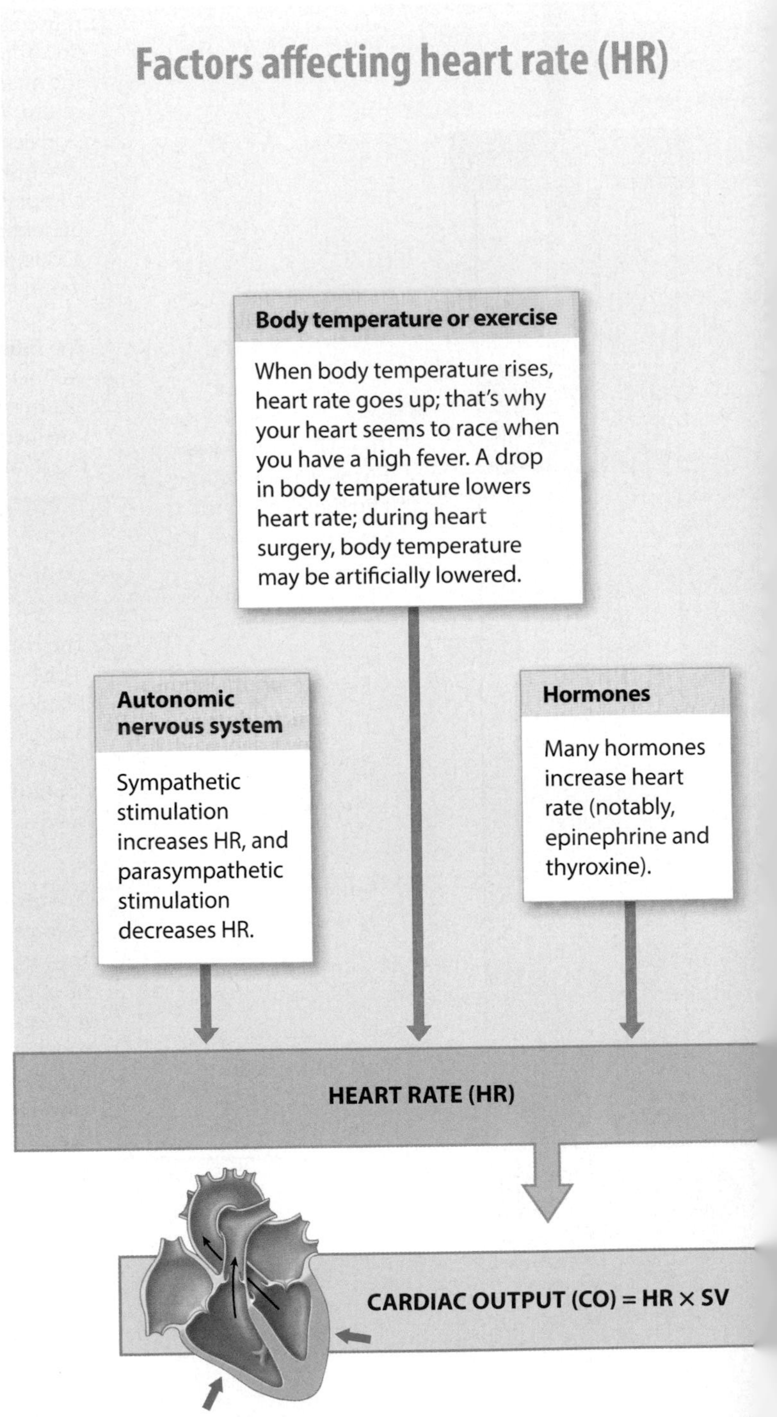

Factors affecting heart rate (HR)
Body temperature or exercise
When body temperature rises, heart rate goes up; that's why your heart seems to race when you have a high fever. A drop in body temperature lowers heart rate; during heart surgery, body temperature may be artificially lowered.
Autonomic nervous system
Sympathetic stimulation increases HR, and parasympathetic stimulation decreases HR.
Hormones
Many hormones increase heart rate (notably, epinephrine and thyroxine).
HEART RATE (HR)
CARDIAC OUTPUT (CO) = HR × SV

Factors affecting stroke volume (SV)

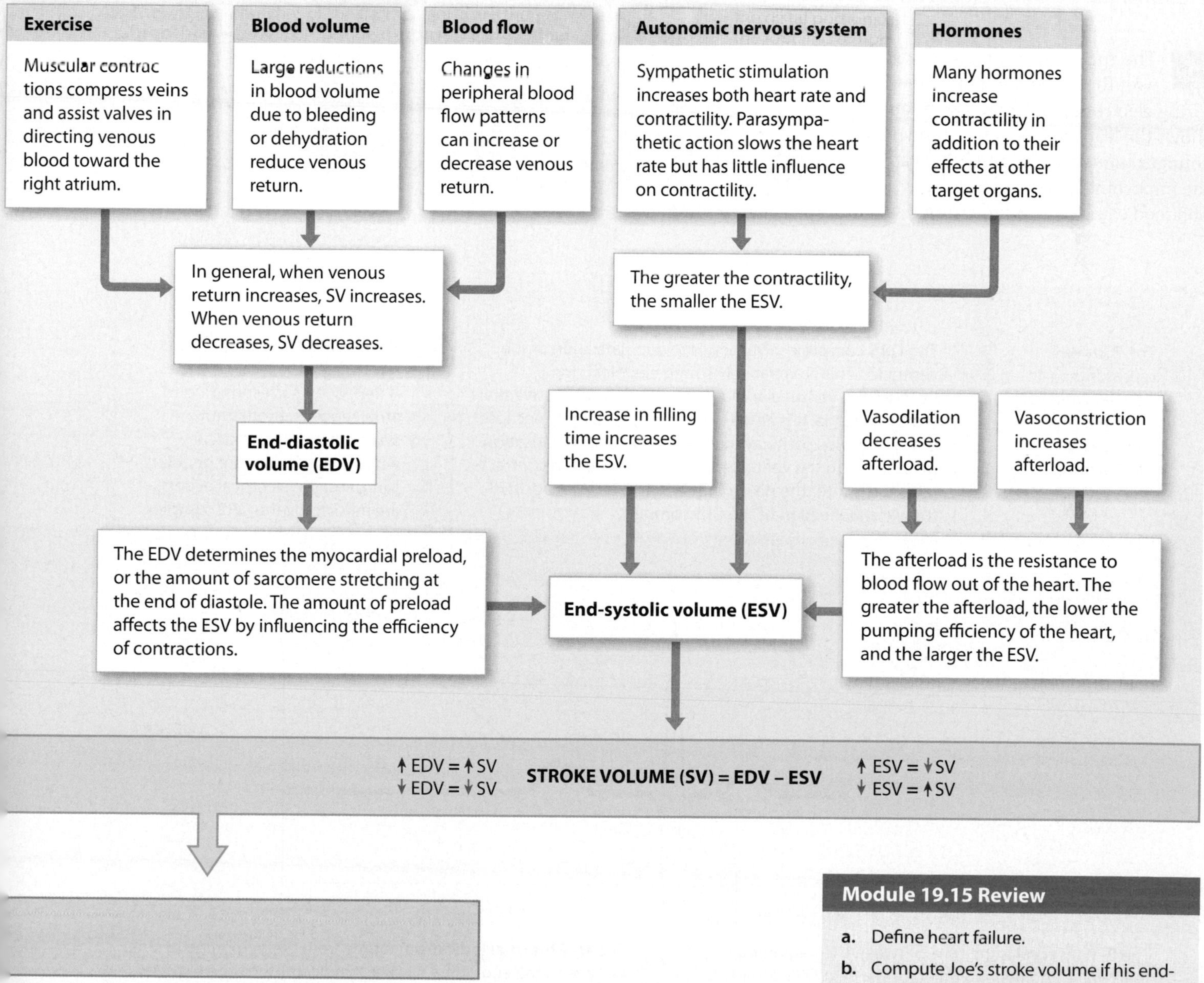
Exercise
Muscular contractions compress veins and assist valves in directing venous blood toward the right atrium.
Blood volume
Large reductions in blood volume due to bleeding or dehydration reduce venous return.
Blood flow
Changes in peripheral blood flow patterns can increase or decrease venous return.
Autonomic nervous system
Sympathetic stimulation increases both heart rate and contractility. Parasympathetic action slows the heart rate but has little influence on contractility.
Hormones
Many hormones increase contractility in addition to their effects at other target organs.
In general, when venous return increases, SV increases. When venous return decreases, SV decreases.
The greater the contractility, the smaller the ESV.
End-diastolic volume (EDV)
Increase in filling time increases the ESV.
Vasodilation decreases afterload.
Vasoconstriction increases afterload.
The EDV determines the myocardial preload, or the amount of sarcomere stretching at the end of diastole. The amount of preload affects the ESV by influencing the efficiency of contractions.
End-systolic volume (ESV)
The afterload is the resistance to blood flow out of the heart. The greater the afterload, the lower the pumping efficiency of the heart, and the larger the ESV.
↑ EDV = ↑ SV
↓ EDV = ↓ SV
STROKE VOLUME (SV) = EDV – ESV
↑ ESV = ↓ SV
↓ ESV = ↑ SV

Module 19.15 Review

a. Define heart failure.

b. Compute Joe's stroke volume if his end-systolic volume (ESV) is 40 mL and his end-diastolic volume (EDV) is 125 mL.

c. Why is it a potential problem if the heart beats too rapidly?

LO 19.15 Explain how stroke volume and cardiac output are coordinated.

Normal and abnormal cardiac activity can be detected in an electrocardiogram

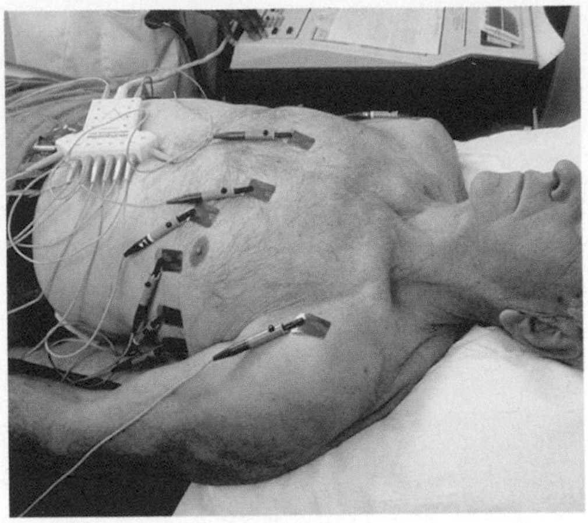

The electrical events occurring in the heart are powerful enough to be detected by electrodes on the surface of the body. A recording of these events over time is an **electrocardiogram** (e-lek-trō-KAR-dē-ō-gram), also called an **ECG** or **EKG**. Clinicians can use an ECG to assess the performance of specific nodal, conducting, and contractile components. When a portion of the heart has been damaged by a heart attack, for example, the ECG will reveal an abnormal pattern of impulse conduction.

1 The appearance of the ECG varies with the placement of the monitoring electrodes, or leads. The photo shows the leads in one of the standard configurations and the graph indicates the important features of an ECG obtained using that configuration.

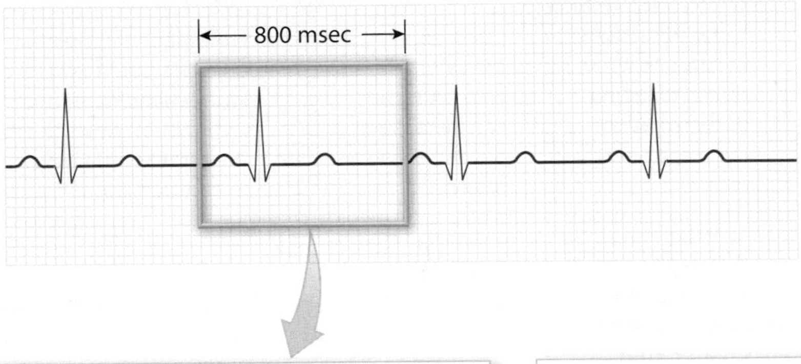
800 msec

The **P wave** represents depolarization of the atria, which causes atrial contraction.

The **QRS complex** corresponds to depolarization of the ventricles. This is a relatively strong electrical signal, because the ventricular muscle is much more massive than that of the atria. It is also a complex signal, largely because of the complex pathway that the spread of depolarization takes through the ventricles. The ventricles begin contracting shortly after the peak of the **R wave**, the first positive (upward) deflection of the QRS complex.

The smaller **T wave** represents repolarization of the ventricles. A deflection corresponding to atrial repolarization is not apparent, because it occurs while the ventricles are depolarizing, and the electrical events are masked by the QRS complex.

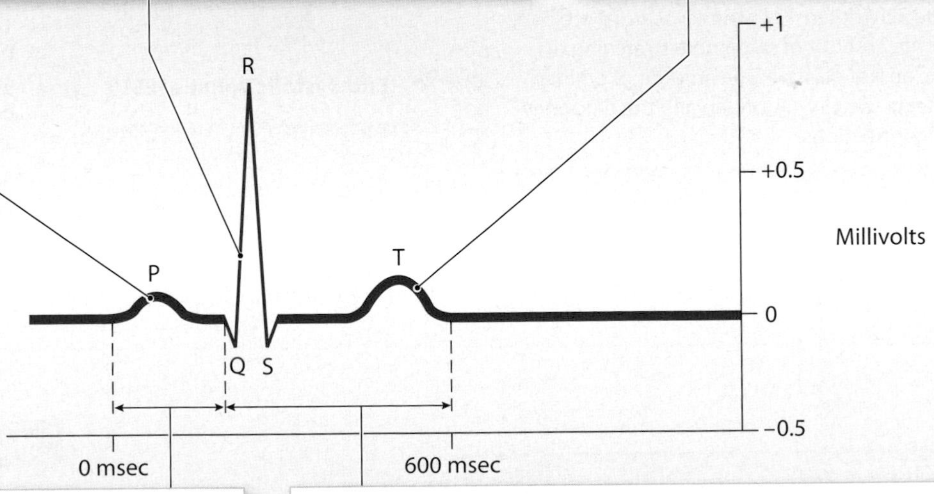
R
P
T
Q
S
+1
+0.5
0
−0.5
Millivolts
0 msec
600 msec

The **P–R interval** is the time between the beginning of the P wave and the beginning of the next QRS complex. It does not extend just to R, because in abnormal ECGs the peak at R can be difficult to determine. Extension of the P–R interval to more than 200 msec can indicate damage to the conducting pathways or AV node.

The **Q–T interval** indicates the time from the beginning of the Q wave to the end of the T wave. It indicates the time required for the ventricles to undergo a single cycle of depolarization and repolarization. The Q–T interval can be lengthened by electrolyte disturbances, some medications, conduction problems, coronary ischemia, or myocardial damage.

2 Despite the variety of sophisticated equipment available to assess or visualize cardiac function, in the majority of cases the ECG provides the most important diagnostic information. ECG analysis is especially useful in detecting and diagnosing **cardiac arrhythmias** (ā-RITH-mē-uz)—abnormal patterns of cardiac electrical activity. Momentary arrhythmias are not inherently dangerous; about 5 percent of healthy people experience a few abnormal heartbeats each day. Clinical problems appear when arrhythmias reduce the pumping efficiency of the heart. Several important examples of arrhythmias are described here.

Premature Atrial Contractions (PACs)

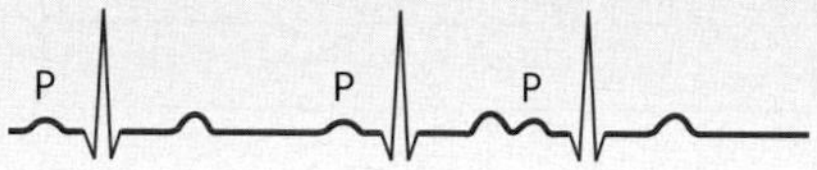
P
P
P

Premature atrial contractions (PACs) often occur in healthy people. In a PAC, the normal atrial rhythm is momentarily interrupted by a "surprise" atrial contraction. Stress, caffeine, and various drugs may increase the incidence of PACs, presumably by increasing the permeabilities of the SA pacemakers. The impulse spreads along the conduction pathway, and a normal ventricular contraction follows the atrial beat.

Paroxysmal Atrial Tachycardia (PAT)

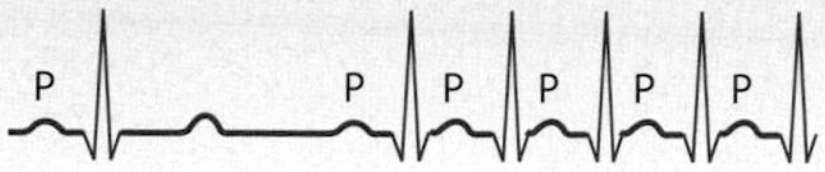
P
P
P
P
P
P

In **paroxysmal** (par-ok-SIZ-mal) **atrial tachycardia**, or **PAT**, a premature atrial contraction triggers a flurry of atrial activity. The ventricles are still able to keep pace, and the heart rate jumps to about 180 bpm.

Atrial Fibrillation (AF)

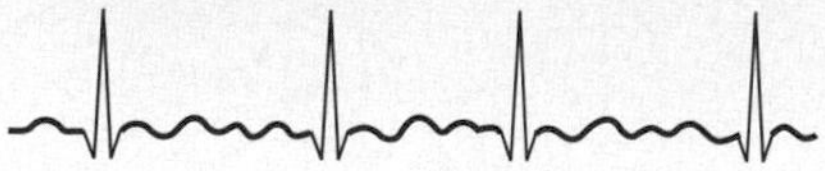

During **atrial fibrillation** (fib-ri-LĀ-shun), the impulses move over the atrial surface at rates of perhaps 500 bpm. The atrial wall quivers instead of producing an organized contraction. The ventricular rate cannot follow the atrial rate and may remain within normal limits. Even though the atria are now nonfunctional, their contribution to ventricular end-diastolic volume is so small that the condition may go unnoticed in older individuals.

Premature Ventricular Contractions (PVCs)

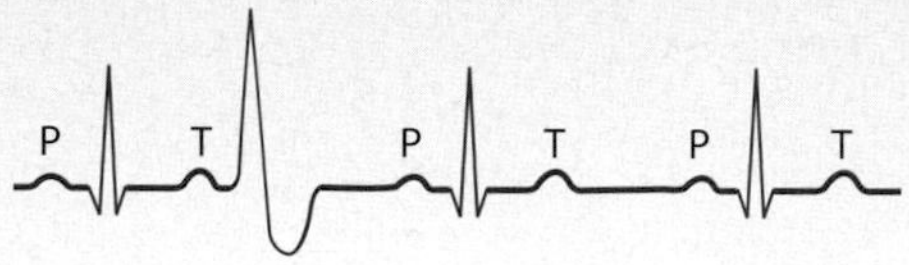
P
T
P
T
P
T

Premature ventricular contractions (PVCs) occur when a Purkinje cell or ventricular myocardial cell depolarizes to threshold and triggers a premature contraction. Single PVCs are common and not dangerous. The cell responsible is called an **ectopic pacemaker**, which is a pacemaker other than the SA node. The frequency of PVCs can be increased by exposure to epinephrine, to other stimulatory drugs, or to ionic changes that depolarize cardiac muscle cell membranes.

Ventricular Tachycardia (VT)

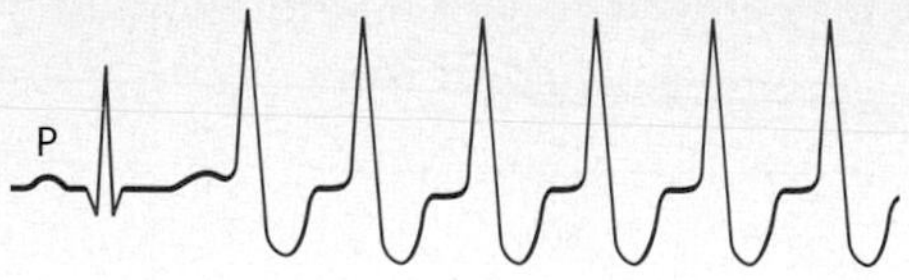
P

Ventricular tachycardia is defined as four or more PVCs without intervening normal beats. It is also known as **VT** or **V-tach**. Multiple PVCs and VT may indicate that serious cardiac problems exist.

Ventricular Fibrillation (VF)

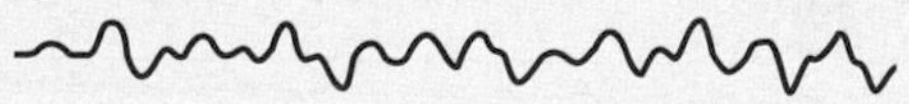

Ventricular fibrillation (VF), or v-fib, is responsible for the condition known as **cardiac arrest**. VF is rapidly fatal, because the ventricles quiver and stop pumping blood. The heart in VF is said to resemble a bag of squirming worms.

Module 19.16 Review

a. Define electrocardiogram.

b. List the important features of the ECG, and indicate what each represents.

c. Why is ventricular fibrillation fatal?

LO 19.16 Identify the electrical events shown on an electrocardiogram.

Matching

Match each lettered term with the most closely related description.

a. P wave	1	AV valves close
b. cardiac output	2	Self-stimulated cardiac muscle contractions
c. automaticity	3	Atrial depolarization
d. "lubb" sound	4	Amount of blood ejected by ventricle during a single beat
e. "dupp" sound	5	HR × SV
f. sympathetic neurons	6	Decrease the heart rate
g. stroke volume	7	Term for slower-than-normal HR
h. tachycardia	8	Increase the heart rate
i. bradycardia	9	Semilunar valve closes
j. parasympathetic neurons	10	Term for faster-than-normal HR

1 ______
2 ______
3 ______
4 ______
5 ______
6 ______
7 ______
8 ______
9 ______
10 ______

Short answer

Refer to the accompanying graph of the cardiac cycle to answer the questions below.

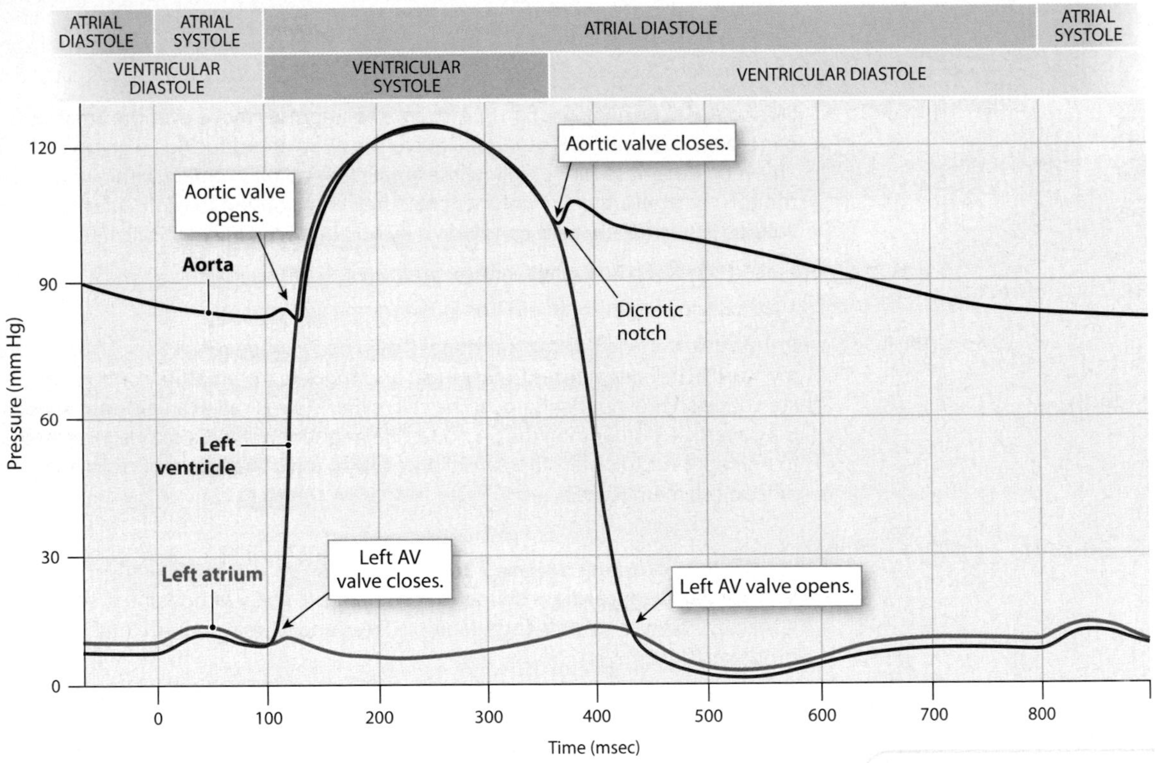
ATRIAL DIASTOLE
ATRIAL SYSTOLE
ATRIAL DIASTOLE
ATRIAL SYSTOLE
VENTRICULAR DIASTOLE
VENTRICULAR SYSTOLE
VENTRICULAR DIASTOLE
Pressure (mm Hg)
120
90
60
30
0
Aortic valve opens.
Aorta
Aortic valve closes.
Dicrotic notch
Left ventricle
Left atrium
Left AV valve closes.
Left AV valve opens.
0
100
200
300
400
500
600
700
800
Time (msec)

11 What event occurs when the pressure in the left ventricle rises above that in the left atrium?

12 During ventricular systole, the blood volume in the **atria** is (increasing or decreasing)?

13 During ventricular systole, the volume in the **ventricles** is (increasing or decreasing)?

14 During most of ventricular diastole, the pressure in the left ventricle is (greater than, the same as, or less than) the pressure in the left atrium.

15 What event occurs when the pressure within the left ventricle becomes greater than the pressure within the aorta?

16 During isovolumetric contraction, where is pressure the highest?

17 During what part of the cardiac cycle is blood pressure highest in the large systemic arteries?

11 ______
12 ______
13 ______
14 ______
15 ______
16 ______
17 ______

Pressure, resistance, and venous return affect cardiac output

You are now familiar with the factors involved in the regulation of cardiac output. Both cardiac output and the distribution of blood within the pulmonary and systemic circuits must constantly be adjusted to meet the demands of active tissues. The sites and mechanisms of cardiovascular regulation are summarized here.

Cardiac Output

To maintain **cardiac output**, the heart must generate enough pressure to force blood through thousands of miles of peripheral capillaries, most no larger in diameter than a single red blood cell.

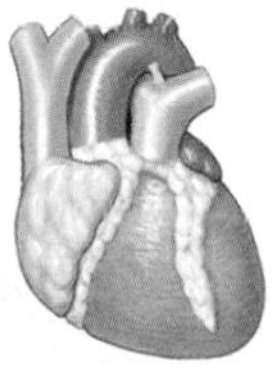

Venous Return

The **venous return** is the amount of blood arriving at the right atrium each minute. On average, it is equal to the cardiac output.

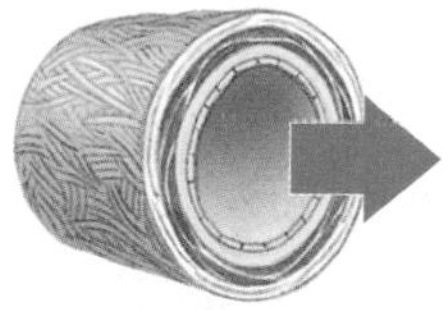

Arterial Blood Pressure

Blood pressure is the pressure within the cardiovascular system as a whole. **Arterial pressure** is much higher than **venous pressure** because it must push blood a greater distance and through smaller and smaller arteries and then through innumerable capillary networks.

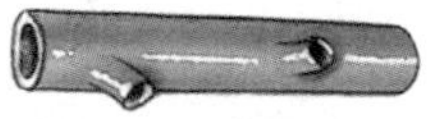

Regulation (Neural and Hormonal)

Neural and hormonal regulation make coordinated adjustments in heart rate, stroke volume, peripheral resistance, and venous pressure so that cardiac output is sufficient to meet the demands of peripheral tissues.

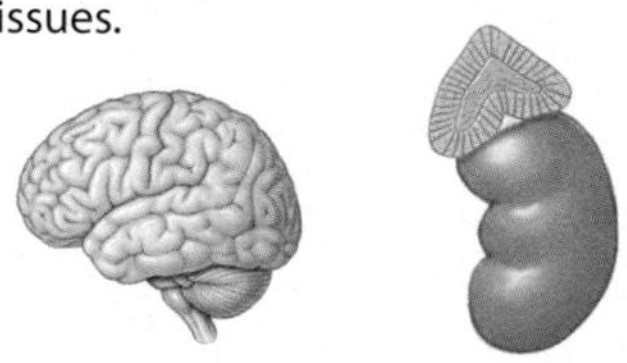

Venous Pressure

Valves and muscular compression of peripheral veins play an important role in maintaining venous pressure and venous blood flow. As blood moves toward the heart, the vessels get larger and larger in diameter, and resistance is continuously decreasing.

Peripheral Resistance

Resistance is a force that opposes movement. The **peripheral resistance** is the resistance of the arterial system as a whole. Resistance increases as the arterial branches get smaller and smaller.

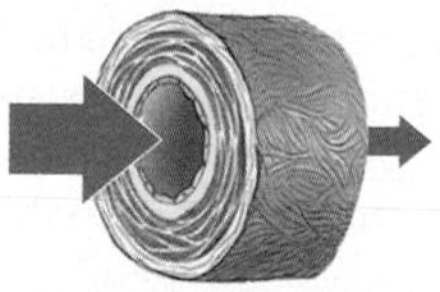

Capillary Pressure

Because capillary pressures are very low, blood flows slowly, allowing plenty of time for diffusion between the blood and the surrounding interstitial fluid. This interplay, called **capillary exchange**, is the primary focus of the entire cardiovascular system.

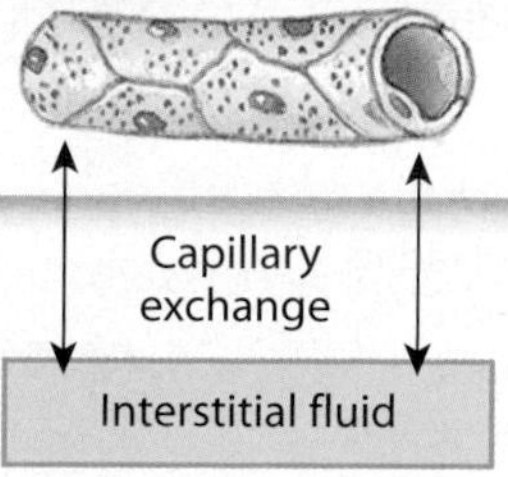
Capillary exchange
Interstitial fluid

In this section we will consider the functioning of the cardiovascular system as a whole, and learn how neural and hormonal mechanisms regulate cardiovascular function.

Module 19.17 Review

a. Neural and hormonal regulation influence which factors?

b. Which is greater: arterial pressure or venous pressure?

c. Why is it beneficial for capillary pressure to be very low?

LO 19.17 Explain the effects of pressure, resistance, and venous return on cardiac output.

Under normal conditions, vessel diameter is the primary source of resistance within the cardiovascular system

For circulation to occur, the heart must develop sufficient pressure to overcome the **total peripheral resistance**—the resistance of the entire cardiovascular system. The total peripheral resistance of the cardiovascular system depends on three factors: vascular resistance, viscosity, and turbulence.

1 **Vascular resistance**, the opposition to blood flow in vessels, is the largest component of total peripheral resistance. Vascular resistance primarily results from friction between blood and the vessel walls. The amount of friction depends on two factors: vessel length and vessel diameter.

Friction and Vessel Length

Friction occurs between the moving blood and the walls of the vessel. The longer the vessel, the greater the surface area in contact with the blood, and the greater the resistance. You can easily blow the water out of a snorkel that is 25 cm (10 in.) long, but you cannot blow the water out of a 15 m garden hose of the same diameter, because the total frictional resistance is too great. The most dramatic changes in blood vessel length occur between birth and maturity, as individuals grow to adult size. In adults, vessel length can be considered constant.

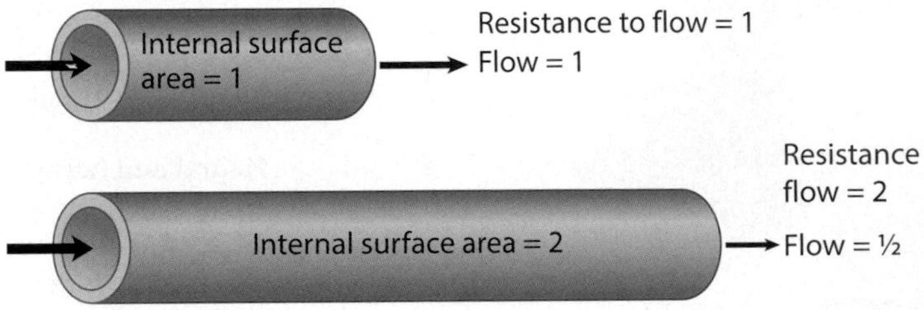

Internal surface area = 1
Resistance to flow = 1
Flow = 1
Internal surface area = 2
Resistance to flow = 2
Flow = ½

Friction and Vessel Diameter

Friction also occurs between layers of fluid moving at different speeds. The layer of blood closest to the vessel wall is slowed down by friction with the endothelial surface. The adjacent layer of blood is slowed down by friction with the more superficial layer. This effect gradually diminishes as the distance from the wall increases. In a small-diameter vessel, all the blood is slowed to some degree, and resistance is relatively high. In a large-diameter vessel the central region is unaffected by events at the periphery, so the resistance is relatively low.

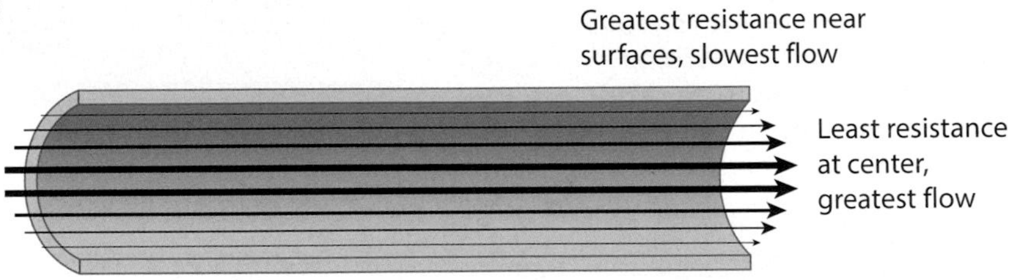

Greatest resistance near surfaces, slowest flow
Least resistance at center, greatest flow

Vessel Length versus Vessel Diameter

Differences in diameter have much more significant effects on resistance than do differences in length. As shown here, if two vessels are equal in diameter (or radius) but one is twice as long as the other, the longer vessel offers twice as much resistance to blood flow. But for two vessels of equal length, one twice the diameter of the other, the narrower one offers 16 times as much resistance to blood flow. This relationship, expressed in terms of the vessel radius (r) and resistance (R), can be summarized as $R = 1/r^4$. This means that a small change in vessel diameter produces a large change in resistance. The vasomotor center controls peripheral resistance and blood flow primarily by altering the diameters of arterioles.

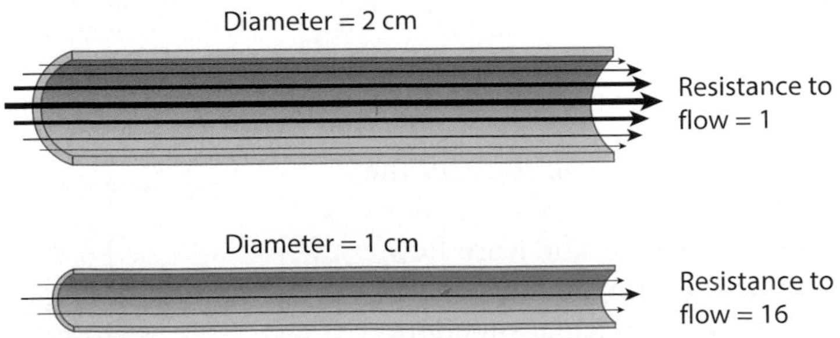

Diameter = 2 cm
Resistance to flow = 1
Diameter = 1 cm
Resistance to flow = 16

2 **Viscosity** is the resistance to flow caused by interactions among molecules and suspended materials in a liquid. Liquids of low viscosity, such as water (viscosity 1.0), flow at low pressures. Thick, syrupy fluids, such as molasses (viscosity 300), flow only under higher pressures. Whole blood has a viscosity about five times that of water, due to the presence of plasma proteins and blood cells. Under normal conditions, the viscosity of blood remains stable, but disorders that affect the hematocrit or plasma composition also change blood viscosity, and thus peripheral resistance.

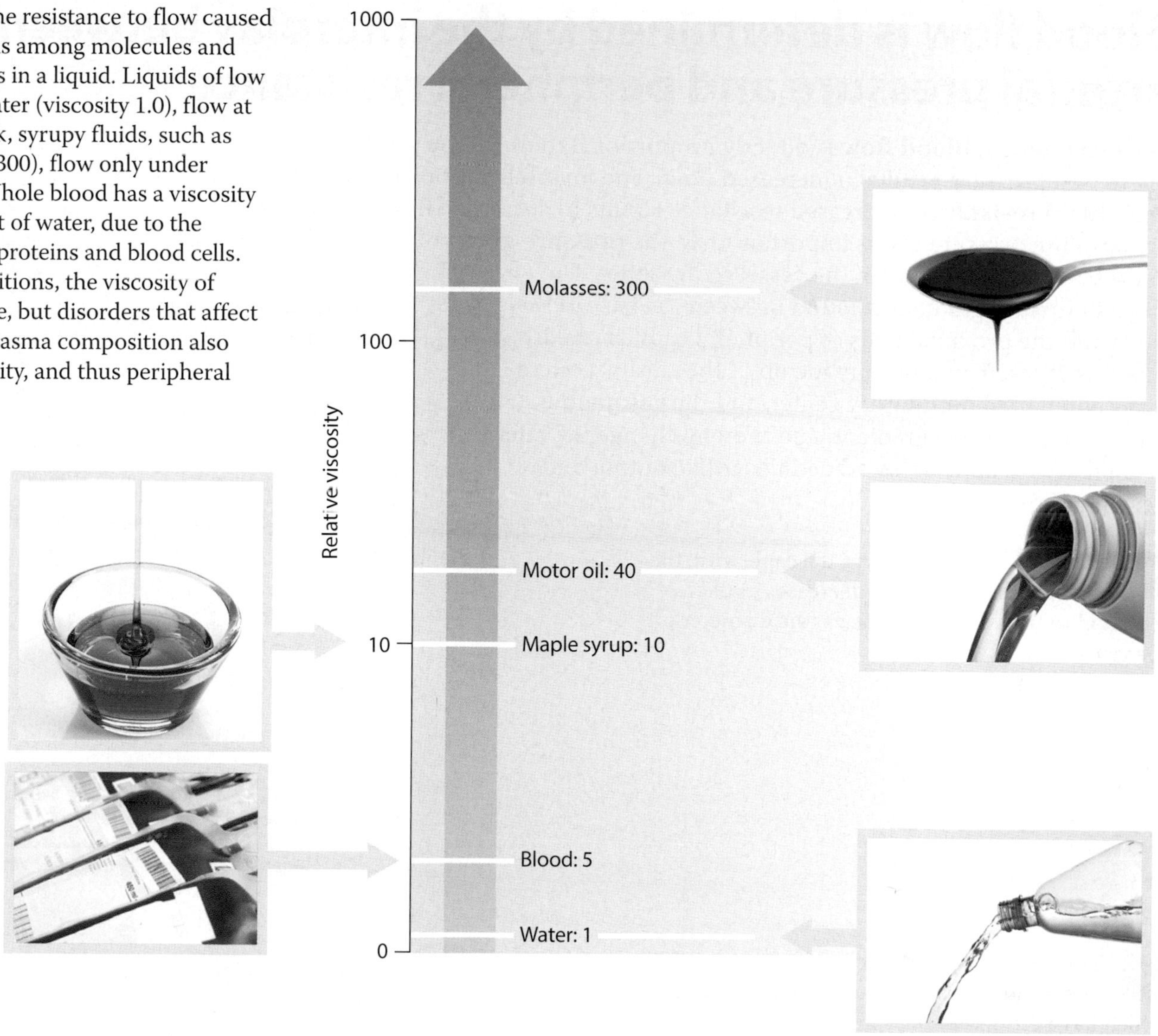

1000
100
10
0
Relative viscosity
Molasses: 300
Motor oil: 40
Maple syrup: 10
Blood: 5
Water: 1

3 High flow rates, irregular surfaces, and sudden changes in vessel diameter upset the smooth flow of blood, creating eddies and swirls. This phenomenon, called **turbulence**, increases resistance and slows blood flow. Here you see the turbulence created by the presence of a plaque within a small artery. Turbulence also occurs in normal individuals when blood flows between the atria and the ventricles, and between the ventricles and the aortic and pulmonary trunks. Turbulent flow is responsible for the production of the third and fourth heart sounds.

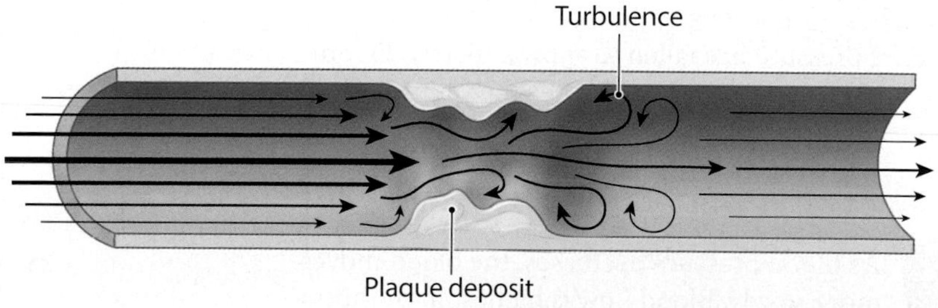

Turbulence
Plaque deposit

Module 19.18 Review

a. List the factors that contribute to total peripheral resistance.

b. Which would reduce peripheral resistance: an increase in vessel length or an increase in vessel diameter?

c. Explain the formula $R = 1/r^4$.

LO **19.18** Describe the factors that influence total peripheral resistance.

Blood flow is determined by the interplay between arterial pressure and peripheral resistance

In general terms, **blood flow** is directly proportional to the **blood pressure** (increased pressure results in increased flow), and inversely proportional to **peripheral resistance** (increased resistance results in decreased flow). However, the absolute pressure is less important than the **pressure gradient**—the difference in pressure from one end of the vessel to the other. The largest pressure gradient is found between the base of the aorta and the proximal ends of peripheral capillary beds. The cardiovascular center (made up of the cardioacceleratory center, cardioinhibitory center, and the vasomotor center) can alter this gradient, and thereby change the rate of capillary blood flow, by adjusting cardiac output and peripheral resistance.

1 As blood proceeds toward the capillaries, the diameter of the arteries decreases markedly. As blood returns toward the heart, the diameter of the veins increases.

2 The heart generates a pressure of around 120 mm Hg as it pushes blood into the aorta, which has a cross-sectional area of roughly 4.5 cm^2. At each branching of the arterial system, the arterial pressure drops as blood is pushed into ever-increasing numbers of smaller and smaller branches. At the start of peripheral capillaries, arterial pressure has fallen to approximately 35 mm Hg, and by the time blood reaches the venules, it has fallen to approximately 18 mm Hg.

3 As blood pressure decreases, the blood moves more slowly. Blood flow is highest in the aorta, where blood pressure is high and the large diameter means that resistance is low. Blood flow is slowest in the capillaries, whose diameter is nearly the same as that of a single red blood cell. This is important because exchange with interstitial fluid occurs only in the capillaries, and diffusion is a relatively slow process. Blood flow then accelerates in the venous system because although pressures are falling, the vessels are merging into large-diameter passages with very low resistance.

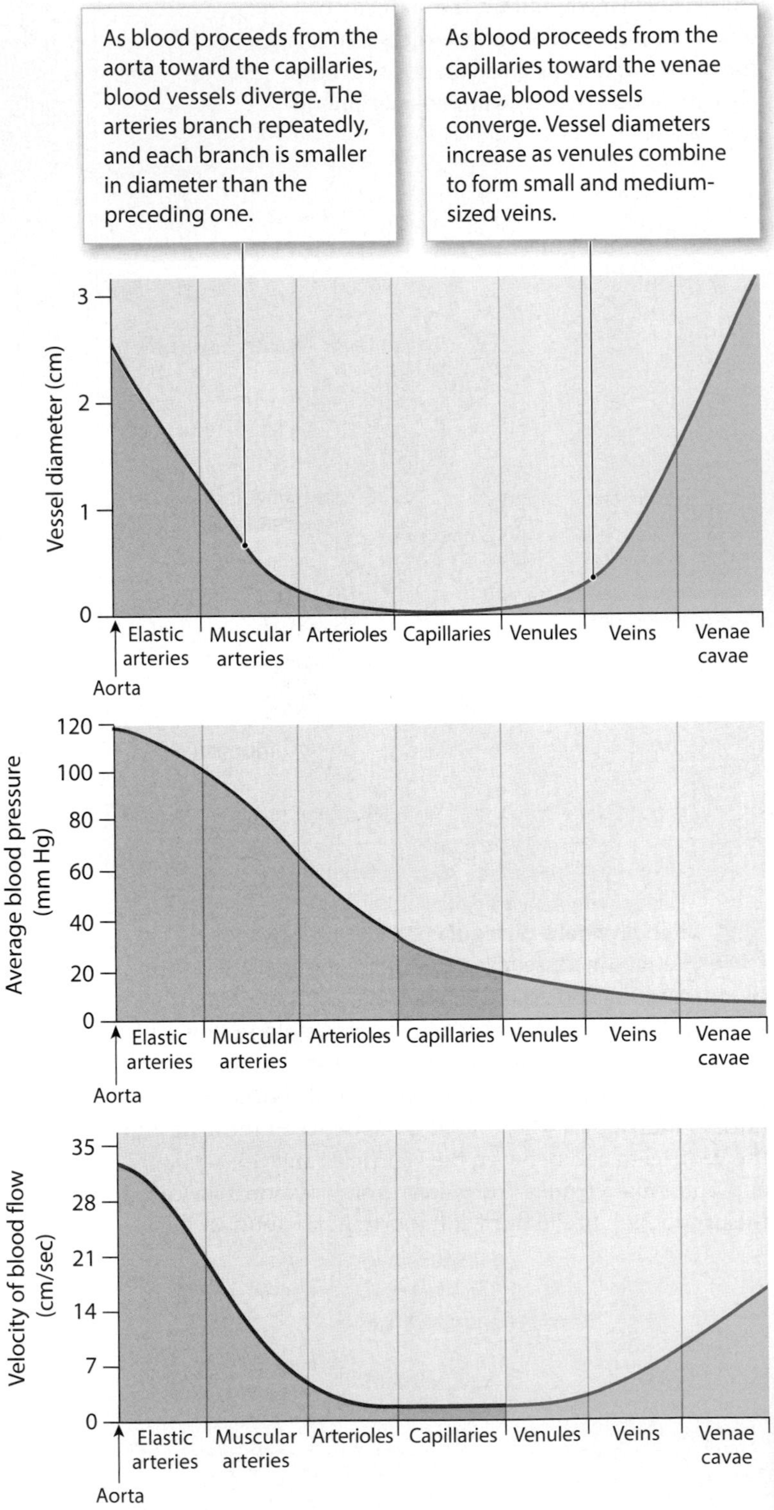
As blood proceeds from the aorta toward the capillaries, blood vessels diverge. The arteries branch repeatedly, and each branch is smaller in diameter than the preceding one.
As blood proceeds from the capillaries toward the venae cavae, blood vessels converge. Vessel diameters increase as venules combine to form small and medium-sized veins.
Vessel diameter (cm)
0
1
2
3
Average blood pressure (mm Hg)
0
20
40
60
80
100
120
Velocity of blood flow (cm/sec)
0
7
14
21
28
35
Elastic arteries
Muscular arteries
Arterioles
Capillaries
Venules
Veins
Venae cavae
Aorta

4 Arterial pressure is not constant. It rises during ventricular systole and falls during ventricular diastole as the elastic arterial walls stretch and recoil. The peak blood pressure measured during ventricular systole is called **systolic pressure**, and the minimum blood pressure at the end of ventricular diastole is called **diastolic pressure**. In recording blood pressure, we separate systolic and diastolic pressures with a slash. Thus the blood pressure illustrated here is written as 120/90 (read as "one-twenty over ninety").

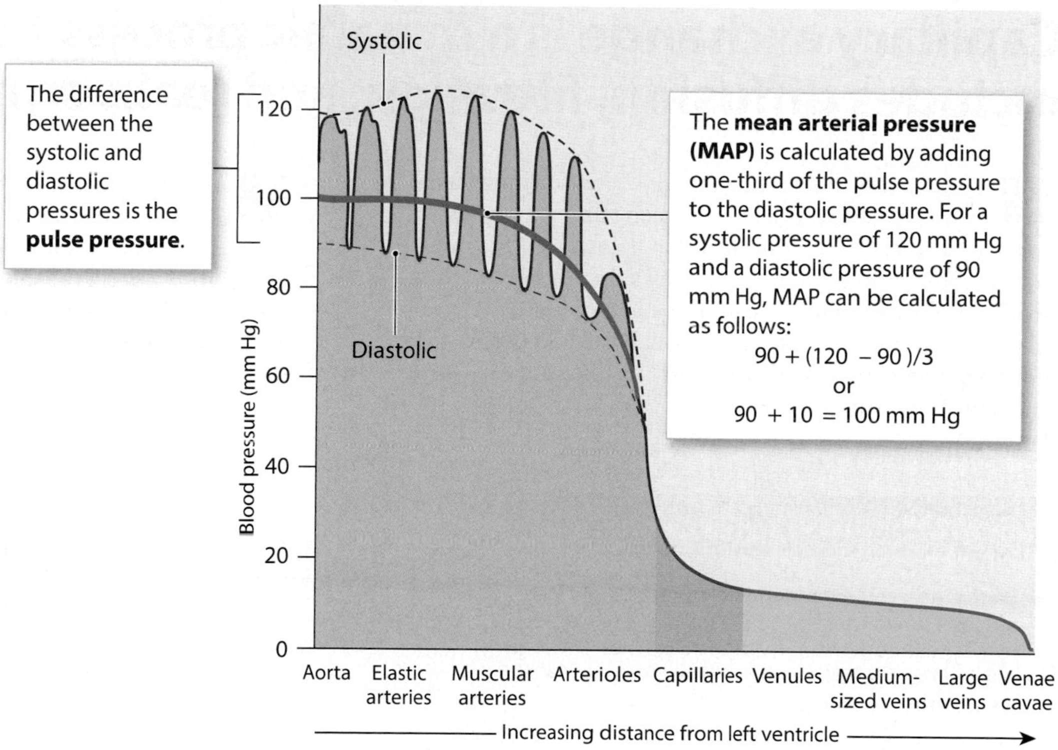

The difference between the systolic and diastolic pressures is the **pulse pressure**.
Systolic
Diastolic
The **mean arterial pressure (MAP)** is calculated by adding one-third of the pulse pressure to the diastolic pressure. For a systolic pressure of 120 mm Hg and a diastolic pressure of 90 mm Hg, MAP can be calculated as follows:
90 + (120 − 90)/3
or
90 + 10 = 100 mm Hg
Blood pressure (mm Hg)
120
100
80
60
40
20
0
Aorta
Elastic arteries
Muscular arteries
Arterioles
Capillaries
Venules
Medium-sized veins
Large veins
Venae cavae
Increasing distance from left ventricle

5 **Capillary exchange** involves a combination of filtration, diffusion, and osmosis. **Capillary hydrostatic pressure (CHP)** is the blood pressure within capillary beds, and it provides the driving force for filtration. Along the length of a typical capillary, CHP pushes water and soluble molecules out of the bloodstream and into the interstitial fluid. Only small solutes can cross the endothelium; larger molecules, such as plasma proteins, remain in the bloodstream. This size-selective process is called **filtration**. We will consider further details about capillary exchange in the next module.

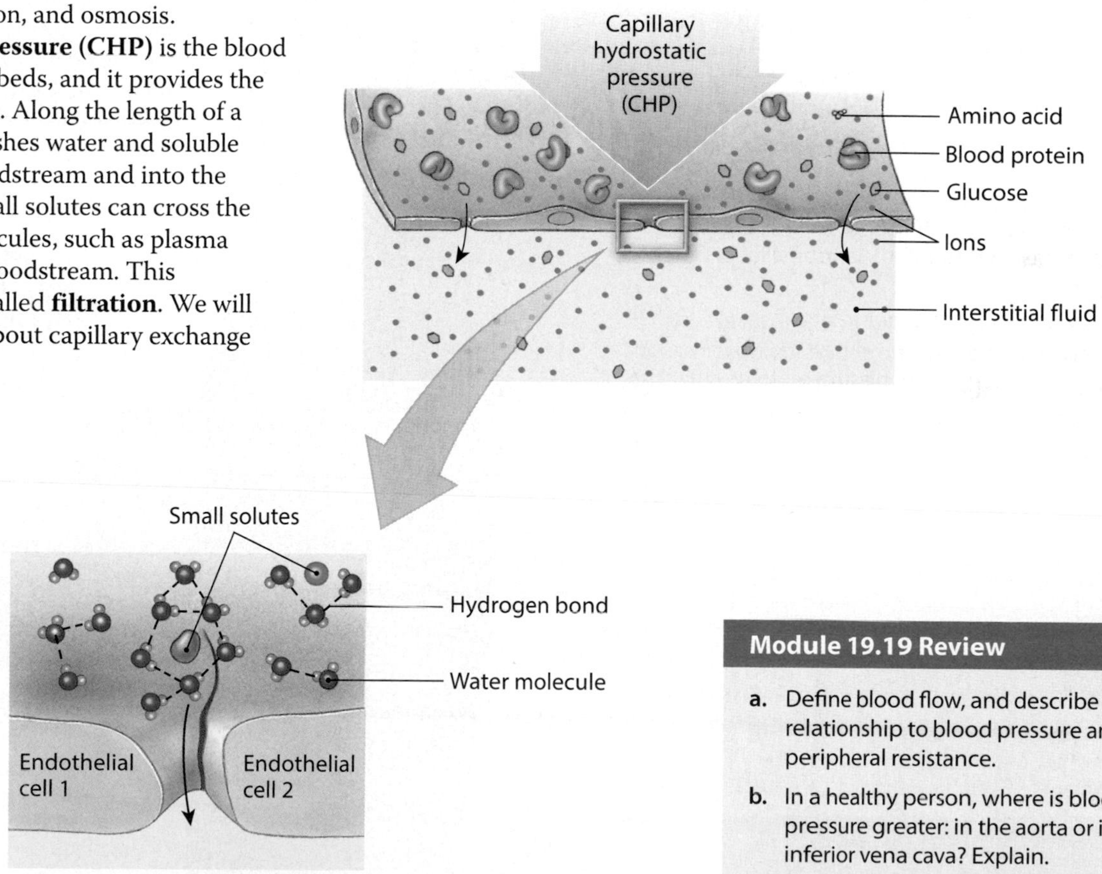

Capillary hydrostatic pressure (CHP)
Amino acid
Blood protein
Glucose
Ions
Interstitial fluid
Small solutes
Hydrogen bond
Water molecule
Endothelial cell 1
Endothelial cell 2

Relatively small solutes are carried along as CHP pushes water into the interstitial spaces.

Module 19.19 Review

a. Define blood flow, and describe its relationship to blood pressure and peripheral resistance.

b. In a healthy person, where is blood pressure greater: in the aorta or in the inferior vena cava? Explain.

c. Calculate the mean arterial pressure for a person whose blood pressure is 125/70.

LO 19.19 Describe the factors that determine blood flow.

Capillary exchange is a dynamic process that includes diffusion, filtration, and reabsorption

1 Diffusion is the net movement of ions or molecules from an area where their concentration is higher to an area where their concentration is lower. Diffusion occurs most rapidly when (1) the distances involved are short, (2) the concentration gradient is large, and (3) the ions or molecules involved are small. Diffusion occurs continuously across capillary walls, but different substances use different routes, as noted in the table below.

Routes of Diffusion across Capillary Walls

- Water, ions, and small organic molecules, such as glucose, amino acids, and urea, can usually enter or leave the bloodstream by diffusion either between adjacent endothelial cells or through the pores of fenestrated capillaries.
- Many ions, including sodium, potassium, calcium, and chloride, can diffuse across endothelial cells by passing through channels in plasma membranes.
- Large water-soluble compounds are unable to enter or leave the bloodstream except at fenestrated capillaries, such as those in the hypothalamus, the kidneys, many endocrine organs, and the intestinal tract.
- Lipids, such as fatty acids and steroids, and lipid-soluble materials, including soluble gases such as oxygen and carbon dioxide, can cross capillary walls by diffusion through the endothelial plasma membranes.
- Plasma proteins are normally unable to cross the endothelial lining anywhere except in sinusoids, such as those of the liver, where plasma proteins enter the bloodstream.

2 In a capillary, blood pressure decreases as blood flows from the arterial end to the venous end. As a result, the rates of filtration and reabsorption gradually change as blood passes along the length of a capillary. The factors involved are diagrammed in the figure below.

Filtration Predominates

Filtration predominates at the arterial end of a capillary, where **capillary hydrostatic pressure (CHP)** is highest. As filtration occurs, the plasma osmolarity, or **blood colloid osmotic pressure (BCOP)**, increases. BCOP rises because water is leaving but the larger solutes, especially soluble plasma proteins, are remaining behind. The difference between CHP (which pushes water out) and BCOP (which draws water in) is known as the **net filtration pressure (NFP)**. At the start of the capillary, NFP is high, but as blood flows along the capillary, CHP decreases as water flows out, and BCOP rises as plasma protein concentrations increase.

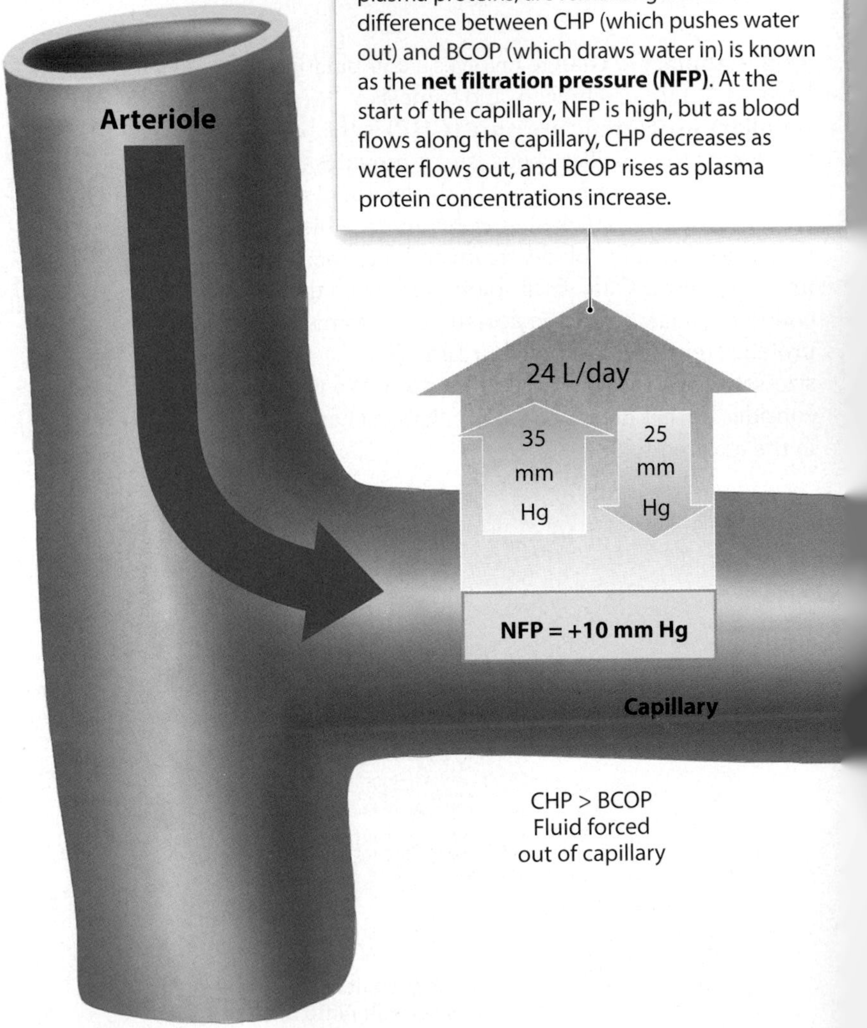
Arteriole
24 L/day
35 mm Hg
25 mm Hg
NFP = +10 mm Hg
Capillary

CHP > BCOP
Fluid forced
out of capillary

3 Any conditions that affect either blood pressure or the osmotic pressures of the blood or interstitial fluid will shift the balance between hydrostatic and osmotic forces. We can then predict the effects on the basis of an understanding of capillary dynamics. Three examples are presented in the table to the right.

Representative Variations in Capillary Exchange

- If hemorrhaging occurs, both blood volume and blood pressure decline. This decrease in CHP lowers the NFP and increases the amount of capillary reabsorption. The result is a reduction in the volume of interstitial fluid and an increase in the circulating plasma volume. This process is known as a **recall of fluids**.
- If dehydration occurs, the plasma volume decreases due to water loss, and the concentration of plasma proteins increases. The increase in BCOP accelerates reabsorption and a recall of fluids that delays the onset and severity of clinical problems caused by low blood volume and blood pressure.
- If the CHP rises or the BCOP decreases, fluid moves out of the blood in capillaries and builds up in peripheral tissues, an abnormal condition called **edema**.

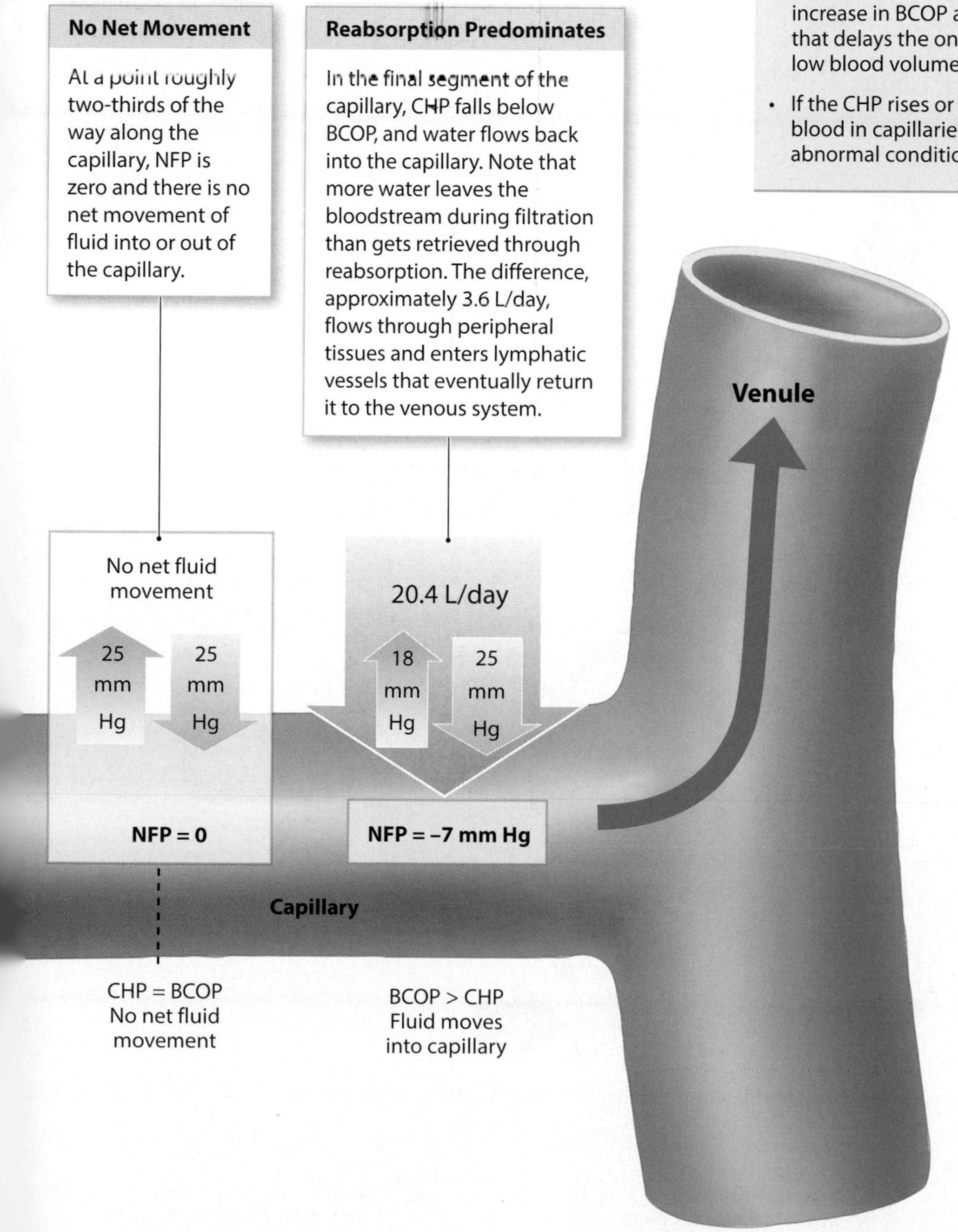
No Net Movement
At a point roughly two-thirds of the way along the capillary, NFP is zero and there is no net movement of fluid into or out of the capillary.
Reabsorption Predominates
In the final segment of the capillary, CHP falls below BCOP, and water flows back into the capillary. Note that more water leaves the bloodstream during filtration than gets retrieved through reabsorption. The difference, approximately 3.6 L/day, flows through peripheral tissues and enters lymphatic vessels that eventually return it to the venous system.
Venule
No net fluid movement
25 mm Hg
25 mm Hg
NFP = 0
20.4 L/day
18 mm Hg
25 mm Hg
NFP = –7 mm Hg
Capillary
CHP = BCOP
No net fluid movement
BCOP > CHP
Fluid moves into capillary

KEY

- **CHP** (Capillary hydrostatic pressure)
- **BCOP** (Blood colloid osmotic pressure)
- **NFP** (Net filtration pressure)

Module 19.20 Review

a. Identify the conditions that would shift the balance between hydrostatic and osmotic forces.

b. Under what general conditions would fluid move into a capillary?

c. Define edema.

LO 19.20 Describe the movement of fluids between capillaries and interstitial spaces.

Cardiovascular regulatory mechanisms respond to changes in blood pressure or blood chemistry

1 Homeostatic mechanisms regulate cardiovascular activity to ensure that **tissue perfusion**—the blood flow through the tissues—meets the tissue demands for oxygen and nutrients. There are two regulatory pathways: one that involves local autoregulation and, if that is ineffective, a second that involves central regulation by neural and endocrine mechanisms.

Central Regulation

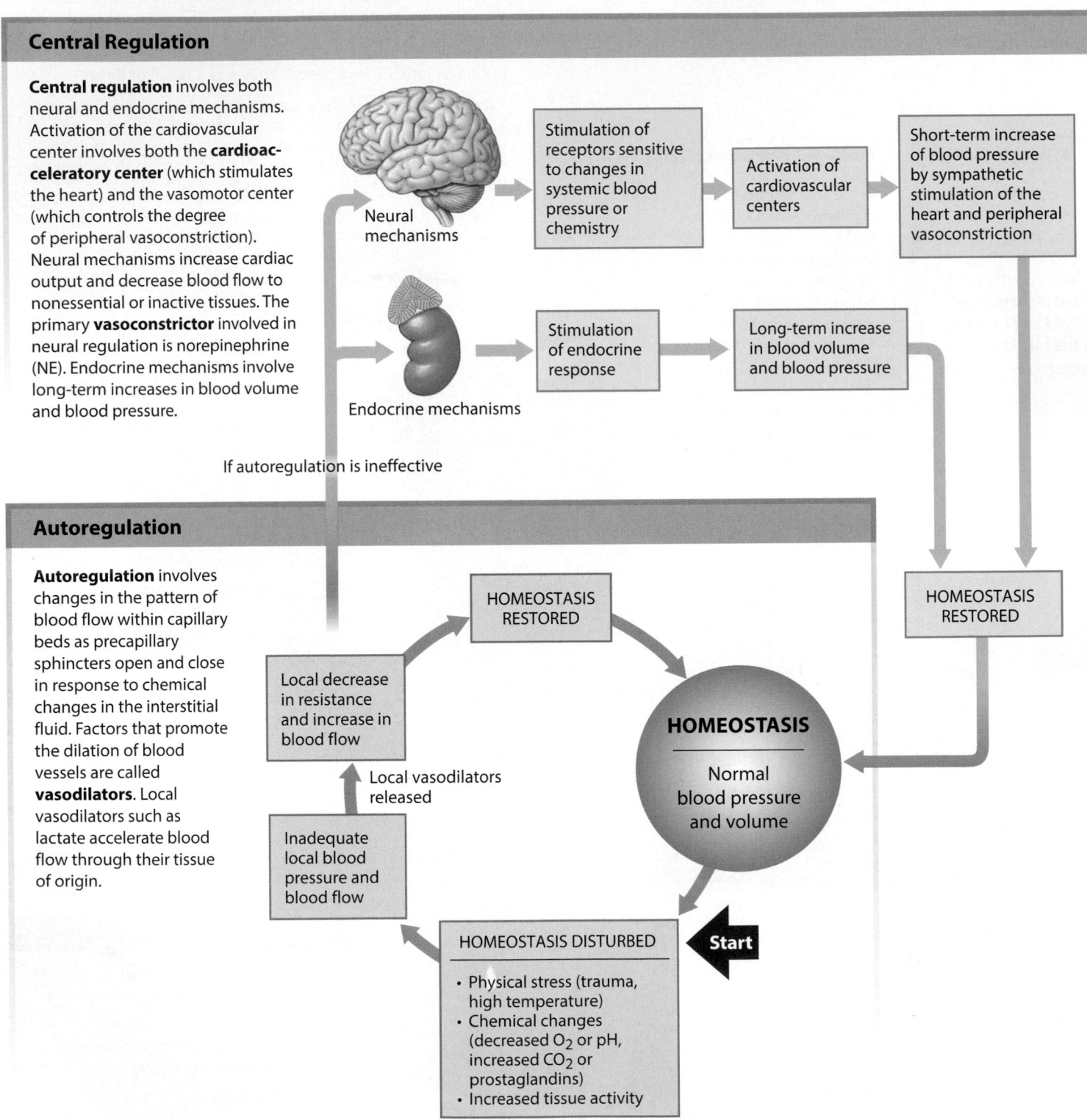

Central regulation involves both neural and endocrine mechanisms. Activation of the cardiovascular center involves both the **cardioacceleratory center** (which stimulates the heart) and the vasomotor center (which controls the degree of peripheral vasoconstriction). Neural mechanisms increase cardiac output and decrease blood flow to nonessential or inactive tissues. The primary **vasoconstrictor** involved in neural regulation is norepinephrine (NE). Endocrine mechanisms involve long-term increases in blood volume and blood pressure.

Autoregulation

Autoregulation involves changes in the pattern of blood flow within capillary beds as precapillary sphincters open and close in response to chemical changes in the interstitial fluid. Factors that promote the dilation of blood vessels are called **vasodilators**. Local vasodilators such as lactate accelerate blood flow through their tissue of origin.

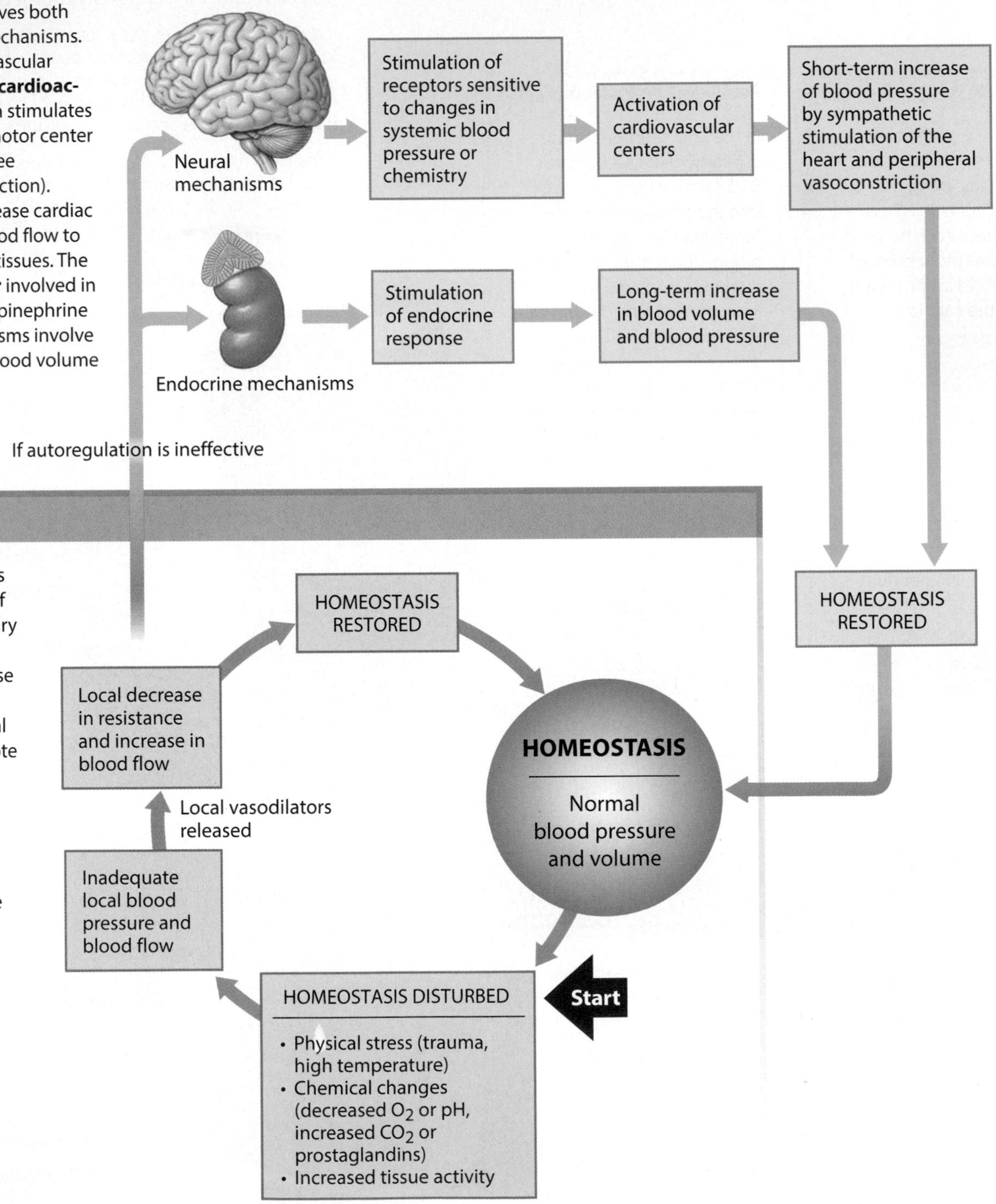
Neural mechanisms
Stimulation of receptors sensitive to changes in systemic blood pressure or chemistry
Activation of cardiovascular centers
Short-term increase of blood pressure by sympathetic stimulation of the heart and peripheral vasoconstriction
Endocrine mechanisms
Stimulation of endocrine response
Long-term increase in blood volume and blood pressure
If autoregulation is ineffective
HOMEOSTASIS RESTORED
HOMEOSTASIS RESTORED
Local decrease in resistance and increase in blood flow
Local vasodilators released
Inadequate local blood pressure and blood flow
HOMEOSTASIS
Normal blood pressure and volume
HOMEOSTASIS DISTURBED
• Physical stress (trauma, high temperature)
• Chemical changes (decreased O_2 or pH, increased CO_2 or prostaglandins)
• Increased tissue activity
Start

2 The **baroreceptor reflexes** (*baro-*, pressure) respond to changes in blood pressure. As noted in Module 14.10 (p. 526), the receptors are located in the walls of (1) the carotid sinuses, expanded chambers near the bases of the internal carotid arteries of the neck; (2) the aortic sinuses, pockets in the walls of the ascending aorta adjacent to the heart; and (3) the right atrium.

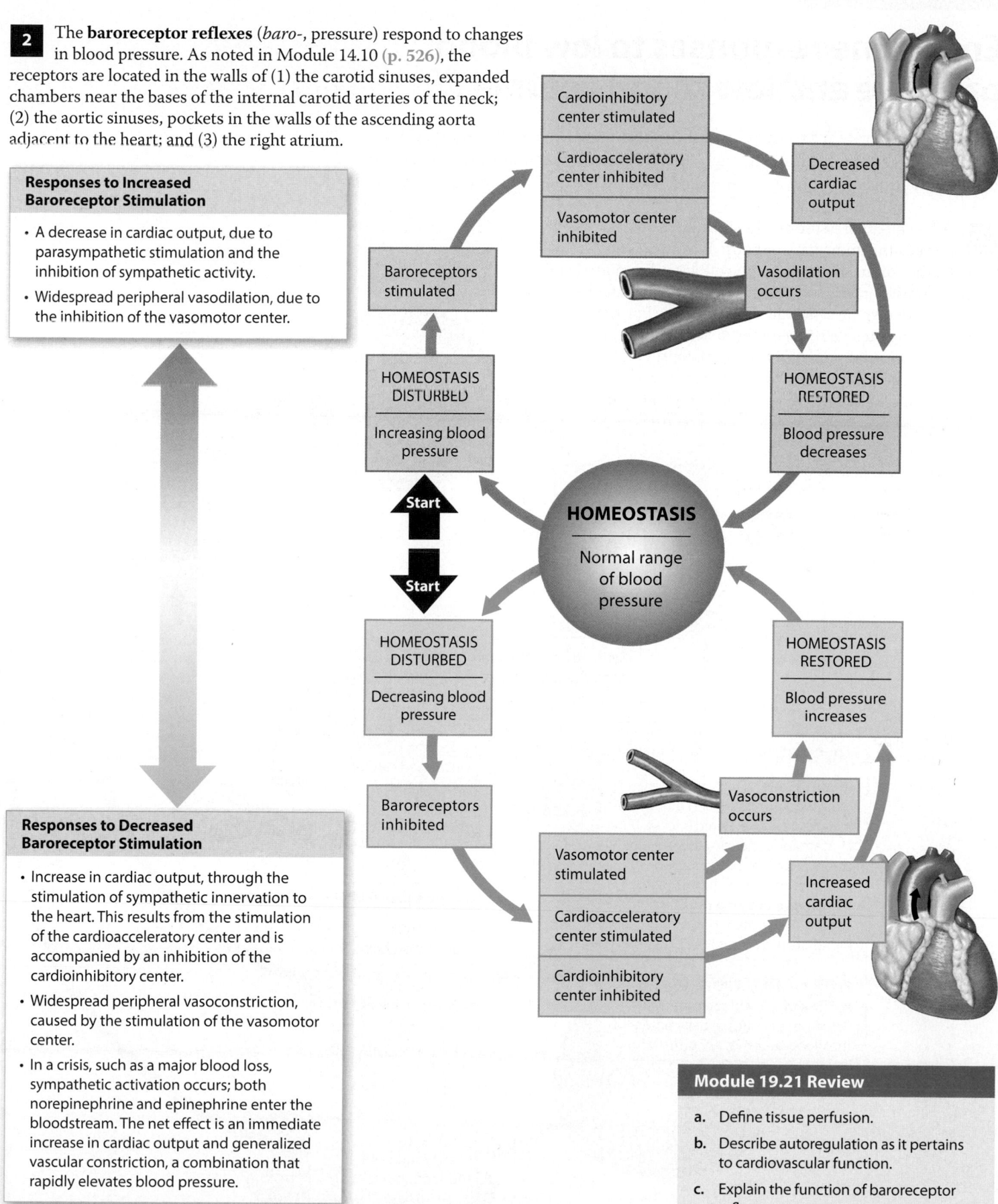

Responses to Increased Baroreceptor Stimulation

- A decrease in cardiac output, due to parasympathetic stimulation and the inhibition of sympathetic activity.
- Widespread peripheral vasodilation, due to the inhibition of the vasomotor center.

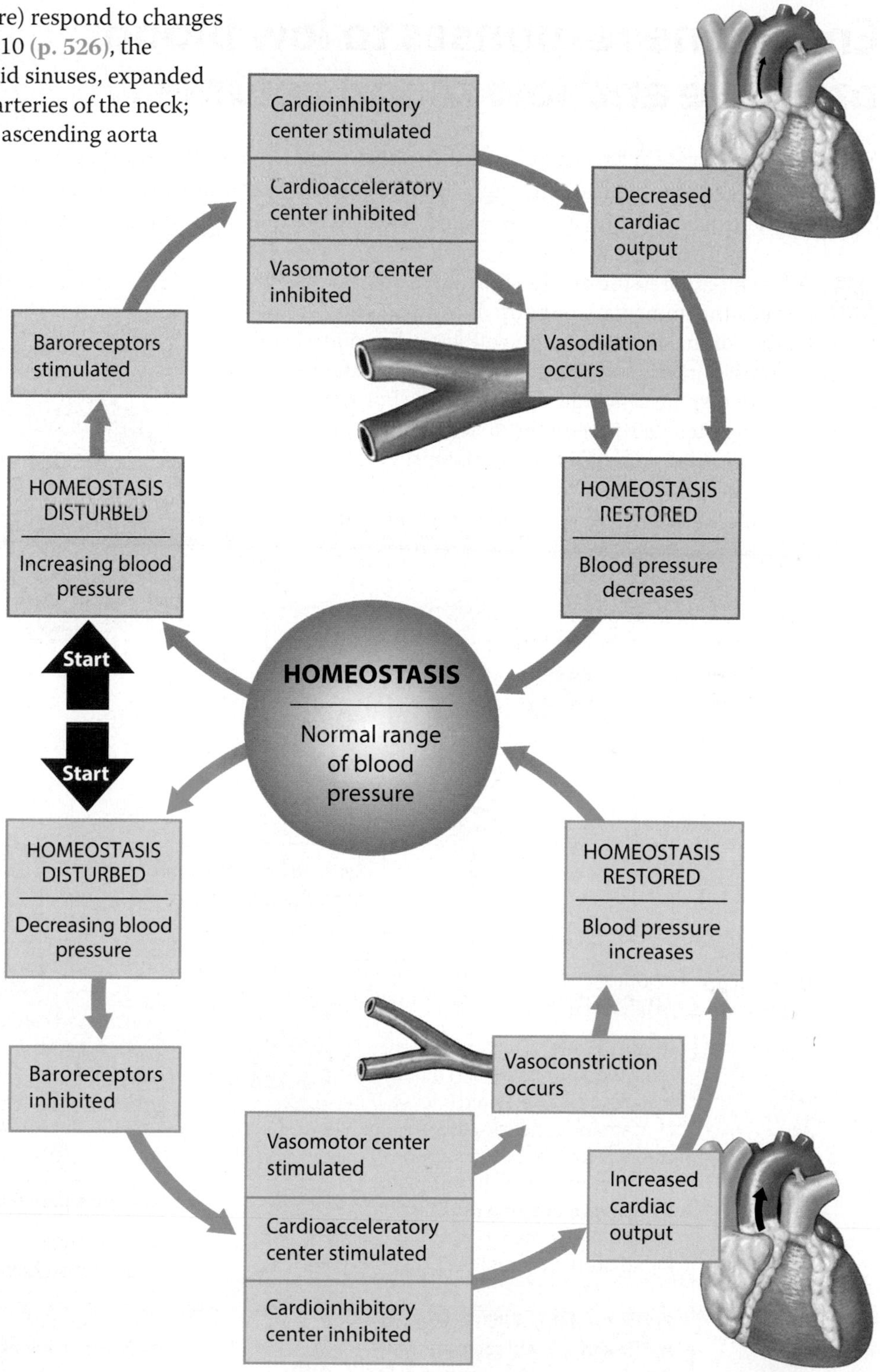

Cardioinhibitory center stimulated
Cardioacceleratory center inhibited
Vasomotor center inhibited
Decreased cardiac output
Vasodilation occurs
Baroreceptors stimulated
HOMEOSTASIS DISTURBED
Increasing blood pressure
HOMEOSTASIS RESTORED
Blood pressure decreases
Start
Start
HOMEOSTASIS
Normal range of blood pressure
HOMEOSTASIS DISTURBED
Decreasing blood pressure
HOMEOSTASIS RESTORED
Blood pressure increases
Baroreceptors inhibited
Vasoconstriction occurs
Vasomotor center stimulated
Cardioacceleratory center stimulated
Cardioinhibitory center inhibited
Increased cardiac output

Responses to Decreased Baroreceptor Stimulation

- Increase in cardiac output, through the stimulation of sympathetic innervation to the heart. This results from the stimulation of the cardioacceleratory center and is accompanied by an inhibition of the cardioinhibitory center.
- Widespread peripheral vasoconstriction, caused by the stimulation of the vasomotor center.
- In a crisis, such as a major blood loss, sympathetic activation occurs; both norepinephrine and epinephrine enter the bloodstream. The net effect is an immediate increase in cardiac output and generalized vascular constriction, a combination that rapidly elevates blood pressure.

Module 19.21 Review

a. Define tissue perfusion.

b. Describe autoregulation as it pertains to cardiovascular function.

c. Explain the function of baroreceptor reflexes.

LO **19.21** Explain central regulation, autoregulation, and baroreceptor reflexes in response to changes in blood pressure and blood composition.

Module 19.22

Endocrine responses to low blood pressure and low blood volume . . .

The endocrine system provides both short-term and long-term regulation of cardiovascular performance through the endocrine functions of the heart and the kidneys, in addition to the actions of antidiuretic hormone from the pituitary gland.

1 When blood pressure and blood volume decrease below normal, the immediate response is the release of E and NE from the adrenal medullae, stimulating cardiac output and peripheral vasoconstriction. Other hormones important in the long-term response include (1) antidiuretic hormone (ADH), (2) angiotensin II, (3) erythropoietin (EPO), and (4) aldosterone. Although ADH and angiotensin II also affect blood pressure, these two hormones are concerned primarily with the long-term regulation of blood volume.

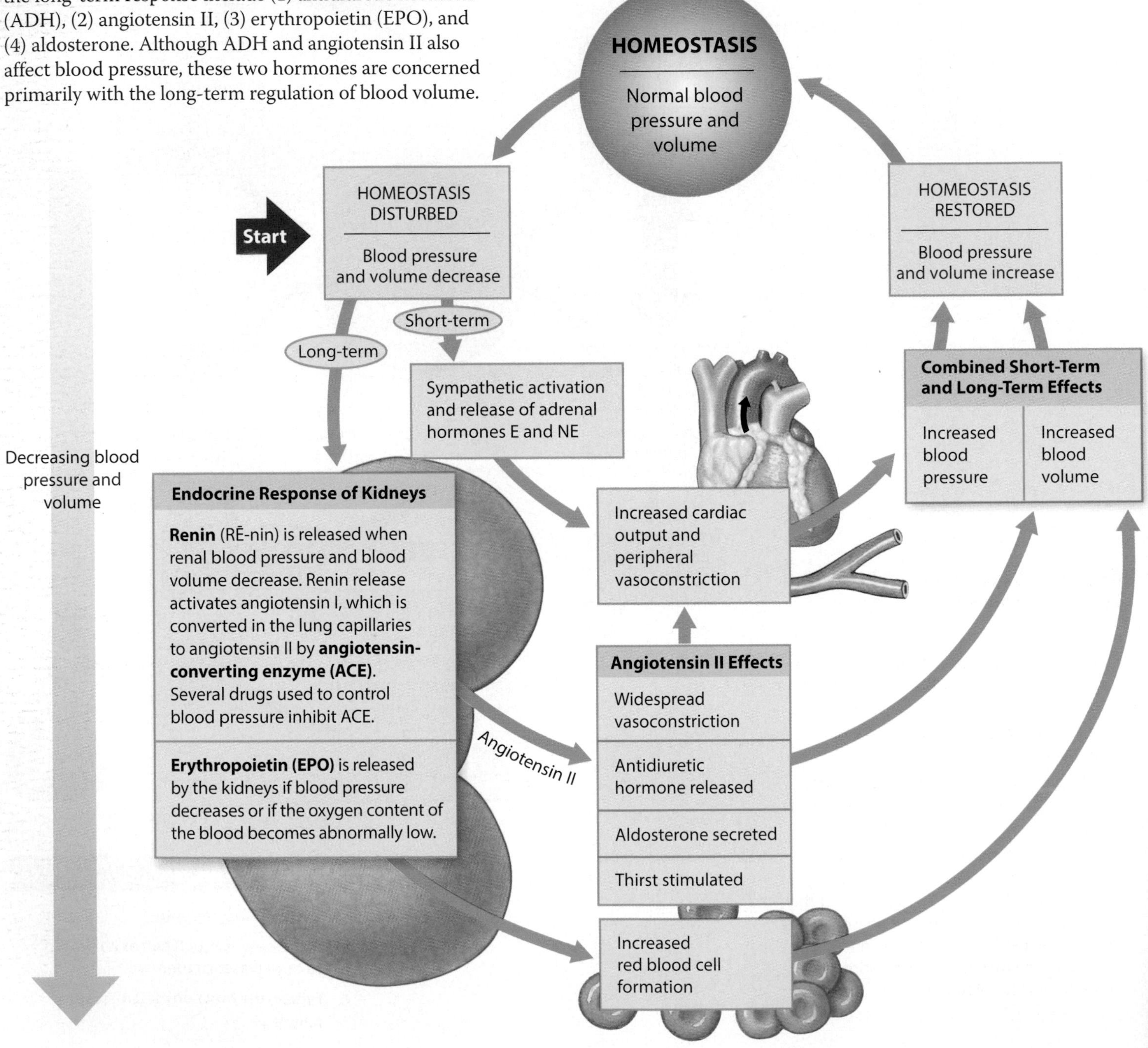
HOMEOSTASIS
Normal blood pressure and volume
Start
HOMEOSTASIS DISTURBED
Blood pressure and volume decrease
Short-term
Long-term
Sympathetic activation and release of adrenal hormones E and NE
Decreasing blood pressure and volume
Endocrine Response of Kidneys
Renin (RĒ-nin) is released when renal blood pressure and blood volume decrease. Renin release activates angiotensin I, which is converted in the lung capillaries to angiotensin II by angiotensin-converting enzyme (ACE). Several drugs used to control blood pressure inhibit ACE.
Erythropoietin (EPO) is released by the kidneys if blood pressure decreases or if the oxygen content of the blood becomes abnormally low.
Angiotensin II
Increased cardiac output and peripheral vasoconstriction
Angiotensin II Effects
Widespread vasoconstriction
Antidiuretic hormone released
Aldosterone secreted
Thirst stimulated
Increased red blood cell formation
HOMEOSTASIS RESTORED
Blood pressure and volume increase
Combined Short-Term and Long-Term Effects
Increased blood pressure
Increased blood volume

. . . are very different from those to high blood pressure and high blood volume

2 Excessive blood volume triggers a response through its effects on the walls of the heart. When the heart walls are abnormally stretched during diastole, cardiac muscle cells release **natriuretic peptides**. **Atrial natriuretic peptide** (nā-trē-ū-RET-ik; *natrium*, sodium + *ouresis*, urination), or **ANP**, is produced by cardiac muscle cells in the wall of the right atrium. A related hormone called **brain natriuretic peptide** (**BNP**) is produced by ventricular muscle cells. The homeostatic responses to ANP and BNP are summarized below. As blood volume and blood pressure decrease, the stresses on the walls of the heart are removed, and natriuretic peptide production ceases.

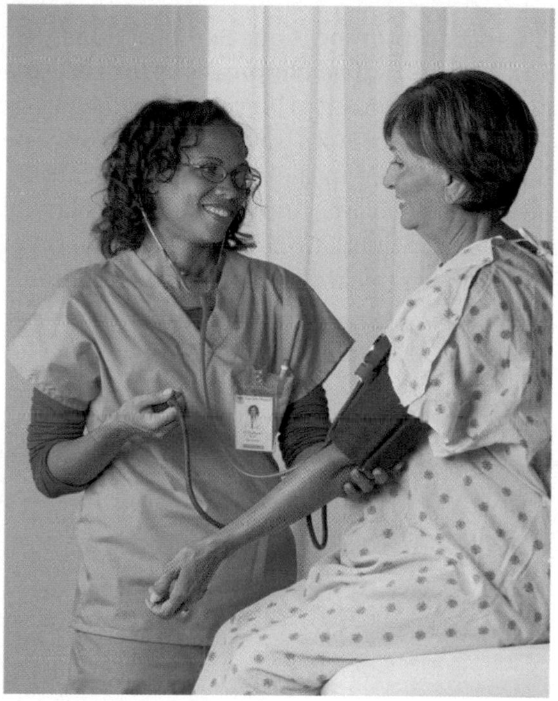

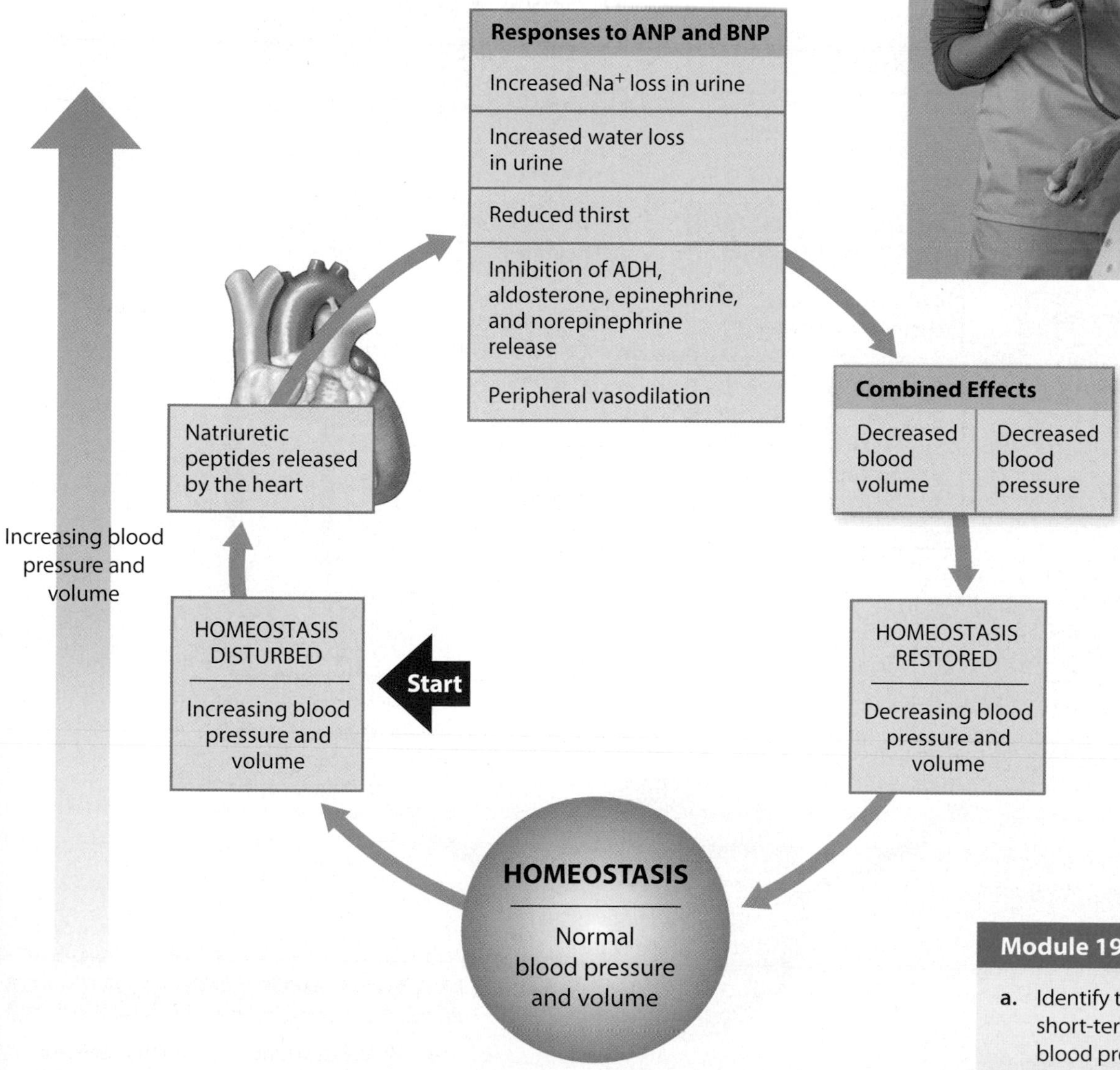
Responses to ANP and BNP
Increased Na+ loss in urine
Increased water loss in urine
Reduced thirst
Inhibition of ADH, aldosterone, epinephrine, and norepinephrine release
Peripheral vasodilation
Combined Effects
Decreased blood volume
Decreased blood pressure
HOMEOSTASIS RESTORED
Decreasing blood pressure and volume
HOMEOSTASIS
Normal blood pressure and volume
HOMEOSTASIS DISTURBED
Increasing blood pressure and volume
Start
Natriuretic peptides released by the heart
Increasing blood pressure and volume

Module 19.22 Review

a. Identify the hormones responsible for short-term regulation of decreasing blood pressure and blood volume.

b. How does the kidney respond to vasoconstriction of the renal artery?

c. Describe the roles of the natriuretic peptides.

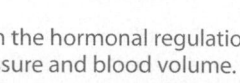

LO 19.22 Explain the hormonal regulation of blood pressure and blood volume.

Module 19.23

Chemoreceptors monitor the chemical composition of the blood and cerebrospinal fluid

Chemoreceptor locations and functions were introduced in Module 14.10 (p. 527). The **chemoreceptor reflexes** respond to changes in carbon dioxide, oxygen, or pH levels in blood and cerebrospinal fluid (CSF). The chemoreceptors involved are sensory neurons located in the **carotid bodies**, situated in the neck near the carotid sinus; in the **aortic bodies**, near the aortic arch; and on the ventrolateral surfaces of the medulla oblongata. Stimulation of these chemoreceptors triggers coordinated adjustments in cardiovascular and respiratory activity.

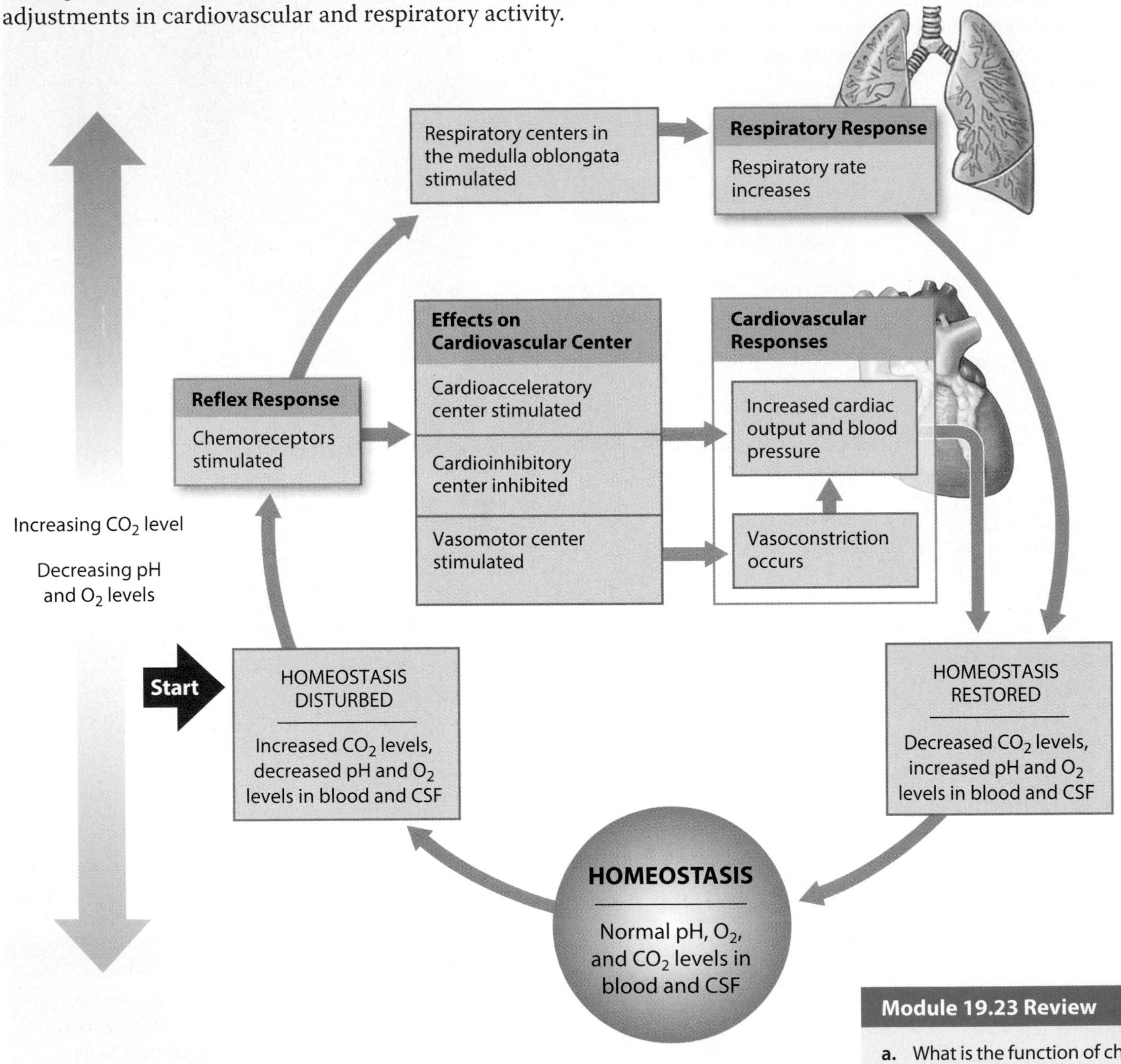

Respiratory centers in the medulla oblongata stimulated
Respiratory Response
Respiratory rate increases
Effects on Cardiovascular Center
Cardioacceleratory center stimulated
Cardioinhibitory center inhibited
Vasomotor center stimulated
Cardiovascular Responses
Increased cardiac output and blood pressure
Vasoconstriction occurs
Reflex Response
Chemoreceptors stimulated
Increasing CO_2 level
Decreasing pH and O_2 levels
Start
HOMEOSTASIS DISTURBED
Increased CO_2 levels, decreased pH and O_2 levels in blood and CSF
HOMEOSTASIS RESTORED
Decreased CO_2 levels, increased pH and O_2 levels in blood and CSF
HOMEOSTASIS
Normal pH, O_2, and CO_2 levels in blood and CSF

Module 19.23 Review

a. What is the function of chemoreceptor reflexes?

b. Cite the locations of chemoreceptors.

c. What effect does an increase in the respiratory rate have on CO_2 levels?

LO **19.23** Describe the role of chemoreceptor reflexes in adjusting cardiovascular activity.

Module 19.24

The cardiovascular center makes extensive adjustments to cardiac output and blood distribution during exercise

At Rest

At rest, cardiac output averages around 5.8 L/min. Note the pattern of blood distribution to major organs, especially the skeletal muscles, brain, and abdominal viscera.

Light Exercise

As you begin light exercise, three interrelated changes take place:

1. Vasodilation occurs, peripheral resistance drops, and blood flow through the capillaries increases.
2. The venous return increases as skeletal muscle contractions squeeze blood along the peripheral veins. At the same time, each inhalation creates a negative pressure in the thoracic cavity that pulls blood into the venae cavae from their branches. This mechanism is called the **respiratory pump**.
3. Cardiac output increases, primarily due to the increased venous return.

Heavy Exercise

At higher levels of exertion, cardiac output increases toward maximal levels. Major changes in the peripheral distribution of blood allow a massive increase in blood flow to skeletal muscles while preventing a potentially disastrous decrease in systemic blood pressure. Under massive sympathetic stimulation, the vasomotor center severely restricts blood flow to "nonessential" organs, such as the digestive viscera. Although blood flow to most tissues is diminished, body temperature rises and skin perfusion increases to promote heat loss. Only the blood supply to the brain remains unchanged.

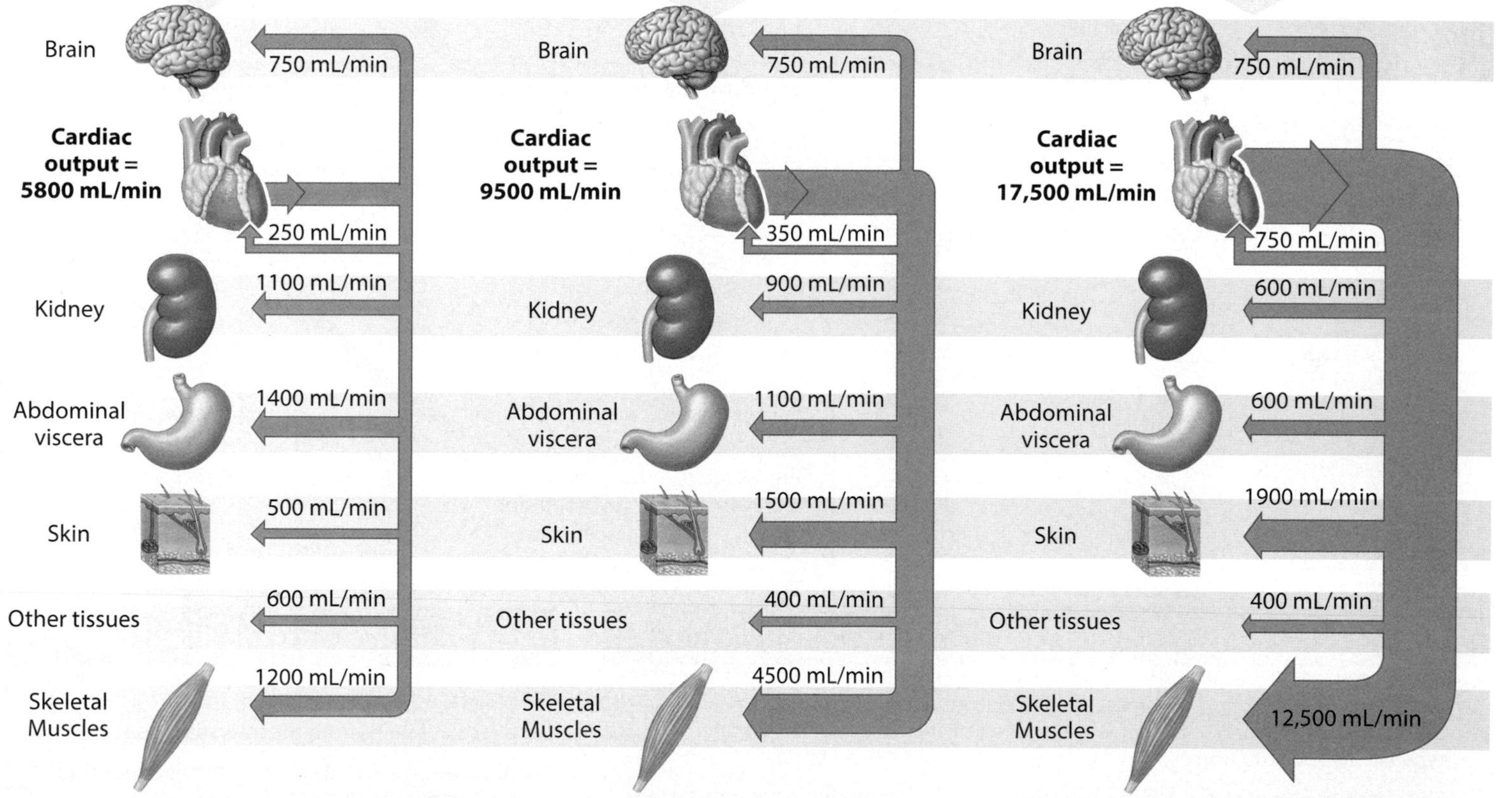

Brain
750 mL/min
Cardiac output = 5800 mL/min
250 mL/min
Kidney
1100 mL/min
Abdominal viscera
1400 mL/min
Skin
500 mL/min
Other tissues
600 mL/min
Skeletal Muscles
1200 mL/min
Brain
750 mL/min
Cardiac output = 9500 mL/min
350 mL/min
Kidney
900 mL/min
Abdominal viscera
1100 mL/min
Skin
1500 mL/min
Other tissues
400 mL/min
Skeletal Muscles
4500 mL/min
Brain
750 mL/min
Cardiac output = 17,500 mL/min
750 mL/min
Kidney
600 mL/min
Abdominal viscera
600 mL/min
Skin
1900 mL/min
Other tissues
400 mL/min
Skeletal Muscles
12,500 mL/min

Cardiovascular performance improves significantly with training. Trained athletes have bigger hearts and larger stroke volumes than do nonathletes. Because cardiac output is equal to the stroke volume times the heart rate, at a given cardiac output, the larger the stroke volume, the slower the heart rate. An athlete at rest can maintain normal blood flow to peripheral tissues at a heart rate as low as 32 beats per minute, and, when necessary, the cardiac output of an athlete in peak condition can increase to levels 50 percent higher than those of nonathletes.

Module 19.24 Review

a. Describe the respiratory pump.

b. Describe the changes in cardiac output and blood flow during exercise.

c. Why must blood flow to visceral organs be reduced during exercise?

19.24 Explain how the cardiovascular system responds to the demands of exercise.

Short-term and long-term mechanisms compensate for a reduction in blood volume

1 When hemostasis fails to prevent significant blood loss, the entire cardiovascular system makes adjustments to maintain blood pressure and restore blood volume. The immediate problem is the maintenance of adequate blood pressure and peripheral blood flow. The long-term problem is the restoration of normal blood volume. These mechanisms can cope with blood losses equivalent to approximately 30 percent of total blood volume.

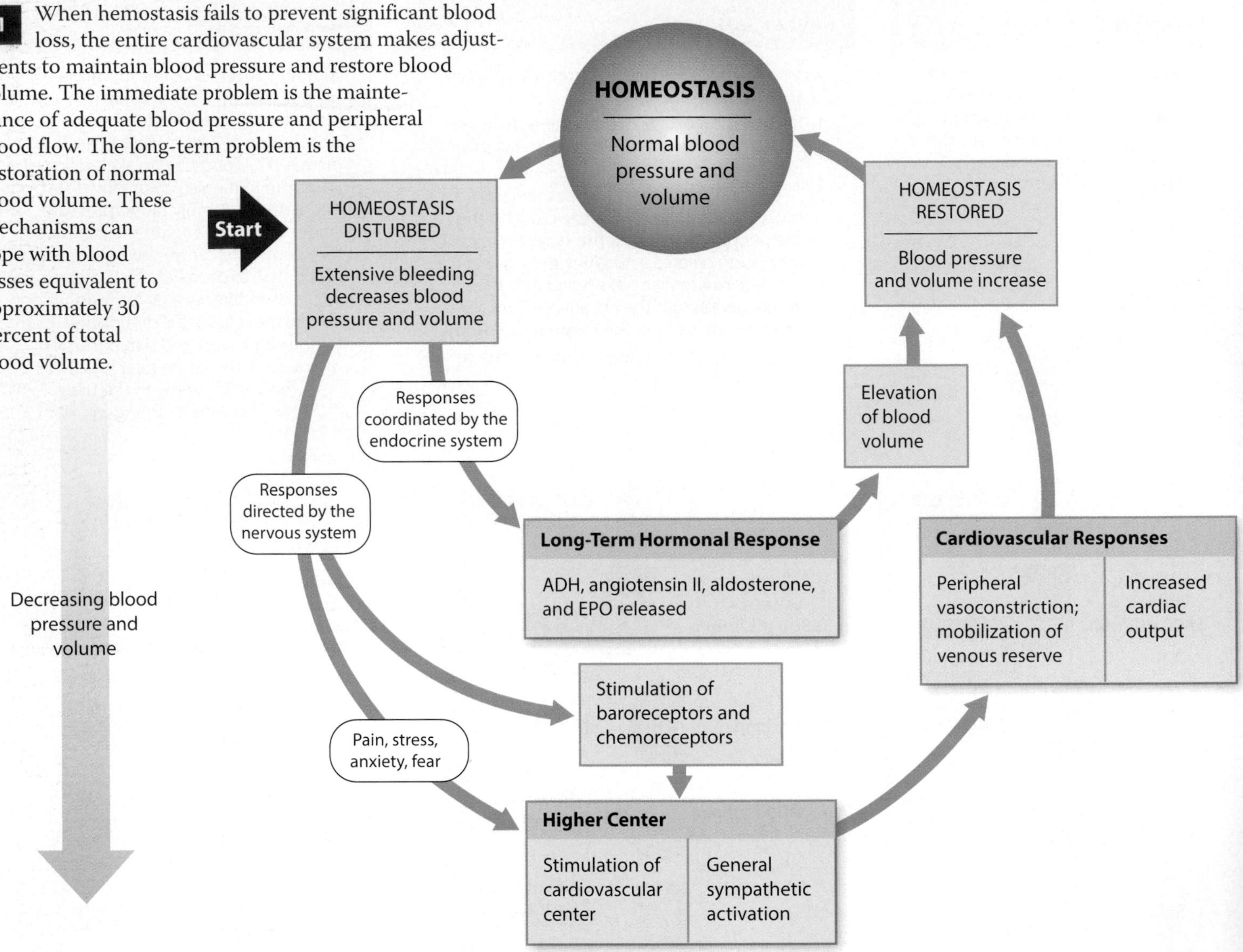
HOMEOSTASIS
Normal blood pressure and volume
Start
HOMEOSTASIS DISTURBED
Extensive bleeding decreases blood pressure and volume
HOMEOSTASIS RESTORED
Blood pressure and volume increase
Responses coordinated by the endocrine system
Responses directed by the nervous system
Elevation of blood volume
Long-Term Hormonal Response
ADH, angiotensin II, aldosterone, and EPO released
Cardiovascular Responses
Peripheral vasoconstriction; mobilization of venous reserve
Increased cardiac output
Decreasing blood pressure and volume
Stimulation of baroreceptors and chemoreceptors
Pain, stress, anxiety, fear
Higher Center
Stimulation of cardiovascular center
General sympathetic activation

Short-Term Responses to Blood Loss

- Carotid and aortic reflexes increase cardiac output and cause peripheral vasoconstriction. Cardiac output is maintained by increasing the heart rate, typically to 180–200 bpm.
- The combination of stress and anxiety stimulates the sympathetic nervous system, which causes a further increase in vasomotor tone, constricting the arterioles and increasing blood pressure. At the same time, venoconstriction, the constriction of large and medium-sized veins, improves venous return and shifts blood into the arterial system.
- Sympathetic activation causes the secretion of E and NE by the adrenal medullae, increasing cardiac output and extending peripheral vasoconstriction. The release of ADH by the neurohypophysis and the production of angiotensin II enhance vasoconstriction; ADH also participates in the long-term response.

Long-Term Responses to Blood Loss

- The decrease in capillary blood pressure triggers a recall of fluids from the interstitial spaces.
- Aldosterone and ADH promote fluid retention and reabsorption by the kidneys.
- Thirst increases, and additional water is obtained by absorption across the digestive tract.
- Erythropoietin stimulates the maturation of red blood cells in the red bone marrow, resulting in increased blood volume and improved oxygen delivery to peripheral tissues.

2 **Shock** is an acute cardiovascular crisis marked by low blood pressure (**hypotension**) and inadequate peripheral blood flow. Severe and potentially fatal signs and symptoms develop as vital tissues become starved for oxygen and nutrients. The most common cardiovascular causes of shock are a drop in cardiac output after hemorrhaging or damage to the heart, as in a heart attack. Shock develops when normal compensation mechanisms fail to maintain adequate cardiac output and peripheral blood flow. **Circulatory shock** involves a series of interlocking positive feedback loops that begin when blood loss exceeds about 35 percent of total blood volume. **Progressive shock** is an initial stage characterized by positive feedback loops that accelerate tissue damage. Unless the positive feedback loops are broken, the person will enter the fatal stage of **irreversible shock**.

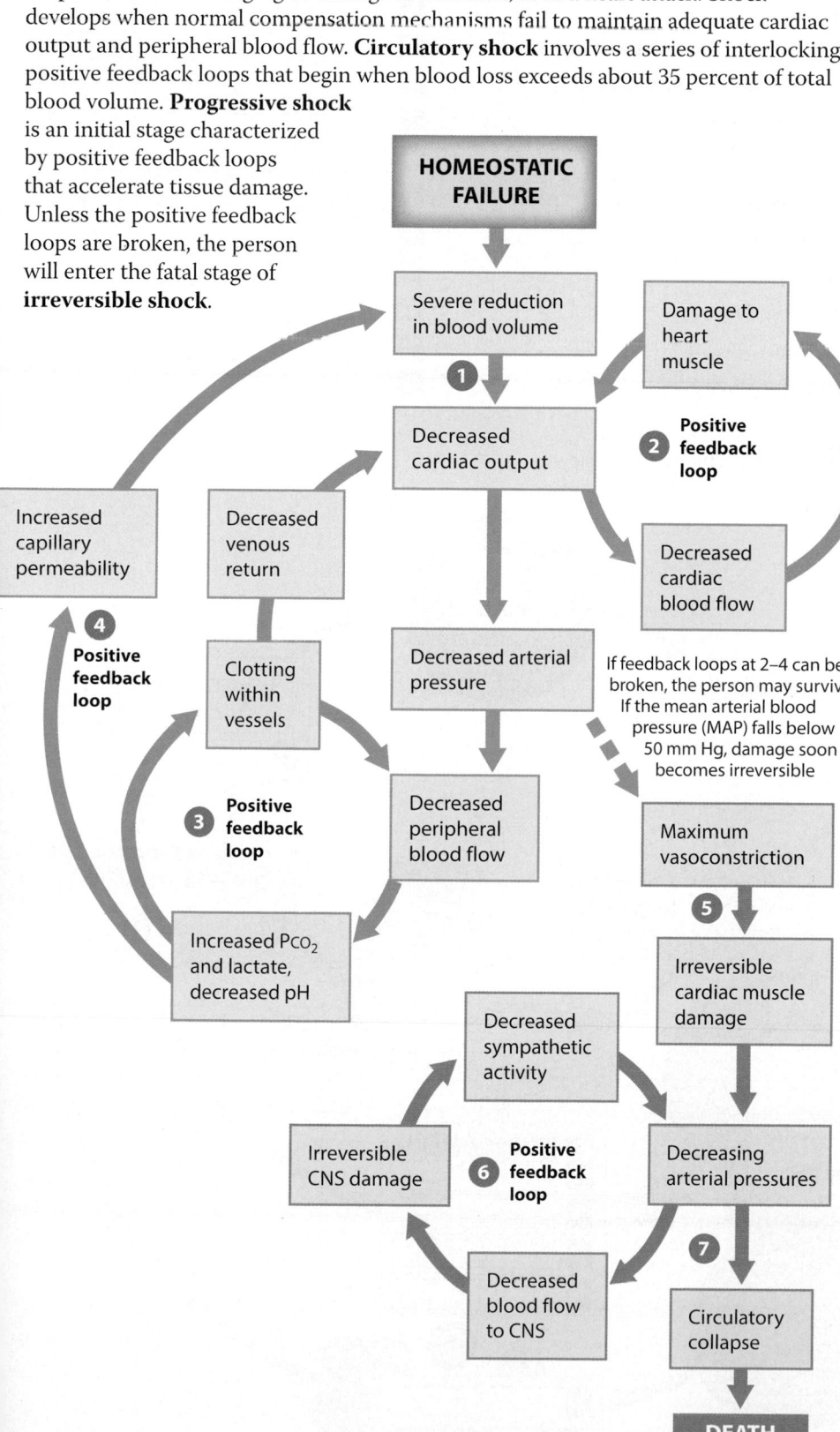

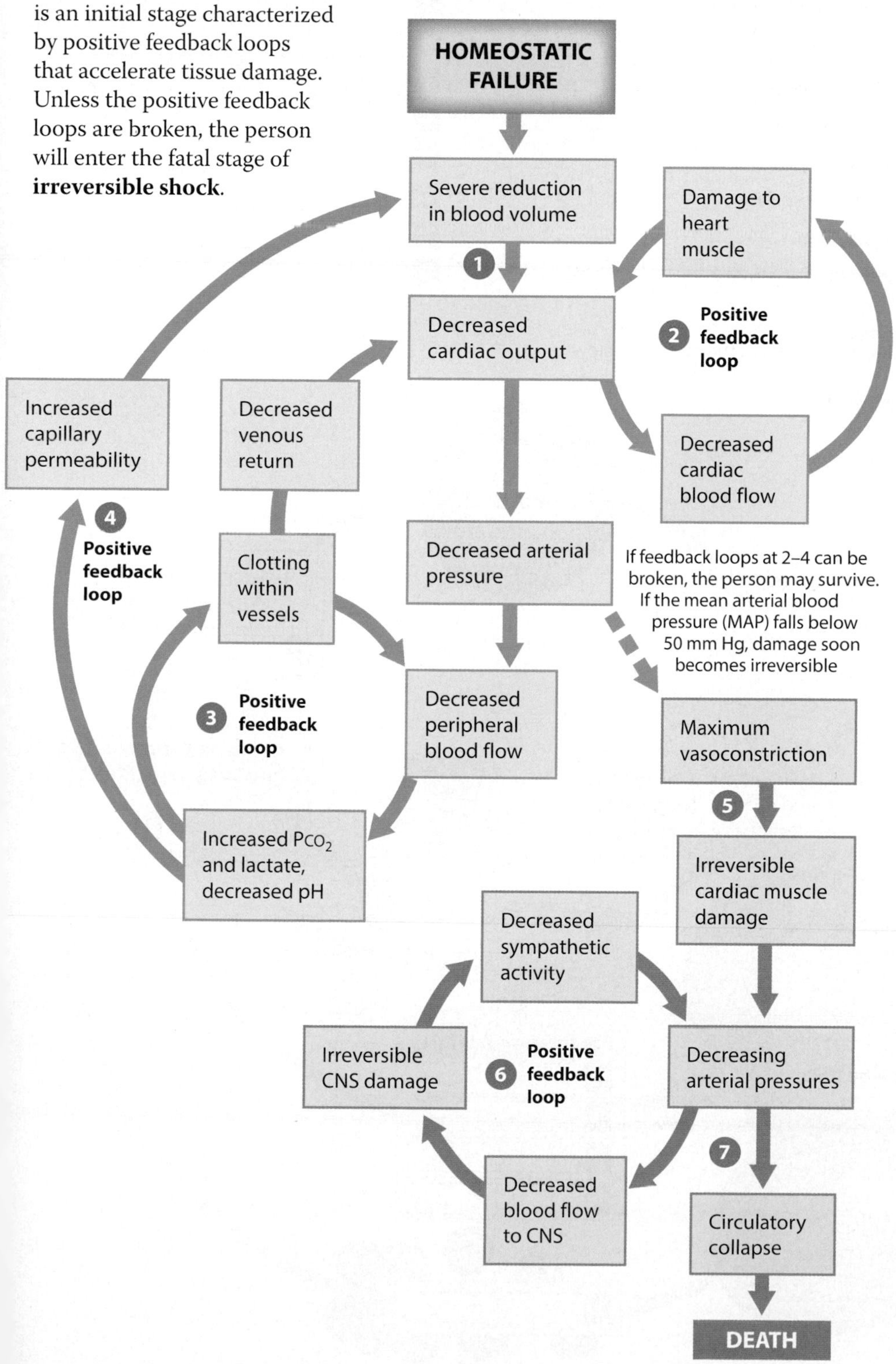

HOMEOSTATIC FAILURE
Severe reduction in blood volume
Damage to heart muscle
1
Decreased cardiac output
2 Positive feedback loop
Increased capillary permeability
Decreased venous return
Decreased cardiac blood flow
4 Positive feedback loop
Clotting within vessels
Decreased arterial pressure
If feedback loops at 2–4 can be broken, the person may survive. If the mean arterial blood pressure (MAP) falls below 50 mm Hg, damage soon becomes irreversible
3 Positive feedback loop
Decreased peripheral blood flow
Maximum vasoconstriction
5
Increased P_{CO_2} and lactate, decreased pH
Irreversible cardiac muscle damage
Decreased sympathetic activity
Irreversible CNS damage
6 Positive feedback loop
Decreasing arterial pressures
7
Decreased blood flow to CNS
Circulatory collapse
DEATH

Progressive Shock

1 When blood volume declines by more than 35 percent, blood pressure remains abnormally low, venous return is reduced, and cardiac output is inadequate despite sustained vasoconstriction and the mobilization of the venous reserve.

2 The first of several dangerous positive feedback loops begins when low cardiac output reduces blood flow to the heart. This damages the myocardium, which leads to a further reduction in cardiac output.

3 Reduced cardiac output accelerates oxygen starvation in peripheral tissues, and the resulting chemical changes promote intravascular clotting that further restricts peripheral blood flow.

4 Local pH changes increase capillary permeability, and this further reduces blood volume.

Irreversible Shock

5 Carotid sinus baroreceptors trigger a massive activation of the vasomotor center. To preserve blood flow to the brain at any cost, the sympathetic output causes a sustained and maximal vasoconstriction. This decreases peripheral circulation to an absolute minimum, but it increases blood pressure to about 70 mm Hg and temporarily improves blood flow to cerebral vessels.

6 Unless prompt treatment is provided, blood pressure will again decrease. By now the heart muscle will be seriously damaged as blood flow to the brain decreases.

7 **Circulatory collapse** occurs when arteriolar smooth muscles and precapillary sphincters become unable to contract, despite the commands of the vasomotor center. The result is widespread peripheral vasodilation, an immediate and fatal decrease in blood pressure, and the cessation of blood flow.

Module 19.25 Review

a. Identify the compensatory mechanisms that respond to blood loss.

b. Name the immediate and long-term problems related to profuse blood loss.

c. Describe circulatory shock, progressive shock, and irreversible shock.

LO **19.25** Explain the body's response to blood loss.

Matching

Match each lettered term with the most closely related description.

a. autoregulation
b. venous return
c. local vasodilators
d. natriuretic peptides
e. chemoreceptors
f. baroreceptors
g. turbulence
h. net hydrostatic pressure
i. medulla oblongata
j. edema
k. viscosity
l. osmotic pressure

1 Detect changes in pressure
2 Vasomotor center
3 Aided by thoracic pressure changes due to breathing
4 Causes immediate, local homeostatic responses
5 Decreased tissue O_2 and increased CO_2
6 Carotid bodies
7 Forces water into a capillary
8 Opposite of smooth blood flow
9 Resistance to flow
10 Forces water out of a capillary
11 Peripheral vasodilation
12 Excess interstitial fluid accumulation

1 ____________________
2 ____________________
3 ____________________
4 ____________________
5 ____________________
6 ____________________
7 ____________________
8 ____________________
9 ____________________
10 ____________________
11 ____________________
12 ____________________

Match each lettered item with the appropriate numbered blank box in the diagram below.

a. increased cardiac output and peripheral vasoconstriction
b. aldosterone secreted
c. increased blood pressure
d. blood pressure and volume decrease
e. thirst stimulated
f. increased blood volume
g. blood pressure and volume increase
h. antidiuretic hormone released
i. increased red blood cell formation
j. widepread vasoconstriction

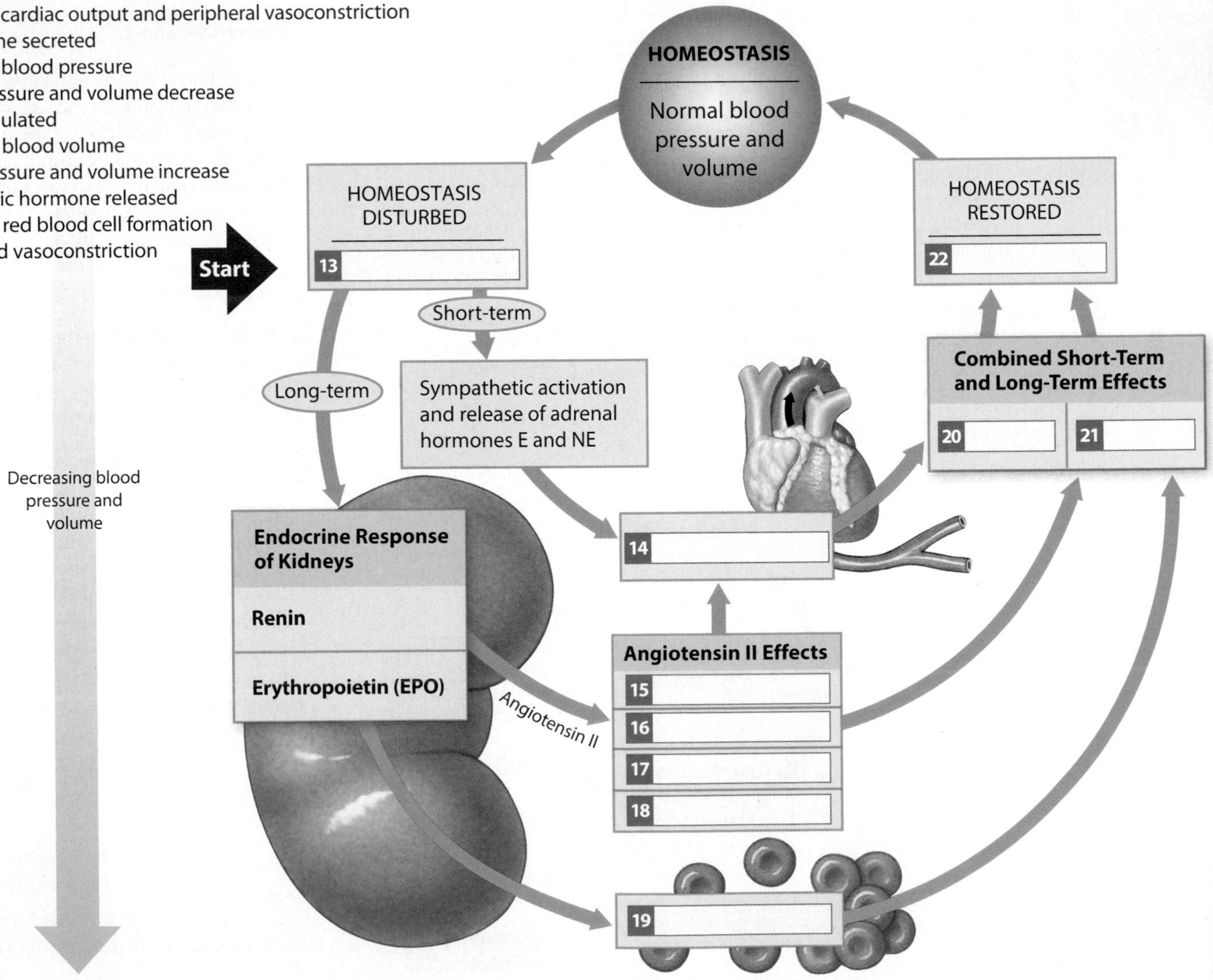
HOMEOSTASIS
Normal blood pressure and volume
HOMEOSTASIS DISTURBED
13
Start
Short-term
Long-term
Sympathetic activation and release of adrenal hormones E and NE
Decreasing blood pressure and volume
Endocrine Response of Kidneys
Renin
Erythropoietin (EPO)
Angiotensin II
14
Angiotensin II Effects
15
16
17
18
19
HOMEOSTASIS RESTORED
22
Combined Short-Term and Long-Term Effects
20
21

Study Outline

SECTION 1 • Structure of the Heart

19.1 The heart has a superior base, an inferior apex, and four borders p. 685

1. The **base** of the heart is at the superior end where the great veins and arteries are attached.
2. The **apex** is a pointed tip at the inferior aspect of the heart.
3. The four borders of the heart are the **superior border**, **inferior border**, **right border**, and **left border**.

19.2 The heart wall contains concentric layers of cardiac muscle tissue p. 686

4. The heart wall contains three layers: **epicardium**, **myocardium**, and **endocardium**.
5. The epicardium, or visceral pericardium, is a serous membrane that covers the surface of the heart. The **parietal pericardium** is a serous membrane that forms the outer wall of the pericardial cavity. It forms part of the **pericardial sac** that surrounds the heart.

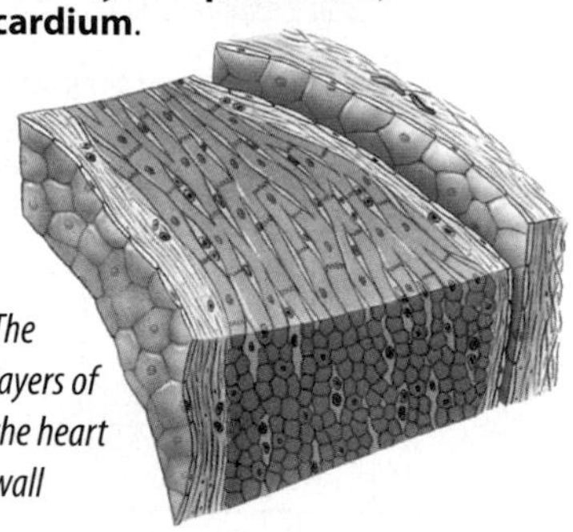

The layers of the heart wall

6. The myocardium contains cardiac muscle tissue, blood vessels, and nerves.
7. The endocardium is a simple squamous epithelium and underlying areolar tissue that covers the inner surface of the heart and heart valves.
8. Cardiac muscle cells are small; contain a single centrally located nucleus, branching interconnections between cells, and specialized intercellular connections known as **intercalated discs**; and are striated in appearance.
9. Cardiac muscle cells are almost totally dependent on aerobic metabolism, and are richly supplied with capillaries. Gap junctions between these cells allow for ions and small molecules to move from one cell to another, and create a direct electrical connection between them. Cardiac muscle is mechanically, chemically, and electrically connected such that it resembles a single, enormous muscle cell, or **functional syncytium**.

19.3 The heart is located in the mediastinum, and enclosed by the pericardial cavity p. 688

10. The heart sits in the anterior portion of the **mediastinum**. The mediastinum also contains the **great vessels**, thymus, esophagus, and trachea.
11. The pericardial cavity is the space between the epicardium and parietal pericardium, and contains 15–50 mL of pericardial fluid.
12. **Pericarditis** is an infection of the pericardium. **Cardiac tamponade** is fluid accumulation within the pericardial cavity.

19.4 The boundaries between the chambers of the heart can be identified on its external surface p. 690

13. The **auricle** of the atrium is an expandable pouch. The **coronary sulcus** is a deep groove between the atria and the ventricles. The **anterior interventricular sulcus** is a depression between the left and right ventricles. The **ligamentum arteriosum** is a remnant of the fetal connection between the aorta and pulmonary trunk.
14. The **coronary sinus** carries blood collected from the coronary veins to the right atrium. The **posterior interventricular sulcus** is a shallow depression between the left and right ventricles.

19.5 The heart has an extensive blood supply p. 692

15. The left and right **coronary arteries** originate at the base of the ascending aorta.
16. The **right coronary artery** follows the coronary sulcus, and supplies blood to the right atrium, portions of both ventricles, and portions of the conducting system of the heart. **Marginal arteries** from the right coronary artery supply the surface of the right ventricle.
17. The **left coronary artery** supplies blood to the left atrium, left ventricle, and interventricular septum. The **anterior interventricular artery** is a branch of the left coronary artery that lies within the anterior interventricular sulcus.
18. The **circumflex artery** is a branch of the left coronary artery that curves to the left around the coronary sulcus. The **posterior interventricular artery** is a branch of the right coronary artery.

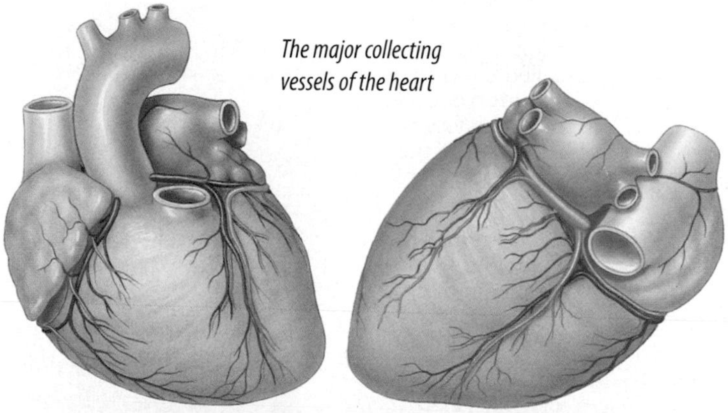

The major collecting vessels of the heart

19. The **anterior cardiac veins** drain the anterior surface of the right ventricle and empty directly into the right atrium. The **great cardiac vein** drains blood from regions supplied by the anterior interventricular artery.
20. The **posterior cardiac vein**, **small cardiac vein**, and **middle cardiac vein** all drain blood from the posterior aspect of the heart and empty into the coronary sinus.

19.6 Internal valves control the direction of blood flow between the heart chambers p. 694

21. The right atrium communicates with the right ventricle, and the left atrium communicates with the left ventricle.

22. The atria are separated by an **interatrial septum**, and a much thicker **interventricular septum** separates the two ventricles.
23. **Atrioventricular** (**AV**) **valves** permit the flow of blood in one direction only: from atria to ventricles. The **right atrioventricular valve** has three **cusps**, and is also known as the **tricuspid valve**. The **left atrioventricular valve** has two cusps, and is also called the **bicuspid valve** or **mitral valve**.
24. The free edge of each AV valve is attached to **chordae tendineae** that originate at **papillary muscles**.
25. Blood leaving the right ventricle passes through the **pulmonary valve** (semilunar) to enter the pulmonary trunk. Blood leaving the left ventricle passes through the **aortic valve** (semilunar) and into the ascending aorta.
26. The atria collect blood returning to the heart and convey it to their attached ventricles.
27. When the right ventricle contracts, it squeezes blood against the left ventricular wall.
28. When the left ventricle contracts, the diameter of the ventricular chamber decreases, and the distance between the apex and base decreases.

19.7 When the heart beats, the AV valves close before the semilunar valves open, and the semilunar valves close before the AV valves open p. 696

29. When the ventricles are relaxed, the chordae tendineae are loose, and the AV valves allow blood flow from the atria into the ventricles.
30. When the ventricles contract, the AV valves close to prevent **regurgitation** into the atria.
31. As the ventricular pressure rises above those in the pulmonary and systemic circuits, the aortic and pulmonary valves open and blood flows out of the ventricles.
32. Interconnected bands of dense connective tissue called the **cardiac skeleton** encircle the heart valves and bases of the pulmonary trunk and aorta.

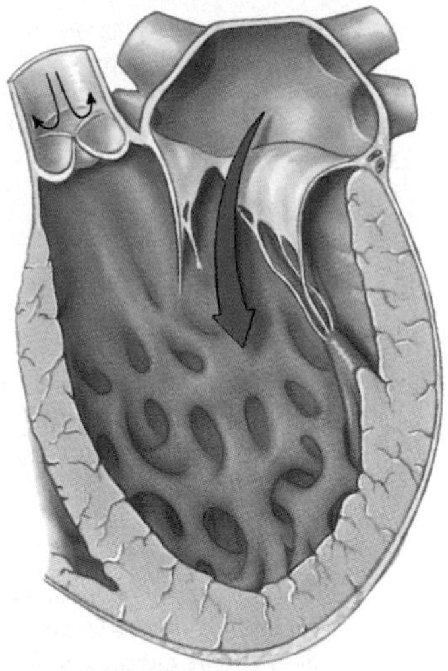

33. The **semilunar valves** do not require muscular braces.
34. A valve with **valvular heart disease (VHD)** cannot maintain adequate blood flow. This can be caused by congenital malformations or **carditis** (inflammation of the heart) and require valve replacement.

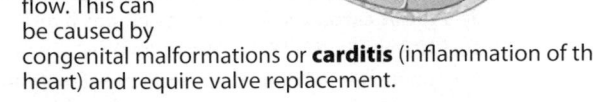

19.8 Arteriosclerosis can lead to coronary artery disease p. 698

35. **Arteriosclerosis** is a thickening and toughening of the arterial walls, and can cause coronary artery disease and stroke.
36. **Atherosclerosis**, deposits of **plaque** in the vessel wall, is the most common form of arteriosclerosis.

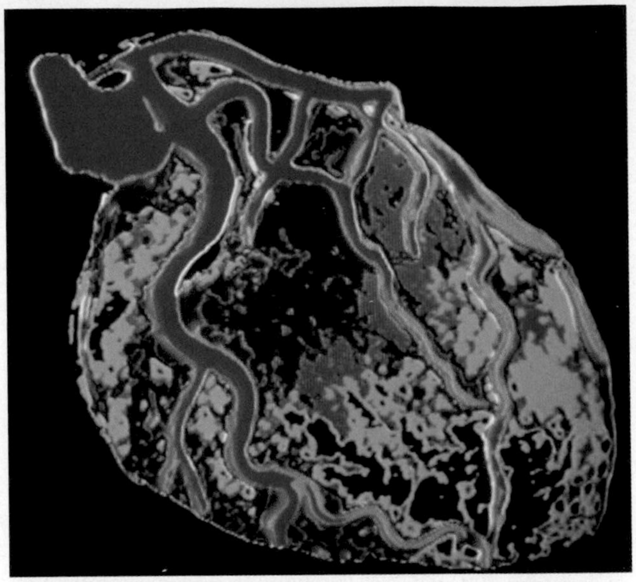

A color-enhanced digital subtraction angiography (DSA) of a normal heart

37. **Coronary artery disease** (**CAD**) is partial or complete blockage of coronary circulation, known as **coronary ischemia**. Blood flow can be blocked by plaque, a **thrombus** (blood clot), and spasms in the vessel.
38. Plaques can be treated by **balloon angioplasty**, or by surgical insertion of a wire-mesh tube called a **stent**.

SECTION 2 • Cardiac Cycle

19.9 The cardiac cycle is a complete round of systole and diastole p. 701

39. The period between the start of one heartbeat and the beginning of the next is a single **cardiac cycle**.
40. All four chambers do not contract at once. Instead, the two atria contract and push the blood into the ventricles, and then the two ventricles contract and push the blood into the pulmonary and systemic circuits.
41. The cardiac cycle can be divided into two phases: **systole** (contraction) and **diastole** (relaxation).
42. A representative heart rate is 75 beats per minute (bpm). A representative cardiac cycle lasts 800 msec.

19.10 The cardiac cycle creates pressure gradients that maintain blood flow p. 702

43. When the cardiac cycle begins, all four chambers are relaxed, and the ventricles are partially filled with blood.
44. During **atrial systole**, the atria contract and fill the ventricles. **Atrial diastole** now begins and continues until the next cardiac cycle.
45. **Ventricular systole** has two phases: **isovolumetric contraction**, when the ventricles contract, the AV valves close, but pressure is not great enough to open the semilunar valves; and **ventricular ejection**, when the pressure is great enough to open the semilunar valves and blood is ejected.
46. **Ventricular diastole—early** is when the pressure drops and blood flows back against the semilunar valve cusps to close them. Blood flows into the relaxed atria, but AV valves are still closed.

47. During **ventricular diastole—late**, all chambers are relaxed and filling. The ventricles fill passively to 70 percent of their final volume.

48. Four heart sounds can be heard with a stethoscope, and they are designated as $\mathbf{S_1}$ to $\mathbf{S_4}$. S_1 is also known as "lubb" and is the sound of the AV valves closing. S_2 is "dupp" and is the sound of the semilunar valves closing. The two other sounds are associated with blood flow, are faint, and are seldom audible in healthy adults.

19.11 The heart rate is established by the sinoatrial node and distributed by the conducting system p. 704

49. **Cardiac output** (**CO**) is the amount of blood ejected by the left ventricle in one minute.

50. CO depends on heart rate (HR), and **stroke volume** (**SV**), which is the amount of blood pumped from the ventricle in a single heartbeat.

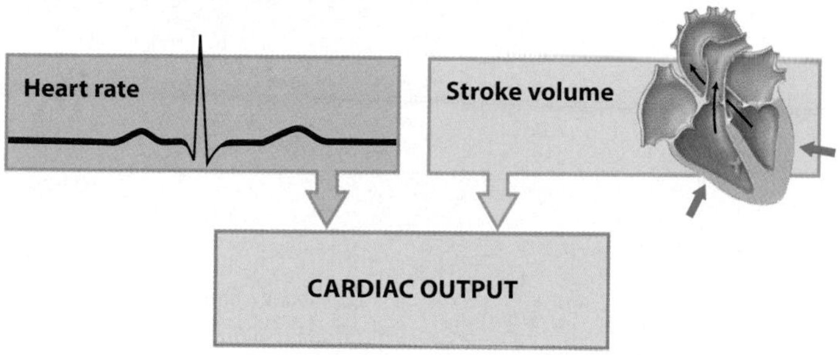

Heart rate
Stroke volume
CARDIAC OUTPUT

51. CO can be calculated as HR × SV. HR: 75 bpm × SV: 80 mL/beat = CO: 6000 mL/min

52. The heart beats independently of neural or hormonal stimulation. This **automaticity** is initiated and distributed by the **conducting system**.

53. Each heartbeat begins with an action potential at the **sinoatrial** (**SA**) **node**. The impulse continues along **internodal pathways**, the **atrioventricular** (**AV**) **node**, the **AV bundle**, **bundle branches**, and **Purkinje fibers**.

19.12 Cardiac muscle cell contractions last longer than skeletal muscle fiber contractions primarily due to differences in membrane permeability p. 706

54. Skeletal muscle fiber action potentials are brief, and end as the muscle twitch begins.

55. Cardiac muscle fiber action potentials are prolonged because calcium ions enter the cell for an extended period. The result is a longer period of active contraction. There is also a long refractory period, which prevents tetanic contractions.

56. There are three stages of cardiac muscle cell action potential: **rapid depolarization**, **plateau**, and **repolarization**.

19.13 The intrinsic heart rate can be altered by autonomic activity p. 708

57. The cells of the SA and AV node cannot maintain a stable resting potential. After repolarization, a gradual, spontaneous depolarization called a **prepotential** or **pacemaker potential** begins. This is fastest at the SA node.

58. Acetylcholine released by the parasympathetic neurons will result in a slower heart rate. Norepinephrine released by sympathetic neurons will increase heart rate.

59. The normal resting heart rate range is 60–100 bpm. Age, health, and physical conditioning influence it.

60. A slower-than-normal heart rate is called **bradycardia**; a faster-than-normal heart rate is called **tachycardia**.

61. The **cardioinhibitory center** of the medulla oblongata controls the parasympathetic neurons that slow heart rate. The **cardioacceleratory center** controls the sympathetic neurons that increase heart rate.

19.14 Stroke volume depends on the relationship between end-diastolic volume and end-systolic volume p. 710

62. The amount of blood in a ventricle at the end of ventricular diastole is **end-diastolic volume** (**EDV**). After ventricular systole, some blood will remain in the ventricle. This is called **end-systolic volume** (**ESV**).

63. EDV is influenced by **venous return** and **filling time**. The Frank-Starling law of the heart describes the relationship between EDV and stroke volume: The greater the EDV, the greater the stroke volume. ESV is influenced by **contractility** and **afterload**.

19.15 Cardiac output is regulated by adjustments in heart rate and stroke volume p. 712

64. **Cardiac output (CO)** varies widely to meet metabolic demands. Important factors that affect **heart rate (HR)** are sympathetic and parasympathetic stimulation, body temperature, and hormones. Important factors that influence **stroke volume (SV)** are exercise, blood volume, blood flow, the autonomic nervous system, and hormones.

19.16 Normal and abnormal cardiac activity can be detected in an electrocardiogram p. 714

65. The electrical events in the heart can be recorded in an **electrocardiogram** (**ECG** or **EKG**).

66. The **P wave** represents atrial depolarization, the **QRS complex** represents ventricular depolarization, and the **T wave** represents ventricular repolarization. Repolarization of the atria is not visible because it is masked by the QRS complex.

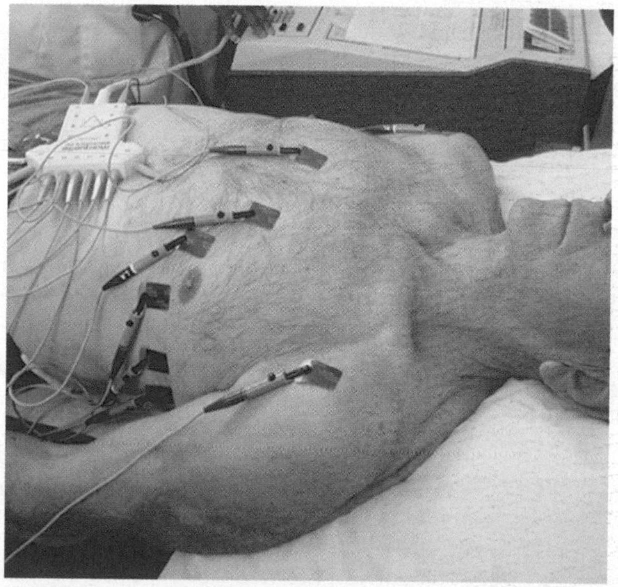

Preparing for an electrocardiogram

67. An ECG can be useful in detecting and diagnosing many **cardiac arrhythmias**.

SECTION 3 • Coordination of Cardiac Output and Blood Flow

19.17 Pressure, resistance, and venous return affect cardiac output p. 717

68. **Cardiac output** and the distribution of blood in the pulmonary and systemic circuits must constantly be adjusted to meet the demands of active tissues.
69. The factors of CO regulation are: **arterial blood pressure**, **peripheral resistance**, **capillary pressure**, **venous pressure**, **venous return**, and **neural and hormonal regulation**.

19.18 Under normal conditions, vessel diameter is the primary source of resistance within the cardiovascular system p. 718

70. The heart must develop enough pressure to overcome **total peripheral resistance**.
71. **Vascular resistance** is the opposition to blood flow in vessels. The longer the vessel length, the greater the resistance due to increased friction. The larger the vessel diameter, the less the resistance due to decreased friction.
72. A small change in vessel diameter produces a large change in resistance.

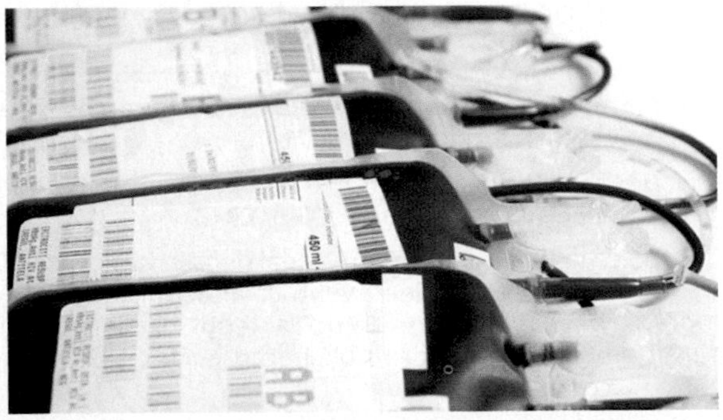

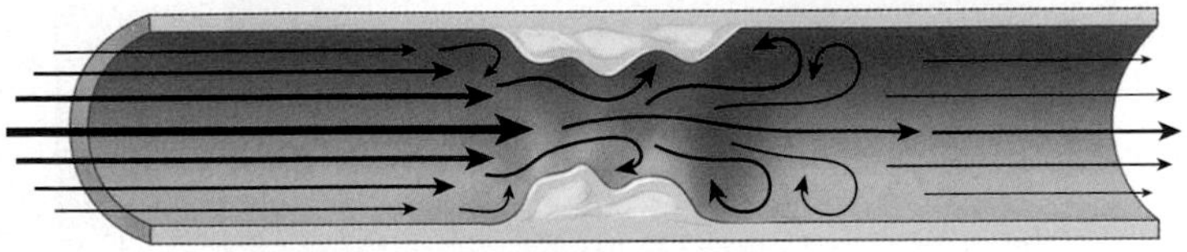

Blood flow turbulence due to plaque

73. **Viscosity** is the resistance to flow. More viscous blood, and turbulence due to plaque or other factors, increase peripheral resistance.

19.19 Blood flow is determined by the interplay between arterial pressure and peripheral resistance p. 720

74. **Blood flow** is directly proportional to **blood pressure**, and inversely proportional to **peripheral resistance**.
75. Absolute pressure is less important than the **pressure gradient**: the difference in pressure from one end of the vessel to the other.
76. As blood moves toward the capillaries, the diameter of the arteries decreases, blood pressure drops, and the blood moves more slowly.
77. As blood returns to the heart, the diameter of the veins increases, blood pressure continues to drop, but blood velocity accelerates.
78. Arterial pressure rises during ventricular systole, and falls during ventricular diastole. The peak pressure measured is called **systolic pressure**, and the minimum pressure measured is called **diastolic pressure**.
79. **Pulse pressure** is the difference between systolic and diastolic pressures. **Mean arterial pressure** (**MAP**) is calculated by adding one-third of pulse pressure to diastolic pressure.

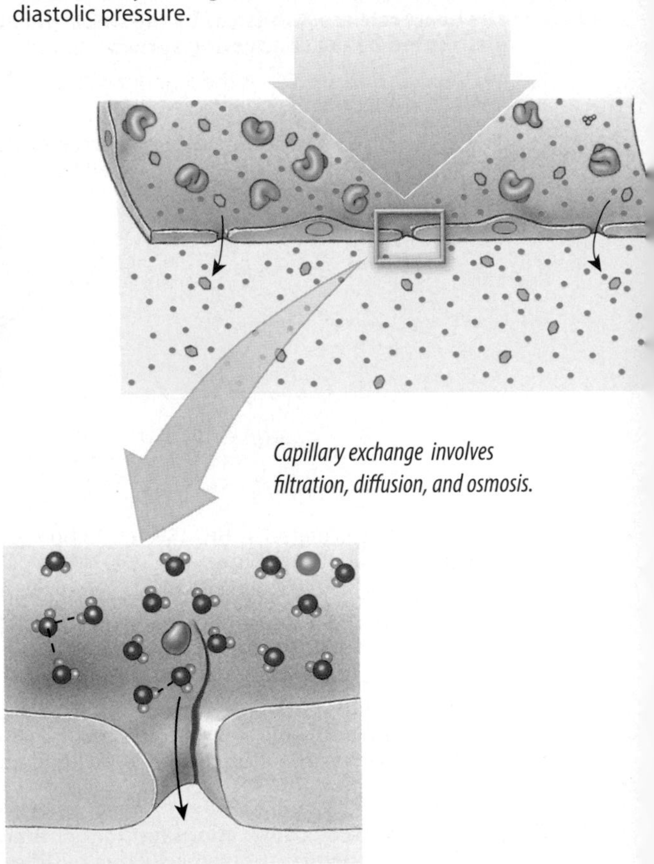

Capillary exchange involves filtration, diffusion, and osmosis.

80. **Capillary exchange** involves a combination of **filtration**, diffusion, and osmosis. **Capillary hydrostatic pressure** is the blood pressure within capillary beds.

19.20 Capillary exchange is a dynamic process that includes diffusion, filtration, and reabsorption p. 722

81. Diffusion occurs most rapidly when the distances are short, the concentration gradient is large, and the ions or molecules are small.
82. In a capillary, blood pressure declines as blood flows from the arterial end to the venous end. Filtration predominates at the arterial end where **capillary hydrostatic pressure** (**CHP**) is highest, and reabsorption predominates at the venous end where **blood colloid osmotic pressure** (**BCOP**) is highest.
83. More water leaves the bloodstream during filtration than is retrieved through reabsorption. The difference is approximately 3.6 L/day and is returned to the venous system by the lymphatic vessels.

19.21 Cardiovascular regulatory mechanisms respond to changes in blood pressure or blood chemistry p. 724

84. Homeostatic mechanisms regulate cardiovascular activity to ensure that **tissue perfusion** (blood flow through tissues) meets the demands for oxygen and nutrients.

85. The two regulatory pathways are autoregulation and central regulation.

86. **Autoregulation** involves changes in capillary blood flow in response to chemical changes in the interstitial fluid. Local **vasodilators** accelerate blood flow.

87. **Central regulation** involves neural **(cardioacceleratory center)** and endocrine mechanisms (norepinephrine).

88. **Baroreceptor reflexes** respond to changes in blood pressure to maintain blood pressure within a normal range.

19.22 Endocrine responses to low blood pressure and low blood volume are very different from those to high blood pressure and high blood volume p. 726

89. When blood pressure and blood volume decrease, epinephrine and norepinephrine stimulate cardiac output and vasoconstriction. Other hormones in long-term response include ADH, angiotensin II, EPO, and aldosterone.

90. Increased blood volume abnormally stretches the heart walls causing cardiac muscle cells to release the **natriuretic peptides**: **atrial natriuretic peptide** and **brain natriuretic peptide**. These hormones reduce blood volume and blood pressure.

19.23 Chemoreceptors monitor the chemical composition of the blood and cerebrospinal fluid p. 728

91. **Chemoreceptor reflexes** stimulated by chemoreceptors in the **carotid bodies**, **aortic bodies**, and medulla oblongata respond to changes in carbon dioxide, oxygen, or pH in blood and CSF.

92. Chemoreceptor reflexes trigger adjustments in cardiovascular and respiratory activity.

19.24 The cardiovascular center makes extensive adjustments to cardiac output and blood distribution during exercise p. 729

93. Cardiac output at rest averages around 5.8 L/min. At light exercise it increases to 9500 L/min, and can increase to 17,500 L/min during heavy exercise.

94. Trained athletes have bigger hearts, larger stroke volumes, and lower resting heart rates than nonathletes. The cardiac output of an athlete in peak condition can be 50 percent higher than that of a nonathlete.

19.25 Short-term and long-term mechanisms compensate for a reduction in blood volume p. 730

95. When homeostasis fails to prevent significant blood loss, the entire cardiovascular system makes adjustments to maintain blood pressure and restore blood volume.

96. **Shock** is an acute cardiovascular crisis marked by low blood pressure (**hypotension**) and inadequate peripheral blood flow. It develops when normal compensation mechanisms fail to maintain adequate cardiac output and peripheral blood flow.

97. **Circulatory shock** begins with **progressive shock** (blood loss exceeding 35 percent) and can lead to the fatal stage of **irreversible shock** and **circulatory collapse**.

Chapter Review Questions

True/False

Indicate whether each statement is true or false.

1 The S_1 or "lubb" sound of the heart is produced by the AV valves closing.

2 The T wave of an electrocardiogram represents repolarization of the atria.

3 The circumflex artery supplies blood to the surface of the right ventricle.

4 Isovolumetric contraction is the first phase of ventricular systole, when pressures close the AV valves, but are not high enough to open the semilunar valves.

5 Tachycardia is a faster-than-normal heart rate, usually above 100 bpm.

1 ______

2 ______

3 ______

4 ______

5 ______

Multiple choice

Select the correct answer from the list provided.

6 Cardiac output (CO) is calculated by which of the following formulas?
- ☐ a) CO = HR × ESV
- ☐ b) CO = HR × EDV
- ☐ c) CO = HR × SV
- ☐ d) CO = HR × MAP

7 During diastole, a chamber of the heart
- ☐ a) relaxes and fills with blood.
- ☐ b) contracts and pushes blood into the adjacent chamber.
- ☐ c) experiences a sharp increase in pressure.
- ☐ d) reaches a pressure of approximately 120 mm Hg.

8 Which of the following is longer?
- ☐ a) the refractory period of a skeletal muscle fiber
- ☐ b) the refractory period of a cardiac muscle cell
- ☐ c) the action potential of a skeletal muscle fiber
- ☐ d) the contraction time of a skeletal muscle fiber

9 If the papillary muscles fail to contract
- ☐ a) the AV valves will not open.
- ☐ b) the semilunar valves will not open.
- ☐ c) the AV valves will not close properly.
- ☐ d) the semilunar valves will not close properly.

10 The right atrium does *not* receive blood from the
- ☐ a) inferior vena cava.
- ☐ b) superior vena cava.
- ☐ c) coronary sinus.
- ☐ d) circumflex artery.

11 Which of the following is a fibrous remnant of a fetal connection between the aorta and pulmonary trunk?
- ☐ a) fossa ovalis
- ☐ b) ligamentum arteriosum
- ☐ c) chordae tendineae
- ☐ d) trabeculae carneae

12 The Frank-Starling law of the heart states that
- ☐ a) the greater the EDV, the greater the stroke volume.
- ☐ b) the greater the ESV, the greater the stroke volume.
- ☐ c) the greater the EDV, the lesser the stroke volume.
- ☐ d) the lesser the ESV, the lesser the stroke volume.

13 Which of the following homeostatic disturbances will cause a chemoreceptor reflex response?
- ☐ a) increased pH in blood
- ☐ b) increased CO_2 in blood
- ☐ c) increased O_2 in blood
- ☐ d) increased O_2 in CSF

14 Which of the following organs receives an unchanging flow of blood as exercise intensity increases?
- ☐ a) brain
- ☐ b) kidney
- ☐ c) skin
- ☐ d) skeletal muscles

15 Abnormal stretch of the right atrium due to excessive blood volume triggers cardiac muscle cells to release
- ☐ a) brain natriuretic peptide (BNP).
- ☐ b) erythropoietin (EPO).
- ☐ c) angiotensin-converting enzyme (ACE).
- ☐ d) atrial natriuretic peptide (ANP).

Short answer

16 What role do the chordae tendineae and papillary muscles play in the normal function of the AV valves?

17 Trace the normal pathway of an electrical impulse through the conducting system of the heart.

18 Describe the three distinct layers that make up the heart wall.

19 What effects do sympathetic and parasympathetic stimulation have on the heart?

20 What are the sources and significance of the four heart sounds?

MasteringA&P®

Access more chapter study tools online in the MasteringA&P Study Area:

- **Chapter Quizzes, Chapter Practice Test, Art-labeling Activities, Animations, MP3 Tutor Sessions, and Clinical Case Studies**

- **Practice Anatomy Lab** PAL™
- **Interactive Physiology** iP®
- **A&P Flix** A&PFlix™
- **PhysioEx** PhysioEx™

Chapter Integration • Applying what you have learned

An abnormal click in a 12-year-old's heart

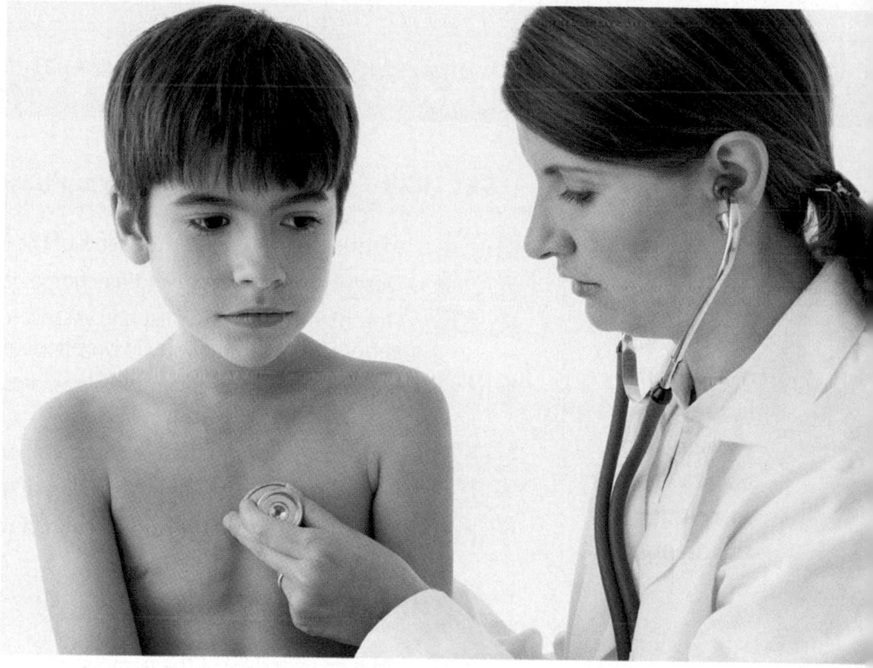

Brian is a 12-year-old boy about to enter the sixth grade. While his pediatrician is listening to his heart during a routine physical examination, she hears an abnormal click during ventricular systole. The doctor informs Brian's mother that she can hear a murmur, or irregular sound, during the beating of his heart, and that the boy most likely has a mitral valve prolapse. She explains that his mitral valve is not functioning perfectly, and that when it closes, it bulges, or prolapses, into the chamber it is attempting to prevent blood from entering. The good news is that she cannot hear any regurgitation.

Brian's mother became noticeably upset at the thought of her son having a heart condition. The doctor calmly explained that his condition is relatively common, is not life-threatening, and often causes no symptoms. She will order an echocardiogram (an ultrasound of the heart) so she can examine his mitral valve and determine the extent of his condition, but that she expects his condition to be relatively minor.

1 **The doctor heard the murmur during ventricular systole. Is the mitral valve closed or open at that time?**

2 **Into which chamber is the valve prolapsing, and why do you think there is no regurgitation?**

3 **Can you predict what the echocardiogram would show to explain Brian's heart murmur?**

20 The Lymphatic System and Immunity

LEARNING OUTCOMES

These Learning Outcomes correspond by number to this chapter's modules and indicate what you should be able to do after completing the chapter.

SECTION 1 • Anatomy of the Lymphatic System

20.1 Identify the various components of the lymphatic system.

20.2 Describe the structure and function of important lymphatic vessels.

20.3 Describe the lymph-collecting vessels, identify the structures returning lymph to the venous system, and explain lymphedema.

20.4 Identify the classes of lymphocytes, discuss their importance, and describe their distribution in the body.

20.5 Describe lymphoid tissues, and trace lymph flow through a lymph node.

20.6 Describe the structure and function of the thymus.

20.7 Describe the structure and function of the spleen, and trace blood flow through it.

SECTION 2 • Innate Immunity

20.8 Give an overview of the components of innate (nonspecific) immunity.

20.9 Explain how physical barriers and phagocytes play a role in innate immunity.

20.10 Describe immune surveillance, and explain the role of NK cells.

20.11 Describe the types of interferons, and explain the pathways of complement activation.

20.12 Explain the significance of inflammation and fever as innate defense mechanisms.

SECTION 3 • Adaptive Immunity

20.13 Describe the types of adaptive immunity and four properties of adaptive immunity.

20.14 Explain how antigens trigger an immune response.

20.15 Explain the events of antigen recognition and the roles of CD markers in T cell differentiation.

20.16 Explain the sensitization and activation of B cells and the role of plasma cells.

20.17 Describe the structure of an antibody, discuss the types of antibodies, and explain the primary and secondary responses to antigen exposure.

20.18 Explain the mechanisms by which antibodies destroy target antigens.

20.19 **CLINICAL MODULE** Explain allergies, anaphylaxis, and the role of antibodies in each.

20.20 Summarize the integration of innate and adaptive immunity.

20.21 **CLINICAL MODULE** Explain autoimmune disorders, graft rejection, and immunodeficiency diseases, and describe age-related changes in the immune response.

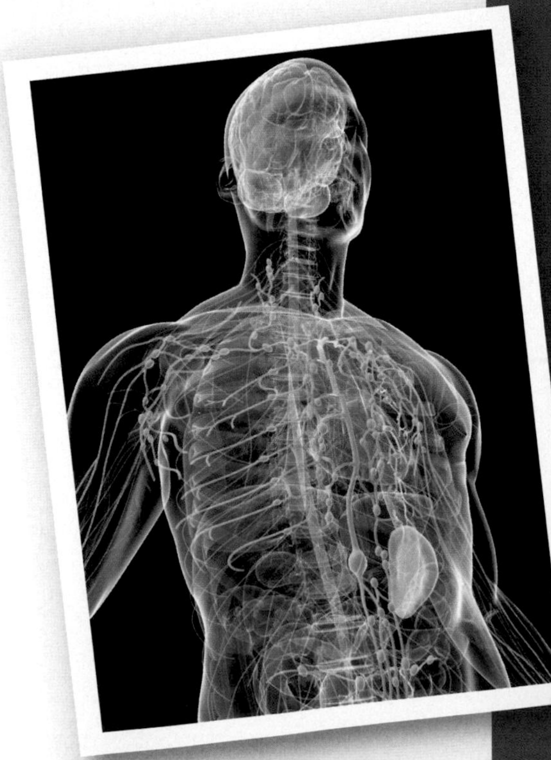

Learning Outcomes are repeated at the bottom of each module.

The lymphatic system consists of lymphatic vessels, nodes, and lymphoid tissue

The lymphatic system includes the cells, tissues, and organs responsible for providing **immunity** (the ability to defend the body against infection, illness, and disease) and returning interstitial fluid to the bloodstream.

1 **Lymphocytes**, the primary cells of the lymphatic system, were introduced in earlier chapters. These cells respond to the presence of invading pathogens (such as bacteria or viruses), abnormal body cells (such as virus-infected cells or cancer cells), and foreign proteins (such as the toxins released by some bacteria). Within the lymphatic system, lymphocytes are surrounded by **lymph**, interstitial fluid that has entered a lymphatic vessel.

2 The lymphatic system includes a network of **lymphatic vessels**, often called **lymphatics**, which begin in peripheral tissues and end at connections to veins, as noted in Module 4.11. The lymphatic system also includes an array of lymphoid tissues and lymphoid organs scattered throughout the body. **Primary lymphoid tissues and organs** are sites where lymphocytes are formed and mature. They include red bone marrow and the thymus gland. Recall that red bone marrow is also where other defense cells, the monocytes and macrophages, are formed. **Secondary lymphoid tissues and organs** are where lymphocytes are activated and cloned (production of identical cellular copies). These structures include lymph nodes, tonsils, MALT, appendix, and the spleen.

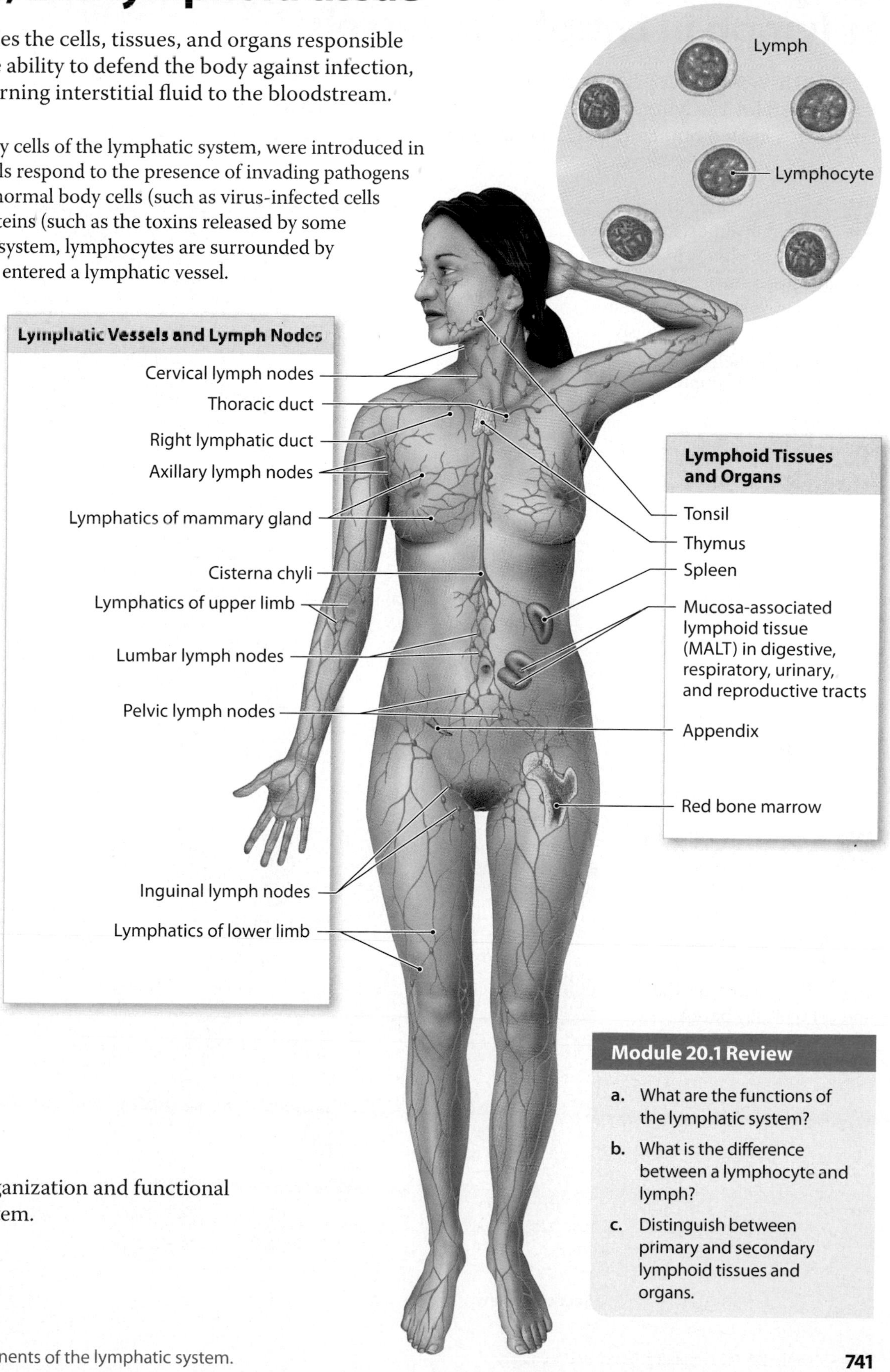
Lymph
Lymphocyte
Lymphatic Vessels and Lymph Nodes
Cervical lymph nodes
Thoracic duct
Right lymphatic duct
Axillary lymph nodes
Lymphatics of mammary gland
Cisterna chyli
Lymphatics of upper limb
Lumbar lymph nodes
Pelvic lymph nodes
Inguinal lymph nodes
Lymphatics of lower limb
Lymphoid Tissues and Organs
Tonsil
Thymus
Spleen
Mucosa-associated lymphoid tissue (MALT) in digestive, respiratory, urinary, and reproductive tracts
Appendix
Red bone marrow

This section considers the organization and functional anatomy of the lymphatic system.

Module 20.1 Review

a. What are the functions of the lymphatic system?

b. What is the difference between a lymphocyte and lymph?

c. Distinguish between primary and secondary lymphoid tissues and organs.

LO 20.1 Identify the various components of the lymphatic system.

Interstitial fluid flows continuously into lymphatic capillaries and exits tissues as lymph in lymphatic vessels

Lymphatic vessels carry lymph from peripheral tissues to the venous system. The lymphatic network begins with **lymphatic capillaries**, which branch through peripheral tissues.

1 Lymphatic capillaries are present in almost every tissue and organ in the body. As shown here, they are closely associated with blood capillary networks within tissues. Interstitial fluid is continuously flowing into lymphatic capillaries, where it is called lymph.

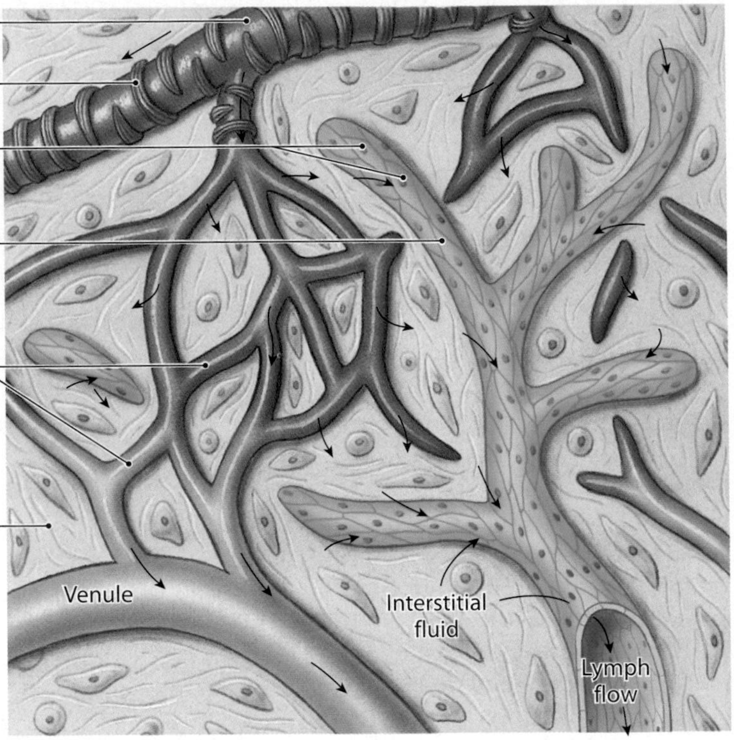
Arteriole
Smooth muscle
Endothelial cells
Lymphatic capillary
Blood capillaries
Loose connective tissue
Venule
Interstitial fluid
Lymph flow

2 Lymphatic capillaries differ from blood capillaries in that they (1) originate as pockets rather than forming continuous tubes, (2) have larger diameters, (3) have thinner walls, and (4) typically have a flattened or irregular outline in sectional view.

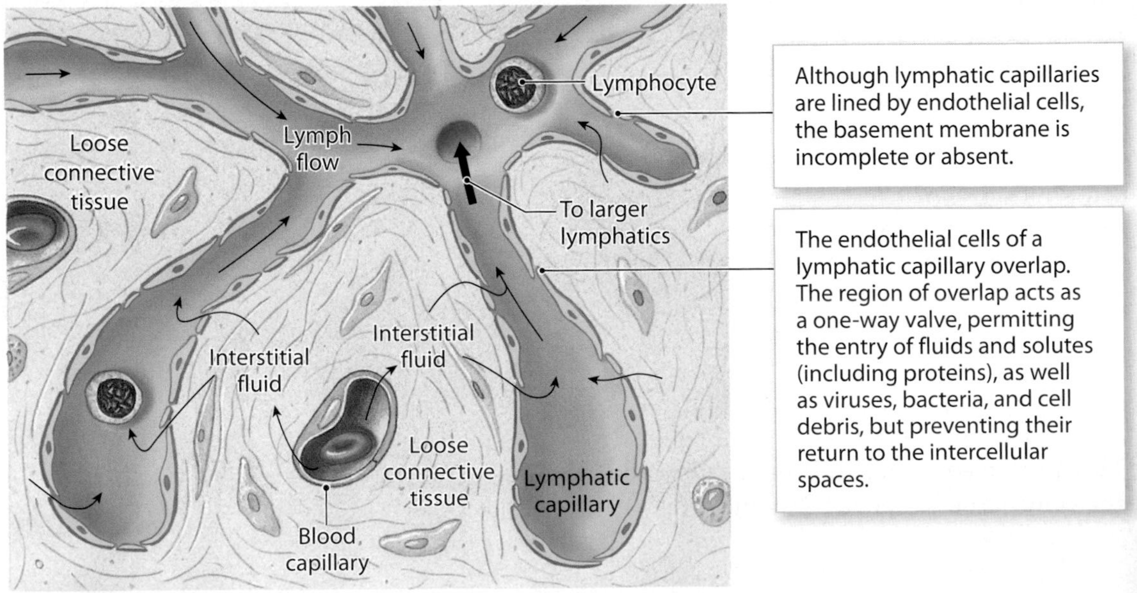
Loose connective tissue
Lymph flow
Lymphocyte
To larger lymphatics
Interstitial fluid
Interstitial fluid
Loose connective tissue
Lymphatic capillary
Blood capillary
Although lymphatic capillaries are lined by endothelial cells, the basement membrane is incomplete or absent.
The endothelial cells of a lymphatic capillary overlap. The region of overlap acts as a one-way valve, permitting the entry of fluids and solutes (including proteins), as well as viruses, bacteria, and cell debris, but preventing their return to the intercellular spaces.

Sectional view

3 From the lymphatic capillaries, lymph flows into larger lymphatic vessels that lead toward the body's trunk. These lymphatic vessels commonly occur in association with blood vessels.

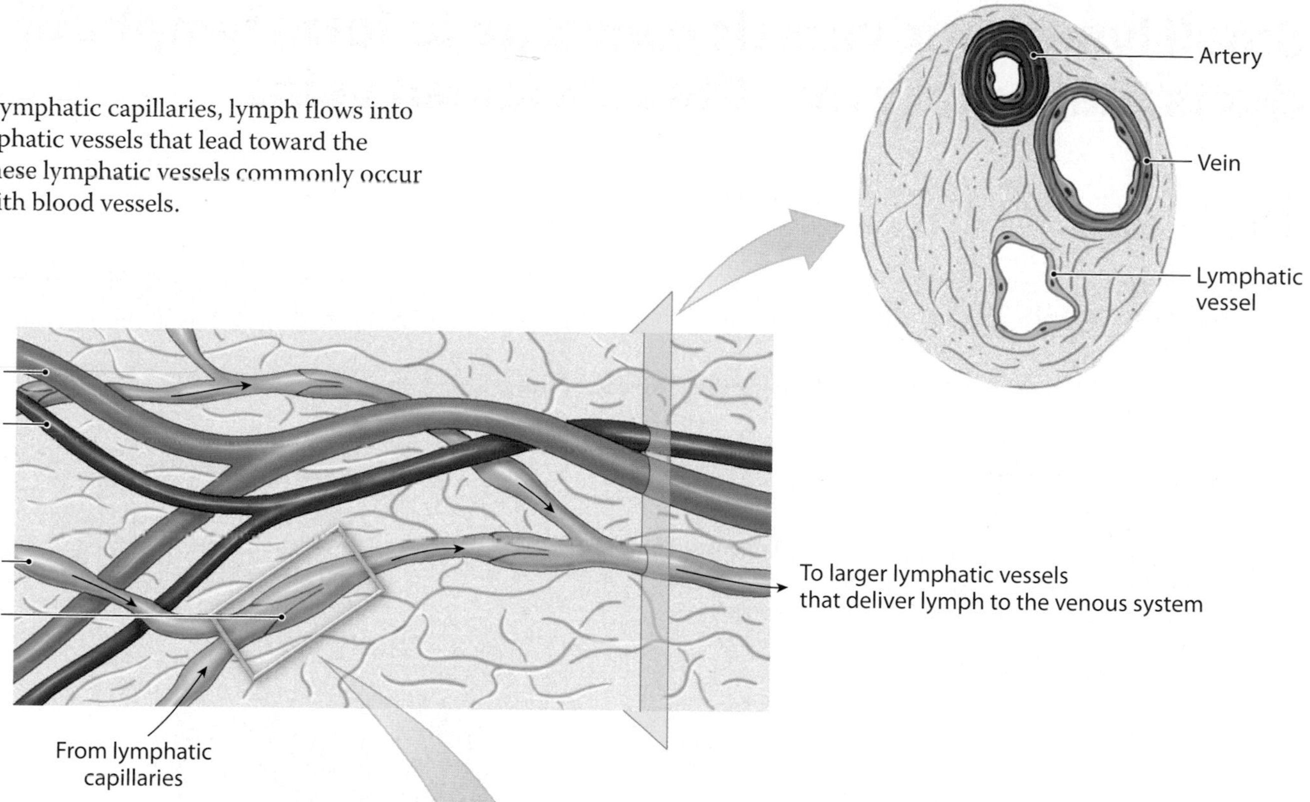

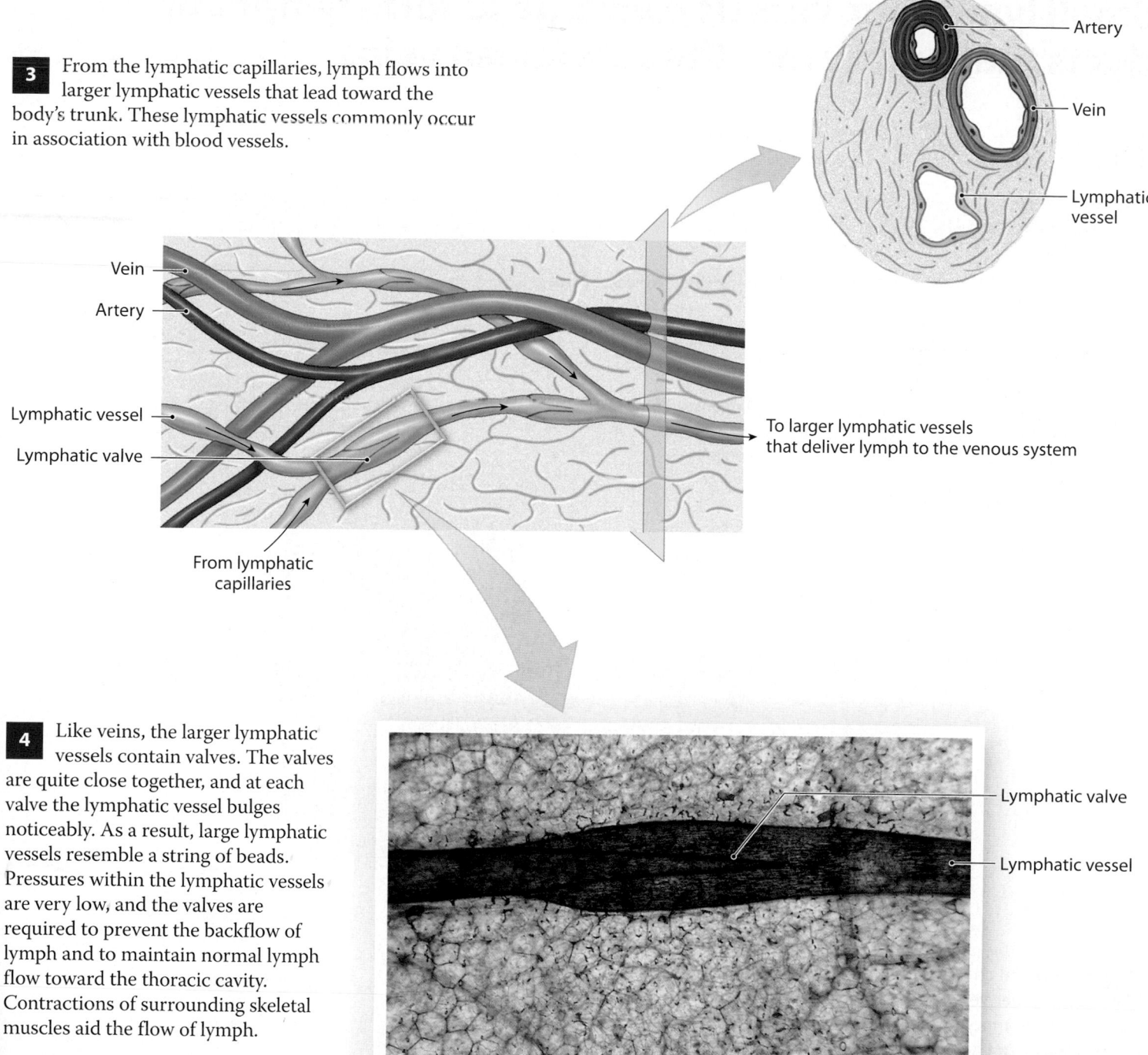

Vein
Artery
Lymphatic vessel
Lymphatic valve
From lymphatic capillaries
To larger lymphatic vessels that deliver lymph to the venous system
Artery
Vein
Lymphatic vessel

4 Like veins, the larger lymphatic vessels contain valves. The valves are quite close together, and at each valve the lymphatic vessel bulges noticeably. As a result, large lymphatic vessels resemble a string of beads. Pressures within the lymphatic vessels are very low, and the valves are required to prevent the backflow of lymph and to maintain normal lymph flow toward the thoracic cavity. Contractions of surrounding skeletal muscles aid the flow of lymph.

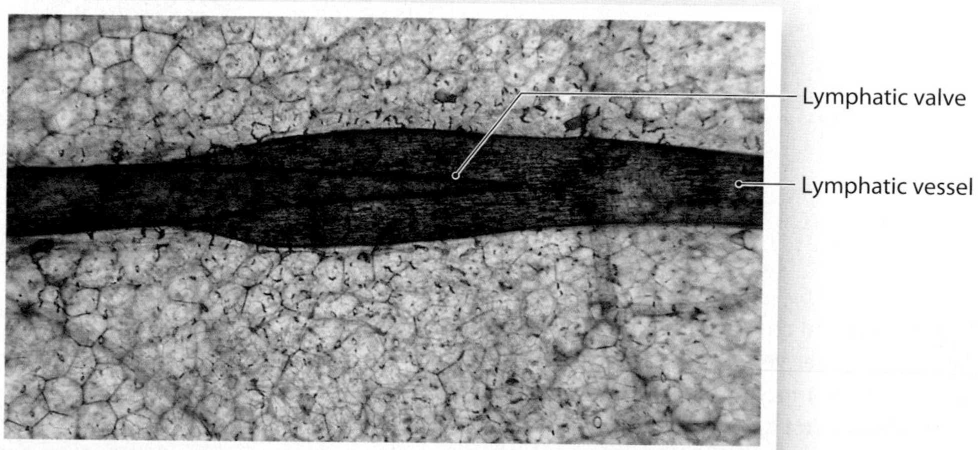

Lymphatic valve
Lymphatic vessel
LM × 65

Valve in lymphatic vessel

Prominent lymphatic capillaries in the small intestine called lacteals are important in the transport of lipids absorbed by the digestive tract. Lymphatic capillaries are absent from areas that lack a blood supply, such as the cornea of the eye. The bone marrow and the central nervous system also lack lymphatic vessels.

Module 20.2 Review

a. What is the function of lymphatic vessels?

b. What is the function of overlapping endothelial cells in lymphatic capillaries?

c. What structure prevents the backflow of lymph in some lymphatic vessels?

LO **20.2** Describe the structure and function of important lymphatic vessels.

Small lymphatic vessels converge to form lymphatic ducts that empty into the subclavian veins

1 The lymph from lymphatic capillaries flows into either superficial or deep lymphatic vessels.

Lymphatic Vessels

Superficial Lymphatics

Superficial lymphatics are located in the subcutaneous layer deep to the skin; in the areolar tissues of the mucous membranes lining the digestive, respiratory, urinary, and reproductive tracts; and in the areolar tissues of the serous membranes lining the pleural, pericardial, and peritoneal cavities.

Deep Lymphatics

Deep lymphatics accompany deep arteries and veins supplying skeletal muscles and other organs of the neck, limbs, and trunk, and the walls of visceral organs.

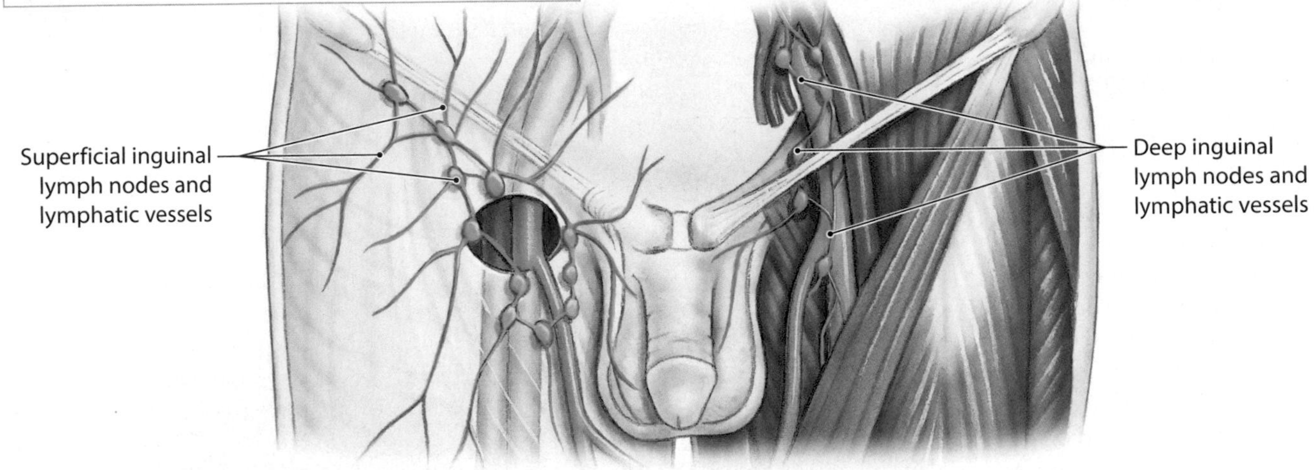
Superficial inguinal lymph nodes and lymphatic vessels
Deep inguinal lymph nodes and lymphatic vessels

2 Superficial and deep lymphatics converge to form larger vessels called **lymphatic trunks**, which in turn empty into two large collecting vessels: the **thoracic duct** and the **right lymphatic duct**. The thoracic duct collects lymph from the body inferior to the diaphragm and from the left side of the body superior to the diaphragm. The smaller right lymphatic duct collects lymph from the right side of the body superior to the diaphragm.

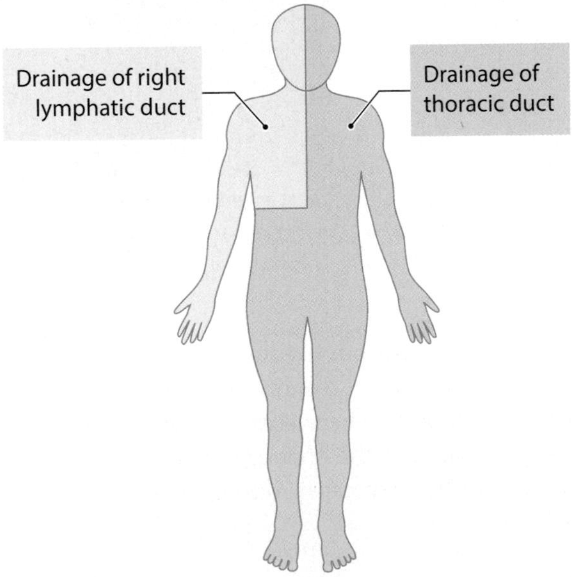
Drainage of right lymphatic duct
Drainage of thoracic duct

3 This illustration shows the relationship between the lymphatic ducts and the venous system. The thoracic duct empties into the left subclavian vein. The right lymphatic duct empties into the right subclavian vein.

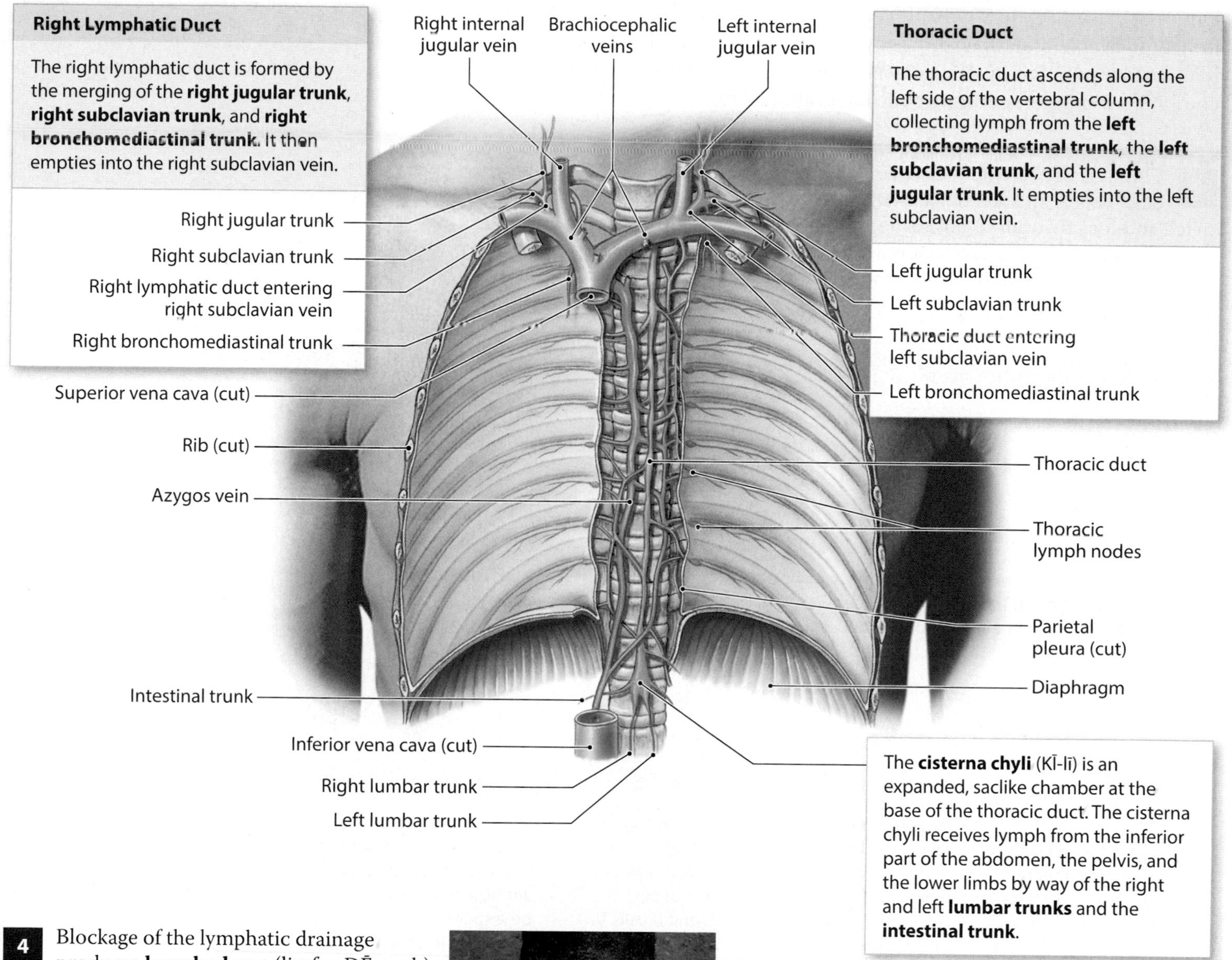

Right Lymphatic Duct
The right lymphatic duct is formed by the merging of the **right jugular trunk**, **right subclavian trunk**, and **right bronchomediastinal trunk**. It then empties into the right subclavian vein.
Right jugular trunk
Right subclavian trunk
Right lymphatic duct entering right subclavian vein
Right bronchomediastinal trunk
Superior vena cava (cut)
Rib (cut)
Azygos vein
Intestinal trunk
Inferior vena cava (cut)
Right lumbar trunk
Left lumbar trunk
Right internal jugular vein
Brachiocephalic veins
Left internal jugular vein
Thoracic Duct
The thoracic duct ascends along the left side of the vertebral column, collecting lymph from the **left bronchomediastinal trunk**, the **left subclavian trunk**, and the **left jugular trunk**. It empties into the left subclavian vein.
Left jugular trunk
Left subclavian trunk
Thoracic duct entering left subclavian vein
Left bronchomediastinal trunk
Thoracic duct
Thoracic lymph nodes
Parietal pleura (cut)
Diaphragm
The **cisterna chyli** (KĪ-lī) is an expanded, saclike chamber at the base of the thoracic duct. The cisterna chyli receives lymph from the inferior part of the abdomen, the pelvis, and the lower limbs by way of the right and left **lumbar trunks** and the **intestinal trunk**.

4 Blockage of the lymphatic drainage produces **lymphedema** (limf-e-DĒ-muh), a condition in which interstitial fluids accumulate and the affected area gradually becomes swollen and grossly distended. Lymphedema most often affects a limb, as in this photo, although it can occur in other locations. If the condition persists, the connective tissues lose their elasticity and the swelling becomes permanent. Because the interstitial fluids are basically stagnant, toxins and pathogens can accumulate and overwhelm local defenses without fully activating the immune system. The immune system is the body's defense against foreign substances.

Module 20.3 Review

a. Name the two large lymphatic vessels into which the lymphatic trunks empty.

b. Describe the drainage of the right lymphatic duct and the thoracic duct.

c. Explain lymphedema.

LO **20.3** Describe the lymph-collecting vessels, identify the structures returning lymph to the venous system, and explain lymphedema.

Module 20.4

Lymphocytes are responsible for the immune functions of the lymphatic system

Lymphocytes make up 20–40 percent of the circulating leukocyte population. However, circulating lymphocytes are only a small fraction of the total lymphocyte population. The body contains some 10^{12} lymphocytes, with a combined weight of more than one kilogram (2.2 lb).

1 Three classes of lymphocytes circulate in blood. While each class has distinctive biochemical and functional characteristics, all are sensitive to specific chemicals called **antigens** that can stimulate the immune system. Most antigens are pathogens, parts or products of pathogens, or other foreign compounds. Most antigens are proteins, but some lipids, polysaccharides, and nucleic acids can also stimulate an immune response that leads to the destruction of the target compound or organism.

Classes of Lymphocytes

T Cells

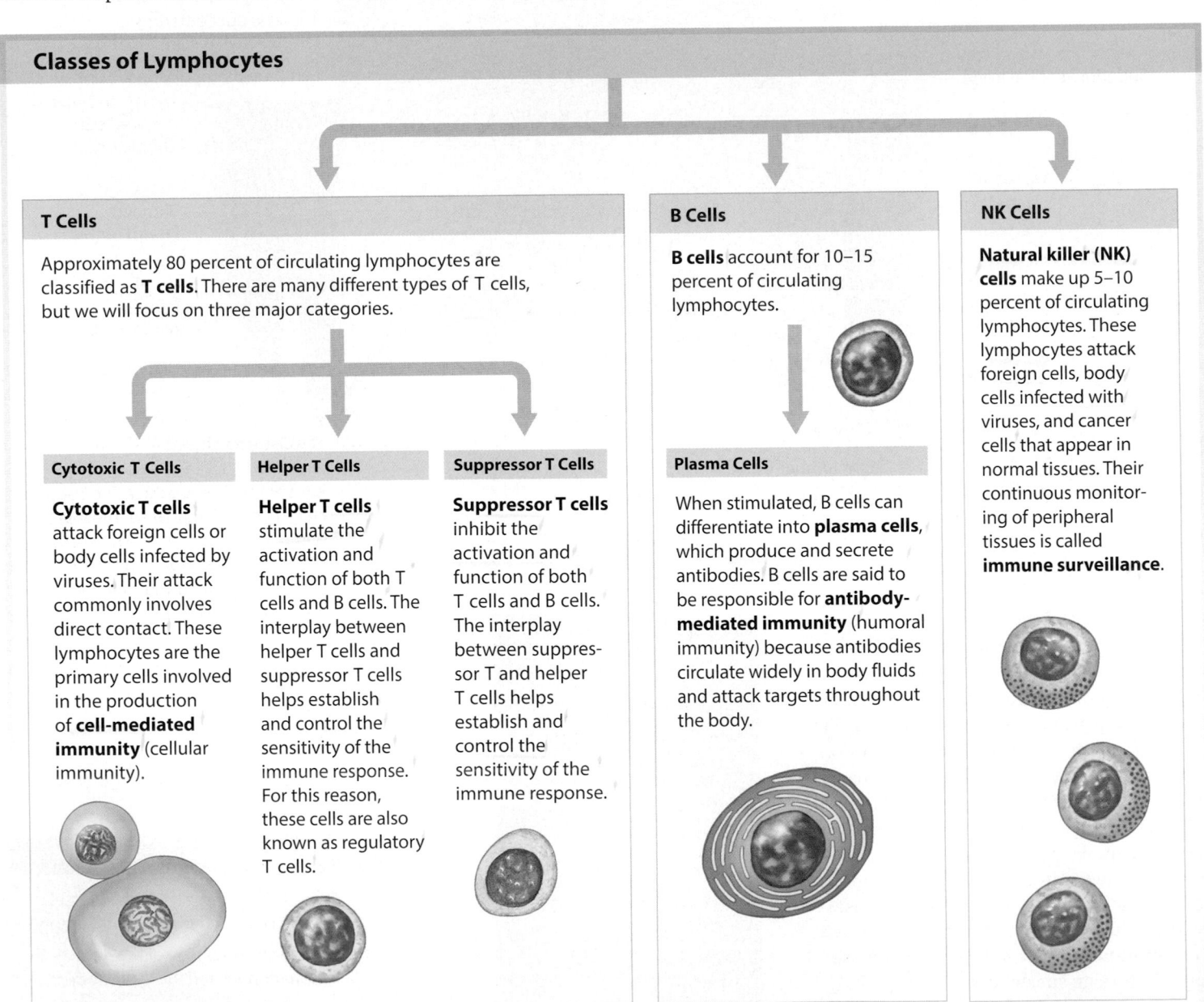

Approximately 80 percent of circulating lymphocytes are classified as **T cells**. There are many different types of T cells, but we will focus on three major categories.

Cytotoxic T Cells

Cytotoxic T cells attack foreign cells or body cells infected by viruses. Their attack commonly involves direct contact. These lymphocytes are the primary cells involved in the production of **cell-mediated immunity** (cellular immunity).

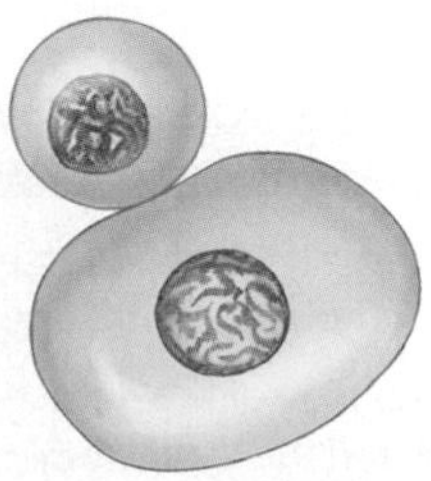

Helper T Cells

Helper T cells stimulate the activation and function of both T cells and B cells. The interplay between helper T cells and suppressor T cells helps establish and control the sensitivity of the immune response. For this reason, these cells are also known as regulatory T cells.

Suppressor T Cells

Suppressor T cells inhibit the activation and function of both T cells and B cells. The interplay between suppressor T and helper T cells helps establish and control the sensitivity of the immune response.

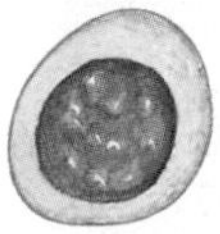

B Cells

B cells account for 10–15 percent of circulating lymphocytes.

Plasma Cells

When stimulated, B cells can differentiate into **plasma cells**, which produce and secrete antibodies. B cells are said to be responsible for **antibody-mediated immunity** (humoral immunity) because antibodies circulate widely in body fluids and attack targets throughout the body.

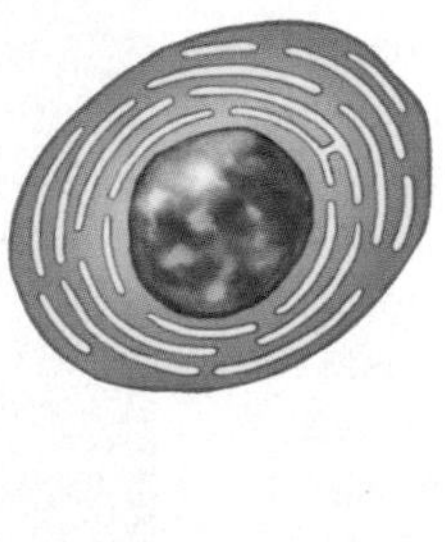

NK Cells

Natural killer (NK) cells make up 5–10 percent of circulating lymphocytes. These lymphocytes attack foreign cells, body cells infected with viruses, and cancer cells that appear in normal tissues. Their continuous monitoring of peripheral tissues is called **immune surveillance**.

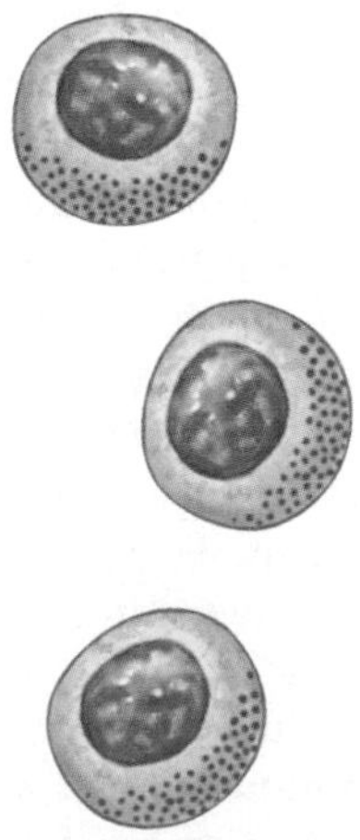

2 In adults, erythropoiesis (red blood cell formation) is normally confined to red bone marrow, but lymphocyte production, or **lymphopoiesis** (lim-fō-poy-Ē-sis), involves the red bone marrow, thymus, and peripheral lymphoid tissues. Red bone marrow plays the primary role in the maintenance of normal lymphocyte populations.

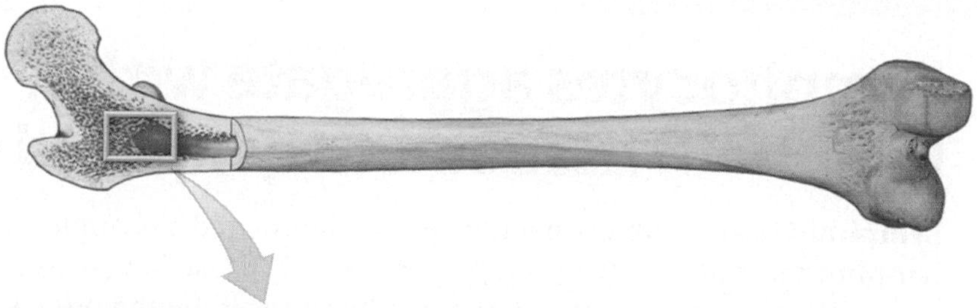

Red Bone Marrow

The divisions of hematopoietic stem cells in the red bone marrow of adults generate the **lymphoid stem cells** that produce all types of lymphocytes.

One group of lymphoid stem cells migrates to the thymus.

A second group of lymphoid stem cells remains in the bone marrow and divides to produce immature B cells and NK cells.

B cells

NK cells

Thymus

These stem cells and their descendants are isolated from the general circulation by the **blood–thymus barrier,** comparable to the blood–brain barrier. Under the influence of thymic hormones, the lymphoid stem cells divide repeatedly, producing the various kinds of T cells. These T cells undergo a selection process to ensure they will not react to the body's own healthy cells and cellular products. During this process up to 98 percent of developing T cells are de-selected and undergo apoptosis.

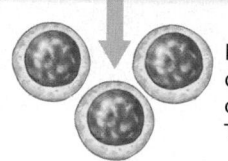

Production and differentiation of surviving T cells

When their maturation is nearing completion, T cells re-enter the bloodstream and travel to peripheral lymphoid tissues and organs.

As they mature, B cells and NK cells enter the bloodstream and migrate to peripheral tissues.

Peripheral Tissues

Cell-mediated immunity

Mature T cells are responsible for cell-mediated immunity.

Antibody-mediated immunity

Most of the B cells move into lymph nodes, the spleen, or other lymphoid tissues.

Immune surveillance

The NK cells migrate throughout the body, moving through peripheral tissues in search of abnormal cells.

The T cells and B cells that migrate from their sites of origin retain the ability to divide. Their divisions produce daughter cells of the same type. For example, a dividing B cell produces other B cells, but not T cells or NK cells. The ability of specific types of lymphocytes to increase in number is crucial to the success of the immune response.

Module 20.4 Review

a. Identify the three main classes of lymphocytes.

b. Which cells are responsible for antibody-mediated immunity?

c. What tissues are involved in lymphopoiesis?

LO **20.4** Identify the classes of lymphocytes, discuss their importance, and describe their distribution in the body.

Lymphocytes aggregate within lymphoid tissues and lymphoid organs

Lymphoid tissues are connective tissues dominated by lymphocytes. In a **lymphoid nodule** the lymphocytes are densely packed in an area of areolar tissue. The boundaries are not distinct, because although nodules may cluster together and form larger masses, no fibrous capsule surrounds the lymphoid tissue. In contrast, **lymphoid organs** (lymph nodes, the thymus, and the spleen) are separated from surrounding tissues by a fibrous connective tissue capsule.

1 Clusters of lymphoid nodules deep to the epithelial lining of the intestine are known as **aggregated lymphoid nodules**, or Peyer's patches. Each nodule often has a central zone called a **germinal center**, which contains dividing lymphocytes. The lymphoid tissues that protect epithelia of the digestive, respiratory, urinary, and reproductive tracts from pathogens and toxins form the **mucosa-associated lymphoid tissue (MALT)**.

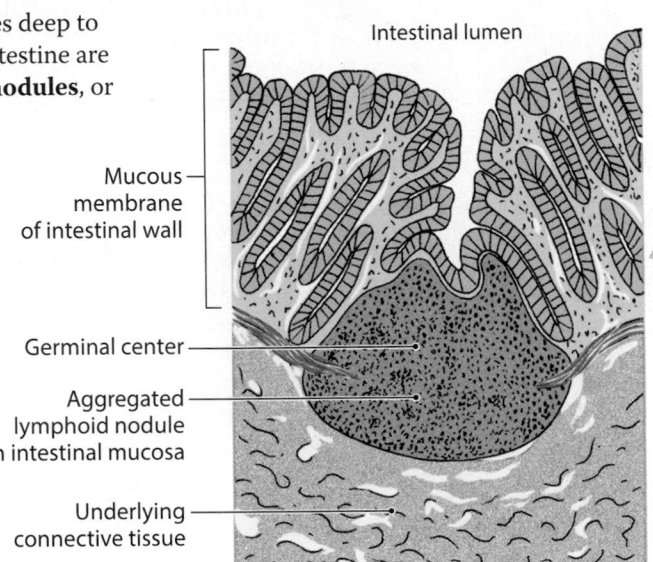

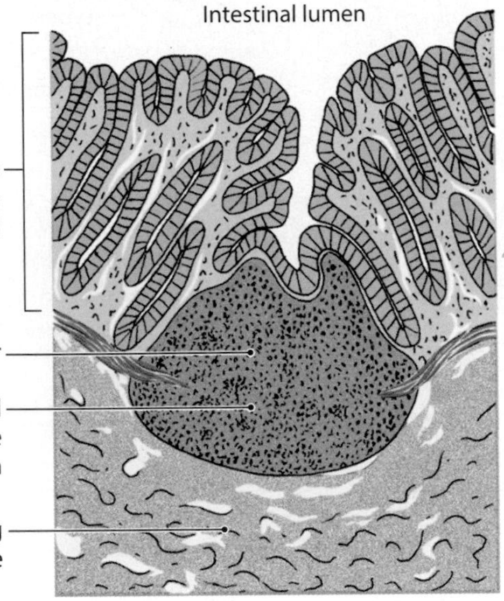
Intestinal lumen
Mucous membrane of intestinal wall
Germinal center
Aggregated lymphoid nodule in intestinal mucosa
Underlying connective tissue

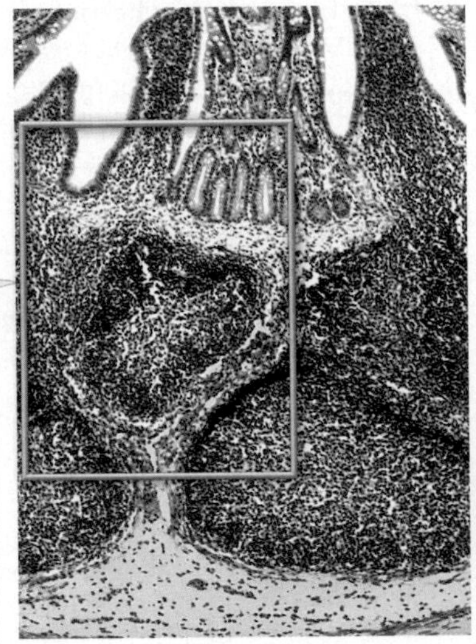

Aggregated lymphoid nodules LM × 20

2 The **tonsils** are large lymphoid nodules in the walls of the pharynx. A single **pharyngeal tonsil**, often called the adenoid, lies in the posterior superior wall of the nasopharynx. Left and right **palatine tonsils** are located at the posterior, inferior margin of the oral cavity, along the boundary of the pharynx. A pair of **lingual tonsils** lie deep to the mucous epithelium covering the base (pharyngeal portion) of the tongue. Because of their location, the lingual tonsils are not usually visible unless they become infected and swollen. **Tonsillitis** is an inflammation of the tonsils (especially the palatines) although the other tonsils may also be affected. Tonsils reach their largest size by puberty and then begin to atrophy.

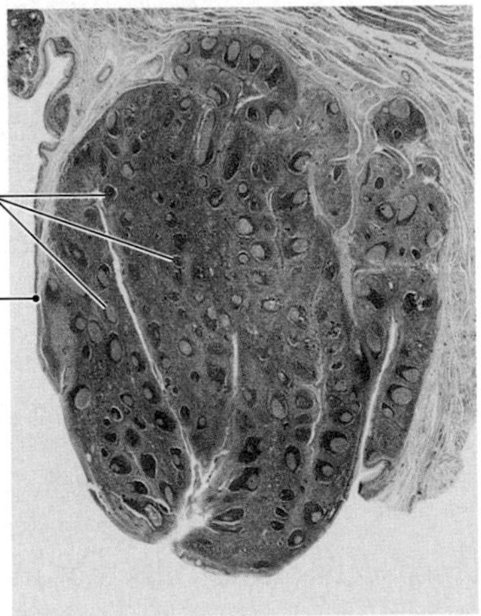
Germinal centers within nodules
Pharyngeal epithelium

Pharyngeal tonsil LM × 40

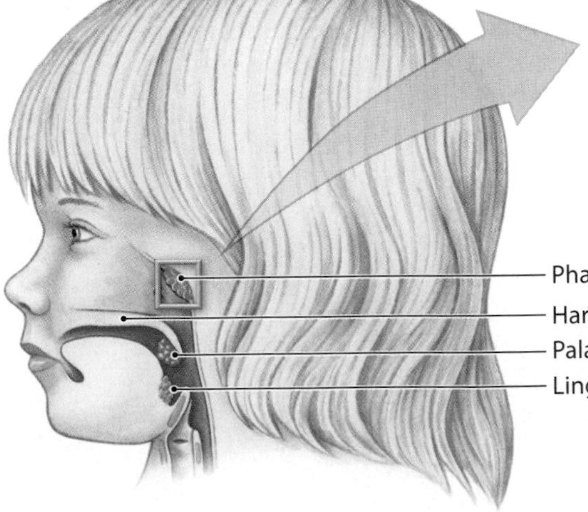
Pharyngeal tonsil
Hard palate
Palatine tonsil
Lingual tonsil

3 **Lymph nodes** are small lymphoid organs ranging in diameter from 1 mm to 25 mm (about 1 in.). The shape of a typical lymph node resembles that of a kidney bean. The largest lymph nodes are located where the peripheral lymphatics from the neck and the limbs connect with the trunk, such as in the groin, the axillae, and the base of the neck. These nodes, often called **lymph glands**, detect pathogens before they reach the vital organs of the trunk. A lymph node functions like a kitchen water filter, purifying lymph before it reaches the venous circulation. As lymph flows through a lymph node, at least 99 percent of the antigens in the lymph are removed, and immune responses are stimulated as needed.

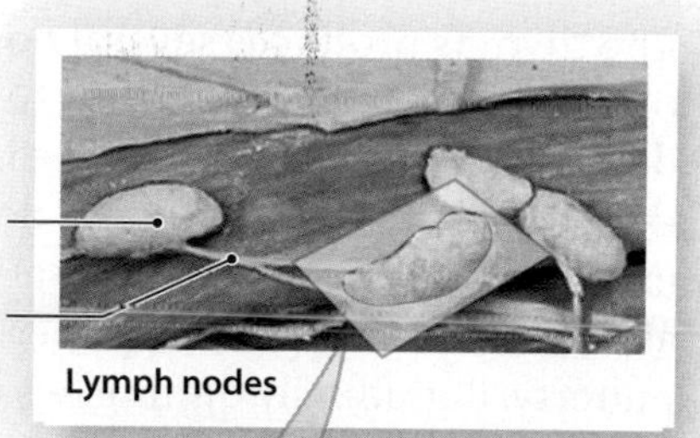
Lymph node
Lymph vessel
Lymph nodes

Path of Lymph Flow through a Lymph Node

6 **Efferent** (*efferens*, to bring out) **lymphatics** leave the lymph node at the hilum. These vessels collect lymph from the medullary sinus and carry it toward the venous circulation.

5 Lymph continues into the **medullary sinus** at the core of the lymph node. This region contains B cells and plasma cells.

4 Lymph then flows through lymph sinuses in the **deep cortex**, which is dominated by T cells.

3 Lymph next flows into the **outer cortex**, which contains B cells within germinal centers that resemble those of lymphoid nodules.

2 The afferent vessels deliver lymph to the subcapsular space, a meshwork of reticular fibers, macrophages, and dendritic cells. **Dendritic cells** are involved in the initiation of the immune response. Some of these cells migrate to a lymph node bearing antigens from peripheral tissues, such as the skin.

1 **Afferent** (*afferens*, to bring to) **lymphatics** carry lymph to the lymph node from peripheral tissues. The afferent lymphatics penetrate the capsule of the lymph node on the side opposite the hilum.

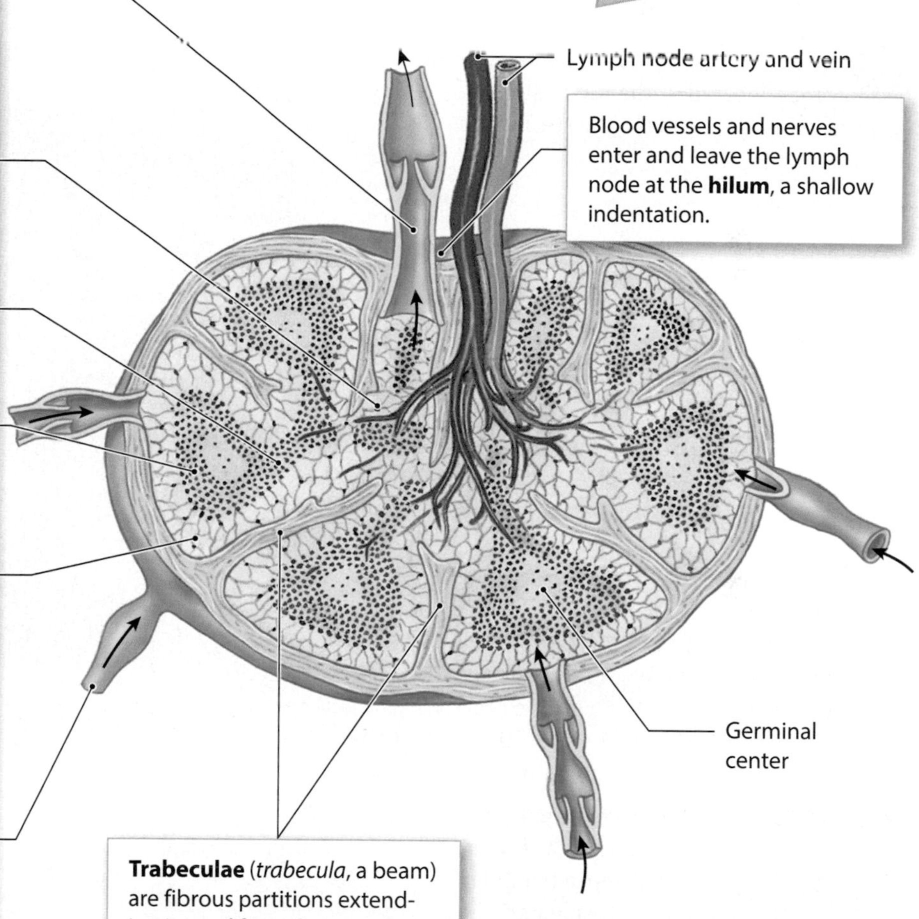
Lymph node artery and vein
Blood vessels and nerves enter and leave the lymph node at the **hilum**, a shallow indentation.
Germinal center
Trabeculae (*trabecula*, a beam) are fibrous partitions extending inward from the capsule.

The digestive, respiratory, urinary, and reproductive tracts communicate with the exterior environment, which contains dangerous pathogens and toxins. MALT, which defends the exposed epithelia, contains much of the lymphoid tissue in the body. A variety of clinical disorders can result from infection and/or inflammation of MALT components. Tonsillitis (see **2**) is one familiar example; another is **appendicitis** (inflammation of the lymphoid tissues of the appendix).

Module 20.5 Review

a. Name the lymphoid tissue that protects epithelia lining the digestive, respiratory, urinary, and reproductive tracts.

b. Define tonsil, and name the five tonsils.

c. Trace the path of lymph through a lymph node, beginning at the afferent lymphatics.

The thymus is a lymphoid organ that produces functional T cells

The thymus produces several hormones that are important to the development of functional T cells, and thus to the maintenance of normal immunological defenses. Thymosin (THĪ-mō-sin) is the name originally given to an extract from the thymus that promotes the development and maturation of lymphocytes. This extract actually contains several complementary hormones collectively known as **thymosins**. Just before puberty, the thymus weighs about 40 g (1.4 oz). After puberty, it gradually diminishes in size and becomes increasingly fibrous, a process called **involution**. By the time a person reaches age 50, the thymus may weigh less than 12 g (0.3 oz). The gradual decrease in the size and secretory abilities of the thymus that occurs with age is correlated with an increased susceptibility to disease.

1 In adults, the thymus is a pink, grainy organ located in the mediastinum, generally just posterior to the sternum.

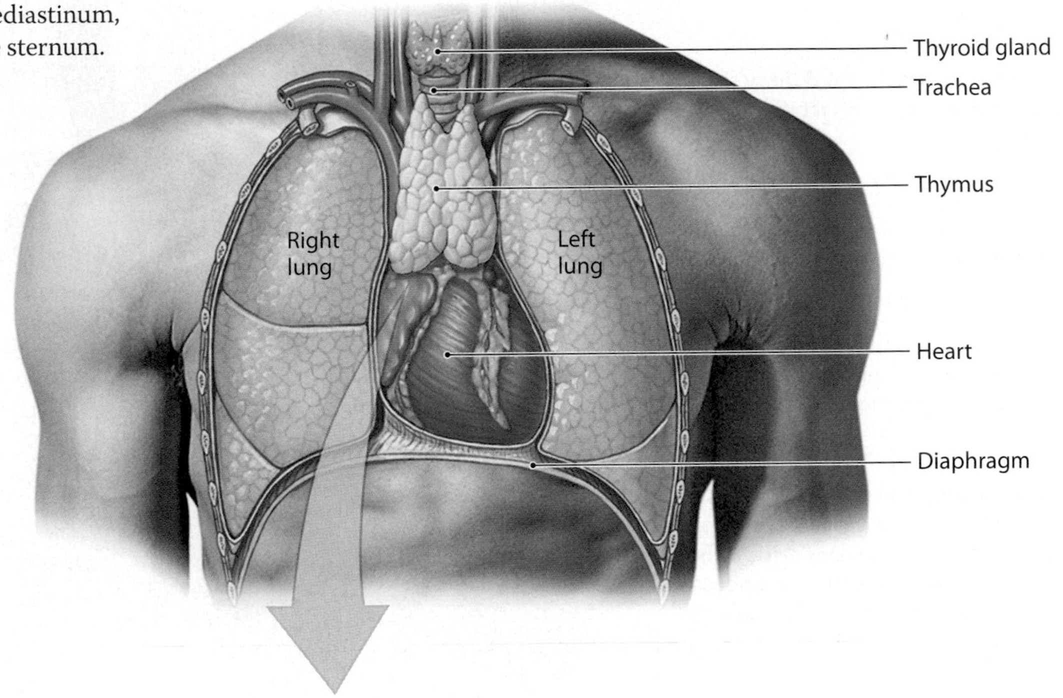
Thyroid gland
Trachea
Thymus
Right lung
Left lung
Heart
Diaphragm

2 The capsule that covers the thymus divides it into left and right **lobes**. Fibrous partitions called **septa** (singular, *septum*) originate at the capsule and divide the lobes into **lobules** averaging 2 mm in diameter.

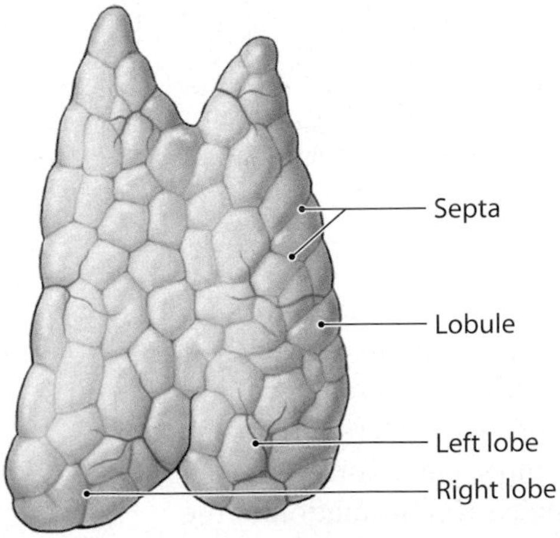
Septa
Lobule
Left lobe
Right lobe

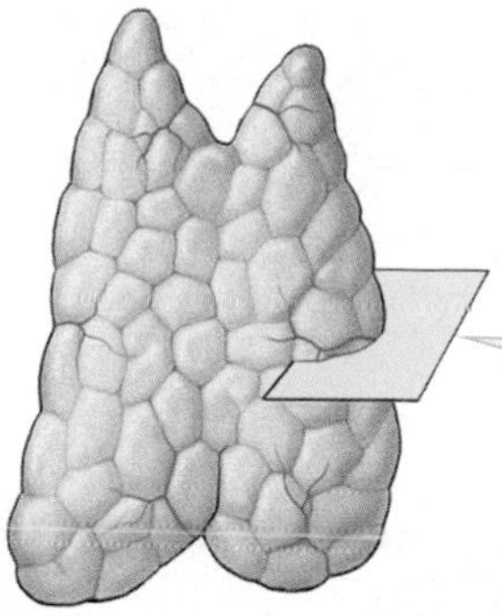

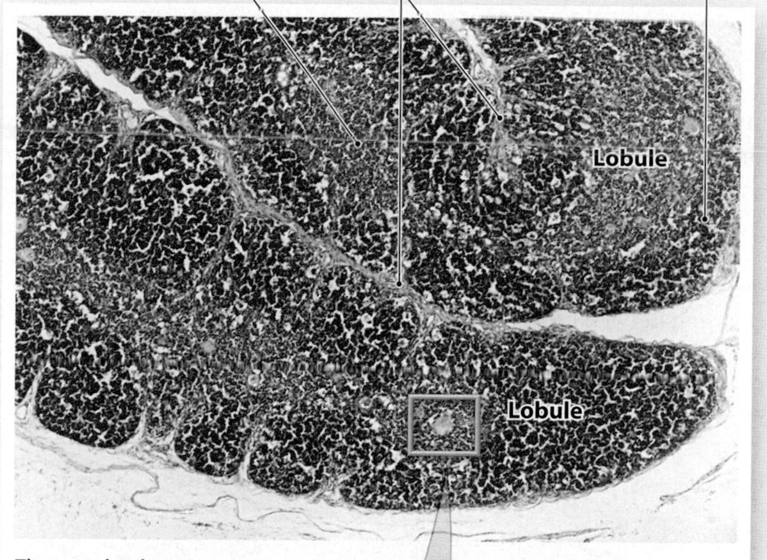
Medulla
Septa
Cortex
Lobule
Lobule
Thymus gland LM × 50

3 Each lobule consists of a dark outer **cortex** and a lighter central **medulla**. Lymphocytes divide and differentiate under controlled conditions in the cortex. Lymphocytes in the cortex are arranged in clusters that are completely surrounded by **thymic epithelial cells (TECs)**. Thymic epithelial cells regulate T cell development and function. Epithelial cells also encircle the blood vessels of the cortex. These cells maintain the **blood–thymus barrier**.

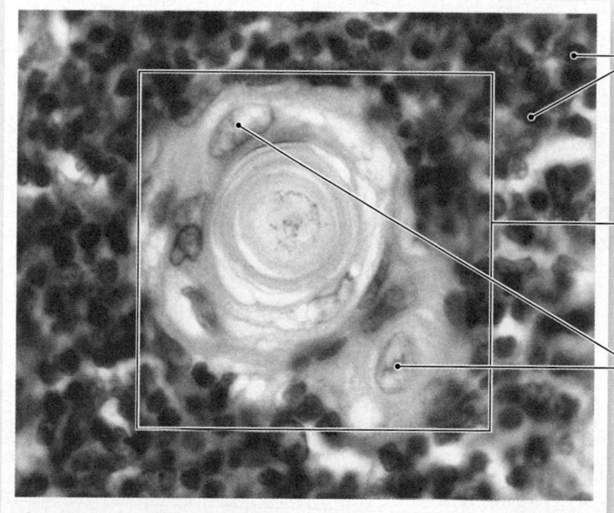
Lymphocytes
Thymic corpuscle
Thymic epithelial cells
Thymic corpuscle LM × 532

4 After roughly 3 weeks, developing T cells leave the cortex and enter the medulla. Because there is no blood–thymus barrier in the medulla, developing T cells may enter a medullary blood vessel. Instead of ensheathing blood vessels as they do in the cortex, the thymic epithelial cells in the medulla cluster together in concentric layers, forming distinctive structures known as **thymic corpuscles**. Within the medulla, developing T cells may also enter a lymphatic vessel and leave the thymus through an efferent lymphatic.

Module 20.6 Review

a. Where is the thymus located?

b. Describe the gross anatomy of the thymus.

c. Which cells constitute and maintain the blood–thymus barrier?

LO 20.6 Describe the structure and function of the thymus.

Module 20.7

The spleen, the largest lymphoid organ, responds to antigens in the bloodstream

The adult **spleen** contains the largest collection of lymphoid tissue in the body. In essence, the spleen performs the same functions for blood that lymph nodes perform for lymph. Functions of the spleen can be summarized as (1) removing abnormal blood cells and other blood components by phagocytosis, (2) storing iron recycled from red blood cells, and (3) initiating immune responses by B cells and T cells in response to antigens in circulating blood.

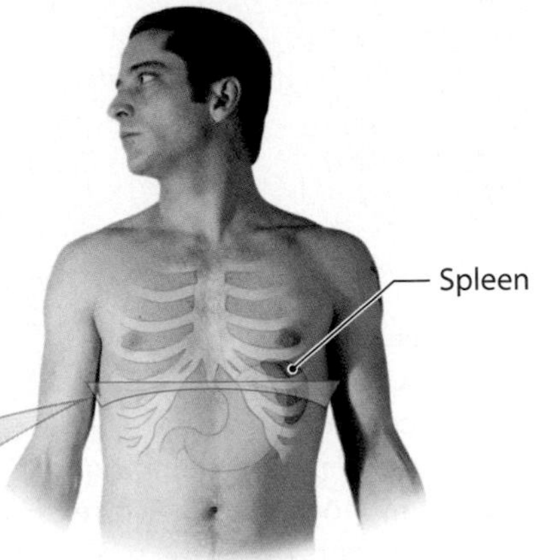
Spleen

1 The spleen lies along the curving lateral border of the stomach, extending between the 9th and 11th ribs on the left side. It is attached to the lateral border of the stomach by the **gastrosplenic ligament**, a broad band of mesentery.

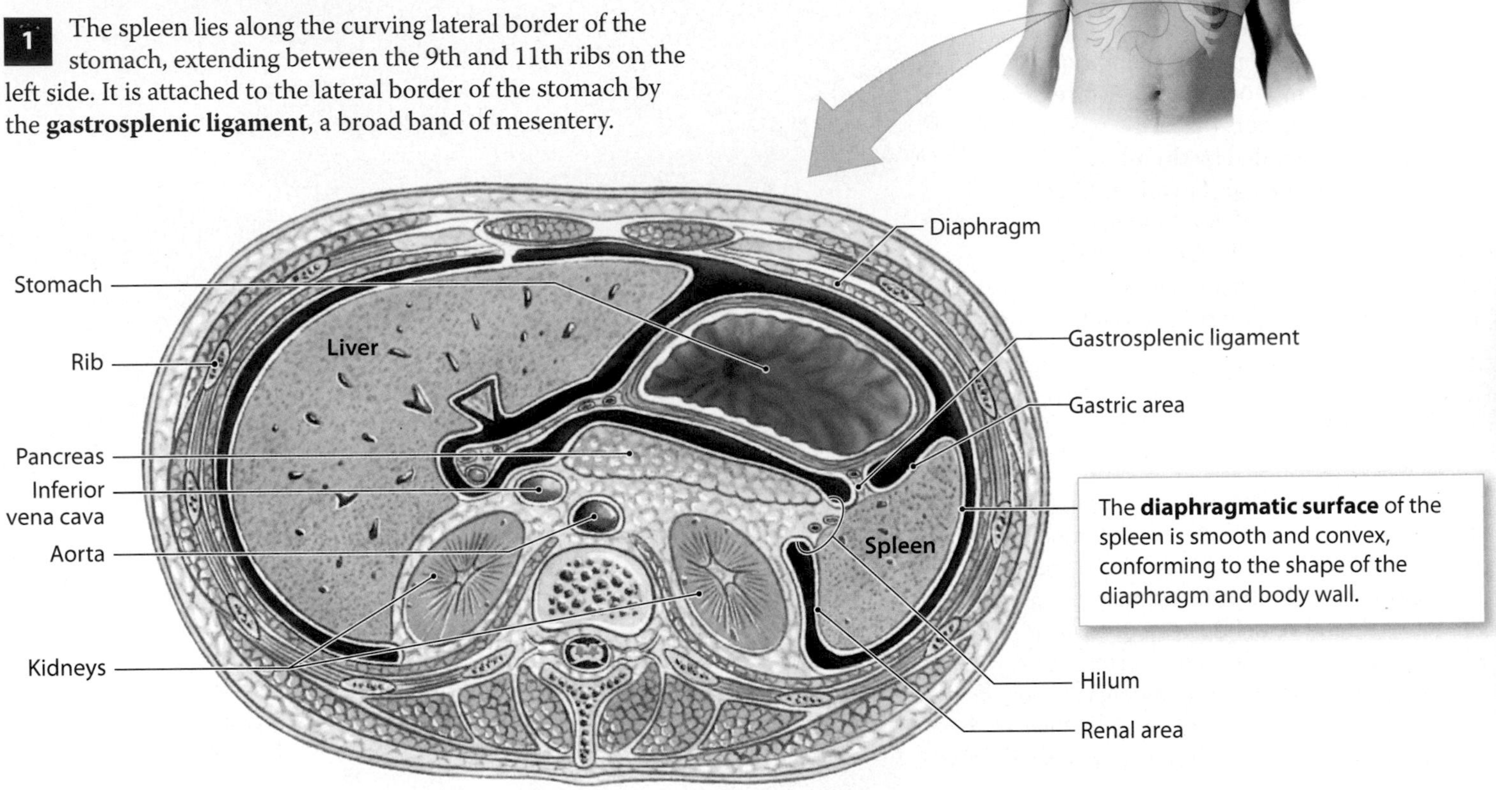
Diaphragm
Stomach
Rib
Liver
Gastrosplenic ligament
Gastric area
Pancreas
Inferior vena cava
Aorta
Spleen
The **diaphragmatic surface** of the spleen is smooth and convex, conforming to the shape of the diaphragm and body wall.
Kidneys
Hilum
Renal area

Inferior view, transverse section

2 The spleen tears so easily that a seemingly minor impact to the left side of the abdomen can rupture the capsule. Because the spleen is relatively fragile, it is very difficult to repair surgically, so a severely ruptured spleen is removed, a process called a **splenectomy** (splē-NEK-tō-mē). A person can survive without a spleen but lives with an increased risk of bacterial infection (particularly involving pneumococcal bacteria).

3 The spleen is about 12 cm (5 in.) long and weighs, on average, nearly 160 g (5.6 oz). In gross dissection, the spleen is deep red because of the blood it contains. The spleen has a soft texture, so its shape primarily reflects its association with the structures around it. The medial, **visceral surface** contains indentations that conform to the shape of the stomach (the **gastric area**) and that of the kidney (the **renal area**).

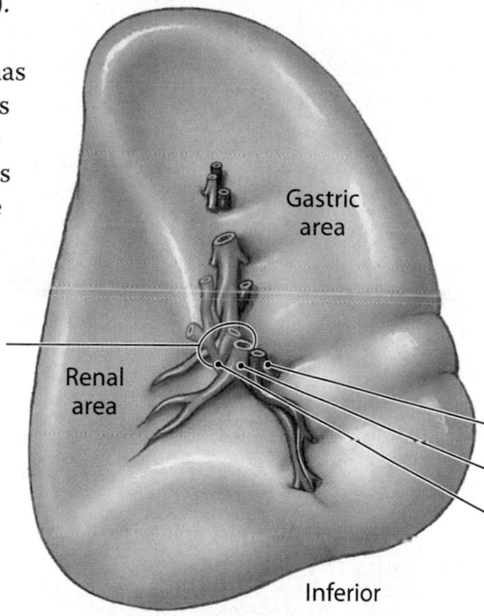
Superior
Gastric area
Hilum
Renal area
Inferior
Splenic artery
Splenic vein
Splenic lymphatic vessels

Landmarks on the visceral surface

Splenic Blood Vessels and Lymphatic Vessels

Splenic blood vessels and lymphatic vessels communicate with the spleen on the visceral surface at the **hilum**, a groove marking the border between the gastric and renal areas.

4 The spleen is surrounded by a capsule containing collagen and elastic fibers. The cellular components within make up the **pulp** of the spleen. **Red pulp** contains large quantities of red blood cells, whereas **white pulp** resembles lymphoid nodules and contains lymphocytes.

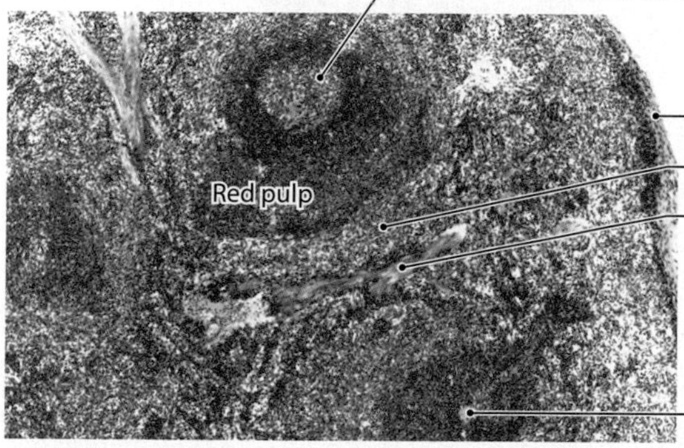
White pulp of splenic nodule
Red pulp
Capsule
Trabecula
Trabecular artery
Central artery in splenic nodule

The histological appearance of the spleen LM × 50

Trabeculae

Fibrous **trabeculae** (partitions) radiate outward toward the capsule through the interior from the hilum. Blood vessels travel within the trabeculae.

Path of Blood Flow through the Spleen

The **trabecular arteries** are branches of the splenic artery. Their finer branches, called **central arteries**, are surrounded by areas of white pulp. → Capillaries discharge the blood into the reticular tissue of the red pulp, which contains free and fixed macrophages. → Blood flows into sinusoids (Module 18.3, p. 655) whose walls contain fixed macrophages. → Blood collects into small veins that in turn merge to form **trabecular veins** that unite at the hilum.

5 The unusual circulatory arrangement within the spleen gives the phagocytes of the spleen an opportunity to identify and engulf any damaged or infected cells in circulating blood. Because macrophages are scattered throughout the red pulp, and lymphocytes and dendritic cells are found within and surrounding the white pulp, any microorganism or other antigen in the blood quickly triggers an immune response.

Module 20.7 Review

a. What is the function of the spleen?

b. Describe red pulp and white pulp found in the spleen.

c. Beginning at the trabecular arteries, trace the path of blood through the spleen.

LO **20.7** Describe the structure and function of the spleen, and trace blood flow through it.

Labeling

Label the structures of the lymphatic system in the accompanying figure.

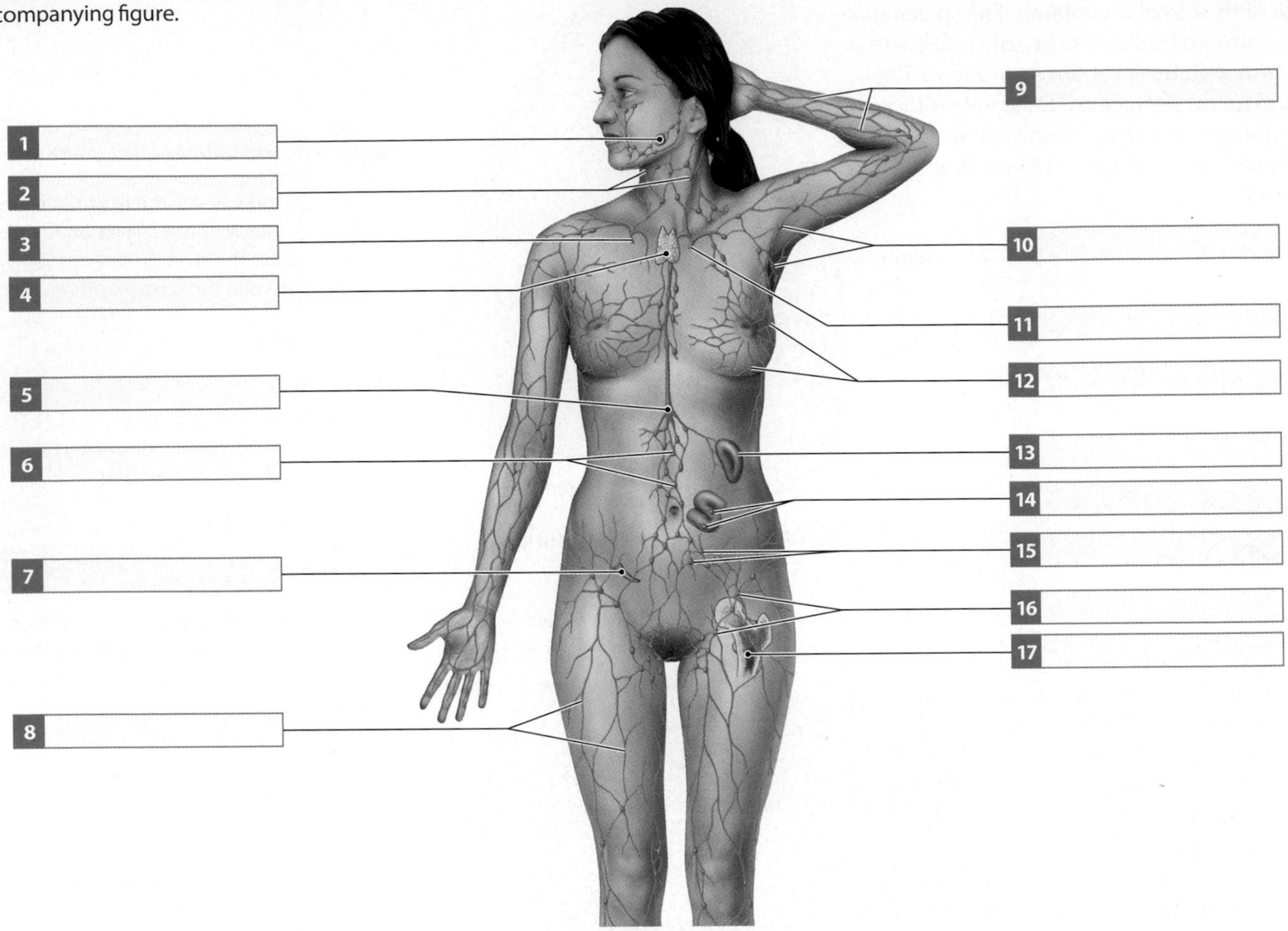
1
2
3
4
5
6
7
8
9
10
11
12
13
14
15
16
17

Matching

Match each lettered term with the most closely related description.

a. B cells
b. spleen
c. lymphatic capillaries
d. cytotoxic T cells
e. epithelial cells
f. tonsils
g. lymphopoiesis
h. thymic corpuscles
i. lymphoid organs
j. helper T cells and suppressor T cells
k. afferent lymphatics
l. right subclavian vein
m. lymph nodes

18 Beginning of lymphatic system
19 Thymus medullary cells
20 Receives lymph from right lymphatic duct
21 Thymus, spleen, and lymph nodes
22 Smallest lymphoid organs
23 Regulate and coordinate the immune response
24 Largest mass of lymphoid tissue in body
25 Maintains blood–thymus barrier
26 Occurs in red bone marrow, thymus, and lymphoid tissues
27 Cell-mediated immunity
28 Lymphoid nodules in walls of pharynx
29 Antibody-mediated immunity
30 Carry lymph to lymph nodes

18 ______
19 ______
20 ______
21 ______
22 ______
23 ______
24 ______
25 ______
26 ______
27 ______
28 ______
29 ______
30 ______

Innate immunity is nonspecfic and is not stimulated by specific antigens

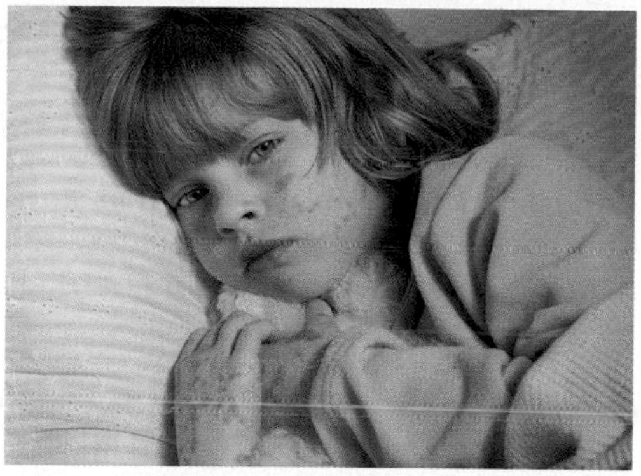

The human body has multiple defense mechanisms that together provide **immunity**—the ability to fight infection, illness, and disease. Two complementary mechanisms are involved, and both must function normally to provide adequate resistance to infection and disease. In this section we consider innate (nonspecific) immunity. Innate immunity is not stimulated by specific antigens and prevents the approach, denies the entry, and limits the spread of microbes or other environmental hazards.

Immunity

Innate (nonspecific) Immunity

Innate defenses do not distinguish one type of threat from another. Their response is the same, regardless of the type of invading agent. These defenses, which are present at birth, provide a defensive capability known as **nonspecific resistance**.

Physical barriers keep hazardous organisms and materials outside the body. For example, a mosquito that lands on your head may be unable to reach the surface of the scalp if you have a full head of hair.

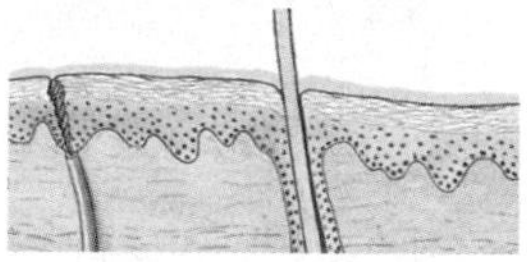

Phagocytes are cells that engulf pathogens and cell debris. Examples of phagocytes are the macrophages of peripheral tissues and the eosinophils and neutrophils of blood.

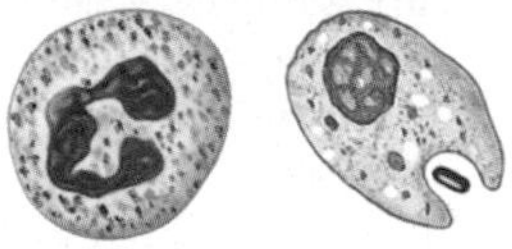

Immune surveillance is the destruction of abnormal cells by NK cells in peripheral tissues.

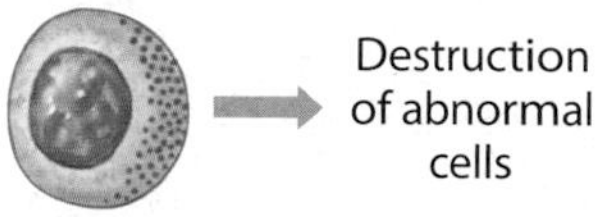
Destruction of abnormal cells

Interferons are chemicals that coordinate the defenses against viral infections.

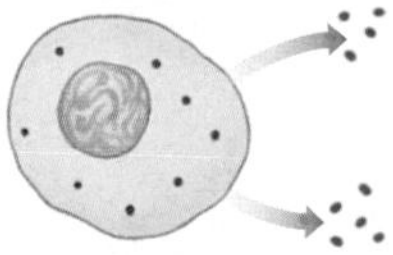

Complement is a system of circulating proteins that assists antibodies in the destruction of pathogens.

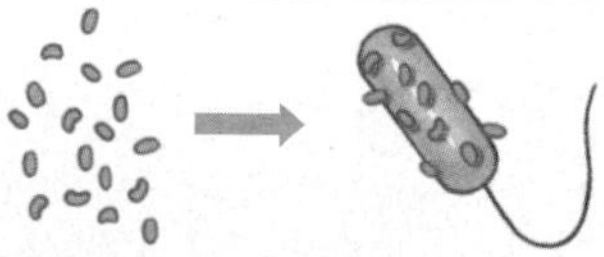

Inflammation is a localized, tissue-level response that tends to limit the spread of an injury or infection.

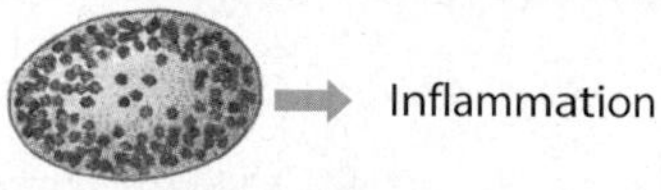
Inflammation

Fever is an elevation of body temperature that accelerates tissue metabolism and defenses.

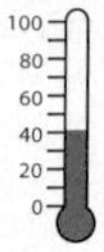
100
80
60
40
20
0

Adaptive (specific) Immunity

Adaptive defenses protect against particular threats. For example, a specific defense may protect against infection by one type of bacterium but be ineffective against other bacteria and viruses. Adaptive immunity depends on the activities of specific lymphocytes. The body's specific defenses produce a state of protection known as **specific resistance**. We discuss adaptive immunity in Section 3.

Module 20.8 Review

a. Distinguish between innate immunity and adaptive immunity.

b. How does innate immunity protect us from disease?

c. A child falls off her bike and skins her knee. Which form of immunity will be activated immediately?

LO **20.8** Give an overview of the components of innate (nonspecific) immunity.

Physical barriers prevent pathogens and toxins from entering body tissues . . .

To cause trouble, an antigenic substance or a pathogen must enter body tissues, which requires crossing an epithelium—either at the skin or across a mucous membrane.

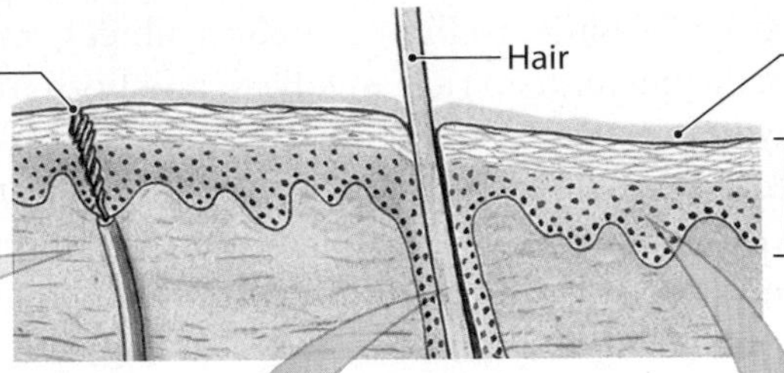
Duct of eccrine sweat gland
Hair
Secretion
Epithelium

1 The integumentary system provides the major physical barrier to the external environment.

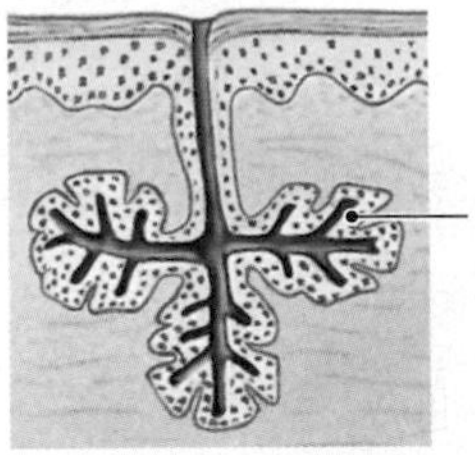
Sebaceous gland

Most epithelia are protected by specialized accessory structures and secretions. The epidermal surface also receives the secretions of sebaceous and sweat glands. These secretions, which flush the surface to wash away microorganisms and chemical agents, may also contain bactericidal chemicals, destructive enzymes (lysozymes), and antibodies.

The hairs on most areas of your body protect against physical abrasion (especially on the scalp), and they often prevent hazardous materials or insects from touching your skin.

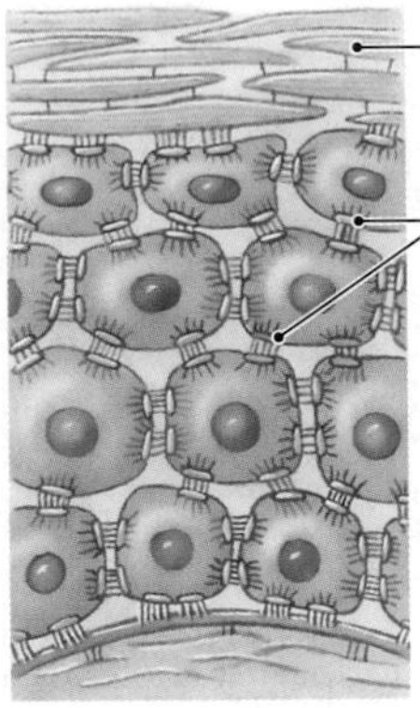
Keratinized cells
Desmosomes

The epithelial covering of the skin has multiple layers, a coating of keratinized cells, and a network of desmosomes that lock adjacent cells together.

2 The epithelia lining the digestive, respiratory, urinary, and reproductive tracts are an important barrier that protects against antigenic substances and pathogens. As noted in Module 20.5, MALT in these locations provides a secondary line of nonspecific defense.

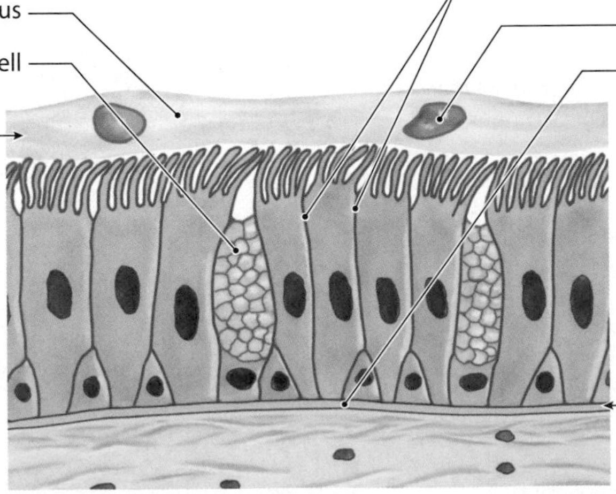
Mucus
Mucous cell
Tight junctions
Entrapped particle
Basement membrane

Mucus bathes most surfaces of your digestive tract, and your stomach contains a powerful acid that can destroy many pathogens. Mucus moves across the lining of the respiratory tract, urine flushes the urinary passageways, and glandular secretions do the same for the reproductive tract. Special enzymes, antibodies, and an acidic pH add to the effectiveness of these secretions.

Along the more delicate internal passageways, epithelial cells are usually tied together by tight junctions and supported by a fibrous basement membrane.

. . . and phagocytes provide the next line of defense

3 **Phagocytes** serve as janitors and police in peripheral tissues. They remove cellular debris and respond to invasion by foreign substances or pathogens. Phagocytes are the "first line of cellular defense" against pathogenic invasion. Many phagocytes attack and remove microorganisms even before lymphocytes detect them. All phagocytic cells function in much the same way, although the target of phagocytosis may differ from one type of phagocyte to another.

Types of Phagocytes

12 μm

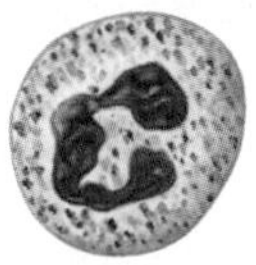

Neutrophils are abundant, mobile, and quick to phagocytize cellular debris or invading bacteria. They circulate in the bloodstream and roam through peripheral tissues, especially at sites of injury or infection.

8–10 μm

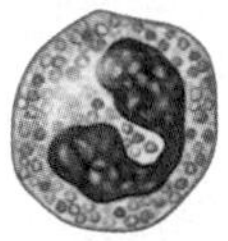

Eosinophils, which are less abundant than neutrophils, phagocytize foreign substances or pathogens that have been coated with antibodies.

There are two major classes of **macrophages** derived from the monocytes of the circulating blood. This collection of phagocytic cells is called the **monocyte–macrophage system**, or the reticuloendothelial system.

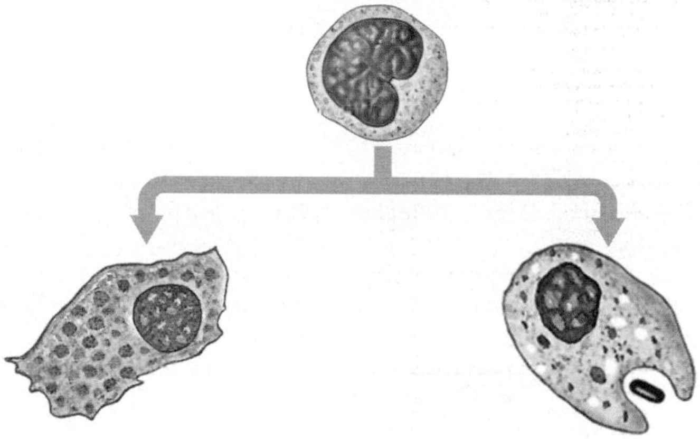

Fixed macrophages are permanent residents of specific tissues and organs and are scattered among connective tissues. They are immobile and stay within these tissues.

Free macrophages travel throughout the body, arriving at the site of an injury by migrating through adjacent tissues or by recruitment from the circulating blood.

4 All phagocytes share several functional characteristics, which are summarized in the table below.

Functional Characteristics of Phagocytes

- Phagocytes can leave capillaries by squeezing between adjacent endothelial cells, a process known as **emigration**, or **diapedesis**.
- Phagocytes may be attracted to or repelled by chemicals in the surrounding fluids, a phenomenon called **chemotaxis**. They are particularly sensitive to chemicals released by either body cells or pathogens.
- Phagocytosis always begins with the attachment of the phagocyte to its target. In this process, receptors on the plasma membrane of the phagocyte bind to the surface of the target.
- After attachment, the phagocyte may either destroy the target itself or promote its destruction by activating specific defenses.

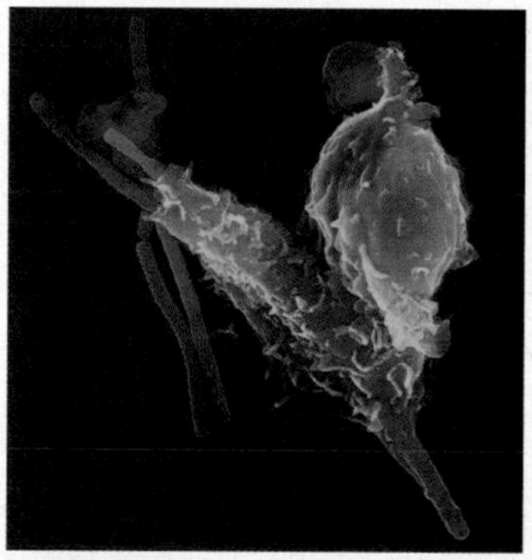

Phagocyte engulfing bacteria SEM × 2900

Module 20.9 Review

a. How does the integumentary system protect the body?
b. Identify the types of phagocytes in the body, and differentiate between fixed macrophages and free macrophages.
c. Define chemotaxis.

LO 20.9 Explain how physical barriers and phagocytes play a role in innate immunity.

NK cells perform immune surveillance, detecting and destroying abnormal cells

The immune system generally ignores the body's own cells unless they become abnormal in some way. **Natural killer (NK) cells** are responsible for recognizing and destroying abnormal cells when they appear in peripheral tissues. The constant monitoring of normal tissues by NK cells is called **immune surveillance**. The plasma membranes of cancer cells generally contain unusual proteins called **tumor-specific antigens**, which NK cells recognize as abnormal. The affected cells are then destroyed, preserving the tissue.

1 NK cells recognize bacteria, foreign cells, cells infected by viruses, and cancer cells. In each case, the steps leading to target cell destruction are similar.

Step 1
If a cell has unusual components in its plasma membrane, an NK cell recognizes that other cell as abnormal. Such recognition activates the NK cell, which then adheres to its target cell.

Step 2
The Golgi apparatus moves around the nucleus until the maturing face points directly toward the abnormal cell. A flood of secretory vesicles is then produced at the Golgi apparatus. These vesicles, which contain proteins called **perforins**, travel through the cytoplasm toward the cell surface.

Step 3
The perforins are released at the cell surface by exocytosis and diffuse across the narrow gap separating the NK cell from its target.

Step 4
As a result of the pores made by perforin molecules, the target cell can no longer maintain its internal environment, and it quickly disintegrates.

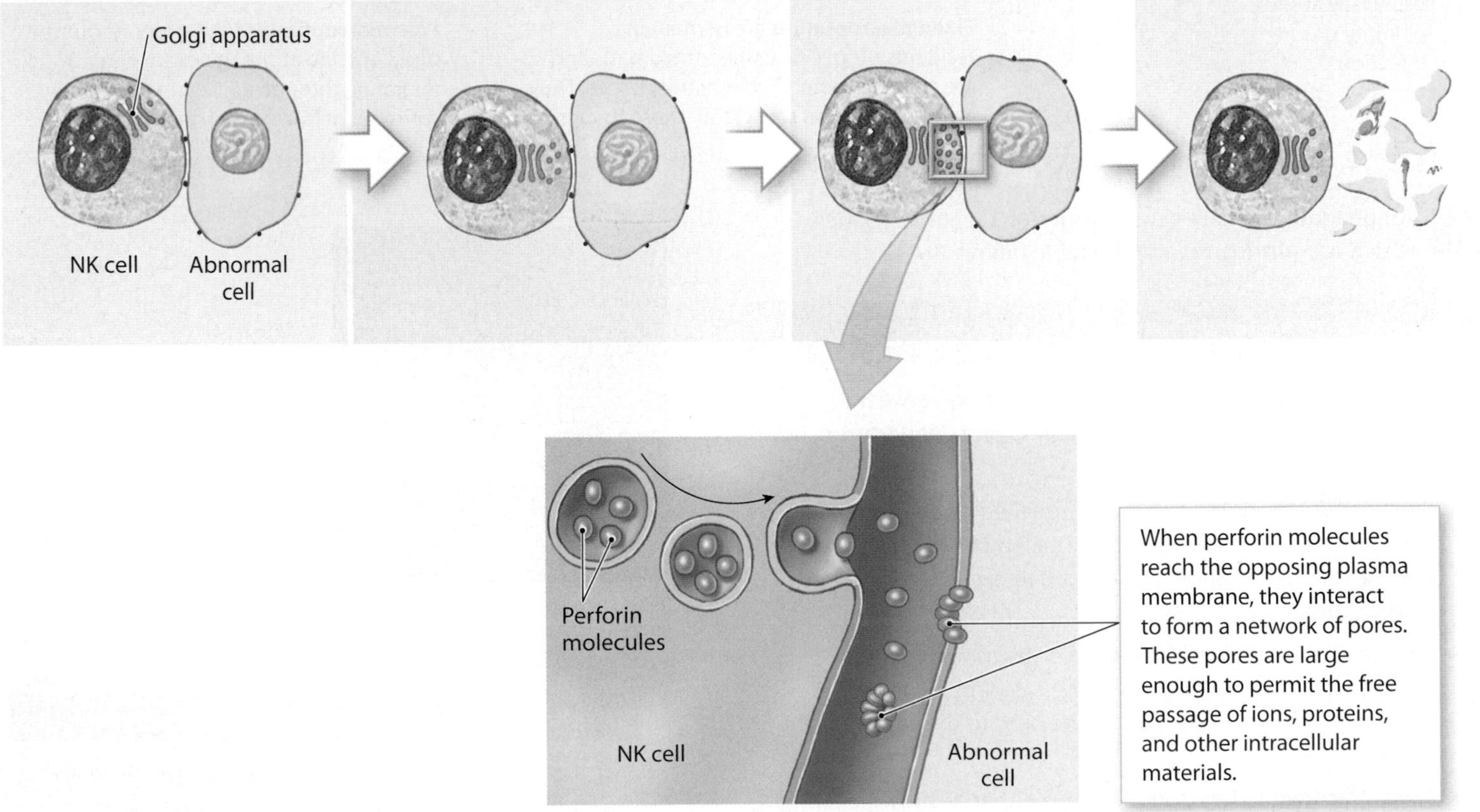
Golgi apparatus
NK cell
Abnormal cell
Perforin molecules
NK cell
Abnormal cell
When perforin molecules reach the opposing plasma membrane, they interact to form a network of pores. These pores are large enough to permit the free passage of ions, proteins, and other intracellular materials.

2 Cell division is a complicated process, and mistakes occasionally occur. The abnormal daughter cells usually cannot survive, and NK cells detect and destroy the few that can still function. That detection is important because a small number of those abnormal cells are potentially lethal cancer cells.

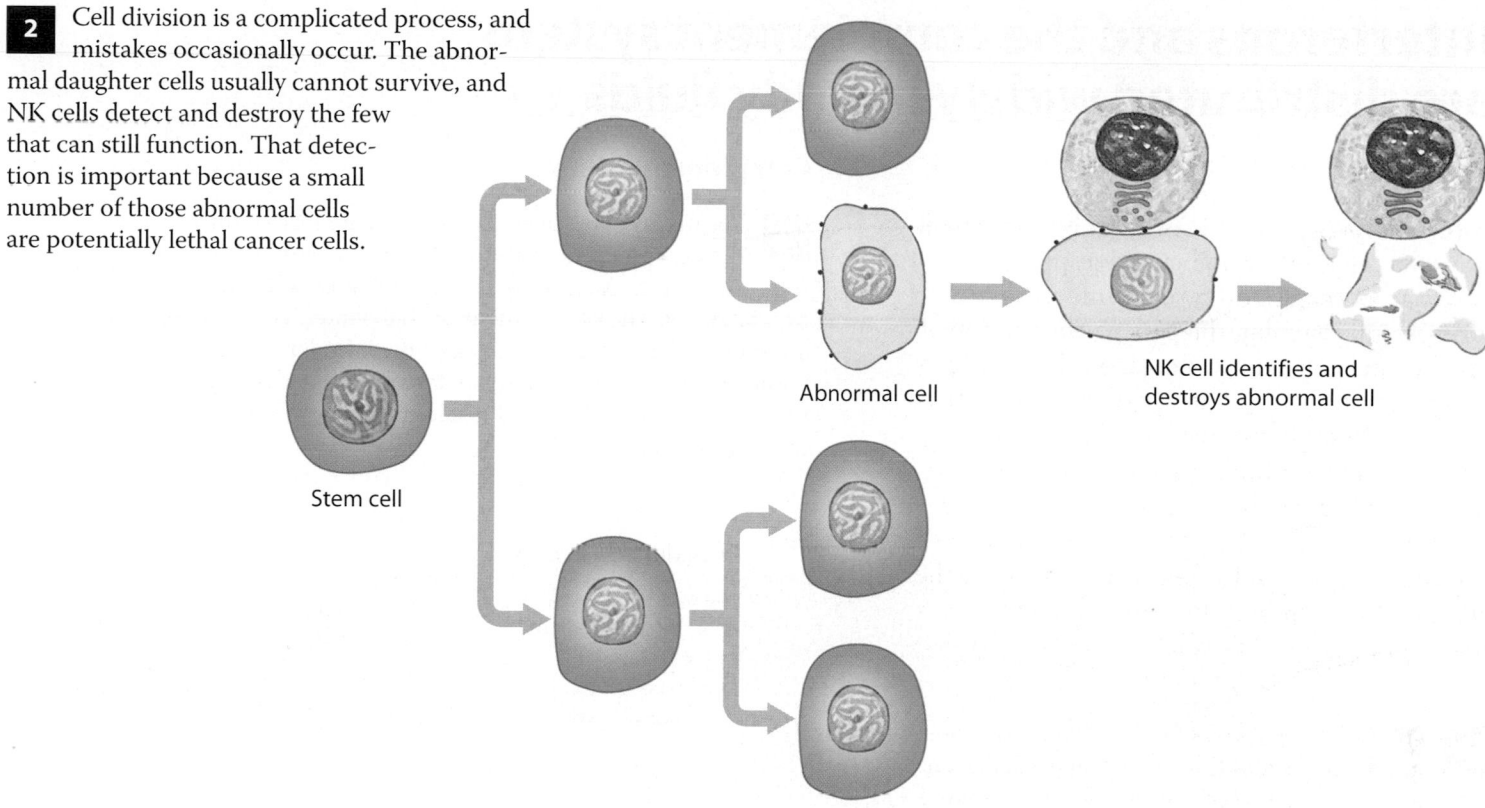

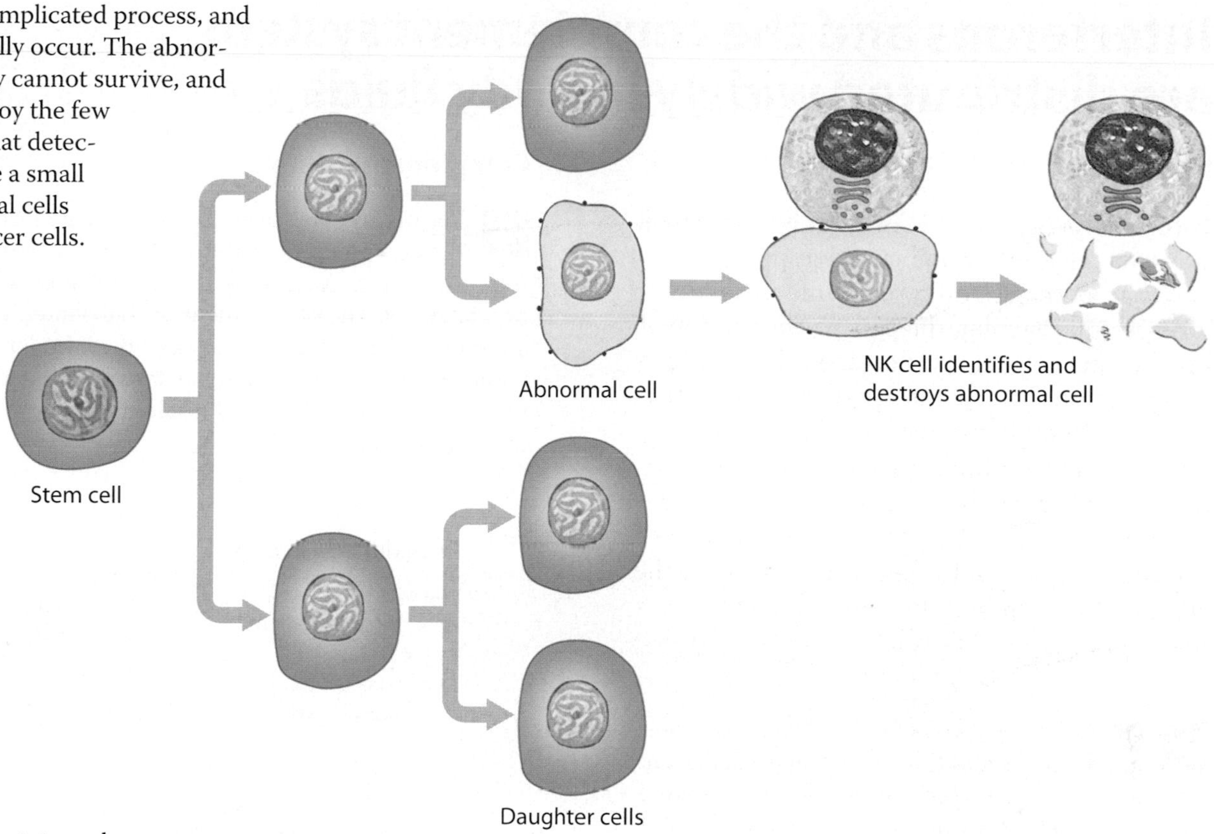
Abnormal cell
NK cell identifies and destroys abnormal cell
Stem cell
Daughter cells

3 Cancer cells often mutate and can sometimes avoid detection by NK cells. This process of avoiding detection or neutralizing body defenses is called **immunological escape**.

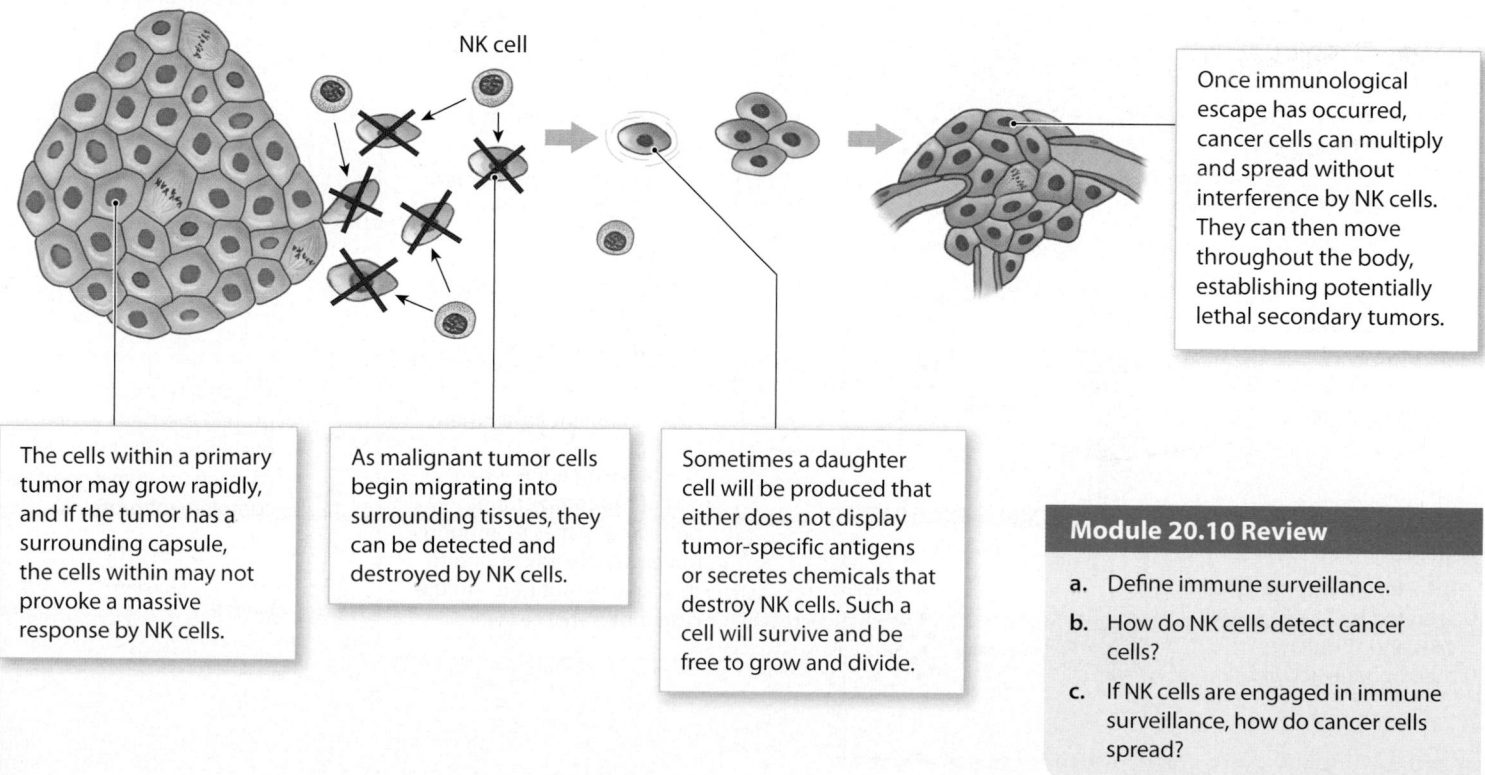

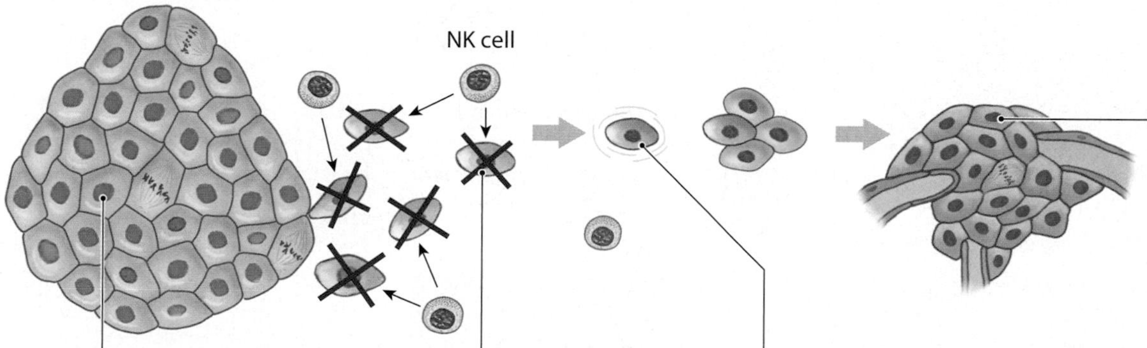
NK cell

Once immunological escape has occurred, cancer cells can multiply and spread without interference by NK cells. They can then move throughout the body, establishing potentially lethal secondary tumors.

The cells within a primary tumor may grow rapidly, and if the tumor has a surrounding capsule, the cells within may not provoke a massive response by NK cells.

As malignant tumor cells begin migrating into surrounding tissues, they can be detected and destroyed by NK cells.

Sometimes a daughter cell will be produced that either does not display tumor-specific antigens or secretes chemicals that destroy NK cells. Such a cell will survive and be free to grow and divide.

Module 20.10 Review

a. Define immune surveillance.

b. How do NK cells detect cancer cells?

c. If NK cells are engaged in immune surveillance, how do cancer cells spread?

LO 20.10 Describe immune surveillance, and explain the role of NK cells.

Interferons and the complement system are distributed widely in body fluids

Interferons

Interferons (in-ter-FĒR-onz) **(IFNs)** are small proteins released by activated lymphocytes and macrophages and by cells infected with viruses. On reaching the plasma membrane of a normal cell, an interferon binds to surface receptors on the cell and, by second messengers, triggers the production of antiviral proteins in the cytoplasm. Antiviral proteins do not interfere with the entry of viruses, but they do interfere with viral replication inside the cell. In addition to their role in slowing the spread of viral infections, IFNs stimulate macrophages and NK cells.

1 At least three types of interferons exist, each of which has additional specialized functions. Interferons are examples of **cytokines** (SĪ-tō-kīnz)—chemicals released by cells to coordinate local activities. Cytokines produced by most cells are used only for cell-to-cell communication within a given tissue. However, cytokines released by phagocytes and lymphocytes can also act as hormones, affecting cells and tissues throughout the body.

1 **Interferon alpha** (α) is produced by cells infected with viruses. IFN alpha attracts and stimulates NK cells and enhances resistance to viral infection.

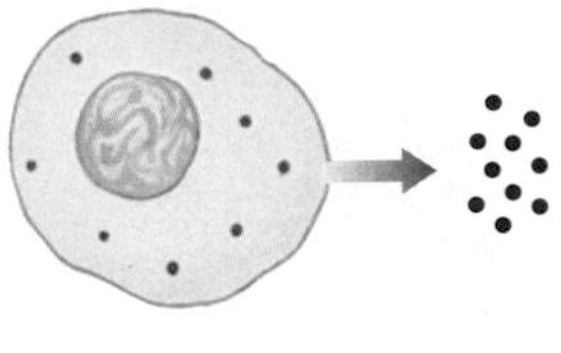

2 **Interferon beta** (β) is secreted by fibroblasts and slows inflammation in a damaged area.

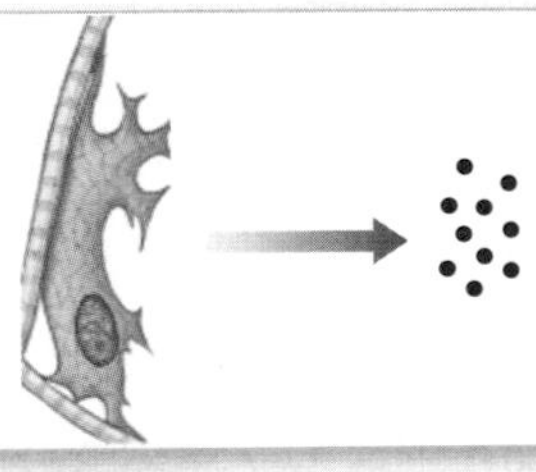

3 **Interferon gamma** (γ) is secreted by T cells and NK cells and stimulates macrophage activity.

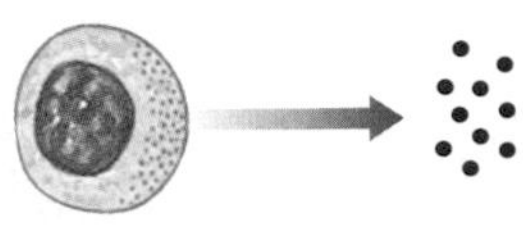

Complement System

2 Plasma contains over 30 special proteins that form the **complement system**. The term *complement* refers to the fact that this system complements the action of antibodies. The complement proteins interact with one another in chain reactions, or cascades, similar to the blood clotting system. There are two possible pathways for complement action: the **classical pathway** and the **alternative pathway**.

Classical Pathway

The most rapid and effective activation of the complement system occurs through the classical pathway.

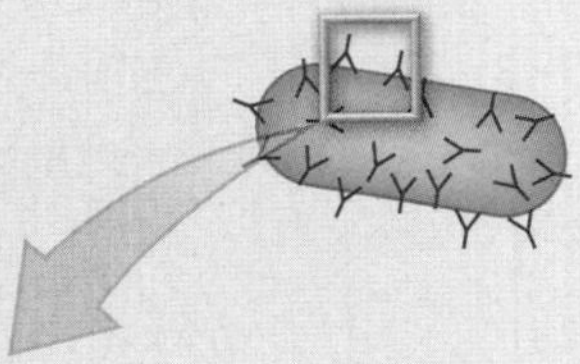

Antibody Binding and Complement Attachment

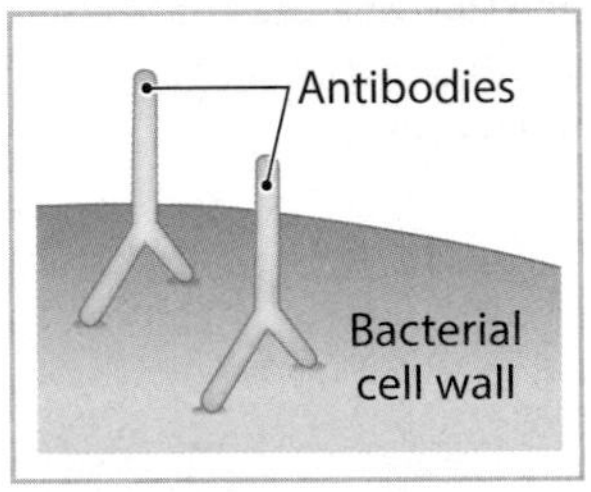
Antibodies
Bacterial cell wall

Antibody binding

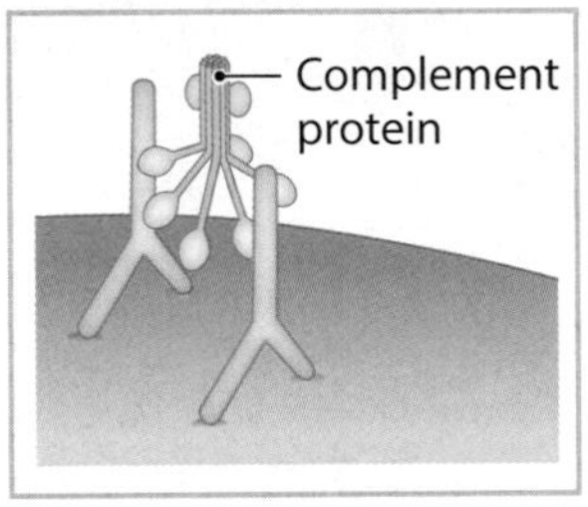
Complement protein

Complement attachment

The process begins when one of the complement proteins attaches to antibody molecules already bound to their specific antigen—in this case, a bacterial cell wall.

Activation and Cascade

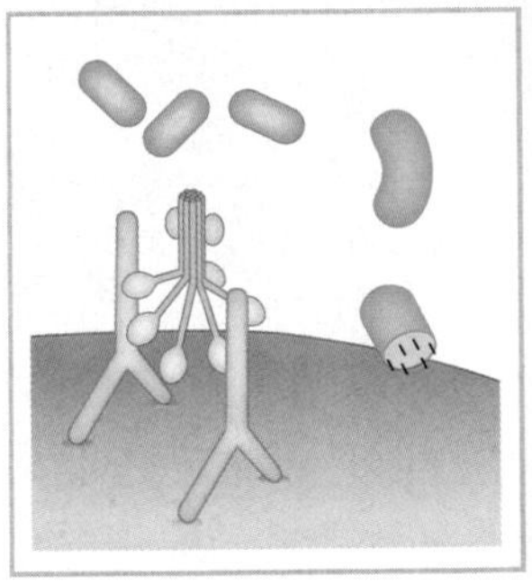

The attached complement protein then acts as an enzyme, catalyzing a series of reactions involving other complement proteins.

Alternative Pathway

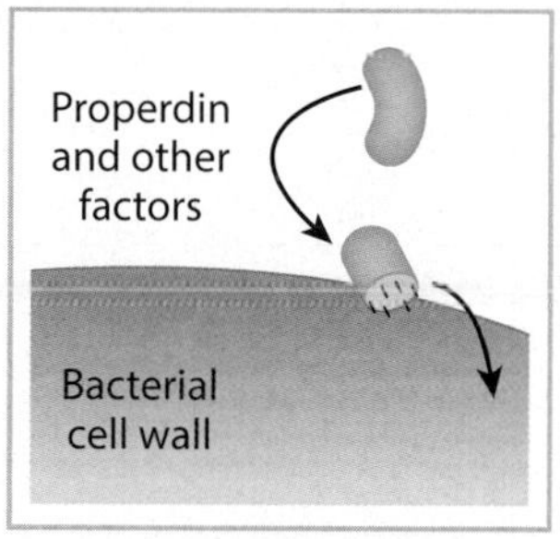

Properdin and other factors
Bacterial cell wall

The alternative pathway begins when several complement proteins, notably **properdin**, interact in the plasma. This interaction can be triggered by exposure to foreign substances, such as the capsule of a bacterium. The end result is the attachment of an activated complement protein to the bacterial cell wall.

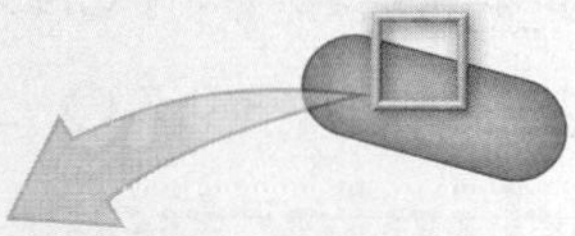

The alternative pathway is important in the defense against bacteria, some parasites, and virus-infected cells.

Complement Attachment (alternative pathway)

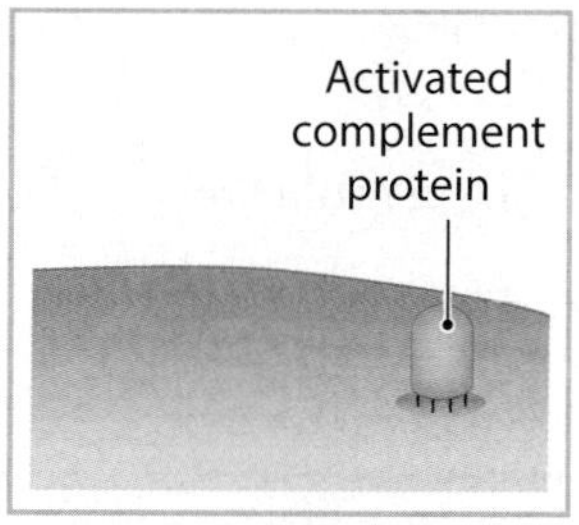

Activated complement protein

Complement Attachment (classical pathway)

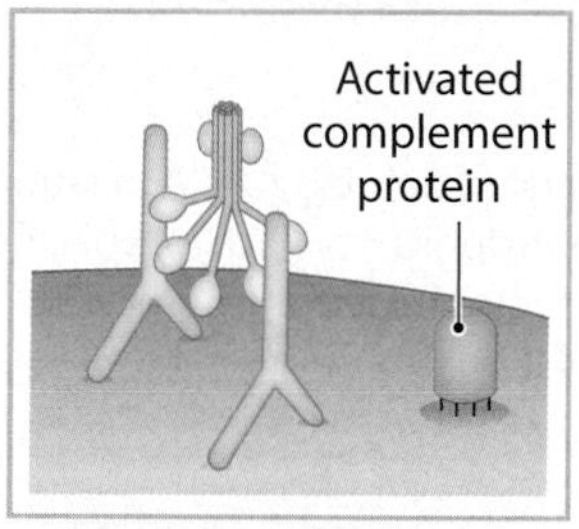

Activated complement protein

The classical pathway ends with the conversion of an inactive complement protein to an activated form that attaches to the cell wall.

Pore Formation and Cell Lysis

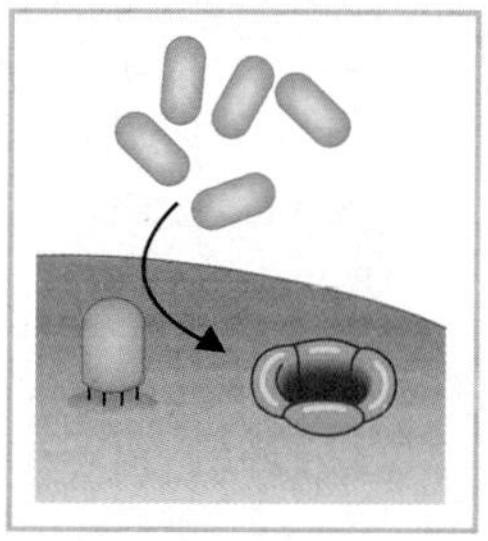

Once an activated complement protein has attached to the cell wall, additional complement proteins form a pore in the membrane that destroys the integrity of the target cell.

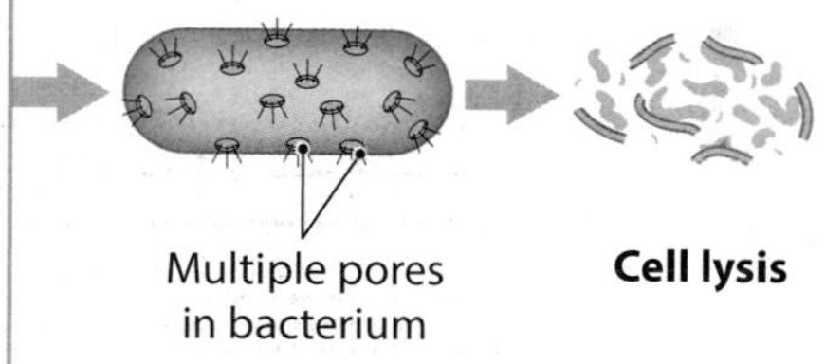

Multiple pores in bacterium
Cell lysis

Enhanced Phagocytosis

A coating of complement proteins and antibodies both attracts phagocytes and makes the target cell easier to engulf. This enhancement of phagocytosis, a process called **opsonization**, occurs because macrophage membranes contain receptors that detect and bind to complement proteins and bound antibodies.

Histamine Release

Release of histamine by mast cells and basophils in tissues increases the degree of local inflammation and accelerates blood flow to the region.

Module 20.11 Review

a. Define interferons.

b. Briefly explain the role of complement proteins.

c. What is the effect of histamine released by complement system activation?

LO 20.11 Describe the types of interferons, and explain the pathways of complement activation.

Inflammation is a localized tissue response to injury; fever is a generalized response to tissue damage and infection

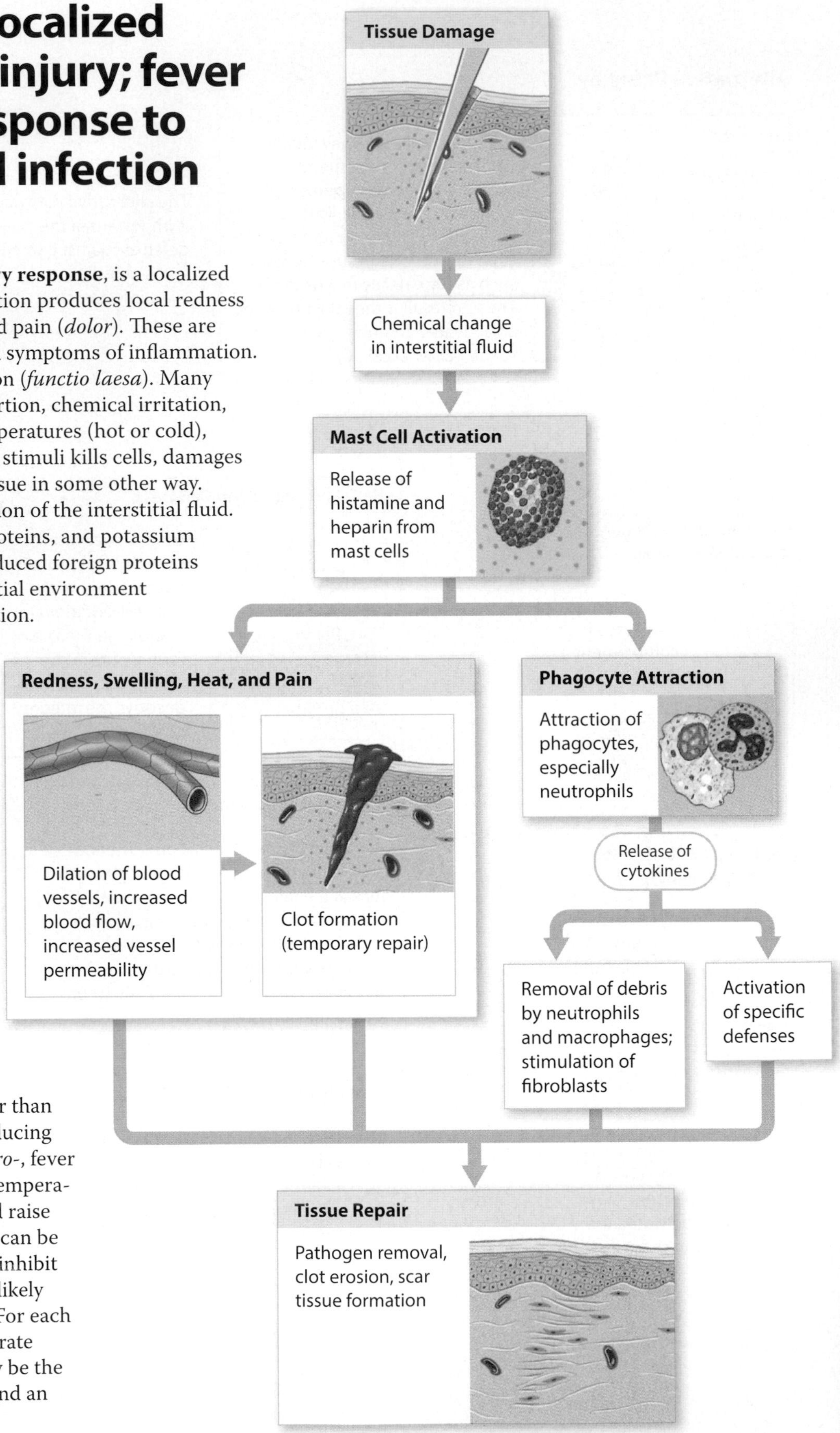
Tissue Damage
Chemical change in interstitial fluid
Mast Cell Activation
Release of histamine and heparin from mast cells
Redness, Swelling, Heat, and Pain
Dilation of blood vessels, increased blood flow, increased vessel permeability
Clot formation (temporary repair)
Phagocyte Attraction
Attraction of phagocytes, especially neutrophils
Release of cytokines
Removal of debris by neutrophils and macrophages; stimulation of fibroblasts
Activation of specific defenses
Tissue Repair
Pathogen removal, clot erosion, scar tissue formation

Inflammation

1 **Inflammation**, or the **inflammatory response**, is a localized tissue response to injury. Inflammation produces local redness (*rubor*), swelling (*tumor*), heat (*calor*), and pain (*dolor*). These are known as the so-called cardinal signs and symptoms of inflammation. Sometimes a fifth is included: lost function (*functio laesa*). Many stimuli, including impact, abrasion, distortion, chemical irritation, infection by pathogens, and extreme temperatures (hot or cold), can produce inflammation. Each of these stimuli kills cells, damages connective tissue fibers, or injures the tissue in some other way. The changes alter the chemical composition of the interstitial fluid. Damaged cells release prostaglandins, proteins, and potassium ions, and the injury itself may have introduced foreign proteins or pathogens. The changes in the interstitial environment trigger the complex process of inflammation.

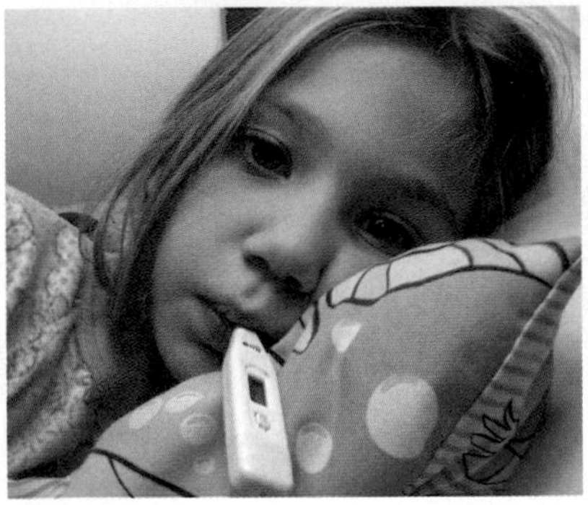

Fever

2 **Fever** is a body temperature greater than 37.2°C (99°F). Circulating fever-inducing proteins called **pyrogens** (PĪ-rō-jenz; *pyro-*, fever or heat + *-gen*, substance) can reset the temperature thermostat in the hypothalamus and raise body temperature. Within limits, a fever can be beneficial. High body temperatures may inhibit some viruses and bacteria, but the most likely beneficial effect is on body metabolism. For each 1°C rise in body temperature, metabolic rate jumps by 10 percent. The net results may be the quicker mobilization of tissue defenses and an accelerated repair process.

Summary of Innate Immunity (Nonspecific Defenses)

3 This table summarizes the information on innate (nonspecific) immunity presented in this section.

Physical Barriers

Prevent approach of and deny access to pathogens

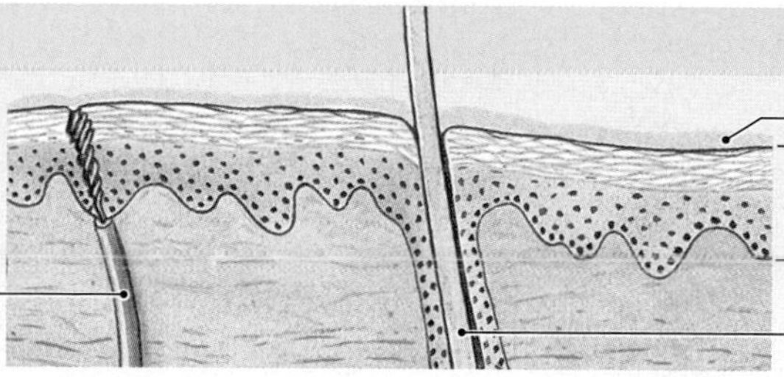

Secretions
Epithelium
Duct of eccrine sweat gland
Hair

Phagocytes

Remove debris and pathogens

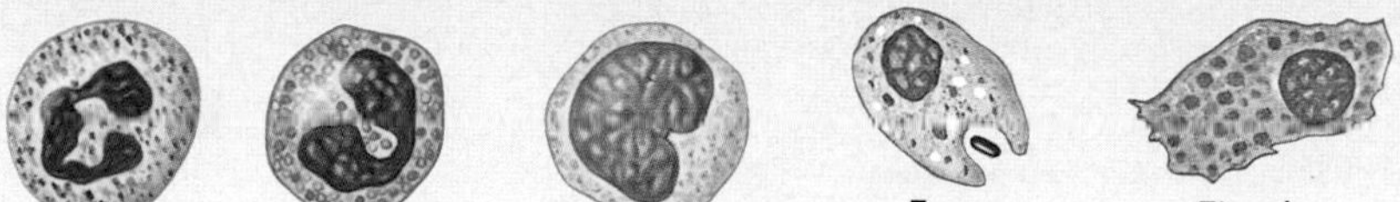

Neutrophil
Eosinophil
Monocyte
Free macrophage
Fixed macrophage

Immune Surveillance

Destroys abnormal cells

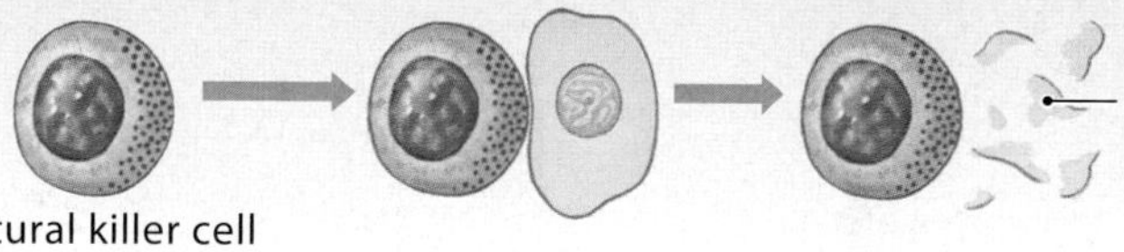

Natural killer cell
Lysed abnormal cell

Interferons

Increase resistance of cells to viral infection; slow the spread of disease

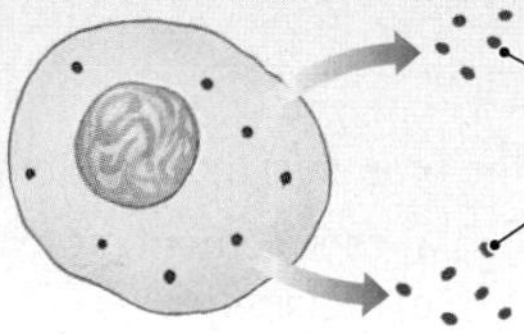

Interferons released by activated lymphocytes, macrophages, or virus-infected cells

Complement System

Attacks and breaks down the surfaces of cells, bacteria, and viruses; attracts phagocytes; stimulates inflammation

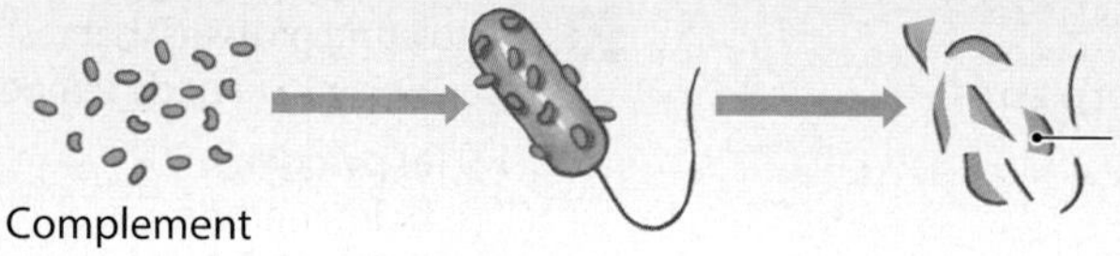

Complement
Lysed pathogen

Inflammation

Multiple effects

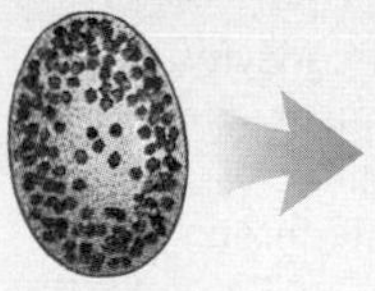

Mast cell

- Blood flow increased
- Phagocytes activated
- Damaged area isolated by clotting reaction
- Capillary permeability increased
- Complement activated
- Regional temperature increased
- Specific defenses activated

Fever

Mobilizes defenses; accelerates repairs; inhibits pathogens

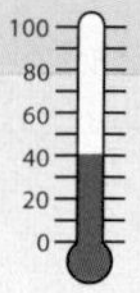

100
80
60
40
20
0

Body temperature rises above 37.2°C in response to pyrogens

Module 20.12 Review

a. Describe inflammation.

b. What effect do pyrogens have in the body?

c. A rise in the level of interferons in the body suggests what kind of infection?

LO **20.12** Explain the significance of inflammation and fever as innate defense mechanisms.

Labeling

Identify the type of innate immunity (nonspecific defense) described below.

1 keep hazardous organisms and materials outside the body. For example, a mosquito that lands on your head may be unable to reach the surface of the scalp if you have a full head of hair.

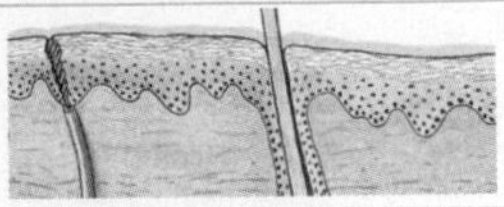

2 are cells that engulf pathogens and cell debris. Examples are the macrophages of peripheral tissues and the eosinophils and neutrophils of blood.

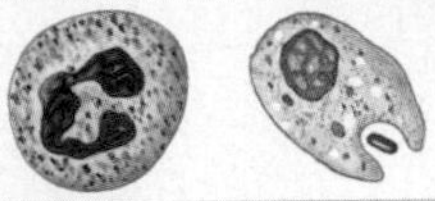

3 is the destruction of abnormal cells by NK cells in peripheral tissues.

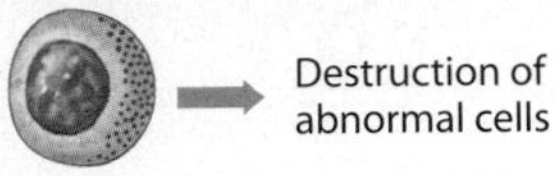
Destruction of abnormal cells

4 are chemical messengers that coordinate the defenses against viral infections.

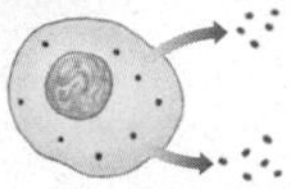

5 is a system of circulating proteins that assists antibodies in the destruction of pathogens.

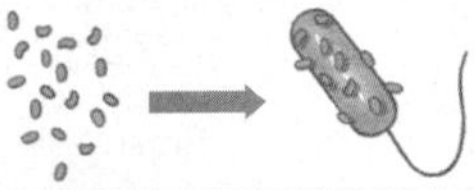

6 is a localized, tissue-level response that tends to limit the spread of an injury or infection.

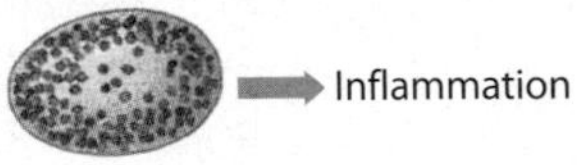
Inflammation

7 is an elevation of body temperature that accelerates tissue metabolism and the activity of defenses.

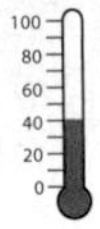
100
80
60
40
20
0

1 ____________________

2 ____________________

3 ____________________

4 ____________________

5 ____________________

6 ____________________

7 ____________________

Multiple choice

Select the correct answer from the list provided.

8 A physical barrier such as the skin provides a nonspecific body defense due to its makeup, which includes

- ☐ a) multiple layers.
- ☐ b) a coating of keratinized cells.
- ☐ c) a network of desmosomes locking adjacent cells together.
- ☐ d) all of these.

9 NK cells sensitive to the presence of abnormal plasma membranes are primarily involved in

- ☐ a) defenses against specific threats.
- ☐ b) phagocytic activity for defense.
- ☐ c) complex, time-consuming defense mechanisms.
- ☐ d) immune surveillance.

10 The nonspecific defense that breaks down cells, attracts phagocytes, and stimulates inflammation is

- ☐ a) the inflammatory response.
- ☐ b) the action of interferons.
- ☐ c) the complement system.
- ☐ d) immune surveillance.

11 The protein(s) that interfere with the replication of viruses is (are)

- ☐ a) complement proteins.
- ☐ b) heparin.
- ☐ c) pyrogens.
- ☐ d) interferons.

12 Circulating proteins that reset the thermostat in the hypothalamus, causing a rise in body temperature, are called

- ☐ a) pyrogens.
- ☐ b) interferons.
- ☐ c) lysosomes.
- ☐ d) complement proteins.

13 The "first line of cellular defense" against pathogenic invasion is

- ☐ a) phagocytes.
- ☐ b) mucus.
- ☐ c) hair.
- ☐ d) interferon.

Short answer

We usually associate a fever with illness or disease. How can a fever be beneficial?

14 ____

Adaptive immunity provides the body's specific defenses

1 **Adaptive (specific) immunity** is coordinated and produced by T cells and B cells. As the flowchart below shows, specific immunity may be acquired in different ways. To prevent disease, the most common way is to induce immunity artificially by vaccination (immunization). **Vaccines** stimulate an immune response by producing antibodies to a specific disease. A vaccine contains either a dead or inactive pathogen, antigens derived from that pathogen, or simulated antigens.

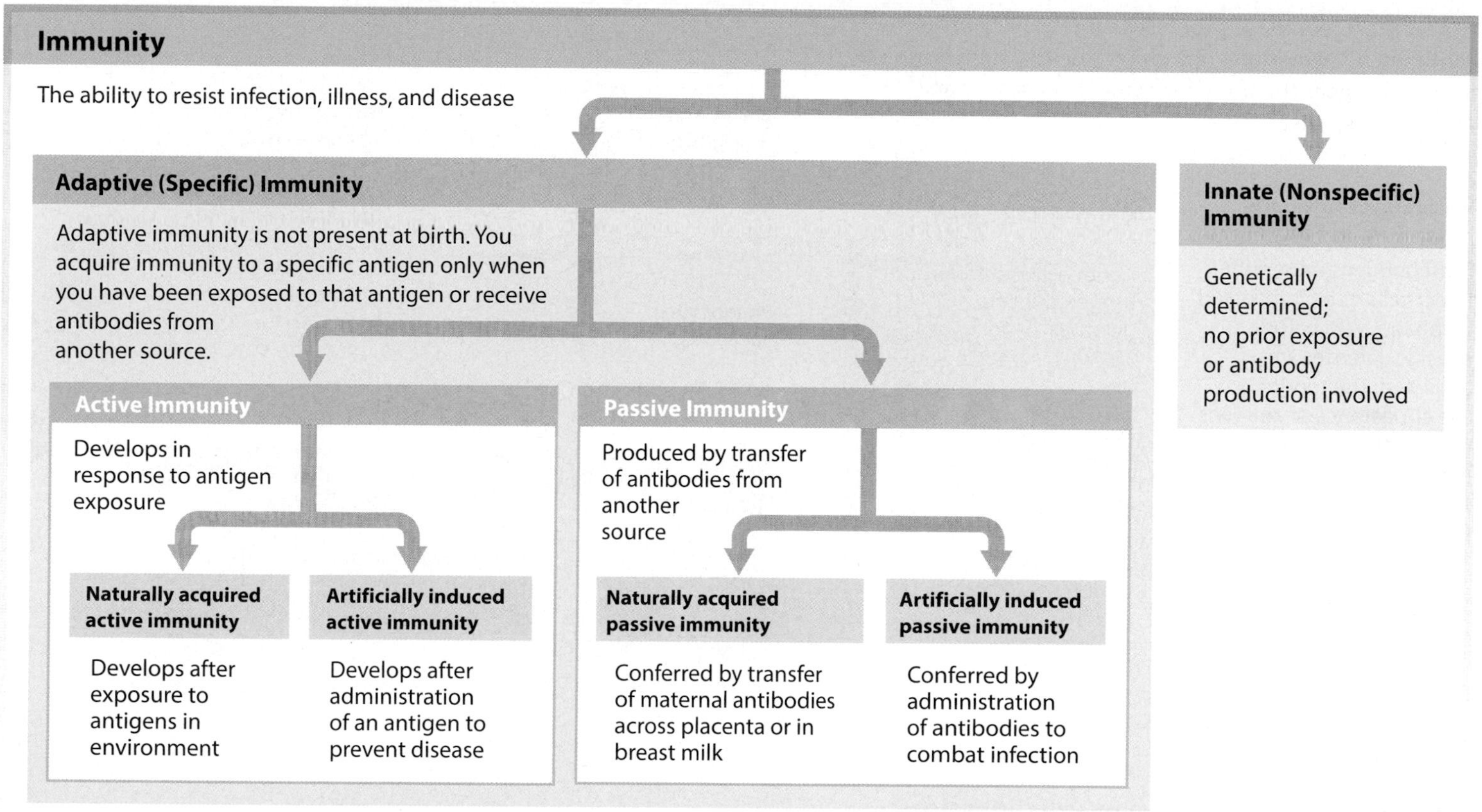
Immunity
The ability to resist infection, illness, and disease
Adaptive (Specific) Immunity
Adaptive immunity is not present at birth. You acquire immunity to a specific antigen only when you have been exposed to that antigen or receive antibodies from another source.
Innate (Nonspecific) Immunity
Genetically determined; no prior exposure or antibody production involved
Active Immunity
Develops in response to antigen exposure
Passive Immunity
Produced by transfer of antibodies from another source
Naturally acquired active immunity
Develops after exposure to antigens in environment
Artificially induced active immunity
Develops after administration of an antigen to prevent disease
Naturally acquired passive immunity
Conferred by transfer of maternal antibodies across placenta or in breast milk
Artificially induced passive immunity
Conferred by administration of antibodies to combat infection

2 Regardless of the form, adaptive immunity exhibits four general properties, which are summarized in the following table.

Properties of Adaptive Immunity

- **Specificity** results from the activation of specific lymphocytes and the production of antibodies with targeted effects. Each T cell or B cell has receptors that bind to one specific antigen, but ignore all others. The response of an activated T cell or B cell is equally specific, and leaves other antigens unaffected.
- **Versatility** results from the large diversity of lymphocytes present in the body. There are millions of different lymphocyte populations, each sensitive to a different antigen. When activated by a specific antigen, a lymphocyte divides, producing more lymphocytes with the same specificity. All the cells produced by the division of an activated lymphocyte constitute a **clone**.
- **Immunologic memory** exists because cell divisions of activated lymphocytes produce two groups of cells: one group that attacks the invader immediately, and another that remains inactive unless it is exposed to the same antigen at a later date. These inactive **memory cells** enable your immune system to "remember" antigens it has previously encountered, and to launch a faster, stronger, and longer-lasting counterattack if such an antigen reappears.
- **Tolerance** exists because the immune response ignores normal ("self") tissues but targets abnormal and foreign cells ("non-self") as well as toxins. Tolerance can also develop over time in response to chronic exposure to an antigen in the environment. Such tolerance generally lasts only as long as the exposure continues.

Module 20.13 Review

a. Which two cells coordinate adaptive immunity?

b. Which type of immunity develops when a child is given the polio vaccine?

c. What are the properties of adaptive immunity?

LO 20.13 Describe the types of adaptive immunity and four properties of adaptive immunity.

Adaptive immunity is triggered by exposure of T cells and B cells to specific antigens

1 This figure provides a "big picture" overview of the immune response. The presentation of specific antigens may stimulate (1) cell-mediated events involving attacks by T cells, and (2) antibody-mediated events involving specific antibodies secreted by activated B cell descendants. We will examine each of these processes more closely in later modules. This module focuses on the first step: how antigens trigger the body's specific defenses, or an immune response.

Antigens or Antigenic Fragments in Body Fluids

Most antigens must either infect cells or be "processed" by phagocytes before specific defenses are activated. The trigger is the appearance of antigens or antigenic fragments in plasma membranes; this is called **antigen presentation**.

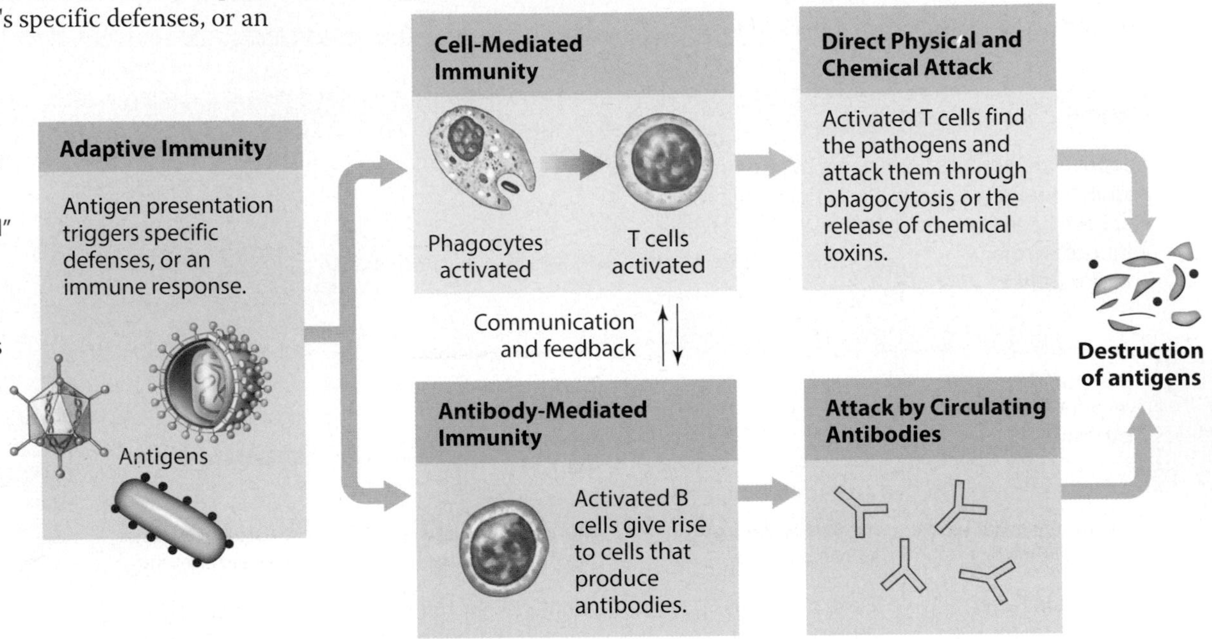

Adaptive Immunity
Antigen presentation triggers specific defenses, or an immune response.
Antigens
Cell-Mediated Immunity
Phagocytes activated
T cells activated
Communication and feedback
Antibody-Mediated Immunity
Activated B cells give rise to cells that produce antibodies.
Direct Physical and Chemical Attack
Activated T cells find the pathogens and attack them through phagocytosis or the release of chemical toxins.
Attack by Circulating Antibodies
Destruction of antigens

MHC Proteins and Antigen Presentation

2 Antigen presentation occurs when an antigen-glycoprotein combination capable of activating T cells appears in a plasma membrane. The structure of these glycoproteins is genetically determined. The genes controlling their synthesis are located along one portion of chromosome 6, in a region called the **major histocompatibility complex (MHC)**. These membrane glycoproteins are called **MHC proteins**. **Class I MHC proteins** are always present in the membranes of all nucleated cells.

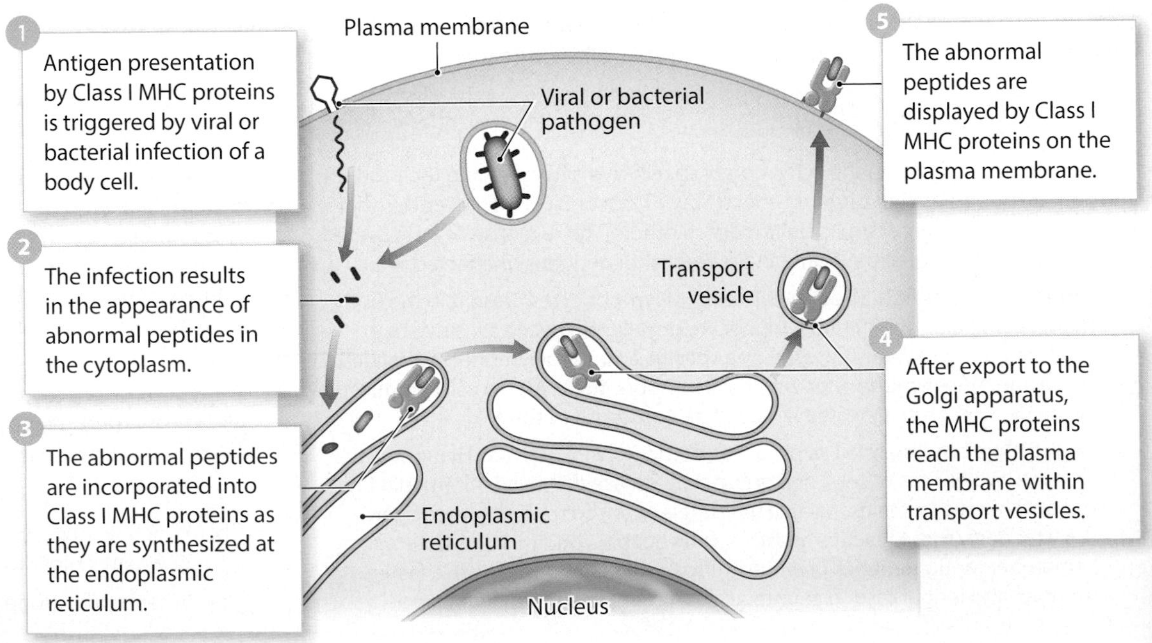

1
Antigen presentation by Class I MHC proteins is triggered by viral or bacterial infection of a body cell.
2
The infection results in the appearance of abnormal peptides in the cytoplasm.
3
The abnormal peptides are incorporated into Class I MHC proteins as they are synthesized at the endoplasmic reticulum.
Plasma membrane
Viral or bacterial pathogen
Transport vesicle
Endoplasmic reticulum
Nucleus
5
The abnormal peptides are displayed by Class I MHC proteins on the plasma membrane.
4
After export to the Golgi apparatus, the MHC proteins reach the plasma membrane within transport vesicles.

3 **Class II MHC proteins** are present only in the membranes of antigen-presenting cells and lymphocytes. **Antigen-presenting cells (APCs)** are specialized cells that include all the phagocytic cells of the monocyte–macrophage group and the dendritic cells of the skin and lymphoid organs. Class II MHC proteins appear in the plasma membrane only when the cell is processing antigens. The process of antigen presentation is shown here for a phagocytic cell.

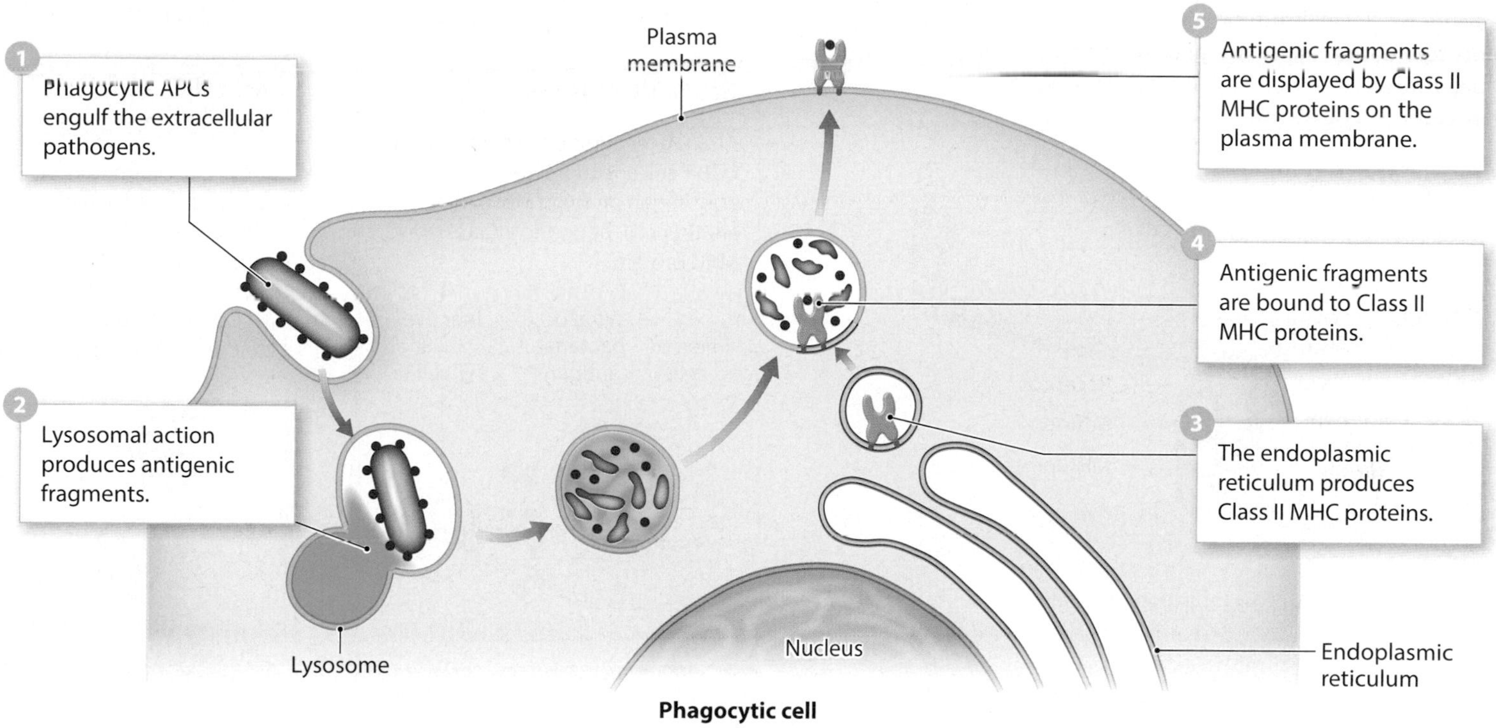
1
Phagocytic APCs engulf the extracellular pathogens.
2
Lysosomal action produces antigenic fragments.
Lysosome
Plasma membrane
Nucleus
Phagocytic cell
5
Antigenic fragments are displayed by Class II MHC proteins on the plasma membrane.
4
Antigenic fragments are bound to Class II MHC proteins.
3
The endoplasmic reticulum produces Class II MHC proteins.
Endoplasmic reticulum

4 In this preparation, cells were exposed to fluorescent-tagged antibodies that bind to specific structures and which glow when the cells are illuminated with fluorescent light. The green cell is an APC and the red cell is a lymphocyte; the bright yellow areas indicate where labeled antigens are being displayed on MHC proteins.

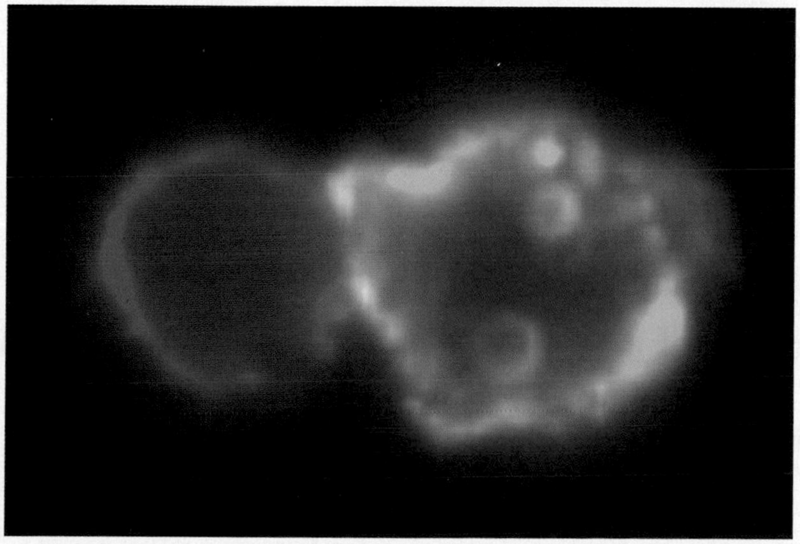

Module 20.14 Review

a. Describe antigen presentation.

b. What is the major histocompatibility complex (MHC)?

c. Where are Class I MHC proteins and Class II MHC proteins found?

LO **20.14** Explain how antigens trigger an immune response.

Infected cells stimulate the formation and division of cytotoxic T cells, memory T_C cells, and suppressor T cells

1 Inactive T cells have receptors that recognize either Class I or Class II MHC proteins on other cells when those proteins are bound to specific antigens. T cell binding will occur only if the MHC protein contains the antigen that the T cell is programmed to detect. This process is called **antigen recognition**.

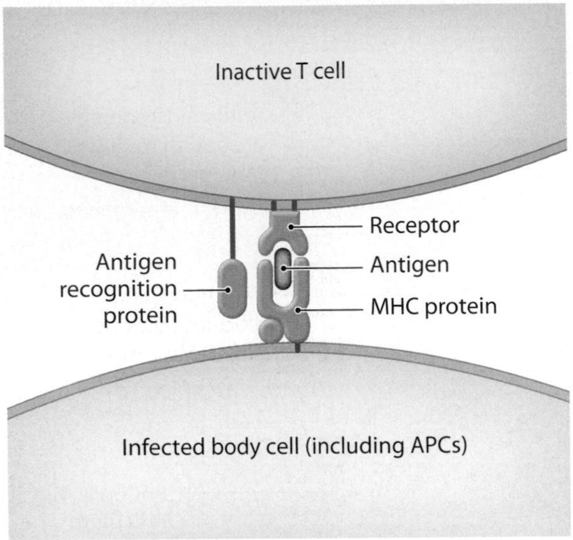
Inactive T cell
Receptor
Antigen
Antigen recognition protein
MHC protein
Infected body cell (including APCs)

2 The membrane proteins involved in antigen recognition are members of a class of proteins called **CD** (cluster of differentiation) **markers**.

CD Markers

There are at least 70 different CD markers, but only two associated with T cells are important to our discussion.

CD8 Markers

CD8 markers are found on **CD8 T cells**. CD8 T cells respond to antigens presented by Class I MHC proteins. These cells provide cell-mediated immunity.

CD4 Markers

CD4 markers are found on **CD4 T cells**. CD4 T cells are discussed further in the next module. They respond to antigens presented by Class II MHC proteins.

3 Antigen recognition prepares a T cell for activation. A CD8 T cell can recognize antigens bound to Class I MHC proteins (1). Full activation requires exposure of the cell to specific physical or chemical stimuli (2). This results in activation and cell division (3).

1 Antigen Recognition

Antigen recognition occurs when a CD8 T cell encounters an appropriate antigen on the surface of another cell, bound to a Class I MHC protein.

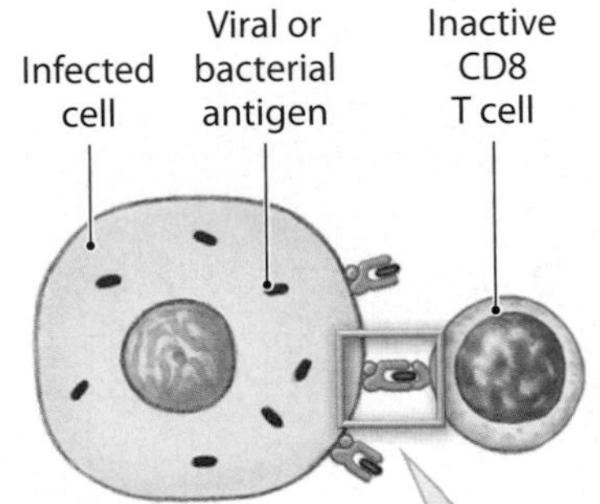
Infected cell
Viral or bacterial antigen
Inactive CD8 T cell

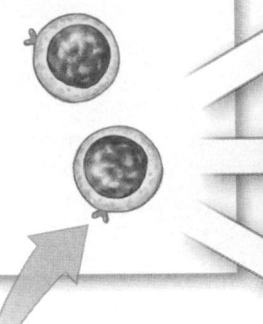

2 Costimulation

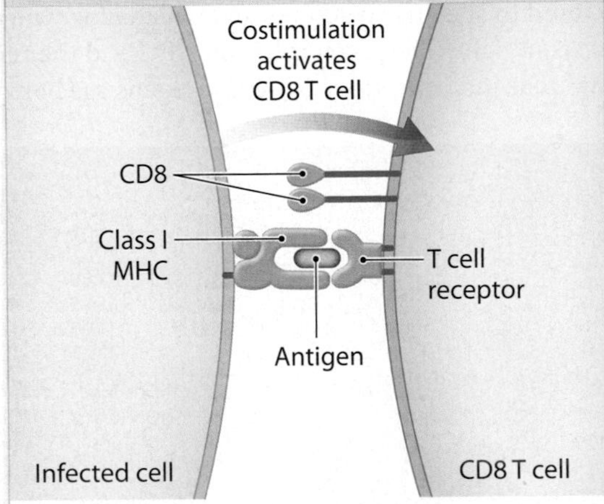
Costimulation activates CD8 T cell
CD8
Class I MHC
T cell receptor
Antigen
Infected cell
CD8 T cell

Before activation can occur, a T cell must be chemically or physically stimulated by the abnormal target cell. This vital secondary binding process, called **costimulation**, confirms the activation signal. Costimulation is like the safety on a gun: It helps prevent T cells from mistakenly attacking normal (self) tissues.

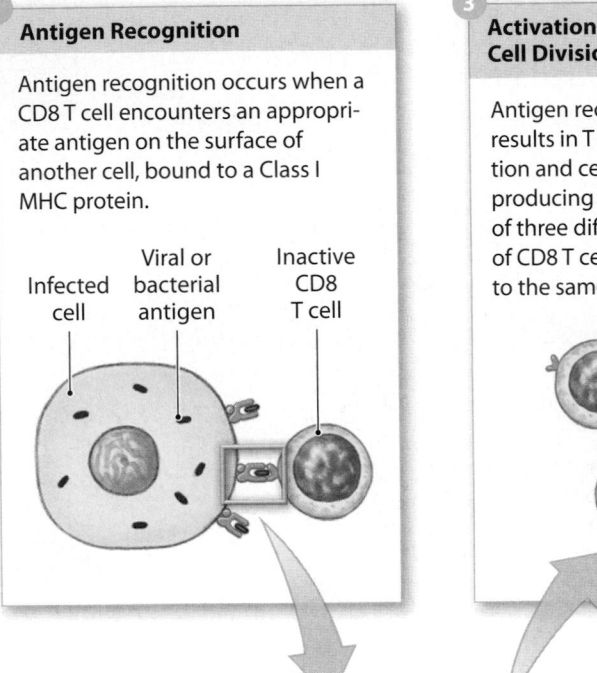

3 Activation and Cell Division

Antigen recognition results in T cell activation and cell division, producing populations of three different types of CD8 T cells sensitive to the same antigen.

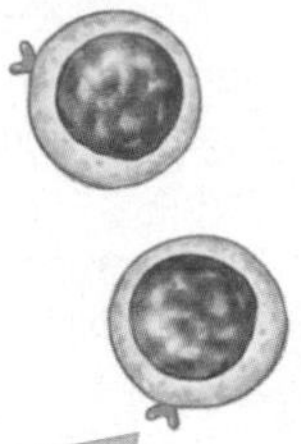

4 Following activation, the cell divides repeatedly, and daughter cells differentiate into three types of CD8 T cells.

Cytotoxic T Cells Seek Out Antigen-Bearing Cells

Cytotoxic T cells, also called **T_C cells**, seek out and destroy abnormal and infected cells. Cytotoxic T cells are highly mobile cells that roam throughout injured tissues. When a T_C cell encounters its target antigens bound to Class I MHC proteins, it attacks the target cell.

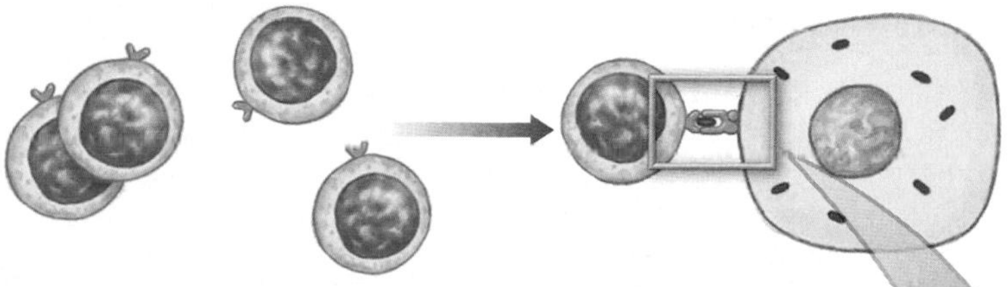

Destruction of Target Cells

The T_C cell destroys the antigen-bearing cell. It may use several different mechanisms to kill the target cell.

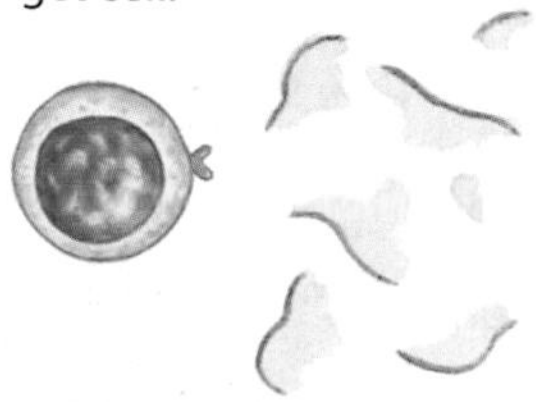

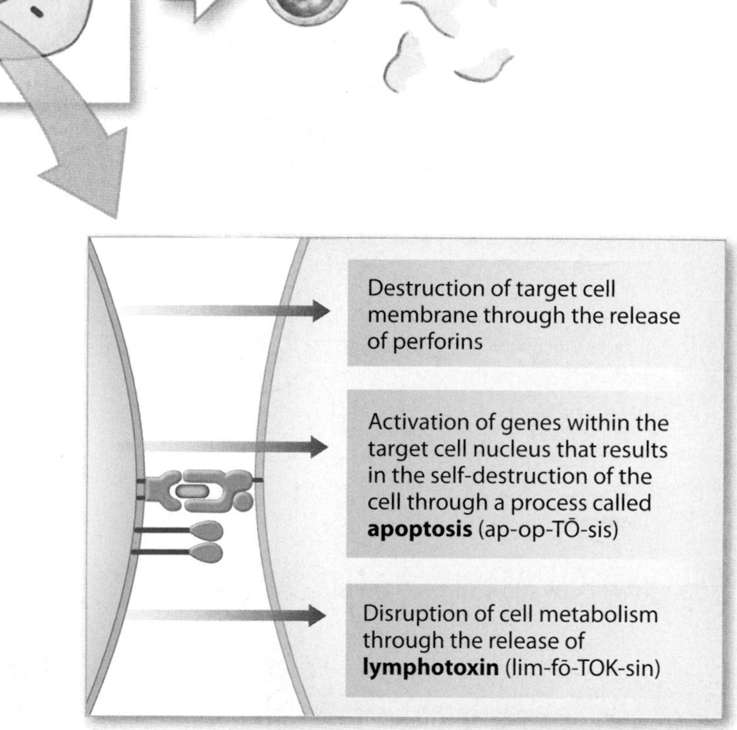
Destruction of target cell membrane through the release of perforins
Activation of genes within the target cell nucleus that results in the self-destruction of the cell through a process called **apoptosis** (ap-op-TŌ-sis)
Disruption of cell metabolism through the release of **lymphotoxin** (lim-fō-TOK-sin)

Memory T_C Cells Are Produced

Memory T_C cells are produced by the same cell divisions that produce cytotoxic T cells. Thousands of these cells are produced, but they do not differentiate further the first time the antigen triggers an immune response.

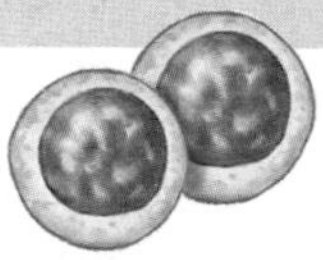

Memory T_C cells (inactive)

Suppressor T Cells Provide a Delayed Suppression

Suppressor T cells (T_S cells) suppress the responses of other T cells and of B cells by secreting **suppression factors** that limit the degree of immune system activation. Suppression does not occur immediately, because suppressor T cell activation takes much longer than the activation of other types of T cells. Suppressor T cells act only after the initial immune response.

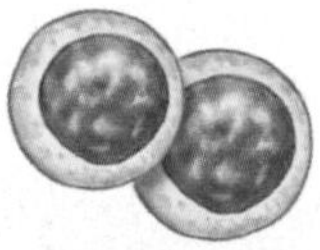

Suppressor T cells

The first time CD8 T cells encounter a specific antigen, the entire sequence of events—from the appearance of the antigen in a tissue to widespread cell destruction by cytotoxic T cells—takes 2 days or more. During this time, the damage or infection may spread, making it more difficult to control. However, if the same antigen appears a second time, memory T_C cells will immediately differentiate into cytotoxic T cells, producing a prompt, effective cellular response that can overwhelm an invading organism before it becomes well established in the tissues.

Module 20.15 Review

a. Describe CD markers.

b. Identify the three major types of T cells activated by Class I MHC proteins.

c. How do abnormal antigens attached to Class I MHC proteins initiate an immune response?

Antigen-presenting cells can stimulate activation of CD4 T cells, producing helper T cells that promote B cell activation and antibody production

1 Before they can initiate antibody-mediated immunity, inactive CD4 T cells must be exposed to antigens that are bound to Class II MHC proteins. Costimulation then completes their activation. Upon activation, CD4 T cells undergo a series of divisions, and daughter cells differentiate into active **helper T cells (T_H cells)** and **memory T_H cells**.

2 When a B cell encounters its specific antigen, it prepares for activation. This preparatory process is called **sensitization**. During sensitization, antigens brought into the cell by endocytosis subsequently appear on the surface of the B cell, bound to Class II MHC proteins.

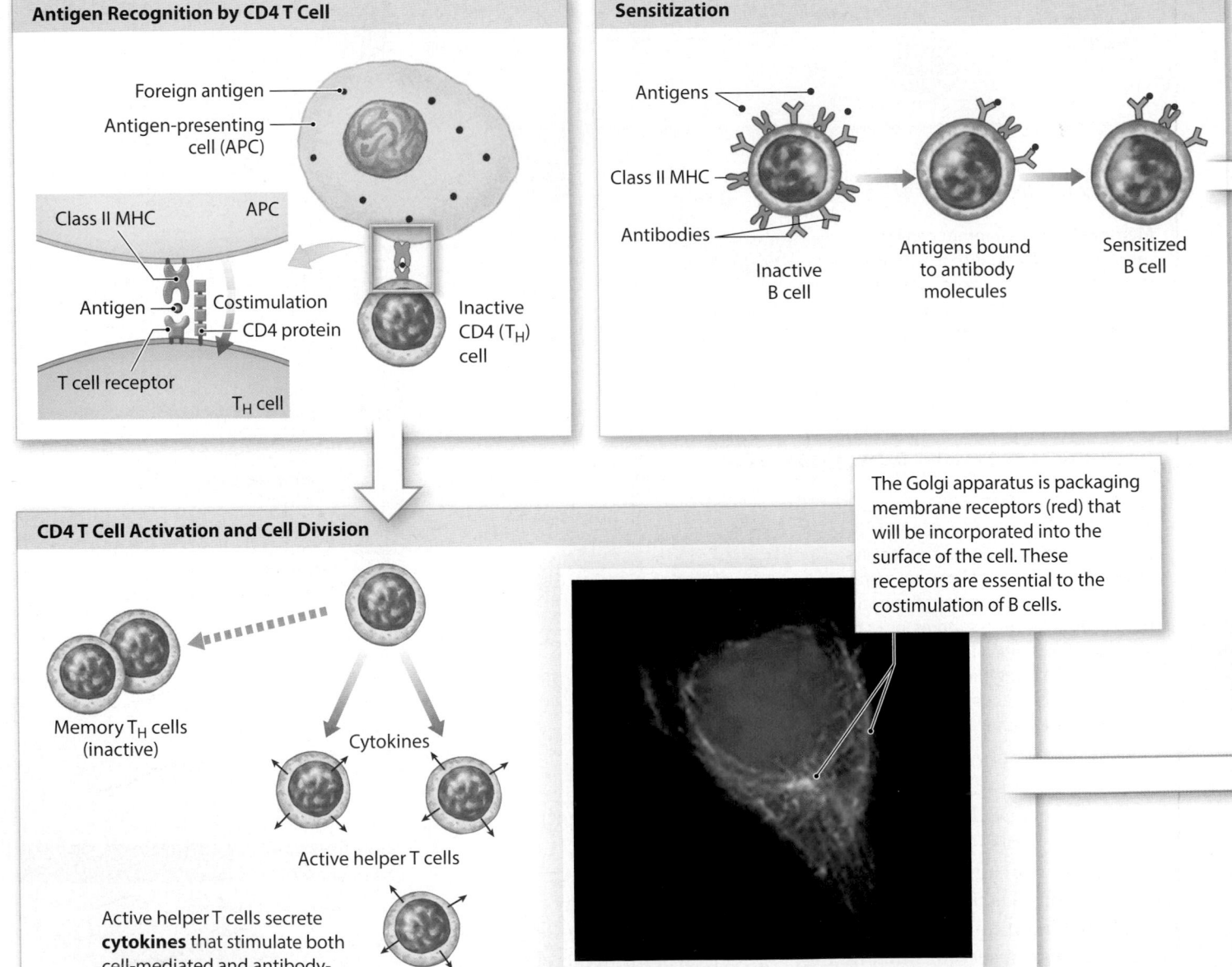

Antigen Recognition by CD4 T Cell

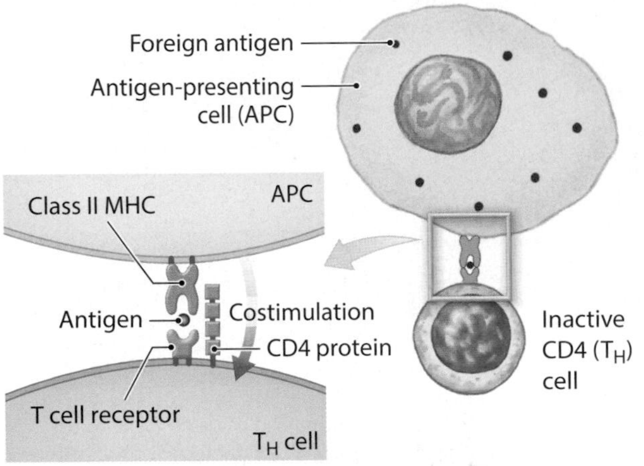
Foreign antigen
Antigen-presenting cell (APC)
Class II MHC
APC
Antigen
Costimulation
CD4 protein
T cell receptor
T_H cell
Inactive CD4 (T_H) cell

Sensitization

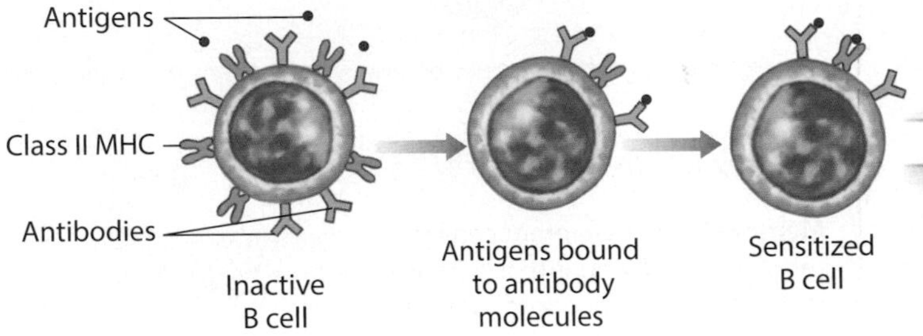
Antigens
Class II MHC
Antibodies
Inactive B cell
Antigens bound to antibody molecules
Sensitized B cell

CD4 T Cell Activation and Cell Division

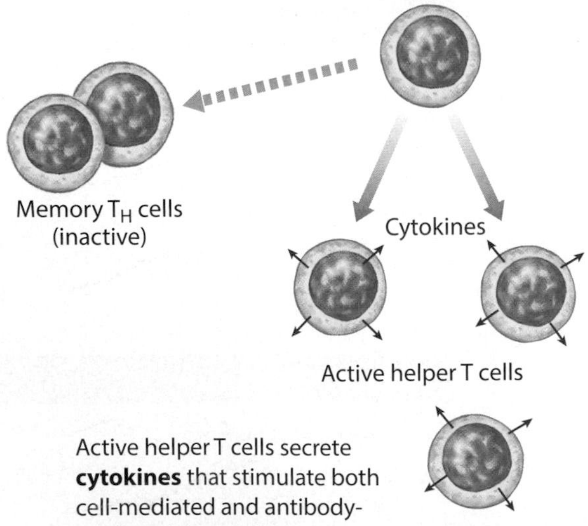
Memory T_H cells (inactive)
Cytokines
Active helper T cells

Active helper T cells secrete **cytokines** that stimulate both cell-mediated and antibody-mediated immunity.

The Golgi apparatus is packaging membrane receptors (red) that will be incorporated into the surface of the cell. These receptors are essential to the costimulation of B cells.

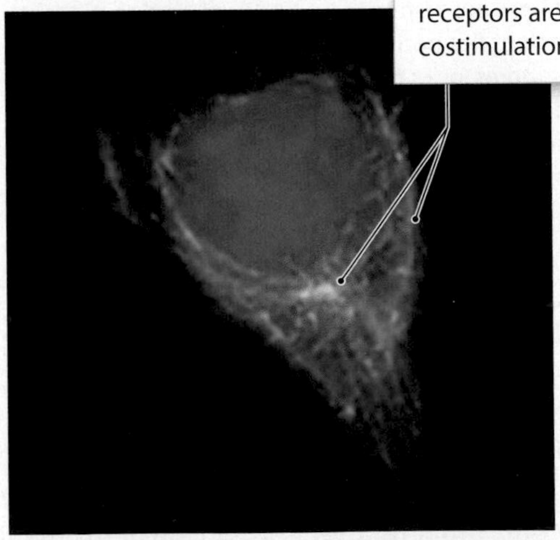

An activated helper T cell Fluorescent LM × 400

3 To become fully activated, a sensitized B cell must encounter a helper T cell that was activated by exposure to the same antigen. When that happens, the helper T cell binds to the MHC complex of the sensitized B cell, recognizes the presence of the antigen, and begins secreting cytokines that promote B cell activation.

4 Under the continued stimulation of cytokines released by active helper T cells, an activated B cell undergoes a series of cell divisions. These divisions yield daughter cells with two different fates.

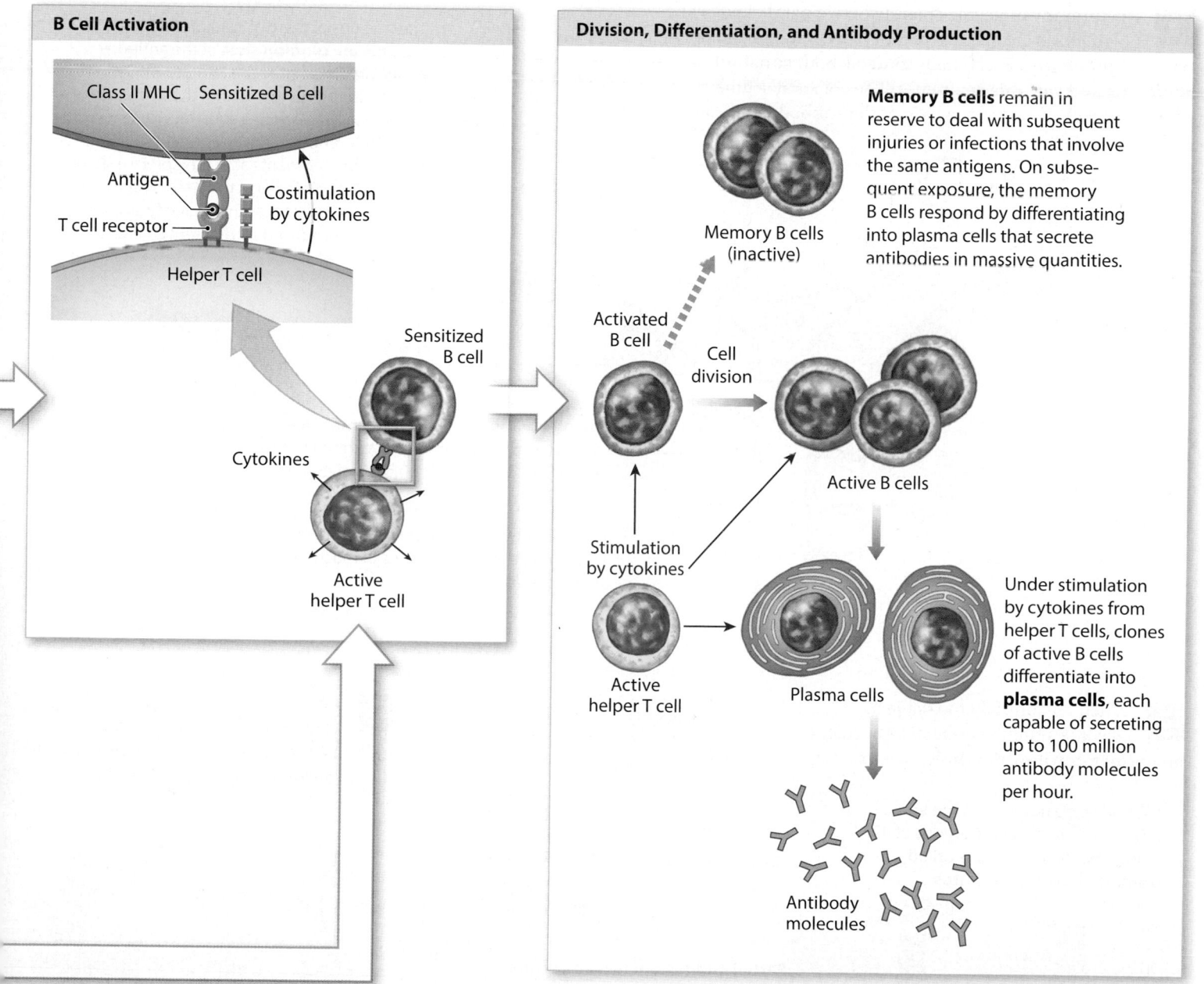

B Cell Activation

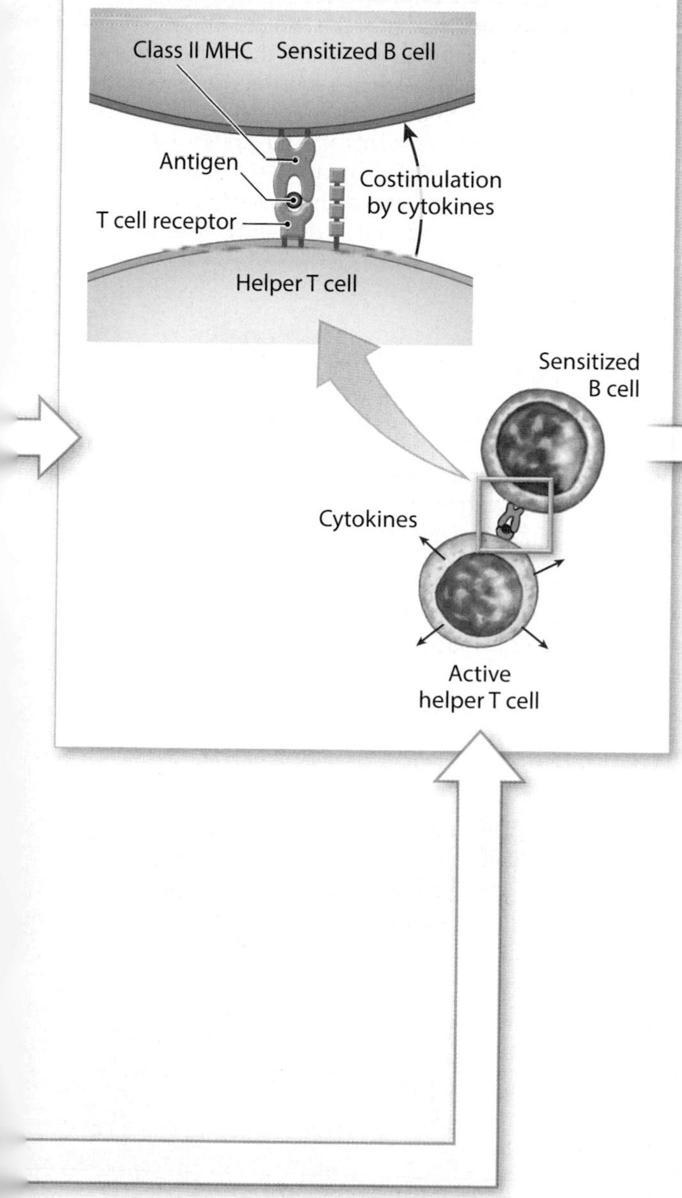
Class II MHC
Sensitized B cell
Antigen
Costimulation by cytokines
T cell receptor
Helper T cell
Sensitized B cell
Cytokines
Active helper T cell

Division, Differentiation, and Antibody Production

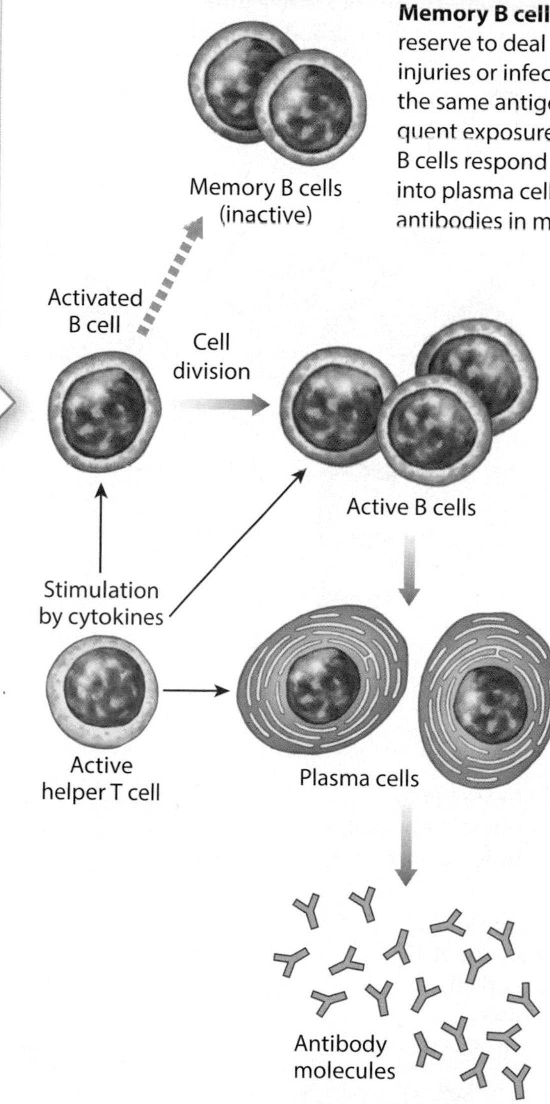
Memory B cells
(inactive)
Activated B cell
Cell division
Active B cells
Stimulation by cytokines
Active helper T cell
Plasma cells
Antibody molecules

Memory B cells remain in reserve to deal with subsequent injuries or infections that involve the same antigens. On subsequent exposure, the memory B cells respond by differentiating into plasma cells that secrete antibodies in massive quantities.

Under stimulation by cytokines from helper T cells, clones of active B cells differentiate into **plasma cells**, each capable of secreting up to 100 million antibody molecules per hour.

Module 20.16 Review

a. Define sensitization.

b. Explain the function of cytokines secreted by helper T cells.

c. If you observed a higher-than-normal number of plasma cells in a sample of lymph, would you expect antibody levels in the blood to be higher or lower than normal?

LO **20.16** Explain the sensitization and activation of B cells and the role of plasma cells.

Antibodies are small soluble proteins that bind to specific antigens; they may inactivate the antigens or trigger another defensive process

1 An antibody molecule consists of two parallel pairs of polypeptide chains: one pair of **heavy chains** and one pair of **light chains**. Each chain contains both **constant segments** and **variable segments**. The constant segments of the heavy chains form the base of the antibody molecule.

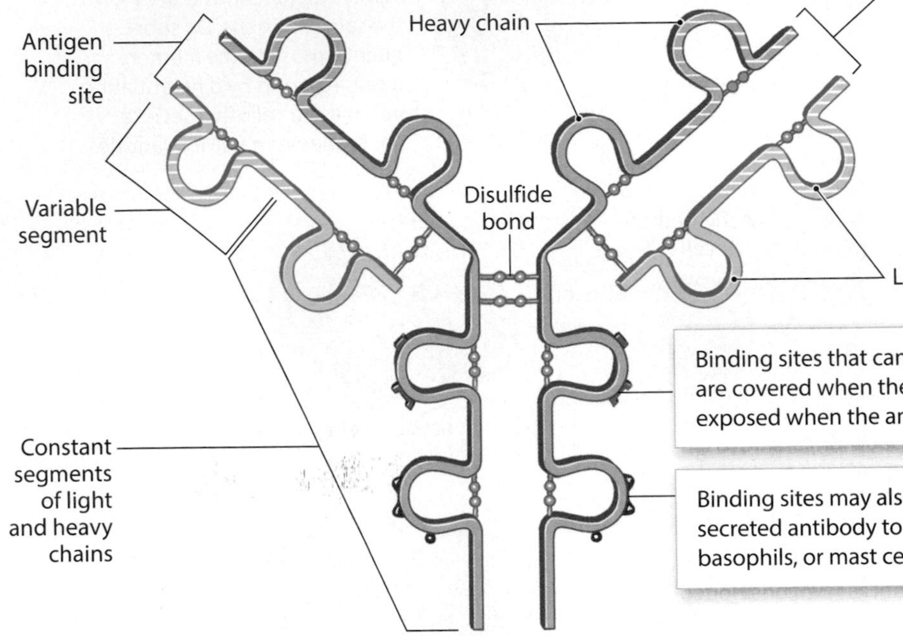

Antigen binding site
Heavy chain
Variable segment
Disulfide bond
Light chain
Constant segments of light and heavy chains

The free tips of the two variable segments form the **antigen binding sites** of the antibody molecule. These sites can interact with an antigen in the same way that the active site of an enzyme interacts with a substrate molecule. Small differences in the amino acid sequence of the variable segments affect the precise shape of the antigen binding site. These differences account for differences in specificity among the antibodies produced by different B cells.

Binding sites that can activate the complement system are covered when the antibody is secreted but become exposed when the antibody binds to an antigen.

Binding sites may also be present that attach the secreted antibody to the surfaces of macrophages, basophils, or mast cells.

2 When an antibody molecule binds to its corresponding antigen molecule, an **antigen-antibody complex** is formed.

1. Antibodies do not bind to the entire antigen, but to specific portions of its exposed surface—regions called **antigenic determinant sites**.

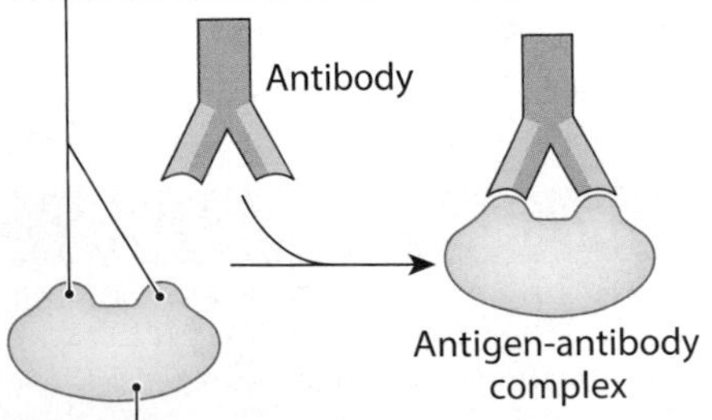

Antibody
Antigen-antibody complex

2. A **complete antigen** is an antigen with at least two antigenic determinant sites, one for each of the antigen binding sites on an antibody molecule.

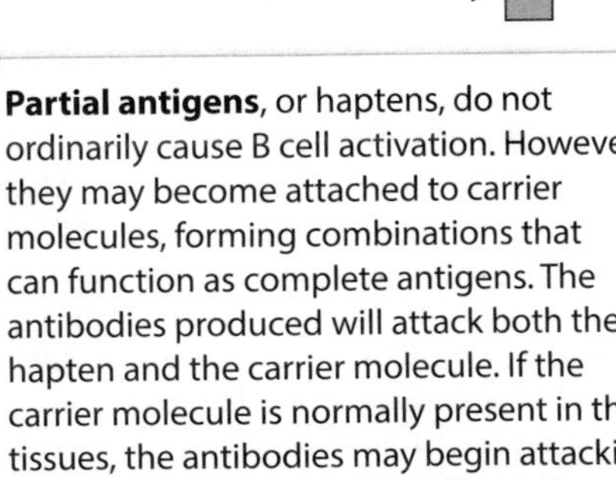

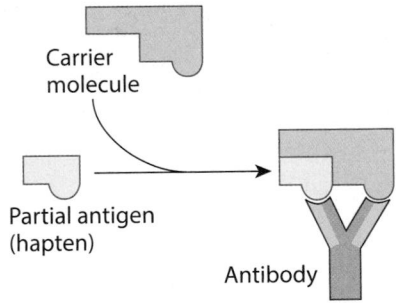

Carrier molecule
Partial antigen (hapten)
Antibody

3. **Partial antigens**, or haptens, do not ordinarily cause B cell activation. However, they may become attached to carrier molecules, forming combinations that can function as complete antigens. The antibodies produced will attack both the hapten and the carrier molecule. If the carrier molecule is normally present in the tissues, the antibodies may begin attacking and destroying normal cells. This is the basis for several drug reactions, including allergies to penicillin.

3 The exposed surface of something as large as a bacterium contains millions of antigenic determinant sites, and it may become carpeted with antibodies.

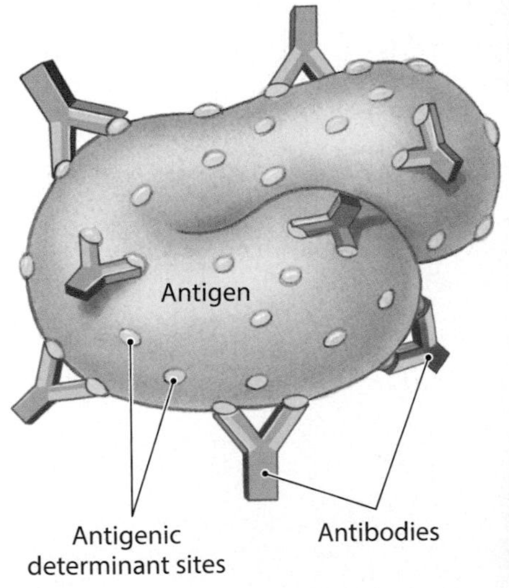

Antigen
Antigenic determinant sites
Antibodies

4 There are five classes of antibodies, or **immunoglobulins (Igs)**. The classes are determined by differences in the structure of the heavy-chain constant segments and so have no effect on the antibody's specificity, which is determined by the antigen binding sites.

Classes of Antibodies

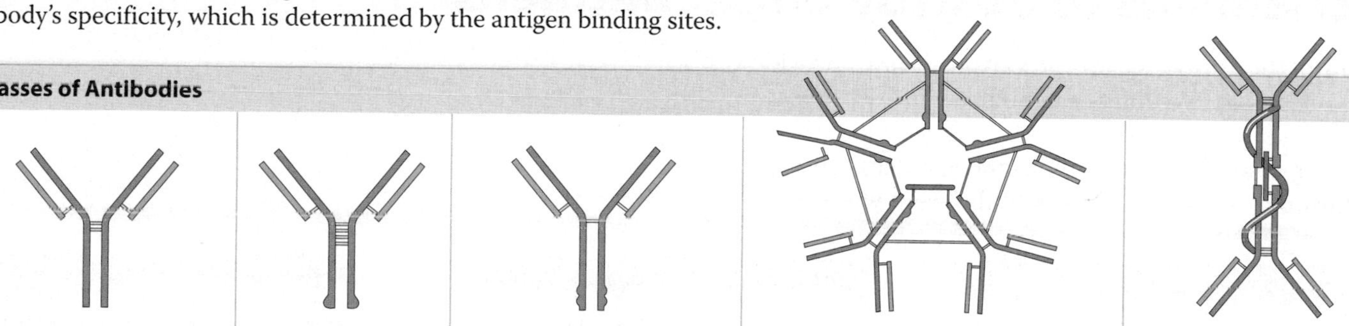

IgG antibodies account for 80 percent of all antibodies. IgG antibodies are responsible for resistance against many viruses, bacteria, and bacterial toxins.

IgE attaches as an individual molecule to the exposed surfaces of basophils and mast cells.

IgD is an individual molecule on the surfaces of B cells, where it can bind antigens in the extracellular fluid. This binding can play a role in the sensitization of the B cell involved.

IgM is the first class of antibody secreted after an antigen is encountered. IgM concentration declines as IgG production accelerates. The anti-A and anti-B antibodies responsible for the agglutination of incompatible blood types are IgM antibodies.

IgA is found primarily in glandular secretions such as mucus, tears, saliva, and semen. These antibodies attack pathogens before they gain access to internal tissues.

5 The initial response to antigen exposure is called the **primary response**. Because the antigen must activate the appropriate B cells, which must then differentiate into antibody-secreting plasma cells, the primary response takes time to develop. During the primary response, the **antibody titer**, which is a laboratory test that measures the level of antibodies in a blood sample, does not peak until 1 or 2 weeks after the initial exposure. If the person is no longer exposed to the antigen, the antibody levels decrease.

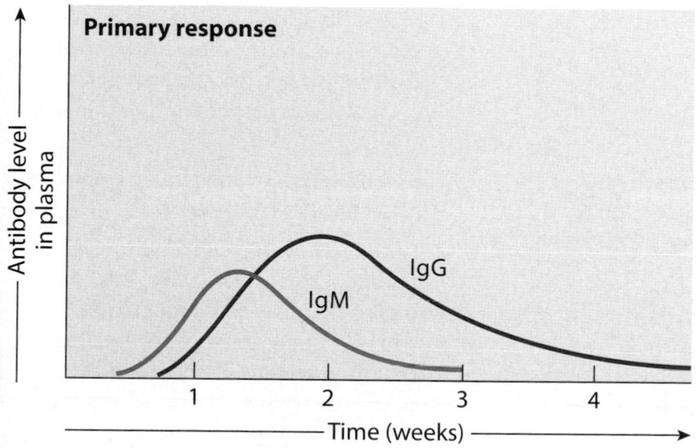
Primary response
Antibody level in plasma
IgM
IgG
1
2
3
4
Time (weeks)

6 When an antigen is encountered a second time, it triggers a more extensive and prolonged **secondary response**. During the secondary response, antibody titers increase more rapidly and reach levels many times higher than they did in the primary response. This reflects the presence of large numbers of memory cells that are already primed for the arrival of the antigen. These memory B cells respond immediately—much faster than the B cells stimulated during the initial exposure. The secondary response appears even if the second exposure occurs years after the first, because memory cells may survive for 20 years or more.

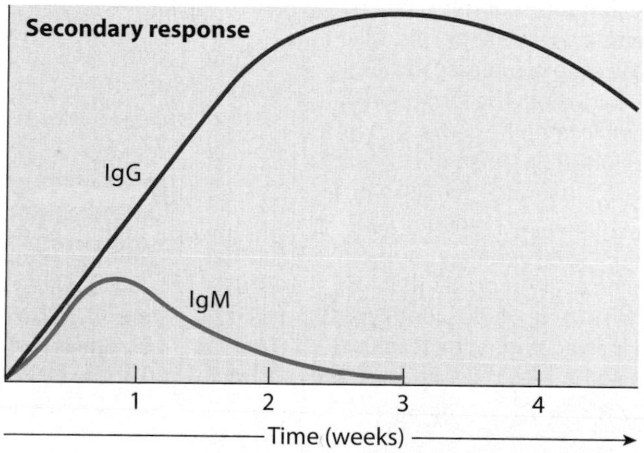
Secondary response
IgG
IgM
1
2
3
4
Time (weeks)

Module 20.17 Review

a. Describe the structure of an antibody.

b. Define antigenic determinant site.

c. Which would be more affected by a lack of memory B cells and memory T cells: the primary response or the secondary response?

LO **20.17** Describe the structure of an antibody, discuss the types of antibodies, and explain the primary and secondary responses to antigen exposure.

Module 20.18

Antibodies use many different mechanisms to destroy target antigens

As a group, the antibodies of the various classes provide a versatile and effective defense against a variety of threats.

1 The formation of an antigen-antibody complex may eliminate the antigen in seven different ways.

Neutralization

Both viruses and bacterial toxins must bind to the plasma membranes of body cells before they can enter or injure those cells. Binding occurs at superficial sites on the bacteria or toxins. Antibodies may bind to those sites, making the virus or toxin incapable of attaching itself to a cell. This mechanism is known as **neutralization**.

Prevention of Pathogen Adhesion

Antibodies dissolved in saliva, mucus, tears, and sweat coat epithelia, providing an additional layer of defense. A covering of antibodies makes it difficult for bacteria or viruses to adhere to and penetrate body surfaces.

Activation of Complement

When bound to an antigen, portions of the antibody molecule change shape, exposing areas that bind complement proteins. The bound complement molecules then activate the complement system, which destroys the antigen.

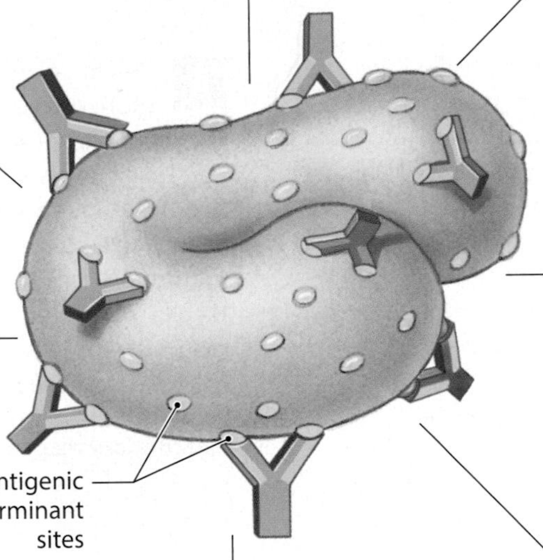

Antigenic determinant sites

Opsonization

Phagocytes can bind more easily to antibodies and complement proteins on the surface of a pathogen than they can to an antibody-free and complement-free surface. As a result, a coating of antibodies and complement proteins increases the effectiveness of phagocytosis. This effect, called **opsonization**, was introduced in Module 20.11. Some bacteria have slick plasma membranes or capsules, and phagocytes must be able to hang onto their prey before they can engulf it.

Precipitation and Agglutination

If antigens are close together, an antibody can bind to antigenic determinant sites on two different antigens. In this way, antibodies can tie large numbers of antigens together, creating an **immune complex**. When the antigen is a soluble molecule, this process may crate complexes too large to remain in solution. The formation of these insoluble immune complexes is called **precipitation**. When the target antigen is on the surface of a cell or a virus, the formation of immune complexes is called **agglutination**. The clumping of erythrocytes that occurs when incompatible blood types are mixed is an agglutination reaction.

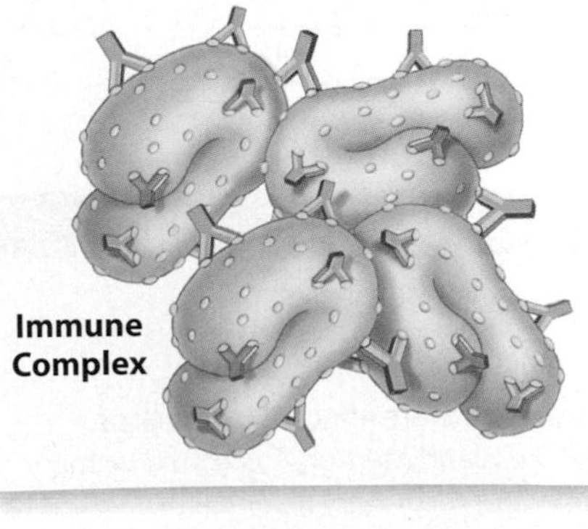

Immune Complex

Stimulation of Inflammation

Antibodies may promote inflammation by stimulating the release of chemicals from basophils and mast cells.

Attraction of Phagocytes

Antigens covered with antibodies attract eosinophils, neutrophils, and macrophages—cells that phagocytize pathogens and destroy foreign or abnormal plasma membranes.

Module 20.18 Review

a. List the ways that antigen-antibody complexes can destroy target antigens.

b. Define opsonization.

c. Which cells are involved in inflammation?

LO 20.18 Explain the mechanisms by which antibodies destroy target antigens.

Allergies and anaphylaxis are antibody responses

Allergies are inappropriate or excessive immune responses to antigens. The sudden increase in cellular activity or antibody titers can have several unpleasant side effects. For example, neutrophils or cytotoxic T cells may destroy normal cells while attacking the antigen, or the antigen-antibody complex may trigger massive inflammation. Antigens that trigger allergic reactions are often called **allergens**.

1 Sensitization to an allergen during the initial exposure leads to the production of large quantities of IgE. The tendency to produce IgE antibodies in response to specific allergens may be genetically determined.

2 **Immediate hypersensitivity** is a rapid and especially severe response to the presence of an antigen. One form, **allergic rhinitis**, includes hay fever and other environmental allergies and may affect 15 percent of the U.S. population. Allergic rhinitis, characterized by inflammation of the nasal membranes, is one example of a **hypersensitivity reaction** that is restricted to the body surface. If the allergen enters the bloodstream, however, the response could be lethal. In **anaphylaxis** (an-a-fi-LAK-sis; *ana-*, again + *phylaxis*, protection), a circulating allergen affects mast cells throughout the body. In severe cases extensive peripheral vasodilation occurs, producing a fall in blood pressure that can lead to a circulatory collapse. This response is called **anaphylactic shock**.

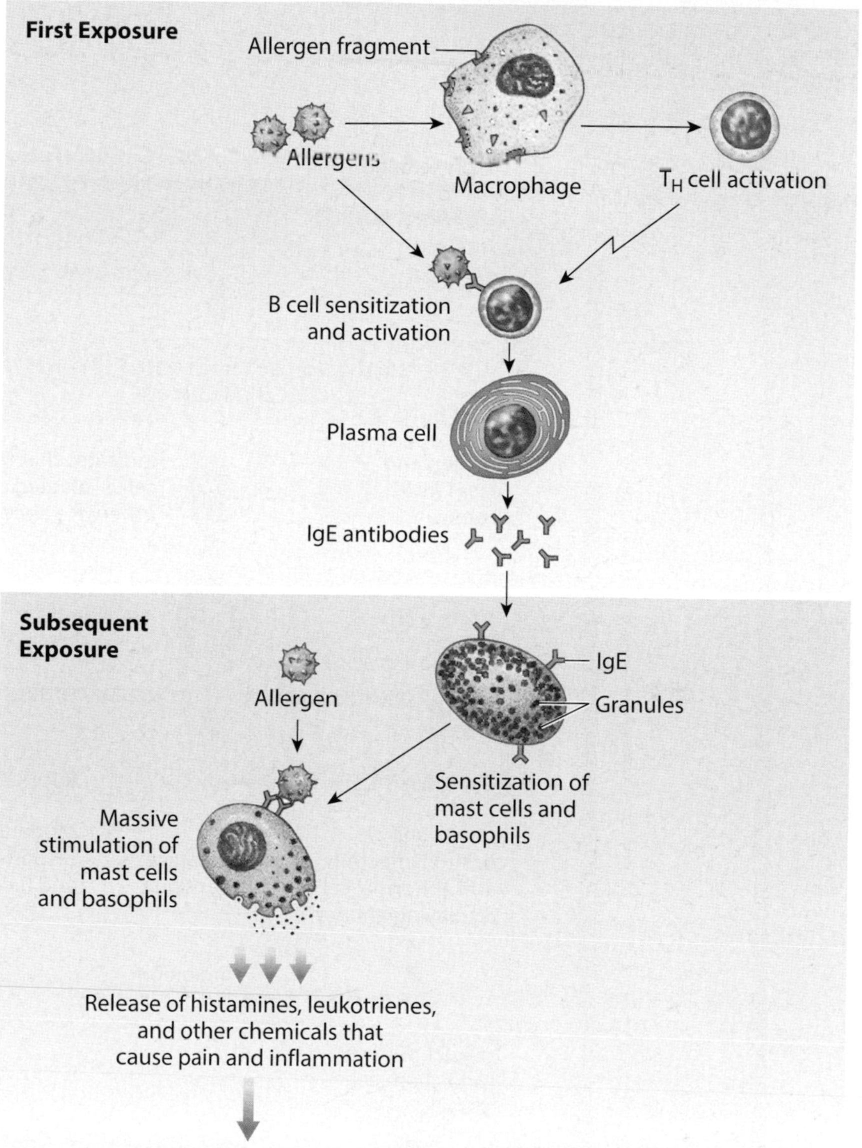
First Exposure
Allergen fragment
Allergens
Macrophage
T_H cell activation
B cell sensitization and activation
Plasma cell
IgE antibodies
Subsequent Exposure
Allergen
IgE
Granules
Sensitization of mast cells and basophils
Massive stimulation of mast cells and basophils
Release of histamines, leukotrienes, and other chemicals that cause pain and inflammation

Localized Allergic Reactions	Systemic Allergic Reactions
If the allergen is at the body surface: localized inflammation, pain, and itching Example: allergic rhinitis	If the allergen is in the bloodstream: itching, swelling, and difficulty breathing (due to airway constriction) Example: anaphylaxis

Module 20.19 Review

a. Define allergy and allergen.

b. What is anaphylaxis?

c. Which chemicals do mast cells and basophils release when stimulated in an allergic reaction?

LO 20.19 Explain allergies, anaphylaxis, and the role of antibodies in each.

Module 20.20

Innate and adaptive immunity work together to defeat pathogens

1 This flowchart depicts the relationships among the responses of innate and adaptive immunity to antigens.

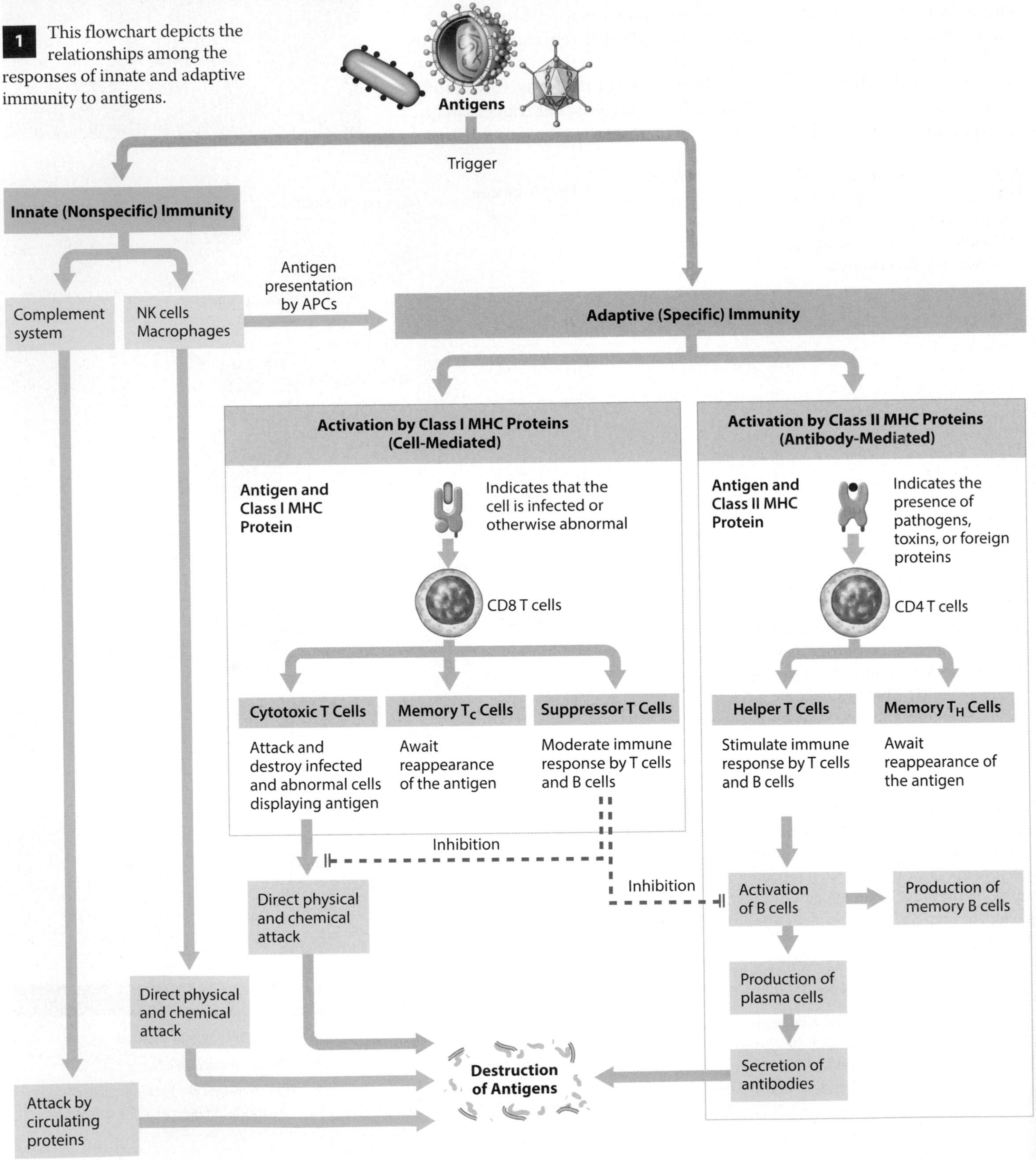

Antigens
Trigger
Innate (Nonspecific) Immunity
Complement system
NK cells Macrophages
Antigen presentation by APCs
Adaptive (Specific) Immunity
Activation by Class I MHC Proteins (Cell-Mediated)
Antigen and Class I MHC Protein
Indicates that the cell is infected or otherwise abnormal
CD8 T cells
Cytotoxic T Cells
Attack and destroy infected and abnormal cells displaying antigen
Memory T_C Cells
Await reappearance of the antigen
Suppressor T Cells
Moderate immune response by T cells and B cells
Activation by Class II MHC Proteins (Antibody-Mediated)
Antigen and Class II MHC Protein
Indicates the presence of pathogens, toxins, or foreign proteins
CD4 T cells
Helper T Cells
Stimulate immune response by T cells and B cells
Memory T_H Cells
Await reappearance of the antigen
Inhibition
Inhibition
Direct physical and chemical attack
Activation of B cells
Production of memory B cells
Production of plasma cells
Secretion of antibodies
Direct physical and chemical attack
Attack by circulating proteins
Destruction of Antigens

2 This illustration provides an overview of the course of events responsible for overcoming a bacterial infection. The most effective defenses against bacteria involve phagocytosis and antigen presentation by APCs.

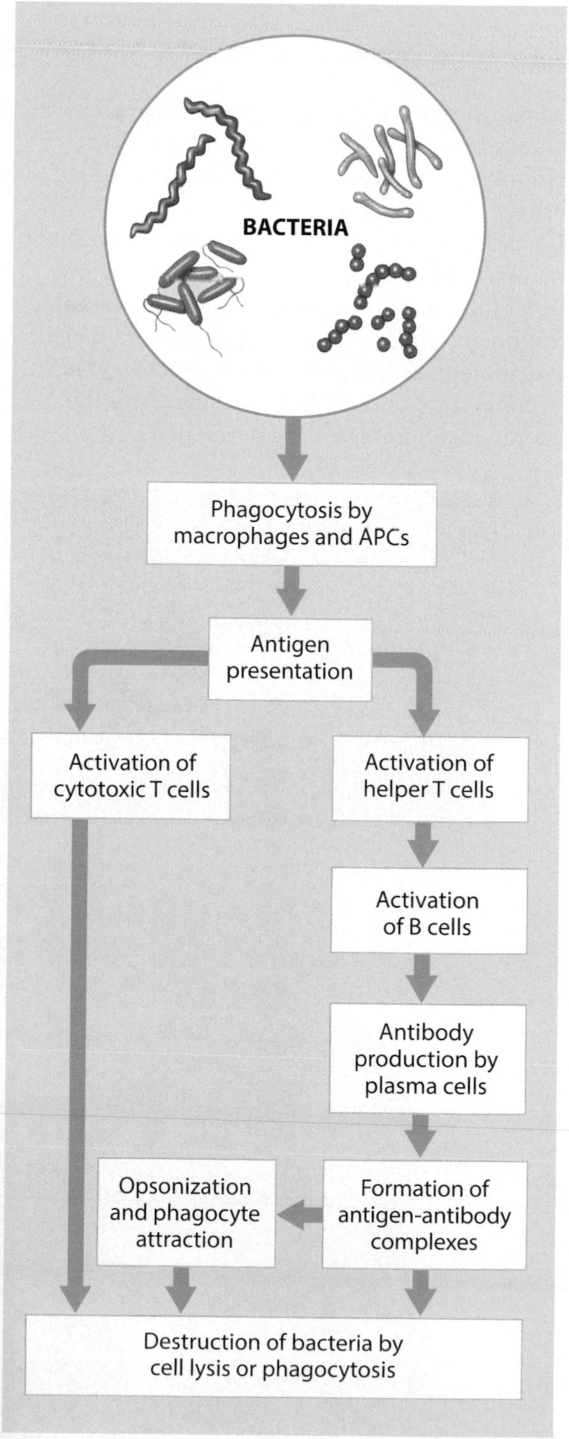
BACTERIA
Phagocytosis by macrophages and APCs
Antigen presentation
Activation of cytotoxic T cells
Activation of helper T cells
Activation of B cells
Antibody production by plasma cells
Formation of antigen-antibody complexes
Opsonization and phagocyte attraction
Destruction of bacteria by cell lysis or phagocytosis

3 The basic sequence of events following a viral infection differs from those of a bacterial infection because cytotoxic T cells and NK cells can be activated by direct contact with virus-infected cells.

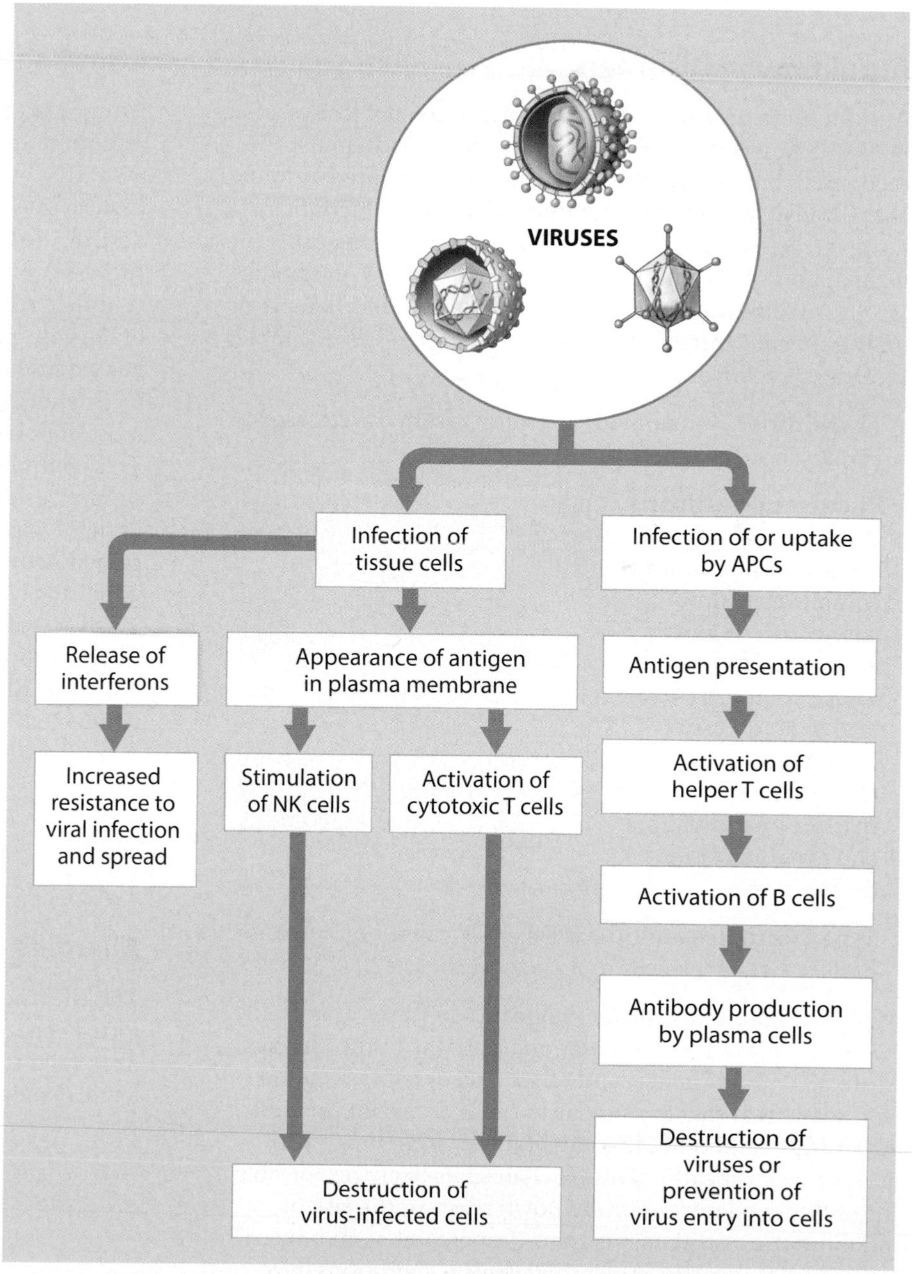
VIRUSES
Infection of tissue cells
Infection of or uptake by APCs
Release of interferons
Appearance of antigen in plasma membrane
Antigen presentation
Increased resistance to viral infection and spread
Stimulation of NK cells
Activation of cytotoxic T cells
Activation of helper T cells
Activation of B cells
Antibody production by plasma cells
Destruction of virus-infected cells
Destruction of viruses or prevention of virus entry into cells

Module 20.20 Review

a. Identify the type of T cell whose plasma membrane contains CD8 markers and the type with CD4 markers.

b. Which cells produce antibodies?

c. Which cells can be activated by direct contact with virus-infected cells?

LO **20.20** Summarize the integration of innate and adaptive immunity.

Immune disorders involving both overactivity and underactivity can be harmful

Excessive or Misdirected Immune Response

Autoimmune Disorders

Autoimmune disorders affect an estimated 5 percent of adults in North America and Europe. The immune system usually recognizes but ignores self-antigens—normal antigens found in the body. When the recognition system malfunctions, however, activated B cells may make antibodies against normal body cells and tissues. These "misguided" antibodies are called **autoantibodies**. The condition produced depends on the specific antigen attacked by autoantibodies. Examples include the following:

- **Thyroiditis** is inflammation resulting from the release of autoantibodies against thyroglobulin.
- **Rheumatoid arthritis** occurs when autoantibodies form immune complexes within connective tissues around the joints. These complexes lead to an excessive immune response characterized by marked inflammation and eventually joint destruction.

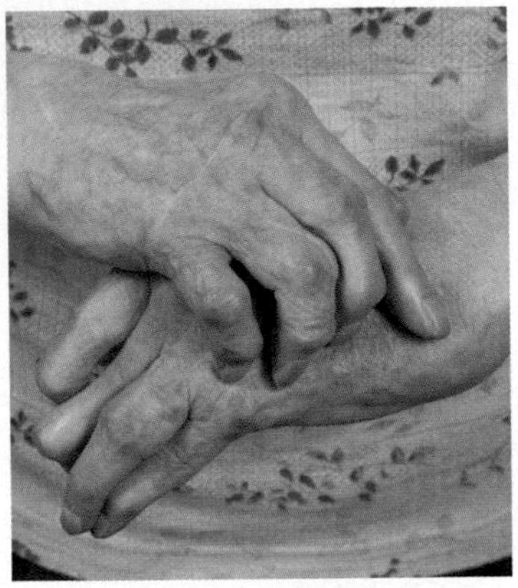

- **Type 1 diabetes mellitus** is generally caused by autoantibodies that attack cells in the pancreatic islets.

Many autoimmune disorders appear to be cases of mistaken identity. For example, proteins associated with the measles, Epstein–Barr, influenza, and other viruses contain amino acid sequences that are similar to those of myelin proteins. As a result, antibodies that target these viruses may also attack myelin sheaths. This mechanism is likely responsible for multiple sclerosis. For unknown reasons, the risk of autoimmune disorders increases if a person has an unusual type of MHC protein. At least 50 clinical conditions have been linked to specific variations in MHC structure.

Graft Rejection

After organ transplant surgery, the major problem is **graft rejection**. In graft rejection, T cells are activated by contact with MHC proteins on plasma membranes in the donated tissues. The cytotoxic T cells that develop then attack and destroy the foreign cells. Significant improvements in transplant success can be made by **immunosuppression**, a partial or complete reduction of the immune response. An understanding of the communication among T cells, macrophages, and B cells has led to the development of drugs with more selective effects. **Cyclosporin A**, a compound derived from a fungus, was the most important immunosuppressive drug developed in the 1980s. This compound suppresses the immune response primarily by inhibiting helper T cell activity while leaving suppressor T cells relatively unaffected.

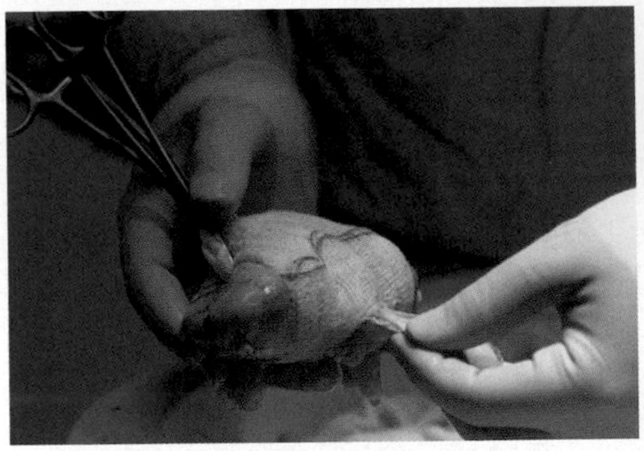

Allergies

The effects of the many forms of allergies range from mild to potentially lethal (Module 20.19).

Inadequate Immune Response

Immunodeficiency Diseases

Immunodeficiency diseases result from (1) problems with the embryonic development of lymphoid organs and tissues; (2) an infection with a virus that depresses immune function; or (3) treatment with, or exposure to, immunosuppressive agents, such as radiation or drugs. **Acquired immunodeficiency syndrome (AIDS)**, the most common immunodeficiency disease, is caused by the **human immunodeficiency virus (HIV)**. The virus binds to CD4 proteins and infects helper T cells. The infected cells begin synthesizing viral proteins, and these new viruses are then shed from the cell surface. Cells infected with HIV are ultimately destroyed by either the virus or immune defenses. The gradual destruction of helper T cells impairs both cell-mediated and antibody-mediated responses to antigens. Making matters worse, suppressor T cells are relatively unaffected by the virus, and over time the excess of suppressing factors "turns off" the normal immune response. Circulating antibody levels decline, cell-mediated immunity is reduced, and the body is left vulnerable to microbial invaders. With immune function suppressed, ordinarily harmless microorganisms can initiate lethal **opportunistic infections**. Because immune surveillance is also depressed, the risk of cancer increases. Infection with HIV occurs through intimate contact with the body fluids of infected individuals. An estimated 33 million people are infected worldwide; 22 million of them are in sub-Saharan Africa, and 1.2 million in North America. AIDS causes about 17,000 deaths each year in the United States, and 2 million deaths worldwide.

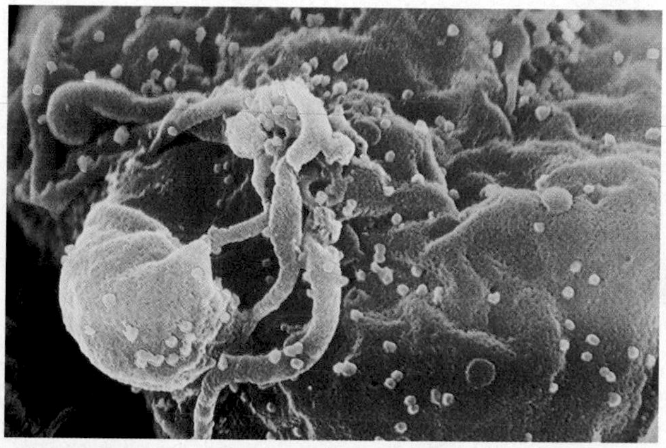

HIV (green) budding from an infected T_H cell SEM × 40,000

Age-Related Reductions in Immune Activity

As we age, the immune system becomes less effective at combating disease. T cells become less responsive to antigens, so fewer cytotoxic T cells respond to an infection. This effect may, at least in part, be associated with the gradual involution of the thymus and a reduction in circulating levels of thymic hormones. Because the number of helper T cells is also reduced, B cells are less responsive, so antibody levels do not rise as quickly after antigen exposure. The net result is an increased susceptibility to viral and bacterial infections. For this reason, vaccinations for acute viral diseases such as the flu (influenza), and for pneumococcal pneumonia, are strongly recommended for elderly people. The increased incidence of cancer in the elderly reflects the fact that immune surveillance declines, so tumor cells are not eliminated as effectively.

Module 20.21 Review

a. Define autoimmune disorders.

b. Describe immunosuppression.

c. Provide a plausible explanation for the increased incidence of cancer in the elderly.

20.21 Explain autoimmune disorders, graft rejection, and immunodeficiency diseases, and describe age-related changes in the immune response.

Matching

Match each lettered term with the most closely related description.

a. opsonization
b. helper T cells
c. antibody
d. Class II MHC
e. costimulation
f. IgM
g. Class I MHC
h. IgG
i. passive immunity
j. anaphylaxis
k. CD4 markers
l. acquired immunity
m. B lymphocytes

1 Two parallel pairs of polypeptide chains
2 Found on helper T cells
3 Active and passive
4 Transfer of antibodies
5 Attacked by HIV
6 Enhances phagocytosis
7 MHC proteins present in the plasma membranes of all nucleated cells
8 Differentiate into memory and plasma cells
9 MHC proteins present in the plasma membranes of all APCs and lymphocytes
10 Antibodies used to determine blood type
11 Secondary binding process required for T cell activation
12 Accounts for 80 percent of all immunoglobulins
13 Circulating allergen stimulates mast cells throughout body

1 ________
2 ________
3 ________
4 ________
5 ________
6 ________
7 ________
8 ________
9 ________
10 ________
11 ________
12 ________
13 ________

Match each lettered term with the most closely related description.

a. cytotoxic T cells
b. viruses
c. B cells
d. antibodies
e. helper T cells
f. macrophages
g. natural killer (NK) cells
h. suppressor T cells
i. memory T cells and B cells

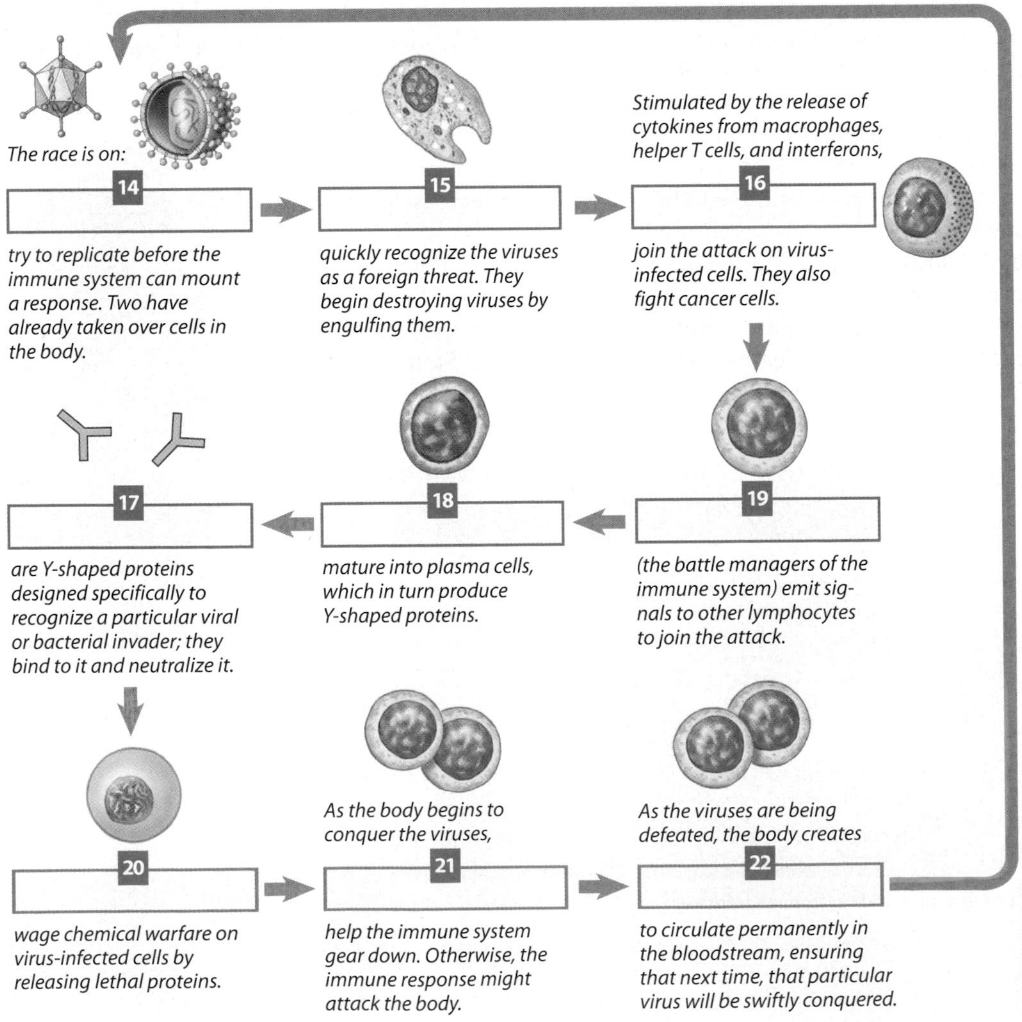

The race is on:
14
try to replicate before the immune system can mount a response. Two have already taken over cells in the body.
15
quickly recognize the viruses as a foreign threat. They begin destroying viruses by engulfing them.
Stimulated by the release of cytokines from macrophages, helper T cells, and interferons,
16
join the attack on virus-infected cells. They also fight cancer cells.
17
are Y-shaped proteins designed specifically to recognize a particular viral or bacterial invader; they bind to it and neutralize it.
18
mature into plasma cells, which in turn produce Y-shaped proteins.
19
(the battle managers of the immune system) emit signals to other lymphocytes to join the attack.
20
wage chemical warfare on virus-infected cells by releasing lethal proteins.
As the body begins to conquer the viruses,
21
help the immune system gear down. Otherwise, the immune response might attack the body.
As the viruses are being defeated, the body creates
22
to circulate permanently in the bloodstream, ensuring that next time, that particular virus will be swiftly conquered.

Study Outline

SECTION 1 • Anatomy of the Lymphatic System

20.1 The lymphatic system consists of lymphatic vessels, nodes, and lymphoid tissue p. 741

1. The lymphatic system includes cells, tissues, and organs that provide **immunity** (the ability to defend the body against infection, illness, and disease) and return interstitial fluid to the bloodstream.
2. **Lymphocytes** are the cells of the lymphatic system. They are surrounded by **lymph**, interstitial fluid that has entered a lymphatic vessel.
3. **Lymphatics (lymphatic vessels)** are vessels that begin in the peripheral tissues and end at veins.
4. **Primary lymphoid tissues and organs** are where lymphocytes form and mature and include the red bone marrow and the thymus gland.
5. **Secondary lymphoid tissues and organs** are where lymphocytes are activated and cloned and include the lymph nodes, tonsils, mucosa-associated lymphoid tissue (MALT), appendix and spleen.

20.2 Interstitial fluid flows continuously into lymphatic capillaries and exits tissues as lymph in lymphatic vessels p. 742

6. **Lymphatic vessels** begin as **lymphatic capillaries** and carry lymph from the peripheral tissues to the venous system.
7. Lymphatic capillaries differ from blood capillaries. They originate at pockets rather than from continuous tubes, are larger in diameter, have thinner walls, and have a flattened or irregular outline in sectional view.

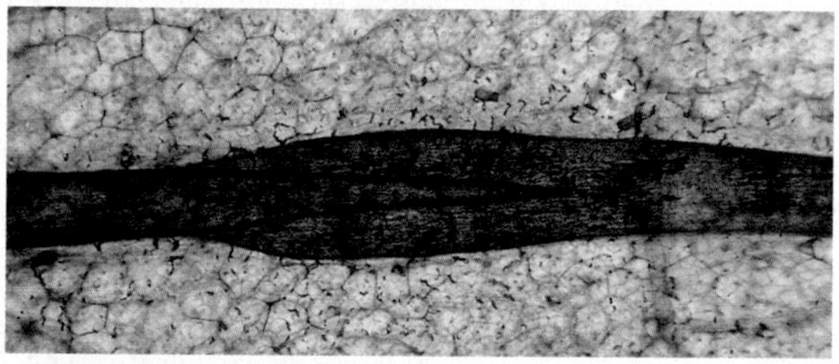

Valve in a lymphatic vessel

8. Like veins, larger lymphatic vessels have valves.
9. Lacteals are prominent lymphatic capillaries in the small intestine.

20.3 Small lymphatic vessels converge to form lymphatic ducts that empty into the subclavian veins p. 744

10. The lymph from lymphatic capillaries flows into either **superficial** or **deep lymphatics** that converge to form **lymphatic trunks**. The lymphatic trunks empty into either the **thoracic duct** or **right lymphatic duct**.
11. The **right jugular trunk**, **right subclavian trunk**, and **right bronchomediastinal trunk** merge to form the right lymphatic duct. It then empties into the right subclavian vein.
12. The **left bronchomediastinal trunk**, the **left subclavian trunk**, and the **left jugular trunk** merge to form the thoracic duct. It then empties into the left subclavian vein.
13. Blockage of lymphatic drainage can cause **lymphedema**.

20.4 Lymphocytes are responsible for the immune functions of the lymphatic system p. 746

14. Lymphocytes make up 20–40 percent of the circulating leukocyte population, yet circulating lymphocytes are only a fraction of the total lymphocyte population.
15. Three classes of lymphocytes circulate in the blood (**T cells**, **B cells**, **NK cells**), and they are all sensitive to **antigens** that can stimulate the immune system.
16. **Lymphopoiesis** (lymphocyte production) involves red bone marrow, thymus, and peripheral lymphoid tissues.

20.5 Lymphocytes aggregate within lymphoid tissues and lymphoid organs p. 748

17. **Lymphoid tissues** are connective tissues dominated by lymphocytes.
18. In a **lymphoid nodule** the lymphocytes are densely packed in areolar tissues. In contrast, **lymphoid organs** (lymph nodes, the thymus, and spleen) are surrounded by a fibrous connective tissue capsule.
19. **Aggregated lymphoid nodules** such as Peyer's patches are found in the small intestine.
20. **Tonsils** are large lymphoid nodules. A single **pharyngeal tonsil** (adenoid) is located in the posterior wall of the nasopharynx. Paired **palatine** and **lingual tonsils** are located in the oral cavity.
21. **Lymph nodes**, or **lymph glands**, are small lymphoid organs shaped like a kidney bean. They detect pathogens before they reach vital organs.
22. **Afferent lymphatics** bring lymph to the lymph node on the side opposite the **hilum**. **Efferent lymphatics** exit at the hilum, along with blood vessels and nerves.

20.6 The thymus is a lymphoid organ that produces functional T cells p. 750

23. The thymus produces a collection of hormones called **thymosins** that are important to the development of T cells.

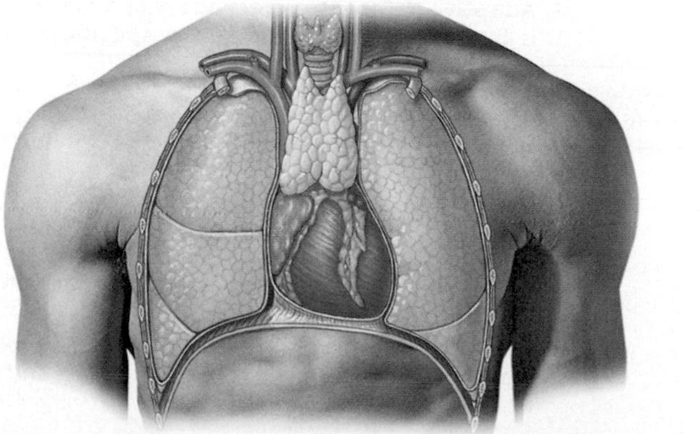

24. The thymus undergoes **involution** with aging. This process is associated with increased susceptibility to disease.
25. The thymus is located in the mediastinum, posterior to the sternum. It has right and left **lobes**, and **lobules** that are divided by **septa**.
26. The thymus has a dark outer **cortex** and lighter central **medulla**. Lymphocytes in the cortex are in clusters and surrounded by **thymic epithelial cells (TECs)**. These cells maintain the blood–thymus barrier.

20.7 The spleen, the largest lymphoid organ, responds to antigens in the bloodstream p. 752

27. The spleen functions to (1) remove abnormal blood cells and other blood components by phagocytosis, (2) store iron recycled from red blood cells, and (3) initiate immune responses by B cells and T cells.
28. The spleen is located on the left side of the body at the level of the 9th and 11th ribs. It is attached to the stomach by the **gastrosplenic ligament**.
29. A **splenectomy** is the removal of the spleen due to injury.
30. The **visceral surface** (medial) contains indentations from the stomach **(gastric area)** and kidney **(renal area)**.
31. **Red pulp** contains red blood cells, and **white pulp** resembles lymphoid nodules and contains lymphocytes.

SECTION 2 • Innate Immunity

20.8 Innate immunity is nonspecific and is not stimulated by specific antigens p. 755

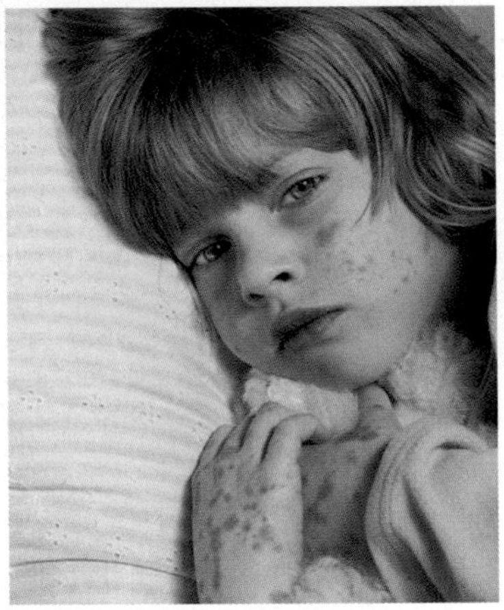

32. Two complementary mechanisms protect the human body from disease: **innate** (nonspecific) **immunity**, and **adaptive** (specific) **immunity**.
33. **Innate defenses** do not distinguish one type of threat from another, and are the following: **physical barriers**, **phagocytes**, **immune surveillance**, **interferons**, **complement**, **inflammation**, and **fever**.
34. **Adaptive defenses** protect against particular threats. This is known as adaptive immunity or **specific resistance**.

20.9 Physical barriers prevent pathogens and toxins from entering body tissues, and phagocytes provide the next line of defense p. 756

35. The integumentary system provides the major physical barrier to the external environment.
36. The epithelia lining the digestive, respiratory, urinary, and reproductive tracts also provide a barrier.
37. **Phagocytes** (**neutrophils**, **eosinophils**, and **fixed** and **free macrophages**) remove cellular debris and respond to foreign substances and pathogens.

20.10 NK cells perform immune surveillance, detecting and destroying abnormal cells p. 758

38. **Natural killer (NK) cells** recognize and destroy abnormal cells. Constant monitoring by NK cells is called **immune surveillance**. Many cancer cells have **tumor-specific antigens** on their plasma membranes that NK cells recognize as abnormal.
39. Cancer cells often mutate and avoid NK cell detection, in a process called **immunological escape**.

20.11 Interferons and the complement system are distributed widely in body fluids p. 760

40. **Interferons (IFNs)** are proteins released by activated lymphocytes and macrophages, and by cells infected with viruses.
41. Interferons bind to surface receptors, and through second messengers, trigger the production of antiviral proteins in the cytoplasm.
42. Interferons are examples of **cytokines**, chemicals released by cells to coordinate local activities.
43. **Interferon alpha** (α) is produced by cells infected with viruses and attracts NK cells. **Interferon beta** (β) is secreted by fibroblasts and slows inflammation. **Interferon gamma** (γ) is secreted by T cells and NK cells and stimulates macrophage activity.
44. The **complement system** complements the actions of antibodies. There are two possible pathways: the **classical pathway** and the **alternative pathway**.

20.12 Inflammation is a localized tissue response to injury; fever is a generalized response to tissue damage and infection p. 762

45. **Inflammation**, or the **inflammatory response**, is a localized tissue response to injury that produces redness, swelling, heat, and pain.
46. **Fever**, a body temperature greater than 37.2°C (99°F), is caused by circulating proteins called **pyrogens** that reset the thermostat in the hypothalamus.
47. Fever can be beneficial in that it inhibits some viruses and bacteria; it also increases metabolic rate.

SECTION 3 • Adaptive Immunity

20.13 Adaptive immunity provides the body's specific defenses p. 765

48. **Adaptive** (**specific**) **immunity** is coordinated and produced by T cells and B cells. **Vaccines** stimulate an immune response by producing antibodies to a specific disease.
49. Two types of adaptive (specific) immunity are **active immunity** (which appears after exposure to an antigen) and **passive immunity** (produced by the transfer of antibodies from another source).
50. Adaptive immunity exhibits four general properties: **specificity**, **versatility**, **immunologic memory**, and **tolerance**.
51. **Memory cells** enable the immune system to "remember" previous target antigens.

20.14 Adaptive immunity is triggered by exposure of T cells and B cells to specific antigens p. 766

52. The first step in triggering an immune response is **antigen presentation**.
53. If a cell-mediated response is triggered, phagocytes and then T cells will be activated. The activated T cells will attack through phagocytosis or chemical toxins.
54. If an antibody-mediated response is triggered, B cells will form antibody-producing cells whose products will destroy the antigen.
55. Antigen presentation occurs when an antigen is attached to a **Class I MHC protein** or **Class II MHC protein** displayed on the plasma membrane of an infected cell or **antigen-presenting cells (APCs)**, respectively.

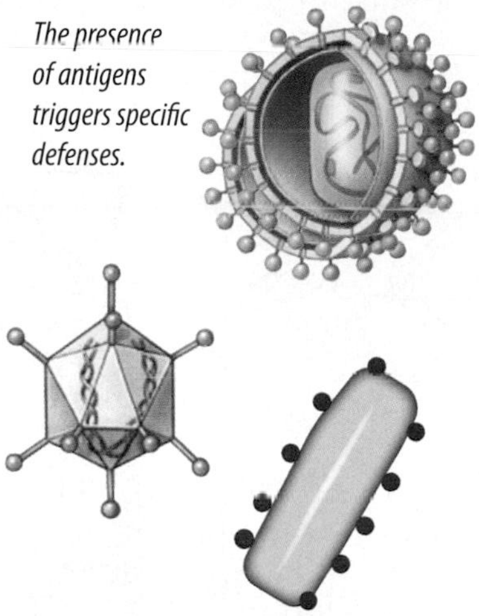

The presence of antigens triggers specific defenses.

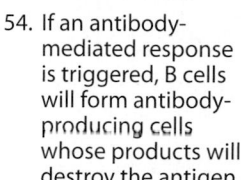

20.15 Infected cells stimulate the formation and division of cytotoxic T cells, memory T_C cells, and suppressor T cells p. 768

56. T cells have receptors for either Class I or Class II MHC proteins. **Antigen recognition** will occur only if the MHC protein presents the antigen to a T cell that can detect it.
57. There are many **CD** (cluster of differentiation) **markers**. **CD8** and **CD4** markers are two important membrane proteins involved in antigen recognition.
58. CD8 markers are found on **CD8 cells.** These cells respond to antigens presented by Class I MHC proteins. They provide cell-mediated immunity.
59. CD4 markers are found on **CD4 cells.** These cells respond to antigens presented by Class II MHC proteins.
60. Before activation can occur, a T cell undergoes a process called **costimulation**, a process that prevents T cells from mistakenly attacking normal body cells.
61. Following activation CD8 T cells will divide and then differentiate into three types of CD8 T cells: **cytotoxic T (T_C) cells**, **memory T_C cells**, and **suppressor T (T_S) cells**.

20.16 Antigen-presenting cells can stimulate activation of CD4 T cells, producing helper T cells that promote B cell activation and antibody production p. 770

62. For CD4 T cells to initiate antibody-mediated immunity, they must be exposed to antigen-bearing Class II MHC proteins. Costimulation then completes their activation.
63. Upon activation, CD4 T cells divide and differentiate into **active helper T cells (T_H cells)** and **memory T_H cells.**
64. B cells prepare for activation by presenting an antigen bound to Class II MHC proteins in a process called **sensitization**.
65. A sensitized B cell must encounter a helper T cell that was activated by the same antigen. When these two cells bind, the active helper T cell secretes **cytokines**.
66. The cytokines continue stimulation of the activated B cell causing it to divide into either **memory B cells** or **plasma cells**. The memory B cells remain in reserve to deal with subsequent infections. The plasma cells secrete antibodies.

20.17 Antibodies are small soluble proteins that bind to specific antigens; they may inactivate the antigens or trigger another defensive process p. 772

67. An antibody molecule contains two parallel pairs of polypeptide chains: one pair of **heavy chains** and one pair of **light chains**. Each chain contains **constant** and **variable segments**.
68. An antibody molecule binds to an antigen at **antigenic determinant sites** forming an **antigen-antibody complex**.
69. There are five different classes of antibodies, or **immunoglobulins (Igs)**, determined by structure. These classes are **IgG, IgE, IgD, IgM,** and **IgA.**
70. During the **primary response**, the concentration of antibodies in the blood **(antibody titer)** does not peak until 1 or 2 weeks after initial exposure.
71. When an antigen is encountered a second time, a more extensive and prolonged **secondary response** occurs. Antibody levels increase more rapidly and reach much higher levels than they did during the primary response.

20.18 Antibodies use many different mechanisms to destroy target antigens p. 774

72. An antigen-antibody complex may eliminate an antigen in seven ways: **neutralization, prevention of pathogen adhesion, activation of complement, opsonization, attraction of phagocytes, stimulation of inflammation,** and **precipitation and agglutination.**

20.19 Allergies and anaphylaxis are antibody responses p. 775

73. **Allergies** are inappropriate or excessive immune responses to antigens. Antigens that trigger allergic reactions are called **allergens.**
74. An allergen entering the bloodstream could be lethal, as in **anaphylaxis**, when an allergen affects mast cells throughout the body. In severe cases, circulatory collapse is possible, in a response called **anaphylactic shock.**

20.20 Innate and adaptive immunity work together to defeat pathogens p. 776

75. Elements of innate and adaptive immunity work together in a coordinated response to antigens.

20.21 Immune disorders involving both overactivity and underactivity can be harmful p. 778

76. Immune disorders of excessive or misdirected response include **autoimmune disorders** (**thyroiditis, rheumatoid arthritis,** and **Type 1 diabetes mellitus**), **graft rejection,** and **allergies.**
77. Immune disorders of inadequate immune response include **immunodeficiency diseases** (such as **acquired immunodeficiency syndrome**, or **AIDS**, which is caused by the **human immunodeficiency virus**, or **HIV**), and age-related reductions in immune activity.

Chapter Review Questions

Labeling

Label the antibody molecule shown here.

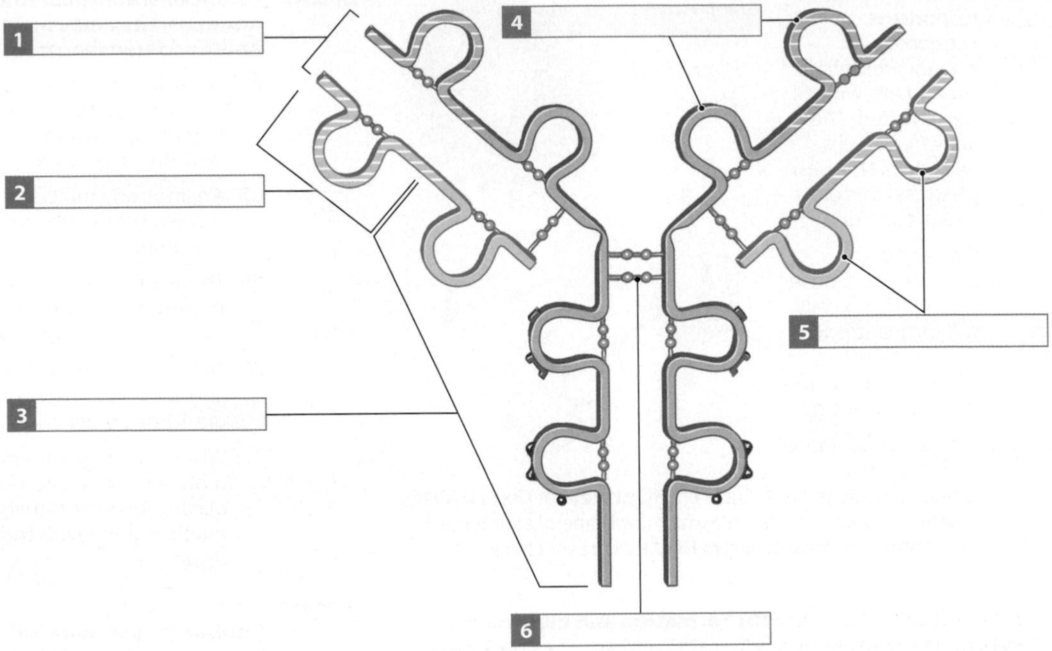
1
2
3
4
5
6

True/False

Indicate whether each statement is true or false.

7 One difference between lymphatic vessels and veins is that lymphatic vessels do not have valves.

8 Lymphopoiesis occurs exclusively in lymphoid tissue.

9 Efferent lymphatics arise from the hilum of a lymph node.

10 A decrease in circulating pyrogens results in an increase in body temperature.

11 Antibody molecules consist of four pairs of heavy chains and four pairs of light chains.

12 Under stimulation from cytokines, B cells can differentiate into plasma cells.

7 ____________________

8 ____________________

9 ____________________

10 ____________________

11 ____________________

12 ____________________

Multiple choice

Select the correct answer from the list provided.

13 Lymph from the left arm, the left half of the head, the left side of the body superior to the diaphragm, and the entire body inferior to the diaphragm is received by the

- ☐ a) thoracic duct.
- ☐ b) cisterna chyli.
- ☐ c) azygos vein.
- ☐ d) left lymphatic duct.

14 CD4 T cells respond to antigens presented by

- ☐ a) Class I MHC.
- ☐ b) Class II MHC.
- ☐ c) cytokines.
- ☐ d) plasma cells.

15 Interferons increase resistance of cells to

- ☐ a) fungal infections.
- ☐ b) bacterial infections.
- ☐ c) viral infections.
- ☐ d) cancer.

16 Which of the following immunoglobulins accounts for 80 percent of all antibodies?

- ☐ a) IgG
- ☐ b) IgE
- ☐ c) IgD
- ☐ d) IgM

17 The human immunodeficiency virus (HIV) infects which of the following cells?
- ☐ a) suppressor T cells
- ☐ b) memory T cells
- ☐ c) helper T cells
- ☐ d) cytotoxic T cells

18 Which of the following is *not* a phagocyte?
- ☐ a) neutrophil
- ☐ b) monocyte
- ☐ c) free macrophage
- ☐ d) natural killer cell

19 The adenoid is also known as the
- ☐ a) pharyngeal tonsil.
- ☐ b) palatine tonsil.
- ☐ c) lingual tonsil.
- ☐ d) Peyer's patch.

20 The gradual decrease in secretions and size of this organ with age is correlated with an increased susceptibility to disease.
- ☐ a) spleen
- ☐ b) thymus
- ☐ c) tonsils
- ☐ d) appendix

Short answer

21 What is graft rejection?

22 How does a cytotoxic T cell (T_C cell) destroy its target cell?

23 Why is it necessary for certain vaccinations, such as hepatitis B, to be administered as a series of injections over a period of months?

Chapter Integration • Applying what you have learned

Catching childhood diseases in adulthood

The varicella-zoster virus (VZV) causes chickenpox (varicella) and shingles (herpes zoster). Chickenpox is a common childhood disease characterized by small, itchy blisters, fever, fatigue, and headache. Shingles is caused by a reactivation of VZV later in life. The Centers for Disease Control and Prevention (CDC) recommends that people aged 19–49 who had chickenpox earlier in life receive the chickenpox vaccine to prevent shingles. The CDC also recommends the shingles vaccine, Zostavax, for people 60 years old and older to prevent shingles.

A 20-year-old student discovers that she has been exposed to chickenpox by her twin nieces and is concerned that she might have contracted the disease. She goes to see her physician, who takes a blood sample and sends it to a lab for antibody titers. The results reveal an elevated level of IgM antibodies to the VZV but very few IgG antibodies to the virus.

1 **Has this student come down with chickenpox?**

2 **Who is susceptible to contracting chickenpox?**

3 **What is the causative agent to chickenpox and for shingles?**

4 **Distinguish between IgM and IgG antibodies.**

5 **Explain how antigenic specificity of antibodies is possible.**

6 **Compare the primary and secondary immune responses as they pertain to IgM and IgG.**

7 **Administration of the chickenpox vaccine confers what type of immunity?**

21 The Respiratory System

LEARNING OUTCOMES

These Learning Outcomes correspond by number to this chapter's modules and indicate what you should be able to do after completing the chapter.

SECTION 1 • Anatomy of the Respiratory System

21.1 Identify the structures of the respiratory system, and list its major functions.

21.2 Explain how the delicate respiratory exchange surfaces are protected from pathogens, debris, and other hazards, and describe cystic fibrosis.

21.3 Identify the organs and structures of the upper respiratory system, and describe their functions.

21.4 Describe the structure of the larynx, and discuss its role in normal breathing and in the production of sound.

21.5 Discuss the structures and functions of the airways outside and inside the lungs.

21.6 Describe the superficial anatomy of the lungs.

21.7 Describe the structure of a pulmonary lobule and the functional anatomy of the alveoli.

SECTION 2 • Respiratory Physiology

21.8 Describe external respiration and internal respiration.

21.9 Summarize the physical principles governing the movement of air into and out of the lungs.

21.10 Name the respiratory muscles, and describe the actions of the muscles responsible for respiratory movements.

21.11 Explain how respiratory rate and tidal volume affect pulmonary and alveolar ventilation.

21.12 Summarize the physical principles governing the diffusion of gases into and out of the blood.

21.13 Discuss the structure and function of hemoglobin, explain the oxygen-hemoglobin saturation curve, and describe the role of 2,3-bisphosphoglycerate.

21.14 Describe how carbon dioxide is transported in the blood, and explain how oxygen is picked up, transported, and released into the bloodstream.

21.15 **CLINICAL MODULE** Explain how pulmonary disease affects compliance and resistance.

21.16 Describe the brain stem structures that influence the control of respiration.

21.17 Identify and discuss reflex respiratory activity in pulmonary ventilation.

21.18 **CLINICAL MODULE** Describe age-related changes to, and the effects of cigarette smoking on, the respiratory system.

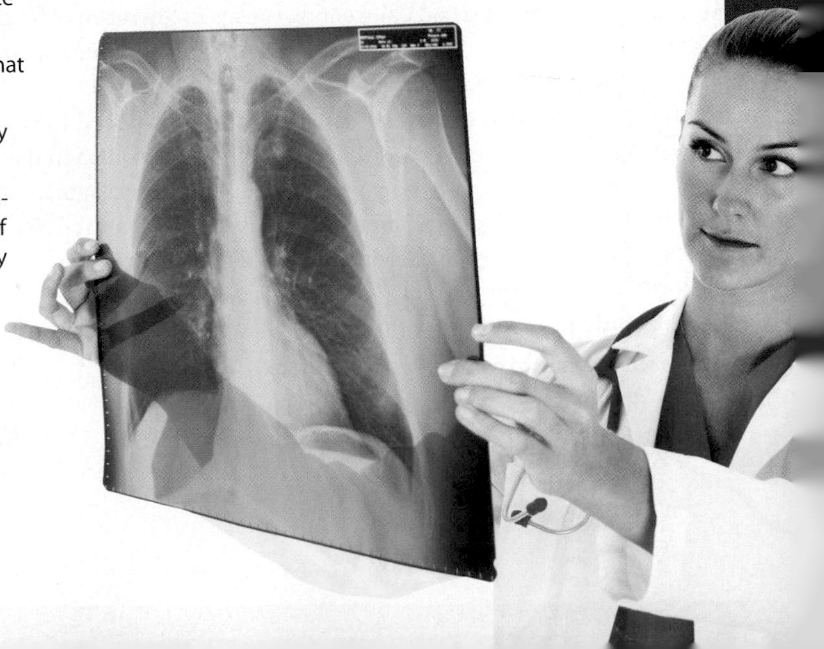

Learning Outcomes are repeated at the bottom of each module.

The respiratory system has an upper and lower respiratory tract with different functions

The **respiratory system** is composed of structures involved in breathing, or **pulmonary ventilation** (airflow to and from the lungs), and gas exchange. In this section we consider this body system's structures and functions.

1 The respiratory system is anatomically divided into an upper respiratory system and a lower respiratory system. The **respiratory tract** is a branching passageway that carries air to and from the gas exchange surfaces of the lungs. The tract is composed of a conducting portion and a respiratory portion. The **conducting portion** begins at the nasal cavity and extends through the pharynx, larynx, trachea, bronchi, and larger bronchioles. The **respiratory portion** includes the smallest, most delicate bronchioles and the air-filled sacs called **alveoli** (al-VĒ-ō-lī). Gas exchange between air and blood occurs at the alveoli.

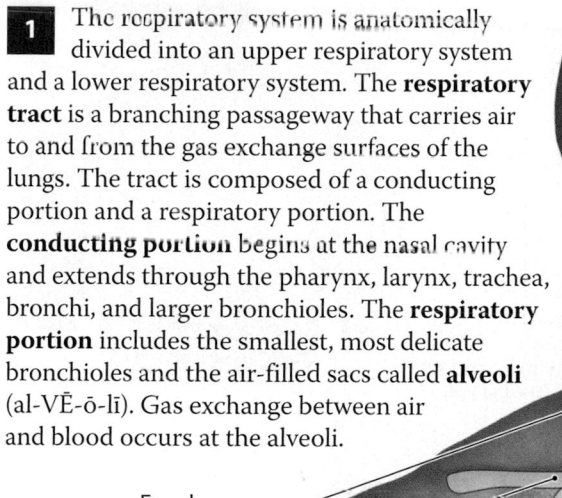

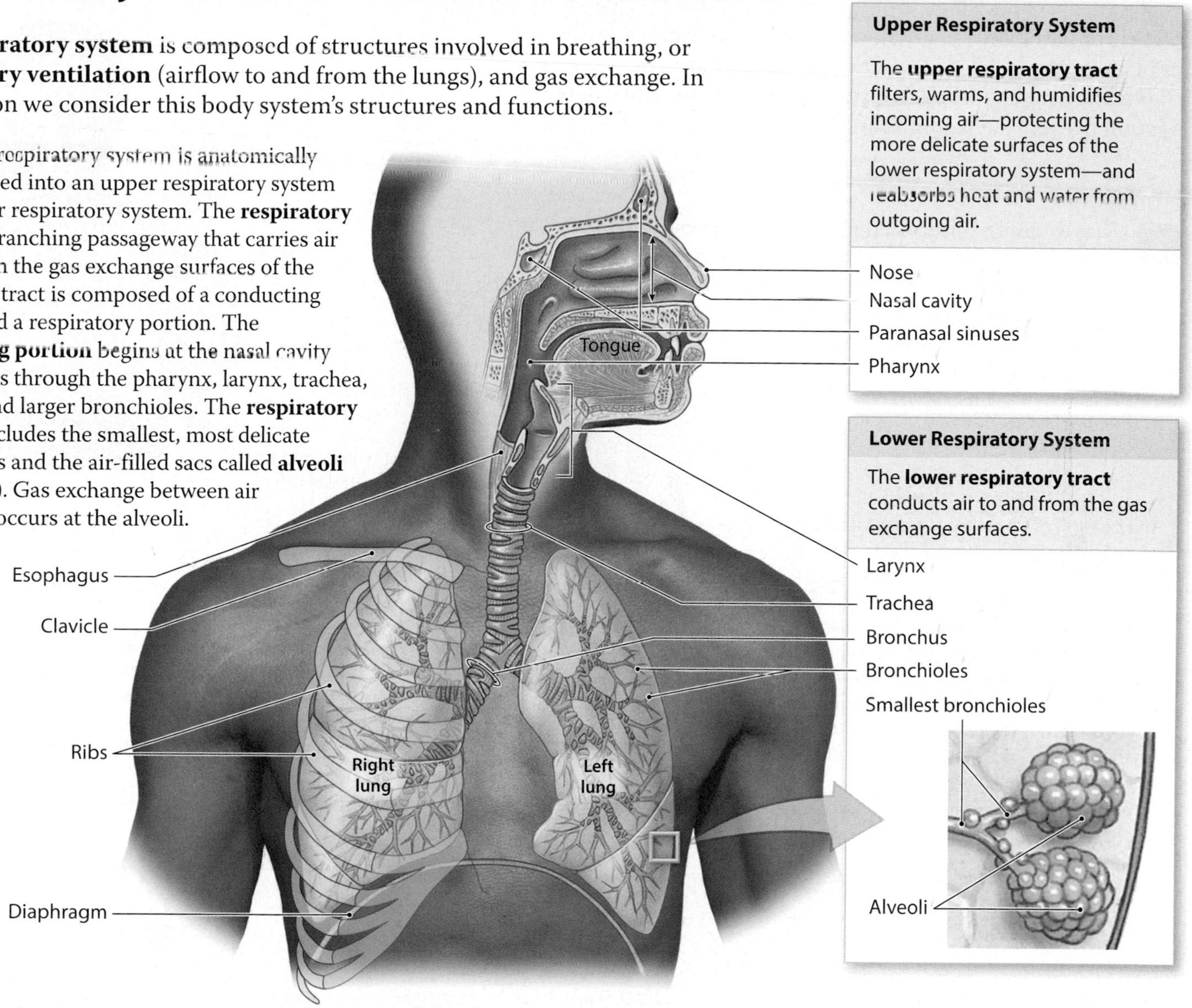
Upper Respiratory System
The **upper respiratory tract** filters, warms, and humidifies incoming air—protecting the more delicate surfaces of the lower respiratory system—and reabsorbs heat and water from outgoing air.
Nose
Nasal cavity
Paranasal sinuses
Pharynx
Tongue
Lower Respiratory System
The **lower respiratory tract** conducts air to and from the gas exchange surfaces.
Larynx
Trachea
Bronchus
Bronchioles
Smallest bronchioles
Alveoli
Esophagus
Clavicle
Ribs
Right lung
Left lung
Diaphragm

2 This table summarizes the major functions of the respiratory system.

Functions of the Respiratory System

- Providing an extensive surface area for gas exchange between air and circulating blood
- Moving air to and from the exchange surfaces of the lungs along the respiratory passageways
- Protecting respiratory surfaces from dehydration, temperature changes, or other environmental variations, and defending the respiratory system and other tissues from invasion by pathogens
- Producing sounds for speaking, singing, and other forms of communication
- Facilitating the detection of olfactory stimuli by olfactory receptors in the superior portions of the nasal cavity

Module 21.1 Review

a. Define pulmonary ventilation.

b. Distinguish between the conducting portion and respiratory portion of the respiratory tract.

c. Where does gas exchange between air and the lungs occur?

LO 21.1 Identify the structures of the respiratory system, and list its major functions.

The respiratory mucosa is protected by the respiratory defense system

Debris or pathogens in inhaled air can severely damage the delicate exchange surfaces of the respiratory system. A series of filtration mechanisms that make up the **respiratory defense system** prevent such contamination.

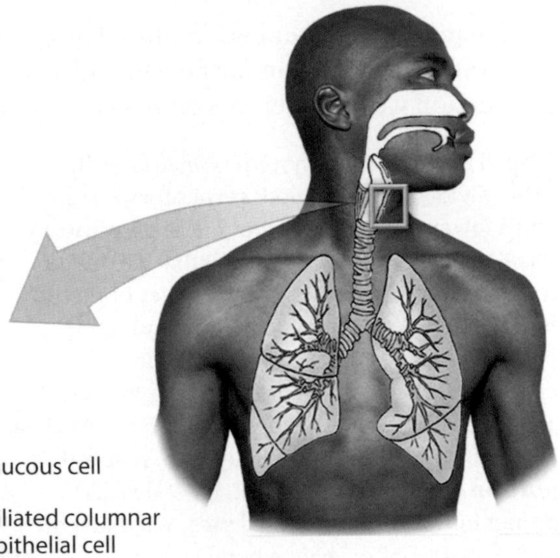

1 The **respiratory mucosa** (mū-KŌ-suh) lines the conducting portion of the respiratory tract. A pseudostratified ciliated columnar epithelium with numerous mucous cells lines the nasal cavity, the superior portion of the pharynx, and the trachea, bronchi, and large bronchioles.

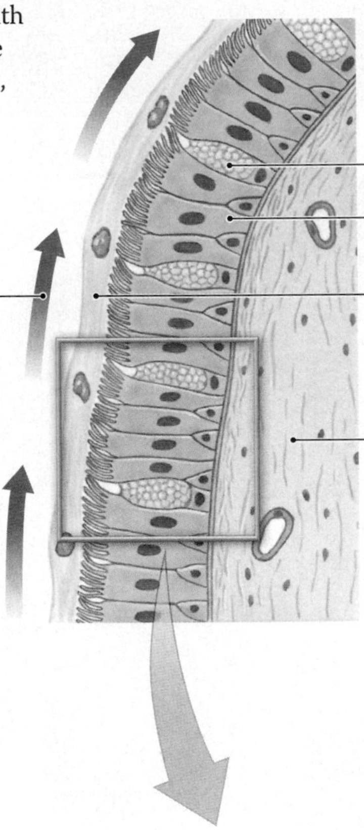

Mucous cell
Ciliated columnar epithelial cell
Mucus layer

The beating of cilia sweeps mucus and any trapped debris or microorganisms toward the pharynx, where they will be coughed out or swallowed and exposed to the acids and enzymes of the stomach. This ciliary movement continuously cleans and protects the respiratory surfaces. The flow of mucus is often described as a **mucus escalator**. By the time air reaches the respiratory portion of the tract, particles larger than around 5 μm have been trapped and removed.

The **lamina propria** (LAM-in-nuh PRŌ-prē-uh) is the underlying layer of areolar tissue that supports the respiratory epithelium. In the upper respiratory system, trachea, and bronchi, the lamina propria contains mucous glands that discharge secretions onto the epithelial surface.

2 This sectional view of the respiratory mucosa shows the histological appearance of the respiratory epithelium.

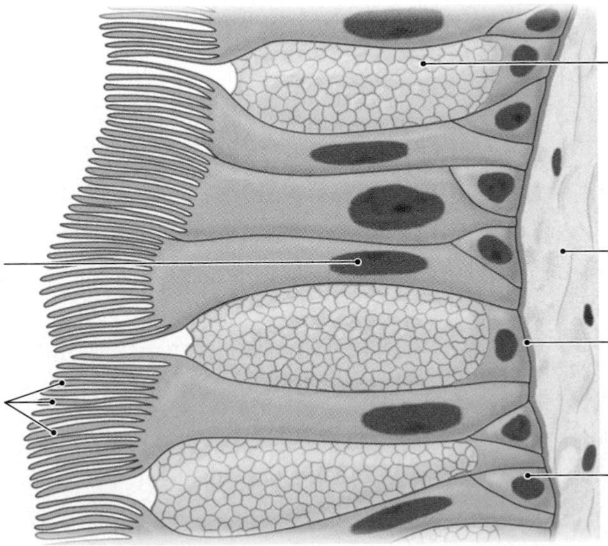

Nucleus of columnar epithelial cell
Cilia
Lamina propria
Basement membrane

Mucous cells in the epithelium and mucous glands in the lamina propria produce a sticky mucus that bathes exposed surfaces.

Stem cells in the epithelium divide to replace damaged or aged cells.

3 The structure of the respiratory epithelium changes markedly along the respiratory tract.

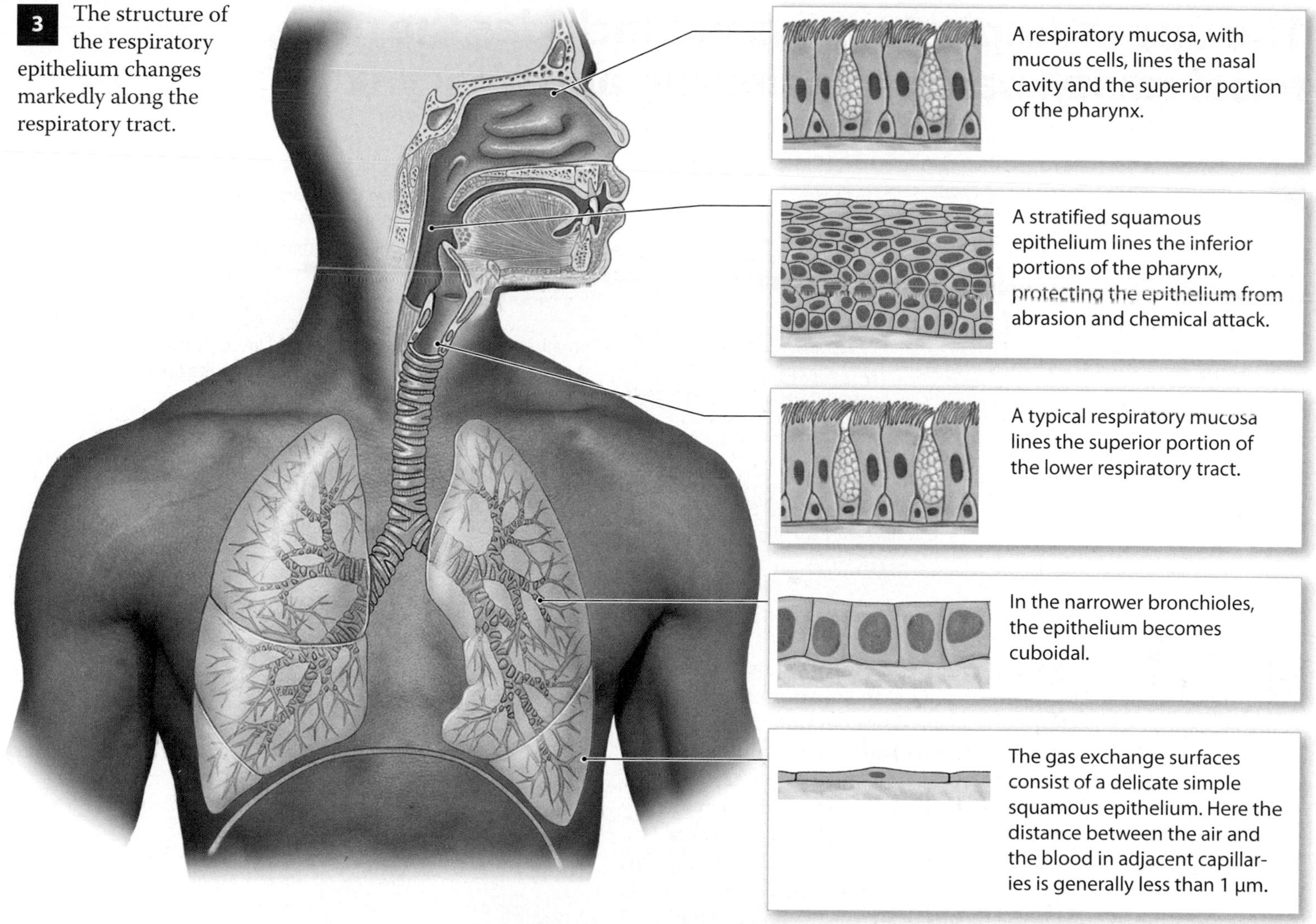
A respiratory mucosa, with mucous cells, lines the nasal cavity and the superior portion of the pharynx.
A stratified squamous epithelium lines the inferior portions of the pharynx, protecting the epithelium from abrasion and chemical attack.
A typical respiratory mucosa lines the superior portion of the lower respiratory tract.
In the narrower bronchioles, the epithelium becomes cuboidal.
The gas exchange surfaces consist of a delicate simple squamous epithelium. Here the distance between the air and the blood in adjacent capillaries is generally less than 1 μm.

4 **Cystic fibrosis (CF)** is the most common lethal inherited disease among Caucasians of Northern European descent. It occurs at a frequency of 1 in 2500 births. The most dangerous signs and symptoms result from the production of abnormally thick and sticky mucus in conducting portions of the respiratory tract. The respiratory defense system cannot transport such dense mucus, and it accumulates, restricting airflow and possibly blocking the smaller respiratory passageways. Potentially lethal infections may develop in the respiratory passageways and lungs if bacteria such as *Pseudomonas aeruginosa* colonize the stagnant mucus. For people who have CF and live in the United States, the median predicted age of survival is the mid-30s. Death generally results from heart failure associated with a massive chronic bacterial infection of the lungs.

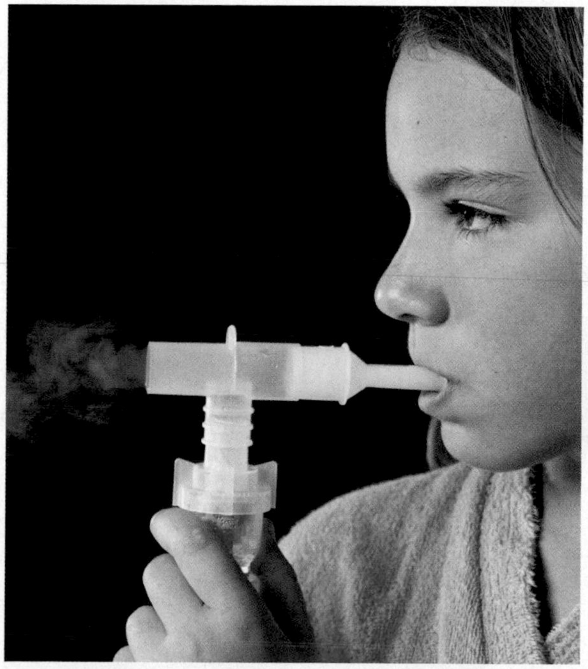

Module 21.2 Review

a. Define respiratory defense system.

b. What membrane lines the conducting portion of the respiratory tract?

c. Why can cystic fibrosis become lethal?

LO **21.2** Explain how the delicate respiratory exchange surfaces are protected from pathogens, debris, and other hazards, and describe cystic fibrosis.

The upper respiratory system includes the nose, nasal cavity, paranasal sinuses, and pharynx

1 The nose is the primary passageway for air entering the respiratory system when you are resting and breathing quietly. This anterior view shows the nasal cartilages and external landmarks.

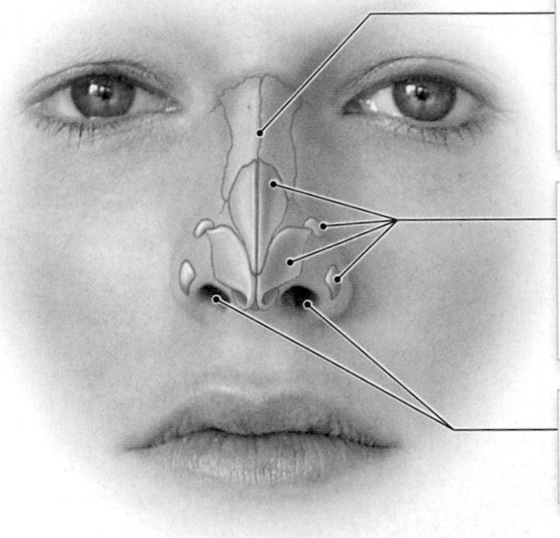

The two nasal bones form the **bridge of the nose** or *dorsum nasi*. This is supported by the anterior portion of the nasal septum, which is formed of hyaline cartilage.

Small, elastic **nasal cartilages** extend laterally from the bridge of the nose. These cartilages help to keep the external nares open and prevent their collapse during a strong inhalation.

Air normally enters through the paired **external nares** (NĀ-res), or nostrils, which open into the nasal cavity.

2 Several important features of the nasal cavity can be best seen in frontal section. The superior, middle, and inferior nasal conchae (singular, *concha*) project toward the nasal septum from the lateral walls of the nasal cavity. (Some references use the terms *superior*, *middle*, and *inferior turbinates*.) To pass from the external nares to the internal nares, air flows between adjacent conchae, through the **superior**, **middle**, and **inferior meatuses** (mē-Ā-tus-ez; *meatus*, a passage). The incoming air bounces off the conchal surfaces and churns like a stream flowing over rocks. As the air swirls, small airborne particles become trapped in the mucus that coats the lining of the nasal cavity. In addition, the turbulence provides extra time for warming and humidifying incoming air. It also creates circular air currents that bring olfactory stimuli to the olfactory receptors.

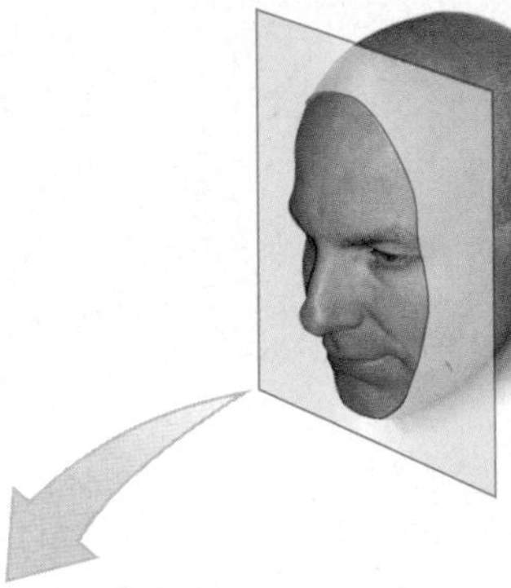

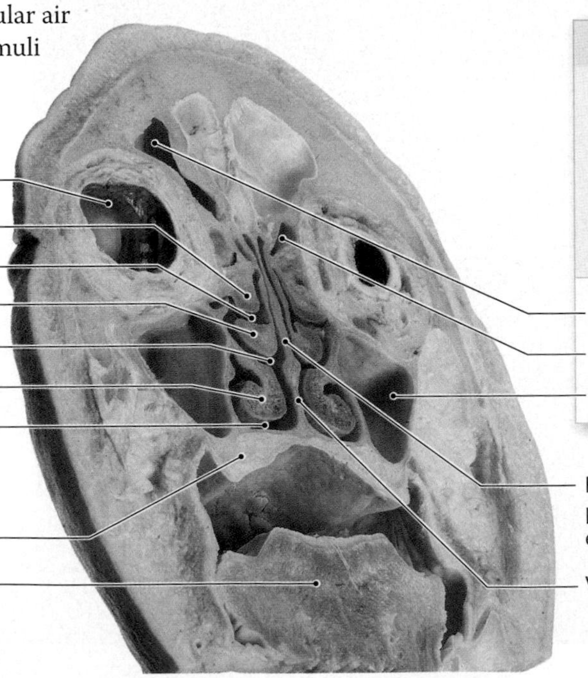
Right eye
Superior nasal concha
Superior meatus
Middle nasal concha
Middle meatus
Inferior nasal concha
Inferior meatus
Hard palate
Tongue
Frontal sinus
Ethmoid air cell
Maxillary sinus
Perpendicular plate of ethmoid bone
Vomer

Paranasal Sinuses

The maxillary, frontal, ethmoid, and sphenoid bones that form the lateral and superior walls of the nasal cavity contain **paranasal sinuses** (Module 7.7, p. 245). The mucus produced in these sinuses, aided by tears draining through the nasolacrimal ducts, keep the surfaces of the nasal cavity moist and clean.

Nasal Septum

The **nasal septum**, formed by the fusion of the perpendicular plate of the ethmoid bone and the vomer, divides the nasal cavity into left and right portions.

3 Other features of the upper respiratory system are best seen in sagittal section. The larynx marks the beginning of the lower respiratory system.

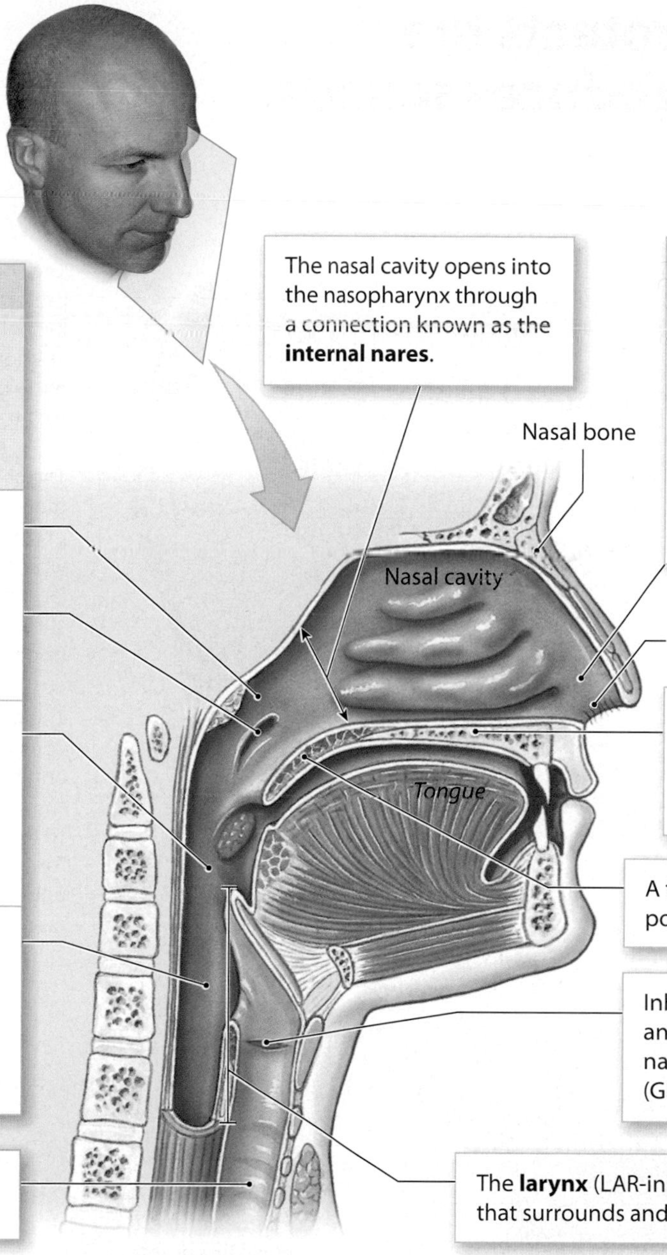
Nasal bone
Nasal cavity
Tongue
External nares

Pharynx

The **pharynx** (FAR-inks) is a chamber shared by the digestive and respiratory systems. The curving superior and posterior walls of the pharynx are closely bound to the axial skeleton, but the lateral walls are flexible and muscular.

The **nasopharynx** (nā-zō-FAR-inks) is the superior portion of the pharynx located between the soft palate and the internal nares. The **nasopharyngeal meatus** is the pharyngeal opening of the auditory tube that leads to the middle ear.

The **oropharynx** (*oris*, mouth) extends between the soft palate and the base of the tongue at the level of the hyoid bone. At the boundary between the nasopharynx and the oropharynx, the epithelium changes from pseudostratified columnar to stratified squamous epithelium.

The narrow **laryngopharynx** (la-rin-gō-FAR-inks) includes that portion of the pharynx between the hyoid bone and the entrance to the larynx and esophagus. Like the oropharynx, the laryngopharynx is lined with a stratified squamous epithelium.

The **trachea**, or windpipe, conducts air toward the lungs.

The nasal cavity opens into the nasopharynx through a connection known as the **internal nares**.

The **nasal vestibule** is the space contained within the flexible tissues of the nose. Its epithelium contains coarse hairs that extend across the external nares. Large airborne particles are trapped in these hairs and are thereby prevented from entering the nasal cavity.

A bony **hard palate** forms the floor of the nasal cavity and separates it from the oral cavity.

A fleshy **soft palate** extends posterior to the hard palate.

Inhaled air leaves the pharynx and enters the larynx through a narrow opening of the **glottis** (GLOT-is).

The **larynx** (LAR-inks) is a cartilaginous structure that surrounds and protects the glottis.

Throughout much of the nasal cavity, the lamina propria contains an extensive network of highly expandable veins that can release heat like a radiator. As cool, dry air passes inward over the exposed surfaces of the nasal cavity, the air warms and water in the mucus evaporates. Air moving from your nasal cavity to your lungs is thus heated almost to body temperature, and it is nearly saturated with water vapor. This protects more delicate respiratory surfaces from chilling or drying out. As air moves out of the respiratory tract, it again passes over the epithelium of the nasal cavity. The air now is warmer and more humid than it was when it entered. The nasal mucosa reabsorbs the heat and water from the outgoing air, thus reducing both heat loss and water loss to the environment. Breathing through your mouth eliminates much of the conditioning of inhaled air and increases heat and water loss at every exhalation.

Module 21.3 Review

a. List the structures of the upper respiratory system.

b. Trace the pathway of air through the upper respiratory system.

c. Why is the vascularization of the nasal cavity important?

LO **21.3** Identify the organs and structures of the upper respiratory system, and describe their functions.

The larynx protects the glottis and produces sounds

1 The **larynx**, sometimes called the "voice box," consists of three large unpaired cartilages and three small paired cartilages.

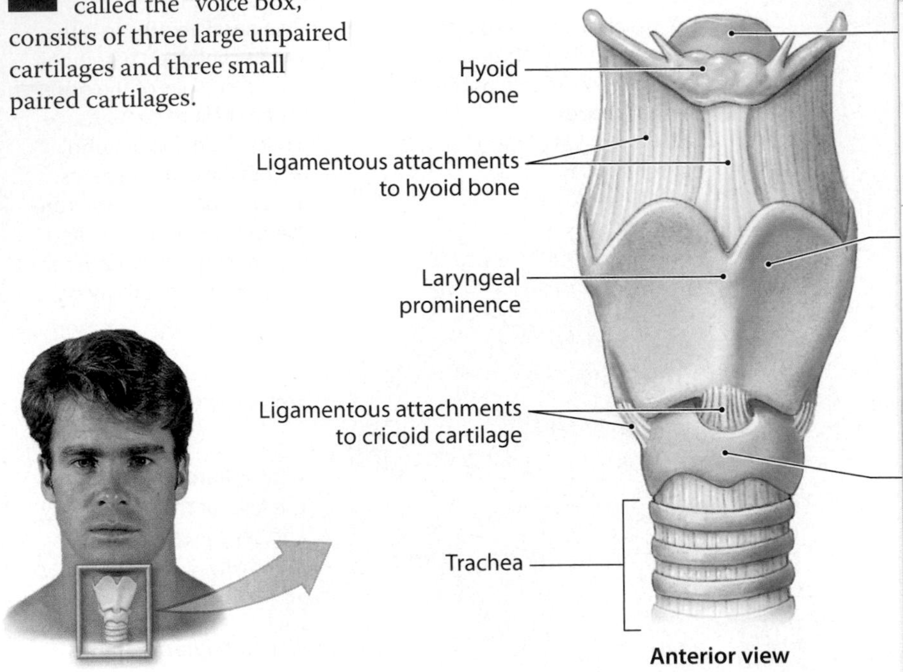

Hyoid bone
Ligamentous attachments to hyoid bone
Laryngeal prominence
Ligamentous attachments to cricoid cartilage
Trachea

Anterior view

Large Cartilages of the Larynx

The shoehorn-shaped **epiglottis** (ep-i-GLOT-is) projects superior to the glottis and forms a lid over it. During swallowing, the larynx is elevated and the epiglottis folds back over the glottis, preventing the entry of both liquids and solid food into the respiratory tract.

The **thyroid** (*thyroid*, shield shaped) **cartilage** forms most of the anterior and lateral walls of the larynx. The prominent anterior surface of the thyroid cartilage, which you can easily see and feel, is called the laryngeal prominence, or Adam's apple. The superior surface has ligamentous attachments to the hyoid bone, the epiglottis, and smaller laryngeal cartilages.

The **cricoid** (KRĪ-koyd; ring shaped) **cartilage** is a complete ring of cartilage with a narrow anterior band and broad posterior portion. Together the cricoid and thyroid cartilages protect the glottis and the entrance to the trachea, and their broad surfaces provide sites for the attachment of important laryngeal muscles and ligaments.

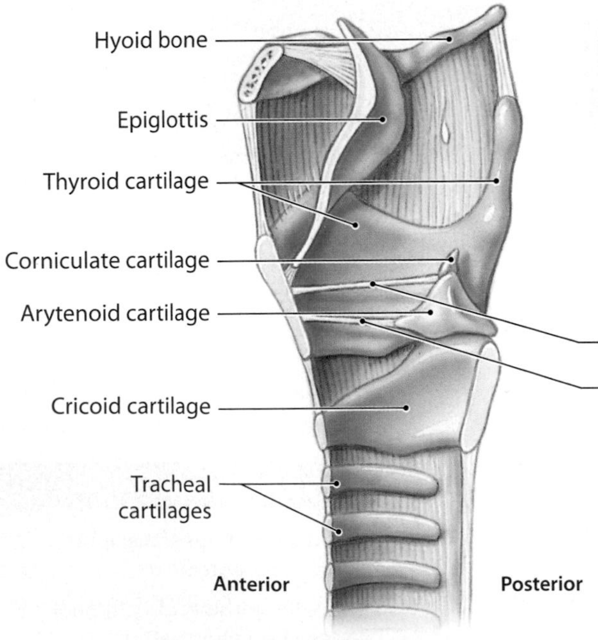

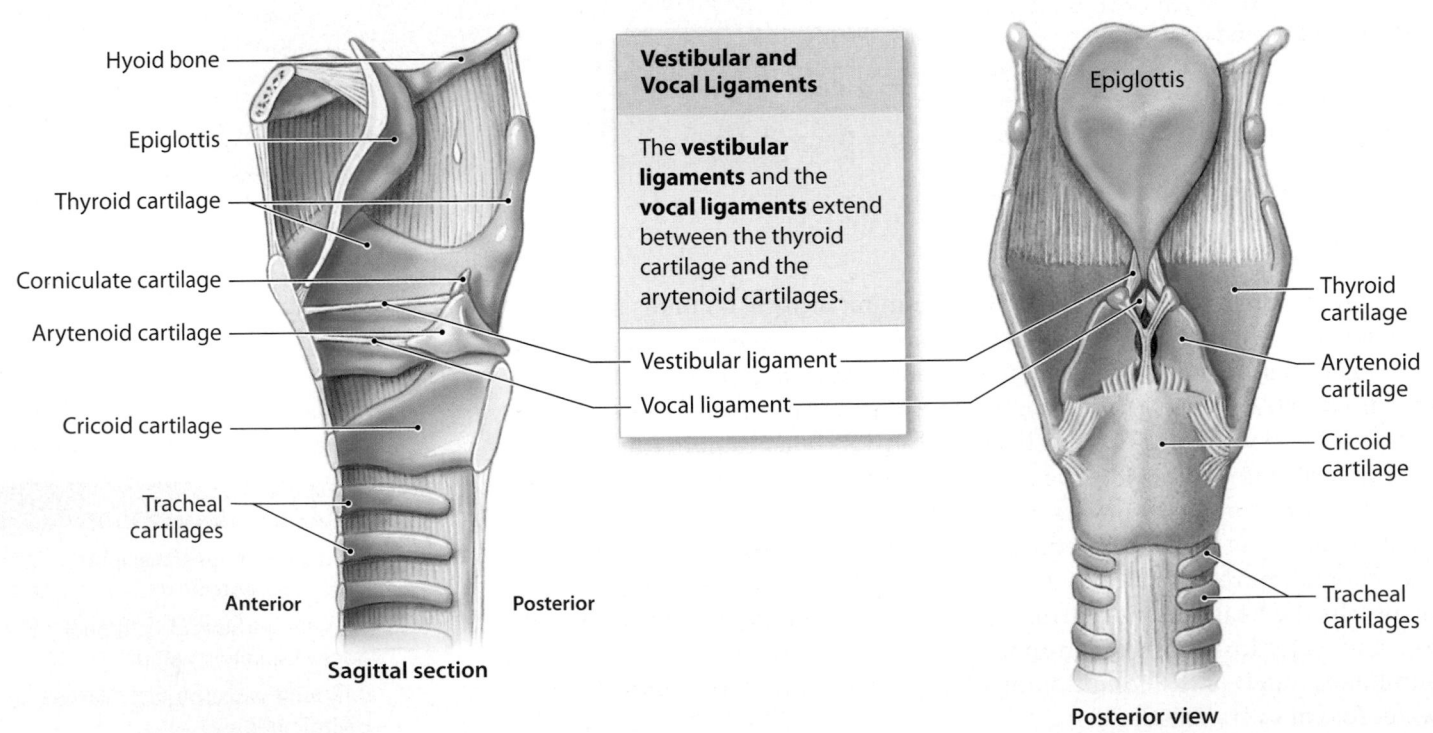

Hyoid bone
Epiglottis
Thyroid cartilage
Corniculate cartilage
Arytenoid cartilage
Cricoid cartilage
Tracheal cartilages
Anterior
Posterior

Sagittal section

Vestibular and Vocal Ligaments

The **vestibular ligaments** and the **vocal ligaments** extend between the thyroid cartilage and the arytenoid cartilages.

Vestibular ligament

Vocal ligament

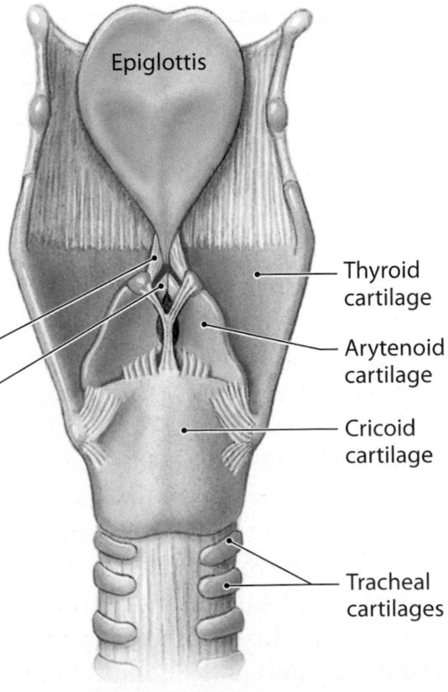

Epiglottis
Thyroid cartilage
Arytenoid cartilage
Cricoid cartilage
Tracheal cartilages

Posterior view

2 The cartilages of the larynx are isolated in this posterior view. Now the small laryngeal cartilages can be seen in relationship to the larger, supporting cartilages.

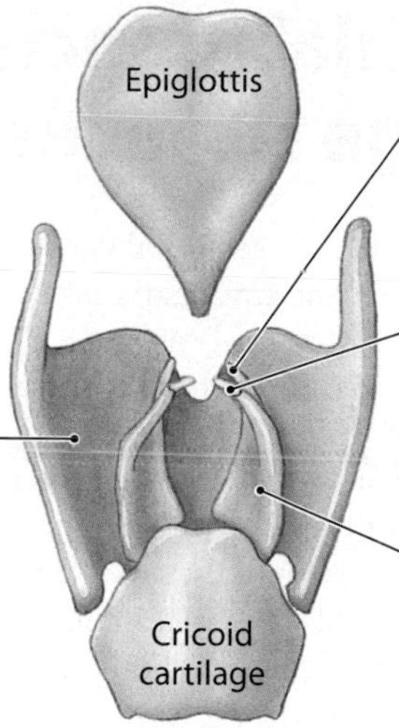
Epiglottis
Thyroid cartilage
Cricoid cartilage

Small Laryngeal Cartilages

The **cuneiform** (kū-NĒ-i-form; wedge shaped) **cartilages** are long and curved, and they lie within folds of tissue that extend between the lateral surface of each arytenoid cartilage and the epiglottis.

The **corniculate** (kor-NIK-ū-lat; horn shaped) **cartilages** articulate with the arytenoid cartilages. The corniculate and arytenoid cartilages function in the opening and closing of the glottis and the production of sound.

The small, paired **arytenoid** (ar-i-TĒ-noyd; ladle shaped) **cartilages** articulate with the superior surface of the cricoid cartilage.

3 Air enters or leaves the larynx through the glottis. The **glottis** (GLOT-is) is made up of the vocal folds and the rima glottidis. The opening or closing of the glottis involves rotational movements of the arytenoid cartilages. When the glottis is open, passing air vibrates its **vocal folds**, tissue folds that contain the elastic **vocal ligaments**. The vibration of the vocal folds produces sound waves, and the pitch of the sound produced depends on the diameter, length, and tension in the vocal folds. The tension is controlled by the contraction of voluntary muscles that reposition the arytenoid cartilages relative to the thyroid cartilage. These diagrammatic superior views show the glottis in the open and closed positions.

Glottis

The **rima glottidis** is the opening between the vocal folds and the arytenoid cartilages.

The vocal folds, which contain the vocal ligaments, lie inferior to the vestibular folds. The vocal folds are involved with the production of sound and are also known as the **vocal cords**.

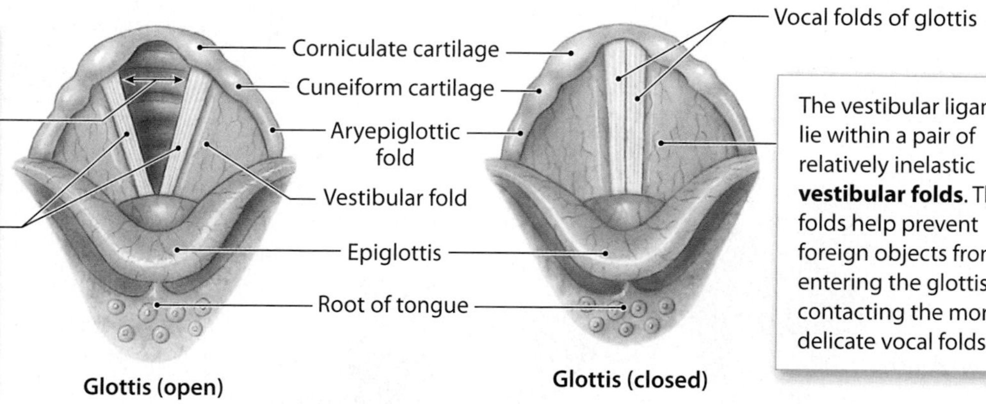
Vocal folds of glottis
Corniculate cartilage
Cuneiform cartilage
Aryepiglottic fold
Vestibular fold
Epiglottis
Root of tongue
Glottis (open)
Glottis (closed)

The vestibular ligaments lie within a pair of relatively inelastic **vestibular folds**. These folds help prevent foreign objects from entering the glottis and contacting the more delicate vocal folds.

4 Sound production at the larynx is called **phonation** (fō-NĀ-shun; *phone*, voice). Phonation is one component of speech. However, clear speech also requires **articulation**, the modification of those sounds by other structures, such as the tongue, teeth, and lips. In a stringed instrument, such as a guitar, the quality of the sound produced does not depend solely on the nature of the vibrating string. Rather, the entire instrument becomes involved as the walls vibrate and the composite sound echoes within the hollow body. Similar amplification and resonance occur within your pharynx, oral cavity, nasal cavity, and paranasal sinuses. The combination determines the particular and distinctive sound of your voice.

Module 21.4 Review

a. Identify the paired and unpaired cartilages that compose the larynx.

b. What are the highly elastic vocal folds of the glottis also called?

c. Distinguish between phonation and articulation.

LO 21.4 Describe the structure of the larynx, and discuss its role in normal breathing and in the production of sound.

Module 21.5

The trachea, bronchi, and bronchial branches convey air to and from lung gas exchange surfaces

1 The **trachea** (TRĀ-kē-uh), or windpipe, is a tough, flexible tube with a diameter of about 2.5 cm (1 in.) and a length of about 11 cm (4.33 in.). It begins anterior to vertebra C_6 and ends in the mediastinum, at the level of vertebra T_5, where it branches to form the right and left **primary bronchi** (BRONG-ki; singular, *bronchus*).

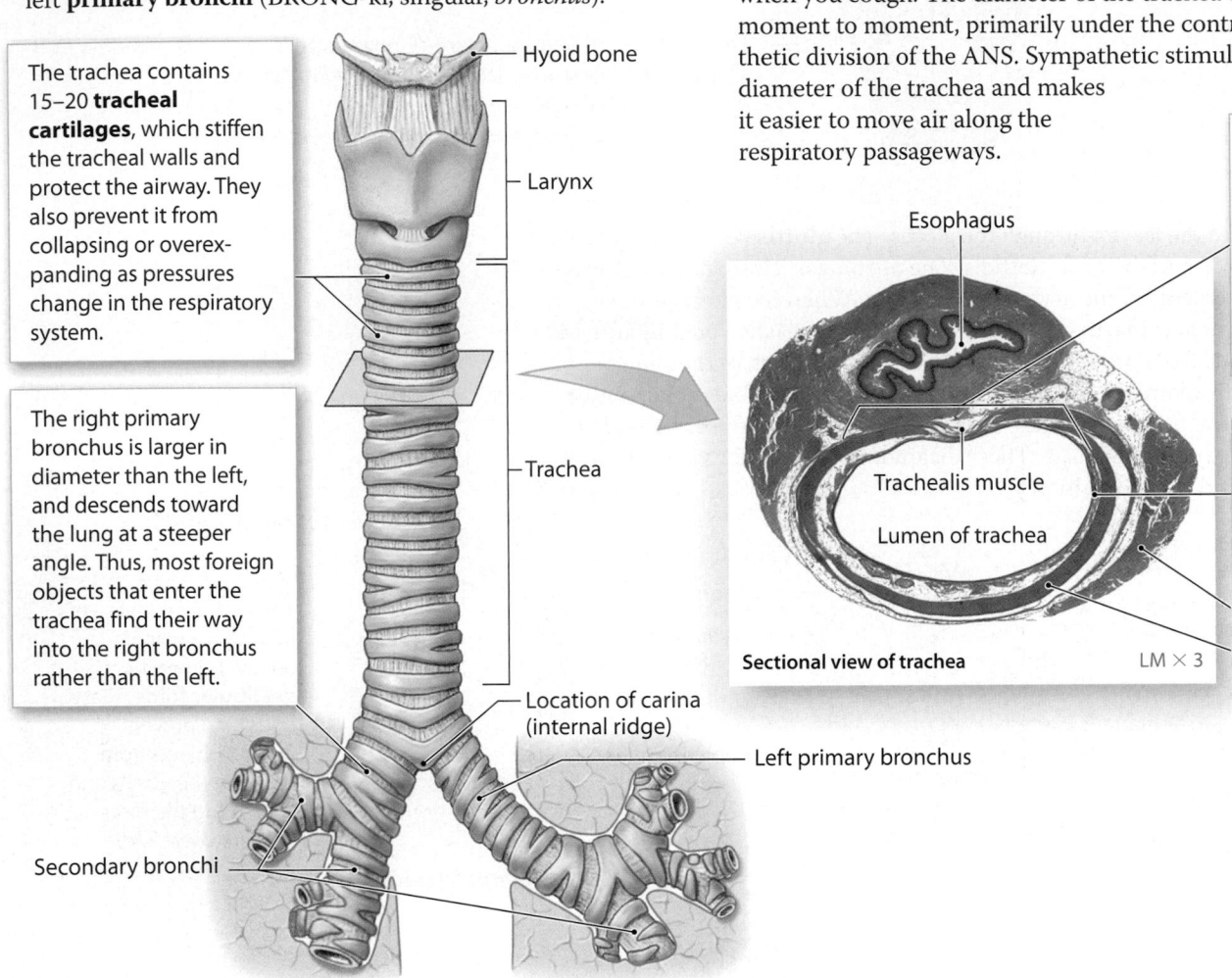

The trachea contains 15–20 **tracheal cartilages**, which stiffen the tracheal walls and protect the airway. They also prevent it from collapsing or overexpanding as pressures change in the respiratory system.

The right primary bronchus is larger in diameter than the left, and descends toward the lung at a steeper angle. Thus, most foreign objects that enter the trachea find their way into the right bronchus rather than the left.

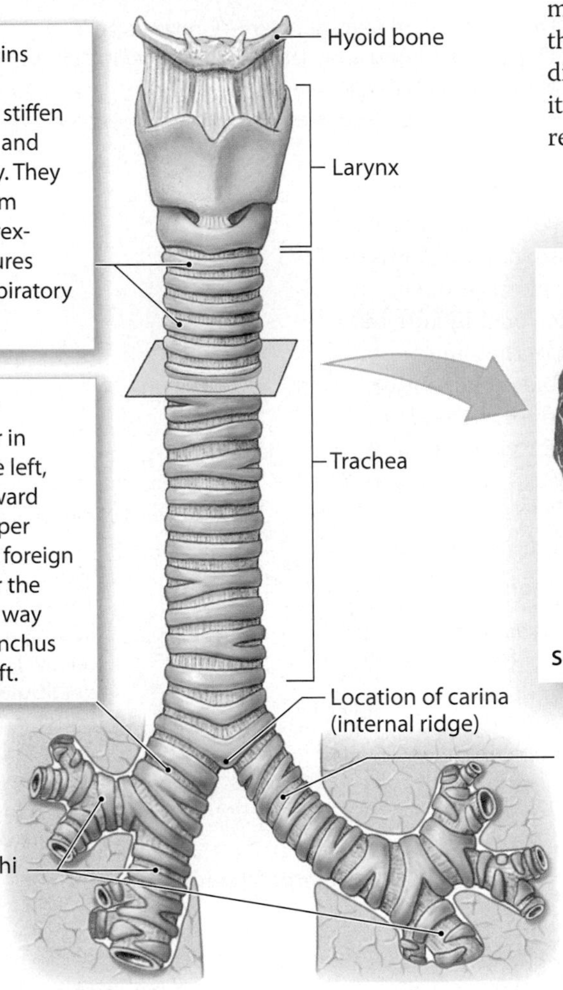

Hyoid bone
Larynx
Trachea
Location of carina (internal ridge)
Left primary bronchus
Secondary bronchi
Right lung
Left lung

2 This is a sectional view of the trachea. An elastic ligament and the **trachealis muscle** connect the ends of each C-shaped tracheal cartilage. Contraction of the trachealis muscle reduces the diameter of the trachea, increasing the resistance to airflow. This allows air to be more forcefully expelled, such as when you cough. The diameter of the trachea changes from moment to moment, primarily under the control of the sympathetic division of the ANS. Sympathetic stimulation increases the diameter of the trachea and makes it easier to move air along the respiratory passageways.

Because the tracheal cartilages are incomplete posteriorly, the posterior tracheal wall can easily distort when large masses of food pass along the esophagus.

The mucosa of the trachea resembles that of the nasal cavity and the nasopharynx.

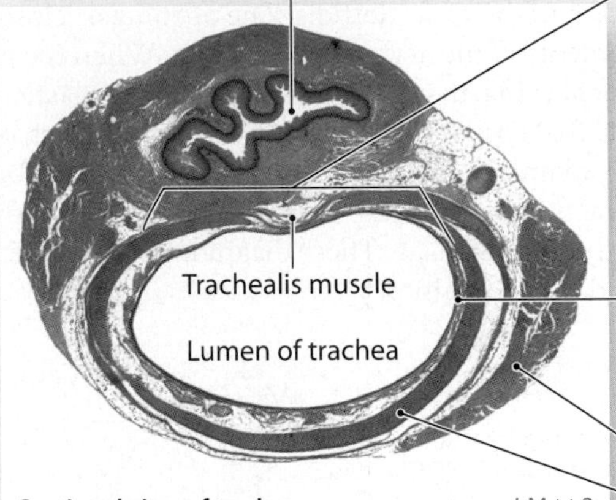

Esophagus
Trachealis muscle
Lumen of trachea
Thyroid gland
Tracheal cartilage
LM × 3

Sectional view of trachea

3 This flowchart indicates the general pattern of airflow in the conducting portion of the respiratory tract.

Air-Conducting Passageways in the Lower Respiratory Tract

The trachea is a single conducting tube, extending from the larynx to the mediastinum. →	The trachea branches to form two primary bronchi, one for each lung. Unlike the C-shaped cartilaginous rings of the trachea, those of the primary bronchi form complete rings. →	After entering the lung, each primary bronchus divides to form **secondary bronchi**. The right lung has three and the left lung has two. The secondary bronchi are supported by small cartilage plates rather than rings. →	Each secondary bronchus divides to form **tertiary bronchi**. The cartilages in the walls of tertiary bronchi resemble those of secondary bronchi. →	Each tertiary bronchus branches several times, giving rise to multiple **bronchioles**. →

4 This diagrammatic view highlights important structural features of bronchi and bronchioles, and the general pattern of airway distribution along the lower respiratory tract. At each new branch, the diameter decreases. Overall, there is a five- to eight-fold decrease in diameter from the trachea to the terminal bronchioles, which have a diameter of 0.3–0.5 mm.

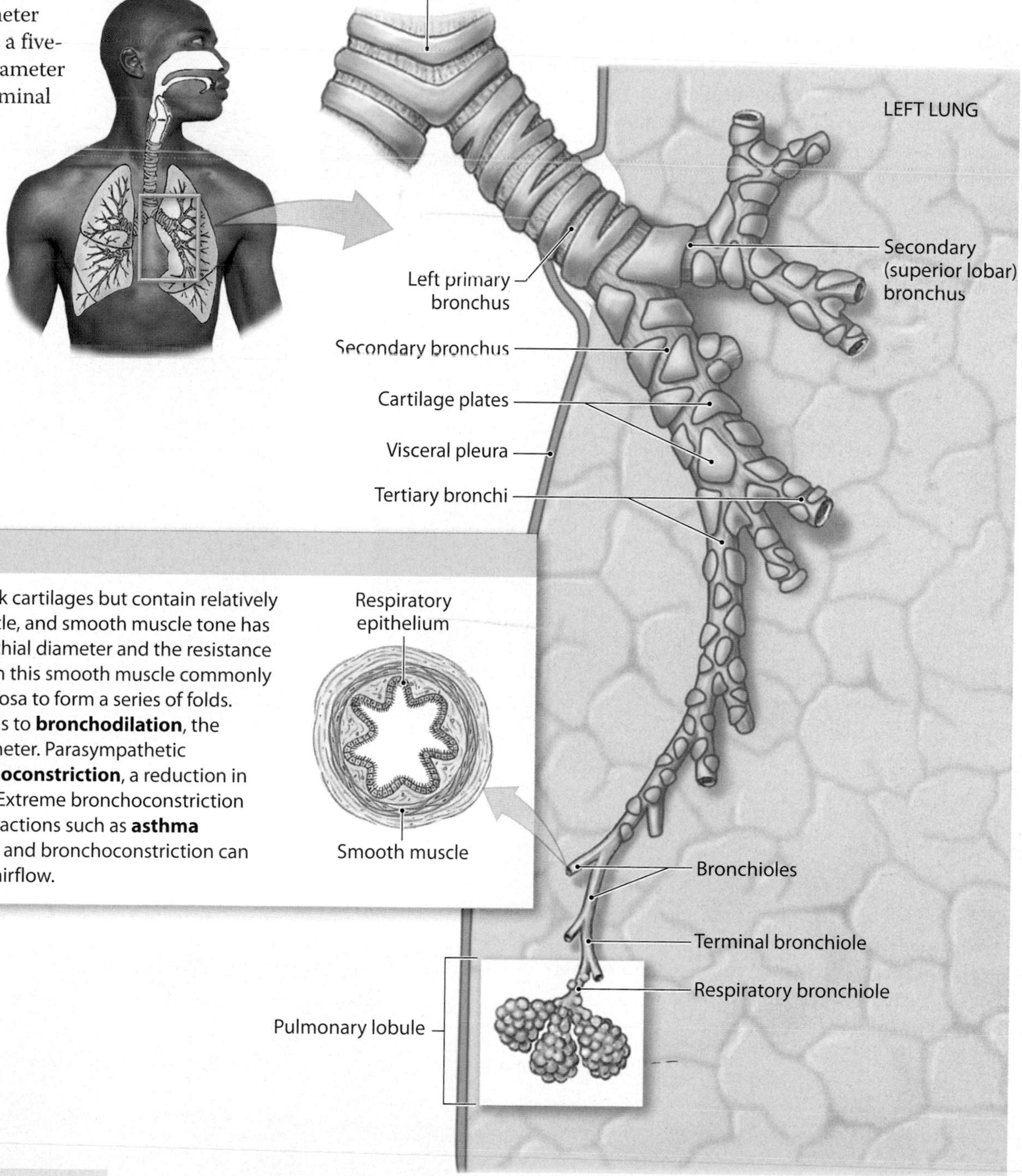

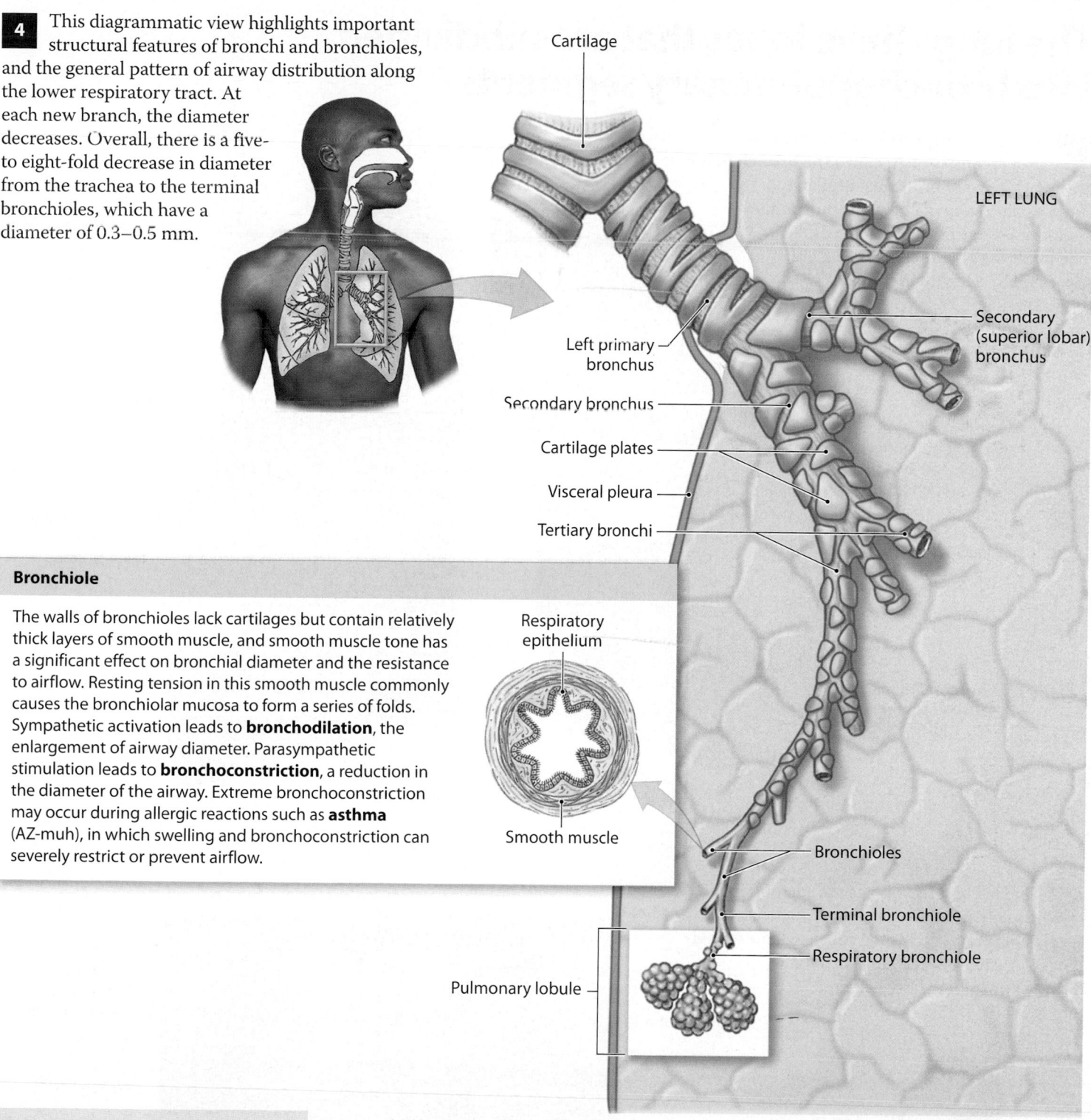

Cartilage
LEFT LUNG
Left primary bronchus
Secondary (superior lobar) bronchus
Secondary bronchus
Cartilage plates
Visceral pleura
Tertiary bronchi
Bronchioles
Terminal bronchiole
Respiratory bronchiole
Pulmonary lobule
Respiratory epithelium
Smooth muscle

Bronchiole

The walls of bronchioles lack cartilages but contain relatively thick layers of smooth muscle, and smooth muscle tone has a significant effect on bronchial diameter and the resistance to airflow. Resting tension in this smooth muscle commonly causes the bronchiolar mucosa to form a series of folds. Sympathetic activation leads to **bronchodilation**, the enlargement of airway diameter. Parasympathetic stimulation leads to **bronchoconstriction**, a reduction in the diameter of the airway. Extreme bronchoconstriction may occur during allergic reactions such as **asthma** (AZ-muh), in which swelling and bronchoconstriction can severely restrict or prevent airflow.

Each bronchiole branches further to form **terminal bronchioles**. Roughly 6500 terminal bronchioles arise from each tertiary bronchus, and each terminal bronchiole supplies a single **pulmonary lobule** where gas exchange occurs within alveoli.

Module 21.5 Review

a. Compare the two primary bronchi.

b. What function do the C-shaped tracheal cartilages allow?

c. Trace the pathway of airflow along the passages of the lower respiratory tract.

The lungs have lobes that are subdivided into bronchopulmonary segments

1 The branching pattern of bronchi and bronchioles is often called the **bronchial tree**. Each tertiary bronchus ultimately supplies air to a single **bronchopulmonary segment**, a specific region of one lung. The lungs have distinct lobes that are separated by deep fissures. Each lobe contains at least a pair of bronchopulmonary segments.

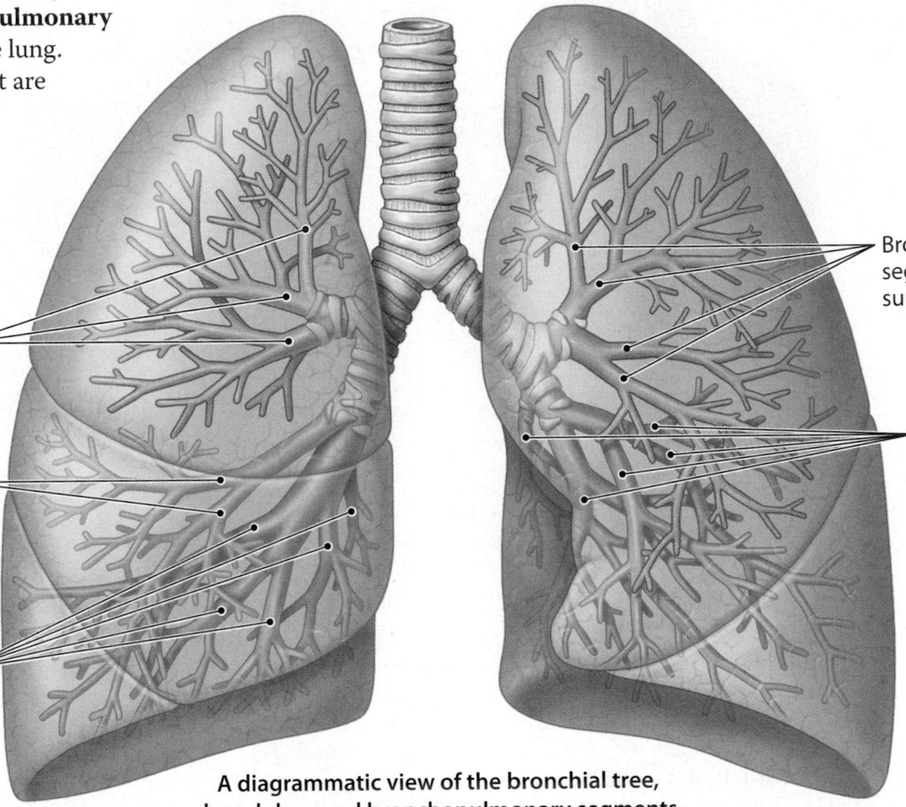
Right
Left
Bronchopulmonary segments of right superior lobe
Bronchopulmonary segments of right middle lobe
Bronchopulmonary segments of right inferior lobe
Bronchopulmonary segments of left superior lobe
Bronchopulmonary segments of left inferior lobe

A diagrammatic view of the bronchial tree, lung lobes, and bronchopulmonary segments

2 The left and right lungs are surrounded by the left and right pleural cavities, respectively. This photo shows the collapsed, preserved lungs in a cadaver. The lungs in a living, healthy nonsmoker have a lighter, pinkish color.

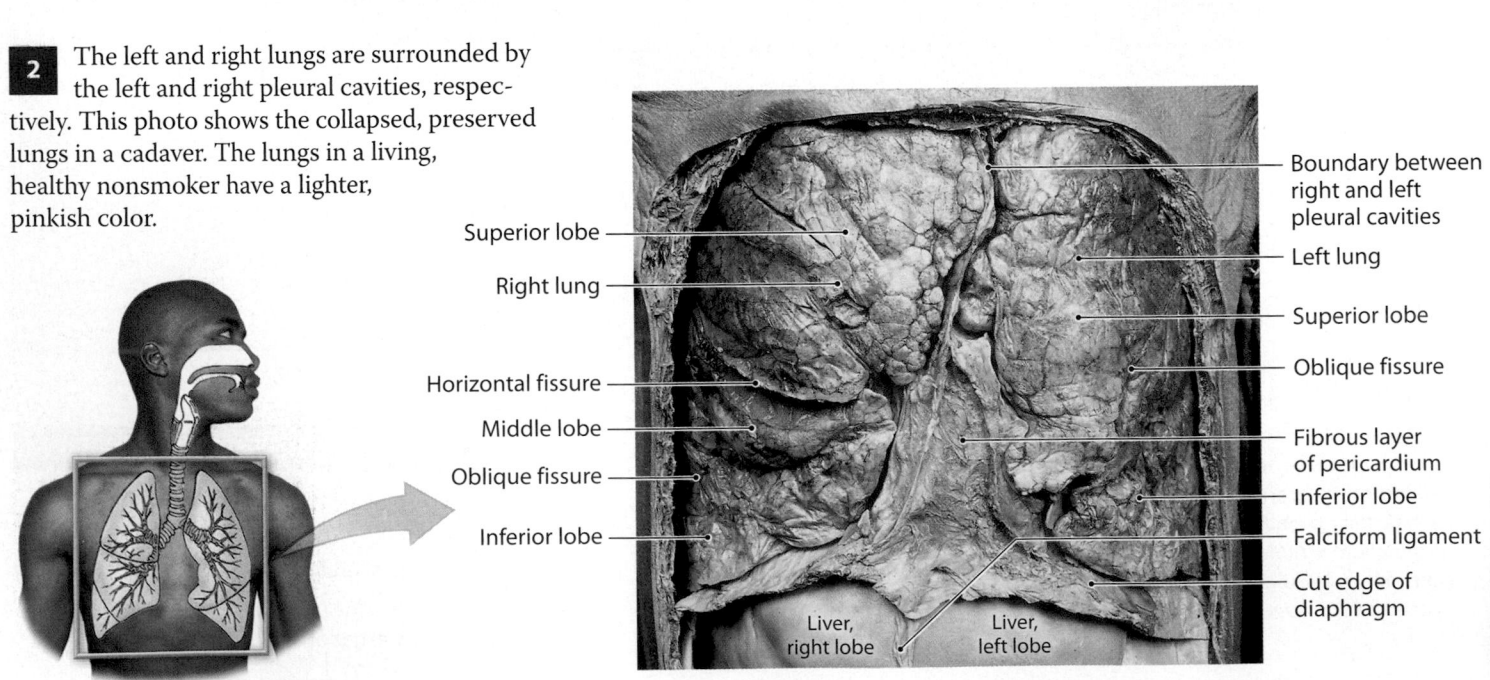

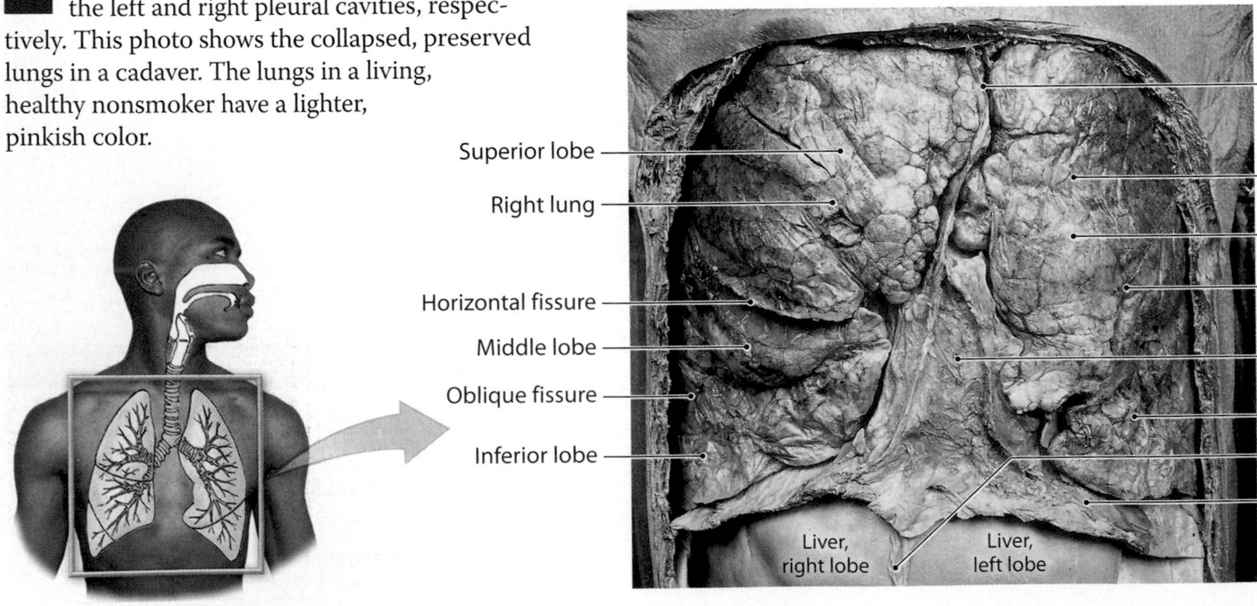
Superior lobe
Right lung
Horizontal fissure
Middle lobe
Oblique fissure
Inferior lobe
Liver, right lobe
Liver, left lobe
Boundary between right and left pleural cavities
Left lung
Superior lobe
Oblique fissure
Fibrous layer of pericardium
Inferior lobe
Falciform ligament
Cut edge of diaphragm

3 Each lung is a blunt cone with a tip extending to the superior border of the first rib. The broad concave inferior portion of each lung rests on the superior surface of the diaphragm. Major superficial landmarks on the left and right lungs can be seen in these lateral and medial views.

Lateral Surfaces

The curving anterior and lateral surfaces of each lung follow the inner contours of the rib cage.

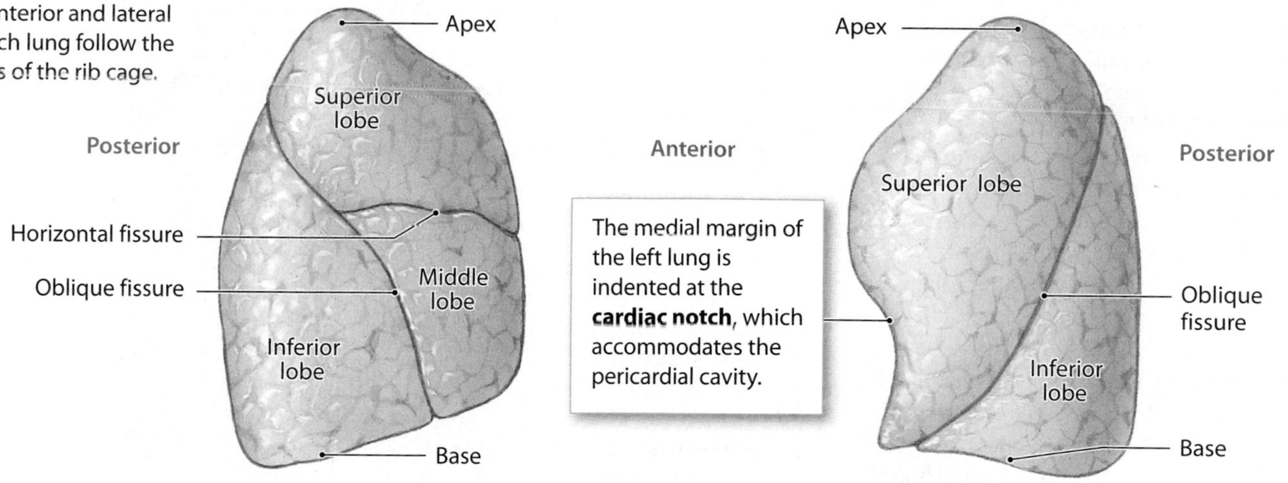

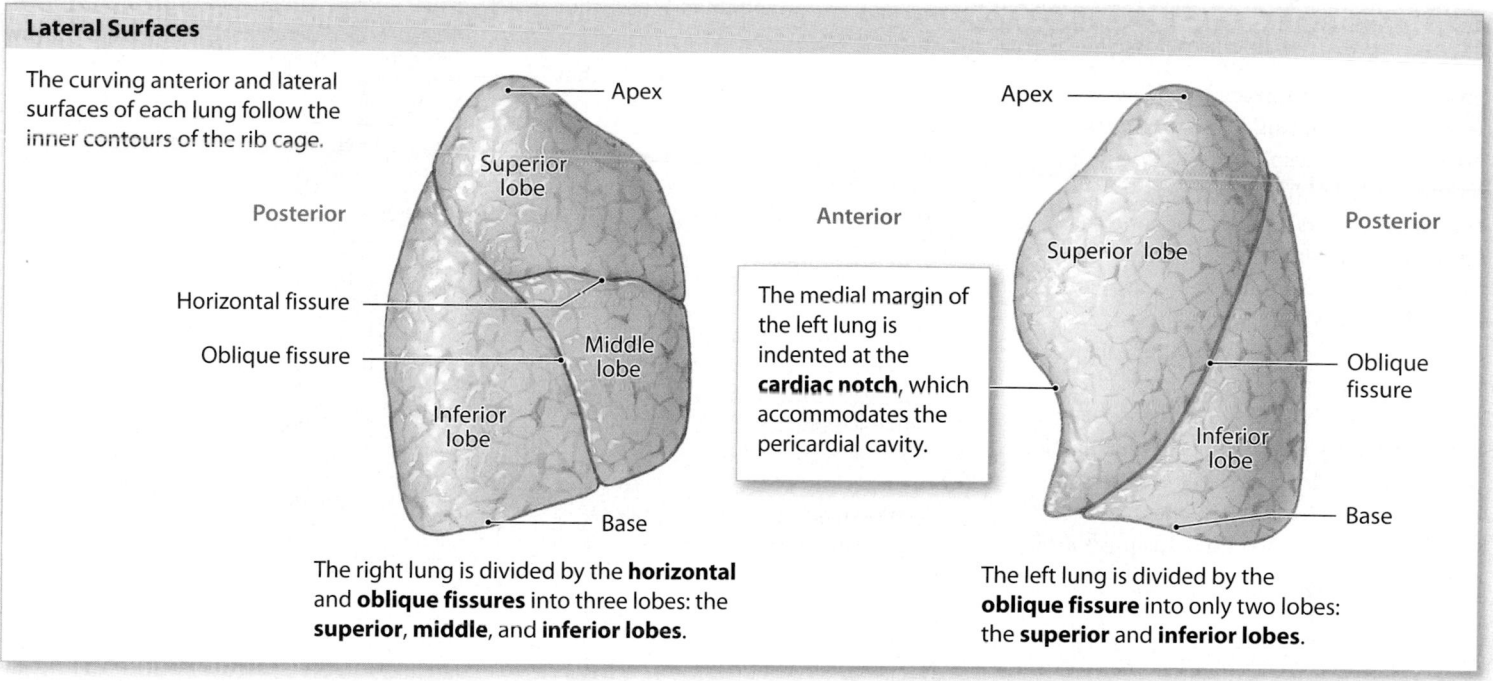

Apex
Superior lobe
Posterior
Horizontal fissure
Oblique fissure
Middle lobe
Inferior lobe
Base
Apex
Anterior
Superior lobe
Posterior
The medial margin of the left lung is indented at the **cardiac notch**, which accommodates the pericardial cavity.
Oblique fissure
Inferior lobe
Base

The right lung is divided by the **horizontal** and **oblique fissures** into three lobes: the **superior**, **middle**, and **inferior lobes**.

The left lung is divided by the **oblique fissure** into only two lobes: the **superior** and **inferior lobes**.

Medial Surfaces

The medial surfaces, which contain the hilum, have more irregular shapes. The medial surfaces of both lungs have grooves that mark the positions of the great vessels and the heart.

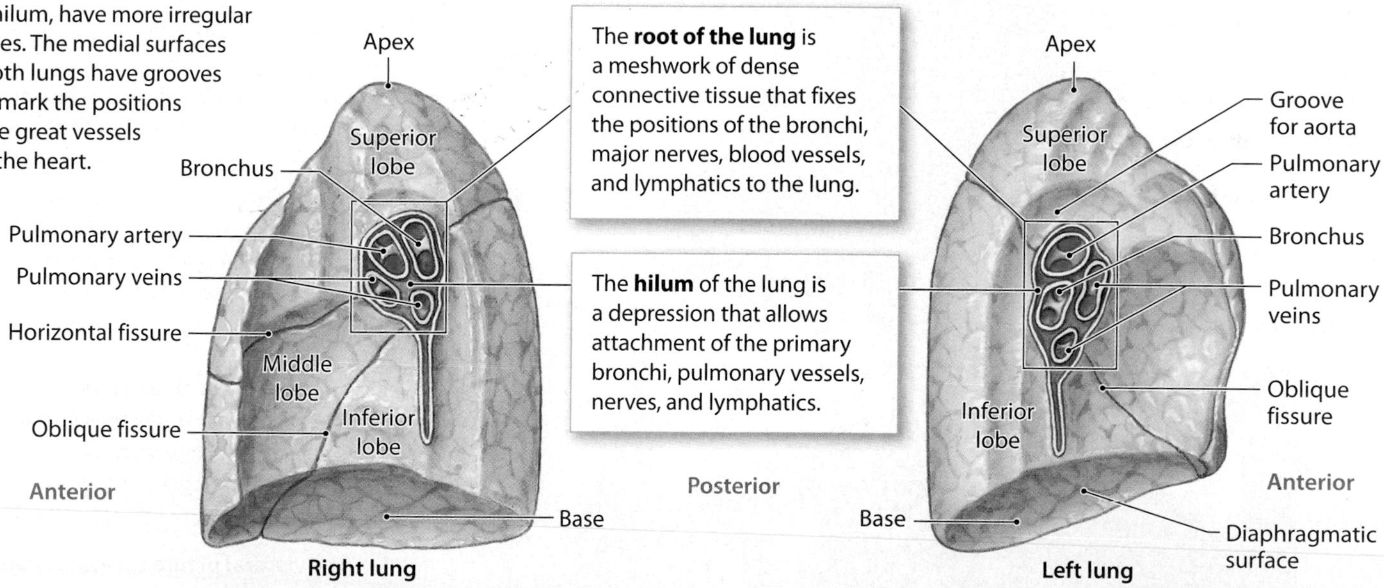

Apex
Superior lobe
Bronchus
Pulmonary artery
Pulmonary veins
Horizontal fissure
Middle lobe
Inferior lobe
Oblique fissure
Anterior
Base
Right lung
The **root of the lung** is a meshwork of dense connective tissue that fixes the positions of the bronchi, major nerves, blood vessels, and lymphatics to the lung.
The **hilum** of the lung is a depression that allows attachment of the primary bronchi, pulmonary vessels, nerves, and lymphatics.
Posterior
Apex
Superior lobe
Groove for aorta
Pulmonary artery
Bronchus
Pulmonary veins
Oblique fissure
Inferior lobe
Anterior
Base
Diaphragmatic surface
Left lung

Module 21.6 Review

a. Define bronchopulmonary segment.

b. Describe the location of the lungs within the thoracic cavity.

c. Name the lobes and fissures of each lung.

LO **21.6** Describe the superficial anatomy of the lungs.

Module 21.7

Pulmonary lobules contain alveoli, where gas exchange occurs

1 The continuous passageways of the conducting and respiratory portions of the respiratory tract end in air sacs called **alveoli** (al-VĒ-ō-lī; singular, *alveolus*). Each lung contains about 150 million alveoli, and they give the lung an open, spongy appearance. Each alveolus is surrounded by an extensive capillary network that receives blood from a branch of a pulmonary artery and discharges blood into a branch of a pulmonary vein. A network of elastic fibers surrounds these capillaries. Recoil of these fibers during exhalation reduces the size of the alveoli and helps push air out of the lungs.

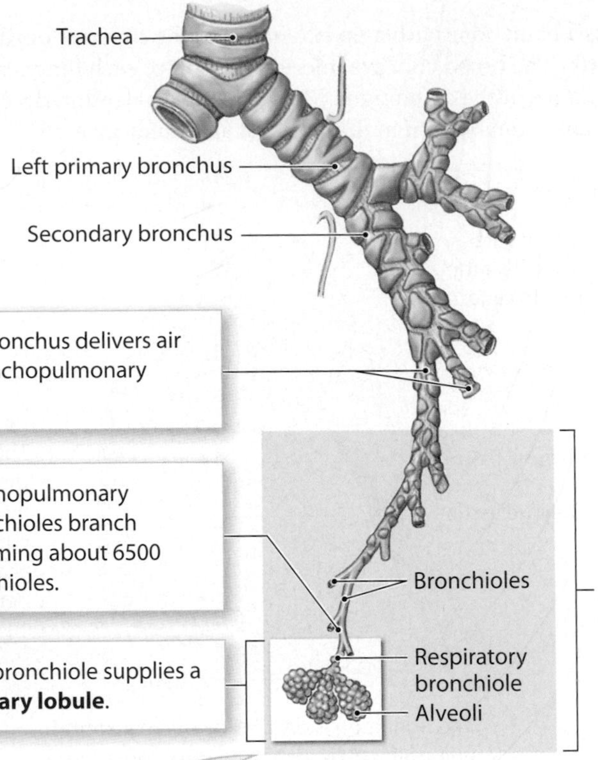

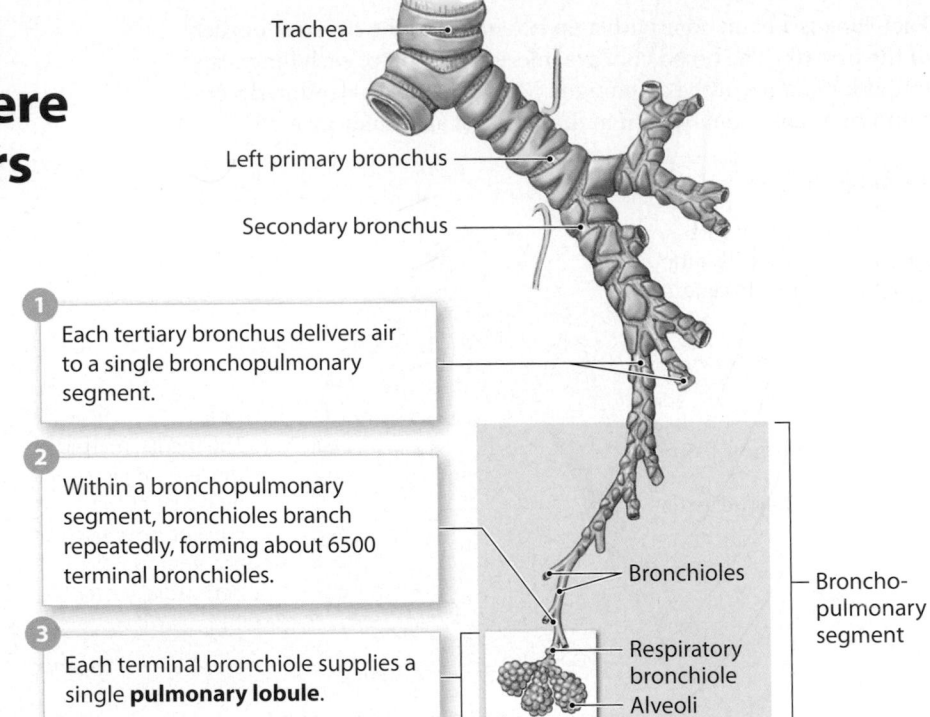

Trachea
Left primary bronchus
Secondary bronchus
Bronchioles
Respiratory bronchiole
Alveoli
Broncho-pulmonary segment

1. Each tertiary bronchus delivers air to a single bronchopulmonary segment.

2. Within a bronchopulmonary segment, bronchioles branch repeatedly, forming about 6500 terminal bronchioles.

3. Each terminal bronchiole supplies a single **pulmonary lobule**.

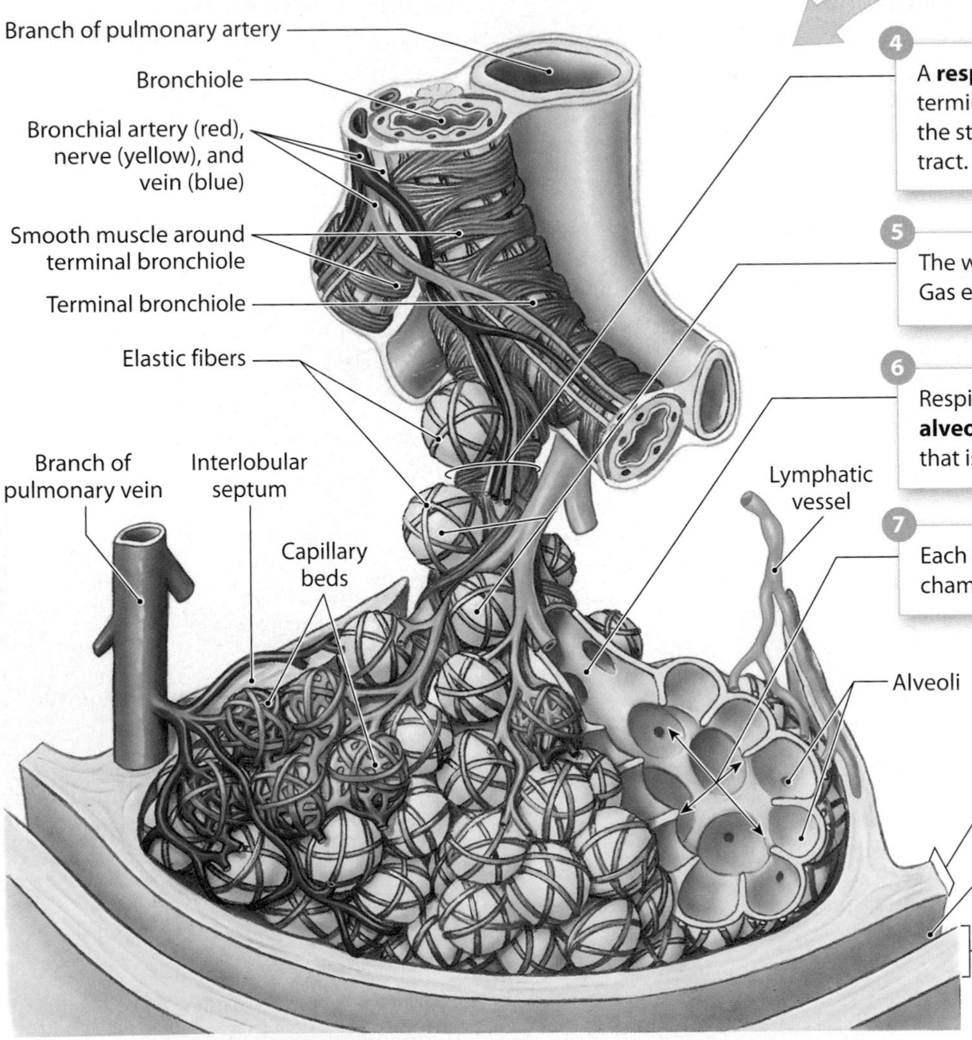

Branch of pulmonary artery
Bronchiole
Bronchial artery (red), nerve (yellow), and vein (blue)
Smooth muscle around terminal bronchiole
Terminal bronchiole
Elastic fibers
Branch of pulmonary vein
Interlobular septum
Capillary beds
Lymphatic vessel
Alveoli

4. A **respiratory bronchiole** forms by the branching of a terminal bronchiole inside a pulmonary lobule. This is the start of the respiratory portion of the respiratory tract.

5. The walls of respiratory bronchioles open into alveoli. Gas exchange occurs across the walls of the alveoli.

6. Respiratory bronchioles open into regions called **alveolar ducts**. An alveolar duct is a common passage that is connected to multiple individual alveoli.

7. Each alveolar duct ends at an **alveolar sac**, a common chamber connected to several individual alveoli.

The Pleura

The **visceral pleura** covers the outer surfaces of the lungs.

The pleural cavity contains a small volume of **pleural fluid** that coats the pleural surfaces and reduces friction.

The **parietal pleura** covers the inner surface of the thoracic wall and extends over the diaphragm and mediastinum.

2 The alveolar epithelium is primarily a simple squamous epithelium. The surrounding capillaries differ functionally from other capillaries; they dilate when alveolar oxygen levels are high, and constrict when alveolar oxygen levels are low. This response directs blood flow to the alveoli containing the most oxygen.

Type II pneumocytes are scattered among the squamous cells. These cells produce **surfactant** (sur-FAK-tant), an oily secretion containing a mixture of phospholipids and proteins that forms a superficial coating over a thin layer of water.

The squamous epithelial cells, called **type I pneumocytes**, are unusually thin and delicate and are the sites of gas diffusion.

Roaming **alveolar macrophages** (or dust cells) patrol the epithelial surface, phagocytizing any particles that have eluded other respiratory defenses and reached the alveoli.

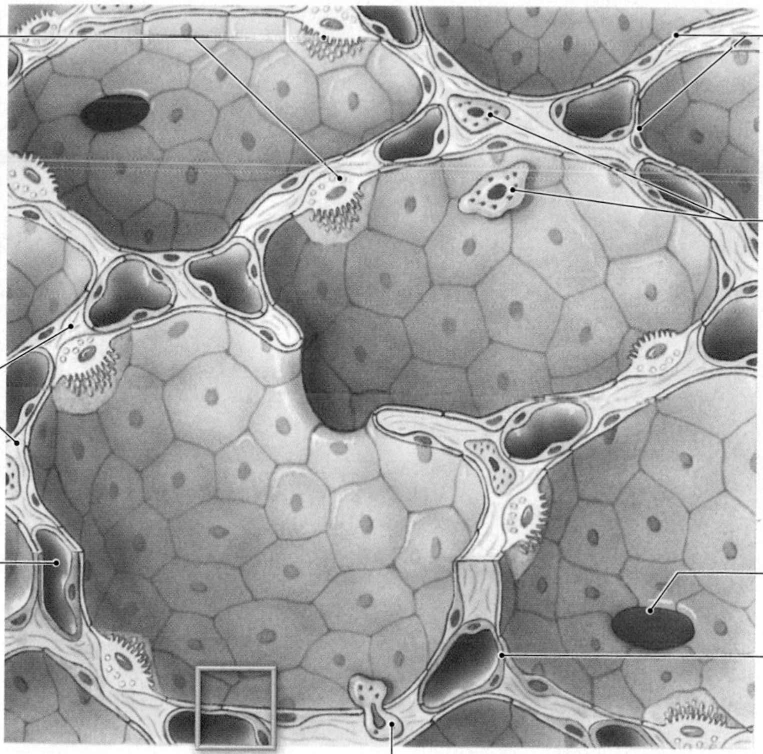
Elastic fibers
Capillary
Interconnection between adjacent alveoli
Endothelial cell of capillary
Alveolar macrophage

3 Gas exchange occurs across the **respiratory membrane** at each alveolus. Diffusion across the respiratory membrane proceeds very rapidly, because the distance is short and both oxygen and carbon dioxide are lipid soluble.

The Respiratory Membrane

The total distance separating alveolar air from blood can be as little as 0.1 μm; it averages about 0.5 μm.

Alveolar epithelium
Fused basement membranes of alveolar epithelium and capillary endothelium
Capillary endothelium

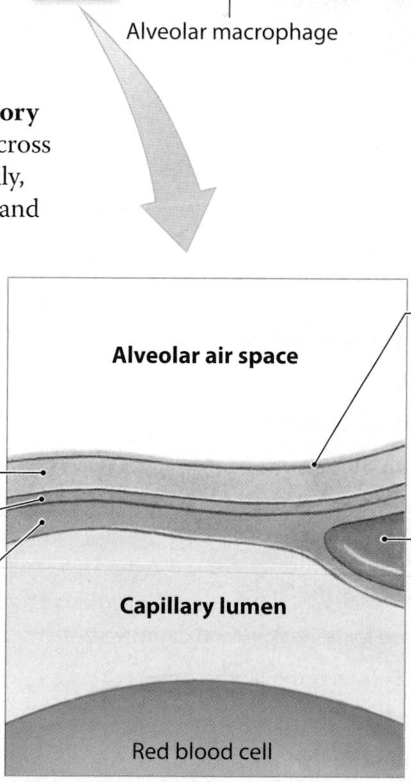
Alveolar air space
Capillary lumen
Red blood cell
Nucleus of endothelial cell

Surfactant plays a key role in keeping the alveoli open. It reduces the surface tension in the liquid coating the alveolar surface. Alveolar walls, like soap bubbles, are very delicate; without surfactant, the surface tension would be so high that the alveoli would collapse.

Module 21.7 Review

a. Define pulmonary lobule.
b. What would happen to the alveoli if surfactant were not produced?
c. Describe the structure and function of the respiratory membrane.

LO 21.7 Describe the structure of a pulmonary lobule and the functional anatomy of the alveoli.

Labeling

Label each of the respiratory system structures in the figure below.

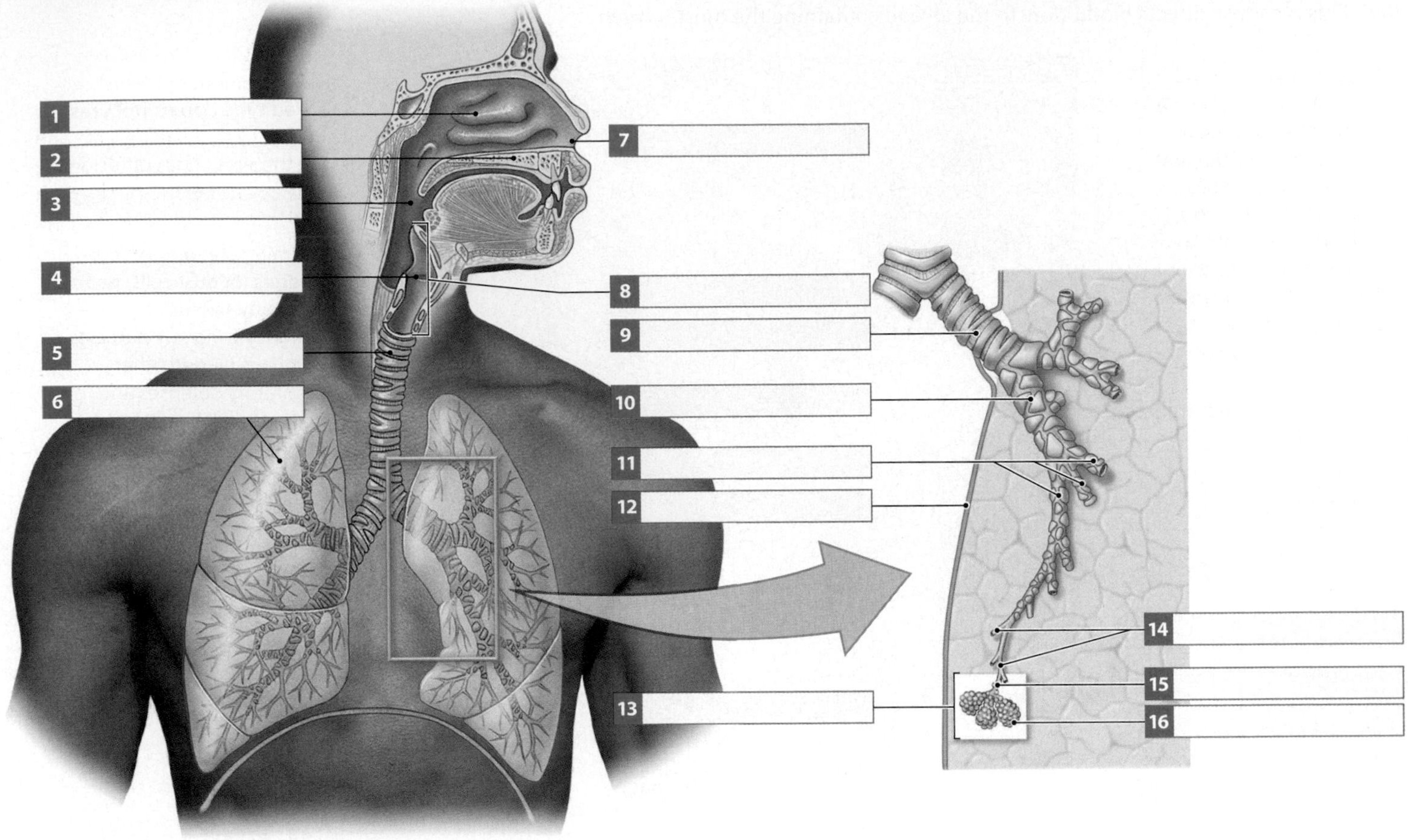
1
2
3
4
5
6
7
8
9
10
11
12
13
14
15
16

Matching

Match each lettered term with the most closely related description.

a. respiratory bronchiole
b. respiratory mucosa
c. phonation
d. bronchodilation
e. terminal bronchiole
f. laryngeal prominence
g. type I pneumocytes
h. type II pneumocytes
i. cystic fibrosis
j. trachea
k. pharynx
l. respiratory membrane
m. larynx
n. bronchoconstriction

17 Produce surfactant
18 Windpipe
19 Simple squamous epithelial cells
20 Sympathetic activation
21 Supplies a pulmonary lobule
22 Parasympathetic activation
23 Start of respiratory portion of respiratory tract
24 Gas exchange
25 Sound production at the larynx
26 Chamber shared by respiratory and digestive systems
27 Surrounds and protects the glottis
28 Lethal inherited respiratory disease
29 Lines the conducting portion of respiratory tract
30 Anterior surface of thyroid cartilage

17 ____________
18 ____________
19 ____________
20 ____________
21 ____________
22 ____________
23 ____________
24 ____________
25 ____________
26 ____________
27 ____________
28 ____________
29 ____________
30 ____________

Respiratory physiology involves external and internal respiration

The general term **respiration** refers to two integrated processes: external respiration and internal respiration.

1 This illustration provides an overview of respiration and shows relationships among external respiration, gas diffusion and transport, and internal respiration.

Respiration

External Respiration

External respiration includes all the processes involved in the exchange of oxygen and carbon dioxide between blood, lungs, and the external environment. The purpose of external respiration, and the primary function of the respiratory system, is meeting the respiratory demands of the body's cells.

Gas diffusion occurs across the respiratory membrane between alveoli and capillaries, and across capillary walls between blood and other tissues.

Internal Respiration

Internal respiration is the absorption of O_2 from blood and the release of CO_2 by tissue cells. We will consider the biochemical pathways responsible for O_2 consumption and CO_2 generation in Chapter 23.

Pulmonary ventilation, or breathing, involves the physical movement of air into and out of the lungs. The primary function of pulmonary ventilation is to maintain adequate **alveolar ventilation**—the movement of air into and out of the alveoli. Alveolar ventilation prevents the buildup of carbon dioxide in the alveoli and ensures a continuous supply of oxygen that keeps pace with its absorption by the bloodstream.

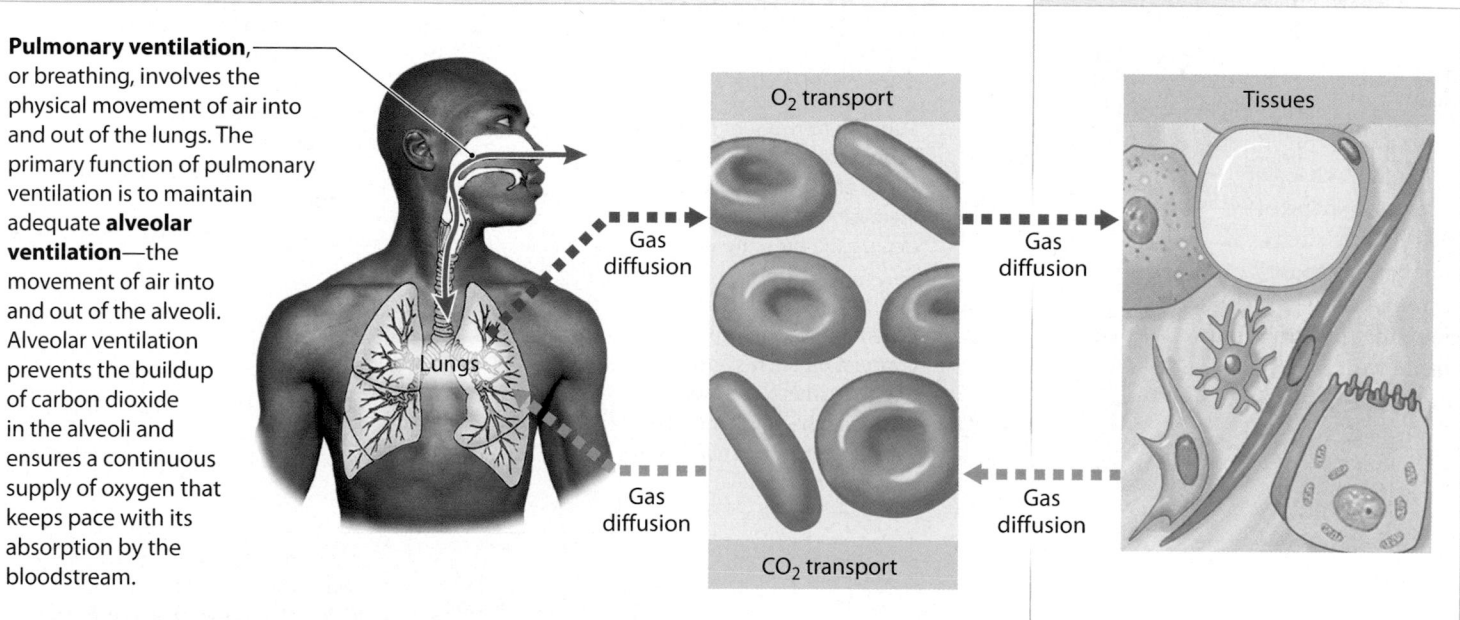

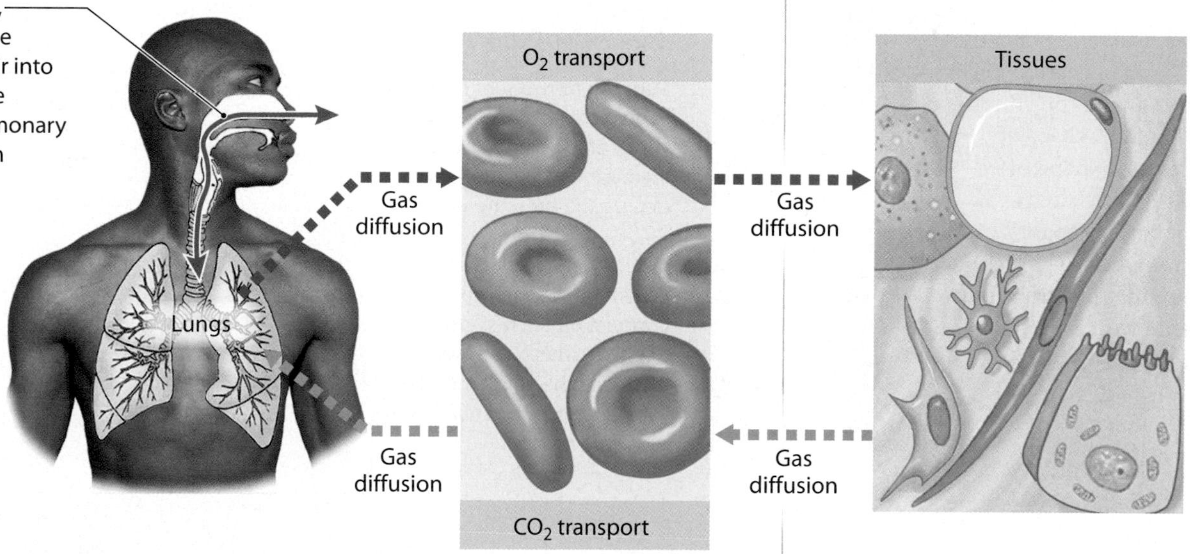
Lungs
Gas diffusion
Gas diffusion
O_2 transport
CO_2 transport
Gas diffusion
Gas diffusion
Tissues

In this section, we will examine the integrated steps involved in external respiration. Abnormalities affecting any of the steps involved in external respiration will ultimately affect the gas concentrations of interstitial fluids, and thus cellular activities as well. If the oxygen content declines, the affected tissues will become starved for oxygen. **Hypoxia**, or low tissue oxygen levels, places severe limits on the metabolic activities of the affected area. If the supply of oxygen is cut off completely, the condition called **anoxia** (an-OK-sē-a; *a-*, without + *ox-*, oxygen) results. Much of the damage caused by strokes and heart attacks is the result of localized anoxia.

Module 21.8 Review

a. Distinguish between external respiration and internal respiration.

b. What is the primary function of pulmonary ventilation?

c. How are hypoxia and anoxia different?

LO 21.8 Describe external respiration and internal respiration.

Pulmonary ventilation is driven by pressure changes within the pleural cavities

In a gas, such as air, the molecules bounce around as independent objects. At normal atmospheric pressures, gas molecules are much farther apart than the molecules in a liquid, so the density of air is relatively low. The pressure exerted by the enclosed gas results from the collision of gas molecules with the walls of the container. The greater the number of collisions, the higher the pressure.

1 These diagrams should help you understand the relationships between pressure and volume in a gas. For a gas in a closed container and at a constant temperature, pressure (P) is inversely proportional to volume (V). That is, if you decrease the volume of a gas, its pressure will increase; if you increase the volume of a gas, its pressure will decrease. The relationship between pressure and volume is reciprocal: If you reduce the volume of a flexible container by half, the pressure within it will double; if you double the volume of the container, the pressure inside it will decline by half. This inverse proportional relationship, first recognized by Robert Boyle in the 1600s, is called **Boyle's law**.

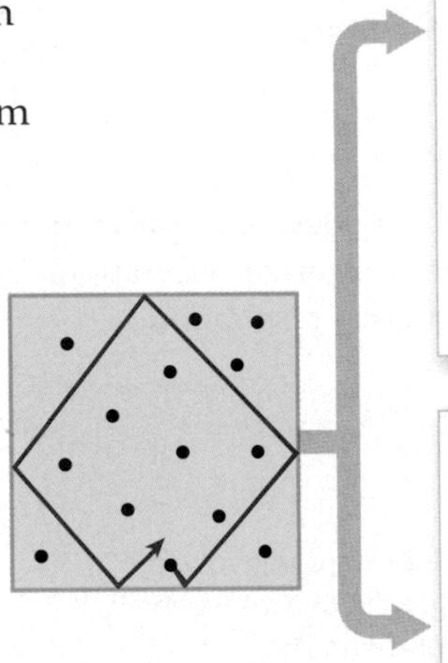

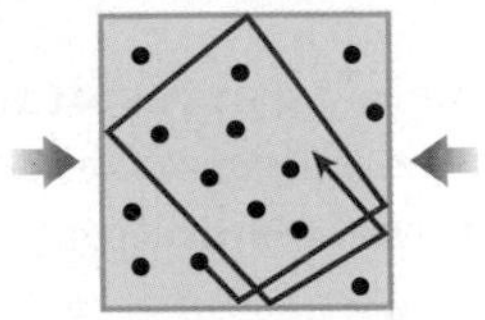

If you decrease the volume of the container, collisions occur more frequently per unit time, increasing the pressure of the gas.

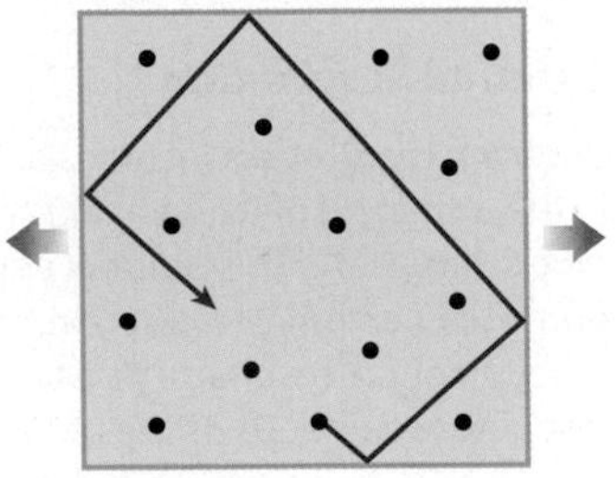

If you increase the volume, fewer collisions occur per unit time, because it takes longer for a gas molecule to travel from one wall to another. As a result, the gas pressure inside the container decreases.

2 Movements of the diaphragm and rib cage change the volume of the thoracic cavity. When the shape of the thoracic cavity changes, it expands or compresses the lungs, changing the air pressure within the respiratory tract.

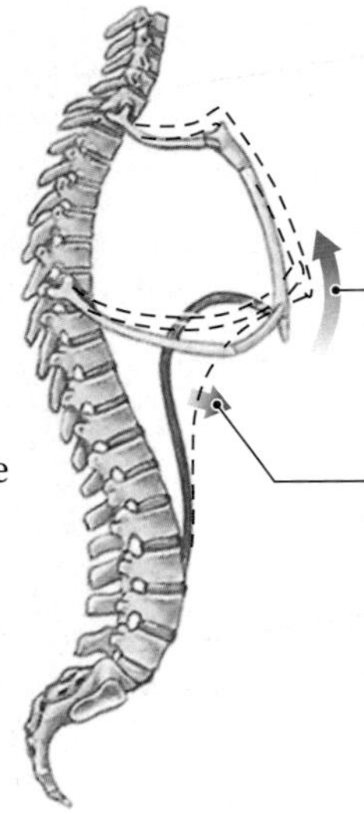

Superior movement of the rib cage increases the depth and width of the thoracic cavity, increasing its volume and decreasing pressure within it.

When the diaphragm contracts, it tenses and moves inferiorly. This movement increases the volume of the thoracic cavity, decreasing the pressure within it.

3 At the start of a breath, pressures inside and outside the thoracic cavity are identical, and no air moves into or out of the lungs.

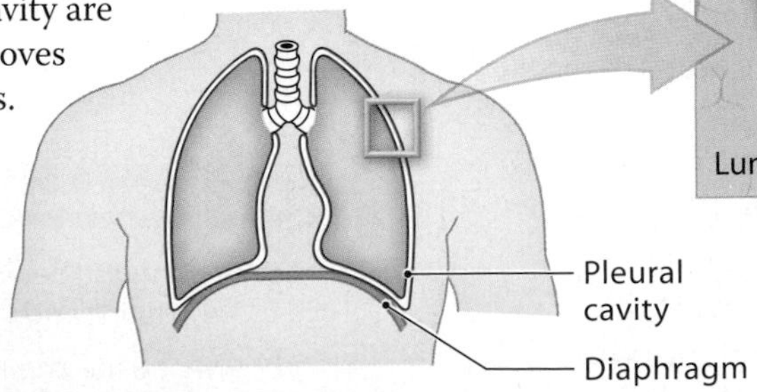
Pleural cavity
Diaphragm

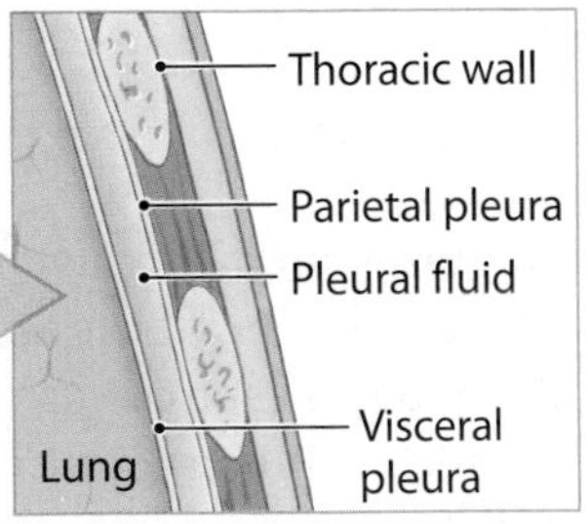
Thoracic wall
Parietal pleura
Pleural fluid
Visceral pleura
Lung

$P_{outside} = P_{inside}$

Pressure outside and inside are equal, so no air movement occurs

Although separated by the pleural cavity, the layer of pleural fluid makes the lungs stick to the inner walls of the thorax. This kind of fluid bond is responsible for making a coaster stick to the bottom of a wet glass. The elastic tissues of the lungs are always trying to recoil and reduce lung volume to about 5 percent of its normal size. This collapse is prevented by the fluid bond between the parietal and visceral pleura. If an injury allows air into the pleural cavity, this bond is broken and the lung collapses. This condition is called **atelectasis** (a-te-LEK-ta-sis).

4 Air will flow from an area of higher pressure to an area of lower pressure. When the thoracic cavity enlarges and increases in volume during inhalation, pressure falls inside the lungs and air flows in. When the thoracic cavity decreases in volume during exhalation, pressure rises inside the lungs, forcing air out of the respiratory tract.

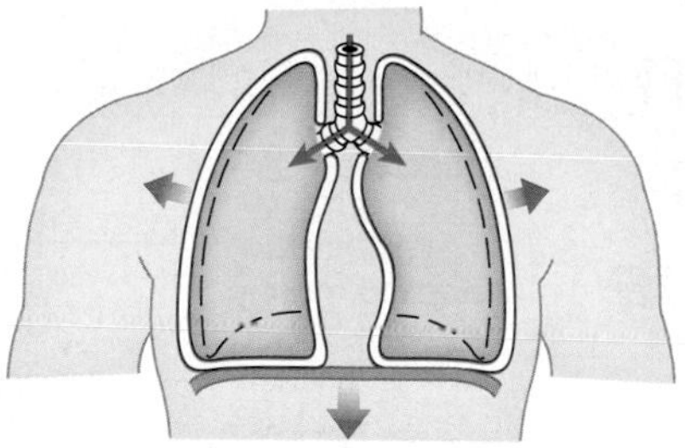

Inhalation: volume increases
$P_{outside} > P_{inside}$
Pressure inside falls, so air flows in

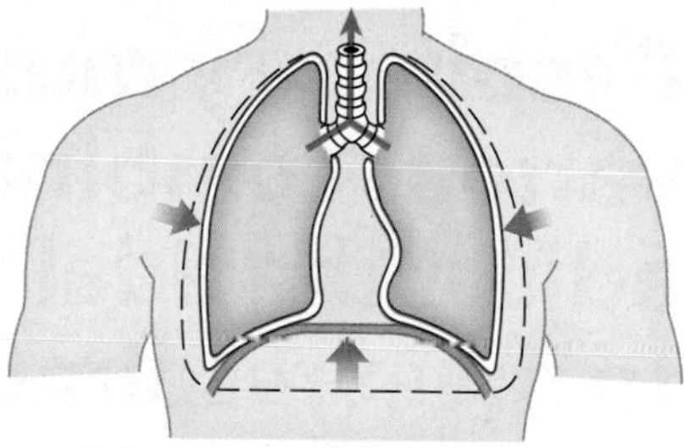

Exhalation: volume decreases
$P_{outside} < P_{inside}$
Pressure inside rises, so air flows out

5 The direction of airflow is determined by the difference between atmospheric pressure and **intrapulmonary** (in-tra-PUL-mo-nār-ē) **pressure**, the pressure inside the respiratory tract (usually measured at the alveoli).

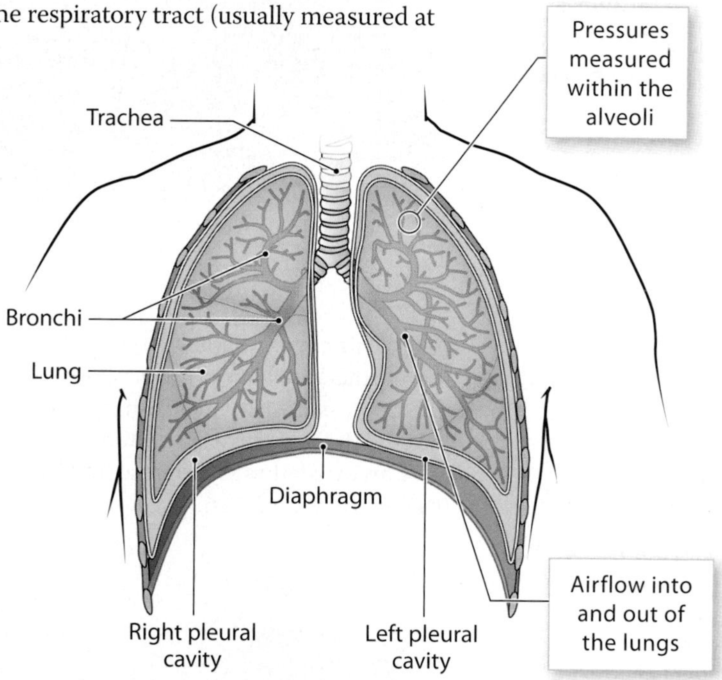

Pressures measured within the alveoli
Trachea
Bronchi
Lung
Diaphragm
Right pleural cavity
Left pleural cavity
Airflow into and out of the lungs

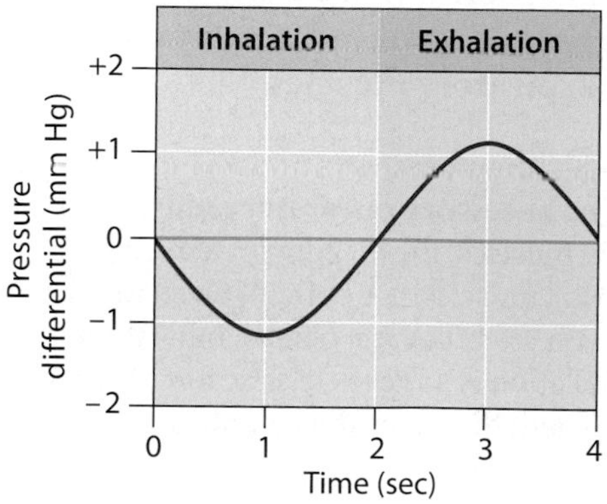

Inhalation
Exhalation
+2
+1
0
−1
−2
Pressure differential (mm Hg)
0
1
2
3
4
Time (sec)

A pressure differential of 0 mm Hg exists when atmospheric and intrapulmonary pressures are equal. Positive intrapulmonary pressures will push air out of the lungs; negative intrapulmonary pressures will pull air into the lungs.

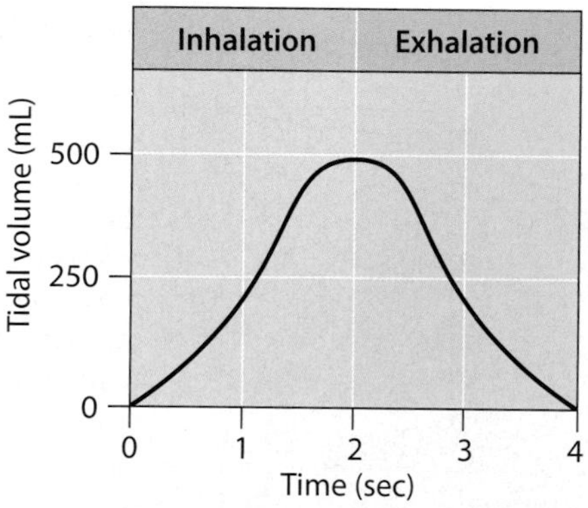

Inhalation
Exhalation
500
250
0
Tidal volume (mL)
0
1
2
3
4
Time (sec)

The **tidal volume (V_T)** is the amount of air moved into the lungs during inhalation and out of the lungs during exhalation. At rest, the tidal volume is approximately 500 mL.

6 Although we will use mm Hg to report pressures, other units are also used in clinical practice. The table below compares several units used to report gas pressures.

Four Common Units for Reporting Gas Pressures

- **Millimeters of mercury (mm Hg)**: This is the most common unit for reporting blood pressure and gas pressures. Normal atmospheric pressure is approximately 760 mm Hg.
- **Torr**: This unit of measurement is preferred by many respiratory therapists; it is also commonly used in Europe and in some technical journals. One torr is equivalent to 1 mm Hg; in other words, normal atmospheric pressure is equal to 760 torr.
- **Centimeters of water (cm H_2O)**: In a hospital setting, anesthetic gas pressures and oxygen pressures are commonly measured in centimeters of water. One cm H_2O is equivalent to 0.735 mm Hg; normal atmospheric pressure is 1033.6 cm H_2O.
- **Pounds per square inch (psi)**: Pressures in compressed gas cylinders and other industrial applications are generally reported in psi. Normal atmospheric pressure at sea level is approximately 15 psi.

Module 21.9 Review

a. Define Boyle's law.

b. What physical changes affect the volume of the lungs?

c. What pressures determine the direction of airflow within the respiratory tract?

LO **21.9** Summarize the physical principles governing the movement of air into and out of the lungs.

Module 21.10

Respiratory muscles in various combinations adjust the tidal volume to meet respiratory demands

Respiratory muscles may be involved with either inhalation or exhalation. Those involved with inhalation are called **inspiratory muscles**; those involved in exhalation are called **expiratory muscles**.

1 This anterior view introduces the **primary** and **accessory respiratory muscles**. The primary muscles, the diaphragm and external intercostal muscles, are both involved in inhalation. When you are breathing quietly, inhalation is active but exhalation is passive—elastic forces and gravity are sufficient to reduce the volume of the lungs.

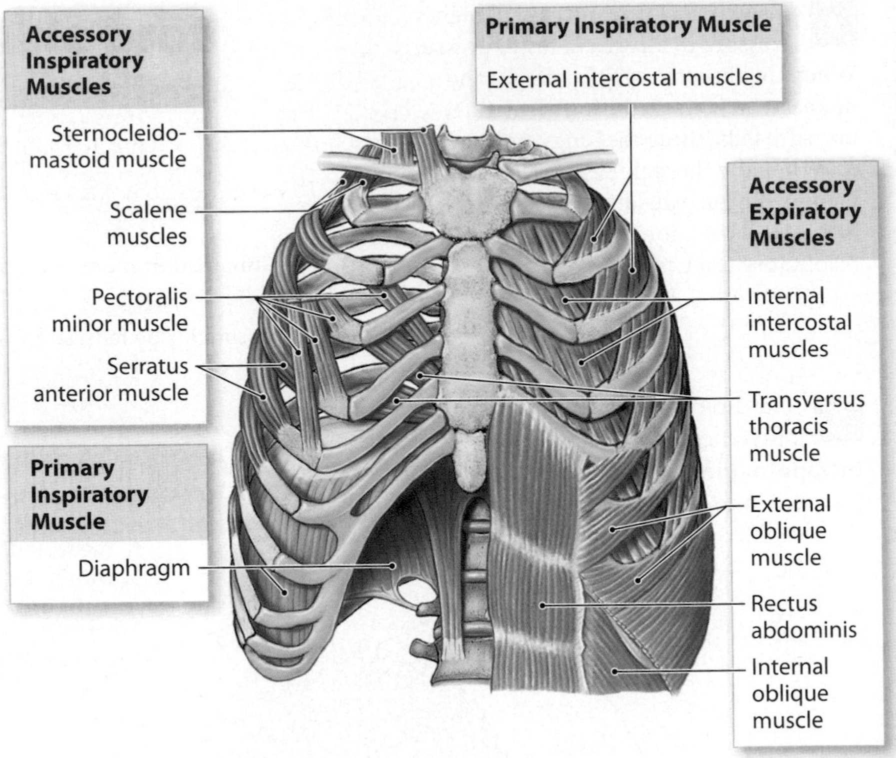

Accessory Inspiratory Muscles
Sternocleido-mastoid muscle
Scalene muscles
Pectoralis minor muscle
Serratus anterior muscle
Primary Inspiratory Muscle
Diaphragm
Primary Inspiratory Muscle
External intercostal muscles
Accessory Expiratory Muscles
Internal intercostal muscles
Transversus thoracis muscle
External oblique muscle
Rectus abdominis
Internal oblique muscle

2 This lateral view during inhalation shows the inspiratory muscles that elevate the ribs and depress the diaphragm to enlarge the thoracic cavity.

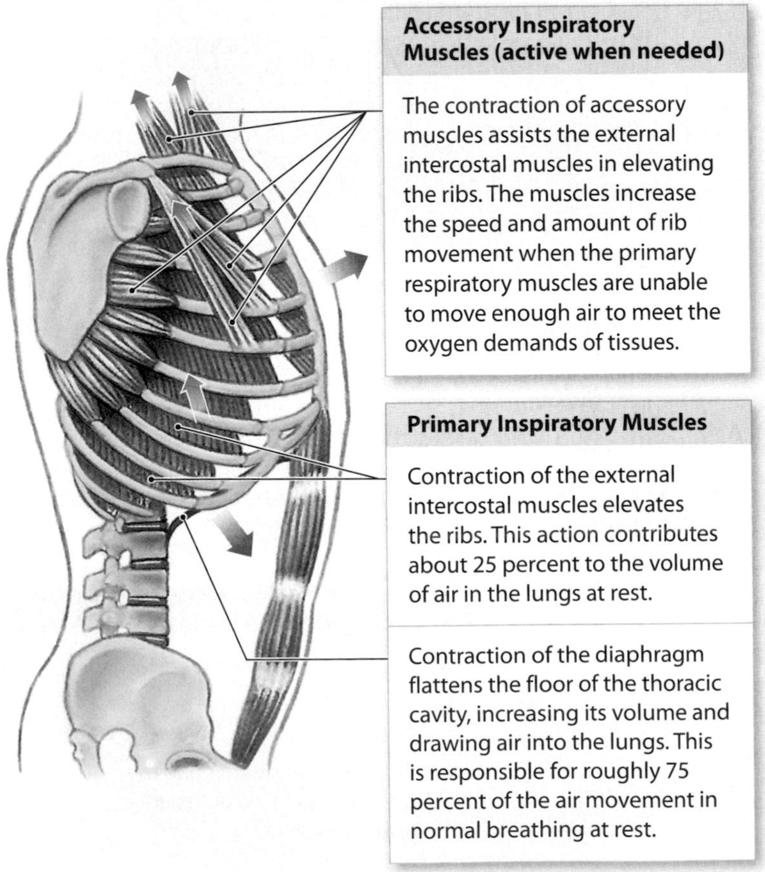

Accessory Inspiratory Muscles (active when needed)
The contraction of accessory muscles assists the external intercostal muscles in elevating the ribs. The muscles increase the speed and amount of rib movement when the primary respiratory muscles are unable to move enough air to meet the oxygen demands of tissues.
Primary Inspiratory Muscles
Contraction of the external intercostal muscles elevates the ribs. This action contributes about 25 percent to the volume of air in the lungs at rest.
Contraction of the diaphragm flattens the floor of the thoracic cavity, increasing its volume and drawing air into the lungs. This is responsible for roughly 75 percent of the air movement in normal breathing at rest.

3 This corresponding lateral view during active exhalation shows the accessory expiratory muscles that depress the ribs and push the relaxed diaphragm into the thoracic cavity. The abdominal muscles that assist in exhalation are represented by a single muscle (the rectus abdominis).

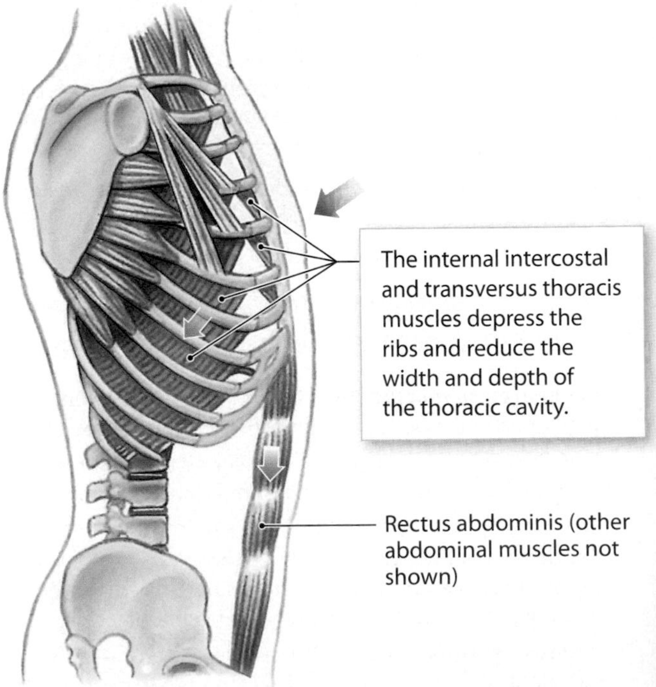

The internal intercostal and transversus thoracis muscles depress the ribs and reduce the width and depth of the thoracic cavity.
Rectus abdominis (other abdominal muscles not shown)

4 Only a small proportion of the air in the lungs is exchanged during a single quiet respiratory cycle (consisting of an inhalation and an exhalation); the tidal volume can be increased by inhaling more vigorously and exhaling more completely. We can divide the total volume of the lungs into a series of **volumes** and **capacities** (each the sum of various volumes), as indicated in this spirogram. The red line indicates the volume of air within the lungs during breathing.

Pulmonary Volumes and Capacities (adult male)

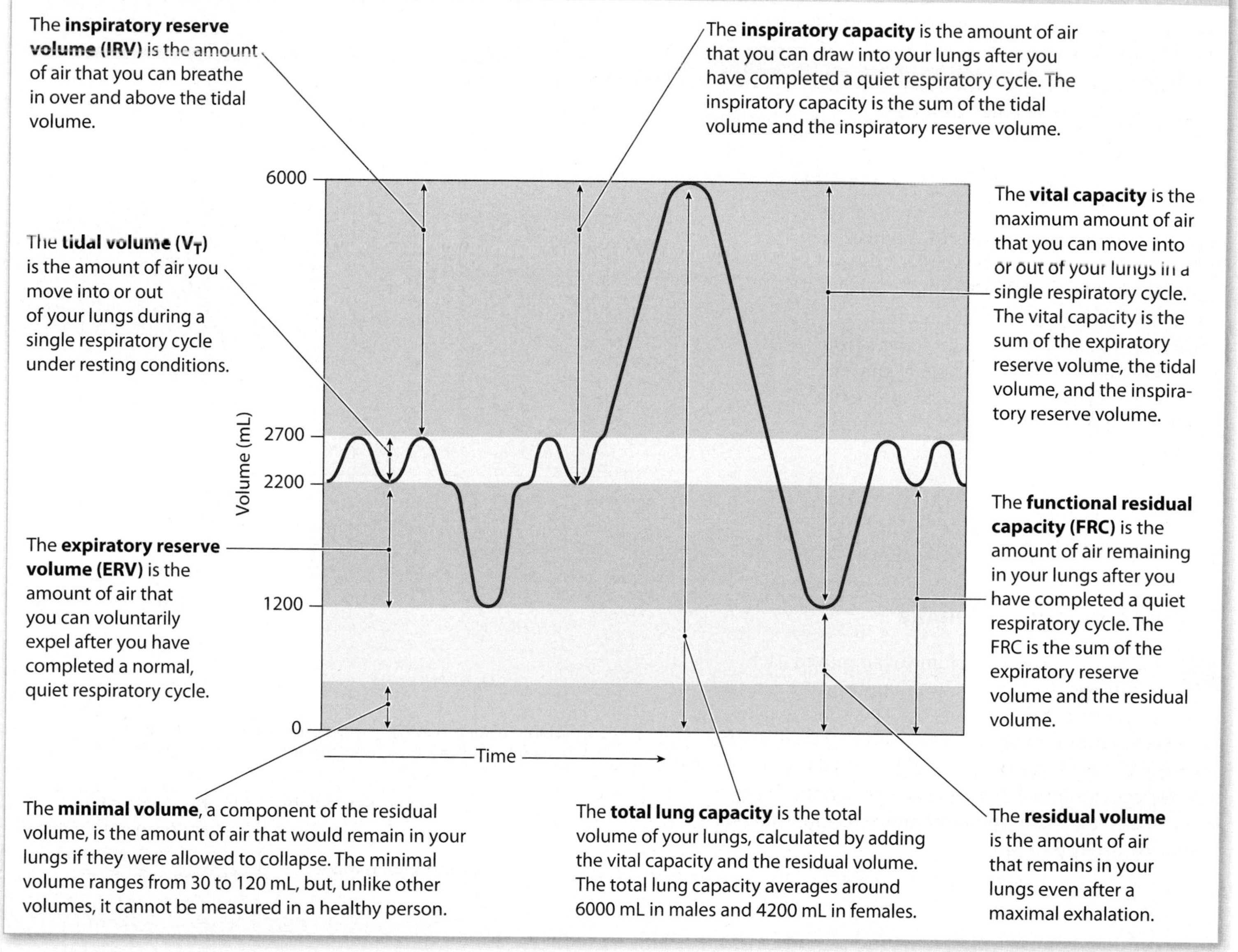

The **inspiratory reserve volume (IRV)** is the amount of air that you can breathe in over and above the tidal volume.
The **inspiratory capacity** is the amount of air that you can draw into your lungs after you have completed a quiet respiratory cycle. The inspiratory capacity is the sum of the tidal volume and the inspiratory reserve volume.
6000
The **tidal volume (V_T)** is the amount of air you move into or out of your lungs during a single respiratory cycle under resting conditions.
The **vital capacity** is the maximum amount of air that you can move into or out of your lungs in a single respiratory cycle. The vital capacity is the sum of the expiratory reserve volume, the tidal volume, and the inspiratory reserve volume.
Volume (mL)
2700
2200
The **expiratory reserve volume (ERV)** is the amount of air that you can voluntarily expel after you have completed a normal, quiet respiratory cycle.
1200
0
The **functional residual capacity (FRC)** is the amount of air remaining in your lungs after you have completed a quiet respiratory cycle. The FRC is the sum of the expiratory reserve volume and the residual volume.
Time
The **minimal volume**, a component of the residual volume, is the amount of air that would remain in your lungs if they were allowed to collapse. The minimal volume ranges from 30 to 120 mL, but, unlike other volumes, it cannot be measured in a healthy person.
The **total lung capacity** is the total volume of your lungs, calculated by adding the vital capacity and the residual volume. The total lung capacity averages around 6000 mL in males and 4200 mL in females.
The **residual volume** is the amount of air that remains in your lungs even after a maximal exhalation.

Pulmonary Volumes

		Males	Females	
Vital capacity	IRV	3300 mL	1900 mL	Inspiratory capacity
	V_T	500 mL	500 mL	
	ERV	1000 mL	700 mL	Functional residual capacity
Residual volume		1200 mL	1100 mL	
Total lung capacity		6000 mL	4200 mL	

Module 21.10 Review

a. Identify the primary inspiratory muscles.

b. When do the accessory respiratory muscles become active?

c. Name the various measurable pulmonary volumes.

LO 21.10 Name the respiratory muscles, and describe the actions of the muscles responsible for respiratory movements.

Pulmonary ventilation must be closely regulated to meet tissue oxygen demands

The respiratory system adjusts pulmonary ventilation over a broad range to meet the oxygen demands of the body. These adjustments involve varying both the number of breaths per minute and the amount of air moved each breath. When you are exercising at peak levels, the amount of air moving into and out of the respiratory tract can be 50 times the amount moved at rest. The factors involved in pulmonary ventilation should remind you of the factors involved in regulating cardiovascular function.

Respiratory Rate

1 Your **respiratory rate** is the number of breaths you take each minute. As you read this, you are probably breathing quietly, with a low respiratory rate. The normal respiratory rate of a resting adult ranges from 12–18 breaths each minute, roughly one for every four heartbeats. Children breathe more rapidly, at rates of about 18–20 breaths per minute.

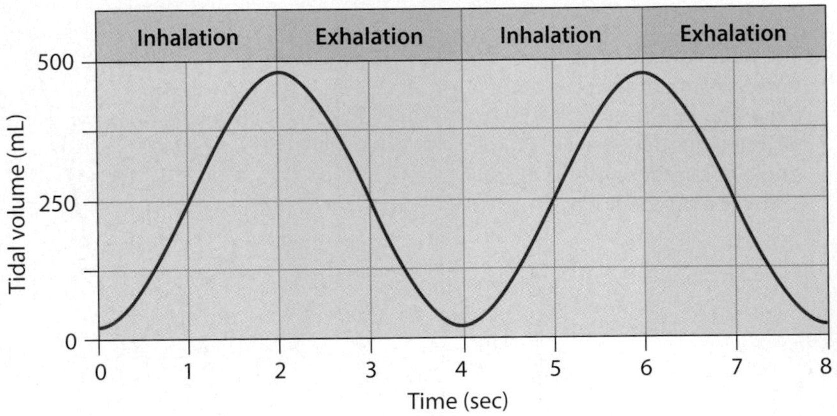
Inhalation
Exhalation
Inhalation
Exhalation
500
250
0
Tidal volume (mL)
0
1
2
3
4
5
6
7
8
Time (sec)

Respiratory Minute Volume

2 We can calculate the volume of air moved each minute, symbolized $\mathbf{V_E}$, by multiplying the respiratory rate, f, by the tidal volume, V_T. This value is called the **respiratory minute volume**. The respiratory rate at rest averages 12 breaths per minute, and the tidal volume at rest averages around 500 mL per breath. On that basis, we can calculate that respiratory minute volume at rest is approximately 6 L/min.

$$V_E = f \times V_T$$

$$\left(\text{Volume of air moved each minute}\right) = \left(\text{Breaths per minute}\right) \times \left(\text{Tidal volume}\right)$$

= 12 × 500 mL per minute
= 6000 mL per minute
= 6.0 liters per minute

In other words, the respiratory minute volume at rest is approximately 6 liters per minute.

3 The factors involved in respiratory minute volume are easily diagrammed, but their functional relationships are complex. Increasing either respiratory rate or tidal volume will increase the respiratory minute volume. But what if the respiratory rate goes up, and the tidal volume goes down? The respiratory minute volume may remain the same, but the effects on respiratory performance are very different. We will now consider the differences and their significance.

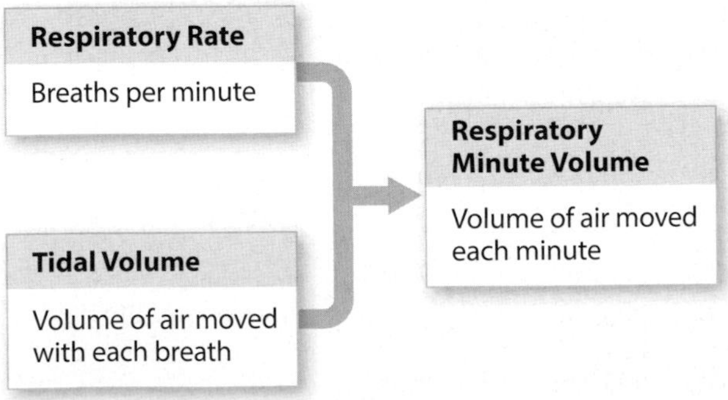
Respiratory Rate
Breaths per minute
Tidal Volume
Volume of air moved with each breath
Respiratory Minute Volume
Volume of air moved each minute

Alveolar Ventilation

4 **Alveolar ventilation**, symbolized V_A, is the amount of air reaching the alveoli each minute. The alveolar ventilation is less than the respiratory minute volume, because some of the air never reaches the alveoli, but remains in the conducting portion of the respiratory system. This is known as the **anatomic dead space (V_D)**, and at rest it amounts to nearly 150 mL of the 500 mL of tidal air.

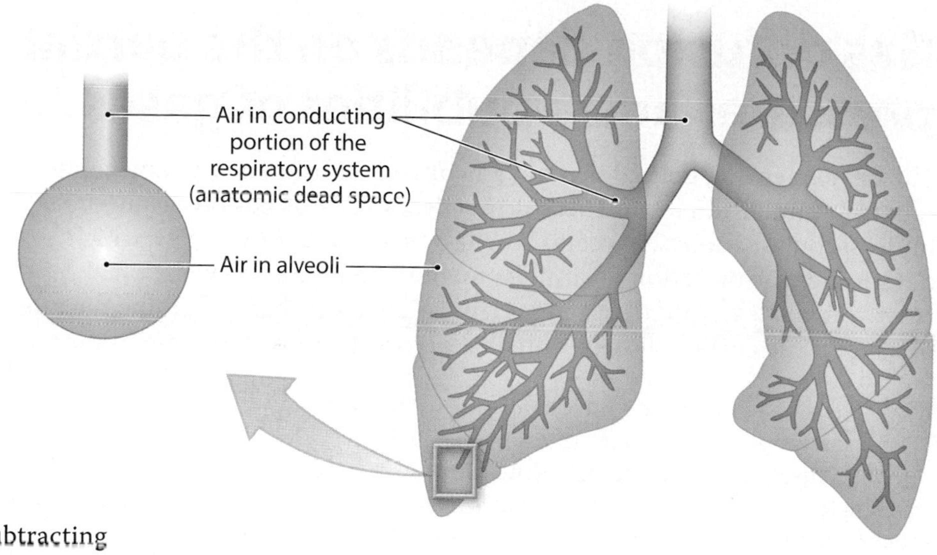

Air in conducting portion of the respiratory system (anatomic dead space)
Air in alveoli

5 We can calculate alveolar ventilation by subtracting the dead space from the tidal volume:

$$V_A = f \times V_T - V_D$$

Alveolar ventilation = Breaths per minute × (Tidal volume − Anatomic dead space)

$$V_A = 12 \times (500\text{ mL} - 150\text{ mL})$$
$$= 12 \times 350\text{ mL}$$
$$= 4200\text{ mL}$$

However, the composition of the gas arriving in the alveoli is significantly different from that of the surrounding atmosphere, because inhaled air always mixes with "used" air in the conducting passageways (the anatomic dead space) on its way to the exchange surfaces. The alveolar air thus contains less oxygen and more carbon dioxide than atmospheric air.

6 Let's return to the effects of altering tidal volume and respiratory rate. If the respiratory rate jumps to 20 breaths per minute but the tidal volume drops to 300 mL, the respiratory minute volume will remain unchanged. However, the alveolar ventilation rate drops dramatically, falling from 4.2 L/min to 3 L/min.

$$V_E = f \times V_T$$
$$= 20 \times 300\text{ mL per minute}$$
$$= 6.0\text{ liters per minute}$$

$$V_A = f \times (V_T - V_D)$$
$$= 20 \times (300\text{ mL} - 150\text{ mL})\text{ per minute}$$
$$= 3.0\text{ liters per minute}$$

This alveolar ventilation rate is almost 30 percent below its original value of 4.2 L/min, and if tissue demands are elevated as well, widespread tissue hypoxia could result. Thus, whenever the demand for oxygen increases, both the tidal volume *and* the respiratory rate must be increased.

Module 21.11 Review

a. Define respiratory rate.

b. How does the respiratory minute volume differ from alveolar ventilation?

c. Which ventilates alveoli more effectively: slow, deep breaths or rapid, shallow breaths? Explain why.

LO **21.11** Explain how respiratory rate and tidal volume affect pulmonary and alveolar ventilation.

Module 21.12

Gas diffusion depends on the partial pressures and solubilities of gases

The principles that govern the movement and diffusion of gas molecules are relatively straightforward. These principles, known as **gas laws**, have been understood for about 250 years. You have read about Boyle's law, which determines the direction of air movement in pulmonary ventilation. Now you will learn about other gas laws and factors that determine the rate of oxygen and carbon dioxide diffusion across the respiratory membrane.

1 The air we breathe is a mixture of gases. Their total pressure at sea level, one atmosphere, equals 760 mm Hg. The **partial pressure (P)** is the pressure exerted by a single gas in a mixture of gases. The partial pressures for atmospheric gases are included in this table. All the partial pressures added together equal the total pressure exerted by the gas mixture; this is known as **Dalton's law**. As soon as air enters the respiratory tract, its characteristics begin to change. In passing through the nasal cavity, inhaled air becomes warmer, and the amount of water vapor increases. When it reaches the alveoli, the incoming air mixes with air remaining in the alveoli from the previous respiratory cycle. During the subsequent exhalation, the departing alveolar air mixes with air in the anatomic dead space, producing yet another mixture that differs from both atmospheric and alveolar air samples.

Atmospheric Pressure = 760 mm Hg

$$597(P_{N_2}) + 159(P_{O_2}) + 0.3(P_{CO_2}) + 3.7(P_{H_2O}) = 760 \text{ mm Hg}$$

Partial Pressures (mm Hg) and Normal Gas Concentrations (%) in Air

Source of Sample	Nitrogen (N_2)	Oxygen (O_2)	Carbon Dioxide (CO_2)	Water Vapor (H_2O)
Inhaled air (dry)	597 (78.6%)	159 (20.9%)	0.3 (0.04%)	3.7 (0.5%)
Alveolar air (saturated)	573 (75.4%)	100 (13.2%)	40 (5.2%)	47 (6.2%)
Exhaled air (saturated)	569 74.8%)	116 (15.3%)	28 (3.7%)	47 (6.2%)

2 At a given temperature, the amount of a particular gas in solution is directly proportional to the partial pressure of that gas. This principle is known as **Henry's law**.

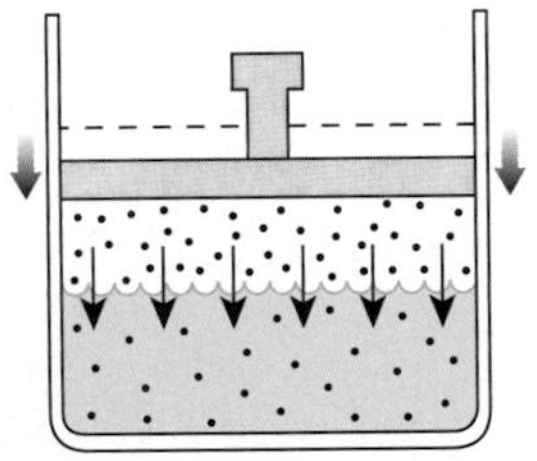

When a gas under pressure contacts a liquid, the pressure tends to force gas molecules into solution. At a given pressure, the number of dissolved gas molecules will increase until an equilibrium is established.

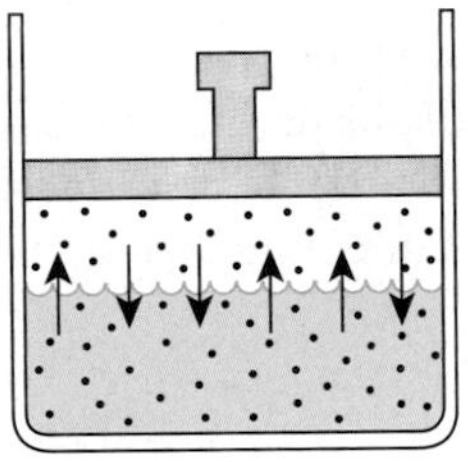

At equilibrium, gas molecules diffuse out of the liquid as quickly as they enter it, so the total number of gas molecules in solution remains constant.

Example:
Soda is put into the can under pressure, and the gas (carbon dioxide) is in solution at equilibrium.

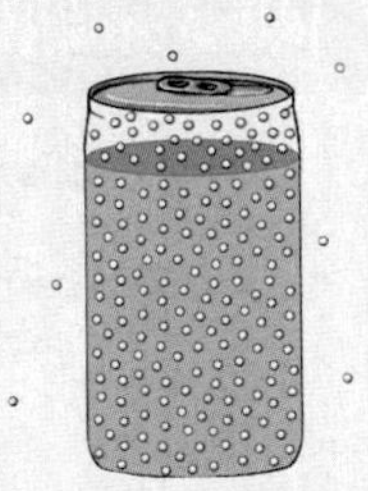

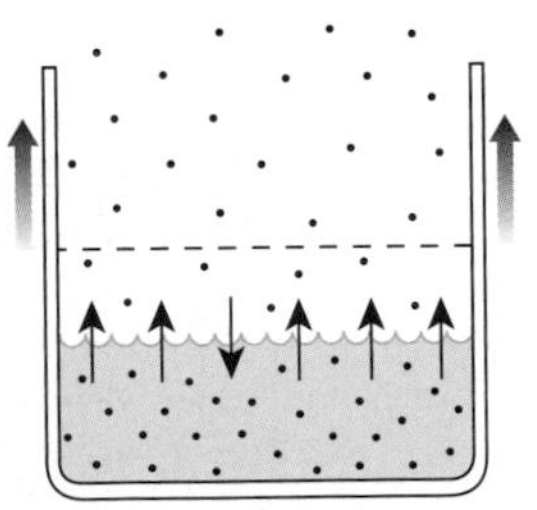

If the partial pressure is decreased, gas molecules will come out of solution.

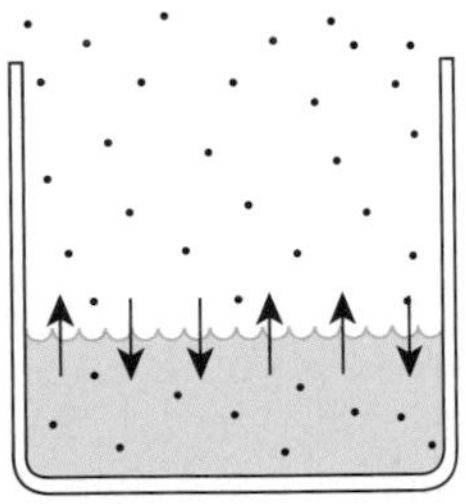

Eventually a new equilibrium will be established.

When you open a soda can, the internal pressure decreases and the gas molecules begin to come out of solution. The volume of the can is so small, and the volume of the atmosphere so great, that within a half hour or so virtually all the carbon dioxide comes out of solution, and you are left with "flat" soda.

3 This illustration shows the partial pressures of oxygen and carbon dioxide during external respiration in the pulmonary circuit and during internal respiration in the systemic circuit.

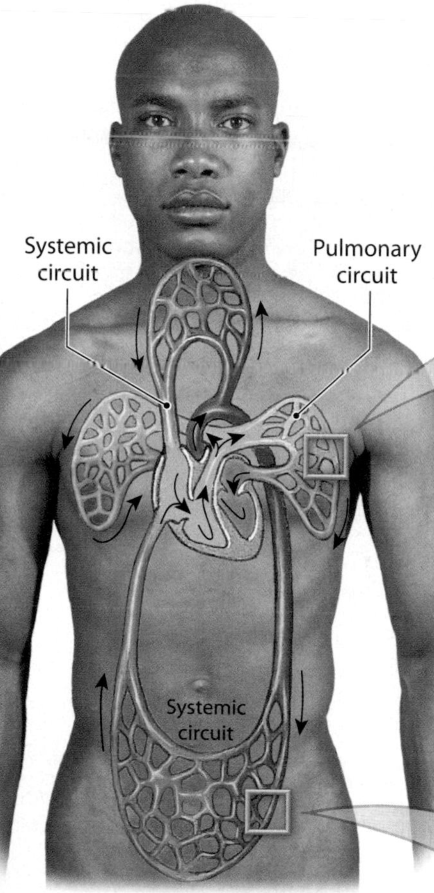

Systemic circuit
Pulmonary circuit
Systemic circuit

External Respiration

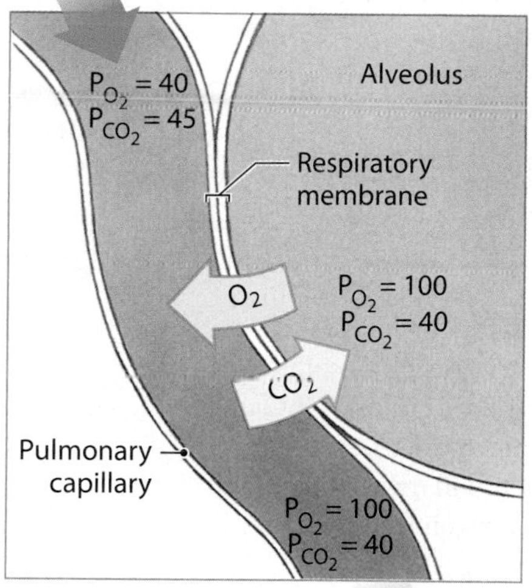

$P_{O_2} = 40$
$P_{CO_2} = 45$
Alveolus
Respiratory membrane
O_2
$P_{O_2} = 100$
$P_{CO_2} = 40$
CO_2
Pulmonary capillary
$P_{O_2} = 100$
$P_{CO_2} = 40$

Blood arriving in the pulmonary arteries has a lower P_{O_2} and a higher P_{CO_2} than does alveolar air. Diffusion between the alveolar mixture and the pulmonary capillaries thus increases blood P_{O_2} while decreasing its P_{CO_2}. By the time the blood enters the pulmonary venules, it has reached equilibrium with the alveolar air. Hence, blood departs the alveoli with a P_{O_2} of about 100 mm Hg and a P_{CO_2} of about 40 mm Hg. This is possible because each gas moves independently.

Internal Respiration

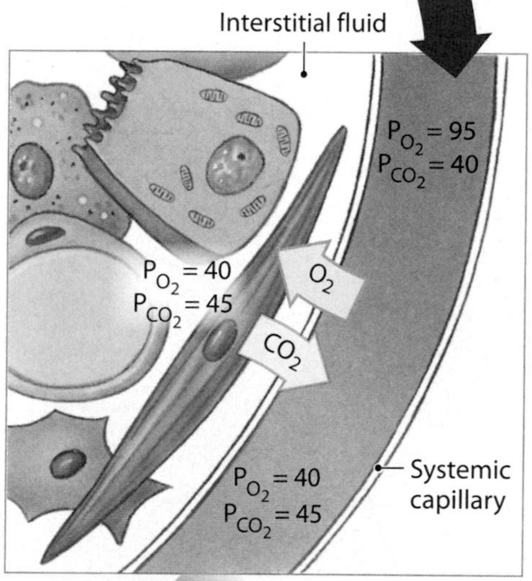

Interstitial fluid
$P_{O_2} = 95$
$P_{CO_2} = 40$
$P_{O_2} = 40$
$P_{CO_2} = 45$
O_2
CO_2
$P_{O_2} = 40$
$P_{CO_2} = 45$
Systemic capillary

The partial pressure of oxygen in the pulmonary veins decreases to about 95 mm Hg as it mixes with venous blood that was distributed to the conducting passageways rather than to the alveoli. This is the P_{O_2} in the blood that arrives at peripheral capillaries. Normal interstitial fluid has a P_{O_2} of 40 mm Hg. As a result, oxygen diffuses out of the capillaries until the capillary P_{O_2} is the same as that in the adjacent tissues. Inactive peripheral tissues normally have a P_{CO_2} of about 45 mm Hg, whereas blood entering peripheral capillaries normally has a P_{CO_2} of 40 mm Hg. As a result, carbon dioxide diffuses into the blood as oxygen diffuses into the tissue.

The system just described is at equilibrium, and the P_{O_2} and P_{CO_2} are stable in the alveoli and in the tissues. Every oxygen molecule entering peripheral tissues is balanced by an oxygen molecule absorbed at the alveoli, and the absorbed oxygen molecule will be replaced in the next respiratory cycle. But if tissue oxygen demand accelerates, that equilibrium is disturbed, and the respiratory rate and tidal volume must then increase.

Module 21.12 Review

a. Define Dalton's law.

b. What is the significance of Henry's law to the process of respiration?

c. Explain the decrease in P_{O_2} from the pulmonary venules to the blood arriving in the peripheral capillaries of the systemic circuit.

LO **21.12** Summarize the physical principles governing the diffusion of gases into and out of the blood.

Almost all the oxygen in the blood is transported bound to hemoglobin within red blood cells

Each 100 mL of blood leaving the alveolar capillaries carries away roughly 20 mL of oxygen. Of this amount, only about 0.3 mL (1.5 percent) consists of oxygen molecules in solution. The rest of the oxygen molecules are bound to hemoglobin (Hb) molecules—specifically, to the iron ions in the center of **heme units.**

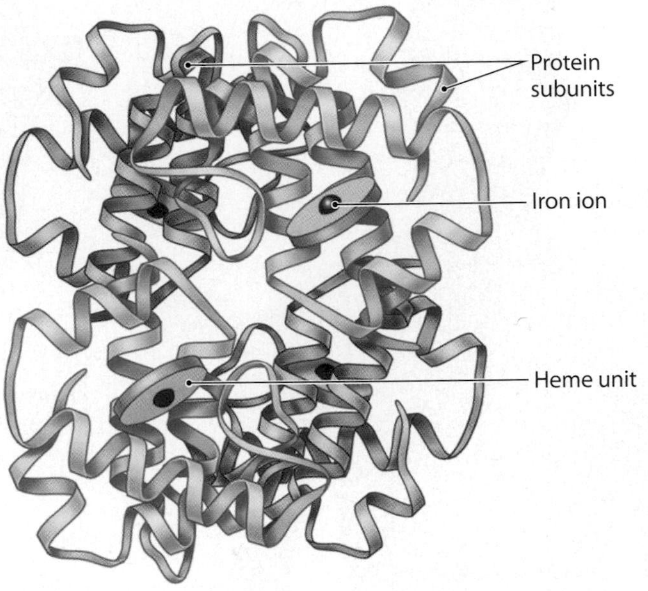
Protein subunits
Iron ion
Heme unit

1 A hemoglobin molecule consists of four globular protein subunits, each containing a heme unit that can bind an oxygen molecule. Thus, each hemoglobin molecule can reversibly bind up to four molecules of oxygen, forming **oxyhemoglobin (HbO_2)**. Carbon monoxide, CO, is a gas released by petroleum-burning engines and heaters. It is dangerous because it will irreversibly bind to heme units, making them unavailable for oxygen transport.

2 The percentage of heme units containing bound oxygen at any given moment is called the **hemoglobin saturation**. If all the Hb molecules in the blood are fully loaded with oxygen, saturation is 100 percent. If, on average, each Hb molecule carries two O_2 molecules, saturation is 50 percent. This graph is an **oxygen-hemoglobin saturation curve**, which shows the saturation of hemoglobin at different partial pressures of oxygen. Notice that hemoglobin will be more than 90 percent saturated if exposed to a P_{O_2} above 60 mm Hg.

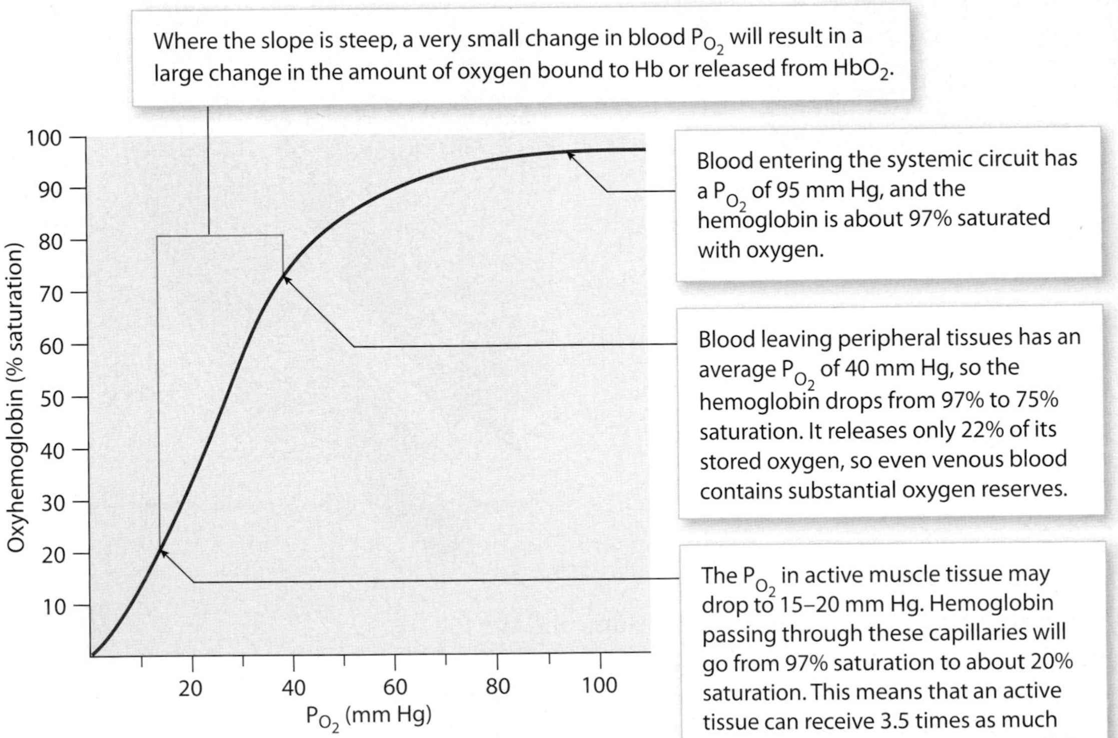
Where the slope is steep, a very small change in blood P_{O_2} will result in a large change in the amount of oxygen bound to Hb or released from HbO_2.
Blood entering the systemic circuit has a P_{O_2} of 95 mm Hg, and the hemoglobin is about 97% saturated with oxygen.
Blood leaving peripheral tissues has an average P_{O_2} of 40 mm Hg, so the hemoglobin drops from 97% to 75% saturation. It releases only 22% of its stored oxygen, so even venous blood contains substantial oxygen reserves.
The P_{O_2} in active muscle tissue may drop to 15–20 mm Hg. Hemoglobin passing through these capillaries will go from 97% saturation to about 20% saturation. This means that an active tissue can receive 3.5 times as much oxygen as an inactive tissue, even if the rate of blood flow remains the same.
Oxyhemoglobin (% saturation)
100
90
80
70
60
50
40
30
20
10
20
40
60
80
100
P_{O_2} (mm Hg)

Note: The curve has this shape because each arriving oxygen molecule increases the affinity of hemoglobin for the next oxygen molecule. Once the first oxygen molecule binds to the hemoglobin, the slope rises steeply until reaching a plateau near 100% saturation.

3 Blood pH has a direct effect on the oxygen-hemoglobin saturation curve because the shape of hemoglobin molecules changes as the number of bound O_2 molecules increases and these changes affect its affinity for oxygen. This change is called the **Bohr effect**. The curves in this figure show how the slope of the oxygen-hemoglobin saturation curve changes when the pH of the blood shifts away from the normal average value of 7.4.

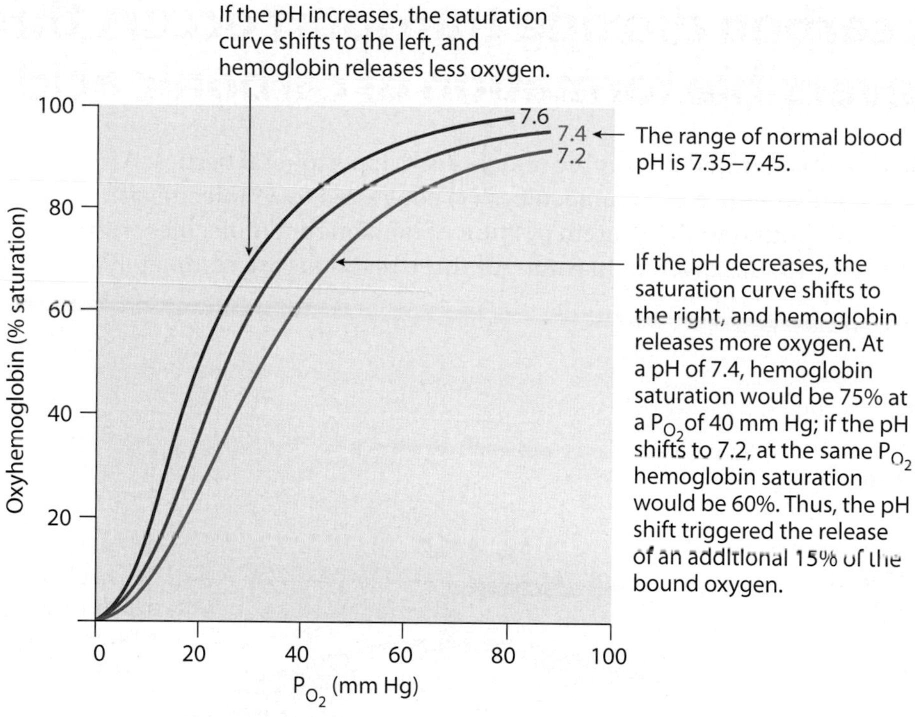

If the pH increases, the saturation curve shifts to the left, and hemoglobin releases less oxygen.
7.6
7.4
7.2
The range of normal blood pH is 7.35–7.45.
If the pH decreases, the saturation curve shifts to the right, and hemoglobin releases more oxygen. At a pH of 7.4, hemoglobin saturation would be 75% at a P_{O_2} of 40 mm Hg; if the pH shifts to 7.2, at the same P_{O_2} hemoglobin saturation would be 60%. Thus, the pH shift triggered the release of an additional 15% of the bound oxygen.
Oxyhemoglobin (% saturation)
100
80
60
40
20
0
20
40
60
80
100
P_{O_2} (mm Hg)

4 Temperature changes also affect the saturation curve. The higher the temperature, the more readily hemoglobin gives up its oxygen reserves. Blood temperature is about 38°C.

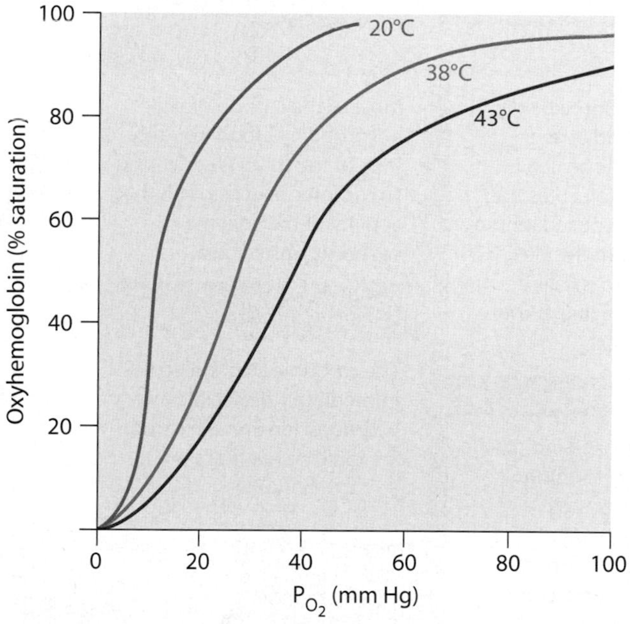

20°C
38°C
43°C
Oxyhemoglobin (% saturation)
100
80
60
40
20
0
20
40
60
80
100
P_{O_2} (mm Hg)

Red blood cells (RBCs) do not contain mitochondria, so they can only generate ATP through glycolysis. The metabolic pathways involved in glycolysis also generate the compound **2,3-bisphosphoglycerate** (biz-fos-fō-GLIS-er-āt), or **BPG**. For any partial pressure of oxygen, the higher the concentration of BPG, the more oxygen will be released by the Hb molecules. BPG production decreases as RBCs age, and levels of BPG can determine how long a blood bank can store fresh whole blood. When BPG levels get too low, hemoglobin becomes firmly bound to the available oxygen. The blood is then useless for transfusions, because the RBCs will no longer release oxygen to peripheral tissues, even at a disastrously low P_{O_2}.

Module 21.13 Review

a. Define oxyhemoglobin.

b. During exercise, hemoglobin releases more oxygen to active skeletal muscles than it does when those muscles are at rest. Why?

c. Explain the relationship among BPG, oxygen, and hemoglobin.

LO 21.13 Discuss the structure and function of hemoglobin, explain the oxygen-hemoglobin saturation curve, and describe the role of 2,3-bisphosphoglycerate.

Most carbon dioxide transport occurs through the reversible formation of carbonic acid

Carbon dioxide is generated by aerobic metabolism in peripheral tissues. After entering the bloodstream, a CO_2 molecule is (1) converted to a molecule of carbonic acid, (2) bound to the protein portion of hemoglobin molecules within red blood cells, or (3) dissolved in plasma. All three reactions are completely reversible.

1 We will consider the events that occur as blood enters peripheral tissues in which the P_{CO_2} is 45 mm Hg.

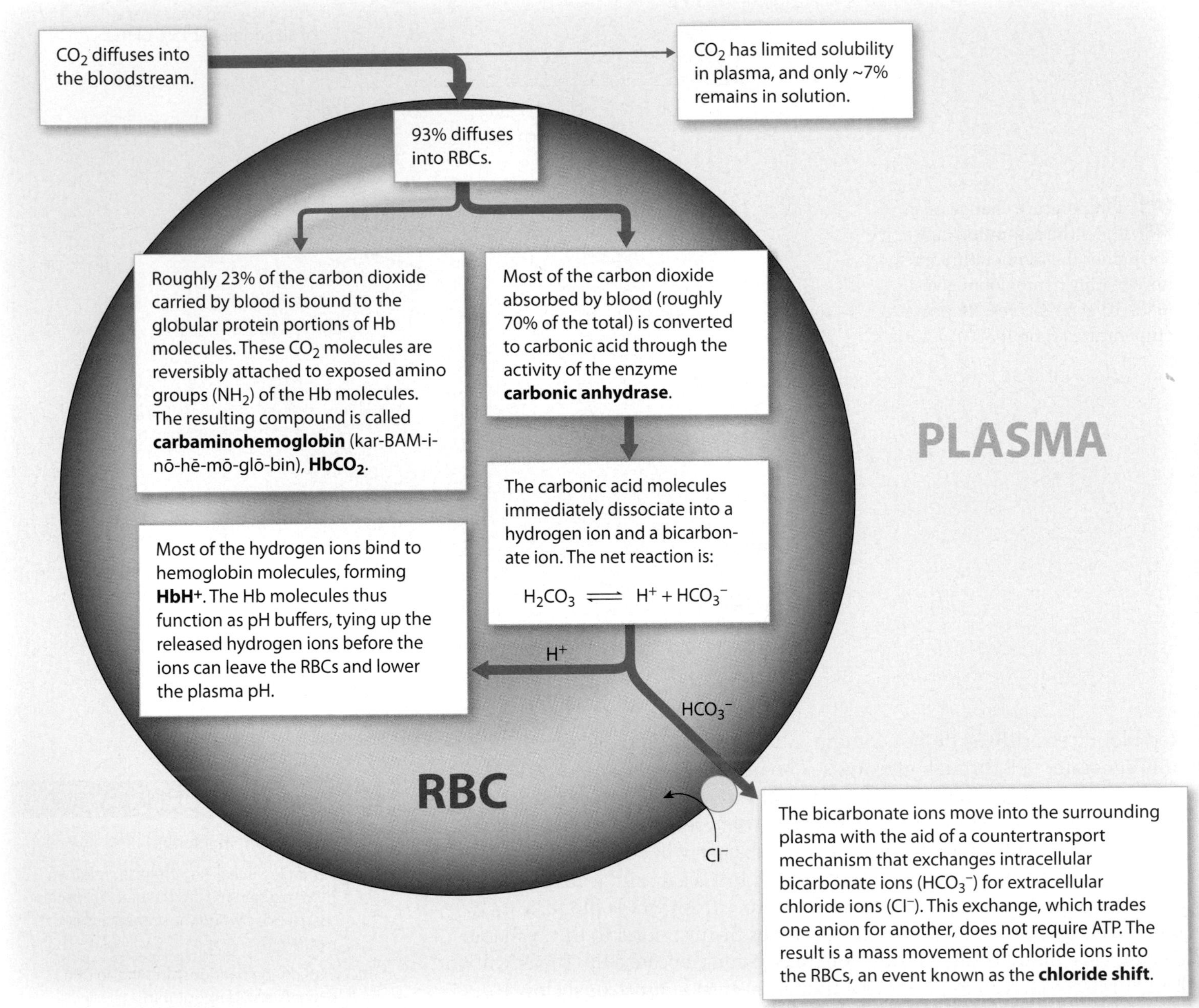

CO_2 diffuses into the bloodstream.
CO_2 has limited solubility in plasma, and only ~7% remains in solution.
93% diffuses into RBCs.
Roughly 23% of the carbon dioxide carried by blood is bound to the globular protein portions of Hb molecules. These CO_2 molecules are reversibly attached to exposed amino groups (NH_2) of the Hb molecules. The resulting compound is called **carbaminohemoglobin** (kar-BAM-i-nō-hē-mō-glō-bin), **$HbCO_2$**.
Most of the carbon dioxide absorbed by blood (roughly 70% of the total) is converted to carbonic acid through the activity of the enzyme **carbonic anhydrase**.
The carbonic acid molecules immediately dissociate into a hydrogen ion and a bicarbonate ion. The net reaction is:
$H_2CO_3 \rightleftharpoons H^+ + HCO_3^-$
Most of the hydrogen ions bind to hemoglobin molecules, forming **HbH^+**. The Hb molecules thus function as pH buffers, tying up the released hydrogen ions before the ions can leave the RBCs and lower the plasma pH.
H^+
HCO_3^-
PLASMA
RBC
Cl^-
The bicarbonate ions move into the surrounding plasma with the aid of a countertransport mechanism that exchanges intracellular bicarbonate ions (HCO_3^-) for extracellular chloride ions (Cl^-). This exchange, which trades one anion for another, does not require ATP. The result is a mass movement of chloride ions into the RBCs, an event known as the **chloride shift**.

2 This illustration summarizes the transport of oxygen and carbon dioxide in the lungs (at left) and in peripheral tissues (at right). This system is at equilibrium, and the partial pressures of oxygen (P_{O_2}) and carbon dioxide (P_{CO_2}) are stable in the alveoli and in the tissues (Module 21.12, p. 809). Every oxygen molecule entering peripheral tissues is balanced by an oxygen molecule absorbed by the alveoli, and the absorbed oxygen molecule will be replaced in the next respiratory cycle.

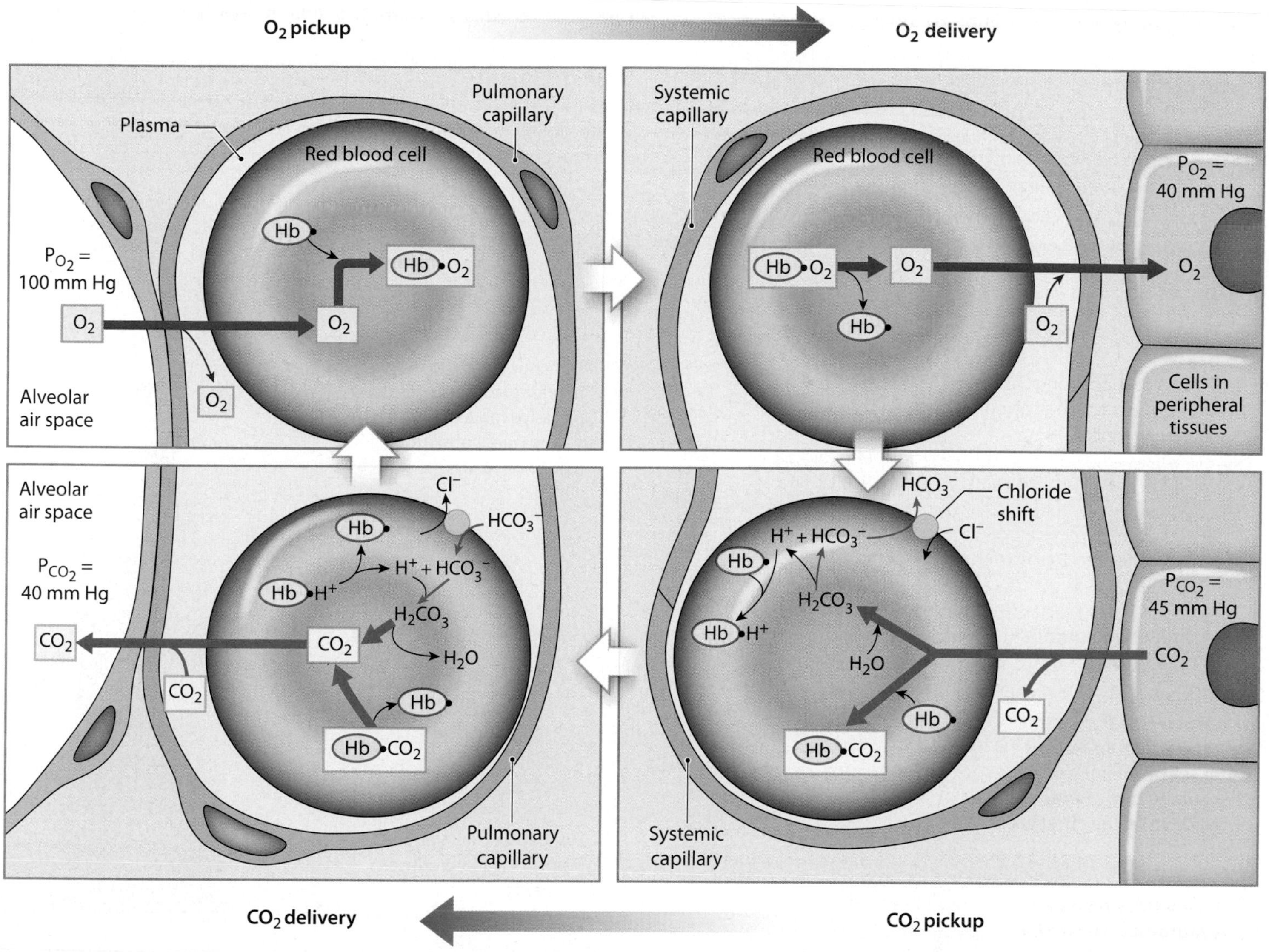

O2 pickup
O2 delivery
Plasma
Pulmonary capillary
Red blood cell
Hb
Hb O2
PO2 = 100 mm Hg
O2
Alveolar air space
Systemic capillary
PO2 = 40 mm Hg
Cells in peripheral tissues
Cl−
HCO3−
Hb H+
H+ + HCO3−
H2CO3
CO2
H2O
Hb CO2
PCO2 = 40 mm Hg
Chloride shift
PCO2 = 45 mm Hg
CO2 delivery
CO2 pickup

The equilibrium between oxygen absorption and oxygen use is disturbed when tissue oxygen demand increases. If the respiratory rate and tidal volume do not increase, the alveolar P_{O_2} will steadily decrease, and the alveolar, blood, and tissue P_{CO_2} will steadily rise. This is a particulary unpleasant combination that can lead to widespread hypoxia and a dangerous decrease in the pH of body fluids. In a clinical setting, pulmonary ventilation is therefore closely regulated by monitoring the P_{O_2}, P_{CO_2}, and pH of body fluids.

Module 21.14 Review

a. Identify three ways that carbon dioxide is transported in the bloodstream.

b. Describe the forces that drive oxygen and carbon dioxide transport between the blood and peripheral tissues.

c. How would blockage of the trachea affect blood pH?

LO 21.14 Describe how carbon dioxide is transported in the blood, and explain how oxygen is picked up, transported, and released into the bloodstream.

Pulmonary disease can affect both lung elasticity and airflow

1 The **compliance** of the lungs is an indication of their expandability, how easily the lungs expand. It is influenced by the internal structure of the lungs (elasticity and resilience) and the flexibility of the chest wall. Compliance is a static measurement determined by monitoring the intrapulmonary pressure at different lung volumes.

2 The **resistance** of the lungs is an indication of how much force is required to inflate or deflate them. At rest, the muscular activity involved in pulmonary ventilation accounts for 3–5 percent of the resting energy demand. If resistance increases, that figure climbs dramatically, and an individual may become exhausted simply trying to breathe.

Compliance

The greater the compliance, the lower the tension in the walls of the lungs at a given volume.

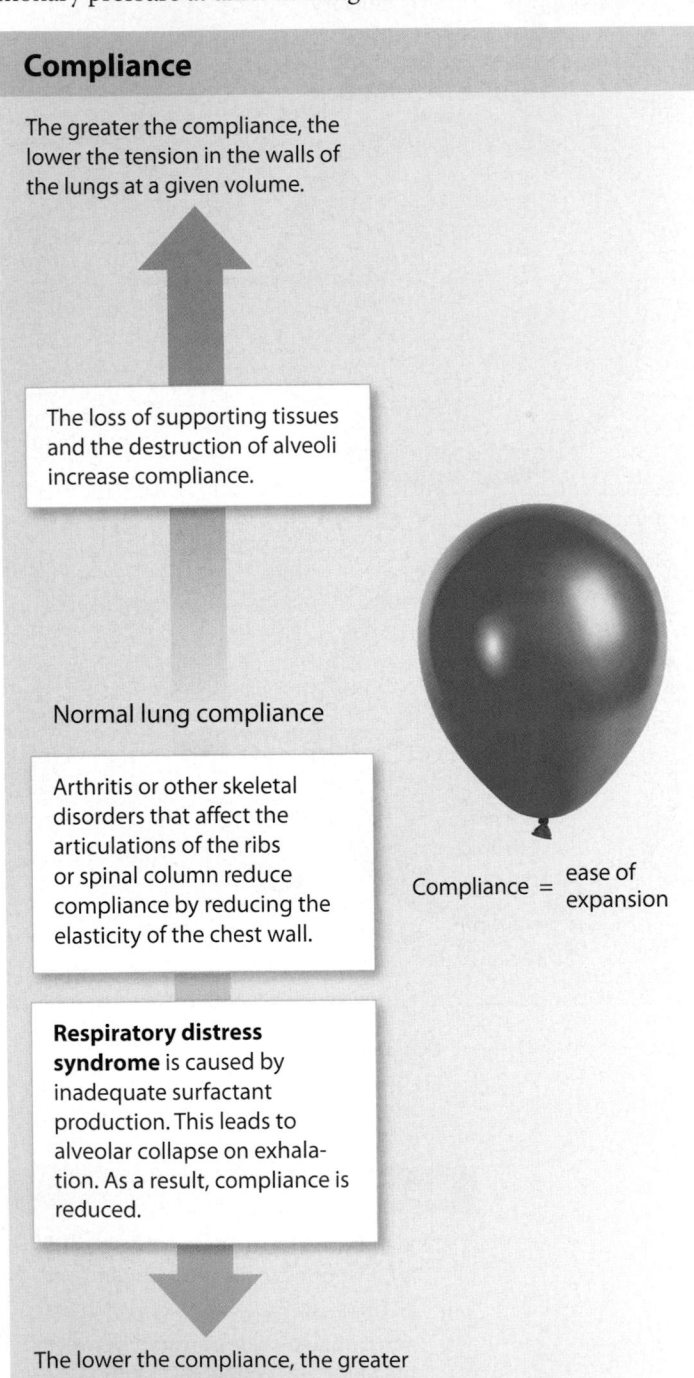

The loss of supporting tissues and the destruction of alveoli increase compliance.

Normal lung compliance

Compliance = ease of expansion

Arthritis or other skeletal disorders that affect the articulations of the ribs or spinal column reduce compliance by reducing the elasticity of the chest wall.

Respiratory distress syndrome is caused by inadequate surfactant production. This leads to alveolar collapse on exhalation. As a result, compliance is reduced.

The lower the compliance, the greater the tension in the walls of the lungs at a given volume, and the harder it is for air to flow along the conducting passages.

Resistance

The higher the resistance, the harder it is to force air along the conducting passages.

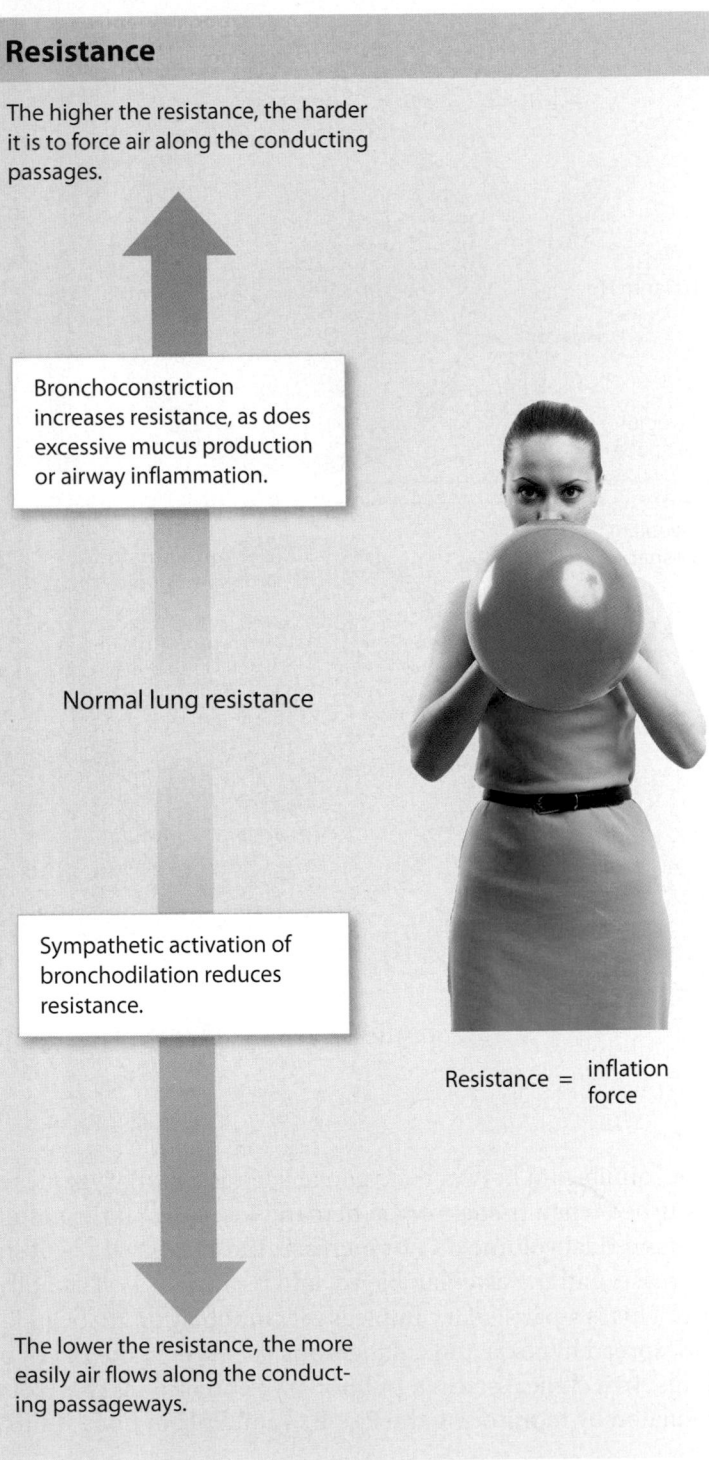

Bronchoconstriction increases resistance, as does excessive mucus production or airway inflammation.

Normal lung resistance

Resistance = inflation force

Sympathetic activation of bronchodilation reduces resistance.

The lower the resistance, the more easily air flows along the conducting passageways.

Chronic Obstructive Pulmonary Disease

3 **Chronic obstructive pulmonary disease (COPD)** is a general term for a progressive disorder of the airways that restricts airflow and reduces alveolar ventilation. We highlight three different COPDs below. Asthma, or asthmatic bronchitis, is the term used when symptoms are acute and intermittent. The terms chronic bronchitis and emphysema are usually applied when signs and symptoms are long lasting and progressive with occasional crises due to infections.

Asthma

Asthma (AZ-muh) is a condition characterized by conducting passageways that are extremely sensitive to irritation. The airways respond to irritation by constricting smooth muscles all along the bronchial tree. This is accompanied by edema and swelling of the mucosa of the respiratory passageways, and the accelerated production of mucus. The combination makes breathing very difficult, and resistance is markedly increased. This can be caused by allergies, toxins, or exercise.

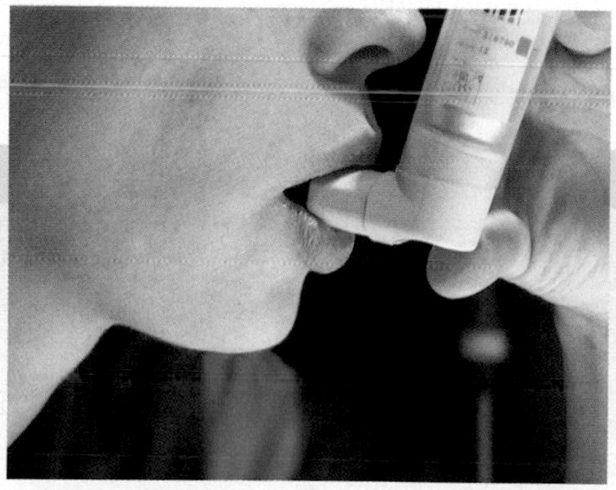

Chronic Bronchitis

Chronic bronchitis (brong-KĪ-tis) is a long-term inflammation and swelling of the bronchial lining, leading to overproduction of mucus. The characteristic sign is frequent coughing with copious sputum production. This condition is most commonly related to cigarette smoking but also results from other environmental irritants, such as chemical vapors. Over time, the increased mucus production can block smaller airways, increasing resistance and reducing respiratory efficiency. Chronic bacterial infections leading to more lung damage are common. People with chronic bronchitis may have signs of heart failure, including widespread edema. Their blood oxygenation is low, and their skin may have a bluish color. The combination of widespread edema and bluish coloration has led to the descriptive term **blue bloaters** for people with this condition.

Emphysema

Emphysema (em-fi-ZĒ-muh) is a chronic, progressive condition characterized by shortness of breath and an inability to tolerate physical exertion. The underlying problem is the destruction of alveolar surfaces and inadequate surface area for oxygen and carbon dioxide exchange. The alveoli gradually expand, and adjacent alveoli merge to form larger air spaces supported by fibrous tissue without alveolar capillary networks. As elastic connective tissues are lost, compliance increases, but the loss of respiratory surface area restricts oxygen absorption, so the person becomes short of breath. The respiratory muscles work hard, and these people, who use a lot of energy just breathing, tend to be thin. Chest x-rays, such as the one at right, show overexpanded lungs. Because their exaggerated respiratory movements typically maintain near normal blood oxygenation, the skin of pale-skinned emphysema patients is usually pink. The combination of heavy breathing and pink coloration has led to the descriptive term **pink puffers** for these people.

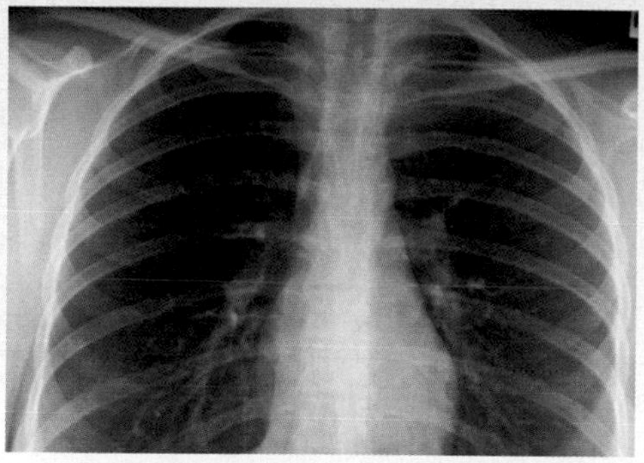

Module 21.15 Review

a. Define compliance and resistance.

b. Identify three chronic obstructive pulmonary diseases (COPDs).

c. Compare chronic bronchitis with emphysema.

LO 21.15 Explain how pulmonary disease affects compliance and resistance.

Module 21.16

Respiratory control mechanisms involve interacting centers in the brain stem

1 Respiratory control involves multiple levels of regulation. Respiratory rate and rhythm are a function of a network of respiratory centers that includes centers in the pons and medulla. Most of the regulatory activities occur outside of our awareness.

Level 3: Higher Centers

Higher centers in the hypothalamus, limbic system, and cerebral cortex can alter the activity of the pneumotaxic centers, but essentially normal respiratory cycles continue even if the brain stem superior to the pons has been severely damaged.

Level 2: Apneustic and Pneumotaxic Centers

The **apneustic** (ap-NŪ-stik) **centers** and the **pneumotaxic** (nū-mō-TAKS-ik) **centers** of the pons are paired nuclei that adjust the output of the respiratory rhythmicity centers.

Start

Level 1: Respiratory Rhythmicity Centers

The most basic level of respiratory control involves pacemaker cells in the medulla oblongata. These neurons generate cycles of contraction and relaxation in the diaphragm. The paired **respiratory rhythmicity centers** establish the pace of respiration by adjusting these pacemakers and coordinating additional respiratory muscle actions. Each rhythmicity center is subdivided into a **dorsal respiratory group (DRG)** and a **ventral respiratory group (VRG)**. The DRG modifies its activities in response to input from chemoreceptors that monitor O_2, CO_2, and pH in the blood and cerebrospinal fluid and from baroreceptors (stretch receptors) that monitor stretching in the walls of the lungs. Current research suggests that it is impossible to attribute respiratory rate and rhythm to a single area, aside from the basic pattern generator in the ventrolateral medulla known as the pre-Bötzinger complex. However, the DRG is mainly concerned with inspiration, and the VRG is primarily associated with expiration.

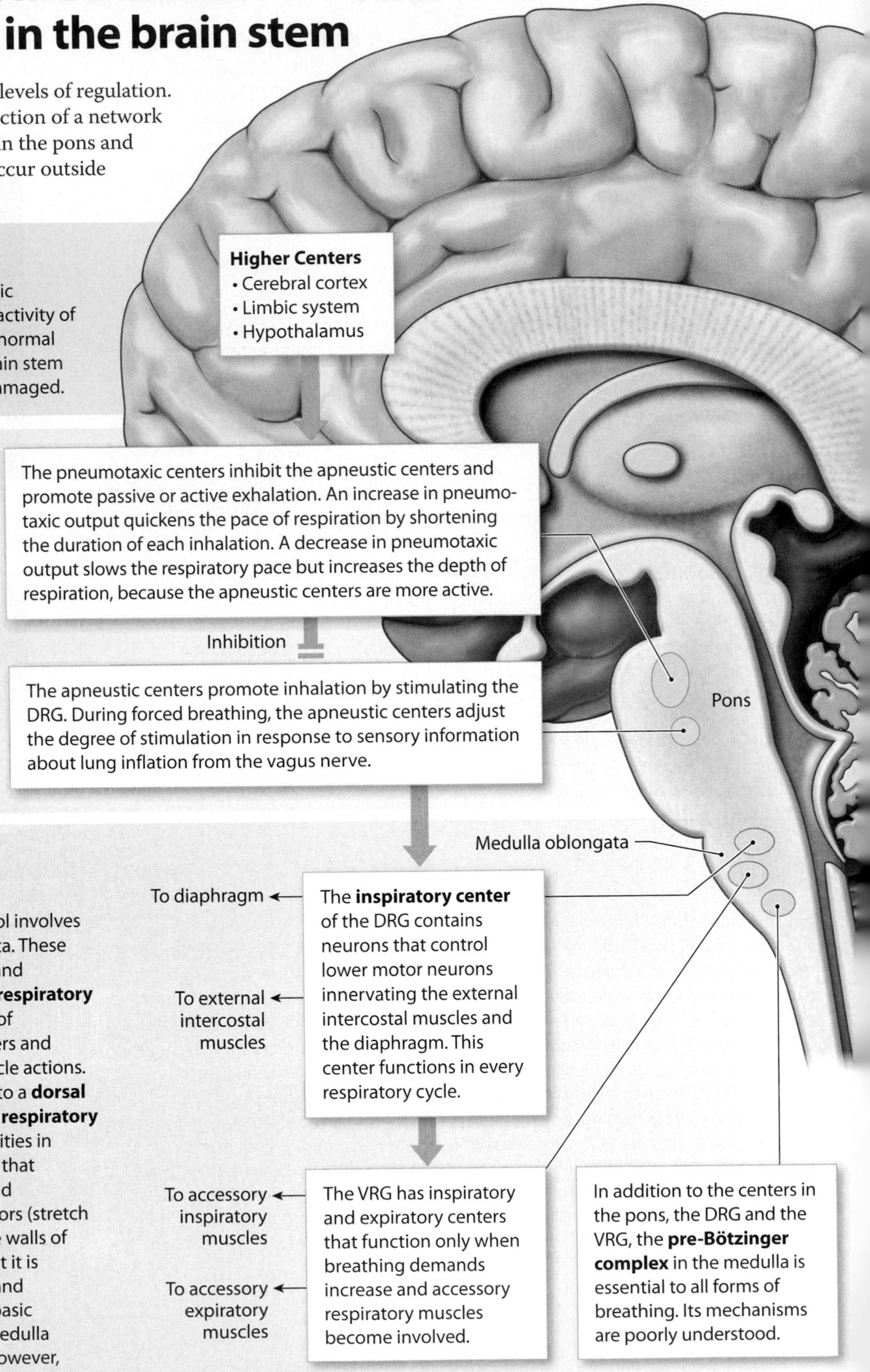

Higher Centers
• Cerebral cortex
• Limbic system
• Hypothalamus
The pneumotaxic centers inhibit the apneustic centers and promote passive or active exhalation. An increase in pneumotaxic output quickens the pace of respiration by shortening the duration of each inhalation. A decrease in pneumotaxic output slows the respiratory pace but increases the depth of respiration, because the apneustic centers are more active.
Inhibition
The apneustic centers promote inhalation by stimulating the DRG. During forced breathing, the apneustic centers adjust the degree of stimulation in response to sensory information about lung inflation from the vagus nerve.
Pons
Medulla oblongata
To diaphragm
To external intercostal muscles
The inspiratory center of the DRG contains neurons that control lower motor neurons innervating the external intercostal muscles and the diaphragm. This center functions in every respiratory cycle.
To accessory inspiratory muscles
To accessory expiratory muscles
The VRG has inspiratory and expiratory centers that function only when breathing demands increase and accessory respiratory muscles become involved.
In addition to the centers in the pons, the DRG and the VRG, the pre-Bötzinger complex in the medulla is essential to all forms of breathing. Its mechanisms are poorly understood.

2 This flowchart shows the events involved in **quiet breathing** in a person at rest.

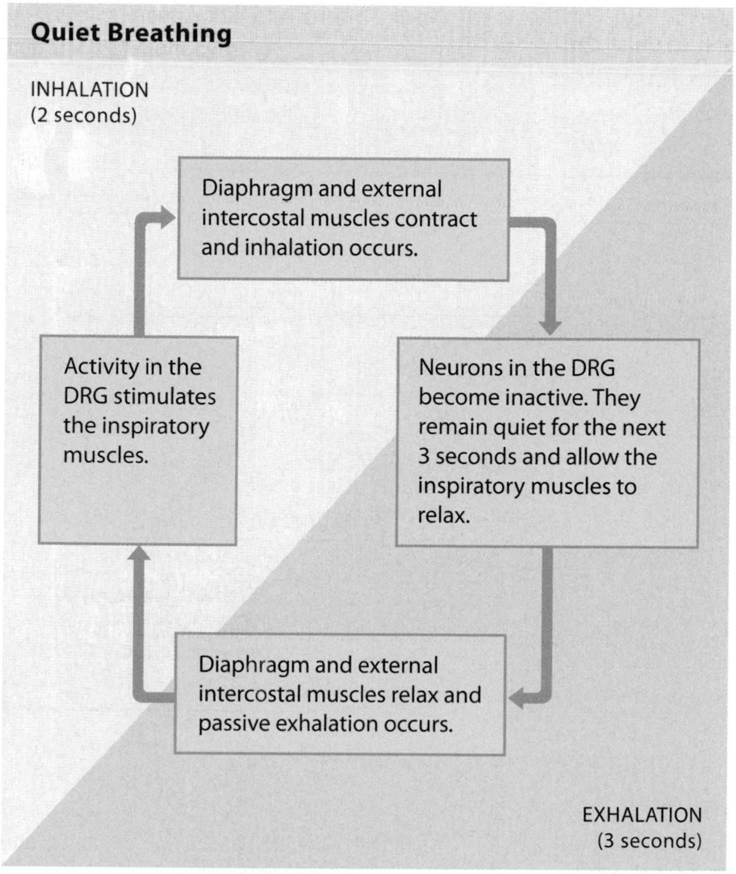

Quiet Breathing
INHALATION
(2 seconds)
Diaphragm and external intercostal muscles contract and inhalation occurs.
Neurons in the DRG become inactive. They remain quiet for the next 3 seconds and allow the inspiratory muscles to relax.
Diaphragm and external intercostal muscles relax and passive exhalation occurs.
Activity in the DRG stimulates the inspiratory muscles.
EXHALATION
(3 seconds)

3 During **forced breathing**, pulmonary ventilation is increased toward maximal levels through the involvement of accessory respiratory muscles.

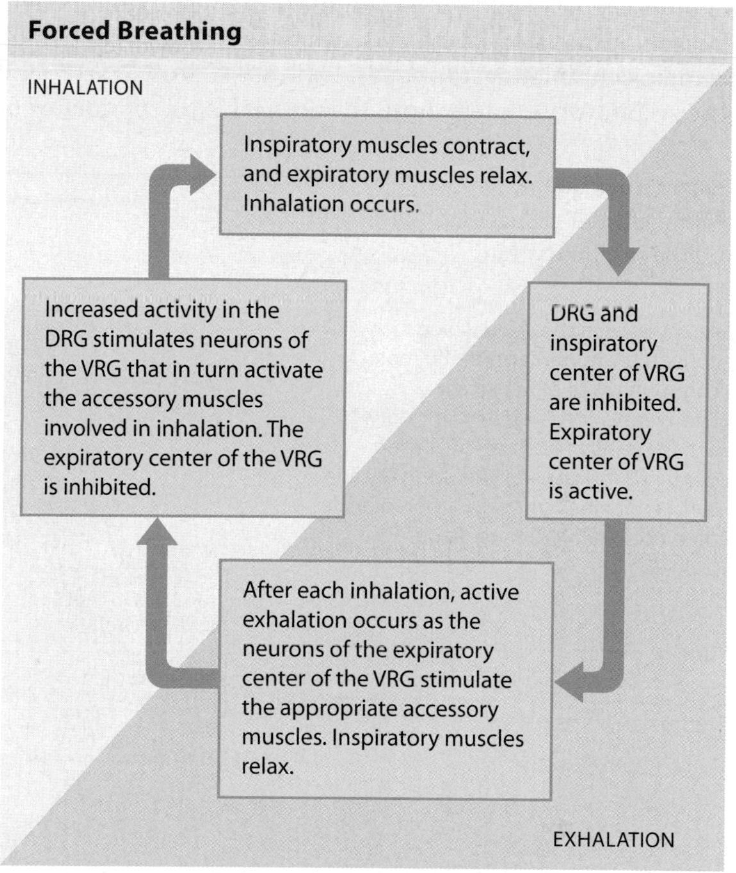

Forced Breathing
INHALATION
Inspiratory muscles contract, and expiratory muscles relax. Inhalation occurs.
DRG and inspiratory center of VRG are inhibited. Expiratory center of VRG is active.
After each inhalation, active exhalation occurs as the neurons of the expiratory center of the VRG stimulate the appropriate accessory muscles. Inspiratory muscles relax.
Increased activity in the DRG stimulates neurons of the VRG that in turn activate the accessory muscles involved in inhalation. The expiratory center of the VRG is inhibited.
EXHALATION

4 This table lists the various sensory stimuli that can modify the activities of the respiratory centers. These stimuli can trigger automatic responses known as **respiratory reflexes**, which are the focus of the next module.

Representative Respiratory Reflexes

- Chemoreceptors sensitive to the pH, P_{O_2}, or P_{CO_2} of the blood or cerebrospinal fluid alter the activities of the respiratory centers.
- Baroreceptors in the aortic or carotid sinuses sensitive to changes in blood pressure alter the activities of the respiratory centers.
- Stretch receptors that respond to changes in the volume of the lungs are responsible for inflation and deflation reflexes.
- Irritating physical or chemical stimuli in the nasal cavity, larynx, or bronchial tree initiate protective reflexes, such as coughing or sneezing.

Respiratory control is an area of intense study, and our understanding of the interactions and mechanisms remains incomplete. Due to the brain stem locations, it is nearly impossible to study this mechanism in humans, and much of the work has been done in cats and newborn rodents.

Module 21.16 Review

a. Name the paired central nervous system nuclei that adjust the pace of respiration.

b. Which brain stem centers generate the respiratory pace?

c. Which chemical factors in blood or cerebrospinal fluid stimulate the respiratory centers?

LO **21.16** Describe the brain stem structures that influence the control of respiration.

Module 21.17

Respiratory reflexes provide rapid automatic adustments in pulmonary ventilation

Chemoreceptor Reflexes

Under normal conditions, the P_{CO_2} is the most important factor stimulating chemoreceptors and thereby influencing respiratory activity. A rise of just 10 percent in the arterial P_{CO_2} causes the respiratory rate to double, even if the P_{O_2} remains completely normal. In contrast, a drop in arterial P_{O_2} has little effect on the respiratory centers, until the arterial P_{O_2} drops below 60 mm Hg.

1 This diagram shows the effect of changes in P_{CO_2} on the respiratory rate.

An increase in the arterial blood P_{CO_2} constitutes **hypercapnia**. The most common cause of hypercapnia is hypoventilation—when respiratory activity is insufficient to meet the demands for tissue oxygen delivery and carbon dioxide removal. Carbon dioxide then accumulates in the blood.

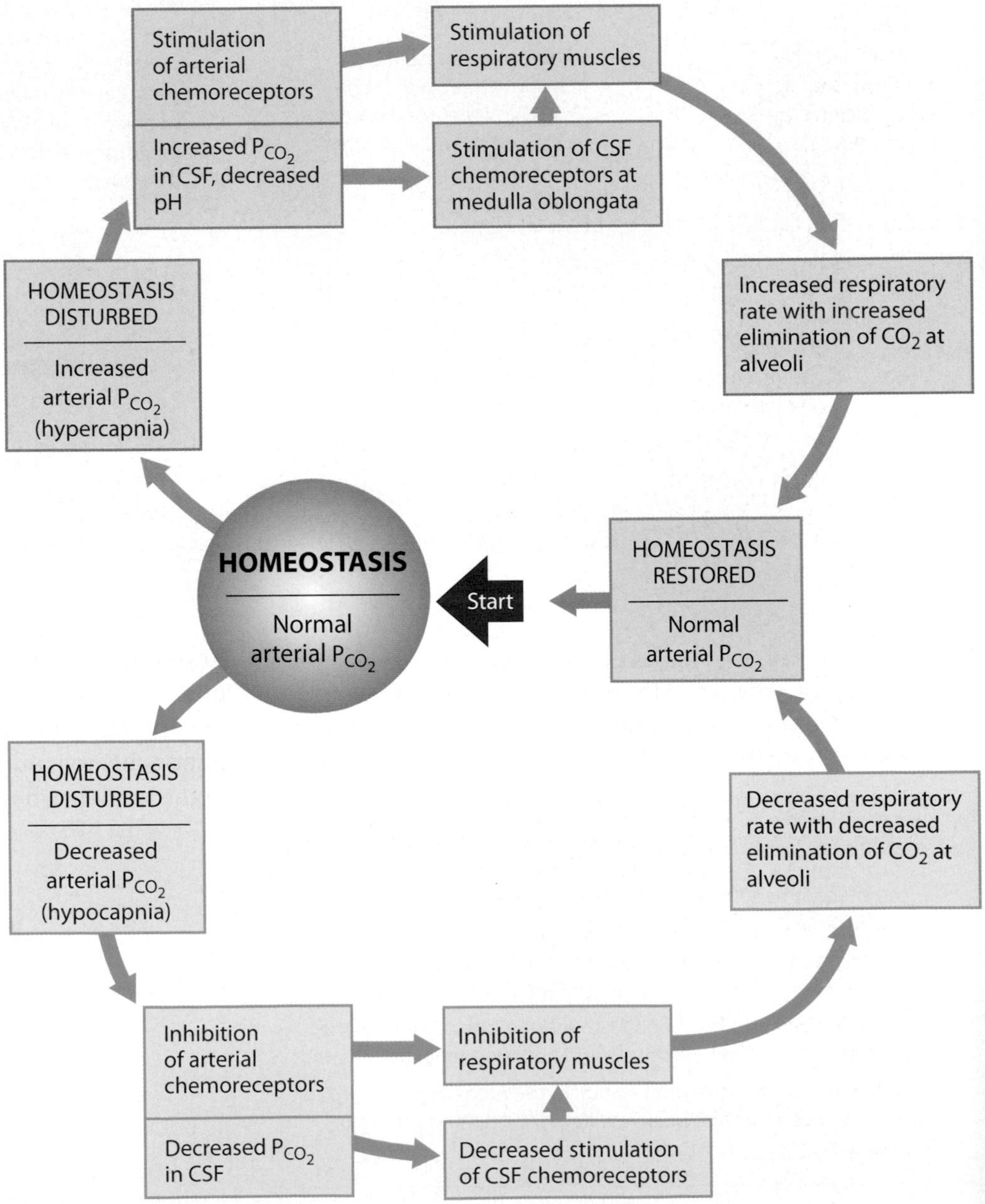

Stimulation of arterial chemoreceptors
Stimulation of respiratory muscles
Increased P_{CO_2} in CSF, decreased pH
Stimulation of CSF chemoreceptors at medulla oblongata
HOMEOSTASIS DISTURBED
Increased arterial P_{CO_2} (hypercapnia)
Increased respiratory rate with increased elimination of CO_2 at alveoli
HOMEOSTASIS
Normal arterial P_{CO_2}
Start
HOMEOSTASIS RESTORED
Normal arterial P_{CO_2}
HOMEOSTASIS DISTURBED
Decreased arterial P_{CO_2} (hypocapnia)
Decreased respiratory rate with decreased elimination of CO_2 at alveoli
Inhibition of arterial chemoreceptors
Inhibition of respiratory muscles
Decreased P_{CO_2} in CSF
Decreased stimulation of CSF chemoreceptors

Hyperventilation, when the rate and depth of respiration exceed the demands for oxygen delivery and carbon dioxide removal, gradually leads to **hypocapnia**, an abnormally low P_{CO_2}. Snorkelers sometimes hyperventilate to extend their time underwater; this works because it is the P_{CO_2} that stimulates respiratory activity. If the P_{CO_2} is driven down too far, a snorkeler may become unconscious from oxygen starvation in the brain without ever feeling the urge to breathe. This condition is called **shallow water blackout**.

Baroreceptor Reflexes

2 Baroreceptors in the carotid and aortic sinuses are monitored by sensory nerves within the glossopharyngeal and vagus nerves, respectively. The cardiovascular effects of their stimulation were considered in Module 19.21 (p. 725). The sensory information is also distributed to the respiratory centers, triggering adjustments in the respiratory minute volume.

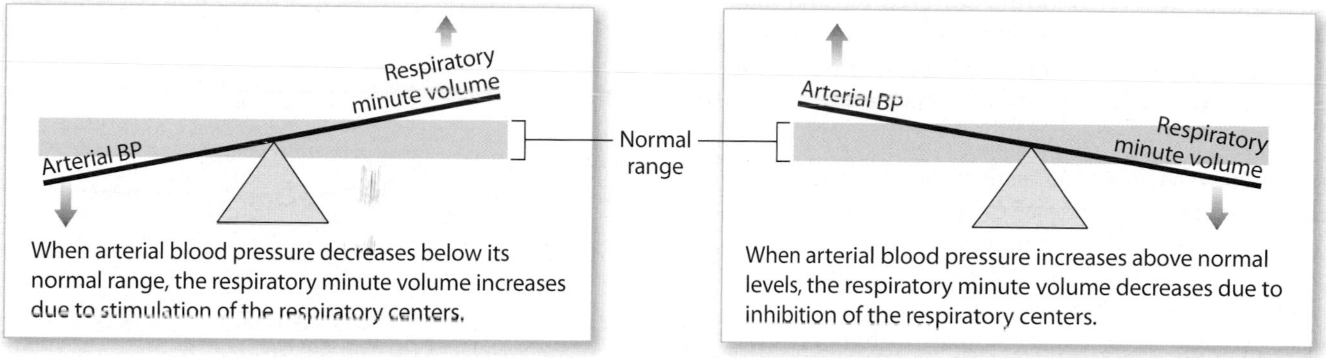
Respiratory minute volume
Arterial BP
Normal range
When arterial blood pressure decreases below its normal range, the respiratory minute volume increases due to stimulation of the respiratory centers.
Arterial BP
Respiratory minute volume
When arterial blood pressure increases above normal levels, the respiratory minute volume decreases due to inhibition of the respiratory centers.

Inflation/Deflation Reflexes

3 Inflation and deflation reflexes are activated by stretch receptors in the lungs during forced breathing when the tidal volume is large (1000 mL or more). This sensory information is distributed to the apneustic centers and the VRG.

The **inflation reflex** prevents overexpansion of the lungs during forced breathing.

The **deflation reflex** inhibits the expiratory centers and stimulates the inspiratory centers when the lungs are deflating.

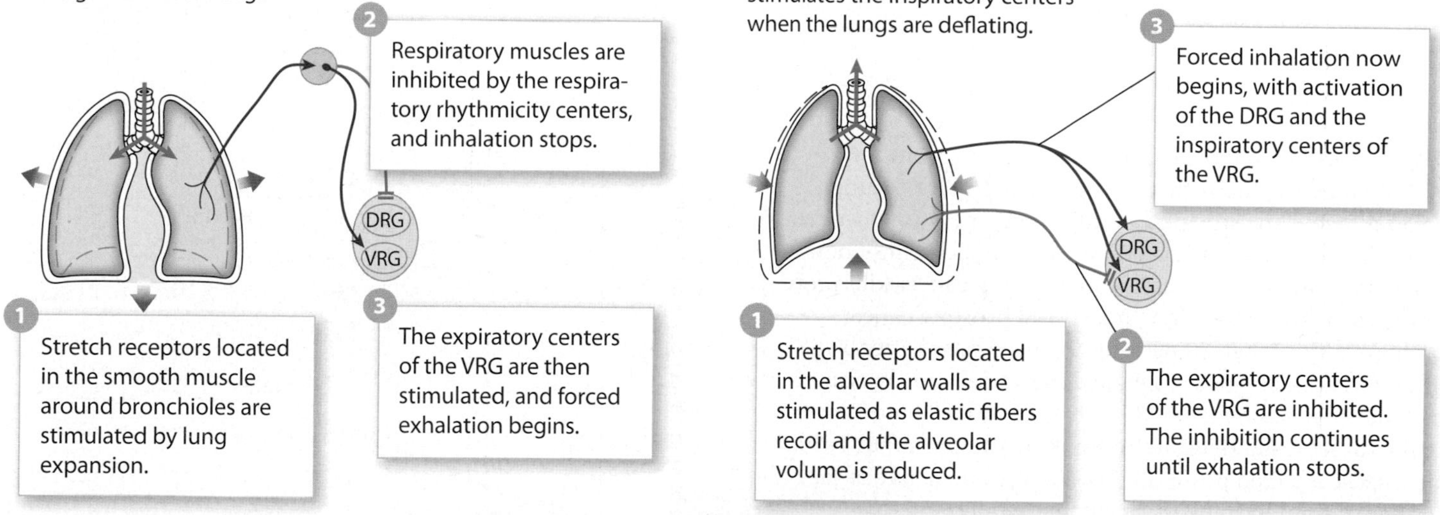
2 Respiratory muscles are inhibited by the respiratory rhythmicity centers, and inhalation stops.
DRG
VRG
1 Stretch receptors located in the smooth muscle around bronchioles are stimulated by lung expansion.
3 The expiratory centers of the VRG are then stimulated, and forced exhalation begins.
3 Forced inhalation now begins, with activation of the DRG and the inspiratory centers of the VRG.
DRG
VRG
1 Stretch receptors located in the alveolar walls are stimulated as elastic fibers recoil and the alveolar volume is reduced.
2 The expiratory centers of the VRG are inhibited. The inhibition continues until exhalation stops.

Protective Reflexes

4 **Protective reflexes** include sneezing and coughing. Sneezing is triggered by an irritation of the nasal cavity wall. Coughing is triggered by an irritation of the larynx, trachea, or bronchi. Both reflexes involve **apnea** (AP-nē-uh), a period in which breathing has stopped, usually followed by a forceful expulsion of air to remove the offending stimulus. Air leaving the larynx can travel at 160 kph (99 mph), carrying mucus, particles, and irritating gases out of the respiratory tract through the nose or mouth.

Module 21.17 Review

a. Define hypercapnia and hypocapnia.

b. Are chemoreceptors more sensitive or less sensitive to plasma levels of CO_2 than they are to plasma levels of O_2?

c. Little Johnny is angry with his mother, so he tells her that he will hold his breath until he turns blue and dies. Should Johnny's mother worry that this will happen?

LO 21.17 Identify and discuss reflex respiratory activity in pulmonary ventilation.

Respiratory function decreases with age; smoking makes matters worse

All aspects of respiratory function are affected by aging. As elastic tissue deteriorates throughout the body, vital capacity decreases. As arthritic changes stiffen rib joints, compliance and maximum respiratory minute volume are decreased. These respiratory limitations make strenuous exercise difficult or impossible.

1 In addition to skeletal and connective tissue changes associated with age, some degree of emphysema is normal in people over age 50. However, the extent varies widely with the lifetime exposure to cigarette smoke and other respiratory irritants. This graph compares the respiratory performance of people who have never smoked with those who have smoked for various periods of time. The message is quite clear: Although some decrease in respiratory performance is inevitable, you can prevent serious respiratory deterioration by stopping smoking or never starting.

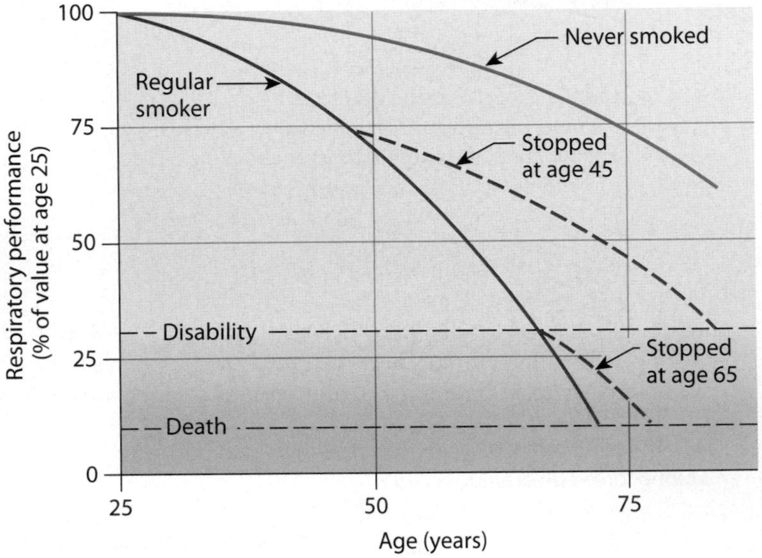
100
75
50
25
0
Respiratory performance (% of value at age 25)
Never smoked
Regular smoker
Stopped at age 45
Disability
Stopped at age 65
Death
25
50
75
Age (years)

Lung Cancer

2 According to the CDC, more people die from **lung cancer** than any other type of cancer. In 2009, lung cancer affected an estimated 110,000 men and 96,000 women in the United States, and accounted for 12.6 percent of new cancer cases in both men and women. Lung cancer kills more people each year than colon, breast, and prostate cancer combined. More than 50 percent of lung cancer patients die within a year of diagnosis. Detailed statistical and experimental evidence has shown that 85–90 percent of all lung cancers are the direct result of cigarette smoking. Before about 1970, this disease affected primarily middle-aged men, but as the number of women smokers has increased (a trend that started in the 1940s), so has the number of women who develop lung cancer. This graph shows that while the incidence of lung cancer among men has gradually declined since 1992, the incidence among women has been increasing. (Incidence refers to new lung cancer cases.)

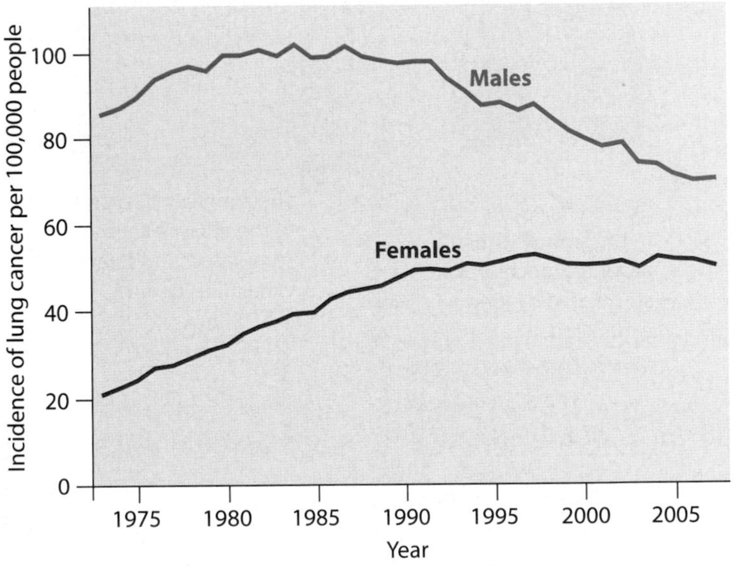
Incidence of lung cancer per 100,000 people
100
80
60
40
20
0
Males
Females
1975
1980
1985
1990
1995
2000
2005
Year

3 Smoke is an irritant, and chemicals found in cigarette smoke include several carcinogens (cancer-causing substances). This figure shows the progression of events leading to lung cancer. The fact that cigarette smoking causes cancer is not surprising, given that the smoke contains carcinogens. What is surprising is that more smokers do not develop lung cancer. Evidence suggests that some smokers have a genetic predisposition to developing at least one form of lung cancer.

Normal Respiratory Epithelium

The normal respiratory epithelium consists of ciliated columnar epithelium with an abundance of mucous cells, which produce mucus that helps clean inhaled air. The irritants and inhaled carcinogens contained in cigarette smoke begin a series of changes to this epithelium.

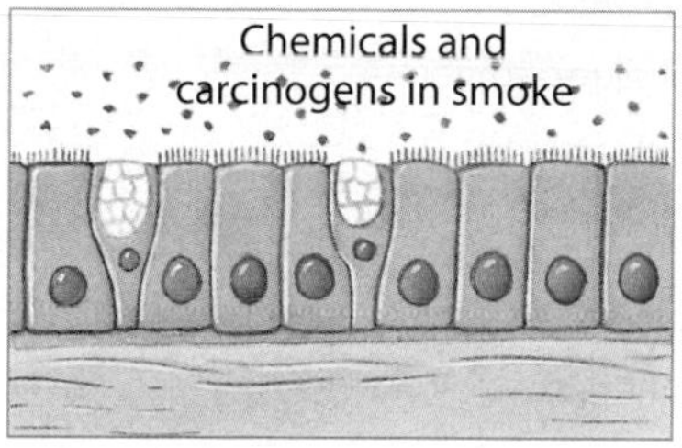
Chemicals and
carcinogens in smoke

Reversible

Dysplasia

In **dysplasia**, cells are damaged and the functional characteristics change. The cilia of respiratory epithelial cells are damaged and paralyzed by exposure to cigarette smoke. These changes cause the local buildup of mucus and reduce the effectiveness of the epithelium in protecting deeper, more delicate portions of the respiratory tract.

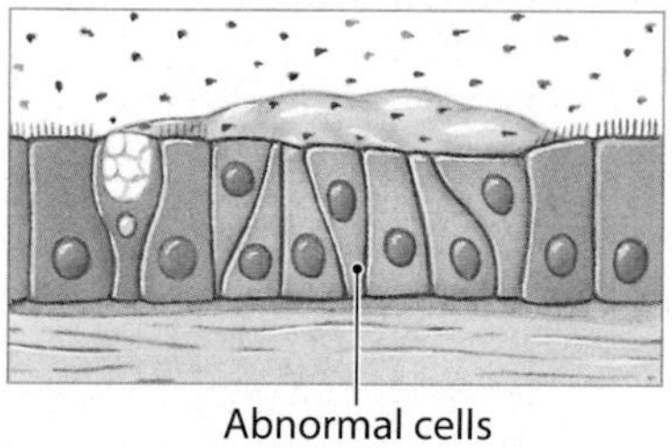
Abnormal cells

Reversible

Metaplasia

In **metaplasia**, a tissue changes its structure, in response to injury or chemical stresses. In this case the stressed respiratory surface converts to a stratified epithelium that protects underlying connective tissue but does nothing for other areas of the respiratory tract. Metaplasia may be reversed if the stressful stimulus is removed before further damage occurs.

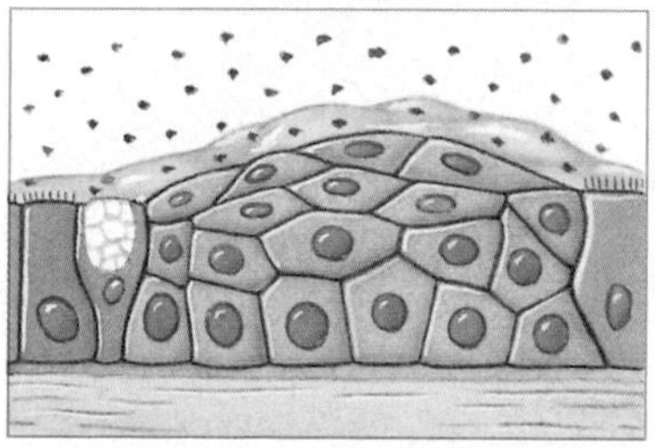

Irreversible

Neoplasia and Anaplasia

In **neoplasia**, the growth of abnormal cells forms a cancerous tumor called a **neoplasm**. In **anaplasia**, the most dangerous stage, the cells become malignant and metastasize (spread) to other parts of the body. Neither neoplasia nor anaplasia is reversible, although the cancers may be treated by chemicals, radiation, and/or surgery.

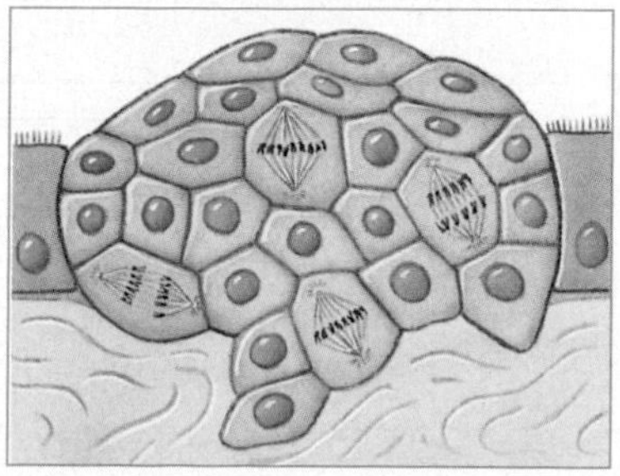

Module 21.18 Review

a. Name several age-related factors that affect the respiratory system.

b. Is the incidence of lung cancer in U.S. males increasing or decreasing?

c. Compare dysplasia, metaplasia, neoplasia, and anaplasia.

LO 21.18 Describe age-related changes to, and the effects of cigarette smoking on, the respiratory system.

Matching

Match each lettered term with the most closely related description.

a. apnea
b. hemoglobin releases more O_2
c. lowers vital capacity
d. external intercostals
e. iron ion
f. Boyle's law
g. compliance
h. bicarbonate ion
i. anoxia
j. atelectasis
k. hypocapnia
l. pneumotaxic centers
m. apneustic centers
n. partial pressure

1 Single gas in a mixture
2 A cause of tissue death
3 Inverse pressure/volume relationship
4 Expandability of lungs
5 CO_2 transport
6 Elastic tissue deterioration
7 Act to elevate ribs
8 Heme unit
9 Blood pH decreases
10 Promotes passive or active exhalation
11 Stimulate DRG and promotes inhalation
12 Hyperventilation
13 Collapsed lung
14 Period of suspended respiration

1 ______
2 ______
3 ______
4 ______
5 ______
6 ______
7 ______
8 ______
9 ______
10 ______
11 ______
12 ______
13 ______
14 ______

Short answer

Identify and describe the various pulmonary volumes and capacities indicated in the spirogram below.

15 ______
16 ______
17 ______
18 ______
19 ______
20 ______
21 ______
22 ______
23 ______

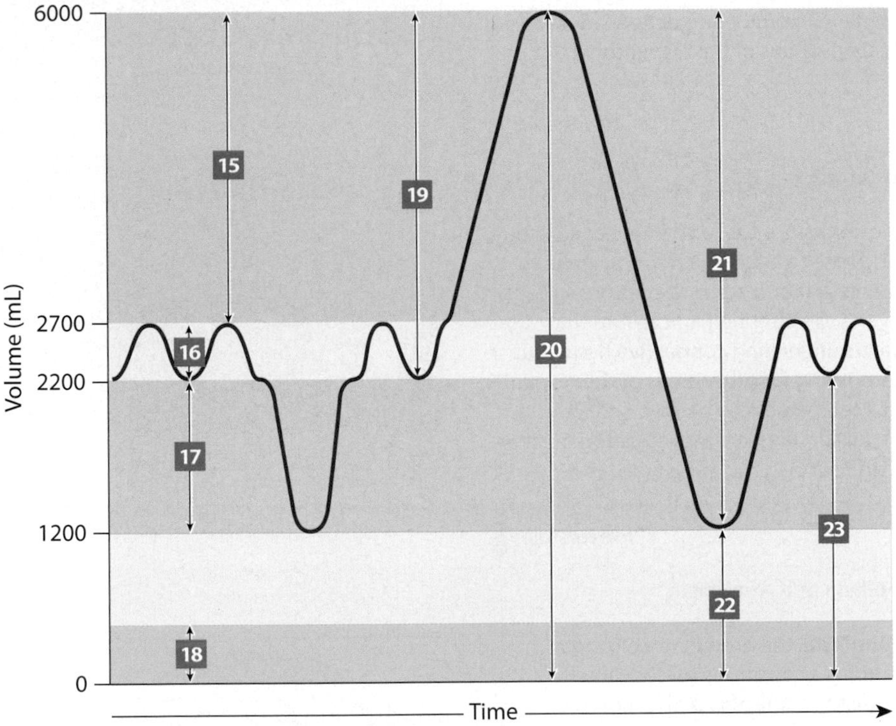
6000
2700
2200
1200
0
Volume (mL)
Time
15
16
17
18
19
20
21
22
23

Section integration

Compare and contrast external respiration, pulmonary ventilation, and internal respiration.

24 ______

Study Outline

SECTION 1 • Anatomy of the Respiratory System

21.1 The respiratory system has an upper and lower respiratory tract with different functions p. 787

1. The **respiratory system** is composed of structures for breathing, or **pulmonary ventilation** (airflow to and from lungs), and gas exchange.
2. The respiratory system is divided into an upper and lower respiratory system.
3. The **respiratory tract** is a passageway that carries air to and from the exchange surfaces of the lungs. It has a **conducting portion** and a **respiratory portion**.

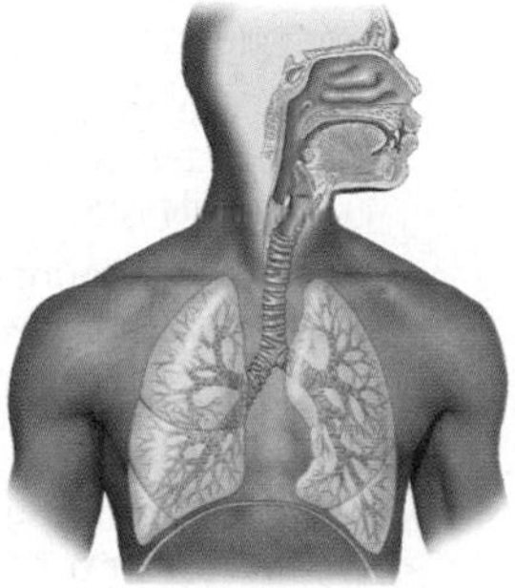

4. The conducting portion extends from the nasal cavity down to the larger bronchioles of the lungs. The respiratory portion includes the smallest bronchioles and **alveoli** (air-filled sacs where gas exchange occurs).
5. The respiratory system functions to provide an extensive surface area for gas exchange between the air and circulating blood, move air to and from the gas exchange surfaces of the lungs, protect the respiratory surfaces, produce sounds for communication, and facilitate olfaction.

21.2 The respiratory mucosa is protected by the respiratory defense system p. 788

6. The **respiratory defense system** lines the conducting portion of the respiratory tract, and prevents contamination of the delicate exchange surfaces by debris or pathogens.
7. The **respiratory mucosa** lines the conducting portion of the respiratory tract.
8. A pseudostratified ciliated columnar epithelium with numerous mucous cells lines the nasal cavity, the superior portion of the pharynx, and the trachea, bronchi, and large bronchioles.
9. The cilia move mucus, debris, and microorganisms toward the pharynx, where they will be swallowed. This is the **mucus escalator**.
10. **Cystic fibrosis (CF)** causes production of a thick and sticky mucus in the conducting portions of the respiratory tract. Potentially lethal infections may result as such dense mucus accumulates because it cannot be transported.

21.3 The upper respiratory system includes the nose, nasal cavity, paranasal sinuses, and pharynx p. 790

11. The nose is the passageway for air entering the respiratory system. It is made of the **bridge of the nose** and **nasal cartilages**. Air enters through the **external nares**, or nostrils.
12. Air then flows past the conchae, causing swirling that can trap particles and assist olfaction.
13. The maxillary, frontal, ethmoid, and sphenoid bones that form the lateral and superior walls of the nasal cavity contain **paranasal sinuses**. The mucus produced in these sinuses drains into the nasal cavity and helps keep it moist.
14. The **nasal septum** divides the nasal cavity, and is formed by the perpendicular plate of the ethmoid bone and vomer.
15. The **pharynx** is divided into the **nasopharynx**, **oropharynx**, and **laryngopharynx**. Other important structures of the upper respiratory system are the **internal nares**, **nasal vestibule**, **hard palate**, and **soft palate**.
16. The extensive network of veins in the nasal cavity delivers body heat to the nasal cavity, so inhaled air is warmed before it leaves the nasal cavity. The heat also evaporates moisture from the epithelium to humidify the incoming air.

21.4 The larynx protects the glottis and produces sounds p. 792

17. The **larynx** is made of three unpaired cartilages: the **epiglottis** (prevents liquids and solid food from entering glottis when swallowing), **thyroid cartilage** (forms most of the anterior and lateral larynx), and **cricoid cartilage** (supports and protects the glottis).
18. The **vestibular ligaments** and the **vocal ligaments** extend between the thyroid cartilage and the **arytenoid cartilages.**
19. When the **glottis** is open, flowing air vibrates its **vocal folds**, or **vocal cords**, that contain the elastic vocal ligaments. Vibration of the vocal folds produces sound waves.
20. Sound production at the larynx is called **phonation**. Clear speech requires **articulation**.

21.5 The trachea, bronchi, and bronchial branches convey air to and from lung gas exchange surfaces p. 794

21. The **trachea** is a flexible airway containing15–20 C-shaped tracheal rings that support and protect its wall. The trachea branches to form left and right **primary bronchi**.

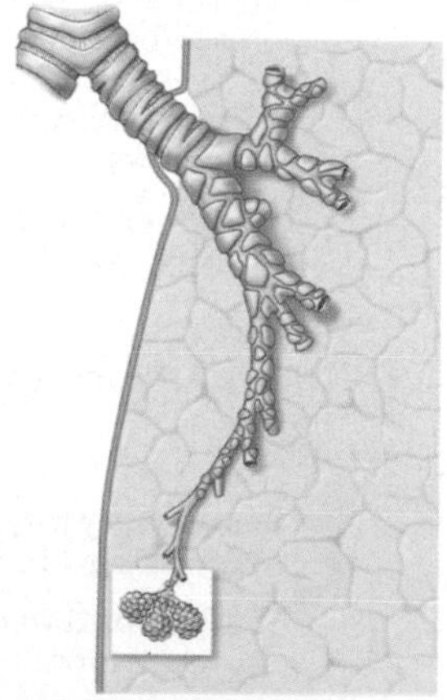

22. The **trachealis muscle** connects the ends of the C-shaped **tracheal cartilages.** Sympathetic stimulation of this muscle increases the diameter of the trachea to allow more airflow.
23. Air-conducting passageways in the lower respiratory tract occur in the following order: trachea, primary bronchi, **secondary bronchi** (three to right lung, three to left lung), **tertiary bronchi**, **bronchioles**, **terminal bronchioles**, and **pulmonary lobules**.
24. Bronchiole walls lack cartilages, but contain smooth muscle. Sympathetic stimulation causes **bronchodilation**, and parasympathetic stimulation causes **bronchoconstriction**. Extreme bronchoconstriction may occur during allergic reactions such as **asthma**.

21.6 The lungs have lobes that are subdivided into bronchopulmonary segments p. 796

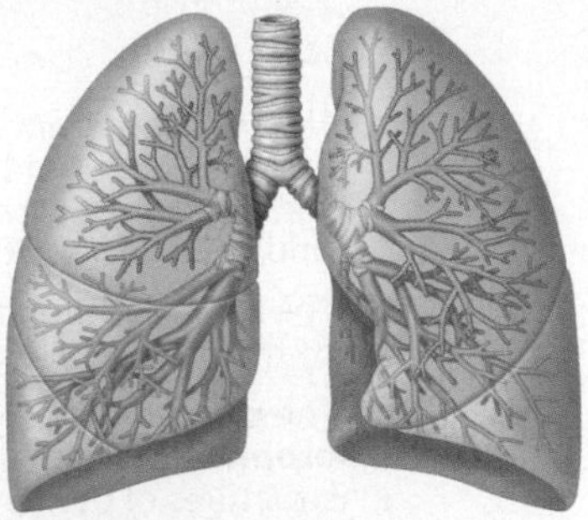

25. The **bronchial tree** is the branching pattern of bronchi and bronchioles.
26. Each tertiary bronchus supplies air to a single **bronchopulmonary segment**. Each lung lobe contains at least two bronchopulmonary segments.
27. The **root of the lung** is a meshwork of dense connective tissue that positions bronchi and other vessels against the lungs.
28. The right lung has three **lobes (superior**, **middle**, and **inferior)**, an **oblique fissure**, and a **horizontal fissure**. The left lung has two **lobes (superior** and **inferior)** and an **oblique fissure**.
29. A **hilum** is on the medial surface of each lung, and is where the bronchi and other vessels enter and exit the lung.

21.7 Pulmonary lobules contain alveoli, where gas exchange occurs p. 798

30. Each lung contains 150 million **alveoli**, or air sacs. They are at the end of the respiratory tract.
31. Each tertiary bronchus delivers air to a single bronchopulmonary segment, where it repeatedly branches into terminal bronchioles.
32. Each terminal bronchiole forms a respiratory bronchiole as it supplies a single **pulmonary lobule**. This is the start of the respiratory portion of the respiratory tract.
33. A **respiratory bronchiole** opens into an **alveolar duct**. Each alveolar duct opens up into an **alveolar sac**. An alveolar sac is connected to many individual alveoli.
34. The alveolar epithelium is primarily a simple squamous epithelium of **type I pneumocytes. Type II pneumocytes** secrete surfactant, an oily secretion that prevents alveoli from collapsing. **Alveolar macrophages** patrol the epithelium.
35. Gas exchange occurs across the **respiratory membrane** at each alveolus.

SECTION 2 • Respiratory Physiology

21.8 Respiratory physiology involves external and internal respiration p. 801

36. **Respiration** refers to external respiration and internal respiration.
37. **External respiration** involves the exchange of oxygen and carbon dioxide between the body and the external environment.
38. **Internal respiration** involves the absorption of oxygen and release of carbon dioxide by tissues.
39. **Pulmonary ventilation**, or breathing, involves the physical movement of air into and out of the lungs. It primarily functions to maintain **alveolar ventilation**, the movement of air into and out of the alveoli.

21.9 Pulmonary ventilation is driven by pressure changes within the pleural cavities p. 802

40. The relationship between pressure (P) and volume (V) is inversely proportional. If the volume of a container is decreased by half, the pressure within it will double; if the volume is doubled, the pressure decreases by half. This is called **Boyle's law**.
41. Movements of the diaphragm and rib cage cause a change in volume of the thoracic cavity. At the start of a breath, no air moves as the pressures inside and outside the thoracic cavity are identical.
42. Air will flow from high pressure to lower pressure. As the thoracic cavity volume enlarges during inhalation, air flows into the lungs. When the thoracic cavity volume decreases in exhalation, increasing air pressure forces air out of the lungs.
43. The direction of airflow is determined by the difference between atmospheric pressure and **intrapulmonary pressure**.
44. The amount of air moved during inhalation and exhalation is called **tidal volume** (V_T). At rest this is approximately 500 mL.

21.10 Respiratory muscles in various combinations adjust the tidal volume to meet respiratory demands p. 804

45. **Inspiratory muscles** are involved with inhalation, and **expiratory muscles** are involved with exhalation.
46. The **primary inspiratory muscles** are the diaphragm and external intercostals. The **accessory inspiratory muscles** are sternocleidomastoid, scalene, pectoralis minor, and serratus anterior.
47. The **accessory expiratory muscles** are internal intercostals, transversus thoracis, external oblique, rectus abdominis, and internal oblique.
48. Pulmonary volumes and capacities include **tidal volume** (**V_T**), **inspiratory reserve volume (IRV)**, **inspiratory capacity**, **vital capacity**, **functional residual capacity (FRC)**, **residual volume**, **total lung capacity**, **minimal volume**, and **expiratory reserve volume (ERV)**.

21.11 Pulmonary ventilation must be closely regulated to meet tissue oxygen demands p. 806

49. The respiratory system adjusts pulmonary ventilation to meet the oxygen demands of the body.
50. **Respiratory rate** is the number of breaths per minute. The normal rate for a resting adult is 12–18 breaths per minute, and for children it is 18–20 breaths per minute.
51. The volume of air moved each minute is called **respiratory minute volume** (V_E). It is calculated by multiplying respiratory rate (f) by tidal volume (V_T). Tidal volume at rest averages 500 mL per breath. Respiratory minute volume at rest is about 6 L/min.
52. **Alveolar ventilation** (**V_A**) is the amount of air reaching the alveoli each minute. Some of the air never reaches the alveoli because it remains in the conducting portion of the lungs. This is the **anatomic dead space** (**V_D**), and at rest is nearly 150 mL of the 500 mL of tidal volume.
53. Alveolar ventilation equals respiratory rate multiplied by tidal volume minus anatomic dead space.
54. Whenever the demand for oxygen increases, both tidal volume and respiratory rate must be increased.

21.12 Gas diffusion depends on the partial pressures and solubilities of gases p. 808

55. The **partial pressure (P)** of a gas is the pressure exerted by a single gas in a mixture of gases.
56. All the partial pressures added together equal the total pressure exerted by the gas mixture; this is known as **Dalton's law**.
57. A gas in solution is directly proportional to the partial pressure of that gas. This principle is known as **Henry's law**.
58. In external respiration, blood arriving at the pulmonary arteries has a lower P_{O_2} and a higher P_{CO_2} than does alveolar air. Hence, diffusion across the respiratory membrane increases blood P_{O_2} and decreases blood P_{CO_2}.
59. The values for pulmonary arterial blood are a P_{O_2} of 40 mm Hg, and a P_{CO_2} of 45 mm Hg. The values for alveolar air are a P_{O_2} of 100 mm Hg, and a P_{CO_2} of 40 mm Hg. Blood reaches equilibrium with alveolar air and departs the alveoli at a P_{O_2} of 100 mm Hg, and a P_{CO_2} of 40 mm Hg.
60. Partial pressure of oxygen drops to 95 mm Hg as it mixes with venous blood from the conducting passageways. Therefore, P_{O_2} in the peripheral capillaries is 95 mm Hg. Normal interstitial fluid has a P_{O_2} of 40 mm Hg.
61. In internal respiration, blood enters systemic capillaries at a P_{O_2} of 95 mm Hg, and a P_{CO_2} of 40 mm Hg. Interstitial fluid is a P_{O_2} of 40 mm Hg, and a P_{CO_2} of 45 mm Hg. Blood reaches equilibrium and departs systemic capillaries with a P_{O_2} of 40 mm Hg, and a P_{CO_2} of 45 mm Hg.

21.13 Almost all the oxygen in the blood is transported bound to hemoglobin within red blood cells p. 810

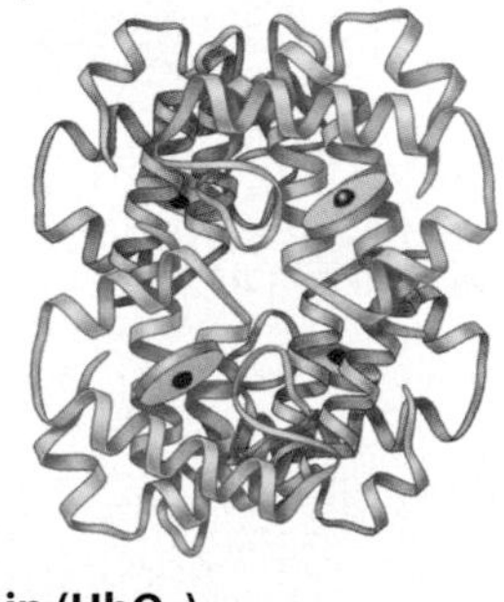

62. In the blood leaving the alveolar capillaries, only 1.5 percent of the oxygen molecules are in solution. The rest are bound to the **heme units** of hemoglobin (Hb) molecules.
63. A hemoglobin molecule has four globular protein subunits, each containing a heme unit. Thus, each hemoglobin molecule can reversibly bind up to four oxygen molecules, forming **oxyhemoglobin (HbO_2)**.
64. **Hemoglobin saturation** is the percentage of heme units containing bound oxygen. An **oxygen-hemoglobin saturation curve** shows the saturation of hemoglobin at different partial pressures of oxygen. Above 60 mm Hg, hemoglobin is 90 percent saturated.
65. The **Bohr effect** is the change of oxygen-hemoglobin saturation due to pH. Normal blood pH is 7.35–7.45. As pH increases, the saturation curve shifts to the left (hemoglobin releases less oxygen); as pH decreases, it shifts to the right (hemoglobin releases more oxygen).
66. The higher the temperature, the more readily hemoglobin releases oxygen.

21.14 Most carbon dioxide transport occurs through the reversible formation of carbonic acid p. 812

67. After entering the bloodstream, a CO_2 molecule is converted to a molecule of carbonic acid (70%), bound to the protein portion of hemoglobin molecules (23%), or dissolved in plasma (7%). All three reactions are reversible.
68. **Carbonic anhydrase** converts 70 percent of the CO_2 in blood to carbonic acid. The carbonic acid molecules immediately dissociate into a hydrogen ion and a bicarbonate ion. The bicarbonate ions move into the plasma in exchange for chloride ions in an event called the **chloride shift**.
69. Some 23 percent of the CO_2 is bound to the globular protein portion of the Hb molecule in **carbaminohemoglobin ($HbCO_2$)**.
70. Only 7 percent of CO_2 is dissolved in plasma.

21.15 Pulmonary disease can affect both lung elasticity and airflow p. 814

71. **Compliance** describes how easily lungs expand. **Resistance** describes how much force is required to inflate or deflate lungs.
72. **Chronic obstructive pulmonary disease (COPD)** is a general term for a disorder restricting airflow and decreasing alveolar ventilation.
73. **Asthma** is the term used when symptoms are acute and intermittent. The terms **chronic bronchitis** and **emphysema** are usually applied when signs and symptoms are chronic and progressive with occasional crises due to infections.

21.16 Respiratory control mechanisms involve interacting centers in the brain stem p. 816

74. Respiratory control involves the **respiratory rhythmicity centers** in the medulla oblongata, the **apneustic** and **pneumotaxic centers** in the pons, and the **higher centers**: the hypothalamus, limbic system, and cerebral cortex.

21.17 Respiratory reflexes provide rapid automatic adjustments in pulmonary ventilation p. 818

75. Normally, P_{CO_2} is the most important factor stimulating chemoreceptors and thereby influencing respiratory factors.
76. An increase in arterial P_{CO_2} **(hypercapnia)** is most commonly caused by hypoventilation. This occurs when respiratory activity is insufficient for CO_2 removal, causing CO_2 to increase in blood.
77. A decrease in arterial P_{CO_2} **(hypocapnia)** occurs when the rate and depth of respiration exceed the demands for O_2 delivery and CO_2 removal, as in **hyperventilation**.
78. Baroreceptors in the carotid and aortic sinuses trigger adjustments in respiratory minute volume.
79. **Inflation reflexes** prevent overexpansion of the lungs. **Deflation reflexes** stimulate the inspiratory centers when the lungs are deflating.
80. **Protective reflexes** include sneezing and coughing. Both reflexes involve **apnea**, a period when breathing has stopped.

21.18 Respiratory function decreases with age; smoking makes matters worse p. 820

81. Respiratory function (vital capacity and compliance) decreases with age. Never having smoked, or stopping smoking, can prevent serious respiratory deterioration.
82. **Lung cancer** kills more people than any other type of cancer. Smoking causes 85–90 percent of all cancers.

Chapter Review Questions

Labeling

Provide the values, in millimeters of mercury (mm Hg), on this illustration of the partial pressures of oxygen and carbon dioxide during external respiration in the pulmonary circuit and internal respiration in the systemic circuit.

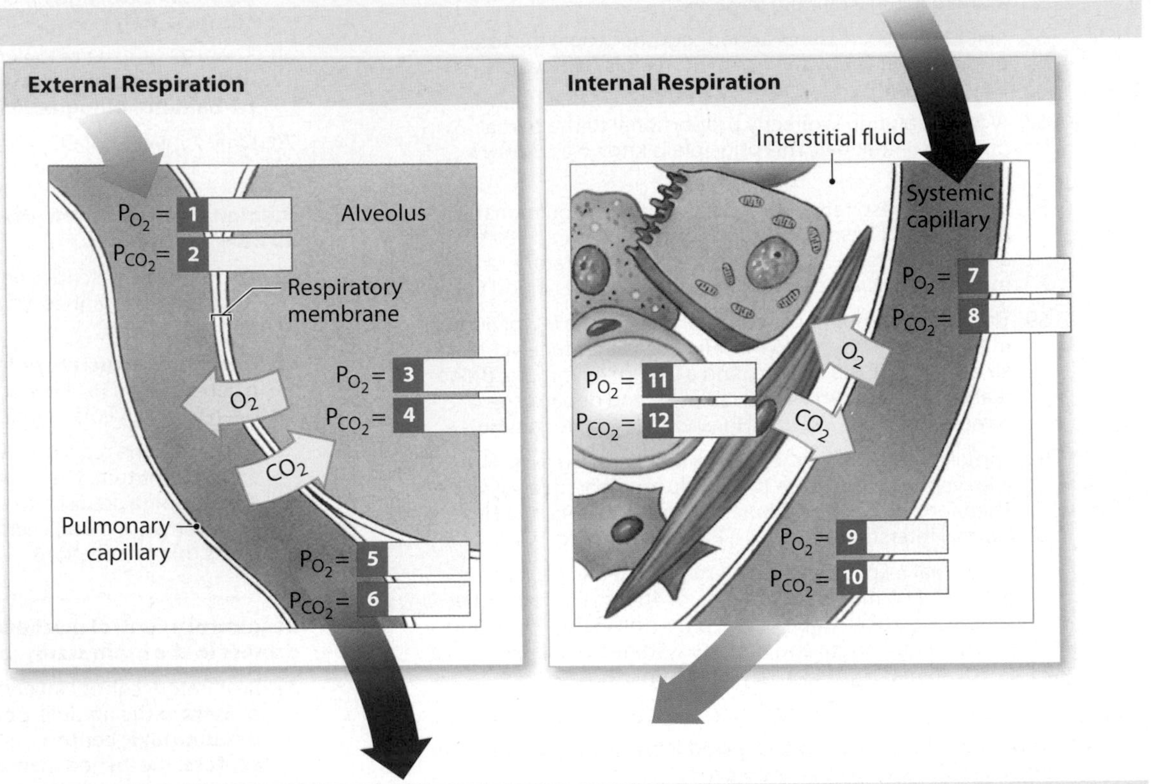
External Respiration
P_{O_2} = 1
P_{CO_2} = 2
Alveolus
Respiratory membrane
P_{O_2} = 3
P_{CO_2} = 4
O_2
CO_2
Pulmonary capillary
P_{O_2} = 5
P_{CO_2} = 6
Internal Respiration
Interstitial fluid
Systemic capillary
P_{O_2} = 7
P_{CO_2} = 8
O_2
CO_2
P_{O_2} = 11
P_{CO_2} = 12
P_{O_2} = 9
P_{CO_2} = 10

Matching

Match each lettered term with the most closely related description.

a. respiratory rate
b. heme unit of Hb molecule
c. left lung
d. increasing pH
e. pulmonary ventilation
f. vital capacity
g. total lung capacity
h. right lung
i. globular unit of Hb molecule
j. respiratory minute volume
k. resistance
l. alveolar ventilation

13 Airflow to and from lungs
14 Volume of air reaching alveoli each minute
15 Has two lobes
16 Binds CO_2
17 Volume of air moved into and out of lungs each minute
18 Binds O_2
19 Number of breaths each minute
20 Hb molecule releases less O_2
21 Force required to inflate/deflate lungs
22 Maximum amount of air moved into and out of lungs in one respiratory cycle
23 Has three lobes
24 Vital capacity plus residual volume

13 ____
14 ____
15 ____
16 ____
17 ____
18 ____
19 ____
20 ____
21 ____
22 ____
23 ____
24 ____

Multiple choice

Select the correct answer from the list provided.

25 Surfactant
- a) protects the outer surface of the lungs.
- b) phagocytizes small particles.
- c) replaces mucus in the alveoli.
- d) helps prevent the alveoli from collapsing.

26 One of the responses to an abnormally low P_{CO_2} would be
- a) decreased respiratory rate.
- b) increased respiratory rate.
- c) hypercapnea.
- d) chronic obstructive pulmonary disease.

27 The narrow opening in the larynx for inhaled air is called the

- [ ] a) epiglottis.
- [ ] b) rima glottidis.
- [ ] c) pharynx.
- [ ] d) internal nares.

28 Which of the following muscle combinations are the primary muscles of inspiration?

- [ ] a) diaphragm and the internal intercostal muscles
- [ ] b) diaphragm and the external intercostal muscles
- [ ] c) abdominal muscles and the internal intercostal muscles
- [ ] d) abdominal muscles and the external intercostal muscles

29 Boyle's law states that

- [ ] a) all the partial pressures in a gas mixture added together equals the total pressure exerted by the gas mixture.
- [ ] b) the amount of a particular gas in solution is directly proportional to the partial pressure of that gas.
- [ ] c) the shape of hemoblogin molecules changes as the number of bound O_2 molecules increases, and these changes affect its affinity for oxygen.
- [ ] d) if you reduce the volume of a flexible container by half, the pressure within it will double.

30 $V_E = f \times V_T$ is the formula used to calculate

- [ ] a) Dalton's law.
- [ ] b) alveolar ventilation.
- [ ] c) respiratory minute volume.
- [ ] d) vital capacity.

Short answer

31 Name the three regions of the pharynx, and identify where each region is located.

32 By what three mechanisms is carbon dioxide transported in the bloodstream?

33 Why is breathing through the nasal cavity more desirable than breathing through the mouth?

MasteringA&P®

Access more chapter study tools online in the MasteringA&P Study Area:

- **Chapter Quizzes, Chapter Practice Test, Art-labeling Activities, Animations, MP3 Tutor Sessions, and Clinical Case Studies**

- **Practice Anatomy Lab** PAL™
- **Interactive Physiology** iP®
- **A&P Flix** A&PFlix™
- **PhysioEx** PhysioEx™

Chapter Integration • Applying what you have learned

Spring break snorkeling danger

Jake is an active 22-year-old college student who enjoys water sports. In fact, he has an athletic scholarship to swim on the university swim team. While on spring break at an oceanfront hotel with both beach and pool access, Jake decides to go snorkeling. The sun is shining, the air temperature is 80°F, and the ocean is a refreshing 72°F. His normal alveolar ventilation rate (AVR) during mild exercise is 6.0 L/min, and his snorkel has a volume of 50 mL.

After snorkeling in the ocean all afternoon, Jake heads to the pool to meet up with some friends. Before diving in, he hyperventilates for several minutes. After he enters and begins swimming underwater, he blacks out and almost drowns. Fortunately, friends notice that something's wrong and pull him out of the pool to safety. Jake recovers within 15 seconds and does not require pulmonary resuscitation or medical treatment.

1 **Assuming a constant tidal volume of 500 mL and an anatomic dead space of 150 mL, what would Jake's respiratory rate have to be for him to maintain an AVR of 6.0 L/min while snorkeling?**

2 **What caused Jake to black out and nearly drown?**

22 The Digestive System

LEARNING OUTCOMES

These Learning Outcomes correspond by number to this chapter's modules and indicate what you should be able to do after completing the chapter.

SECTION 1 • Organization of the Digestive System

- 22.1 Name the major and accessory organs of the digestive system.
- 22.2 Describe the functional histology of the digestive tract.
- 22.3 Describe the structural and functional features of smooth muscle tissue.
- 22.4 Explain the processes by which materials move through the digestive tract.

SECTION 2 • Digestive Tract

- 22.5 Name the structures and primary functions of the digestive tract organs.
- 22.6 Describe the anatomy of the oral cavity, and discuss the functions of its structures.
- 22.7 Describe the types of teeth, and differentiate between the primary and secondary dentition.
- 22.8 Describe the anatomy and functions of the pharynx and esophagus, and explain the swallowing process.
- 22.9 Explain the embryonic development of the mesenteries, and describe the mesenteries that remain in adulthood.
- 22.10 Describe the anatomy of the stomach and its histological features.
- 22.11 Describe the anatomy of the stomach relating to its role in digestion and absorption.
- 22.12 Describe the anatomy of the intestinal tract and its histological features.
- 22.13 Describe the anatomy and physiology of the small intestine.
- 22.14 Discuss the major digestive hormones and their primary effects.
- 22.15 Explain the regulation of gastric activity by central and local mechanisms.
- 22.16 Describe the gross anatomy of the three segments of the large intestine.
- 22.17 Describe the large intestine's histology and role in fecal compaction, and explain the defecation reflex.

SECTION 3 • Accessory Digestive Organs

- 22.18 Describe the functions of the accessory organs of the digestive system.
- 22.19 Discuss the structure and functions of the salivary glands.
- 22.20 Describe the anatomy and location of the liver and gallbladder.
- 22.21 Describe the histological features of liver tissue.
- 22.22 Describe the structure, functions, and regulatory activities of the gallbladder and pancreas.
- 22.23 **CLINICAL MODULE** Briefly describe several digestive system disorders.

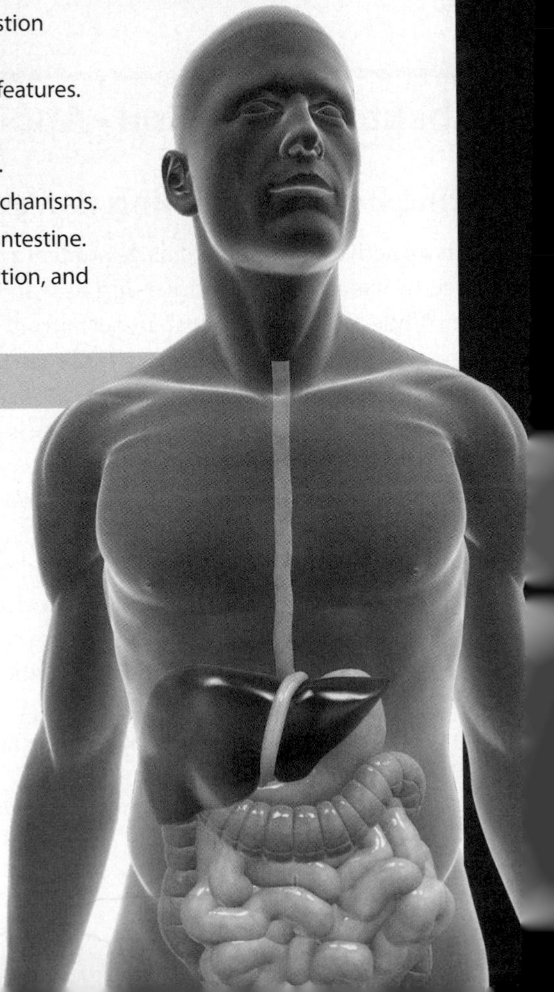

Learning Outcomes are repeated at the bottom of each module.

The digestive system consists of the digestive tract and accessory organs

The **digestive system** consists of a muscular tube, the digestive tract—also called the gastrointestinal (GI) tract, or alimentary canal—and various accessory organs. Food enters the mouth and passes along the length of the digestive tract. On the way, accessory organs mechanically process the food and produce secretions containing water, enzymes, buffers, and other components that assist in preparing organic and inorganic nutrients for absorption.

1 The digestive system works with other systems to support tissues that have no direct connection with the outside environment and no other means of obtaining nutrients.

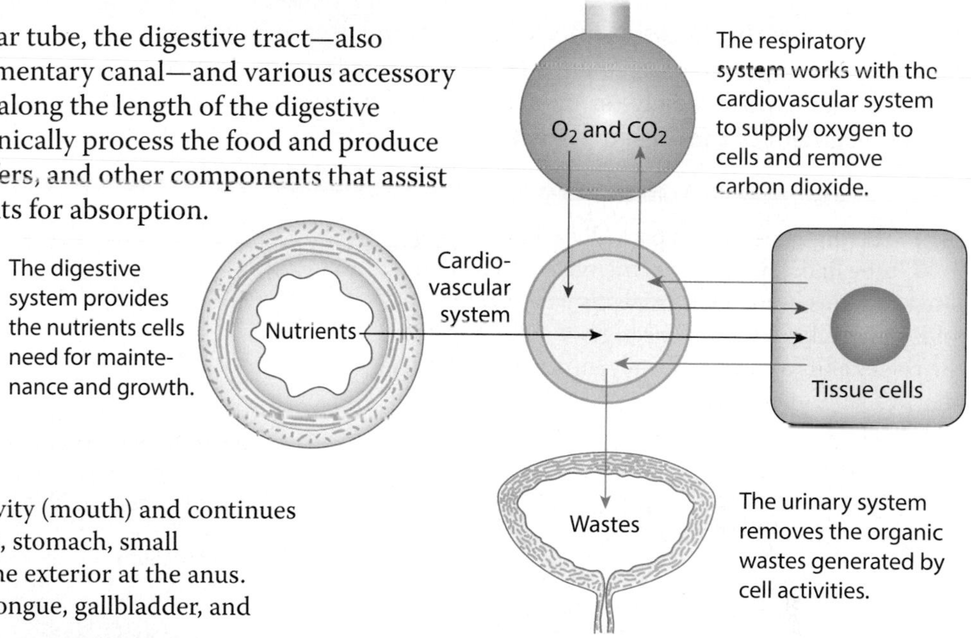
The digestive system provides the nutrients cells need for maintenance and growth.
Nutrients
Cardio-vascular system
O_2 and CO_2
The respiratory system works with the cardiovascular system to supply oxygen to cells and remove carbon dioxide.
Tissue cells
Wastes
The urinary system removes the organic wastes generated by cell activities.

2 The **digestive tract** begins at the oral cavity (mouth) and continues through the pharynx (throat), esophagus, stomach, small intestine, and large intestine, which opens to the exterior at the anus. Accessory digestive organs include the teeth, tongue, gallbladder, and *glandular organs*, such as the salivary glands, liver, and pancreas.

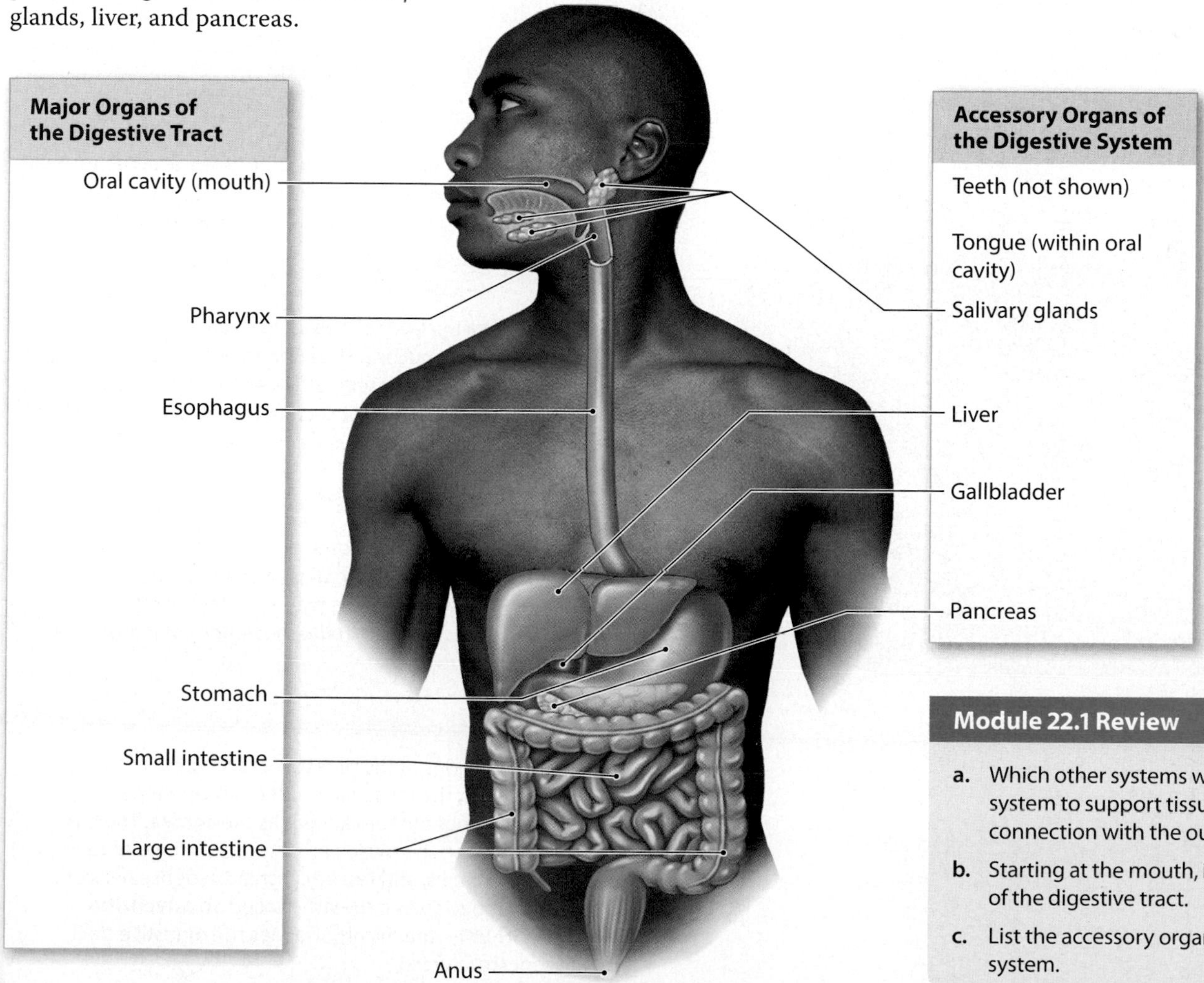
Major Organs of the Digestive Tract
Oral cavity (mouth)
Pharynx
Esophagus
Stomach
Small intestine
Large intestine
Anus
Accessory Organs of the Digestive System
Teeth (not shown)
Tongue (within oral cavity)
Salivary glands
Liver
Gallbladder
Pancreas

Module 22.1 Review

a. Which other systems work with the digestive system to support tissues that have no direct connection with the outside environment?

b. Starting at the mouth, identify the organs of the digestive tract.

c. List the accessory organs of the digestive system.

LO 22.1 Name the major and accessory organs of the digestive system.

Module 22.2

The digestive tract is a muscular tube lined by a mucous epithelium

The structure of the digestive tract varies to some degree from one region of the digestive tract to another. The composite view shown in this module most closely resembles the small intestine, the longest segment of the digestive tract.

1 The digestive tract is basically a long muscular tube lined by a mucous membrane. The lining often contains permanent ridges as well as temporary folds that disappear as the passageway fills. Together the ridges and folds dramatically increase the surface area available for absorption.

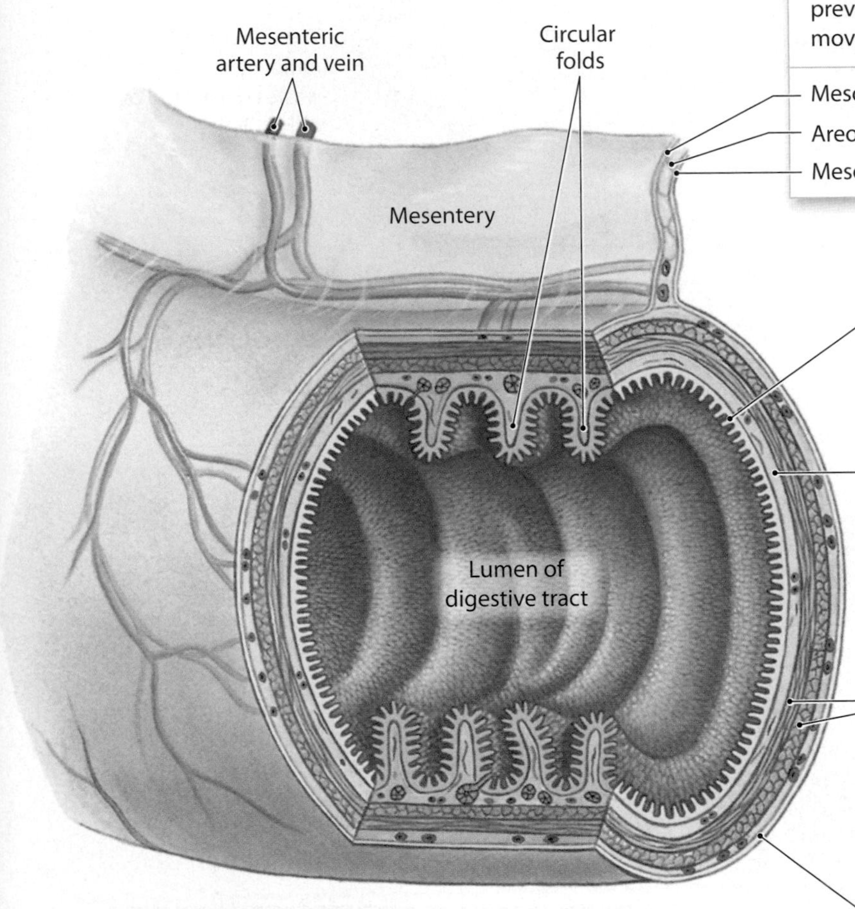

Mesenteric artery and vein
Circular folds
Mesentery
Lumen of digestive tract

Mesentery

A **mesentery** is a double sheet of peritoneal membrane. The areolar tissue between the mesothelial surfaces provides an access route for the passage of blood vessels, nerves, and lymphatic vessels to and from the digestive tract. Mesenteries also stabilize the positions of the attached organs and prevent the intestines from becoming entangled during digestive movements or sudden changes in body position.

Mesothelium
Areolar tissue
Mesothelium

Major Layers of the Digestive Tract

Mucosa

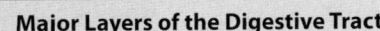

The inner lining, or **mucosa**, of the digestive tract is a mucous membrane consisting of an epithelium, which is moistened by glandular secretions, and a lamina propria of areolar tissue.

Submucosa

The **submucosa** is a layer of dense irregular connective tissue that surrounds the mucosa. The submucosa has large blood vessels and lymphatic vessels, and in some regions it also contains exocrine glands that secrete buffers and enzymes into the lumen of the digestive tract.

Muscularis Externa

The **muscularis externa** is dominated by smooth muscle cells in two layers, an inner circular layer and an outer longitudinal layer. These layers play an essential role in mechanical processing and in the movement of materials along the digestive tract.

Serosa

Along most portions of the digestive tract within the peritoneal cavity, the muscularis externa is covered by a layer of visceral peritoneum known as the **serosa**. There is no serosa covering the muscularis externa of the oral cavity, pharynx, esophagus, and rectum. Instead, a dense network of collagen fibers forms a sheath—called an **adventitia** (ad-ven-TISH-uh)—that firmly attaches the digestive tract to adjacent structures.

2 Each layer in the wall of the digestive tract has a different set of functions. The mucosa has several structural features that enhance its abilities to absorb nutrients from the lumen of the tract. Those features are shown in this diagrammatic view.

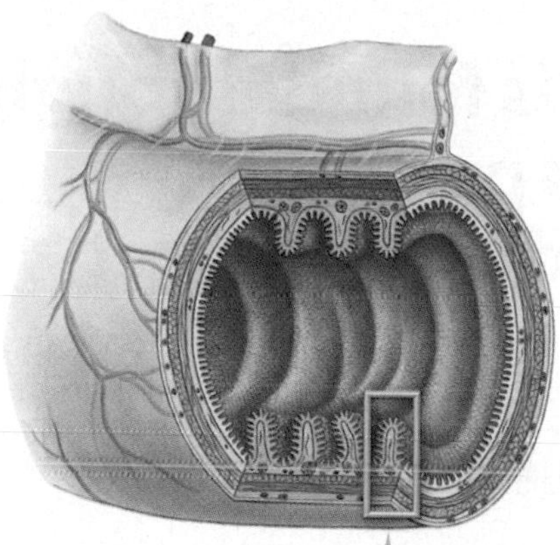

Secretory Glands

The secretions of gland cells located in the mucosa and submucosa—or in accessory glandular organs—are carried to the epithelial surfaces by ducts.

Circular folds (plicae circulares) are permanent transverse folds in the intestinal lining.

Components of the Mucosa

The mucosal epithelium of the oral cavity, pharynx, and esophagus are lined by a stratified squamous epithelium. The stomach, small intestine, and almost the entire length of the large intestine are lined by a simple columnar epithelium that contains mucous cells.

Villi (singular, *villus*) are small mucosal projections, like tiny fingers, that stick into the lumen of the small intestine. These are permanent features that further increase the surface area available for absorption.

The **lamina propria** consists of a layer of areolar tissue that also contains blood vessels, sensory nerve endings, lymphatic vessels, smooth muscle cells, scattered areas of lymphoid tissue, and, in some regions, mucous glands.

The **muscularis** (mus-kū-LAIR-is) **mucosae** (mū-KŌ-sē) consists of two concentric layers of smooth muscle. The inner layer encircles the lumen (the circular muscle layer), and the outer layer contains muscle cells oriented parallel to the long axis of the tract (the longitudinal muscle layer). Contractions in these layers alter the shape of the lumen and move the circular folds and villi.

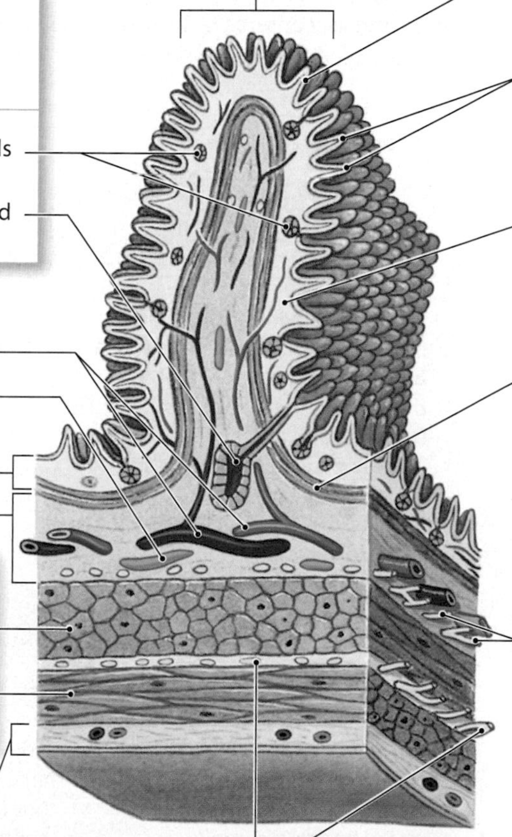

Mucosal glands
Submucosal gland
Artery and vein
Lymphatic vessel
Mucosa
Submucosa
Muscularis Externa
Circular muscle layer
Longitudinal muscle layer
Serosa

Superficial to the muscularis externa, the submucosa contains the **submucosal plexus**, a nerve network that contains sensory neurons, parasympathetic ganglionic neurons, and sympathetic postganglionic fibers that innervate the mucosa and submucosa.

The muscularis externa contains the **myenteric** (mī-en-TER-ik) **plexus** (*mys*, muscle + *enteron*, intestine). This network of parasympathetic ganglia, sensory neurons, interneurons, and sympathetic postganglionic fibers lies sandwiched between the circular and longitudinal muscle layers. The myenteric plexus and the submucosal plexus contain neurons of the enteric nervous system, and most digestive activities are locally controlled. In general, parasympathetic stimulation increases muscle tone and activity; sympathetic stimulation decreases muscle tone and activity.

Module 22.2 Review

a. What is the importance of the mesenteries?

b. Name the four layers of the digestive tract beginning from the lumen of the digestive tract.

c. Compare the submucosal plexus with the myenteric plexus.

LO 22.2 Describe the functional histology of the digestive tract.

Module 22.3

Smooth muscle tissue is found throughout the body, but it plays a particularly prominent role in the digestive tract

Smooth muscle tissue forms sheets, bundles, or sheaths around other tissues in almost every organ. Smooth muscles around blood vessels regulate blood flow through vital organs. In the digestive and urinary systems, smooth muscle sphincters regulate the movement of materials along internal passageways.

1 Smooth muscle cells are relatively long and slender, ranging from 5 to 10 μm in diameter and from 30 to 200 μm in length. Wherever smooth muscle tissue forms layers, the cells are aligned parallel to one another. In the digestive tract, there is usually an inner circular layer and an outer longitudinal layer. In a longitudinal section of the digestive tract, the muscle cells in the circular layer of the muscularis externa look like little round balls, whereas those in the longitudinal layer look like long spindles.

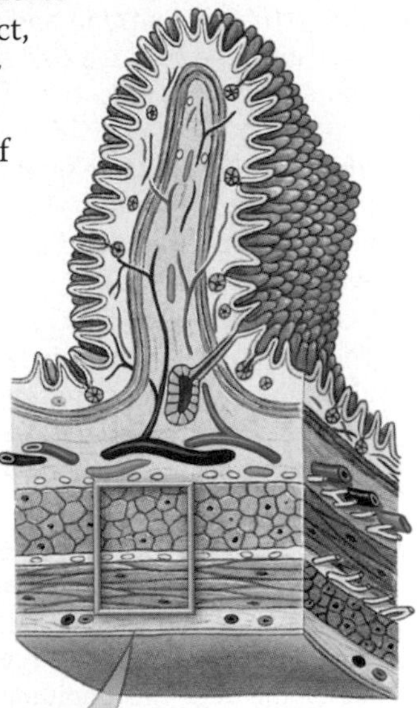

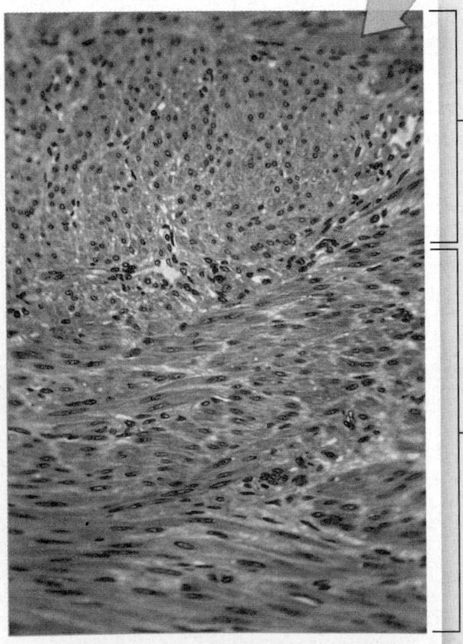
Circular muscle layer
Longitudinal muscle layer

Smooth muscle LM × 65

2 Like skeletal and cardiac muscle cells, actin and myosin filaments are involved in smooth muscle contractions. However, the organization of those filaments differs from that of skeletal or cardiac muscle cells. A smooth muscle fiber has no T tubules, and the sarcoplasmic reticulum forms a loose network throughout the sarcoplasm. Smooth muscle cells also lack myofibrils and sarcomeres. As a result, this tissue also has no striations and is called nonstriated muscle.

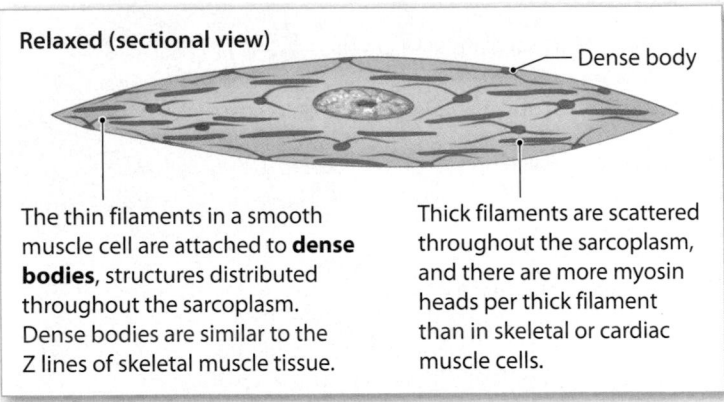

Relaxed (sectional view)

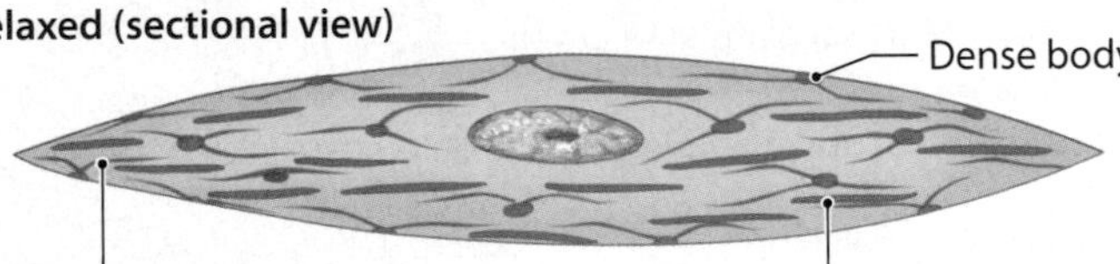
Dense body

The thin filaments in a smooth muscle cell are attached to **dense bodies**, structures distributed throughout the sarcoplasm. Dense bodies are similar to the Z lines of skeletal muscle tissue.

Thick filaments are scattered throughout the sarcoplasm, and there are more myosin heads per thick filament than in skeletal or cardiac muscle cells.

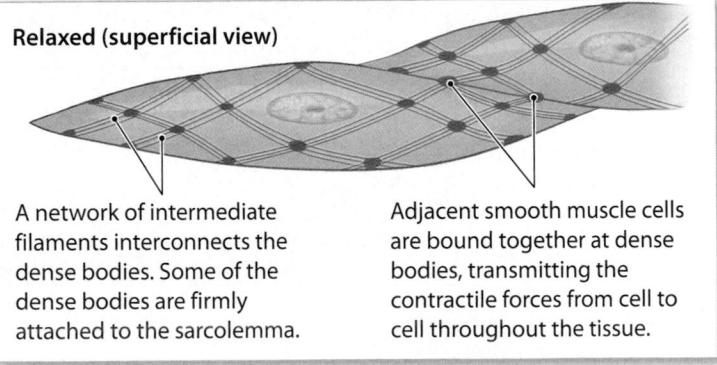

Relaxed (superficial view)

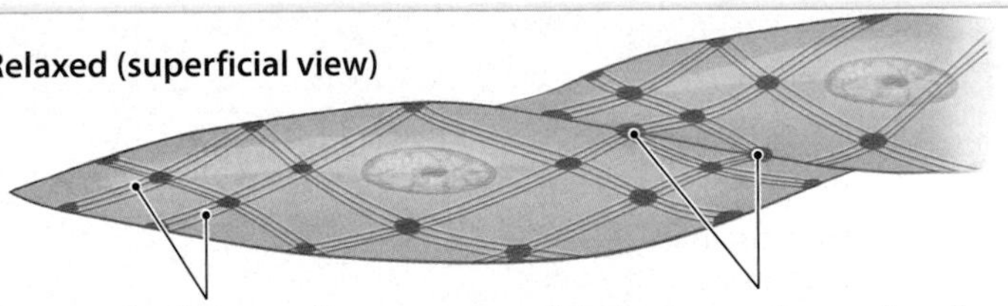

A network of intermediate filaments interconnects the dense bodies. Some of the dense bodies are firmly attached to the sarcolemma.

Adjacent smooth muscle cells are bound together at dense bodies, transmitting the contractile forces from cell to cell throughout the tissue.

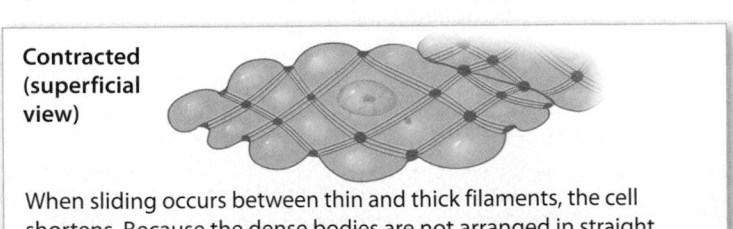

Contracted (superficial view)

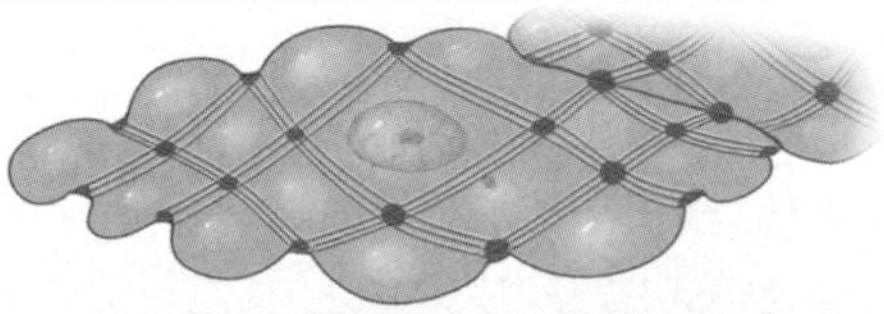

When sliding occurs between thin and thick filaments, the cell shortens. Because the dense bodies are not arranged in straight lines, the muscle cell twists like a corkscrew as it contracts.

3 Smooth muscle tissue can be divided into two subtypes based on distinctively different methods of innervation and control.

Types of Smooth Muscle

Multi-unit Smooth Muscle

Multi-unit smooth muscle cells are innervated in motor units comparable to those of skeletal muscles, but each smooth muscle cell may be connected to more than one motor neuron. Multi-unit smooth muscle tissue is located in the iris of the eye, where it regulates the diameter of the pupil; along portions of the male reproductive tract; within the walls of large arteries; and in the arrector pili muscles of the skin.

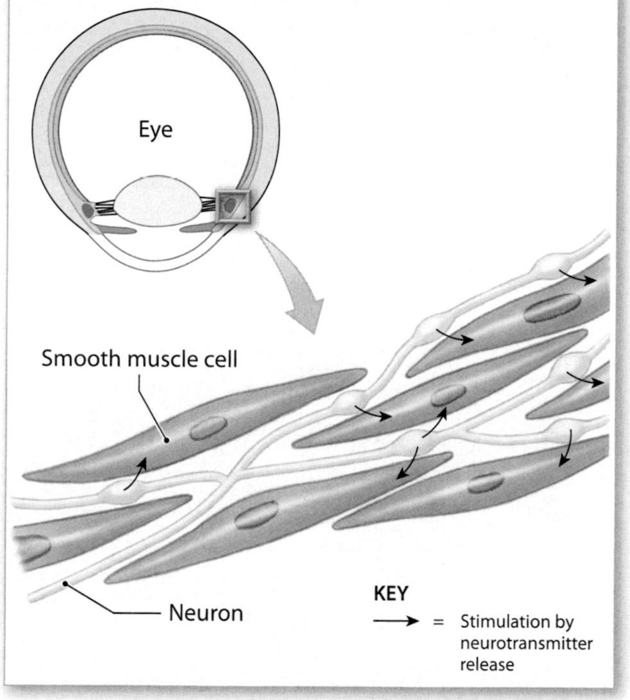

Eye
Smooth muscle cell
Neuron
KEY
= Stimulation by neurotransmitter release

Visceral Smooth Muscle

Most **visceral smooth muscle cells** lack a direct contact with any motor neuron. These muscle cells are arranged in sheets or layers, with adjacent muscle cells electrically connected by gap junctions and mechanically connected by dense bodies. As a result, whenever one muscle cell contracts, the stimulus for contraction can travel to adjacent smooth muscle cells, and the contraction spreads in a wave throughout the layer. A contraction can occur in response to neural, hormonal, or chemical stimuli. In addition, many visceral smooth muscle networks show rhythmic cycles of activity triggered by **pacesetter cells** that contract spontaneously at regular intervals. Visceral smooth muscle cells are located in the walls of the digestive tract, the gallbladder, the urinary bladder, and many other internal organs.

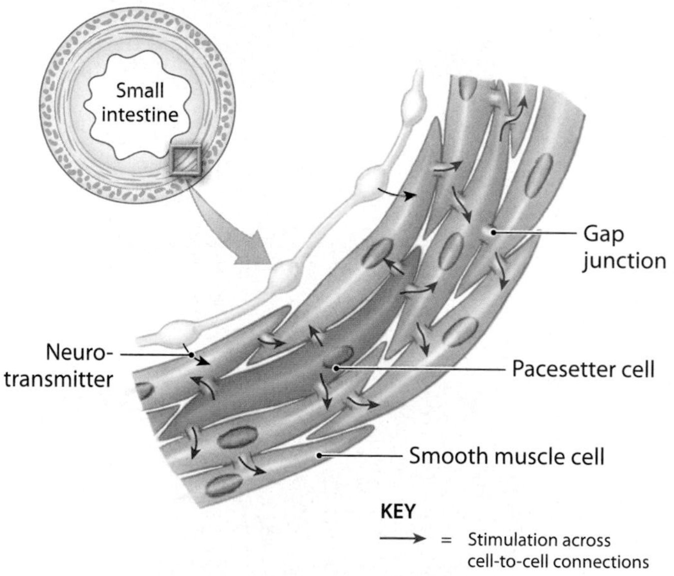

Small intestine
Gap junction
Neuro-transmitter
Pacesetter cell
Smooth muscle cell
KEY
= Stimulation across cell-to-cell connections

Because the thick and thin filaments of smooth muscle are scattered and are not organized into sarcomeres, tension development and resting length are not directly related. A stretched smooth muscle soon adapts to its new length and retains the ability to contract on demand. This ability to function over a wide range of lengths is called **plasticity**. Smooth muscle can contract over a range of lengths four times greater than that of skeletal muscle. This is extremely important in organs like the stomach, the intestines, the urinary bladder, or the uterus, which must undergo major changes in size and shape. These smooth muscle tissues have a normal background level of activity known as **smooth muscle tone**. Neural, hormonal, or local chemical factors can increase or decrease smooth muscle tone and alter the degree of tension in the wall of a muscular organ.

Module 22.3 Review

a. Describe the orientation of smooth muscle fibers in the muscularis externa of the digestive tract.

b. Identify the structural characteristics of smooth muscle fibers.

c. Why can smooth muscle contract over a wider range of resting lengths than skeletal muscle?

LO 22.3 Describe the structural and functional features of smooth muscle tissue.

Smooth muscle contractions mix the contents of the digestive tract and propel materials along its length

The coordinated contractions of the smooth muscle layers in the muscularis externa play a vital role in both moving material along the digestive tract (through peristalsis) and in mechanical processing (through segmentation).

Peristalsis

1 Food enters the digestive tract as a moist, compact mass known as a **bolus** (BŌ-lus). The muscularis externa propels materials from one portion of the digestive tract to another by contractions known as **peristalsis** (per-i-STAL-sis). During peristalsis, a wave of contraction in the circular muscles forces the bolus forward.

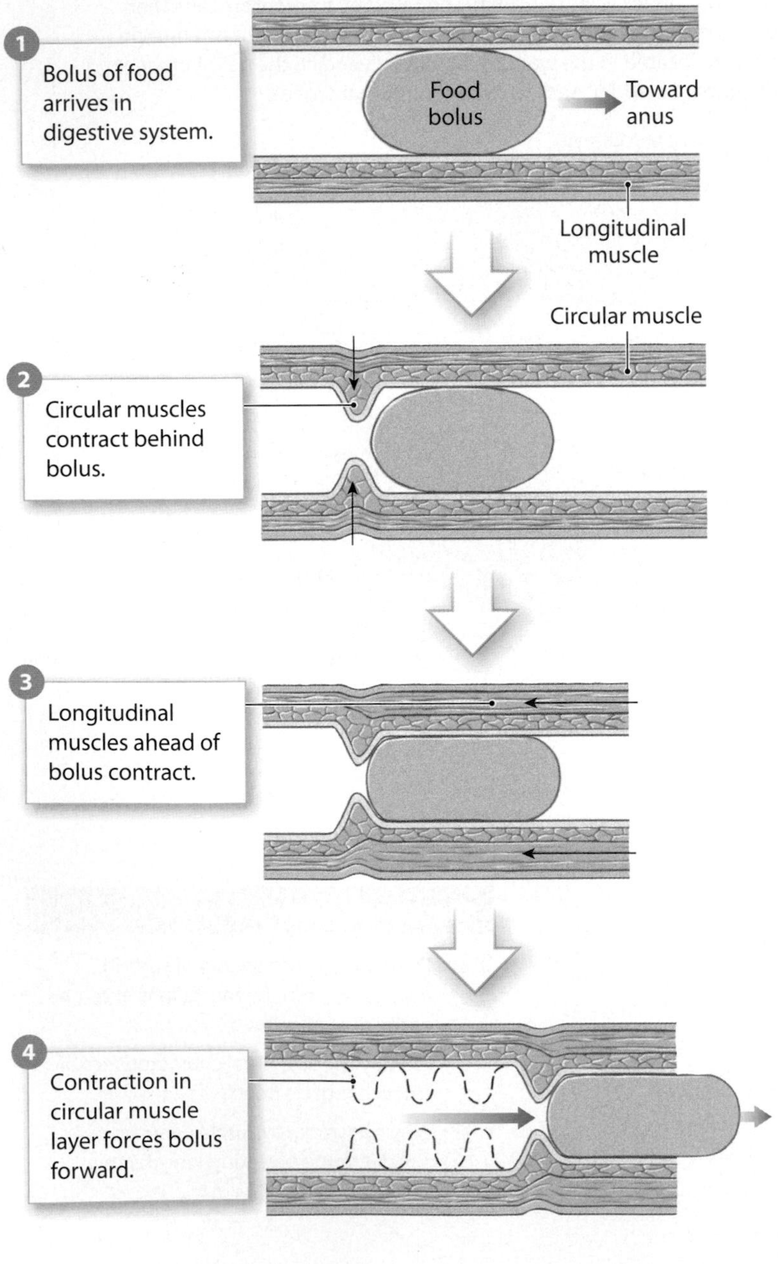

1
Bolus of food arrives in digestive system.
Food bolus
Toward anus
Longitudinal muscle
Circular muscle
2
Circular muscles contract behind bolus.
3
Longitudinal muscles ahead of bolus contract.
4
Contraction in circular muscle layer forces bolus forward.

Segmentation

2 Most areas of the small intestine and some portions of the large intestine undergo cycles of contraction that churn and fragment the bolus, mixing the contents with intestinal secretions. This activity is called **segmentation**. These rhythmic cycles of contraction do not follow a set pattern and thus do not push materials along the tract in any one direction.

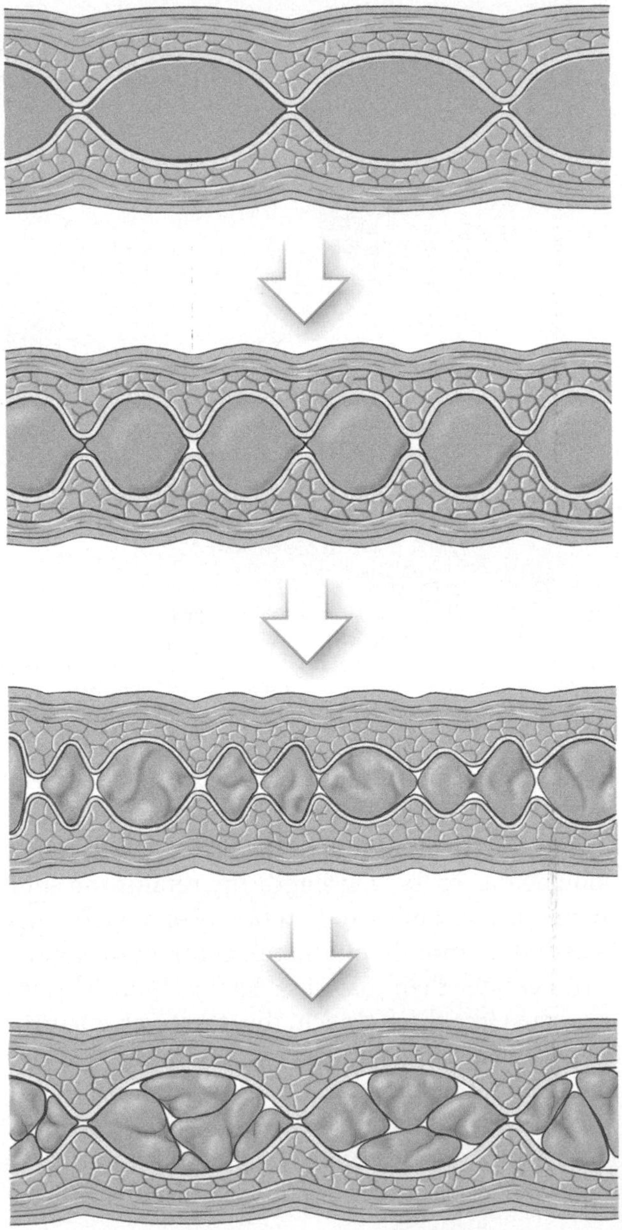

3 Three major mechanisms regulate and control digestive activities. The primary stimuli involved are local factors that may in turn activate neural or hormonal control mechanisms.

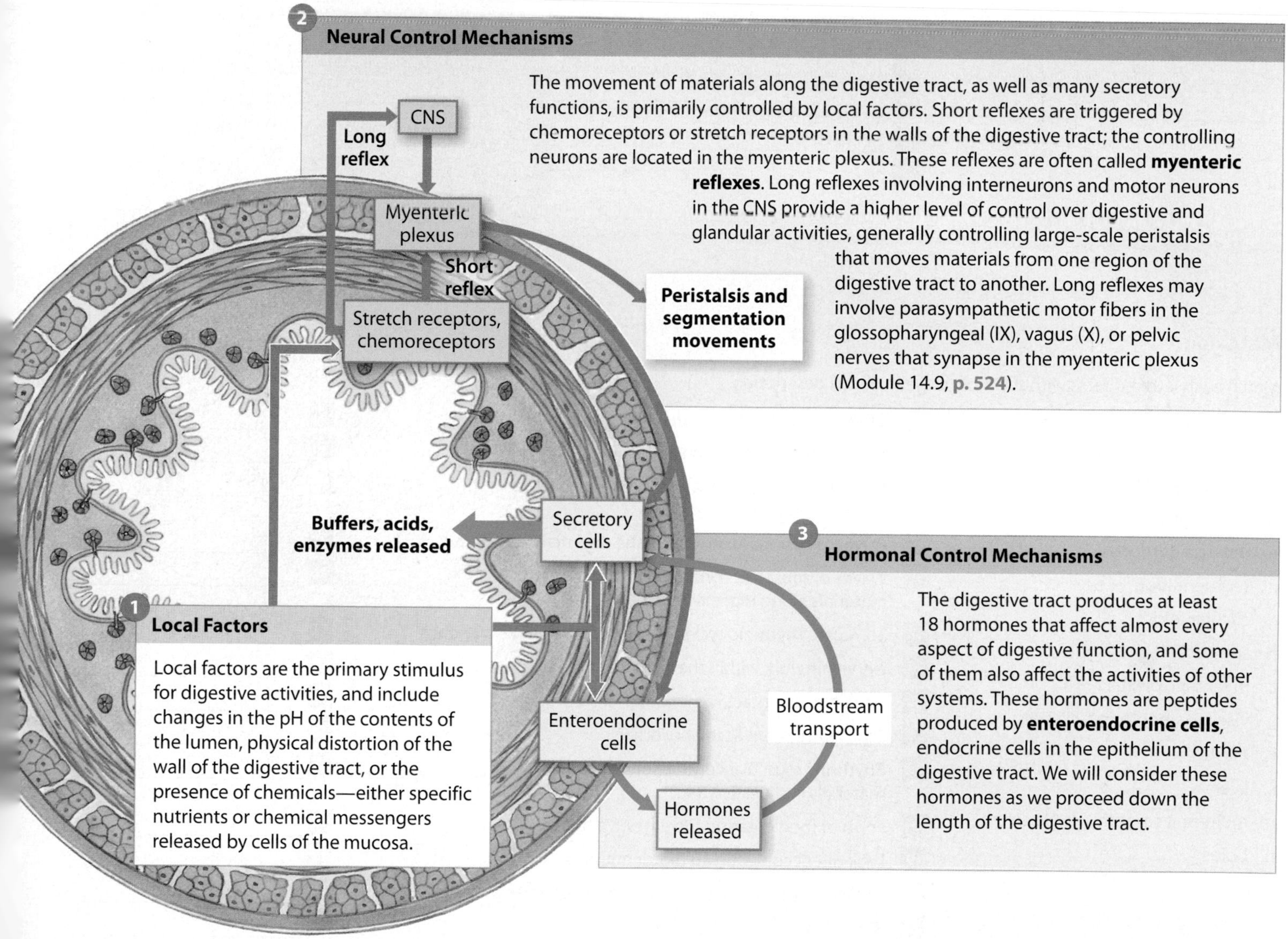

CNS
Long reflex
Myenteric plexus
Short reflex
Stretch receptors, chemoreceptors
Peristalsis and segmentation movements
Buffers, acids, enzymes released
Secretory cells
Enteroendocrine cells
Bloodstream transport
Hormones released

2 **Neural Control Mechanisms**

The movement of materials along the digestive tract, as well as many secretory functions, is primarily controlled by local factors. Short reflexes are triggered by chemoreceptors or stretch receptors in the walls of the digestive tract; the controlling neurons are located in the myenteric plexus. These reflexes are often called **myenteric reflexes**. Long reflexes involving interneurons and motor neurons in the CNS provide a higher level of control over digestive and glandular activities, generally controlling large-scale peristalsis that moves materials from one region of the digestive tract to another. Long reflexes may involve parasympathetic motor fibers in the glossopharyngeal (IX), vagus (X), or pelvic nerves that synapse in the myenteric plexus (Module 14.9, p. 524).

1 **Local Factors**

Local factors are the primary stimulus for digestive activities, and include changes in the pH of the contents of the lumen, physical distortion of the wall of the digestive tract, or the presence of chemicals—either specific nutrients or chemical messengers released by cells of the mucosa.

3 **Hormonal Control Mechanisms**

The digestive tract produces at least 18 hormones that affect almost every aspect of digestive function, and some of them also affect the activities of other systems. These hormones are peptides produced by **enteroendocrine cells**, endocrine cells in the epithelium of the digestive tract. We will consider these hormones as we proceed down the length of the digestive tract.

Module 22.4 Review

a. Which is more efficient in propelling intestinal contents along the digestive tract: peristalsis or segmentation? Why?

b. Cite the major mechanisms that regulate and control digestive activities.

c. Describe enteroendocrine cells.

Labeling

Label the structures of the digestive tract in the accompanying figure.

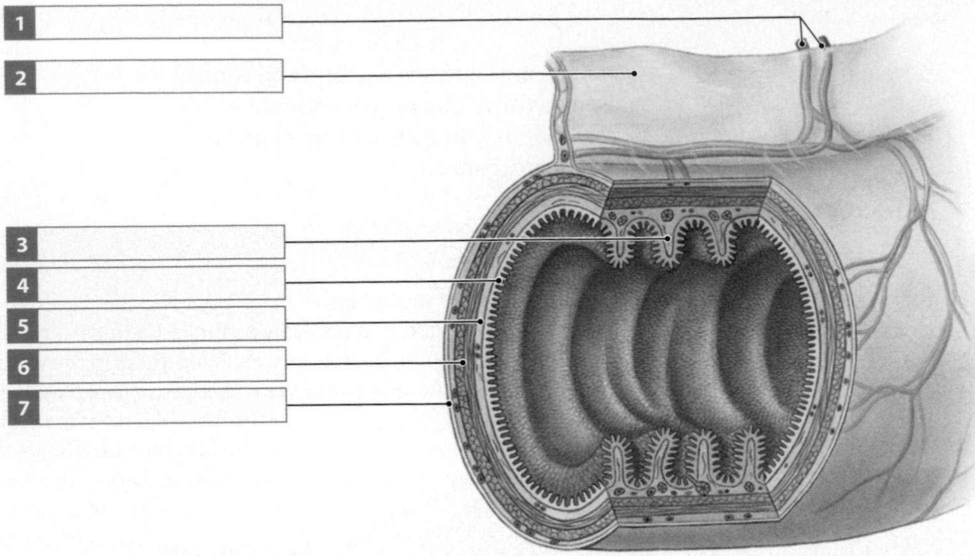
1
2
3
4
5
6
7

Matching

Match each lettered term with the most closely related description.

a. lamina propria
b. peristalsis
c. pacesetter cells
d. esophagus
e. muscularis mucosa
f. segmentation
g. circular folds
h. sphincter
i. myenteric plexus
j. visceral smooth muscle cells
k. plasticity
l. liver
m. multi-unit smooth muscle cells
n. bolus

8 Digestive tube between the pharynx and stomach
9 Moves circular folds and villi
10 Areolar tissue layer containing blood vessels, nerve endings, and lymphatics
11 Permanent transverse folds in the digestive tract lining
12 Waves of muscular contractions that propel materials along digestive tract
13 Stimulate rhythmic cycles of activity along digestive tract
14 Nerve network within the muscularis externa
15 Have direct contact with motor neurons
16 Digestive system accessory organ
17 Rhythmic muscular contractions that mix materials in digestive tract
18 Form of food entering the digestive tract
19 Lack direct contact with motor neurons
20 Ability of smooth muscle cells to function over varied lengths
21 Ring of muscle tissue

8 ______
9 ______
10 ______
11 ______
12 ______
13 ______
14 ______
15 ______
16 ______
17 ______
18 ______
19 ______
20 ______
21 ______

Section integration

How would a decrease in smooth muscle tone affect the digestive processes and possibly cause constipation (infrequent bowel movement)?

22 ______

The digestive tract begins with the mouth and ends with the anus

1 The digestive tract is a muscular tube approximately 10 m (33 ft) long. It can be divided into regions that differ in histological structure and functional properties. In this section we consider each of the major organs of the digestive tract, which are summarized below.

Major Organs of the Digestive Tract
Oral Cavity (Mouth)
Mechanical processing with accessory organs (teeth and tongue), moistening, mixing with salivary secretions
Pharynx
Muscular propulsion of materials into the esophagus
Esophagus
Transport of materials to the stomach
Stomach
Chemical breakdown of materials by acid and enzymes; mechanical processing through muscular contractions
Small Intestine
Enzymatic digestion and absorption of water, organic substrates, vitamins, and ions
Large Intestine
Dehydration and compaction of indigestible materials in preparation for elimination

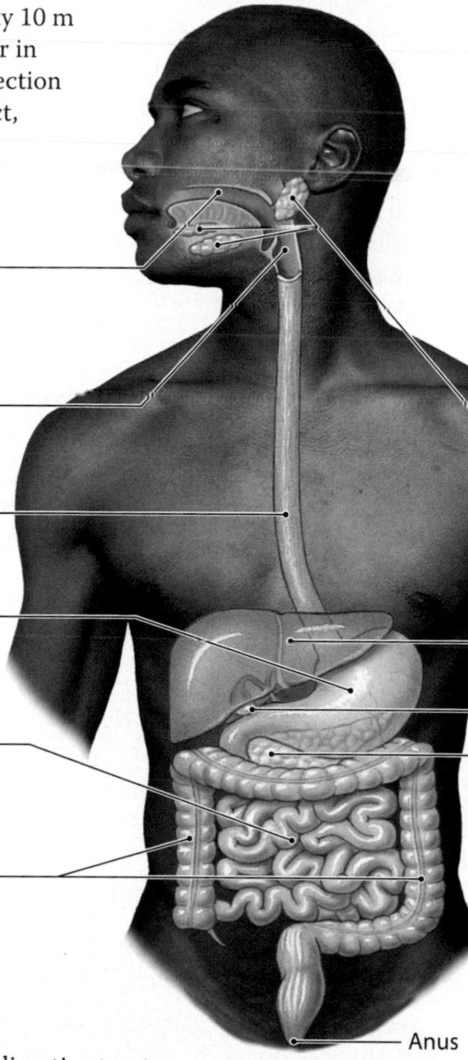
Anus

Accessory Organs of the Digestive System
Teeth
Tongue
Salivary glands
Liver
Gallbladder
Pancreas

2 This table summarizes the general functions of the digestive tract. The lining of the digestive tract also protects surrounding tissues against digestive acids and enzymes, abrasion, and pathogens within the digestive tract.

Functions of the Digestive Tract

- **Ingestion** occurs when solid food and liquid enter the oral cavity of the digestive tract.
- **Mechanical processing** of food occurs either before it is swallowed or in the proximal portions of the digestive tract.
- **Digestion** is the chemical and enzymatic breakdown of food into small organic molecules that can be absorbed by the digestive epithelium.
- **Secretion** is the release of water, acids, enzymes, buffers, and salts by the epithelium of the digestive tract and by accessory digestive organs.
- **Absorption** is the movement of organic molecules, electrolytes, vitamins, and water across the digestive epithelium and into the interstitial fluid of the digestive tract.
- **Compaction** is the progressive dehydration of indigestible materials and organic wastes prior to elimination from the body. The compacted material is called **feces**. The discharge of feces from the body is called **defecation** (def-e-KĀ-shun).

Module 22.5 Review

a. Define ingestion.

b. Distinguish between digestion and absorption.

c. Define compaction.

LO 22.5 Name the structures and primary functions of the digestive tract organs.

The oral cavity is a space that contains the tongue, teeth, and gums

The **oral cavity** (mouth) is lined by the **oral mucosa**, which has a stratified squamous epithelium. A layer of keratinized cells covers regions exposed to severe abrasion, such as the superior surface of the tongue and the opposing surface of the hard palate. The epithelial lining of the cheeks, lips, and inferior surface of the tongue is relatively thin and nonkeratinized. Nutrients are not absorbed in the oral cavity, but digestion of carbohydrates and lipids begins here. The mucosa inferior to the tongue is thin enough and vascular enough to allow rapid absorption of lipid-soluble drugs, such as nitroglycerin, used to treat angina attacks.

1 This sagittal section introduces the major structures forming the boundaries of the oral cavity (mouth).

Superior Boundary of the Oral Cavity

The **hard palate** is formed by the palatine processes of the maxillary bones and the horizontal plates of the palatine bones.

The muscular **soft palate** lies posterior to the hard palate.

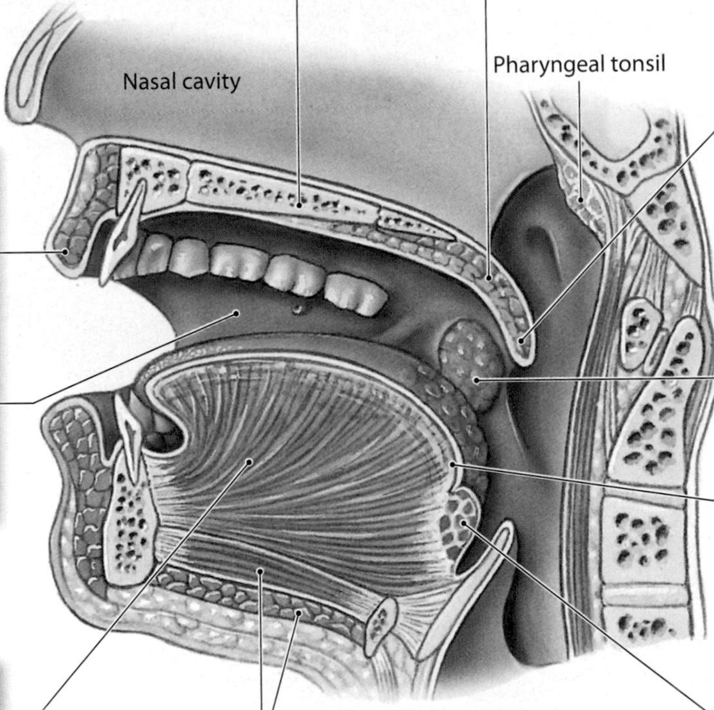
Nasal cavity
Pharyngeal tonsil

Anterior and Lateral Boundary of the Oral Cavity

Anteriorly, the mucosa of each cheek is continuous with that of the lips, or **labia** (LĀ-bē-uh; singular, *labium*).

The mucosae of the **cheeks**, or lateral walls of the oral cavity, are supported by pads of fat and the buccinator muscles.

The **body** of the tongue is a mobile organ that forms part of the oral cavity floor.

Posterior Boundary of the Oral Cavity

The posterior margin of the soft palate supports the **uvula** (Ū-vū-luh), a dangling process that helps prevent food from entering the pharynx prematurely and swings upwards during swallowing to prevent food from entering the nasopharynx.

A palatine tonsil lies on either side of the entrance to the oropharynx.

The **root** of the tongue is the fixed portion that projects into the oropharynx. A V-shaped line of vallate papillae marks the boundary between the root and the body of the tongue.

A pair of lingual tonsils are embedded in the root of the tongue.

Inferior Boundary of the Oral Cavity

The floor of the mouth inferior to the tongue receives extra support from the geniohyoid and mylohyoid muscles.

2 An anterior view of the oral cavity shows additional details not visible in the sagittal section.

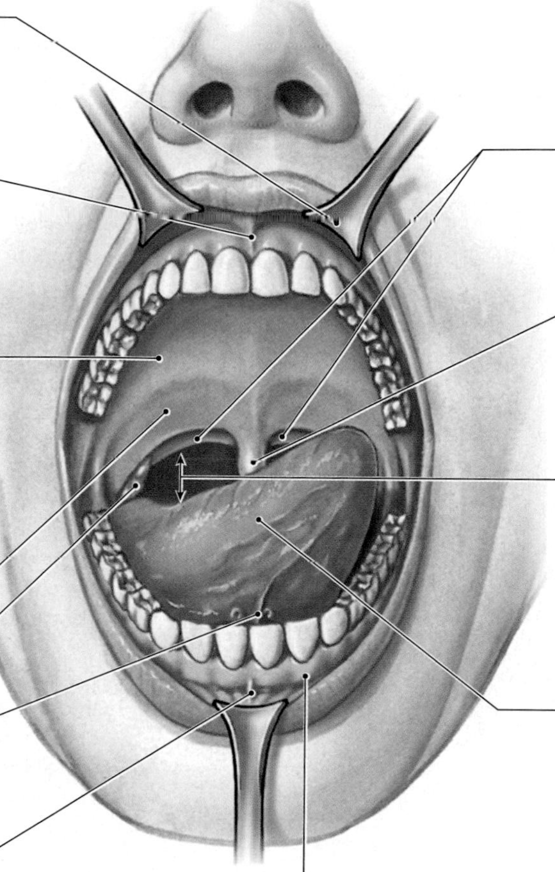

The retractors are placed in the **vestibule**, the space between the cheeks (or lips) and the teeth.
Anteriorly, a delicate **labial frenulum** (FREN-ū-lum; *frenulum*, a small bridle) attaches the upper and lower lips to the gums.
The mucosa covering the hard palate is thick and has complex ridges. When your tongue compresses food against the hard palate, these ridges provide traction.
Soft palate
Palatine tonsil
The **lingual frenulum** attaches the body of the tongue to the floor of the mouth.
Frenulum of lower lip
On either side of the uvula are two pairs of muscular **pharyngeal arches**.
Uvula
On each side, the ridge that connects the more anterior pharyngeal arch with the uvula forms the superior margin of the **fauces** (FAW-sēz), the connection between the oral cavity and the oropharynx. (The root of the tongue forms the inferior margin.)
The **tongue** manipulates materials inside the mouth and may occasionally be used to bring foods into the oral cavity, as when licking an ice cream cone. The tongue's surface is flushed by the secretions of small glands that extend into the underlying lamina propria. These secretions contain water, mucins, and the enzyme **lingual lipase**, which starts digesting lipids even before they leave the oral cavity.
The **gingivae** (JIN-ji-vē), or gums, are ridges of oral mucosa that surround the base of each tooth on the alveolar processes of the maxillary bones and mandible. In most regions, the gingivae are firmly bound to the periostea of the underlying bones.

Three pairs of salivary glands secrete saliva into the oral cavity. These accessory glands, which begin the process of carbohydrate digestion, will be considered in Module 22.19.

Module 22.6 Review

a. The oral cavity is lined by which type of epithelium?

b. Name the structure that forms the roof of the mouth.

c. Describe the location of the fauces.

LO 22.6 Describe the anatomy of the oral cavity, and discuss the functions of its structures.

Module 22.7

Teeth in different regions of the jaws vary in size, shape, and function

1 This is a sectional view through a representative adult tooth. The bulk of each tooth consists of a mineralized matrix similar to that of bone. This material, called **dentin**, differs from bone in that it does not contain cells. Instead, cytoplasmic processes extend into the dentin from cells making up the pulp in the central **pulp cavity**.

The **crown** of the tooth projects into the oral cavity from the surface of the gums.

The **neck** of the tooth marks the boundary between the crown and the root.

The **root** of each tooth sits in a bony cavity called an **alveolus**, or tooth socket.

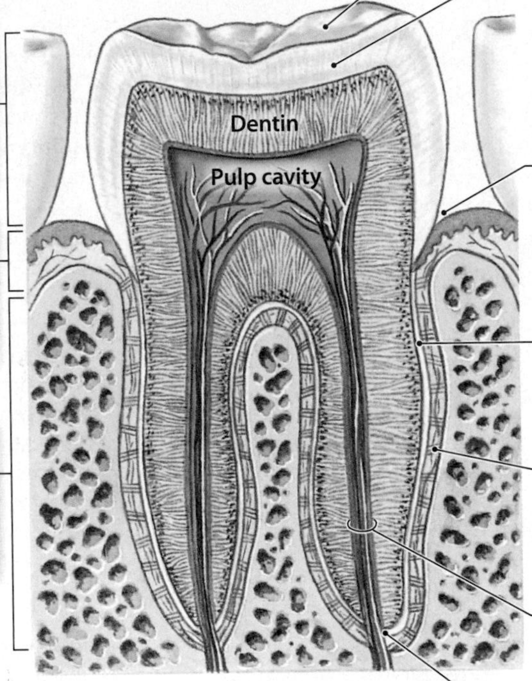

Dentin
Pulp cavity

Components of a Tooth
The portion of the crown that is used for crushing, slicing, or chewing is the **occlusal surface** of the tooth.
A layer of **enamel** covers the dentin of the crown. Enamel, which contains calcium phosphate in a crystalline form, is the hardest biologically manufactured substance. Adequate amounts of calcium, phosphates, and vitamin D during childhood are essential if the enamel coating is to be complete and resistant to decay.
A shallow groove called the **gingival** (JIN-ji-val) **sulcus** surrounds the neck of each tooth. The epithelium of the gingiva is bound to the tooth at the base of the sulcus. This epithelial attachment prevents bacteria from entering the deeper tissues around the root.
A layer of **cementum** (se-MEN-tum) covers the dentin of the root. Cementum is less resistant to erosion than is dentin.
Collagen fibers of the **periodontal ligament** extend from the dentin of the root to the compact bone of the alveolus, creating a strong articulation known as a **gomphosis** (Module 8.1, p. 281).
The pulp cavity receives blood vessels and nerves through the **root canal**, a narrow tunnel within the root of the tooth. The opening into the root canal is called the **apical foramen**.

2 There are four different types of teeth, each with a distinctive shape and root pattern.

Types of Teeth

Incisors (in-SĪ-zerz) are blade-shaped teeth located at the front of the mouth. Incisors are useful for clipping or cutting. These teeth have a single root.	The **cuspids** (KUS-pidz), or canine teeth, are conical, with a sharp ridgeline and a pointed tip. They are used for tearing or slashing. Cuspids have a single root.	**Bicuspids** (bī-KUS-pidz), or premolars, have flattened crowns with prominent ridges. They crush, mash, and grind. Bicuspids have one or two roots.	**Molars** have very large, flattened crowns with prominent ridges adapted for crushing and grinding. Molars in the upper jaw typically have three roots, while those in the lower jaw usually have two roots.

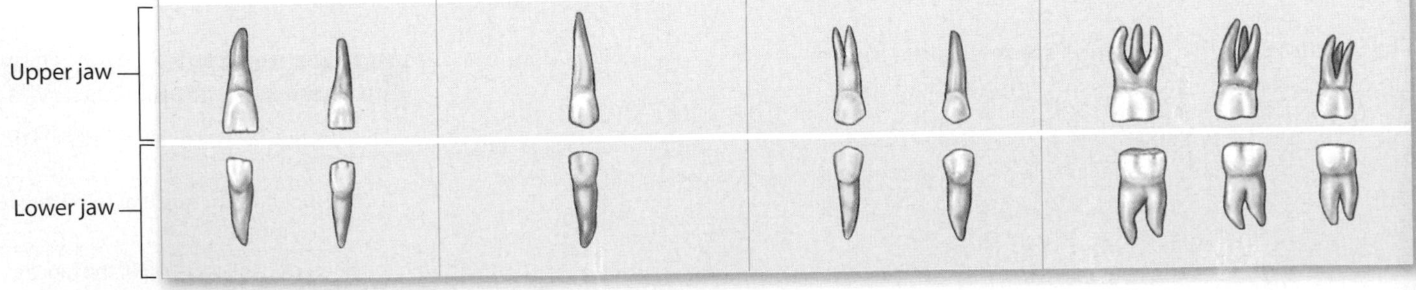

Upper jaw
Lower jaw

3 Two sets of teeth form during embryonic development. The first to erupt through the gums are the **deciduous** (de-SID-ū-us; *deciduus*, falling off) **teeth**, the temporary teeth of the **primary dentition**. Deciduous teeth are also called primary teeth, milk teeth, or baby teeth. At 2 years of age, children have 20 deciduous teeth—five on each side of the upper and lower jaws: two incisors, one cuspid, and a pair of deciduous molars.

4 As the jaws grow larger, the primary dentition is gradually replaced by the **secondary dentition**. Three additional molars appear on each side of the upper and lower jaws as the person ages, bringing the count of permanent teeth to 32. As replacement proceeds, the periodontal ligaments and roots of the primary teeth erode until the deciduous teeth either fall out or are pushed aside by the eruption (emergence) of the secondary teeth.

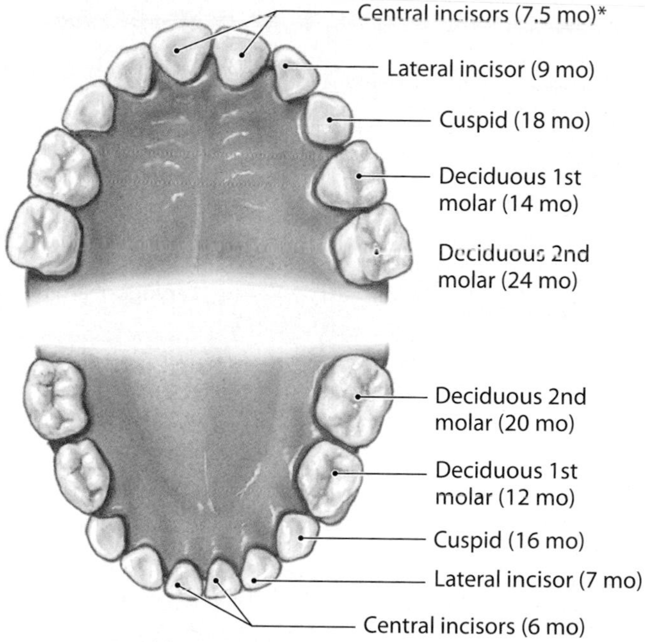

Central incisors (7.5 mo)*
Lateral incisor (9 mo)
Cuspid (18 mo)
Deciduous 1st molar (14 mo)
Deciduous 2nd molar (24 mo)
Deciduous 2nd molar (20 mo)
Deciduous 1st molar (12 mo)
Cuspid (16 mo)
Lateral incisor (7 mo)
Central incisors (6 mo)

Deciduous teeth

*Indicates month of eruption

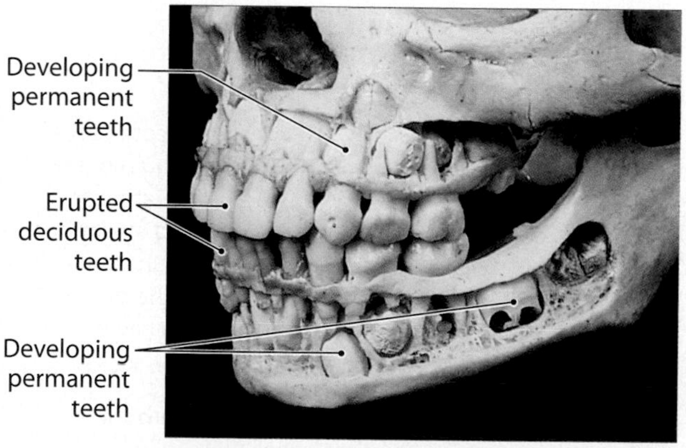

Developing permanent teeth
Erupted deciduous teeth
Developing permanent teeth

Mandible and maxilla exposed to show developing permanent teeth in the skull of a child

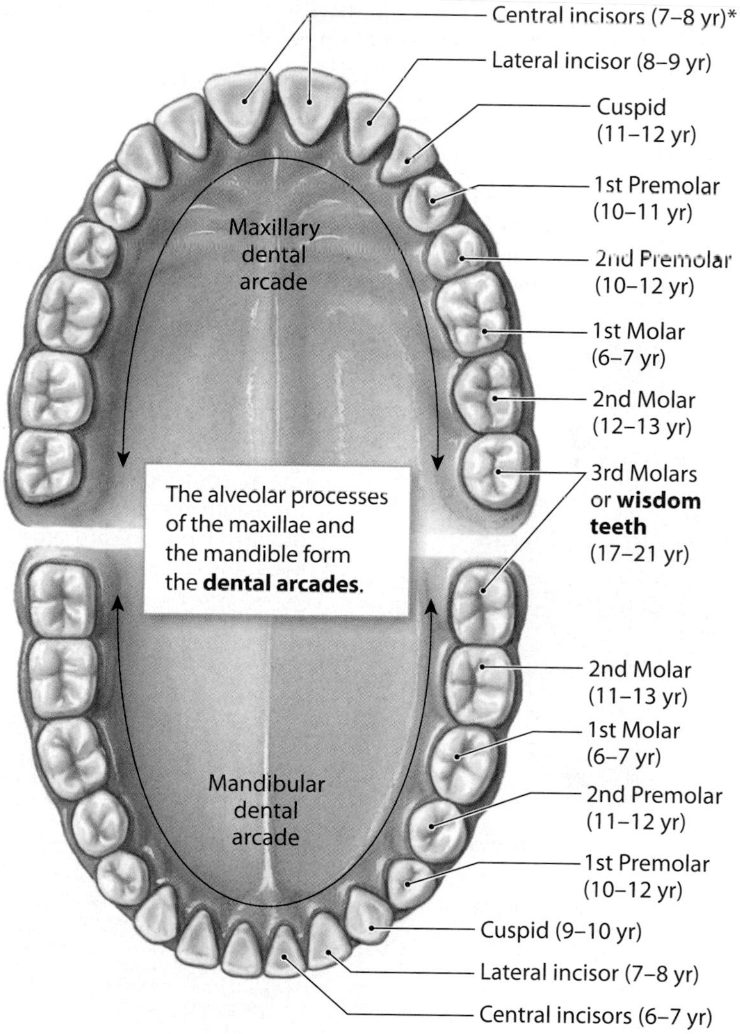

Central incisors (7–8 yr)*
Lateral incisor (8–9 yr)
Cuspid (11–12 yr)
1st Premolar (10–11 yr)
2nd Premolar (10–12 yr)
1st Molar (6–7 yr)
2nd Molar (12–13 yr)
3rd Molars or wisdom teeth (17–21 yr)
Maxillary dental arcade
The alveolar processes of the maxillae and the mandible form the dental arcades.
Mandibular dental arcade
2nd Molar (11–13 yr)
1st Molar (6–7 yr)
2nd Premolar (11–12 yr)
1st Premolar (10–12 yr)
Cuspid (9–10 yr)
Lateral incisor (7–8 yr)
Central incisors (6–7 yr)

Adult teeth, upper and lower jaws

*Indicates year of eruption

When you brush and massage your gums, the stimulation strengthens the epithelial attachment at the gingival sulcus. A condition called **gingivitis**, an inflammation of the gingivae, can occur if the gingival attachment weakens. Severe gingivitis, usually caused by bacterial infection, causes erosion of the gums, root damage, and eventual tooth loss. **Tooth decay** generally results from the action of bacteria that normally inhabit your mouth. Bacteria adhering to the surfaces of the teeth produce a sticky matrix that traps food particles and creates deposits known as **dental plaque**.

Module 22.7 Review

a. Name the four types of teeth and the three main parts of a typical tooth.

b. Differentiate between the primary dentition and the secondary dentition.

c. What is the name sometimes given to the third set of molars?

LO 22.7 Describe the types of teeth, and differentiate between the primary and secondary dentition.

The muscular walls of the pharynx and esophagus play a key role in swallowing

1 The **pharynx** (FAR-ingks), an anatomical space, is a common passageway for solid food, liquids, and air. Food passes through the oropharynx and laryngopharynx on its way to the esophagus. This illustration reviews the major landmarks and boundaries of the pharynx, or throat (Module 21.3, p. 791).

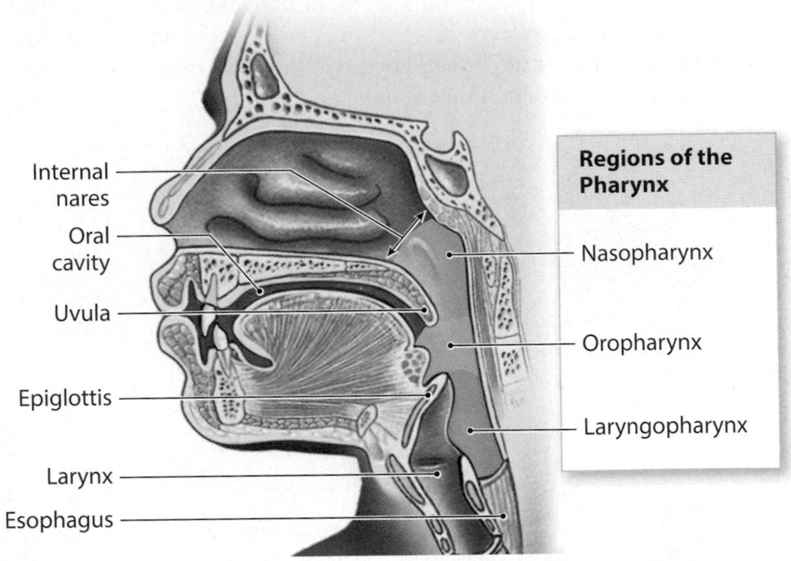
Internal nares
Oral cavity
Uvula
Epiglottis
Larynx
Esophagus
Regions of the Pharynx
Nasopharynx
Oropharynx
Laryngopharynx

2 The **esophagus** is a hollow muscular tube approximately 25 cm (10 in.) long and about 2 cm (0.80 in.) in diameter at its widest point. The primary function of the esophagus is to convey solid food and liquids to the stomach.

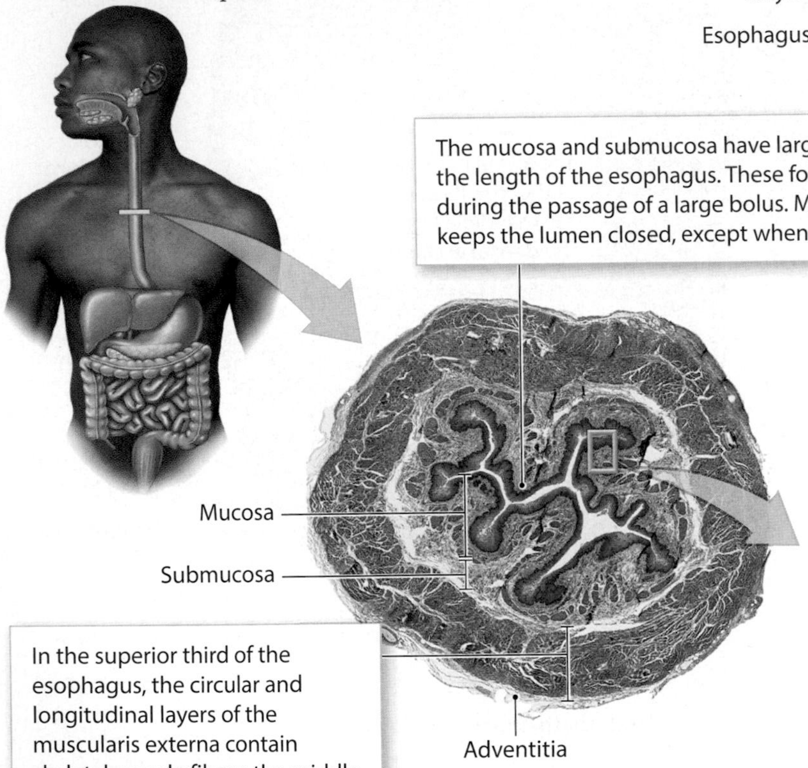
The mucosa and submucosa have large folds that extend the length of the esophagus. These folds allow for expansion during the passage of a large bolus. Muscle tone in the walls keeps the lumen closed, except when you swallow.
Mucosa
Submucosa
Adventitia
In the superior third of the esophagus, the circular and longitudinal layers of the muscularis externa contain skeletal muscle fibers; the middle third contains a mixture of skeletal and smooth muscle tissue; and along the inferior third, only smooth muscle occurs.

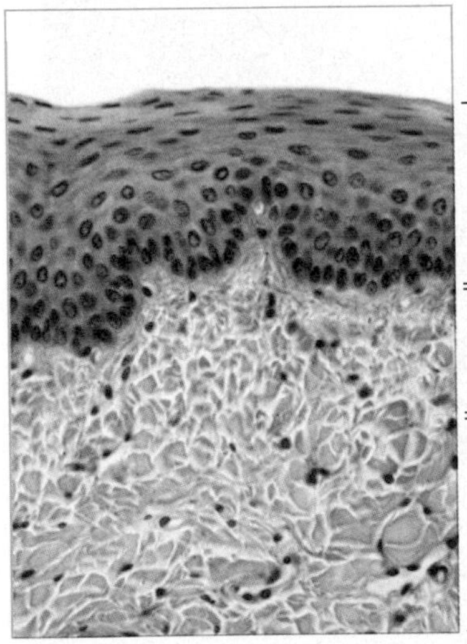
The mucosa of the esophagus contains a nonkeratinized, stratified squamous epithelium similar to that of the pharynx and oral cavity.
Lamina propria
The muscularis mucosae consists of an irregular layer of smooth muscle.

Esophageal mucosa LM × 300

3 The layers of the esophageal wall are comparable with those in other portions of the digestive tract. However, the shape of the lumen and the structure of the muscularis externa are unique to the esophagus. There is no serosa, but an **adventitia** of connective tissue outside the muscularis externa anchors the esophagus to the posterior body wall.

4 This light micrograph illustrates the extreme thickness of the epithelium making up the esophageal mucosa layer.

5 Swallowing, or **deglutition** (dē-glū-TISH-un), is a complex process that can be started voluntarily but proceeds automatically once it begins. Although you take conscious control over swallowing when you eat or drink, swallowing is also controlled at the subconscious level. Each day you swallow approximately 2400 times. We divide swallowing into three phases.

Buccal Phase

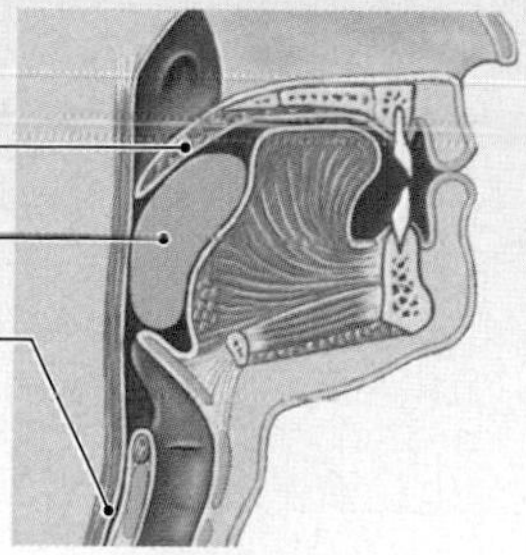

Soft palate
Bolus
Esophagus

The **buccal phase** begins with the compression of the bolus against the hard palate. Retraction of the tongue then forces the bolus into the oropharynx and assists in the elevation of the soft palate, thereby sealing off the nasopharynx. The buccal phase is strictly voluntary. Once the bolus enters the oropharynx, reflex responses are initiated and the bolus is moved toward the stomach.

Pharyngeal Phase

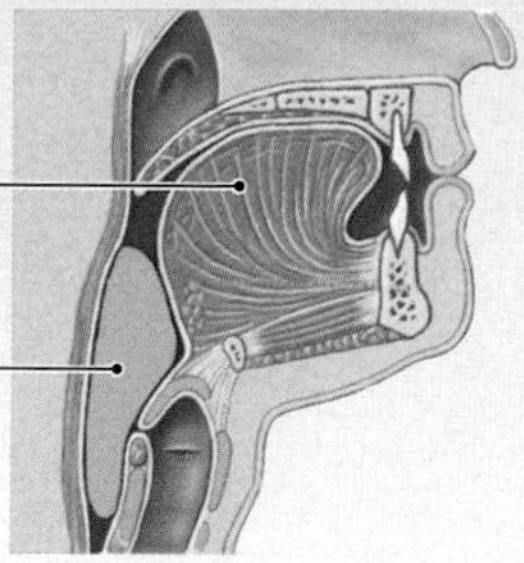

Tongue
Bolus

The **pharyngeal phase** begins when tactile receptors on the palatal arches and uvula are stimulated. In response, motor commands from the swallowing center in the medulla oblongata then direct a coordinated pattern of muscle contraction in the pharyngeal muscles. Elevation of the larynx and folding of the epiglottis results from contractions of the pharyngeal muscles, while the palatal muscles elevate the uvula and soft palate to block the entrance to the nasopharynx. Pharyngeal constrictors then force the bolus through the pharynx, past the closed glottis, and into the esophagus.

Esophageal Phase

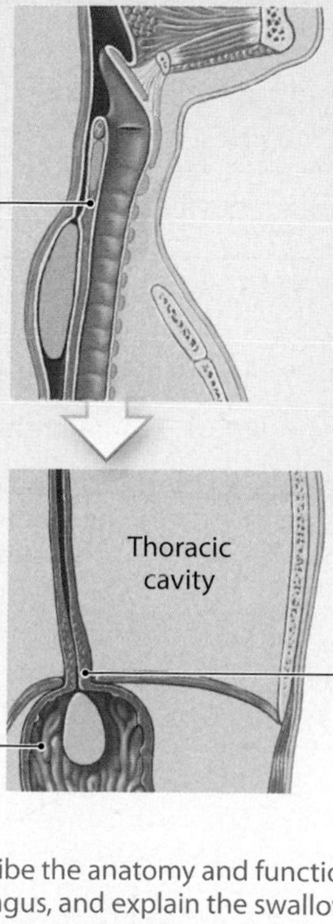

Peristalsis
Thoracic cavity
Lower esophageal sphincter
Stomach

The **esophageal phase** of swallowing begins as the contraction of pharyngeal muscles forces the bolus through the entrance to the esophagus. Once in the esophagus, the bolus is pushed toward the stomach by peristalsis. The approach of the bolus triggers the opening of the lower esophageal sphincter, and the bolus then continues into the stomach. For a typical bolus, the entire trip takes about 9 seconds. Liquids may travel faster, flowing ahead of the peristaltic contractions with the assistance of gravity. A dry or poorly lubricated bolus travels much more slowly, and a series of local reflexes, called **secondary peristaltic waves**, may be required to push it all the way to the stomach.

The esophagus begins posterior to the cricoid cartilage. From this point, where it is at its narrowest, the esophagus descends toward the thoracic cavity posterior to the trachea. It passes inferiorly along the posterior wall of the mediastinum and enters the abdominopelvic cavity through the **esophageal hiatus** (hī-Ā-tus), an opening in the diaphragm. The esophagus is innervated by parasympathetic and sympathetic fibers from the esophageal plexus. Resting muscle tone in the circular muscle layer in the superior 3 cm of the esophagus normally prevents air from entering the esophagus. The band of smooth muscle involved functions as an **upper esophageal sphincter**. A comparable area of smooth muscle at the inferior end of the esophagus forms the **lower esophageal sphincter** (cardiac sphincter). It normally remains in a state of active contraction, preventing the backflow of materials from the stomach into the esophagus.

Module 22.8 Review

a. Describe the structure and function of the pharynx.
b. Name the structure connecting the pharynx to the stomach.
c. Describe the major event in each of the three phases of swallowing.

LO 22.8 Describe the anatomy and functions of the pharynx and esophagus, and explain the swallowing process.

The stomach and most of the intestinal tract are suspended by mesenteries and enclosed by the peritoneal cavity

The **peritoneal cavity** is lined by a serous membrane called the peritoneum (Module 1.17, p. 35). We can divide the serous membrane into the serosa or **visceral peritoneum**, which covers organs enclosed by the peritoneal cavity, and the **parietal peritoneum**, which lines the inner surfaces of the peritoneal cavity. Recall that a mesentery is a double sheet of peritoneal membrane (p. 830).

1 These diagrams are cross-sectional views from early embryos. During embryonic development, the digestive tract and accessory organs are enclosed by the peritoneal cavity and suspended by **dorsal** and **ventral mesenteries** (MEZ-en-ter-ēz).

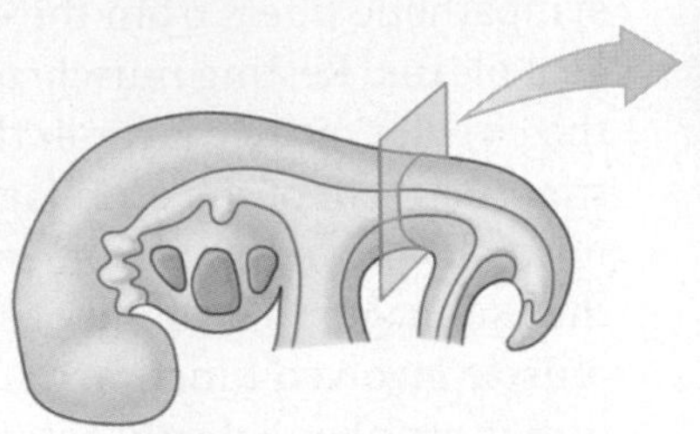

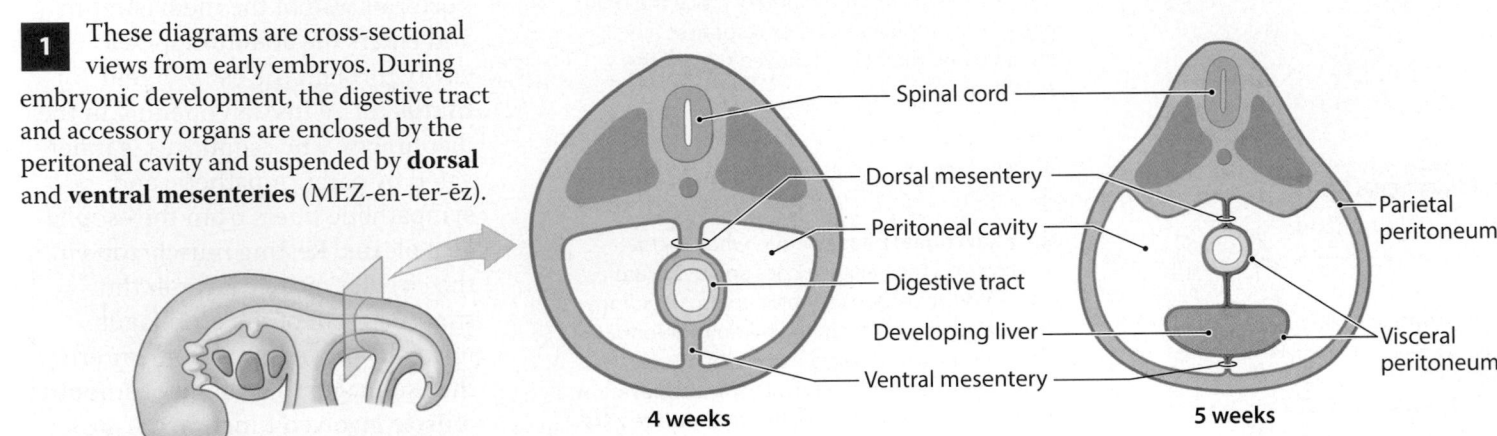

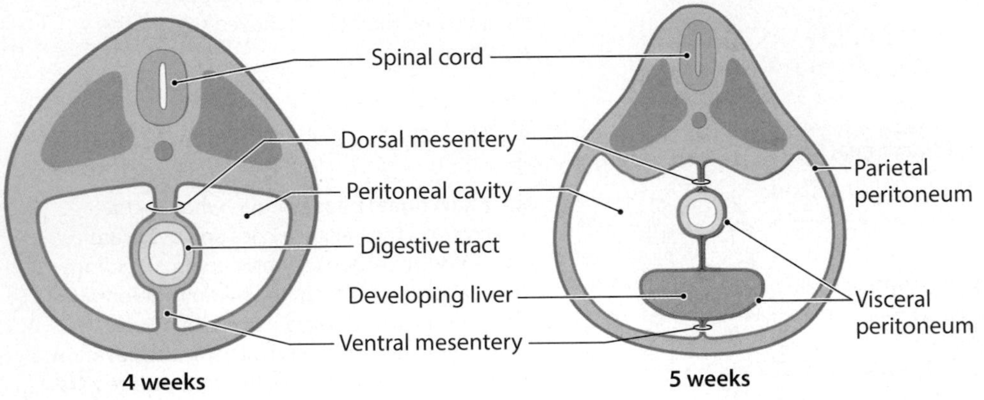
Spinal cord
Dorsal mesentery
Peritoneal cavity
Digestive tract
Developing liver
Ventral mesentery
Parietal peritoneum
Visceral peritoneum
4 weeks
5 weeks

2 The diagrammatic cross-sectional view below and the lateral view to the right show the mesenteries supporting the digestive tract and accessory organs early in development. The ventral mesentery later disappears along most of the digestive tract, remaining in adults in only two places: between the stomach and the liver (the lesser omentum), and between the liver and the anterior abdominal wall (the falciform ligament).

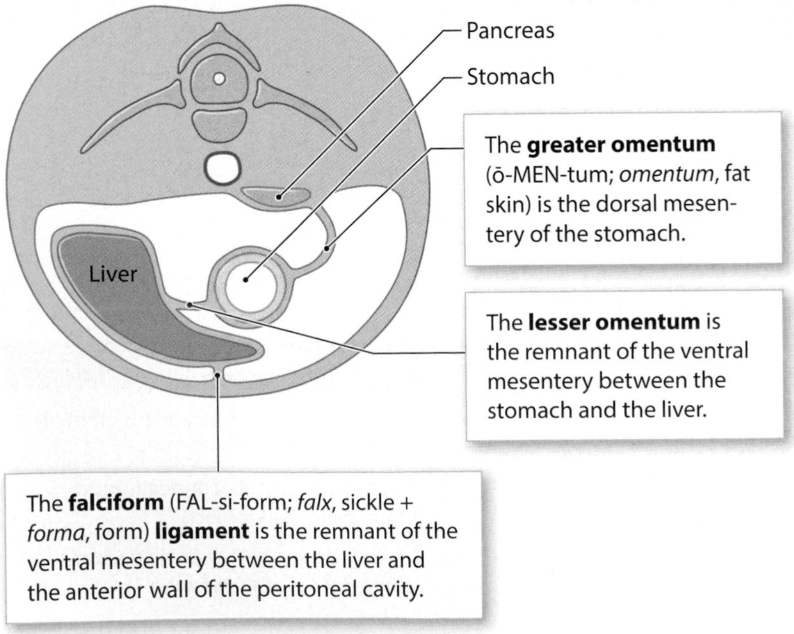
Pancreas
Stomach
Liver
The **greater omentum** (ō-MEN-tum; *omentum*, fat skin) is the dorsal mesentery of the stomach.
The **lesser omentum** is the remnant of the ventral mesentery between the stomach and the liver.
The **falciform** (FAL-si-form; *falx*, sickle + *forma*, form) **ligament** is the remnant of the ventral mesentery between the liver and the anterior wall of the peritoneal cavity.

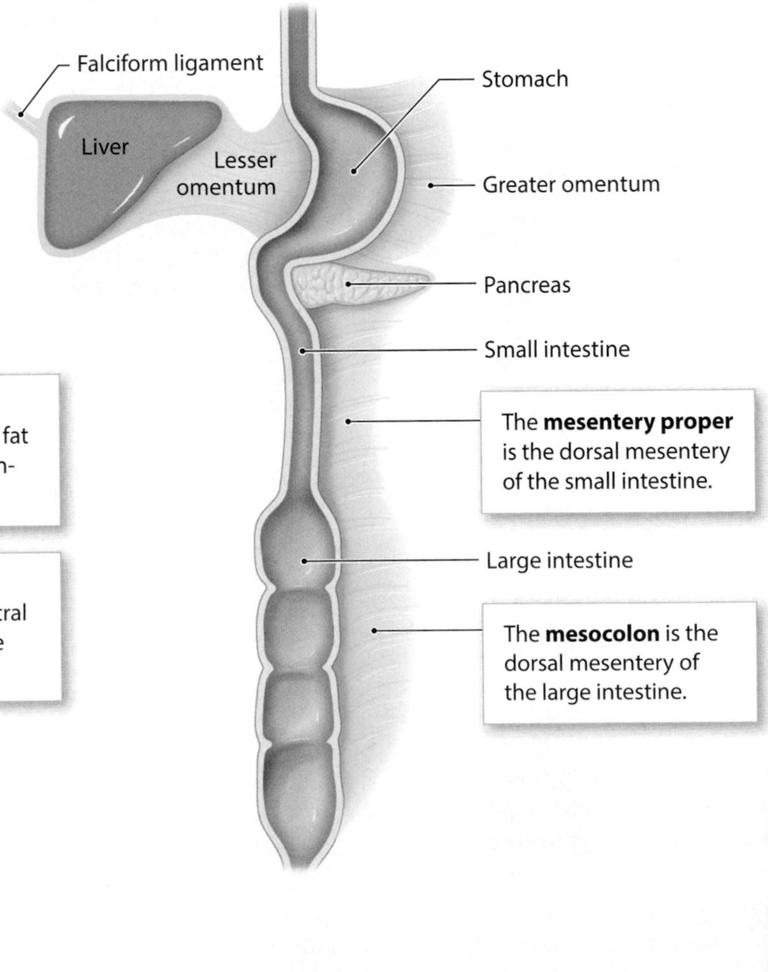
Falciform ligament
Liver
Lesser omentum
Stomach
Greater omentum
Pancreas
Small intestine
The **mesentery proper** is the dorsal mesentery of the small intestine.
Large intestine
The **mesocolon** is the dorsal mesentery of the large intestine.

3 As the digestive tract elongates, the position of the mesenteries changes. Some segments of the tract contact the posterior wall and become fixed in position. As the small intestine coils and increases in length, segments of the mesentery proper come into contact and fuse together.

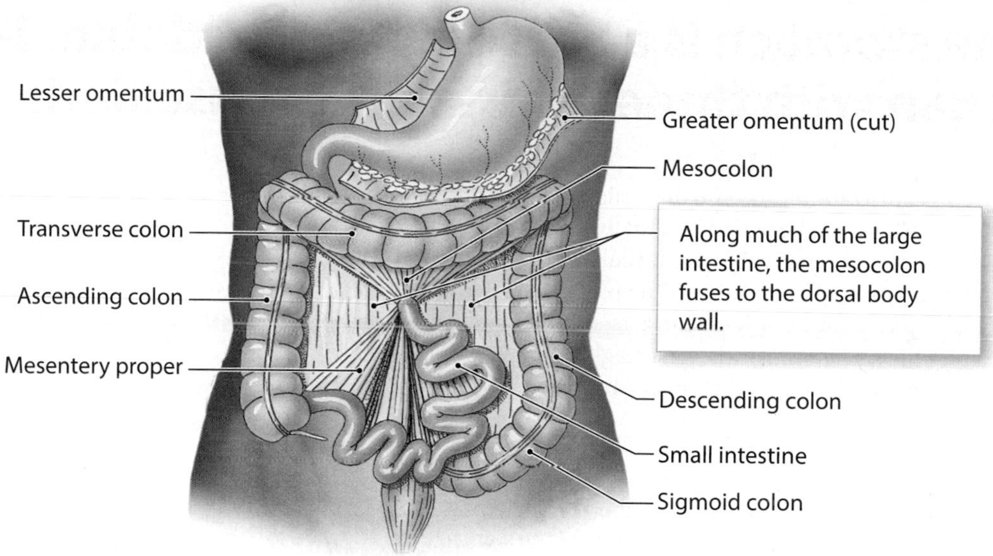

Lesser omentum
Greater omentum (cut)
Mesocolon
Transverse colon
Along much of the large intestine, the mesocolon fuses to the dorsal body wall.
Ascending colon
Mesentery proper
Descending colon
Small intestine
Sigmoid colon

4 A sagittal section reveals the orientation of the mesenteries in an adult. The coils of the small intestine are suspended by the mesentery proper. The transverse colon and a small section near the end of the large intestine are suspended by remnants of the original mesocolon.

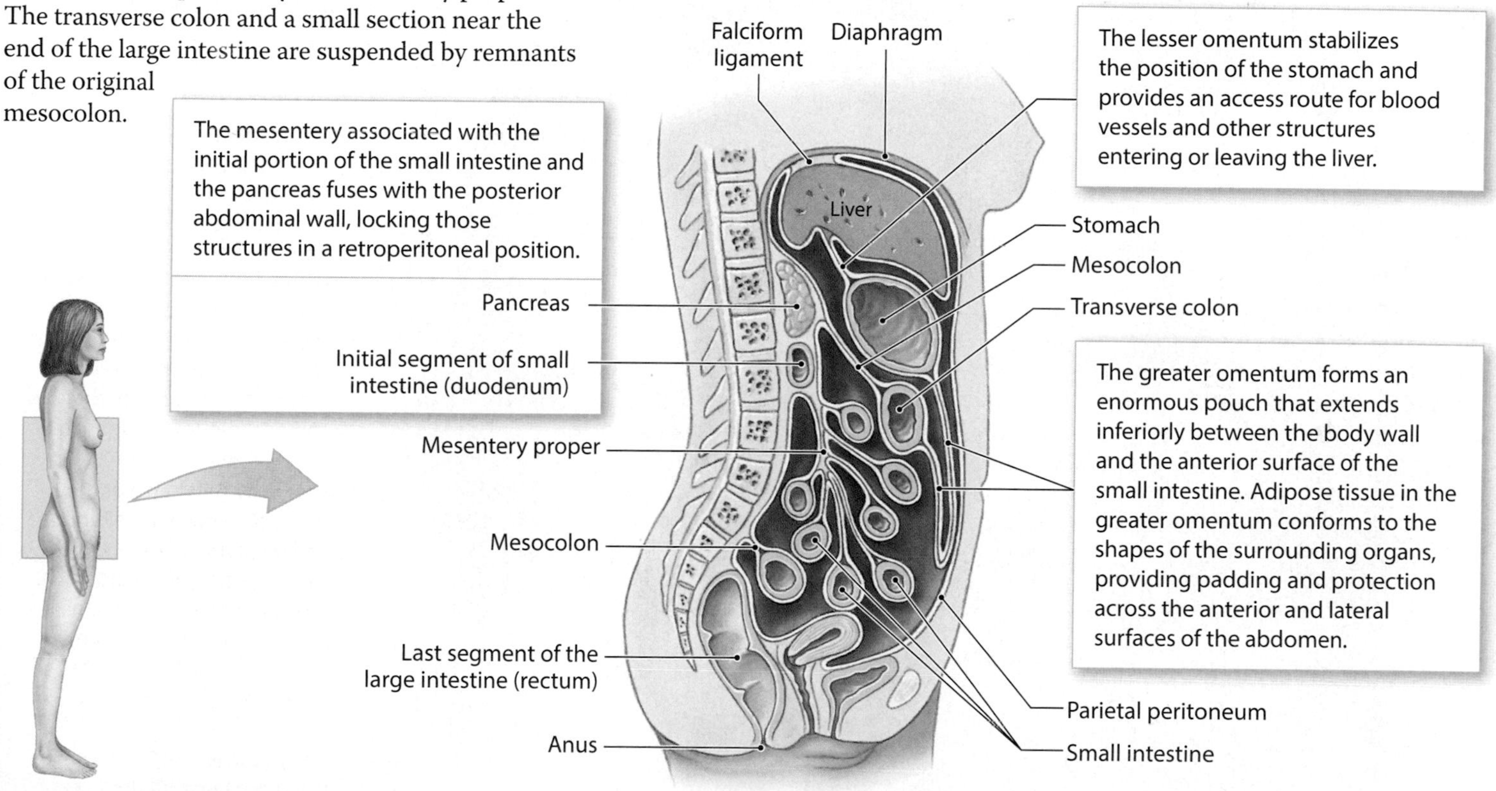

Falciform ligament
Diaphragm
The lesser omentum stabilizes the position of the stomach and provides an access route for blood vessels and other structures entering or leaving the liver.
The mesentery associated with the initial portion of the small intestine and the pancreas fuses with the posterior abdominal wall, locking those structures in a retroperitoneal position.
Liver
Stomach
Mesocolon
Transverse colon
Pancreas
Initial segment of small intestine (duodenum)
The greater omentum forms an enormous pouch that extends inferiorly between the body wall and the anterior surface of the small intestine. Adipose tissue in the greater omentum conforms to the shapes of the surrounding organs, providing padding and protection across the anterior and lateral surfaces of the abdomen.
Mesentery proper
Mesocolon
Last segment of the large intestine (rectum)
Anus
Parietal peritoneum
Small intestine

The serous membrane lining the peritoneal cavity continuously produces peritoneal fluid, which provides essential lubrication. Because a thin layer of peritoneal fluid separates the parietal and visceral surfaces, sliding movements can occur without friction and resulting irritation. About 7 liters of fluid are secreted and reabsorbed each day, although the volume within the peritoneal cavity at any one time is about 50 mL. Liver disease, kidney disease, and heart failure can accelerate the rate at which fluids move into the peritoneal cavity. The accumulation of peritoneal fluid creates a characteristic abdominal swelling called **ascites** (a-SĪ-tēz).

Module 22.9 Review

a. What is the falciform ligament?

b. What is the function of the lesser omentum?

c. Explain the significance of peritoneal fluid.

LO 22.9 Explain the embryonic development of the mesenteries, and describe the mesenteries that remain in adulthood.

The stomach is a muscular, expandable, J-shaped organ with three layers in the muscularis externa

1 This illustration presents the major surfaces and regions of the stomach. The shape is actually highly variable. When empty, the stomach resembles a muscular tube with a narrow, constricted lumen. When full, it can contain 1–1.5 liters of material. That material, which consists of food combined with saliva and the secretions of the gastric glands, is a viscous, highly acidic, soupy mixture called **chyme** (kīm).

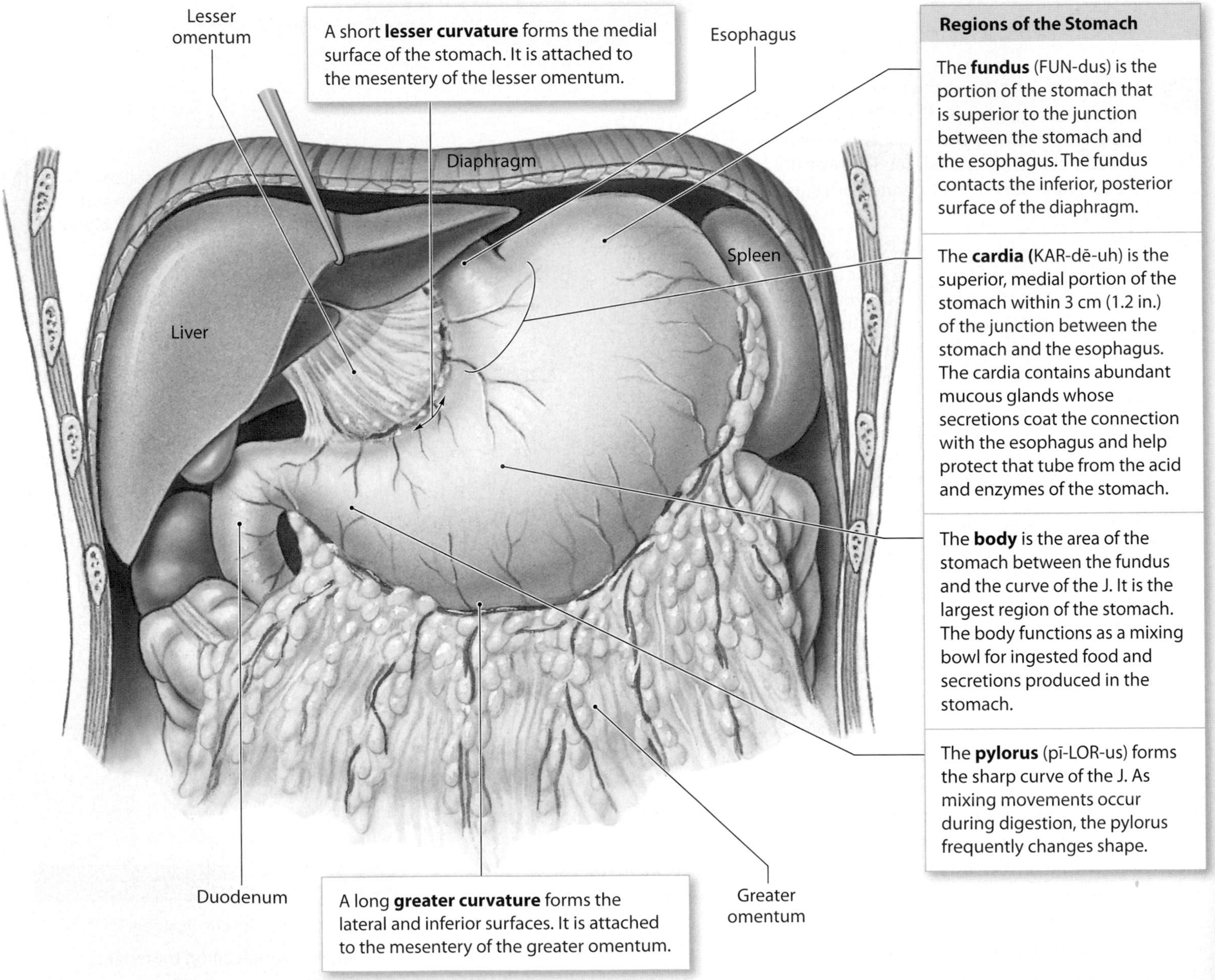
Lesser omentum
A short **lesser curvature** forms the medial surface of the stomach. It is attached to the mesentery of the lesser omentum.
Esophagus
Diaphragm
Spleen
Liver
Duodenum
A long **greater curvature** forms the lateral and inferior surfaces. It is attached to the mesentery of the greater omentum.
Greater omentum
Regions of the Stomach
The **fundus** (FUN-dus) is the portion of the stomach that is superior to the junction between the stomach and the esophagus. The fundus contacts the inferior, posterior surface of the diaphragm.
The **cardia** (KAR-dē-uh) is the superior, medial portion of the stomach within 3 cm (1.2 in.) of the junction between the stomach and the esophagus. The cardia contains abundant mucous glands whose secretions coat the connection with the esophagus and help protect that tube from the acid and enzymes of the stomach.
The **body** is the area of the stomach between the fundus and the curve of the J. It is the largest region of the stomach. The body functions as a mixing bowl for ingested food and secretions produced in the stomach.
The **pylorus** (pī-LOR-us) forms the sharp curve of the J. As mixing movements occur during digestion, the pylorus frequently changes shape.

2 This cadaver dissection exposes the stomach and greater omentum.

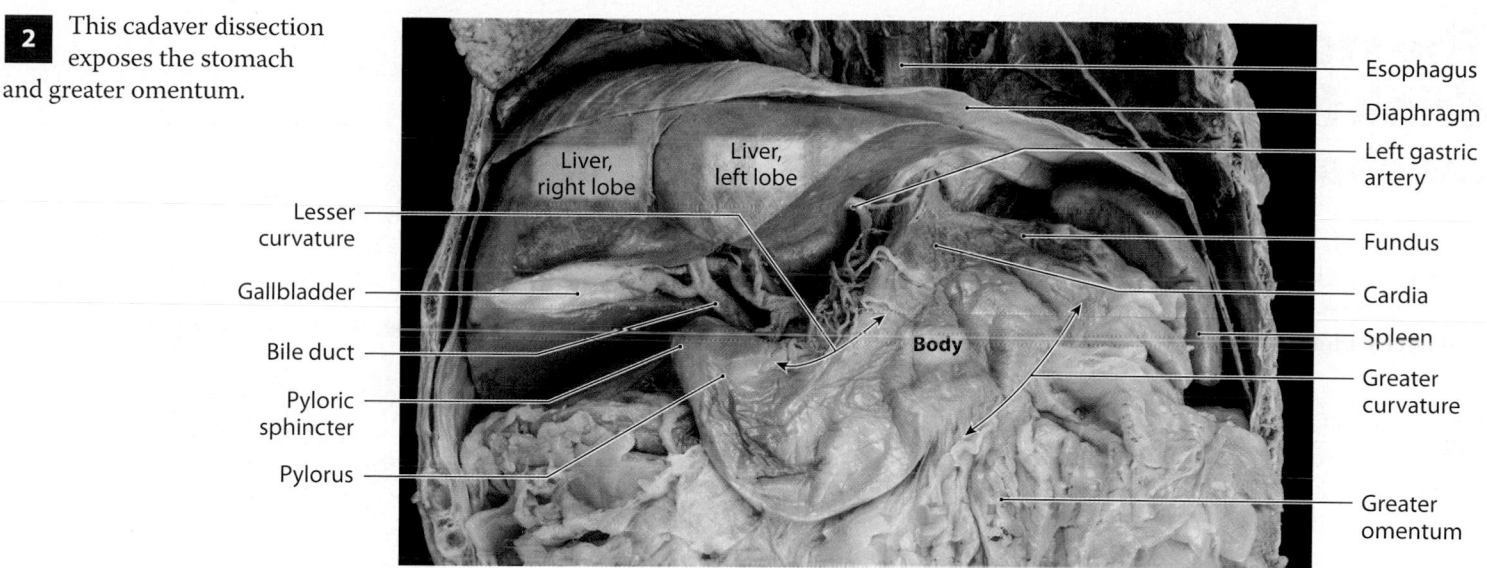

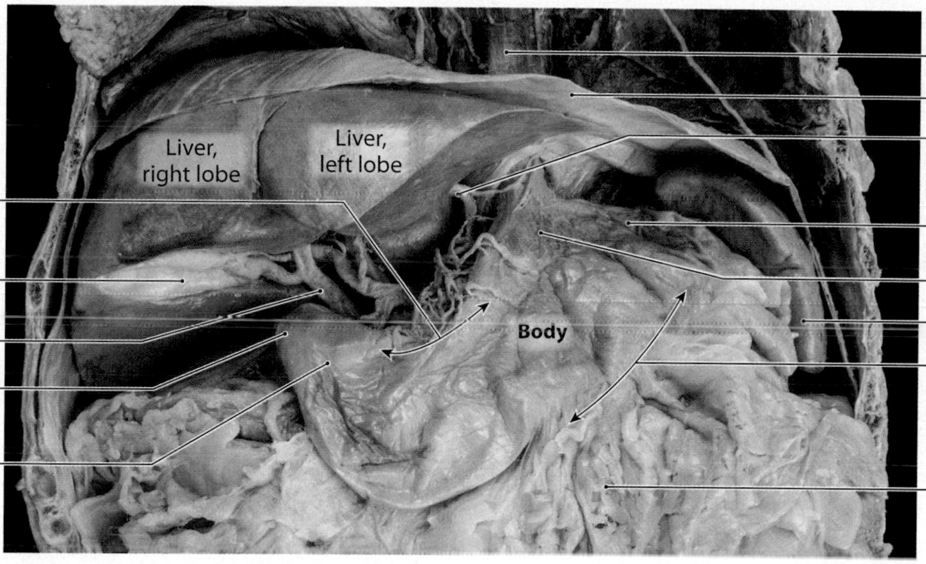
Esophagus
Diaphragm
Left gastric artery
Fundus
Cardia
Spleen
Greater curvature
Greater omentum
Liver, right lobe
Liver, left lobe
Body
Lesser curvature
Gallbladder
Bile duct
Pyloric sphincter
Pylorus

3 A partially sectioned and dissected stomach reveals additional details about the internal structure of this organ.

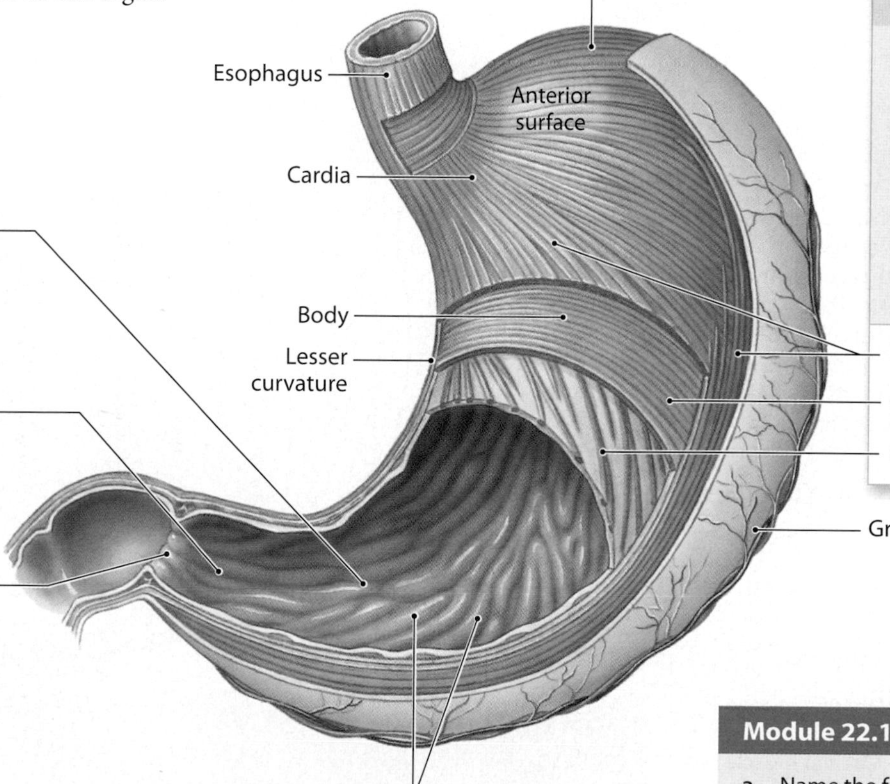
Fundus
Esophagus
Anterior surface
Cardia
Body
Lesser curvature
Longitudinal muscle layer
Circular muscle layer
Oblique muscle layer
Greater curvature

Layers of the Muscularis Externa

The muscularis externa of the stomach contains an inner, oblique layer of smooth muscle. This additional layer of smooth muscle strengthens the stomach wall and assists in the mixing and churning to form chyme.

The Pylorus

The **pyloric antrum** (*antron*, cavity) is the portion of the pylorus connected to the body of the stomach.

The **pyloric canal** empties into the duodenum, the proximal segment of the small intestine.

A muscular **pyloric sphincter** regulates the release of chyme into the duodenum.

Rugae (RŪ-gē; wrinkles) are prominent but temporary mucosal folds that allow the gastric lumen to expand. As the stomach fills, the rugae gradually flatten out until, at maximum distension, they almost disappear.

Module 22.10 Review

a. Name the four major regions of the stomach in order from its connection with the esophagus to the small intestine.

b. What anatomical feature of the stomach allows the organ to form chyme?

c. Describe the inner lining of the stomach.

The stomach receives food and drink from the esophagus and aids in mechanical and chemical digestion

The **stomach** has four major functions: (1) storage of ingested food, (2) mechanical breakdown of ingested food, (3) disruption of chemical bonds in food through the action of acid and enzymes, and (4) production of intrinsic factor.

1 The wall of the stomach is relatively thick and muscular, and its mucosa has deep folds that form gastric glands.

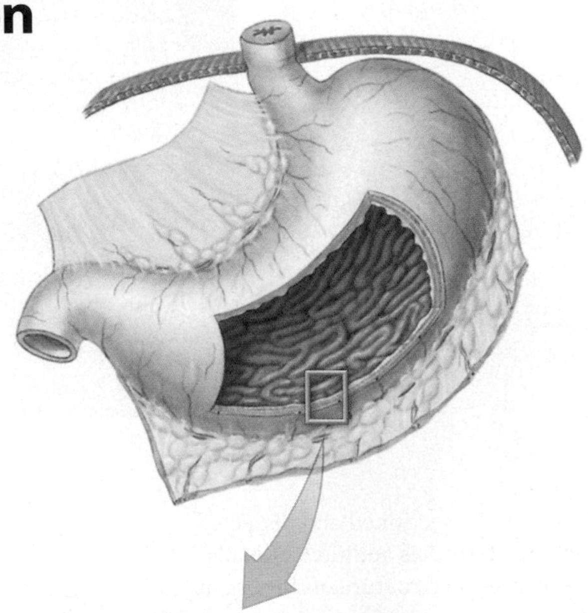

Layers of the Stomach Wall

Mucosa

A simple columnar epithelium lines all portions of the stomach. This epithelium produces a carpet of mucus that covers the interior surfaces of the stomach. The alkaline mucous layer protects epithelial cells against the acid and enzymes in the gastric lumen. Still, the environment is harsh, and a typical gastric epithelial cell has a life span of only 3 to 7 days.

Lamina propria

Muscularis mucosae

Submucosa

Muscularis Externa

Oblique muscle

Circular muscle

Longitudinal muscle

Serosa

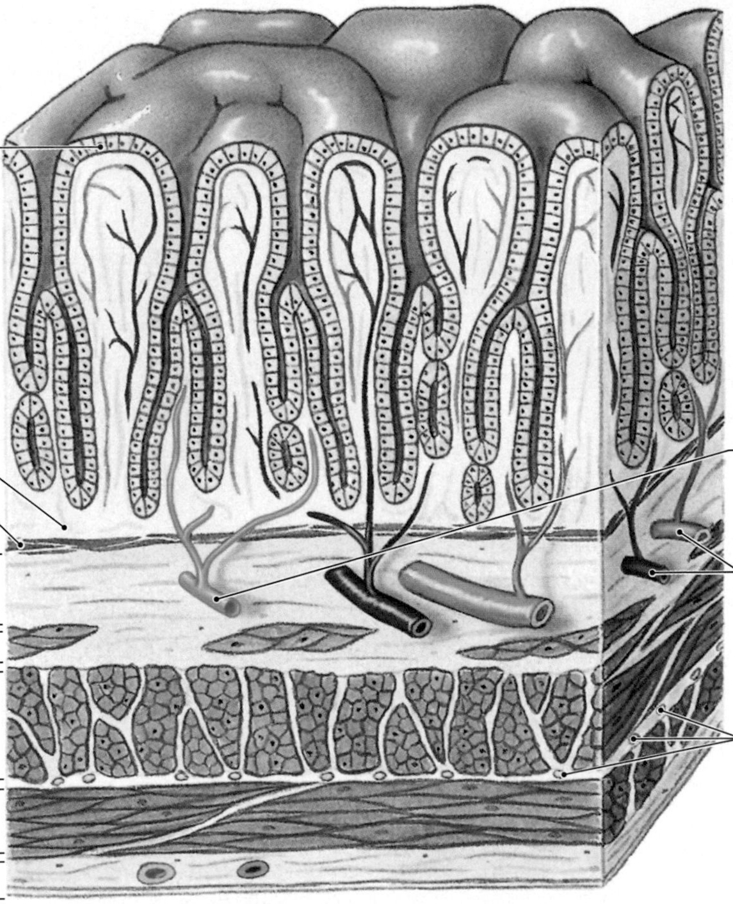
Lymphatic vessel
Artery and vein
Myenteric plexus

2 **Gastric glands** in the fundus and body secrete most of the acid and enzymes involved in gastric digestion. The gastric glands in these areas are dominated by **parietal cells** and **chief** (zymogenic) **cells**. Together, they secrete about 1500 mL of gastric juice each day. Gastric glands in the pylorus secrete mucus and hormones involved in the coordination and control of digestive activity.

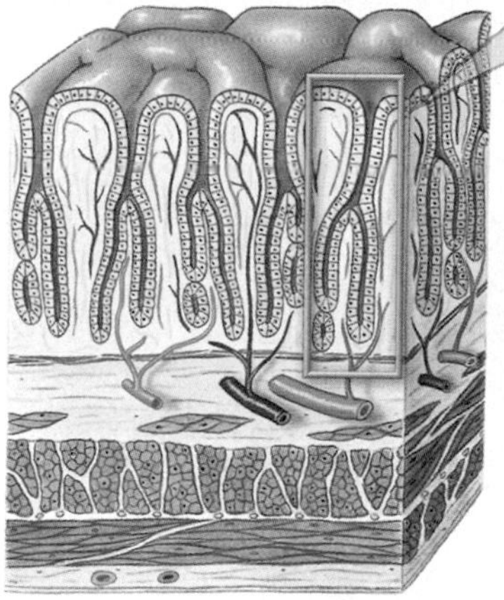

Shallow depressions called **gastric pits** open onto the gastric surface. Stem cells at the base, or neck, of each gastric pit actively divide, replacing superficial cells that are shed into the chyme.

Each gastric pit communicates with several gastric glands that extend deep into the lamina propria.

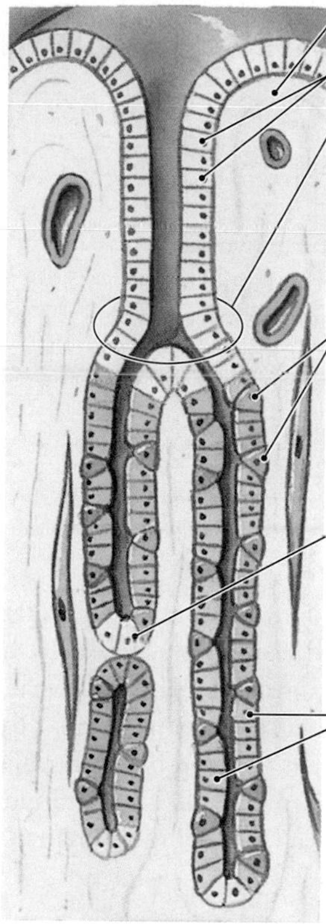

Lamina propria
Mucous epithelial cells
Neck

Cells of Gastric Glands

Parietal cells secrete **intrinsic factor**, a glycoprotein that facilitates the absorption of vitamin B_{12} across the intestinal lining. This vitamin is necessary for erythropoiesis (Module 17.11, p. 642). Parietal cells also secrete hydrochloric acid (HCl).

G cells are enteroendocrine cells that produce a variety of hormones. These hormones and their functions will be considered in Module 22.14.

Chief cells secrete **pepsinogen** (pep-SIN-ō-jen), an inactive proenzyme. Acid (HCl) in the gastric lumen converts pepsinogen to **pepsin**, an active proteolytic (protein-digesting) enzyme. In addition, the stomachs of newborn infants (but not of adults) produce **rennin** and **gastric lipase**, enzymes important for the digestion of milk.

3 In addition to intrinsic factor, parietal cells secrete HCl to keep the stomach contents at pH 1.5–2.0. However, parietal cells do not produce this strong acid in the cytoplasm, because it would erode secretory vesicles and destroy the cell. Instead, H^+ and Cl^- are transported and secreted independently, as diagrammed below. When gastric glands are actively secreting, enough bicarbonate ions enter the bloodstream to increase the pH of the blood significantly. This sudden influx of bicarbonate ions has been called the **alkaline tide**.

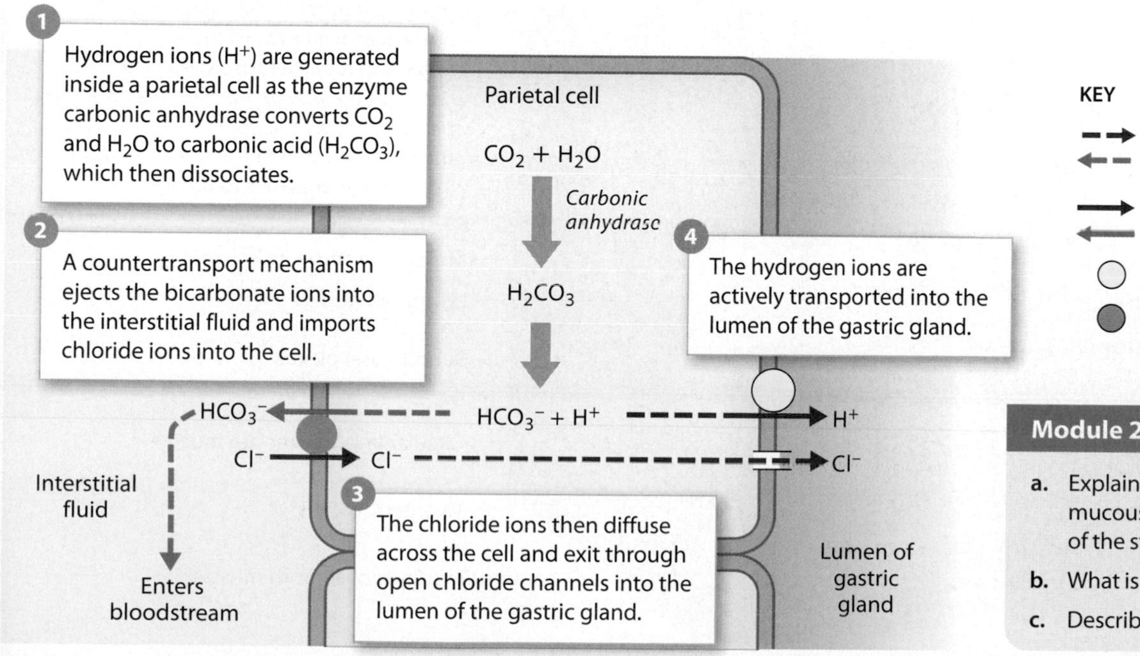

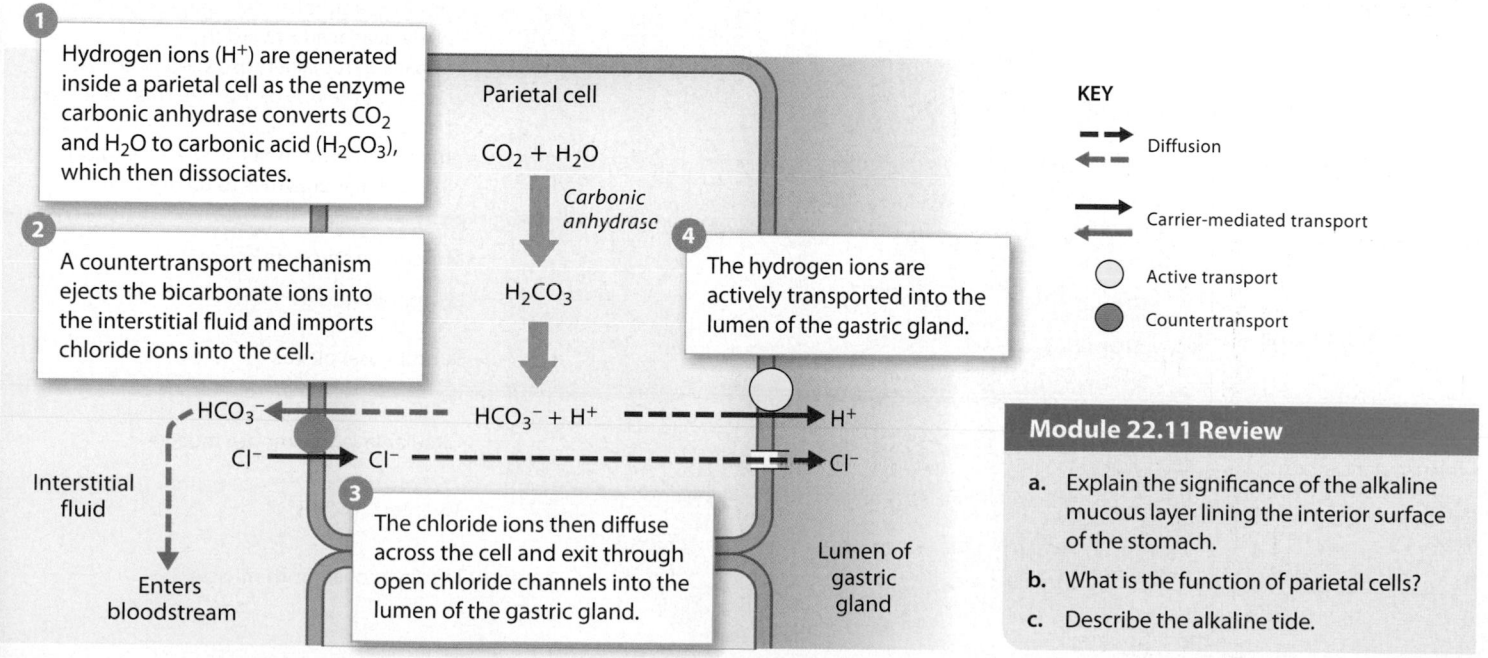

1 Hydrogen ions (H+) are generated inside a parietal cell as the enzyme carbonic anhydrase converts CO2 and H2O to carbonic acid (H2CO3), which then dissociates.
2 A countertransport mechanism ejects the bicarbonate ions into the interstitial fluid and imports chloride ions into the cell.
3 The chloride ions then diffuse across the cell and exit through open chloride channels into the lumen of the gastric gland.
4 The hydrogen ions are actively transported into the lumen of the gastric gland.
Parietal cell
CO2 + H2O
Carbonic anhydrase
H2CO3
HCO3− + H+
HCO3−
Cl−
Cl−
H+
Cl−
Interstitial fluid
Enters bloodstream
Lumen of gastric gland
KEY
Diffusion
Carrier-mediated transport
Active transport
Countertransport

Module 22.11 Review

a. Explain the significance of the alkaline mucous layer lining the interior surface of the stomach.

b. What is the function of parietal cells?

c. Describe the alkaline tide.

22.11 Describe the anatomy of the stomach relating to its role in digestion and absorption.

Module 22.12

The intestinal tract is specialized to absorb nutrients

1 The intestinal lining has a series of transverse folds called **circular folds**. Unlike the rugae in the stomach, these circular folds are permanent features that do not disappear when the small intestine fills. The small intestine contains roughly 800 circular folds, most of them within the jejunum. Their presence greatly increases the surface area available for absorption.

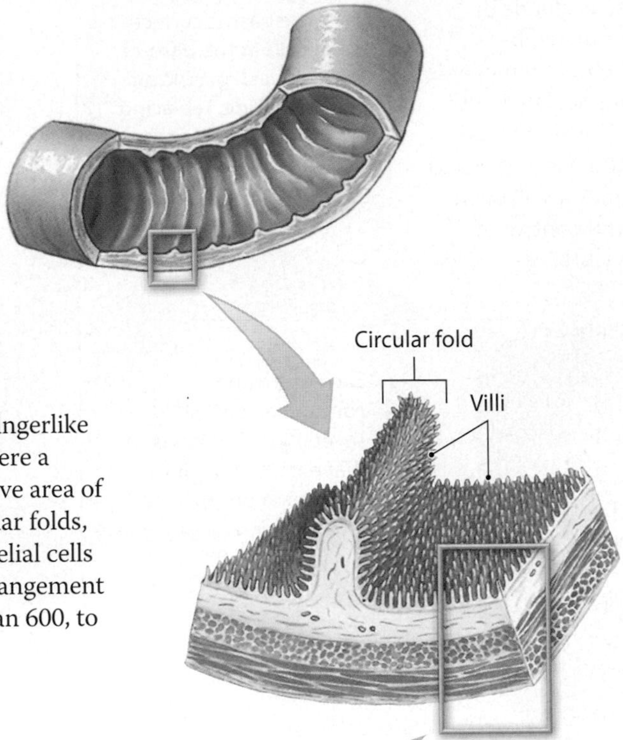

Circular fold
Villi

2 The mucosa of the small intestine contains a series of fingerlike projections, the **intestinal villi**. If the small intestine were a simple tube with smooth walls, it would have a total absorptive area of nearly 3300 cm^2 (3.6 ft^2). Instead, the mucosa contains circular folds, which support a forest of villi. Each villus is covered by epithelial cells whose exposed surfaces are covered with microvilli. This arrangement increases the total area for absorption by a factor of more than 600, to approximately 2 million cm^2 (more than 2200 ft^2).

3 This diagrammatic sectional view of the intestinal wall shows features common to all segments of the small intestine.

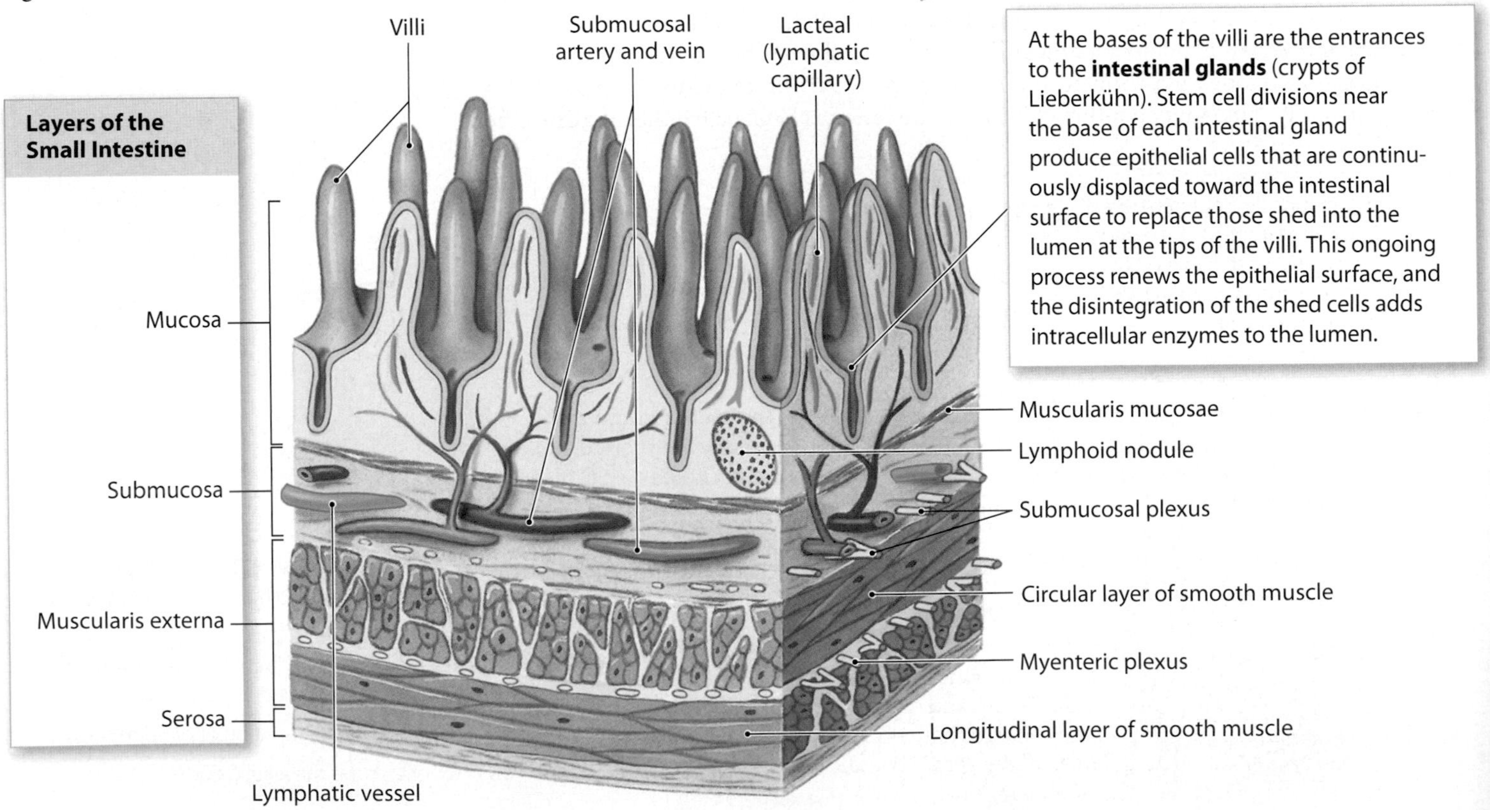

Villi
Submucosal artery and vein
Lacteal (lymphatic capillary)
Layers of the Small Intestine
Mucosa
Submucosa
Muscularis externa
Serosa
Lymphatic vessel
At the bases of the villi are the entrances to the **intestinal glands** (crypts of Lieberkühn). Stem cell divisions near the base of each intestinal gland produce epithelial cells that are continuously displaced toward the intestinal surface to replace those shed into the lumen at the tips of the villi. This ongoing process renews the epithelial surface, and the disintegration of the shed cells adds intracellular enzymes to the lumen.
Muscularis mucosae
Lymphoid nodule
Submucosal plexus
Circular layer of smooth muscle
Myenteric plexus
Longitudinal layer of smooth muscle

4 Each villus has a complex internal structure. The lamina propria of each villus contains an extensive network of capillaries that originate in a vascular network within the submucosa. These capillaries carry absorbed nutrients to the hepatic portal system for delivery to the liver, which adjusts the nutrient concentrations in blood before the blood reaches the general systemic circulation.

5 The surface of each villus consists of a simple columnar epithelium that is carpeted with microvilli. Because the microvilli project from the epithelium like the bristles on a brush, these cells are said to have a **brush border**. Brush border enzymes are integral membrane proteins located on the surfaces of intestinal microvilli. These enzymes break down materials that come in contact with the brush border. The epithelial cells then absorb the breakdown products.

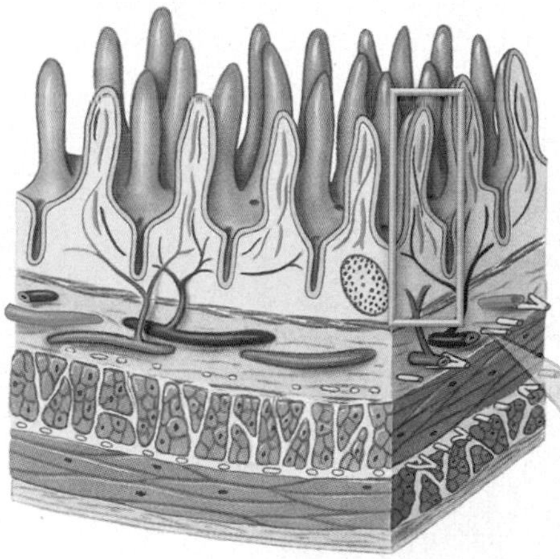

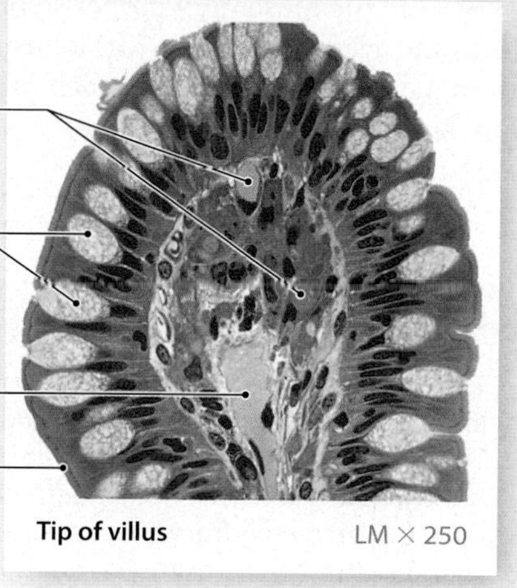

Capillaries
Lacteal
Brush border
Tip of villus
LM × 250

Mucous cells between the columnar epithelial cells eject mucins onto the intestinal surfaces.

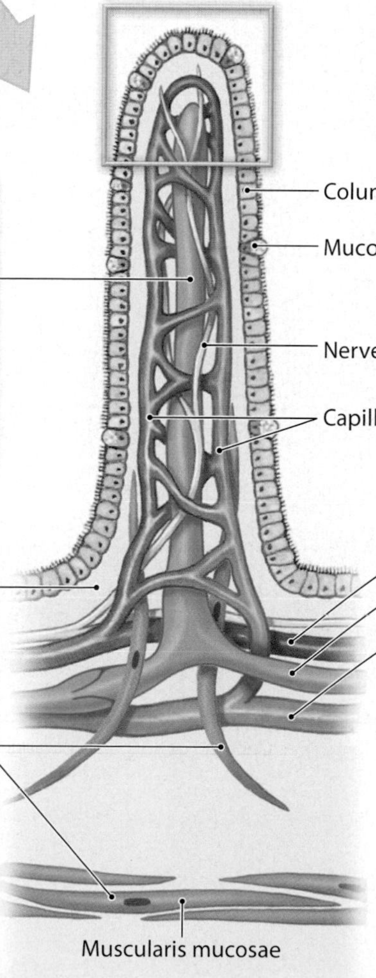

Columnar epithelial cell
Mucous cell
Nerve
Capillary network
Lamina propria
Arteriole
Lymphatic vessel
Venule
Muscularis mucosae

Each villus contains a lymphatic capillary called a **lacteal** (LAK-tē-ul; *lacteus*, milky). Lacteals transport materials that cannot enter blood capillaries. For example, absorbed fatty acids are assembled into protein–lipid packages that are too large to diffuse into the bloodstream. These lipid-rich packets reach the venous circulation as the thoracic duct delivers lymph to the left subclavian vein.

Contractions of the muscularis mucosae and smooth muscle cells within the intestinal villi move the villi back and forth, exposing the epithelial surfaces to the liquefied intestinal contents. This movement improves the efficiency of absorption by quickly eliminating local differences in nutrient concentration. Movements of the villi also squeeze the lacteals, which assist moving lymph out of the villi.

Module 22.12 Review

a. Name the layers of the small intestine from superficial to deep.

b. Describe the anatomy of the intestinal mucosa.

c. Explain the function of lacteals.

LO **22.12** Describe the anatomy of the intestinal tract and its histological features.

Module 22.13

The small intestine is divided into the duodenum, jejunum, and ileum

The small intestine plays the key role in nutrient digestion and absorption. Ninety percent of nutrient absorption occurs in the small intestine; most of the rest occurs in the large intestine.

1 The small intestine fills much of the abdominopelvic cavity, and the mesentery proper stabilizes its position. The small intestine averages 6 m (19.7 ft) long and has a diameter ranging from 4 cm (1.6 in.) at the stomach to about 2.5 cm (1 in.) at the junction with the large intestine.

Segments of the Small Intestine

The **duodenum** (dū-ō-DĒ-num), 25 cm (10 in.) long, is the segment closest to the stomach. This portion of the small intestine is a "mixing bowl." It receives chyme from the stomach and digestive secretions from the pancreas and liver. Except for the proximal 2.5 cm (1 in.), the duodenum is in a retroperitoneal position, firmly attached to the posterior body wall.

The **jejunum** (je-JŪ-num) is the segment between the duodenum and the ileum and is marked by a sharp bend at its beginning. At this junction, the small intestine reenters the peritoneal cavity, supported by a sheet of mesentery. The jejunum is about 2.5 meters (8.2 ft) long. The bulk of chemical digestion and nutrient absorption occurs in the jejunum.

The **ileum** (IL-ē-um), the final segment of the small intestine, is also the longest, averaging 3.5 meters (11.5 ft) in length. The ileum ends at the **ileocecal** (il-ē-o-SĒ-kal) **valve**, a sphincter that controls the flow of material from the ileum into the cecum of the large intestine. To help you remember the order of the small intestine segments, use this mnemonic: ***d***on't ***j***ump ***i***n—***d***uodenum, ***j***ejunum, and ***i***leum.

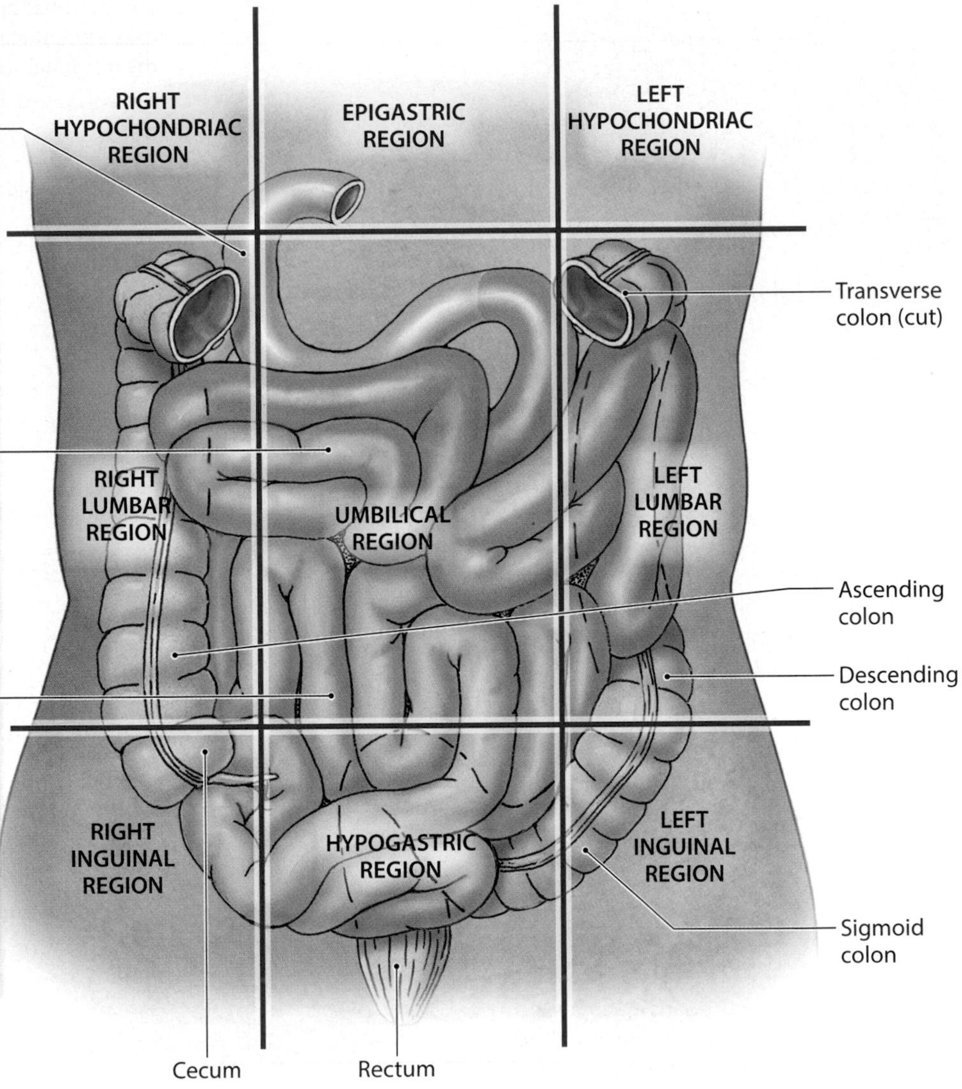

RIGHT HYPOCHONDRIAC REGION
EPIGASTRIC REGION
LEFT HYPOCHONDRIAC REGION
Transverse colon (cut)
RIGHT LUMBAR REGION
UMBILICAL REGION
LEFT LUMBAR REGION
Ascending colon
Descending colon
RIGHT INGUINAL REGION
HYPOGASTRIC REGION
LEFT INGUINAL REGION
Sigmoid colon
Cecum
Rectum

2 Each segment of the small intestine has characteristic features related to its primary functions. The transition from one region to another is gradual, and the boundaries are indistinct.

Duodenum

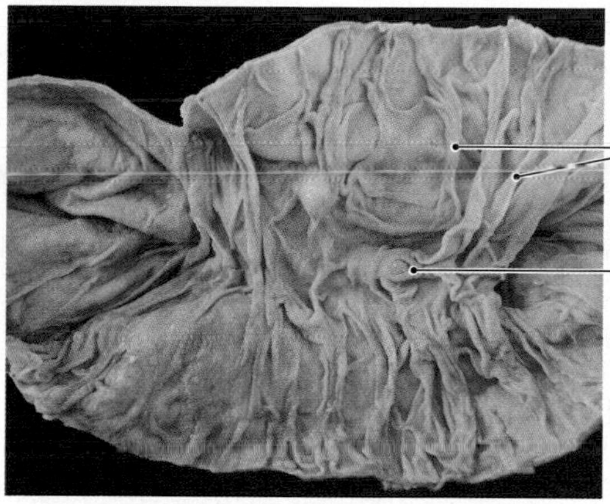

Circular folds
Duodenal ampulla

The duodenum has few circular folds, and their villi are small. The primary function of the duodenum is to receive chyme from the stomach and neutralize its acids before they can damage the absorptive surfaces of the small intestine.

Jejunum

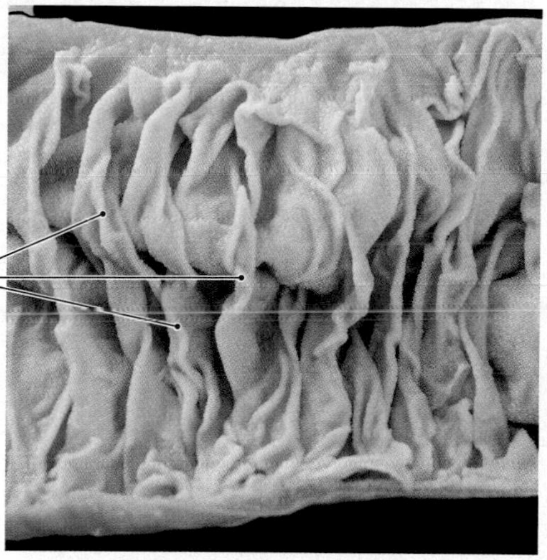

Circular folds

The jejunum has numerous circular folds and the villi are abundant and very long.

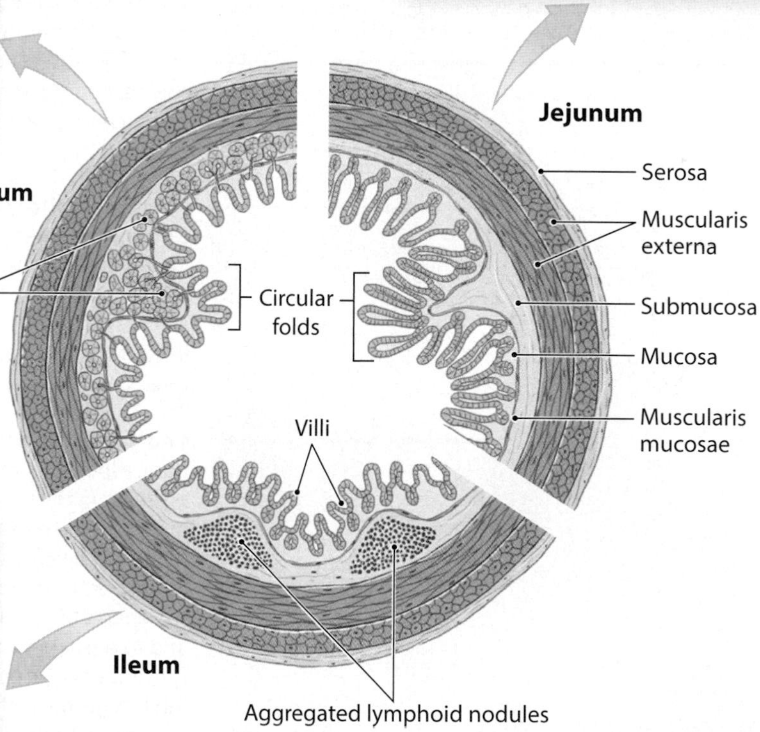

Duodenum
Jejunum
Ileum
Serosa
Muscularis externa
Submucosa
Mucosa
Muscularis mucosae
Circular folds
Villi
Aggregated lymphoid nodules

The submucosa of the duodenum is dominated by **duodenal glands** that produce mucous secretions.

Ileum

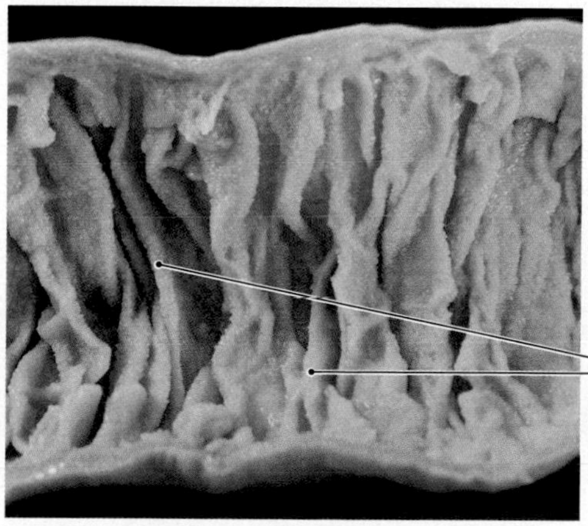

Circular folds

The ileum has fewer circular folds than the jejunum and those disappear distally. Its villi are relatively stumpy. The submucosa contains aggregated lymphoid nodules.

Module 22.13 Review

a. Name the three segments of the small intestine from proximal to distal.

b. Identify the segment of the small intestine found within the epigastric region.

c. What is the primary function of the duodenum?

22.13 Describe the anatomy and physiology of the small intestine.

Module 22.14

Five hormones are involved in the regulation of digestive activities

1 Four of the five major hormones involved in regulating digestion are produced by the duodenum, which receives partially digested materials from the stomach. The duodenum adjusts gastric activity and coordinates the secretions of accessory digestive organs according to the characteristics of the arriving chyme.

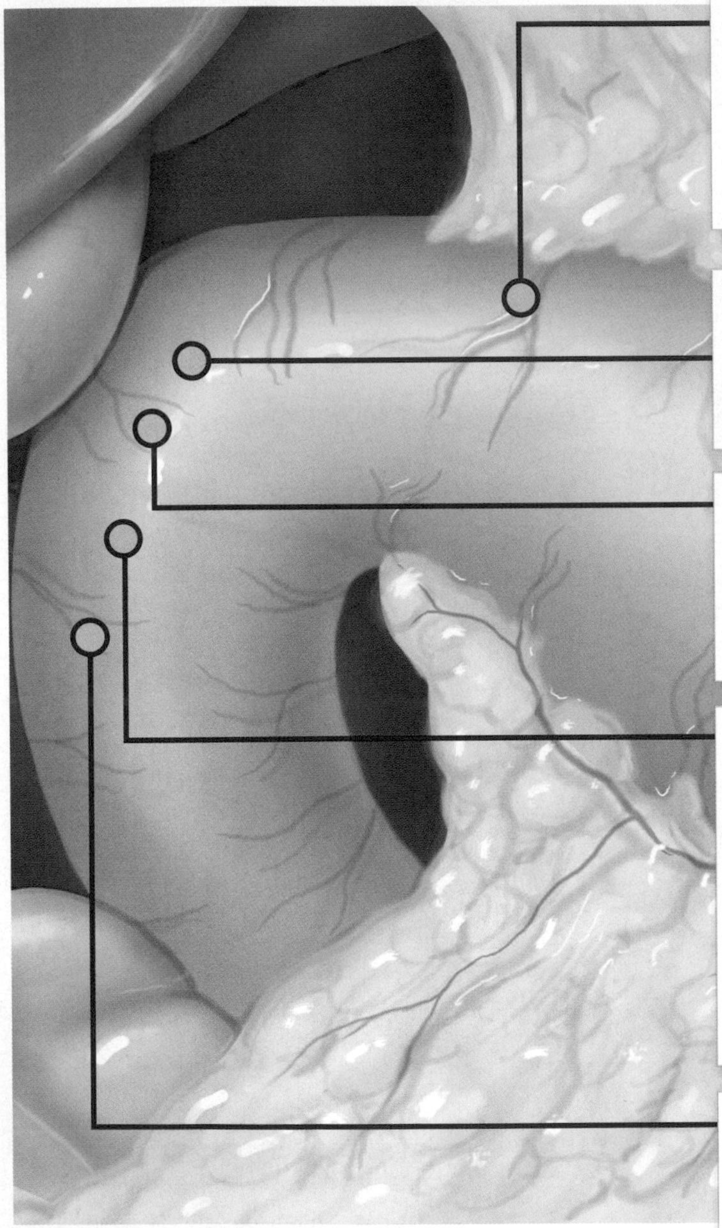

Gastrin is secreted by G cells in the pyloric antrum and enteroendocrine cells in the duodenum. The pyloric antrum secretes gastrin when stimulated by the vagus nerves or when food arrives in the stomach. Duodenal cells release gastrin when they are exposed to large quantities of incompletely digested proteins. The functions of gastrin include increasing stomach motility and stimulating gastric acid and enzyme production.

Secretin is released when chyme arrives in the duodenum. Secretin's primary effect is an increase in the secretion of bile (by the liver) and buffers (by the pancreas) which in turn act to increase the pH of the chyme. Among its secondary effects, secretin reduces gastric motility and secretory rates.

Gastric inhibitory peptide (GIP) is secreted when fats and carbohydrates—especially glucose—enter the small intestine. The inhibition of gastric activity is accompanied by the stimulation of insulin release at the pancreatic islets. GIP has several secondary effects, including stimulating duodenal gland activity, stimulating lipid synthesis in adipose tissue, and increasing glucose use by skeletal muscles.

Cholecystokinin (CCK) is secreted when chyme arrives in the duodenum, especially when the chyme contains lipids and partially digested proteins. In the pancreas, CCK accelerates the production and secretion of all types of digestive enzymes. It also causes a relaxation of the hepatopancreatic sphincter and contraction of the gallbladder, resulting in the ejection of bile and pancreatic juice into the duodenum. Thus, the net effects of CCK are to increase the secretion of pancreatic enzymes and to push pancreatic secretions and bile into the duodenum. The presence of CCK in high concentrations has two additional effects: It inhibits gastric activity, and it appears to have CNS effects that reduce the sensation of hunger.

Vasoactive intestinal peptide (VIP) stimulates the secretion of intestinal glands, dilates regional capillaries, and inhibits acid production in the stomach. By dilating capillaries in active areas of the intestinal tract, VIP provides an efficient mechanism for removing absorbed nutrients.

2 This flowchart summarizes the pattern of hormone release and the effects of those hormones within the digestive system.

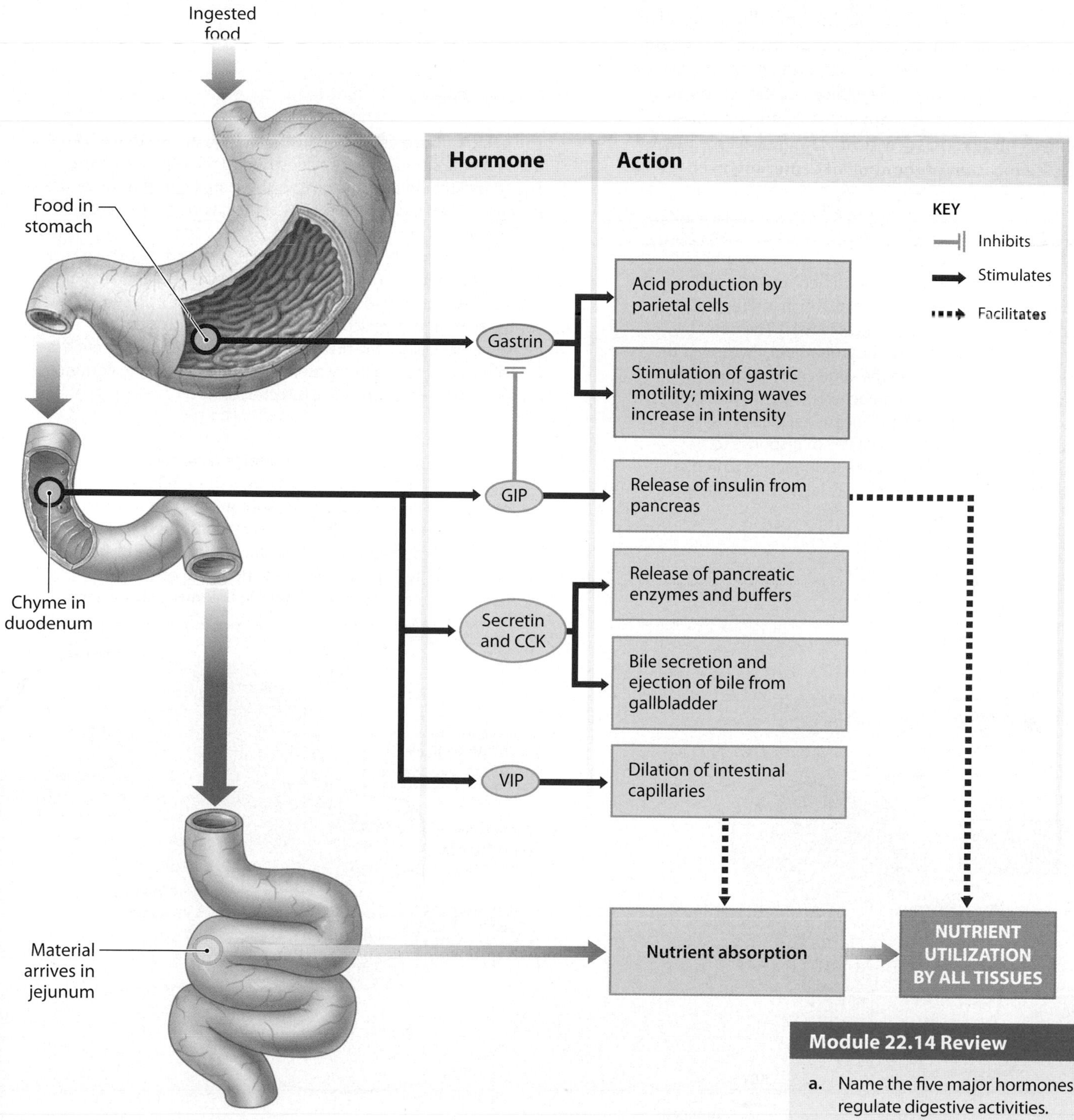

Ingested food
Food in stomach
Chyme in duodenum
Material arrives in jejunum
Hormone
Action
KEY
Inhibits
Stimulates
Facilitates
Gastrin
GIP
Secretin and CCK
VIP
Acid production by parietal cells
Stimulation of gastric motility; mixing waves increase in intensity
Release of insulin from pancreas
Release of pancreatic enzymes and buffers
Bile secretion and ejection of bile from gallbladder
Dilation of intestinal capillaries
Nutrient absorption
NUTRIENT UTILIZATION BY ALL TISSUES

Module 22.14 Review

a. Name the five major hormones that regulate digestive activities.

b. How would the pH of the intestinal contents be affected if the small intestine did not produce secretin?

c. Does a high-fat meal raise or lower the level of cholecystokinin (CCK) in the blood?

Module 22.15

Central and local mechanisms coordinate gastric and intestinal activities

1 The duodenum plays a key role in controlling digestive function because it monitors the contents of the chyme and adjusts the activities of the stomach and accessory glands to protect the delicate absorptive surfaces of the jejunum. This pivotal role of the duodenum is apparent when you consider the three **phases of gastric secretion**. The phases are named according to the location of the control center involved.

Cephalic Phase

The **cephalic phase** of gastric secretion begins when you see, smell, taste, or think of food. This phase, which is directed by the CNS, prepares the stomach to receive food. The neural output proceeds by way of the parasympathetic division of the autonomic nervous system, and the vagus nerves innervate the submucosal plexus of the stomach. Next, postganglionic parasympathetic fibers innervate mucous cells, chief cells, parietal cells, and G cells of the stomach. In response to stimulation, gastric juice production accelerates, reaching rates of about 500 mL/h. This phase generally lasts only minutes.

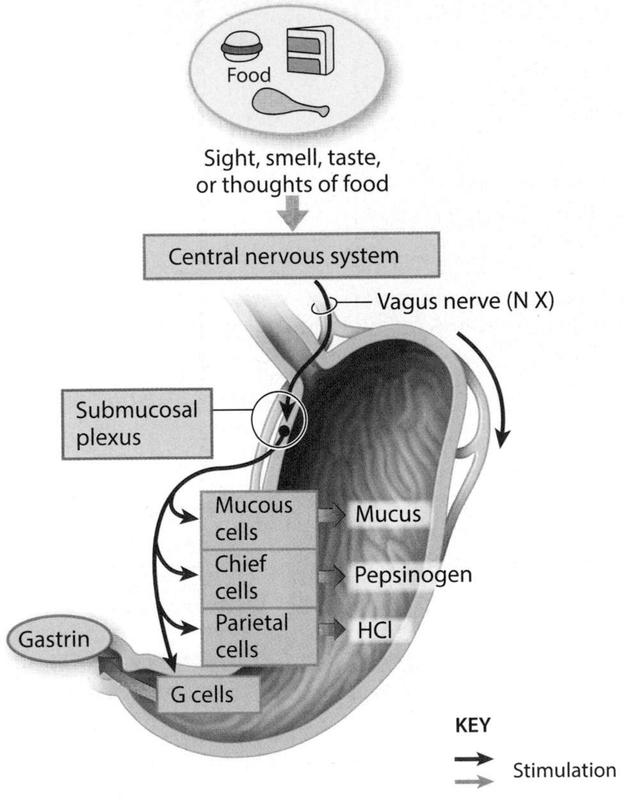

Food
Sight, smell, taste, or thoughts of food
Central nervous system
Vagus nerve (N X)
Submucosal plexus
Mucous cells
Mucus
Chief cells
Pepsinogen
Parietal cells
HCl
Gastrin
G cells
KEY
Stimulation

Gastric Phase

The **gastric phase** begins when food arrives in the stomach and builds on the stimulation provided during the cephalic phase. The stimuli that initiate the gastric phase are (1) distension of the stomach, (2) an increase in the pH of the gastric contents (due to dilution of acid during mixing), and (3) the presence of undigested food in the stomach, especially proteins and peptides. The gastric phase may continue for 3 to 4 hours while the acid and enzymes process the ingested food. During this period, gastrin stimulates contractions in the muscularis externa of the stomach and intestinal tract. After the first hour, the material in the stomach is churning like clothing in a washing machine. As mixing continues, a large volume of gastric juice is secreted.

The stimulation of stretch receptors and chemoreceptors triggers short reflexes coordinated in the submucosal and myenteric plexuses. This in turn activates the stomach's secretory cells. The stimulation of the myenteric plexus produces powerful contractions called **mixing waves** in the muscularis externa.

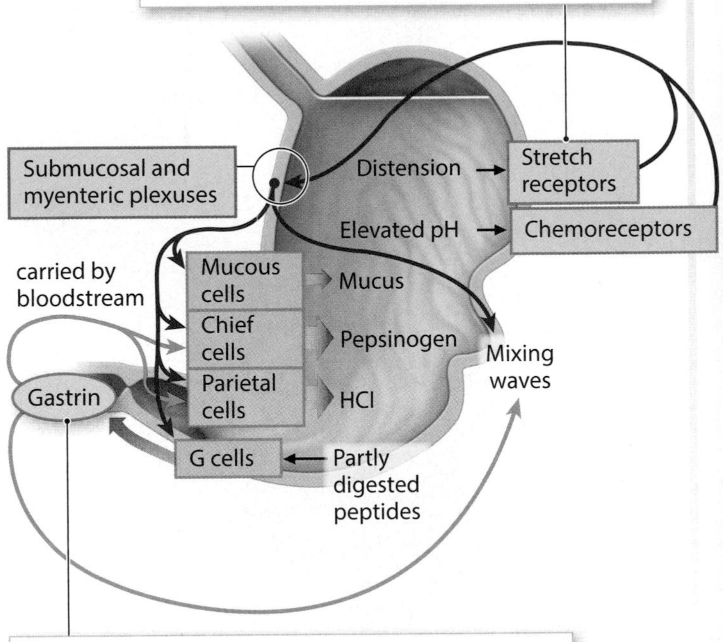

Submucosal and myenteric plexuses
Distension
Stretch receptors
Elevated pH
Chemoreceptors
carried by bloodstream
Mucous cells
Mucus
Chief cells
Pepsinogen
Parietal cells
HCl
Mixing waves
Gastrin
G cells
Partly digested peptides

Neural stimulation and the presence of peptides and amino acids in chyme stimulate the secretion of the hormone gastrin. Gastrin travels by the bloodstream to parietal and chief cells, whose increased secretions reduce the pH of the gastric juice. In addition, gastrin also stimulates gastric motility.

Intestinal Phase

The **intestinal phase** of gastric secretion begins when chyme first enters the small intestine, usually after several hours of mixing contractions. The function of the intestinal phase is controlling the rate of gastric emptying to ensure that the secretory, digestive, and absorptive functions of the small intestine can proceed with reasonable efficiency. Although here we consider the intestinal phase as it affects stomach activity, the arrival of chyme in the small intestine also triggers other neural and hormonal events that coordinate the activities of the intestinal tract and the pancreas, liver, and gallbladder.

Chyme leaving the stomach decreases the distension in the stomach, thereby reducing the stimulation of stretch receptors. Distension of the duodenum by chyme stimulates stretch receptors and chemoreceptors that trigger the **enterogastric reflex.** This reflex inhibits both gastrin production and gastric contractions and stimulates the contraction of the pyloric sphincter, which prevents further discharge of chyme. At the same time, local reflexes at the duodenum stimulate mucus production, which helps protect the duodenal lining from the arriving acid and enzymes.

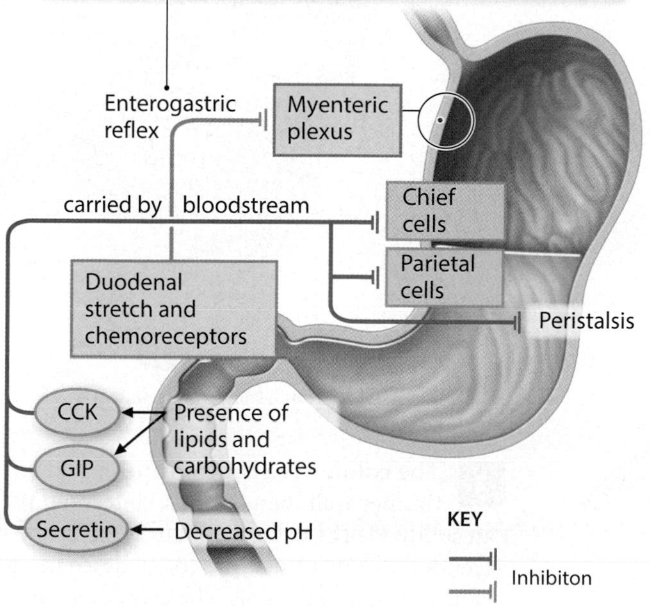
Enterogastric reflex
Myenteric plexus
carried by bloodstream
Chief cells
Parietal cells
Duodenal stretch and chemoreceptors
Peristalsis
CCK
GIP
Presence of lipids and carbohydrates
Secretin
Decreased pH
KEY
Inhibiton

2 Two central reflexes are triggered by the stimulation of stretch receptors in the stomach wall as it fills. These reflexes accelerate movement along the small intestine while the enterogastric reflex controls the rate of chyme entry into the duodenum.

Central Gastric Reflexes

The **gastroenteric reflex** stimulates motility and secretion along the entire small intestine.

The **gastroileal** (gas-trō-IL-ē-al) **reflex** triggers the opening of the ileocecal valve, allowing materials to pass from the small intestine into the large intestine.

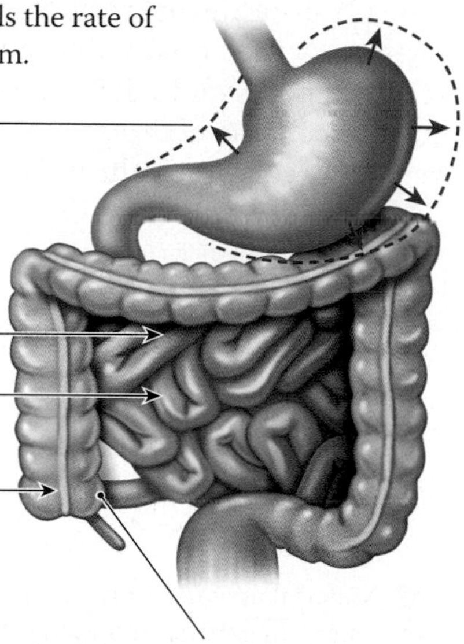

The **ileocecal valve** regulates the passage of materials into the large intestine.

In general, the rate of chyme movement into the small intestine is fastest when the stomach is greatly distended and the meal contains little protein. A big meal that contains small amounts of protein, large amounts of carbohydrates, alcohol, or caffeine will leave the stomach very quickly. One reason for this is that both alcohol and caffeine stimulate gastric secretion and motility.

Module 22.15 Review

a. Name and briefly describe an important characteristic of each of the three phases of gastric secretion.

b. Why might severing the branches of the vagus nerves that supply the stomach provide relief for a person who suffers from chronic gastric ulcers (sores on the stomach lining)?

c. Describe two central reflexes triggered by stimulation of the stretch receptors in the stomach wall.

Module 22.16

The large intestine stores and concentrates fecal material

The **large intestine**, also known as the large bowel, has an average length of about 1.5 meters (4.9 ft) and a diameter of 7.5 cm (3 in.). The major functions of the large intestine include (1) reabsorbing water and compacting the intestinal contents into feces, (2) absorbing important vitamins generated by bacterial action, and (3) storing fecal material prior to defecation. The large intestine consists of three segments: the cecum, the colon, and the rectum.

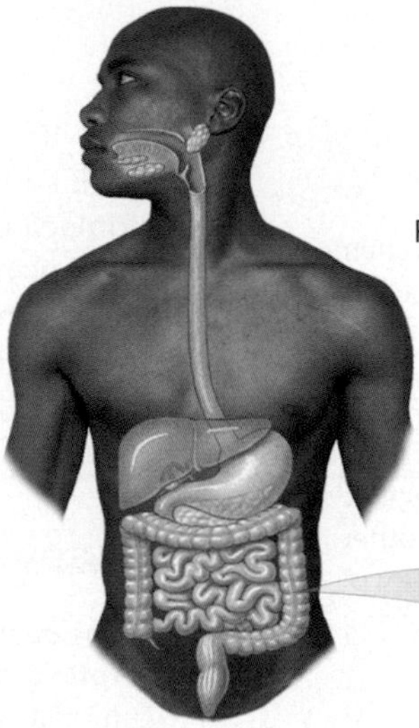

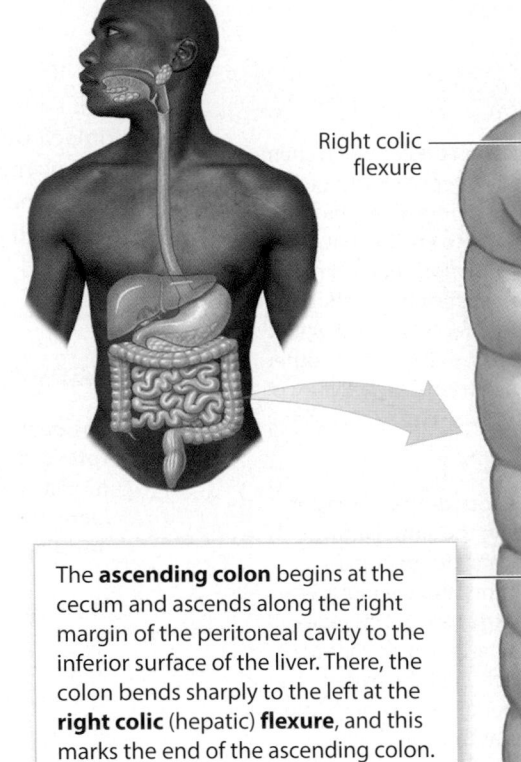

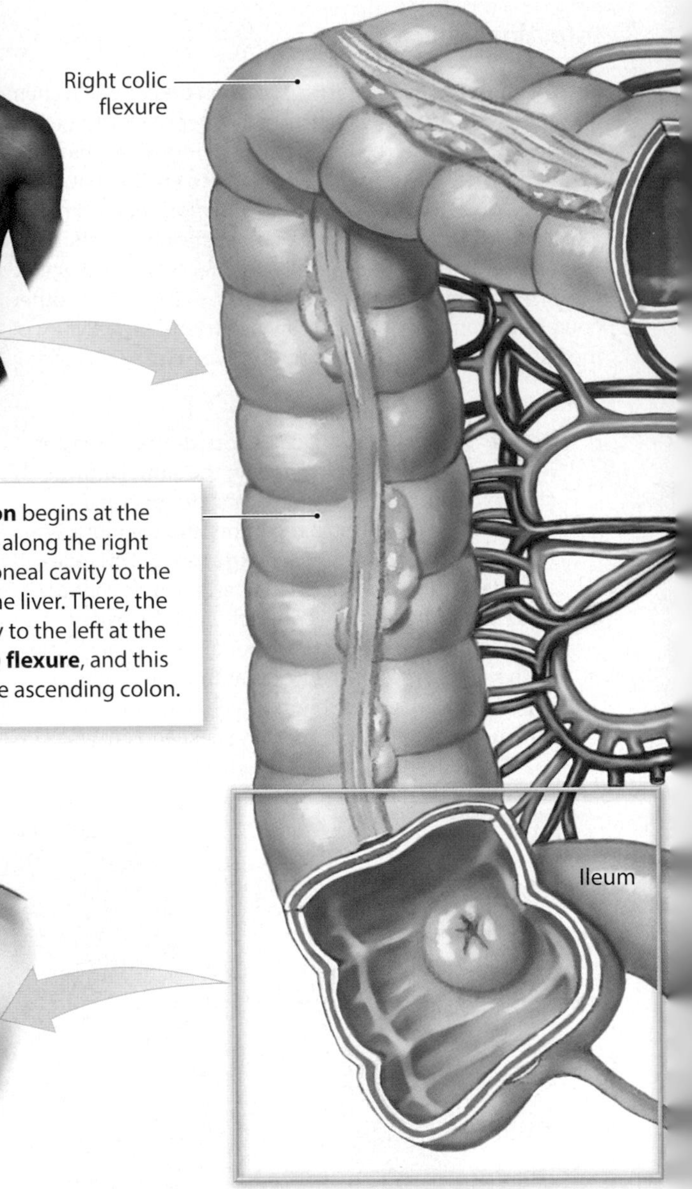
Right colic flexure
Ileum

The **ascending colon** begins at the cecum and ascends along the right margin of the peritoneal cavity to the inferior surface of the liver. There, the colon bends sharply to the left at the **right colic** (hepatic) **flexure**, and this marks the end of the ascending colon.

1 Material arriving from the ileum first enters an expanded pouch called the **cecum** (SĒ-kum). The cecum collects and stores materials from the ileum and begins the process of **compaction** (the forming of feces by compression).

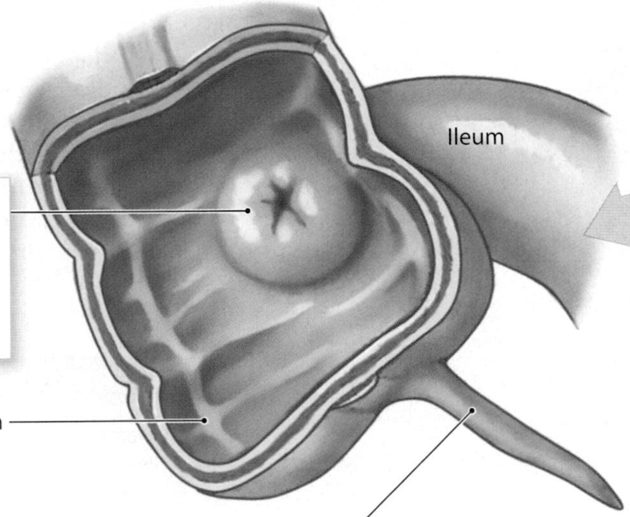
Ileum
Cecum

The ileum attaches to the medial surface of the cecum and opens into the cecum at the **ileocecal valve**.

The slender, hollow **appendix** (also called the vermiform appendix) is attached to the cecum. The appendix is generally about 9 cm (3.6 in.) long, but its size and shape are quite variable. The mucosa and submucosa of the appendix are dominated by lymphoid nodules, and the appendix functions primarily as an organ of the lymphatic system. Inflammation of the appendix is known as **appendicitis**.

2 The **colon** has a larger diameter and a thinner wall than the small intestine. We can subdivide the colon into four regions: the ascending colon, transverse colon, descending colon, and sigmoid colon. The ascending and descending colon are retroperitoneal and firmly attached to the abdominal wall. The transverse colon and sigmoid colon are suspended by remnants of the embryonic mesocolon.

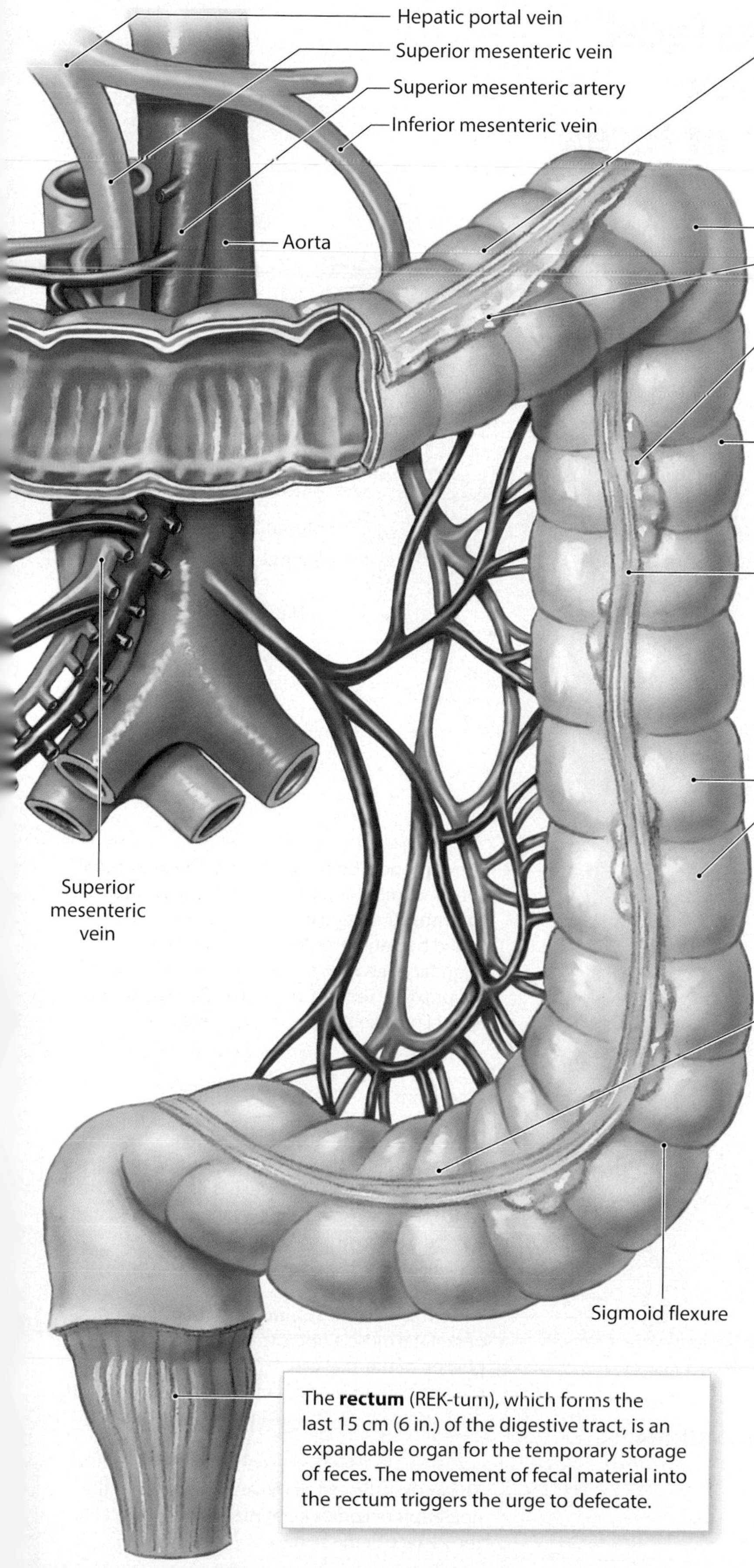
Hepatic portal vein
Superior mesenteric vein
Superior mesenteric artery
Inferior mesenteric vein
Aorta
Left colic flexure
Greater omentum (cut)
Superior mesenteric vein
Sigmoid flexure

The **transverse colon** crosses the abdomen from right to left. It is supported by the transverse mesocolon and is separated from the anterior abdominal wall by the layers of the greater omentum. As the transverse colon reaches the left side of the body, the colon makes a 90° turn at the **left colic** (splenic) **flexure**.

The serosa of the colon contains numerous teardrop-shaped sacs of fat called **omental** (fatty) **appendices**.

The **descending colon** proceeds inferiorly along the body's left side until reaching the iliac fossa. At the iliac fossa, the descending colon ends at the **sigmoid flexure**.

Three separate longitudinal bands of smooth muscle—called the **teniae coli** (TĒ-nē-ē KŌ-lē; singular, *tenia*)—run along the outer surfaces of the colon just deep to the serosa. These bands correspond to the outer layer of the muscularis externa in other portions of the digestive tract.

Muscle tone within the teniae coli is what creates **haustra** (HAWS-truh), a series of pouches in the wall of the colon. Cutting into the intestinal lumen reveals that the creases between the haustra affect the mucosal lining as well, producing a series of internal folds. Haustra permit the expansion and elongation of the colon, rather like the bellows that allow an accordion to lengthen.

The sigmoid flexure is the start of the **sigmoid** (SIG-moyd; *sigmeidos*, Greek letter S) **colon**, an S-shaped segment that is about 15 cm (6 in.) long and empties into the rectum.

The **rectum** (REK-tum), which forms the last 15 cm (6 in.) of the digestive tract, is an expandable organ for the temporary storage of feces. The movement of fecal material into the rectum triggers the urge to defecate.

Powerful peristaltic contractions called **mass movements** occur a few times each day in response to distension of the stomach and duodenum. These contractions begin at the transverse colon and push materials along the distal portion of the large intestine.

Module 22.16 Review

a. Name the major functions of the large intestine.

b. Identify the segments of the large intestine and the four regions of the colon.

c. Describe mass movements.

LO 22.16 Describe the gross anatomy of the three segments of the large intestine.

The large intestine compacts fecal material; the defecation reflex coordinates the elimination of feces

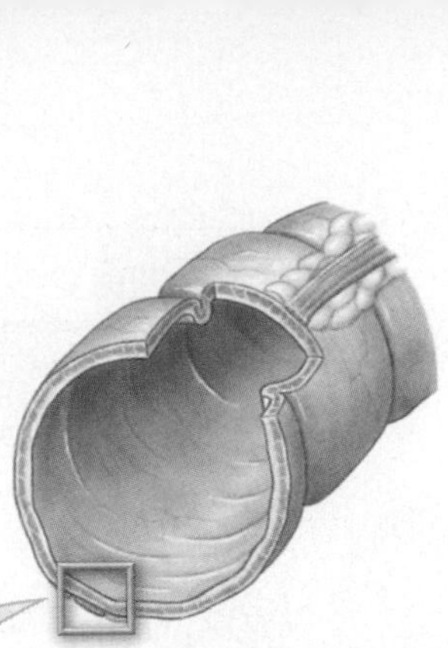

1 The major characteristics of the wall of the large intestine are the lack of villi and the presence of distinctive intestinal glands dominated by mucous cells. The mucosa of the large intestine does not produce enzymes; any digestion that occurs results from enzymes secreted into the small intestine or from bacterial action. The mucus lubricates the feces as it becomes drier and more compact.

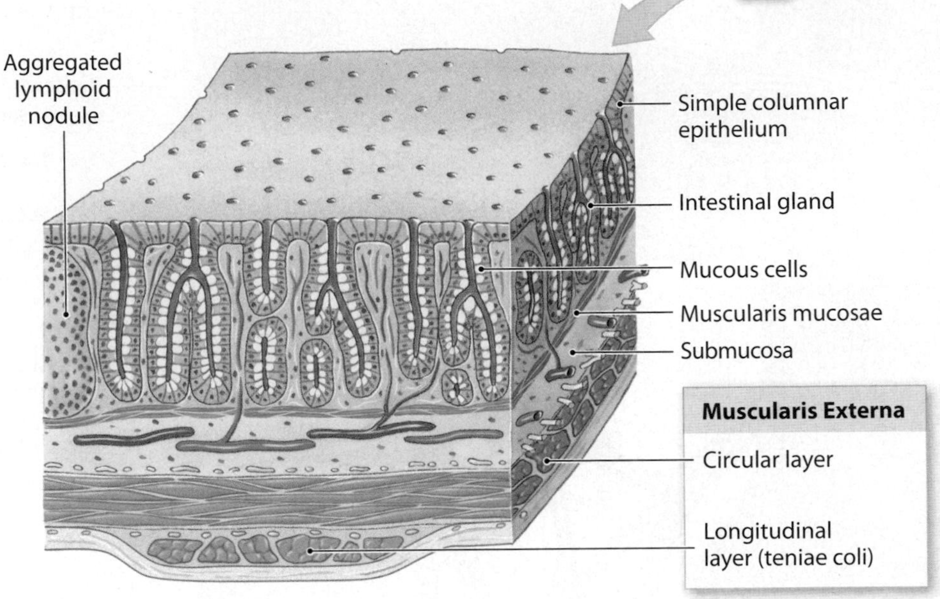
Aggregated lymphoid nodule
Simple columnar epithelium
Intestinal gland
Mucous cells
Muscularis mucosae
Submucosa
Muscularis Externa
Circular layer
Longitudinal layer (teniae coli)

2 This illustration shows the characteristic features of the rectum, the last segment of the digestive tract. The lamina propria and submucosa of the distal portion of the rectum contain a network of veins. If venous pressures there rise too high due to straining during defecation or pregnancy, the veins can become distended, producing **hemorrhoids**.

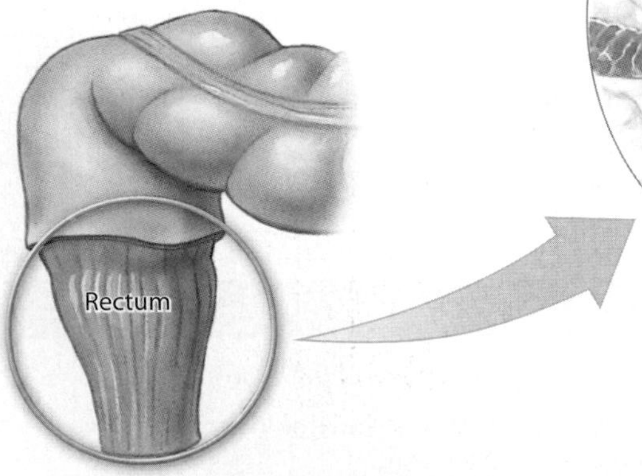
Rectum

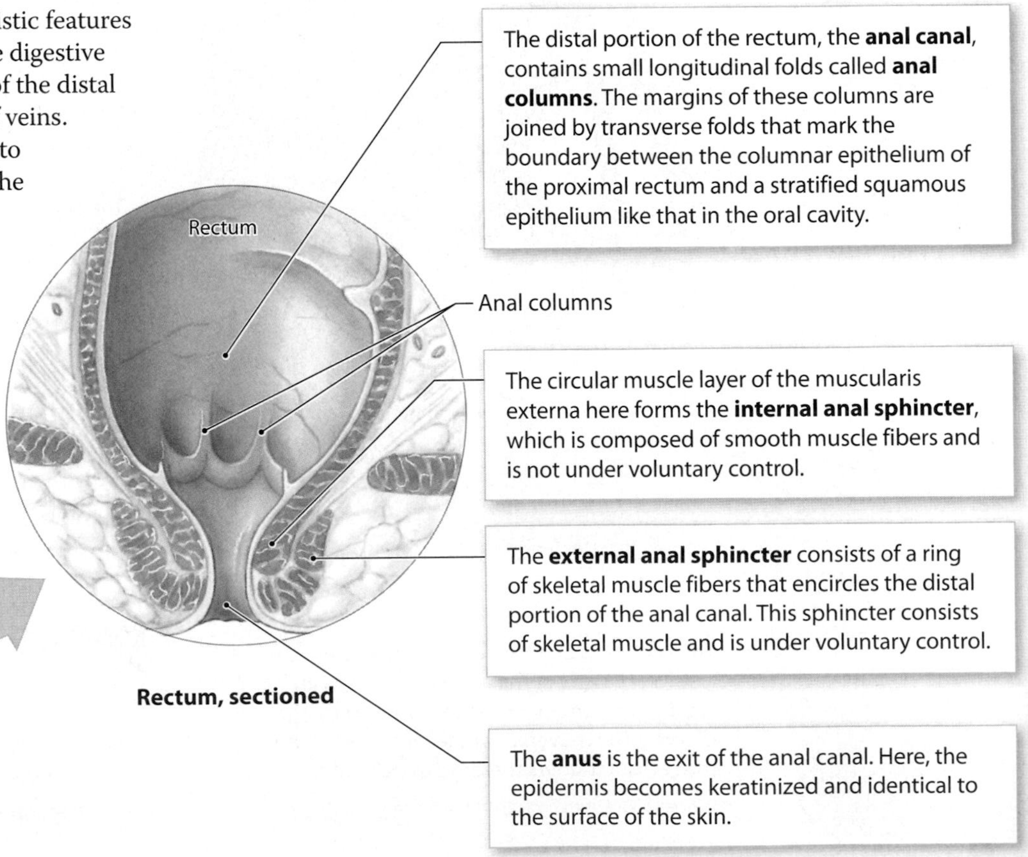
Rectum
The distal portion of the rectum, the **anal canal**, contains small longitudinal folds called **anal columns**. The margins of these columns are joined by transverse folds that mark the boundary between the columnar epithelium of the proximal rectum and a stratified squamous epithelium like that in the oral cavity.
Anal columns
The circular muscle layer of the muscularis externa here forms the **internal anal sphincter**, which is composed of smooth muscle fibers and is not under voluntary control.
The **external anal sphincter** consists of a ring of skeletal muscle fibers that encircles the distal portion of the anal canal. This sphincter consists of skeletal muscle and is under voluntary control.
The **anus** is the exit of the anal canal. Here, the epidermis becomes keratinized and identical to the surface of the skin.

Rectum, sectioned

3 Less than 10 percent of the nutrient absorption within the digestive tract occurs in the large intestine. Nevertheless, absorption in this segment of the digestive tract is very important. In addition to preventing dehydration by reabsorbing water, the epithelium absorbs three vitamins produced by the normal bacteria living in the colon.

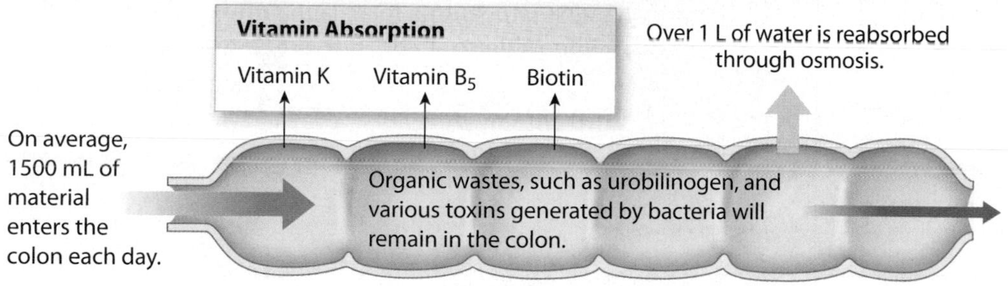

Vitamin Absorption
Vitamin K
Vitamin B_5
Biotin
Over 1 L of water is reabsorbed through osmosis.
On average, 1500 mL of material enters the colon each day.
Organic wastes, such as urobilinogen, and various toxins generated by bacteria will remain in the colon.

Only 200 mL of feces is ejected. Fecal material is 75% water, 5% bacteria, and the rest a mixture of indigestible materials, inorganic matter, and the remains of epithelial cells. Bacteria produce several compounds that contribute to the odor of feces, including ammonia, **indole** and **skatole** (two nitrogen-containing compounds), and **hydrogen sulfide** (H_2S), a gas that produces a "rotten egg" odor.

4 The rectal chamber is usually empty, except when a powerful peristaltic contraction forces feces out of the sigmoid colon. Distension of the rectal wall then starts the **defecation reflex**, which involves two positive feedback loops, triggered by the stimulation of stretch receptors in the walls of the rectum.

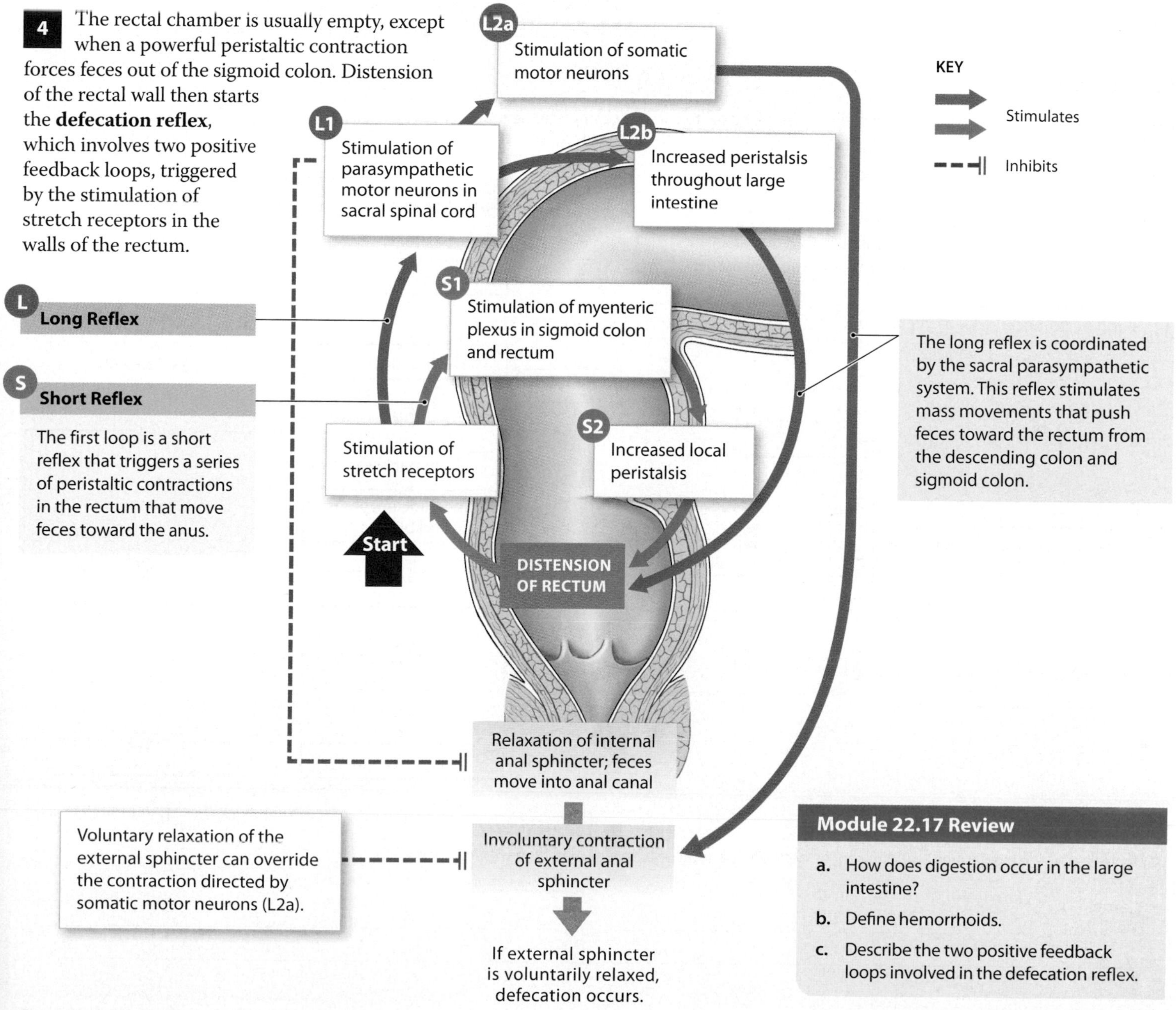

L2a Stimulation of somatic motor neurons
L1 Stimulation of parasympathetic motor neurons in sacral spinal cord
L2b Increased peristalsis throughout large intestine
KEY
Stimulates
Inhibits
L Long Reflex
S Short Reflex
The first loop is a short reflex that triggers a series of peristaltic contractions in the rectum that move feces toward the anus.
S1 Stimulation of myenteric plexus in sigmoid colon and rectum
Stimulation of stretch receptors
S2 Increased local peristalsis
Start
DISTENSION OF RECTUM
The long reflex is coordinated by the sacral parasympathetic system. This reflex stimulates mass movements that push feces toward the rectum from the descending colon and sigmoid colon.
Relaxation of internal anal sphincter; feces move into anal canal
Voluntary relaxation of the external sphincter can override the contraction directed by somatic motor neurons (L2a).
Involuntary contraction of external anal sphincter
If external sphincter is voluntarily relaxed, defecation occurs.

Module 22.17 Review

a. How does digestion occur in the large intestine?

b. Define hemorrhoids.

c. Describe the two positive feedback loops involved in the defecation reflex.

LO 22.17 Describe the large intestine's histology and role in fecal compaction, and explain the defecation reflex.

Labeling

Label the structures of a typical tooth in the accompanying figure.

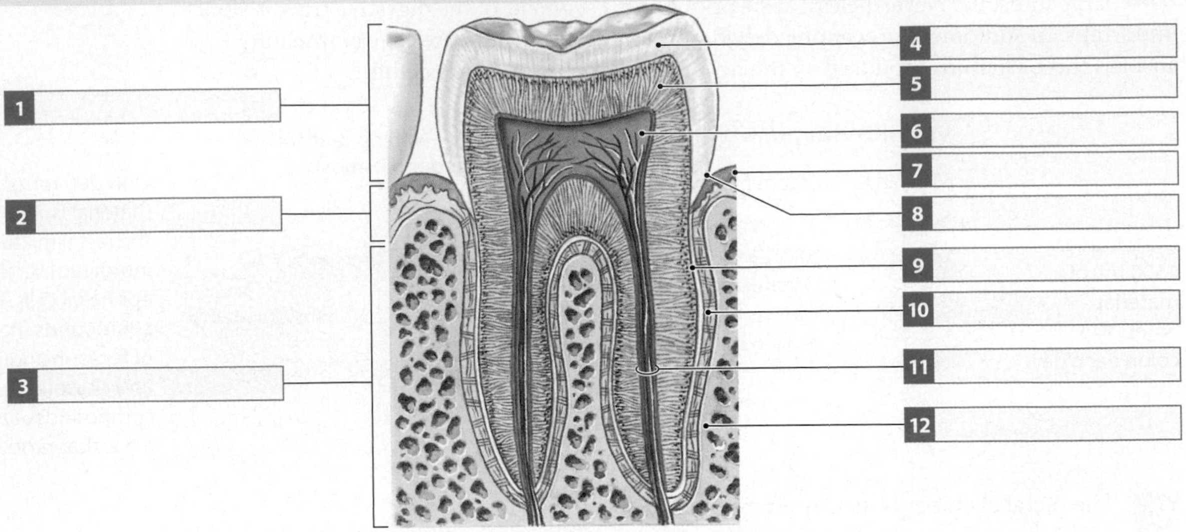
1
2
3
4
5
6
7
8
9
10
11
12

Concept map

Use each of the following terms once to fill in the blank boxes to correctly complete the map.

- acid production
- gastrin
- VIP
- insulin
- intestinal capillaries
- GIP
- material in jejunum
- gallbladder
- bile
- inhibits
- secretin and CCK
- nutrient utilization by tissues

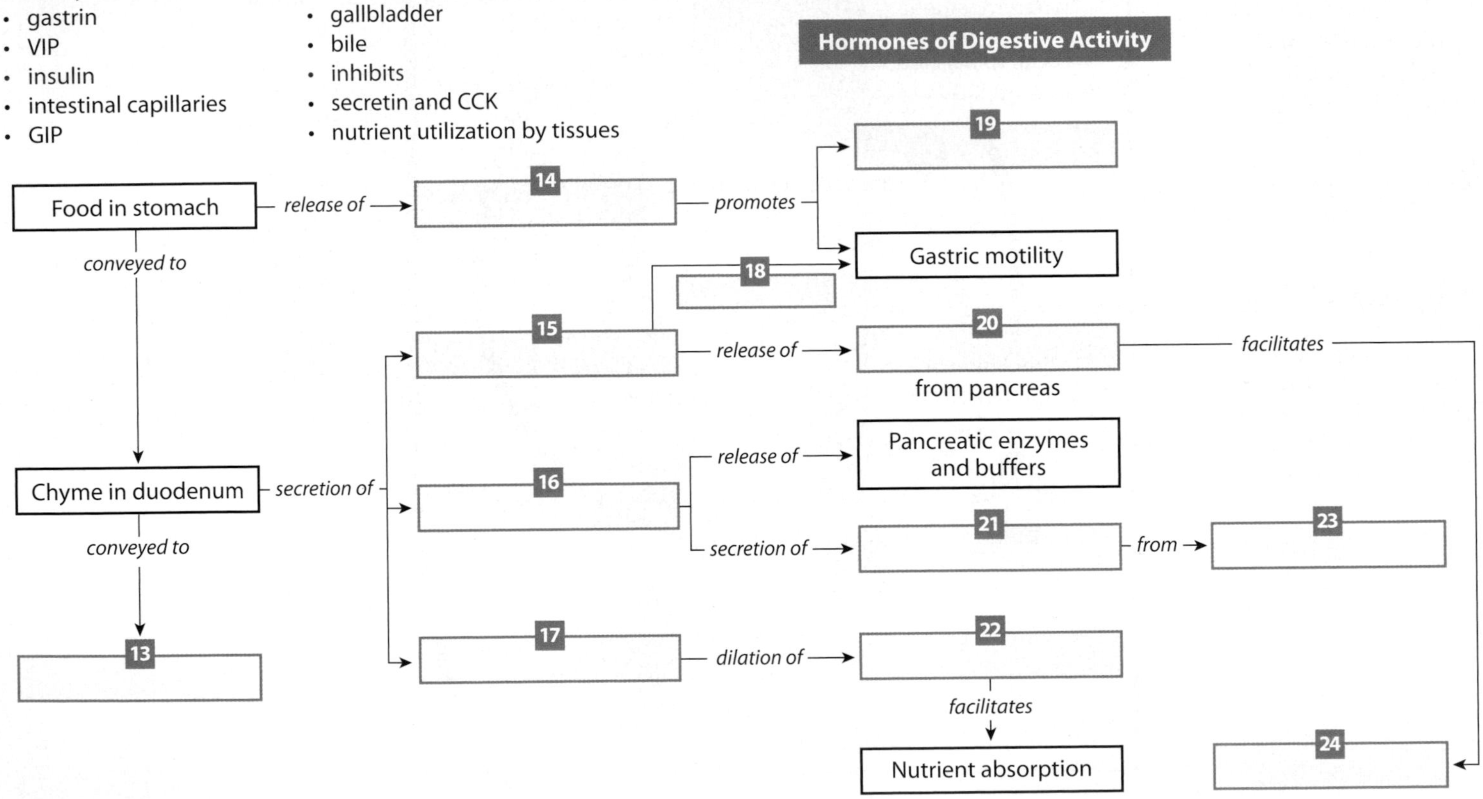
Hormones of Digestive Activity
Food in stomach
release of
14
promotes
19
Gastric motility
conveyed to
18
15
release of
20
from pancreas
facilitates
Chyme in duodenum
secretion of
release of
Pancreatic enzymes and buffers
16
secretion of
21
from
23
conveyed to
13
17
dilation of
22
facilitates
Nutrient absorption
24

Short answer

Briefly describe the similarities and differences between parietal cells and chief cells in the stomach wall.

25 ____

____

Some accessory digestive organs have secretory functions

The accessory digestive organs are the salivary glands, the gallbladder, the pancreas, and the liver. The salivary glands and pancreas produce and store enzymes and buffers that are essential to normal digestive function. In addition to their roles in digestion, the salivary glands, liver, and pancreas have vital metabolic and endocrine functions.

Accessory Digestive Organs

Salivary Glands

Three pairs of salivary glands produce secretions that contain mucins and enzymes.

Gallbladder

The gallbladder stores and concentrates bile secreted by the liver.

Pancreas

Exocrine cells secrete buffers and digestive enzymes; endocrine cells secrete insulin, glucagon, pancreatic polypeptide, and GH–IH, hormones introduced in Module 16.10 (p. 604).

Liver

The liver has almost 200 known functions. Some of the most important are listed in the table below.

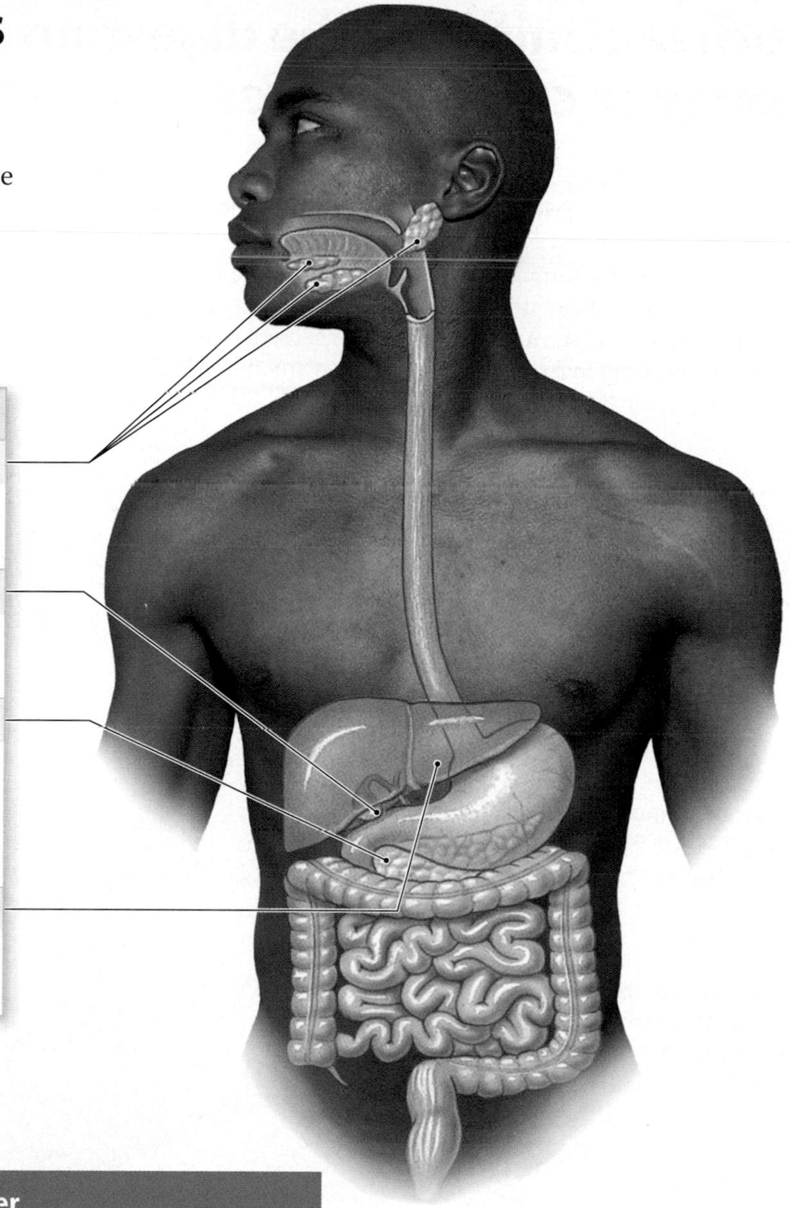

Digestive and Metabolic Functions of the Liver

- Synthesizing and secreting bile
- Storing glycogen and lipids
- Maintaining normal concentrations of glucose, amino acids, and fatty acids in the bloodstream
- Synthesizing and interconverting nutrient types (such as the conversion of carbohydrates to lipids)
- Synthesizing and releasing cholesterol bound to transport proteins
- Inactivating toxins
- Storing iron
- Storing fat-soluble vitamins

Other Major Functions

- Synthesizing plasma proteins
- Synthesizing clotting factors
- Phagocytizing damaged red blood cells (by Kupffer cells)
- Storing blood
- Absorbing and breaking down circulating hormones and immunoglobulins
- Absorbing and inactivating lipid-soluble drugs

Module 22.18 Review

a. What is the function of the salivary glands?

b. Distinguish between the exocrine and endocrine secretions of the pancreas.

c. Which accessory organ of the digestive system is responsible for almost 200 known functions?

LO 22.18 Describe the functions of the accessory organs of the digestive system.

Module 22.19

Saliva lubricates and moistens the mouth and initiates the digestion of complex carbohydrates

1 Three pairs of salivary glands secrete into the oral cavity. Each pair has a distinctive cellular organization and produces saliva with slightly different properties. Any object in your mouth can trigger a salivary reflex by stimulating receptors monitored by the trigeminal nerve (V) or taste buds innervated by cranial nerves VII, IX, or X. Parasympathetic stimulation accelerates secretion by all the salivary glands, resulting in the production of large amounts of saliva.

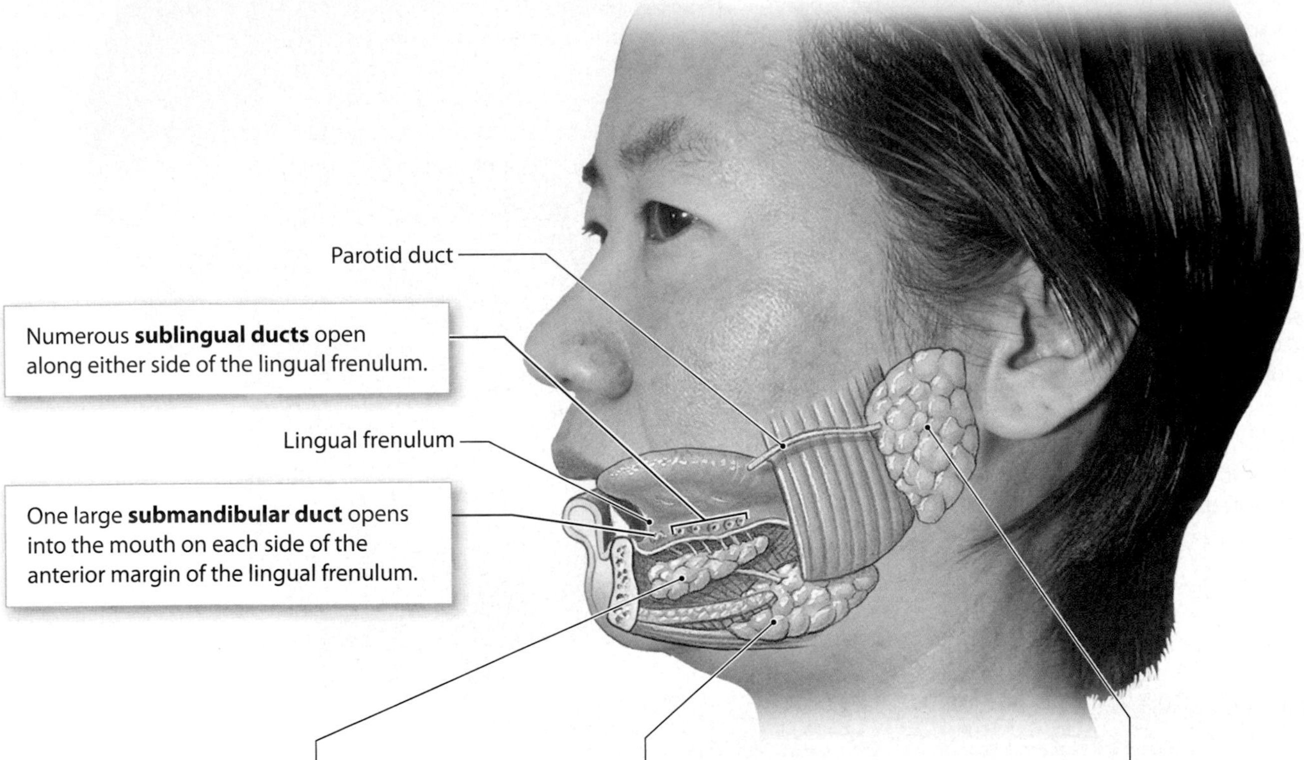

Parotid duct
Numerous **sublingual ducts** open along either side of the lingual frenulum.
Lingual frenulum
One large **submandibular duct** opens into the mouth on each side of the anterior margin of the lingual frenulum.

The Salivary Glands		
The **sublingual** (sub-LING-gwal) **salivary glands** lie under either side of the tongue, covered by the mucous membrane of the floor of the mouth. These glands produce a mucous secretion that acts as a buffer and lubricant.	The **submandibular salivary glands** lie along the inner surface of the mandible within the mandibular groove. Cells of the submandibular glands secrete a mixture of buffers, mucins, and **salivary amylase**, an enzyme that breaks down starches (complex carbohydrates). The gland cells also transport antibodies (IgA) into the saliva, to provide additional protection against pathogens in food.	The large **parotid** (pa-ROT-id) **salivary glands** lie inferior to the zygomatic arch deep to the skin covering the lateral and posterior surface of the mandible. Each gland has an irregular shape, extending from the mastoid process of the temporal bone across the outer surface of the masseter muscle. The parotid salivary glands produce a serous secretion containing large amounts of salivary amylase. The secretions of each parotid gland are drained by a **parotid duct**, which empties into the vestibule at the level of the second upper molar.

2 Each submandibular salivary gland contains a mixture of secretory cells, some specialized for mucous secretion and others specialized for enzyme production. The **saliva** in the mouth is a mixture of glandular secretions; about 70 percent of the saliva originates in the submandibular salivary glands, 25 percent from the parotid salivary glands, and 5 percent from the sublingual salivary glands. Collectively the salivary glands produce 1.0–1.5 L of saliva each day, and 99.4 percent of that volume is water.

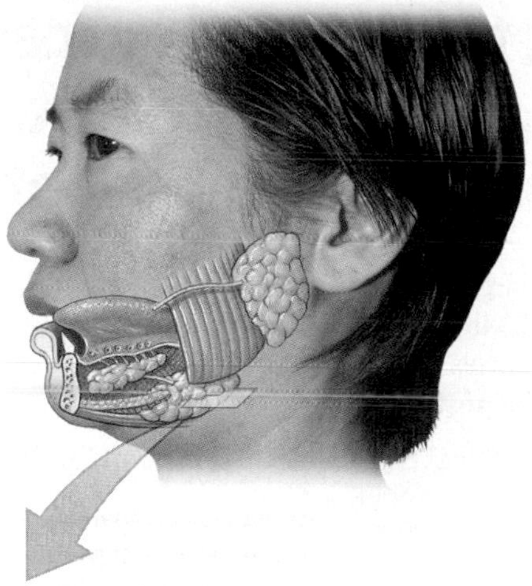

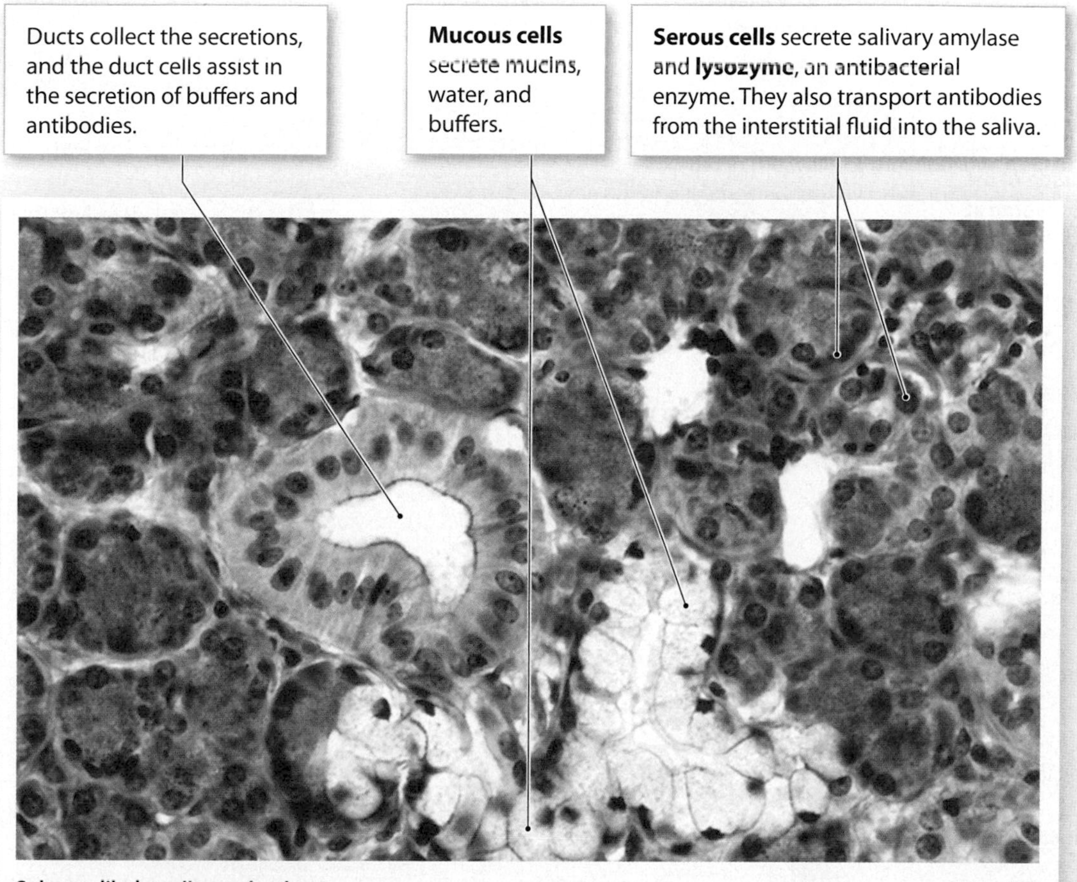
Ducts collect the secretions, and the duct cells assist in the secretion of buffers and antibodies.
Mucous cells secrete mucins, water, and buffers.
Serous cells secrete salivary amylase and **lysozyme**, an antibacterial enzyme. They also transport antibodies from the interstitial fluid into the saliva.

Submandibular salivary gland LM × 600

A continuous background level of saliva secretion flushes the oral surfaces, helping keep them clean. Buffers in the saliva keep the pH of your mouth near 7.0 and prevent the buildup of acids produced by bacteria. In addition, saliva contains antibodies (IgA) and lysozyme, which help control populations of oral bacteria. Food usually remains in the mouth long enough for chewing (mastication) to mix it with saliva and break the combination into a relatively homogeneous, pulpy mass. This is compacted by the tongue to form a bolus that can be easily swallowed.

Module 22.19 Review

a. Name the three pairs of salivary glands.

b. The digestion of which nutrient would be affected by damage to the parotid salivary glands?

c. Which glandular secretions contribute least to saliva production?

LO **22.19** Discuss the structure and functions of the salivary glands.

The liver, the largest visceral organ, is divided into left, right, caudate, and quadrate lobes

1 The **liver**, the largest visceral organ, weighs about 1.5 kg (3.3 lb). These two horizontal sections at vertebral levels T_{11} and T_{12} shown below give you an idea of the position of the liver relative to other visceral organs.

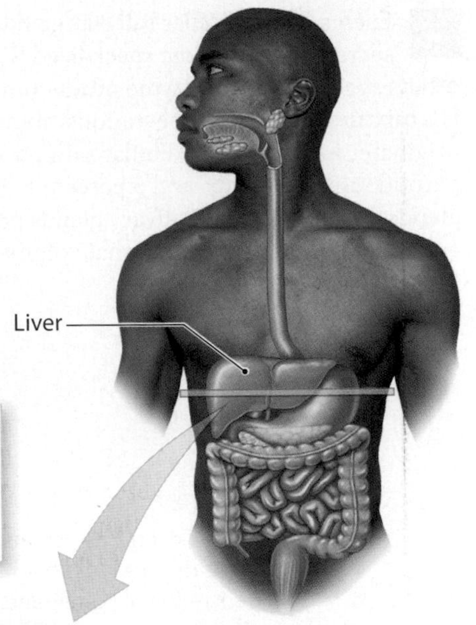
Liver

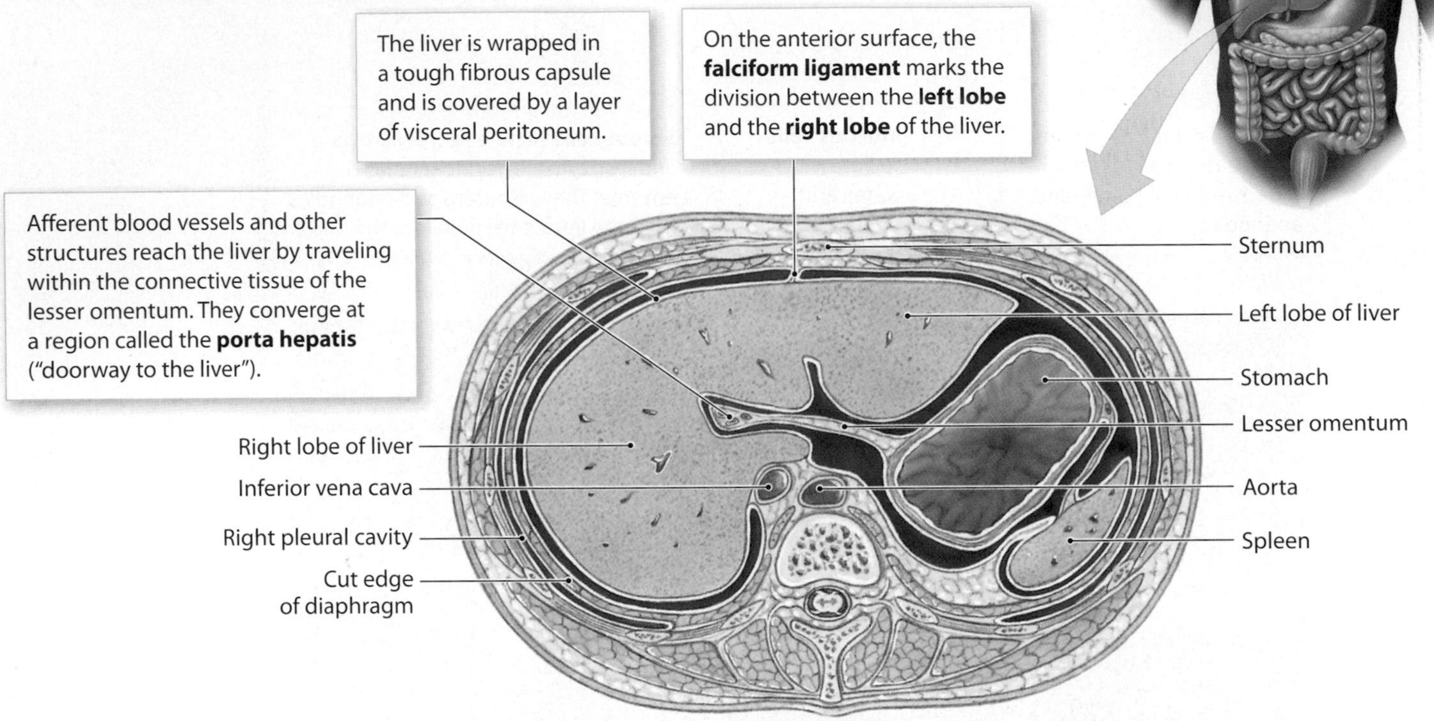
The liver is wrapped in a tough fibrous capsule and is covered by a layer of visceral peritoneum.
On the anterior surface, the **falciform ligament** marks the division between the **left lobe** and the **right lobe** of the liver.
Afferent blood vessels and other structures reach the liver by traveling within the connective tissue of the lesser omentum. They converge at a region called the **porta hepatis** ("doorway to the liver").
Sternum
Left lobe of liver
Stomach
Lesser omentum
Aorta
Spleen
Right lobe of liver
Inferior vena cava
Right pleural cavity
Cut edge of diaphragm

Horizontal section at the level of vertebra T_{11} (diagrammatic view)

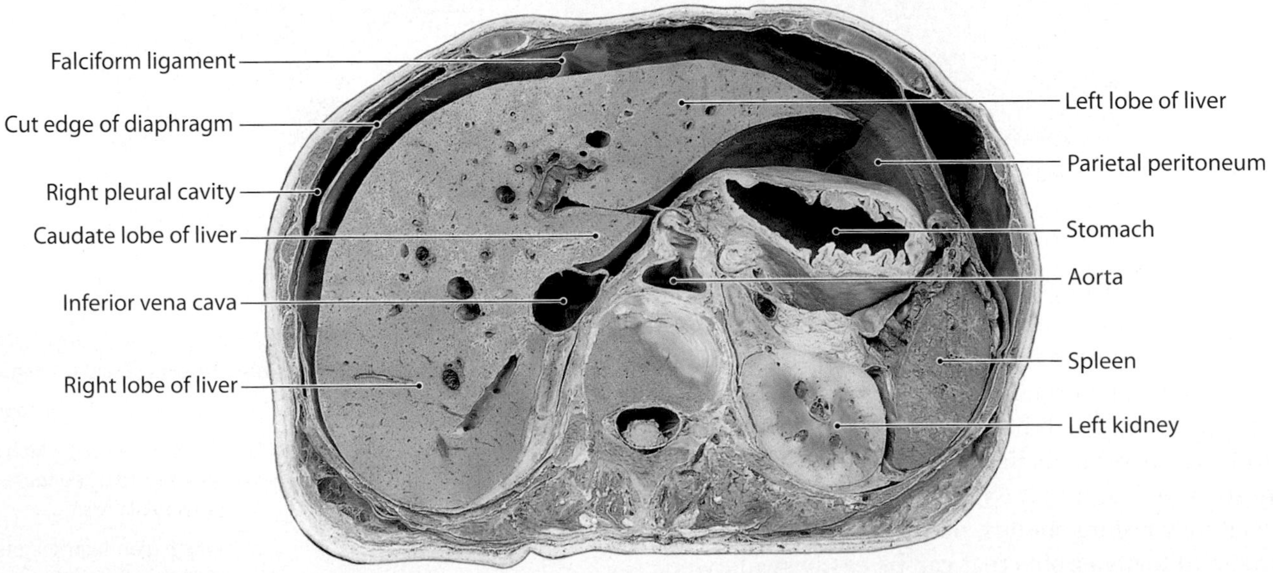
Falciform ligament
Cut edge of diaphragm
Right pleural cavity
Caudate lobe of liver
Inferior vena cava
Right lobe of liver
Left lobe of liver
Parietal peritoneum
Stomach
Aorta
Spleen
Left kidney

Horizontal section at the level of vertebra T_{12} (cadaver)

2 The major anatomical landmarks and the four liver lobes are shown on these anterior and posterior views of the isolated liver.

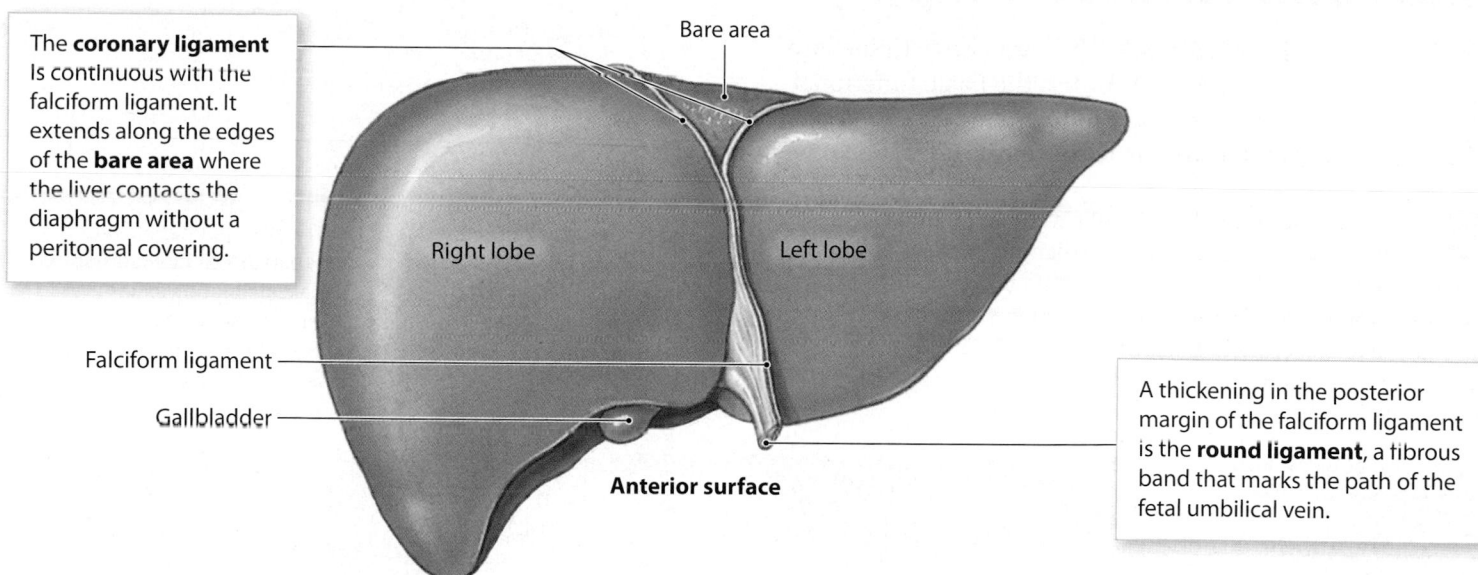

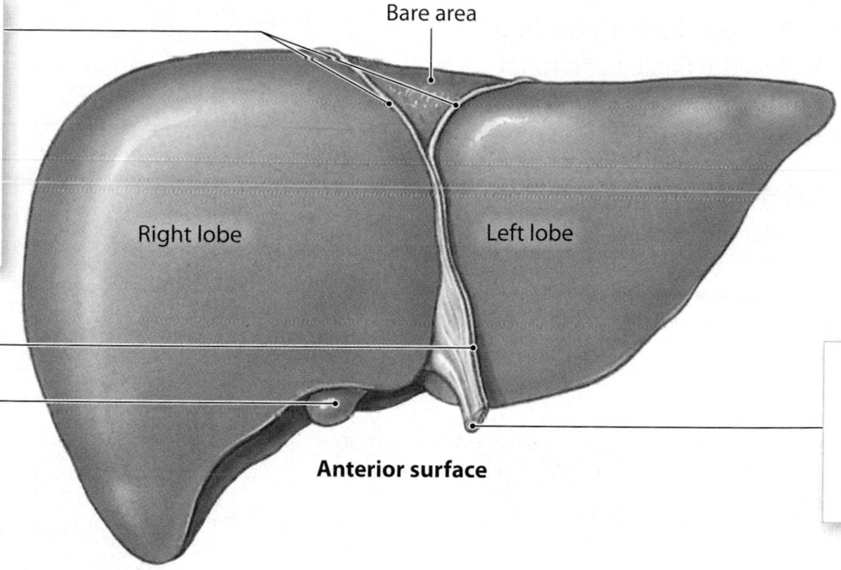
The **coronary ligament** is continuous with the falciform ligament. It extends along the edges of the **bare area** where the liver contacts the diaphragm without a peritoneal covering.
Bare area
Right lobe
Left lobe
Falciform ligament
Gallbladder
A thickening in the posterior margin of the falciform ligament is the **round ligament**, a fibrous band that marks the path of the fetal umbilical vein.
Anterior surface

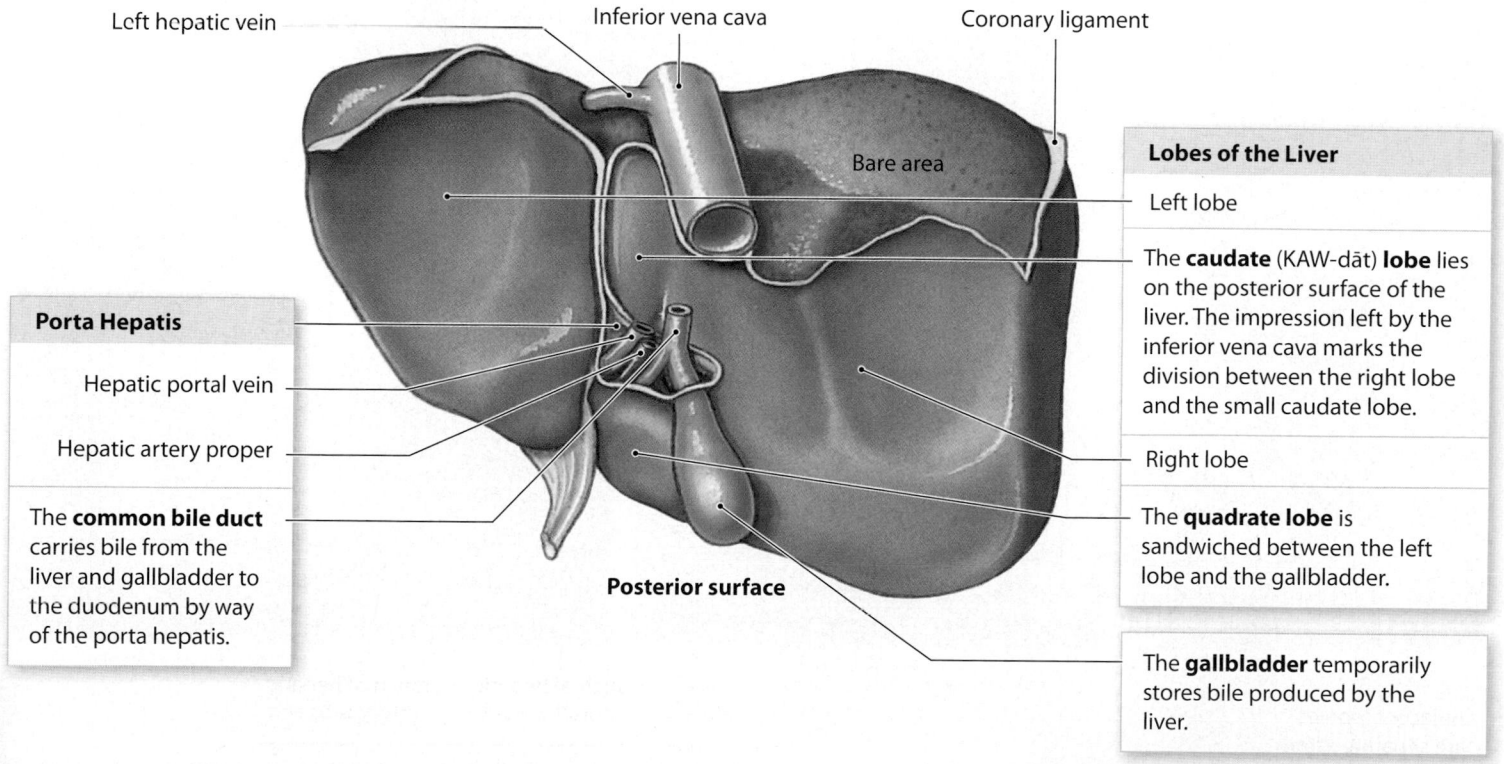

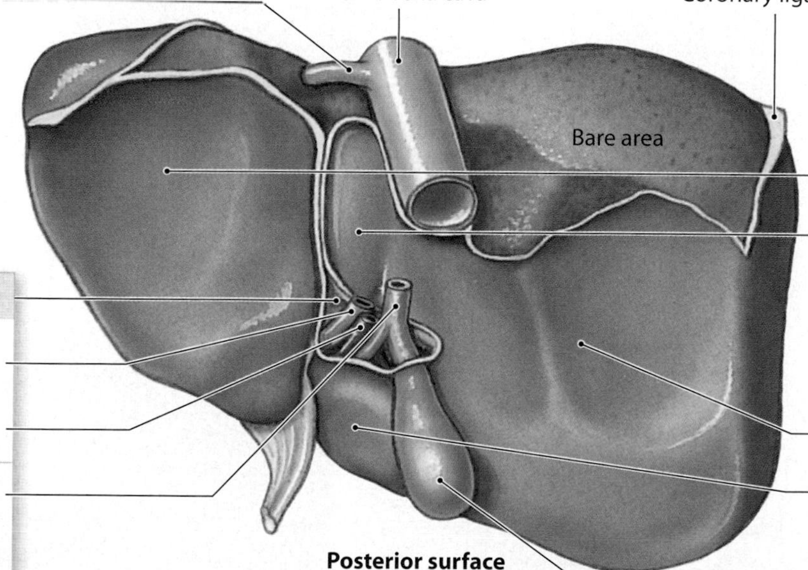
Left hepatic vein
Inferior vena cava
Coronary ligament
Bare area
Porta Hepatis
Hepatic portal vein
Hepatic artery proper
The **common bile duct** carries bile from the liver and gallbladder to the duodenum by way of the porta hepatis.
Lobes of the Liver
Left lobe
The **caudate** (KAW-dāt) **lobe** lies on the posterior surface of the liver. The impression left by the inferior vena cava marks the division between the right lobe and the small caudate lobe.
Right lobe
The **quadrate lobe** is sandwiched between the left lobe and the gallbladder.
The **gallbladder** temporarily stores bile produced by the liver.
Posterior surface

Module 22.20 Review

a. Name the lobes of the liver.

b. What structure marks the division between the left lobe and right lobe of the liver?

c. What is the function of the gallbladder?

LO **22.20** Describe the anatomy and location of the liver and gallbladder.

Module 22.21

The liver tissues have an extensive and complex blood supply

The lobes of the liver are divided by connective tissue into approximately 100,000 **liver lobules**, the basic functional units of the liver. The histological organization and structure of a typical liver lobule are the focus of this module.

1 This is a diagrammatic view of several liver lobules. Each lobule is roughly 1 mm in diameter. Liver cells, called **hepatocytes** (HEP-a-tō-sīts), within these lobules adjust circulating levels of nutrients through selective absorption and secretion. In cross section, a typical liver lobule has a hexagonal shape and is surrounded by six **portal areas**, one at each corner of the lobule.

The hepatocytes in a liver lobule form a series of irregular plates arranged like the spokes of a wheel. The plates are only one cell thick, and exposed hepatocyte surfaces are covered with short microvilli.

The plates of hepatocytes are separated by **liver sinusoids**, delicate blood vessels that lack a basement membrane but otherwise resemble large fenestrated capillaries.

Bile duct
Branch of hepatic portal vein
Branch of hepatic artery proper

Adjacent lobules are separated from each other by an **interlobular septum**.

Portal Area

A portal area, or **portal triad**, contains three structures: (1) a branch of the hepatic portal vein, (2) a branch of the hepatic artery proper, and (3) a bile duct. Branches from the arteries and veins of each portal area deliver blood to the liver sinusoids, or hepatic sinusoids, of adjacent liver lobules. The **hepatic portal system** is a venous portal system in which the hepatic portal vein receives blood from the capillaries of most of the abdominal viscera and delivers it to the hepatic sinusoids (Module 18.12, p. 673).

2 This figure provides an enlarged view of one portion of a single liver lobule, and shows additional details of its functional anatomy.

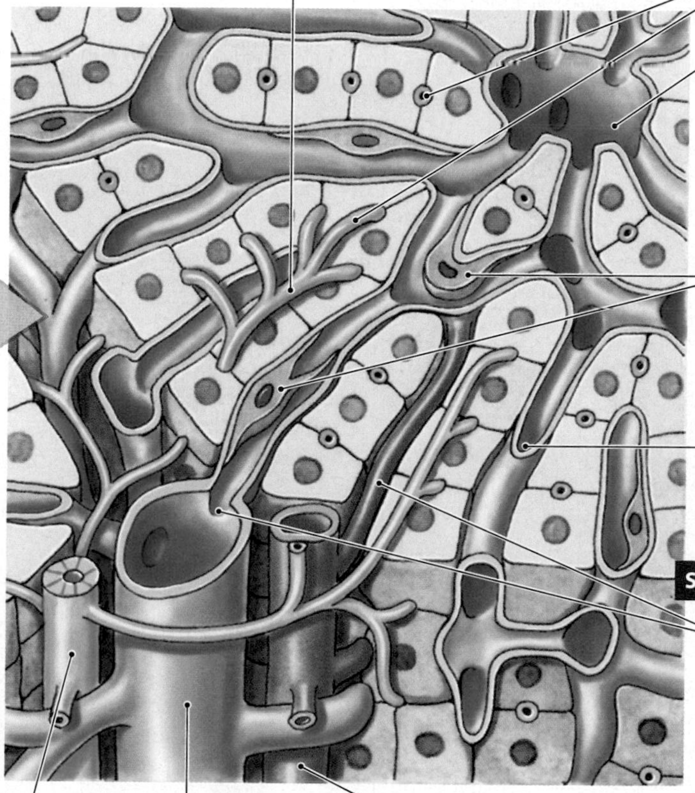

Start
Bile duct
Branch of hepatic portal vein
Branch of hepatic artery proper

1 Blood enters the liver sinusoids from small branches of the hepatic portal vein and hepatic artery proper. About one-third of the blood supply to the liver is arterial blood from the hepatic artery proper. The rest is venous blood from the hepatic portal vein, which begins in the capillaries of the esophagus, stomach, small intestine, and most of the large intestine.

2 As blood flows through the liver sinusoids, hepatocytes adjacent to them regulate solute and nutrient levels and absorb or secrete molecules as large as plasma proteins.

3 In addition to containing typical endothelial cells, the sinusoidal lining includes a large number of **Kupffer** (KOOP-fer) **cells**. These phagocytic cells, part of the monocyte–macrophage system, engulf pathogens, cell debris, and damaged blood cells. Kupffer cells are also responsible for storing iron, some lipids, and heavy metals (such as tin or mercury) that are absorbed by the digestive tract.

4 The **central vein** collects blood from the sinusoids of the lobule. The central veins of all the lobules ultimately merge to form the hepatic veins, which then empty into the inferior vena cava.

5 The hepatocytes secrete a fluid called **bile** into a network of narrow channels between the opposing membranes of adjacent liver cells. These passageways, called **bile canaliculi**, extend outward, away from the central vein. Bile contains acidic bile salts.

6 Bile canaliculi merge to form **bile ductules** (DUK-tūlz), which carry bile to **bile ducts** in the nearest portal area. Bile plays an important role in the digestion of fats in the small intestine.

Liver diseases (such as the various forms of **viral hepatitis**) and conditions such as alcoholism can lead to degenerative changes in the liver tissue and constriction of blood flow. Pressures in the hepatic portal system are usually low, averaging 10 mm Hg or less. This pressure can increase markedly, however, if blood flow through the liver becomes restricted as a result of a blood clot or damage to the organ. Such a rise in portal pressure is called **portal hypertension**. As pressures rise, small peripheral veins and capillaries in the portal system become distended; if they rupture, potentially fatal bleeding can occur. Portal hypertension can also force fluid into the peritoneal cavity across the serosal surfaces of the liver and viscera, producing ascites.

Module 22.21 Review

- **a.** Define hepatocyte.
- **b.** Describe a portal area.
- **c.** Define Kupffer cells, and indicate their functions.

LO **22.21** Describe the histological features of liver tissue.

The gallbladder stores and concentrates bile . . .

The **gallbladder** is a hollow, pear-shaped organ that stores and concentrates bile prior to its ejection into the small intestine. This muscular sac is located in a recess in the posterior surface of the liver's right lobe. The gallbladder is divided into three regions: the **fundus**, the **body**, and the **neck**.

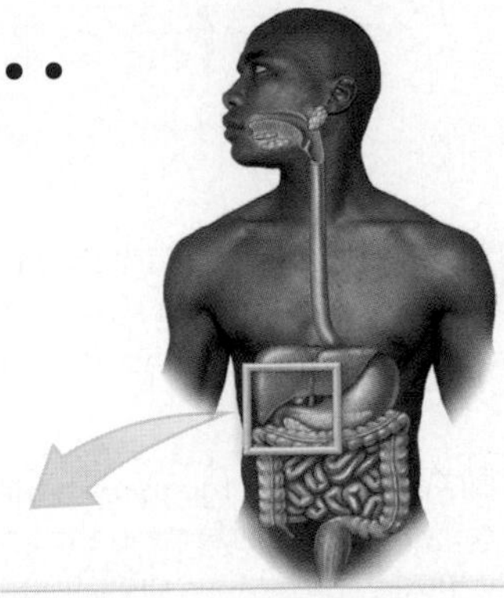

1 In this view, the liver has been pulled upward to show the gallbladder and its associated ducts.

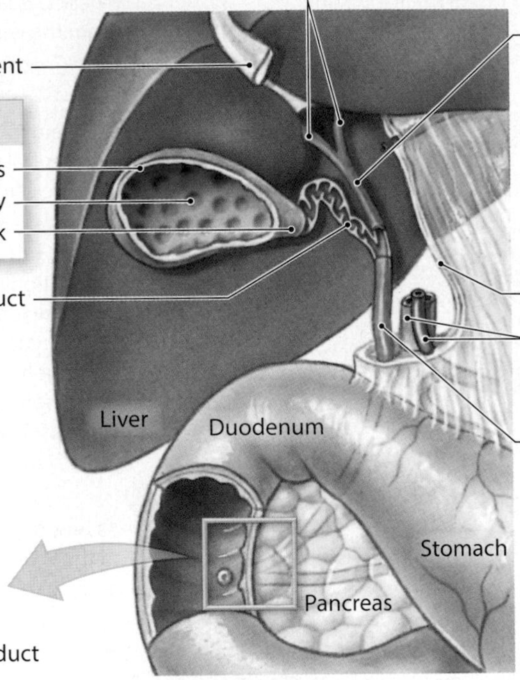
The **right** and **left hepatic ducts** collect bile from all the bile ducts of the liver lobes.
Round ligament
Gallbladder
Fundus
Body
Neck
Cystic duct
The hepatic ducts unite to form the **common hepatic duct**. Bile in the common hepatic duct either flows into the common bile duct or enters the **cystic duct**, which leads to the gallbladder.
Cut edge of lesser omentum
Hepatic portal vein and hepatic artery proper
Liver
Duodenum
Stomach
Pancreas
The **common bile duct** is formed by the union of the cystic duct and the common hepatic duct. It empties into the duodenum.

2 The common bile duct passes within the lesser omentum toward the stomach, turns, and penetrates the wall of the duodenum to meet the pancreatic duct at the entrance to the **duodenal ampulla** (am-PUL-a), a chamber within the **duodenal papilla** that projects into the intestinal lumen.

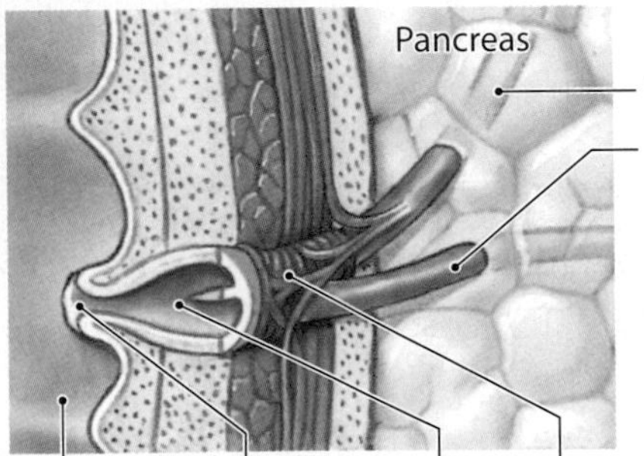
Pancreas
Common bile duct
Pancreatic duct
Intestinal lumen
Duodenal papilla
Duodenal ampulla
The muscular **hepatopancreatic sphincter** encircles the lumen of the common bile duct, pancreatic duct, and duodenal ampulla. Resting tension in the sphincter prevents bile flow into the duodenum except at mealtimes.

3 This diagram shows the functional relationships involved in the storage and ejection of bile. Bile can enter the duodenum only when the hepatopancreatic sphincter is open. So for most of the day, bile flows into the gallbladder instead.

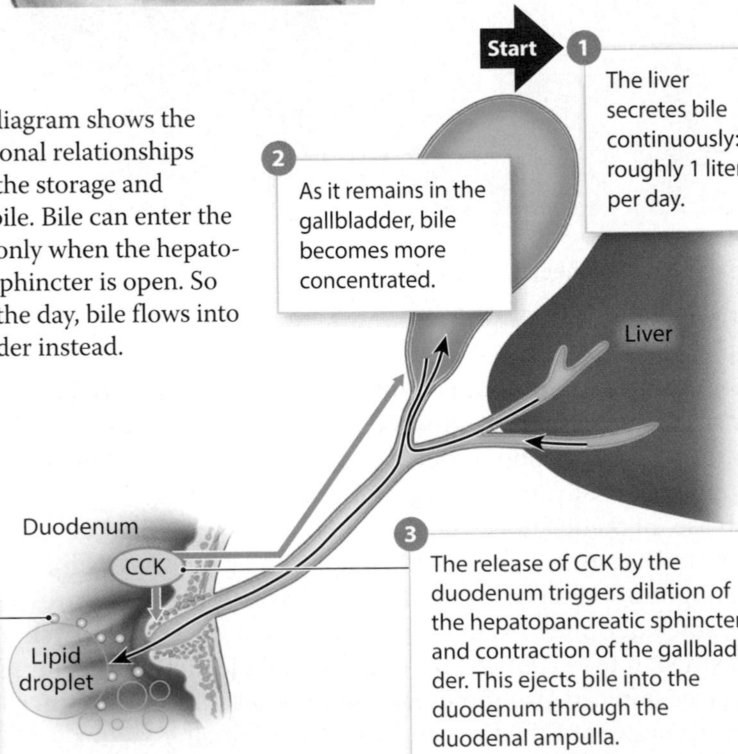
Start
1 The liver secretes bile continuously: roughly 1 liter per day.
2 As it remains in the gallbladder, bile becomes more concentrated.
Liver
Duodenum
CCK
Lipid droplet
3 The release of CCK by the duodenum triggers dilation of the hepatopancreatic sphincter and contraction of the gallbladder. This ejects bile into the duodenum through the duodenal ampulla.
In the lumen of the digestive tract, bile salts break the lipid droplets apart in a process called **emulsification** (ē-mul-si-fi-KĀ-shun). This increases the surface area accessible to enzymes.

. . . and the pancreas has vital endocrine and exocrine functions

The **pancreas** lies posterior to the stomach, extending laterally from the duodenum toward the spleen. The pancreas is a slender, pinkish-gray organ about 15 cm (6 in.) long and weighing about 80 g (3 oz).

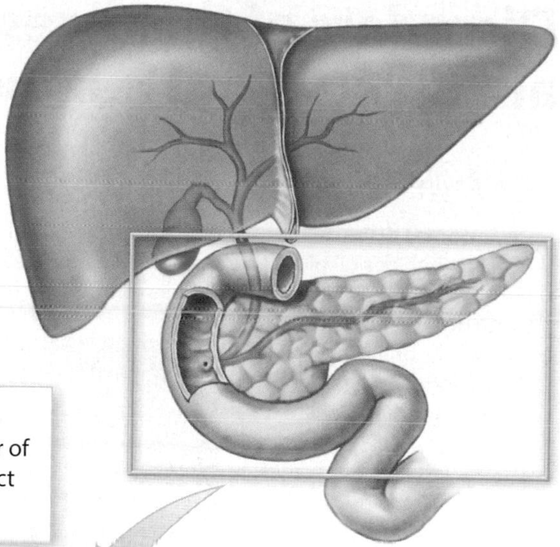

4 The broad **head** of the pancreas lies within the loop formed by the duodenum as it leaves the pylorus. Like the duodenum, the pancreas is retroperitoneal and firmly bound to the posterior wall of the abdominal cavity.

The large **pancreatic duct** delivers the exocrine secretions to the duodenum. The epithelial cells lining the duct and its smaller branches secrete water and ions that mix with the secretions of the exocrine gland cells to form a watery **pancreatic juice**.

Partitions of connective tissue divide the interior of the pancreas into distinct **pancreatic lobules**.

In 3–10 percent of the population, a small **accessory pancreatic duct** (Santorini duct) branches from the pancreatic duct and empties separately into the duodenum.

The pancreatic duct meets the common bile duct at the entrance to the duodenal ampulla.

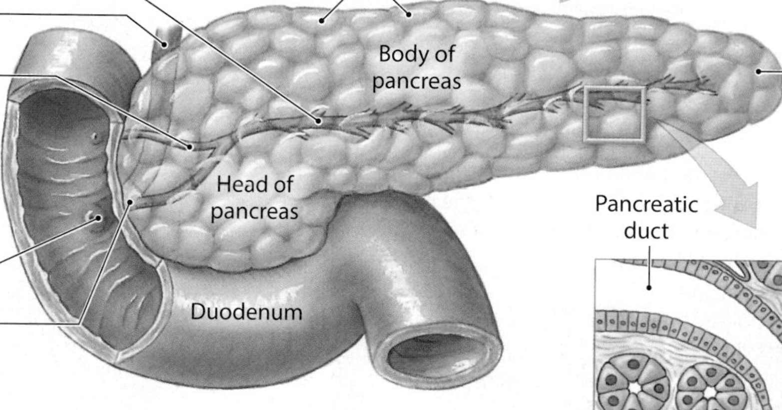

Common bile duct
Body of pancreas
Tail of pancreas
Head of pancreas
Duodenal papilla
Duodenum

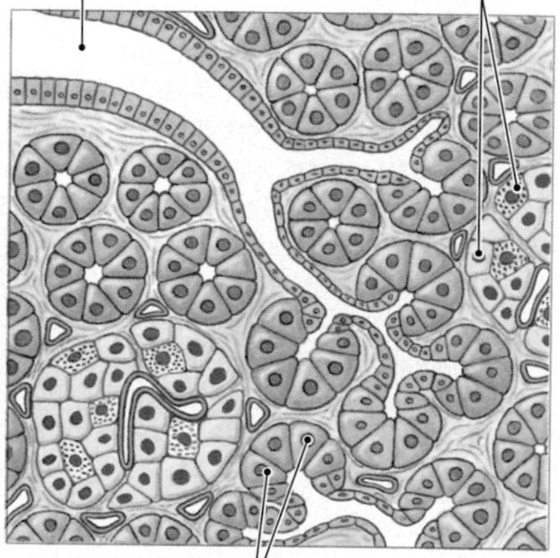

Pancreatic duct
Endocrine cells in pancreatic islet

5 The pancreas is primarily an exocrine organ, and pancreatic tissue is dominated by the **pancreatic acini** (AS-i-nī; *berry*), which produce digestive enzymes and buffers.

Pancreatic acinar cells secrete pancreatic enzymes that do most of the digestive work in the small intestine, breaking down ingested materials into small molecules suitable for absorption.

6 Each day, the pancreas secretes about 1000 mL (1 qt) of pancreatic juice containing a variety of enzymes and a watery buffer solution. This table introduces the primary pancreatic enzymes produced; their functions will be considered further in the next chapter.

Major Pancreatic Enzymes

- **Pancreatic alpha-amylase** is a carbohydrase (kar-bō-HĪ-drās)—an enzyme that breaks down certain starches. Pancreatic alpha-amylase is almost identical to salivary amylase.
- **Pancreatic lipase** breaks down certain complex lipids, releasing products (such as fatty acids) that can be easily absorbed.
- **Nucleases** break down RNA or DNA.
- **Proteolytic enzymes** break proteins apart. They are secreted as inactive proenzymes that become active once they are in the duodenal lumen. The active enzymes include trypsin, chymotrypsin, carboxypeptidase, and elastase. Together, they break down proteins into a mixture of dipeptides, tripeptides, and amino acids.

Module 22.22 Review

a. Define emulsification.

b. Trace a drop of bile from the hepatic ducts to the duodenal lumen.

c. What is the primary digestive function of the pancreas?

LO 22.22 Describe the structure, functions, and regulatory activities of the gallbladder and pancreas.

Disorders of the digestive system are diverse and relatively common

Oral Cavity

Periodontal disease, the most common cause for the loss of teeth, occurs when dental plaque forms in the area between the gums and teeth. The bacterial activity may cause **gingivitis** (shown here), tooth decay, and, eventually, breakdown of the periodontal ligament and surrounding bone.

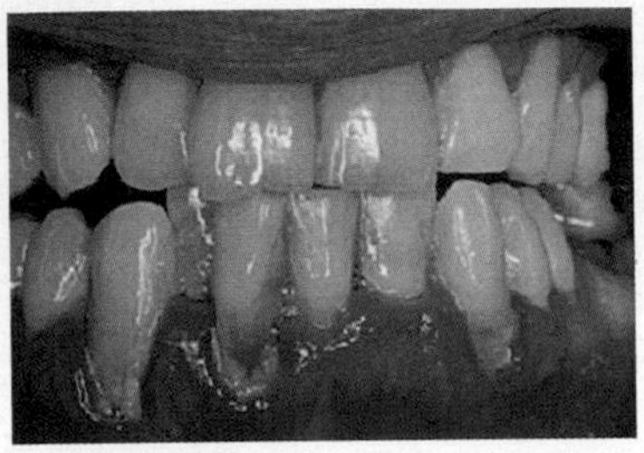

Periodontal disease

Salivary Glands

The **mumps virus** causes an infection of the salivary glands that is called **mumps**. The infection most often occurs in the parotid salivary gland, as seen here, but it may also infect other salivary glands and other organs, including the gonads and the meninges. Infection typically occurs at 5 to 9 years of age. In postadolescent males, the mumps virus infecting the testes may cause sterility. An effective mumps vaccine became available in 1967. The vaccine is usually combined with measles and rubella vaccines (to form the MMR vaccine), and is administered to infants after the age of 15 months.

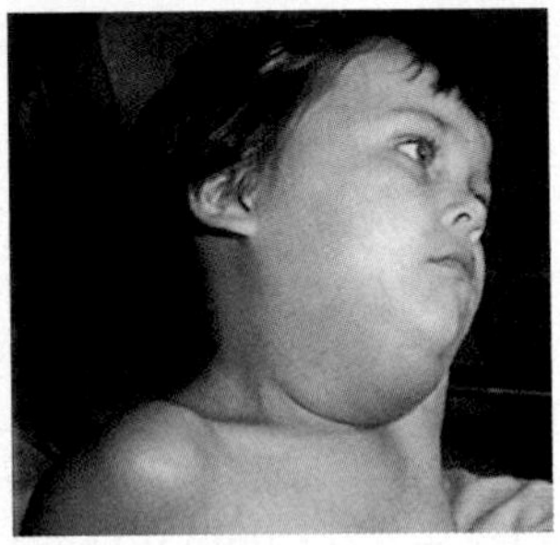

Mumps in parotid gland

Esophagus

Esophagitis (ē-sof-a-JĪ-tis) is an inflammation of the esophagus. This painful condition, which can be seen in this endoscopic view of the esophagus, usually results from the presence of stomach acids that leak through a weakened or permanently relaxed lower esophageal sphincter. Such backflow, or **gastroesophageal reflux**, is responsible for the symptoms of heartburn.

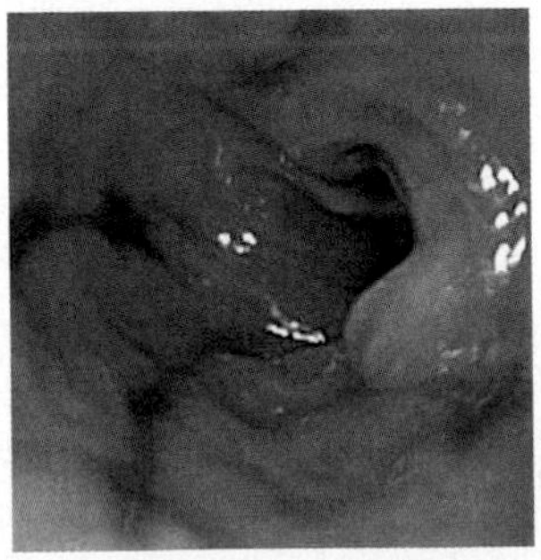

Esophagitis seen on endoscopy

Liver

Any condition that severely damages the liver is a threat to life. **Hepatitis**, an inflammation of the liver, can be caused by alcohol abuse, drugs, or infection. **Cirrhosis** (sir-RŌ-sis) is a form of hepatitis characterized by the degeneration of liver cells and their replacement with fibrous connective tissue (a process called scarring). The surviving liver cells divide, but the fibrous tissue prevents the reestablishment of normal tissues. As a result, liver function declines and a variety of other complications develop. There are many different forms of viral hepatitis: The most common are **hepatitis A**, **B,** and **C**. The hepatitis viruses disrupt liver function by attacking and destroying liver cells. An infected person may develop a high fever, and the liver may become inflamed and tender. In a condition called **jaundice** (JAWN-dis), the skin and eyes develop a yellow color because the bilirubin normally excreted in the bile is accumulating in body fluids.

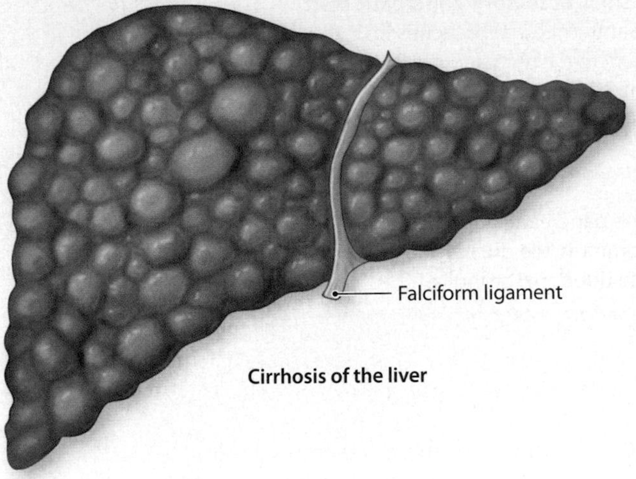
Falciform ligament

Cirrhosis of the liver

Gallbladder

If bile becomes too concentrated, crystals of insoluble minerals and salts called **gallstones** form. Small gallstones are not a problem if they can be flushed through the bile duct and excreted. In **cholecystitis** (kō-lē-sis-TĪ-tis; *chole*, bile + *kystis*, bladder + *itis*, inflammation), the gallstones are so large that they damage the wall of the gallbladder or block the cystic duct or common bile duct. In that case, the gallbladder may need to be surgically removed. This removal does not seriously impair digestion, because bile production continues at normal levels.

Gallstones

Stomach

Inflammation of the mucous membrane lining the stomach is called **gastritis** (gas-TRĪ-tis). This condition may develop after ingesting drugs, including aspirin and alcohol. It may also appear after severe emotional or physical stress, bacterial infection of the gastric wall, or the ingestion of strong chemicals. Gastritis may lead to ulcer formation. A **peptic ulcer** develops when gastric enzymes and acids erode through the stomach or duodenal lining. Specifically, a peptic ulcer located in the stomach is termed a **gastric ulcer**, and one located in the duodenum is called a **duodenal ulcer**. It is now known that infection by the bacterium *Helicobacter pylori* (HE-li-kō-bak-ter pī-LŌR-ī) is responsible for over 80 percent of peptic ulcers. Treatment for ulcers involves the administration of drugs, such as **cimetidine** (si-MET-i-dēn) (Tagamet), that inhibit acid production by gastric glands, combined with antibiotics if *Helicobacter pylori* is present.

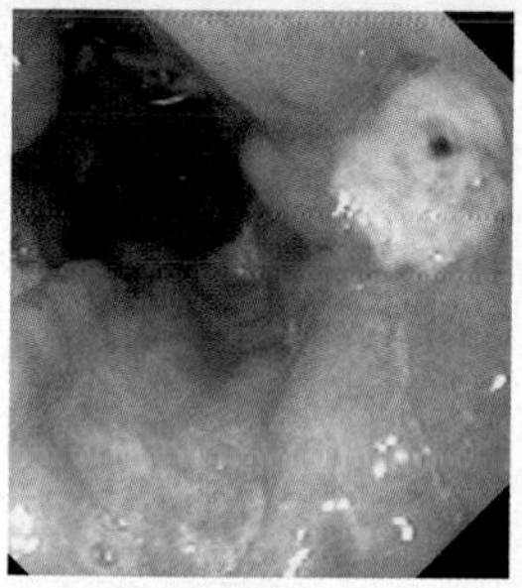

Gastric ulcer

Pancreas

Pancreatitis (pan-krē-a-TĪ-tis) is an inflammation of the pancreas. Factors that may cause this condition include blockage of the excretory ducts by gallstones, viral infections, and toxic drugs, such as alcohol. Any of these stimuli may begin to injure exocrine cells in a portion of the organ. Lysosomes then activate digestive enzymes within the cells, which begin to break down. In about one-eighth of the cases, death results when the process does not stop, and the released lysosomal enzymes destroy the pancreas.

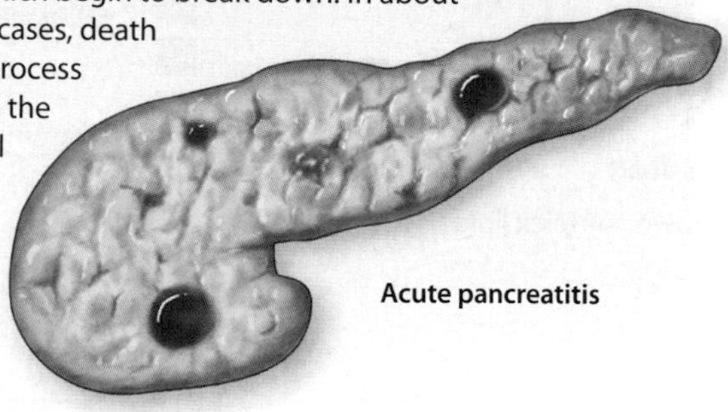

Acute pancreatitis

Small Intestine

Enteritis is inflammation of the intestine (usually applied to the small intestine). Enteritis typically causes watery bowel movements, or **diarrhea** (dī-a-RĒ-uh). One cause of diarrhea due to enteritis is the protozoan *Giardia lamblia* (shown here). **Dysentery** (dis-en-TER-ē) is inflammation of the small and large intestine that usually produces diarrhea containing blood and mucus. **Gastroenteritis** is inflammation of the stomach and the intestines due to bacterial, viral, protozoan, or parasitic worm infections. Most of these conditions are prevalent in areas that have poor sanitation and low water quality.

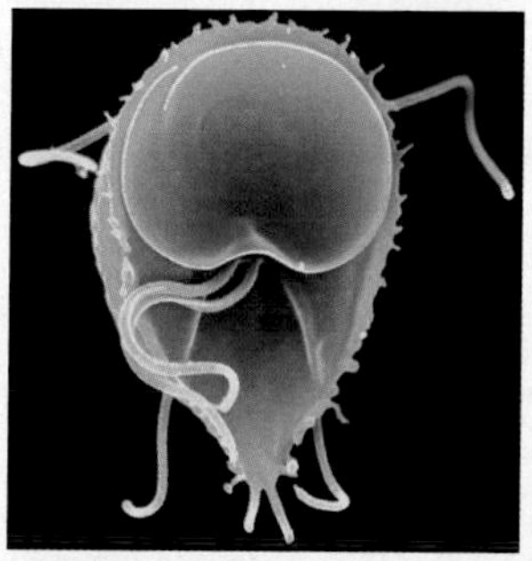

Giardia lamblia SEM × 4000

Large Intestine

Colitis (kō-LĪ-tis) is a general term referring to inflammation of the colon, often involving diarrhea or constipation. Diarrhea results when the lining of the colon is unable to reabsorb water normally, or when so much fluid enters the colon that its water reabsorption capacity is exceeded. **Constipation** is infrequent bowel movement (or defecation), generally involving dry, hard feces. It results when fecal material moves through the colon so slowly that excessive water reabsorption occurs.

Colorectal cancer is relatively common in the United States. Aside from skin cancers, colorectal cancer is the third most common cancer in the United States, affecting both men and women. According to the American Cancer Society's estimates for 2013, there were 102,480 new cases of colon cancer and 40,340 new cases of rectal cancer. The death rate has declined over the past 20 years for both men and women. It is most common among persons over 50 years of age. Primary risk factors for colorectal cancer include a diet rich in animal fats and low in fiber. There are also a number of inherited disorders that promote epithelial tumor formation along the intestines. It is believed that most colorectal cancers begin as small, localized tumors, or **polyps** (POL-ips), that grow from the mucosa lining the intestinal wall. The prognosis improves dramatically if cancerous polyps are removed before metastasis has occurred.

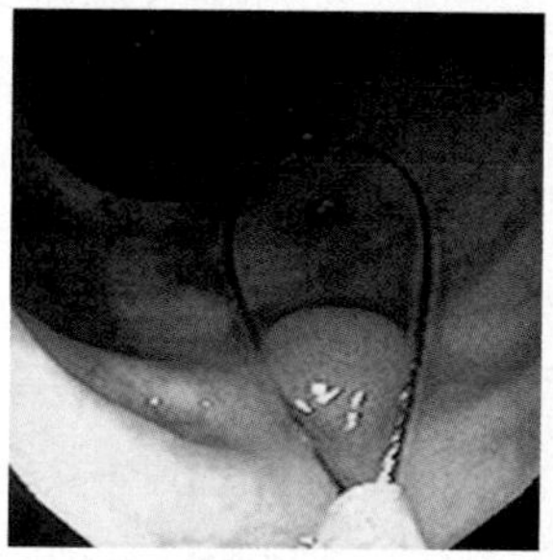

Colon polyp seen on colonoscopy

Module 22.23 Review

a. Describe periodontal disease.
b. Describe cholecystitis.
c. What bacterium is responsible for most peptic ulcers?

LO 22.23 Briefly describe several digestive system disorders.

Labeling

Label the structures of a liver lobule in the accompanying figure.

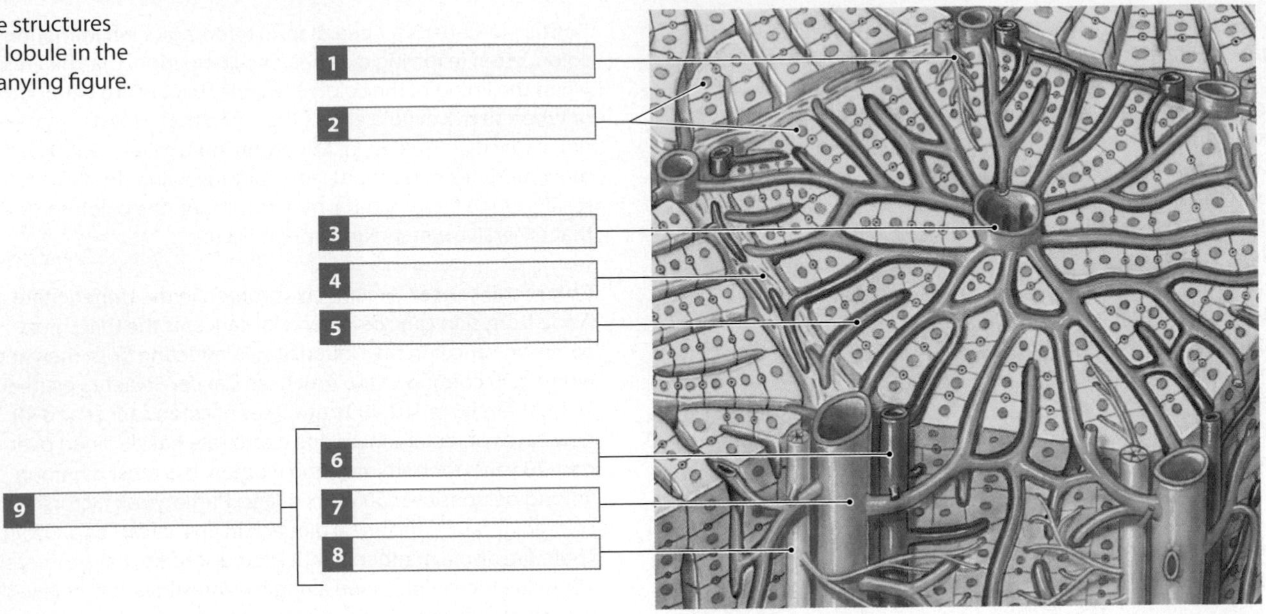
1
2
3
4
5
6
7
8
9

Matching

Match each lettered term with the most closely related description.

a. lysozyme
b. emulsification
c. gallstones
d. Kupffer cells
e. pancreatic lipase
f. liver
g. starch
h. pancreas
i. submandibular glands
j. hepatocytes
k. gallbladder
l. mumps
m. common bile duct
n. peptic ulcer

10 Pancreatic alpha-amylase substrate
11 Retroperitoneal organ
12 Drains liver and gallbladder
13 Bile-secreting cells
14 Viral infection of salivary glands
15 Digestive epithelial damage by acids
16 Process of breaking lipid droplets apart
17 Pancreatic enzyme that breaks down complex lipids
18 Organ that secretes bile continuously
19 Antibacterial enzyme
20 Greatest producer of saliva
21 Phagocytize and store iron
22 Stores bile
23 Cholecystitis

10 ______
11 ______
12 ______
13 ______
14 ______
15 ______
16 ______
17 ______
18 ______
19 ______
20 ______
21 ______
22 ______
23 ______

Short answer

Describe the beneficial roles of saliva.

24 ______

Section integration

Predict the consequences of a blockage of the duodenal ampulla by a tumor.

25 ______

Study Outline

SECTION 1 • Organization of the Digestive System

22.1 The digestive system consists of the digestive tract and accessory organs p. 829

1. The **digestive system** consists of the digestive tract plus accessory organs.
2. The **digestive tract** begins at the oral cavity (mouth) and continues through the pharynx (throat), esophagus, stomach, small intestine, and large intestine. It ends with the large intestine, which opens to the exterior at the anus.
3. The accessory organs of the digestive system are the teeth, tongue, salivary glands, liver, gallbladder, and pancreas.
4. The digestive system works with other systems to support tissues that have no direct connection with the outside environment.

22.2 The digestive tract is a muscular tube lined by a mucous epithelium p. 830

5. A double sheet of peritoneal membrane called **mesentery** stabilizes the positions of the digestive tract.
6. The major layers of the digestive tract are the **mucosa** (inner lining), **submucosa** (surrounds the mucosa), **muscularis externa** (circular and longitudinal layers of smooth muscle cells), and **serosa** (visceral peritoneum).

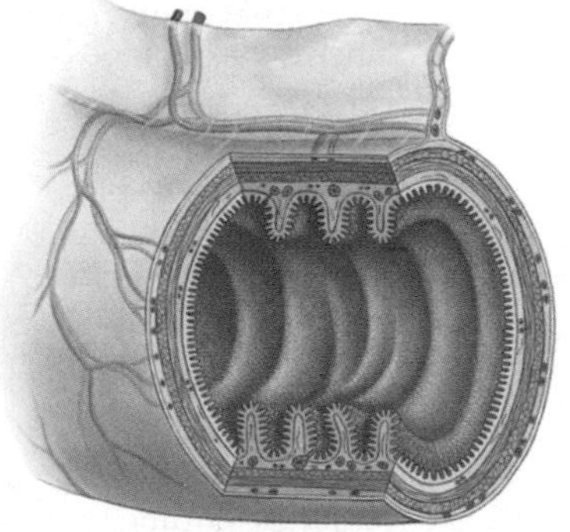

Section of digestive tract

7. Secretions from gland cells located in the mucosa and submucosa, or in accessory glandular organs, are carried to the epithelial surfaces by ducts.
8. **Circular folds** (plicae circulares) are permanent transverse folds in the intestinal lining. **Villi** are small mucosal projections into the lumen of the small intestine.

22.3 Smooth muscle tissue is found throughout the body, but it plays a particularly prominent role in the digestive tract p. 832

9. In the digestive tract, the muscularis externa usually contains an inner circular layer and an outer longitudinal layer of smooth muscle.
10. **Multi-unit smooth muscle cells** are innervated by motor units, and are located in the iris of the eye, the male reproductive tract, large arteries, and the arrector pili muscles of the skin.
11. **Visceral smooth muscle cells** lack a direct contact with motor neurons. They are arranged in sheets, and the layers contract in waves triggered by **pacesetter cells**.
12. **Plasticity** is the ability of a stretched smooth muscle to function over a wide range of lengths. **Smooth muscle tone** is the normal background level of activity.

22.4 Smooth muscle contractions mix the contents of the digestive tract and propel materials along its length p. 834

13. Food enters the digestive tract as a moist, compact mass called a **bolus**.

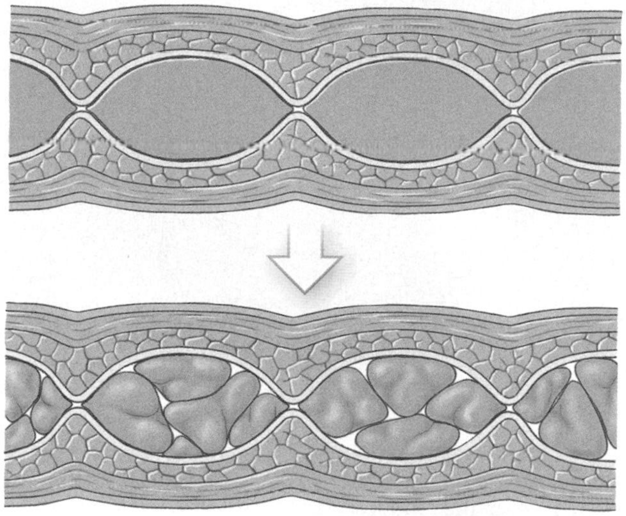

14. The muscularis externa propels the bolus with wavelike contractions called **peristalsis**. **Segmentation** churns and fragments a bolus without any directional movement.
15. The three major mechanisms that regulate and control digestive activities are **local factors**, **neural control mechanisms**, and **hormonal control mechanisms**.

SECTION 2 • Digestive Tract

22.5 The digestive tract begins with the mouth and ends with the anus p. 837

16. The major organs of the digestive tract are the **oral cavity**, **pharynx**, **esophagus**, **stomach**, **small intestine**, and **large intestine**. The accessory organs are the teeth, tongue, salivary glands, liver, gallbladder, and pancreas.
17. The general functions of the digestive tract are **ingestion**, **mechanical processing**, **digestion**, **secretion**, **absorption**, and **compaction**.

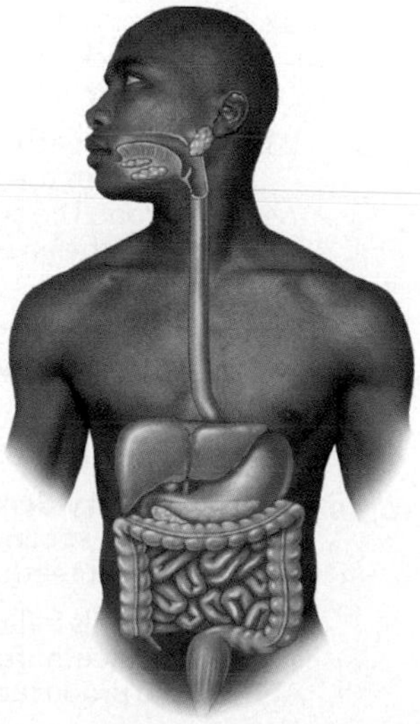

22.6 The oral cavity is a space that contains the tongue, teeth, and gums p. 838

18. The superior boundary of the **oral cavity** is formed by the **hard palate** (palatine processes of the maxillary bones and the horizontal plates of the palatine bones), and the muscular **soft palate.**

Sagittal secton of the oral cavity

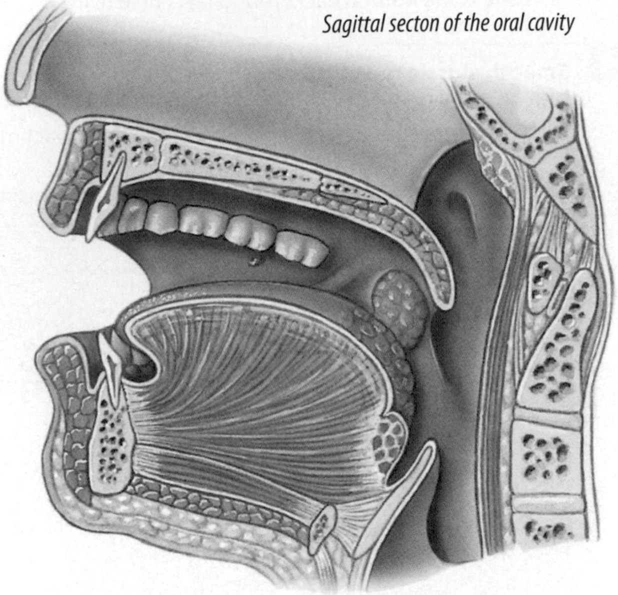

19. The posterior boundary of the oral cavity contains the **uvula**, the palatine tonsils, the lingual tonsils, and the **root** of the tongue.
20. The tongue forms the inferior boundary of the oral cavity. The **labia** (lips) and **cheeks** form the anterior and lateral boundaries.
21. The **vestibule** is the space between the cheeks and teeth. The **labial frenulum** attaches the lips to the **gingivae**, or gums. The **lingual frenulum** attaches the tongue to the floor of the mouth. The **pharyngeal arches** are on either side of the uvula. The **fauces** is the connection between the oral cavity and the oropharynx.

22.7 Teeth in different regions of the jaws vary in size, shape, and function p. 840

22. The components of a tooth are the **crown** (above the gums), **neck** (boundary between the crown and root), and **root** (within a bony socket, or alveolus).
23. The bulk of each tooth is made of **dentin**. **Enamel** covers the dentin of the crown. **Cementum** covers the dentin of the root. The **periodontal ligament** extends from the dentin to the alveolus to form an articulation called a **gomphosis**.
24. The pulp cavity receives blood vessels and nerves through the **apical foramen** into the **root canal**.
25. The four different types of teeth are **incisors**, **cuspids**, **bicuspids**, and **molars**.
26. The **primary dentition** is composed of 20 **deciduous teeth**. The **secondary dentition** is composed of 32 permanent teeth.
27. **Gingivitis** is inflammation of the gingivae, and can damage teeth. **Tooth decay** can result from the **dental plaque** produced by bacteria.

22.8 The muscular walls of the pharynx and esophagus play a key role in swallowing p. 842

28. The **pharynx** is a passageway for solid food, drinks, and air. Food passes through the oropharynx and laryngopharynx on its way to the **esophagus**, a hollow muscular organ leading to the stomach.
29. The shape of the lumen and structure of the muscularis externa of the esophagus are unique to the rest of the digestive tract. There is no serosa, but an **adventitia** of connective tissue anchors it to the posterior body wall.
30. Swallowing, or **deglutition**, involves three phases: the **buccal phase**, the **pharyngeal phase**, and the **esophageal phase**.
31. The esophagus enters the abdominopelvic cavity through the **esophageal hiatus** of the diaphragm. The **upper esophageal sphincter** prevents air from entering the esophagus, and the **lower esophageal sphincter** prevents the backflow of food from the stomach into the esophagus.

22.9 The stomach and most of the intestinal tract are suspended by mesenteries and enclosed by the peritoneal cavity p. 844

32. **Visceral peritoneum** covers the organs enclosed by the **peritoneal cavity**. The **parietal peritoneum** lines the inner surfaces of the body wall.
33. **Dorsal** and **ventral mesenteries** suspend the digestive tract and accessory organs during embryonic development.

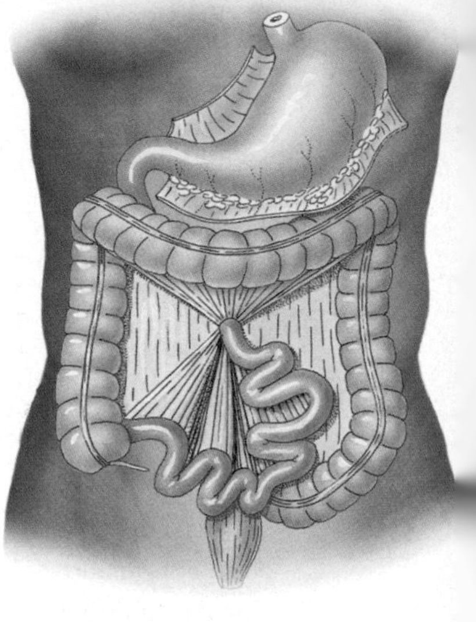

34. The ventral mesentery remains in the adult as the **lesser omentum** and the **falciform ligament**.
35. The **greater omentum** is the dorsal mesentery of the stomach, the **mesentery proper** is the dorsal mesentery of the small intestine, and the **mesocolon** is the dorsal mesentery of the large intestine.

22.10 The stomach is a muscular, expandable, J-shaped organ with three layers in the muscularis externa p. 846

36. The regions of the stomach are the **fundus**, **cardia**, **body**, and **pylorus**.
37. The medial surface of the stomach forms the **lesser curvature**, and the lateral and inferior surfaces form the **greater curvature**.
38. The pylorus is composed of the **pyloric antrum**, **pyloric canal**, and **pyloric sphincter**.
39. The muscularis externa of the stomach has an inner oblique layer of smooth muscle and mucosal folds called **rugae**.

22.11 The stomach receives food and drink from the esophagus and aids in mechanical and chemical digestion
p. 848

40. The **stomach** stores, breaks down, and disrupts chemical bonds in food and produces intrinsic factor.
41. The lumen of the stomach is lined with shallow depressions called **gastric pits**, which communicate with several gastric glands. **Gastric glands** secrete most of the acid and enzymes involved in gastric digestion.

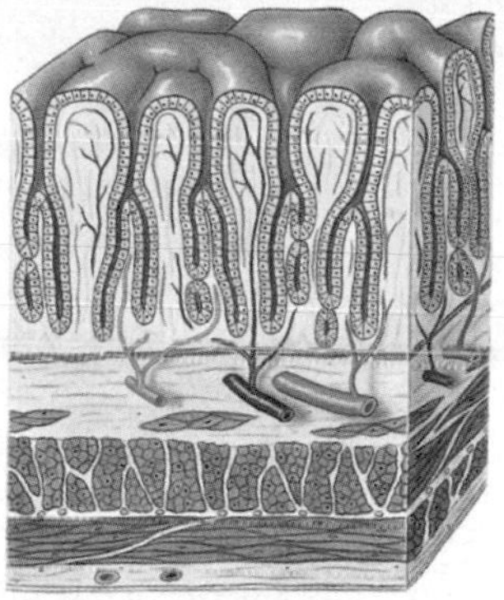

Layers of the stomach wall

42. The cells of gastric glands are **parietal cells** that secrete **intrinsic factor**, which aids vitamin B_{12} absorption in the small intestine, and HCl; **G cells** (enteroendocrine cells); and **chief cells** that secrete **pepsinogen**.
43. Pepsinogen is converted to **pepsin** by acid in the gastric lumen.

22.12 The intestinal tract is specialized to absorb nutrients
p. 850

44. The intestinal lining has transverse folds called **circular folds** that greatly increase the surface area for absorption.
45. The mucosa of the small intestine contains fingerlike projections called **intestinal villi**. At the base of the villi are the entrances into the **intestinal glands**.
46. Each villus has a complex internal structure with a blood capillary network that carries absorbed nutrients to the liver via hepatic portal circulation, and a lymphatic capillary called a **lacteal** that absorbs materials that cannot enter the blood capillaries.
47. The microvilli of the simple columnar epithelium on the surface of each villus project to form a **brush border**.

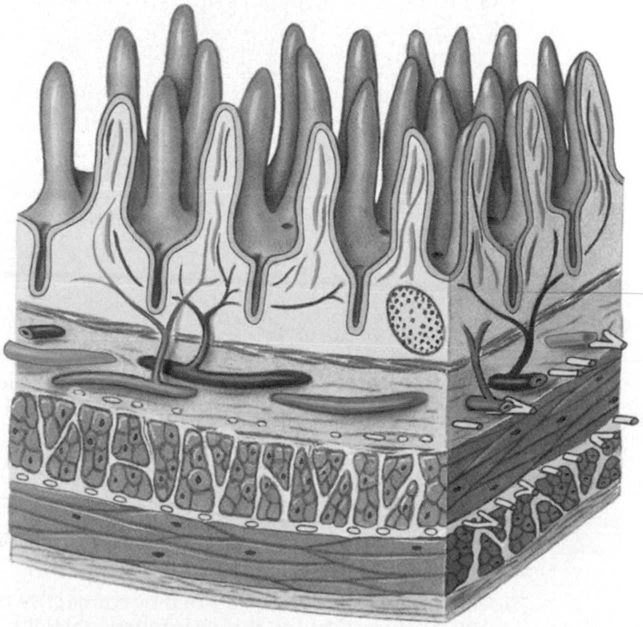

Layers of the small intestine

22.13 The small intestine is divided into the duodenum, jejunum, and ileum
p. 852

48. The small intestines account for 90 percent of nutrient absorption.
49. The regions of the small intestine are the **duodenum**, **jejunum**, and **ileum**. The ileum ends at the **ileocecal valve**.
50. Each segment of the small intestine has characteristic features. The transition from one region to another is gradual, and the boundaries are indistinct.

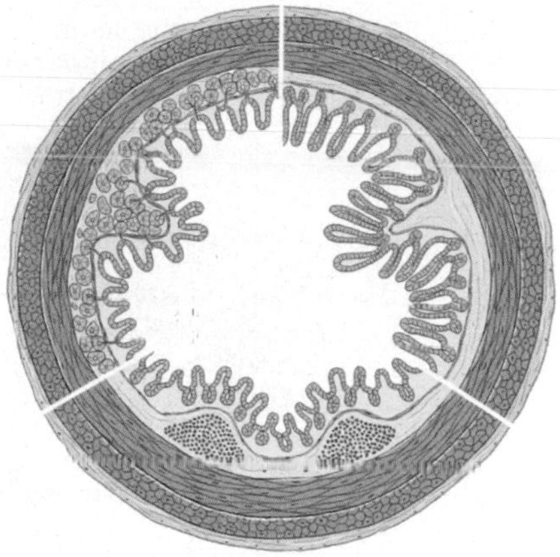

22.14 Five hormones are involved in the regulation of digestive activities
p. 854

51. The duodenum produces four of the five hormones that regulate digestion.
52. The hormones are **gastrin** (secreted by G cells in stomach; stimulates gastric motility), **secretin** (release of pancreatic enzymes and bile secretion), **gastric inhibitory peptide** (release of insulin from pancreas), **cholecystokinin** (release of pancreatic enzymes and bile secretion), and **vasoactive intestinal peptide** (dilation of intestinal capillaries).

22.15 Central and local mechanisms coordinate gastric and intestinal activities
p. 856

53. The duodenum plays a key role in controlling the three **phases of gastric secretion**: cephalic phase, gastric phase, and intestinal phase.
54. The **cephalic phase** of gastric secretion begins when you see, smell, taste, or think of food. The CNS is preparing the stomach to receive food.
55. The **gastric phase** is when food arrives in the stomach. It includes distension of the stomach, an increase in pH, and the presence of undigested materials in the stomach.
56. The **intestinal phase** is when chyme first enters the small intestine.
57. Two **central gastric reflexes** are stimulated by stretch receptors in the stomach. The **gastroenteric reflex** stimulates motility and secretion, and the **gastroileal reflex** opens the ileocecal valve.

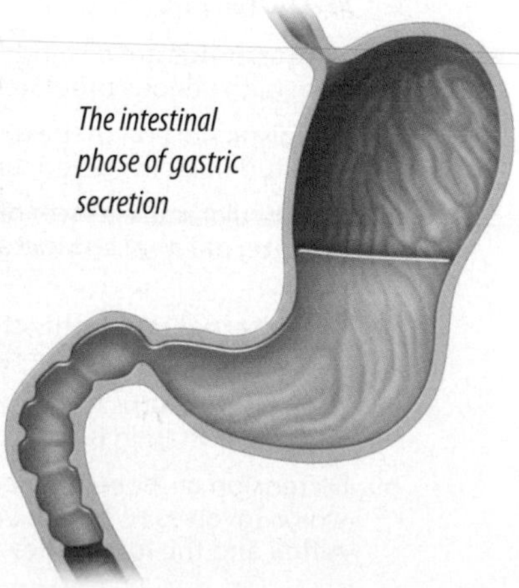

The intestinal phase of gastric secretion

22.16 The large intestine stores and concentrates fecal material p. 858

58. The functions of the **large intestine** include reabsorbing water and compacting the intestinal contents into feces, absorbing important vitamins generated by bacteria, and storing fecal material prior to defecation.
59. Material from the ileum enters the **cecum** through the **ileocecal valve**. **Compaction** begins in the cecum. The slender **appendix** is attached to the cecum.

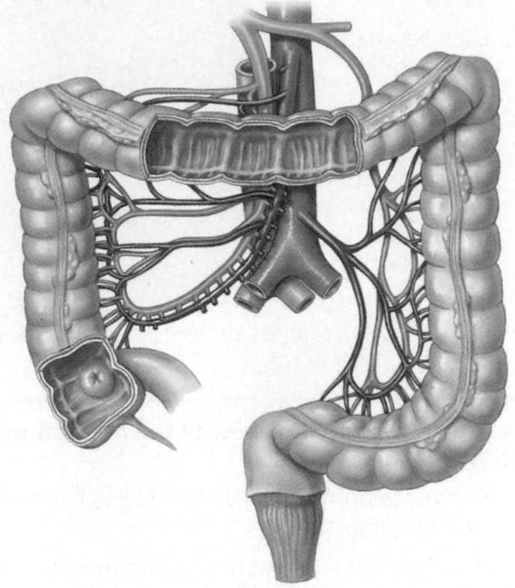

60. The three segments of the large intestine are the cecum, the **colon,** and the rectum. The anatomy of the colon begins at the cecum and continues to the **ascending colon**, **right colic flexure**, **transverse colon**, **left colic flexure**, **descending colon**, **sigmoid flexure**, **sigmoid colon**, and **rectum**.
61. Along the length of the large intestine are pouches called **haustra**, and longitudinal bands of smooth muscle called **teniae coli**.
62. Powerful peristaltic contractions called **mass movements** push materials toward the distal portion of the large intestine.

22.17 The large intestine compacts fecal material; the defecation reflex coordinates the elimination of feces p. 860

63. The large intestine lacks villi, but has distinct intestinal glands dominated by mucous cells. Digestion in the small intestine is by enzymes secreted into the small intestine, and by bacteria.
64. **Hemorrhoids** are distended veins in the rectum that result from high venous pressures.
65. The distal portion of the **anal canal** contains small longitudinal folds called **anal columns**.
66. The circular muscle layer of the muscularis externa forms the **internal anal sphincter**, which is not under voluntary control.
67. The **external anal sphincter** is a ring of skeletal muscle fibers, and is under voluntary control.
68. The colon absorbs three bacteria-produced vitamins: vitamin K, vitamin B_5, and biotin.
69. Distension of the rectal walls starts the **defecation reflex**, which involves two positive feedback loops: the **short reflex** and the **long reflex**.

SECTION 3 • Accessory Digestive Organs

22.18 Some accessory digestive organs have secretory functions p. 863

70. The accessory digestive organs are the **salivary glands**, **gallbladder**, **pancreas**, and **liver**. The liver has almost 200 known functions.

22.19 Saliva lubricates and moistens the mouth and initiates the digestion of complex carbohydrates p. 864

71. Three pairs of salivary glands secrete into the oral cavity: the **sublingual salivary glands**, the **submandibular salivary glands**, and the **parotid salivary glands**.
72. **Saliva** is a mixture of glandular secretions: 70 percent submandibular salivary glands, 25 percent parotid salivary glands, 5 percent sublingual salivary glands.
73. 99.4 percent of the 1.0–1.5 L of saliva produced each day is water.
74. **Mucous cells** secrete mucins, water, and buffers. **Serous cells** secrete salivary amylase and **lysozyme**.

22.20 The liver, the largest visceral organ, is divided into left, right, caudate, and quadrate lobes p. 866

75. The **liver** is wrapped in a tough fibrous capsule and covered by visceral peritoneum.
76. The **falciform ligament** marks the division between the **right lobe** and **left lobe**. A thickening of the posterior margin of the falciform ligament forms the **round ligament**.

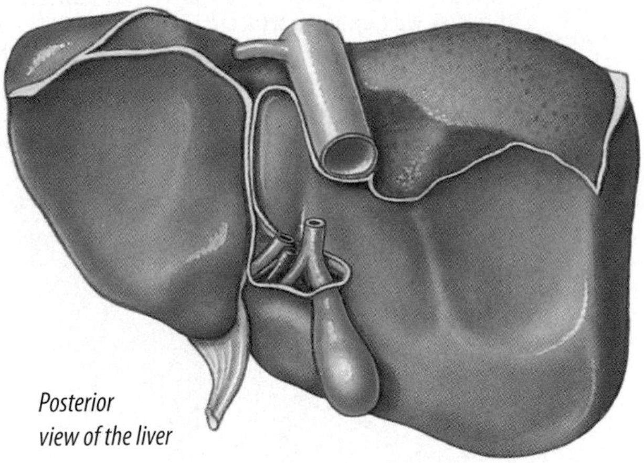

Posterior
view of the liver

77. The inferior vena cava lies between the **caudate lobe** and right lobe. The **quadrate lobe** lies between the left lobe and gallbladder.
78. The **gallbladder** stores bile. The **common bile duct** carries bile from the liver and gallbladder to the duodenum.

22.21 The liver tissues have an extensive and complex blood supply p. 868

79. The lobes of the liver are divided by connective tissue into 100,000 **liver lobules**, the basic functional units of the liver. A typical lobule is hexagonal and surrounded by six **portal areas**, one at each corner.

80. A portal area, or **portal triad**, contains three structures: a branch of the hepatic portal vein, a branch of the hepatic artery proper, and a small branch of bile duct.

81. Liver cells, or **hepatocytes**, within the lobules adjust circulating levels of nutrients. **Liver sinusoids** separate plates of hepatocytes. Hepatocytes secrete bile into **bile canaliculi**. Bile canaliculi merge to form **bile ductules**, which carry bile to the **bile ducts**.

82. The **central vein** collects blood from the sinusoids of the lobule. The central veins of all the lobules merge to form the hepatic veins, which empty into the inferior vena cava.

83. Sinusoids contain **Kupffer cells**, phagocytic cells that are part of the monocyte–macrophage system that engulf pathogens, cell debris, and damaged blood cells.

84. **Viral hepatitis**, alcoholism, and **portal hypertension** can damage the liver.

22.22 The gallbladder stores and concentrates bile and the pancreas has vital endocrine and exocrine functions p. 870

85. The **gallbladder** is located in a recess in the posterior surface of the liver's right lobe. The three regions of the gallbladder are the **fundus**, **body**, and **neck**.

86. The **right** and **left hepatic ducts** collect bile from the liver, and transport the bile to the **common hepatic duct**. Bile enters either the common bile duct or the **cystic duct**, which leads to the gallbladder. The **common bile duct** is formed by the union of the cystic duct and common hepatic duct, and enters the duodenum.

87. The common bile duct penetrates the wall of the duodenum and meets with the pancreatic duct at the **duodenal ampulla**, a chamber within the **duodenal papilla** that projects into the intestinal lumen. The **hepatopancreatic sphincter** prevents bile flow into the duodenum except at mealtimes.

88. The **head** of the **pancreas** lies within the loop formed by the duodenum. The **pancreatic duct** delivers **pancreatic juice** to the duodenum.

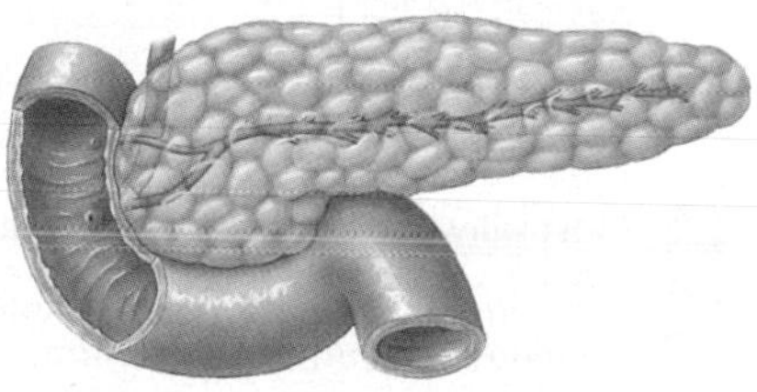

89. The pancreas is primarily an exocrine organ. **Pancreatic acini** produce digestive enzymes that break down starches, lipids, RNA, DNA, and proteins. Pancreatic acini also secrete buffers.

22.23 Disorders of the digestive system are diverse and relatively common p. 872

90. **Periodontal disease**, **mumps**, and **gastroesophageal reflux** are common disorders of the digestive system. Other disorders include **hepatitis**, **gallstones**, **pancreatitis**, **gastritis**, and **enteritis**. **Colorectal cancer** is the third most common cancer in the United States, affecting both men and women.

Chapter Review Questions

Labeling

Label the figure shown here.

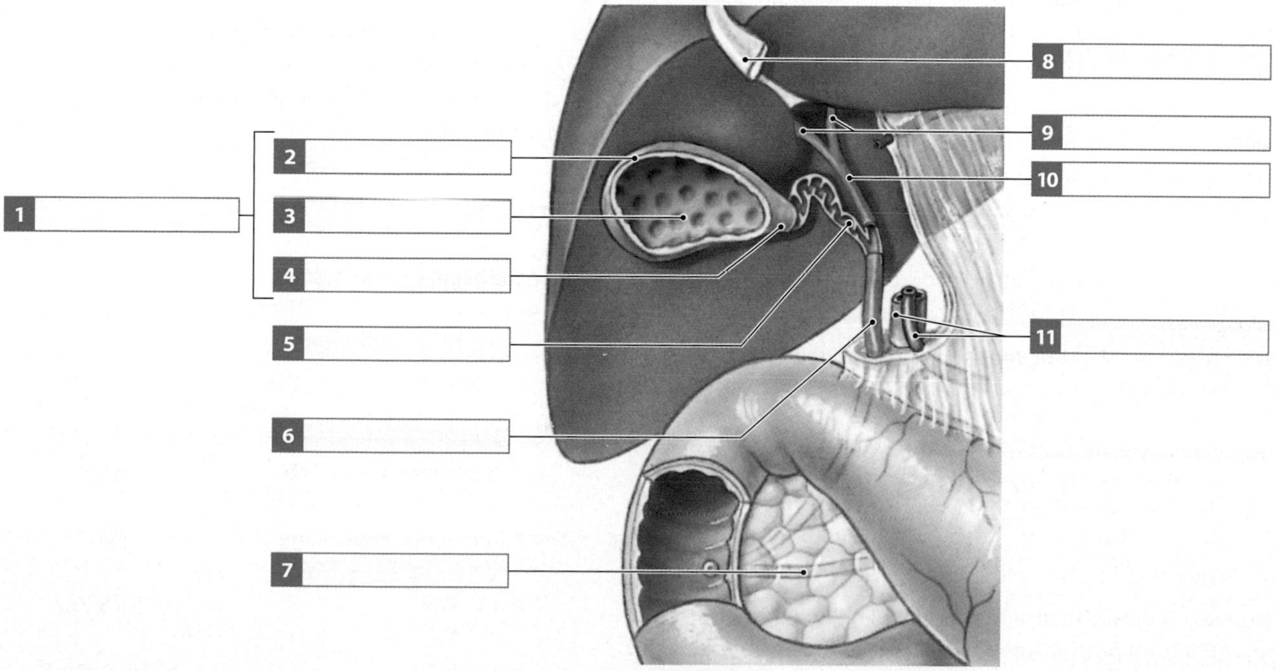

1
2
3
4
5
6
7
8
9
10
11

True/False

Indicate whether each statement is true or false.

12 Kupffer cells are phagocytic cells within liver sinusoids.

13 The gallbladder produces almost as much bile on a daily basis as the liver.

14 Pancreatic islets produce the exocrine secretions for digestion.

15 Most saliva originates in the submandibular glands.

16 The correct order of small intestine segments is: duodenum, ileum, and jejunum.

17 Haustra are a series of pouches along the length of the colon.

12 ______________

13 ______________

14 ______________

15 ______________

16 ______________

17 ______________

Multiple choice

Select the correct answer from the list provided.

18 The release of pancreatic secretions and bile into the duodenum is caused by

- [ ] a) cholecystokinin.
- [ ] b) gastric inhibitory peptide.
- [ ] c) gastrin.
- [ ] d) vasoactive intestinal peptide.

19 Which of the following is *not* a major function of the large intestine?

- [ ] a) reabsorbing water and compacting the intestinal contents into feces
- [ ] b) producing enzymes that assist pancreatic juice
- [ ] c) absorbing vitamins produced by bacterial action
- [ ] d) storing fecal material prior to defecation

20 Fingerlike projections into the lumen of the small intestine are called

- [ ] a) microvilli.
- [ ] b) villi.
- [ ] c) circular folds.
- [ ] d) rugae.

21 A mesentery that forms a pouch between the body wall and the anterior surface of the small intestine is the

- [ ] a) greater omentum.
- [ ] b) lesser omentum.
- [ ] c) mesocolon.
- [ ] d) mesentery proper.

22 Pepsin is formed by a reaction between

- [ ] a) gastric lipase and acid.
- [ ] b) pepsinogen and intrinsic factors.
- [ ] c) bile and pancreatic juice.
- [ ] d) pepsinogen and acid.

23 Which of the following structures is *not* a component of a portal area?

- [ ] a) branch of the hepatic portal vein
- [ ] b) branch of the hepatic vein
- [ ] c) branch of the hepatic artery proper
- [ ] d) branch of a bile duct

Short answer

24 Name the four different types of teeth. What is the number of deciduous teeth and permanent teeth?

25 Identify the three phases of gastric secretion, and provide a brief description of each phase.

26 Describe the hepatic portal system.

Chapter Integration • Applying what you have learned

Three forms of weight control surgery

Obesity is a medical condition in which excess body fat adversely affects the quality of life, leading to increased health problems and decreased life expectancy. According to 2010 statistics reported by the Centers for Disease Control and Prevention (CDC), 35.7 percent of American adults and 16.9 percent of children and adolescents are obese. Moreover, despite public awareness and health initiatives, not one of the 50 states met the government-sponsored Healthy People 2010 goal of reducing to 15 percent the percentage of the population that is obese.

Because obesity is a leading preventable cause of death, more people have turned to surgery to help them lose weight. One form of weight control surgery involves gastric stapling. In this procedure, a large portion of the gastric lumen is stapled shut, leaving only a small pouch in contact with the esophagus and duodenum.

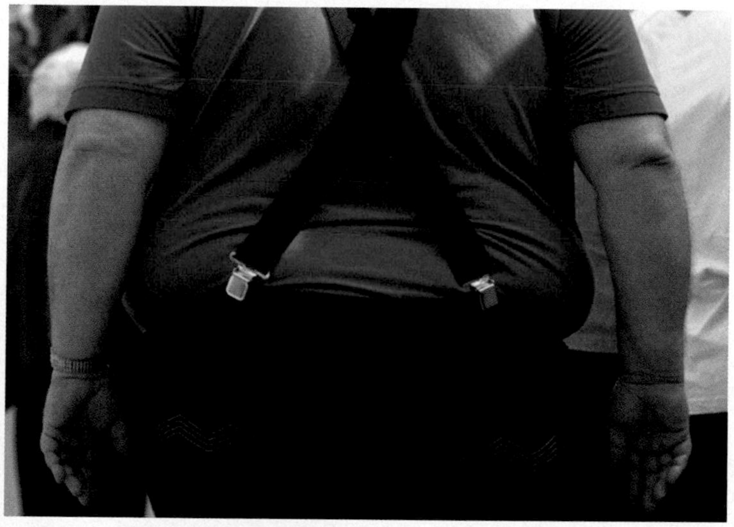

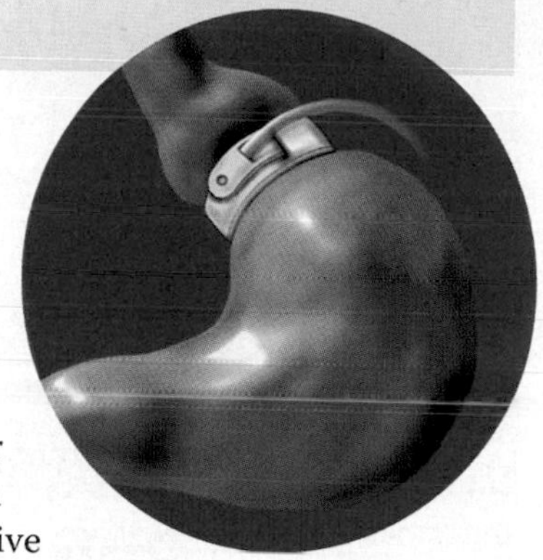

Gastric bypass is another surgical procedure to induce weight loss. In this surgery, the proximal small intestine (duodenum) is connected to a small pouch formed by a superior portion of the stomach. Although this procedure seems more effective than gastric stapling, it involves more complicated surgery.

A third procedure, adjustable gastric band surgery (also known as lap-band surgery), is shown above. It involves placing an inflatable silicone device around the superior aspect of the stomach by laparoscopic surgery. This procedure is performed on obese patients with a body mass index (BMI) of 35–40 or greater. (The body mass index is a ratio calculated by dividing one's weight in kilograms by the square of one's height in meters; it is used to identify those people who are overweight or underweight.)

After each type of surgery, the stomach will hold approximately 110–220 grams of food at each meal, compared to 1500 grams for a normal, distended stomach. Although surgical intervention is successful in many cases, the risks for potential complications are also high.

1. **How would gastric stapling result in weight loss?**
2. **Explain the roles of the stretch receptors and the feeling of fullness after the gastric stapling procedure.**
3. **Using your knowledge of smooth muscle tissue, what will happen to the gastric muscularis externa over time after gastric stapling?**
4. **How do gastric bypass surgery and lap-band surgery help achieve the goal of weight loss?**

23 Metabolism and Energetics

LEARNING OUTCOMES

These Learning Outcomes correspond by number to this chapter's modules and indicate what you should be able to do after completing the chapter.

SECTION 1 • Introduction to Cellular Metabolism

23.1 Define metabolism, catabolism, and anabolism, and give an overview of cellular metabolism.

23.2 Describe the role of the nutrient pool in cellular metabolism.

23.3 Describe the basic steps in the citric acid cycle.

23.4 Describe the basic steps in the electron transport system.

SECTION 2 • Digestion and Metabolism of Organic Nutrients

23.5 Outline the steps involved in digestion, and list the nutrients used by the body.

23.6 Describe carbohydrate metabolism.

23.7 Describe the fate of glucose in glycolysis.

23.8 Describe the mechanisms of lipid transport and distribution.

23.9 Describe the fate of fatty acids in lipid metabolism.

23.10 Summarize the main features of protein metabolism and the use of proteins as an energy source.

23.11 Differentiate between the absorptive and postabsorptive metabolic states and summarize the characteristics of each.

23.12 Explain the role of fat-soluble vitamins and water-soluble vitamins in metabolic pathways.

23.13 Explain what constitutes a balanced diet and why such a diet is important.

23.14 + **CLINICAL MODULE** Describe several metabolic disorders resulting from nutritional or biochemical problems.

SECTION 3 • Energetics and Thermoregulation

23.15 Explain energetics and the role of thermoregulation in maintaining homeostasis.

23.16 Describe the roles of the satiety center and the feeding center in the regulation of food intake.

23.17 Discuss the mechanisms involved in heat gain and heat loss.

23.18 Discuss the homeostatic mechanisms that maintain a constant body temperature.

LO Learning Outcomes are repeated at the bottom of each module.

Metabolism refers to all the chemical reactions in the body

The term **metabolism** (me-TAB-ō-lizm) refers to all the chemical reactions that occur in an organism. Chemical reactions within cells, collectively known as **cellular metabolism**, provide the energy needed to maintain homeostasis and to perform essential functions. As noted in Chapter 2, **catabolism** is the breakdown of organic substrates in the body, whereas **anabolism** is the synthesis of new organic molecules.

An Overview of Cellular Metabolism

In the process of **metabolic turnover**, cells continuously break down and replace all their organic components except DNA. The catabolic reactions involved provide very little ATP to the cell. Cells continuously absorb organic molecules from the surrounding interstitial fluids, adding to those released through catabolic reactions in metabolic turnover.

All of the cell's organic building blocks form a **nutrient pool**—an accessible source of organic substrates. The components of the nutrient pool can be either used for anabolism or broken down further for ATP production. This section considers the origins and fate of the nutrient pool, and ends with an overview of the catabolic pathways that provide ATP.

In mitochondria, the catabolic reactions of aerobic metabolism release significant amounts of energy. About 40 percent of the energy released in these catabolic reactions is captured and used to convert ADP to ATP. The other 60 percent escapes as heat that warms the interior of the cell and the surrounding tissues.

The ATP produced by mitochondria provides energy to support both anabolism and other cell functions.

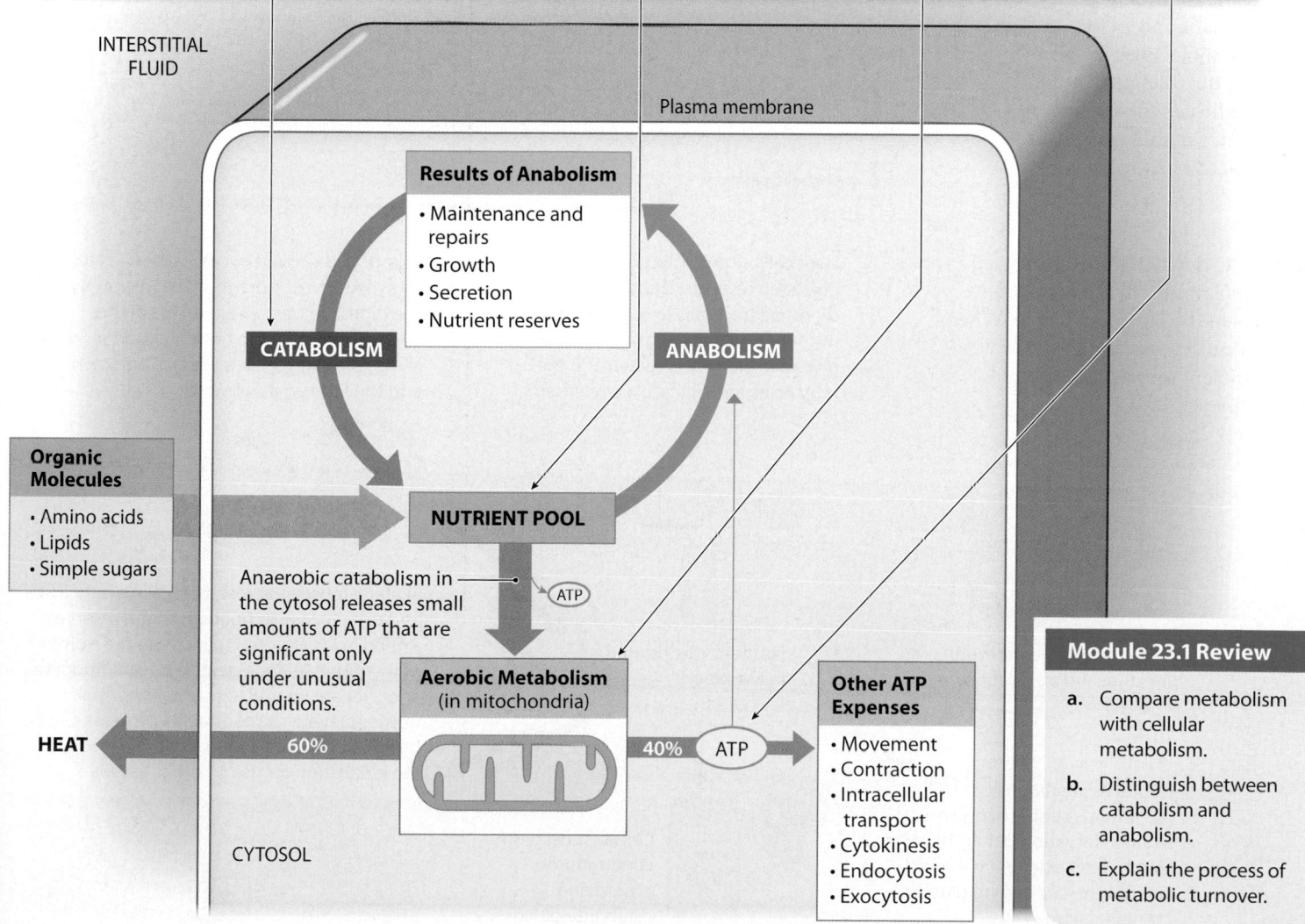

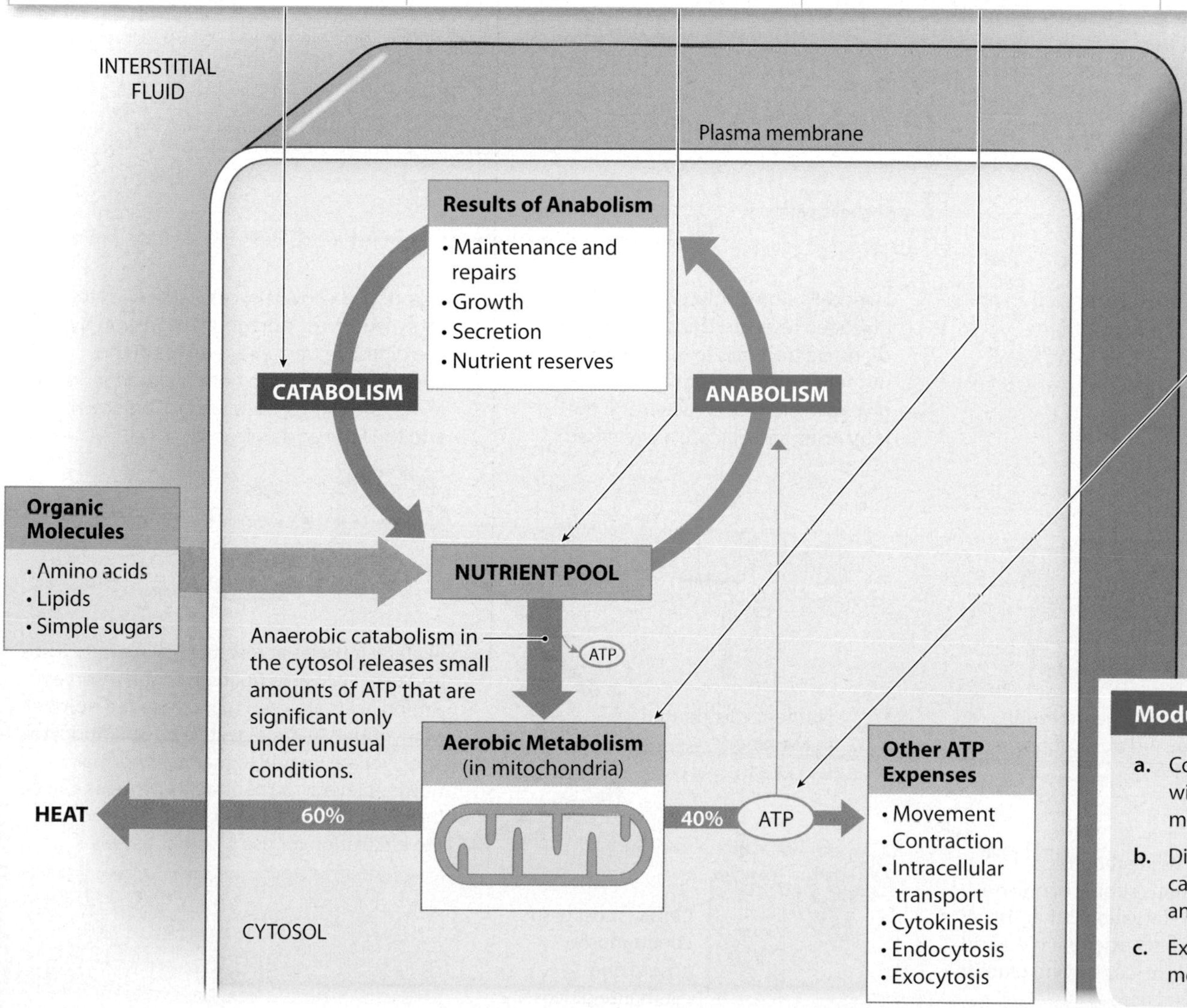
INTERSTITIAL FLUID
Plasma membrane
Results of Anabolism
• Maintenance and repairs
• Growth
• Secretion
• Nutrient reserves
CATABOLISM
ANABOLISM
Organic Molecules
• Amino acids
• Lipids
• Simple sugars
NUTRIENT POOL
Anaerobic catabolism in the cytosol releases small amounts of ATP that are significant only under unusual conditions.
ATP
Aerobic Metabolism (in mitochondria)
HEAT
60%
40%
ATP
Other ATP Expenses
• Movement
• Contraction
• Intracellular transport
• Cytokinesis
• Endocytosis
• Exocytosis
CYTOSOL

Module 23.1 Review

a. Compare metabolism with cellular metabolism.

b. Distinguish between catabolism and anabolism.

c. Explain the process of metabolic turnover.

23.1 Define metabolism, catabolism, and anabolism, and give an overview of cellular metabolism.

Cells can break down any available substrate from the nutrient pool to obtain the energy they need

The **nutrient pool** is the source of the organic substrates for both catabolism and anabolism. Cells tend to conserve the materials needed to build new compounds and break down the excess.

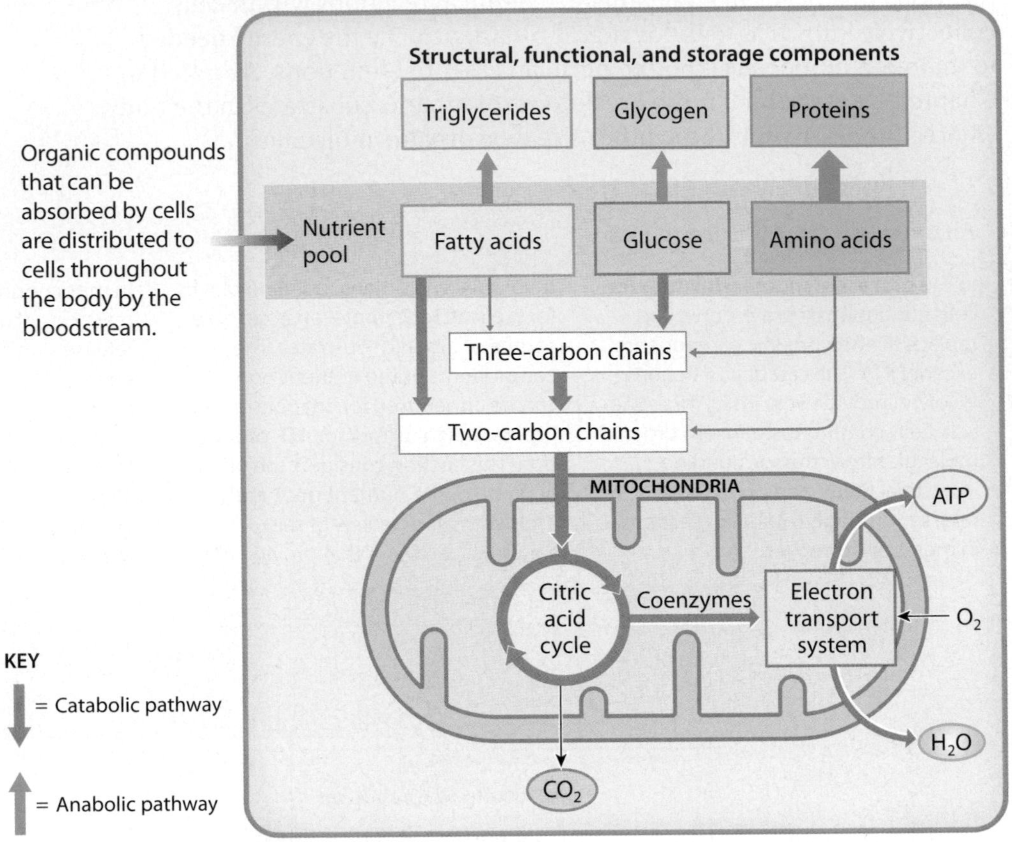
Organic compounds that can be absorbed by cells are distributed to cells throughout the body by the bloodstream.
Structural, functional, and storage components
Triglycerides
Glycogen
Proteins
Nutrient pool
Fatty acids
Glucose
Amino acids
Three-carbon chains
Two-carbon chains
MITOCHONDRIA
Citric acid cycle
Coenzymes
Electron transport system
ATP
O_2
H_2O
CO_2
KEY
= Catabolic pathway
= Anabolic pathway

1 The nutrient pool is the key to the cell's survival because it contains the organic materials required for both anabolism and catabolism. Cells are continuously replacing membranes, organelles, enzymes, and structural proteins as well as performing their specialized functions. All of these functions require ATP, so the cell's mitochondria must have a continuous supply of two-carbon substrate molecules. In this diagram, the thickness of the arrow indicates the relative importance of that pathway in an inactive cell.

2 The nutrient pool of each cell contributes to the metabolic reserves of the body as a whole. When absorption across the digestive tract is insufficient to maintain normal nutrient levels in the blood, those reserves can be mobilized.

Liver cells store triglycerides and glycogen reserves. If absorption by the digestive tract fails to maintain normal nutrient levels, the triglycerides and glycogen are broken down and the fatty acids and glucose are released.

Adipocytes convert excess fatty acids to triglycerides for storage. If absorption by the digestive tract and reserves in the liver cannot maintain normal nutrient levels, the triglycerides are broken down and the fatty acids released.

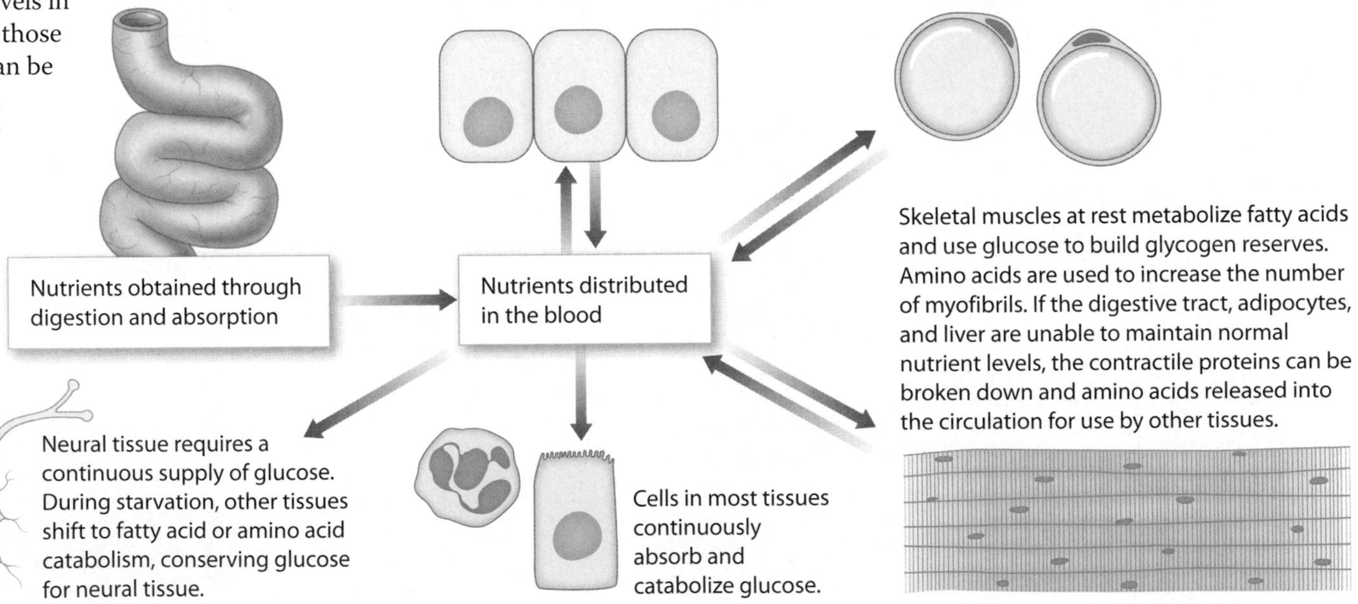
Nutrients obtained through digestion and absorption
Nutrients distributed in the blood
Skeletal muscles at rest metabolize fatty acids and use glucose to build glycogen reserves. Amino acids are used to increase the number of myofibrils. If the digestive tract, adipocytes, and liver are unable to maintain normal nutrient levels, the contractile proteins can be broken down and amino acids released into the circulation for use by other tissues.
Neural tissue requires a continuous supply of glucose. During starvation, other tissues shift to fatty acid or amino acid catabolism, conserving glucose for neural tissue.
Cells in most tissues continuously absorb and catabolize glucose.

3 Neither the diet nor the nutrient pool provides everything needed to build every protein, carbohydrate, and lipid a cell might require. With few exceptions, this is not a problem because the cell also contains the enzymes necessary to synthesize what it needs from available substrates. This diagram is a general overview of the catabolic and anabolic pathways involved; these pathways will be the focus of the next section. Several modules in earlier chapters provided an overview of some of the key steps and pathways (Modules 2.8, 2.14, 2.15, 2.17, 2.19, and 9.12). You should take the time to go back and review that material before proceeding.

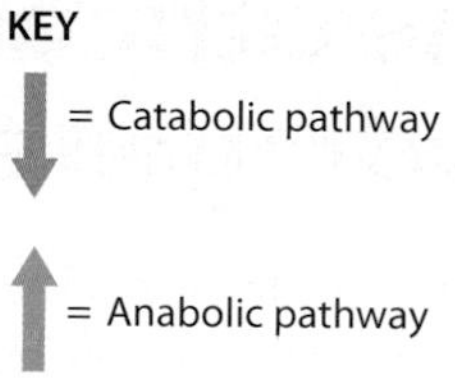
KEY
= Catabolic pathway
= Anabolic pathway

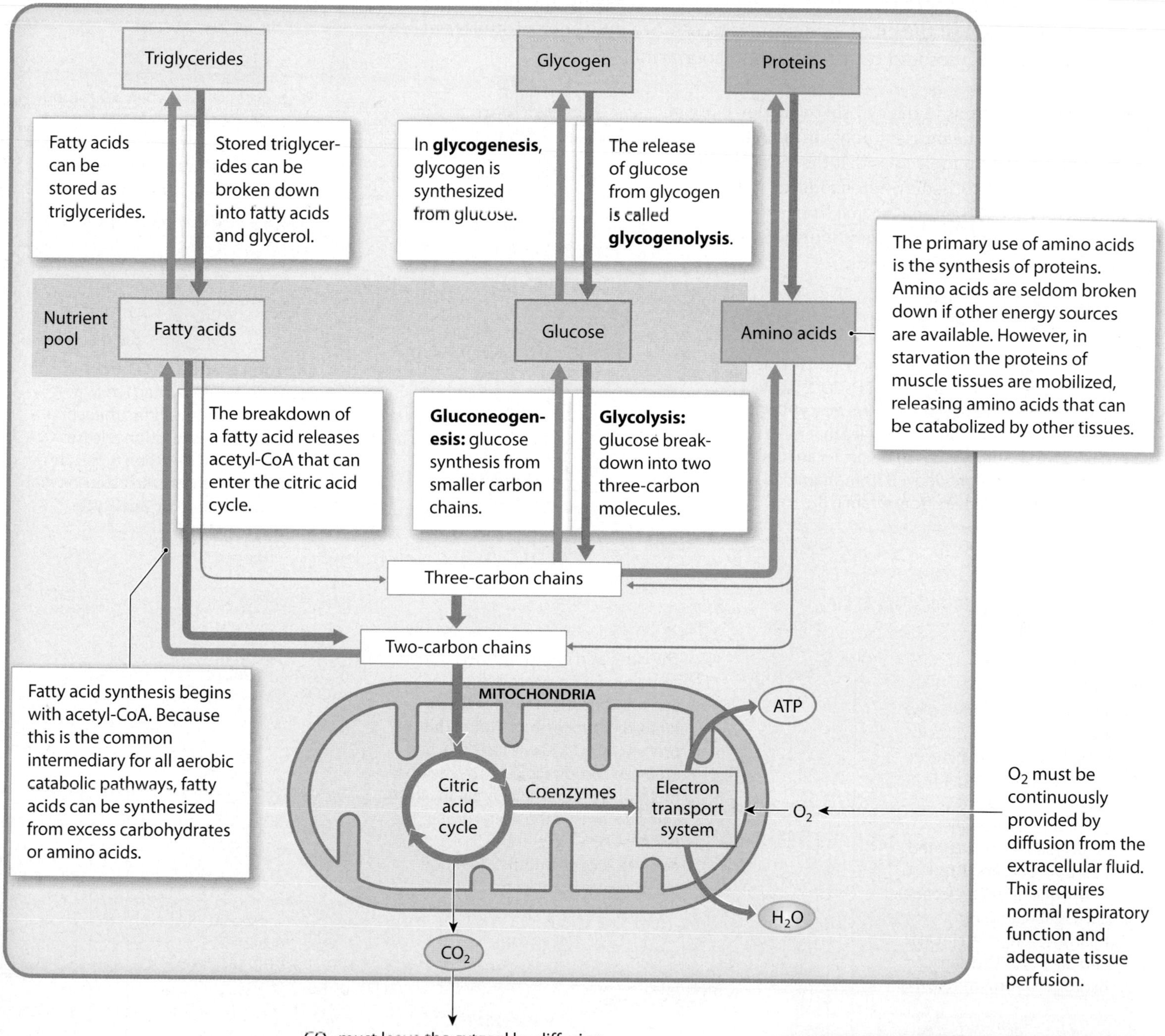
Triglycerides
Glycogen
Proteins
Fatty acids can be stored as triglycerides.
Stored triglycerides can be broken down into fatty acids and glycerol.
In **glycogenesis**, glycogen is synthesized from glucose.
The release of glucose from glycogen is called **glycogenolysis**.
The primary use of amino acids is the synthesis of proteins. Amino acids are seldom broken down if other energy sources are available. However, in starvation the proteins of muscle tissues are mobilized, releasing amino acids that can be catabolized by other tissues.
Nutrient pool
Fatty acids
Glucose
Amino acids
The breakdown of a fatty acid releases acetyl-CoA that can enter the citric acid cycle.
Gluconeogenesis: glucose synthesis from smaller carbon chains.
Glycolysis: glucose breakdown into two three-carbon molecules.
Three-carbon chains
Two-carbon chains
Fatty acid synthesis begins with acetyl-CoA. Because this is the common intermediary for all aerobic catabolic pathways, fatty acids can be synthesized from excess carbohydrates or amino acids.
MITOCHONDRIA
Citric acid cycle
Coenzymes
Electron transport system
ATP
O_2
H_2O
CO_2
O_2 must be continuously provided by diffusion from the extracellular fluid. This requires normal respiratory function and adequate tissue perfusion.

CO_2 must leave the cytosol by diffusion into the extracellular fluid. The bloodstream must continuously absorb CO_2 in peripheral tissues and eliminate it at the lungs to prevent potentially dangerous changes in body fluid pH.

Module 23.2 Review

a. Define nutrient pool.

b. Why do cells engage in catabolism?

c. Why do cells make new compounds?

LO 23.2 Describe the role of the nutrient pool in cellular metabolism.

The citric acid cycle transfers hydrogen atoms to coenzymes

Mitochondria provide almost all of the energy that supports cellular operations. The cell "feeds" its mitochondria from its nutrient pool, and in return, the cell gets the ATP it needs. However, mitochondria are picky eaters: They will only accept specific organic molecules for processing and producing energy. The function of the citric acid cycle is to remove hydrogen atoms from those organic molecules and transfer them to coenzymes.

1 Chemical reactions in the cytosol break down organic nutrients from the nutrient pool into smaller 3-carbon and 2-carbon molecules that the mitochondria can process. Once absorbed by mitochondria, it doesn't matter whether these fragments came from a carbohydrate, a lipid, or an amino acid. Everything is broken down through the catabolic pathway.

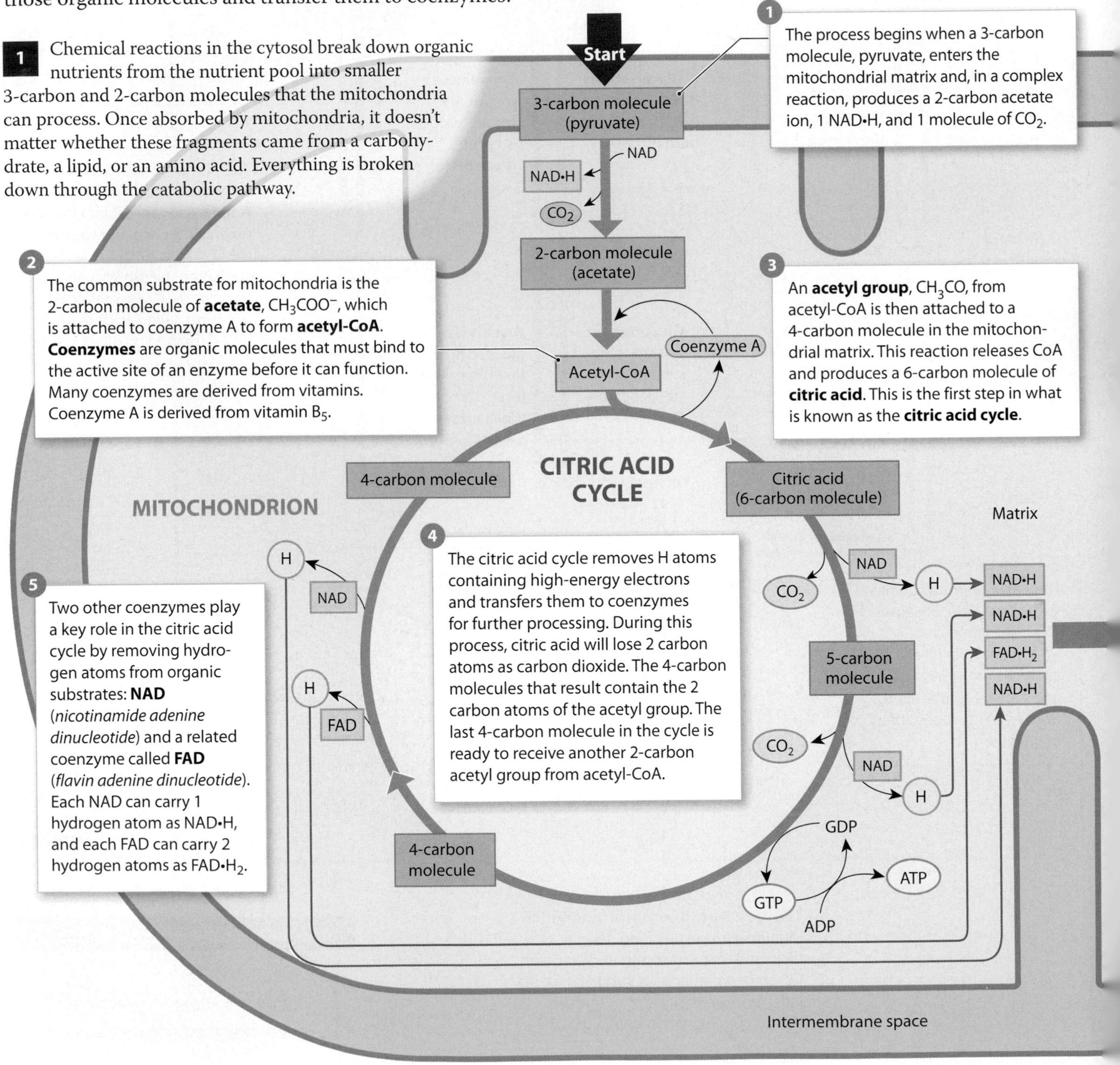

Start
3-carbon molecule (pyruvate)
NAD
NAD•H
CO_2
2-carbon molecule (acetate)
Coenzyme A
Acetyl-CoA
CITRIC ACID CYCLE
4-carbon molecule
Citric acid (6-carbon molecule)
MITOCHONDRION
Matrix
H
NAD
FAD
NAD•H
NAD•H
$FAD•H_2$
NAD•H
5-carbon molecule
CO_2
CO_2
4-carbon molecule
GDP
GTP
ATP
ADP
Intermembrane space
1 The process begins when a 3-carbon molecule, pyruvate, enters the mitochondrial matrix and, in a complex reaction, produces a 2-carbon acetate ion, 1 NAD•H, and 1 molecule of CO_2.
2 The common substrate for mitochondria is the 2-carbon molecule of **acetate**, CH_3COO^-, which is attached to coenzyme A to form **acetyl-CoA**. **Coenzymes** are organic molecules that must bind to the active site of an enzyme before it can function. Many coenzymes are derived from vitamins. Coenzyme A is derived from vitamin B_5.
3 An **acetyl group**, CH_3CO, from acetyl-CoA is then attached to a 4-carbon molecule in the mitochondrial matrix. This reaction releases CoA and produces a 6-carbon molecule of **citric acid**. This is the first step in what is known as the **citric acid cycle**.
4 The citric acid cycle removes H atoms containing high-energy electrons and transfers them to coenzymes for further processing. During this process, citric acid will lose 2 carbon atoms as carbon dioxide. The 4-carbon molecules that result contain the 2 carbon atoms of the acetyl group. The last 4-carbon molecule in the cycle is ready to receive another 2-carbon acetyl group from acetyl-CoA.
5 Two other coenzymes play a key role in the citric acid cycle by removing hydrogen atoms from organic substrates: **NAD** (*nicotinamide adenine dinucleotide*) and a related coenzyme called **FAD** (*flavin adenine dinucleotide*). Each NAD can carry 1 hydrogen atom as NAD•H, and each FAD can carry 2 hydrogen atoms as $FAD•H_2$.

2 This figure is a more detailed view of the citric acid cycle showing the changes in the carbon chain as it works its way through the cycle. For each acetyl-CoA molecule that enters the citric acid cycle, 5 hydrogen atoms are removed and transferred to coenzymes, 2 molecules of CO_2 are produced, and 2 molecules of water are consumed. Note that the net energy gain from each acetyl group entering the citric acid cycle is 1 ATP molecule.

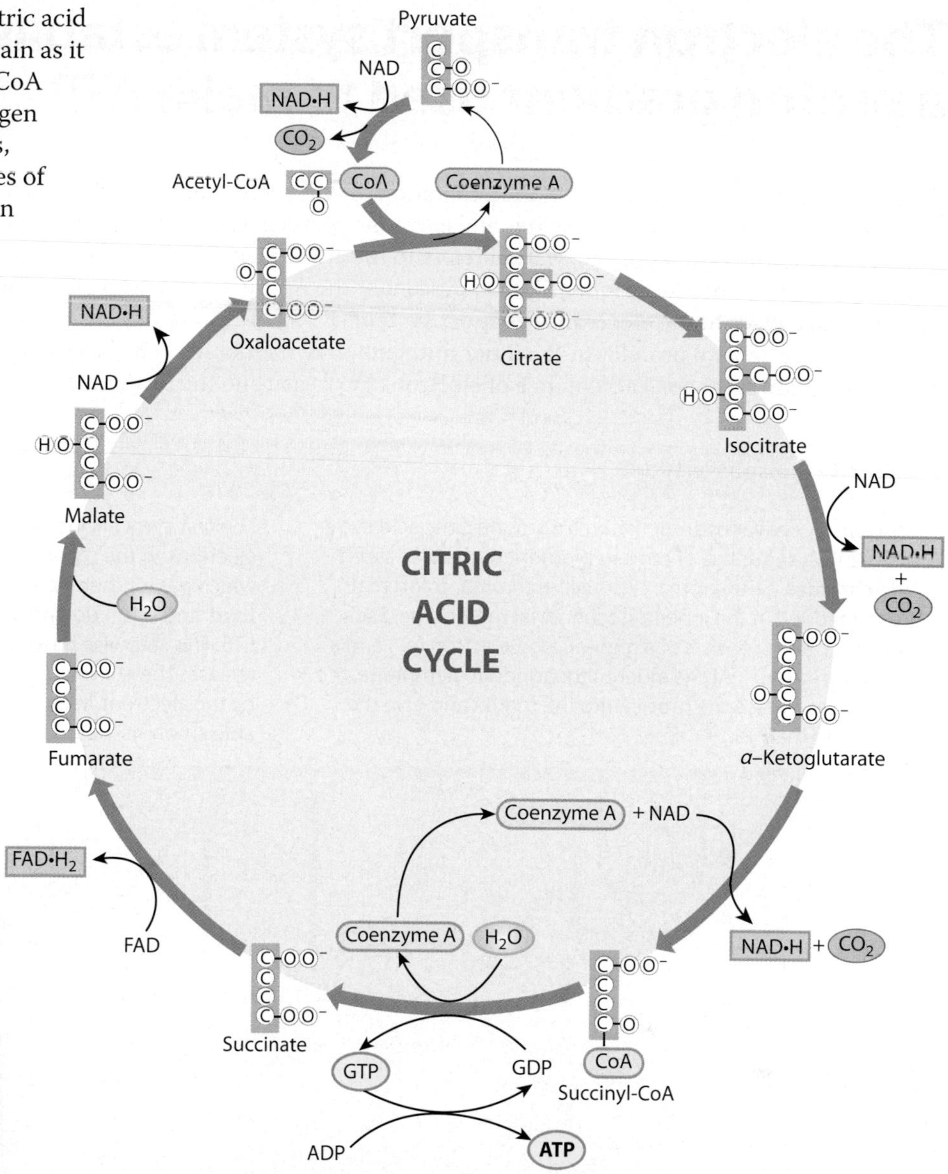

Pyruvate
NAD
NAD•H
CO_2
Acetyl-CoA
CoA
Coenzyme A
Oxaloacetate
Citrate
Isocitrate
NAD
NAD•H
+
CO_2
α–Ketoglutarate
CITRIC ACID CYCLE
Coenzyme A + NAD
NAD•H + CO_2
Succinyl-CoA
Coenzyme A
H_2O
GDP
GTP
ADP
ATP
Succinate
FAD
$FAD•H_2$
Fumarate
H_2O
Malate
NAD
NAD•H

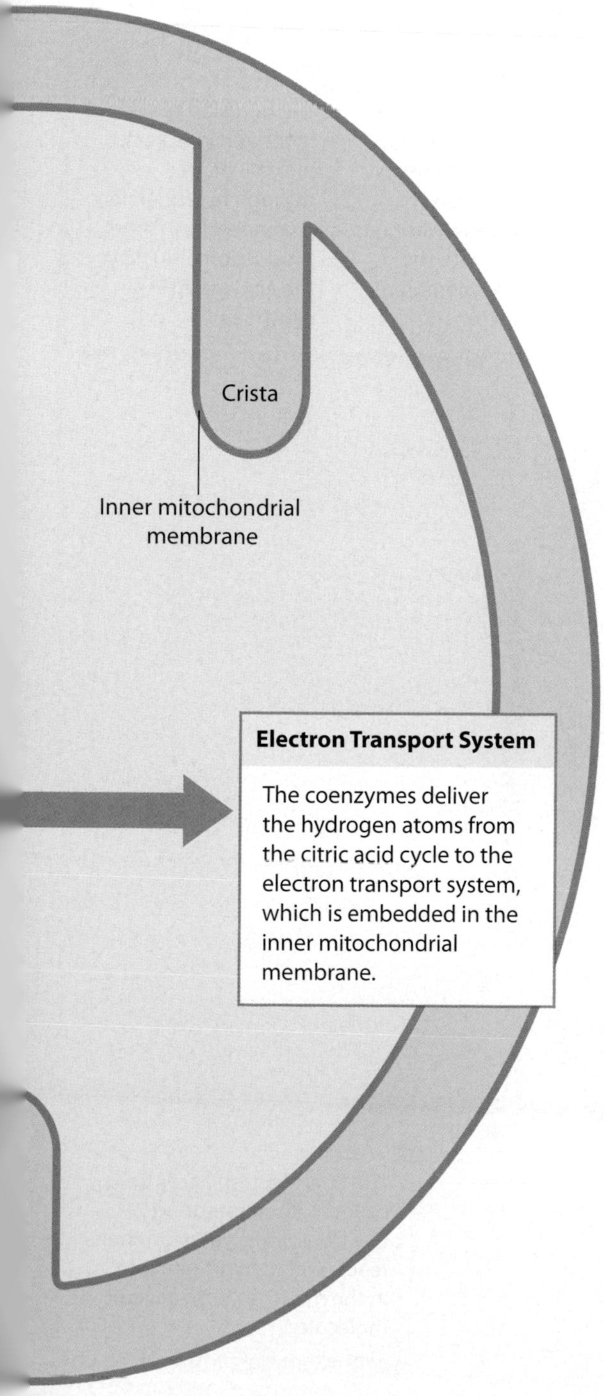

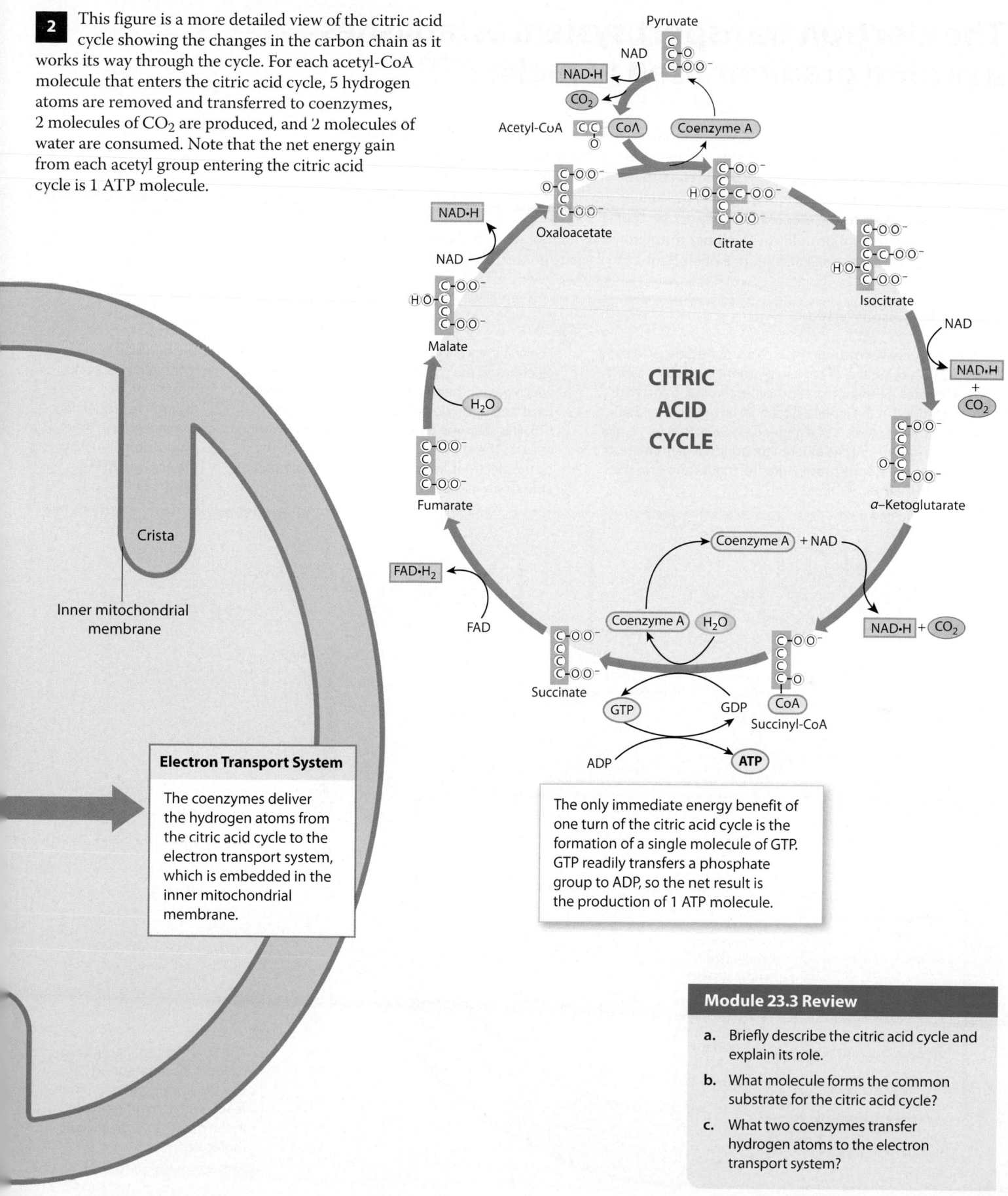

Crista
Inner mitochondrial membrane

Electron Transport System

The coenzymes deliver the hydrogen atoms from the citric acid cycle to the electron transport system, which is embedded in the inner mitochondrial membrane.

The only immediate energy benefit of one turn of the citric acid cycle is the formation of a single molecule of GTP. GTP readily transfers a phosphate group to ADP, so the net result is the production of 1 ATP molecule.

Module 23.3 Review

a. Briefly describe the citric acid cycle and explain its role.

b. What molecule forms the common substrate for the citric acid cycle?

c. What two coenzymes transfer hydrogen atoms to the electron transport system?

LO 23.3 Describe the basic steps in the citric acid cycle.

Module 23.4

The electron transport system establishes a proton gradient used to make ATP

1 **Oxidative phosphorylation** is the generation of ATP within mitochondria in a reaction sequence that requires coenzymes and consumes oxygen. Oxidation refers to the transfer of electrons and phosphorylation refers to the attachment of a high-energy phosphate group. The process produces more than 90 percent of the ATP used by body cells. The key reactions take place in the **electron transport system (ETS),** or respiratory chain, a series of integral and peripheral proteins in the inner mitochondrial membrane. The basis of oxidative phosphorylation is the final acceptance of electrons by oxygen and the formation of water.

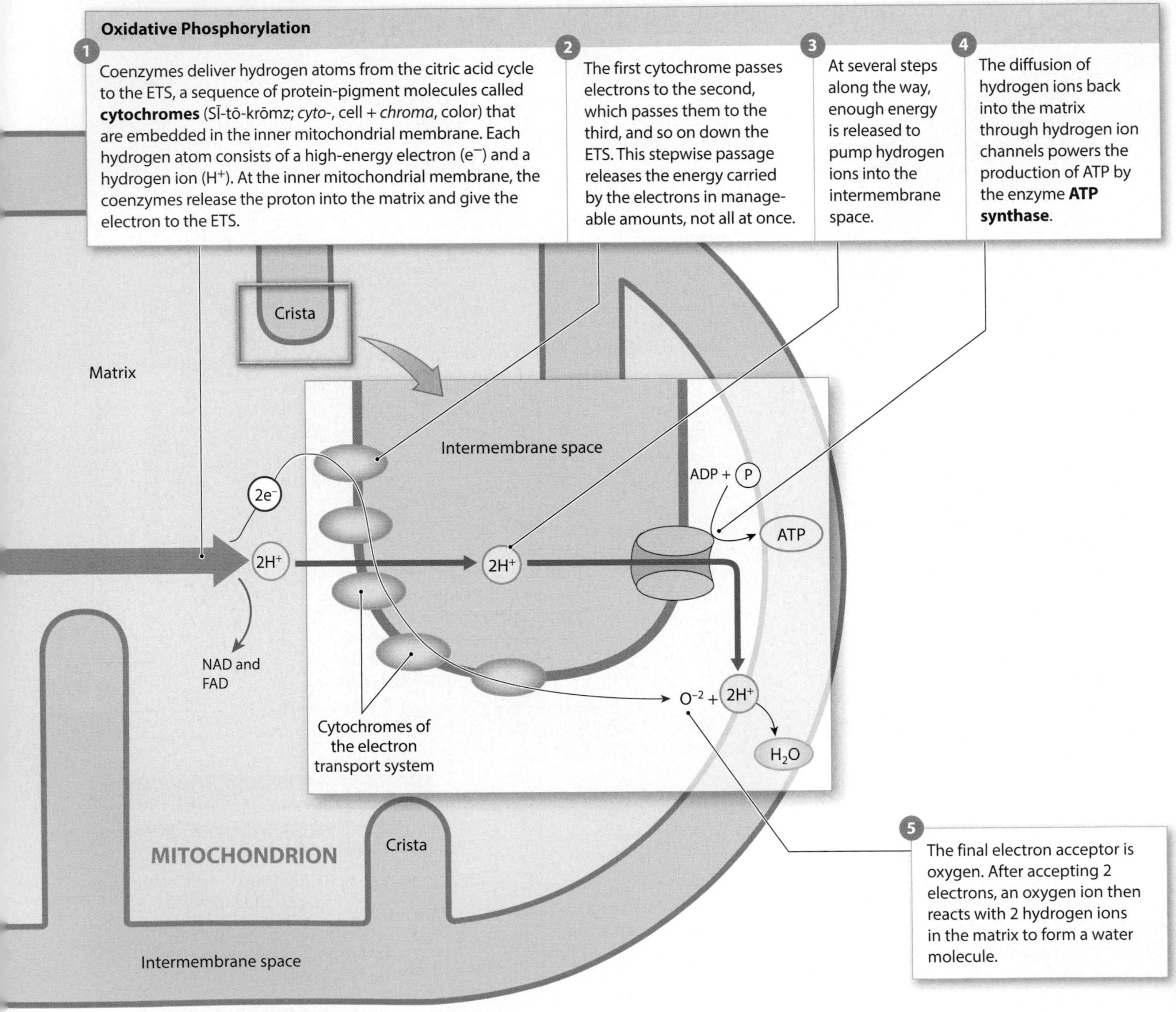

Oxidative Phosphorylation
1 Coenzymes deliver hydrogen atoms from the citric acid cycle to the ETS, a sequence of protein-pigment molecules called cytochromes (SĪ-tō-krōmz; cyto-, cell + chroma, color) that are embedded in the inner mitochondrial membrane. Each hydrogen atom consists of a high-energy electron (e⁻) and a hydrogen ion (H⁺). At the inner mitochondrial membrane, the coenzymes release the proton into the matrix and give the electron to the ETS.
2 The first cytochrome passes electrons to the second, which passes them to the third, and so on down the ETS. This stepwise passage releases the energy carried by the electrons in manageable amounts, not all at once.
3 At several steps along the way, enough energy is released to pump hydrogen ions into the intermembrane space.
4 The diffusion of hydrogen ions back into the matrix through hydrogen ion channels powers the production of ATP by the enzyme ATP synthase.
Crista
Matrix
Intermembrane space
ADP + P
ATP
2e⁻
2H⁺
2H⁺
NAD and FAD
Cytochromes of the electron transport system
O⁻² + 2H⁺
H₂O
MITOCHONDRION
Crista
Intermembrane space
5 The final electron acceptor is oxygen. After accepting 2 electrons, an oxygen ion then reacts with 2 hydrogen ions in the matrix to form a water molecule.

2 The transfer of electrons in the ETS and and their relative loss of energy is summarized in the figure below.

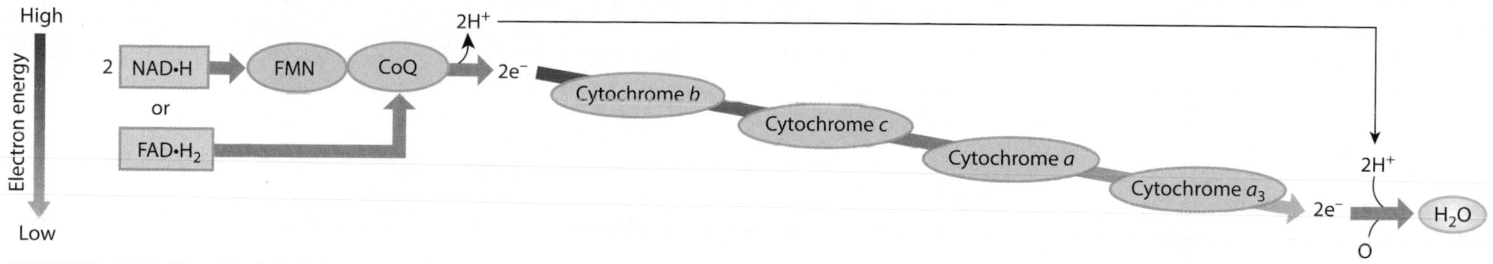
High
Electron energy
Low
2 NAD•H
or
FAD•H2
FMN
CoQ
2H+
2e−
Cytochrome b
Cytochrome c
Cytochrome a
Cytochrome a3
2H+
2e−
O
H2O

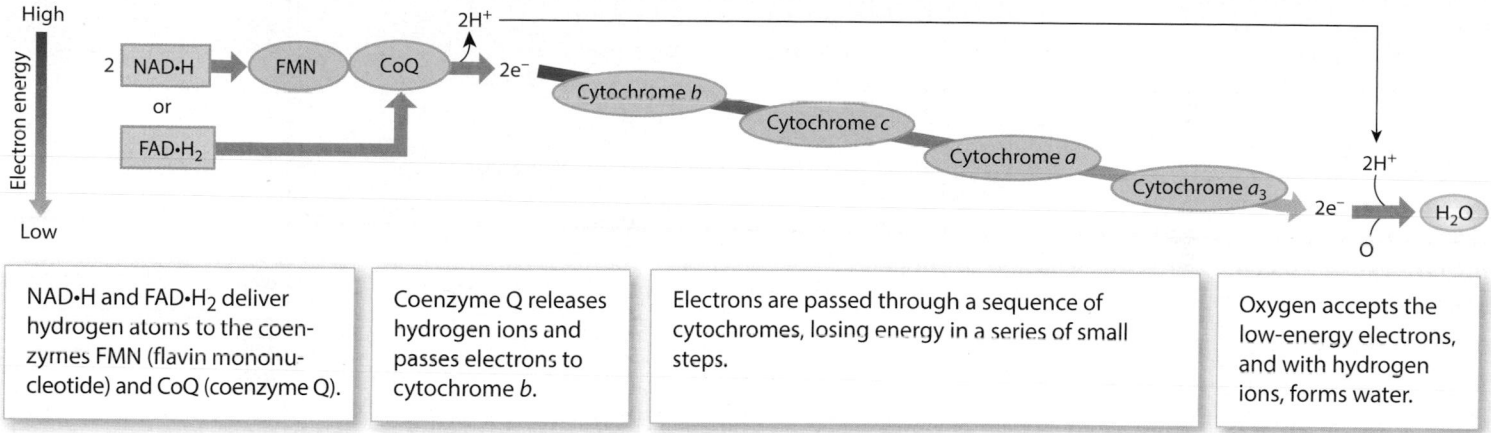

NAD•H and FAD•H_2 deliver hydrogen atoms to the coenzymes FMN (flavin mononucleotide) and CoQ (coenzyme Q).

Coenzyme Q releases hydrogen ions and passes electrons to cytochrome *b*.

Electrons are passed through a sequence of cytochromes, losing energy in a series of small steps.

Oxygen accepts the low-energy electrons, and with hydrogen ions, forms water.

3 A more detailed look at the electron transport system shows where the reactions occur in the mitochondrion. Notice the sites where hydrogen ions are pumped into the intermembrane space, providing the concentration gradient essential for the generation of ATP. The red line indicates the path taken by the electrons.

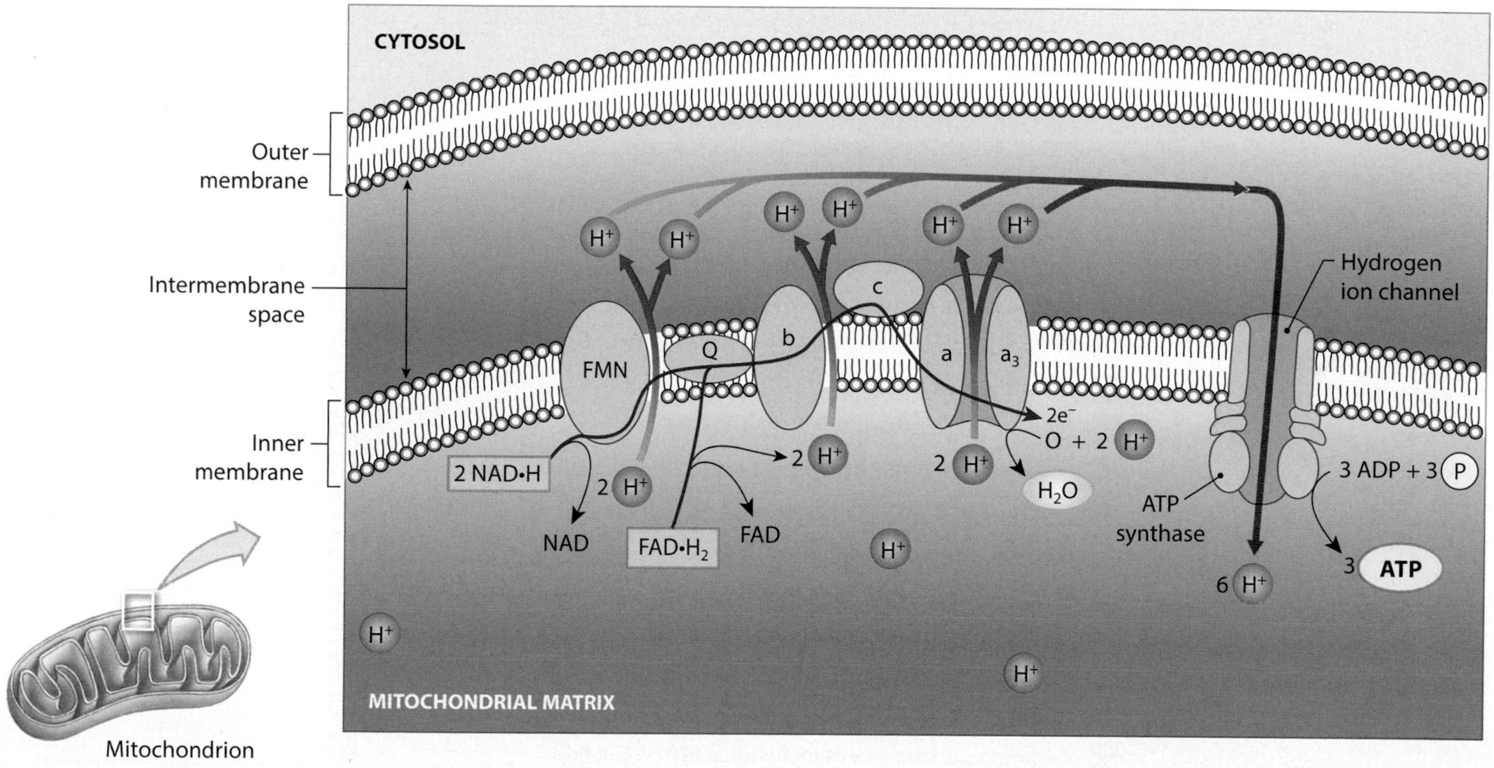
CYTOSOL
Outer membrane
Intermembrane space
Inner membrane
H+
FMN
Q
b
c
a
a3
2 NAD•H
2 H+
NAD
FAD•H2
FAD
2 H+
2e−
O + 2 H+
H2O
Hydrogen ion channel
ATP synthase
3 ADP + 3 P
3 ATP
6 H+
MITOCHONDRIAL MATRIX
Mitochondrion

Mitochondria are very efficient, but aerobic metabolism is dependent on the availability of oxygen that must be absorbed first by the cell and then by the mitochondria. If that supply of oxygen is severely restricted, the cytochromes cannot pass on their electrons and the ETS stops working. When the ETS stops, NAD and FAD can't drop off their hydrogen atoms, so the citric acid cycle halts as well. If the problem persists, the cell will die because it lacks the ATP to support vital functions. Poisons like cyanide (used as a pesticide to kill rats or mice) bind to cytochromes and prevent the transfer of electrons to oxygen. As a result, cells die from energy starvation.

Module 23.4 Review

a. Define oxidative phosphorylation.

b. Where are the cytochromes located in a mitochondrion?

c. Describe the role that hydrogen ion channels play in the generation of ATP.

LO 23.4 Describe the basic steps in the electron transport system.

Matching

Match each lettered term with the appropriate numbered blank in the cellular metabolism figure to the right.

a. glucose
b. electron transport system
c. O_2
d. fatty acids
e. proteins
f. citric acid cycle
g. ATP
h. CO_2
i. H_2O
j. coenzymes
k. two-carbon chains

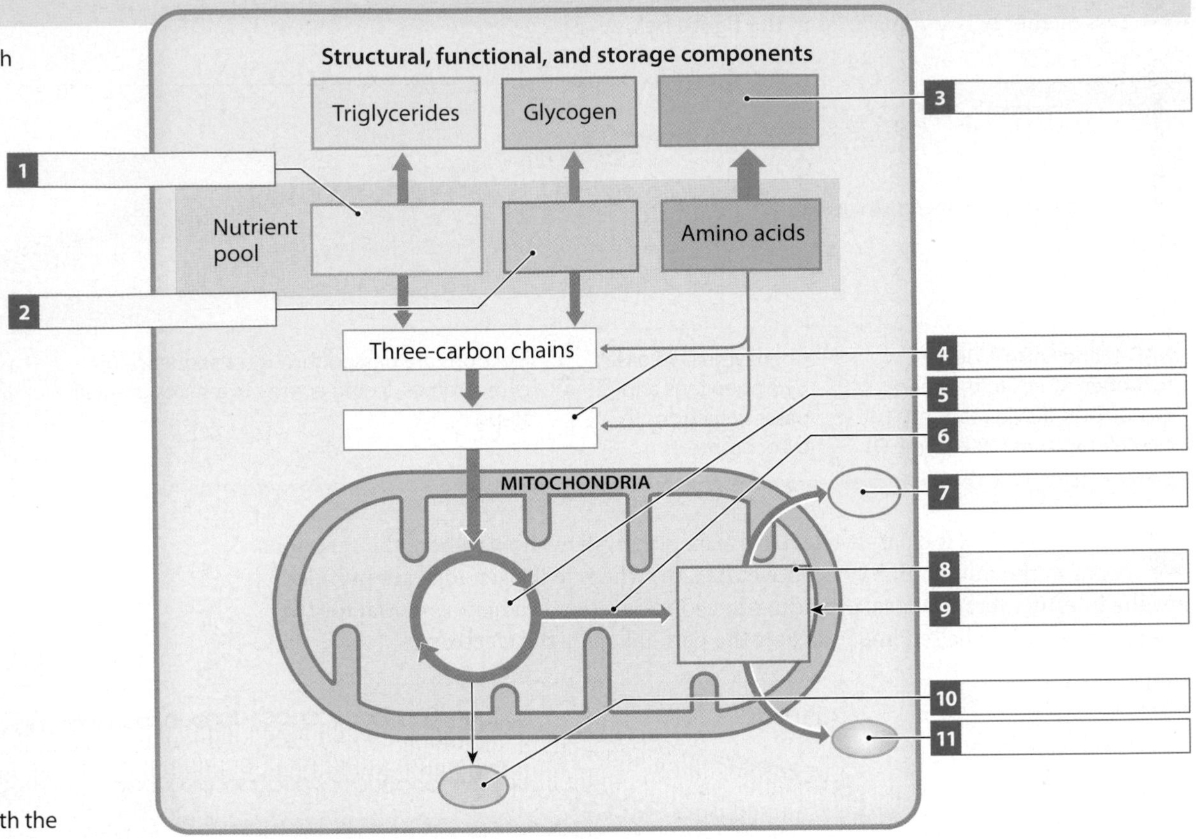
Structural, functional, and storage components
Triglycerides
Glycogen
Nutrient pool
Amino acids
Three-carbon chains
MITOCHONDRIA
1
2
3
4
5
6
7
8
9
10
11

Match each lettered term with the most closely related description.

a. coenzymes
b. cytochromes
c. citric acid
d. nutrients scarce
e. nutrient pool
f. anabolism
g. ATP
h. water
i. acetate
j. oxygen
k. citric acid cycle
l. catabolism
m. oxidative phosphorylation
n. nutrients abundant

12 6-carbon molecule
13 Collection of all the cell's organic substances
14 Synthesis of new organic molecules
15 Process that produces over 90 percent of ATP used by body cells
16 ETS proteins
17 Carry hydrogen atoms to the ETS
18 Final acceptor of electrons from the ETS
19 Breakdown of organic molecules
20 Condition when cells preferentially break down carbohydrates
21 Product of hydrogen ion diffusion within mitochondria
22 Source of mitochondrial CO_2 production
23 ETS byproduct
24 Condition when cells preferentially break down lipids
25 Common substrate for mitochondrial ATP production

12 ______
13 ______
14 ______
15 ______
16 ______
17 ______
18 ______
19 ______
20 ______
21 ______
22 ______
23 ______
24 ______
25 ______

Short answer

Neural tissue requires a constant supply of glucose. What general shifts in cellular metabolism occur during fasting or starvation to meet that requirement?

26 ______

Digestion involves a series of steps to make nutrients available to the body

1 The food we eat has an organized physical structure, and the organic compounds it contains are large and often insoluble. During digestion the physical structure is broken down, and a combination of chemical and enzymatic attack breaks the complex organic compounds into simpler components that can be absorbed by the digestive tract, and then distributed by the bloodstream. Cells throughout the body rely on these organic molecules for energy production and to replenish the intracellular nutrient pool.

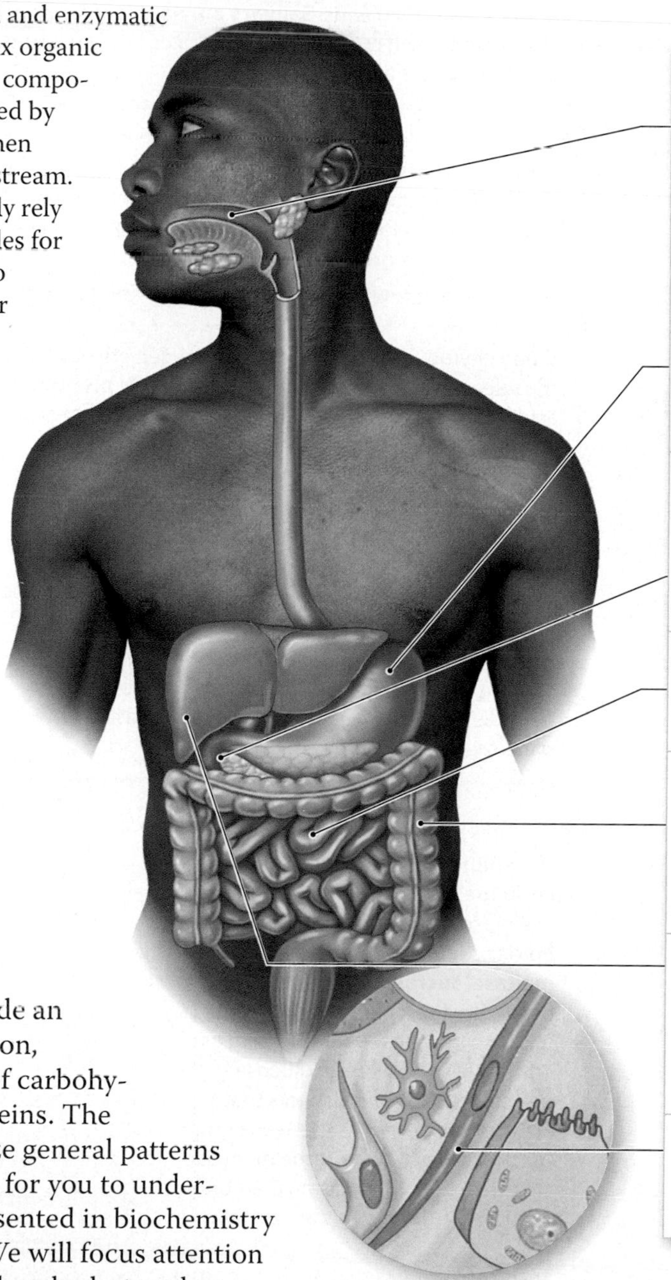

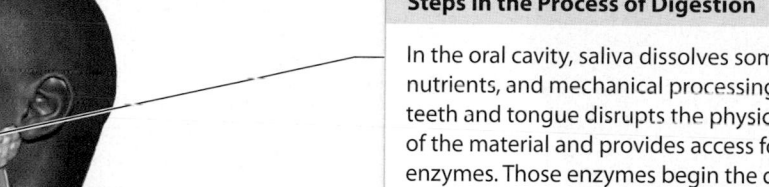

Steps in the Process of Digestion

In the oral cavity, saliva dissolves some organic nutrients, and mechanical processing with the teeth and tongue disrupts the physical structure of the material and provides access for digestive enzymes. Those enzymes begin the digestion of complex carbohydrates (polysaccharides) and lipids.

In the stomach, the food is further broken down physically and chemically by stomach acid and by enzymes that can operate at an extremely low pH.

In the duodenum, buffers from the pancreas and liver moderate the pH of the arriving chyme. The pancreas also secretes various digestive enzymes that catalyze the catabolism of carbohydrates, lipids, proteins, and nucleic acids.

Nutrient absorption then occurs in the small intestine, primarily in the jejunum, and the nutrients enter the bloodstream.

Indigestible food and wastes enter the large intestine, where water is reabsorbed and bacteria generate both organic nutrients and vitamins. These organic products are absorbed before the residue is excreted as feces.

Most of the nutrients absorbed by the digestive tract are transported in a branch of the hepatic portal vein that ends at the liver. The liver absorbs nutrients as needed to maintain normal levels in the bloodstream.

Within peripheral tissues, cells absorb the nutrients needed to maintain their nutrient pool and ongoing operations.

This section will provide an overview of the digestion, absorption, and fates of carbohydrates, lipids, and proteins. The modules will emphasize general patterns that will make it easier for you to understand the specifics presented in biochemistry or nutrition courses. We will focus attention on how nutrients are absorbed, stored or interconverted, or catabolized to yield the energy needed to support vital activities. We will not consider the major anabolic pathways here because Chapter 2 discussed the synthesis of carbohydrates, lipids, and proteins in sufficient detail.

Module 23.5 Review

a. Why is digestion important?

b. Where does most nutrient absorption occur?

c. Most of the absorbed nutrients enter into which blood vessel?

LO 23.5 Outline the steps involved in digestion, and list the nutrients used by the body.

Carbohydrates are usually the preferred substrates for catabolism and ATP production under resting conditions

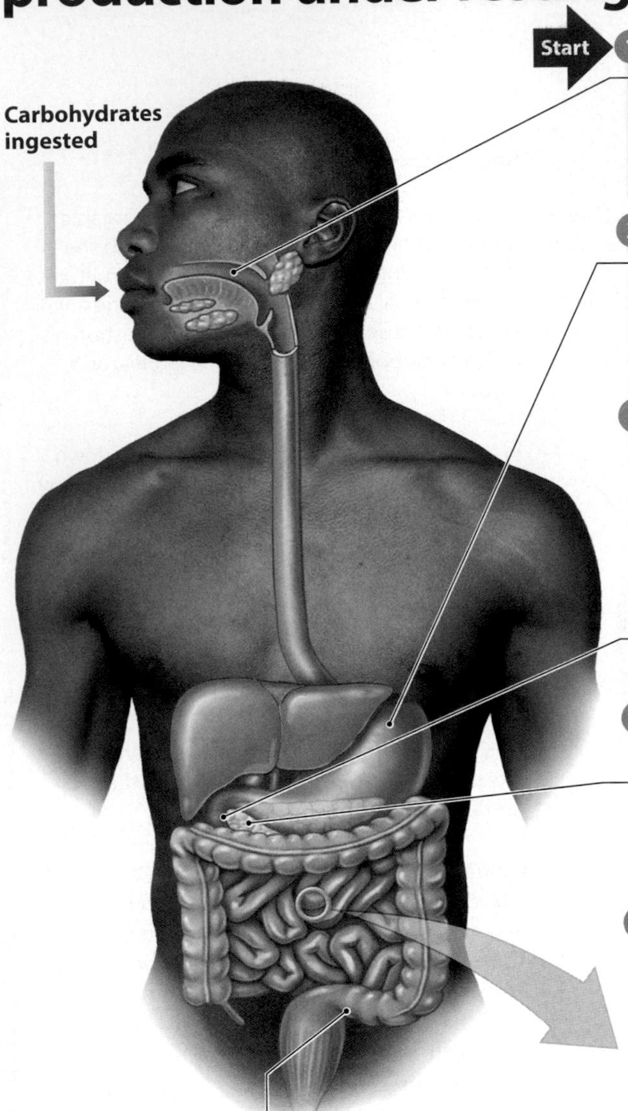
Carbohydrates ingested
Start

1 In the mouth, chewing saturates the bolus with secretions from the salivary glands. **Salivary amylase** is an enzyme that breaks down complex carbohydrates into a mixture of disaccharides and trisaccharides.

2 Salivary amylase continues to digest carbohydrates from the meal until the pH throughout the contents of the stomach falls below 4.5. This enzyme generally remains active for 1–2 hours after a meal.

3 When chyme arrives in the duodenum, secretin stimulates the release of buffers that shift the duodenal pH from acidic to alkaline (all intestinal enzymes require an alkaline pH). At the same time, cholecystokinin (CCK) release triggers the secretion of pancreatic buffers and enzymes, including **pancreatic alpha-amylase**. This enzyme has the same functions as salivary amylase, which was deactivated by denaturation in the stomach.

4 The arrival of chyme containing large amounts of carbohydrates triggers the release of gastric inhibitory peptide (GIP), which stimulates insulin release by the pancreas (Module 22.14, p. 854).

5 The epithelial cells lining the jejunum finish the digestion of carbohydrates. The plasma membrane at the brush border contains the enzymes **maltase**, **sucrase**, and **lactase**, which break the disaccharides maltose (glucose + glucose), sucrose (glucose + fructose), and lactose (glucose + galactose) into simple sugars that are then absorbed. As these enzymes function, they transport the monosaccharides across the plasma membrane and release them into the cytosol.

The simple sugars that are transported into the cell at its apical surface diffuse through the cytosol and reach the interstitial fluid by facilitated diffusion across the basolateral surfaces. These monosaccharides then diffuse into the capillaries of the intestinal villi for eventual transport to the liver in the hepatic portal vein.

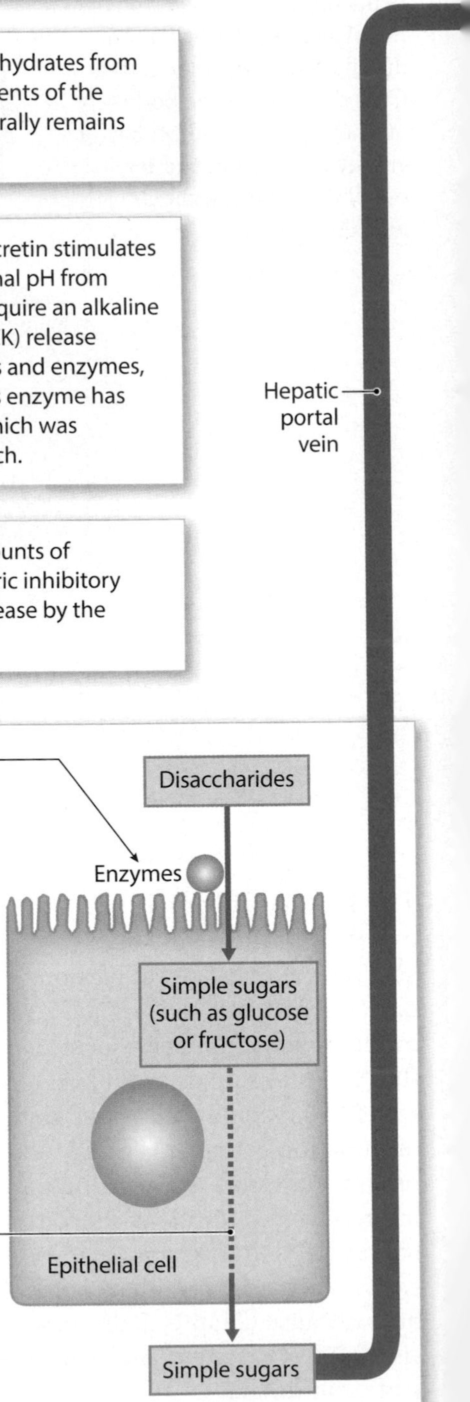
Disaccharides
Enzymes
Simple sugars (such as glucose or fructose)
Epithelial cell
Simple sugars
Hepatic portal vein

6 Intestinal enzymes do not alter indigestible carbohydrates such as cellulose, so they arrive in the colon virtually intact. These carbohydrates provide a reliable nutrient source for bacteria in the colon, whose metabolic activities generate small quantities of **flatus**, or intestinal gas. Foods containing large amounts of indigestible carbohydrates (such as beans) stimulate bacterial gas production, leading to distension of the colon, cramps, and the frequent discharge of intestinal gases.

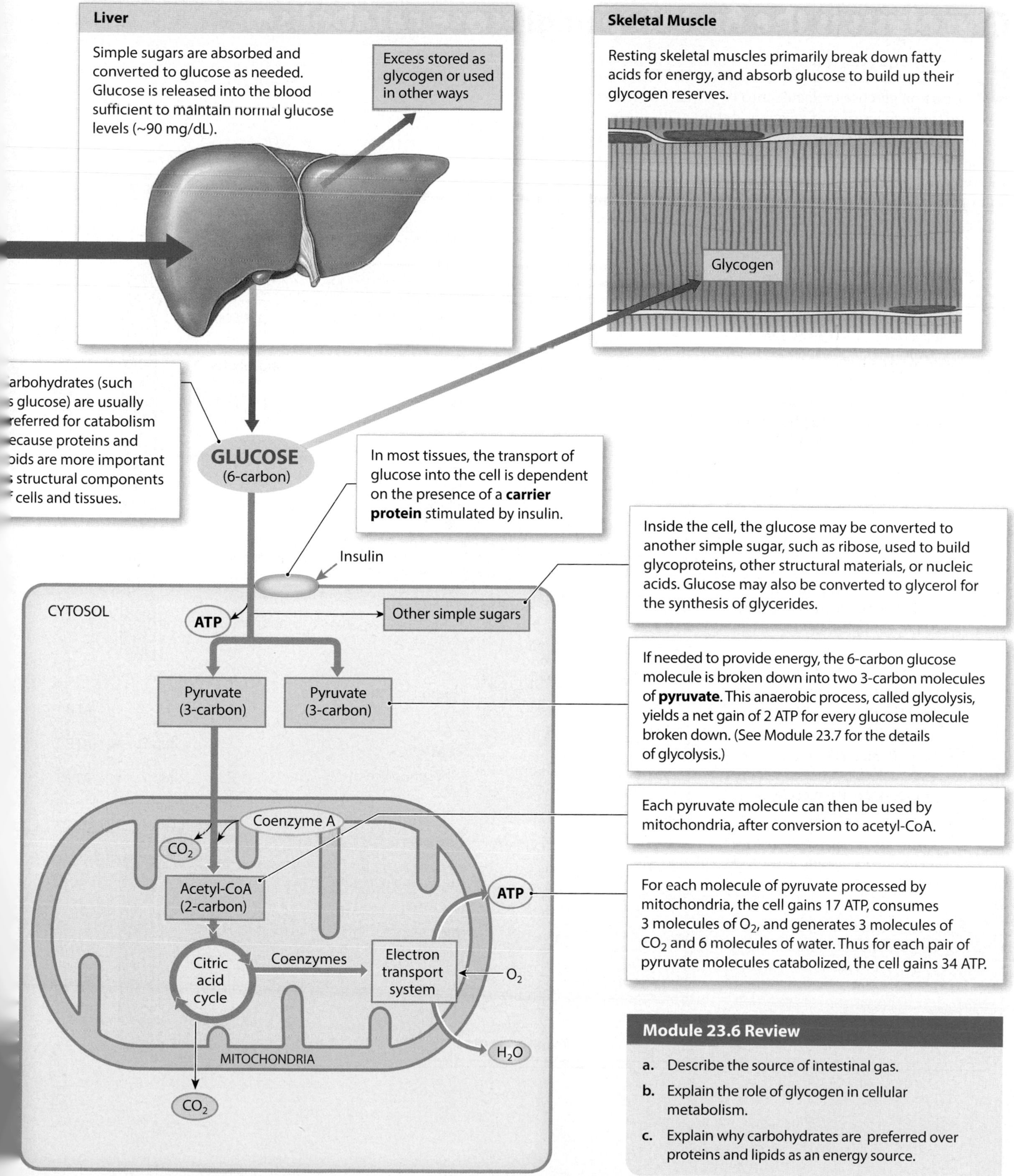
Liver
Simple sugars are absorbed and converted to glucose as needed. Glucose is released into the blood sufficient to maintain normal glucose levels (~90 mg/dL).
Excess stored as glycogen or used in other ways
Skeletal Muscle
Resting skeletal muscles primarily break down fatty acids for energy, and absorb glucose to build up their glycogen reserves.
Glycogen
arbohydrates (such s glucose) are usually referred for catabolism ecause proteins and pids are more important s structural components cells and tissues.
GLUCOSE (6-carbon)
In most tissues, the transport of glucose into the cell is dependent on the presence of a carrier protein stimulated by insulin.
Insulin
Inside the cell, the glucose may be converted to another simple sugar, such as ribose, used to build glycoproteins, other structural materials, or nucleic acids. Glucose may also be converted to glycerol for the synthesis of glycerides.
CYTOSOL
ATP
Other simple sugars
Pyruvate (3-carbon)
Pyruvate (3-carbon)
If needed to provide energy, the 6-carbon glucose molecule is broken down into two 3-carbon molecules of pyruvate. This anaerobic process, called glycolysis, yields a net gain of 2 ATP for every glucose molecule broken down. (See Module 23.7 for the details of glycolysis.)
Coenzyme A
CO_2
Each pyruvate molecule can then be used by mitochondria, after conversion to acetyl-CoA.
Acetyl-CoA (2-carbon)
ATP
For each molecule of pyruvate processed by mitochondria, the cell gains 17 ATP, consumes 3 molecules of O_2, and generates 3 molecules of CO_2 and 6 molecules of water. Thus for each pair of pyruvate molecules catabolized, the cell gains 34 ATP.
Citric acid cycle
Coenzymes
Electron transport system
O_2
MITOCHONDRIA
H_2O
CO_2

Module 23.6 Review

a. Describe the source of intestinal gas.

b. Explain the role of glycogen in cellular metabolism.

c. Explain why carbohydrates are preferred over proteins and lipids as an energy source.

LO 23.6 Describe carbohydrate metabolism.

Glycolysis is the first step in glucose catabolism

1 **Glycolysis** is an anaerobic process that breaks down a 6-carbon glucose molecule into two 3-carbon molecules of pyruvate through a series of enzymatic steps. This diagram follows the fate of the carbon chain. There is a net gain of 2 ATP molecules for each glucose molecule converted to 2 molecules of pyruvate. In addition, NAD accepts hydrogen atoms that can be transferred to mitochondria for use in the ETS. The further catabolism of pyruvate begins with its entry into a mitochondrion, as summarized in **2**.

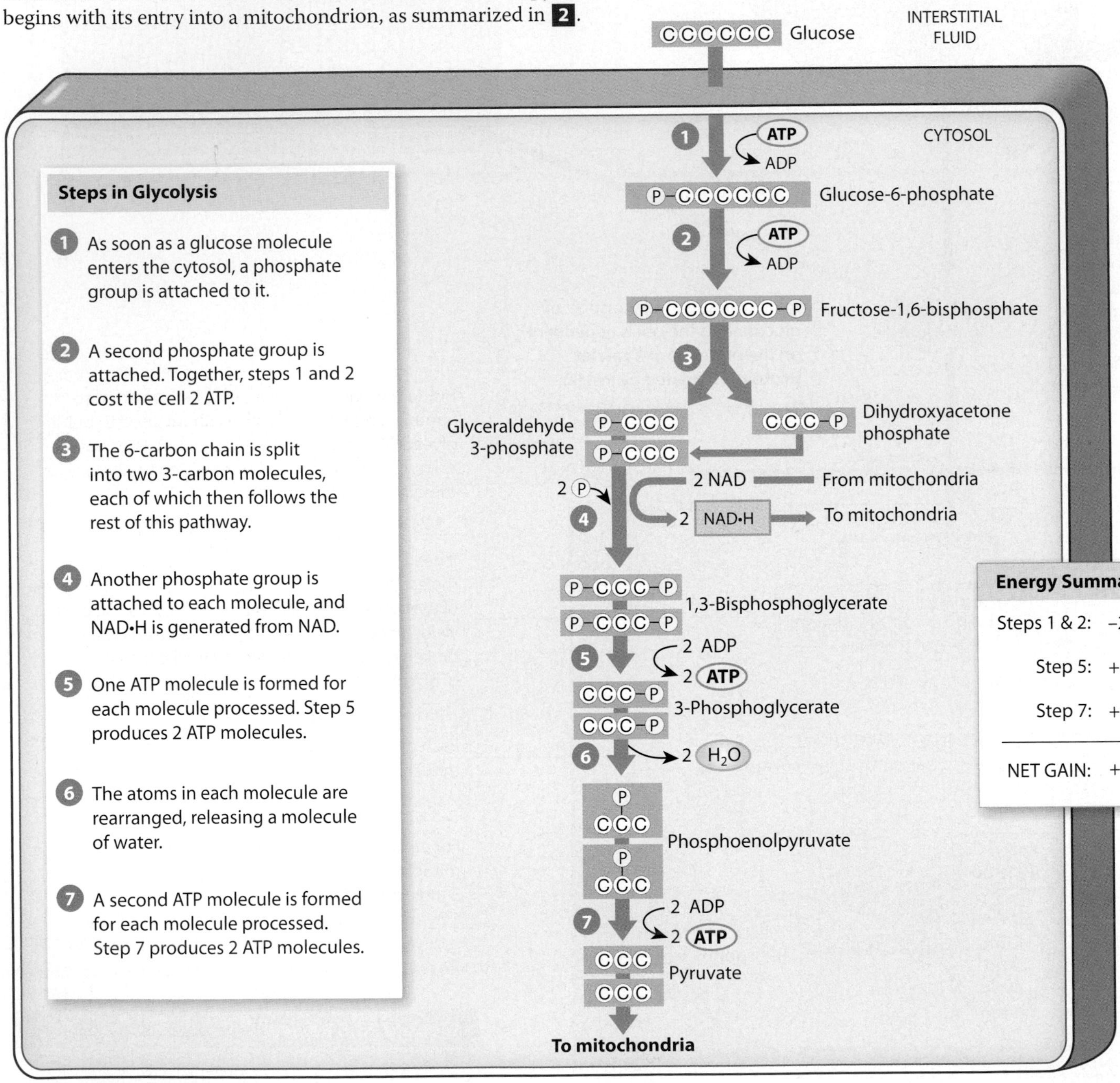

CCCCCC Glucose
INTERSTITIAL FLUID
CYTOSOL
1 ATP ADP
P–CCCCCC Glucose-6-phosphate
2 ATP ADP
P–CCCCCC–P Fructose-1,6-bisphosphate
3
Glyceraldehyde 3-phosphate
P–CCC
P–CCC
CCC–P Dihydroxyacetone phosphate
2 P
4
2 NAD From mitochondria
2 NAD•H To mitochondria
P–CCC–P
P–CCC–P
1,3-Bisphosphoglycerate
5
2 ADP
2 ATP
CCC–P
CCC–P
3-Phosphoglycerate
6
2 H2O
P CCC
P CCC
Phosphoenolpyruvate
7
2 ADP
2 ATP
CCC
CCC
Pyruvate
To mitochondria

Steps in Glycolysis

1. As soon as a glucose molecule enters the cytosol, a phosphate group is attached to it.
2. A second phosphate group is attached. Together, steps 1 and 2 cost the cell 2 ATP.
3. The 6-carbon chain is split into two 3-carbon molecules, each of which then follows the rest of this pathway.
4. Another phosphate group is attached to each molecule, and NAD•H is generated from NAD.
5. One ATP molecule is formed for each molecule processed. Step 5 produces 2 ATP molecules.
6. The atoms in each molecule are rearranged, releasing a molecule of water.
7. A second ATP molecule is formed for each molecule processed. Step 7 produces 2 ATP molecules.

Energy Summary

Steps 1 & 2:	−2 ATP
Step 5:	+2 ATP
Step 7:	+2 ATP
NET GAIN:	+2 ATP

2 This diagram summarizes the sources of ATP produced during the complete catabolism of a single glucose molecule. Two ATP are gained through glycolysis, whereas 34 are gained through aerobic metabolism in mitochondria. Note that the energy yields of NAD•H and FAD•H_2 are different; this is because they are transferred to different coenzymes in the ETS.

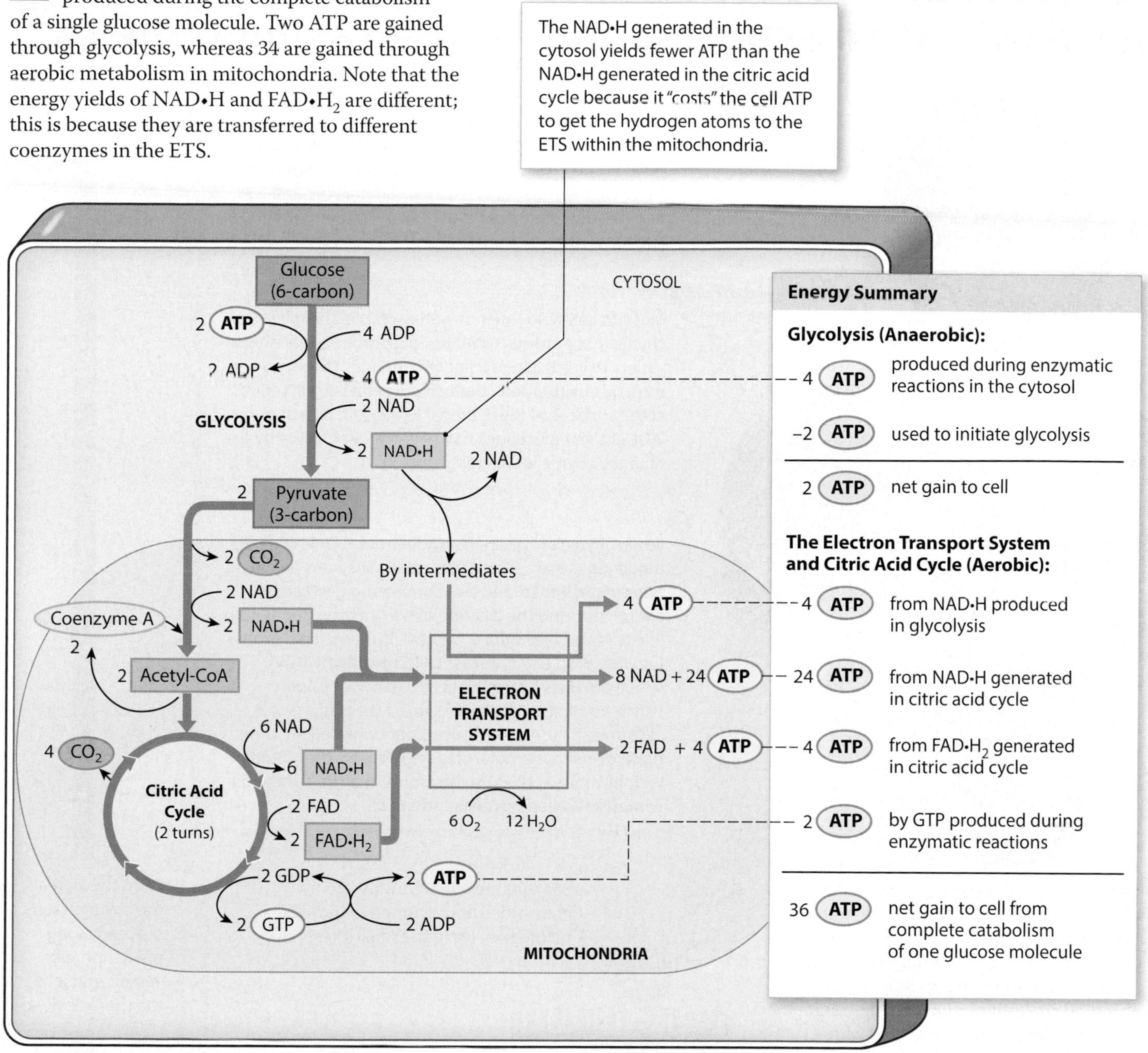

The NAD•H generated in the cytosol yields fewer ATP than the NAD•H generated in the citric acid cycle because it "costs" the cell ATP to get the hydrogen atoms to the ETS within the mitochondria.
Glucose (6-carbon)
CYTOSOL
2 ATP
4 ADP
2 ADP
4 ATP
GLYCOLYSIS
2 NAD
2 NAD•H
2 NAD
2 Pyruvate (3-carbon)
2 CO2
By intermediates
2 NAD
Coenzyme A
2 NAD•H
2
2 Acetyl-CoA
4 ATP
8 NAD + 24 ATP
ELECTRON TRANSPORT SYSTEM
6 NAD
4 CO2
6 NAD•H
Citric Acid Cycle
(2 turns)
2 FAD
2 FAD•H2
2 FAD + 4 ATP
6 O2
12 H2O
2 GDP
2 ATP
2 GTP
2 ADP
MITOCHONDRIA
Energy Summary
Glycolysis (Anaerobic):
4 ATP produced during enzymatic reactions in the cytosol
−2 ATP used to initiate glycolysis
2 ATP net gain to cell
The Electron Transport System and Citric Acid Cycle (Aerobic):
4 ATP from NAD•H produced in glycolysis
24 ATP from NAD•H generated in citric acid cycle
4 ATP from FAD•H2 generated in citric acid cycle
2 ATP by GTP produced during enzymatic reactions
36 ATP net gain to cell from complete catabolism of one glucose molecule

Reasons Why Glucose Is the Primary Energy Source for Cells

- Glucose is a small, soluble molecule that is easily distributed through body fluids.
- Glucose can provide ATP anaerobically through glycolysis. Although only a small amount of ATP is produced, glycolysis is important during peak levels of physical activity, in red blood cells, or when a tissue is temporarily deprived of oxygen.
- Glucose can be stored as glycogen, which forms compact, insoluble granules.
- Glucose can be easily mobilized because the breakdown of glycogen (glycogenolysis) occurs very quickly and involves only a single enzymatic step. Mobilization of other intracellular reserves involves much more complex pathways and takes considerably more time.

Module 23.7 Review

a. List the molecular products from a glucose molecule after glycolysis.

b. Identify when most of the CO_2 is released during the complete catabolism of glucose.

c. Explain when glycolysis is important in cellular metabolism.

Module 23.8

Lipids reach the bloodstream in chylomicrons; the cholesterol is then extracted and released as lipoproteins

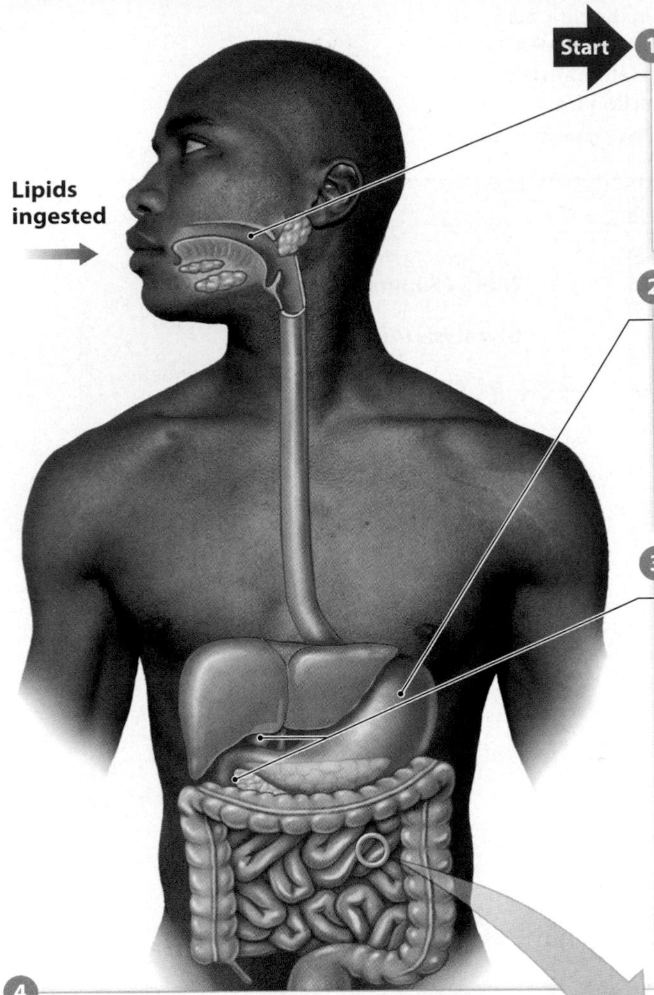

Lipids ingested
Start

1 Chewing in the mouth breaks material into smaller chunks and disrupts connective tissue organization. As this is under way, the bolus becomes saturated with saliva containing **lingual lipase**. This enzyme attacks triglycerides, breaking them down into monoglycerides and fatty acids.

2 Most dietary lipids are not water soluble. The mixing of chyme in the stomach creates large drops containing a variety of lipids. Lingual lipase continues to function in the acid environment, but can only attack triglycerides at the surfaces of these drops. As a result, only about 20% of the triglycerides have been broken down by the time the chyme leaves the stomach.

3 When chyme arrives in the duodenum, CCK is released, triggering the secretion of pancreatic enzymes, including **pancreatic lipase**, and also stimulating gallbladder contraction and the ejection of bile into the duodenum. Bile contains **bile salts** that break the large lipid drops into tiny droplets, a process called **emulsification**, which provides better access for pancreatic lipase. Pancreatic lipase then breaks apart the triglycerides to form a mixture of fatty acids, monoglycerides, and glycerol. As these molecules are released, they interact with bile salts in the lumen to form small lipid–bile salt complexes called **micelles** (mī-SELZ).

4 When a micelle contacts the intestinal epithelium, the lipids diffuse across the plasma membrane and enter the cytosol. The intestinal cells synthesize new triglycerides from the monoglycerides, fatty acids, and glycerol. These triglycerides, in company with absorbed cholesterol, phospholipids, and other lipid-soluble materials, are then coated with proteins, creating complexes known as **chylomicrons** (kī-lō-MĪ-kronz; *chylos*, milky lymph + *mikros*, small). Chylomicrons are **lipoproteins**—lipid-protein complexes that contain insoluble lipids. The superficial coating of phospholipids and proteins makes the entire complex water soluble.

Bile salts are released. Most are reabsorbed and returned to the liver before they leave the small intestine.

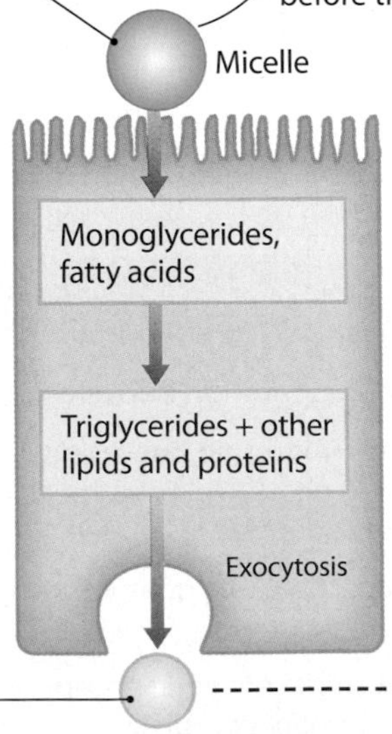

Micelle
Monoglycerides, fatty acids
Triglycerides + other lipids and proteins
Exocytosis

The intestinal cells then secrete the chylomicrons into interstitial fluid by exocytosis. The superficial protein coating of the chylomicrons keeps them suspended in the interstitial fluid, but their size generally prevents them from diffusing into capillaries.

Most of the chylomicrons released diffuse into the intestinal lacteals of the lymphatic system, which lack basement membranes and have large gaps between adjacent endothelial cells.

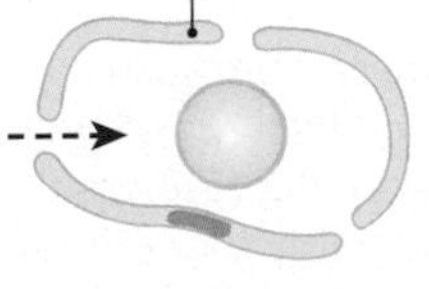

5 From the lacteals, the chylomicrons proceed along the lymphatic vessels and into the thoracic duct.

Thoracic duct

6 The chylomicrons enter the bloodstream at the left subclavian vein, then pass through the pulmonary circuit before entering the systemic circuit.

Capillary walls contain the enzyme **lipoprotein lipase**, which breaks down the chylomicrons and releases fatty acids and monoglycerides that can diffuse into the interstitial fluid.

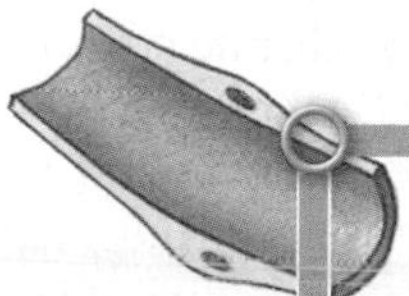

Resting skeletal muscles absorb fatty acids and break them down, using the ATP provided both to power the contractions that maintain muscle tone and to convert glucose to glycogen.

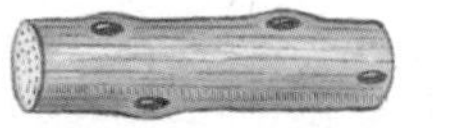

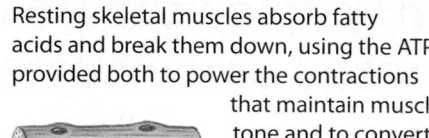

Adipocytes absorb the monoglycerides and fatty acids, and use them to synthesize triglycerides for storage.

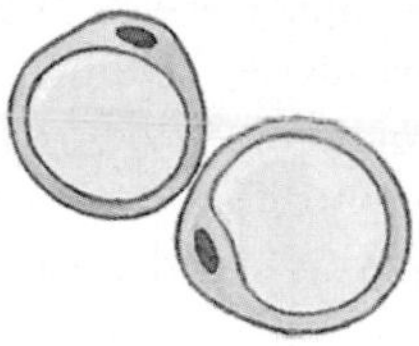

Lipoproteins and Lipid Transport and Distribution

7 The liver absorbs chylomicrons and creates **low density lipoproteins (LDLs)** and **very low density lipoproteins (VLDLs)**. To make LDLs, triclycerides are removed from the chylomicrons, cholesterol is added, and the surface proteins are altered. VLDLs contain triglycerides manufactured by the liver, plus small amounts of phospholipids and cholesterol. Some of the cholesterol is used by the liver to synthesize bile salts; excess cholesterol is excreted in the bile.

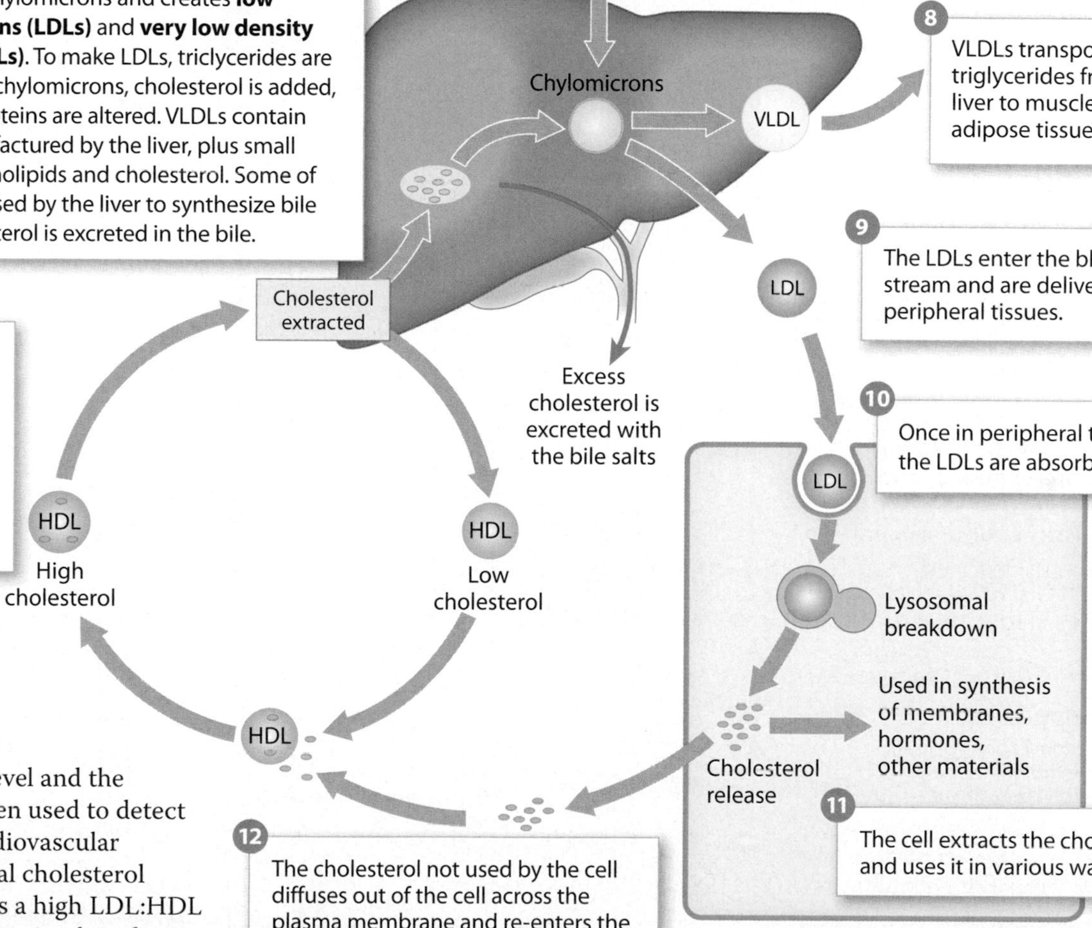

Chylomicrons
VLDL
LDL
Cholesterol extracted
Excess cholesterol is excreted with the bile salts
LDL
HDL
High cholesterol
HDL
Low cholesterol
HDL
Lysosomal breakdown
Cholesterol release
Used in synthesis of membranes, hormones, other materials

8 VLDLs transport triglycerides from the liver to muscle and adipose tissue.

9 The LDLs enter the bloodstream and are delivered to peripheral tissues.

10 Once in peripheral tissues, the LDLs are absorbed.

11 The cell extracts the cholesterol and uses it in various ways.

12 The cholesterol not used by the cell diffuses out of the cell across the plasma membrane and re-enters the bloodstream. **High density lipoproteins (HDLs)**, proteins released by the liver, absorb excess cholesterol and transport it back to the liver for storage or excretion in the bile.

13 The HDLs return the cholesterol to the liver, where it is extracted and packaged in new LDLs and VLDLs or excreted with bile salts in bile.

The total cholesterol level and the LDL:HDL ratio are often used to detect potentially serious cardiovascular problems. Elevated total cholesterol (above 200 mg/dL) plus a high LDL:HDL ratio means that (1) there is a lot of cholesterol in the circulation and (2) most of it is going out to the tissues (in LDLs and VLDLs) and staying there, rather than returning to the liver (in HDLs) for recycling or excretion. This is a potential problem because excess cholesterol in peripheral tissues tends to accumulate in atherosclerotic plaques (Module 19.8) that can cause heart attacks and strokes.

Module 23.8 Review

a. What is the difference between a micelle and a chylomicron?

b. What does the liver do with the chylomicrons it receives?

c. Describe the roles of LDL and HDL.

LO 23.8 Describe the mechanisms of lipid transport and distribution.

Fatty acids can be broken down to provide energy or converted to other lipids

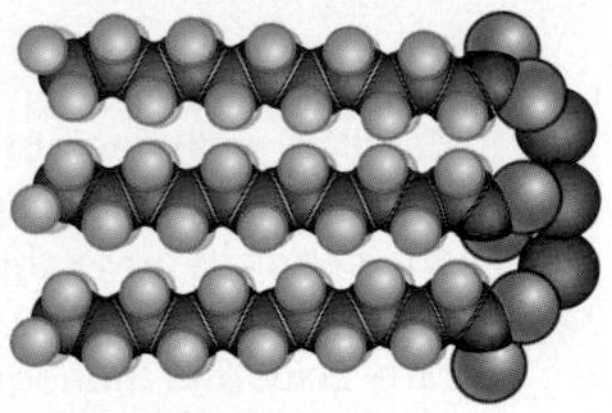

1 During lipid catabolism, or **lipolysis**, lipids are broken down into pieces that can be either converted to pyruvate or channeled directly into the citric acid cycle. We will consider triglyceride catabolism, as the principles apply to other lipids as well. A triglyceride is first split into its component parts by hydrolysis, yielding 1 molecule of glycerol and 3 fatty acid molecules. Glycerol is converted to pyruvate. The catabolism of fatty acids involves a completely different set of enzymes that generate acetyl-CoA directly. This process is called **beta-oxidation**.

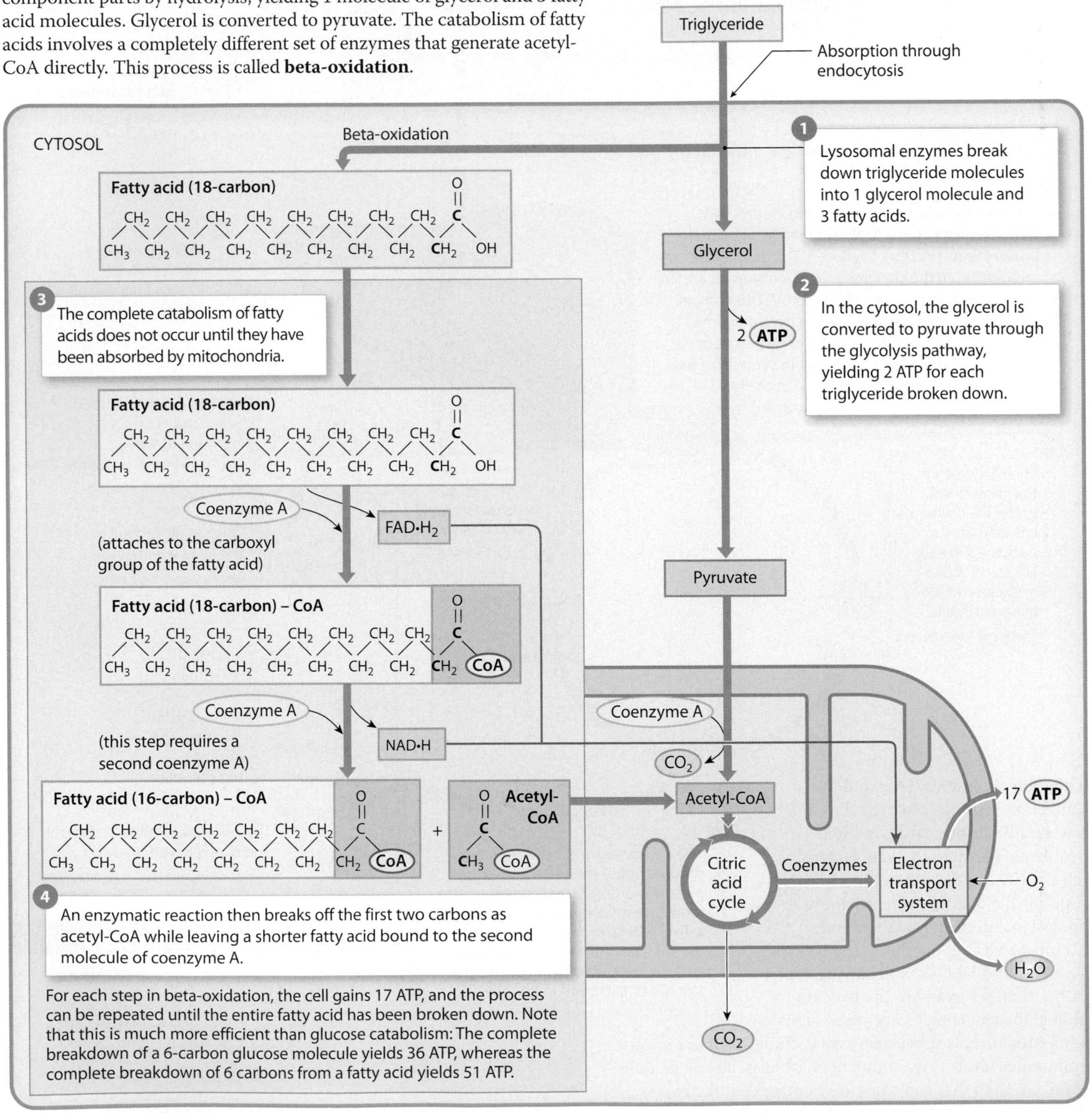

Triglyceride
Absorption through endocytosis
CYTOSOL
Beta-oxidation
1 Lysosomal enzymes break down triglyceride molecules into 1 glycerol molecule and 3 fatty acids.
Fatty acid (18-carbon)
Glycerol
2 ATP
2 In the cytosol, the glycerol is converted to pyruvate through the glycolysis pathway, yielding 2 ATP for each triglyceride broken down.
3 The complete catabolism of fatty acids does not occur until they have been absorbed by mitochondria.
Fatty acid (18-carbon)
Coenzyme A
FAD•H2
(attaches to the carboxyl group of the fatty acid)
Pyruvate
Fatty acid (18-carbon) – CoA
Coenzyme A
(this step requires a second coenzyme A)
NAD•H
Coenzyme A
CO2
Fatty acid (16-carbon) – CoA
Acetyl-CoA
Acetyl-CoA
17 ATP
Citric acid cycle
Coenzymes
Electron transport system
O2
H2O
4 An enzymatic reaction then breaks off the first two carbons as acetyl-CoA while leaving a shorter fatty acid bound to the second molecule of coenzyme A.
For each step in beta-oxidation, the cell gains 17 ATP, and the process can be repeated until the entire fatty acid has been broken down. Note that this is much more efficient than glucose catabolism: The complete breakdown of a 6-carbon glucose molecule yields 36 ATP, whereas the complete breakdown of 6 carbons from a fatty acid yields 51 ATP.
CO2

4 The glycerol required for triglyceride production is synthesized from one of the intermediate products of glycolysis.

3 All of the other structural and functional lipids can be synthesized from fatty acids.

2 Fatty acid synthesis involves a reaction sequence quite distinct from that of beta-oxidation. As a result, body cells cannot build every fatty acid they can break down. For example, our cells lack the enzymes to insert double bonds in the proper locations to synthesize two 18-carbon fatty acids synthesized by plants: **linolenic acid** (an **omega-3** fatty acid) or **linoleic acid** (an **omega-6** fatty acid). However, these fatty acids are needed to synthesize prostaglandins and some of the phospholipids found in plasma membranes throughout the body. They are therefore called **essential fatty acids**, because they must be included in your diet.

Start **1** The synthesis of most types of lipids, including nonessential fatty acids and steroids, begins with acetyl-CoA. Lipogenesis can use almost any organic substrate, because lipids, amino acids, and carbohydrates can be converted to acetyl-CoA.

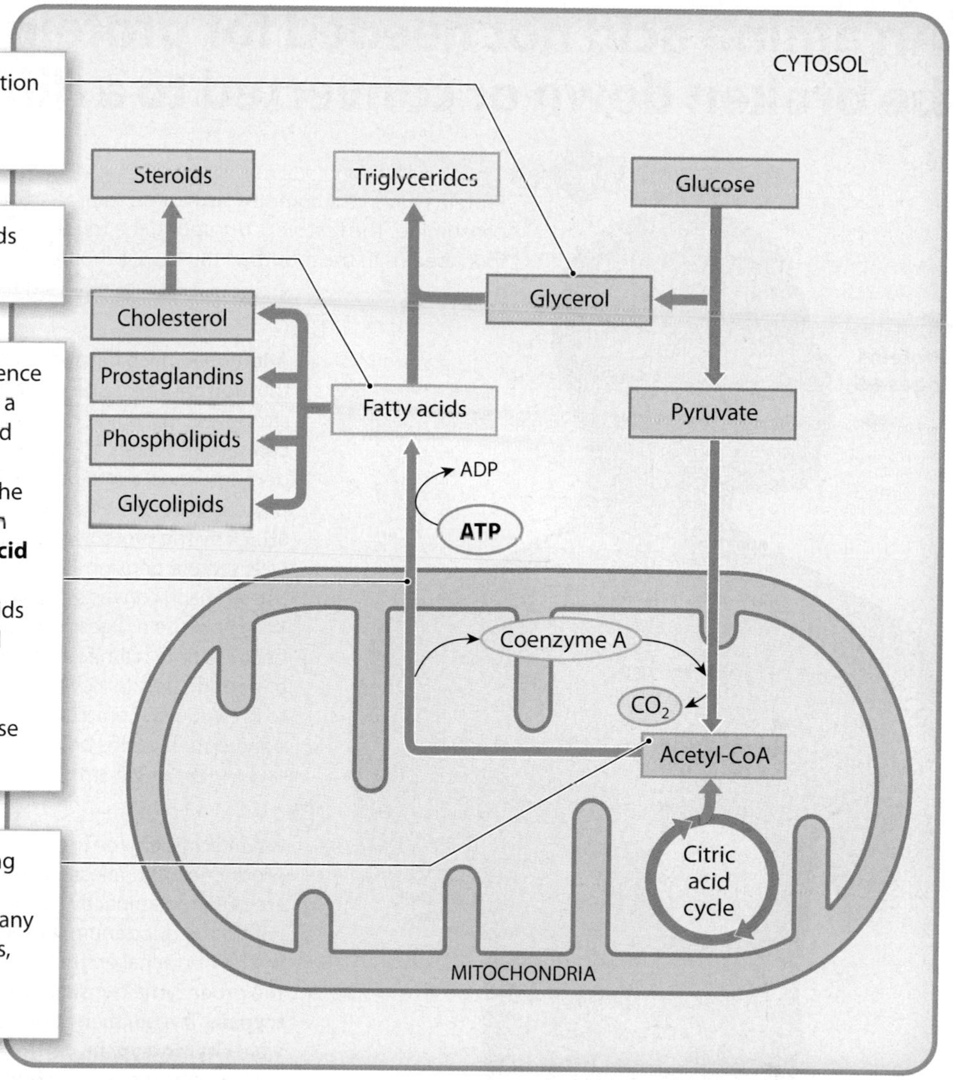
CYTOSOL
Steroids
Triglycerides
Glucose
Glycerol
Cholesterol
Prostaglandins
Fatty acids
Pyruvate
Phospholipids
Glycolipids
ADP
ATP
Coenzyme A
CO_2
Acetyl-CoA
Citric acid cycle
MITOCHONDRIA

2 The synthesis of lipids is known as **lipogenesis** (lip-ō-JEN-e-sis). Because most catabolic pathways are reversible, the cell can also synthesize most—but not all—lipids on demand. This diagram shows the major pathways for lipogenesis.

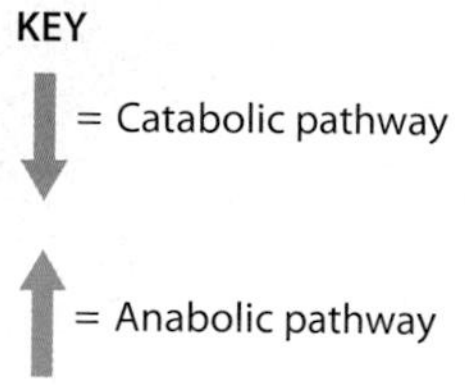
KEY
= Catabolic pathway
= Anabolic pathway

Lipids are useful as an energy reserve primarily because beta-oxidation is so efficient, and excess lipids can be stored as triglycerides. The problem is that triglyceride reserves form insoluble droplets that are difficult for water-soluble enzymes to access. As a result, stored lipids cannot provide large amounts of ATP very quickly—that's why carbohydrate catabolism is so important. Lipids, however, can be stored in much larger quantities, and they are well suited for meeting chronic energy demands during periods of stress or starvation.

Module 23.9 Review

a. Define beta-oxidation.

b. What molecule plays a key reactant role in both ATP production from fatty acids and lipogenesis?

c. Identify the fates of fatty acids.

LO 23.9 Describe the fate of fatty acids in lipid metabolism.

An amino acid not needed for protein synthesis may be broken down or converted to a different amino acid

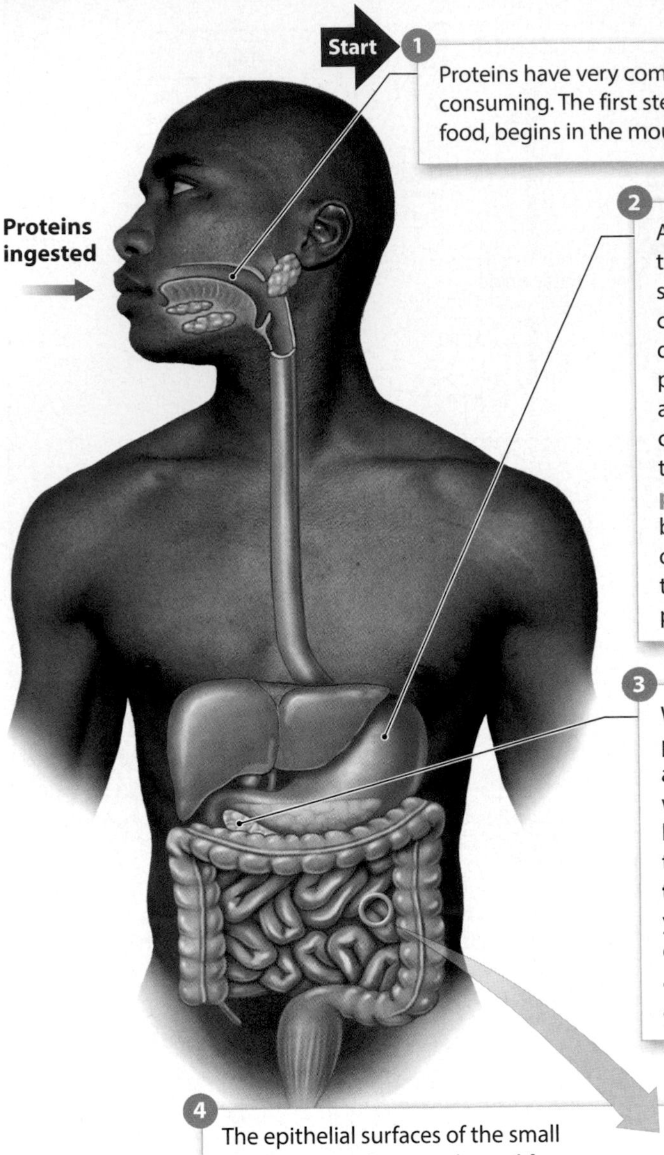
Start
Proteins ingested

1 Proteins have very complex structures, so protein digestion is both complex and time-consuming. The first step, disrupting the tough three-dimensional organization of the food, begins in the mouth as the food is thoroughly chewed and mixed with saliva.

2 Additional mechanical processing occurs in the stomach through churning and mixing. Exposure of the bolus to a strongly acidic environment kills pathogens and breaks down connective tissues and plant cell walls. Stomach acids also denature most proteins and disrupt tertiary and secondary protein structure, exposing peptide bonds to enzymatic attack by the proteolytic enzyme **pepsin**. Recall that chief cells secrete pepsinogen, an inactive proenzyme, but HCl in the stomach converts pepsinogen to pepsin (Module 22.11, p. 849). Protein digestion is not completed in the stomach, because time is limited and pepsin attacks only specific types of peptide bonds. However, pepsin generally has enough time to break down complex proteins into smaller peptide and polypeptide chains before the chyme enters the duodenum.

3 When acidic chyme arrives in the duodenum, CCK stimulates production and release of pancreatic enzymes. These enzymes are secreted as inactive proenzymes that are then activated within the duodenum. **Enteropeptidase**, an enzyme released by the duodenal epithelium, begins the process, converting the proenzyme trypsinogen to the proteolytic enzyme **trypsin**. Trypsin then converts the other proenzymes to yield **chymotrypsin, carboxypeptidase**, and **elastase**. Each enzyme attacks peptide bonds linking specific amino acids and ignores others. Together, they break down proteins into a mixture of dipeptides, tripeptides, and amino acids.

4 The epithelial surfaces of the small intestine contain several **peptidases**—enzymes that break peptide bonds—notably **dipeptidases** that break dipeptides apart and release individual amino acids. These amino acids, as well as those released by the action of pancreatic enzymes, are absorbed through both facilitated diffusion and cotransport mechanisms.

After diffusing to the basal surface of the cell, the amino acids are released into interstitial fluid by facilitated diffusion and cotransport. Once in the interstitial fluid, the amino acids enter intestinal capillaries for transport to the liver in the hepatic portal vein.

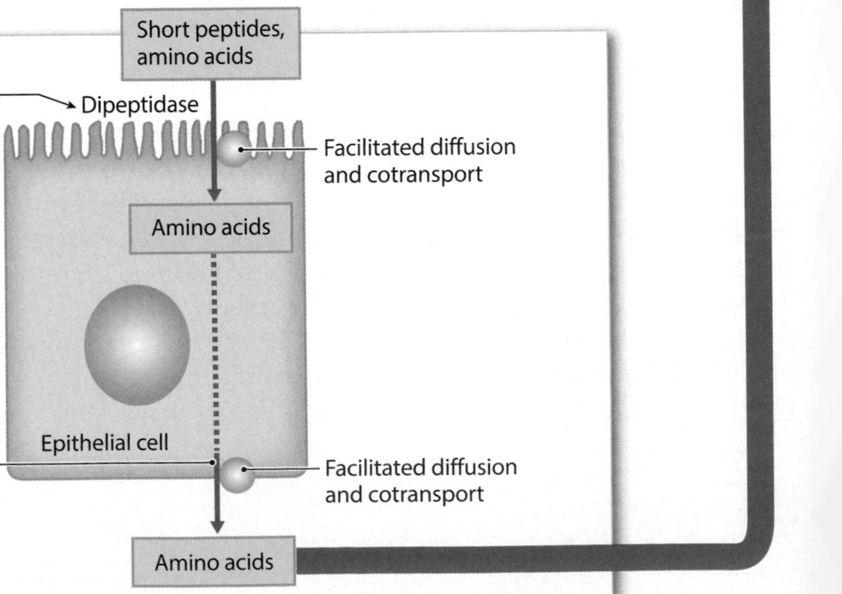
Short peptides, amino acids
Dipeptidase
Facilitated diffusion and cotransport
Amino acids
Epithelial cell
Facilitated diffusion and cotransport
Amino acids
Hepatic portal vein

5

The liver does not control circulating levels of amino acids as precisely as it does glucose concentrations. Blood amino acid levels normally range between 35 and 65 mg/dL, but they may become elevated after a protein-rich meal. The liver itself uses many amino acids for synthesizing plasma proteins, and it has all of the enzymes needed to synthesize, convert, or catabolize amino acids. In addition, amino acids that can be broken down to 3-carbon molecules can be used for gluconeogenesis when other sources of glucose are unavailable.

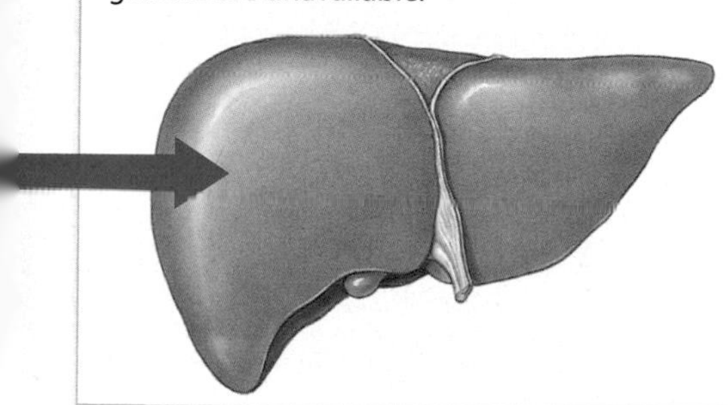

Amino Acid Synthesis

Liver cells and other body cells can readily synthesize the carbon frameworks of about half of the amino acids needed to synthesize proteins. There are 10 **essential amino acids** that the body either cannot synthesize or that cannot be produced in amounts sufficient for growing children.

In **amination**, an ammonium ion (NH_4^+) is used to form an amino group that is attached to a molecule, yielding an amino acid.

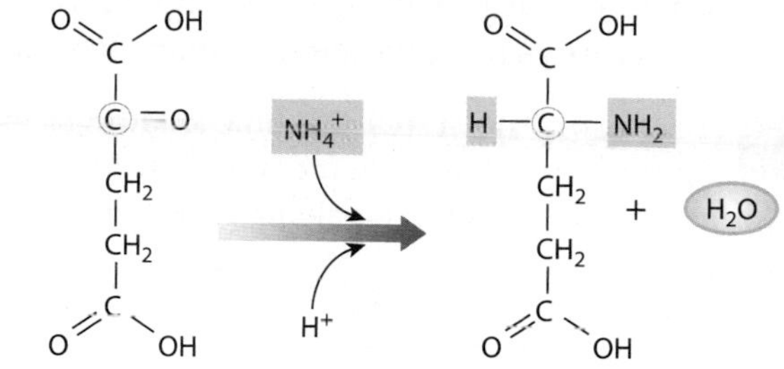
NH_4^+
H^+
H_2O

α–Ketoglutarate   Glutamic acid

In **transamination**, the amino group of one amino acid gets transferred to another molecule, yielding a different amino acid. The remaining carbon chain can then be broken down or used in other ways.

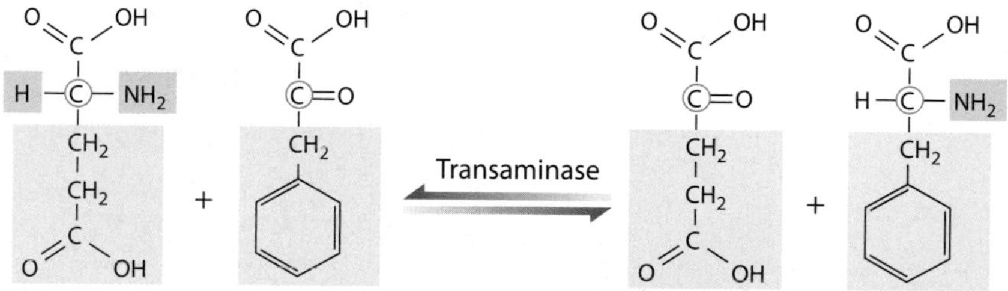
Transaminase

Glutamic acid   Organic acid 1   Organic acid 2   Tyrosine

Note: Alanine aminotransferase (ALT) and aspartate aminotransferase (AST) are transaminases that leak from injured hepatocytes into the circulation. High levels in a blood test indicate liver disease or inflammation.

Amino Acid Catabolism

The first step in amino acid catabolism is the removal of the amino group, leaving a carbon chain that can be converted to pyruvate, acetyl-CoA, or an intermediary in the citric acid cycle.

In **deamination**, the amino group is removed and an ammonium ion is released.

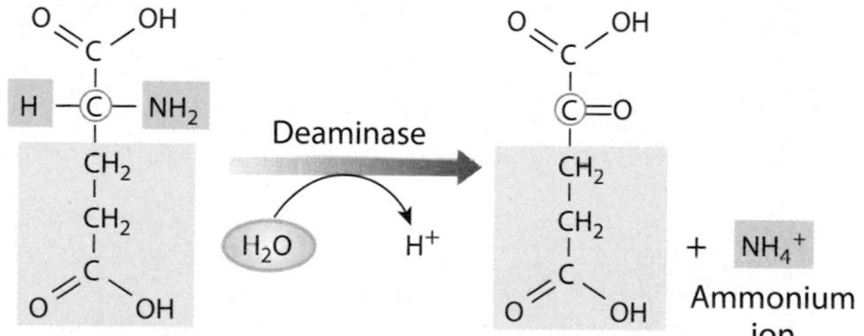
Deaminase
H_2O
H^+
NH_4^+
Ammonium ion

Glutamic acid   Organic acid

Ammonium ions are highly toxic, even in low concentrations. Liver cells, the primary sites of deamination, have enzymes that use ammonium ions to synthesize **urea**, a relatively harmless water-soluble compound that is excreted in urine. The **urea cycle** is the reaction sequence responsible for the production of urea.

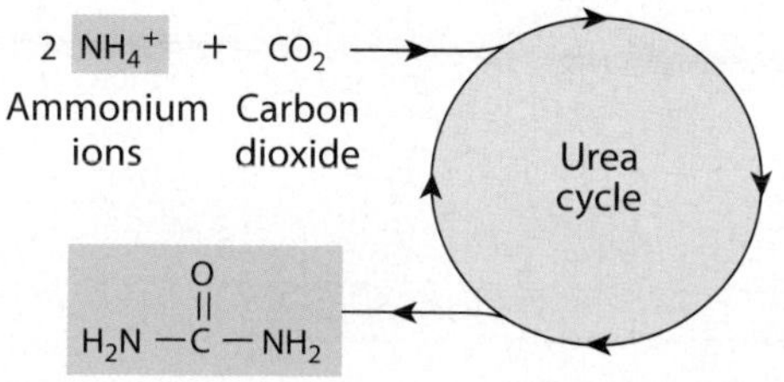
2 NH_4^+ + CO_2
Ammonium ions
Carbon dioxide
Urea cycle

Urea

When broken down in the mitochondria, the energy yield of amino acids is comparable with that of carbohydrates. However, a cell surviving by amino acid catabolism is like someone in a winter cabin burning the walls to stay warm. It can help temporarily, but it isn't a permanent solution, and in the long run it only makes matters worse.

Module 23.10 Review

a. Describe the role of CCK release and its effects on proteins.

b. Identify the processes by which the amino group is removed.

c. What happens to the ammonium ions that are removed from amino acids during deamination?

LO **23.10** Summarize the main features of protein metabolism and the use of proteins as an energy source.

Module 23.11

There are two general patterns of metabolic activity: the absorptive and postabsorptive states

Metabolic activity changes from moment to moment, but there are broad patterns that follow a daily cycle. These cycles are maintained by many different hormones. The degree of response differs from tissue to tissue, but we will consider a representative cell that shares characteristics with all body cells.

1 The **absorptive state** is the time following a meal, when nutrient absorption is under way. After a typical meal, the absorptive state continues for about 4 hours. If you are fortunate enough to eat three meals a day, you spend 12 out of every 24 hours in the absorptive state. Insulin is the primary hormone of the absorptive state, although various other hormones stimulate amino acid uptake (growth hormone) and protein synthesis (growth hormone, androgens, and estrogens).

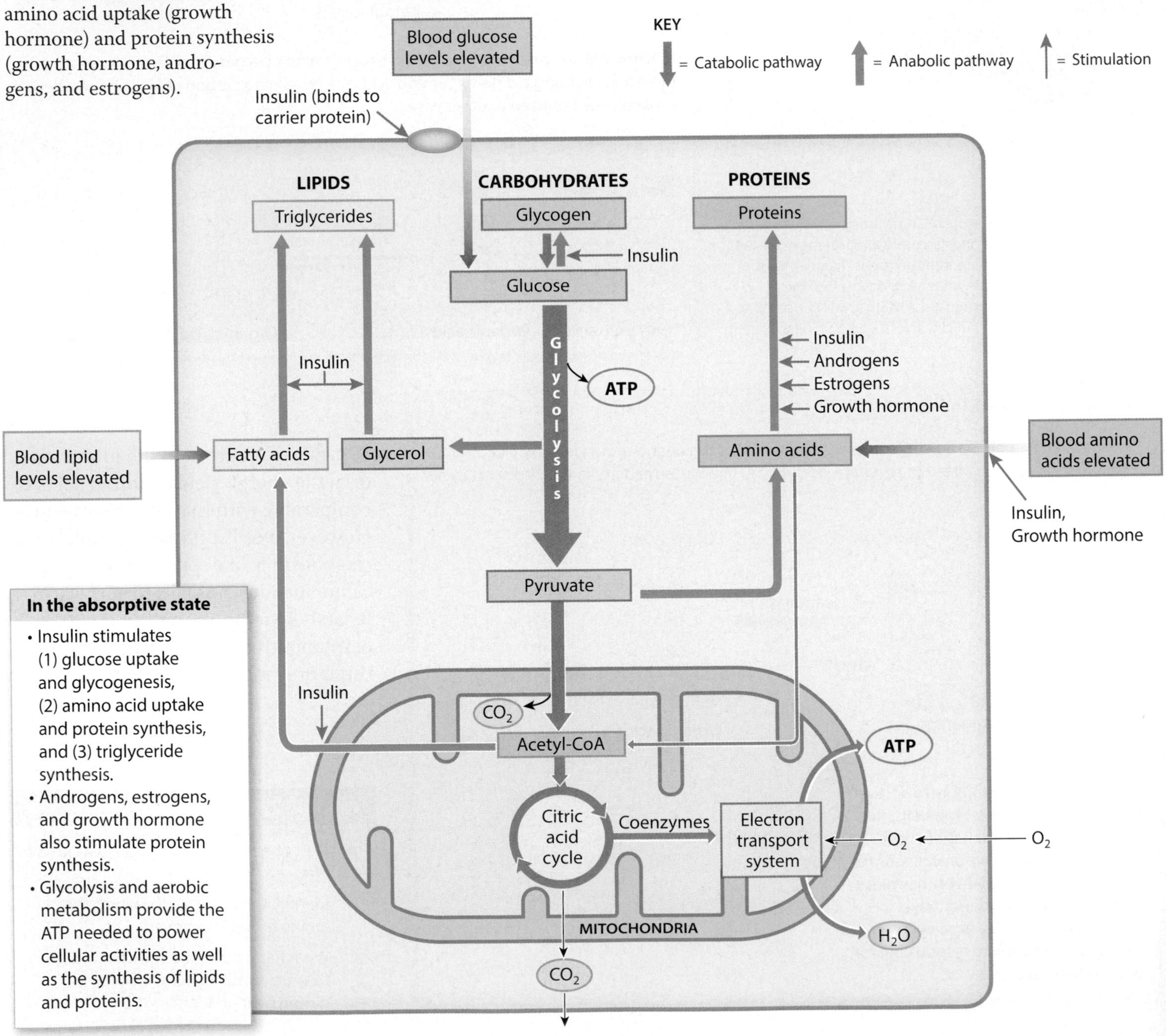

Blood glucose levels elevated
KEY
= Catabolic pathway
= Anabolic pathway
= Stimulation
Insulin (binds to carrier protein)
LIPIDS
CARBOHYDRATES
PROTEINS
Triglycerides
Glycogen
Proteins
Insulin
Glucose
Insulin
Insulin
Androgens
Estrogens
Growth hormone
Glycolysis
ATP
Blood lipid levels elevated
Fatty acids
Glycerol
Amino acids
Blood amino acids elevated
Insulin, Growth hormone
Pyruvate
In the absorptive state
• Insulin stimulates (1) glucose uptake and glycogenesis, (2) amino acid uptake and protein synthesis, and (3) triglyceride synthesis.
• Androgens, estrogens, and growth hormone also stimulate protein synthesis.
• Glycolysis and aerobic metabolism provide the ATP needed to power cellular activities as well as the synthesis of lipids and proteins.
Insulin
CO_2
Acetyl-CoA
ATP
Citric acid cycle
Coenzymes
Electron transport system
O_2
O_2
MITOCHONDRIA
H_2O
CO_2

2 The **postabsorptive state** is the time when nutrient absorption is not under way and your body must rely on internal energy reserves to continue meeting its energy demands. You spend about 12 hours each day in the postabsorptive state, although a person who is skipping meals can extend that time considerably. Metabolic activity in the postabsorptive state is focused on the mobilization of energy reserves and the maintenance of normal blood glucose levels. These activities are coordinated by several hormones, including glucagon, epinephrine, glucocorticoids, and growth hormone.

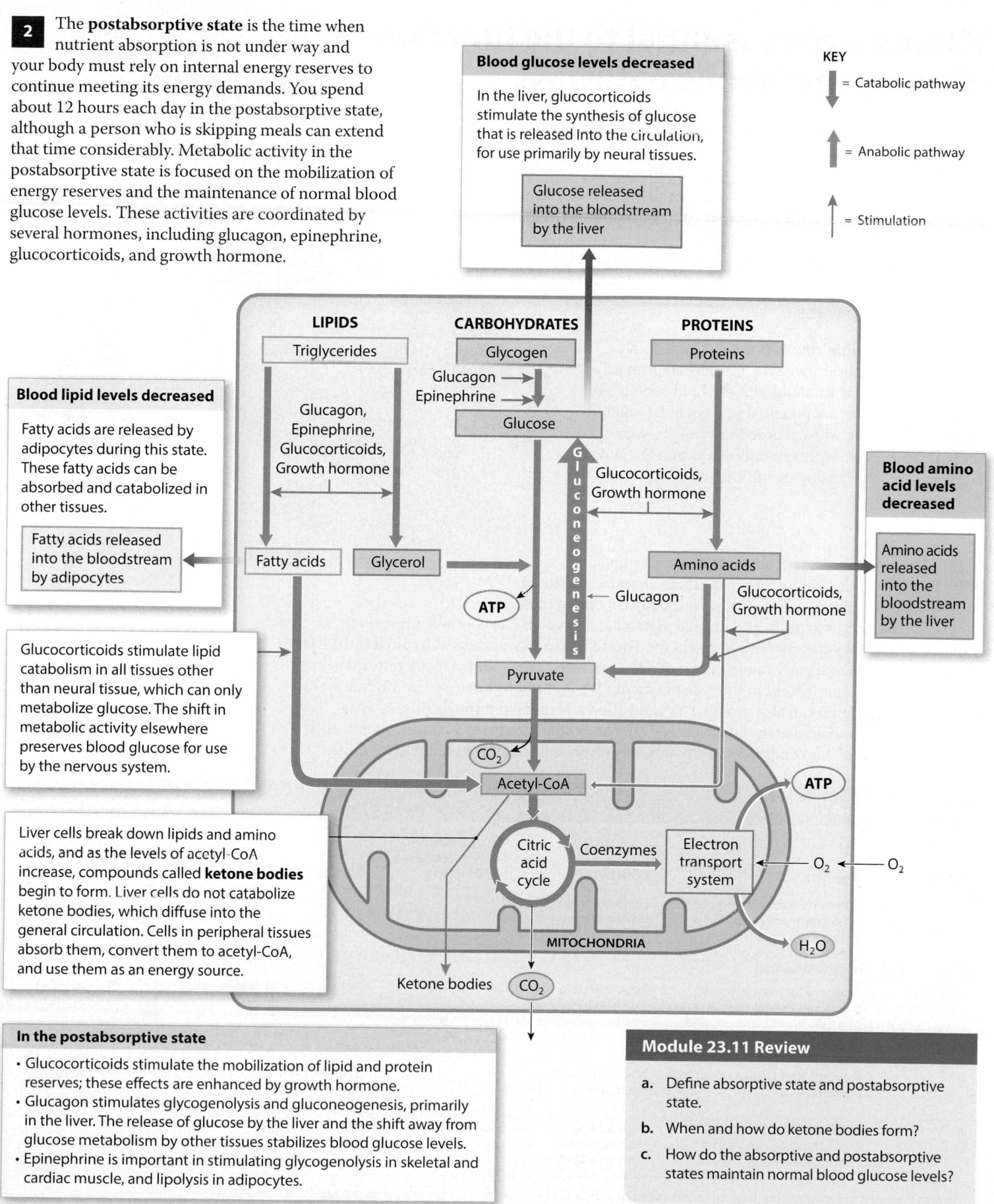

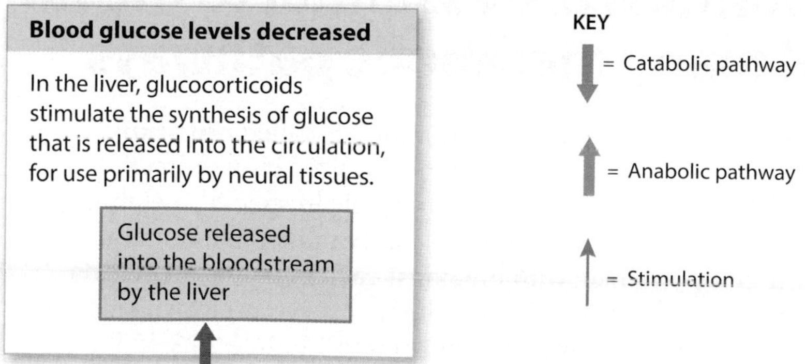

Blood glucose levels decreased
In the liver, glucocorticoids stimulate the synthesis of glucose that is released into the circulation, for use primarily by neural tissues.
Glucose released into the bloodstream by the liver
KEY
= Catabolic pathway
= Anabolic pathway
= Stimulation

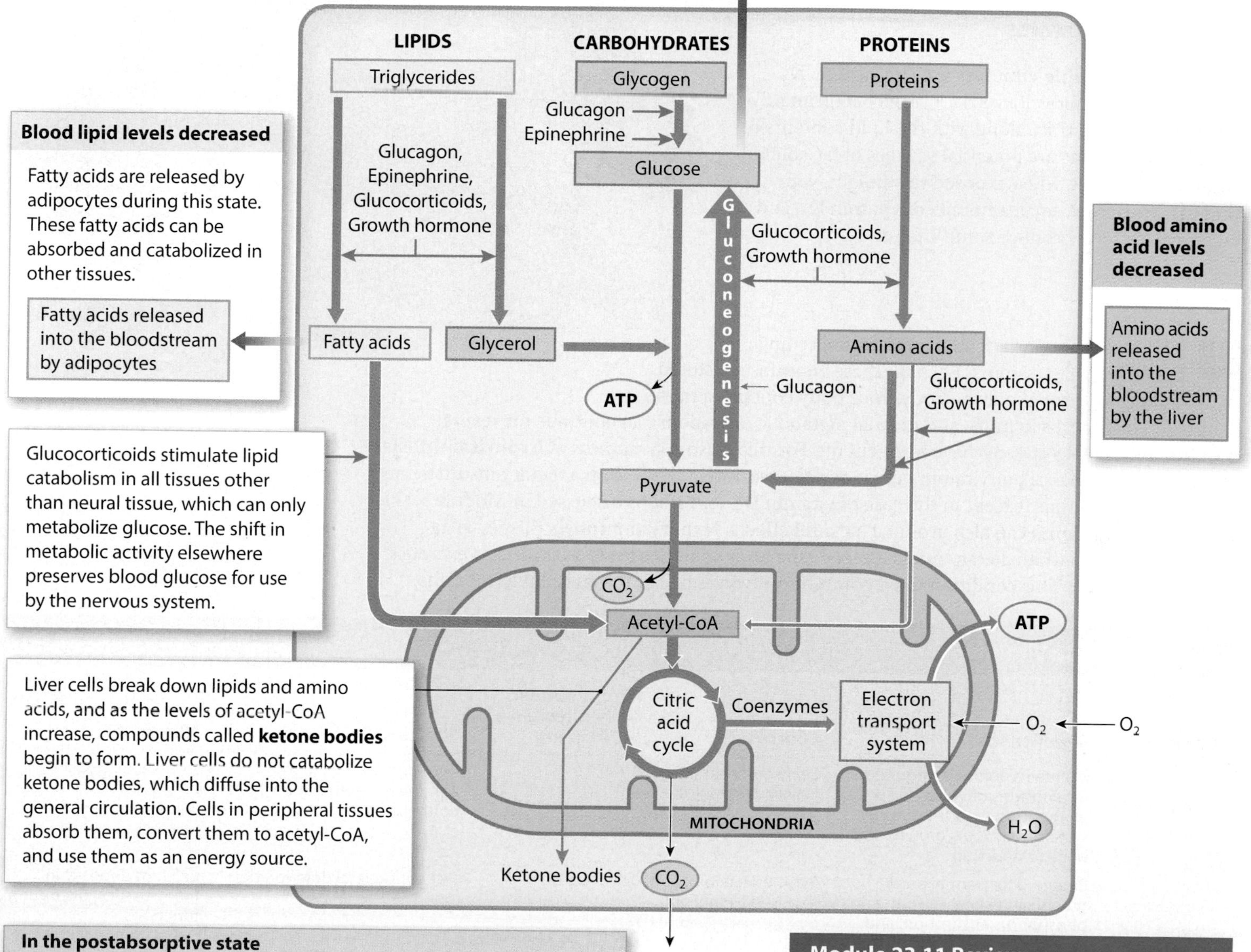

LIPIDS
Triglycerides
Glucagon, Epinephrine, Glucocorticoids, Growth hormone
Fatty acids
Glycerol
CARBOHYDRATES
Glycogen
Glucagon
Epinephrine
Glucose
Gluconeogenesis
PROTEINS
Proteins
Glucocorticoids, Growth hormone
Amino acids
Glucagon
Glucocorticoids, Growth hormone
ATP
Pyruvate
Blood lipid levels decreased
Fatty acids are released by adipocytes during this state. These fatty acids can be absorbed and catabolized in other tissues.
Fatty acids released into the bloodstream by adipocytes
Blood amino acid levels decreased
Amino acids released into the bloodstream by the liver
Glucocorticoids stimulate lipid catabolism in all tissues other than neural tissue, which can only metabolize glucose. The shift in metabolic activity elsewhere preserves blood glucose for use by the nervous system.
CO_2
Acetyl-CoA
ATP
Liver cells break down lipids and amino acids, and as the levels of acetyl-CoA increase, compounds called ketone bodies begin to form. Liver cells do not catabolize ketone bodies, which diffuse into the general circulation. Cells in peripheral tissues absorb them, convert them to acetyl-CoA, and use them as an energy source.
Citric acid cycle
Coenzymes
Electron transport system
O_2
O_2
MITOCHONDRIA
H_2O
Ketone bodies
CO_2

In the postabsorptive state

- Glucocorticoids stimulate the mobilization of lipid and protein reserves; these effects are enhanced by growth hormone.
- Glucagon stimulates glycogenolysis and gluconeogenesis, primarily in the liver. The release of glucose by the liver and the shift away from glucose metabolism by other tissues stabilizes blood glucose levels.
- Epinephrine is important in stimulating glycogenolysis in skeletal and cardiac muscle, and lipolysis in adipocytes.

Module 23.11 Review

a. Define absorptive state and postabsorptive state.

b. When and how do ketone bodies form?

c. How do the absorptive and postabsorptive states maintain normal blood glucose levels?

LO 23.11 Differentiate between the absorptive and postabsorptive metabolic states and summarize the characteristics of each.

Module 23.12

Vitamins are essential to the function of many metabolic pathways

The absorption of nutrients from food is called **nutrition**. **Vitamins** are organic compounds required in very small quantities but that play an essential role in specific metabolic pathways. However, many people only think of vitamins as something they take with breakfast each morning. **Minerals** are important inorganic nutrients that serve as electrolytes in body fluids. We discuss their roles in Module 25.3.

Fat-Soluble Vitamins

1 The **fat-soluble vitamins** are vitamins A, D_3, E, and K. These vitamins are absorbed primarily from the digestive tract along with the lipid contents of micelles. Vegetables are potential sources of fat-soluble vitamins. However, when exposed to sunlight, your skin can synthesize small amounts of vitamin D_3, and intestinal bacteria produce some vitamin K.

2 This table summarizes information concerning the fat-soluble vitamins. Because these vitamins are stored in lipid deposits throughout the body, your body contains a significant reserve of these vitamins, and normal metabolic operations can continue for several months after dietary sources have been cut off. For this reason, symptoms of **hypovitaminosis** (hī-pō-vī-ta-min-Ō-sis), or vitamin deficiency disease, rarely result from a dietary insufficiency of fat-soluble vitamins (except in the case of vitamin D_3, for reasons discussed in Module 5.11). Too much of a vitamin can also produce harmful effects. **Hypervitaminosis** (hī-per-vī-ta-min-Ō-sis) occurs when dietary intake exceeds the body's abilities to store, utilize, or excrete a particular vitamin. This condition most commonly involves one of the fat-soluble vitamins.

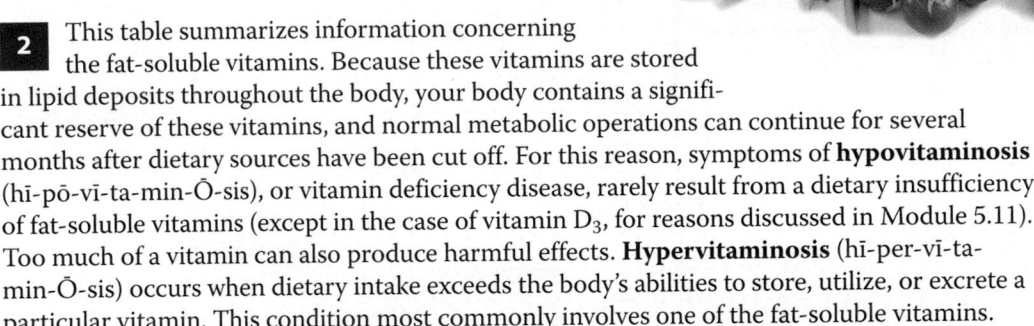

The Fat-Soluble Vitamins

Vitamin	Significance	Sources	Recommended Daily Allowance (RDA) in mg	Effects of Deficiency	Effects of Excess
A	Maintains epithelia; required for synthesis of visual pigments; supports immune system; promotes growth and bone remodeling	Leafy green and yellow vegetables	0.7–0.9	Retarded growth, night blindness, deterioration of epithelial membranes	Liver damage, skin paling, CNS effects (nausea, anorexia)
D_3	Required for normal bone growth, intestinal calcium and phosphorus absorption, and retention of these ions by the kidneys	Synthesized in skin exposed to sunlight	0.005–0.015*	Rickets, skeletal deterioration	Calcium deposits in many tissues, disrupting functions
E	Prevents breakdown of vitamin A and fatty acids	Meat, milk, vegetables	15	Anemia, other problems suspected	Nausea, stomach cramps, blurred vision, fatigue
K	Essential for liver synthesis of prothrombin and other clotting factors	Vegetables; production by intestinal bacteria	0.09–0.12	Bleeding disorders	Liver dysfunction, jaundice

*Unless exposure to sunlight is inadequate for extended periods and alternative sources (fortified milk products) are unavailable.

Water-Soluble Vitamins

3 The **water-soluble vitamins** are the B vitamins and vitamin C. Most of them are components of coenzymes. The B vitamins are found in meat, eggs, and dairy products, while vitamin C is found in citrus fruits.

4 This table summarizes information about water-soluble vitamins. These water-soluble vitamins are rapidly exchanged between the fluid compartments of the digestive tract and the circulating blood, and excessive amounts are readily excreted in urine. For this reason, hypervitaminosis involving water-soluble vitamins is unlikely unless you are taking megadoses of vitamin supplements.

The Water-Soluble Vitamins

Vitamin	Component or Precursor of	Sources	Recommended Daily Allowance (RDA) in mg	Effects of Deficiency	Effects of Excess
B_1 (thiamine)	Coenzyme in many pathways	Milk, meat, bread	1.1–1.2	Muscle weakness, CNS and cardiovascular problems, including heart disease; called *beriberi*	Hypotension
B_2 (riboflavin)	Part of FAD, involved in multiple pathways, including glycolysis and citric acid cycle	Milk, meat, eggs and cheese	1.1–1.3	Epithelial and mucosal deterioration	Itching, tingling
B_3 (niacin)	Part of NAD, involved in multiple pathways	Meat, bread, potatoes	14–16	CNS, GI, epithelial, and mucosal deterioration; called *pellagra*	Itching, burning; vasodilation; death after large dose
B_5 (pantothenic acid)	Coenzyme A, in multiple pathways	Milk, meat	10	Retarded growth, CNS disturbances	None reported
B_6 (pyridoxine)	Coenzyme in amino acid and lipid metabolism	Meat, whole grains, vegetables, orange juice, cheese and milk	1.3–1.7	Retarded growth, anemia, convulsions, epithelial changes	CNS alterations, perhaps fatal
B_9 (folic acid)	Coenzyme in amino acid and nucleic acid metabolism	Leafy vegetables, some fruits, liver, cereal and bread	0.2–0.4	Retarded growth, anemia, gastrointestinal disorders, developmental abnormalities	Few noted, except at massive doses
B_{12} (cobalamin)	Coenzyme in nucleic acid metabolism	Milk, meat	0.0024	Impaired RBC production, causing pernicious anemia	Polycythemia
B_7 (biotin)	Coenzyme in many pathways	Eggs, meat, vegetables	0.03	Fatigue, muscular pain, nausea, dermatitis	None reported
C (ascorbic acid)	Coenzyme in many pathways	Citrus fruits	75–90; smokers add 35 mg	Epithelial and mucosal deterioration; called *scurvy*	Kidney stones

Only vitamins B_{12} and C are stored in significant quantities, so insufficient intake of other water-soluble vitamins can lead to initial signs and symptoms of vitamin deficiency within a period of days to weeks. The bacteria living in your intestines help prevent deficiency diseases by producing small amounts of four of the nine water-soluble vitamins (B_5, B_7, B_9, and B_{12}). The intestinal epithelium can easily absorb all the water-soluble vitamins except B_{12}. The B_{12} molecule is large, and it must be bound to **intrinsic factor** synthesized by the gastric mucosa before absorption can occur.

Module 23.12 Review

a. Define nutrition.

b. Identify the two classes of vitamins.

c. If vitamins do not provide a source of energy, what is their role in nutrition?

LO **23.12** Explain the role of fat-soluble vitamins and water-soluble vitamins in metabolic pathways.

Module 23.13

Proper nutrition depends on eating a balanced diet

A **balanced diet** contains all the ingredients needed to maintain homeostasis, including adequate substrates to produce ATP, essential amino acids and fatty acids, and vitamins. In addition, the diet must include minerals (electrolytes) and enough water to replace losses in urine, feces, and evaporation. A balanced diet prevents **malnutrition**, an unhealthy state resulting from inadequate or excessive absorption of one or more nutrients.

Know the Limits on Fats, Sugars, and Salt (Sodium)

- Get most of your fat from fish, nuts, and vegetable oils, which are key sources of essential fatty acids.
- Limit solid fats like butter, shortening, and lard that are high in saturated fats and cholesterol.
- Check the Nutrition Facts label to keep saturated fats, trans fats, and sodium low.
- Choose food and beverages low in added sugars. Added sugars contribute to excessive caloric intake.

1 How do you know if you're eating a balanced diet? Follow the updated food recommendations on the Website www.choosemyplate.gov. The United States Department of Agriculture created this diagram to offer personalized eating plans and food assessments based on the current Dietary Guidelines for Americans. The color-coded food group slices indicate the proportions of food we should consume from each of the five basic food groups: grains, vegetables, fruits, dairy, and proteins.
In addition, use oils sparingly.

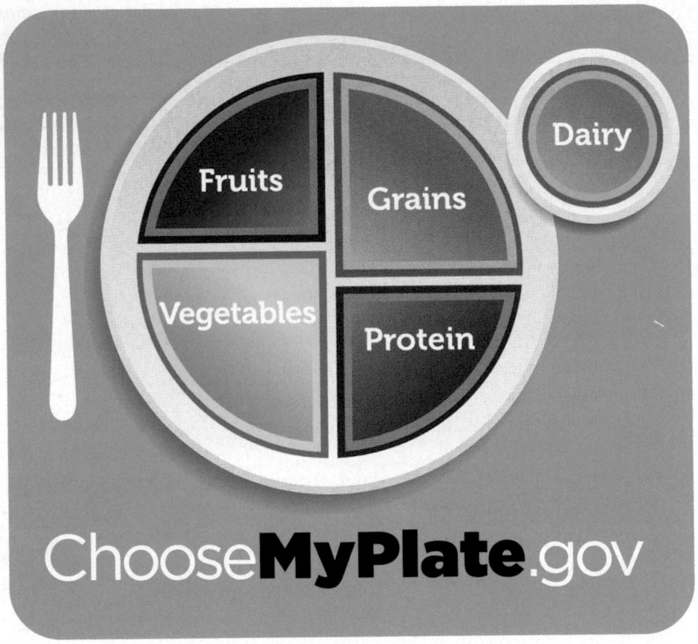

Fruits
Grains
Dairy
Vegetables
Protein
ChooseMyPlate.gov

Find Your Balance between Food and Physical Activity

- Stay within your daily calorie needs.
- Adults should be physically active for at least 30 minutes most days of the week.
- About 60 minutes a day of physical activity may be needed to prevent weight gain.
- To sustain weight loss, at least 60 to 90 minutes a day of physical activity may be required.
- Children and teenagers should be physically active for 60 minutes most days.

GRAINS	VEGETABLES	FRUITS	OILS	DAIRY	PROTEINS
Make half your grains whole	Vary your veggies	Focus on fruits		Get your calcium-rich foods	Go lean with proteins

2 In addition to considering the types of food you eat, think about how much energy that food contains relative to how much energy you use on a daily basis. We express the energy content of food in **calories**. One calorie is the energy needed to raise the temperature of 1 g of water by 1°C. We use the term **kilocalorie** (KIL-ō-kal-o-rē) (kcal) or **Calorie** (Cal) when talking about the metabolism of the entire body. One Calorie is the amount of energy needed to raise the temperature of 1 kilogram of water 1°C. The energy yield of the components of your diet differ; the values are 4.18 Cal/g for carbohydrates, 4.32 Cal/g for proteins, and 9.46 Cal/g for lipids. Depending on the level of activity, the average adult needs between 2000 to 3000 Cal each day to maintain a stable weight. That's a lot of mitochondrial activity—1 Cal is roughly equivalent to 83 million trillion ATP.

Food name	Serving	Cal
Breakfast bar	1 bar	368
Long-grain rice	1.5 cup	308
Bread, whole wheat	4 slices	277
Butter	0.5 tbsp	51
Beer, regular	12 fl oz	160
Cola, regular	12 fl oz	140

3 Here is additional information about the various food groups. The food groups in themselves are less important than making intelligent choices about what (and how much) you eat. Poor choices can lead to malnutrition even if all five groups are represented.

Five Basic Food Groups and Their Effects on Health

Nutrient Group	Provides	Health Effects
Grains (recommended: at least half of total eaten should be whole grains)	Carbohydrates; vitamins E, thiamine, niacin, folate; calcium; phosphorus; iron; sodium; dietary fiber	Whole grains prevent rapid rise in blood glucose levels, and consequent rapid rise in insulin levels
Vegetables (recommended: especially dark-green and orange vegetables)	Carbohydrates; vitamins A, C, E, folate; dietary fiber; potassium	Reduce risk of cardiovascular disease; protect against colon cancer (folate) and prostate cancer (lycopene in tomatoes)
Fruits (recommended: a variety of fruit each day)	Carbohydrates; vitamins A, C, E, folate; dietary fiber; potassium	Reduce risk of cardiovascular disease; protect against colon cancer (folate)
Dairy (recommended: low-fat or fat-free milk, yogurt, and cheese)	Complete proteins; fats; carbohydrates; calcium; potassium; magnesium; sodium; phosphorus; vitamins A, B_{12}, pantothenic acid, thiamine, riboflavin	Good source of calcium, which strengthens bones. Whole milk: High in calories, may cause weight gain; saturated fats associated with heart disease
Meat and Beans (recommended: lean meats, fish, poultry, eggs, dry beans, nuts, legumes)	Complete proteins; fats; calcium; potassium; phosphorus; iron; zinc; vitamins E, thiamine, B_6	Fish and poultry lower risk of heart disease and colon cancer (compared to red meat). Consumption of up to one egg per day does not appear to increase incidence of heart disease; nuts and legumes improve blood cholesterol ratios, lower risk of heart disease and diabetes

4 Some members of the dairy products and meat and beans groups—for example milk, yogurt, beef, fish, poultry, and eggs—provide all the essential amino acids in sufficient quantities. They are said to contain **complete proteins**. (Recall that essential amino acids must come from the diet because they cannot be synthesized by the body. Nonessential amino acids are those the body can make on demand.)

5 Many plants supply adequate *amounts* of protein but they are **incomplete proteins**, which lack one or more of the essential amino acids. Vegetarians, who largely restrict themselves to grains, vegetables, and fruits groups (with or without dairy products), must become adept at varying their food choices to include combinations of ingredients that meet all their amino acid requirements. Even with a proper balance of amino acids, vegans, who avoid all animal products, face a significant problem, because vitamin B_{12} is obtained only from animal products or from fortified foods.

Module 23.13 Review

a. Define balanced diet.

b. Of these three—carbohydrates, lipids, or proteins—which releases the greatest amount of energy per gram during catabolism?

c. Distinguish between a complete protein and an incomplete protein.

LO **23.13** Explain what constitutes a balanced diet and why such a diet is important.

Module 23.14 CLINICAL MODULE

Metabolic disorders may result from nutritional or biochemical problems

Disorders Related to Diet and Digestion

Eating Disorders

Eating disorders are psychological problems that result in either inadequate or excessive food consumption. There are many different forms; a common thread in these conditions is an obsessive concern about food and body weight.

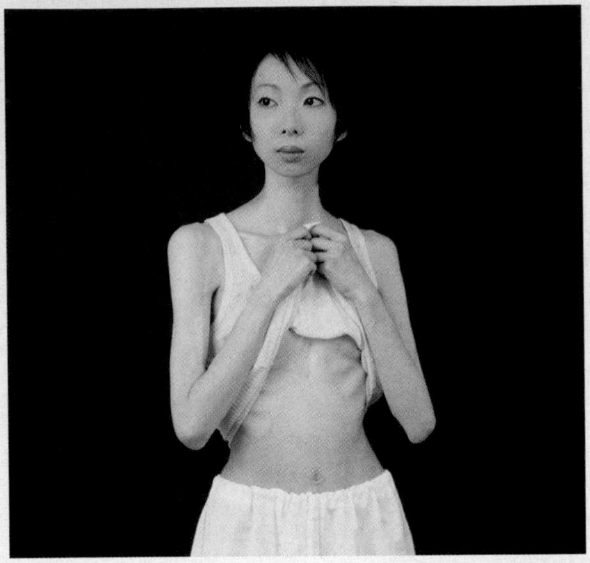

Anorexia Nervosa

Anorexia (an-ō-REK-sē-uh) is the lack or loss of appetite. It may also accompany disorders that involve other systems. **Anorexia nervosa** is a form of self-induced starvation that appears to be the result of severe psychological problems. It is most common in adolescent Caucasian females whose weight is nearly 30 percent below normal levels. Although very obviously underweight, patients are convinced that they are too fat and refuse to eat normal amounts of food. Death rates from severe cases range from 10 to 15 percent.

Bulimia

In **bulimia** (bū-LĒM-ē-uh) a person goes on an "eating binge" that may involve a meal that lasts 1–2 hours and may include 20,000 or more calories. The meal is followed by induced vomiting, usually along with the use of laxatives (to promote the movement of material through the digestive tract), and diuretics (drugs that cause excessive urination). Bulimia is more common than anorexia nervosa, and generally affects adolescent females. The health risks of bulimia result from (1) cumulative damage to the stomach, esophagus, oral cavity, and teeth from repeated exposure to stomach acids; (2) electrolyte imbalances resulting from the loss of sodium and potassium ions in gastric juices, diarrhea, and urine; (3) edema; and (4) cardiac arrhythmias.

Obesity

Obesity is defined as a condition of being 20 percent over ideal weight. It is at this point that serious health risks appear, such as diabetes, hypertension, and hypercholesterolemia. On that basis, the U.S. Centers for Disease Control and Prevention estimate that more than one-third of U.S. adults (35.7 percent) and about 17 percent of children aged 2–19 are obese. Basically, obese people take in more food energy than they are using.

There are two major categories of obesity: regulatory obesity and metabolic obesity. **Regulatory obesity**, the most common form, results from a failure to regulate food intake so that appetite, diet, and activity are in balance. In **metabolic obesity**, the condition is secondary to some underlying bodily malfunction that affects cell and tissue metabolism. Cases of metabolic obesity are relatively rare.

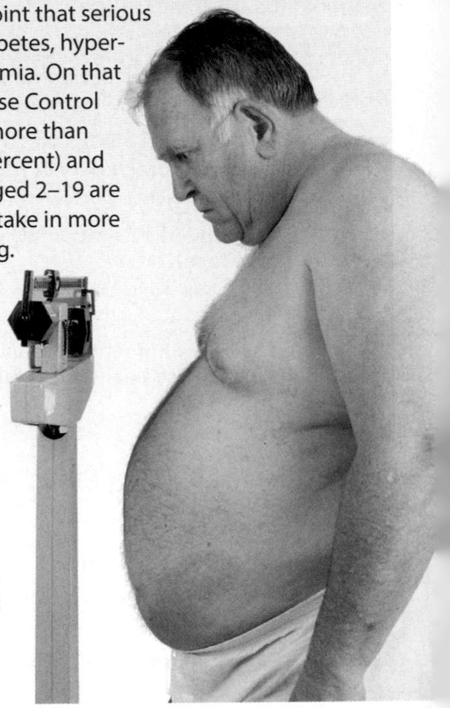

Elevated Cholesterol Levels

Earlier chapters have noted that elevated cholesterol levels are associated with the development of atherosclerosis and coronary artery disease (Module 19.8, p. 698). Nutritionists currently recommend that you reduce cholesterol intake to under 300 mg per day. This amount represents a 40 percent reduction for the average American adult. In fasting people, triglycerides are usually present at levels of 40–150 mg/dL. A high total cholesterol value linked to a high LDL level spells trouble. In effect, an excessive amount of cholesterol is being exported to peripheral tissues. Problems can also exist in people with low HDL levels (below 35 mg/dL). In such cases, excess cholesterol delivered to the tissues cannot easily be returned to the liver for excretion. In either event, the amount of cholesterol in peripheral tissues—and especially in arterial walls—is likely to increase.

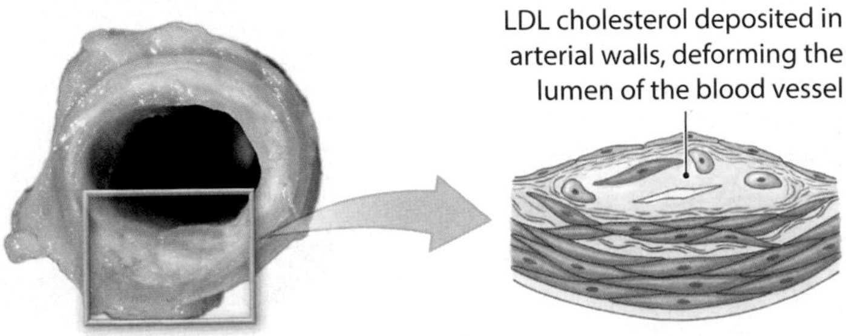
LDL cholesterol deposited in arterial walls, deforming the lumen of the blood vessel

Nutritional/Metabolic Disorders

Phenylketonuria

Several inherited metabolic disorders result from an inability to produce specific enzymes involved in amino acid metabolism. People with **phenylketonuria** (fen-il-kē-tō-NŪ-rē-uh), or **PKU**, for example, cannot convert phenylalanine to tyrosine. This reaction is an essential step in the synthesis of norepinephrine, epinephrine, dopamine, and melanin. If PKU is not detected in infancy, central nervous system development is inhibited, and severe brain damage results. The condition is common enough that a warning is printed on the packaging of products, such as diet drinks, that contain phenylalanine.

Diet
Coke

Ketoacidosis

When glucose supplies are limited, the breakdown of fatty acids and some amino acids in liver cells elevates acetyl-CoA levels and results in the production of small organic acids called **ketone bodies**. Most of these compounds diffuse out of the liver and accumulate in the bloodstream. One of the ketone bodies, acetone, has a "fruity" aroma that can be smelled in the breath of someone who has skipped one or more meals. This condition is called **ketosis** (ke-TŌ-sis), and over time ketone bodies can lower the pH of blood. This acidification of the blood is called **ketoacidosis** (kē-tō-as-i-DŌ-sis). In severe cases, the blood pH may drop below 7.05, and this may cause coma, cardiac arrhythmias, and death. A person with poorly controlled or undiagnosed diabetes mellitus is at serious risk of developing ketoacidosis because the liver responds as if the person is starving, catabolizing proteins and lipids and dumping ketone bodies into the circulation.

Gout

When RNA is recycled as part of metabolic turnover, the purines (adenine and guanine) cannot be catabolized. Instead, they are deaminated and excreted as **uric acid**. Like urea, uric acid is a relatively nontoxic waste product, but it is far less soluble than urea. Urea and uric acid are called **nitrogenous wastes**, because they are waste products that contain nitrogen atoms. Normal uric acid concentrations in plasma average 2.7–7.4 mg/dL, depending on gender and age. At concentrations over 7.4 mg/dL, body fluids become saturated with uric acid and insoluble uric acid crystals may begin to form. The condition that then develops is called **gout**. Initially, the joints of the limbs, especially the metatarsal–phalangeal joint of the great toe, are likely to be affected. This intensely painful condition, called **gouty arthritis**, may persist for several days and then disappear for a period of days to years.

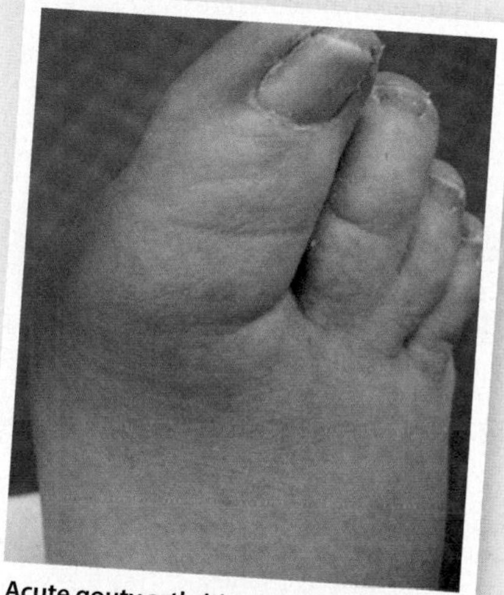

Acute gouty arthritis of the great toe

Protein Deficiency Diseases

Regardless of the energy content of the diet, if it is deficient in essential amino acids, the person will be malnourished to some degree. In a **protein deficiency disease**, protein synthesis decreases throughout the body. As protein synthesis in the liver fails to keep pace with the breakdown of plasma proteins, plasma osmolarity falls. This reduced osmolarity results in a fluid shift as more water moves out of the capillaries and into interstitial spaces, the peritoneal cavity, or both. The longer the person remains in this state, the more severe the ascites and edema that result. It is estimated that more than 100 million children worldwide suffer from protein deficiency diseases.

Kwashiorkor (kwash-ē-OR-kor) occurs in children whose protein intake is inadequate, even if the caloric intake is acceptable. In each case, additional complications include damage to the developing brain. The term is from the Ghana language and its literal translation means "first-second", describing the development of the disease in an older child who had been weaned from his mother's breast when the younger sibling was born. Protein deficiency diseases are rare in the United States.

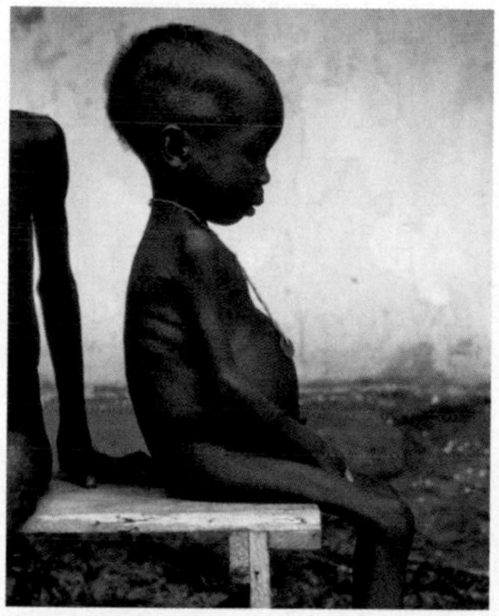

Kwashiorkor

Module 23.14 Review

a. Identify and briefly define two eating disorders.

b. Briefly describe phenylketonuria (PKU).

c. Define protein deficiency disease, and cite an example.

LO 23.14 Describe several metabolic disorders resulting from nutritional or biochemical problems.

Matching

Match each lettered term with the most closely related description.

a. lipogenesis	1	Absorptive state hormone	1 ______
b. anorexia	2	Glycogen reserves	2 ______
c. lipolysis	3	Water-soluble vitamins	3 ______
d. A, D, E, K	4	Fat catabolism	4 ______
e. absorptive state	5	Lipid synthesis	5 ______
f. deamination	6	Amino acid catabolism	6 ______
g. ketone bodies	7	Fat-soluble vitamins	7 ______
h. calorie	8	Lipid transport	8 ______
i. uric acid	9	Removal of amino group	9 ______
j. B complex and C	10	Lipid breakdown	10 ______
k. urea formation	11	Gout	11 ______
l. lipoproteins	12	Unit of energy	12 ______
m. insulin	13	Period following a meal	13 ______
n. skeletal muscle	14	Lack or loss of appetite	14 ______

Multiple choice

Select the correct answer from the list provided.

15 Intestinal absorption of nutrients occurs in the
- ☐ a) duodenum.
- ☐ b) ileocecum.
- ☐ c) ileum.
- ☐ d) jejunum.

16 When blood glucose concentrations are elevated, the glucose molecules are
- ☐ a) catabolized for energy.
- ☐ b) used to build protein.
- ☐ c) used for tissue repair.
- ☐ d) all of these

17 Most of the lipids absorbed by the digestive tract are immediately transferred to the
- ☐ a) liver.
- ☐ b) red blood cells.
- ☐ c) hepatocytes for storage.
- ☐ d) venous circulation by the thoracic duct.

18 Hypervitaminosis involving water-soluble vitamins is relatively uncommon because
- ☐ a) the excess amount is stored in adipose tissue.
- ☐ b) the excess amount is readily excreted in the urine.
- ☐ c) the excess amount is stored in the bones.
- ☐ d) excess amounts are readily absorbed by skeletal muscle tissue.

Short answer

19 What is the difference between an essential amino acid and a non-essential amino acid?

19 ______

20 Describe four reasons why protein catabolism is an impractical source of quick energy.

20 ______

21 What is the primary difference between the absorptive and postabsorptive states?

21 ______

22 Why is the liver the focal point for metabolic regulation and control?

22 ______

Section integration

Claudia suffers from anorexia nervosa. One afternoon she is rushed to the emergency room because of cardiac arrhythmias. Her breath smells fruity, and her blood and urine samples contain high levels of ketone bodies. Why does her breath have a fruity smell, and what would be the pH of her blood and urine?

23 ______

Energetics is the study of energy changes, and thermoregulation involves heat balance

The amount of energy needed to support ongoing activities varies from moment to moment. The study of the flow of energy and its change(s) from one form to another is called **energetics**. A common benchmark in energetics studies is the **basal metabolic rate (BMR)**, the minimum resting energy expenditure of an awake, alert person.

1 A direct method of determining the BMR involves monitoring respiratory activity. If we assume that average amounts and proportions of carbohydrates, lipids, and proteins are being catabolized, 4.825 Cal are expended per liter of oxygen consumed.

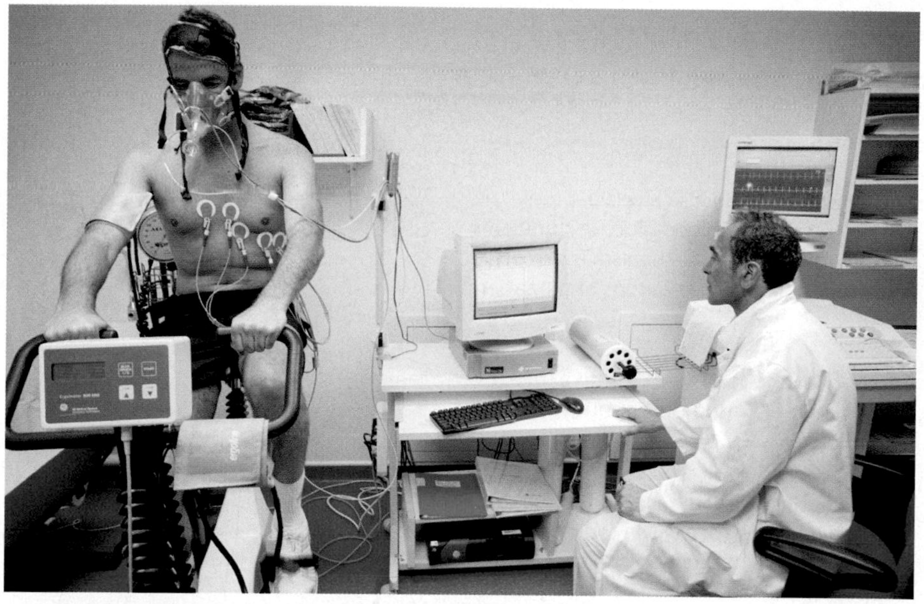

2 An average person has a BMR of 70 Cal per hour, or about 1680 Cal per day. The actual energy consumption per day may be several times that amount, depending on a person's size, weight, and level of physical activity. This graph shows the approximate number of Calories expended per hour at various levels of physical exertion.

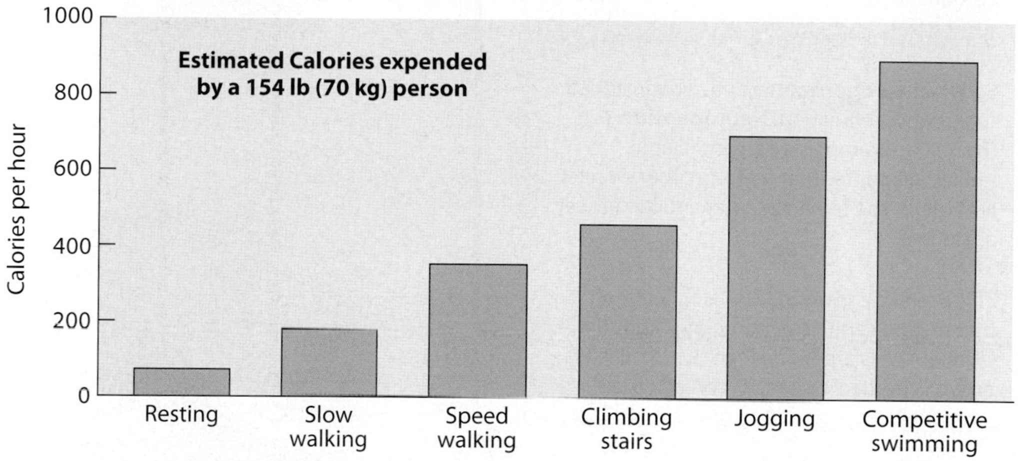
Estimated Calories expended by a 154 lb (70 kg) person
Calories per hour
1000
800
600
400
200
0
Resting
Slow walking
Speed walking
Climbing stairs
Jogging
Competitive swimming

To maintain energy balance, food intake must be adequate to support the activities under way. This section begins with a look at the control of appetite. All of the reactions that generate ATP also generate heat; only about 40 percent of the energy released through catabolism can be used to form ATP, and the rest warms the surrounding cytoplasm. When activity levels increase, ATP production accelerates, and more heat is generated. Enzymes will only function normally over a relatively narrow range of temperatures, and metabolic pathways are at risk unless heat is lost as quickly as it is produced. So this section ends with the topic of **thermoregulation**—the homeostatic control of body temperature.

Module 23.15 Review

a. Define energetics.

b. What is basal metabolic rate?

c. Define thermoregulation.

LO 23.15 Explain energetics and the role of thermoregulation in maintaining homeostasis.

Module 23.16

The control of appetite is complex and involves both short-term and long-term mechanisms

The **feeding center** (involved with hunger) and the **satiety center** (involved with food satisfaction) are hypothalamic nuclei involved with controlling appetite. Multiple factors influence these centers, and the most important satiety factors are diagrammed here. In addition, social factors, psychological pressures, and dietary habits can play a role, although the mechanisms and pathways involved remain to be determined.

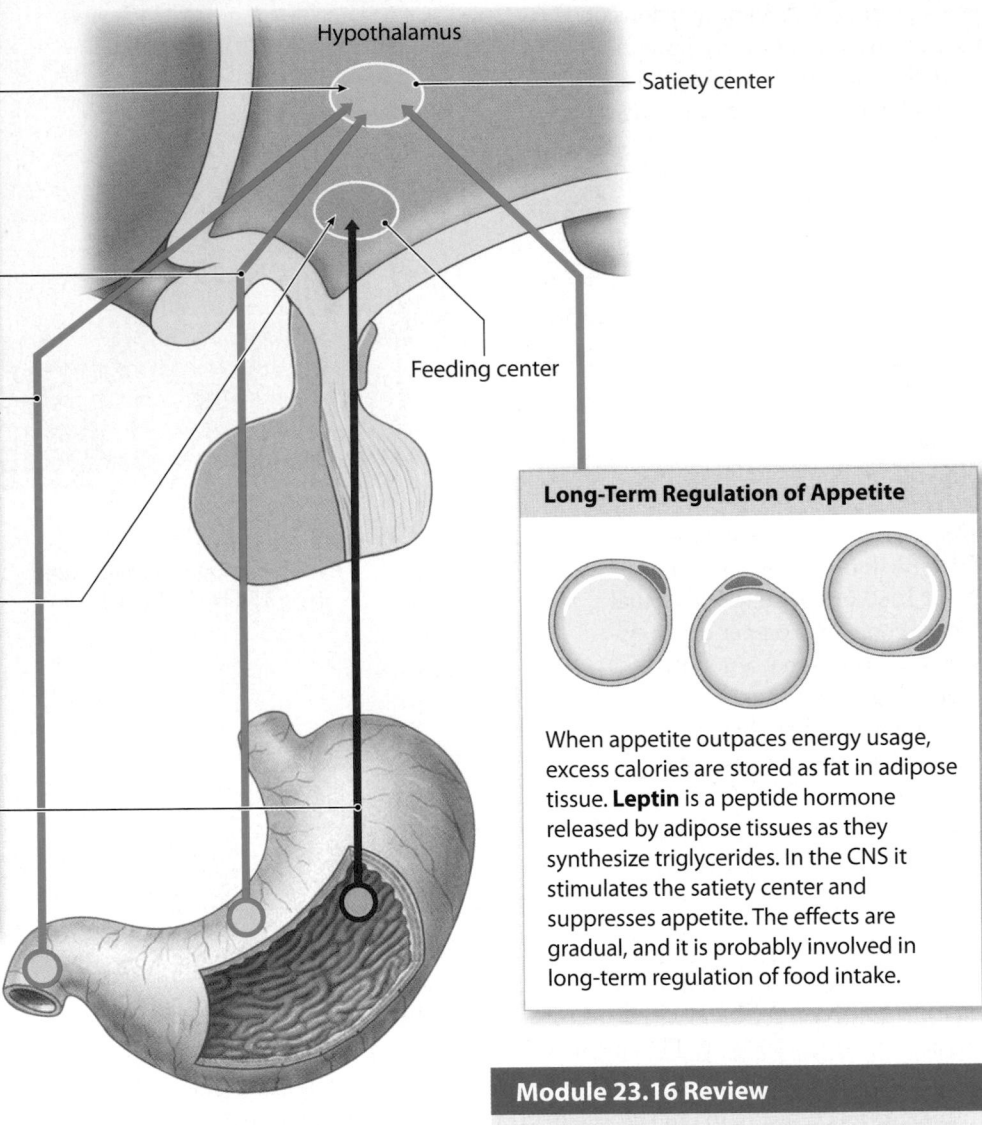

Short-Term Regulation of Appetite

Stimulation of Satiety Center →

Elevated blood glucose levels depress appetite, and low blood glucose stimulates appetite. The likely mechanism is glucose entry stimulating the neurons of the satiety center.

Several hormones of the digestive tract, including CCK, suppress appetite during the absorptive state.

Stimulation of stretch receptors along the digestive tract, especially in the stomach, causes a sense of satiation and suppresses appetite.

Stimulation of Feeding Center →

Several neurotransmitters have been linked to appetite regulation. **Neuropeptide Y (NPY)**, for example, is a hypothalamic neurotransmitter that (among other effects) stimulates the feeding center and increases appetite.

The hormone **ghrelin** (GREL-in), secreted by the gastric mucosa, stimulates appetite. Ghrelin levels are high when the stomach is empty, and decline as the stomach fills.

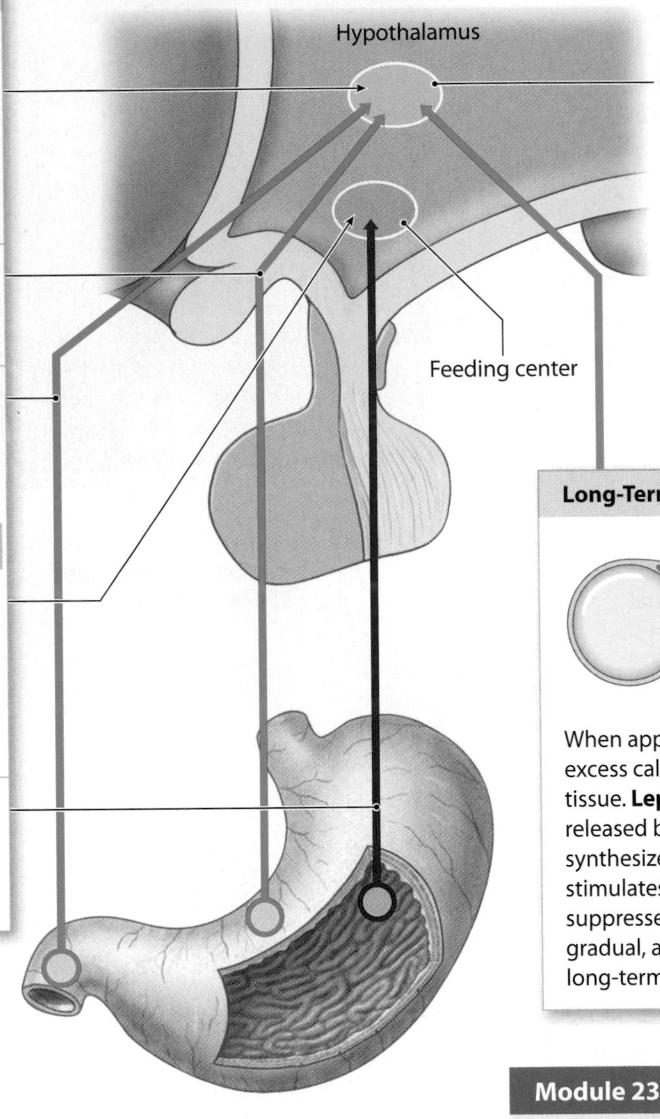

Hypothalamus
Satiety center
Feeding center

Long-Term Regulation of Appetite

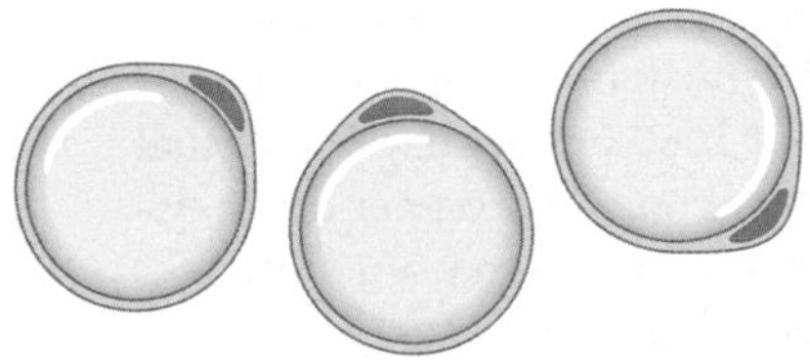

When appetite outpaces energy usage, excess calories are stored as fat in adipose tissue. **Leptin** is a peptide hormone released by adipose tissues as they synthesize triglycerides. In the CNS it stimulates the satiety center and suppresses appetite. The effects are gradual, and it is probably involved in long-term regulation of food intake.

In general, activation of the satiety center causes inhibition of the feeding center, and vice versa. In addition, factors that primarily stimulate one center often have a secondary, inhibitory effect on the other.

Module 23.16 Review

a. How might a lack of neuropeptide Y in the hypothalamus affect the control of appetite?

b. What hormone inhibits the satiety center and stimulates appetite in the short-term?

c. Describe leptin and its effect on appetite.

LO **23.16** Describe the roles of the satiety center and the feeding center in the regulation of food intake.

Module 23.17

To maintain a constant body temperature, heat gain and heat loss must be in balance

Only about 40 percent of energy released by catabolism can be captured as ATP, and the rest is lost as heat that warms surrounding tissues. If body temperature is to remain constant, heat production and heat loss must be kept in balance despite wide variations in activity levels and environmental conditions.

1 This illustration introduces the primary mechanisms of heat transfer between the body and the surrounding environment.

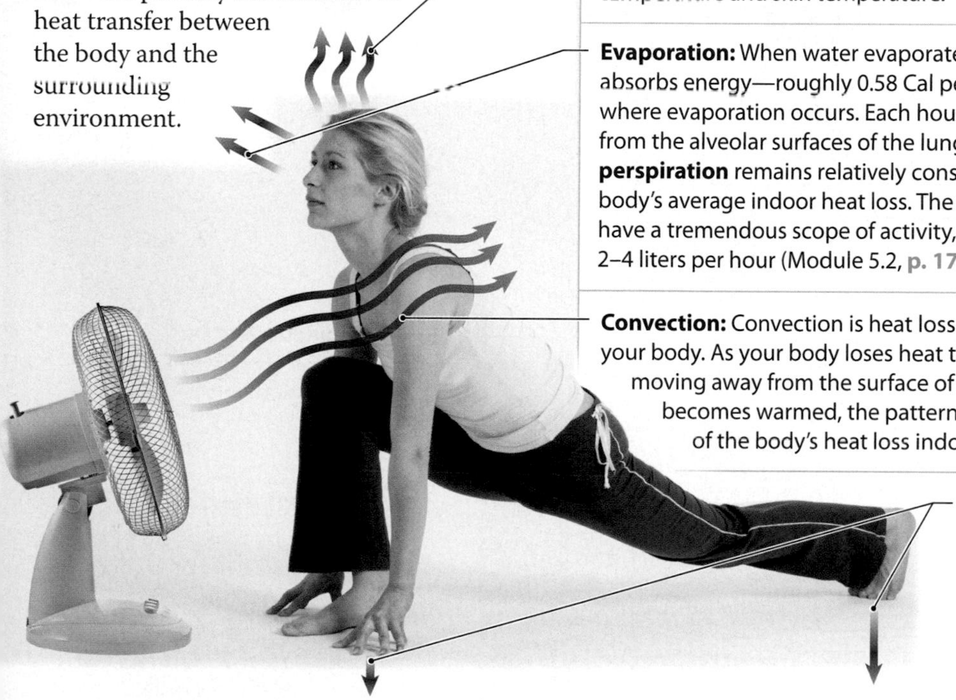

Primary Mechanisms of Heat Transfer

Radiation: Objects warmer than the environment lose heat energy as infrared radiation. When you feel the heat from the sun, you are detecting that radiation. Your body loses heat the same way, but in proportionately smaller amounts. More than 50 percent of the heat you lose indoors is attributable to radiation; the exact amount varies with both body temperature and skin temperature.

Evaporation: When water evaporates, it changes from a liquid to a vapor. Evaporation absorbs energy—roughly 0.58 Cal per gram of water evaporated—and cools the surface where evaporation occurs. Each hour, 20–25 mL of water crosses epithelia and evaporates from the alveolar surfaces of the lungs and the surface of the skin. This **insensible perspiration** remains relatively constant; at rest, it accounts for about 20 percent of your body's average indoor heat loss. The sweat glands responsible for **sensible perspiration** have a tremendous scope of activity, ranging from virtual inactivity to secretory rates of 2–4 liters per hour (Module 5.2, p. 177).

Convection: Convection is heat loss to the cooler air that moves across the surface of your body. As your body loses heat to the air next to your skin, that air warms and rises, moving away from the surface of the skin. Cooler air replaces it, and as this air in turn becomes warmed, the pattern repeats. Convection accounts for nearly 15 percent of the body's heat loss indoors.

Conduction: Conduction is the direct transfer of energy through physical contact. Conduction is generally not an effective way for gaining or losing heat, and its impact depends on the temperature of the object and the amount of skin area it contacts. When you are standing, conductive losses are negligible.

Underlying physical or environmental condition	°F	°C	Thermoregulatory capabilities	Major physiological effects
CNS damage Heat stroke	114 110 106	44 42	Severely impaired Impaired	Death Proteins denature Convulsions Cell damage
Disease-related fevers Severe exercise Active children	102	40 38	Effective	Disorientation
Normal range (oral)	98	36		Systems normal
Early mornings in cold weather	94	34	Impaired	Disorientation
Severe exposure	90 86	32 30	Severely impaired	Loss of muscle control Loss of consciousness
Hypothermia for open heart surgery	82 78 74	28 26 24	Lost	Cardiac arrest Death

2 The physiological effects of a failure to control body temperature are indicated in the chart to the left.

Module 23.17 Review

a. Define insensible perspiration.

b. What heat transfer process accounts for about one-half of a person's heat loss when indoors?

c. How is heat loss different between conduction and convection?

LO 23.17 Discuss the mechanisms involved in heat gain and heat loss.

Thermoregulatory centers in the hypothalamus adjust heat loss and heat gain

Heat loss and heat gain involve the activities of many systems. Those activities are coordinated by the **heat-loss center** and **heat-gain center**, respectively, in the preoptic area of the hypothalamus.

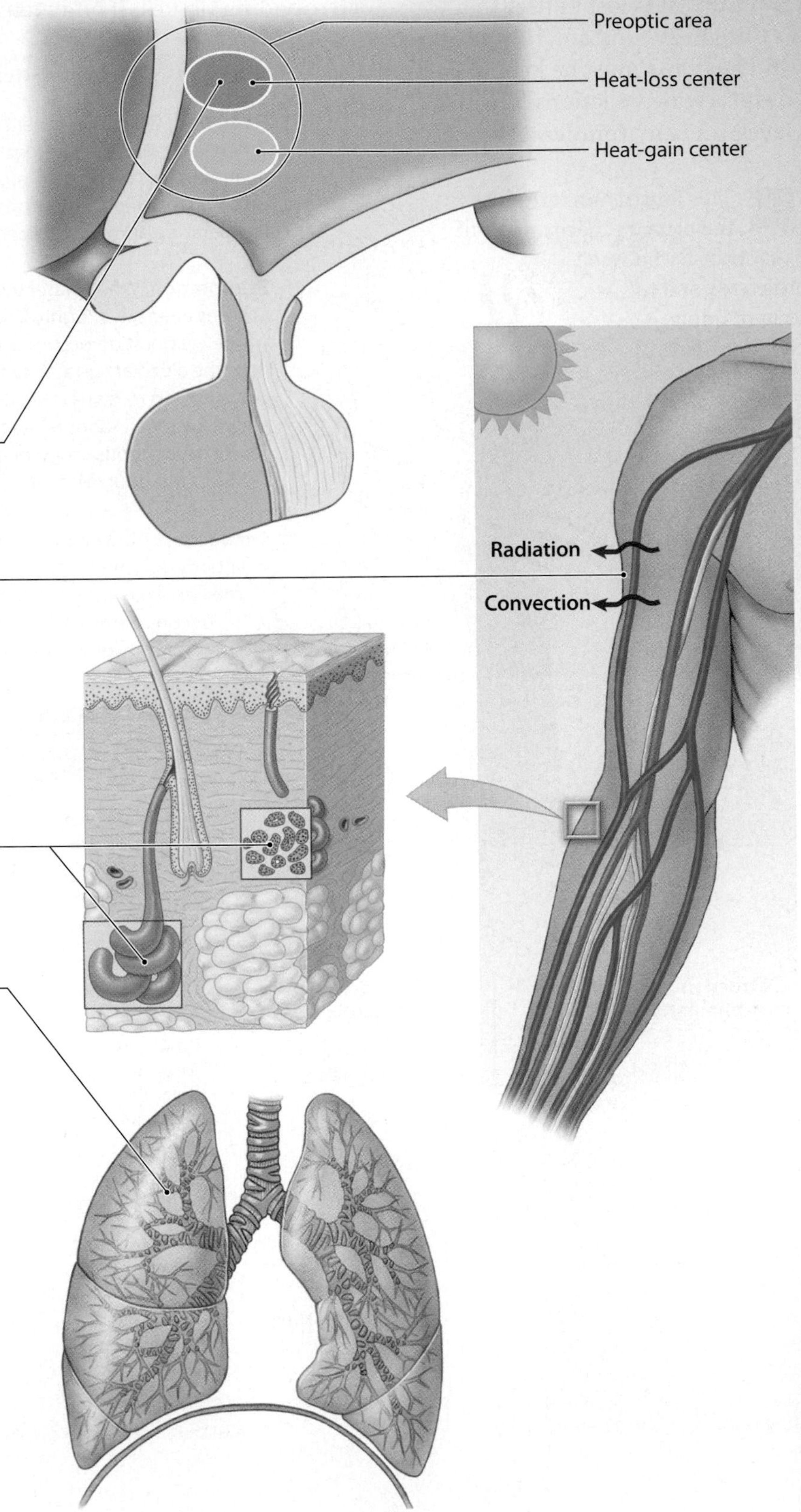
Preoptic area
Heat-loss center
Heat-gain center
Radiation
Convection

Responses to High Body Temperature Coordinated by the Heat-Loss Center

Behavioral Changes: A sense of discomfort leads to behavioral responses—getting into the shade, going into the water, or taking other steps that reduce body temperature.

Vasodilation and Shunting of Blood to Skin Surface: The inhibition of the vasomotor center causes peripheral vasodilation, and warm blood flows to the surface of the body. The skin takes on a reddish color, skin temperatures rise, and radiational and convective losses increase.

Sweat Production: As blood flow to the skin increases, sweat glands are stimulated to increase sweating. The perspiration flows across the body surface, and evaporative heat losses accelerate. Maximal secretion, if completely evaporated, would remove 2320 Cal per hour.

Respiratory Heat Loss: The respiratory centers are stimulated, and the depth of respiration increases. Often, the person begins breathing through an open mouth rather than through the nasal passageways, increasing evaporative heat losses through the lungs.

When Body Temperature Falls

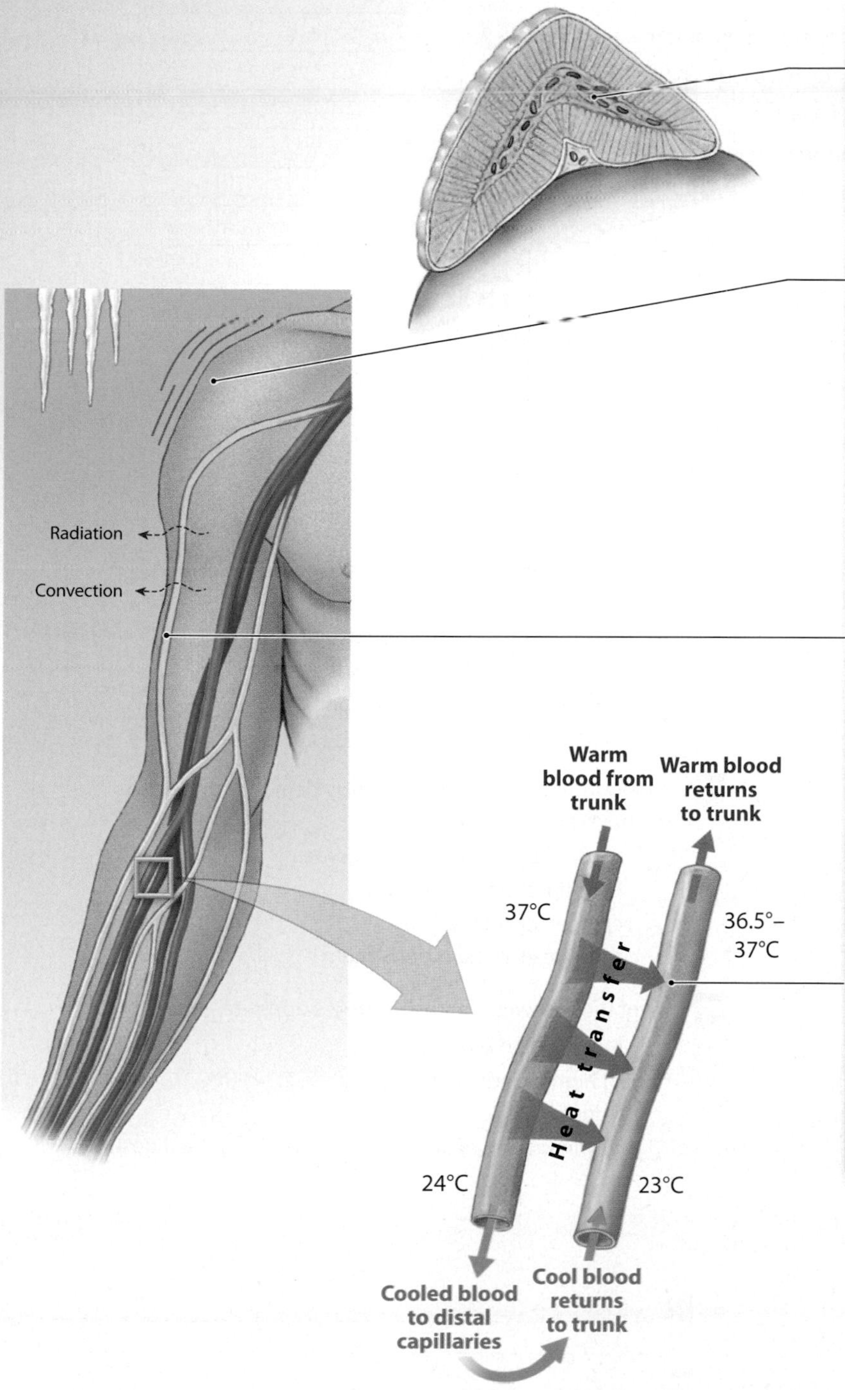
Radiation
Convection
Warm blood from trunk
Warm blood returns to trunk
37°C
36.5°–37°C
Heat transfer
24°C
23°C
Cooled blood to distal capillaries
Cool blood returns to trunk

Responses to Low Body Temperature Coordinated by the Heat-Gain Center

The heat-gain center responds to low body temperature in two ways:

Increased Generation of Body Heat

Nonshivering thermogenesis (ther-mō-JEN-e-sis) involves the release of hormones that increase the metabolic activity of all tissues. Sympathetic stimulation of the adrenal medullae releases epinephrine, which quickly increases the rate of glycogenolysis in the liver and skeletal muscle and the metabolic rate of most tissues.*

In **shivering thermogenesis**, a gradual increase in muscle tone increases the energy consumption of skeletal muscle tissue throughout your body. Both agonists and antagonists are involved, and muscle tone gradually increases to the point at which stretch receptor stimulation will produce brief, oscillatory contractions of antagonistic muscles. In other words, you begin to shiver. Shivering can elevate body temperature quite effectively, increasing the rate of heat generation by as much as 400 percent.

Conservation of Body Heat

The vasomotor center decreases blood flow to the dermis, thereby reducing losses by radiation and convection. The skin cools, and with blood flow restricted, it may take on a bluish or pale color. The epithelial cells are not damaged, because they can tolerate extended periods at temperatures as low as 25°C (77°F) or as high as 49°C (120°F).

The deep veins lie alongside the deep arteries, and heat is conducted from the warm blood flowing outward to the limbs to the cooler blood returning from the periphery. This arrangement traps the heat close to the body core and dramatically reduces heat loss. The transfer of heat, water, or solutes between fluids moving in opposite directions is called **countercurrent exchange**. The intermittent cycle of constriction and release when capillaries in our hands are exposed to cold temperature is called the *Lewis Wave* or the *hunter's response*. This is an autonomic response to reduce heat loss at the extremities.

* Note: In children, but not usually in adults, the heat-gain center can increase the rate of thyrotropin-releasing hormone (TRH) release by the hypothalamus when body temperature is below normal. This stimulates the release of thyroid-stimulating hormone (TSH) by the anterior lobe of the pituitary gland, and as the rate of thyroid hormone release increases, so does the rate of catabolism throughout the body (Module 16.7, p. 599).

Module 23.18 Review

a. Predict the effect of peripheral vasodilation on a person's body temperature.

b. Describe the role of nonshivering thermogenesis in regulating body temperature.

c. Name the heat conservation mechanism that conducts heat from deep arteries to adjacent deep veins in the limbs.

LO **23.18** Discuss the homeostatic mechanisms that maintain a constant body temperature.

Matching

Match each lettered term with the most closely related description.

a. ghrelin
b. basal metabolic rate
c. 40 percent
d. leptin
e. insensible perspiration
f. inhibits feeding center
g. shivering thermogenesis
h. peripheral vasoconstriction
i. thermoregulation
j. sensible perspiration
k. neuropeptide Y
l. 60 percent
m. nonshivering thermogenesis
n. peripheral vasodilation

1 Sweat gland activity; heat loss
2 Adipose tissue hormone
3 Homeostatic control of body temperature
4 General role of satiety center
5 Release of hormones; increased metabolism
6 Percent of catabolic energy released as heat
7 Stimulation of vasomotor center
8 Appetite-regulating neurotransmitter
9 Resting energy expenditure
10 Stomach hormone
11 Percent of catabolic energy captured as ATP
12 Inhibition of vasomotor center
13 Lung and epithelial water loss
14 Result of increased skeletal muscle tone

1 ______
2 ______
3 ______
4 ______
5 ______
6 ______
7 ______
8 ______
9 ______
10 ______
11 ______
12 ______
13 ______
14 ______

Multiple choice

Select the correct answer from the list provided.

15 A person's BMR is influenced by their
- a) gender.
- b) body weight.
- c) age.
- d) all of these

16 The processes involved in heat transfer between the body and the environment are
- a) sensible, insensible, heat loss, and heat gain.
- b) radiation, conduction, convection, and evaporation.
- c) physiological responses and behavioral modifications.
- d) sensible, insensible, hormones, and heat conservation.

17 The primary mechanisms for increasing heat loss from the body include
- a) vasomotor and respiratory.
- b) sensible and insensible.
- c) physiological responses and behavioral modifications.
- d) acclimatization and vasomotor.

18 All of the following are responses to an increase in body temperature, except
- a) stimulation of the respiratory centers.
- b) stimulation of sweat glands.
- c) peripheral vasoconstriction.
- d) peripheral vasodilation.

19 If daily intake exceeds total energy demands, the excess energy is stored primarily as
- a) triglycerides in adipose tissue.
- b) lipoproteins in the liver.
- c) glycogen in the liver.
- d) glucose in the bloodstream.

20 All of the following factors suppress appetite, except
- a) low blood glucose levels.
- b) high blood glucose levels.
- c) leptin.
- d) stimulation of stretch receptors along the digestive tract.

Short answer

21 How can a person's energy expenditure at rest be estimated by monitoring their oxygen consumption?

22 Describe the responses generated by the heat-gain center.

23 Describe the heat-gain mechanisms involved in nonshivering thermogenesis.

21 ______
______

22 ______
______

23 ______
______
______
______

Study Outline

SECTION 1 • Introduction to Cellular Metabolism

23.1 Metabolism refers to all the chemical reactions in the body p. 883

1. **Metabolism** is all the chemical reactions in the body. **Cellular metabolism** is the chemical reactions within cells.
2. **Catabolism** is the breakdown of organic substances; **anabolism** is the synthesis of new organic molecules.
3. In **metabolic turnover**, cells continuously break down and replace all their organic components except DNA.
4. All of the cell's organic building blocks form a **nutrient pool**.
5. In mitochondria, the catabolic reactions of aerobic metabolism release significant amounts of energy. Forty percent of this energy is captured and used to convert ADP to ATP. The remainder is lost as heat that warms the cell interior and surrounding tissues.

23.2 Cells can break down any available substrate from the nutrient pool to obtain the energy they need p. 884

6. The **nutrient pool** is the source of substrates for both catabolism and anabolism.
7. Cellular activity requires ATP, so the cell's mitochondria must have a continuous supply of 2-carbon substrate molecules.
8. The nutrient pool of each cell contributes to the metabolic reserves of the body as a whole.
9. If absorption in the digestive tract does not provide adequate nutrient levels, liver cells, adipocytes, and skeletal muscles can mobilize nutrient reserves for use by other cells.
10. Neither the diet nor the nutrient pool provides everything needed to build every protein, carbohydrate, and lipid a cell might require. Cells can synthesize what they need from available substrates.

23.3 The citric acid cycle transfers hydrogen atoms to coenzymes p. 886

11. Mitochondria provide almost all of the energy that supports cellular operations in the form of ATP by absorbing specific organic molecules, removing the hydrogen atoms from those molecules, and transferring them to **coenzymes**.
12. Organic nutrients from the nutrient pool are broken down into smaller 3-carbon and 2-carbon molecules that the mitochondria can process.
13. The process begins when pyruvate enters the mitochondrial matrix, and produces a 2-carbon **acetate** ion, 1 NAD•H, and 1 molecule of CO_2.
14. Acetate attaches to coenzyme A to form **acetyl-CoA**. The acetyl group from the acetyl-CoA attaches to a 4-carbon molecule in the mitochondrial matrix, and produces a 6-carbon molecule of **citric acid**. This is the first step of the **citric acid cycle**.
15. The citric acid cycle removes H atoms containing high-energy electrons and transfers them to coenzymes for further processing.
16. **NAD** and **FAD** are coenzymes that play a key role in the citric acid cycle by removing H atoms and delivering them to the electron transport system.
17. The energy gain from each acetyl group entering the citric acid cycle is 1 ATP molecule.

23.4 The electron transport system establishes a proton gradient used to make ATP p. 888

18. **Oxidative phosphorylation** is the generation of ATP within mitochondria in a reaction sequence that requires coenzymes and consumes oxygen. This process produces 90 percent of the ATP used by cells. The key reactions take place in the **electron transport system (ETS)**.
19. Coenzymes deliver hydrogen atoms from the citric acid cycle to a sequence of molecules called **cytochromes**. The first cytochrome passes electrons to the second, which passes them down the ETS. This passage gradually releases the energy carried by the electrons.
20. Along the way, hydrogen ions are pumped into the intermembrane space. The diffusion of hydrogen ions back into the matrix powers the production of ATP by the enzyme **ATP synthase**.
21. The final electron acceptor is oxygen. After accepting 2 electrons, an oxygen ion then reacts with a pair of hydrogen ions in the matrix to form a water molecule.
22. Aerobic metabolism is dependent on the availability of oxygen. If oxygen is severely restricted, the cytochromes cannot pass on their electrons and the ETS stops working.

SECTION 2 • Digestion and Metabolism of Organic Nutrients

23.5 Digestion involves a series of steps to make nutrients available to the body p. 891

23. During digestion, a combination of chemical and enzymatic attack breaks the complex organic compounds of food into simpler components that can be absorbed by the digestive tract and then distributed by the bloodstream.
24. Cells throughout the body rely on these organic molecules for energy production and to replenish the intracellular nutrient pool.

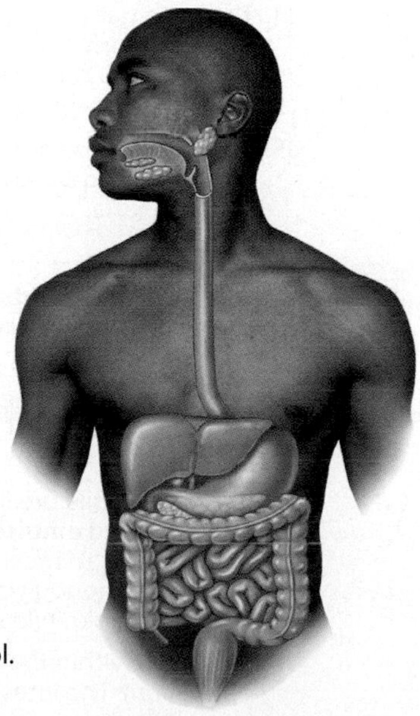

23.6 Carbohydrates are usually the preferred substrates for catabolism and ATP production under resting conditions p. 892

25. **Salivary amylase** in saliva is an enzyme that breaks down complex carbohydrates into disaccharides and trisaccharides. It remains active until pH falls below 4.5 in the stomach.
26. In the duodenum, buffers shift pH to alkaline. **Pancreatic alpha-amylase** now functions as salivary amylase did. Chyme with carbohydrates causes the release of gastric inhibitory peptide (GIP), which causes the release of insulin.
27. Epithelial cells of the jejunum finish the digestion of carbohydrates. The plasma membrane of the brush border contains the enzymes **maltase**, **sucrase**, and **lactase**, which break down the disaccharides into simple sugars that are absorbed.
28. Indigestible carbohydrates such as cellulose provide a nutrient source for bacteria in the colon. These bacteria produce intestinal gas, or **flatus**.
29. Simple sugars are absorbed in the liver and converted to glucose. Excess glucose is stored as glycogen in the liver and skeletal muscles.
30. Glucose is transported into cells by a **carrier protein** stimulated by insulin. Inside the cell, glucose is used to build structural materials or nucleic acids. If energy is needed, it is broken down into two molecules of **pyruvate** in an anaerobic process called glycolysis.
31. Glycolysis yields 2 ATP for every glucose molecule broken down. Each pyruvate molecule can be used by the mitochondria after conversion to acetyl-CoA.
32. For each molecule of pyruvate processed by mitochondria, the cell gains 17 ATP, consumes 3 molecules of O_2, and generates 3 molecules of CO_2 and 6 molecules of water. Thus for each pair of pyruvate molecules catabolized, the cell gains 34 ATP.

23.7 Glycolysis is the first step in glucose catabolism p. 894

33. **Glycolysis** is an anaerobic process that breaks down a 6-carbon glucose molecule into two 3-carbon molecules of pyruvate through a series of enzymatic steps.
34. There is a net gain of 2 ATP molecules for each glucose molecule converted to 2 molecules of pyruvate. In addition, NAD accepts hydrogen atoms that can be transferred to mitochondria for use in the ETS.
35. A single glucose molecule yields 2 ATP through glycolysis, and 34 ATP through aerobic metabolism in the mitochondria.

23.8 Lipids reach the bloodstream in chylomicrons; the cholesterol is then extracted and released as lipoproteins p. 896

36. **Lingual lipase** in the saliva breaks down triglycerides into monoglycerides and fatty acids. Large drops containing a variety of lipids form in the stomach.
37. **Bile salts** in bile break the large drops into tiny droplets in a process called **emulsification**. This allows **pancreatic lipase** better access to break apart the triglycerides to a mixture of fatty acids, monoglycerides, and glycerol. These molecules form small lipid–bile salt complexes called **micelles**.
38. The lipids within the micelle diffuse across the intestinal epithelium. The intestinal cells synthesize triglycerides and other lipids that are coated with proteins in a complex known as **chylomicrons**, which are **lipoproteins**.
39. Most of the chylomicrons diffuse into the intestinal lacteals, and proceed along the lymphatic vessels, the thoracic duct, the left subclavian vein, the pulmonary circuit, and then into the systemic circuit.
40. The liver absorbs chylomicrons and creates **low density lipoproteins (LDLs)** and **very low density lipoproteins (VLDLs)**. VLDLs transport triglycerides from the liver to muscle and adipose tissue. The LDLs enter the bloodstream and are delivered to peripheral tissues.
41. The cholesterol not used by the cell re-enters the bloodstream. **High density lipoproteins (HDLs)** absorb excess cholesterol and transport it back to the liver for storage or excretion in the bile.

23.9 Fatty acids can be broken down to provide energy or converted to other lipids p. 898

42. During lipid catabolism, or **lipolysis**, lipids are broken down into pieces that can be either converted to pyruvate or channeled directly into the citric acid cycle in a process called **beta-oxidation**.

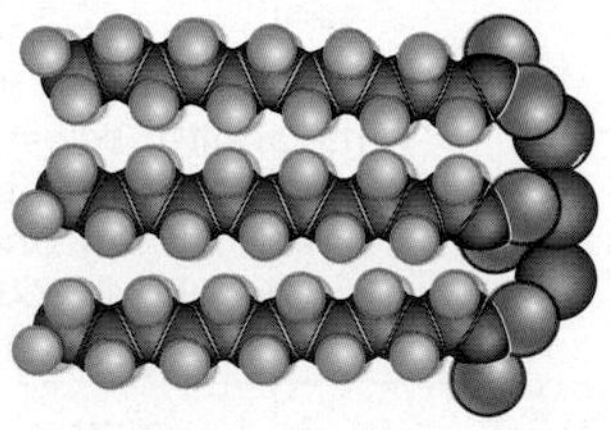

The principles of triglyceride catabolism apply to other lipids as well.

43. The synthesis of lipids, known as **lipogenesis**, begins with acetyl-CoA, and results in fatty acids that can be used to synthesize other structural and functional lipids.

23.10 An amino acid not needed for protein synthesis may be broken down or converted to a different amino acid p. 900

44. Protein digestion begins when food is chewed and mixed with saliva, and continues through the churning and mixing in the stomach. Stomach acids also denature most proteins, exposing them to enzymatic attack by **pepsin**.
45. A variety of enzymes attack specific peptide bonds to break down proteins into a mixture of dipeptides, tripeptides, and amino acids.
46. The epithelial surfaces of the small intestine contain **peptidases** that break dipeptides apart and release individual amino acids, which are absorbed into the cell. Amino acids are then released into interstitial fluid, intestinal capillaries, hepatic portal vein, and then the liver.
47. The liver uses amino acids for synthesizing plasma proteins, and it has all of the enzymes needed to synthesize, convert, or catabolize amino acids. **Deamination** releases toxic ammonium ions, which liver cells use to synthesize relatively harmless **urea** that is excreted in urine.

23.11 There are two general patterns of metabolic activity: the absorptive and postabsorptive states p. 902

48. The **absorptive state** is the time following a meal, when nutrient absorption is under way. After a typical meal, the absorptive state continues for about 4 hours.
49. In the absorptive state, insulin stimulates glucose uptake and glycogenesis, protein synthesis is stimulated, and glycolysis and aerobic metabolism provide the ATP needed for cellular activities.

50. The **postabsorptive state** is when nutrient absorption is not under way and your body must rely on internal energy reserves to continue meeting its energy demands.
51. In the postabsorptive state, lipid and protein reserves are mobilized, blood glucose levels are stabilized by the release of glucose by the liver, skeletal and cardiac muscle stimulate glycogenolysis, and adipocytes stimulate lipolysis.

23.12 Vitamins are essential to the function of many metabolic pathways p. 904

52. **Nutrition** is the absorption of nutrients from food.
53. **Vitamins** are organic compounds required in very small quantities that play an essential role in specific metabolic pathways. The **fat-soluble vitamins** are vitamins A, D_3, E, and K. They are absorbed from the digestive tract along with the lipid contents of micelles. Vegetables are potential sources of fat-soluble vitamins.
54. The **water-soluble vitamins** are the B vitamins and vitamin C. The B vitamins are found in meat, eggs, and dairy products; vitamin C is found in citrus fruits.

23.13 Proper nutrition depends on eating a balanced diet p. 906

55. A **balanced diet** contains everything needed to maintain homeostasis, including adequate substrates to produce ATP, essential amino acids and fatty acids, vitamins, electrolytes, and water. A balanced diet helps prevent **malnutrition**.
56. In addition to considering the types of food we eat, we should consider the energy that food contains. The energy content of food is expressed in **calories**.
57. One calorie is the energy needed to raise the temperature of 1 g of water by 1°C. We use the term **kilocalorie** (kcal) or **Calorie** (Cal) when talking about the metabolism of the entire body. One Calorie is the amount of energy needed to raise the temperature of 1 kilogram of water 1°C.
58. The energy yield is different for components of the diet. The values are 4.18 Cal/g for carbohydrates, 4.32 Cal/g for proteins, and 9.46 Cal/g for lipids. Depending on activity level, the average adult needs between 2000 and 3000 Cal each day.
59. The five basic food groups are grains, vegetables, fruits, dairy, and meat and beans.
60. Some members of the dairy and meat and beans groups provide all the essential amino acids and are said to be **complete proteins**. Many plants supply adequate amounts of protein, but they are **incomplete proteins** because they lack one or more of the essential amino acids.

23.14 Metabolic disorders may result from nutritional or biochemical problems p. 908

61. **Eating disorders** are psychological problems that result in either inadequate or excessive food consumption. These include self-induced starvation called **anorexia nervosa**, or binge eating and purging, called **bulimia**.
62. **Obesity** is defined as being 20 percent over ideal weight. Elevated cholesterol levels are associated with atherosclerosis and coronary artery disease.
63. **Phenylketonuria** is an inherited inability to produce the enzymes involved in amino acid metabolism and can result in brain damage if not detected in infancy.
64. **Kwashiorkor** is a **protein deficiency disease** in children whose protein intake is inadequate.
65. When glucose is limited, the breakdown of fatty acids and some amino acids result in the production of **ketone bodies. Ketosis** is an accumulation of ketone bodies in the bloodstream; it can often be detected on the breath because it has a fruity aroma. This condition can lower blood pH in a condition called **ketoacidosis**. This can occur in poorly controlled or undiagnosed diabetes mellitus, and in cases of starvation.
66. An accumulation of the **nitrogenous waste**, **uric acid**, can cause the formation of uric acid crystals. This condition is called **gout**, and can lead to **gouty arthritis** in the joints of the limbs.

SECTION 3 • Energetics and Thermoregulation

23.15 Energetics is the study of energy changes, and thermoregulation involves heat balance p. 911

67. The study of the flow of energy and its change(s) from one form to another is called **energetics**. A common benchmark in energetics studies is the **basal metabolic rate (BMR)**, the minimum resting energy expenditure of an awake, alert person.

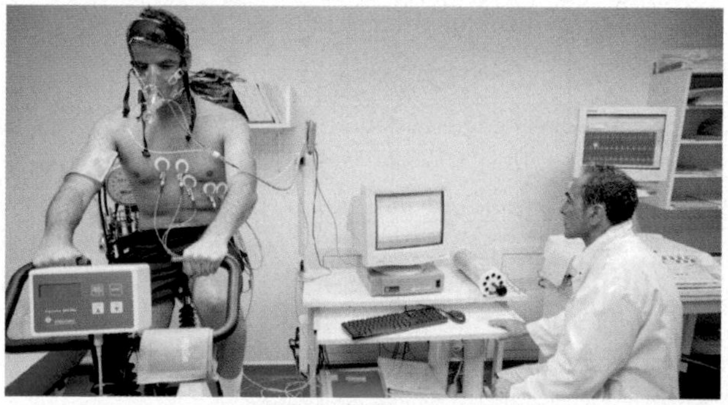

68. BMR may be determined by measuring respiratory activity and assuming that 4.825 Calories are expended per liter of oxygen consumed. An average person has a BMR of 70 Cal per hour, or about 1680 Cal per day.
69. All of the reactions that generate ATP also generate heat; only about 40 percent of the energy released through catabolism can be used to form ATP, and the rest warms the surrounding cytoplasm.
70. The homeostatic control of body temperature is **thermoregulation**.

23.16 The control of appetite is complex and involves both short-term and long-term mechanisms p. 912

71. The **feeding center** (involved with hunger) and the **satiety center** (involved with food satisfaction) are hypothalamic nuclei involved with the control of appetite.
72. Elevated blood glucose levels depress appetite, and low blood glucose stimulates appetite. Hormones of the digestive tract and stimulation of stretch receptors in the digestive tract also suppress appetite.
73. The feeding center is stimulated by **neuropeptide Y (NPY)** and **ghrelin**.
74. Excess calories are stored as fat in adipocytes. **Leptin**, a hormone released by adipose tissues as they synthesize triglycerides, suppresses appetite.

23.17 **To maintain a constant body temperature, heat gain and heat loss must be in balance** p. 913

75. If body temperature is to remain constant, heat production and heat loss must be kept in balance despite wide variations in activity levels and environmental conditions.

76. The primary mechanisms for heat transfer are **radiation**, **evaporation**, **convection**, and **conduction**.

23.18 **Thermoregulatory centers in the hypothalamus adjust heat loss and heat gain** p. 914

77. Heat loss and heat gain are regulated by the **heat-loss center** and the **heat-gain center**, respectively, in the preoptic area of the hypothalamus.
78. Responses to high body temperature are coordinated by the heat-loss center. These responses include **behavioral change**, **vasodilation and shunting of blood to the skin surface**, **sweat production**, and **respiratory heat loss**.
79. Responses to low body temperature are coordinated by the heat-gain center. These responses include **nonshivering thermogenesis**, **shivering thermogenesis**, and the **conservation of body heat** using the **countercurrent exchange** mechanism as blood flow decreases to the surface.

Chapter Review Questions

True/False

Indicate whether each statement is true or false.

1. The chemical reactions within cells are collectively known as cellular anabolism.
2. The breakdown of glucose into two 3-carbon molecules is called glycolysis.
3. The energy yield for carbohydrates is 4.18 Cal/g.
4. Kwashiorkor occurs in adults when uric acid crystals form in body fluids, causing a specific type of arthritis.
5. Leptin is a hormone secreted by the gastric mucosa that stimulates appetite.
6. The average person has a BMR of 70 Cal/hour, or about 1680 Cal/day.

1 ____________
2 ____________
3 ____________
4 ____________
5 ____________
6 ____________

Multiple choice

Select the correct answer from the list provided.

7. Catabolism refers to
   - ☐ a) the creation of a nutrient pool.
   - ☐ b) the sum total of all chemical reactions in the body.
   - ☐ c) the production of organic compounds.
   - ☐ d) the breakdown of organic substrates.

8. The breakdown of glucose to pyruvate is
   - ☐ a) anaerobic.
   - ☐ b) aerobic.
   - ☐ c) lipogenic.
   - ☐ d) anabolic.

9. The process that produces 90 percent of the ATP used by our cells is
   - ☐ a) glycolysis.
   - ☐ b) the citric acid cycle.
   - ☐ c) oxidative phosphorylation.
   - ☐ d) thermoregulation.

10. Glucose synthesis from smaller carbon chains is known as
    - ☐ a) glycolysis.
    - ☐ b) gluconeogenesis.
    - ☐ c) glycogenesis.
    - ☐ d) glycogenolysis.

11. The net result of one turn of the citric acid cycle is
    - ☐ a) 1 ATP.
    - ☐ b) 2 ATP.
    - ☐ c) 17 ATP.
    - ☐ d) 34 ATP.

12. Which of the following is a protein deficiency disease?
    - ☐ a) phenylketonuria
    - ☐ b) kwashiorkor
    - ☐ c) ketoacidosis
    - ☐ d) gouty arthritis

13. The enzyme that breaks down complex carbohydrates into a mixture of disaccharides and trisaccharides in the mouth is
    - ☐ a) pepsin.
    - ☐ b) lingual lipase.
    - ☐ c) sucrase.
    - ☐ d) salivary amylase.

14. Which of the following is a water-soluble vitamin?
    - ☐ a) vitamin A
    - ☐ b) vitamin C
    - ☐ c) vitamin D_3
    - ☐ d) vitamin K

15 The hormone released by adipose tissues as they synthesize triglycerides is

- a) leptin.
- b) ghrelin.
- c) neuropeptide Y.
- d) insulin.

16 More than 50 percent of the heat you lose indoors is attributable to

- a) radiation.
- b) sensible perspiration.
- c) convection.
- d) conduction.

Short answer

17 Explain how catabolism and anabolism are "linked" by ATP.

18 What level of total cholesterol is considered elevated, and what are the health risks of elevated cholesterol?

19 What are the risks of too much vitamin intake, and which vitamins are most commonly involved?

20 Distinguish between a calorie, a kilocalorie, and a Calorie.

Chapter Integration • Applying what you have learned

Finding balance in foods

Diet has a profound influence on a person's general health. Too many or too few nutrients, hypervitaminosis (too many vitamins) or hypovitaminosis (too few vitamins), and above-normal or below-normal concentrations of minerals (inorganic nutrients) can adversely affect health. Subtle, long-term problems can occur when the diet includes the wrong proportions or combinations of nutrients. The average diet in the United States contains too much sodium and too many calories in general, and lipids (particularly saturated fats) provide too great a proportion of those calories. Poor diets increase the incidence of obesity, heart disease, atherosclerosis, hypertension, and diabetes in the U.S. population.

Mrs. Henderson, who has a family history of cardiovascular disease, is concerned about her health. A comprehensive blood test reveals elevated levels of LDL (so-called "bad cholesterol") and decreased levels of HDL (so-called "good cholesterol"), normal glucose, and adequate vitamin levels except vitamin B_6. The physician orders a consultation with the registered dietician, who discovers that Mrs. Henderson's semi-vegetarian diet is high in carbohydrates and fats.

1 **Why are vitamins and minerals essential components of the diet?**

2 **Explain the labeling of HDL as "good cholesterol" and LDL as "bad cholesterol."**

3 **Why are HDLs considered beneficial?**

4 **What process in the liver increases after a high-carbohydrate meal?**

5 **What is the significance of a pyridoxine (vitamin B_6) deficiency?**

24 The Urinary System

LEARNING OUTCOMES

These Learning Outcomes correspond by number to this chapter's modules and indicate what you should be able to do after completing the chapter.

SECTION 1 • Anatomy of the Urinary System

- **24.1** Identify the organs of the urinary system, and cite a primary function of each.
- **24.2** Describe the location and structural features of the kidneys.
- **24.3** Describe the gross structural features of the kidney, and distinguish between cortical and juxtamedullary nephrons.
- **24.4** Describe the segments of the nephron and collecting system, including their general functions and histological appearance.
- **24.5** Trace the pathway of blood flow through a kidney, and compare the pattern of blood flow in cortical and juxtamedullary nephrons.

SECTION 2 • Overview of Renal Physiology

- **24.6** Briefly describe how the kidneys maintain homeostasis and produce urine.
- **24.7** Describe filtration, reabsorption, and secretion along each segment of the nephron and collecting system.
- **24.8** Describe the structural features of a renal corpuscle, and explain the functions of the filtration membrane components.
- **24.9** Describe the factors that influence filtration pressure and the glomerular filtration rate.
- **24.10** Identify the types of transport mechanisms along the proximal and distal convoluted tubules of the nephron.
- **24.11** Explain the role of countercurrent multiplication in the formation of a concentration gradient in the renal medulla.
- **24.12** Describe how antidiuretic hormone influences the volume and concentration of urine.
- **24.13** Summarize the major steps involved in water reabsorption and urine production.
- **24.14** **CLINICAL MODULE** Compare and contrast chronic and acute renal failure, and explain the process of hemodialysis.

SECTION 3 • Urine Storage and Elimination

- **24.15** Name the organs responsible for the transport, storage, and elimination of urine.
- **24.16** Describe the structures and functions of the ureters, urinary bladder, and urethra.
- **24.17** Discuss the roles of local and central pathways in urination, and describe the micturition reflex.
- **24.18** **CLINICAL MODULE** Describe common urinary disorders related to output and frequency.

Learning Outcomes are repeated at the bottom of each module.

The urinary system organs are the kidneys, ureters, urinary bladder, and urethra

In this section we consider the functional anatomy of the components of the urinary system. Later sections consider how the kidneys remove metabolic wastes from the bloodstream to produce urine, how the process is regulated, and how urine is eliminated through the passageways of the **urinary tract** (the ureters, urinary bladder, and urethra).

1 The **urinary system** eliminates excess water, salts, and physiological wastes through the production of urine.

The **kidneys** are metabolically very active. At rest they receive about 25 percent of the cardiac output. The two kidneys perform the excretory functions of the urinary system. They produce **urine**, a fluid containing water, ions, and small soluble substances.

The **ureters** receive the urine from the kidneys and conduct it to the urinary bladder. Urine movement involves a combination of gravity and the peristaltic contractions of smooth muscle in the walls of the ureters.

The **urinary bladder** receives and stores urine prior to its elimination from the body. Urine elimination, or **urination**, is driven by the contraction of smooth muscle layers in the walls of the urinary bladder.

The **urethra** is a passageway that conducts urine from the urinary bladder to the exterior.

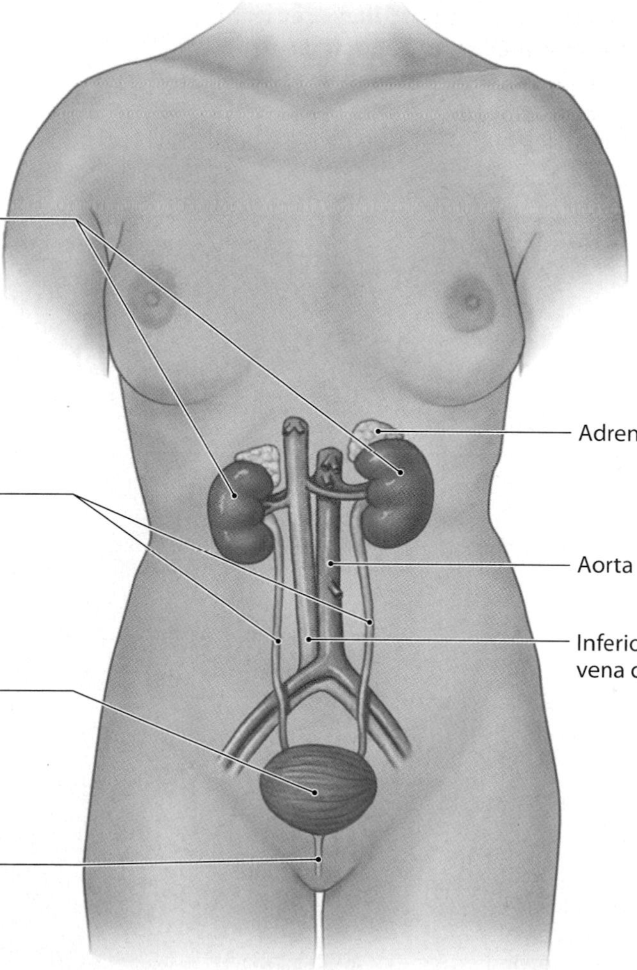
Adrenal gland
Aorta
Inferior vena cava

2 The urinary system removes wastes generated by cells throughout the body, but it has several other essential homeostatic functions. These functions are summarized in the following table.

Functions of the Urinary System

- Adjusting blood volume and blood pressure
- Regulating blood plasma concentrations of sodium, potassium, chloride, and other ions
- Stabilizing blood pH
- Conserving valuable nutrients by preventing their loss in urine
- Removing drugs and toxins from the bloodstream

Module 24.1 Review

a. Which organ performs the excretory functions of the urinary system?

b. The urinary tract is composed of which structures?

c. Describe the functions of the urinary system.

LO 24.1 Identify the organs of the urinary system, and cite a primary function of each.

The kidneys are paired retroperitoneal organs

In this module you will learn the location of the kidneys and their placement in relation to the axial skeleton and the organs of the abdominopelvic cavity. You will also see how the kidneys are supported, protected, and stabilized in position.

1 The kidneys, ureters, urinary bladder, and the associated blood vessels are shown in this anterior view. Because the kidneys are in a retroperitoneal position, they are clearly visible in an anterior view only after other abdominal organs have been removed. To help you remember the organs positioned retroperitoneally, use the mnemonic **SAD PUCKER** for **S**uprarenal (adrenal) glands, **A**orta and inferior vena cava, **D**uodenum, **P**ancreas, **U**reters, **C**olon, **K**idneys, **E**sophagus, **R**ectum.

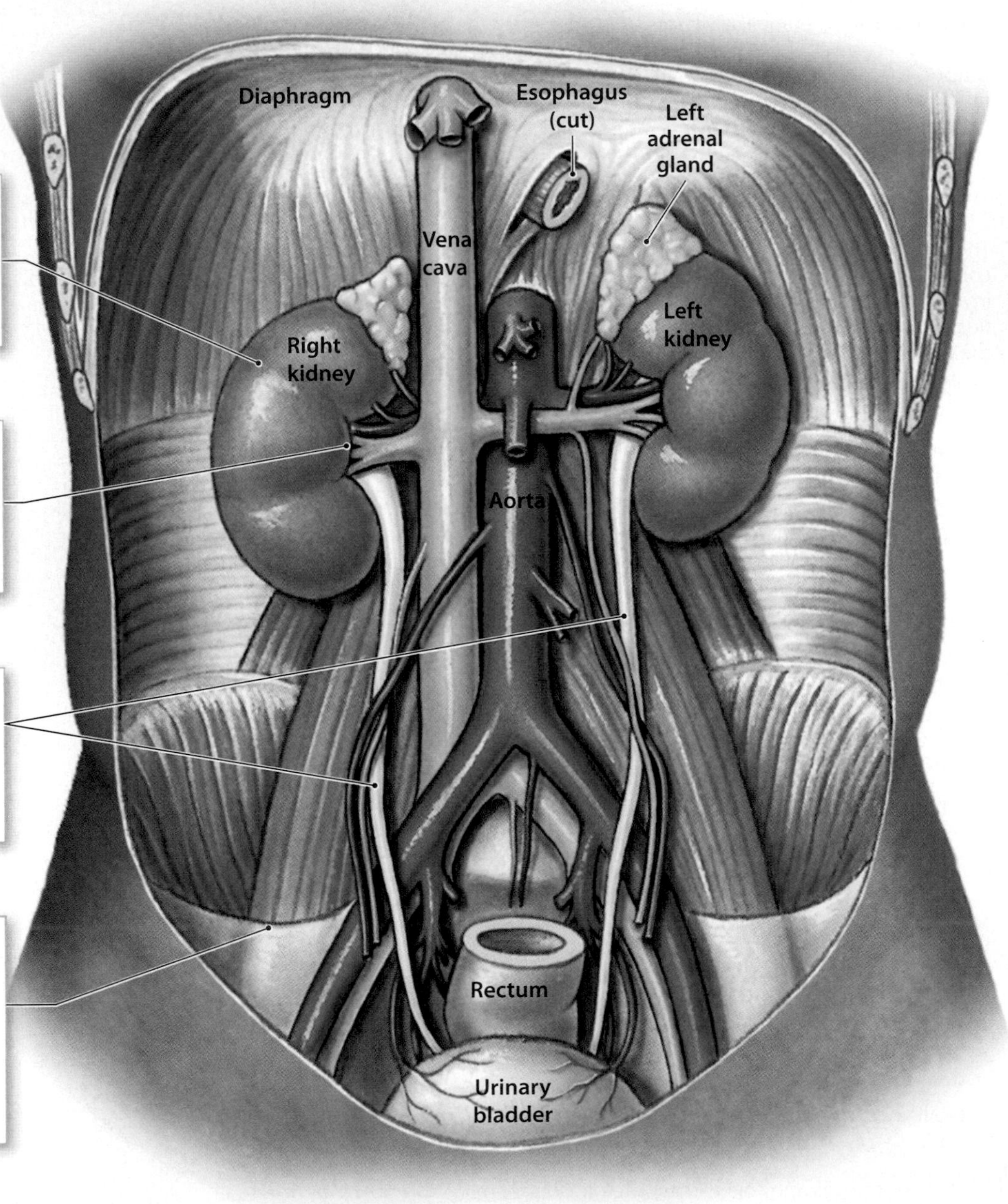

Diaphragm
Esophagus (cut)
Left adrenal gland
Vena cava
Left kidney
Right kidney
Aorta
Rectum
Urinary bladder

A typical adult **kidney** is reddish-brown and about 10 cm (4 in.) long, 5.5 cm (2.2 in.) wide, and 3 cm (1.2 in.) thick. Each kidney weighs about 150 g (5.25 oz).

The **hilum**, a prominent medial indentation, is the point of entry for the renal artery and renal nerves, and the point of exit for the renal vein and the ureter.

The **ureters** pass inferiorly and cross the anterior surfaces of the external iliac artery and vein before emptying into the posterior, inferior surface of the urinary bladder.

This is the cut edge of the posterior peritoneum, which has been removed. The kidneys, adrenal glands, and ureters lie between the muscles of the posterior body wall and the parietal peritoneum, in a **retroperitoneal** position.

2 In this posterior view, you can see that the kidneys are located on either side of the vertebral column, between vertebrae T_{12} and L_3. In this position the kidneys are protected by the visceral organs (anteriorly), the musculature of the body wall, and the 11th and 12th ribs (posteriorly and laterally). The left kidney lies slightly superior to the right kidney.

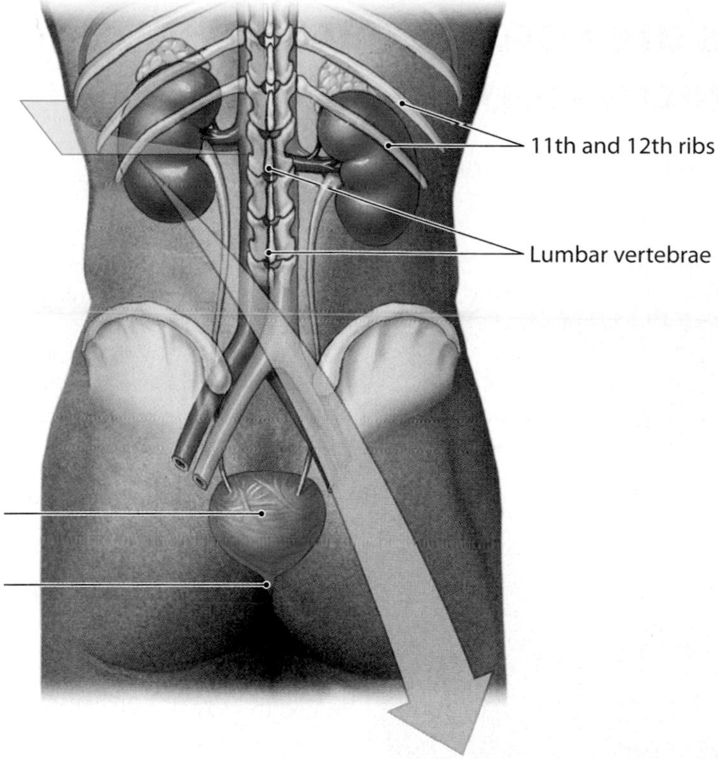
11th and 12th ribs
Lumbar vertebrae
Urinary bladder
Urethra

3 The position of the kidneys in the abdominal cavity is maintained by (1) the overlying peritoneum, (2) contact with adjacent visceral organs, and (3) supporting connective tissues. As seen in this sectional view, three concentric layers of connective tissue protect and stabilize each kidney.

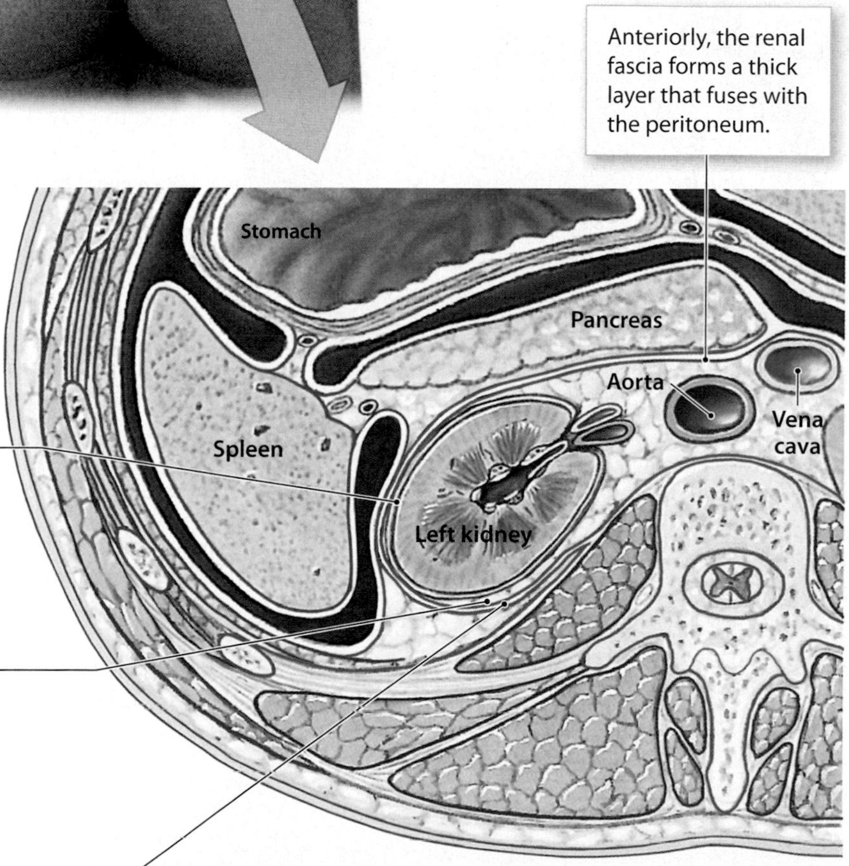
Anteriorly, the renal fascia forms a thick layer that fuses with the peritoneum.
Stomach
Pancreas
Aorta
Vena cava
Spleen
Left kidney

Fibrous Capsule

The **fibrous capsule** is a layer of collagen fibers that covers the outer surface of the entire organ.

Perinephric Fat

The **perinephric fat** (perinephric fat capsule) is a thick layer of adipose tissue that surrounds the fibrous capsule.

Renal Fascia

The **renal fascia** is a dense, fibrous outer layer that anchors the kidney to surrounding structures. Collagen fibers extend outward from the fibrous capsule through the perinephric fat to this layer. Posteriorly, the renal fascia fuses with the deep fascia surrounding the muscles of the body wall.

Module 24.2 Review

a. List the main structures composing the urinary system.

b. Describe the concentric layers of connective tissue that protect and anchor the kidney.

c. What would happen to a kidney's position if the perinephric fat layer were depleted and the collagen fibers of the fibrous capsule were to become detached?

LO **24.2** Describe the location and structural features of the kidneys.

Module 24.3

The kidneys are complex at the gross and microscopic levels

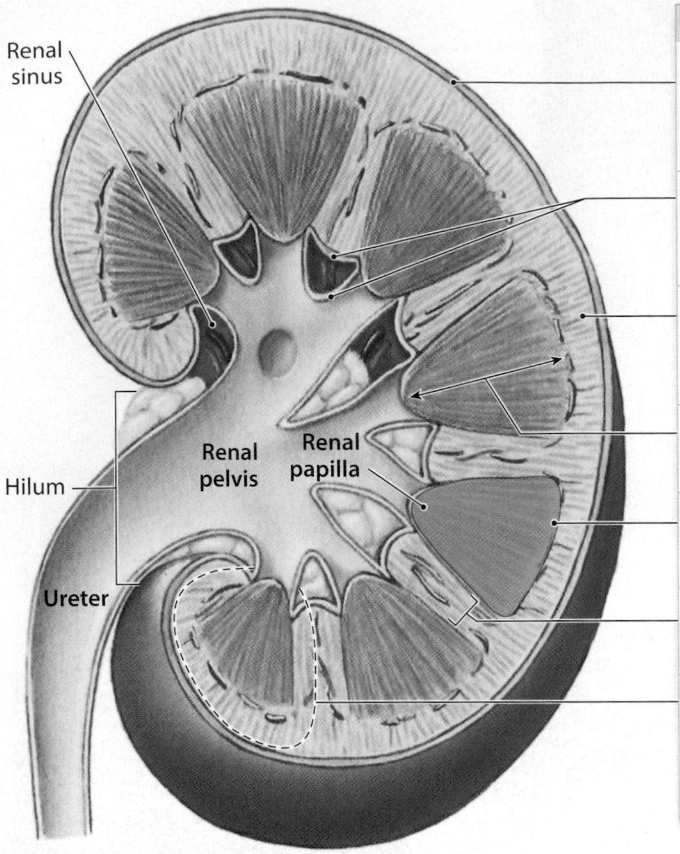

Renal sinus
Hilum
Ureter
Renal pelvis
Renal papilla

Major Structural Landmarks of the Kidney
The **fibrous capsule** covering the outer surface of the kidney also lines the **renal sinus**, an internal cavity within the kidney. The outer and inner linings are continuous at the hilum.
Within the renal sinus, the fibrous capsule stabilizes the positions of the ureter, the renal blood vessels, and renal nerves.
The **renal cortex** is the superficial portion of the kidney, in contact with the fibrous capsule. The cortex is reddish brown and granular.
The **renal medulla** extends from the renal cortex to the renal sinus.
A **renal pyramid** is a conical structure extending from the cortex to a tip called the **renal papilla**.
A **renal column** is a band of granular tissue that separates adjacent pyramids.
A **kidney lobe** consists of a renal pyramid, the overlying area of renal cortex, and adjacent tissues of the renal columns. Each kidney contains 6–18 kidney lobes. Urine production occurs in the kidney lobes.

1 Here is a diagrammatic view of a sectioned kidney, showing the major landmarks and features. The blood vessels servicing the kidney enter through the **hilum**.

A **minor calyx** collects the urine produced by a single kidney lobe.

A **major calyx** forms through the fusion of 4–5 minor calyces (KĀ-li-sēz).

The **renal pelvis** is a large, funnel-shaped structure that collects urine from the major calyces. It is continuous with the ureter.

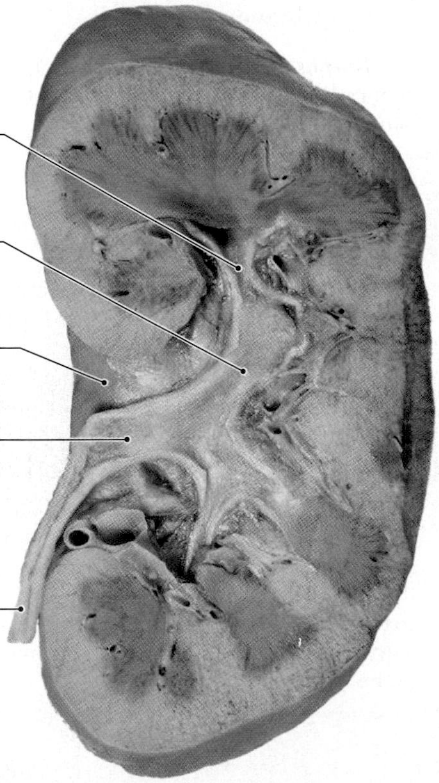

Hilum
Ureter

2 Here is a frontal section of a human kidney. The labels follow the path of urine within the collecting system (discussed later) of the renal sinus, which communicates with the ureter at the hilum.

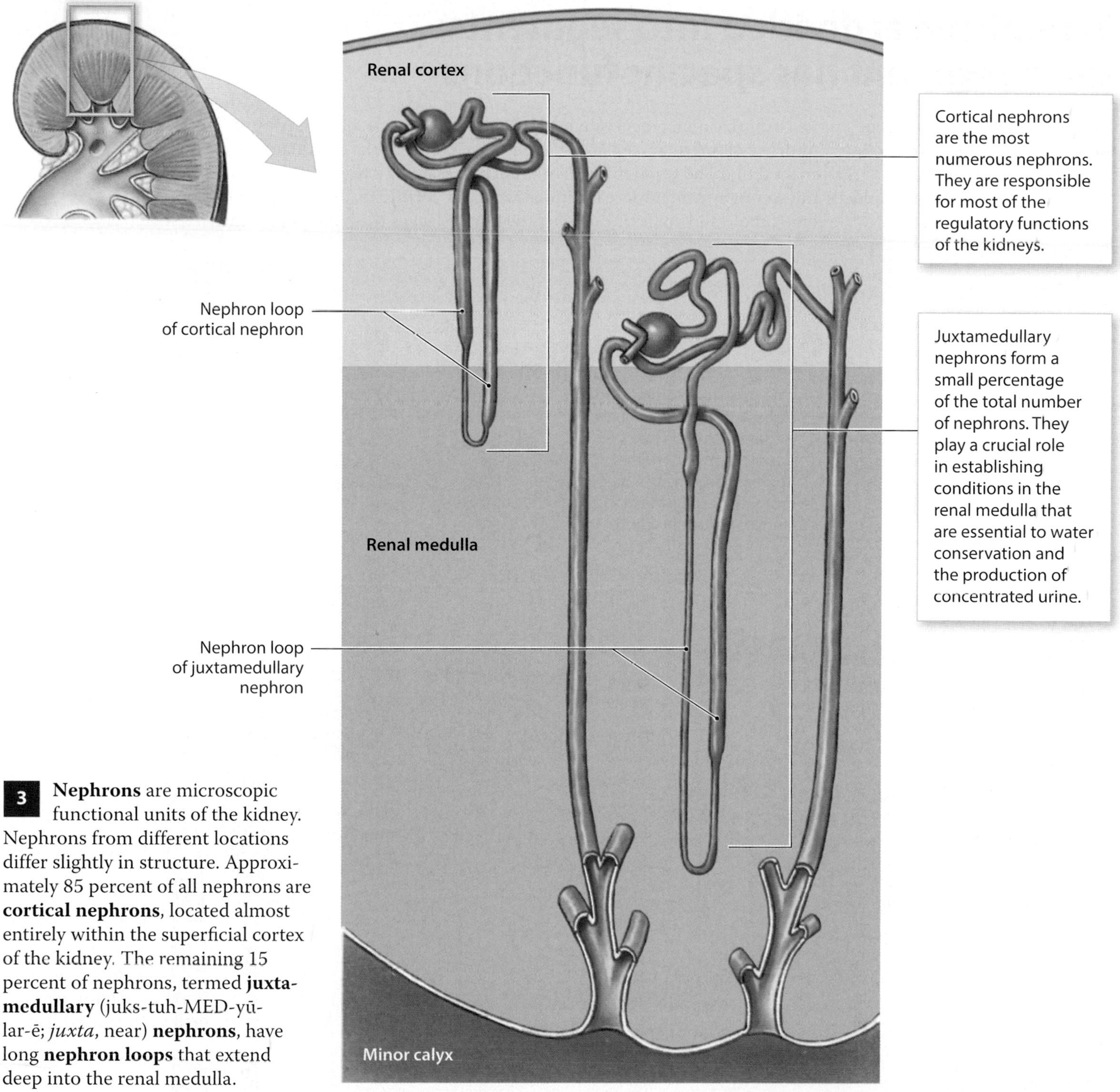

Renal cortex
Cortical nephrons are the most numerous nephrons. They are responsible for most of the regulatory functions of the kidneys.
Nephron loop of cortical nephron
Juxtamedullary nephrons form a small percentage of the total number of nephrons. They play a crucial role in establishing conditions in the renal medulla that are essential to water conservation and the production of concentrated urine.
Renal medulla
Nephron loop of juxtamedullary nephron
Minor calyx

3 **Nephrons** are microscopic functional units of the kidney. Nephrons from different locations differ slightly in structure. Approximately 85 percent of all nephrons are **cortical nephrons**, located almost entirely within the superficial cortex of the kidney. The remaining 15 percent of nephrons, termed **juxtamedullary** (juks-tuh-MED-yū-lar-ē; *juxta*, near) **nephrons**, have long **nephron loops** that extend deep into the renal medulla.

Module 24.3 Review

a. Which structure is a conical mass within the renal medulla that ends at the papilla?

b. Describe the renal papilla.

c. Which type of nephron is essential for the conservation of water and the production of concentrated urine?

LO 24.3 Describe the gross structural features of the kidney, and distinguish between cortical and juxtamedullary nephrons.

Module 24.4

A nephron is divided into segments; each segment has specific functions

1 The nephron consists of a renal corpuscle and a renal tubule. At the renal corpuscle, blood pressure forces water and dissolved solutes out of the glomerular capillaries and into a chamber—the **capsular space**—that is continuous with the lumen of the renal tubule. The **renal tubule** is a tubular passageway which may be 50 mm (1.97 in.) long. Filtration produces an essentially protein-free solution, known as a **filtrate**, that is similar to blood plasma. After modification by the renal tubule and collecting system, the filtrate leaves the kidneys as urine.

Nephron

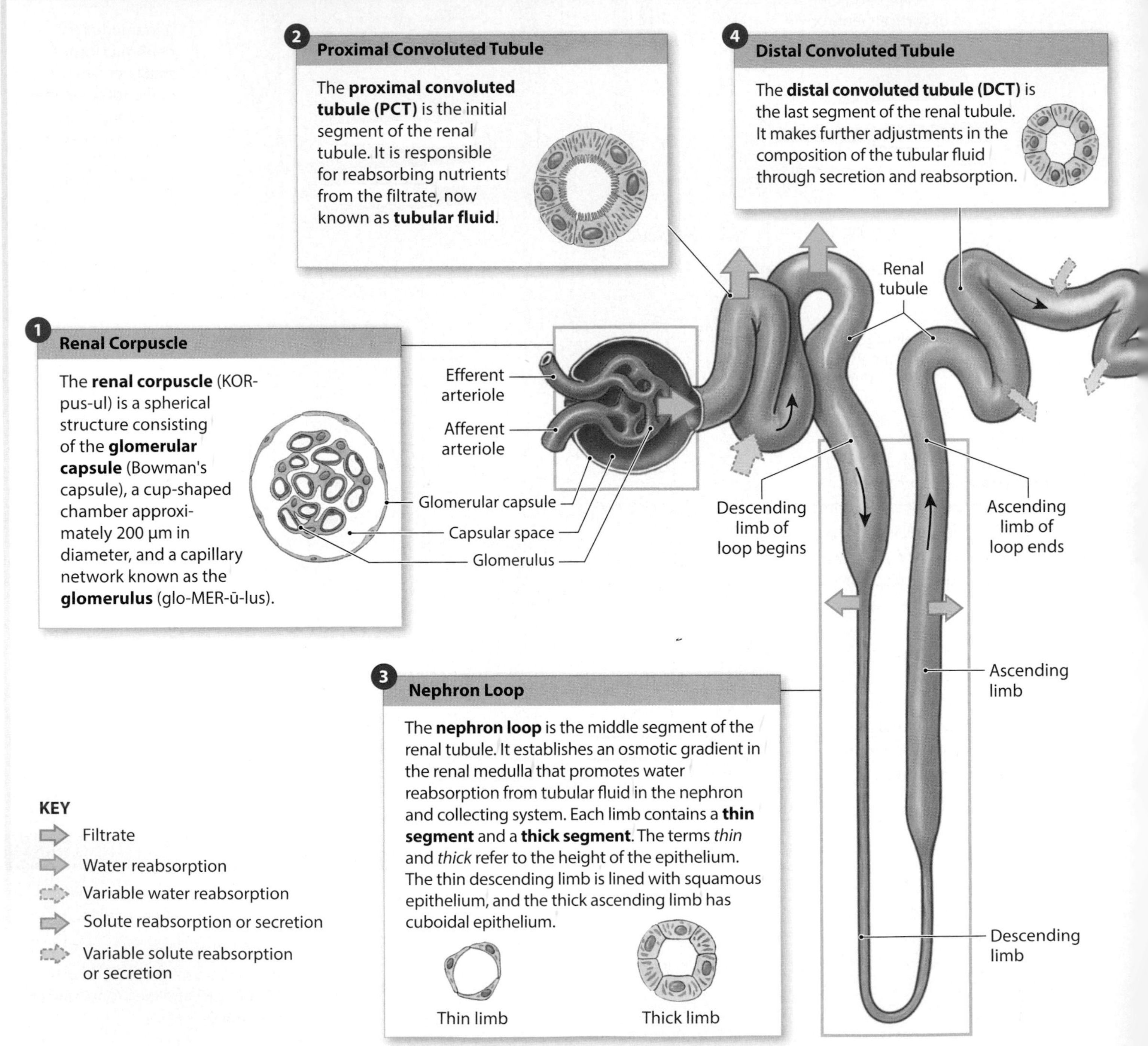

Efferent arteriole
Afferent arteriole
Glomerular capsule
Capsular space
Glomerulus
Renal tubule
Descending limb of loop begins
Ascending limb of loop ends
Ascending limb
Descending limb
Thin limb
Thick limb
KEY
Filtrate
Water reabsorption
Variable water reabsorption
Solute reabsorption or secretion
Variable solute reabsorption or secretion

1 Renal Corpuscle

The **renal corpuscle** (KOR-pus-ul) is a spherical structure consisting of the **glomerular capsule** (Bowman's capsule), a cup-shaped chamber approximately 200 μm in diameter, and a capillary network known as the **glomerulus** (glo-MER-ū-lus).

2 Proximal Convoluted Tubule

The **proximal convoluted tubule (PCT)** is the initial segment of the renal tubule. It is responsible for reabsorbing nutrients from the filtrate, now known as **tubular fluid**.

3 Nephron Loop

The **nephron loop** is the middle segment of the renal tubule. It establishes an osmotic gradient in the renal medulla that promotes water reabsorption from tubular fluid in the nephron and collecting system. Each limb contains a **thin segment** and a **thick segment**. The terms *thin* and *thick* refer to the height of the epithelium. The thin descending limb is lined with squamous epithelium, and the thick ascending limb has cuboidal epithelium.

4 Distal Convoluted Tubule

The **distal convoluted tubule (DCT)** is the last segment of the renal tubule. It makes further adjustments in the composition of the tubular fluid through secretion and reabsorption.

2 Each nephron empties into the **collecting system**, a series of tubes that carry tubular fluid away from the nephron. Collecting ducts receive this fluid from many nephrons. Each collecting duct begins in the renal cortex and descends into the renal medulla, carrying fluid to a papillary duct that drains into a minor calyx.

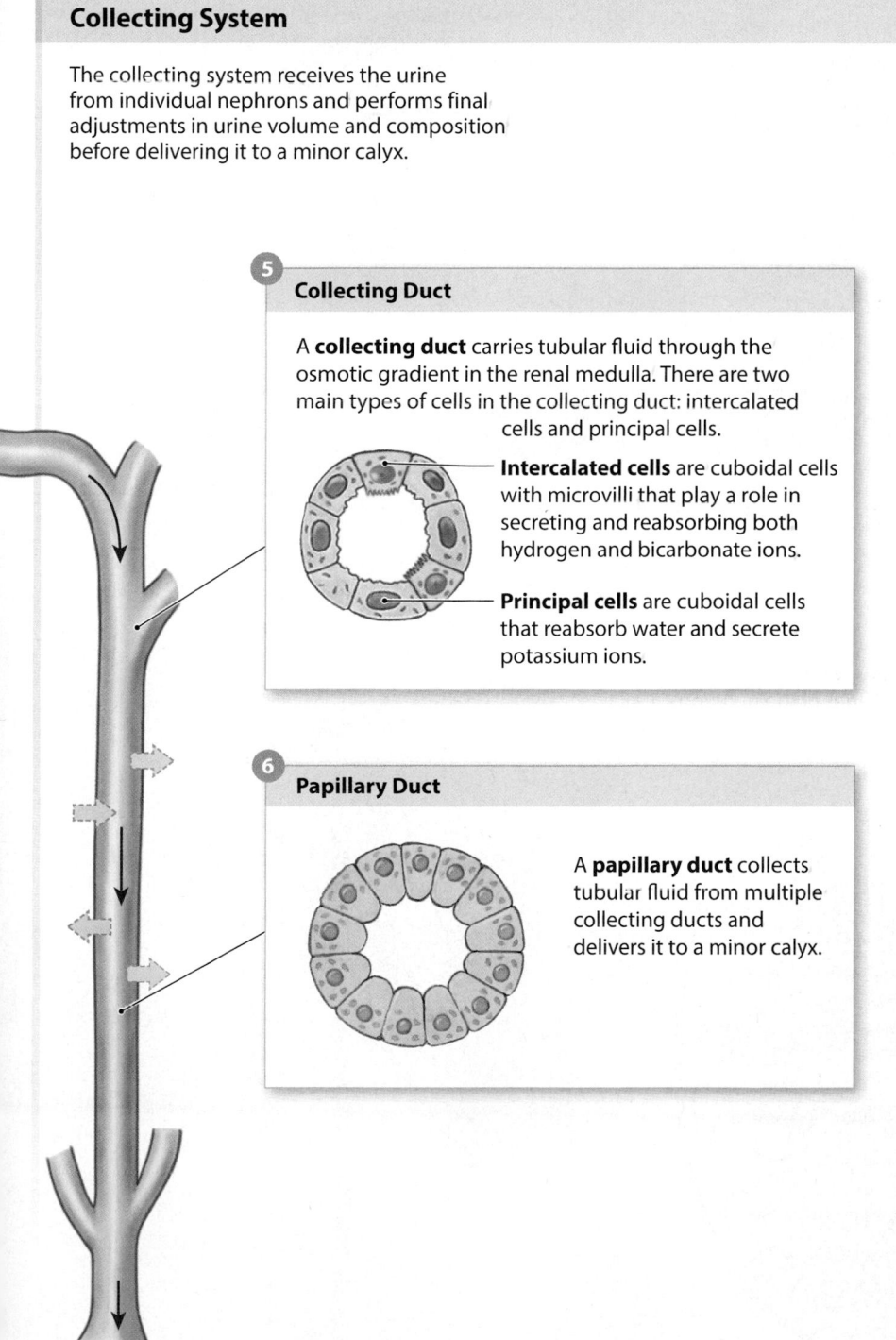

Collecting System

The collecting system receives the urine from individual nephrons and performs final adjustments in urine volume and composition before delivering it to a minor calyx.

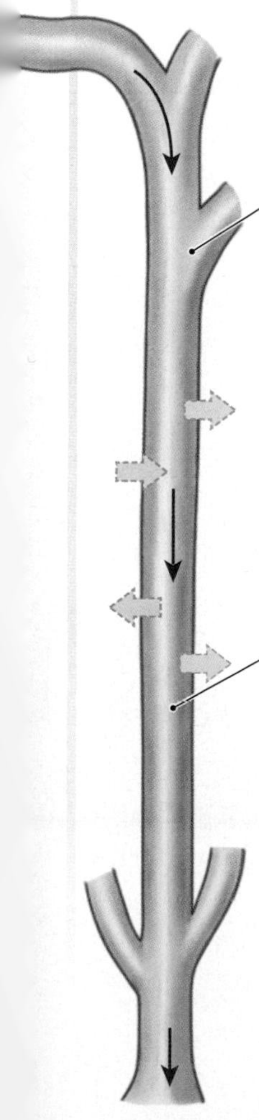

5 Collecting Duct

A **collecting duct** carries tubular fluid through the osmotic gradient in the renal medulla. There are two main types of cells in the collecting duct: intercalated cells and principal cells.

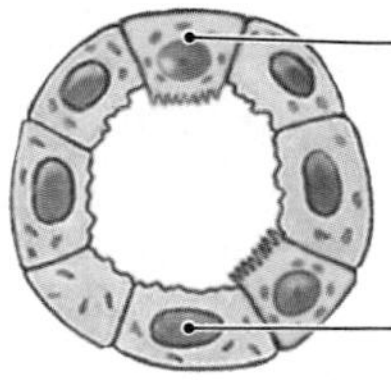

Intercalated cells are cuboidal cells with microvilli that play a role in secreting and reabsorbing both hydrogen and bicarbonate ions.

Principal cells are cuboidal cells that reabsorb water and secrete potassium ions.

6 Papillary Duct

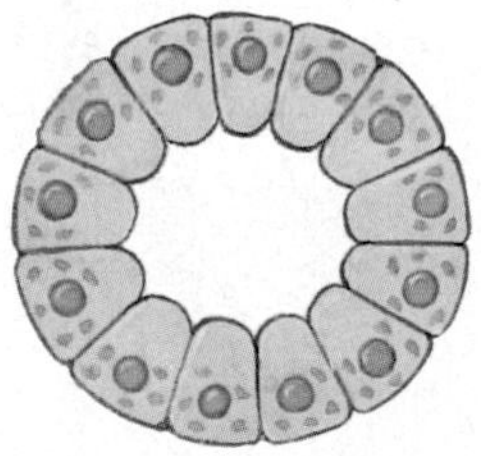

A **papillary duct** collects tubular fluid from multiple collecting ducts and delivers it to a minor calyx.

As it travels along the renal tubule, tubular fluid gradually changes in composition. The characteristics of the urine that enters the minor calyx vary from moment to moment depending on the activities under way in each segment of the nephron and collecting system.

Module 24.4 Review

a. Describe filtrate.

b. List the primary structures of the nephron and the collecting system.

c. Identify the components of the renal corpuscle.

LO 24.4 Describe the segments of the nephron and collecting system, including their general functions and histological appearance.

Module 24.5

The kidneys are highly vascular, and the circulation patterns are complex

1 Here are diagrammatic views of the **arterial** system supplying the kidney (top) and the **venous** system draining the kidney (bottom).

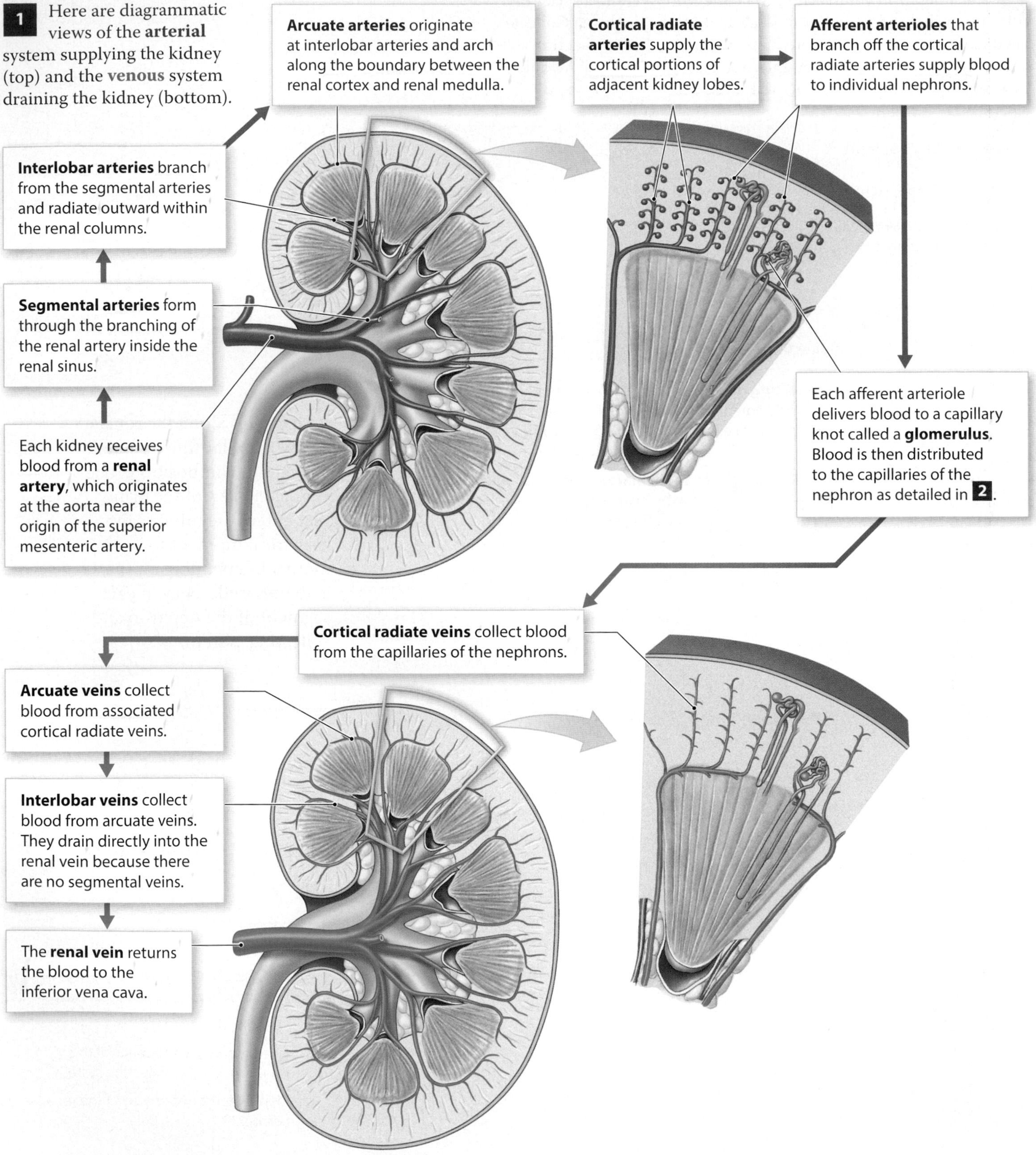

Arcuate arteries originate at interlobar arteries and arch along the boundary between the renal cortex and renal medulla.
Cortical radiate arteries supply the cortical portions of adjacent kidney lobes.
Afferent arterioles that branch off the cortical radiate arteries supply blood to individual nephrons.
Interlobar arteries branch from the segmental arteries and radiate outward within the renal columns.
Segmental arteries form through the branching of the renal artery inside the renal sinus.
Each kidney receives blood from a **renal artery**, which originates at the aorta near the origin of the superior mesenteric artery.
Each afferent arteriole delivers blood to a capillary knot called a **glomerulus**. Blood is then distributed to the capillaries of the nephron as detailed in 2.
Cortical radiate veins collect blood from the capillaries of the nephrons.
Arcuate veins collect blood from associated cortical radiate veins.
Interlobar veins collect blood from arcuate veins. They drain directly into the renal vein because there are no segmental veins.
The **renal vein** returns the blood to the inferior vena cava.

2 In a cortical nephron, the nephron loop is relatively short, and the **efferent arteriole** delivers blood to a network of **peritubular capillaries**, which surround the entire renal tubule. Both the nephrons and the peritubular capillaries are surrounded by interstitial fluid called **peritubular fluid**. These capillaries drain into small venules that carry blood to the cortical radiate veins.

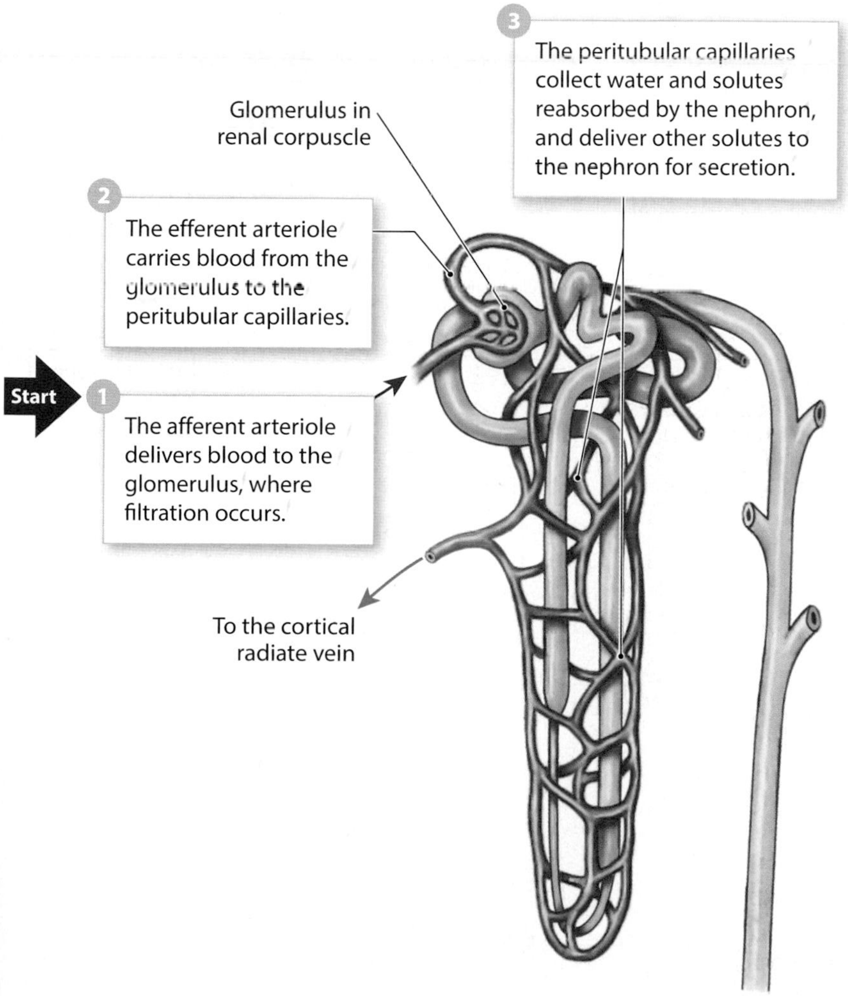

Glomerulus in renal corpuscle
3 The peritubular capillaries collect water and solutes reabsorbed by the nephron, and deliver other solutes to the nephron for secretion.
2 The efferent arteriole carries blood from the glomerulus to the peritubular capillaries.
Start
1 The afferent arteriole delivers blood to the glomerulus, where filtration occurs.
To the cortical radiate vein

3 In a juxtamedullary nephron, the peritubular capillaries are connected to the **vasa recta** (*vasa*, vessels + *recta*, straight)—long, straight capillaries that parallel the nephron loop.

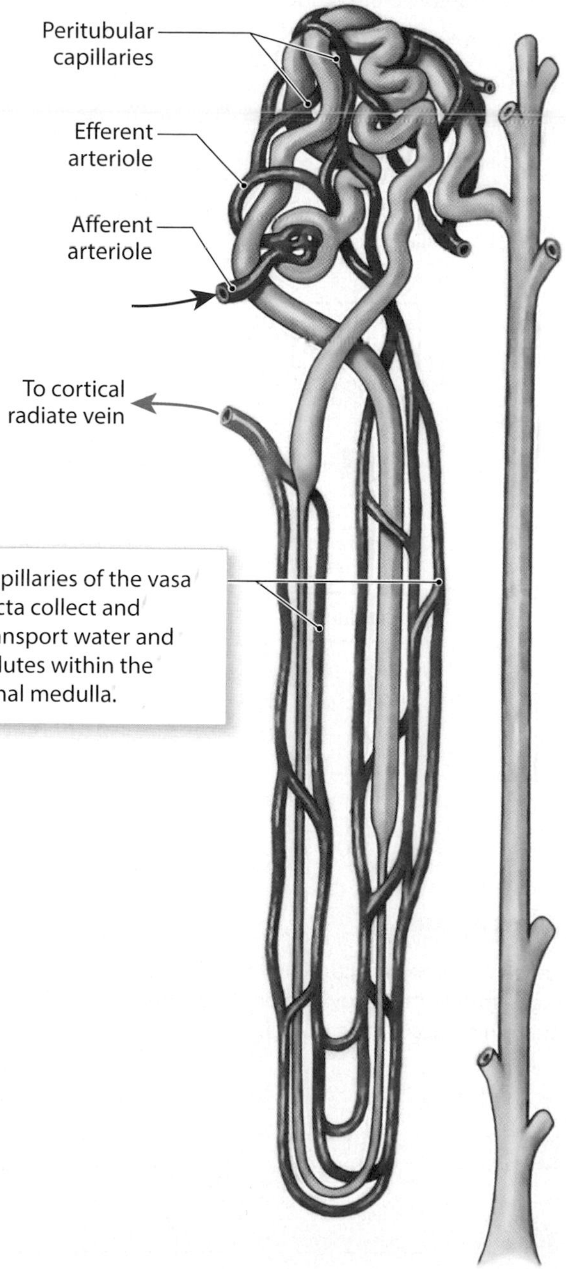

Peritubular capillaries
Efferent arteriole
Afferent arteriole
To cortical radiate vein
Capillaries of the vasa recta collect and transport water and solutes within the renal medulla.

Each kidney has approximately 1.25 million nephrons, with a combined length of about 145 km (85 miles). Both the cortical and the juxtamedullary nephrons are innervated by renal nerves that enter at the hilum and follow the branches of the renal arteries. Most of the nerve fibers involved are sympathetic postganglionic fibers from the celiac plexus and the inferior splanchnic nerves. Sympathetic innervation adjusts blood flow and blood pressure at the glomeruli, and stimulates the release of renin.

Module 24.5 Review

a. Trace the pathway of blood from the renal artery to the renal vein.

b. Describe how blood enters and leaves the glomerulus.

c. What are the vasa recta?

LO **24.5** Trace the pathway of blood flow through a kidney, and compare the pattern of blood flow in cortical and juxtamedullary nephrons.

Concept map

Use each of the following terms once to fill in the blank boxes to correctly complete the map.

- ureter
- proximal convoluted tubule
- glomerulus
- urinary bladder
- renal tubules
- papillary ducts
- major calyces
- renal medulla
- renal sinus
- nephrons

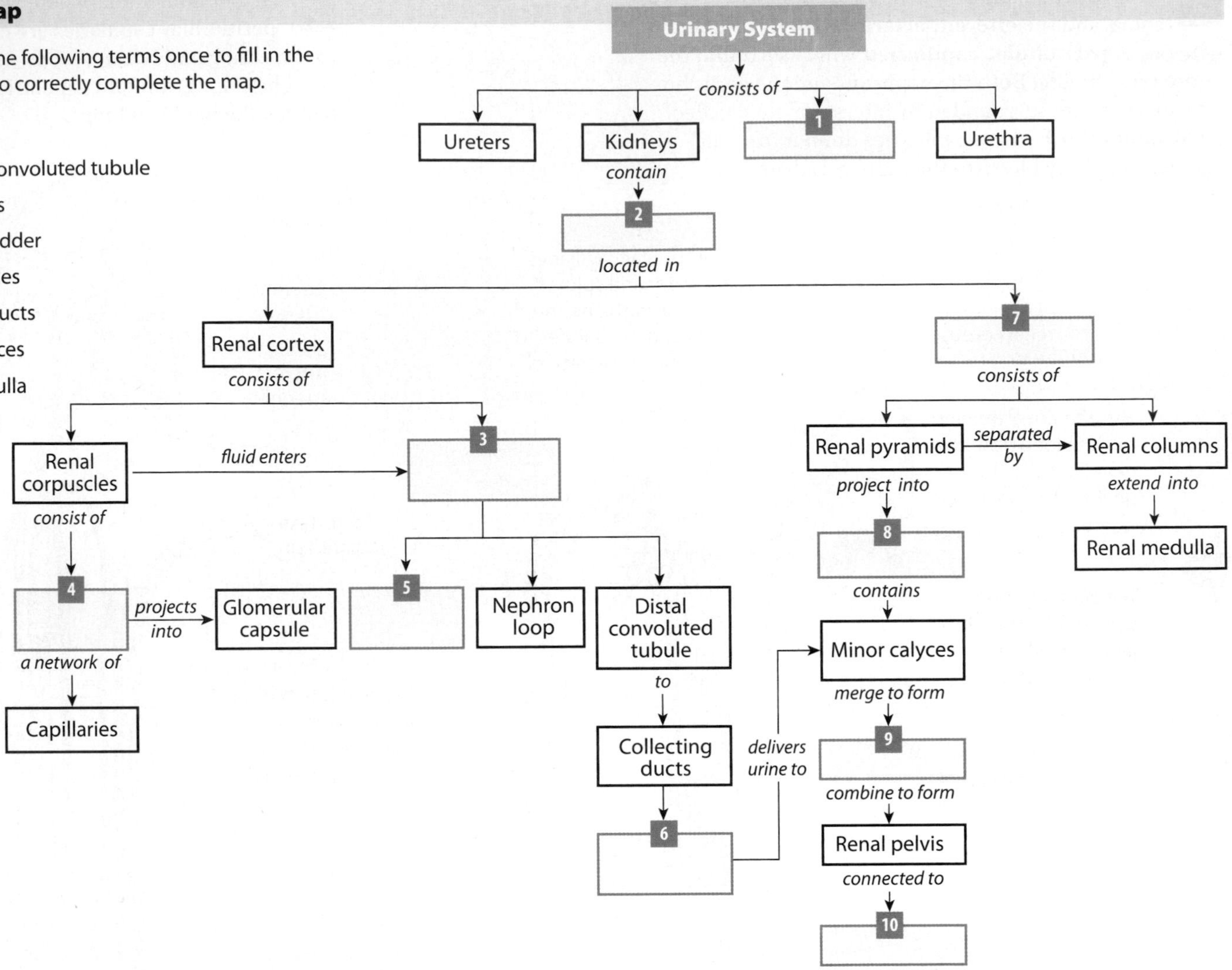
Urinary System
consists of
Ureters
Kidneys
1
Urethra
contain
2
located in
Renal cortex
consists of
7
consists of
Renal corpuscles
fluid enters
3
Renal pyramids
separated by
Renal columns
consist of
project into
extend into
4
projects into
Glomerular capsule
5
Nephron loop
Distal convoluted tubule
8
Renal medulla
a network of
contains
Capillaries
to
Minor calyces
Collecting ducts
delivers urine to
merge to form
9
6
combine to form
Renal pelvis
connected to
10

Short answer

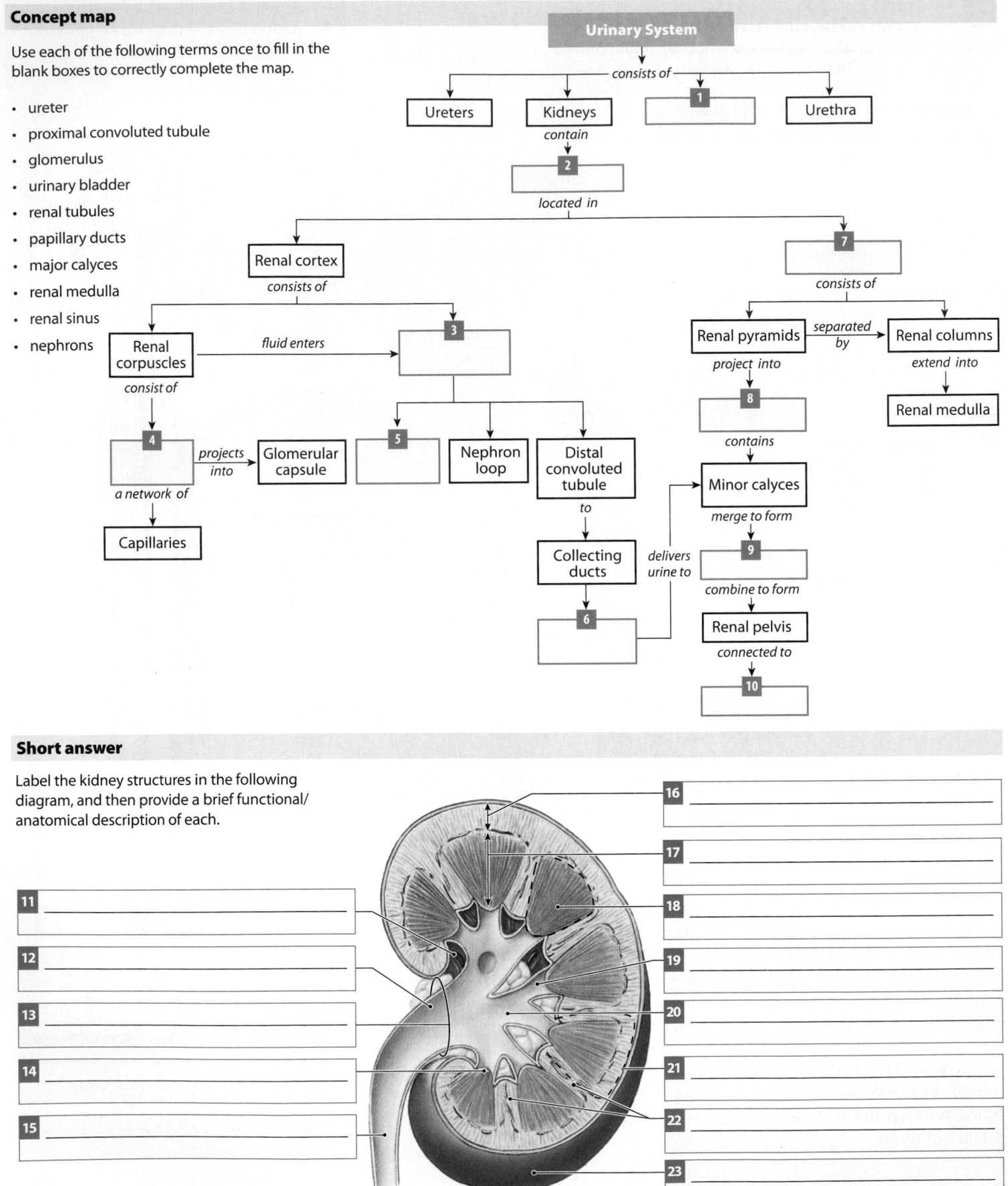

Label the kidney structures in the following diagram, and then provide a brief functional/anatomical description of each.

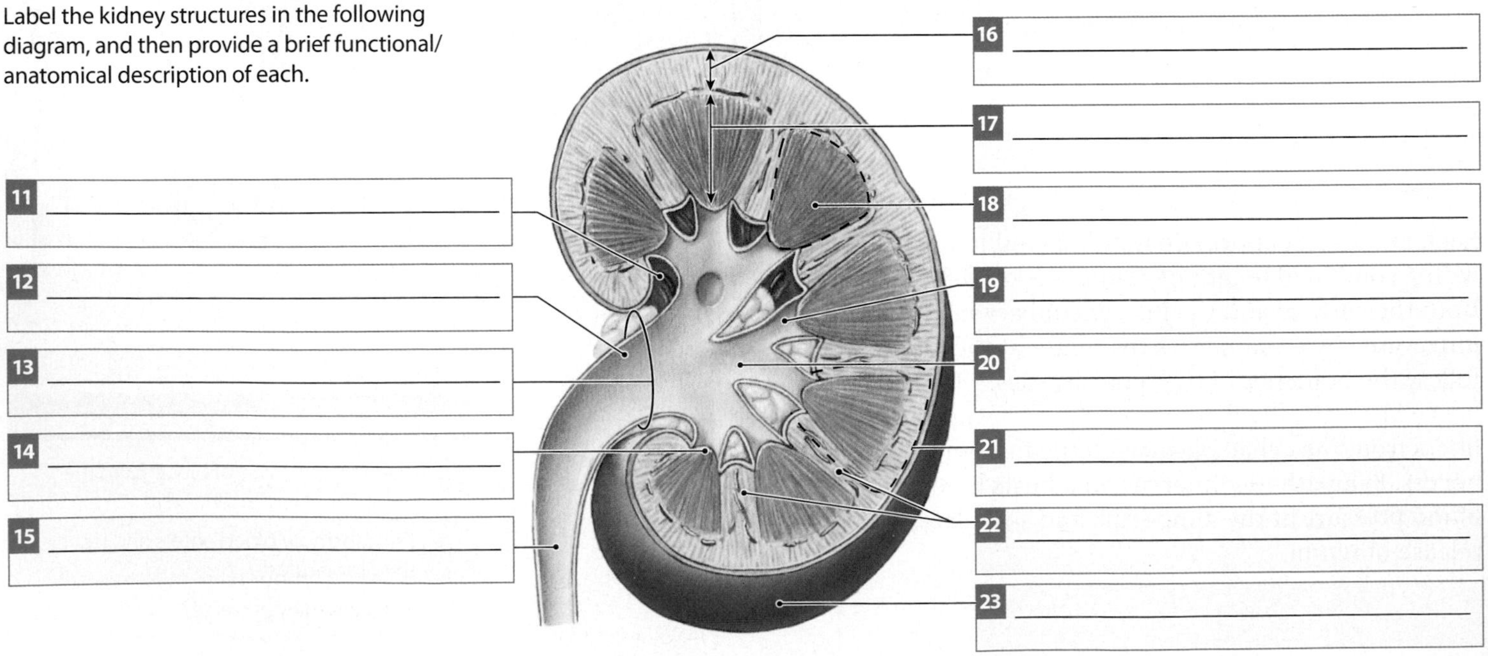
11
12
13
14
15
16
17
18
19
20
21
22
23

The kidneys maintain homeostasis by removing wastes and producing urine

The goal of urine production is to maintain homeostasis by regulating the volume and composition of blood. This process involves the excretion of solutes—specifically, metabolic wastes. Three organic wastes are noteworthy:

- **Urea**, the most abundant organic waste, is a by-product of the breakdown of amino acids in the liver.
- **Creatinine** is generated in skeletal muscle tissue through the breakdown of creatine phosphate, a high-energy compound that plays an important role in muscle contraction.
- **Uric acid** is a waste formed during the recycling of the nitrogenous bases of RNA molecules.

The kidneys are usually capable of producing concentrated urine with an osmotic concentration of 855–1335 mOsm/L, more than four times that of blood plasma. In a clinical setting, the osmolarity of urine is often reported as osmolality, which is milliosmoles per kilogram of water (mOsm/kg H_2O). When discussing osmotic concentrations in body fluids, these terms are often used interchangeably.

1 The kidneys produce fluid (urine) that is different from other body fluids. This table shows renal efficiency by comparing concentrations of typical substances found in blood plasma and urine.

Normal Laboratory Values for Solutes in Plasma and Urine

Solute	Blood Plasma	Urine
Ions (mEq/L)		
Sodium (Na^+)	135–145	40–220
Potassium (K^+)	3.5–5.0	25–100
Bicarbonate (HCO_3^-)	20–28	1.9
Metabolites and Nutrients (mg/dL)		
Glucose	70–110	0.009
Lipids	450–1000	0.002
Proteins	6.0–8.0 g/dL	0.000
Nitrogenous Wastes (mg/dL)		
Urea	8–25	1800
Creatinine	0.6–1.5	150
Uric acid	2–6	40
Ammonia	<0.1	60

Note: The values indicated are typical ranges; specific numbers vary depending on the laboratory and testing methods used.

2 Kidney function requires three distinct physiological processes.

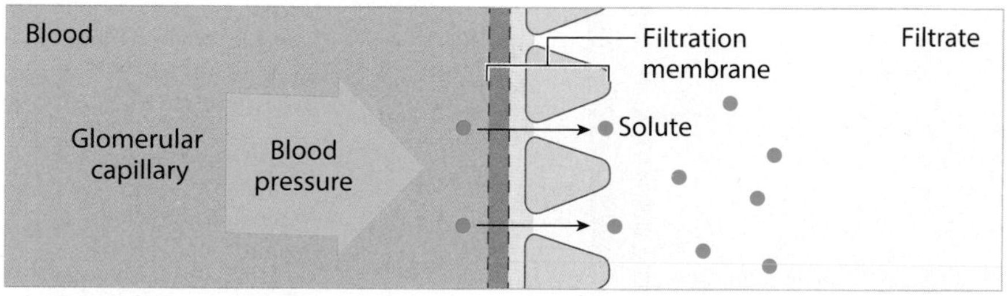
Blood
Filtration membrane
Filtrate
Glomerular capillary
Blood pressure
Solute

In **filtration**, blood pressure forces water and solutes across the membranes of the glomerular capillaries and into the capsular space. Solute molecules small enough to pass through the filtration membrane are carried by the surrounding water molecules.

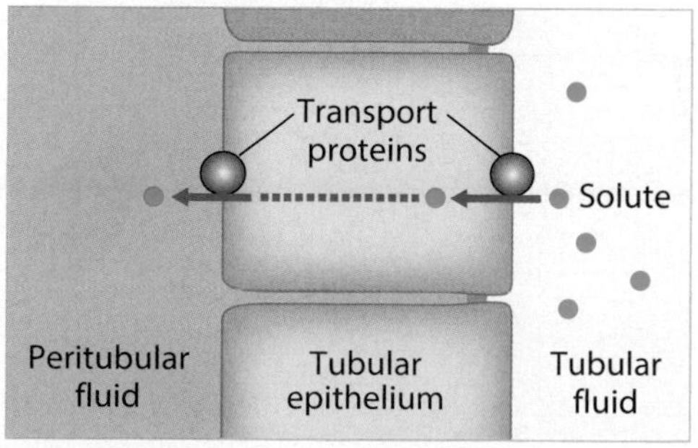
Transport proteins
Solute
Peritubular fluid
Tubular epithelium
Tubular fluid

Reabsorption is the transport of water and solutes from the tubular fluid, across the tubular epithelium, and into the peritubular fluid.

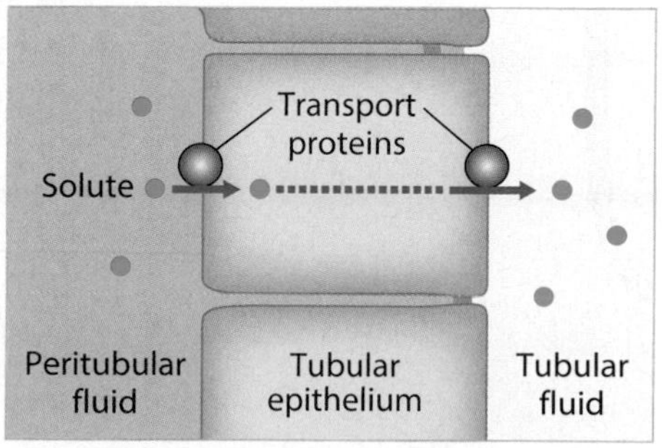
Transport proteins
Solute
Peritubular fluid
Tubular epithelium
Tubular fluid

Secretion is the transport of solutes from the peritubular fluid, across the tubular epithelium, and into the tubular fluid.

Module 24.6 Review

a. Identify three solutes excreted in urine, and describe how they are generated.

b. In which direction do fluids and solutes move in each of the three kidney processes?

c. Blood pressure is required for which kidney function?

LO 24.6 Briefly describe how the kidneys maintain homeostasis and produce urine.

Filtration, reabsorption, and secretion occur in specific segments of the nephron and collecting system

1 This diagram summarizes the general functions of the various segments of the nephron and collecting system in the formation of urine. Most segments perform a combination of reabsorption and secretion, but the balance between the two processes varies from one segment to another. Regulation of the final volume and solute concentration of the urine results from the interaction between the collecting system and the nephron loops, especially the long loops of the juxtamedullary nephrons.

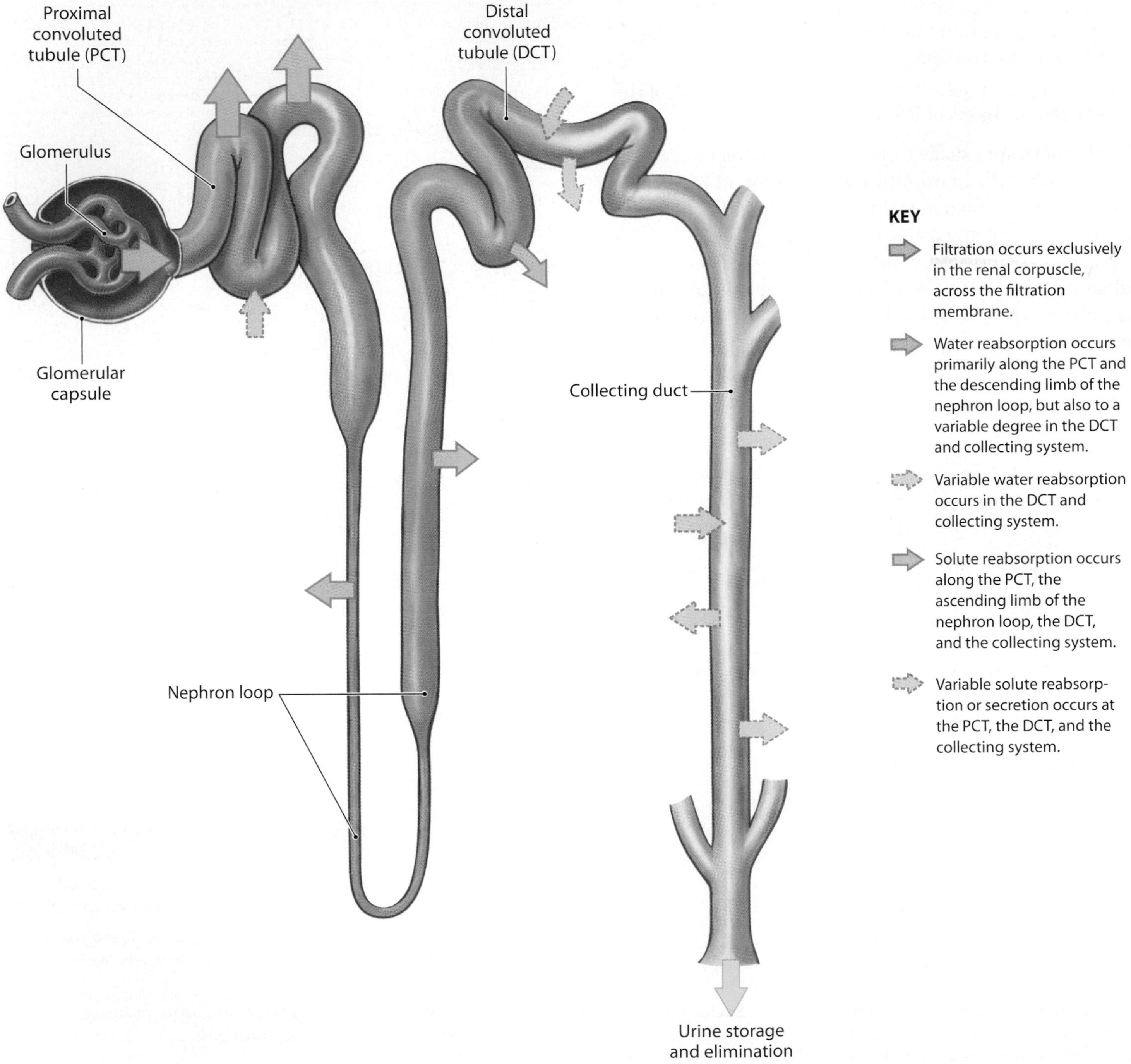
Proximal convoluted tubule (PCT)
Distal convoluted tubule (DCT)
Glomerulus
Glomerular capsule
Collecting duct
Nephron loop
Urine storage and elimination
KEY
Filtration occurs exclusively in the renal corpuscle, across the filtration membrane.
Water reabsorption occurs primarily along the PCT and the descending limb of the nephron loop, but also to a variable degree in the DCT and collecting system.
Variable water reabsorption occurs in the DCT and collecting system.
Solute reabsorption occurs along the PCT, the ascending limb of the nephron loop, the DCT, and the collecting system.
Variable solute reabsorption or secretion occurs at the PCT, the DCT, and the collecting system.

2 The table below gives an overview of the functions of the various parts of the nephron and collecting system. We will consider the roles of these nephron segments and surrounding blood vessels in greater detail in subsequent modules.

Renal Structures and Their Functions

Segment	General Functions	Specific Functions
Renal corpuscle	*Filtration* of blood; generates approximately 180 L/day of filtrate similar in composition to blood plasma but without plasma proteins	*Filtration* Water and inorganic and organic solutes from plasma *Retention* Plasma proteins and blood cells
Proximal convoluted tubule (PCT)	*Reabsorption* of 60–70% of the water (108–116 L/day), 99–100% of the organic substrates, and 60–70% of the sodium and chloride ions in the original filtrate	*Active reabsorption* Glucose, other simple sugars, amino acids, vitamins, ions (including sodium, potassium, calcium, magnesium, phosphate, and bicarbonate) *Passive reabsorption* Urea, chloride ions, lipid-soluble materials, water *Secretion* Hydrogen ions, ammonium ions, creatinine, drugs, and toxins
Nephron loop	*Reabsorption* of 25% of the water (45 L/day) and 20–25% of the sodium and chloride ions in the original filtrate; creation of the concentration gradient in the renal medulla	*Reabsorption* Sodium and chloride ions, water
Distal convoluted tubule (DCT)	*Reabsorption* of a variable amount of water (usually 5%, or 9 L/day) under antidiuretic hormone stimulation, and a variable amount of sodium ions under aldosterone stimulation	*Reabsorption* Sodium and chloride ions, sodium ions (variable), calcium ions (variable), water (variable) *Secretion* Hydrogen ions, ammonium ions, creatinine, drugs, and toxins
Collecting system	*Reabsorption* of a variable amount of water (usually 9.3%, or 16.8 L/day) under antidiuretic hormone stimulation, and a variable amount of sodium ions under aldosterone stimulation	*Reabsorption* Sodium ions (variable), bicarbonate ions (variable), water (variable) *Secretion* Potassium and hydrogen ions (variable)
Blood Vessels		
Peritubular capillaries	*Redistribution* of water and solutes reabsorbed in the renal cortex	Return of water and solutes from the peritubular fluid to the general circulation
Vasa recta	*Redistribution* of water and solutes reabsorbed in the renal medulla, and stabilization of the concentration gradient of the renal medulla	Return of water and solutes from the peritubular fluid to the general circulation

Module 24.7 Review

a. Identify the three distinct processes of urine formation in the kidney.

b. Where does filtration exclusively occur in the kidney?

c. Which segment of the nephron is solely involved in the reabsorption of water and sodium and chloride ions?

LO **24.7** Describe filtration, reabsorption, and secretion along each segment of the nephron and collecting system.

Filtration occurs at the renal corpuscle

The renal corpuscle, the start of the nephron, filters blood. This is the vital first step in the formation of urine.

1 At the renal corpuscle, the capillary knot of the glomerulus projects into the capsular space like the heart projects into the pericardial cavity. Like the pericardium, the glomerular capsule has an outer parietal layer and an inner visceral layer.

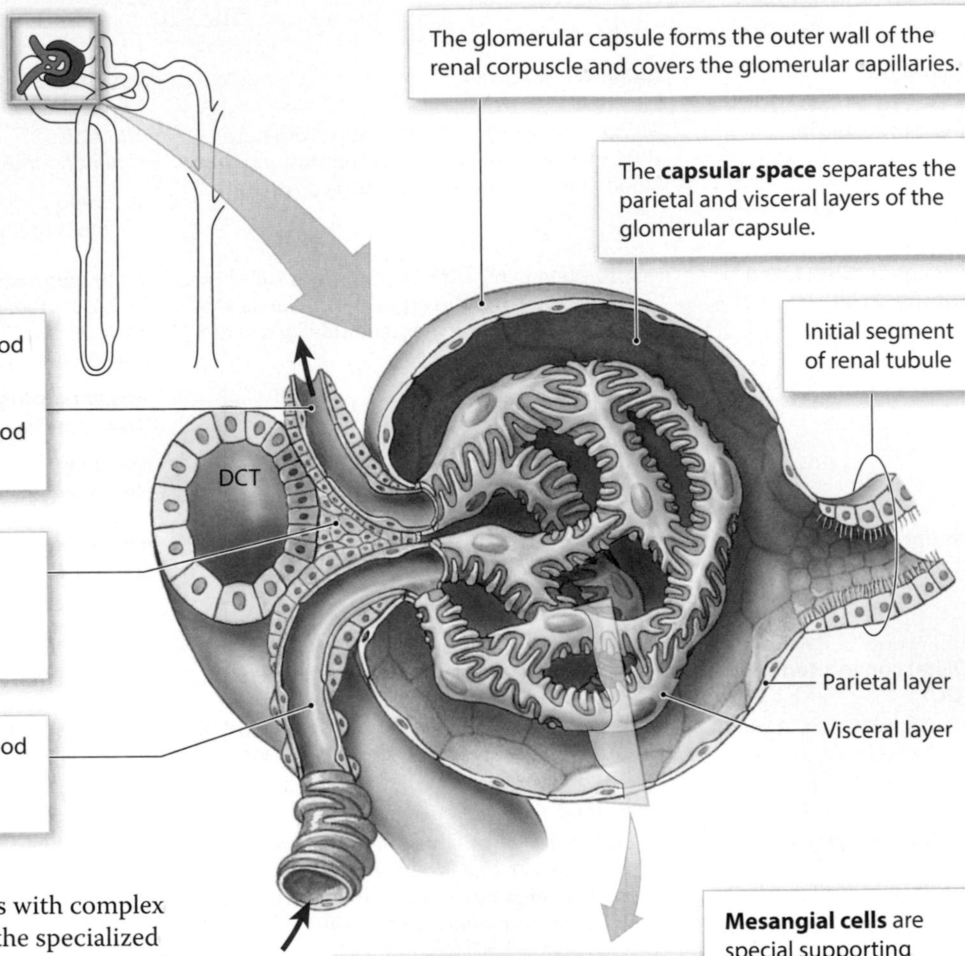
The glomerular capsule forms the outer wall of the renal corpuscle and covers the glomerular capillaries.
The **capsular space** separates the parietal and visceral layers of the glomerular capsule.
Initial segment of renal tubule
The **efferent arteriole** delivers blood to peritubular capillaries. It has a smaller diameter than the afferent arteriole, and this increases the blood pressure within the glomerulus.
DCT
The **juxtaglomerular complex** consists of specialized cells that secrete renin when glomerular blood pressure decreases.
The **afferent arteriole** delivers blood to the glomerulus from a cortical radiate artery.
Parietal layer
Visceral layer

2 The visceral layer consists of large cells with complex processes, or "feet," that wrap around the specialized dense layer of the glomerular capillaries. These unusual cells are called **podocytes** (PŌ-dō-sīts; *podos*, foot + *-cyte*, cell), and their feet are known as **pedicels**. Substances passing out of the blood at the glomerulus must be small enough to pass between the narrow gaps, or **filtration slits**, between adjacent pedicels.

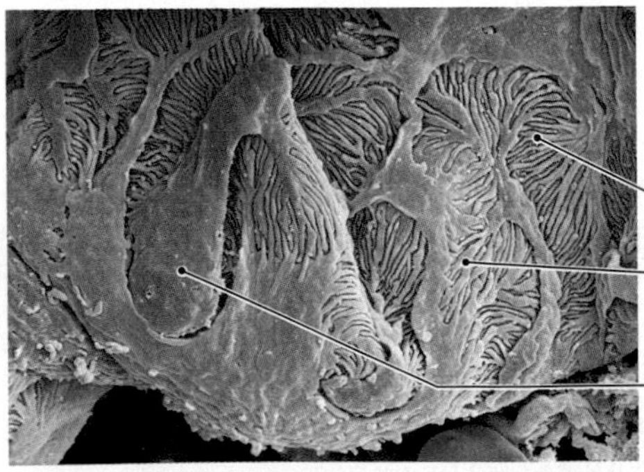

A podocyte SEM × 2100

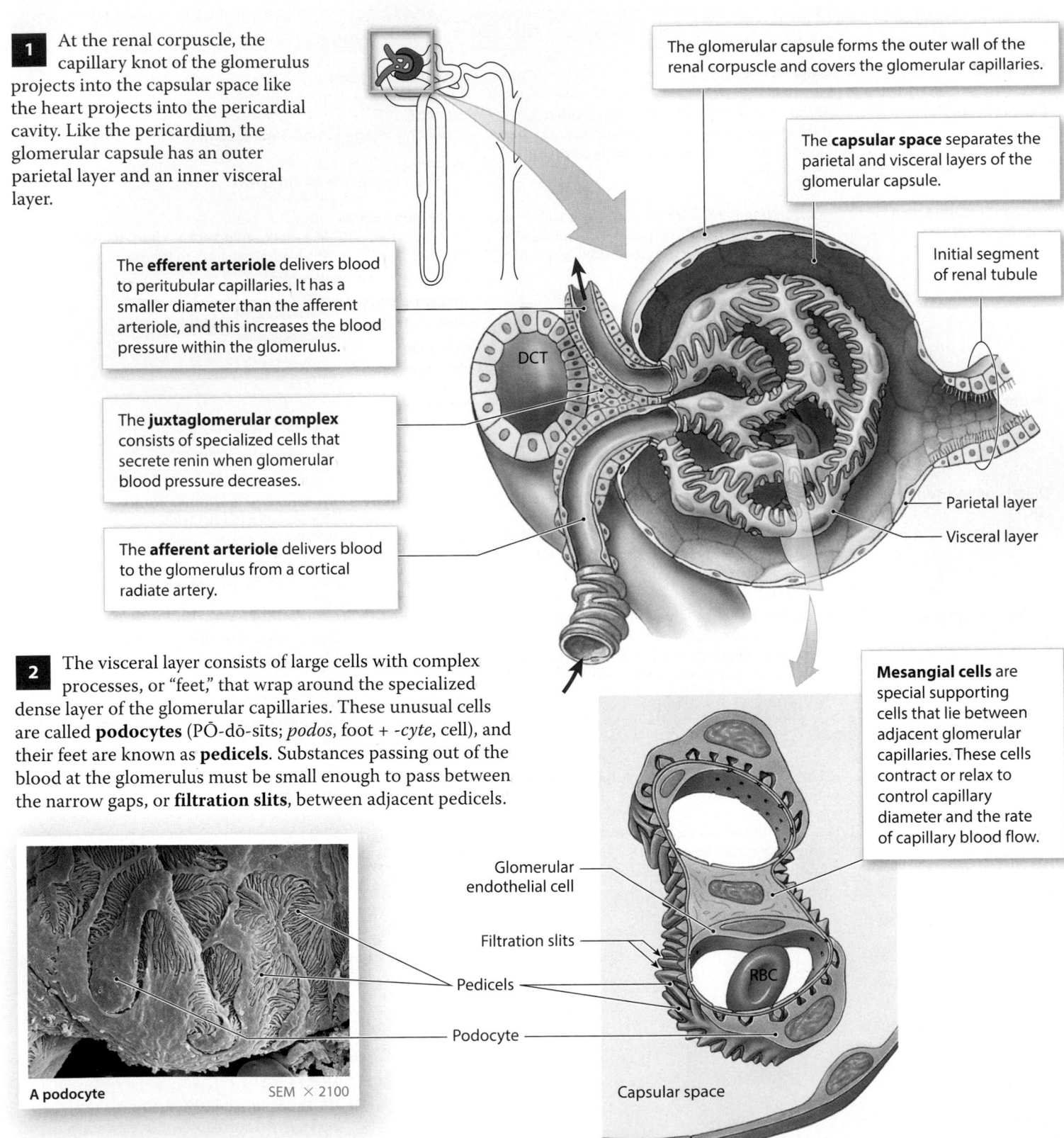

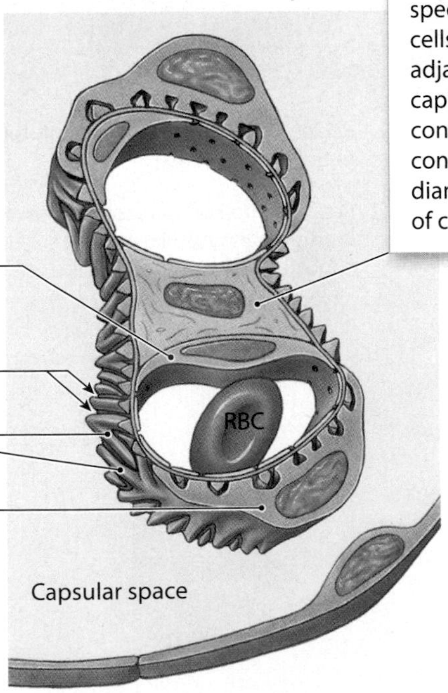
Mesangial cells are special supporting cells that lie between adjacent glomerular capillaries. These cells contract or relax to control capillary diameter and the rate of capillary blood flow.
Glomerular endothelial cell
Filtration slits
Pedicels
Podocyte
RBC
Capsular space

3 The glomerular capillaries are fenestrated capillaries containing large-diameter pores (Module 18.3, p. 654). The endothelium covers a **dense layer**, a specialized type of basement membrane. Together, the fenestrated endothelium, the dense layer, and the filtration slits form the **filtration membrane**. Under normal circumstances only a few plasma proteins—such as albumin molecules, with an average diameter of 7 nm—can cross the filtration membrane and enter the capsular space.

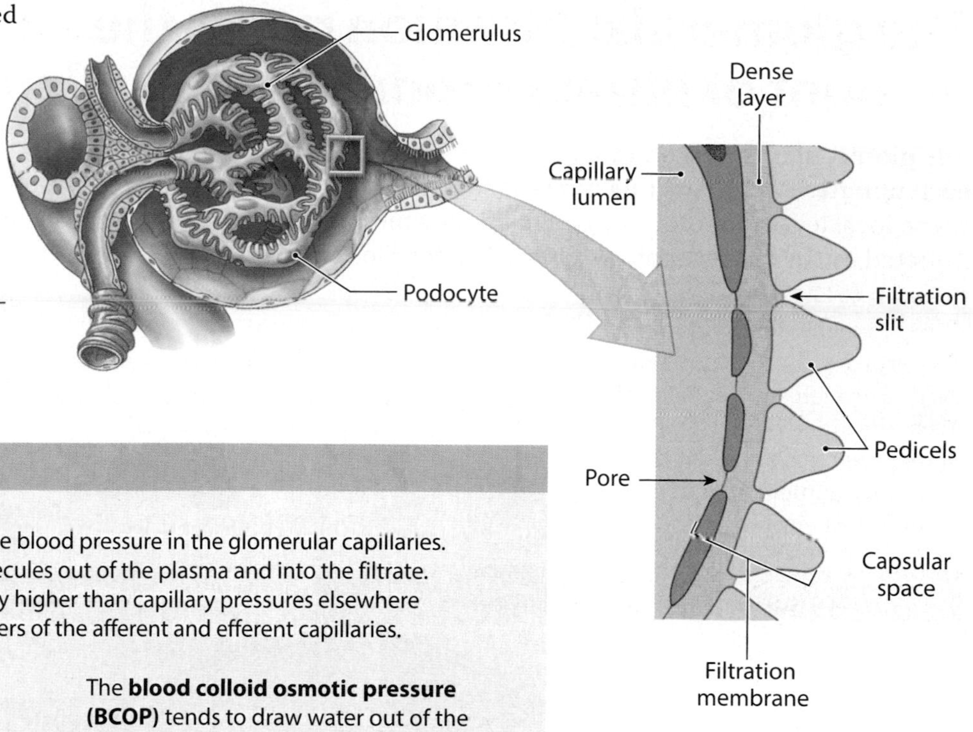
Glomerulus
Podocyte
Dense layer
Capillary lumen
Filtration slit
Pedicels
Pore
Capsular space
Filtration membrane

Factors Controlling Glomerular Filtration

The **glomerular hydrostatic pressure (GHP)** is the blood pressure in the glomerular capillaries. This pressure tends to push water and solute molecules out of the plasma and into the filtrate. The GHP, which averages 50 mm Hg, is significantly higher than capillary pressures elsewhere in the systemic circuit, due to the different diameters of the afferent and efferent capillaries.

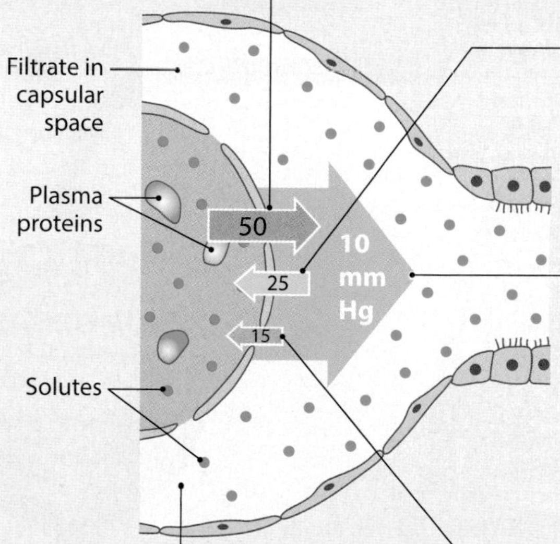
Filtrate in capsular space
Plasma proteins
Solutes
50
25
15
10 mm Hg

The **blood colloid osmotic pressure (BCOP)** tends to draw water out of the filtrate and into the plasma; it thus opposes filtration. Over the entire length of the glomerular capillary bed, the BCOP averages about 25 mm Hg.

The **net filtration pressure (NFP)** is the pressure acting across the glomerular capillaries. It represents the sum of the hydrostatic pressures and the colloid osmotic pressures. Under normal circumstances, the net filtration pressure is approximately 10 mm Hg. This is the average pressure forcing water and dissolved substances out of the glomerular capillaries and into the capsular space.

The **capsular colloid osmotic pressure** is usually 0 because few, if any, plasma proteins enter the capsular space.

Capsular hydrostatic pressure (CsHP) opposes GHP. CsHP, which tends to push water and solutes out of the filtrate and into the plasma, results from the resistance of filtrate already present in the nephron that must be pushed toward the renal pelvis.

4 The primary factor involved in glomerular filtration is basically the same as that governing fluid and solute movement across capillaries throughout the body: the balance between **hydrostatic pressure** (fluid pressure) and **colloid osmotic pressure** (pressure due to materials in solution) on either side of the capillary membrane.

Module 24.8 Review

a. The capsular space separates which layers of the glomerular capsule?

b. Explain why blood pressure is higher in glomerular capillaries than in other systemic capillaries.

c. Blood colloidal osmotic pressure tends to draw water out of the filtrate and into the plasma. Why does this occur?

LO 24.8 Describe the structural features of a renal corpuscle, and explain the functions of the filtration membrane components.

Module 24.9

The glomerular filtration rate is the amount of filtrate produced each minute

The **glomerular filtration rate (GFR)** is the amount of filtrate the kidneys produce each minute. Two interacting levels of control stabilize GFR: (1) autoregulation at the local level, and (2) central regulation, which has an endocrine component initiated by the kidneys and an autonomic component involving the sympathetic division of the ANS.

1 Through autoregulation the kidneys adjust GFR in response to changes in the local environment at and around the nephrons.

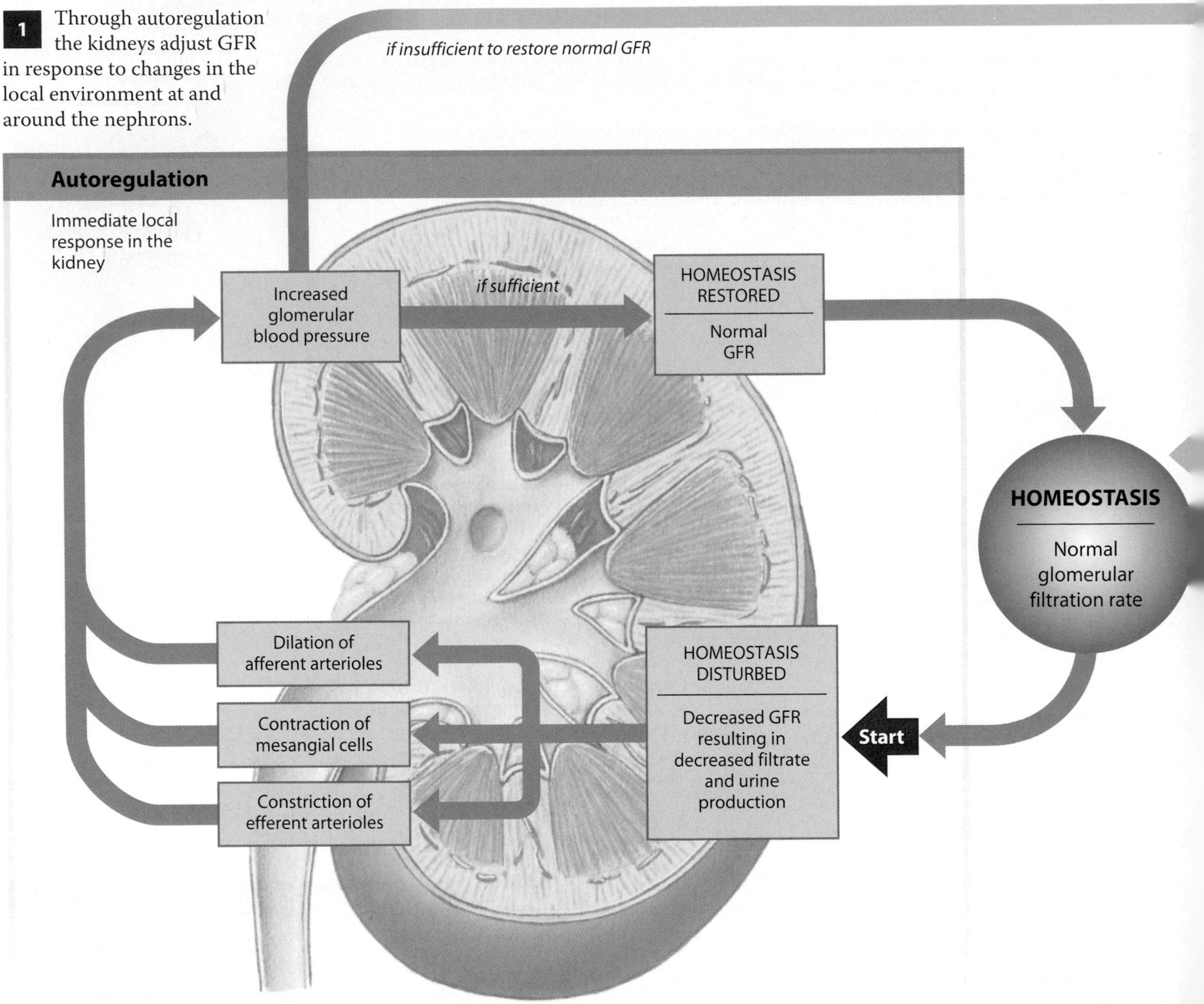

if insufficient to restore normal GFR
Autoregulation
Immediate local response in the kidney
Increased glomerular blood pressure
if sufficient
HOMEOSTASIS RESTORED
Normal GFR
HOMEOSTASIS
Normal glomerular filtration rate
Dilation of afferent arterioles
Contraction of mesangial cells
Constriction of efferent arterioles
HOMEOSTASIS DISTURBED
Decreased GFR resulting in decreased filtrate and urine production
Start

2 In essence, the kidneys call for a central response that increases GFR if autoregulation is ineffective. This response involves multiple systems and mechanisms.

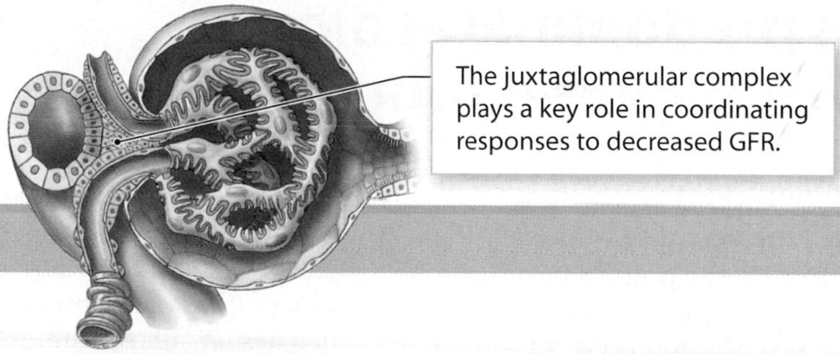

The juxtaglomerular complex plays a key role in coordinating responses to decreased GFR.

Central Regulation

Integrated endocrine and neural mechanisms activated

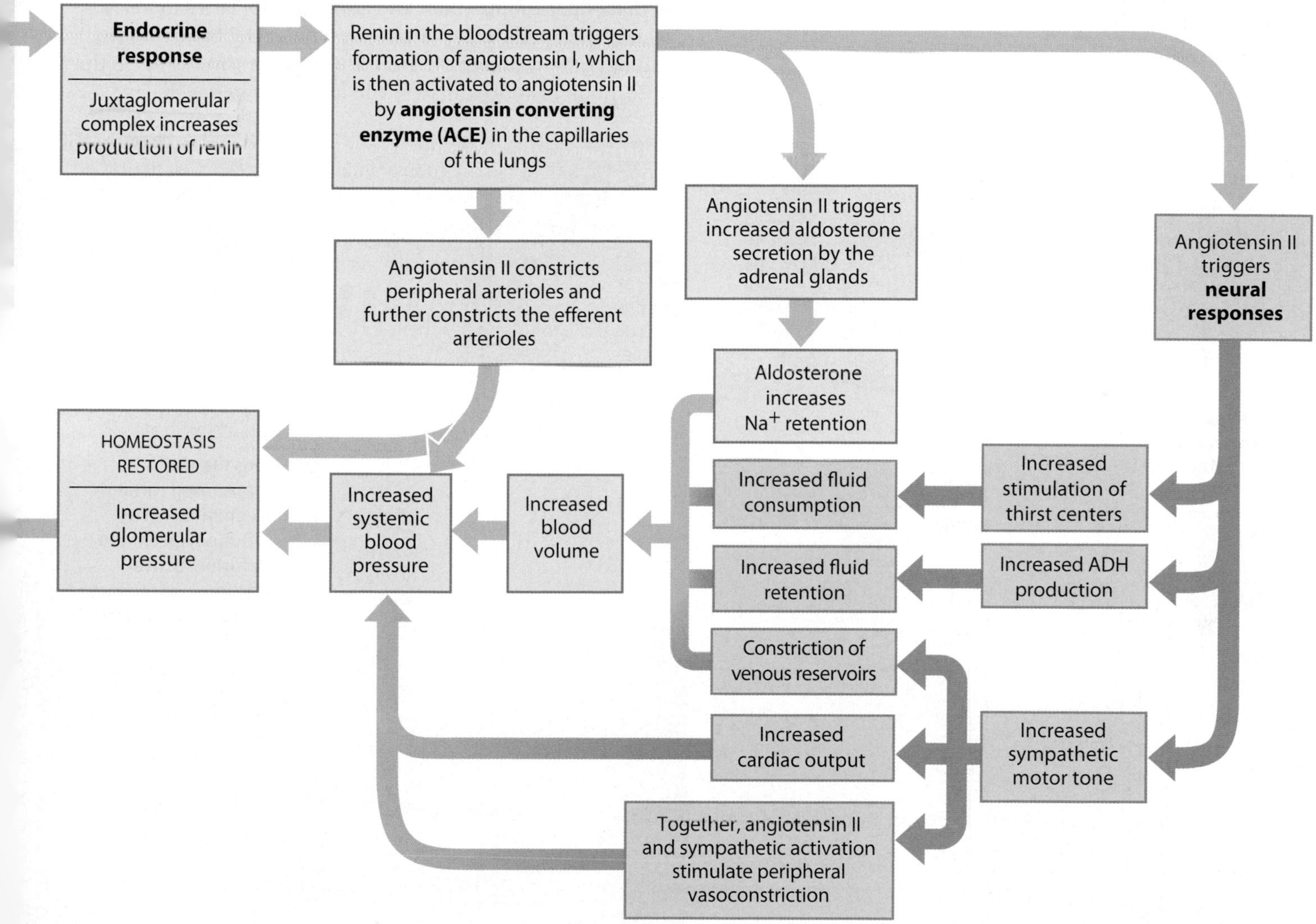

Endocrine response
Juxtaglomerular complex increases production of renin
Renin in the bloodstream triggers formation of angiotensin I, which is then activated to angiotensin II by **angiotensin converting enzyme (ACE)** in the capillaries of the lungs
Angiotensin II constricts peripheral arterioles and further constricts the efferent arterioles
Angiotensin II triggers increased aldosterone secretion by the adrenal glands
Aldosterone increases Na^+ retention
Angiotensin II triggers **neural responses**
Increased stimulation of thirst centers
Increased ADH production
Increased sympathetic motor tone
Increased fluid consumption
Increased fluid retention
Constriction of venous reservoirs
Increased cardiac output
Together, angiotensin II and sympathetic activation stimulate peripheral vasoconstriction
Increased blood volume
Increased systemic blood pressure
HOMEOSTASIS RESTORED
Increased glomerular pressure

Each kidney contains about 6 m^2—some 64 square feet—of filtration surface, and the GFR averages an astounding 125 mL per minute. This means that about 10 percent of the fluid delivered to the kidneys by the renal arteries leaves the bloodstream and enters the capsular spaces. In the course of a single day, the glomeruli generate about 180 liters (48 gal) of filtrate, about 70 times the total plasma volume. But as filtrate passes through the renal tubules, about 99 percent of it is reabsorbed.

Module 24.9 Review

a. In response to decreased filtration pressure, the juxtaglomerular complex does what?

b. Angiotensin II has what effect on nephrons?

c. Angiotensin II has what effect on the CNS?

LO 24.9 Describe the factors that influence filtration pressure and the glomerular filtration rate.

Reabsorption predominates along the proximal convoluted tubule . . .

The proximal convoluted tubule (PCT) is the first segment of the renal tubule. The entrance to the PCT lies almost directly opposite the point where the afferent and efferent arterioles connect to the glomerulus. The distal convoluted tubule (DCT), which forms the last segment of the nephron, makes final adjustments to the solute composition of the tubular fluid.

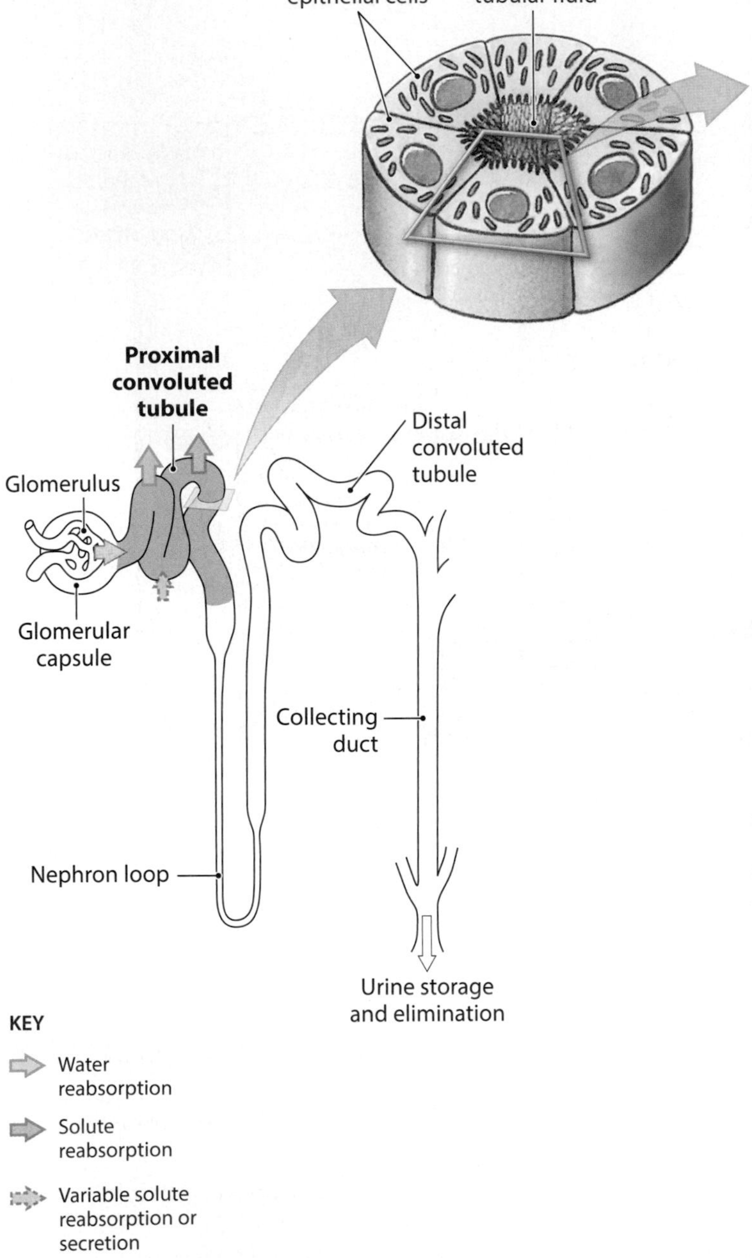
Cuboidal epithelial cells
Lumen containing tubular fluid
Proximal convoluted tubule
Glomerulus
Glomerular capsule
Distal convoluted tubule
Collecting duct
Nephron loop
Urine storage and elimination
KEY
Water reabsorption
Solute reabsorption
Variable solute reabsorption or secretion

1 Normally, before the tubular fluid enters the nephron loop, the PCT reabsorbs more than 99 percent of the glucose, amino acids, and other organic nutrients in the fluid. The PCT also reabsorbs sodium, potassium, bicarbonate, magnesium, phosphate, and sulfate ions. As reabsorption occurs, the solute concentration of tubular fluid decreases, and that of peritubular fluid and adjacent capillaries increases. Osmosis then pulls water out of the tubular fluid and into the peritubular fluid. Along the PCT, this mechanism results in the reabsorption of approximately 108 liters of water each day.

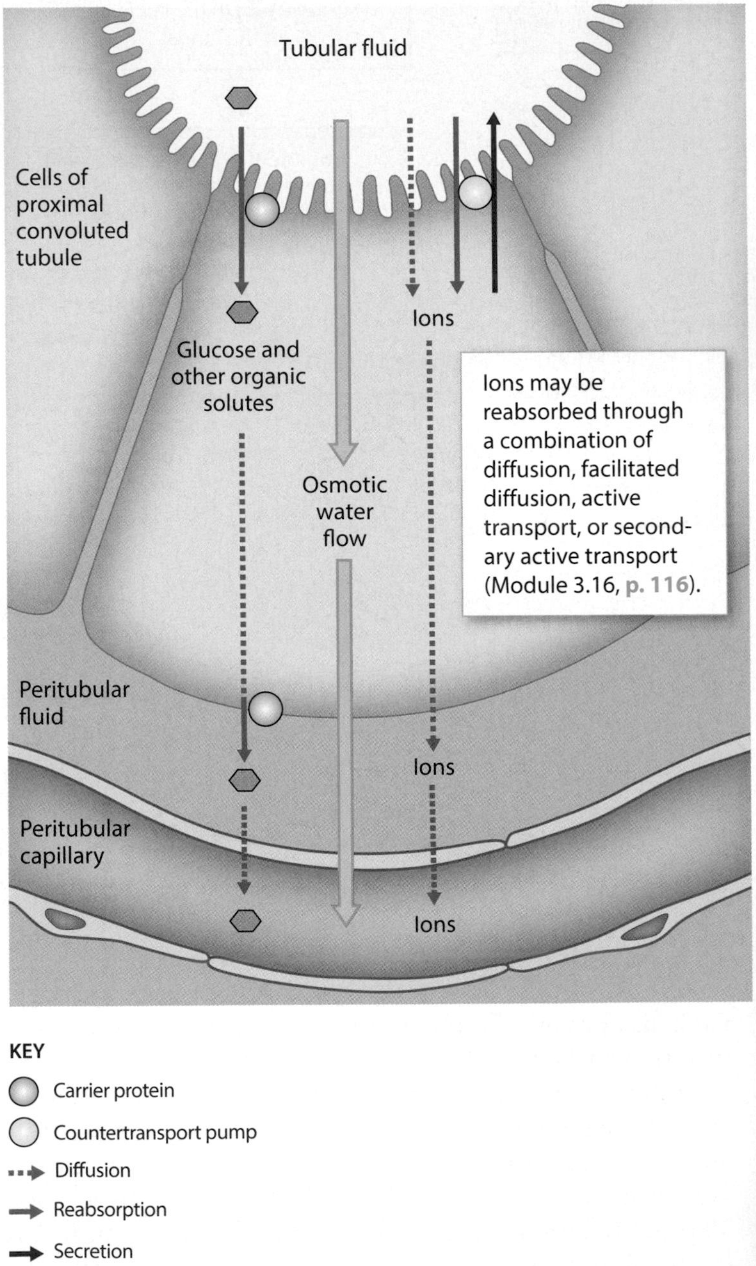
Tubular fluid
Cells of proximal convoluted tubule
Glucose and other organic solutes
Osmotic water flow
Ions
Ions may be reabsorbed through a combination of diffusion, facilitated diffusion, active transport, or secondary active transport (Module 3.16, p. 116).
Peritubular fluid
Ions
Peritubular capillary
Ions
KEY
Carrier protein
Countertransport pump
Diffusion
Reabsorption
Secretion

. . . whereas reabsorption and secretion are often linked along the distal convoluted tubule

2 Only 15–20 percent of the initial filtrate volume reaches the DCT, and the concentrations of electrolytes and organic wastes in the arriving tubular fluid no longer resemble the concentrations in blood plasma. In the DCT, a combination of secretion and reabsorption further alters the solute composition of the tubular fluid.

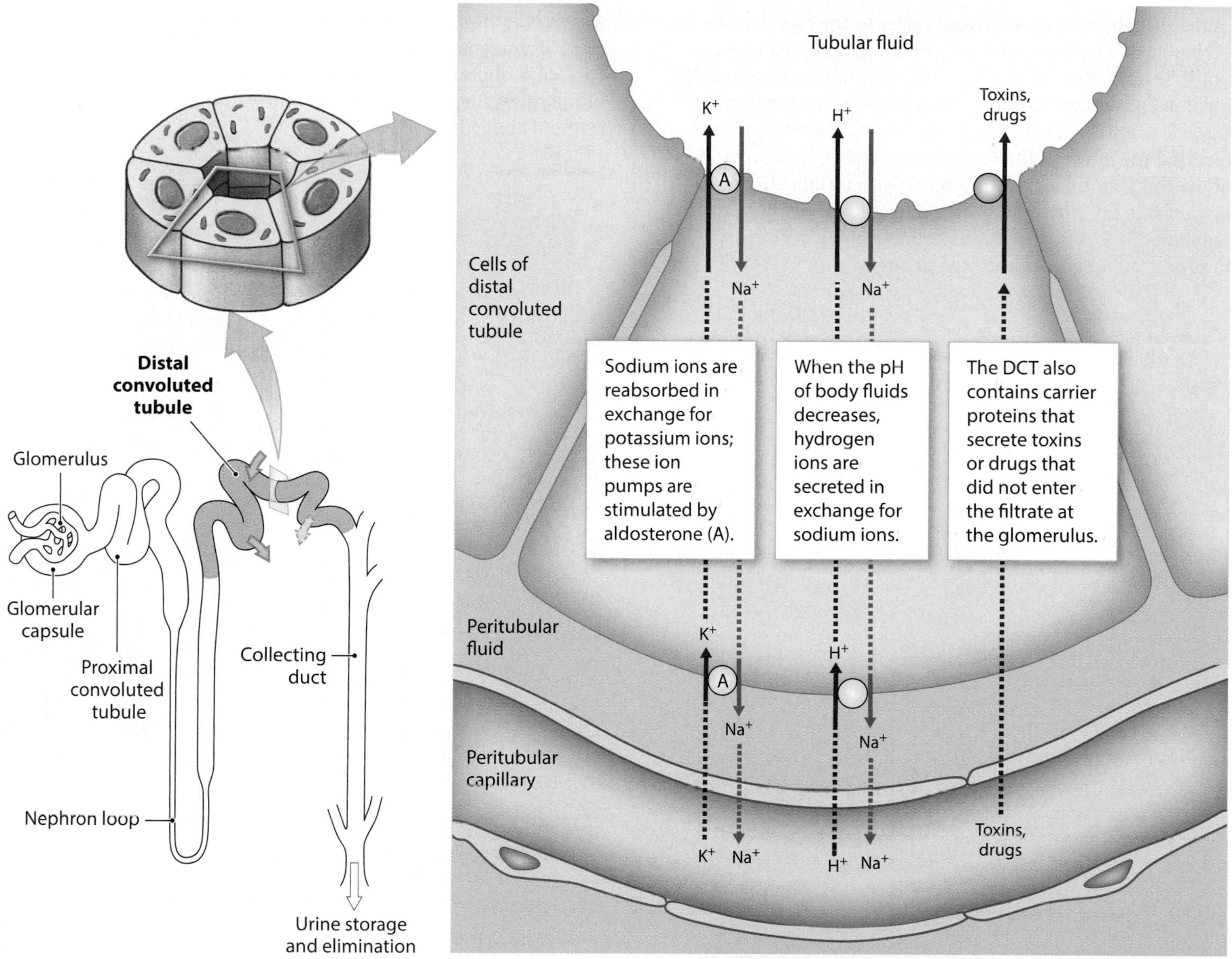

Distal convoluted tubule
Glomerulus
Glomerular capsule
Proximal convoluted tubule
Collecting duct
Nephron loop
Urine storage and elimination
Tubular fluid
K+
H+
Toxins, drugs
A
Na+
Cells of distal convoluted tubule
Sodium ions are reabsorbed in exchange for potassium ions; these ion pumps are stimulated by aldosterone (A).
When the pH of body fluids decreases, hydrogen ions are secreted in exchange for sodium ions.
The DCT also contains carrier proteins that secrete toxins or drugs that did not enter the filtrate at the glomerulus.
Peritubular fluid
Peritubular capillary
K+ Na+
H+ Na+
Toxins, drugs

KEY

- Carrier protein
- Countertransport pump
- Diffusion
- Reabsorption
- Secretion

Module 24.10 Review

a. Identify the segment of the nephron that makes final adjustments to the composition of tubular fluid.

b. What effect would increased amounts of aldosterone have on the K^+ concentration in urine?

c. What effect would a decrease in the Na^+ concentration of filtrate have on the pH of tubular fluid?

24.10 Identify the types of transport mechanisms along the proximal and distal convoluted tubules of the nephron.

Module 24.11

Exchange between the limbs of the nephron loop creates an osmotic concentration gradient in the renal medulla

1 The thin descending limb and the thick ascending limb of the nephron loop are very close together, separated only by peritubular fluid. The exchange that occurs between these segments is called **countercurrent multiplication**. *Countercurrent* refers to the fact that the exchange occurs between fluids moving in opposite directions: Tubular fluid in the descending limb flows toward the renal pelvis, whereas tubular fluid in the ascending limb flows toward the renal cortex. *Multiplication* refers to the fact that the effect of the exchange increases as movement of the fluid continues. Countercurrent multiplication is responsible for creating the concentration gradient in the renal medulla. It is this gradient that enables the kidney to produce highly concentrated urine.

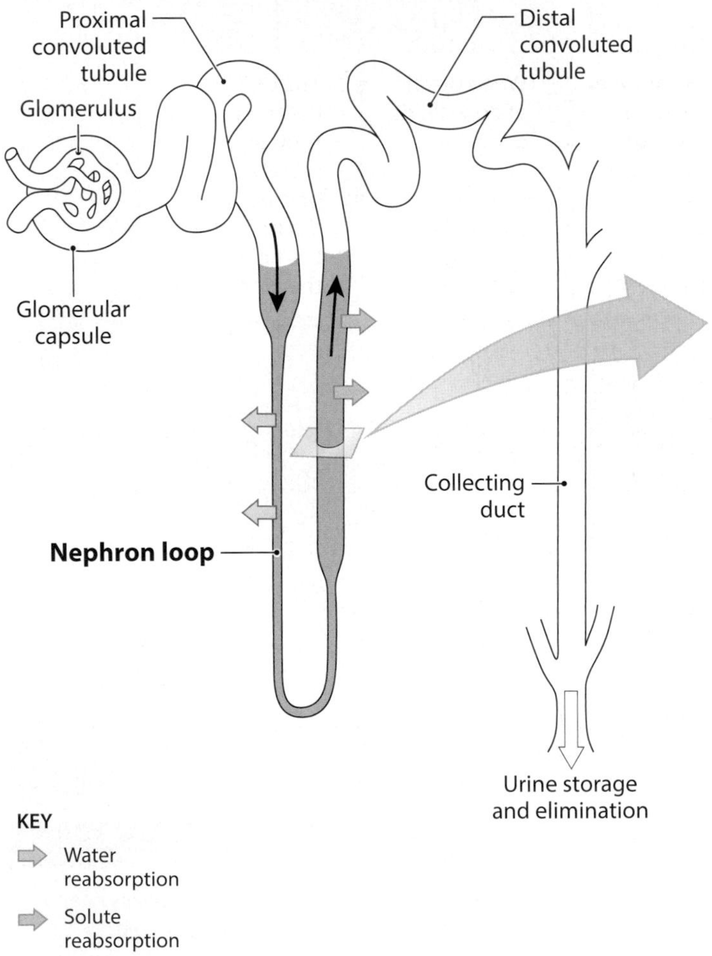
Proximal convoluted tubule
Glomerulus
Glomerular capsule
Distal convoluted tubule
Collecting duct
Nephron loop
Urine storage and elimination
KEY
Water reabsorption
Solute reabsorption

2 The key to this process is the transport activity performed by the cells of the thick ascending limb. Active transport at the apical surface moves sodium and chloride ions out of the tubular fluid, and they then enter the peritubular fluid of the renal medulla. Because the apical surface is impermeable to water, the transport of ions does not result in an osmotic water flow as it did along the PCT.

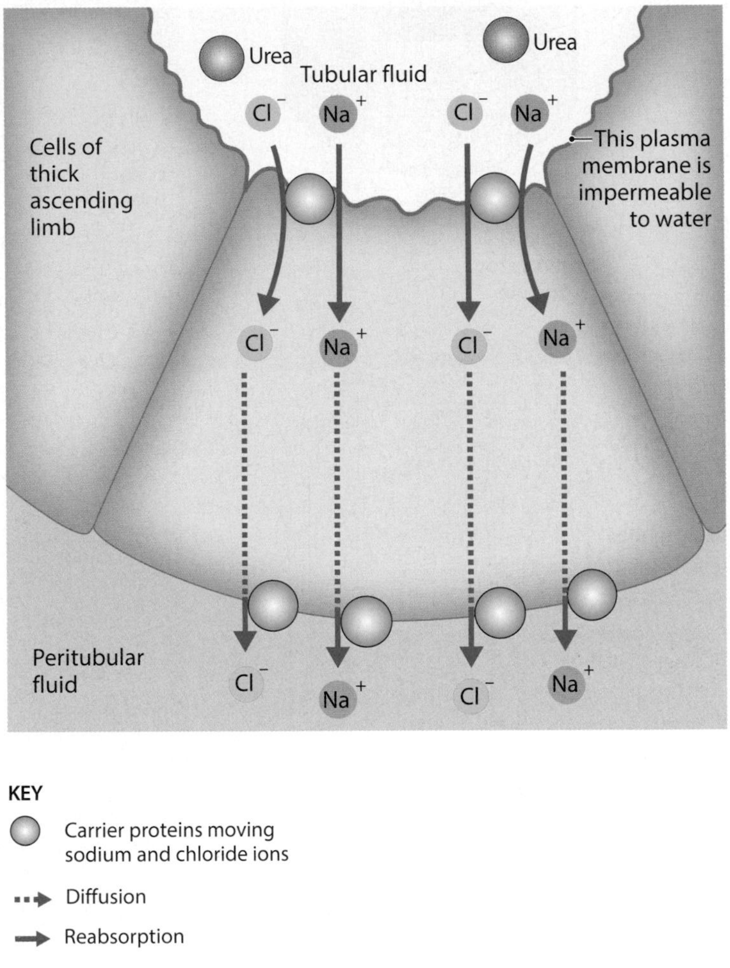
Urea
Tubular fluid
Urea
Cl^-
Na^+
Cl^-
Na^+
Cells of thick ascending limb
This plasma membrane is impermeable to water
Cl^-
Na^+
Cl^-
Na^+
Peritubular fluid
Cl^-
Na^+
Cl^-
Na^+
KEY
Carrier proteins moving sodium and chloride ions
Diffusion
Reabsorption

3 The removal of sodium and chloride ions from the tubular fluid in the ascending limb increases the osmotic concentration of the peritubular fluid around the thin descending limb. The thin descending limb is permeable to water but impermeable to solutes; as tubular fluid travels deeper into the renal medulla along the thin descending limb, osmosis moves water into the peritubular fluid.

The thin descending limb is permeable to water, but impermeable to solutes. Solutes remain behind as water leaves, so the tubular fluid reaching the turn of the nephron loop has a higher osmotic concentration than it did at the start.

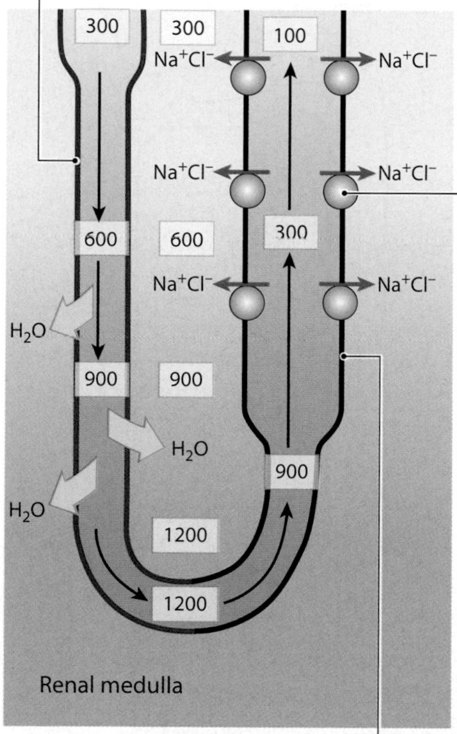

300
300
100
Na^+Cl^-
Na^+Cl^-
Na^+Cl^-
Na^+Cl^-
Na^+Cl^-
Na^+Cl^-
600
600
300
H_2O
H_2O
H_2O
900
900
900
1200
1200
Renal medulla

The pumping mechanism of the thick ascending limb is highly effective: Almost two-thirds of the sodium and chloride ions that enter it are pumped out of the tubular fluid before that fluid reaches the DCT.

The thick ascending limb is impermeable to water and other solutes and selectively permeable to Na^+ and Cl^-. In other tissues, differences in solute concentration are quickly resolved by osmosis. But since osmosis cannot occur across the impermeable wall of the thick ascending limb, the solute concentration in the tubular fluid decreases as Na^+ and Cl^- are removed.

4 Urea is an important solute that is eliminated in the urine. As water is reabsorbed along the DCT and collecting duct, the concentration of urea gradually increases in the tubular fluid. The tubular fluid reaching the papillary duct typically contains urea at a concentration of about 450 mOsm/L.

1 The thin descending limb is permeable to water, but impermeable to solutes, including urea.

2 The thick ascending limb of the nephron loop is impermeable to water and solutes.

3 The DCT and the collecting ducts are impermeable to urea but have variable permeability to water.

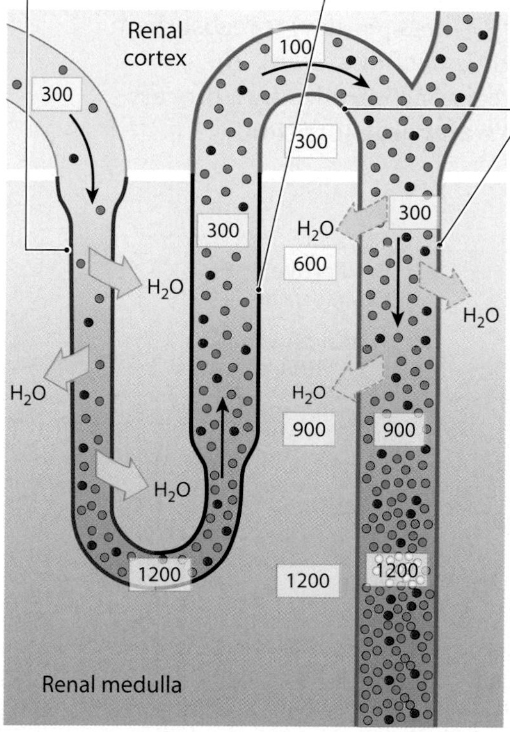

Renal cortex
300
100
300
300
300
H_2O
600
H_2O
H_2O
H_2O
H_2O
H_2O
900
900
1200
1200
1200
Renal medulla

KEY
- ● Na^+
- ● Cl^-
- ● Urea

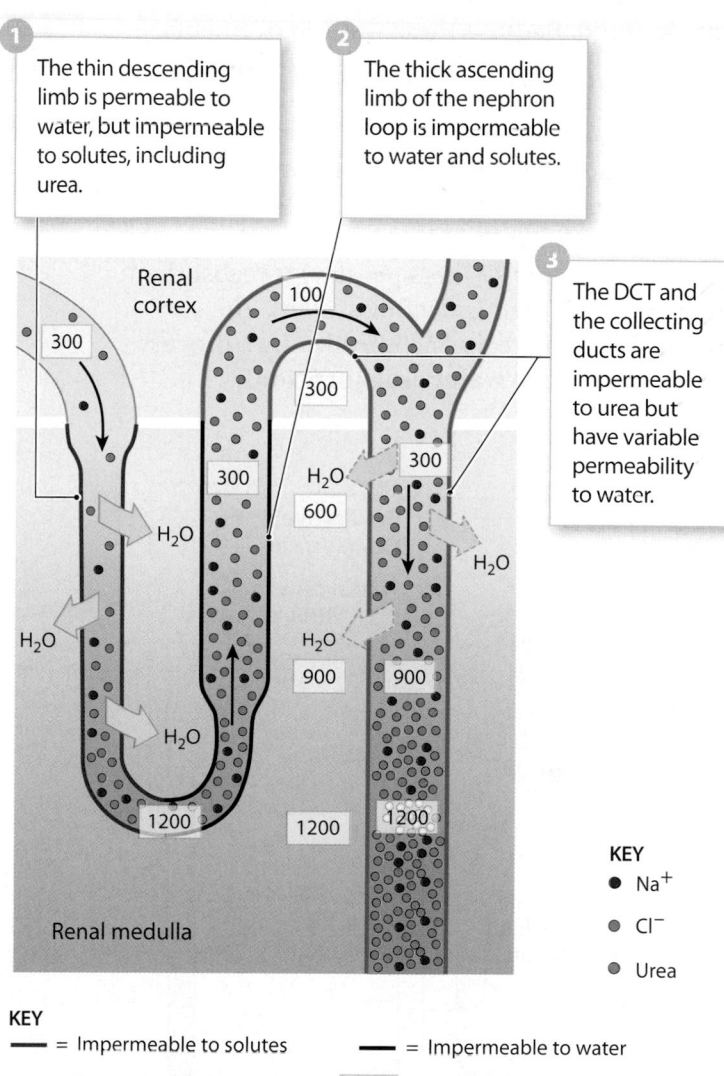

KEY
- ― = Impermeable to solutes
- ― = Impermeable to water
- ― = Impermeable to urea; variable permeability to water
- 300 = Fluid osmotic concentration (mOsm/L)

Module 24.11 Review

a. Define countercurrent multiplication as it occurs in the kidneys.

b. The thick ascending limb of the nephron loop actively pumps what substances into the peritubular fluid?

c. An increase in sodium and chloride ions in the peritubular fluid affects the thin descending limb in what way?

LO 24.11 Explain the role of countercurrent multiplication in the formation of a concentration gradient in the renal medulla.

Urine volume and concentration are hormonally regulated

Urine volume and osmotic concentration are regulated through the hormonal control of water reabsorption by antidiuretic hormone (ADH). The water permeabilities of the PCT and descending limb of the nephron loop cannot be adjusted, and water reabsorption occurs whenever the osmotic concentration of the peritubular fluid exceeds that of the tubular fluid. Because these water movements cannot be prevented, they represent **obligatory water reabsorption**. Obligatory reabsorption usually recovers 85 percent of the volume of filtrate produced. The volume of water lost in urine depends on how much of the water in the remaining tubular fluid (15 percent of the filtrate volume, or nearly 27 liters per day) is reabsorbed along the DCT and collecting system. The amount can be precisely controlled by a process called **facultative water reabsorption**.

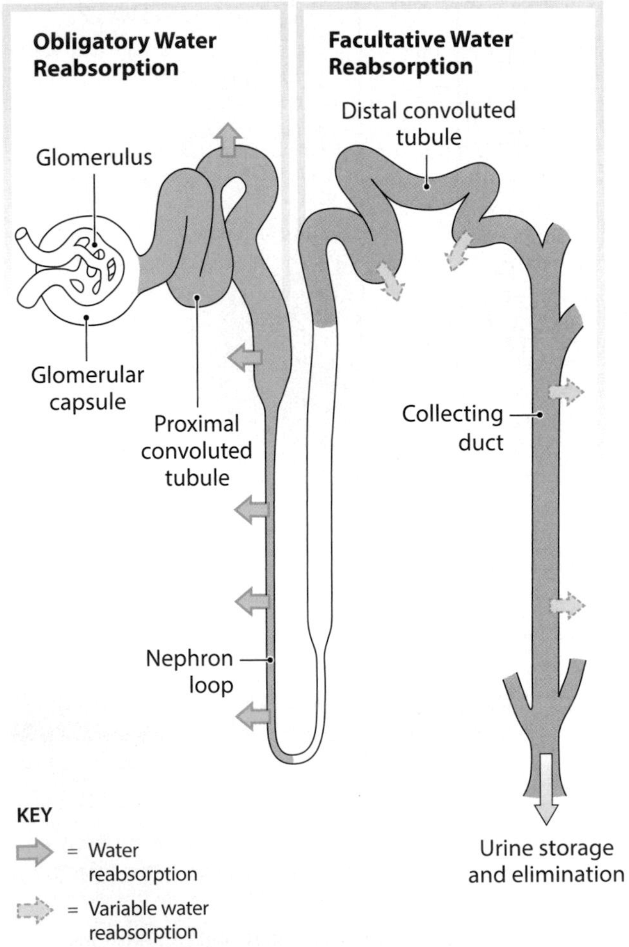

Obligatory Water Reabsorption
Glomerulus
Glomerular capsule
Proximal convoluted tubule
Nephron loop
Facultative Water Reabsorption
Distal convoluted tubule
Collecting duct
Urine storage and elimination
KEY
= Water reabsorption
= Variable water reabsorption

1 Without antidiuretic hormone (ADH), water is not reabsorbed in the distal collecting tubule or the collecting duct, so all the fluid reaching the DCT is lost in the urine. No facultative water reabsorption occurs, and the person then produces large amounts of very dilute urine.

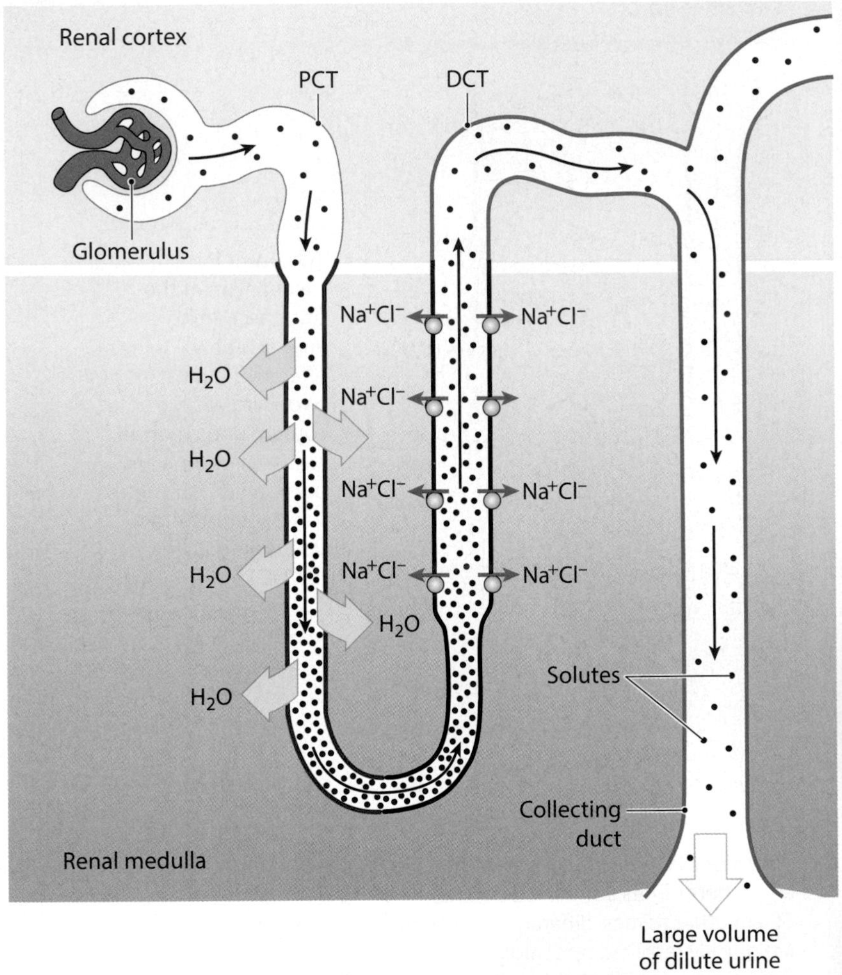

Renal cortex
PCT
DCT
Glomerulus
Na+Cl−
H2O
Solutes
Collecting duct
Renal medulla
Large volume of dilute urine

2 ADH causes the appearance of special water channels, called **aquaporins**, in the apical plasma membranes lining the DCT and collecting duct. The aquaporins dramatically increase the rate of osmotic water movement. As ADH levels increase, the DCT and collecting system become more permeable to water, the amount of water reabsorbed increases, and the urine osmotic concentration increases. Under maximum ADH stimulation, the DCT and collecting system become so permeable to water that the osmotic concentration of the urine is equal to that of the peritubular fluid in the deepest portion of the renal medulla.

3 A healthy adult typically produces 1200 mL of urine per day (about 0.6 percent of the filtrate volume), with an osmotic concentration of about 1000 mOsm/L. However, normal values differ from person to person and from day to day, as the kidneys alter their function to maintain homeostatic conditions within body fluids.

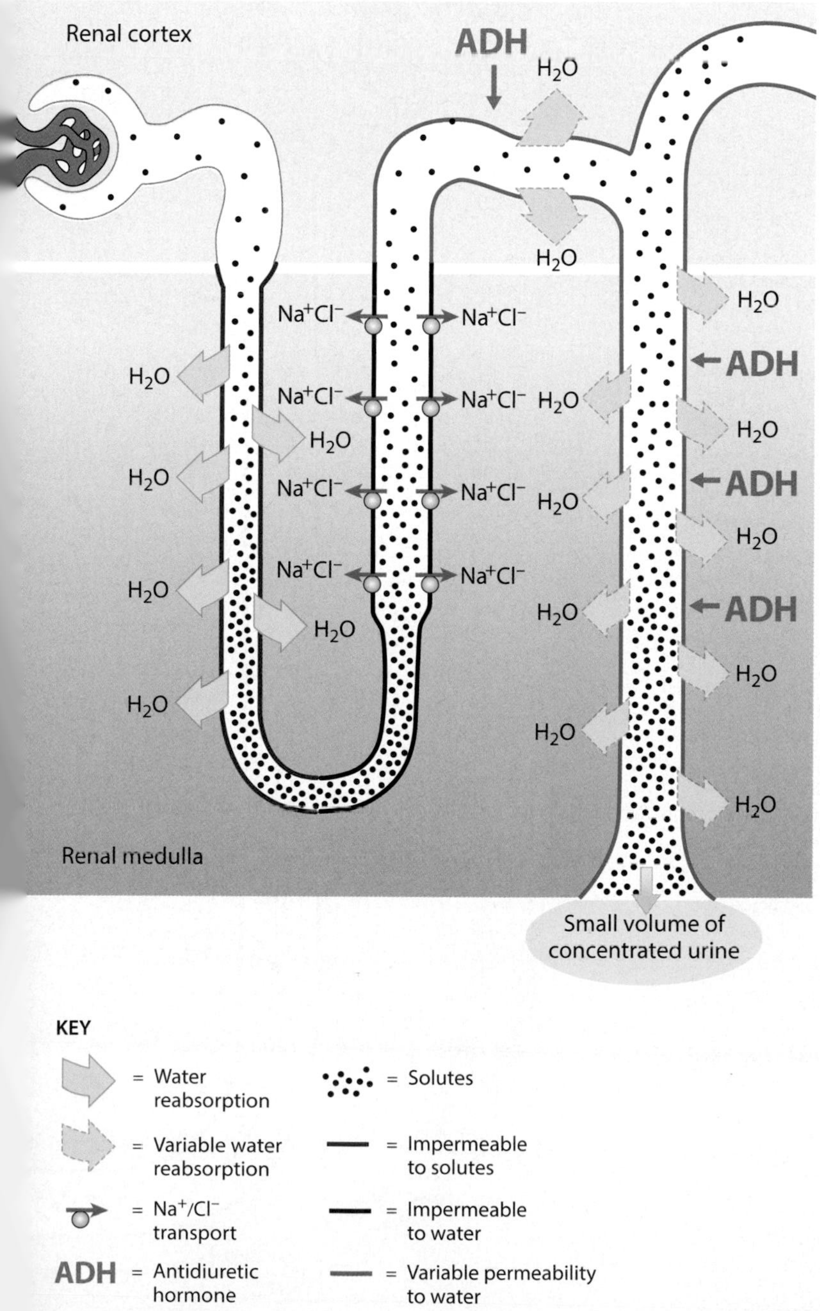

Renal cortex
ADH
H_2O
Na^+Cl^-
ADH
Renal medulla
Small volume of concentrated urine
KEY
= Water reabsorption
= Variable water reabsorption
= Na^+/Cl^- transport
ADH = Antidiuretic hormone
= Solutes
= Impermeable to solutes
= Impermeable to water
= Variable permeability to water

General Characteristics of Normal Urine

Characteristic	Normal Range
pH	4.5–8 (average: 6.0)
Specific gravity	1.003–1.030
Osmotic concentration (osmolarity)	855–1335 mOsm/L
Water content	93–97%
Volume	700–2000 mL/day
Color	Clear yellow
Odor	Varies with composition
Bacterial content	None (sterile)

Module 24.12 Review

a. Can the water permeability of the PCT or DCT ever change? Explain.

b. What effect does an increase in ADH levels have on the DCT?

c. When ADH levels in the DCT decrease, what happens to the urine osmotic concentration?

LO 24.12 Describe how antidiuretic hormone influences the volume and concentration of urine.

Module 24.13

Renal function is an integrative process involving filtration, reabsorption, and secretion

1 The filtrate produced by the renal corpuscle has the same osmotic concentration as plasma—about 300 mOsm/L. It has the same composition as plasma but does not contain plasma proteins.

2 In the proximal convoluted tubule (PCT), the active removal of ions and organic nutrients produces a continuous osmotic flow of water out of the tubular fluid. This decreases the volume of filtrate but keeps the tubular fluid inside and the peritubular fluid outside isotonic.

3 In the PCT and descending limb of the nephron loop, water moves into the surrounding peritubular fluid, leaving a small volume of highly concentrated tubular fluid. This reduction occurs by obligatory water reabsorption.

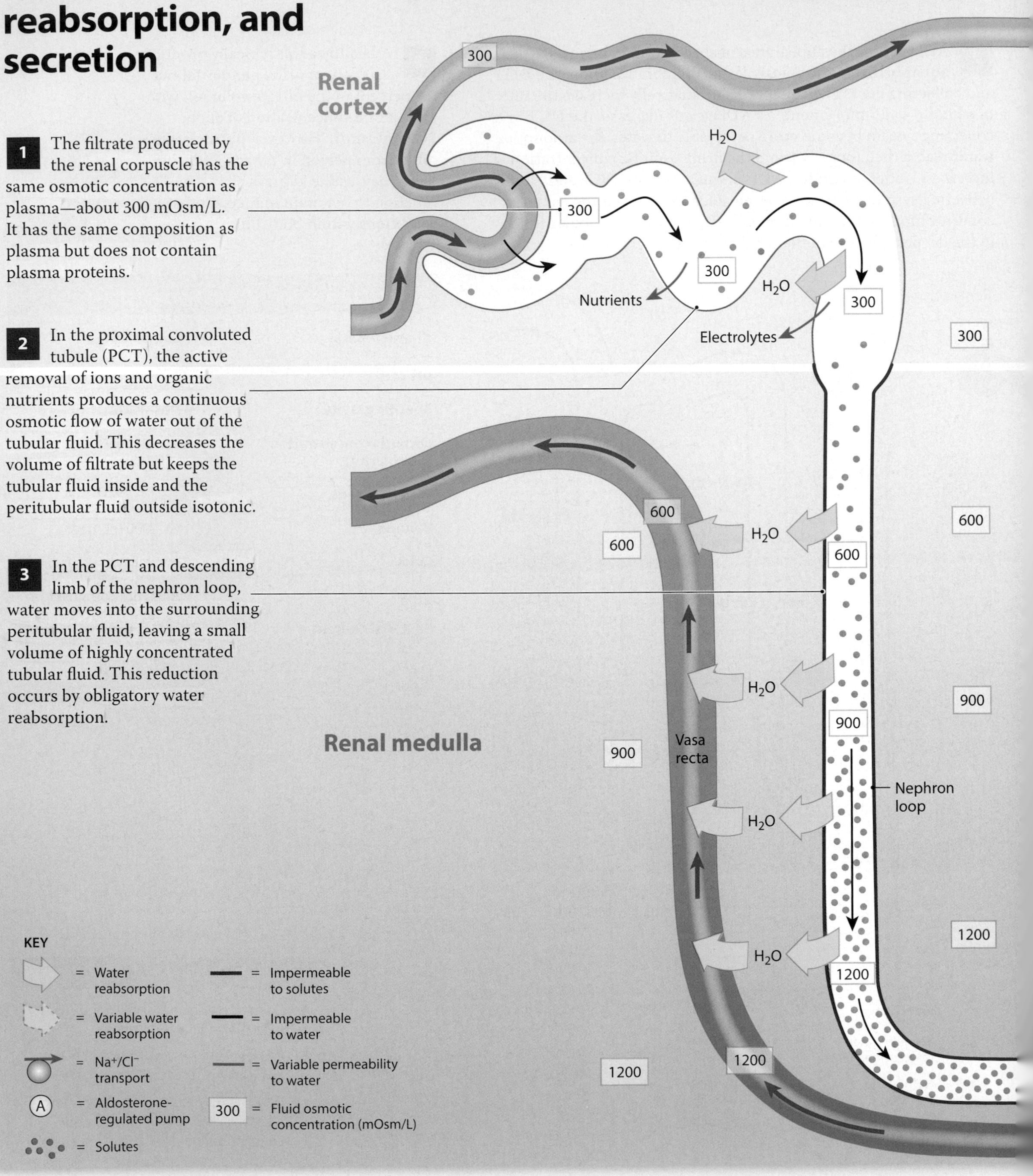

Renal cortex
300
H_2O
Nutrients
Electrolytes
Renal medulla
600
900
1200
Vasa recta
Nephron loop
KEY
= Water reabsorption
= Variable water reabsorption
= Na^+/Cl^- transport
= Aldosterone-regulated pump
= Solutes
= Impermeable to solutes
= Impermeable to water
= Variable permeability to water
300 = Fluid osmotic concentration (mOsm/L)

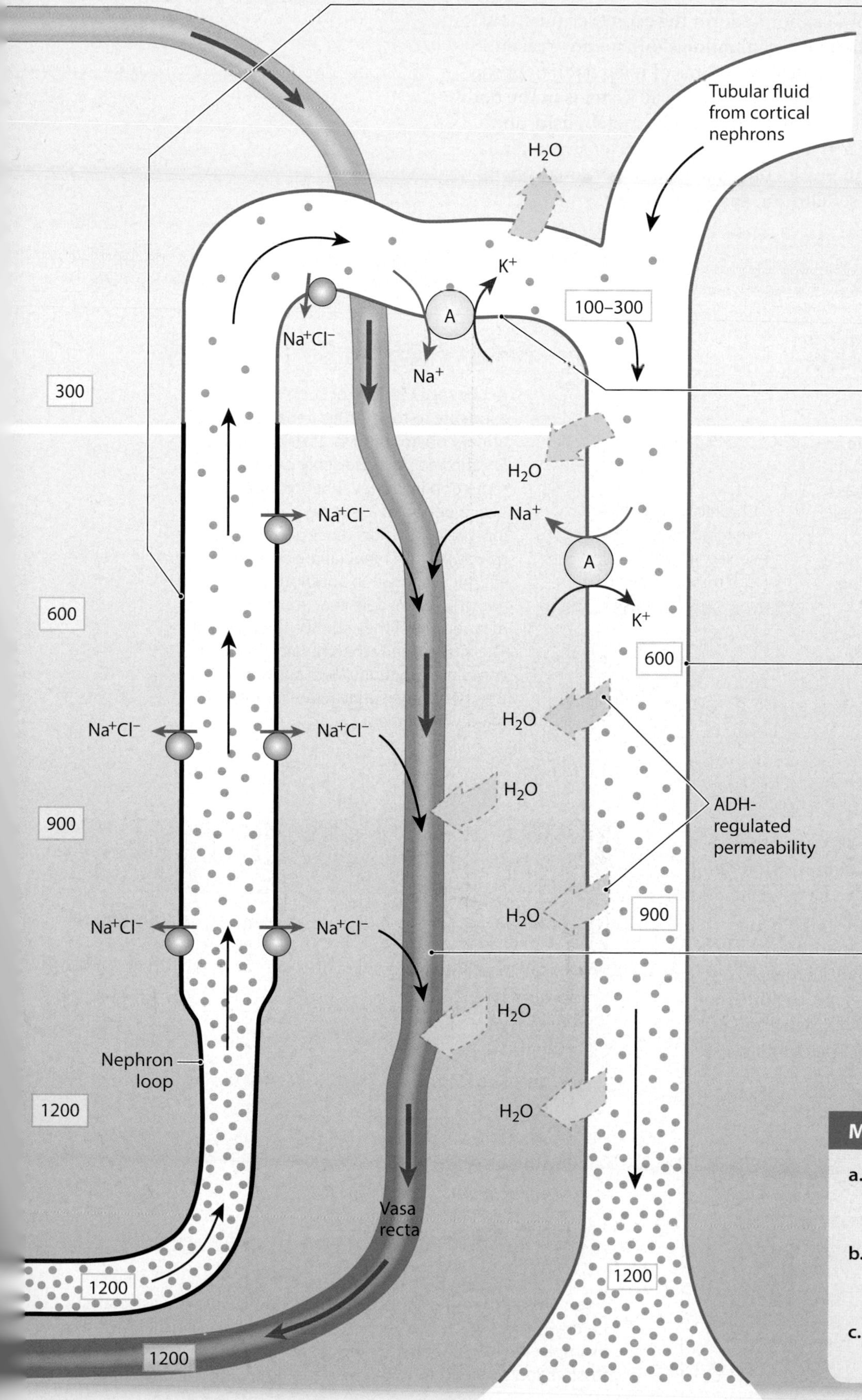

Tubular fluid from cortical nephrons
H_2O
K^+
A
Na^+Cl^-
Na^+
100–300
300
H_2O
Na^+Cl^-
Na^+
A
600
K^+
600
Na^+Cl^-
Na^+Cl^-
H_2O
H_2O
900
ADH-regulated permeability
Na^+Cl^-
Na^+Cl^-
H_2O
900
H_2O
Nephron loop
1200
H_2O
Vasa recta
1200
1200
1200

4 The ascending limb is impermeable to water and solutes. The tubule cells of the thick ascending limb actively transport Na^+ and Cl^- out of the tubule, thereby decreasing the osmotic concentration of the tubular fluid. Because just Na^+ and Cl^- are removed, urea makes up a higher proportion of the total osmotic concentration at the end of the nephron loop.

5 Further adjustments in the composition of the tubular fluid occur in the DCT and the collecting system. The osmotic concentration of the tubular fluid can be adjusted through active transport (reabsorption or secretion).

6 The final adjustments in the volume and osmotic concentration of the tubular fluid are made by controlling the water permeabilities of the distal portions of the DCT and the collecting system. The level of ADH determines the final urine concentration.

7 The vasa recta absorb the solutes and water reabsorbed by the nephron loop and the collecting ducts. By transporting these solutes and water into the systemic circuit, the vasa recta maintain the concentration gradient of the renal medulla.

Module 24.13 Review

a. The filtrate produced at the renal corpuscle has the same osmotic concentration as ________.

b. In the PCT, ions and organic substrates are actively removed, thus causing what to occur?

c. How is the concentration gradient of the renal medulla maintained?

Renal failure is a life-threatening condition

1 **Renal failure** occurs when the kidneys cannot perform the excretory functions needed to maintain homeostasis. When kidney filtration slows for any reason, urine production declines. As the decline continues, symptoms of renal failure appear because water, ions, and metabolic wastes are retained. Virtually all systems in the body are affected. For example, fluid balance, pH, muscular contraction, metabolism, and digestive function are disturbed. The person generally becomes hypertensive, anemia develops due to a decrease in erythropoietin production, and central nervous system problems can lead to sleeplessness, seizures, delirium, and even coma.

Renal Failure

Chronic Renal Failure

In **chronic renal failure**, kidney function deteriorates gradually, and the associated problems accumulate over time. The management of chronic renal failure typically involves restricting water and salt intake and minimizing protein intake. This combination reduces strain on the urinary system by (1) minimizing the volume of urine produced and (2) preventing the generation of large quantities of nitrogenous wastes. Acidosis, a common problem in persons with renal failure, can be countered by ingesting bicarbonate ions.

Chronic renal failure generally cannot be reversed; its progression can only be slowed.

Acute Renal Failure

Acute renal failure occurs when exposure to toxic drugs, renal ischemia, urinary obstruction, or trauma causes filtration to slow suddenly or stop. The reduction in kidney function occurs over a period of a few days and lasts for weeks. Sensitized people can also develop acute renal failure after an allergic response to antibiotics or anesthetics. People in acute renal failure may recover if they survive the incident. The kidneys may then regain partial or complete function. (With supportive treatment, the survival rate is approximately 50 percent.)

2 In **hemodialysis** (hē-mō-dī-AL-i-sis), an artificial membrane is used to regulate the composition of blood by means of a dialysis machine. The basic principle involved in this process, called **dialysis**, is passive diffusion across a selectively permeable membrane. The patient's blood flows past an artificial dialysis membrane, which contains pores large enough to permit the diffusion of ions, nutrients, and organic wastes, but small enough to prevent the loss of plasma proteins. A special dialysis fluid flows on the other side of the membrane.

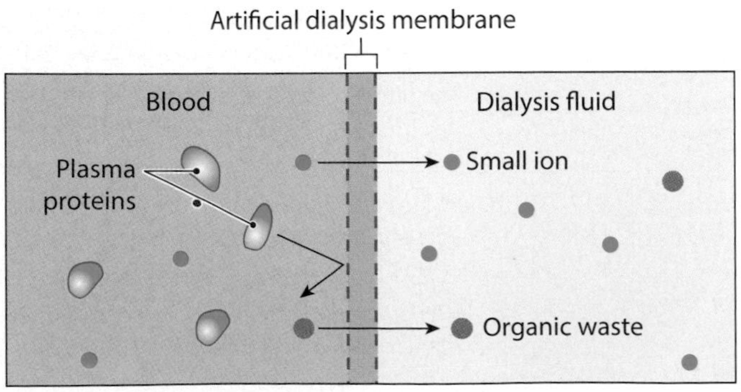
Artificial dialysis membrane
Blood
Dialysis fluid
Plasma proteins
Small ion
Organic waste

Comparative Composition of Plasma and Dialysis Fluid

Component	Plasma	Dialysis Fluid
Electrolytes (mEq/L)		
Sodium (Na^+)	135–145	136–140
Potassium (K^+)	3.5–5.0	0–3.0
Calcium (Ca^{2+})	4.3–5.3	1.5
Magnesium (Mg^{2+})	1.4–2.0	0.5–1.0
Chloride (Cl^-)	100–108	99–110
Bicarbonate (HCO_3^-)	21–28	27–39
Phosphate (PO_4^{2-})	3	0
Sulfate (SO_4^{2-})	1	0
Nutrients (mg/dL)		
Glucose	70–110	100

Note: Although these values are representative, the precise dialysis fluid composition can be tailored to meet specific clinical needs. For example, if plasma potassium levels are too low, the dialysis fluid potassium ion concentration can be elevated to remedy the situation.

As diffusion takes place across the dialysis membrane, the composition of the blood changes. Potassium ions, phosphate ions, sulfate ions, urea, creatinine, and uric acid diffuse across the membrane into the dialysis fluid. Bicarbonate ions and glucose diffuse into the bloodstream. In effect, diffusion across the dialysis membrane takes the place of normal glomerular filtration, and the characteristics of the dialysis fluid ensure that important metabolites remain in the bloodstream rather than diffusing across the membrane.

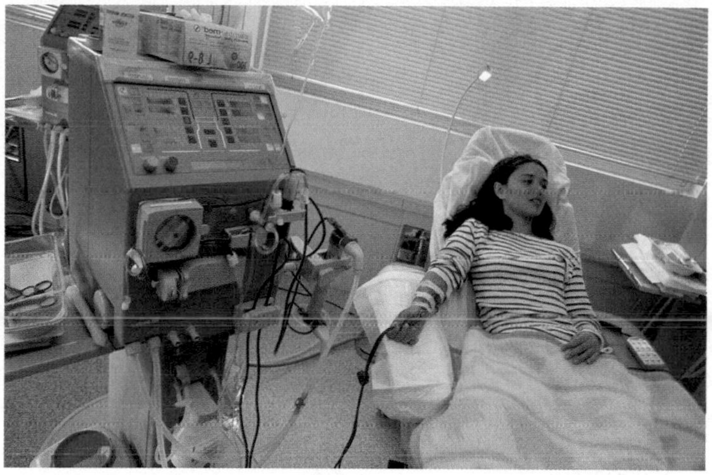

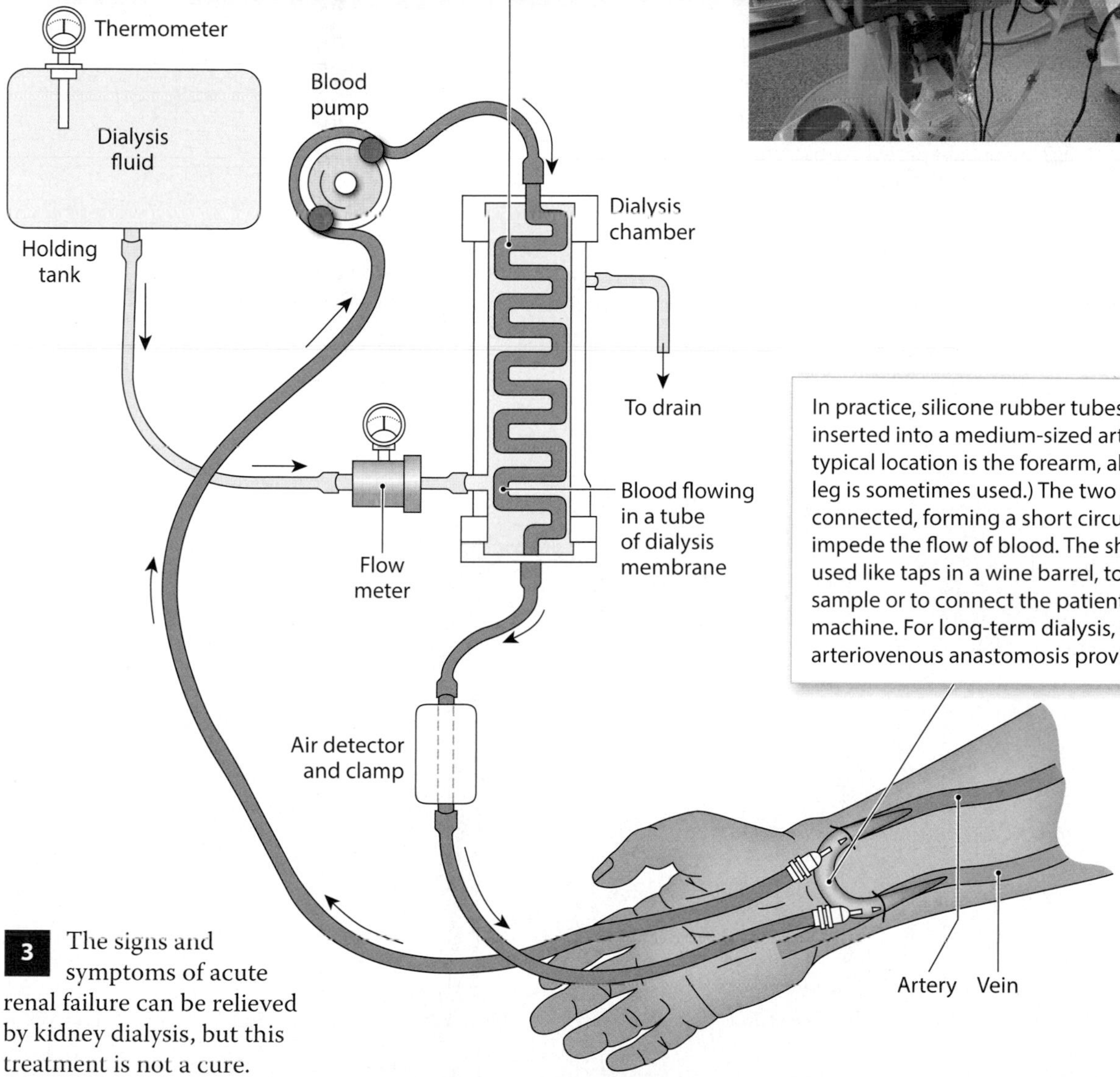
Thermometer
Dialysis fluid
Holding tank
Blood pump
Dialysis chamber
To drain
Flow meter
Blood flowing in a tube of dialysis membrane
Air detector and clamp
Artery
Vein

In practice, silicone rubber tubes called **shunts** are inserted into a medium-sized artery and vein. (The typical location is the forearm, although the lower leg is sometimes used.) The two shunts are then connected, forming a short circuit that does not impede the flow of blood. The shunts can then be used like taps in a wine barrel, to draw a blood sample or to connect the patient to a dialysis machine. For long-term dialysis, a surgically created arteriovenous anastomosis provides access.

3 The signs and symptoms of acute renal failure can be relieved by kidney dialysis, but this treatment is not a cure.

The only real cure for severe renal failure is a kidney transplant. This procedure involves implanting a new kidney obtained from a living donor or from a deceased donor. The recipient's nonfunctioning kidney(s) may be removed, especially if an infection is present. Patient survival is more than 90 percent at 2 years after the transplant. The use of kidneys from close relatives significantly improves the chances that the transplant will succeed. Immunosuppressive drugs are given to reduce tissue rejection. Unfortunately, this treatment also lowers the recipient's resistance to infection or cancer.

Module 24.14 Review

a. Briefly explain the difference between chronic renal failure and acute renal failure.

b. Define dialysis.

c. Explain why patients on dialysis often receive Epogen or Procrit, a synthetic form of erythropoietin.

LO **24.14** Compare and contrast chronic and acute renal failure, and explain the process of hemodialysis.

Matching

Match each lettered term with the most closely related description.

a. aquaporins
b. ADH
c. aldosterone
d. PCT
e. secretion
f. renal corpuscle
g. nephron loop
h. filtrate
i. BCOP
j. podocytes

1 Site of plasma filtration
2 Glomerular epithelium
3 Protein-free solution
4 Opposes filtration
5 Countercurrent multiplication
6 Water channels
7 Primary method for eliminating drugs or toxins
8 Stimulates ion pump—Na^+ reabsorbed
9 Primary site of nutrient reabsorption in the nephron
10 Regulates passive reabsorption of water from urine in the collecting system

1 __________
2 __________
3 __________
4 __________
5 __________
6 __________
7 __________
8 __________
9 __________
10 __________

Short answer

Identify the structures of the representative nephron and collecting system in the following diagram, and describe the functions of each.

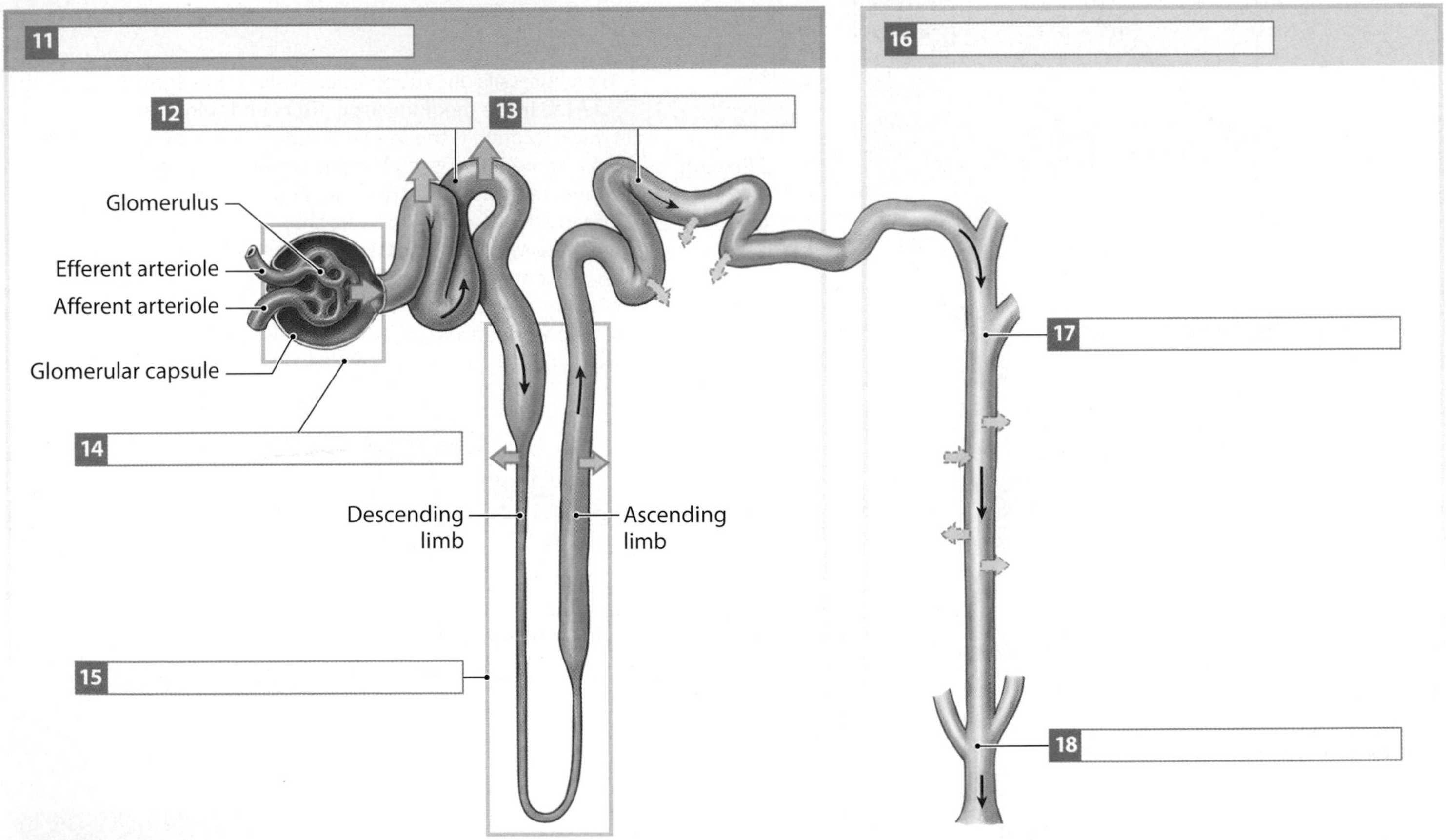
11
16
12
13
Glomerulus
Efferent arteriole
Afferent arteriole
Glomerular capsule
14
17
Descending limb
Ascending limb
15
18

Section integration

Marissa has had a urinalysis that detected large amounts of plasma proteins and white blood cells in her urine. What condition might be responsible, and what effects would it have on her urine output?

19 __________

The urinary tract transports, stores, and eliminates urine

Filtrate modification and urine production end when the fluid enters the renal pelvis. The **urinary tract** (the ureters, urinary bladder, and urethra) is responsible for the transport, storage, and elimination of urine.

1 A **pyelogram** (PĪ-el-ō-gram) is an image of the urinary system. It is obtained by taking an x ray of the kidneys after a radiopaque dye has been administered intravenously. Such an image provides an orientation to the relative sizes and positions of the system's main structures.

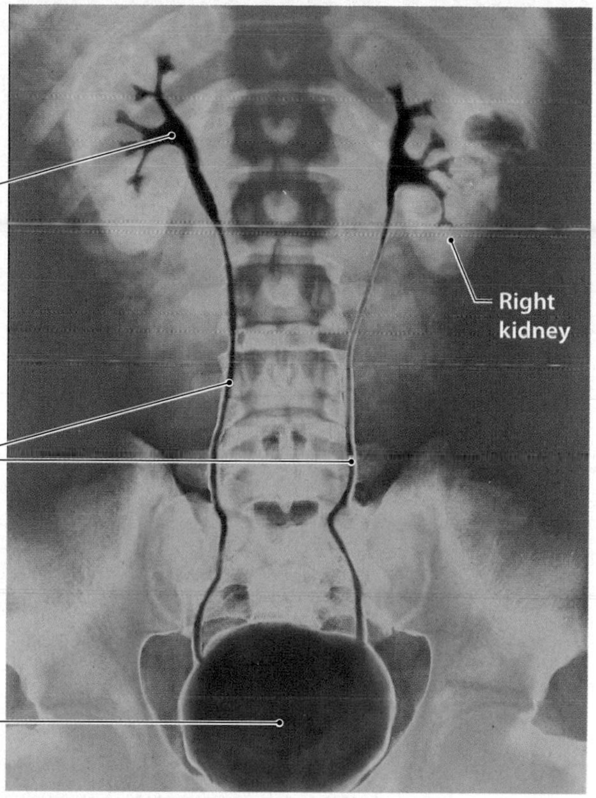
Renal pelvis
Right kidney

The **ureters** are a pair of muscular tubes that extend from the kidneys to the urinary bladder—a distance of about 30 cm (12 in.). The ureters are retroperitoneal and are firmly attached to the posterior abdominal wall. The paths taken by the ureters in men and women are different, due to variations in the nature, size, and position of the reproductive organs. In males, the base of the urinary bladder lies between the rectum and the pubic symphysis. In females, the base of the urinary bladder sits inferior to the uterus and anterior to the vagina.

The **urinary bladder** is a hollow, muscular organ that serves as a temporary reservoir for urine. A full urinary bladder can contain as much as a liter of urine.

2 This is a sectional view of a male pelvis showing the locations of the lower components of the urinary tract.

3 This is a sectional view of a female pelvis showing the locations of the lower components of the urinary tract.

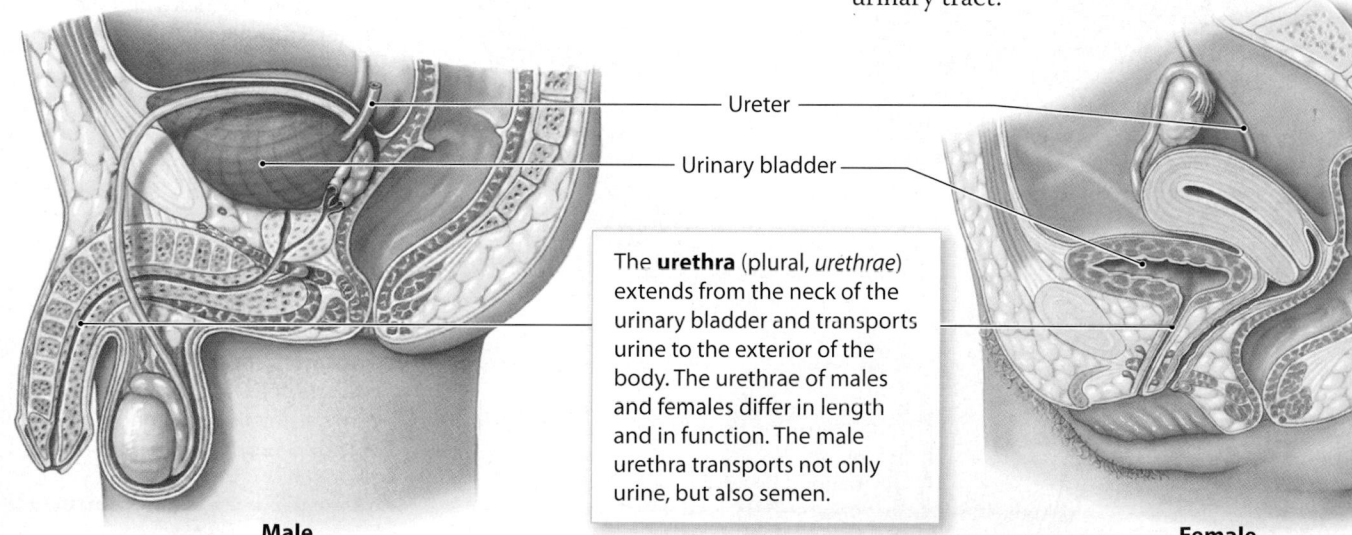

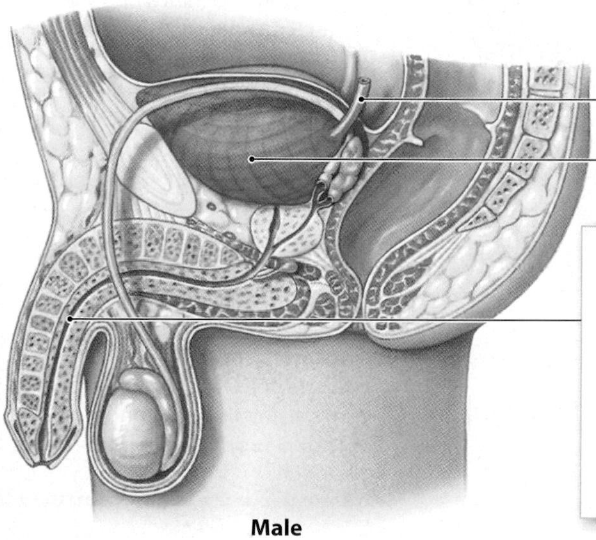
Ureter
Urinary bladder

Male

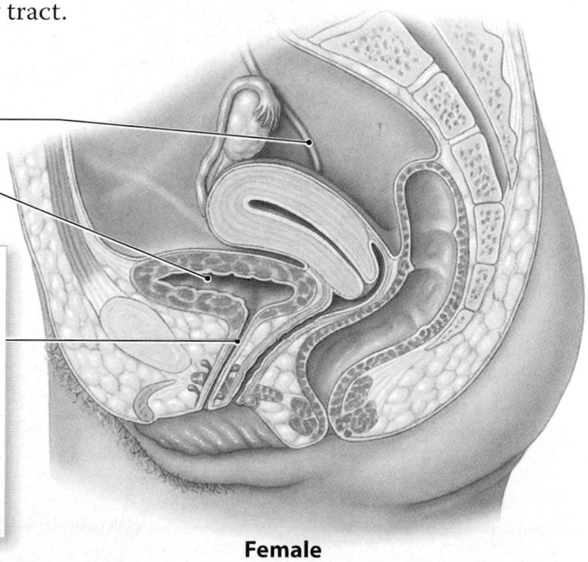

Female

The **urethra** (plural, *urethrae*) extends from the neck of the urinary bladder and transports urine to the exterior of the body. The urethrae of males and females differ in length and in function. The male urethra transports not only urine, but also semen.

Module 24.15 Review

a. When does filtrate modification and urine production end?

b. What is a pyelogram?

c. How does the urethra differ between males and females?

The ureters, urinary bladder, and urethra are specialized to conduct urine

1 The urinary bladder is filled by the ureters and drained by the urethra. Urinary bladder dimensions vary with its state of distension. The bladder's posterior, inferior, and anterior surfaces lie outside the peritoneal cavity. In these areas, tough ligamentous bands anchor it to the pelvic and pubic bones.

Supporting Ligaments

The **lateral umbilical ligaments** pass along the sides of the bladder to the umbilicus (navel). These fibrous cords are the vestiges of the two umbilical arteries, which supplied blood to the placenta during embryonic and fetal development.

The **middle umbilical ligament** extends from the anterior, superior border toward the umbilicus.

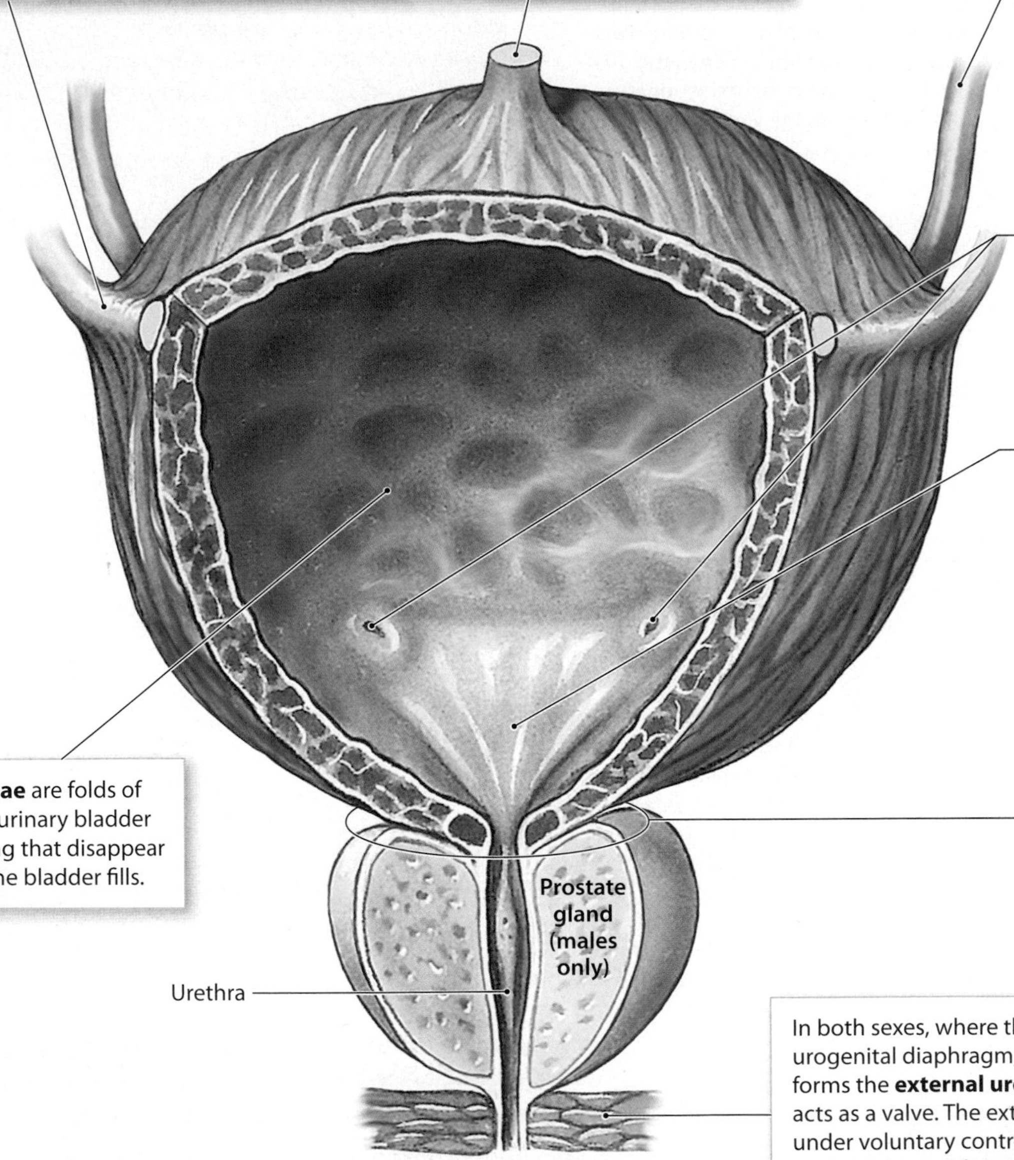
Prostate gland (males only)
Urethra

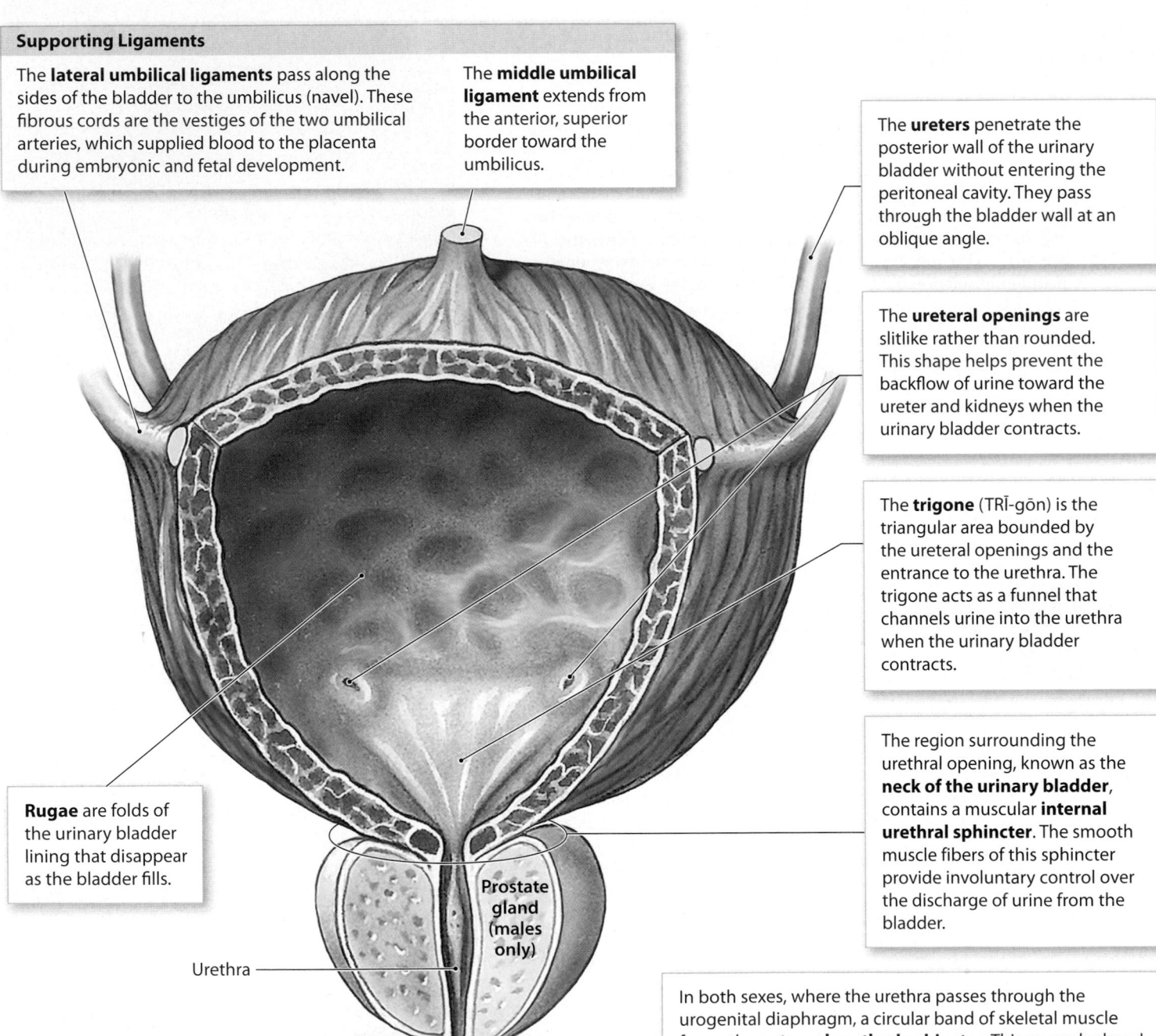

The **ureters** penetrate the posterior wall of the urinary bladder without entering the peritoneal cavity. They pass through the bladder wall at an oblique angle.

The **ureteral openings** are slitlike rather than rounded. This shape helps prevent the backflow of urine toward the ureter and kidneys when the urinary bladder contracts.

The **trigone** (TRĪ-gōn) is the triangular area bounded by the ureteral openings and the entrance to the urethra. The trigone acts as a funnel that channels urine into the urethra when the urinary bladder contracts.

Rugae are folds of the urinary bladder lining that disappear as the bladder fills.

The region surrounding the urethral opening, known as the **neck of the urinary bladder**, contains a muscular **internal urethral sphincter**. The smooth muscle fibers of this sphincter provide involuntary control over the discharge of urine from the bladder.

In both sexes, where the urethra passes through the urogenital diaphragm, a circular band of skeletal muscle forms the **external urethral sphincter**. This muscular band acts as a valve. The external urethral sphincter, which is under voluntary control by the perineal branch of the pudendal nerve, has a resting muscle tone and must be voluntarily relaxed to permit urination.

2 The wall of each ureter consists of three layers: (1) an inner mucosa, comprising a transitional epithelium and the surrounding lamina propria; (2) a middle muscular layer made up of longitudinal and circular bands of smooth muscle; and (3) an outer connective tissue layer that is continuous with the fibrous capsule and peritoneum. About every 30 seconds, a peristaltic contraction begins at the renal pelvis and sweeps along the ureter, forcing urine toward the urinary bladder.

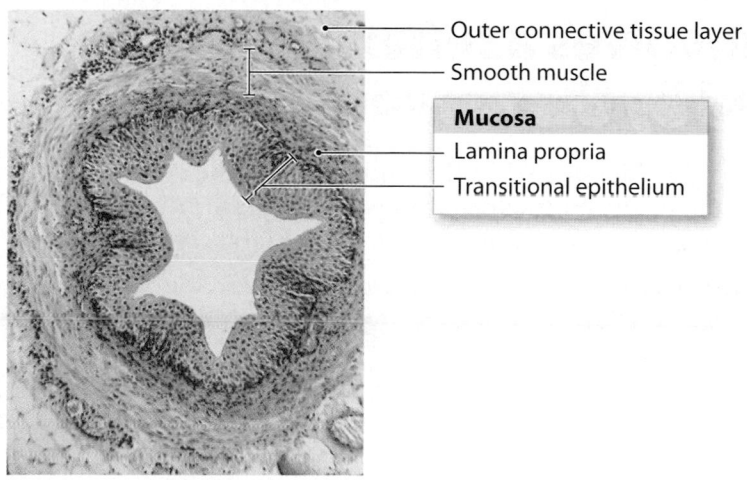

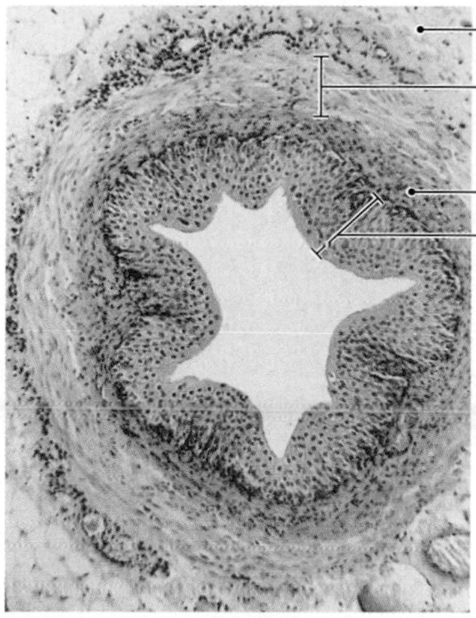

Outer connective tissue layer
Smooth muscle
Mucosa
Lamina propria
Transitional epithelium

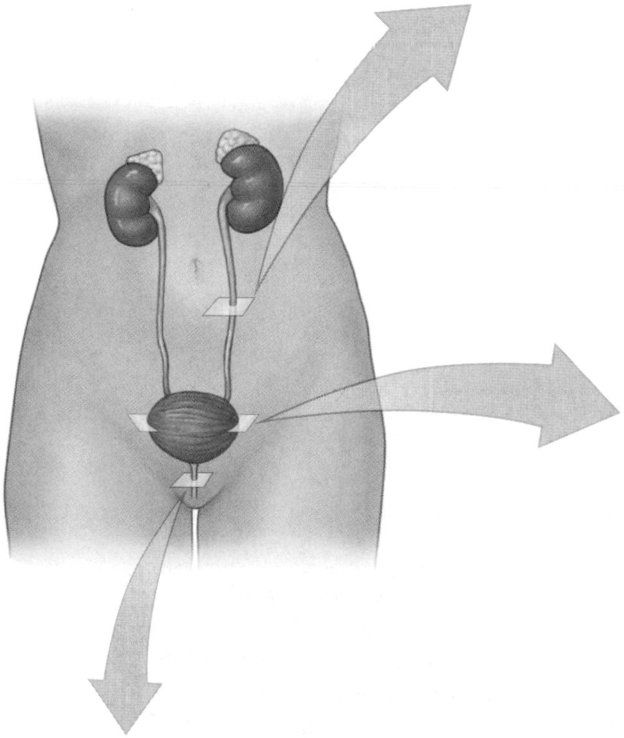

3 The wall of the urinary bladder contains mucosa, submucosa, and muscularis layers. The muscularis layer consists of inner and outer layers of longitudinal smooth muscle, with a circular layer between the two. Collectively, these layers form the powerful **detrusor** (dē-TRŪ-sor) **muscle** of the urinary bladder. Contraction of this muscle compresses the urinary bladder and expels its contents into the urethra.

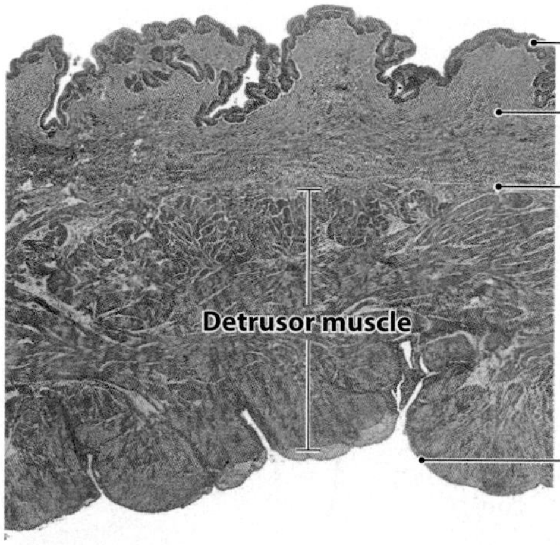

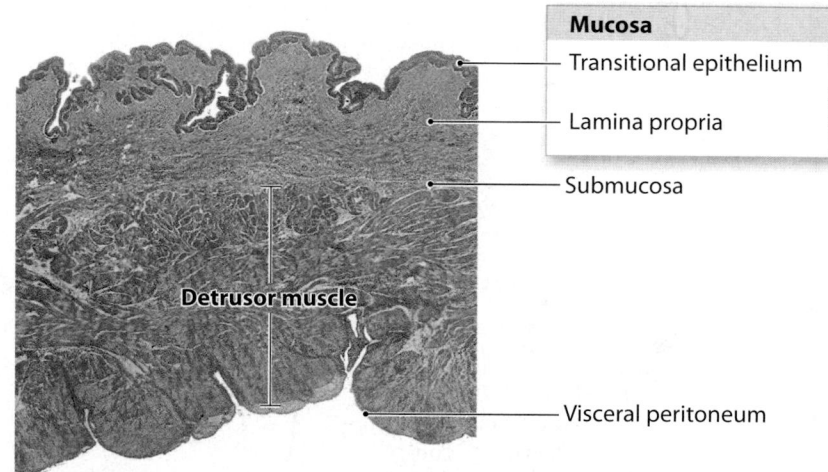

Mucosa
Transitional epithelium
Lamina propria
Submucosa
Detrusor muscle
Visceral peritoneum

4 The urethral lining consists of a stratified epithelium that varies from transitional at the neck of the urinary bladder, to stratified columnar at the midpoint, to stratified squamous near the external urethral orifice. The lamina propria is thick and elastic, and the mucous membrane has longitudinal folds. Mucin-secreting cells are located in the epithelial pockets. Connective tissues of the lamina propria anchor the urethra to surrounding structures.

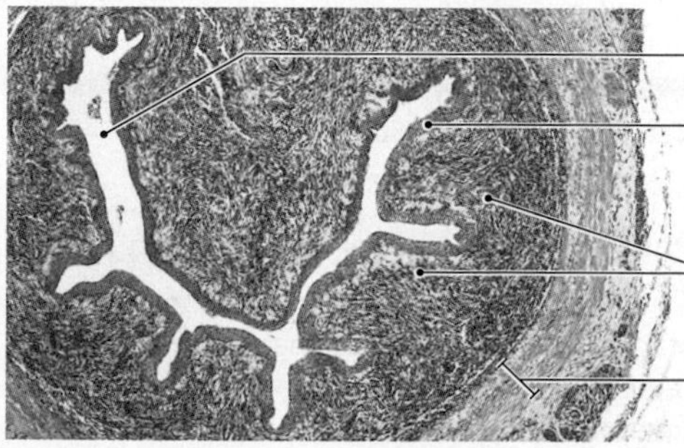

Lumen of urethra
Stratified squamous epithelium of mucosa
Lamina propria containing mucous epithelial glands
Smooth muscle

Module 24.16 Review

a. Urine is transported by the __________, stored within the __________, and eliminated through the __________.

b. What has to happen to the external urethral sphincter to allow urination?

c. The wall of the urinary bladder is composed of a specialized smooth muscle. Name it and describe its physiological role.

Module 24.17

Urination involves a reflex coordinated by the nervous system

Urine reaches the urinary bladder by peristaltic contractions of the ureters. The process of urination is coordinated by the **micturition reflex**, a complex process involving both a local reflex pathway and a central pathway through the cerebral cortex.

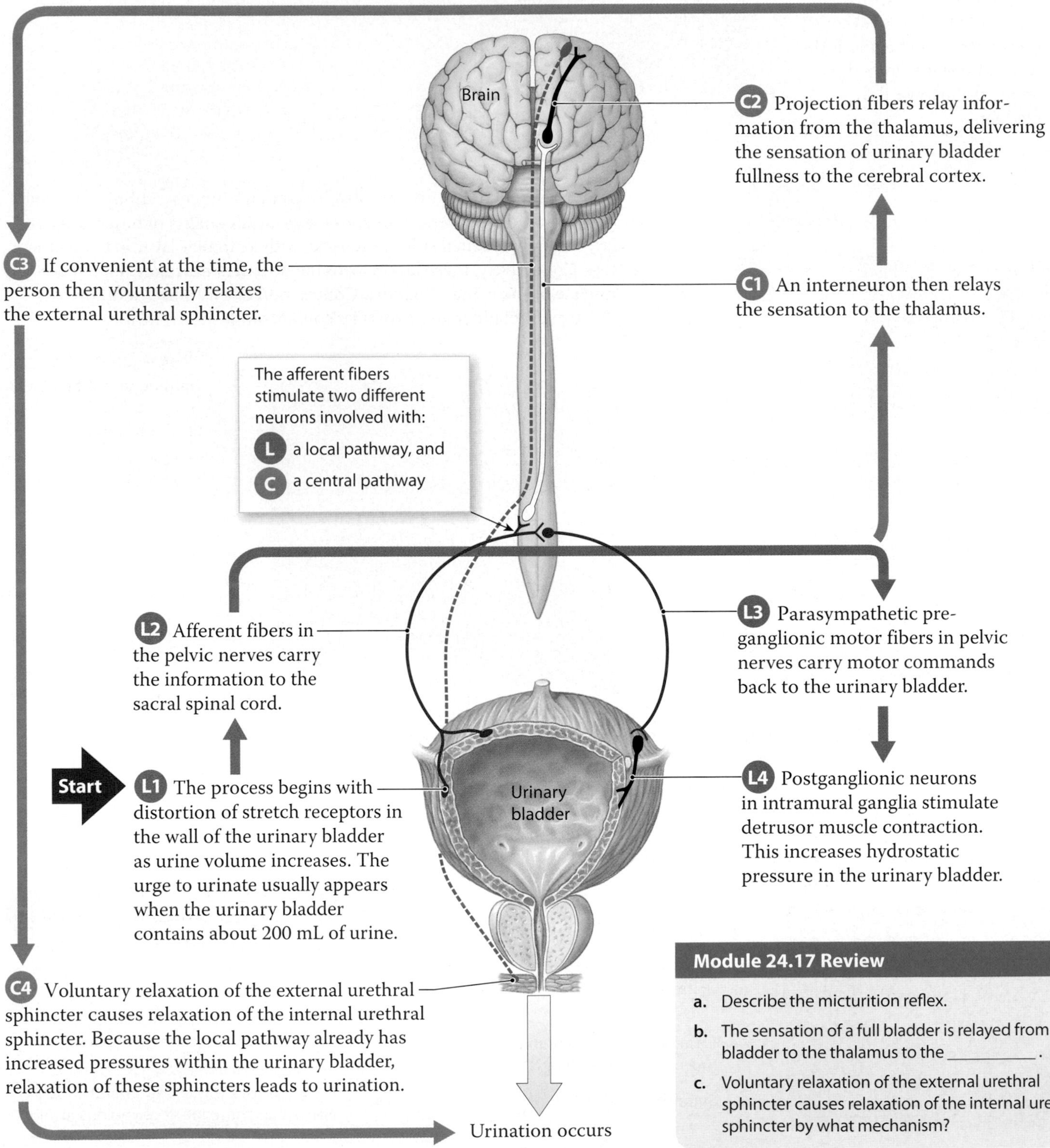

Brain
C2 Projection fibers relay information from the thalamus, delivering the sensation of urinary bladder fullness to the cerebral cortex.
C3 If convenient at the time, the person then voluntarily relaxes the external urethral sphincter.
C1 An interneuron then relays the sensation to the thalamus.
The afferent fibers stimulate two different neurons involved with:
L a local pathway, and
C a central pathway
L2 Afferent fibers in the pelvic nerves carry the information to the sacral spinal cord.
L3 Parasympathetic preganglionic motor fibers in pelvic nerves carry motor commands back to the urinary bladder.
Start
L1 The process begins with distortion of stretch receptors in the wall of the urinary bladder as urine volume increases. The urge to urinate usually appears when the urinary bladder contains about 200 mL of urine.
Urinary bladder
L4 Postganglionic neurons in intramural ganglia stimulate detrusor muscle contraction. This increases hydrostatic pressure in the urinary bladder.
C4 Voluntary relaxation of the external urethral sphincter causes relaxation of the internal urethral sphincter. Because the local pathway already has increased pressures within the urinary bladder, relaxation of these sphincters leads to urination.
Urination occurs

Module 24.17 Review

a. Describe the micturition reflex.

b. The sensation of a full bladder is relayed from the bladder to the thalamus to the ___________.

c. Voluntary relaxation of the external urethral sphincter causes relaxation of the internal urethral sphincter by what mechanism?

LO 24.17 Discuss the roles of local and central pathways in urination, and describe the micturition reflex.

Urinary disorders can often be detected by physical examinations and laboratory tests

1 The primary signs and symptoms of urinary system disorders include changes in volume and appearance of urine, frequency of urination, and pain. The nature and location of the pain can provide clues to the source.

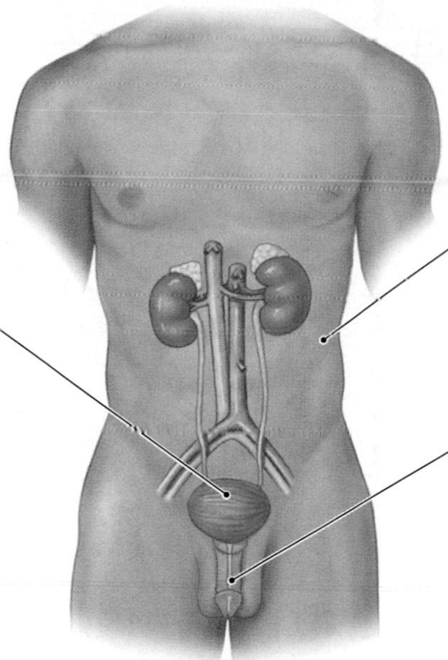

Pain in the superior pubic region may be associated with urinary bladder disorders.
Pain in the superior lumbar region or in the flank that radiates to the right upper quadrant or left upper quadrant can be caused by kidney infections such as **pyelonephritis**, or by **renal calculi** (kidney stones).
Dysuria (painful or difficult urination) can occur with cystitis or urethritis, or with urinary obstructions. In males, enlargement of the prostate gland can compress the urethra and lead to dysuria.

2 Characteristic changes in urinary output or frequency give clues to underlying problems with this system or other systems.

Abnormal Urine Output and Frequency

Increased Urgency or Increased Frequency

An irritation of the lining of the ureters or urinary bladder can lead to the strong desire to urinate (urgency) with the increased need to urinate more often (frequency), although the total amount of urine produced each day remains normal.

Changes in Urinary Output

Changes in the volume of urine produced by a person with average fluid intake indicate problems either with the kidneys or with the control of renal function. **Polyuria**, the production of excessive amounts of urine, results from hormonal or metabolic problems, such as those associated with diabetes or glomerulonephritis. **Oliguria** (a urine volume of 50–500 mL/day) and **anuria** (0–50 mL/day) are conditions that indicate serious kidney problems and potential renal failure.

Incontinence

Incontinence, an inability to control urination voluntarily, may involve periodic involuntary leakage (stress incontinence), or inability to delay urination (urge incontinence)—or a continual, slow trickle of urine from a bladder that is always full (overflow incontinence).

Urinary Retention

In **urinary retention**, renal function is normal, at least initially, but urination does not occur. Urinary retention in males commonly results from enlargement of the prostate gland and compression of the prostatic urethra.

Important clinical signs of urinary system disorders include the following:

- **Edema**. Renal disorders often lead to **proteinuria** (protein in the urine), and if severe, result in generalized edema in peripheral tissues. Facial swelling, especially around the eyes, is common.
- **Fever.** A fever commonly develops when the urinary system is infected by pathogens. **Cystitis** (urinary bladder inflammation), usually caused by infection, may result in a low-grade fever; kidney infections, such as pyelonephritis, can produce very high fevers.

Module 24.18 Review

a. What is the term for painful or difficult urination?

b. Obstruction of a ureter by a kidney stone would interfere with the flow of urine between which two points?

c. Why is urinary obstruction at the urethra more dangerous than at the ureter?

LO 24.18 Describe common urinary disorders related to output and frequency.

Labeling

Use the following descriptions to fill in the boxes in the micturition reflex diagram.

- sensation relayed to thalamus
- person relaxes external urethral sphincter
- afferent fibers carry information to sacral spinal cord
- sensation of bladder fullness delivered to cerebral cortex
- stretch receptors stimulated
- detrusor muscle contraction stimulated
- parasympathetic preganglionic fibers carry motor commands
- internal urethral sphincter relaxes

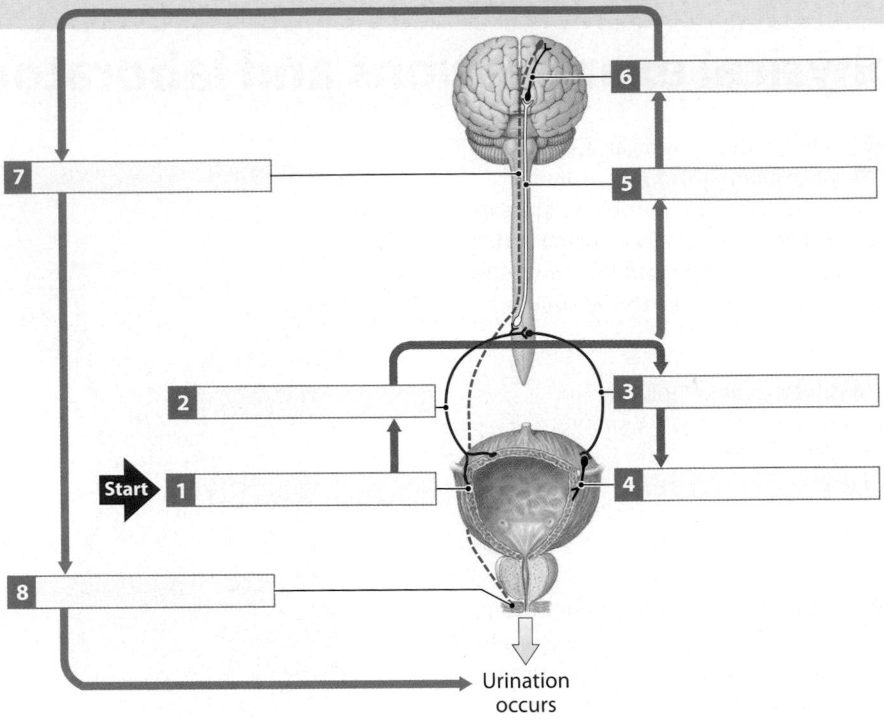
6
7
5
2
3
Start
1
4
8
Urination occurs

Matching

Match each lettered term with the most closely related description.

a. urethra
b. external urethral sphincter
c. detrusor
d. rugae
e. internal urethral sphincter
f. trigone
g. transitional epithelium
h. micturition
i. stratified squamous epithelium
j. external urethral orifice
k. middle umbilical ligament

9 The ring of smooth muscle in the neck of the urinary bladder
10 Triangular area within the urinary bladder
11 Relaxation of this muscle leads to urination
12 The external opening of the urethra
13 Folds lining the surface of the empty urinary bladder
14 Superior, supporting fibrous cord of the urinary bladder
15 Contraction of this smooth muscle compresses the urinary bladder
16 The type of epithelium that lines the ureters
17 Tube that transports urine to the exterior
18 Term for urination
19 Epithelium that lines the urethra

9 ________
10 ________
11 ________
12 ________
13 ________
14 ________
15 ________
16 ________
17 ________
18 ________
19 ________

Short answer

List four primary signs and symptoms of urinary disorders.

20 ________

Briefly describe the similarities and differences in the following pairs of terms.

21 cystitis/pyelonephritis
22 stress incontinence/overflow incontinence
23 polyuria/proteinuria

21 ________
22 ________
23 ________

Study Outline

SECTION 1 • Anatomy of the Urinary System

24.1 The urinary system organs are the kidneys, ureters, urinary bladder, and urethra p. 923

1. The **urinary system** eliminates excess water, salts, and physiological wastes through urine.
2. The **kidneys** perform the excretory functions of the urinary system. Urine is eliminated by the **urinary tract** (**ureters**, **urinary bladder**, and **urethra**).
3. The functions of the urinary system are adjusting blood volume and pressure, regulating blood plasma concentrations of ions, stabilizing blood pH, conserving nutrients by preventing their loss in urine, and removing drugs and toxins from the bloodstream.

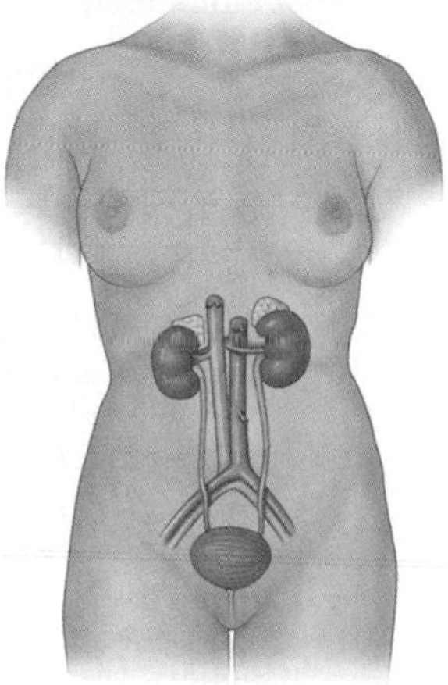

24.2 The kidneys are paired retroperitoneal organs p. 924

4. The **hilum** is a medial indentation on the kidney. It is the entry point for the renal artery and renal nerves, and an exit point for the renal vein and ureters.
5. The **ureters** pass inferiorly and cross the anterior surfaces of the external iliac artery and vein before emptying into the posterior, inferior surface of the urinary bladder.
6. The **kidneys**, adrenal glands, and ureters lie between the muscles of the posterior body wall and the parietal peritoneum, in a **retroperitoneal** position.
7. Visceral organs, the body wall musculature, and the 11th and 12th ribs protect the kidneys. The left kidney lies slightly superior to the right kidney.
8. A **fibrous capsule** of collagen fibers covers the outer surface of the kidney.
9. **Perinephric fat** surrounds the fibrous capsule. **Renal fascia** anchors the kidney to surrounding structures.

24.3 The kidneys are complex at the gross and microscopic levels p. 926

10. The **fibrous capsule** also lines the **renal sinus**, an internal cavity within the kidney.
11. The **renal cortex** is the superficial portion of the kidney. The **renal medulla** extends from the renal cortex to the renal sinus.
12. A **renal pyramid** is a conical structure extending from the cortex to a tip called the **renal papilla**. A **renal column** separates adjacent pyramids.
13. A **minor calyx** collects the urine produced by a single **kidney lobe**. A **major calyx** is a fusion of 4–5 minor calyces. The **renal pelvis** collects urine from the major calyces and is continuous with the ureters.
14. **Nephrons** are microscopic functional units of the kidney. Approximately 85 percent of nephrons are **cortical nephrons**. The remaining nephrons are **juxtamedullary nephrons** that have long **nephron loops**.

24.4 A nephron is divided into segments; each segment has specific functions p. 928

15. A nephron consists of a **renal corpuscle** and a **renal tubule**.
16. At the renal corpuscle (which consists of the **glomerular capsule** and **glomerulus**), blood pressure forces water and dissolved solutes out of the glomerular capillaries and into the **capsular space**. Filtration produces **filtrate** that is similar to blood plasma.
17. After modification by the renal tubule and collecting system, the filtrate leaves the kidneys as urine.
18. The **proximal convoluted tubule** (**PCT**) reabsorbs nutrients from the **tubular fluid**. The **distal convoluted tubule** (**DCT**) further adjusts the tubular fluid.
19. The **nephron loop** establishes an osmotic gradient in the renal medulla that promotes water reabsorption from tubular fluid, permitting the production of concentrated urine.

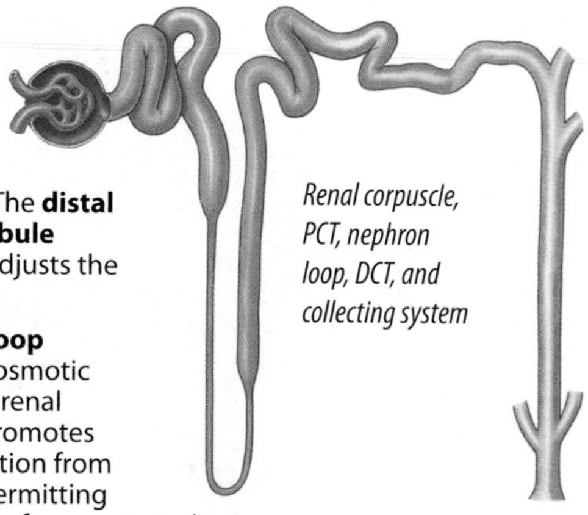
Renal corpuscle, PCT, nephron loop, DCT, and collecting system

20. Each nephron empties into a **collecting system** of tubes (**collecting duct** and **papillary duct**) that performs final adjustments to urine volume and composition.

24.5 The kidneys are highly vascular, and the circulation patterns are complex p. 930

21. The pathway of renal blood flow is: **renal artery** → **segmental arteries** → **interlobar arteries** → **arcuate arteries** → **cortical radiate arteries** → **afferent arterioles** → **glomerulus** → **efferent arterioles** → **peritubular capillaries** → **cortical radiate veins** → **arcuate veins** → **interlobar veins** → **renal vein**.

SECTION 2 • Overview of Renal Physiology

24.6 The kidneys maintain homeostasis by removing wastes and producing urine p. 933

22. Three solutes excreted in urine are **urea** (from liver breakdown of amino acids), **creatinine** (from the breakdown of creatine phosphate), and **uric acid** (from the recycling of the nitrogenous bases of RNA).
23. The three distinct processes of kidney function are **filtration**, **reabsorption**, and **secretion**.

24. The kidneys produce a fluid (urine) very different from other body fluids. Laboratory tests can measure the concentration of substances within the urine.

24.7 Filtration, reabsorption, and secretion occur in specific segments of the nephron and collecting system p. 934

25. Filtration occurs at the renal corpuscle.
26. Most segments of the nephron perform a combination of reabsorption and secretion, but the balance between the two processes varies from one segment to another.
27. Final volume and solute concentration of the urine result from the interaction between the collecting system and the nephron loops.

24.8 Filtration occurs at the renal corpuscle p. 936

28. Filtration is the vital first step in urine formation.
29. The **afferent arteriole** delivers blood from a cortical radiate artery, while the **efferent arteriole** delivers blood to peritubular capillaries. The efferent arteriole has a smaller diameter than the afferent arteriole, and this elevates the blood pressure within the glomerulus.

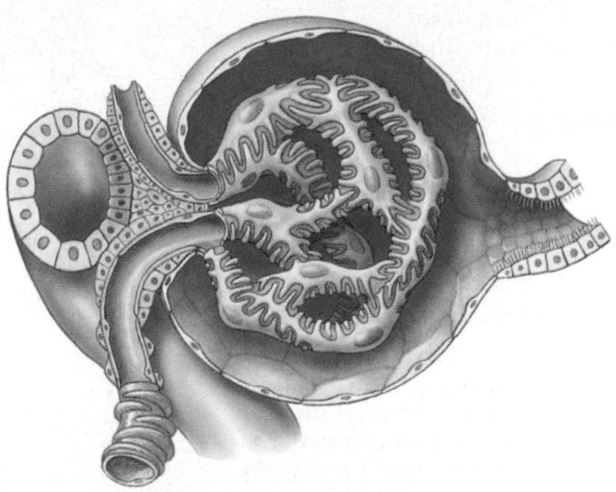

30. The **juxtaglomerular complex** secretes renin when glomerular blood pressure decreases.
31. Substances passing out of the glomerulus must be small enough to pass through the **filtration slits** between the **pedicels** of the **podocytes**.
32. Glomerular capillaries are fenestrated capillaries. The endothelium covers a **dense layer** that along with the filtration slits forms the **filtration membrane**.
33. The primary factor involved in glomerular filtration is the balance between **hydrostatic pressure** and **colloid osmotic pressure**.

24.9 The glomerular filtration rate is the amount of filtrate produced each minute p. 938

34. The **glomerular filtration rate (GFR)** is the amount of filtrate the kidneys produce each minute.
35. GFR is controlled by autoregulation at the local level, and central regulation, which has an endocrine component initiated by the kidneys and an autonomic component involving the sympathetic division of the ANS.
36. Through **autoregulation**, the kidneys adjust GFR when filtration rate decreases by dilation of the afferent arterioles, contraction of the mesangial cells, and constriction of the efferent arterioles. The result is an increase in glomerular blood pressure and normal GFR.
37. If autoregulation is ineffective, **central regulation** responds by activating endocrine and neural mechanisms. If GFR does not return to normal, the juxtaglomerular complex increases renin production.
38. Renin triggers the formation of angiotensin I, which then is activated to angiotensin II by **angiotensin converting enzyme** (**ACE**) in the capillaries of the lungs.
39. Angiotensin II triggers a series of **neural responses** to increase glomerular pressure, including stimulation of thirst centers, increased ADH and aldosterone production, increased cardiac output, and increased sympathetic motor tone.

24.10 Reabsorption predominates along the proximal convoluted tubule, whereas reabsorption and secretion are often linked along the distal convoluted tubule p. 940

40. The proximal convoluted tubule (PCT) reabsorbs more than 99 percent of the glucose, amino acids, and other organic nutrients in the fluid. The PCT also reabsorbs sodium, potassium, bicarbonate, magnesium, phosphate, and sulfate ions.
41. Only 15 to 20 percent of the initial filtrate volume reaches the distal convoluted tubule (DCT). In the DCT, sodium ions are reabsorbed and exchanged for potassium ions by ion pumps that are stimulated by aldosterone. When pH decreases, hydrogen ions are exchanged for sodium ions, and the DCT also secretes toxins and drugs.

24.11 Exchange between the limbs of the nephron loop creates an osmotic concentration gradient in the renal medulla p. 942

42. The exchange that occurs between the thin descending limb and the thick ascending limb of the nephron loop is called **countercurrent multiplication**.
43. *Countercurrent* refers to the fact that tubular fluid in the descending limb flows toward the renal pelvis, whereas tubular fluid in the ascending limb flows toward the renal cortex.
44. *Multiplication* refers to the fact that the effect of the exchange increases as fluid movement continues.
45. Active transport in the thick ascending limb moves sodium and chloride ions out of the tubular fluid and into the peritubular fluid of the renal medulla.
46. The thin descending limb is permeable to water but impermeable to solutes.
47. As water is reabsorbed along the DCT and collecting duct, the concentration of urea gradually increases in the tubular fluid.

24.12 Urine volume and concentration are hormonally regulated p. 944

48. Water reabsorption may occur passively or by hormonal control. **Obligatory water reabsorption**, when water movements cannot be prevented, occurs in the PCT and descending limb of the nephron loop.
49. **Facultative water reabsorption**, which is controlled by antidiuretic hormone (ADH), occurs in the DCT and collecting system. **Aquaporins** in the apical plasma membranes lining the DCT and collecting duct enhance the rate of osmotic water movement.
50. An adult typically produces 1200 mL of urine per day.

24.13 Renal function is an integrative process involving filtration, reabsorption, and secretion p. 946

51. The filtrate produced by the renal corpuscle has the same osmotic concentration as plasma, about 300 mOsm/L.
52. In the PCT, the active removal of ions and organic nutrients produces a continuous osmotic flow of water out of the tubular fluid. Obligatory water reabsorption in the PCT and descending limb moves water into the peritubular fluid leaving highly concentrated tubular fluid.
53. The thick ascending limb of the nephron loop is impermeable to water and solutes. The tubule cells actively transport Na^+ and Cl^- out of the tubule.
54. Further adjustments in the composition of the tubular fluid occur in the DCT and the collecting system.
55. Final adjustments of the tubular fluid are made in the DCT and the collecting system where the level of ADH determines the final urine concentration.
56. The vasa recta absorb the solutes and water reabsorbed by the nephron loop, DCT, and the collecting ducts.

24.14 Renal failure is a life-threatening condition p. 948

57. **Renal failure** occurs when the kidneys cannot perform the excretory functions needed to maintain homeostasis.
58. In **chronic renal failure**, kidney function deteriorates gradually, and the associated problems accumulate over time.
59. **Acute renal failure** occurs when exposure to toxic drugs, renal ischemia, urinary obstruction, or trauma causes filtration to slow suddenly or stop.
60. In **hemodialysis** an artificial membrane regulates the composition of blood by means of a dialysis machine. The basic principle in this process, called **dialysis**, is passive diffusion across a selectively permeable membrane. Diffusion across the dialysis membrane takes the place of normal glomerular filtration.

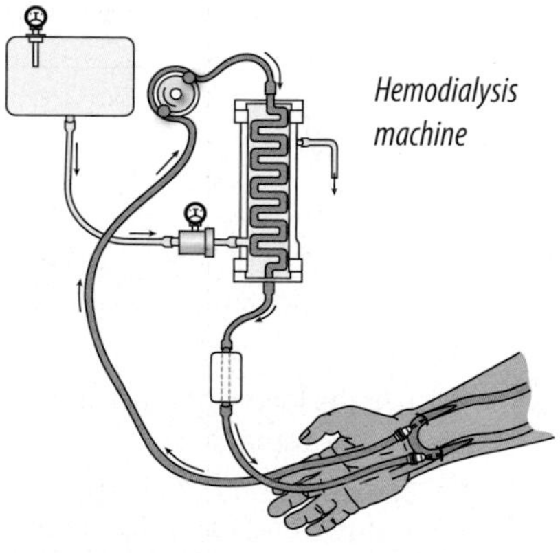

Hemodialysis machine

SECTION 3 • Urine Storage and Elimination

24.15 The urinary tract transports, stores, and eliminates urine p. 951

61. The **urinary tract** is composed of the ureters, urinary bladder, and urethra. It transports, stores, and eliminates urine.
62. The **ureters** are a pair of muscular tubes that extend from the kidneys to the urinary bladder.
63. The **urinary bladder** is a hollow, muscular organ that serves as a temporary reservoir for urine.
64. The **urethra** extends from the neck of the urinary bladder and transports urine to the exterior of the body. The male urethra is longer, and transports semen as well as urine.

24.16 The ureters, urinary bladder, and urethra are specialized to conduct urine p. 952

65. The **lateral umbilical ligaments** and the **middle umbilical ligament** support the urinary bladder.
66. The **ureters** penetrate the posterior wall of the urinary bladder at an oblique angle without entering the peritoneal cavity. The **urethral openings** are slitlike to help prevent the backflow of urine as the bladder contracts.
67. The **trigone** is a triangular area bounded by the urethral openings and the entrance to the urethra. It acts like a funnel that channels urine toward the urethra. **Rugae** are folds of the urinary bladder lining that disappear as the bladder fills.
68. The **neck of the urinary bladder** is the region surrounding the urethral opening. It contains the smooth muscle **internal urethral sphincter**.

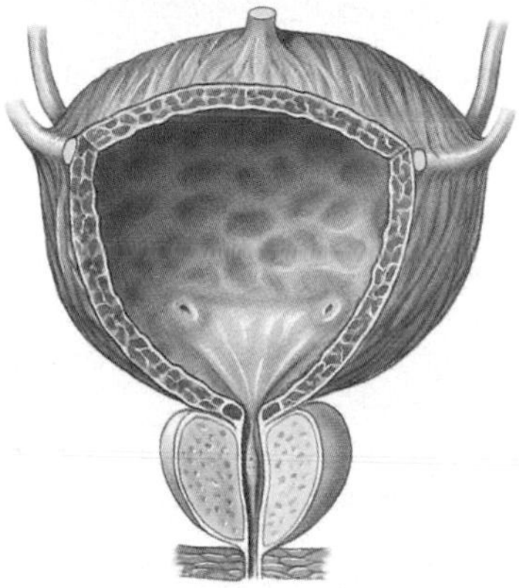

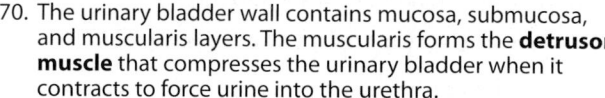

69. Each ureter consists of three layers: (1) an inner mucosa, comprising a transitional epithelium and the surrounding lamina propria; (2) a middle muscular layer made up of longitudinal and circular bands of smooth muscle; and (3) an outer connective tissue layer. Peristalsis moves urine toward the urinary bladder.
70. The urinary bladder wall contains mucosa, submucosa, and muscularis layers. The muscularis forms the **detrusor muscle** that compresses the urinary bladder when it contracts to force urine into the urethra.
71. The urethral lining consists of a stratified epithelium that varies from transitional at the neck of the urinary bladder, to stratified columnar at the midpoint, to stratified squamous near the external urethral orifice.

24.17 Urination involves a reflex coordinated by the nervous system p. 954

72. The **micturition reflex** coordinates the process of urination.
73. Voluntary relaxation of the external urethral sphincter causes relaxation of the internal urethral sphincter. Because the local pathway already has increased pressures within the urinary bladder, relaxation of these sphincters leads to urination.

24.18 Urinary disorders can often be detected by physical examinations and laboratory tests p. 955

74. The primary signs and symptoms of urinary system disorders include changes in volume and appearance of urine, frequency of urination, and pain.
75. **Pyelonephritis** (kidney infection) or **renal calculi** (kidney stones) can cause pain in the superior lumbar region. **Dysuria** (painful or difficult urination) can be caused by a variety of reasons.
76. **Proteinuria** (protein in the urine), **edema**, **fever**, and **cystitis** (urinary bladder inflammation) are all important clinical signs of urinary system disorders.

Chapter Review Questions

Labeling

Label the image of a renal corpuscle and associated structures shown here.

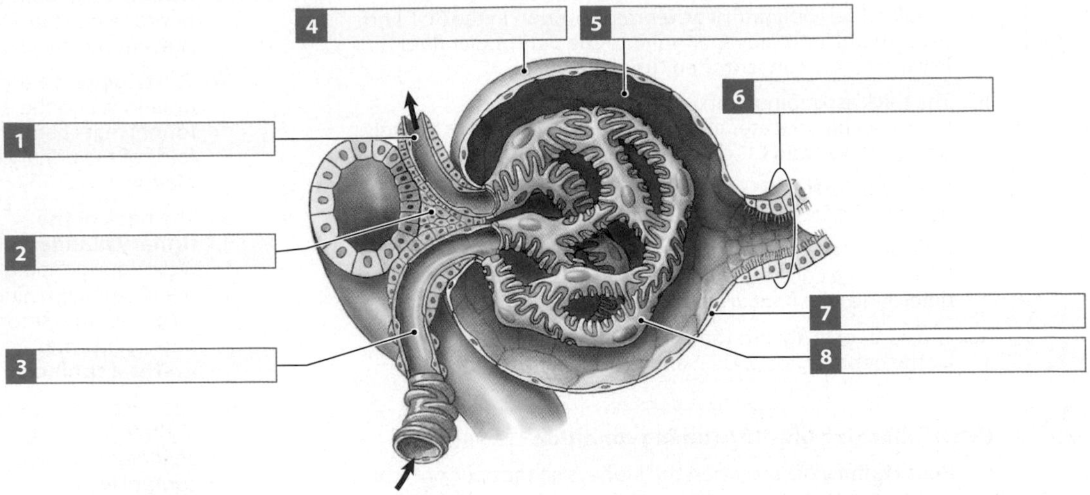
1
2
3
4
5
6
7
8

True/False

Indicate whether each statement is true or false.

9 The urinary tract includes the kidneys, ureters, urinary bladder, and urethra.

10 The left kidney lies slightly superior to the right kidney.

11 Most nephrons are cortical nephrons.

12 Tubular fluid is no longer modified beyond the DCT.

13 Contraction of the detrusor muscle moves urine along the ureters, toward the urinary bladder.

14 The process of urination is coordinated by the micturition reflex.

9 ____________
10 ____________
11 ____________
12 ____________
13 ____________
14 ____________

Multiple choice

Select the correct answer from the list provided.

15 The basic functional unit of the kidney is the
- [ ] a) filtration unit.
- [ ] b) nephron loop.
- [ ] c) glomerulus.
- [ ] d) nephron.

16 When ADH levels increase,
- [ ] a) the amount of water reabsorbed increases.
- [ ] b) the amount of water reabsorbed decreases.
- [ ] c) the DCT becomes impermeable to water.
- [ ] d) sodium ions are exchanged for potassium ions.

17 Which of the following conditions would cause an increase in the glomerular filtration rate?
- [ ] a) constriction of the afferent arteriole
- [ ] b) constriction of the efferent arteriole
- [ ] c) decrease in systemic blood pressure
- [ ] d) decrease in aldosterone and ADH production

18 Which of the following statements best describes the action of aldosterone at the DCT?
- [ ] a) Aldosterone stimulates actions that cause the secretion of drugs and toxins from the peritubular fluid.
- [ ] b) Sodium ions are reabsorbed in exchange for potassium ions by ion pumps stimulated by aldosterone.
- [ ] c) Aldosterone increases water secretion.
- [ ] d) Aldosterone causes glucose to be reabsorbed into the peritubular fluid.

19 Water reabsorption occurs primarily along the
- [ ] a) entire length of the nephron loop.
- [ ] b) ascending limb of the nephron loop and the DCT.
- [ ] c) PCT and descending limb of the nephron loop.
- [ ] d) DCT and collecting system.

20 In central regulation, decreased GFR is coordinated by the
- [ ] a) renal corpuscle.
- [ ] b) juxtamedullary complex.
- [ ] c) renal pelvis.
- [ ] d) cerebral cortex.

21 Urea is a byproduct of the breakdown of

- ☐ a) amino acids by the liver.
- ☐ b) creatine phosphate in skeletal muscle.
- ☐ c) the recycling of the nitrogenous bases in RNA molecules.
- ☐ d) glycogen to glucose.

22 Which of the following is *not* a response to angiotensin II?

- ☐ a) decreased ADH production
- ☐ b) increased stimulation of the thirst centers
- ☐ c) increased aldosterone secretion by the adrenal glands
- ☐ d) constriction of peripheral arteries and further constriction of the efferent arterioles

Short answer

23 Identify and describe the three distinct processes in urine production.

24 Describe the primary process involved in glomerular filtration.

25 What is countercurrent multiplication?

MasteringA&P®

Access more chapter study tools online in the MasteringA&P Study Area:

- **Chapter Quizzes, Chapter Practice Test, Art-labeling Activities, Animations, MP3 Tutor Sessions, and Clinical Case Studies**

- **Practice Anatomy Lab** PAL™
- **Interactive Physiology** iP®
- **A&P Flix** A&PFlix™
- **PhysioEx** PhysioEx™

Chapter Integration • Applying what you have learned

Melamine contamination and food product safety

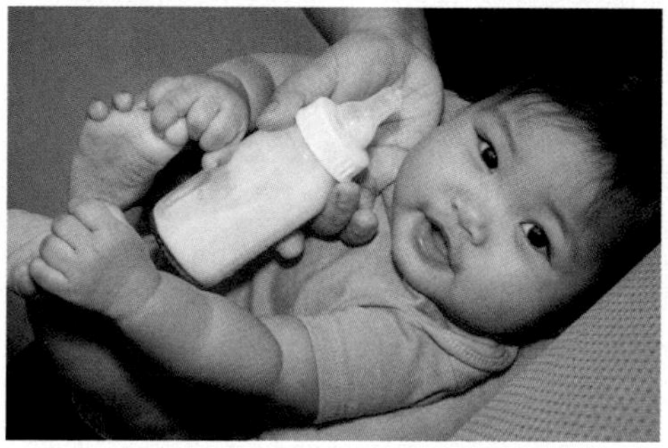

For consumers, some troubling developments have turned the purity of food and food products, for humans and animals alike, from a matter of trust into a matter of concern.

In 2007, the unintentional adulteration of pet food made in China with the nitrogen-containing industrial chemical melamine caused pet illnesses and deaths in the United States and resulted in massive recalls of contaminated products. In 2008, milk intentionally contaminated with melamine—put there to skew laboratory tests that measure nitrogen as an index of the protein content in a food or drink powder—sickened at least 64,000 children in China, several of whom died. Clearly, vigilance concerning our food supply is crucial, even while standards and regulations are in place.

In the body, melamine contamination has two major effects: (1) the formation of crystalline masses in the filtrate and/or urine, and (2) acidification of the tubular fluid. Resulting clinical problems range from blood in the urine, to acid-base and electrolyte disorders, to urinary obstruction and (in severe cases) kidney failure.

1 **Propose a linkage between the stated effects of melamine poisoning and the clinical problems observed.**

2 **Primary therapeutic options for melamine poisoning include infusion of fluids, dialysis, and medication. Explain how these options address specific problems caused by melamine poisoning, and suggest possible follow-up tests.**

25 Fluid, Electrolyte, and Acid–Base Balance

LEARNING OUTCOMES

These Learning Outcomes correspond by number to this chapter's modules and indicate what you should be able to do after completing the chapter.

SECTION 1 • Fluid and Electrolyte Balance

25.1 Name the body's fluid compartments, identify the solid components, and summarize their contents.

25.2 Explain what is meant by fluid balance, and discuss its importance for homeostasis.

25.3 Explain what is meant by mineral balance, and discuss its importance for homeostasis.

25.4 Summarize the relationship between sodium and water in maintaining fluid and electrolyte balance.

25.5 **CLINICAL MODULE** Explain factors that control potassium balance, and discuss hypokalemia and hyperkalemia.

SECTION 2 • Acid–Base Balance

25.6 Describe the three classes of acids in the body.

25.7 Explain the role of buffer systems in maintaining acid–base balance and pH.

25.8 Explain the role of buffer systems in regulating the pH of the intracellular fluid and the extracellular fluid.

25.9 Describe the compensatory mechanisms involved in the maintenance of acid–base balance.

25.10 **CLINICAL MODULE** Describe respiratory acidosis and respiratory alkalosis.

LO Learning Outcomes are repeated at the bottom of each module.

Body composition may be viewed in terms of solids and two fluid compartments

Chapter 23 considered the metabolism of the organic components of the body. This chapter takes a broader look at the composition of the body as a whole. We will focus on the inorganic components: water and minerals. **Minerals** are the inorganic substances that dissociate in body fluids to form ions called electrolytes.

1 These pie charts compare the total body composition of adult males and females. The greatest variation is in the intracellular fluid (ICF), or cytosol, as a result of differences in the intracellular water content of fat (10 percent) versus skeletal muscle (75 percent). Less striking differences occur in the extracellular fluid (ECF) values, due to variations in the interstitial fluid volume of various tissues and the larger blood volume in males versus females. (For simplification, plasma refers to blood plasma.) The ECF and ICF are called **fluid compartments**, because they commonly behave as distinct entities. Because cells have a plasma membrane, and active transport occurs at the membrane surface, cells are able to maintain internal environments quite distinct from that of the ECF.

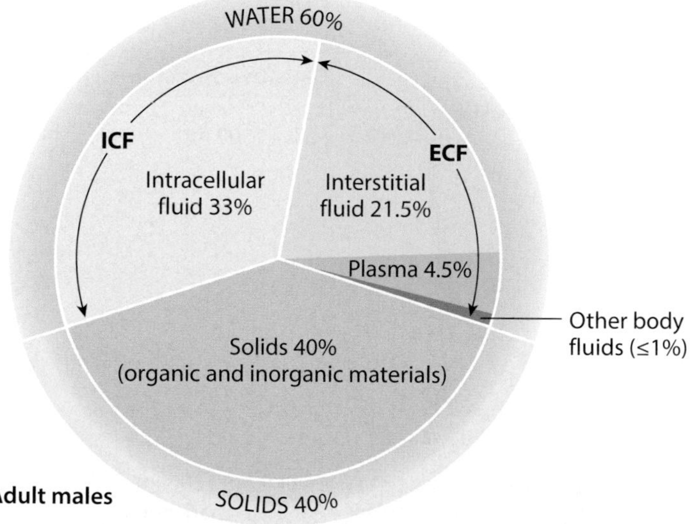
WATER 60%
ICF
ECF
Intracellular fluid 33%
Interstitial fluid 21.5%
Plasma 4.5%
Other body fluids (≤1%)
Solids 40% (organic and inorganic materials)
SOLIDS 40%
Adult males

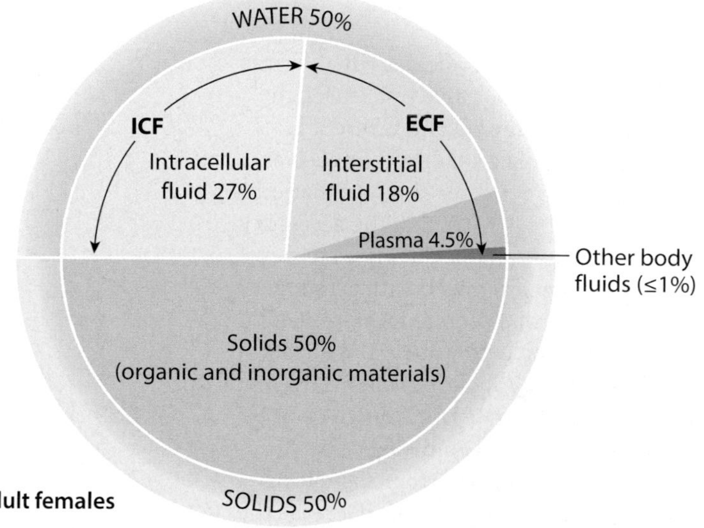
WATER 50%
ICF
ECF
Intracellular fluid 27%
Interstitial fluid 18%
Plasma 4.5%
Other body fluids (≤1%)
Solids 50% (organic and inorganic materials)
SOLIDS 50%
Adult females

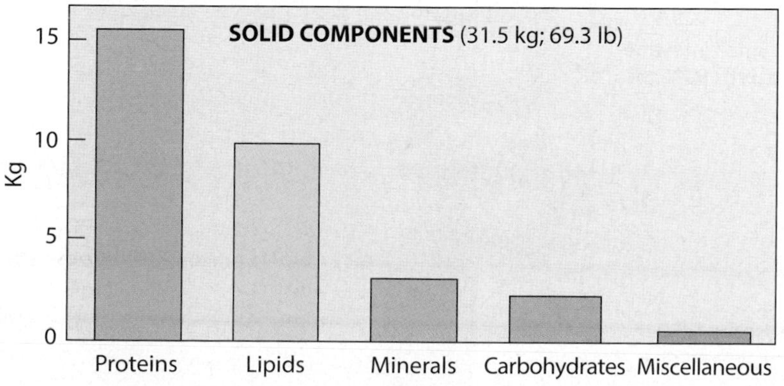
SOLID COMPONENTS (31.5 kg; 69.3 lb)
Kg
15
10
5
0
Proteins
Lipids
Minerals
Carbohydrates
Miscellaneous

2 Solid components make up only 40–50 percent of the body mass. This bar graph presents an overview of the solid components of a 70-kg (154-pound) person with a minimum of body fat. The distribution was obtained by averaging values for males and females ages 18–40 years.

In this section we will consider the exchange of water and electrolytes between the ECF and ICF, and between the body and the external environment.

Module 25.1 Review

a. What are minerals?

b. Name the two fluid compartments.

c. How is it that cells are able to maintain their internal environments?

LO 25.1 Name the body's fluid compartments, identify the solid components, and summarize their contents.

Fluid balance exists when water gains equal water losses

1 Your body is in **fluid balance** when its water content remains stable over time. Water gains occur primarily in the digestive tract. Water losses occur through many routes, but almost half of daily water loss occurs through urination.

Fluid Balance

Source	Daily Input (mL)
Water content of food	1000
Water consumed as liquid	1200
Metabolic water produced during catabolism	300
Total	**2500**
Method of Elimination	**Daily Output (mL)**
Urination	1200
Evaporation at skin	750
Evaporation at lungs	400
Loss in feces	150
Total	**2500**

2 This diagram indicates where water enters the digestive tract through ingestion or secretion, and where it is reabsorbed. Only a small amount leaves the digestive tract in feces. The situation is complicated by the fact that the accessory digestive glands are producing watery secretions that are mixed with arriving food. Most of that secreted water must be recovered along with water gained from food and drink. All of this water movement involves passive water flow down osmotic gradients. Intestinal epithelial cells continuously absorb nutrients and ions, and these activities gradually decrease the solute concentration in the lumen of the digestive tract and increase the solute concentration in the interstitial fluid of the lamina propria. As the solute concentration decreases in the lumen, water moves across the epithelium and into the interstitial fluid, maintaining osmotic equilibrium. Once within the interstitial fluid, the absorbed water is rapidly distributed throughout the ECF.

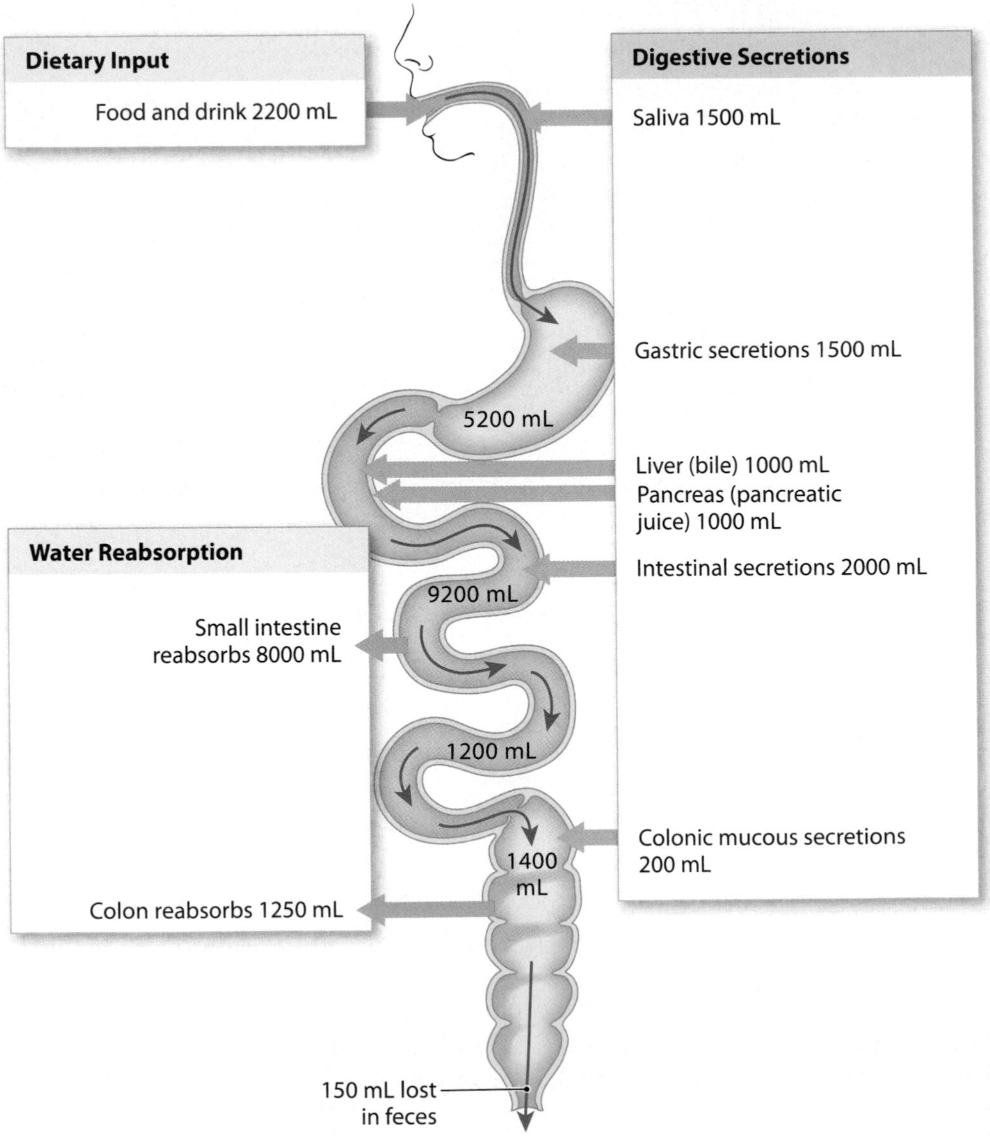
Dietary Input
Food and drink 2200 mL
Digestive Secretions
Saliva 1500 mL
Gastric secretions 1500 mL
5200 mL
Liver (bile) 1000 mL
Pancreas (pancreatic juice) 1000 mL
Intestinal secretions 2000 mL
Water Reabsorption
9200 mL
Small intestine reabsorbs 8000 mL
1200 mL
Colonic mucous secretions 200 mL
1400 mL
Colon reabsorbs 1250 mL
150 mL lost in feces

3 This diagram illustrates the major factors that affect ECF volume. Although the composition of the ECF and ICF are very different, the two are at osmotic equilibrium. Note that the volume of the ICF is larger than that of the ECF. The volume of water held within cells represents a significant reserve that can prevent sudden changes in the solute and water concentrations in the ECF. A rapid water movement between the ECF and the ICF in response to an osmotic gradient is called a **fluid shift**. Fluid shifts occur rapidly in response to changes in the osmotic concentration of the ECF and reach equilibrium within minutes to hours. This can be an important factor when water intake is restricted but water losses are severe.

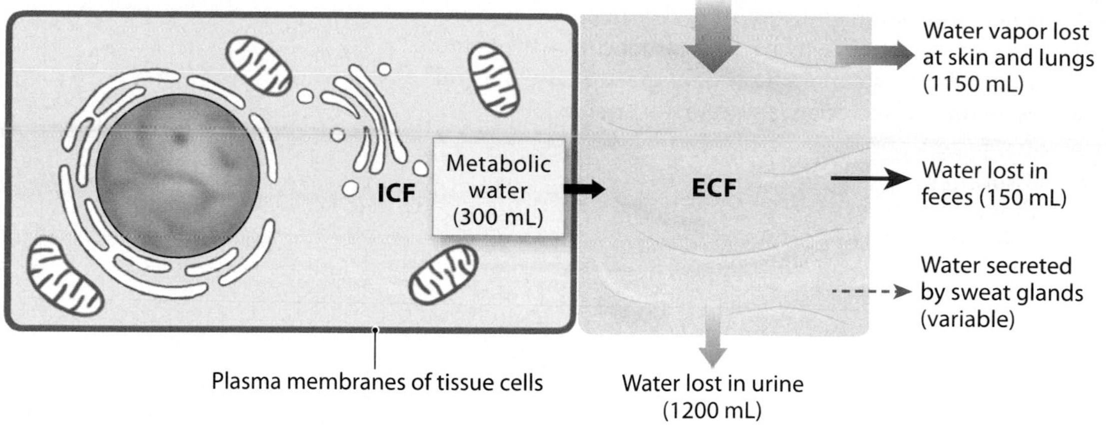

Water absorbed across digestive epithelium (2200 mL)
Water vapor lost at skin and lungs (1150 mL)
ICF
Metabolic water (300 mL)
ECF
Water lost in feces (150 mL)
Water secreted by sweat glands (variable)
Plasma membranes of tissue cells
Water lost in urine (1200 mL)

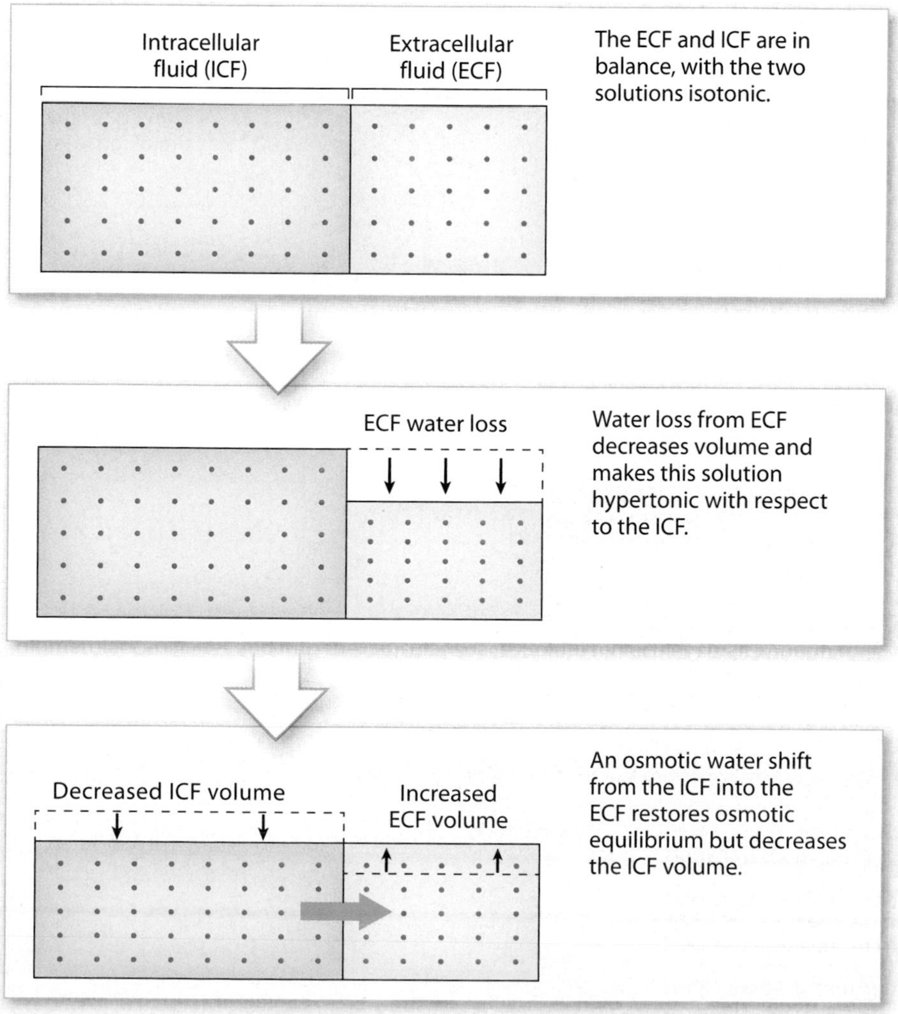

Intracellular fluid (ICF)
Extracellular fluid (ECF)
The ECF and ICF are in balance, with the two solutions isotonic.
ECF water loss
Water loss from ECF decreases volume and makes this solution hypertonic with respect to the ICF.
Decreased ICF volume
Increased ECF volume
An osmotic water shift from the ICF into the ECF restores osmotic equilibrium but decreases the ICF volume.

4 **Dehydration** develops when water losses outpace water gains. When you lose water but retain electrolytes, the osmotic concentration of the ECF increases. Osmosis then moves water out of the ICF and into the ECF until the two solutions are again isotonic. At that point, both the ECF and ICF are somewhat more concentrated than normal, and both volumes are lower than they were before the fluid loss. Because the ICF is considerably larger than the ECF, the ICF is an effective water reserve. However, if the fluid imbalance continues unchecked, the loss of water from the ICF will produce severe thirst, dryness, and wrinkling of the skin. Eventually blood volume and blood pressure may drop to the point at which circulatory shock develops.

Module 25.2 Review

a. Identify routes of fluid loss from the body.

b. Describe a fluid shift.

c. Explain dehydration and its effect on the osmotic concentration of blood.

Module 25.3

Mineral balance involves balancing electrolyte gains and losses

1 **Mineral balance** is the balance between ion absorption, which occurs across the lining of the small intestine and colon, and ion excretion, which occurs primarily by the kidneys. Sweat glands are a potential source of both water and mineral loss, but the rate of secretion is extremely variable.

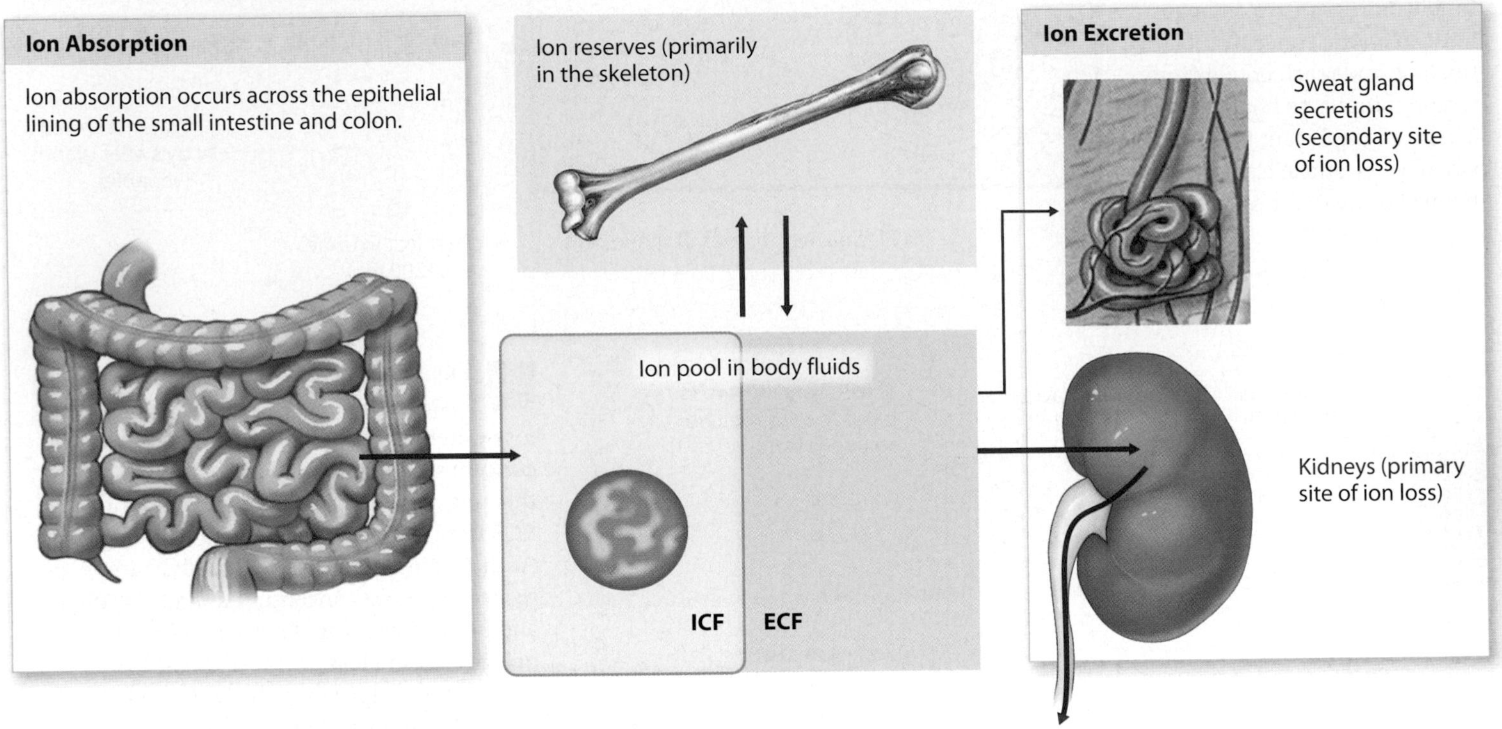

Ion Absorption
Ion absorption occurs across the epithelial lining of the small intestine and colon.
Ion reserves (primarily in the skeleton)
Ion pool in body fluids
ICF
ECF
Ion Excretion
Sweat gland secretions (secondary site of ion loss)
Kidneys (primary site of ion loss)

2 This table lists the mechanisms involved in the absorption of major electrolytes along the digestive tract.

Mineral Absorption

Electrolyte	Mechanism(s)
Na^+	Channel-mediated diffusion, cotransport, or active transport
Ca^{2+}	Active transport
K^+	Channel-mediated diffusion
Mg^{2+}	Active transport
Fe^{2+}	Active transport
Cl^-	Channel-mediated diffusion or carrier-mediated transport
I^-	Channel-mediated diffusion or carrier-mediated transport
HCO_3^-	Channel-mediated diffusion or carrier-mediated transport
NO_3^-	Channel-mediated diffusion or carrier-mediated transport
PO_4^{3-}	Active transport
SO_4^{2-}	Active transport

3 The body contains substantial reserves of key minerals. This table summarizes the functions of the various minerals and indicates the primary routes of ion excretion. The amount lost each day must equal the daily intake if the person is to stay in mineral balance.

Minerals and Mineral Reserves

Mineral	Functions	Total Body Content	Primary Route of Excretion	Recommended Dietary Allowance (RDA) (in mg)
Bulk Minerals				
Sodium	Major cation in body fluids; essential for normal membrane function	110 g, primarily in body fluids	Urine, sweat, feces	1500
Potassium	Major cation in cytosol; essential for normal membrane function	140 g, primarily in cytosol	Urine	4700
Chloride	Major anion in body fluids; functions in forming HCl	89 g, primarily in body fluids	Urine, sweat	2300
Calcium	Essential for normal muscle and neuron function and normal bone structure	1.36 kg, primarily in skeleton	Urine, feces	1000–1200
Phosphorus	In high-energy compounds, nucleic acids, and bone matrix (as phosphate)	744 g, primarily in skeleton	Urine, feces	700
Magnesium	Cofactor of enzymes, required for normal membrane functions	29 g (skeleton, 17 g; cytosol and body fluids, 12 g)	Urine	310–400
Trace Minerals				
Iron	Component of hemoglobin, myoglobin, and cytochromes	3.9 g (1.6 g stored as ferritin or hemosiderin)	Urine (traces)	8–18
Zinc	Cofactor of enzyme systems, notably carbonic anhydrase	2 g	Urine, hair (traces)	8–11
Copper	Required as cofactor for hemoglobin synthesis	127 mg	Urine, feces (traces)	0.9
Manganese	Cofactor for some enzymes	11 mg	Feces, urine (traces)	1.8–2.3
Cobalt	Cofactor for transaminations; mineral in vitamin B_{12} (cobalamin)	1.1 g	Feces, urine	0.0001

Module 25.3 Review

a. Define mineral balance.

b. Identify the ions absorbed by active transport.

c. Explain the significance of two important body minerals: sodium and calcium.

LO **25.3** Explain what is meant by mineral balance, and discuss its importance for homeostasis.

Water balance depends on sodium balance, and the two are regulated simultaneously

Response to Changes in Sodium Levels

1 **Sodium balance** exists when sodium gains equal sodium losses. The regulatory mechanism involved changes the ECF volume but keeps the Na^+ concentration relatively stable. When sodium gains exceed losses, the ECF volume increases; when sodium losses exceed gains, the volume of the ECF decreases. The changes in ECF volume occur without a significant change in the osmotic concentration of the ECF. These adjustments cause minor changes in ECF volume that do not cause adverse physiological effects.

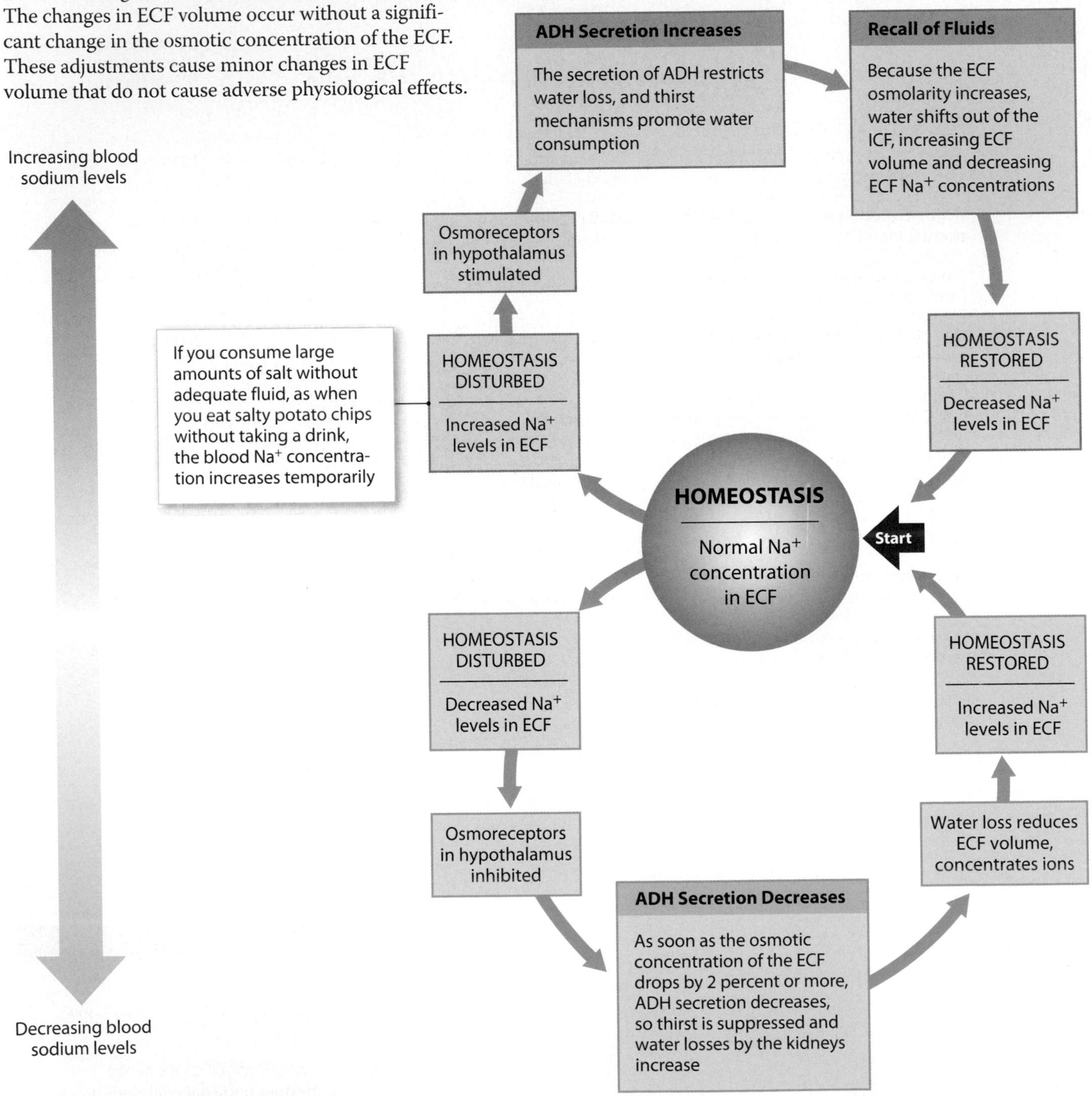

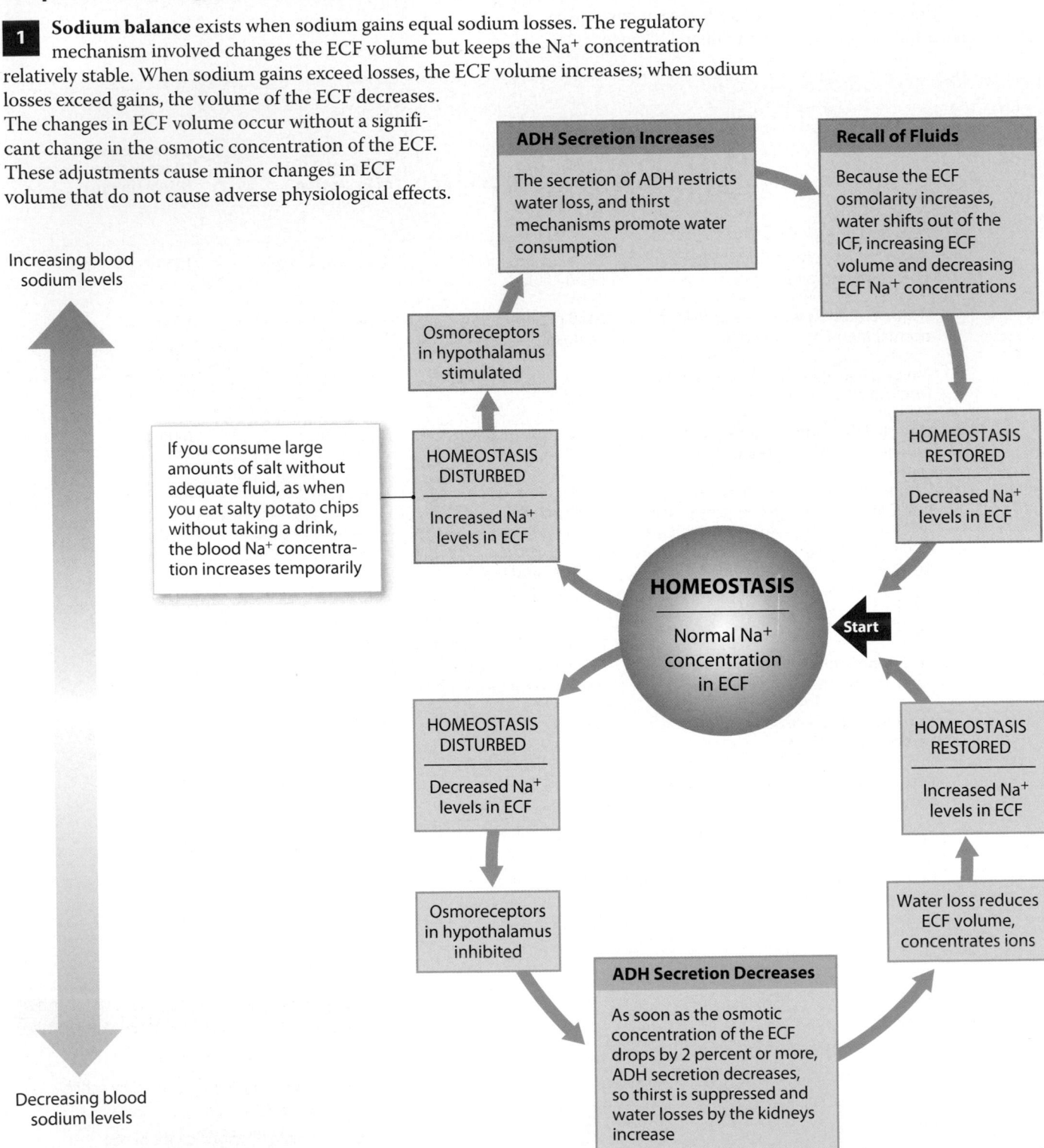

Increasing blood sodium levels
Decreasing blood sodium levels
If you consume large amounts of salt without adequate fluid, as when you eat salty potato chips without taking a drink, the blood Na+ concentration increases temporarily
HOMEOSTASIS DISTURBED
Increased Na+ levels in ECF
Osmoreceptors in hypothalamus stimulated
ADH Secretion Increases
The secretion of ADH restricts water loss, and thirst mechanisms promote water consumption
Recall of Fluids
Because the ECF osmolarity increases, water shifts out of the ICF, increasing ECF volume and decreasing ECF Na+ concentrations
HOMEOSTASIS RESTORED
Decreased Na+ levels in ECF
HOMEOSTASIS
Normal Na+ concentration in ECF
Start
HOMEOSTASIS DISTURBED
Decreased Na+ levels in ECF
Osmoreceptors in hypothalamus inhibited
ADH Secretion Decreases
As soon as the osmotic concentration of the ECF drops by 2 percent or more, ADH secretion decreases, so thirst is suppressed and water losses by the kidneys increase
Water loss reduces ECF volume, concentrates ions
HOMEOSTASIS RESTORED
Increased Na+ levels in ECF

Response to Changes in ECF Volume

2 If the changes in ECF volume caused by disturbances in sodium balance are extreme, the homeostatic mechanisms responsible for regulating blood volume and blood pressure will be activated. This is the case because when ECF volume changes, so does plasma volume and, in turn, blood volume. If ECF volume rises, blood volume goes up; if ECF volume drops, blood volume goes down.

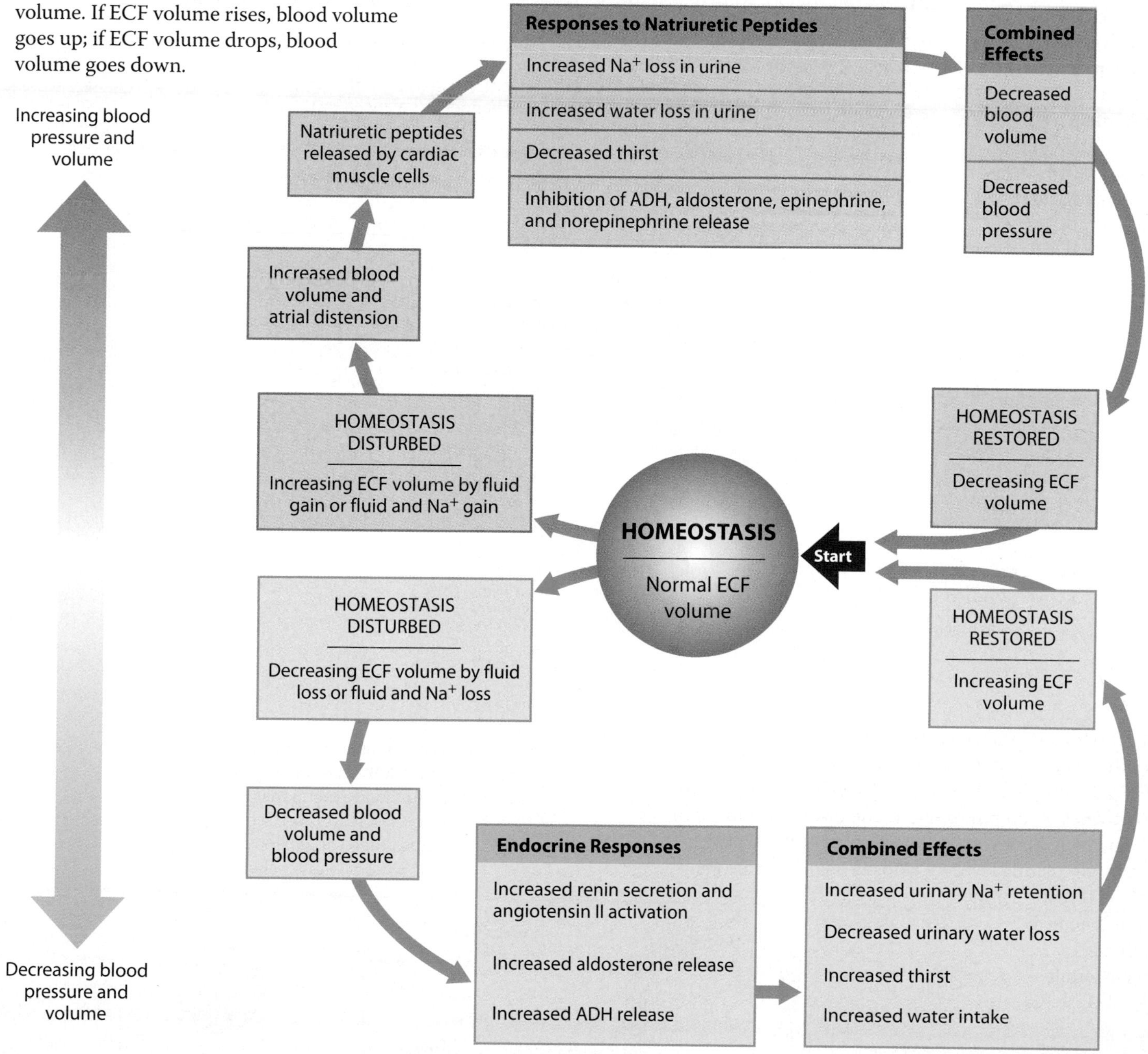

Increasing blood pressure and volume
Natriuretic peptides released by cardiac muscle cells
Responses to Natriuretic Peptides
Increased Na+ loss in urine
Increased water loss in urine
Decreased thirst
Inhibition of ADH, aldosterone, epinephrine, and norepinephrine release
Combined Effects
Decreased blood volume
Decreased blood pressure
Increased blood volume and atrial distension
HOMEOSTASIS DISTURBED
Increasing ECF volume by fluid gain or fluid and Na+ gain
HOMEOSTASIS RESTORED
Decreasing ECF volume
HOMEOSTASIS
Normal ECF volume
Start
HOMEOSTASIS DISTURBED
Decreasing ECF volume by fluid loss or fluid and Na+ loss
HOMEOSTASIS RESTORED
Increasing ECF volume
Decreased blood volume and blood pressure
Endocrine Responses
Increased renin secretion and angiotensin II activation
Increased aldosterone release
Increased ADH release
Combined Effects
Increased urinary Na+ retention
Decreased urinary water loss
Increased thirst
Increased water intake
Decreasing blood pressure and volume

Sustained abnormalities in the Na^+ concentration in the ECF occur only when there are severe problems with fluid balance. When the Na^+ concentration of the ECF falls below 136 mEq/L, a state of **hyponatremia** (*natrium*, sodium) exists; this can be caused by excessive water intake (overhydration) or inadequate salt intake. When body water content declines, the Na^+ concentration rises; when that concentration exceeds 145 mEq/L, **hypernatremia** exists; dehydration is the most common cause. Both conditions are serious and potentially life-threatening.

Module 25.4 Review

a. What effect does inhibition of osmoreceptors have on ADH secretion and thirst?

b. What effect does aldosterone have on sodium ion concentration in the ECF?

c. Briefly summarize the relationship between sodium ion concentration and the ECF.

LO **25.4** Summarize the relationship between sodium and water in maintaining fluid and electrolyte balance.

Disturbances of potassium balance are uncommon but extremely dangerous

1 This diagram indicates the major factors involved in **potassium balance**. The key factors in maintaining potassium balance are (1) the rate of K^+ entry across the digestive epithelium and (2) the rate of K^+ loss into urine.

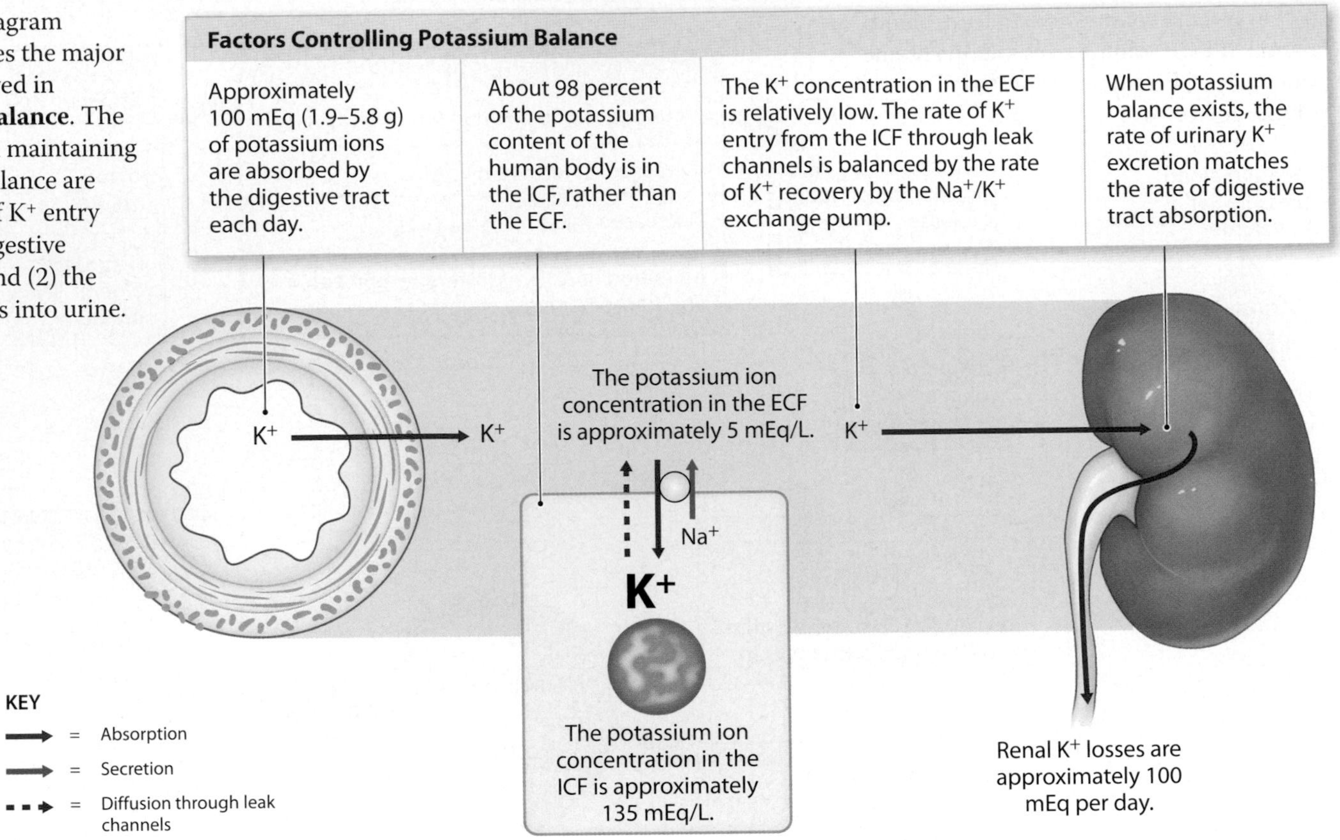

Factors Controlling Potassium Balance
Approximately 100 mEq (1.9–5.8 g) of potassium ions are absorbed by the digestive tract each day.
About 98 percent of the potassium content of the human body is in the ICF, rather than the ECF.
The K+ concentration in the ECF is relatively low. The rate of K+ entry from the ICF through leak channels is balanced by the rate of K+ recovery by the Na+/K+ exchange pump.
When potassium balance exists, the rate of urinary K+ excretion matches the rate of digestive tract absorption.
K+
K+
The potassium ion concentration in the ECF is approximately 5 mEq/L.
K+
Na+
K+
The potassium ion concentration in the ICF is approximately 135 mEq/L.
Renal K+ losses are approximately 100 mEq per day.
KEY
= Absorption
= Secretion
= Diffusion through leak channels

2 Because dietary intake is relatively constant, the kidneys are the main factor determining the K^+ concentration in the ECF. Potassium loss by the kidneys is regulated by controlling the activities of ion pumps along the distal portions of the nephron and collecting system. The activity of these pumps is regulated by circulating levels of aldosterone. Aldosterone stimulates Na^+ reabsorption while simultaneously accelerating K^+ excretion (Module 24.10, p. 941).

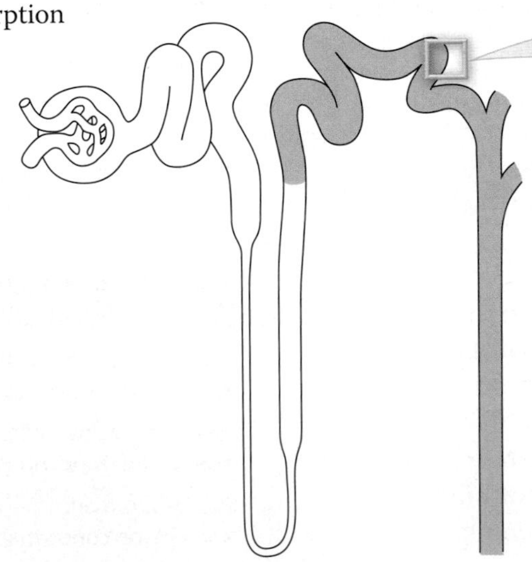

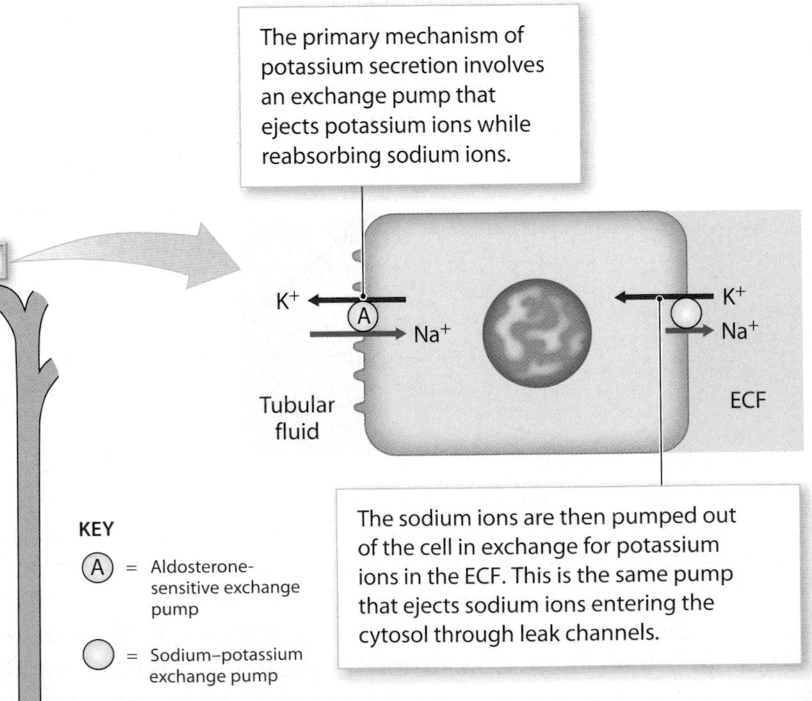

The primary mechanism of potassium secretion involves an exchange pump that ejects potassium ions while reabsorbing sodium ions.
K+
A
Na+
K+
Na+
Tubular fluid
ECF
The sodium ions are then pumped out of the cell in exchange for potassium ions in the ECF. This is the same pump that ejects sodium ions entering the cytosol through leak channels.
KEY
A = Aldosterone-sensitive exchange pump
= Sodium–potassium exchange pump

3 Potassium balance does not occur in isolation. For example, aldosterone secretion increases in response to either high blood K^+ concentrations or a decline in ECF sodium levels. However, disturbances of acid–base balance can disrupt potassium balance by affecting the aldosterone-sensitive pumps (Module 24.13, p. 947).

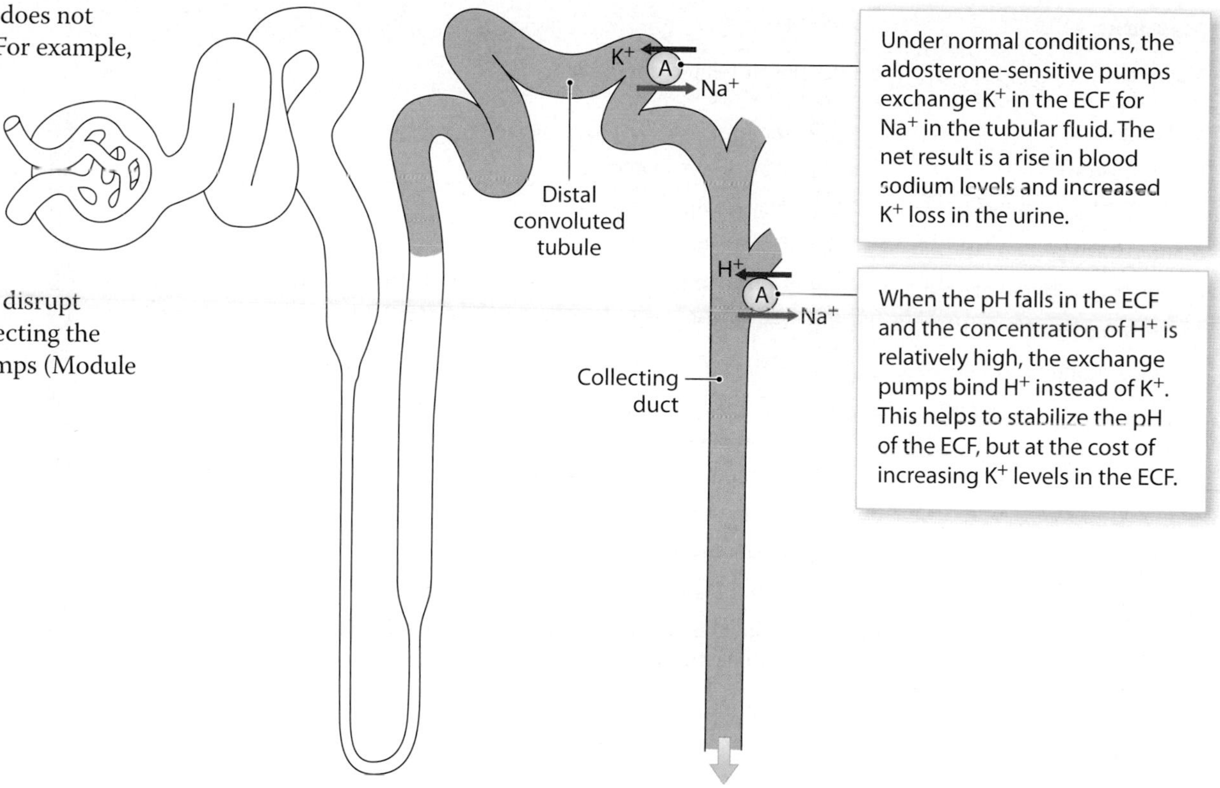

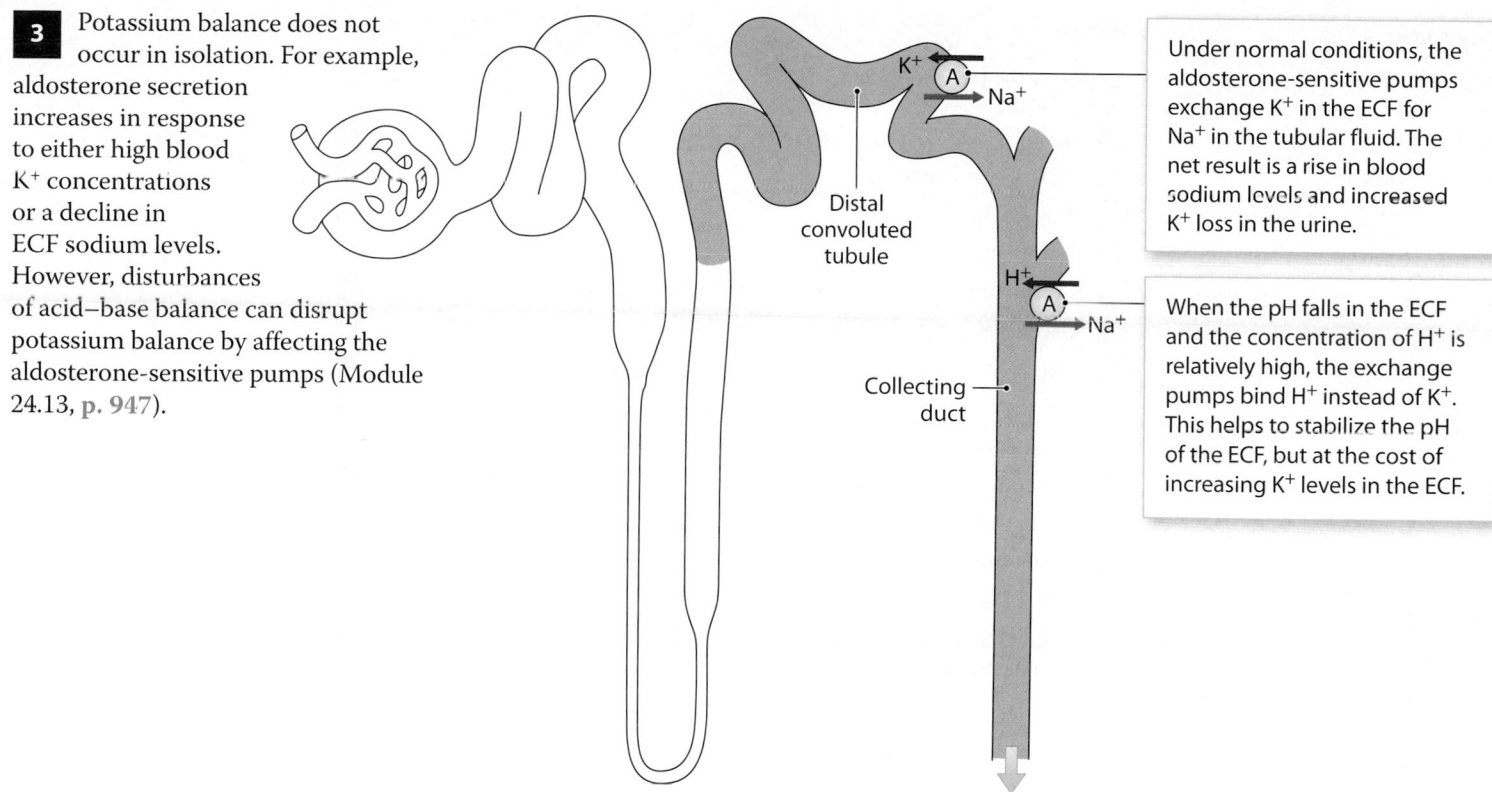

K+
A
Na+
Distal convoluted tubule
H+
A
Na+
Collecting duct
Under normal conditions, the aldosterone-sensitive pumps exchange K+ in the ECF for Na+ in the tubular fluid. The net result is a rise in blood sodium levels and increased K+ loss in the urine.
When the pH falls in the ECF and the concentration of H+ is relatively high, the exchange pumps bind H+ instead of K+. This helps to stabilize the pH of the ECF, but at the cost of increasing K+ levels in the ECF.

4 The diagram below indicates major factors involved in disturbances of potassium balance. The normal range of K^+ concentration in the ECF is 3.5–5.0 mEq/L. Variations outside of that range are unusual and potentially dangerous.

When the blood concentration of potassium falls below 2 mEq/L, extensive muscular weakness develops, followed by eventual paralysis. This condition, called severe **hypokalemia** (*kalium*, potassium), is potentially lethal due to its effects on the heart.	Normal potassium levels in blood (3.5–5.0 mEq/L)	High K^+ concentrations in the blood produce an equally dangerous condition known as **hyperkalemia**. Severe hyperkalemia occurs when K^+ concentration exceeds 7 mEq/L, which results in cardiac arrhythmias.

Factors Promoting Hypokalemia	
Several diuretics can produce hypokalemia by increasing the volume of urine produced.	The endocrine disorder called **aldosteronism**, characterized by excessive aldosterone secretion, results in hypokalemia by overstimulating sodium retention and potassium loss.

Factors Promoting Hyperkalemia		
Chronically low blood pH promotes hyperkalemia by interfering with K^+ excretion by the kidneys.	Kidney failure due to damage or disease will prevent normal K^+ secretion and thereby produce hyperkalemia.	Several drugs promote urination by blocking Na^+ reabsorption by the kidneys. When sodium reabsorption slows down, so does potassium secretion, and hyperkalemia can result.

Treatment for hypokalemia generally includes increasing dietary intake of potassium by salting food with potassium salts (KCl) or by taking potassium tablets, such as Slow-K. Treatment for hyperkalemia typically includes diluting the ECF with a solution low in K^+, stimulating K^+ loss in urine by using diuretics such as *Lasix*, adjusting the pH of the ECF, and restricting dietary K^+ intake. In cases resulting from renal failure, kidney dialysis may also be required.

Module 25.5 Review

a. What organs are primarily responsible for regulating the potassium ion concentration in the ECF?

b. Identify factors that cause potassium excretion.

c. Define hypokalemia and hyperkalemia.

LO 25.5 Explain factors that control potassium balance, and discuss hypokalemia and hyperkalemia.

Matching

Match each lettered term with the most closely related description.

a. kidneys
b. potassium
c. fluid compartments
d. fluid balance
e. hypertonic blood plasma
f. dehydration
g. aldosterone
h. plasma, interstitial fluid
i. osmoreceptors
j. fluid shift
k. hypokalemia
l. ADH
m. hyponatremia
n. sodium

1 Monitor blood osmotic concentration
2 Water gain = water loss
3 Major components of ECF
4 Dominant cation in ECF
5 Hormone that restricts water loss and stimulates thirst
6 Caused by overhydration
7 Dominant cation in ICF
8 ICF and ECF
9 Most important sites of sodium ion regulation
10 Water movement between ECF and ICF
11 Water moves from cells into ECF
12 Result of aldosteronism
13 Water losses greater than water gains
14 Regulates sodium ion absorption along distal convoluted tubule and collecting system

1 ______
2 ______
3 ______
4 ______
5 ______
6 ______
7 ______
8 ______
9 ______
10 ______
11 ______
12 ______
13 ______
14 ______

Multiple choice

Select the correct answer from the list provided.

15 Nearly two-thirds of the total body water content is
- a) extracellular fluid (ECF).
- b) intracellular fluid (ICF).
- c) tissue fluid.
- d) interstitial fluid.

16 Electrolyte balance involves balancing the rates of absorption across the digestive tract with rates of loss at the
- a) heart and lungs.
- b) stomach and liver.
- c) kidneys and sweat glands.
- d) pancreas and gallbladder.

17 If the ECF is hypertonic with respect to the ICF, water will move
- a) from the ECF into cells until osmotic equilibrium is restored.
- b) from cells into the ECF until osmotic equilibrium is restored.
- c) in both directions until osmotic equilibrium is restored.
- d) in response to the sodium–potassium exchange pump.

18 When pure water is consumed, the ECF
- a) becomes hypotonic with respect to the ICF.
- b) becomes hypertonic with respect to the ICF.
- c) becomes isotonic with respect to the ICF.
- d) electrolytes become more concentrated.

19 Physiological adjustments affecting fluid and electrolyte balance are mediated primarily by
- a) antidiuretic hormone.
- b) aldosterone.
- c) natriuretic peptides.
- d) all of the above

20 When water is lost but electrolytes are retained, the osmolarity of the ECF rises, and osmosis then moves water
- a) out of the ECF and into the ICF.
- b) back and forth between the ICF and ECF.
- c) out of the ICF and into the ECF.
- d) none of the above

Section integration

Malia, a nursing student, has been caring for burn patients. She notices that they consistently show elevated levels of potassium in their urine and wonders why. What would you tell her?

21 ______

There are three classes of acids in the body

Your body is in **acid–base balance** when the production of hydrogen ions is precisely offset by their loss, and when the pH of body fluids remains within normal limits.

1 This diagram shows the major factors involved in maintaining acid–base balance. The primary challenge to homeostasis is that your body generates a variety of acids during normal metabolic operations, and a significant decrease in body fluid pH must be prevented. The respiratory and urinary systems, and buffer systems in the ICF and ECF, play important homeostatic roles in acid–base balance.

Active tissues continuously generate carbon dioxide (CO_2), which in solution forms carbonic acid (H_2CO_3) that immediately dissociates into a hydrogen ion and a bicarbonate ion ($H^+ + HCO_3^-$). Additional acids, such as lactic acid, are produced in the course of normal metabolic operations.

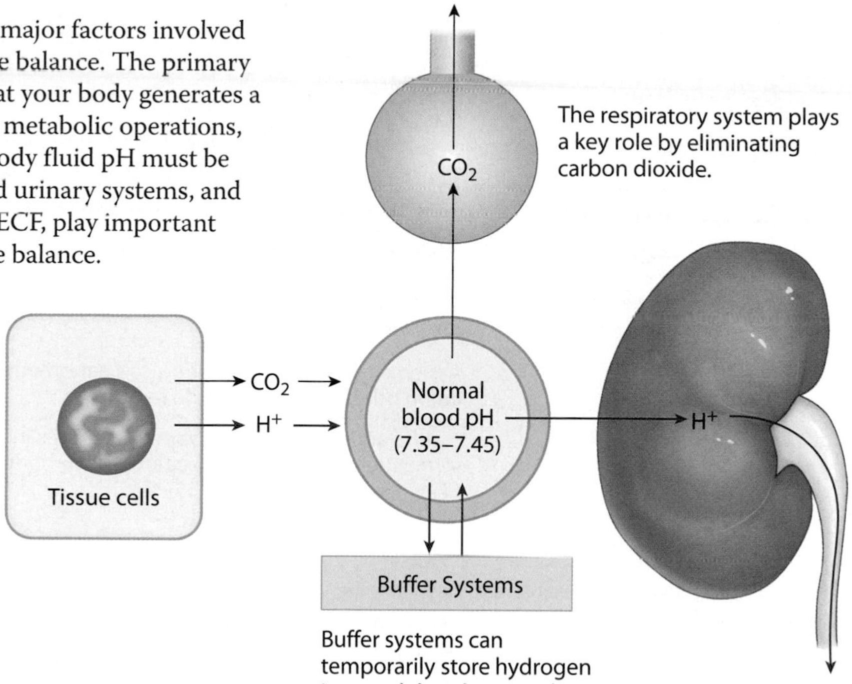
CO_2
The respiratory system plays a key role by eliminating carbon dioxide.
Tissue cells
CO_2
H^+
Normal blood pH (7.35–7.45)
H^+
Buffer Systems

The kidneys play a major role by secreting hydrogen ions into the urine and generating buffers that enter the bloodstream. The rate of excretion rises and falls as needed to maintain normal blood pH. As a result, the normal pH of urine varies widely but averages 6.0—slightly acidic.

Buffer systems can temporarily store hydrogen ions and thereby provide short-term pH stability.

2 There are three classes of acids that can threaten pH balance.

Classes of Acids

Fixed Acids

Fixed acids are acids that do not leave solution; once produced, they remain in body fluids until they are eliminated at the kidneys. Sulfuric acid and phosphoric acid are the most important fixed acids in the body. They are generated in small amounts during the catabolism of amino acids and compounds that contain phosphate groups, including phospholipids and nucleic acids.

Organic Acids

Organic acids are acid participants in, or by-products of, cellular metabolism. Important organic acids include lactic acid (produced by the anaerobic metabolism of pyruvate) and ketone bodies (synthesized from acetyl-CoA). Under normal conditions, most organic acids are metabolized rapidly, so significant accumulations do not occur.

Volatile Acids

Volatile acids can leave the body by entering the atmosphere at the lungs. Carbonic acid (H_2CO_3) is a volatile acid that forms through the interaction of water and carbon dioxide.

Module 25.6 Review

a. When is your body in acid–base balance?

b. What is the primary challenge to acid–base homeostasis?

c. Name the three classes of acids that threaten pH balance.

LO 25.6 Describe the three classes of acids in the body.

Potentially dangerous disturbances in acid–base balance are opposed by buffer systems

1 The topic of pH and the chemical nature of acids, bases, and buffers was introduced in Module 2.12 (p. 64). This table reviews key terms important to the discussion that follows.

A Review of Important Terms Relating to Acid–Base Balance

pH	The negative exponent (negative logarithm) of the hydrogen ion concentration $[H^+]$ in a solution
Neutral	A solution with a pH of 7; the solution contains equal numbers of hydrogen ions (H^+) and hydroxide ions (OH^-)
Acidic	A solution with a pH below 7; in this solution, hydrogen ions predominate
Basic (alkaline)	A solution with a pH above 7; in this solution, hydroxide ions predominate
Acid	A substance that dissociates to release hydrogen ions, decreasing pH
Base	A substance that dissociates to release hydroxide ions or to remove hydrogen ions, increasing pH
Salt	An ionic compound consisting of a cation other than a hydrogen ion and an anion other than a hydroxide ion
Buffer	A substance that tends to oppose changes in the pH of a solution by removing or replacing hydrogen ions; in body fluids, buffers maintain blood pH within normal limits (7.35–7.45)

2 The pH of the ECF normally remains within relatively narrow limits, usually 7.35–7.45. Any deviation from the normal range is extremely dangerous, because changes in H^+ concentrations disrupt the stability of plasma membranes, alter the structure of proteins, and change the activities of important enzymes. You could not survive for long with an ECF pH below 6.8 or above 7.7. In practice, decreases in pH are much more common than increases, because several acids, including carbonic acid, are generated by normal cellular activities. Any shift in pH affects virtually all body systems, but the nervous and cardiovascular systems are particularly sensitive to pH fluctuations.

The pH of the ECF normally ranges from 7.35 to 7.45.

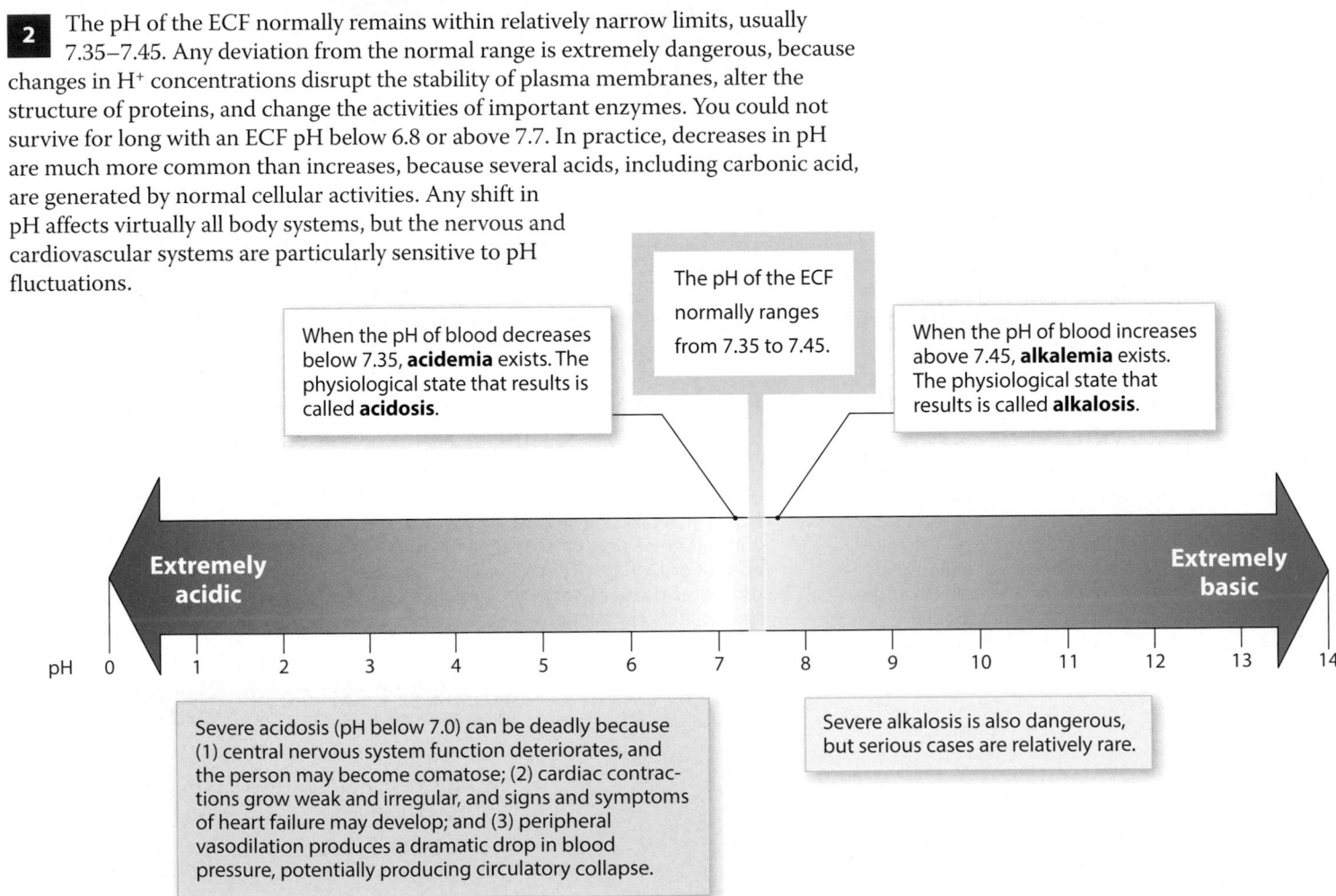

When the pH of blood decreases below 7.35, **acidemia** exists. The physiological state that results is called **acidosis**.

When the pH of blood increases above 7.45, **alkalemia** exists. The physiological state that results is called **alkalosis**.

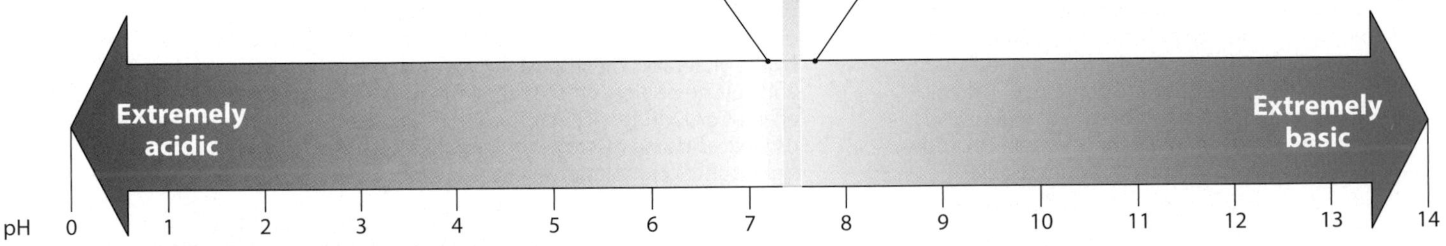
Extremely acidic
Extremely basic
pH
0
1
2
3
4
5
6
7
8
9
10
11
12
13
14

Severe acidosis (pH below 7.0) can be deadly because (1) central nervous system function deteriorates, and the person may become comatose; (2) cardiac contractions grow weak and irregular, and signs and symptoms of heart failure may develop; and (3) peripheral vasodilation produces a dramatic drop in blood pressure, potentially producing circulatory collapse.

Severe alkalosis is also dangerous, but serious cases are relatively rare.

3 The partial pressure of carbon dioxide (P_{CO_2}) in blood is the most important factor affecting the pH of body tissues. However, its effect on pH is indirect. This is because carbon dioxide first combines with water to form **carbonic acid** (H_2CO_3). Most of the carbon dioxide in solution is converted to carbonic acid by carbonic anhydrase, and most of the carbonic acid dissociates into hydrogen ions and bicarbonate ions. For this reason, there is an inverse relationship between the P_{CO_2} and pH.

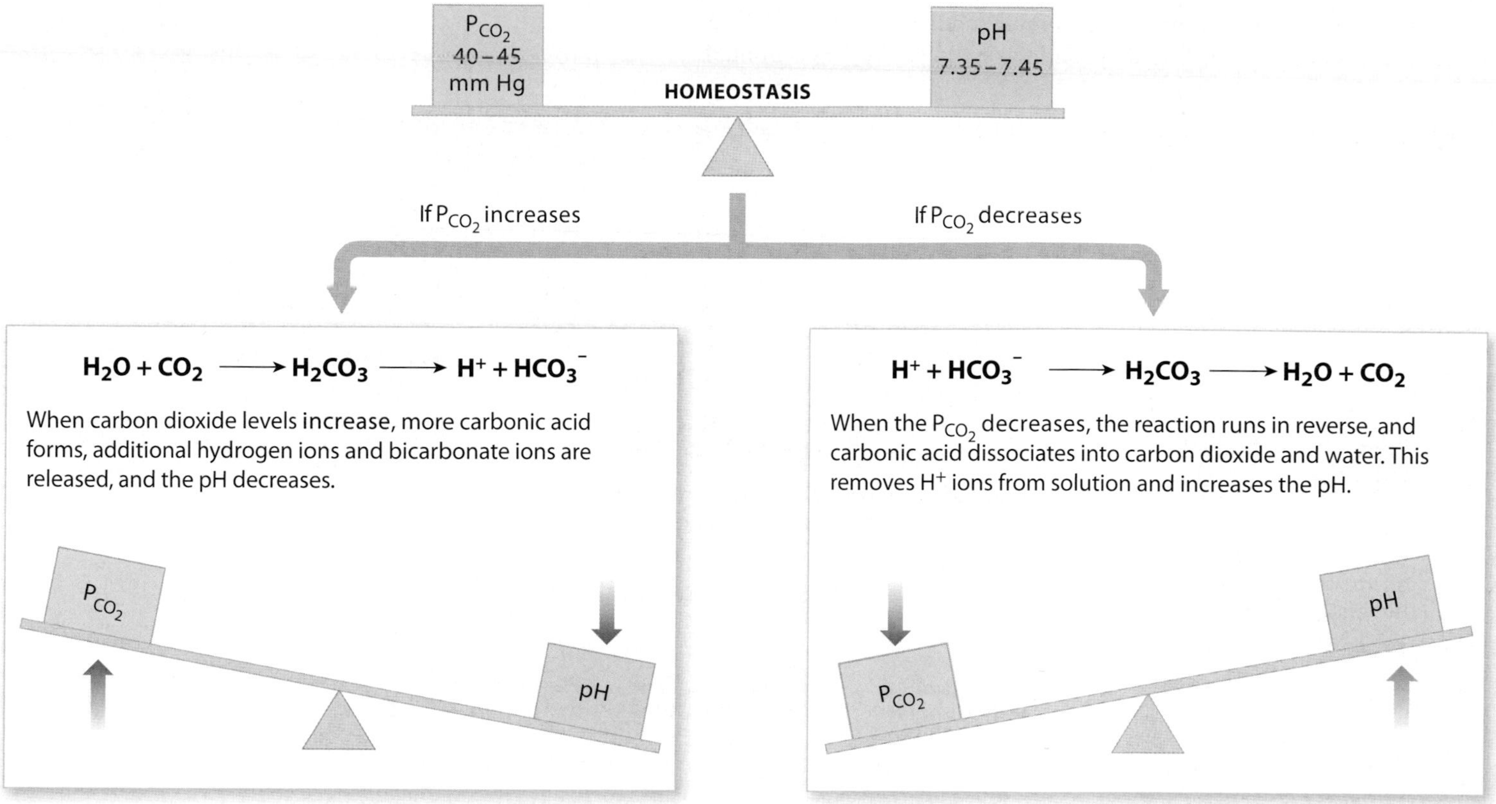

P_{CO_2} 40–45 mm Hg
pH 7.35–7.45
HOMEOSTASIS
If P_{CO_2} increases
If P_{CO_2} decreases
$H_2O + CO_2 \longrightarrow H_2CO_3 \longrightarrow H^+ + HCO_3^-$
When carbon dioxide levels increase, more carbonic acid forms, additional hydrogen ions and bicarbonate ions are released, and the pH decreases.
P_{CO_2}
pH
$H^+ + HCO_3^- \longrightarrow H_2CO_3 \longrightarrow H_2O + CO_2$
When the P_{CO_2} decreases, the reaction runs in reverse, and carbonic acid dissociates into carbon dioxide and water. This removes H^+ ions from solution and increases the pH.
P_{CO_2}
pH

4 A **buffer system** in body fluids generally consists of a combination of a weak acid (HY) and the anion (Y^-) released by its dissociation. The anion functions as a weak base. In solution, molecules of the weak acid exist in equilibrium with its dissociation products. In chemical notation, this relationship is represented as:

$$HY \rightleftharpoons H^+ + Y^-$$

Adding H^+ to the solution upsets the equilibrium and results in the formation of additional molecules of the weak acid.

$$H^+ + Y^- \xrightarrow{H^+} H^+ + HY$$

Removing H^+ from the solution also upsets the equilibrium and results in the dissociation of additional molecules of HY. This releases H^+.

$$H^+ + HY \longrightarrow H^+ + Y^- \quad (H^+ \text{ released})$$

Module 25.7 Review

a. Define acidemia and alkalemia.

b. What intermediate compound formed from water and carbon dioxide directly affects the pH of the ECF?

c. Summarize the relationship between P_{CO_2} levels and pH.

LO 25.7 Explain the role of buffer systems in maintaining acid–base balance and pH.

Module 25.8

Buffer systems can delay but not prevent pH shifts in the ICF and ECF

1 The body has three major buffer systems, each with slightly different characteristics and distributions. Although buffer systems can bind excess H^+, they provide only a temporary solution to an acid–base imbalance. The hydrogen ions are not eliminated, but merely rendered harmless. In the process, a buffer molecule is tied up, and the supply of buffers is limited.

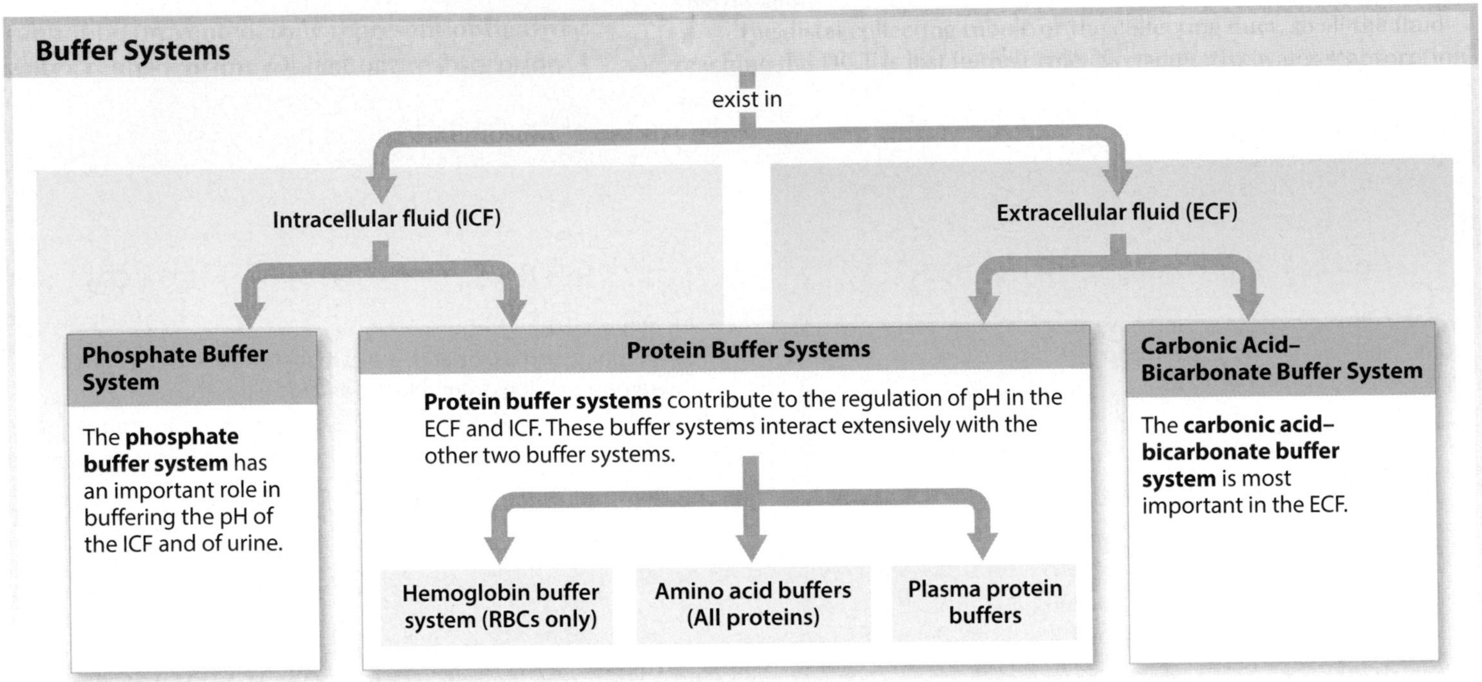

Buffer Systems
exist in
Intracellular fluid (ICF)
Extracellular fluid (ECF)
Phosphate Buffer System
The phosphate buffer system has an important role in buffering the pH of the ICF and of urine.
Protein Buffer Systems
Protein buffer systems contribute to the regulation of pH in the ECF and ICF. These buffer systems interact extensively with the other two buffer systems.
Hemoglobin buffer system (RBCs only)
Amino acid buffers (All proteins)
Plasma protein buffers
Carbonic Acid–Bicarbonate Buffer System
The carbonic acid–bicarbonate buffer system is most important in the ECF.

2 The **hemoglobin buffer system** is the only intracellular buffer system that can have an immediate effect on the pH of body fluids. In the tissues, red blood cells absorb carbon dioxide from the plasma and convert it to carbonic acid. As the carbonic acid dissociates, the hydrogen ions are buffered by hemoglobin proteins. At the lungs, the entire reaction sequence proceeds in reverse and the CO_2 diffuses into the alveoli for exhalation (Module 21.14, p. 812).

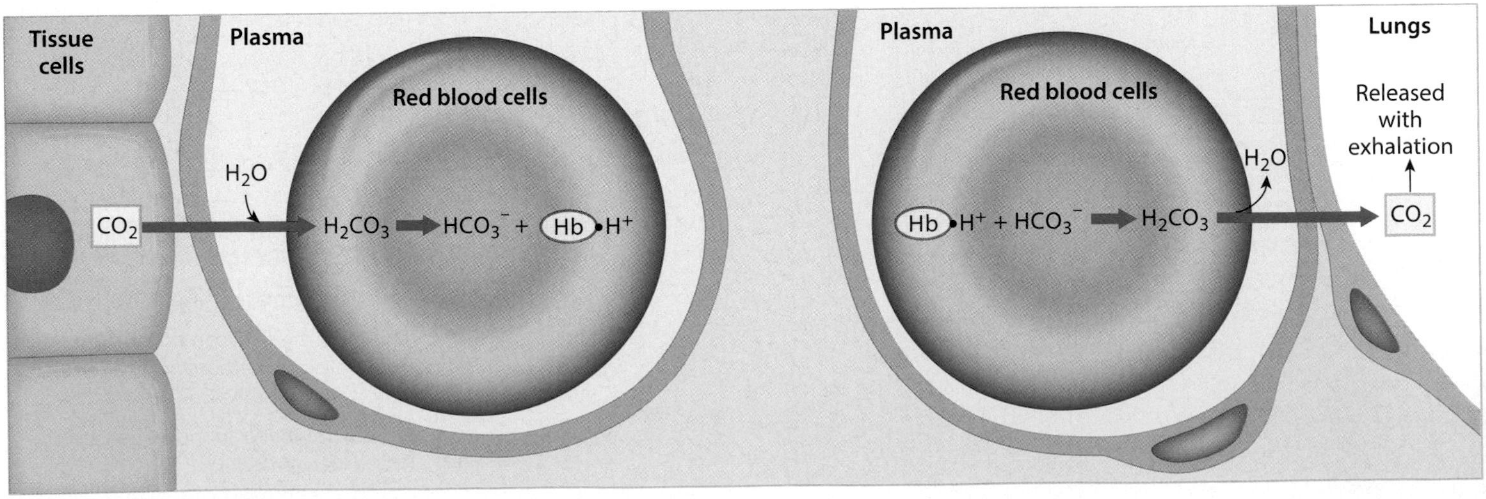

Tissue cells
Plasma
Red blood cells
H2O
CO2
H2CO3 → HCO3− + Hb•H+
Plasma
Red blood cells
Hb•H+ + HCO3− → H2CO3
H2O
Lungs
Released with exhalation
CO2

3 Protein buffer systems prevent significant pH change, usually by binding excess hydrogen ions (H^+). These buffer systems depend on the ability of amino acids to respond to pH changes by accepting or releasing H^+. Each amino acid has a specific pH (usually less than 7) at which the carboxyl group has released a hydrogen ion and the amino group has bound one. This form, which has no net ionic charge, is called a *zwitterion*. In solutions above pH 7, the amino group (NH_2), carboxylate group (COO^-), and side group (R) of most free amino acids can act as buffers. The underlying mechanism is shown here. However, in a protein, most of the carboxylate and amino groups in the main chain are tied up in peptide bonds. Only the exposed amino group and carboxylate group at either end of a protein are available as buffers. Thus, most of the buffering capacity of proteins is provided by the R-groups of the component amino acids.

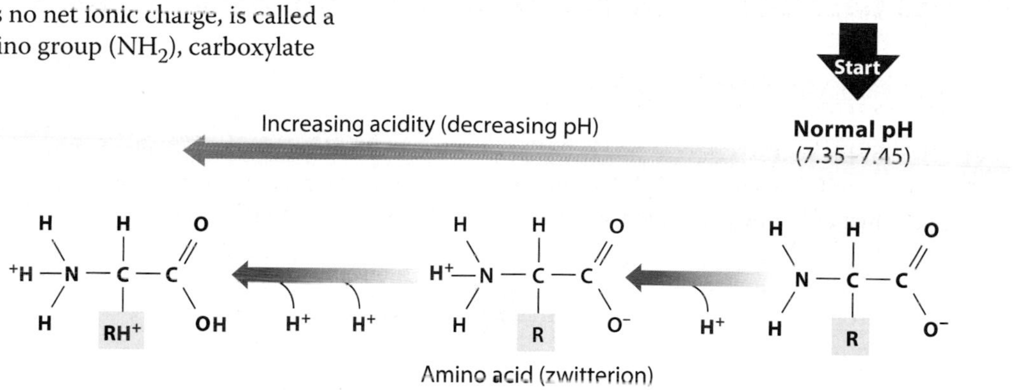

Start
Increasing acidity (decreasing pH)
Normal pH
(7.35–7.45)
RH+
OH
H+
H+
Amino acid (zwitterion)
O−
H+
R

If pH decreases, the carboxylate group (COO^-) and the amino group ($-NH_2$) of an amino acid can act as weak bases and accept additional hydrogen ions, forming a carboxyl group (–COOH) and an amino ion ($-NH_3^+$), respectively. Many of the R-groups can also accept hydrogen ions, forming RH^+.

At the normal pH of body fluids (7.35–7.45), neither the carboxylate groups (COO^-) nor the amino groups ($-NH_2$) of most amino acids are bound to hydrogen ions.

4 The carbonic acid–bicarbonate buffer system involves freely reversible reactions. A change in the concentration of any participant affects the concentrations of all other participants and shifts the direction of the reactions under way.

BICARBONATE RESERVE

Body fluids contain a large reserve of HCO_3^-, primarily in the form of dissolved molecules of the weak base sodium bicarbonate ($NaHCO_3$). This readily available supply of HCO_3^- is known as the **bicarbonate reserve**.

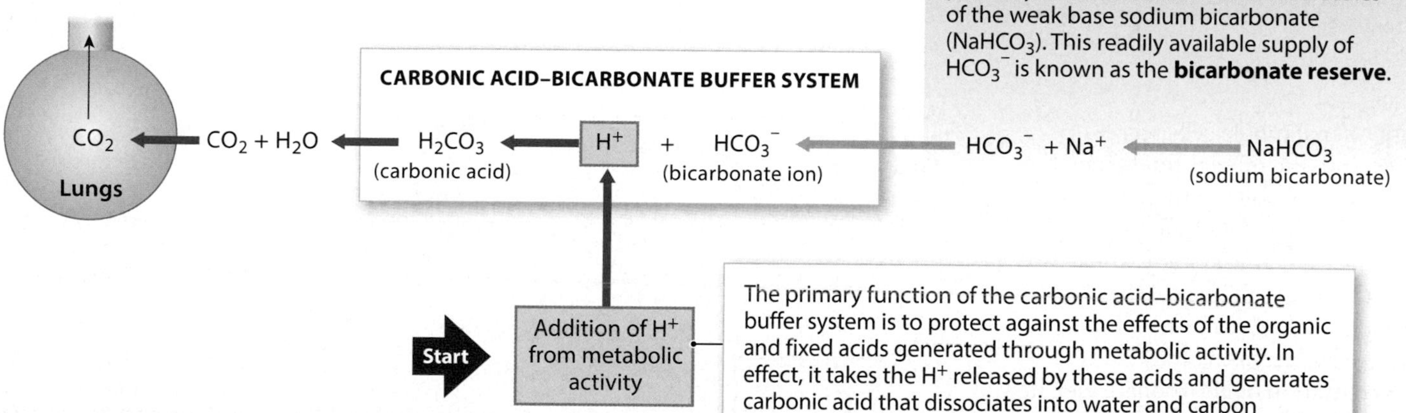

CARBONIC ACID–BICARBONATE BUFFER SYSTEM
CO2
Lungs
CO2 + H2O
H2CO3
(carbonic acid)
H+
+ HCO3−
(bicarbonate ion)
HCO3− + Na+
NaHCO3
(sodium bicarbonate)
Start
Addition of H+ from metabolic activity

The primary function of the carbonic acid–bicarbonate buffer system is to protect against the effects of the organic and fixed acids generated through metabolic activity. In effect, it takes the H^+ released by these acids and generates carbonic acid that dissociates into water and carbon dioxide, which can easily be eliminated by the lungs.

Metabolic acid–base disorders result from the production or loss of excessive amounts of fixed or organic acids. The primary role of the carbonic acid–bicarbonate buffer system is to protect against such disorders. **Respiratory acid–base disorders** result from an imbalance between the rate of CO_2 generation and the rate of CO_2 elimination at the lungs. The carbonic acid–bicarbonate buffer system cannot protect against respiratory disorders; the imbalances must be corrected by reflexive changes in the depth and rate of respiration.

Module 25.8 Review

a. Identify the body's three major buffer systems.

b. Which fluids are buffered by the phosphate buffer system?

c. Describe the carbonic acid–bicarbonate buffer system.

LO **25.8** Explain the role of buffer systems in regulating the pH of the intracellular fluid and the extracellular fluid.

The homeostatic responses to metabolic acidosis and alkalosis involve respiratory and renal mechanisms as well as buffer systems

Responses to Metabolic Acidosis

1 **Metabolic acidosis** develops when large numbers of hydrogen ions are released by organic or fixed acids, and the pH decreases. For homeostasis to be preserved, the excess H^+ must either be permanently bound through the formation of water (linked to the formation of CO_2 that can be eliminated by the lungs), or removed from body fluids through secretion by the kidneys.

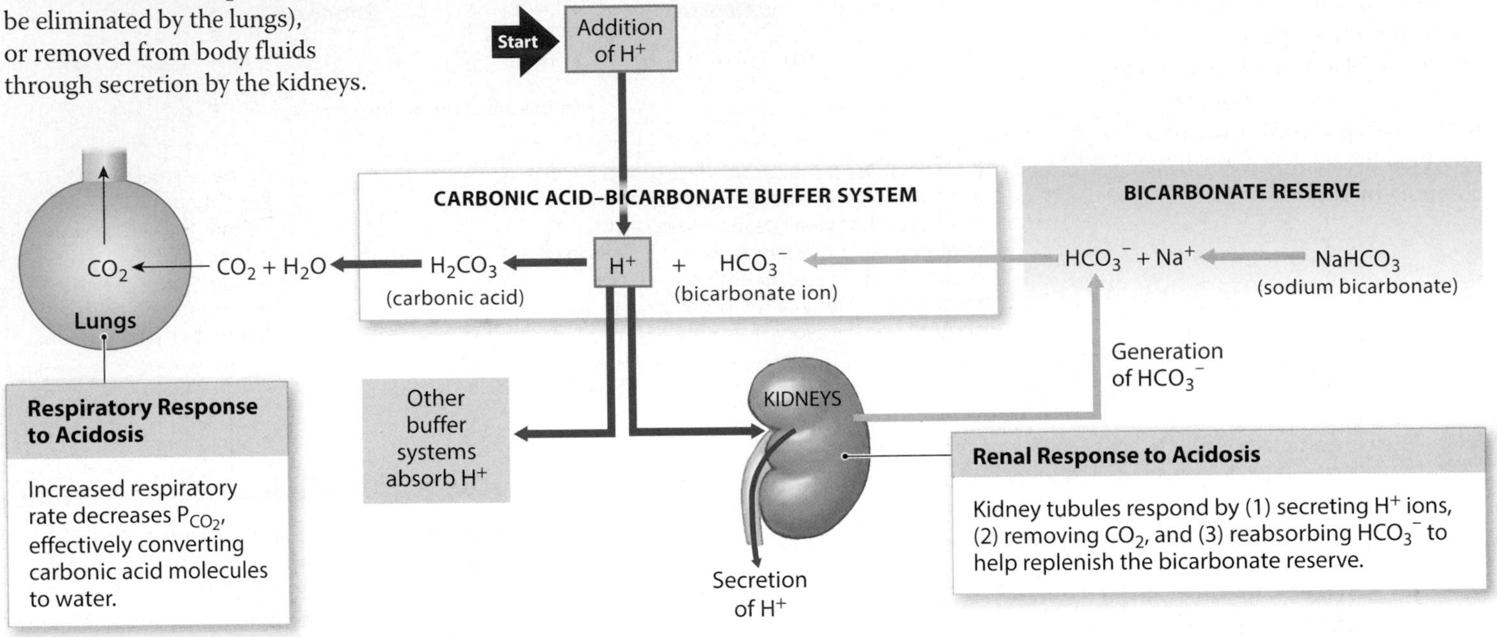

Start
Addition of H^+
CARBONIC ACID–BICARBONATE BUFFER SYSTEM
CO_2
Lungs
$CO_2 + H_2O$
H_2CO_3 (carbonic acid)
H^+ + HCO_3^- (bicarbonate ion)
BICARBONATE RESERVE
$HCO_3^- + Na^+$
$NaHCO_3$ (sodium bicarbonate)
Generation of HCO_3^-
Other buffer systems absorb H^+
KIDNEYS
Secretion of H^+
Respiratory Response to Acidosis
Increased respiratory rate decreases P_{CO_2}, effectively converting carbonic acid molecules to water.
Renal Response to Acidosis
Kidney tubules respond by (1) secreting H^+ ions, (2) removing CO_2, and (3) reabsorbing HCO_3^- to help replenish the bicarbonate reserve.

2 Renal tubule cells secrete H^+ into the tubular fluid along the proximal convoluted tubule (PCT), the distal convoluted tubule (DCT), and the collecting system. They also bolster the capabilities of the carbonic acid–bicarbonate buffer system by increasing the concentration of bicarbonate ions in the ECF, replacing those pulled from the bicarbonate reserve.

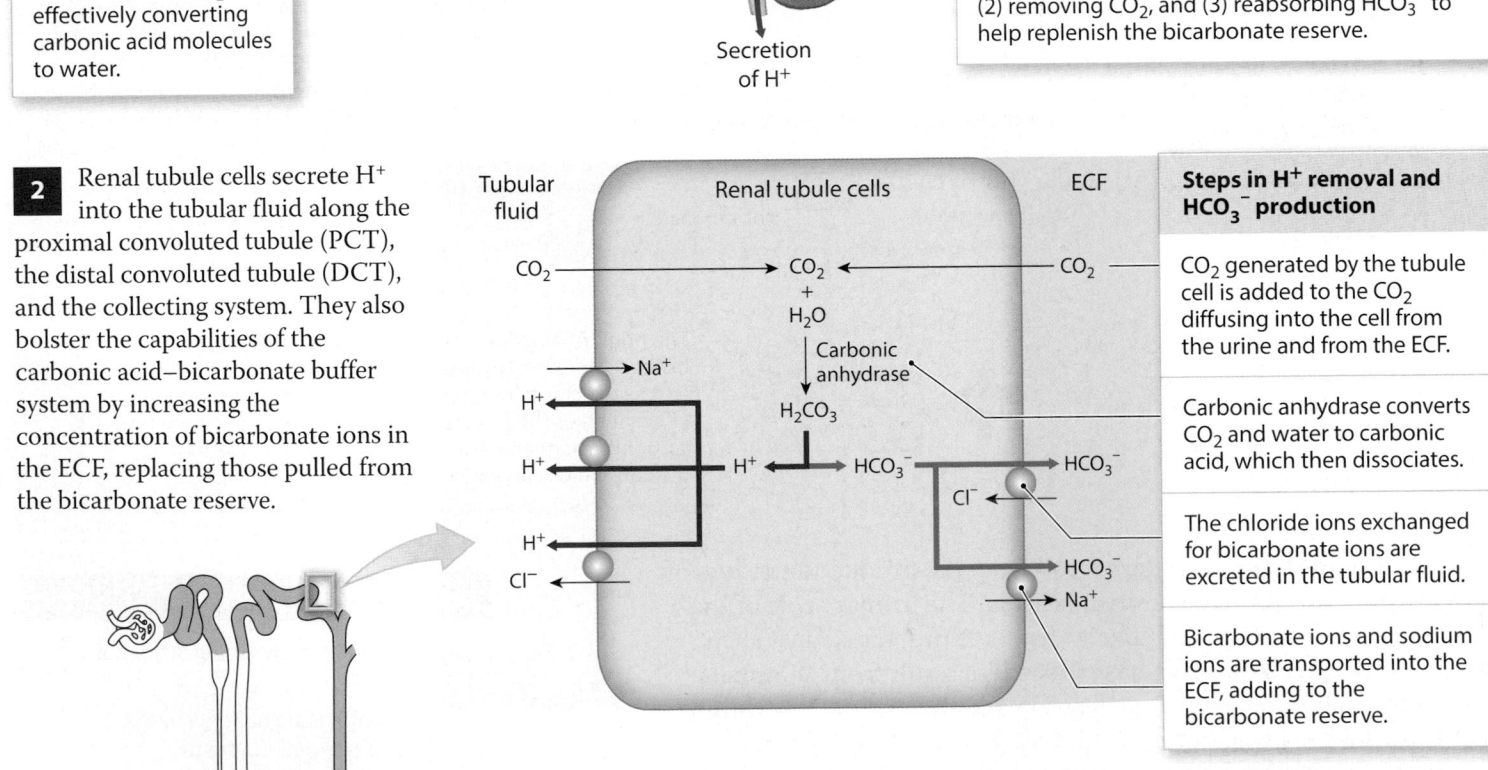

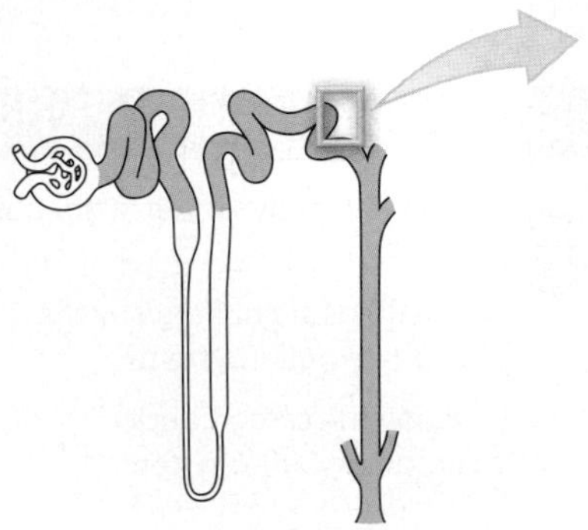

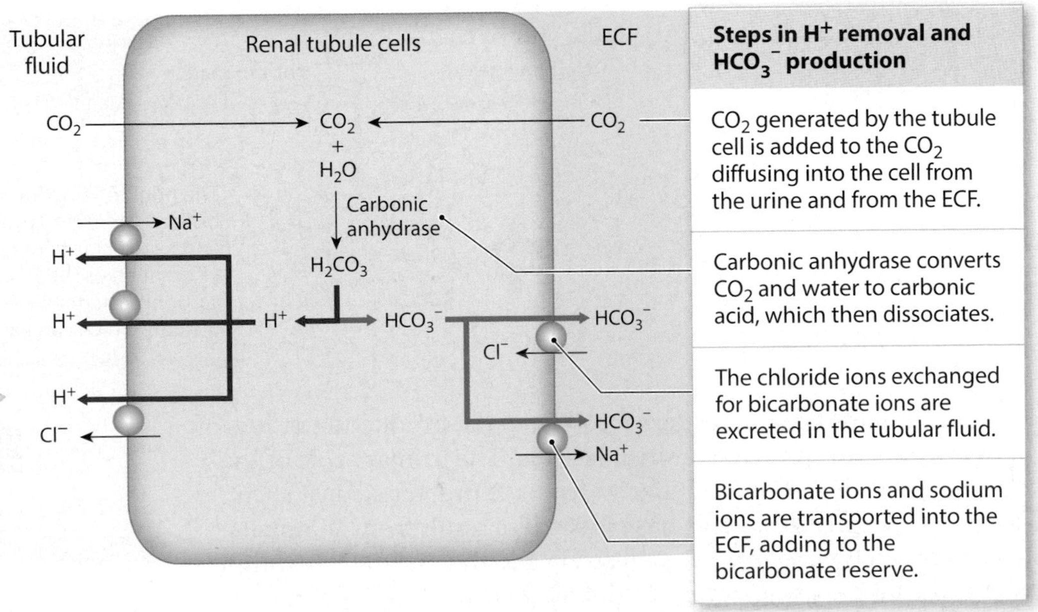

Tubular fluid
Renal tubule cells
ECF
CO_2
CO_2 + H_2O
Carbonic anhydrase
H_2CO_3
H^+
HCO_3^-
Na^+
Cl^-
Steps in H^+ removal and HCO_3^- production
CO_2 generated by the tubule cell is added to the CO_2 diffusing into the cell from the urine and from the ECF.
Carbonic anhydrase converts CO_2 and water to carbonic acid, which then dissociates.
The chloride ions exchanged for bicarbonate ions are excreted in the tubular fluid.
Bicarbonate ions and sodium ions are transported into the ECF, adding to the bicarbonate reserve.

Responses to Metabolic Alkalosis

3 **Metabolic alkalosis** develops when large numbers of hydrogen ions are removed from body fluids, resulting in a rise in pH. When this occurs, (1) the rate of H^+ secretion by the kidneys declines, (2) tubule cells do not reclaim the bicarbonates in tubular fluid, and (3) the collecting system transports HCO_3^- into tubular fluid while releasing a strong acid (HCl) into the ECF.

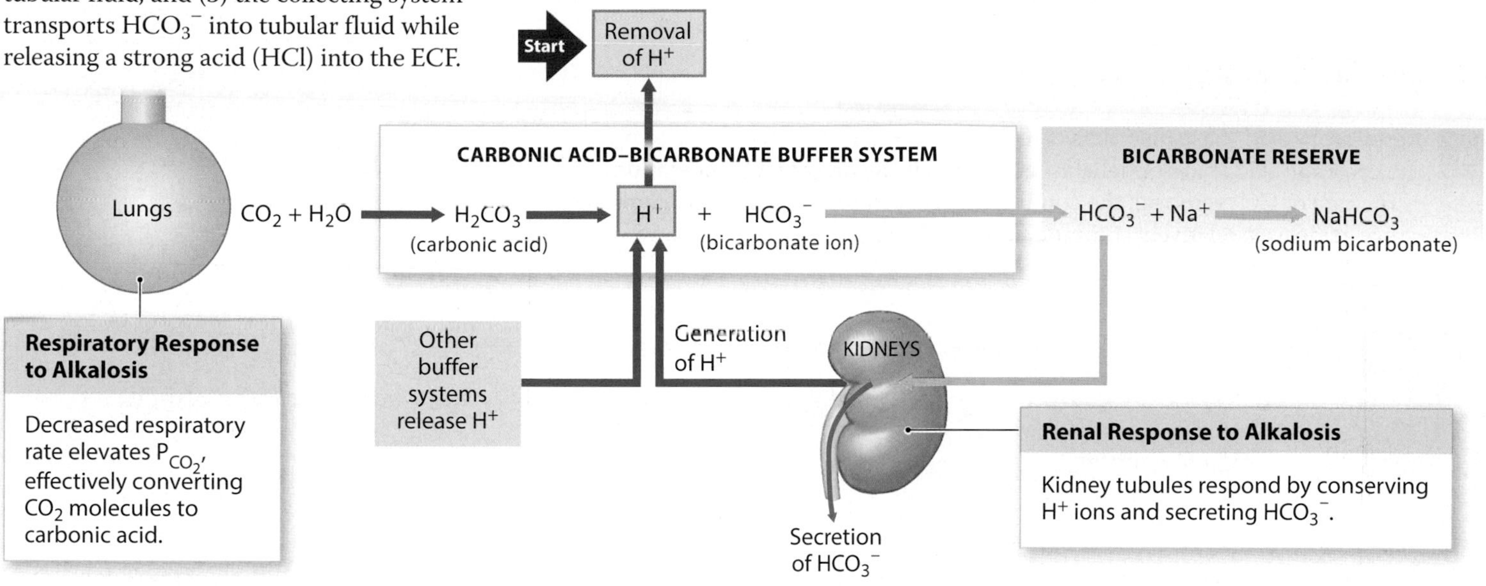

Start
Removal of H+
CARBONIC ACID–BICARBONATE BUFFER SYSTEM
Lungs
CO2 + H2O → H2CO3 (carbonic acid) → H+ + HCO3− (bicarbonate ion)
BICARBONATE RESERVE
HCO3− + Na+ → NaHCO3 (sodium bicarbonate)
Respiratory Response to Alkalosis
Decreased respiratory rate elevates PCO2, effectively converting CO2 molecules to carbonic acid.
Other buffer systems release H+
Generation of H+
KIDNEYS
Renal Response to Alkalosis
Kidney tubules respond by conserving H+ ions and secreting HCO3−.
Secretion of HCO3−

4 Tubule cells secrete bicarbonate ions into the tubular fluid along the proximal convoluted tubule (PCT), the distal convoluted tubule (DCT), and the collecting system. They also move H^+ ions into the ECF.

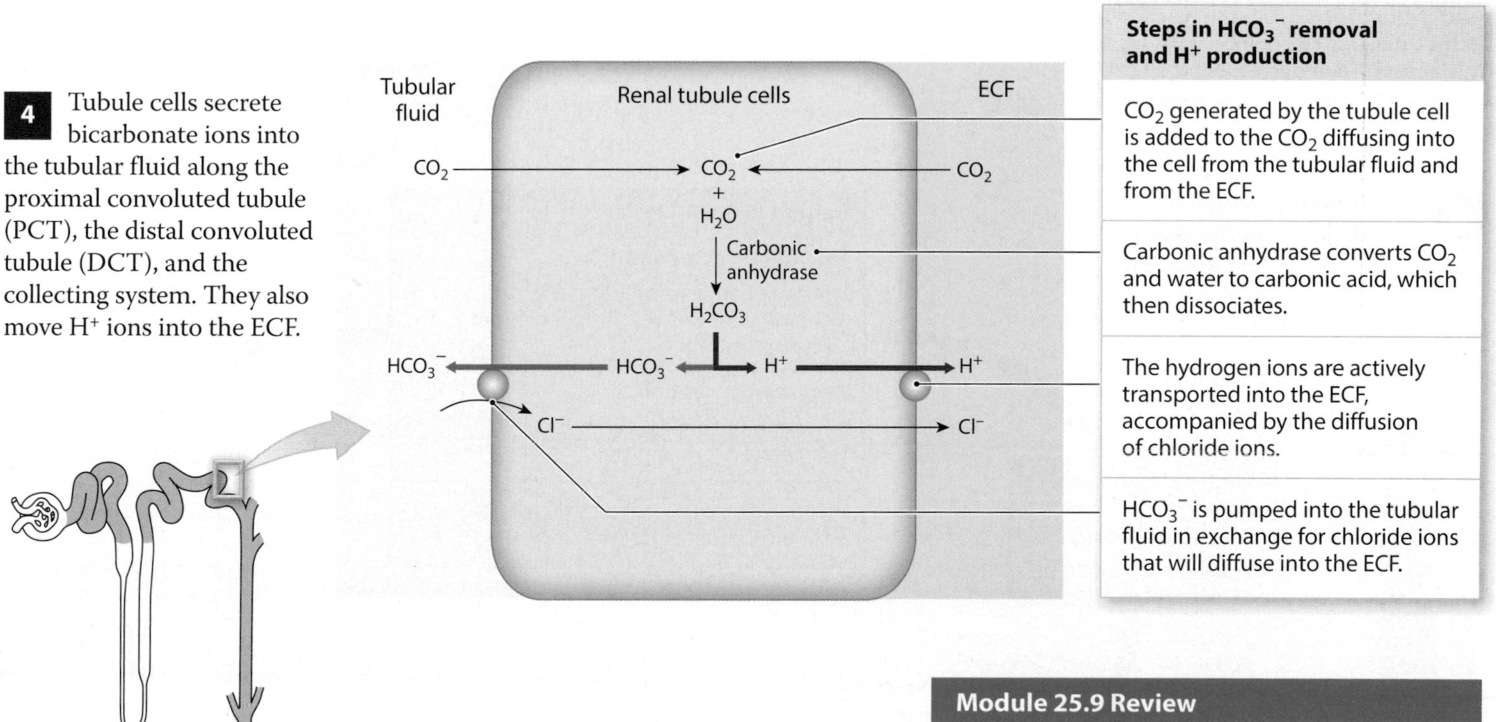

Tubular fluid
Renal tubule cells
ECF
CO2
CO2 + H2O
Carbonic anhydrase
H2CO3
HCO3−
H+
Cl−
Steps in HCO3− removal and H+ production
CO2 generated by the tubule cell is added to the CO2 diffusing into the cell from the tubular fluid and from the ECF.
Carbonic anhydrase converts CO2 and water to carbonic acid, which then dissociates.
The hydrogen ions are actively transported into the ECF, accompanied by the diffusion of chloride ions.
HCO3− is pumped into the tubular fluid in exchange for chloride ions that will diffuse into the ECF.

Module 25.9 Review

a. Describe metabolic acidosis.

b. Describe metabolic alkalosis.

c. If the kidneys are conserving HCO_3^- and eliminating H^+ in acidic urine, which is occurring: metabolic alkalosis or metabolic acidosis?

LO **25.9** Describe the compensatory mechanisms involved in the maintenance of acid–base balance.

Respiratory acid–base disorders are the most common challenges to acid–base balance

Respiratory acid–base disorders result from an imbalance between the rate of CO_2 generation in body tissues and the rate of CO_2 elimination at the lungs.

Respiratory Acidosis

1 Respiratory acid–base disorders cannot be corrected, even temporarily, by the carbonic acid–bicarbonate buffer system. If the rate of CO_2 generation exceeds the rate of CO_2 removal, the condition of **respiratory acidosis** develops. Respiratory acidosis is relatively common and may be life-threatening.

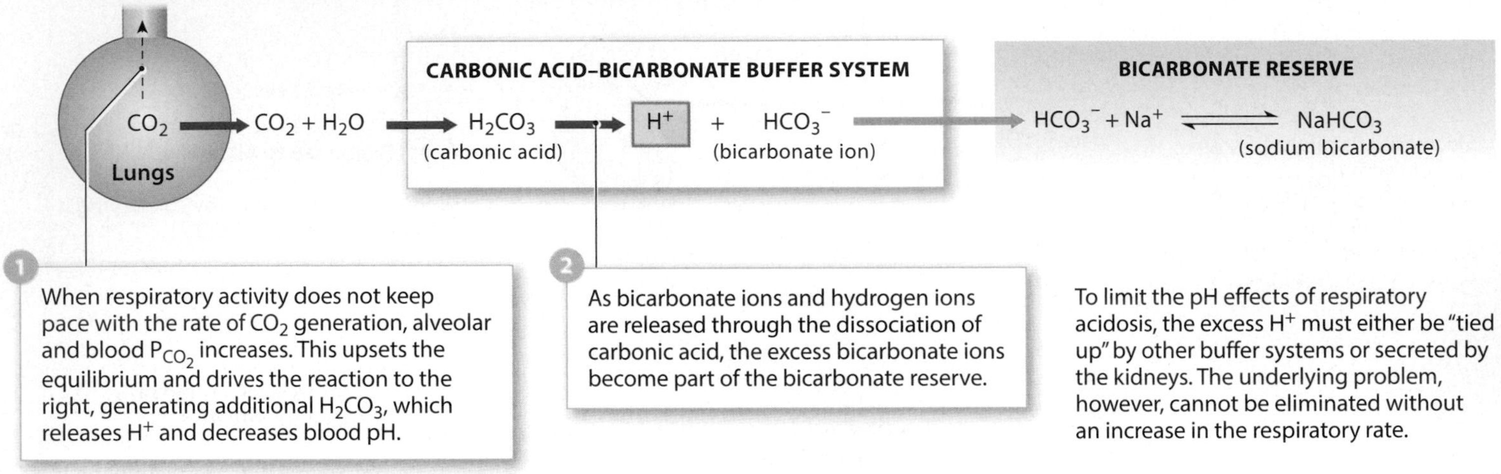

CO_2
Lungs
$CO_2 + H_2O$
CARBONIC ACID–BICARBONATE BUFFER SYSTEM
H_2CO_3 (carbonic acid)
H^+ + HCO_3^- (bicarbonate ion)
BICARBONATE RESERVE
$HCO_3^- + Na^+ \rightleftharpoons NaHCO_3$ (sodium bicarbonate)
1 When respiratory activity does not keep pace with the rate of CO_2 generation, alveolar and blood P_{CO_2} increases. This upsets the equilibrium and drives the reaction to the right, generating additional H_2CO_3, which releases H^+ and decreases blood pH.
2 As bicarbonate ions and hydrogen ions are released through the dissociation of carbonic acid, the excess bicarbonate ions become part of the bicarbonate reserve.
To limit the pH effects of respiratory acidosis, the excess H^+ must either be "tied up" by other buffer systems or secreted by the kidneys. The underlying problem, however, cannot be eliminated without an increase in the respiratory rate.

2 This flowchart summarizes the integrated homeostatic responses to respiratory acidosis.

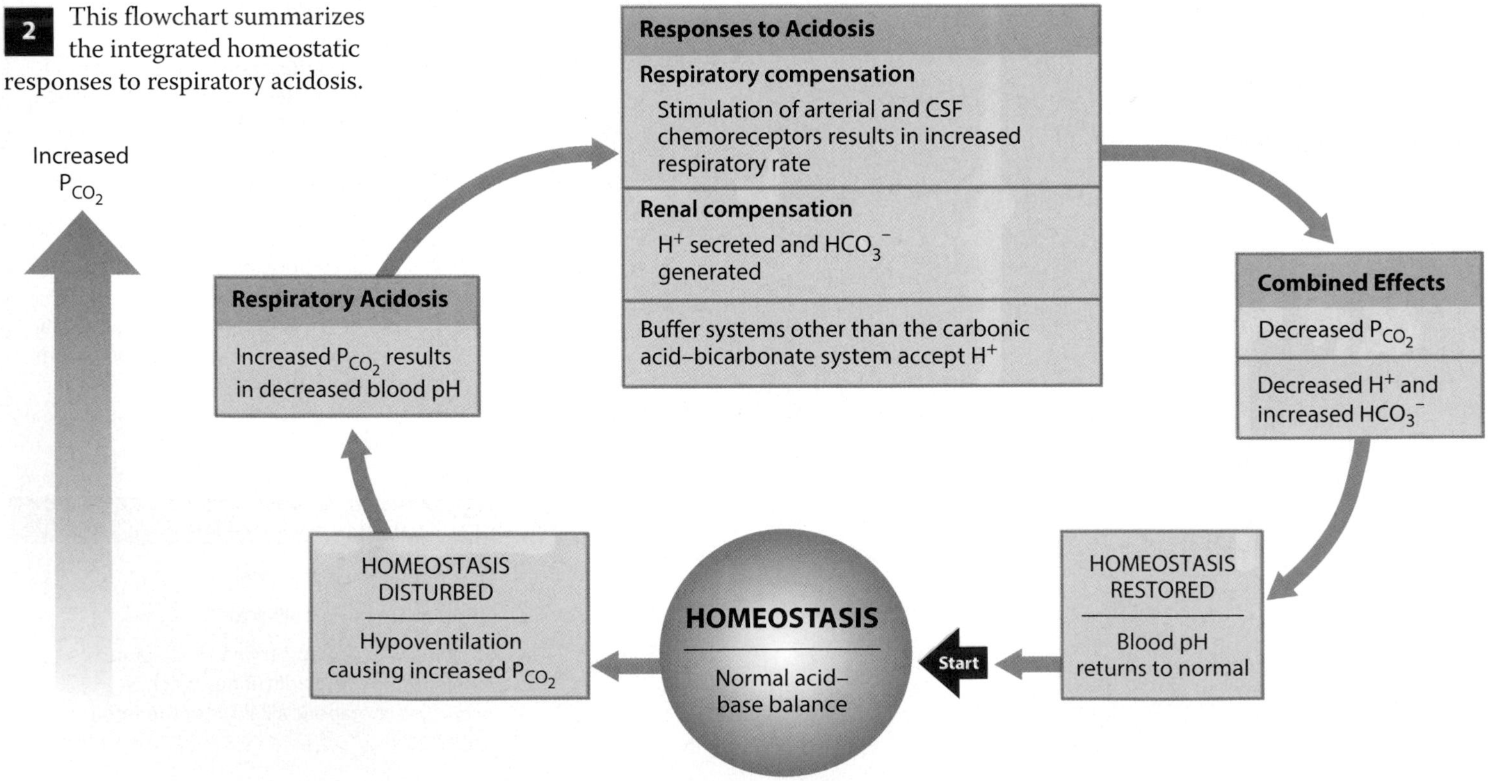

Increased P_{CO_2}
Responses to Acidosis
Respiratory compensation
Stimulation of arterial and CSF chemoreceptors results in increased respiratory rate
Renal compensation
H^+ secreted and HCO_3^- generated
Buffer systems other than the carbonic acid–bicarbonate system accept H^+
Respiratory Acidosis
Increased P_{CO_2} results in decreased blood pH
Combined Effects
Decreased P_{CO_2}
Decreased H^+ and increased HCO_3^-
HOMEOSTASIS DISTURBED
Hypoventilation causing increased P_{CO_2}
HOMEOSTASIS
Normal acid–base balance
Start
HOMEOSTASIS RESTORED
Blood pH returns to normal

Respiratory Alkalosis

3 If the rate of CO_2 elimination exceeds the rate of CO_2 generation, the condition of **respiratory alkalosis** develops. Respiratory alkalosis is relatively uncommon and rarely severe.

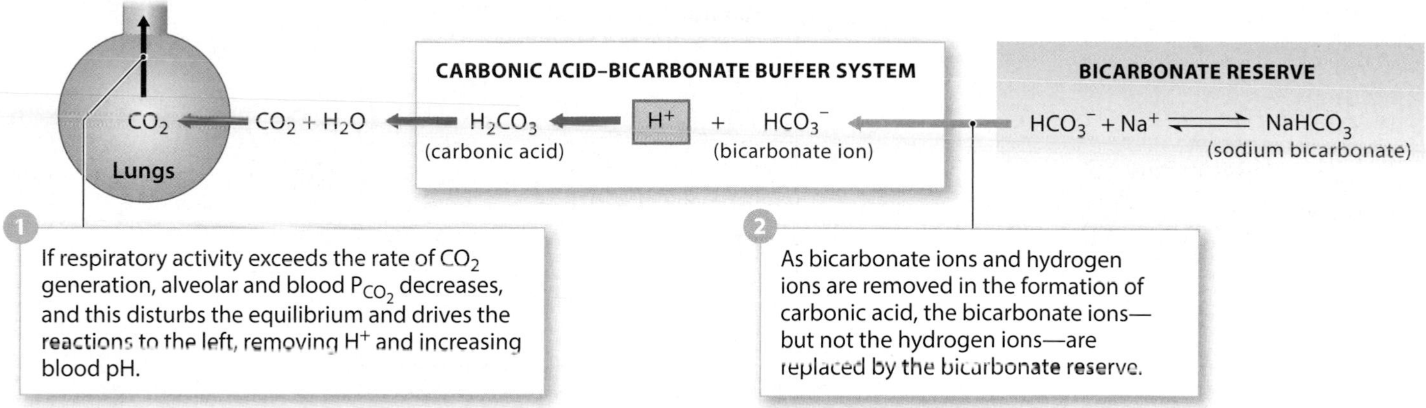

CARBONIC ACID–BICARBONATE BUFFER SYSTEM
BICARBONATE RESERVE
CO_2
Lungs
$CO_2 + H_2O$
H_2CO_3
(carbonic acid)
H^+
+
HCO_3^-
(bicarbonate ion)
$HCO_3^- + Na^+ \rightleftharpoons NaHCO_3$
(sodium bicarbonate)
1 If respiratory activity exceeds the rate of CO_2 generation, alveolar and blood P_{CO_2} decreases, and this disturbs the equilibrium and drives the reactions to the left, removing H^+ and increasing blood pH.
2 As bicarbonate ions and hydrogen ions are removed in the formation of carbonic acid, the bicarbonate ions—but not the hydrogen ions—are replaced by the bicarbonate reserve.

4 The homeostatic responses to respiratory alkalosis are summarized in this flowchart. Most cases are related to anxiety, and the resulting hyperventilation is self-limiting—the person often faints and the respiratory rate then decreases to normal levels. One common treatment is to have the person breathe in and out of a paper bag; the rising CO_2 level in the recycled air increases blood P_{CO_2} and eliminates the alkalosis.

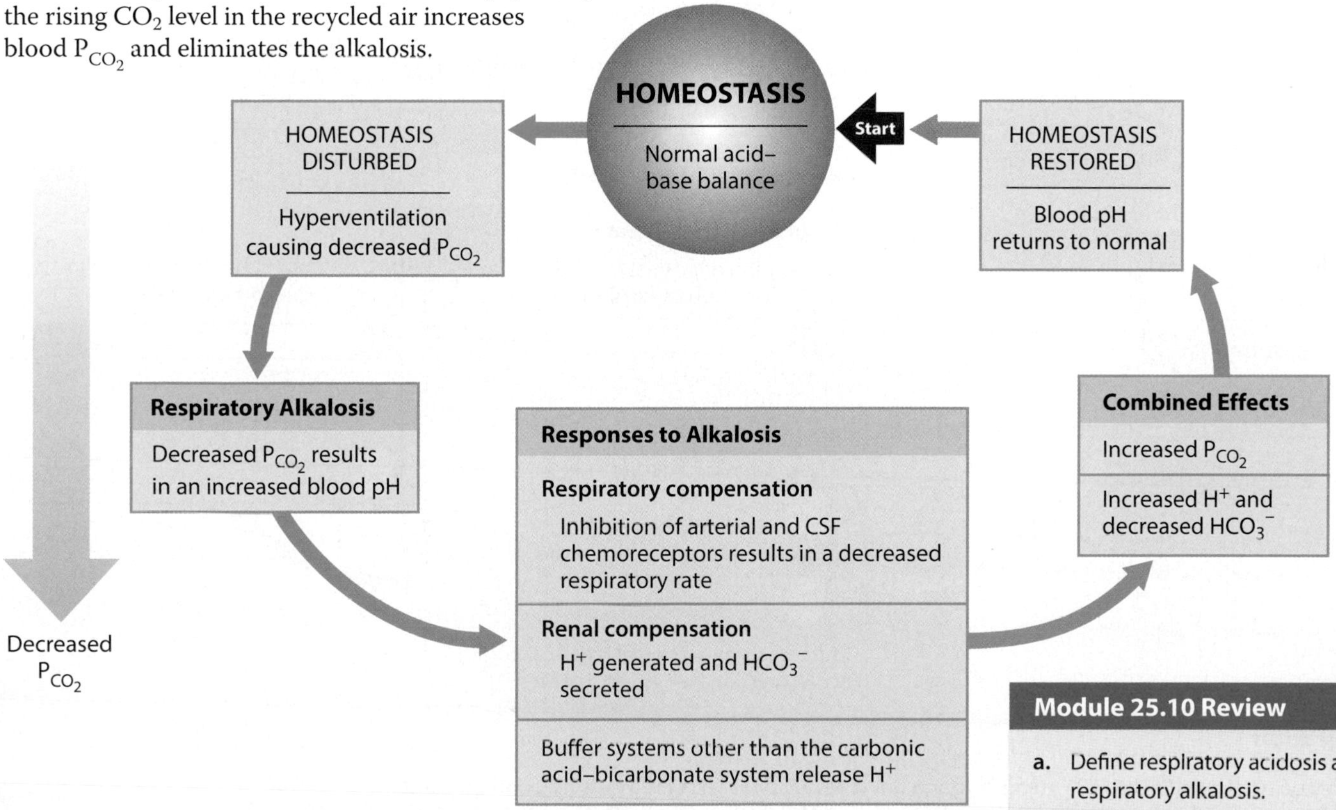

HOMEOSTASIS
Normal acid–base balance
Start
HOMEOSTASIS DISTURBED
Hyperventilation causing decreased P_{CO_2}
HOMEOSTASIS RESTORED
Blood pH returns to normal
Decreased P_{CO_2}
Respiratory Alkalosis
Decreased P_{CO_2} results in an increased blood pH
Responses to Alkalosis
Respiratory compensation
Inhibition of arterial and CSF chemoreceptors results in a decreased respiratory rate
Renal compensation
H^+ generated and HCO_3^- secreted
Buffer systems other than the carbonic acid–bicarbonate system release H^+
Combined Effects
Increased P_{CO_2}
Increased H^+ and decreased HCO_3^-

Module 25.10 Review

a. Define respiratory acidosis and respiratory alkalosis.

b. What would happen to the blood P_{CO_2} of a patient who has an airway obstruction?

c. How would a decrease in the pH of body fluids affect the respiratory rate?

LO 25.10 Describe respiratory acidosis and respiratory alkalosis.

Labeling

Use the following terms to label the boxes in the two flowcharts. Terms may be used more than once.

- blood pH decrease
- blood pH increase
- increased P_{CO_2}
- decreased P_{CO_2}
- increased
- decreased
- alkalosis
- acidosis
- generated
- secreted

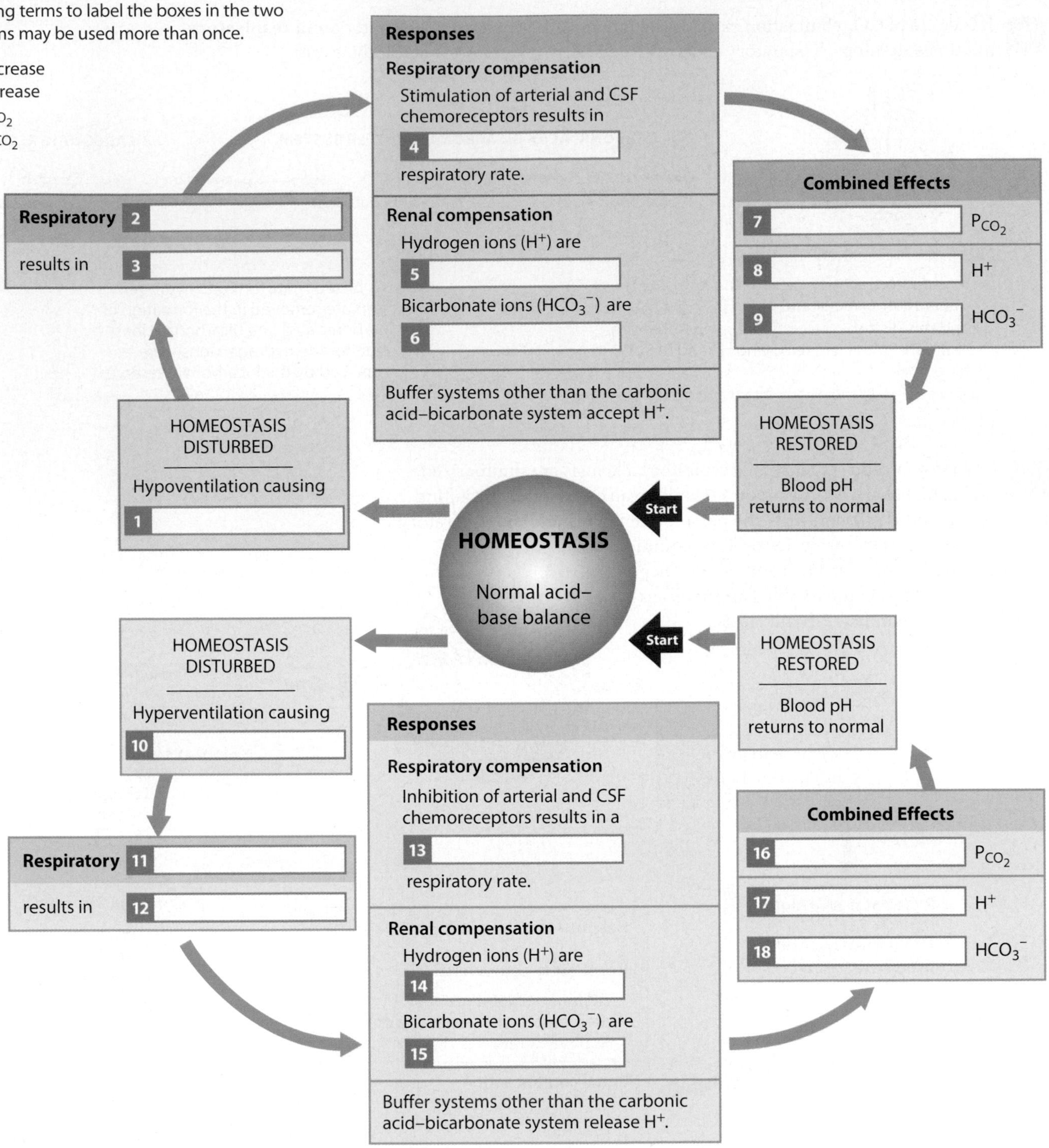
Responses
Respiratory compensation
Stimulation of arterial and CSF chemoreceptors results in
4
respiratory rate.
Renal compensation
Hydrogen ions (H+) are
5
Bicarbonate ions (HCO3−) are
6
Buffer systems other than the carbonic acid–bicarbonate system accept H+.
Combined Effects
7 PCO2
8 H+
9 HCO3−
Respiratory 2
results in 3
HOMEOSTASIS DISTURBED
Hypoventilation causing
1
HOMEOSTASIS RESTORED
Blood pH returns to normal
Start
HOMEOSTASIS
Normal acid–base balance
Start
HOMEOSTASIS DISTURBED
Hyperventilation causing
10
HOMEOSTASIS RESTORED
Blood pH returns to normal
Responses
Respiratory compensation
Inhibition of arterial and CSF chemoreceptors results in a
13
respiratory rate.
Renal compensation
Hydrogen ions (H+) are
14
Bicarbonate ions (HCO3−) are
15
Buffer systems other than the carbonic acid–bicarbonate system release H+.
Respiratory 11
results in 12
Combined Effects
16 PCO2
17 H+
18 HCO3−

Section integration

After falling into a deep lake and nearly drowning, a young boy is rescued. His rescuers assess his condition and find that his body fluids have high P_{CO_2} and lactate levels, and low P_{O_2} levels. Identify the underlying problem, and recommend the necessary treatment to restore homeostatic conditions.

19 ______

Study Outline

SECTION 1 • Fluid and Electrolyte Balance

25.1 Body composition may be viewed in terms of solids and two fluid compartments p. 963

1. **Minerals** are the inorganic substances that dissociate in body fluids to form ions called electrolytes.
2. Intracellular fluid (ICF) and extracellular fluid (ECF) are called **fluid compartments**.
3. Cells are able to maintain internal environments quite distinct from that of the ECF because they have a plasma membrane and active transport at the membrane surface.
4. Solid components make up only 40–50 percent of body mass.

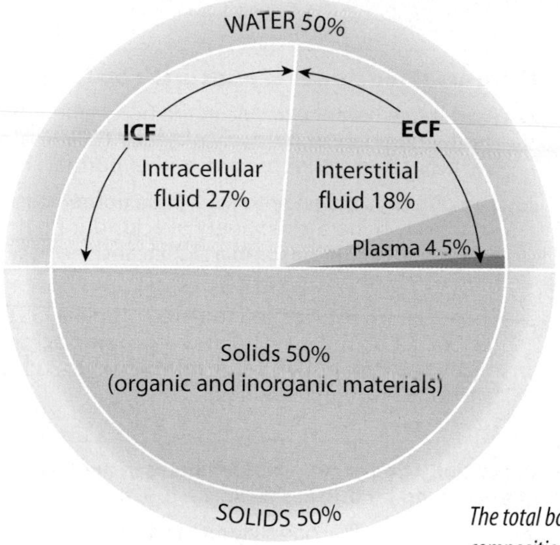
WATER 50%
ICF
ECF
Intracellular fluid 27%
Interstitial fluid 18%
Plasma 4.5%
Solids 50% (organic and inorganic materials)
SOLIDS 50%
The total body composition of an adult female

25.2 Fluid balance exists when water gains equal water losses p. 964

5. **Fluid balance** in the body occurs when water content remains stable over time.
6. Water gains occur primarily in the digestive tract. Almost half of water loss occurs though urination.
7. Daily water input from water content in food, consumed liquids, and metabolic water produced during catabolism is approximately 2500 mL. A total of 2500 mL of water is eliminated through urination, evaporation at the skin and lungs, loss through feces, and some secretion by sweat glands.

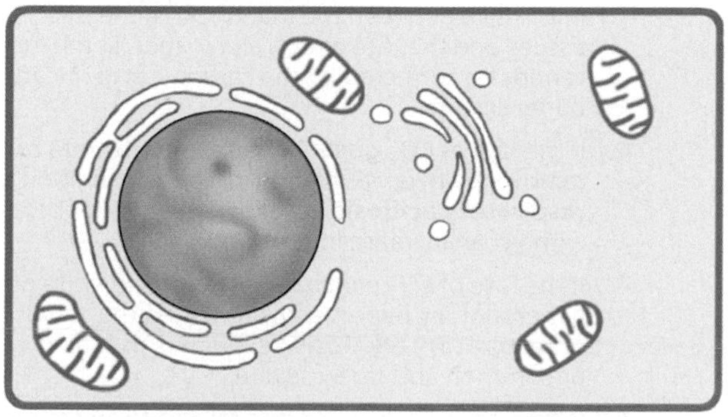

8. A rapid water movement between the ECF and the ICF in response to an osmotic gradient is called a **fluid shift**.
9. **Dehydration** develops when water losses outpace water gains.

25.3 Mineral balance involves balancing electrolyte gains and losses p. 966

10. **Mineral balance** is the balance between ion absorption, which occurs across the lining of the small intestine and colon, and ion excretion, which occurs primarily by the kidneys.
11. The body contains substantial reserves of key minerals. The amount lost each day must equal the daily intake if the person is to stay in mineral balance.

25.4 Water balance depends on sodium balance, and the two are regulated simultaneously p. 968

12. **Sodium balance** exists when sodium gains equal sodium losses.
13. When sodium gains exceed losses, the ECF volume increases; when sodium losses exceed gains, the volume of the ECF decreases.
14. If ECF volume rises, blood volume goes up. If ECF volume drops, blood volume goes down
15. When the Na^+ concentration in the ECF falls below 136 mEq/L, a state of **hyponatremia** exists. This can be caused by excessive water intake (overhydration) or inadequate salt intake.
16. When body water content declines, the Na^+ concentration rises; when that concentration exceeds 145 mEq/L, **hypernatremia** exists; dehydration is the most common cause.

25.5 Disturbances of potassium balance are uncommon but extremely dangerous p. 970

17. The key factors in maintaining **potassium balance** are the rate of K^+ entry across the digestive epithelium and the rate of K^+ loss into urine.
18. The kidneys are the main factor determining the K^+ concentration in the ECF.
19. Potassium loss is regulated by the kidneys through aldosterone stimulating Na^+ reabsorption while simultaneously accelerating K^+ excretion.
20. The endocrine disorder **aldosteronism** (excessive aldosterone secretion), results in **hypokalemia** (low potassium) from excess sodium retention and potassium loss. Diuretics, such as *Lasix,* can cause hypokalemia by increasing the volume of urine produced.

21. **Hyperkalemia** (excess potassium) can be caused by low blood pH, kidney failure, and diuretics that block Na^+ reabsorption (which also causes slowed potassium secretion).

SECTION 2 • Acid–Base Balance

25.6 There are three classes of acids in the body p. 973

22. **Acid–base balance** occurs in the body when the production of H^+ is precisely offset by their loss, and when the pH of body fluids remains within normal limits.
23. The primary challenge to acid–base homeostasis is that your body generates a variety of acids during normal metabolic operations, and a significant decrease in body fluid pH must be prevented.
24. Three classes of acids can threaten pH balance: **fixed acids** (acids that do not leave solution), **organic acids** (products of cellular metabolism), and **volatile acids** (acids leaving the body through the lungs).

25.7 Potentially dangerous disturbances in acid–base balance are opposed by buffer systems p. 974

25. The pH of the ECF normally remains between 7.35 and 7.45.
26. Changes in H^+ concentrations from the normal range disrupt the stability of plasma membranes, alter the structure of proteins, and change the activities of important enzymes.
27. When the pH of blood decreases below 7.35, **acidemia** exists, resulting in **acidosis**. When the pH of blood increases above 7.45, **alkalemia** exists, resulting in **alkalosis**.
28. The most important factor affecting the pH of body tissues is P_{CO_2} because carbon dioxide combines with water to form carbonic acid that dissociates into a H^+ and a bicarbonate ion (HCO_3^-) in the reaction: $H_2O + CO_2 \rightarrow H_2CO_3 \rightarrow H^+ + HCO_3^-$. There is an inverse relationship between P_{CO_2} and pH.
29. A **buffer system** in body fluids generally consists of a combination of a weak acid (HY) and the anion (Y^-) released by its dissociation.

25.8 Buffer systems can delay but not prevent pH shifts in the ICF and ECF p. 976

30. There are three major buffer systems: the phosphate buffer system, the protein buffer systems, and the carbonic acid–bicarbonate buffer system.
31. The **phosphate buffer system** plays an important role in buffering the pH of the ICF and urine.
32. The **protein buffer systems** regulate the pH of the ECF and ICF, and interact with the two other buffer systems. Protein buffer systems prevent significant pH change by binding excess H^+. This involves the amino, carboxylate, and R (side) groups of free amino acids or, in proteins, mostly exposed R-groups.
33. The **carbonic acid–bicarbonate buffer system** is most important in the ECF. It takes the H^+ released by metabolic activity and generates carbonic acid that dissociates into water and carbon dioxide, which can easily be eliminated by the lungs.

25.9 The homeostatic responses to metabolic acidosis and alkalosis involve respiratory and renal mechanisms as well as buffer systems p. 978

34. **Metabolic acidosis** develops when organic or fixed acids release large numbers of H^+, and the pH decreases.
35. The excess H^+ must either be permanently bound up through the formation of water, or removed from body fluids through secretion by the kidneys.
36. The respiratory response to acidosis is increased respiratory rate to decrease P_{CO_2}, resulting in carbonic acid and water. The renal response is for tubule cells to secrete H^+ into the tubular fluid along the PCT, the DCT, and the collecting system.
37. **Metabolic alkalosis** develops when large numbers of H^+ are removed from body fluids, resulting in a rise in pH.
38. The respiratory response to metabolic alkalosis is decreased respiratory rate. This elevates P_{CO_2}, resulting in CO_2 molecules converting into carbonic acid and adding H^+. The renal response is for the kidney tubules to conserve H^+ and secrete HCO_3^-.

25.10 Respiratory acid–base disorders are the most common challenges to acid–base balance p. 980

39. **Respiratory acid–base disorders** result from an imbalance between the rate of CO_2 generation in body tissues and the rate of CO_2 elimination at the lungs. This cannot be corrected by the carbonic acid–bicarbonate buffer system.
40. If the rate of CO_2 generation exceeds the rate of CO_2 removal by hypoventilation, then the condition of **respiratory acidosis** develops. This cannot be eliminated without an increase in respiratory rate.
41. If the rate of CO_2 elimination exceeds the rate of CO_2 generation by hyperventilation, then the condition of **respiratory alkalosis** develops. This is relatively uncommon and rarely severe.
42. Respiratory alkalosis is often related to anxiety-induced hyperventilation, and the person faints, resulting in normal ventilation rates. One common treatment is to have the person breathe into a paper bag. The rising CO_2 level in the recycled air increases blood P_{CO_2} and eliminates the alkalosis.

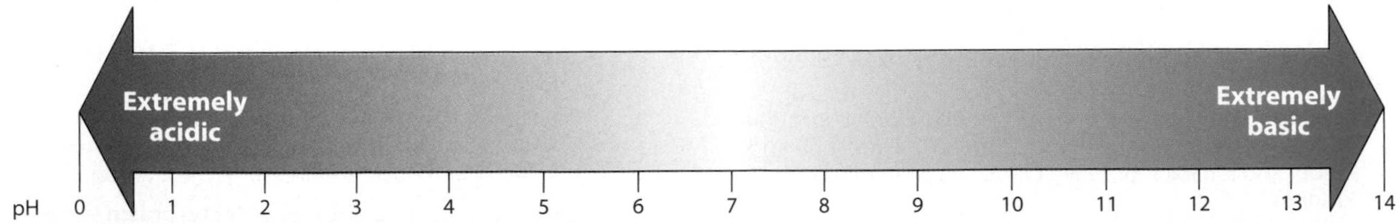
Extremely acidic
Extremely basic
pH 0 1 2 3 4 5 6 7 8 9 10 11 12 13 14

Severe acidosis and severe alkalosis are both dangerous to the human body.

Chapter Review Questions

True/False

Indicate whether each statement is true or false.

1. Metabolic acidosis develops when large numbers of hydrogen ions are removed from body fluids, resulting in an increase in pH.
2. Proteins are the most abundant solid components of body mass.
3. About half of daily water loss occurs through evaporation at the lungs.
4. Respiratory alkalosis can be treated by simply having the person breathe in and out of a paper bag.
5. Hypokalemia is caused by inadequate aldosterone secretion.
6. Iron is classified as a bulk mineral.
7. Carbonic acid is a volatile acid that forms through the interaction of water and carbon dioxide.
8. The pH of ECF normally ranges from 7.0 to 7.30.
9. The total body composition of adult males is 60 percent water and 40 percent solids, while for females, total body composition is 50 percent water and 50 percent solids.
10. About 98 percent of the potassium content of the human body is in the ICF, rather than the ECF.

1 ______
2 ______
3 ______
4 ______
5 ______
6 ______
7 ______
8 ______
9 ______
10 ______

Matching

Match each lettered term with the most closely related description.

a. buffer
b. acid
c. basic, or alkaline
d. pH
e. acidic
f. base
g. neutral
h. salt

11. The negative exponent of the hydrogen ion concentration in a solution
12. A substance that dissociates to release hydroxide ions or to remove hydrogen ions, increasing pH
13. A solution with a pH of 7; it contains equal numbers of hydrogen ions and hydroxide ions
14. A substance that tends to oppose changes in the pH of a solution by removing or replacing hydrogen ions
15. A solution with a pH above 7; hydroxide ions predominate
16. A substance that dissociates to release hydrogen ions, decreasing pH
17. An ionic compound consisting of a cation other than a hydrogen ion and an anion other than a hydroxide ion
18. A solution with a pH below 7; hydrogen ions predominate

11 ______
12 ______
13 ______
14 ______
15 ______
16 ______
17 ______
18 ______

Multiple choice

Select the correct answer from the list provided.

19 The most important factor affecting the pH of body tissues is the

- ☐ a) hydrochloric acid.
- ☐ b) ketone bodies.
- ☐ c) lactic acid.
- ☐ d) partial pressure of carbon dioxide.

20 Respiratory acidosis develops when the pH of blood is

- ☐ a) increased due to decreased P_{CO_2} level.
- ☐ b) decreased due to increased P_{CO_2} level.
- ☐ c) increased due to increased P_{CO_2} level.
- ☐ d) decreased due to decreased P_{CO_2} level.

21 Which of the following is the correct chemical equation of the reaction between water and carbon dioxide?

- ☐ a) $H_2CO_3 + H^+ \longrightarrow HCO_3^- + H_2O \longrightarrow CO_2$
- ☐ b) $H_2O + CO_2 \longrightarrow H_2CO_3 \longrightarrow H^+ + HCO_3^-$
- ☐ c) $H_2O + CO_2 \longrightarrow HCO_3^- \longrightarrow H_2CO_3 \longrightarrow H^+$
- ☐ d) $H_2O + CO_2 + H^+ \longrightarrow H_2CO_3 \longrightarrow HCO_3^-$

22 Most of the buffering capacity of proteins is provided by the

- ☐ a) R-groups of the component amino acids.
- ☐ b) peptide bonds between adjacent amino acids.
- ☐ c) amino and carboxyl groups of the component amino acids.
- ☐ d) central carbon group of most amino acids.

23 Which of the following is *not* an endocrine response to decreased blood volume and blood pressure?

- ☐ a) increased renin secretion and angiotensin II activation
- ☐ b) increased aldosterone release
- ☐ c) increased ADH release
- ☐ d) increased Na^+ in urine

24 The condition of below normal Na^+ concentration in the blood is called

- ☐ a) hyponatremia.
- ☐ b) hypernatremia.
- ☐ c) hypokalemia.
- ☐ d) hyperkalemia.

25 Which of the following minerals functions as a component of myoglobin, hemoglobin, and cytochromes?

- ☐ a) zinc
- ☐ b) copper
- ☐ c) iron
- ☐ d) cobalt

26 The primary site of ion loss is at the

- ☐ a) sweat glands.
- ☐ b) kidneys.
- ☐ c) intestines.
- ☐ d) lungs.

Short answer

27 Differentiate between fluid balance, mineral balance, and acid–base balance.

28 What would happen if a dehydrated patient accidentally received an intravenous solution of hypertonic saline?

Chapter Integration • Applying what you have learned

Intestinal trouble in paradise

While visiting a foreign country with poor-quality potable water, Beth and Tom each had a drink with ice during dinner. Many travelers avoid drinking the water in foreign countries but forget that ice is often made with local tap water! As a result, both Beth and Tom contracted an intestinal disease that causes severe diarrhea. (Diarrhea is characterized by loose watery bowel movements that, if left untreated, can cause serious health complications.)

It is estimated that each year such "traveler's diarrhea" affects as many as 10 million Americans. High-risk destinations include the lesser-developed countries of Latin America, Africa, the Middle East, and regions of Asia. The Centers for Disease Control and Prevention (CDC) maintain a Website to advise travelers.

After 2 days of experiencing uncontrolled bouts of diarrhea, the couple sought medical attention at a nearby medical clinic. Various tests revealed that the culprit was the bacterium *Escherichia coli.* Further tests revealed that both Beth and Tom were dehydrated, so intravenous fluids were prescribed. Unfortunately, the attending nurse became distracted and erroneously gave Tom a hypertonic glucose solution instead of the normal saline that Beth received.

Thailand
Africa
THE ROUGH GUIDE to
Mexico

1 **How would you expect Beth's and Tom's dehydrated states to affect their blood pH, urine pH, and breathing pattern?**

2 **What effect will the intravenous hypertonic glucose solution have on Tom's blood ADH levels and urine volume?**

26 The Reproductive System

LEARNING OUTCOMES

These Learning Outcomes correspond by number to this chapter's modules and indicate what you should be able to do after completing the chapter.

SECTION 1 • Male Reproductive System

- **26.1** Identify the structures of the male reproductive system, and cite their functions.
- **26.2** Describe the structures of the testes, and outline their fetal development.
- **26.3** Summarize the events of meiosis in the production of spermatozoa, and describe the functional anatomy of a mature spermatozoon.
- **26.4** Explain meiosis and early spermiogenesis within the seminiferous tubules.
- **26.5** Explain the roles played by the male reproductive tract and accessory glands in the functional maturation, nourishment, storage, and transport of spermatozoa.
- **26.6** Describe the structure and function of the penis.
- **26.7** Explain the roles of regulatory hormones and testosterone in the establishment and maintenance of male sexual function.

SECTION 2 • Female Reproductive System

- **26.8** Identify the structures of the female reproductive system, and cite their functions.
- **26.9** Describe the anatomy of the ovaries, uterus, and associated structures.
- **26.10** Outline the processes of meiosis and oogenesis in the ovaries.
- **26.11** Describe the structure, histology, and functions of the uterine tubes and uterus.
- **26.12** Identify the phases and events of the uterine cycle.
- **26.13** Describe the structure, histology, and functions of the vagina.
- **26.14** Discuss the structure and function of the mammary glands.
- **26.15** Summarize the hormonal regulation of the female reproductive cycles.
- **26.16** + **CLINICAL MODULE** Discuss various birth control strategies and their associated risks.
- **26.17** + **CLINICAL MODULE** Discuss several common reproductive system disorders.

LO Learning Outcomes are repeated at the bottom of each module.

Male reproductive structures are gonads, accessory organs, and external genitalia

The reproductive system of both sexes includes the following basic structures:

- **Gonads** (GŌ-nadz; *gone*, seed), or reproductive organs, which produce gametes and hormones.
- Accessory glands and organs that secrete fluids into the ducts of the reproductive system or into other excretory ducts.
- Perineal structures collectively known as the **external genitalia** (jen-i-TĀ-lē-uh).

1 This figure provides an overview of the structures of the male reproductive system. The male gonads are called the **testes**, or **testicles**, and they produce gametes called **spermatozoa** (sper-ma-tō-ZŌ-uh; singular, *spermatozoon*), or sperm. Mature spermatozoa travel along the male duct system, or **male reproductive tract**. As they proceed, they are mixed with the secretions of accessory glands to form a fluid known as **semen** (SĒ-men).

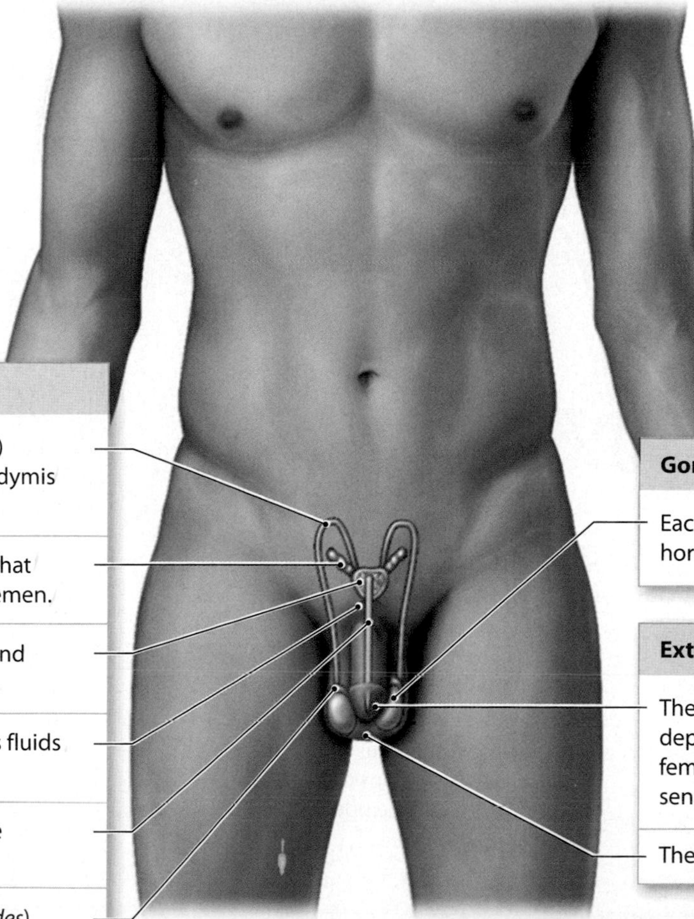
Accessory Organs
Each **ductus deferens** (sperm duct) conducts sperm between the epididymis and prostate gland.
Each **seminal gland** secretes fluid that makes up much of the volume of semen.
The **prostate gland** secretes fluid and enzymes.
Each **bulbourethral gland** secretes fluids that lubricate the tip of the penis
The **urethra** conducts semen to the exterior.
Each **epididymis** (plural, *epididymides*) is the site of sperm maturation.
Gonads
Each testis produces sperm and hormones.
External Genitalia
The **penis** contains erectile tissue, deposits sperm in the vagina of the female, and produces pleasurable sensations during sexual activities.
The **scrotum** surrounds the testes.

Module 26.1 Review

a. What is the function of the testes?

b. List the accessory organs of the male reproductive system.

c. Identify the structures of the male external genitalia.

LO **26.1** Identify the structures of the male reproductive system, and cite their functions.

Module 26.2

The coiled seminiferous tubules of the testes are connected to the male reproductive tract

1 The main structures of the male reproductive system are shown in this sagittal section. Proceeding from a testis, the spermatozoa travel within the **epididymis** (ep-i-DID-i-mus), along the **ductus deferens** (DUK-tus DEF-e-renz), and then along the **ejaculatory duct** and the urethra before leaving the body. Accessory organs—the **seminal** (SEM-i-nal) **glands**, the **prostate** (PROS-tāt) **gland**, and the **bulbourethral** (bul-bō-ū-RĒ-thral) **glands**—secrete various fluids into the ejaculatory ducts and urethra. The external genitalia consist of the **scrotum** (SKRŌ-tum), which encloses the testes, and the **penis** (PĒ-nis), an erectile organ through which the distal portion of the urethra passes.

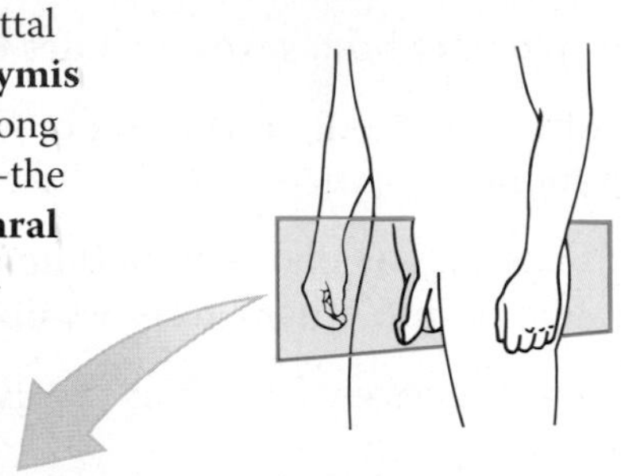

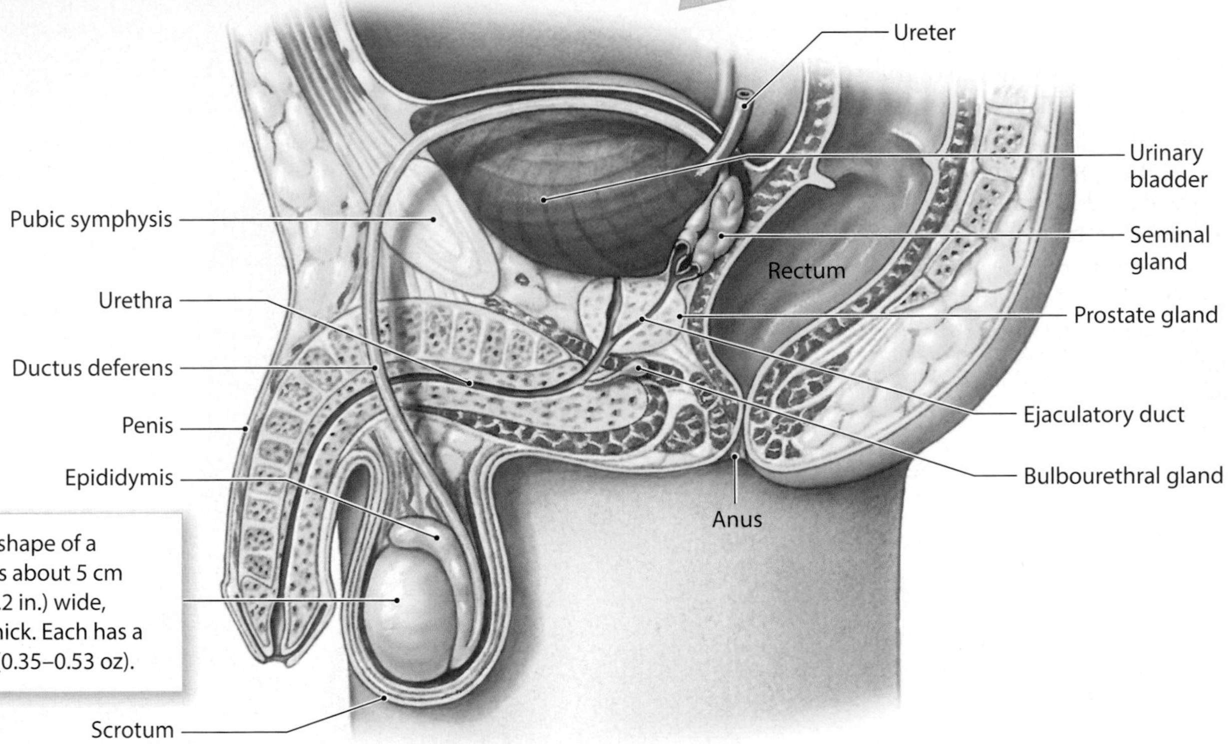
Ureter
Urinary bladder
Pubic symphysis
Seminal gland
Rectum
Urethra
Prostate gland
Ductus deferens
Ejaculatory duct
Penis
Bulbourethral gland
Epididymis
Anus
Each testis has the shape of a flattened egg and is about 5 cm (2 in.) long, 3 cm (1.2 in.) wide, and 2.5 cm (1 in.) thick. Each has a weight of 10–15 g (0.35–0.53 oz).
Scrotum

2 These sagittal sectional views illustrate the positional changes involved in the fetal descent of a testis.

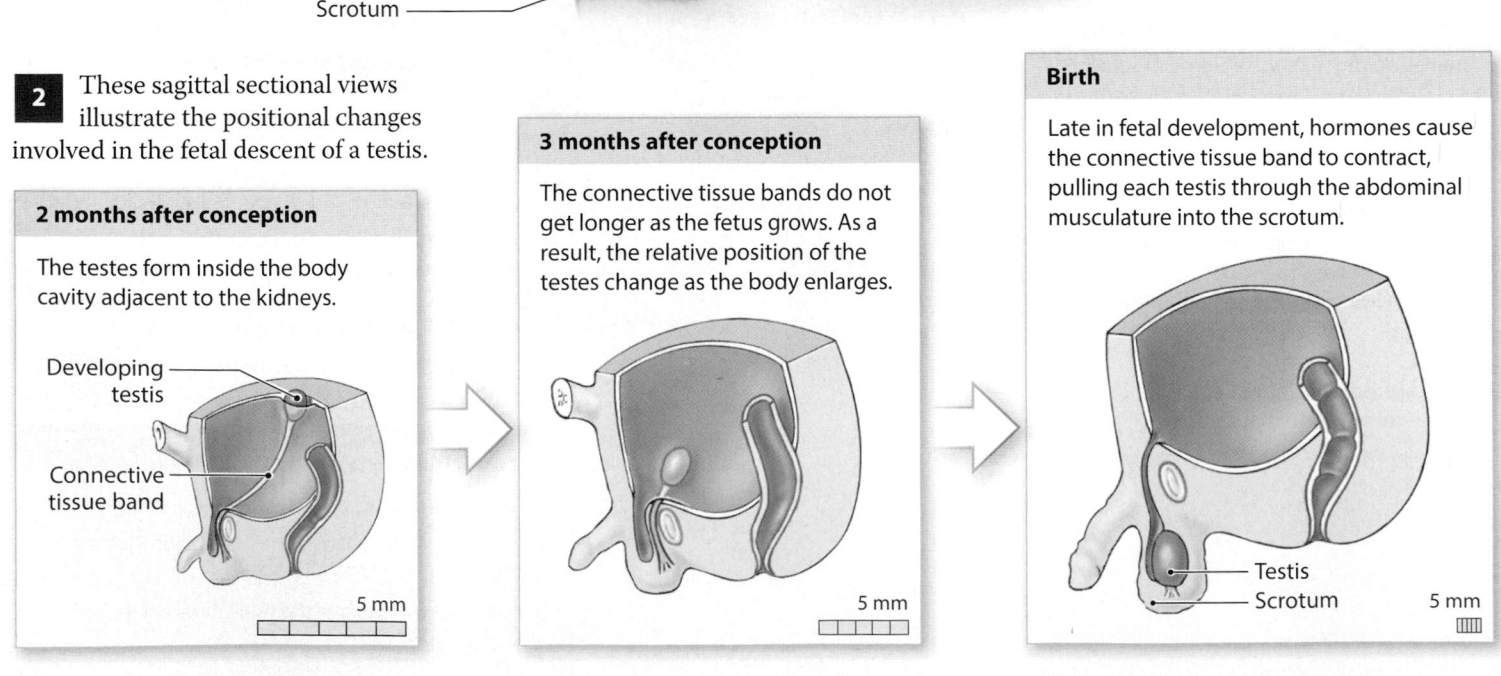

2 months after conception

The testes form inside the body cavity adjacent to the kidneys.

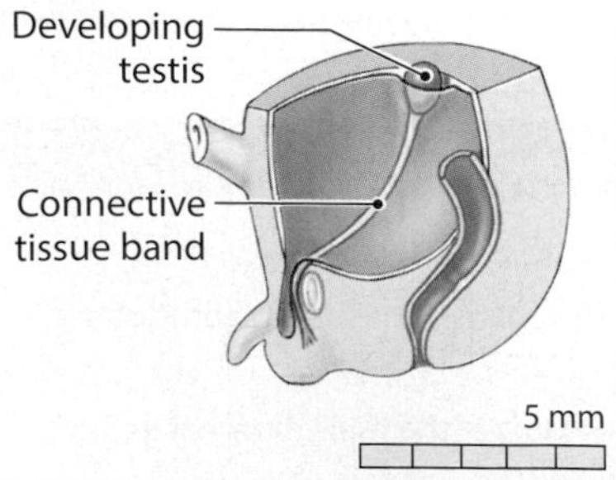
Developing testis
Connective tissue band
5 mm

3 months after conception

The connective tissue bands do not get longer as the fetus grows. As a result, the relative position of the testes change as the body enlarges.

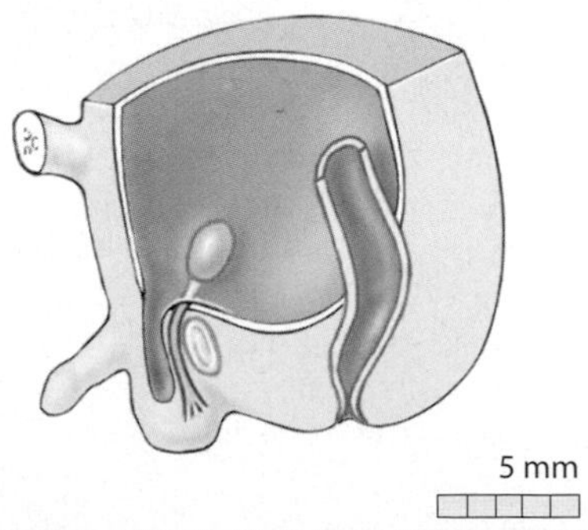
5 mm

Birth

Late in fetal development, hormones cause the connective tissue band to contract, pulling each testis through the abdominal musculature into the scrotum.

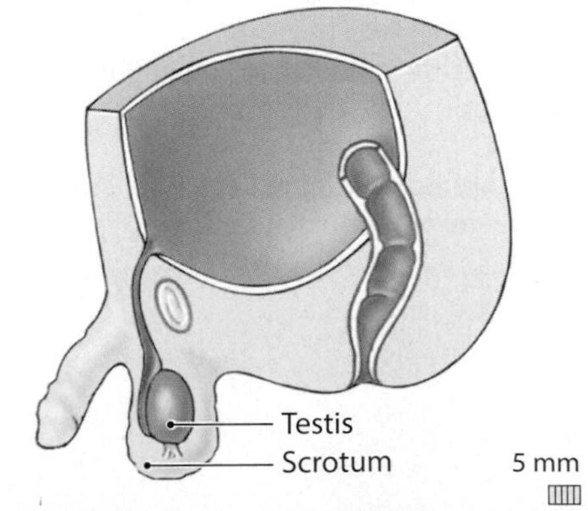
Testis
Scrotum
5 mm

3 This dissection view shows superficial and deeper features of the scrotum, testes, and related structures.

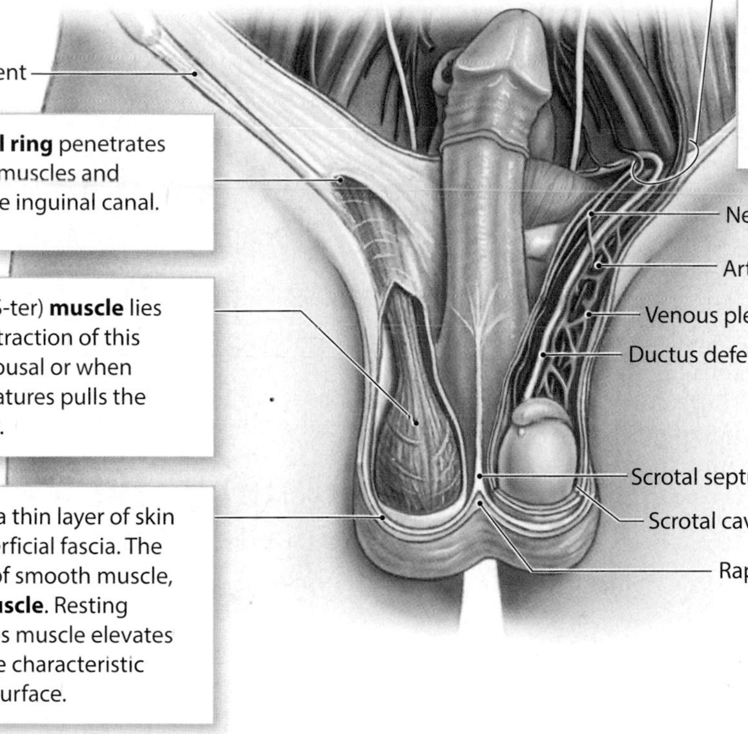

Inguinal ligament
Nerve
Artery
Venous plexus
Ductus deferens
Scrotal septum
Scrotal cavity
Raphe

The **superficial inguinal ring** penetrates the layers of abdominal muscles and forms the entrance to the inguinal canal.

The **cremaster** (krē-MAS-ter) **muscle** lies deep to the dermis. Contraction of this muscle during sexual arousal or when exposed to cool temperatures pulls the testes closer to the body.

The scrotum consists of a thin layer of skin and the underlying superficial fascia. The dermis contains a layer of smooth muscle, the **dartos** (DAR-tōs) **muscle**. Resting muscle tone in the dartos muscle elevates the testes and causes the characteristic wrinkling of the scrotal surface.

The **inguinal canal** extends from the inguinal ring to the scrotal cavity. In normal adult males, the inguinal canals are closed, but the presence of the spermatic cords creates weak points in the abdominal wall that remain throughout life. As a result, **inguinal hernias**—protrusions of visceral tissues or organs into the inguinal canal—are relatively common in males.

The **spermatic cords** extend through the inguinal canals between the abdominopelvic cavity and the testes. Each spermatic cord consists of layers of fascia and muscle enclosing the ductus deferens and the blood vessels, nerves, and lymphatic vessels that supply the testes.

The left and right **scrotal cavities** are separated by the **scrotal septum**. The partition between the two is marked by a raised thickening in the scrotal surface known as the **raphe** (RĀ-fē).

4 This horizontal section through the scrotum shows the internal organization of the testes.

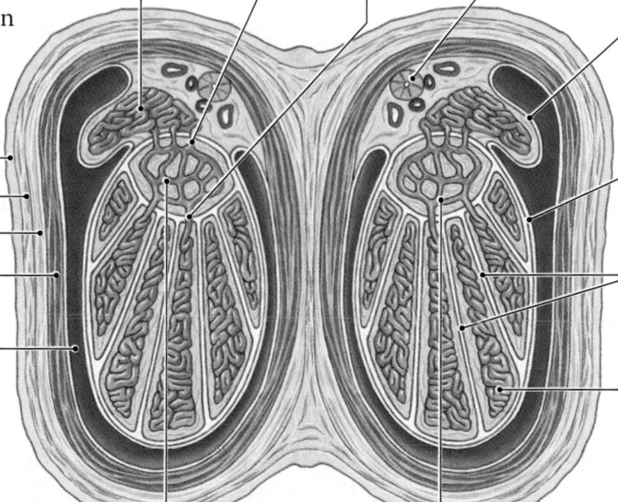

Epididymis
Efferent ductule
Straight tubule
Ductus deferens
Scrotum
Skin
Dartos muscle
Superficial fascia
Cremaster muscle
Scrotal cavity
Septa

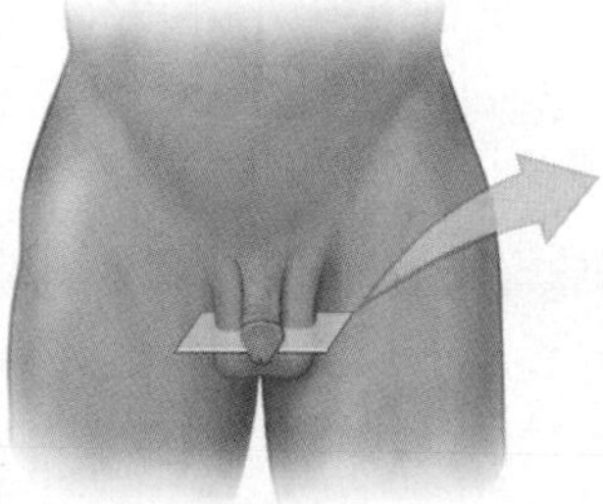

Mesothelium lines the scrotal cavity and reduces friction between opposing surfaces.

The **tunica albuginea** (al-bū-JIN-ē-uh), a tough fibrous capsule, covers the testis. This capsule is continuous with septa that subdivide the interior of the testis into separate **lobules**.

Each lobule contains several coiled **seminiferous tubules**. Each averages about 80 cm (32 in.) in length. A typical testis contains nearly one-half mile of seminiferous tubules. Sperm production occurs within these tubules.

The seminiferous tubules merge into straight tubules, which are connected to a maze of passageways known as the **rete** (RĒ-tē; *rete*, a net) **testis**. Fifteen to 20 large **efferent ductules** connect the rete testis to the epididymis.

Module 26.2 Review

a. Describe the fetal descent of the testes.

b. Identify the complex network of channels that is connected to the seminiferous tubules.

c. On a warm day, would the cremaster muscle be contracted or relaxed? Why?

LO **26.2** Describe the structures of the testes, and outline their fetal development.

Meiosis in the testes produces haploid spermatids that mature into spermatozoa

Spermatogenesis, or sperm production, involves (1) **mitosis** and cell division (cytokinesis), (2) **meiosis** (mī-Ō-sis), a special form of cell division involved in gamete production, and (3) **spermiogenesis**, the differentiation of immature male gametes into physically mature spermatozoa.

1 As you may recall from Chapter 3, somatic cells contain 23 pairs of chromosomes. Mitosis is part of the process of somatic cell division, producing two daughter cells each containing identical pairs of chromosomes. This diagram follows the fates of three representative chromosomes during mitosis and cell division. Because the daughter cells each contain 23 pairs of chromosomes, or two sets of chromosomes, they are called **diploid** (DIP-loyd; *diplo*, double) (2n) cells. Each mitotic division of a stem cell within the seminiferous tubules produces one daughter cell that remains a stem cell, and one daughter cell that enters meiosis.

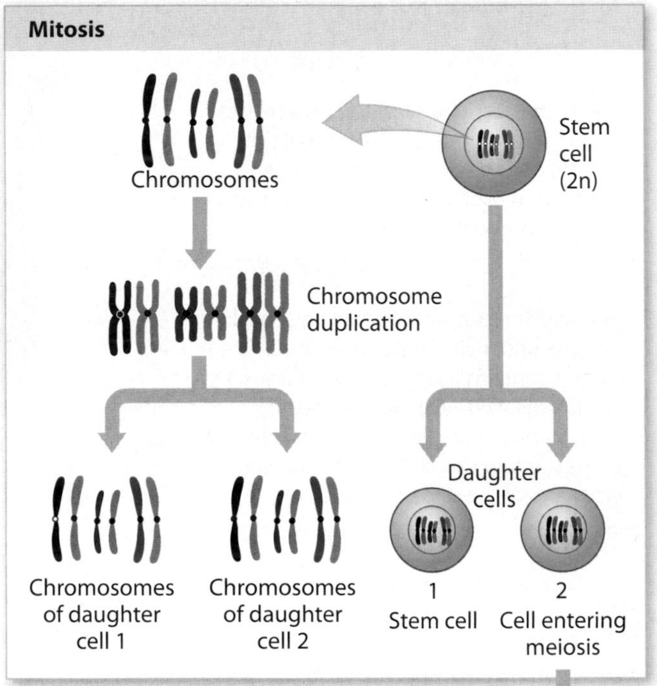
Mitosis
Chromosomes
Stem cell (2n)
Chromosome duplication
Daughter cells
Chromosomes of daughter cell 1
Chromosomes of daughter cell 2
1
Stem cell
2
Cell entering meiosis

2 Meiosis involves two cycles of cell division (**meiosis I** and **meiosis II**) and produces four cells, each of which contains 23 individual chromosomes, or one set of chromosomes. Because these reproductive cells (gametes) contain only one member of each pair of chromosomes, they are called **haploid** (HAP-loyd; *haplo*, single) (n) cells. The events in the nucleus are the same for the formation of spermatozoa or oocytes (female gametes). In **synapsis**, corresponding maternal and paternal chromosomes associate to form 23 chromosome pairs. Each member of a pair consists of two duplicate chromatids, and the set of four chromatids is called a **tetrad** (TET-rad; *tetras*, four).

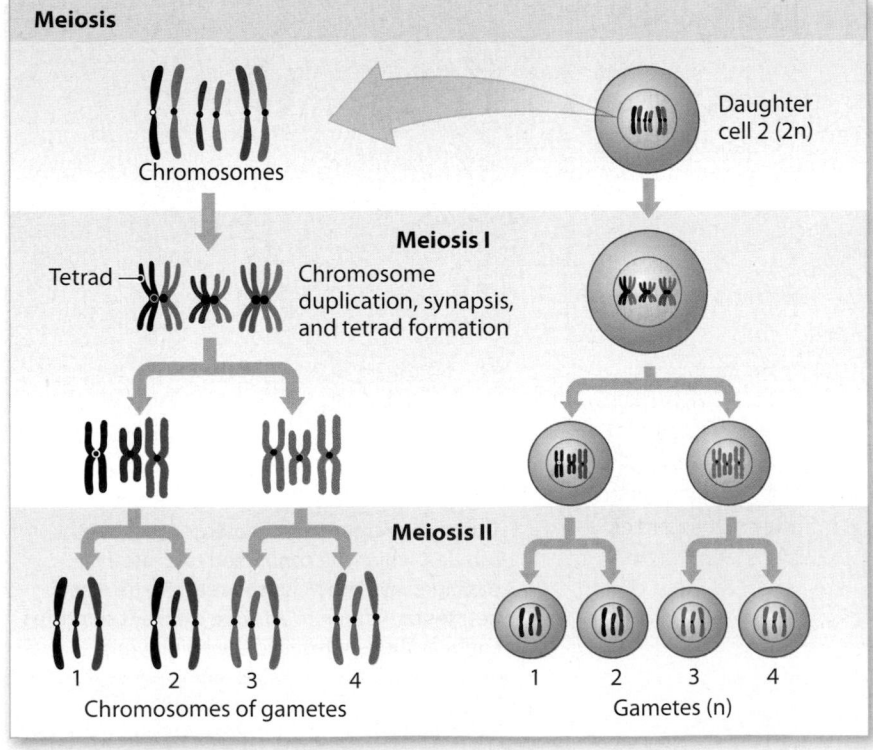

Meiosis
Chromosomes
Daughter cell 2 (2n)
Meiosis I
Tetrad
Chromosome duplication, synapsis, and tetrad formation
Meiosis II
1
2
3
4
Chromosomes of gametes
1
2
3
4
Gametes (n)

3 **Spermatogonia** (singular, *spermatogonium*) are the stem cells in the seminiferous tubules that divide to produce spermatozoa in a process called **spermatogenesis**. The complete process of spermatogenesis occurs in four phases and takes about 64 days.

Spermatogenesis

Mitosis of spermatogonium

Each division of a diploid spermatogonium produces two daughter cells. One is a spermatogonium that remains in contact with the basement membrane of the tubule, and the other is a **primary spermatocyte** that is displaced toward the lumen. These events from spermatogonium to primary spermatocyte take 16 days.

Meiosis I

As meiosis I begins, each primary spermatocyte contains 46 individual chromosomes. At the end of meiosis I, the daughter cells are called **secondary spermatocytes**. Every secondary spermatocyte contains 23 chromosomes, each with a pair of duplicate chromatids. This phase of spermatogenesis takes about 24 days.

Meiosis II

The secondary spermatocytes soon enter meiosis II, which yields four haploid **spermatids**, each containing 23 chromosomes. For each primary spermatocyte that enters meiosis, four spermatids are produced. This phase lasts only a few hours.

Spermiogenesis (physical maturation)

In spermiogenesis, the last step of spermatogenesis, each spermatid matures into a single spermatozoon, or sperm. The process of spermiogenesis—from spermatids to spermatozoa—takes 24 days.

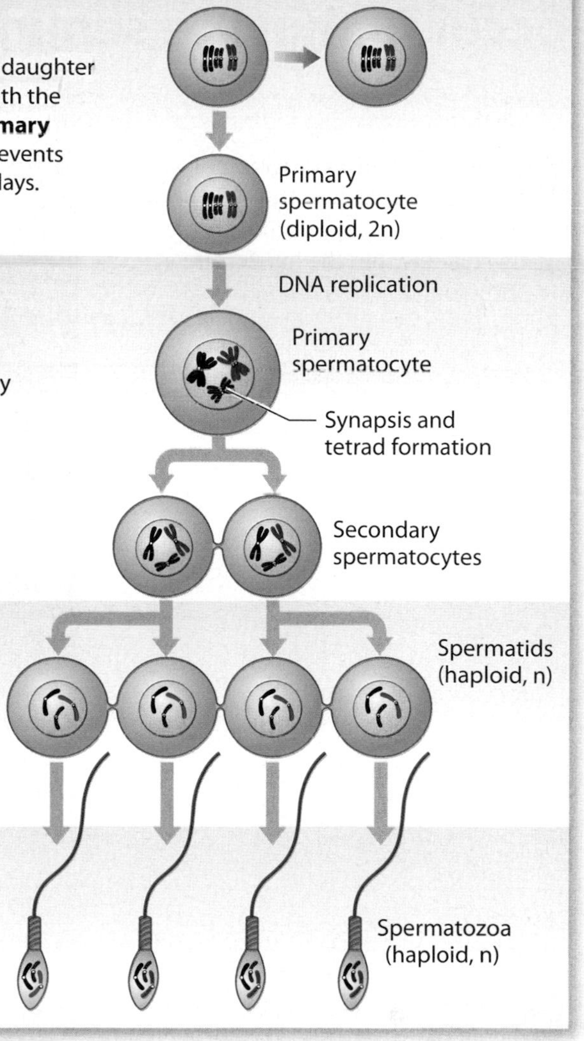

Primary spermatocyte (diploid, 2n)
DNA replication
Primary spermatocyte
Synapsis and tetrad formation
Secondary spermatocytes
Spermatids (haploid, n)
Spermatozoa (haploid, n)

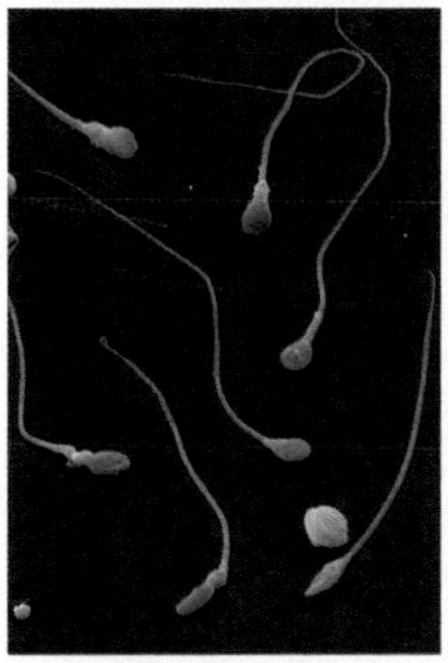
Mature spermatozoa

4 Here you can see the distinctive, specialized features of a **spermatozoon**. Unlike other, less specialized cells, a mature spermatozoon lacks an endoplasmic reticulum, a Golgi apparatus, lysosomes, peroxisomes, inclusions, and many other intracellular structures. The loss of these organelles reduces the cell's size and mass; it is essentially a mobile carrier for the enclosed chromosomes, and extra weight would slow it down.

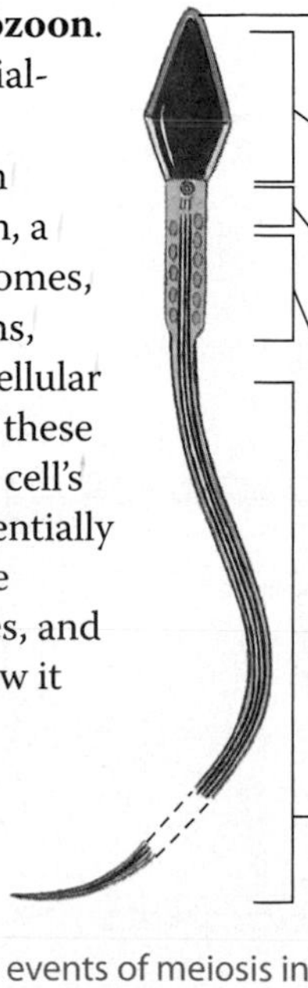

Structure of a Spermatozoon

The **acrosome** (AK-rō-sōm), or acrosomal cap, is a membranous compartment containing enzymes essential to fertilization.

The **head** is a flattened ellipse containing a nucleus with densely packed chromosomes.

The **neck** contains both centrioles of the original spermatid. The microtubules of the distal centriole are continuous with those of the middle piece and tail.

The **middle piece** contains mitochondria arranged in a spiral around the microtubules. Mitochondrial activity provides the ATP required to move the tail.

The **tail** is a **flagellum**, a whiplike organelle that moves the sperm.

The tail of a spermatozoon is the only flagellum in the human body. Whereas cilia beat in a predictable, wavelike fashion, the flagellum of a spermatozoon has a complex, corkscrew motion.

Module 26.3 Review

a. Define spermatogenesis.

b. How many spermatozoa will eventually be produced from each primary spermatocyte?

c. Describe the functional anatomy of a typical spermatozoon.

LO 26.3 Summarize the events of meiosis in the production of spermatozoa, and describe the functional anatomy of a mature spermatozoon.

Meiosis and early spermiogenesis occur within the seminiferous tubules

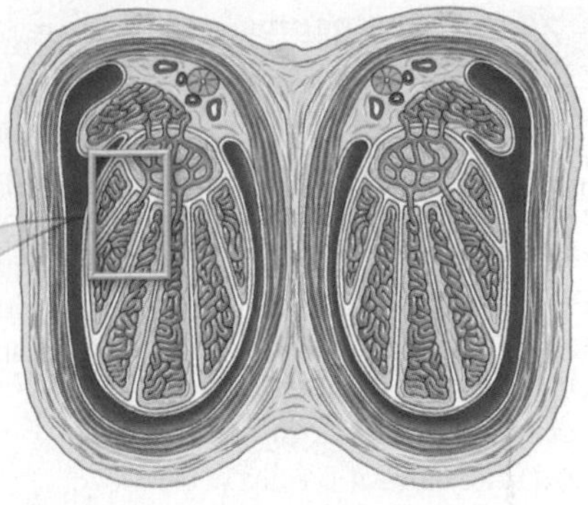

1 This is a light micrograph of a horizontal section through a testis. Because the seminiferous tubules are tightly coiled within the lobules, they are often seen in cross section.

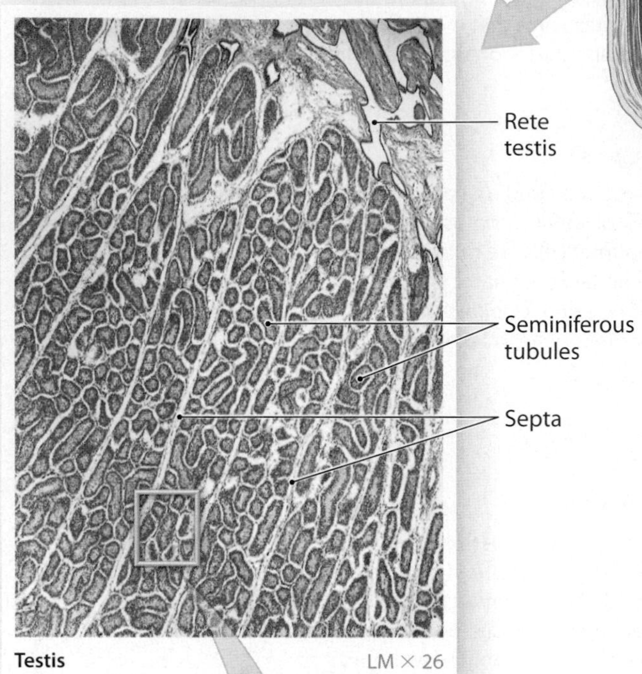
Rete testis
Seminiferous tubules
Septa

Testis LM × 26

2 Here is a higher-magnification view of several sections through one or more seminiferous tubules. Spermatogenesis and spermiogenesis together take approximately 9 weeks. If spermatogenesis were synchronized along the entire length of a tubule, spermatozoa would be released once every nine weeks. Instead, each segment of a seminiferous tubule is at a different phase of the process of spermatogenesis. As a result, every seminiferous tubule is continuously producing spermatozoa.

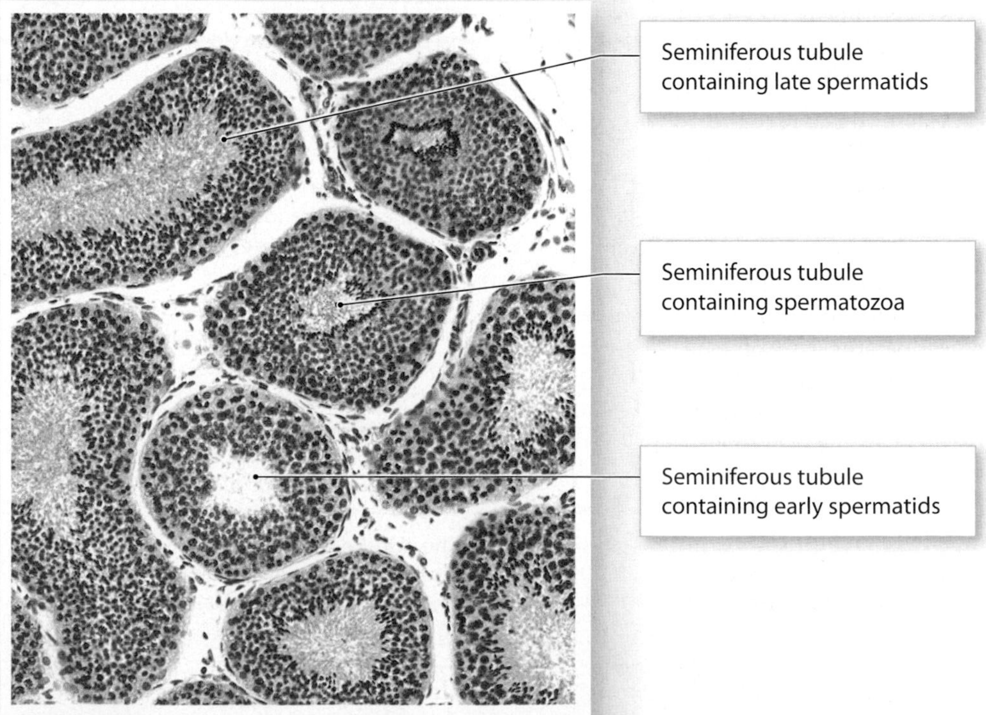
Seminiferous tubule containing late spermatids
Seminiferous tubule containing spermatozoa
Seminiferous tubule containing early spermatids

Seminiferous tubules LM × 75

3 Each tubule is surrounded by a delicate connective tissue capsule, and areolar tissue fills the spaces between the tubules. Within those spaces are numerous blood vessels and large interstitial cells. **Interstitial cells** (Leydig cells) produce androgens, such as testosterone and androstenedione, the dominant sex hormones in males.

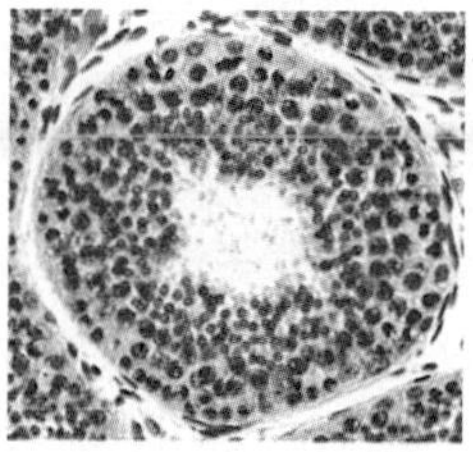

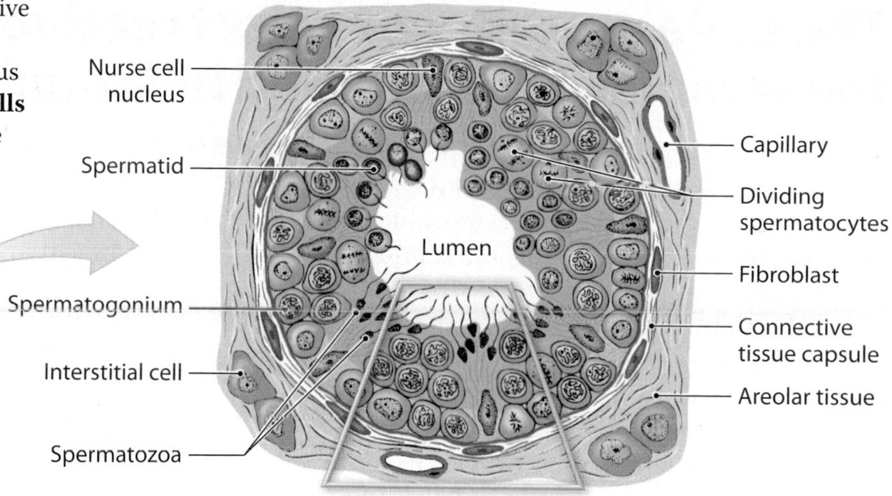

Nurse cell nucleus
Spermatid
Spermatogonium
Interstitial cell
Spermatozoa
Lumen
Capillary
Dividing spermatocytes
Fibroblast
Connective tissue capsule
Areolar tissue

4 Each seminiferous tubule contains spermatogonia, spermatocytes at various stages of meiosis, spermatids, spermatozoa, and large **nurse cells** (Sertoli cells). Nurse cells are attached to the tubular capsule and extend to the lumen between the other types of cells. Developing spermatocytes undergoing meiosis, and spermatids undergoing spermiogenesis, are not free in the seminiferous tubules. Instead, they are surrounded by the cytoplasm of the nurse cells. As spermiogenesis proceeds, the spermatids gradually develop into mature spermatozoa. At **spermiation**, a spermatozoon loses its attachment to the nurse cell and enters the lumen of the seminiferous tubule.

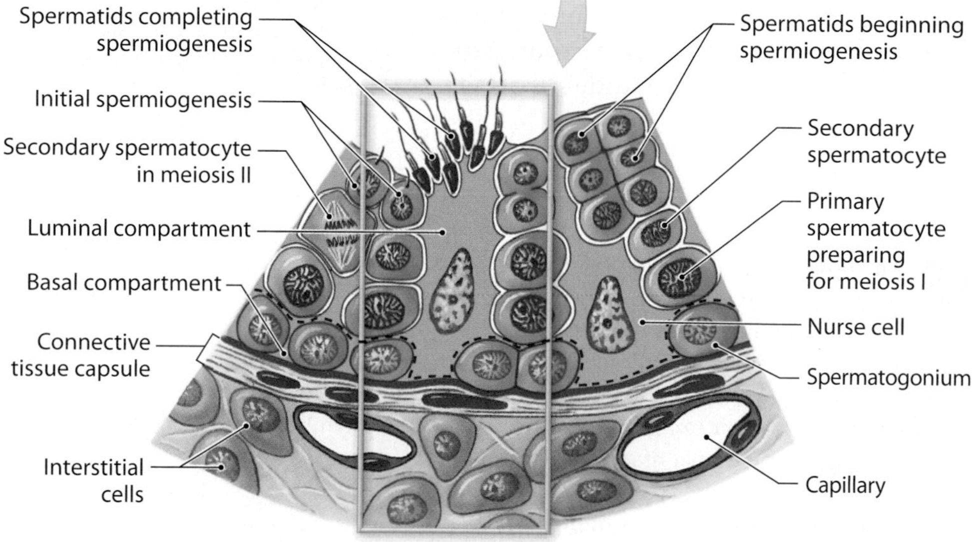

Spermatids completing spermiogenesis
Initial spermiogenesis
Secondary spermatocyte in meiosis II
Luminal compartment
Basal compartment
Connective tissue capsule
Interstitial cells
Spermatids beginning spermiogenesis
Secondary spermatocyte
Primary spermatocyte preparing for meiosis I
Nurse cell
Spermatogonium
Capillary

5 Nurse cells are not only essential to the process of spermatogenesis, but as you will see in Module 26.7, they are involved in its hormonal regulation. One of their many important functions is to isolate the seminiferous tubules from the general circulation by creating a **blood–testis barrier** comparable in function to the blood–brain barrier. Nurse cells are joined by tight junctions, forming a layer that divides the seminiferous tubule into an outer **basal compartment**, which contains the spermatogonia, and an inner **luminal compartment**, where meiosis and spermiogenesis occur.

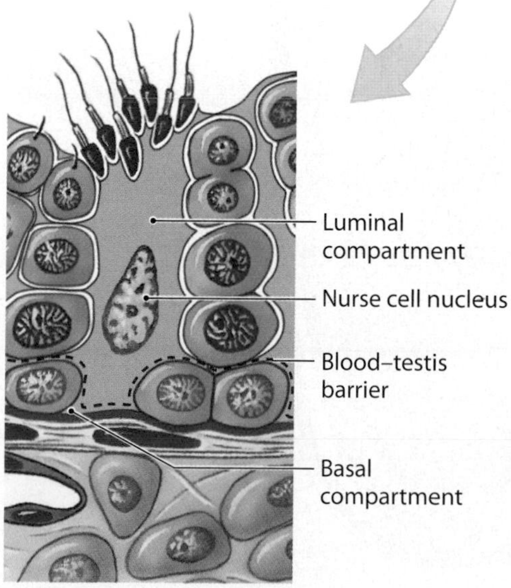

Luminal compartment
Nurse cell nucleus
Blood–testis barrier
Basal compartment

Module 26.4 Review

a. What is the function of interstitial cells?

b. What is the role of nurse cells?

c. Describe the process of spermiation.

LO 26.4 Explain meiosis and early spermiogenesis within the seminiferous tubules.

The male reproductive tract receives secretions from the seminal, prostate, and bulbourethral glands

The testes produce physically mature spermatozoa. The other portions of the male reproductive system are responsible for the functional maturation, nourishment, storage, and transport of spermatozoa. The spermatozoa leaving the testes are physically mature, but immobile and incapable of fertilizing an oocyte. To become motile (actively swimming) and fully functional, spermatozoa must undergo a process called **capacitation**. Capacitation normally occurs in two steps: (1) Spermatozoa become motile when they are mixed with secretions of the seminal glands, and (2) they become capable of successful fertilization when exposed to conditions in the female reproductive tract.

1 This diagrammatic posterior view shows the urinary bladder, prostate gland, and other structures of the male reproductive system.

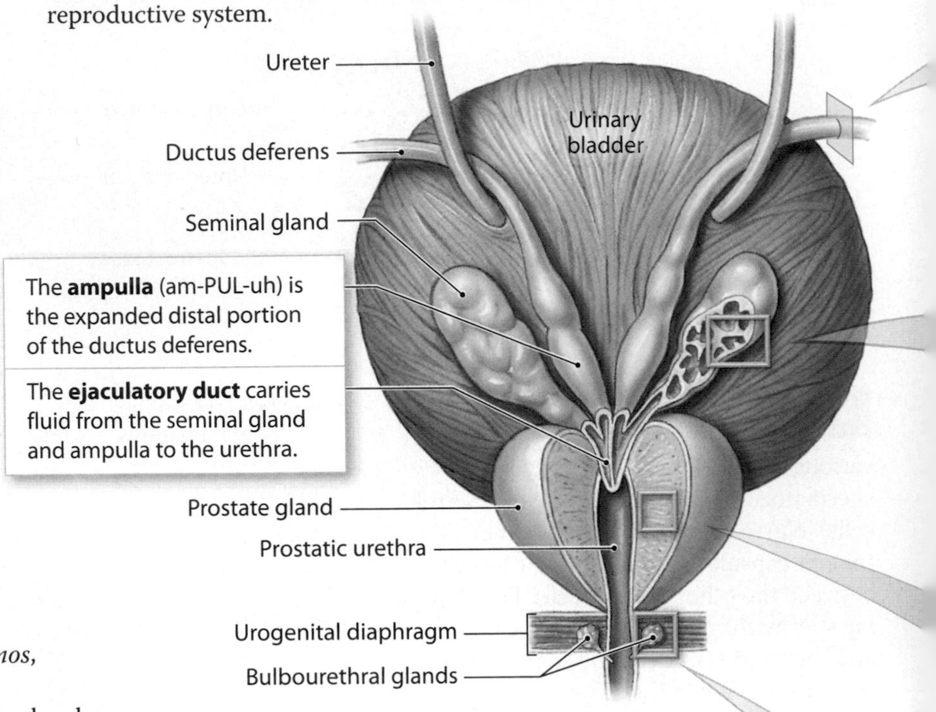

Ureter
Ductus deferens
Urinary bladder
Seminal gland
The **ampulla** (am-PUL-uh) is the expanded distal portion of the ductus deferens.
The **ejaculatory duct** carries fluid from the seminal gland and ampulla to the urethra.
Prostate gland
Prostatic urethra
Urogenital diaphragm
Bulbourethral glands

2 The **epididymis** (ep-i-DID-i-mis; *epi*, on + *didymos*, twin; plural, *epididymides*), the start of the male reproductive tract, is a coiled tube bound to the posterior border of each testis. A tubule almost 7 m (23 ft) long, the epididymis is coiled and twisted so as to take up very little space. It takes up to two weeks for a spermatozoon to pass through the epididymis and complete its functional maturation. During this time, spermatozoa exist in a sheltered environment that is precisely regulated by the surrounding epithelial cells.

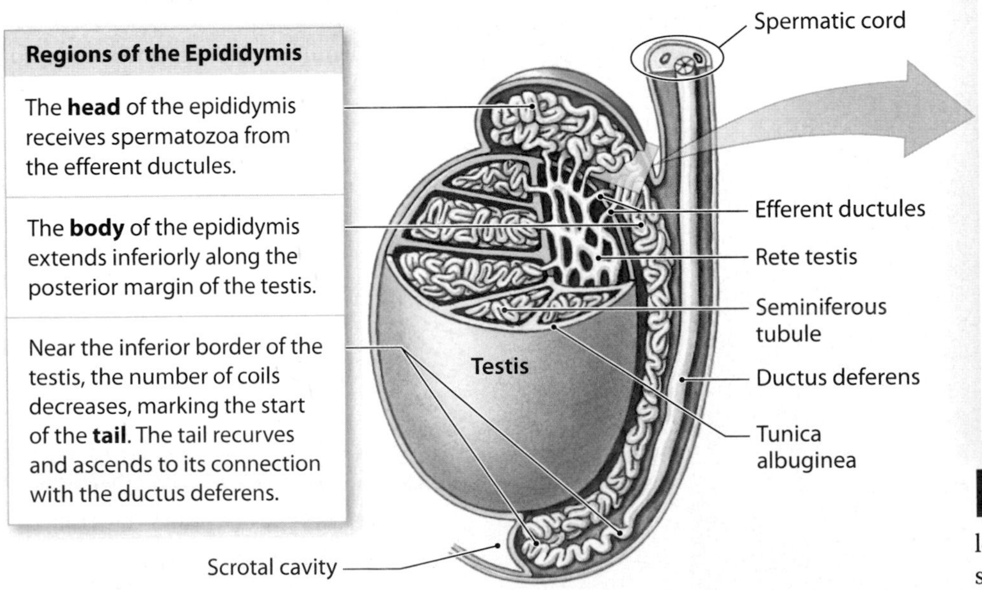

Regions of the Epididymis
The **head** of the epididymis receives spermatozoa from the efferent ductules.
The **body** of the epididymis extends inferiorly along the posterior margin of the testis.
Near the inferior border of the testis, the number of coils decreases, marking the start of the **tail**. The tail recurves and ascends to its connection with the ductus deferens.
Spermatic cord
Efferent ductules
Rete testis
Seminiferous tubule
Ductus deferens
Tunica albuginea
Testis
Scrotal cavity

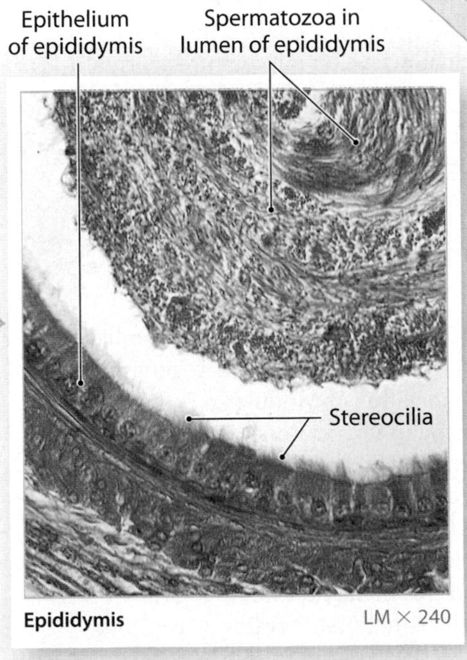

Epithelium of epididymis
Spermatozoa in lumen of epididymis
Stereocilia

Epididymis LM × 240

3 The pseudostratified columnar epithelial lining of the epididymis has extremely long **stereocilia**. These processes increase the surface area available for absorption from, and secretion into, the fluid in the lumen.

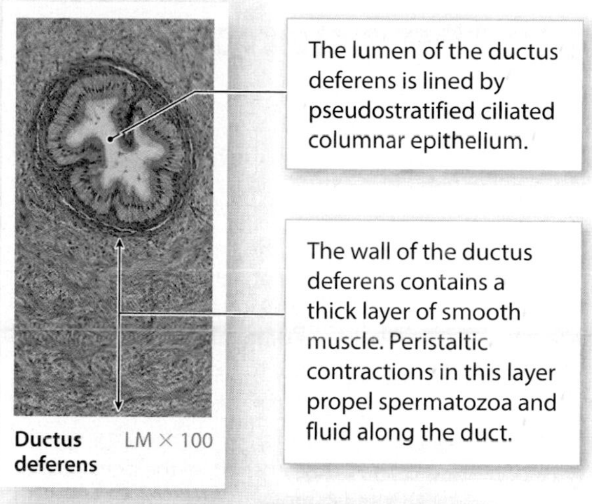
The lumen of the ductus deferens is lined by pseudostratified ciliated columnar epithelium.
The wall of the ductus deferens contains a thick layer of smooth muscle. Peristaltic contractions in this layer propel spermatozoa and fluid along the duct.
Ductus deferens
LM × 100

4 Each **ductus deferens**, or *vas deferens*, is 40–45 cm (16–18 in.) long. As part of the spermatic cord, it ascends through the inguinal canal, enters the abdominal cavity, and passes posterior to the urinary bladder to reach the prostate gland. In addition to transporting spermatozoa from the epididymis, the ductus deferens can store spermatozoa for several months. During this time, the spermatozoa remain in a state of suspended animation (a temporary state with stoppage of functions) and have low metabolic rates.

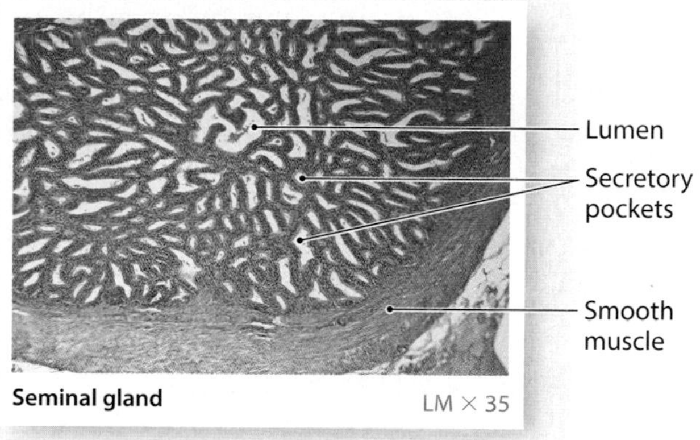
Lumen
Secretory pockets
Smooth muscle
Seminal gland
LM × 35

5 The **seminal glands**, also called the *seminal vesicles*, are embedded in connective tissue on either side of the midline, sandwiched between the posterior wall of the urinary bladder and the rectum. The seminal glands produce about 60 percent of the volume of **semen**. When mixed with the secretions of the seminal glands, previously inactive but functional spermatozoa undergo the first step in capacitation and begin beating their flagella, becoming highly motile. The secretions are ejected by smooth muscle contractions controlled by the sympathetic division of the ANS. About 2–5 mL of semen is released in a typical ejaculation. Semen contains spermatozoa and **seminal fluid**. In addition to seminal gland secretions, semen also contains secretions from nurse cells, the epididymis, the prostate gland, and the bulbourethral glands.

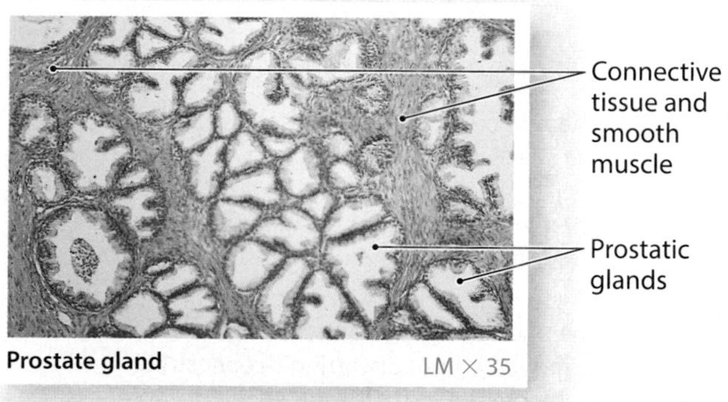
Connective tissue and smooth muscle
Prostatic glands
Prostate gland
LM × 35

6 The **prostate gland** is a small, muscular, rounded organ about 4 cm (1.6 in.) in diameter. The prostate gland encircles the proximal portion of the urethra as it leaves the urinary bladder. The glandular tissue of the prostate is surrounded by and wrapped in a thick blanket of smooth muscle fibers. The prostate gland produces 20–30 percent of the volume of semen. In addition to several other substances of uncertain significance, prostatic secretions contain **seminalplasmin** (sem-i-nal-PLAZ-min), an antibiotic protein that may help prevent urinary tract infections in males. These secretions are ejected into the prostatic urethra by peristaltic contractions of the muscular prostate wall.

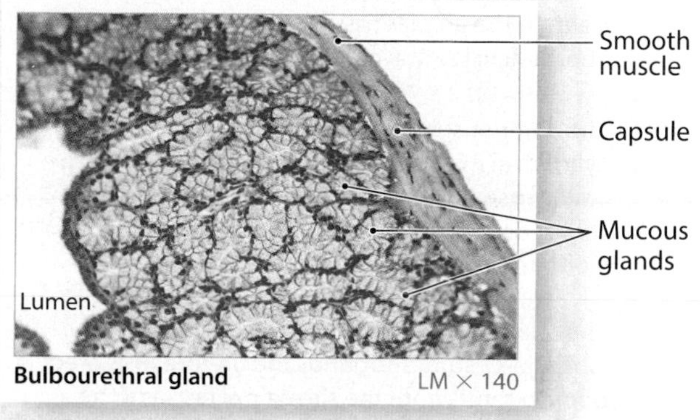
Smooth muscle
Capsule
Mucous glands
Lumen
Bulbourethral gland
LM × 140

7 The paired **bulbourethral glands** (Cowper's glands) are located at the base of the penis, covered by the fascia of the urogenital diaphragm. The duct of each gland empties into the urethra. These glands secrete a thick, alkaline mucus that helps neutralize any urinary acids that may remain in the urethra, and it also lubricates the tip of the penis.

Module 26.5 Review

a. Define semen.

b. What are the functions of the secretion of the bulbourethral glands?

c. Trace the ductal pathway from the epididymis to the urethra.

LO 26.5 Explain the roles played by the male reproductive tract and accessory glands in the functional maturation, nourishment, storage, and transport of spermatozoa.

The penis conducts urine and semen to the exterior

1 The penis has two functions. It conducts urine to the exterior and introduces semen into the female's vagina during sexual intercourse.

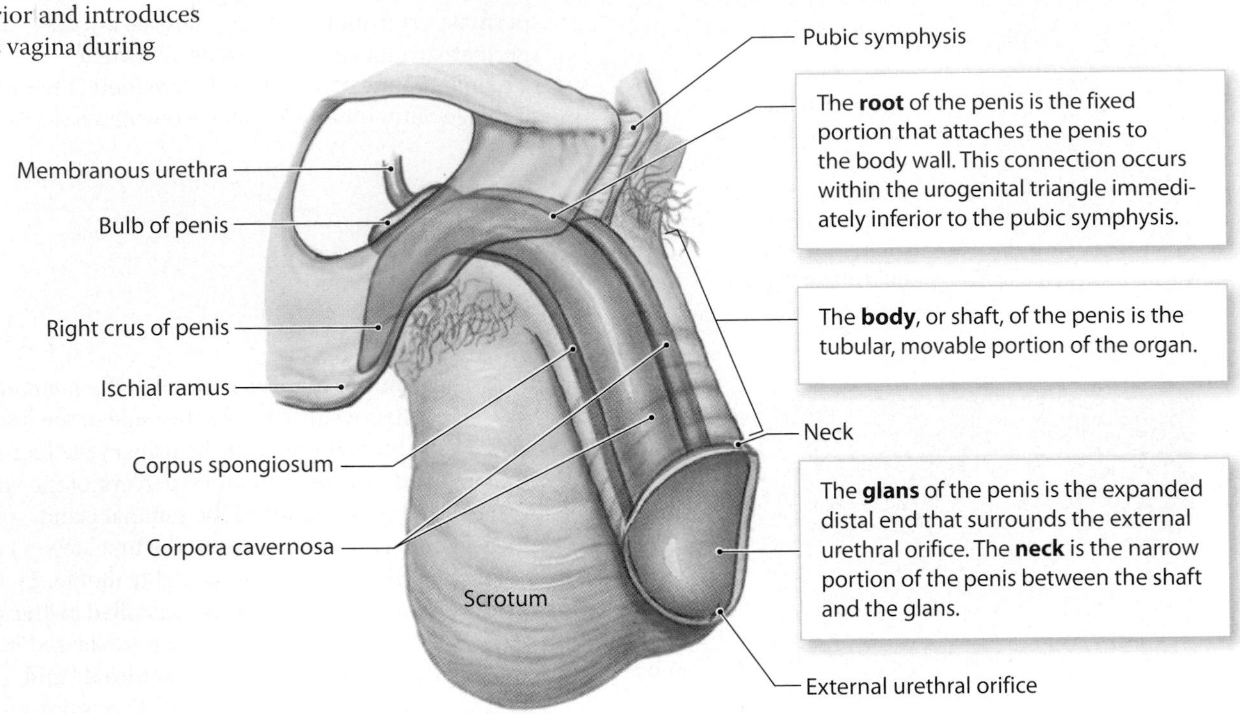
Pubic symphysis
The **root** of the penis is the fixed portion that attaches the penis to the body wall. This connection occurs within the urogenital triangle immediately inferior to the pubic symphysis.
Membranous urethra
Bulb of penis
The **body**, or shaft, of the penis is the tubular, movable portion of the organ.
Right crus of penis
Ischial ramus
Neck
Corpus spongiosum
The **glans** of the penis is the expanded distal end that surrounds the external urethral orifice. The **neck** is the narrow portion of the penis between the shaft and the glans.
Corpora cavernosa
Scrotum
External urethral orifice

2 This cross section shows that most of the body of the penis consists of three cylindrical columns of vascularized **erectile tissue**. Erectile tissue consists of a three-dimensional network with vascular spaces incompletely separated by partitions of elastic connective tissue and smooth muscle fibers. In the resting state, the arterial branches are constricted, and the muscular partitions are tense. This combination restricts blood flow into the erectile tissue.

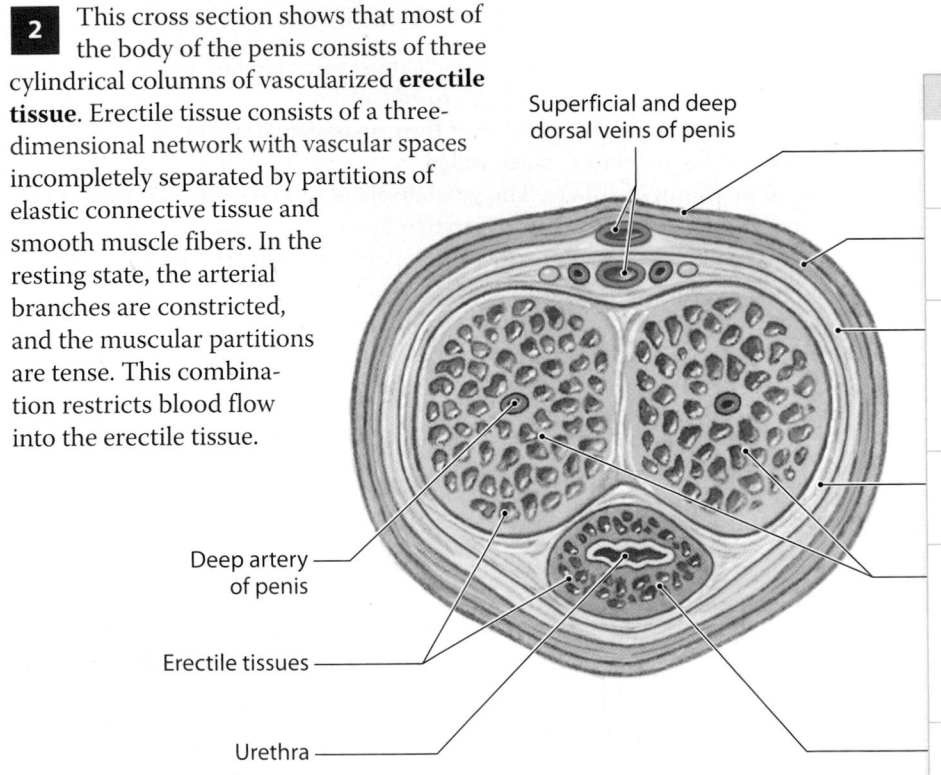

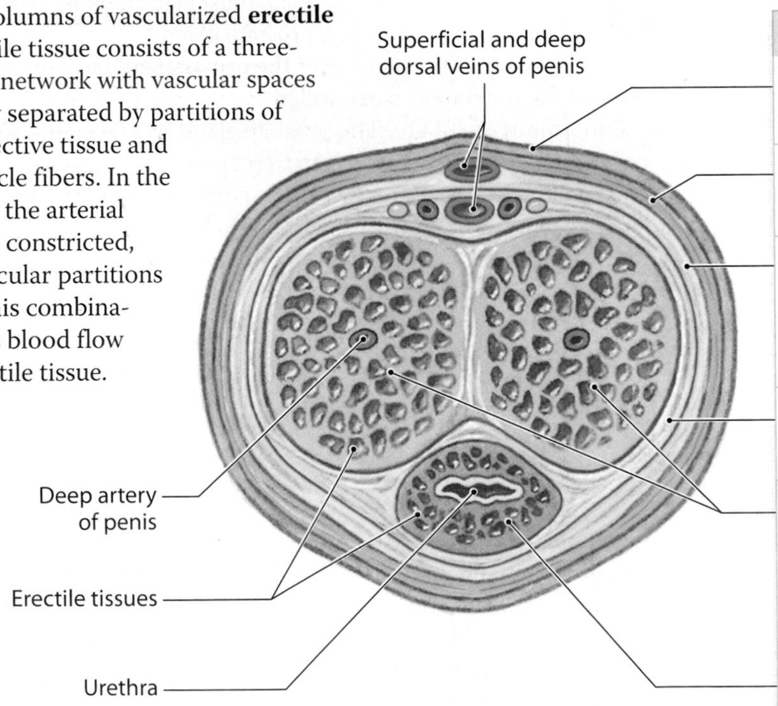
Superficial and deep dorsal veins of penis
Deep artery of penis
Erectile tissues
Urethra

Tissue Layers of the Penis

The skin overlying the penis resembles that of the the skin on the scrotum.

The dermis has a layer of smooth muscle that is a continuation of the dartos muscle of the scrotum.

The underlying areolar tissue allows the thin skin to move without distorting deeper structures. The areolar tissue also contains superficial arteries, veins, and lymphatic vessels.

Deep to the areolar tissue, a dense network of elastic fibers encircles the internal structures of the penis.

The anterior surface of the flaccid penis covers two cylindrical masses of erectile tissue: the **corpora cavernosa** (KOR-por-a ka-ver-NŌ-suh; singular, *corpus cavernosum*). The corpora cavernosa extend along the length of the penis as far as its neck.

The relatively slender **corpus spongiosum** (spon-jē-Ō-sum) surrounds the urethra. This erectile body extends from the superficial fascia of the urogenital diaphragm to the tip of the penis, where it expands to form the glans.

3 This diagrammatic view of the male reproductive tract provides a three-dimensional perspective on the relationships among the glands and passageways. The numbers will be useful as you follow the phases of sexual response in **4**.

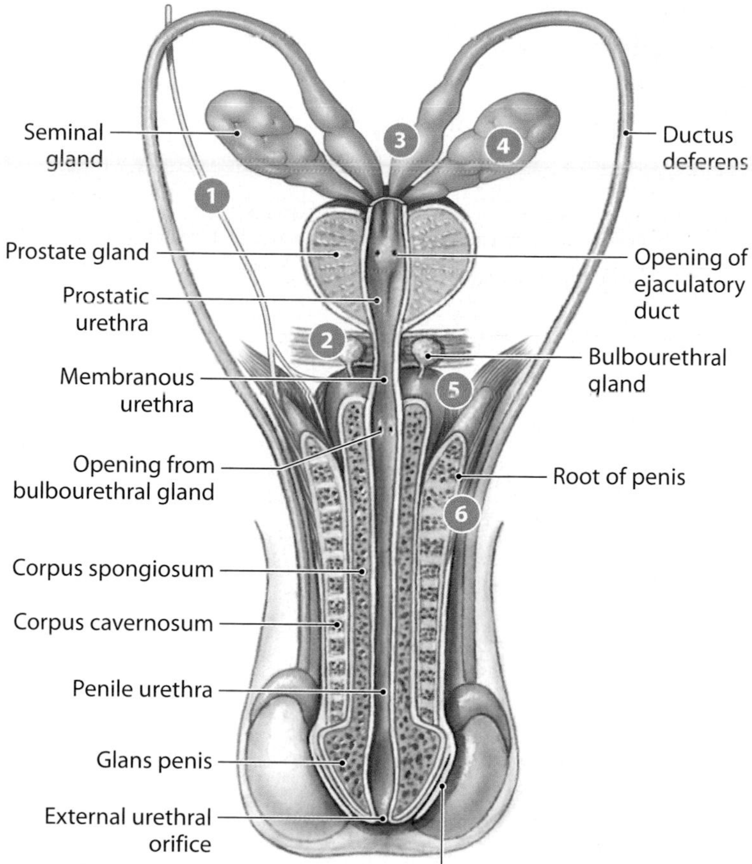

Seminal gland
Prostate gland
Prostatic urethra
Membranous urethra
Opening from bulbourethral gland
Corpus spongiosum
Corpus cavernosum
Penile urethra
Glans penis
External urethral orifice
Ductus deferens
Opening of ejaculatory duct
Bulbourethral gland
Root of penis
1
2
3
4
5
6

A fold of skin called the **prepuce** (PRĒ-pūs), or foreskin, surrounds the tip of the penis. The prepuce attaches to the relatively narrow neck of the penis and continues over the glans penis. Glands in the skin of the neck and the inner surface of the prepuce secrete a waxy material known as **smegma** (SMEG-ma).

Impotence, or **erectile dysfunction (ED)**, is an inability to achieve or maintain an erection. There may be a variety of physical causes, because erection involves vascular changes as well as neural commands. For example, low blood pressure in the arteries of the penis, due to a cardiovascular blockage such as a plaque, can impair the ability to achieve an erection. Psychological factors such as depression or anxiety can also result in impotence. The erection pathway activated by nitric oxide (NO) is opposed by enzymes that cause constriction of the arteries supplying the erectile tissues. The drugs Viagra and Cialis temporarily inactivate these enzymes, so that even small levels of NO are sufficient to produce an erection.

4 The three physiological phases in the male sexual response are summarized here.

Arousal

During **arousal**, erotic thoughts or stimulation of sensory nerves in the genital region leads to an increase in the parasympathetic outflow over the pelvic nerves.

1. The parasympathetic innervation of the penile arteries involves neurons that release **nitric oxide** (NO) at their axon terminals. The smooth muscles in the arterial walls relax when NO is released, at which time the vessels dilate, blood flow increases, the vascular channels become engorged with blood, and **erection** of the penis occurs.

2. Arousal also stimulates the bulbourethral glands. Their secretion lubricates the tip of the glans penis.

Emission

Further stimulation leads to sympathetic activation that causes **emission**.

3. Emission begins with peristaltic contractions in the ampullae of the ductus deferens. This pushes spermatozoa into the prostatic urethra.

4. Contractions then begin in the walls of the seminal glands and the prostate gland, and their secretions now enter the urethra and mix with the spermatozoa introduced by ampullary contractions to form **semen**.

Ejaculation

Ejaculation occurs as powerful, rhythmic contractions take place in the ischiocavernosus and bulbocavernosus muscles. These contractions are controlled by somatic motor neurons in the lower lumbar and upper sacral segments of the spinal cord. These contractions are associated with the pleasurable sensations known as **male orgasm**.

5. The bulbocavernosus muscles wrap around the base of the penis, and their contractions push semen toward the external urethral orifice.

6. The ischiocavernosus muscles insert along the sides of the penis, and their contractions serve primarily to stiffen the erect penis.

Module 26.6 Review

a. Name the three columns of erectile tissue in the penis.
b. List the phases of the male sexual response.
c. An inability to contract the ischiocavernosus and bulbospongiosus muscles would interfere with which phase of the male sexual response?

LO 26.6 Describe the structure and function of the penis.

Testosterone plays a key role in establishing and maintaining male sexual function

Testosterone is produced primarily by the interstitial cells of the testes. (Small amounts are also produced by the zona reticularis of the adrenal glands in both sexes [Module 16.9, p. 602].) This flowchart diagrams the hormonal interactions that regulate male reproductive function.

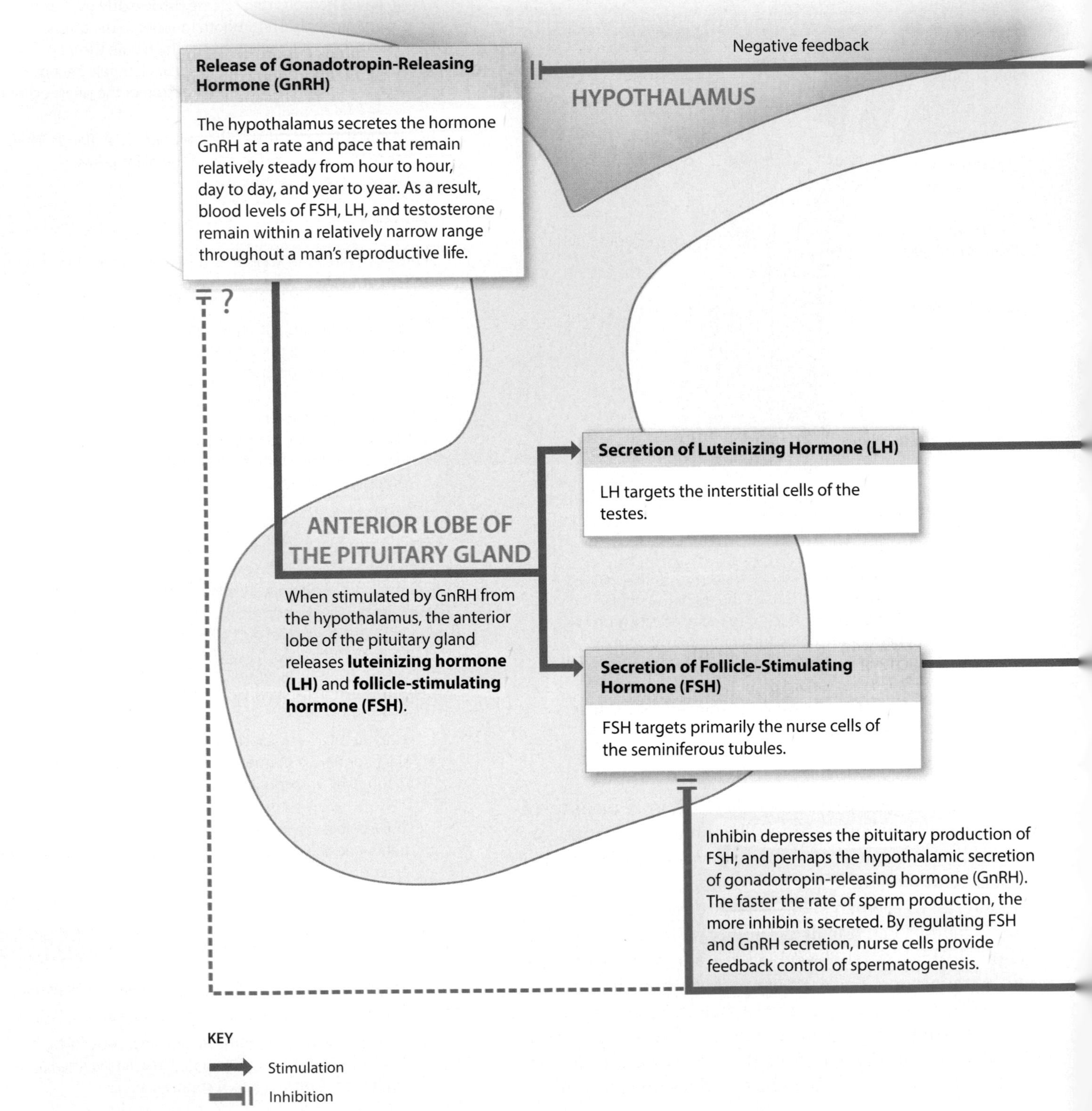
Negative feedback
Release of Gonadotropin-Releasing Hormone (GnRH)
The hypothalamus secretes the hormone GnRH at a rate and pace that remain relatively steady from hour to hour, day to day, and year to year. As a result, blood levels of FSH, LH, and testosterone remain within a relatively narrow range throughout a man's reproductive life.
HYPOTHALAMUS
?
ANTERIOR LOBE OF THE PITUITARY GLAND
When stimulated by GnRH from the hypothalamus, the anterior lobe of the pituitary gland releases **luteinizing hormone (LH)** and **follicle-stimulating hormone (FSH)**.
Secretion of Luteinizing Hormone (LH)
LH targets the interstitial cells of the testes.
Secretion of Follicle-Stimulating Hormone (FSH)
FSH targets primarily the nurse cells of the seminiferous tubules.
Inhibin depresses the pituitary production of FSH, and perhaps the hypothalamic secretion of gonadotropin-releasing hormone (GnRH). The faster the rate of sperm production, the more inhibin is secreted. By regulating FSH and GnRH secretion, nurse cells provide feedback control of spermatogenesis.
KEY
Stimulation
Inhibition

In many target tissues, some of the arriving testosterone is converted to **dihydrotestosterone (DHT)**. A small amount of DHT diffuses back out of the cell and into the bloodstream, and DHT levels are usually about 10 percent of circulating testosterone levels. Dihydrotestosterone can also enter peripheral cells and bind to the same hormone receptors targeted by testosterone. In addition, some tissues (notably those of the external genitalia) respond to DHT rather than to testosterone, and other tissues (including the prostate gland) are more sensitive to DHT than to testosterone.

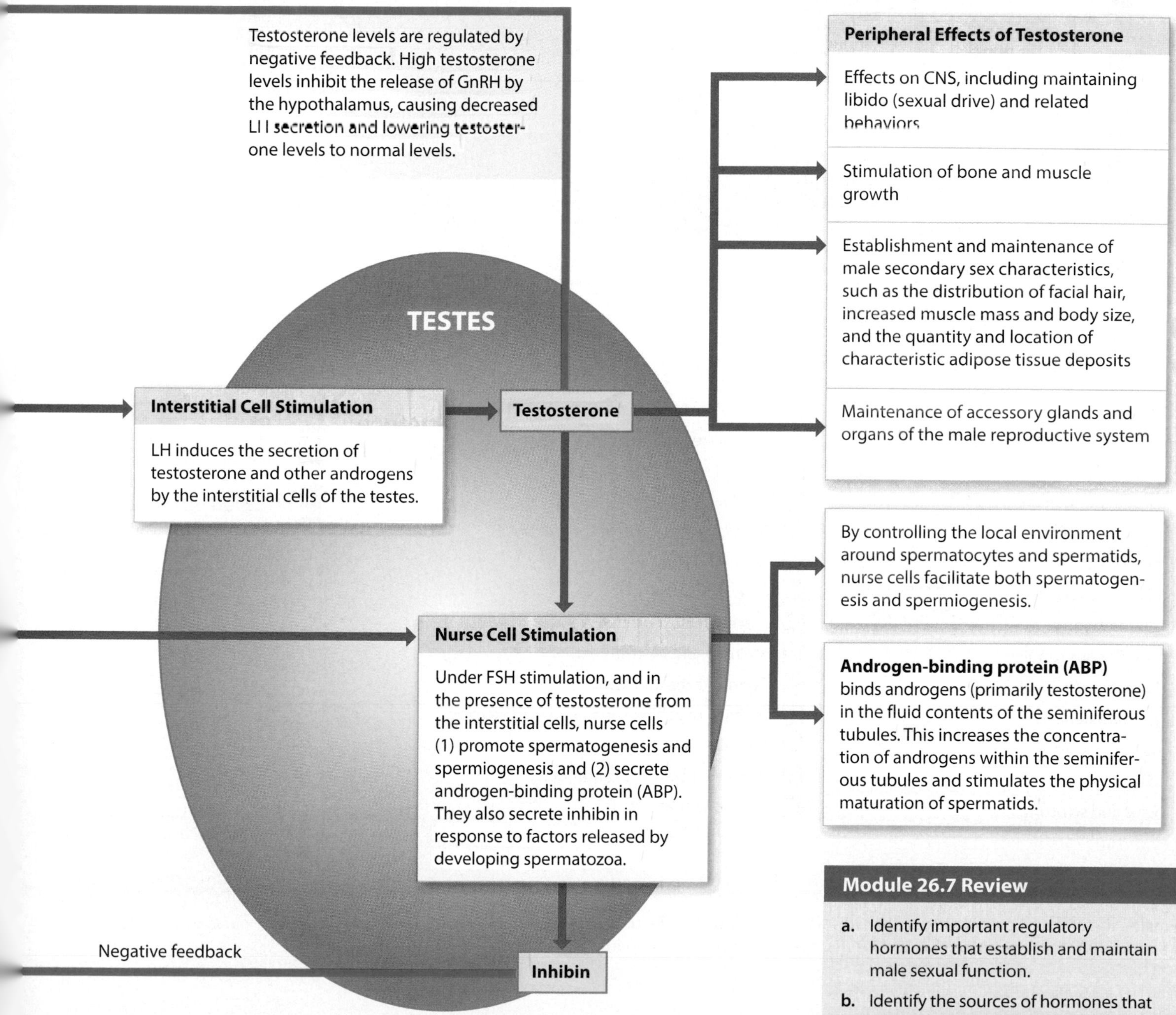

Testosterone levels are regulated by negative feedback. High testosterone levels inhibit the release of GnRH by the hypothalamus, causing decreased LH secretion and lowering testosterone levels to normal levels.
TESTES
Interstitial Cell Stimulation
LH induces the secretion of testosterone and other androgens by the interstitial cells of the testes.
Testosterone
Peripheral Effects of Testosterone
Effects on CNS, including maintaining libido (sexual drive) and related behaviors
Stimulation of bone and muscle growth
Establishment and maintenance of male secondary sex characteristics, such as the distribution of facial hair, increased muscle mass and body size, and the quantity and location of characteristic adipose tissue deposits
Maintenance of accessory glands and organs of the male reproductive system
Nurse Cell Stimulation
Under FSH stimulation, and in the presence of testosterone from the interstitial cells, nurse cells (1) promote spermatogenesis and spermiogenesis and (2) secrete androgen-binding protein (ABP). They also secrete inhibin in response to factors released by developing spermatozoa.
By controlling the local environment around spermatocytes and spermatids, nurse cells facilitate both spermatogenesis and spermiogenesis.
Androgen-binding protein (ABP) binds androgens (primarily testosterone) in the fluid contents of the seminiferous tubules. This increases the concentration of androgens within the seminiferous tubules and stimulates the physical maturation of spermatids.
Negative feedback
Inhibin

Module 26.7 Review

a. Identify important regulatory hormones that establish and maintain male sexual function.

b. Identify the sources of hormones that control male reproductive functions.

c. What effect would low FSH levels have on sperm production?

26.7 Explain the roles of regulatory hormones and testosterone in the establishment and maintenance of male sexual function.

Labeling

Label the structures of the male reproductive system in the accompanying diagram.

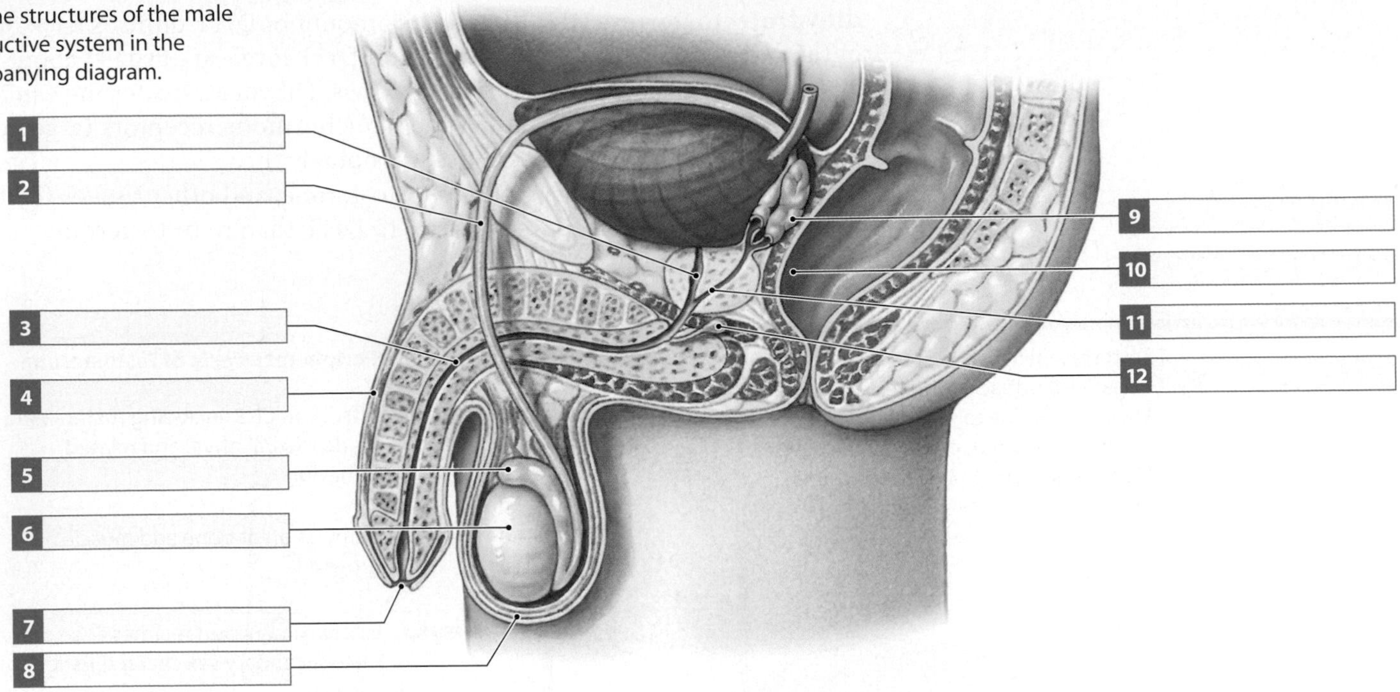
1
2
3
4
5
6
7
8
9
10
11
12

Matching

Match each lettered term with the most closely related description.

a. semen
b. epididymis
c. nurse cells
d. corpus spongiosum
e. luteinizing hormone (LH)
f. impotence
g. follicle-stimulating hormone (FSH)
h. spermatogonia
i. seminiferous tubules
j. dartos muscle
k. spermatogenesis
l. interstitial cells
m. spermiogenesis
n. penis and scrotum

13 Scrotal smooth muscle
14 Sperm stem cells
15 Sites of sperm production
16 Produce testosterone
17 Physical maturation of spermatids
18 Sperm production
19 External genitalia
20 Start of male reproductive tract
21 Maintain blood–testis barrier
22 Spermatozoa and seminal fluid
23 Inability to achieve or maintain an erection
24 Erectile tissue surrounding the urethra
25 Induces secretion of androgens
26 Hormone that targets nurse cells

13 ______
14 ______
15 ______
16 ______
17 ______
18 ______
19 ______
20 ______
21 ______
22 ______
23 ______
24 ______
25 ______
26 ______

Section integration

In males, the endocrine disorder hypogonadism is primarily due to the underproduction of testosterone or the lack of tissue sensitivity to testosterone, and results in sterility. What are five primary functions of testosterone in males?

27 ______

Female reproductive structures are mammary glands, gonads, external genitalia, and the reproductive tract

A woman's reproductive system produces sex hormones and functional gametes, and it must also be able to protect and support a developing embryo, maintain a growing fetus, and nourish a newborn infant. The main organs of the female reproductive system are the ovaries, uterine tubes, uterus, vagina, and components of the external genitalia. The female reproductive system also includes accessory organs—the mammary glands—and a variety of smaller accessory glands that secrete their products into the female reproductive tract.

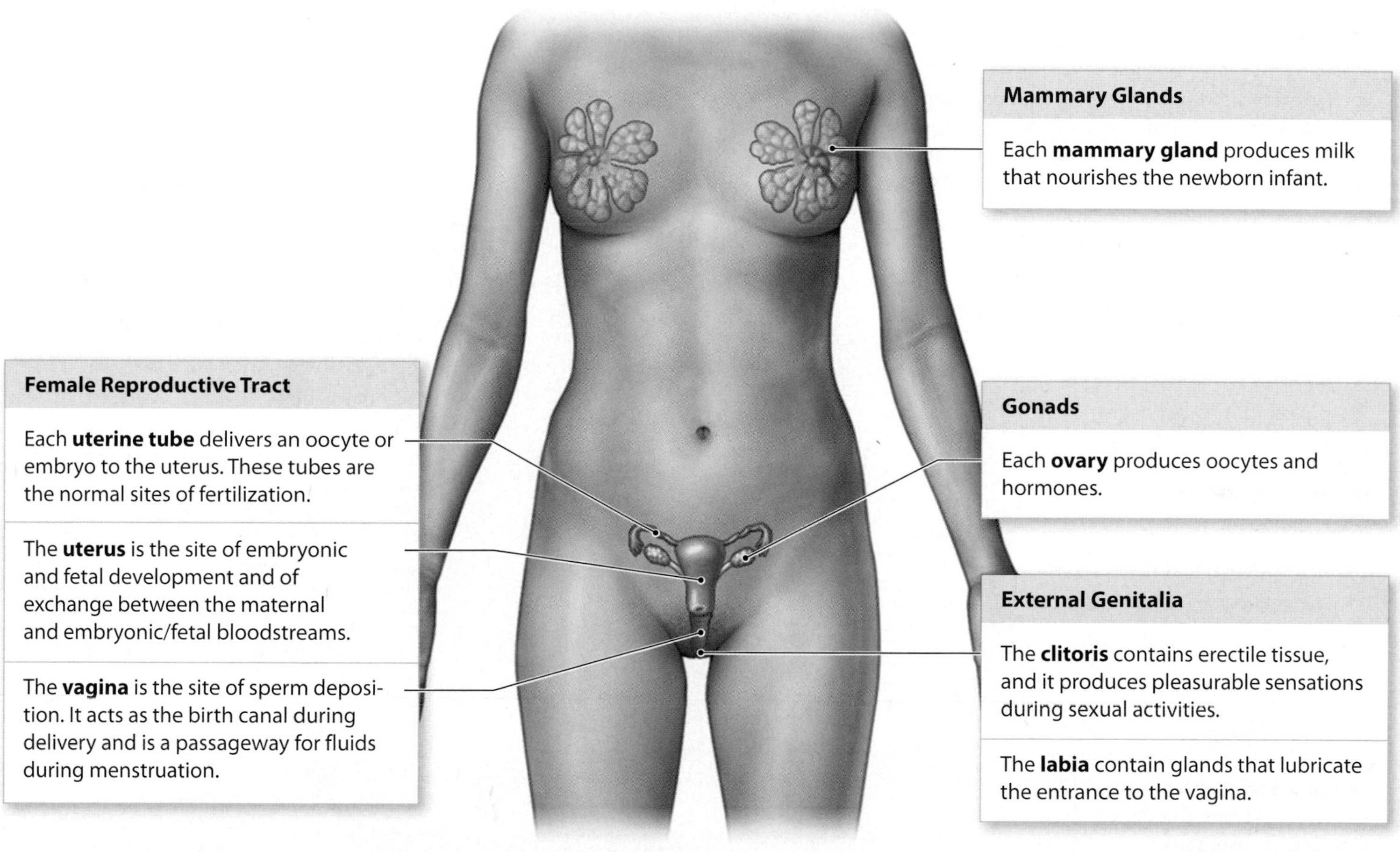
Mammary Glands
Each **mammary gland** produces milk that nourishes the newborn infant.
Female Reproductive Tract
Each **uterine tube** delivers an oocyte or embryo to the uterus. These tubes are the normal sites of fertilization.
The **uterus** is the site of embryonic and fetal development and of exchange between the maternal and embryonic/fetal bloodstreams.
The **vagina** is the site of sperm deposition. It acts as the birth canal during delivery and is a passageway for fluids during menstruation.
Gonads
Each **ovary** produces oocytes and hormones.
External Genitalia
The **clitoris** contains erectile tissue, and it produces pleasurable sensations during sexual activities.
The **labia** contain glands that lubricate the entrance to the vagina.

This figure gives an overview of the structures of the female reproductive system. The gonads in females, called ovaries (singular, *ovary*), produce immature female gametes called **oocytes** (Ō-ō-sīts), which later mature into **ova** (singular, *ovum*). Oocytes leave the ovary and then travel along the female duct system, or **female reproductive tract**. If fertilization occurs, it will normally occur in the uterine tubes and further embryonic development will occur within the uterus.

Module 26.8 Review

a. Identify the main organs of the female reproductive system.

b. Name the structures of the female external genitalia.

c. Where does fertilization normally occur?

LO **26.8** Identify the structures of the female reproductive system, and cite their functions.

Module 26.9

The ovaries and the female reproductive tract are in close proximity but are not directly connected

1 This sagittal section through the pelvic cavity shows the location and orientation of the female reproductive organs.

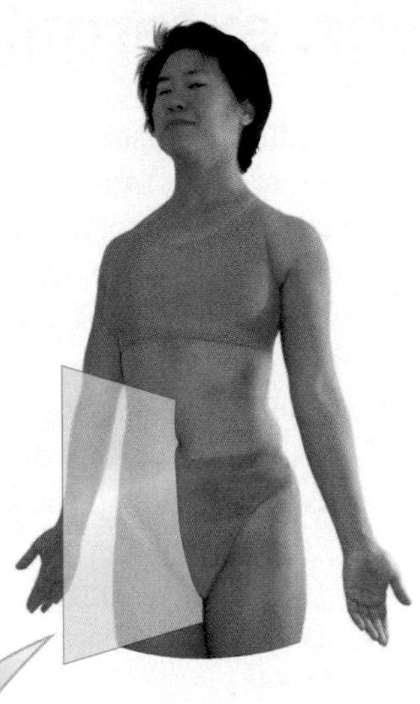

The paired **ovaries** are small, lumpy, almond-shaped organs near the lateral walls of the pelvic cavity. The ovaries have three main functions: (1) production of immature female gametes, or **oocytes**; (2) secretion of female sex hormones, including **estrogens** and **progesterone**; and (3) secretion of **inhibin**, involved in the feedback control of pituitary FSH production. (Estrogens include estradiol, estriol, and estrone; estradiol is the most abundant of the three types.)

Each **uterine tube** begins with an expanded funnel, called an **infundibulum**, that is open into the pelvic cavity along the medial surface of the ovary. The other end of the uterine tube opens into the uterine cavity.

The **uterus** sits inferior to the ovaries, usually angled anteriorly above the urinary bladder.

The pocket formed between the uterus and the posterior wall of the urinary bladder is the **vesicouterine** (ves-i-kō-Ū-ter-in) **pouch**.

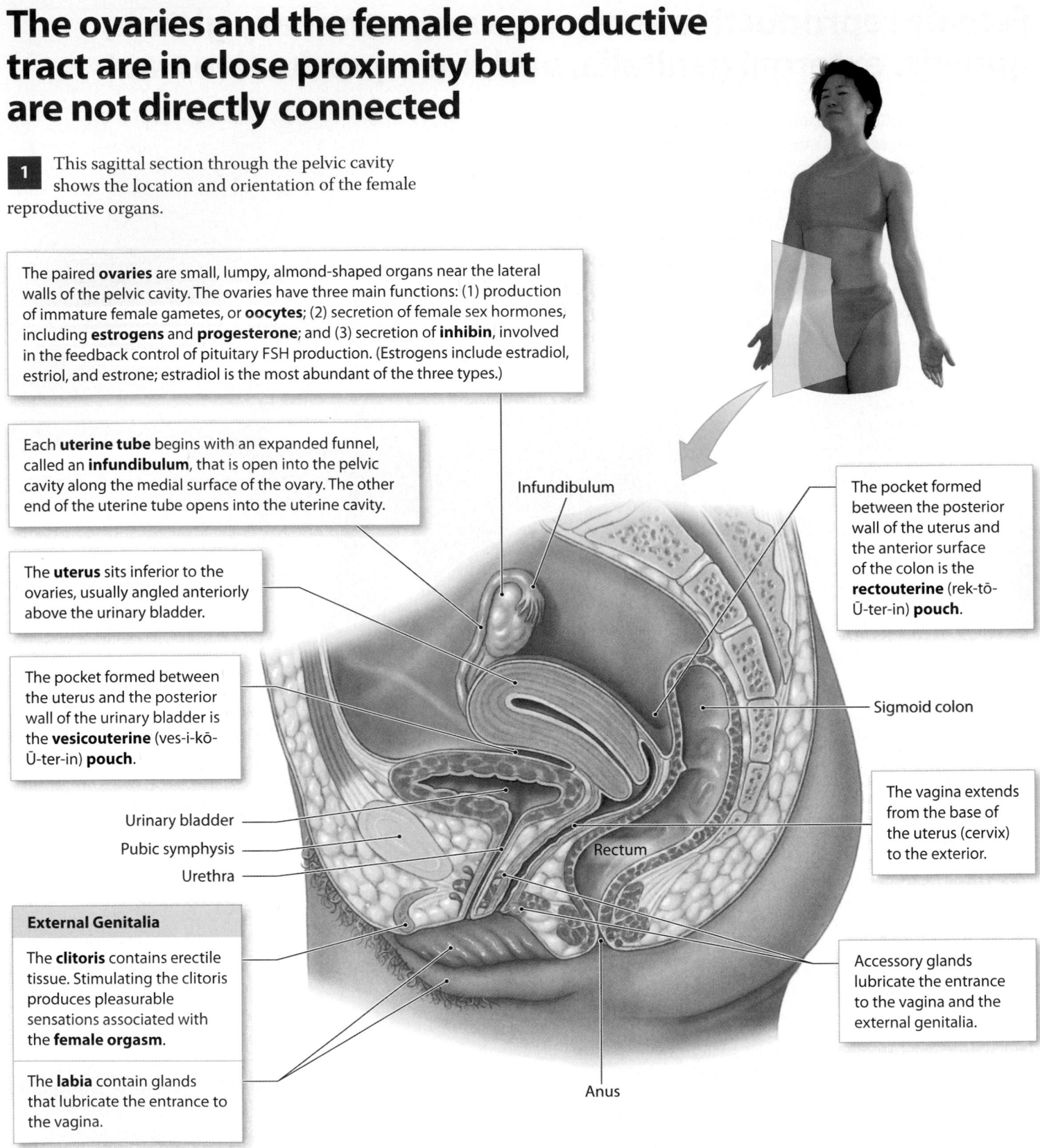

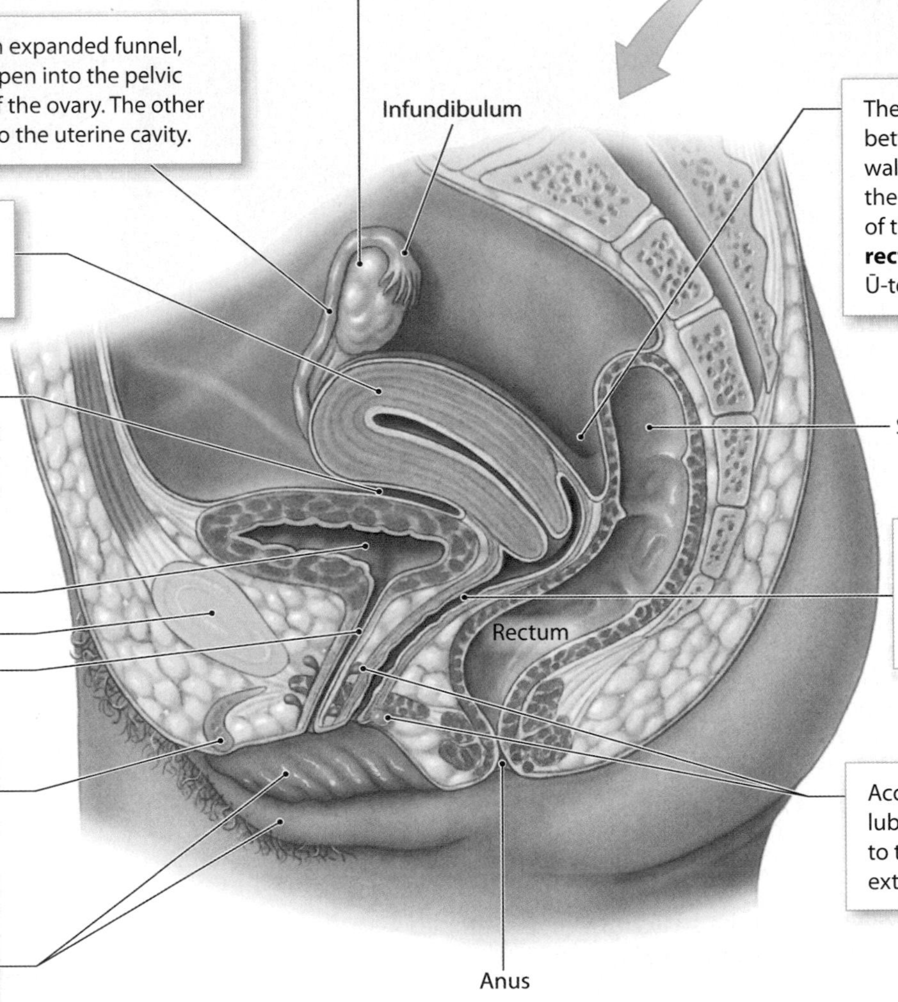

Infundibulum
Sigmoid colon
Urinary bladder
Pubic symphysis
Urethra
Rectum
Anus

The pocket formed between the posterior wall of the uterus and the anterior surface of the colon is the **rectouterine** (rek-tō-Ū-ter-in) **pouch**.

The vagina extends from the base of the uterus (cervix) to the exterior.

Accessory glands lubricate the entrance to the vagina and the external genitalia.

External Genitalia

The **clitoris** contains erectile tissue. Stimulating the clitoris produces pleasurable sensations associated with the **female orgasm**.

The **labia** contain glands that lubricate the entrance to the vagina.

2 The position of each ovary is stabilized by several thickened peritoneal folds that are called *ligaments*. This is a view from above and behind, with the left uterine tube pulled away from the ovary to show the ligaments more clearly.

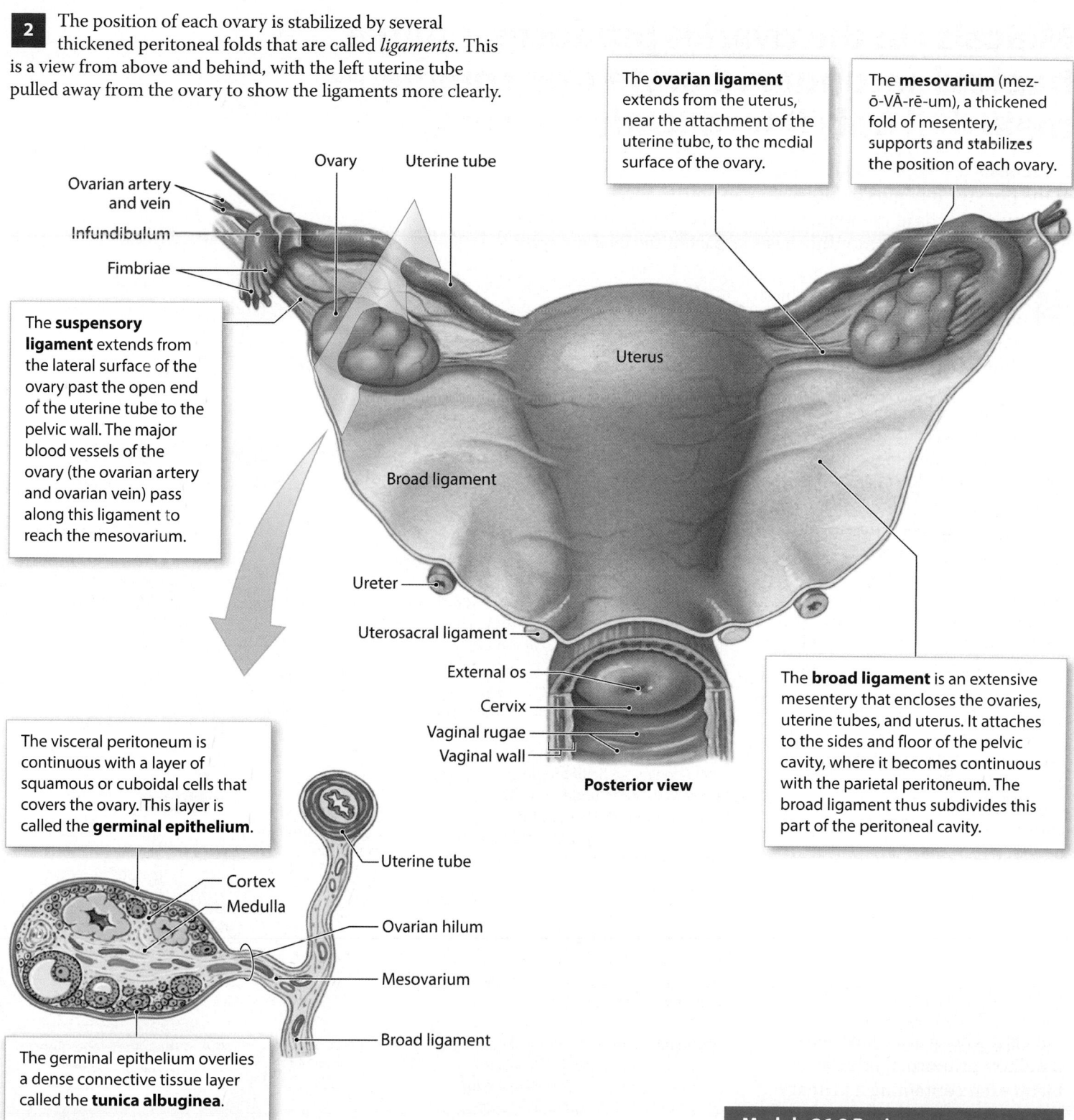

The **ovarian ligament** extends from the uterus, near the attachment of the uterine tube, to the medial surface of the ovary.

The **mesovarium** (mez-ō-VĀ-rē-um), a thickened fold of mesentery, supports and stabilizes the position of each ovary.

The **suspensory ligament** extends from the lateral surface of the ovary past the open end of the uterine tube to the pelvic wall. The major blood vessels of the ovary (the ovarian artery and ovarian vein) pass along this ligament to reach the mesovarium.

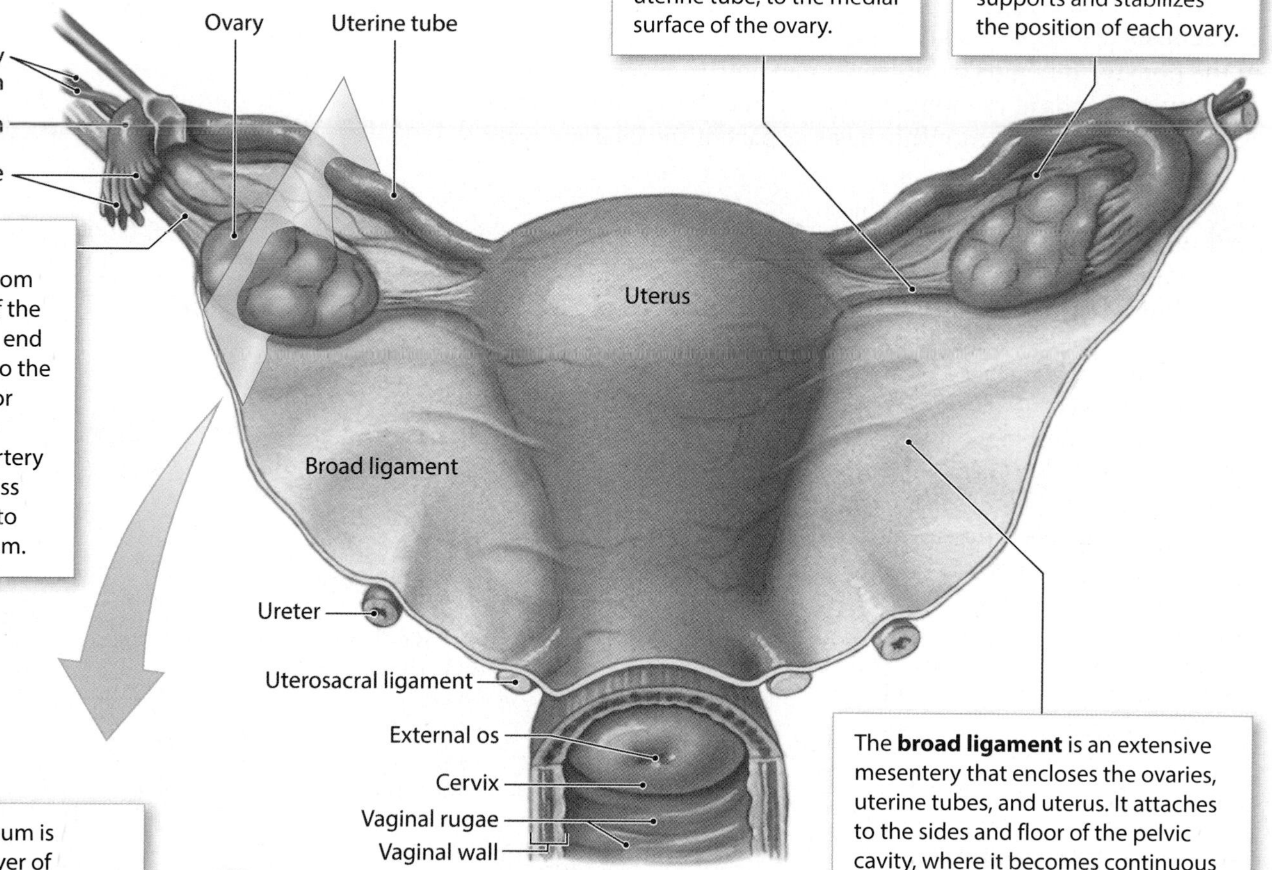
Ovary
Uterine tube
Ovarian artery and vein
Infundibulum
Fimbriae
Uterus
Broad ligament
Ureter
Uterosacral ligament
External os
Cervix
Vaginal rugae
Vaginal wall
Posterior view

The **broad ligament** is an extensive mesentery that encloses the ovaries, uterine tubes, and uterus. It attaches to the sides and floor of the pelvic cavity, where it becomes continuous with the parietal peritoneum. The broad ligament thus subdivides this part of the peritoneal cavity.

The visceral peritoneum is continuous with a layer of squamous or cuboidal cells that covers the ovary. This layer is called the **germinal epithelium**.

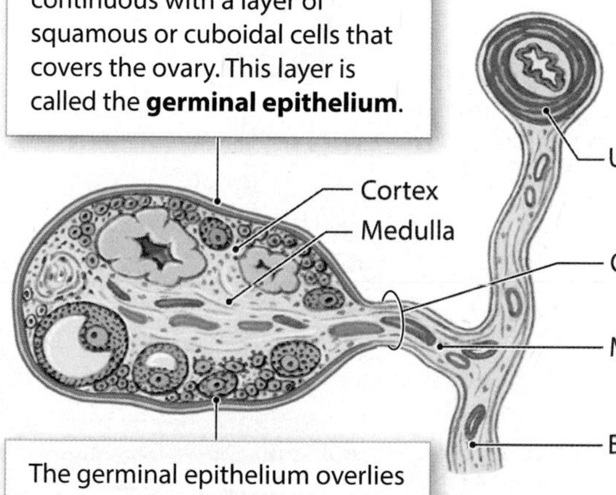
Cortex
Medulla
Uterine tube
Ovarian hilum
Mesovarium
Broad ligament

The germinal epithelium overlies a dense connective tissue layer called the **tunica albuginea**.

3 This is a cross section taken through the mesovarium and the broad ligament. A typical ovary is about 5 cm long, 2.5 cm wide, and 8 mm thick (2 in. by 1 in. by 0.33 in.) and weighs 6–8 g (roughly 0.25 oz). Blood vessels enter and leave the ovary at the ovarian hilum, where the ovary attaches to the mesovarium. The interior of the ovary is divided into a superficial **cortex** and a deeper **medulla**. Gametes are produced in the cortex.

Module 26.9 Review

a. What roles do the ovaries perform?

b. Distinguish between the vesicouterine and rectouterine pouches.

c. Name the structures enclosed by the broad ligament, and cite the function of the mesovarium.

LO 26.9 Describe the anatomy of the ovaries, uterus, and associated structures.

Module 26.10

Meiosis I in the ovaries produces a single haploid secondary oocyte that completes meiosis II only if fertilization occurs

Oogenesis (ō-ō-JEN-e-sis) is the formation and development of the oocyte. It begins before a woman's birth, accelerates at puberty, and ends at menopause. Between puberty and menopause, oogenesis occurs on a monthly basis as part of the ovarian cycle.

1 Although the nuclear events in the ovaries during meiosis are the same as those in the testes, the cytoplasm of the **primary oocyte** is unevenly distributed during the two meiotic divisions. Oogenesis produces one functional **secondary oocyte**, which contains most of the original cytoplasm, and two or three **polar bodies**, nonfunctional cells that later disintegrate. Another difference is that the ovary releases a secondary oocyte rather than a mature ovum. The secondary oocyte is suspended in metaphase of meiosis II. Meiosis will not be completed unless and until fertilization occurs.

Why is the secondary oocyte considered haploid even though it has the same amount of DNA as a diploid primary oocyte? It is haploid because each set of duplicate chromatids originated from only *one* member of a chromosome pair in the primary oocyte.

Not all primary oocytes produced during development survive until puberty. The ovaries have about 2 million primordial follicles at birth, each containing a primary oocyte. Primordial follicles exist at the earliest stages of development. By the time of puberty, the number has dropped to about 400,000. The rest of the primordial follicles degenerate in a process called **atresia** (a-TRĒ-zē-uh).

Oogenesis

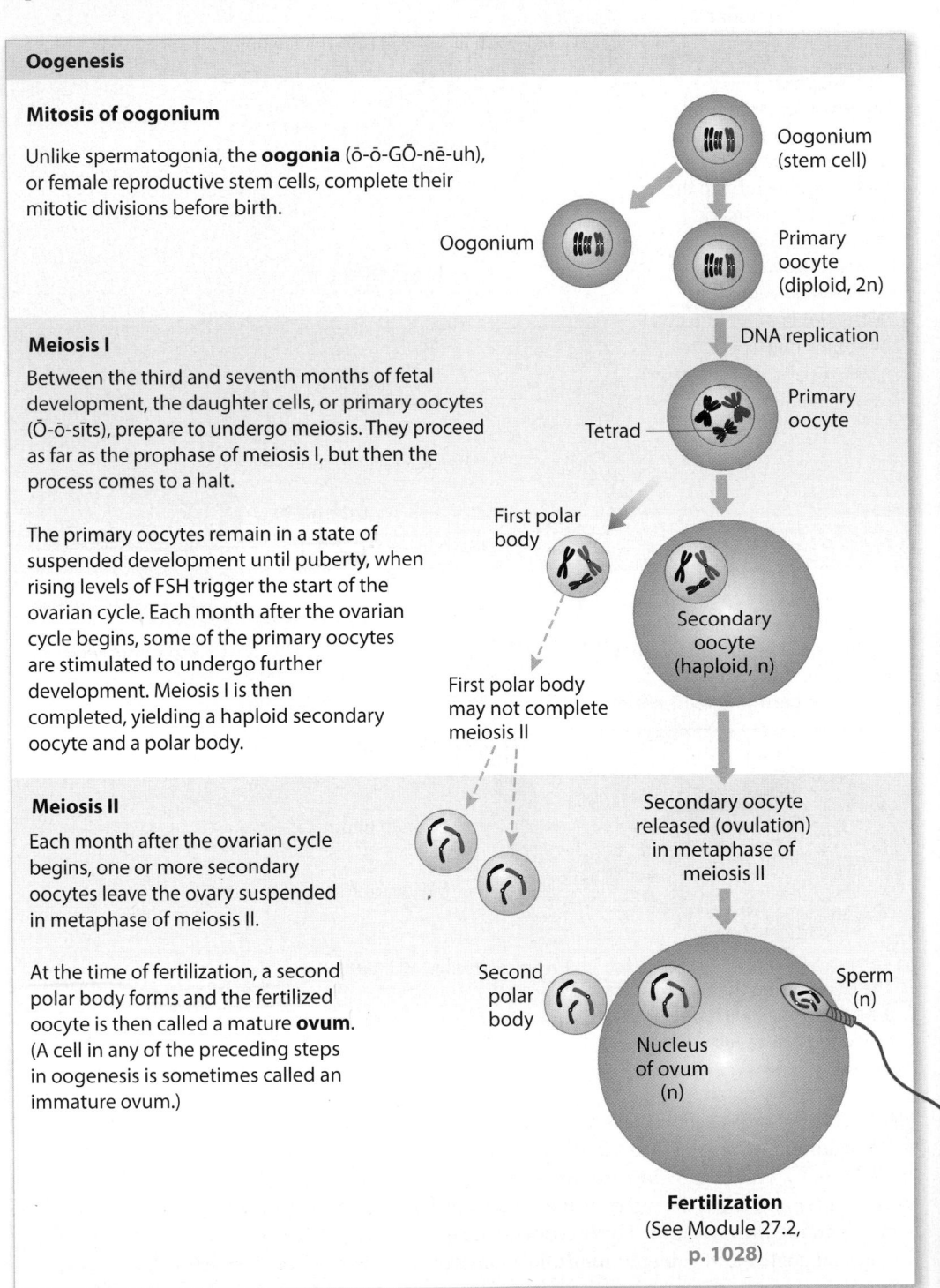

Mitosis of oogonium

Unlike spermatogonia, the **oogonia** (ō-ō-GŌ-nē-uh), or female reproductive stem cells, complete their mitotic divisions before birth.

Meiosis I

Between the third and seventh months of fetal development, the daughter cells, or primary oocytes (Ō-ō-sīts), prepare to undergo meiosis. They proceed as far as the prophase of meiosis I, but then the process comes to a halt.

The primary oocytes remain in a state of suspended development until puberty, when rising levels of FSH trigger the start of the ovarian cycle. Each month after the ovarian cycle begins, some of the primary oocytes are stimulated to undergo further development. Meiosis I is then completed, yielding a haploid secondary oocyte and a polar body.

Meiosis II

Each month after the ovarian cycle begins, one or more secondary oocytes leave the ovary suspended in metaphase of meiosis II.

At the time of fertilization, a second polar body forms and the fertilized oocyte is then called a mature **ovum**. (A cell in any of the preceding steps in oogenesis is sometimes called an immature ovum.)

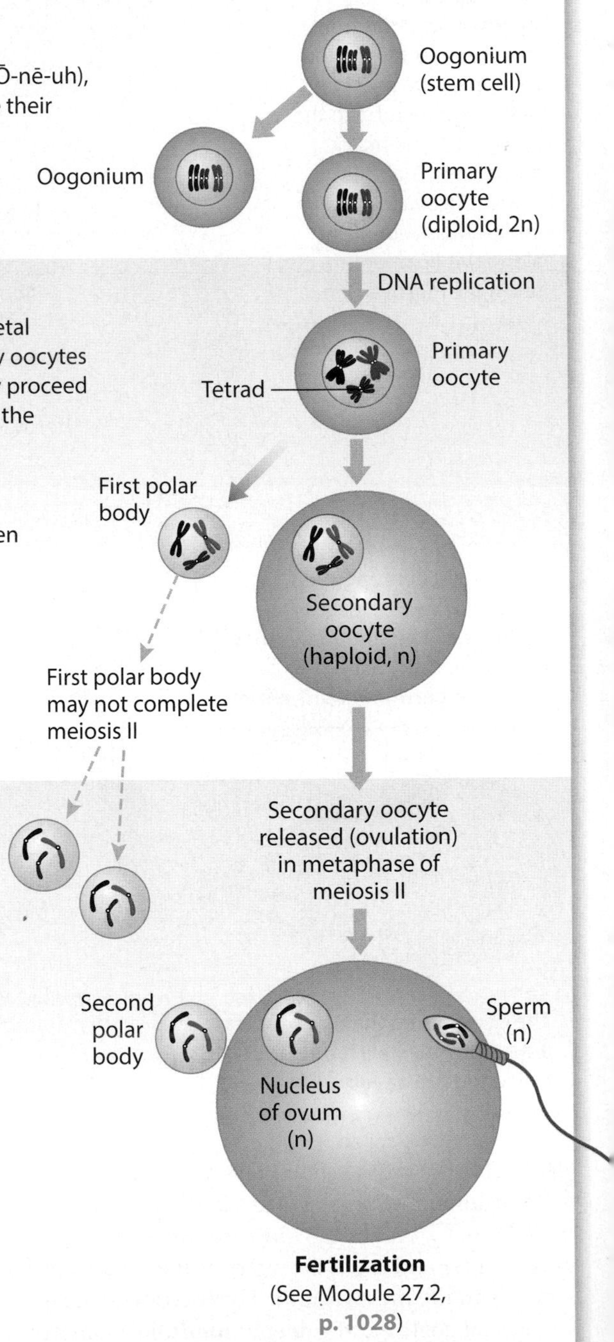
Oogonium (stem cell)
Oogonium
Primary oocyte (diploid, 2n)
DNA replication
Primary oocyte
Tetrad
First polar body
Secondary oocyte (haploid, n)
First polar body may not complete meiosis II
Secondary oocyte released (ovulation) in metaphase of meiosis II
Second polar body
Sperm (n)
Nucleus of ovum (n)
Fertilization
(See Module 27.2, p. 1028)

2 **Ovarian follicles** are structures where oocyte growth and meiosis I occur. During an ovarian cycle, the follicle gradually changes. Follicles do not physically move around the cortex, and this figure shows the relative sizes and sequence of events.

Primordial Follicles in Egg Nest

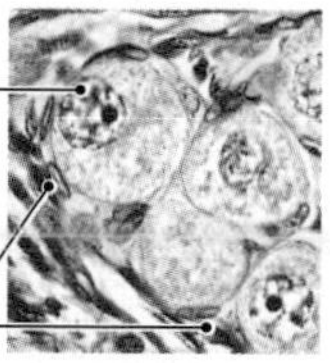
Primary oocyte
Follicle cells

Primary oocytes are located in the outer portion of the ovarian cortex, near the tunica albuginea, in clusters called **egg nests**. An inactive primary oocyte is surrounded by a simple squamous layer of follicle cells, forming a **primordial follicle**.

Formation of Primary Follicles

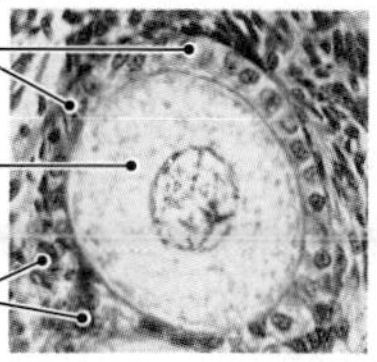
Granulosa cells
Primary oocyte
Thecal cells

Follicular cells enlarge, divide, and form several layers of cells around an activated primary oocyte. These follicle cells are now called **granulosa cells**. A region around the oocyte develops, called the **zona pellucida** (ZŌ-na pe-LŪ-si-duh; *pellucidus*, translucent). As the granulosa cells enlarge and multiply, a layer of **thecal cells** (*theca*, a box) forms around the follicle. Thecal cells and granulosa cells work together to produce estrogens.

Formation of Secondary Follicles

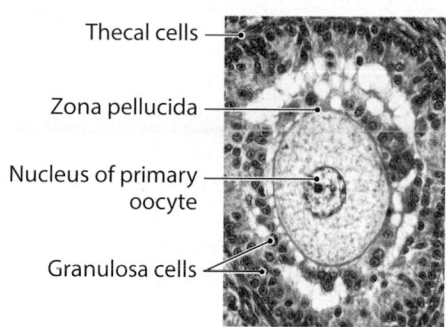

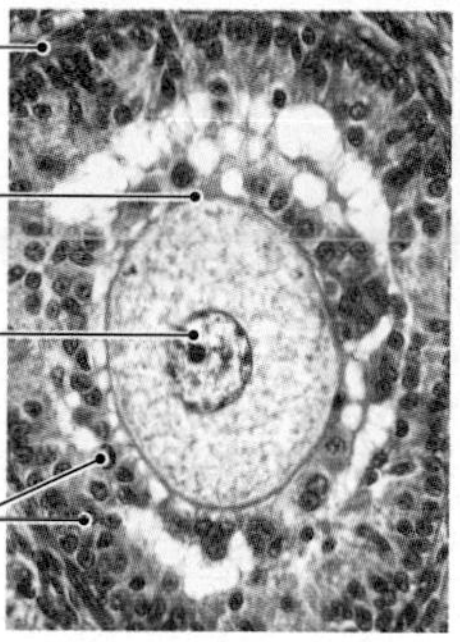
Thecal cells
Zona pellucida
Nucleus of primary oocyte
Granulosa cells

Secondary follicles develop as the wall of the follicle thickens and the deeper follicular cells begin secreting fluid that accumulates in small pockets. These pockets gradually expand and separate the inner and outer layers of the follicle.

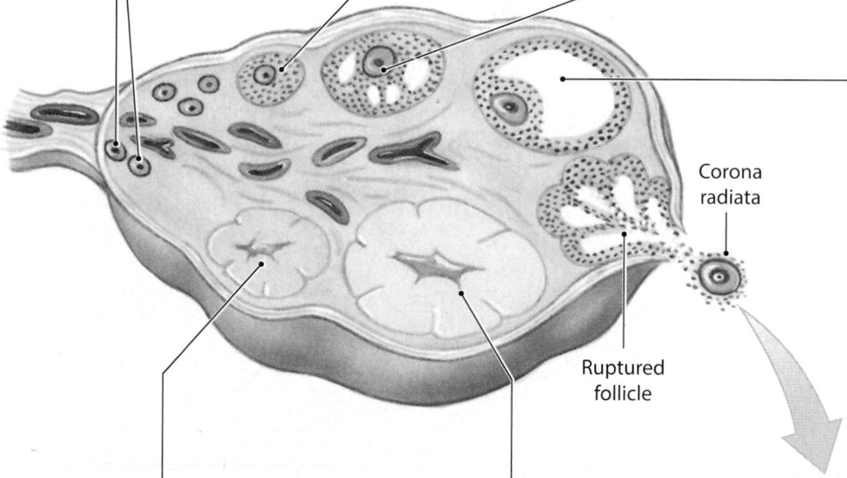
Corona radiata
Ruptured follicle

Formation of a Tertiary Follicle

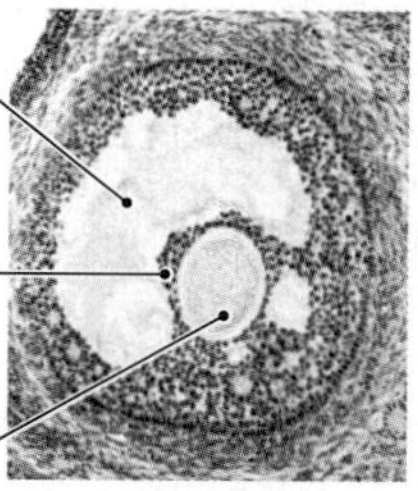
Antrum containing follicular fluid
Corona radiata
Secondary oocyte

By days 10–14 of the cycle, usually only one secondary follicle has become a **tertiary follicle**, or mature graafian (GRAH-fē-an) follicle, roughly 15 mm in diameter. The oocyte projects into the **antrum** (AN-trum), or expanded central chamber of the follicle. The granulosa cells associated with the secondary oocyte form a protective layer known as the **corona radiata** (kō-RŌ-nuh rā-dē-AH-tuh).

Ovulation

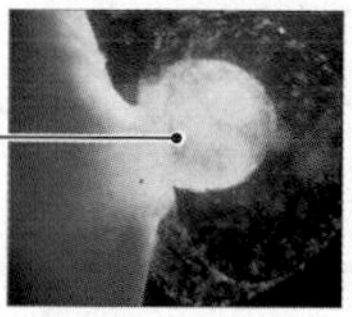
Secondary oocyte

At **ovulation**, the tertiary follicle releases the secondary oocyte and corona radiata into the pelvic cavity. Ovulation marks the end of the **follicular phase** of the ovarian cycle and the start of the **luteal phase**.

Formation of Corpus Luteum

The empty tertiary follicle initially collapses, and under LH stimulation the remaining granulosa cells proliferate to create the **corpus luteum** (LŪ-tē-um; *lutea*, yellow), which secretes progesterone (prō-JES-ter-on) and estrogens. Progesterone prepares the uterus for pregnancy by stimulating the maturation of the uterine lining and the secretions of uterine glands.

Formation of Corpus Albicans

If fertilization does not occur, after 12 days progesterone and estrogen levels fall markedly. Fibroblasts invade the nonfunctional corpus luteum, producing a knot of pale scar tissue called a **corpus albicans** (AL-bi-kanz). The disintegration, or involution, of the corpus luteum marks the end of the ovarian cycle. A new ovarian cycle then begins with the activation of another group of primordial follicles.

Module 26.10 Review

a. Define oocyte.

b. What are the main differences in gamete production between males and females?

c. List the important events in the ovarian cycle.

LO 26.10 Outline the processes of meiosis and oogenesis in the ovaries.

Module 26.11

The uterine tubes are connected to the uterus, a hollow organ with thick muscular walls

1 Each **uterine tube** (fallopian tube) is a hollow, muscular structure measuring approximately 13 cm (5.2 in.) in length. The distal portion of each uterine tube connects to the uterus. The **uterus** is a hollow, muscular organ that is about 7.5 cm (3 in.) long with a maximum diameter of 5 cm (2 in.). It weighs 30–40 g (1–1.4 oz). The sectional illustration below shows the internal structure of the uterine tube and the connection between the lumen of the uterine tube and the large uterine cavity within the uterus.

The thickness of the smooth muscle layers in the wall of the **ampulla**, the middle segment of the uterine tube, gradually increases as the tube approaches the uterus.

The ampulla leads to the **isthmus** (IS-mus) of the uterine tube, a short segment connected to the uterine wall.

The **infundibulum** is a funnel-like expansion of the uterine tube. It has numerous fingerlike projections that extend into the pelvic cavity. The projections are called **fimbriae** (FIM-brē-ē). Fimbriae drape over the surface of the ovary, but there is no physical connection between the two structures. The inner surface of the infundibulum are lined with cilia that beat toward the lumen of the uterine tube.

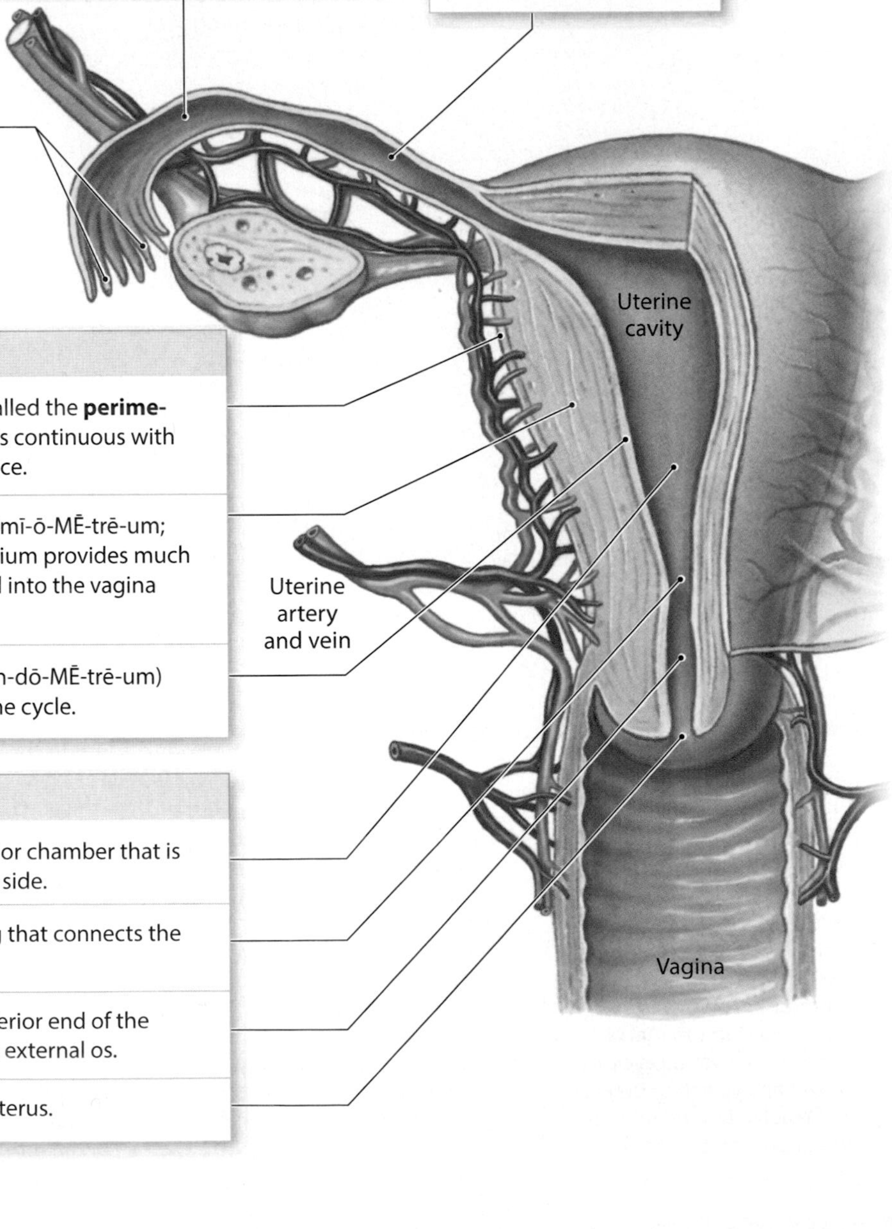

Uterine cavity
Uterine artery and vein
Vagina

Layers of the Uterine Wall

The outer surface of the uterus is an incomplete serosa called the **perimetrium** (per-i-mē-trē-um; *peri-*, around + *metra*, uterus). It is continuous with the peritoneal lining and covers most of the uterine surface.

The perimetrium covers a thick, muscular **myometrium** (mī-ō-MĒ-trē-um; *myo-*, muscle). The smooth muscle tissue of the myometrium provides much of the force needed to move a fetus out of the uterus and into the vagina during delivery.

The inner lining consists of a glandular **endometrium** (en-dō-MĒ-trē-um) whose tissue changes in the course of the monthly uterine cycle.

The Uterine Cavity

The **uterine cavity**, or **uterine lumen**, is the large, superior chamber that is continuous with the isthmus of the uterine tube on each side.

The **internal os** (*os*, an opening or mouth) is the opening that connects the uterine cavity to the cervical canal.

The **cervical canal** is a constricted passageway at the inferior end of the uterine cavity. It begins at the internal os and ends at the external os.

The **external os** is the curved vaginal opening into the uterus.

2 Concentric layers of smooth muscle surround the mucosa of the uterine tube. Oocyte transport along the tube involves a combination of ciliary movement and peristaltic contraction stimulated by autonomic nerves.

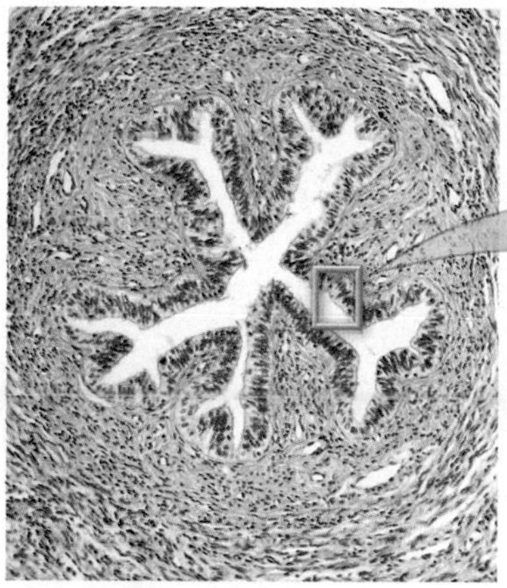

Isthmus of uterine tube LM × 100

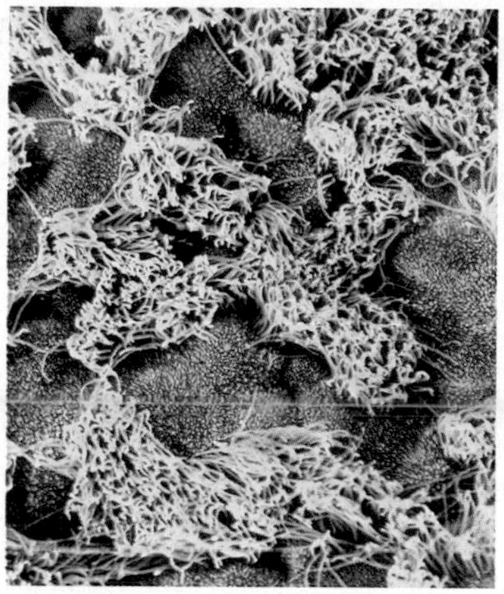

Epithelial surface SEM × 4000

3 This colorized SEM shows the ciliated epithelium of the uterine tube. The cilia (colored yellow-green) beat toward the uterine cavity, establishing fluid currents that help collect and transport the secondary oocyte after ovulation.

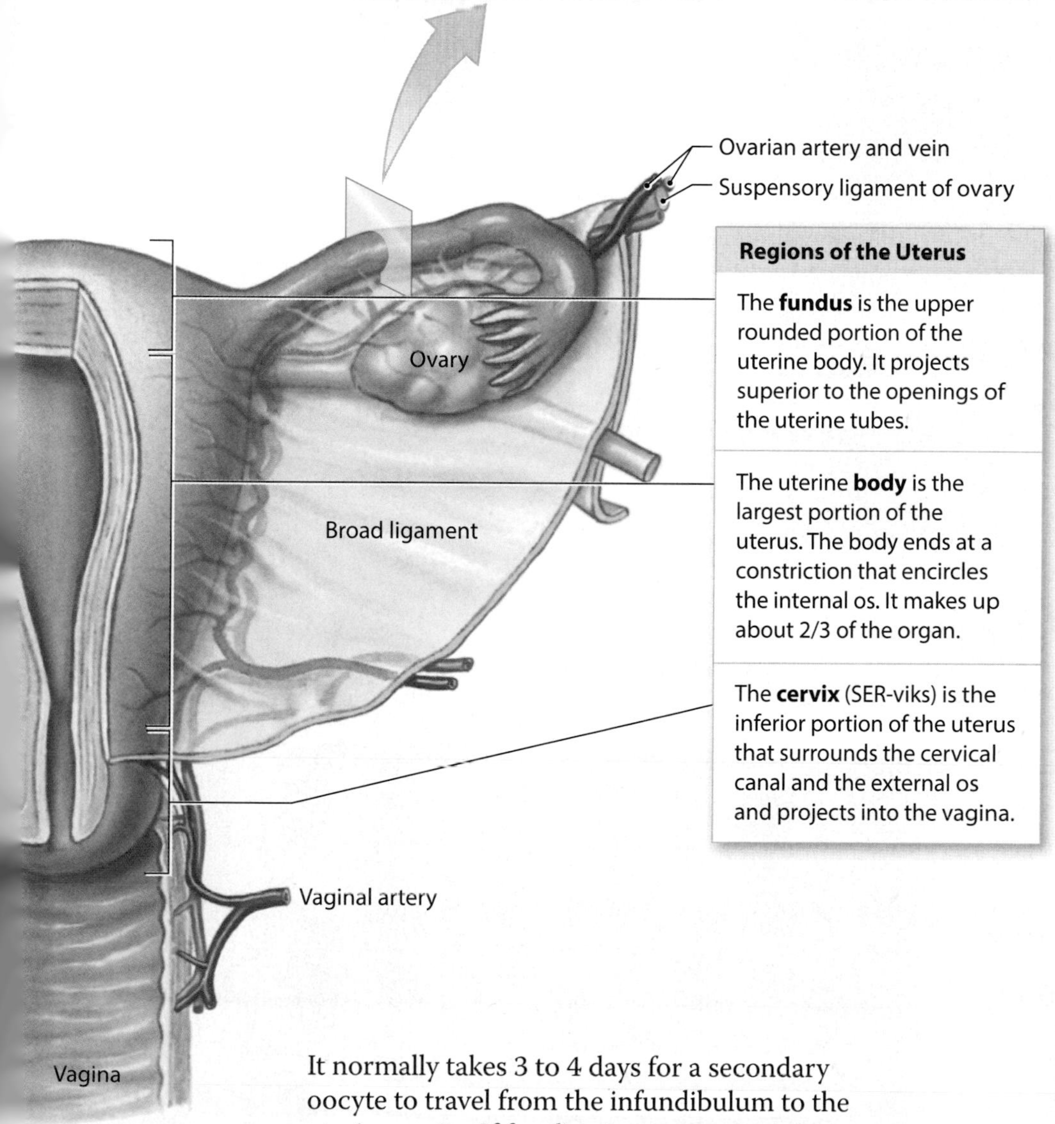
Ovarian artery and vein
Suspensory ligament of ovary
Ovary
Broad ligament
Vaginal artery
Vagina
Regions of the Uterus
The fundus is the upper rounded portion of the uterine body. It projects superior to the openings of the uterine tubes.
The uterine body is the largest portion of the uterus. The body ends at a constriction that encircles the internal os. It makes up about 2/3 of the organ.
The cervix (SER-viks) is the inferior portion of the uterus that surrounds the cervical canal and the external os and projects into the vagina.

4 The uterus is divided into three anatomical regions, as shown here. The uterus, which is capable of great changes in size and shape, provides mechanical protection, nutritional support, and waste removal for the developing **embryo** (weeks 1–8) and **fetus** (week 9 through delivery). In addition, contractions in the muscular wall of the uterus are important in delivering the fetus at birth.

It normally takes 3 to 4 days for a secondary oocyte to travel from the infundibulum to the uterine cavity. If fertilization is to occur, the secondary oocyte must encounter spermatozoa during the first 12–24 hours of its passage along the uterine tube.

Module 26.11 Review

a. How do recently released secondary oocytes reach the uterine tube?

b. Describe the three layers of the uterine wall.

c. Name the regions of the uterus.

Module 26.12

The uterine cycle involves changes in the functional zone of the endometrium

1 Within the myometrium, branches of the uterine arteries form **arcuate arteries**, which encircle the endometrium. From the arcuate arteries, **radial arteries** supply the endometrium.

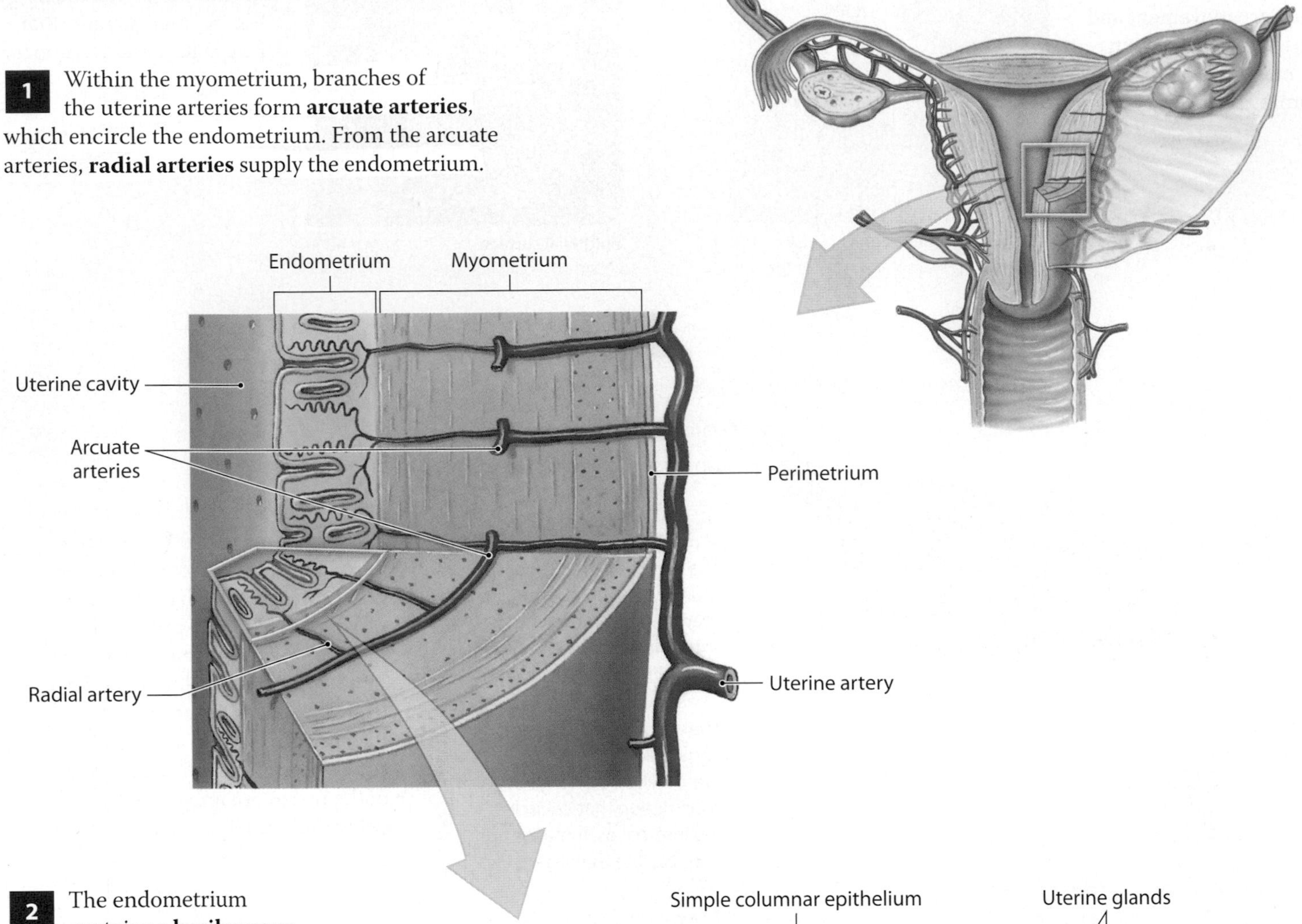

Endometrium
Myometrium
Uterine cavity
Arcuate arteries
Perimetrium
Radial artery
Uterine artery

2 The endometrium contains a **basilar zone** adjacent to the myometrium, and a **functional zone**, the region closest to the uterine cavity. **Straight arteries** deliver blood to the basilar zone, and **spiral arteries** supply the functional zone. The functional zone of the endometrium contains large tubular uterine glands.

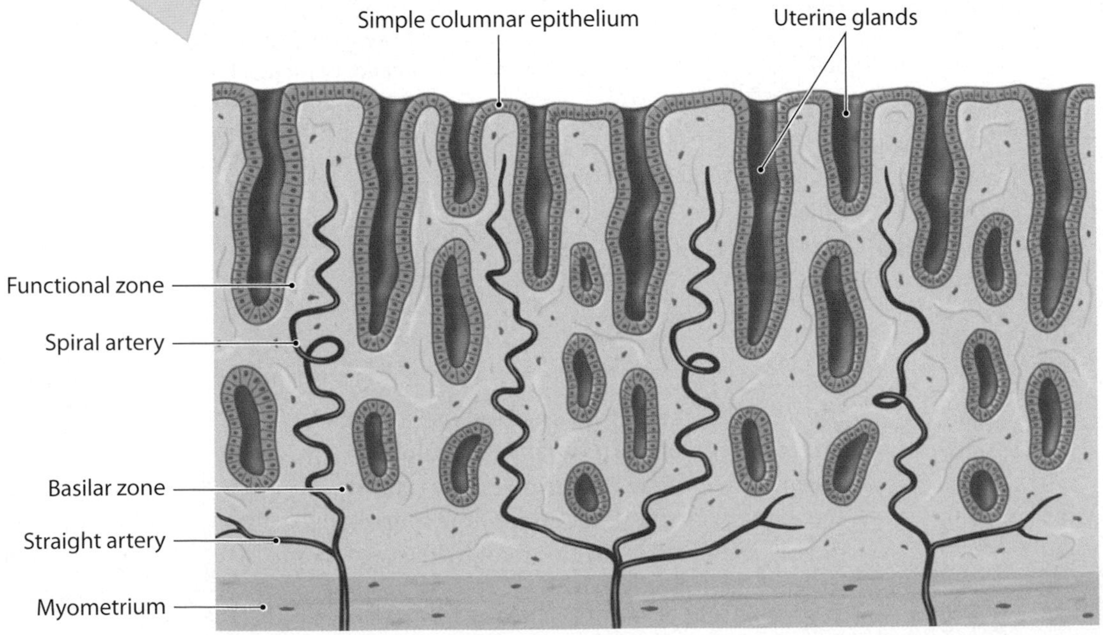

Simple columnar epithelium
Uterine glands
Functional zone
Spiral artery
Basilar zone
Straight artery
Myometrium

3 The structure of the basilar zone remains relatively constant over time, but that of the functional zone undergoes cyclical changes in response to sex hormone levels. These cyclical changes produce the characteristic histological features of the **uterine cycle**. The uterine cycle averages 28 days in length, but it can range from 21 to 35 days in healthy women of reproductive age.

Menses

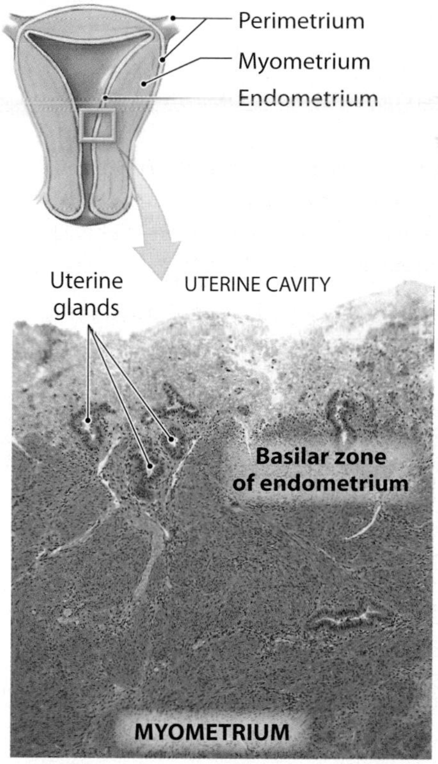
Perimetrium
Myometrium
Endometrium
Uterine glands
UTERINE CAVITY
Basilar zone of endometrium
MYOMETRIUM

The uterine cycle begins with the onset of **menses** (MEN-sēz), an interval marked by the degeneration of the functional zone of the endometrium. This degeneration is caused by constriction of the spiral arteries, which reduces endometrial blood flow. Eventually, the weakened arterial walls rupture, and blood pours into the connective tissues of the functional zone. Blood cells and degenerating tissues then break away and enter the uterine lumen, to be lost by passage through the external os and into the vagina. The process of endometrial shedding, called **menstruation** (men-strū-Ā-shun), generally lasts from one to seven days. Over this period roughly 35 to 50 mL of blood are lost.

Proliferative Phase

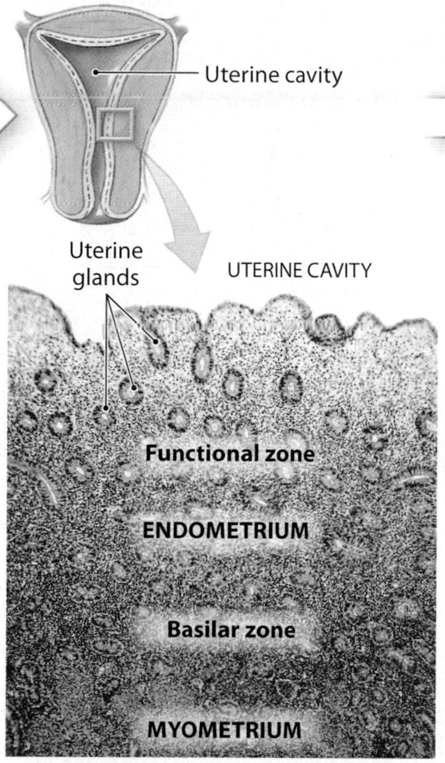
Uterine cavity
Uterine glands
UTERINE CAVITY
Functional zone
ENDOMETRIUM
Basilar zone
MYOMETRIUM

The basilar zone and the deepest uterine glands survive menses intact. The epithelial cells of the uterine glands then multiply and spread across the endometrial surface, restoring the integrity of the uterine epithelium. As this reorganization proceeds, the endometrium is in the **proliferative phase**. Restoration is stimulated and sustained by estrogens secreted by the developing ovarian follicles. By the time ovulation occurs, the functional zone is several millimeters thick, and prominent mucous glands extend to the border with the basilar zone. At this time, the uterine glands are manufacturing a glycogen-rich mucus that can be metabolized by an early embryo.

Secretory Phase

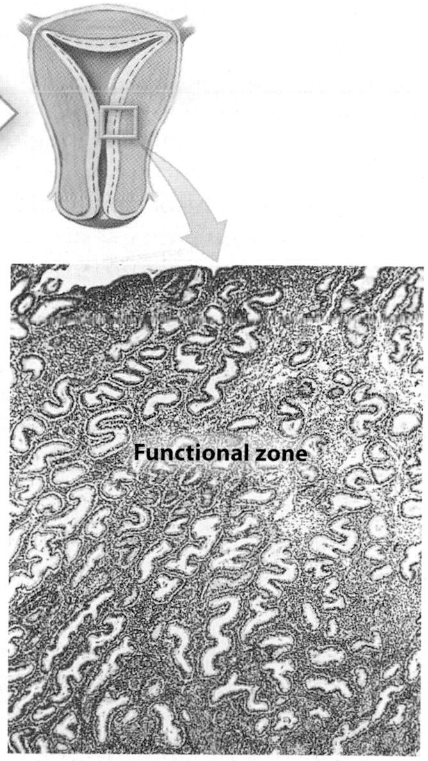
Functional zone

During the **secretory phase**, the uterine glands enlarge, accelerating their rate of secretion, and the arteries that supply the uterine wall elongate and spiral through the tissues of the functional zone. This activity occurs under the combined stimulatory effects of progesterone and estrogens from the corpus luteum. The secretory phase begins at the time of ovulation and lasts as long as the corpus luteum remains intact. When the corpus luteum stops producing stimulatory hormones, a new uterine cycle begins with the onset of menses and the disintegration of the functional zone.

The uterine cycle, or **menstrual cycle**, begins at puberty. The first cycle, known as **menarche** (me-NAR-kē; *men*, month + *arche*, beginning), typically occurs at age 11–12. The cycles continue until **menopause** (MEN-ō-pawz), the termination of the uterine cycle, at age 45–55. Over the interim, the regular appearance of uterine cycles is interrupted only by circumstances such as illness, stress, starvation, or pregnancy.

Module 26.12 Review

a. Name the zones of the endometrium.

b. Differentiate between menses and menstruation.

c. Describe the phases of the uterine cycle.

The entrance to the vagina is enclosed by external genitalia

1 The **vagina** is an elastic, muscular tube extending between the cervix and the **vestibule**, a space bordered by the female external genitalia. The vagina is typically 7.5–9 cm (3–3.6 in.) long, but its diameter varies because it is highly distensible. The internal passageway is called the **vaginal canal**. The vagina (1) serves as a passageway for the elimination of menstrual fluids; (2) receives the penis during sexual intercourse, and holds spermatozoa prior to their passage into the uterus; and (3) forms the inferior portion of the birth canal, through which the fetus passes during delivery.

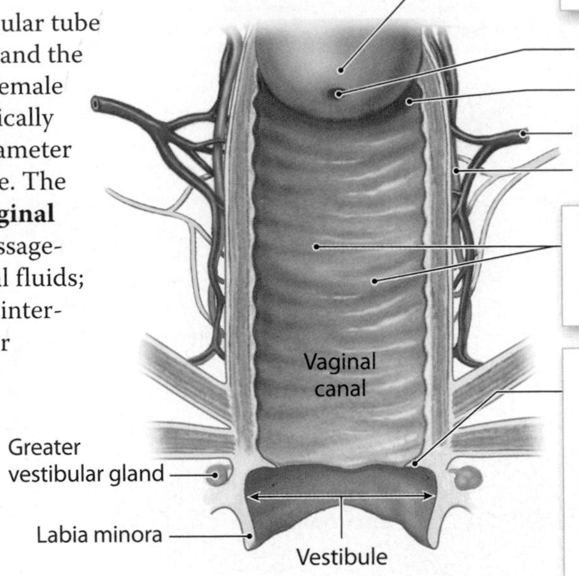
At the proximal end of the vagina, the **cervix** projects into the vaginal canal. The shallow recess in the vagina surrounding the tip of the cervix is known as the **fornix** (FOR-niks).
External os
Fornix
Vaginal artery
Vaginal vein
In the relaxed state, the vaginal lining forms folds called **rugae**. The vaginal canal is lined by a nonkeratinized stratified squamous epithelium.
Vaginal canal
Throughout childhood the vagina and vestibule are usually separated by the **hymen** (HĪ-men), an elastic epithelial fold of variable size that partially blocks the entrance to the vagina. An intact hymen is typically stretched or torn during sexual intercourse, tampon use, or heavy physical exercise. It is frequently absent, and its condition is not an indicator of sexual virginity.
Greater vestibular gland
Labia minora
Vestibule

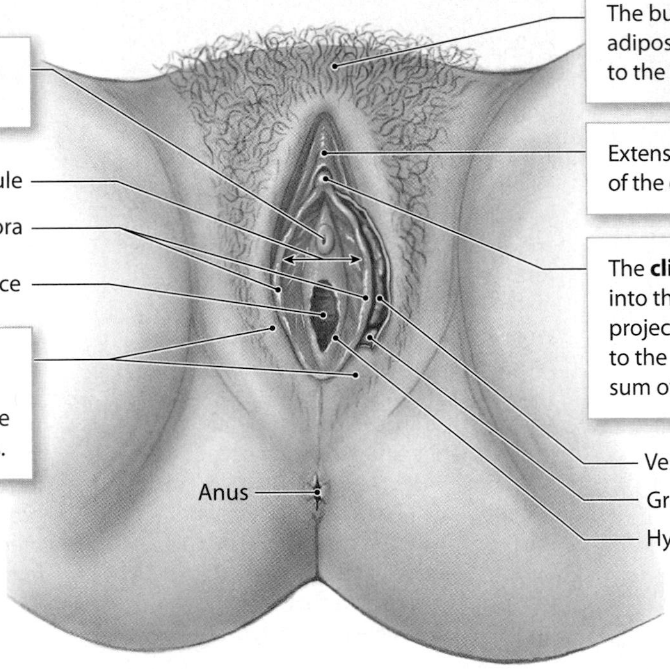
The urethra opens into the vestibule just anterior to the vaginal entrance.
The bulge of the **mons pubis** is created by adipose tissue deep to the skin and superficial to the pubic symphysis.
Extensions of the labia minora encircle the body of the clitoris, forming its **prepuce**, or hood.
Vestibule
Labia minora
Vaginal entrance
The **clitoris** (KLIT-ō-ris or kli-TŌR-is) projects into the vestibule. This small, rounded tissue projection contains erectile tissue comparable to the corpora cavernosa and corpus spongiosum of the penis.
The **labia majora** (singular, *labium majus*) are prominent folds of skin that encircle and partially conceal the labia minora and adjacent structures.
Anus
Vestibular bulb
Greater vestibular gland
Hymen (torn)

2 The area containing the female external genitalia is the **vulva** (VUL-vuh), or **pudendum** (pū-DEN-dum). The vagina opens into the vestibule, a central space bounded by small folds known as the **labia minora** (LĀ-be-uh mi-NOR-uh; singular, *labium minus*). A variable number of small **lesser vestibular glands** discharge their secretions onto the exposed surface of the vestibule, keeping it moist. During sexual arousal, a pair of ducts discharges the secretions of the **greater vestibular glands** (Bartholin's glands) into the vestibule. These mucous glands have the same embryonic origins as the bulbourethral glands of males. The **vestibular bulbs** are masses of erectile tissue on either side of the vaginal entrance. The vestibular bulbs have the same embryonic origins as the corpus spongiosum of the penis in males.

Module 26.13 Review

a. List the functions of the vagina.

b. Describe the anatomy of the vagina.

c. Cite the similarities that exist between certain structures in the reproductive systems of females and males.

LO 26.13 Describe the structure, histology, and functions of the vagina.

The mammary glands nourish the infant after delivery

A newborn infant cannot fend for itself, and several of its key systems are not fully developed. Over the initial period of adjustment to an independent life, the infant can gain nourishment from the milk secreted by the maternal **mammary glands**. These organs are controlled mainly by hormones released by the reproductive system and the placenta, a temporary structure that provides the embryo and fetus with nutrients. The interaction of these hormones results in milk production, or **lactation** (lak-TĀ-shun).

1 The mammary gland lies directly over the pectoralis major muscle. This illustration shows the internal organization of the mammary tissue and its supporting structures.

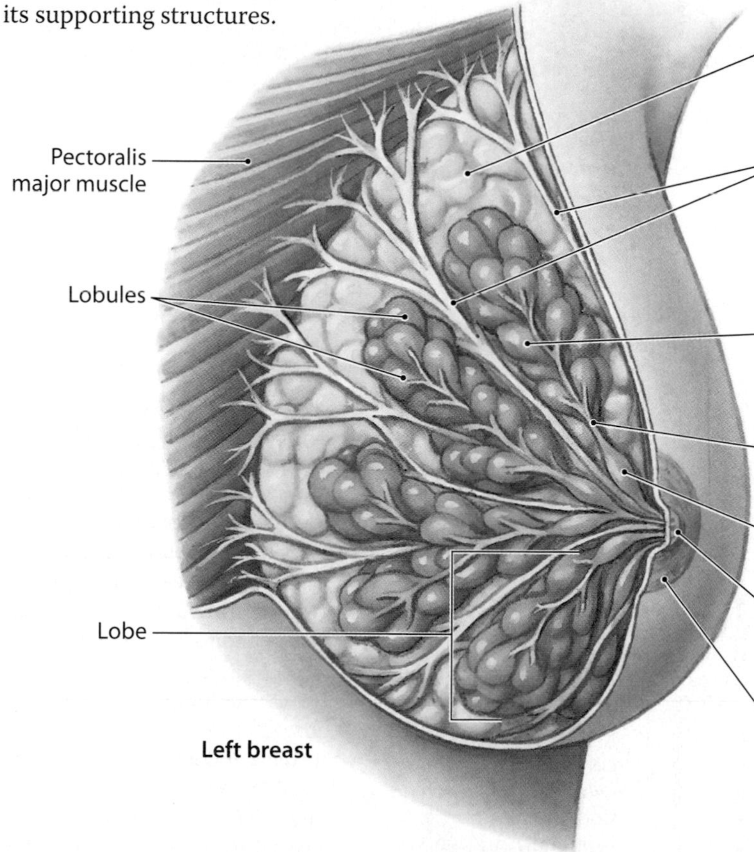
Pectoralis major muscle
Lobules
Lobe
Left breast

The Structure of a Mammary Gland

On each side, a mammary gland lies in the subcutaneous tissue of the **pectoral fat pad** deep to the skin of the chest.

Dense connective tissue surrounds the duct system and forms partitions that extend between the lobes and the lobules. These bands of connective tissue, the **suspensory ligaments of the breast**, originate in the dermis of the overlying skin.

The glandular tissue of the breast consists of separate **lobes**, each containing several secretory **lobules**. Each lobule is composed of many **secretory alveoli**.

Ducts leaving the lobules converge, giving rise to a single **lactiferous** (lak-TIF-er-us) **duct** in each lobe.

Near the nipple, each lactiferous duct enlarges, forming an expanded chamber called a **lactiferous sinus**.

Each breast has a **nipple**, a small conical projection where 15–20 lactiferous sinuses open onto the body surface.

The reddish-brown skin around each nipple is the **areola** (a-RĒ-ō-luh). Large sebaceous glands deep to the areolar surface give it a grainy texture.

Module 26.14 Review

a. Define lactation.

b. Explain whether the blockage of a single lactiferous sinus would or would not interfere with the delivery of milk to the nipple.

c. Trace the route of milk from its site of production to the body surface.

LO **26.14** Discuss the structure and function of the mammary glands.

Module 26.15

The ovarian and uterine cycles are regulated by hormones of the hypothalamus, pituitary gland, and ovaries

The ovarian and uterine cycles must operate in synchrony to ensure proper reproductive function. If the two cycles are not properly coordinated, infertility results. A female who doesn't ovulate cannot conceive, even if her uterus is perfectly normal. A female who ovulates normally, but whose uterus is not ready to support an embryo, will also be infertile.

1 As in males, GnRH from the hypothalamus regulates reproductive function in females. However, in females, GnRH levels change throughout the course of the ovarian cycle.

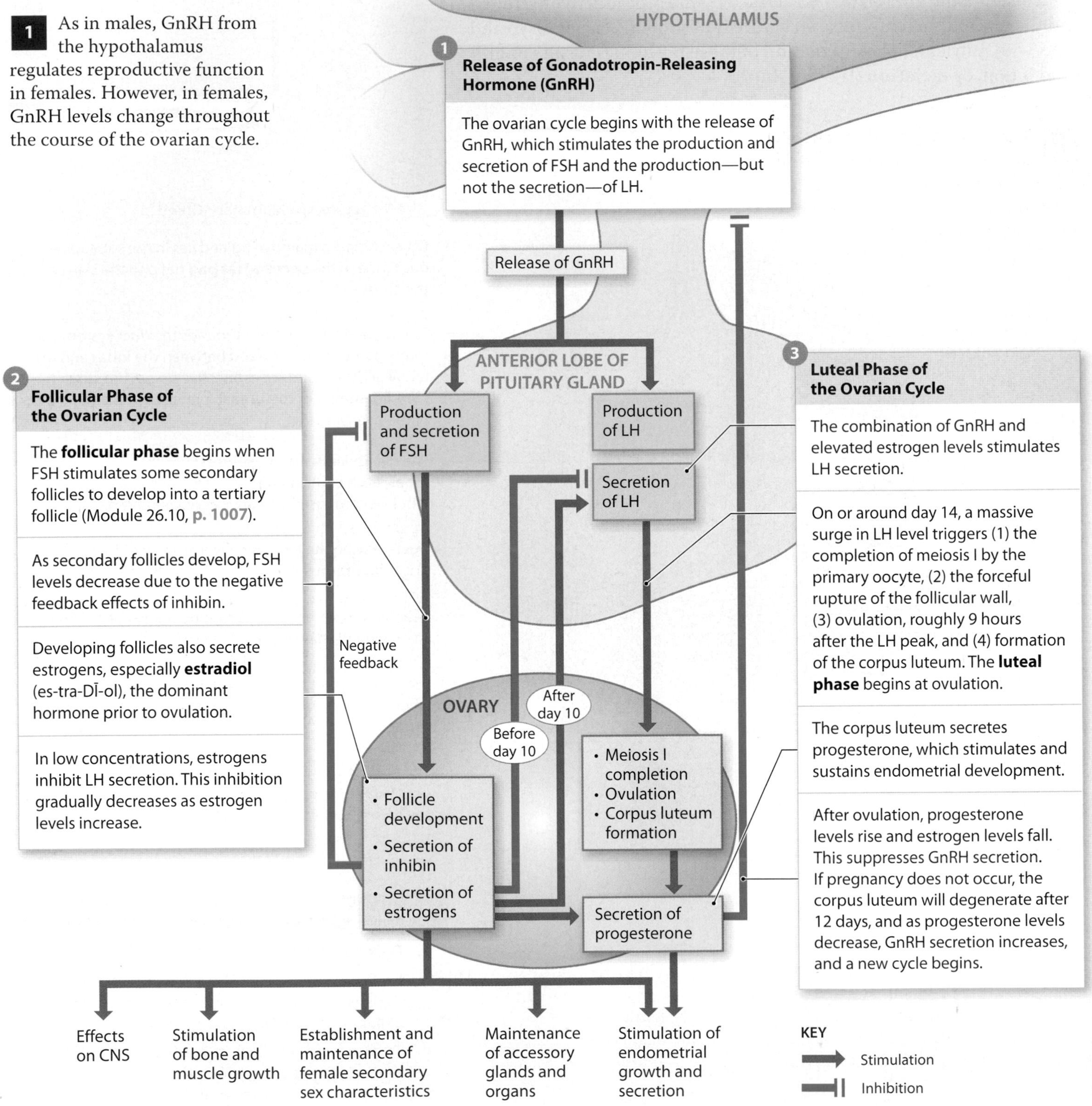

HYPOTHALAMUS
1
Release of Gonadotropin-Releasing Hormone (GnRH)
The ovarian cycle begins with the release of GnRH, which stimulates the production and secretion of FSH and the production—but not the secretion—of LH.
Release of GnRH
ANTERIOR LOBE OF PITUITARY GLAND
Production and secretion of FSH
Production of LH
Secretion of LH
2
Follicular Phase of the Ovarian Cycle
The follicular phase begins when FSH stimulates some secondary follicles to develop into a tertiary follicle (Module 26.10, p. 1007).
As secondary follicles develop, FSH levels decrease due to the negative feedback effects of inhibin.
Developing follicles also secrete estrogens, especially estradiol (es-tra-DĪ-ol), the dominant hormone prior to ovulation.
In low concentrations, estrogens inhibit LH secretion. This inhibition gradually decreases as estrogen levels increase.
Negative feedback
OVARY
After day 10
Before day 10
• Follicle development
• Secretion of inhibin
• Secretion of estrogens
• Meiosis I completion
• Ovulation
• Corpus luteum formation
Secretion of progesterone
3
Luteal Phase of the Ovarian Cycle
The combination of GnRH and elevated estrogen levels stimulates LH secretion.
On or around day 14, a massive surge in LH level triggers (1) the completion of meiosis I by the primary oocyte, (2) the forceful rupture of the follicular wall, (3) ovulation, roughly 9 hours after the LH peak, and (4) formation of the corpus luteum. The luteal phase begins at ovulation.
The corpus luteum secretes progesterone, which stimulates and sustains endometrial development.
After ovulation, progesterone levels rise and estrogen levels fall. This suppresses GnRH secretion. If pregnancy does not occur, the corpus luteum will degenerate after 12 days, and as progesterone levels decrease, GnRH secretion increases, and a new cycle begins.
Effects on CNS
Stimulation of bone and muscle growth
Establishment and maintenance of female secondary sex characteristics
Maintenance of accessory glands and organs
Stimulation of endometrial growth and secretion
KEY
Stimulation
Inhibition

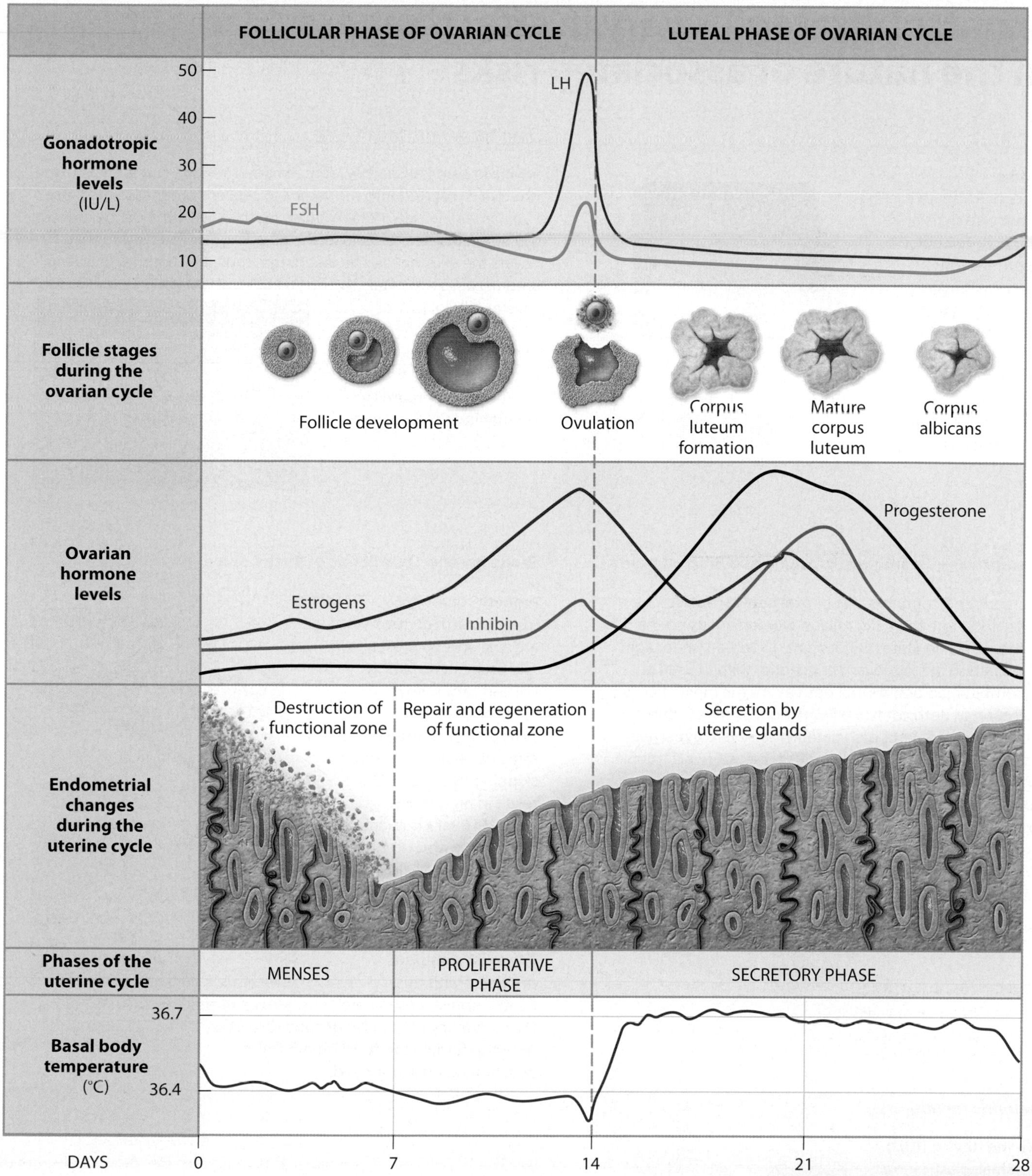

FOLLICULAR PHASE OF OVARIAN CYCLE
LUTEAL PHASE OF OVARIAN CYCLE
Gonadotropic hormone levels (IU/L)
50
40
30
20
10
LH
FSH
Follicle stages during the ovarian cycle
Follicle development
Ovulation
Corpus luteum formation
Mature corpus luteum
Corpus albicans
Ovarian hormone levels
Progesterone
Estrogens
Inhibin
Endometrial changes during the uterine cycle
Destruction of functional zone
Repair and regeneration of functional zone
Secretion by uterine glands
Phases of the uterine cycle
MENSES
PROLIFERATIVE PHASE
SECRETORY PHASE
Basal body temperature (°C)
36.7
36.4
DAYS
0
7
14
21
28

2 This illustration combines the key events in the ovarian and uterine cycles. The monthly hormonal fluctuations cause physiological changes that affect core body temperature. During the follicular phase—when estrogens are the dominant hormones—the **basal body temperature**, or the resting body temperature measured upon awakening in the morning, is about 0.3°C (0.5°F) lower than it is during the luteal phase, when progesterone dominates.

Module 26.15 Review

a. What uterine cycle event occurs when estrogen and progesterone decrease?

b. What ovarian cycle changes would result if the LH surge did not occur?

c. Summarize the roles of the hormones in the ovarian and uterine cycles.

 CLINICAL MODULE

Birth control strategies vary in effectiveness and in the nature of associated risks

Male Condom

Male condoms (prophylactics or "rubbers") cover the glans and shaft of the penis during intercourse and keep spermatozoa from reaching the female reproductive tract. Of all the strategies described in this module, only latex condoms protect against **sexually transmitted diseases (STDs)**, such as syphilis, gonorrhea, human papillomavirus (HPV), and AIDS.

Diaphragm with Spermicide

A **diaphragm** is a shallow, dome-shaped silicone cup with a flexible rim that is inserted into the vagina to prevent pregnancy. Because vaginas vary in size, women choosing this method must be individually fitted. Before intercourse, the diaphragm is inserted so that it covers the external os. The diaphragm must be coated with a small amount of spermicidal (sperm-killing) jelly or cream to be an effective contraceptive. The failure rate of a properly fitted and used diaphragm is estimated at 5–6 percent.

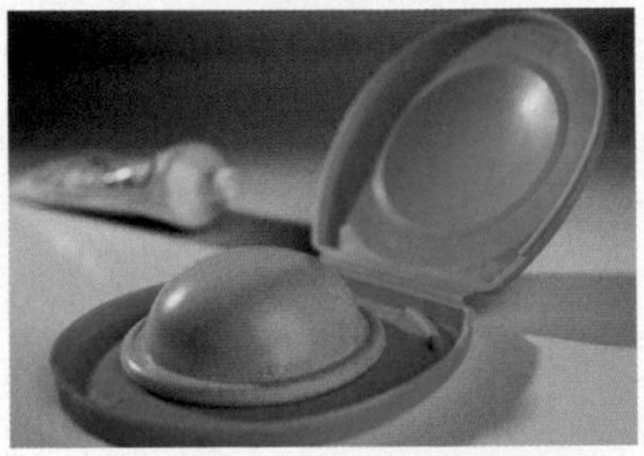

Oral Contraceptives—Combined (Estrogen and Progesterone)

Numerous brands and combinations of **oral contraceptives** are now available, and more than 200 million women are using them worldwide. In the United States, 33 percent of women under age 45 use a combination pill to prevent conception. When used as prescribed, combination oral contraceptives are the most effective form of birth control. Birth control pills are not risk free: Combination pills can worsen problems associated with severe hypertension, diabetes mellitus, epilepsy, gallbladder disease, heart trouble, and acne. Women taking oral contraceptives are also at increased risk of venous thrombosis, strokes, pulmonary embolism, and (for women over 35) heart disease. However, pregnancy itself has similar or higher risks.

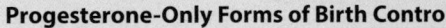

Progesterone-Only Forms of Birth Control

Progesterone-only forms of birth control are available by prescripton: The progesterone-only pill and Depo-Provera are examples. The **progesterone-only pill** must be taken daily, and skipping even one pill may result in pregnancy. **Depo-Provera** is injected every 3 months. Uterine cycles are initially irregular and eventually cease in roughly 50 percent of women using this product. The most common problems with this contraceptive method are (1) a tendency to gain weight and (2) a slow return to fertility (up to 18 months) after injections are discontinued.

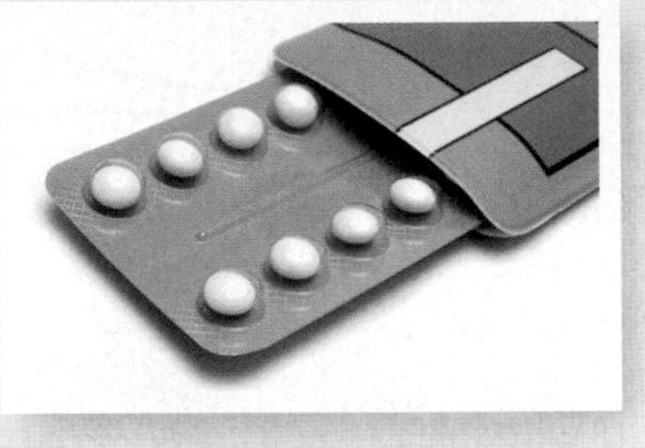

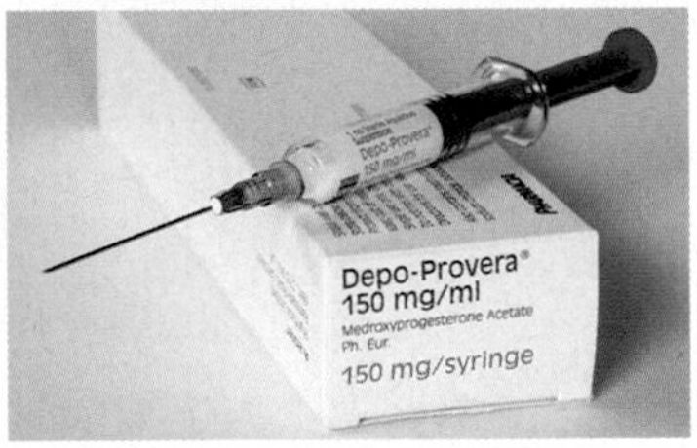
Depo-Provera®
150 mg/ml
150 mg/syringe

IUD (Intrauterine Device)

An **intrauterine device (IUD)** consists of a small plastic loop or a T that is inserted into the uterine cavity. The mechanism of action remains unclear, but IUDs are known to stimulate prostaglandin production in the uterus, and they are effective for up to 12 years after insertion.

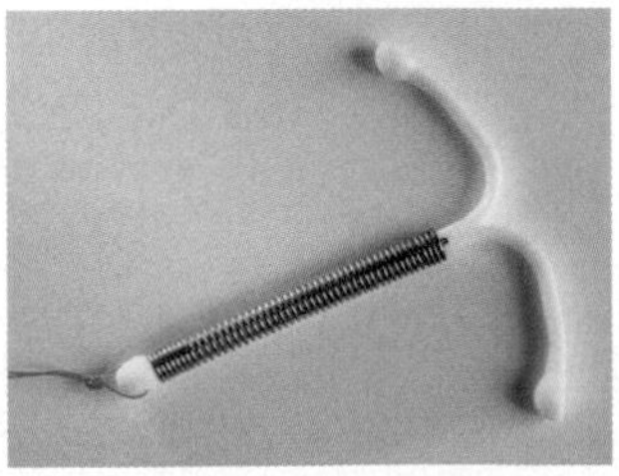

The Rhythm Method

"Natural Family Planning," also called the **rhythm method** or fertility awareness-based methods, involves abstaining from sexual intercourse on the days ovulation might be occurring. The timinq is estimated on the basis of previous patterns of menstruation; monitoring changes in indications of ovulation, including basal body temperature and cervical mucus texture; and, for some, urine tests for LH. Because of the irregularity of many women's uterine cycles, this method of contraception has a failure rate estimated to be 13–20 percent.

Post-Coital Contraceptives (Plan B)

Hormonal post-coital contraception, or the emergency "morning after" pill, involves taking levonorgestrel contraceptive pills up to 5 days (120 hours) after unprotected intercourse. Emergency contraception prevents a woman's ovary from releasing an oocyte for longer than usual, thereby preventing an oocyte from joining a sperm. Particularly useful when barrier methods malfunction or coerced intercourse occurs, it reduces expected pregnancy rates by up to 89 percent when taken within 72 hours of unprotected sex. Its effectiveness decreases with time, up to 120 hours. Anyone—regardless of age—can purchase Plan B One-Step brand as an over-the-counter product and a prescription is not required.

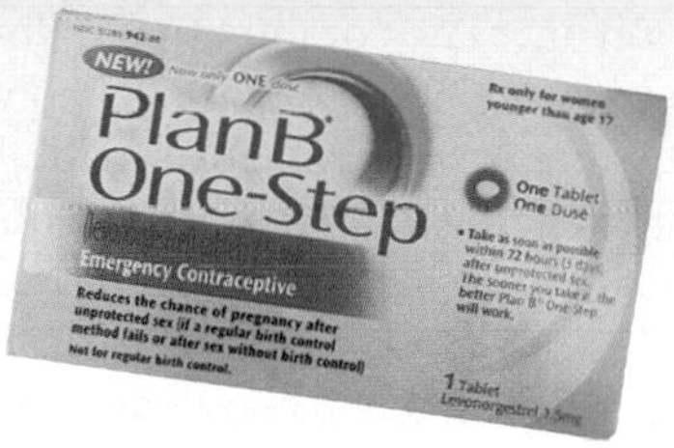
NEW!
Plan B
One-Step
Emergency Contraceptive
One Tablet
One Dose

Surgical Sterilization—Male

In a **vasectomy** (va-SEK-tō-mē), each ductus deferens is cut and either a segment is removed (and the ends tied or cauterized) or silicone plugs are inserted (which makes it relatively easy to reverse the procedure). After a vasectomy, spermatozoa cannot pass from the epididymides to the distal portions of the reproductive tract. The surgery can be performed in a physician's office in a matter of minutes; the failure rate (due to incomplete closure or blockage of either ductus deferens) is extremely low. After vasectomy, men experience normal sexual function, because the secretions of the epididymides and testes normally account for only about 5 percent of the volume of semen. Spermatozoa continue to develop, but they remain inactive and eventually degenerate.

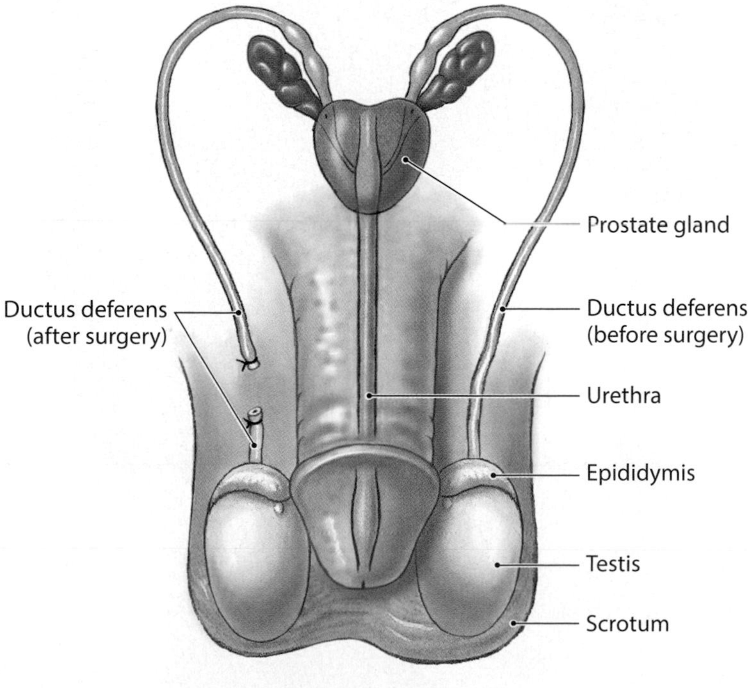
Prostate gland
Ductus deferens (after surgery)
Ductus deferens (before surgery)
Urethra
Epididymis
Testis
Scrotum

Surgical Sterilization—Female

The uterine tubes can be blocked by a surgical procedure known as a **tubal ligation**. The failure rate for this procedure is also extremely low. Because the surgery requires that the abdominopelvic cavity be entered, most commonly by laparoscopy, a general anesthetic is required and complications are more likely than with a vasectomy.

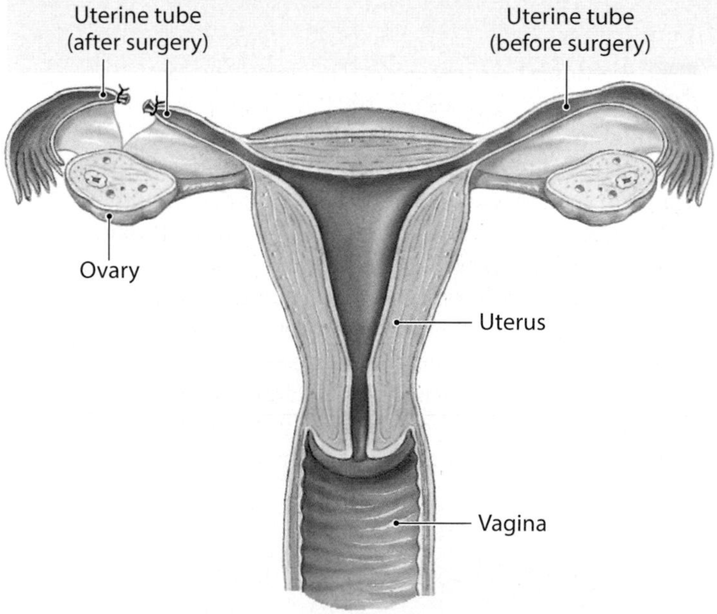
Uterine tube (after surgery)
Uterine tube (before surgery)
Ovary
Uterus
Vagina

Module 26.16 Review

a. Which birth control method provides some protection against sexually transmitted diseases?

b. The use of which birth control method often results in the cessation of the uterine cycle?

c. Define vasectomy.

LO 26.16 Discuss various birth control strategies and their associated risks.

Reproductive system disorders are relatively common and often deadly

1 Enlargement of the prostate gland, or **benign prostatic hypertrophy (BPH)**, typically occurs spontaneously in men over age 50. The increase in size occurs as testosterone production by the interstitial cells decreases. At the same time, the interstitial cells begin releasing small quantities of estrogens into the bloodstream. The combination of lower testosterone levels and the presence of estrogens probably stimulates prostatic growth. In severe cases, prostatic swelling constricts and blocks the urethra, producing urinary obstruction. **Prostate cancer**, a malignancy of the prostate gland, is the second most common cause of cancer deaths in males. The American Cancer Society estimates that approximately 239,000 new prostate cancer cases in 2013 will result in about 30,000 deaths. Blood tests are often used for screening purposes. The most sensitive is a blood test for **prostate-specific antigen (PSA)**. Elevated levels of this antigen, normally present in low concentrations, may indicate the presence of prostate cancer. Screening with periodic PSA tests is now being recommended for men over age 50. Treatment of localized prostate cancer often involves radiation therapy or surgical removal of the prostate gland—a **prostatectomy** (pros-ta-TEK-tō-mē).

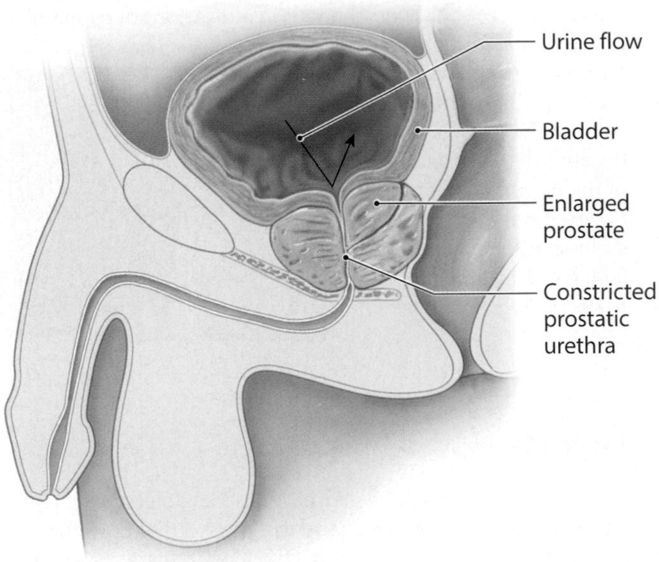
Urine flow
Bladder
Enlarged prostate
Constricted prostatic urethra

2 **Testicular cancer** occurs at a relatively low rate: about 3 cases per 100,000 males per year. Although only about 7900 new cases are reported each year in the United States, with less than 400 deaths, testicular cancer is the most common cancer among males aged 15–35. More than 95 percent of testicular cancers result from abnormal spermatogonia or spermatocytes, rather than abnormal nurse cells, interstitial cells, or other testicular cells. Treatment generally consists of a combination of orchiectomy (removal of the testes) and chemotherapy. The survival rate is now near 95 percent, primarily as a result of earlier diagnosis and improved treatment protocols.

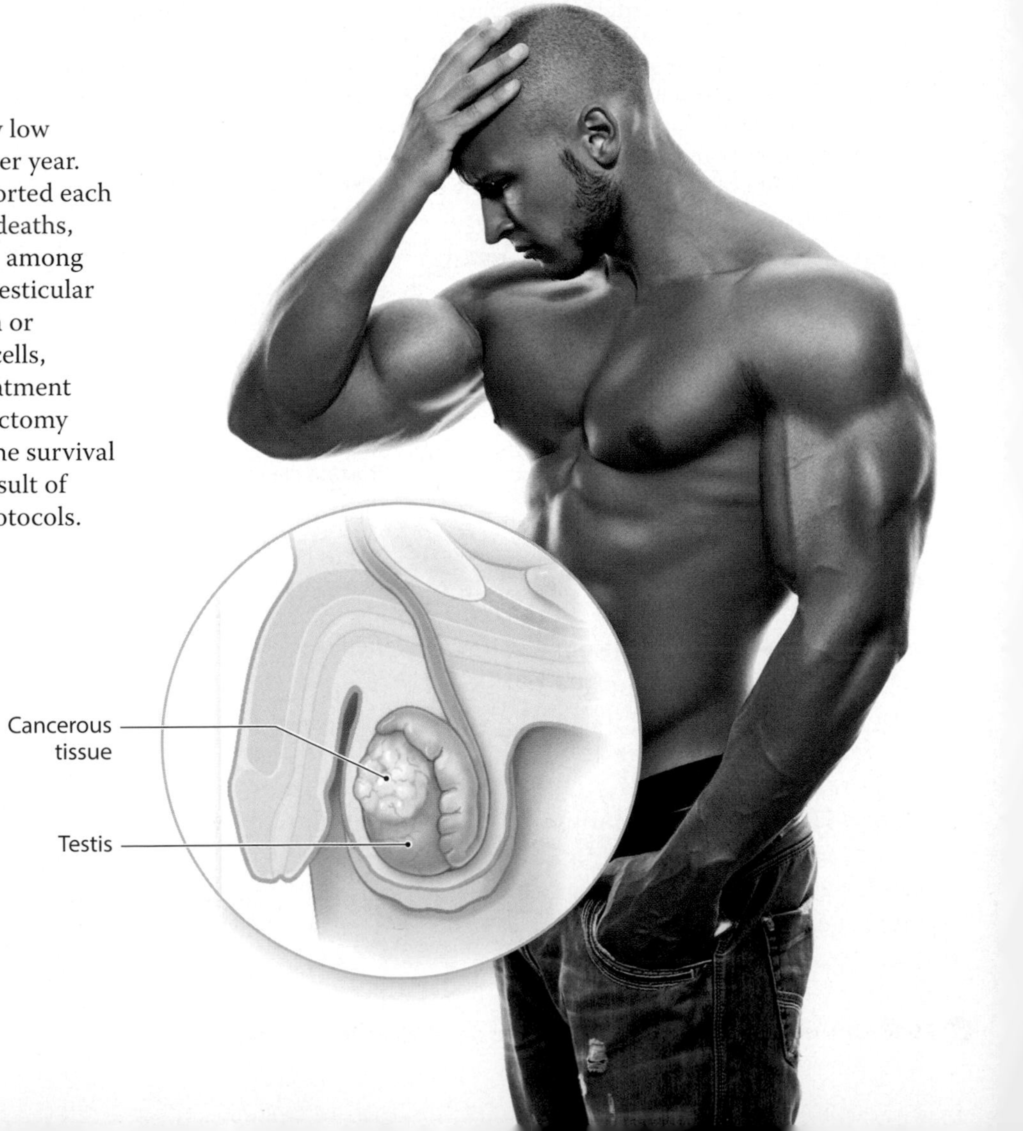
Cancerous tissue
Testis

3 The mammary glands are stimulated by the changing levels of circulating reproductive hormones that accompany the uterine cycle, and, late in the ovarian cycle, occasional discomfort or even inflammation of mammary gland tissues can occur. If inflamed lobules become walled off by scar tissue, **cysts** are created. Clusters of cysts can be felt in the breast as discrete masses, a condition known as **fibrocystic disease**. Biopsies may be needed to distinguish between this benign condition and **breast cancer**. Breast cancer, a malignant, metastasizing tumor of the mammary gland, is the leading cause of death in women between ages 35 and 45, but it is most common in women over age 50. An estimated 12.4 percent of U.S. women will develop breast cancer at some point in their lifetime. Notable risk factors include (1) a family history of breast cancer, (2) a first pregnancy after age 30, and (3) early menarche or late menopause. Treatment of breast cancer includes surgical removal of the tumor. Because in many cases cancer cells begin to spread before the condition is diagnosed, part or all of the affected mammary gland is surgically removed and usually the axillary lymph nodes on that side are biopsied to detect signs of metastasis. A combination of chemotherapy, radiation treatments, and hormone treatments may be used to supplement the surgical procedures.

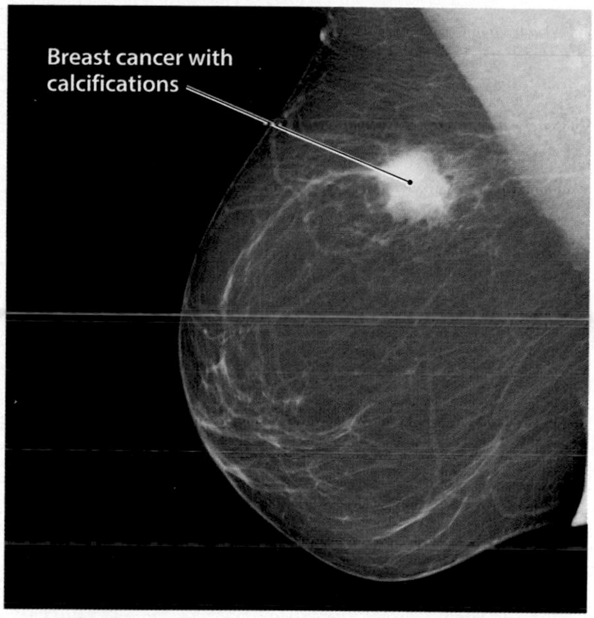
Breast cancer with calcifications

4 A woman in the United States has a 1-in-70 chance of developing **ovarian cancer** in her lifetime. In 2013, there were an estimated 22,240 new cases, and an estimated 14,030 deaths. Although ovarian cancer is the third most common reproductive cancer among women, it is the most dangerous because it is seldom diagnosed in its early stages. The prognosis is relatively good for cancers that originate in the general ovarian tissues or from abnormal oocytes. These cancers respond well to some combination of chemotherapy, radiation, and surgery. However, 85 percent of ovarian cancers are **carcinomas** (cancers derived from epithelial cells), and sustained remission can be obtained in only about one-third of the cases of this type.

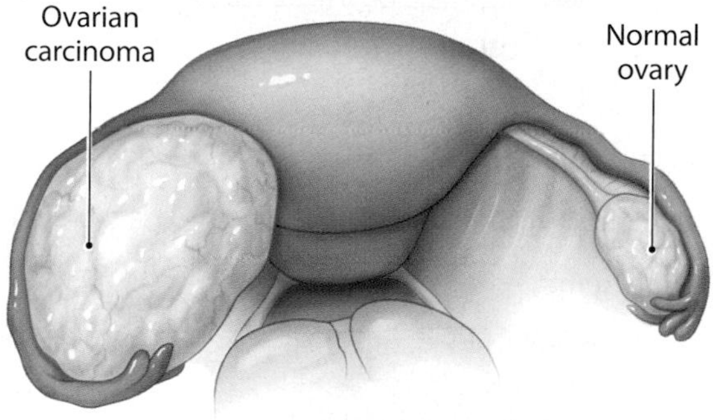
Ovarian carcinoma
Normal ovary

5 **Cervical cancer** is the most common cancer of the reproductive system in women ages 15–34. Each year roughly 12,340 U.S. women are diagnosed with invasive cervical cancer, and approximately one-third of them eventually die from the condition. Another 35,000 women are diagnosed with a less aggressive form of cervical cancer. *Gardasil* is a new vaccine that protects against four types of **human papillomavirus (HPV)**, which cause about 75 percent of cervical cancers.

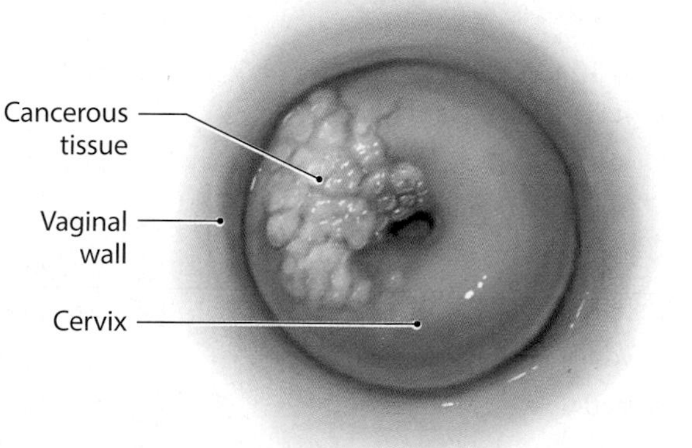
Cancerous tissue
Vaginal wall
Cervix

Sexual activity carries with it the risk of infection with a variety of microorganisms. The consequences of such an infection may range from merely inconvenient to potentially lethal. **Sexually transmitted diseases (STDs)/sexually transmitted infections (STIs)** are transferred from person to person, primarily or exclusively by sexual intercourse. At least two dozen bacterial, viral, and fungal infections are currently recognized as STDs. The bacterium *Chlamydia* can cause pelvic inflammatory disease (PID) and infertility; AIDS, caused by HIV, is a deadly viral disease. The incidence of STDs has been increasing in the United States since 1984; an estimated 20 million new cases occur each year—almost 50 percent in persons aged 15–24. Poverty, intravenous drug use, prostitution, and the appearance of drug-resistant pathogens all contribute to the problem.

Module 26.17 Review

a. From which cell type does ovarian cancer usually arise?

b. Which pathogen is associated with most cases of cervical cancer?

c. Define sexually transmitted disease.

Labeling

Label the structures of the female reproductive system in the accompanying diagram.

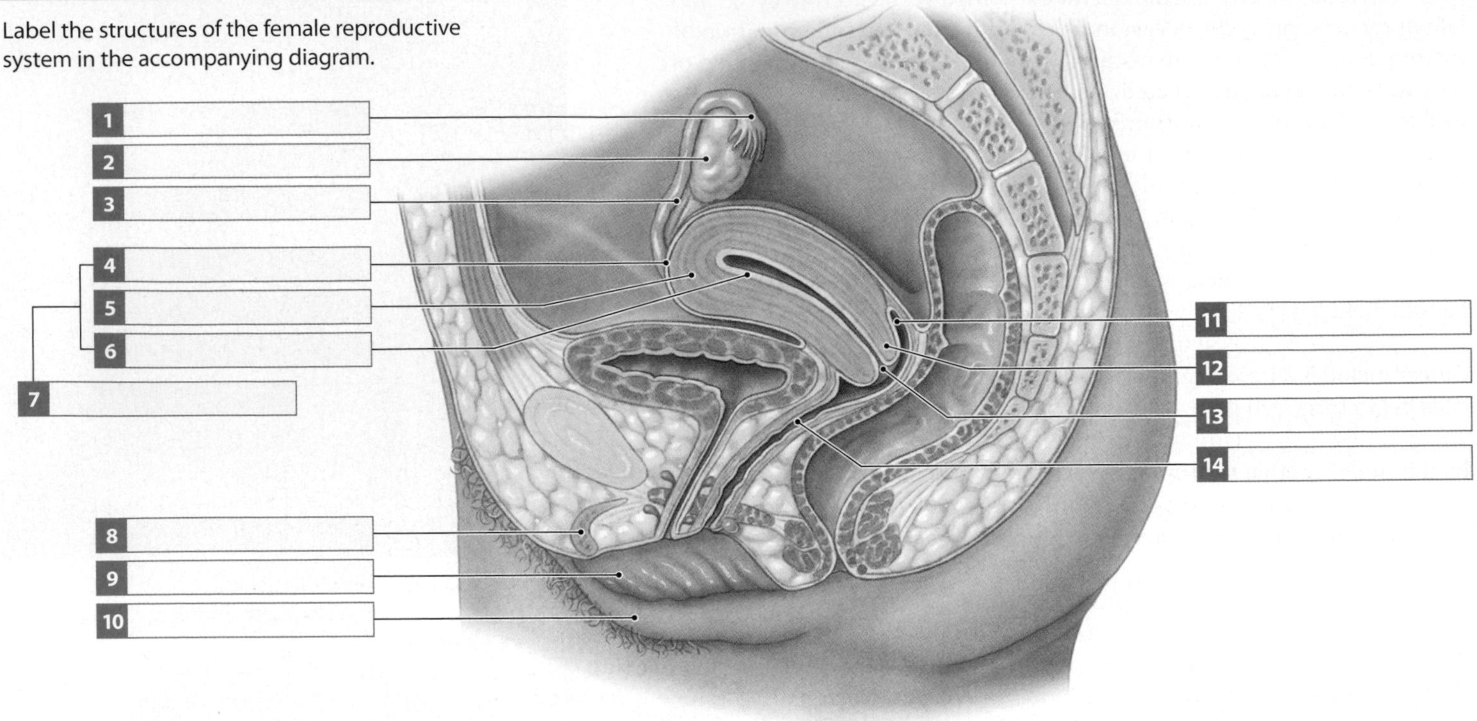
1
2
3
4
5
6
7
8
9
10
11
12
13
14

Matching

Match each lettered term with the most closely related description.

a. LH surge
b. rectouterine pouch
c. tubal ligation
d. menarche
e. ovaries
f. corpus luteum
g. vulva
h. cervix
i. broad ligament
j. oocytes
k. GnRH
l. vesicouterine pouch
m. lactation
n. uterine cycle

15 Immature female gametes
16 First menstrual cycle
17 Pocket anterior to the uterus
18 Encloses the ovaries, uterine tubes, and uterus
19 Pocket posterior to the uterus
20 Endocrine structure
21 Averages 28 days
22 Milk production
23 Oocyte and hormone production
24 Triggers ovulation
25 Inferior portion of the uterus
26 Female surgical sterilization
27 Stimulates FSH production and secretion
28 Contains female external genitalia

15 ____________
16 ____________
17 ____________
18 ____________
19 ____________
20 ____________
21 ____________
22 ____________
23 ____________
24 ____________
25 ____________
26 ____________
27 ____________
28 ____________

Section integration

In a condition known as endometriosis, endometrial cells migrate from the body of the uterus either into the uterine tubes or through the uterine tubes and into the peritoneal cavity, where they become established. Explain why periodic pain is a major symptom of endometriosis.

29 ____

Study Outline

SECTION 1 • Male Reproductive System

26.1 Male reproductive structures are gonads, accessory organs, and external genitalia p. 989

1. The reproductive system of both sexes includes **gonads**, accessory glands and organs, and external genitalia.
2. The male gonads are the **testes (testicles)**, and they produce **spermatozoa** that travel along the **male reproductive tract**.
3. The spermatozoa mix with **semen**.
4. The male **accessory organs** include a **ductus deferens**, **seminal gland**, **bulbourethral gland**, and **epididymis** on each side, and the **prostate gland** and **urethra**.
5. The **penis** and **scrotum** are structures of the male **external genitalia**.

26.2 The coiled seminiferous tubules of the testes are connected to the male reproductive tract p. 990

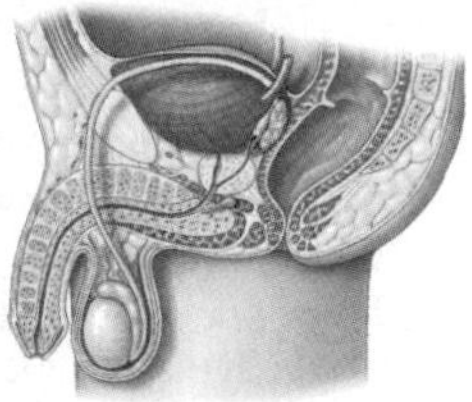

6. Starting from the **epididymis**, spermatozoa travel the **ductus deferens**, **ejaculatory duct**, and urethra before leaving the body.
7. The accessory organs (**seminal glands**, **prostate**, and **bulbourethral glands**) secrete fluids into the ejaculatory ducts and urethra.
8. The testes descend into the **scrotum**.
9. The **superficial inguinal ring** forms the entrance to the **inguinal canal**. The **spermatic cord** lies within the inguinal canal and encloses the ductus deferens, nerves, and blood and lymphatic vessels.
10. The **dartos muscle** is the smooth muscle of the scrotum. The two **scrotal cavities** are separated by the **scrotal septum**.
11. The **tunica albuginea** cover the testes, and are continuous with the septa that separate the **lobules**. Each lobule contains coiled **seminiferous tubules** where sperm production occurs.
12. The seminiferous tubules are connected to the **rete testis**, and **efferent ductules** connect to the epididymis.

26.3 Meiosis in the testes produces haploid spermatids that mature into spermatozoa p. 992

13. **Spermatogenesis** involves **mitosis**, **meiosis**, and **spermiogenesis**.
14. Each mitotic division of a stem cell produces one daughter cell that remains a stem cell, and one daughter cell that enters meiosis. These **diploid** (2n) cells contain 23 pairs of chromosomes.
15. Meiosis involves two cycles of cell division (**meiosis I** and **meiosis II**) and produces four **haploid** (n) cells that each contain 23 individual chromosomes.
16. In **synapsis**, maternal and paternal chromosomes form 23 chromosome pairs. Each member of a pair consists of two duplicate chromatids, and the set of four chromatids is called a **tetrad**.
17. **Spermatogenesis** begins with the division of a **spermatogonium** (stem cell) into a daughter stem cell and **primary spermatocyte**.
18. In meiosis I, **secondary spermatocytes** form from a primary spermatocyte. Meiosis II yields four haploid **spermatids**.
19. In spermiogenesis, each spermatid matures into a single **spermatozoon**, or sperm, that contains an **acrosome**, **head**, **neck**, **middle piece**, and **tail**, or **flagellum**.

26.4 Meiosis and early spermiogenesis occur within the seminiferous tubules p. 994

20. Every seminiferous tubule is continuously producing spermatozoa.
21. **Interstitial cells** (Leydig cells) produce testosterone and androstenedione, the dominant sex hormones in males.
22. **Nurse cells** (Sertoli cells) surround developing spermatocytes and spermatids. At **spermiation**, a spermatozoon loses its attachment and enters the lumen of the seminiferous tubule.
23. Nurse cells are joined by tight junctions to isolate the seminiferous tubules from the general circulation by creating a **blood–testis barrier**.

26.5 The male reproductive tract receives secretions from the seminal, prostate, and bulbourethral glands p. 996

24. Spermatozoa become motile and capable of fertilization only after undergoing **capacitation**.
25. The **epididymis** consists of a **head**, **body**, and **tail**. It is here that spermatozoa complete maturation.
26. Each **ductus deferens** is part of the spermatic cord. It transports and stores spermatozoa.
27. The **seminal glands** produce about 60 percent of the volume of **semen**. Semen includes spermatozoa and **seminal fluid**.
28. The **prostate gland** encircles the proximal portion of the urethra as it leaves the urinary bladder, and produces 20–30 percent of the volume of semen.
29. The paired **bulbourethral glands** are located at the base of the penis, and secrete alkaline mucus that helps neutralize urinary acids that may remain in the urethra, and lubricates the tip of the penis.

26.6 The penis conducts urine and semen to the exterior p. 998

30. The penis conducts urine to the exterior and introduces semen into the female's vagina during sexual intercourse.
31. The penis is attached to the body wall by the **root**. The **body**, or shaft, of the penis is the tubular portion of the organ. The **glans** is the expanded distal end that surrounds the external urethral orifice. The **neck** is the narrow portion of the penis between the shaft and the glans. The **prepuce** surrounds the tip of the penis.
32. The penis consists of three cylindrical columns of **erectile tissue**. The two **corpora cavernosa** extend along the length of the penis as far as its neck. The **corpus spongiosum** surrounds the urethra.
33. The three phases of male sexual response are **arousal**, **emission**, and **ejaculation**.

34. **Impotence**, or **erectile dysfunction** (**ED**), is the inability to achieve or maintain an erection due to physical or psychological causes.

26.7 Testosterone plays a key role in establishing and maintaining male sexual function p. 1000

35. The hypothalamus releases gonadotropin-releasing hormone (GnRH), which causes the anterior pituitary to release **luteinizing hormone** (**LH**) and **follicle-stimulating hormone** (**FSH**). LH stimulates the interstitial cells of the testes to secrete **testosterone**. FSH stimulates nurse cell activity.
36. Through negative feedback, testosterone decreases GnRH secretion, and inhibin secretion by nurse cells depresses FSH production.
37. The peripheral effects of testosterone are to maintain libido, stimulate bone and muscle growth, maintain male secondary sex characteristics, and maintain organs of the male reproductive system.

SECTION 2 • Female Reproductive System

26.8 Female reproductive structures are mammary glands, gonads, external genitalia, and the reproductive tract p. 1003

38. The main organs of the female reproductive system are the ovaries, uterine tubes, uterus, vagina, and the external genitalia (clitoris and labia).
39. **Ovaries** are the female gonads, and they produce **oocytes** that mature into **ova**.
40. The **female reproductive tract** is made up of the **uterine tubes**, **uterus**, and **vagina**.
41. **Mammary glands** produce milk that nourishes the newborn infant.

26.9 The ovaries and the female reproductive tract are in close proximity but are not directly connected p. 1004

42. The **ovaries** have three main functions: **oocyte** production, secretion of **estrogens** and **progesterone**, and secretion of **inhibin**.
43. Each **uterine tube** begins with an expanded funnel called an **infundibulum**, and ends at the uterine cavity.
44. The pocket formed between the uterus and the posterior wall of the urinary bladder is the **vesicouterine pouch**. The pocket formed between the posterior wall of the uterus and the anterior surface of the colon is the **rectouterine pouch**.
45. The vagina extends from the base of the uterus to the exterior.
46. The **external genitalia** include the **clitoris** and the **labia**.
47. Each ovary is stabilized by an **ovarian ligament** and **suspensory ligament**, as well as the **mesovarium**.

26.10 Meiosis I in the ovaries produces a single haploid secondary oocyte that completes meiosis II only if fertilization occurs p. 1006

48. **Oogenesis** is the formation and development of the oocyte. Between puberty and menopause, it occurs on a monthly basis as part of the ovarian cycle.
49. Oogenesis produces one functional **secondary oocyte** and two or three **polar bodies** (nonfunctional cells that later disintegrate).
50. The **oogonia** complete their mitotic divisions before birth. Between the third and seventh months of fetal development primary oocytes undergo meiosis. The process stops at prophase of meiosis I. The meiotic process begins again at puberty when rising FSH triggers the **ovarian cycle**. Meiosis I is then completed, yielding a secondary oocyte and a polar body.
51. Each month after the ovarian cycle begins, one or more secondary oocytes leave the ovary suspended in metaphase of meiosis II.
52. At the time of fertilization, a second polar body forms and the fertilized oocyte is then called a mature **ovum**.
53. **Ovarian follicles** are specialized structures in the cortex of the ovaries where both oocyte growth and meiosis I occur.
54. An inactive primary oocyte is surrounded by a layer of follicle cells, forming a **primordial follicle**. Follicular cells are called **granulosa cells** when they enlarge, divide, and form several layers of cells around an activated primary oocyte.
55. A **zona pellucida** forms around the oocyte, and **thecal cells** form around the follicle.
56. Together, thecal cells and granulosa cells produce estrogens.
57. **Secondary follicles** develop, and then by days 10–14 of the ovarian cycle, one secondary follicle develops into a **tertiary follicle**.
58. The oocyte projects into the **antrum**. The granulosa cells associated with the secondary oocyte form a protective layer known as the **corona radiata**.
59. At **ovulation**, the tertiary follicle releases the secondary oocyte with its corona radiata into the pelvic cavity. Ovulation marks the end of the **follicular phase** of the ovarian cycle and the start of the **luteal phase**.
60. The empty tertiary follicle initially collapses, and under LH stimulation forms the **corpus luteum**, which secretes progesterone. If fertilization does not occur, the nonfunctional corpus luteum forms a pale scar called a **corpus albicans**.

26.11 The uterine tubes are connected to the uterus, a hollow organ with thick muscular walls p. 1008

61. The regions of the **uterine tube** are the funnel-like **infundibulum**, the fingerlike **fimbriae**, the **ampulla**, and **isthmus**.
62. The **layers of the uterine wall** are an outer **perimetrium**, a muscular **myometrium**, and an inner glandular **endometrium**.
63. The **uterine cavity**, or **uterine lumen**, is continuous with the **internal os** that leads to the **cervical canal**, and the **external os** that opens into the vagina.
64. The **fundus** is the rounded upper portion of the uterine body. The **body** is the largest portion of the uterus. The **cervix** is the inferior portion of the uterus.

26.12 The uterine cycle involves changes in the functional zone of the endometrium p. 1010

65. Within the myometrium, branches of the uterine arteries form **arcuate arteries**. From the arcuate arteries, **radial arteries** supply the endometrium.
66. The endometrium contains a **basilar zone** and a **functional zone**. **Straight arteries** deliver blood to the basilar zone, and **spiral arteries** supply the functional zone.

67. The basilar zone remains constant over time, but the functional zone changes in response to sex hormone levels.
68. The **uterine cycle** averages 28 days in length, but can range from 21 to 35 days.
69. The uterine cycle begins when the functional zone of the endometrium degenerates, the weakened arterial walls rupture, and blood pours into the connective tissues of the functional zone. This is called **menses**. The process of endometrial shedding is called **menstruation** and lasts from 1 to 7 days.
70. Restoration of the endometrium occurs in the **proliferative phase**. During the **secretory phase** the uterine glands enlarge, and the arteries that supply the uterine wall elongate and spiral through the functional zone. The secretory phase begins at the time of ovulation and lasts as long as the corpus luteum remains intact.
71. The uterine cycle, or **menstrual cycle**, begins at puberty. The first cycle is known as **menarche**. The cycles continue until **menopause**.

26.13 The entrance to the vagina is enclosed by external genitalia p. 1012

72. The **vagina** is an elastic, muscular tube extending between the cervix and the **vestibule**. The internal passageway is called the **vaginal canal**.
73. The vagina serves as a passageway for menstrual fluids, receives the penis during sexual intercourse, and forms the inferior portion of the birth canal.
74. The shallow recess in the vagina surrounding the tip of the **cervix** is known as the **fornix**. The vaginal lining forms folds called **rugae**.
75. Throughout childhood the vagina and vestibule are usually separated by the **hymen**, but the hymen is typically stretched or torn during sexual intercourse, tampon use, or heavy physical exercise.
76. The **mons pubis** is a bulge of adipose tissue superficial to the pubic symphysis.
77. The **labia majora** are prominent folds of skin that encircle and partially conceal the **labia minora** and adjacent structures. Extensions of the labia minora encircle the body of the **clitoris**, forming its **prepuce**.
78. The area containing the female external genitalia is called the **vulva**, or **pudendum**.
79. **Lesser vestibular glands** discharge onto the exposed surface of the vestibule, keeping it moist. During sexual arousal the **greater vestibular glands** secrete into the vestibule.

26.14 The mammary glands nourish the infant after delivery p. 1013

80. On each side, a **mammary gland** lies directly over the pectoralis major muscle within the subcutaneous tissue of the **pectoral fat pad**.
81. The **suspensory ligaments of the breast** surround the duct system and form partitions between the **lobes** and **lobules**.
82. The glandular tissue of the breast consists of lobes that contain lobules. Each lobule is composed of many **secretory alveoli**.

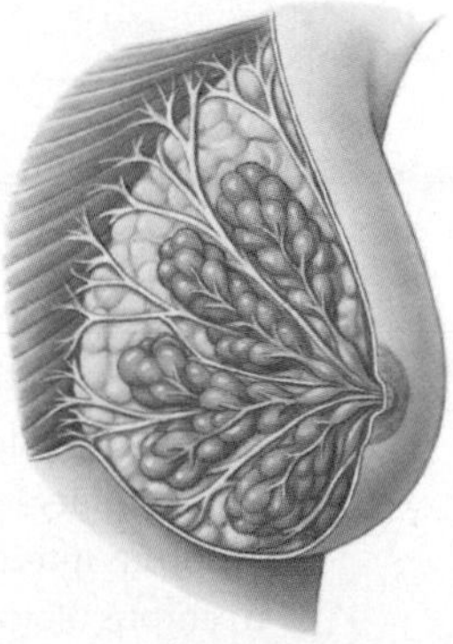

83. Ducts leaving the lobules converge to a single **lactiferous duct** that enlarges to a **lactiferous sinus**. At the **nipple**, 15 to 20 lactiferous sinuses open onto the body surface.
84. The reddish-brown skin around each nipple is the **areola**.

26.15 The ovarian and uterine cycles are regulated by hormones of the hypothalamus, pituitary gland, and ovaries p. 1014

85. The ovarian and uterine cycles must operate in synchrony to ensure proper reproductive function.
86. GnRH levels from the hypothalamus change throughout the ovarian cycle. The ovarian cycle begins with the release of GnRH, which stimulates the release of FSH from the anterior lobe of the pituitary gland.
87. The **follicular phase** of the ovarian cycle begins when FSH stimulates secondary follicles to develop into tertiary follicles. Follicles release inhibin and estrogens (especially **estradiol**, the dominant hormone prior to ovulation).
88. Inhibin decreases FSH. Estrogens affect the CNS, stimulate bone and muscle growth, maintain female secondary sex characteristics, maintain accessory glands, and stimulate endometrial growth and secretion.
89. GnRH and elevated estrogen levels stimulate LH secretion from the anterior pituitary. LH causes the completion of meiosis I and ovulation. The **luteal phase** of the ovarian cycle begins at ovulation.
90. The corpus luteum secretes progesterone, which stimulates and sustains endometrial development.
91. After ovulation GnRH decreases because progesterone levels increase and estrogens decrease. If there is no pregnancy, the corpus luteum degenerates, progesterone decreases, GnRH increases, and a new cycle begins.

26.16 Birth control strategies vary in effectiveness and in the nature of associated risks p. 1016

92. Birth control methods include **male condoms**, the **diaphragm**, **combined** and **progesterone-only oral contraceptives**, **Depo-Provera** injections, the **intrauterine device**, the **rhythm method**, **hormonal post-coital contraceptives**, and **surgical sterilization** (**vasectomy** and **tubal ligation**).

26.17 Reproductive system disorders are relatively common and often deadly p. 1018

93. **Benign prostatic hypertrophy** (**BPH**) occurs in men over age 50 as testosterone decreases. It can cause urinary obstruction. **Prostate cancer** is the second most common cause of cancer deaths in men, and can be screened for with the **prostate-specific antigen** (**PSA**) blood test.
94. Although **testicular cancer** is rare, it is the most common cancer among males aged 15–35. It has a near 95 percent survival rate.
95. Benign **fibrocystic disease** and **breast cancer** are distinguished with a biopsy.
96. **Ovarian cancer** is dangerous because it is seldom diagnosed in its early stages.
97. **Cervical cancer** is the most common cancer of the reproductive system in women ages 15–34. The *Gardasil* vaccine protects against four types of **human papillomavirus** (**HPV**), which cause about 75 percent of cervical cancers.
98. At least two dozen **sexually transmitted diseases** (**STDs**)/**sexually transmitted infections** (**STIs**) can be acquired primarily or exclusively by sexual intercourse.

Chapter Review Questions

Labeling

Label the male and female reproductive structures shown here.

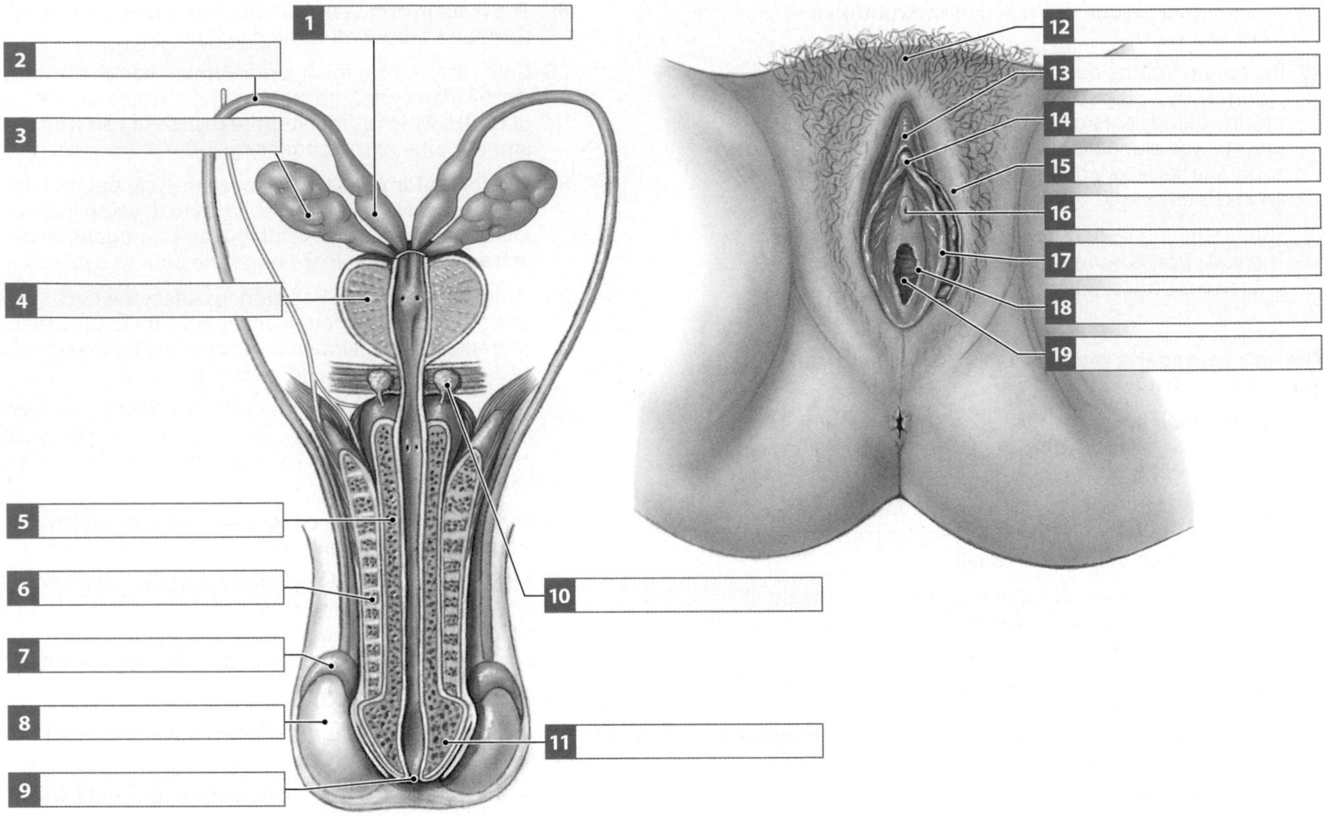
1
2
3
4
5
6
7
8
9
10
11
12
13
14
15
16
17
18
19

True/False

Indicate whether each statement is true or false.

20 The prostate gland is the site of sperm maturation.

21 Secondary spermatocytes contain 23 chromosomes.

22 In the male, LH induces the secretion of testosterone and other androgens by the interstitial cells of the testes.

23 At the end of meiosis II in females, four functional oocytes are produced.

24 A massive surge of LH on or around day 14 of the menstrual cycle triggers ovulation and formation of the corpus luteum.

25 *Gardasil* is a vaccine that protects against the virus that causes about 75 percent of ovarian cancers.

20 ______________

21 ______________

22 ______________

23 ______________

24 ______________

25 ______________

Multiple choice

Select the correct answer from the list provided.

26 In females, meiosis II is not completed until
- ☐ a) birth.
- ☐ b) ovulation.
- ☐ c) fertilization.
- ☐ d) uterine implantation.

27 Progesterone secretion is performed by the
- ☐ a) primary follicle.
- ☐ b) secondary follicle.
- ☐ c) corpus luteum.
- ☐ d) corpus albicans.

28 Spermatogenesis begins with the division of spermatogonia in the
- ☐ a) epididymis.
- ☐ b) seminiferous tubules.
- ☐ c) ductus deferens.
- ☐ d) bulbourethral glands.

29 Erection of the penis is a result of
- ☐ a) parasympathetic neurons releasing nitric oxide at their axon terminals, causing the penile arteries to dilate, increasing blood flow to the penis.
- ☐ b) sympathetic neurons releasing nitric oxide at their axon terminals, causing the penile arteries to dilate, increasing blood flow to the penis.
- ☐ c) parasympathetic neurons releasing nitric oxide at their axon terminals, causing the penile veins to constrict, decreasing blood flow from the the penis.
- ☐ d) sympathetic neurons releasing nitric oxide at their axon terminals, causing the penile veins to constrict, decreasing blood flow from the penis.

30 Estrogens are produced by
- ☐ a) the zona pellucida and the corona radiata.
- ☐ b) primordial follicles and the zona pellucida.
- ☐ c) primary oocytes and cells of the corpus luteum.
- ☐ d) thecal cells and granulosa cells.

31 About 60 percent of the volume of semen is produced by the
- ☐ a) seminal glands.
- ☐ b) prostate gland.
- ☐ c) bulbourethral glands.
- ☐ d) ductus deferens.

Short answer

32 Explain how oral contraceptives that contain estrogen and progesterone, or only progesterone, prevent ovulation.

33 What happens to the tertiary follicle after ovulation?

34 Identify the sterilization procedures for men and women. How is sexual function affected as a result of each?

MasteringA&P®

Access more chapter study tools online in the MasteringA&P Study Area:

- **Chapter Quizzes, Chapter Practice Test, Art-labeling Activities, Animations, MP3 Tutor Sessions, and Clinical Case Studies**

- **Practice Anatomy Lab** PAL™
- **Interactive Physiology** iP®
- **A&P Flix** A&PFlix™
- **PhysioEx** PhysioEx™

Chapter Integration • Applying what you have learned

Exercise and the absence of menstruation

Exercise-induced amenorrhea, the abnormal absence of menstruation, occurs in 5–25 percent of women athletes. The variability depends on the level of competition and type of sport. For example, female bodybuilders and ballet dancers, both of whom are likely to have low body fat, commonly experience amenorrhea. Among well-nourished female athletes, hard training and exercise may cause the release of stress hormones, which then interfere with the pituitary gland's production of the hormones necessary for maintaining the menstrual cycle.

Although temporary cessation of menstrual cycles itself is not dangerous, there are long-term health consequences for prolonged exercise-induced amenorrhea in premenopausal women. As a result of long-term exercise-induced amenorrhea, women become estrogen deficient, which can lead to other health-related consequences.

Once the diagnosis is confirmed, treatment involves increasing caloric intake and restoring estrogen levels to the normal range. In most cases, exercise-induced amenorrhea is reversible with treatment.

1 **What does amenorrhea in female athletes suggest about the relationship between body fat and menstruation?**

2 **How might exercise-induced amenorrhea affect pregnancy?**

3 **What are some health-related consequences of estrogen deficiency?**

27 Development and Inheritance

LEARNING OUTCOMES

These Learning Outcomes correspond by number to this chapter's modules and indicate what you should be able to do after completing the chapter.

SECTION 1 • Overview of Development

- **27.1** Describe the various stages of gestation and development.
- **27.2** Describe the process of fertilization, and explain the significance of multiple sperm to ensuring its success.
- **27.3** Discuss cleavage, blastocyst formation, and implantation of the blastocyst in the uterine wall.
- **27.4** Describe gastrulation and the formation of the three germ layers.
- **27.5** Identify and describe the formation, location, and functions of the extra-embryonic membranes.
- **27.6** Discuss the roles of the extra-embryonic membranes in embryonic development and placenta formation.
- **27.7** Discuss the importance of the placenta to the fetus and as an endocrine organ.
- **27.8** Describe organogenesis and its role in the developing fetus.
- **27.9** Describe the interplay between maternal organ systems and the developing fetus.
- **27.10** List and discuss the events that occur during labor and delivery.
- **27.11** Identify the features of and the functions associated with the various life stages.
- **27.12** Explain the roles of hormones in males and females at puberty.

SECTION 2 • Genetics and Inheritance

- **27.13** Describe genetics and inheritance and the relationship between genotype and phenotype.
- **27.14** Relate the basic principles of genetics to the inheritance of human traits.
- **27.15** Describe the relationships among the various forms of inheritance, and give examples of representative phenotypic characters, both normal and abnormal.
- **27.16** **CLINICAL MODULE** Identify several chromosomal disorders, and describe the human genome and epigenome.

LO Learning Outcomes are repeated at the bottom of each module.

Gestation and development are marked by various stages

1 **Gestation** (jes-TĀ-shun) is the time spent in prenatal development within the womb. **Development** is the gradual modification of anatomical structures and physiological characteristics during the period from fertilization to maturity. The changes that occur during development are truly remarkable. In a mere 9 months, all the tissues, organs, and organ systems we have studied so far take shape and begin to function. What begins as a single cell slightly larger than the period at the end of this sentence becomes an individual whose body contains trillions of cells organized into a complex array of highly specialized structures.

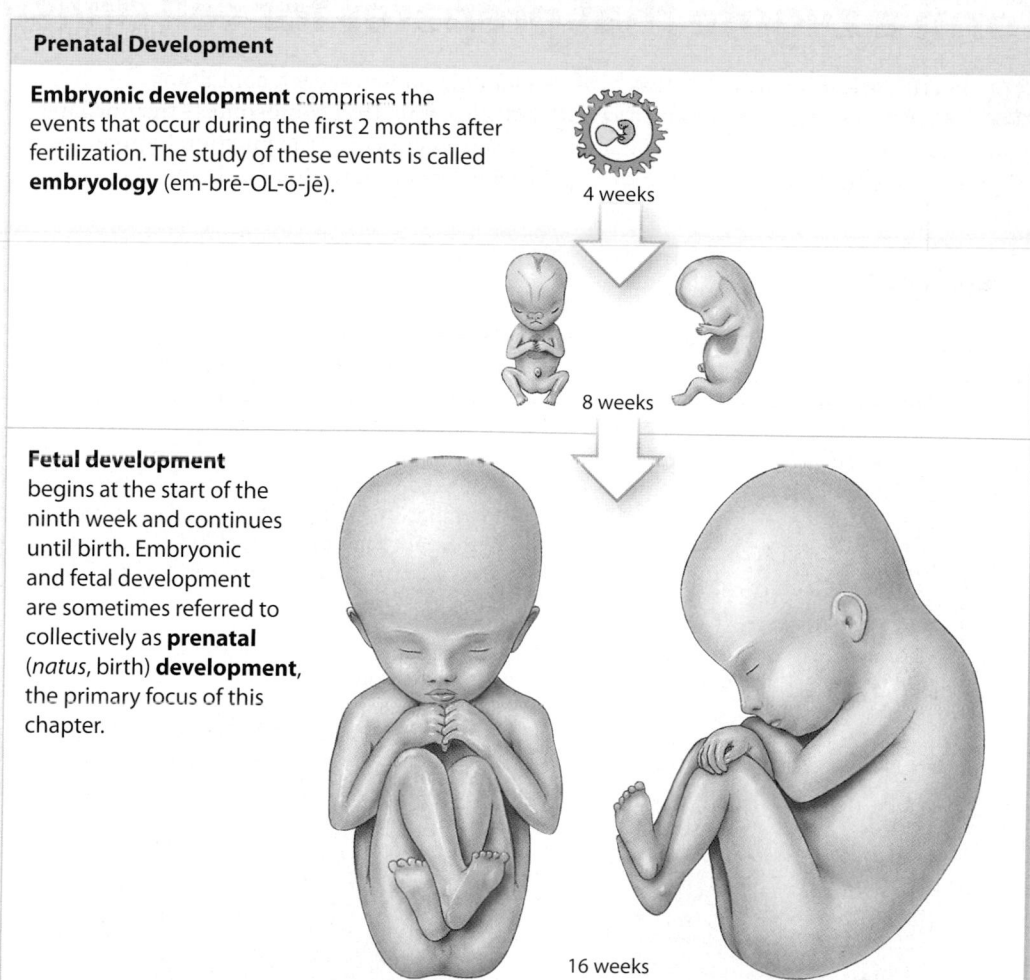

Prenatal Development

Embryonic development comprises the events that occur during the first 2 months after fertilization. The study of these events is called **embryology** (em-brē-OL-ō-jē).

4 weeks

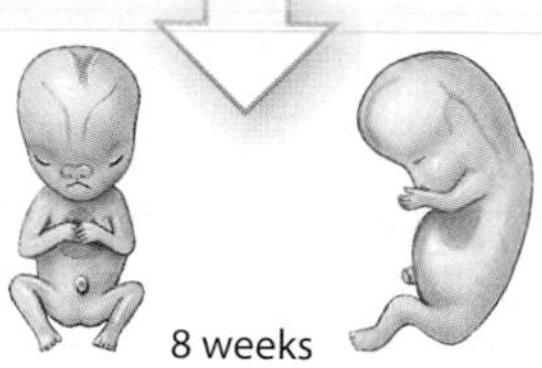
8 weeks

Fetal development begins at the start of the ninth week and continues until birth. Embryonic and fetal development are sometimes referred to collectively as **prenatal** (*natus*, birth) **development**, the primary focus of this chapter.

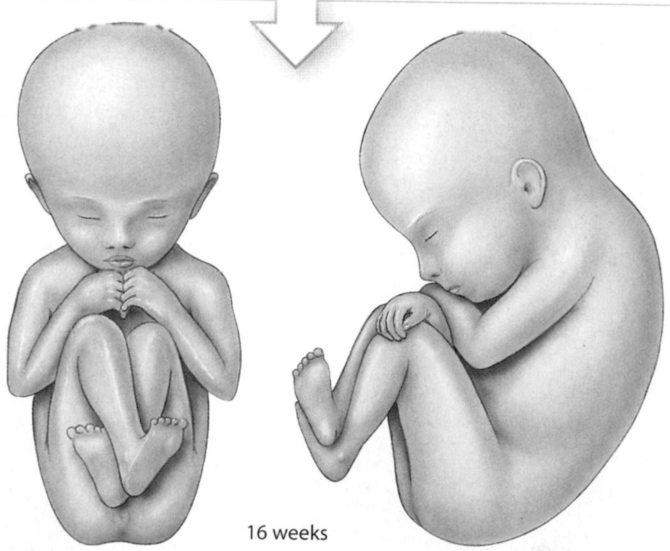
16 weeks

Gestation		
First Trimester	**Second Trimester**	**Third Trimester**
The **first trimester** is the time of embryonic and early fetal development. During this time, the rudiments (beginning structures) of all the major organ systems appear.	The **second trimester** is dominated by the development of organs and organ systems, a process that nears completion by the end of the sixth month. During this time, body shape and proportions change. By the end of this trimester, the fetus looks distinctively human.	The **third trimester** is characterized by the largest gain in fetal weight. Early in the third trimester, most of the fetus's major organ systems become fully functional. An infant born 1 month or even 2 months prematurely has a reasonable chance of survival if appropriate medical care is available.

Postnatal development begins at birth and continues to **maturity**, the state of full development or completed growth. A basic understanding of prenatal and postnatal development gives important insights into anatomical structures. In addition, many of the mechanisms of development and growth are similar to those responsible for tissue repair following injury.

2 For convenience, we usually think of the gestation period as three integrated trimesters, each 3 months long.

Module 27.1 Review

a. Define gestation.

b. Distinguish among embryonic, fetal, and prenatal development.

c. What is postnatal development?

LO 27.1 Describe the various stages of gestation and development.

At fertilization, an ovum and a spermatozoon form a zygote that prepares for cell division

1 **Fertilization** involves the fusion of two haploid gametes, each containing 23 chromosomes, producing a **zygote** (ZĪ-gōt) with 46 chromosomes—the normal complement in a somatic cell. Fertilization typically occurs near the junction between the ampulla and isthmus of the uterine tube, generally within a day after ovulation.

Fertilization

Step 1
Oocyte at Ovulation
Ovulation releases a secondary oocyte and the first polar body; both are surrounded by the corona radiata. The oocyte is suspended in metaphase of meiosis II (Module 26.10, p. 1006).

Step 2
Fertilization and Oocyte Activation
Acrosomal enzymes from multiple spermatozoa create gaps between the cells of the corona radiata. A single spermatozoon then makes contact with the oocyte membrane, and membrane fusion occurs, triggering **oocyte activation*** and the completion of meiosis. The secondary oocyte is now an **ovum**.

Step 3
Pronucleus Formation Begins
The sperm is absorbed into the cytoplasm, and the female nuclear material within the ovum reorganizes as the **female pronucleus**.

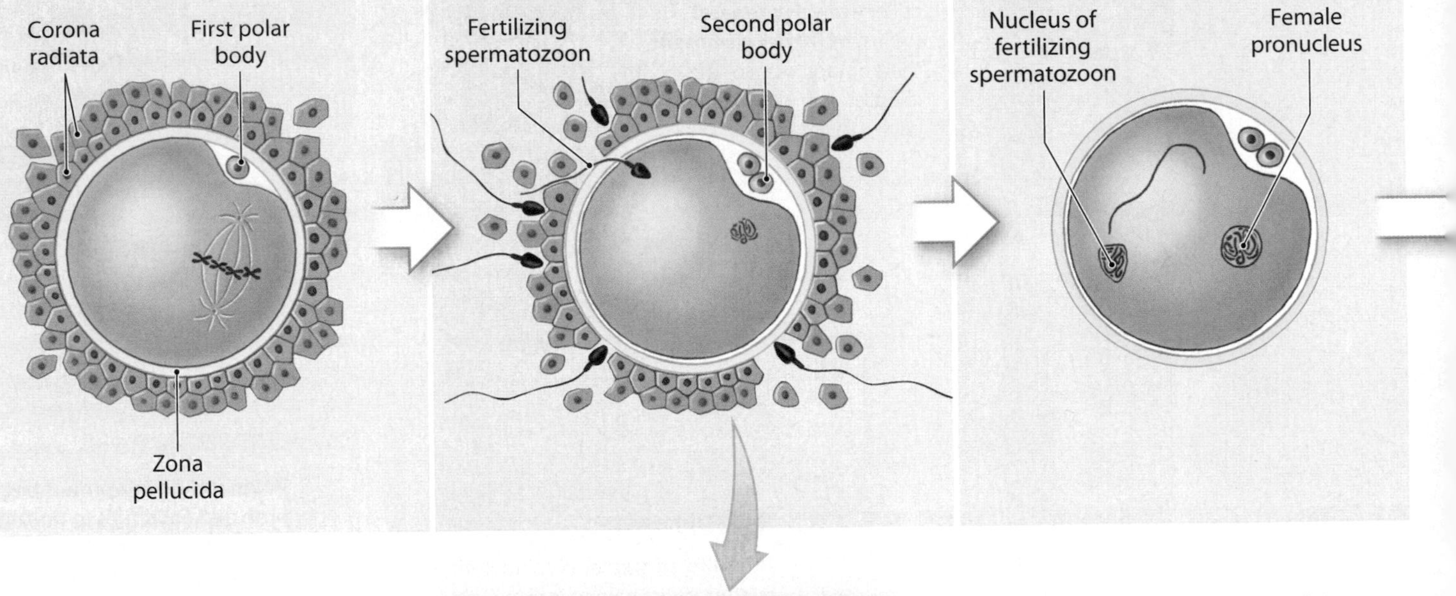
Corona radiata
First polar body
Zona pellucida
Fertilizing spermatozoon
Second polar body
Nucleus of fertilizing spermatozoon
Female pronucleus

2 This is a photograph of a secondary oocyte surrounded by spermatozoa. The function of a spermatozoon is to deliver the paternal (father's) chromosomes to the secondary oocyte and then facilitate the process of fertilization. In contrast, the secondary oocyte provides all the cellular organelles and inclusions, nourishment, and genetic programming necessary to support embryo development for nearly a week after conception. The volume of the secondary oocyte is now 2000 times greater than that of a spermatozoon.

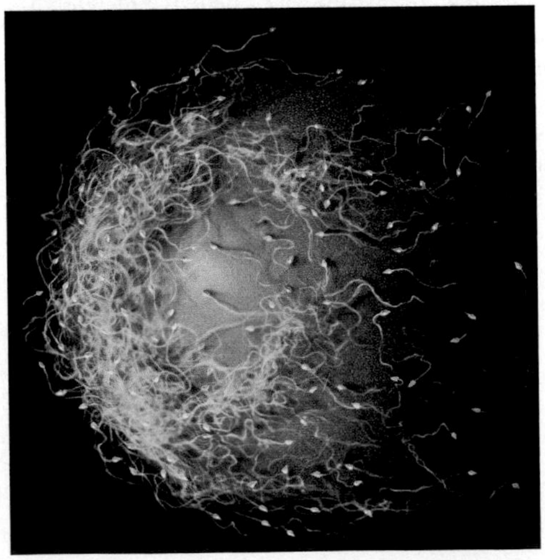

*Note: Oocyte activation is a series of changes in the metabolic activity of the oocyte leading to Ca^{2+} release from the smooth ER. The rise in Ca^{2+} has three effects: (1) Enzymes are released that prevent fertilization by more than one sperm, (2) completion of meiosis II and formation of the second polar body, and (3) activation of enzymes that cause a rapid increase in the cell's metabolic rate.

Step 4
Spindle Formation and Cleavage Preparation
The **male pronucleus** develops, and spindle fibers appear in preparation for cell division. This is the start of the process of **cleavage**, a series of cell divisions that produces an increasing number of smaller and smaller daughter cells.

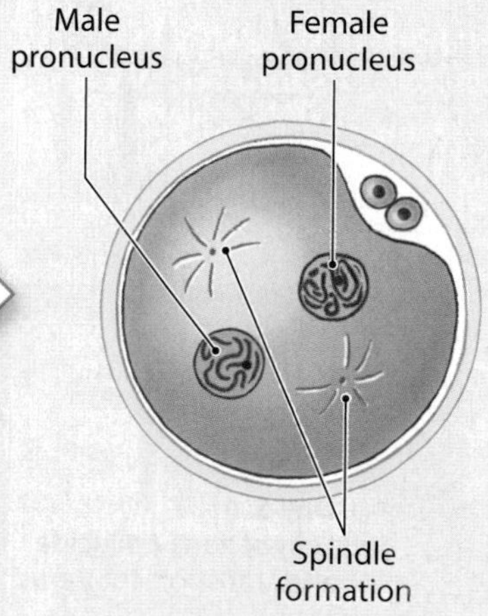
Male pronucleus
Female pronucleus
Spindle formation

Step 5
Amphimixis Occurs and Cleavage Begins
The male pronucleus migrates toward the center of the ovum, where the spindle fibers are forming. The two pronuclei then fuse in a process called **amphimixis** (am-fi-MIK-sis). The cell is now a zygote that contains the normal complement of 46 chromosomes. Fertilization is complete.

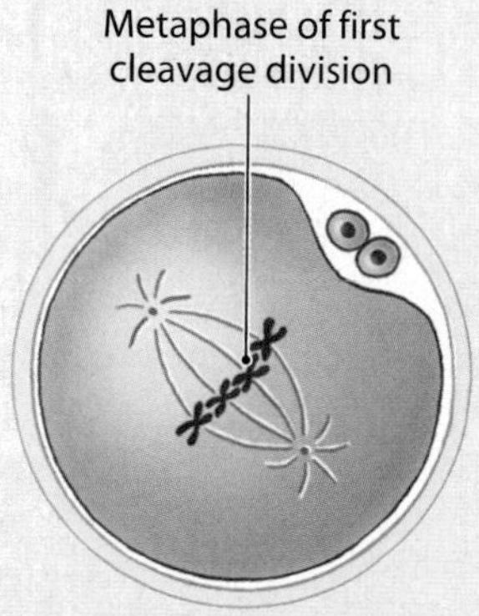
Metaphase of first cleavage division

This is the "moment of conception." Almost immediately the chromosomes line up along a metaphase plate, and the cell prepares to divide.

Step 6
Cytokinesis Begins
The first cleavage division nears completion about 30 hours after fertilization. This division produces two daughter cells, each one-half the size of the original zygote. These cells are called **blastomeres** (BLAS-tō-mērz).

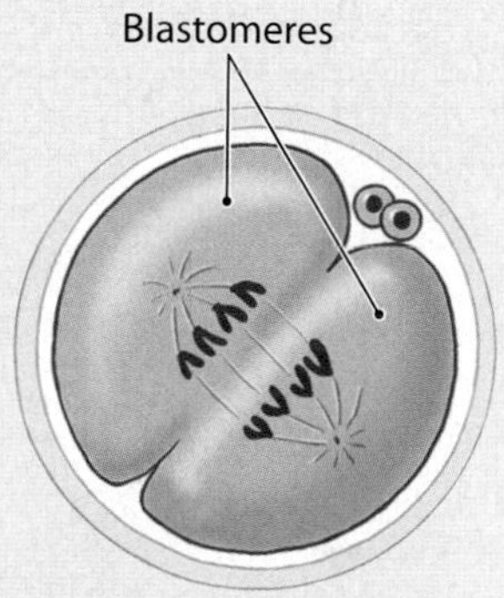
Blastomeres

Of the approximately 200 million spermatozoa introduced into the vagina in a typical ejaculation, only about 10,000 enter a uterine tube, and fewer than 100 reach the isthmus. In general, a male with a sperm count below 20 million per milliliter is functionally sterile because too few spermatozoa survive to reach and fertilize an oocyte. While it is true that only one spermatozoon fertilizes an oocyte, dozens of spermatozoa are required for successful fertilization. The additional sperm are essential because one sperm does not contain enough acrosomal enzymes to disrupt the corona radiata that surrounds the secondary oocyte.

Module 27.2 Review

a. Define fertilization.
b. How many chromosomes are contained within a human zygote?
c. Why are numerous spermatozoa required to fertilize a secondary oocyte?

LO **27.2** Describe the process of fertilization, and explain the significance of multiple sperm to ensuring its success.

Cleavage continues until the blastocyst implants in the uterine wall

During cleavage, the cytoplasm of the zygote becomes subdivided among an increasing number of progressively smaller blastomeres. A group of blastomeres created by cleavage divisions is called a **pre-embryo**.

1 Cleavage lasts about 7 days. During that time the pre-embryo travels the length of the uterine tube. After 3 days, the pre-embryo is a solid ball of cells known as a **morula** (MOR-ū-la; *morus*, mulberry). The morula typically reaches the uterus on day 4. Over the next 2 days, the blastomeres form a **blastocyst** (BLAS-tō-sist), a hollow ball with an inner cavity known as the **blastocoele** (BLAS-tō-sēl). At this stage the blastomeres are no longer identical in size and shape.

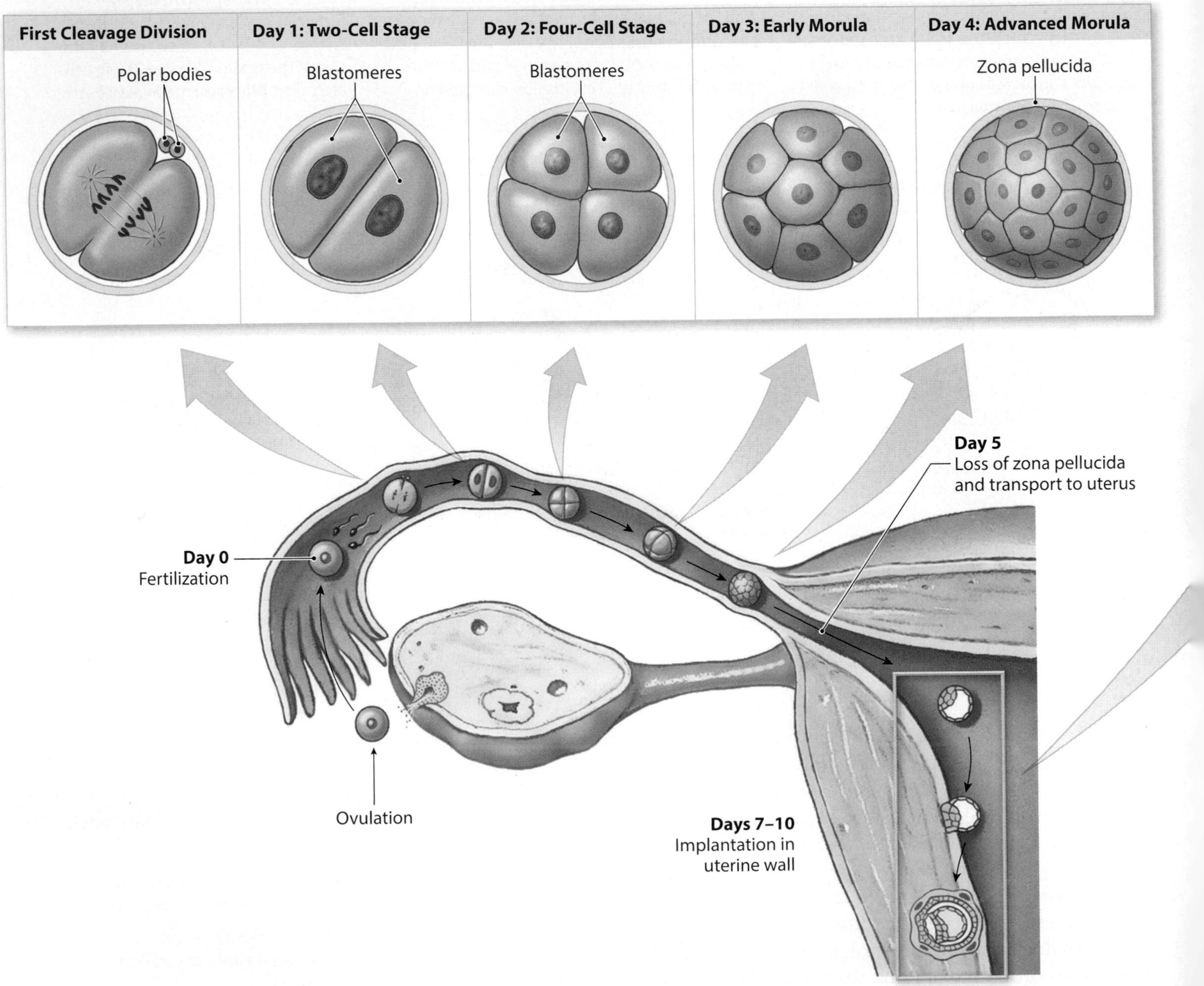
First Cleavage Division
Polar bodies
Day 1: Two-Cell Stage
Blastomeres
Day 2: Four-Cell Stage
Blastomeres
Day 3: Early Morula
Day 4: Advanced Morula
Zona pellucida
Day 5
Loss of zona pellucida and transport to uterus
Day 0
Fertilization
Ovulation
Days 7–10
Implantation in uterine wall

Day 6: Blastocyst

The blastocyst is freely exposed to the fluid contents of the uterine cavity with the loss of the zona pellucida. The uterine cavity contains the glycogen-rich secretions of the uterine glands. The rate of growth and cell division now accelerates, and the blastocyst enlarges rapidly.

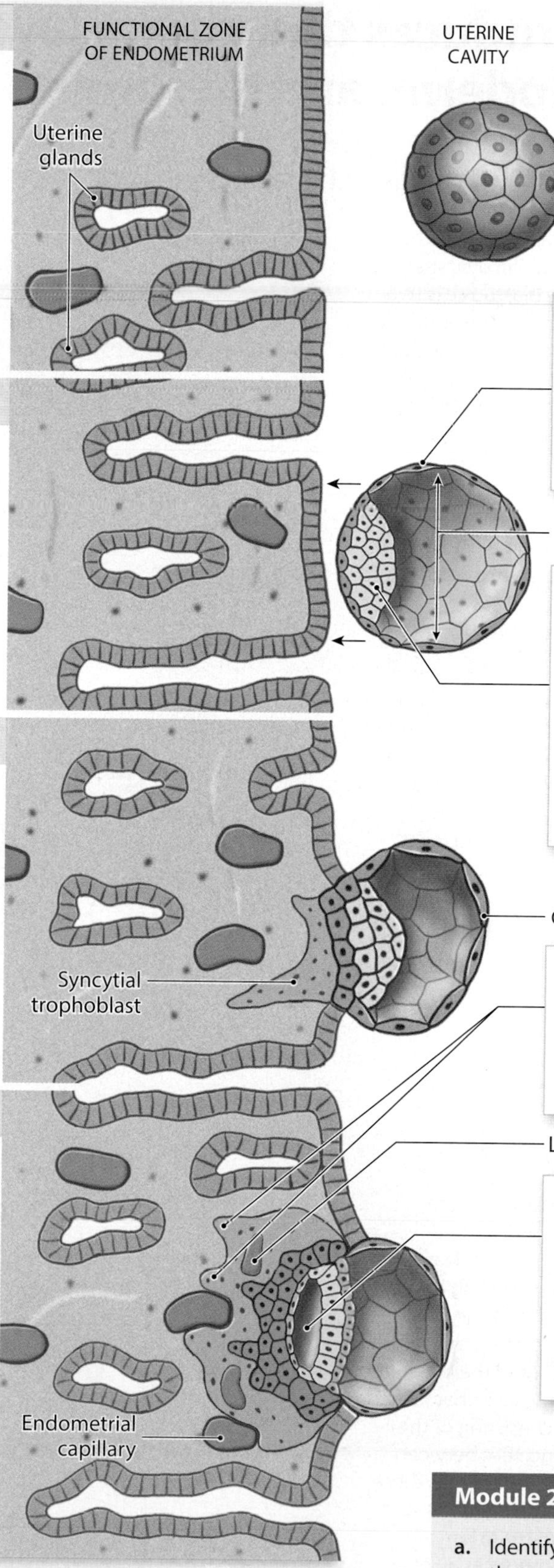

FUNCTIONAL ZONE OF ENDOMETRIUM
UTERINE CAVITY
Uterine glands
Blastocyst
Blastocoele
Cellular trophoblast
Syncytial trophoblast
Lacuna
Endometrial capillary

The **trophoblast** (TRŌ-fō-blast; *trophos*, food + *blast*, precursor) is the cell layer surrounding the blastocyst. The cells in this layer nourish the embryo and later form part of the placenta.

Day 7: Implantation

When fully formed, the blastocyst contacts the endometrium. **Implantation** begins with the attachment of the blastocyst to the endometrium of the uterus. Implantation proceeds as the blastocyst erodes the endometrial lining and becomes enclosed within the endometrium by day 10.

The **inner cell mass** lies clustered at one end of the blastocyst. These cells are exposed to the blastocoele but are insulated from contact with the intrauterine environment by the trophoblast. In time, the inner cell mass will form the embryo.

Day 8: Trophoblast Development

At the point of contact, the trophoblast cells divide rapidly, making the trophoblast several layers thick. The cells closest to the blastocoele remain intact, forming a layer of **cellular trophoblast**. Near the endometrial wall, the plasma membranes separating the trophoblast cells disappear, creating a layer of cytoplasm containing multiple nuclei. This layer is called the **syncytial** (sin-SISH-ul) **trophoblast**.

Fingerlike **villi** extend away from the trophoblast into the surrounding endometrium, gradually increasing in size and complexity.

Day 9: Formation of Amniotic Cavity

As implantation proceeds, the syncytial trophoblast continues to enlarge and erode the surrounding endometrium. Nutrients released by the eroding uterine glands are absorbed by the syncytial trophoblast and distributed by diffusion through the underlying cellular trophoblast to the inner cell mass. These nutrients provide the energy needed to support the early stages of embryo formation. Trophoblastic extensions (villi) grow around endometrial capillaries. As the capillary walls are destroyed, maternal blood begins to percolate through trophoblastic channels known as **lacunae**.

At the time of implantation, the inner cell mass has separated from the trophoblast. The separation gradually increases, creating a fluid-filled chamber called the **amniotic** (am-nē-OT-ik) **cavity**.

Module 27.3 Review

a. Identify the final stage of development that results from cleavage.

b. Describe the trophoblast.

c. Describe the blastocyst and its role in implantation.

LO 27.3 Discuss cleavage, blastocyst formation, and implantation of the blastocyst in the uterine wall.

Module 27.4

Gastrulation produces three germ layers: ectoderm, endoderm, and mesoderm

Day 9: Formation of Amniotic Cavity (continued)

When the amniotic cavity first appears, the cells of the inner cell mass are organized into an oval sheet known as the **blastodisc**. The early blastodisc is two layers thick: a superficial layer that faces the amniotic cavity, and a deeper layer that is exposed to the fluid contents of the blastocoele. Cells of the superficial layer migrate along the walls of the amniotic cavity and separate the amniotic cavity from the trophoblast. This is the first step in the formation of the **amnion**, one of four **extra-embryonic membranes** we will consider further in Module 27.5. At this stage nutrients released into the amnion and blastocoele by the advancing trophoblast are absorbed directly by the cells of the blastodisc.

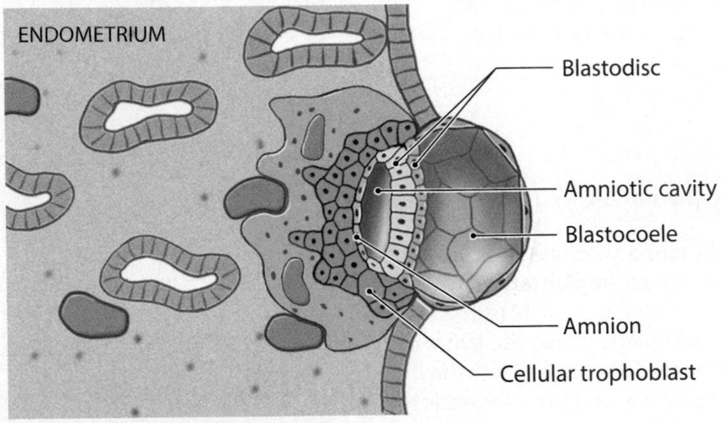
ENDOMETRIUM
Blastodisc
Amniotic cavity
Blastocoele
Amnion
Cellular trophoblast

Day 10: Yolk Sac Formation

While cells from the superficial layer of the inner cell mass migrate around the amniotic cavity, forming the amnion, cells from the deeper layer migrate around the outer edges of the blastocoele. This is the first step in the formation of the **yolk sac**, a second extra-embryonic membrane. For roughly the next 2 weeks, the yolk sac is the primary nutrient source for the inner cell mass. It absorbs and distributes nutrients released into the blastocoele by the trophoblast.

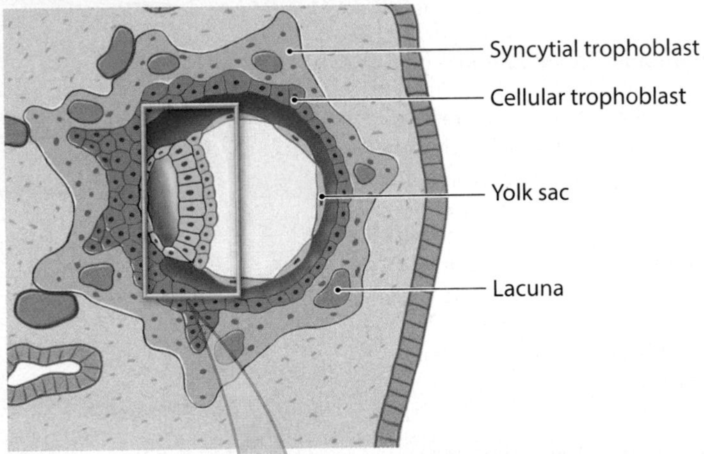
Syncytial trophoblast
Cellular trophoblast
Yolk sac
Lacuna

Day 12: Gastrulation

By day 12, superficial cells of the blastodisc are migrating toward a central line known as the **primitive streak**. Here, the migrating cells leave the surface and move between the two existing layers. This movement creates three distinct embryonic layers: (1) the **ectoderm**, consisting of superficial cells that did not migrate into the interior of the blastodisc; (2) the **endoderm**, consisting of the cells that face the yolk sac; and (3) the **mesoderm**, consisting of the poorly organized layer of migrating cells between the ectoderm and the endoderm. Collectively, these three embryonic layers are called **germ layers**, and the migration process is called **gastrulation** (gas-troo-LĀ-shun). Gastrulation produces an oval, three-layered sheet known as the **embryonic disc**. This disc will form the body of the embryo, whereas all other cells of the blastocyst will be part of the extra-embryonic membranes.

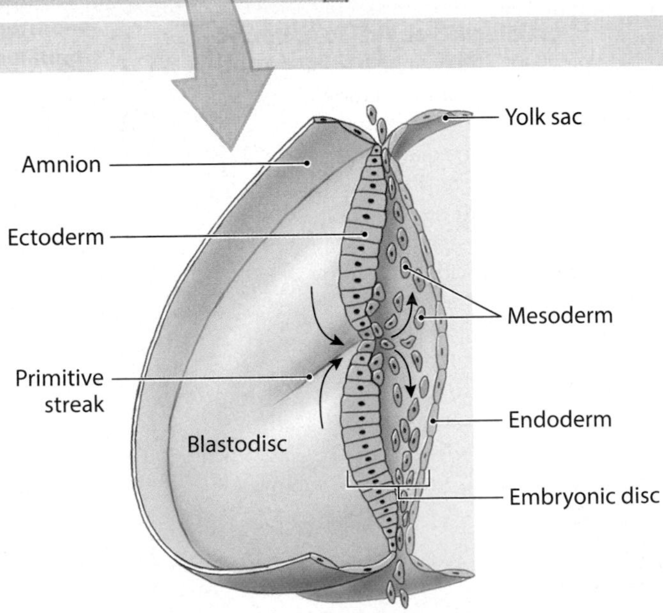
Yolk sac
Amnion
Ectoderm
Mesoderm
Primitive streak
Endoderm
Blastodisc
Embryonic disc

The panels to the left illustrate the formation of the three germ layers, and the table below summarizes their importance to later development.

The Fates of the Germ Layers

Body System	Ectodermal Contributions
Integumentary system	Epidermis, hair follicles and hairs, nails, and glands communicating with the skin (sweat glands, mammary glands, and sebaceous glands)
Skeletal system	Pharyngeal cartilages and their derivatives in adults (portion of sphenoid, the auditory ossicles, the styloid processes of the temporal bones, the cornu and superior rim of the hyoid bone)
Nervous system	All neural tissue, including brain and spinal cord
Endocrine system	Pituitary gland and adrenal medullae
Respiratory system	Mucous epithelium of nasal passageways
Digestive system	Mucous epithelium of mouth and anus, salivary glands
	Mesodermal Contributions
Integumentary system	Dermis and hypodermis
Skeletal system	All structures except some pharyngeal derivatives
Muscular system	All structures
Endocrine system	Adrenal cortex, endocrine tissues of heart, kidneys, and gonads
Cardiovascular system	All structures
Lymphatic system	All structures
Urinary system	The kidneys, including the nephrons and the initial portions of the collecting system
Reproductive system	The gonads and the adjacent portions of the duct systems
Miscellaneous	The lining of the pleural, pericardial, and peritoneal cavities and the connective tissues that support all organ systems
	Endodermal Contributions
Endocrine system	Thymus, thyroid gland, and pancreas
Respiratory system	Respiratory epithelium (except nasal passageways) and associated mucous glands
Digestive system	Mucous epithelium (except mouth and anus), exocrine glands (except salivary glands), liver, and pancreas
Urinary system	Urinary bladder and distal portions of the duct system
Reproductive system	Distal portions of the duct system, stem cells that produce gametes

The trophoblast undergoes repeated nuclear divisions, shows extensive and rapid growth, has a very high demand for energy, invades and spreads through adjacent tissues, yet fails to activate the maternal immune system—in short, the trophoblast has many of the characteristics of cancer cells. In about 0.1 percent of pregnancies, something goes wrong with the regulatory mechanisms, and instead of developing normally, the syncytial trophoblast behaves like a tumor. This condition is called **gestational trophoblastic neoplasia**. Approximately 20 percent of gestational trophoblastic neoplasias metastasize to other tissues, with potentially fatal results. Consequently, prompt surgical removal of the mass is essential, and the surgery is sometimes followed by chemotherapy.

Module 27.4 Review

a. Describe gastrulation and the formation of the germ layers.

b. What germ layer gives rise to nearly all body systems except the nervous and respiratory systems?

c. Define gestational trophoblastic neoplasia.

LO **27.4** Describe gastrulation and the formation of the three germ layers.

The extra-embryonic membranes form the placenta that supports fetal growth and development

Germ layers form four extra-embryonic membranes: the yolk sac (endoderm + mesoderm), the amnion (ectoderm + mesoderm), the allantois (endoderm + mesoderm), and the chorion (mesoderm + trophoblast). Although these membranes support embryonic and fetal development, few traces of their existence remain in adult systems.

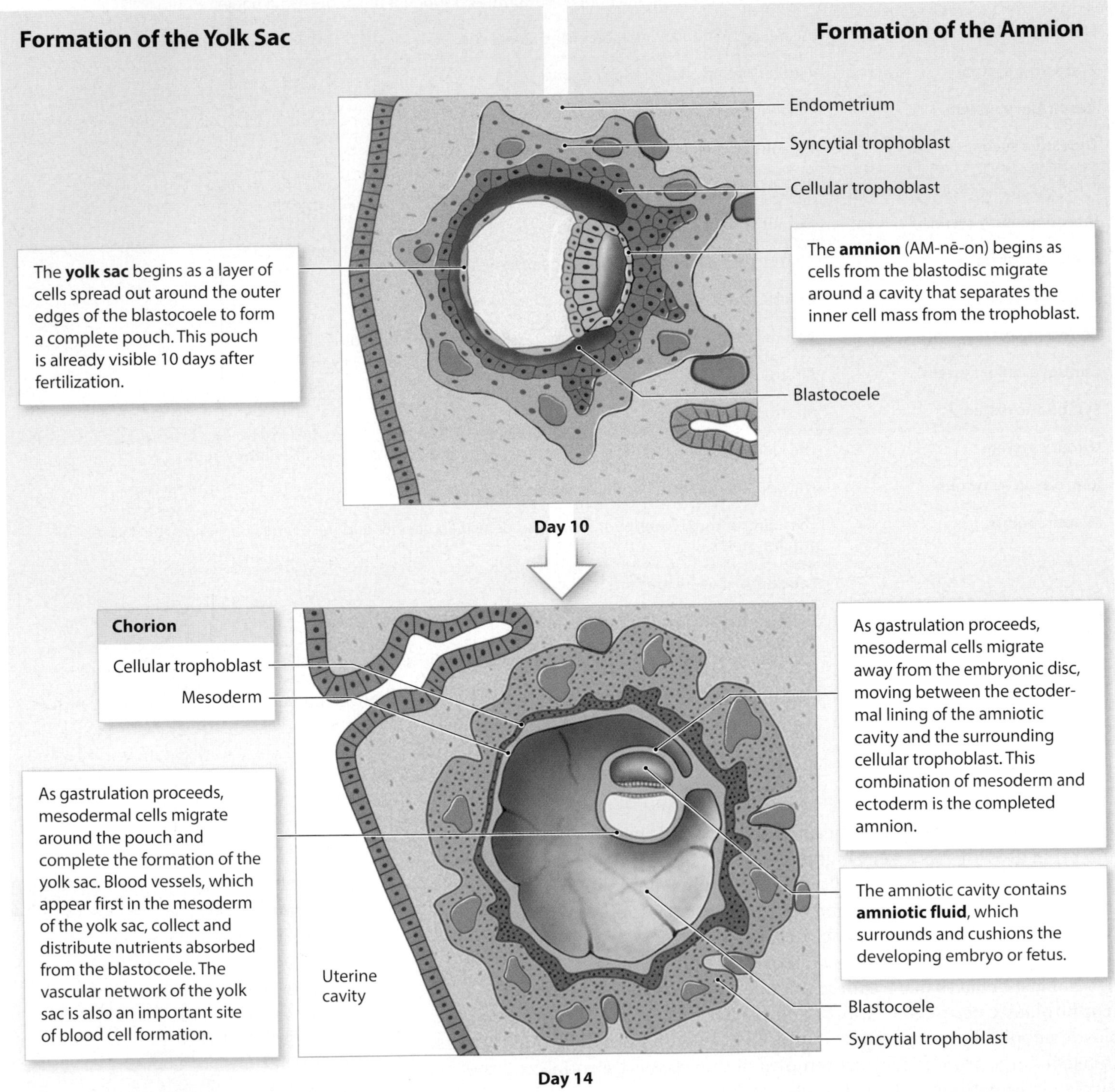
Formation of the Yolk Sac
Formation of the Amnion
Endometrium
Syncytial trophoblast
Cellular trophoblast
The **amnion** (AM-nē-on) begins as cells from the blastodisc migrate around a cavity that separates the inner cell mass from the trophoblast.
The **yolk sac** begins as a layer of cells spread out around the outer edges of the blastocoele to form a complete pouch. This pouch is already visible 10 days after fertilization.
Blastocoele
Day 10
Chorion
Cellular trophoblast
Mesoderm
As gastrulation proceeds, mesodermal cells migrate away from the embryonic disc, moving between the ectodermal lining of the amniotic cavity and the surrounding cellular trophoblast. This combination of mesoderm and ectoderm is the completed amnion.
As gastrulation proceeds, mesodermal cells migrate around the pouch and complete the formation of the yolk sac. Blood vessels, which appear first in the mesoderm of the yolk sac, collect and distribute nutrients absorbed from the blastocoele. The vascular network of the yolk sac is also an important site of blood cell formation.
The amniotic cavity contains **amniotic fluid**, which surrounds and cushions the developing embryo or fetus.
Uterine cavity
Blastocoele
Syncytial trophoblast
Day 14

Formation of the Allantois

Formation of the Chorion

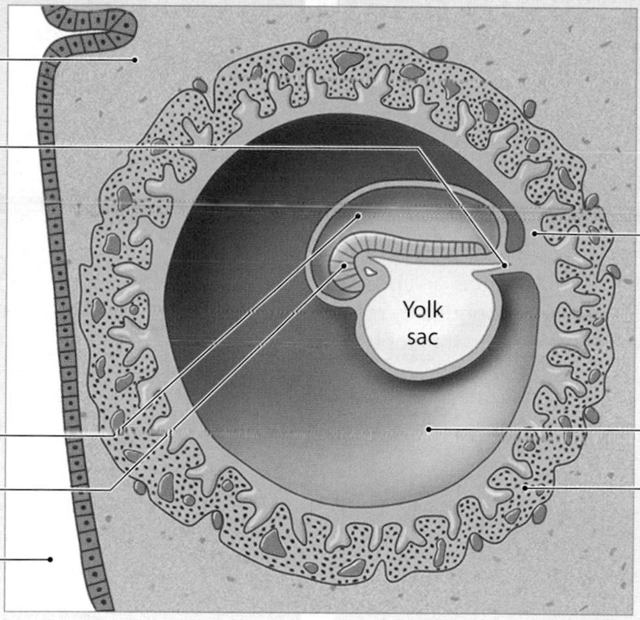
Endometrium
Yolk sac
Amniotic cavity
Embryo
Uterine cavity
Blastocoele
Syncytial trophoblast

The **allantois** (a-LAN-tō-is) begins as an outpocketing of the endoderm near the base of the yolk sac. The free endodermal tip then grows toward the wall of the blastocyst, surrounded by a mass of mesodermal cells.

The mesoderm associated with the allantois spreads around the blastocyst, separating the cellular trophoblast from the blastocoele. This combination of mesoderm and trophoblast is the **chorion** (KŌ-rē-on).

Week 3

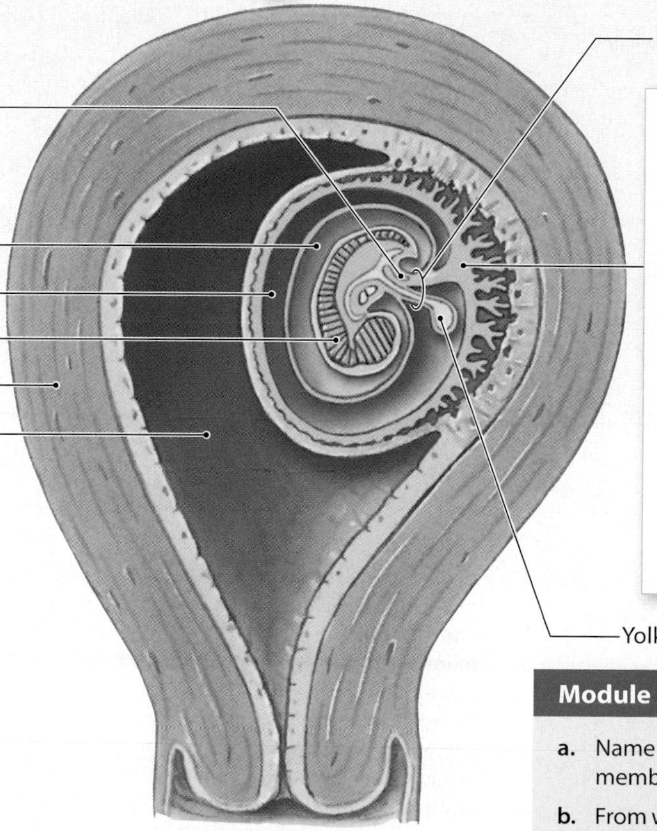
Umbilical stalk
Amniotic cavity
Blastocoele
Embryo
Uterus
Uterine cavity
Yolk sac

The allantois extends partway into the umbilical stalk. The base of the allantois will form the urinary bladder.

The **placenta** forms the interface between embryonic/fetal and maternal systems. It develops as villi of the chorion invade the endometrium and break down maternal blood vessels. The placenta becomes the primary support structure for the developing embryo (and later, the fetus). The placenta is the site where oxygen and nutrients are absorbed from the maternal bloodstream and exchanged for carbon dioxide and wastes.

Week 5

Module 27.5 Review

a. Name the four extra-embryonic membranes.

b. From which germ layers do the extra-embryonic membranes form, and what are each membrane's functions?

c. Describe the placenta.

27.5 Identify and describe the formation, location, and functions of the extra-embryonic membranes.

The formation of extra-embryonic membranes is associated with major changes in the shape and complexity of the embryo

Week 2

Migration of mesoderm around the inner surface of the trophoblast creates the chorion. Mesodermal migration around the outside of the amniotic cavity, between the ectodermal cells and the trophoblast, forms the amnion. Mesodermal migration around the endodermal pouch creates the yolk sac.

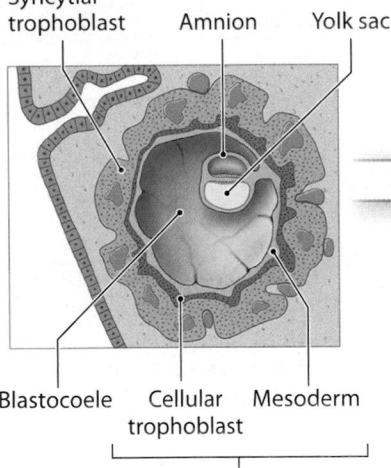

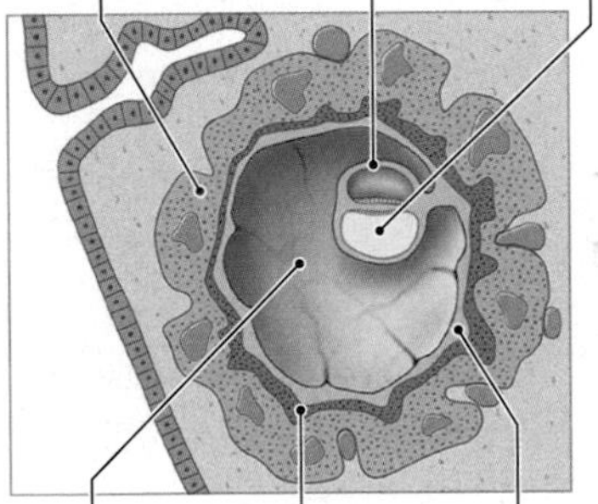
Syncytial trophoblast
Amnion
Yolk sac
Blastocoele
Cellular trophoblast
Mesoderm
Chorion

Week 3

The embryonic disc bulges into the amniotic cavity at the **head fold**. The allantois, an endodermal extension surrounded by mesoderm, extends toward the trophoblast.

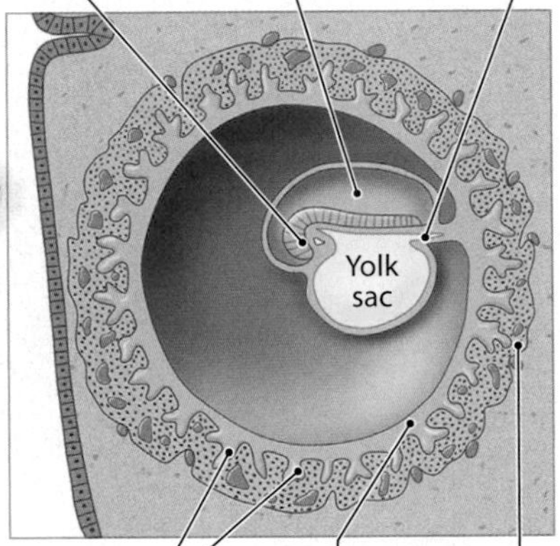
Head fold of embryo
Amniotic cavity (containing amniotic fluid)
Allantois
Yolk sac
Chorion
Syncytial trophoblast

Mesoderm extends along the core of each trophoblastic villus, forming **chorionic villi** in contact with maternal tissues. Embryonic blood vessels develop within each villus. Blood flow through those chorionic vessels begins early in the third week of development, when the embryonic heart starts beating.

Week 4

The embryo now has a **tail fold** as well as a head fold. The anterior/posterior, left/right, and superior/inferior axes of the developing embryo are now clearly established. The connections between the embryo and the surrounding trophoblast begin to constrict.

The **body stalk**, the connection between embryo and chorion, contains the distal portions of the allantois and blood vessels that carry blood to and from the placenta.

The narrow connection between the endoderm of the embryo and the yolk sac is called the **yolk stalk**.

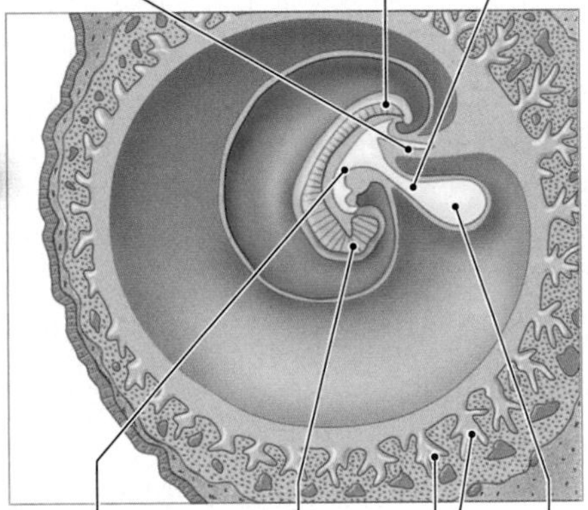
Tail fold
Embryonic gut
Head fold
Yolk sac

As the chorionic villi enlarge, more maternal blood vessels are eroded. Maternal blood now moves slowly through complex lacunae lined by the syncytial trophoblast. Chorionic blood vessels pass close by, and gases and nutrients diffuse between the embryonic and maternal circulations across the layers of the trophoblast.

Week 5

The developing embryo and extra-embryonic membranes bulge into the uterine cavity. The trophoblast pushing out into the uterine cavity remains covered by endometrium but no longer participates in nutrient absorption and embryo support. The embryo moves away from the placenta, and the body stalk and yolk stalk fuse to form an **umbilical stalk**.

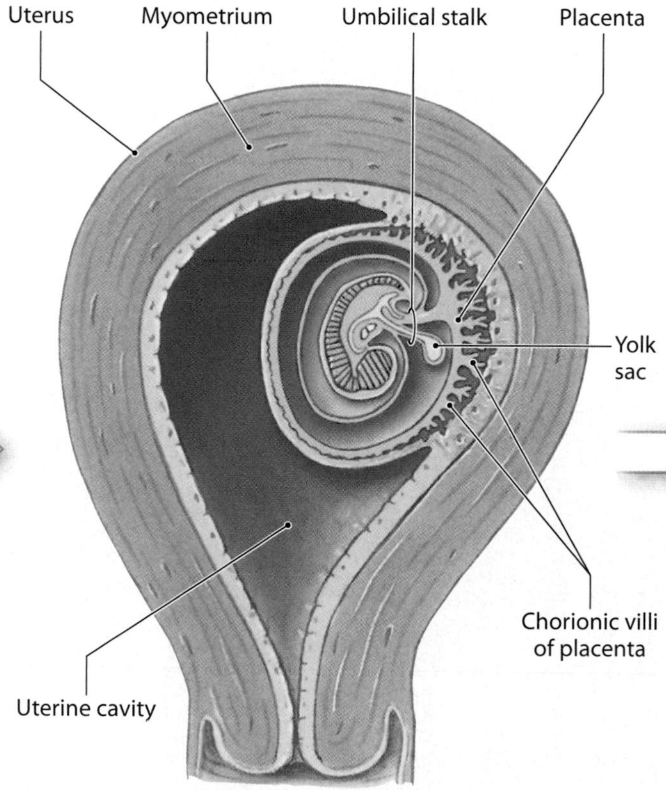

Uterus
Myometrium
Umbilical stalk
Placenta
Yolk sac
Chorionic villi of placenta
Uterine cavity

Week 10

The amnion has expanded greatly, filling the uterine cavity. The fetus is connected to the placenta by an elongated **umbilical cord** that contains a portion of the allantois, blood vessels, and the remnants of the yolk stalk.

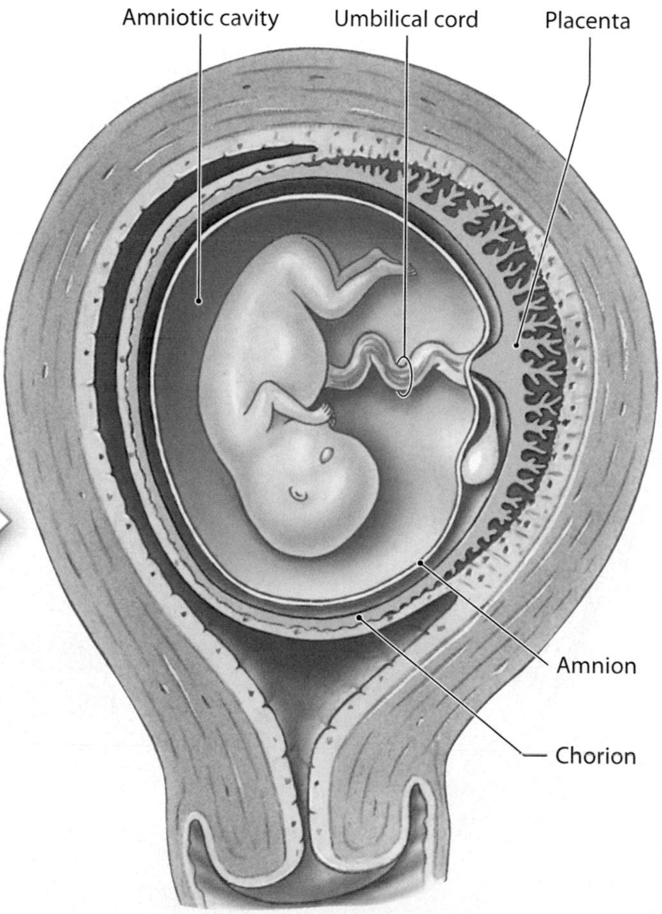

Amniotic cavity
Umbilical cord
Placenta
Amnion
Chorion

Module 27.6 Review

a. Describe the chorionic villi.

b. Compare the body stalk with the yolk stalk.

c. Identify the structure connecting the fetus to the placenta, and name the extra-embryonic membrane from which it is derived.

LO **27.6** Discuss the roles of the extra-embryonic membranes in embryonic development and placenta formation.

The placenta performs many vital functions during prenatal development

1 This illustration gives a closer look at the structure of the placenta. The chorionic villi provide the surface area for active and passive exchange of gases, nutrients, and wastes between the fetal and maternal bloodstreams. Blood flowing to the placenta through the paired **umbilical arteries** is deoxygenated (blue-colored) and contains wastes generated by fetal tissues. At the placenta, oxygen supplies are replenished, organic nutrients are added, and carbon dioxide and other organic wastes are removed. Oxygenated blood (red-colored) then returns to the fetus within a single **umbilical vein**. Recall that arteries carry blood away from the heart, and veins carry blood toward the heart (Module 18.2, p. 652).

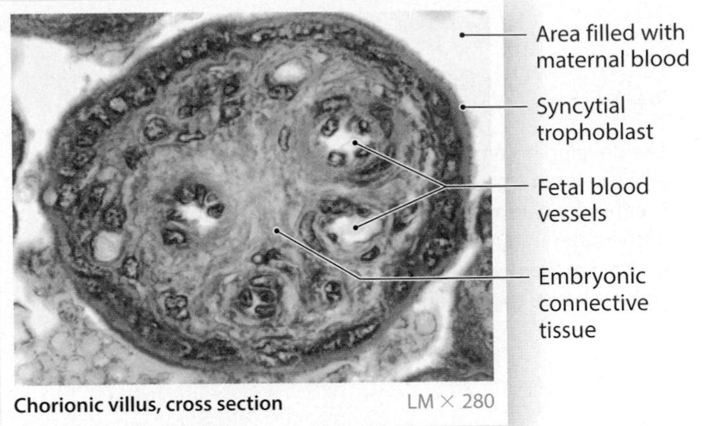
Area filled with maternal blood
Syncytial trophoblast
Fetal blood vessels
Embryonic connective tissue
LM × 280
Chorionic villus, cross section

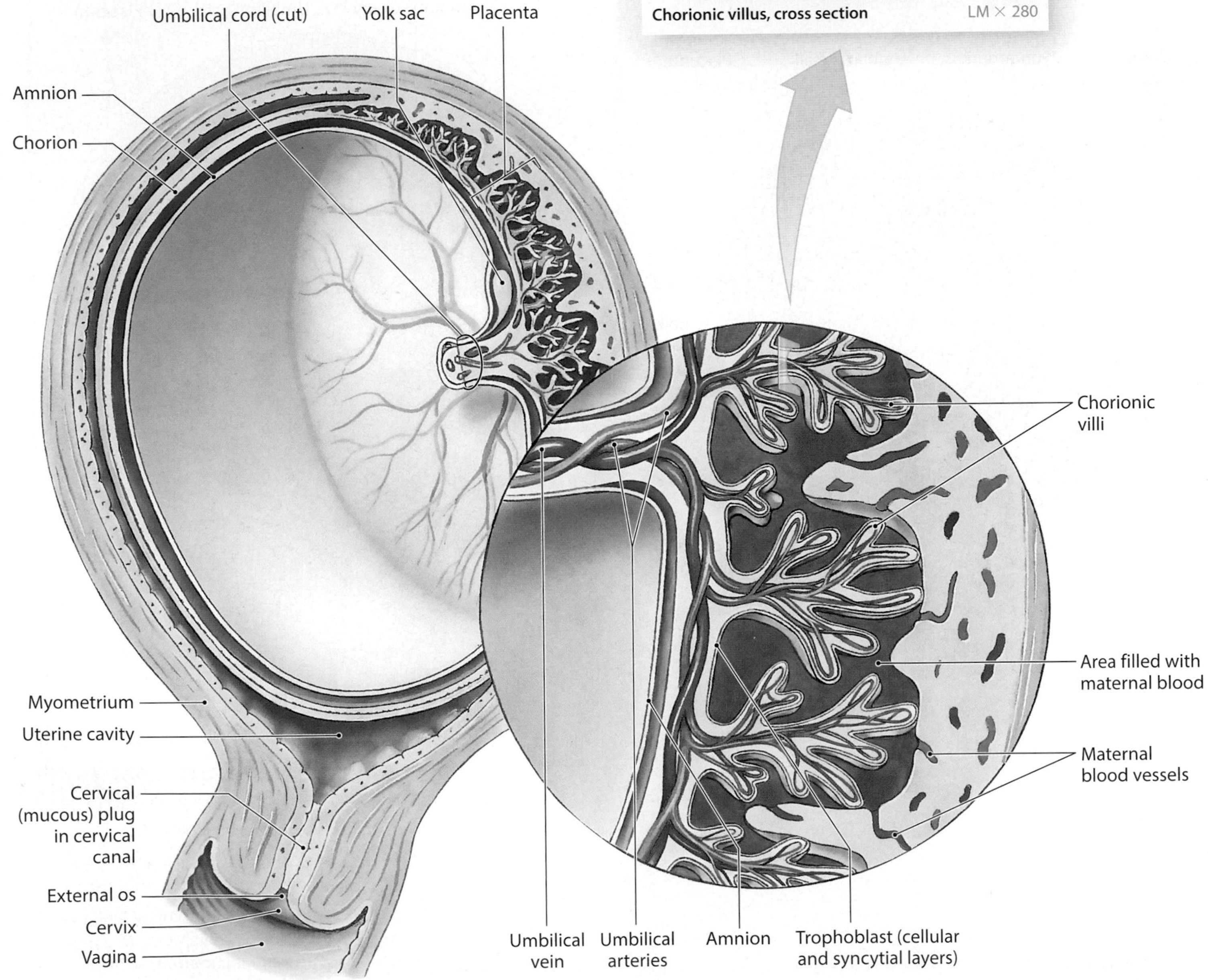
Umbilical cord (cut)
Yolk sac
Placenta
Amnion
Chorion
Myometrium
Uterine cavity
Cervical (mucous) plug in cervical canal
External os
Cervix
Vagina
Chorionic villi
Area filled with maternal blood
Maternal blood vessels
Umbilical vein
Umbilical arteries
Amnion
Trophoblast (cellular and syncytial layers)

2 In addition to supplying nutrients to the fetus, the placenta acts as an endocrine organ. Several hormones are synthesized by the syncytial trophoblast and released into the maternal bloodstream.

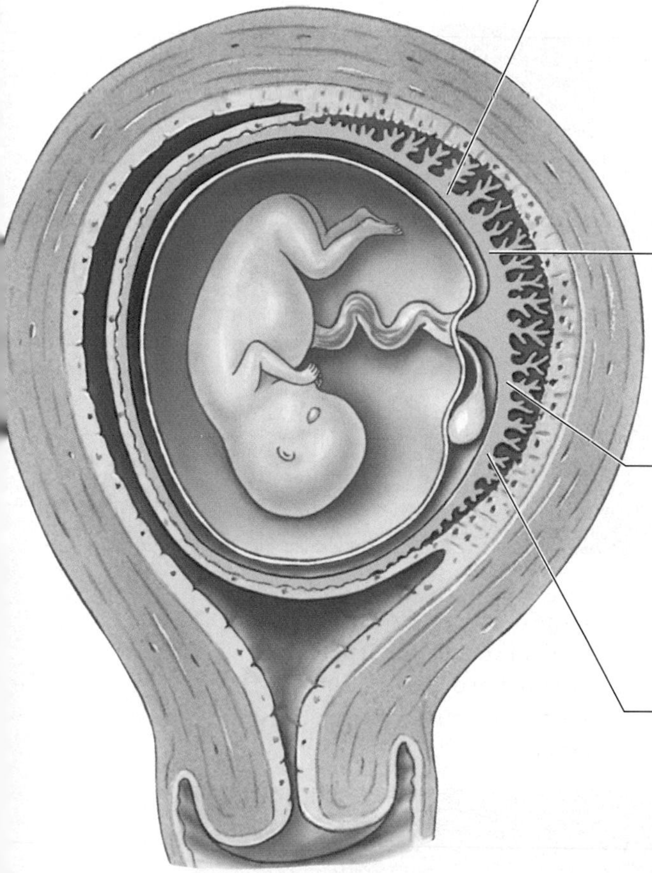

Placental Hormones

Human Chorionic Gonadotropin

Human chorionic gonadotropin (hCG) appears in the maternal bloodstream soon after implantation has occurred. The presence of hCG in blood or urine samples provides a reliable indication of pregnancy. Over the counter kits sold for the early detection of pregnancy are sensitive to the presence of this hormone in urine. Functionally, hCG resembles LH, because it maintains the integrity of the corpus luteum and promotes the continued secretion of progesterone. As a result, in pregnancy the endometrial lining remains perfectly functional, and menses does not occur. In the presence of hCG, the corpus luteum persists for 3–4 months before gradually decreasing in size and secretory function. The decline in luteal function does not trigger the return of uterine cycles, because by the end of the first trimester, the placenta is secreting both estrogens and progesterone.

Human Placental Lactogen

Human placental lactogen (hPL) helps prepare the mammary glands for milk production. The mammary glands convert from inactive to active status when stimulated by placental hormones (hPL, estrogens, and progesterone) and several maternal hormones (GH, prolactin, and thyroid hormones).

Relaxin

Relaxin is a peptide hormone that is secreted by the placenta and the corpus luteum during pregnancy. Relaxin (1) increases the flexibility of the pubic symphysis, permitting the pelvis to expand during delivery; (2) causes dilation of the cervix, making it easier for the fetus to enter the vaginal canal; and (3) delays the onset of labor contractions until late in the pregnancy.

Progesterone and Estrogens

After the first trimester, the placenta produces sufficient amounts of progesterone to maintain the endometrial lining and continue the pregnancy. As the end of the third trimester approaches, estrogen production by the placenta accelerates. As we will see in a later module, the rising estrogen levels play a role in stimulating labor and delivery.

Module 27.7 Review

a. Name the hormones synthesized by the syncytial trophoblast.

b. The presence of which hormone in the urine provides a reliable indicator of pregnancy in home pregnancy tests?

c. When does the placenta become sufficiently functional to continue the pregnancy?

LO 27.7 Discuss the importance of the placenta to the fetus and as an endocrine organ.

Organ systems form in the first trimester and become functional in the second and third trimesters

The first trimester is a critical period for development, because events in the first 12 weeks establish the basis for **organogenesis**, the process of organ formation. Over the next two trimesters the fetus grows larger and the organ systems increase in complexity to the stage at which they are capable of normal function.

1 This is a scanning electron micrograph of an embryo in the second week of development. The CNS is forming as a deep groove develops in a thick ectodermal band that lies along the posterior midline of the embryo.

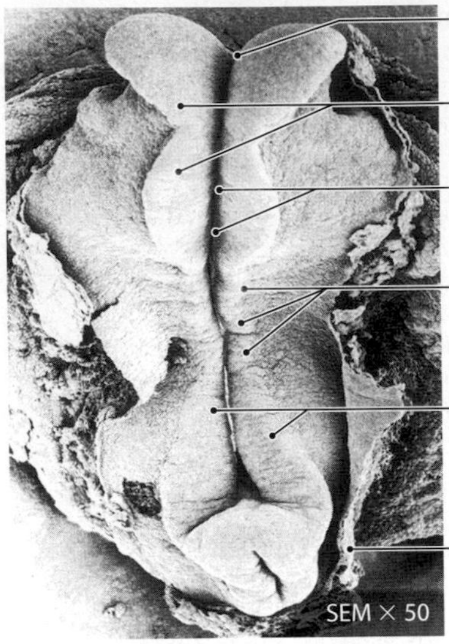
Future head of embryo
Thickened neural plate (will form brain)
Central canal of future spinal cord
Somites (mesodermal blocks that will form muscles and vertebrae)
Neural folds (fuse to enclose brain ventricles and central canal of spinal cord)
Cut wall of amnion
SEM × 50

2 This photograph of a 4-week-old embryo shows many features you can probably recognize easily. The heart is beating, pushing blood to and from the placenta, providing the nutrients needed to promote additional growth and development.

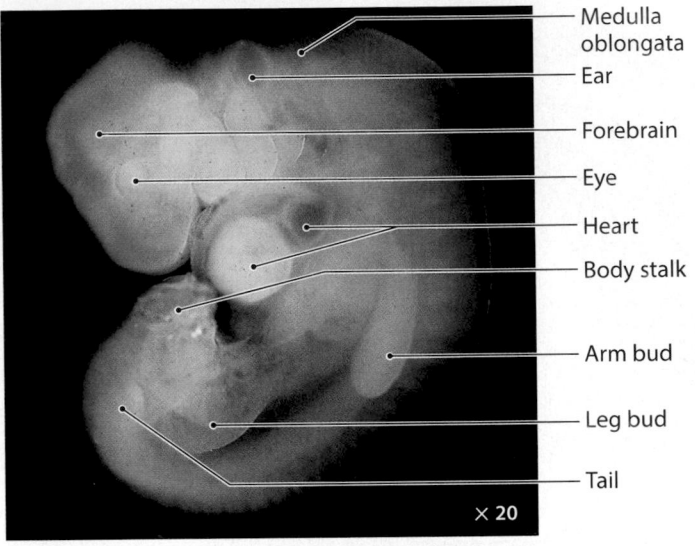

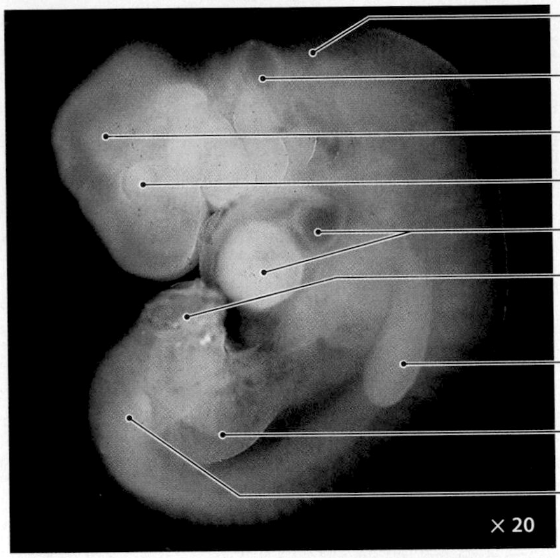
Medulla oblongata
Ear
Forebrain
Eye
Heart
Body stalk
Arm bud
Leg bud
Tail
× 20

3 By week 6, the placenta has formed and the embryo floats within the amniotic cavity. Body proportions are changing, the limbs are growing longer, and the skull bones are beginning to organize around the already-formed brain and eyes.

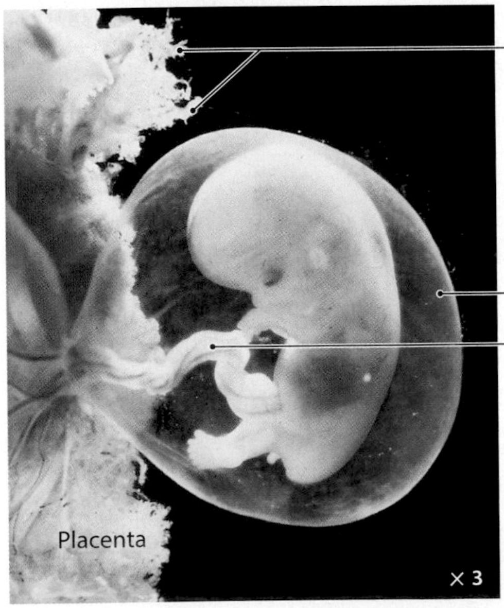
Chorionic villi
Amnion
Umbilical cord
Placenta
× 3

4 At the end of the first trimester, the fetus is considerably larger and its human features better defined. The axial and appendicular muscles are forming, and fetal movements will soon begin.

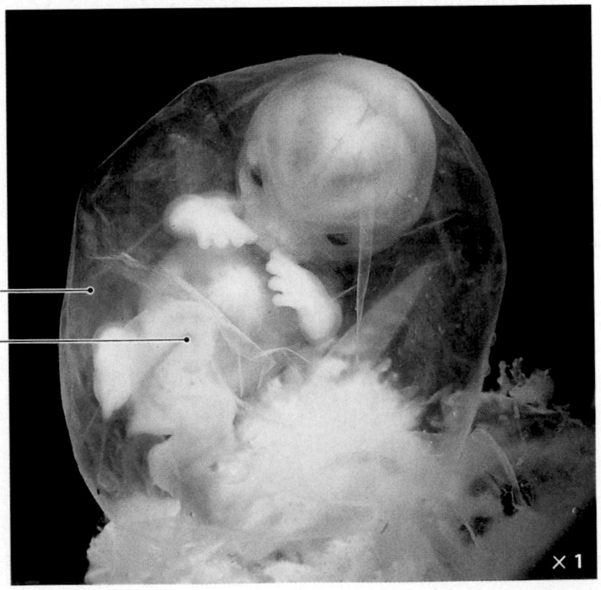
× 1

5 This is a fetus after 4 months of gestation. The face and palate have their proper form; the cerebral hemispheres are rapidly enlarging. Hair follicles are present, and hair growth begins. Peripheral nerves have formed, sensory receptors are developing, and the fetus moves frequently. The first 8 weeks of fetal growth are the most rapid, with the fetus increasing in weight some twenty-five-fold. By the end of the second trimester, the fetus will have grown to a weight of about 0.64 kg (1.4 lb).

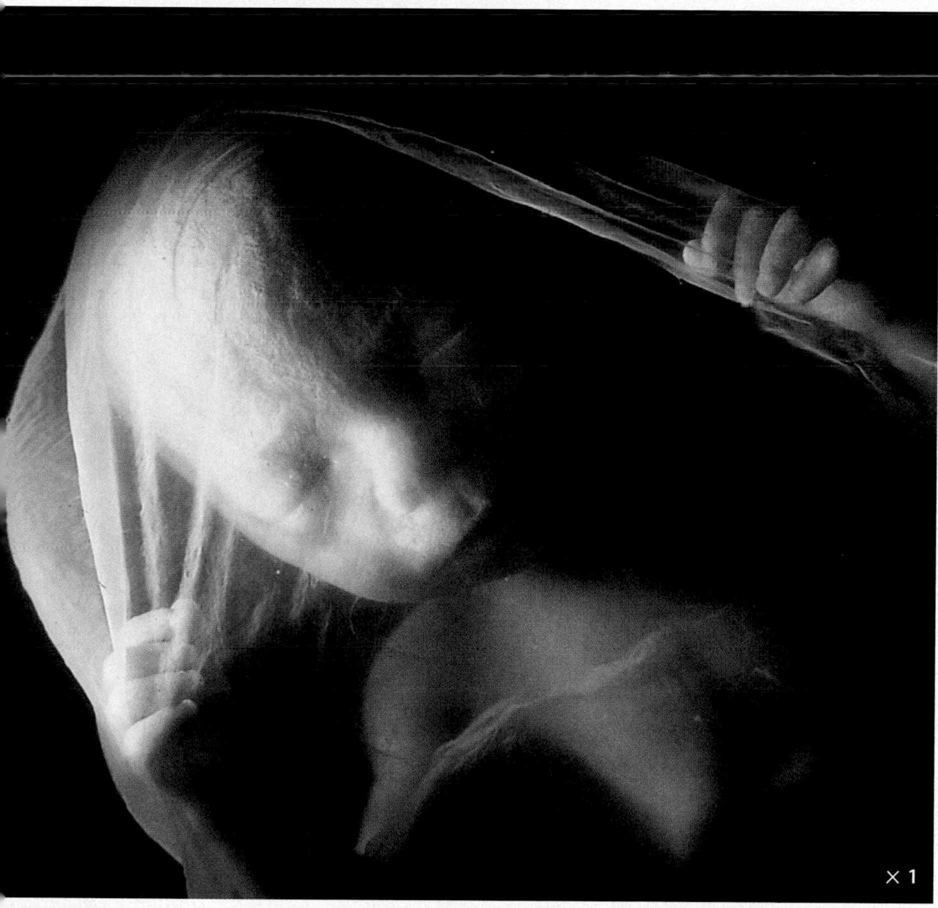
× 1

6 This is an ultrasound of a fetus after 6 months of gestation. During the third trimester, most of the organ systems become ready to function normally without maternal assistance. The rate of growth starts to slow, but in absolute terms the largest weight gain occurs in this trimester. In the final 3 months of gestation, the fetus gains about 2.6 kg (5.7 lb), reaching a full-term weight of approximately 3.2 kg (7 lb).

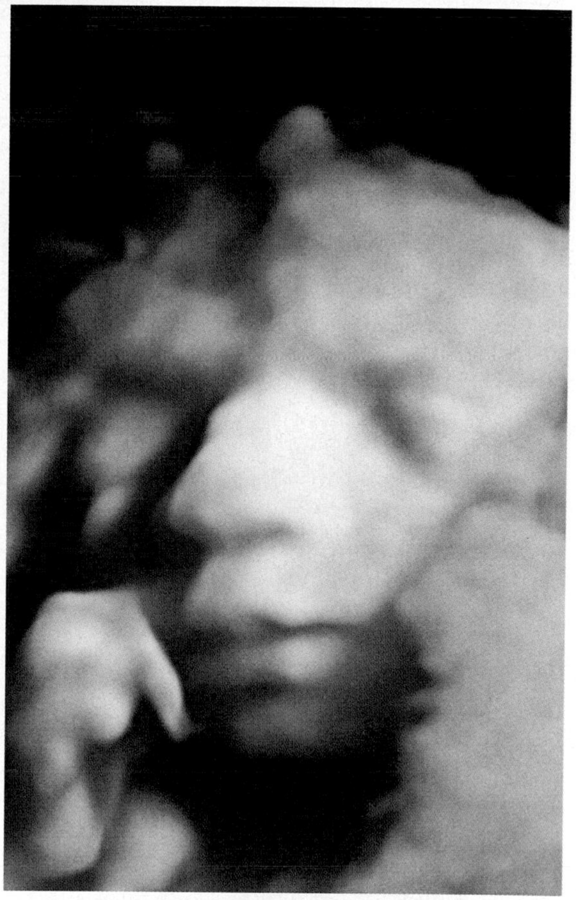

7 Multiple births (twins, triplets, and so forth) can take place for several reasons. The ratio of twin births to single births in the U.S. population is about 1:89. "Fraternal," or **dizygotic** (dī-zī-GOT-ick), twins develop when two separate oocytes are ovulated and fertilized. Seventy percent of twins are dizygotic.

"Identical," or **monozygotic**, twins result either from the separation of blastomeres early in cleavage or from the splitting of the inner cell mass before gastrulation. In either event, the genetic makeup of the twins is identical because both formed from the same pair of gametes.

Multiple births can result from multiple ovulations, blastomere splitting, or some combination of the two. Infertility treatments may also lead to multiple births.

Module 27.8 Review

a. Define organogenesis.
b. Identify the general events of fetal development during the second and third trimesters.
c. During which trimester does the fetus undergo its largest absolute weight gain?

LO 27.8 Describe organogenesis and its role in the developing fetus.

Module 27.9

Pregnancy places anatomical and physiological stresses on maternal systems

Pregnancy places tremendous strains on the mother. The developing fetus is totally dependent on maternal organ systems for nourishment, respiration, and waste removal. Maternal systems perform these functions in addition to their normal operations. For example, the mother must absorb enough oxygen, nutrients, and vitamins for herself and for her fetus, and she must eliminate all the wastes that are generated. Although this is not a burden over the initial weeks of gestation, the demands placed on the mother become significant as the fetus grows.

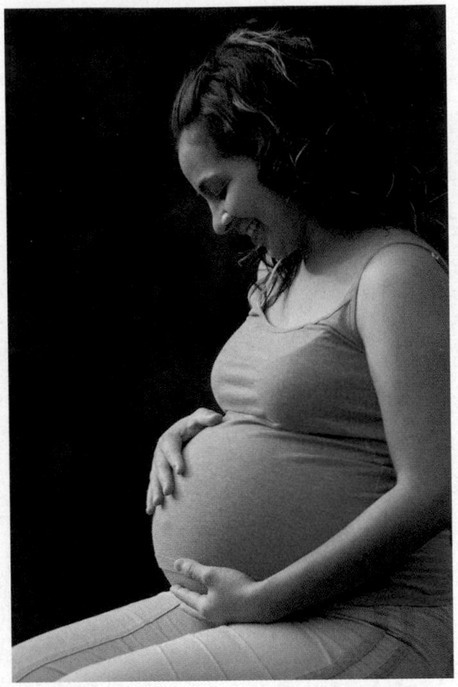

1 The physical strains of pregnancy are considerable. It is not unusual for a woman to gain 11–16 kg (25–35 lb) in weight during pregnancy, and the weight is not aligned with the body axis. This means that moving and maintaining balance use additional energy. The sectional views below compare the organ positions in nonpregnant and pregnant women. When the pregnancy is at full term, the uterus and fetus are so large that they push many of the maternal abdominal organs out of their normal positions.

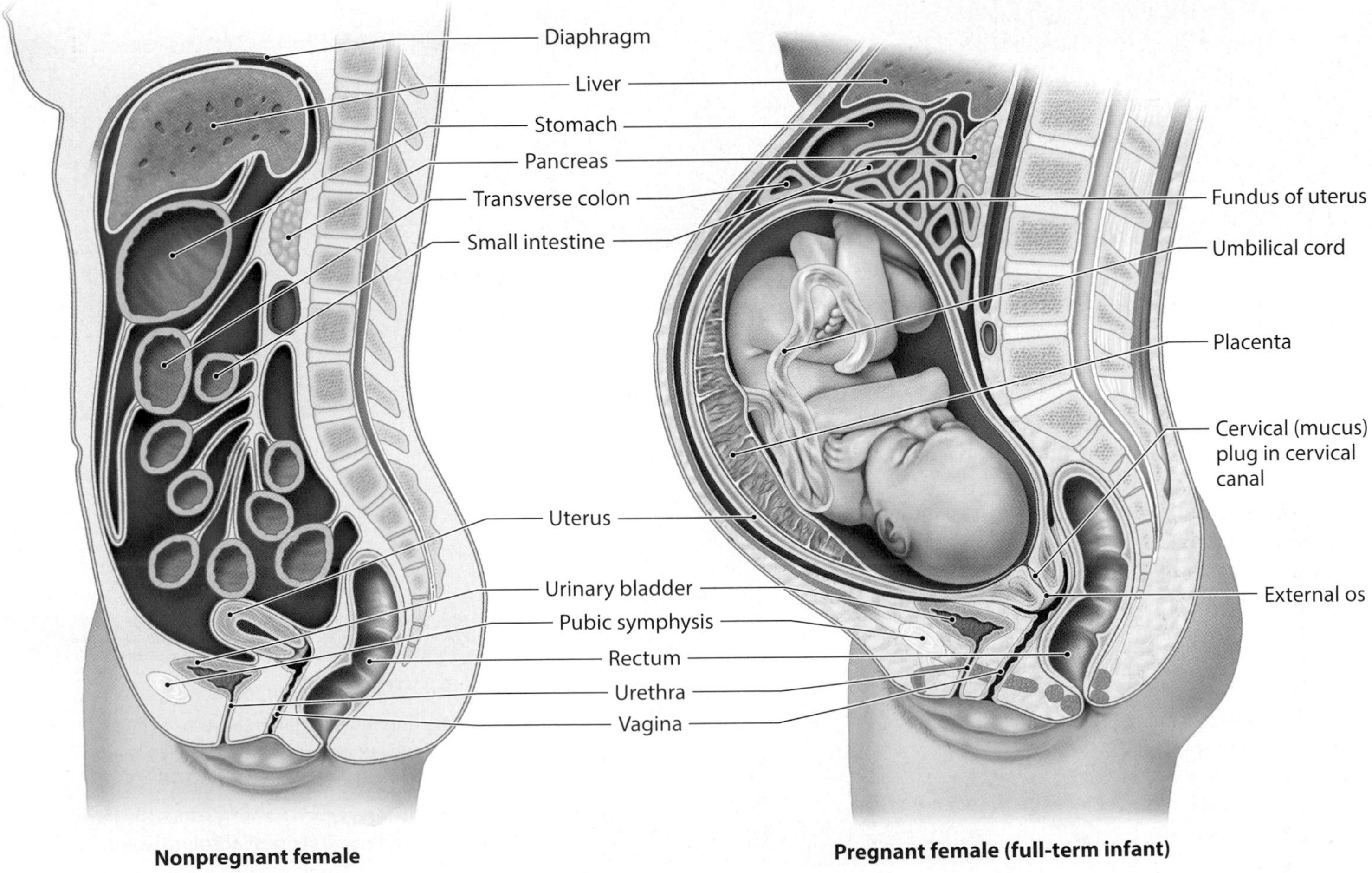

Diaphragm
Liver
Stomach
Pancreas
Transverse colon
Small intestine
Fundus of uterus
Umbilical cord
Placenta
Cervical (mucus) plug in cervical canal
Uterus
Urinary bladder
Pubic symphysis
Rectum
Urethra
Vagina
External os
Nonpregnant female
Pregnant female (full-term infant)

2 The physiological stresses are even more extreme than the physical ones. This is a summary of the physiological changes that have occurred in maternal systems by the end of the third trimester.

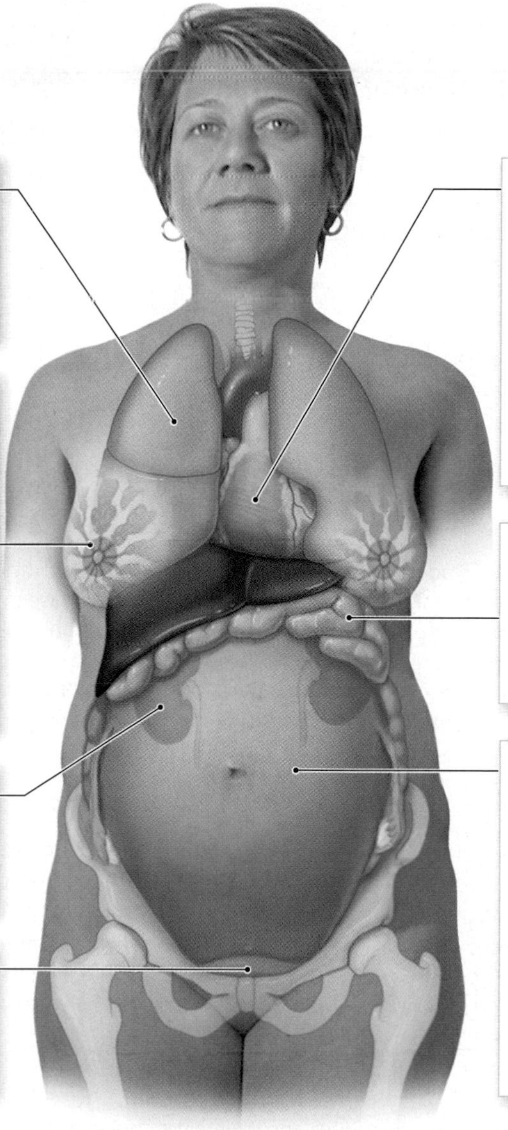

The mother's lungs must deliver the extra oxygen required, and remove the excess carbon dioxide generated, by the fetus. As a result, the maternal respiratory rate increases and tidal volume increases.
Mammary gland development requires a combination of hormones, including human placental lactogen and placental prolactin (PRL), estrogens, progesterone, GH, and thyroxine from the maternal endocrine organs. By the end of the sixth month of pregnancy, the mammary glands are fully developed and begin to produce clear secretions that are stored in the duct system of those glands and may be expressed from the nipple.
The kidneys must eliminate the wastes produced by maternal and fetal systems. As a result, the maternal glomerular filtration rate (GFR) increases by nearly 50 percent.
Because the volume of urine produced increases and the weight of the uterus presses down on the urinary bladder, pregnant women need to urinate frequently.
The volume of blood flowing into the placenta decreases the volume in the systemic circuit of the mother. At the same time, fetal metabolic activity "steals" maternal oxygen and increases maternal CO_2 levels. This combination stimulates the production of renin and erythropoietin, leading to an increase in maternal blood volume. By the end of gestation, maternal blood volume has increased by almost 50 percent.
Pregnant women must nourish both themselves and their fetus and so they tend to have increased hunger sensations. Maternal requirements for nutrients can increase by up to 30 percent above normal.
At the end of gestation, a typical uterus has grown from 7.5 cm (3 in.) in length and 30–40 g (1–1.4 oz) in weight to 30 cm (12 in.) in length and 1100 g (2.4 lb) in weight. The uterus may then contain 2 liters of fluid, plus fetus and placenta, for a total weight of roughly 6–7 kg (13–15 lb). This remarkable expansion occurs through the enlargement (hypertrophy) of existing cells, especially smooth muscle fibers, rather than by an increase in the total number of cells.

Although pregnancy is a natural phenomenon, the physical and physiological demands on maternal systems make it potentially dangerous. At any age, the risks associated with pregnancy are significantly greater than those associated with the use of oral contraceptives. (The notable exception involves women who both take the pill and smoke.) For pregnant women over age 35 the chances of dying from pregnancy-related complications are almost twice as great as the chances of being killed in an automobile accident.

Module 27.9 Review

a. Based on the illustrations showing the locations of the internal organs in nonpregnant and pregnant women, explain why some women experience difficulty breathing while pregnant.

b. List the major changes that occur in maternal systems during pregnancy.

c. Why does a mother's blood volume increase during pregnancy?

LO **27.9** Describe the interplay between maternal organ systems and the developing fetus.

Multiple factors initiate and accelerate labor

The tremendous stretching of the uterus is associated with a gradual increase in the rate of spontaneous smooth muscle contractions in the myometrium. In the early stages of pregnancy, progesterone released by the placenta inhibits the uterine smooth muscle, preventing powerful contractions. Late in pregnancy, some women experience occasional spasms in the uterine musculature, but these contractions are neither regular nor persistent. Such contractions are called **false labor**. **True labor** begins when biochemical and mechanical factors reach a point of no return.

1 This sagittal section shows the position of the fetus at the onset of true labor.

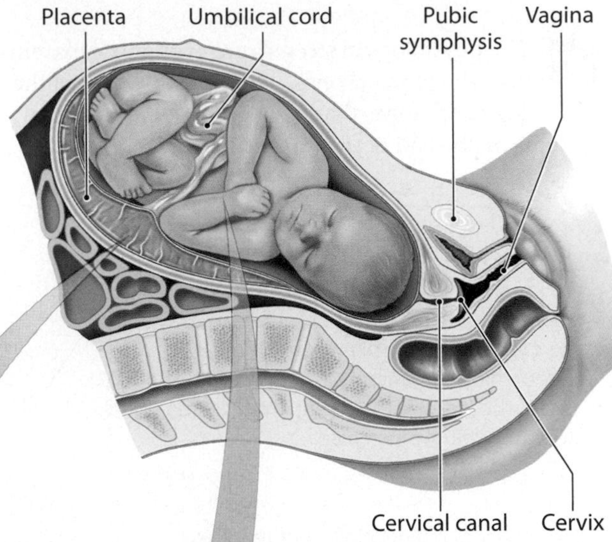
Placenta
Umbilical cord
Pubic symphysis
Vagina
Cervical canal
Cervix

2 After 9 months of gestation, multiple factors interact to initiate true labor. Once labor contractions have begun in the myometrium, positive feedback ensures that they will continue until delivery has been completed.

Placental Factors

Placental estrogens increase the sensitivity of the smooth muscle cells of the myometrium and make contractions more likely. As delivery approaches, the production of estrogens accelerates. Estrogens also increase the sensitivity of smooth muscle fibers to oxytocin.

Relaxin produced by the placenta relaxes the pelvic articulations and dilates the cervix.

Fetal Factors

Growth and the increase in fetal weight stretch and distort the myometrium.

The fetal pituitary gland releases oxytocin in response to estrogens.

Distortion of Stretched Myometrium

Distortion of the myometrium increases the sensitivity of the smooth muscle layers, promoting spontaneous contractions that get stronger and more frequent as the pregnancy advances.

Maternal Oxytocin Release

Maternal oxytocin release is stimulated by high estrogen levels.

Prostaglandin Production

Estrogens and oxytocin stimulate the production of prostaglandins in the endometrium. These prostaglandins further stimulate smooth muscle contractions.

Increased Excitability of the Myometrium

Oxytocin and prostaglandins both stimulate the myometrium. In addition, the sensitivity of the uterus to oxytocin increases dramatically. The smooth muscle in a late-term uterus is 100 times more sensitive to oxytocin than the smooth muscle in a nonpregnant uterus.

LABOR CONTRACTIONS OCCUR

Labor contractions move the fetus and further distort the myometrium. This distortion stimulates additional oxytocin and prostaglandin release. This **positive feedback** continues until delivery is completed.

3 The goal of labor is **parturition** (par-tū-RISH-un), or childbirth, the forcible expulsion of the fetus, followed by the placenta. Labor is divided into three stages: the dilation stage, the expulsion stage, and the placental stage.

Dilation Stage

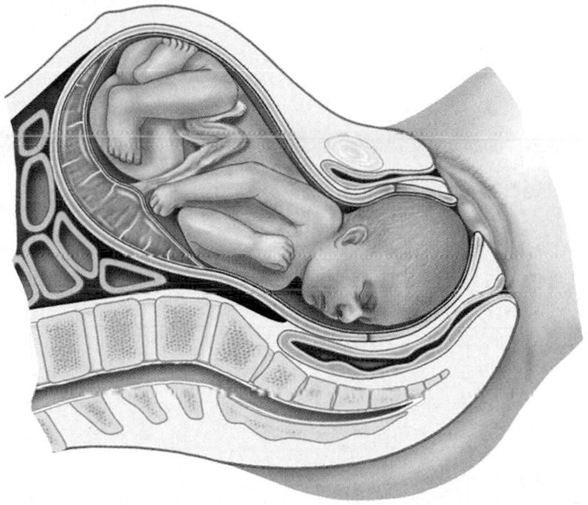

The **dilation stage** begins with the onset of true labor, as the cervix dilates and the fetus shifts toward the cervical canal, moved by gravity and uterine contractions. This stage typically lasts 8 or more hours. At the start of the dilation stage, labor contractions last up to half a minute and occur once every 10–30 minutes; their frequency increases steadily. Late in this stage, the amnion ruptures, an event sometimes referred to as "having one's water break."

Expulsion Stage

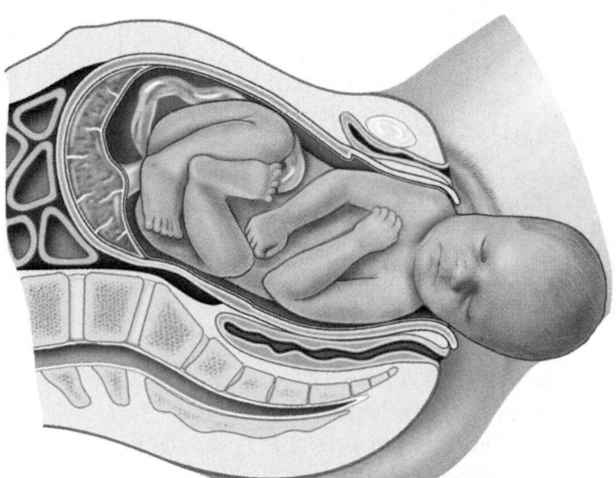

The **expulsion stage** begins as the cervix, pushed open by the approaching fetus, completes dilation (about 10 cm). In this stage, contractions reach maximum intensity, occurring at perhaps 2- or 3-minute intervals and lasting a full minute. Expulsion continues until the fetus has emerged from the vagina; in most cases, the expulsion stage lasts less than 2 hours. The arrival of the newborn infant into the outside world is **delivery**, or birth.

Placental Stage

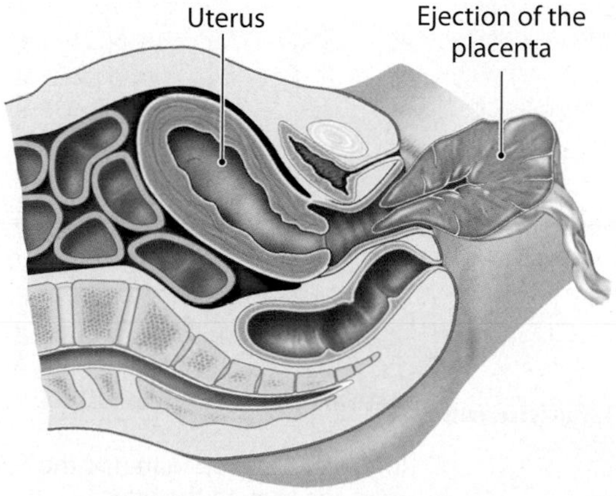
Uterus
Ejection of the placenta

During the **placental stage** of labor, muscle tension builds in the walls of the partially empty uterus, which gradually decreases in size. This uterine contraction tears the connections between the endometrium and the placenta, and the placenta, or **afterbirth**, is ejected. The disruption of the placenta is accompanied by a loss of blood, but associated uterine contractions compress the uterine vessels and usually restrict this flow.

Premature labor occurs when true labor begins before the fetus has completed normal development. The newborn's chances of surviving are directly related to its body weight at delivery. Even with massive supportive efforts, newborns weighing less than 400 g (14 oz) at birth will not survive, primarily because their respiratory, cardiovascular, and urinary systems are unable to support life without aid from maternal systems. Most fetuses born at 25–27 weeks of gestation (a birth weight under 600 g or 21.1 oz) die despite intensive neonatal care, and survivors have a high risk of developmental abnormalities. **Premature delivery** usually refers to birth at 28–36 weeks (a birth weight over 1 kg or 2.2 lb). With care, these newborns have a good chance of surviving and developing normally.

Module 27.10 Review

a. List and describe the factors involved in initiating labor contractions.

b. What chemicals are primarily responsible for initiating contractions of true labor?

c. Name the three stages of labor, and describe the events that characterize each stage.

LO 27.10 List and discuss the events that occur during labor and delivery.

Module 27.11

After delivery, development initially requires nourishment by maternal systems

Developmental processes do not end at delivery, because newborns have few of the anatomical, functional, or physiological characteristics of mature adults. During an initial **neonatal period**, the newborn is dependent on the mother for nourishment, and protection from environmental hazards such as extreme changes in temperature. The neonatal period lasts from birth through the first 28 days of life.

1 Milk is provided to infants through the **milk let-down reflex** (milk-ejection reflex), which is diagrammed here. By the end of the sixth month of pregnancy, the mammary glands are fully developed, and the gland cells begin to produce a secretion known as **colostrum** (kō-LOS-trum). Colostrum contains antibodies that may help the infant ward off infections until the infant's own immune system becomes fully functional. After the first few days of nursing, the mammary glands produce breast milk, which has a much higher fat content than colostrum but still contains antibodies. It also contains **lysozyme**, an enzyme with antibiotic properties.

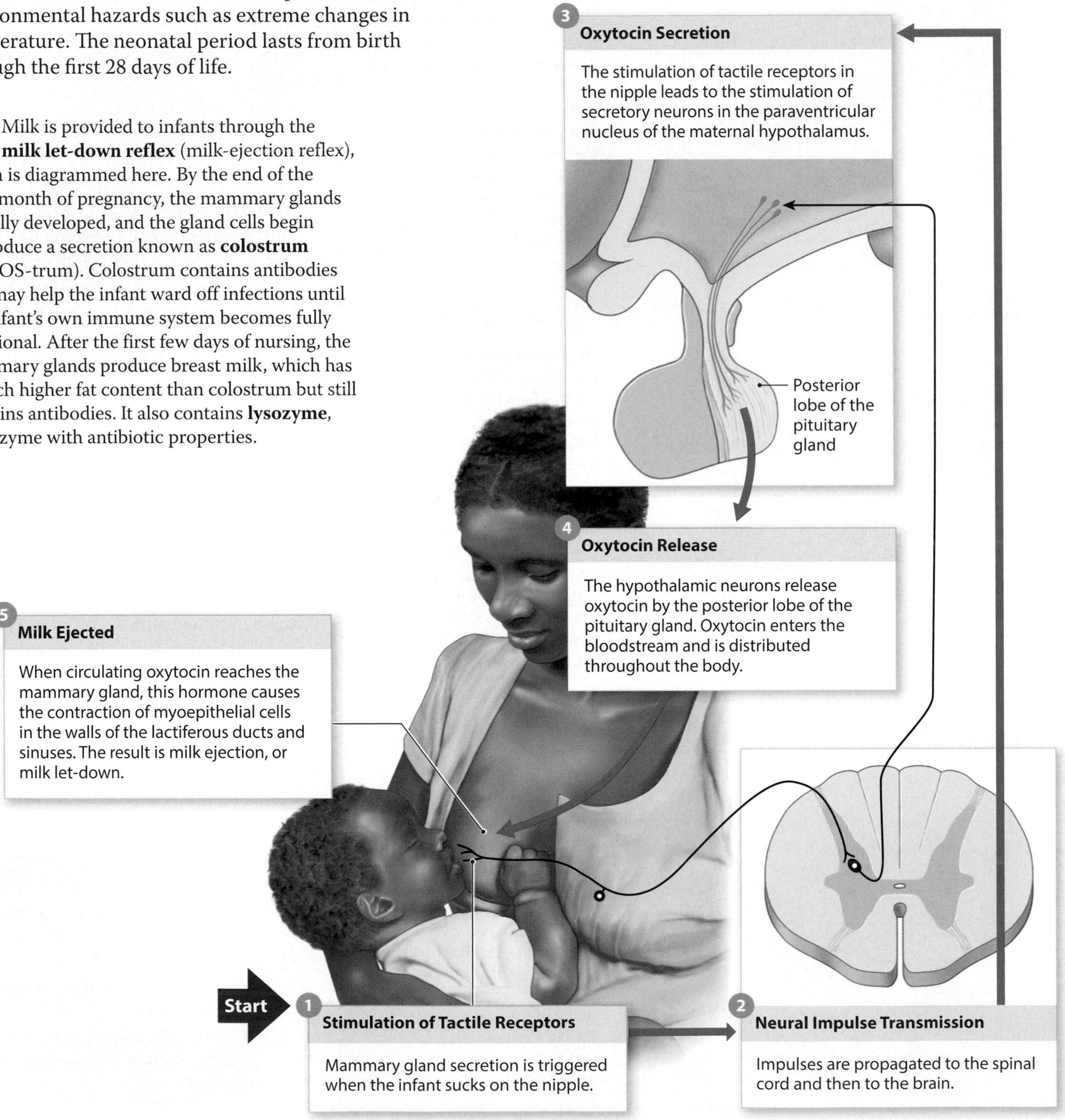

3 Oxytocin Secretion
The stimulation of tactile receptors in the nipple leads to the stimulation of secretory neurons in the paraventricular nucleus of the maternal hypothalamus.
Posterior lobe of the pituitary gland
4 Oxytocin Release
The hypothalamic neurons release oxytocin by the posterior lobe of the pituitary gland. Oxytocin enters the bloodstream and is distributed throughout the body.
5 Milk Ejected
When circulating oxytocin reaches the mammary gland, this hormone causes the contraction of myoepithelial cells in the walls of the lactiferous ducts and sinuses. The result is milk ejection, or milk let-down.
Start
1 Stimulation of Tactile Receptors
Mammary gland secretion is triggered when the infant sucks on the nipple.
2 Neural Impulse Transmission
Impulses are propagated to the spinal cord and then to the brain.

2 Postnatal development includes five **life stages**: (1) the neonatal period, (2) infancy, (3) childhood, (4) adolescence, and (5) maturity. Once maturity has been reached, the individual is subject to the gradual changes that accompany **senescence** (*senesco*, to grow old), or aging. Growth during infancy and childhood occurs under the direction of circulating hormones, notably growth hormone, adrenal steroids, and thyroid hormones. These hormones affect each tissue and organ in specific ways, depending on cellular sensitivity.

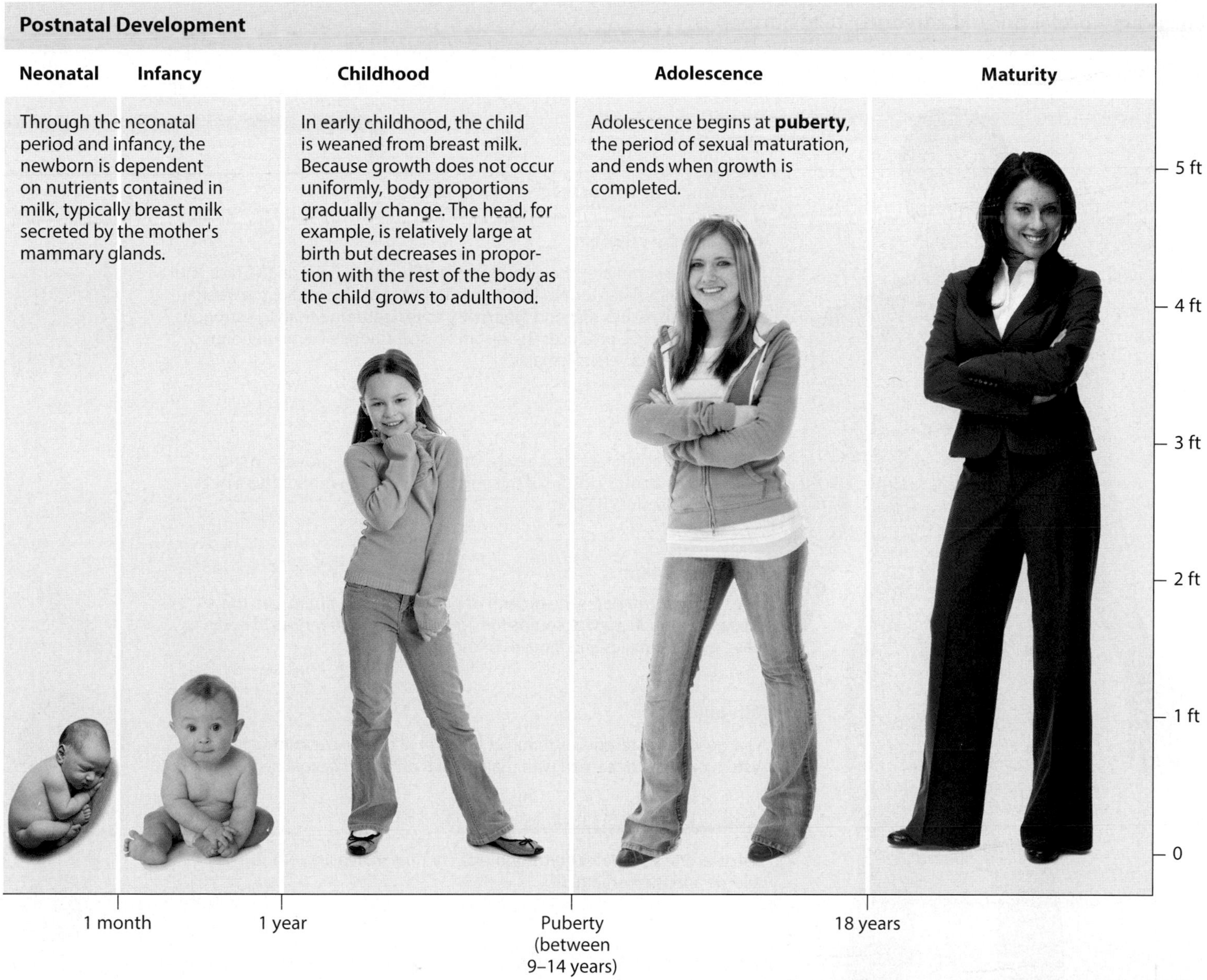

Postnatal Development
Neonatal
Infancy
Childhood
Adolescence
Maturity
Through the neonatal period and infancy, the newborn is dependent on nutrients contained in milk, typically breast milk secreted by the mother's mammary glands.
In early childhood, the child is weaned from breast milk. Because growth does not occur uniformly, body proportions gradually change. The head, for example, is relatively large at birth but decreases in proportion with the rest of the body as the child grows to adulthood.
Adolescence begins at **puberty**, the period of sexual maturation, and ends when growth is completed.
5 ft
4 ft
3 ft
2 ft
1 ft
0
1 month
1 year
Puberty (between 9–14 years)
18 years

Module 27.11 Review

a. What hormone causes the milk let-down reflex?

b. Explain the difference between colostrum and breast milk.

c. Name the stages of postnatal development, and describe the time frame involved for each of the stages.

LO 27.11 Identify the features of and the functions associated with the various life stages.

Module 27.12

At puberty, male and female sex hormones have differing effects on most body systems

Many body systems alter their activities in response to changes in circulating levels of sex hormones at **puberty**. The most important sex-related changes are summarized here. The effects of testosterone and estrogens are facilitated and enhanced by growth hormone, thyroid hormones, prolactin, and adrenocortical hormones.

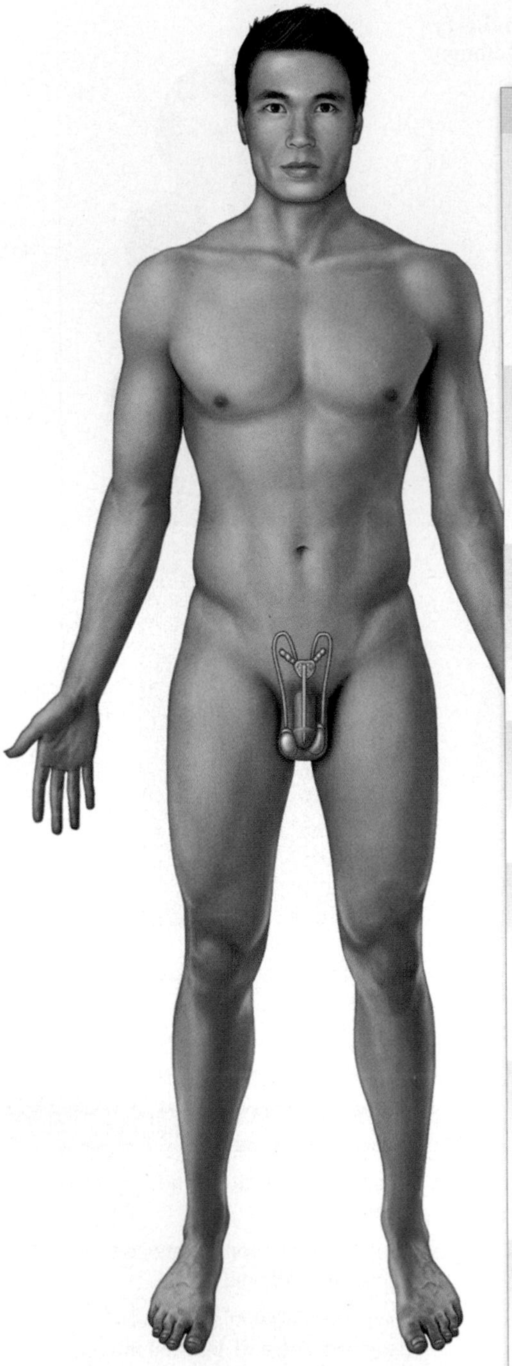

Responses to Testosterone in Males

Integumentary System

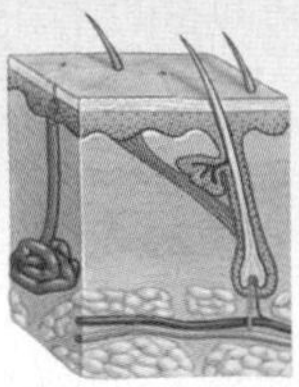

Testosterone stimulates the development of terminal hairs on the face and chest, and stimulates terminal hair growth in the axillae and in the genital area. Adipose tissues respond differently to testosterone than to estrogens, and this difference produces the distinct distributions of subcutaneous body fat in males versus females.

Skeletal System

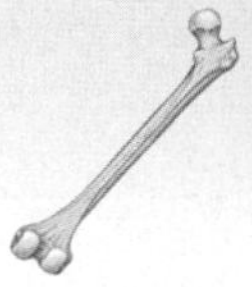

Testosterone accelerates bone deposition and skeletal growth. In the process, it promotes closure of the epiphyseal cartilages and thus places a limit on growth in height.

Muscular System

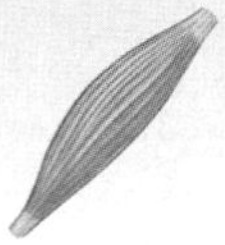

Testosterone stimulates the growth of skeletal muscle fibers, and the increased muscle mass accounts for significant sex differences in body mass, even for males and females of the same height.

Nervous System

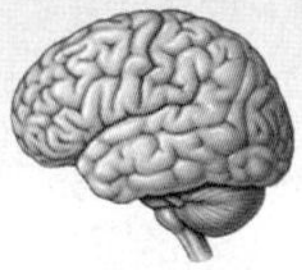

A surge in testosterone secretion at puberty activates the central nervous system centers concerned with male sexual drive and sexual behaviors.

Cardiovascular System

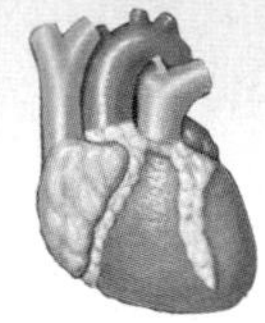

Testosterone stimulates erythropoiesis, thereby increasing blood volume and the hematocrit.

Respiratory System

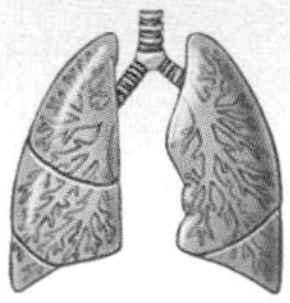

Testosterone stimulates growth of the larynx and a thickening and lengthening of the vocal cords. These changes cause a gradual deepening of the voice in males.

Reproductive System

Testosterone stimulates the functional development of the accessory reproductive glands, such as the prostate gland and seminal glands, and helps promote spermatogenesis.

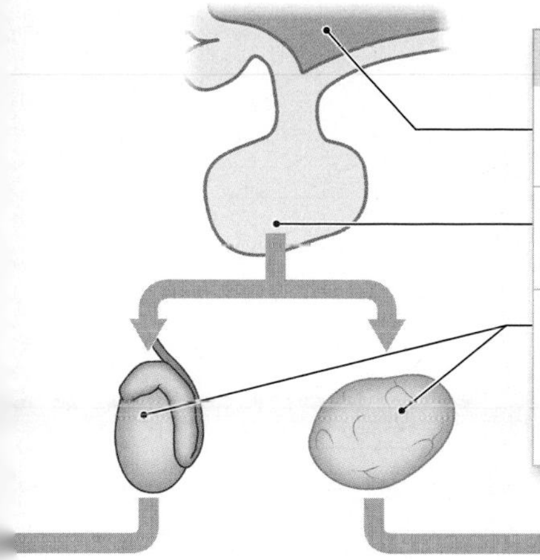

At Puberty

The hypothalamus increases its production of gonadotropin-releasing hormone (GnRH).

Endocrine cells in the anterior lobe of the pituitary gland become more sensitive to GnRH, and circulating levels of FSH and LH rise rapidly.

Testicular or ovarian cells become more sensitive to FSH and LH, initiating (1) gamete production; (2) the secretion of sex hormones, which stimulate the appearance of secondary sex characteristics and behaviors; and (3) a sudden acceleration in the growth rate, ending with closure of the epiphyseal cartilages.

Responses to Estrogens in Females

Integumentary System

Estrogens stimulate the hair follicles to continue to produce fine vellus hairs and stimulate terminal hair growth in the axillae and in the genital area. The combination of estrogens, prolactin, growth hormone, and thyroid hormones promotes the initial development of the mammary glands.

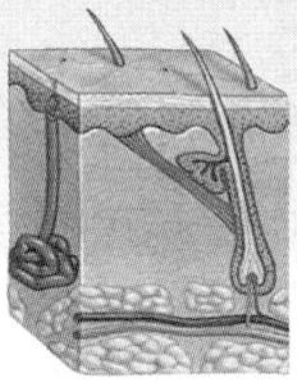

Skeletal System

Estrogens cause more rapid epiphyseal closure than does testosterone. In addition, the period of skeletal growth ends at an earlier age in girls than in boys, and so females generally do not grow as tall as males.

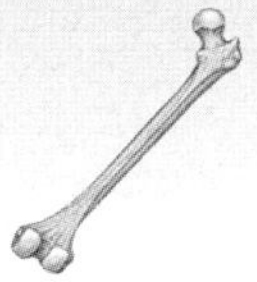

Muscular System

Estrogens stimulate the growth of skeletal muscle fibers, but not to the extent that testosterone does in males.

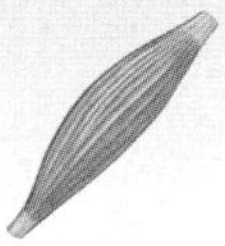

Nervous System

A surge in estrogen secretion at puberty activates central nervous system centers involved in female sexual drive and sexual behaviors.

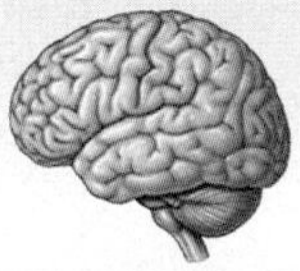

Cardiovascular System

The iron loss associated with menses increases the risk of developing iron-deficiency anemia. Estrogens decrease plasma cholesterol levels and slow the formation of plaque within arteries. As a result, premenopausal women have a lower risk of atherosclerosis than do adult men.

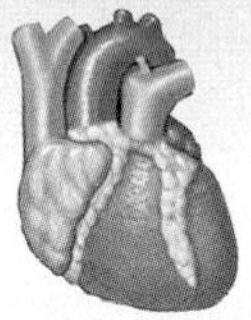

Respiratory System

Estrogens do not cause excessive growth of the larynx and vocal cords, so females typically have higher-pitched voices than males.

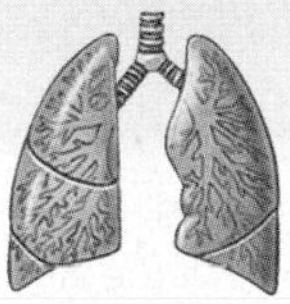

Reproductive System

Estrogens target the uterus, promoting a thickening of the myometrium and increasing blood flow to the endometrium. Estrogens also promote the functional development of accessory reproductive structures in females.

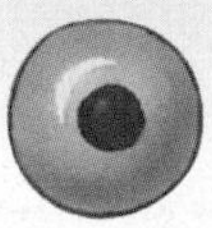

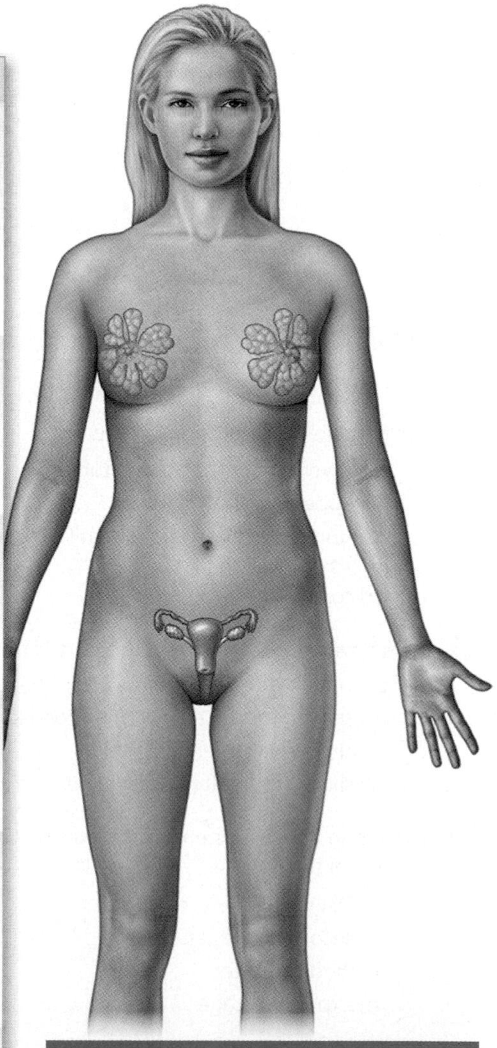

Module 27.12 Review

a. Name the three major interacting hormonal events associated with the onset of puberty.

b. Why does a man have a deeper voice and a larger larynx than a woman?

c. Why are premenopausal women at lesser risk of atherosclerosis than men?

27.12 Explain the roles of hormones in males and females at puberty.

SECTION 1 Review

Matching

Match each lettered term with the most closely related description.

a. hCG
b. conception
c. chorion
d. syncytial trophoblast
e. colostrum
f. amnion
g. amphimixis
h. morula
i. embryonic disc
j. neonate
k. gestation
l. inner cell mass
m. relaxin
n. blastocyst

1 Fertilization
2 Newborn infant
3 Pronuclei fuse
4 Period of prenatal development
5 Pregnancy test
6 Softens pubic symphysis
7 Mammary gland secretion
8 Mesoderm and ectoderm
9 Hollow ball of cells
10 Mesoderm and trophoblast
11 Forms the embryo
12 Cytoplasm with many nuclei
13 Solid ball of cells
14 Gastrulation product

1 ____
2 ____
3 ____
4 ____
5 ____
6 ____
7 ____
8 ____
9 ____
10 ____
11 ____
12 ____
13 ____
14 ____

Multiple choice

Select the correct answer from the list provided.

15 Fertilization typically occurs in the
- a) lower part of the uterine tube.
- b) upper part of the uterus.
- c) junction between the ampulla and isthmus of the uterine tube.
- d) cervix.

16 Fetal development begins at the start of the
- a) implantation process.
- b) second month after fertilization.
- c) ninth week after fertilization.
- d) sixth month after fertilization.

17 Organs and organ systems complete most of their development by the end of the
- a) first trimester.
- b) second trimester.
- c) third trimester.
- d) expulsion stage.

18 The four general processes that occur during the first trimester include
- a) blastomere, blastocyst, morula, and trophoblast.
- b) cleavage, implantation, placentation, and embryogenesis.
- c) placentation, dilation, expulsion, and organogenesis.
- d) yolk sac, amnion, allantois, and chorion.

19 The most dangerous period in prenatal or neonatal life is the
- a) first trimester.
- b) second trimester.
- c) third trimester.
- d) expulsion stage.

20 The systems that were relatively nonfunctional during the fetal period that must become functional at birth are the
- a) cardiovascular, muscular, and skeletal systems.
- b) integumentary, reproductive, and nervous systems.
- c) respiratory, digestive, and urinary systems.
- d) endocrine, nervous, and digestive systems.

Section integration

Tina gives birth to a baby with a congenital deformity of the stomach. Tina thinks her baby's condition resulted from a viral infection she had during her third trimester. Explain if the viral infection likely caused the baby's condition.

21 ____

A person may be described in terms of their genotype and phenotype

Although everyone goes through the same developmental stages, differences in both genetic structure and local environments produce distinctive individual characteristics. Traditionally, **inheritance** or **heredity** refers to the transfer of genetically determined characteristics from generation to generation. The study of the mechanisms responsible for inheritance is called **genetics**. We begin our study of genetics by examining two important concepts: *genotype* and *phenotype*.

1 One way to understand genotype and phenotype is to compare them to the architecture of a house. The house plan is the genotype and how the finished house looks is its phenotype.

Every nucleated somatic cell in your body carries copies of the original 46 chromosomes present when you were a zygote. Those chromosomes and their component genes constitute your **genotype** (JĒN-ō-tīp; *geno*, gene). In architectural terms, the genotype is a set of plans, like the blueprints for a house.

The collective expression of your genes determines the anatomical and physiological characteristics that make up your **phenotype** (FĒ-nō-tīp; *phaino*, to display). The field of genetics recently expanded with the understanding that not all of a person's traits are determined solely by their DNA sequence. **Epigenetics** is the study of gene expression due to alterations in the reading of the DNA sequences.

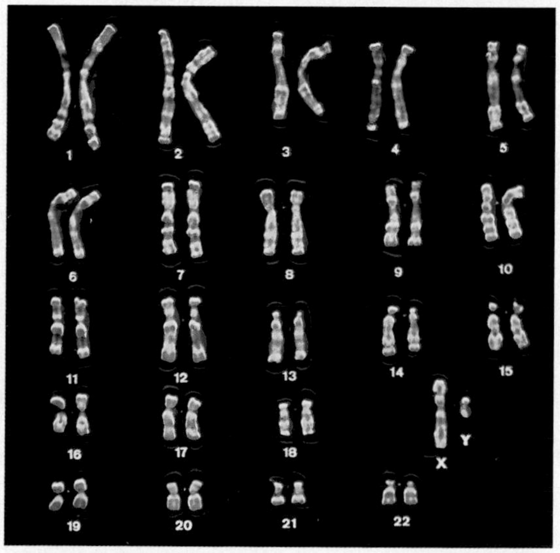
1
2
3
4
5
6
7
8
9
10
11
12
13
14
15
16
17
18
X
Y
19
20
21
22

2 This photograph shows the **karyotype** (*karyon*, nucleus + *typos*, mark), or entire set of chromosomes, of a normal male. At amphimixis, one member of each of the 23 chromosome pairs was contributed by the spermatozoon, and the other by the oocyte. The two members of each pair are known as **homologous** (huh-MOL-ō-gus) **chromosomes**. Twenty-two of those pairs are called **autosomal** (aw-tō-SŌ-mul) **chromosomes**. Most of the genes of the autosomal chromosomes affect somatic characteristics, such as hair color and skin pigmentation. The chromosomes of the 23rd pair are called the **sex chromosomes**; one of their functions is to determine whether the individual is genetically male or female. The sex chromosomes of a male consist of an **X chromosome** and a shorter **Y chromosome**, whereas females have two X chromosomes.

In this section we consider basic genetics as it applies to inherited characteristics, such as sex, hair color, and various disorders.

Module 27.13 Review

a. Distinguish between genotype and phenotype.

b. How are autosomal and sex chromosomes different?

c. How can you tell that the karyotype shown here is male?

LO 27.13 Describe genetics and inheritance and the relationship between genotype and phenotype.

Genes and chromosomes determine patterns of inheritance

1 The two chromosomes in a homologous autosomal pair have the same structure and carry genes that affect the same traits. The two chromosomes in a pair may carry the same form or different forms of each gene. The various forms of a given gene are called **alleles** (uh-LĒLZ). The genes are also located at equivalent positions on their respective chromosomes. A gene's position on a chromosome is called a **locus** (LŌ-kus; plural, *loci*).

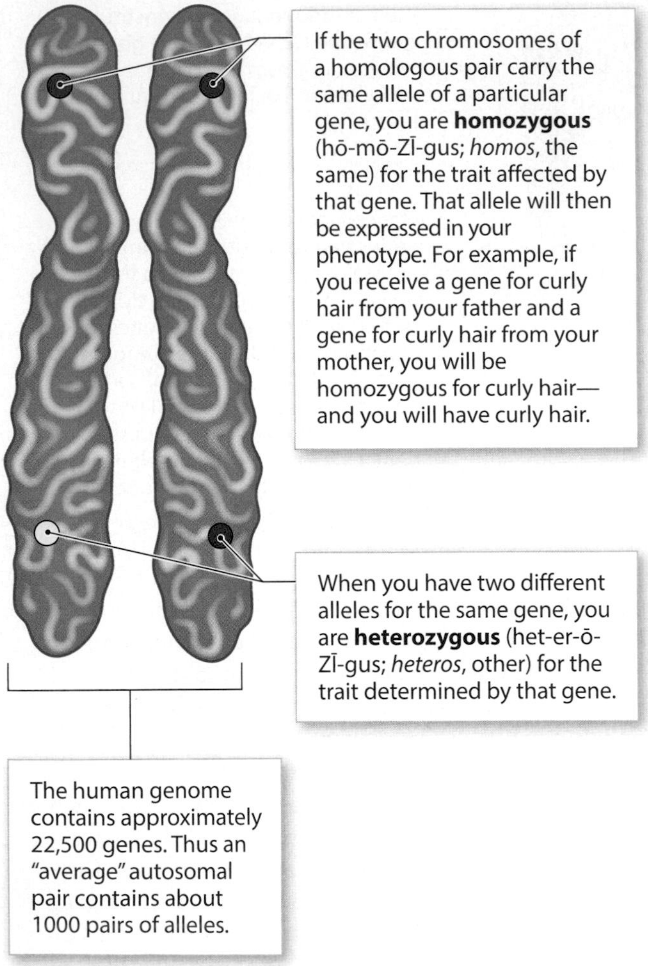
If the two chromosomes of a homologous pair carry the same allele of a particular gene, you are **homozygous** (hō-mō-ZĪ-gus; *homos*, the same) for the trait affected by that gene. That allele will then be expressed in your phenotype. For example, if you receive a gene for curly hair from your father and a gene for curly hair from your mother, you will be homozygous for curly hair—and you will have curly hair.
When you have two different alleles for the same gene, you are **heterozygous** (het-er-ō-ZĪ-gus; *heteros*, other) for the trait determined by that gene.
The human genome contains approximately 22,500 genes. Thus an "average" autosomal pair contains about 1000 pairs of alleles.

2 The phenotype that results from a heterozygous genotype depends on the nature of the interaction between the corresponding alleles. The most common form of interaction among autosomal genes is called **simple inheritance**. In one kind of simple inheritance—**strict dominance**—any dominant allele will be expressed in the phenotype, regardless of any conflicting instructions carried by the other allele. For instance, a person with only one allele for freckles will have freckles, because that allele is dominant over the "nonfreckle" allele. An allele that is **recessive** will be expressed in the phenotype only if that same allele is present on both chromosomes of a homologous pair.

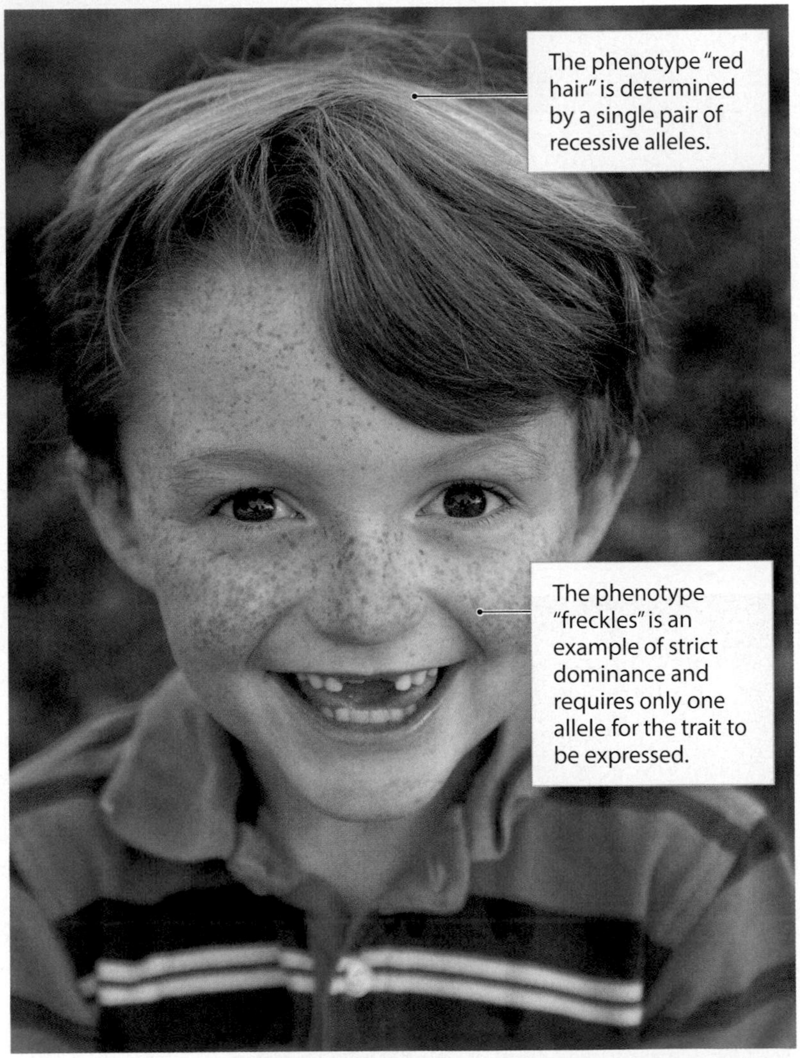
The phenotype "red hair" is determined by a single pair of recessive alleles.
The phenotype "freckles" is an example of strict dominance and requires only one allele for the trait to be expressed.

3 By convention, dominant alleles are indicated by capital letters, and recessive alleles by lowercase letters. Thus for a given trait *T*, the possible genotypes are *TT* (homozygous dominant), *Tt* (heterozygous), and *tt* (homozygous recessive). The **Punnett squares** shown in this figure indicate the possible offspring of a mother with albinism (no skin pigmentation) and a father with normal skin pigmentation. Because albinism is a recessive trait, the maternal alleles are designated *aa*. The father has normal pigmentation, a dominant trait, so his genotype could be either *AA* or *Aa*.

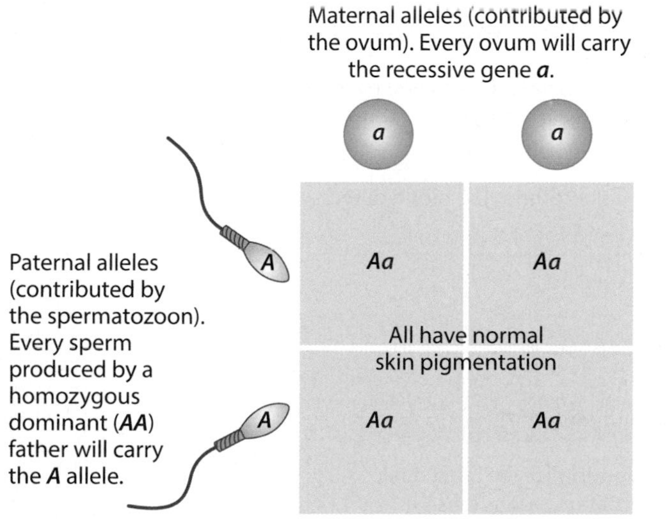

Maternal alleles (contributed by the ovum). Every ovum will carry the recessive gene ***a***.
a
a
Paternal alleles (contributed by the spermatozoon). Every sperm produced by a homozygous dominant (***AA***) father will carry the ***A*** allele.
A
A
Aa
Aa
All have normal skin pigmentation
Aa
Aa

If the father is homozygous for normal pigmentation, all of the children will have the genotype *Aa*, and all will have normal skin pigmentation.

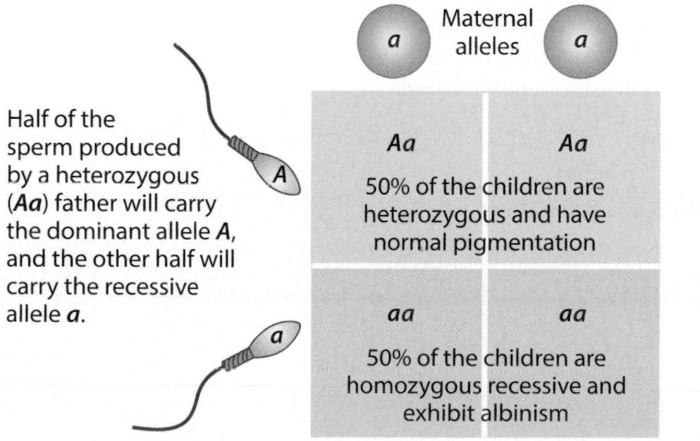

Maternal alleles
a
a
Half of the sperm produced by a heterozygous (***Aa***) father will carry the dominant allele ***A***, and the other half will carry the recessive allele ***a***.
A
a
Aa
Aa
50% of the children are heterozygous and have normal pigmentation
aa
aa
50% of the children are homozygous recessive and exhibit albinism

If the father is heterozygous for normal skin pigmentation, the probability that a child will have normal pigmentation is reduced to 50%.

4 Many phenotypic characteristics are determined by interactions among the alleles of several genes. Such interactions constitute **polygenic inheritance**. The resulting phenotype depends not only on the nature of the alleles, but also on how those alleles interact. For this reason, you cannot predict the presence or absence of phenotypic characters using a simple Punnett square. An example of polygenic inheritance is brown or black hair color.

Module 27.14 Review

a. Describe homozygous and heterozygous.

b. Differentiate between simple inheritance and polygenic inheritance.

c. The trait "freckles" operates through strict dominance. What would be the phenotype of a person who is heterozygous for this trait?

LO **27.14** Relate the basic principles of genetics to the inheritance of human traits.

Module 27.15

There are several different patterns of inheritance

1 Autosomal chromosomes and sex chromosomes show different patterns of inheritance. This flowchart summarizes the major patterns of inheritance.

Major Patterns of Inheritance

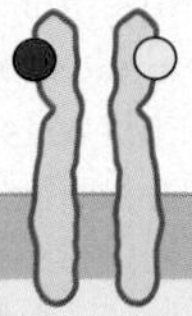

Inheritance Involving Autosomal Chromosomes

Simple Inheritance

The phenotype is determined by a single pair of alleles. Approximately 80% of your genotype falls within this category.

Polygenic Inheritance

The phenotype is determined by interactions among the alleles of several genes.

Examples

- Hair color (other than blond or red, which are recessive traits)
- Skin color
- Eye color
- Height

Strict Dominance

One allele dominates the other allele and determines the phenotype

Examples of dominant traits

- Normal skin pigmentation
- Freckles
- Nearsightedness
- Farsightedness
- Astigmatism
- Free earlobes
- Tongue-rolling
- Rh factor
- Type A or B blood
- Huntington's disease

Examples of recessive traits

- Albino pigmentation
- Absence of freckles
- Normal vision
- Attached earlobes
- Inability to roll tongue
- Rh factor absent
- Type O blood
- Sickle cell anemia
- Cystic fibrosis
- Tay–Sachs disease
- Phenylketonuria

Codominance

In **codominance**, an individual who is heterozygous (has different alleles) for a given trait exhibits both of the phenotypes for that trait. Blood type in humans is determined by codominance. The alleles for type A and type B blood are dominant over the allele for type O blood, but a person with one type A allele and one type B allele has type AB blood.

Examples

- Type AB blood
- Structure of albumins

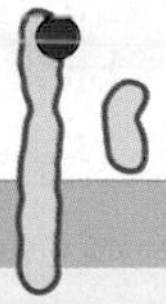

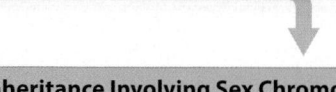

Inheritance Involving Sex Chromosomes

Sex-linked inheritance involves genes on the sex chromosomes. Most examples involve the X chromosome. An X-linked allele determines the phenotype in males since there is no corresponding allele on the Y chromosome to mask or affect its expression. Most known cases involve alleles that are recessive in females.

Examples

- Red–green color blindness
- Hemophilia (some forms)
- Duchenne muscular dystrophy

2 Sex-linked genes are carried by either sex chromosome. But, because the X chromosome is much larger than the Y chromosome, it carries many more genes. Only a few of these sex-linked genes are homologous. Genes that are found on the X chromosome but not on the Y chromosome are called **X-linked**. In contrast, those genes found only on the Y chromosome are called **Y-linked**. A Y-linked example is the SRY gene for "maleness." It triggers male development in the embryo. A number of clinical disorders are X-linked traits.

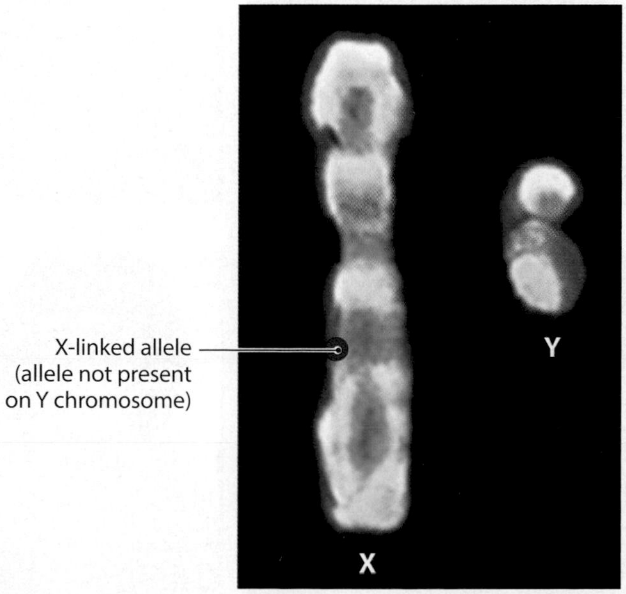

X-linked allele (allele not present on Y chromosome)
Y
X

3 The inheritance of color blindness exemplifies the differences between sex-linked inheritance and autosomal inheritance. In males, the presence of a dominant allele, ***C***, on the X chromosome (X^C) results in normal color vision; a recessive allele, ***c***, on the X chromosome (X^c) results in red–green color blindness. This Punnett square reveals that each son of a father with normal vision and a heterozygous (carrier) mother has a 50 percent chance of being red–green color blind, whereas any daughters will have normal color vision.

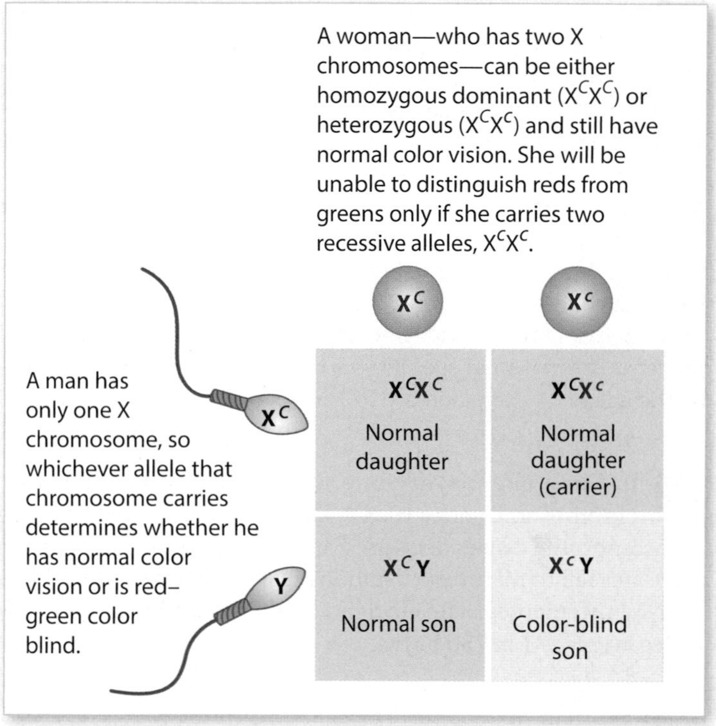

A woman—who has two X chromosomes—can be either homozygous dominant (X^CX^C) or heterozygous (X^CX^c) and still have normal color vision. She will be unable to distinguish reds from greens only if she carries two recessive alleles, X^cX^c.
X^C
X^c
A man has only one X chromosome, so whichever allele that chromosome carries determines whether he has normal color vision or is red–green color blind.
X^C
Y
X^CX^C
Normal daughter
X^CX^c
Normal daughter (carrier)
X^CY
Normal son
X^cY
Color-blind son

Module 27.15 Review

a. Compare strict dominance with codominance.

b. Why are sex-linked traits expressed more frequently in males than in females?

c. Indicate the type of inheritance involved in each of the following situations: (1) children who exhibit the trait have at least one parent who also exhibits it; (2) children exhibit the trait even though neither parent exhibits it; and (3) the trait is expressed equally in daughters and sons.

LO 27.15 Describe the relationships among the various forms of inheritance, and give examples of representative phenotypic characters, both normal and abnormal.

Many clinical disorders are linked to the human genome and epigenome

Chromosomal abnormalities can involve thousands of genes, and as a result they are usually lethal. Most of the time the embryo or fetus dies before delivery. However, there are a few autosomal chromosome abnormalities that do not invariably result in prenatal mortality (death). In contrast, variations in the structure of individual genes are relatively common. Although more than 99 percent of human nucleotide bases are the same in all people, there are about 1.4 million single-base differences, or **single nucleotide polymorphisms (SNPs)**. Some of these SNPs are inconsequential, but others are associated with specific diseases.

1 **Trisomy 21**, or **Down's syndrome**, is the most common viable chromosomal abnormality. Affected individuals have a third chromosome on the 21st pair, and exhibit intellectual disability and physical malformations, including a characteristic facial appearance. The degree of intellectual disability ranges from moderate to severe, and anatomical problems affecting the cardiovascular system often prove fatal during childhood or early adulthood. Although some people survive to moderate old age, many develop Alzheimer's disease while still relatively young (before age 40). For unknown reasons, there is a direct correlation between maternal age and the risk of having a child with trisomy 21. For a maternal age below 25, the incidence of Down's syndrome approaches 1 in 2000 births, or 0.05 percent. For maternal ages 30–34, the odds increase to 1 in 900, and by the age of 44 they increase to 1 in 46, or more than 2 percent.

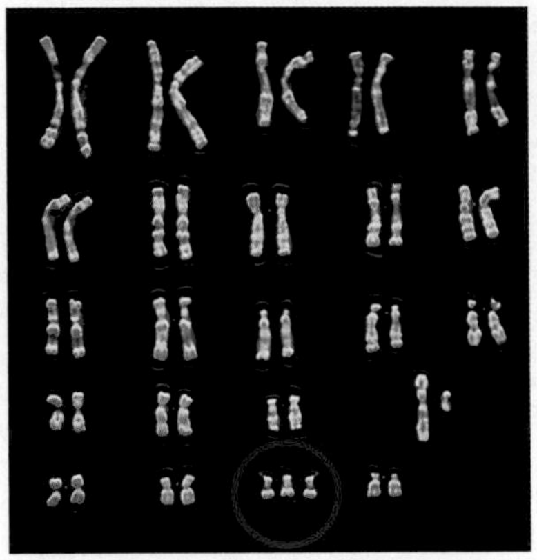

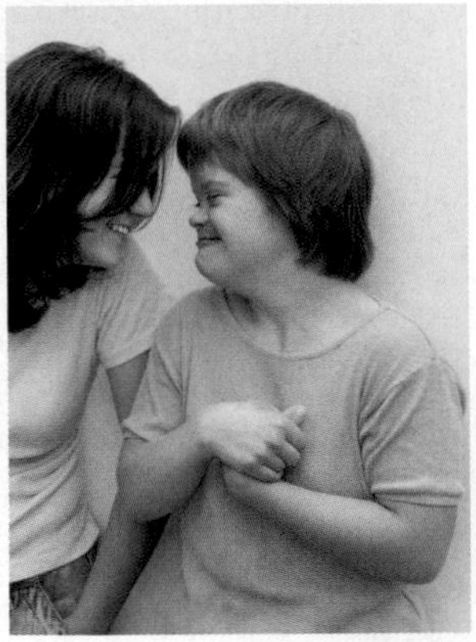

2 In **Klinefelter's syndrome**, the individual carries the sex chromosome pattern XXY. The phenotype is male, but the extra X chromosome causes decreased androgen production. As a result, the testes fail to mature so the individuals are sterile, and the breasts are slightly enlarged. The incidence of this condition among newborn males averages 1 in 750 births.

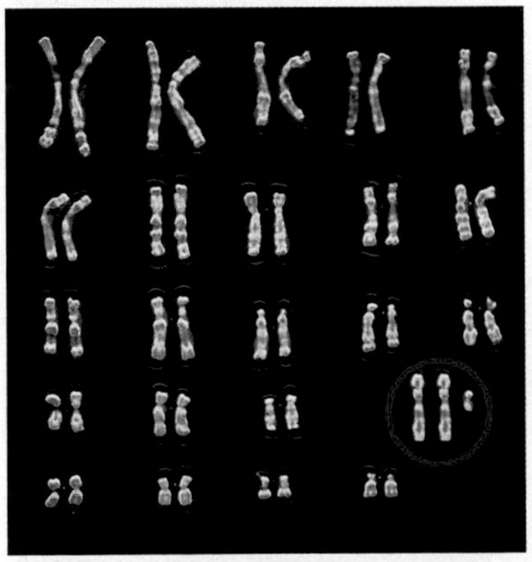

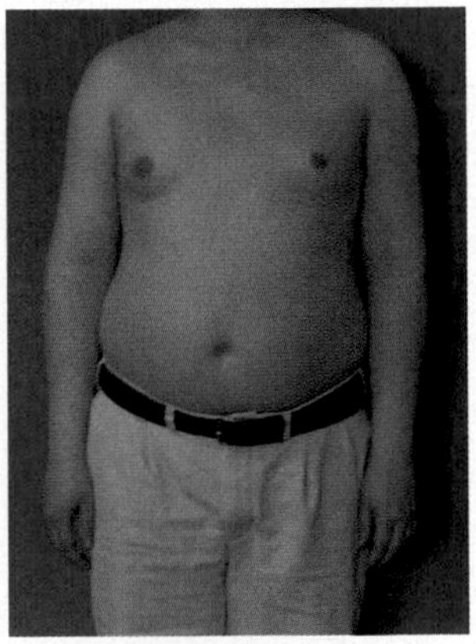

3 Individuals with **Turner's syndrome** have only a single, female sex chromosome; their sex chromosome complement is designated XO. This kind of chromosomal deletion is known as **monosomy**. The incidence of this condition at delivery has been estimated as 1 in 10,000 live births. The condition may not be recognized at birth, because the phenotype is normal female. But maturational changes do not appear at puberty. The ovaries are nonfunctional, and estrogen production occurs at negligible levels.

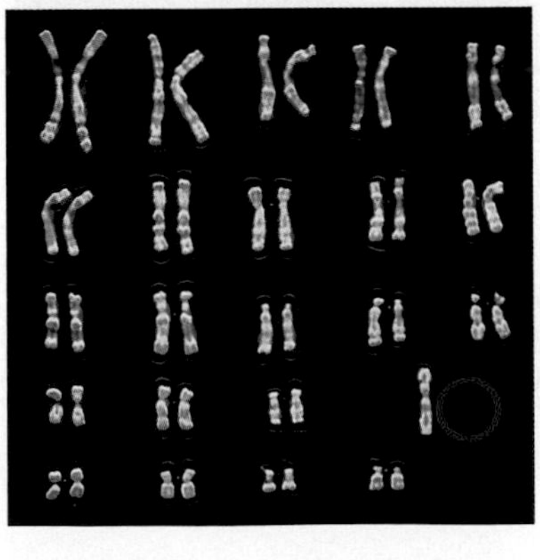

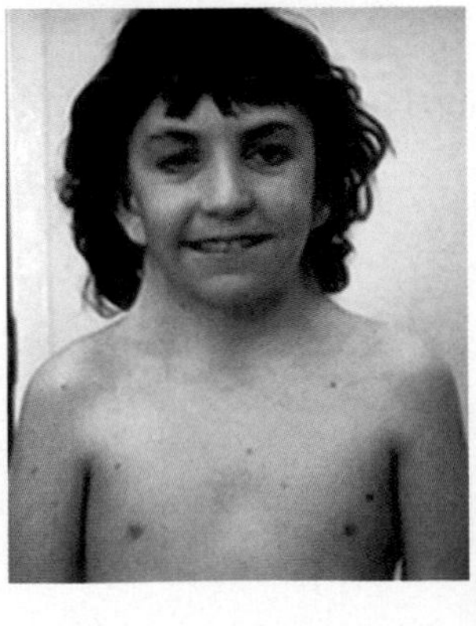

4 The **human genome**—the full set of genetic material (DNA) in our chromosomes, listed nucleotide by nucleotide—contains an estimated 20,000–25,000 genes. The sequence of nucleotides along all of the chromosomes has been determined. Roughly 10,000 different single gene disorders have been described. Most are very rare, but collectively they may affect 1 in every 200 births. Over 900 of these disorders have been mapped on the genome, and several examples are included in this chromosome map. Many of these conditions were noted in earlier chapters.

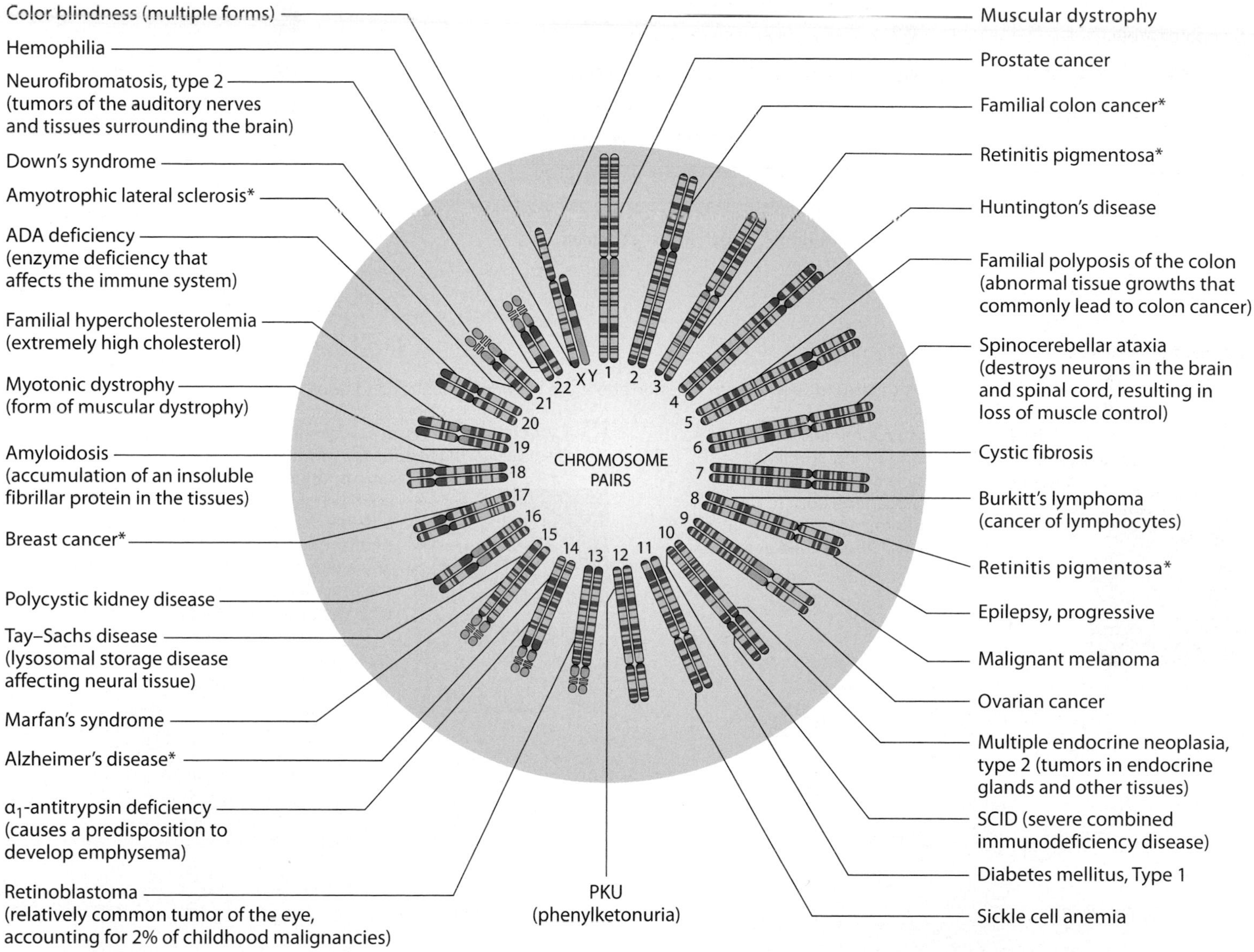
Color blindness (multiple forms)
Hemophilia
Neurofibromatosis, type 2 (tumors of the auditory nerves and tissues surrounding the brain)
Down's syndrome
Amyotrophic lateral sclerosis*
ADA deficiency (enzyme deficiency that affects the immune system)
Familial hypercholesterolemia (extremely high cholesterol)
Myotonic dystrophy (form of muscular dystrophy)
Amyloidosis (accumulation of an insoluble fibrillar protein in the tissues)
Breast cancer*
Polycystic kidney disease
Tay–Sachs disease (lysosomal storage disease affecting neural tissue)
Marfan's syndrome
Alzheimer's disease*
α_1-antitrypsin deficiency (causes a predisposition to develop emphysema)
Retinoblastoma (relatively common tumor of the eye, accounting for 2% of childhood malignancies)
PKU (phenylketonuria)
Muscular dystrophy
Prostate cancer
Familial colon cancer*
Retinitis pigmentosa*
Huntington's disease
Familial polyposis of the colon (abnormal tissue growths that commonly lead to colon cancer)
Spinocerebellar ataxia (destroys neurons in the brain and spinal cord, resulting in loss of muscle control)
Cystic fibrosis
Burkitt's lymphoma (cancer of lymphocytes)
Retinitis pigmentosa*
Epilepsy, progressive
Malignant melanoma
Ovarian cancer
Multiple endocrine neoplasia, type 2 (tumors in endocrine glands and other tissues)
SCID (severe combined immunodeficiency disease)
Diabetes mellitus, Type 1
Sickle cell anemia
CHROMOSOME PAIRS
X Y 1 2 3 4 5 6 7 8 9 10 11 12 13 14 15 16 17 18 19 20 21 22

* One form of the disease

Understanding the role of epigenetics in human development and disease is an area of active study. Researchers are piecing together the human **epigenome**, that is, all the chemicals that mark our genome and affect its activities. Although the epigenetic marks are not part of the DNA, they can be passed on and inherited when cells divide.

Module 27.16 Review

a. Define single nucleotide polymorphism.

b. Name the disorder characterized by each of the following chromosome patterns: (1) XO and (2) XXY.

c. Identify the chromosome involved in each of the following disorders: (1) ovarian cancer, (2) Tay–Sachs disease, and (3) spinocerebellar ataxia.

LO 27.16 Identify several chromosomal disorders, and describe the human genome and epigenome.

Matching

Match each lettered term with the most closely related description.

a. genotype
b. heterozygous
c. locus
d. autosomes
e. simple inheritance
f. homozygous
g. alleles
h. polygenic inheritance
i. homologous
j. genetics
k. karyotype
l. phenotype

1. Visible characteristics
2. Alternate forms of a gene
3. Refers to the two members of a pair of chromosomes
4. Array of the entire set of chromosomes in a cell
5. An individual's chromosomes and genes
6. Gene's position on a chromosome
7. Two different alleles for the same gene
8. Study of the mechanisms of inheritance
9. Two identical alleles for the same gene
10. Interactions between alleles on several genes
11. Phenotype determined by a single pair of alleles
12. Chromosomes affecting somatic characteristics

1 ____________
2 ____________
3 ____________
4 ____________
5 ____________
6 ____________
7 ____________
8 ____________
9 ____________
10 ____________
11 ____________
12 ____________

Section integration

Using the templates below, draw Punnett squares to answer each question about the following genetic conditions.

13 Tongue-rolling is inherited as a strictly dominant trait. Using *T* for tongue-rolling and *t* for non-tongue-rolling, determine the possible genotypic and phenotypic ratios of offspring born to two heterozygous tongue-rolling parents. What is the probability they will have a child unable to roll his/her tongue?

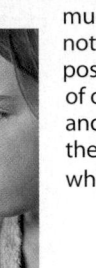

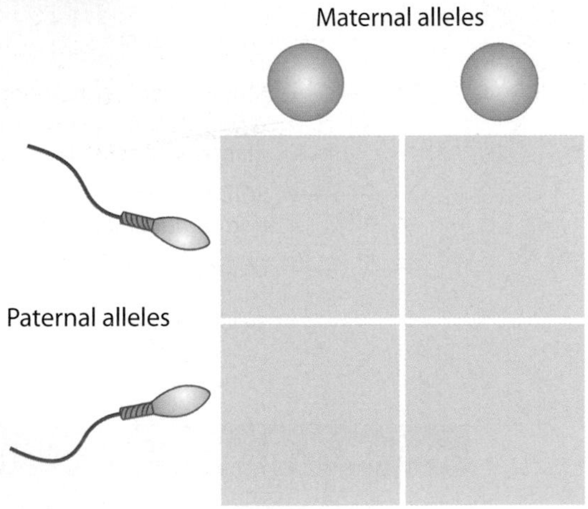
Maternal alleles
Paternal alleles

14 Cystic fibrosis (CF) is an autosomal homozygous recessive disorder (*cc*) that causes the production of excessively thick mucus. Using *c* for the CF trait and *C* for not having the CF trait, determine the possible genotypic and phenotypic ratios of offspring produced by a mother with CF and a father who is a carrier for CF. What is the probability that they will have a child who will be a carrier for CF?

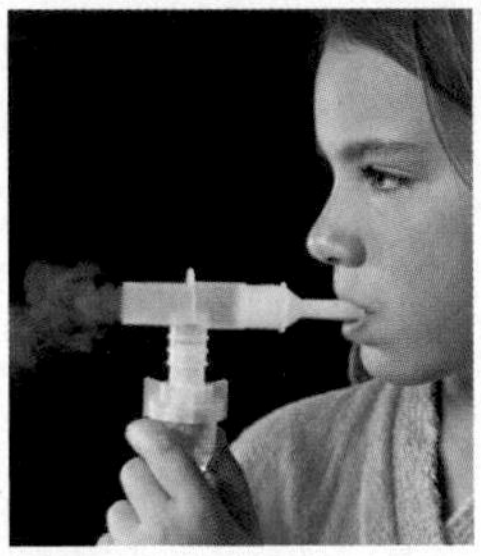

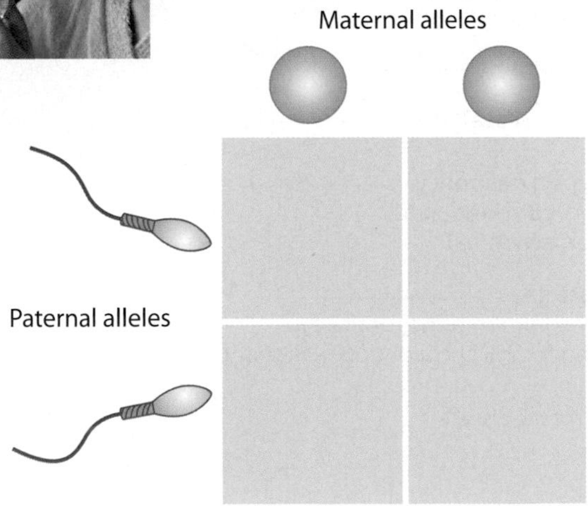
Maternal alleles
Paternal alleles

13 ____

____

____

14 ____

____

____

Study Outline

SECTION 1 • Overview of Development

27.1 Gestation and development are marked by various stages p. 1027

1. **Embryonic development** occurs during the first 2 months after fertilization.
2. **Fetal development** begins at the start of the ninth week and continues until birth.
3. **Prenatal development** comprises both embryonic and fetal development.
4. **Gestation** is the time spent in prenatal development, and occurs during three integrated trimesters, each 3 months long.
5. **Postnatal development** begins at birth and continues to **maturity**.

27.2 At fertilization, an ovum and a spermatozoon form a zygote that prepares for cell division p. 1028

6. **Fertilization** is the fusion of two haploid gametes, each containing 23 chromosomes, producing a **zygote** with 46 chromosomes. This usually occurs in the uterine tube within a day of ovulation.
7. **Oocyte activation** occurs when a single sperm makes contact with the oocyte membrane, and membrane fusion takes place. Meiosis II is completed and the secondary oocyte is now an **ovum**.

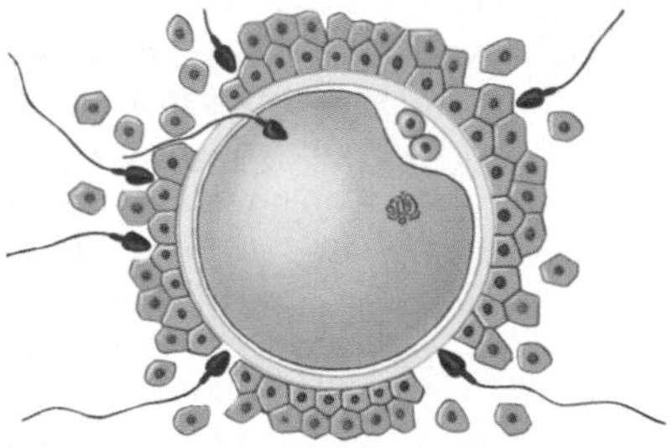

8. The female nuclear material within the ovum reorganizes as the **female pronucleus**.
9. The **male pronucleus** develops and spindle fibers appear in preparation for cell division. This is the start of **cleavage**, in which cell divisions produce increasing numbers of smaller and smaller daughter cells.
10. The two pronuclei then fuse in a process called **amphimixis**. The cell is now a zygote that contains the normal complement of 46 chromosomes. Fertilization is complete.
11. Cell division of the zygote produces cells called **blastomeres**.

27.3 Cleavage continues until the blastocyst implants in the uterine wall p. 1030

12. A group of blastomeres created by cleavage divisions is called a **pre-embryo**.
13. After 3 days, the pre-embryo is a solid ball of cells known as a **morula**.
14. By day 6, the blastomeres form a **blastocyst**, a hollow ball with an inner cavity (**blastocoele**), a surrounding cell layer (**trophoblast**), and an inner cluster of cells (**inner cell mass**).
15. **Implantation** (day 7) begins with the attachment of the blastocyst to the uterine endometrium.
16. The cells closest to the blastocoele remain intact, by day 8 forming the **cellular trophoblast**. The **syncytial trophoblast** forms from the cells near the endometrial wall.
17. **Villi** extend from the trophoblast into the endometrium. The **amniotic cavity** forms by day 9 between the inner cell mass and the cellular trophoblast.

27.4 Gastrulation produces three germ layers: ectoderm, endoderm, and mesoderm p. 1032

18. Cells of the inner cell mass are organized into an oval sheet known as the **blastodisc**. Cells of the superficial layer migrate and then separate the amniotic cavity from the trophoblast in the first step of **amnion** formation.
19. The **yolk sac** forms by day 10 from the outer edges of the blastocoele. For the next 2 weeks, the yolk sac is the primary nutrient source for the inner cell mass.
20. By day 12, superficial cells of the blastodisc have begun to migrate toward a central line known as the **primitive streak**, and migrate between the existing cell layers. This creates three embryonic layers that are collectively known as **germ layers**: **ectoderm**, **endoderm**, and **mesoderm**.
21. **Gastrulation** produces the three-layered **embryonic disc**. This will form the body of the embryo, whereas all other cells of the blastocyst will be part of the **extra-embryonic membranes**.

27.5 The extra-embryonic membranes form the placenta that supports fetal growth and development p. 1034

22. Germ layers form four extra-embryonic membranes: the yolk sac, the amnion, the allantois, and the chorion.
23. The **yolk sac** begins as a layer of cells spread out around the outer edges of the blastocoele to form a complete pouch.
24. The **amnion** begins as cells from the blastodisc migrate around a cavity that separates the inner cell mass from the trophoblast. The amniotic cavity contains **amniotic fluid** that cushions the developing embryo or fetus.
25. The **allantois** begins as an outpocketing of the endoderm near the base of the yolk sac by week 3.
26. The mesoderm associated with the allantois spreads around the blastocyst, separating the cellular trophoblast from the blastocoele. By day 14 this combination of mesoderm and trophoblast is the **chorion**.

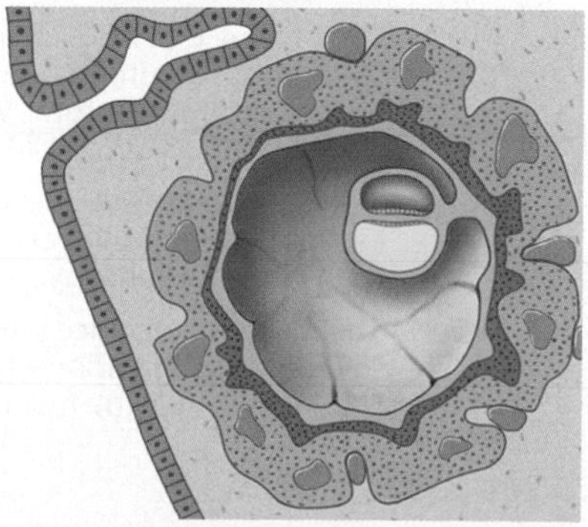

27. The **placenta** develops by week 5 as villi of the chorion invade the endometrium and break down maternal blood vessels. It becomes the primary support structure for the developing embryo.

27.6 The formation of extra-embryonic membranes is associated with major changes in the shape and complexity of the embryo p. 1036

28. By week 3, the embryonic disc bulges into the amniotic cavity at the **head fold**. By week 4 the embryo also has a **tail fold**.
29. By week 5 the embryo moves away from the placenta, and the **body stalk** and **yolk stalk** fuse to form an **umbilical stalk**.
30. By week 10 the fetus is connected to the placenta by an elongated **umbilical cord** that contains a portion of the allantois, blood vessels, and the remnants of the yolk stalk.

27.7 The placenta performs many vital functions during prenatal development p. 1038

31. Blood flows from the fetus to the placenta through paired **umbilical arteries**.

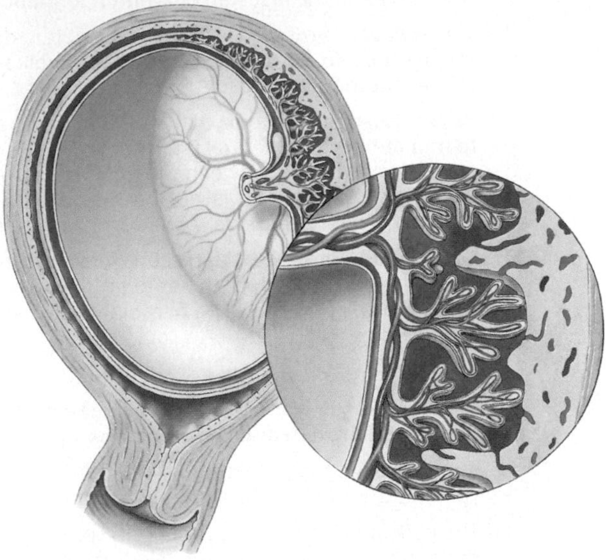

32. At the placenta, oxygen supplies are replenished, organic nutrients are added, and carbon dioxide and other organic wastes are removed.
33. Blood returns to the fetus within a single **umbilical vein**.
34. The placenta also acts as an endocrine gland, secreting **human chorionic gonadotropin** (**hCG**), **human placental lactogen** (**hPL**), **relaxin**, and **progesterone** and **estrogens**.
35. The presence of hCG, an indicator of pregnancy, causes the corpus luteum to persist for 3–4 months and continue to secrete progesterone, while hPL helps prepare the mammary glands for milk production.
36. Relaxin increases the flexibility of the pubic symphysis, causes dilation of the cervix, and delays labor contractions until late in pregnancy.
37. The placenta produces sufficient amounts of progesterone to maintain the endometrial lining and continue the pregnancy after the first trimester. Toward the end of the third trimester, estrogen production by the placenta accelerates and assists in stimulating labor and delivery.

27.8 Organ systems form in the first trimester and become functional in the second and third trimesters p. 1040

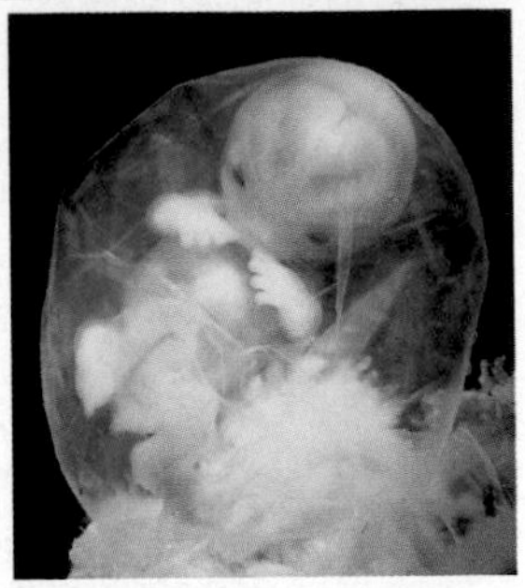

38. Events in the first 12 weeks establish **organogenesis**, the process of organ formation.
39. The heart is beating by week 4. By week 6, the placenta has formed and the embryo floats within the amniotic cavity.
40. By the end of the first trimester, the axial and appendicular muscles are forming, and fetal movements will soon begin.
41. By the end of the second trimester, the fetus will have grown to a weight of about 0.64 kg (1.4 lb).
42. During the third trimester, most of the organ systems become ready to function normally without maternal assistance.
43. In the final 3 months of gestation, the fetus gains about 2.6 kg (5.7 lb), reaching a full-term weight of approximately 3.2 kg (7 lb).
44. **Dizygotic** twins develop when two separate oocytes are ovulated and fertilized. Seventy percent of twins are dizygotic. **Monozygotic** twins result either from the separation of blastomeres early in cleavage or from the splitting of the inner cell mass before gastrulation.

27.9 Pregnancy places anatomical and physiological stresses on maternal systems p. 1042

45. The physical strains of pregnancy are considerable. The developing fetus is totally dependent on maternal organ systems that must support the mother as well as the fetus.
46. At full term, the uterus and fetus are so large that they push many of the maternal abdominal organs out of their normal positions.
47. By the end of the third trimester, many physiological stressors related to nourishment, oxygen supply, and waste removal occur in the maternal system.
48. The physical and physiological demands on maternal systems make pregnancy potentially dangerous.

27.10 Multiple factors initiate and accelerate labor p. 1044

49. Late in pregnancy, some women experience occasional spasms in the uterine musculature called **false labor**. **True labor** begins when biochemical and mechanical factors reach a point of no return.
50. **Placental factors** and **fetal factors** interact to initiate true labor. **Positive feedback** ensures that contractions of the myometrium will continue until delivery has been completed.
51. Placental factors include placental estrogens that increase the sensitivity of the smooth muscle cells of the myometrium and make contractions more likely, as well as relaxin that dilates the cervix.
52. Fetal factors include fetal stretching of the myometrium that promotes uterine contractions and fetal secretion of oxytocin.

53. **Parturition**, or childbirth, is the forcible expulsion of the fetus and placenta. Labor is divided into three stages: the dilation, expulsion, and placental stages.

54. The **dilation stage** begins with the start of true labor as the cervix dilates. It can last 8 or more hours.

55. The **expulsion stage** begins when the cervix completes dilating, and continues until the fetus has emerged from the vagina (**delivery**). This usually lasts less than 2 hours.

56. The **placental stage** is the ejection of the placenta, or **afterbirth**.

57. **Premature labor** occurs when true labor begins before the fetus has completed normal development. **Premature delivery** usually refers to birth at 28–36 weeks.

27.11 After delivery, development initially requires nourishment by maternal systems p. 1046

58. The **neonatal period** lasts from birth through the first 28 days of life. At this time the newborn is dependent on the mother for nourishment.

59. Mammary gland cells produce protein-rich **colostrum** during the neonate's first few days of life; then the gland cells convert to milk production. These secretions are released as a result of the **milk let-down** (milk-ejection) **reflex**.

60. Postnatal development includes five **life stages**: the **neonatal** period, **infancy**, **childhood**, **adolescence**, and **maturity**.

27.12 At puberty, male and female sex hormones have differing effects on most body systems p. 1048

61. At **puberty**, the hypothalamus increases its production of gonadotropin-releasing hormone (GnRH).

62. Endocrine cells in the anterior lobe of the pituitary gland become more sensitive to GnRH, and circulating levels of FSH and LH rise rapidly.

63. Testicular or ovarian cells become more sensitive to FSH and LH. The following responses are initiated: gamete production; the secretion of sex hormones, which stimulate the appearance of secondary sex characteristics; and a sudden acceleration in the growth rate, ending with closure of the epiphyseal cartilages.

SECTION 2 • Genetics and Inheritance

27.13 A person may be described in terms of their genotype and phenotype p. 1051

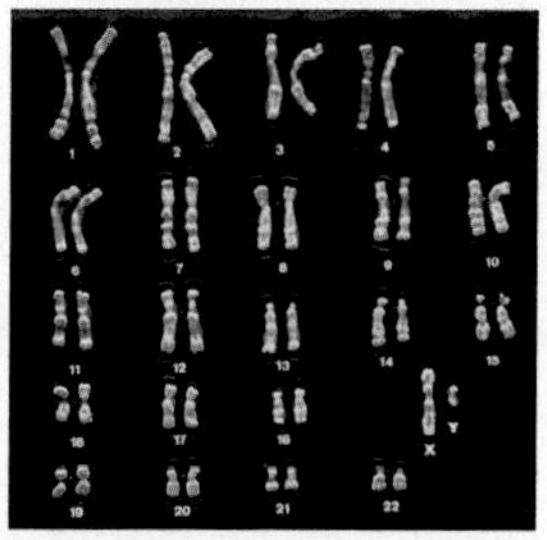

64. **Genetics** is the study of **inheritance**, or **heredity**, the transfer of genetically determined characteristics from generation to generation.

65. Somatic cells contain copies of the original 46 chromosomes from the zygote. These chromosomes and their genes constitute a person's **genotype**. Anatomical and physiological characteristics make up the **phenotype. Epigenetics** examines how phenotypes may be affected by alterations in the reading of DNA sequences. A **karyotype** is a complete set of a person's chromosomes.

66. The two members of each pair of chromosomes are known as **homologous chromosomes**. Twenty-two of those pairs are called **autosomal chromosomes**. The chromosomes of the 23rd pair are called the **sex chromosomes**.

67. The sex chromosomes of a male consist of an **X chromosome** and a shorter **Y chromosome**, whereas females have two X chromosomes.

27.14 Genes and chromosomes determine patterns of inheritance p. 1052

"Red hair" phenotype is determined by a single pair of recessive alleles. "Freckles" phenotype follows strict dominance.

68. The two chromosomes in a homologous autosomal pair have the same structure and carry genes that affect the same traits.

69. The various forms of a given gene are called **alleles**. A gene's position on a chromosome is called a **locus**.

70. If the two chromosomes of a homologous pair carry the same allele of a particular gene, you are **homozygous** for the trait affected by that gene.

71. When you have two different alleles for the same gene, you are **heterozygous** for the trait determined by that gene.

72. The most common form of interaction among autosomal genes is **simple inheritance**. In one kind of simple inheritance called **strict dominance**, any dominant allele will be expressed in the phenotype.

73. An allele that is **recessive** will be expressed in the phenotype only if that same allele is present on both chromosomes of a homologous pair.

74. In a **Punnett square**, dominant alleles are indicated by capital letters, and recessive alleles by lowercase letters.

75. Many phenotypic characteristics involve **polygenic inheritance**, and are determined by interactions among the alleles of several genes. In these situations, a Punnett square cannot predict the presence or absence of phenotypic characteristics.

27.15 There are several different patterns of inheritance p. 1054

76. Autosomal chromosomes and sex chromosomes show different patterns of inheritance.

77. Autosomal chromosomes involve simple inheritance. This includes strict dominance and **codominance**. Autosomal chromosomes also involve polygenic inheritance.

78. **Sex-linked inheritance** involves genes on the sex chromosomes. Most examples involve the X chromosome. An **X-linked** allele determines the phenotype in males since there is no corresponding allele on the Y chromosome to mask its expression. Most known cases involve alleles that are recessive in females.

27.16 **Many clinical disorders are linked to the human genome and epigenome** p. 1056

79. Chromosome abnormalities and **single nucleotide polymorphisms** (**SNPs**) have been linked with various disorders and diseases.
80. **Trisomy 21**, or **Down's syndrome**, is the most common viable chromosomal abnormality. Affected individuals exhibit intellectual disability and physical malformations, including a characteristic facial appearance.

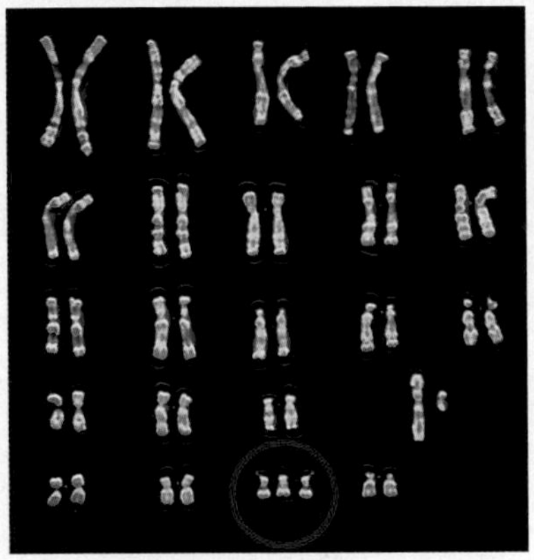

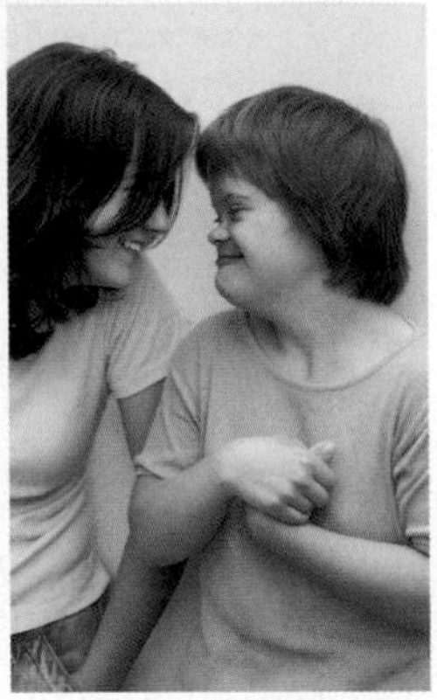

81. Individuals with **Klinefelter's syndrome** have the sex chromosome pattern XXY. The phenotype is male, but the extra X chromosome causes decreased androgen production, resulting in immature testes, sterility, and enlarged breasts.
82. Individuals with **Turner's syndrome** have only a single, female sex chromosome, so their sex chromosome complement is designated XO (**monosomy**). Maturational changes do not appear at puberty; effects include nonfunctional ovaries and negligible estrogen production.
83. The full set of genetic material (DNA) in our chromosomes is called the **human genome**, and contains an estimated 20,000–25,000 genes. Over 900 single gene disorders have been mapped on the genome.
84. Knowing the human **epigenome**, that is, all the chemicals that mark our genome and affect its activities, is expected to improve our understanding of human development and disease.

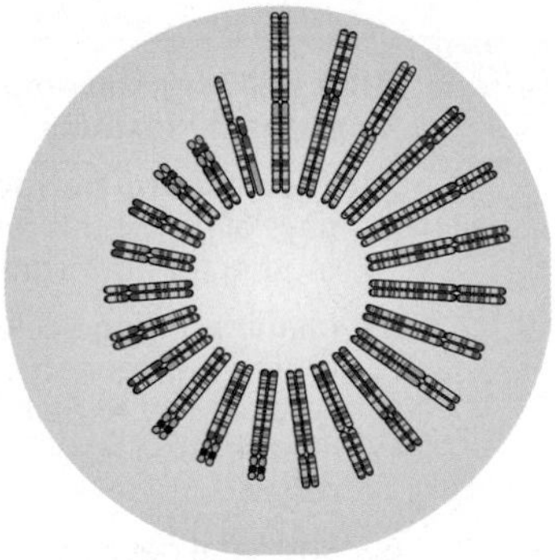

Chapter Review Questions

True/False

Indicate whether each statement is true or false.

1 Fertilization typically occurs within the uterus.

2 In the 2 weeks after its development, the yolk sac is the primary nutrient source for the inner cell mass.

3 Dizygotic twins are also called identical twins.

4 Relaxin is a hormone released by the placenta as well as the corpus luteum.

5 There is one umbilical artery and two umbilical veins.

6 The four extra-embryonic membranes are the yolk sac, amnion, allantois, and chorion.

7 The structure that implants in the uterine endometrium is an advanced morula.

8 The health risks associated with oral contraceptives are far greater than the risks associated with pregnancy.

1 ______
2 ______
3 ______
4 ______
5 ______
6 ______
7 ______
8 ______

Multiple choice

Select the correct answer from the list provided.

9 The hormone detected in pregnancy kits is

- ☐ a) relaxin.
- ☐ b) human chorionic gonadotropin (hCG).
- ☐ c) human placental lactogen (hPL).
- ☐ d) gonadotropin-releasing hormone (GnRH).

10 Concerning sex-linked inheritance, X-linked recessive traits

- ☐ a) are more likely to be expressed in females.
- ☐ b) are more likely to be expressed in males.
- ☐ c) are never expressed in females.
- ☐ d) are never expressed in males.

11 Amphimixis is when

- ☐ a) the blastocyst implants into the uterine endometrium.
- ☐ b) a single spermatozoa contacts the oocyte membrane.
- ☐ c) the male pronucleus develops and spindle fibers appear in preparation for cell division.
- ☐ d) the male and female pronuclei fuse.

12 For a given trait *T*, how would a homozygous dominant genotype be indicated?

- ☐ a) *TT*
- ☐ b) *Tt*
- ☐ c) *tt*
- ☐ d) *tT*

13 If an allele must be present on both the maternal and paternal chromosomes to affect the phenotype, the allele is said to be

- ☐ a) dominant.
- ☐ b) recessive.
- ☐ c) heterozygous.
- ☐ d) polygenic.

14 Which of the following sex chromosome patterns represents Klinefelter's syndrome?

- ☐ a) YO
- ☐ b) XO
- ☐ c) YYX
- ☐ d) XXY

15 The hormone responsible for milk ejection is

- ☐ a) prolactin.
- ☐ b) estrogen.
- ☐ c) oxytocin.
- ☐ d) progesterone.

16 The phases of labor in the correct order are

- ☐ a) dilation, placental, expulsion.
- ☐ b) placental, dilation, expulsion.
- ☐ c) placental, expulsion, dilation.
- ☐ d) dilation, expulsion, placental.

Short answer

17 Identify the five life stages of postnatal development, and identify the approximate ages for each stage.

18 What is trisomy 21, and what are the risk factors for having a baby with this condition?

MasteringA&P®

Access more chapter study tools online in the MasteringA&P Study Area:

- **Chapter Quizzes, Chapter Practice Test, Art-labeling Activities, Animations, MP3 Tutor Sessions, and Clinical Case Studies**

- **Practice Anatomy Lab** PAL™
- **Interactive Physiology** iP®
- **A&P Flix** A&PFlix™
- **PhysioEx** PhysioEx™

Chapter Integration • Applying what you have learned

The blocks that build a family

Joe and Jane have not been successful in having a child although they have been trying for 2 years. They finally consult with a physician and undergo complete physical exams and genetics testing. They discover that Joe's genotype is a normal XY, but he suffers from oligospermia (low sperm count). Jane is a carrier (XX^h) for hemophilia A, an X-linked recessive trait in which the blood does not clot properly. The doctor informs them that Joe's low sperm count is interfering with their ability to have children, and that there is a risk that any child who is born will have hemophilia A.

For most couples, the only sign that the male has a low sperm count is the inability to conceive a child after 1 year of regular intercourse. In most cases, the cause is unknown, but other reasons include damaged sperm ducts, infections, antisperm antibodies, or Klinefelter's syndrome, which results in a complete lack of sperm in the semen (azoospermia).

Many people with hemophilia maintain active, productive lives. Signs and symptoms vary and include unexplained bruising, nosebleeds with unknown cause, and painful headaches. Hemophilia A is the most common type of hemophilia and is caused by insufficient amounts of clotting factor VIII.

1 **If it only takes one sperm to fertilize a secondary oocyte, why is Joe's oligospermia problematic for achieving pregnancy?**

2 **Regarding hemophilia A, what is the probability that the couple will have daughters with hemophilia? What is the probability that the couple will have sons with hemophilia?**

Periodic Table of Elements

The **periodic table** presents the known elements in order of their atomic weights. Each horizontal row represents a single electron shell. The number of elements in that row is determined by the maximum number of electrons that can be stored at that energy level. The element at the left end of each row contains a single electron in its outermost electron shell; the element at the right end of the row has a filled outer electron shell. Organizing the elements in this fashion highlights similarities that reflect the composition of the outer electron shell. These similarities are evident when you examine the vertical columns. All the gases of the right-most column—helium, neon, argon, krypton, xenon, and radon—have full electron shells; each is a gas at normal atmospheric temperature and pressure, and none reacts readily with other elements. These elements, highlighted in blue, are known as the *noble*, or *inert*, *gases*. In contrast, the elements of the left-most column below hydrogen—lithium, sodium, potassium, rubidium, caesium, and francium—are silvery, soft metals that are so highly reactive that pure forms cannot be found in nature. The fourth and fifth electron levels can hold up to 18 electrons. Table inserts are used for the so-called *lanthanoids* and *actinoids* to save space, as higher levels can store up to 32 electrons. Elements of particular importance to our discussion of human anatomy and physiology are highlighted in pink.

Atomic number — 1 H — Chemical symbol
Hydrogen — Element name
Atomic weight — 1.01

1 H Hydrogen 1.01																	2 He Helium 4.00
3 Li Lithium 6.94	4 Be Beryllium 9.01											5 B Boron 10.81	6 C Carbon 12.01	7 N Nitrogen 14.01	8 O Oxygen 16.00	9 F Fluorine 19.00	10 Ne Neon 20.18
11 Na Sodium 22.99	12 Mg Magnesium 24.31											13 Al Aluminum 26.98	14 Si Silicon 28.09	15 P Phosphorus 30.97	16 S Sulfur 32.07	17 Cl Chlorine 35.45	18 Ar Argon 39.95
19 K Potassium 39.10	20 Ca Calcium 40.08	21 Sc Scandium 44.96	22 Ti Titanium 47.87	23 V Vanadium 50.94	24 Cr Chromium 52.00	25 Mn Manganese 54.94	26 Fe Iron 55.85	27 Co Cobalt 58.93	28 Ni Nickel 58.69	29 Cu Copper 63.55	30 Zn Zinc 65.38	31 Ga Gallium 69.72	32 Ge Germanium 72.63	33 As Arsenic 74.92	34 Se Selenium 78.96	35 Br Bromine 79.90	36 Kr Krypton 83.80
37 Rb Rubidium 85.47	38 Sr Strontium 87.62	39 Y Yttrium 88.91	40 Zr Zirconium 91.22	41 Nb Niobium 92.91	42 Mo Molybdenum 95.96	43 Tc Technetium	44 Ru Ruthenium 101.1	45 Rh Rhodium 102.9	46 Pd Palladium 106.4	47 Ag Silver 107.9	48 Cd Cadmium 112.4	49 In Indium 114.8	50 Sn Tin 118.7	51 Sb Antimony 121.8	52 Te Tellurium 127.6	53 I Iodine 126.9	54 Xe Xenon 131.3
55 Cs Caesium 132.9	56 Ba Barium 137.3	57–71 Lanthanoids	72 Hf Hafnium 178.5	73 Ta Tantalum 180.9	74 W Tungsten 183.8	75 Re Rhenium 186.2	76 Os Osmium 190.2	77 Ir Iridium 192.2	78 Pt Platinum 195.1	79 Au Gold 197.0	80 Hg Mercury 200.6	81 Tl Thallium 204.3	82 Pb Lead 207.2	83 Bi Bismuth 209.0	84 Po Polonium	85 At Astatine	86 Rn Radon
87 Fr Francium	88 Ra Radium	89–103 Actinoids	104 Rf Rutherfordium	105 Db Dubnium	106 Sg Seaborgium	107 Bh Bohrium	108 Hs Hassium	109 Mt Meitnerium	110 Ds Darmstadtium	111 Rg Roentgenium	112 Cn Copernicium		114 Fl Flerovium		116 Lv Livermorium		

57 La Lanthanum 138.9	58 Ce Cerium 140.1	59 Pr Praseodymium 140.9	60 Nd Neodymium 144.2	61 Pm Promethium	62 Sm Samarium 150.4	63 Eu Europium 152.0	64 Gd Gadolinium 157.3	65 Tb Terbium 158.9	66 Dy Dysprosium 162.5	67 Ho Holmium 164.9	68 Er Erbium 167.3	69 Tm Thulium 168.9	70 Yb Ytterbium 173.1	71 Lu Lutetium 175.0
89 Ac Actinium	90 Th Thorium 232.0	91 Pa Protactinium 231.0	92 U Uranium 238.0	93 Np Neptunium	94 Pu Plutonium	95 Am Americium	96 Cm Curium	97 Bk Berkelium	98 Cf Californium	99 Es Einsteinium	100 Fm Fermium	101 Md Mendelevium	102 No Nobelium	103 Lr Lawrencium

Normal Physiological Values

Tables 1 and 2 present normal averages or ranges for the chemical composition of body fluids. These values are approximations rather than absolute values, because test results vary from laboratory to laboratory due to differences in procedures, equipment, normal solutions, and so forth. Blanks in the tabular data appear where data are not available. The following modules in the text contain additional information about body fluid analysis:

- Module 17.2 (p. 624) discusses the formed elements of whole blood.
- Module 17.4 (p. 629) provides an example of a complete blood count (CBC) and examples of red blood cell (RBC) tests.
- Module 17.9 (p. 638) presents data on the white blood cells (WBCs) in whole blood.
- Module 24.6 (p. 933) compares the normal values for solutes in blood, plasma, and urine.
- Module 24.12 (p. 945) lists the general characteristics of normal urine.
- Module 24.14 (p. 948) compares the composition of plasma and dialysis fluid.

Table 1 The Composition of Minor Body Fluids

	Normal Averages or Ranges					
Test	**Perilymph**	**Endolymph**	**Synovial Fluid**	**Sweat**	**Saliva**	**Semen**
pH			7.4	4–6.8	6.4*	7.19
Specific gravity			1.008–1.015	1.001–1.008	1.007	1.028
Electrolytes (mEq/L)						
Potassium	5.5–6.3	140–160	4.0	4.3–14.2	21	31.3
Sodium	143–150	12–16	136.1	0–104	14*	117
Calcium	1.3–1.6	0.05	2.3–4.7	0.2–6	3	12.4
Magnesium	1.7	0.02		0.03–4	0.6	11.5
Bicarbonate	17.8–18.6	20.4–21.4	19.3–30.6		6*	24
Chloride	121.5	107.1	107.1	34.3	17	42.8
Proteins (total) (mg/dL)	200	150	1.72 g/dL	7.7	386†	4.5 g/dL
Metabolites (mg/dL)						
Amino acids				47.6	40	1.26 g/dL
Glucose	104		70–110	3.0	11	224 (fructose)
Urea				26–122	20	72
Lipids (total)	12		20.9	‡	25–500§	188

*Increases under salivary stimulation.
†Primarily alpha-amylase, with some lysozymes.
‡Not present in merocrine (eccrine) secretions.
§Cholesterol.

APPENDIX

Table 2 The Chemistry of Blood, Cerebrospinal Fluid (CSF), and Urine			
	Normal Averages or Ranges		
Test	**Blood***	**CSF**	**Urine**
pH	S: 7.35–7.45	7.31–7.34	4.5–8.0
Osmolarity (mOsm/L)	S: 280–295	292–297	855–1335
Electrolytes	(mEq/L unless noted)		(urinary loss, mEq per 24-hour period†)
Bicarbonate	P: 20–28	20–24	0
Calcium	S: 8.5–10.3	2.1–3.0	6.5–16.5
Chloride	P: 97–107	100–108	110–250
Iron	S: 50–150 μg/L	23–52 μg/L	40–150 μg
Magnesium	S: 1.4–2.1	2–2.5	6.0–10.0
Phosphorus	S: 1.8–2.9	1.2–2.0	0.4–1.3 g
Potassium	P: 3.5–5.0	2.7–3.9	25–125
Sodium	P: 135–145	137–145	40–220
Sulfate	S: 0.2–1.3		1.07–1.3 g
Metabolites	(mg/dL unless noted)		(urinary loss, mg per 24-hour period‡)
Amino acids	P/S: 2.3–5.0	10.0–14.7	41–133
Ammonia	P: 20–150 μg/dL	25–80 μg/dL	340–1200
Bilirubin	S: 0.5–1.0	<0.2	0
Creatinine	P/S: 0.6–1.5	0.5–1.9	770–1800
Glucose	P/S: 70–110	40–70	0
Ketone bodies	S: 0.3–2.0	1.3–1.6	10–100
Lactic acid	WB: 0.7–2.5 mEq/L§	10–20	100–600
Lipids (total)	P: 450–1000	0.8–1.7	0.002
Cholesterol (total)	S: 150–300	0.2–0.8	1.2–3.8
Triglycerides	S: 40–150	0–0.9	0
Urea	P: 8–25	12.0	1800
Uric acid	P: 2.0–6.0	0.2–1.5	250–750
Proteins	(g/dL)	(mg/dL)	(urinary loss, mg per 24-hour period‡)
Total	P: 6.0–8.0	2.0–4.5	0–8
Albumin	S: 3.2–4.5	10.6–32.4	0–3.5
Globulins (total)	S: 2.3–3.5	2.8–15.5	7.3
Immunoglobulins	S: 1.0–2.2	1.1–1.7	3.1
Fibrinogen	P: 0.2–0.4	0.65	0

*S = serum, P = plasma, WB = whole blood.
†Because urinary output averages just over 1 liter per day, these electrolyte values are comparable to mEq/L.
‡Because urinary metabolite and protein data approximate mg/L or g/L, these data must be divided by 10 for comparison with CSF or blood concentrations.
§Venous blood sample.

Codon Chart

A codon is a sequence of three consecutive nucleotides in mRNA that codes for a particular amino acid or signals to stop protein synthesis. Because each mRNA codon consists of three nucleotides, the four nucleotides in mRNA (A, U, G, and C) can produce 64 different combinations. Of these, 61 codons correspond to amino acids and three act as stop signals of protein synthesis. There are only 20 different amino acids, so most amino acids are represented by more than one codon. One codon, AUG, has a dual role. It codes for the amino acid methionine and also as the start signal for protein synthesis. It is always the first codon in a strand of mRNA.

Genetic Code (mRNA codons)

First nucleotide	Second nucleotide: U	Second nucleotide: C	Second nucleotide: A	Second nucleotide: G	Third nucleotide
U	UUU Phenylalanine	UCU Serine	UAU Tyrosine	UGU Cysteine	U
U	UUC Phenylalanine	UCC Serine	UAC Tyrosine	UGC Cysteine	C
U	UUA Leucine	UCA Serine	UAA **Stop**	UGA **Stop**	A
U	UUG Leucine	UCG Serine	UAG **Stop**	UGG Tryptophan	G
C	CUU Leucine	CCU Proline	CAU Histidine	CGU Arginine	U
C	CUC Leucine	CCC Proline	CAC Histidine	CGC Arginine	C
C	CUA Leucine	CCA Proline	CAA Glutamine	CGA Arginine	A
C	CUG Leucine	CCG Proline	CAG Glutamine	CGG Arginine	G
A	AUU Isoleucine	ACU Threonine	AAU Asparagine	AGU Serine	U
A	AUC Isoleucine	ACC Threonine	AAC Asparagine	AGC Serine	C
A	AUA Isoleucine	ACA Threonine	AAA Lysine	AGA Arginine	A
A	AUG **Start** *or* Met*	ACG Threonine	AAG Lysine	AGG Arginine	G
G	GUU Valine	GCU Alanine	GAU Aspartic acid	GGU Glycine	U
G	GUC Valine	GCC Alanine	GAC Aspartic acid	GGC Glycine	C
G	GUA Valine	GCA Alanine	GAA Glutamic acid	GGA Glycine	A
G	GUG Valine	GCG Alanine	GAG Glutamic acid	GGG Glycine	G

*Abbreviation for methionine.

Answers

CHAPTER 1

Module Reviews

Module 1.1 Review
a. Several strategies for success in this course include approaching the information in ways that include teaching fellow students, asking questions in class, having a set study schedule with a devoted block of time, mastering memorization, avoiding shortcuts, regularly attending and participating in all class sessions, reading material before class, avoiding procrastination, and asking for help immediately if you have any problems. **b.** The learning outcomes indicate what you should be able to do after completing each chapter. **c.** A "Black Box" refers to missing knowledge. The more you learn about the body, the more you will realize how many other things you don't know.

Module 1.2 Review
a. Biology is the study of life. **b.** The basic functions shared by all living things are: responsiveness, adaptability, growth and development, reproduction, and movement. **c.** Most animals have an internal circulation network system to transport the inputs and/or products of digestion, respiration, and excretion throughout the body or to excretory sites.

Module 1.3 Review
a. Anatomy is the study of internal and external body structures; physiology is the study of how living organisms perform their vital functions. **b.** Gross anatomy (also called macroscopic anatomy) is the study of body structures that can be seen with the unaided eye; microscopic anatomy is the study of body structures that cannot be seen without magnification. **c.** The structures of body parts (anatomy) are closely related to their functions (physiology)—that is, function follows form.

Module 1.4 Review
a. All specific functions are performed by specific structures, and the link between the two is always present, but not always understood. **b.** The elbow joint moves in a single plane, like the opening and closing of a door on a hinge. **c.** If a structure's anatomy were altered, the structure's function would likely be impaired or perhaps eliminated.

Module 1.5 Review
a. An organ is two or more tissues working to perform several functions. **b.** The lowest level of organization that includes the smallest living units in the body (cells) is the cellular level. **c.** The levels of organization between cells and organisms are: tissue, organ, and organ system.

Module 1.6 Review
a. The unit used to measure cell size is the micrometer, which is equal to one-millionth of a meter. **b.** The cell theory holds that (1) cells are the structural building blocks of all plants and animals, (2) new cells are produced through the division of pre-existing cells, and (3) cells are the smallest structural units that perform all the vital functions of life. **c.** A fat cell's large volume relative to its surface area makes it ideal for storing fat, whereas the extensive branching of a neuron permits communication with a large number of other cells.

Module 1.7 Review
a. Histology is the study of tissues. **b.** The body's four primary tissue types that form all body structures are: epithelial tissue, connective tissue, muscle tissue, and neural tissue. **c.** Epithelial tissue covers external and internal surfaces and produces secretions; connective tissue fills internal spaces, provides support, and stores energy; muscle tissue is specialized to contract and produce movement; and neural tissue transmits information.

Module 1.8 Review
a. The body's 11 organ systems are the integumentary, skeletal, muscular, nervous, endocrine, cardiovascular, lymphatic, respiratory, digestive, urinary, and reproductive systems. **b.** The digestive system provides nutrients and minerals for the skeletal system, which in turn protects soft tissues and organs of the digestive system. **c.** Your answer may differ, but could include the following: Falling down a flight of stairs could cause a compound fracture (a broken bone that protrudes through the skin), which could affect (1) the skeletal system (a broken bone), (2) the integumentary system (disruption of skin integrity), (3) the muscular system (broken bone tearing through a muscle), (4) the cardiovascular system (blood loss at the site of injury), (5) the lymphatic system (mobilization of specialized cells to defend against infection), and (6) the nervous system (pain and nerve injury as a result of the trauma).

Module 1.9 Review
a. Major organs of the integumentary system are: skin, hair, nails, associated glands and sensory receptors, and the subcutaneous layer. Major organs of the skeletal system are bones, cartilages, joints, and bone marrow. Major organs of the muscular system are the skeletal muscles, tendons, and aponeuroses. Major organs of the nervous system are the brain, spinal cord, organs of the special senses, and the peripheral nerves. **b.** The functions of the integumentary system are to cover and protect the body, secrete lipid coating for hair and skin, provide perspiration for cooling, and provide for sensation. The functions of the skeletal system are to support and protect soft tissues and organs, allow for movement, store minerals, produce blood cells, and store energy. The functions of the muscular system are to provide skeletal movement, control entrances into the body, and produce heat. The functions of the nervous system are to act as a control and processing center, relay information to and from the brain, and provide sensory input. **c.** Your answer may differ, but could include the following: A disorder of the nervous system could affect the muscular system by preventing a patient from being able to walk properly if she had a spinal cord injury, or causing tremors if she had Parkinson's disease.

Module 1.10 Review
a. Major organs of the endocrine system are the endocrine glands: pineal, pituitary, thyroid, parathyroids, thymus, and adrenals; as well as other organs that have endocrine function such as the kidneys, pancreas, and gonads. Major organs of the cardiovascular system are the heart, blood vessels, and blood. Major organs of the lymphatic system are the lymphatic vessels, lymph nodes, spleen, and thymus. Major organs of the respiratory system are the nasal cavities, paranasal sinuses, pharynx, larynx, trachea, bronchi, and lungs. **b.** The functions of the endocrine system are to control and regulate many physiological processes such as reproduction, growth and development, and homeostasis of blood pressure and composition. The functions of the cardiovascular system are to propel and distribute blood, maintain blood pressure, and transport gases, nutrients, and wastes. The functions of the lymphatic system are to transport lymph, defend against disease, and produce immune cells. The functions of the respiratory system are to detect, smell, warm, intake, and filter air, and to exchange gases between the lungs and blood. **c.** Your answer may differ, but could include the following: A lymphatic system disease affects the cardiovascular system when a patient has an asthma attack and cannot fully oxygenate blood due to difficulty in breathing.

Module 1.11 Review
a. Major organs of the digestive system are the mouth, salivary glands, pharynx, esophagus, stomach, small intestine, liver, gallbladder, pancreas, and large intestine. Major organs of the urinary system are the kidneys,

ureters, urinary bladder, and urethra. Major organs of the female reproductive system are the ovaries, uterine tubes, uterus, vagina, clitoris, labia, and mammary glands. Major organs of the male reproductive system are the testes, epididymis, ductus deferens, seminal glands, prostate gland, urethra, penis, and scrotum. **b.** The functions of the digestive system are to ingest food, mechanically and enzymatically digest the food, and prepare the remaining wastes for elimination. The functions of the urinary system are to form and concentrate urine, and to store urine for eventual elimination. The reproductive system produces cells and sex hormones in both sexes. The female reproductive system also supports embryonic development from fertilization to birth. **c.** Your answer may differ, but could include the following: A reproductive system disorder such as a sexually transmitted disease that is acquired through the urethra could affect the urinary system if the bacterium travels up to the urinary bladder and kidneys, causing a more serious infection.

Module 1.12 Review
a. Homeostasis is the presence of a stable internal environment. **b.** Homeostatic regulation is important because failure to maintain homeostasis soon leads to illness or death. **c.** The three parts necessary for homeostatic regulation are a receptor or sensor, which senses an environmental change or stimulus; a control or integration center, which receives and processes the information from the receptor; and an effector, which responds by opposing the stimulus.

Module 1.13 Review
a. An example of negative feedback homeostatic regulation in the body is temperature regulation. When temperature receptors in the skin and brain detect a rise in body temperature, a negative feedback regulation is initiated. The control centers in the brain receive this input and send commands to effectors causing blood vessels in the skin to dilate and sweat glands to secrete. The result is the body is cooled, and homeostasis is achieved. **b.** Negative feedback systems maintain homeostasis (and provide long-term control over the body's internal conditions and systems) by counteracting any stimulus that moves conditions outside their normal range. **c.** Positive feedback is useful in processes such as blood clotting, which, once begun, must move quickly to completion. It is harmful in situations where stable conditions must be maintained, because it tends to exaggerate any departure from the desired condition. Thus, positive feedback in the regulation of body temperature would cause a slight fever to spiral out of control, with fatal results.

Module 1.14 Review
a. Greek and Latin are the source of many modern anatomical terms. **b.** An eponym is an anatomical structure or clinical condition named after a person. **c.** Cadaver-based anatomy was established as a discipline studied by medical professionals in Italy at the University of Bologna.

Module 1.15 Review
a. A person in the anatomical position is standing erect, facing the observer, arms at the sides with the palms facing forward, and the feet together. **b.** Clinicians base their descriptions on four abdominopelvic quadrants (determined by the intersection of two imaginary perpendicular lines that cross at the navel, or umbilicus), whereas anatomists use nine abdominopelvic regions. **c.** A person lying face down in the anatomical position is prone.

Module 1.16 Review
a. The purpose of directional and sectional terms is to provide a standardized frame of reference for describing the human body. **b.** In the anatomical position, an anterior view shows the subject's face, whereas a posterior view shows the subject's back. **c.** A midsagittal section would separate the two eyes.

Module 1.17 Review
a. Body cavities (1) protect internal organs and cushion them from shocks that occur during activity, and (2) allow organs within them to change size and shape without disrupting the activities of nearby organs. **b.** The body cavities of the trunk are the thoracic cavity (which contains the pleural and pericardial cavities) and the abdominopelvic cavity (consisting of the peritoneal, abdominal, and pelvic cavities). **c.** The incision would open the peritoneal portion of the abdominopelvic cavity, which is just inferior to the diaphragm.

Section Reviews

Section 1 Review
1. respiration; **2.** growth and development; **3.** adaptability; **4.** circulation; **5.** excretion; **6.** digestion; **7.** movement; **8.** responsiveness; **9.** Right atrium, Myocardium, Left ventricle, Endocardium, Superior vena cava; **10.** Valve to aorta opens, Valve between left atrium and left ventricle closes, Pressure in left atrium, Electrocardiogram, Heartbeat; **11.** Both keys and messenger molecules have specific three-dimensional shapes, so both can "fit" into and function with their complementary structures (a lock or a receptor protein) only if the shapes match closely enough. Because messenger molecules need not be flat in one dimension (as keys tend to be), they can be very complex in shape, and thus may be able to bind with a variety of complementary receptor proteins. Moreover, although a lock has moving parts, it does not actually change shape, whereas a receptor protein bound by a messenger can change shape, which in turn affects its function. **12.** Larger organisms simply cannot absorb the amount of needed materials or excrete wastes rapidly enough across their body surfaces as well as very small organisms can. Their food must be processed, or digested, before being absorbed, and because the processes of absorption, respiration, and excretion occur in different parts of the organism, an efficient, internal circulation network for their materials is necessary for survival.

Section 2 Review
1. cells; **2.** organs; **3.** organ systems; **4.** epithelial tissue; **5.** external and internal surfaces; **6.** glandular secretions; **7.** connective tissue; **8.** matrix; **9.** protein fibers; **10.** ground substance; **11.** muscle tissue; **12.** movement; **13.** bones of the skeleton; **14.** blood; **15.** materials within digestive tract; **16.** neural tissue; **17.** neuroglia; **18.** In order from simplest to most complex, the correct levels of organization are: chemical, cellular, tissue, organ, organ system, and organism. **19.** integumentary: protection from environmental hazards, temperature control; **20.** skeletal: provides support, protects soft tissues, stores minerals, forms blood cells; **21.** muscular: provides skeletal movement, guards entrances and exits to body, produces heat, supports skeleton, protects soft tissues; **22.** nervous: directs immediate responses to stimuli, coordinates the activities of other organ systems; **23.** endocrine: controls and regulates growth, development, day-night rhythms, fluid balance, calcium, glucose and water balance, blood cell production, metabolic rate; **24.** cardiovascular: propels and distributes blood, maintains blood pressure, transports gases and nutrients, eliminates wastes, assists in temperature regulation and defense against disease; **25.** lymphatic: transports lymph and lymphocytes, controls immune response and immune cells; **26.** respiratory: filters, warms, and humidifies air; detects smell; conducts air into body; gas exchange between air and blood; **27.** digestive: processing

of food and absorption of nutrients, minerals, vitamins, and water; **28.** urinary: eliminates excess water, salts, and waste products; controls blood pH; **29.** reproductive: produces sperm or oocytes, produces hormones, provides the necessary structure and function for sexual intercourse, fetal growth and development, and nourishment of the newborn. **30.** cardiovascular system: blood cells; digestive system: smooth muscle cells; reproductive system: sperm cells (male) and oocytes (female); skeletal system: bone cells (osteocytes); nervous system: nerve cells (neurons). Other cells could be listed for other systems.

Section 3 Review

1. positive feedback; **2.** homeostatic regulation; **3.** homeostasis; **4.** sensor; **5.** negative feedback; **6.** negative feedback; **7.** positive feedback; **8.** negative feedback; **9.** positive feedback; **10.** Blood flow to skin increases, sweating increases, body surface cools, and temperature declines. **11.** Blood flow to skin decreases, shivering occurs, body heat is conserved, and temperature rises. **12.** One reason your body temperature may have dropped is that your body may be losing heat faster than it is being produced. (This is more likely to occur on a cool day.) Perhaps hormones have caused a decrease in your metabolic rate, so your body is not producing as much heat as it normally would. Or, you may have an infection that has temporarily reset the set point of the body's "thermostat" to a value higher than normal. The last possibility is the most likely explanation given the circumstances.

Section 4 Review

1. superior; **2.** inferior; **3.** cranial; **4.** caudal; **5.** posterior or dorsal; **6.** anterior or ventral; **7.** lateral; **8.** medial; **9.** proximal; **10.** distal; **11.** proximal; **12.** distal; **13.** thoracic cavity; **14.** mediastinum; **15.** left lung; **16.** trachea, esophagus; **17.** heart; **18.** diaphragm; **19.** abdominopelvic cavity; **20.** peritoneal cavity; **21.** digestive glands and organs; **22.** pelvic cavity; **23.** reproductive organs

Chapter Review Questions

1. right pleural cavity encloses the right lung; **2.** pelvic cavity contains urinary bladder, reproductive organs, and the last portion of the digestive tract; **3.** pericardial cavity encloses the heart; **4.** left pleural cavity encloses the left lung; **5.** diaphragm is a muscular sheet that separates the thoracic cavity and abdominopelvic cavity; **6.** abdominal cavity contains many digestive glands and organs enclosed by the peritoneum; **7.** c; **8.** e; **9.** k; **10.** b; **11.** j; **12.** h; **13.** a; **14.** i; **15.** n; **16.** d; **17.** f; **18.** m; **19.** g; **20.** l; **21.** b; **22.** d; **23.** b; **24.** b; **25.** c; **26.** d; **27.** c; **28.** b; **29.** Anatomy is the study of structure. Physiology is the study of function. **30.** The three basic principles of the cell theory are: 1. Cells are the structural building blocks of all plants and animals. 2. Cells are produced by the divisions of pre-existing cells. 3. Cells are the smallest structural units that perform all vital functions. **31.** The heart is enclosed by the pericardial cavity; the small intestine is enclosed by the peritoneum within the abdominal cavity; the large intestine is enclosed by the peritoneum within the abdominal cavity, but with the last portion in the pelvic cavity (infraperitoneal); the lungs are enclosed by the pleural cavities; the kidneys are retroperitoneal. **32.** The four main tissue types are epithelial tissue, connective tissue, muscle tissue, and neural tissue. Module 1.7 identifies a comprehensive list of where these tissues are found in the body. **33.** Calcitonin is released when calcium levels are elevated. This hormone should bring about a decrease in blood calcium levels, thus decreasing the stimulus for its release. **34.** A stroke causes damage to the brain, the organ responsible for controlling both voluntary and autonomic activities. Therefore, a patient could lose control of voluntary activities such as walking or speech. Autonomic activities such as bladder control could also be lost.

Chapter Integration

1. The bullet entered the anterior body approximately 2 centimeters inferior to the umbilicus, within the hypogastric (pelvic) abdominopelvic region. **2.** Organs found within the abdominal cavity include the liver, stomach, spleen, small intestine, and most of the large intestine. The abdominal aorta and inferior vena cava are two large blood vessels situated posteriorly. If the bullet entered at this location, these structures may have been affected. **3.** It is likely that a massive internal hemorrhage led to this man's death. **4.** Aaron's endocrine system is failing. If he is not treated, long-term high blood glucose could cause damage to the circulatory system resulting in high blood pressure, which could also damage the urinary system's kidneys, and cause damage to the nervous system by depriving the retina of blood flow, causing blindness. **5.** Normally, cellular receptors would detect high levels of blood glucose. Control centers would stimulate the pancreas to secrete insulin. Insulin would have an effect on the tissues to absorb blood glucose. The failing mechanism is likely the response of the pancreas in that it is not producing insulin. **6.** Dietary interventions could include controlling Aaron's diet so that his insulin does not rise rapidly. This could be achieved by eating smaller meals more regularly. Medical interventions could be insulin injections when Aaron's blood glucose is high.

CHAPTER 2

Module Reviews

Module 2.1 Review

a. Atoms are the smallest stable units of matter. They are composed of subatomic particles. **b.** Protons have a positive charge, and neutrons are electrically neutral. Electrons have a negative charge and are much smaller than either protons or neutrons. **c.** Electrons are subatomic particles not in the nucleus. They whirl around the nucleus creating an electron cloud.

Module 2.2 Review

a. The number of protons in an atom is known as the atomic number. Mass number is the total number of protons and neutrons. **b.** Isotopes are atoms with the same number of protons, but different numbers of neutrons. Hydrogen has three isotopes: hydrogen-1, with a mass number of 1; deuterium, with a mass number of 2; and tritium, with a mass number of 3. The mass number is greater in each isotope because the atoms contain an increasing number of neutrons. The heavier hydrogen sample must contain a higher proportion of one or both of the heavier isotopes. **c.** Trace elements, found in very small amounts in the body, are chemical elements required for normal growth and maintenance.

Module 2.3 Review

a. The maximum number of electrons that can occupy an atom's first three energy levels (electron shells) is 2, 8, and 8, respectively. **b.** Atoms of inert elements are nonreactive because their outermost electron shell contains the maximum number of electrons possible. **c.** A cation is formed when an atom loses one or more electrons from its outermost electron shell; it has an overall positive charge because it contains more protons than electrons. An anion is formed when an atom gains one or more electrons in its outermost electron shell; it has an overall negative charge because it contains more electrons than protons.

Module 2.4 Review

a. The two most common types of chemical bonds are ionic bonds, which result from the attraction of oppositely charged atoms (ions), and covalent bonds, which result from the sharing of electrons.

b. The atoms in a water molecule are held together by polar covalent bonds, in which electrons are shared unequally. **c.** The term *molecule* refers only to chemical structures held together by covalent bonds. Table salt is an ionic compound whose components—sodium ions and chloride ions—are held together by ionic bonds.

Module 2.5 Review

a. Solids have a fixed volume and shape, liquids have a constant volume but no fixed shape, and gases have neither a constant volume nor a fixed shape. **b.** Water molecules are attracted to each other by hydrogen bonds. **c.** The attraction between water molecules at the water's surface creates surface tension, which prevents small objects from penetrating into the water.

Module 2.6 Review

a. Cells are chemical factories because they use complex chemical reactions to provide the energy they need to maintain homeostasis and to perform essential functions. **b.** *Work* is the movement of an object or a change in the physical structure of matter. *Energy* is the capacity to perform work. *Potential energy* is stored energy that has the potential (capability) to do work. *Kinetic energy* is the energy of motion. **c.** Cells do *work* when they synthesize complex molecules and move materials into and out of cells. Muscle contraction requires *energy*. Molecules inside muscle cells store the *potential energy* of contraction. The potential energy of contraction is converted into *kinetic energy* when a muscle contracts.

Module 2.7 Review

a. The chemical shorthand used to describe chemical compounds and reactions effectively is known as chemical notation. **b.** The molecular formula for glucose, a compound composed of 6 carbon atoms, 12 hydrogen atoms, and 6 oxygen atoms, is $C_6H_{12}O_6$. **c.** The weight of 1 mole of glucose is 180 grams. (Add the atomic weights of 6 C = 72 g, 12 H = 12 g, and 6 O = 96 g.)

Module 2.8 Review

a. Three types of chemical reactions important in human physiology are (1) decomposition reactions, in which a molecule is broken down into smaller fragments; (2) synthesis reactions, in which small molecules are assembled into larger ones; and (3) exchange reactions, in which parts of the reacting molecules are shuffled around to produce new products. **b.** In hydrolysis reactions, water is a reactant, whereas in dehydration synthesis reactions water is a product. **c.** The source of the released energy is the potential energy stored in the covalent bonds of the glucose molecule. Energy was released when some of the covalent bonds were broken during this catabolic reaction.

Module 2.9 Review

a. An enzyme is a protein that lowers the activation energy, which is the amount of energy required to start a chemical reaction. **b.** Enzymes promote chemical reactions by lowering the activation energy requirements and making it possible for chemical reactions to proceed under conditions compatible with life. **c.** Metabolites are molecules that can be synthesized or broken down by chemical reactions inside our bodies. Nutrients are essential metabolites normally obtained from the diet.

Module 2.10 Review

An exercising student's body could use water to: **a.** lubricate joints for easy movement. **b.** function in the dehydration synthesis and hydrolysis reactions of muscle contraction. **c.** form perspiration that would evaporate from the skin, thereby cooling the body, and to dissolve the waste products generated by exercise.

Module 2.11 Review

a. Sodium chloride dissociates in water as the positive poles of water molecules are attracted to the negatively charged chloride ions, and the negative poles of water molecules are attracted to the positively charged sodium ions. The ions stay dissolved in solution because a layer of surrounding water molecules, or hydration sphere, separates them from each other. **b.** Electrolytes are soluble inorganic molecules whose ions will conduct an electrical current in solution. **c.** Hydrophilic molecules are attracted to water molecules, whereas hydrophobic molecules do not interact with water molecules.

Module 2.12 Review

a. pH is a measure of the hydrogen ion concentration in a solution, defined as the negative logarithm of the hydrogen ion concentration, expressed in moles per liter. On the pH scale, 7 represents neutrality, values below 7 indicate acidic solutions, and values above 7 indicate alkaline (basic) solutions. **b.** The pH of various body fluids must remain relatively constant if the body is to maintain homeostasis and remain healthy. **c.** An acid is a compound whose dissociation in solution releases a hydrogen ion (H^+) and an anion; a base is a compound whose dissociation releases a hydroxide ion (OH^-) into the solution or removes a hydrogen ion (H^+) from the solution; and a salt is an ionic compound consisting of a cation other than H^+ and an anion other than OH^-.

Module 2.13 Review

a. Organic compounds always contain the elements carbon and hydrogen, and generally oxygen as well. **b.** A functional group is a grouping of atoms that confer specific chemical properties to the rest of a molecule to which it is attached. **c.** Amino groups are the functional groups of amino acids. Carboxyl groups are functional groups of fatty acids and amino acids. Hydroxyl groups are the functional groups that link molecules by dehydration synthesis and affect a molecule's solubility. Phosphate groups are functional groups found in nucleic acids and high-energy compounds.

Module 2.14 Review

a. Compounds with a C:H:O ratio of 1:2:1 are carbohydrates and are used in the body mainly as energy sources. **b.** The three structural classes of carbohydrates are monosaccharides (glucose), disaccharides (sucrose), and polysaccharides (starch). **c.** Muscle cells make (synthesize) glycogen by linking numerous glucose molecules in a series of dehydration synthesis reactions.

Module 2.15 Review

a. Lipids are a diverse group of water-insoluble organic compounds that contain carbon, hydrogen, and oxygen in a ratio that does not approximate 1:2:1 (they contain less oxygen than do carbohydrates). Examples are fatty acids, glycerides, eicosanoids, steroids, phospholipids, and glycolipids. **b.** All fatty acids consist of a hydrocarbon chain and a carboxyl group. In saturated fatty acids, each carbon atom in the hydrocarbon chain has four single covalent bonds that bind the maximum number of hydrogen atoms possible. In unsaturated fatty acids, one or more of the carbon atoms in the hydrocarbon chain has double covalent bonds, so fewer hydrogen atoms are bonded. **c.** In the hydrolysis of a triglyceride, the reactants are a triglyceride and three water molecules. The products are a glycerol molecule and three fatty acids.

Module 2.16 Review

a. Eicosanoids function primarily as chemical messengers; steroids function as both components of cellular membranes (cholesterol) and sex hormones (androgens and estrogens); and phospholipids and glycolipids are important components of cellular membranes. **b.** Cholesterol is a component of plasma membranes and is important for cell growth and division. **c.** When phospholipids and glycolipids form a micelle, the hydrophobic tails of both molecules are inside the micelle, and their hydrophilic heads are outside the micelle.

Module 2.17 Review
a. Proteins are organic compounds formed from amino acids. Each amino acid contains a carbon atom, a hydrogen atom, an amino group ($-NH_2$), a carboxyl group (–COOH), and an R group (a variable side chain of one or more atoms). **b.** During the dehydration synthesis of two amino acids, a peptide bond links the amino group of one amino acid with the carboxyl group of the other amino acid. **c.** The heat of boiling breaks bonds that maintain the protein's tertiary structure, quaternary structure, or both. The resulting change in shape affects the ability of the protein molecule to perform its normal biological functions. These changes are known as denaturation.

Module 2.18 Review
a. The active site is the location on an enzyme where substrate binding occurs. **b.** The reactants in an enzymatic reaction are called substrates. **c.** An enzyme's specificity results from the unique shape of its active site, which permits only a substrate with a complementary shape to bind.

Module 2.19 Review
a. Cells obtain energy from the high-energy bonds of compounds such as ATP. **b.** ATP (adenosine triphosphate) is a compound consisting of adenosine to which three phosphate groups are attached; high-energy bonds attach the second and third phosphates. **c.** Adenosine monophosphate (AMP) is a nucleotide consisting of adenosine plus a phosphate group (PO_4^{3-}), and adenosine diphosphate (ADP) is a compound consisting of adenosine with two phosphate groups attached. Adding a phosphate group to AMP creates ADP. Breaking one phosphate linkage of ATP, to form ADP, provides the primary source of energy for physiological processes.

Module 2.20 Review
a. Nucleic acids are large organic molecules composed of carbon, hydrogen, oxygen, nitrogen, and phosphorus that regulate the synthesis of proteins and make up the genetic material in cells. **b.** The nucleic acid RNA (ribonucleic acid) contains the sugar ribose. The nucleic acid DNA (deoxyribonucleic acid) contains the sugar deoxyribose; both contain nitrogenous bases and phosphate groups. **c.** The complementary strands of DNA are held together through complementary base pairing between adenine and thymine, and between guanine and cytosine.

Section Reviews

Section 1 Review
1. f; **2.** d; **3.** k; **4.** a; **5.** l; **6.** h; **7.** i; **8.** j; **9.** c; **10.** e; **11.** g; **12.** b; **13.** 2; **14.** 4; **15.** 1; **16.** 0; **17.** 6; **18.** 12; **19.** 7; **20.** 7; **21.** 20; **22.** 20; **23.** molecule; **24.** compound; **25.** molecule; **26.** compound; **27.** The outer energy level of inert elements is filled with electrons, whereas the outer energy level of reactive elements is not filled with electrons. Atoms with unfilled outer energy levels will react with other atoms. Atoms with filled outer energy levels will not react with other atoms. **28.** Both polar and nonpolar molecules are held together by covalent bonds. However, in a polar molecule the electrons are not shared equally, so it carries small, or partial, positive and negative charges on its surface; in a nonpolar molecule the electrons are shared equally, so it is electrically neutral. **29.** Both covalent bonds and ionic bonds bind atoms together, but covalent bonds involve the sharing of electrons between atoms, whereas ionic bonds involve the electrical attraction of oppositely charged atoms (ions).

Section 2 Review
1. h; **2.** g; **3.** e; **4.** f; **5.** d; **6.** c; **7.** a; **8.** b; **9.** H_2; **10.** 2 H; **11.** 6 H_2O; **12.** $C_{12}H_{22}O_{11}$; **13.** $C_6H_{12}O_6 + 6\ O_2 \rightarrow 6\ CO_2 + 6\ H_2O$; **14.** hydrolysis reaction; **15.** dehydration synthesis reaction; **16.** A decreased amount of enzyme at the second step would limit the amount of the intermediate products in the next two steps. This would cause a decrease in the amount of the final product.

Section 3 Review
1. e; **2.** h; **3.** i; **4.** d; **5.** a; **6.** j; **7.** c; **8.** g; **9.** f; **10.** b; **11.** effective lubrication (as between bony surfaces in a joint); **12.** reactivity (participates in chemical reactions); **13.** high heat capacity (readily absorbs and retains heat); **14.** solubility (is a solvent for many substances); **15.** acidic; **16.** neutral; **17.** alkaline; **18.** The pH 3 solution is 1000 times more acidic than the pH 6 solution—it contains a one-thousandfold (10^3) increase in the concentration of hydrogen ions (H^+). **19.** Three negative effects of abnormal fluctuations in pH are cell and tissue damage (due to broken bonds), changes in the shapes of proteins, and altered cellular functions. **20.** Table salt dissociates (dissolves) in pure water but because it does not release either hydrogen ions (H^+) or hydroxide ions (OH^-), no change in pH occurs.

Section 4 Review
1. carbohydrates; **2.** polysaccharides; **3.** disaccharides; **4.** monosaccharides; **5.** lipids; **6.** fatty acids; **7.** glycerol; **8.** proteins; **9.** amino acids; **10.** nucleic acids; **11.** RNA; **12.** DNA; **13.** nucleotides; **14.** ATP; **15.** phosphate groups; **16.** polysaccharide, polyunsaturated, polypeptide; **17.** triglyceride; **18.** disaccharide, diglyceride, dipeptide; **19.** glycogen, glycolipids; **20.** g; **21.** f; **22.** c; **23.** h; **24.** d; **25.** e; **26.** b; **27.** j; **28.** k; **29.** i; **30.** a

Chapter Review Questions

1. c; **2.** c; **3.** b; **4.** e; **5.** c; **6.** a; **7.** c; **8.** b; **9.** d; **10.** a; **11.** b; **12.** d; **13.** c; **14.** b; **15.** d; **16.** a; **17.** b; **18.** d; **19.** protons, neutrons, and electrons; **20.** carbohydrates, lipids, proteins, and nucleic acids; **21.** Lipids perform many functions in the body including: energy reserve, insulation and heat conservation, organ protection, chemical messengers, structural components of cell membranes, and digestive secretions. **22.** Proteins are long chains of amino acids bound together by peptide bonds. Progressive folding and pleating of the chain into more complex shapes modifies the primary linear arrangement of amino acids. The increasing levels of complexity are primary, secondary, tertiary, and quaternary structures. **23.** **(a)** DNA: deoxyribose, phosphate, and nitrogenous bases (A, T, C, G); **(b)** RNA: ribose, phosphate, and nitrogenous bases (A, U, C, G); **24.** Enzymes are specialized protein catalysts that lower the activation energy for chemical reactions. Enzymes speed up chemical reactions but are not used up or changed in the process. **25.** A salt is an ionic compound consisting of any cations other than hydrogen ions and any anions other than hydroxide ions. Acids dissociate and release hydrogen ions, whereas bases remove hydrogen ions from solution (usually by releasing hydroxide ions). **26.** Nonpolar covalent bonds involve an equal sharing of electrons. Polar covalent bonds involve an unequal sharing of electrons. Ionic bonds result from the attraction of oppositely charged ions. (Ions are atoms which have either lost or gained electrons.) **27.** The molecule is a nucleic acid. Carbohydrates and lipids do not contain nitrogen. Although both proteins and nucleic acids contain nitrogen, only nucleic acids normally contain phosphorus. **28.** The insect can walk across the surface of the pond because its small mass is not sufficient to break the surface tension created by the hydrogen bonds between the water molecules. **29.** All of the foods and drink that the student consumed have a pH below 7 and are therefore acidic. These acidic foods, in conjunction with the hydrochloric acid in his stomach necessary for digestion, are likely the cause of his upset stomach. Consuming foods that are slightly alkaline could buffer his stomach contents and relieve his symptoms.

30. (a)

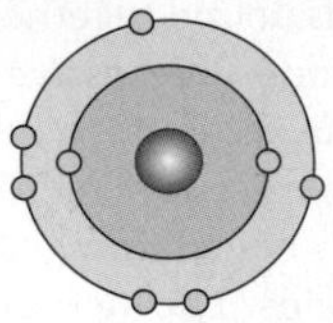

Oxygen atom

(b) Two more electrons can fit into the outermost energy level of an oxygen atom.

Chapter Integration

1. Digestion involves breaking down molecules into smaller fragments; therefore, the *digestivo* would be promoting decomposition reactions. **2.** The decomposition reactions occurring in Claire's digestive tract are releasing some of the potential energy stored in the covalent bonds of the food she just consumed. This energy can be used to do work, but some of that energy is lost to heat, which could result in her feeling warmer and wanting to remove her jacket. **3.** Nutrients, the essential metabolites obtained from our diet, are digested by enzymes. If the *digestivo* is truly aiding in digestion, it would have to be rich in digestive enzymes. **4.** The polysaccharides would be broken down, or catabolized, into monosaccharides. Lipids such as glycerides would be broken down into simpler molecules such as diglycerides, monoglycerides, and glycerol. The proteins would be broken down into amino acids.

CHAPTER 3

Module Reviews

Module 3.1 Review

a. The cell theory states that cells are the building blocks of all plants and animals, all new cells come from the division of pre-existing cells, and cells are the smallest units that carry out all vital physiological functions. **b.** All cells are descendants of the fertilized ovum. **c.** Differentiation is the gradual specialization of cells due to regional differences in the cytoplasm of the fertilized ovum.

Module 3.2 Review

a. Cytoplasm is the material between the plasma membrane and the nuclear membrane; cytosol is the fluid portion of the cytoplasm. **b.** Membranous organelles and their functions are as follows: Endoplasmic reticulum synthesizes secretory products and participates in intracellular storage and transport; the nucleus controls metabolism, stores and processes genetic information, and controls protein synthesis; rough ER modifies and packages newly synthesized proteins; smooth ER synthesizes lipids and carbohydrates; Golgi apparatus stores, alters, and packages secretory products and lysosomal enzymes; lysosomes remove damaged organelles or intracellular pathogens; mitochondria produce 95 percent of the ATP required by the cell; and peroxisomes neutralize toxic compounds. **c.** The cytoskeleton strengthens and supports the cells and enables movement of cellular structures and materials.

Module 3.3 Review

a. The general functions of the plasma membrane include physical isolation of the cell from its environment, regulation of exchange with the environment, sensitivity to the environment, and structural support. **b.** The phospholipid bilayer of the plasma membrane is mostly responsible for isolating a cell from its external environment. **c.** Channel proteins are integral proteins that allow water and small ions to pass through the plasma membrane.

Module 3.4 Review

a. The three basic components of the cytoskeleton are microfilaments, intermediate filaments, and microtubules. **b.** Microtubules are common to both centrioles and cilia. **c.** Motile cilia propel materials across cell surfaces.

Module 3.5 Review

a. Newly synthesized proteins from free ribosomes enter the cytosol; those from fixed ribosomes enter the ER. **b.** Smooth endoplasmic reticulum (SER) lacks ribosomes, and its cisternae are tubular. **c.** The SER functions in the synthesis of lipids such as steroids. Ovaries and testes produce large amounts of steroid hormones, which are lipids, and thus contain large amounts of SER.

Module 3.6 Review

a. The Golgi apparatus (1) modifies and packages cellular secretions, such as hormones or enzymes; (2) renews or modifies the plasma membrane; and (3) packages special enzymes within vesicles for use within the cell. **b.** Lysosomes contain digestive enzymes. **c.** Lysosomes may (1) break down other intracellular organelles to enhance recycling and membrane flow; (2) fuse with vesicles containing fluids or solids from the external environment to obtain nutrients or destroy pathogens; and (3) break down and release their digestive enzymes, resulting in the destruction of the cell (autolysis).

Module 3.7 Review

a. A double membrane encloses a mitochondrion; the outer membrane surrounds the organelle, whereas the inner membrane contains folds called cristae and encloses a fluid, enzyme-filled matrix. **b.** Mitochondria require oxygen to produce ATP. **c.** Mitochondria produce energy in the form of ATP molecules. A great number of mitochondria in a cell indicates a high demand for energy.

Module 3.8 Review

a. The nucleus functions as a control center in a cell. It stores all the information needed to direct the synthesis of the more than 100,000 proteins in the human body. It controls which proteins are synthesized and when. **b.** Most cells contain a single nucleus, but there are exceptions. Skeletal muscle cells have many nuclei; red blood cells have none. **c.** The genetic information in the cell is coded in the sequence of DNA nucleotides.

Module 3.9 Review

a. The nucleus is a cellular organelle containing DNA, RNA, and proteins. Surrounding the nucleus is the double-membraned nuclear envelope; the gap within this double membrane is the perinuclear space. Nuclear pores allow for chemical communication between the nucleus and the cytosol. **b.** The DNA in the nucleus stores the cell's instructions for synthesizing proteins. **c.** The nuclei of human somatic cells contain 23 pairs of chromosomes.

Module 3.10 Review

a. A gene is a portion of a DNA strand that functions as a hereditary unit and codes for a specific protein. **b.** The genetic code is described as a triplet code because a sequence of three nitrogenous bases—a triplet—specifies the identity of a specific amino acid. **c.** The three types of RNA involved in protein synthesis are ribosomal RNA (rRNA), messenger RNA (mRNA), and transfer RNA (tRNA).

Module 3.11 Review

a. Transcription is the encoding of the genetic instructions in a segment of DNA onto a strand of RNA (typically mRNA). **b.** The DNA template strand is the strand of DNA that will be used to synthesize RNA. **c.** A cell that lacked the enzyme RNA polymerase would not be able to transcribe RNA from DNA.

Module 3.12 Review

a. Translation is the synthesis of a protein using the information provided by the sequence of codons along an mRNA strand. **b.** The second codon is GCA, and its complementary anticodon sequence is CGU. **c.** Deletion of a base during transcription will change the makeup of all subsequent codons in the mRNA base sequence; the altered mRNA sequence will then result in the incorporation of a different series of amino acids into the protein during translation. Almost certainly the protein so produced will not be functional.

Module 3.13 Review

a. Permeability is the plasma membrane property that precisely determines which substances can enter or leave the cytoplasm. **b.** Three different types of membranes based on permeability are freely permeable membranes, which permit any substance to pass without difficulty; selectively permeable membranes, such as plasma membranes, which permit some substances to pass, but prevent other substances from passing, and impermeable membranes, which permit nothing to pass through. **c.** Passive processes do not require ATP for a substance to pass through the plasma membrane. Active processes require ATP for a substance to pass through the plasma membrane.

Module 3.14 Review

a. Diffusion is the passive movement of molecules from an area of higher concentration to an area of lower concentration until equilibrium is reached. **b.** Factors that influence diffusion rates include distance, molecule size, temperature, concentration gradient size, electrical forces, and the presence or absence of membrane channel proteins. **c.** Diffusion is driven by a concentration gradient. The higher the concentration gradient, the faster the rate of diffusion; the lower the concentration gradient, the slower the rate of diffusion. If the concentration of oxygen in the lungs were to decrease, the concentration gradient between oxygen in the lungs and oxygen in the blood would decrease (as long as the oxygen level of the blood remained constant). Thus, oxygen would diffuse more slowly into the blood.

Module 3.15 Review

a. Osmosis is the passive movement of water across a selectively permeable membrane from a solution with a lower solute concentration to a solution with a higher solute concentration. **b.** A hypotonic solution would cause an osmotic flow into the cell, leading to swelling and eventual bursting of the red blood cell (hemolysis). A hypertonic solution would cause an osmotic flow out of the cell, resulting in a shriveled or collapsed red blood cell (crenation). **c.** The 10 percent salt solution is hypertonic with respect to the cells lining the nasal cavity (the solution contains a higher salt concentration than do the cells). The hypertonic solution would draw water out of the cells (which would shrink), adding water to (diluting) the mucus. Thinner mucus would help relieve the congestion.

Module 3.16 Review

a. In carrier-mediated transport, integral proteins bind specific ions or molecules and transport them across the plasma membrane. **b.** In both facilitated diffusion and active transport, a carrier molecule transports materials across the plasma membrane. **c.** Transporting hydrogen ions against their concentration gradient—that is, from a site where they are less concentrated (the cells lining the stomach, which produce the hydrogen ions) to a region where they are more concentrated (the digesting stomach contents)—requires energy. Thus, an active transport process must be involved.

Module 3.17 Review

a. Endocytosis is the process in which a membranous vesicle forms at the cell surface, encloses a relatively large volume of extracellular material, and then moves into the cytoplasm. Types of endocytosis are pinocytosis, the vesicle-mediated movement into the cytoplasm of extracellular fluid and its contents, and phagocytosis, the vesicle-mediated movement into the cytoplasm of extracellular solids, especially bacteria and debris. **b.** The engulfment of bacteria and their transport into the cell is called phagocytosis. **c.** Exocytosis is the ejection of cytoplasmic materials after a membranous vesicle fuses with the plasma membrane.

Module 3.18 Review

a. Cell division is important because it is essential to survival. Aging and environmental stressors damage cells. Cellular division replaces old and damaged cells. **b.** Apoptosis is the genetically programmed death of a cell. **c.** Mitosis produces two daughter cells, each containing 46 chromosomes. Meiosis produces sex cells (sperm or oocytes) containing only 23 chromosomes.

Module 3.19 Review

a. Interphase is the portion of a cell's life cycle during which the chromosomes are uncoiled and all normal cellular functions except mitosis are under way. Its stages are G_1, S, G_2, and G_0. **b.** Important enzymes for DNA replication are DNA helicases, DNA polymerase, and DNA ligases. **c.** This cell is likely in the G_1 phase of interphase.

Module 3.20 Review

a. Mitosis is the division of a cell nucleus into two identical daughter cell nuclei, an essential step in cell division. Its four stages are prophase, metaphase, anaphase, and telophase. **b.** A chromatid is a copy of a duplicated chromosome. Human cells normally contain 23 pairs of chromosomes, so during mitosis a human cell would contain 92 chromatids. **c.** If spindle fibers failed to form during mitosis, the chromosome would not be able to separate into two sets. If cytokinesis occurred, the result would be one cell with two sets of chromosomes, and one cell with none.

Module 3.21 Review

a. Cancer is an illness characterized by mutations that disrupt normal cell control mechanisms and produce malignant cells. **b.** A tumor (or neoplasm) is a mass or swelling produced by abnormal growth and division of cells. The cells of a benign tumor typically remain within their tissue of origin and seldom threaten life. **c.** Metastasis is the spread of cancer cells from one site to another, producing secondary tumors.

Section Reviews

Section 1 Review

1. glycolysis; **2.** aerobic (metabolism); **3.** microvilli; **4.** lysosome; **5.** microvilli: increase surface area to facilitate absorption of extracellular materials; **6.** Golgi apparatus: stores, alters, and packages newly synthesized proteins; **7.** lysosome: intracellular removal of pathogens or damaged organelles; **8.** mitochondrion: produces 95 percent of the ATP required by the cell; **9.** peroxisome: neutralizes toxic compounds; **10.** nucleus: controls metabolism, stores and processes genetic information, controls protein synthesis; **11.** rough endoplasmic reticulum: has ribosomes bound to the membranes, modifies and packages newly synthesized proteins; **12.** ribosomes: protein synthesis; **13.** cytoskeleton: provides strength and support, enables movement of cellular structures and materials; **14.** Similar to the role of a plasma membrane around a cell, an organelle membrane physically isolates the organelle's contents from the cytosol, regulates exchange with the cytosol, and provides structural support.

Section 2 Review

1. d; **2.** h; **3.** l; **4.** a; **5.** k; **6.** n; **7.** m; **8.** b; **9.** j; **10.** e; **11.** i; **12.** c; **13.** g; **14.** f; **15.** AUG/UUU/UGU/GCC/GCC/UUA; **16.** UAC/AAA/ACA/CGG/CGG/AAU; **17.** Methionine-Phenylalanine-Cysteine-Alanine-Alanine-Leucine; **18.** The nucleus contains the information for synthesizing proteins in the nucleotide sequence of its DNA. Changes in the extracellular fluid can affect cells through the binding of molecules to plasma membrane receptors or by the diffusion of molecules through membrane channels. Such stimuli may result in alterations of genetic activity in the nucleus. These alterations may change biochemical processes and metabolic pathways through the synthesis of additional, fewer, or different enzymes. Altered genetic activity may also change the physical structure of the cell by synthesizing additional, fewer, or different structural proteins.

Section 3 Review

1. diffusion; **2.** facilitated diffusion; **3.** molecular size; **4.** net diffusion of water; **5.** active transport; **6.** specificity; **7.** vesicular transport; **8.** exocytosis; **9.** pinocytosis; **10.** "cell eating"; **11.** diffusion; **12.** neither; **13.** osmosis; **14.** diffusion; **15.** osmosis

Section 4 Review

1. somatic cells; **2.** G_1 phase; **3.** G_2 phase; **4.** DNA replication; **5.** mitosis; **6.** metaphase; **7.** telophase; **8.** cytokinesis; **9.** telophase; **10.** prophase; **11.** centromere; **12.** A cell that undergoes repeated rounds of the cell cycle without cytokinesis could result in a large, multinucleated cell.

Chapter Review Questions

1. b; **2.** c; **3.** b; **4.** b; **5.** a; **6.** b; **7.** a; **8.** b; **9.** d; **10.** b; **11.** c; **12.** c; **13.** anchoring proteins, recognition proteins, receptor proteins, carrier proteins, and channels; **14.** Mitosis specifically refers to the division and duplication of the cell's nucleus. Cytokinesis is the division of the cytoplasm into two distinct new cells. **15.** Diffusion is the movement of solutes by concentration differences. Osmosis is the net diffusion of water across a membrane. **16.** Mitochondria contain their own DNA called mtDNA. It codes for small numbers of RNA and polypeptide molecules. The polypeptides are used in enzymes required for energy production. **17.** The nonmembranous organelles are the cytoskeleton, microvilli, centrioles, cilia, flagella, and ribosomes. The membranous organelles are the mitochondria, nucleus, endoplasmic reticulum, Golgi apparatus, lysosomes, and peroxisomes. **18.** Prophase: chromosomes coil forming visible chromosomes, each copy called a chromatid. Metaphase: chromatids move to the metaphase plate. Anaphase: centromere of each chromatid pair splits and the chromatids separate. Telophase: each new cell prepares to return to interphase, nuclear membranes re-form, nuclei enlarge, and chromosomes uncoil; **19.** If a cell had microvilli on its plasma membrane it is likely to be actively engaged in absorbing materials from the extracellular fluid. **20.** The cell would swell up like a balloon and may eventually burst, releasing its contents. This event is called hemolysis. **21.** e, b, a, g, f, c, d; **22.** Phagocytosis begins when cytoplasmic extensions called pseudopodia surround targeted material. The pseudopodia fuse to form a phagosome containing the targeted material. This vesicle fuses with lysosomes. Lysosomal enzymes digest the targeted material. Nutrients are released from the vesicle, and the residue is ejected from the cell through exocytosis. **23.** All somatic cells contain 46 chromosomes. Since mitosis produces two identical nuclei, the resulting cells will each have 46 chromosomes. Meiosis produces sex cells (sperm or oocytes) containing 23 chromosomes, which can unite to form a new somatic cell of 46 chromosomes. **24.** Lipid molecules such as steroid hormones are lipid soluble and therefore easily diffuse across the lipid bilayer of the plasma membrane. **25.** Malignant tumors divide rapidly and release chemicals that stimulate the growth of blood vessels in a process called angiogenesis. The availability of additional nutrients from these new vessels accelerates tumor growth and metastasis.

Chapter Integration

1. They would want to read in the lab report that the tumor is benign. This would mean the tumor is localized and likely to be easily removed surgically. They would not want to read the word malignant because these types of cells divide rapidly and are more likely to threaten life. **2.** The tumor is likely caused by a mutation in a single cell. This mutated cell continued to undergo mitosis producing more abnormal cells, resulting in a tumor, or mass of abnormal cells. **3.** If this tumor is malignant, it is likely to undergo metastasis and spread to other organs. **4.** If they could, Vito and his doctor would like to switch on apoptosis, or programmed cell suicide, in the tumor cells. **5.** Mitochondria contain RNA, DNA, and the necessary enzymes to synthesize proteins. Mitochondrial DNA is different from nuclear DNA and allows each mitochondrion to control its own maintenance, growth, and reproduction. Serious exercisers have high rates of energy consumption, and over time their mitochondria respond to this high demand for energy by dividing and increasing in number. **6.** Sweating during exercise results in losses of water from the extracellular fluid. As the extracellular fluid osmolarity increases, osmosis moves water out of the cells, which begin to dehydrate. Energy drinks contain a solution of water, electrolytes (especially sodium chloride, which is lost in sweat), and glucose. As the sodium ions and glucose are absorbed by the digestive tract, water follows by osmosis. As the water influx restores normal ECF volume and osmolarity, water moves back into the cells, rehydrating them.

CHAPTER 4

Module Reviews

Module 4.1 Review

a. Histology is the study of tissues. **b.** The four basic tissue types are epithelial tissue, connective tissue, muscle tissue, and neural tissue. **c.** Epithelial tissue covers exposed surfaces, lines passageways, and forms secretory glands. Connective tissue fills spaces, provides support, and stores energy. Muscle tissue contracts to produce movement. Neural tissue conducts impulses and carries information.

Module 4.2 Review

a. Early microscopes used only light, and could only magnify 10 or 20 times. Today light microscopes can magnify up to 1000 times, while modern microscopes using beams of electrons instead of light can magnify over 1 million times. **b.** A light microscope (LM) passes visible light through a thin section of tissue. A transmission electron microscope (TEM) uses magnets to direct a beam of electrons through a finely sectioned object. A scanning electron microscope (SEM) creates a three-dimensional image by first coating a surface with electron-dense material, then bombarding it with electron beams. **c.** The microscope in your lab is most likely a compound light microscope. (Answers may vary.)

Module 4.3 Review

a. Epithelial tissue provides physical protection, controls permeability, provides sensation, and produces specialized secretions. **b.** The presence of many cilia on the free surface of epithelial cells aids the movement of substances over the epithelial surface. **c.** The three characteristic shapes of epithelial cells are squamous (flat), cuboidal (cube-shaped), and columnar (appearing tall and rectangular). A single layer of epithelial cells is a simple epithelium, whereas multiple layers of epithelial cells constitute a stratified epithelium.

Module 4.4 Review

a. Epithelial intercellular connections are occluding junctions, adhesion belts, gap junctions, desmosomes, and hemidesmosomes. **b.** Gap junctions help coordinate the functioning of adjacent cells by allowing small molecules and ions to pass from cell to cell. In epithelial cells, gap junctions help coordinate functions such as cilia movement. In cardiac and smooth muscle tissues, they are essential in achieving coordinated muscle cell contractions. **c.** Epithelia rely on blood vessels in underlying tissues to supply needed nutrients.

Module 4.5 Review

a. No. A simple squamous epithelium provides too little protection against infection, abrasion, or dehydration to function effectively on the skin surface. **b.** Because all these sites are subject to mechanical stresses and abrasion: by food (pharynx and esophagus), by feces (anus), or during intercourse or childbirth (vagina). **c.** Keratinized epithelia are both tough (have mechanical strength) and water resistant.

Module 4.6 Review
a. In a sectional view, simple cuboidal epithelial cells are square and have central nuclei, and the distance between adjacent nuclei is roughly equal to the height of the epithelium. **b.** Stratified cuboidal epithelia are associated with the ducts of sweat glands, mammary glands, and other exocrine glands. **c.** The epithelium that lines the urinary bladder and changes in appearance during stretching is a transitional epithelium.

Module 4.7 Review
a. In a sectional view, simple columnar epithelial cells appear as tall, slender rectangles containing elongated nuclei close to the basal lamina. Cell height is several times the distance between adjacent nuclei. **b.** A pseudostratified columnar epithelium is not truly stratified—even though its nuclei make it appear so—because all of its epithelial cells contact the basal lamina. **c.** The columnar epithelium lining the intestine typically has microvilli on its apical surface.

Module 4.8 Review
a. The two primary types of glands are endocrine glands and exocrine glands. **b.** A gland that lacks ducts and releases its secretions directly into the interstitial fluid is an endocrine gland. **c.** The mode of secretion in which secretory cells fill with secretions and then rupture is holocrine secretion.

Module 4.9 Review
a. The three basic components of connective tissue are specialized cells, extracellular protein fibers, and a fluid known as ground substance. **b.** Connective tissue provides a structural framework for the body, transports fluids and dissolved materials, protects organs, supports and interconnects other tissue types, stores energy, and defends against invading microorganisms. **c.** Connective tissue proper consists of connective tissues with many types of cells and extracellular fibers in a viscous ground substance. Fluid connective tissues have distinct cells in a watery matrix containing dissolved proteins. Supporting connective tissues have less diverse cells and matrix with more densely packed fibers.

Module 4.10 Review
a. Cells found in connective tissue proper are melanocytes, fixed macrophages, mast cells, fibroblasts, adipocytes (fat cells), plasma cells, free macrophages, mesenchymal cells, neutrophils, eosinophils, and lymphocytes. **b.** Fibroblasts secrete the proteins that form collagen, elastic fibers, and reticular fibers in the matrix, and fibroblasts also maintain these connective tissue fibers. **c.** The connective tissue that contains primarily lipids is adipose (fat) tissue.

Module 4.11 Review
a. A vitamin C deficiency would impair the production of collagen fibers, which add strength to connective tissue. The deficiency would result in tissue that is weak and prone to damage. **b.** The two types of connective tissue that contain a liquid matrix (that is, are fluid connective tissues) are blood and lymph. **c.** Continuous recirculation of extracellular fluid (plasma, interstitial fluid, and lymph) helps eliminate local differences in the levels of nutrients, wastes, or toxins; maintains blood volume; and alerts the immune system to infectious agents or infections.

Module 4.12 Review
a. Mature cartilage cells are called chondrocytes. **b.** The fibers characteristic of cartilage supporting the ear are elastic fibers. **c.** The type of cartilage in intervertebral discs is fibrocartilage.

Module 4.13 Review
a. Whereas bone contains only collagen fibers, cartilage may contain collagen, elastic, and reticular fibers. **b.** Mature bone cells in lacunae are called osteocytes. **c.** The functional unit of compact bone is an osteon.

Module 4.14 Review
a. The four types of membranes found in the body are mucous membranes, serous membranes, the cutaneous membrane, and synovial membranes. **b.** The body cavities lined by serous membranes are the pleural, peritoneal, and pericardial cavities. **c.** The three layers of fascia are superficial fascia (areolar and adipose tissue), deep fascia (dense irregular connective tissue), and subserous fascia (areolar tissue).

Module 4.15 Review
a. The relative body weight of each of the four tissue types is muscle tissue 50 percent, connective tissue 45 percent, epithelial tissue 3 percent, and neural tissue 2 percent. **b.** The three classifications of muscle tissue are skeletal muscle tissue, cardiac muscle tissue, and smooth muscle tissue. **c.** Skeletal muscle tissue functions to pull on bones for movement; cardiac muscle tissue moves blood through blood vessels; and smooth muscle tissue moves substances through the digestive tract and regulates the diameter of blood vessels, among other functions.

Module 4.16 Review
a. Skeletal muscle tissue features long and cylindrical cells with striations and multiple nuclei. Cardiac muscle tissue features branched cells with striations, intercalated discs, and a single nucleus. Smooth muscle tissue features spindle-shaped cells, with no striations, and a single nucleus. **b.** Smooth muscle tissue regulates blood vessel diameter. **c.** Neurons transfer and process information. Neuroglia protect, support, and repair neural tissue.

Module 4.17 Review
a. The two processes in the response to tissue injury are inflammation and regeneration. **b.** Inflammation produces swelling, redness, warmth, and pain. **c.** Inflammation can occur in any organ in the body because all organs have connective tissues.

Section Reviews

Section 1 Review
1. simple squamous epithelium; **2.** simple cuboidal epithelium; **3.** simple columnar epithelium; **4.** stratified squamous epithelium; **5.** stratified cuboidal epithelium; **6.** stratified columnar epithelium; **7.** mucous cells; **8.** mucin; **9.** mucus; **10.** ducts; **11.** epithelial surfaces; **12.** exocrine glands; **13.** merocrine secretion; **14.** apocrine secretion; **15.** holocrine secretion; **16.** interstitial fluid; **17.** endocrine glands; **18.** avascular; **19.** alveolar (acinar) gland; **20.** transitional epithelium; **21.** occluding junction; **22.** basal lamina; **23.** simple gland; **24.** mesothelium (simple squamous epithelium); **25.** pseudostratified columnar epithelium; **26.** urinary bladder, ureters, urine-collecting chambers in the kidney; **27.** stratified squamous epithelium; **28.** simple columnar epithelium; **29.** lining of the peritoneum and pericardium, exchange surfaces (alveoli) within the lungs; **30.** lining of exocrine glands and ducts, kidney tubules; **31.** stratified cuboidal epithelium

Section 2 Review
1. loose connective tissue; **2.** adipose; **3.** regular; **4.** tendons; **5.** ligaments; **6.** fluid connective tissue; **7.** blood; **8.** hyaline; **9.** chondrocytes in lacunae; **10.** bone; **11.** perichondrium, periosteum, peritoneum, pericardium; **12.** periosteum, osteocytes, osteon; **13.** chondrocyte, perichondrium; **14.** interstitial growth; **15.** lacunae; **16.** chondrocyte; **17.** osseous tissue; **18.** fibrocartilage; **19.** adipocytes; **20.** synovial membrane; **21.** cutaneous membrane; **22.** perichondrium

Section 3 Review
1. cardiac; **2.** smooth; **3.** nonstriated; **4.** multinucleate; **5.** cell body; **6.** dendrites; **7.** neuroglia; **8.** maintain physical structure of neural tissue; **9.** repair neural tissue framework after injury; **10.** perform phagocytosis; **11.** provide nutrients to neurons;

12. regulate the composition of interstitial fluid surrounding neurons; **13.** axon; **14.** intercalated disc; **15.** neuroglia; **16.** skeletal muscle tissue; **17.** smooth muscle tissue; **18.** regeneration; **19.** inflammation; **20.** Increased blood flow and blood vessel permeability enhance the delivery of oxygen and nutrients and the migration of additional phagocytes into the area, and the removal of toxins and waste products from the area.

Chapter Review Questions

1. hyaline cartilage; **2.** elastic cartilage; **3.** fibrocartilage; **4.** cardiac muscle tissue; **5.** smooth muscle tissue; **6.** skeletal muscle tissue; **7.** bone or osseous tissue; **8.** adipose tissue; **9.** dense regular connective tissue; **10.** reticular tissue; **11.** areolar tissue; **12.** fluid connective tissues (blood and lymph); **13.** c; **14.** b; **15.** c; **16.** a; **17.** d; **18.** b; **19.** b; **20.** a; **21.** b; **22.** d; **23.** The skin and the lining of the mouth and throat are areas regularly exposed to severe mechanical or chemical stresses. The superficial cells, therefore, are continually worn away due to these stressors and must be replaced by cells in deeper layers. Stratified squamous epithelium contains a series of layers that allow for deeper cells to replace the lost superficial cells. **24.** Endocrine glands secrete hormones onto the surface of the gland or directly into the surrounding fluid. Exocrine glands secrete directly into ducts. **25.** Neural tissue contains (1) neurons, which transmit electrical impulses, and (2) neuroglia, which protect, support, and repair neural tissue and maintain the nutrient supply to neurons. **26.** Normally the extensive interlocking connections between the epithelial skin cells protect from infection by blocking pathogen access. After a severe skin burn or abrasion, this mechanism can no longer provide protection and pathogens easily enter the deeper tissues, resulting in an infection. **27.** Underneath a light microscope cardiac muscle tissue would demonstrate branched cells, striations, a single nucleus, and intercalated discs. **28.** Skeletal muscle tissue is made up of densely packed fibers running parallel that demonstrate striations, but because muscle fibers are composed of cells, they also have many visible nuclei. Jason is probably looking at a tendon (dense regular connective tissue). The small nuclei would be those of fibroblasts. **29.** The skin in the injured area will become red, warm, swollen, and painful. These changes occur as a result of inflammation, the body's first response to injury. Injury to the epithelium and underlying connective tissue will trigger the release of chemicals such as histamine and heparin from mast cells in the area. These chemicals, in turn, initiate the changes identified above.

Chapter Integration

1. The reason for the discrepancy between her microscope and her text is that her microscope is focused at 40×, the typical lowest power on light microscopes found in the common anatomy and physiology lab, but the image in her text is focused at 350×. That is a difference of almost 10 times, so the images will be greatly different. That's like the difference in the view from an airplane window at 1000 feet versus 10,000 feet. **2.** The best way to help Alexis is to advise her to continue to focus her slide with progressively greater magnification to get as close as possible to the image in the book. Most microscopes can magnify at 400×, which is a good power for her to use because that will allow for the two images to look very similar. The best microscope technique is to first focus an image at the lowest objective, and then focus at progressively greater power. In most cases, you will be able to focus an image that is very close to the magnification identified in the text.

CHAPTER 5

Module Reviews

Module 5.1 Review

a. The two major components of the cutaneous membrane are the epidermis and the dermis. The hypodermis is not part of the integument. **b.** The various accessory structures of the integument are hairs, nails, and exocrine glands (sebaceous glands and sweat glands). Supporting structures include blood vessels and nerve fibers. **c.** The major functions of the integumentary system are to protect underlying tissues and organs, excrete wastes by integumentary glands, maintain body temperature, produce melanin and keratin, synthesize vitamin D_3, store lipids, and detect touch, pressure, pain, and temperature.

Module 5.2 Review

a. The layers of the epidermis from deep to superficial are the stratum basale, stratum spinosum, stratum granulosum, stratum lucidum, and stratum corneum. **b.** Dandruff consists of cells from the stratum corneum. **c.** A splinter that penetrates to the third epidermal layer of the palm is lodged in the stratum granulosum.

Module 5.3 Review

a. The two pigments in the epidermis are carotene (an orange-yellow pigment) and melanin (a brown, yellow-brown, or black pigment). **b.** Exposure to sunlight or sunlamps darkens skin because the ultraviolet radiation they emit stimulates melanocytes in the epidermis and dermis to synthesize the pigment melanin. **c.** When skin gets warm, arriving oxygenated blood is diverted to the superficial dermis for the purpose of eliminating heat. The oxygenated blood imparts a reddish coloration to the skin.

Module 5.4 Review

a. The dermis (a connective tissue layer) lies between the epidermis and the hypodermis. **b.** The presence of elastic fibers allows the dermis to undergo repeated cycles of stretching and recoil (returning to its original shape). **c.** The capillaries and sensory neurons that supply the epidermis are located in the papillary layer of the dermis.

Module 5.5 Review

a. A first-degree burn affects only the surface of the skin. This would include most sunburns. A second-degree burn damages the entire epidermis and perhaps some of the dermis. Although blistering occurs, accessory structures such as hair follicles and glands are generally not affected. Third-degree burns, also called full-thickness burns, destroy the epidermis and dermis, and extend into the hypodermis. **b.** Third-degree burns often require skin grafting because the tissue damage is so great it cannot repair itself. **c.** In an autograft, the undamaged skin of a patient is used as a graft source. An allograft uses skin from a donor, and a xenograft uses skin from an animal such as a pig. Autografts are best because they are not rejected by the patient's immune system.

Module 5.6 Review

a. Epidermal derivatives are accessory structures that originate from the epidermis during embryological development. They include hair follicles, sebaceous and sweat glands, and nails. **b.** Sweat glands and sebaceous glands are exocrine glands in the integument. **c.** The accessory structures protect the integument (as in the case of hair and nails), provide sensations, assist in thermoregulation, and excrete wastes.

Module 5.7 Review

a. A typical hair is a keratinous strand produced by basal cells within a hair follicle. **b.** When an arrector pili muscle contracts, it pulls the hair follicle erect. The overall effect is known as "goose bumps" or "chicken skin." **c.** Pulling a hair is painful because its root is attached deep within the hair follicle, the base of which is surrounded by a root hair plexus consisting of sensory nerves. Cutting a hair is painless because a hair shaft contains no sensory nerves.

Module 5.8 Review

a. Two types of exocrine glands found in the skin are sebaceous (oil) glands and sweat glands. **b.** Sebaceous secretions (sebum) lubricate and protect the keratin of the hair shaft, lubricate and condition the surrounding skin, and inhibit the growth of bacteria. **c.** Deodorants are used to mask the odor of apocrine sweat gland secretions.

Module 5.9 Review

a. A fingernail is a keratinous structure that is produced by epithelial cells of the nail root and protects the underlying fingertip. Structures of the nail include a distal free edge, lateral nail fold, lunula, proximal nail fold, eponychium, nail root, nail body, and hyponychium. **b.** Nail production occurs at the nail root, an epidermal fold that is not visible from the surface. **c.** The hyponychium is the thickened stratum corneum underlying the free edge of a nail.

Module 5.10 Review

a. Common effects of the aging process on the skin include epidermal thinning due to declining basal cell activity, fewer melanocytes, reduced sebaceous gland secretion, declining dendritic cell numbers, reduced vitamin D3 production, declining glandular activity, reduced blood flow to the dermis, cessation of hair follicle functioning, and slower skin repair. **b.** With advancing age, melanocyte activity decreases, leading to gray or white hair. **c.** As a person ages, the blood supply to the dermis decreases, and merocrine sweat glands become less active. Both changes make it more difficult for the elderly to cool themselves in hot weather.

Module 5.11 Review

a. Some hormones that are necessary for maintaining healthy skin are growth hormone, sex hormones, growth factors (including epidermal growth factor [EGF]), steroid hormones (glucocorticoids), and thyroid hormones. **b.** In the presence of UV radiation in sunlight, epidermal cells in the stratum spinosum and stratum basale convert a cholesterol-related steroid into cholecalciferol, also known as vitamin D_3. **c.** The hormone cholecalciferol (vitamin D_3) is needed to form strong bones and teeth. When the body surface is covered, UV radiation cannot reach the skin to stimulate cholecalciferol (vitamin D_3) production, so fragile bones can develop.

Module 5.12 Review

a. The first step in tissue repair is inflammation. Inflammation produces swelling, redness, heat, and pain. **b.** Granulation tissue is the combination of fibrin clots, fibroblasts, and the extensive network of capillaries in healing tissue. **c.** Skin can regenerate effectively even after undergoing considerable damage because stem cells persist in both the epithelial and connective tissue components of skin. In response to injury, cells of the stratum basale replace epithelial cells while mesenchymal cells replace lost dermal cells.

Section Reviews

Section 1 Review

1. accessory structures; **2.** epidermis; **3.** granulosum; **4.** papillary layer; **5.** nerves; **6.** reticular layer; **7.** collagen; **8.** hypodermis; **9.** connective; **10.** fat; **11.** dermis: the connective tissue layer beneath the epidermis; **12.** epidermis: the protective epithelium covering the surface of the skin; **13.** papillary layer: vascularized areolar tissue containing capillaries, lymphatic vessels, and sensory neurons that supply the skin surface; **14.** reticular layer: interwoven meshwork of dense irregular connective tissue containing collagen fibers and elastic fibers; **15.** hypodermis (subcutaneous layer or superficial fascia): layer of loose connective tissue below the dermis; **16.** Malignant melanoma is often fatal because melanocytes are located close to the dermal layer, so if they become malignant, they can easily metastasize through the blood vessels and lymphatic vessels in nearby connective tissues; **17.** The firefighter's comment that his burns are not painful should concern the doctors because it indicates that these are third-degree burns. This type of burn is full-thickness and can destroy the sensory neurons, which is why the firefighter could not detect pain, despite being seriously burned.

Section 2 Review

1. free edge; **2.** lateral nail fold; **3.** nail body; **4.** lunula; **5.** proximal nail fold; **6.** eponychium; **7.** eponychium; **8.** proximal nail fold; **9.** nail root; **10.** lunula; **11.** nail body; **12.** hyponychium; **13.** phalanx; **14.** dermis; **15.** epidermis; **16.** hair shaft; **17.** sebaceous gland; **18.** arrector pili muscle; **19.** connective tissue sheath of hair bulb; **20.** root hair plexus; **21.** h; **22.** c; **23.** e; **24.** g; **25.** i; **26.** b; **27.** d; **28.** a; **29.** j; **30.** f; **31.** The chemicals in hair dyes break the protective covering of the cortex, allowing the dyes to stain the medulla of the shaft. This is not permanent because the cortex remains damaged, allowing shampoo and UV rays from the sun to enter the medulla and affect the color. Also, the viable portion of the hair remains unaffected, so that when the shaft is replaced, the color or any change in curl will be lost.

Chapter Review Questions

1. stratum corneum; **2.** stratum lucidum; **3.** stratum granulosum; **4.** stratum spinosum; **5.** stratum basale; **6.** false; **7.** true; **8.** false; **9.** false; **10.** true; **11.** true; **12.** false; **13.** true; **14.** b; **15.** b; **16.** b; **17.** d; **18.** a; **19.** b; **20.** c; **21.** a; **22.** epidermis, dermis; **23.** subcutaneous; **24.** Apocrine; **25.** basal cell carcinoma; **26.** melanin; **27.** Epidermal cell division occurs in the stratum basale. **28.** These smooth muscles cause hairs to stand erect when stimulated. **29.** Variations in hair color are due to differences in the pigment produced by the melanocytes in the hair papilla. With age, pigment production decreases, and hair color lightens. White hair is due to the combination of no pigment and air bubbles in the hair shaft medulla. **30.** A lack of vitamin D_3 can result in rickets in children. In adults, it can lead to decreased bone density, and a greater risk for bone fractures. **31.** Compared to the thin skin of the rest of the body, the thick skin of the palms of the hands has a much thicker stratum corneum, and also contains a stratum lucidum. The heat from the campfire, therefore, has less tissue to penetrate on the thin skin of your face before reaching the sensory neurons. **32.** A cut parallel to the cleavage lines will usually remain closed and heal with little scarring. A cut at right angles to a cleavage line will be pulled open as severed elastic fibers recoil and will result in greater scarring. **33.** Using the rule of nines, the right leg is 19 percent of surface area, the right arm is 9 percent of surface area, and the back of the trunk is 18 percent of body area. Summing these numbers estimates that this patient has burned 46 percent of her body.

Chapter Integration

1. The tattoo ink stays in place because it is deposited into the dermis. The dermal layer is composed mostly of connective tissues with widely scattered cells. The dermal layer does not contain cells that are shed, as they are in the epidermis. Also, the ink particles are too large for them to be removed by the body's normal defense mechanisms. **2.** A tattoo can fade over time for a variety of reasons including the tendency of the ink to disperse, and the effect of sunlight on the ink. **3.** Tattoos can be painful because the papillary layer of the dermis contains many sensory neurons. The tattoo needle would be damaging to these neurons, causing pain. **4.** The reticular and papillary layers of the dermis contain networks of blood vessels and lymphatics. If the tattoo artist is not hygienic in his practice, the risk for infection is substantial. **5.** Many years ago tattoos were removed by a variety of methods that breached the epidermis to reach the ink. These methods included abrasion, freezing (cryosurgery), or physically removing the tattoo by cutting it out. These methods often resulted in significant scarring. Today, tattoo removal involves the use of lasers that heat up the ink, causing it to break into smaller particles. Macrophages then successfully remove these smaller particles.

CHAPTER 6

Module Reviews

Module 6.1 Review

a. The axial skeleton is composed of the bones of the skull, thorax, and vertebral column. It contains a total of 80 bones. **b.** The appendicular skeleton includes the bones of the limbs, and the pectoral and pelvic girdles. It contains a total of 126 bones. **c.** The skeletal system functions to support the body, store minerals and lipids, produce blood cells, protect delicate organs and tissues, and function as levers for movement.

Module 6.2 Review

a. The six broad categories for classifying bones according to shape are flat bones, irregular bones, long bones, sesamoid bones, short bones, and sutural (Wormian) bones. **b.** A surface marking is a characteristic of a bone's surface that has a certain function, such as forming a joint, serving as a site of muscle attachment, or allowing the passage of nerves and blood vessels. **c.** A tubercle is a small, rounded projection on a bone, whereas a tuberosity is a small, rough projection that may occupy a broad area on the bone's surface.

Module 6.3 Review

a. The major parts of a long bone are the epiphysis, diaphysis, metaphysis, and medullary cavity. **b.** The medullary cavity—the space within a bone—contains the red bone marrow, the site of blood cell production, and the yellow bone marrow, which is adipose tissue that is an important site for energy reserves. **c.** Articular cartilage is nourished by diffusion from the synovial fluid within the joint.

Module 6.4 Review

a. Osteocytes are cells responsible for the maintenance and turnover of the mineral content of bone; osteoblasts are cells that produce the fibers and matrix of bone; osteogenic cells are stem cells that differentiate into osteoblasts; and osteoclasts are cells that dissolve the fibers and matrix of bone. **b.** If the activity of osteoclasts (which demineralize bone) exceeded osteoblast activity (production of new bone), then the bone's mineral content (and thus its mass) would decline, making it weaker. **c.** If the ratio of collagen to hydroxyapatite in a bone increased, the bone's compressive strength would decrease, and it would also become more flexible.

Module 6.5 Review

a. An osteon is the basic functional unit of mature compact bone; it consists of osteocytes organized around a central canal and separated by concentric lamellae. **b.** Compact bone, which lies over spongy bone and makes up most of a bone's diaphysis, consists of compactly arranged osteons (Haversian systems); it protects, supports, and resists stress. Spongy bone makes up most of the mass of short, flat, and irregular bones and is also found at the epiphyses of long bones; it stores marrow and provides some support. **c.** The sample is from an epiphysis. The presence of lamellae that are not arranged in osteons is indicative of spongy bone, which occurs in epiphyses.

Module 6.6 Review

a. Appositional growth is bone enlargement by the addition of bone matrix at its surface. **b.** As a bone increases in diameter, the medullary cavity also increases in diameter. **c.** The periosteum is the layer that surrounds a bone. It consists of an outer fibrous region and an inner cellular region. The endosteum is an incomplete cellular lining on the inner (medullary) surfaces of bones.

Module 6.7 Review

a. Endochondral ossification is the replacement of a cartilaginous model with bone. **b.** In endochondral ossification, the source of osteoblasts is the differentiation of cells in the inner layer of the perichondrium. **c.** X-rays of long bones, such as the femur, can reveal the presence or absence of the epiphyseal cartilage, which separates the epiphysis from the diaphysis so long as the bone is still lengthening. If the epiphyseal cartilage is still present, growth is still occurring; if it is not, the bone has reached its full length.

Module 6.8 Review

a. Intramembranous ossification is bone formation within connective tissue without the prior development of a cartilaginous model. **b.** During intramembranous ossification, mesenchymal cells or fibrous connective tissue is replaced by bone. **c.** The principal difference between the two types of ossification is that in intramembranous ossification, bone develops from mesenchymal cells or fibrous connective tissue, whereas in endochondral ossification, bone develops from a cartilage model.

Module 6.9 Review

a. Pituitary dwarfism is less common today in the United States because children can be treated with synthetic growth hormone, which provides adequate or near adequate amounts for normal growth and development. **b.** Marfan's syndrome is a hereditary disorder of connective tissue that results in abnormally long and thin limbs and digits. The condition usually causes life-threatening cardiovascular problems. **c.** Gigantism results from overproduction of growth hormone *before* puberty, causing extreme height, whereas acromegaly results from overproduction of growth hormone *after* puberty, causing abnormally thick bones.

Module 6.10 Review

a. Bone is 33 percent organic compounds and 67 percent inorganic compounds, a ratio of 1:2. **b.** The three organ systems that coordinate blood calcium levels are the skeletal system, the digestive system, and the urinary system. **c.** If blood calcium were seriously reduced, a patient's neuron and muscle cell functions would be disrupted. Skeletal muscles might not function properly, so the patient's movement would be altered. Cardiac muscle could also be affected, so heart rate is likely to be irregular.

Module 6.11 Review

a. The hormones involved in stimulating and inhibiting the release of calcium ions from bone matrix are parathyroid hormone (PTH), which increases blood calcium levels by indirectly causing osteoclasts to release stored calcium from bone; calcitonin, which decreases blood calcium levels by inhibiting osteoclasts and causing osteoblasts to continue depositing calcium in bone; and calcitriol, which increases intestinal calcium absorption, stimulates calcium resorption from bone, aids the effect of PTH on bone resorption, and increases kidney reabsorption of calcium so fewer calcium ions are lost in urine. **b.** Increased PTH secretion would increase blood calcium levels by stimulating osteoblasts to secrete a factor leading to the maturation of osteoclasts, which are then capable of eroding bone and releasing stored calcium ions. **c.** Calcitonin lowers blood calcium levels by inhibiting osteoclast activity and increasing the rate of calcium excretion at the kidneys.

Module 6.12 Review

a. Immediately after a fracture, extensive bleeding occurs at the injury site. After several hours, a large blood clot called a fracture hematoma develops. Next, an internal callus forms as a network of spongy bone unites the inner edges, and an external callus of cartilage and bone stabilizes the outer edges. The cartilaginous external callus is eventually replaced by bone, and the struts of spongy bone then unite the broken ends. With time, the swelling that initially marked the location of the fracture subsides and the fracture site is remodeled, leaving little evidence that a break occurred. **b.** An external callus forms early in the healing process, when cells from the endosteum and periosteum migrate to the area of the fracture. These cells form an enlarged collar (external callus) that encircles the bone in the area of the fracture. **c.** An open fracture (also called a compound fracture) is a break in the bone in which bone pierces the skin; a closed fracture (also called a simple fracture) is a break in the bone in which no bone breaks the skin.

Section Reviews

Section 1 Review

1. intramembranous ossification; **2.** collagen; **3.** osteocytes; **4.** lacunae; **5.** hyaline cartilage; **6.** periosteum; **7.** compact bone; **8.** irregular bones; **9.** epiphyses; **10.** fossa; **11.** medullary cavity; **12.** trabeculae; **13.** osteoclasts; **14.** sesamoid bones; **15.** ossification or osteogenesis; **16.** osteon; **17.** appositional growth; **18.** endochondral ossification; **19.** The fracture might have damaged the epiphyseal cartilage in Rebecca's right leg. Even though the bone healed properly, the damaged leg did not produce as much cartilage as did the undamaged leg. The result would be a shorter bone on the side of the injury.

Section 2 Review

1. calcitonin; **2.** $\downarrow Ca^{2+}$ concentration in blood; **3.** $\downarrow Ca^{2+}$ level; **4.** parathyroid glands; **5.** $\uparrow Ca^{2+}$ concentration in blood; **6.** release of stored Ca^{2+} from bone; **7.** homeostasis; **8.** spiral fracture; **9.** transverse fracture; **10.** greenstick fracture; **11.** comminuted fracture; **12.** compression fracture; **13.** Colles fracture

Chapter Review Questions

1. epiphysis; **2.** metaphysis; **3.** diaphysis; **4.** epiphysis; **5.** spongy bone or cancellous bone or trabecular bone; **6.** compact bone; **7.** medullary cavity; **8.** true; **9.** false; **10.** true; **11.** true; **12.** false; **13.** true; **14.** false; **15.** true; **16.** b; **17.** c; **18.** a; **19.** b; **20.** d; **21.** c; **22.** a; **23.** b; **24.** c; **25.** 80, 126; **26.** Sesamoid; **27.** sinus; **28.** osteon or Haversian system; **29.** Osteoclasts; **30.** (1) support, (2) storage of minerals and lipids, (3) blood cell production, (4) protection, (5) leverage; **31.** The twisting forces of a pirouette are likely to cause a spiral fracture. **32.** Pituitary growth failure causes reduced epiphyseal cartilage activity and therefore abnormally short bones. This is rare because children can be treated with synthetic growth hormone. Achondroplasia is caused by reduced epiphyseal activity resulting in short, stocky limbs. The trunk, however, is of normal size. **33.** In intramembranous ossification, bone replaces mesenchymal cells or fibrous connective tissue. In endochondral ossification, bone replaces a cartilage model. **34.** An epiphyseal fracture is of particular concern because the cartilage can stop growing as it heals. This could result in the fractured limb being shorter than the opposite limb. **35.** Since the osteons of compact bone run parallel to the long axis of the shaft, impacts perpendicular to the length of the bone are more likely to cause a fracture than stresses parallel to the length of the bone. **36.** Gigantism is overproduction of growth hormone (GH) before puberty. People with gigantism can be very tall, exceeding 2.7 m (8 ft 11 in.). In acromegaly, GH rises after puberty, so the person does not grow taller, but the bones become thicker. Facial contours change in a characteristic manner.

Chapter Integration

1. The worker forcefully fell on his buttocks in a seated position, which would have resulted in vertical forces on his vertebrae causing a compression fracture. As confirmed by his coworkers, no bones were protruding through the skin; therefore, he has closed fractures. **2.** When the doctor shaves the surface of the worker's hipbone, she will be collecting the osteoblasts located in the cellular layer of the periosteum. As these cells are deposited, they will continue ossification (osteogenesis) and will produce new bone. The grafted material that fills the space between the damaged and healthy vertebrae will continue the ossification process until it forms into a single fused bone. **3.** The region from which the graft is removed will undergo a healing process similar to that which occurs in a fracture. A hematoma will form, cells will form a callus, new bone will be deposited, and the graft region will be remodeled with osteoblasts and osteoclasts until it is completely healed.

CHAPTER 7

Module Reviews

Module 7.1 Review

a. The skull is made up of 22 bones (eight cranial bones, and 14 facial bones). There are seven associated bones (six auditory ossicles and one hyoid bone). Therefore, there are 29 bones in the skull. **b.** The axial skeleton provides a framework that supports and protects the brain, the spinal cord, and the organs in the body cavities of the trunk. It also provides surface area for the attachment of muscles. **c.** The muscles that attach to the axial skeleton function to position the head, neck, and trunk, perform respiratory movements, and stabilize or position the limbs.

Module 7.2 Review

a. The bones of the cranium are the occipital bone, frontal bone, sphenoid, ethmoid, and the paired parietal and temporal bones. **b.** The facial bones protect and support the entrances to the digestive and respiratory tracts. These bones also provide attachment points for the muscles of facial expression. **c.** A suture is a fibrous connective tissue joint between the flat bones of the skull.

Module 7.3 Review

a. The facial bones are the paired maxillae, palatine, nasal, inferior nasal conchae, zygomatic, lacrimal, and the unpaired vomer and mandible. **b.** The right parietal bone is fractured. **c.** Vomer: facial; ethmoid: cranial; sphenoid: cranial; temporal: cranial; inferior nasal conchae: facial.

Module 7.4 Review

a. Both the external acoustic meatus and the internal acoustic meatus are found in the temporal bone. **b.** The alveolar processes support the upper teeth in the maxillae and the lower teeth in the mandible. **c.** The internal acoustic meatus carries blood vessels and nerves to the internal ear and serves as a passageway for the facial nerve and vestibulocochlear nerves.

Module 7.5 Review

a. The temporal bone contains the carotid canal, through which passes the internal carotid artery supplying the brain. **b.** The foramen ovale provides a passageway for nerves innervating the jaw. **c.** The foramen magnum is located in the occipital bone. This opening surrounds the connection between the brain and spinal cord.

ANSWERS

Module 7.6 Review

a. The optic canal is found in the sphenoid. The optic nerve and ophthalmic artery pass through this opening. **b.** The sphenoid bone contains the sella turcica. The hypophyseal fossa within the sella turcica encloses the pituitary gland. **c.** The ethmoid contains the cribriform plate. In addition to forming the floor of the cranium and the roof of the nasal cavity, olfactory foramina in this structure serve as passageways for olfactory nerves.

Module 7.7 Review

a. The bones of the orbital complex are the frontal, sphenoid, zygomatic, palatine, maxilla, lacrimal, and ethmoid. **b.** Frontal: both; maxilla: both; palatine: both; nasal: nasal. **c.** The frontal sinuses are cavities within the frontal bone that usually appear after age 6.

Module 7.8 Review

a. The foramina of the mandible are the two mental foramina and the two mandibular foramina. **b.** Three auditory ossicles are located in each middle ear cavity, found within the petrous part of the temporal bone. The ossicles play a key role in hearing by conducting vibrations produced by sound waves arriving at the tympanic membrane to the internal ear. **c.** Your lab partner is correct. The hyoid bone does not directly attach to any other bone; instead, it supports the larynx and is the attachment site for muscles of the larynx, pharynx, and tongue.

Module 7.9 Review

a. The major fontanelles are the anterior fontanelle, occipital fontanelle, sphenoidal fontanelle, and mastoid fontanelle. **b.** Because fontanelles are not ossified at birth, they permit flexibility of the skull during childbirth, and allow for growth of the brain during infancy and early childhood. **c.** An infant's cranial bones are larger relative to the facial bones to accommodate the size of the brain. The facial bones develop after birth.

Module 7.10 Review

a. The secondary curves of the spine allow us to balance our body weight to permit an upright posture with minimal muscular effort. Without the secondary curves, we would not be able to stand upright for extended periods of time. **b.** The major components of a typical vertebra are the vertebral body, articular processes, and the vertebral arch. The vertebral arch is composed of a spinous process, laminae, transverse processes, and pedicles. **c.** The intervertebral discs attach to the body of the vertebra.

Module 7.11 Review

a. The dens is part of the axis, or second cervical vertebra, which is located in the cervical (neck) region of the vertebral column. **b.** The presence of transverse foramina indicates that this vertebra is a cervical vertebra. **c.** When you run your finger down a person's spine, you can feel the spinous processes of the vertebrae.

Module 7.12 Review

a. There are five vertebrae in the lumbar region and five fused vertebrae in the sacrum. **b.** The lumbar vertebrae must support a great deal more weight than do vertebrae that are superior in the spinal column. The large vertebral bodies allow the weight to be distributed over a larger area. **c.** The sacrum.

Module 7.13 Review

a. Vertebrosternal ribs are attached directly to the sternum by their own individual costal cartilage. Vertebrochondral ribs do not attach directly to the sternum but by means of a shared costal cartilage. **b.** Improper chest compressions during CPR can—and commonly do—result in fractures of the sternum or ribs. **c.** In addition to the ribs and sternum, the 12 thoracic vertebrae make up the thoracic cage.

Module 7.14 Review

a. There are 126 bones in the appendicular skeleton. **b.** The pectoral girdles attach the upper limbs to the axial skeleton. The pelvic girdle attaches the lower limbs to the axial skeleton. **c.** There are 16 carpal bones in the wrist and only 14 tarsal bones in the ankles. The metacarpal and metatarsal bones are the same in number (10 each), and the fingers and toes are also the same in number (28 each).

Module 7.15 Review

a. The bones of the pectoral girdles are two clavicles (collarbones) and two scapulae (shoulder blades). **b.** In attaching the scapula to the sternum, the clavicle restricts the scapula's range of movement. A broken clavicle thus gives the scapula a greater range of movement, but makes it less stable. **c.** The humerus articulates with the scapula at the glenoid cavity.

Module 7.16 Review

a. The bone of the arm is the humerus, and the bones of the forearm are the radius and ulna. **b.** The two rounded projections on either side of the elbow are the lateral and medial epicondyles of the humerus. **c.** The radius is positioned laterally when the forearm is in the anatomical position.

Module 7.17 Review

a. Phalanges are bones of the fingers (or toes). **b.** The carpal bones are the scaphoid, lunate, triquetrum, pisiform, trapezium, trapezoid, capitate, and hamate. **c.** Bill has broken the tip of his thumb, also known as the pollex.

Module 7.18 Review

a. The three bones that fuse to make up a hip bone (coxal bone) are the ilium, ischium, and pubis. **b.** When you are seated, your body weight is borne by the ischial tuberosities. **c.** The acetabulum is the concave socket (fossa) on the lateral aspect of the pelvis that articulates with the head of the femur.

Module 7.19 Review

a. The bones of the pelvis are the two hip (coxal) bones, the sacrum, and the coccyx. **b.** The two pubic bones are joined anteriorly by the pubic symphysis. **c.** The female pelvis is adapted to support the weight of the developing fetus and to enable the newborn to pass through the pelvic outlet during delivery. Compared with the male pelvis, the female pelvis is smoother and lighter, broader and shallower, and has less prominent markings, an enlarged pelvic outlet, a sacrum and coccyx with less curvature, a wider and more circular pelvic inlet, ilia that project farther laterally, and an inferior angle between the pubic bones that is greater than 100° (as opposed to 90° or less for males).

Module 7.20 Review

a. The bones of the lower limb are the femur (thigh), patella (kneecap), tibia and fibula (leg), tarsal bones, metatarsal bones, and phalanges. **b.** The head of the femur articulates with the acetabulum. **c.** The fibula both stabilizes the ankle joint and is an important point of attachment for muscles that move the foot and toes. When the fibula is fractured, those muscles cannot function properly, so walking becomes difficult—and painful.

Module 7.21 Review

a. The tarsal bones are the talus, calcaneus, cuboid, navicular, medial cuneiform, intermediate cuneiform, and lateral cuneiform. **b.** The talus transmits the weight of the body from the tibia toward the toes. **c.** Joey most likely fractured the calcaneus (heel bone).

Section Reviews

Section 1 Review

1. frontal; **2.** sphenoid; **3.** zygomatic; **4.** ethmoid; **5.** lacrimal; **6.** palatine; **7.** maxilla; **8.** sphenoidal fontanelle; **9.** squamous suture; **10.** lambdoid suture; **11.** mastoid fontanelle; **12.** anterior fontanelle; **13.** sagittal suture; **14.** occipital fontanelle; **15.** Region: thoracic; **16.** Characteristics: found in chest, heart-shaped body, smaller vertebral foramen, long slender spinous process that points inferiorly, all but two have facets for rib articulations; **17.** Region: cervical; **18.** Characteristics: found in neck, small oval body, large vertebral foramen, bifid spinous process, transverse foramina; **19.** Region: lumbar; **20.** Characteristics: found in inferior back, massive vertebral body, smallest vertebral foramen, spinous process blunt and projects posteriorly, no articular facets for ribs or transverse foramina.

Section 2 Review

1. clavicle; **2.** scapula; **3.** humerus; **4.** radius; **5.** ulna; **6.** carpal bones; **7.** metacarpal bones; **8.** phalanges; **9.** hip bone (coxal bone); **10.** femur; **11.** patella; **12.** tibia; **13.** fibula; **14.** tarsal bones; **15.** metatarsal bones; **16.** phalanges; **17.** Sex: female; **18.** Differences: smoother, lighter, less prominent markings, enlarged pelvic outlet, pubic angle greater than 100°, less curvature of sacrum and coccyx, ilia project more laterally; **19.** Sex: male; **20.** Differences: rougher, heavier, more prominent markings, smaller pelvic outlet, pubic angle less than 100°, more curvature of sacrum and coccyx, ilia project more superiorly.

Chapter Review Questions

1. occipital bone; **2.** parietal bone; **3.** frontal bone; **4.** temporal bone; **5.** sphenoid **6.** ethmoid; **7.** vomer; **8.** mandible; **9.** lacrimal bone; **10.** nasal bone; **11.** zygomatic bone; **12.** maxilla; **13.** true; **14.** false; **15.** false; **16.** true; **17.** c; **18.** c; **19.** b; **20.** d; **21.** b; **22.** b; **23.** b; **24.** a; **25.** Fontanelles, which are fibrous connections between cranial bones in the infant, permit the skull to distort without damage during delivery, helping to ease the infant through the birth canal. **26.** Movement of the ribs affects the width and depth of the thoracic cage, increasing or decreasing its volume accordingly. This change in volume assists in breathing by increasing and decreasing the pressure inside the thoracic cavity. **27.** The three bones that fuse to form the hip bone (or coxal bone) are the ilium, ischium, and pubis. They meet at the acetabulum. **28.** Primary curves develop before birth, and the secondary curves develop after birth. The thoracic curve and sacral curves are primary curves. The cervical curve is a secondary curve that forms as an infant begins to balance the weight of the head. The lumbar curve is a secondary curve that develops with the ability to stand. **29.** Arches assist with transferring the weight of the body to the feet and eventually to the ground. The longitudinal arch runs from the calcaneus to the distal metatarsals. The transverse arch runs perpendicular to the longitudinal arch and is the result of the change in the degree of longitudinal arch curvature from the medial border to the lateral border of the foot.

Chapter Integration

1. Donny is likely experiencing difficulty breathing because he has two fractured ribs. Breathing entails rib movements that change the volume of the thoracic cavity; broken ribs cause difficulty breathing because the pain hampers full expansion of the rib cage. **2.** Donny probably dislocated his shoulder, a common injury due to the weak nature of the shoulder joint (that is, the articulation between the head of the humerus and the glenoid cavity of the scapula). **3.** Both the appendicular and axial skeleton have been affected in Donny's injuries. The ribs are part of the axial skeleton, and the shoulder joint is part of the appendicular skeleton. **4.** The bones that do not properly form a suture in cleft palate are the horizontal plates of the palatine bones, and maxillae. These bones are paired bones, so there would be a total of four bones involved. **5.** If you were to look into the mouth of a baby with cleft palate, you would see a visible gap in the hard palate. If there is a gap or cleft between these bones, there is no single hard palate in the roof of the mouth. It is therefore possible to see into the nasal cavity because it is immediately superior to the oral cavity. **6.** Babies with cleft palate are likely to have problems with nursing and eating. Later on these babies could also have speech difficulties. The teeth could be misaligned, requiring orthodontia. If the lip is involved, the condition is called cleft lip.

CHAPTER 8

Module Reviews

Module 8.1 Review

a. The amount of movement at a joint is known as range of motion (ROM). **b.** A synarthrosis is a joint with no movement. An amphiarthrosis is a joint with little movement. **c.** The greatest ROM occurs in a diarthrosis. This category of joint allows for free movement.

Module 8.2 Review

a. Components of a synovial joint are a fibrous joint (articular) capsule, which surrounds the joint; articular cartilages, which resemble hyaline cartilages and cover the articulating bone surfaces; and a synovial membrane, which lines the articular capsule and secretes synovial fluid that provides lubrication, distributes nutrients, and absorbs shocks. Accessory structures include bursae, which are pockets filled with synovial fluid, that reduce friction and absorb shocks; fat pads, which protect the articular cartilages; menisci, which are fibrocartilage articular discs that allow for variation in the shapes of the articulating surfaces; ligaments, which are cords of fibrous tissue that support, strengthen, and reinforce the joint; and tendons, which pass across or around a joint, limit the range of motion, and provide mechanical support. **b.** Articular cartilages lack a blood supply, and thus rely on synovial fluid to supply nutrients and remove wastes. If the circulation of synovial fluid were impaired, the cartilages would no longer receive nutrients, and wastes would accumulate. This could cause the cartilages to degenerate, and cells in the tissue may die. **c.** In a joint dislocation (luxation), the articulating surfaces of a joint are forced out of position.

Module 8.3 Review

a. Based on the shapes of the articulating surfaces, synovial joints are classified as gliding, hinge, pivot, condylar, saddle, and ball-and-socket joints. **b.** A ball-and-socket joint permits the widest ROM. **c.** Shoulder: ball-and-socket; elbow: hinge; ankle: gliding; thumb: saddle.

Module 8.4 Review

a. When doing jumping jacks, both the upper and lower limbs must perform abduction (when the limbs are spread apart) and adduction (when they are brought back together again). **b.** Hinge joints perform flexion and extension. **c.** Dorsiflexion is upward movement of the foot through flexion at the ankle, whereas plantar flexion is ankle extension, as when pointing the toes.

Module 8.5 Review

a. Pronation and supination of the hand are made possible by the rotation of the radius head. **b.** Protraction, supination, and pronation occur while wriggling into tight-fitting gloves. **c.** Snapping your fingers involves opposition of the thumb.

Module 8.6 Review

a. Joints cannot be both strong and highly mobile. As a result, the more movable a joint is, the weaker it is. **b.** Typically, the joints of the appendicular skeleton have a more extensive ROM than those of the axial skeleton. **c.** The upper limb is attached to the axial skeleton by the sternoclavicular joint.

Module 8.7 Review

a. The primary vertebral ligaments are the ligamentum flavum, posterior longitudinal ligament, interspinous ligament, supraspinous ligament, and anterior longitudinal ligament. **b.** The nucleus pulposus is the gelatinous central region of an intervertebral disc. The anulus fibrosus is the tough layer of fibrocartilage encircling the nucleus pulposus. **c.** A bulging disc is a vertebral disc that is displaced or partly protruding as a result of a compressed nucleus pulposus distorting the anulus fibrosus. In a herniated disc, the nucleus pulposus breaks through the anulus fibrosus, causing it to protrude into the vertebral canal.

Module 8.8 Review

a. Ligaments and muscles provide most of the stability for the shoulder joint. **b.** The iliofemoral, pubofemoral, and ischiofemoral ligaments are at the hip joint. **c.** The upward force of the humeral head could cause partial or complete dislocation of the acromioclavicular joint. This is called a shoulder separation. The bones involved include the clavicle, scapula, and humerus; and the stabilizing ligaments involved are the coracoclavicular, acromioclavicular, coracoacromial, coracohumeral, and glenohumeral ligaments.

Module 8.9 Review

a. Menisci are found in the knee joint. **b.** Damage to the menisci of the knee joint decreases the joint's lateral stability, so the person would have a difficult time locking the knee in place while standing and would have to use muscle contractions to stabilize the joint. If the person had to stand for a long time, the muscles would fatigue and the knee would "give out." It is also likely that the person would feel pain. **c.** A severely hyperextended knee would damage the ACL.

Module 8.10 Review

a. Rheumatism is a general term describing any painful condition of joints, muscles, or both that is not caused by infection or injury. Osteoarthritis is a form of rheumatism characterized by degeneration of the joint cartilage and the underlying bone. Osteoarthritis results from cumulative wear and tear or genetic factors affecting collagen formation. **b.** An arthroscope is an instrument that uses thin, flexible optical fibers and a tiny camera to view the interior structures of a joint. This instrument can also be modified to perform surgical procedures without the trauma of major surgery. **c.** A person can slow the progression of arthritis by engaging in regular exercise, doing physical therapy, and taking anti-inflammatory drugs.

Section Reviews

Section 1 Review

1. medullary cavity; **2.** spongy bone; **3.** periosteum; **4.** synovial membrane; **5.** articular cartilage; **6.** joint cavity (containing synovial fluid); **7.** joint capsule; **8.** compact bone; **9.** flexion; **10.** extension; **11.** hyperextension; **12.** flexion; **13.** hyperextension; **14.** abduction; **15.** adduction; **16.** head rotation; **17.** pronation; **18.** abduction; **19.** adduction; **20.** opposition; **21.** e; **22.** d; **23.** f; **24.** g; **25.** b; **26.** c; **27.** h; **28.** a

Section 2 Review

1. coracoclavicular ligaments; **2.** acromioclavicular ligament; **3.** tendon of supraspinatus muscle; **4.** acromion; **5.** articular capsule; **6.** subdeltoid bursa; **7.** synovial membrane; **8.** humerus; **9.** clavicle; **10.** coracoacromial ligament; **11.** coracoid process; **12.** scapula; **13.** articular cartilages; **14.** joint cavity; **15.** glenoid labrum; **16.** patellar surface of femur; **17.** fibular collateral ligament or lateral collateral ligament; **18.** lateral condyle; **19.** lateral meniscus; **20.** tibia; **21.** fibula; **22.** posterior cruciate ligament (PCL); **23.** medial condyle; **24.** tibial collateral ligament or medial collateral ligament; **25.** medial meniscus; **26.** anterior cruciate ligament (ACL); **27.** b; **28.** e; **29.** h; **30.** g; **31.** c; **32.** a; **33.** f; **34.** d; **35.** The sternoclavicular joints are the only joints between the pectoral girdles and the axial skeleton. The sacroiliac joints are the joints between the pelvic girdle and the axial skeleton.

Chapter Review Questions

1. true; **2.** false; **3.** false; **4.** true; **5.** true; **6.** false; **7.** d; **8.** c; **9.** c; **10.** d; **11.** d; **12.** c; **13.** b; **14.** b; **15.** b; **16.** b; **17.** d; **18.** a; **19.** b; **20.** c; **21.** In a bulging disc, the nucleus pulposus does not extrude. In a herniated disc, the nucleus pulposus breaks through the anulus fibrosus. **22.** Articular cartilage lacks a perichondrium, and its matrix contains more water than does the matrix of other cartilages. **23.** (1) gliding joint: sternoclavicular joint; (2) hinge joint: elbow joint; (3) pivot joint: proximal radioulnar joint; (4) condylar joint: radiocarpal joints; (5) saddle joint: first carpometacarpal joint; (6) ball-and-socket joint: shoulder joint; **24.** Menisci may subdivide a synovial cavity, channel the flow of synovial fluid, or allow for variations in the shapes of articular surfaces to assist with joint stability and cushioning. **25.** A loss of bone mass as a result of aging is called osteopenia. When the reduction of bone mass is sufficient to compromise normal function, the condition is called osteoporosis.

Chapter Integration

1. The bones composing the knee joint are the femur, patella, and tibia. The fibula is not part of the joint. **2.** Physical therapy may improve ROM at the joint and increase blood flow to the tissue. **3.** The knee joint is a diarthrotic synovial joint that functions as a hinge; as such, it allows for flexion and extension at the knee. **4.** People who have undergone a total knee arthroplasty may be restricted from running/jogging, high-impact aerobics, and jumping—that is, any activity that places sudden, extreme loads on the artificial joint. **5.** In this scenario, a downward force onto the arm would force the head of the humerus inferiorly out of its articulation with the glenoid cavity. Therefore, Tommy suffered a dislocation of the shoulder joint (glenohumeral joint). A shoulder separation involves the acromioclavicular joint, which is not involved in Tommy's injury. **6.** The affected bones would be the humerus and scapula. The affected ligaments would be the coracohumeral ligament and the glenohumeral ligaments. The affected joint structures would be the articular capsule and the glenoid labrum. The tendons of the muscles surrounding the humeral head are also likely to be involved. **7.** While a full recovery is possible, it is more likely that he will have long-term joint instability if the ligaments are permanently stretched or ruptured. If the glenoid labrum is damaged, it will no longer help to hold the humeral head in the glenoid cavity. Physical therapy to strengthen the shoulder joint muscles is critical to his recovery. Surgery to tighten the ruptured ligaments may also be required.

CHAPTER 9

Module Reviews

Module 9.1 Review

a. The three types of muscle tissue are skeletal, cardiac, and smooth. Skeletal muscle is directly or indirectly attached to bones and allows for movement of the body. Cardiac muscle forms the heart and propels blood. Smooth muscle is found throughout the body, and, among other functions, moves substances along the digestive tract and regulates the diameter of small arteries. **b.** Skeletal muscle tissue is voluntary. Cardiac and smooth muscle tissue are involuntary. **c.** The functions of skeletal muscle tissue include producing movement of the skeleton, maintaining body posture and position, supporting soft tissues such as the abdominal wall and pelvic cavity floor, guarding entrances and exits of the body, maintaining body temperature, and providing nutrient reserves.

Module 9.2 Review

a. A tendon is a collagenous bundle that connects a skeletal muscle to a bone, whereas an aponeurosis is a broad collagenous sheet that takes the place of tendons and connects a skeletal muscle to a wider area of bone or more than one bone. **b.** The epimysium is a dense layer of collagen fibers that surrounds the entire muscle; the perimysium divides the skeletal muscle into a series of compartments, each containing a bundle of muscle fibers called a fascicle; and the endomysium surrounds individual skeletal muscle cells (fibers). The collagen fibers of the epimysium, perimysium, and endomysium come together to form either bundles known as tendons or broad sheets called aponeuroses. Tendons and aponeuroses attach skeletal muscles to bones. **c.** Because tendons attach muscles to bones, severing the tendon would disconnect the muscle from the bone, so the muscle could not move a body part.

Module 9.3 Review

a. Sarcomeres, the smallest contractile units of a striated muscle cell, are segments of myofibrils. Each sarcomere has dark A bands and light I bands. The A band contains the M line, the H band, and the zone of overlap. Each I band contains thin filaments, but not thick filaments. Z lines mark the boundaries between adjacent sarcomeres. **b.** Transverse tubules are tubular extensions of the sarcolemma that extend deep into the sarcoplasm, contacting cisternae of the sarcoplasmic reticulum. **c.** You would expect the greatest concentration of calcium ions in a resting skeletal muscle to be in the cisternae of the sarcoplasmic reticulum.

Module 9.4 Review

a. Thin filaments consist of actin, troponin, and tropomyosin; thick filaments are composed of myosin surrounding a core of titin. **b.** The zone of overlap is important because it is there that the free head of the myosin molecule can interact with the thin filaments. **c.** The sliding filament theory describes the process of sarcomere shortening caused by the sliding of thin and thick filaments past one another.

Module 9.5 Review

a. In a resting (polarized) cell, the inside of its plasma membrane has a slight negative charge with respect to its outside surface. In undisturbed neurons and skeletal muscle fibers, typical resting potentials are -70 mV and -85 mV, respectively. **b.** Open leak channels in the plasma membrane of cells allow ions to move down their concentration gradients. Sodium–potassium ion pumps maintain the cell's resting potential by exporting three Na^+ from the cell, in exchange for two K^+. **c.** A cell is depolarized when a threshold potential is reached (-55 mV for skeletal muscle) and Na^+ rush into the cell. The depolarization peaks at $+30$ mV. Na^+ inflow then abruptly stops and repolarization begins as K^+ rush out of the cell. This loss of positive ions causes the membrane potential to become negative again.

Module 9.6 Review

a. The neuromuscular junction is a specialized intercellular connection that enables a motor neuron to communicate with a skeletal muscle fiber. **b.** Acetylcholine release is necessary for skeletal muscle contraction, because it serves as the first step in the process. A drug that blocks acetylcholine release would prevent ACh from binding with receptors on the motor end plate, so sodium ions would not rush into the muscle fiber's sarcoplasm, and no action potential would be generated in the sarcolemma. As a result, muscle contraction could not occur. **c.** Without AChE, the motor end plate would be continuously stimulated by acetylcholine, locking the muscle in a state of contraction.

Module 9.7 Review

a. ATP is the molecule that supplies the energy for a muscle contraction. **b.** Once the contraction process has begun, the steps that occur are (1) active sites exposed, (2) cross-bridges form, (3) myosin heads pivot (power stroke), (4) cross-bridges detach, and (5) myosin reactivates ("recocks"). **c.** The breakdown of ATP into ADP + P enables myosin reactivation because the energy released during this process is used to "recock" the myosin heads.

Module 9.8 Review

a. ACh is released into the synaptic cleft from the axon terminal. This begins the action potential in the sarcolemma. **b.** The action potential in the sarcolemma travels along the T tubules to the triads, where it triggers the release of Ca^{2+} from the terminal cisternae of the sarcoplasmic reticulum. **c.** The contraction cycle will continue as long as there is ATP available and action potentials are still produced at the motor end plate.

Module 9.9 Review

a. The length of individual sarcomeres is a factor that affects the amount of tension produced when a skeletal muscle fiber contracts. **b.** According to the sarcomere length–tension relationship, (1) the greater the zone of overlap in the sarcomere, the greater the tension the muscle can develop; and (2) there is an optimum range of actin and myosin overlap that will produce the greatest amount of tension. **c.** In an initial latent period (after the stimulus arrives and before tension begins to increase), an action potential generated in the muscle fiber triggers the release of calcium ions from the SR. In the contraction phase, calcium binds to troponin (cross-bridges form), and tension begins to increase. In the relaxation phase, tension drops because cross-bridges have detached and because calcium levels have fallen; the active sites are once again covered by the troponin–tropomyosin complex.

Module 9.10 Review

a. Incomplete tetanus refers to a muscle producing near-peak tension during rapid cycles of contraction and relaxation. Wave summation refers to the addition of one twitch to another. **b.** A motor unit is all of the muscle fibers controlled by a single motor neuron. **c.** The finer and more precise the movement produced by a particular muscle, the fewer the number of muscle fibers in the motor unit.

Module 9.11 Review

a. In an isotonic contraction, tension rises and the skeletal muscle's length changes. In an isometric contraction, tension rises but the muscle's length does not change, and the load does not move. **b.** The heavier the load on a muscle, the longer it will take for the muscle to begin to shorten and the less the muscle will shorten. **c.** Yes, a skeletal muscle can contract without shortening, as occurs during an isometric contraction. Whether a contracting muscle shortens (a concentric isotonic contraction), elongates (an eccentric isotonic contraction), or remains the same length (an isometric contraction) depends on the relationship between the resistance and the tension produced by actin–myosin interactions.

Module 9.12 Review

a. Three sources of energy utilized by muscle fibers are ATP, creatine phosphate, and glycogen. **b.** Muscle cells continuously synthesize ATP by utilizing creatine phosphate and by metabolizing glycogen and fatty acids. **c.** Muscle fibers produce lactic acid under conditions of anaerobic metabolism (when there is a lack of oxygen). These conditions occur at peak levels of muscle activity.

Module 9.13 Review

a. After moderate activity, it may take several hours for a muscle fiber to recover. After sustained activity at higher levels, complete recovery can take a week. **b.** Much of the lactate produced during peak exertion diffuses out of the muscle fibers and into the bloodstream. The liver absorbs this lactate and begins converting it into pyruvate. **c.** Oxygen debt (or excess postexercise oxygen consumption) is the amount of oxygen intake required after strenuous activity to produce the ATP needed to restore normal, pre-exertion conditions in the body.

Module 9.14 Review

a. The three types of skeletal muscle fibers are (1) fast fibers—also called white muscle fibers, fast-twitch glycolytic fibers, Type II-B fibers, and fast fatigue (FF) fibers; (2) intermediate fibers—also called fast-twitch oxidative fibers, Type II-A fibers, and fast resistant (FR) fibers; and (3) slow fibers—also called red muscle fibers, slow-twitch oxidative fibers, Type I fibers, and slow oxidative (SO) fibers. **b.** A sprinter requires large amounts of energy for a short burst of activity. To supply this energy, the sprinter's muscles rely on anaerobic metabolism. Anaerobic metabolism is less efficient in producing energy than aerobic metabolism, and the process also produces acidic waste products; this combination contributes to muscle fatigue. Conversely, marathon runners derive most of their energy from aerobic metabolism, which is more efficient and produces fewer waste products than anaerobic metabolism does. **c.** Individuals who excel at endurance activities have a higher-than-normal percentage of slow fibers. Slow fibers are physiologically better adapted to this type of activity than are fast fibers, which are less vascular and fatigue faster.

Module 9.15 Review

a. Muscle hypertrophy is enlargement of fiber size without cell division (and thus of the entire muscle) stemming from increases in myofibrils, mitochondria, glycolytic enzymes, and glycogen reserves, often in response to activities such as bodybuilding. Muscle atrophy is the wasting away of tissues from lack of use, ischemia, or nutritional abnormalities. **b.** While Fred's leg was immobilized, its muscles did not receive sufficient neural stimulation to maintain normal mass, tone, and strength—that is, his leg muscles atrophied—and the muscles were thus unable to support his weight. **c.** A murder victim's time of death can be estimated according to the body's flexibility or rigidity because rigor mortis typically begins a few hours after death, reaches maximum rigidity some 2–7 hours after death, and subsides about 1–6 days later or when decomposition begins. Thus, for example, a victim whose body lacks any signs of rigor mortis likely died within the past few hours. At the molecular level, the membranes of the dead cells are no longer selectively permeable and the SR is no longer able to retain calcium ions. As calcium ions enter the sarcoplasm, a sustained contraction develops, making the body extremely stiff. Contraction persists because the dead muscle cells can no longer make the ATP required for cross-bridge detachment from the active sites. Rigor mortis lasts until the lysosomal enzymes released by autolysis break down the myofilaments.

Section Reviews

Section 1 Review

1. mitochondrion; **2.** sarcolemma; **3.** myofibril; **4.** thin filament; **5.** thick filament; **6.** triad; **7.** sarcoplasmic reticulum; **8.** T tubule; **9.** terminal cisterna; **10.** sarcoplasm; **11.** myofibril; **12.** I band; **13.** A band; **14.** H band; **15.** Z line; **16.** titin; **17.** zone of overlap; **18.** M line; **19.** thin filament; **20.** thick filament; **21.** sarcomere; **22.** skeletal muscle; **23.** muscle fascicle; **24.** muscle fiber; **25.** myofibril; **26.** myofilament; **27.** sarcomere

Section 2 Review

1. fatty acids; **2.** O_2; **3.** glucose; **4.** glycogen; **5.** CP; **6.** creatine; **7.** b; **8.** c; **9.** a; **10.** d; **11.** large; **12.** small; **13.** intermediate; **14.** white; **15.** pink; **16.** low; **17.** high; **18.** dense; **19.** intermediate; **20.** few; **21.** many; **22.** intermediate; **23.** rapid; **24.** medium; **25.** fast; **26.** slow; **27.** fast; **28.** low; **29.** high; **30.** low; **31.** high

Chapter Review Questions

1. true; **2.** false; **3.** true; **4.** false; **5.** b; **6.** g; **7.** i; **8.** c; **9.** a; **10.** h; **11.** j; **12.** f; **13.** e; **14.** d; **15.** d; **16.** b; **17.** a; **18.** b; **19.** d; **20.** a; **21.** b; **22.** c; **23.** a; **24.** d; **25.** The T tubules conduct action potentials into the interior of the cell. **26.** When contracting to lift the coffee cup, the muscles of elbow joint flexion are recruiting few motor units. The greater tension required for curling a dumbbell would be achieved by recruitment of more motor units including motor units of more powerful muscle fibers. **27.** Throughout the recovery period, oxygen demand remains above normal resting levels as the liver absorbs and recycles the lactate released by muscle fibers during intense exercise. The liver converts the lactate to glucose, which is released into the bloodstream. The transport of lactate to the liver and of glucose back to muscle cells is called the Cori cycle. The additional oxygen consumed by the liver for glucose synthesis is called the oxygen debt, or excess postexercise oxygen consumption. The muscle fibers use the glucose to restore ATP, creatine phosphate, and glycogen concentrations to pre-exertion levels. **28.** Aerobic metabolism and glycolysis generate ATP from glucose in muscle cells. **29.** When a muscle experiences repeated, exhaustive stimulation, as in exercise, muscle fibers develop more mitochondria, a higher concentration of glycolytic enzymes, larger glycogen reserves, and more myofibrils. Each myofibril contains more thick and thin filaments. The net effect is hypertrophy, or enlargement of the stimulated muscle.

Chapter Integration

1. Sean's muscles are likely rich in slow fibers (type I, red, slow-twitch oxidative fibers). These fibers help him excel because they have a high resistance to fatigue due to their high myoglobin content, dense capillary supply, and high number of mitochondria. This fiber type is excellent for endurance activities like soccer, but is not well suited for sprinting from home plate to first base. **2.** Stuart's skeletal muscles are probably rich in fast fibers (Type II-B, white, fast-twitch glycolytic fibers). Fast fibers provide Stuart with the speed necessary to get him from home plate to first base because they have a large diameter, and they rapidly reach peak tension, contract fast, and are high in glycolytic enzymes. This fiber type is not well suited for endurance activities and probably explains why he does not perform well in a sport like soccer. **3.** The percentage of fast versus slow fibers in an individual is genetically determined. Endurance training, however, can cause fast fibers to take on the appearance and characteristics of intermediate fibers (Type II-A, fast resistant, fast-twitch oxidative fibers). Sean the soccer player is more likely to have experienced this training benefit.

CHAPTER 10

Module Reviews

Module 10.1 Review

a. The muscular system can be divided into axial and appendicular divisions. Axial muscles support and position the axial skeleton. Appendicular muscles support and move the limbs. **b.** There are approximately 700 skeletal muscles in the body. **c.** Factors that influence the performance of a skeletal muscle are muscle fiber organization and how the muscle attaches to the bone.

Module 10.2 Review

a. A lever is a rigid structure—such as a board, a crowbar, or a bone—that moves on a fixed point called the fulcrum. There are three classes of levers: In a first-class lever, the fulcrum lies between the applied force and the load; in a second-class lever, the load lies between the fulcrum and applied force; and in a third-class lever, the applied force is between the fulcrum and the load. Third-class levers are the most common type in the body. **b.** The joint between the occipital bone and the first cervical vertebra is part of a first-class lever system. The joint between the two bones (the fulcrum) lies between the skull (which provides the load) and the neck muscles (which provide the applied force). **c.** Contraction of a pennate muscle generates more tension than would contraction of a parallel muscle of the same size because a pennate muscle contains more muscle fibers, and thus more myofibrils and sarcomeres.

Module 10.3 Review

a. A synergist is a muscle that helps a larger agonist (or prime mover—a muscle that is responsible for a specific movement) perform its actions more efficiently. **b.** Muscles A and B are antagonists because they perform opposite actions. **c.** The name *flexor carpi radialis longus* tells you that this muscle is a long (*longus*) muscle that lies next to the radius (*radialis*) and flexes (*flexor*) the wrist (*carpi*).

Module 10.4 Review

a. Axial muscles arise on the axial skeleton and position the head and spinal column and also move the rib cage. This arrangement makes breathing possible.
b. biceps brachii = appendicular; external oblique = axial; temporalis = axial; vastus medialis = appendicular. **c.** The following structures labeled on these figures are not muscles: linea alba, flexor retinaculum, iliotibial tract, patella, tibia, clavicle, sternum, superior extensor retinaculum, inferior extensor retinaculum, lateral malleolus of fibula, medial malleolus of tibia, calcaneal tendon, calcaneus.

Module 10.5 Review

a. The axial musculature stabilizes and positions the head, neck, and trunk. **b.** The first axial muscle group includes the muscles of the head and neck that are not associated with the vertebral column. The second axial muscle group is the muscles of the vertebral column. **c.** The third axial muscle group is the oblique and rectus muscles of the trunk that form the muscular walls of the thoracic and abdominopelvic cavities. The fourth axial muscle group is the muscles of the pelvic floor.

Module 10.6 Review

a. The muscles associated with the mouth include the buccinator, depressor labii inferioris, levator labii superioris, levator anguli oris, mentalis, orbicularis oris, risorius, depressor anguli oris, zygomaticus major, and zygomaticus minor. The orbicularis oris would be involved in kissing or whistling.
b. Buccinator = mouth; corrugator supercilii = eye; mentalis = mouth; nasalis = nose; platysma = neck; procerus = nose; risorius = mouth.
c. A person is able to consciously move the skin on the scalp because the fibers of the epimysium are woven into those of the superficial fascia and the dermis, whereas in the femoral region, muscle fibers (epimysium) insert into bone or tendons, not skin tissue.

Module 10.7 Review

a. The extrinsic eye muscles are the inferior rectus, medial rectus, superior rectus, lateral rectus, inferior oblique, and superior oblique. **b.** The medial and lateral pterygoids originate on the lateral pterygoid plates and insert at the medial surface of the mandibular ramus. **c.** Contracting the masseter muscle elevates the mandible and relaxing this muscle depresses the mandible. So you would probably be eating or chewing something.

Module 10.8 Review

a. The muscles of the tongue include the genioglossus, hyoglossus, palatoglossus, and styloglossus. **b.** The muscles that elevate the soft palate are the levator veli palatini and tensor veli palatini. **c.** Muscles associated with the hyoid that form the floor of the mouth include the mylohyoid, geniohyoid, and digastric.

Module 10.9 Review

a. The spinal flexor muscles include the longus capitis, longus colli, and quadratus lumborum.
b. The splenius capitis muscles enable you to extend your neck.
c. The vertebral column does not need a massive series of flexor muscles because (1) many of the large trunk muscles flex the vertebral column when they contract, and (2) since most of the body weight lies anterior to the vertebral column, gravity tends to flex the spine.

Module 10.10 Review

a. The transversus abdominis forms the deepest layer of the abdominal wall muscles. **b.** The rectus abdominis connects the ribs and sternum to the pubic bones.
c. The external oblique muscle tenses the abdominal wall and compresses the contents of the abdominal cavity, depresses the ribs, and flexes or bends the spine.

Module 10.11 Review

a. The external urethral sphincter and deep transverse perineal muscle make up the urogenital diaphragm. **b.** The bulbospongiosus muscle compresses and stiffens the clitoris and narrows the vaginal opening. **c.** The coccygeus muscle extends from the sacrum and coccyx to the ischial spine.

Module 10.12 Review

a. The appendicular muscles stabilize, position, and support the limbs. **b.** Pectoral girdle muscles originate on the axial skeleton.
c. A muscle that inserts on the femur likely originates from the pelvic region.

Module 10.13 Review

a. The rectus abdominis is the axial muscle that is known as the "six-pack" in physically fit people.
b. As you move proximally to distally, the muscles become smaller and more numerous, enabling precise movements.
c. deltoid = appendicular; external oblique =axial; gluteus maximus = appendicular; pectoralis major = appendicular; platysma = axial; rectus femoris = appendicular.

Module 10.14 Review

a. The largest of the superficial muscles that position the pectoral girdle is the trapezius. **b.** The levator scapulae muscles enable you to shrug your shoulders.
c. The subclavius originates on the first rib and inserts on the inferior border of the clavicle.

Module 10.15 Review

a. The line of action is the line of force produced when a muscle contracts. **b.** The deltoid muscle abducts the upper arm. **c.** The subscapularis muscle originates on the anterior surface of the scapula and inserts on the lesser tubercle of the humerus.

Module 10.16 Review

a. A retinaculum is a wide band of connective tissue that stabilizes tendons. **b.** The wrist extensors are located on the posterior surface of the forearm.
c. Supinator and pronator muscles are involved in turning a doorknob.

Module 10.17 Review

a. Muscles that extend the fingers are the extensor digitorum and extensor digiti minimi. **b.** The abductor pollicis longus, extensor pollicis brevis, and extensor pollicis longus are muscles that abduct the wrist. **c.** The names of muscles associated with the thumb frequently include the term *pollicis*; pollex is the anatomical term for the thumb.

Module 10.18 Review

a. The intrinsic muscles of the thumb are the adductor pollicis, flexor pollicis brevis, opponens pollicis, and abductor pollicis brevis. **b.** None. No muscles originate on the phalanges.
c. We are able to move our fingers because tendons from forearm muscles extend across the distal finger joints. When the forearm muscles contract, our fingers move.

Module 10.19 Review

a. The muscles of the gluteal group are the gluteus maximus, gluteus medius, gluteus minimus, and tensor fasciae latae. **b.** The muscle whose origin is the lateral border of the ischial tuberosity and insertion is the intertrochanteric crest of the femur is the quadratus femoris. **c.** Injury to the obturator muscles would impair lateral rotation at the hip.

Module 10.20 Review

a. The biceps femoris, semimembranosus, semitendinosus, sartorius, and popliteus flex the knee. **b.** The quadriceps muscles are the rectus femoris, vastus intermedius, vastus lateralis, and vastus medialis. **c.** The muscle whose origin is on the lateral condyle of the femur is the popliteus.

Module 10.21 Review

a. The muscles that extend the ankle are the plantaris, gastrocnemius, soleus, tibialis posterior, fibularis longus, and fibularis brevis. **b.** A torn calcaneal tendon would make plantar flexion difficult, because this tendon attaches the soleus and gastrocnemius muscles to the calcaneus (heel bone). **c.** The muscles involved in flexing the toes are the flexor digitorum longus and flexor hallucis longus.

Module 10.22 Review

a. The flexor hallucis brevis muscle flexes the great toe. **b.** The retinacula stabilize the positions of the tendons descending from the leg. **c.** The lumbrical muscles originate on the tendons of the flexor digitorum, and insert on the tendons of the extensor digitorum longus of toes two to five. Their contraction results in flexion at the proximal metatarsophalangeal joints and extension at the distal interphalangeal joints.

Module 10.23 Review

a. The eight muscle compartments of the limbs are the lateral compartment, medial compartment, anterior compartment, superficial anterior compartment, deep anterior compartment, posterior compartment, superficial posterior compartment, and deep posterior compartment. **b.** Compartment syndrome is a condition in which increased pressure in the confined anatomical space adversely affects circulation. **c.** Compartment syndrome can be life threatening because prolonged ischemia leads to tissue death.

Section Reviews

Section 1 Review

1. unipennate; **2.** convergent; **3.** bipennate; **4.** circular or sphincter; **5.** multipennate; **6.** sternocleidomastoid; **7.** deltoid; **8.** biceps brachii; **9.** external oblique; **10.** pronator teres; **11.** brachioradialis; **12.** flexor carpi radialis; **13.** rectus femoris; **14.** vastus lateralis; **15.** vastus medialis; **16.** gastrocnemius; **17.** soleus; **18.** pectoralis major; **19.** rectus abdominis; **20.** iliopsoas; **21.** tensor fascia latae; **22.** gracilis; **23.** sartorius; **24.** tibialis anterior; **25.** extensor digitorum longus

Section 2 Review

1. occipitofrontalis (frontal belly); **2.** temporalis; **3.** orbicularis oculi; **4.** levator labii superioris; **5.** zygomaticus minor; **6.** zygomaticus major; **7.** buccinator; **8.** orbicularis oris; **9.** risorius; **10.** depressor labii inferioris; **11.** depressor anguli oris; **12.** masseter; **13.** mylohyoid; **14.** digastric; **15.** geniohyoid; **16.** omohyoid; **17.** stylohyoid; **18.** thyrohyoid; **19.** sternothyroid; **20.** sternohyoid; **21.** sternocleidomastoid

Section 3 Review

1. triceps brachii, long head; **2.** anconeus; **3.** extensor carpi ulnaris; **4.** extensor carpi radialis longus; **5.** extensor digitorum; **6.** flexor carpi ulnaris; **7.** gluteus medius; **8.** tensor fasciae latae; **9.** gluteus maximus; **10.** adductor magnus; **11.** gracilis; **12.** biceps femoris; **13.** semitendinosus; **14.** semimembranosus; **15.** sartorius; **16.** popliteus; **17.** gastrocnemius; **18.** tibialis anterior; **19.** fibularis longus; **20.** soleus; **21.** extensor digitorum longus; **22.** fibularis brevis; **23.** superior extensor retinaculum; **24.** inferior extensor retinaculum; **25.** calcaneal tendon

Chapter Review Questions

1. second-class lever; **2.** first-class lever; **3.** third-class lever; **4.** true; **5.** false; **6.** true; **7.** true; **8.** false; **9.** j; **10.** h; **11.** f; **12.** a; **13.** g; **14.** b; **15.** e; **16.** c; **17.** d; **18.** i; **19.** b; **20.** b; **21.** c; **22.** a; **23.** a; **24.** c; **25.** d; **26.** a; **27.** c; **28.** b; **29.** The muscles of the pelvic floor support the organs of the pelvic cavity, flex the coccygeal joints, and control the movement of materials through the anus and urethra. **30.** The muscles of the rotator cuff are supraspinatus, infraspinatus, teres minor, and subscapularis. **31.** The muscles of the quadriceps muscle group are rectus femoris, vastus intermedius, vastus medialis, and vastus lateralis. These muscles produce extension at the knee. Rectus femoris also produces flexion at the hip.

Chapter Integration

1. The triceps brachii is an antagonist to the biceps brachii muscle. This means that the triceps muscle opposes the movement of the biceps muscle. To achieve muscle hypertrophy and tone of both muscles, the countering exercises must be performed. **2.** Exercises to tone the abdominal muscle groups include flexing and twisting the trunk. The muscles that the person should target include the rectus abdominis muscles, the external and internal oblique muscles, and the latissimus dorsi muscles. Sit-ups and twisting movements will stimulate these muscles. Placing a weight on the chest while doing sit-ups will produce faster results because the rectus abdominis muscle would be working against a greater load. **3.** When jumping, the hip and knee joints are performing extension, and the ankle joint is conducting plantar flexion. **4.** The primary muscles for hip joint extension are gluteus maximus, biceps femoris, semitendinosus, and semimembranosus. The quadriceps femoris muscles (rectus femoris, vastus intermedius, vastus lateralis, and vastus medialis) extend the knee. Plantar flexion is performed by the gastrocnemius, soleus, fibularis longus, and fibularis brevis muscles. **5.** Weight room activities that perform the joint actions in jumping are best to help Jennifer increase her strength. Such exercises include squats, knee extensions, knee curls, and heel raises.

CHAPTER 11

Module Reviews

Module 11.1 Review

a. The central nervous system (CNS) consists of the brain and spinal cord. The CNS integrates, processes, and coordinates sensory data and motor commands. The peripheral nervous system (PNS) includes all the neural tissue outside the CNS. **b.** The sensory division of the PNS brings information to the CNS from peripheral receptors. **c.** Smooth muscle, cardiac muscle, glands, and adipose tissue are target organs of the ANS.

Module 11.2 Review

a. Structural components of a typical neuron include a cell body (including the nucleus and perikaryon, plus neurofilaments and neurotubules), an axon (including the axon hillock, axoplasm, axolemma, telodendria, collateral branches, and axon terminals), and dendrites (including dendritic spines). **b.** A synapse is the site of communication between a neuron and some other cell. If the other cell is a muscle cell, not a neuron, the term neuromuscular junction is used; if the other cell is a gland cell, the communication is often called a neuroglandular synapse. **c.** Most CNS neurons lack centrioles, which organize the microtubules of the spindle apparatus during mitosis, so these cells cannot divide and replace themselves.

Module 11.3 Review

a. According to structure, neurons are classified as anaxonic, bipolar, unipolar, or multipolar. **b.** According to function, neurons are classified as sensory neurons, interneurons, and motor neurons. **c.** Most sensory neurons of the PNS are unipolar, so these neurons are more likely to be sensory neurons. Motor neurons, which are multipolar, are found in the CNS.

Module 11.4 Review

a. Central nervous system neuroglia are ependymal cells, astrocytes, oligodendrocytes, and microglia. **b.** Astrocytes protect the CNS from circulating chemicals and hormones by maintaining the blood–brain barrier. **c.** Microglia are the phagocytic cells that occur in increased numbers in infected (and damaged) areas of the CNS.

Module 11.5 Review

a. Neuroglia of the PNS are satellite cells and Schwann cells. **b.** The neurilemma is the outer surface of a Schwann cell encircling an axon. **c.** Wallerian degeneration occurs in the PNS, where Schwann cells participate in the repair of damaged nerves.

Module 11.6 Review

a. The membrane potential, or transmembrane potential, is the unequal charge distribution between the inner and outer surfaces of the plasma membrane. **b.** Graded potentials are temporary, localized changes in resting potentials that decrease with distance away from the stimulus. In response to sufficiently large graded potentials, action potentials occur. Action potentials begin in one location and spread along the surface of an axon toward the axon terminals. **c.** Information processing is the integration of stimuli by an individual cell.

Module 11.7 Review

a. The resting membrane potential, or resting potential, is the membrane potential of an undisturbed (nonstimulated) cell. **b.** The sodium–potassium exchange pump maintains the cell's resting potential by ejecting three sodium ions from the cell for every two potassium ions it recovers from the extracellular fluid. **c.** Decreasing the concentration of extracellular potassium ions would cause more potassium to leave the cell, which would make the membrane potential of the neuron more negative.

Module 11.8 Review

a. Gated channels are active channels in the plasma membrane that typically open in response to specific stimuli. **b.** Chemically gated channels operate when they bind specific chemicals (such as ACh); voltage-gated channels operate in response to changes in the membrane potential; and mechanically gated channels operate in response to physical distortion of the membrane surface. **c.** If the voltage-gated sodium ion channels in a neuron's plasma membrane could not open, sodium ions could not rush into the cell, and its membrane potential would not change.

Module 11.9 Review

a. A graded potential (also called a local potential) is a change in the membrane potential that cannot spread far from the site of stimulation. **b.** Depolarization is a shift from the resting potential in which the membrane potential becomes less negative. Repolarization is the return of the membrane potential to the resting potential after the membrane has been depolarized. Hyperpolarization is a shift from the resting potential in which the membrane potential becomes more negative. **c.** Movement of sodium ions parallel to the inner and outer surfaces of the plasma membrane—after passing through open chemically gated sodium channels—accounts for the local currents associated with graded potentials.

Module 11.10 Review

a. An action potential is a propagated change in the membrane potential of excitable cells, initiated by a change in the plasma membrane's permeability to sodium ions. **b.** The events involved in the generation of action potentials are (1) depolarization to threshold, (2) activation of sodium channels and rapid depolarization, (3) inactivation of sodium channels and activation of potassium channels, and (4) closing of potassium channels. **c.** The refractory period is the time between the initiation of an action potential and the restoration of the normal resting potential. The absolute refractory period is the portion of the refractory period during which the membrane cannot respond to further stimulation, no matter its magnitude. The relative refractory period is the time during which the membrane can respond only to a larger-than-normal stimulus.

Module 11.11 Review

a. Continuous propagation is the propagation of an action potential along an unmyelinated axon, where the action potential affects every portion of the membrane surface. Saltatory propagation is the relatively rapid propagation of an action potential between successive nodes of a myelinated axon. **b.** The presence of myelin greatly increases the propagation speed of action potentials.

Module 11.12 Review

a. The parts of a chemical synapse—the site where a neuron communicates with another neuron or with a cell of a different type—are a presynaptic cell and a postsynaptic cell, whose plasma membranes are separated by a narrow gap called the synaptic cleft. **b.** Synaptic fatigue occurs in an axon terminal when intensive stimulation exceeds its ability to keep pace with the demand for neurotransmitter. It is reversed and eliminated by resynthesis of the neurotransmitter. **c.** In chemical synapses, a neurotransmitter crosses a narrow synaptic cleft, whereas in electrical synapses the membranes of the presynaptic and postsynaptic cells are joined together by gap junctions.

Module 11.13 Review

a. An excitatory postsynaptic potential (EPSP) is a graded depolarization of a postsynaptic membrane by a neurotransmitter released by a presynaptic cell. An inhibitory postsynaptic potential (IPSP) is a graded hyperpolarization of a postsynaptic membrane after the arrival of a neurotransmitter. **b.** No action potential will be generated, because the depolarization did not reach threshold. **c.** Temporal summation is the addition of a rapid succession of stimuli occurring at a single synapse. Spatial summation involves the addition of simultaneous stimuli applied at different locations; that is, it involves multiple synapses that are active all at the same time.

Module 11.14 Review

a. Regulatory neurons facilitate or inhibit the activities of presynaptic neurons by affecting the plasma membrane of the cell body, or by altering the sensitivity of axon terminals. **b.** The degree of sustained depolarization at the axon hillock determines the frequency of action potential generation. **c.** The greater the degree of sustained depolarization at the axon hillock, the *higher* the frequency of generation of action potentials.

Section Reviews

Section 1 Review

1. dendrite; **2.** Nissl bodies; **3.** mitochondrion; **4.** nucleus; **5.** nucleolus; **6.** cell body; **7.** axon hillock; **8.** initial segment; **9.** axolemma; **10.** axon; **11.** telodendrion; **12.** axon terminal; **13.** multipolar; **14.** unipolar; **15.** anaxonic; **16.** bipolar; **17.** neuron, neuroglia, neurofilaments, neurofibrils, neurotubules, neurilemma; **18.** dendrite, dendritic spines, telodendria, oligodendrocytes; **19.** efferent fibers; **20.** afferent fibers

Section 2 Review

1. action potential; **2.** electrical synapse; **3.** resting potential; **4.** gated channels; **5.** cholinergic synapses; **6.** hyperpolarization; **7.** local current; **8.** depolarization; **9.** acetylcholine (ACh); **10.** calcium ions (Ca^{2+}); **11.** synaptic vesicle; **12.** acetylcholinesterase (AChE); **13.** ACh receptor; **14.** sodium ions (Na^{+}); **15.** An action potential depolarizes the axon terminal. **16.** Calcium ions enter the cytosol of the axon terminal. **17.** ACh is released through exocytosis. **18.** ACh binds to sodium channel receptors on the postsynaptic membrane, producing a graded depolarization. **19.** The depolarization ends as ACh is broken down into acetate and choline by AChE. **20.** The axon terminal reabsorbs choline from the synaptic cleft and uses it to synthesize new molecules of ACh. **21.** In myelinated fibers, saltatory propagation transmits nerve impulses to the neuromuscular junctions rapidly enough to initiate muscle contractions and promote normal movements. In axons that have become demyelinated, nerve impulses cannot be propagated, so the muscles are not stimulated to contract. Eventually, the muscles atrophy from lack of stimulation (a condition termed disuse atrophy).

Chapter Review Questions

1. ependymal cell; **2.** astrocyte; **3.** node (node of Ranvier); **4.** axon; **5.** oligodendrocyte; **6.** myelinated axon; **7.** neuron cell body; **8.** microglia; **9.** false; **10.** true; **11.** false; **12.** true; **13.** true; **14.** c; **15.** e; **16.** a; **17.** b; **18.** f; **19.** g; **20.** h; **21.** d; **22.** c; **23.** b; **24.** b; **25.** c; **26.** d; **27.** a; **28.** b; **29.** The central nervous system (CNS) is composed of the brain and spinal cord. The peripheral nervous system (PNS) is made up of all the neural tissue outside the CNS and is divided into the sensory and motor divisions. **30.** The three functional classes of neurons in the nervous system are sensory neurons, interneurons, and motor neurons. Sensory neurons transmit impulses from the PNS to the CNS. Interneurons analyze sensory inputs and coordinate motor outputs. Motor neurons transmit impulses from the CNS to the PNS. **31.** In continuous propagation, which occurs in unmyelinated axons, an action potential appears to move across the membrane surface in a series of tiny steps. In saltatory propagation, which occurs in myelinated axons, only the nodes can respond to a depolarizing stimulus. Saltatory propagation is much faster than continuous propagation. **32.** Localized changes in membrane potential that cannot spread far from the site of stimulation are called graded potentials. Action potentials occur when threshold is reached, causing a depolarization that spreads along the axon, resulting in synaptic activity at the axon terminal. **33.** Temporal summation is the addition of stimuli that arrive at a single synapse in rapid succession. Spatial summation occurs when simultaneous stimuli at multiple synapses have a cumulative effect on membrane potential.

Chapter Integration

1. Demyelination is the loss or destruction of the myelin sheaths that insulate nerve fibers. **2.** The loss of myelin slows nerve impulse propagation, so within the CNS, information about limb movement and body position moves slowly; and motor commands move slowly and erratically. **3.** The glial cells affected by MS are oligodendrocytes, which form the myelin sheaths around axons in the CNS.

CHAPTER 12

Module Reviews

Module 12.1 Review

a. Sensory input travels toward the spinal cord, while motor commands travel away from the spinal cord. **b.** A reflex is a rapid, automatic response triggered by specific stimuli. **c.** Spinal reflexes are reflexes that are controlled in the spinal cord, and can function without any input from the brain.

Module 12.2 Review

a. A typical spinal cord has 31 pairs of spinal nerves, and the spinal cord ends at the level of lumbar vertebra 1 or 2 (L_1 or L_2). **b.** The gray matter of the spinal cord is composed of the cell bodies of neurons, neuroglia, and unmyelinated axons. **c.** Gross anatomical features of the cross-sectioned spinal cord include the anterior median fissure (a deep groove along the anterior or ventral surface); the posterior median sulcus (a shallow longitudinal groove); white matter (composed of myelinated and unmyelinated axons); gray matter (composed of cell bodies of neurons, neuroglia, and unmyelinated axons); the central canal (a passageway containing cerebrospinal fluid); a dorsal root of each spinal nerve (axons of neurons whose cell bodies are in the dorsal root ganglion); a ventral root of each spinal nerve (the axons of motor neurons that extend into the periphery to control somatic and visceral effectors); dorsal root ganglia (contain cell bodies of sensory neurons); and spinal nerves (contain the axons of sensory and motor neurons).

Module 12.3 Review

a. The three spinal meninges are the dura mater (the outermost component of the cranial and spinal meninges), arachnoid mater (the middle layer that encloses cerebrospinal fluid), and pia mater (the innermost layer of the meninges bound to the underlying neural tissue). **b.** Cerebrospinal fluid is found in the subarachnoid space, which lies deep to the epithelium of the arachnoid mater and superficial to the pia mater. **c.** The lumbar puncture needle would penetrate the epidermis, dermis, subcutaneous layer (hypodermis), and then skeletal muscle before reaching the protective spinal coverings: the dura mater, then the arachnoid matter, and finally the subarachnoid space, which contains cerebrospinal fluid.

Module 12.4 Review

a. Sensory nuclei receive and relay sensory information from peripheral receptors; motor nuclei issue motor commands to peripheral effectors. **b.** The poliovirus-infected neurons would be in the anterior gray horns of the spinal cord, where the cell bodies of somatic motor neurons are located. **c.** A disease that damages myelin sheaths would affect the white matter columns of the spinal cord, which are composed of bundles of myelinated axons.

Module 12.5 Review

a. The three layers of connective tissue of a spinal nerve are the outer epineurium, middle perineurium, and inner endoneurium; the major peripheral branches of a spinal nerve are the dorsal rami, the ventral rami, and the communicating rami. **b.** A dermatome is a specific bilateral sensory region monitored by a single pair of spinal nerves. **c.** Shingles is caused by a reactivation of the varicella-zoster virus (VZV), the same herpes virus that causes chickenpox. Once reactivated, the virus stimulates painful inflammation of the nerve ganglia and causes skin eruptions in a pattern that corresponds to the affected dermatome.

Module 12.6 Review

a. The gray ramus is a bundle of postganglionic sympathetic nerve fibers that are distributed to effectors in the body wall, skin, and limbs by way of a spinal nerve; the white ramus is a nerve bundle containing the myelinated preganglionic axons of sympathetic motor neurons en route to sympathetic ganglia. **b.** 1) = white rami communicantes, 2) = gray rami communicantes;

c. The dorsal ramus of each thoracic or superior lumbar spinal nerve innervates the skin and skeletal muscles of the back.

Module 12.7 Review

a. A nerve plexus is a complex, interwoven network of nerves. The major plexuses are the cervical, brachial, lumbar, and sacral plexuses. **b.** An anesthetic that blocks the function of the ventral rami of the cervical spinal nerves would affect the skin and muscles of the back of the neck and of the shoulders. **c.** Damage to the cervical plexus—more specifically, to the phrenic nerves, which originate in this plexus and innervate the diaphragm—would interfere with the ability to breathe.

Module 12.8 Review

a. The brachial plexus is a network of nerves formed by branches of spinal nerve segments C_4–T_1, en route to innervating the upper limb. **b.** The major nerves associated with the brachial plexus are the dorsal scapular, long thoracic, suprascapular, medial and lateral pectoral, subscapular, thoracodorsal, axillary, medial antebrachial cutaneous, radial, musculocutaneous, median, and ulnar nerves. **c.** A nerve plexus trunk is a large bundle of axons contributed by several spinal nerves; a nerve plexus cord is a smaller branch of nerves that originates at a trunk.

Module 12.9 Review

a. The lumbar plexus is a nerve network formed by axons from the ventral rami of spinal nerve segments T_{12}–L_4; the sacral plexus is a nerve network formed by the ventral rami of spinal nerve segments L_4–S_4. **b.** The major nerves of the sacral plexus are the superior and inferior gluteal nerves; the posterior femoral cutaneous nerve; the sciatic nerve, which branches into the tibial nerve and the common fibular nerve; and the pudendal nerve. **c.** The sciatic nerve divides into the tibial nerve and the common fibular nerve.

Module 12.10 Review

a. In divergent neural circuits, information spreads from one neuron to several neurons. In convergent neural circuits several neurons synapse with a single postsynaptic neuron. **b.** In serial processing, information is processed in a stepwise fashion, one neuron to another. **c.** The most complex neural processing occurs in the brain.

Module 12.11 Review

a. All reflex arcs include a receptor, a sensory neuron, a motor neuron, and a peripheral effector; interneurons may or may not be present as well. **b.** All reflexes are rapid, unconscious patterned responses to a physical stimulus, which restore or maintain homeostasis. **c.** Neural reflexes are classified according to their development (innate reflexes vs. acquired reflexes), the nature of the resulting motor response (somatic reflexes vs. visceral reflexes), the complexity of the neural circuit involved (polysynaptic reflexes vs. monosynaptic reflexes), and the site of information processing (spinal reflexes vs. cranial reflexes).

Module 12.12 Review

a. A stretch reflex is a monosynaptic reflex that provides automatic regulation of skeletal muscle length. **b.** In the patellar reflex, the response observed is leg extension, and the effectors involved are the quadriceps femoris muscles. **c.** When stretch receptors are stimulated by gamma motor neurons, the muscle spindles become more sensitive. As a result, little (if any) stretching stimulus is needed to stimulate the contraction of the quadriceps muscles. Thus, the reflex response would occur more quickly.

Module 12.13 Review

a. All polysynaptic reflexes involve pools of interneurons, are intersegmental in distribution, involve reciprocal inhibition, and have reverberating circuits. **b.** The flexor reflex is an example of a withdrawal reflex that contracts the flexor muscles of a limb in response to a painful stimulus; so, it has a protective function. **c.** During a withdrawal reflex, the limb on the opposite side is extended. This response is called a crossed extensor reflex.

Module 12.14 Review

a. Reinforcement is an enhancement of a spinal reflex through the facilitation of motor neurons involved in reflexes. **b.** Reflex testing provides information about the nervous system's functional status. **c.** A positive Babinski reflex is abnormal in adults; it indicates possible damage of descending tracts in the spinal cord.

Section Reviews

Section 1 Review

1. white matter; **2.** dorsal root ganglion; **3.** lateral white column; **4.** posterior gray horn; **5.** lateral gray horn; **6.** anterior gray horn; **7.** posterior median sulcus; **8.** central canal; **9.** sensory nuclei; **10.** motor nuclei; **11.** anterior gray commissure; **12.** ventral root; **13.** anterior white commissure; **14.** anterior median fissure; **15.** anterior view; **16.** radial nerve; **17.** ulnar nerve; **18.** median nerve; **19.** posterior view; **20.** columns; **21.** conus medullaris; **22.** nerves; **23.** meninges; **24.** cauda equina; **25.** brachial plexus; **26.** dura mater; **27.** perineurium; **28.** gray ramus

Section 2 Review

1. divergence; **2.** convergence; **3.** serial processing; **4.** parallel processing; **5.** reverberation; **6.** receptor; **7.** sensory neuron; **8.** interneuron; **9.** spinal cord (CNS); **10.** motor neuron; **11.** effector; **12.** ipsilateral reflex; **13.** withdrawal reflexes; **14.** gamma motor neuron; **15.** flexor reflex; **16.** visceral reflexes; **17.** acquired reflexes; **18.** contralateral reflex; **19.** reciprocal inhibition; **20.** reinforcement; **21.** The withdrawal reflex illustrated in the reflex arc illustration is an innate, somatic, polysynaptic, spinal reflex.

Chapter Review Questions

1. gray matter; **2.** white matter; **3.** ventral root; **4.** spinal nerve; **5.** dorsal root ganglion; **6.** dorsal root; **7.** sympathetic ganglion; **8.** dorsal ramus; **9.** pia mater; **10.** arachnoid; **11.** dura mater; **12.** true; **13.** false; **14.** false; **15.** true; **16.** false; **17.** c; **18.** c; **19.** b; **20.** d; **21.** d; **22.** a; **23.** b; **24.** b; **25.** b; **26.** a; **27.** Shingles is caused by the varicella-zoster virus (VZV), the same virus that causes chickenpox. Symptoms include a painful rash and blisters whose distribution corresponds to the affected sensory nerve and its associated dermatome. **28.** The student bumped her ulnar nerve as it passed behind the medial epicondyle of the humerus. She is experiencing pain in the fourth and fifth finger, as well as the medial hand. **29.** Spinal nerve C_1 exits superior to vertebra C_1. Spinal nerve C_8 exits inferior to vertebra C_7. Therefore, there are eight cervical nerves but only seven cervical vertebrae. **30.** Cerebrospinal fluid is located in the subarachnoid space. It functions as a shock absorber and a diffusion medium for dissolved gases, nutrients, chemical messengers, and wastes.

Chapter Integration

1. A head that snaps to the left would cause a stretching of the spinal nerves on the right side of the neck. The majority of these nerves travel into the right upper limb, which is why the pain would extend only into the right side. The pain was in his neck as well because these nerves originate from the spinal cord in the cervical region. **2.** The spinal nerves involved in Dominic's stinger would be those of the brachial plexus, which are spinal nerves C_4–T_1. **3.** The burning, numbing pain is the result of the stretch placed on the brachial plexus as Dominic's head snapped to the opposite side. These nerves innervate the entire upper limb, which is why the pain extended all the way to his fingertips. **4.** The anterior horn in the lumbar region of the spinal cord contains somatic motor neurons that direct the activity of skeletal muscles of the hip, lower limb, and foot. As a result of the injury, Karen would be expected to have poor control of most muscles of the lower limbs, causing difficulty walking (if she could walk at all) and problems maintaining balance (if she could stand). **5.** The person would still exhibit defecation (bowel) and urination (urinary bladder) reflexes because these spinal reflexes are processed at the level of the spinal cord. Afferent impulses from the organs would stimulate specific interneurons in the sacral region that synapse with the motor neurons controlling the sphincters, thus bringing about emptying when the organs began to fill. However, a person with a spinal cord transected at L_1 would lose voluntary control of the bowel and urinary bladder, because these functions rely on impulses carried by motor neurons in the brain that must travel down the spinal cord and synapse with the interneurons and motor neurons involved in the reflex. **6.** The effects of a transection of the spinal cord at L_1 in an adult are the same as the situation in newborns, in whom the descending tracts required for conscious control of urination have not yet fully developed.

CHAPTER 13

Module Reviews

Module 13.1 Review

a. The three primary brain vesicles are the prosencephalon, mesencephalon, and rhombencephalon. **b.** The diencephalon and telencephalon are secondary brain vesicles forming from the prosencephalon. The metencephalon and myelencephalon are secondary brain vesicles forming from the rhombencephalon. **c.** The cerebrum is the largest region of the adult brain.

Module 13.2 Review

a. The six major regions of the brain are the cerebrum, diencephalon, cerebellum, midbrain, pons, and medulla oblongata. The midbrain, pons, and medulla oblongata make up the brain stem. The thalamus and hypothalamus make up the diencephalon. **b.** The medulla oblongata (the most caudal of the brain regions) relays sensory information to other parts of the brain stem and to the thalamus. It also contains centers that regulate autonomic function, such as heart rate and blood pressure. **c.** The corpus callosum is a tract of white matter that links the left and right cerebral hemispheres, whereas the septum pellucidum is a partition that separates the two lateral ventricles.

Module 13.3 Review

a. The layers of the cranial meninges, from superficial to deep, are the outer dura mater, the middle arachnoid mater, and the inner pia mater. **b.** If the normal movement of CSF were blocked, CSF would continue to be produced at the choroid plexuses in each ventricle, but the fluid would remain there, causing the ventricles to swell. **c.** Decreased diffusion across the arachnoid granulations would increase the volume of CSF in the ventricles, because less CSF would re-enter the bloodstream. The increased pressure within the brain due to accumulated CSF could damage the brain.

Module 13.4 Review

a. The ascending and descending tracts of white matter in the medulla oblongata link the brain with the spinal cord. **b.** The nucleus gracilis and nucleus cuneatus are components of the medulla oblongata that relay somatic sensory information to the thalamus. **c.** The pyramids contain tracts of motor fibers that originate at the cerebral cortex. Some of the motor fibers cross to opposite sides of the medulla oblongata, and that crossing over is called a decussation.

Module 13.5 Review

a. The components of the cerebellar gray matter are the cerebellar cortex and the cerebellar nuclei. **b.** The arbor vitae is the white matter of the cerebellum, and it connects the cerebellar cortex and nuclei with the cerebellar peduncles. **c.** Ataxia is the failure of muscular coordination that can result from damage to the cerebellum from trauma, stroke, or certain drugs, including alcohol.

Module 13.6 Review

a. The two pairs of sensory nuclei contained within the corpora quadrigemina are the superior colliculi and inferior colliculi. **b.** The superior colliculi of the midbrain control reflexive movements of the eyes, head, and neck. **c.** Cranial nerves III to XII arise from the brain stem (midbrain, pons, and medulla oblongata).

Module 13.7 Review

a. The main components of the diencephalon are the epithalamus, thalamus, and hypothalamus. **b.** Damage to the lateral geniculate nuclei of the thalamus would interfere with the flow of visual information and thus affect the sense of sight. **c.** The preoptic area of the hypothalamus, a component of the diencephalon, is stimulated by changes in body temperature.

Module 13.8 Review

a. The limbic system establishes emotional states; links the conscious, intellectual functions of the cerebral cortex with the unconscious and autonomic functions of the brain stem; and facilitates memory storage and retrieval. **b.** The hippocampus is important in the storage and retrieval of long-term memories. **c.** The amygdaloid body plays a role in the regulation of heart rate, in the control of the “fight or flight” response, and in linking emotions with specific memories.

Module 13.9 Review

a. The basal nuclei are masses of cerebral gray matter that function in the subconscious control of skeletal muscle activity. **b.** The caudate nucleus is one of the basal nuclei involved with the subconscious control of skeletal muscular activity. **c.** Damage to the basal nuclei would result in decreased muscle tone and the loss of coordination of learned movement patterns.

Module 13.10 Review

a. The lobes of the cerebrum—the frontal lobe, parietal lobe, occipital lobe, and temporal lobe—are named for the overlying bones of the skull. **b.** The insula is an island of cortex located medial to the lateral sulcus. **c.** Damage to the left postcentral gyrus would interfere with the awareness of sensory information from the right side of the body.

Module 13.11 Review

a. The primary motor cortex is located in the precentral gyrus of the frontal lobe of the cerebrum. **b.** Damage to the temporal lobes of the cerebrum would interfere with the processing of olfactory (smell) and auditory (sound) sensations. **c.** The stroke has damaged the speech center, located in the frontal lobe.

Module 13.12 Review

a. The axons in the cerebral white matter are called association fibers, commissural fibers, and projection fibers. **b.** The longitudinal fasciculi connect the frontal lobe to the other lobes of the same hemisphere. **c.** The fibers carrying information between the cerebral cortex and the spinal cord are called projection fibers, and they pass through the diencephalon, brain stem, and cerebellum.

Module 13.13 Review

a. An electroencephalogram (EEG) is a printed record or graph of the electrical activity of the brain. **b.** The four wave types associated with an EEG are alpha waves (characteristic of normal resting adults), beta waves (characteristic of a person who is concentrating), theta waves (observed in children and frustrated adults), and delta waves (found in a person who is sleeping deeply, in infants, or in people with damaged portions of the brain). **c.** A seizure is a temporary cerebral disorder accompanied by abnormal movements, unusual sensations, inappropriate behavior, or some combination of these signs and symptoms. Epilepsy is a clinical condition characterized by seizures.

Module 13.14 Review
a. The cranial nerves are the olfactory (I), optic (II), oculomotor (III), trochlear (IV), trigeminal (V), abducens (VI), facial (VII), vestibulocochlear (VIII), glossopharyngeal (IX), vagus (X), accessory (XI), and hypoglossal (XII) nerves. **b.** The cranial nerves with motor functions only are the oculomotor (III), trochlear (IV), abducens (VI), accessory (XI), and hypoglossal (XII). **c.** The cranial nerves with mixed functions are the trigeminal (V), facial (VII), glossopharyngeal (IX), and vagus (X) nerves.

Module 13.15 Review
a. The term general senses is used to describe our sensitivity to temperature, pain, touch, pressure, vibration, and proprioception. **b.** You are able to distinguish the exact spot of touch more precisely in certain areas due to the size of the receptive field. The smaller the receptive field, the greater your ability to localize an area of stimulus. **c.** The events of a sensory pathway are the following: arrival of a stimulus, depolarization of receptor, action potential generation, propagation over labeled line, and CNS processing that can result in a voluntary or involuntary response.

Module 13.16 Review
a. The four types of general sensory receptors (and the stimuli that excite them) are nociceptors (pain), thermoreceptors (temperature), mechanoreceptors (physical distortion), and chemoreceptors (chemicals dissolved in body fluids). **b.** The three classes of mechanoreceptors are tactile receptors, which respond to the sense of touch; baroreceptors, which detect changes in pressure; and proprioceptors, which monitor the positions of bones, joints, and muscles. **c.** Adaptation is a decrease in receptor sensitivity in the presence of constant stimulation. Peripheral adaptation reduces the amount of information from receptors that reach the central nervous system. In central adaptation, awareness of a stimulus virtually disappears, even though sensory neurons in the CNS remain active.

Module 13.17 Review
a. The six types of tactile receptors are free nerve endings (sensitive to touch and pressure), the root hair plexuses (monitor distortions of and movements across the body surface), tactile discs (detect fine touch and pressure), tactile corpuscles (detect fine touch and pressure, and low frequency vibration), lamellated corpuscles (sensitive to pulsing or vibrating stimuli, such as deep pressure), and Ruffini corpuscles (sensitive to pressure and distortion of the skin). **b.** Tactile receptors found only in the dermis are tactile corpuscles, lamellated corpuscles, and Ruffini corpuscles. **c.** A Ruffini corpuscle is more sensitive to continuous deep pressure because, unlike a lamellated corpuscle, it undergoes little adaptation.

Module 13.18 Review
a. A sensory homunculus is a functional map of the primary sensory cortex. **b.** The lateral spinothalamic tracts carry action potentials generated by nociceptors. **c.** The left cerebral hemisphere (specifically, the primary sensory cortex in that hemisphere) receives impulses conducted by the right fasciculus gracilis.

Module 13.19 Review
a. Corticospinal tracts are descending tracts that carry motor commands from the cerebral cortex to the anterior gray horns of the spinal cord. **b.** The corticobulbar tracts (which are descending tracts) carry information or commands from the cerebral cortex to nuclei and centers in the brain stem. **c.** Increased stimulation of the motor neurons of the red nucleus would increase stimulation of the skeletal muscles in the upper limbs, thereby increasing their muscle tone.

Module 13.20 Review
a. The basic motor patterns related to eating and drinking are controlled by the hypothalamus. **b.** The thalamus and midbrain control reflexes in response to visual and auditory stimuli experienced while viewing a movie. **c.** As you decide to hit the ball, the motor association areas receive information from the frontal lobes and then relay that information to the basal nuclei and cerebellum. As the hitting movement begins, the motor association areas send additional information to the primary motor cortex.

Module 13.21 Review
a. Referred pain is a sensation felt in a part of the body other than its actual source. **b.** Rabies is contracted by a bite from a rabid animal. The bite injects the rabies virus into peripheral tissues, and the virus infects axons. Retrograde flow then carries the viral particles into the CNS. **c.** Amyotrophic lateral sclerosis (ALS), commonly called Lou Gehrig's disease, is a progressive degeneration of the motor neurons of the CNS, leading to muscle atrophy and eventual paralysis.

Section Reviews

Section 1 Review
1. precentral gyrus; **2.** frontal lobe; **3.** lateral sulcus; **4.** temporal lobe; **5.** pons; **6.** central sulcus; **7.** postcentral gyrus; **8.** parietal lobe; **9.** occipital lobe; **10.** cerebellum; **11.** medulla oblongata; **12.** olfactory bulb (associated with cranial nerve I, olfactory), S; **13.** oculomotor (III), M; **14.** trigeminal (V), B; **15.** facial (VII), B; **16.** glossopharyngeal (IX), B; **17.** vagus (X), B; **18.** optic (II), S; **19.** trochlear (IV), M; **20.** abducens (VI), M; **21.** vestibulocochlear (VIII), S; **22.** hypoglossal (XII), M; **23.** accessory (XI), M; **24.** thalamus; **25.** arcuate fibers; **26.** fornix; **27.** commissural fibers; **28.** basal nuclei; **29.** The sensory innervation of the nasal lining, or nasal mucosa, is by way of the maxillary branch of the trigeminal nerve (V). Irritation of the nasal lining increases the frequency of action potentials along the maxillary branch of the trigeminal nerve through the semilunar ganglion to reach centers in the midbrain, which in turn excite the neurons of the reticular activating system (RAS). Increased activity by the RAS can raise the cerebrum back to consciousness.

Section 2 Review
1. free nerve ending; **2.** root hair plexus; **3.** tactile discs; **4.** tactile corpuscle; **5.** Ruffini corpuscle; **6.** lamellated corpuscle; **7.** anterior; **8.** posterior; **9.** lateral corticospinal tract of corticospinal pathway (conscious control of skeletal muscles throughout the body); **10.** rubrospinal tract of lateral pathway (subconscious regulation of muscle tone and movement of distal limb muscles); **11.** reticulospinal tract of medial pathway (subconscious regulation of muscle tone, and movements of the neck, trunk, and proximal limb muscles); **12.** vestibulospinal tract of medial pathway (subconscious regulation of muscle tone, and movements of the neck, trunk, and proximal limb muscles); **13.** tectospinal tract of medial pathway (subconscious regulation of muscle tone, and movements of the neck, trunk, and proximal limb muscles in response to bright lights, sudden movements, and loud noises); **14.** anterior corticospinal tract of corticospinal pathway (conscious control of skeletal muscles throughout the body); **15.** posterior column pathway (carries sensations of fine touch, pressure, vibration, and proprioception); **16.** posterior spinocerebellar tract of spinocerebellar pathway (carries proprioceptive information about the position of skeletal muscles, tendons, and joints); **17.** lateral spinothalamic tract of spinothalamic pathway (carries pain and temperature sensations); **18.** anterior spinocerebellar tract of

spinocerebellar pathway (carries proprioceptive information about the position of skeletal muscles, tendons, and joints); **19.** anterior spinothalamic tract of spinothalamic pathway (carries crude touch and pressure sensations); **20.** Injuries to the primary motor cortex eliminate the ability to exert fine control over motor units, but gross movements may still be produced by cerebral nuclei using the reticulospinal or rubrospinal tracts.

Chapter Review Questions

1. frontal lobe; **2.** corpus callosum; **3.** thalamus; **4.** hypothalamus; **5.** optic chiasm; **6.** mammillary body; **7.** midbrain; **8.** pons; **9.** precentral gyrus; **10.** central sulcus; **11.** postcentral gyrus; **12.** parietal lobe; **13.** parieto-occipital sulcus; **14.** occipital lobe; **15.** pineal gland; **16.** cerebral aqueduct; **17.** fourth ventricle; **18.** medulla oblongata; **19.** false; **20.** true; **21.** false; **22.** false; **23.** true; **24.** b; **25.** b; **26.** a; **27.** d; **28.** b; **29.** a; **30.** b; **31.** d; **32.** In any inflamed tissue, swelling occurs in the area of the inflammation. The accumulation of fluid in the subarachnoid space can cause damage by pressing against neurons. If the intracranial pressure is excessive, brain damage can occur, and if the pressure involves vital autonomic reflex areas, death could occur. **33.** The nerve that is blocked when a dentist works on a tooth in the bottom jaw would be the mandibular branch of the trigeminal nerve (V). This is a mixed nerve. Other sensory areas that could be affected by the injection are the inferior gums, lips, and portions of the tongue and palate. Motor functions that could be affected are the muscles of mastication.

Chapter Integration

1. The cranial nerve involved in Bell's palsy is the facial nerve (VII) on the left side of his face. **2.** The facial nerve is a mixed nerve, so John is likely to be experiencing both sensory and motor symptoms. His entire face is likely to be experiencing unusual sensations, including the loss of taste on the anterior two-thirds of his tongue on the left side. Besides being unable to move his facial muscles, he probably can't close his eye, make tears (from his lacrimal gland), or produce saliva (from his submandibular and sublingual salivary glands). **3.** Because John is not producing tears and he is unable to close his eye, he should probably take care to keep his eye moist. Using artificial tears (eye drops) and wearing an eye patch while sleeping is a good idea. He may also want to drink with a straw since he is probably having difficulty in controlling his lips to successfully drink from a glass.

CHAPTER 14

Module Reviews

Module 14.1 Review

a. The SNS provides conscious and subconscious control over skeletal muscles. The ANS controls the visceral effectors: smooth muscle, glands, cardiac muscle, and adipocytes. **b.** Preganglionic neurons are part of the visceral reflex arcs. They are involved mostly in direct reflex responses, rather than responses to commands from the hypothalamus. **c.** Ganglionic neurons of autonomic ganglia innervate visceral effectors.

Module 14.2 Review

a. The major divisions of the ANS are the sympathetic division, the parasympathetic division, and the enteric division. **b.** The sympathetic division of the ANS is responsible for the physiological changes that occur when you are startled by a loud noise. **c.** In the sympathetic division, axons emerge from thoracic and lumbar segments of the spinal cord and innervate ganglia relatively close to the spinal cord, whereas in the parasympathetic division, axons emerge from the brain stem and sacral segments of the spinal cord and innervate ganglia very close to (or within) target organs.

Module 14.3 Review

a. General responses to increased sympathetic activity include heightened mental alertness, increased metabolic rate, reduced digestive and urinary functions, activation of energy reserves, increased respiratory rate and dilation of respiratory passageways, elevated heart rate and blood pressure, and activation of sweat glands. General responses to increased parasympathetic activity include decreased metabolic rate, decreased heart rate and blood pressure, increased secretion by salivary and digestive glands, increased motility and blood flow in the digestive tract, and stimulation of urination and defecation. **b.** preganglionic neurons (T_1–L_5) → collateral ganglia → ganglionic neurons (postganglionic fibers) → visceral effectors in abdominopelvic cavity; **c.** An intramural ganglion is a group of neurons embedded in the tissue of a target organ.

Module 14.4 Review

a. Splanchnic nerves are three groups of nerves—cardiopulmonary, abdominopelvic (greater, lesser, and lumbar), and pelvic (sacral)—that supply the viscera. **b.** The vagus nerve innervates the cardiac plexus, celiac plexus, inferior mesenteric plexus, and the hypogastric plexus. **c.** The vagus nerve (X) carries most of the outflow of the parasympathetic division of the ANS.

Module 14.5 Review

a. Both alpha receptors and beta receptors are adrenergic receptors on the membranes of target cells. Alpha receptors are more sensitive to NE than beta receptors, but both receptors are stimulated by E. Alpha-1 receptor stimulation typically excites the target cell, and alpha-2 receptor stimulation generally inhibits the target cell. Stimulation of beta receptors may result in the excitation or inhibition of the target cell. **b.** Blocking the beta receptors on cells would decrease (or prevent) sympathetic stimulation of tissues containing those cells. As a result, heart rate, force of cardiac muscle contraction, and contraction of the smooth muscle in blood vessel walls would decrease, lowering blood pressure. **c.** Nicotinic receptors are acetylcholine (ACh) receptors on the surfaces of sympathetic and parasympathetic ganglionic cells; muscarinic receptors are ACh membrane receptors that are located at all parasympathetic neuromuscular and neuroglandular junctions, and at a few sympathetic neuromuscular and neuroglandular junctions.

Module 14.6 Review

a. In tense (or anxious) people, increased sympathetic stimulation typically causes some or all of the following changes: dry mouth; increased heart rate, blood pressure, and respiration rate; cold sweats; an urge to urinate or defecate; changes in digestive tract motility (for example, "butterflies in the stomach"); and dilated pupils. **b.** Acetylcholine (ACh) is released by all parasympathetic neurons. **c.** The parasympathetic division is sometimes referred to as the anabolic system because parasympathetic stimulation leads to a general increase in the nutrient content of the blood. Cells throughout the body respond to the increase by absorbing the nutrients and using them to support growth and other anabolic activities.

Module 14.7 Review

a. The ANS adjusts the activities of virtually every body system, maintaining homeostasis without instructions or interference from the conscious mind. **b.** The two types of motor pathways are somatic and visceral. **c.** The somatic effector is skeletal muscle. The visceral effectors are smooth muscles, glands, cardiac muscle, and adipocytes.

Module 14.8 Review

a. Dual innervation is the situation in which a given body structure receives instructions from both the sympathetic and parasympathetic divisions of the ANS. **b.** Autonomic tone is the background level of activity in sympathetic or parasympathetic motor neurons under resting conditions. It provides a mechanism for fine control of visceral function because a resting neuron may be *less active* or *more active* rather than simply switching from off to on. This is particularly important when only one division innervates a visceral organ, or when an organ must be precisely controlled over a broad range of activity levels (e.g., the heart). **c.** The blood vessels of the skin receive only sympathetic innervation. When you go outside into the cold, sympathetic neurons release NE and cause vasoconstriction of superficial blood vessels through stimulation of alpha receptors. When you get angry, sympathetic activation occurs and large amounts of epinephrine enter the circulation. This stimulates beta receptors in the superficial blood vessels, dilating those vessels, and stimulates the heart, increasing blood pressure and blood flow. As a result, your skin—and most obviously your face—turns red.

Module 14.9 Review

a. A visceral reflex is an autonomic motor response that can be modified, facilitated, or inhibited by higher centers, especially those of the hypothalamus. **b.** Short reflexes are autonomic responses that bypass the CNS, whereas long reflexes involve interneurons within the CNS and autonomic delivery of motor commands to the effectors. **c.** The solitary nucleus is a large mass of gray matter in the medulla oblongata that serves as a processing center and sorting center for visceral sensory information.

Module 14.10 Review

a. Baroreceptors are receptors that detect changes in pressure; chemoreceptors are receptors that detect changes in the concentrations of specific chemicals or compounds. **b.** Baroreceptors are located along the digestive tract (stomach, intestines, and colon), within the walls of the urinary bladder, in the carotid and aortic sinuses, and in the lungs. **c.** Chemoreceptors are sensitive to changes in blood pH and levels of carbon dioxide and oxygen.

Module 14.11 Review

a. The hypothalamus is the brain structure considered to be the headquarters for the ANS. **b.** A brain tumor pressing on the hypothalamus could interfere with autonomic function because the hypothalamus receives visceral sensory information and controls both sympathetic and parasympathetic functions. **c.** The thalamus relays somatic sensory information.

Section Reviews

Section 1 Review

1. cervical sympathetic ganglia; **2.** sympathetic chain ganglia; **3.** coccygeal ganglia; **4.** cardiopulmonary splanchnic nerves; **5.** cardiac and pulmonary plexuses; **6.** celiac ganglion; **7.** superior mesenteric ganglion; **8.** splanchnic nerves; **9.** inferior mesenteric ganglion; **10.** sympathetic division; **11.** thoracolumbar division; **12.** thoracic nerves; **13.** lumbar nerves; **14.** parasympathetic division; **15.** craniosacral division; **16.** cranial nerves III, VII, IX, X; **17.** sacral nerves; **18.** enteric nervous system; **19.** f; **20.** g; **21.** h; **22.** b; **23.** a; **24.** c; **25.** e; **26.** d

Section 2 Review

1. limbic system and thalamus; **2.** hypothalamus; **3.** pons; **4.** spinal cord T_1–L_2; **5.** complex visceral reflexes; **6.** vasomotor; **7.** coughing; **8.** respiratory; **9.** sympathetic visceral reflexes; **10.** parasympathetic visceral reflexes; **11.** P; **12.** P; **13.** S; **14.** P; **15.** S; **16.** S; **17.** P; **18.** S; **19.** S; **20.** S; **21.** P; **22.** S; **23.** P; **24.** Even though most sympathetic postganglionic fibers are adrenergic, releasing norepinephrine, a few are cholinergic, releasing acetylcholine. This distribution of the cholinergic fibers by the sympathetic division provides a method of regulating sweat gland secretion and selectively controlling blood flow to skeletal muscles while reducing the flow to other tissues in a body wall to maintain homeostasis.

Chapter Review Questions

1. preganglionic neurons; **2.** preganglionic fibers; **3.** sympathetic ganglion; **4.** ganglionic neurons; **5.** postganglionic fibers; **6.** parasympathetic ganglion; **7.** acetylcholine; **8.** norepinephrine; **9.** epinephrine; **10.** false; **11.** true; **12.** false; **13.** false; **14.** false; **15.** true; **16.** c; **17.** d; **18.** a; **19.** c; **20.** a; **21.** c; **22.** b; **23.** a; **24.** b; **25.** c; **26.** a; **27.** a; **28.** Visceral reflex arcs include a receptor, a sensory neuron, a processing center of one or more interneurons, and two or more visceral motor neurons. Short reflexes bypass the CNS entirely, and control very simple motor responses in one small part of a target organ. Long reflexes deliver information to the CNS along the dorsal roots of spinal nerves, within the sensory branches of cranial nerves, and within the autonomic nerves that innervate visceral effectors. Long reflexes predominate and are responsible for coordinating responses involving multiple organ systems. **29.** There are many differences between the somatic and autonomic nervous systems. Three of the differences are the following: the somatic nervous system exerts control only over skeletal muscle, while the autonomic nervous system controls smooth muscle, glands, cardiac muscle, and adipocytes; somatic nervous system effector activity involves an upper motor neuron in the CNS and a lower motor neuron in the PNS, while autonomic nervous system effector activity involves an upper motor neuron in the CNS, and two lower motor neurons in the PNS; and the somatic nervous system uses only ACh at its target organ, while the autonomic nervous system uses ACh, epinephrine, and norepinephrine. **30.** The cranial nerves associated with the parasympathetic division are cranial nerves III, VII, IX, and X. The sacral nerves associated with the parasympathetic division are S_2, S_3, and S_4.

Chapter Integration

1. The parasympathetic division of the autonomic nervous system would be affected if muscarinic receptors were blocked. **2.** Blocking muscarinic receptors would block the neurotransmitter acetylcholine (ACh). **3.** Muscarinic receptors are cholinergic receptors because they bind with ACh. If scopolamine blocks ACh, it would therefore be described as an anticholinergic drug. **4.** ACh is the solitary neurotransmitter of the parasympathetic nervous system. If ACh is blocked while wearing this patch, Andrew is likely to experience a decrease in the "rest and digest" activities of the PNS. He is likely to experience dry eyes as a result of decreased lacrimal gland secretion, a dry mouth from decreased salivary gland secretion, dilated pupils due to pupil constriction inhibition, and blurred near vision as accommodation of the lens for close vision is decreased. **5.** Epinephrine would be more effective because it binds to the beta-2 receptors of the smooth muscles surrounding the airways, resulting in their relaxation and airway dilation, thus making it easier for Gregor to breathe. **6.** A decrease in blood pressure would stimulate baroreceptors in the carotid sinus and aortic sinus to relay information to the cardiac and vasomotor centers in the medulla oblongata. These centers respond by increasing sympathetic impulses that initiate two complementary sympathetic visceral reflexes. The cardiac centers increase the heart rate and force of contraction, and the vasomotor center changes the diameter of peripheral blood vessels to increase blood pressure.

CHAPTER 15

Module Reviews

Module 15.1 Review

a. The special senses originate at receptor cells that may be neurons or specialized receptor cells that communicate with sensory neurons. **b.** A generator potential is the depolarization of specialized sensory neurons. **c.** The origin of the other senses differs from olfaction in that their receptors have inexcitable membranes that generate graded potentials, not action potentials.

Module 15.2 Review

a. Olfaction—the sense of smell—involves olfactory receptors in paired olfactory organs responding to chemical stimuli. **b.** The neurons associated with olfaction that are continually regenerated are olfactory receptor cells. New receptor cells are produced by the division and differentiation of basal (stem) cells. **c.** Axons from the olfactory epithelium collect into bundles that synapse in the olfactory bulb. Axons leaving the olfactory bulb then travel along the olfactory tract to the olfactory cortex, hypothalamus, and portions of the limbic system.

Module 15.3 Review

a. Gustation is the sense of taste, provided by taste receptors responding to chemical stimuli. **b.** Filiform papillae are slender conical projections on the superior surface of the tongue. They provide friction for the tongue to move objects in the mouth, but they do not contain taste buds. **c.** Taste receptors respond more readily to unpleasant than to pleasant stimuli. Therefore we are more sensitive to bitter and sour sensations than to sweet and salty sensations. Such sensitivity has survival value by helping us avoid substances such as acids that may harm mucous membranes, and bitter-tasting biological toxins.

Module 15.4 Review

a. Gustducins are G proteins—protein complexes that use second messengers to produce effects—that are associated with sweet, bitter, and umami sensations. **b.** The cranial nerves that carry gustatory information are the facial nerve (VII), glossopharyngeal nerve (IX), and vagus nerve (X). **c.** The gustatory pathway: taste receptors → facial nerve (VII), glossopharyngeal nerve (IX), and vagus nerve (X) → synapse in solitary nucleus of medulla oblongata → medial lemniscus → synapse in thalamus → primary sensory cortex.

Module 15.5 Review

a. In olfaction and gustation the receptor cells are exposed to the external environment and communicate with the CNS. By contrast, in equilibrium and hearing the receptors are isolated within the internal ear and protected from the external environment. **b.** The internal ear receptors are called hair cells because their free surfaces are covered with specialized processes similar to cilia and microvilli. Certain hair cells contain a kinocilium (a single large cilium) and many stereocilia. **c.** The internal ear can respond to gravity and acceleration, rotation, or sound, depending on which region is stimulated.

Module 15.6 Review

a. External ear infections are relatively uncommon because the skin in the external acoustic meatus contains glands that secrete cerumen (ear wax), which inhibits microbial growth. **b.** The bones in the middle ear are the malleus, incus, and stapes. **c.** The auditory tube (also called the pharyngotympanic tube or eustachian tube) connects the nasopharynx with the middle ear and permits pressure equalization on either side of the tympanic membrane.

Module 15.7 Review

a. The bony labyrinth is composed of semicircular canals enclosing the semicircular ducts, the vestibule, and the cochlea. **b.** The semicircular canals are part of the bony labyrinth and the canals surround the semicircular ducts, which are part of the membranous labyrinth. **c.** Within the membranous labyrinth, receptors in the vestibule respond to gravity or linear acceleration, receptors in the semicircular ducts respond only to rotation, and receptors in the cochlear duct respond only to sound.

Module 15.8 Review

a. Damage to the cupula of the lateral semicircular duct would interfere with the perception of horizontal rotation of the head. **b.** Otoliths are densely packed calcium carbonate crystals that sit upon the gelatinous otolithic membrane in the maculae. **c.** Receptors in the saccule and utricle provide sensations of gravity and linear acceleration.

Module 15.9 Review

a. The organ of Corti is located in the cochlea of the internal ear. **b.** Perilymph fills the scala vestibuli and the scala tympani, and endolymph fills the cochlear duct. **c.** When the basilar membrane moves in response to pressure changes in the perilymph, the hair cells of the organ of Corti press against the tectorial membrane.

Module 15.10 Review

a. A decibel is the unit of measurement for the intensity of sound. **b.** Sound waves enter the external acoustic meatus → tympanic membrane → auditory ossicles → oval window → basilar membrane → round window of scala tympani → hair cells vibrate against tectorial membrane → information relayed to CNS by cochlear branch of cranial nerve VIII. **c.** If the round window could not move, the perilymph would not be moved by the vibration of the stapes at the oval window; this would reduce or eliminate the perception of sound.

Module 15.11 Review

a. The hair cell receptors for equilibrium are in the vestibule and the semicircular ducts. **b.** Cranial nerves III, IV, VI, and XI are involved with eye, head, and neck movements. **c.** The reflexive response to a loud noise is to turn your head and eyes toward the source of the noise.

Module 15.12 Review

a. The first structures that form in eye development are the optic vesicles. **b.** The retina develops from the inner and outer layers of the optic cup. **c.** The ependymal cells of the outer layer of the optic cup develop into the photoreceptors.

Module 15.13 Review

a. Accessory structures associated with the eye include the eyelids (palpebrae), eyelashes, medial canthus, cornea, lateral canthus, lacrimal caruncle, conjunctiva, tarsal glands, and lacrimal apparatus. **b.** The conjunctiva is the first layer of the eye that would be affected by inadequate tear production. **c.** Conjunctivitis, or pinkeye, is inflammation of the conjunctiva, generally caused by a pathogenic infection or by physical, allergic, or chemical irritation of the conjunctival surface. The most obvious sign of conjunctivitis, redness, results from the dilation of blood vessels deep to the conjunctival epithelium.

Module 15.14 Review

a. The layers (tunics) of the eye are the fibrous layer, the vascular layer (uvea), and the inner layer (retina). **b.** The color of eyes is largely determined by the density and distribution of melanocytes in the iris. **c.** Aqueous humor is found in the anterior cavity, between the cornea and the lens, and in the posterior cavity where it is called vitreous humor.

Module 15.15 Review

a. The cornea does not contain blood vessels. **b.** Light passes through the cornea, aqueous humor, pupil of the iris, lens, and the gelatinous vitreous body (containing vitreous humor) before arriving at the retina. **c.** When light intensity decreases, sympathetic stimulation causes the pupillary dilator muscles to contract, resulting in dilated (enlarged) pupils.

Module 15.16 Review

a. The focal point is the point at which the light rays from an object intersect on the retina. **b.** When the ciliary muscles are relaxed, you are viewing something in the distance. **c.** The near point of vision typically increases with age because the elasticity of the lens tends to decrease with age.

Module 15.17 Review

a. Rods are more numerous than cones, are responsible for vision in dim light, and provide poorer visual acuity than cones. Cones provide color vision and require more intense light than rods. **b.** When you enter a dimly lit room, you are unlikely to be able to see at all. The low-intensity light in the room would be focused on the fovea, which contains only cones, which cannot be stimulated by low-intensity light. **c.** If you had been born without cones, you would still be able to see—so long as you had functioning rods—but you would see in black and white only, and with reduced visual acuity.

Module 15.18 Review

a. A photoreceptor is made up of an outer segment and inner segment. The tip of the outer segment is enclosed by pigment epithelial cells. The outer segment contains flattened membranous discs containing visual pigment molecules sensitive to photons of light. The inner segment contains the major organelles of the cell and it synapses with a bipolar cell. **b.** A dietary vitamin A deficiency would reduce the quantity of retinal the body could produce, thereby interfering with night vision. **c.** The three types of cones are blue cones, green cones, and red cones.

Module 15.19 Review

a. The visual pigments undergo activation, bleaching, and reassembly during photoreception. **b.** The two configurations of retinal are the 11-*cis* form and more linear 11-*trans* form. **c.** ATP is required for converting 11-*trans* retinal back to its original 11-*cis* form.

Module 15.20 Review

a. Optic radiation refers to the bundles of projection fibers linking the lateral geniculate nuclei of the thalamus with the visual cortex in each cerebral hemisphere. **b.** Visual images are perceived in the visual cortex of the occipital lobes of the cerebrum. **c.** The visual pathway: photoreceptors in the retina → bipolar cell → ganglion cell → axons from the population of ganglion cells converge on optic disc → optic nerve (II) → optic chiasm → optic tract → lateral geniculate nucleus → collateral fibers to diencephalon and brain stem, other collateral fibers to superior colliculus, other collateral fibers to occipital cortex of cerebral hemisphere.

Module 15.21 Review

a. Emmetropia is the term for normal vision. **b.** A converging lens (one with at least one convex surface) is used to correct hyperopia. **c.** Two surgical procedures for correcting myopia and hyperopia are photorefractive keratectomy (PRK) and laser-assisted in-situ keratomileusis (LASIK). These procedures—both types of refractive surgery—use lasers to slice the corneal epithelium, thereby permanently reshaping the cornea.

Module 15.22 Review

a. Cranial nerves VII (facial), IX (glossopharyngeal), and X (vagus) provide taste sensations from the tongue. **b.** Vertigo is caused by any condition that alters the function of the internal ear receptor complex, the vestibular branch of the vestibulocochlear nerve, or sensory nuclei and pathways in the central nervous system. Common causes include motion sickness, excessive alcohol consumption, and exposure to certain drugs. **c.** Two common classes of hearing-related disorders are conductive deafness and nerve deafness.

Section Reviews

Section 1 Review

1. umami; **2.** sour; **3.** bitter; **4.** salty; **5.** sweet; **6.** vallate papillae; **7.** foliate papillae; **8.** fungiform papillae; **9.** filiform papillae; **10.** e; **11.** i; **12.** f; **13.** m; **14.** a; **15.** c; **16.** k; **17.** j; **18.** l; **19.** h; **20.** g; **21.** b; **22.** d; **23.** The olfactory sensory receptor cells are specialized neurons whose cilia-shaped dendrites contain receptor proteins. The binding of odorant molecules to the receptor proteins results in a depolarization of the receptor cell and the production of action potentials. In contrast, the membranes of the sensory receptor cells for taste, vision, equilibrium, and hearing are inexcitable and do not generate action potentials. These cells all form synapses with the processes of sensory neurons, which depolarize and produce action potentials when stimulated by chemical transmitters (neurotransmitters).

Section 2 Review

1. external ear; **2.** middle ear; **3.** internal ear; **4.** auricle; **5.** external acoustic meatus; **6.** elastic cartilage; **7.** tympanic membrane; **8.** auditory ossicles (malleus, incus, and stapes); **9.** tympanic cavity; **10.** petrous part of temporal bone; **11.** vestibulocochlear nerve (VIII); **12.** cochlea; **13.** auditory tube; **14.** scala vestibuli; **15.** vestibular membrane; **16.** cochlear duct; **17.** organ of Corti; **18.** basilar membrane; **19.** scala tympani; **20.** spiral ganglion; **21.** The rapid descent in the elevator causes the otoliths in the macula of the saccule of each vestibule to slide upward, producing the sensation of downward vertical motion. When the elevator abruptly stops, the otoliths do not. It takes a few seconds for them to come to rest in the normal position. As long as the otoliths are displaced, you will perceive movement.

Section 3 Review

1. posterior cavity; **2.** choroid; **3.** fovea; **4.** optic nerve; **5.** optic disc; **6.** retina; **7.** sclera; **8.** fornix; **9.** palpebral conjunctiva; **10.** ocular conjunctiva; **11.** ciliary body; **12.** iris; **13.** lens; **14.** cornea; **15.** ciliary zonule (suspensory ligaments); **16.** ora serrata; **17.** l; **18.** k; **19.** g; **20.** h; **21.** o; **22.** i; **23.** f; **24.** d; **25.** n; **26.** m; **27.** j; **28.** b; **29.** a; **30.** c; **31.** e; **32.** Light falling on the eye passes through the cornea and strikes the photoreceptors of the retina, bleaching (breaking down) many molecules of the pigment rhodopsin into retinal and opsin. After an intense exposure to light, a photoreceptor cannot respond to further stimulation until its rhodopsin molecules have been regenerated by the conversion of retinal molecules to their original shape and recombination with opsin molecules. The "ghost" image remains until the rhodopsin molecules are regenerated.

Chapter Review Questions

1. pigmented part of the retina; **2.** central retinal artery; **3.** optic nerve; **4.** central retinal vein; **5.** neural part of the retina; **6.** optic disc (blind spot); **7.** ganglion cell; **8.** sclera; **9.** choroid; **10.** false; **11.** false; **12.** true; **13.** false; **14.** true; **15.** true; **16.** c; **17.** b; **18.** a; **19.** c; **20.** b; **21.** The four primary taste sensations are salt, sweet, sour, and bitter. The two other taste sensations that have been identified are umami and water. **22.** A cataract is a condition in which the lens loses its transparency. This can be caused by injury, radiation, or a reaction to drugs. Senile cataracts are a natural consequence of aging and are the most common form. Cataracts may be treated with surgery, which involves removing the lens and then replacing it with an artificial substitute. Vision is then refined with glasses or contact lenses. **23.** Otitis media is a middle ear infection caused by microorganisms traveling along the auditory tube from the nasopharynx to the middle ear.

ANSWERS

Chapter Integration

1. Myopia is the medical term for nearsightedness; people with myopia are able to see nearby objects clearly, but distant objects appear blurry or fuzzy. Hence, Mr. Drummond would have difficulty seeing while driving. **2.** Mr. Drummond will likely have a follow-up visit with an optometrist or ophthalmologist, who will prescribe spectacle lenses (glasses) or contact lenses to alleviate the myopia and restore his distance sight to normal, or near-normal, vision. **3.** Vertigo is another term for dizziness and is generally associated with a problem in the internal ear. The movement of Mr. Drummond's arms while standing still provided evidence that there may be a problem with his equilibrium centers (saccule and utricle). Often, inflammation due to infection or a cold affects the internal ear (or cranial nerve VIII) and causes vertigo. **4.** When Mr. Drummond closes his eyes, he does not have any visual cues, so his brain must rely solely on proprioception (the sense of relative position of body parts) and information from the equilibrium centers of the internal ear to maintain normal posture. Because either the internal ear receptors or the sensory nerves are not functioning normally, he is unstable. **5.** The most likely reason for Mr. Drummond's arms to drift to the left is that he is getting inappropriate sensations from equilibrium receptors, either at the maculae (affecting his ability to determine which way is "down") or at one of the horizontal semicircular ducts (making him attempt to compensate for a perceived roll to the right).
6. Daniel is experiencing nerve deafness caused by his chronic exposure to loud sounds. This type of deafness can be caused by a single exposure to an extremely loud sound (as in a bomb blast), or by long-term exposure to sounds above 90 dB. **7.** Within Daniel's cochlea, the stereocilia in the hair cells of the organ of Corti are most likely damaged. While he is likely to be conducting sound waves from his tympanic membrane, auditory ossicles, and perilymph, the hair cells are not generating action potentials along the spiral ganglion to the cochlear branch of cranial nerve VIII. It is also possible that the nerve itself has been damaged as a result of Daniel's exposure to loud sounds. **8.** It is not likely that Daniel will ever regain his hearing because the hair cells of the organ of Corti, the spiral ganglion, and cranial nerve VIII do not have the capacity to regenerate.

CHAPTER 16

Module Reviews

Module 16.1 Review

a. Paracrine communication is by the release of paracrine factors into the extracellular fluid, and is limited to a local area. Endocrine communication is by the release of hormones into the bloodstream, and it can target cells beyond the local area. **b.** The nervous system communicates with neurotransmitters, whereas the endocrine system communicates with hormones. **c.** The common goal of both the nervous and endocrine systems is to preserve homeostasis by coordinating and regulating the activities of other cells, tissues, organs, and organ systems.

Module 16.2 Review

a. The structural classes of hormones are (1) amino acid derivatives (thyroid hormones, catecholamines, and tryptophan derivatives); (2) peptide hormones (glycoproteins or short polypeptide chains), which are chains of amino acids that are synthesized as prohormones; and (3) lipid derivatives (eicosanoids and steroid hormones), which contain carbon rings and side chains that are built from fatty acids or cholesterol. **b.** The endocrine system is one of several body systems that includes organs whose primary function is the production of hormones or paracrine factors, which are chemical secretions that are transported by the extracellular fluid or bloodstream to target cells in other sites within the body. **c.** Organs of the endocrine system are the hypothalamus, pituitary gland, thyroid gland, adrenal glands, pancreas (pancreatic islets), pineal gland, and parathyroid glands. Organs of other systems that have endocrine functions are the heart, thymus, digestive tract, kidneys, and gonads.

Module 16.3 Review

a. A hormone receptor is a protein molecule, located either on the plasma membrane or inside the cell, that binds to a specific hormone. **b.** A first messenger is a hormone whose binding to a protein receptor in the plasma membrane gives rise to a second messenger in the cytoplasm. A second messenger changes the rate of various metabolic reactions by acting as an enzyme activator, an enzyme inhibitor, or a cofactor. **c.** Steroid hormones diffuse across the plasma membrane and bind to receptors in the cytoplasm.

Module 16.4 Review

a. A regulatory hormone is a hormone secreted by the hypothalamus that controls endocrine cells in the anterior lobe of the pituitary gland. **b.** The three mechanisms of hypothalamic integration of neural and endocrine function are (1) secretion of antidiuretic hormone (ADH) and oxytocin (OXT), (2) secretion of regulatory hormones that control activity of the anterior lobe of the pituitary gland, and (3) neural (sympathetic) control over the endocrine cells of the adrenal medullae. **c.** The blood vessels of the hypophyseal portal system link the hypothalamus and anterior lobe of the pituitary gland. Unusually permeable fenestrated capillary beds in each structure are connected by portal vessels. This arrangement ensures that hypothalamic regulatory hormones reach the "downstream" endocrine cells of the anterior lobe directly, before mixing with, and being diluted by, the general circulation.

Module 16.5 Review

a. The two lobes of the pituitary gland are the anterior lobe and the posterior lobe. **b.** The hormones produced and released by the anterior lobe of the pituitary gland are (1) thyroid-stimulating hormone (TSH), which targets the thyroid gland; (2) adrenocorticotropic hormone (ACTH), which targets the adrenal cortex; (3) follicle-stimulating hormone (FSH) and (4) luteinizing hormone (LH), which target the testes in males and the ovaries in females; (5) growth hormone (GH), which targets liver cells (which respond by synthesizing somatomedins); (6) prolactin (PRL), which targets mammary glands in females; and (7) melanocyte-stimulating hormone (MSH), which targets melanocytes in the skin. Hormones released by the posterior lobe of the pituitary gland are (8) oxytocin (OXT), which targets the uterus and mammary glands; and (9) antidiuretic hormone (ADH), which targets the kidneys. **c.** In a person who is dehydrated, the amount of ADH released by the posterior lobe of the pituitary gland increases in response to increased blood osmotic pressure resulting from an increase in solute concentration.

Module 16.6 Review

a. The hypothalamic releasing hormones are corticotropin-releasing hormone (CRH), thyrotropin-releasing hormone (TRH), growth hormone–releasing hormone (GH–RH), prolactin-releasing factor (PRF), and gonadotropin-releasing hormone (GnRH). **b.** Somatomedins mediate the action of growth hormone (GH). Increased levels of GH typically accompany increased levels of somatomedins. **c.** Increased circulating levels of glucocorticoids inhibit the release of CRH by the hypothalamus.

The lack of CRH decreases the secretion of ACTH from the pituitary gland, so ACTH levels would decrease.

Module 16.7 Review

a. The hormones of the thyroid gland are thyroxine (T_4), triiodothyronine (T_3), and calcitonin. **b.** Calcitonin aids in calcium regulation. **c.** Most of the body's reserves of the thyroid hormone, thyroxine (T_4), are bound to transport proteins in the bloodstream called thyroid-binding globulins. Because these compounds represent such a large reserve of thyroxine, it takes several days after removal of the thyroid gland for blood levels of thyroxine to decrease.

Module 16.8 Review

a. The parathyroid glands are embedded in the posterior surfaces of the lateral lobes of the thyroid gland. **b.** Parathyroid hormone (PTH) increases blood calcium levels by decreasing calcium deposition in bones, by increasing reabsorption of calcium from the blood by the kidneys, and by increasing the production of calcitriol by the kidneys. **c.** Decreased blood calcium levels result in increased secretion of PTH.

Module 16.9 Review

a. The two regions of an adrenal gland are the cortex and medulla. The cortex secretes mineralocorticoids, primarily aldosterone; glucocorticoids, mainly cortisol (or hydrocortisone) and corticosterone; and androgens. The medulla secretes epinephrine and norepinephrine. **b.** The three zones of the adrenal cortex are the zona glomerulosa, zona fasciculata, and zona reticularis. **c.** Increased cortisol levels would result in increased blood glucose levels, because cortisol decreases the use of glucose by cells while increasing both the available glucose (by promoting the breakdown of glycogen) and the conversion rate of amino acids to carbohydrates.

Module 16.10 Review

a. The types of cells in the pancreatic islets (and their hormones) are alpha cells (glucagon), beta cells (insulin), delta cells (GH–IH), and F cells (pancreatic polypeptide, or PP). **b.** Insulin secretion lowers blood glucose concentrations. **c.** Increased levels of glucagon stimulate the conversion of glycogen to glucose in the liver, which in turn decreases the amount of glycogen stored in the liver.

Module 16.11 Review

a. The hormone-secreting cells of the pineal gland are pinealocytes. **b.** Melatonin secretion is influenced by circadian rhythms, the daily changes in physiological processes that follow a regular day/night pattern. Increased amounts of light would inhibit the production (and release) of melatonin from the pineal gland, which receives neural input concerning the presence of light or darkness from visual pathway collaterals. **c.** In humans, melatonin may affect the timing of sexual maturation, protect against free radical damage, and maintain circadian rhythms.

Module 16.12 Review

a. Diabetes mellitus is an endocrine disorder characterized by elevated blood glucose levels resulting from inadequate insulin production or diminished cell sensitivity to insulin. **b.** The two types of diabetes mellitus are Type 1, characterized by inadequate insulin production by the pancreatic beta cells, and Type 2, characterized by insulin resistance (failure of the body to use insulin properly). **c.** Some clinical problems associated with diabetes mellitus are diabetic retinopathy, blockages in heart circulation and increased risk for heart attack, diabetic nephropathy, diabetic neuropathy, and decreased blood flow to the distal portions of the limbs.

Module 16.13 Review

a. Antagonistic effects are involved in a negative feedback response. **b.** A synergistic effect is when two hormones have an additive effect so that the net result is greater than the effect each would produce alone. **c.** Hormones producing different but complementary effects are described as having integrative effects.

Module 16.14 Review

a. The heart secretes natriuretic peptides, and the kidneys release erythropoietin. **b.** When renin is released into the bloodstream, it functions as an enzyme that activates the renin-angiotensin-aldosterone system, which ultimately causes blood pressure to increase. **c.** Hormones necessary for normal growth and development include GH, thyroid hormones, insulin, PTH, calcitriol, and reproductive hormones.

Module 16.15 Review

a. The three phases of the stress response are the alarm phase, the resistance phase, and the exhaustion phase. **b.** The resistance phase is characterized by long-term metabolic adjustments, including mobilization of remaining energy reserves, conservation of glucose, increased blood glucose concentrations, and conservation of salts and water coupled with the loss of K^+ and H^+. **c.** The collapse of vital systems occurs during the exhaustion phase of the stress response (general adaptation syndrome).

Module 16.16 Review

a. The prefix *hyper-* refers to excessive hormone production, whereas *hypo-* refers to inadequate hormone production. **b.** Three common causes of hormone hyposecretion are metabolic factors, physical damage, and congenital disorders. **c.** Aldosteronism is characterized by increased body weight due to Na^+ and water retention and a low blood K^+ concentration.

Section Reviews

Section 1 Review

1. catecholamines; **2.** thyroid hormones; **3.** tryptophan derivatives; **4.** peptide hormones; **5.** short polypeptides; **6.** glycoproteins; **7.** small proteins; **8.** lipid derivatives; **9.** eicosanoids; **10.** steroid hormones; **11.** transport proteins; **12.** c; **13.** e; **14.** f; **15.** g; **16.** j; **17.** h; **18.** b; **19.** i; **20.** a; **21.** d; **22.** thymus; **23.** pineal gland; **24.** pancreatic islet; **25.** hypothalamus; **26.** kidney; **27.** adrenal gland; **28.** pituitary gland; **29.** gonad

Section 2 Review

1. release of natriuretic peptides; **2.** suppression of thirst; **3.** Na^+ and H_2O loss from kidneys; **4.** decreased blood pressure; **5.** increased fluid loss; **6.** decreasing blood pressure and volume; **7.** erythropoietin released; **8.** renin released; **9.** increased red blood cell production; **10.** aldosterone secreted; **11.** ADH secreted; **12.** increasing blood pressure and volume; **13.** c; **14.** g; **15.** a; **16.** e; **17.** j; **18.** d; **19.** i; **20.** b; **21.** f; **22.** h; **23.** (1) The two hormones may have opposing or antagonistic effects, such as occurs between insulin (decreases blood glucose levels) and glucagon (increases blood glucose levels). (2) The two hormones may have an additive or synergistic effect, in which the net result is greater than the sum of each acting alone. An example is the enhanced glucose-sparing action of GH in the presence of glucocorticoids. (3) One hormone may have a permissive effect on another, in which the first hormone is needed for the second hormone to produce its effect. For example, epinephrine cannot alter the rate of tissue energy consumption without the presence of thyroid hormones. (4) The hormones may have integrative effects, in which the hormones may produce different but complementary results in specific tissues and organs. An example is the differing effects of calcitriol and parathyroid hormone (PTH) on tissues involved in calcium metabolism; calcitriol increases calcium ion absorption by the intestinal tract, and PTH inhibits osteoblast activity and enhances calcium ion reabsorption by the kidneys.

Chapter Review Questions

1. true; **2.** false; **3.** true; **4.** false; **5.** false; **6.** corticotropin-releasing hormone (CRH); **7.** thyrotropin-releasing hormone (TRH); **8.** growth hormone–releasing hormone (GH–RH); **9.** growth hormone–inhibiting hormone (GH–IH); **10.** prolactin-releasing factor (PRF); **11.** prolactin-inhibiting hormone (PIH); **12.** gonadotropin-releasing hormone (GnRH); **13.** adrenocorticotropic hormone (ACTH); **14.** glucocorticoids (steroid hormones); **15.** thyroid-stimulating hormone (TSH); **16.** thyroid hormones; **17.** growth hormone (GH); **18.** somatomedins; **19.** prolactin (PRL); **20.** follicle-stimulating hormone (FSH); **21.** luteinizing hormone (LH); **22.** inhibin; **23.** testosterone; **24.** estrogen; **25.** progesterone; **26.** inhibin; **27.** melanocyte-stimulating hormone; **28.** oxytocin (OXT); **29.** antidiuretic hormone (ADH); **30.** b; **31.** d; **32.** d; **33.** a; **34.** b; **35.** d; **36.** The kidney releases hormones that function to regulate red blood cell production and the rate of calcium and phosphate absorption by the intestinal tract. **37.** Calcitonin decreases blood calcium levels. Parathyroid hormone increases blood calcium levels.

Chapter Integration

1. Sherry's physician suspected hyperthyroidism because she exhibited the classic signs and symptoms of the condition: restlessness, anxiety, irritability, difficulty sleeping, diarrhea, weight loss, rapid heart rate, and tremors. **2.** Hyperthyroidism is the excess production of T_3, T_4, or both, by the thyroid gland. Excess thyroid hormone results in a rapid heartbeat and an increased metabolic rate, among other clinical signs and symptoms. **3.** Sherry's physician could order blood tests to assay the levels of TSH, T_3, and T_4. Results confirming hyperthyroidism are lower-than-normal TSH levels and elevated T_3 and T_4 levels. Her physician could also determine whether her condition is primary hyperthyroidism (a problem with the thyroid gland) or secondary hyperthyroidism (a problem with hypothalamic-pituitary control of the thyroid gland). **4.** Sherry's signs and symptoms indicating nervous system involvement include tremors and anxiety.

CHAPTER 17

Module Reviews

Module 17.1 Review

a. The components of the cardiovascular system are the heart, blood vessels, and blood. **b.** Arteries carry blood away from the heart to the capillaries. Capillaries allow diffusion between blood and the interstitial fluids, and veins carry blood from capillaries to the heart. **c.** Blood functions to transport dissolved gases, nutrients, hormones, and metabolic wastes; regulate pH and ion composition of interstitial fluids; restrict fluid loss at injury sites; defend against toxins and pathogens; and stabilize body temperature.

Module 17.2 Review

a. Whole blood is composed of plasma (which contains albumins, globulins, fibrinogen, electrolytes, organic nutrients, and organic wastes) and formed elements (which are platelets, white blood cells, and red blood cells). White blood cells include neutrophils, eosinophils, basophils, lymphocytes, and monocytes. **b.** The hematocrit, also called the packed cell volume, is the percentage of formed elements in a sample of whole blood. **c.** During an infection, you would expect the level of immunoglobulins (antibodies) in the blood to be elevated.

Module 17.3 Review

a. A hematopoietic stem cell (hemocytoblast) is a multipotent stem cell whose divisions produce lymphoid and myeloid stem cells, which divide to form each of the various populations of blood cells. **b.** Platelets develop from megakaryocytes. These large cells shed their cytoplasm in small, membrane-enclosed packets. These packets are the platelets that enter the bloodstream. **c.** Lymphoid stem cells originate in the red bone marrow and give rise to lymphocytes; these stem cells also produce lymphocytes in the thymus, spleen, and lymph nodes. Myeloid stem cells are cells in red bone marrow that give rise to all the formed elements except lymphocytes.

Module 17.4 Review

a. Hematology is the medical study of blood, blood-producing organs, and blood disorders. **b.** A complete blood count (CBC) is a diagnostic blood test used to determine underlying medical conditions. A CBC includes the RBC count, WBC count, erythrocyte indices (such as hemoglobin content), hematocrit, platelet count, and WBC differential count. **c.** A patient with a depressed hematocrit level would have anemia.

Module 17.5 Review

a. Rouleaux are stacks of red blood cells. **b.** Hemoglobin is a protein—composed of four globular subunits, each bound to a heme molecule—that gives RBCs the ability to transport oxygen in the blood. **c.** Oxyhemoglobin is hemoglobin whose iron has bound oxygen; it is bright red. Deoxyhemoglobin is hemoglobin whose iron has not bound oxygen; it is dark red.

Module 17.6 Review

a. Hemolysis is the rupture of red blood cells; it results in the release of hemoglobin. **b.** After the removal of iron within macrophages, heme is converted into biliverdin, which is then converted to bilirubin. In the large intestine, bilirubin is converted to either stercobilins, which are eliminated in the feces, or urobilins, which are eliminated in the feces or in urine. **c.** Bilirubin would accumulate in the blood and produce jaundice. This condition occurs because diseases that damage the liver impair its ability to excrete bilirubin in the bile.

Module 17.7 Review

a. Surface antigens on RBCs are glycolipids in the plasma membrane; they determine blood type. **b.** Only type O^- blood can be safely transfused into a person whose blood type is O^-. **c.** A person with type A blood also has anti-B antibodies, so if they received a transfusion of type B blood, the transfused red blood cells would clump, or agglutinate, potentially blocking blood flow to various organs and tissues.

Module 17.8 Review

a. Hemolytic disease of the newborn (HDN) is a condition in which maternal antibodies attack and destroy fetal red blood cells, resulting in fetal anemia; it occurs in a sensitized Rh^- mother who is carrying an Rh^+ fetus. **b.** When RhoGAM (which contains anti-Rh antibodies) is injected into a pregnant Rh^- woman, the anti-Rh antibodies circulate in the mother's bloodstream, where they destroy any fetal RBCs there. This prevents the mother's immune system from making antibodies against the developing fetus's red blood cells. **c.** An Rh^+ mother carrying an Rh^- fetus does not require a RhoGAM injection because the fetus is not at risk of Rh incompatibility. The fetus is not at risk because its RBCs lack Rh surface antigens, and the mother's plasma lacks anti-Rh antibodies.

Module 17.9 Review

a. The five types of white blood cells are neutrophils, eosinophils, basophils, monocytes, and lymphocytes. **b.** An infected cut would contain a large number of neutrophils, phagocytic white blood cells that are generally the first to arrive at the site of an injury. **c.** Basophils enter damaged tissues and release a variety of chemicals, including histamine, which promotes inflammation.

Module 17.10 Review

a. Hemostasis is the stoppage of blood flow. It involves three phases: the vascular phase, the platelet phase, and the coagulation phase. **b.** During the vascular phase, local blood vessel constriction (vascular spasm) occurs at the injury site. In the platelet phase, platelets are activated, aggregate at the site, and adhere to damaged blood vessel surfaces. In the coagulation phase, factors released by platelets and endothelial cells interact with clotting factors (through either the extrinsic pathway, the intrinsic pathway, or the common pathway) to form a blood clot, a process involving the conversion of soluble fibrinogen to insoluble fibers of fibrin. **c.** The correct sequence for the events in hemostasis and clot dissolution is (1) vascular spasm, (2) platelet phase, (3) coagulation, (4) retraction, and (5) fibrinolysis.

Module 17.11 Review

a. Venipuncture is the piercing of a vein to obtain a blood sample. **b.** Pernicious anemia is insufficient red blood cell production that results from a lack of vitamin B_{12}; the blood cells that do develop tend to be macrocytic (abnormally large) and abnormally shaped. Iron deficiency anemia results when the dietary intake or absorption of iron is insufficient, impairing normal hemoglobin synthesis; these blood cells are microcytic (abnormally small). **c.** The two types of leukemia are myeloid leukemia and lymphoid leukemia.

Section Reviews

Section 1 Review

1. f; **2.** l; **3.** j; **4.** o; **5.** b; **6.** g; **7.** a; **8.** i; **9.** k; **10.** m; **11.** n; **12.** c; **13.** e; **14.** h; **15.** d

Section 2 Review

1. plasma; **2.** water; **3.** solutes; **4.** proteins; **5.** electrolytes, glucose, urea; **6.** albumins; **7.** globulins; **8.** fibrinogen; **9.** formed elements; **10.** erythrocytes; **11.** leukocytes; **12.** platelets; **13.** neutrophils; **14.** basophils; **15.** monocytes; **16.** eosinophils; **17.** lymphocytes; **18.** e; **19.** f; **20.** a; **21.** k; **22.** b; **23.** j; **24.** l; **25.** c; **26.** d; **27.** g; **28.** i; **29.** h; **30.** During differentiation, the red blood cells of humans (and other mammals) lose most of their organelles, including nuclei and ribosomes. As a result, mature circulating RBCs can neither divide nor synthesize the structural proteins and enzymes required for cellular repairs.

Chapter Review Questions

1. monocyte; **2.** neutrophil; **3.** erythrocyte, or red blood cell; **4.** lymphocyte; **5.** basophil; **6.** eosinophil; **7.** false; **8.** false; **9.** true; **10.** false; **11.** true; **12.** a; **13.** c; **14.** a; **15.** d; **16.** b; **17.** b; **18.** b; **19.** The two types of leukemias are characterized by elevated levels of circulating WBCs. Myeloid leukemia is characterized by the presence of abnormal granulocytes (neutrophils, eosinophils, and basophils) or other cells of the bone marrow. Lymphoid leukemia involves lymphocytes and their stem cells. **20.** Blood stabilizes and maintains body temperature by absorbing and redistributing the heat produced by active skeletal muscles. If body temperature is high, the heat will be lost across the surface of the skin. If body temperature is low, the warm blood is directed to the brain and to other temperature-sensitive organs. **21.** Type A blood has surface antigen A and anti-B plasma antibody; type B blood has surface antigen B and anti-A plasma antibody; type AB blood has both A and B surface antigens and neither anti-A nor anti-B plasma antibodies; and type O blood is lacking A and B surface antigens, and has both anti-A and anti-B plasma antibodies.

Chapter Integration

1. The drug that Piero is considering performs like the hormone erythropoietin (EPO). **2.** EPO is carried to red bone marrow, where it stimulates stem cells and developing RBCs. With more RBCs in circulation, Piero will be carrying more oxygen in his blood. This will allow his cells to deliver more oxygen to his working muscles, and perform at a higher level. **3.** Serious side effects from taking EPO include blood clots, heart attack, and stroke from the increased blood viscosity due to the increased levels of erythrocytes in blood. **4.** Piero could consider moving from Miami to the mountains so he could train at a higher altitude. Strenuous exercise at a higher altitude would result in Piero being more hypoxic than at his training at sea level. This could result in his body releasing more EPO, and stimulating erythropoiesis in a natural way, and within healthy limits. **5.** Normal red blood cells are shaped like biconcave discs. Hemolysis is the rupture or destruction of red blood cells. In most cases, hemolysis occurs outside the bloodstream in the spleen, liver, and bone marrow. **6.** Some functions of blood include transporting heat throughout the body and oxygen to cells. Because Ursula's oxygen-carrying red blood cells were being actively destroyed, her heart had to pump faster to supply tissues with the oxygen they need. The resulting increase in circulation brought more warm blood to her body surface, where greater-than-normal heat loss lowered her body temperature. **7.** Anemia and the accompanying low hemoglobin levels reduce the blood's ability to carry oxygen and nutrients to the body's vital organs and tissues. As a result, Ursula experienced fatigue and weakness.

CHAPTER 18

Module Reviews

Module 18.1 Review

a. The pulmonary circuit carries blood through the arteries, capillaries, and veins of the lungs from the right ventricle to the left atrium. **b.** The systemic circuit transports blood through the arteries, capillaries, and veins of the body from the left ventricle to the right atrium. Blood returning to the heart from the systemic circuit must complete the pulmonary circuit before it re-enters the systemic circuit. **c.** The right atrium receives blood from the systemic circuit.

Module 18.2 Review

a. The five general classes of blood vessels are arteries, arterioles, capillaries, venules, and veins. **b.** A capillary is a small blood vessel, located between an arteriole and a venule, whose thin wall permits the diffusion of gases, nutrients, and wastes between blood and interstitial fluids. **c.** These blood vessels are veins. Arteries and arterioles have a large amount of smooth muscle tissue in a thick, well-developed tunica media.

Module 18.3 Review

a. The two types of capillaries are continuous capillaries and fenestrated capillaries. **b.** Fenestrated capillaries are located where solutes as large as small peptides move freely into and out of the blood, including endocrine glands, the choroid plexus of the brain, absorptive areas of the intestine, and filtration areas of the kidneys. **c.** Capillary walls are thin, so distances for diffusion are short. Continuous capillaries have small gaps between adjacent endothelial cells that permit the diffusion of water and small solutes into the surrounding interstitial fluid but prevent the loss of blood cells and plasma proteins. Fenestrated capillaries contain pores that permit very rapid exchange of fluids and solutes between interstitial fluid and blood. The walls of arteries and veins are several cell layers thick and are not specialized for diffusion.

Module 18.4 Review

a. In the arterial system, pressures are high enough to keep the blood

ANSWERS

moving away from the heart. In the venous system, blood pressure is too low to keep the blood moving back toward the heart. Valves in veins prevent blood from flowing backward whenever the venous pressure drops. **b.** Assisted by the presence of valves in the veins, which prevent backflow of the blood, the contraction of the surrounding skeletal muscles squeezes venous blood toward the heart. **c.** Varicose veins are sagging, swollen veins distorted by the pooling of blood resulting from gravity and the failure of venous valves.

Module 18.5 Review

a. Vasculogenesis refers to the formation of the first blood vessels. Angiogenesis is the growth of new blood vessels from pre-existing vessels. **b.** Blood islands are aggregated groups of cells scattered in the yolk sac. They are formed from early precursor cells called angioblasts. **c.** Angioblasts remodel blood islands first into capillary networks, and then into larger arterial and venous networks.

Module 18.6 Review

a. The two circulatory circuits of the cardiovascular system are the pulmonary circuit and the systemic circuit. The pulmonary circuit carries deoxygenated blood from the right ventricle to the lungs and returns oxygenated blood to the left atrium. The systemic circuit carries oxygenated blood to the organs and tissues of the body and returns deoxygenated blood to the right atrium. **b.** The three general patterns of blood vessel organization are the following: (1) The peripheral distributions of arteries and veins on the body's left and right sides are generally identical, except near the heart, where the largest vessels connect to the atria or ventricles; (2) a single vessel may have several names as it crosses specific anatomical boundaries, making accurate anatomical descriptions possible; and (3) tissues and organs are usually serviced by several arteries and veins. **c.** right ventricle → right and left pulmonary arteries → pulmonary arterioles → alveolar capillaries → pulmonary venules → pulmonary veins → left atrium.

Module 18.7 Review

a. The largest artery in the body is the aorta. **b.** The two large veins that collect blood from the systemic circuit are the superior vena cava and the inferior vena cava. **c.** A major anatomical difference between the arterial and venous systems is the existence of dual venous drainage in the neck and limbs.

Module 18.8 Review

a. The two arteries formed by the division of the brachiocephalic trunk are the right common carotid artery and the right subclavian artery. **b.** A blockage of the left subclavian artery would interfere with blood flow to the left arm. **c.** Thor's bulging vein is his external jugular vein.

Module 18.9 Review

a. The arterial structure in the neck region that contains baroreceptors is the carotid sinus. **b.** The branches of the external carotid artery are the superficial temporal, maxillary, occipital, facial, and lingual arteries. **c.** The veins that combine to form the brachiocephalic vein are the external jugular, internal jugular, vertebral, and subclavian veins.

Module 18.10 Review

a. The three branches of the internal carotid artery are the ophthalmic, anterior cerebral, and middle cerebral arteries. **b.** The cerebral arterial circle (also known as the circle of Willis) is a ring-shaped anastomosis that encircles the infundibulum of the pituitary gland. Its anatomical arrangement creates alternate pathways in the cerebral circulation, so that if blood flow is interrupted in one area, other blood vessels can continue to perfuse the entire brain with blood. **c.** The internal jugular veins drain the dural sinuses of the brain.

Module 18.11 Review

a. Rupturing the celiac trunk would most directly affect the stomach, inferior portion of the esophagus, spleen, liver, gallbladder, and proximal portion of the small intestine. **b.** The inferior vena cava collects most of the venous blood inferior to the diaphragm. **c.** The major branches of the inferior vena cava are the lumbar, gonadal, hepatic, renal, adrenal, and phrenic veins.

Module 18.12 Review

a. The unpaired branches of the abdominal aorta that supply blood to the visceral organs are the celiac trunk, superior mesenteric artery, and inferior mesenteric artery. **b.** The three veins that merge to form the hepatic portal vein are the superior mesenteric, inferior mesenteric, and splenic veins. **c.** The left and right gastroepiploic veins carry blood away from the stomach.

Module 18.13 Review

a. The first two branches of the common iliac artery are the internal iliac artery and the external iliac artery. **b.** A blockage of the popliteal vein would interfere with blood flow in the tibial and fibular (peroneal) veins (which form the popliteal vein) and the small saphenous vein (which joins the popliteal vein). **c.** The plantar venous arch delivers blood to the anterior tibial, posterior tibial, and fibular (peroneal) veins.

Module 18.14 Review

a. Deoxygenated blood flows from the fetus to the placenta through a pair of umbilical arteries, and oxygenated blood returns from the placenta in a single umbilical vein. The umbilical vein then drains into the ductus venosus within the fetal liver. **b.** The six necessary structures in the fetal circulation are two umbilical arteries, one umbilical vein, the ductus venosus, the foramen ovale, and the ductus arteriosus. After birth, the foramen ovale closes and persists as the fossa ovalis, a shallow depression; the ductus arteriosus persists as the ligamentum arteriosum, a fibrous cord; and the umbilical vessels and ductus venosus persist throughout life as fibrous cords. **c.** Ventricular septal defects are abnormal openings between the left and right ventricles. Tetralogy of Fallot includes a ventricular septal defect plus three other heart defects: a narrowing of the pulmonary trunk, a displaced aorta, and an enlarged right ventricle with corresponding thickened right and left ventricles.

Section Reviews

Section 1 Review

1. artery; **2.** vein; **3.** smooth muscle; **4.** internal elastic membrane; **5.** endothelium; **6.** tunica intima; **7.** tunica media; **8.** tunica externa, or tunica adventitia; **9.** e; **10.** i; **11.** j; **12.** g; **13.** a; **14.** k; **15.** b; **16.** h; **17.** l; **18.** c; **19.** d; **20.** f; **21.** Contraction or relaxation of the smooth muscle cells in precapillary sphincters control the blood flow into capillaries. During exercise, certain precapillary sphincters will relax to allow more blood flow into the working muscles to meet their metabolic needs, and into the skin to help with cooling. Other precapillary sphincters will contract to decrease blood flow into nonessential organs such as the digestive viscera. In cold temperatures, precapillary sphincters in the periphery and skin will contract to decrease blood flow to these areas and conserve heat.

Section 2 Review

1. common carotid; **2.** subclavian; **3.** brachiocephalic trunk; **4.** brachial; **5.** radial; **6.** popliteal; **7.** fibular (peroneal); **8.** aortic arch; **9.** celiac trunk; **10.** renal; **11.** common iliac; **12.** external iliac; **13.** femoral; **14.** anterior tibial **15.** vertebral; **16.** internal jugular; **17.** brachiocephalic; **18.** axillary; **19.** cephalic; **20.** median antebrachial; **21.** ulnar; **22.** great saphenous; **23.** fibular (peroneal); **24.** superior vena cava;

25. inferior vena cava; **26.** internal iliac; **27.** deep femoral; **28.** posterior tibial

Chapter Review Questions

1. internal carotid artery; **2.** basilar artery; **3.** vertebral artery; **4.** anterior communicating artery; **5.** anterior cerebral artery; **6.** posterior communicating artery; **7.** posterior cerebral artery; **8.** cerebral arterial circle, or circle of Willis; **9.** d; **10.** b; **11.** d; **12.** c; **13.** a; **14.** d; **15.** a; **16.** b; **17.** A brachiocephalic artery is not needed on the left side of the aortic arch because the majority of the heart lies to the left of midline; this allows the left common carotid artery and left subclavian artery to have direct access to the head and left upper limb, respectively. **18.** Blood volume is distributed throughout the body as follows: 64 percent in the systemic venous system, 9 percent in the pulmonary circuit, 7 percent in the heart, 13 percent in the systemic arterial circuit, and 7 percent in systemic capillaries. **19.** The hepatic portal system directs blood with absorbed nutrients from the digestive system to the liver for processing.

Chapter Integration

1. The catheter would be inserted into Mr. Samuel's femoral artery. It would be guided then to his external iliac artery, common iliac artery, abdominal aorta, and thoracic aorta to the aortic arch. **2.** The dye would be released into the aortic arch at the base of the brachiocephalic trunk and left common carotid artery. This is the best place because the dye would then travel to the brain by the two common carotid arteries, and the two vertebral arteries. **3.** Mr. Samuel was lucky to have had his blockage on one of his vertebral arteries because this vessel joins with the vertebral artery from the opposite side to form the basilar artery. Because the cerebral arterial circle (circle of Willis) that is distal to the basilar artery ensured even distribution of blood, Mr. Samuel did not experience a complete interruption of blood flow to his brain. A blockage in the basilar artery would be more serious because it would block the blood from both vertebral arteries, and a blockage in the middle cerebral artery would be serious because it is distal to the cerebral arterial circle.

CHAPTER 19

Module Reviews

Module 19.1 Review

a. The anterior view of the heart is dominated by the right atrium and right ventricle. **b.** The great veins and arteries of the heart are attached to the base of the heart. **c.** The apex is located on the inferior aspect of the heart.

Module 19.2 Review

a. From superficial to deep, the layers of the heart wall are the epicardium, myocardium, and endocardium. **b.** The epicardium consists of an outer mesothelium and an underlying layer of areolar tissue that attaches directly to the myocardium. **c.** Cardiac tissue is metabolically active and dependent on mitochondrial activity for ATP and local capillaries for obtaining oxygen and nutrients.

Module 19.3 Review

a. The mediastinum is the region between the two pleural cavities that contains the heart along with the great vessels (large arteries and veins attached to the heart), thymus, esophagus, and trachea. **b.** The heart is surrounded by the pericardial sac in the anterior mediastinum, deep to the sternum and superior to the diaphragm. **c.** Cardiac tamponade can be a life-threatening condition because the accumulating fluid within the pericardial cavity restricts heart movement.

Module 19.4 Review

a. The four cardiac chambers are the left atrium, right atrium, left ventricle, and right ventricle. **b.** The anterior interventricular sulcus marks the boundary between the left and right ventricles on the heart's anterior surface; the shallower posterior interventricular sulcus marks the boundary between the left and right ventricles on the posterior surface; and the coronary sulcus is a deep groove that marks the border between the atria and the ventricles. **c.** Coronary veins collect blood from the myocardium and carry it to the right atrium.

Module 19.5 Review

a. Arteries: left coronary artery, anterior interventricular artery, right coronary artery, marginal arteries, circumflex artery, and posterior interventricular artery. Veins: great cardiac vein, anterior cardiac veins, posterior cardiac vein, middle cardiac vein, and small cardiac vein. **b.** The anterior cardiac veins drain the anterior surface of the right ventricle and empty into the right atrium. The posterior cardiac vein drains the area (posterior surface of the left ventricle) supplied by the circumflex artery. **c.** During elastic rebound, some blood in the aorta is driven forward into the systemic circuit, and some is forced back toward the left ventricle and into the coronary arteries.

Module 19.6 Review

a. Damage to the semilunar valve on the right side of the heart would affect blood flow to the pulmonary trunk. **b.** Contraction of the papillary muscles pulls on the chordae tendineae, which prevent the AV valves from swinging into the atria. **c.** The more muscular left ventricle must generate enough force to propel blood throughout the body (except the lungs), whereas the right ventricle must generate only enough force to propel blood the short distance to the lungs.

Module 19.7 Review

a. Cardiac regurgitation is the abnormal backflow of blood into the atria when the ventricles contract, and it is prevented by the chordae tendineae and papillary muscles. **b.** The tricuspid valve is composed of three relatively large flaps (cusps); the pulmonary valve is made up of three smaller half-moon–shaped cusps. **c.** Semilunar valves prevent the backflow of blood into the ventricles.

Module 19.8 Review

a. Arteriosclerosis is any thickening and toughening of arterial walls; atherosclerosis is a type of arteriosclerosis characterized by changes in the endothelial lining and the formation of fatty deposits (plaque) in the tunica media. **b.** Coronary ischemia is a condition in which the blood supply of the coronary arteries is reduced. **c.** Stents are wire-mesh tubes that prop open the natural blood vessel, creating a conduit to restore blood flow. Without adequate blood flow to the cardiac muscle, the tissue would die.

Module 19.9 Review

a. The cardiac cycle is the period between the start of one heartbeat and the beginning of the next. **b.** The alternate term for heart contraction is systole, and the term for heart relaxation is diastole. **c.** When a chamber is relaxed, it is in the diastole phase of the cardiac cycle.

Module 19.10 Review

a. The two phases of ventricular systole are isovolumetric contraction and ventricular ejection. **b.** Atrial systole, atrial diastole, ventricular systole, and ventricular diastole. **c.** No. When pressure in the left ventricle first rises, the heart is contracting but blood is not leaving the heart. During this initial phase of contraction, called the period of isovolumetric contraction, both the AV valves and the semilunar valves are closed. The increase in pressure is the result of the cardiac muscle contracting. When the pressure in the ventricle exceeds that in the aorta, the aortic semilunar valves are forced open, and blood is rapidly ejected from the ventricle.

Module 19.11 Review

a. Automaticity is the ability of cardiac muscle tissue to contract without neural or hormonal stimulation. **b.** If the cells of the SA node failed to function, the heart would continue to beat, but at a slower rate; the AV node would act as the pacemaker. **c.** If the impulses from the atria were not delayed at the AV node, they would be conducted through the ventricles so quickly by the bundle branches and Purkinje cells that the ventricles would begin contracting before the atria had finished contracting. As a result, the ventricles would not be as full of blood as they could be, and the pumping action of the heart would be less efficient.

Module 19.12 Review

a. Cardiac muscle has a long refractory period that continues until relaxation is well under way. As a result, another action potential cannot arrive quickly enough for summation to occur, and thus tetany cannot occur. **b.** The three stages of an action potential in a cardiac muscle cell are rapid depolarization, plateau, and repolarization. **c.** Slow calcium channels are voltage-gated calcium channels that open slowly and remain open for a relatively long period—about 175 msec. When they are open, the entry of calcium ions into the cell roughly balances the loss of sodium ions through the active transport of sodium ions. As a result, the membrane potential remains near 0 mV for an extended period.

Module 19.13 Review

a. Like NE, caffeine acts directly on the conducting system and contractile cells of the heart, increasing the rate at which they depolarize. Drinking large amounts of caffeinated beverages would therefore increase the heart rate. **b.** Bradycardia is a heart rate below 60 beats per minute; tachycardia is a heart rate above 100 beats per minute. **c.** The cardioacceleratory center in the medulla oblongata activates sympathetic neurons to increase heart rate; the cardioinhibitory center (also in the medulla oblongata) controls the parasympathetic neurons that slow heart rate.

Module 19.14 Review

a. The end-diastolic volume (EDV) is the amount of blood a ventricle contains at the end of diastole, just before a contraction begins; the end-systolic volume (ESV) is the amount of blood that remains in the ventricle at the end of ventricular systole. **b.** An increase in venous return would stretch the heart muscle. The more the heart muscle is stretched, the more forcefully it will contract (to a point). The more forceful the contraction, the more blood the heart will eject with each beat (stroke volume). Therefore, increased venous return would increase the stroke volume (if all other factors are constant). **c.** An increase in sympathetic stimulation of the heart would increase heart rate and force of contraction. The end-systolic volume (ESV) is the amount of blood that remains in a ventricle after a contraction (systole). The more forcefully the heart contracts, the more blood it ejects. Therefore, increased sympathetic stimulation results in a lower ESV.

Module 19.15 Review

a. Heart failure is a condition in which the heart can no longer meet the oxygen and nutrient demands of peripheral tissues. **b.** SV = EDV − ESV, so SV = 125 mL − 40 mL = 85 mL. **c.** The amount of blood that the heart pumps is proportional to the amount of blood that enters it. A heart that is beating too rapidly does not have adequate filling time, and it pumps less blood; peripheral tissues can be damaged by inadequate blood flow.

Module 19.16 Review

a. An electrocardiogram (ECG or EKG) is a recording of the electrical activities of the heart over time. **b.** The important features of an ECG are the P wave (atrial depolarization), the QRS complex (ventricular depolarization), and the T wave (ventricular repolarization). **c.** Ventricular fibrillation, which causes the condition known as cardiac arrest, is fatal because the ventricles merely quiver and do not pump blood into the systemic circulation.

Module 19.17 Review

a. Neural and hormonal regulation influence heart rate, stroke volume, peripheral resistance, and venous pressure. **b.** Arterial pressure is much higher than venous pressure because it must push blood a greater distance and through progressively smaller vessels. **c.** It is beneficial for capillary pressure to be very low so blood can flow slowly and allow time for diffusion between the blood and the surrounding interstitial fluid.

Module 19.18 Review

a. Total peripheral resistance reflects a combination of vascular resistance, vessel length, vessel diameter, blood viscosity, and turbulence. **b.** An increase in vessel diameter would reduce peripheral resistance. (An increase in vessel length would increase peripheral resistance.) **c.** The formula $R = 1/r^4$ states that resistance (R) is inversely proportional to the fourth power of the vessel radius (r). This means that a small change in vessel diameter results in a large change in resistance.

Module 19.19 Review

a. Blood flow is the volume of blood flowing per unit of time through a vessel or group of vessels; it is directly proportional to blood pressure and inversely proportional to peripheral resistance. **b.** In a healthy person, blood pressure is greater in the aorta than in the inferior vena cava. If the pressure were higher in the inferior vena cava than in the aorta, blood would flow in the reverse direction. **c.** Using the formula MAP = diastolic pressure + (pulse pressure)/3, MAP equals 70 + (125 − 70)/3, which equals 70 + 18.3, or 88.3 mm Hg.

Module 19.20 Review

a. Any condition that affects either blood pressure or osmotic pressures in the blood or interstitial fluid will shift the balance between hydrostatic and osmotic forces. **b.** Fluid moves into a capillary whenever blood colloid osmotic pressure (BCOP) is greater than capillary hydrostatic pressure (CHP). **c.** Edema is an abnormal accumulation of interstitial fluid in peripheral tissues.

Module 19.21 Review

a. Tissue perfusion is blood flow to tissues that is sufficient to deliver adequate oxygen and nutrients. **b.** Cardiovascular autoregulation involves local factors changing the pattern of blood flow within capillary beds in response to chemical changes in interstitial fluids. **c.** Baroreceptor reflexes respond to changes in blood pressure. The baroreceptors—located in the walls of the carotid sinuses, aortic sinuses, and right atrium—monitor the degree of stretch at those sites.

Module 19.22 Review

a. Epinephrine and norepinephrine from the adrenal medullae provide short-term regulation of decreasing blood pressure and blood volume. **b.** Vasoconstriction of the renal artery would decrease both blood flow and blood pressure at the kidney. In response, the kidney would increase the amount of renin it releases, which in turn would increase the level of angiotensin II. The angiotensin II would bring about increased blood pressure and increased blood volume. **c.** Excessive stretching of the right atrium during diastole causes the release of atrial natriuretic peptide (ANP). Excessive stretching of the ventricles during diastole causes the release of brain natriuretic peptide (BNP). The roles of these peptides are to trigger responses whose combined effects act to decrease blood volume and blood pressure. As blood volume and blood pressure decrease, natriuretic peptide production stops.

Module 19.23 Review
a. Chemoreceptor reflexes respond to decreasing pH and oxygen levels, and increasing CO_2 levels in the blood and cerebrospinal fluid (CSF). These reflexes stimulate responses by the cardiovascular centers to increase vasoconstriction, cardiac output and blood pressure that increase pH and oxygen levels and decrease CO_2 levels. **b.** Chemoreceptors are located in the carotid bodies, in the aortic bodies, and on the ventrolateral surfaces of the medulla oblongata. **c.** An increase in the respiratory rate reduces CO_2 levels.

Module 19.24 Review
a. The respiratory pump is a mechanism by which a reduction of pressure in the thoracic cavity during inhalation assists venous return to the heart. **b.** During exercise, cardiac output increases, and blood flow to skeletal muscles increases at the expense of blood flow to less essential organs. **c.** Unless compensatory vasoconstriction occurs in "nonessential" organs, such as those of the digestive system, vasodilation in skeletal muscles would cause a potentially dangerous decrease in blood pressure and blood flow throughout the body during exercise.

Module 19.25 Review
a. Compensatory mechanisms that respond to blood loss include an increase in cardiac output, a mobilization of venous reserves, peripheral vasoconstriction, and the release of hormones that promote the retention of fluids and the maturation of erythrocytes. **b.** The immediate, short-term, problem during hemorrhaging is maintaining adequate blood pressure and peripheral blood flow; the long-term problem is restoring normal blood volume. **c.** Circulatory shock occurs when blood loss exceeds about 35 percent of the total blood volume. Circulatory shock involves a series of positive feedback loops that are initiated after homeostasis has been disrupted. Progressive shock is the next stage after circulatory shock. It, too, is a series of positive feedback loops that accelerate tissue damage. Irreversible shock is the fatal stage that occurs if the positive feedback loops initiated during progressive shock are not broken.

Section Reviews

Section 1 Review
1. aortic arch; **2.** superior vena cava; **3.** right pulmonary arteries; **4.** ascending aorta; **5.** fossa ovalis; **6.** opening of coronary sinus; **7.** right atrium; **8.** pectinate muscles; **9.** tricuspid valve cusp; **10.** chordae tendineae; **11.** papillary muscle; **12.** right ventricle; **13.** inferior vena cava; **14.** pulmonary trunk; **15.** pulmonary valve; **16.** left pulmonary arteries; **17.** left pulmonary veins; **18.** left atrium; **19.** aortic valve; **20.** bicuspid valve cusp; **21.** left ventricle; **22.** interventricular septum; **23.** trabeculae carneae; **24.** moderator band; **25.** g; **26.** h; **27.** j; **28.** a; **29.** d; **30.** i; **31.** b; **32.** c; **33.** f; **34.** e; **35.** (a) deoxygenated blood flow: right atrium → right atrioventricular valve (tricuspid valve) → right ventricle → pulmonary semilunar valve. (b) oxygenated blood flow: left atrium → left atrioventricular valve (bicuspid valve) → left ventricle → aortic semilunar valve.

Section 2 Review
1. d; **2.** c; **3.** a; **4.** g; **5.** b; **6.** j; **7.** i; **8.** f; **9.** e; **10.** h; **11.** left AV valve closes; **12.** increasing; **13.** decreasing; **14.** less than; **15.** aortic valve is forced open; **16.** aorta; **17.** ventricular systole

Section 3 Review
1. f; **2.** i; **3.** b; **4.** a; **5.** c; **6.** e; **7.** l; **8.** g; **9.** k; **10.** h; **11.** d; **12.** j; **13.** d; **14.** a; **15.** j; **16.** h; **17.** b; **18.** e; **19.** i; **20.** c; **21.** f; **22.** g

Chapter Review Questions

1. true; **2.** false; **3.** false; **4.** true; **5.** true; **6.** c; **7.** a; **8.** b; **9.** c; **10.** d; **11.** b; **12.** a; **13.** b; **14.** a; **15.** d; **16.** The regurgitation of blood into the atria as the ventricles contract is prevented as the papillary muscles contract and pull on the chordae tendineae. **17.** SA node → internodal pathways → AV node → AV bundle → right and left bundle branches → Purkinje fibers; **18.** The epicardium, or visceral pericardium, is a mesothelial serous membrane that covers the outer surface of the heart. The myocardium is the middle layer of the heart and contains cardiac muscle tissue, blood vessels, and nerves. The endocardium is made of simple squamous epithelium (endothelium) and lines the inner surface of the heart. **19.** Sympathetic stimulation increases heart rate and force of contraction (contractility). Parasympathetic stimulation decreases heart rate and force of contraction. **20.** The first heart sound is "lubb" (S_1) and marks the start of ventricular contraction. The AV valves closing produce this sound. The second sound is "dupp" (S_2) and marks the start of ventricular diastole. This sound is produced by the semilunar valves closing. The third heart sound (S_3) is associated with blood flow into the atria. The fourth heart sound (S_4) is associated with atrial contraction.

Chapter Integration

1. The mitral valve is closed during ventricular systole. **2.** The mitral valve is prolapsing into the left atrium. There is no regurgitation because, despite the prolapse, the two cusps of the mitral valve are still in contact and preventing the backflow of blood. **3.** The echocardiogram could reveal abnormally shaped mitral valve cusps, or chordae tendineae that are elongated and unable to restrain the mitral valve cusps during ventricular systole.

CHAPTER 20

Module Reviews

Module 20.1 Review
a. The functions of the lymphatic system are to provide immunity (the ability to defend the body against infection, illness, and disease) and to return tissue fluid to the bloodstream. **b.** Lymphocytes are the primary cells of the lymphatic system. These cells respond to pathogens, abnormal body cells, and foreign proteins such as toxins released by some bacteria. Lymph is the interstitial fluid that has entered a lymphatic vessel and surrounds lymphocytes. **c.** Primary lymphoid tissues and organs are sites where lymphocytes are formed and mature. Secondary lymphoid tissues and organs are sites where lymphocytes are activated and cloned.

Module 20.2 Review
a. Lymphatic vessels transport lymph from peripheral tissues to the venous system. **b.** Overlapping endothelial cells in lymphatic capillaries act as one-way valves that permit the entry of fluids and solutes but prevent their return to the intercellular spaces. **c.** Lymphatic valves prevent the backflow of lymph in some lymphatic vessels.

Module 20.3 Review
a. The lymphatic trunks empty into the thoracic duct and the right lymphatic duct. **b.** The right lymphatic duct collects lymph from the right side of the body superior to the diaphragm; the thoracic duct collects lymph from the body inferior to the diaphragm and from the left side of the body superior to the diaphragm. **c.** Lymphedema is the accumulation of interstitial fluids that results from blocked lymphatic drainage. If the condition persists, connective tissues lose their elasticity, and the swelling becomes permanent.

Module 20.4 Review
a. The three main classes of lymphocytes are T cells, B cells, and natural killer (NK) cells. **b.** B cells are responsible for antibody-mediated immunity. **c.** The red bone marrow, thymus, and peripheral lymphoid tissues are involved in lymphopoiesis.

Module 20.5 Review
a. Mucosa-associated lymphoid tissue (MALT) is the collection of lymphoid tissue that protects epithelia lining the digestive, respiratory, urinary, and reproductive tracts. **b.** Tonsils are large lymphoid nodules in the walls of the pharynx. The five tonsils are the left and right palatine tonsils, a single pharyngeal tonsil (adenoid), and a pair of lingual tonsils. **c.** Lymph flow through a lymph node: afferent lymphatics → subcapsular space → outer cortex → deep cortex → medullary sinus → efferent lymphatics.

Module 20.6 Review
a. The thymus is located in the anterior mediastinum, posterior to the sternum. **b.** The thymus is a pink, grainy organ ranging in weight from 40 g at puberty to less than 12 g by age 50. A capsule covers the thymus and divides it into two lobes, and fibrous partitions called septa divide the lobes into lobules. **c.** Thymic epithelial cells in the cortex maintain the blood–thymus barrier.

Module 20.7 Review
a. The spleen filters the blood; the phagocytes it contains identify and engulf pathogens or infected cells circulating in the blood.
b. Red pulp contains large numbers of red blood cells; white pulp resembles lymphoid nodules and contains lymphocytes.
c. Beginning with the trabecular arteries, blood then flows to the central arteries → capillaries → reticular tissue of red pulp → sinusoids → trabecular veins.

Module 20.8 Review
a. Innate immunity is nonspecific and does not distinguish one type of threat from another. Adaptive immunity is specific and protects against particular threats.
b. Innate immunity protects us using physical barriers such as skin, phagocytes that engulf pathogens, immune surveillance of abnormal cells, interferons to defend from viruses, complement to assist antibodies, inflammation to limit the spread of infection, and fever.
c. Innate (nonspecific) immunity will be activated immediately in the child with a skinned knee.

Module 20.9 Review
a. The integumentary system provides a physical barrier that is the first line of defense in preventing pathogens and toxins from entering body tissues. Skin secretions flush the surface, hair protects against physical abrasion, and the multiple layers of the skin's epithelium create an interlocking barrier. **b.** The body's phagocytes are neutrophils, eosinophils, and macrophages. Fixed macrophages are scattered among connective tissues and do not move; free macrophages are mobile and reach injury sites by migrating through adjacent tissues or traveling in the bloodstream. **c.** Chemotaxis is phagocyte movement in response to attraction to or repulsion from chemical stimuli.

Module 20.10 Review
a. Immune surveillance is the constant monitoring of normal tissues by NK cells sensitive to abnormal antigens on the surfaces of cells. **b.** NK cells recognize unusual proteins, called tumor-specific antigens, on the plasma membranes of cancer cells. When these antigens are detected, the NK cells then destroy the abnormal cells. **c.** Cancer cells can mutate such that either they do not display tumor-specific antigens, or they secrete chemicals that can destroy NK cells. This ability to escape detection is called immunological escape.

Module 20.11 Review
a. Interferons are small proteins that are released by activated lymphocytes, macrophages, and cells infected with viruses; they trigger the production of antiviral proteins that interfere with viral replication within tissue cells.
b. The complement proteins of the complement system interact with each other in chain reactions that ultimately produce activated forms that target bacterial cell walls and plasma membranes, stimulate inflammation, attract phagocytes, or enhance phagocytosis.
c. Histamine release by mast cells and basophils in tissues increases local inflammation, thereby accelerating blood flow to the region.

Module 20.12 Review
a. Inflammation, also called the inflammatory response, is a localized response to injury in the body. It is characterized by redness (*rubor*), swelling (*tumor*), heat (*calor*), and pain (*dolor*). Lost function (*functio laesa*) is also commonly associated with inflammation. **b.** Pyrogens increase body temperature (produce a fever), which can mobilize defenses, accelerate repairs, and inhibit pathogens.
c. A rise in the level of interferons suggests a viral infection.

Module 20.13 Review
a. The two cells responsible for coordinating adaptive immunity are T cells and B cells. **b.** A child who receives the polio vaccine develops artificially induced active immunity. **c.** The properties of adaptive immunity are specificity, versatility, immunologic memory, and tolerance.

Module 20.14 Review
a. Antigen presentation occurs when an antigen-glycoprotein, or antigen-MHC protein, combination capable of activating T cells appears in a plasma membrane (typically that of a macrophage). T cells sensitive to this antigen are activated if they contact the antigen on the plasma membrane of the antigen-presenting cell.
b. The major histocompatibility complex (MHC) is a portion of chromosome 6 containing genes that control the synthesis of membrane glycoproteins. **c.** Class I MHC proteins are in the plasma membranes of all nucleated body cells. Class II MHC proteins are only in the plasma membranes of antigen-presenting cells (APCs) and lymphocytes.

Module 20.15 Review
a. T cell plasma membranes contain proteins called CD (cluster of differentiation) markers. Cells with CD8 markers respond to antigens presented by Class I MHC proteins and are on cytotoxic T cells, memory T_C cells, and suppressor T cells. Cells with CD4 markers respond to antigens presented by Class II MHC proteins. **b.** The three major types of T cells activated by Class I MHC proteins are cytotoxic T cells, memory T_C cells, and suppressor T cells. **c.** Abnormal antigens attached to Class I MHC proteins of an infected cell are displayed on the surface of the cell's plasma membrane. The recognition of such antigens by CD8 T cells initiates an immune response.

Module 20.16 Review
a. Sensitization is the process by which a B cell prepares to undergo activation after encountering a specific antigen. During sensitization, the specific antigens are brought into the cell where they become bound to Class II MHC proteins. Together, they then appear at the cell surface.
b. Cytokines secreted by activated T cells aid in coordinating specific and nonspecific defenses and regulate cell-mediated and antibody-mediated immunity.
c. Plasma cells produce and secrete antibodies, so observing an elevated number of plasma cells in the lymph would lead you to expect higher-than-normal antibody levels in the blood.

Module 20.17 Review
a. An antibody molecule consists of two parallel pairs of polypeptide chains: a pair of long, heavy chains and a pair of short, light chains. Each chain contains both constant segments and variable segments. The constant segments of the heavy chains form the base of the antibody molecule; the free tips of the two variable segments form the antigen binding sites.
b. An antigenic determinant site is the part of an antigen molecule to which an antibody molecule binds. **c.** The secondary response would be more affected by a lack of memory cells, which are produced in response to an initial exposure to an antigen during the primary response.

Module 20.18 Review
a. Antigen-antibody complexes help to destroy antigens through seven processes: neutralization, prevention of bacterial and viral adhesion, activation of complement, opsonization, attraction of phagocytes, stimulation of inflammation, and precipitation and agglutination. **b.** Opsonization is the process by which the coating of pathogens with antibodies and complement proteins makes the pathogens more susceptible to phagocytosis. **c.** Basophils and mast cells are involved in inflammation.

Module 20.19 Review
a. An allergy is an inappropriate or excessive immune response to an allergen, which is an antigen that triggers an allergic reaction. **b.** In anaphylaxis, an immune response to a circulating antigen stimulates mast cells throughout the body to release chemicals that prompt the inflammatory response. **c.** Histamines, leukotrienes, and other chemicals that cause pain and inflammation are released when mast cells and basophils are stimulated in an allergic reaction.

Module 20.20 Review
a. CD8 markers are found on cytotoxic T cells, memory T_C cells, and suppressor T cells; CD4 markers are on all helper T cells. **b.** Plasma cells produce antibodies. **c.** Cytotoxic T cells and NK cells can be activated by direct contact with virus-infected cells.

Module 20.21 Review
a. Autoimmune disorders are diseases that result from the production of antibodies (called autoantibodies) directed against normal substances in the body (self-antigens). **b.** Immunosuppression is the partial or complete reduction of the immune response in a person. It is also induced to enhance the survival of organ transplant recipients. **c.** The increased incidence of cancer in the elderly may result from a decline in immune surveillance, which results in reduced elimination of tumor cells as they arise.

Section Reviews

Section 1 Review
1. tonsil; **2.** cervical lymph nodes; **3.** right lymphatic duct; **4.** thymus; **5.** cisterna chyli; **6.** lumbar lymph nodes; **7.** appendix; **8.** lymphatics of lower limb; **9.** lymphatics of upper limb; **10.** axillary lymph nodes; **11.** thoracic duct; **12.** lymphatics of mammary gland; **13.** spleen; **14.** mucosa-associated lymphoid tissue (MALT); **15.** pelvic lymph nodes; **16.** inguinal lymph nodes; **17.** red bone marrow; **18.** c; **19.** h; **20.** l; **21.** i; **22.** m; **23.** j; **24.** b; **25.** e; **26.** g; **27.** d; **28.** f; **29.** a; **30.** k

Section 2 Review
1. physical barriers; **2.** phagocytes; **3.** immune surveillance; **4.** interferons; **5.** complement system; **6.** inflammation; **7.** fever; **8.** d; **9.** d; **10.** c; **11.** d; **12.** a; **13.** a; **14.** The high body temperatures of a fever may inhibit some viruses and bacteria, or increase their reproductive rate so that the disease runs its course more quickly. High body temperatures also accelerate the body's metabolic processes, which may help to mobilize tissue defenses and speed the repair process.

Section 3 Review
1. c; **2.** k; **3.** l; **4.** i; **5.** b; **6.** a; **7.** g; **8.** m; **9.** d; **10.** f; **11.** e; **12.** h; **13.** j; **14.** b; **15.** f; **16.** g; **17.** e; **18.** c; **19.** d; **20.** a; **21.** h; **22.** i

Chapter Review Questions

1. antigen binding site; **2.** variable segment of light chain; **3.** constant segments of light and heavy chains; **4.** heavy chain; **5.** light chain; **6.** disulfide bond; **7.** false; **8.** false; **9.** true; **10.** false; **11.** false; **12.** true; **13.** a; **14.** b; **15.** c; **16.** a; **17.** c; **18.** d; **19.** a; **20.** b; **21.** In graft rejection, T cells are activated by contact with MHC proteins on plasma membranes in the donated tissues. The cytotoxic T cells that develop then attack and destroy the foreign cells. **22.** A cytotoxic T cell (T_C cell) destroys its target cell by first encountering its target antigen bound to a Class I MHC protein. It then attacks it by any one of the following mechanisms: (1) destroying the target cell's membrane through the release of perforins; (2) activating apoptosis in the target cell; (3) disrupting the target cell's metabolism through the release of lymphotoxin. **23.** Multiple injections of certain vaccines are timed to produce primary and then secondary responses of the immune system. In the primary response to the vaccine injection, B cells produce daughter cells that differentiate into plasma cells and memory B cells. The plasma cells produce antibodies, which represent the primary response, but the primary response does not maintain elevated antibody levels for long periods. Subsequent vaccine injections are necessary to trigger secondary responses, when memory B cells differentiate into plasma cells that produce antibody concentrations that remain high much longer.

Chapter Integration

1. The student cannot yet know whether she will come down with the chickenpox. Her elevated blood IgM levels indicate that she is in the early stages of a primary response to the chickenpox virus. If her immune response proves unable to control and then eliminate the virus, she will develop chickenpox. **2.** Because chickenpox is highly contagious, people who have not had chickenpox previously and have not been vaccinated are susceptible to this highly contagious disease. **3.** The causative agent for chickenpox and shingles is the varicella-zoster virus (VZV). **4.** IgM antibodies are the first class of immunoglobulins secreted after an antigen is encountered. IgG antibodies are the most common class of immunoglobulins. **5.** Antigenic specificity results from the arrangement of genes controlling the amino acid sequences of the variable segments of antibodies. **6.** Both responses involve the production of antibodies as stimulated by antigens. The primary response develops relatively slowly and levels of IgG rise more slowly than do levels of IgM. The secondary response develops much more rapidly, is more prolonged, and produces much more IgG. **7.** Any vaccine, including that for chickenpox, stimulates artificially induced active immunity.

CHAPTER 21

Module Reviews

Module 21.1 Review
a. Pulmonary ventilation is airflow to and from the lungs. **b.** The conducting portion of the respiratory tract begins at the nasal cavity and extends through the pharynx, larynx, trachea, bronchi, and larger bronchioles. The respiratory portion includes the smallest bronchioles and alveoli. **c.** Gas exchange between air and the lungs occurs in the alveoli.

Module 21.2 Review
a. The respiratory defense system is a series of filtration mechanisms that prevent airway contamination by debris and pathogens. **b.** The respiratory mucosa lines the conducting portion of the respiratory tract. **c.** Cystic fibrosis is a lethal, inherited disease that results from the production of dense mucus that restricts respiratory passages and accumulates in the lungs. Harmful bacterial infection of the lungs may also develop, leading to death.

Module 21.3 Review
a. The structures of the upper respiratory system are the nose, nasal cavity, paranasal sinuses, and pharynx. **b.** Pathway of air through the upper respiratory system: external nares → nasal vestibule (guarded by hairs that

screen out large particles) → nasal cavity → superior, middle, and inferior meatuses (air bounces off the conchal surfaces) → internal nares (the connections between the nasal cavity and nasopharynx) → nasopharynx → oropharynx → laryngopharynx. **c.** The rich vascularization of the nasal cavity by the expandable veins in the lamina propria radiates body heat, so inhaled air is warmed before it leaves the nasal cavity. The heat also evaporates moisture from the epithelium to humidify the incoming air.

Module 21.4 Review

a. The paired cartilages are the arytenoid cartilages, corniculate cartilages, and cuneiform cartilages. The unpaired laryngeal cartilages are the thyroid cartilage, cricoid cartilage, and epiglottis. **b.** The highly elastic vocal folds of the glottis are also called the vocal cords. **c.** Phonation is the production of sound, and one component of speech; articulation is the modification of sound by the tongue, teeth, and lips for clear speech.

Module 21.5 Review

a. The right primary bronchus is larger in diameter than the left primary bronchus, and it descends toward the lung at a steeper angle than the left primary bronchus. **b.** The C-shaped tracheal cartilages allow room for the esophagus to expand when food or liquids are swallowed. **c.** Pathway of airflow along the lower respiratory tract: trachea → primary bronchi → secondary bronchi → tertiary bronchi → terminal bronchioles → pulmonary lobule.

Module 21.6 Review

a. A bronchopulmonary segment is a specific region of a lung supplied by a tertiary bronchus. **b.** Within the thoracic cavity, the left and right lungs are surrounded by the left and right pleural cavities, respectively. The apex of each lung extends superiorly to the first rib, and the base of each lung rests on the superior surface of the diaphragm. **c.** The left lung is divided into a superior lobe and an inferior lobe by the oblique fissure; in the right lung, the horizontal fissure separates the superior lobe from the middle lobe, while the oblique fissure separates the superior and middle lobes from the inferior lobe.

Module 21.7 Review

a. Pulmonary lobules are the smallest subdivisions of the lungs; branches of the pulmonary arteries, pulmonary veins, and tertiary bronchi supply each lobule. **b.** Without surfactant, the alveoli would collapse due to the high surface tension in the thin layer of water that moistens the alveolar surfaces. **c.** The respiratory membrane is made up of the fused basement membranes of the alveolar epithelium and capillary endothelium. Because it is very thin and oxygen and carbon dioxide are lipid soluble, diffusion occurs rapidly across the membrane.

Module 21.8 Review

a. External respiration is all the processes involved in the exchange of oxygen and carbon dioxide between the blood, lungs, and the external environment. Internal respiration is oxygen absorption from the blood and carbon dioxide release by body tissues. **b.** The primary function of pulmonary ventilation is to maintain adequate alveolar ventilation, the movement of air into and out of the alveoli. **c.** Hypoxia is low tissue oxygen levels. Anoxia is when oxygen is cut off completely.

Module 21.9 Review

a. Boyle's law states that at a constant temperature, the pressure of a gas is inversely proportional to its volume. **b.** The movements of the diaphragm and rib cage affect the volume of the lungs. **c.** The intrapulmonary pressure (the pressure inside the respiratory tract) and the atmospheric pressure (the pressure outside the respiratory tract) determine the direction of airflow. Air moves from the area with the higher pressure to the area with the lower pressure.

Module 21.10 Review

a. The primary inspiratory muscles are the diaphragm and the external intercostal muscles. **b.** The accessory respiratory muscles become active whenever the primary respiratory muscles are unable to move enough air to meet the oxygen demands of tissues. **c.** The measurable pulmonary volumes are the tidal volume, expiratory reserve volume (ERV), residual volume, and inspiratory reserve volume (IRV).

Module 21.11 Review

a. The respiratory rate is the number of breaths taken each minute. **b.** Respiratory minute volume is the amount of air moved into and out of the respiratory tract each minute, whereas alveolar ventilation is the amount of air reaching the alveoli each minute. Because some of the air never reaches the alveoli but instead remains in the anatomic dead space, alveolar ventilation is lower than respiratory minute volume. **c.** Slow, deep breaths ventilate alveoli more effectively, because a smaller amount of the tidal volume of each breath is spent moving air into and out of the anatomic dead space of the lungs.

Module 21.12 Review

a. Dalton's law states that in a mixture of gases, the individual gases exert a pressure proportional to their abundance in the mixture. **b.** Henry's law states that, at a given temperature, the amount of a particular gas that dissolves in a liquid is directly proportional to the partial pressure of that gas. Henry's law underlies the diffusion of gases between capillaries and alveoli, and between capillaries and interstitial fluid. **c.** The P_{O_2} decreases from about 100 mm Hg to 95 mm Hg in the pulmonary veins is due to mixing with venous blood from the conducting passageways. The blood arriving at the peripheral capillaries has a P_{O_2} of 95 mm Hg.

Module 21.13 Review

a. Oxyhemoglobin is hemoglobin to which oxygen molecules have bound. **b.** Both increased temperature and decreased pH (from heat and acidic wastes generated by active skeletal muscles) cause hemoglobin to release more oxygen during exercise than when the muscles are at rest. **c.** BPG (2,3-bisphosphoglycerate) is a compound that decreases hemoglobin's affinity for oxygen. For any partial pressure of oxygen, if the concentration of BPG increases, the amount of oxygen released by hemoglobin will increase.

Module 21.14 Review

a. Carbon dioxide is transported in the bloodstream as bicarbonate ions, bound to hemoglobin, or dissolved in the plasma. **b.** Driven by differences in partial pressure, oxygen enters the blood at the lungs and leaves it in peripheral tissues; similar forces drive carbon dioxide into the blood at the tissues and into the alveoli at the lungs. **c.** Blockage of the trachea would interfere with the body's ability to take in oxygen and eliminate carbon dioxide. Because most carbon dioxide is transported in blood as bicarbonate ions formed from the dissociation of carbonic acid, an inability to eliminate carbon dioxide would result in a buildup of excess hydrogen ions, which would decrease blood pH.

Module 21.15 Review

a. Compliance is the ease with which the lungs expand and recoil. Resistance is an indication of how much force is required to inflate or deflate the lungs. **b.** Three COPDs are asthma, chronic bronchitis, and emphysema. **c.** Chronic bronchitis is long-term inflammation of the mucous membranes in the bronchial tubes; emphysema is a condition in which the alveolar surfaces of the lungs are destroyed and alveoli merge, which reduces respiratory surface area and oxygen absorption, causing breathlessness. People who have chronic bronchitis are sometimes called "blue bloaters"; those with emphysema are called "pink puffers."

Module 21.16 Review
a. The pairs of central nervous system nuclei that adjust the pace of respiration are the apneustic centers and the pneumotaxic centers in the pons. **b.** The respiratory rhythmicity centers in the medulla oblongata generate the respiratory pace. **c.** The pH, P_{O_2}, and P_{CO_2} in blood and cerebrospinal fluid stimulate the respiratory centers.

Module 21.17 Review
a. Hypercapnia is an increase in the P_{CO_2} of arterial blood above the normal range. Hypocapnia is an abnormally low arterial P_{CO_2}. **b.** Chemoreceptors are more sensitive to carbon dioxide levels than they are to oxygen levels. **c.** Johnny's mother should not worry. When Johnny holds his breath, carbon dioxide levels in his blood increase, causing increased stimulation of the inspiratory centers, forcing him to breathe again.

Module 21.18 Review
a. Aging results in deterioration of elastic tissue, arthritic changes that stiffen rib articulations, decreased flexibility at costal cartilages, decreased vital capacity, and some degree of emphysema. **b.** The incidence of lung cancer in U.S. males is decreasing. **c.** Dysplasia is the development of abnormal cells; metaplasia is the development of abnormal changes in tissue structure; neoplasia is the conversion of normal cells to tumor (cancerous) cells; and, in anaplasia, the malignant cells spread (metastasize) throughout the body.

Section Reviews

Section 1 Review
1. nasal cavity; **2.** hard palate; **3.** pharynx; **4.** glottis; **5.** trachea; **6.** right lung; **7.** external nares; **8.** larynx; **9.** primary bronchus; **10.** secondary bronchus; **11.** tertiary bronchi; **12.** visceral pleura; **13.** pulmonary lobule; **14.** bronchioles; **15.** terminal bronchiole; **16.** alveolus; **17.** h; **18.** j; **19.** g; **20.** d; **21.** e; **22.** n; **23.** a; **24.** l; **25.** c; **26.** k; **27.** m; **28.** i; **29.** b; **30.** f

Section 2 Review
1. n; **2.** i; **3.** f; **4.** g; **5.** h; **6.** c; **7.** d; **8.** e; **9.** b; **10.** l; **11.** m; **12.** k; **13.** j; **14.** a; **15.** inspiratory reserve volume (IRV): the amount of air that can be taken in above the tidal volume; **16.** tidal volume (V_T): the amount of air inhaled and exhaled during a single respiratory cycle while resting; **17.** expiratory reserve volume (ERV): the amount of air that can be expelled after a completely normal, quiet respiratory cycle; **18.** minimal volume: the amount of air remaining in the lungs if they were to collapse; **19.** inspiratory capacity: the amount of air that can be drawn into the lungs after completing a quiet respiratory cycle; **20.** total lung capacity: the total volume of the lungs; **21.** vital capacity: the maximum amount of air that can be moved into or out of the lungs in a single respiratory cycle; **22.** residual volume: the amount of air remaining in the lungs after a maximal exhalation; **23.** functional residual capacity (FRC): the amount of air that remains in the lungs after completing a quiet respiratory cycle; **24.** External respiration includes all the processes involved in the exchange of oxygen and carbon dioxide between blood, lungs, and the external environment. Pulmonary ventilation, or breathing, is a process of external respiration that involves the physical movement of air into and out of the lungs. Internal respiration is the absorption of oxygen and the release of carbon dioxide by tissue cells.

Chapter Review Questions

1. 40 mm Hg; **2.** 45 mm Hg; **3.** 100 mm Hg; **4.** 40 mm Hg; **5.** 100 mm Hg; **6.** 40 mm Hg; **7.** 95 mm Hg; **8.** 40 mm Hg; **9.** 40 mm Hg; **10.** 45 mm Hg; **11.** 40 mm Hg; **12.** 45 mm Hg; **13.** e; **14.** l; **15.** c; **16.** i; **17.** j; **18.** b; **19.** a; **20.** d; **21.** k; **22.** f; **23.** h; **24.** g; **25.** d; **26.** a; **27.** b; **28.** b; **29.** d; **30.** c; **31.** The regions of the pharynx are (1) the nasopharynx, located between the soft palate and the internal nares; (2) the oropharynx, located between the soft palate and the base of the tongue at the level of the hyoid bone; and (3) the laryngopharynx, which is the portion of the pharynx located between the hyoid bone and entrance to the larynx and esophagus. **32.** The three ways that carbon dioxide is transported in blood are (1) as carbonic acid, (2) bound to the protein portion of hemoglobin molecules within the RBC, or (3) dissolved in plasma as CO_2. **33.** Breathing through the nasal cavity is more desirable than breathing through the mouth because the nasal cavity cleanses, moistens, and warms inhaled air, whereas the mouth does not. Breathing through the mouth eliminates much of the conditioning of inhaled air and increases heat and water loss at every exhalation.

Chapter Integration

1. Alveolar ventilation rate (AVR) = respiratory rate × (tidal volume − anatomic dead space). In this case, the dead space is 200 mL (the anatomic dead space plus the volume of the snorkel); therefore, AVR = respiratory rate × (500 − 200). To maintain an AVR of 6.0 L/min, or 6000 mL/minute, the respiratory rate must be 6000/(500 − 200), or 20 breaths per minute. **2.** Jake's hyperventilation resulted in abnormally low P_{CO_2}. This reduced his urge to breathe, so he stayed underwater longer, unaware that his P_{CO_2} was so low that he would lose consciousness.

CHAPTER 22

Module Reviews

Module 22.1 Review
a. The respiratory, cardiovascular, and urinary systems all work with the digestive system to support tissues that have no direct connection with the outside environment. **b.** The organs of the digestive tract are the oral cavity (mouth), pharynx (throat), esophagus, stomach, small intestine, and large intestine. **c.** The accessory organs of the digestive system are the teeth, tongue, salivary glands, liver, gallbladder, and pancreas.

Module 22.2 Review
a. The mesenteries—sheets consisting of two layers of serous membrane (visceral peritoneum) separated by loose connective tissue—support and stabilize the organs in the abdominal cavity and provide a route for the passage of associated blood vessels, nerves, and lymphatic vessels. **b.** The four layers of the digestive tract beginning from the lumen are the mucosa (adjacent to the lumen), submucosa, muscularis externa, and serosa. **c.** The submucosal plexus is a nerve network that contains sensory neurons, parasympathetic ganglionic neurons, and sympathetic postganglionic fibers that innervate the mucosa and submucosa. The deeper myenteric plexus is a network of parasympathetic neurons, interneurons, and sympathetic postganglionic fibers that lies between the circular and longitudinal muscle layers of the muscularis externa.

Module 22.3 Review
a. Smooth muscle fibers in either the circular or longitudinal layers of the muscularis externa lie parallel to each other. In a longitudinal section of the digestive tract, the fibers of the superficial circular layer appear as little round balls, whereas the fibers of the deeper longitudinal layer are spindle shaped. **b.** Smooth muscle fibers are spindle shaped; lack T tubules, myofibrils, and sarcomeres; and the sarcoplasmic reticulum forms a loose network throughout the sarcoplasm. Because the tissue lacks sarcomeres, it is nonstriated.

Additionally, the thin filaments are anchored to dense bodies. **c.** Smooth muscle can contract over a wider range of resting lengths compared with skeletal muscle because of the looser organization of actin and myosin filaments in smooth muscle.

Module 22.4 Review

a. The waves of contractions that constitute peristalsis are more efficient in propelling intestinal contents along the digestive tract than segmentation, which is basically a churning action that mixes intestinal contents with digestive fluids. **b.** The major mechanisms that regulate and control digestive activities are local factors (stimuli for digestive activities, such as changes in pH or distortion of the intestinal lumen), neural mechanisms (myenteric reflexes), and hormonal mechanisms (involving neuroendocrine cells). **c.** Enteroendocrine cells are endocrine cells scattered among the epithelial cells lining the digestive tract that secrete peptide hormones important to digestion.

Module 22.5 Review

a. Ingestion is the entry of foods and liquids into the oral cavity of the digestive tract. **b.** Digestion is the chemical and enzymatic breakdown of food into small molecules. Absorption is the movement of molecules and other substances across the digestive epithelium and into the interstitial fluid of the digestive tract. **c.** Compaction is the dehydration of indigestible materials and organic wastes prior to elimination from the body.

Module 22.6 Review

a. The oral cavity is lined by stratified squamous epithelium, which provides protection against friction and abrasion. **b.** The hard palate forms the roof of the mouth. **c.** The fauces is the arched opening between the oral cavity and the oropharynx.

Module 22.7 Review

a. The four types of teeth are incisors, cuspids (canines), bicuspids (premolars), and molars. A typical tooth has a crown, a neck, and a root. **b.** The primary dentition is typically composed of 20 deciduous teeth (also called primary teeth, milk teeth, or baby teeth), which are temporary and the first teeth to appear in children. The secondary dentition is typically composed of 32 permanent teeth that appear subsequent to the primary dentition. **c.** The third set of molars is sometimes called the wisdom teeth.

Module 22.8 Review

a. The pharynx is an anatomical space that serves as a common passageway in the digestive and respiratory tracts; in its digestive function, it receives a food bolus or liquids and passes them to the esophagus as part of the swallowing process. **b.** The esophagus is the structure connecting the pharynx to the stomach. **c.** During the buccal phase, food is formed into a bolus; during the pharyngeal phase, the bolus contacts the palatal arches and moves into the esophagus; during the esophageal phase, swallowing begins as pharyngeal muscles contract and the bolus is moved toward the stomach by peristaltic waves.

Module 22.9 Review

a. The falciform ligament is a sheet of mesentery that is a remnant of the ventral mesentery between the liver and the anterior wall of the peritoneal cavity. **b.** The lesser omentum stabilizes the position of the stomach and provides an access route for blood vessels (and other structures) entering or leaving the liver. **c.** Peritoneal fluid separates the parietal and visceral surfaces of the peritoneal cavity and prevents friction and subsequent irritation during sliding movements of organs enclosed by the peritoneal cavity.

Module 22.10 Review

a. The four major regions of the stomach are the cardia, fundus, body, and pylorus. **b.** The muscularis externa of the stomach has an additional inner, oblique layer of smooth muscle. The longitudinal, circular, and oblique orientations of the muscle fibers in the three layers of the muscularis externa allow for the mixing and churning actions necessary for chyme formation. **c.** The inner lining of the stomach contains rugae, mucosal folds in the lining of the empty stomach that disappear as gastric distension occurs.

Module 22.11 Review

a. The alkaline mucous layer protects epithelial cells against the acid and enzymes in the gastric lumen. **b.** Parietal cells secrete intrinsic factor and hydrochloric acid (HCl). **c.** The alkaline tide is a sudden influx of bicarbonate ions into the bloodstream from parietal cells that are actively secreting HCl; it causes a temporary increase in blood pH.

Module 22.12 Review

a. The layers of the small intestine from superficial to deep are the mucosa, submucosa, muscularis externa, and serosa. **b.** The intestinal mucosa has transverse folds called circular folds that have small projections called intestinal villi. These folds and projections increase the surface area available for absorption. Each villus contains a terminal lymphatic capillary called a lacteal. Between the bases of the villi are intestinal glands lined by enteroendocrine, mucous, and stem cells. **c.** Lacteals are lymphatic capillaries in the intestinal villi that transport materials that cannot enter blood capillaries, such as absorbed fatty acids combined with proteins.

Module 22.13 Review

a. The three segments of the small intestine are the duodenum, jejunum, and ileum. **b.** The proximal portion of the duodenum is found within the epigastric region. **c.** The duodenum is the intestinal "mixing bowl." Its primary function is to receive chyme from the stomach and neutralize its acids to avoid damaging the absorptive surfaces of the remaining regions of the small intestine. It also receives digestive secretions from the pancreas and liver.

Module 22.14 Review

a. The five major hormones that regulate digestive activities are gastrin, secretin, gastric inhibitory peptide (GIP), cholecystokinin (CCK), and vasoactive intestinal peptide (VIP). **b.** If the small intestine did not produce secretin, the pH of the intestinal contents would be lower (more acidic) than normal, because secretin stimulates the pancreas to release a fluid high in buffers that neutralizes the acidic chyme entering the duodenum from the stomach. **c.** A high-fat meal raises the cholecystokinin (CCK) level in the blood.

Module 22.15 Review

a. The three phases of gastric secretion (named according to the location of the control center involved) are the cephalic phase, which prepares the stomach to receive ingested materials; the gastric phase, which begins with the arrival of food in the stomach; and the intestinal phase, which controls the rate of gastric emptying. **b.** Severing the branches of the vagus nerves that supply the stomach would interrupt parasympathetic stimulation of gastric secretions, and the consequent reduction in acid secretions would provide some relief from the pain of gastric ulcers. **c.** The two central reflexes triggered by stimulation of stretch receptors in the stomach wall are the gastroenteric reflex, which stimulates motility and secretion along the entire small intestine, and the gastroileal reflex, which triggers the opening of the ileocecal valve to allow passage of materials from the small intestine into the large intestine.

Module 22.16 Review

a. The major functions of the large intestine are (1) reabsorbing water and compacting material into feces, (2) absorbing vitamins produced by bacteria, and (3) storing fecal material prior to defecation. **b.** The segments of the large intestine are the cecum, colon, and rectum. The four regions of the colon are the ascending colon, transverse colon, descending colon, and sigmoid colon. **c.** Mass movements are powerful peristaltic contractions that occur a few times daily in response to distension of the stomach and duodenum.

Module 22.17 Review

a. Since the large intestine does not produce any enzymes, any digestion that occurs results from enzymes secreted into the small intestine or from bacterial action. However, the large intestine compacts fecal material, prevents dehydration by reabsorbing water, and absorbs vitamins produced by the normal bacteria that live in the colon. **b.** Hemorrhoids are distended (swollen) veins in the distal portion of the rectum that may result from straining during defecation. **c.** The two positive feedback loops in the defecation reflex are (1) the short reflex, whereby stretch receptors in the rectal walls promote a series of peristaltic contractions in the colon and rectum, moving feces into the anal canal; and (2) the long reflex, whereby parasympathetic motor neurons in the sacral spinal cord, also activated by stretch receptors, stimulate peristalsis by motor commands distributed by somatic motor neurons.

Module 22.18 Review

a. The salivary glands produce secretions that contain mucins and enzymes that are essential to normal digestive function. **b.** The exocrine secretions of the pancreas are buffers and digestive enzymes. The endocrine secretions are insulin, glucagon, pancreatic polypeptide, and GH–IH hormones. **c.** The liver is the accessory organ of the digestive system that has almost 200 known functions.

Module 22.19 Review

a. The three pairs of salivary glands are the parotid, sublingual, and submandibular salivary glands. **b.** Damage to the parotid salivary glands, which secrete the enzyme salivary amylase, would interfere with the digestion of starches (complex carbohydrates). **c.** The sublingual salivary glands contribute least (about 5 percent) to the secretions that make up saliva.

Module 22.20 Review

a. The liver is divided into left, right, caudate, and quadrate lobes. **b.** The falciform ligament marks the division between the left lobe and the right lobe of the liver. **c.** The gallbladder temporarily stores bile produced by the liver.

Module 22.21 Review

a. A hepatocyte is a liver cell. **b.** A portal area is located at each of the six corners of a liver lobule; each portal area contains (1) a branch of the hepatic portal vein, (2) a branch of the hepatic artery proper, and (3) a small branch of the bile duct. **c.** Kupffer cells are liver macrophages that engulf pathogens, cell debris, and damaged blood cells.

Module 22.22 Review

a. Emulsification is the breaking apart of lipid droplets by bile salts. **b.** The pathway of a drop of bile: hepatic ducts → common hepatic duct → common bile duct → duodenal ampulla and papilla → duodenal lumen; **c.** The pancreas produces pancreatic juice that contains buffers and several digestive enzymes necessary for the breakdown of starches, lipids, nucleic acids, and proteins.

Module 22.23 Review

a. Periodontal disease is characterized by a loosening of the teeth within the alveolar sockets due to erosion of the periodontal ligaments by acids produced by bacterial action. **b.** Cholecystitis is inflammation of the gallbladder, usually resulting from a blockage of the cystic duct or the common bile duct by gallstones. **c.** The bacterium responsible for most peptic ulcers is *Helicobacter pylori.*

Section Reviews

Section 1 Review

1. mesenteric artery and vein; **2.** mesentery; **3.** circular fold; **4.** mucosa; **5.** submucosa; **6.** muscularis externa; **7.** serosa; **8.** d; **9.** e; **10.** a; **11.** g; **12.** b; **13.** c; **14.** i; **15.** m; **16.** l; **17.** f; **18.** n; **19.** j; **20.** k; **21.** h; **22.** With a decrease in smooth muscle tone, general motility along the digestive tract decreases, and peristaltic contractions are weaker.

Section 2 Review

1. crown; **2.** neck; **3.** root; **4.** enamel; **5.** dentin; **6.** pulp cavity; **7.** gingiva; **8.** gingival sulcus; **9.** cementum; **10.** periodontal ligament; **11.** root canal; **12.** bone of alveolus; **13.** material in jejunum; **14.** gastrin; **15.** GIP; **16.** secretin and CCK; **17.** VIP; **18.** inhibits; **19.** acid production; **20.** insulin; **21.** bile; **22.** intestinal capillaries; **23.** gallbladder; **24.** nutrient utilization by tissues; **25.** Both parietal cells and chief cells are secretory cells found in the gastric glands of the wall of the stomach. However, parietal cells secrete intrinsic factor and hydrochloric acid, whereas chief cells secrete pepsinogen, an inactive proenzyme.

Section 3 Review

1. bile ductule; **2.** hepatocytes; **3.** central vein; **4.** interlobular septum; **5.** sinusoid; **6.** branch of hepatic artery proper; **7.** branch of hepatic portal vein; **8.** bile duct; **9.** portal area (portal triad); **10.** g; **11.** h; **12.** m; **13.** j; **14.** l; **15.** n; **16.** b; **17.** e; **18.** f; **19.** a; **20.** i; **21.** d; **22.** k; **23.** c; **24.** Saliva (1) continuously flushes and cleans oral surfaces, (2) contains buffers that prevent the buildup of acids produced by bacterial action, and (3) contains antibodies (IgA) and lysozyme, which help control the growth of oral bacterial populations. **25.** Such a blockage would interfere with the release of secretions into the duodenum by the pancreas, gallbladder, and liver. The pancreas normally secretes about 1 liter of pancreatic juice, a mixture of a variety of digestive enzymes and buffer solution. The blockage of pancreatic juice would lead to pancreatitis, an inflammation of the pancreas. Extensive damage to exocrine cells by the blocked digestive enzymes would lead to autolysis that could destroy the pancreas and result in the person's death. Blockage of bile secretion from the common bile duct could lead to damage of the wall of the gallbladder by the formation of gallstones and to jaundice because bilirubin from the liver would not be excreted in the bile and, instead, would accumulate in body fluids.

Chapter Review Questions

1. gallbladder; **2.** fundus; **3.** body; **4.** neck; **5.** cystic duct; **6.** common bile duct; **7.** pancreatic duct; **8.** round ligament; **9.** left and right hepatic ducts; **10.** common hepatic duct; **11.** hepatic portal vein and hepatic artery proper; **12.** true; **13.** false; **14.** false; **15.** true; **16.** false; **17.** true; **18.** a; **19.** b; **20.** b; **21.** a; **22.** d; **23.** b; **24.** The four different types of teeth are incisors, cuspids, bicuspids, and molars. There are 20 deciduous teeth and 32 permanent teeth. **25.** The three phases of gastric secretion are the cephalic phase, the gastric phase, and the intestinal phase. The cephalic phase of gastric secretion begins when you see, smell, taste, or think of food. The CNS is preparing the stomach to receive food. The gastric phase is when food arrives in the stomach. It includes distension of the stomach, an increase in pH, and the presence of undigested materials. The intestinal phase is when chyme first enters the small intestine. **26.** The hepatic portal system is a venous portal system in which the hepatic portal vein receives blood from capillaries of most of the abdominal viscera and delivers it to the hepatic sinusoids.

Chapter Integration

1. Gastric stapling would result in weight loss because the volume of food (and thus the amount of calories) consumed is reduced because the person feels full after eating only a small amount. **2.** After gastric stapling surgery, the person can eat only a small amount of food before the stretch receptors in the gastric wall become stimulated and a feeling of fullness results. **3.** The smooth muscle in the gastric muscularis externa of the functional portion of the stomach gradually becomes increasingly tolerant of distention, and the operation may have to be repeated to achieve the same results. **4.** Gastric bypass and lap-band surgeries, like gastric stapling, reduce gastric capacity, thereby decreasing the amount of food a person can ingest before feeling full.

CHAPTER 23

Module Reviews

Module 23.1 Review
a. Metabolism refers to all the chemical reactions that occur in an organism. Cellular metabolism refers to the chemical reactions within cells. **b.** Catabolism is the breakdown of organic substrates in the body, whereas anabolism is the synthesis of new organic molecules. **c.** Metabolic turnover is the process in which cells continuously break down and replace all their organic components except DNA.

Module 23.2 Review
a. The nutrient pool within a cell includes all the organic substrates in the cytosol that are available for anabolism or catabolism. **b.** Cells carry out catabolism to release energy for use in cell growth, cell division, and tissue-specific activities. **c.** Cells make new compounds to maintain and repair structures, to support growth, and to build up nutrient reserves.

Module 23.3 Review
a. The citric acid cycle is the reaction sequence that occurs in the matrix of mitochondria. In the process, organic molecules are broken down, carbon dioxide molecules are released, and hydrogen atoms are transferred to coenzymes that deliver them to the electron transport system. **b.** The common substrate for the citric acid cycle in mitochondria is the 2-carbon molecule of acetate, CH_3COO^-, which is attached to coenzyme A to form acetyl-CoA. **c.** Hydrogen atoms are transferred to the electron transport system by way of the coenzymes NAD and FAD.

Module 23.4 Review
a. Oxidative phosphorylation is the generation of ATP within mitochondria in a reaction sequence that requires coenzymes and consumes oxygen. **b.** Cytochromes are embedded in the inner mitochondrial membrane. **c.** Hydrogen ion channels are the passageways for the diffusion of hydrogen ions from the inner membrane space of the mitochondria to the matrix. This movement of hydrogen ions powers the production of ATP by ATP synthase.

Module 23.5 Review
a. Digestion is important because cells throughout the body rely on the organic molecules from the food we eat for energy production and to replenish the intracellular nutrient pool. **b.** Most nutrient absorption occurs in the small intestine, primarily in the jejunum. **c.** Most nutrients enter into a branch of the hepatic portal vein and are transported to the liver.

Module 23.6 Review
a. Intestinal gas, or flatus, is generated by bacterial activities in the colon when indigestible carbohydrates stimulate bacterial gas production. **b.** Glycogen is synthesized from excess glucose molecules by liver and muscle cells, and serves as an intracellular glucose reserve. **c.** Carbohydrates are the preferred energy source because proteins and lipids are more important as structural components of cells and tissues.

Module 23.7 Review
a. The products are two molecules each of pyruvate, ATP, and NAD•H. **b.** Most (two-thirds) of the CO_2 released in the complete catabolism of glucose occurs during the citric acid cycle. **c.** ATP produced anaerobically through glycolysis is important during peak levels of physical activity, in red blood cells, or when a tissue is temporarily deprived of oxygen.

Module 23.8 Review
a. Micelles are lipid–bile salt complexes (containing fatty acids, glycerol, and monoglycerides) formed in the intestinal lumen. Chylomicrons are lipoproteins formed in intestinal epithelial cells and contain newly synthesized triglycerides, cholesterol, and other lipids surrounded by phospholipids and proteins. **b.** The liver absorbs chylomicrons, removes the triglycerides, combines the cholesterol from the chylomicron with recycled cholesterol, and alters the surface proteins. Newly synthesized complexes are released into the bloodstream as low density lipoproteins (LDLs) or very low density lipoproteins (VLDLs). **c.** Low density lipoproteins (LDLs) deliver cholesterol to body tissues, and high density lipoproteins (HDLs) absorb unused cholesterol from body tissues, returning it to the liver where it may be packaged into new LDLs or excreted with bile salts in bile.

Module 23.9 Review
a. Beta-oxidation is fatty acid catabolism that produces molecules of acetyl-CoA. **b.** Acetyl-CoA is a reactant molecule in ATP production and in the synthesis of most types of lipids. **c.** Fatty acids may become a source of energy or a component of triglycerides, glycolipids, phospholipids, prostaglandins, cholesterol, and steroids.

Module 23.10 Review
a. The arrival of acidic chyme in the duodenum triggers the release of CCK, which stimulates the production and release of inactive pancreatic proenzymes. Enteropeptidase, released from the duodenum, converts the pancreatic proenzyme trypsinogen into the proteolytic enzyme, trypsin. Trypsin then converts other proenzymes to yield chymotrypsin, carboxypeptidase, and elastase. Each of these enzymes attacks peptide bonds that link specific amino acids while ignoring others. As a result, they break down proteins into a mixture of dipeptides, tripeptides, and amino acids. **b.** The amino group is removed by deamination or transamination. **c.** The ammonium ions combine with carbon dioxide to form urea (in the urea cycle), which is ultimately excreted in the urine.

Module 23.11 Review
a. The absorptive state, lasting about 4 hours, is the period following a meal, when nutrient absorption is under way. The postabsorptive state, lasting about 12 hours, is the period when nutrient absorption is not under way and the body relies on internal energy reserves to meet demands. **b.** Ketone bodies form during the postabsorptive state when lipids and amino acids are broken down in the liver. The increased concentration of acetyl-CoA that results from their breakdown forms ketone bodies. Ketone bodies are not catabolized by liver cells, and they diffuse into the circulation. **c.** During the absorptive state, insulin prevents a large surge in blood glucose after a meal by stimulating the liver to remove glucose from the circulation. During the postabsorptive state, blood glucose begins to decrease, triggering the release of glucagon, which stimulates the liver to release glucose into the circulation.

Module 23.12 Review
a. Nutrition is the absorption of nutrients from food. **b.** The two classes of vitamins are fat-soluble vitamins and water-soluble vitamins. **c.** Vitamins play an important role in metabolic pathways by serving as coenzymes.

Module 23.13 Review
a. A balanced diet contains all the ingredients needed to maintain homeostasis and prevent malnutrition. **b.** The catabolism of lipids releases the greatest amount of energy per gram. **c.** A complete protein meets the body's amino acid requirements; an incomplete protein is deficient in one or more of the essential amino acids.

Module 23.14 Review
a. Eating disorders are psychological problems that result in inadequate food consumption (anorexia nervosa) or excessive food consumption followed by purging (bulimia). **b.** Phenylketonuria (PKU) is an inherited metabolic disorder resulting from an inability to convert phenylalanine to tyrosine. **c.** Protein deficiency diseases are nutritional disorders resulting from a lack of one or more essential amino acids. Kwashiorkor is an example of a protein deficiency disease.

Module 23.15 Review
a. Energetics is the study of the flow of energy and its change from one form to another. **b.** Basal metabolic rate is the minimum resting energy expenditure of an awake, alert person. **c.** Thermoregulation is the homeostatic control of body temperature.

Module 23.16 Review
a. A lack of neuropeptide Y, a hypothalamic neurotransmitter, would probably decrease appetite because it normally stimulates the feeding center. **b.** Ghrelin, a hormone secreted by the gastric mucosa when the stomach is not full, inhibits the satiety center and stimulates appetite. **c.** Leptin is a peptide hormone produced by adipose tissue during the synthesis of triglycerides. It stimulates the satiety center and suppresses appetite.

Module 23.17 Review
a. Insensible perspiration is the evaporation of water from the skin and alveolar surfaces of the lungs. **b.** Radiation accounts for about one-half of a person's heat loss indoors. **c.** Conduction is the direct transfer of heat through physical contact. Convection is heat loss to the cooler air in contact with the skin. The air warmed by the skin rises and it is repeatedly replaced by cooler air until there is no difference in temperature.

Module 22.18 Review
a. The vasodilation of peripheral vessels would increase blood flow to the skin and thus the amount of heat the body can lose. As a result, body temperature would decrease. **b.** Nonshivering thermogenesis involves the release of hormones that increase the metabolic activity of all tissues, resulting in an increase in body temperature. **c.** Countercurrent exchange is the heat conservation mechanism that results in the conduction of heat from deep arteries to adjacent deep veins in the limbs.

Section Reviews

Section 1 Review
1. d; **2.** a; **3.** e; **4.** k; **5.** f; **6.** j; **7.** g; **8.** b; **9.** c; **10.** h; **11.** i; **12.** c; **13.** e; **14.** f; **15.** m; **16.** b; **17.** a; **18.** j; **19.** l; **20.** n; **21.** g; **22.** k; **23.** h; **24.** d; **25.** i; **26.** During fasting or starvation, other tissues shift to fatty acid catabolism or amino acid catabolism, to conserve glucose for neural tissue.

Section 2 Review
1. m; **2.** n; **3.** j; **4.** g; **5.** a; **6.** k; **7.** d; **8.** l; **9.** f; **10.** c; **11.** i; **12.** h; **13.** e; **14.** b; **15.** d; **16.** a; **17.** d; **18.** b; **19.** Essential amino acids are necessary in the diet because the body cannot synthesize them. The body can synthesize nonessential amino acids on demand. **20.** (1) Proteins are difficult to break apart because of their complex three-dimensional structure. (2) The energy yield of proteins (4.32 Cal/g) is less than that of lipids (9.46 Cal/g). (3) The byproducts of protein or amino acid catabolism are ammonium ions, a toxin that can damage cells. (4) Proteins form the most important structural and functional components of cells. Excessive protein catabolism would threaten homeostasis at the cellular to system levels of organization. **21.** During the absorptive state, the intestinal mucosa is absorbing nutrients from the digested food. In the postabsorptive state, metabolic activity centers on the mobilization of energy reserves and the maintenance of normal blood glucose levels. **22.** Liver cells can break down or synthesize most carbohydrates, lipids, and amino acids. The liver has an extensive blood supply and thus can easily monitor and regulate the blood levels of these nutrients. The liver also stores energy in the form of glycogen. **23.** It appears that Claudia is suffering from ketoacidosis as a consequence of her anorexia. Because she is literally starving herself, her body is metabolizing large amounts of fatty acids and amino acids to provide energy and in the process is producing large quantities of ketone bodies (normal metabolites from these catabolic processes). One of the ketones formed is acetone, which can be eliminated through the lungs. Acetone has a fruity aroma, so her breath would also smell fruity. The ketones are organic acids and when they accumulate in the blood, they would decrease the blood pH, and when the ketone bodies spill over into the urine, they would decrease the pH in urine. (In severe cases of ketoacidosis, the ketone bodies may lower blood pH below 7.05, which may cause coma, cardiac arrhythmias, and death.)

Section 3 Review
1. j; **2.** d; **3.** i; **4.** f; **5.** m; **6.** l; **7.** h; **8.** k; **9.** b; **10.** a; **11.** c; **12.** n; **13.** e; **14.** g; **15.** d; **16.** b; **17.** c; **18.** c; **19.** a; **20.** a; **21.** The amount of energy used or expended at rest is powered by mitochondrial energy production, which requires oxygen. Using the relationship of 4.825 Calories/L of oxygen, a measurement of the oxygen consumed provides an estimate of the energy expenditure. **22.** The heat-gain center functions in preventing hypothermia, or below-normal body temperature, by conserving body heat and increasing the rate of heat production by the body. **23.** Nonshivering thermogenesis increases the metabolic rate of most tissues through the actions of two hormones, epinephrine and thyroid-stimulating hormone (TSH). In the short term, the heat-gain center stimulates the adrenal medullae to release epinephrine by the sympathetic division of the ANS. Epinephrine quickly increases the breakdown of glycogen (glycogenolysis) in the liver and skeletal muscle, and it increases the metabolic rate of most tissues. The long-term increase in metabolism occurs primarily in children as the heat-gain center adjusts the rate of thyrotropin-releasing hormone (TRH) release by the hypothalamus. When body temperature is low, additional TRH is released, which stimulates the release of TSH by the anterior lobe of the pituitary gland. The thyroid gland then increases its rate of thyroid hormone release, and these hormones increase the rate of catabolism throughout the body.

Chapter Review Questions

1. false; **2.** true; **3.** true; **4.** false; **5.** false; **6.** true; **7.** d; **8.** a; **9.** c; **10.** b; **11.** a; **12.** b; **13.** d; **14.** b; **15.** a; **16.** a; **17.** Some of the energy released during catabolism is captured as ATP, and this ATP can be used as the energy source for anabolism. **18.** A cholesterol level above 200 mg/dL is considered elevated. Elevated cholesterol plus a high LDL:HDL ratio can result in atherosclerotic plaques that can cause heart attacks and strokes. **19.** Too much of a vitamin can produce harmful effects in a condition called hypervitaminosis. This condition most commonly involves one of the fat-soluble vitamins. **20.** A calorie is the energy needed to raise the temperature of 1 g of water 1°C. A kilocalorie is the amount of energy needed to raise the temperature of 1 kilogram of water 1°C. A Calorie is equal to a kilocalorie.

Chapter Integration

1. Vitamins and minerals are essential components of the diet because the body cannot synthesize most of the vitamins and minerals it requires. **2.** These terms refer to the lipoproteins in the blood that transport cholesterol. So-called "good cholesterol" (high density lipoproteins, or HDLs) transports excess cholesterol to the liver for storage and breakdown, whereas "bad cholesterol" (low density lipoproteins, or LDLs) transports cholesterol to peripheral tissues, which includes the arteries. The buildup of cholesterol in arterial walls is linked to cardiovascular disease. **3.** HDLs are considered beneficial because they reduce the amount of cholesterol in the bloodstream by transporting it to the liver for storage or excretion in bile. **4.** Glycogenesis (the synthesis of glycogen from glucose) in the liver increases after a high-carbohydrate meal. **5.** A diet deficient in pyridoxine (vitamin B_6) would interfere with the body's ability to metabolize proteins. Subsequent effects of the deficiency include retarded growth, anemia, convulsions, and epithelial changes.

CHAPTER 24

Module Reviews

Module 24.1 Review

a. The kidneys perform the excretory functions of the urinary system. **b.** The urinary tract is composed of the ureters, urinary bladder, and urethra. **c.** The functions of the urinary system include adjusting blood volume and pressure, regulating blood plasma concentrations of ions, stabilizing blood pH, conserving nutrients by preventing their loss in urine, and removing drugs and toxins from the bloodstream.

Module 24.2 Review

a. The main structures of the urinary system are the kidneys, ureters, urinary bladder, and urethra. **b.** The layers of connective tissue that protect and anchor the kidney are the fibrous capsule, the perinephric fat (perinephric fat capsule), and the renal fascia. The fibrous capsule covers the surface of the kidney, the perinephric fat is a thick layer of adipose tissue surrounding the fibrous capsule, and the renal fascia is a dense, fibrous outer layer that anchors the kidney to surrounding structures. **c.** If the perinephric fat, which helps hold the kidneys in position against the posterior body wall, were depleted, the kidneys could drop, which may cause pain or distortion of other structures.

Module 24.3 Review

a. The renal pyramid is a conical mass within the renal medulla that ends at the papilla. **b.** The renal papilla is the tip of the conically shaped renal pyramid; it empties formed urine into the renal pelvis. **c.** Juxtamedullary nephrons are essential for the conservation of water and the production of concentrated urine.

Module 24.4 Review

a. Filtrate is similar to blood plasma, but filtrate does not contain proteins. **b.** The primary structures of the nephron are the renal corpuscle and the renal tubule, which is divided into the proximal convoluted tubule, nephron loop, and distal convoluted tubule. The main structures of the collecting system are the collecting duct and papillary duct. **c.** The renal corpuscle consists of the glomerular capsule and the glomerulus.

Module 24.5 Review

a. Pathway of renal blood flow: renal artery → segmental arteries → interlobar arteries → arcuate arteries → cortical radiate arteries → afferent arterioles → glomerulus → efferent arterioles →peritubular capillaries → cortical radiate veins → arcuate veins → interlobar veins → renal vein. **b.** Blood enters the glomerulus by the afferent arteriole and leaves by the efferent arteriole. **c.** The vasa recta are long, straight peritubular capillaries that parallel the nephron loop of a juxtamedullary nephron.

Module 24.6 Review

a. Three solutes excreted in urine are urea (from liver breakdown of amino acids), creatinine (from the breakdown of creatine phosphate), and uric acid (from the recycling of the nitrogenous bases of RNA). **b.** During filtration, fluids and solutes move from the glomerular capillaries and into the capsular space. During reabsorption, fluids and solutes move from the tubular fluid and into the peritubular fluid. During secretion, fluids and solutes move from the peritubular fluid and into the tubular fluid. **c.** Blood pressure is required for filtration.

Module 24.7 Review

a. The three distinct processes of urine formation in the kidney are filtration, reabsorption, and secretion. **b.** Filtration exclusively occurs across the filtration membrane in the renal corpuscle. **c.** The nephron loop is solely involved in the reabsorption of water and sodium and chloride ions.

Module 24.8 Review

a. The capsular space separates the parietal and visceral layers of the glomerular capsule. **b.** Blood pressure is higher in glomerular capillaries than systemic capillaries because the efferent arteriole has a smaller diameter than the afferent arteriole. **c.** Blood colloidal pressure tends to draw water out of the filtrate and into the plasma because the solute concentration within the blood exceeds that within the filtrate. The "advantage" that results when blood colloidal pressure draws water out of the filtrate and into the plasma is that this action helps conserve body water.

Module 24.9 Review

a. In response to decreased filtration pressure, the juxtaglomerular complex increases renin production and release into the bloodstream. **b.** Angiotensin II causes efferent arteriole constriction, which increases glomerular blood pressure. **c.** Angiotensin II triggers CNS neural responses, including enhancing the sensation of thirst and increasing antidiuretic hormone (ADH) production.

Module 24.10 Review

a. The distal convoluted tubule (DCT) makes final adjustments to the composition of tubular fluid. **b.** Increased amounts of aldosterone, which promotes Na^+ retention and K^+ secretion by the kidneys, would increase the K^+ concentration of urine. **c.** If the concentration of Na^+ in the filtrate decreased, fewer hydrogen ions could be secreted by the countertransport mechanism involving these two ions. As a result, the pH of the tubular fluid would increase.

Module 24.11 Review

a. Countercurrent multiplication in the kidneys is the exchange of substances between two adjacent nephron loop limbs containing tubular fluid moving in opposite directions; the process enables the kidney tubules to concentrate urine. **b.** The thick ascending limb actively pumps sodium and chloride ions into the peritubular fluid. **c.** An increase of sodium and chloride ions in the peritubular fluid increases the osmotic concentrations around the thin descending limb, resulting in an osmotic outflow of water.

Module 24.12 Review

a. No, the permeability of the PCT cannot change, and water reabsorption occurs whenever the osmotic concentration of the peritubular fluid exceeds that of the tubular fluid. However, the permeability of the DCT to water increases in response to antidiuretic hormone (ADH). **b.** Increased ADH levels cause the appearance of more water channels, or aquaporins, in the DCT; as a result, more water is reabsorbed into the peritubular fluid, which decreases the volume of water in the urine. **c.** Decreased ADH levels in the DCT reduce the urine osmotic concentration because there is more water in the urine; the result is a larger volume of more dilute urine.

Module 24.13 Review

a. The filtrate produced at the renal corpuscle has the same osmotic concentration as plasma. **b.** The active removal of ions and organic substrates from the PCT causes a continuous osmotic flow of water out of the tubular fluid. **c.** The concentration gradient of the renal medulla is maintained by the removal of solutes and water from the area by the vasa recta, which transport them to the circulatory system.

Module 24.14 Review
a. Chronic renal failure is a gradual loss of renal function, whereas acute renal failure is a sudden loss of renal function. **b.** Dialysis is the process of using an artificial semipermeable membrane to remove wastes and retain plasma proteins from the blood of a person whose kidneys are not functioning properly. **c.** Patients on dialysis are often given Epogen or Procrit (a synthetic form of erythropoietin) to treat anemia, which occurs because their malfunctioning kidneys produce too little erythropoietin. Erythropoietin is the hormone that stimulates the development of red blood cells in the red bone marrow.

Module 24.15 Review
a. Filtrate modification and urine production end when the fluid enters the renal pelvis. **b.** A pyelogram is an image of the urinary system that is obtained by taking an x-ray after radiopaque dye has been administered intravenously. **c.** The male urethra is longer, and transports not only urine, but also semen.

Module 24.16 Review
a. Urine is transported by the ureters, stored within the urinary bladder, and eliminated through the urethra. **b.** The external urethral sphincter must be consciously relaxed to allow urination. **c.** The specialized smooth muscle in the wall of the urinary bladder is the detrusor muscle; its contraction compresses (squeezes) the urinary bladder and expels urine into the urethra.

Module 24.17 Review
a. The urge to urinate usually appears when the urinary bladder contains about 200 mL of urine. The micturition reflex begins when the stretch receptors in the urinary bladder wall have provided adequate stimulation to the parasympathetic motor neurons. The activity in the motor neurons generates action potentials that reach the smooth muscle in the wall of the urinary bladder. These efferent impulses travel over the pelvic nerves, producing a sustained contraction of the urinary bladder. **b.** The sensation of a full bladder is relayed from the bladder to the thalamus to the cerebral cortex. **c.** Voluntary relaxation of the external urethral sphincter causes relaxation of the internal urethral sphincter by the micturition reflex.

Module 24.18 Review
a. Dysuria is the term for painful or difficult urination. **b.** Obstruction of a ureter by a kidney stone would interfere with the flow of urine between the kidney and the urinary bladder. **c.** Urinary obstruction at the urethra is more dangerous than urinary obstruction in the ureter because urine would be prevented from exiting the body and would build up in the urinary bladder, leading to its rupture. Obstruction of only one of the two ureters is less hazardous than obstruction of the lone urethra because in the former case, the other kidney could still excrete wastes that could be voided.

Section Reviews

Section 1 Review
1. urinary bladder; **2.** nephrons; **3.** renal tubules; **4.** glomerulus; **5.** proximal convoluted tubule; **6.** papillary ducts; **7.** renal medulla; **8.** renal sinus; **9.** major calyces; **10.** ureter; **11.** renal sinus: cavity within kidney that contains calyces, pelvis of the ureter, and segmental vessels; **12.** renal pelvis: funnel-shaped expansion of the superior portion of ureter; **13.** hilum: depression on the medial border of kidney, and site of the apex of the renal pelvis and the passage of segmental renal vessels and renal nerves; **14.** renal papilla: tip of the renal pyramid that projects into a minor calyx; **15.** ureter: tube that conducts urine from the renal pelvis to the urinary bladder; **16.** renal cortex: the outer portion of the kidney containing renal lobules, renal columns (extensions between the pyramids), renal corpuscles, and the proximal and distal convoluted tubules; **17.** renal medulla: the inner, darker portion of the kidney that contains the renal pyramids; **18.** renal pyramid: conical mass of the kidney projecting into the medullary region containing part of the secreting tubules and collecting tubules; **19.** minor calyx: subdivision of major calyces into which urine enters from the renal papillae; **20.** major calyx: primary subdivision of renal pelvis formed from the merging of four or five minor calyces, **21.** kidney lobe: portion of kidney consisting of a renal pyramid and its associated cortical tissue; **22.** renal columns: cortical tissue separating renal pyramids; **23.** fibrous capsule (outer layer): covering of the kidney's outer surface and lining of the renal sinus

Section 2 Review
1. f; **2.** j; **3.** h; **4.** i; **5.** g; **6.** a; **7.** e; **8.** c; **9.** d; **10.** b; **11.** nephron: functional unit of the kidney that filters and excretes waste materials from the blood and forms urine; **12.** proximal convoluted tubule: reabsorbs water, ions, and all organic nutrients; **13.** distal convoluted tubule: important site of active secretion; **14.** renal corpuscle: expanded chamber that encloses the glomerulus; **15.** nephron loop: portion of the nephron that produces the concentration gradient in the renal medulla; **16.** collecting system: series of tubes that carry tubular fluid away from the nephron; **17.** collecting duct: portion of collecting system that receives fluid from many nephrons and performs variable reabsorption of water and reabsorption or secretion of sodium, potassium, hydrogen, and bicarbonate ions; **18.** papillary duct: delivers urine to the minor calyx; **19.** The presence of plasma proteins and numerous WBCs in Marissa's urine indicates an increased permeability of the filtration membrane. This condition usually results from inflammation of the filtration membrane within the renal corpuscle. If the condition is temporary, it is probably an acute glomerular nephritis usually associated with a bacterial infection (such as streptococcal sore throat). If the condition is long term, resulting in a nonfunctional kidney, it is referred to as chronic glomerular nephritis. The urine volume would be greater than normal because the plasma proteins increase the osmolarity of the filtrate.

Section 3 Review
1. stretch receptors stimulated; **2.** afferent fibers carry information to sacral spinal cord; **3.** parasympathetic preganglionic fibers carry motor commands; **4.** detrusor muscle contraction stimulated; **5.** sensation relayed to thalamus; **6.** sensation of bladder fullness delivered to cerebral cortex; **7.** person relaxes external urethral sphincter; **8.** internal urethral sphincter relaxes; **9.** e; **10.** f; **11.** b; **12.** j; **13.** d; **14.** k; **15.** c; **16.** g; **17.** a; **18.** h; **19.** i; **20.** Four primary signs and symptoms of urinary disorders are (1) changes in the volume of urine, (2) changes in the appearance of urine, (3) changes in the frequency of urination, and (4) pain. **21.** cystitis/pyelonephritis: Both conditions involve inflammation and infections of the urinary system, but cystitis refers to the urinary bladder, whereas pyelonephritis refers to the kidney. **22.** stress incontinence/overflow incontinence: Both conditions involve an inability to control urination, but stress incontinence involves periodic involuntary leakage, whereas overflow incontinence involves a continual, slow trickle of urine. **23.** polyuria/proteinuria: Both are abnormal urine conditions, but polyuria is the production of excessive amounts of urine, whereas proteinuria refers to the presence of protein in the urine.

Chapter Review Questions

1. efferent arteriole; **2.** juxtamedullary complex;

3. afferent arteriole; **4.** glomerular capsule; **5.** capsular space; **6.** initial segment of renal tubule; **7.** parietal layer of glomerular capsule; **8.** visceral layer of glomerular capsule; **9.** false; **10.** true; **11.** true; **12.** false; **13.** false; **14.** true; **15.** d; **16.** a; **17.** b; **18.** b; **19.** c; **20.** b; **21.** a; **22.** a; **23.** The processes of urine production are (1) filtration: blood pressure forces water and solutes across the membranes of the glomerular capillaries and into the capsular space; (2) reabsorption: the transport of water and solutes from the tubular fluid, across the tubular epithelium, and into the peritubular fluid; and (3) secretion: the transport of solutes from the peritubular fluid, across the tubular epithelium, and into the tubular fluid. **24.** The primary process involved in glomerular filtration is basically the same as that regulating fluid and solute movement across capillaries throughout the body: the balance between hydrostatic pressure (fluid pressure) and colloid osmotic pressure (pressure due to materials in solution) on either side of the capillary membrane. **25.** The thin descending limb and the thick ascending limb of the loop of the nephron are very close together and the tubular fluid flows in opposite directions. The exchange that occurs between these segments, called countercurrent multiplication, results in a higher osmotic concentration in the peritubular fluid compared to the tubular fluid in the ascending limb. This process is responsible for creating the concentration gradient in the renal medulla that enables the kidney to produce highly concentrated urine.

Chapter Integration

1. The formation of crystalline masses can lead to kidney stones, which can cause urinary obstruction. The kidney stones can cause tiny ruptures along the urinary tract, leading to blood in the urine (hematuria). The metabolism of nitrogen-containing melamine releases ammonia, which may cause electrolyte imbalance and renal failure. **2.** Administration of intravenous fluids and dialysis would flush the body of the harmful chemical. Medication to control blood pressure, nausea, and anemia might also be necessary. Frequent blood tests and urinalysis are also performed to monitor kidney function.

CHAPTER 25

Module Reviews

Module 25.1 Review

a. Minerals are inorganic substances that dissociate in body fluids to form ions called electrolytes. **b.** Intracellular fluid (ICF), or cytosol, and extracellular fluid (ECF) are the two fluid compartments. **c.** Cells have a plasma membrane and active transport at the membrane surface that allows them to maintain their internal environments.

Module 25.2 Review

a. The major routes of fluid loss are urination, evaporation at the skin and at the lungs, and water lost in feces. Varying amounts of water are secreted by sweat glands. **b.** A fluid shift is a rapid movement of water between the ECF and ICF in response to an osmotic gradient. **c.** Dehydration is a reduction in the water content of the body that develops when water losses outpace water gains. In dehydration, the osmotic concentration of blood plasma increases.

Module 25.3 Review

a. Mineral balance is when the body's ion gains and losses are equal. **b.** The ions absorbed by active transport are sodium, calcium, magnesium, iron, phosphate, and sulfate. **c.** Sodium is a major cation that is essential for normal membrane function. Calcium is a cation that is essential for normal muscle and neuron function and for normal bone structure.

Module 25.4 Review

a. When osmoreceptors are inhibited, ADH release is decreased, and thirst is suppressed. **b.** Aldosterone causes increased urinary sodium retention and thus increases the sodium ion concentration in the ECF. **c.** Shifts in sodium balance result in expansion or contraction of the ECF. Large variations in ECF volume are corrected by homeostatic mechanisms triggered by changes in blood volume. If the blood volume becomes too low, ADH and aldosterone are secreted, increasing the sodium ion concentration in the ECF; if the volume becomes too high, natriuretic peptides are secreted.

Module 25.5 Review

a. The kidneys are primarily responsible for regulating the potassium ion concentration in the ECF. **b.** Potassium excretion increases as potassium concentrations rise in the ECF, under aldosterone stimulation, and when the ECF pH rises. **c.** Hypokalemia is a condition characterized by blood K^+ levels below 3.5 mEq/L. Hyperkalemia is a condition characterized by blood K^+ levels above 5.0 mEq/L.

Module 25.6 Review

a. Your body is in acid–base balance when the production of hydrogen ions is precisely offset by their loss, and when the pH of body fluids remains within normal limits. **b.** The primary challenge to acid–base homeostasis is that your body generates a variety of acids during normal metabolic operations, and a significant decrease in body fluid pH must be prevented. **c.** The three classes of acids that threaten pH balance are fixed acids, organic acids, and volatile acids.

Module 25.7 Review

a. Acidemia is the condition in which blood pH decreases below 7.35. Alkalemia exists when blood pH is above 7.45. **b.** Carbonic acid and its dissociation into hydrogen ions and bicarbonate ions directly affects the pH of the ECF. **c.** An inverse relationship exists between P_{CO_2} levels and pH.

Module 25.8 Review

a. The body's three major buffer systems are the phosphate buffer system, the protein buffer systems, and the carbonic acid–bicarbonate buffer system. **b.** The phosphate buffer system plays an important role in buffering the pH of the ICF and the urine. **c.** The carbonic acid–bicarbonate buffer system prevents pH changes caused by organic acids and fixed acids generated by metabolic activity. It uses the H^+ released by these acids to generate carbonic acid, which dissociates into H_2O and CO_2, the latter of which is exhaled from the lungs.

Module 25.9 Review

a. Metabolic acidosis results from the production of large numbers of fixed and organic acids, from the depletion of the bicarbonate reserve, caused by an inability to excrete hydrogen ions at the kidneys, or bicarbonate loss. **b.** Metabolic alkalosis results from the removal of large numbers of hydrogen ions from body fluids when bicarbonate ion concentrations become elevated. **c.** The kidneys conserve HCO_3^- and eliminate H^+ in the urine during metabolic acidosis.

Module 25.10 Review

a. Respiratory acidosis is lowered blood pH resulting from inadequate respiratory activity and is characterized by elevated levels of carbon dioxide. Respiratory alkalosis is elevated blood pH due to excessive respiratory activity, which decreases carbon dioxide levels and increases the pH of body fluids. **b.** The blood P_{CO_2} of a patient with an airway obstruction would increase, resulting in respiratory acidosis. **c.** A decrease in the pH of body fluids would cause an increase in the respiratory rate.

Section Reviews

Section 1 Review

1. i; **2.** d; **3.** h; **4.** n; **5.** l; **6.** m; **7.** b; **8.** c; **9.** a; **10.** j; **11.** e;

12. k; **13.** f; **14.** g; **15.** b; **16.** c; **17.** b; **18.** a; **19.** d; **20.** c; **21.** When tissues are burned, cells are destroyed and the contents of their cytoplasm leak into the interstitial fluid and then move into the blood. Since potassium ions are normally found within cells, damage to a large number of cells releases relatively large amounts of potassium ions into the blood. The elevated potassium level would stimulate cells of the adrenal cortex to produce aldosterone. The elevated levels of aldosterone would promote sodium retention and potassium secretion by the kidneys, thereby accounting for the elevated potassium levels in the patients' urine.

Section 2 Review

1. increased P_{CO_2}; **2.** acidosis; **3.** blood pH decrease; **4.** increased; **5.** secreted; **6.** generated; **7.** decreased; **8.** decreased; **9.** increased; **10.** decreased P_{CO_2}; **11.** alkalosis; **12.** blood pH increase; **13.** decreased; **14.** generated; **15.** secreted; **16.** increased; **17.** increased; **18.** decreased; **19.** The young boy has metabolic and respiratory acidosis. The metabolic acidosis resulted primarily from the large amounts of lactic acid generated by the boy's muscles as he struggled in the water. (The dissociation of lactic acid releases hydrogen ions and lactate ions.) Sustained hypoventilation during drowning contributed to both tissue hypoxia and respiratory acidosis. Respiratory acidosis developed as the P_{CO_2} increased in the ECF, increasing the production of carbonic acid and its dissociation into H^+ and $HCO3^-$. Prompt emergency treatment is essential; the usual procedure involves some form of artificial or mechanical respiratory assistance (to increase the respiratory rate and decrease the P_{CO_2} in the ECF) coupled with the intravenous infusion of a buffered isotonic solution that would absorb the hydrogen ions in the ECF and increase body fluid pH.

Chapter Review Questions

1. false; **2.** true; **3.** false; **4.** true; **5.** false; **6.** false; **7.** true; **8.** false; **9.** true; **10.** true; **11.** d; **12.** f; **13.** g; **14.** a; **15.** c; **16.** b; **17.** h; **18.** e; **19.** d; **20.** b; **21.** b; **22.** a; **23.** d; **24.** c; **25.** c; **26.** b; **27.** Fluid balance is when water content in the body remains stable over time. Water gains from the digestive tract must balance with water losses through urination, evaporation at the skin and lungs, feces, and varying degrees of secretion from sweat glands. Mineral balance is the balance between ion absorption, which occurs across the lining of the small intestine and colon, and ion excretion, which is done primarily by the kidneys. Acid–base balance is when the production of hydrogen ions is precisely offset by their loss, and when the pH of body fluids remains within normal limits. **28.** The hypertonic saline solution would cause fluid to move from the ICF to the ECF, further aggravating the patient's dehydration.

Chapter Integration

1. Digestive secretions contain high levels of bicarbonate, so people with diarrhea can lose significant amounts of this important ion, leading to acidosis. We would expect the blood pH to be lower than 7.35, and the urine pH to be low (due to increased renal excretion of hydrogen ions). We would also expect an increase in the rate and depth of breathing as the respiratory system tries to compensate for the lowered pH by eliminating carbon dioxide. **2.** The hypertonic solution will cause fluid to move from the ICF to the ECF, further aggravating Tom's dehydration. The slight increase in pressure and osmolarity of the blood should lead to an increase in ADH, even though ADH levels are probably quite high already. Despite the high ADH levels, urine volume would probably increase, because the kidneys could not reabsorb much of the glucose. The remaining glucose would increase the osmolarity of the tubular fluid, decreasing water reabsorption and increasing urine volume.

CHAPTER 26

Module Reviews

Module 26.1 Review

a. The function of the testes is to produce hormones and the male gametes, called spermatozoa. **b.** The accessory organs of the male reproductive system are the ductus deferens (paired), seminal glands, prostate gland, bulbourethral glands, urethra, and epididymis (paired). **c.** The structures of the male external genitalia are the penis and scrotum.

Module 26.2 Review

a. The testes form during fetal development inside the body cavity adjacent to the kidneys. As the fetus grows, connective tissue bands connected to the developing testes do not grow. So, the relative position of the testes change as the body grows. Late in fetal development, hormones cause the connective tissue bands to contract and pull each testis through the abdominal musculature and into the scrotum. **b.** The complex network of channels that is connected to the seminiferous tubules is the rete testis. **c.** On a warm day, the cremaster muscle (as well as the dartos muscle) would be relaxed so that the scrotum could descend away from the warmth of the body, thereby cooling the testes.

Module 26.3 Review

a. Spermatogenesis is the production of spermatozoa and involves mitosis, meiosis, and spermiogenesis. **b.** Four haploid spermatozoa will be produced from each diploid primary spermatocyte. **c.** A typical spermatozoon has an acrosome (acrosomal cap) that contains enzymes essential to fertilization; a head that is packed with chromosomes; a neck that contains centrioles; a middle piece containing mitochondria to provide ATP for propulsion; and a flagellum, the whiplike structure that moves the cell.

Module 26.4 Review

a. Interstitial cells produce male sex hormones, or androgens, including testosterone and androstenedione, the dominant sex hormones in males. **b.** Nurse cells provide nutrients to the developing sperm and form the blood–testis barrier that isolates sperm from the blood. **c.** During spermiation, a spermatozoon loses its attachment to the nurse cell and enters the lumen of the seminiferous tubule.

Module 26.5 Review

a. Semen is sperm plus seminal fluid, which includes the secretions from nurse cells; the epididymis; and the seminal, prostate, and bulbourethral glands. **b.** The secretion of the bulbourethral glands lubricates the tip of the penis and neutralizes any urinary acids that may remain in the urethra. **c.** The tail of each epididymis connects with the ductus deferens, which passes through the inguinal canal as part of the spermatic cord. Near the prostate gland, each ductus deferens enlarges to form an ampulla. The junction of the base of the seminal gland and the ampulla creates the ejaculatory duct, which empties into the urethra.

Module 26.6 Review

a. The three columns of erectile tissue are the corpus spongiosum and the paired corpora cavernosa. **b.** The phases of the male sexual response are arousal, emission, and ejaculation. **c.** An inability to contract the ischiocavernosus and bulbospongiosus muscles would interfere with the ejaculation phase, including the contractions that are part of the male orgasm.

Module 26.7 Review

a. Important regulatory hormones that establish and maintain male sexual function are follicle-stimulating hormone (FSH), luteinizing hormone (LH), and gonadotropin-releasing hormone

(GnRH). Testosterone is the most important androgen, although some tissues, such as the external genitalia and prostate gland, are more responsive or sensitive to dihydrotestosterone (DHT). **b.** The testes, hypothalamus, and the anterior lobe of the pituitary gland secrete the hormones that control male reproductive functions. **c.** Low FSH levels would lead to low levels of testosterone in the seminiferous tubules, decreasing both the sperm production rate and sperm count.

Module 26.8 Review

a. The main organs of the female reproductive system are the ovaries, uterine tubes, uterus, vagina, and external genitalia. **b.** The female external genitalia include the clitoris and labia. **c.** Fertilization normally occurs in the uterine tubes.

Module 26.9 Review

a. The ovaries produce immature female gametes called oocytes; secrete female sex hormones, including estrogens and progesterone; and secrete inhibin, which is involved in the feedback control of FSH production. **b.** The vesicouterine pouch is the pocket between the uterus and posterior wall of the urinary bladder. The rectouterine pouch is the pocket between the uterus and the anterior surface of the colon. **c.** The ovaries, uterine tubes, and uterus are enclosed within the broad ligament. The mesovarium supports and stabilizes each ovary.

Module 26.10 Review

a. An oocyte is an immature female gamete whose meiotic divisions will produce a single ovum and two or three polar bodies. **b.** Males produce gametes from puberty until death; females produce gametes only from puberty to menopause. Males produce many gametes at a time; females typically produce one or two per 28-day cycle. Males release mature gametes that have completed meiosis; females release secondary oocytes suspended in metaphase of meiosis II. **c.** Important events in the ovarian cycle are (1) formation of primary follicles from primordial follicles in an egg nest, (2) formation of secondary follicles, (3) formation of a tertiary follicle, (4) ovulation, and (5) formation and degeneration of the corpus luteum. A corpus albicans forms if fertilization does not occur.

Module 26.11 Review

a. Recently released secondary oocytes reach the uterine tube with the aid of the beating action of cilia on the inner surfaces of the fimbriae of the infundibulum. **b.** The perimetrium is the outer, incomplete serosal layer; the myometrium is the middle, muscular layer; and the endometrium is the inner, glandular layer. **c.** The regions of the uterus are the fundus (upper, rounded portion), body (ending at the internal os), and cervix (inferior portion).

Module 26.12 Review

a. The endometrium contains the deeper basilar zone and the more superficial functional zone. **b.** Menses is the first phase of the uterine cycle and is marked by degeneration of the functional zone of the uterus. Menstruation is the process of endometrial shedding that occurs during menses. **c.** The uterine cycle begins with menses, the destruction of the functional zone. After menses, the proliferative phase begins, during which the functional zone undergoes repair and thickens. Following the proliferative phase is the secretory phase, during which uterine (endometrial) glands enlarge.

Module 26.13 Review

a. The vagina (1) serves as a passageway for the elimination of menstrual fluids; (2) receives the penis during sexual intercourse, and holds spermatozoa prior to their passage into the uterus; and (3) forms the inferior portion of the birth canal, through which the fetus passes during delivery. **b.** The vagina is a muscular tube extending between the uterus and external genitalia; its lining forms folds called rugae. The proximal portion of the vagina is marked by the cervix, which dips into the vaginal canal, and the shallow recess known as the fornix. The hymen, a thin epithelial fold, partially blocks the entrance to the vagina until physical distortion ruptures it. **c.** The greater vestibular glands in females are similar to the bulbourethral glands in males; and both the male penis and the female clitoris have erectile tissue.

Module 26.14 Review

a. Lactation is the secretion of milk by the mammary glands. **b.** Blockage of a single lactiferous sinus would not interfere with the delivery of milk to the nipple, because each breast generally has several lactiferous sinuses. **c.** Route of milk flow: secretory alveoli of the secretory lobules → ducts within a lobe → lactiferous duct of the lobe → lactiferous sinus → surface of the nipple.

Module 26.15 Review

a. Completion of the decrease in the levels of estrogens and progesterone signals the beginning of menses and the start of a new uterine cycle. **b.** If the LH surge did not occur during an ovarian cycle, ovulation and corpus luteum formation could not occur. **c.** The hypothalamic secretion of GnRH triggers the pituitary secretion of FSH and LH. FSH initiates follicular development, and activated follicles and ovarian interstitial cells produce estrogens. High estrogen levels stimulate LH secretion and increase anterior pituitary gland sensitivity to GnRH, causing the release of LH. Progesterone is the main hormone of the luteal phase. Changes in estrogen and progesterone levels are responsible for maintaining the uterine cycle.

Module 26.16 Review

a. Condoms provide some protection against sexually transmitted diseases. **b.** Depo-Provera injections (a progesterone-only form of birth control) result in the cessation of the uterine cycle in 50 percent of women using this product. (Uterine cycles eventually resume after use of the product is discontinued.) **c.** A vasectomy is the surgical removal of a segment of each ductus deferens and the tying or cauterizing of the cut ends, preventing spermatozoa from reaching the distal portions of the male reproductive tract.

Module 26.17 Review

a. Ovarian cancer usually arises from epithelial cells. **b.** The human papillomavirus (HPV) causes most cases of cervical cancer. **c.** A sexually transmitted disease is a disease that is transferred from one person to another primarily or exclusively through sexual contact.

Section Reviews

Section 1 Review

1. prostatic urethra; **2.** ductus deferens; **3.** penile urethra; **4.** penis; **5.** epididymis; **6.** testis; **7.** external urethral orifice; **8.** scrotum; **9.** seminal gland; **10.** prostate gland; **11.** ejaculatory duct; **12.** bulbourethral gland; **13.** j; **14.** h; **15.** i; **16.** l; **17.** m; **18.** k; **19.** n; **20.** b; **21.** c; **22.** a; **23.** f; **24.** d; **25.** e; **26.** g; **27.** Normal levels of testosterone (1) promote the functional maturation of spermatozoa, (2) maintain the accessory organs of the male reproductive tract, (3) are responsible for establishing and maintaining male secondary sex characteristics, (4) stimulate bone and muscle growth, and (5) stimulate sexual behaviors and sexual drive (libido).

Section 2 Review

1. infundibulum; **2.** ovary; **3.** uterine tube; **4.** perimetrium; **5.** myometrium; **6.** endometrium; **7.** uterus; **8.** clitoris; **9.** labium minus; **10.** labium majus; **11.** fornix; **12.** cervix; **13.** external os; **14.** vagina; **15.** j; **16.** d; **17.** l; **18.** i; **19.** b; **20.** f; **21.** n; **22.** m; **23.** e; **24.** a; **25.** h; **26.** c; **27.** k; **28.** g; **29.** The endometrial cells have receptors for estrogens and progesterone and respond to these hormones as if the cells were still in the body of the uterus. Under the influence of estrogens, the endometrial cells proliferate at the beginning of the uterine (menstrual) cycle and begin to develop glands and blood vessels, which then further develop under the control of progesterone. This dramatic increase in tissue size exerts pressure on neighboring tissues or in some other way interferes with their function. It is the recurring expansion of tissue in an abnormal location that causes periodic pain.

Chapter Review Questions

1. ampulla; **2.** ductus deferens; **3.** seminal gland; **4.** prostate gland; **5.** corpus spongiosum; **6.** corpus cavernosum; **7.** epididymis; **8.** testis; **9.** external urethral orifice; **10.** bulbourethral gland; **11.** glans penis; **12.** mons pubis; **13.** prepuce; **14.** clitoris; **15.** labium majus; **16.** urethral opening; **17.** labium minus; **18.** hymen; **19.** vaginal entrance; **20.** false; **21.** true; **22.** true; **23.** false; **24.** true; **25.** false; **26.** c; **27.** c; **28.** b; **29.** a; **30.** d; **31.** a; **32.** Oral contraceptives that contain estrogen and progesterone or only progesterone inhibit GnRH release at the hypothalamus and thus FSH and LH release from the anterior lobe of the pituitary gland through negative feedback. Without FSH, primordial follicles do not begin to develop, and levels of estrogen remain low. An LH surge, triggered by the peaking of estrogen, is necessary for ovulation to occur. If the level of estrogen is not allowed to rise above the critical level, the LH surge will not occur, and thus ovulation will not occur, even if a follicle managed to develop to a stage at which it could ovulate. Any mature follicles would ultimately degenerate, and no new follicles would mature to take their place. Although the ovarian cycle is interrupted, the level of hormones is still adequate to regulate a normal menstrual cycle. **33.** After ovulation the tertiary follicle initially collapses, and the remaining granulosa cells proliferate to create the corpus luteum that secretes progesterone. **34.** The sterilization procedure for men is vasectomy; for women it is tubal ligation. After vasectomy, male sexual function is not affected. Spermatozoa continue to develop, but they are not present in semen because the ductus deferens is now blocked. A man continues to experience an erection as he did before. After tubal ligation, a woman still experiences the menstrual cycle, but the spermatozoa are unable to contact the secondary oocyte because the uterine tubes are now blocked.

Chapter Integration

1. The presence of exercise-induced amenorrhea suggests that a certain amount of body fat is necessary for menstrual cycles to occur. If body fat levels fall below some set point, menstruation ceases. **2.** Because a woman lacking adequate body fat might not have the energy reserves needed to have a successful pregnancy, compensatory mechanisms in her body prevent pregnancy by shutting down the ovarian cycle, and thus the menstrual cycle. When her body subsequently accumulates sufficient energy reserves in body fat, the cycles begin again. **3.** Estrogen deficiency can lead to infertility, vaginal and breast atrophy, and osteoporosis. It may also increase the risk of heart attacks later in life.

CHAPTER 27

Module Reviews

Module 27.1 Review

a. Gestation is the time spent in prenatal development, and the gestation period is the three integrated trimesters, each 3 months long. **b.** Embryonic development refers to the events that occur during the first 2 months after fertilization. Fetal development begins at the start of the ninth week and continues until birth. Prenatal development comprises both embryonic and fetal development. **c.** Postnatal development begins at birth and continues until maturity.

Module 27.2 Review

a. Fertilization is the fusion of a secondary oocyte and a spermatozoon to form a zygote. **b.** A normal human zygote contains 46 chromosomes. **c.** Many spermatozoa are needed to achieve fertilization because one spermatozoon (sperm) does not contain enough acrosomal enzymes to erode the corona radiata surrounding the secondary oocyte.

Module 27.3 Review

a. The morula is the final stage of development that results from cleavage. **b.** The trophoblast is the cell layer surrounding the blastocyst. These cells nourish the embryo and later form part of the placenta. **c.** The blastocyst consists of an outer trophoblast and an inner cell mass. Implantation begins about 7 days after fertilization when the blastocyst adheres to the uterine lining. The trophoblast cells in contact with the uterine lining divide and form a syncytial trophoblast that erodes the endometrial lining, and the blastocyst becomes enclosed within the endometrium by about 10 days after fertilization.

Module 27.4 Review

a. Gastrulation is the formation of the primary germ layers—the endoderm, ectoderm, and mesoderm—from the embryonic disc. It is from these germ layers that the body systems differentiate. **b.** The mesoderm gives rise to nearly all body systems except the nervous and respiratory systems. **c.** Gestational trophoblastic neoplasia is a tumor formed by undifferentiated, rapid growth of the syncytial trophoblast; if untreated, the neoplasm may become malignant.

Module 27.5 Review

a. The four extra-embryonic membranes are the yolk sac, amnion, allantois, and chorion. **b.** The yolk sac forms from the endoderm and mesoderm; it is an important site of blood cell formation. The amnion forms from the ectoderm and mesoderm; it encloses the fluid that surrounds and cushions the developing embryo and fetus. The allantois forms from the endoderm and mesoderm; its base gives rise to the urinary bladder. The chorion forms from the mesoderm and trophoblast; it surrounds the blastocoele. **c.** The placenta forms the interface between embryonic/fetal system and the maternal system. It becomes the primary support structure for the developing embryo and fetus. The placenta is also the site where oxygen and nutrients are absorbed from the maternal bloodstream and exchanged for carbon dioxide and wastes.

Module 27.6 Review

a. The chorionic villi are structures that extend outward into the maternal tissues, forming an intricate, branching network through which maternal blood flows. Embryonic blood vessels extend into each chorionic villus. **b.** The body stalk is the connection between the embryo and the chorion; it contains portions of the allantois and blood vessels that carry blood to and from the placenta. The yolk stalk is the connection between the endoderm of the embryo and the yolk sac. **c.** The umbilical cord connects the fetus to the placenta, and it is derived from the allantois.

Module 27.7 Review

a. The hormones synthesized by the syncytial trophoblast are human chorionic gonadotropin (hCG), human placental lactogen (hPL), relaxin, progesterone, and estrogens. **b.** The presence of human chorionic gonadotropin (hCG) in the urine provides a reliable indicator of pregnancy. **c.** After the first trimester, the placenta is sufficiently functional to maintain the pregnancy.

Module 27.8 Review

a. Organogenesis is the process of organ formation. **b.** During the second and third trimesters the fetus grows larger, and the organ systems increase in complexity and approach normal functional capabilities. **c.** During the third trimester, the fetus undergoes its largest absolute weight gain.

Module 27.9 Review
a. Difficulty breathing in pregnant women results from the enlarged uterus pressing against the diaphragm and crowding the lungs. **b.** The major changes that occur in maternal systems during pregnancy are increases in respiratory rate and tidal volume, blood volume, nutrient requirements, glomerular filtration rate (GFR), and the size of the uterus and mammary glands. **c.** A mother's blood volume increases during pregnancy to compensate for the reduction in maternal blood volume resulting from blood flow through the placenta.

Module 27.10 Review
a. Relaxin, produced by the placenta, softens the pubic symphysis and dilates the cervix, and the weight of the fetus distorts the external os of the uterus. Distortion of the cervix and rising estrogen levels promote the release of oxytocin, and the already stretched smooth muscles of the myometrium become even more excitable. **b.** Estrogens and oxytocin stimulate the production of prostaglandins, which are then primarily responsible for initiating true labor. **c.** The dilation stage begins with the onset of true labor, as the cervix dilates and the fetus begins to move toward the cervical canal; late in this stage, the amnion ruptures. The expulsion stage begins as the cervix dilates completely and continues until the fetus has completely emerged from the vagina (delivery). In the placental stage, the uterus gradually contracts, tearing the connections between the endometrium and the placenta and ejecting the placenta (also referred to as the afterbirth).

Module 27.11 Review
a. Oxytocin causes the milk let-down (milk-ejection) reflex. **b.** Colostrum contains antibodies and is produced by the mammary glands from the end of the sixth month of pregnancy until a few days after delivery. After that, the glands begin producing breast milk, which contains antibodies and lysozyme but has a higher fat content than colostrum. **c.** The postnatal stages of development are the neonatal period, from birth to 1 month; infancy, from 1 month to age 2 years; childhood, from age 2 years until sexual maturation begins; adolescence, which begins with the onset of sexual maturation (puberty) between ages 9–14 years and ends when growth in body size ends (around 18 years); and maturity, which includes the rest of the person's life. A final stage called senescence, or aging, overlaps with maturity.

Module 27.12 Review
a. The three interacting hormonal events associated with the onset of puberty are (1) increased GnRH production by the hypothalamus; (2) increased sensitivity to GnRH by the anterior lobe of the pituitary gland, and a rapid increase in circulating levels of FSH and LH; and (3) increased sensitivity to FSH and LH by ovarian and testicular cells. **b.** A man has a deeper voice and a larger larynx than a woman because testosterone stimulates laryngeal development in males to a greater extent than estrogen stimulates laryngeal development in females. **c.** The higher estrogen levels in premenopausal women (compared with adult men) decrease plasma cholesterol levels, thereby slowing plaque formation and lowering the risk of atherosclerosis.

Module 27.13 Review
a. Chromosomes and their associated component genes constitute the genotype. Anatomical and physiological characteristics from the expression of genes make up the phenotype. **b.** Autosomal chromosomes are the 22 pairs that affect somatic characteristics. The chromosomes of the 23rd pair are the sex chromosomes that determine whether an individual is male (XY) or female (XX). **c.** The karyotype shown is male because it has an XY for the 23rd pair of chromosomes.

Module 27.14 Review
a. Homozygous means that homologous chromosomes carry the same allele of a given gene. Heterozygous means that homologous chromosomes carry different alleles of a given gene. **b.** In simple inheritance, phenotypic characteristics are determined by interactions between a single pair of alleles. Polygenic inheritance involves interactions among alleles of several genes. **c.** The phenotype of a person who is heterozygous for "freckles"—that is, a person with one dominant allele and one recessive allele for that trait—would be "freckles."

Module 27.15 Review
a. In strict dominance, one allele dominates the other, so an individual who is heterozygous for a given trait exhibits the dominant phenotype. In codominance, an individual who is heterozygous for a given trait exhibits both of the phenotypes for that trait. **b.** Sex-linked traits are expressed more frequently in males because the small Y chromosome of males contains few genes that correspond to those on the larger X chromosome. **c.** (1) simple inheritance (dominant); (2) simple inheritance (recessive); (3) autosomal inheritance.

Module 27.16 Review
a. A single nucleotide polymorphism is a variation in a single base pair in a DNA sequence. **b.** (1) XO: Turner's syndrome; (2) XXY: Klinefelter's syndrome. **c.** (1) ovarian cancer: chromosome 9; (2) Tay–Sachs disease: chromosome 15; (3) spinocerebellar ataxia: chromosome 6.

Section Reviews

Section 1 Review
1. b; **2.** j; **3.** g; **4.** k; **5.** a; **6.** m; **7.** e; **8.** f; **9.** n; **10.** c; **11.** l; **12.** d; **13.** h; **14.** i; **15.** c; **16.** c; **17.** b; **18.** b; **19.** a; **20.** c; **21.** It is very unlikely that the baby's condition is the result of a viral infection contracted during the third trimester. The development of organ systems occurs during the first trimester, and by the end of the second trimester, most organ systems are fully formed. During the third trimester, the fetus undergoes tremendous growth, but very little new organ formation occurs.

Section 2 Review

1. l; **2.** g; **3.** i; **4.** k; **5.** a; **6.** c; **7.** b; **8.** j; **9.** f; **10.** h; **11.** e; **12.** d;
13. Tongue-rolling

		Maternal alleles	Maternal alleles
		T	*t*
Paternal alleles	*T*	*TT*	*Tt*
Paternal alleles	*t*	*Tt*	*tt*

Mom = *Tt*
Dad = *Tt*
The possible genotypic and phenotypic ratios of offspring are the following:
1 *TT* = tongue-roller; 1 in 4 = 25%
2 *Tt* = tongue-rollers; 2 in 4 = 50%
1 *tt* = non-tongue-roller; 1 in 4 = 25%
The probability that they will have a child unable to roll his/her tongue is 1 in 4 or 25%.
14. Cystic fibrosis

		Maternal alleles	Maternal alleles
		c	*c*
Paternal alleles	*C*	*Cc*	*Cc*
Paternal alleles	*c*	*cc*	*cc*

Mom = *cc*
Dad = *Cc*
The possible genotypic and phenotypic ratios of offspring are the following:
2 *Cc* = normal carriers; 2 in 4 = 50%
2 *cc* = cystic fibrosis; 2 in 4 = 50%
The probability that they will have a child who will be a carrier for CF is 2 in 4 or 50%.
(A child has to inherit two copies of the faulty allele to be born with CF.)

Chapter Review Questions

1. false; **2.** true; **3.** false; **4.** true; **5.** false; **6.** true; **7.** false; **8.** false; **9.** b; **10.** b; **11.** d; **12.** a; **13.** b; **14.** d; **15.** c; **16.** d; **17.** The five stages of postnatal development are neonatal (1 month), infancy (1 year), childhood (up to 9 years), adolescence (between 9 and 14 years), and maturity (18 years). **18.** Trisomy 21, or Down's syndrome, is when the affected person has a third chromosome on the 21st pair. Affected people exhibit intellectual disability and physical malformations, including a characteristic facial appearance. There is a direct correlation between maternal age and the risk of having a child with trisomy 21. For a maternal age below 25, the incidence of Down's syndrome approaches 1 in 2000 births, or 0.05 percent. For maternal ages 30–34, the odds increase to 1 in 900, and during ages 35–44 they increase to 1 in 46, or more than 2 percent.

Chapter Integration

1. Although technically it takes only one sperm to fertilize a secondary oocyte, Joe's oligospermia means that most sperm entering the female reproductive tract are killed or disabled before they reach the uterus. Thus, too few sperm reach the secondary oocyte in a uterine tube to produce sufficient acrosomal enzymes to enable one sperm to penetrate the corona radiata and fuse with the oocyte membrane. **2.** Hemophilia A is an X-linked recessive trait. The probability that this couple's daughters will have hemophilia is zero, because each daughter will receive a dominant normal allele from her father. There is a 50 percent chance that a son will have hemophilia, because each son has a 50 percent chance of receiving the mother's normal allele, and a 50 percent chance of receiving the mother's recessive allele.

A

abdomen: The region of the trunk between the inferior margin of the rib cage and the superior margin of the pelvis.
abdominopelvic cavity: The term used to refer to the general region bounded by the abdominal wall and the pelvis; it contains the peritoneal cavity and visceral organs.
abducens: Cranial nerve VI, which innervates the lateral rectus muscle of the eye.
abduction: Movement away from the midline of the body, as viewed in the anatomical position.
abscess: A localized collection of pus within a damaged tissue.
absorption: The active or passive uptake of gases, fluids, or solutes.
accommodation: An alteration in the curvature of the lens of the eye to focus an image on the retina.
acetabulum: The fossa on the lateral aspect of the pelvis that accommodates the head of the femur.
acetylcholine (ACh): A chemical neurotransmitter in the brain and peripheral nervous system; the dominant neurotransmitter in the peripheral nervous system, released at neuromuscular junctions and synapses of the parasympathetic division.
acetylcholinesterase (AChE): An enzyme found in the synaptic cleft, bound to the postsynaptic membrane, and in tissue fluids; breaks down and inactivates acetylcholine molecules.
acetyl-CoA: An acetyl group bound to coenzyme A, a participant in the anabolic and catabolic pathways for carbohydrates, lipids, and many amino acids.
acetyl group: $—CH_3CO$.
Achilles tendon: The large tendon that inserts on the calcaneus; tension on this tendon produces extension (plantar flexion) of the foot; also called *calcaneal tendon.*
acid: A compound whose dissociation in solution releases a hydrogen ion and an anion; an acidic solution has a pH below 7.0 and contains an excess of hydrogen ions.
acidosis: An abnormal physiological state characterized by a blood pH below 7.35.
acinus/acini: A histological term referring to a blind pocket, pouch, or sac.
acoustic: Pertaining to sound or the sense of hearing.
acromion: A continuation of the scapular spine that projects superior to the capsule of the shoulder joint.
acrosomal cap: A membranous sac at the tip of a spermatozoon that contains hyaluronidase.
actin: The protein component of microfilaments that forms thin filaments in skeletal muscles and produces contractions of all muscles through interaction with thick (myosin) filaments.
action potential: A propagated change in the membrane potential of excitable cells, initiated by a change in the membrane permeability to sodium ions.
active transport: The ATP-dependent absorption or secretion of solutes across a plasma membrane.
acute: Sudden in onset, severe in intensity, and brief in duration.
adaptation: A change in pupillary size in response to changes in light intensity; a decrease in receptor sensitivity or perception after chronic stimulation.
adaptive immunity: Type of defense that is not present at birth and is related to previous antigen exposure; also called *specific immunity.*
Addison's disease: A condition resulting from the hyposecretion of glucocorticoids; characterized by lethargy, weakness, hypotension, and increased skin pigmentation.
adduction: Movement toward the axis or midline of the body, as viewed in the anatomical position.
adenine: A purine; one of the nitrogenous bases in the nucleic acids RNA and DNA.
adenohypophysis: The anterior lobe of the pituitary gland.
adenosine: A compound consisting of adenine and ribose.
adenosine diphosphate (ADP): A compound consisting of adenosine with two phosphate groups attached.
adenosine monophosphate (AMP): A nucleotide consisting of adenosine plus a phosphate group (PO_4^{3-}); also called *adenosine phosphate.*
adenosine triphosphate (ATP): A high-energy compound consisting of adenosine with three phosphate groups attached; the third is attached by a high-energy bond.
adenylate cyclase: An enzyme bound to the inner surfaces of plasma membranes that can convert ATP to cyclic AMP; formerly called *adenylyl cyclase.*
adipocyte: A fat cell.
adipose tissue: Loose connective tissue dominated by adipocytes.
adrenal cortex: The superficial portion of the adrenal gland that produces steroid hormones; also called *suprarenal cortex.*
adrenal gland: A small endocrine gland that secretes steroids and catecholamines and is located superior to each kidney; also called *suprarenal gland.*
adrenal medulla: The core of the adrenal gland; a modified sympathetic ganglion that secretes catecholamines into the blood during sympathetic activation; also called *suprarenal medulla.*
adrenergic: An axon terminal that, when stimulated, releases norepinephrine.
adrenocortical hormone: Any steroid produced by the adrenal cortex.
adrenocorticotropic hormone (ACTH): The hormone that stimulates the production and secretion of glucocorticoids by the zona fasciculata of the adrenal cortex; released by the anterior lobe of the pituitary gland in response to corticotropin-releasing hormone.
adventitia: The superficial layer of connective tissue surrounding an internal organ; fibers are continuous with those of surrounding tissues, providing support and stabilization.
aerobic: Requiring the presence of oxygen.
aerobic metabolism: The complete breakdown of organic substrates into carbon dioxide and water, via pyruvate; a process that yields large amounts of ATP but requires mitochondria and oxygen.
afferent: Toward a center.
afferent arteriole: An arteriole that carries blood to a glomerulus of the kidney.
afferent fiber: An axon that carries sensory information to the central nervous system.
agglutination: The aggregation of red blood cells due to interactions between surface antigens and plasma antibodies.
aggregated lymphoid nodules: Lymphoid nodules beneath the epithelium of the small intestine; also called *Peyer's patches.*
agonist: A muscle responsible for a specific movement; also called a *prime mover.*
agranular: Without granules; *agranular leukocytes* are monocytes and lymphocytes.
alba: White.
albicans: White.
albuginea: White.
aldosterone: A mineralocorticoid produced by the zona glomerulosa of the adrenal cortex; stimulates sodium and water conservation at the kidneys; secreted in response to the presence of angiotensin II.
alkalosis: The condition characterized by blood pH greater than 7.45; associated with a relative deficiency of hydrogen ions or an excess of bicarbonate ions.
allograft: Tissue transplant between members of the same species.
alpha receptors: Membrane receptors sensitive to norepinephrine or epinephrine; stimulation normally results in the excitation of the target cell.
alveolar sac: An air-filled chamber that supplies air to several alveoli.
alveolus/alveoli: Blind pockets at the end of the respiratory tree, lined by a simple squamous epithelium and surrounded by a capillary network; sites of gas exchange with the blood; a bony socket that holds the root of a tooth.
Alzheimer's disease: A disorder resulting from degenerative changes in populations of neurons in the cerebrum, causing dementia characterized by problems with attention, short-term memory, and emotions.
amination: The attachment of an amino group to a carbon chain; performed by a variety of cells and important in the synthesis of amino acids.
amino acid: Organic compound made up of carbon, hydrogen, oxygen, and nitrogen; building blocks of protein.
amino group: $—NH_2$.
amnion: One of the four extraembryonic membranes; surrounds the developing embryo or fetus.
amniotic fluid: Fluid that fills the amniotic cavity; cushions and supports the embryo or fetus.
amphiarthrosis: A joint that permits a small degree of independent movement; *see* **interosseous membrane** and **pubic symphysis.**
amphipathic: Molecule containing both hydrophobic and hydrophilic portions.
ampulla/ampullae: A localized dilation in the lumen of a canal or passageway.
amygdaloid body: A basal nucleus that is a component of the limbic system and acts as an interface between that system, the cerebrum, and sensory systems.
amylase: An enzyme that breaks down polysaccharides; produced by the salivary glands and pancreas.
anabolism: The synthesis of complex organic compounds from simpler precursors.
anaerobic: Without oxygen.
anal triangle: The posterior subdivision of the perineum.
anaphase: The mitotic stage in which the paired chromatids separate and move toward opposite ends of the spindle apparatus.
anaphylaxis: A hypersensitivity reaction due to the binding of antigens to immunoglobulins (IgE) on the surfaces of mast cells; the release of histamine, serotonin, and prostaglandins by mast cells then causes widespread inflammation; a sudden decrease in blood pressure may occur, producing anaphylactic shock.
anastomosis: The joining of two tubes, usually referring to a connection between two peripheral vessels without an intervening capillary bed.
anatomical position: An anatomical reference position; the body viewed from the anterior surface with the palms facing forward.
anatomy: The study of the structure of the body.
androgen: A steroid sex hormone primarily produced by the interstitial cells of the testis and manufactured in small quantities by the adrenal cortex in both sexes.
anemia: The condition marked by a decrease in the hematocrit, the hemoglobin content of the blood, or both.
angiogenesis: Formation of new blood vessels.
angiotensin I: The hormone produced by the activation of angiotensinogen by renin; angiotensin-converting enzyme (ACE) converts angiotensin I into angiotensin II in lung capillaries.

angiotensin II: A hormone that causes an increase in systemic blood pressure, stimulates the secretion of aldosterone, promotes thirst, and causes the release of antidiuretic hormone; angiotensin-converting enzyme in lung capillaries converts angiotensin I into angiotensin II.
angiotensinogen: The blood protein produced by the liver that is converted to angiotensin I by the enzyme renin.
anion: An ion having a negative charge.
anoxia: Tissue oxygen deprivation.
antagonist: A muscle that opposes the movement of an agonist.
antebrachium: The forearm.
anterior: On or near the front, or ventral surface, of the body.
antibiotic: A chemical agent that selectively kills pathogens, primarily bacteria.
antibody: A globular protein produced by plasma cells that will bind to specific antigens and promote their destruction or removal from the body.
antibody-mediated immunity: The form of immunity resulting from the presence of circulating antibodies produced by plasma cells; also called *humoral immunity*.
anticodon: Three nitrogenous bases on a tRNA molecule that interact with a complementary codon on a strand of mRNA.
antidiuretic hormone (ADH): A hormone synthesized in the hypothalamus and secreted by the posterior lobe of the pituitary gland (neurohypophysis); causes water retention by the kidneys and an increase of blood pressure.
antigen: A substance capable of inducing the production of antibodies.
antigen-antibody complex: The combination of an antigen and a specific antibody.
antigenic determinant site: A portion of an antigen that can interact with an antibody molecule.
antigen-presenting cell (APC): A cell that processes antigens and displays them, bound to Class II MHC proteins; essential to the initiation of a normal immune response.
antihistamine: A chemical agent that blocks the action of histamine on peripheral tissues.
antrum: A chamber or pocket.
anulus: A cartilage or bone shaped like a ring; also spelled *annulus*.
anus: The external opening of the anal canal.
aorta: The large, elastic artery that carries blood away from the left ventricle and into the systemic circuit.
apocrine secretion: A mode of secretion in which the glandular cell sheds portions of its cytoplasm.
aponeurosis/aponeuroses: Broad tendinous sheet(s) that may serve as the origin or insertion of a skeletal muscle.
appendicular: Pertaining to the upper or lower limbs.
appendix: A blind sac connected to the cecum of the large intestine.
appositional growth: The enlargement of a cartilage or bone by the addition of cartilage or bony matrix at its surface.
aqueous humor: A fluid similar to perilymph or cerebrospinal fluid that fills the anterior chamber of the eye.
arachidonic acid: One of the essential fatty acids.
arachnoid granulations: Processes of the arachnoid mater that project into the superior sagittal sinus; sites where cerebrospinal fluid enters the venous circulation.
arachnoid mater: The middle meninx that encloses cerebrospinal fluid and protects the central nervous system.
arbor vitae: The central, branching mass of white matter inside the cerebellum.
arcuate: Curving.
areolar: Containing minute spaces, as in areolar tissue.
areolar tissue: Loose connective tissue with an open framework.
arrector pili: Smooth muscles whose contractions force hairs to stand erect.
arrhythmias: Abnormal patterns of cardiac contractions.
arteriole: A small arterial branch that delivers blood to a capillary network.
artery: A blood vessel that carries blood away from the heart and toward a peripheral capillary.
articular: Pertaining to a joint.
articular capsule: The dense collagen fiber sleeve that surrounds a joint and provides protection and stabilization.
articular cartilage: The cartilage pad that covers the surface of a bone inside a joint cavity.
articulation: A joint.
ascending tract: A tract carrying information from the spinal cord to the brain.
association areas: Cortical areas of the cerebrum that are responsible for the integration of sensory inputs and/or motor commands.
astrocyte: One of the four types of neuroglia in the central nervous system; responsible for maintaining the blood–brain barrier by the stimulation of endothelial cells.
atherosclerosis: The formation of fatty plaques in the walls of arteries, restricting blood flow to deep tissues.
atom: The smallest stable unit of matter.
atomic number: The number of protons in the nucleus of an atom.
atomic weight: The average of the different atomic masses and isotope proportions of a particular element.
atria: Thin-walled chambers of the heart that receive venous blood from the pulmonary or systemic circuit.
atrial natriuretic peptide (ANP): *See* **natriuretic peptides**.
atrioventricular (AV) node: Specialized cardiocytes that relay the contractile stimulus to the bundle of His, the bundle branches, the Purkinje fibers, and the ventricular myocardium; located at the boundary between the atria and ventricles.
atrioventricular (AV) valve: One of the valves that prevents backflow into the atria during ventricular systole (contraction).
atrophy: The wasting away of tissues from a lack of use, ischemia, or nutritional abnormalities.
auditory: Pertaining to the sense of hearing.
auditory ossicles: The bones of the middle ear: malleus, incus, and stapes.
auditory tube: A passageway that connects the nasopharynx with the middle ear cavity.
auricle: A broad, flattened process that resembles the external ear; in the ear, the expanded, projecting portion that surrounds the external auditory meatus, also called *pinna*; in the heart, the externally visible flap formed by the collapse of the outer wall of a relaxed atrium.
autoantibodies: Antibodies that react with antigens on the surfaces of a person's own cells and tissues.
autograft: Tissue transplant from the same person.
autoimmunity: The immune system's sensitivity to normal cells and tissues, resulting in the production of autoantibodies.
autolysis: The destruction of a cell due to the rupture of lysosomal membranes in its cytoplasm.
automaticity: The spontaneous depolarization to threshold, characteristic of cardiac pacemaker cells.
autonomic ganglion: A collection of visceral motor neurons outside the central nervous system.
autonomic nerve: A peripheral nerve consisting of preganglionic or postganglionic autonomic fibers.
autonomic nervous system (ANS): Centers, nuclei, tracts, ganglia, and nerves involved in the unconscious regulation of visceral functions; includes components of the central nervous system and the peripheral nervous system.
autopsy: The detailed examination of a body after death.
autoregulation: Changes in activity that maintain homeostasis in direct response to changes in the local environment; does not require neural or endocrine control.
autosomal: Chromosomes other than the X or Y sex chromosome.
avascular: Without blood vessels.
axilla: The armpit.
axolemma: The plasma membrane of an axon, continuous with the plasma membrane of the cell body and dendrites and distinct from any neuroglial coverings.
axon: The elongated extension of a neuron that conducts an action potential.
axon hillock: In a multipolar neuron, the portion of the cell body adjacent to the initial segment.
axon terminal: The club-shaped ending of an axon that contains neurotransmitters and makes synaptic contact with an effector; also called a *synaptic terminal*.
axoplasm: The cytoplasm within an axon.

B

bacteria: Single-celled microorganisms, some pathogenic, that are common in the environment and in and on the body.
baroreception: The ability to detect changes in pressure.
baroreceptor reflex: A reflexive change in cardiac activity in response to changes in blood pressure.
baroreceptors: The receptors responsible for baroreception.
basal lamina: A layer of filaments and fibers that attaches an epithelium to the underlying connective tissue; also called the *basement membrane*.
basal nuclei: Nuclei of the cerebrum that are important in the subconscious control of skeletal muscles.
base: A compound whose dissociation releases a hydroxide ion (OH^-) or removes a hydrogen ion (H^+) from the solution.
basement membrane: A layer of filaments and fibers that attaches an epithelium to the underlying connective tissue; also called the *basal lamina*.
basophils: Circulating granulocytes (white blood cells) similar in size and function to tissue mast cells.
B cells: Lymphocytes capable of differentiating into plasma cells, which produce antibodies.
benign: Not malignant.
beta cells: Cells of the pancreatic islets that secrete insulin in response to increased blood sugar concentrations.
beta oxidation: Fatty acid catabolism that produces molecules of acetyl-CoA.
beta receptors: Membrane receptors sensitive to epinephrine; stimulation may result in the excitation or inhibition of the target cell.
bicarbonate ion: HCO_3^-; anion component of the carbonic acid–bicarbonate buffer system.
bicuspid: Having two cusps or points; refers to a premolar tooth, which has two roots, or to the left AV valve, which has two cusps.
bicuspid valve: The left atrioventricular (AV) valve, also called *mitral valve*.
bifurcate: To branch into two parts.

bile: The exocrine secretion of the liver; stored in the gallbladder and ejected into the duodenum.
bile salts: Steroid derivatives in bile; responsible for the emulsification of ingested lipids.
bilirubin: A pigment that is the byproduct of hemoglobin catabolism.
biopsy: The removal of a small tissue sample for pathological analysis.
bladder: A muscular sac that distends as fluid is stored and whose contraction ejects the fluid at an appropriate time; used alone, the term usually refers to the urinary bladder.
blastocyst: An early stage in the developing embryo, consisting of an outer trophoblast and an inner cell mass.
blood–brain barrier: The isolation of the central nervous system from the general circulation; primarily the result of astrocyte regulation of capillary permeabilities.
blood–CSF barrier: The isolation of the cerebrospinal fluid from the capillaries of the choroid plexus; primarily the result of specialized ependymal cells.
blood pressure: A force exerted against vessel walls by the blood in the vessels, due to the push exerted by cardiac contraction and the elasticity of the vessel walls; usually measured along one of the muscular arteries, with systolic pressure measured during ventricular systole (contraction) and diastolic pressure during ventricular diastole (relaxation).
blood–testis barrier: The isolation of the interior of the seminiferous tubules from the general circulation, due to the activities of the nurse (sustentacular) cells.
Bohr effect: The increased oxygen release by hemoglobin due to increased carbon dioxide levels.
bolus: A compact mass; usually refers to compacted ingested material (food) on its way to the stomach.
bone: *See* **osseous tissue**.
bowel: The intestinal tract.
brachial: Pertaining to the arm.
brachial plexus: A network formed by branches of spinal nerves C_5–T_1 en route to innervating the upper limb.
brachium: The arm.
bradycardia: An abnormally slow heart rate, usually below 50 bpm.
brain natriuretic peptide (BNP): *See* **natriuretic peptides**.
brain stem: The brain minus the cerebrum, diencephalon, and cerebellum.
brevis: Short.
bronchial tree: The trachea, bronchi, and bronchioles.
bronchodilation: The dilation of the bronchial passages; can be caused by sympathetic stimulation.
bronchus/bronchi: Branch/branches of the bronchial tree between the trachea and bronchioles.
buccal: Pertaining to the cheeks.
buffer: A compound that stabilizes the pH of a solution by removing or releasing hydrogen ions.
buffer system: Interacting compounds that prevent increases or decreases in the pH of body fluids; includes the carbonic acid–bicarbonate buffer system, the phosphate buffer system, and the protein buffer system.
bulbar: Pertaining to the brain stem.
bulbourethral glands: Mucous glands at the base of the penis that secrete into the penile urethra; the equivalent of the greater vestibular glands of females; also called *Cowper's glands.*
bundle branches: Specialized conducting cells in the ventricles that carry the contractile stimulus from the atrioventricular bundle (bundle of His) to the Purkinje fibers.
bundle of His: Specialized conducting cells in the interventricular septum that carry the contracting stimulus from the AV node to bundle branches and then to Purkinje fibers.
bursa: A small sac filled with synovial fluid that cushions adjacent structures and reduces friction.

C

calcaneal tendon: The large tendon that inserts on the calcaneus; tension on this tendon produces extension (plantar flexion) of the foot; also called *Achilles tendon.*
calcaneus: The heel bone, the largest of the tarsal bones.
calcification: The deposition of calcium salts within a tissue.
calcitonin: The hormone secreted by C cells of the thyroid when calcium ion concentrations are abnormally high; restores homeostasis by increasing the rate of bone deposition and the rate of calcium loss by the kidneys.
calculus/calculi: Solid mass/masses of insoluble materials that form within body fluids, especially the gallbladder, kidneys, or urinary bladder.
callus: A localized thickening of the epidermis due to chronic mechanical stresses; a thickened area that forms at the site of a bone break as part of the repair process.
canaliculi: Microscopic passageways between cells; bile canaliculi carry bile to bile ducts in the liver; in bone, canaliculi permit the diffusion of nutrients and wastes to and from osteocytes.
cancellous bone: Spongy bone, composed of a network of bony struts.
cancer: An illness caused by mutations leading to the uncontrolled growth and replication of the affected cells.
cannula: A tube that can be inserted into the body; commonly placed in blood vessels prior to transfusion or dialysis.
capacitation: The activation process that must occur before a spermatozoon can successfully fertilize an oocyte; occurs in the vagina after ejaculation.
capillary: A small blood vessel, located between an arteriole and a venule, whose thin wall permits the diffusion of gases, nutrients, and wastes between plasma and interstitial fluids.
capitulum: A general term for a small, elevated articular process; refers to the rounded distal surface of the humerus that articulates with the head of the radius.
caput: The head.
carbaminohemoglobin: Hemoglobin bound to carbon dioxide molecules.
carbohydrase: An enzyme that breaks down carbohydrate molecules.
carbohydrate: An organic compound containing carbon, hydrogen, and oxygen in a ratio that approximates 1:2:1.
carbon dioxide: CO_2; a compound produced by the citric acid cycle reactions of aerobic metabolism.
carbonic anhydrase: An enzyme that catalyzes the reaction $H_2O + CO_2 \rightarrow H_2CO_3$; important in carbon dioxide transport, gastric acid secretion, and renal pH regulation.
carcinogenic: Stimulating cancer formation in affected tissues.
cardia: The area of the stomach surrounding its connection with the esophagus.
cardiac: Pertaining to the heart.
cardiac cycle: One complete heartbeat, including atrial and ventricular systole and diastole.
cardiac output: The amount of blood ejected by the left ventricle each minute.
cardiac reserve: The potential percentage increase in cardiac output above resting levels.
cardiac tamponade: A compression of the heart due to fluid accumulation in the pericardial cavity.
cardiocyte: A cardiac muscle cell.
cardiovascular: Pertaining to the heart, blood, and blood vessels.
cardiovascular center: Poorly localized area in the reticular formation of the medulla oblongata of the brain; includes cardioacceleratory, cardioinhibitory, and vasomotor centers.
cardium: The heart.
carotene: A yellow-orange pigment, found in carrots and in green and orange leafy vegetables, that the body can convert to vitamin A.
carotid artery: The principal artery of the neck, servicing cervical and cranial structures; one branch, the internal carotid, provides a major blood supply to the brain.
carotid body: A group of receptors, adjacent to the carotid sinus, that are sensitive to changes in the carbon dioxide levels, pH, and oxygen concentrations of arterial blood.
carotid sinus: A dilated segment at the base of the internal carotid artery whose walls contain baroreceptors sensitive to changes in blood pressure.
carotid sinus reflex: Reflexive changes in blood pressure that maintain homeostatic pressures at the carotid sinus, stabilizing blood flow to the brain.
carpus/carpal: The wrist.
cartilage: A connective tissue with a gelatinous matrix that contains an abundance of fibers.
catabolism: The breakdown of complex organic molecules into simpler components, accompanied by the release of energy.
catalyst: A substance that accelerates a specific chemical reaction but that is not altered by the reaction.
catecholamines: Epinephrine, norepinephrine, dopamine, and related compounds.
catheter: A tube surgically inserted into a body cavity or along a blood vessel or excretory passageway for the collection of body fluids, monitoring of blood pressure, or introduction of medications or radiographic dyes.
cation: An ion that has a positive charge.
cauda equina: Spinal nerve roots distal to the tip of the adult spinal cord; they extend caudally inside the vertebral canal en route to lumbar and sacral segments.
caudal/caudally: Closest to or toward the tail (coccyx).
caudate nucleus: One of the basal nuclei involved with the subconscious control of skeletal muscles.
cell: The smallest living structural unit in the human body.
cell-mediated immunity: Resistance to disease by sensitized T cells that destroy antigen-bearing cells by direct contact or through the release of lymphotoxins; also called *cellular immunity.*
center of ossification: The site in a connective tissue where bone formation begins.
central canal: Longitudinal canal in the center of an osteon that contains blood vessels and nerves, a passageway along the longitudinal axis of the spinal cord that contains cerebrospinal fluid.
central nervous system (CNS): The brain and spinal cord.
centriole: A cylindrical intracellular organelle composed of nine groups of microtubules, three in each group; functions in mitosis or meiosis by organizing the microtubules of the spindle apparatus.
centromere: The localized region where two chromatids remain connected after the chromosomes have replicated; site of spindle fiber attachment.
centrosome: A region of cytoplasm that contains a pair of centrioles oriented at right angles to one another.

cephalic: Pertaining to the head.
cerebellum: The posterior portion of the metencephalon, containing the cerebellar hemispheres; includes the arbor vitae, cerebellar nuclei, and cerebellar cortex.
cerebral cortex: An extensive area of neural cortex covering the surfaces of the cerebral hemispheres.
cerebral hemispheres: A pair of expanded portions of the cerebrum covered in neural cortex.
cerebrospinal fluid (CSF): Fluid bathing the internal and external surfaces of the central nervous system; secreted by the choroid plexus.
cerebrovascular accident (CVA): The occlusion of a blood vessel that supplies a portion of the brain, resulting in damage to the dependent neurons; also called *stroke*.
cerebrum: The largest portion of the brain, composed of the cerebral hemispheres; includes the cerebral cortex, the basal nuclei, and the internal capsule.
cerumen: The waxy secretion of the ceruminous glands along the external acoustic meatus.
ceruminous glands: Integumentary glands that secrete cerumen.
cervix: The inferior portion of the uterus.
chemoreception: The detection of changes in the concentrations of dissolved compounds or gases.
chemotaxis: The attraction of phagocytes to the source of abnormal chemicals in tissue fluids.
chloride shift: The movement of plasma chloride ions into red blood cells in exchange for bicarbonate ions generated by the intracellular dissociation of carbonic acid.
cholecystokinin (CCK): A duodenal hormone that stimulates the contraction of the gallbladder and the secretion of enzymes by the exocrine pancreas.
cholesterol: A steroid component of plasma membranes and a substrate for the synthesis of steroid hormones and bile salts.
choline: A breakdown product or precursor of acetylcholine.
cholinergic synapse: A synapse where the presynaptic membrane releases acetylcholine on stimulation.
cholinesterase: The enzyme that breaks down and inactivates acetylcholine.
chondrocyte: A cartilage cell.
chondroitin sulfate: The predominant proteoglycan in cartilage, responsible for the gelatinous consistency of the matrix.
chordae tendineae: Fibrous cords that stabilize the position of the AV valves in the heart, preventing backflow during ventricular systole.
chorion/chorionic: An extraembryonic membrane, consisting of the trophoblast and underlying mesoderm, that forms the placenta.
choroid: The middle, vascular layer in the wall of the eye.
choroid plexus: The vascular complex in the roof of the third and fourth ventricles of the brain, responsible for the production of cerebrospinal fluid.
chromatid: One complete copy of a DNA strand and its associated nucleoproteins.
chromatin: A histological term referring to the grainy material visible in cell nuclei during interphase; the appearance of the DNA content of the nucleus when the chromosomes are uncoiled.
chromosomes: Dense structures, composed of tightly coiled DNA strands and associated histones, that become visible in the nucleus when a cell prepares to undergo mitosis or meiosis; normal human somatic cells each contain 46 chromosomes.
chronic: Habitual or long term.
chylomicrons: Relatively large droplets that may contain triglycerides, phospholipids, and cholesterol in association with proteins; synthesized and released by intestinal cells and transported to the venous blood by the lymphatic system.
ciliary body: A thickened region of the choroid that encircles the lens of the eye; includes the ciliary muscle and the ciliary processes that support the suspensory ligaments of the lens.
cilium/cilia: Slender organelle(s) extending above the free surface of an epithelial cell and generally undergoes cycles of movement; composed of a basal body and microtubules in a $9 + 2$ array.
circulatory system: The network of blood vessels and lymphatic vessels that facilitate the distribution and circulation of extracellular fluid.
circumduction: A movement at a synovial joint in which the distal end of the bone moves in a circular direction, but the shaft does not rotate.
cisterna: An expanded or flattened chamber derived from and associated with the endoplasmc reticulum.
citric acid cycle: The reaction sequence that occurs in the matrix of mitochondria; in the process, organic molecules are broken down, carbon dioxide molecules are released, and hydrogen atoms are transferred to coenzymes that deliver them to the electron transport system; also called the *Krebs cycle*.
clot: A network of fibrin fibers and trapped blood cells; also called a *thrombus* if it occurs within the cardiovascular system.
clotting factors: Plasma proteins, synthesized by the liver, that are essential to the clotting response.
clotting response: The series of events that result in the formation of a clot.
coccygeal ligament: The fibrous extension of the dura mater and filum terminale; provides longitudinal stabilization to the spinal cord.
coccyx: The terminal portion of the spinal column, consisting of relatively tiny, fused vertebrae.
cochlea: The spiral portion of the bony labyrinth of the internal ear that surrounds the organ of hearing.
cochlear duct: The central membranous tube within the cochlea that is filled with endolymph and contains the organ of Corti; also called *scala media*.
codon: A sequence of three nitrogenous bases along an mRNA strand that will specify the location of a single amino acid in a peptide chain.
coelom: The ventral body cavity, lined by a serous membrane and subdivided during fetal development into the pleural, pericardial, and abdominopelvic (peritoneal) cavities.
coenzymes: Complex organic cofactors; most are structurally related to vitamins.
cofactor: Ions or molecules that must be attached to the active site before an enzyme can function; examples include mineral ions and several vitamins.
collagen: A strong, insoluble protein fiber common in connective tissues.
collateral ganglion: A sympathetic ganglion located anterior to the spinal column and separate from the sympathetic chain.
colliculus/colliculi: Little mound(s); in the brain, refers to one of the thickenings in the roof of the mesencephalon; the superior colliculi are associated with the visual system, and the inferior colliculi with the auditory system.
colloid/colloidal suspension: A solution containing large organic molecules in suspension.
colon: The large intestine.
comminuted: Broken or crushed into small pieces.
commissure: A crossing over from one side to another.
common bile duct: The duct formed by the union of the cystic duct from the gallbladder and the bile ducts from the liver; terminates at the duodenal ampulla, where it meets the pancreatic duct.
compact bone: Dense bone that contains parallel osteons.
complement: A group of distinct plasma proteins that interact in a chain reaction after exposure to activated antibodies or the surfaces of certain pathogens; complement proteins promote cell lysis, phagocytosis, and other defense mechanisms.
compliance: Distensibility; the ability of certain organs to tolerate changes in volume; indicates the presence of elastic fibers and smooth muscles.
compound: A pure chemical substance made up of atoms of two or more different elements in a fixed proportion, regardless of the type of chemical bond joining them.
concentration: The amount (in grams) or number of atoms, ions, or molecules (in moles) per unit volume.
concentration gradient: Regional differences in the concentration of a particular substance.
conception: Fertilization.
concha/conchae: Three pairs of thin, scroll-like bones that project into the nasal cavities; the superior and middle conchae are part of the ethmoid, and the inferior conchae articulate with the ethmoid, lacrimal, maxilla, and palatine bones.
condyle: A rounded articular projection on the surface of a bone.
congenital: Present at birth.
congestive heart failure (CHF): The failure to maintain adequate cardiac output due to cardiovascular problems or myocardial damage.
conjunctiva: A layer of stratified squamous epithelium that covers the inner surfaces of the eyelids and the anterior surface of the eye to the edges of the cornea.
connective tissue: One of the four primary tissue types; provides a structural framework that stabilizes the relative positions of the other tissue types; includes connective tissue proper, cartilage, bone, and blood; contains cell products, cells, and ground substance.
continuous propagation: The propagation of an action potential along an unmyelinated axon or a muscle plasma membrane, wherein the action potential affects every portion of the membrane surface.
contractility: The ability to contract; possessed by skeletal, smooth, and cardiac muscle cells.
contralateral reflex: A reflex that affects the opposite side of the body from the stimulus.
conus medullaris: The conical tip of the spinal cord that gives rise to the filum terminale.
convergence: In the nervous system, the innervation of a single neuron by axons from several neurons; most common along motor pathways.
coracoid process: A hook-shaped process of the scapula that projects above the anterior surface of the capsule of the shoulder joint.
Cori cycle: The metabolic exchange of lactate from skeletal muscle for glucose from the liver; performed during the recovery period after muscular exertion.
cornea: The transparent portion of the fibrous layer of the anterior surface of the eye.
corniculate cartilages: A pair of small laryngeal cartilages.
cornu: Horn-shaped.
coronoid: Hooked or curved.
corpora quadrigemina: The superior and inferior colliculi of the mesencephalic tectum (roof) in the brain.
corpus/corpora: Body/bodies.
corpus callosum: A large bundle of axons that links centers in the left and right cerebral hemispheres.
corpus luteum: The progesterone-secreting mass of follicle cells that develops in the ovary after ovulation.

cortex: The outer layer or region of an organ.
corticobulbar tracts: Descending tracts that carry information or commands from the cerebral cortex to nuclei and centers in the brain stem.
corticospinal tracts: Descending tracts that carry motor commands from the cerebral cortex to the anterior gray horns of the spinal cord.
corticosteroid: A steroid hormone produced by the adrenal cortex.
corticosterone: A corticosteroid secreted by the zona fasciculata of the adrenal cortex; a glucocorticoid.
corticotropin-releasing hormone (CRH): The releasing hormone, secreted by the hypothalamus, that stimulates secretion of adrenocorticotropic hormone (ACTH) by the anterior lobe of the pituitary gland (adenohypophysis).
cortisol: A corticosteroid secreted by the zona fasciculata of the adrenal cortex; a glucocorticoid.
costa/costae: Rib(s).
cotransport: The membrane transport of a nutrient, such as glucose, in company with the movement of an ion, normally sodium; transport requires a carrier protein but does not involve direct ATP expenditure and can occur regardless of the concentration gradient for the nutrient.
countercurrent exchange: The transfer of heat, water, or solutes between two fluids that travel in opposite directions.
countercurrent multiplication: Active transport between two limbs of a loop that contains a fluid moving in one direction; responsible for the concentration of urine in the kidney tubules.
covalent bond: A chemical bond between atoms that involves the sharing of electrons.
coxal bone: Hip bone.
cranial: Pertaining to the head.
cranial nerves: Peripheral nerves originating at the brain.
craniosacral division: *See* **parasympathetic division**.
cranium: The braincase; the skull bones that surround and protect the brain.
creatine: A nitrogenous compound, synthesized in the body, that can form a high-energy bond by connecting to a phosphate group and that serves as an energy reserve.
creatine phosphate: A high-energy compound in muscle cells; during muscle activity, the phosphate group is donated to ADP, regenerating ATP; also called *phosphorylcreatine*.
creatinine: A breakdown product of creatine metabolism.
crenation: Cellular shrinkage due to an osmotic movement of water out of the cytoplasm.
cribriform plate: A portion of the ethmoid bone that contains the foramina used by the axons of olfactory receptors en route to the olfactory bulbs of the cerebrum.
cricoid cartilage: A ring-shaped cartilage that forms the inferior margin of the larynx.
crista/cristae: A ridge-shaped collection of hair cells in the ampulla of a semicircular duct; the crista and cupula form a receptor complex sensitive to movement along the plane of the semicircular canal.
cross-bridge: A myosin head that projects from the surface of a thick filament and that can bind to an active site of a thin filament in the presence of calcium ions.
cuneiform cartilages: A pair of small cartilages in the larynx.
cupula: A gelatinous mass that is located in the ampulla of a semicircular duct in the internal ear and whose movement stimulates the hair cells of the crista.
Cushing's disease: A condition caused by the oversecretion of adrenal steroids.
cutaneous membrane: The epidermis and papillary layer of the dermis.
cuticle: The layer of dead, keratinized cells that surrounds the shaft of a hair; for nails, *see* **eponychium**.
cyanosis: An abnormal bluish coloration of the skin due to the presence of deoxygenated blood in vessels near the body surface.
cystic duct: A duct that carries bile between the gallbladder and the common bile duct.
cytochrome: A pigment component of the electron transport system; a structural relative of heme.
cytokinesis: The cytoplasmic movement that separates two daughter cells at the completion of mitosis.
cytology: The study of cells.
cytoplasm: The material between the plasma membrane and the nuclear membrane; cell contents.
cytosine: A pyrimidine; one of the nitrogenous bases in the nucleic acids RNA and DNA.
cytoskeleton: A network of microtubules and microfilaments in the cytoplasm.
cytosol: The fluid portion of the cytoplasm.
cytotoxic: Poisonous to cells.
cytotoxic T cells: Lymphocytes involved in cell-mediated immunity that kill target cells by direct contact or by the secretion of lymphotoxins; also called *killer T cells* and T_C *cells*.

D

daughter cells: Genetically identical cells produced by somatic cell division.
deamination: The removal of an amino group from an amino acid.
decomposition reaction: A chemical reaction that breaks a molecule into smaller fragments.
decussate: To cross over to the opposite side, usually referring to the crossover of the descending tracts of the corticospinal pathway on the ventral surface of the medulla oblongata.
defecation: The elimination of fecal wastes.
degradation: Breakdown, catabolism.
dehydration: A reduction in the water content of the body that threatens homeostasis.
dehydration synthesis: The joining of two molecules associated with the removal of a water molecule.
demyelination: The loss of the myelin sheath of an axon, normally due to chemical or physical damage to Schwann cells or oligodendrocytes.
denaturation: A temporary or permanent change in the three-dimensional structure of a protein.
dendrite: A sensory process of a neuron.
deoxyribonucleic acid (DNA): A nucleic acid consisting of a double chain of nucleotides that contains the sugar deoxyribose and the nitrogenous bases adenine, guanine, cytosine, and thymine.
deoxyribose: A five-carbon sugar resembling ribose but lacking an oxygen atom.
depolarization: A change in the membrane potential from a negative value toward 0 mV.
depression: Inferior (downward) movement of a body part.
dermatome: A sensory region monitored by the dorsal rami of a single spinal segment.
dermis: The connective tissue layer beneath the epidermis of the skin.
detrusor muscle: Collectively, the three layers of smooth muscle in the wall of the urinary bladder.
development: Growth and the acquisition of increasing structural and functional complexity; includes the period from conception to maturity.
diabetes mellitus: Polyuria and glycosuria, most commonly due to the inadequate production or diminished sensitivity to insulin with a resulting increase of blood glucose levels.
diapedesis: The movement of white blood cells through the walls of blood vessels by migration between adjacent endothelial cells.
diaphragm: Any muscular partition; the respiratory muscle that separates the thoracic cavity from the abdominopelvic cavity.
diaphysis: The shaft of a long bone.
diarthrosis: A synovial joint.
diastolic pressure: Pressure measured in the walls of a muscular artery when the left ventricle is in diastole (relaxation).
diencephalon: A division of the brain that includes the epithalamus, thalamus, and hypothalamus.
differential count: The determination of the relative abundance of each type of white blood cell on the basis of a random sampling of 100 white blood cells.
differentiation: The gradual appearance of characteristic cellular specializations during development as the result of gene activation or repression.
diffusion: Passive molecular movement from an area of higher concentration to an area of lower concentration.
digestion: The chemical breakdown of ingested food into simple molecules that can be absorbed by the cells of the digestive tract.
digestive system: The digestive tract and associated glands.
digestive tract: An internal passageway that begins at the oral cavity (mouth), ends at the anus, and is lined by a mucous membrane; also called *gastrointestinal tract*.
dilate: To increase in diameter; to enlarge or expand.
disaccharide: A compound formed by the joining of two simple sugars by dehydration synthesis.
dissociation: *See* **ionization**.
distal: A direction away from the point of attachment or origin; for a limb, away from its attachment to the trunk.
distal convoluted tubule (DCT): The segment of the nephron closest to the connecting tubules and collecting duct; an important site of active secretion.
diuresis: Fluid loss by the kidneys; the production of unusually large volumes of urine.
divergence: In neural tissue, the spread of information from one neuron to many neurons; an organizational pattern common along sensory pathways of the central nervous system.
diverticulum: A sac or pouch in the wall of the colon or other organ.
DNA molecule: Two DNA strands wound in a double helix and held together by hydrogen bonds between complementary nitrogenous base pairs.
dopamine: An important neurotransmitter in the central nervous system.
dorsal: Toward the back, posterior.
dorsal root ganglion: A peripheral nervous system ganglion containing the cell bodies of sensory neurons.
dorsiflexion: Upward movement of the foot through flexion at the ankle.
Down's syndrome: A genetic abnormality resulting from the presence of three copies of chromosome 21; people with this condition have characteristic physical and intellectual deficits; also called *trisomy 21*.
duct: A passageway that delivers exocrine secretions to an epithelial surface.
ductus arteriosus: A vascular connection between the pulmonary trunk and the aorta that functions throughout fetal life; normally closes at birth or

shortly thereafter and persists as the ligamentum arteriosum.

ductus deferens: A passageway that carries spermatozoa from the epididymis to the ejaculatory duct; also called the *vas deferens*.

duodenal ampulla: A chamber that receives bile from the common bile duct and pancreatic secretions from the pancreatic duct.

duodenal papilla: A conical projection from the inner surface of the duodenum that contains the opening of the duodenal ampulla.

duodenum: The proximal 25 cm (9.8 in.) of the small intestine that contains short villi and submucosal glands.

dura mater: The outermost layer of the cranial and spinal meninges.

E

eccrine glands: Sweat glands of the skin that produce a watery secretion.

ectoderm: One of the three primary germ layers; covers the surface of the embryo and gives rise to the nervous system, the epidermis and associated glands, and a variety of other structures.

ectopic: Outside the normal location.

effector: A peripheral gland or muscle cell innervated by a motor neuron.

efferent: Conducting a fluid or nerve impulse away from an organ or structure.

efferent arteriole: An arteriole carrying blood away from a glomerulus of the kidney.

efferent fiber: An axon that carries impulses away from the central nervous system.

ejaculation: The ejection of semen from the penis as the result of muscular contractions of the bulbospongiosus and ischiocavernosus muscles.

ejaculatory ducts: Short ducts that pass within the walls of the prostate gland and connect the ductus deferens with the prostatic urethra.

elastin: Connective tissue fibers that stretch and recoil, providing elasticity to connective tissues.

electrical coupling: A connection between adjacent cells that permits the movement of ions and the transfer of graded or propagated changes in the membrane potential from cell to cell.

electrocardiogram (ECG, EKG): A graphic record of the electrical activities of the heart, as monitored at specific locations on the body surface.

electroencephalogram (EEG): A graphic record of the electrical activities of the brain.

electrolytes: Soluble inorganic compounds whose ions will conduct an electrical current in solution.

electron: One of the three fundamental subatomic particles; has a negative charge and normally orbits the protons of the nucleus.

electron transport system (ETS): The cytochrome system responsible for aerobic energy production in cells; a complex bound to the inner mitochondrial membrane.

element: All the atoms with the same atomic number.

elevation: Movement in a superior, or upward, direction.

embryo: The developmental stage beginning at fertilization and ending at the start of the third developmental month.

embryology: The study of embryonic development, focusing on the first two months after fertilization.

endocardium: The simple squamous epithelium that lines the heart and is continuous with the endothelium of the great vessels.

endochondral ossification: The replacement of a cartilaginous model by bone; the characteristic mode of formation for skeletal elements other than the bones of the cranium, the clavicles, and sesamoid bones.

endocrine gland: A gland that secretes hormones into the blood.

endocrine system: The endocrine (ductless) glands and organs of the body.

endocytosis: The movement of relatively large volumes of extracellular material into the cytoplasm by the formation of a membranous vesicle at the cell surface; includes pinocytosis and phagocytosis.

endoderm: One of the three primary germ layers; the layer on the undersurface of the embryonic disc; gives rise to the epithelia and glands of the digestive system, the respiratory system, and portions of the urinary system.

endogenous: Produced within the body.

endolymph: The fluid contents of the membranous labyrinth (the saccule, utricle, semicircular ducts, and cochlear duct) of the internal ear.

endometrium: The mucous membrane lining the uterus.

endomysium: A delicate network of connective tissue fibers that surrounds individual muscle cells.

endoneurium: A delicate network of connective tissue fibers that surrounds individual nerve fibers.

endoplasmic reticulum: A network of membranous channels in the cytoplasm of a cell that function in intracellular transport, synthesis, storage, packaging, and secretion.

endosteum: An incomplete cellular lining on the inner (medullary) surfaces of bones.

endothelium: The simple squamous epithelial cells that line blood and lymphatic vessels.

enteroendocrine cells: Endocrine cells scattered among the epithelial cells that line the digestive tract.

enterogastric reflex: The reflexive inhibition of gastric secretion; initiated by the arrival of chyme in the small intestine.

enterokinase: An enzyme in the lumen of the small intestine that activates the proenzymes secreted by the pancreas.

enzyme: A protein that catalyzes a specific biochemical reaction.

eosinophil: A type of white blood cell with a lobed nucleus and red-staining granules; participates in the immune response and is especially important during allergic reactions.

ependyma: The layer of cells lining the ventricles and central canal of the central nervous system.

epicardium: A serous membrane covering the outer surface of the heart; also called *visceral pericardium*.

epidermis: The epithelium covering the surface of the skin.

epididymis: A coiled duct that connects the rete testis to the ductus deferens; site of functional maturation of spermatozoa.

epidural space: The space between the spinal dura mater and the walls of the vertebral foramen; contains blood vessels and adipose tissue; a common site of injection for regional anesthesia.

epigenetics: Study of gene expression due to alterations in the reading of DNA sequences.

epiglottis: A blade-shaped flap of tissue, reinforced by cartilage, that is attached to the dorsal and superior surface of the thyroid cartilage; folds over the entrance to the larynx during swallowing.

epimysium: A dense layer of collagen fibers that surrounds a skeletal muscle and is continuous with the tendons/aponeuroses of the muscle and with the perimysium.

epineurium: A dense layer of collagen fibers that surrounds a peripheral nerve.

epiphyseal cartilage: The cartilaginous region between the epiphysis and diaphysis of a growing bone.

epiphysis: The head of a long bone.

epithelium: One of the four primary tissue types; a layer of cells that forms a superficial covering or an internal lining of a body cavity or vessel.

eponychium: A narrow zone of stratum corneum that extends across the surface of a nail at its exposed base; also called the *cuticle*.

equilibrium: A dynamic state in which two opposing forces or processes are in balance.

erection: The stiffening of the penis due to the engorgement of the erectile tissues of the corpora cavernosa and corpus spongiosum.

erythema: Redness and inflammation at the surface of the skin.

erythrocyte: A red blood cell; has no nucleus and contains large quantities of hemoglobin.

erythropoietin: A hormone released by most tissues, and especially by the kidneys, when oxygen levels decrease; stimulates erythropoiesis (red blood cell formation) in red bone marrow.

Escherichia coli: A normal bacterial resident of the large intestine.

esophagus: A muscular tube that connects the pharynx to the stomach.

essential amino acids: Amino acids that cannot be synthesized in the body in adequate amounts and must be obtained from the diet.

essential fatty acids: Fatty acids that cannot be synthesized in the body and must be obtained from the diet.

estrogens: A class of steroid sex hormones that includes estradiol.

evaporation: A movement of molecules from the liquid state to the gaseous state.

eversion: A turning outward.

excitable membranes: Membranes that propagate action potentials, a characteristic of muscle cells and nerve cells.

excitatory postsynaptic potential (EPSP): The depolarization of a postsynaptic membrane by a chemical neurotransmitter released by the presynaptic cell.

excretion: The removal of wastes from the blood, tissues, or organs.

exocrine gland: A gland that secretes onto the body surface or into a passageway connected to the exterior.

exocytosis: The ejection of cytoplasmic materials by the fusion of a membranous vesicle with the plasma membrane.

expiration: Exhalation; breathing out.

extension: An increase in the angle between two articulating bones; the opposite of flexion.

external acoustic meatus: A passageway in the temporal bone that leads to the tympanic membrane of the internal ear.

external ear: The auricle, external acoustic meatus, and tympanic membrane.

external nares: The nostrils; the external openings into the nasal cavity.

external respiration: The diffusion of gases between the alveolar air and the alveolar capillaries and between the systemic capillaries and peripheral tissues.

exteroceptors: General sensory receptors in the skin, mucous membranes, and special sense organs that provide information about the external environment and about our position within it.

extracellular fluid: All body fluids other than that contained within cells; includes plasma and interstitial fluid.

extraembryonic membranes: The yolk sac, amnion, chorion, and allantois.

extrinsic pathway: A clotting pathway that begins with damage to blood vessels or surrounding tissues and ends with the formation of tissue factor complex.

F

facilitated: Brought closer to threshold, as in the depolarization of a nerve plasma membrane toward threshold; making the cell more sensitive to depolarizing stimuli.

facilitated diffusion: The passive movement of a substance across a plasma membrane by means of a protein carrier.

falciform ligament: A sheet of mesentery that contains the ligamentum teres, the fibrous remains of the umbilical vein of the fetus.

falx: Sickle-shaped.

falx cerebri: The curving sheet of dura mater that extends between the two cerebral hemispheres; encloses the superior sagittal sinus.

fasciae: Connective tissue fibers, primarily collagen, that form sheets or bands beneath the skin to attach, stabilize, enclose, and separate muscles and other internal organs.

fasciculus: A small bundle; usually refers to a collection of nerve axons or muscle fibers.

fatty acids: Hydrocarbon chains that end in a carboxyl group.

fauces: The passage from the mouth to the pharynx, bounded by the palatal arches, the soft palate, and the uvula.

feces: Waste products eliminated by the digestive tract at the anus; contains indigestible residue, bacteria, mucus, and epithelial cells.

fenestra/fenestrae: Opening/openings.

fertilization: The fusion of a secondary oocyte and a spermatozoon to form a zygote.

fetus: The developmental stage lasting from the start of the third developmental month to delivery.

fibrin: Insoluble protein fiber that forms the basic framework of a blood clot.

fibrinogen: A plasma protein that is the soluble precursor of the insoluble protein fibrin.

fibroblasts: Cells of connective tissue proper that produce extracellular fibers and secrete the organic compounds of the extracellular matrix.

fibrous cartilage: Cartilage containing an abundance of collagen fibers; located around the edges of joints, in the intervertebral discs, and the menisci of the knee.

fibrous layer: The outermost layer of the eye, composed of the sclera and cornea.

fibula: The lateral, slender bone of the leg.

filiform papillae: Slender conical projections from the dorsal surface of the anterior two-thirds of the tongue.

filtrate: The fluid produced by filtration at a glomerulus in the kidney.

filtration: The movement of a fluid across a membrane whose pores restrict the passage of solutes on the basis of size.

filtration pressure: The hydrostatic pressure responsible for filtration.

filum terminale: A fibrous extension of the spinal cord, from the conus medullaris to the coccygeal ligament.

fimbriae: Fringes; the fingerlike processes that surround the entrance to the uterine tube.

first-degree burn: Burn involving only the epidermis and causing redness and swelling; type of partial-thickness burn.

fissure: An elongated groove or opening.

flagellum/flagella: An organelle that is structurally similar to a cilium but is used to propel a cell through a fluid; found on spermatozoa.

flatus: Intestinal gas.

flexion: A movement that reduces the angle between two articulating bones; the opposite of extension.

flexor: A muscle that produces flexion.

flexor reflex: A reflex contraction of the flexor muscles of a limb in response to a painful stimulus.

flexure: A bending.

folia: Leaflike folds; the slender folds in the surface of the cerebellar cortex.

foliate papillae: Taste buds located on the lateral margins of the posterior region of the tongue.

follicle: A small secretory sac or gland.

follicle-stimulating hormone (FSH): A hormone secreted by the anterior lobe of the pituitary gland (adenohypophysis); stimulates oogenesis (female) and spermatogenesis (male).

fontanelle: A relatively soft, flexible, fibrous region between two flat bones in the developing skull; also spelled *fontanel.*

foramen/foramina: Opening(s) or passage(s) through a bone.

forearm: The distal portion of the upper limb between the elbow and wrist.

forebrain: The cerebrum.

fornix: An arch or the space bounded by an arch; in the brain, an arching tract that connects the hippocampus with the mammillary bodies; in the eye, a slender pocket situated where the epithelium of the ocular conjunctiva folds back on itself as the palpebral conjunctiva; in the vagina, the shallow recess surrounding the protrusion of the cervix.

fossa: A shallow depression or furrow in the surface of a bone.

fourth ventricle: An elongated ventricle of the metencephalon (pons and cerebellum) and the myelencephalon (medulla oblongata) of the brain; the roof contains a region of choroid plexus.

fovea: The portion of the retina that provides the sharpest vision because it has the highest concentration of cones; also called *fovea centralis.*

fracture: A break or crack in a bone.

frenulum: A bridle; usually referring to a band of tissue that restricts movement (e.g., *lingual frenulum*).

frontal plane: A sectional plane that divides the body into an anterior portion and a posterior portion; also called *coronal plane.*

fructose: A hexose (six-carbon simple sugar) in foods and in semen.

fundus: The base of an organ such as the stomach, uterus, or gallbladder.

G

gallbladder: The pear-shaped reservoir for bile after it is secreted by the liver.

gametes: Reproductive cells (spermatozoa or oocytes) that contain half the normal chromosome complement.

gametogenesis: The formation of gametes.

gamma aminobutyric acid (GABA): A neurotransmitter of the central nervous system whose effects are generally inhibitory.

gamma motor neurons: Motor neurons that adjust the sensitivities of muscle spindles (intrafusal fibers).

ganglion/ganglia: A collection of neuron cell bodies in the peripheral nervous system.

gap junctions: Connections between cells that permit electrical coupling.

gaster: The stomach; the body, or belly, of a skeletal muscle.

gastric: Pertaining to the stomach.

gastric glands: The tubular glands of the stomach whose cells produce acid, enzymes, intrinsic factor, and hormones.

gastrointestinal (GI) tract: *See* **digestive tract**.

gene: A portion of a DNA strand that functions as a hereditary unit, is located at a particular site on a specific chromosome, and codes for a specific protein or polypeptide.

genetics: The study of mechanisms of heredity.

geniculate: Like a little knee; the medial geniculate nuclei and lateral geniculate nuclei are in the thalamus of the brain.

genitalia: The reproductive organs.

germinal centers: Pale regions in the interior of lymphoid tissues or lymphoid nodules, where cell divisions occur that produce additional lymphocytes.

gestation: The period of intrauterine development; pregnancy.

gland: Cells that produce exocrine or endocrine secretions.

glenoid cavity: A rounded depression that forms the articular surface of the scapula at the shoulder joint.

glial cells: *See* **neuroglia**.

globular proteins: Proteins whose tertiary structure makes them rounded and compact.

glomerular capsule: The expanded initial portion of the nephron that surrounds the glomerulus.

glomerular filtration rate: The rate of filtrate formation at the glomerulus.

glomerulus: A ball or knot; in the kidneys, a knot of capillaries that projects into the enlarged, proximal end of a nephron; the site of filtration, the first step in the production of urine.

glossopharyngeal nerve: Cranial nerve IX.

glucagon: A hormone secreted by the alpha cells of the pancreatic islets; increases blood glucose concentrations.

glucocorticoids: Hormones secreted by the zona fasciculata of the adrenal cortex to modify glucose metabolism; cortisol and corticosterone are important examples.

gluconeogenesis: The synthesis of glucose from protein or lipid precursors.

glucose: A six-carbon sugar, $C_6H_{12}O_6$; the preferred energy source for most cells and normally the only energy source for neurons.

glycerides: Lipids composed of glycerol bound to fatty acids.

glycogen: A polysaccharide that is an important energy reserve; a polymer consisting of a long chain of glucose molecules.

glycogenesis: The synthesis of glycogen from glucose molecules.

glycogenolysis: Glycogen breakdown and the liberation of glucose molecules.

glycolipids: Compounds created by the combination of carbohydrate and lipid components.

glycolysis: The anaerobic cytosolic breakdown of glucose into two 3-carbon molecules of pyruvate, with a net gain of two ATP molecules.

glycoprotein: A compound containing a relatively small carbohydrate group attached to a large protein.

glycosuria: The presence of glucose in urine; also called *glucosuria.*

Golgi apparatus: A cellular organelle consisting of a series of membranous plates that give rise to lysosomes and secretory vesicles.

gomphosis: A fibrous synarthrosis that binds a tooth to the bone of the jaw.

gonadotropin-releasing hormone (GnRH): A hypothalamic releasing hormone that causes the secretion of both follicle-stimulating hormone and luteinizing hormone by the anterior lobe of the pituitary gland (adenohypophysis).

gonadotropins: Follicle-stimulating hormone and luteinizing hormone; hormones that stimulate gamete development and sex hormone secretion.

gonads: Reproductive organs that produce gametes and sex hormones.

granulocytes: White blood cells containing granules that are visible with the light microscope; includes eosinophils, basophils, and neutrophils; also called *granular leukocytes*.

gray matter: Areas in the central nervous system that are dominated by neuron cell bodies, neuroglia, and unmyelinated axons.

gray ramus: A bundle of postganglionic sympathetic nerve fibers that are distributed to effectors in the body wall, skin, and limbs by way of a spinal nerve.

greater omentum: A large fold of the dorsal mesentery of the stomach; hangs anterior to the intestines.

groin: The inguinal region.

gross anatomy: The study of the structural features of the body without the aid of a microscope.

growth hormone (GH): An anterior lobe of the pituitary gland (adenohypophysis) hormone that stimulates tissue growth and anabolism when nutrients are abundant and restricts tissue glucose dependence when nutrients are in short supply.

growth hormone–inhibiting hormone (GH–IH): A hypothalamic regulatory hormone that inhibits growth hormone secretion by the anterior lobe of the pituitary gland.

guanine: A purine; one of the nitrogenous bases in the nucleic acids RNA and DNA.

gustation: The sense of taste.

gyrus: A prominent fold or ridge of neural cortex on the surfaces of the cerebral hemispheres.

H

hair: A keratinous strand produced by epithelial cells of the hair follicle.

hair cells: Sensory cells of the internal ear.

hair follicle: An accessory structure of the integument; a tube lined by a stratified squamous epithelium that begins at the surface of the skin and ends at the hair papilla.

hallux: The big toe; also called the *great toe.*

haploid: Possessing half the normal number of chromosomes; a characteristic of gametes.

hard palate: The bony roof of the oral cavity, formed by the maxillae and palatine bones.

helper T cells: Lymphocytes whose secretions and other activities coordinate cell-mediated and antibody-mediated immunity; also called T_H *cells*.

hematocrit: The percentage of formed elements in a sample of blood; also called *volume of packed red cells (VPRC)* or *packed cell volume (PCV)*.

hematoma: An abnormal collection of clotted or partially clotted blood outside a blood vessel.

hematopoietic stem cells: Stem cells whose divisions produce each of the various populations of blood cells; also called *hemocytoblasts*.

hematuria: The abnormal presence of red blood cells in urine.

heme: A porphyrin ring containing a central iron atom that can reversibly bind oxygen molecules; a component of the hemoglobin molecule.

hemoglobin: A protein composed of four globular subunits, each bound to a heme molecule; gives red blood cells the ability to transport oxygen in the blood.

hemolysis: The breakdown of red blood cells.

hemopoiesis: Blood cell formation and differentiation.

hemorrhage: Blood loss; to bleed.

hemostasis: The stoppage of bleeding.

heparin: An anticoagulant released by activated basophils and mast cells.

hepatic duct: The duct that carries bile away from the liver lobes and toward the union with the cystic duct.

hepatic portal vein: The vessel that carries blood between the intestinal capillaries and the sinusoids of the liver.

hepatocyte: A liver cell.

heterozygous: Having two different alleles at corresponding sites on a chromosome pair.

hiatus: A gap, cleft, or opening.

high density lipoprotein (HDL): A lipoprotein with a relatively small lipid content; responsible for the movement of cholesterol from peripheral tissues to the liver.

hilum: A localized region where blood vessels, lymphatic vessels, nerves, and/or other anatomical structures are attached to an organ.

hippocampus: A region, deep to the floor of a lateral ventricle, involved with emotional states and the conversion of short-term to long-term memories.

histamine: The chemical released by stimulated mast cells or basophils to initiate or enhance inflammation.

histology: The study of tissues.

histones: Proteins associated with the DNA of the nucleus; the DNA strands are wound around them.

holocrine: A form of exocrine secretion in which the secretory cell becomes swollen with vesicles and then ruptures.

homeostasis: The maintenance of a relatively constant internal environment.

homozygous: Having identical alleles at corresponding sites on a chromosome pair.

hormone: A compound that is secreted by one cell and travels through the bloodstream to affect the activities of cells in another part of the body.

human chorionic gonadotropin (hCG): The placental hormone that maintains the corpus luteum for the first three months of pregnancy.

human immunodeficiency virus (HIV): The infectious agent that causes acquired immunodeficiency syndrome (AIDS).

human leukocyte antigen (HLA): *See* **MHC protein**.

human placental lactogen (hPL): The placental hormone that stimulates the functional development of the mammary glands.

humoral immunity: *See* **antibody-mediated immunity**.

hyaluronan: A carbohydrate component of proteoglycans in the matrix of many connective tissues.

hyaluronidase: An enzyme that breaks down the bonds between adjacent follicle cells; produced by some bacteria and found in the acrosome of a spermatozoon.

hydrogen bond: A weak interaction between the hydrogen atom on one molecule and a negatively charged portion of another molecule.

hydrolysis: The breakage of a chemical bond through the addition of a water molecule; the reverse of dehydration synthesis.

hydrophilic: Freely associating with water; readily entering into solution; water-loving.

hydrophobic: Incapable of freely associating with water molecules; insoluble; water-fearing.

hydrostatic pressure: Fluid pressure.

hydroxide ion: OH^-.

hypercapnia: High blood carbon dioxide concentrations, commonly as a result of hypoventilation or inadequate tissue perfusion.

hyperpolarization: The movement of the membrane potential away from the normal resting potential and farther from 0 mV.

hypersecretion: The overactivity of glands that produce exocrine or endocrine secretions.

hypertension: Abnormally high blood pressure.

hypertonic: In comparing two solutions, the solution with the higher osmolarity.

hypertrophy: An increase in tissue size without cell division.

hyperventilation: A rate of respiration sufficient to decrease blood P_{CO_2} to levels below normal.

hypocapnia: An abnormally low blood P_{CO_2} concentration commonly as a result of hyperventilation.

hypodermic needle: A needle inserted through the skin to introduce drugs into the subcutaneous layer.

hypodermis: The layer of loose connective tissue below the dermis; also called *subcutaneous layer* or *superficial fascia*.

hypophyseal portal system: The network of vessels that carries blood from capillaries in the hypothalamus to capillaries in the anterior lobe of the pituitary gland.

hypophysis: The pituitary gland.

hyposecretion: Abnormally low exocrine or endocrine secretion.

hypothalamus: The floor of the diencephalon; the region of the brain containing centers involved with the subconscious regulation of visceral functions, emotions, drives, and the coordination of neural and endocrine functions.

hypotonic: In comparing two solutions, the solution with the lower osmolarity.

hypoventilation: A respiratory rate that is insufficient to keep blood P_{CO_2} within normal levels.

hypoxia: A low tissue oxygen concentration.

I

ileum: The distal 2.5 m of the small intestine.

ilium: The largest of the three bones whose fusion creates a coxal bone.

immunity: Resistance to injuries and diseases caused by foreign substances, toxins, or pathogens.

immunization: The production of immunity by the deliberate exposure to antigens under conditions that prevent the development of illness but stimulate the production of memory B cells.

immunoglobulin (Ig): A circulating antibody.

implantation: The attachment of a blastocyst into the endometrium of the uterine wall.

inclusions: Aggregations of insoluble pigments, nutrients, or other materials in cytoplasm.

incus: The central auditory ossicle, located between the malleus and the stapes in the middle ear cavity.

inexcitable: Incapable of propagating an action potential.

infarct: An area of dead cells that results from an interruption of blood flow.

infection: The invasion and colonization of body tissues by pathogens.

inferior: Below, in reference to a particular structure, with the body in the anatomical position.

inferior vena cava: The vein that carries blood from the parts of the body inferior to the heart to the right atrium.

infertility: The inability to conceive; also called *sterility*.

inflammation: A nonspecific defense mechanism that operates at the tissue level; characterized by swelling, redness, heat (warmth), pain, and sometimes loss of function.

infundibulum: A tapering, funnel-shaped structure; in the brain, the connection between the pituitary gland and the hypothalamus; in the uterine tube, the entrance bounded by fimbriae that receives the oocyte at ovulation.

ingestion: The introduction of food into the digestive tract by way of the mouth; eating.

inguinal canal: A passage through the abdominal wall that marks the path of testicular descent and that contains the testicular arteries, veins, and ductus deferens.
inguinal region: The area of the abdominal wall near the junction of the trunk and the thighs that contains the external genitalia; the groin.
inhibin: A hormone, produced by nurse (sustentacular) cells of the testes and follicular cells of the ovaries, that inhibits the secretion of follicle-stimulating hormone by the anterior lobe of the pituitary gland (adenohypophysis).
inhibitory postsynaptic potential (IPSP): A hyperpolarization of the postsynaptic membrane after the arrival of a neurotransmitter.
initial segment: The proximal portion of the axon where an action potential first appears.
injection: The forcing of fluid into a body part or organ.
innate immunity: Type of defense that is genetically determined, present at birth, and is not related to previous exposure to a particular antigen; also called *nonspecific immunity*.
inner cell mass: Cells of the blastocyst that will form the body of the embryo.
innervation: The distribution of sensory and motor nerves to a specific region or organ.
insensible perspiration: Evaporative water loss by diffusion across the epithelium of the skin or evaporation across the alveolar surfaces of the lungs.
insertion: A point of attachment of a muscle; the end that is easily movable.
insoluble: Incapable of dissolving in solution.
inspiration: Inhalation; the movement of air into the respiratory system.
insulin: A hormone secreted by beta cells of the pancreatic islets; causes a decrease in blood glucose concentrations.
integument: The skin.
intercalated discs: Regions where adjacent cardiocytes interlock and where gap junctions permit electrical coupling between the cells.
intercellular fluid: *See* **interstitial fluid**.
interferons: Peptides released by virus-infected cells, especially lymphocytes, that slow viral replication and make other cells more resistant to viral infection.
interleukins: Peptides, released by activated monocytes and lymphocytes, that assist in the coordination of cell-mediated and antibody-mediated immunities.
internal capsule: The collection of afferent and efferent fibers of the white matter of the cerebral hemispheres, visible on gross dissection of the brain.
internal ear: The membranous labyrinth that contains the organs of hearing and equilibrium.
internal nares: The entrance to the nasopharynx from the nasal cavity.
internal respiration: The diffusion of gases between interstitial fluid and cytoplasm.
interneuron: An association neuron; central nervous system neurons that are between sensory and motor neurons.
interoceptors: Sensory receptors monitoring the functions and status of internal organs and systems.
interosseous membrane: The fibrous connective tissue membrane between the shafts of the tibia and fibula and between the radius and ulna; an example of a fibrous amphiarthrosis.
interphase: The stage in the life cycle of a cell during which the chromosomes are uncoiled and all normal cellular functions except mitosis are under way.
intersegmental reflex: A reflex that involves several segments of the spinal cord.
interstitial fluid: The fluid in the tissues that fills the spaces between cells.
interstitial growth: Type of cartilage growth that occurs during development in which the cartilage expands from within as chondrocytes divide and daughter cells secrete additional matrix.
interventricular foramen: The opening that permits fluid movement between the lateral and third ventricles of the brain.
intervertebral disc: A fibrocartilage pad between the bodies of successive vertebrae that absorbs shocks.
intestinal crypt: A tubular epithelial pocket that is lined by secretory cells and opens into the lumen of the digestive tract; also called *intestinal gland*.
intestine: The tubular organ of the digestive tract.
intracellular fluid: The cytosol.
intramembranous ossification: The formation of bone within a connective tissue without the prior development of a cartilaginous model.
intrinsic factor: A glycoprotein, secreted by the parietal cells of the stomach, that facilitates the intestinal absorption of vitamin B_{12}.
intrinsic pathway: A pathway of the clotting system that begins with the activation of platelets and ends with the formation of factor X activator complex.
inversion: A turning inward.
in vitro: Outside the body, in an artificial environment.
in vivo: In the living body.
involuntary: Not under conscious control.
ion: An atom or molecule having a positive or negative charge due to the loss or gain, respectively, of one or more electrons.
ionic bond: A molecular bond created by the attraction between ions with opposite charges.
ionization: Dissociation; the breakdown of a molecule in solution to form ions.
ipsilateral: A reflex response that affects the same side as the stimulus.
iris: A contractile structure, made up of smooth muscle, that forms the colored portion of the eye.
ischemia: An inadequate blood supply to a region of the body.
ischium: One of the three bones whose fusion creates a coxal bone.
islets of Langerhans: *See* **pancreatic islets**.
isotonic: A solution with an osmolarity that does not result in water movement across plasma membranes.
isotopes: Forms of an element whose atoms contain the same number of protons but different numbers of neutrons (and thus differ in atomic mass).
isthmus: A narrow band of tissue connecting two larger masses.

J

jejunum: The middle part of the small intestine.
joint: An area where adjacent bones interact; also called *articulation*.
juxtaglomerular complex: Specialized cells in between the walls of the distal convoluted tubule and afferent and efferent arterioles adjacent to the glomerulus; a complex responsible for the release of renin and erythropoietin.

K

keratin: The tough, fibrous protein component of nails, hair, calluses, and the general integumentary surface.
keto acid: The carbon chain that remains after the deamination or transamination of an amino acid.
ketoacidosis: A condition characterized by a decrease in blood pH due to the presence of large numbers of ketone bodies.
ketone bodies: Keto acids produced during the catabolism of lipids and some amino acids; specifically, acetone, acetoacetate, and beta-hydroxybutyrate.
kidney: A component of the urinary system; an organ functioning in the regulation of blood composition, including the excretion of wastes and the maintenance of normal fluid and electrolyte balances.
killer T cells: *See* **cytotoxic T cells**.
Krebs cycle: *See* **citric acid cycle**.
Kupffer cells: Phagocytic cells of the liver sinusoids.

L

labium/labia: Lip(s); the labia majora and labia minora are structures of the female external genitalia.
labrum: A lip or rim.
labyrinth: A maze of passageways; the structures of the internal ear.
lacrimal gland: A tear gland on the dorsolateral surface of the eye.
lactase: An enzyme that breaks down the milk sugar lactose.
lactate: An anion released by the dissociation of lactic acid, produced from pyruvate under anaerobic conditions.
lactation: The production of milk by the mammary glands.
lacteal: A terminal lymphatic within an intestinal villus.
lacuna: A small pit or cavity.
lambdoid suture: The synarthrosis between the parietal and occipital bones of the cranium.
lamellae: Concentric layers; the concentric layers of bone within an osteon.
lamellated corpuscle: A receptor sensitive to vibration.
lamina: A thin sheet or layer.
lamina propria: The reticular tissue that underlies a mucous epithelium and forms part of a mucous membrane.
Langerhans cells: Cells in the epithelium of the skin and digestive tract that participate in the immune response by presenting antigens to T cells; also called *dendritic cells*.
large intestine: The terminal portion of the intestinal tract, consisting of the colon, the rectum, and the anal canal.
laryngopharynx: The division of the pharynx that is inferior to the epiglottis and superior to the esophagus.
larynx: A complex cartilaginous structure that surrounds and protects the glottis and vocal cords; the superior margin is bound to the hyoid bone, and the inferior margin is bound to the trachea.
latent period: The time between the stimulation of a muscle and the start of the contraction phase.
lateral apertures: Openings in the roof of the fourth ventricle that permit the circulation of cerebrospinal fluid into the subarachnoid space.
lateral ventricle: A fluid-filled chamber within a cerebral hemisphere.
lateral: Pertaining to the side.
lens: The transparent refractive structure that is between the iris and the vitreous humor.
lesser omentum: A double-layered peritoneal fold formed by the mesentery that connects the lesser curvature of the stomach to the liver.

leukocyte: A white blood cell.
ligament: A dense band of connective tissue fibers that attaches one bone to another.
ligamentum arteriosum: The fibrous strand in adults that is the remnant of the ductus arteriosus of the fetal stage.
ligamentum nuchae: An elastic ligament between the vertebra prominens and the occipital bone.
ligamentum teres: The fibrous strand in the falciform ligament of adults that is the remnant of the umbilical vein; also called the *round ligament of liver*.
ligate: To tie off.
limbic system: The group of nuclei and centers in the cerebrum and diencephalon that are involved with emotional states, memories, and behavioral drives.
lingual: Pertaining to the tongue.
lipid: An organic compound containing carbon, hydrogen, and oxygen in a ratio that does not approximate 1:2:1; includes fats, oils, and waxes.
lipogenesis: The synthesis of lipids from nonlipid precursors.
lipolysis: The catabolism of lipids as a source of energy.
lipoprotein: A compound containing a relatively small lipid bound to a protein.
liver: An organ of the digestive system that has varied and vital functions, including the production of plasma proteins, the excretion of bile, the storage of energy reserves, the detoxification of poisons, and the interconversion of nutrients.
lobule: Histologically, the basic organizational unit of the liver.
local hormone: *See* **prostaglandin**.
loop of Henle: *See* **nephron loop**.
loose connective tissue: A loosely organized, easily distorted connective tissue that contains several fiber types, a varied population of cells, and a viscous ground substance.
low density lipoprotein (LDL): Form of lipoprotein that transports cholesterol in the blood to tissues other than the liver.
lumbar: Pertaining to the lower back.
lumen: The central space within a duct or other internal passageway.
lungs: The paired organs of breathing located in the pleural cavities.
luteinizing hormone (LH): A hormone produced by the anterior lobe of the pituitary gland (adenohypophysis). In females, it assists FSH in follicle stimulation, triggers ovulation, and promotes the maintenance and secretion of endometrial glands. In males, it stimulates testosterone secretion by the interstitial cells of the testes.
lymph: The fluid contents of lymphatic vessels, similar in composition to interstitial fluid.
lymphatic vessels: The vessels of the lymphatic system; also called *lymphatics*.
lymph nodes: Lymphoid organs that monitor the composition of lymph.
lymphocyte: A cell of the lymphatic system that plays a role in the immune response.
lymphokines: Chemicals secreted by activated lymphocytes.
lymphopoiesis: The production of lymphocytes from lymphoid stem cells.
lymphotoxin: A secretion of lymphocytes that kills the target cells.
lysis: The destruction of a cell through the rupture of its plasma membrane.
lysosome: An intracellular vesicle containing digestive enzymes.
lysozyme: An enzyme, present in some exocrine secretions that has antibiotic properties.

M

macrophage: A phagocytic cell of the monocyte–macrophage system.
macula: A receptor complex, located in the saccule or utricle of the internal ear, that responds to linear acceleration or gravity.
macula (of retina): An oval area in the retina whose center (the fovea) is the region of sharpest vision; also called *macula lutea*.
major histocompatibility complex: *See* **MHC protein**.
malignant tumor: A form of cancer characterized by rapid cell growth and the spread of cancer cells throughout the body.
malleus: The first auditory ossicle, bound to the tympanic membrane and the incus.
malnutrition: An unhealthy state produced by inadequate dietary intake or absorption of nutrients.
mammary glands: Milk-producing glands of the female breasts.
mammillary bodies: Nuclei in the hypothalamus that affect eating reflexes and behaviors; a component of the limbic system.
manus: The hand; manual.
marrow: A tissue that fills the internal cavities in bone; dominated by hemopoietic cells (red bone marrow) or by adipose tissue (yellow bone marrow).
mast cell: A connective tissue cell that, when stimulated, releases histamine, serotonin, and heparin, initiating the inflammatory response.
mastication: Chewing.
mastoid sinus: Air-filled spaces in the mastoid process of the temporal bone.
matrix: The extracellular fibers and ground substance of a connective tissue.
maxillary sinus: One of the paranasal sinuses; an air-filled chamber lined by a respiratory epithelium that is located in a maxilla and opens into the nasal cavity.
meatus: An opening or entrance into a passageway.
mechanoreception: The detection of mechanical stimuli, such as touch, pressure, or vibration.
medial: Toward the midline of the body.
mediastinum: The central tissue mass that divides the thoracic cavity into two pleural cavities.
medulla: The inner layer or core of an organ.
medulla oblongata: The most caudal of the brain regions, also called the *myelencephalon*.
medullary cavity: The space within a bone that contains the marrow.
medullary rhythmicity center: The center in the medulla oblongata that sets the background pace of respiration; includes inspiratory and expiratory centers.
megakaryocytes: Bone marrow cells responsible for the formation of platelets.
meiosis: Cell division that produces gametes with half the normal somatic chromosome number.
melanin: The yellow-brown pigment produced by the melanocytes of the skin.
melanocyte: A specialized cell in the deeper layers of the stratified squamous epithelium of the skin; responsible for the production of melanin.
melanocyte-stimulating hormone (MSH): A hormone, produced by the pars intermedia of the anterior lobe of the pituitary gland (adenohypophysis) that stimulates melanin production.
melatonin: A hormone secreted by the pineal gland.
membrane: Any sheet or partition; a layer consisting of an epithelium and the underlying connective tissue.
membrane flow: The movement of sections of membrane surface to and from the cell surface and components of the endoplasmic reticulum, the Golgi apparatus, and vesicles.
membrane potential: The potential difference, measured across a plasma membrane and expressed in millivolts, that results from the uneven distribution of positive and negative ions across the plasma membrane.
membranous labyrinth: Endolymph-filled tubes that enclose the receptors of the internal ear.
meninges: Three membranes that surround the surfaces of the central nervous system; the dura mater, the pia mater, and the arachnoid.
meniscus: A fibrocartilage pad between opposing surfaces in a joint.
menses: The first part of the uterine cycle in which the endometrial lining sloughs away; menstrual period.
merocrine: A method of secretion in which the cell ejects materials from secretory vesicles through exocytosis.
mesencephalon: The midbrain; the region between the diencephalon and pons.
mesenchyme: Embryonic or fetal connective tissue.
mesentery: A double layer of serous membrane that supports and stabilizes the position of an organ in the abdominopelvic cavity and provides a route for the associated blood vessels, nerves, and lymphatic vessels.
mesoderm: The middle germ layer, between the ectoderm and endoderm of the embryo.
mesothelium: A simple squamous epithelium that lines the body cavities enclosing the lungs, heart, and abdominal organs.
messenger RNA (mRNA): RNA formed at transcription to direct protein synthesis in the cytoplasm.
metabolic turnover: The continuous breakdown and replacement of organic materials within cells.
metabolism: The sum of all biochemical processes under way within the human body at any moment; includes anabolism and catabolism.
metabolites: Compounds produced in the body as a result of metabolic reactions.
metacarpal bones: The five bones of the palm of the hand.
metaphase: The stage of mitosis in which the chromosomes line up along the equatorial plane of the cell.
metaphysis: The region of a long bone between the epiphysis and diaphysis, corresponding to the location of the epiphyseal cartilage of the developing bone.
metarteriole: A vessel that connects an arteriole to a venule and that provides blood to a capillary plexus.
metastasis: The spread of cancer cells from one organ to another, leading to the establishment of secondary tumors.
metatarsal bone: One of the five bones of the foot that articulate with the tarsal bones (proximally) and the phalanges (distally).
metencephalon: The pons and cerebellum of the brain.
MHC protein: A surface antigen that is important to the recognition of foreign antigens and that plays a role in the coordination and activation of the immune response; also called *major histocompatibility complex (MHC) protein* or *human leukocyte antigen (HLA)*.
micelle: A droplet with hydrophilic portions on the outside; a spherical aggregation of bile salts, monoglycerides, and fatty acids in the lumen of the intestinal tract.
microfilaments: Fine protein filaments visible with the electron microscope; components of the cytoskeleton.
microglia: Phagocytic neuroglia in the central nervous system.
microtubules: Microscopic tubules that are part of the cytoskeleton and are a component in cilia, flagella, the centrioles, and spindle fibers.

microvilli: Small, fingerlike extensions of the exposed plasma membrane of an epithelial cell.
micturition: Urination.
midbrain: The mesencephalon.
middle ear: The space between the external and internal ears that contains auditory ossicles.
midsagittal plane: A plane passing through the midline of the body that divides it into left and right halves.
mineralocorticoids: Corticosteroids produced by the zona glomerulosa of the adrenal cortex; steroids such as aldosterone that affect mineral metabolism.
mitochondrion: An intracellular organelle responsible for generating most of the ATP required for cellular operations.
mitosis: The division of a single cell nucleus that produces two identical daughter cell nuclei; an essential step in cell division.
mitral valve: *See* **bicuspid valve**.
mixed gland: A gland that contains exocrine and endocrine cells, or an exocrine gland that produces serous and mucous secretions.
mixed nerve: A peripheral nerve that contains sensory and motor fibers.
mole: A quantity of an element or compound having a mass in grams equal to the element's atomic weight or to the compound's molecular weight.
molecular weight: The sum of the atomic weights of all the atoms in a molecule or compound.
molecule: A chemical structure containing two or more atoms that are held together by covalent chemical bonds.
monocytes: Phagocytic agranulocytes (white blood cells) in the circulating blood.
monoglyceride: A lipid consisting of a single fatty acid bound to a molecule of glycerol.
monosaccharide: A simple sugar, such as glucose or ribose.
monosynaptic reflex: A reflex in which the sensory afferent neuron synapses directly on the motor efferent neuron.
motor unit: All of the muscle cells controlled by a single motor neuron.
mucins: Proteoglycans responsible for the lubricating properties of mucus.
mucosa: A mucous membrane; the epithelium plus the lamina propria.
mucosa-associated lymphoid tissue (MALT): The extensive collection of lymphoid tissues linked with the epithelia of the digestive, respiratory, urinary, and reproductive tracts.
mucous (adjective)**:** Indicating the presence or production of mucus.
mucous cell: A goblet-shaped, mucus-producing, unicellular gland in certain epithelia of the digestive and respiratory tracts; also called *goblet cells*.
mucus (noun)**:** A lubricating fluid that is composed of water and mucins and is produced by unicellular and multicellular glands along the digestive, respiratory, urinary, and reproductive tracts.
multipolar neuron: A neuron with many dendrites and a single axon; the typical form of a motor neuron.
multi-unit smooth muscle: A smooth muscle tissue whose muscle cells are innervated in motor units.
muscarinic receptors: Membrane receptors sensitive to acetylcholine and to muscarine, a toxin produced by certain mushrooms; located at all parasympathetic neuromuscular and neuroglandular junctions and at a few sympathetic neuromuscular and neuroglandular junctions.
muscle: A contractile organ composed of muscle tissue, blood vessels, nerves, connective tissues, and lymphatic vessels.
muscle tissue: A tissue characterized by the presence of cells capable of contraction; includes skeletal, cardiac, and smooth muscle tissues.
muscularis externa: Concentric layers of smooth muscle responsible for peristalsis.
muscularis mucosae: The layer of smooth muscle beneath the lamina propria; responsible for moving the mucosal surface.
mutagens: Chemical agents that induce mutations and may be carcinogenic.
mutation: A change in the nucleotide sequence of the DNA in a cell.
myelencephalon: *See* **medulla oblongata**.
myelin: An insulating sheath around an axon; consists of multiple layers of neuroglial membrane; significantly increases the impulse propagation rate along the axon.
myelination: The formation of myelin.
myenteric plexus: Parasympathetic motor neurons and sympathetic postganglionic fibers located between the circular and longitudinal layers of the muscularis externa.
myocardial infarction: A heart attack; damage to the heart muscle due to an interruption of regional coronary circulation.
myocardium: The cardiac muscle tissue of the heart.
myofibril: Organized collections of myofilaments in skeletal and cardiac muscle cells.
myofilaments: Fine protein filaments composed primarily of the proteins actin (thin filaments) and myosin (thick filaments).
myoglobin: An oxygen-binding pigment that is especially common in slow skeletal muscle fibers and cardiac muscle cells.
myogram: A recording of the tension produced by muscle fibers when stimulated.
myometrium: The thick layer of smooth muscle in the wall of the uterus.
myosin: The protein component of thick filaments.

N

nail: A keratinous structure produced by epithelial cells of the nail root.
nares, external: The entrance from the exterior to the nasal cavity.
nares, internal: The entrance from the nasal cavity to the nasopharynx.
nasal cavity: A chamber in the skull that is bounded by the internal and external nares.
nasolacrimal duct: The passageway that transports tears from the nasolacrimal sac to the nasal cavity.
nasolacrimal sac: A chamber that receives tears from the lacrimal ducts.
nasopharynx: A region that is posterior to the internal nares and superior to the soft palate and ends at the oropharynx.
natriuretic peptides (NP): Hormones released by specialized cardiocytes when they are stretched by an abnormally large venous return; promote fluid loss and reductions in blood pressure and in venous return. Include atrial natriuretic peptide (ANP) and brain natriuretic peptide (BNP).
necrosis: The death of cells or tissues from disease or injury.
negative feedback: A corrective mechanism that opposes or negates a variation from normal limits.
neonate: A newborn infant, or baby.
neoplasm: A tumor, or mass of abnormal tissue.
nephron: The basic functional unit of the kidney.
nephron loop: The portion of the nephron that creates the concentration gradient in the renal medulla; also called *loop of Henle*.
nerve impulse: An action potential in a neuron plasma membrane.
neural cortex: An area of gray matter at the surface of the central nervous system.
neurilemma: The outer surface of a Schwann cell that encircles an axon in the peripheral nervous system (PNS).
neurofibrils: Microfibrils in the cytoplasm of a neuron.
neurofilaments: Microfilaments in the cytoplasm of a neuron.
neuroglandular junction: A cell junction at which a neuron controls or regulates the activity of a secretory (gland) cell.
neuroglia: Cells of the central nervous system and peripheral nervous system that support and protect neurons; also called *glial cells*.
neurohypophysis: The posterior lobe of the pituitary gland; contains the axons of hypothalamic neurons, and these axons release OXT and ADH.
neuromuscular junction: A synapse between a neuron and a muscle cell.
neuron: A cell in neural tissue that is specialized for intercellular communication through (1) changes in membrane potential and (2) synaptic connections.
neurotransmitter: A chemical compound released by one neuron to affect the membrane potential of another.
neurotubules: Microtubules in the cytoplasm of a neuron.
neutron: A fundamental particle that does not carry a positive or a negative charge.
neutrophil: A white blood cell that is very numerous and normally the first of the mobile phagocytic cells to arrive at an area of injury or infection.
nicotinic receptors: Acetylcholine receptors on the surfaces of sympathetic and parasympathetic ganglion cells; respond to the compound nicotine.
nipple: An elevated epithelial projection on the surface of the breast; contains the openings of the lactiferous sinuses.
Nissl bodies: The ribosomes, Golgi apparatus, rough endoplasmic reticulum, and mitochondria of the perikaryon of a typical neuron.
nitrogenous wastes: Organic waste products of metabolism that contain nitrogen, such as urea, uric acid, and creatinine.
nociception: Pain perception.
node: A gap in the myelin sheath of an axon between adjacent Schwann cells or oligodendrocytes; also called *node of Ranvier*.
nodose ganglion: A sensory ganglion of cranial nerve X; also called *inferior ganglion*.
noradrenaline: *See* **norepinephrine**.
norepinephrine (NE): A catecholamine neurotransmitter in the peripheral nervous system and central nervous system, released at most sympathetic neuromuscular and neuroglandular junctions, and a hormone secreted by the adrenal medulla; also called *noradrenaline*.
nucleic acid: A polymer of nucleotides that contains a pentose sugar, a phosphate group, and one of four nitrogenous bases that regulate the synthesis of proteins and make up the genetic material in cells.
nucleolus: The dense region in the nucleus that is the site of ribosomal RNA synthesis.
nucleoplasm: The fluid content of the nucleus.
nucleoproteins: Proteins of the nucleus that are generally associated with DNA.
nucleotide: A compound consisting of a nitrogenous base, a simple sugar, and a phosphate group.
nucleus: A cellular organelle that contains DNA, RNA, and proteins; in the central nervous system, a mass of gray matter.
nucleus pulposus: The gelatinous central region of an intervertebral disc.
nurse cells: Supporting cells of the seminiferous tubules of the testis; responsible for the differentiation of spermatids, the maintenance of the blood–testis barrier, and the secretion of

inhibin, and androgen-binding protein; also called *sustentacular cells.*
nutrient: An inorganic or organic substance that can be used by the body as a cofactor or to produce energy.

O

occlusal surface: The opposing surfaces of the teeth that come into contact when chewing food.
ocular: Pertaining to the eye.
oculomotor nerve: Cranial nerve III, which controls the extra-ocular muscles other than the superior oblique and the lateral rectus muscles.
olecranon: The proximal end of the ulna that forms the prominent point of the elbow.
olfaction: The sense of smell.
olfactory bulb: The expanded ends of the olfactory tracts; the sites where the axons of the first cranial nerves (I) synapse on central nervous system interneurons that lie inferior to the frontal lobes of the cerebrum.
oligodendrocytes: Central nervous system neuroglia that maintain cellular organization within gray matter and provide a myelin sheath in areas of white matter.
oocyte: A cell whose meiotic divisions will produce a single ovum and three polar bodies.
oogenesis: Formation and development of an oocyte.
opsonization: An effect of coating an object with antibodies; the attraction and enhancement of phagocytosis.
optic chiasm: The crossing point of the optic nerves.
optic nerve: The second cranial nerve (II), which carries signals from the retina of the eye to the optic chiasm.
optic tract: The tract over which nerve impulses from the retina are transmitted between the optic chiasm and the thalamus.
orbit: The bony recess of the skull that contains the eyeball.
organelle: An intracellular structure with a specific function or group of functions.
organic compound: A compound containing carbon, hydrogen, and in most cases oxygen.
organogenesis: The formation of organs during embryonic and fetal development.
organs: Combinations of tissues that perform complex functions.
origin: In a skeletal muscle, the point of attachment that does not change position when the muscle contracts; usually defined in terms of movements from the anatomical position.
oropharynx: The middle portion of the pharynx, bounded superiorly by the nasopharynx, anteriorly by the oral cavity, and inferiorly by the laryngopharynx.
osmolarity: The total concentration of dissolved materials in a solution, regardless of their specific identities, expressed in moles; also called *osmotic concentration.*
osmoreceptor: A receptor sensitive to changes in the osmolarity of plasma.
osmosis: The movement of water across a selectively permeable membrane from one solution to another solution that contains a higher solute concentration.
osmotic pressure: The force of osmotic water movement; the pressure that must be applied to prevent osmosis across a membrane.
osseous tissue: A strong connective tissue containing specialized cells and a mineralized matrix of crystalline calcium phosphate and calcium carbonate; also called *bone.*
ossicles: Small bones.
ossification: The formation of bone; osteogenesis.
osteoblast: A cell that produces the fibers and matrix of bone.
osteoclast: A cell that dissolves the fibers and matrix of bone.
osteocyte: A bone cell responsible for the maintenance and turnover of the mineral content of the surrounding bone.
osteogenic layer: The inner, cellular layer of the periosteum that aids in bone growth and repair.
osteolysis: The breakdown of the mineral matrix of bone.
osteon: The basic histological unit of compact bone, consisting of osteocytes organized around a central canal and separated by concentric lamellae.
otic: Pertaining to the ear.
otoliths: Calcium carbonate crystals embedded in a gelatinous matrix; located on each macula of the vestibule.
oval window: An opening in the bony labyrinth where the stapes attaches to the membranous wall of the scala vestibuli (vestibular duct).
ovarian cycle: The monthly chain of events that leads to ovulation.
ovary: The female reproductive organ that produces oocytes (gametes).
ovulation: The release of a secondary oocyte, surrounded by cells of the corona radiata, after the rupture of the wall of a tertiary follicle; in females, the periodic release of an oocyte from an ovary.
ovum/ova: The functional product of meiosis II, produced after the fertilization of a secondary oocyte.
oxytocin (OXT): A hormone produced by hypothalamic cells and secreted into capillaries at the posterior lobe of the pituitary gland (neurohypophysis); stimulates smooth muscle contractions of the uterus or mammary glands in females and the prostate gland in males.

P

pacemaker cells: Cells of the sinoatrial node that set the pace of cardiac contraction.
palate: The horizontal partition separating the oral cavity from the nasal cavity and nasopharynx; divided into an anterior bony (hard) palate and a posterior fleshy (soft) palate.
palatine: Pertaining to the palate.
palpate: To examine by touch.
palpebrae: Eyelids.
pancreas: A digestive organ containing exocrine and endocrine tissues; the exocrine portion secretes pancreatic juice, and the endocrine portion secretes hormones, including insulin and glucagon.
pancreatic duct: A tubular duct that carries pancreatic juice from the pancreas to the duodenum.
pancreatic islets: Aggregations of endocrine cells in the pancreas; also called *islets of Langerhans.*
pancreatic juice: A mixture of buffers and digestive enzymes that is discharged into the duodenum under the stimulation of the enzymes secretin and cholecystokinin.
papilla: A small, conical projection.
paralysis: The loss of voluntary motor control over a portion of the body.
paranasal sinuses: Bony chambers, lined by respiratory epithelium, that open into the nasal cavity; the frontal, ethmoidal, sphenoidal, and maxillary sinuses.
parasagittal: A section or plane that parallels the midsagittal plane but that does not pass along the midline.
parasympathetic division: One of the two divisions of the autonomic nervous system; generally responsible for activities that conserve energy and lower the metabolic rate; also called *craniosacral division.*
parathyroid glands: Four small glands embedded in the posterior surface of the thyroid gland; secrete parathyroid hormone.
parathyroid hormone (PTH): A hormone secreted by the parathyroid glands when blood calcium levels decrease below the normal range; causes increased osteoclast activity, increased intestinal calcium uptake, and decreased calcium ion loss at the kidneys.
parenchyma: The cells of a tissue or organ that are responsible for fulfilling its functional role; distinguished from the stroma of that tissue or organ.
parietal: Relating to the parietal bone; referring to the wall of a cavity.
parietal cells: Cells of the gastric glands that secrete hydrochloric acid and intrinsic factor.
Parkinson's disease: A neurological disorder resulting from dopamine deficiency characterized by rhythmic muscular tremors, rigidity, and a mask-like appearance.
parotid salivary glands: Large salivary glands that secrete a saliva containing high concentrations of salivary (alpha) amylase.
patella: The sesamoid bone of the knee; also called the *kneecap.*
pathogen: A disease-causing organism.
pathogenic: Disease-causing.
pathologist: A physician specializing in the identification of diseases on the basis of characteristic structural and functional changes in tissues and organs.
pelvic cavity: The inferior subdivision of the abdominopelvic cavity; encloses the urinary bladder, the sigmoid colon and rectum, and male or female reproductive organs.
pelvis: A bony complex created by the articulations among the coxal bones, the sacrum, and the coccyx.
penis: A component of the male external genitalia; a copulatory organ that surrounds the urethra and introduces semen into the female vagina; the developmental equivalent of the female clitoris.
peptide: A chain of amino acids linked by peptide bonds.
peptide bond: A covalent bond between the amino group of one amino acid and the carboxyl group of another.
pericardial cavity: The space between the parietal pericardium and the epicardium (visceral pericardium) that covers the outer surface of the heart.
pericardium: The fibrous sac that surrounds the heart; its inner, serous lining is continuous with the epicardium.
perichondrium: The layer that surrounds a cartilage, consisting of an outer fibrous region and an inner cellular region.
perikaryon: The cytoplasm that surrounds the nucleus in the cell body of a neuron.
perilymph: A fluid similar in composition to cerebrospinal fluid; located in the spaces between the bony labyrinth and the membranous labyrinth of the internal ear.
perimysium: A connective tissue partition that separates adjacent fasciculi in a skeletal muscle.
perineum: The pelvic floor and its associated structures.
perineurium: A connective tissue partition that separates adjacent bundles of nerve fibers in a peripheral nerve.
periodontal ligament: Collagen fibers that bind the cementum of a tooth to the periosteum of the surrounding alveolus.
periosteum: The layer that surrounds a bone, consisting of an outer fibrous region and inner cellular region.

peripheral nervous system (PNS): All neural tissue outside the central nervous system.
peripheral resistance: The resistance to blood flow; primarily caused by friction with the vessel walls.
peristalsis: A wave of smooth muscle contractions that propels materials along the lumen of a tube such as the digestive tract, the ureters, or the ductus deferens.
peritoneum: The serous membrane that lines the peritoneal cavity.
peritubular capillaries: A network of capillaries that surrounds the proximal and distal convoluted tubules of the kidneys.
permeability: The ease with which dissolved materials can cross a membrane; if the membrane is freely permeable, any molecule can cross it; if impermeable, nothing can cross; most biological membranes are selectively permeable.
peroxisome: A membranous vesicle containing enzymes that break down hydrogen peroxide (H_2O_2).
pes: The foot.
petrosal ganglion: A sensory ganglion of the glossopharyngeal nerve (N IX).
petrous: Stony; usually refers to the thickened portion of the temporal bone that encloses the internal ear.
pH: The negative exponent (negative logarithm) of the hydrogen ion concentration, expressed in moles per liter.
phagocyte: A cell that performs phagocytosis.
phagocytosis: The engulfing of extracellular materials or pathogens; the movement of extracellular materials into the cytoplasm by enclosure in a membranous vesicle.
phalanx/phalanges: Bone(s) of the finger(s) or toe(s).
pharmacology: The study of drugs, their physiological effects, and their clinical uses.
pharynx: The throat; a muscular passageway shared by the digestive and respiratory tracts.
phenotype: Physical characteristics that are genetically determined.
phosphate group: PO_4^{3-}; a functional group that can be attached to an organic molecule; required for the formation of high-energy bonds.
phospholipid: An important membrane lipid whose structure includes both hydrophilic and hydrophobic regions.
phosphorylation: The addition of a high-energy phosphate group to a molecule.
photoreception: Sensitivity to light.
physiology: The study of function; deals with the ways organisms perform vital activities.
pia mater: The innermost layer of the meninges bound to the underlying neural tissue.
pineal gland: Neural tissue in the posterior portion of the roof of the diencephalon; secretes melatonin.
pinna: *See* **auricle**.
pinocytosis: The introduction of fluids into the cytoplasm by enclosing them in membranous vesicles at the cell surface.
pituitary gland: An endocrine organ that is located in the sella turcica of the sphenoid and is connected to the hypothalamus by the infundibulum; includes the posterior lobe (neurohypophysis) and the anterior lobe (adenohypophysis); also called the *hypophysis*.
placenta: A temporary structure in the uterine wall that permits diffusion between the fetal and maternal circulatory systems.
plantar: Referring to the sole of the foot.
plantar flexion: Ankle extension; toe pointing.
plasma: The fluid ground substance of whole blood; what remains after the cells have been removed from a sample of whole blood.
plasma cell: An activated B cell that secretes antibodies.
plasma membrane: A cell membrane; plasmalemma.
platelets: Small packets of cytoplasm that contain enzymes important in the clotting response; manufactured in bone marrow by megakaryocytes.
pleura: The serous membrane that lines the pleural cavities.
pleural cavities: Body cavities of the thoracic region that surround the lungs.
plexus: A network or braid.
polar body: A nonfunctional packet of cytoplasm that contains chromosomes eliminated from an oocyte during meiosis.
polar bond: A covalent bond in which electrons are shared unequally.
polarized: Referring to cells that have regional differences in organelle distribution or cytoplasmic composition along a specific axis, such as between the basement membrane and free surface of an epithelial cell.
pollex: The thumb.
polypeptide: A chain of amino acids strung together by peptide bonds; those containing more than 100 peptides are called *proteins*.
polysaccharide: A complex sugar, such as glycogen or a starch.
polysynaptic reflex: A reflex in which interneurons are interposed between the sensory fiber and the motor neuron(s).
polyunsaturated fats: Fatty acids containing carbon atoms that are linked by double bonds.
pons: The portion of the metencephalon that is anterior to the cerebellum.
popliteal: Pertaining to the back of the knee.
positive feedback: A mechanism that increases a deviation from normal limits after an initial stimulus.
postcentral gyrus: The primary sensory cortex, where touch, vibration, pain, temperature, and taste sensations arrive and are consciously perceived.
posterior: Toward the back; dorsal.
postganglionic neuron: An autonomic neuron in a peripheral ganglion, whose activities control peripheral effectors.
postsynaptic membrane: The portion of the plasma membrane of a postsynaptic cell that is part of a synapse.
potential difference: The separation of opposite charges; requires a barrier that prevents ion migration.
precentral gyrus: The primary motor cortex of a cerebral hemisphere, located anterior to the central sulcus.
prefrontal cortex: The anterior portion of each cerebral hemisphere; thought to be involved with higher intellectual functions, predictions, and calculations.
preganglionic neuron: A visceral motor neuron in the central nervous system whose output controls one or more ganglionic motor neurons in the peripheral nervous system.
premotor cortex: The motor association area between the precentral gyrus and the prefrontal area.
preoptic nucleus: The hypothalamic nucleus that coordinates thermoregulatory activities.
presynaptic membrane: The synaptic surface where neurotransmitter release occurs.
prevertebral ganglion: *See* **collateral ganglion**.
prime mover: A muscle that performs a specific action.
proenzyme: An inactive enzyme secreted by an epithelial cell.
progesterone: The most important hormone secreted by the corpus luteum after ovulation.
prognosis: A prediction about the possible course or outcome from a specific disease.
projection fibers: Axons carrying information from the thalamus to the cerebral cortex.
prolactin (PRL): The hormone that stimulates functional development of the mammary glands in females; a secretion of the anterior lobe of the pituitary gland (adenohypophysis).
pronation: The rotation of the forearm that makes the palm face posteriorly.
prone: Lying face down with the palms facing the floor.
pronucleus: An enlarged ovum or spermatozoon nucleus that forms after fertilization but before amphimixis.
prophase: The initial phase of mitosis; characterized by the appearance of chromosomes, the breakdown of the nuclear membrane, and the formation of the spindle apparatus.
proprioception: The awareness of the positions of bones, joints, and muscles.
prostaglandin: A fatty acid secreted by one cell that alters the metabolic activities or sensitivities of adjacent cells; also called *local hormone*.
prostate gland: An accessory gland of the male reproductive tract, contributing about one-third of the volume of semen.
prosthesis: An artificial substitute for a body part.
protease: *See* **proteinase**.
protein: A large polypeptide with a complex structure.
proteinase: An enzyme that breaks down proteins into peptides and amino acids.
proteoglycan: A substance containing a large polysaccharide complex attached to a relatively small protein; examples include hyaluronan and chondroitin sulfate.
proton: A fundamental particle having a positive charge.
protraction: Movement anteriorly in the horizontal plane.
proximal: Toward the attached base of an organ or structure.
proximal convoluted tubule (PCT): The segment of the nephron that is located between the glomerular capsule (Bowman's capsule) and the nephron loop; the major site of active reabsorption from filtrate.
pseudopodia: Temporary cytoplasmic extensions typical of mobile or phagocytic cells.
pseudostratified epithelium: An epithelium that contains several layers of nuclei but whose cells are all in contact with the underlying basement membrane.
puberty: A period of rapid growth, sexual maturation, and the appearance of secondary sexual characteristics; normally occurs at ages 10–15 years.
pubic symphysis: The fibrocartilaginous amphiarthrosis between the pubic bones of the coxal bones.
pubis: The anterior, inferior component of the hip bone.
pudendum: The external genitalia.
pulmonary circuit: Blood vessels between the pulmonary semilunar valve of the right ventricle and the entrance to the left atrium; the blood flow through the lungs.
pulmonary ventilation: The movement of air into and out of the lungs.
pulvinar nucleus: The thalamic nucleus involved in the integration of sensory information prior to projection to the cerebral hemispheres.
pupil: The opening in the center of the iris through which light enters the eye.

purine: A nitrogen compound with a double ring-shaped structure; examples include adenine and guanine, two nitrogenous bases that are common in nucleic acids.

Purkinje cell: A large, branching neuron of the cerebellar cortex.

Purkinje fibers: Specialized conducting cardiocytes in the ventricles of the heart.

pyloric sphincter: A ring of smooth muscle that regulates the passage of chyme from the stomach to the duodenum.

pylorus: The gastric region between the body of the stomach and the duodenum; includes the pyloric sphincter.

pyrimidine: A nitrogen compound with a single ring-shaped structure; examples include cytosine, thymine, and uracil, nitrogenous bases that are common in nucleic acids.

pyruvate: The anion formed by the dissociation of pyruvic acid, a three-carbon compound produced by glycolysis.

Q

quaternary structure: The three-dimensional protein structure produced by interactions between protein subunits.

R

rami communicantes: Axon bundles that link the spinal nerves with the ganglia of the sympathetic chain.

ramus/rami: A branch/branches.

raphe: A seam.

receptive field: The area monitored by a single sensory receptor.

rectum: The inferior 15 cm (6 in.) of the digestive tract.

rectus: Straight.

red blood cell (RBC): *See* **erythrocyte**.

reductional division: The first meiotic division, which reduces the chromosome number from 46 to 23.

reflex: A rapid, automatic response to a stimulus.

reflex arc: The receptor, sensory neuron, motor neuron, and effector involved in a particular reflex; interneurons may be present, depending on the reflex considered.

refractory period: The period between the initiation of an action potential and the restoration of the normal resting membrane potential; during this period, the membrane will not respond normally to stimulation.

relaxation phase: The period after a contraction when the tension in the muscle fiber returns to resting levels.

relaxin: A hormone that loosens the pubic symphysis; secreted by the placenta.

renal: Pertaining to the kidneys.

renal corpuscle: The initial segment of the nephron, consisting of an expanded chamber that encloses the glomerulus.

renin: The enzyme released by cells of the juxtaglomerular complex when renal blood flow decreases; converts angiotensinogen to angiotensin I.

repolarization: The movement of the membrane potential away from a positive value and toward the resting potential.

respiration: The exchange of gases between cells and the environment; includes pulmonary ventilation, external respiration, internal respiration, and cellular respiration.

respiratory minute volume (V_E): The amount of air moved into and out of the respiratory system each minute.

respiratory pump: A mechanism by which changes in the intrapleural pressures during the respiratory cycle assist the venous return to the heart.

resting membrane potential: The membrane potential of an undisturbed cell, or simply the *resting potential*.

resting potential: The membrane potential of a normal cell under homeostatic conditions.

rete: An interwoven network of blood vessels or passageways.

reticular activating system (RAS): The mesencephalic portion of the reticular formation; responsible for arousal and the maintenance of consciousness.

reticular formation: A diffuse network of gray matter that extends the entire length of the brain stem.

reticulospinal tracts: Descending tracts of the medial pathway that carry involuntary motor commands issued by neurons of the reticular formation.

retina: The innermost layer of the eye, lining the vitreous chamber; also called *neural layer*.

retinal: A visual pigment derived from vitamin A.

retraction: Movement posteriorly in the horizontal plane.

retroperitoneal: Behind or outside the peritoneal cavity.

reverberation: A positive feedback along a chain of neurons such that they remain active once stimulated.

rheumatism: A general term used to describe pain in muscles, tendons, bones, or joints.

Rh factor: A surface antigen that may be present (Rh-positive) or absent (Rh-negative) from the surfaces of red blood cells.

rhodopsin: The visual pigment in the membrane discs of the distal segments of rods.

rhythmicity center: A medullary center responsible for the pace of respiration; includes inspiratory and expiratory centers.

ribonucleic acid: A nucleic acid consisting of a chain of nucleotides that contain the sugar ribose and the nitrogenous bases adenine, guanine, cytosine, and uracil.

ribose: A five-carbon sugar that is a structural component of RNA.

ribosome: An organelle that contains rRNA and proteins and is essential to mRNA translation and protein synthesis.

rod: A photoreceptor responsible for vision in dim lighting.

rough endoplasmic reticulum (RER): A membranous organelle that is a site of protein synthesis and storage.

round window: An opening in the bony labyrinth of the internal ear that exposes the membranous wall of the scala tympani (tympanic duct) to the air of the middle ear cavity.

rubrospinal tracts: Descending tracts of the lateral pathway that carry involuntary motor commands issued by the red nucleus of the mesencephalon.

rugae: Mucosal folds in the lining of the empty stomach that disappear as gastric distension occurs; folds in the urinary bladder.

rule of nines: Method used to calculate the total surface area involved in burns in which the body surface of an adult is divided into multiples of 9.

S

saccule: A portion of the vestibular apparatus of the internal ear; contains a macula important for providing sensations of gravity and linear acceleration in a vertical dimension.

sagittal plane: A sectional plane that divides the body into left and right portions.

salt: An inorganic compound consisting of a cation other than H^+ and an anion other than OH^-.

saltatory propagation: The relatively rapid propagation of an action potential between successive nodes of a myelinated axon.

sarcomere: The smallest contractile unit of a striated muscle cell.

sarcoplasm: The cytoplasm of a muscle cell.

scala media: *See* **cochlear duct.**

scala tympani: The perilymph-filled chamber of the internal ear, adjacent to the basilar membrane; pressure changes there distort the round window; also called *tympanic duct*.

scala vestibuli: A coiled tube filled with perilymph that lies within the bony labyrinth; it is continous with the scala tympani (tympanic duct) at the tip of the cochlear spiral; also called *vestibular duct*.

scar tissue: The thick, collagenous tissue that forms at an injury site.

Schwann cells: Neuroglia responsible for the neurilemma that surrounds axons in the peripheral nervous system.

sciatic nerve: A nerve innervating the posteromedial portions of the thigh and leg.

sclera: The fibrous, outer layer of the eye that forms the white area of the anterior surface; a portion of the fibrous layer of the eye.

sclerosis: A hardening and thickening that commonly occurs secondary to tissue inflammation.

scrotum: The loose-fitting, fleshy pouch that encloses the testes of the male.

sebaceous glands: Glands that secrete sebum; normally associated with hair follicles.

sebum: A waxy secretion that coats the surfaces of hairs.

secondary sex characteristics: Physical characteristics that appear at puberty in response to sex hormones.

second-degree burn: Burn involving the epidermis and dermis that usually forms blisters; type of partial-thickness burn.

secretin: A hormone, secreted by the duodenum, that stimulates the production of buffers by the pancreas and inhibits gastric activity.

semen: The fluid ejaculate that contains spermatozoa and the secretions of accessory glands of the male reproductive tract.

semicircular ducts: The tubular components of the membranous labyrinth of the internal ear; responsible for dynamic equilibrium.

semilunar valve: A three-cusped valve guarding the exit from one of the cardiac ventricles; the pulmonary and aortic valves.

seminal glands: Glands of the male reproductive tract that produce roughly 60 percent of the volume of semen; also called *seminal vesicles*.

seminiferous tubules: Coiled tubules where spermatozoon production occurs in the testes.

senescence: Aging.

sensible perspiration: Water loss due to secretion by sweat glands.

septa: Partitions that subdivide an organ.

serous cell: A cell that produces a serous secretion.

serous membrane: A squamous epithelium and the underlying loose connective tissue; the lining of the pericardial, pleural, and peritoneal cavities; also called a *serosa*.

serous secretion: A watery secretion that contains high concentrations of enzymes.

serum: The ground substance of blood plasma from which clotting agents have been removed.

sesamoid bone: A bone that forms within a tendon.

sigmoid colon: The S-shaped segment of the colon between the descending colon and the rectum.

sign: The visible, objective evidence of the presence of a disease.

simple epithelium: An epithelium containing a single layer of cells above the basal lamina.

sinoatrial (SA) node: The natural pacemaker of the heart; situated in the wall of the right atrium.
sinus: A chamber or hollow in a tissue; a large, dilated vein.
sinusoid: An exchange vessel that is similar in general structure to a fenestrated capillary. The two differ in size (sinusoids are larger and more irregular in cross section), continuity (sinusoids have gaps between endothelial cells), and support (sinusoids have thin basement membranes, if present at all).
skeletal muscle: A contractile organ of the muscular system.
skeletal muscle tissue: A contractile tissue dominated by skeletal muscle fibers; characterized as striated, voluntary muscle.
sliding filament theory: The concept that a sarcomere shortens as the thick and thin filaments slide past one another.
small intestine: The duodenum, jejunum, and ileum; the digestive tract between the stomach and the large intestine.
smooth endoplasmic reticulum (SER): A membranous organelle in which lipid and carbohydrate synthesis and storage occur.
smooth muscle tissue: Muscle tissue in the walls of many visceral organs; characterized as nonstriated, involuntary muscle.
soft palate: The fleshy posterior extension of the hard palate, separating the nasopharynx from the oral cavity.
solute: Any materials dissolved in a solution.
solution: A fluid containing dissolved materials.
somatic: Pertaining to the body.
somatic nervous system (SNS): The efferent division of the nervous system that innervates skeletal muscles.
somatomedins: Substances stimulating tissue growth; released by the liver after the secretion of growth hormone; also called *insulin-like growth factors*.
sperm: *See* **spermatozoon**.
spermatic cord: Collectively, the spermatic vessels, nerves, lymphatic vessels, and the ductus deferens, extending between the testes and the proximal end of the inguinal canal.
spermatocyte: A cell of the seminiferous tubules that is engaged in meiosis.
spermatogenesis: Spermatozoon production.
spermatozoon/spermatozoa: Male gamete(s); also called *sperm*.
sphincter: A muscular ring that contracts to close the entrance or exit of an internal passageway.
spinal nerve: One of 31 pairs of nerves that originate on the spinal cord from anterior and posterior roots.
spindle apparatus: Microtubule-based structure that distributes duplicated chromosomes to opposite ends of a dividing cell during mitosis.
spinocerebellar tracts: Ascending tracts that carry sensory information to the cerebellum.
spinothalamic tracts: Ascending tracts that carry poorly localized touch, pressure, pain, vibration, and temperature sensations to the thalamus.
spinous process: The prominent posterior projection of a vertebra; formed by the fusion of two laminae.
spiral organ: A receptor complex in the scala media of the cochlea that includes the inner and outer hair cells, supporting cells and structures, and the tectorial membrane; provides the sensation of hearing; also called the *organ of Corti*.
spleen: A lymphoid organ important for the phagocytosis of red blood cells, the immune response, and lymphocyte production.
squama: A broad, flat surface.
squamous: Flattened.
squamous epithelium: An epithelium whose superficial cells are flattened and platelike.
stapes: The auditory ossicle attached to the tympanic membrane.
stenosis: A constriction or narrowing of a passageway.
stereocilia: Elongate microvilli characteristic of the epithelium of the epididymis, portions of the ductus deferens, and the internal ear.
steroid: A ring-shaped lipid structurally related to cholesterol.
stimulus: An environmental change that produces a change in cellular activities; often used to refer to events that alter the membrane potentials of excitable cells.
stratified: Containing several layers.
stratum: A layer.
stretch receptors: Sensory receptors that respond to stretching of the surrounding tissues.
stroma: The connective tissue framework of an organ; distinguished from the functional cells (parenchyma) of that organ.
subarachnoid space: A meningeal space containing cerebrospinal fluid; the area between the arachnoid membrane and the pia mater.
subclavian: Pertaining to the region immediately posterior and inferior to the clavicle.
subcutaneous layer: *See* **hypodermis**.
submucosa: The region between the muscularis mucosae and the muscularis externa.
subserous fascia: The loose connective tissue layer beneath the serous membrane that lines the ventral body cavity.
substrate: A participant (product or reactant) in an enzyme-catalyzed reaction.
sulcus: A groove or furrow.
summation: The temporal or spatial addition of contractile force or neural stimuli.
superior: Above, in reference to a portion of the body in the anatomical position.
superior vena cava (SVC): The vein that carries blood to the right atrium from parts of the body that are superior to the heart.
supination: The rotation of the forearm such that the palm faces anteriorly.
supine: Lying face up, with palms facing anteriorly.
suppressor T cells: Lymphocytes that inhibit B cell activation and the secretion of antibodies by plasma cells.
surfactant: A lipid secretion that coats the alveolar surfaces of the lungs and prevents their collapse.
sustentacular cells: *See* **nurse cells**.
sutural bones: Irregular bones that form in fibrous tissue between the flat bones of the developing cranium; also called *Wormian bones*.
suture: A fibrous joint between flat bones of the skull.
sympathetic division: The division of the autonomic nervous system that is responsible for "fight or flight" reactions; primarily concerned with the elevation of metabolic rate and increased alertness; also called *thoracolumbar division*.
symphysis: A fibrous amphiarthrosis, such as that between adjacent vertebrae or between the pubic bones of the coxal bones.
symptom: An abnormality of function as a result of disease; subjective experience of patient.
synapse: The site of communication between a nerve cell and some other cell; if the other cell is not a neuron, the term *neuromuscular junction* or *neuroglandular junction* is often used.
synaptic delay: The period between the arrival of an impulse at the presynaptic membrane and the initiation of an action potential in the postsynaptic membrane.
syncytium: A multinucleate mass of cytoplasm, produced by the fusion of cells or repeated mitoses without cytokinesis.
syndrome: A discrete set of signs and symptoms that occur together.
synergist: A muscle that assists a prime mover in performing its primary action.
synovial cavity: A fluid-filled chamber in a synovial joint.
synovial fluid: The substance secreted by synovial membranes that lubricates joints.
synovial joint: A freely movable joint where the opposing bone surfaces are separated by synovial fluid; a diarthrosis.
synovial membrane: An incomplete layer of fibroblasts facing the synovial cavity, plus the underlying loose connective tissue.
synthesis: Manufacture; anabolism.
system: An interacting group of organs that performs one or more specific functions.
systemic circuit: The vessels between the aortic valve and the entrance to the right atrium; the system other than the vessels of the pulmonary circuit.
systole: A period of contraction in a chamber of the heart, as part of the cardiac cycle.
systolic pressure: The peak arterial pressure measured during ventricular systole.

T

tarsus: The ankle.
tactile: Pertaining to the sense of touch.
tarsal bones: The bones of the ankle (the talus, calcaneus, navicular, and cuneiform bones).
T cells: Lymphocytes responsible for cell-mediated immunity and for the coordination and regulation of the immune response; includes regulatory T cells (helpers and suppressors) and cytotoxic (killer) T cells.
tectospinal tracts: Descending tracts of the medial pathway that carry involuntary motor commands issued by the colliculi.
telodendria: Terminal axonal branches that end in axon terminals.
telophase: The final stage of mitosis, characterized by the disappearance of the spindle apparatus, the reappearance of the nuclear membrane, the disappearance of the chromosomes, and the completion of cytokinesis.
temporal: Pertaining to time (temporal summation) or to the temples (temporal bone).
tendon: A collagenous band that connects a skeletal muscle to an element of the skeleton.
teres: Round.
terminal: Toward the end.
tertiary structure: The protein structure that results from interactions among distant portions of the same molecule; complex coiling and folding.
testes: The male gonads, sites of gamete production and hormone secretion.
testosterone: The main androgen produced by the interstitial cells of the testes.
tetraiodothyronine: T_4, or thyroxine, a thyroid hormone.
thalamus: The walls of the diencephalon.
therapy: The treatment of disease.
thermoreception: Sensitivity to temperature changes.
thermoregulation: Homeostatic maintenance of body temperature.
thick filament: A cytoskeletal filament in a skeletal or cardiac muscle cell; composed of myosin, with a core of titin.
thin filament: A cytoskeletal filament in a skeletal or cardiac muscle cell; consists of actin, troponin, and tropomyosin.
third-degree burn: Burn involving the entire thickness of the skin and extends into the subcutaneous tissue, muscle, and bone causing scarring; full-thickness burn.

thoracolumbar division: The sympathetic division of the autonomic nervous system.

thorax: The chest.

threshold: The membrane potential at which an action potential begins.

thrombin: The enzyme that converts (soluble) fibrinogen to (insoluble) fibrin.

thymine: A pyrimidine; one of the nitrogenous bases in the nucleic acid DNA.

thymosins: Thymic hormones essential to the development and differentiation of T cells.

thymus: A lymphoid organ, the site of T cell development and maturation.

thyroglobulin: A circulating transport globulin that binds thyroid hormones.

thyroid gland: An endocrine gland whose lobes are lateral to the thyroid cartilage of the larynx.

thyroid hormones: Thyroxine (T_4) and T_3, hormones of the thyroid gland; stimulate tissue metabolism, energy utilization, and growth.

thyroid-stimulating hormone (TSH): The hormone, produced by the anterior lobe of the pituitary gland, that triggers the secretion of thyroid hormones by the thyroid gland.

thyroxine: A thyroid hormone; also called T_4 or *tetraiodothyronine.*

tidal volume: The volume of air moved into and out of the lungs during a normal quiet respiratory cycle.

tissue: A collection of specialized cells and cell products that performs a specific function.

titin: Fibrous protein that connects thick myosin filaments to the z lines in the sarcomere.

tonsil: A lymphoid nodule in the wall of the pharynx; the palatine, pharyngeal, and lingual tonsils.

topical: Applied to the body surface.

toxic: Poisonous.

trabecula: A connective tissue partition that subdivides an organ.

trachea: The windpipe; an airway extending from the larynx to the primary bronchi.

tract: A bundle of axons in the central nervous system.

transcription: The encoding of genetic instructions on a strand of mRNA.

transection: The severing or cutting of an object in the transverse plane.

translation: The process of peptide formation from the instructions carried by an mRNA strand.

transudate: A fluid that diffuses across a serous membrane and lubricates opposing surfaces.

transverse tubules: The transverse, tubular extensions of the sarcolemma that extend deep into the sarcoplasm, contacting cisternae of the sarcoplasmic reticulum; also called *T tubules*.

tricuspid valve: The right atrioventricular valve, which prevents the backflow of blood into the right atrium during ventricular systole.

trigeminal nerve: Cranial nerve V, which provides sensory information from the lower portions of the face (including the upper and lower jaws) and delivers motor commands to the muscles of mastication.

triglyceride: A lipid that is composed of a molecule of glycerol attached to three fatty acids.

trochanter: Large process near the head of the femur.

trochlea: A pulley; the spool-shaped medial portion of the condyle of the humerus.

trochlear nerve: Cranial nerve IV, controlling the superior oblique muscle of the eye.

trunk: The thoracic and abdominopelvic regions; a major arterial branch.

T tubules: *See* **transverse tubules**.

tuberculum: A small, localized elevation on a bony surface.

tuberosity: A large, roughened elevation on a bony surface.

tumor: A tissue mass formed by the abnormal growth and replication of cells.

tunica: A layer or covering.

twitch: A single stimulus–contraction–relaxation cycle in a skeletal muscle.

tympanic membrane: The membrane that separates the external acoustic meatus from the middle ear; the membrane whose vibrations are transferred to the auditory ossicles and ultimately to the oval window; also called *eardrum* or *tympanum.*

U

umbilical: Pertaining to the navel.

umbilical cord: The connecting stalk between the fetus and the placenta; contains the allantois, the umbilical arteries, and the umbilical vein.

unipolar neuron: A sensory neuron whose cell body is in a dorsal root ganglion or a sensory ganglion of a cranial nerve.

unmyelinated axon: An axon whose neurilemma does not contain myelin and across which continuous propagation occurs.

uracil: A pyrimidine; one of the nitrogenous bases in the nucleic acid RNA.

ureters: Muscular tubes, lined by transitional epithelium, that carry urine from the renal pelvis to the urinary bladder.

urethra: A muscular tube that carries urine from the urinary bladder to the exterior.

urinary bladder: The muscular, distensible sac that stores urine prior to micturition.

urination: The voiding of urine; micturition.

uterus: The muscular organ of the female reproductive tract in which implantation, placenta formation, and fetal development occur.

utricle: The largest chamber of the vestibular apparatus of the internal ear; contains a macula important for static equilibrium.

V

vagina: A muscular tube extending between the uterus and the vestibule.

vallate papilla: One of the large, dome-shaped papillae on the superior surface of the tongue that forms a V, separating the body of the tongue from the root.

vascular: Pertaining to blood vessels.

vasculogenesis: Formation of the vascular system.

vasoconstriction: A decrease in the diameter of arterioles due to the contraction of smooth muscles in the tunica media; increases peripheral resistance; may occur in response to local factors, through the action of hormones, or from the stimulation of the vasomotor center.

vasodilation: An increase in the diameter of arterioles due to the relaxation of smooth muscles in the tunica media; decreases peripheral resistance; may occur in response to local factors, through the action of hormones, or after decreased stimulation of the vasomotor center.

vasomotion: Rhythmic changes in the pattern of blood flow through a capillary bed due to the alternate contraction and relaxation of precapillary sphincters.

vasomotor center: The center in the medulla oblongata whose stimulation produces vasoconstriction and an increase of peripheral resistance.

vein: A blood vessel carrying blood from a capillary bed toward the heart.

vena cava: One of the major veins delivering systemic blood to the right atrium; superior and inferior venae cavae.

ventilation: Air movement into and out of the lungs.

ventral: Pertaining to the anterior surface.

ventricle: A fluid-filled chamber; in the heart, one of the large chambers discharging blood into the pulmonary or systemic circuits; in the brain, one of four fluid-filled interior chambers.

venule: Thin-walled veins that receive blood from capillaries.

vertebral canal: The passageway that encloses the spinal cord; a tunnel bounded by the neural arches of adjacent vertebrae.

vertebral column: The cervical, thoracic, and lumbar vertebrae, the sacrum, and the coccyx.

very low density lipoprotein (VLDL): Form of lipoprotein that transports triglycerides from the intestine and liver to muscle and adipose tissue.

vesicle: A membranous sac in the cytoplasm of a cell.

vestibular nucleus: The processing center for sensations that arrive from the vestibular apparatus of the internal ear, located near the border between the pons and the medulla oblongata.

vestibulospinal tracts: Descending tracts of the medial pathway that carry involuntary motor commands issued by the vestibular nucleus to stabilize the position of the head.

villus/villi: Slender, finger-shaped projection(s) of a mucous membrane.

virus: A noncellular pathogen.

viscera: Internal organs of the thoracic and abdominopelvic cavities.

visceral: Pertaining to viscera (internal organs) or their outer coverings.

visceral smooth muscle: A smooth muscle tissue that forms sheets or layers in the walls of visceral organs; the cells may not be innervated, and the layers often show automaticity (rhythmic contractions).

viscosity: The resistance to flow that a fluid exhibits as a result of molecular interactions within the fluid.

viscous: Thick, syrupy.

vitamin: An essential organic nutrient that functions as a coenzyme in vital enzymatic reactions.

vitreous humor: The fluid portion of the vitreous body, the gelatinous mass in the posterior cavity (vitreous chamber) of the eye.

voluntary: Controlled by conscious thought processes.

W

white blood cells (WBCs): The granulocytes and agranulocytes of whole blood.

white matter: Regions in the central nervous system that are dominated by myelinated axons.

white ramus: A nerve bundle containing the myelinated preganglionic axons of sympathetic motor neurons en route to the sympathetic chain or to a collateral ganglion.

Wormian bones: *See* **sutural bones**.

X

xenograft: Tissue transplant from another species, for example, a pig heart valve is transplanted into a human.

xiphoid process: The slender, inferior extension of the sternum.

Y

Y chromosome: The sex chromosome whose presence indicates that the individual is a genetic male.

Z

zona fasciculata: The region of the adrenal cortex that secretes glucocorticoids.

zona glomerulosa: The region of the adrenal cortex that secretes mineralocorticoids.

zona reticularis: The region of the adrenal cortex that secretes androgens.

zygote: The fertilized ovum, prior to the start of cleavage.

Credits

Illustration Credits

Unless otherwise stated, all illustrations are © Pearson Education. Illustration on p. 906 courtesy of U.S. Department of Agriculture.

Photo Credits

Front Matter p. iv (top to bottom) PK Martini. Amy Heflinger. Cesere Brothers Photography. Tim Mantoani; p. v (top to bottom) William C. Ober and Claire E. Ober. William C. Ober and Claire E. Ober. PK Martini. Ralph T. Hutchings.

Chapter 1 Chapter opener Jacom Stephens/Getty Images; p. 3 1.1.1 Darren Baker/Fotolia; p. 3 1.1.2 Larry Mulvehill/Science Source; p. 4 1.2.1 Susan Schmitz/Shutterstock; p. 4 1.2.2 ooyoo/iStockphoto; p. 4 1.2.3 Nick Byrne/iStockphoto; p. 4 1.2.4 William C. Ober; p. 8 1.4 Ralph T. Hutchings; p. 26 1.13-SA Pearson Education; p. 29 1.14 The Granger Collection; p. 30 1.15.1 Ed Fisher/The Cartoon Bank; p. 31 1.15.2 Custom Medical Stock Photo; p. 34 1.17SA-1 Pearson Education; p. 35 1.17SA-2 Pearson Education; p. 35 1.17SA-3 Pearson Education; p. 35 1.17SA-4 Pearson Education; p. 38 1-SO The Granger Collection; p. 39 1-CR-SA Pearson Education; p. 41 1-CI-1 Michelle Del Guercio/Science Source; p. 41 1-CI-2 ben44/Shutterstock.

Chapter 2 Chapter opener Africa Studio/Shutterstock; p. 48 2.4 Vidura Luis Barrios/Alamy; p. 50 2.5.1 pixelrobot/Fotolia; p. 50 2.5.2 pabijan/Fotolia; p. 50 2.5.3 Martina I. Meyer/iStockphoto; p. 50 2.5.4a meunierd/Shutterstock; p. 50 2.5.4b EpicStockMedia/Shutterstock; p. 50 2.5.4c Maria Pavlova/iStockphoto; p. 51 2.5.6 Markus Gayda; p. 51 2.5.7 Donald Clever/seelevel.com; p. 53 2.6 Valua Vitaly/iStockphoto; p. 59 2.9.3a Serg64/Shutterstock; p. 59 2.9.3b Aaron Amat/Shutterstock; p. 61 2.10.1 Brandon Perry/Shutterstock; p. 61 2.10.2 Trevor Chriss/Alamy; p. 85 02-CI Phil Date/Shutterstock.

Chapter 3 Chapter opener iLexx/iStockphoto; p. 92 3.4 Fawcett/Hirokawa/Heuser/Science Source; p. 96 3.6 Dr. Birgit H. Satir; p. 98 3.7 CNRI/Science Source; p. 102 3.9 Don W. Fawcett/Science Source; p. 115 3.15.1 Steve Gschmeissner/Science Source; p. 115 3.15.2 David M. Phillips/Science Source; p. 115 3.15.3 Steve Gschmeissner/Science Source; p. 119 3.17 Don W. Fawcett/Science Source; p. 124 3.20.1 Ed Reschke; p. 124 3.20.2 Ed Reschke; p. 124 3.20.3 Ed Reschke; p. 125 3.20.4 Ed Reschke; p. 125 3.20.5 Ed Reschke; p. 125 3.20.6 Ed Reschke; p. 125 3.20.7 Ed Reschke; p. 127 3.21 The International Journal of Surgery Image Library (theijs.com); p. 133 3-CI-1 epstock/Shutterstock; p. 133 3-CI-2 Darren Hubley/Shutterstock.

Chapter 4 Chapter opener BIOPHOTO ASSOCIATES/Getty Images; Chapter opener Ed Reschke/Getty Images; Chapter opener Garry DeLong/Getty Images; p. 136 4.2.1 Alexander Raths/Shutterstock; p. 136 4.2.2 Ariwasabi/Shutterstock; p. 136 4.2.3 O.Bellini/Shutterstock; p. 136 4.2.4 Brian Maudsley/Shutterstock; p. 136 4.2.5 iStockphoto; p. 136 4.2.6 FEI; p. 137 4.2.7 Robert B. Tallitsch, Pearson Education; p. 137 4.2.8 Biophoto Associates/Science Source; p. 137 4.2.9 Custom Medical Stock Photo, Inc.; p. 137 4.2.10 Brian Maudsley/Shutterstock; p. 142 4.5.1 Robert B. Tallitsch, Pearson Education; p. 143 4.5.2 Frederic H. Martini; p. 144 4.6.1 Robert B. Tallitsch, Pearson Education; p. 144 4.6.3 Gregory N. Fuller, M.D., Ph.D.; p. 145 4.6.4a Robert B. Tallitsch, Pearson Education; p. 145 4.6.4b Robert B. Tallitsch, Pearson Education; p. 146 4.7.1 Frederic H. Martini; p. 146 4.7-SA-1 Pearson Education; p. 146 4.7-SA-3 William C. Ober; p. 147 4.7.2 Robert B. Tallitsch, Pearson Education; p. 147 4.7.3 Gregory N. Fuller, M.D., Ph.D.; p. 147 4.7-SA-2 Pearson Education; p. 147 4.7-SA-3 William C. Ober; p. 148 4.8 Dr. H. Jastrow; p. 153 4.10.1 Robert B. Tallitsch, Pearson Education; p. 153 4.10.2 Ward's Natural Science Establishment; p. 153 4.10-SA Pearson Education; p. 154 4.11.1 Robert B. Tallitsch, Pearson Education; p. 154 4.11.2 Frederic H. Martini; p. 154 4.11.3 Robert B. Tallitsch, Pearson Education; p. 154 4.11-SA Pearson Education; p. 156 4.12.2 Science Source; p. 156 4.12.3 Robert B. Tallitsch, Pearson Education; p. 156 4.12.4 Robert B. Tallitsch, Pearson Education; p. 156 4.12-SA-1 Pearson Education; p. 156 4.12-SA-2 William C. Ober; p. 159 4.13 Robert B. Tallitsch, Pearson Education; p. 161 4.14-SA Pearson Education; p. 163 4.15 Ward's Natural Science Establishment; p. 164 4.16.1 Robert B. Tallitsch, Pearson Education; p. 164 4.16.2 Robert B. Tallitsch, Pearson Education; p. 164 4.16.3 Robert B. Tallitsch, Pearson Education; p. 164 4.16-SA Pearson Education; p. 166 4.17.1 Pixlmaker/Fotolia; p. 166 4.17.2 Leah-Anne Thompson/Shutterstock; p. 171 4-CR-1 Robert B. Tallitsch, Pearson Education; p. 171 4-CR-2 Robert B. Tallitsch, Pearson Education; p. 171 4-CR-3 Robert B. Tallitsch, Pearson Education; p. 171 4-CR-4 Robert B. Tallitsch, Pearson Education; p. 171 4-CR-5 Robert B. Tallitsch, Pearson Education; p. 171 4-CR-6 Robert B. Tallitsch, Pearson Education; p. 171 4-CR-7 Robert B. Tallitsch, Pearson Education; p. 171 4-CR-8 Robert B. Tallitsch, Pearson Education; p. 171 4-CR-9 Robert B. Tallitsch, Pearson Education; p. 171 4-CR-10 Robert B. Tallitsch, Pearson Education; p. 173 4-CI-1 Frederic H. Martini; p. 173 4-CI-2 Dan Trubovitz and Kevin Petti, Pearson Education; p. 173 4-CI-3 Superstock.

Chapter 5 Chapter opener Valua Vitaly/Shutterstock; p. 176 5.2.1 Robert B. Tallitsch, Pearson Education; p. 176 5.2.2 Robert B. Tallitsch, Pearson Education; p. 177 5.2.3 Robert B. Tallitsch, Pearson Education; p. 177 5.2.4 Cloud Hill Imaging; p. 178 5.3.1 Robert B. Tallitsch, Pearson Education; p. 179 5.3.2 Elizabeth A. Abel, M.D.; p. 179 5.3.3 Elizabeth A. Abel, M.D.; p. 180 5.4.1 Steve Gschmeissner/Science Source; p. 180 5.4.2 Prof. P. Motta/Science Photo Library/Science Source; p. 183 5.5.1 U.S. Air Force photo by Senior Airman Nancy Hooks; p. 189 5.8.1 Frederic H. Martini; p. 189 5.8.2 Frederic H. Martini; p. 190 5.9 Jim Thomas, M.D.; p. 193 5.11 Biophoto Associates/Science Source; p. 194 5.12.1 William C. Ober; p. 195 5.12.2 Kristine Hannon; p. 199 5-CR-1 Robert B. Tallitsch, Pearson Education; p. 201 5-CI-1 LEDPIX / Alamy; p. 201 5-CI-2 SuperflyImages/iStockphoto.

Chapter 6 Chapter opener Dorling Kindersley/Getty Images; p. 206 6.3 Ralph T. Hutchings; p. 209 6.4 Donald Clever/Seelevel.com; p. 210 6.5.1 Robert B. Tallitsch, Pearson Education; p. 211 6.5.2 Ralph T. Hutchings; p. 215 6.7 Science Photo Library/Alamy; p. 216 6.8.1 Frederic H. Martini; p. 217 6.8.2 Ralph T. Hutchings; p. 217 6.8.3 Ralph T. Hutchings; p. 218 6.9.1 Dennis MacDonald/Alamy; p. 218 6.9.2 Gary Parker Photography; p. 218 6.9.3 Joeff Davis; p. 218 6.9.4 Jim Stevenson/Science Source; p. 219 6.9.5 Bettmann/Corbis; p. 219 6.9.6 Frederick Kaplan MD/From: Overexpression of an Osteogenic Morphogen in Fibrodysplasia Ossificans Progressiva. Adam B. Shafritz et al. New England Journal of Medicine. 1996 Aug 22;335(8):555-561. Courtesy of the Mütter Museum, College of Physicians of Philadelphia. Copyright ©1996 Massachusetts Medical Society. All rights reserved.; p. 219 6.9.7 From: Mosby's Dental Dictionary, 2nd edition. 2008.; p. 224 6.12.1 kameel/Fotolia; p. 225 6.12.2 Kathleen Welch; p. 225 6.12.3 SIU BIOMED COM/Custom Medical Stock Photo; p. 225 6.12.4 Custom Medical Stock Photo; p. 225 6.12.5 Frederic H. Martini; p. 225 6.12.6 Frederic H. Martini; p. 225 6.12.7 Mark Aiken (chezlark.com); p. 225 6.12.8 Dan Zika; p. 225 6.12.9 LIVING ART ENTERPRISES, LLC/Science Source; p. 225 6.12-10 Scott Camazine/Science Source; p. 226 6-SR2-1 SIU BIOMED COM/Custom Medical Stock Photo; p. 226 6-SR2-2 Kathleen Welch; p. 226 6-SR2-3 Frederic H. Martini; p. 226 6-SR2-4 Mark Aiken (chezlark.com); p. 226 6-SR2-6 Scott Camazine/Science Source; p. 228 6-SO-1 Ralph T. Hutchings; p. 228 6-SO-2 Ralph T. Hutchings; p. 228 6-SO-3 kameel/Fotolia; p. 229 6-CR Ralph T. Hutchings; p. 231 6-CI Jabejon/iStockphoto.

Chapter 7 Chapter opener Alexander Potapov/Fotolia; p. 235 7.2.1 Ralph T. Hutchings; p. 235 7.2.2 Ralph T. Hutchings; p. 243 7.6.1 Ralph T. Hutchings; p. 243 7.6.2 Ralph T. Hutchings; p. 246 7.8.1 Ralph T. Hutchings; p. 246 7.8.2 Ralph T. Hutchings; p. 247 7.8.3 Ralph T. Hutchings; p. 247 7.8.4 Based on Life-size Auditory Ossicles, model LT-E13 by 3B Scientific; p. 247 7.8-SA Pearson Education; p. 248 7.9.1 Michael J. Timmons; p. 248 7.9.2 Ralph T. Hutchings; p. 250 7.10-SA Pearson Education; p. 252 7.11.1 Ralph T. Hutchings; p. 252 7.11.2 Ralph T. Hutchings; p. 253 7.11.3 Ralph T. Hutchings; p. 253 7.11.4 Ralph T. Hutchings; p. 254 7.12.1 Ralph T. Hutchings; p. 254 7.12.2 Ralph T. Hutchings; p. 254 7.12.3 Ralph T. Hutchings; p. 255 7.12.4 Ralph T. Hutchings; p. 255 7.12.5 Ralph T. Hutchings; p. 256 7.13.1 Ralph T. Hutchings; p. 256 7.13.2 Ralph T. Hutchings; p. 257 7.13.3 Ralph T. Hutchings; p. 258 7-SR1-1 Ralph T. Hutchings; p. 258 7-SR1-2 Ralph T. Hutchings; p. 258 7-SR1-3 Ralph T. Hutchings; p. 260 7.15.1 Ralph T. Hutchings; p. 260 7.15.2 Ralph T. Hutchings; p. 261 7.15.3 Ralph T. Hutchings; p. 261 7.15.4 Ralph T. Hutchings; p. 261 7.15.5 Ralph T. Hutchings; p. 262 7.16.1 Ralph T. Hutchings; p. 262 7.16.2 Ralph T. Hutchings; p. 263 7.16.3 Ralph T. Hutchings; p. 263 7.16.4 Ralph T. Hutchings; p. 264 7.17.1 Ralph T. Hutchings; p. 265 7.17.2 Ralph T. Hutchings; p. 266 7.18.1 Ralph T. Hutchings; p. 267 7.18.2 Ralph T. Hutchings; p. 268 7.19 Ralph T. Hutchings; p. 270 7.20.1 Ralph T. Hutchings; p. 270

7.20.2 Ralph T. Hutchings; p. 271 7.20.3 Ralph T. Hutchings; p. 271 7.20.4 Ralph T. Hutchings; p. 271 7.20.5 Ralph T. Hutchings; p. 271 7.20.6 Ralph T. Hutchings; p. 272 7.21.1 Ralph T. Hutchings; p. 273 7.21.2 Ralph T. Hutchings; p. 273 7.21.3 Ralph T. Hutchings; p. 273 7.21.4 Scott Camazine/Science Source; p. 279 7-CI-1 Pressmaster/Shutterstock; p. 279 7-CI-2 SHOUT/Alamy.

Chapter 8 Chapter opener michaeljung/Shutterstock; p. 286 8.4-SA-1 Pearson Education; p. 286 8.4-SA-2 Ralph T. Hutchings; p. 286 8.4-SA-3 Ralph T. Hutchings; p. 287 8.4-SA-4 Pearson Education; p. 287 8.4-SA-5 Ralph T. Hutchings; p. 287 8.4-SA-6 Ralph T. Hutchings; p. 287 8.4-SA-7 Ralph T. Hutchings; p. 288 8.5-SA-1 Pearson Education; p. 288 8.5-SA-2 Pearson Education; p. 289 8.5-SA-3 Ralph T. Hutchings; p. 289 8.5-SA-4 Ralph T. Hutchings; p. 289 8.5-SA-5 Ralph T. Hutchings; p. 289 8.5-SA-6 Ralph T. Hutchings; p. 290 8-SR1-SA-1 Pearson Education; p. 290 8-SR1-SA-2 Pearson Education; p. 290 8-SR1-SA-3 Ralph T. Hutchings; p. 290 8-SR1-SA-4 Ralph T. Hutchings; p. 291 8.6.1 Ralph T. Hutchings; p. 291 8.6.2 Ralph T. Hutchings; p. 292 8.7.1 Ralph T. Hutchings; p. 293 8.7.2 Prof. P. Motta/Science Photo Library/Science Source; p. 293 8.7.3 Prof. P. Motta/Science Photo Library/Science Source; p. 293 8.7.4 Michael J. Timmons; p. 296 8.9.1 Ralph T. Hutchings; p. 296 8.9.2 Ralph T. Hutchings; p. 297 8.9.3 Ralph T. Hutchings; p. 297 8.9.4 Ralph T. Hutchings; p. 298 8.10.1 MILA Zed/Shutterstock; p. 298 8.10.2 Mukundan Attur, Ph.D.; p. 298 8.10.3 Jari Sanjiv; p. 298 8.10.4 New York University School of Medicine; p. 298 8.10.5 Stanley S. Tao; p. 299 8.10.6 Mark D. Miller, M.D.; p. 299 8.10.7 Frederic H. Martini; p. 299 8.10.8 Smith & Nephew, Inc.; p. 299 8.10.9 Medical RF/SuperStock; p. 299 8.10.10 SIU BIOMED COMM/Custom Medical Stock Photo; p. 300 8-SR-2 Ralph T. Hutchings; p. 301 8-SO-SA Pearson Education; p. 302 8-SO-1 Prof. P. Motta/Science Photo Library/Science Source; p. 302 8-SO-2 Prof. P. Motta/Science Photo Library/Science Source; p. 302 8-SO-2 MILA Zed/Shutterstock; p. 302 8-SO-4 Stanley S. Tao; p. 305 8-CI-1 Apogee/Science Source; p. 305 8-CI-2 Dan Kosmayer/Shutterstock; p. 305 8-CI-3 Nagel Photography/Shutterstock.

Chapter 9 Chapter opener Yuri Arcurs/Fotolia; p. 308 9.2-SA Pearson Education; p. 325 9.10-SA Pearson Education; p. 326 9.11 photobank.kiev.ua/Shutterstock; p. 332 9.14.1 Frederic H. Martini; p. 332 9.14.2 Frederic H. Martini; p. 332 9.14.3 Comack, D. (ed): HAM'S HISTOLOGY, 9th ed. Philadelphia: J. B. Lippincott, 1987. By Permission./Lippincott Williams & wilkins/Books; p. 334 9.15.1 Yuri Arcurs/Fotolia; p. 334 9.15.2 Science Photo Library/Custom Medical Stock Photo; p. 334 9.15.3 Fotosearch/age fotostock; p. 341 9-CI-1 Andrey_Popov/Shutterstock; p. 341 9-CI-2 Nicholas Piccillo/Shutterstock.

Chapter 10 Chapter opener Aleksandr Doodko/Shutterstock; p. 356 10.8.1 Ralph T. Hutchings; p. 356 10.8-SA-1 William C. Ober; p. 357 10.8-SA-2 William C. Ober; p. 360 10.10-SA Pearson Education; p. 364 10-SR2-SA William C. Ober; p. 366 10.13-SA-1 Pearson Education; p. 367 10.13-SA-2 Pearson Education; p. 368 10.14-SA-1 Pearson Education; p. 369 10.14-SA-2 Pearson Education; p. 370 10.15-SA-1 Pearson Education; p. 370 10.15-SA-2 Pearson Education; p. 372 10.16-SA-1 Pearson Education; p. 372 10.16-SA-2 Pearson Education; p. 374 10.17-SA-1 Pearson Education; p. 374 10.17-SA-2 Pearson Education; p. 375 10.17-SA-3 Pearson Education; p. 375 10.17-SA-4 Pearson Education; p. 388 10-SR3-SA Pearson Education; p. 390 10-SO-SA Pearson Education; p. 393 10-CI-1 Felix Mizioznikov/Shutterstock; p. 393 10-CI-2 Valery Bareta/Shutterstock.

Chapter 11 Chapter opener SARANS/Shutterstock; Chapter opener SARANS/Shutterstock; Chapter opener SARANS/Shutterstock; p. 395 11.1-SA Pearson Education; p. 427 11-CI Dan Kollmann/Shutterstock.

Chapter 12 p. 433 12.3.1 Ralph T. Hutchings; p. 433 12.3.2 Hinerfeld/Custom Medical Stock Photo, Inc; p. 433 12.3.3 Ralph T. Hutchings; p. 434 12.4 Michael J. Timmons; p. 437 12.5 William C. Ober; p. 440 12.7-SA-1 Pearson Education; p. 441 12.7-SA-2 Pearson Education; p. 442 12.8-SA Pearson Education; p. 458 12-SO-SA Pearson Education; p. 461 12-CI-2 image100/age fotostock.

Chapter 13 Chapter opener Lightspring/Shutterstock; p. 470 13.5 Ward's Natural Science Establishment; p. 472 13.6 Pearson Education; p. 480 13.10.1 Ralph T. Hutchings; p. 480 13.10.2 Ralph T. Hutchings; p. 485 13.13 Jodi Jacobson/iStockphoto; p. 490 13.16.1 iStockphoto; p. 490 13.16.2 Anneka/Shutterstock; p. 500 13.21.1 Lisa F. Young/Shutterstock; p. 500 13.21.2 Ralph F. Józefowicz, MD, University of Rochester Department of Pathology; p. 500 13.21.3 Svet-lana/Shutterstock; p. 501 13.21.4 Cecilia Magill/Science Source; p. 501 13.21.5 Bettmann/CORBIS; p. 501 13.21.6 Gunilla Elam/Science Source; p. 505 13-SO Gunilla Elam/Science Source; p. 507 13-SI-1 Dr. P. Marazzi/Science Source.

Chapter 14 Chapter opener Sebastian Kaulitzki/Shutterstock; p. 511 14.2-SA Pearson Education; p. 517 14.5.1 Snaprender/Shutterstock; p. 517 14.5.2 objectsforall/Shutterstock; p. 525 14.9-SA Pearson Education; p. 527 14.10 Frederic H. Martini; p. 531 14-SO-1 objectsforall/Shutterstock; p. 532 14-SO-2 Frederic H. Martini; p. 535 14-CI-1 Sam Chadwick/Shutterstock; p. 535 14-CI-2 irin-k/Shutterstock; p. 535 14-CI-3 Simone van den Berg/Shutterstock.

Chapter 15 Chapter opener Image Source/Alamy; p. 541 15.3.1 Robert B. Tallitsch; p. 541 15.3.2 Biophoto Associates/Science Source; p. 541 15.3.3 Monkey Business Images/Shutterstock; p. 551 15.8-SA-1 Pearson Education; p. 551 15.8-SA-2 Pearson Education; p. 552 15.9 Michael J. Timmons; p. 558 15-SR2 Michael J. Timmons; p. 560 15.13.1 Ralph T. Hutchings; p. 561 15.13.2 William C. Ober; p. 561 15.13.3 Medical-on-Line/Alamy; p. 565 15.15.1 1988 Diane Hirsch/Fundamental Photographs; p. 565 15.15.2 1988 Diane Hirsch/Fundamental Photographs; p. 568 15.17 Keith/Custom Medical Stock Photo; p. 575 15.21 ZUMA Press, Inc./Alamy; p. 576 15.22.1 absolut/Shutterstock; p. 576 15.22.2 Blend Images/Alamy; p. 576 15.22.3 O.Bellini/Shutterstock; p. 576 15.22.4 Biophoto Associates/Science Source; p. 577 15.22.5 ernstc/Shutterstock; p. 577 15.22.6 Helen McArdle/Science Source; p. 583 15-SO ernstc/Shutterstock; p. 585 15-CI-1 Lisa F. Young/Shutterstock; p. 585 15-CI-2 Derek Gordon/Shutterstock.

Chapter 16 Chapter opener Carlos Caetano/Shutterstock; p. 598 16.7 Pearson Education; p. 598 16.7-SA William C. Ober; p. 600 16.8.1 Frederic H. Martini; p. 600 16.8.2 Pearson Education; p. 600 16.8-SA Pearson Education; p. 602 16.9 Ward's Natural Science Establishment; p. 602 16.9-SA Pearson Education; p. 604 16.10-SA Pearson Education; p. 606 16.11.1 Lutz Slomianka; p. 607 16.12 Kokhanchikov/Shutterstock; p. 612 16.15.1 violetblue/Shutterstock; p. 612 16.15.2 Ruhani Kaur, UNICEF India Country Office; p. 614 16.16.1 Mosby's Dental Dictionary, 2nd edition. 2008. Copyright Elsevier/Laura Gould; p. 614 16.16.2 Camazine Scott/Getty Images; p. 614 16.16.3 Medical-on-Line/Alamy; p. 615 16.16.4 Custom Medical Stock Photo, Inc.; p. 615 16.16.5 Biophoto Associates/Science Source; p. 618 16-SO-1 Kokhanchikov/Shutterstock; p. 619 16-SO-2 violetblue/Shutterstock; p. 621 16-CI Adam Gregor/Shutterstock.

Chapter 17 Chapter opener martan/Shutterstock; p. 629 17.4 Eugene Sim/Shutterstock; p. 630 17.5.1 Frederic H. Martini; p. 630 17.5.2 Cheryl Power/Science Source; p. 630 17.5.3 Ed Reschke/Getty Images; p. 635 17.7 Karen E. Petersen; p. 638 17.9 Pearson Education; p. 641 17.10 SPL/Custom Medical Stock Photo, Inc.; p. 642 17.11.1 Shevelev Vladimir/Shutterstock; p. 642 17.11.2 Monkey Business Images/Shutterstock; p. 642 17.11.3 Eye of Science/Science Source; p. 643 17.11.4 National Medical Slide Bank/Custom Medical Stock Photo, Inc.; p. 643 17.11.5 Paulo Henrique Orlandi Mourao; p. 643 17.11.6 claffra/Shutterstock; p. 647 17-SO SPL/Custom Medical Stock Photo, Inc.; p. 649 17-CI-1 Stefan Schurr/Shutterstock; p. 649 17-CI-3 Susumu Inoue MD; p. 649 17-CI-3 Rudchenko Liliia /Shutterstock.

Chapter 18 Chapter opener Alex Mit/Shutterstock; p. 651 18.1 Michelangelus/Shutterstock; p. 651 18.1-SA Pearson Education; p. 652 18.2 Biophoto Associates/Science Source; p. 656 18.4 Timothey Kosachev/Shutterstock; p. 658 18-SR-1 Biophoto Associates/Science Source; p. 660 18.6-SA Pearson Education; p. 664 18.8-SA-1 Pearson Education; p. 665 18.8-SA-2 Pearson Education; p. 666 18.9-SA-1 William C. Ober; p. 666 18.9-SA-2 William C. Ober; p. 670 18.11-SA-1 Pearson Education; p. 670 18.11-SA-2 Pearson Education; p. 672 18.12-SA-1 Pearson Education; p. 672 18.12-SA-2 Pearson Education; p. 679 18-SO Timothey Kosachev/Shutterstock; p. 680 18-SO-SA-1 Pearson Education; p. 680 18-SO-SA-2 Pearson Education; p. 680 18-SO-SA-3 William C. Ober; p. 683 18-CI-1 Helder Almeida/Shutterstock; p. 683 18-CI-2 Medical Body Scans / Science Source.

Chapter 19 Chapter opener dream designs/ Shutterstock; p. 685 19.1-SA Pearson Education; p. 687 19.2 Pearson Education; p. 688 19.3-SA Pearson Education; p. 691 19.4 Pearson Education; p. 697 19.7.1 Science Photo Library/Science Source; p. 697 19.7.2 Biophoto Associates/Science Source; p. 697 19.7.3 Margaret Grimes; p. 697 19.7.4 Patty Quinehan; p. 698 19.8.1 Ed Reschke//Getty Images; p. 698 19.8.2 B & B Photos/Custom Medical Stock Photo, Inc.; p. 698 19.8.3 Southern Illinois University/Science Source; p. 699 19.8.4 Howard Sochurek/ CORBIS- NY; p. 699 19.8.5 Howard Sochurek/ CORBIS- NY; p. 699 19.8.6 ICVI-CCN/Science Source; p. 703 19.10 ayzek/Shutterstock; p. 712 19.15 William C. Ober; p. 714 19.16 Larry Mulvehill/ Science Source; p. 719 19.18.1 vladm/Shutterstock; p. 719 19.18.1a Mara Zemgaliete; p. 719 19.18.2 Shutterstock; p. 719 19.18.3 Shutterstock; p. 719 19.18.4 Shutterstock; p. 727 19.22 Terry Vine/Corbis; p. 734 19-SO-1 Howard Sochurek/CORBIS- NY; p. 735 19-SO-2 Larry Mulvehill/Science Source; p. 736 19-SO-3 Shutterstock; p. 736 19-SO-4 Terry Vine/Corbis; p. 739 19-CI © Jose Luis Pelaez/Corbis.

Chapter 20 Chapter opener Sebastian Kaulitzki/ Shutterstock; p. 743 20.2 Frederic H. Martini; p. 745 20.3 Barkley Fahnestock/iStockphoto; p. 745 20.3-SA Pearson Education; p. 748 20.5.1 Pearson Education; p. 748 20.5.2 Biophoto Associates/ Science Source; p. 749 20.5.3 Ralph T. Hutchings; p. 751 20.6.1 Pearson Education; p. 751 20.6.2 Pearson Education; p. 752 20.7.1 iStockphoto; p. 752 20.7-SA Pearson Education; p. 753 20.7.2 Pearson Education; p. 755 20.8 iStockphoto; p. 756 20.9.1 iStockphoto; p. 757 20.9.2 CC-BY-SA Photo:Extrge, Wikipedia, The Free Encyclopedia; p. 762 20.12 Ronald Bloom/iStockphoto; p. 767 20.14 ImageStream image courtesy of Annis Corp.; p. 770 20.16 James P. Allison; p. 775 20.19 Monika Wisnievska/iStockphoto; p. 778 20.21.1 Stan Rohrer/iStockphoto; p. 778 20.21.2 Keith Bedford/ Reuters; p. 778 20.21.3 Ramona Heim/iStockphoto; p. 779 20.21.4 Centers for Disease Control and Prevention (CDC); p. 779 20.21.5 Steve Code/ iStockphoto; p. 780 20-SO-1 Frederic H. Martini; p. 781 20-SO-SA Pearson Education; p. 782 20-SO-2 iStockphoto; p. 785 20-CI NatUlrich/Shutterstock.

Chapter 21 Chapter opener Minerva Studio/ Shutterstock; p. 787 21.1-SA Pearson Education; p. 788 21.2-SA-1 Pearson Education; p. 789 21.2 Juan Monino/iStockphoto; p. 789 21.2-SA-2 Pearson Education; p. 790 21.3 Ralph T. Hutchings; p. 790 21.3-SA-1 William C. Ober; p. 791 21.3-SA-2 William C. Ober; p. 791 21.3-SA-3 William C. Ober; p. 792 21.4-SA Pearson Education; p. 793 21.4 Jason Stitt/iStockphoto; p. 794 21.5 Lester V. Bergman/ Corbis; p. 795 21.5-SA Pearson Education; p. 796 21.6 Ralph T. Hutchings; p. 796 21.6-SA Pearson Education; p. 800 21-SR-1-SA Pearson Education; p. 801 21.8-SA Pearson Education; p. 809 21.12-SA Pearson Education; p. 814 21.15.1 iStockphoto; p. 814 21.15.2 George Mayer/iStockphoto; p. 815 21.15.3 Dana Spiropoulou/iStockphoto; p. 815 21.15.4 Jaren Wicklund/iStockphoto; p. 815 21.15.5 Yen Teoh/iStockphoto; p. 819 21.17 George Mayer/ iStockphoto; p. 821 21.18 knape/iStockphoto; p. 823 21-SO-SA Pearson Education; p. 827 21-CI-1 Mike Tan C. T./Shutterstock; p. 827 21-CI-2 Dudarev Mikhail/Shutterstock.

Chapter 22 Chapter opener Sebastian Kaulitzki/ Shutterstock; p. 829 22.1-SA Pearson Education; p. 832 22.3 Frederic H. Martini; p. 837 22.5-SA Pearson Education; p. 841 22.7 Ralph T. Hutchings; p. 842 22.8.1 Alfred Pasieka/Getty Images; p. 842 22.8.2 Frederic H. Martini; p. 842 22.8-SA-1 Pearson Education; p. 842 22.8-SA-2 Pearson Education; p. 845 22.9SA Pearson Education; p. 847 22.10 Karen Krabbenhoft, Pearson Education; p. 851 22.12 M.I. Walker/Photo Researchers, Inc.; p. 853 22.13.1 Karen Krabbenhoft, Pearson Education; p. 853 22.13.2 Karen Krabbenhoft, Pearson Education; p. 853 22.13.3 Karen Krabbenhoft, Pearson Education; p. 858 22.16-SA Pearson Education; p. 863 22.18-SA Pearson Education; p. 864 22.19-SA-1 Pearson Education; p. 865 22.19 Robert B. Tallitsch, Pearson Education; p. 865 22.19-SA-2 Pearson Education; p. 866 22.20 Ralph T. Hutchings; p. 866 22.20-SA Pearson Education; p. 870 22.22-SA Pearson Education; p. 872 22.23.1 Dr Richard Nejat; p. 872 22.23.2 Barbara Rice/ Centers for Disease Control and Prevention (CDC); p. 872 22.23.3 Gastrolab/Science Source; p. 872 22.23.4 Stell98; p. 873 22.23.5 David M. Martin, MD/Science Source; p. 873 22.23.6 Ronald Bleday M.D.; p. 873 22.23.7 Joel Mancuso; p. 875 22-SO-SA-1 Pearson Education; p. 881 22-CI-1 Joanna Pecha/Thinkstock; p. 881 22-CI-2 James P. Gray.

Chapter 23 Chapter opener MidoSemsem/ Shutterstock; p. 891 23.5-SA Pearson Education; p. 892 23.6-SA Pearson Education; p. 896 23.8-SA Pearson Education; p. 900 23.10-SA Pearson Education; p. 904 23.12.1 Suharjoto/iStockphoto; p. 904 23.12.2 Olga Lyubkina/iStockphoto; p. 905 23.12.3 Ilya Genkin/iStockphoto; p. 907 23.13.1 Joop Hoek/iStockphoto; p. 907 23.13.2 Kelly Cline/ iStockphoto; p. 908 23.14.1 Hisayoshi Osawa/Getty Images; p. 908 23.14.2 David Gaylor/Shutterstock; p. 908 23.14.3 B & B Photos/Custom Medical Stock Photo, Inc.; p. 909 23.14.4 William C. Ober; p. 909 23.14.5 Lyle Conrad/Centers for Disease Control and Prevention (CDC); p. 909 23.14.6 Thalia Oster COE; p. 911 23.15 Phanie Agency/Science Source; p. 913 23.17 Fotandy/Dreamstime LLC; p. 913 23.17a Madlen/Shutterstock; p. 917 23-SO-SA Pearson Education; p. 919 23-SO-1 Phanie Agency/ Science Source; p. 920 23-SO-2 Fotandy/Dreamstime LLC; p. 920 23-SO-2a Madlen/Shutterstock; p. 921 23-CI moodboard/Fotolia.

Chapter 24 Chapter opener Andrey Popov/ Shutterstock; p. 924 24.2-SA-1 Pearson Education; p. 925 24.2-SA-2 Pearson Education; p. 926 24.3 Ralph T. Hutchings; p. 936 24.8 Steve Gschmeissner/ Science Source; p. 949 24.14 Beranger/Science Source; p. 951 24.15 Science Source; p. 953 24.16.1 Ward's Natural Science Establishment; p. 953 24.16.2 Frederic H. Martini; p. 953 24.16.3 Pearson Education; p. 961 24-CI-1 Thinkstock; p. 961 24-CI-2 WilleeCole/Shutterstock.

Chapter 25 Chapter opener Rido/Shutterstock; Chapter opener Alan Poulson/Shutterstock; p. 987 25-CI-1 Brand X Pictures/Thinkstock; p. 987 25-CI-2 haveseen/Shutterstock; p. 987 25-CI-3 Seelevel, Pearson Education.

Chapter 26 Chapter opener arek_malang/ Shutterstock; p. 993 26.3 Eye of Science/Science Source; p. 994 26.4.1 Frederic H- Martini; p. 994 26.4.2 Frederic H. Martini; p. 994 26.4.3 Robert B. Tallitsch, Pearson Education; p. 996 26.5.1 Frederic H. Martini; p. 997 26.5.2 Ward's Natural Science Establishment; p. 997 26.5.3 Frederic H. Martini; p. 997 26.5.4 Frederic H. Martini; p. 997 26.5.5 Frederic H. Martini; p. 1004 26.9-SA Pearson Education; p. 1007 26.10.1 Frederic H. Martini; p. 1007 26.10.2 Frederic H. Martini; p. 1007 26.10.3 Frederic H. Martini; p. 1007 26.10.4 Frederic H. Martini; p. 1007 26.10.5 C. Edelmann/Science Source; p. 1009 26.11.1 Frederic H. Martini; p. 1009 26.11.2 SPL/Custom Medical Stock Photo, Inc.; p. 1011 26.12.1 Frederic H. Martini; p. 1011 26.12.2 Frederic H. Martini; p. 1011 26.12.3 Michael J. Timmons; p. 1016 26.16.1 Kim Gunkel/iStockphoto; p. 1016 26.16.2 Getty Images; p. 1016 26.16.3 Brent Melton/iStockphoto; p. 1016 26.16.4 Marc Dietrich/ Shutterstock; p. 1016 26.16.5 Kumar Sriskandan/ Alamy Images; p. 1016 26.16.6 Ottfried Schreiter/ Alamy; p. 1017 26.16.7 Jennifer Trechard/ iStockphoto; p. 1017 26.16.8 Maureen Spuhler, Pearson Education; p. 1018 26.17.1 AlexSutula/ Shutterstock; p. 1019 26.17.2 ksass/iStockphoto; p. 1025 26-CI Igor Klimov/Shutterstock.

Chapter 27 Chapter opener Vitalinka/Shutterstock; p. 1028 27.2 Francis Leroy/Science Source; p. 1038 27.7 Frederic H. Martini; p. 1040 27.8.1 Dr. Arnold Tamarin; p. 1040 27.8.2 Lennart Nilsson/Tidningarnas Telelgrambyra AB; p. 1040 27.8.3 Lennart Nilsson/Tidningarnas Telelgrambyra AB; p. 1040 27.8.4 Lennart Nilsson/Tidningarnas Telelgrambyra AB; p. 1041 27.8.5 Lennart Nilsson/Tidningarnas Telelgrambyra AB; p. 1041 27.8.6 Molly Ward; p. 1041 27.8.7 Hannamariah/ Shutterstock; p. 1042 27.9 Aldo Murillo/iStockphoto; p. 1043 27.9-SA Pearson Education; p. 1047 27.11.1 Victoria Penafiel/Getty Images; p. 1047 27.11.2 Jaroslav Wojcik/iStockphoto; p. 1047 27.11.3 Jaroslav Wojcik/iStockphoto; p. 1047 27.11.4 Justin Horrocks/iStockphoto; p. 1047 27.11.5 Jacob Wackerhausen/iStockphoto; p. 1051 27.13.1 Casarsa/iStockphoto; p. 1051 27.13.2 Bob Ainsworth/iStockphoto; p. 1051 27.13.3 Science Photo Library/Science Source; p. 1052 27.14.1 Lisa McCorkle/iStockphoto; p. 1053 27.14.2 Lisa F. Young/Shutterstock; p. 1055 27.15 Science Photo Library/Science Source; p. 1056 27.16.1 Science Photo Library/Science Source; p. 1056 27.16.2 Phil Idor/Fotolia; p. 1056 27.16.3 Science Photo Library/ Science Source; p. 1056 27.16.4 Malkolm Gin/ Wikimedia Commons; p. 1056 27.16.5 Science Photo Library/Science Source; p. 1056 27.16.6 Kumar Vikram; p. 1058 27-SR2-1 Allen Johnson/ iStockphoto; p. 1058 27-SR2-2 Juanmonino/Getty Images; p. 1060 27-SO-1 Lennart Nilsson/ Tidningarnas Telelgrambyra AB; p. 1061 27-SO-2 Science Photo Library/Science Source; p. 1061 27-SO-3 Lisa McCorkle/iStockphoto; p. 1062 27-SO-4 Science Photo Library/Science Source; p. 1062 27-SO-5 Phil Idor/Fotolia; p. 1063 27-CI Alexander Raths/Shutterstock.

Index

A

INDEX

INDEX

INDEX

INDEX

G

H

INDEX

INDEX

M

INDEX

INDEX

O

INDEX

P

INDEX

S

INDEX

INDEX

INDEX

INDEX

U

V

INDEX

X

Y

Z

COMMON ABBREVIATIONS USED IN SCIENCE

ACh	acetylcholine
AChE	acetylcholinesterase
ACTH	adrenocorticotropic hormone
ADH	antidiuretic hormone
ADP	adenosine diphosphate
AIDS	acquired immunodeficiency syndrome
ALS	amyotrophic lateral sclerosis
AMP	adenosine monophosphate
ANP	atrial natriuretic peptide
ANS	autonomic nervous system
AP	arterial pressure
ARDS	adult respiratory distress syndrome
atm	atmospheric pressure
ATP	adenosine triphosphate
ATPase	adenosine triphosphatase
AV	atrioventricular
AVP	arginine vasopressin
BCOP	blood colloid osmotic pressure
BMR	basal metabolic rate
BPG	bisphosphoglycerate
bpm	beats per minute
BUN	blood urea nitrogen
C	large calorie; Celsius
CABG	coronary artery bypass graft
CAD	coronary artery disease
cAMP	cyclic-AMP
CAPD	continuous ambulatory peritoneal dialysis
CCK	cholecystokinin
CD	cluster of differentiation
CF	cystic fibrosis
CHF	congestive heart failure
CHP	capillary hydrostatic pressure
CsHP	capsular hydrostatic pressure
CNS	central nervous system
CO	cardiac output; carbon monoxide
CoA	coenzyme A
COMT	catechol-O-methyltransferase
COPD	chronic obstructive pulmonary disease
CP	creatine phosphate
CPK, CK	creatine phosphokinase
CPM	continuous passive motion
CPR	cardiopulmonary resuscitation
CRF	chronic renal failure
CRH	corticotropin-releasing hormone
CSF	cerebrospinal fluid; colony-stimulating factors
CT	computerized tomography; calcitonin
CVA	cerebrovascular accident
CVS	cardiovascular system
DAG	diacylglycerol
DC	Doctor of Chiropractic
DCT	distal convoluted tubule
DDST	Denver Developmental Screening Test
DIC	disseminated intravascular coagulation
DJD	degenerative joint disease
DMD	Duchenne's muscular dystrophy
DNA	deoxyribonucleic acid
DO	Doctor of Osteopathy
DPM	Doctor of Podiatric Medicine
DSA	digital subtraction angiography
E	epinephrine
ECF	extracellular fluid
ECG	electrocardiogram
EDV	end-diastolic volume
EEG	electroencephalogram
EKG	electrocardiogram
ELISA	enzyme-linked immunosorbent assay
EPSP	excitatory postsynaptic potential
ERV	expiratory reserve volume
ESV	end-systolic volume
ETS	electron transport system
FAD	flavin adenine dinucleotide
FAS	fetal alcohol syndrome
FES	functional electrical stimulation
FMN	flavin mononucleotide
FRC	functional residual capacity
FSH	follicle-stimulating hormone
GABA	gamma aminobutyric acid
GAS	general adaptation syndrome
GC	glucocorticoid
GFR	glomerular filtration rate
GH	growth hormone
GH–IH	growth hormone–inhibiting hormone
GHP	glomerular hydrostatic pressure
GH–RH	growth hormone–releasing hormone
GIP	gastric inhibitory peptide
GnRH	gonadotropin releasing hormone
GTP	guanosine triphosphate
Hb	hemoglobin
hCG	human chorionic gonadotropin
HCl	hydrochloric acid
HDL	high-density lipoprotein
HDN	hemolytic disease of the newborn
hGH	human growth hormone
HIV	human immunodeficiency virus
HLA	human leukocyte antigen
HMD	hyaline membrane disease
HP	hydrostatic pressure
hPL	human placental lactogen
HR	heart rate
Hz	Hertz
ICF	intracellular fluid
ICOP	interstitial fluid colloid osmotic pressure
IGF	insulin-like growth factor
IH	inhibiting hormone
IM	intramuscular
IP_3	inositol triphosphate
IPSP	inhibitory postsynaptic potential
IRV	inspiratory reserve volume
ISF	interstitial fluid
IUD	intrauterine device
IVC	inferior vena cava
IVF	in vitro fertilization
kcal	kilocalorie
LDH	lactate dehydrogenase
LDL	low density lipoprotein
L-DOPA	levodopa
LH	luteinizing hormone
LLQ	left lower quadrant
LM	light micrograph
LSD	lysergic acid diethylamide
LUQ	left upper quadrant
MAO	monoamine oxidase
MAP	mean arterial pressure
MC	mineralocorticoid
MD	Doctor of Medicine
mEq	millequivalent; (10^{-3})
MHC	major histocompatibility complex
MI	myocardial infarction
mm Hg	millimeters of mercury
mmol	millimole
mOsm	milliosmole
MRI	magnetic resonance imaging
mRNA	messenger RNA
MS	multiple sclerosis
MSH	melanocyte-stimulating hormone
MSH–IH	melanocyte-stimulating hormone–inhibiting hormone
NAD	nicotinamide adenine dinucleotide
NE	norepinephrine
NFP	net filtration pressure
NHP	net hydrostatic pressure
NO	nitric oxide
NRDS	neonatal respiratory distress syndrome
OP	osmotic pressure
Osm	osmoles
OXT	oxytocin
PAC	premature atrial contraction
PAT	paroxysmal atrial tachycardia
PCT	proximal convoluted tubule
PCV	packed cell volume
PEEP	positive end-expiratory pressure
PET	positron emission tomography
PFC	perfluorochemical emulsion
PG	prostaglandin
PID	pelvic inflammatory disease
PIH	prolactin-inhibiting hormone
PIP	phosphatidylinositol
PKC	protein kinase C
PKU	phenylketonuria
PLC	phospholipase C
PMN	polymorphonuclear leukocyte
PNS	peripheral nervous system
PR	peripheral resistance
PRF	prolactin-releasing factor
PRL	prolactin
psi	pounds per square inch
PT	prothrombin time
PTA	post-traumatic amnesia; plasma thromboplastin antecedent
PTC	phenylthiocarbamide
PTH	parathyroid hormone
PVC	premature ventricular contraction
RAS	reticular activating system
RBC	red blood cell
RDA	recommended daily allowance
RDS	respiratory distress syndrome
REM	rapid eye movement
RER	rough endoplasmic reticulum
RH	releasing hormone
RHD	rheumatic heart disease
RLQ	right lower quadrant
RNA	ribonucleic acid
rRNA	ribosomal RNA
RUQ	right upper quadrant
SA	sinoatrial
SCA	sickle cell anemia
SCID	severe combined immunodeficiency disease
SEM	scanning electron micrograph
SER	smooth endoplasmic reticulum
SGOT	serum glutamic oxaloacetic transaminase
SIADH	syndrome of inappropriate ADH secretion
SIDS	sudden infant death syndrome
SLE	systemic lupus erythematosus
SNS	somatic nervous system
STD	sexually transmitted disease
SV	stroke volume
SVC	superior vena cava
T_3	triiodothyronine
T_4	tetraiodothyronine, or thyroxine
TB	tuberculosis
TBG	thyroid-binding globulin
TEM	transmission electron micrograph
TIA	transient ischemic attack
T_m	transport (tubular) maximum
TMJ	temporomandibular joint
t-PA	tissue plasminogen activator
TRH	thyrotropin-releasing hormone
tRNA	transfer RNA
TSH	thyroid-stimulating hormone
TSS	toxic shock syndrome
U.S.	United States
UTI	urinary tract infection
UTP	uridine triphosphate
UV	ultraviolet
$\dot{V}_A$	alveolar ventilation
V_D	anatomic dead space
$\dot{V}_E$	respiratory minute volume
V_T	tidal volume
VF	ventricular fibrillation
VLDL	very low-density lipoprotein
VPRC	volume of packed red cells
VT	ventricular tachycardia
WBC	white blood cell

FOREIGN WORD ROOTS, PREFIXES, SUFFIXES, AND COMBINING FORMS

Each entry starts with the commonly used form or forms of the prefix, suffix, or combining form followed by the word root (shown in italics) and its English translation. One example is also given to illustrate the use of each entry.

a-, *a-,* without: avascular
ab-, *ab,* from: abduct
-ac, *-akos,* pertaining to: cardiac
acr-, *akron,* extremity: acromegaly
ad-, *ad,* to, toward: adduct
aden-, adeno-, *adenos,* gland: adenoid
adip-, *adipos,* fat: adipocytes
aer-, *aeros,* air: aerobic metabolism
-al, *-alis,* pertaining to: brachial
alb-, *albicans,* white: albino
-algia, *algos,* pain: neuralgia
allo-, *allos,* other: allograft
ana-, *ana,* up, back: anaphase
andro-, *andros,* male: androgen
angio-, *angeion,* vessel: angiogram
ante-, *ante,* before: antebrachial
anti-, ant-, *anti,* against: antibiotic
apo-, *apo,* from: apocrine
arachn-, *arachne,* spider: arachnoid
arter-, *arteria,* artery: arterial
arthro-, *arthros,* joint: arthroscopy
astro-, *aster,* star: astrocyte
atel-, *ateles,* imperfect: atelectasis
aur-, *auris,* ear: auricle
auto-, *auto,* self: autonomic
baro-, *baros,* pressure: baroreceptor
bi-, *bi-* two: bifurcate
bio-, *bios,* life: biology
-blast, *blastos,* precursor: osteoblast
brachi-, *brachium,* arm: brachiocephalic
brachy-, *brachys,* short: brachydactyly
brady-, *bradys,* slow: bradycardia
bronch-, *bronchus,* windpipe, airway: bronchial
carcin-, *karkinos,* cancer: carcinoma
cardi-, cardio-, *kardia,* heart: cardiac
-cele, *kele,* tumor, hernia, or swelling: blastocele
-centesis, *kentesis,* puncture: thoracocentesis
cephal-, *cephalos,* head: brachiocephalic
cerebr-, *cerebrum,* brain: cerebral hemispheres
cerebro-, *cerebros,* brain: cerebrospinal fluid
cervic-, *cervicis,* neck: cervical vertebrae
chole-, *chole,* bile: cholecystitis
-chondrion, *chondrion,* granule: mitochondrion
chondro-, *chondros,* cartilage: chondrocyte
chrom-, chromo-, *chroma,* color: chromatin
circum-, *circum,* around: circumduction
-clast, *klastos,* broken: osteoclast
coel-, -coel, *koila,* cavity: coelom
colo-, *kolon,* colon: colonoscopy
contra-, *contra,* against: contralateral
corp-, *corpus,* body: corpuscle
cortic-, *cortex,* rind or bark: corticospinal
cost-, *costa,* rib: costal
cranio-, *cranium,* skull: craniosacral
cribr-, *cribrum,* sieve: cribriform
-crine, *krinein,* to separate: endocrine
cut-, *cutis,* skin: cutaneous
cyan-, *kyanos,* blue: cyanosis
cyst-, -cyst, *kystis,* sac: blastocyst
cyt-, *cyto-, kytos,* a hollow cell: cytology
de-, *de,* from, away: deactivation
dendr-, *dendron,* tree: dendrite
dent-, *dentes,* teeth: dentition
derm-, *derma,* skin: dermatome
desmo-, *desmos,* band: desmosome
di-, *dis,* twice: disaccharide
dia-, *dia,* through: diameter
digit-, *digit,* a finger or toe: digital
dipl-, *diploos,* double: diploid
dis-, *dis,* apart, away from: disability
diure-, *diourein,* to urinate: diuresis
dys-, *dys,* painful: dysmenorrhea
-ectasis, *ektasis,* expansion: atelectasis
ecto-, *ektos,* outside: ectoderm
-ectomy, *ektome,* excision: appendectomy
ef-, *ex,* away from: efferent
emmetro-, *emmetros,* in proper measure: emmetropia
encephalo-, *enkephalos,* brain: encephalitis
end-, endo-, *endon,* within: endometrium
entero-, *enteron,* intestine: enteric
epi-, *epi,* upon: epimysium
erythema-, *erythema,* flushed (skin): erythematosis
erythro-, *erythros,* red: erythrocyte
ex-, *ex,* out of, away from: exocytosis
extra-, *exter,* outside of, beyond, in addition: extracellular
ferr-, *ferrum,* iron: transferrin
fil-, *filum,* thread: filament
-form, *-formis,* shape: fusiform
gastr-, *gaster,* stomach: gastrointestinal
-gen, -genic, *gennan,* to produce: mutagen
genicula-, *geniculum,* kneelike structure: geniculates
genio-, *geneion,* chin: geniohyoid
gest-, *gesto,* to bear: gestation
glosso-, -glossus, *glossus,* tongue: hypoglossal
glyco-, *glykys,* sugar: glycogen
-gram, *gramma,* record: myogram
gran-, *granulum,* grain: granulocyte
-graph, -graphia, *graphein,* to write, record: electroencephalograph
gyne-, gyno-, *gynaikos,* woman: gynecologist
hem-, hemo-, *haima,* blood: hemopoiesis
hemi-, *hemi-,* one half: hemisphere
hepato-, *hepaticus,* liver: hepatocyte
hetero-, *heteros,* other: heterozygous
histo-, *histos,* tissue: histology
holo-, *holos,* entire: holocrine
homeo-, *homoios,* similar: homeostasis
homo-, *homos,* same: homozygous
hyal-, hyalo-, *hyalos,* glass: hyaline
hydro-, *hydros,* water: hydrolysis
hyo-, *hyoeides,* U-shaped: hyoid bone
hyper-, *hyper,* above: hyperpolarization
hypo-, *hypo,* under: hypothyroid
hyster-, *hystera,* uterus: hysterectomy
-ia, *-ia,* state or condition: insomnia
idi-, *idios,* one's own: idiopathic
in-, *in-,* in, within, or denoting negative effect: inactivate
infra-, *infra,* beneath: infraorbital
inter-, *inter,* between: interventricular
intra-, *intra,* within: intracapsular
ipsi-, *ipse,* itself: ipsilateral
iso-, *isos,* equal: isotonic
-itis, *-itis,* inflammation: dermatitis